W9-AKV-046

HELENA CURTIS **BIOLOGY**

HELENA CURTIS

BIOLOGY

FOURTH EDITION

BIOLOGY, FOURTH EDITION

PRINTED IN THE UNITED STATES OF AMERICA

LIBRARY OF CONGRESS CATALOG CARD NO. 82-083895

ISBN: 0-87901-186-6

FIRST PRINTING, MARCH 1983

EDITOR: SALLY ANDERSON

ILLUSTRATOR: SHIRLEY BATY

PICTURE EDITOR: ANNE FELDMAN

PRODUCTION: GEORGE TOULOUMES

DESIGN: MALCOLM GREAR DESIGNERS

TYPOGRAPHER: NEW ENGLAND TYPOGRAPHIC SERVICE, INC.

PRINTING AND BINDING: R. R. DONNELLEY & SONS

FRONTISPIECE: FOOTPRINTS MADE 3.8 MILLION YEARS AGO,

UNCOVERED AT LAETOLI, TANZANIA

WORTH PUBLISHERS, INC.

444 PARK AVENUE SOUTH

NEW YORK, NEW YORK 10016

This book is dedicated to C. P. Rhoads, M.D.

Contents in Brief

A Darwin's finch

Contents

Photosynthetic membranes

Transposon

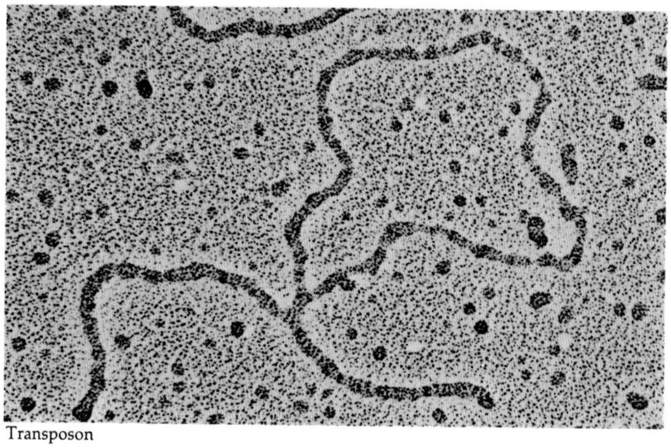

Leopard

Fern spores

Wolf spider

Blueberry flowers

Penguins

Nerve fibers

Elephant seals

Preface

> This book is the result of an unusually broad cooperative venture. There were several essential factors in this venture. My own contribution was my long experience as a professional writer in the field of biology and a desire to communicate my interests and enthusiasms about the subject. Second, and more important, was the arrival on the publishing scene of a vigorous young company with fresh ideas and initiative. The third and by far the most important factor was the willingness of a large number of teachers and experts in all fields of biology to lend their assistance at every stage of the planning, writing, and innumerable revisions of the manuscript and illustrations.

The paragraph just quoted was written fifteen years ago in the preface to the first edition of *Biology.* There have been many changes since that time. For one thing, neither Worth Publishers nor I are quite so young, though we are, we hasten to add, just as vigorous. In retrospect, our youthful inexperience was one of our chief assets. Had we been less innocent, I do not think we would have dared as newcomers to undertake so vast a project as an introductory biology text. I still wonder how we did it.

Another change of slightly more importance has been in the general nature of biology textbooks. As a writer, I felt strongly that any book, textbook or not, had an obligation to be well written, interesting, and even, on occasion, entertaining. Biology, for me, is a source of pleasure and excitement, as it is for most of the men and women engaged in biological research. I did not understand why textbooks found it necessary to keep that information a trade secret. Robert Worth not only agreed with me but went one step further. He maintained that a book, even a textbook, should also be well crafted, handsomely illustrated, and even beautiful. Now there are many biology textbooks that are handsome and readable as well as serving their pedagogical functions. Bob and I are proud to have been part of this general trend.

The most important changes in these fifteen years have been, of course, in biology itself. There has been not only a flood of new information, but also of new ideas and unifying concepts. Moreover, the rate of change is itself increasing; the changes in the last four years have been enormous. In molecular genetics, in particular, there has been an absolutely unprecedented explosion of knowledge, made even more dramatic by the unexpected nature of much of what has been discovered. In such cases, where a subject is out-of-date before the printer's ink dries, the best one can hope to do is to establish a basis for the understanding not only of what is happening now, but what may happen a year from now.

Finally, there has been an extrabiological development that must affect all biology textbooks: the resurgence of creationism in a new but not very different form. The recent public debates were foreshadowed for me by letters from students who were using *Biology* and were obviously having difficulty reconciling these new (to them) ideas about evolution with their religious beliefs. While we may be impatient with those who attempt to impose creationist beliefs upon others, those students deserve our attention and respect. In the first edition of *Biology,* I simply tried to present current thinking about how evolution occurs, without taking enough time to explain the reasons why biologists so overwhelmingly accept—and have done so for considerably more than a hundred years—the fact that evolution has occurred. ("'Shut up,' he explained," will always remain for me one of the greatest lines of modern literature.) This time I do present the evidence in what I hope is a quiet and nonbelligerent way. (Many young people do not like being told what to think, and I don't blame them.)

Despite these changes, this edition follows the same general pattern as the previous ones. The Introduction focuses on evolution, still the major theme of this text, as it must be of all modern biology texts. The chapter closes with a brief discussion of the controversies now going on. The purpose of this discussion is to make clear that there are two types of disagreement—one between scientists and nonscientists and the other between groups of scientists with opposing viewpoints. The former is not the business of science (though, of course, of concern to scientists), whereas the latter is not only the business of science but indeed its life's blood. Following this thread elsewhere in the book, I have identified areas of disagreement among scientists and instances where prevailing opinion has proved to be wrong. This tactic elicited criticism from some reviewers who feel it trivializes science and distracts the student from the more important business of learning the fundamentals of biology. As I look back over the changes in biological thought in just the last fifteen years, it seems to me that perhaps one of the most useful things a student should learn is a healthy skepticism about what he or she reads, even in textbooks. Scientific progress is a process that takes place only within the confines of the skull of human beings, which makes it both marvelous and fallible.

After the Introduction, this edition, as was the case with previous ones, follows the levels-of-organization approach. Part I deals with life at the subcellular and cellular level, Part II with organisms, and Part III with populations, ending with the evolution of the human species. Each part is divided into two or three sections. We have been tempted from time to time to reverse the order and begin with ecology, but the great majority of you who were surveyed were in favor of the present organization.

Those familiar with previous editions will notice many changes within the sections, however. In Section 1, a notable addition is an expanded presentation on techniques of microscopy. This addition reflects our decision, made early in the planning stages of the present edition, to include more information about how biologists know what they know and how scientists in general go about their business. Also new in Section 1—although not new to the book—is the discussion of the origin of life. The new placement of this material provides a logical bridge from biochemistry to cytology and also offers a good opportunity to reiterate the evolutionary theme.

A major innovation in Section 2, Energetics, is the greatly expanded discussion of the chemiosmotic theory. A separate section on energetics first made its appearance in the second edition of *Invitation to Biology,* which has often served as a proving ground for changes that have later showed up in the parent text. Since that time the section has grown considerably in size and strength. In this edition, following a precedent established in the third edition of *Invitation to Biology,* almost all of the discussion of enzyme structure and function has been incorporated into this section, making possible a much more dynamic view of cellular activities.

Section 3, Genetics, has of course been greatly revised. We have ventured to lead students right up to the frontier of what is happening now, which is going to demand some work on their parts (and yours too), but the rewards, it seems to me, are tremendous. Not very long ago, at least one leader in the field of molecular genetics announced that all the fun was over and all that remained was a little mopping up. Those of real talent, it was announced, were switching to other fields of greater promise. As one who has followed the history of molecular genetics almost from its birth, albeit as a bystander, my sadness on reading this obituary was tempered only by a natural skepticism. I am very glad that this particular prophecy was dead wrong.

Part II, The Biology of Organisms, has also undergone many changes. The first chapter of Section 4, on the classification of organisms, has been given a thorough renovation. New taxonomic tools are discussed, particularly the sequencing of macromolecules and the existence (or nonexistence) of molecular clocks. Also, the dispute among cladists, numerical pheneticists, and more traditional systematists has reached a point where it becomes of general interest, not only because of the subject itself, but because of what it reveals about the different ways to think about relationships among the earth's biota.

The following chapters of this section, The Diversity of Organisms, are considerably expanded. Always fighting the battle of the bulge, we embarked on this expansion after a great deal of debate. Of all the sections in the book, this one has always been to me the least satisfactory. The reviewers did not like it either. We had selected the wrong examples, omitted the most interesting organisms, and left out whole phyla, including, of course, somebody's favorite. But no matter how we framed the question, reviewers refused to let us omit the section entirely. So we took the alternative route and made it longer, by about 25 percent. This increase has made it possible to include all the major modern taxa, to add some more examples where they were lacking, and to weave the section into a more coherent whole.

The second section of Part II, Plants, has been enriched not only by new materials but by new illustrations, many of them borrowed from the latest edition of Raven, Evert, and Curtis: *Biology of Plants.* I am no longer an active coauthor of that book, but I am still very grateful for the experience of collaborating with Peter Raven and Ray Evert and for the appreciation of the plant world that I received from these two excellent teachers. This experience provided me with the fortitude to continue to resist an integrated approach to plant and animal physiology. This route seems a tempting one, especially to someone seeking unifying principles. But it results, inevitably, it seems to me, in students seeing plants as slow, dull animals rather than as highly successful well-integrated organisms with methods of survival equal to those of animals, though quite different.

In the third section of Part II, we have retained our focus on human physiology. However, the internal organization of this section has been considerably revised in a way that increases the emphasis on the problem-solving approach; this strengthens its biological foundations and opens the way for more comparative examples. We have also increased the emphasis throughout the section on control mechanisms. Finally, we have added a chapter on immunology. Immunology is an excellent example of how proliferation of knowledge can quite suddenly result in a whole lot of details falling into place. The section now ends with the sequence on human embryological development and birth.

Part III, the Biology of Populations, begins with a new chapter, previously mentioned, that presents some of the traditional evidence for evolution. The evolution section ends with a chapter on the origin of species and the question of gradualism versus punctuated equilibria. Ecology, which follows evolution in this edition, begins with a new chapter on single-species ecology, or, as it is more popularly known, life-history strategies. Now the flow of the ecology section, like that of the rest of the book, moves from smaller to larger, ending with the planet earth and its biomes. This section has been thoroughly revised throughout: ecology is a field in which yesterday's paradigm is today's pitfall.

The next to the last chapter in this book concerns social behavior, focusing particularly on matters of kinship, altruism, selfish genes, and conflicts of interest. This is a chapter students should enjoy, both because the examples are fun and because these studies and speculations introduce a novel way of looking at something familiar—and it is always exciting to have your mind stretched in this way.

The book ends, as in previous editions, with a chapter on human evolution. This chapter is a fitting ending for the book, it seems to me, and not just because it brings matters up to the present moment. For one thing, the evidence for human evolution is so overwhelming that this particular story serves as a resounding reprise of biology's major theme. For another, the view of human evolution has changed completely in the lifetime of this book. The story we told in the first edition of a single line of human descent is, very simply, quite wrong. I am not mentioning this by way of apology, but rather to point out what a dynamic process biology is.

Each section ends with suggestions for further reading. Scientifically speaking, the selections are arbitrary. They were chosen not as documentation for statements in the book or as fuller presentations of difficult subjects, but rather because of their accessibility to students. Our hope is that at least some students will continue reading on their own. What I would really like them to read are reports not yet published about discoveries just now being dreamed of.

In the Preface to the first edition, I noted that the most important of the several ingredients of *Biology* was the help and cooperation we received from many different experts in biology research and teaching. Their number has grown, as can be seen from the following list, and so has my appreciation of their role. I continue to be impressed by their range of knowledge and their sharpness of wit, but what I am really grateful for is their patience, goodwill, and generosity. Many of them I have never even spoken to on the telephone (an instrument I avoid at all costs), much less met, but I nevertheless feel a strong bond of friendship and shared experience with them.

VERNON AHMADJIAN, Clark University
HARRY BERNHEIM, Tufts University
WILLIAM BISCHOFF, University of Toledo
JACQUES M. CHILLER, Scripps Clinic and Research Foundation
EDWARD I. COHER, Southampton College of Long Island University
JOHN O. CORLISS, University of Maryland
JERRY DAVIS, University of Wisconsin, La Crosse
MARK DUBIN, University of Colorado, Boulder
MARTIN DWORKIN, University of Minnesota
NILES ELDREDGE, American Museum of Natural History
RAY F. EVERT, University of Wisconsin
BEN T. FEESE, Centre College of Kentucky
DOUGLAS FUTUYMA, State University of New York, Stony Brook
ARTHUR GALSTON, Yale University
MICHAEL T. GHISELIN, University of Utah
PETER GIEBEL, Virginia Commonwealth University
URSULA GOODENOUGH, Washington University
JEAN B. HARRISON, University of California, Los Angeles
H. CRAIG HELLER, Stanford University
HENRY HORN, Princeton University
GERALD KARP, University of Florida
JOHN KIRSCH, Harvard University
JAMES LLOYD, University of Florida
DOROTHY LUCIANO, formerly of the University of Michigan
ALLEN MACNEILL, Cornell University
R. WILLIAM MARKS, Villanova University
HAROLD MARTINSON, University of California, Los Angeles
RICHARD MARZOLF, Kansas State University
ROGER MILKMAN, University of Iowa
DOUGLAS W. MORRISON, Rutgers University, Newark
BETTE NICOTRI, University of Washington
MAX OESCHGER, Johns Hopkins University
ROBERT ORNDUFF, University of California, Berkeley
VICKI PEARSE, University of California, Santa Cruz
FRANK PRICE, Hamilton College
ANTHONY RUSSELL, University of Calgary
KATHLEEN SCOTT, Rutgers University, Piscataway
IAN TATTERSALL, American Museum of Natural History
IAIN TAYLOR, University of British Columbia
JOHN TAYLOR, University of California, Berkeley
JOHN YARNALL, Humboldt State University

I wish to acknowledge a relatively new ingredient in the person of N. Sue Barnes. Sue first joined us in the course of preparing the third edition of *Biology.* She was a major factor in the successful revision of the early chapters of that book in which we made the jump from the Bohr model of the atom to a more modern level of organization. She rapidly became so invaluable that, at my request, she did most of the work on the third edition of *Invitation to Biology,* on which she is acknowledged as coauthor. In this present edition, she has worked particularly on

Sections 1 and 2 and on the expanded version of Diversity, but the entire book has passed under her scrutiny. She assures me that she is now a permanent member of the team, which is good because she has become the fourth essential component.

Penultimately, I would like to thank Shirley Baty and Anne Feldman, good friends who have been with me since the beginning; Sally Anderson and George Touloumes, for their extraordinary editorial and production talents; and all the others at Worth Publishers who have put up with my bad handwriting, my messy manuscripts, and my innumerable revisions. We have done it once again!

Finally, I want to thank the students who have written to me, some with criticisms, some with questions, and some—bless them—just because they enjoyed the book. Their letters serve to remind me how privileged I am to be writing for young people; I like their curiosity, their energies, their imaginativeness, and their dislike of the pompous and pedantic. I hope I serve them well.

East Hampton, New York
January, 1983

Helena Curtis

HELENA CURTIS **BIOLOGY**

Introduction

In 1831, the young Charles Darwin set sail from England on what was to prove the most consequential voyage in the history of biology. Not yet 23, Darwin had already abandoned a proposed career in medicine—he describes himself as fleeing a surgical theater in which an operation was being performed on an unanesthetized child—and was a reluctant candidate for the clergy, a profession deemed suitable for the younger son of an English gentleman. An indifferent student, Darwin was an ardent hunter and horseman, a collector of beetles, mollusks, and shells, and an amateur botanist and geologist. When the captain of the surveying ship H.M.S. *Beagle,* himself only a little older than Darwin, offered passage to any young man who would volunteer to go without pay as a naturalist, Darwin eagerly seized the opportunity to escape from Cambridge. This voyage, which lasted five years, shaped the course of Darwin's future work. He returned to an inherited fortune, an estate in the English countryside, and a lifetime of independent work and study that radically changed mankind's view of life and of our place in the living world.

THE ROAD TO EVOLUTIONARY THEORY

That Darwin was the founder of the modern theory of evolution is well known. In order to understand the meaning of his theory, however, it is useful to look briefly at the intellectual climate in which it was formulated. Aristotle (384–322 B.C.), the first great biologist, believed that all living things could be arranged in a hierarchy. This hierarchy became known as the *Scala Naturae,* or ladder of nature, in which the simplest creatures had a humble position on the bottommost rung, mankind occupied the top, and all other organisms had their proper places between. Until the end of the last century, many biologists believed in such a natural hierarchy. But whereas to Aristotle living organisms had always existed, the later biologists (at least those of the Occidental world) believed, in harmony with the teachings of the Old Testament, that all living things were the products of a divine creation. They believed, moreover, that most were created for the service or pleasure of mankind. Indeed, it was pointed out, even the lengths of day and night were planned to coincide with the human need for sleep.

I–1
"Afterwards, on becoming very intimate with Fitz Roy [the captain of the Beagle*], I heard that I had run a very narrow risk of being rejected on account of the shape of my nose! He . . . was convinced that he could judge of a man's character by the outline of his features; and he doubted whether anyone with my nose could possess sufficient energy and determination for the voyage. But I think he was afterwards well satisfied that my nose had spoken falsely." (Charles Darwin,* The Voyage of the Beagle.*)*

That each type of living thing came into existence in its present form—specially and specifically created—was a compelling idea. How else could one explain the astonishing extent to which every living thing was adapted to its environment and

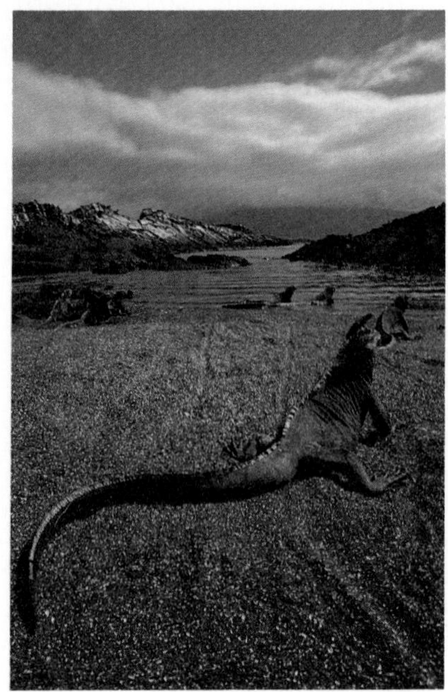

I–2
The voyage of the Beagle *took young Charles Darwin to the Galapagos islands, off the western coast of South America. The archipelago consists of 13 volcanic islands that pushed up from the sea about a million years ago. Craters still covered with black basaltic lava rise to over 1,000 meters. The major vegetation is a dreary grayish-brown thornbush, making up a wasteland of dense leafless thicket, and a few tall tree cactuses—"what we might imagine the cultivated parts of the Infernal regions to be," young Darwin wrote in his diary. The photo shows marine iguanas on Narborough, one of the Galapagos.*

to its role in nature? It was not only the authority of the church but also, so it seemed, the evidence before one's own eyes that gave such strength to the concept of special creation.

Among those who believed in divine creation was Carolus Linnaeus (1707–1778), the great eighteenth-century Swedish naturalist, who developed our present system of nomenclature for biological species. In 1753, Linnaeus published *Species Plantarum,* which described, in two volumes, every species of plant known at that time. Even as Linnaeus was at work on these encyclopedic volumes, explorers were returning to Europe from Africa and the New World with previously undescribed plants and animals and even, apparently, new kinds of human beings. Linnaeus revised edition after edition to accommodate these findings, but he did not change his opinion that all species now in existence were created by the sixth day of God's labor and have remained fixed ever since. During Linnaeus's time, however, it became clear that the pattern of creation was far more complex than had been originally envisioned.

Evolution before Darwin

The French scientist Georges-Louis Leclerc de Buffon (1707–1788) was among the first to suggest that species might undergo some changes in the course of time. Buffon believed that these changes took place by a process of degeneration. He suggested that, in addition to the numerous creatures that were produced by divine creation at the beginning of the world, "there are lesser families conceived by Nature and produced by Time." In fact, as he summed it up, ". . . improvement and degeneration are the same thing, for both imply an alteration of the original constitution." Buffon's hypothesis, although vague as to mechanism, did attempt to explain the bewildering variety of creatures in the modern world.

Another early doubter of the fixity of species was Erasmus Darwin (1731–1802), Charles Darwin's grandfather. Erasmus Darwin was a physician, a gentleman naturalist, and a prolific writer, often in verse, on both botany and zoology. Erasmus Darwin suggested, largely in asides and footnotes, that species have historical connections with one another, that animals may change in response to their environment, and that their offspring may inherit these changes. He maintained, for instance, that a polar bear is an "ordinary" bear that, by living in the Arctic, became modified and passed the modifications along to its cubs. These ideas were never clearly formulated but are interesting because of their possible effects on Charles Darwin, although the latter, born after his grandfather had died, did not profess to hold his grandfather's views in high esteem.

The Age of the Earth

It was geologists, more than biologists, who paved the way for evolutionary theory. One of the most influential of these was James Hutton (1726–1797). Hutton proposed that the earth had been molded not by sudden, violent events but by slow and gradual processes—wind, weather, and the flow of water—the same processes that can be seen at work in the world today. This theory of Hutton's, which was known as uniformitarianism, was important for three reasons. First, it implied that the earth has a long history. This was a new idea. Christian theologians, by counting the successive generations since Adam (as recorded in the Bible), had calculated the maximum age of the earth at about 6,000 years. No one, as far as we know, had ever thought in terms of a longer period. Yet, 6,000 years is far too short

I–3
While the Beagle *sailed up the west coast of South America, Darwin explored the Andes on foot and horseback. He saw geological strata such as these, discovered fossil sea shells at 12,000 feet, and was witness to the upheaval of the earth produced by a major earthquake that occurred while he was there. In 1846, he published a book on his geological observations in South America. Strata are now seen as pages in evolutionary history.*

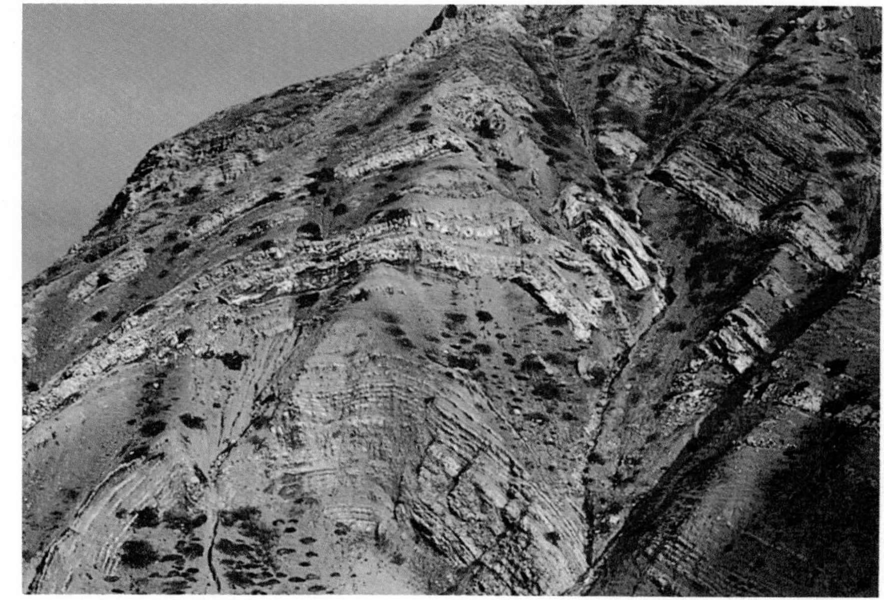

I–4
Particular strata, even though on different continents, have characteristic assemblages of fossils. These fossil gastropods from the Potomac River in Virginia are characteristic of the Paleocene, now dated at 53 to 65 million years ago.

for major evolutionary changes to take place, by any theory. Second, the theory of uniformitarianism stated that change is itself the normal course of events, as opposed to a static system interrupted by an occasional unusual event, such as an earthquake. Third, although this was never explicit, uniformitarianism suggested that there might be alternatives to the literal interpretation of the Bible.

The Fossil Record

During the latter part of the eighteenth century, there was a revival of interest in fossils. In previous centuries, fossils had been collected as curiosities, but they had generally been regarded either as accidents of nature—stones that somehow looked like shells—or as evidence of great catastrophes, such as Noah's Flood. The English surveyor William Smith (1769–1839) was among the first to study the distribution of fossils scientifically. Whenever his work took him down into a mine or along canals or cross-country, he carefully noted the order of the different layers of rock, which are called strata, and collected the fossils from each layer. He eventually established that each stratum, no matter where he came across it in England, contained characteristic kinds of fossils and that these fossils were actually the best way to identify a particular stratum in a number of different geographic locations. (The use of fossil distribution to identify strata is still widely practiced, for instance, by geologists looking for oil.) Smith did not interpret his findings, but the implication that the present surface of the earth had been formed layer by layer over the course of time was an unavoidable one.

Like Hutton's world, the world seen and described by William Smith was clearly a very ancient one. A revolution in geology was beginning; earth science was becoming a study of time and change rather than a mere cataloging of types of rocks. As a consequence, the history of the earth became inseparable from the history of living organisms, as revealed in the fossil record.

I–5

A drawing by Georges Cuvier of a mastodon. Although Cuvier was one of the world's experts in reconstructing extinct animals from their fossil remains, he was a powerful opponent of evolutionary theories.

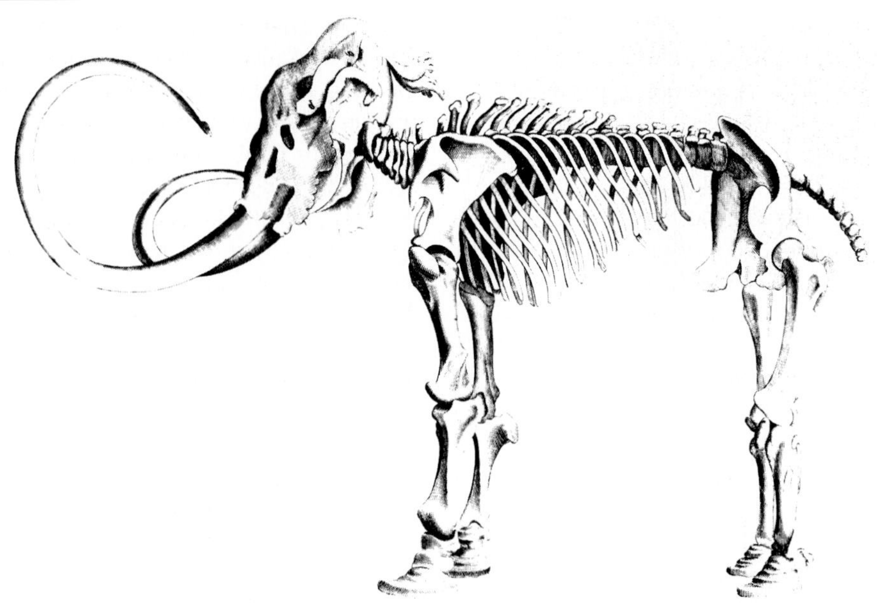

Catastrophism

Although the way to evolutionary theory was being prepared by the revolution in geology, the time was not yet ripe for a parallel revolution in biology. The dominating force in European science in the early nineteenth century was Georges Cuvier (1769–1832). Cuvier was the founder of vertebrate paleontology, the scientific study of the fossil record. An expert in anatomy and zoology, he applied his special knowledge of the way in which animals are constructed to the study of fossil animals and was able to make brilliant deductions about the form of an entire animal from a few fragments of bone. We think of paleontology and evolution as so closely connected that it is surprising to learn that Cuvier was a staunch and powerful opponent of evolutionary theories. He recognized the fact that many species that had once existed no longer did. (In fact, according to modern estimates, considerably less than one percent of all species that have ever lived are represented on the earth today.) Cuvier explained the extinction of species by postulating a series of catastrophes. After each catastrophe, the most recent of which was Noah's Flood, new species filled the vacancies.

Louis Agassiz (1807–1873), America's leading nineteenth-century biologist, was also a major opponent of evolution. According to Agassiz, the fossil record revealed 50 to 80 total extinctions of life and an equal number of separate creations.

The Theories of Lamarck

The first scientist to work out a systematic theory of evolution was Jean Baptiste Lamarck (1744–1829). "This justly celebrated naturalist," as Darwin himself referred to him, boldly proposed in 1801 that all species, including *Homo sapiens,* are descended from other species. Lamarck, unlike most of the other zoologists of his time, was particularly interested in one-celled organisms and invertebrates. Undoubtedly it was his long study of these forms of life that led him to think of living things in terms of constantly increasing complexity, each species derived from an earlier, less complex one.

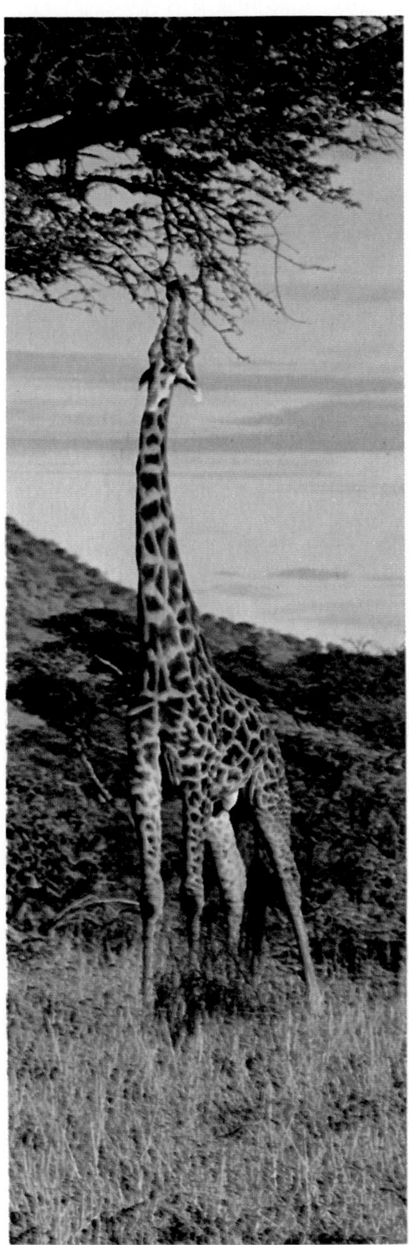

I–6
According to Lamarck's now-discredited hypothesis, the necks of giraffes became longer when they stretched to reach high branches, and this acquired characteristic was transmitted to their offspring.

Like Cuvier and others, Lamarck noted that older rocks generally contained fossils of simpler forms of life. Unlike Cuvier, however, Lamarck interpreted this as meaning that the higher forms had arisen from the simpler forms by a kind of progression. According to his hypothesis, this progression, or "evolution," to use the modern term, is dependent on two main forces. The first is the inheritance of acquired characteristics. Organs in animals become stronger or weaker, more or less important, through use or disuse, and these changes, according to Lamarck's theory, are transmitted from the parents to the progeny. His most famous example, and the one that Cuvier used most often to ridicule him, was the evolution of the giraffe. According to Lamarck, the modern giraffe evolved from ancestors that stretched their necks longer and longer to reach leaves on high branches. These ancestors transmitted their longer necks to their offspring, which stretched their necks even longer, and so on.

The second, equally important factor in Lamarck's theory of evolution was a universal creative principle, an unconscious striving upward on the *Scala Naturae* that moved every living creature toward greater complexity. Every amoeba was on its way to man. Some might get waylaid—the orangutan, for instance, had been diverted from its course by being caught in an unfavorable environment—but the will was always present. Life in its simplest forms was constantly emerging by spontaneous generation to fill the void left at the bottom of the ladder. In Lamarck's formulation, the ladder of nature of Aristotle had been transformed into a steadily ascending escalator powered by a universal will.

Lamarck's contemporaries did not object to his ideas about the inheritance of acquired characteristics, as we would today with our present knowledge of the mechanisms of heredity. Nor did they criticize his belief in a metaphysical force, which was actually a common element in many of the theories of the time. But these vague, untestable postulates provided a very shaky foundation for the radical proposal that "higher" forms evolved from simpler ones, and Lamarck personally was no match for the brilliant and witty Cuvier. As a result, Lamarck's career was ruined, and both scientists and the public became even less prepared to accept any evolutionary doctrine.

DEVELOPMENT OF DARWIN'S THEORY

The Earth Has a History

The person who most influenced Darwin, it is generally agreed, was Charles Lyell (1797–1875), a geologist who was Darwin's senior by 12 years. One of the books that Darwin took with him on his voyage was the first volume of Lyell's newly published *Principles of Geology,* and the second volume was sent to him while he was on the *Beagle.* On the basis of his own observations and those of his predecessors, Lyell opposed the theory of catastrophes. Instead, he produced new evidence in support of Hutton's earlier theory of uniformitarianism. According to Lyell, the slow, steady, and cumulative effect of natural forces had produced continuous change in the course of the earth's history. Since this process is demonstrably slow, its results being barely visible in a single lifetime, it must have been going on for a very long time. Lyell himself was an antievolutionist until 1862. However, what Darwin's theory needed was time, and it was time that Lyell gave him. In the words of Ernst Mayr of Harvard University, the discovery that the earth was ancient "was the snowball that started the whole avalanche."

The Voyage of the *Beagle*

This, then, was the intellectual climate in which Charles Darwin set sail from England. As the *Beagle* moved down the Atlantic coast of South America, through the Strait of Magellan, and up the Pacific coast, Darwin traveled the interior. He explored the rich fossil beds of South America (with the theories of Lyell fresh in his mind) and collected specimens of the many new kinds of plant and animal life he encountered. He was impressed most strongly during his long, slow trip down the coast and up again by the constantly changing varieties of organisms he saw. The birds and other animals on the west coast, for example, were very different from those on the east coast, and even as he moved slowly up the western coast, one species would give way to another.

Most interesting to Darwin were the animals and plants that inhabited a small, barren group of islands, the Galapagos, which lie some 950 kilometers off the coast of Ecuador. The Galapagos were named after the islands' most striking inhabitants, the tortoises (*galápagos* in Spanish), some of which weigh 100 kilograms or more. Each island has its own type of tortoise; sailors who took these tortoises on board and kept them as convenient sources of fresh meat on their sea voyages could readily tell which island any particular tortoise had come from. Then there was a group of finchlike birds, 13 species in all, that differed from one another in the sizes and shapes of their bodies and beaks, and particularly in the type of food they ate. In fact, although clearly finches, they had many characteristics seen only in completely different types of birds on the mainland. One finch, for example, feeds by routing insects out of the bark of trees. It is not fully equipped for this, however, lacking the long tongue with which the woodpecker flicks out insects from under the bark. Instead, the woodpecker finch uses a small stick or cactus spine to pry the insects loose.

From his knowledge of geology, Darwin knew that these islands, clearly of volcanic origin, were much younger than the mainland. Yet the plants and animals of the islands were different from those of the mainland, and in fact the inhabitants of different islands in the archipelago differed from one another. Were the living things on each island the product of a separate special creation? "One might really fancy," Darwin mused at a later date, "that from an original paucity of birds in this archipelago one species had been taken and modified for different ends." This problem continued, in his own word, to "haunt" him.

I–7

Cutaway view of the Beagle. *Only 28 meters in length, this "good little vessel" set sail on its five-year voyage with 74 people aboard. Darwin shared the poop cabin with a midshipman and 22 chronometers belonging to Captain Fitz Roy, who had a passion for exactness. His sleeping space was so confined that he had to remove a drawer from a locker to make room for his feet.*

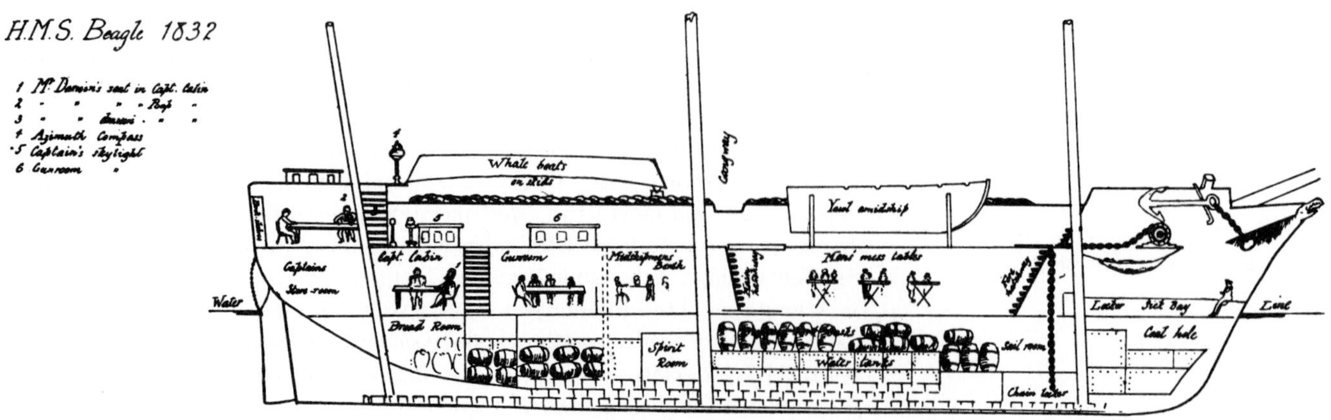

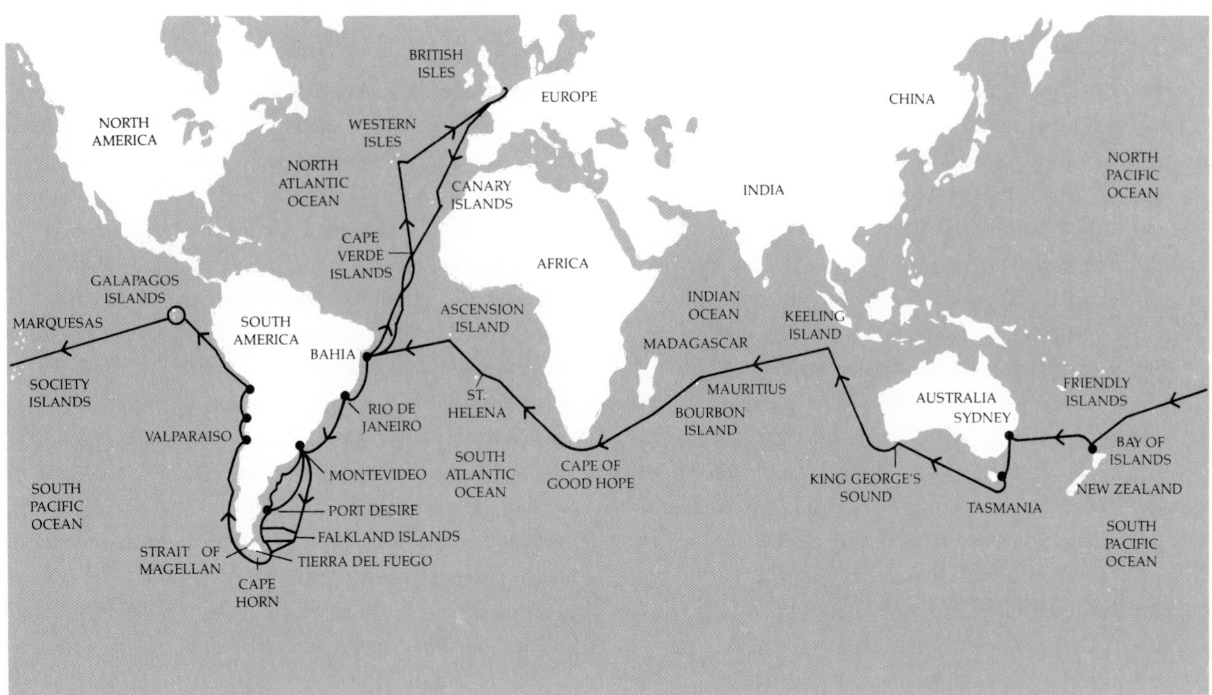

I–8
The Beagle's *voyage.*

The Darwinian Theory

Darwin was an assiduous and voracious reader. Not long after his return, he came across a short but much talked about sociological treatise by the Reverend Thomas Malthus, which first appeared in 1798. In this book, Malthus warned, as economists have warned frequently ever since, that the human population was increasing so rapidly that it would soon be impossible to feed all the earth's inhabitants. Darwin saw that Malthus's conclusion—that food supply and other factors hold populations in check—is true for all species, not just the human one. For example, Darwin calculated that a single breeding pair of elephants, which are among the slowest breeders of all animals, would, if all their progeny lived and reproduced the normal number of offspring over a normal life span, produce a standing population of 19 million elephants in 750 years, yet the average number of elephants generally remains the same over the years. So, although this single breeding pair could have, in time, produced 19 million elephants, it did, in fact, produce only two. But why these particular two? The process by which the two survivors are "chosen" Darwin called natural selection.

Natural selection, according to Darwin, was a process analogous to the type of selection exercised by breeders of cattle, horses, or dogs. In the case of artificial selection, we humans choose individual specimens of plants or animals for breeding on the basis of characteristics that seem to us to be desirable. In the case of natural selection, the environment takes the place of human choice. As individuals with certain hereditary characteristics survive and reproduce and individuals with other hereditary characteristics are eliminated, the population will slowly change. If some horses were swifter than others, for example, these individuals would be more likely to survive, and their progeny, in turn, might be swifter, and so on.

DARWIN'S LONG DELAY

Darwin returned to England with the Beagle *in 1836. Two years later, he read the essay by Malthus, and in 1842 he wrote a preliminary sketch of his theory, which he revised in 1844. On completing the revision, he wrote a formal letter to his wife requesting her, in the event of his death, to publish the manuscript (which was some 230 pages long). Then, with the manuscript and letter in safekeeping, he turned to other work, including a four-volume treatise on barnacles. For more than 20 years following his return from the Galapagos, Darwin mentioned his ideas on evolution only in his private notebooks and in letters to his scientific colleagues.*

In 1856, urged on by his friends Charles Lyell and botanist Joseph Hooker, Darwin set slowly to work preparing a manuscript for publication. In 1858, some 10 chapters later, Darwin received a letter from the Malay archipelago from another English naturalist, Alfred Russel Wallace, who had corresponded with Darwin on several previous occasions. Wallace presented a theory of evolution that exactly paralleled Darwin's own. Like Darwin, Wallace had traveled extensively and also had read Malthus's essay. Wallace, tossing in bed one night with a fever, had a sudden flash of insight. "Then I saw at once," Wallace recollected, "that the ever-present variability of all living things would furnish the material from which, by the mere weeding out of those less adapted to the actual conditions, the fittest alone would continue the race." Within two days, Wallace's 20-page manuscript was completed and in the mail.

When Darwin received Wallace's letter, he turned to his friends for advice, and Lyell and Hooker, taking matters into their own hands, presented the theory of Darwin and Wallace at a scientific meeting just one month later. (Darwin described Wallace as "noble and generous," as indeed he was.) Lyell and Hooker read four papers from Darwin's notes of 1844, excerpts from two letters written by Darwin, and Wallace's manuscript. Their presentation received little attention, but for Darwin the floodgates were opened. He finished his long treatise in little more than a year, and the book was finally published. The first printing was a mere 1,250 copies, but they were sold out the same day.

Why Darwin's long delay? His own writings, voluminous though they are, shed little light on this question. But perhaps his background does. He came from a conventionally devout family, and he himself had been a divinity student preparing for a career as "a country parson." Perhaps most important, his wife, to whom he was deeply devoted, was extremely religious. It is difficult to avoid the speculation that Darwin, as has been the case with others, found the implications of his theory difficult to confront.

Alfred Russel Wallace (1823–1913). As a young man, Wallace explored the Malay archipelago for eight years, covering 14,000 miles by foot and native canoe. During his stay there, he collected 125,000 specimens of plants and animals, many of them previously unknown. His book about his Malay travels bears this inscription: "To Charles Darwin, Author of 'The Origin of Species,' I dedicate this book, not only as a token of personal esteem and friendship but also to express my deep admiration for his genius and his works."

According to Darwin's theory, these variations among individuals, which occur in every natural population, are a matter of chance. They are not produced by the environment, by a "creative force," or by the unconscious striving of the organism. In themselves, they have no goal or direction. It is the operation of natural selection over a series of generations that gives direction to evolution. A variation that gives an animal even a slight advantage makes that animal more likely to leave surviving offspring. Thus, to return to Lamarck's giraffe, an animal with a slightly longer neck has an advantage in feeding and so is likely to leave more offspring than one with a shorter neck. If the longer neck is an inherited trait, some of these offspring will also have long necks, and if the long-necked animals in this generation have an advantage, the next generation will include more long-necked individuals. Finally, the population of short-necked giraffes will have become a population of longer-necked ones (although there will still be variations in neck length).

As you can see, the essential difference between Darwin's formulation and that of any of his predecessors is the central role he gave to variation. Others had thought of variations as mere disturbances in the overall design, whereas Darwin conceived of variations among individuals as the real fabric of the evolutionary process. Species arise, he proposed, when differences among individuals within a group are gradually converted into differences between groups as the groups become separated in space and time.

The Origin of Species, which Darwin pondered for more than 20 years after his return to England, is, in his own words, "one long argument." Fact after fact, observation after observation, culled from the most remote Pacific island to a neighbor's pasture, is recorded, analyzed, and commented upon. Every objection is weighed, anticipated, and countered. *The Origin of Species* was published on November 24, 1859, and the Western world has not been the same since.

Acceptance of Darwin's argument revolutionized the science of biology. "The theory of evolution," in the words of Ernst Mayr, "is quite rightly called the greatest unifying theory in biology." It is the thread that links together all the diverse phenomena of the living world. (This statement will be amply supported in the chapters that follow.)

It also deeply influenced our way of thinking about ourselves. With the possible exception of the storm that raged around Copernicus and Galileo, no revolution in scientific thought has had as much effect on human culture as this one. One reason is, of course, that evolution is in contradiction to the literal, fundamentalist interpretation of the Bible.

Another, perhaps more deep-seated, difficulty has been that it seems to diminish human significance. The new astronomy had made it clear that the earth is not the center of the universe or even our own solar system. Now the new biology asks us to accept the proposition that, like all other organisms, we too, as far as science can show, are not created for any special purpose or as part of any universal design.

CHALLENGES TO EVOLUTIONARY THEORY

Today, with almost no exceptions, modern biologists are convinced by the vast body of accumulated evidence that the earth has a long history and that all living organisms, including ourselves, arose in the course of this history from earlier, more primitive forms.

Yet, as everyone who reads a newspaper knows, evolutionary theory is once again a matter of lively controversy. Moreover, the advocates of special creation—which means simply the "theory" that each species was created separately—seek to lend strength to their arguments from the fact that scientists, too, are asking questions about evolution. They point out that, even among scientists, evolution is "only a theory," and that even leading scientists do not agree on this "theory."

How do we resolve this apparent paradox? As we shall see, much of the confusion surrounding this controversy stems from the phrase "theory of evolution," and, indeed, from the very definition of the word "theory," and from a misunderstanding of the scientific process.

The Nature of Science

Science, biological and other, is a way of seeking principles of order. Art is another, as are religion and philosophy. Science differs from these others in that it limits its search to the natural world, the physical universe. Also, and perhaps even more significant, it differs from them in the central value it gives to observation (particularly to that structured kind of observation called experimentation). Scientists begin their search by accumulating data and trying to fit these data into systems of order, conceptual schemes that organize the data in some meaningful way. These are not two steps, the accumulating and the ordering; they go on simultaneously. Or to put it another way, the accumulation of data is undertaken by scientists as a way of answering a question, or supporting or rejecting a hunch or hypothesis. (This is what Einstein meant when he said that it is the theory that decides what we can observe.) The data may be generated by deliberate, planned experiments or may be gleaned retrospectively from one's earlier systematic observation or from verifiable information recorded by others. (Darwin made copious use of all three.)

When a scientist has collected sufficient data to support a particular conclusion, he or she then reports the results to other scientists; today such a report usually takes place at a scientific meeting (such as the meeting at which Darwin's and Wallace's papers were read), or in a scientific publication *(The Origin of Species)*. If the data are sufficiently interesting or the conclusion important, the observations or experiments will be repeated in an attempt to confirm, deny, or extend them. Hence, scientists always also report the methods that they used as well as their results.

When a hypothesis has been sufficiently tested in this way, it is generally referred to as a theory. Thus, a theory in science has a somewhat different meaning from the word "theory" in common usage, in which "just a theory" carries with it the implication of a flight of fancy or an abstract notion, rather than a carefully formulated, well-tested proposition. Because the theory of evolution pervades all of modern biology, it has been tested and retested, directly and indirectly, for the past 125 years.

The Mendelian Challenge

The most important new evidence bearing on evolutionary theory has been in the field of genetics. The concept of the gene, set forth by Mendel but unknown to Darwin, made it possible to understand, as Darwin could not, how variations could originate, be preserved, and be transmitted from one generation to another. In fact, Mendelian genetics, because it implied complete fidelity in reproduction, seemed to some to make evolution virtually impossible. This problem was resolved by the

SOME COMMENTS ON SCIENCE AND SCIENTISTS

Science and Theory

On principle, it is quite wrong to try founding a theory on observable magnitudes alone. It is the theory which decides what we can observe.

A. Einstein, from J. Bernstein, "The Secrets of the Old Ones, II," New Yorker, *March 17, 1973.*

The Scientific Method

Indeed, scientists are in the position of a primitive tribe which has undertaken to duplicate the Empire State Building, room for room, without ever seeing the original building or even a photograph. Their own working plans, of necessity, are only a crude approximation of the real thing, conceived on the basis of miscellaneous reports volunteered by interested travelers and often in apparent conflict on points of detail. In order to start the building at all, some information must be ignored as erroneous or impossible, and the first constructions are little more than large grass shacks. Increasing sophistication, combined with methodical accumulation of data, make it necessary to tear down the earlier replicas (each time after violent arguments), replacing them successively with more up-to-date versions. We may easily doubt that the version current after only 300 years of effort is a very adequate restoration of the Empire State Building; yet, in the absence of clear knowledge to the contrary, the tribe must regard it as such (and ignore odd travelers' tales that cannot be made to fit).

E. J. DuPraw, Cell and Molecular Biology, *Academic Press, Inc., New York, 1968.*

Scientists at Work

Scientists at work have the look of creatures following genetic instructions; they seem to be under the influence of a deeply placed human instinct. They are, despite their efforts at dignity, rather like young animals engaged in savage play. When they are near to an answer their hair stands on end, they sweat, they are awash in their own adrenalin. To grab the answer, and grab it first, is for them a more powerful drive than feeding or breeding or protecting themselves against the elements.

It sometimes looks like a solitary activity, but it is as much the opposite of solitary as human behavior can be. There is nothing so social, so communal, so interdependent. An active field of science is like an immense intellectual anthill; the individual almost vanishes into the mass of minds tumbling over each other, carrying information from place to place, passing it around at the speed of light.

There are special kinds of information that seem to be chemotactic. As soon as a trace is released, receptors at the back of the neck are caused to tremble, there is a massive convergence of motile minds flying upwind on a gradient of surprise, crowding around the source. It is an infiltration of intellects, an inflammation.

There is nothing to touch the spectacle. In the midst of what seems a collective derangement of minds in total disorder, with bits of information being scattered about, torn to shreds, disintegrated, reconstituted, engulfed, in a kind of activity that seems as random and agitated as that of bees in a disturbed part of the hive, there suddenly emerges, with the purity of a slow phrase of music, a single new piece of truth about nature. . . .

There is something like aggression in the activity, but it differs from other forms of aggressive behavior in having no sort of destruction as the objective. While it is going on, it looks and feels like aggression: get at it, uncover it, bring it out, grab it, halloo! It is like a primitive running hunt, but there is nothing at the end of it to be injured. More probably, the end is a sigh. But then, if the air is right and the science is going well, the sigh is immediately interrupted, there is a yawping new question, and the wild, tumbling activity begins once more, out of control all over again.

Lewis Thomas, The Lives of a Cell: Notes of a Biology Watcher, *Viking Press, Inc., New York, 1974.*

I think we realized almost immediately that we had stumbled onto something important. According to Jim, I went into the Eagle, the pub across the road where we lunched every day, and told everyone that we'd discovered the secret of life. Of that I have no recollection, but I do recall going home and telling my wife Odile that we seemed to have made a big discovery. Years later she told me that she hadn't believed a word of it. "You were always coming home and saying things like that," she said, "so naturally I thought nothing of it."

"Crick Looks Back on DNA," Science, *vol. 206, page 667, November 9, 1979.*

I have long felt that biology ought to seem as exciting as a mystery story, for a mystery story is exactly what biology is.

Richard Dawkins, The Selfish Gene, *Oxford University Press, New York, 1976.*

THE EVOLUTIONARY PARADIGM

A paradigm is, essentially, a way of looking at the world. Because we are members of the western European culture, we view the world in a particular framework. We assume, for example, that the universe can be studied rationally. We assume that what we perceive by our senses is an accurate representation of the "real world." We are just beginning to abandon our long-held assumption that all other organisms on earth are placed here for our pleasure and benefit.

Science has its own paradigms, its basic assumptions. Atomic theory is a paradigm for physicists and chemists, as is cell theory for biologists. Similarly, the concept that the different species of organisms are related to one another and share a common ancestor—the theory of evolution—has become a paradigm of modern biology, although the exact details of how evolution occurred may be under question as are, for example, the details of subatomic structure. The greatest revolutions take place in science when one paradigm replaces another, as when astronomers established that the sun and not the earth is the center of our planetary system, or when the Newtonian view of the physical world was replaced by Einstein's theory of relativity. Once a voyager relinquishes the concept that the earth is flat, travel is never the same again.

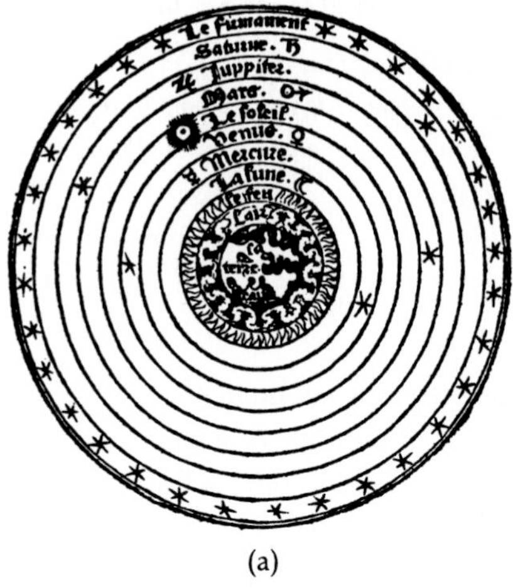

(a)

(a) *A view of the universe first proposed by the early Greeks and accepted throughout the Middle Ages. In this drawing, dated 1528, earth is in the center of the universe, surrounded by concentric circles representing spheres of air, fire, the moon, the sun, five planets, and fixed stars.* (b) *The solar system, as proposed by Nicholas Copernicus. In 1543, Copernicus set forth in* De Revolutionibus *the new concept that the*

concept of mutation—an abrupt, discontinuous genetic change. The discovery of the nature of mutations brought with it new challenges to the Darwinian theory: evolution was conceived, by some, as occurring in sudden mutational leaps, rather than by the slow accumulation of variation envisaged by Darwin. This controversy was resolved in the 1930s by the formulation of the synthetic ("neo-Darwinian") theory of evolution, so called because it offered a synthesis of what was known at the time about genetics and evolutionary theory. As we shall see in Section 7, the synthetic theory, which has dominated evolutionary thinking for half a century, has been enormously productive as a source of new experiments and new ideas.

The synthetic theory is now itself undergoing challenges, some of them stemming from new evidence arising from studies in molecular biology, some from a reinterpretation of the fossil record. One of the current controversies involves how large a role is played by chance in the direction of evolutionary change. Another controversy concerns the rate of major evolutionary change and whether or not natural selection operates on species as well as on individuals. Note that it is not evolution itself that is being questioned, it is the mechanism of how it occurred.

As you read the chapters that follow, you will see that evolutionary theory is not alone in being subjected to constant challenge and change. Atomic theory has gone

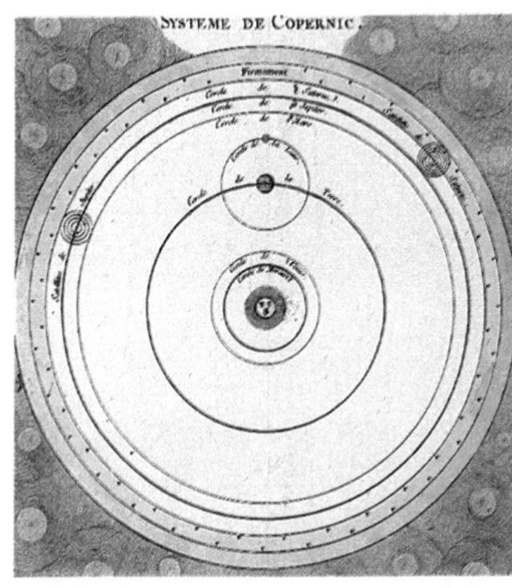

(b)

sun, not the earth, is the center of the solar system. His theory was supported by the German astronomer Johannes Kepler (1571–1630), who discovered the laws of planetary motion, and by the Italian Galileo Galilei (1564–1642). The latter spent the last ten years of his life confined to his home for heresy because of his advocacy of Copernican beliefs.

through even more drastic revisions, yet no one has interpreted these as meaning that scientists have rejected the concept of the atom. Similarly, developments in molecular genetics have been so rapid in the last decade that geneticists can no longer reach any agreement on a simple definition of a gene, yet no one denies such an entity exists.

All of this is evidence that biology as a science is alive and well and that scientists are doing what they are supposed to do—asking questions. In the words of Steven M. Stanley, one of the chief critics of the neo-Darwinian synthesis, "the theory of evolution is not just getting older, it is getting better."

SCIENCE AS PROCESS

You are fortunate to be studying biology at this time of controversy and rapid change. New ideas and unexpected discoveries have opened up exciting frontiers in many different areas of biology—genetics, evolution, ecology, and energetics, to mention just a few. Because there is so much to tell, most texts, and this one is no exception, tend to stress what is known at the present time, rather than what is not known or how we came to know what we do. This tendency, although understandable, distorts the nature of biology and, indeed, of science in general. Modern science is not a static accumulation of facts organized in a particular way but a process taking place in the minds of living scientists. In our enthusiasm for telling you all that biology has discovered, do not let us convince you that all is known. Many questions are still unanswered. More important, many good questions have not been asked. Perhaps you may be the one to ask them.

You may have been persuaded to study biology because of the environmental problems now confronting us or because of a desire to know more about the mechanisms of your own body or an interest in the "green revolution" or genetic engineering or a career in medicine—in short, because it is "relevant." The study of biology is, indeed, pertinent to many aspects of our day-to-day existence, but do not make this your main reason for the study of biology. Above all other considerations, study biology because it is "irrelevant"—that is, study it for its own sake, because, like art and music and literature, it is an adventure for the mind and nourishment for the spirit.

QUESTIONS

1. What is the essential difference between Darwin's theory of evolution and that of Lamarck?

2. The chief predator of an English species of snail is the song thrush. Snails that inhabit woodland floors have dark shells, whereas those that live on grass have yellow shells, which are less clearly visible against the lighter background. Explain, in terms of Darwinian principles.

3. The phrase "chance and necessity" has been used to describe the Darwinian theory of evolution. Relate this to the fact that snails living on grass do not have green shells, but there are, for example, green frogs and green insects.

4. When scientists report new findings, they are expected to reveal their methods and raw data as well as their results. Why is such reporting considered essential?

SUGGESTIONS FOR FURTHER READING

Books

BATES, MARSTON, and PHILIP S. HUMPHREY, eds.: *The Darwin Reader,* Charles Scribner's Sons, New York, 1956.*

A collection of Darwin's writings, including The Autobiography, *and excerpts from* The Voyage of the Beagle, The Origin of Species, The Descent of Man, *and* The Expression of the Emotions. *Darwin was a fine writer, and you can discover here the wide range of his interests and concerns at different periods of his life.*

BRONOWSKI, J.: *The Ascent of Man,* Little, Brown and Company, Boston, 1973.*

An informal and illuminating history of the sciences, originally prepared as a television series. The emphasis is on science's relation to human culture. Well designed and illustrated.

DARWIN, CHARLES: *The Origin of Species by Means of Natural Selection, or The Preservation of Favored Races in the Struggle for Life,* Doubleday & Company, Inc., Garden City, N.Y., 1960.*

Darwin's "long argument." Every student of biology should, at the very least, browse through this book to catch its special flavor and to begin to understand its extraordinary force.

DARWIN, CHARLES: *The Voyage of the Beagle,* Natural History Press, Garden City, N.Y., 1962.*

Darwin's own chronicle of the expedition on which he made the discoveries and observations that eventually led him to his theory of evolution. The sensitive, eager, young Darwin that emerges from these pages is very unlike the solemn image many of us have formed of him from his later portraits.

GHISELIN, MICHAEL T.: *The Triumph of the Darwinian Method,* University of California Press, Berkeley, Calif., 1969.*

A popular view of Darwin is that he set aside his work on evolution and distracted himself with less consequential matters, such as the classification of barnacles and the pollination of flowers. In this case study of Darwin's works, Ghiselin argues that all of Darwin's writings constitute a coherent body of work, all dealing with various aspects of his theory of evolution.

MOREHEAD, ALAN: *Darwin and the Beagle,* Harper & Row, Publishers, Inc., New York, 1969.*

A delightful narrative of Darwin's journey, beautifully illustrated with contemporary or near contemporary drawings, paintings, and lithographs.

TOULMIN, STEPHEN, and JUNE GOODFIELD: *The Discovery of Time,* Harper & Row, Publishers, Inc., New York, 1965.*

The historical development of our concepts of time as they relate to nature, human nature, and human society.

Articles

GODFREY, LAURIE R.: "The Flood of Antievolutionism," *Natural History,* June 1981, pages 4–10.

MAYR, ERNST: "Darwin and Natural Selection," *American Scientist,* vol. 65, pages 321–327, 1977.

MAYR, ERNST: "The Nature of the Darwinian Revolution," *Science,* vol. 176, pages 981–989, 1972.

---

* Available in paperback.

PART I

Biology of Cells

SECTION 1

The Unity of Life

CHAPTER 1

Atoms and Molecules

The universe began, according to current theory, with an explosion that filled all space, with every particle of matter hurled away from every other particle. The temperature at the time of the explosion—some 18 billion years ago—was about 100,000,000,000 degrees Celsius (10^{11} °C). At this temperature, not even atoms could hold together; all matter was in the form of subatomic, elementary particles. Moving at enormous velocities, even these particles had fleeting lives. Colliding with great force, they annihilated one another, creating new particles and releasing more energy.

As the universe cooled, two types of stable particles, previously present only in relatively small amounts, began to assemble. (By this time, 100,000 years after the "big bang" is believed to have taken place, the temperature had dropped to a mere 2500°C, about the temperature of a white-hot wire in an incandescent light bulb.) These particles—protons and neutrons—are very heavy as subatomic particles go. Held together by forces that are still incompletely understood, they formed the central cores, or nuclei, of atoms.

These nuclei, with their positively charged protons, attracted small, light, negatively charged particles—electrons—which moved rapidly around them. Thus, atoms came into being. (Perhaps atoms existed before the explosion, but how will one ever know?)

It is from these atoms—blown apart, formed, and re-formed over the course of some 18 billion years—that all the stars and planets of the universe are formed, including our particular star and planet. And it is from the atoms present on this planet that living systems assembled themselves and evolved. Each atom in our own bodies had its origin in this explosion 18 billion years ago. You and I are flesh and blood, but we are also stardust.

This text begins where life began, with the atom. At first, the universe aside, it might appear that lifeless atoms have little to do with biology. Bear with us, however. A closer look reveals that the activities we associate with being alive depend on combinations and exchanges between atoms, and the force that binds the electron to the atomic nucleus stores the energy that powers living systems.

1–1
The Serpens Nebula, shown here, is composed of clouds of gases and dust. In the small, dark regions of the photograph, new clusters of stars are forming from dense concentrations of gas. The stars and planets of the universe have their origins in nebulae such as this.

THE SIGNS OF LIFE

What do we mean when we speak of "the evolution of life," or "life on other planets," or "when life began"? Actually, there is no simple definition. Life does not exist in the abstract; there is no "life," only living things. Moreover, there is no single, simple way to draw a sharp line between the living and the nonliving. There are, however, certain properties that, taken together, distinguish animate (that is, living) objects from inanimate ones.

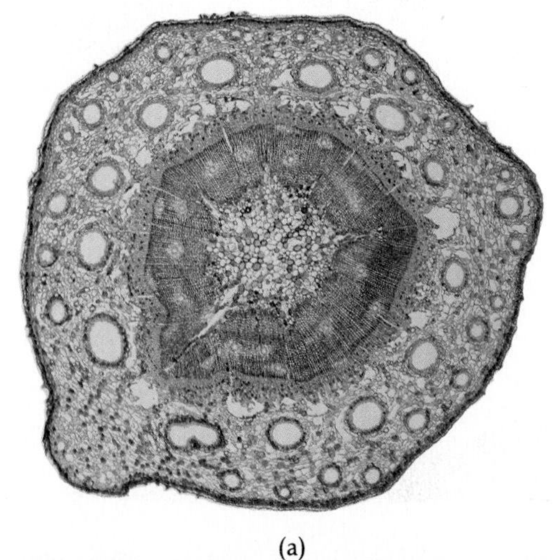

(a) *Living things are highly organized, as in this cross section of a one-year-old pine stem. It reflects the complicated organization of many different kinds of atoms into molecules and of molecules into complex structures. Such complexity of form is not found in inanimate objects.*

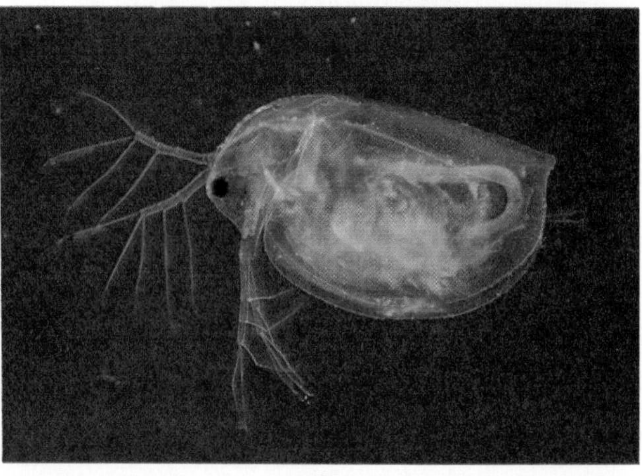

(b) *Living organisms are homeostatic, which means simply "staying the same." Although they constantly exchange materials with the outside world, they maintain a relatively stable internal environment quite unlike that of their surroundings. Even this tiny, apparently fragile animal, a water flea, has a chemical composition that differs from its changing environment.*

(c) *Living things reproduce themselves. They make more of themselves, generation after generation, with astonishing fidelity (and yet, as we shall see, with just enough variation to provide the raw material for evolution).*

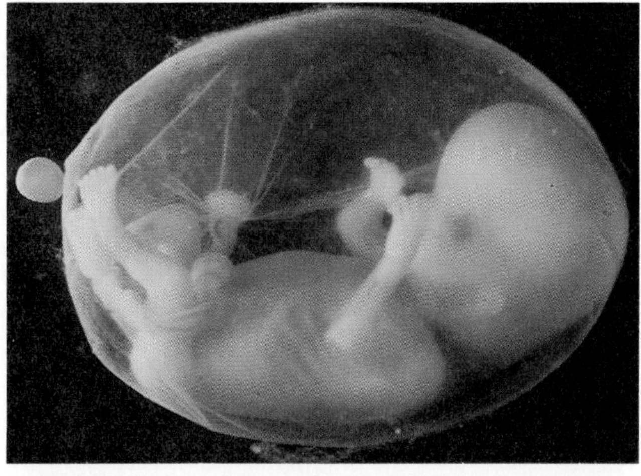
(d)

(e)

(f)

(g)

(d) Living organisms grow and develop. Growth and development are the processes by which, for example, a single living cell, the fertilized egg, becomes a tree, or an elephant, or as shown here, a human fetus.

(e) Living things take energy from the environment and change it from one form to another. They are highly specialized at energy conversion. This young female Alaskan brown bear has just converted chemical energy stored in her body to energy of motion used in catching a salmon. After she has eaten and digested the salmon, the chemical energy stored in its body will be available for her use.

(f) Living organisms respond to stimuli. For example, web-building spiders, such as this garden spider, are sensitive to the slightest vibrations of their webs. They can distinguish between vibrations caused by the wind and those caused by an intruder such as the grasshopper at the left. When the grasshopper became entangled in the web, the spider responded promptly, injecting it with venom and wrapping it in silk.

(g) Living things are adapted. Moles, for instance, live underground in tunnels shoveled out by their large forepaws. Their eyes are small and almost sightless. Their noses, with which they sense the worms and other small invertebrates that make up their diet, are fleshy and enlarged.

ATOMS

All matter, including the most complex living organisms, is made up of combinations of *elements,* which are substances that cannot be broken down by ordinary chemical means. The smallest particle of an element is an *atom.* There are 92 naturally occurring elements, each differing from the others in the structure of its atoms (Table 1-1).

Table 1-1 Atomic Structure of Some Familiar Elements

		NUCLEUS		
ELEMENT	SYMBOL	NUMBER OF PROTONS	NUMBER OF NEUTRONS*	NUMBER OF ELECTRONS
Hydrogen	H	1	0	1
Helium	He	2	2	2
Carbon	C	6	6	6
Nitrogen	N	7	7	7
Oxygen	O	8	8	8
Sodium	Na	11	12	11
Phosphorus	P	15	16	15
Sulfur	S	16	16	16
Chlorine	Cl	17	18	17
Calcium	Ca	20	20	20

* In most common isotope.

The atoms of each different element have a characteristic number of positively charged particles, called protons, in their nuclei. For example, an atom of hydrogen, the lightest of the elements, has 1 proton in its nucleus; an atom of the heaviest naturally occurring element, uranium, has 92 protons in its nucleus. The number of protons in the nucleus of a particular atom is called the *atomic number.*

Outside the nucleus of an atom are negatively charged particles, the electrons, which are attracted by the positive charge of the protons. The number of electrons in an atom equals the number of protons in its nucleus. The electrons determine the chemical properties of atoms, and chemical reactions involve changes in the numbers and energy of these electrons.

Atoms also contain neutrons, which are uncharged particles of about the same weight as protons. These, too, are found in the nucleus of the atom, where they seem to have a stabilizing effect. The *atomic weight* of an element is essentially equal to the number of protons plus neutrons in the nuclei of its atoms. (Electrons are so light by comparison that their weight is usually disregarded. When you weigh yourself, only about 30 grams—approximately 1 ounce—of your total weight is made up of electrons.)

Isotopes

All atoms of a particular element have the same number of protons in their nuclei. Sometimes, however, atoms of the same element contain different numbers of neutrons. These atoms, which differ from one another in their atomic weights but not in their atomic numbers, are known as *isotopes* of the element. For example, three isotopes of hydrogen exist (Table 1-2). The common form of hydrogen, with its one proton, has an atomic weight of 1 and is symbolized as ^{1}H, or simply H. A second isotope of hydrogen, known as deuterium, contains one proton and one neutron and so has an atomic weight of 2; this isotope is symbolized as ^{2}H. Tritium, ^{3}H, a third, extremely rare isotope, has one proton and two neutrons and so has an atomic weight of 3. The chemical behavior of the two heavier isotopes is essentially the same as that of ordinary hydrogen—all three isotopes have only one electron each, and it is the electrons that determine chemical properties.

Table 1-2 Isotopes of Hydrogen

ISOTOPE							
NAME	SYMBOL	ATOMIC NUMBER	ATOMIC WEIGHT	NUMBER OF PROTONS	NUMBER OF NEUTRONS	NUMBER OF ELECTRONS
Hydrogen	^{1}H	1	1	1	0	1
Deuterium	^{2}H	1	2	1	1	1
Tritium	^{3}H	1	3	1	2	1

Most elements have several isotopic forms. The differences in weight, although very small, are sufficiently great that they can be detected with modern laboratory apparatus. Moreover, many, but not all, of the less common isotopes are radioactive. This means that the nucleus of the atom is unstable and emits energy as it changes to a more stable form. The energy given off by a radioactive isotope can be detected with a Geiger counter or on photographic film.

Isotopes have a number of important uses in biological research and in medicine. They can be used, for example, to determine the age of fossils and of the rocks in which fossils are found (Figure 1-2). Each type of radioactive isotope emits energy and changes into another kind of isotope at a characteristic and fixed rate. As a result, the relative proportions of different isotopes in a rock sample give a good indication of how long ago that rock was formed. Another use of radioactive isotopes is as "tracers." Since isotopes of the same element all have the same chemical properties, a radioactive isotope will behave in an organism just as its more common nonradioactive isotope does. As a result, biologists have been able to use isotopes of a number of elements—especially carbon, hydrogen, and oxygen—to trace the course of many essential processes in living organisms.

1-2
The dating of fossils using radioactive isotopes has established that the oldest known direct ancestors of human beings lived about 3.5 million years ago. This fossil footprint, one of a trail of five found in Tanzania in east Africa, may have been left by one of those ancestors.

Models of Atomic Structure

The concept of the atom as the indivisible unit of the elements is almost 200 years old; however, our ideas about its structure have undergone many changes and may well undergo further changes in the future. These ideas, or hypotheses, are usually presented in the form of models, as are many other scientific hypotheses. (As we shall see, a hypothesis may be stated verbally, mathematically, or in the form of a model. Until it can be so stated, it is useless, scientifically speaking.)

The earliest model of the atom, emphasizing its indivisibility, portrayed the atom as a sphere like a billiard ball. When it was realized that electrons could be removed from the atom, the billiard-ball model gave way to the plum-pudding model, in which the atom was represented as a solid, positively charged mass with

1-3
Two models of the carbon atom: (a) *a planetary model and* (b) *a Bohr model.*

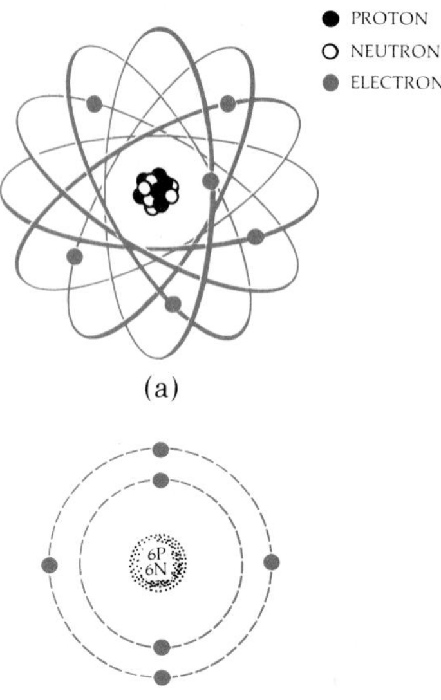

negatively charged particles, the electrons, embedded in it. Subsequently, however, physicists found that an atom is, in fact, mostly empty space. The distance from electron to nucleus, experiments indicated, is about 1,000 times greater than the diameter of the nucleus; the electrons are so exceedingly small that the space is almost entirely empty. Thus the more familiar planetary model of the atom came into being, in which the electrons were depicted as moving in orbits around the nucleus (Figure 1-3a). Later, the Bohr model (named after physicist Niels Bohr) became the most popular one (Figure 1-3b). It emphasized the fact that different electrons of an atom have different amounts of energy and are at different distances from the nucleus. As we shall see, neither the planetary model nor the Bohr model gives an accurate "picture" of an atom, and they have been superseded by another model, the orbital model (Figure 1-6, page 24). However, the Bohr model can help us to understand certain properties of atoms that are of great importance in the chemistry of living systems.

ELECTRONS AND ENERGY

The distance of an electron from the nucleus is determined by the amount of *potential energy* (often called "energy of position") the electron possesses. The greater the amount of energy possessed by the electron, the farther it will be from the nucleus. Thus, an electron with a relatively small amount of energy is found close to the nucleus and is said to be at a low *energy level;* an electron with more energy is farther from the nucleus, at a higher energy level.

An analogy may be useful. A boulder on flat ground may be said to have no energy. If you change its position by pushing it up a hill, you give it potential energy. As long as it sits on the peak of the hill, the rock neither gains nor loses energy. If it rolls down the hill, however, its potential energy is released (Figure 1-4a). Similarly, water that has been pumped up to a water tank for storage has potential energy that will be released when the water runs back down.

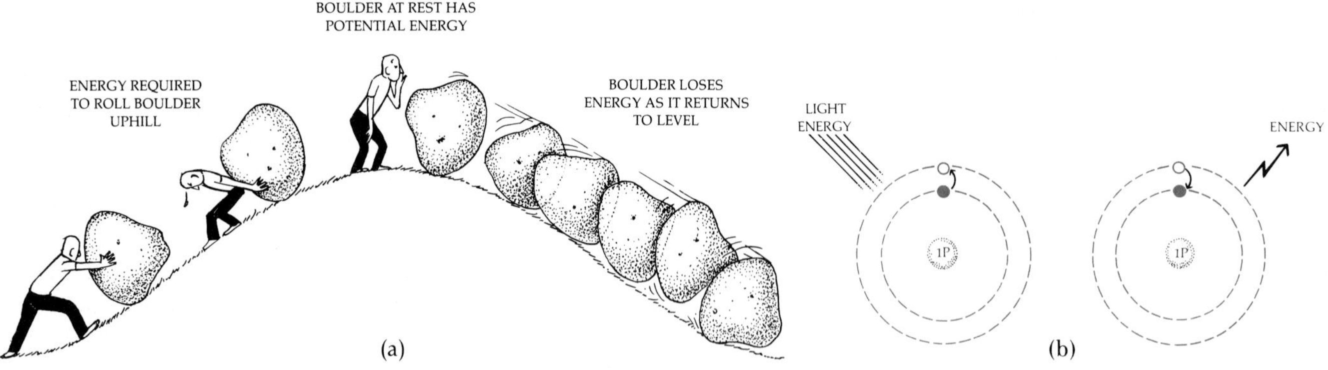

1-4
(a) *The energy used to push a boulder to the top of a hill (less the heat energy produced by friction between the boulder and hill) becomes potential energy, stored in the boulder as it rests at the top of the hill. This potential energy is converted to kinetic energy (energy of motion) as the boulder rolls downhill.*

(b) *When an atom, such as the hydrogen atom diagrammed here, receives an input of energy, an electron may be boosted to a higher energy level. The electron thus gains potential energy, which is released when the electron returns to its previous energy level.*

1–5
The leaves of these corn plants contain a pigment known as chlorophyll, which gives them their green color. When a packet of light energy—a photon—strikes a molecule of chlorophyll, electrons in the molecule are raised to higher energy levels. As each electron returns to its previous energy level, the energy released is captured in the bonds of carbon-containing compounds.

The electron is like the boulder, or the water, in that an input of energy can move it to a higher energy level—farther away from the nucleus. As long as it remains at the higher energy level, it possesses the added energy. And, just as the rock is likely to roll, and the water to run, downhill, the electron also tends to go to its lowest possible energy level.

It takes energy to move a negatively charged electron farther away from a positively charged nucleus, just as it takes energy to push a rock up a hill. However, unlike the rock on the hill, the electron cannot be pushed partway up. With an input of energy, an electron can move from a lower energy level to any one of several higher energy levels, but it cannot move to an energy state somewhere in between. For an electron to move from one energy level to a higher one, it must absorb a discrete amount of energy, equal to the difference between the two particular energy levels. When the electron returns to its original energy level, that same amount of energy is released (Figure 1–4b). The discrete amount of energy involved in the transition between two energy levels is known as a quantum. Thus the study of electron movements is known as quantum mechanics, and the term "quantum jump," which has invaded our everyday discourse, refers to an abrupt, discontinuous movement from one level to another.

In the green cells of plants and algae, the radiant energy of sunlight raises electrons to a higher energy level. In a series of reactions, which will be described in Chapter 10, these electrons are passed "downhill" from one energy level to another until they return to their original energy level. During these transitions, the radiant energy of sunlight is transformed into the chemical energy on which all life on earth depends.

The Arrangement of Electrons

At a given energy level, an electron moves around the nucleus at almost the speed of light. The electron is so small and moves so rapidly that it is theoretically impossible to determine, at any given moment, both its precise location and the exact amount of energy it possesses. As a result of this difficulty, the current model of atomic structure describes the pattern of the electron's motion rather than its position. The volume of space in which the electron will be found 90 percent of the time is defined as its *orbital*.

In any atom, the electrons at the lowest energy level—the first energy level—occupy a single spherical orbital, which can contain a maximum of two electrons (Figure 1–6a, on the next page). Thus, for instance, hydrogen's single electron moves about the nucleus—90 percent of the time—within this single spherical orbital. Similarly, the two electrons of helium (atomic number 2) move within the single spherical orbital of the first energy level.

Atoms of higher atomic number than helium have more than two electrons. Since the first energy level is filled, these additional electrons must occupy higher energy levels, farther from the nucleus. At the second energy level, there are four orbitals, each of which can hold a maximum of two electrons (Figure 1–6b). Thus, the second energy level can contain a total of eight electrons—and so can the third.

The way an atom reacts chemically is determined by the number and arrangement of its electrons. An atom is most stable when all of its electrons are at their lowest possible energy levels. Therefore, the electrons of an atom fill the energy levels in order—the first is filled before the second, the second before the third, and so on. Moreover, an atom in which the outermost energy level is completely filled

FIRST ENERGY LEVEL:

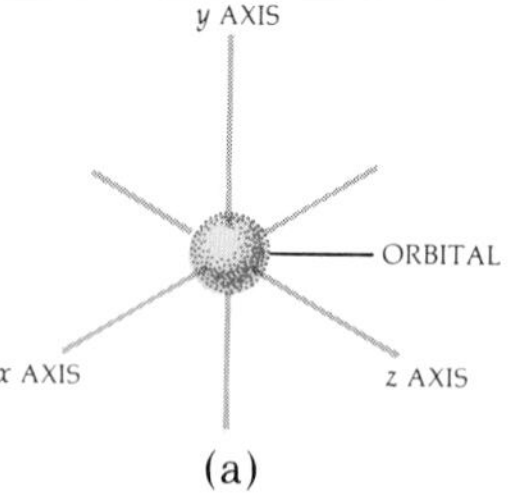

(a)

SECOND ENERGY LEVEL:

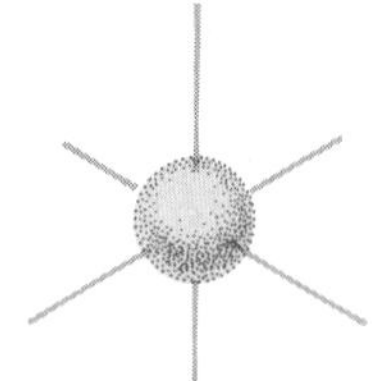
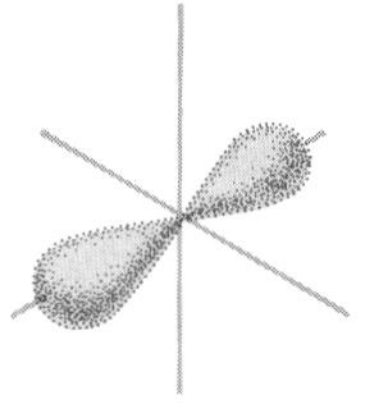
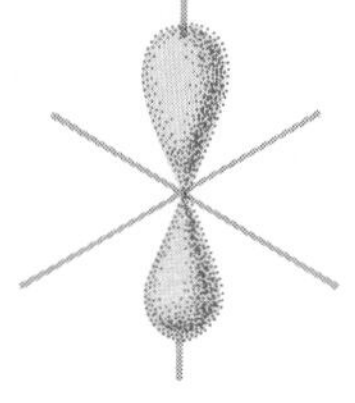
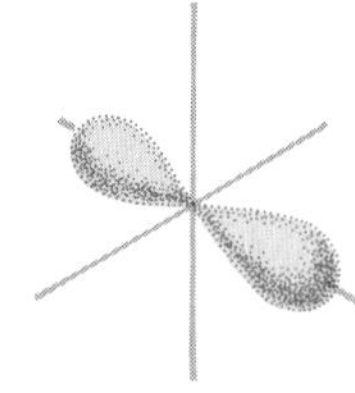

(b)

1-6
The most accurate representation of our current knowledge of atomic structure is provided by orbital models. (a) *The two electrons at the first energy level of an atom occupy a single spherical orbital. The nucleus is at the intersection of the axes (indicated by the gray lines).*

(b) *At the second energy level, there are four orbitals, each containing two electrons. One of these orbitals is spherical and the other three are dumbbell-shaped. The axes of the dumbbell-shaped orbitals are perpendicular to one another.*

For clarity, the orbitals are shown individually in this diagram. In reality, however, the spherical orbital of the second energy level surrounds the orbital of the first energy level, and portions of the dumbbell-shaped orbitals pass through the two spherical orbitals. The orbitals influence one another and determine the overall shape of the atom.

with electrons is more stable than one in which the outer energy level is only partially filled. For example, helium (atomic number 2) has two electrons at the first energy level, which means that its outer energy level (in this case, also its lowest energy level) is completely filled. Helium is therefore extremely stable and tends to be unreactive. Similarly, neon (atomic number 10) has two electrons at the first energy level and eight at the second energy level; both energy levels are completely filled, and neon is unreactive. Helium, neon, and argon (atomic number 18) are called the "noble" gases because of their disdain for reacting with other elements.

In the atoms of most elements, however, the outer energy level is only partially filled (Table 1-3). These atoms tend to interact with other atoms in such a way that after the reaction both atoms have completely filled outer energy levels. Some atoms lose electrons; others gain electrons; and, in many of the most important chemical reactions that occur in living systems, atoms share their electrons with each other.

Table 1-3 Electron Arrangements in Some Familiar Elements

ELEMENT	ATOMIC NUMBER	NUMBER OF ELECTRONS IN EACH ENERGY LEVEL*		
		FIRST	SECOND	THIRD
Hydrogen (H)	1	1	—	—
Helium (He)	2	2	—	—
Carbon (C)	6	2	4	—
Nitrogen (N)	7	2	5	—
Oxygen (O)	8	2	6	—
Neon (Ne)	10	2	8	—
Sodium (Na)	11	2	8	1
Phosphorus (P)	15	2	8	5
Sulfur (S)	16	2	8	6
Chlorine (Cl)	17	2	8	7

* The first energy level can hold a maximum of 2 electrons; the second energy level can hold a maximum of 8 electrons, as can the third energy level of the elements through calcium (atomic number 20). In elements of higher atomic number, the third energy level has additional orbitals that can hold a maximum of 10 more electrons.

BONDS AND MOLECULES

When atoms interact with one another, resulting in filled outer energy levels, new, larger particles are formed. These particles, consisting of two or more atoms, are known as *molecules*, and the forces that hold them together are known as *bonds*. There are two principal types of bonds: ionic and covalent.

Ionic Bonds

For many atoms, the simplest way to attain a completely filled outer energy level is either to gain or to lose one or two electrons. For example, chlorine (atomic number 17) needs one electron to complete its outer energy level (see Table 1–3). By contrast, sodium (atomic number 11) has a single electron in its outer energy level. This electron is strongly attracted by the chlorine atom and jumps from the sodium to the chlorine. As a result of this transfer, both atoms have completed outer energy levels and all the electrons are at the lowest possible energy levels. In the process, however, the original atoms have become electrically charged. Such charged atoms are known as *ions*. The chlorine atom, having accepted an electron from sodium, now has one more electron than proton and is a negatively charged chloride ion: Cl^-. Conversely, the sodium ion has one less electron than proton and is positively charged: Na^+.

Because of their charges, positive and negative ions attract one another. Thus the sodium ion (Na^+) with its single positive charge is attracted to the chloride ion (Cl^-) with its single negative charge. The resulting substance, sodium chloride (NaCl), is ordinary table salt (Figure 1–7). Similarly, when a calcium atom (atomic number 20) loses two electrons, the resulting calcium ion (Ca^{2+}) can attract and hold two Cl^- ions. Calcium chloride is identified in chemical shorthand as $CaCl_2$, with the subscript 2 indicating that two chloride ions are present for each ion of calcium.

Bonds that involve the mutual attraction of ions of opposite charge are known as *ionic bonds*. Such bonds can be quite strong, but, as we shall see in the next chapter,

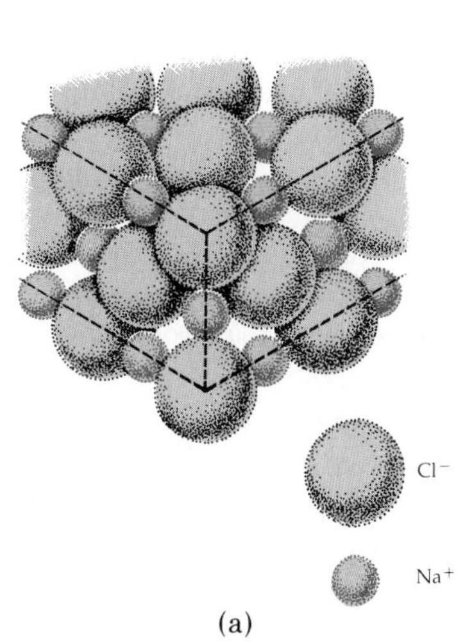

(a)

(b)

1–7
(a) *Oppositely charged ions, such as the sodium and chloride ions depicted here as spheres, attract one another. Table salt is crystalline NaCl, a latticework of alternating Na^+ and Cl^- ions held together by their opposite charges. Such bonds between oppositely charged ions are known as ionic bonds.*

(b) *The regularity of the latticework is reflected in the structure of salt crystals, magnified here about 14 times.*

1-8
This electron micrograph shows a nerve fiber resting in (and, at the right, pulled away from) a groove in a muscle fiber. The fibers are magnified about 1,900 times.

The nerve fiber transmits impulses to the muscle that cause it to contract. Sodium and potassium ions are involved in producing and propagating impulses along the nerve fiber, while calcium ions are required for the contraction of the muscle fiber.

many ionic substances break apart easily in water, producing free ions. Small ions such as Na^+ and Cl^- make up less than 1 percent of the weight of most living matter, but they play crucial roles. Potassium ion (K^+) is the principal positively charged ion in most organisms, and many essential biological processes occur only in its presence. Both Na^+ and K^+ are involved in the production and propagation of the nerve impulse. Calcium ion (Ca^{2+}) is required for the contraction of muscles, and magnesium ion (Mg^{2+}) forms a part of the chlorophyll molecule, the molecule in green plants and algae that traps radiant energy from the sun.

Covalent Bonds

Another way for an atom to complete its outer energy level is by sharing electrons with another atom. Bonds formed by shared pairs of electrons are known as *covalent bonds.* In a covalent bond, the shared pair of electrons forms a new orbital (called a molecular orbital) that envelops the nuclei of both atoms (Figure 1–9). In such a bond, each electron spends part of its time around one nucleus and part of its time around the other. Thus the sharing of electrons both completes the outer energy level and neutralizes the nuclear charge.

Atoms that need to gain electrons to achieve a filled, and therefore stable, outer energy level have a strong tendency to form covalent bonds. Thus, for example, a hydrogen atom forms a single covalent bond with another hydrogen atom. It can also form a covalent bond with any other atom that needs to gain an electron to complete its outer energy level.

Of extraordinary importance in living systems is the capacity of carbon atoms to form covalent bonds. A carbon atom has four electrons in its outer energy level (see Table 1–3). It can share each of those electrons with another atom, forming covalent bonds with as many as four other atoms (Figure 1–10). The covalent bonds formed by a carbon atom may be with different atoms (most frequently hydrogen, oxygen, and nitrogen) or with other carbon atoms. As we shall see in Chapter 3, this tendency of carbon atoms to form covalent bonds with other carbon atoms gives rise to the large molecules that form the structures of living organisms and that participate in essential life processes.

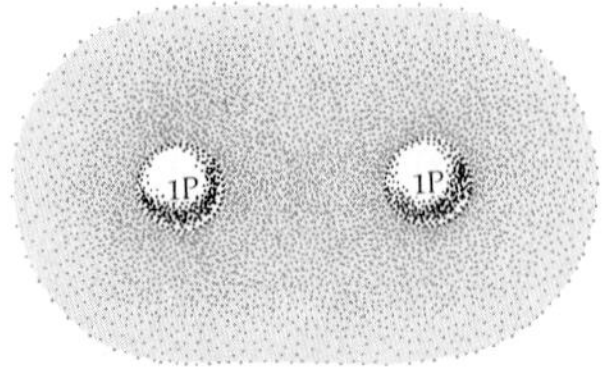

HYDROGEN MOLECULE (H_2)

1-9
In a molecule of hydrogen, each atom shares its single electron with the other atom. As a result, both atoms effectively have a filled first energy level, containing two electrons—a highly stable arrangement. This type of bond, in which electrons are shared, is known as a covalent bond.

1-10

(a) When a carbon atom forms covalent bonds with four other atoms, the electrons in its outer energy level form new orbitals. These new orbitals, which are all the same shape, are oriented toward the four corners of a tetrahedron. Thus the four orbitals are separated as far as possible. (b) When a carbon atom reacts with four hydrogen atoms, each of the electrons in its outer energy level forms a covalent bond with the single electron of one hydrogen atom, producing a methane molecule (c). Each pair of electrons moves in a new, molecular orbital. The molecule has the shape of a tetrahedron.

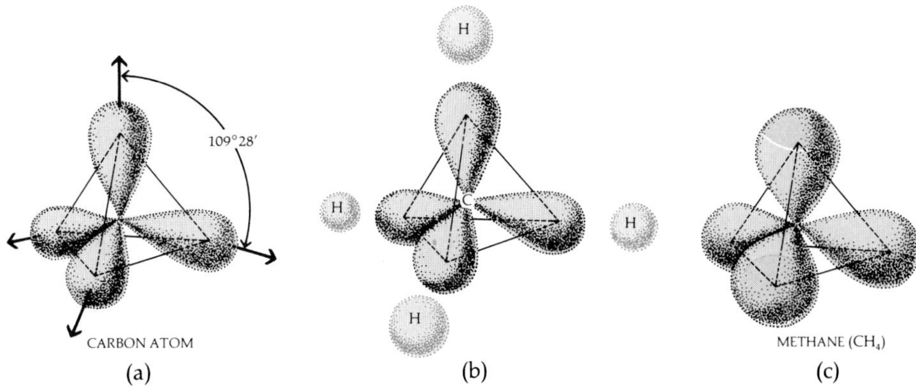

Polar Covalent Bonds

The atomic nuclei of different elements have different degrees of attraction for electrons. Factors that determine the strength with which a nucleus attracts its outer electrons include the number of protons it contains, the closeness of the outer electrons to the nucleus, and the number of other, "shielding" electrons between the nucleus and the outer electrons. In covalent bonds formed between atoms of different elements, the electrons are not shared equally between the atoms involved—instead, the shared electrons tend to spend more time around the nucleus with the greater attraction. The atom around which the electrons spend more time has a slightly negative charge, whereas the other atom has a slightly positive charge, since its nuclear charge is not entirely neutralized (Figure 1-11).

Covalent bonds in which electrons are shared unequally are known as *polar covalent bonds,* and the molecules containing these bonds are said to be *polar molecules.* Such molecules often contain oxygen atoms, to which electrons are strongly attracted. The polar properties of many oxygen-containing molecules have very important consequences for living things. For example, many of the special properties of water (H_2O), upon which life depends, derive largely from its polar nature, as we shall see in the next chapter.

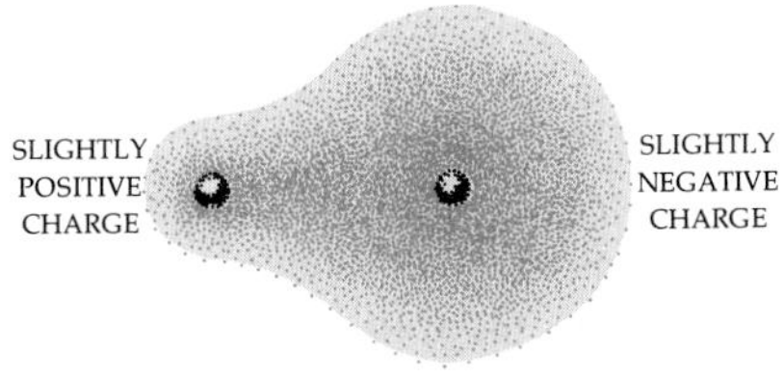

1-11

In a polar molecule, such as hydrogen chloride (HCl), the shared electrons tend to spend more time around one atom, in this case, the chlorine atom, than around the other atom. As a result, the atom that attracts the electrons more strongly (chlorine) has a slightly negative charge, and the atom that attracts the electrons less strongly (hydrogen) has a slightly positive charge.

Ionic, polar covalent, and covalent bonds actually may be considered different versions of the same type of bond. The differences depend on the differing attractions of the combining atoms for electrons. In a wholly nonpolar covalent bond, the electrons are shared equally; such bonds can exist only between identical atoms, H_2, Cl_2, O_2, and N_2, for example. In polar covalent bonds, electrons are shared unequally, and in ionic bonds, there is an electrostatic attraction between the negatively and positively charged ions as a result of their having previously gained or lost electrons.

Double and Triple Bonds

There are various ways in which atoms can participate in covalent bonds and fill their outer energy levels. Oxygen, for example, has six electrons in its outer energy level. Four of these electrons are grouped into two pairs and are generally unavailable for covalent bonding; the other two electrons are unpaired, and each can be shared with another atom in a covalent bond. In the water molecule (H_2O), one of these electrons participates in a covalent bond with one hydrogen atom, and the other in a covalent bond with a different hydrogen atom. Two single bonds are formed, and all three atoms have filled outer energy levels.

The bonding situation is different in another familiar substance, carbon dioxide (CO_2). In this molecule, the two available electrons of each oxygen atom participate with two electrons of a *single* carbon atom in the formation of *two* covalent bonds. Each oxygen atom is joined to the central carbon atom by two pairs of electrons (four electrons). Such bonds are called *double bonds*, and they are symbolized in a structural formula by two lines connecting the atomic symbols: O═C═O. Carbon atoms can form double or even triple bonds (in which three pairs of electrons are shared) with each other as well as with other atoms, and so the variety of kinds of molecules that carbon can form is very large.

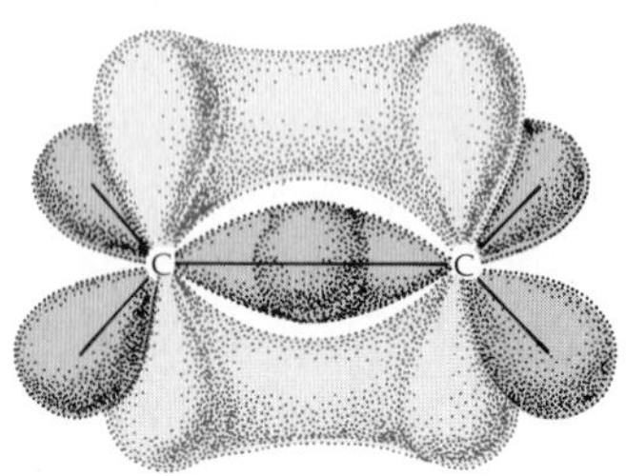

1–12
An orbital model of a carbon-carbon double bond. One pair of electrons occupies the inner orbital between the two carbon atoms. The other pair of electrons occupies the outer orbital, which has two phases, one above the plane and one below. This creates a rigid bond, about which the atoms cannot rotate. The two smaller orbitals extending from each carbon atom each contain one electron and can form covalent bonds with other atoms.

Electrons shared in double and triple bonds form orbitals that differ in shape from the orbitals filled by single electron pairs. For instance, when four single bonds satisfy the electron requirements of carbon, they will be directed toward the four corners of a tetrahedron that has the carbon atom at its center, as we saw in Figure 1–10. When two bonds are replaced by a double bond, the remaining single bonds form the arms of a Y with the double bond as its leg (Figure 1–12). When two double bonds are made by a single carbon atom, as in carbon dioxide, the three bonded atoms lie in a straight line.

Single bonds are flexible, leaving atoms free to rotate in relation to one another. Double and triple bonds hold the atoms relatively rigid in relation to one another. The presence of double bonds in a molecule can make a significant difference in its properties. For example, both fats and oils are composed of carbon and hydrogen atoms covalently bonded together, but in fats the bonds are all single and in oils some of the bonds are double. The rigidity caused by these double bonds prevents the oil molecules from packing together, and, as a consequence, oils are liquids at room temperature. In fats, by contrast, the molecules can bend and twist, fitting closely together in a solid structure at room temperature.

CHEMICAL REACTIONS

Chemical reactions—exchanges of electrons among atoms—can be compactly described by chemical equations. For example, the equation for the formation of sodium chloride is

$$Na^+ + Cl^- \longrightarrow NaCl$$

The arrow in the equation designates "forms" or "yields" and shows the direction of chemical change. Like algebraic equations, chemical equations "balance"; that is, there are the same number of atoms in the products of the reaction as in the original reactants. To take a slightly more complex example, hydrogen gas can combine with oxygen gas to produce water. Hydrogen gas is H_2, and oxygen gas is O_2. However, we know that each molecule of water contains two atoms of hydrogen and one of oxygen, and therefore the proportions must be two to one:

$$2H_2 + O_2 \longrightarrow 2H_2O$$

Two molecules of H_2 plus one molecule of O_2 yield two molecules of water. The equation for a chemical reaction thus tells the kinds of atoms that are present, their proportions, and the direction of the reaction.

LEVELS OF ORGANIZATION

Notice that by beginning the study of biology with an examination of how atoms and molecules are put together and how they interact, we have made a very fundamental assumption. We have said, in effect, that living things are made up of the same chemical and physical components as nonliving things and that they obey the same chemical and physical laws. Yet, as we have seen, there are recognizable differences between living and nonliving systems. What then is the basis of these differences?

A proton is a positively charged particle, an electron a negatively charged particle. Put them together and you have a hydrogen atom, an entity with quite different properties from those of an electron or a proton. Combine hydrogen atoms with each other in hydrogen molecules, and they form a colorless, highly flammable gas. Combine hydrogen atoms with the atoms of oxygen (whose molecules form a different colorless gas), and liquid water is produced—not a gas at all and with very different and remarkable properties, not a bit like those of isolated electrons and protons or of elemental hydrogen and oxygen. Take just a few more kinds of atoms—carbon, nitrogen, a little sulfur—and put them together in a certain way, and you have the contractile machinery of a muscle cell, the basis for all the voluntary movements of our bodies. Add a little phosphorus, combine the atoms in a different way, and you can spell out the genetic code. Put together enough of these atoms in the right way, enclose them in a membrane (made up of the same kinds of atoms), and they form a living cell.

The key to these enormous qualitative differences, as you have guessed by now, is organization. Each of the states just described represents another level of organization. The characteristics of each different level are not simply a combination of the original characteristics of its components but are, in fact, totally different. At each new level of organization, new and different properties emerge.

The cell is the level of organization at which life can unarguably be said to appear as a new, emergent property. Organize cells in a particular way and a structure with new, extraordinary properties can take shape—a brain, for instance. The human brain represents the highest degree of organizational complexity on this planet. Yet it is, in turn, only a part of a larger entity whose characteristics are different from those of the brain, although they depend upon them.

Nor is the living organism the ultimate level of biological order. One would know comparatively little, for example, about bees had one only a single bee to study. Or, as the primatologist Robert Yerkes once said, "One chimpanzee is no chimpanzee." One of the exciting new developments in biology is the discovery of some underlying principles governing the cohesion of social groups and the behavior of individual animals toward one another (sociobiology)—in other words, the organization of societies.

Finally, groups of living organisms are themselves part of an even vaster system of organization involving not only the great diversity of plants and animals and microorganisms and their interactions with each other but also the physical characteristics of the environment and, ultimately, the planet earth itself.

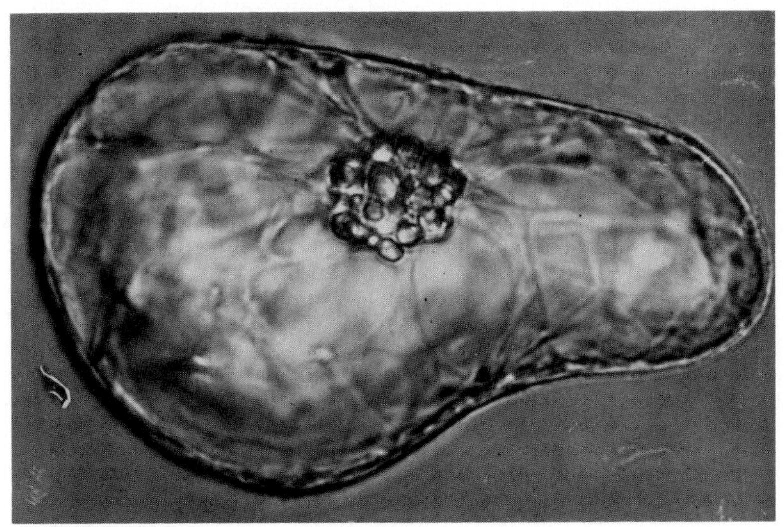

A single living cell from a pea plant. Made up mostly (more than 95 percent) of only four kinds of atoms, it is nevertheless capable of carrying out a wide variety of intricate chemical reactions. This cell's capacity to exhibit the properties, such as energy conversion, homeostasis, reproduction, and growth, that are characteristic of living organisms depends on the complex organization of its constituent parts.

Types of Reactions

The multitude of chemical reactions that occur in both the animate and the inanimate worlds can be classified into a few general types. One type can be represented by the expression:

$$A + B \longrightarrow AB$$

Examples of this sort of reaction are the combination of sodium ions and chloride ions to form sodium chloride and the combination of hydrogen gas with oxygen gas to produce water.

A reaction may also take the form of a dissociation:

$$AB \longrightarrow A + B$$

For example, the earlier equation showing the formation of water can be reversed:

$$2H_2O \longrightarrow 2H_2 + O_2$$

This means that water molecules yield hydrogen and oxygen gases.

A reaction may also involve an exchange, taking the form:

$$AB + CD \longrightarrow AD + CB$$

An example of such a reaction is the combination of sodium hydroxide (NaOH) and hydrochloric acid (HCl) to make table salt and water:

$$NaOH + HCl \longrightarrow NaCl + H_2O$$

As we pursue our study of living organisms, we shall encounter numerous examples of these three general types of chemical reactions.

THE BIOLOGICALLY IMPORTANT ELEMENTS

Of the 92 naturally occurring elements, only six make up some 99 percent of all living tissue (Table 1-4). These six elements are carbon, hydrogen, nitrogen, oxygen, phosphorus, and sulfur, conveniently remembered as CHNOPS. These are not the most abundant of the elements of the earth's surface. Why, as life assembled and evolved from stardust, were these of such importance? One clue is that the atoms of all of these elements need to gain electrons to complete their outer energy levels (see Table 1-3). Thus they generally form covalent bonds.

Table 1-4 Atomic Composition of Three Representative Organisms

ELEMENT	HUMAN	ALFALFA	BACTERIUM
Carbon	19.37%	11.34%	12.14%
Hydrogen	9.31	8.72	9.94
Nitrogen	5.14	0.83	3.04
Oxygen	62.81	77.90	73.68
Phosphorus	0.63	0.71	0.60
Sulfur	0.64	0.10	0.32
CHNOPS total:	97.90%	99.60%	99.72%

Because these atoms are small, the shared electrons in the bonds are held closely to the nuclei, producing very stable molecules. Moreover, with the exception of hydrogen, atoms of these elements can all form bonds with two or more atoms, making possible the formation of the large and complex molecules essential for the structures and functions of living systems.

SUMMARY

Matter is composed of atoms, the smallest units of chemical elements. Atoms are made up of smaller particles. The nucleus of an atom contains positively charged protons and (except for hydrogen, ^{1}H) neutrons, which have no charge. The atomic number of an atom is equal to the number of protons in its nucleus. The atomic weight of an atom is the sum of the number of protons and neutrons in its nucleus. The chemical properties of an atom are determined by its electrons—small, negatively charged particles found outside the nucleus. The number of electrons in an atom equals the number of protons and thus the atomic number.

The nuclei of different isotopes of the same element contain the same number of protons but different numbers of neutrons. Thus the isotopes of an element have the same atomic number but different atomic weights.

The electrons of an atom have differing amounts of energy. Electrons closer to the nucleus have less energy than those farther from the nucleus and thus are at a lower energy level. An electron tends to move to the lowest unoccupied energy level, but with an input of energy, it can be boosted to a higher energy level. When the electron returns to a lower energy level, energy is released.

The chemical behavior of an atom is determined by the number and arrangement of its electrons. An atom is most stable when all of its electrons are at their lowest possible energy levels and those energy levels are completely filled with electrons. The first energy level can hold two electrons, and the second and third energy levels can each hold eight electrons. Chemical reactions between atoms result from the tendency of atoms to reach the most stable electron arrangement possible.

Particles consisting of two or more atoms are known as molecules, which are held together by chemical bonds. Two common types of bonds are ionic and covalent. Ionic bonds are formed by the mutual attraction of particles of opposite electric charge; such particles, formed when an electron jumps from one atom to another, are known as ions. In covalent bonds, pairs of electrons are shared between atoms; in some covalent bonds, known as polar covalent bonds, pairs of electrons are shared unequally, giving the molecule regions of positive and negative charge. Covalent bonds in which two atoms share two pairs of electrons (four electrons) are known as double bonds, and those in which they share three pairs of electrons (six electrons) are known as triple bonds.

Chemical reactions—exchanges of electrons among atoms—can be represented by chemical equations. Three general types of chemical reactions are (1) the combination of two or more substances to form a different substance, (2) the dissociation of a substance into two or more substances, and (3) the mutual exchange of atoms among two or more substances.

Six elements (CHNOPS) make up 99 percent of all living matter. The atoms of all of these elements are small and form tight, stable covalent bonds. With the exception of hydrogen, they can all form covalent bonds with two or more atoms, giving rise to the complex molecules that characterize living systems.

QUESTIONS

1. Describe the three types of particles of which atoms are composed. What is the atomic number of an atom? The atomic weight?

2. For each of the following isotopes, determine the number of protons and neutrons in the nucleus: (a) ^{11}C, ^{12}C, ^{14}C; (b) ^{31}P, ^{32}P, ^{33}P; (c) ^{32}S, ^{35}S, ^{38}S.

3. Although no model of the atom gives us an exact "picture," different models can help us to understand important characteristics of atoms. What characteristic of the atom was stressed by the planetary model? What important characteristic of electrons is emphasized by the Bohr model? What additional information about electrons is provided by the orbital model?

4. The street lights in many cities contain bulbs filled with sodium vapor. When electricity is passed through the bulb, a brilliant yellow light is given off. What is happening to the sodium atoms to cause this?

5. What is the difference between an energy level and an orbital? How many electrons can the first energy level of an atom hold? The second energy level? The third energy level?

6. Determine the number of protons, the number of neutrons, the number of energy levels, and the number of electrons in the outermost energy level in each of the following atoms: oxygen, nitrogen, carbon, sulfur, phosphorus, chlorine, potassium, and calcium.

7. How many electrons does each of the atoms in Question 6 need to share, gain, or lose to acquire a completed outer energy level?

8. Magnesium has an atomic number of 12. How many electrons are in its first energy level? Its second energy level? Its third energy level? How would you expect magnesium and chlorine to interact? Write the formula for magnesium chloride.

9. Explain the differences between ionic, covalent, and polar covalent bonds. What tendency of atoms causes them to interact with each other, forming bonds?

10. Knowing that chemical reactions have to be balanced, fill the appropriate numbers into the underlined spaces (*hint:* from 1 to 6 in all cases):

(a) $\_\_H_2CO_3 \longrightarrow \_\_H_2O + \_\_CO_2$
CARBONIC ACID

(b) $\_\_H_2 + \_\_N_2 \longrightarrow \_\_NH_3$
AMMONIA

(c) $\_\_NaOH + \_\_H_2CO_3 \longrightarrow \_\_Na_2CO_3 + \_\_H_2O$
SODIUM HYDROXIDE

(d) $\_\_CH_3OH + \_\_O_2 \longrightarrow \_\_CO_2 + \_\_H_2O$
METHYL ALCOHOL

(e) $\_\_O_2 + \_\_C_6H_{12}O_6 \longrightarrow \_\_H_2O + \_\_CO_2$
SUGAR

11. What six elements make up the bulk of living tissue? What characteristics do the atoms of these six elements share?

CHAPTER 2

Water

In this chapter and the next, we are going to examine the molecules of which living things are composed. By far the most abundant of these molecules is water, which makes up 50 to 95 percent of the weight of any functioning living system.

Life on this planet began in water, and today, wherever liquid water is found, life is also present. There are one-celled organisms that eke out their entire existence in no more water than can cling to a grain of sand. Some kinds of algae are found only on the melting undersurfaces of polar ice floes. Certain bacteria can tolerate the near-boiling water of hot springs. In the desert, plants race through an entire life cycle—seed to flower to seed—following a single rainfall. In the rain forest, the water cupped in the leaves of a tropical plant forms a microcosm in which a myriad of small organisms are born, spawn, and die.

Water is the most common liquid on earth. Three-fourths of the surface of the earth is covered by water. In fact, if the earth's land surface were absolutely smooth, all of it would be 2.5 kilometers* under water. But do not mistake "common" for "ordinary"; water is not in the least an ordinary liquid. Compared with other liquids it is, in fact, quite extraordinary. If it were not, it is unlikely that life on earth could ever have evolved.

---

* About 1.5 miles. A metric table with English equivalents is provided in Appendix A.

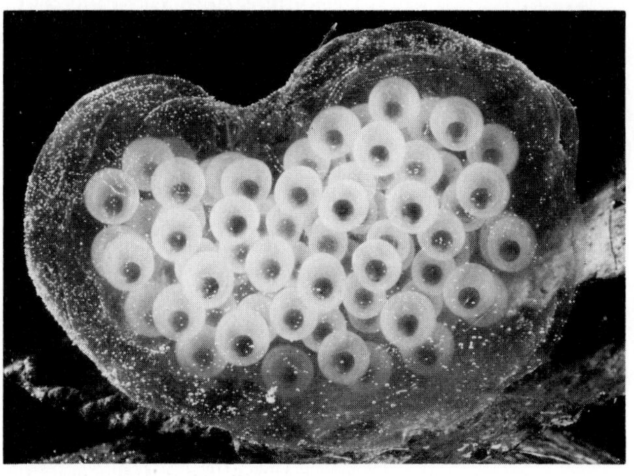

2–1
The first living systems came into being, according to present hypotheses, in the warm primitive seas, and for many organisms, ourselves included, each new individual begins life bathed and cradled in water. These are frog eggs, encased in a jellylike sac.

WATER AND THE HYDROGEN BOND

In order to understand why water is so extraordinary and how, as a consequence, it can play its unique and crucial role in relation to living systems, we must look again at its molecular structure. Each water molecule is made up of two atoms of hydrogen and one atom of oxygen (Figure 2–2). Each of the hydrogen atoms is linked to the oxygen atom by a covalent bond; that is, the single electron of each hydrogen atom is shared with the oxygen atom, which also contributes an electron to each bond.

The water molecule as a whole is neutral in charge, having an equal number of electrons and protons. However, the molecule is polar (page 27). Because of the very strong attraction of the oxygen nucleus for electrons, the shared electrons of the covalent bonds spend more time around the oxygen nucleus than they do around the hydrogen nuclei. As a consequence, the region near each hydrogen nucleus is a weakly positive zone. Moreover, the oxygen atom has four additional electrons in its outer energy level. These electrons are paired in two orbitals that are not involved in covalent bonding to hydrogen. Each of these orbitals is a weakly negative zone. Thus, the water molecule, in terms of its polarity, is four-cornered, with two positively charged "corners" and two negatively charged ones (Figure 2–3a).

When one of these charged regions comes close to an oppositely charged region of another water molecule, the force of attraction forms a bond between them, which is known as a *hydrogen bond*. Hydrogen bonds are not found only in water. A hydrogen bond can form between any hydrogen atom that is covalently bonded to an atom that has a strong attraction for electrons—usually oxygen or nitrogen—and the oxygen or nitrogen atom of another molecule. In water, a hydrogen bond forms between a negative "corner" of one water molecule and a positive "corner" of another. Every water molecule can establish hydrogen bonds with four other water molecules (Figure 2–3b).

Any single hydrogen bond is relatively weak and has an exceedingly short lifetime; on an average, each hydrogen bond in liquid water lasts approximately 1/100,000,000,000th of a second. But, as one is broken, another is made. All together, the hydrogen bonds have considerable strength, causing the water molecules to cling together under ordinary conditions of temperature and pressure.

Now let us look at some of the consequences of these attractions among water molecules, especially as they affect living systems.

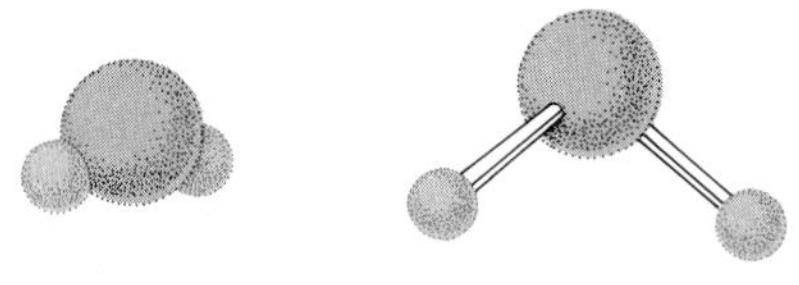

(a) (b)

2–2
The structure of the water molecule (H_2O) can be depicted in several different ways. (a) *In the space-filling model, the oxygen atom is represented by the gray sphere and the hydrogen atoms by the colored spheres. Because of its simplicity, this model is often used as a convenient symbol of the water molecule.*

(b) *The ball-and-stick model emphasizes that the atoms are joined by covalent bonds; it also gives some indication of the geometry of the molecule. A more accurate description of the molecule's shape is provided by the orbital model in Figure 2–3a.*

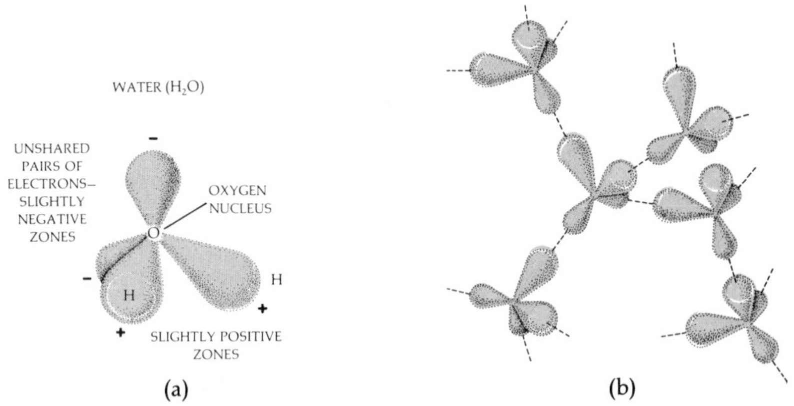

(a) (b)

2–3
The polarity of the water molecule and its consequences. (a) *As shown in this model, four orbitals branch off from the oxygen nucleus of a water molecule. Two of the orbitals are formed by the shared electrons bonding the hydrogen atoms to the oxygen atom. They have a slightly positive charge. The other two orbitals have a slightly negative charge.*

(b) *As a result of these positive and negative zones, each water molecule can form hydrogen bonds (dashed lines) with four other water molecules. Under ordinary conditions of pressure and temperature, the hydrogen bonds are continually breaking and re-forming in a shifting pattern. Thus water is a liquid.*

2–4
A water strider's legs dimple the surface of a pond. The water strider, which lives on this surface, has specialized hairs on its first and third pairs of legs that enable it to rest upon the surface film, depressing it but not penetrating. The second pair of legs, which penetrate the film, serve as propelling oars.

Surface Tension

Look at water dripping from a faucet. Each drop clings to the rim and dangles for a moment by a thread of water; then just as the tug of gravity breaks it loose, its outer surface is drawn taut, to form a sphere as the drop falls free. Gently place a needle or a razor blade flat on the surface of the water in a glass. Although the metal is denser than water, it floats. Look at a pond in spring or summer; you will see water striders and other insects walking on its surface almost as if it were solid. These phenomena are all the result of *surface tension*. Surface tension is the result of the cohesion, or clinging together, of the water molecules. (*Cohesion* is, by definition, the holding together of molecules of the same substance. *Adhesion* is the holding together of molecules of different substances.)

The only liquid with a surface tension greater than that of water is mercury. Atoms of mercury are so greatly attracted to one another that they tend not to adhere to anything else. Water, however, because of its negative and positive charges, adheres strongly to any other charged molecules and to charged surfaces. The "wetting" capacity of water—that is, its ability to coat a surface—results from its polar structure, as does its cohesiveness.

Capillary Action and Imbibition

If you hold two dry glass slides together and dip one corner in water, the combination of cohesion and adhesion will cause water to spread upward between the two slides. This is *capillary action*. Capillary action similarly causes water to rise in very fine glass tubes, to creep up a piece of blotting paper, or to move slowly through the micropores of the soil and so become available to the roots of plants.

Imbibition ("drinking up") is the capillary movement of water molecules into substances such as wood or gelatin, which swell as a result. The pressures developed by imbibition can be astonishingly great. It is said that stone for the ancient Egyptian pyramids was quarried by driving wooden pegs into holes drilled in the rock face and then soaking the pegs with water. The swelling of the wood created a force great enough to break the stone slab free. Seeds imbibe water as they begin to germinate, swelling and bursting their seed coats (Figure 2–5).

2–5
The germination of seeds begins with changes in the seed coat that permit a massive uptake of water. The embryo and surrounding structures then swell, bursting the seed coat. In these acorns, photographed on the forest floor, the embryonic roots have emerged through the tough outer layers of the fruits.

Specific Heat of Water

Table 2-1 Comparative Specific Heats (The quantity of heat, in calories, required to raise the temperature of 1 gram through 1°C)

SUBSTANCE	SPECIFIC HEAT
Water	1.00
Lead	0.03
Iron	0.10
Salt (NaCl)	0.21
Glass	0.20
Sugar (sucrose)	0.30
Liquid ammonia	1.23
Chloroform	0.24
Ethyl alcohol (ethanol)	0.60

The amount of heat a given amount of a substance requires for a given increase in temperature is its *specific heat* (also called heat capacity). One calorie* is defined as the amount of heat that will raise the temperature of 1 gram (1 milliliter or 1 cubic centimeter) of water 1°C. The specific heat of water is about twice the specific heat of oil or alcohol; that is, approximately 0.5 calorie will raise the temperature of 1 gram of oil or alcohol 1°C. It is four times the specific heat of air or aluminum and 10 times that of iron. Only liquid ammonia has a higher specific heat (Table 2-1). In other words, it requires a high input of energy to raise the temperature of water.

Heat is a form of energy—the *kinetic energy*, or energy of movement, of molecules. Molecules are always in motion; they vibrate, rotate, and shift position in relation to other molecules. Note that heat and temperature are not the same. Temperature, which is measured in degrees, reflects the *average* kinetic energy of the molecules. Heat, which is measured in calories, reflects the *total* kinetic energy in a collection of molecules; it includes both the magnitude of the molecular movements and the mass and number of moving molecules present. For example, a lake may have a lower temperature than does a bird flying over it, but the lake contains more heat because it has many more molecules in motion.

The high specific heat of water is a consequence of hydrogen bonding. The hydrogen bonds in water tend to restrict the movement of the molecules. In order for the kinetic energy of water molecules to increase sufficiently for the temperature to rise 1°C, it is necessary first to rupture a number of the hydrogen bonds holding the molecules together. When you heat a pot of water, much of the heat energy added to the water is used in breaking the hydrogen bonds between the water molecules. Only a relatively small amount of heat energy is therefore available to increase molecular movement.

What does the high specific heat of water mean in biological terms? It means that for a given rate of heat input, the temperature of water will rise more slowly than the temperature of almost any other material. Conversely, the temperature will drop more slowly as heat is removed. Because so much heat input or heat loss is required to raise or lower the temperature of water, organisms that live in the oceans or large bodies of fresh water live in an environment where the temperature is relatively constant. Also, the high water content of terrestrial plants and animals helps them to maintain a relatively constant internal temperature. This constancy of temperature is critical because biologically important chemical reactions take place only within a narrow temperature range.

Table 2-2 Comparative Heats of Vaporization (The quantity of heat, in calories, required to convert 1 gram of liquid to 1 gram of gas)

LIQUID	HEAT REQUIRED
Water (at 0°C)	596
Water (at 100°C)	540
Ammonia	295
Chlorine	67.4
Hydrofluoric acid	360
Nitric acid	115
Carbon dioxide	72.2
Ethyl alcohol (ethanol)	236.5
Ether	9.4

Heat of Vaporization

Vaporization—or evaporation, as it is more commonly called—is the change from a liquid to a gas. Water has a high *heat of vaporization*. It takes more than 500 calories to change a gram of liquid water into vapor, 60 times as much as for ether and almost twice as much as for ammonia (Table 2-2).

Hydrogen bonding is also responsible for water's high heat of vaporization. Vaporization comes about because some of the most rapidly moving molecules of a

* Nutritional "calories" are actually kilocalories (kcal); 1,000 calories equal 1 kilocalorie.

liquid break loose from the surface and enter the air. The hotter the liquid, the more rapid the movement of its molecules and, hence, the more rapid the rate of evaporation. But, whatever the temperature, so long as a liquid is exposed to air that is less than 100 percent saturated with the vapor of that liquid, evaporation will take place, right down to the last drop.

In order for a water molecule to break loose from its fellow molecules—that is, to vaporize—the hydrogen bonds have to be broken. This requires heat energy. At water's boiling point (100°C at a pressure of 1 atmosphere), 540 calories are needed to change 1 gram of liquid water into vapor. As a consequence, when water evaporates, as from the surface of your skin or a leaf, the escaping molecules carry a great deal of heat away with them. Thus evaporation has a cooling effect. Evaporation from the surface of a land-dwelling plant or animal is one of the principal ways in which these organisms "unload" excess heat and so stabilize their temperatures.

Freezing

Water exhibits another peculiarity when it undergoes the transition from a liquid to a solid (ice). In most liquids, the density—that is, the weight of the material in a given volume—increases as the temperature drops. This greater density occurs because the individual molecules are moving more slowly and so the spaces between them decrease. The density of water also increases as the temperature drops, until it nears 4°C. Then the water molecules come so close together and are moving so slowly that every one of them can maintain hydrogen bonds simultaneously with four other molecules. However, because of the geometry of the water molecule, the formation of four simultaneous hydrogen bonds requires that the molecules move apart from each other. This creates an open latticework (Figure 2–7a, on the next page) that is the most stable structure for an ice crystal. Thus water as a solid takes up more volume than water as a liquid. Ice is less dense than liquid water and therefore floats in it.

2–6

Ammonia is very similar to water in its molecular structure, and biologists have speculated about whether it might substitute for water in life processes. The ammonia molecule (NH_3) is made up of hydrogen atoms covalently bonded to nitrogen, which, like the oxygen in the water molecule, retains a slight negative charge. But because there are three hydrogens with slight positive charges to one nitrogen, ammonia does not have the cohesive power of water and evaporates much more quickly. Perhaps this is why no form of life based on ammonia has been found, although NH_3 may have been very common in the primitive atmosphere.

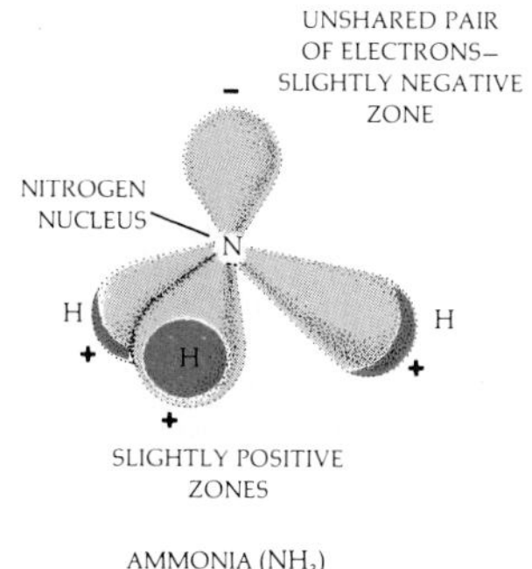

AMMONIA (NH_3)

"Ammonia! Ammonia!"

[*Drawing by R. Grossman, © 1962 The New Yorker Magazine, Inc.*]

2–7
(a) *In the crystalline structure of ice, each water molecule is hydrogen-bonded to four other water molecules in a three-dimensional open latticework. The bond angles in some of the water molecules are distorted as they link up in a hexagonal arrangement. This arrangement, shown here in a small section of the latticework, is repeated throughout the crystal and is responsible for the beautiful patterns seen in snowflakes and frost. The water molecules are actually farther apart in ice than they are in liquid water.*

(b) *When water freezes in the cracks and crevices of rock, the force created by its expansion splits the rock. Over long periods of time, this process breaks up masses of rock and contributes to the formation of soil.*

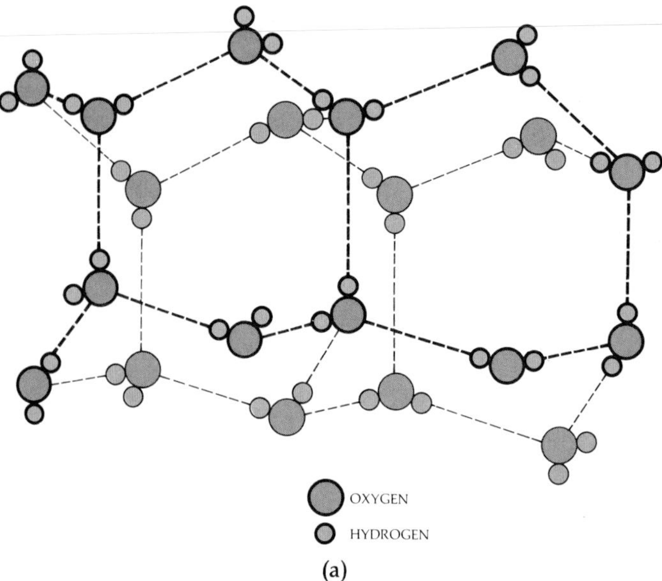

(a)

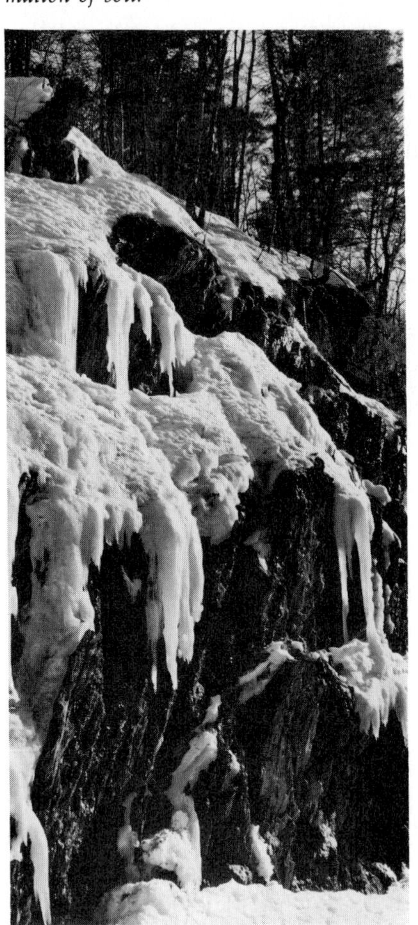
(b)

This increase in volume has occasional disastrous effects on water pipes but, on the whole, turns out to be enormously beneficial for life forms. If water continued to contract as it froze, ice would be heavier than liquid water. As a result, lakes and ponds and other bodies of water would freeze from the bottom up. Once ice began to accumulate on the bottom, it would tend not to melt, season after season. Spring and summer might stop the freezing process, but laboratory experiments have shown that if ice is held to the bottom of even a relatively shallow tank, water can be boiled on the top without melting the ice. Thus if water did not expand when it froze, it would continue to freeze from the bottom up, year after year, and never melt again. Eventually, the body of water would freeze solid and any life in it would be destroyed. By contrast, the layer of floating ice that actually forms tends to protect the organisms in the water. The ice layer effectively insulates the liquid water beneath it, keeping its temperature at or above the freezing point of water.

The freezing point of water is 0°C, as is the melting point. To make the transition from a solid to a liquid, water requires 79.7 calories per gram, a quantity known as the *heat of fusion*. As ice melts, it draws this much heat from its surroundings, thereby cooling the surroundings. The heat energy absorbed by the ice breaks the hydrogen bonds of the latticework. Conversely, as water freezes, it releases the same amount of heat into its surroundings. In this way, ice and snow also serve as temperature stabilizers, particularly during the transition periods of fall and spring. Moderation of sudden changes in temperature gives organisms time to make seasonal adjustments essential to survival.

The presence of dissolved substances in water reduces the temperature at which water freezes, which is why salt is thrown on icy sidewalks and used in ice-cream freezers. The "hardening" process in several species of winter-hardy plants, by which they prepare themselves for cold weather, includes the breakdown of starch (which is insoluble) into simple sugars (which are soluble). Freshwater fish, whose body fluids are salty compared to the pond or lake in which they live, do not freeze when the temperature of water is at or near 0°C. However, logically speaking, saltwater fish, whose body fluids are less salty than the ocean water surrounding

THE SEASONAL CYCLE OF A LAKE

As we have seen, water increases in density as its temperature drops, until it reaches 4° C, the temperature of maximum density. Water either colder or warmer than 4° C is less dense and floats above water at 4° C. As a result, the water of temperate-zone lakes is stratified in the summer and winter but undergoes considerable mixing in the fall and spring. The stratifications of summer and winter enable lake-dwelling organisms to avoid life-threatening temperature extremes, while the mixing that occurs in fall and spring provides nutrients and oxygen to organisms at all levels of the lake.

In the summer, the top layer of water, called the epilimnion, is heated by the sun and the surrounding air, becoming warmer than the lower layers. Since it becomes less dense as it becomes warmer, this water remains at the surface. Only the water in the epilimnion circulates. In the middle layer, there is an abrupt drop in temperature, known as the thermocline. Since the water in this layer is progressively more dense, it does not mix with the lighter water above. The water of the middle layer effectively cuts off the circulation of oxygen from the surface into the third layer, the hypolimnion. As the organisms of the hypolimnion gradually use up the available oxygen, the summer stagnation results (a).

In the fall, the temperature of the epilimnion drops until it is the same as that of the hypolimnion. The warmer water in the middle layer then rises to the surface, producing the fall overturn (b). *Aided by the fall winds, water begins to circulate throughout the lake* (c); *oxygen is returned to the depths, and nutrients released by the activities of bottom-dwelling bacteria are carried to the upper layers of the lake.*

As winter deepens, the surface water cools below 4° C, becoming lighter as it expands. This water remains on the surface and, in many areas, freezes. The result is winter stratification (d).

In the spring, as the ice melts and the water on the surface warms to 4° C, it sinks to the bottom, producing the spring overturn (e) *and another thorough mixing of the water in the lake* (f).

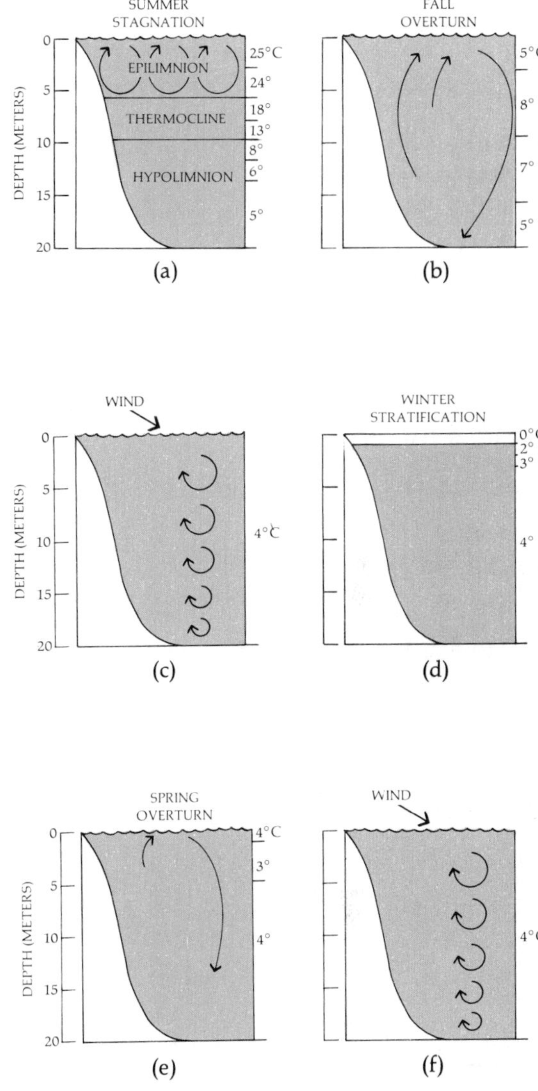

them, should freeze at the below-zero temperatures of Arctic water. They do not, however, and animal physiologists investigating this phenomenon have discovered that at least one species, the ghost fish, produces a complex protein named, appropriately, antifreeze protein. This protein, which is secreted into the bloodstream, appears to interfere with formation of the crystalline structure of ice. Recently, studies of several species of terrestrial frogs that spend the winter hibernating beneath leaf litter have revealed that their body fluids contain a high concentration of glycerol, one of the ingredients sometimes used in automobile antifreeze.

2–8
Because of the polarity of water molecules, water can serve as a solvent for ionic substances and polar molecules. This diagram shows sodium chloride (NaCl) dissolving in water as the water molecules cluster around the individual sodium and chloride ions, separating them from each other.

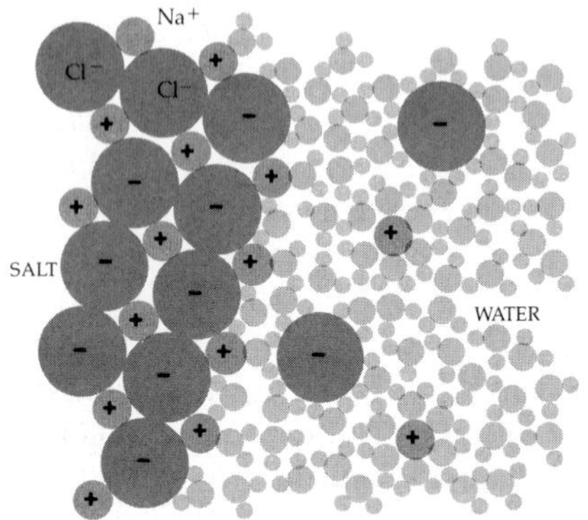

WATER AS A SOLVENT

Many substances within living systems are found in solution. (A *solution* is a uniform mixture of the molecules of two or more substances. The substance present in the greatest amount—usually a liquid—is called the solvent, and the substances present in lesser amounts are called solutes.) The polarity of water molecules is responsible for water's capacity as a solvent. The polar water molecules tend to separate ionic substances, such as sodium chloride (NaCl), into their constituent ions. As shown in Figure 2–8, the water molecules cluster around and segregate the charged ions. Many of the molecules important in living systems—such as sugars—also have areas of positive and negative charge. (Such polar regions arise, as you might expect, in the neighborhood of covalently bonded atoms whose nuclei exert differing degrees of attraction for electrons.) These molecules therefore attract water molecules and so dissolve in water.

Polar molecules that readily dissolve in water are often called *hydrophilic* ("water-loving"). Such molecules slip into aqueous solution easily because their partially charged regions attract water molecules as much as or more than they attract each other. The polar water molecules thus compete with the attraction between the solute molecules themselves.

Molecules, such as fats, that lack polar regions tend to be very insoluble in water. The hydrogen bonding between the water molecules acts as a force to exclude the nonpolar molecules. As a result of this exclusion, nonpolar molecules tend to cluster together in water, just as droplets of fats tend to coalesce, for example, on the surface of chicken soup. Such molecules are said to be *hydrophobic* ("water-fearing"), and the clusterings are known as hydrophobic interactions.

We will encounter these properties of hydrophobic and hydrophilic molecules again in later chapters. These weak forces—hydrogen bonds and hydrophobic forces—play very important roles in shaping the architecture of large, biologically important molecules and, as a consequence, in determining their properties.

IONIZATION OF WATER: ACIDS AND BASES

In liquid water, there is a slight tendency for a hydrogen atom to jump from the oxygen atom to which it is covalently bonded to the oxygen atom to which it is hydrogen-bonded (Figure 2–9). In this reaction, two ions are produced: the hydronium ion (H_3O^+) and the hydroxide ion (OH^-). In any given volume of pure water, a small but constant number of water molecules will be ionized in this way. The number is constant because the tendency of water to ionize is exactly offset by the tendency of the ions to reunite; thus even as some molecules are ionizing, an equal number of others are forming, a state known as dynamic *equilibrium*.

Although the positively charged ion formed when water ionizes is the hydronium ion (H_3O^+), rather than the hydrogen ion (H^+), by convention the ionization of water is expressed by the equation:

$$HOH \rightleftharpoons H^+ + OH^-$$

The arrows indicate that the reaction goes in both directions. The fact that the arrow pointing toward HOH is longer indicates that, at equilibrium, most of the H_2O is not ionized. As a consequence, in any sample of pure water, only a small fraction exists in ionized form.

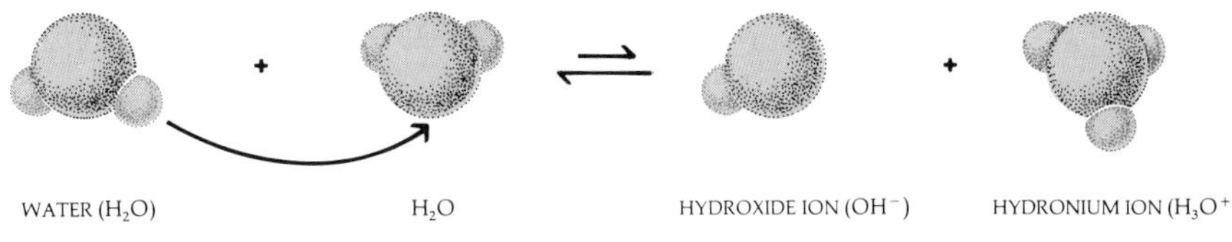

2–9
When water ionizes, a hydrogen nucleus (that is, a proton) shifts from the oxygen atom to which it is covalently bonded to the oxygen atom to which it is hydrogen-bonded. The resulting ions are the negatively charged hydroxide ion and the positively charged hydronium ion. In this diagram, the large spheres represent oxygen and the small spheres represent hydrogen.

In pure water, the number of H^+ ions exactly equals the number of OH^- ions. This is necessarily the case since neither ion can be formed without the other when only H_2O molecules are present. However, when an ionic substance or a substance with polar molecules is dissolved in water, it may change the relative numbers of H^+ and OH^- ions. For example, when hydrochloric acid (HCl) dissolves in water, it is almost completely ionized into H^+ and Cl^- ions; as a result, an HCl solution contains more H^+ ions than OH^- ions. Conversely, when sodium hydroxide (NaOH) dissolves in water, it forms Na^+ and OH^- ions; thus, in a solution of sodium hydroxide in water, there are more OH^- ions than H^+ ions.

A solution acquires the properties we recognize as acidic when the number of H^+ ions exceeds the number of OH^- ions; conversely, a solution is basic (alkaline) when the number of OH^- ions exceeds the number of H^+ ions. Thus, an *acid* is a substance that causes an increase in the relative number of H^+ ions in a solution, and a *base* is a substance that causes an increase in the relative number of OH^- ions.

Strong and Weak Acids and Bases

Strong acids and bases are substances, like HCl and NaOH, that ionize almost completely in water, resulting in relatively large increases in the concentrations of H^+ and OH^- ions, respectively. Weak acids and bases, by contrast, are those that ionize only slightly, resulting in relatively small increases in the concentration of H^+ or OH^- ions.

Because of the strong tendency of H^+ and OH^- ions to combine and the weak tendency of water to ionize, the concentration of OH^- ions will always decrease as the concentration of H^+ ions increases (as, for example, when HCl is added to water), and vice versa. If HCl is added to a solution containing NaOH, the following reaction will take place:

$$H^+ + Cl^- + Na^+ + OH^- \longrightarrow H_2O + Na^+ + Cl^-$$

In other words, if an acid and a base of comparable strength are added in equivalent amounts, the solution will not have an excess of either H^+ or OH^- ions.

Many of the acids important in living systems owe their acidic properties to a group of atoms called the carboxyl group, which includes one carbon atom, two oxygen atoms, and a hydrogen atom (symbolized as —COOH). When a substance containing a carboxyl group is dissolved in water, some of the —COOH groups dissociate to yield hydrogen ions:

$$\text{—COOH} \rightleftharpoons \text{—COO}^- + H^+$$

Thus compounds containing carboxyl groups are hydrogen-ion donors, or acids. They are weak acids, however, because, as indicated by the arrows, the —COOH ionizes only slightly.

Among the most important bases in living systems are compounds that contain the amino group ($—NH_2$). This group has a weak tendency to accept hydrogen ions, thereby forming $—NH_3^+$:

$$—NH_2 + H^+ \rightleftharpoons —NH_3^+$$

As hydrogen ions are removed from solution by the amino group, the relative concentration of H^+ ions decreases and the relative concentration of OH^- ions increases. Groups, such as $—NH_2$, that are weak hydrogen-ion acceptors are thus weak bases.

The pH Scale

Chemists express degrees of acidity by means of the *pH scale*. The symbol "pH" stands for the negative logarithm of the concentration of hydrogen ions in moles per liter. Although this sounds complicated, in practice it is relatively simple. As you may recall from your mathematics courses, the logarithm is the exponential power to which a specified number (usually 10) must be raised to equal a given number—in this case, the concentration of hydrogen ions expressed in moles per liter.

A *mole* is the amount of an element equivalent to its atomic weight expressed in grams, or the amount of a substance equivalent to its molecular weight expressed in grams. (The *molecular weight* of a substance is the sum of the atomic weights of the atoms constituting the molecule.) Thus, a mole of atomic hydrogen (atomic weight 1) is 1 gram of hydrogen atoms; a mole of atomic oxygen (atomic weight 16) is 16 grams of oxygen atoms; and a mole of water (molecular weight 18) is 18 grams of water molecules. The most interesting thing about the mole is that a mole—of any substance—contains the same number of particles as any other mole. This number, known as Avogadro's number, is 6.02×10^{23}. Thus, a mole of water molecules (18 grams) contains exactly the same number of molecules as a mole of hydrochloric acid molecules (36.5 grams). Use of the mole in specifying quantities

Table 2-3 The pH Scale

	CONCENTRATION OF H^+ IONS (MOLES PER LITER)	pH	CONCENTRATION OF OH^- IONS (MOLES PER LITER)
	1.0 $= 10^0$	0	10^{-14}
	0.1 $= 10^{-1}$	1	10^{-13}
	0.01 $= 10^{-2}$	2	10^{-12}
Acidic	0.001 $= 10^{-3}$	3	10^{-11}
	0.0001 $= 10^{-4}$	4	10^{-10}
	0.00001 $= 10^{-5}$	5	10^{-9}
	0.000001 $= 10^{-6}$	6	10^{-8}
Neutral	0.0000001 $= 10^{-7}$	7	10^{-7} = 0.0000001
	10^{-8}	8	10^{-6} = 0.000001
	10^{-9}	9	10^{-5} = 0.00001
	10^{-10}	10	10^{-4} = 0.0001
Basic	10^{-11}	11	10^{-3} = 0.001
	10^{-12}	12	10^{-2} = 0.01
	10^{-13}	13	10^{-1} = 0.1
	10^{-14}	14	10^0 = 1.0

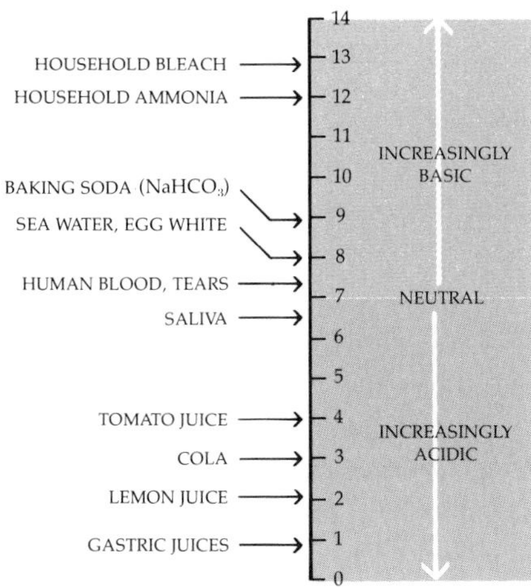

2–10
pH values of various common solutions. A difference of one pH unit reflects a tenfold difference in the H^+ ion concentration. Cola, for instance, is 10 times as acidic as tomato juice, and gastric juices are about 100 times more acidic than cola drinks.

of substances involved in chemical reactions makes it possible for us to consider comparable numbers of reacting particles.

The ionization that occurs in a liter of pure water results in the formation, at equilibrium, of 1/10,000,000 mole of hydrogen ions (and, as we noted earlier, of exactly the same quantity of hydroxide ions). For convenience, this concentration of hydrogen ions is written exponentially as 10^{-7} mole per liter. The logarithm is the exponent, -7, and the negative logarithm is 7; in terms of the pH scale, it is referred to simply as pH 7 (see Table 2–3). At pH 7, the concentrations of free H^+ and OH^- are exactly the same, and thus pure water is said to be neutral. Any pH below 7 is acidic, and any pH above 7 is basic. The lower the pH number, the higher the concentration of hydrogen ions. Thus pH 2 means 10^{-2} mole of hydrogen ions per liter of water, or 1/100 mole per liter—which is, of course, a much larger figure than 1/10,000,000. A difference of one pH unit represents a tenfold difference in the concentration of hydrogen ions.

We can now define "acid" and "base" more fully:

1. An acid is a substance that causes an increase in the number of H^+ ions and a decrease in the number of OH^- ions in a solution. Most acids are hydrogen-ion donors, but some acids function by removing OH^- ions from the solution. A solution with a pH below 7 (with more than 10^{-7} mole of H^+ ions per liter) is acidic.
2. A base is a substance that causes a decrease in the number of H^+ ions and an increase in the number of OH^- ions in a solution. Some bases, such as NaOH, donate OH^- ions to the solution; others, such as the $—NH_2$ group, are hydrogen-ion acceptors, removing H^+ ions from the solution. A solution with a pH above 7 (with less than 10^{-7} mole of H^+ ions per liter) is basic.

2–11
Surface of the lining of the stomach, as shown in a scanning electron micrograph (magnified approximately 185 times). The numerous indentations are the openings into gastric pits, in which acid-secreting cells are located. Mucus, also secreted by cells of the stomach lining, coats the surface of the stomach and protects it from the acid.

Buffers

Solutions more acidic than pH 1 or more basic than pH 14 are possible, but these are not included in the scale because they are almost never encountered in biological systems (Figure 2–10). In fact, almost all the chemistry of living things takes place at pH's between 6 and 8. Notable exceptions are the chemical processes in the stomach of humans and other animals, which take place at a pH of about 2 (Figure 2–11). Human blood, for instance, maintains an almost constant pH of 7.4 despite the fact that it is the vehicle for a large number and variety of nutrients and other chemicals being delivered to the cells, as well as for the removal of wastes, many of which are acids and bases.

The maintenance of a constant pH—an example of homeostasis (see page 18)—is important because the pH greatly influences the rate of chemical reactions. Organisms resist strong, sudden changes in the pH of blood and other fluids by means of *buffers*, which are combinations of H^+-donor and H^+-acceptor forms of weak acids or bases.

Buffers help maintain constant pH by their tendency to combine with H^+ ions and thus remove them from solution as the H^+ ion concentration begins to rise and to release them as it falls. The capacity of a buffer system to resist changes in pH is greatest when the concentrations of its H^+-donor and H^+-acceptor forms are equal. As the concentration of one form increases and that of the other decreases, the buffer becomes less effective. A variety of buffers function in living systems, each most effective at the particular pH at which its H^+-donor and H^+-acceptor concentrations are equal.

ACID RAIN

The average pH of normal rainfall is about 5.6 (mildly acidic), a result of the combination of carbon dioxide with water vapor to produce carbonic acid. In the 1920s, however, the pH of rain and snow in Scandinavia began to drop, and by the 1950s, similar phenomena were observed elsewhere in Europe and in the northeastern United States. As more data were collected, it was found that, in certain geographic areas, the average annual pH of precipitation was between 4.0 and 4.5. Occasional storms would release rain with a pH as low as 2.1, which is extremely acidic.

The low pH was traced to several acids found in the rainfall: sulfurous (H_2SO_3), sulfuric (H_2SO_4), nitrous (HNO_2), and nitric (HNO_3), all of which ionize almost completely in aqueous solution, producing hydrogen ions. These acids are formed when the gaseous oxides of sulfur and nitrogen dissolve in water. Sulfur and nitrogen oxides are released into the atmosphere by some natural processes (for example, volcanic eruptions), but far greater quantities are released as a result of human activities. Sulfur oxides are produced by the combustion of high-sulfur coal and oil and by the smelting of sulfur-containing ores. Nitrogen oxides are by-products of gasoline combustion in automobile engines and of some generating processes for electricity.

That sulfur oxides could be damaging to vegetation was evident as early as the turn of the century, when a large copper smelter was opened in a mountainous area in Tennessee. Within a few years, all vegetation had been killed in the formerly luxuriant forest surrounding the smelter. The solution devised for this problem—still used today—was to build very tall smokestacks, so the wind would carry the pollutants away from the immediate area. It was assumed that they would be so widely dispersed that they would be rendered harmless. By the 1960s, accumulating evidence indicated that sulfur oxides released from tall smokestacks are transported hundreds or thousands of miles by the prevailing winds and then return to the earth in rain and snow. Nitrogen oxides released from automobiles are also carried off by the wind. What was once a local problem has become an international problem, in which the pollutants respect no boundaries.

The biological consequences of acid rain depend in part on the characteristics of the soil and underlying rock on which it falls. In areas where the principal rock is limestone (calcium carbonate), the buffering action of the H_2CO_3-HCO_3^- system can generally prevent acidification of the soil, lakes, and streams. In other areas, where the soil and bodies of water do not contain such a natural buffer, the pH drops gradually but steadily as a result of repeated additions of acid rain. The drop in pH is often quite sudden and extreme when the spring melt brings an infusion of acid accumulated in the winter snows. Although the low pH resulting from the spring melt is usually temporary, it can be particularly devastating to salamanders and frogs, many of which lay their eggs in small ponds and puddles formed by the melt water.

Lakes in mountainous regions are especially vulnerable to acid rain. A 1977 Cornell University study of the lakes at high elevations (above 600 meters) in the western Adirondack Mountains of New York found that 51 percent had a pH below 5.0, of which 90 percent were devoid of fish life. By contrast, a similar study performed between 1929 and 1937 found that only 4 percent of the lakes were acidic and without fish. More recent studies in the Adirondacks indicate increasingly acidic conditions in lakes at lower levels, which were previously unaffected. The effects of low pH on fish include the depletion of calcium in their bodies, leading to weakened and deformed bones; the failure of many eggs to hatch and deformed fish from those that do hatch; and the clogging of the gills by aluminum, which is released from the soil by acid.

The major buffer system in the human bloodstream is the acid-base pair H_2CO_3-HCO_3^-. The weak acid H_2CO_3 (carbonic acid) dissociates into H^+ and bicarbonate ions as shown in the equation below.

$$\underset{H^+ \text{ DONOR}}{H_2CO_3} \rightleftharpoons H^+ + \underset{H^+ \text{ ACCEPTOR}}{HCO_3^-}$$

The H_2CO_3-HCO_3^- buffer system resists the changes in pH that might result from the addition of small amounts of acid or base by "soaking up" the acid or base. For example, if a small amount of H^+ is added to the system, it combines with the H^+ acceptor HCO_3^- to form H_2CO_3. This reaction removes the added H^+ and main-

This map, based on estimates made by the Environmental Protection Agency, shows the sensitivity of different areas of the continental United States to acid rain. It takes into account such factors as major sources of sulfur and nitrogen oxides, weather patterns, altitude, and soil characteristics.

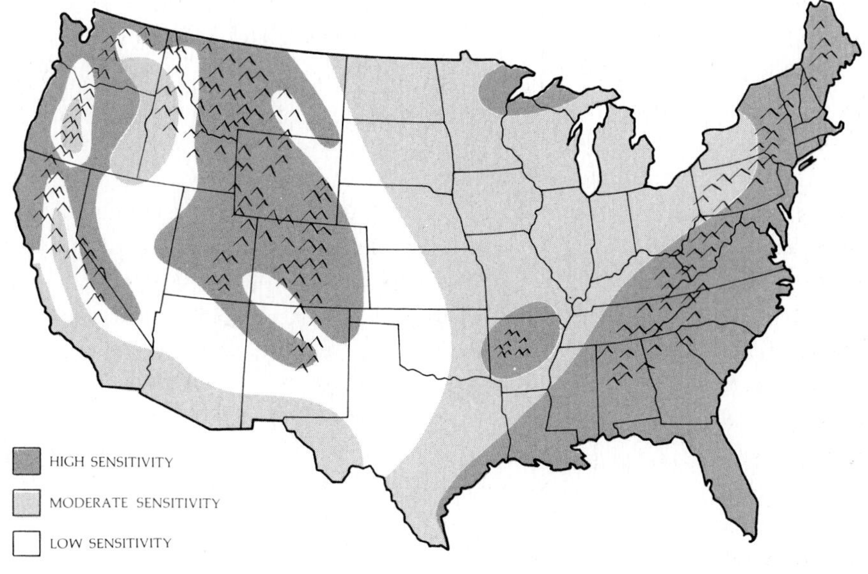

The effects of acid rain on plants depend on both the species and the soil conditions. Among the observed effects are reduced germination of seeds, a decrease in the number of seedlings that mature, reduced photosynthesis, and lowered resistance to disease. If the soil is not adequately buffered, essential nutrients are leached from it and are thus unavailable to the plants.

Many people believe that acid rain is the most serious worldwide pollution problem confronting us today. The potential consequences of its effects on biological systems are immense: lowered crop yields, decreased timber production, the loss of important freshwater fishing areas, and the need for greater amounts of increasingly expensive fertilizer to compensate for nutrient losses from leaching. The monetary and social costs of allowing the conditions that create acid rain to continue (or even to increase) are potentially very great, as are the costs of available processes to remove the sulfur and nitrogen oxides at the source, before they enter the air.

In the last few years, scientists from many fields have begun new research projects to gain a greater understanding of the causes and effects of acid rain and the likely consequences of proposed solutions. Although scientists can provide information on which decisions can be based, the choices that lie ahead are essentially social and economic, to be made through political processes.

tains the pH near its original value. If a small amount of OH^- is added, it combines with the H^+ to form H_2O; more H_2CO_3 tends to ionize to replace the H^+ as it is used.

Control of the pH of the blood is rendered even "tighter" by the fact that the H_2CO_3 is in equilibrium with dissolved carbon dioxide (CO_2) in the blood:

$$H_2O + CO_2 \rightleftharpoons H_2CO_3$$

As the arrows indicate, the two reactions are in equilibrium and the equilibrium favors the formation of CO_2; in fact, the ratio is about 100 to 1 in favor of CO_2 formation.

Dissolved CO_2 in the blood is, in turn, in equilibrium with the CO_2 in the lungs. By changing your rate of breathing, you can change the HCO_3^- concentration in the blood and thus adjust the pH of your internal fluids.

Obviously, if the blood should be flooded with a very large excess of acid or base, the buffer would fail, but normally it is able to adjust continuously and very rapidly to the constant small additions of acid or base that normally occur in body fluids.

THE WATER CYCLE

Most of the water on earth—almost 98 percent—is in liquid form, in the oceans, lakes, and streams. Of the remaining 2 percent, some is frozen in polar ice and glaciers, some is in the soil, some is in the atmosphere in the form of vapor, and some is in the bodies of living organisms.

Water is made available to land organisms by processes powered by the sun. Solar energy evaporates water from the oceans, leaving the salt behind. Water is also evaporated, but in much smaller amounts, from moist soil surfaces, from the leaves of plants, and from the bodies of other organisms. These molecules—now water vapor—are carried up into the atmosphere by air currents. Eventually they fall to the earth's surface again as snow or rain. Most of the water falls on the oceans, since these cover most of the earth's surface. The water that falls on land is pulled back to the oceans by the force of gravity. Some of it, reaching low ground, forms ponds or lakes and streams or rivers, which pour water back into the oceans.

Some of the water that falls on the land percolates down through the soil until it reaches a zone of saturation. In the zone of saturation, all pores and cracks in the

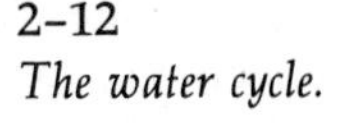

2–12
The water cycle.

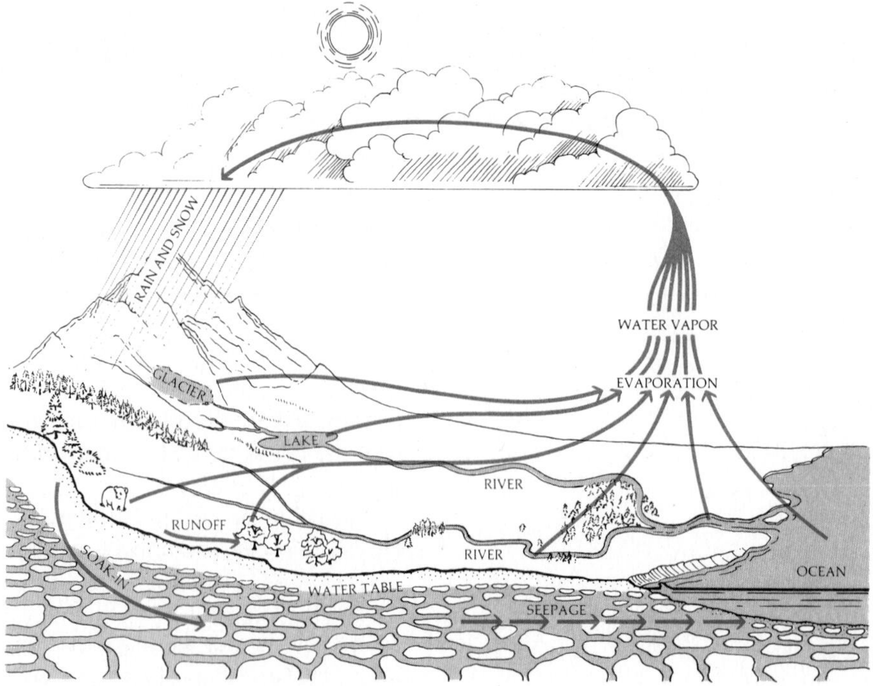

rock are filled with water (groundwater). The upper surface of the zone of saturation is known as the water table. Below the zone of saturation is solid rock, through which the water cannot penetrate. The deep groundwater, moving extremely slowly, eventually also reaches the ocean, thereby completing the water cycle.

As we have seen in this chapter, water, essential for life, is a most extraordinary substance. The earth's supply of water is the permanent possession of our planet, held to its surface by the force of gravity. Through the movements of the water cycle, it is perpetually available to living organisms.

SUMMARY

Water, the most common liquid on the earth's surface and the major component, by weight, of all living things, has a number of remarkable properties. These properties are a consequence of its molecular structure and are responsible for water's "fitness" for its roles in living systems.

Water is made up of two hydrogen atoms and one oxygen atom held together by covalent bonds. The water molecule is polar, with two weakly negative zones and two weakly positive zones. As a consequence, weak bonds form between water molecules. Such bonds, which link a somewhat positively charged hydrogen atom that is part of one molecule to a somewhat negatively charged oxygen atom that is part of another molecule, are known as hydrogen bonds. Each water molecule can form hydrogen bonds with four other water molecules. Although individual bonds are weak and constantly shifting, the total strength of the bonds holding the molecules together is very great.

Because of the hydrogen bonds holding the water molecules together (cohesion), water has a high surface tension and a high specific heat (the amount of heat that a given amount of the substance requires for a given increase in temperature). It also has a high heat of vaporization (the heat required to change a liquid to a gas) and a high heat of fusion (the heat required to change a solid to a liquid). Just before it freezes, water expands; thus ice has a lower density and a larger volume than liquid water. As a result, ice floats on water.

The polarity of the water molecule is responsible for water's adhesion to other polar substances and hence its tendency for capillary movement. Similarly, water's polarity makes it a good solvent for ions and polar molecules. Molecules that dissolve readily in water are known as hydrophilic. Water molecules, as a consequence of their polarity, actively exclude nonpolar molecules from solution. Molecules that are excluded from water solution are known as hydrophobic.

Water has a slight tendency to ionize, that is, to separate into H^+ ions (actually H_3O^+, hydronium ions) and OH^- ions. In pure water, the number of H^+ ions and OH^- ions is equal at 10^{-7} mole per liter. A solution that contains more H^+ ions than OH^- ions is acidic; one that contains more OH^- ions than H^+ ions is basic. The pH scale reflects the proportion of H^+ ions to OH^- ions. An acidic solution has a pH less than 7.0; a basic solution has a pH more than 7.0. Almost all of the chemical reactions of living systems take place within a narrow range of pH around neutrality. Organisms maintain this narrow pH range by means of buffers, which are combinations of H^+-donor and H^+-acceptor forms of weak acids or bases.

Through the water cycle, the water above, on, and below the earth's surface is recirculated. As a result, it is continuously available to living organisms.

QUESTIONS

1. (a) Sketch the water molecule and label the areas of positive and negative charge. (b) What are the major consequences of the polarity of the water molecule? (c) How are these effects important to living systems?

2. The trick with the razor blade works better if the blade is a little greasy. Why?

3. Surfaces such as glass or raincoat cloth can be made "nonwettable" by application of silicone oils or other substances that cause water to bead up instead of spread flat. What do you suppose is happening, in molecular terms, when a surface becomes nonwettable?

4. Generally, coastal areas have more moderate temperatures (not as cold in the winter, nor as hot in the summer) than inland areas at the same latitude. What reasonable explanation can you give for this phenomenon?

5. What is vaporization? Describe the changes that take place in water as it vaporizes. What is heat of vaporization? Why does water have an unusually high heat of vaporization?

6. Can you explain maple sugar production in terms of its value to the sugar maple tree?

7. As we have seen, the digestive processes in the human stomach take place at a pH of about 2. When the food being digested reaches the small intestine, sodium bicarbonate ($NaHCO_3$) is released from the pancreas into the small intestine. What effect would you expect this to have on the pH of the food?

8. Describe the properties of water that are exhibited and suggested by the photograph at the left.

CHAPTER 3

Organic Molecules

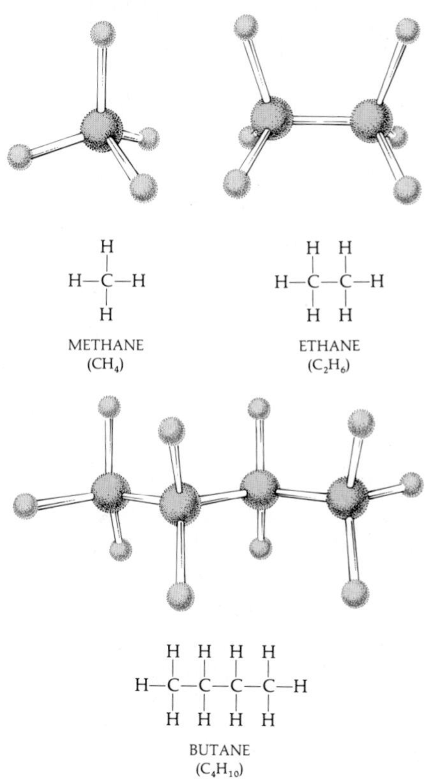

3–1
Ball-and-stick models and structural formulas of methane, ethane, and butane. In the models, the gray spheres represent carbon atoms and the smaller colored spheres represent hydrogen atoms. The sticks in the models—and the lines in the structural formulas—represent covalent bonds, each of which consists of a pair of shared electrons.

In this chapter, we present some of the types of *organic molecules*—that is, molecules containing carbon—that are found in living things. As you will see, the molecular drama is a grand spectacular with, literally, a cast of thousands; a single bacterial cell contains some 5,000 different kinds of molecules, and an animal or plant cell has about twice that many. These thousands of molecules, however, are composed of relatively few elements. Similarly, relatively few kinds of molecules play the major roles in living systems. If you bear with us, you will find that you readily learn to recognize the players and their roles and to distinguish the stars from the members of the chorus. Consider this, if you will, an introduction to the principal characters; the plot begins to unfold in Chapter 4.

THE CENTRAL ROLE OF CARBON

As we noted previously, water makes up from 50 to 95 percent of a living system, and ions such as K^+, Na^+, and Ca^{2+} account for no more than 1 percent. Almost all the rest, chemically speaking, is composed of organic molecules.

Four different kinds of organic molecules are found in large quantities in organisms. These four are carbohydrates (composed of sugars), lipids (including fats and waxes), proteins (composed of amino acids), and nucleotides (complex molecules that can combine to form very large molecules—DNA and RNA—known as nucleic acids). In this chapter, we shall consider the first three of these. Nucleotides will be taken up in later chapters when we discuss their key roles in energy exchanges and in genetic systems.

All of these molecules—carbohydrates, lipids, proteins, and nucleotides—contain carbon, hydrogen, and oxygen. In addition, proteins contain nitrogen and sulfur; and nucleotides, as well as some lipids, contain nitrogen and phosphorus.

The Carbon Backbone

As you will recall from Chapter 1, a carbon atom has six protons and six electrons, two electrons in its first energy level and four in its second energy level. Thus carbon can form four covalent bonds with as many as four different atoms. Methane (CH_4), which is natural gas, is an example (Figure 1–10, page 27). Even more important, in terms of carbon's biological role, carbon atoms can form bonds with each other. Ethane contains two carbons; propane, three; butane, four; and so on, forming long chains (Figure 3–1). In general, an organic molecule derives its overall

WHY NOT SILICON?

Silicon, which has an atomic number of 14, is more abundant than carbon, which has an atomic number of 6. As you can tell from its atomic number, silicon also requires four electrons to complete its outer energy level. Why then is it found so rarely in living systems? Because silicon is larger, the distance between two silicon atoms is much greater than the distance between two carbon atoms. As a result, the bonds between the more tightly held carbon atoms are almost twice as strong as those between silicon atoms. Thus carbon can form long stable chains and silicon cannot.

Carbon's capacity to form double bonds is also crucial to its central role in biology. Carbon combines with oxygen by means of two double bonds, and the carbon dioxide molecule, free and independent, all its electron requirements satisfied, floats in air as a gas. It also dissolves readily in water and so is available to living systems. In SiO_2, by contrast, silicon is joined to oxygen by single bonds, leaving two unpaired electrons on the silicon atom and one on each oxygen. Unable to participate in multiple bonds, these unpaired electrons pair with the unpaired electrons on neighboring SiO_2 molecules, forming, eventually, grains of sand, rocks, or with biological intervention, the shells of microscopic marine organisms.

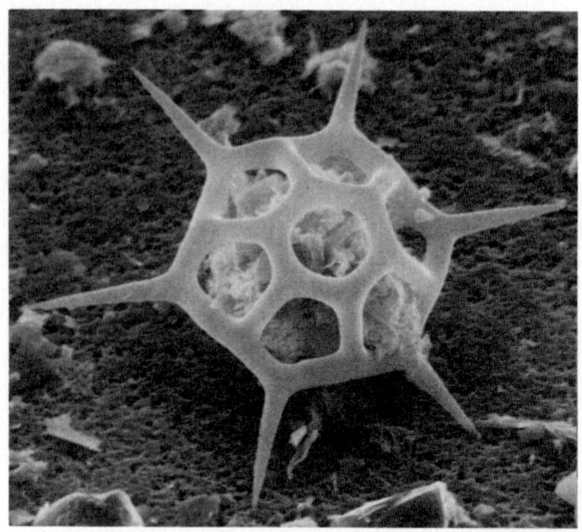

This delicate skeleton of a microorganism, from Narragansett Bay, Rhode Island, is composed of silicon dioxide. The material within the skeleton is organic debris.

shape from the arrangement of the carbon atoms that form the backbone, or skeleton, of the molecule. The shape of the molecule, in turn, determines many of its properties and its function within living systems.

In the molecules shown in Figure 3–1, every carbon bond that is not occupied by another carbon atom is taken up by a hydrogen atom. Such compounds, consisting of only carbon and hydrogen, are known as *hydrocarbons*. Structurally, they are the simplest kind of organic compounds. Although most hydrocarbons are derived from the remains of organisms that died millions of years ago (see page 466), they are relatively unimportant in living organisms. They are, however, of great economic importance; the liquid fuels upon which we depend—gasoline, diesel fuel, and heating oil—are all hydrocarbons.

Functional Groups

The specific chemical properties of a molecule derive principally from groups of atoms, known as *functional groups*, that are attached to the carbon skeleton. An OH (hydroxyl) group is an example.* When one hydrogen and one oxygen are bonded covalently, one outer electron of the oxygen is left over for sharing. A compound with a hydroxyl group in place of one or more of the hydrogens in a hydrocarbon is known as an alcohol. Thus methane (CH_4), with the replacement of one hydrogen atom by a hydroxyl group, becomes methanol, or wood alcohol (CH_3OH), a pleasant-smelling, poisonous compound noted for its ability to cause blindness

* —OH, the functional group, is called hydroxyl; OH^-, the ion, is called hydroxide.

ALPHA-GLUCOSE

GLUCOSE, STRAIGHT-CHAIN FORM

BETA-GLUCOSE

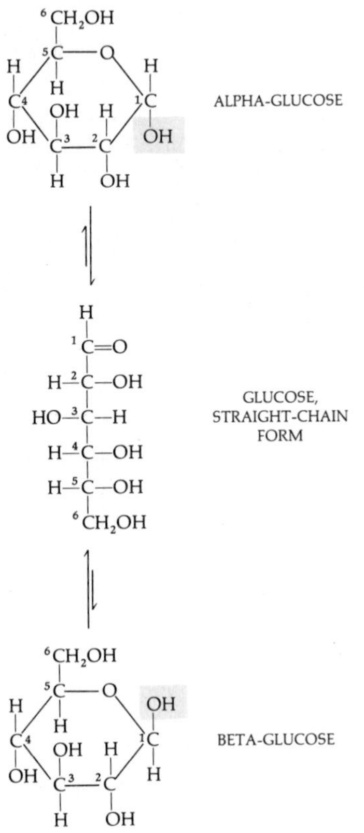

3–2
In aqueous solution, the six-carbon sugar glucose exists in two different ring structures, alpha and beta, that are in equilibrium with each other. The molecules pass through the straight-chain form to get from one structure to the other. The sole difference in the two ring structures is the position of the hydroxyl group attached to carbon atom 1; in the alpha form it is below the plane of the ring, and in the beta form it is above the plane. As we shall see, this small difference can lead to very significant differences in the properties of larger molecules formed by living systems from glucose.

Table 3–1 Some Biologically Important Functional Groups

GROUP	NAME	BIOLOGICAL SIGNIFICANCE
—OH	Hydroxyl	Polar, thus water-soluble; forms hydrogen bonds
—COOH	Carboxyl	Weak acid (hydrogen donor); becomes negatively charged when it loses a hydrogen: $-COO^- + H^+$
$-NH_2$	Amino	Weak base (hydrogen acceptor); becomes positively charged when it accepts a hydrogen: $-NH_3^+$
—CHO	Aldehyde	Polar, thus water-soluble; characterizes some sugars
>C=O	Ketone (or carbonyl)	Polar, thus water-soluble; characterizes other sugars
$-CH_3$	Methyl	Hydrophobic (insoluble in water)
$-PO(OH)_2$	Phosphate	Acid (hydrogen donor); usually negatively charged: $-PO_3^{2-} + 2H^+$

and death. Ethane similarly becomes ethanol, or grain alcohol (C_2H_5OH), which is present in all alcoholic beverages. Glycerol, $C_3H_5(OH)_3$, contains, as its formula indicates, three carbon atoms, five hydrogen atoms, and three hydroxyl groups. The carboxyl group (COOH), mentioned in the previous chapter, is a functional group that gives a compound the properties of an acid.

Many functional groups are polar and so have a partial positive or negative charge in aqueous solution. Some of these tend to become fully ionized, depending on the pH of the solution. Thus these groups confer water-solubility and local electric charge to the molecules that contain them. Many functional groups participate directly in chemical reactions.

Functional groups are essential for recognizing a particular molecule and helping predict its properties. Alcohols, with their polar hydroxyl groups, tend to be soluble in water, for instance, whereas hydrocarbons, such as butane, with only nonpolar functional groups (such as methyl groups), are highly insoluble in water. Sugars contain hydroxyl groups plus an aldehyde group or a ketone group; when a sugar molecule that contains five or more carbon atoms dissolves in water, the aldehyde or ketone group often reacts with one of the hydroxyl groups, resulting in a ring structure (Figure 3–2). Aldehyde groups are often associated with pungent odors and tastes. Smaller molecules with aldehyde groups, such as formaldehyde, have unpleasant odors, whereas larger ones, such as the chemicals that give vanilla, apples, cherries, and almonds their distinctive flavors, tend to be pleasing to the human sensory apparatus.

Four Representative Molecules

Figure 3–3 shows four kinds of molecules found in living systems, most often as constituents of larger molecules. The first (a), as you will recognize, has a long hydrocarbon chain; this chain, being nonpolar, is hydrophobic. Because of the COOH, you know that it is an acid. It is a *fatty acid,* a major component of lipids such as oils, fats, and waxes.

The second molecule (b) is a *sugar.* It is recognizable as a sugar because it consists of carbon, hydrogen, and oxygen in the ratio of one carbon to two hydrogens to one oxygen (CH_2O is the general formula for a sugar). The molecule shown has a ketone group characteristic of a sugar, and like glucose (Figure 3–2), it forms a ring in aqueous solution. You can also tell that it is water-soluble because of the polar OH (alcohol) groups. This particular sugar is fructose.

The third molecule (c) is an *amino acid.* An amino acid consists of a carbon atom bonded to one hydrogen atom and to three functional groups. One of these groups is always a carboxyl group, which gives the molecule its acidic properties. Because of the amino group, an amino acid is also a weak base. The third functional group varies from amino acid to amino acid; in this case it is a methyl (CH_3) group, and the amino acid is alanine.

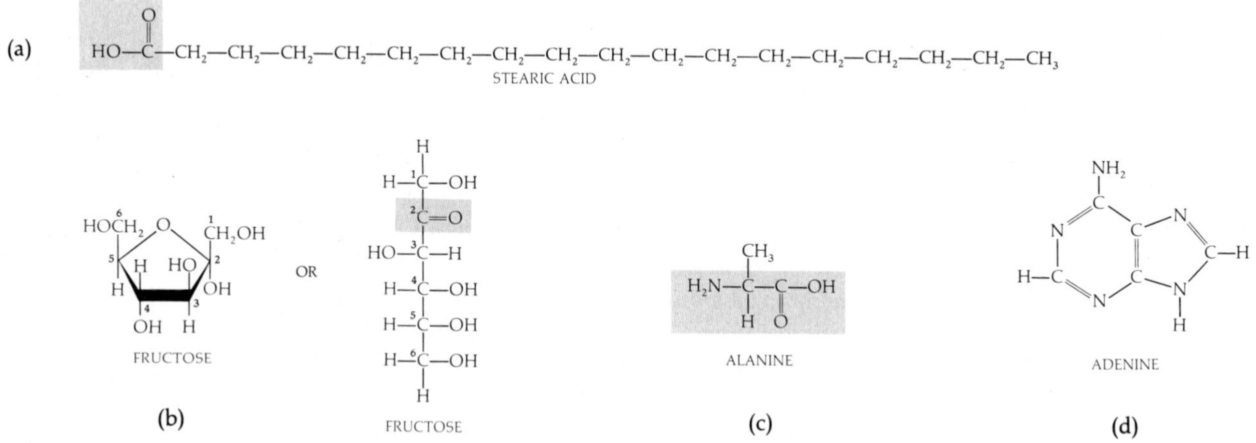

(a) STEARIC ACID

(b) FRUCTOSE OR FRUCTOSE

(c) ALANINE

(d) ADENINE

3–3
(a) *A fatty acid is a long hydrocarbon chain terminating in a carboxyl group (indicated in color). Fatty acids are major components of many lipids.* (b) *The sugar fructose, which tends to assume the ring form in aqueous solution. In the chain form, the ketone group is indicated by color.* (c) *Every amino acid contains an amino group (NH_2) and a carboxyl group (COOH) bonded to a central carbon atom. A hydrogen atom and a side group are also bonded to the same carbon atom. This basic structure is the same in all amino acids. In this amino acid, called alanine, the side group is CH_3 (methyl), one of the functional groups shown in Table 3–1.* (d) *A nitrogenous base, adenine. Adenine is an important component of DNA, the hereditary material, and of a number of other compounds as well. Note the two nitrogen-containing rings.*

3–4
A chemical bond is a force holding atoms together. The strength of the bond is measured in terms of the energy required to break it. The figures at the left indicate the number of kilocalories that will break the bonds between the pairs of atoms shown. The lines connecting the atoms represent the bonds, and the figures above them the characteristic center-to-center distances between them, expressed in nanometers (1 nanometer, abbreviated nm, equals 10^{-9} meter). Double lines indicate double bonds, which, as you can see, hold the atoms closer together and are stronger.

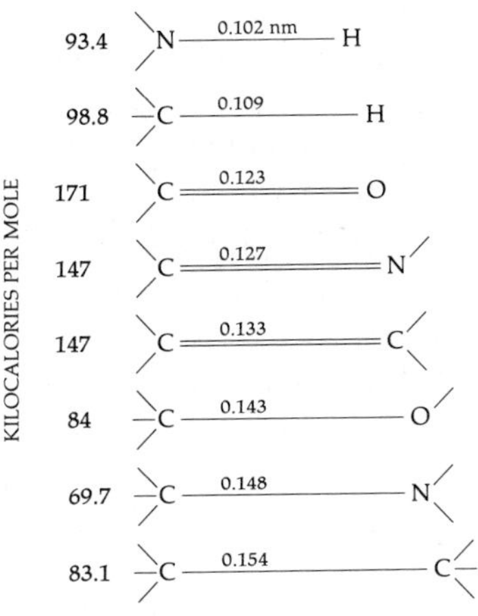

The fourth molecule (d) is a *nitrogenous base*. It is called nitrogenous because the rings of the molecule contain not only carbon but also nitrogen. It is a base because of the amino group. Nitrogenous bases are important constituents of nucleotides. This nitrogenous base is called adenine and, as we shall see, it is definitely one of the stars in the biological drama.

It has been said that it is necessary only to be able to recognize about 30 molecules for a working knowledge of the biochemistry of cells. One of these is a fatty acid; two others are the sugars glucose and ribose; five are nitrogenous bases such as adenine; and 20 are the biologically important amino acids. So, as you can see, the cast of characters is nearly complete.

The Energy Factor

Covalent bonds—the bonds commonly found in organic molecules—are strong, stable bonds consisting of electrons moving in orbitals about two or more different atomic nuclei. These bonds have different characteristic strengths, depending on the configurations of these orbitals. Bond strengths are conventionally measured in terms of the energy, expressed in kilocalories per mole, required to break the bond (Figure 3–4). Chemical reactions always involve a change in electron configurations and therefore in bond strengths. Depending on the relative strengths of the bonds broken and the bonds formed in the course of a chemical reaction, energy will either be released from the system or will be taken up by it from the surroundings.

Consider, for example, the burning of methane, represented by the following equation:

$$CH_4 + 2O_2 \longrightarrow CO_2 + 2H_2O$$

This reaction can be set in motion by a spark, which is what causes explosions in coal mines, and it releases energy in the form of heat. The amount of energy released can be measured quite precisely (Figure 3–5). It turns out to be 213 kilocalories per mole of methane. This can be expressed by a simple equation:

$$\Delta H^\circ = -213 \text{ kcal/mole}$$

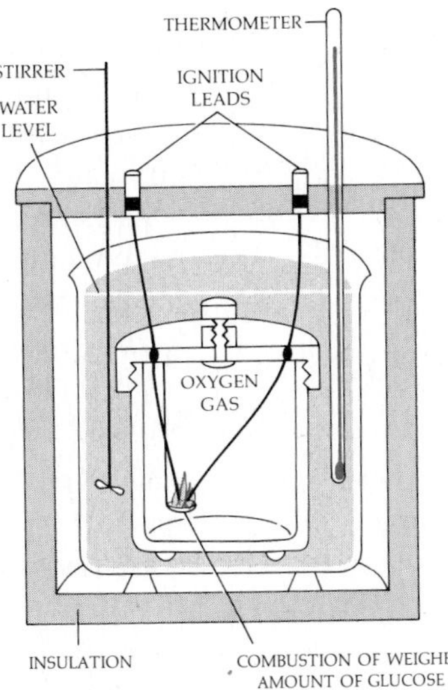

3–5
A calorimeter. A known quantity of glucose or some other material is ignited electrically. As it burns, the rise in the temperature of the surrounding water is measured. Using the specific heat of water and the known weight of water in the calorimeter, one can then calculate the number of calories released by the burning of the sample.

MOLECULAR MODELS

As we saw in Chapters 1 and 2, chemists have developed various models to represent the structures of atoms and molecules. Each of these models is a way of organizing a particular set of scientific data and of focusing attention on particular characteristics of atoms and molecules.

Because the properties of a molecule depend on its three-dimensional characteristics, physical models are often the most useful. For example, ball-and-stick models of the kind shown in Figure 3–1 emphasize the geometry of a molecule and, in particular, the bonds between atoms. But these models fail to suggest the overall shape of the molecule created by the movement of electrons within their orbitals.

A closer approximation of molecular shape is provided by space-filling models. Each atom is represented by the edge of its outermost orbitals. Space-filling models are misleading, however, in that molecules do not fill space in the same way that we think of a table or a rock as filling space. The atoms that make up molecules consist mostly of empty space. If the outer orbitals of the electrons in an oxygen atom were the size of the perimeter of the Astrodome in Houston, the nucleus would be a ping-pong ball in the center of the stadium. What "fills" the space in molecules are regions of charge, associated with the movements of the electrons around the nuclei. One molecule "sees" another molecule in terms of those regions of charge. As a consequence, for instance, a protein that transports glucose molecules into the living cell will not transport fructose molecules because of the differences in the shape of those regions of charge. All the intricate biochemistry that goes on in the cell is based on this ability of molecules to "recognize" one another.

Ball-and-stick and space-filling models are often used in the laboratory, but they are less useful on paper because it is necessary to see them from all angles to see all of the atoms and their bonds. The most accurate two-dimensional representations of molecular structure are orbital models, such as those shown in Figure 2–3 (page 34). For molecules containing more than a few atoms, however, orbital models become extremely complicated. Thus, when representing complex molecules, such as those found in living systems, chemists usually use molecular formulas or structural formulas. A molecular formula indicates the number of atoms of each kind within the molecule, while a structural formula shows how the atoms are bonded to one another.

Glucose, for example, has 6 carbon atoms, 12 hydrogen atoms, and 6 oxygen atoms. Its molecular formula is $C_6H_{12}O_6$. *However, fructose also contains 6 carbons, 12 hydrogens, and 6 oxygens and has a similar structure—a chain of carbon atoms to which hydrogen and oxygen atoms are attached. The differences between glucose and fructose are determined by which carbon atoms the other atoms are attached to. The molecules can therefore be distinguished by their structural formulas, as shown at the right.*

The Greek letter delta (**Δ**) stands for change, *H* for heat, and the superscript ° indicates that the reaction occurs under certain standard conditions of temperature and pressure. The minus sign indicates that energy has been released.

Similarly, changes in energy occur in the chemical reactions that take place in organisms. However, as we shall see in Section 2, living systems have evolved strategies for minimizing the energy required to set a reaction in motion and also the proportion of energy released as heat. These strategies involve, among other things, specialized protein molecules known as *enzymes*, which are essential participants in the chemical reactions of living systems. (The word "strategy" in its ordinary meaning is a deliberate plan to achieve a specified goal. Biologists use it to mean a group of related traits, evolved by organisms under the influence of natural selection, that solves particular problems encountered by living systems.)

CARBOHYDRATES: SUGARS AND POLYMERS OF SUGARS

Carbohydrates are the principal energy-storage molecules in most living things. In addition, they form a variety of structural components of living cells; the walls of

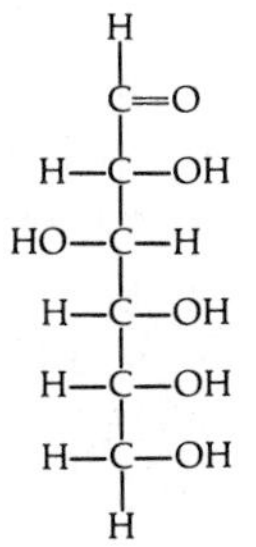

GLUCOSE (CHAIN)

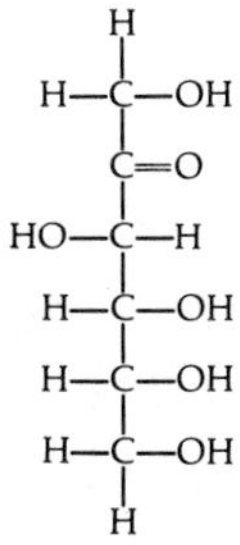

FRUCTOSE (CHAIN)

Or, in the ring form:

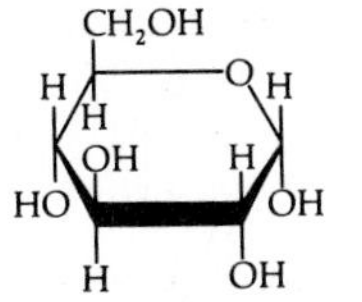

GLUCOSE (RING)

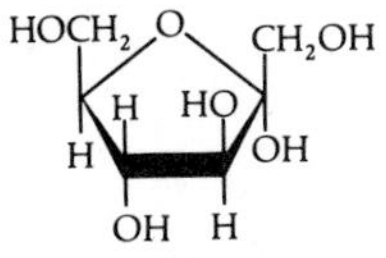

FRUCTOSE (RING)

The lower edges of the rings are made thicker to hint at a three-dimensional structure. By convention, the carbon atoms at the intersections of the links in an organic ring structure are "understood" and not labeled.

GLUCOSE

FRUCTOSE

Space-filling models of the sugars glucose and fructose. The darkest spheres, almost completely hidden at the center of each molecule, represent the carbon atoms. The large dark spheres at the surface of each molecule represent oxygen atoms, while the white spheres represent hydrogen atoms.

Although structural formulas give us less information than physical models, they are a convenient tool as we examine the molecules involved in the structures and processes of living systems.

young plant cells, for example, are about 40 percent cellulose, which is the most common organic compound in the biosphere.

Carbohydrates are classified according to the number of sugar molecules they contain. *Monosaccharides* ("single sugars"), such as ribose, glucose, and fructose, contain only one sugar molecule. *Disaccharides* consist of two sugar molecules linked covalently, while *polysaccharides* contain many sugar molecules. Large molecules, such as polysaccharides, that are made up of similar or identical subunits are known as *polymers* ("many parts") and the subunits are called *monomers* ("single parts").

Monosaccharides: Ready Energy for Living Systems

As we noted previously, monosaccharides (sugars) are characterized by hydroxyl groups and an aldehyde or ketone group. They can be described by the formula $(CH_2O)_n$, where n may be as small as 3, as in $C_3H_6O_3$, or as large as 8, as in $C_8H_{16}O_8$ (Figure 3-6). These proportions gave rise to the term carbohydrate ("hydrate of carbon") for sugars and the larger molecules formed from sugar subunits.

3–6

Two different ways of classifying monosaccharides: according to the number of carbon atoms and according to the functional groups (indicated in color). Glyceraldehyde, ribose, and glucose contain, in addition to hydroxyl groups, an aldehyde group. Dihydroxyacetone, ribulose, and fructose each contain a ketone group.

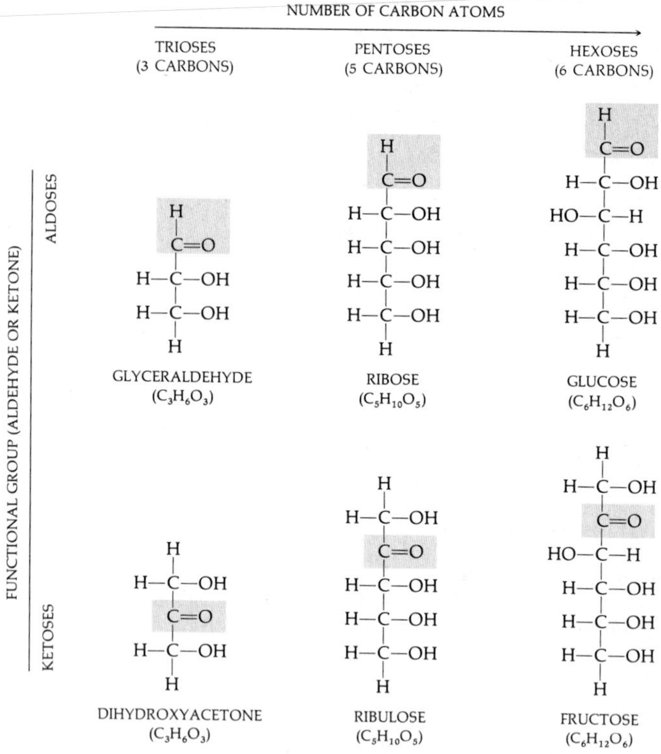

NUMBER OF CARBON ATOMS

TRIOSES (3 CARBONS) | PENTOSES (5 CARBONS) | HEXOSES (6 CARBONS)

FUNCTIONAL GROUP (ALDEHYDE OR KETONE)

ALDOSES

GLYCERALDEHYDE ($C_3H_6O_3$)

RIBOSE ($C_5H_{10}O_5$)

GLUCOSE ($C_6H_{12}O_6$)

KETOSES

DIHYDROXYACETONE ($C_3H_6O_3$)

RIBULOSE ($C_5H_{10}O_5$)

FRUCTOSE ($C_6H_{12}O_6$)

3–7

Sucrose is a disaccharide made up of two monosaccharides, glucose and fructose. Note that the fructose has been rotated 180° and joins to glucose in what is known as a 1⟶2 linkage (the bonding between the two rings involves the 1-carbon of glucose and the 2-carbon of fructose). The formation of sucrose involves the removal of a molecule of water (condensation). Splitting sucrose requires, conversely, the addition of a water molecule (hydrolysis).

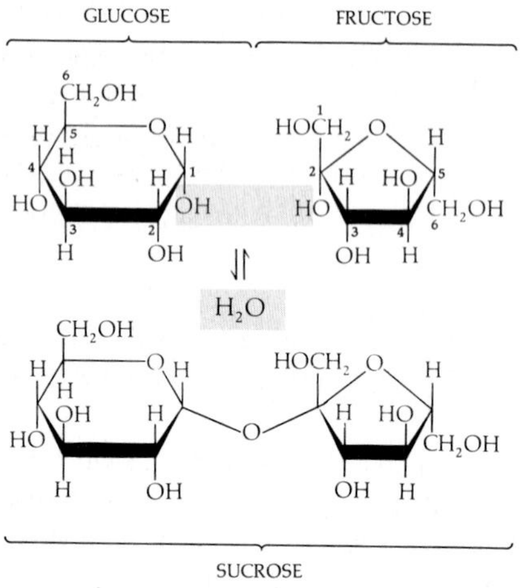

GLUCOSE | FRUCTOSE

H_2O

SUCROSE

Like hydrocarbons, carbohydrates can be burned, or oxidized, to yield carbon dioxide and water:

$$(CH_2O)_n + nO_2 \longrightarrow (CO_2)_n + (H_2O)_n$$

This reaction is also energy-yielding, and the amount of energy released as heat can be calculated by burning sugar molecules in a calorimeter. The same amount of energy is released—although not nearly so wastefully—when the equivalent amount of carbohydrate is oxidized in a living cell as when it is burned in a calorimeter.

This statement comparing the oxidation of food molecules with that of fuel molecules is not a metaphor; it is a fact. For example, the cost of transporting a kilogram the distance of a kilometer is 0.95 kilocalories for a pigeon, 0.73 for a person, and 0.83 for a Cadillac.

A principal energy source for vertebrates is the monosaccharide glucose. It is in this form that sugar is generally transported in the animal body. A patient receiving an intravenous feeding in a hospital is getting glucose dissolved in water. This glucose is carried through the bloodstream to the cells of the body where the energy-yielding reactions are carried out. As measured in a calorimeter, the oxidation of a mole of glucose yields 673 kilocalories:

$$C_6H_{12}O_6 + 6O_2 \longrightarrow 6CO_2 + 6H_2O$$

$$\Delta H^\circ = -673 \text{ kcal}$$

WHY IS SUGAR SWEET?

Actually it is not. Sugars have a number of objectively definable properties, such as molecular weight, melting point, and calorie content. Sweetness is not such a property. Just as beauty is in the eye of the beholder, sweetness resides not in the molecule itself but in the perception systems that have evolved to detect it.

Houseflies, for instance, have sugar detectors in their feet. If a housefly lights on a droplet of weak sugar solution, its proboscis (a tubular mouthpart through which it feeds) will automatically be extended. This useful response evolved over the millennia as a food-detecting mechanism. We similarly have sensory receptors tuned for the detection of sugar, although our receptors, in keeping with our eating habits, are in our tongues. Given sugar's value as an energy source, it is not surprising that many other organisms have sugar-detecting mechanisms. A housefly reacts positively to about the same types of sugars that we do, although its detection mechanism, as gauged by the proboscis extension test, is about 10 million times more sensitive.

The capacity to detect sugars, and the fact that the sensation is pleasurable, probably evolved in insects and other animals because it promoted their ingestion of these energy-rich food molecules and, by extension, their survival. This taste for sugar has been exploited by plants, particularly the flowering plants. They have evolved nectaries, dripping with sugary syrup, by which they lure pollinators to their flowers, and fruits that turn sweet just as the seeds become mature and are ready to be transported to a germination site. Animals that consume these sugary products obtain not only energy-rich sugar molecules but also other essential nutrients, such as plant proteins and lipids, vitamins, and minerals.

Today, our taste for sugar is being further exploited by manufacturers of prepared foods. Many dry cereals are more than 50 percent sugar, and sugar is an added hidden ingredient in countless other products. We consume increasing amounts of sugar in place of other, equally essential nutrients. Sugar has thus become what behavioral scientists call a "supernormal stimulus." Baby cuckoos provide an example of a supernormal stimulus. Hatched in the nests of other birds, they gape so widely at the sight of food that the adult birds, all their parental attention focused on this one oversized craw, let their own offspring starve. As another example, herring gulls abandon their own eggs for an artificial egg 1.5 times its size. And following a cue designed to lead us to a nutritious, vitamin-rich food supply, we let our children and ourselves grow fat, toothless, and malnourished, and die before our time.

Disaccharides: Transport Forms

Although glucose is the common transport sugar for vertebrates, sugars are often transported in other organisms as disaccharides. Sucrose, commonly called cane sugar, is the form in which sugar is transported in plants from the photosynthetic cells (mostly in the leaf), where it is produced, to other parts of the plant body. Sucrose is composed of the monosaccharides glucose and fructose. As is the case with all disaccharides and, indeed, with most organic polymers, a molecule of water is removed in the course of bond formation, a reaction known as *condensation*. Thus, only the monomers of carbohydrates actually have a CH_2O ratio because of the removal of two atoms of hydrogen and one of oxygen every time such a bond is formed.

When the sucrose molecule is split into glucose and fructose, as it is when it is used as an energy source, the molecule of water is added again. This splitting requires the addition of a water molecule and so is known as *hydrolysis*, from *hydro*, meaning "water," and *lysis*, meaning "breaking apart."

Another common disaccharide is lactose, a sugar that occurs only in milk. Lactose is made up of glucose combined with another monosaccharide, galactose. Sugar is transported through the blood of many insects in the form of another disaccharide, trehalose, which consists of two glucose units linked together.

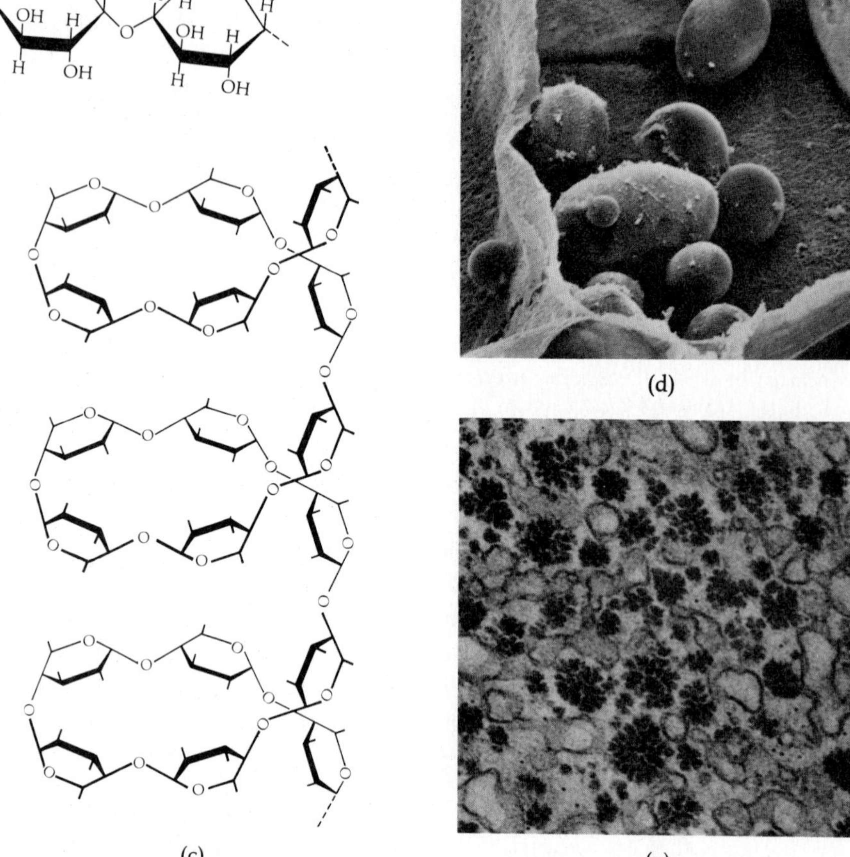

3–8
In plants, sugars are stored in the form of starch. Starch is composed of two different types of polysaccharides, amylose (a) *and amylopectin* (b). *A single molecule of amylose may contain 1,000 or more glucose units with the first carbon of one glucose ring linked to the fourth carbon of the next in a long unbranched chain, which winds to form a helix* (c). *A molecule of amylopectin may contain from 1,000 to 6,000 glucose units; short chains containing about 24 to 36 glucose units periodically branch off from the main chain.*

(d) *Starch molecules, perhaps because of their helical nature, tend to cluster into granules. In this scanning electron micrograph of a single storage cell of a potato, the spherical and egg-shaped objects are starch granules.*

(e) *Glycogen, which is the common storage form for sugar in vertebrates, resembles amylopectin in its general structure except that each branch contains only 16 to 24 glucose units. The dark granules in this liver cell are glycogen. When glucose is needed, it is provided by the conversion of glycogen.*

Storage Polysaccharides

Polysaccharides are made up of monosaccharides linked together in long chains. Some of them are storage forms of sugar. *Starch*, for instance, is the principal food storage form in most plants. A potato, for example, contains starch produced from the sugar formed in the green leaves of the plant; the sugar is transported underground and accumulated there in a form suitable for winter storage, after which it will provide for new growth in the spring. Starch occurs in two forms, amylose and amylopectin. Both consist of glucose units linked together (Figure 3–8).

Glycogen is the principal storage form for sugar in higher animals. Glycogen has a structure much like that of amylopectin except that it is more highly branched, with branches occurring every eight to ten glucose units. In vertebrates, glycogen is stored principally in the liver and in muscle tissue. When there is an excess of glucose in the bloodstream, the liver forms glycogen. When the concentration of glucose in the blood drops, the hormone glucagon, produced by the pancreas, is released into the bloodstream; glucagon stimulates the liver to hydrolyze glycogen to glucose, which then enters the bloodstream.

Formation of polysaccharides from monosaccharides requires energy. However, when the cell needs energy, these polysaccharides can be hydrolyzed, releasing monosaccharides that can, in turn, be oxidized to provide energy for cellular work.

Structural Polysaccharides

A major function of molecules in living systems is to form the structural components of cells and tissues. The principal structural molecule in plants is *cellulose*. In fact, half of all of the organic carbon in the biosphere is contained in cellulose. Wood is about 50 percent cellulose, and cotton is nearly pure cellulose.

Cellulose molecules form the fibrous part of the plant cell wall. The cellulose fibers, embedded in a matrix of other kinds of polysaccharides, form an external envelope around the plant cell. When the cell is young, this envelope is flexible and stretches as the cell grows, but it becomes thicker and more rigid as the cell matures. In some plant tissues, such as the tissues that form wood and bark, the cells eventually die, leaving only their tough outer walls.

Cellulose is a polymer composed of monomers of glucose, just as starch and glycogen are. Starch and glycogen can be readily utilized as fuels by almost all kinds of living systems, but only a few microorganisms—certain bacteria, protozoa, and fungi—can hydrolyze cellulose. Cows and other ruminants, termites, and cockroaches can use cellulose for energy only because of microorganisms that inhabit their digestive tracts.

To understand the differences between structural polysaccharides, such as cellulose, and energy-storage polysaccharides, such as starch or glycogen, we have to look briefly once again at the glucose molecule. You will remember that the molecule is basically a chain of carbon atoms and that when it is in solution, as it is in the cell, it assumes a ring form. The ring may close in either of two ways (see Figure 3–2). One ring form is known as alpha, and the other as beta. The alpha and beta forms are in equilibrium, with a certain number of the molecules changing from one to the other all of the time, going through the open-chain structure to reach the other form. Starch and glycogen are both made up entirely of alpha units. Cellulose consists entirely of beta units (see Figure 3–9a). This slight difference has a profound effect on the three-dimensional structure of the molecules, which align in parallel, forming crystalline cellulose microfibrils. As a result, cellulose is impervious to the enzymes that so successfully break down the storage polysaccharides.

3–9

(a) *Cellulose consists of beta-glucose monomers, whereas starch (Figure 3–8) consists of alpha-glucose monomers. In cellulose, the OH groups (indicated in color), which project from both sides of the chain, form hydrogen bonds with neighboring OH groups, resulting in the formation of bundles of cross-linked parallel chains* (b). *By contrast, in the starch molecule, most of the OH groups capable of forming hydrogen bonds face toward the exterior of the helix, making it more readily soluble in the surrounding water.*

(c) *The plant cell wall is composed largely of cellulose. Each of the microfibrils you can see here (magnified about 30,000 times) is a bundle of hundreds of cellulose strands, and each strand is a chain of glucose units* (a). *The microfibrils, as strong as an equivalent amount of steel, are embedded in other polysaccharides, one of which is pectin.*

(a)

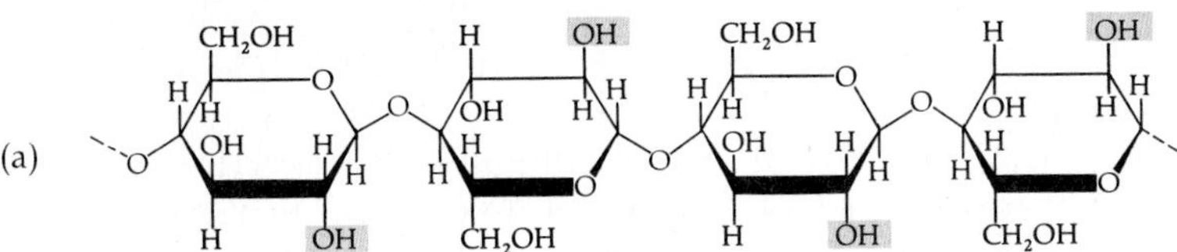

CELLULOSE MOLECULE

(b)

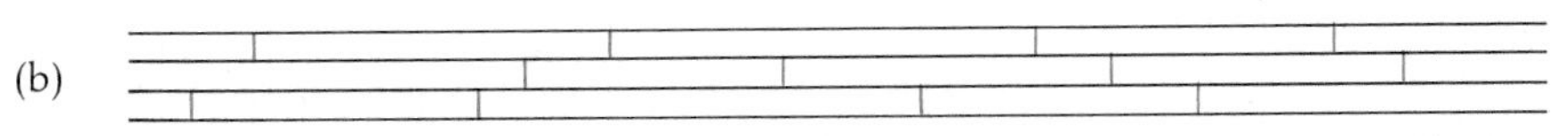

MODEL OF CROSS-LINKED CELLULOSE MOLECULES

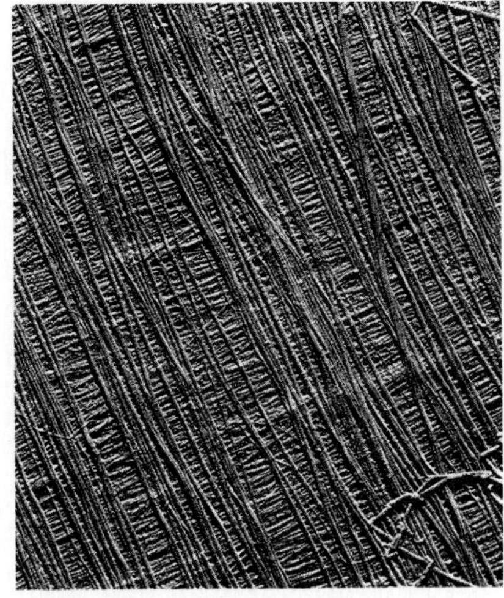

(c)

3-10

(a) *Chitin is a polymer consisting of repeated modified monosaccharides. As you can see, the monomer is a six-carbon sugar, like glucose, to which a nitrogen-containing group has been added.* (b) *Green darner molting. The relatively hard outer coverings, or exoskeletons, of insects contain chitin. Some types of insects, after molting, recycle their sugars and nitrogen by thriftily eating their discarded exoskeletons.*

(a)

CHITIN

(b)

Chitin, which is a major component of the exoskeletons of arthropods, such as insects and crustaceans, and also of the cell walls of many fungi, is a tough, resistant, modified polysaccharide (Figure 3-10). At least 900,000 different species of organisms can synthesize chitin, and it has been estimated that the individuals belonging to a single species of crab produce several million tons of chitin a year.

LIPIDS

Lipids are a general group of organic substances that are insoluble in polar solvents, such as water, but that dissolve readily in nonpolar organic solvents, such as chloroform, ether, and benzene. Many, although not all, contain fatty acids as major structural components. Lipids serve both as energy-storage forms—usually in the form of fats or oils—and for structural purposes, as in the case of phospholipids and waxes.

Fats and Oils: Energy in Storage

Unlike many plants, such as the potato, animals have only a limited capacity to store carbohydrates. In vertebrates, sugars in excess of what can be stored as glycogen are converted into fats. Some plants also store food energy as oils, especially in seeds and fruits. Fats and oils contain a higher proportion of energy-rich carbon-hydrogen bonds (see Figure 3-4) than carbohydrates do and, as a consequence, contain more chemical energy. On the average, fats yield about 9.3 kilocalories per gram* as compared to 3.79 kcal per gram of carbohydrate, or 3.12 kcal per gram of protein. Also, because they are hydrophobic, they do not attract water molecules and hence are not "weighted down" by them, as glycogen is. Taking into account the water factor, fats store six times as much energy, gram for gram, as glycogen, which is undoubtedly why in the course of evolution they came to play a major role in energy storage.

For instance, a male ruby-throated hummingbird has a fat-free weight of 2.5 grams (about 1/10 ounce). It migrates every fall from Florida to Yucatan, some 2,000 kilometers. Before doing so, it accumulates 2.0 grams of body fat, an amount almost equal to its original weight. However, if it were to carry the same energy reserves in the form of glycogen, it would have to carry 5 grams, twice its own fat-free weight.

---

* 1,000 grams = 1 kilogram = 2.2 pounds, so oxidation of a pound of fat would yield about 4,200 kilocalories, more than the 24-hour requirement for moderately active adults.

A neutral fat molecule consists of three molecules of fatty acid joined to one glycerol molecule. (Note the similarities of glycerol and glyceraldehyde, as shown in Figure 3–11a). Fatty acids, which are seldom found in cells in a free state (that is, not as part of another molecule), typically consist of chains between 14 and 22 carbon atoms long. About 70 different fatty acids are known. They differ in their chain lengths, in whether the chain contains any double bonds (as in oleic acid) or not (as in stearic acid), and in the position in the chain of any double bonds (Figure 3–11b). A fatty acid, such as stearic acid, in which there are no double bonds is said to be saturated because the bonding possibilities are complete for all the carbon atoms of the chain. A fatty acid, such as oleic acid, that contains carbon atoms joined by double bonds is said to be unsaturated because those carbon atoms are able to form additional bonds with other atoms. Unsaturated fats, which tend to be oily liquids, are more common in plants than in animals; examples are olive oil, peanut oil, and corn oil.

Sugars, Fats, and Calories

As we noted earlier, when carbohydrates are taken into the body in excess of the body's energy requirements, they are stored temporarily as glycogen or, more permanently, as fats. Conversely, when the energy requirements of the body are not met by its immediate intake of food, glycogen and, subsequently, fat are broken down to fill these requirements. Whether or not the body uses up its own storage molecules has nothing to do with the molecular form in which the energy comes into the body. It is simply a matter of whether these molecules, as they are broken down, release sufficient numbers of calories.

Insulators and Cushions

In general, fat stored in fat cells can be mobilized for energy when caloric intake is less than caloric expenditures. Some types of fat, however, seem to be protected from such mobilization. Large masses of fatty tissue, for example, surround mammalian kidneys and serve to protect these precious organs from physical shock. For reasons that are not understood, these fat deposits remain intact even at times of starvation. Another mammalian characteristic is a layer of fat under the skin, which serves as thermal insulation. This layer is particularly well developed in seagoing mammals (Figure 3–12).

3–11

(a) *Glycerol differs from the monosaccharide glyceraldehyde by only two hydrogen atoms. The hydroxyl (OH) groups of glycerol are characteristic of an alcohol. Glycerol is one of the building blocks of fats.*

(b) *A fat molecule consists of three fatty acids joined to a glycerol molecule. The long hydrocarbon chains of which the fatty acids are composed terminate in carboxyl groups, which become covalently bonded to the glycerol molecule. Each bond is formed when a molecule of water (color) is removed (condensation). The physical properties of the fat—such as its melting point—are determined by the lengths of the chains and by whether its component fatty acids are saturated or unsaturated. Three different fatty acids are shown here. Stearic acid and palmitic acid are saturated, and oleic acid is unsaturated, as you can see by the double bond in its structure.*

(a)

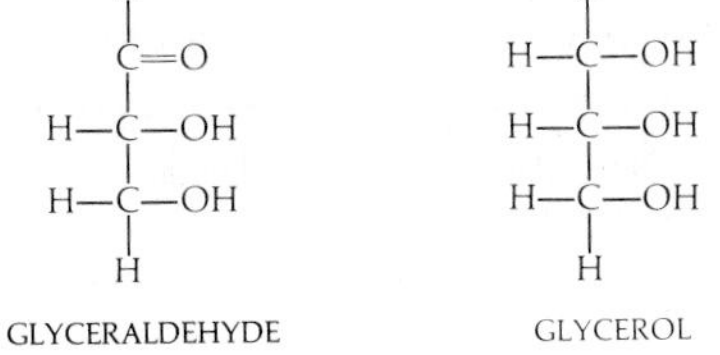

GLYCERALDEHYDE: H—C(=O)—C(H)(OH)—C(H)(OH)—H

GLYCEROL: H—C(H)(OH)—C(H)(OH)—C(H)(OH)—H

(b)

$HO—C(=O)—CH_2—CH_2—CH_2—CH_2—CH_2—CH_2—CH_2—CH_2—CH_2—CH_2—CH_2—CH_2—CH_2—CH_2—CH_2—CH_2—CH_3$

STEARIC ACID

$HO—C(=O)—CH_2—CH_2—CH_2—CH_2—CH_2—CH_2—CH_2—CH{=}CH—CH_2—CH_2—CH_2—CH_2—CH_2—CH_2—CH_2—CH_3$

OLEIC ACID

$HO—C(=O)—CH_2—CH_2—CH_2—CH_2—CH_2—CH_2—CH_2—CH_2—CH_2—CH_2—CH_2—CH_2—CH_2—CH_2—CH_3$

PALMITIC ACID

CARBOXYL GROUP

GLYCEROL

FATTY ACID

FAT MOLECULE

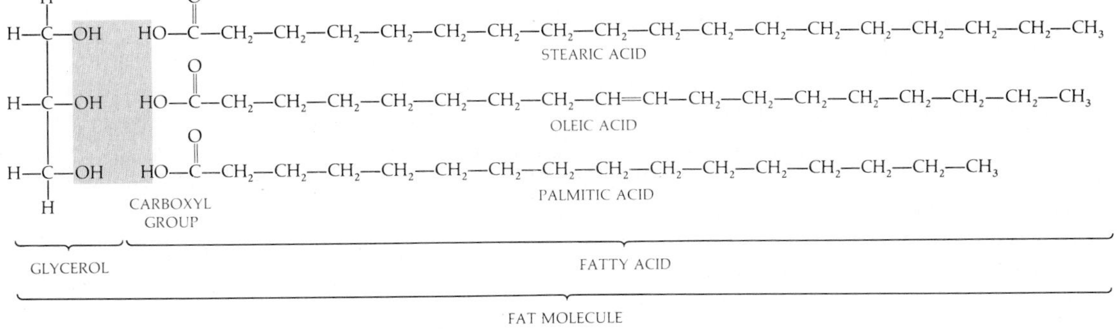

3–12
This Weddell seal, enjoying the Antarctic spring, is well insulated by a thick layer of fat under the skin, which serves the same function that a wet suit serves for a diver.

Among humans, females characteristically have a thicker layer of subdermal ("under-the-skin") fat than males. This capacity to store fat, although not much admired in our present culture, was undoubtedly very valuable 10,000 or more years ago. At that time, as far as we know, there was no other reserve food supply, and this extra fat not only nourished the woman but, more important, the unborn child and the nursing infant, whose ability to fast without damage is much less than that of the adult. Thus many of us are strenuously dieting off what millennia of evolution have given us the capacity to accumulate.

Phospholipids

Lipids also play extremely important structural roles. The lipids most important for structural purposes are phospholipids. Like fats, the phospholipids are composed of fatty acid chains attached to a glycerol backbone. In the phospholipids, however, the third carbon of the glycerol molecule is occupied not by a fatty acid but by a phosphate group (Figure 3–13) to which another polar group is usually attached. Phosphate groups are negatively charged. As a result, the phosphate end of the molecule is hydrophilic and the fatty acid portions are not. The consequences are shown in Figure 3–14. As we shall see, according to the present model, this arrangement of phospholipid molecules, with their hydrophilic heads extended and their hydrophobic tails clustered together, forms the structural basis of the cell membrane. We shall examine this structure further in Chapter 6.

3–13
A phospholipid molecule consists of two fatty acids linked to a glycerol molecule, as in a fat, and a phosphate group (indicated by color) linked to the glycerol's third carbon. It also usually contains an additional chemical group, indicated by the letter R. The fatty acid "tails" are nonpolar and therefore insoluble in water (hydrophobic); the polar "head" containing the phosphate and R groups is soluble (hydrophilic).

POLAR HEAD — NONPOLAR TAILS

$$R{-}O{-}\overset{O^-}{\underset{O}{\overset{|}{\underset{\|}{P}}}}{-}O{-}\overset{3}{C}H_2$$

$$H{-}\overset{2}{C}{-}O{-}\overset{O}{\overset{\|}{C}}{-}CH_2CH_2CH_2CH_2CH_2CH_2CH_2CH{=}CHCH_2CH_2CH_2CH_2CH_2CH_2CH_2CH_3$$

$$H{-}\overset{1}{C}{-}O{-}\overset{O}{\overset{\|}{C}}{-}CH_2CH_2CH_2CH_2CH_2CH_2CH_2CH_2CH_2CH_2CH_2CH_2CH_2CH_2CH_2CH_2CH_3$$

$$\underset{|}{}H$$

GLYCEROL

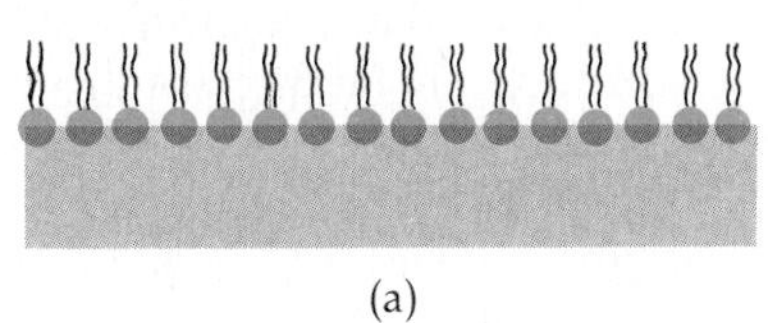

(a)

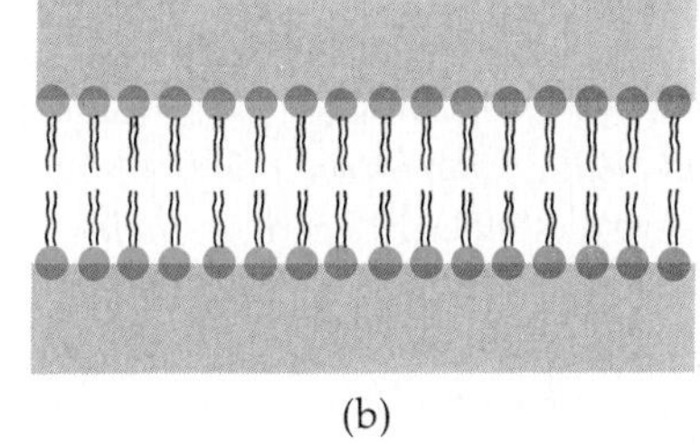

(b)

3–14
(a) *Because phospholipids have water-soluble heads and water-insoluble tails, they tend to form a thin film on a water surface with their tails extending above the water.* (b) *Surrounded by water, they spontaneously arrange themselves in two layers with their heads extending outward and their hydrophobic (water-fearing) tails inward. This arrangement is important in the structure of the cell membrane.*

Waxes

Waxes are also a form of structural lipid. They form protective coatings on skin, fur, feathers, on the leaves and fruits of land plants (Figure 3–15), and on the exoskeletons of many insects.

Cholesterol and Other Steroids

Cholesterol belongs to a group of compounds known as the *steroids* (Figure 3–16). Although steroids do not resemble the other lipids structurally, they are grouped with the lipids because they are insoluble in water. All the steroids have four linked carbon rings, like cholesterol, and several of them, like cholesterol, have a tail. In addition, many of them have the OH functional group, which classifies them as alcohols.

Cholesterol is found in cell membranes (with the exception of bacterial cells); about 25 percent (by dry weight) of the membranes of red blood cells is cholesterol. It is also a major component of the myelin sheath, the lipid membrane that wraps around fast-conducting nerve fibers, speeding the nerve impulse.

In some older people, cholesterol forms fatty deposits on the inner lining of the blood vessels. This blocks the vessels and decreases their elasticity, making such people more susceptible to high blood pressure, heart attacks, and strokes. Attempts have been made to reduce the incidence and extent of such deposits by reducing the dietary intake of high-cholesterol foods, such as eggs and cheese. But since many cells of the human body are able to synthesize cholesterol, it is not clear that a reduction of dietary cholesterol will solve the problem.

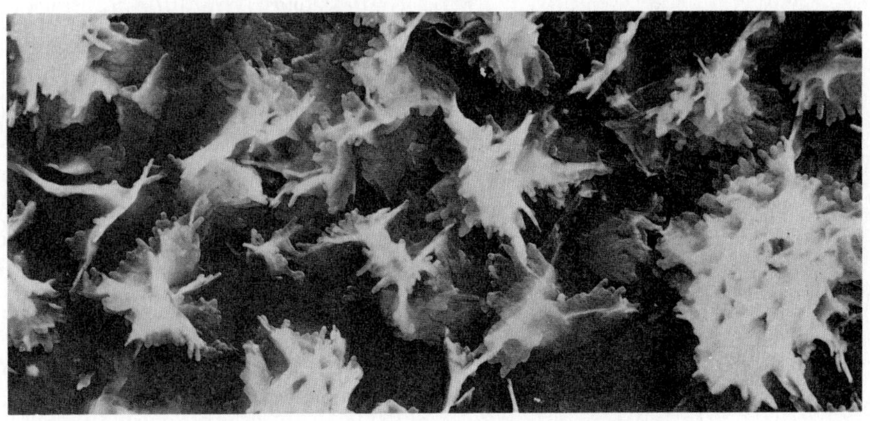

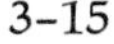
3–15
Waxes are also lipids. This electron micrograph shows waxy deposits on the upper surface of a eucalyptus leaf. The deposits are magnified 10,800 times. All groups of land plants synthesize waxes, which protect exposed plant surfaces from water loss.

3–16

Two examples of steroids. (a) *The cholesterol molecule consists of four carbon rings and a hydrocarbon chain.* (b) *Testosterone, a male sex hormone synthesized from cholesterol by cells in the testes, also has the characteristic four-ring structure but lacks the hydrocarbon tail.*

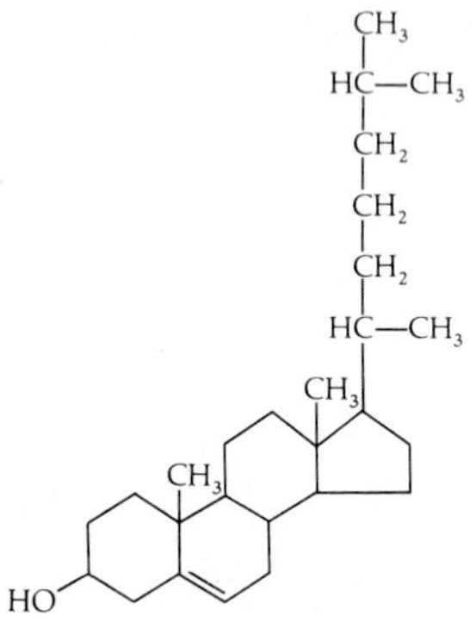

CHOLESTEROL

(a)

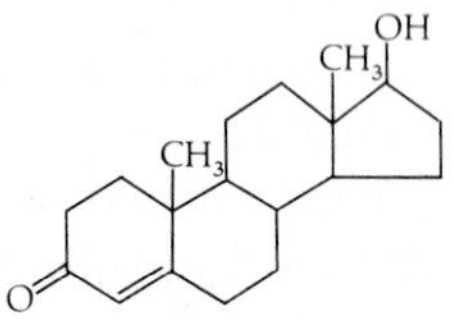

TESTOSTERONE

(b)

Sex hormones and the hormones of the adrenal cortex are also steroids; they are formed from cholesterol in the ovaries, testes, and other glands that produce them. Prostaglandins are a recently discovered group of chemicals with hormonelike actions, which are derived from fatty acids. Both the steroid hormones and the prostaglandins will be discussed more fully in Section 6.

PROTEINS

Proteins are among the most abundant organic molecules; in most living systems they make up 50 percent or more of the dry weight. Only plants, with their high cellulose content, are less than half protein. There are many different protein molecules: enzymes; hormones; storage proteins, such as those in the eggs of birds and reptiles and in seeds; transport proteins, such as hemoglobin; contractile proteins of the sort found in muscle; immunoglobulins (antibodies); membrane proteins; and many different types of structural proteins (Table 3–2). In their functions, their diversity is overwhelming. In their structure, however, they all follow the same simple blueprint: they are all polymers of amino acids, arranged in a linear sequence.

Table 3–2 Biological Functions of Proteins

TYPES OF PROTEINS*	EXAMPLES
Structural proteins	Collagen, silk, virus coats, microtubules
Regulatory proteins	Insulin, ACTH, growth hormones
Contractile proteins	Actin, myosin
Transport proteins	Hemoglobin, myoglobin
Storage proteins	Egg white, seed protein
Protective proteins in vertebrate blood	Antibodies, complement
Membrane proteins	Membrane-transport proteins, antigens
Toxins	Botulism toxin, diphtheria toxin
Enzymes	Sucrase, pepsin

* Many of the proteins listed here will be discussed in other sections of the book, particularly Section 6.

Amino Acids: The Building Blocks of Proteins

As we saw earlier, every amino acid has the same fundamental structure, which consists of a central carbon atom bonded to an amino group (NH_2), to a carboxyl group (COOH), and to a hydrogen atom. In every amino acid there is also another atom or group of atoms bonded to the central carbon. As shown in Figure 3–17, this side (R) group can be a hydrogen atom, in which case the amino acid is glycine; a CH_3 group, in which case the amino acid is alanine; and so on. The side group, depending on the atom or atoms that compose it, may have a positive charge, have a negative charge, be polar (with negative and positive zones), or have no charge at all (in which case it is hydrophobic).

A large variety of different amino acids is theoretically possible, but only 20 different kinds are used to build proteins. And it is always the same 20, whether in a bacterial cell, a plant cell, or a cell in your own body.

3–17

(a) *Every amino acid contains an amino group (NH_2) and a carboxyl group (COOH) bonded to a central carbon atom. A hydrogen atom and a side group are also bonded to the same carbon atom. This basic structure is the same in all amino acids. The "R" stands for the side group, which is different in each kind of amino acid.* (b) *The 20 kinds of amino acids used in making proteins. As you can see, their basic structures are the same, but they differ in their side groups. The side groups may be nonpolar (with no difference in charge between one zone and another), polar but with the two charges balancing one another out so that the side group as a whole is uncharged, negatively charged, or positively charged. The nonpolar side groups are not soluble in water, whereas the charged and polar side groups are water-soluble.*

(a)

R
H₂N—C—C—OH
H O

(b)

NONPOLAR

ALANINE (ala)
VALINE (val)
LEUCINE (leu)
ISOLEUCINE (ile)
PROLINE (pro)
PHENYLALANINE (phe)
TRYPTOPHAN (trp)
METHIONINE (met)

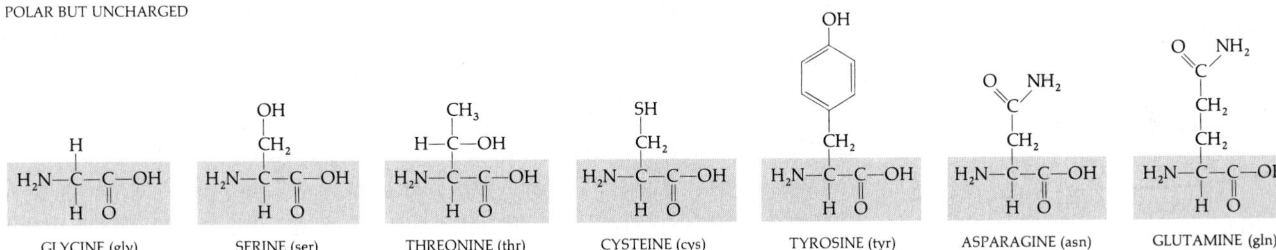

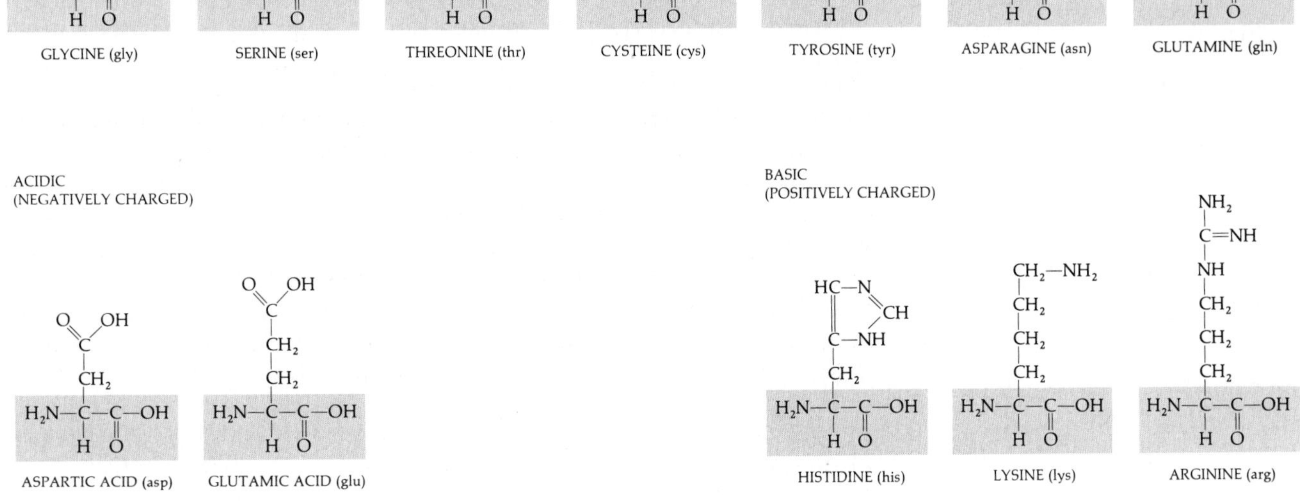

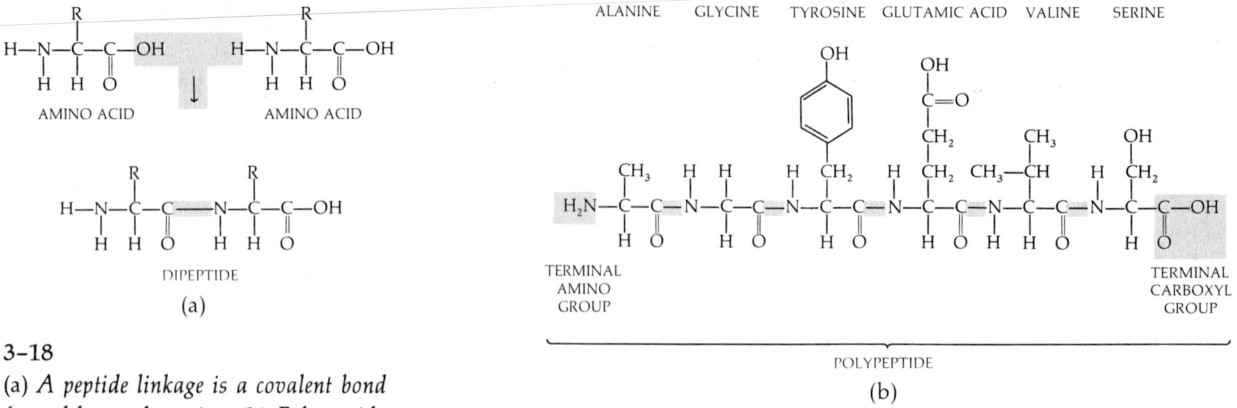

3–18

(a) *A peptide linkage is a covalent bond formed by condensation.* (b) *Polypeptides are polymers of amino acids linked together by peptide bonds, with the amino group of one acid joining the carboxyl group of its neighbor. The polypeptide chain shown here contains six different amino acids, but some chains may contain as many as 300 linked amino acid monomers.*

In proteins, amino acids are joined together by *peptide bonds*, which form as the result of condensation reactions (Figure 3–18). The sequence of amino acids in the chains determines the biological character of the protein molecule; even one small variation in the sequence may alter or destroy the way in which the protein functions.

Protein molecules are large, often containing several hundred amino acids. Thus the number of different amino acid sequences, and therefore the possible variety of protein molecules, is enormous—about as enormous as the number of different sentences that can be written with our own 26-letter alphabet. Organisms, however, have only a very small fraction of the possible proteins. The single-celled bacterium *Escherichia coli,** for example, contains 600 to 800 different kinds of proteins at any one time, and the cell of a plant or animal has several times that number. In a complex organism, there are at least several thousand different proteins, each with a special function and each, by its unique chemical nature, specifically fitted for that function.

The Levels of Protein Organization

Proteins are assembled in living systems with the amino group of one amino acid linked to the carbonyl† of another, like a line of boxcars. Such chains of covalently bonded amino acids are called *polypeptides*. The linear sequence of amino acids, which is dictated by the hereditary information in the cell for that particular protein, is known as the *primary structure* of the protein. Each different protein has a different primary structure. The primary structure of one protein is shown in Figure 3–19.

---

* Biologists use a binomial ("two-name") system for designating organisms. Every different kind of organism has a unique two-part name. The first part of the name refers to the genus (plural, genera) to which the organism belongs, and the second part refers to the species, a subdivision of the genus category. In this name, for example, *Escherichia* denotes the genus, while *coli* designates a particular kind, or species, of *Escherichia,* distinguished from all others by certain characteristics.

† When a peptide bond forms, the OH of the carboxyl group and an H of the amino group split out to form a water molecule. All that remains of the carboxyl group is the $\rangle C = O$ group, which we refer to as carbonyl. (See Table 3–1.)

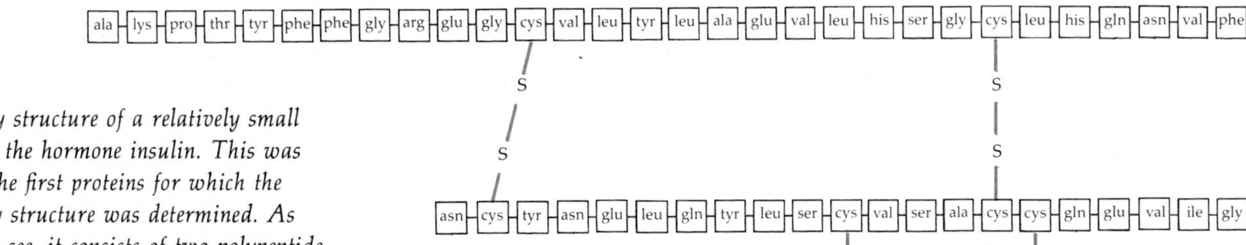

3–19
Primary structure of a relatively small protein, the hormone insulin. This was one of the first proteins for which the primary structure was determined. As you can see, it consists of two polypeptide chains held together by disulfide bridges.

The protein is assembled in a long chain, one amino acid at a time (a process to be described in some detail in Chapter 15). As it is assembled, interactions begin to take place among the various amino acids along the chain. Linus Pauling and coworker Robert Corey discovered that hydrogen bonds could form between the slightly positive amino hydrogen of one amino acid and the slightly negative carbonyl oxygen of another amino acid. They elucidated two structures that could result from these hydrogen bonds. One of these they called the alpha helix, because it was the first to be discovered, and the second, the beta pleated sheet. These structures are shown in Figure 3–20. Biochemists refer to the regular, repeated configurations caused by hydrogen bonding between atoms of the polypeptide backbone as the *secondary structure* of a protein. Proteins that exist for most of their length in a helical or pleated-sheet form are known as *fibrous proteins*, and they play important structural roles in organisms.

Other forces, which involve the nature of the R groups in the individual amino acids, are also at work on the polypeptide chain, and these counteract the formation of the hydrogen bonds just described. For instance, an R group such as that of isoleucine is so bulky that it interrupts the turn of the helix, making the hydrogen bonding impossible. Wherever a cysteine encounters another cysteine, a covalent bond may form, making a disulfide bridge that locks the molecule in that position. R groups with unlike charges are attracted to each other and those with like charges are mutually repelled. As the molecule is twisted and turned in solution, the hydrophobic R groups become clustered together in the interior of the molecule and the hydrophilic R groups extend outward into the aqueous solution. Hydrogen

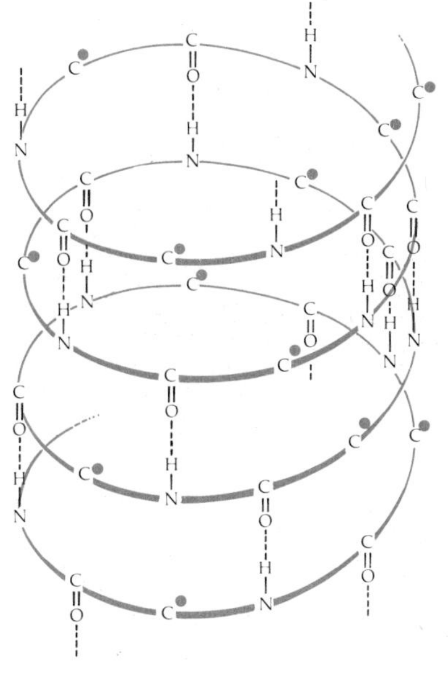
(a)

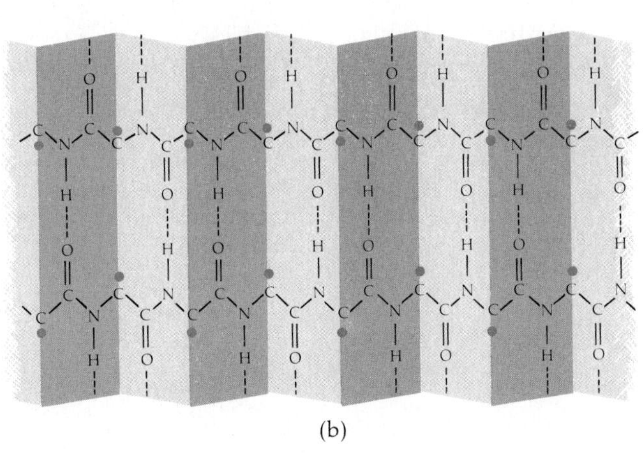
(b)

3–20
(a) *The alpha helix. The helix is held in shape by hydrogen bonds, indicated by the dashed lines. The bonds form between the oxygen atom in the carbonyl group in one amino acid and the amino group in another amino acid that occurs four amino acids farther along the chain. The R groups, not shown in this diagram, are attached to the carbons indicated by the colored dots. The groups extend out from the helix.* (b) *The beta-pleated sheet structure of proteins. The pleats are formed by hydrogen bonding between atoms of the backbone of the polypeptide; the R groups, which are attached to the carbons indicated by the colored dots, extend above and below the folds of the pleat.*

3–21

(a) *Types of bonds that stabilize the tertiary structure of a protein molecule.* (b) *Space-filling model of the digestive enzyme chymotrypsin. The intricate folding of the tertiary structure produces a globular molecule.*

● = AMINO ACID

N = NONPOLAR (HYDROPHOBIC) INTERACTIONS

S—S = DISULFIDE BRIDGES

---- = HYDROGEN BONDS

P+, P− = CHARGED (HYDROPHILIC) GROUPS

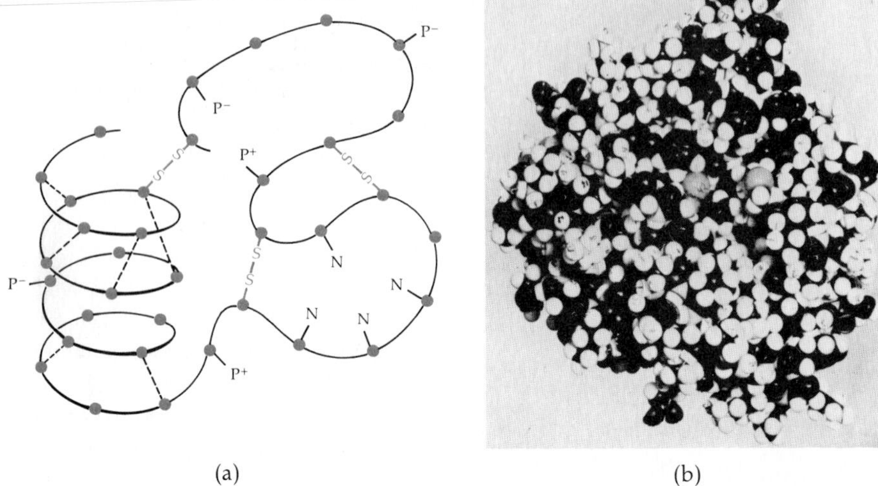

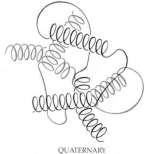

(a) (b)

bonds form, linking together segments of the amino acid backbone. The intricate three-dimensional structure that results from these interactions among R groups is called the *tertiary structure* of a protein. Figure 3–21 shows the various types of bonds that are involved in forming the tertiary structure.

In many proteins, the tertiary structure produces an intricately folded, globular shape for the molecule as a whole; these proteins are called *globular proteins*. Enzymes—proteins that regulate chemical reactions in living systems—are globular proteins. As we shall see in Chapter 8, the three-dimensional structure of an enzyme is the critical factor in determining its biological function. Antibodies, important components of the immune system, are also globular proteins.

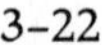

3–22

(a) *The primary structure of a protein is the linear sequence of its amino acids.* (b) *Because of interactions among these amino acids, the molecule spontaneously forms a secondary structure, such as the alpha helix shown here, and* (c) *a tertiary structure, such as a globule.* (d) *Many globular proteins, including hemoglobin and some enzymes, are made up of more than one amino acid chain. This structure is known as a quaternary structure.*

Many proteins are composed of more than one polypeptide chain. The polypeptide chains are held together by hydrogen bonds, disulfide bridges, attractions between positive and negative charges, and hydrophobic forces. Such proteins are often called multimeric; a protein containing four polypeptide chains is termed a tetramer. Hemoglobin, for example, is a tetramer; it is composed of four polypeptide chains, two of one kind (alpha) and two of another (beta). This level of organization of proteins, which involves interaction of two or more polypeptides, is called a *quaternary structure*.

Note that the secondary, tertiary, and quaternary structures all depend on the primary structure—the sequence of amino acids—and on the local chemical environment.

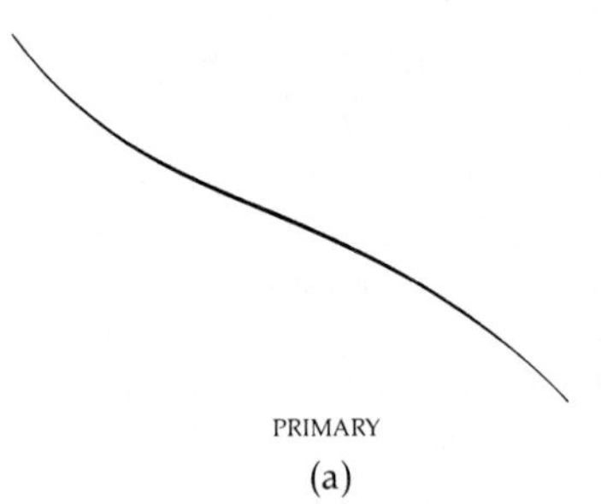

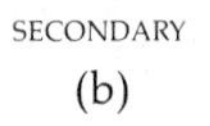

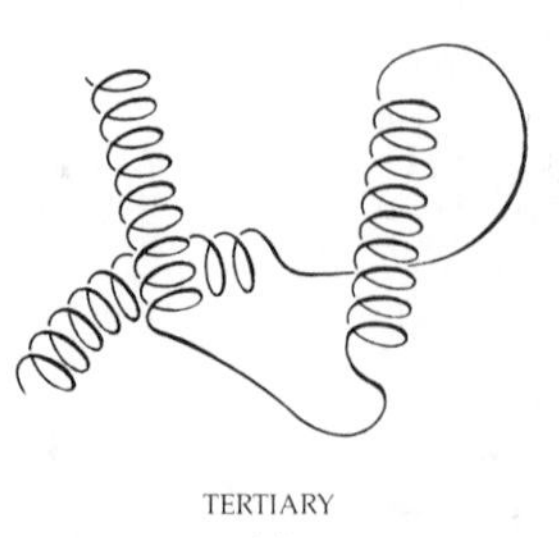

PRIMARY (a) SECONDARY (b) TERTIARY (c) QUATERNARY (d)

Structural Uses of Proteins

Fibrous Proteins

In general, fibrous proteins have a regular, repeated sequence of amino acids and therefore a regular, repetitious structure. An example is collagen, which makes up about one-third of all the protein in vertebrates. The basic collagen molecule is composed of three very long polypeptides—about 1,000 amino acids per chain. These three polypeptides, which are made up of repeating groups of amino acids, are held together by hydrogen bonds linking amino acids of different chains into a tight coil. The molecules can coil so tightly because every third amino acid is glycine, the smallest of the amino acids. The collagen molecules are packed together to form fibrils (Figure 3-23), which are, in turn, associated into larger fibers.

Collagen is actually a family of proteins. Different types of collagen molecules contain polypeptides with slightly different amino acid sequences. The larger structures formed from the different types of molecules perform a variety of functions in the body. Consider a cow: Tendons, which link muscle to bone, are made up of collagen fibers in parallel bundles; thus arranged, they are very strong but do not stretch. The cow's hide, by contrast, is made up of collagen fibrils arranged in an interlacing network laid down in sheets. Even its corneas—the transparent coverings of the eyeballs—are composed of collagen. When collagen is boiled in water, the polymers are dispersed into shorter chains, which we know as gelatin.

Other fibrous proteins include keratin (Figure 3-24), silk, and elastin, present in the elastic tissue of ligaments.

3-23
Collagen molecules are packed together in fibrils that are a major constituent of skin, tendon, ligament, cartilage, and bone. Within an individual fibril (a), *the collagen molecules are arranged in a staggered pattern, with gaps between individual molecules. This arrangement strengthens the fibrils, making them resistant to shearing forces. In electron micrographs of collagen fibrils, such as* (b), *a striated pattern is observed because the stain used in preparing the specimen is concentrated in the gaps between the molecules, causing the regions with no gaps to appear as lighter bands. These fibrils are magnified 23,500 times.*

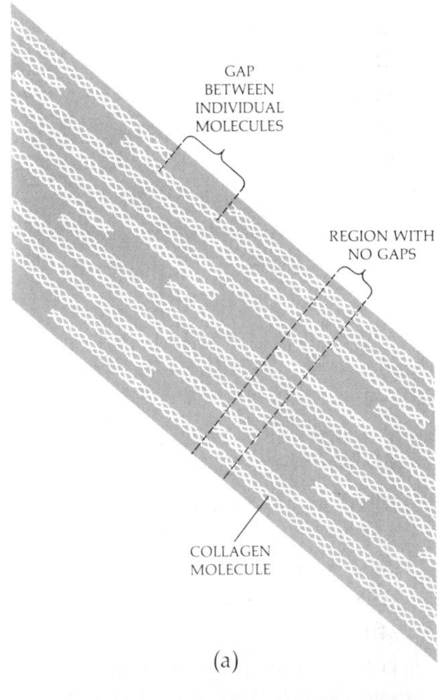

(a)

(b)

(a)

(b)

3-24
The structural protein keratin is found in all vertebrates. It is the chief component of scales, wool, nails, and feathers. (a) *These are cross sections of human hairs, magnified 160 times.* (b) *A feather is made up of a shaft to which thousands of barbs—each with many tiny barbules—are attached. The barbules on the lower half of each feather have tiny hooks on them that catch on the barbules of the adjacent feather, forming a solid, though flexible, structure for flight. When a bird preens itself, it is "zipping" its feathers back together. This feather, from a hummingbird, is magnified 56 times.*

3–25

(a) *Microtubules are hollow tubes, so small that they cannot be visualized by a light microscope. They are composed of subunits, each of which is a globular protein. The subunits are of two types, alpha tubulin and beta tubulin, which first come together to form a soluble dimer. The dimers then self-assemble into insoluble hollow tubules.* (b) *Among their many functions, microtubules make up the internal structure of cilia, the small, hairlike appendages visible on this* Paramecium. *The organism is magnified 452 times.*

(a)

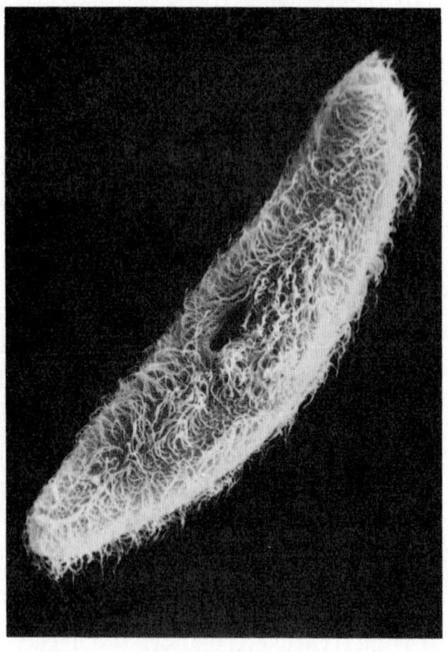

(b)

Globular Proteins

Some structural proteins are globular. For example, microtubules, which function in a variety of ways inside the cell, are made up of globular proteins. They are long, hollow tubes—so long that their entire length can seldom be traced in a single microscopic section. Microtubules play a critical role in cell division, as we shall see in Chapter 7. They also may serve as tracks along which substances can move inside the cell. Microtubules participate in the internal skeleton that stiffens parts of the cell body and also seem to function as a kind of scaffolding for cellular construction projects. For example, the formation of a new cell wall in a plant can be predicted by the appearance at the site of large numbers of microtubules; when cellulose fibrils are being laid down outside a plant cell membrane, as a cell wall forms or grows, it is possible to detect microtubules inside the cell aligned in the same direction as the fibrils outside.

Chemical analysis shows that each microtubule consists of a very large number of subunits. There are two types of subunits, each of which is a globular protein formed from one polypeptide chain. Because of their complementary configurations, the two subunits fit together, forming approximately dumbbell-shaped dimers. The dimers assemble themselves into tubules (Figure 3–25), adding on length as required. When their job is over, they separate. The way in which the cell controls the assembly and disassembly of microtubules is the subject of a great deal of current research.

Hemoglobin: An Example of Specificity

Fibrous proteins, like structural polysaccharides, are usually molecules with a relatively small variety of monomers in a repetitive sequence. Many globular proteins, by contrast, have extremely complex, irregular amino acid sequences, as complex and irregular as the sequence of letters in a sentence on this page. Just as these sentences make sense (if they do) because the letters are the right ones and in the right order, the proteins make sense, biologically speaking, because their amino acids are the right ones in the right order.

Hemoglobin, for example, has a quaternary structure that consists of four polypeptide chains, each of which is combined with an iron-containing molecule known as *heme*. In heme, an iron atom is held by nitrogen atoms that are part of a larger structure known as a porphyrin ring (Figure 3–26). Hemoglobin has two identical alpha chains and two identical beta chains, each with a unique primary structure containing about 150 amino acids, for a total of about 600 amino acids in all. In humans, hemoglobin molecules are manufactured and carried in the red blood cells. A mature red blood cell contains about 265 million molecules of hemoglobin. These molecules possess the special property of being able to combine loosely with oxygen so that they can collect oxygen in the lungs and release it in the tissues.

Sickle cell anemia is a disease in which the hemoglobin molecules are defective. When oxygen is removed, the defective molecules change shape and combine with one another to form stiffened rodlike structures. Red cells containing large proportions of these molecules become stiff and deformed, taking on the characteristic sickle shape (Figure 3–28). The deformed cells may clog the smallest blood vessels (capillaries). This causes often painful blood clots and deprives vital organs of their full supply of blood, resulting in intermittent illness and often a shortened life span.

3–26

The heme group of hemoglobin. It contains an iron atom (Fe) held in a porphyrin ring. The porphyrin ring consists of four nitrogen-containing rings, which are numbered in the diagram. Each heme group is attached to a long polypeptide chain that wraps around it. The oxygen molecule is held flat against the heme.

3–27

The hemoglobin molecule consists of four heme groups with their polypeptide chains intertwined in a quaternary structure. The outside of the molecule and the hole through the middle are lined by charged amino acids, and the uncharged amino acids are packed inside. Each molecule can hold up to four oxygen molecules, one bound to each heme group. The sequence of the amino acids in each chain is its primary structure. The helical form assumed by any part of the chain as a consequence of hydrogen bonding between nearby C=O and NH groups is its secondary structure. The folding of the chains in three-dimensional shapes is the tertiary structure, and the combination of the four chains into a single functional molecule is the quaternary structure. (Adapted with permission from R. E. Dickerson and I. Geis, The Structure and Action of Proteins, *W. A. Benjamin, Inc., Menlo Park, Calif., 1969. Copyright 1969 by Dickerson and Geis.)*

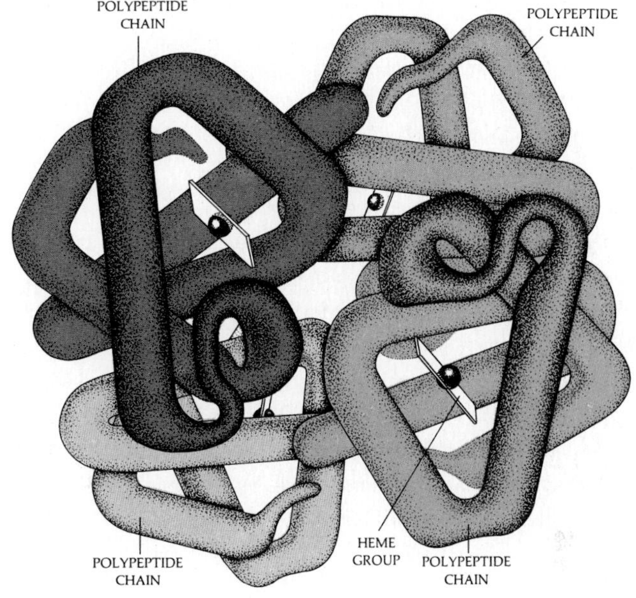

3–28

Scanning electron micrographs of (a) *red blood cells containing normal hemoglobin, and* (b) *red blood cell containing the abnormal hemoglobin associated with sickle cell anemia. When the oxygen concentration in the blood is low, the abnormal hemoglobin molecules stick together, distorting the shape of the cells. As a result, the cells cannot pass readily through the capillaries. These cells are magnified about 7,000 times.*

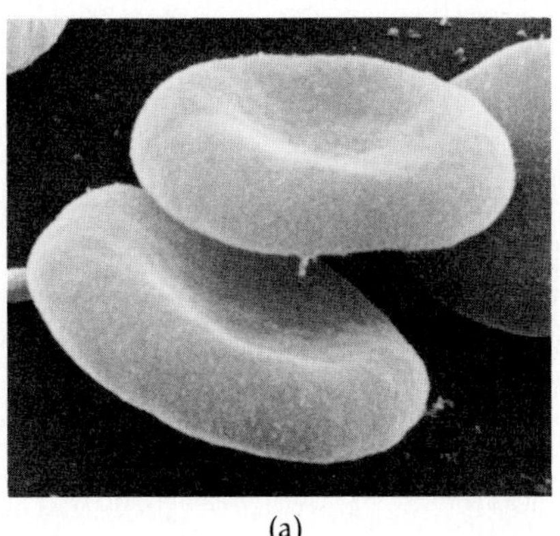

(a)

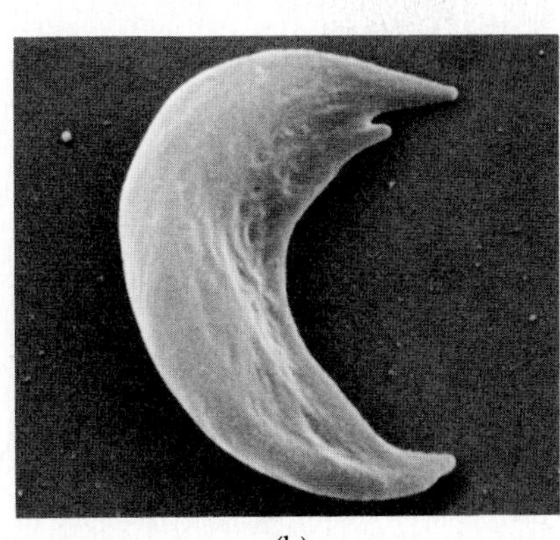

(b)

DETERMINING THE AMINO ACID SEQUENCE OF A PROTEIN

In 1953, Frederick Sanger of Cambridge University in England reported, for the first time, the complete amino acid sequence of a protein. He chose the hormone insulin because it is one of the smallest proteins known. Many different chemical procedures (and several years of work) were required. First, the hormone was heated in an acidic solution, which hydrolyzed all the peptide bonds, releasing the amino acids. The number and identity of the individual amino acids were determined by a newly perfected technique called ion-exchange chromatography. Ion-exchange chromatography depends on the fact that different amino acids have different charges and so will adhere to oppositely charged surfaces with different degrees of intensity. This first step gave Sanger the amino acid composition but told him nothing about the sequence.

The next step was to divide the protein into smaller, uniform pieces. For example, the chemical cyanogen bromide, it was discovered, cleaves a polypeptide chain only on the carboxyl side of the amino acid methionine. Thus a polypeptide that contains seven methionines will generally be cleaved into eight fragments by such a treatment. The digestive enzyme trypsin will cleave a polypeptide only on the carboxyl side of arginine and of lysine, permitting the breakdown of the polypeptide into still smaller but still uniform fragments. These fragments were then separated by chromatography.

Working with these smaller fragments, a method was developed by which the amino acid at the amino terminal end of the polypeptide chain could be labeled (by attaching a specific chemical group to it), and then released and chemically identified. Thus, by removing one amino acid at a time, Sanger and his coworkers were able to painstakingly identify the sequence of amino acids in a number of short fragments. However, they still could not tell how the fragments fit together to make up the complete protein.

The next step was to find other reagents that cleaved the polypeptide chain at completely different sites. For instance, hydroxylamine breaks only those peptide bonds between asparagine and glycine, so treatment with this reagent gives a totally different set of uniform fragments, which, again, could be separated and analyzed. Finally the results of both operations were compared by analysis of overlapping regions, fitting the whole thing together like a puzzle. Sanger compared his method to trying to determine how a car is put together by looking at the pieces in a junkyard. You might first find an axle joined to a wheel and then, in another pile of debris, an axle joined to a steering column, and so one could piece together the sequence: steering column–axle–wheel.

The development of the methodology for analyzing proteins was of great importance. With minor modifications, Sanger's method is still in use and has been employed in the amino acid sequencing of several hundred proteins. More important, however, it was an enormous conceptual breakthrough. It established for the first time that a given protein had its own highly specific amino acid sequence and so, conversely, opened the way to the idea that it is the amino acid sequence of the protein that gives the protein its specificity—in other words, that complex macromolecules can carry large amounts of information. The science of molecular biology is founded on this concept.

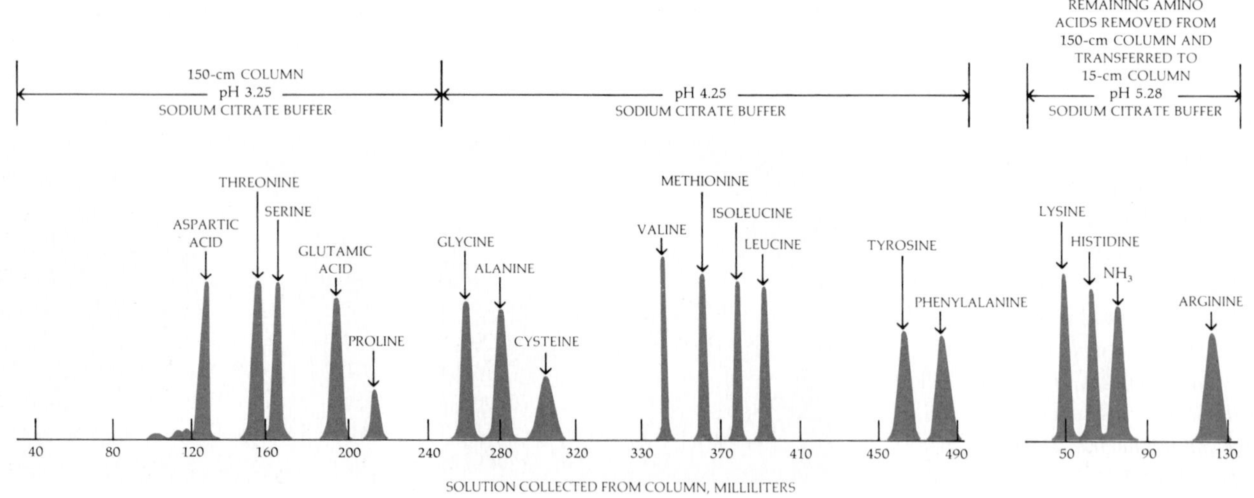

3–29

An example of the remarkable precision of the "language" of proteins. Portions of the beta chains of the hemoglobin A (normal) molecule and the hemoglobin S (sickle cell) molecule are shown. The hemoglobin molecule is composed of two identical alpha chains and two identical beta chains, each chain consisting of about 150 amino acids, or a total of 600 amino acids in the molecule. The entire structural difference between the normal molecule and the sickle cell molecule (literally, a life-and-death difference) consists of one change in the sequence of each beta chain: one glutamic acid is replaced by one valine.

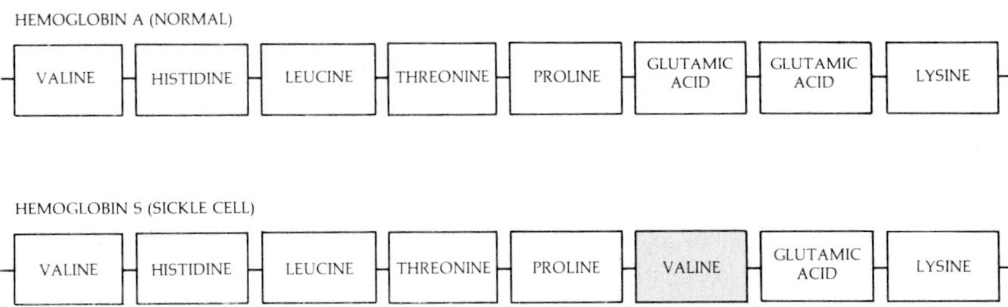

The chromatographic analysis of the amino acids in a protein. All of the peptide bonds are first hydrolyzed, producing a mixture of the individual amino acids. This mixture is slowly added to a column filled with a resin that contains fixed charged groups. The strength with which the different amino acids are bound to the resin depends on whether they are positively or negatively charged, which is, in turn, determined by the pH of the solution. After the amino acids are bound to the resin, buffers of increasing pH are passed through the column; the solution leaving the column is collected in samples of a few milliliters each. Each sample is analyzed to determine the particular amino acid it contains and the amount, a procedure that can now be performed by an automated "amino acid analyzer." The result is a chart, such as this, in which the area under each peak is proportional to the amount of each amino acid in the original protein.

The size of the column and the particular resin, buffer, and pH range that will give the best separation for a particular protein are determined experimentally.

Analysis of the hemoglobin molecules of patients with sickle cell anemia reveals that the only difference between normal and sickle cell hemoglobin is that in a precise location in each beta chain, one glutamic acid is replaced by one valine. When one considers that this difference of two amino acids in a total of almost 600 can be the difference between life and death, one begins to get an idea of the meaning of specificity and of the precision and the importance of the arrangement of amino acids in a particular sequence in a protein. In living systems, which must perform many different activities simultaneously, the specificity of function that results from the structural precision of different protein molecules is of crucial importance.

SUMMARY

The chemistry of living organisms is, in essence, the chemistry of compounds containing carbon. Carbon is uniquely suited to this central role by the fact that it is the lightest atom capable of forming multiple covalent bonds. Because of this capacity, carbon can combine with carbon and other atoms to form a great variety of strong and stable chain and ring compounds. Organic molecules derive their three-dimensional shapes primarily from their carbon skeletons. Many of their specific properties, however, are dependent on functional groups. Among the major types of organic molecules are (1) fatty acids, which have long hydrocarbon chains containing only carbon and hydrogen; (2) carbohydrates, whose monomers contain carbon, hydrogen, and oxygen in the ratio CH_2O; (3) amino acids; and (4) nitrogenous bases, such as adenine.

A general characteristic of all organic compounds is that they yield energy when oxidized. Sugars serve as a primary source of chemical energy for living systems. The simplest sugars are the monosaccharides ("single sugars"), such as glucose and fructose. Monosaccharides can be combined to form disaccharides ("two sugars"), such as sucrose, which is the form in which sugars are transported through the plant body.

Polysaccharides (chains of many monosaccharides) such as starch and glycogen are storage forms for sugars. These molecules can be broken apart by hydrolysis. Other polysaccharides form important structural elements. The chief structural polysaccharide is cellulose, found in plant cell walls.

Lipids are hydrophobic organic molecules that, like carbohydrates, play important roles in energy storage and as structural components. Fats are the chief energy-storing lipids. A fat molecule consists of one molecule of glycerol bonded to three fatty acids. Fats are designated as unsaturated or saturated depending on whether or not their fatty acids contain any double bonds. Unsaturated fats, which tend to be oily liquids, are more commonly found in plants.

Phospholipids are major structural components of cellular membranes. Phospholipids consist of one unit of glycerol, two fatty acids (instead of the three fatty acids present in fats), and a phosphate group to which another polar group may be attached. Because of their hydrophilic "heads" and hydrophobic "tails," phospholipids spontaneously orient in water to form films and clusters that are the basis of membrane structure.

Proteins are very large molecules composed of long chains of amino acids. The 20 different amino acids used in making proteins vary according to the properties of their side groups. From these an extremely large variety of different kinds of protein molecules can be synthesized, each of which has a highly specific function in living systems.

The sequence of amino acids is known as the primary structure of the protein. Depending on the amino acid sequence, the molecule may take on any of a variety of forms. Hydrogen bonds between C=O and NH groups tend to fold the chain into a repeating secondary structure, such as the alpha helix or the beta pleated sheet. Interactions between the side groups of the amino acids may result in further folding into a tertiary structure, which is often an intricate, globular form. Two or more polypeptides may interact to form a quaternary structure.

In fibrous proteins, the long molecules interact with other similar, or identical, long polypeptide chains to form cables or sheets. Collagen, hair, silk, wool, horns, nails, and feathers are all made up of fibrous proteins. Globular proteins may also serve structural purposes. Microtubules, which are important cell components, are composed of repeating units of globular proteins assembled helically into a hollow tubule.

Because of the variety of amino acids, proteins can have a high degree of specificity. An example is hemoglobin, the oxygen-carrying molecule of the blood, which is composed of four (two pairs of) polypeptide chains, each attached to an iron-containing (heme) group. Substitution of one amino acid for another in one of the pairs of chains results in a serious and sometimes fatal form of anemia known as sickle cell anemia.

QUESTIONS

1. Distinguish among the following: hydrocarbon/carbohydrate; glucose/fructose/sucrose; monomer/polymer; glycogen/starch/cellulose; saturated/unsaturated; polysaccharide/polypeptide; peptide bond/disulfide bridge/hydrophobic interaction; primary structure/secondary structure/tertiary structure/quaternary structure; heme/hemoglobin.

(a) CH_3COOH

MAJOR COMPONENT OF VINEGAR

(b) $H-\overset{\displaystyle H}{\overset{|}{C}}=O$

PRESERVATIVE USED FOR BIOLOGICAL SPECIMENS

(c) HCOOH

ACTIVE INGREDIENT IN AN ANT'S STING

(d) $CH_3-\underset{\displaystyle O}{\underset{\|}{C}}-CH_3$

NAIL-POLISH REMOVER

(e) $\underset{\displaystyle OH}{\underset{|}{CH_2}}-\underset{\displaystyle OH}{\underset{|}{CH_2}}$

AUTOMOBILE ANTIFREEZE

(f) NH_2

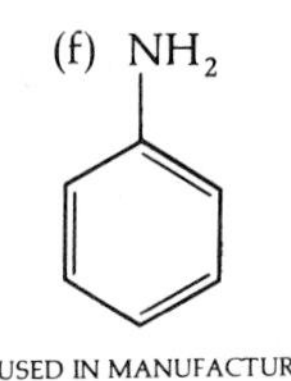

USED IN MANUFACTURE OF COMMERCIAL DYES

2. Identify the functional groups in the compounds at the left. Which of these is hydrophilic? Hydrophobic?

3. Draw a structural formula for (a) a carbohydrate; (b) a fatty acid; (c) an amino acid.

4. Butyric acid, $CH_3CH_2CH_2COOH$, gives rancid butter its odor and flavor. Draw its structural formula.

5. Many of the synthetic reactions in living systems take place by condensation. What is a condensation reaction? What types of molecules undergo condensation reactions to form disaccharides and polysaccharides? To form fats? To form proteins?

6. Disaccharides and polysaccharides, as well as lipids and proteins, can be broken down by hydrolysis. What is hydrolysis? What two types of products are released when a polysaccharide such as starch is hydrolyzed? How are these products important for the living cell?

7. What do we mean when we say that some polysaccharides are "energy-storage" molecules and that others are "structural" molecules? Give an example of each. In what sense should any polysaccharide be regarded as an "energy-storage" molecule?

8. Plants usually store energy reserves as polysaccharides, whereas, in most animals, lipids are the principal form of energy storage. Why is it advantageous for animals to have their energy reserves stored as lipids rather than as polysaccharides? (Think about the differences in "life style" between plants and animals.) What kinds of storage materials would you expect to find in seeds?

9. Sketch the arrangement of phospholipids when they are surrounded by water.

10. In pioneer days, soap was made by boiling animal fat with lye (potassium hydroxide). The bonds linking the fatty acids to the glycerol molecule were hydrolyzed, and the potassium hydroxide reacted with the fatty acid to produce soap. A typical soap available today is sodium stearate. In water, it ionizes to produce sodium ions (Na^+) and stearate ions:

$$CH_3CH_2CH_2CH_2CH_2CH_2CH_2CH_2CH_2CH_2CH_2CH_2CH_2CH_2CH_2CH_2CH_2C\begin{matrix} \diagup\!\!\!\!= O \\ \diagdown O^- \end{matrix}$$

Explain how soap functions to trap and remove particles of dirt and grease.

11. Silk is a protein in which polypeptide chains are arranged in a beta pleated sheet. In these chains, the peptide sequence glycine-serine-glycine-alanine-glycine-alanine occurs repeatedly. (a) Draw the structural formula for this hexapeptide, and show the peptide bonds in color. (b) Explain how a peptide bond is formed.

CHAPTER 4

Cells: An Introduction

In the last three chapters, we have progressed from subatomic particles through atoms and molecules to large proteins, among the most complex of all molecules. Although the neutrons, protons, and electrons of which atoms are composed are all the same, the elements differ greatly from one another. Mercury is a heavy metallic liquid, chlorine is a green gas, sulfur is a yellow powder, pure carbon can take the form of a hard solid—a diamond—and so on. The differences lie not in the nature of the subatomic particles but, rather, in their number and arrangement—that is, in their organization.

Just as subatomic particles combine to form atoms, atoms combine to form molecules. At each level of organization, new properties appear. Water, for instance, as we have seen, is not the sum of the properties of elemental hydrogen and oxygen; it is something more and also something different. In proteins, amino acids become organized into polypeptides, and polypeptide chains are arranged in a new level of organization, the tertiary or quaternary structure of the complete protein molecule. Only at this level of organization do the complex properties of the protein emerge, and only then can the molecule assume its function.

Living systems, as we noted earlier, obey the laws of physics and chemistry, and yet there are striking differences between "life" and "nonlife." According to biologists, the basis of these differences is to be found not in the atoms and molecules themselves, but in their organization. The characteristics of living systems—like those of atoms or of molecules—do not emerge gradually as the degree of organization increases. They appear quite suddenly and specifically, in the form of the living cell—something that is more than and different from its constituent atoms and molecules.

THE CELL THEORY

The word "cell" was first used in a biological sense some 300 years ago. In the seventeenth century, Robert Hooke, using a microscope of his own construction, noticed that cork and other plant tissues are made up of small cavities separated by walls. He called these cavities "cells," meaning "little rooms." However, the word did not take on its present meaning, as the basic unit of which all living systems are composed, for more than 150 years.

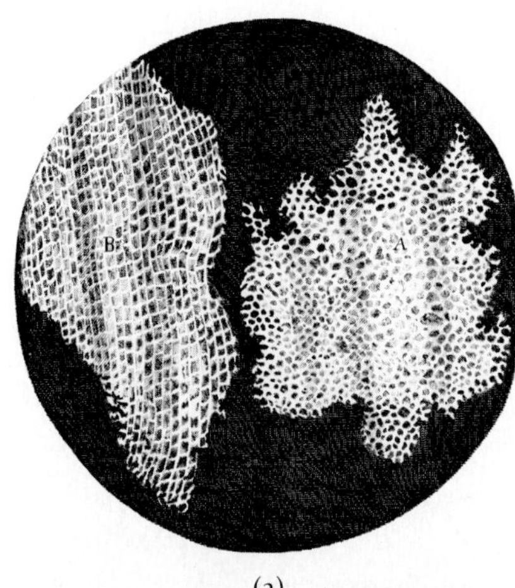

(a)

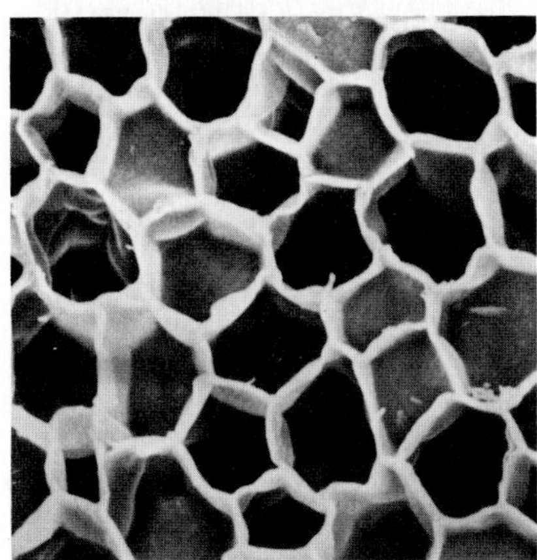

(b)

4–1
(a) *Robert Hooke's drawings of two slices of a piece of cork, reproduced from his* Micrographia, *published in 1665, and* (b) *a scanning electron micrograph of a slice of cork. Hooke was the first to use the word "cells" to describe the tiny compartments that together make up an organism. The cells in these pieces of cork have died—all that remain are the outer walls. As we shall see, the living cell is filled with a variety of substances, organized into distinct structures and carrying out a multitude of essential processes.*

In 1838, Matthias Schleiden, a German botanist, came to the conclusion that all plant tissues are organized in the form of cells. In the following year, zoologist Theodor Schwann extended Schleiden's observations to animal tissues and proposed a cellular basis for all life. This notion was of tremendous importance because it emphasized the basic sameness of all living systems and so brought an underlying unity to widely varied studies involving many different kinds of organisms.

In 1858, the idea that all living organisms are composed of one or more cells took on an even broader significance when the great pathologist Rudolf Virchow generalized that cells can arise only from preexisting cells: "Where a cell exists, there must have been a preexisting cell, just as the animal arises only from an animal and the plant only from a plant. . . . Throughout the whole series of living forms, whether entire animal or plant organisms or their component parts, there rules an eternal law of continuous development."

The *cell theory,* like the theory of evolution, is a major unifying concept in biology. In its modern form, this theory states simply that (1) living matter is composed of cells; (2) the chemical reactions of a living organism, including its energy-yielding processes and its biosynthetic reactions, take place within cells; (3) cells arise from other cells; and (4) cells contain the hereditary information of the organisms of which they are a part, and this information is passed from parent cell to daughter cell.

In the broad perspective provided by the theory of evolution, Virchow's concept that cells arise only from preexisting cells makes clear that there is an unbroken continuity between modern cells—and the organisms they compose—and the earliest cells that appeared on earth. The realization of that continuity, however, required the disproof of long-held notions.

THE DISPROOF OF SPONTANEOUS GENERATION

Most of the early biologists, from the time of Aristotle, believed that simple living things, such as worms, beetles, frogs, and salamanders, could originate spontaneously in dust or mud, that rodents formed from moist grain, and that plant lice condensed from a dewdrop. In the seventeenth century, Francesco Redi performed a famous experiment in which he put out decaying meat in a group of wide-mouthed jars—some with lids, some covered by a fine veil, and some open—and proved that maggots arose only where flies were able to lay their eggs.

By the nineteenth century, no scientist continued to believe that complex organisms arise spontaneously. The advent of microscopy, however, led not only to the formulation of the cell theory but also to a vigorous renewal of belief in the spontaneous generation of very simple organisms. It was necessary only to put decomposing substances in a warm place for a short time and tiny "live beasts" appeared under the lens, before one's very eyes. By 1860, the controversy had become so spirited that the Paris Academy of Sciences offered a prize for experiments that would throw new light on the question. The prize was claimed in 1864 by Louis Pasteur, who devised experiments to show that microorganisms appeared only as contaminants from the air and not "spontaneously" as his opponents claimed. In his experiments he used swan-necked flasks (Figure 4–2), which permitted the entrance of oxygen, thought to be necessary for life, but which, in their long, curving necks, trapped bacteria, fungal spores, and other microbial life and

4–2
Pasteur's swan-necked flasks, which he used to counter the argument that spontaneous generation failed to occur in sealed vessels because air was excluded. These flasks permitted the entrance of oxygen, thought to be essential for life, but their long, curving necks trapped spores of microorganisms and thereby protected the liquids in the flasks from contamination.

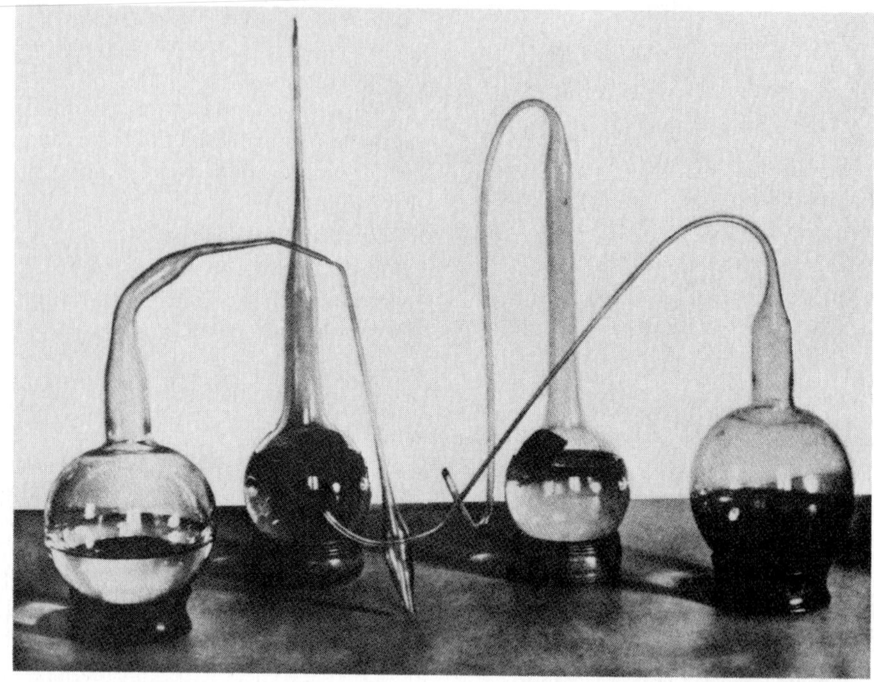

thereby protected the contents of the flasks from contamination. He showed that if the liquid in the flask was boiled (which killed microorganisms already present) and the neck of the flask was allowed to remain intact, no microorganisms would appear. Some of his original flasks, still sterile, remain on display. Only if the curved neck of the flask was broken off, permitting contaminants to enter the flask, did microorganisms appear.

"Life is a germ, and a germ is Life," Pasteur proclaimed at a brilliant "scientific evening" at the Sorbonne before the social elite of Paris. "Never will the doctrine of spontaneous generation recover from the mortal blow of this simple experiment."

In retrospect, Pasteur's well-planned experiments were so decisive because the broad question of whether or not spontaneous generation had *ever* occurred was reduced to the simpler question of whether or not it occurred under the specific conditions claimed for it. Pasteur's experiments answered *only* the latter question, but the results were so dramatic that for many years very few scientists were able to entertain the possibility that under quite different conditions, when the earth was very young, some form of "spontaneous generation" might indeed have taken place. The question of the origin of the first cells, from which all subsequent cells are descended, remained unasked until well into the twentieth century.

THE BEGINNING OF LIFE

Until very recently, the earliest fossil organisms known were from the Cambrian period, a mere 600 million years ago, and for a long time after the publication of *The Origin of Species,* biologists regarded the earliest events in the history of life as chapters that would probably remain forever closed to scientific investigation.

THE ORIGIN OF THE EARTH

The universe was perhaps 10 billion years old, cosmologists calculate, when the star that is our sun came into being. According to current hypotheses, it formed, like other stars, from an accumulation of particles of dust and hydrogen and helium gases whirling in space among the older stars.

The immense cloud that was to become the sun condensed gradually as the hydrogen and helium atoms were pulled toward one another by the force of gravity, falling into the center of the cloud and gathering speed as they fell. As the cluster grew denser, the atoms moved more rapidly. More atoms collided with each other, and the gas in the cloud became hotter and hotter. As the temperature rose, the collisions became increasingly violent until the hydrogen atoms collided with such force that their nuclei fused, forming additional helium atoms and releasing nuclear energy. This thermonuclear reaction is still going on at the heart of the sun and is the source of the energy radiated from its glowing surface.

The planets, according to current theory, formed from the remaining gas and dust moving around the newly formed star. At first, particles would have collected at random, but as each mass grew larger, other particles began to be attracted by the gravity of the largest masses. The whirling dust and forming spheres continued to revolve around the sun until finally each planet had swept its own path clean, picking up loose matter like a giant snowball. The orbit nearest the sun was swept by Mercury, the next by Venus, the third by earth, the fourth by Mars, and so on out to Neptune and Pluto, the most distant of the planets. The planets, including earth, are calculated to have come into being about 4.6 billion years ago.

During the time earth and the other planets were being formed, the release of energy from radioactive materials kept their interiors very hot. When earth was still so hot that it was mostly liquid, the heavier materials collected in a dense core whose diameter is about half that of the planet. As soon as the supply of stellar dust, stones, and larger rocks was exhausted, the planet ceased to grow. As earth's surface cooled, an outer crust, a skin as thin by comparison as the skin of an apple, was formed. The oldest known rocks in this layer have been dated by isotopic methods as about 3.98 billion years old.

Only 50 kilometers below its surface, the earth is still hot—a small fraction of it is even still molten. We see evidence of this in the occasional volcanic eruption that forces lava (molten rock) through weak points in the earth's skin, or in the geyser, which spews up boiling water that has trickled down to the earth's interior.

The biosphere is the part of the planet within which life exists. It forms a thin film on the outermost layer, extending only about 8 or 10 kilometers up into the atmosphere and about as far down into the depths of the sea.

The tremendous amounts of energy released by the thermonuclear reactions at the heart of the sun give rise to an envelope of extremely hot gases surrounding its surface. These glowing gases are normally invisible, but during a solar eclipse they become strikingly apparent. The layer of gases may extend as far as 64,000 kilometers from the surface of the sun—a distance about five times greater than the diameter of the earth.

Two developments, however, have greatly improved our long-distance vision. The first was the formulation of a testable hypothesis about the events preceding life's origins. This hypothesis generated questions for which answers could be sought experimentally. The results of the initial experimental tests led to the formulation of further hypotheses and to additional experiments, a process that continues today as scientists in many laboratories explore the question of life's origins. The second development was the discovery of fossilized cells more than 3 billion years old.

The testable hypothesis was offered by the Russian biochemist A. I. Oparin. According to Oparin, the appearance of life was preceded by a long period of what is sometimes called chemical evolution. The identity of the substances, particularly gases, present in the primitive atmosphere and in the seas during this period is a matter of controversy. There is general agreement, however, on two critical issues: (1) Little or no free oxygen was present, and (2) the four elements—hydrogen, oxygen, carbon, and nitrogen—that make up more than 95 percent of living tissues were available in some form in the atmosphere and waters of the primitive earth.

In addition to these raw materials, energy abounded on the young earth. There was heat energy, both boiling (moist) heat and baking (dry) heat. Water vapor spewed out of the primitive seas, cooled in the upper atmosphere, collected into clouds, fell back on the crust of the earth, and steamed up again. Violent rainstorms were accompanied by lightning, which provided electrical energy. The sun bombarded the earth's surface with high-energy particles and ultraviolet light, another form of energy. Radioactive elements within the earth released their energy into the atmosphere. Oparin hypothesized that under such conditions organic molecules were formed from the atmospheric gases and collected in a thin soup in the earth's seas and lakes. Because there was no free oxygen present to react with and degrade these organic molecules to simple substances such as carbon dioxide (as would happen today), they tended to persist. Some of these molecules might have become locally more concentrated by the drying up of a lake or by the adhesion of the molecules to a solid surface.

4–3
Bolts of lightning in the steam and other gases boiling up from a volcanic crater. This lightning results from static electricity generated by colliding atoms and molecules in the gases. Such sources of energy, present on the primitive earth, might have contributed to the formation of organic molecules. This photograph, taken in 1963, shows the birth of the island of Surtsey off the coast of Iceland.

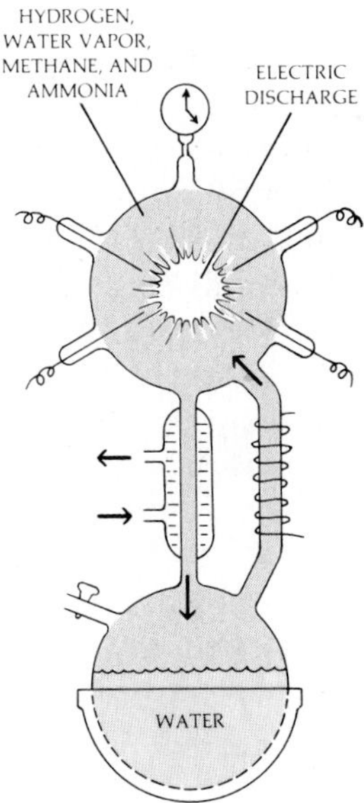

4–4
Miller's experiment. Conditions believed to have existed on the primitive earth were simulated in the apparatus diagrammed here. Methane and ammonia were continuously circulated between a lower "ocean," which was heated, and an upper "atmosphere," through which an electric discharge was transmitted. At the end of 24 hours, about half of the carbon originally present in the methane gas was converted to amino acids and other organic molecules. This was the first test of Oparin's hypothesis.

Oparin published his small monograph in 1922, but at that time biochemists were so convinced by Pasteur's demonstration disproving spontaneous generation that the scientific community ignored his ideas. In the 1950s, the first test of Oparin's hypothesis was performed by Stanley Miller, then a graduate student at the University of Chicago (Figure 4–4). Experiments of this sort, now repeated many times, have shown that almost any source of energy—lightning, ultraviolet radiation, or hot volcanic ash—would have converted molecules believed to have been present on the earth's surface into a variety of complex organic compounds. With various modifications in the experimental conditions and in the mixture of gases placed in the reaction vessel, almost all of the common amino acids have been produced and also nitrogenous bases, ribose, and nucleotides—all essential components of the hereditary material.

These experiments have not proven that such organic compounds were formed spontaneously on the primitive earth, only that they could have formed. The accumulated evidence is nevertheless very great, and most biochemists now believe that, given the conditions existing on the young earth, chemical reactions producing amino acids and other organic compounds were inevitable.

As the concentrations of such molecules increased, bringing them into closer proximity to each other, they would have been subject to the same chemical forces that act on organic molecules today. As we saw in the last chapter, small organic molecules react with each other to form larger molecules, and such forces as hydrogen bonds and hydrophobic interactions cause these molecules to assemble themselves into more complex aggregations. In modern chemical systems—either in the laboratory or in the living organism—the more stable molecules and aggregations tend to survive and the least stable are transitory. Similarly, the compounds and aggregations that had the greatest chemical stability under the prevailing conditions on the primitive earth would have tended to survive. Hence a form of natural selection played a role in chemical evolution as well as in the biological evolution that was to follow.

The First Cells

In the last chapter, we noted that hydrophobic molecules have a tendency to cluster together, sometimes forming boundaries between themselves and the surrounding solution, as when oil droplets form in water. Although the formation of such droplets is most familiar in the case of lipids, it can also occur with protein molecules, many of which have sizable hydrophobic regions. It is hypothesized that in the course of chemical evolution, similar processes occurred and a boundary or membrane formed around an aggregate of macromolecules, separating it from the external environment. In experiments simulating conditions during the earth's first billion years, Sidney W. Fox and his coworkers at the University of Miami have shown that proteinoid microspheres (Figure 4–5) can be formed and can carry out a few chemical reactions analogous to those of living cells. The microspheres grow slowly by the addition of proteinoid material from the solution and eventually bud off smaller microspheres.

Proteinoid microspheres are not living cells. Their formation suggests, however, the kinds of processes that could have given rise to the first tiny living cells capable of carrying out the chemical reactions necessary to sustain their physical and chemical integrity and capable of reproducing themselves. It is not known when the first living cells appeared on earth, but we can establish some sort of time scale.

4-5
When dry mixtures of amino acids are heated at moderate temperatures, polymers known as thermal proteinoids are formed. Each of these polymers may contain as many as 200 amino acid monomers. When the polymers are placed in aqueous solution and maintained under suitable conditions, they spontaneously form proteinoid microspheres, shown here. The microspheres are separated from the surrounding solution by a membrane that appears to be two-layered.

The short, straight line at the bottom of this micrograph and those that follow provides a reference for size; a micrometer, abbreviated μm, is 1/10,000 centimeter. The same system is used to indicate distances on a road map.

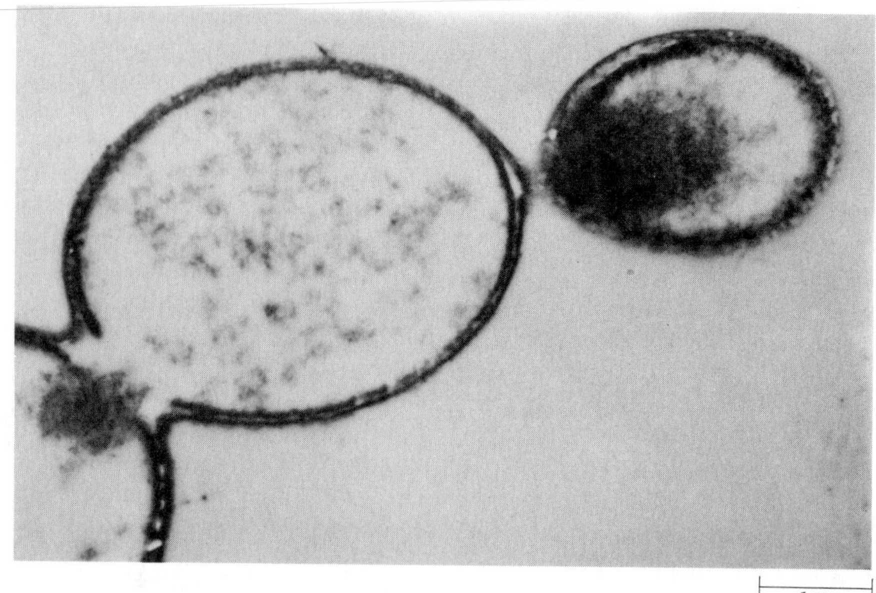

1 μm

In the mid-1960s, microfossils were discovered in flintlike rocks, called chert, in South Africa. The fossils, whose structure is visible only by electron microscopy, resemble present-day bacteria and cyanobacteria (blue-green "algae"). The rocks in which they were found, according to radioactive dating, are 3.4 billion years old. In 1980, the discovery in Australia of still older microfossils was announced. These microfossils, which consist of filaments of bacteria-like cells, have been dated at 3.5 billion years—about 1.1 billion years after the formation of the earth itself. The fossilized cells are sufficiently complex that it is clear that some little aggregation of chemicals had moved through the twilight zone separating the living from the nonliving millions of years before these cells made their appearance on the earth's surface.

Why on Earth?

On the basis of astronomical studies and the explorations carried out by unmanned space vehicles thus far, it appears that earth alone among the planets of our solar system supports life. The conditions on earth are ideal for living systems based on carbon-containing molecules. A major factor is that earth is neither too close to nor too distant from the sun. The chemical reactions on which life—at least as we know it—depends virtually cease at very low temperatures. At high temperatures, complex chemical compounds are too unstable for life to form or survive.

Earth's size and mass are also important factors. Planets much smaller than earth do not have enough gravitational pull to hold a protective atmosphere, and any planet much larger than earth is likely to have so dense an atmosphere that light from the sun cannot reach its surface. The earth's atmosphere blocks out the most energetic radiations from the sun, which are capable of breaking the covalent bonds between carbon molecules. It does, however, permit the passage of visible light, which made possible one of the most significant steps in the evolution of complex living systems.

4-6

(a) This bacterium-like microfossil was found in South Africa in a deposit of a flintlike rock called black chert. It is about 3.4 billion years old. (b) A filament of bacteria-like cells from black chert in Western Australia. The arrows indicate the cross walls between the individual cells. Dated at 3.5 billion years of age, these are the oldest fossils now known.

(a)

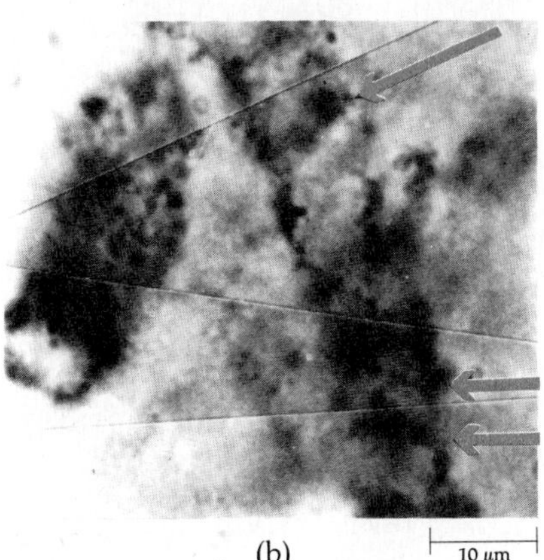

(b)

HETEROTROPHS AND AUTOTROPHS

The energy that produced the first organic molecules came from a variety of sources on the primitive earth and in its atmosphere—heat, ultraviolet radiations, and electrical disturbances. When the first primitive cells or cell-like structures evolved, they required a continuing supply of energy to maintain themselves, to grow, and to reproduce. The manner in which these cells obtained energy is currently the subject of lively discussion.

Modern organisms—and the cells of which they are composed—can meet their energy needs in one of two ways. *Heterotrophs* are organisms that are dependent upon outside sources of organic molecules for both their energy and their small building-block molecules. (*Hetero* comes from the Greek word meaning "other," and *troph* comes from *trophos,* "one that feeds.") All animals and fungi, as well as many single-celled organisms, are heterotrophs. *Autotrophs,* by contrast, are "self-feeders." They do not require organic molecules to use as sources of energy or as small building-block molecules; they are, instead, able to synthesize their own energy-rich organic molecules from simple inorganic substances. Most autotrophs, including plants and several different types of single-celled organisms, are *phototrophs* ("light-eaters"), meaning that their energy source is the sun. Certain groups of bacteria, however, are *chemoautotrophs;* these organisms are able to capture the energy released by specific inorganic reactions to power their life processes, including the synthesis of needed organic molecules.

Although both heterotrophs and autotrophs are represented among the earliest microfossils, it is logical to postulate that the first living cell was an extreme heterotroph. As the primitive heterotrophs increased in number, according to this hypothesis, they began to use up the complex molecules on which their existence depended and which had taken millions of years to accumulate. As the supply of these molecules decreased, competition began. Under the pressure of this competition, cells that could make efficient use of the limited energy sources now available were more likely to survive than cells that could not. In the course of time, cells evolved that were able to synthesize organic molecules out of simple inorganic materials. Without the evolution of autotrophs, life on earth would soon have come to an end.

In recent years, scientists have discovered several different groups of chemoautotrophic bacteria that would have been well-suited to the conditions prevailing on the young earth. Some of these bacteria are the inhabitants of swamps, while others were found in deep ocean trenches in areas where gases escape from fissures in the earth's crust. There is evidence (to be discussed in Chapter 20) that these bacteria are the surviving representatives of very ancient groups of unicellular organisms. Their discovery has raised the possibility that the earliest cells may have been chemoautotrophs, rather than heterotrophs, and that both heterotrophic and phototrophic forms developed subsequently. Continuing studies should shed more light on this question.

In the more than 3.5 billion years since life first appeared on earth, the most successful autotrophs (that is, those that have left the most offspring and diverged into the greatest variety of forms) have been the phototrophs—those organisms that evolved a system for making direct use of the sun's energy in the process of photosynthesis. With the advent of photosynthesis, the flow of energy in the biosphere came to assume its modern form: radiant energy from the sun channeled through photosynthetic autotrophs to heterotrophic forms of life.

PROKARYOTES AND EUKARYOTES

As we noted earlier, there is an unbroken continuity between modern cells—and the organisms they compose—and the primitive cells that first appeared on earth. One essential feature of all cells is an outer membrane, the *cell membrane* (sometimes called the *plasma membrane*), which separates the cell from its external environment. Another is the genetic material—the hereditary information—that directs a cell's activities and enables it to reproduce, passing on its characteristics to its offspring. The other contents of a cell (that is, everything within the cell membrane except the genetic material) are known as the *cytoplasm*. The cytoplasm contains a large variety of molecules as well as formed bodies called *organelles*. These specialized structures carry out particular functions within the cell such as, for example, assembling protein molecules.

There are, however, two fundamentally distinct kinds of cells, *prokaryotes* and *eukaryotes*, which differ most notably in the organization of their genetic material. In prokaryotic cells, the genetic material is in the form of a large, single molecule of a chemical known as DNA (deoxyribonucleic acid); in eukaryotic cells, the DNA is associated with proteins in complex structures known as *chromosomes*. Moreover, in eukaryotes, the chromosomes are surrounded by a double membrane, the *nuclear envelope*, that separates them from the cytoplasm in a distinct *nucleus* (hence their name, *eu*, meaning "true," and *karyon*, meaning "nucleus"or "kernel"). By contrast, in prokaryotes ("before a nucleus"), the DNA is not contained within a membrane-bound nucleus.

Modern prokaryotes include the bacteria (Figure 4–7) and the cyanobacteria (Figure 4–8), a group of photosynthetic prokaryotes that were formerly known as the blue-green algae. (The Latin term *alga*, plural *algae*, means "seaweed." Algae is

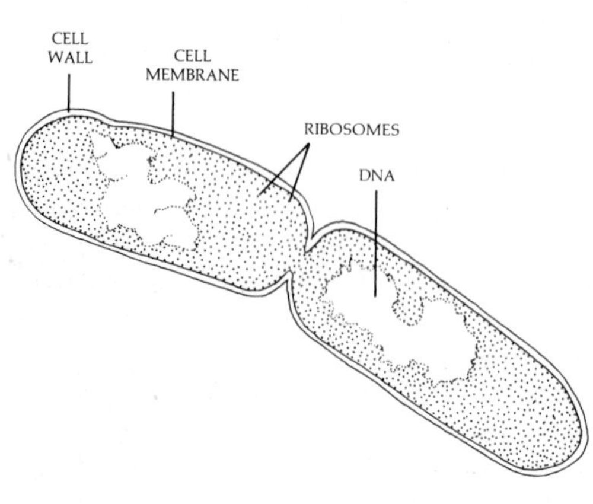

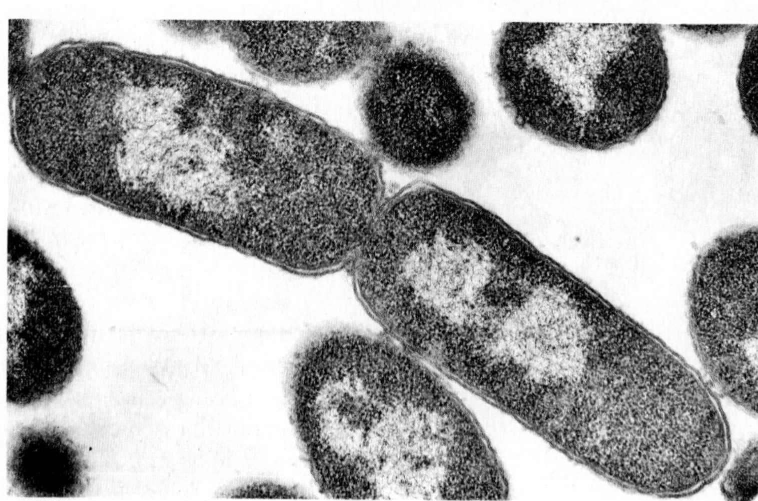

4–7
Cells of Escherichia coli, *a heterotrophic prokaryote that is a common inhabitant of the human digestive tract. The hereditary material (DNA) is in the less dense (lighter-appearing) area in the center of each cell. The small, dense bodies in the cytoplasm are ribosomes. The two cells in the center have just finished dividing and have not yet separated completely.* Escherichia coli *is the most intensively studied and best understood of all living organisms.*

4-8

Electron micrograph and diagram of a photosynthetic prokaryotic cell, the cyanobacterium Anabaena azollae. *In addition to the hereditary material, this cell contains a series of membranes in which chlorophyll and other photosynthetic pigments are embedded. This phototroph synthesizes its own energy-rich organic compounds in chemical reactions powered by the radiant energy of the sun.*

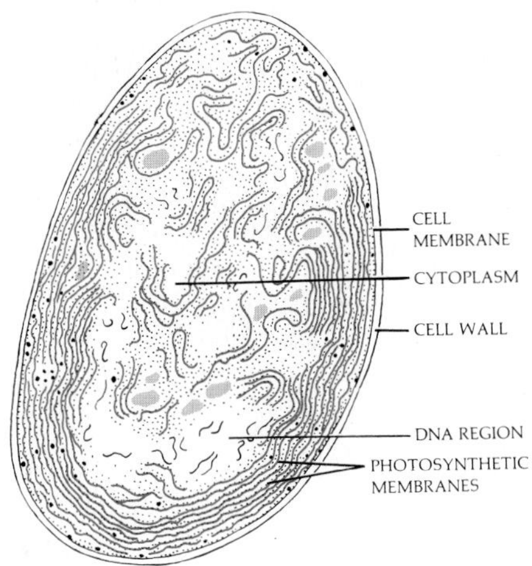

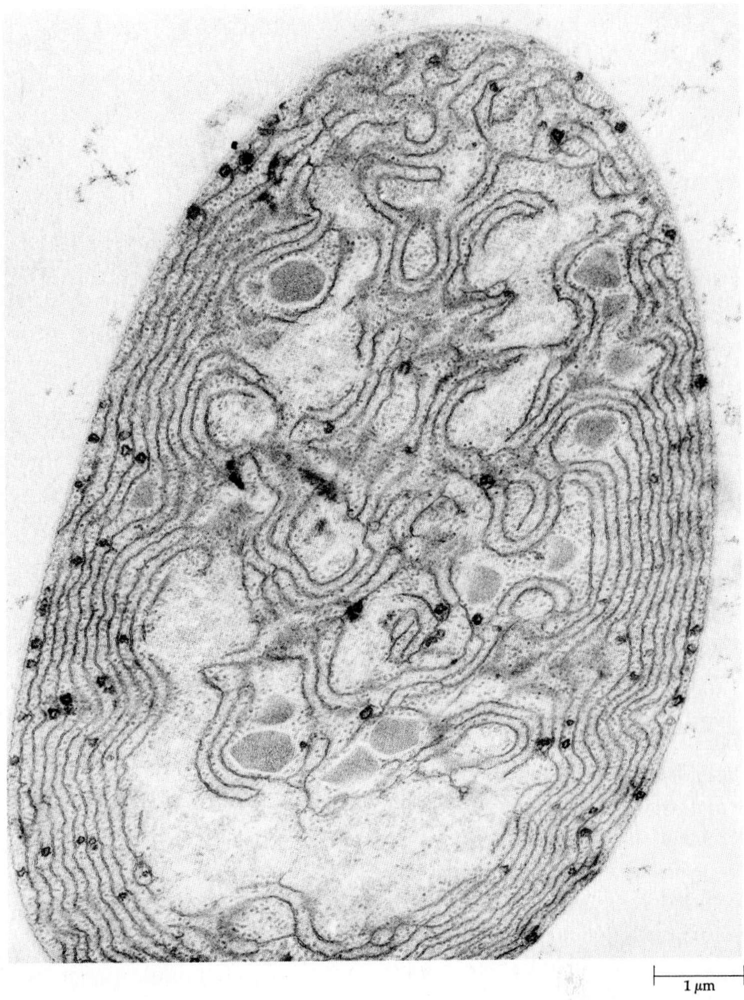

a general term applied to eukaryotic single-celled photosynthetic organisms and to many simple multicellular forms. Until recently, it was also applied to some prokaryotes.) The cell membrane of prokaryotes is surrounded by an outer *cell wall* that is manufactured by the cell itself. The cytoplasm contains very small organelles called *ribosomes*, on which protein molecules are assembled.

According to the fossil record, the earliest living organisms were comparatively simple cells, resembling present-day prokaryotes. Prokaryotes were the only forms of life on this planet for almost 2 billion years until eukaryotes evolved. (The evolutionary relationship between prokaryotes and eukaryotes—quite an interesting subject—will be explored in Chapter 21.) Eukaryotic cells are usually larger than prokaryotic cells, and their organelles are more numerous and more complex, often enclosed within membranes. Some eukaryotic cells, including plant cells and fungi, have a cell wall; others, including the cells of our own bodies and those of other animals, do not. All multicellular organisms are made up of eukaryotic cells.

Figure 4-9 gives an example of a single-celled photosynthetic eukaryote, the alga *Chlamydomonas*. It is a common inhabitant of freshwater ponds and aquariums.

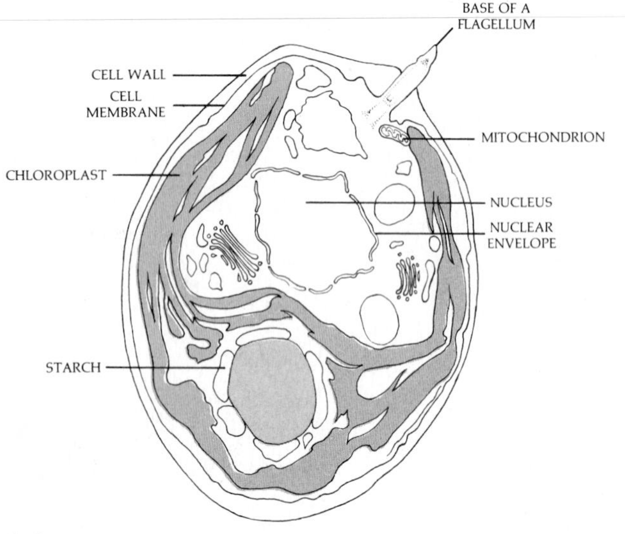

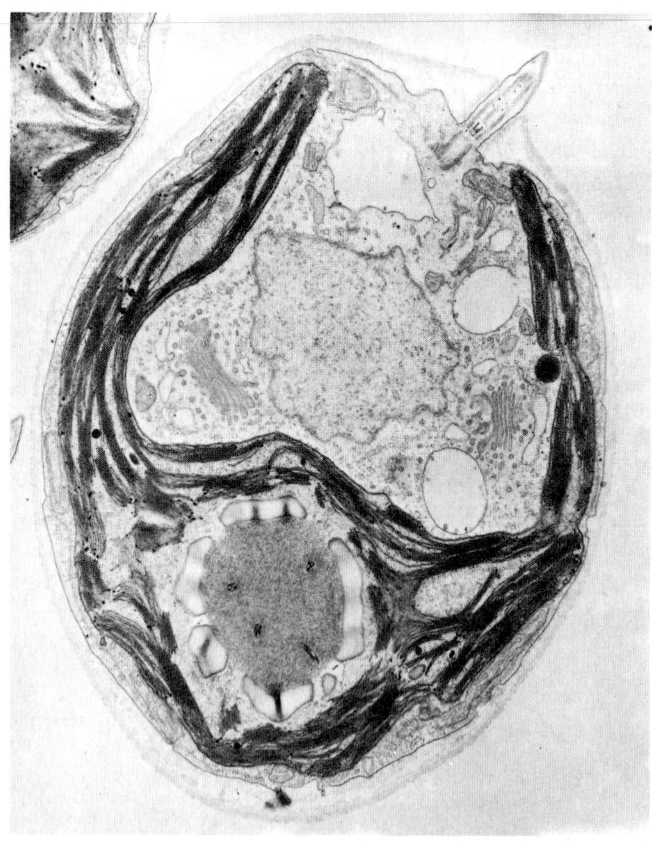

4–9
Electron micrograph of Chlamydomonas, *a photosynthetic eukaryotic cell, which contains a membrane-bound ("true") nucleus and numerous organelles. The most prominent organelle is the single, irregularly shaped chloroplast that fills most of the cell. It is surrounded by a double membrane and is the site of photosynthesis. Other membrane-bound organelles, the mitochondria, provide energy for cellular functions, including the flicking movements of the two flagella (one of which is visible in the micrograph). These movements propel the cell through the water. The organism's food reserves are in the form of starch granules. The cytoplasm is surrounded by a cell membrane, outside of which is a cell wall composed of cellulose and other polysaccharides.*

These organisms are small, bright green (because of their chlorophyll), and move very quickly with a characteristic darting motion. Being photosynthetic, they are usually found near the water's surface. *Chlamydomonas* is believed to resemble the sort of cell from which the plants evolved.

The Origins of Multicellularity

The first multicellular organisms, as far as can be told by the fossil record, made their appearance a mere 750 million years ago (Figure 4–10). The various major groups of multicellular organisms—such as the plants, the animals, and the fungi—presumably evolved from different types of single-celled eukaryotes.

The cells of modern multicellular organisms closely resemble those of single-celled organisms. They are bound by a cell membrane identical in appearance to the cell membrane of single-celled organisms. Their organelles are constructed according to the same design. The cells of multicellular organisms differ from single-celled organisms in that each type of cell is specialized to carry out a relatively limited function in the life of the organism. However, each remains a remarkably self-sustaining unit.

Notice how similar a cell from the leaf of a corn plant (Figure 4–11) is to *Chlamydomonas.* This plant cell is also photosynthetic, supplying its own energy needs from sunlight. However, unlike the alga, it is part of a multicellular organism and depends on other cells for water, minerals, protection from desiccation (drying out), and other necessities.

4-10

The clockface of biological time. Life first appears relatively early in the earth's history, before 6:00 A.M. on a 24-hour time scale. The first multicellular organisms do not appear until the early evening of that 24-hour day, and Homo sapiens *is a late arrival—at about 30 seconds to midnight.*

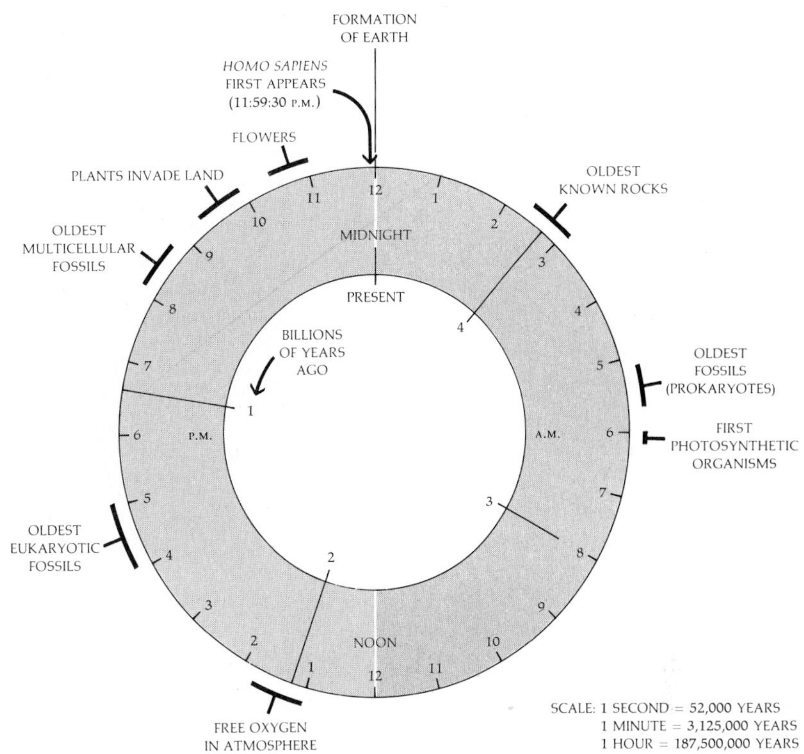

4-11

Electron micrograph of cells from the leaf of a corn plant. The nucleus can be seen on the right side of the central cell. The granular material within the nucleus is chromatin; it contains DNA associated with protein. Note the many mitochondria and chloroplasts, all enclosed by membranes. The vacuole and cell wall are characteristic of plant cells but are generally not found in animal cells. As you can see, this cell closely resembles Chlamydomonas, *shown in Figure 4-9.*

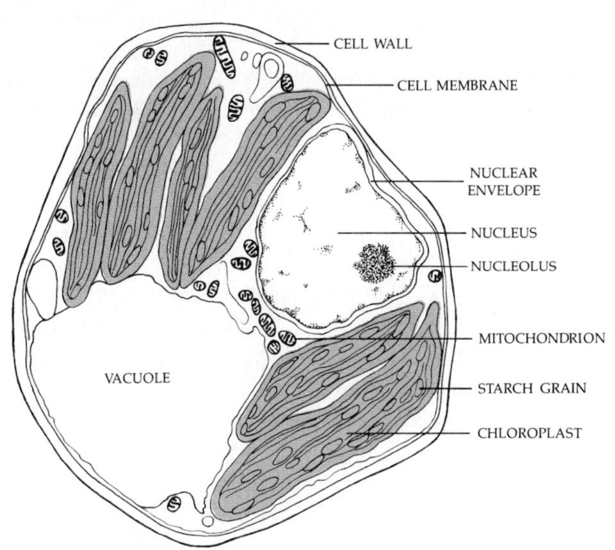

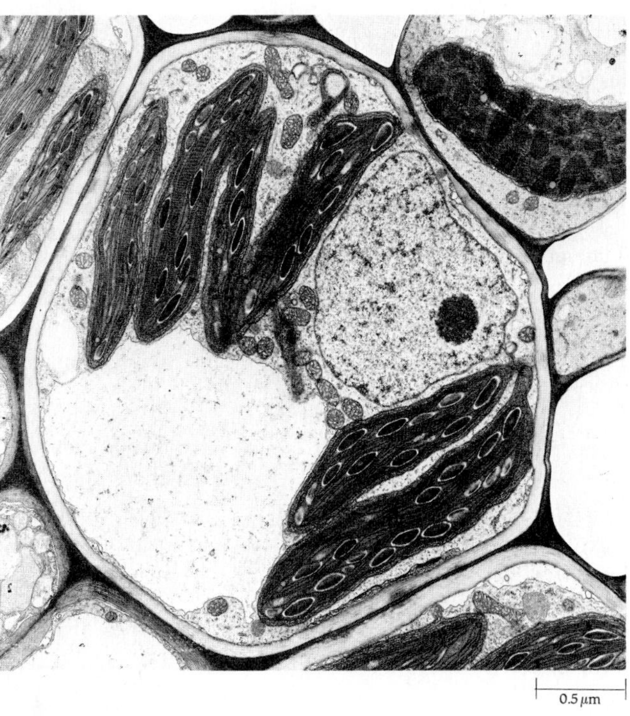

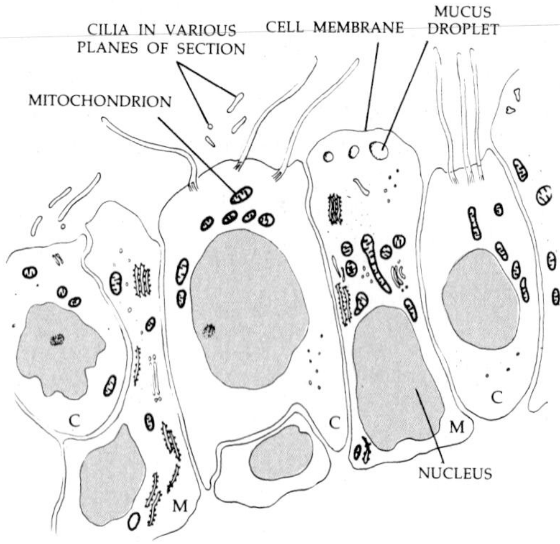

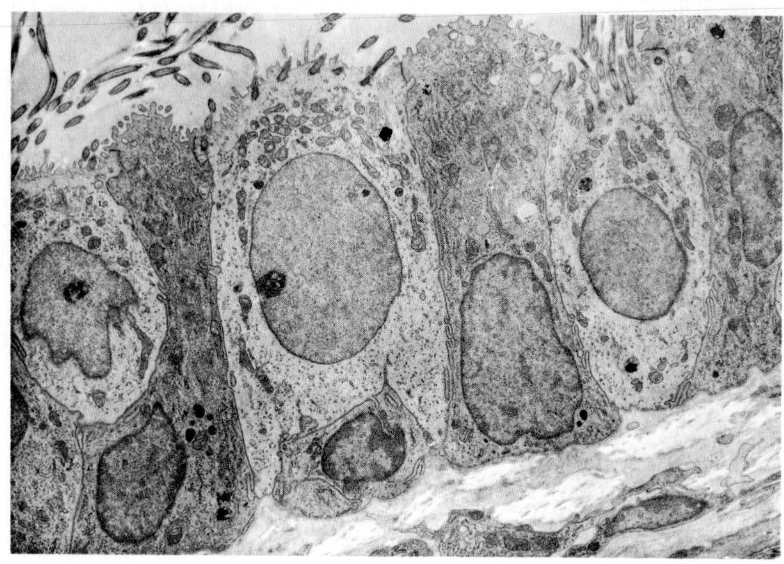

4–12
Cells from the surface of the trachea (windpipe) of a bat. The free surface of the larger cell is covered with cilia, which are exactly the same in structure as the flagella of Chlamydomonas. *(When they are fewer and longer, they are usually called flagella, whereas when shorter and more numerous they are called cilia.) Next to the ciliated cells are cells that secrete mucus onto the cell surface. Mucus currents, swept by cilia, remove foreign particles from the surface of the trachea. Note the mitochondria, clustered at the base of the cilia.*

The human body, made up of trillions of individual cells, is composed of at least 100 different types of cells, each specialized for its particular function. Figure 4–12 shows cells from an animal trachea (windpipe). These cells are epithelial cells, the sort of cells that line all the internal and external body surfaces of animals. The epithelial cells of the trachea are part of an elaborate organ system involved in delivering oxygen to other cells in the body.

Prokaryotes, protists, fungi, plants, and animals constitute the five different kingdoms of organisms recognized in this text. The prokaryotes are all unicellular and include chemoautotrophs, phototrophs, and heterotrophs. Protists are eukaryotic one-celled organisms and some simple multicellular ones; this kingdom includes both phototrophs and heterotrophs. The fungi, plants, and animals are all multicellular eukaryotic organisms. All animals and fungi are heterotrophs, whereas all plants, with a few curious exceptions (such as Indian pipe and dodder, which are parasites) are phototrophs. Within the multicellular plant body, however, some of the cells are phototrophic, such as the cells of a leaf, and some are heterotrophic, such as the cells of a root. The phototrophic cells supply the heterotrophic cells with sucrose produced from photosynthesis.

VIEWING THE CELLULAR WORLD

In the three centuries since Robert Hooke first observed the structure of cork through his simple microscope, a wealth of knowledge has been accumulated both about the structure of cells and their component parts and about the dynamic processes that characterize the living cell. This knowledge, which we shall begin to examine in the next chapter, has generally come in bursts, following the development of new and better techniques for studying the cell and its contents.

Types of Microscopes

Unaided, the human eye has a resolving power of about 1/10 millimeter, or 100 micrometers (Table 4–1). This means that if you look at two lines that are less than

100 micrometers apart, they merge into a single line. Similarly, two dots less than 100 micrometers apart look like a single blurry dot. Most eukaryotic cells are between 10 and 30 micrometers in diameter—some 3 to 10 times below the resolving power of the human eye—and prokaryotic cells are smaller still. In order to distinguish individual cells, to say nothing of examining the structures of which they are composed, we must use instruments that provide greater resolution. Most of our current knowledge of cell structure has been gained with the assistance of three different types of instruments: the light microscope, the transmission electron microscope, and the scanning electron microscope (Figure 4–13).

Table 4–1 Measurements Used in Microscopy

1 centimeter (cm) = 1/100 meter = 0.4 inch*
1 millimeter (mm) = 1/1,000 meter = 1/10 cm
1 micrometer (μm)† = 1/1,000,000 meter = 1/10,000 cm
1 nanometer (nm) = 1/1,000,000,000 meter = 1/10,000,000 cm
1 angstrom (Å)‡ = 1/10,000,000,000 meter = 1/100,000,000 cm
or
1 meter = 10^2 cm = 10^3 mm = 10^6 μm = 10^9 nm = 10^{10} Å

* A metric-to-English conversion table is found in Appendix A.

† Micrometers were formerly known as microns (μ), and nanometers as millimicrons (mμ).

‡ The angstrom is not an accepted measurement in the International System of Units; in the past, however, it was widely used in microscopy, and you will occasionally encounter it in your reading.

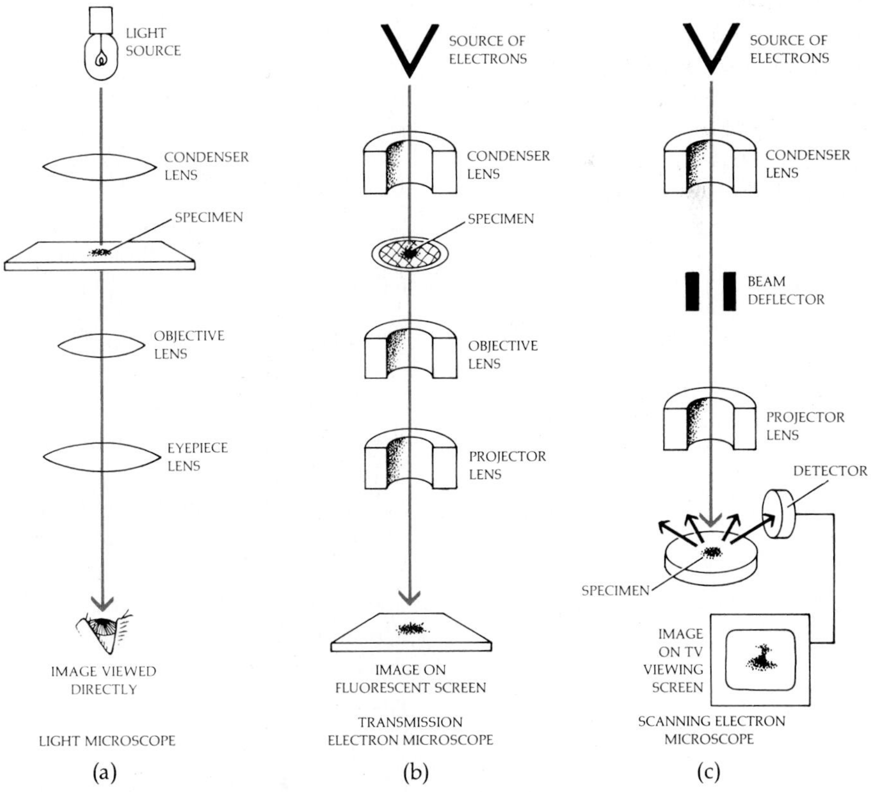

4–13

A comparison of (a) *the light microscope,* (b) *the transmission electron microscope, and* (c) *the scanning electron microscope. The light microscope is shown upside-down, to emphasize its similarities with the electron microscopes. The focusing lenses in the light microscope are glass or quartz; those in the electron microscopes are magnetic coils. In both the light microscope and the transmission electron microscope, the illuminating beam passes through the specimen; in the scanning electron microscope, it is deflected from the surface.*

LIGHT SOURCE
CONDENSER LENS
SPECIMEN
OBJECTIVE LENS
EYEPIECE LENS
IMAGE VIEWED DIRECTLY
LIGHT MICROSCOPE
(a)

SOURCE OF ELECTRONS
CONDENSER LENS
SPECIMEN
OBJECTIVE LENS
PROJECTOR LENS
IMAGE ON FLUORESCENT SCREEN
TRANSMISSION ELECTRON MICROSCOPE
(b)

SOURCE OF ELECTRONS
CONDENSER LENS
BEAM DEFLECTOR
PROJECTOR LENS
DETECTOR
SPECIMEN
IMAGE ON TV VIEWING SCREEN
SCANNING ELECTRON MICROSCOPE
(c)

The best light microscopes have a resolving power of 0.2 micrometer, or 200 nanometers, and so improve on the naked eye about 500 times. It is theoretically impossible to build a light microscope that will do better than this. The limiting factor is the wavelength of light, which ranges from about 0.4 micrometer for violet light to about 0.7 micrometer for red light. With the light microscope, we can distinguish the larger structures within eukaryotic cells and can also distinguish individual prokaryotic cells. We cannot, however, visualize the internal structure of prokaryotic cells or distinguish between the finer structures of eukaryotic cells.

Notice that resolving power and magnification are two different things. If you take a picture through the best light microscope of two lines that are less than 0.2 micrometer, or 200 nanometers, apart, you can enlarge that photograph indefinitely, but the two lines will continue to blur together. By using more powerful lenses, you can increase magnification, but this will not improve resolution.

With the transmission electron microscope, resolving power has been increased about 400 times over that provided by the light microscope. This is achieved by using "illumination" of a much shorter wavelength, consisting of electron beams instead of light rays. Areas in the specimen that permit the transmission of more electrons—"electron-transparent" regions—show up bright, and areas that scatter electrons away from the image—"electron-opaque" regions—are dark. Transmission electron microscopy at present affords a resolving power of about 0.5 nanometer, roughly 200,000 times greater than that of the human eye. This is about five times the diameter of a hydrogen atom.

Although the resolving power of the scanning electron microscope is only about 10 nanometers, this instrument has become an extraordinarily valuable tool. In scanning electron microscopy, the electrons whose imprints are recorded come from the surface of the specimen, rather than from a section through the specimen. The electron beam is focused into a fine probe, which is rapidly passed back and forth over the specimen in a precise pattern. Complete scanning from top to bottom usually takes a few seconds. In addition to the illuminating electrons scattered from the surface of the specimen, the specimen itself emits low-energy secondary electrons as a result of the electron bombardment from the probe. Variations in the surface of the specimen alter the number of secondary electrons emitted; holes and fissures appear dark, and knobs and ridges are light. The scattered electrons plus the secondary electrons are amplified and transmitted to a television screen, which is scanned in synchrony with the probe. Scanning electron microscopy provides vivid three-dimensional representations of cells and cellular structures (Figure 4–16) that more than compensate for its limited resolution.

Preparation of Specimens

In both the light microscope and the transmission electron microscope, the formation of an image with perceptible contrast requires that different parts of the cell differ in their transparency to the beam of illumination—either light rays or electrons. Parts of the specimen that readily permit the passage of light or of electrons appear bright, whereas parts that block the passage of the illuminating beam appear dark. In the scanning electron microscope, areas that appear light are those that emit secondary electrons and deflect electrons back into the image; parts that appear dark are those that are not well illuminated by the electron beam or that deflect electrons away from the detector.

4–14
Rabbit sperm cells, as seen in a light micrograph.

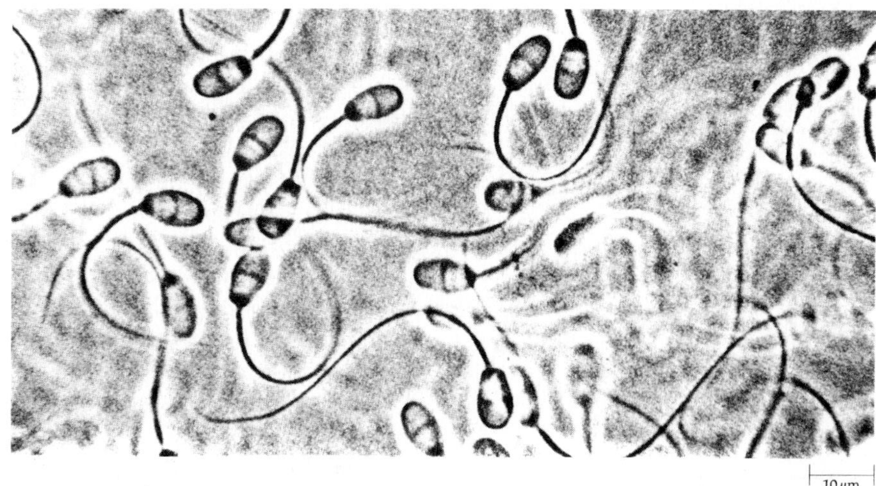

4–15
A rabbit sperm cell, as seen in a transmission electron micrograph. Note the dramatic increase in the resolution of ultrastructural detail.

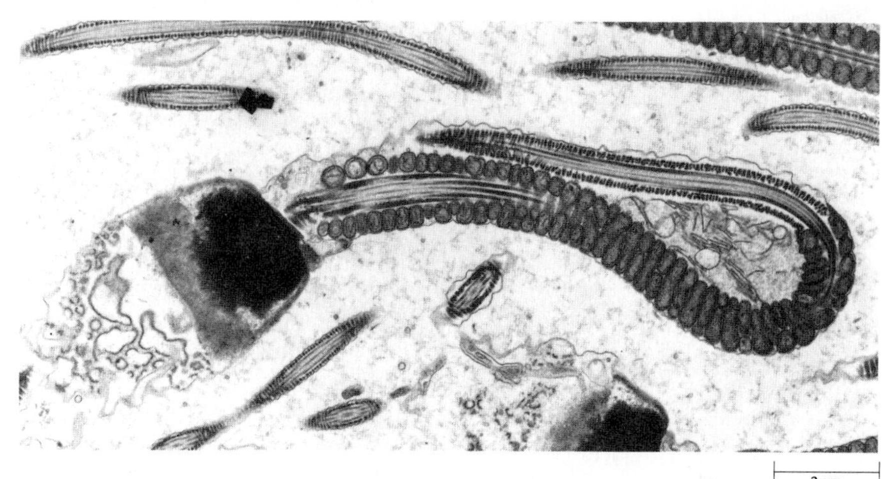

4–16
Rabbit sperm cells, as seen in a scanning electron micrograph.

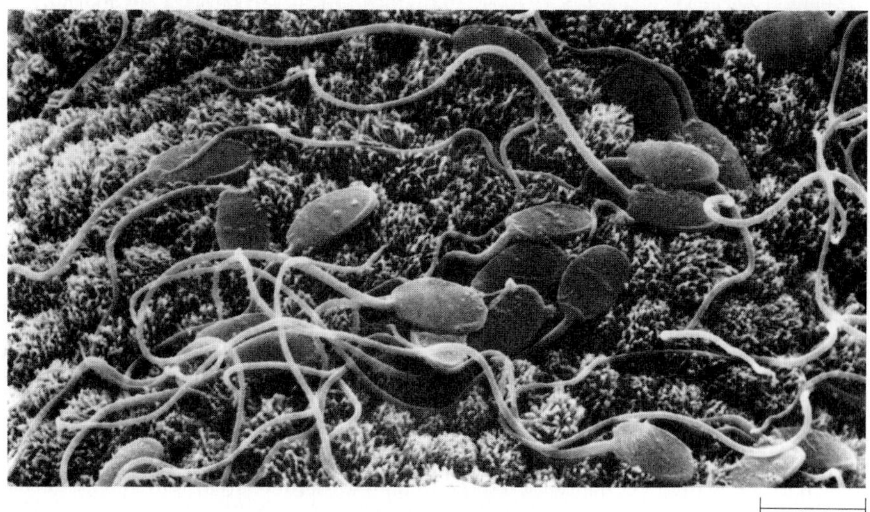

Living cells and their component parts are, however, almost completely transparent to light. By weight, cells are about 70 percent water, through which light passes easily. Moreover, water and the much larger molecules that form cellular structures are composed of small atoms of low atomic weight (CHNOPS). These atoms are relatively transparent to electrons, which are strongly deflected only by atoms of high atomic weight, such as those of heavy metals.

To create sufficient contrast for the light microscope, cells must be treated with dyes or other substances that differentially adhere to or react with specific subcellular components, producing regions of differing opacity. For the electron microscope, the specimens are similarly treated with compounds of heavy metals. In the transmission electron micrograph of collagen fibrils in the previous chapter (page 69), for example, the dark bands are deposits of a metallic compound that has filled the gaps between individual collagen molecules. The light bands are areas filled with tightly packed collagen molecules, which excluded the metallic compound.

After a specimen has been treated with a staining substance, all of the stain that has not adhered to specific structures must be washed away. Cells, however, are quite fragile and any kind of rough treatment disrupts their structure. To solve this problem, biological specimens are generally "fixed" before staining. This procedure involves treatment with compounds that bind cell structures in place, usually through the formation of additional covalent bonds between molecules. Aldehydes, for instance, react with the terminal amino groups of protein molecules, linking adjacent protein molecules together in a fairly rigid structure. Osmium tetroxide, a compound frequently used in preparing specimens for electron microscopy, interacts with lipids, binding the molecules together. Fixation has the added advantage of making cells more permeable to staining substances and to the solutions used to wash away excess stain.

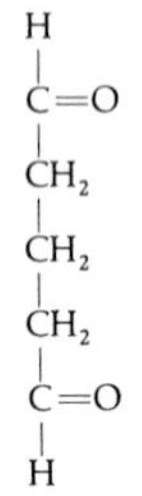

H–C(=O)–CH_2–CH_2–CH_2–C(=O)–H

GLUTARALDEHYDE

(a)

OsO_4 (O=Os(=O)(=O)=O)

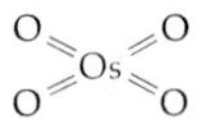

OSMIUM TETROXIDE

(b)

4–17

(a) *Glutaraldehyde and* (b) *osmium tetroxide, two compounds frequently used as fixatives in preparing specimens for microscopic examination. Note that glutaraldehyde has two aldehyde functional groups; each of these groups can react with a different protein molecule, linking the molecules together. The oxygen atoms of osmium tetroxide react with lipids, leading to a similar cross-linking.*

Fixation and staining procedures are usually carried out with groups of cells as in, for example, a piece of liver tissue. Such specimens are not transparent to an illuminating beam—they are simply too thick to allow the passage of light rays or electrons. Before examination under the light microscope or the transmission electron microscope, they must be sliced into sections so thin that the unstained regions are transparent. The specimens are usually embedded in wax or a plastic resin to provide enough firmness to allow a clean slicing.

With the scanning electron microscope, as we noted earlier, electrons do not pass through the sample but are instead deflected from its surface. Usually, the surface of the specimen is first coated with metal, a process known as shadowing (Figure 4–18). A filament of metal is heated to evaporation at one side of the sample; the evaporated atoms then fall on the sample, creating a coating of metal that is thicker on one side of the raised surfaces and thinner on the other. Often, the organic material of the original specimen is removed by chemical treatment, leaving only a metallic replica of the surface, which is then reinforced with a carbon film. Depending on its thickness, the replica can be examined under either the transmission electron microscope or the scanning electron microscope.

In addition to these rather drastic treatments, specimens for the electron microscope must also be dehydrated, either by chemical methods or by freeze-dry methods similar to those used in preparing freeze-dried coffee. This step is necessitated by the properties of the electrons forming the illuminating beam. If they pass through a chamber containing gaseous molecules, the electrons are deflected

4-18

"Shadowing" and the preparation of a replica. (a) *The specimen is placed on a support and* (b) *is "shadowed" by heavy metal atoms evaporated from a heated filament at the side. Because the atoms are deposited from an angle, the metal coating is thicker on raised areas of the specimen.* (c) *A uniform film of carbon atoms is deposited from above, reinforcing and strengthening the replica,* (d) *which is then floated to the surface of a solvent that dissolves away the organic material. The completed replica* (e) *is washed and picked up on a copper grid.*

If the replica is thin enough, it can be examined under the transmission electron microscope as well as under the scanning electron microscope.

(a)

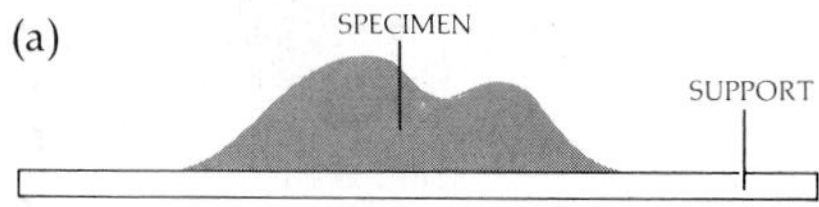

(b)

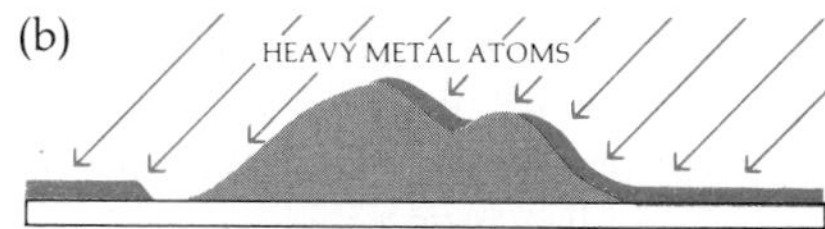

(c)

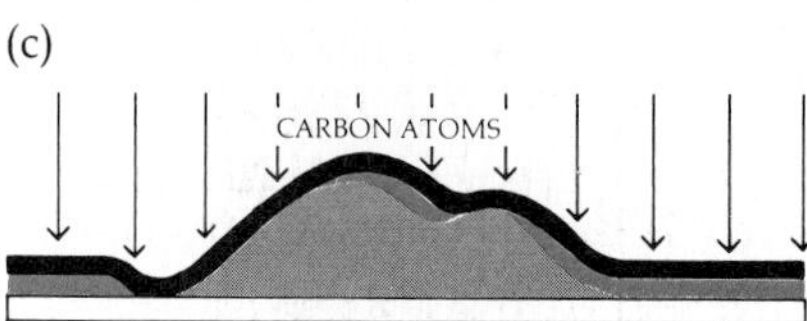

(d)

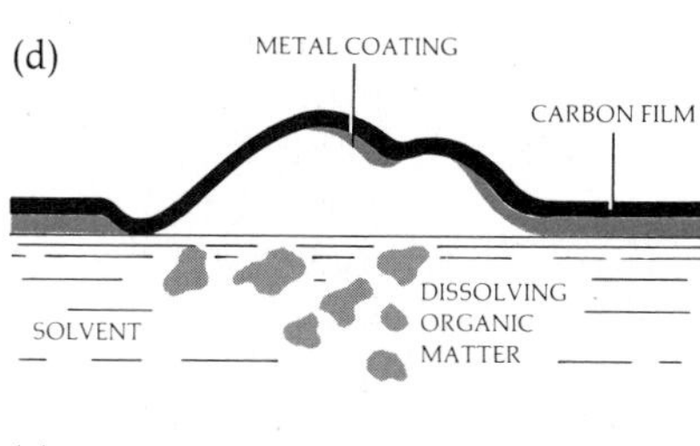

(e)

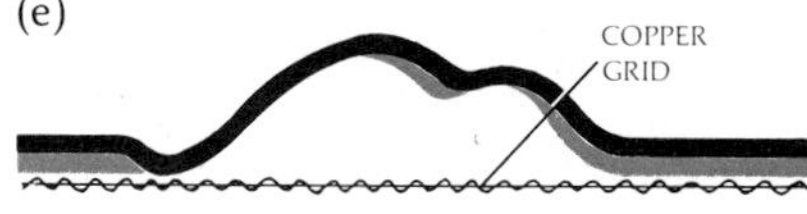

by the molecules and cannot be focused into a beam. Thus all air must be evacuated from the inner, working chamber of an electron microscope, creating a vacuum. If the sample were not first dehydrated, water molecules would evaporate from the sample into the chamber, destroying the vacuum and the focused electron beam.

The procedures required to prepare most stained specimens for the light microscope and to prepare all specimens for the electron microscope usually result in the death of the constituent cells. Moreover, they raise serious questions about whether the structures we see in micrographs are "real" or are distortions introduced by the preparation process. One way of reducing the likelihood that microscopic observations are in error is to prepare similar samples using different techniques. If a feature appears repeatedly in different types of preparation, the probability is great that it exists within the living cell. Another approach has been the development of a variety of new techniques for viewing living cells.

Observation of Living Cells

When light waves are emitted from a coherent source (such as a laser), the waves are in phase, that is, the peaks and troughs of the waves match (Figure 4-19a). This has the effect of reinforcing the waves and creating a greater amplitude, which we perceive as increased brightness. When light waves are out of phase, however, they interfere with one another, reducing amplitude and brightness (Figure 4-19b).

As light waves pass from one material through another, they are bent, or diffracted, and their paths are slightly changed. This diffraction also alters the phase relationships of the light waves, resulting in varying amounts of interference. The amount of interference produced when light passes through the different struc-

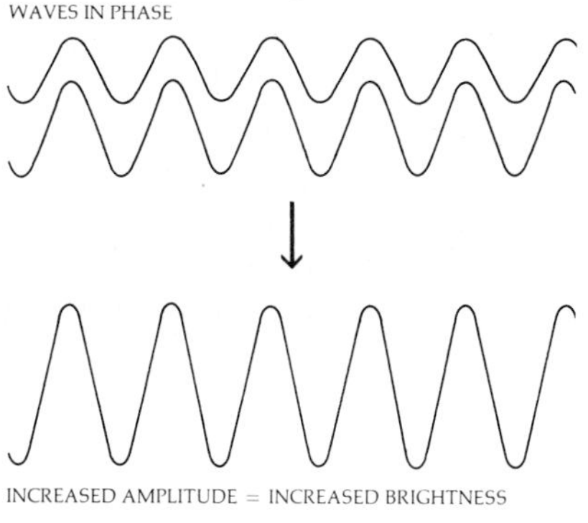

(a)

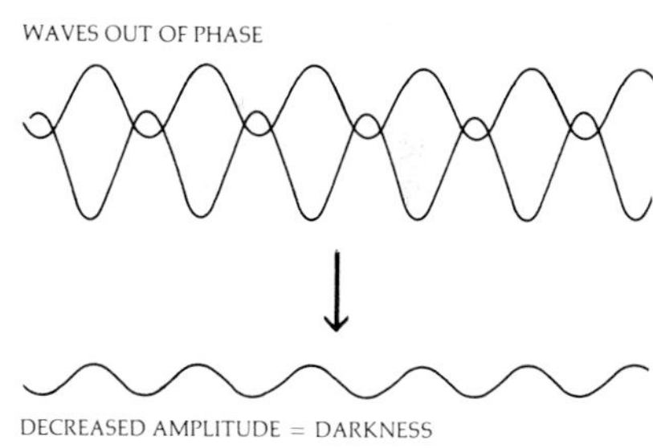

(b)

4-19

The brightness of light depends on the amplitude of the light waves. (a) *When two light waves are in phase, they reinforce one another, resulting in greater amplitude and brightness.* (b) *When the waves are completely out of phase, as shown here, they cancel each other out and no light is perceived.*

Waves that have passed through different structures within the unstained cell are partially out of phase. The resulting interference produces slight differences in contrast that can be enhanced by special optical systems, such as those of phase-contrast and differential-interference microscopes.

4–20

Four views of a living fibroblast, a type of connective tissue cell that can be propagated in the laboratory in culture dishes. This cell has been photographed through (a) *a conventional light microscope,* (b) *a phase-contrast microscope,* (c) *a differential-interference microscope, and* (d) *a dark-field microscope.*

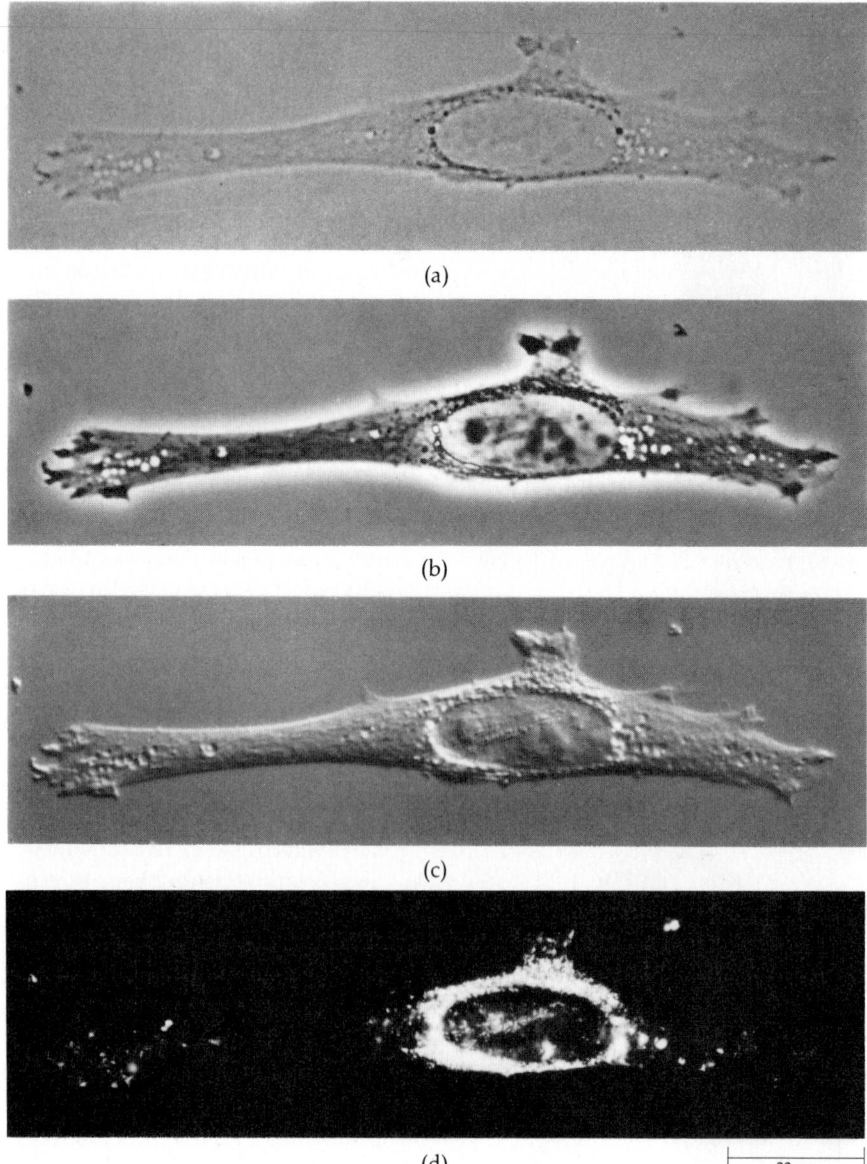

tures of a cell is not great and provides little detectable contrast when the cell is viewed through an ordinary light microscope—which is, of course, why cells must be stained. In phase-contrast and differential-interference microscopes, however, specially designed optical systems enhance the small amounts of interference, providing greater contrast. The resolution of these microscopes is limited, as in ordinary light microscopes, but they do provide a different perspective on the living cell, revealing features difficult to detect with other systems.

Another technique frequently used with living cells is dark-field microscopy. The illuminating beam strikes the specimen from the side and the lens system detects light reflected from the sample, which appears as a bright object against a dark background. Features of the cell that are invisible in other micrographs often come into sharp relief in dark-field micrographs.

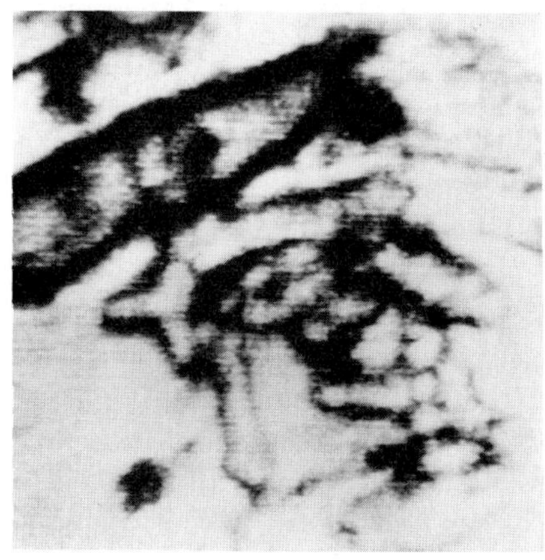

(a)

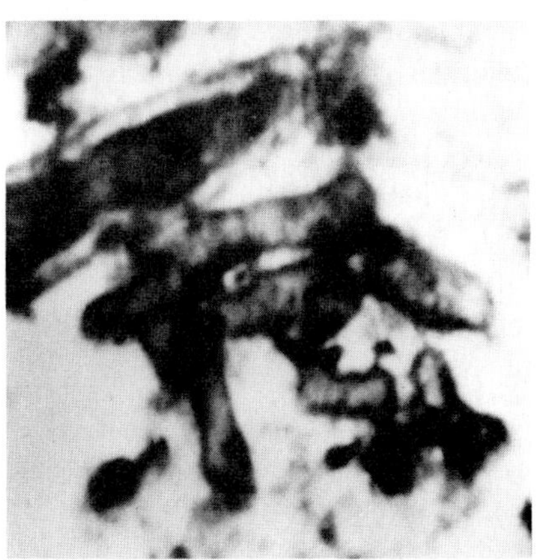

(b)

4–21
When specimens, such as these cells of the freshwater alga Hydrodictyon reticulatum, *are studied with the acoustic microscope, both* (a) *acoustic and* (b) *optical images are produced simultaneously. A particular feature of the specimen may transmit strikingly different amounts of ultrasound and of light. For example, the walls of these cells, rich in pectin, are opaque to ultrasound but relatively transparent to light.*

At the present time, work is progressing rapidly on other microscopic techniques. For example, video cameras are being coupled with light microscopes, producing a display on a screen and a videotape record. By adjusting the controls, as you do on a television set, the background "noise" can be reduced, contrast improved, and particular features enhanced. Video techniques, as applied to the study of the living cell, are in their infancy, but they are generating great excitement as they reveal previously unseen processes within the cell. Another new tool is the acoustic microscope, which uses a beam of sound waves as the source of "illumination." Ultrasound techniques have proved a valuable diagnostic tool in medicine, and the acoustic microscope, which uses even shorter wavelengths of sound, may prove equally valuable in cell biology.

SUMMARY

The properties associated with biological systems emerge at the cellular level of organization. One of the unifying concepts of biology is the cell theory, which states that (1) all living organisms consist of one or more cells; (2) the chemical reactions of a living organism, including energy-yielding processes and biosynthetic reactions, take place within cells; (3) cells arise from other cells; and (4) cells contain the hereditary information of the organisms of which they are a part, and this information is passed from parent cell to daughter cell.

In the 1860s, Pasteur proved decisively that living organisms could not be spontaneously generated from nonliving matter under the conditions now prevailing on the earth. In the 1920s, Oparin hypothesized that under different conditions, which could be tested experimentally, life could have arisen spontaneously.

The primitive atmosphere held the raw materials of living matter—hydrogen, oxygen, carbon, and nitrogen—combined in water vapor and gases. The energy required to break apart the simple gases in the atmosphere and re-form them into more complex molecules was present in heat, lightning, radioactive elements, and high-energy radiation from the sun. Laboratory experiments testing Oparin's hypothesis have shown that the types of organic molecules characteristic of living systems can be formed under the conditions that probably prevailed on the young earth. Other experiments have suggested the kinds of processes by which aggregations of organic molecules could have formed cell-like structures, separated from their environment by a membrane and capable of maintaining their structural and chemical integrity.

The age of the earth is estimated at 4.6 billion years. Microfossils of bacteria-like cells have been discovered that are 3.5 billion years old. The complexity of these cells suggests that the first primitive cells arose very early in the earth's existence—sometime during the first billion years.

The earliest cells may have been heterotrophs (organisms that depend on outside sources for their energy-rich organic molecules) or chemoautotrophs (organisms that capture the energy released by specific inorganic reactions and use it to synthesize their own organic molecules). Phototrophs (organisms that use the sun's energy to power their synthetic reactions) evolved subsequently. With the advent of photosynthesis, the flow of energy through the biosphere assumed its dominant modern form—radiant energy from the sun is captured by photosynthetic organisms and channeled through them to heterotrophic organisms.

There are two fundamentally distinct types of cellular organization—prokaryotes, which include only the bacteria and cyanobacteria, and eukaryotes, which include the protists, fungi, plants, and animals. Prokaryotic cells lack membrane-bound nuclei and most of the organelles found in eukaryotic cells.

Multicellular organisms evolved comparatively recently—only about 750 million years ago. They are composed of eukaryotic cells, specialized to perform particular functions.

Because of the small size of cells and the limited resolving power of the human eye, microscopes are required to visualize cells and subcellular structures. The three principal types are the light microscope, the transmission electron microscope, and the scanning electron microscope. The light microscope has a resolving power of 0.2 micrometer, whereas the transmission electron microscope has a resolving power of about 0.5 nanometer. The resolving power of the scanning electron microscope is more limited, but it provides unique views of the three-dimensional structure of cells and their component parts.

Specimens to be studied using a conventional light microscope or a transmission electron microscope must be fixed, stained, dehydrated (for the electron microscope), embedded, and sliced into thin sections. Surface replicas are generally prepared for study with the scanning electron microscope. Concern about the distortions that may be introduced by these preparation procedures has led to the development of other microscopic techniques. The special optical systems of phase-contrast, differential-interference, and dark-field microscopes make it possible to study living cells. Other new developments are the use of video cameras with microscopes and the acoustic microscope.

QUESTIONS

1. Distinguish among the following: heterotroph/autotroph; chemoautotroph/phototroph; prokaryote/eukaryote; light microscope/transmission electron microscope/scanning electron microscope.

2. Why would energy sources have been necessary for the synthesis of simple organic molecules on the primitive earth?

3. Although there is some uncertainty as to the exact mixture of gases that constituted the early atmosphere, there is general agreement that free oxygen was not present. What properties of oxygen would have made chemical evolution unlikely in an atmosphere containing O_2?

4. A key event in the origin of life was the formation of a membrane that separated the contents of primitive cells from their surroundings. Why was this so critical?

5. Some scientists think that other planets in our galaxy may well contain some form of life. Suppose you were seeking such a planet, what characteristics would you look for?

6. Return to Chapter 2 and add approximate scale markers to Figures 2–1, 2–4, 2–5, and 2–11.

7. What are the advantages and disadvantages of studying cells with the transmission electron microscope and the scanning electron microscope? With special optical microscopes such as phase-contrast, differential-interference, and dark-field?

CHAPTER 5

How Cells Are Organized

There are many, many different kinds of cells. Our own tissues and organs are constructed of about 200 different and distinct types of somatic ("body") cells. In a drop of pond water, you are likely to find several different kinds of protists, and in even a small pond, there are probably several hundred different kinds of protists, plus a variety of prokaryotes. Plants are composed of cells superficially quite different from those of our own bodies, and insects have many cells of kinds not found either in plants or in vertebrates. Thus, the first remarkable fact about cells is their diversity.

The second, even more remarkable, fact is their similarity. Every cell is a self-contained and at least partially self-sufficient unit, surrounded by a membrane that controls the passage of materials into and out of the cell. This makes it possible for the cell to differ biochemically and structurally from its surroundings. All eukaryotic cells also, at least at some time in their existence, contain nuclei, which serve as their information and control centers. Many have a variety of internal structures, the organelles, which are similar or identical from one cell to another throughout a wide range of cell types. And, all are composed of the same remarkably few kinds of atoms and molecules.

CELL SIZE AND SHAPE

Most of the cells that make up a plant or animal body are between 10 and 30 micrometers in diameter. A principal restriction on cell size is that imposed by the relationship between volume and surface area. As Figure 5–1 (page 98) shows, as volume increases, surface area decreases rapidly in proportion to volume. Materials—such as oxygen, carbon dioxide, ions, food molecules, and waste products—entering and leaving the cell must move through its membrane-bound surface. The more active the cell's metabolism,* the more rapidly these materials must be exchanged with the environment if the cell is to continue to function. Moreover, the larger the cell, the greater the quantity of materials that must be exchanged to meet the needs of the larger volume of living matter. In small cells, however, the proportion of surface area to volume is greater than in large cells; thus proportionately greater quantities of materials can move into, out of, and through small cells in a given period of time.

---

* Metabolism is simply the total of all of the chemical activities of a living system.

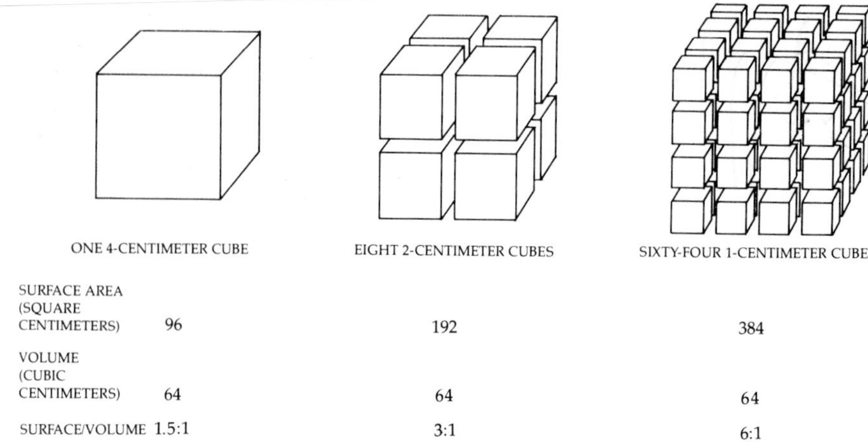

	ONE 4-CENTIMETER CUBE	EIGHT 2-CENTIMETER CUBES	SIXTY-FOUR 1-CENTIMETER CUBES
SURFACE AREA (SQUARE CENTIMETERS)	96	192	384
VOLUME (CUBIC CENTIMETERS)	64	64	64
SURFACE/VOLUME	1.5:1	3:1	6:1

5–1
The single 4-centimeter cube, the eight 2-centimeter cubes, and the sixty-four 1-centimeter cubes all have the same total volume. As the volume is divided up into smaller units, however, the amount of surface area increases (for example, the single 4-centimeter cube has only one-fourth the total surface area of the sixty-four 1-centimeter cubes). Similarly, smaller cells have more surface area (more membrane surface through which materials can move into or out of the cell) and less volume (less living matter to be serviced and shorter distances through which materials must move within the cell).

A second limitation on cell size appears to involve the capacity of the nucleus, the cell's control center, to provide enough copies of the information needed to regulate the processes occurring in a large, metabolically active cell. The exceptions seem to "prove" the rule. In certain large, complex one-celled organisms—the ciliates, of which *Paramecium* is an example—each cell has two or more nuclei, the additional ones apparently copies of the original. Other organisms, such as the slime molds, are made up, in effect, of one giant cell but have thousands of nuclei. Such organisms are also, frequently, very thin and spread out, thereby avoiding the surface/volume problem.

It is not surprising, therefore, that the most metabolically active cells are usually small. The relationship between cell size and metabolic activity is nicely illustrated by egg cells. Many egg cells are very large. A frog's egg, for instance, is 1,500 micrometers in diameter. Some egg cells are several centimeters across—for example, the egg cell, or yolk, of a chicken's egg. Most of this mass consists of stored nutrients for the developing embryo. When the egg cell is fertilized and begins to be active metabolically, many nuclear divisions occur and the cell divides many times before there is any actual increase in volume or mass. Thus the total mass is cut into cellular units small enough for efficient transfer and control processes.

Like drops of water and soap bubbles, cells have a tendency to be spherical. Cells take other shapes because of cell walls, found in plants, fungi, and many one-celled organisms; because of attachments to and pressure from other neighboring cells or surfaces (as in intestinal epithelial cells); or because of arrays of microtubules and other structural filaments within the cell (Figure 5–2).

SUBCELLULAR ORGANIZATION

Antony van Leeuwenhoek discovered the protists some 300 years ago. "This was for me," he wrote, "among all the marvels that I have discovered in nature, the most marvelous of all." As Leeuwenhoek and his successors observed the thousands of living creatures that they found "all alive in a drop of water," they were able to see, but only barely, structures within them, which they interpreted as being miniature hearts and stomachs and lungs, in other words, tiny organs, or organelles.

Modern microscopic techniques have confirmed that eukaryotic cells do indeed contain a multitude of structures. They are not, of course, organs such as those found in multicellular organisms, but they are in some ways comparable: they are specialized in form and function to carry out particular activities required in the cellular economy. Just as the organs of multicellular animals work together in organ systems, the organelles of cells engage in a number of cooperative and interdependent functions.

Every cell must carry out essentially the same processes—acquire and assimilate food, eliminate wastes, synthesize new cellular materials, and in many cases, be able to move and to reproduce. Just as the various organs of your body have a form or design that suits them for the specific functions they carry out (the kidney for elimination of wastes from the blood, the intestine for food absorption, and so on), so all cells have an internal architecture that includes organelles suited to the functions they perform. It is important to realize that a cell is not a random assortment of parts but a dynamic, integrated entity.

Also, although we can look at only one function at a time, remember that most activities of a cell go on simultaneously and influence one another. *Chlamydomonas*, for instance, is swimming, photosynthesizing, absorbing nutrients from the water, building its cell wall, making proteins, converting sugars to starch (or vice versa), and oxidizing food molecules for energy, all at the same time. It is also likely to be orienting itself in the sunlight, it is probably preparing to divide, it is possibly "looking" for a mate, and it is undoubtedly carrying out at least a dozen or more other important activities, many of which may be still unknown.

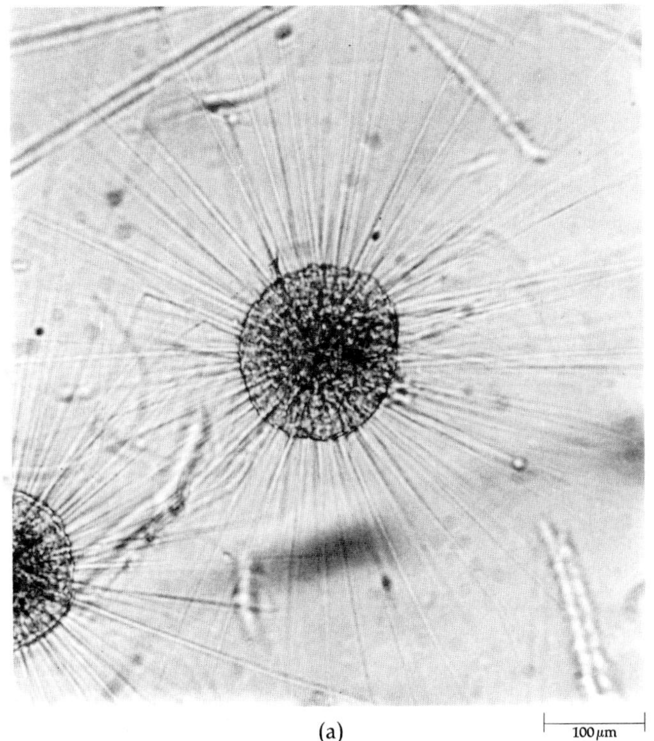

(a)

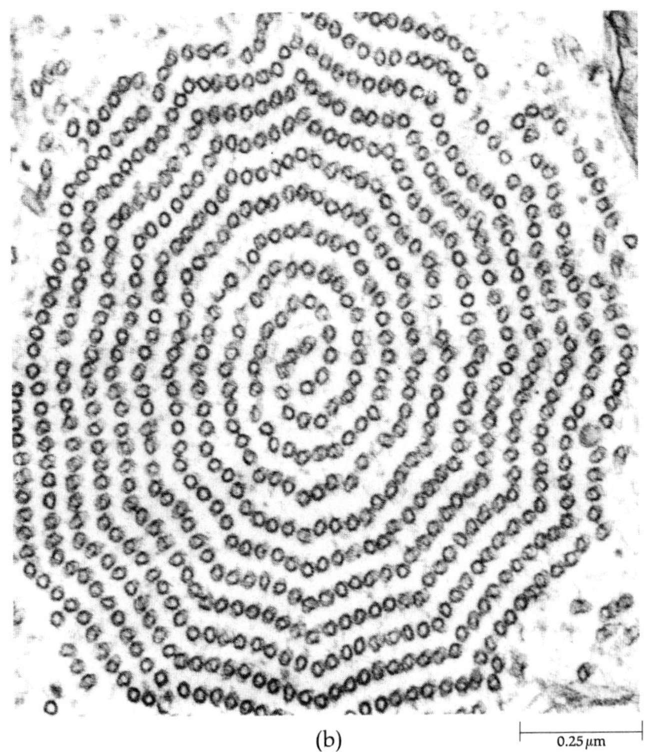

(b)

5–2
(a) *Numerous fine strands of cytoplasm (axopods) extend radially from the body of the protist* Actinosphaerium. *Each axopod is bounded by an extension of the cell membrane and contains many microtubules, arranged longitudinally, which stiffen and extend the axopods.* (b) *As seen in cross section, the microtubules are arranged in two interlocking spirals that form a twelvefold pattern.*

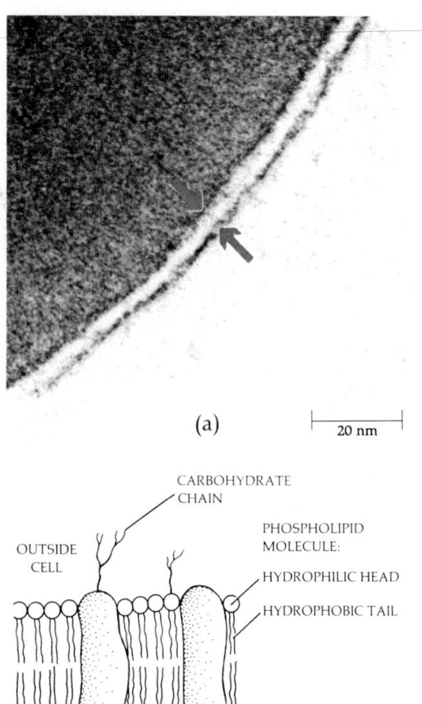

5–3

(a) Electron micrograph showing a cross section of the cell membrane of a red blood cell. The "molecular sandwich" structure of the membrane is believed to consist of two electron-opaque (dark) layers of phospholipid molecules arranged with their hydrophobic tails pointing inward, forming the electron-transparent (light) inner "filling," with globular proteins embedded throughout. The darker material at the left of the micrograph is hemoglobin, which fills the red blood cell.

(b) *Model of the cell membrane showing the two layers of phospholipids within which the large globular proteins are embedded. Short carbohydrate chains are attached to some of the outside layer of phospholipid and protein molecules. These serve as receptors for hormones, antibodies, and viruses, and are also involved with adhesion between cells.*

THE ARCHITECTURE OF THE CELL

All cells, as we stated previously, are basically very similar, containing many of the same structures, the same types of enzymes, and the same kind of genetic material. An outer membrane, apparently conforming to the same general biochemical design in all cells, surrounds the cell. The living materials bounded by the membrane consist, in eukaryotes, of the nucleus and the cytoplasm, which contains the organelles.

The Cell Membrane

A cell can exist as a distinct entity because of the cell membrane, which regulates the passage of materials into and out of the cell. The cell membrane (also called, as we noted previously, the plasma membrane) is only about 7 to 9 nanometers thick and cannot be resolved in the light microscope. Now, with the electron microscope, it can be visualized as a continuous, thin double line (Figure 5–3). According to current hypotheses, the eukaryotic cell membrane consists essentially of phospholipid and cholesterol molecules arranged so that their hydrophobic tails are clustered inward, forming the filling of a molecular sandwich, with globular proteins embedded throughout. Short carbohydrate chains are attached to some of the protein and phospholipid molecules on the side of the membrane facing the external environment, and additional proteins are attached on the side facing the cytoplasm.

All the membranes of a cell, including those surrounding the various organelles, have this same structure. There are, however, local differences in the types of lipids and, particularly, in the number and types of proteins and carbohydrates. These differences give the membranes of different types of cells and of the different organelles unique properties that can be correlated with differences in membrane function. In Chapter 6, we shall examine the molecular structure of cell membranes in more detail and shall consider the ways in which they perform their essential tasks.

The cell membrane of bacterial cells is much the same in basic composition as the membranes of eukaryotes, except that, with a few exceptions, bacterial cell membranes do not contain cholesterol.

The Cell Wall

A principal distinction between plant and animal cells is that the former are surrounded by a cell wall. The wall is outside the membrane and is constructed by the cell. As a plant cell divides, a thin layer of gluey material forms between the two new cells; this becomes the *middle lamella* (Figure 5–4a). Composed of pectins (the compounds that make jellies gel) and other polysaccharides, it holds adjacent cells together. Next, on either side of the middle lamella, the two plant cells construct their primary cell walls. The primary wall contains cellulose molecules bundled together in microfibrils that are laid down in a matrix of gluey polymers. As you can see in Figure 3–9c on page 59, successive layers of microfibrils are oriented at right angles to one another in the completed cell wall. (Those of you familiar with building materials will note that the cellulose cell wall thus combines the structural features of both fiberglass and plywood.)

In plants, growth takes place largely by cell elongation. Studies have shown that the cell adds new materials to its walls throughout this elongation process. The cell,

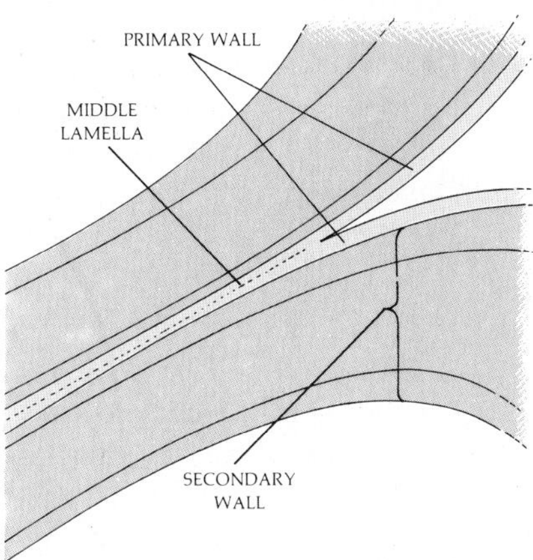

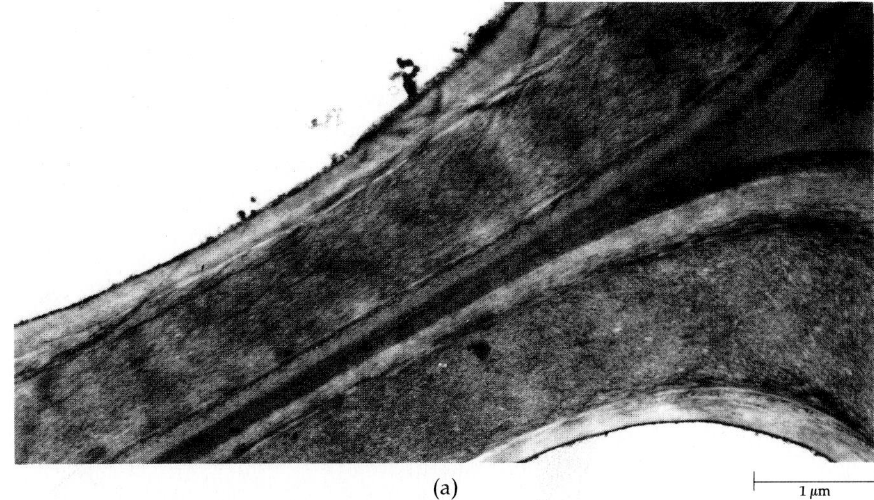

(a)

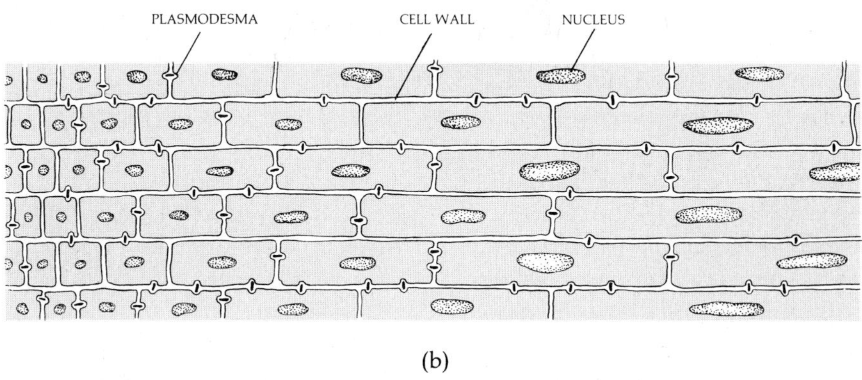

(b)

5–4

(a) *Electron micrograph of two adjacent cell walls of tracheids, cells through which water is conducted in plants. You can see the middle lamella, the primary walls, and the layered secondary walls, deposited inside the primary wall. The cells, which are from the wood of a ground hemlock, have died.*

(b) *Growth of plant cells is limited by the rate at which the cell walls expand. The walls control both the rate of growth and its direction; they do not expand in all directions but elongate in a single dimension. Cells at the left are newly formed; cells farther to the right are older and have started to elongate. Plasmodesmata (singular, plasmodesma) are channels connecting adjacent cells.*

however, does not simply expand in all directions; its final shape is determined by the structure of its cell wall (Figure 5–4b).

As the cell matures, a secondary wall may be constructed. This wall is not capable of expansion, as is the primary wall. It often contains other molecules, such as lignin, that have stiffening properties. In such cells, the living material of the cell often dies, leaving only the outer wall, a monument to the cell's architectural abilities.

Cellulose-containing cell walls are also found in many algae; fungi and prokaryotes also have cell walls, but they usually do not contain cellulose.

The Nucleus

In eukaryotic cells, the nucleus is a large, often spherical body, usually the most prominent structure within the cell. It is surrounded by two lipoprotein membranes, which together make up the nuclear envelope. These two membranes are separated by a gap of about 20 to 40 nanometers. At frequent intervals, however, they are fused together to create pores through which materials pass between the nucleus and cytoplasm (Figure 5–5). The pores, which are surrounded by large protein-containing granules arranged in an octagonal pattern, form a narrow channel through the fused lipid bilayers.

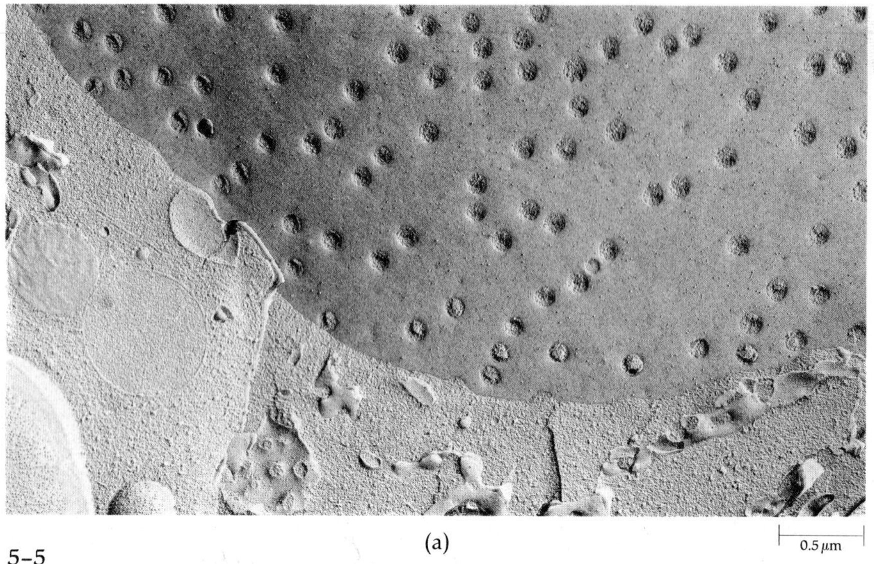

(a)

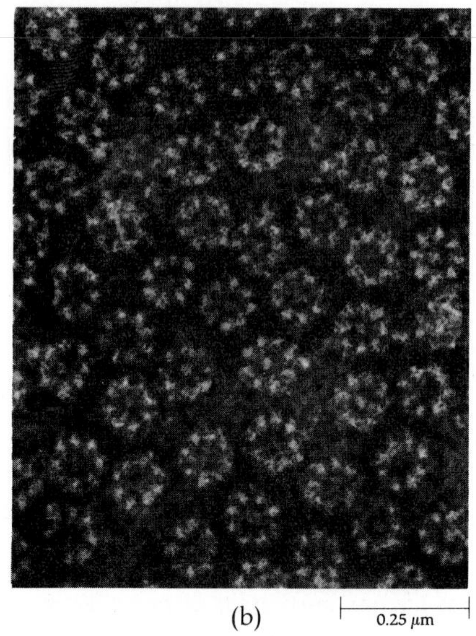

(b)

5–5

(a) *The cell nucleus fills the upper half of this electron micrograph. What you see here is the outer surface of the nuclear envelope. Clearly visible on this surface are pores, through which the interior of the nucleus can be seen. It is believed that the nucleus and the cytoplasm are able to "communicate" through the pores.*

(b) *A section in the plane of the membrane reveals the eight protein-containing granules that surround each nuclear pore.*

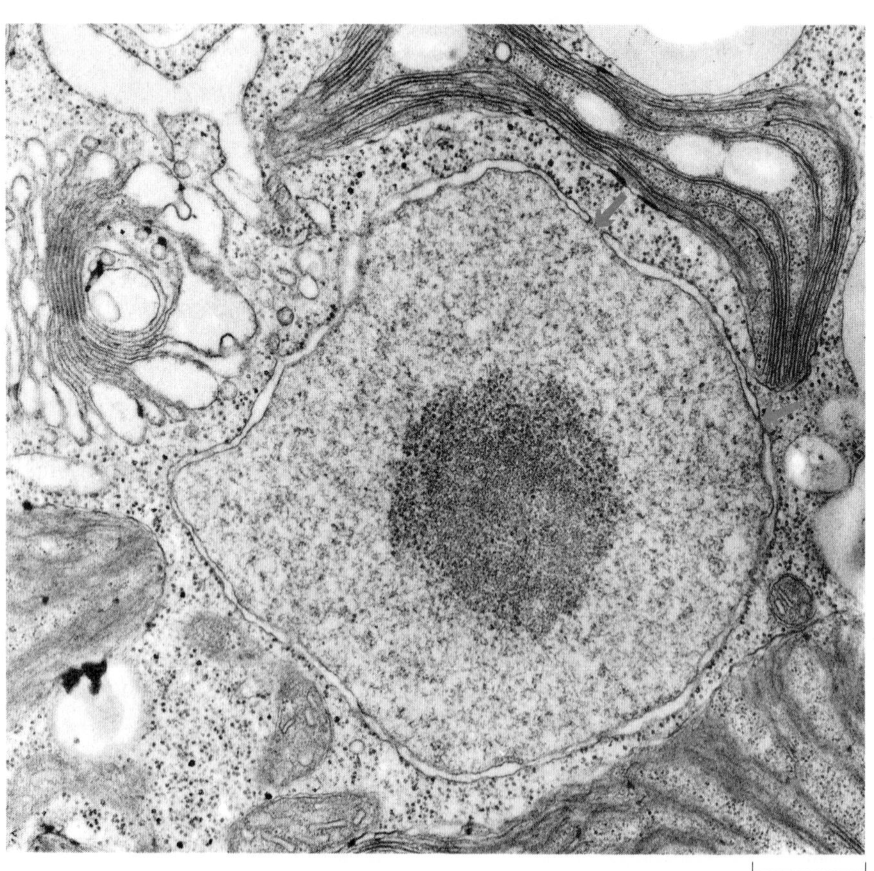

5–6

Electron micrograph of the nucleus of the alga Chlamydomonas. *The dark body in the center of the nucleus is the nucleolus. The major components of ribosomes are produced in the nucleolus; you can see partially formed ribosomes around its periphery. Notice also the nuclear envelope with its two nuclear pores, indicated by arrows. Above the nucleus is a portion of a chloroplast containing starch grains. A Golgi body is to the upper left of the nucleus, and mitochondria are visible below.*

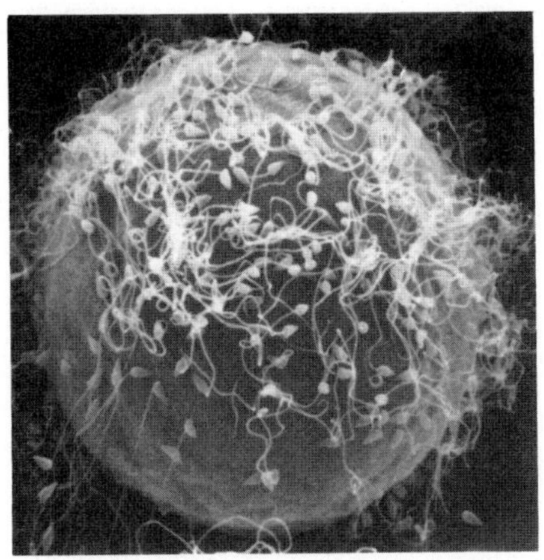

5–7
This egg from a sea urchin is surrounded by sperm cells. Despite the great differences in size of egg and sperm, both contribute equally to the hereditary characteristics of the individual. Since the nucleus is approximately the same size in both cells, the early microscopists postulated that this part of the cell must be the carrier of the hereditary information. Sea urchin eggs and sperm cells have been used in many studies because sea urchins are relatively easy to obtain and fertilization, which is external, can be easily observed in the laboratory.

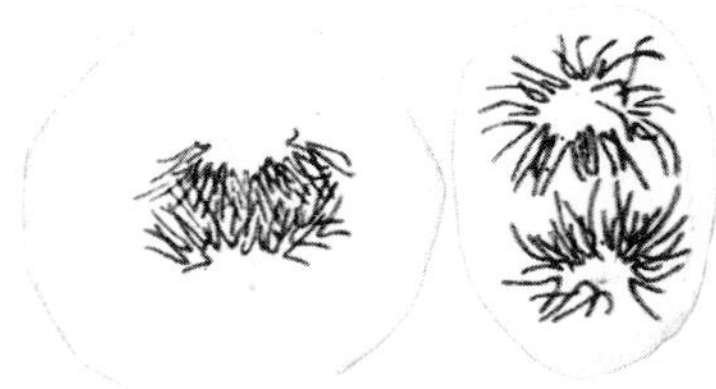

5–8
Drawings made by Walther Flemming in 1882 of chromosomes in dividing cells of salamander larvae. Flemming's observations were dependent upon the development of new cytological staining techniques.

The chromosomes are found within the nucleus. When the cell is not dividing, the chromosomes are visible only as a tangle of fine threads, called *chromatin*. The most conspicuous body within the nucleus is the *nucleolus*. There are usually two nucleoli per nucleus, although often only one is visible in a micrograph. The nucleolus is the site at which ribosomal subunits are constructed (see page 328). Viewed with the electron microscope, the nucleolus appears to be a collection of fine granules and tiny fibers (Figure 5–6). These are thought to be parts of ribosomes.

The Functions of the Nucleus

Our current understanding of the role of the nucleus in the life of the cell began with some early microscopic observations. One of the most important of these observations was made more than a hundred years ago by a German embryologist, Oscar Hertwig, who was observing the eggs and sperm of sea urchins. Sea urchins produce eggs and sperm in great numbers. The eggs are relatively large and so are easy to observe. They are fertilized in the open water, rather than internally, as is the case with land-dwelling vertebrates such as ourselves. Watching the eggs being fertilized under his microscope, Hertwig observed that only a single sperm cell was required. Further, when the sperm cell penetrated the egg, its nucleus was released and fused with the nucleus of the egg. This observation, confirmed by other scientists and in other kinds of organisms, was important in establishing the fact that the nucleus is the carrier of the hereditary information: the only link between father and offspring is the nucleus of the sperm.

Another clue to the importance of the nucleus came about as the result of the observations of Walther Flemming, also about 100 years ago. Flemming observed the "dance of the chromosomes" that takes place when eukaryotic cells divide (a process to be described in Chapter 7), and he painstakingly pieced together the sequence of events. (The fact that Hertwig and Flemming made their observations at about the same time was no coincidence; enormous improvements had just occurred in the light microscope and in techniques of microscopy.)

Since Flemming's time, a number of experiments have explored the role of the nucleus in the cell. In one simple experiment, the nucleus was removed from an amoeba by microsurgery. The amoeba stopped dividing and, in a few days, it died. If, however, a nucleus from another amoeba was implanted within 24 hours after the original one was removed, the cell survived and divided normally.

In the early 1930s, Joachim Hämmerling studied the comparative roles of the nucleus and the cytoplasm by taking advantage of some unusual properties of the marine alga *Acetabularia*. The body of *Acetabularia* consists of a single huge cell 2 to 5 centimeters in height. Individuals have a cap, a stalk, and a "foot," all of which are differentiated portions of the single cell. If the cap is removed, the cell will rapidly regenerate a new one. Different species of *Acetabularia* have different kinds of caps. *Acetabularia mediterranea*, for example, has a compact umbrella-shaped cap, and *Acetabularia crenulata* has a cap of petal-like structures.

Hämmerling took the "foot," which contains the nucleus, from a cell of *A. crenulata** and grafted it onto a cell of *A. mediterranea*, from which he had first

* By convention, with the second mention of a scientific binomial, it is permissible to abbreviate the first (genus) name. This is fortunate, particularly when dealing with such names as *Acetabularia*.

5–9

(a) One species of Acetabularia *has an umbrella-shaped cap, and (b) another has a ragged, petal-like cap. If the caps are removed, new caps form, similar in appearance to the amputated one. However, if the "foot" (containing the nucleus) is removed at the same time as the cap and a new nucleus from the other species is transplanted, the cap (c) that forms will have a structure with characteristics of both species. If this cap is removed, the next cap (d) that grows will be characteristic of the cell that donated the nucleus, not of the cell that donated the cytoplasm.*

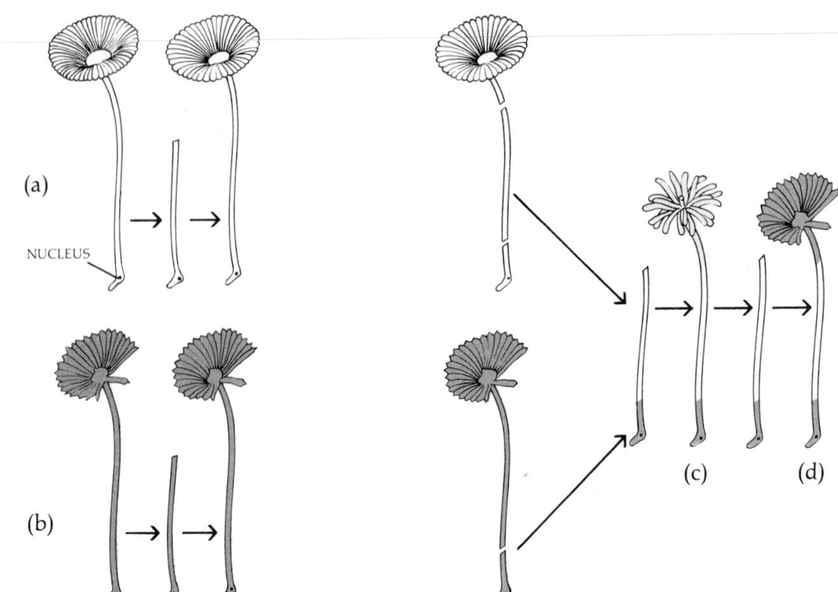

removed the "foot" and the cap. The cap that then formed had a shape intermediate between those of the two species. When this cap was removed, the next cap that formed was completely characteristic of *A. crenulata.*

Hämmerling interpreted these results as meaning that certain cap-determining substances are produced under the direction of the nucleus. These substances accumulate in the cytoplasm, which is why the first cap that formed after nuclear transplantation was of an intermediate type. By the time the second cap formed, however, the cap-determining substances present in the cytoplasm before the transplant had been exhausted, and the form of the cap was completely under the control of the new nucleus.

We can see from these experiments that the nucleus performs two crucial functions for the cell. First, it carries the hereditary information that determines whether a particular cell will develop into (or be a part of) a *Paramecium,* an oak, or a human—and not just any *Paramecium,* oak, or human, but one that resembles the parent or parents of that particular, unique organism. Each time a cell divides, this information is passed on to the two new cells. Second, as Hämmerling's work indicated, the nucleus exerts a continuing influence over the ongoing activities of the cell, ensuring that the complex molecules that the cell requires are synthesized in the number and of the kind needed. The way in which the nucleus performs these functions will be described in Section 3.

The Cytoplasm

Not long ago, the cell was visualized as a bag of fluid containing enzymes and other dissolved molecules along with the nucleus, a few mitochondria, and occasional other organelles that could be seen by special microscopic techniques. Within the last three decades, however, with the development of electron microscopy, an increasing number of structures have been identified within the cytoplasm, which is now known to be highly structured and crowded with organelles. A representative array of organelles is seen in Figure 5–10.

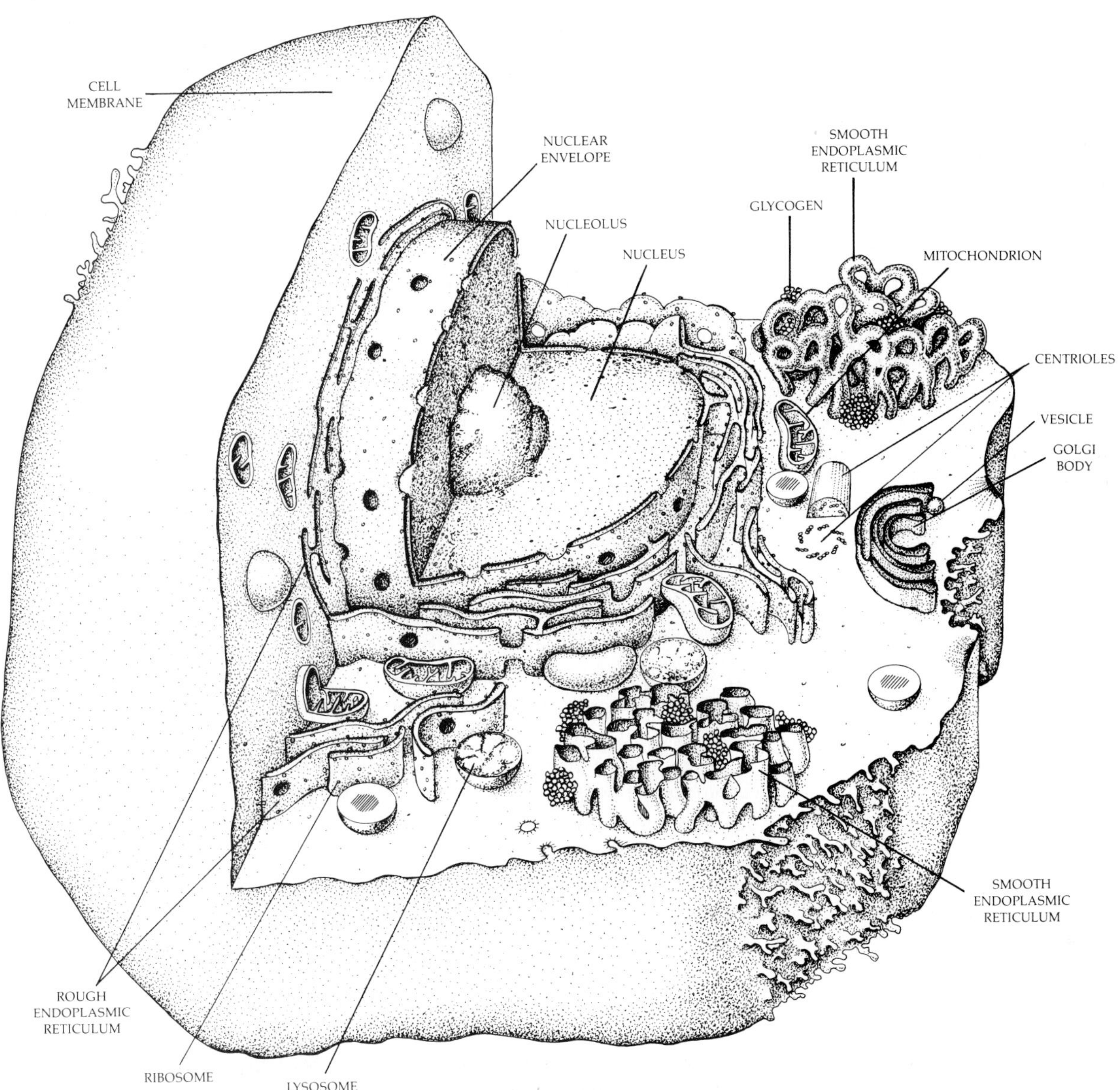

5–10

An animal cell, as interpreted from electron micrographs. Like all cells, this one is bounded by an outer cell membrane, which acts as a selectively permeable barrier to the surrounding environment. All materials that enter or leave the cell, including food, wastes, and chemical messages, must pass through this barrier.

Within the membrane is found the cytoplasm, which contains the enzymes and other solutes of the cell. The cytoplasm is traversed and subdivided by an elaborate system of membranes, the endoplasmic reticulum, a portion of which is shown here. In some areas, the endoplasmic reticulum is covered with ribosomes, the special structures on which amino acids are assembled into proteins. Ribosomes are also found elsewhere in the cytoplasm.

Golgi bodies are packaging centers for molecules synthesized within the cell. The mitochondria are the sites of the chemical reactions that provide energy for cellular activities.

The largest body in the cell is the nucleus. It is surrounded by a double membrane, the nuclear envelope, which is continuous with the endoplasmic reticulum. Within the nuclear envelope is a nucleolus, the site where the ribosomes are formed, and the chromatin, which is the material of the chromosomes in an extended form.

Not shown in this diagram is the cytoskeleton, an elaborate, highly structured network of protein filaments. These filaments pervade the cytoplasm, anchoring the organelles, maintaining the cell's shape, and directing the intracellular molecular traffic.

The Cytoskeleton

As we noted in the previous chapter, extremely thin sections are required for study with the transmission electron microscope. With the recent development, however, of the high-voltage electron microscope, which produces a beam of electrons with greater penetration, it has become possible to use much thicker specimens—in some cases, whole cells. The resulting visualization of the interior of the cell in three dimensions has revealed previously unsuspected interconnections among filamentous protein structures within the cytoplasm. These structures form an internal *cytoskeleton* that maintains the shape of the cell, enables it to move, anchors its organelles, and directs its traffic. Four different types of structure have been identified as participants in the cytoskeleton: *microtubules*, *microfilaments*, *intermediate fibers*, and *microtrabeculae* (from the Latin *trabecula*, meaning "little beam").

Microtubules (described on page 70) are 20 to 25 nanometers in diameter and, in many cells, seem to extend out from the cell center, ending near the cell surface. As noted previously, they play an important role in cell division and seem to provide a temporary scaffolding for the construction of other cellular structures. They are also the key components of cilia and flagella, permanent structures used for locomotion by many types of cells.

Microfilaments are fine protein threads, only 3 to 6 nanometers in diameter, consisting of molecules of a globular protein known as actin. Like microtubules, they can be readily assembled and disassembled by the cell. Microfilaments are involved in cell motility. In cells that move by gradual changes of shape, such as amoebas, they are found concentrated in bundles or a meshwork near the moving edge.

Intermediate fibers, as their name implies, are intermediate in size between microtubules and microfilaments, with a diameter of 7 to 10 nanometers. Unlike microtubules and microfilaments, which consist of globular protein subunits, intermediate fibers are composed of fibrous proteins and cannot be easily disassembled by the cell once they are formed. Intermediate fibers are thought to have a ropelike structure, similar to that of collagen, and they are found in the greatest density in cells subject to mechanical stress.

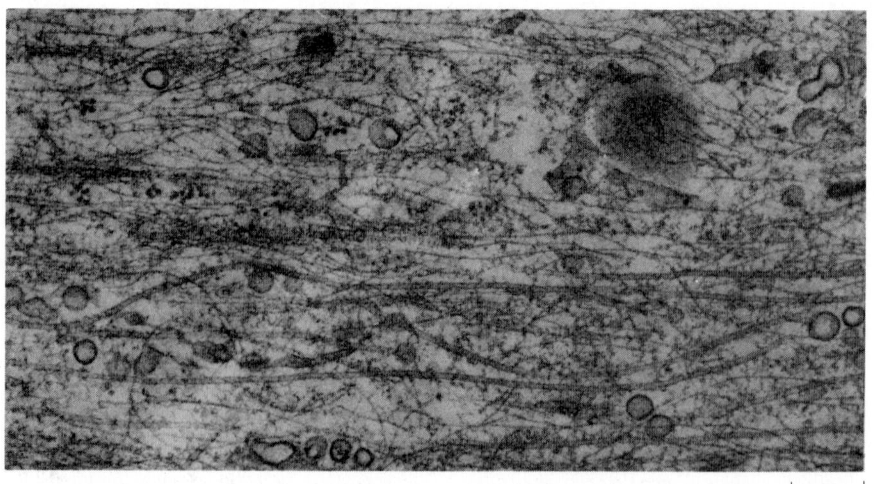

5–11
Elements of the cytoskeleton visible in this electron micrograph of a human fibroblast include microtubules (running horizontally across the lower portion of the micrograph), microfilaments (concentrated in bundles toward the left), and intermediate fibers (the thin strands in the upper half of the micrograph).

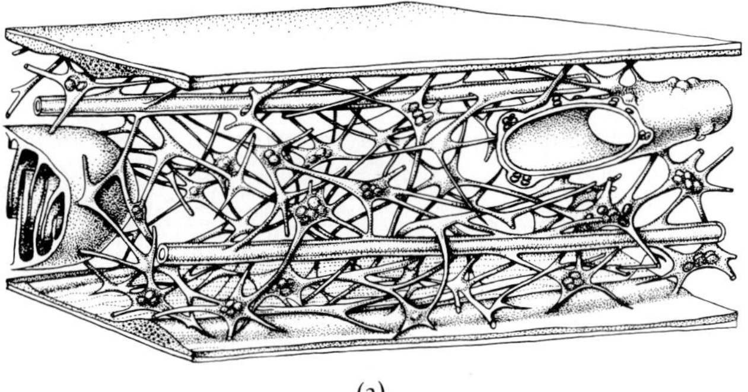

(a)

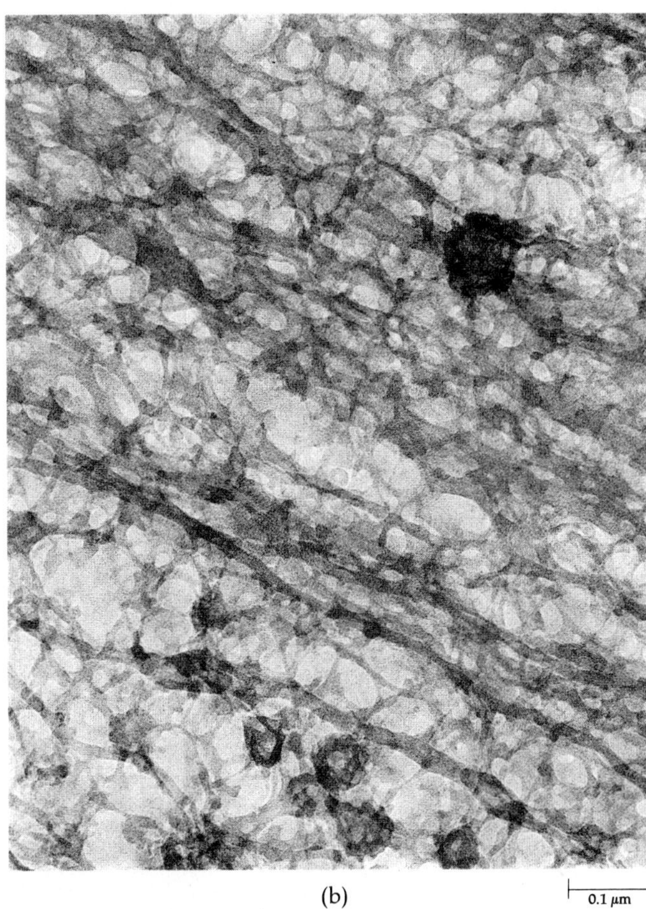

(b) 0.1 μm

5–12

(a) *A model of the cytoskeleton. Microtrabeculae form the bulk of the network, interconnecting the other cytoskeletal elements and the organelles. The two long rods running longitudinally are microtubules. The microfilaments and intermediate fibers are concentrated near the cell membrane, at the top and bottom of the model. A portion of a mitochondrion is at the left, and a section of endoplasmic reticulum is at the right. Ribosomes are found on the endoplasmic reticulum and at some of the intersections of microtrabeculae.*

(b) *This high-voltage electron micrograph of a small section of the cytoplasm of a rat cell reveals the microtrabecular network. The microtrabeculae are the wisplike fibers. The thicker, darker fibers running diagonally from the top left are microtubules. The spherical objects at the intersections of the microtrabeculae in the lower portion of the micrograph are ribosomes.*

The most recently discovered elements of the cytoskeleton are the microtrabeculae, wisplike fibers that form a dense network interconnecting all of the other structures within the cytoplasm (Figure 5–12). Their chemical composition is not yet known, but they are thought to contain proteins.

Although the cytoskeleton gives the cell a highly ordered three-dimensional structure, it is neither rigid nor permanent. It is a dynamic framework, changing and shifting according to the activities of the cell.

Vacuoles

In addition to organelles and the cytoskeleton, the cytoplasm of many cells, especially plant cells, contains *vacuoles.* A vacuole is a space in the cytoplasm filled with water and solutes, and surrounded by a single membrane. Immature plant cells characteristically have many vacuoles, but as the plant cells mature, the numerous smaller vacuoles coalesce into one large, central, fluid-filled vacuole that then becomes a major supporting element of the cell (see Figure 4–11, page 87). The vacuole also increases the size of the cell, including the amount of surface exposed to the environment, with a minimal investment in structural materials by the cell.

Vesicles, more common in animal cells, have the same general structure as vacuoles. They are distinguished by size: vesicles are usually less than 100 nanometers in diameter, whereas vacuoles are larger.

Ribosomes

Ribosomes are the most numerous of the cell's many organelles. A rapidly growing *E. coli* cell has approximately 15,000 ribosomes, and a eukaryotic cell may have many times more. As we noted in Chapter 3, proteins are chains of amino acids, assembled in a specific sequence. The ribosomes are the sites at which this assembly takes place, a process that will be explored in some detail in Chapter 15. The more protein a cell is making, the more ribosomes it has.

The way in which ribosomes are distributed in the cell seems to be related to the kinds of proteins it is making. Some proteins—enzymes or hemoglobin, for example—are used within the cell. Others, such as collagen, digestive enzymes, hormones, or mucus, are released outside the cell, sometimes carrying out their functions at a great distance, on a cellular scale, from their source. In cells that are making proteins for their own use, such as embryonic cells, ribosomes tend to be distributed in the cytoplasm. As the high-voltage electron microscope has revealed, these ribosomes are not floating free in the cytoplasm but are instead clustered at interstices of the microtrabecular network. In cells that are making digestive enzymes or other proteins for export, ribosomes are also found attached to a complex system of internal membranes, the endoplasmic reticulum.

Endoplasmic Reticulum

The *endoplasmic reticulum* is a network of interconnecting flattened sacs, tubes, and channels found in eukaryotic cells. The amount of endoplasmic reticulum in a cell is not fixed but increases or decreases depending on the cell's activity.

There are two types of endoplasmic reticulum, rough (with ribosomes attached) and smooth (without ribosomes), which are, however, continuous with each other. Rough endoplasmic reticulum predominates in cells making large amounts of proteins for export. It is continuous with the outer membrane of the nuclear envelope, which also has ribosomes attached. Rough endoplasmic reticulum often includes large, flattened sacs called cisternae. If cells engaged in protein synthesis

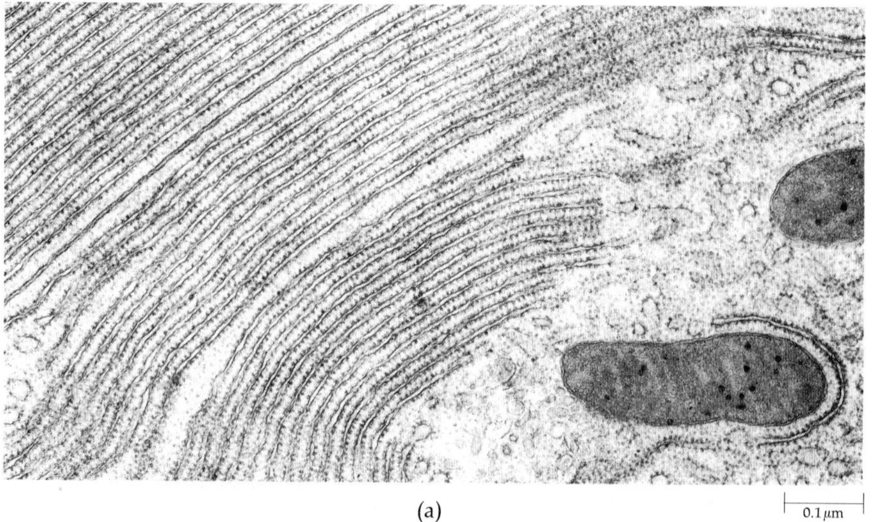

(a)

5–13
(a) *Rough endoplasmic reticulum, which fills most of this micrograph, is a system of membranes that separates the cell into channels and compartments and provides surfaces on which chemical activities take place. The dense objects on the membrane surfaces are ribosomes. This cell is from a pancreas, an organ extremely active in the synthesis of digestive enzymes, which are "exported" to the upper intestine, where most digestion takes place. In the lower right-hand corner of the micrograph are one mitochondrion and, above it, a portion of another.* (b) *Rough endoplasmic reticulum at higher magnification, showing the individual ribosomes.* (c) *An interpretation of the rough endoplasmic reticulum based on electron micrographs.*

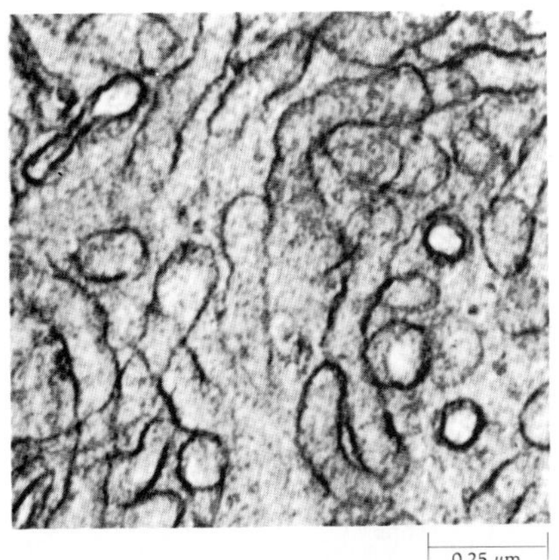

5–14
Smooth endoplasmic reticulum from the testicle of an opossum. These membranes participate in the synthesis of the hormone testosterone.

are permitted to take up radioactive amino acids, the radioactive labels are first detected at the membrane of the rough endoplasmic reticulum and then, slightly later, within its cisternae. The initial portion of proteins synthesized on the rough endoplasmic reticulum consists of a "leader" of hydrophobic amino acids that is thought to assist in the transport of the protein through the lipid bilayer to the interior of the endoplasmic reticulum. The newly synthesized protein molecule moves from the rough endoplasmic reticulum through the smooth endoplasmic reticulum and then to the Golgi bodies. In the course of this progression from one organelle to another, the molecule undergoes further processing, including cleavage of the initial sequence of hydrophobic amino acids and, often, the addition of carbohydrate groups to the protein.

Cells concerned with the synthesis of lipids—such as the gland cells that make steroid hormones—have large amounts of smooth endoplasmic reticulum. Smooth endoplasmic reticulum is also found in liver cells, where it appears to be involved in various detoxification processes (one of the many functions of the liver). For instance, in experimental animals fed large amounts of phenobarbital, the amount of smooth endoplasmic reticulum in the liver cells increases severalfold. Smooth endoplasmic reticulum also seems to be active in the liver's breakdown of glycogen to glucose. It also appears, as we have noted, to serve as a passageway for material moving from the rough endoplasmic reticulum to the Golgi bodies. As more of its functions are discovered, it seems likely that the smooth endoplasmic reticulum system actually represents a number of quite different specializations of endoplasmic reticulum, resembling one another only in their lack of ribosomes.

Golgi Bodies

Each *Golgi body* consists of flattened, membrane-bound sacs stacked loosely on one another and surrounded by tubules and vesicles (Figure 5-15, on the next page). The function of the Golgi body is to accept vesicles from the endoplasmic reticulum, to modify the membranes of the vesicles, to further process their contents, and to distribute the finished product to other parts of the cell and, especially, to the cell surface. Thus they serve as packaging and distribution centers.

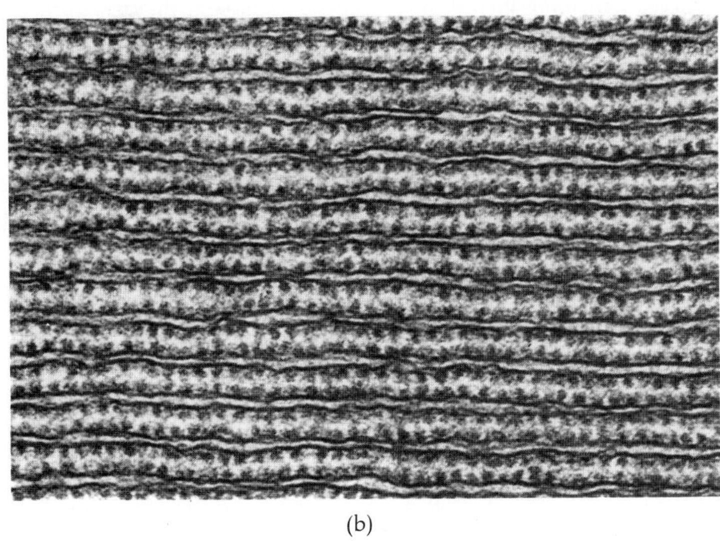

(b)

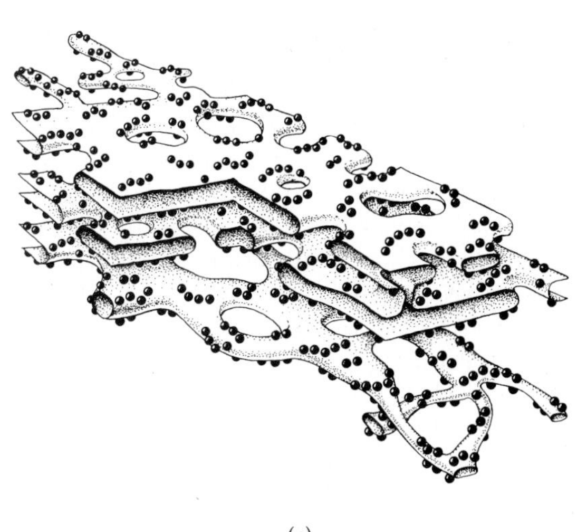

(c)

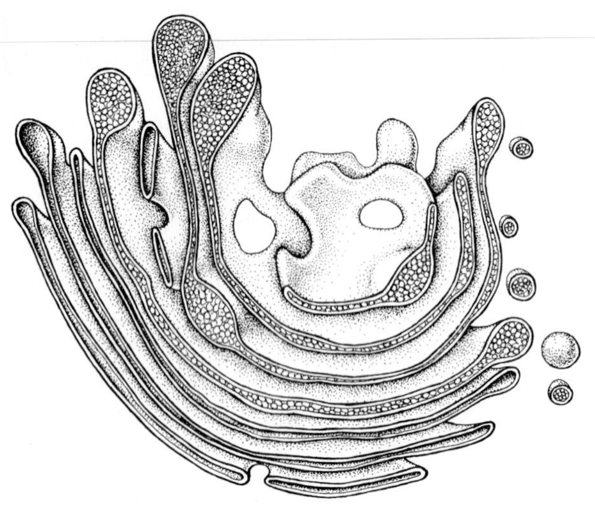

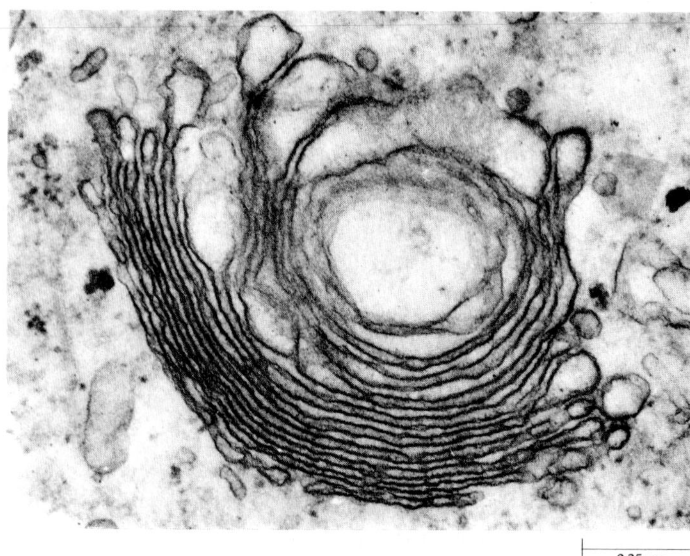

5–15
Graphic interpretation and electron micrograph of a Golgi body. Golgi bodies are composed of special arrangements of membranes. Materials are packaged in membrane-enclosed vesicles at the Golgi bodies and are distributed within the cell or shipped to the cell surface. Note the vesicles pinching off from the edges of the flattened sacs.

The final assembly of the proteins and carbohydrates (glycoproteins) that are found on the surfaces of cell membranes occurs in the Golgi body. In plant cells, Golgi bodies also bring together some of the components of the cell walls and export them to the cell surface where they are assembled. Golgi bodies are found in most types of eukaryotic cells. Animal cells usually contain 10 to 20 Golgi bodies, and plant cells may have several hundred.

Figure 5–16 summarizes the ways in which the rough and smooth endoplasmic reticulum and the Golgi body and its vesicles interact to produce macromolecules for export.

Lysosomes

One type of small vesicle commonly formed in the Golgi body is the *lysosome.* Lysosomes are essentially bags of hydrolytic enzymes enclosed in membranes that separate them from the rest of the cell; they are involved in breaking down proteins, polysaccharides, and lipids. If the lysosomes break open, the cell itself is destroyed, since the enzymes they carry are capable of breaking down all the major compounds found in a living cell. The tenderness and inflammation associated with rheumatoid arthritis and gout appear to be related to the escape of hydrolytic enzymes from lysosomes.

An example of the function of lysosomes is given by white blood cells, which engulf bacteria in the human body. As the bacteria are taken up by the cell, they are wrapped in a membrane-enclosed sac, a vacuole. (This process is known as phagocytosis; we shall discuss it further in the next chapter.) When this occurs, the lysosomes within the cell fuse with the vacuoles containing the bacteria and release their hydrolytic enzymes. The bacteria are quickly digested. In similar fashion, the lysosomes of single-celled organisms, such as *Paramecium, Didinium,* and amoebas, fuse with the phagocytic vacuoles containing food organisms. Hydrolytic enzymes released by the lysosomes into the vacuoles digest the contents. Why the enzymes do not destroy the membranes of the lysosomes that carry them is a pertinent question yet to be answered.

5–16
Diagram illustrating the interaction of cytoplasmic organelles in the synthesis of proteins and in the packaging of macromolecules for export. As they are synthesized, proteins are fed into the rough endoplasmic reticulum. They then move through the smooth endoplasmic reticulum and are released in vesicles that fuse with the sacs of the Golgi body. In the Golgi body, carbohydrates are added to some of the proteins, producing glycoproteins; lipids are added to other proteins, producing lipoproteins. Both glycoproteins and lipoproteins are common components of membranes. Vesicles containing the finished macromolecules are released from the Golgi body and move to other locations within the cell or to its exterior surface.

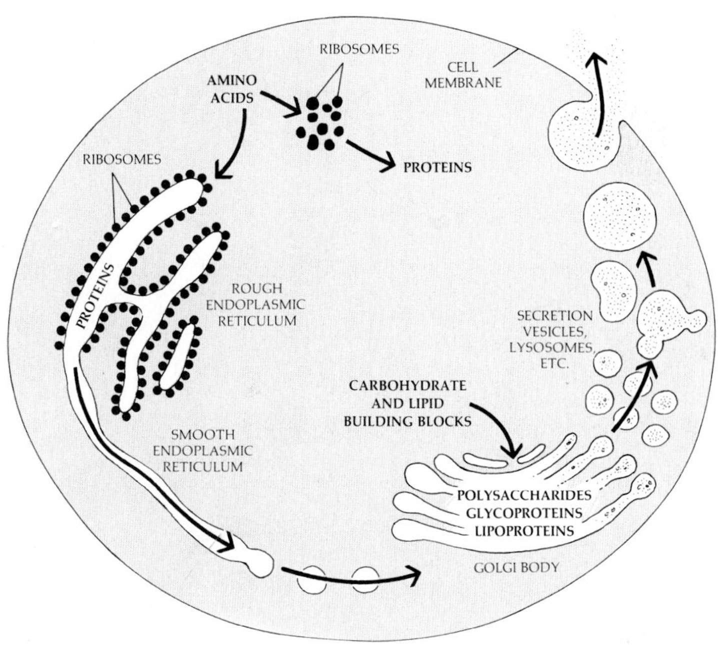

Peroxisomes

Another type of vesicle containing destructive enzymes is the *peroxisome*. Peroxisomes are vesicles in which purines (one of two major categories of nitrogenous bases) are broken down by the cell. In plants, peroxisomes are also the site of a series of reactions that occur in sunlight when the cell contains relatively high concentrations of oxygen. These reactions and the breakdown of purines both produce hydrogen peroxide (H_2O_2), a compound that is extremely toxic to living cells. The peroxisomes, however, contain another enzyme that immediately breaks hydrogen peroxide into water and oxygen, preventing any damage to the cell.

5–17
Lysosomes and peroxisomes are vesicles in which cells break down different types of molecules. (a) *In this portion of a cell from the adrenal gland, the dark oval bodies are lysosomes. They contain 40 or more different hydrolytic enzymes.* (b) *A peroxisome. The crystalline material in the center is a peroxide-producing enzyme involved in the breakdown of purines. Another enzyme destroys the peroxide, preventing its escape into the cytoplasm.*

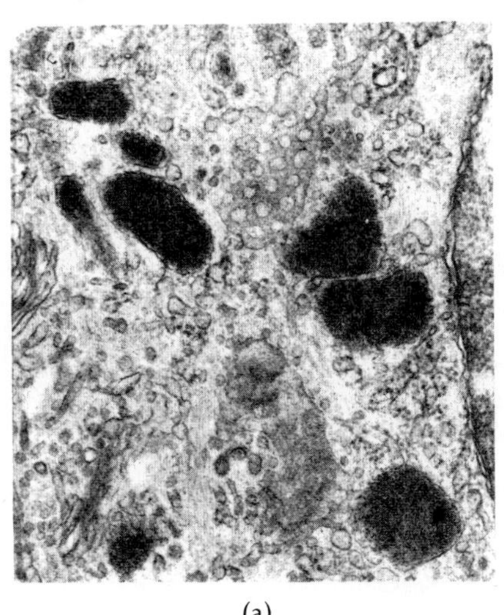

(a)

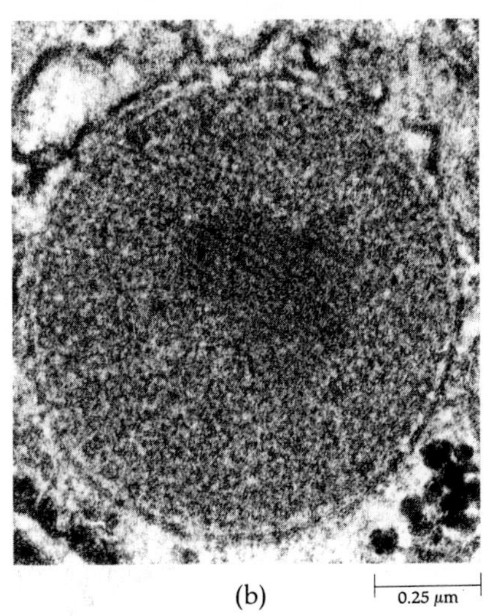

(b)

0.25 μm

5–18
Mitochondria in a fiber of flight muscle from a wasp.

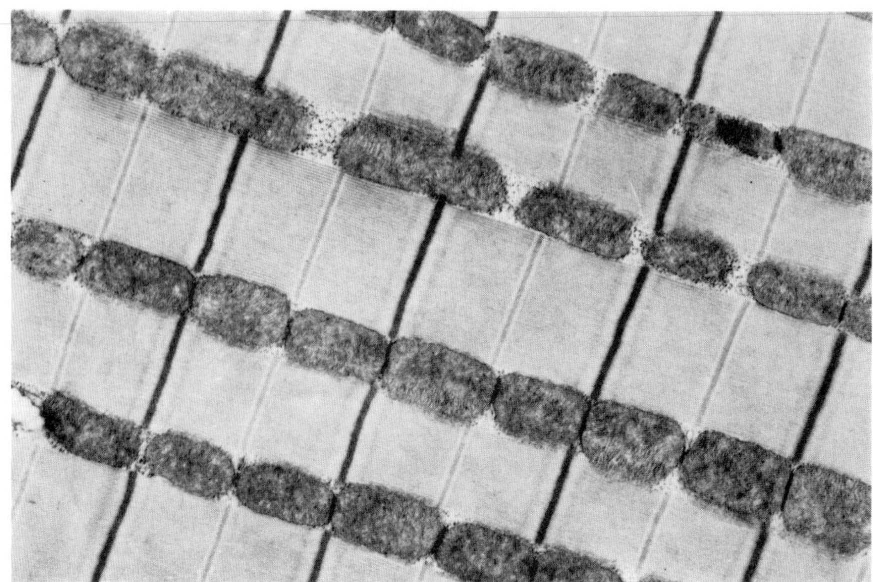

Mitochondria

Mitochondria (singular, mitochondrion) are among the largest organelles in a cell. In the mitochondria, energy-yielding organic molecules are broken down and their energy is repackaged into smaller units, convenient for most cellular processes. The higher the energy requirements of a particular eukaryotic cell, the more or larger mitochondria it is likely to have. A liver cell, for example, which has modest energy requirements, has about 2,500, making up about 25 percent of the volume of the cell, whereas a heart muscle cell has several times as many very large mitochondria. Mitochondria are often found clustered in areas in the cell where energy requirements are high.

Mitochondria vary in shape from almost spherical, to potato-shaped, to greatly elongated cylinders. They are always surrounded by two membranes, the inner one of which folds inward; these folds, known as cristae, are working surfaces for mitochondrial reactions. The more active a mitochondrion, the more cristae it is likely to have. In Chapter 9, we shall examine in more detail the structure of mitochondria and the processes that occur within these important organelles.

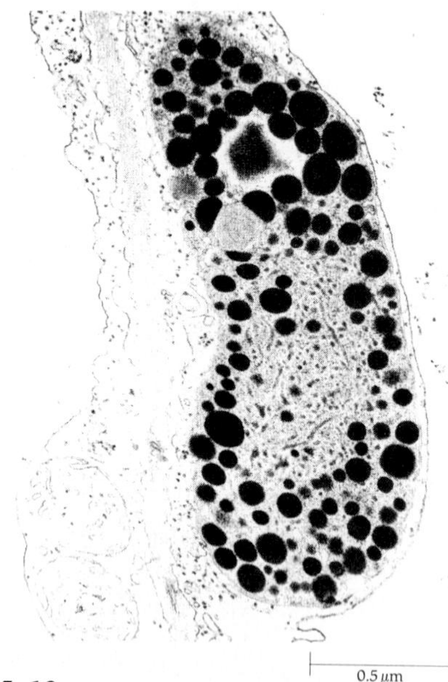

5–19
Chromoplast from a forsythia petal. The large, dark granules contain the orange and yellow pigments characteristic of certain flowers and fall leaves. To the left of the chromoplast are the faintly visible cell wall and two mitochondria. To the left and right of the micrograph are portions of two vacuoles.

Plastids

Plastids are membrane-bound organelles found only in the cells of plants and algae. They are surrounded by two membranes, like mitochondria, and have an internal membrane system that may be folded intricately. Mature plastids are of three types: leucoplasts, chromoplasts, and chloroplasts.

Leucoplasts (*leuco* means "white") store starch or, sometimes, proteins or oils. Leucoplasts are likely to be numerous in storage organs such as roots, as in a turnip, or tubers, as in a potato (see page 58).

Chromoplasts (*chromo* means "color") contain pigments and are associated with the bright orange and yellow colors of fruits, flowers, fall leaves, and carrots.

Chloroplasts (*chloro* means "green") are the chlorophyll-containing plastids in which photosynthesis takes place. Their structure will be described in Chapter 10.

5–20

Chloroplast development. (a) *The immature plastid contains small crystalline structures (top).* (b) *In the presence of light, these structures begin to break up into elongated vesicles.* (c) *The vesicles flatten into stacks of membranes.* (d) *A mature chloroplast. Chromoplasts and leucoplasts develop from plastids similar to* (a).

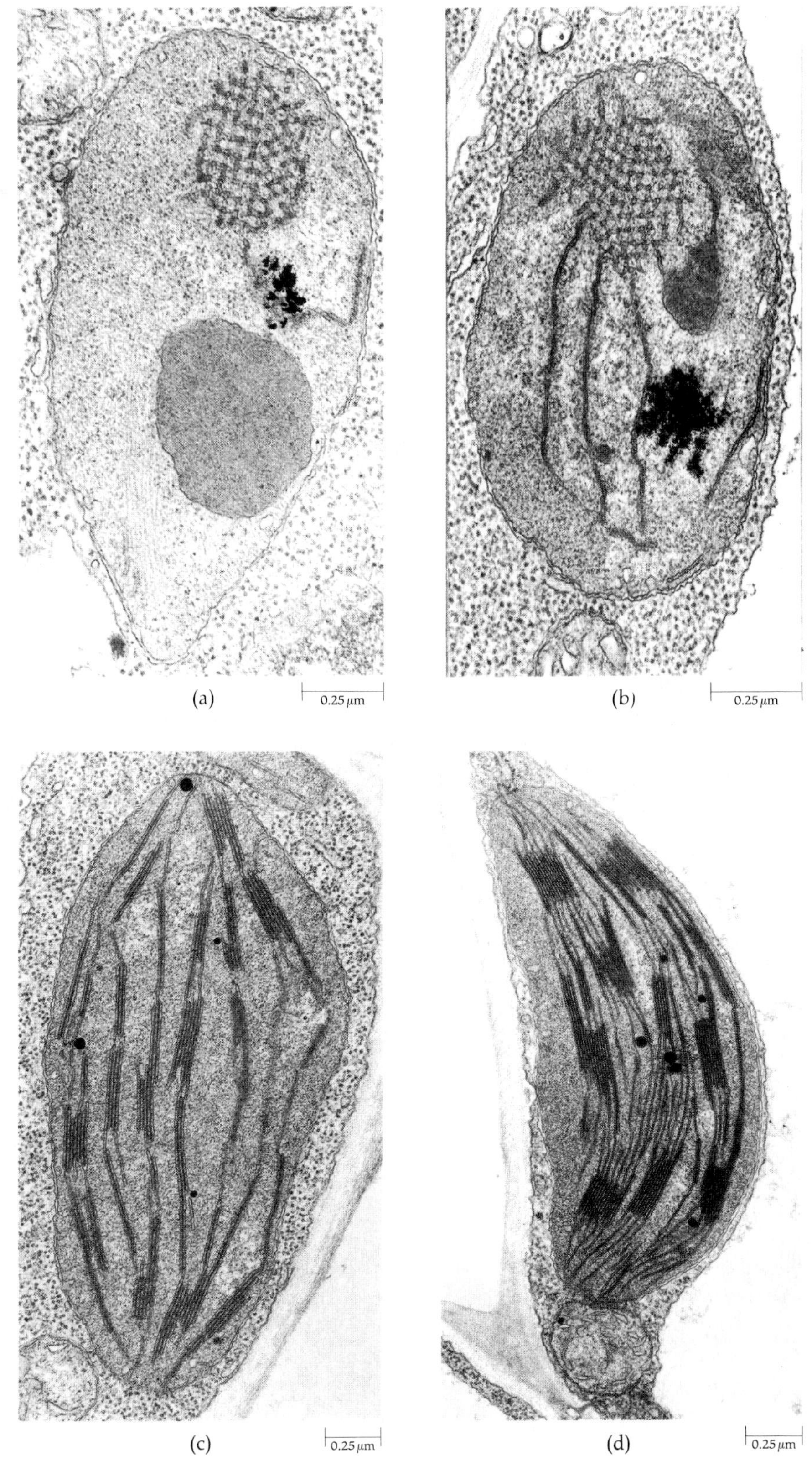

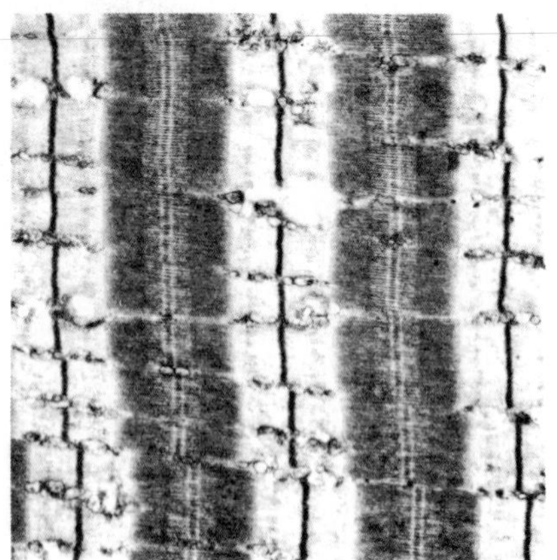

5–21
Contractile assemblies of protein filaments in a vertebrate skeletal muscle. Each unit is known as a sarcomere, approximately 18 of which are visible in this electron micrograph. When the muscle contracts, the distance from left to right in this micrograph shortens.

HOW CELLS MOVE

All cells exhibit some form of movement. Even plant cells, encased in a rigid cell wall, exhibit active cytoplasmic streaming (movement of cytoplasm within the cell) as well as chromosomal movements and changes in shape during cell division. As we saw in Figure 4–12 (page 88), cilia beat along the surface of the tracheal cells of animals. Embryonic cells migrate in the course of development. Differentiating and regenerating nerve cells send out axons, which are long, slender extensions that may be a meter or more in length. Amoebas pursue and engulf their prey. Even little *Chlamydomonas* cells dart toward a light source. Two different molecular mechanisms of cellular movement have been identified: (1) assemblies of protein filaments, usually referred to as muscle protein, and (2) the assemblies of microtubules in cilia and flagella.

Muscle Protein

The skeletal muscles of vertebrates contain elaborate contractile assemblies consisting of protein filaments (Figure 5–21). The two principal types of protein involved are known as actin and myosin. (The contractile mechanism will be described in detail in Chapter 42.) These are commonly thought of as muscle proteins, since it is in muscle tissue that they were first identified and have been studied most intensively. However, it has now been found that actin, in particular (which is easier to isolate chemically), is present in a great variety of cells, including plant cells. As you will recall, the microfilaments of the cytoskeleton are composed of actin subunits. It is in the form of these microfilaments that actin appears to participate in both cell motility and the internal movement of cellular contents.

Contractile proteins are involved in a variety of different cellular processes. For example, they are found in the large, multinucleate organisms known as slime molds, which move like giant amoebas and exhibit vigorous cytoplasmic streaming. Actin has also been found in a true amoeba, *Amoeba proteus.* In algal cells, actin microfilaments occur in bundles wherever cytoplasmic streaming is taking place. Such microfilaments have also been shown to act as a sort of "purse string" in animal cells during cell division, pinching off the cytoplasm to separate the two daughter cells. Cytologists are now coming to the conclusion that "muscle" pro-

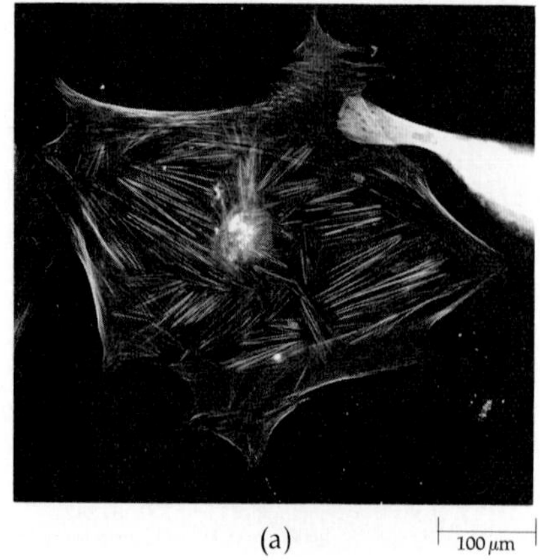

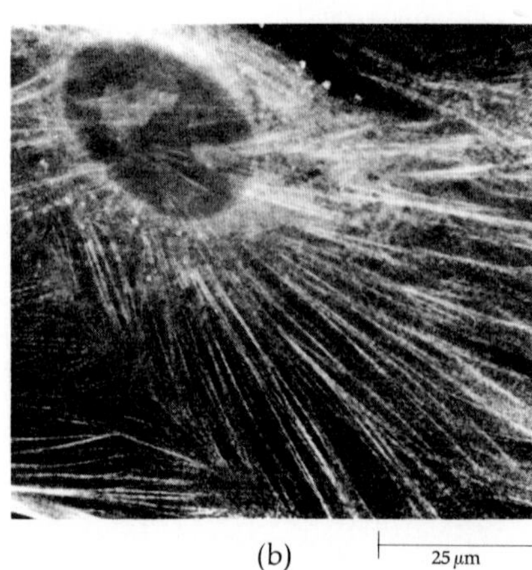

(a) 100 μm (b) 25 μm

5–22
Contractile proteins in nonmuscle cells are revealed by immunofluorescence microscopy. The cell is treated with specially prepared fluorescent antibodies to the protein of interest. These antibodies attach to the protein, and the pattern of their fluorescence indicates the location of the protein. These micrographs of human fibroblast cells reveal (a) *actin in microfilament bundles and* (b) *myosin filaments. These proteins are of the same type as found in skeletal muscle.*

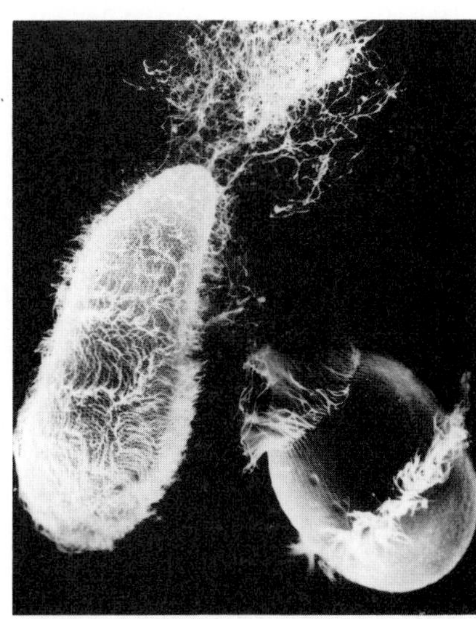

5–23
Two ciliates, one-celled organisms distinguished by their many cilia. On the left, Paramecium; *on the right,* Didinium. Didinium *is stalking* Paramecium. Paramecium, *in defense, has discharged a barrage of barbs (visible as a cloud at the top of the micrograph).* Didinium *is about to eject a bundle of slender, poisonous strands (not visible), which will paralyze* Paramecium *in a matter of seconds. In* Paramecium, *the cilia are distributed fairly evenly over the cell surface. In* Didinium, *they form two wreaths that circle the organism's barrel-shaped body.*

teins are common to all cells and that the elaborate contractile machinery of skeletal muscle cells is a recent evolutionary specialization. The way in which these proteins bring about cytoplasmic streaming and amoeboid movement is currently under active investigation.

Cilia and Flagella

Cilia and *flagella* are long, thin (0.2 micrometer) structures extending from the surface of many types of eukaryotic cells. They are the same except for length (the names were given before their basic similarity was realized). When they are shorter and occur in larger numbers, they are more likely to be called cilia; when they are longer and fewer, they are usually called flagella. Prokaryotic cells also have flagella, but they are quite different in structure.

In one-celled eukaryotes and some small animals (such as flatworms), cilia and flagella are associated with movement of the organism. For example, one type of *Paramecium* has approximately 17,000 cilia, each about 10 micrometers long, which propel it through the water by beating in a coordinated fashion. Other one-celled organisms, such as the members of the genus *Chlamydomonas,* have only two whiplike flagella, which protrude from the anterior end of the organism and move it through the water. The motile power of the human sperm cell comes from its single powerful flagellum, or "tail."

Many of the cells that line the surfaces within our bodies are also ciliated. These cilia do not move the cells but, rather, serve to sweep substances from the environment across the cell surface. For example, cilia on the surface of cells of the respiratory tract beat upward, propelling a current of mucus that sweeps bits of soot, dust, pollen, tobacco tar—whatever foreign substances we have inhaled either accidentally or on purpose—to our throats, where they can be removed by swallowing. Human egg cells are propelled down the oviducts by the beating of cilia that line the inner surfaces of these tubes. Cilia and flagella are found extensively throughout the living world, on the cells of invertebrates, vertebrates, the sex cells of ferns and other plants, as well as on one-celled organisms. Only a few large groups of organisms, such as the red algae, higher fungi, and flowering plants, have no cilia or flagella in any cells.

Almost all eukaryotic cilia and flagella, whether on a *Paramecium* or a sperm cell, have the same structure. Nine pairs of fused microtubules form a ring that surrounds two additional, solitary microtubules in the center (Figure 5–24, on the next page). Microtubules, you will recall, are composed of identical globular protein units assembled in a hollow helix. The movement of cilia and flagella comes from within the structures themselves; if cilia are removed from cells and placed in a medium containing energy-yielding chemicals, they twitch. The movement, according to the generally accepted hypothesis, is caused by each outer pair moving tractor-fashion with respect to its nearest neighbor. The two "arms" that you can see on one of each pair of outer tubules (Figure 5–24a) have been shown to be enzymes involved in energy-releasing chemical reactions. At least two other proteins are involved in the formation of spokes connecting the nine pairs of outer microtubules to the central pair, and still another protein forms more widely spaced bands, rather like the hoops of a barrel, connecting the nine outer pairs to each other. The spokes are thought to play a role in coordinating the tractorlike movements of the microtubules, whereas the bands limit the amount of sliding possible and thus convert it into a bending motion.

5–24

(a) *Diagram of a cilium with its underlying basal body. Virtually all eukaryotic cilia and flagella, whether they are found on one-celled organisms or on the surfaces of cells within our own bodies, have this same internal structure, which consists of an outer ring of nine pairs of microtubules surrounding two additional microtubules in the center. The "arms," the radial spokes, and the connecting bands are formed from different types of protein. The basal bodies from which cilia and flagella arise have nine outer triplets, with no microtubules in the center. The "hub" of the wheel in the basal body is not a microtubule, although it has about the same diameter.*

(b) *Cross section of flagella. These are from* Trichonympha, *a one-celled organism that lives in the guts of termites, where it breaks down cellulose.*

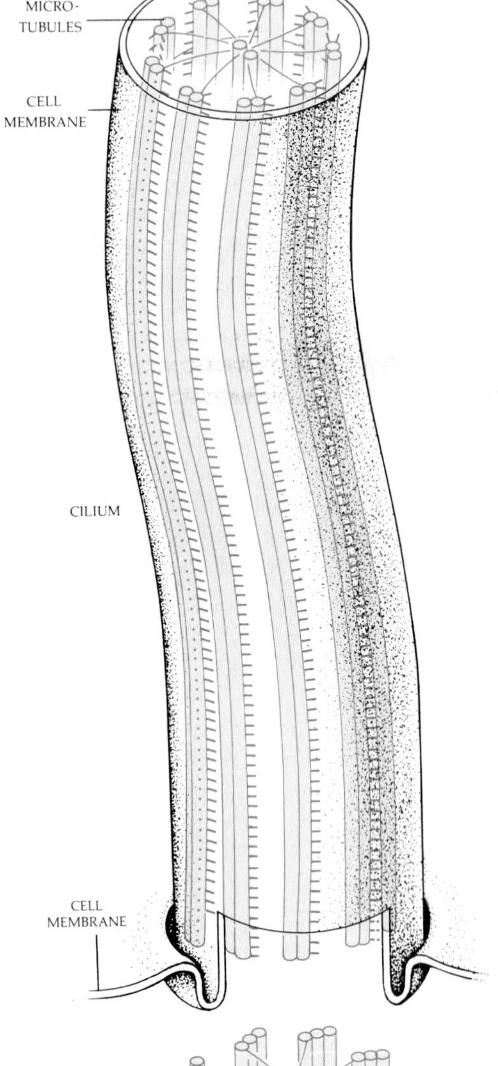

(a)

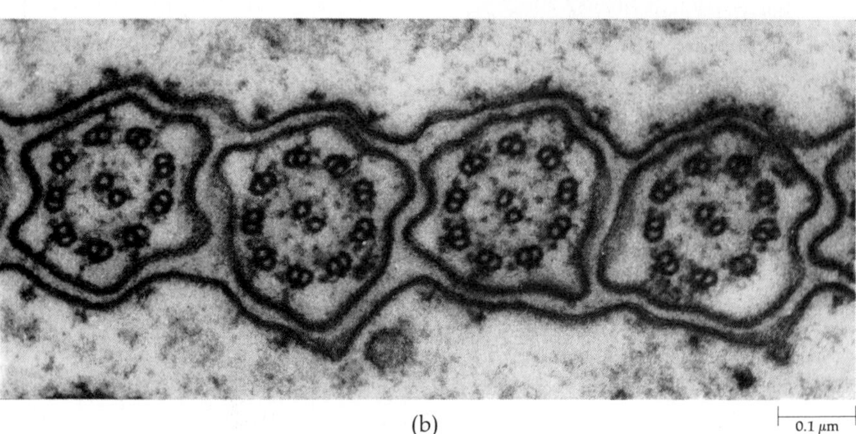

(b)

Basal Bodies and Centrioles

Underlying each cilium is a structure known as a *basal body*, which has the same diameter as a cilium, about 0.2 micrometer. It consists of microtubules arranged in nine triplets around the periphery. Unlike the cilium, it has no microtubules in the center, and none of the microtubules in the basal body have arms. Cilia and flagella arise from basal bodies. For instance, as a sperm cell takes form, a basal body moves near the cell membrane, and the sperm's flagellum arises from it through the assembly of microtubules. Basal bodies also apparently transmit fuel molecules, and perhaps other substances as well, to the cilia and flagella.

Many types of eukaryotic cells contain *centrioles*. Centrioles, which typically occur in pairs, are small cylinders, about 0.2 micrometer in diameter, containing nine microtubule triplets (Figure 5–25). Their structure is identical to that of basal bodies; however, their distribution in the cell is different. Until recently, it appeared that their function was also different, which is why they are called by different names even though electron microscopy has revealed their identical structures. Centrioles usually lie in pairs with their long axes at right angles to one another in the cytoplasm, near the nuclear envelope. They are found only in those groups of organisms that also have cilia or flagella (and, therefore, basal bodies). There is evidence that centrioles play a role in organizing a structure known as the *spindle*, which appears at the time of cell division and is involved in chromosome movements. The spindle, it has been shown, also contains numerous microtubules; thus centrioles and basal bodies both appear to be organizers of microtubules. However, as we shall see in Chapter 7, cells that have no centrioles—such as the cells of the flowering plants—are also able to organize microtubules into a spindle.

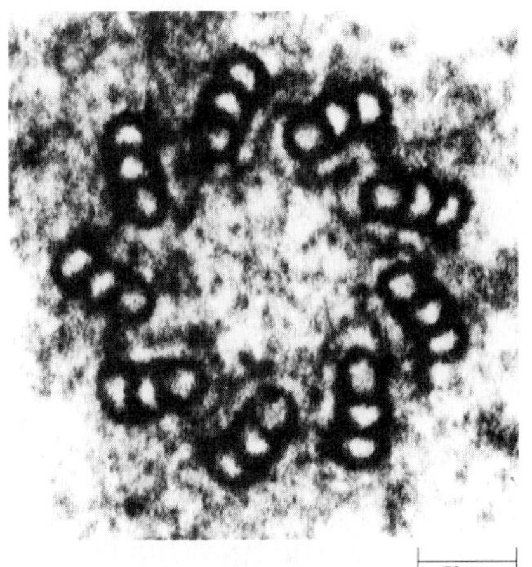

5-25
Centrioles are structurally identical to basal bodies. The one shown here in cross section is from a mouse embryo cell.

Bacterial Flagella

Some bacteria also move by means of flagella, although these flagella are so different in construction from those of eukaryotes that it would be useful if they had a different name. Each bacterial flagellum is made up of repeating units of a small globular protein, flagellin, assembled into chains that are wound in a triple helix (three chains) with a hollow core. These flagella grow from the tip; the individual flagellin molecules pass down a channel in the center of the helix and are assembled onto the end of the chains.

The bacterial flagellum is not enclosed within the cell membrane, as eukaryotic flagella are, but protrudes from the cell as a naked, corkscrew-shaped protein filament. It is anchored in the cell membrane by a complex baseplate (to be described in Chapter 20) and rotates, at about 40 revolutions per second (the rate varies with the species). When it rotates in a counterclockwise direction, it pushes the cell forward. Clockwise rotation pulls the cell backward or, in some species, causes it to tumble through the water. The motor for this flagellar motion is thought to be within the anchoring baseplate. *Escherichia coli* moves by means of six flagella. This bacterial cell, a heterotroph, can feed on glucose. If glucose is present, *E. coli* has no flagella. In the absence of glucose (or another abundant food source), the flagella regrow and propel the little cell into greener pastures.

HOW CELLS COMMUNICATE

In multicellular organisms a yet higher level of organization is found where the specialized cells cooperate to form a functional system. Cells are organized into *tissues*, groups of cells with common functions: muscle, nerve, connective, and epithelial (covering). Tissues are further organized in concert to form an *organ*, such as the heart, brain, or kidney, which, like the subcellular organelles, has a design that suits it for a specific function.

5-26
A bacterial cell (Proteus mirabilis) *with numerous flagella—176, to be exact.*

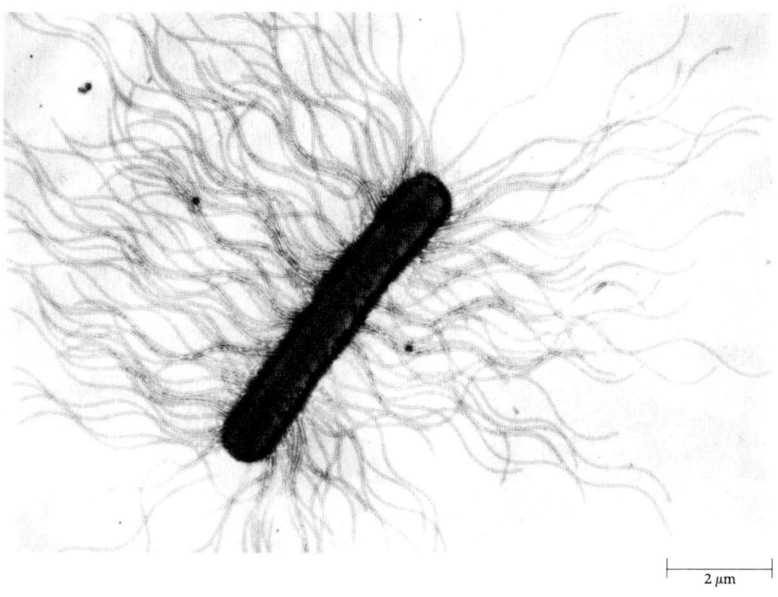

5–27

Plasmodesmata connecting two leaf cells from a corn plant. The wide gray area running horizontally in this electron micrograph is the cell wall, which is traversed by the plasmodesmata. The dark line extending through the center of each plasmodesma is a desmotubule.

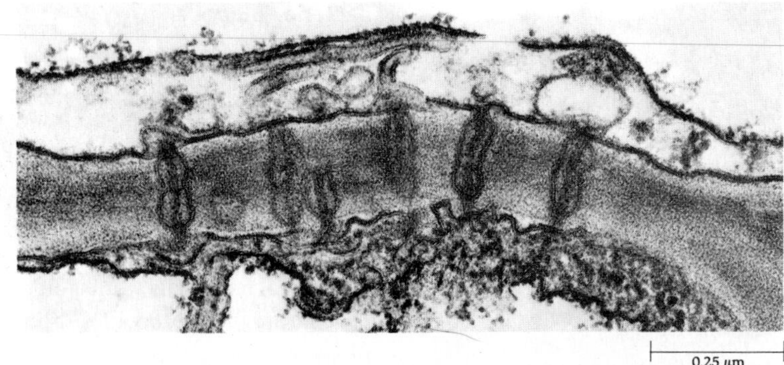

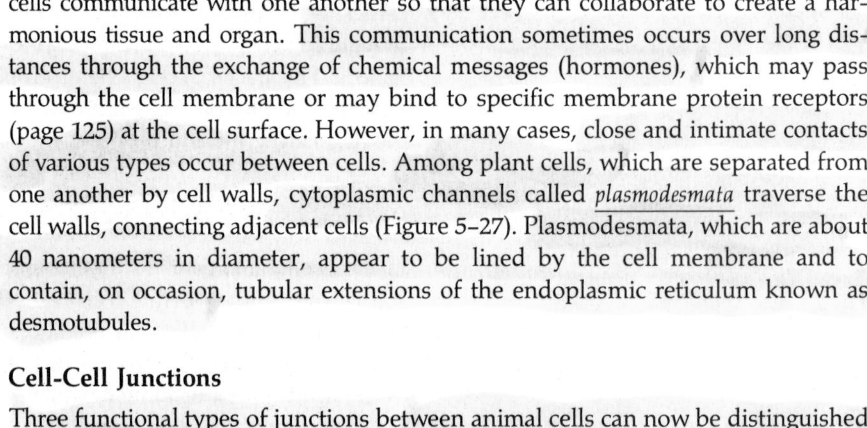

As you might imagine, in multicellular organisms it is essential that individual cells communicate with one another so that they can collaborate to create a harmonious tissue and organ. This communication sometimes occurs over long distances through the exchange of chemical messages (hormones), which may pass through the cell membrane or may bind to specific membrane protein receptors (page 125) at the cell surface. However, in many cases, close and intimate contacts of various types occur between cells. Among plant cells, which are separated from one another by cell walls, cytoplasmic channels called *plasmodesmata* traverse the cell walls, connecting adjacent cells (Figure 5–27). Plasmodesmata, which are about 40 nanometers in diameter, appear to be lined by the cell membrane and to contain, on occasion, tubular extensions of the endoplasmic reticulum known as desmotubules.

5–28

Junctions between cells. (a) *Desmosomes cement cells to adjoining cells. Here three desmosomes connect the cell membranes of two adjacent skin cells of a salamander larva.* (b) *A tight junction (arrow) seals the spaces between cells. These cells are from the surface layer (epithelium) of a rat intestine.* (c) *A portion of a gap junction between two liver cells, as seen in cross section. An ordered arrangement of protein molecules surrounds each of the narrow channels through which small molecules can pass between the cells. In this electron micrograph the channels are filled with an electron-opaque staining substance and appear as black dots.*

Cell-Cell Junctions

Three functional types of junctions between animal cells can now be distinguished by the electron microscope: (1) adhering junctions that hold together adjacent cells in a tissue; *desmosomes* are the principal type of adhering junction; (2) junctions between cells that keep materials from leaking through tissues; in vertebrates these are called *tight junctions,* whereas slightly different structures in invertebrates are known as septate junctions; and (3) junctions through which cells can exchange nutrient molecules and molecular communications; these are called *gap junctions.*

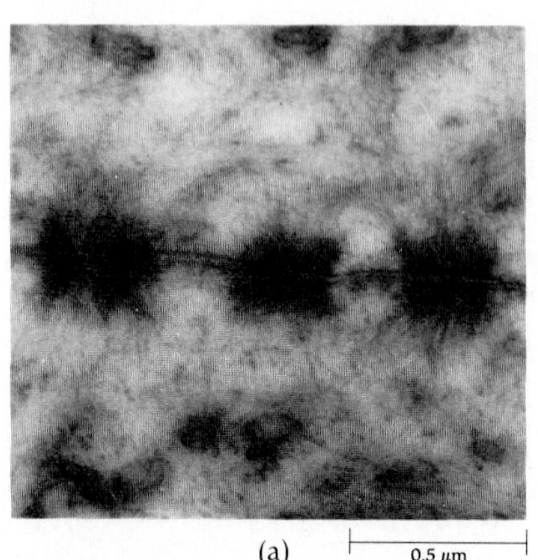

(a)

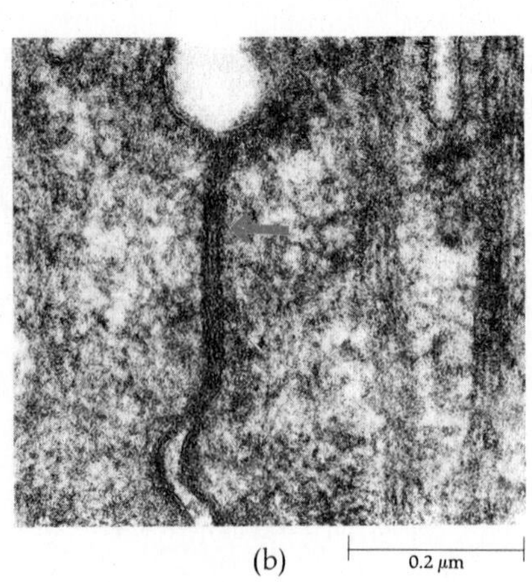

(b)

(c)

Desmosomes have often been compared to spot welds between cells. They consist of plaques of dense fibrous material between cells with clusters of filaments from the cytoplasm of the neighboring cells looping in and out of them. They attach cells to one another and give tissues mechanical strength. They are found in especially large numbers in tissues subjected to severe mechanical stress, such as the skin.

Tight junctions form a continuous ring around each cell in a layer of tissue, preventing leakage between cells. For example, intestinal epithelial cells are surrounded by tight junctions that keep the intestinal contents from seeping between the cells. Similar arrangements are found in all other cells that form linings of internal organs, such as the bladder and the kidney. Tight junctions seem to involve the fusion of adjacent membranes.

Gap junctions permit the passage of materials between cells. They appear as clusters of very small channels (about 2 nanometers in diameter) surrounded by an ordered array of proteins. Experiments with radioactively labeled molecules have shown that small messenger molecules can pass through these channels. The junctions also serve to transmit electrical signals. For example, gap junctions synchronize the contractions of muscle cells of the heart.

Chemical Communication

Cells send one another chemical messages. Nerve impulses are transmitted from neuron to neuron, or from neuron to muscle or gland, by means of chemical signals, as we shall see in Chapter 40. Cells in the body of a plant or animal release hormones that travel over a distance and affect other cells in the same organism. Embryonic cells in the course of development release chemicals that direct the differentiation of neighboring cells into organs and tissues. If normal cells from vertebrates (including humans) are isolated from one another and grown in a nutrient medium on a smooth glass surface, they move, amoeba-like, ruffling their cell borders until they encounter another cell, at which time they stop, apparently in response to some signal. More significantly, they multiply until each cell is touching another cell; then they cease to multiply. This phenomenon is known as contact inhibition. (Cancer cells do not show contact inhibition; they pile on top of one another, moving, multiplying, crowding each other until all of the nutrients are used up.)

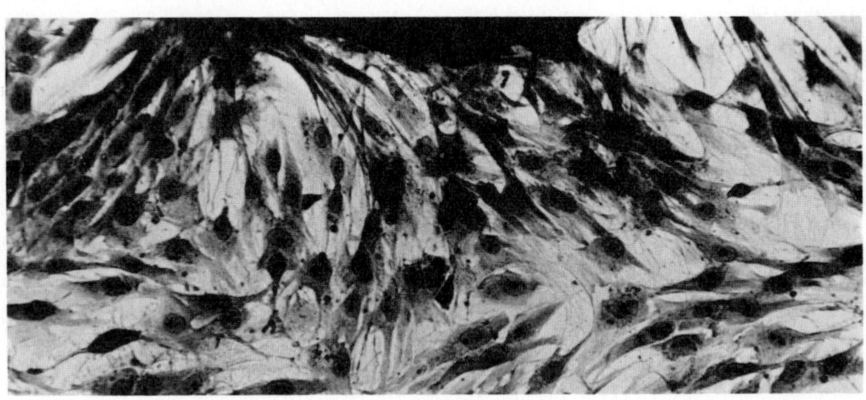

5–29
Fibroblasts from a baboon growing in tissue culture on a glass plate. When the cells touch each other, they stop their amoeba-like movement. They will multiply only until they fill the empty space.

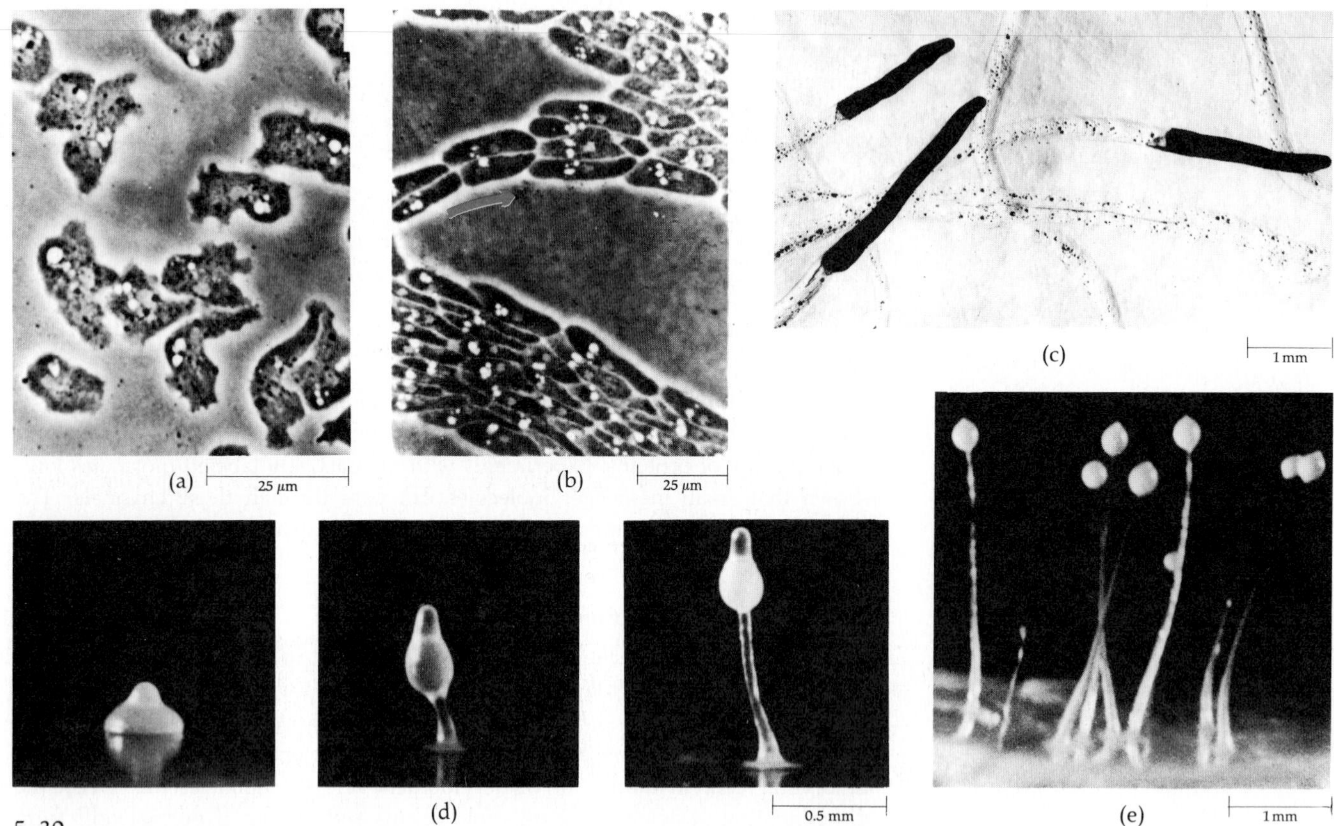

5-30
Life history of a cellular slime mold, Dictyostelium discoideum. (a) *Amoebas in feeding stage. The light gray area in the center of each cell is the nucleus, and the white areas are contractile vacuoles.* (b) *Amoebas aggregating. The direction in which the stream is moving is indicated by the arrow.* (c) *Migrating sluglike masses, each of which deposits a thick slime sheath that collapses behind it.* (d) *At the end of the migration, the mass gathers together and begins to rise vertically, differentiating into a stalk and fruiting body* (e).

A cellular communication system of particular interest to biologists, because of the comparative ease with which it can be studied, is seen among a group of organisms known as the cellular slime molds. The slime mold *Dictyostelium discoideum* is an example. At one stage in its life cycle, it exists as a swarm of small individual amoebas, which divide and grow and feed, amoeba-fashion, until their food supply (mostly bacteria) gives out. At this point, the cells alter both their shape and behavior: they become sausage-shaped and begin to migrate toward the center of the group. Eventually, they pile up in a heap; the heap gradually takes on the form of a multicellular mass somewhat resembling a garden slug. The sluglike mass soon stops its migration, gathers itself into a mound, and sends up a long stalk at the tip of which a small, shimmering fruiting body forms. The fruiting body matures and eventually bursts open, releasing a new swarm of tiny amoebas, and the cycle begins again. The chemical that spreads from cell to cell to initiate this remarkable sequence of events was first called acrasin, after Acrasia, the cruel witch in Spenser's *Faerie Queene* who attracted men and turned them into beasts. Acrasin was later identified as the chemical compound cyclic AMP (adenosine monophosphate). Cyclic AMP has now been shown to have an important role in human physiology, which will be described in Chapter 41.

SUMMARY

Cells are the basic units of biological structure and function. The size of cells is limited by proportions of surface to volume; the greater a cell's surface area in proportion to its volume, the more readily substances can get into and out of the cell. Cell size is also limited by the capacity of the nucleus to regulate the cellular activities of very large cells. Cells that are active metabolically are likely to be small.

Cells are separated from their environment by a cell membrane that restricts the passage of materials into and out of the cell and so protects the cell's structural and functional integrity. Plant cells are also usually enclosed in a cell wall.

The nucleus of eukaryotic cells is separated from the cytoplasm by a double membrane, the nuclear envelope, which contains pores through which molecules pass to and from the cytoplasm. The nucleus contains the hereditary material, the chromosomes, which, when the cell is not dividing, exist in an extended form called chromatin. The nucleolus, visible within the nucleus, is the site of formation of the ribosomes. Interacting with the cytoplasm, the nucleus helps to regulate the cell's ongoing activities.

The cytoplasm of the cell is a concentrated aqueous solution containing enzymes, many other dissolved molecules and ions, and also, in the case of eukaryotic cells, a variety of membrane-bound organelles with specialized functions in the life of the cell. The eukaryotic cytoplasm has a supporting cytoskeleton that includes at least four different types of structures: microtubules, microfilaments, intermediate fibers, and microtrabeculae. The cytoskeleton maintains the shape of the cell, enables it to move, anchors its organelles, and directs its traffic. Vacuoles and vesicles (small vacuoles) are also present in the cytoplasm of many cells, particularly plant cells. They are bounded by a single membrane.

Eukaryotic cells contain many organelles, most of which are not found in prokaryotic cells (see Table 5–1, page 122). The most numerous organelles (in both prokaryotes and eukaryotes) are the ribosomes, the sites of assembly of proteins. The endoplasmic reticulum is an interconnected system of membrane-lined sacs, tubes, and channels. It serves as a work surface for many of the cell's biochemical activities and also as a system of passageways for moving materials through the cell. In eukaryotic cells, ribosomes may be bound to the surface of the endoplasmic reticulum or clustered at interstices of the microtrabecular network. Rough endoplasmic reticulum (with ribosomes attached) is characteristically found in cells producing proteins for export. Smooth endoplasmic reticulum, which lacks ribosomes, is involved with lipid synthesis. Golgi bodies, also composed of membranes, are packaging centers for materials being moved through and out of the cell. Lysosomes, which contain hydrolytic enzymes, are involved in intracellular digestive activities in some cells. The enzymes for cellular reactions that produce hydrogen peroxide as a by-product are sequestered in the peroxisomes, along with an enzyme that breaks the toxic peroxide down into water and oxygen.

Mitochondria are membrane-bound organelles in which energy-yielding organic molecules are broken down and the released energy repackaged into smaller units. Plastids are membrane-bound organelles found only in photosynthetic organisms. Leucoplasts are storage compartments, chromoplasts contain pigments, and chloroplasts are the sites of photosynthesis in eukaryotes.

Contractile, or "muscle," proteins, principally actin and myosin, are associated with internal cellular movement, while cilia and flagella are associated with the

Table 5-1 Comparison of Prokaryotic, Animal, and Plant Cells

	PROKARYOTE	ANIMAL	PLANT
Cell membrane	Present	Present	Present
Cell wall	Present (noncellulose polysaccharide plus protein)	Absent	Present (contains cellulose)
Nucleus	No nuclear envelope	Surrounded by nuclear envelope	Surrounded by nuclear envelope
Chromosomes	Single, continuous DNA molecule	Multiple, consisting of DNA and protein	Multiple, consisting of DNA and protein
Endoplasmic reticulum	Absent	Usually present	Usually present
Mitochondria	Absent	Present	Present
Plastids	Absent	Absent	Present in many cell types; chloroplasts in photosynthetic cells
Ribosomes	Present (smaller)	Present	Present
Golgi bodies	Absent	Present	Present
Lysosomes	Absent	Often present	Similar structures (lysosomal compartments) present
Peroxisomes	Absent	Often present	Often present
Vacuoles	Absent	Small or absent	Usually large single vacuole in mature cell
9 + 2 cilia or flagella	Absent	Often present	Absent (in higher plants)
Centrioles	Absent	Present	Absent (in higher plants)

external movement of cells or the movement of materials along cell surfaces. These hairlike appendages are found on the surface (yet within the cell membrane) of many types of eukaryotic cells. They have a highly characteristic 9 + 2 structure, with nine pairs of microtubules forming a ring surrounding two central microtubules. One of each pair of outer microtubules contains enzymes involved in the chemical reactions that release energy for ciliary motion.

Cilia and flagella arise from basal bodies, which are cylindrical structures containing nine microtubule triplets with no inner pair. Centrioles have the same internal structure as basal bodies and are found only in those groups of organisms that have cilia or flagella. They typically occur in pairs, lying near the nuclear envelope, and may play a role in the formation of the spindle during cell division.

Bacterial flagella, which have a structure different from that of eukaryotic flagella, are not enclosed within the cell membrane. They are, however, anchored in the membrane by a complex baseplate, which is thought to contain the motor that powers their rotation.

Cells have a variety of systems for communicating with one another, including direct physical contact through plasmodesmata and gap junctions. Other types of cell-cell junctions include desmosomes, which "weld" cells together in tissues subject to mechanical stress, and tight junctions, which prevent leakage between cells in lining tissues. Cells also communicate with one another by chemical agents that either pass through the cell membrane or interact with receptors in its surface.

QUESTIONS

1. Distinguish among the following: cell membrane/cell wall; nucleus/nucleolus; rough endoplasmic reticulum/smooth endoplasmic reticulum; lysosomes/peroxisomes; chloroplasts/mitochondria; cilia/flagella; basal body/centriole; plasmodesma/desmotubule; desmosomes/tight junctions/gap junctions.

2. (a) Sketch an animal cell. Include the principal organelles and label them. (b) Prepare a similar, labeled sketch of a plant cell. (c) What are the major differences between the animal cell and the plant cell?

3. Why is the secondary wall of a plant cell *inside* the primary cell wall? Where is the cell membrane in relation to the two cell walls?

4. What are the functions of the cytoskeleton? Describe the similarities and differences between microtubules, microfilaments, intermediate fibers, and microtrabeculae.

5. Explain the functions of each of the following structures: endoplasmic reticulum, ribosomes, Golgi bodies. How do they interact in the export of proteins from the cell?

6. Use a ruler and the scale marker at the bottom of each micrograph on pages 100 and 116 to determine: (a) the thickness (roughly) of a cell membrane, (b) the diameter of a flagellum, and (c) the diameter of a microtubule within a flagellum. (This is how the sizes of cellular components are determined by microscopists.) Would a flagellum be resolvable in a light microscope (that is, is its diameter more than 0.2 μm)?

7. (a) Sketch a cross section of a cilium. (b) Sketch a cross section of the basal body of a cilium. (c) What are the differences between the two structures?

8. In what three major ways do bacterial flagella differ from eukaryotic cilia and flagella?

9. On the basis of what you know of the functions of each of the structures in Table 5–1, what components would you expect to find most prominently in each of the following cell types: muscle cells, sperm cells, green leaf cells, red blood cells, white blood cells?

10. Two brothers were under medical treatment for infertility. Microscopic examination of their semen showed that the sperm were immotile and that the little "arms" were missing from the microtubular arrays. The brothers also had chronic bronchitis and other respiratory difficulties. Can you explain why?

CHAPTER 6

How Things Get into and out of Cells

6–1
The relative concentrations of different ions in pond water (gray bars) and in the cytoplasm of the green alga Nitella *(colored bars). Differences such as these indicate that cells regulate the passage of materials across their membranes.*

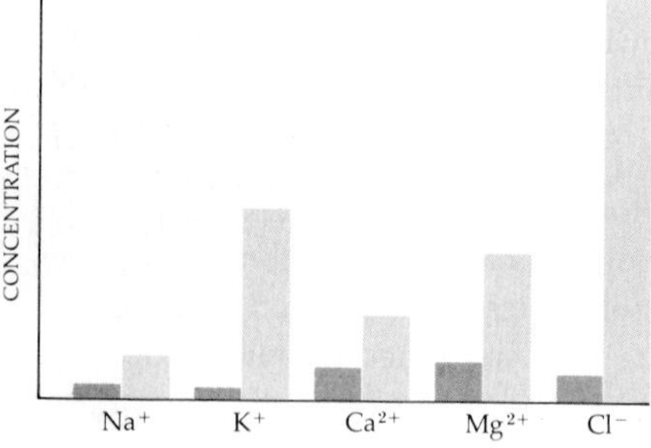

Cells are able to regulate the passage of materials across cell membranes. This is an important capacity. One of the criteria by which we identify living systems is that living matter, although surrounded on all sides by nonliving matter, is different from it in the kinds and amounts of chemical substances it contains (Figure 6–1). Without this difference, of course, living matter would be unable to maintain the organization and structure on which its existence depends.

The cell membrane is not simply an impenetrable barrier, however. Living matter constantly exchanges substances with the nonliving world around it. Control of these exchanges is essential in order to protect the cell's integrity and to maintain those very narrow conditions of pH and ionic concentrations at which metabolic activities can take place. Cell membranes thus have a complex double function of keeping certain things out and letting others in. In addition to the cell membrane, which controls the passage of material from outside the cell, internal membranes, such as those surrounding mitochondria, chloroplasts, and the nucleus, regulate the passage of material among intracellular compartments and so regulate the internal environment (Figure 6–2).

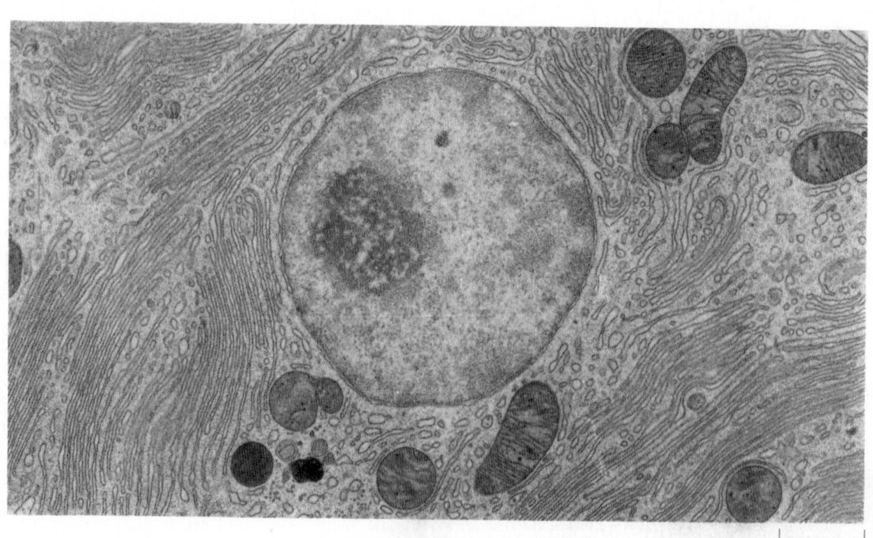

6–2
Electron micrograph of a portion of a cell from the pancreas, an organ of the digestive system. The cell has a large, central nucleus with scattered chromatin, many mitochondria, large quantities of rough endoplasmic reticulum, and many small vesicles. Not only is the cell itself surrounded by a membrane and the nucleus by a double membrane system (the nuclear envelope), but its organelles are also surrounded by membranes. The membranes of the endoplasmic reticulum further divide the cell into membrane-bound channels. Collectively, all of these different membranes regulate the movements of substances into and out of cells and restrict their passage from one part of the cell to another.

Control of these exchanges across membranes depends on physical and chemical properties of the membranes and of the molecules that move through them. Let us first look at the membrane itself.

THE STRUCTURE OF THE CELL MEMBRANE

Figure 6–3 shows a current model of the structure of the cell membrane. As we saw in the last chapter, the membrane is composed of a phospholipid bilayer in which are suspended numerous large globular proteins. These proteins, known as integral proteins, generally span the membrane and protrude on either side; the portions embedded in the bilayer have hydrophobic surfaces, whereas the surfaces of those portions that extend beyond the bilayer are hydrophilic. Although the evidence remains inconclusive, there may be hydrophilic pores running through some of the integral proteins. All the membranes of a cell, including those surrounding the various organelles, have this same basic structure. There are, however, local differences in the types of lipids and, particularly, in the number and types of proteins. Most membranes are about 40 percent lipid and 60 percent protein, though there is considerable variation.

The two surfaces of the cell membrane differ considerably in chemical composition. The two layers of the bilayer generally have differing concentrations of specific types of lipid molecules. Moreover, the integral proteins have a definite orientation within the bilayer and the portions extending on either side have differing amino acid compositions and tertiary structures. On the cytoplasmic side of the membrane, additional protein molecules, known as peripheral proteins, are attached to some of the integral proteins protruding from the bilayer. On the outside of the membrane, short carbohydrate chains are attached to the protruding proteins and to some of the lipid molecules. These form a carbohydrate coat on the outer surface of the membranes of some eukaryotic cells. The carbohydrates are involved in recognition and adhesion processes—between cells and cells, for instance, as well as cells and antibodies, cells and hormones, and cells and viruses.

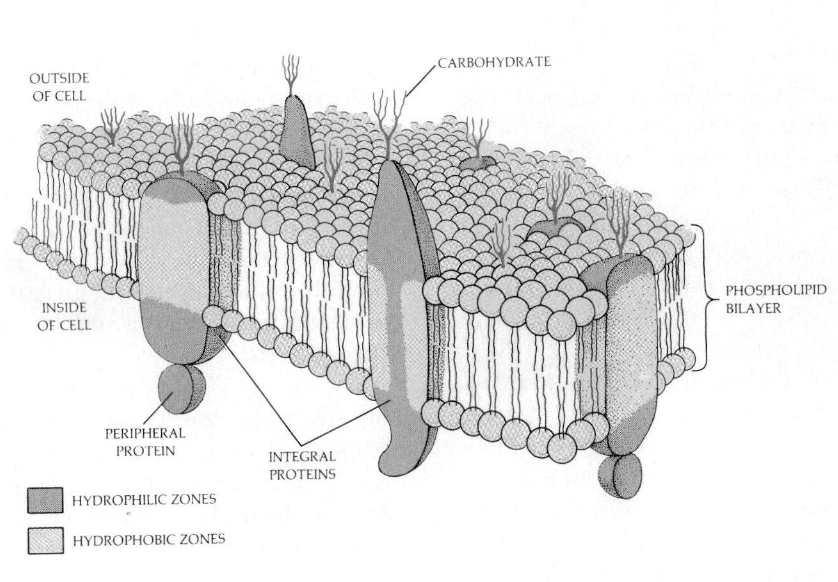

6–3
Model of a cell membrane, the basic features of which were first proposed by S. J. Singer and G. L. Nicolson on the basis of electron micrographs and biochemical data. The membrane is composed of phospholipid molecules and large protein molecules. The phospholipid molecules are arranged in a bilayer with their hydrophobic tails pointing inward and their hydrophilic phosphate heads pointing outward. The proteins embedded in the bilayer are known as integral proteins; on the cytoplasmic side of the membrane, peripheral proteins are attached to some of the integral proteins. The portion of a protein molecule's surface that is within the lipid bilayer is hydrophobic; the portion of the surface that is exposed outside the bilayer is hydrophilic. It is believed that pores with hydrophilic surfaces may pass through some of the protein molecules. The short carbohydrate chains attached to the outside of the membrane are involved in the adhesion of cells to each other and with the "recognition" of molecules at the membrane surface.

The proteins vary from membrane to membrane, depending on the function of the cell (or of the organelle, in the case of intracellular membranes), and also from place to place on the same membrane. Some of the proteins are enzymes, which regulate particular chemical reactions, while others are carriers involved in the transport of molecules across the membrane. Although many of the integral proteins appear to be anchored in place, either by peripheral proteins or by elements of the cytoskeleton that are concentrated near the cytoplasmic surface of the membrane, the structure of the bilayer is generally quite fluid. The lipid molecules and at least some of the protein molecules can move laterally within it, forming different patterns that vary from time to time and place to place. Consequently, this widely accepted model of membrane structure is known as the *fluid-mosaic model.*

THE MOVEMENT OF WATER AND SOLUTES

6–4
Water at the top of a falls, like a boulder on a hilltop, has potential energy. The movement of water molecules as a group, as from the top of the falls to the bottom, is referred to as bulk flow.

Of the many kinds of molecules moving into and out of cells, by far the most important is water. Further, the other ions and molecules moving in and out are dissolved in water. Let us therefore look again at water, focusing our attention this time on how water moves.

Water molecules move from one place to another because of differences in potential energy, usually referred to as the *water potential.* Water moves from a region where water potential is greater to a region where water potential is lower, regardless of the reason for the water potential. A simple example is water running downhill in response to gravity. Water at the top of a hill has more potential energy (that is, a greater water potential) than water at the bottom of a hill. As the water runs downhill, its potential energy is converted to kinetic energy, which can be converted, in turn, to mechanical energy doing useful work if, for example, a water mill is placed in the path of the moving water.

Pressure is another source of water potential. If we fill a rubber bulb with water and then squeeze, water will squirt out of the nozzle. Like water at the top of a hill, this water has been given a high water potential and will move to a lower one. Can we make the water that is running downhill run uphill by means of pressure? Obviously we can. But only so long as the water potential produced by the pressure exceeds the water potential produced by gravity.

In solutions, water potential is affected by the concentration of dissolved particles (solutes). As the concentration of solute particles (that is, the number of solute particles per unit volume of solution) increases, the concentration of water molecules (that is, the number of water molecules per unit volume of solution) decreases, and vice versa. The water potential is directly related to the concentration of water molecules—the higher the concentration of water molecules, the greater the water potential. Conversely, the higher the concentration of solute particles, the lower the water potential. Individual water molecules move from regions of high water potential to regions of low water potential, a fact of great importance for living systems.

The concept of water potential is a useful one because it enables us to predict the way that water will move under various combinations of circumstances. Measurements of water potential are usually made in terms of the pressure required to stop the movement of water—that is, the hydrostatic (water-stopping) pressure—under the particular circumstances. The unit usually used to measure this pressure is the atmosphere. One atmosphere is the average pressure of the air at sea level, about 1 kg/cm^2 (or 15 lb/in^2).

Two mechanisms are involved in the movement of water and of substances dissolved in it: bulk flow and diffusion. In living systems, bulk flow moves water and solutes from one part of a multicellular organism to another part, whereas diffusion moves molecules and ions into, out of, and through cells. A particular instance of diffusion—that of water across a membrane that separates solutions of different composition—is known as osmosis.

Bulk Flow

Bulk flow is the overall movement of a fluid. The molecules move all together and in the same direction. For example, water runs downhill by bulk flow in response to the differences in water potential at the top and the bottom of a hill. Blood moves through your body by bulk flow as a result of the water potential (blood pressure) created by the pumping of your heart. Sap—a concentrated solution of sucrose in water—moves by bulk flow from the leaves of a plant to other parts of the plant body.

Diffusion

Diffusion is a familiar phenomenon. If you sprinkle a few drops of perfume in one corner of a room, the scent will eventually permeate the entire room even if the air is still. If you put a few drops of dye in one end of a glass tank full of water, the dye molecules will slowly become evenly distributed throughout the tank. The process may take a day or more, depending on the size of the tank, the temperature, and the relative size of the molecules.

Why do the dye molecules move apart? If you could observe the individual dye molecules in the tank (Figure 6–5), you would see that each one of them moves individually and at random. Looking at any single molecule—at either its rate of motion or its direction of motion—gives you no clue at all about where the molecule is located with respect to the others. So how do the molecules get from one end of the tank to the other? Imagine a thin section through the tank, running from top to bottom. Dye molecules will move in and out of this section, some moving in one direction, some moving in the other. But you will see more dye molecules

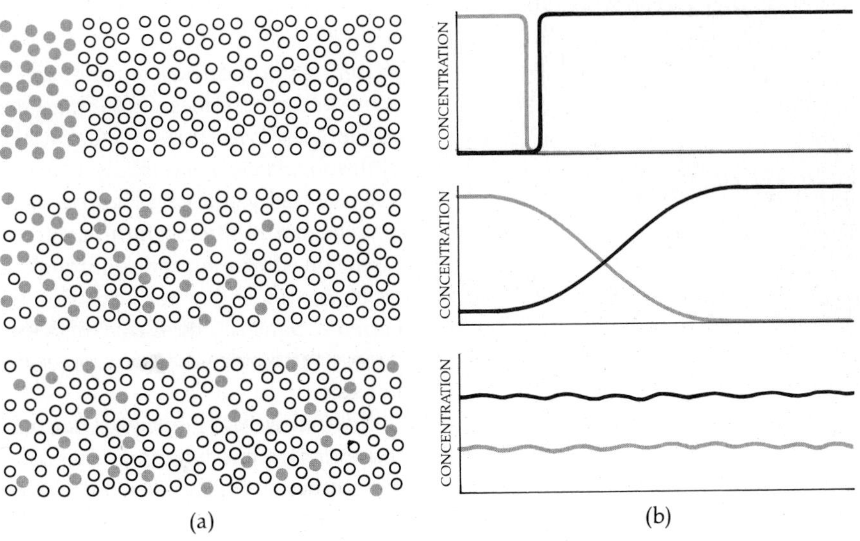

6–5
(a) Diagram of the diffusion process. Diffusion is the result of the random movement of individual molecules, which produces a net movement from a more concentrated to a less concentrated area. Notice that as one type of molecule (indicated by color) diffuses to the right, the other diffuses in the opposite direction. The result will be an even distribution of both types of molecules. Can you see why the net movement of molecules will slow down as equilibrium is reached? (b) *Graphs showing the concentration gradients of dye and water.*

moving from the side of greater dye concentration. Why? Simply because there are more dye molecules at that end of the tank. Since there are more dye molecules on the left, more dye molecules, moving at random, will move to the right, even though there is an equal probability that any one molecule of dye will move from right to left. Consequently, the *net* movement of dye molecules will be from left to right. Similarly, if you could see the movement of the individual water molecules in the tank, you would see that their *net* movement is from right to left.

Substances that are moving from a region of higher concentration of their own molecules to a region of lower concentration are said to be moving *down a gradient.* (A substance moving in the opposite direction, toward a higher concentration of its own molecules, moves *against a gradient,* which is analogous to being pushed uphill.) Diffusion occurs only down a gradient. The steeper the downhill gradient—that is, the larger the difference in concentration—the more rapid the diffusion. In our imaginary tank, there are two gradients; the dye molecules are moving down one of them, and the water molecules are moving down the other in the opposite direction.

What happens when all the molecules are distributed evenly throughout the tank? The even distribution does not affect the behavior of the molecules as individuals; they still move at random. And, since the movements are random, just as many molecules go to the left as to the right. But because there are now as many molecules of dye and as many molecules of water on one side of the tank as on the other, there is no net movement of either. There is, however, just as much random motion as before, provided the temperature has not changed. When the molecules have reached such a state of equal distribution, that is, when there are no more gradients, they are said to be in dynamic equilibrium.

An aqueous solution in which diffusion is occurring is another example of water potential. The region of the tank in which there is pure water has a higher water potential than the region containing water plus dye or some other solute. Thus the water molecules are moving from a region of higher water potential to a region of lower water potential.

The essential characteristics of diffusion are (1) that each molecule moves independently of the others and (2) that these movements are random. The net result of diffusion is that the diffusing substance becomes evenly distributed.

Cells and Diffusion

Water, oxygen, carbon dioxide, and a few other simple molecules diffuse freely across cell membranes. Diffusion is also a principal way in which substances move within cells. One of the major factors limiting cell size is this dependence upon diffusion, which is essentially a slow process, except over very short distances. As you can see by studying Figure 6–5, the process becomes increasingly slower and less efficient as the distance "covered" by the diffusing molecules increases. The rapid spread of a substance through a large volume, such as perfume through the air of a room, is due not primarily to diffusion but rather to the circulation of air currents. Similarly, in many cells, the transport of materials is speeded by active streaming of the cytoplasm, a process in which elements of the cytoskeleton play a key role.

Efficient diffusion requires not only a relatively small volume but also a steep concentration gradient. Cells maintain such gradients by their metabolic activities, thereby hastening diffusion. For example, carbon dioxide is constantly produced as

6–6

(a) Micrograph of a gill filament from a trout, showing the densely packed lamellae, which contain the blood vessels. Water, carrying dissolved oxygen, flows between the lamellae in one direction; blood flows through them in the opposite direction. Thus the blood carrying the most oxygen (that is, the blood leaving the lamellae) meets the water carrying the most oxygen, and the blood carrying the least oxygen (the blood entering the lamellae) meets the water carrying the least oxygen. (b) In this way, a constant concentration gradient is maintained along the lamella, and the transfer of oxygen to blood by diffusion takes place all across its surface.

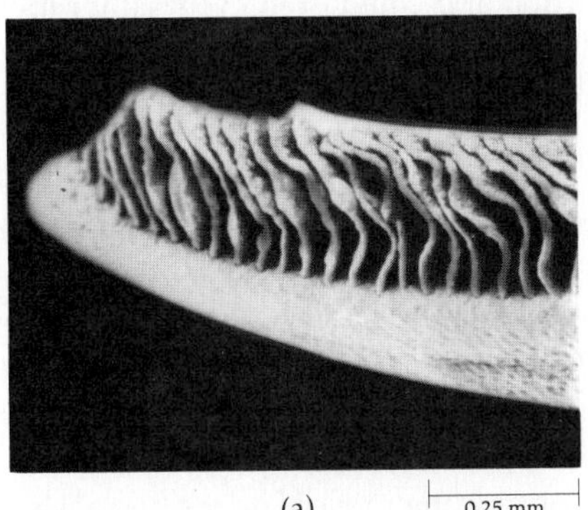

(a)

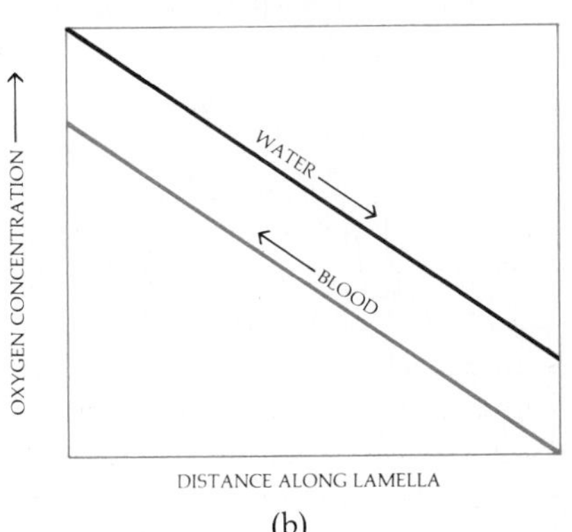

(b)

the cell oxidizes fuel molecules for energy. As a result, there is a higher concentration of carbon dioxide inside the cell than out. Thus a gradient is maintained between the inside of the cell and the outside, and carbon dioxide diffuses out of the cell down this gradient. Conversely, oxygen is used up by the cell in the course of its internal activities, so oxygen present in air or water or blood tends to move into the cells by diffusion, again down a gradient. Similarly, within a cell, materials are often produced at one place and used at another. Thus a concentration gradient is established between the two areas, and the material diffuses down the gradient from the site of production to the site of use.

Countercurrent Exchange

Although diffusion is efficient only over short distances, it nevertheless plays a key role in the transport of substances into and out of multicellular organisms, as well as between different compartments within the organism. For example, oxygen enters the bloodstream of an animal by diffusing through cells that are in contact with the environment. Unless the animal is quite small or leads a sedentary existence, a high rate of diffusion is required to supply it with adequate oxygen. Organisms have developed a number of anatomical arrangements that maintain steep concentration gradients and thus maximize diffusion rates. One of the most common types of arrangement is found in fish gills. Gills are divided into filaments, which are further divided into a number of flat, densely packed lamellae—platelike structures containing many blood vessels (Figure 6–6). Water, containing dissolved oxygen, flows over the lamellae, and oxygen moves by diffusion into the blood vessels just below the thin surfaces. In the lamellae, the direction of blood flow is opposite the direction of water flow. This countercurrent arrangement maintains a constant concentration gradient between the bloodstream and the water, and oxygen can diffuse in through the entire surface of the lamellae. As we shall see in Section 6, this principle of countercurrent exchange is used in several different anatomical systems.

Osmosis

A membrane that permits the passage of some substances, while blocking the passage of others, is said to be *selectively permeable*. The movement of water molecules through such a membrane is a special case of diffusion, known as *osmosis*. Osmosis results in a net transfer of water from a solution that has higher water potential to a solution that has lower water potential. In the absence of other factors that influence water potential (such as pressure), the movement of water in osmosis will be from a region of lower solute concentration (and therefore of higher water concentration) into a region of higher solute concentration (lower water concentration). The presence of solute decreases the water potential and so creates a gradient of water potential along which water diffuses.

The diffusion of water is not affected by *what* is dissolved in the water, only by *how much* is dissolved—that is, the concentration of particles of solute (molecules or ions) in the water. The word *isotonic* was coined to describe two or more solutions that have equal numbers of dissolved particles per unit volume and therefore the same water potential. There is no net movement of water across a membrane separating two solutions that are isotonic to one another, unless, of course, pressure is exerted on one side. In comparing solutions of different concentration, the solution that has less solute (and therefore a higher water potential) is known as *hypotonic*, and the one that has more solute (a lower water potential) is known as *hypertonic*. (Note that *iso* means "the same"; *hyper* means "more"—in this case, more particles of solute; and *hypo* means "less"—in this case, fewer particles of solute.) In osmosis, water molecules diffuse from a hypotonic solution (or from pure water), through a selectively permeable membrane, into a hypertonic solution.

Osmosis and Living Organisms

The osmotic movement of water across the selectively permeable cell membrane causes some crucial problems for living systems. These problems vary according to whether the cell or organism is hypotonic, isotonic, or hypertonic in relation to its environment. One-celled organisms that live in salt water, for example, are usually isotonic with the medium they inhabit, which is one way of solving the problem. Similarly, the cells of higher animals are isotonic with the blood and lymph that constitute the watery medium in which they live.

Many types of cells live in a hypotonic environment. In all single-celled organisms that live in fresh water, such as *Paramecium*, the interior of the cell is hypertonic to the surrounding water; consequently, water tends to move into the cell by osmosis. If too much water were to move into the cell, it could dilute the cell contents to the point of interfering with function and could even eventually rupture the cell membrane. This is prevented by a specialized organelle known as a contractile vacuole, which collects water from various parts of the cell and pumps it out with rhythmic contractions (Figure 6–7). As you might expect, this bulk transport process requires energy.

Osmotic Potential

The water potential on the two sides of a selectively permeable membrane will become equal if enough water moves from the hypotonic solution into the hypertonic solution to equalize the water concentrations (and therefore the solute concentrations)—that is, to make the solutions isotonic. If, however, physical barriers prevent the expansion of the hypertonic solution as water moves into it by osmosis,

6–7
A Paramecium *is hypertonic in relation to its environment, and hence water tends to move into the cell by osmosis. Excess water is expelled through its contractile vacuoles.* (a) *Photomicrograph of a* Paramecium *showing the position of its two rosette-like contractile vacuoles.* (b) *Collecting tubules converge toward the vacuole, filling it.* (c) *Then it contracts, emptying outside the cell membrane by way of a small central pore.*

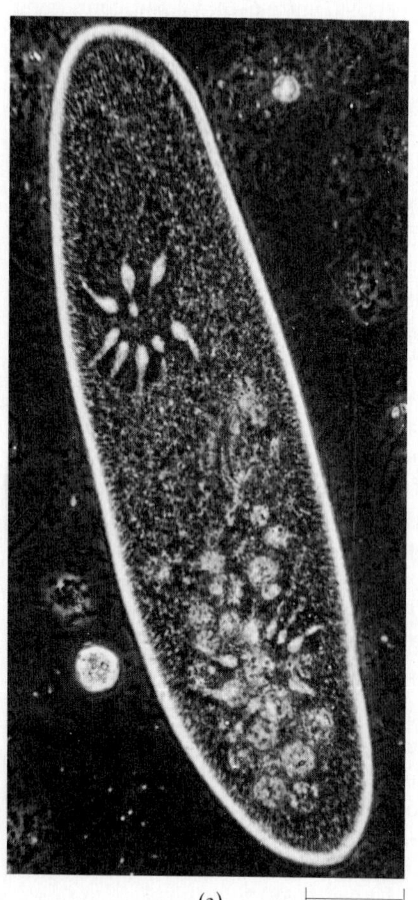

(a) 25 μm

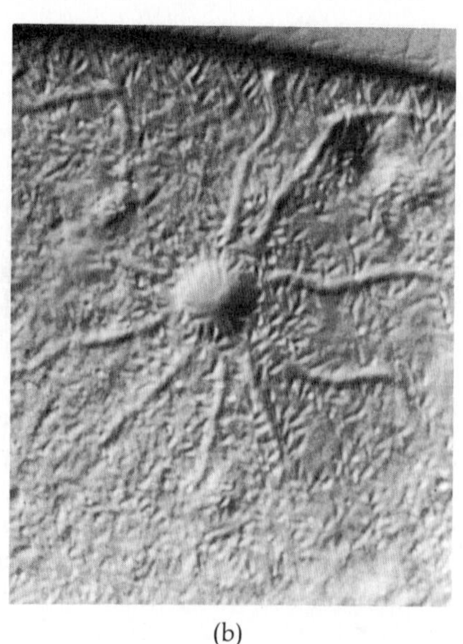

(b)

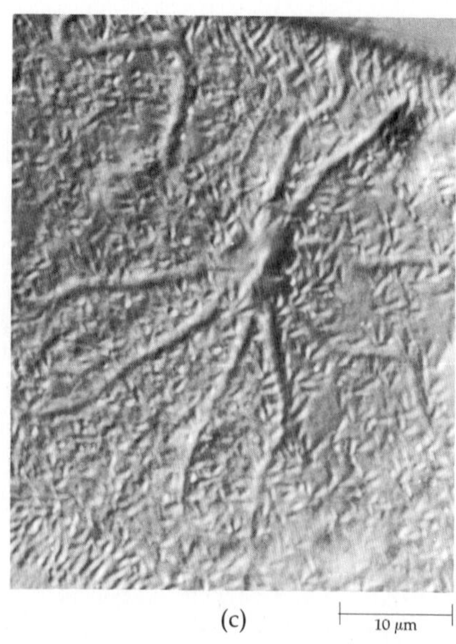

(c) 10 μm

6-8

Osmosis and the measurement of osmotic potential. (a) *The tube contains a solution and the beaker contains distilled water.* (b) *The selectively permeable membrane permits the passage of water but not of solute. The diffusion of water into the solution causes the solution to rise in the tube until the tendency of water to move into a region of lower water concentration is counterbalanced by the pressure resulting from the force of gravity acting on the column of solution. This pressure is proportional to the height,* h, *and density of the column of solution.* (c) *The pressure that must be applied to the piston to force the column of solution back to the level of the water in the beaker provides a quantitative measure of the osmotic potential.*

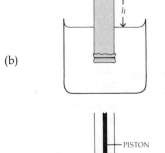

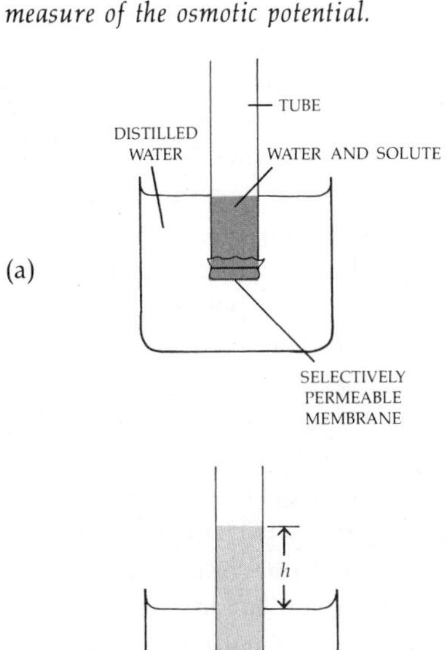

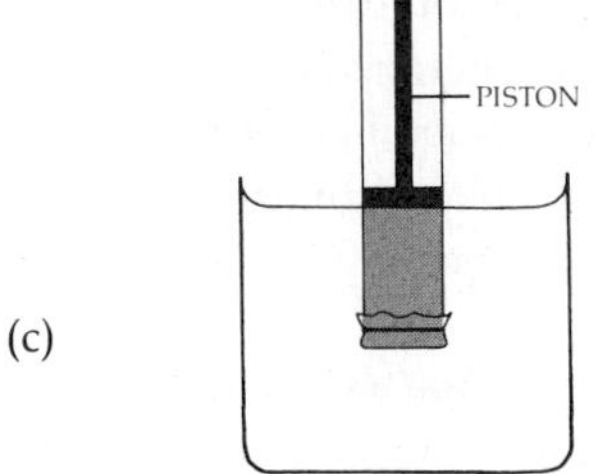

there will be a buildup of pressure as water molecules continue to move across the membrane. This pressure gradually increases the water potential of the hypertonic solution, decreasing the gradient of water potential between the two solutions. As the pressure increases, the net flow of water molecules will slow and then cease as the gradient of water potential disappears. (Individual water molecules, of course, continue to move back and forth across the membrane, but these movements are in equilibrium and there is no net movement of water.) The pressure that is required to stop the osmotic movement of water into a solution is called the osmotic pressure. It is a measure of the *osmotic potential* of the solution—that is, of the tendency of water to move across a membrane into the solution.

The measurement of the osmotic potential of a solution is illustrated in Figure 6-8. The beaker contains distilled water, and within the tube is the solution. Across the mouth of the tube is a selectively permeable membrane; such a membrane is freely permeable to water but not to the particles (ions or molecules) in the solution. Water moving from the beaker through the membrane into the solution causes the solution to rise in the tube. As it does so, the pressure created by the force of gravity acting on the column of solution gradually increases, thus increasing the water potential of the solution. The solution rises in the tube until equilibrium is reached—that is, until the water potential on both sides of the membrane is equal. The amount of pressure that must then be applied to the piston to force the solution in the tube back to the level of the water in the beaker provides a quantitative measure of the osmotic potential.

The lower the water potential of a solution, the greater the tendency of water molecules to move into it by osmosis and, therefore, the greater its osmotic potential. Since solutes decrease the water potential of a solution, a higher solute concentration means a greater osmotic potential.

As we shall see in Chapter 37, the osmotic potential of solutions within the vertebrate kidney plays an important part in determining the composition of the excreted urine.

Turgor

Plant cells are usually hypertonic to their surrounding environment, and so water tends to diffuse into them. This movement of water into the cell creates pressure within the cell against the cell wall. The pressure causes the cell wall to expand and the cell to enlarge. The elongation that occurs as a plant cell matures (Figure 5-4b) is a direct result of the osmotic movement of water into the cell.

As the plant cell matures, the cell wall stops growing. Moreover, mature plant cells typically have large central vacuoles that contain solutions of salts and other materials. (In citrus fruits, for example, they contain the acids that give the fruits their characteristic sour taste.) Because of these concentrated solutions, plant cells have a high osmotic potential—that is, water has a strong tendency to move into the cells. In the mature cell, however, the cell wall does not expand further. Its resistance to expansion results in an inward directed pressure, known as the wall pressure, which prevents the net movement of additional water into the cell. Consequently, equilibrium of salt concentration is not reached and water continues to "try" to move into the cell, maintaining a constant pressure on the cell wall from the inside (Figure 6-9, on the next page). This internal pressure on the cell wall is known as *turgor,* and it keeps the cell walls stiff and the plant body crisp. When turgor is reduced, as a consequence of water loss, the plant wilts.

6–9

(a) *A turgid plant cell. The central vacuole is hypertonic in relation to the fluid surrounding it and so gains water. The expansion of the cell is held in check by the cell wall.* (b) *A plant cell begins to wilt if it is placed in an isotonic solution, so that water pressure no longer builds up within the vacuole.* (c) *A plant cell in a hypertonic solution loses water to the surrounding fluid and so collapses, with its membrane pulling away from the cell wall. Such a cell is said to be plasmolyzed.*

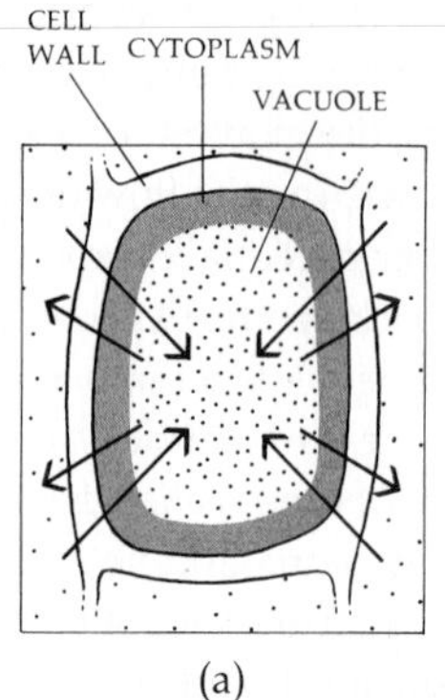

(a)

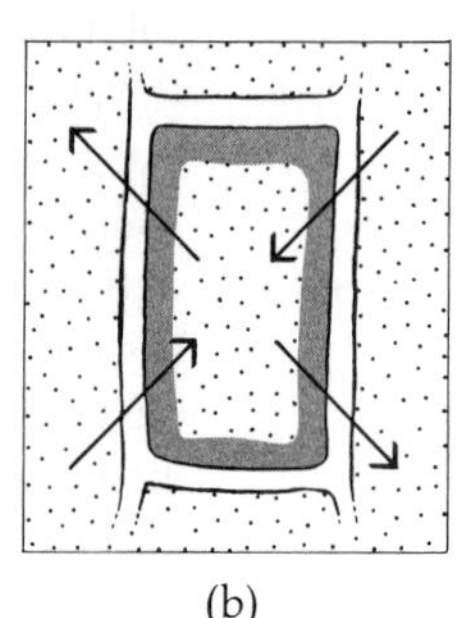

(b)

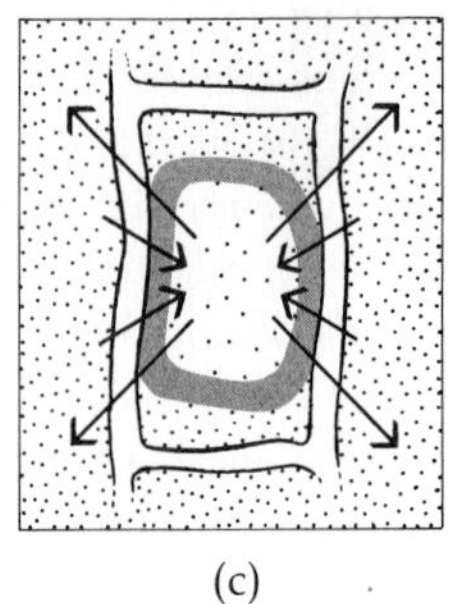

(c)

TRANSPORT ACROSS CELL MEMBRANES

In order to get into or out of a cell (or a membrane-bound organelle, such as a mitochondrion), substances must pass through a membrane. However, as we noted in Chapter 2, water and other polar or charged (hydrophilic) molecules exclude lipids and other hydrophobic molecules. Conversely, hydrophobic molecules exclude hydrophilic ones. Thus the cell membrane's lipid bilayer would be expected to allow easy passage of hydrophobic molecules but to exclude a large number of ions and hydrophilic molecules essential to the cell's existence. (It was in fact the observation that hydrophobic molecules diffused readily across cell membranes that provided the first evidence of the lipid nature of the membrane.) Carbon dioxide and oxygen, which are both nonpolar, are soluble in lipids and move freely through the membrane. Most organic molecules of biological importance, however, have polar functional groups and are therefore hydrophilic; they cannot freely diffuse through the lipid barrier. A number of mechanisms exist for getting hydrophilic molecules and ions through the membrane.

First, apparently there are apertures in the membrane through which water molecules can diffuse. (Water is the smallest polar molecule of biological importance.) These may be either permanent pores created by the tertiary structure of some of the integral proteins or momentary openings resulting from movements of the lipid molecules of the membrane. Other polar molecules, provided they are small enough, also pass through these openings. The permeability of the membrane to these solutes varies inversely with the size of the molecules, indicating that the apertures are small and that the membrane acts like a sieve in this respect.

Second, some of the integral membrane proteins act as carriers, ferrying back and forth molecules that could not readily cross the membrane by diffusion because of their size or polarity. These transport proteins are highly selective; a particular carrier may accept one molecule while it excludes a nearly identical one. Also, the protein molecule is not permanently altered in the transport process. In these respects, as we shall see in Chapter 8, carrier molecules are like enzymes, and to emphasize this, they have been named permeases. Unlike enzymes, however, they do not necessarily produce chemical change in the molecules with which they interact.

Some carrier proteins can transport substances across the membrane only if there is a favorable concentration gradient; such carrier-assisted transport is known as *facilitated diffusion*. It is driven by the concentration gradient and so moves molecules from a region of higher concentration to a region of lower concentration. Other carriers can move molecules against the concentration gradient, an energy-requiring process known as *active transport.*

Depending on the situation, the transport of a particular molecule into a cell may involve either facilitated diffusion or active transport. For example, several lines of evidence support the conclusion that the entry of glucose into most cells takes place by facilitated diffusion. First, because glucose is rapidly broken down when it enters a cell, a steep concentration gradient is maintained. Second, if large numbers of glucose molecules are outside the cell, the rate of entry does not increase as it would if glucose were moving by simple diffusion; it reaches a peak and then levels off. This limitation on the rate of entry is attributed to there being a limited number of carrier molecules. Third, some molecules shaped almost like glucose compete with glucose for transport across the membrane. Thus, although the exact carrier molecule has not been isolated or identified, it is generally agreed that it exists.

In certain situations, glucose can also be actively transported against a concentration gradient, presumably by different carrier molecules. For example, glucose is transported into liver cells, where it is stored as glycogen, even though the concentration of glucose is higher inside the liver cells than in the bloodstream.

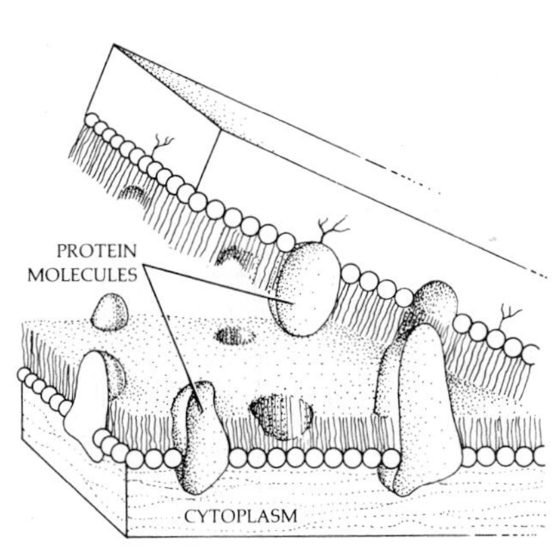

(a)

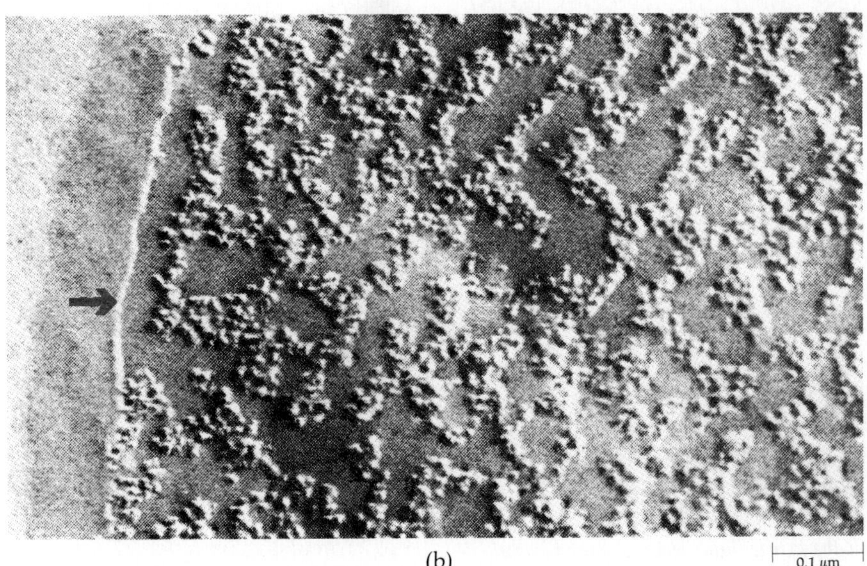

(b)

6–10
(a) *The internal surface of cellular membranes can be revealed when specimens are prepared for the electron microscope by the freeze-fracture technique. In this procedure, the cell is frozen and then fractured with a sharp blow. The fracture line generally runs between the two lipid layers of a membrane, exposing the embedded proteins. A replica of the exposed surface is prepared for examination, using the procedures described on page 92.*

(b) *The internal surface of the membrane of a red blood cell, prepared by the freeze-fracture technique. The arrow indicates the edge of the fracture line. The numerous particulate structures visible in the micrograph are integral membrane proteins, many of which are thought to be permeases.*

6–11

Mitochondria clustered near the surface of kidney cells. These cells are concerned with pumping out sodium against the concentration gradient. The mitochondria, presumably, provide the energy for this active-transport process.

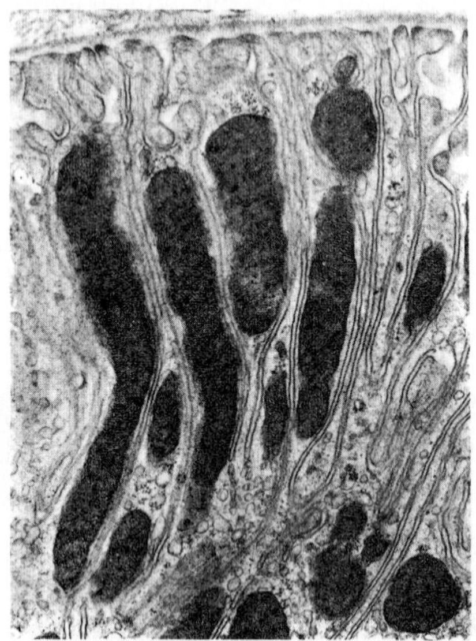

The Sodium-Potassium Pump

One of the most important and best-understood active-transport systems is the sodium-potassium pump. Most cells maintain a differential concentration gradient of sodium ions (Na^+) and potassium ions (K^+) across the cell membrane: Na^+ is maintained at a low concentration inside the cell, and K^+ is kept at a high concentration. This concentration gradient is exploited by nerve cells to propagate electrical impulses, as you will learn in Chapter 40. The sodium-potassium pump requires energy made available by a molecule known as ATP (adenosine triphosphate), which is the form in which most of a cell's ready energy currency is carried. A measure of the importance of the sodium-potassium pump to the organism is that more than a third of the ATP used by a resting animal is consumed by this one ion-pumping mechanism.

The transport of Na^+ and K^+ ions is accomplished by a carrier protein thought to exist in two alternative configurations. One configuration has a cavity opening to the inside of the cell into which an Na^+ ion can fit; the other has a cavity opening to the outside, into which a K^+ ion fits. As shown in Figure 6–12, Na^+ within the cell binds to the transport protein. At the same time, an energy-releasing reaction involving ATP results in the attachment of a phosphate group to the protein. This triggers its shift to the alternative configuration and the release of the Na^+ to the outside of the membrane. The transport protein is now ready to pick up a K^+ ion, which results in the release of the phosphate group from the protein, thus causing it to return to the first configuration and to release the K^+ ion to the inside of the cell. As you can see, this process will generate a gradient of Na^+ and K^+ ions across the membrane.

Many ingenious models have been proposed to show how carrier molecules might accept and eject their passengers. One of the first suggested that the carrier rotated, like a revolving door. More recent evidence indicates that although mem-

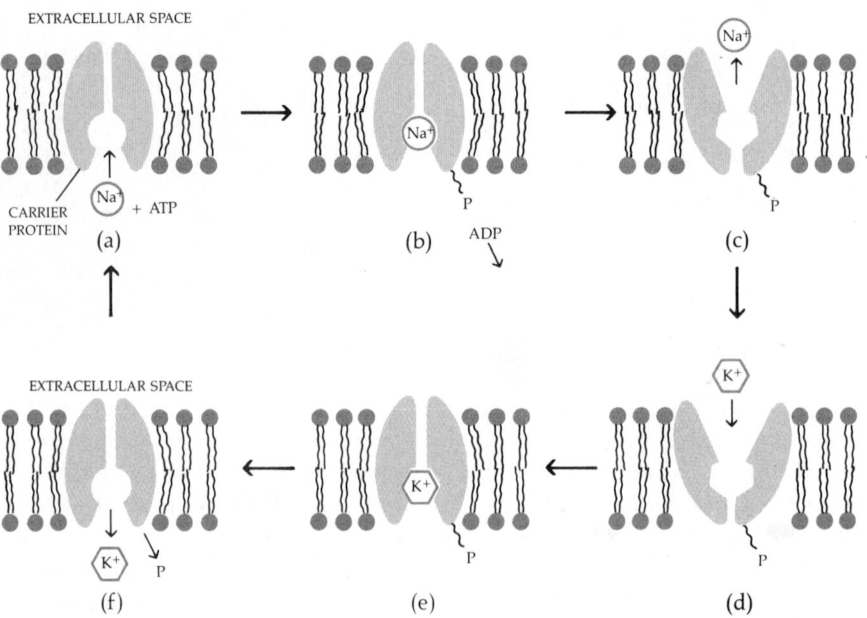

6–12

A model of the sodium-potassium pump. (a) An Na^+ ion in the cytoplasm fits precisely into the carrier protein. (b) A chemical reaction involving ATP then attaches a phosphate group (P) to the protein, releasing ADP (adenosine diphosphate). This process results in (c) a change of shape that causes the Na^+ to be released outside the cell. (d) A K^+ ion in the extracellular space is bound to the carrier protein (e), which in this form provides a better fit for K^+ than for Na^+. (f) The phosphate group is then released from the protein, inducing conversion back to the other shape, and the K^+ ion is released into the cytoplasm. The carrier protein is now ready to transport another Na^+ ion out of the cell.

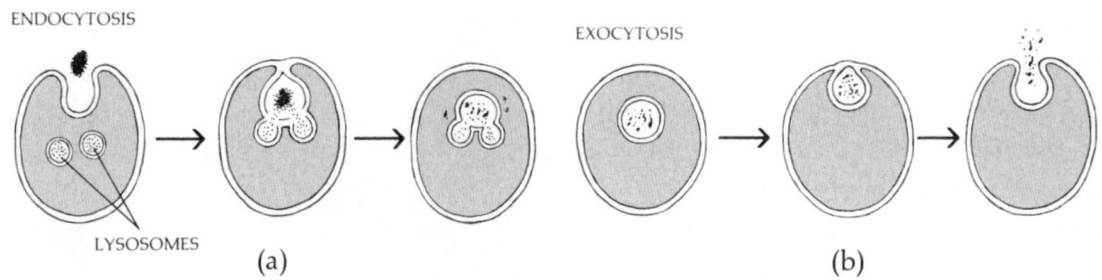

6–13
(a) *Endocytosis. Material to be taken into the cell is enveloped in a portion of the cell membrane, which pinches off to become a separate vacuole. If the material is a food item, lysosomes fuse with the vacuole, spilling their digestive enzymes into it.*

(b) *Exocytosis. Material is transported out of the cell as the vacuole membrane fuses with the cell membrane. This process is used for the secretion of substances synthesized by the cell for export, as well as for the elimination of the indigestible remains in a food vacuole.*

brane proteins can move laterally, they are not free to flip-flop through the membrane as that model required. A current model hypothesizes that the carrier molecule has a hydrophilic core, which the transported molecule is squeezed through, propelled by changes in the configuration of the protein. It is likely that cell membranes contain a large variety of carrier proteins, and they may well function in different ways.

ENDOCYTOSIS AND EXOCYTOSIS

Other types of transport processes involve vacuoles that are formed from or that fuse with the cell membrane. In *endocytosis* (Figure 6–13a), material to be taken into the cell attaches to special areas on the cell membrane and induces the membrane to bulge inward, producing a little pouch or vacuole enclosing the substance. This vacuole is released into the cytoplasm. The process can also work in reverse (Figure 6–13b). For example, many substances are exported from cells in vesicles or vacuoles formed by the Golgi bodies. The vacuoles move to the surface of the cell. When the vacuole reaches the cell surface, its membrane fuses with the membrane of the cell, thus expelling its contents to the outside. This process is *exocytosis.*

As you can see by studying Figure 6–13, the surface of the membrane facing the interior of a vacuole is equivalent to the surface facing the exterior of the cell; similarly, the surface of the vacuole membrane facing the cytoplasm is equivalent to the cytoplasmic surface of the cell membrane. Material needed for expansion of the cell membrane as a cell grows is thought to be transported, ready-made, from the Golgi bodies to the membrane by a process similar to exocytosis. There is also evidence that the portions of the cell membrane used in forming endocytotic vacuoles are returned to the membrane in exocytosis, thus recycling the membrane lipids and proteins.

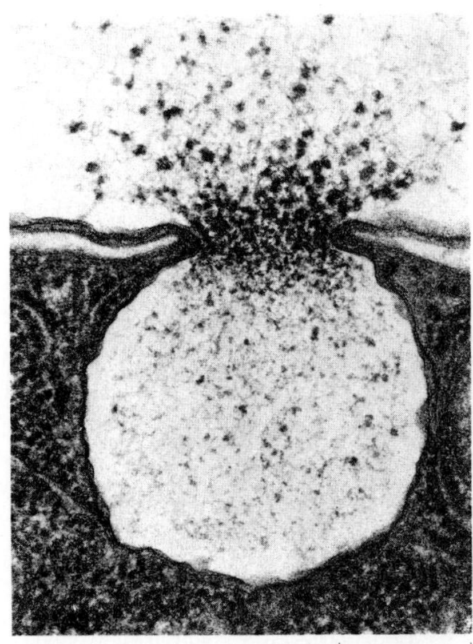

6–14
Exocytosis. A secretory vesicle, formed within the cell, discharges its contents. Notice how the membrane enclosing the vesicle has fused with the cell membrane.

When the substance to be taken into the cell in endocytosis is a solid, such as a bacterial cell, the process is usually called *phagocytosis*, from the Greek word *phage,* "to eat." Many one-celled organisms, such as amoebas, feed in this way, and white blood cells in our own bloodstreams engulf bacteria and other invaders in phagocytic vacuoles. Often lysosomes fuse with these vacuoles, emptying their enzymes into them and so digesting or destroying their contents.

The taking in of dissolved molecules, as distinct from particulate matter, is sometimes given the special name of *pinocytosis*, although it is the same in principle as phagocytosis. Pinocytosis occurs not only in single-celled organisms but also in multicellular animals. One type of cell in which it has been frequently observed is the human egg cell. As the egg cell matures in the ovary of the female, it is surrounded by "nurse cells," which apparently transmit nutrients to the egg cell, which takes them in by pinocytosis.

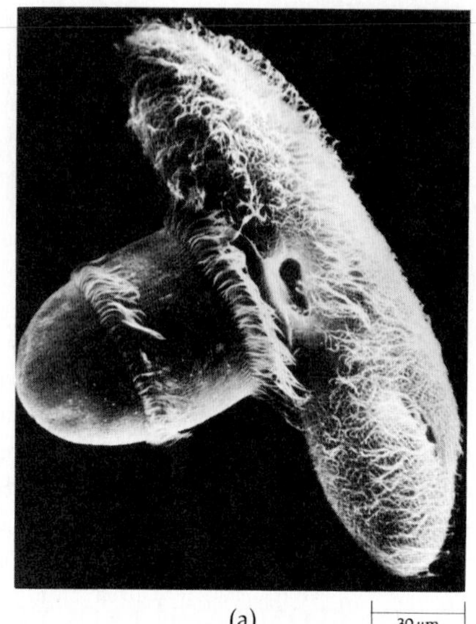

(a)

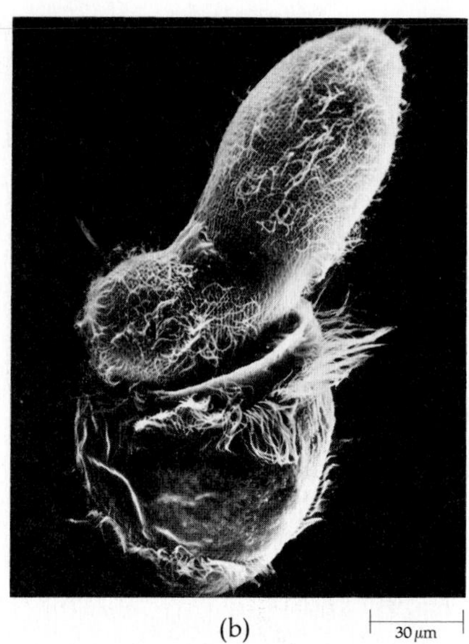

(b)

6–15
Phagocytosis of Paramecium *by* Didinium. *(See Figure 5–23 for the preamble to their encounter.)* (a) *Ingestion of the* Paramecium *has begun. The concave area just above the oral rim of the* Didinium *is the oral groove of the* Paramecium. Paramecium, *a heterotroph, feeds largely on bacteria.* (b) *Because the* Paramecium *is larger than the* Didinium, *folding helps.* (c) *The* Paramecium *is half "swallowed"; the part that is within the* Didinium *is surrounded by a membrane composed of the cell membrane of the* Didinium. *The process of compression has begun, as you can see in the tip of the* Paramecium *protruding from the oral rim of the* Didinium. *This compression is largely a matter of squeezing out water.* (d) *Once the* Paramecium *is completely inside, the cell membrane of the* Didinium *will fuse over it, forming a food vacuole. A* Didinium *can eat a dozen* Paramecium, *each larger than itself, in a single day. Moreover, the* Paramecium *must also provide the means for its own demolition: the* Didinium *apparently lacks a crucial digestive enzyme that the* Paramecium *supplies.*

Phagocytosis and pinocytosis appear superficially different from membrane-transport systems involving carrier molecules. They are fundamentally similar, however, in that all depend on the capacity of the membrane to "recognize" particular molecules. This capacity is, of course, the result of billions of years of an evolutionary process that began, as far as we are able to discern, with the formation of a fragile film around a few organic molecules, thus separating the molecules from their external environment and permitting them to maintain the particular kind of organization that we recognize as life. It is one of the many critical capacities transmitted from parent to offspring each time a cell divides, a process we shall examine in the next chapter.

SUMMARY

The cell membrane regulates the passage of materials into and out of the cell, a function that makes it possible for the cell to maintain its structural and functional integrity. This regulation depends on interaction between the membrane and the materials that pass through it.

According to the fluid-mosaic model of the membrane, cell membranes are phospholipid bilayers in which globular proteins are suspended. Some of these integral proteins act as carriers, ferrying molecules through the membrane, while others are enzymes. The two faces of the membrane differ in chemical composition. The cytoplasmic face is characterized by peripheral protein molecules, attached to the integral proteins embedded in the bilayer; the exterior face of the membrane is characterized by short carbohydrate chains attached to the protruding proteins and to some of the lipid molecules.

One of the principal substances passing into and out of cells is water. Water potential determines the direction in which water moves; that is, water moves from where the water potential is higher to where it is lower. Water movement takes place by bulk flow and diffusion.

(c) 30 μm

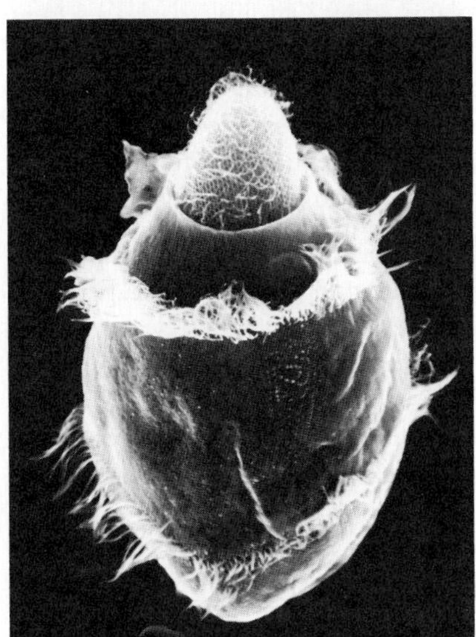

(d) 30 μm

Bulk flow is the overall movement of water molecules as a group, as when water flows in response to gravity or pressure. The circulation of blood through the human body is an example of bulk flow.

Diffusion involves the random movement of molecules and results in net movement down a concentration gradient. Diffusion is most efficient when the surface area is large in relation to volume, when the distance involved is small, and when the concentration gradient is steep. By their metabolic activities, cells maintain steep concentration gradients of many substances. The rate of movement of substances within cells is also increased by cytoplasmic streaming. In several anatomical systems of multicellular organisms, countercurrent exchange maintains a steep concentration gradient over a large surface area, thus maximizing the rate of diffusion.

Osmosis is the diffusion of water through a membrane that permits the passage of water but inhibits the movement of most solutes; such a membrane is called a selectively permeable membrane. In the absence of other forces, the net movement of water in osmosis is from a region of lower solute concentration (a hypotonic medium), and therefore of higher water potential, to one of higher solute concentration (a hypertonic medium), and so of lower water potential. Turgor in plant cells is a consequence of osmosis.

Molecules cross the cell membrane by diffusion or are transported by carrier proteins embedded in the membrane. If the movement is driven by the concentration gradient, the process is known as facilitated diffusion. If the movement requires the expenditure of energy by the cell, it is known as active transport. Active transport can move substances against their concentration gradients. One of the most important active-transport systems is the sodium-potassium pump, which maintains sodium ions at a low concentration and potassium ions at a high concentration in the cytoplasm.

Controlled movement into and out of a cell may also occur by endocytosis or exocytosis, in which substances are transported in vacuoles composed of portions of the cell membrane. Endocytosis of solids is called phagocytosis and of dissolved molecules, pinocytosis.

QUESTIONS

1. Distinguish among the following: bulk flow/diffusion/osmosis; water potential/hydrostatic pressure/osmotic potential; hypotonic/hypertonic/isotonic; facilitated diffusion/active transport; endocytosis/exocytosis; phagocytosis/pinocytosis.

2. Describe the structure of the cell membrane. How do the two faces of the membrane differ? What is the functional significance of these differences?

3. What is a concentration gradient? How does a concentration gradient affect diffusion? How does a concentration gradient affect osmosis?

4. When diffusion of dye molecules in a tank of water is complete, random movement of molecules continues (as long as the temperature remains the same). However, net movement stops. How do you reconcile these two facts?

5. Why is diffusion more rapid in gases than in liquids? Why is it more rapid at higher temperatures than at lower temperatures?

6. How does countercurrent exchange affect diffusion?

7. Three funnels have been placed in a beaker containing a solution (see the figure below). What is the concentration of the solution? Explain your answer.

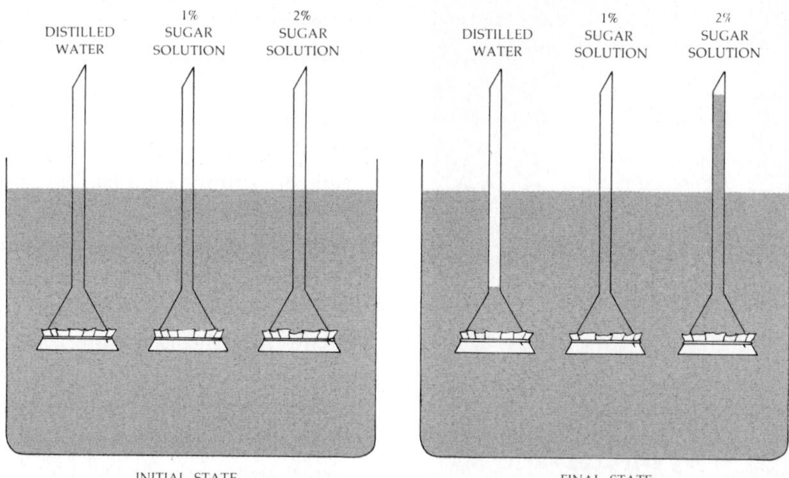

8. Imagine a pouch with a selectively permeable membrane containing a saltwater solution. It is immersed in a dish of fresh water. Which way will the water move? If you add salt to the water in the dish, how will this affect water movement? What living systems exist under analogous conditions? How do you think they maintain water balance?

9. When you forget to water your house plants, they wilt and the leaves (and sometimes the stems) become very limp. What has happened to the plants to cause this change in appearance and texture? Within a few hours after you remember to water your plants, they resume their normal, healthy appearance. What has occurred within the plants to cause this restoration? Sometimes, if you wait too long to water your plants, they never revive. What do you suppose has happened?

10. What limits the passage of water and other polar molecules and ions through the cell membrane? How do such molecules get into and out of the cell? Describe three possible routes.

11. In what three ways does active transport differ from diffusion?

12. Justify the conclusion that differences in ion concentration between cells and their surroundings (see Figure 6–1) indicate that cells regulate the passage of materials across membranes.

13. In Figure 6–15d, the *Paramecium* is sinking below the oral rim of *Didinium*. What will happen next? Complete the scenario, giving as many details as possible. (You might want to end your account with the fact that *Didinium* divides once for every two *Paramecium* consumed.)

CHAPTER 7

How Cells Divide

Cell division is the process by which cellular material is divided between two new daughter cells. In one-celled organisms, it increases the number of individuals in the population. In many-celled plants and animals, it is the means by which the organism grows, starting from one single cell, and also by which injured or worn-out tissues are replaced and repaired. A cell first grows by assimilating materials from its environment and synthesizing these materials into new structural and functional molecules. When a cell reaches a certain critical size and metabolic state, it divides. The two daughter cells, each of which has received about half of the mass of the parent cell, then begin growing again. A bacterial cell may divide every 6 minutes. In a one-celled eukaryote such as *Paramecium,* cell division may occur every few hours.

The new cells produced are structurally and functionally similar both to the parent cell and to one another. They are similar, in part, because each new cell receives about half of the parent cell's cytoplasm and organelles. More important, in terms of structure and function, each new cell inherits an exact replica of the hereditary information of the parent cell.

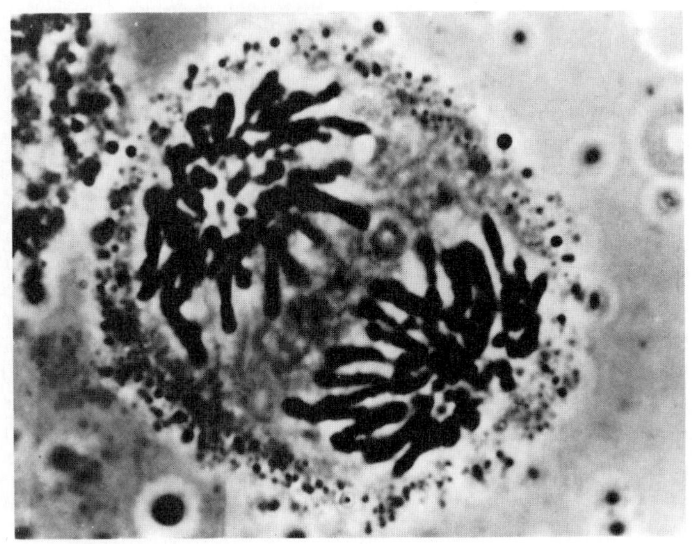

7–1
A human cell divides. The long, dark bodies are chromosomes, the carriers of the hereditary information. The chromosomes have replicated and moved apart. Each set of chromosomes is an exact copy of the other. Thus the two new cells, the daughter cells, will contain the same hereditary material.

7–2

Single-celled eukaryotic organisms typically reproduce by simple cell division. In this scanning electron micrograph of the ciliated protozoan Opisthonecta, *the separation of the two daughter cells is almost complete. Each cell has received not only an exact replica of the parent cell's hereditary information but also approximately half of its organelles and cytoplasm.*

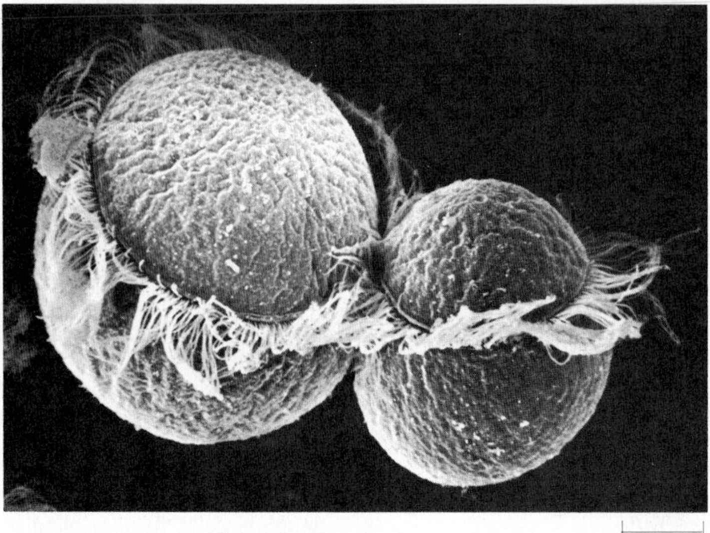

15 μm

CELL DIVISION IN PROKARYOTES

The distribution of exact replicas of the hereditary information is comparatively simple in prokaryotic cells. In such cells, the hereditary material is in the form of a single, long, circular molecule of DNA (deoxyribonucleic acid, about which we shall be hearing considerably more in Section 3). This molecule, the cell's chromosome, is replicated before cell division. According to present evidence, each of the two daughter chromosomes is attached to a different spot on the interior of the cell membrane. As the membrane elongates, the chromosomes move apart (Figure 7–3). When the cell has approximately doubled in size and the chromosomes are separated, the cell membrane pinches inward, and a new cell wall forms that separates the two new cells and their chromosome replicas.

7–3

(a) *A diagram of the attachment of bacterial chromosomes, indicating the possible role of the mesosome (an inward fold of the cell membrane) in ensuring the distribution of the chromosomes in dividing cells.* (b) *A bacterial chromosome attached to a mesosome on the bacterial cell membrane.*

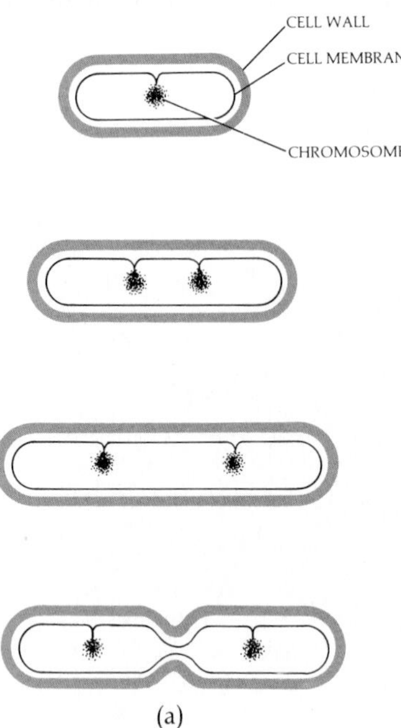

(a)

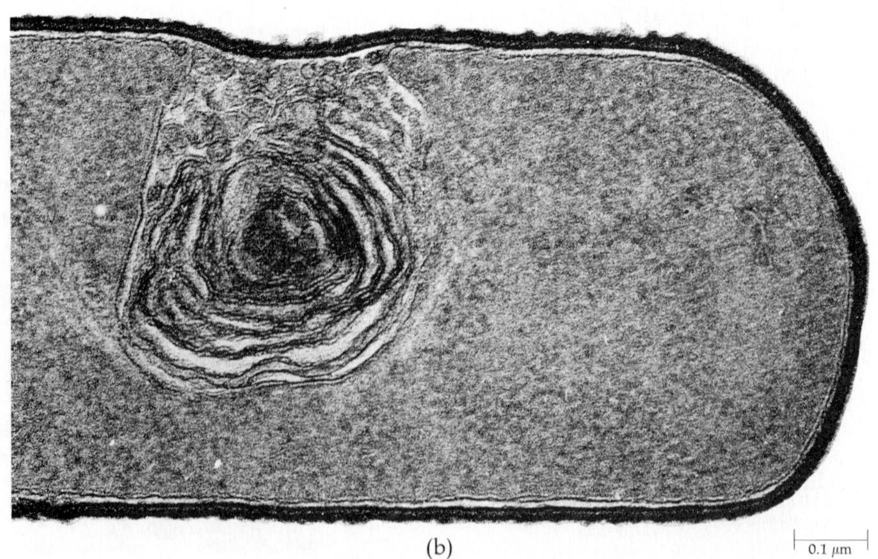

(b)

0.1 μm

CELL DIVISION IN EUKARYOTES

In eukaryotic cells, the problem of exactly dividing the genetic material is much more complex. A typical eukaryotic cell contains about a thousand times more DNA than a prokaryotic cell, and this DNA is associated with protein, forming a number of distinct chromosomes. For instance, human cells have 46 chromosomes, each different from the others; at cell division, each daughter cell has to receive one, and only one, of each of the 46. Moreover, eukaryotic cells contain a variety of organelles that must also be apportioned between the daughter cells.

The solutions to these problems are, as you will see, ingenious and elaborate. In a series of steps called, collectively, *mitosis*, a complete set of chromosomes is allocated to each of the daughter cells. Mitosis is usually followed by *cytokinesis*, a process that separates the two daughter cells, each of which contains not only a full chromosome complement but also approximately half of the cytoplasm and organelles of the parent cell.

Although mitosis and cytokinesis are the culminating events of cell division in eukaryotes, they represent only two stages of a larger process.

THE CELL CYCLE

Dividing cells pass through a regular sequence of cell growth and division, known as the *cell cycle* (Figure 7–4). The cycle consists of five major phases: G_1, S, G_2, mitosis, and cytokinesis. Completion of the cycle requires varying periods of time from a few hours to several days, depending on both the type of cell and external factors, such as temperature or available nutrients.

Before a cell can begin mitosis and actually divide, it must replicate its DNA, synthesize more of the proteins associated with the DNA in the chromosomes, produce a supply of organelles adequate for two daughter cells, and synthesize the structures needed to carry out mitosis and cytokinesis. These preparatory processes occur during the G_1, S, and G_2 phases of the cell cycle, which are known collectively as *interphase*.

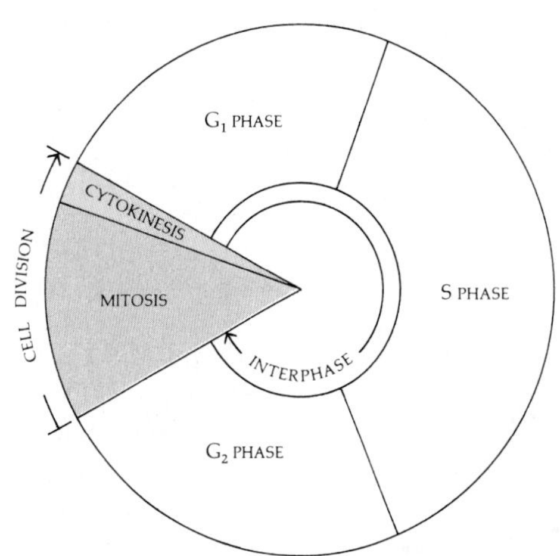

7–4
The cell cycle. Cell division, which consists of mitosis (shown in color) and cytokinesis (gray), takes place after the completion of the three preparatory phases that constitute interphase. During the S (synthesis) phase, the chromosomal material is replicated. Separating cell division and the S phase are two G phases. The first of these (G_1) is a period of general growth and replication of cytoplasmic organelles. During the second (G_2), structures directly associated with mitosis are assembled. After the G_2 phase comes mitosis (the division of the nucleus), which is usually followed immediately by cytokinesis (the division of the cytoplasm). In cells of different species or of different tissues within the same organism, the different phases occupy different proportions of the total cycle.

The key process of DNA replication occurs during the S (synthesis) phase of the cell cycle, a time in which many of the DNA-associated proteins are also synthesized. G (gap) phases precede and follow the S phase; during the G phases, no DNA synthesis can be detected in the cell.

The G_1 phase, which follows cytokinesis and precedes the S phase, is a period of intensive biochemical activity. The cell doubles in size, and its enzymes, ribosomes, mitochondria, and other cytoplasmic molecules and structures approximately double in number. Some of the cellular structures can be synthesized entirely *de novo* ("from scratch") by the cell; these include microtubules, microfilaments, and ribosomes, all of which are composed, at least in part, of protein subunits. Membranous structures, such as the nuclear envelope, Golgi bodies, lysosomes, vacuoles, and vesicles, are all apparently derived from the endoplasmic reticulum, which is renewed and enlarged by the addition of phospholipid and protein molecules. Mitochondria and chloroplasts are produced only from previously existing mitochondria and chloroplasts or plastids (page 113). Each of these organelles also has its own chromosome, which is organized much like the single chromosome of the bacterial cell. (These are two of the reasons that many biologists hypothesize that these organelles originated as separate organisms and then took up a new way of life inside early eukaryotic cells, more than a billion years ago. This question will be discussed further in Chapter 21.)

7–5
Cross section of the tip of an onion root. The cells in this region undergo repeated divisions, providing new cells for the growth of the root. Different cells in this tissue section are in different phases of the cell cycle. The three cells that are significantly larger than the others and that contain dark, rodlike structures are in mitosis. One cell, near the center of the micrograph, is about midway through cytokinesis, and the two cells below it have almost completed cytokinesis.

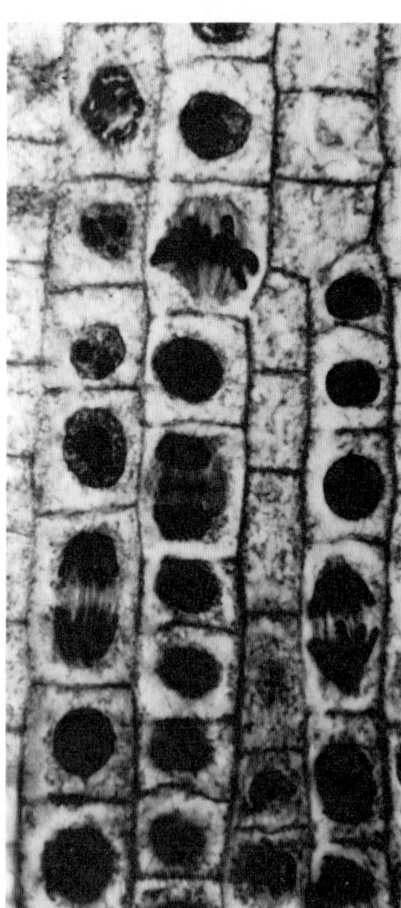

When cells stop growing, as a result, for example, of depletion of nutrients or contact inhibition (page 119), they stop in the G_1 phase. Therefore, it is concluded that substances are synthesized during the G_1 phase that either inhibit or stimulate the S phase and the rest of the cell cycle, thus determining whether or not cell division will occur. Further knowledge of the control mechanisms involved would not only be interesting biologically but also might be of great importance in the control of cancer. As we noted previously, cancer cells differ from normal cells largely in that they keep on dividing at the expense of the host tissues.

Once the synthesis of DNA in the S phase has begun, the cell moves steadily through the remaining phases of the cell cycle. During the G_2 phase, which follows the S phase and precedes mitosis, the cell assembles the special structures required for the separation of the chromosomes during mitosis and of the two daughter cells during cytokinesis.

Dividing cells of different species show characteristic variations in the general pattern of the cell cycle. In the common bean, for example, the complete cycle requires about 19 hours, of which 7 hours are taken up by the S phase; G_1 and G_2 are of equal length (about 5 hours each), and mitosis lasts 2 hours. By contrast, in mouse fibroblast cells, the cell cycle is approximately 22 hours, of which mitosis is less than 1 hour, S is almost 10 hours, G_1 is 9 hours, and G_2 is a little more than 2 hours.

Some cell types pass through successive cell cycles throughout the life of the organism. This group includes the one-celled organisms and certain cells in growth centers of both plants and animals. An example is the cells in the human bone marrow that give rise to red blood cells. The average red blood cell lives only about 120 days, and there are about 25 trillion (2.5×10^{13}) of them in an adult. To maintain this number, about 2.5 million new red blood cells must be produced by cell division each second. At the other extreme, some highly specialized cells, such as nerve cells, lose their capacity to replicate once they are mature. A third group of cells retains the capacity to divide but does so only under special circumstances.

7-6
A condensed chromosome. The chromosomal material was replicated during the S phase of the cell cycle, and each chromosome now consists of two identical parts, called chromatids. The centromere, the constricted area at the center, is the site of attachment of the two chromatids.

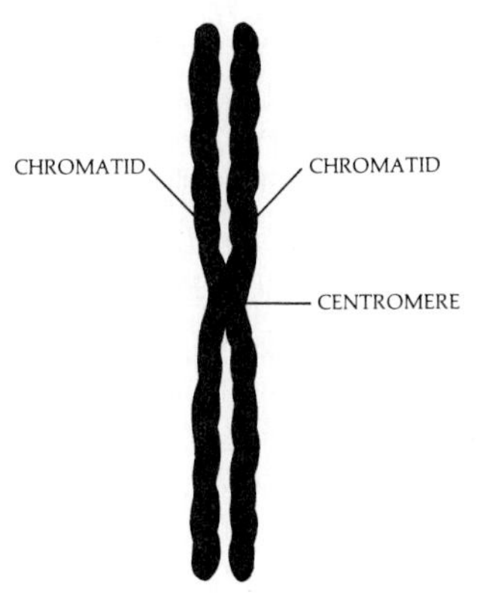

Cells in the human liver, for example, do not ordinarily divide, but if a portion of the liver is removed surgically, the remaining cells (even if as few as a third of the total remain) continue to replicate themselves until the liver reaches its former size. Then they stop. All told, about 2 trillion (2×10^{12}) cell divisions occur in an adult human every 24 hours, in other words, about 25 million per second.

MITOSIS

When cells are in the interphase portions of the cell cycle, the chromosomal material is dispersed and is visible, if at all, only as threadlike strands, the chromatin. As mitosis begins, the chromatin slowly coils and condenses into a compact form; this condensation appears to be necessary for the complex movements and separation of the chromosomes during mitosis.

When the condensed chromosomes become visible under the light microscope, each consists of two replicas, called *chromatids* (Figure 7-6), and the two identical chromatids remain joined together like a pair of Siamese twins. Their point of attachment is a constricted area common to both chromatids, the *centromere*. Within this constricted area are disk-shaped protein structures, the *kinetochores*.

The Phases of Mitosis

The function of mitosis is to maneuver the replicated chromosomes so that each new cell gets a full complement—one of each.

The process of mitosis is conventionally divided into four phases: prophase, metaphase, anaphase, and telophase; of these, prophase is usually by far the longest. If a mitotic division takes 10 minutes (which is about the minimum time required), during about six of these minutes the cell will be in prophase. The schematic drawings that follow show mitosis as it takes place in an animal cell.

During interphase, little can be seen in the nucleus. In early *prophase*, however, the chromatin condenses, and the individual chromosomes begin to become visible. Each chromosome consists of two duplicate chromatids pressed closely together longitudinally and connected at the centromere. In the cells of most organisms (higher plants are the principal exception), two centriole pairs can be seen at one side of the nucleus, outside the nuclear envelope. Each pair consists of one mature centriole and a smaller, newly formed centriole lying at a right angle to the first. The cell becomes more spheroid and the cytoplasm more viscous at this stage.

7-7
Early prophase in an onion cell. The chromosomes have begun to condense. Because of the threadlike appearance of the chromosomes at this stage, the process of mitosis derives its name from mitos, *the Greek word for "thread."*

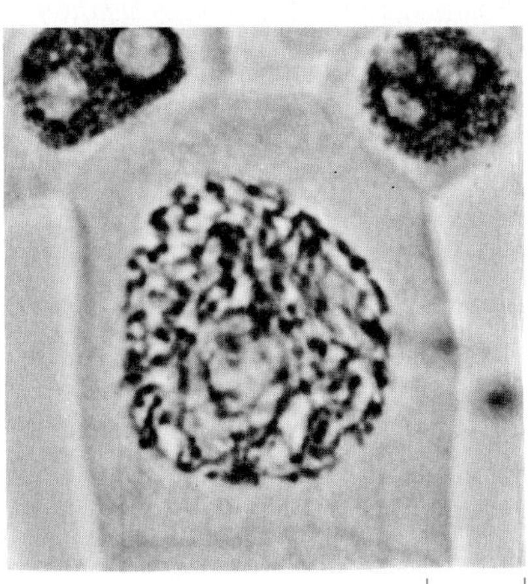

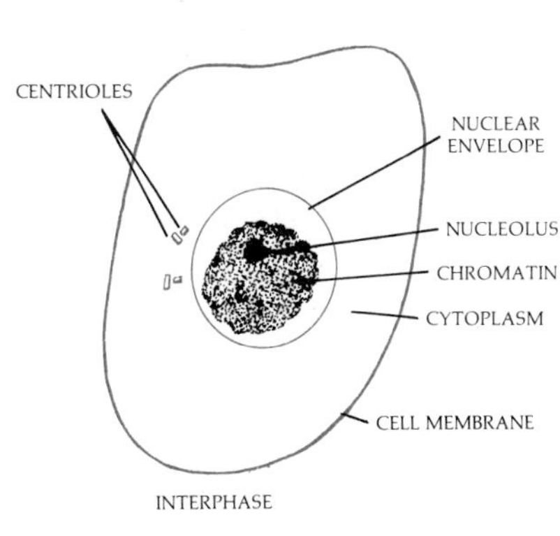

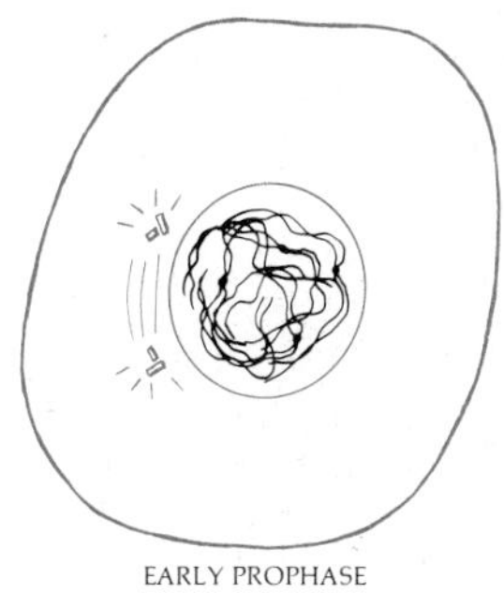

During prophase, the centriole pairs move apart. Between the centriole pairs, forming as they separate (or perhaps separating them as they form), are *spindle fibers* made up of microtubules (see page 70) and other proteins. In those cells that have centrioles, additional fibers, known collectively as the *aster*, radiate outward from the centrioles. By this time the nucleoli usually have disappeared from view. The nuclear envelope breaks down as the chromosomes condense. At the end of prophase, the chromosomes are fully condensed and are no longer separated from the cytoplasm.

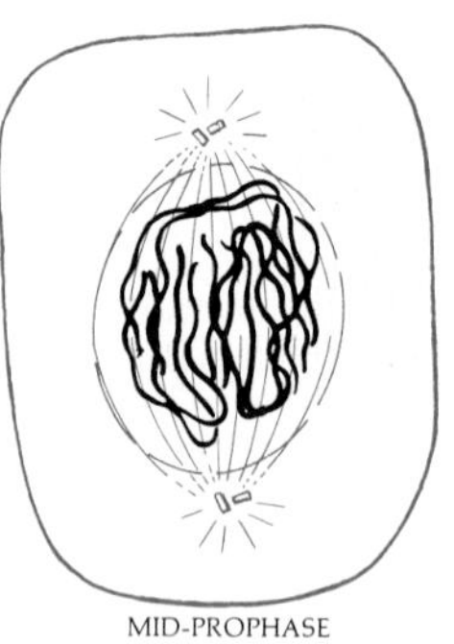

MID-PROPHASE

By the end of prophase, the centriole pairs are at opposite ends of the cell, and the members of each pair are of equal size. The spindle is fully formed. It is a three-dimensional football-shaped structure consisting of three groups of microtubules: (1) astral rays, (2) continuous fibers reaching from pole to pole, and (3) shorter fibers that are attached to the kinetochores of each sister chromatid.

During early *metaphase* the chromatid pairs move back and forth within the spindle, apparently maneuvered by the spindle fibers, as if they were being tugged first toward one pole and then the other. Finally they become arranged precisely at the midplane (equator) of the cell. This marks the end of metaphase.

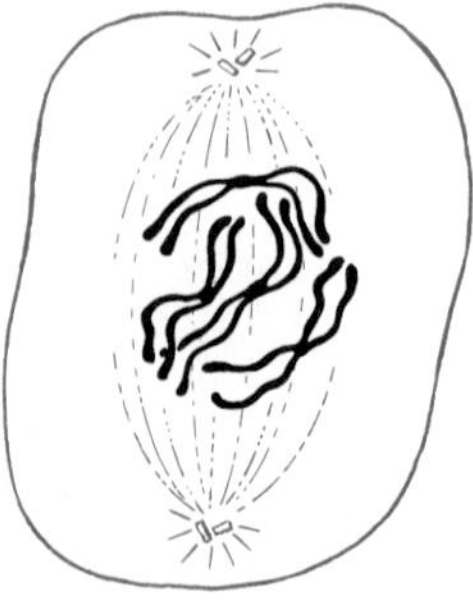

EARLY METAPHASE

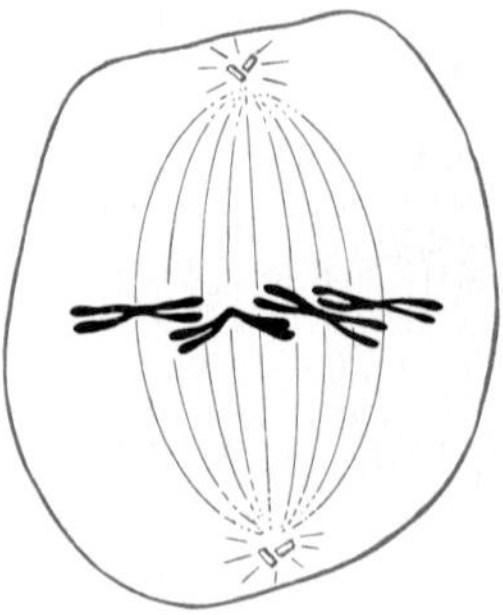

METAPHASE

7-8
A differential-interference micrograph of middle anaphase in a cell from the seed of a South African lily. The chromosomes have moved halfway toward the poles.

At the beginning of *anaphase*, the centromeres separate simultaneously in all the chromatid pairs. The chromatids of each pair then move apart, each chromatid becoming a separate chromosome, each apparently drawn toward the opposite pole by the spindle fibers. The centromeres move first, while the arms of the chromosomes seem to drag behind. As anaphase continues, the two identical sets of newly separated chromosomes move toward the opposite poles of the spindle. Anaphase is the most rapid portion of mitosis.

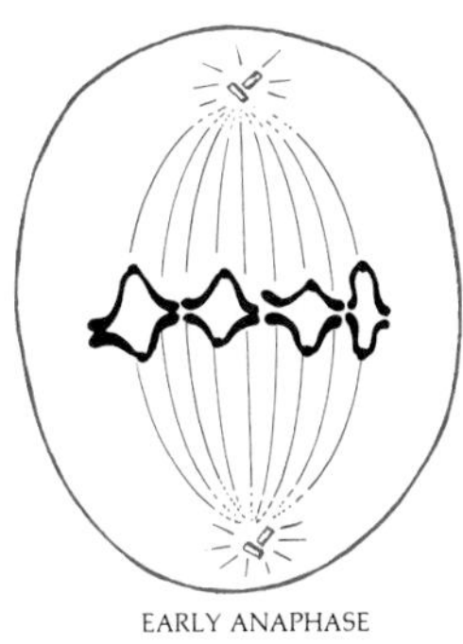

EARLY ANAPHASE

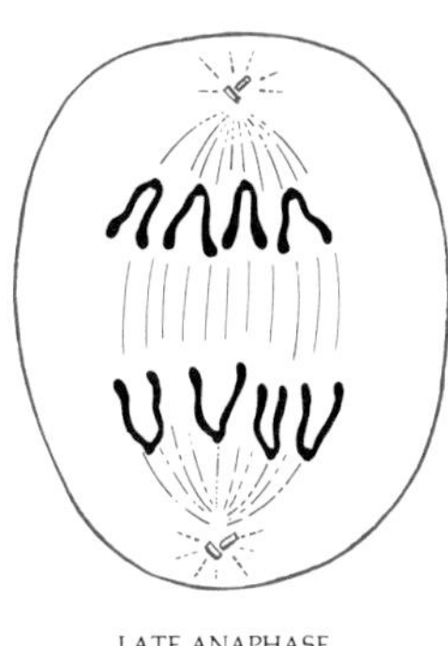

LATE ANAPHASE

By the beginning of *telophase*, the chromosomes have reached the opposite poles and the spindle disperses into tubulin dimers, subunits of the globular protein building blocks that make up its microtubules (see Figure 3-25a on page 70). During late telophase, nuclear envelopes form around the two sets of chromosomes, which once more become diffuse. In each nucleus, the nucleoli reappear. Often, a new centriole begins to form adjacent to each of the previous ones. Replication of the centrioles continues during the subsequent cell cycle, so that each daughter cell has two centriole pairs by prophase of the next mitotic division.

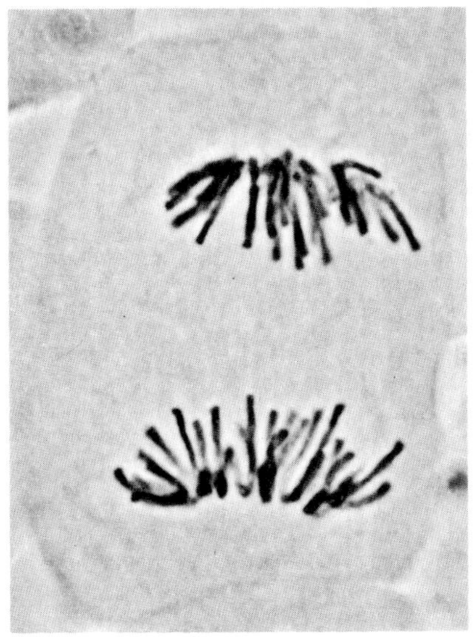

7-9
Early telophase in an onion cell.

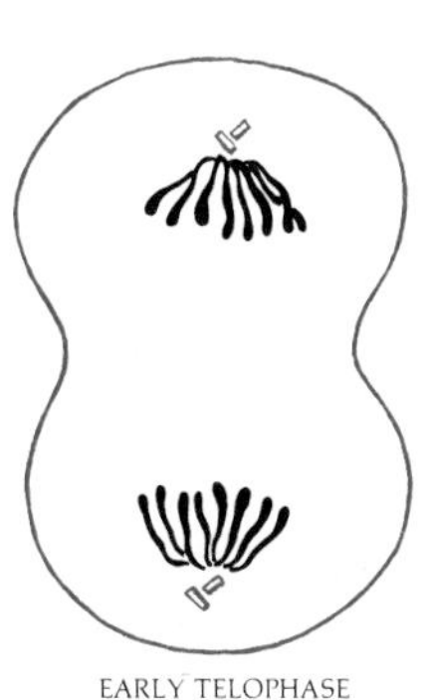

EARLY TELOPHASE

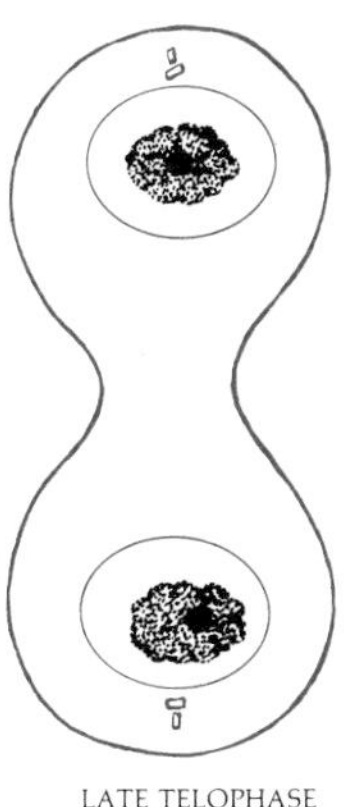

LATE TELOPHASE

7-10

Mitosis in the embryonic cells of a whitefish. (a) Prophase. The chromosomes have become visible, the nuclear envelope has broken down, and the spindle apparatus has formed. (b) Metaphase. The chromatid pairs, perhaps guided by the spindle fibers, are lined up at the equator of the cell. Some of them appear to have begun to separate. (c) Anaphase. The two sets of chromosomes are moving apart. (d) Telophase. The chromosomes are completely separated, the spindle apparatus is disappearing, and a new cell membrane is forming that will complete the separation of the two daughter cells.

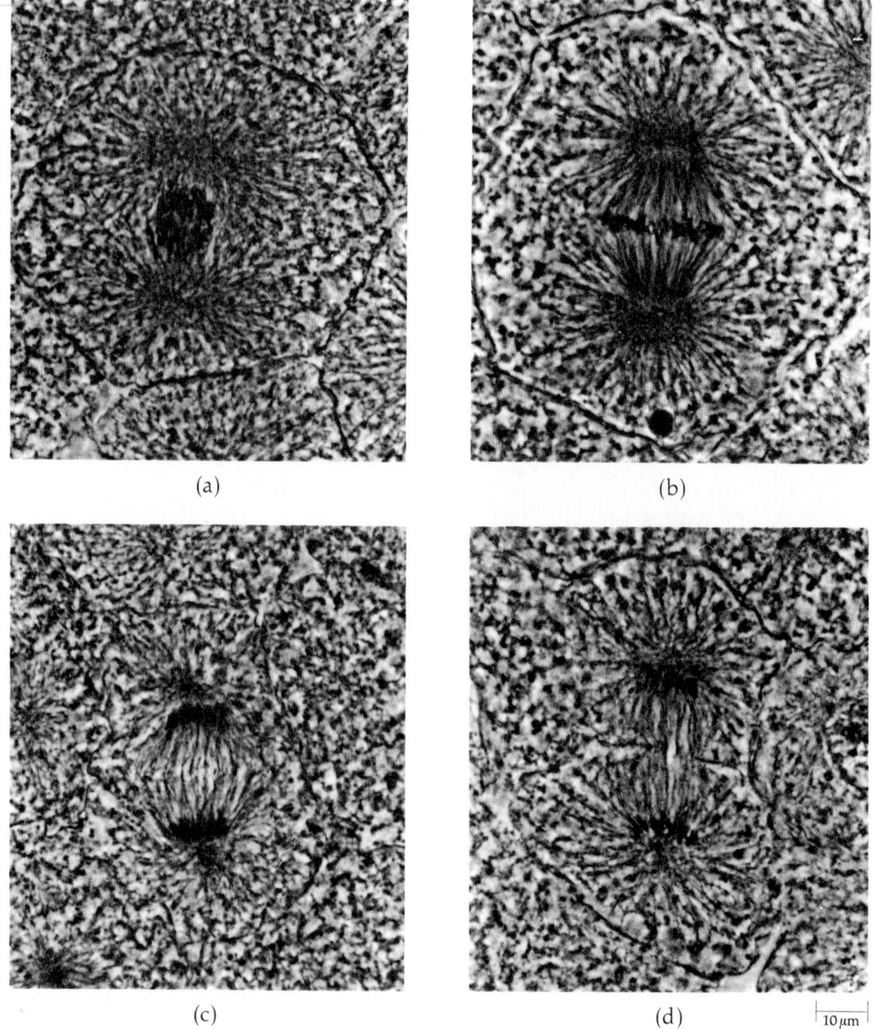

The Spindle

The spindle (Figure 7-11a), as we noted previously, is composed largely of microtubules, some of which stretch from pole to pole and some of which are attached to the chromosomes at their kinetochores (Figure 7-11b). Although the spindle fibers attached to the kinetochores lengthen during prophase and then shorten during anaphase, they do not appear to get thinner or thicker. This suggests that they do not contract but that material is added to or removed from the spindle fiber as the spindle changes shape. This hypothesis is supported by the observation that the cytoskeletal microtubules that radiate outward from the nucleus of a nondividing cell are disassembled at the beginning of mitosis, providing tubulin for the spindle. Following mitosis and the dispersion of the spindle, the cytoskeletal microtubules reassemble. This borrowing from the cytoskeleton is probably responsible for the characteristic rounded appearance of cells during mitosis.

7-11

(a) *This micrograph of the spindle in a wheat cell illustrates the spindle's three-dimensional quality. The gray streaks are the spindle fibers. The chromosomes are the large dark bodies near the equator of the spindle.*

(b) *In this electron micrograph, spindle fibers can be seen extending to the kinetochores of a green alga. The dark material is the metaphase chromosome, most of which is out of the plane of the thin section prepared for this micrograph. The kinetochores are the two disk-shaped areas at either side of the chromosome.*

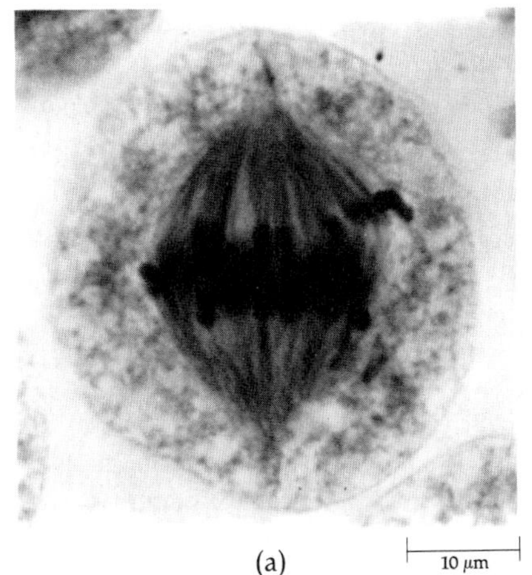

(a) 10 μm

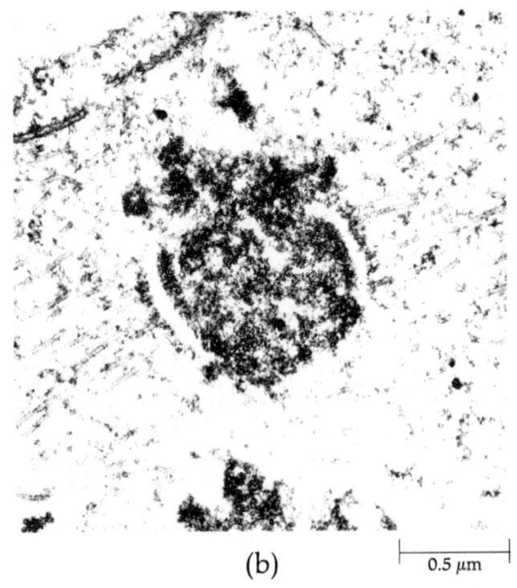

(b) 0.5 μm

The Centrioles

Basal bodies and centrioles are the same structure used, perhaps, for different purposes. Basal bodies organize the microtubules of flagella and cilia, and centrioles have long been thought to play a role in organizing the microtubules of the spindle fibers—spinning them out somewhat as a spider spins out silk. A striking example of the interchangeability of basal bodies and centrioles is provided by little *Chlamydomonas.* At the beginning of mitosis, its two flagella are reabsorbed by the cell and the basal bodies move near the nucleus; during mitosis, they behave exactly like centrioles, appearing to organize the spindle. When mitosis is complete, they migrate to the anterior ends of the daughter cells, giving rise to new flagella. Despite this evidence for the role of the centrioles in spindle formation, plant cells that do not have centrioles or basal bodies also form spindles with microtubules. No asters appear, however, and the spindles are not as sharply focused at the poles as they are in cells with centrioles. In some animal cells, it is possible to remove the centrioles from the cells—yet spindle formation proceeds normally.

The explanation for these seemingly contradictory observations may lie in a densely staining region seen around the centrioles in many electron micrographs.

7-12

(a) *Cross section of a centriole. Centrioles and basal bodies are structurally identical. It is thought that basal bodies organize microtubules into cilia and flagella and that, similarly, centrioles may organize the microtubules of spindle fibers. However, plant cells that do not have centrioles are also able to form spindles.* (b) *A centriole pair. Centrioles are formed either from preexisting centrioles, with the newly formed centriole appearing at right angles to the previously existing one, or from basal bodies.*

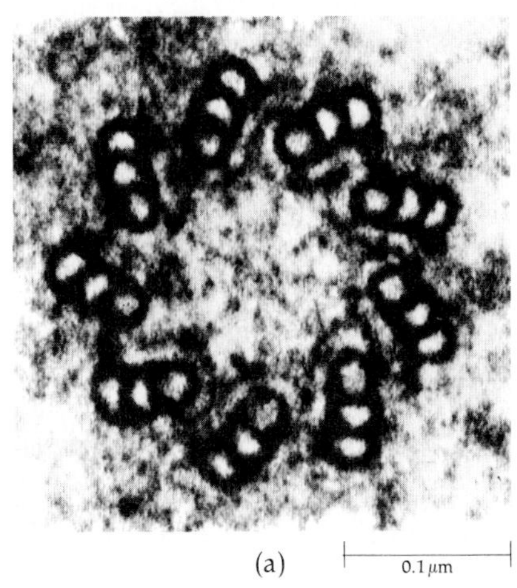

(a) 0.1 μm

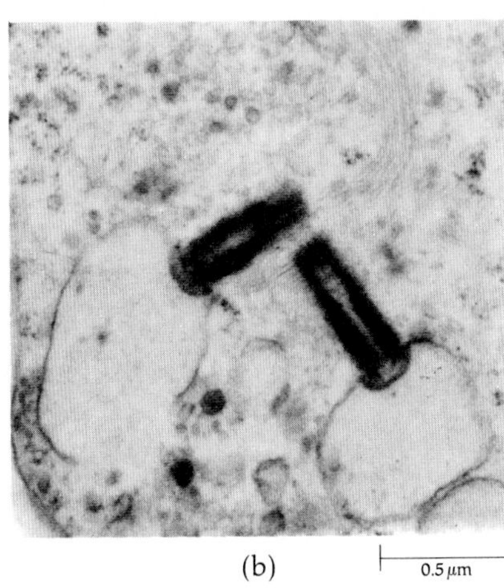

(b) 0.5 μm

Such a densely staining region is also present in cells without centrioles and is the area from which the spindle fibers develop. The material in this region, rather than the centrioles themselves, may organize the microtubules of the spindle fibers. It has also been suggested that the spindles, instead of forming from the centrioles, serve to separate the centrioles, pushing them apart and so ensuring that each daughter cell receives an adequate supply of basal bodies from which to construct flagella or cilia.

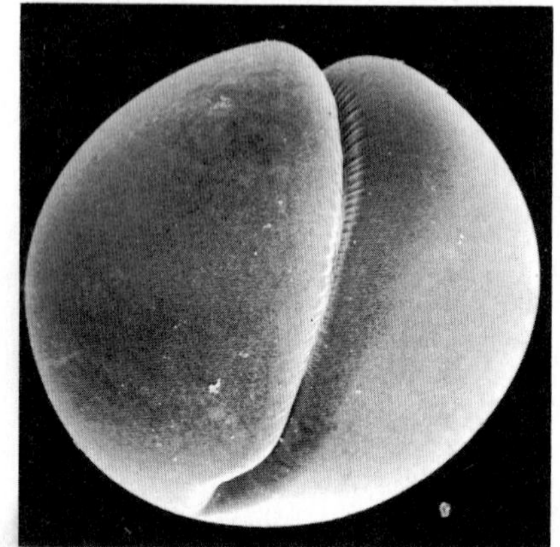

(a)

(b)

7-13

Cytokinesis in an animal cell, the egg of a frog. (a) *The egg divides in two.* (b) *Close-up showing the constriction furrows.*

CYTOKINESIS

Cytokinesis, the division of the cytoplasm, usually but not always accompanies mitosis, the division of the nucleus. The visible process of cytokinesis generally begins during telophase of mitosis, and it usually divides the cell into two nearly equal parts. The cleavage always occurs at the midline of the spindle, although the spindle does not seem to be directly involved in the division of the cytoplasm. There is evidence, however, that the microtubules of the spindle play a role in positioning the other structures that are responsible for the actual division of the cytoplasm. If, for example, the spindle is pushed to the side of the cell shortly after its formation is complete—so that its equator extends only halfway across the cell—only the cytoplasm of that half of the cell is ultimately divided in cytokinesis. The result is a two-lobed cell, with a daughter nucleus in each lobe.

Cytokinesis differs in some respects in plant and animal cells. In animal cells, during early telophase, the cell membrane begins to constrict along the circumference of the cell in the area where the equator of the spindle used to be. At first, a furrow appears on the surface, and this gradually deepens into a groove (Figure 7-13). Eventually the connection between the daughter cells dwindles to a slender thread, which soon parts. Actin microfilaments (page 114), seen in large numbers near the furrows, are thought to play a role in the constriction. They are believed to act as a sort of "purse string," gathering in the membrane of the parent cell at its midline, thus pinching apart the two daughter cells.

In plant cells, the cytoplasm is divided at the midline by the formation of a *cell plate*, which is formed from a series of vesicles produced from Golgi bodies (Figure 7-14). The vesicles eventually fuse to form a flat, membrane-bound space, the cell plate. As more vesicles fuse, the edges of the growing plate fuse with the membrane of the cell. In this way, a space is established between the two daughter cells, completing their separation. This space ultimately becomes impregnated with pectins, forming the middle lamella (see page 100). Each new cell then constructs its own cell wall, laying down cellulose and other polysaccharides against its membrane.

When cell division is complete, two daughter cells are produced, smaller than the parent cell but otherwise indistinguishable from it and from each other.

Although we have focused in this chapter on the processes directly involved in cell division, it is important to realize that the other activities characteristic of the living cell are also occurring throughout the cell cycle. The cell is synthesizing the macromolecules necessary to maintain its structure, it is degrading other molecules, it is regulating both the internal movement of substances and the movement of substances between the cell interior and the exterior environment, and it is responding to a variety of stimuli. All of these activities—as well as cell division itself—require a steady expenditure of energy by the cell. In the next section, we shall see how cells provide themselves with this essential energy.

7–14

In plants, the final separation of the two cells is effected by the formation of a structure known as the cell plate. Vesicles appear across the equatorial plane of the cell and gradually fuse, forming a flat membrane-bound space, the cell plate, which extends outward until it reaches the wall of the dividing cell. The large, dark forms on either side of the micrograph are the chromosomes.

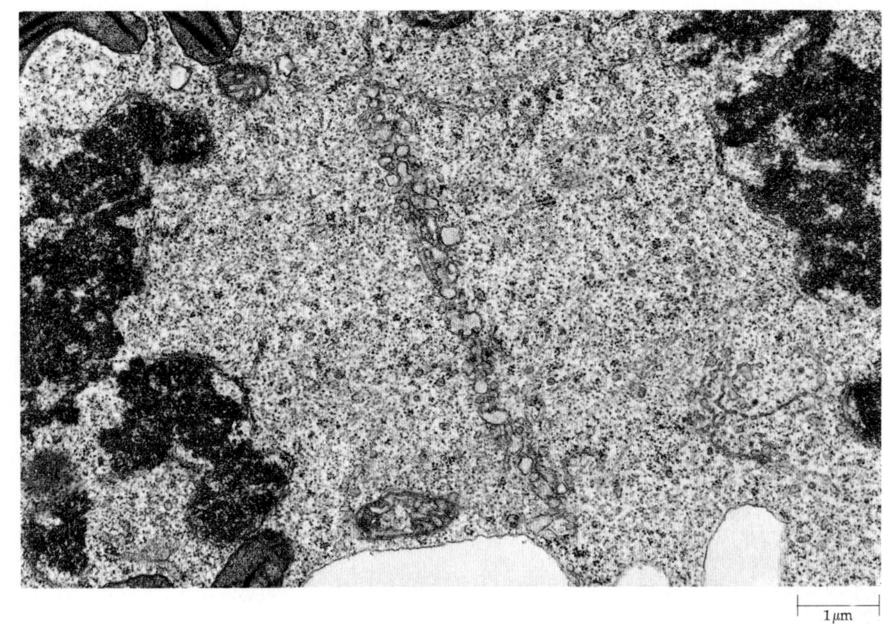

7–15

Mitosis in a plant cell with four chromosomes. Note that a spindle forms, although no centrioles or asters are visible. Cytokinesis takes place with the formation of a cell plate.

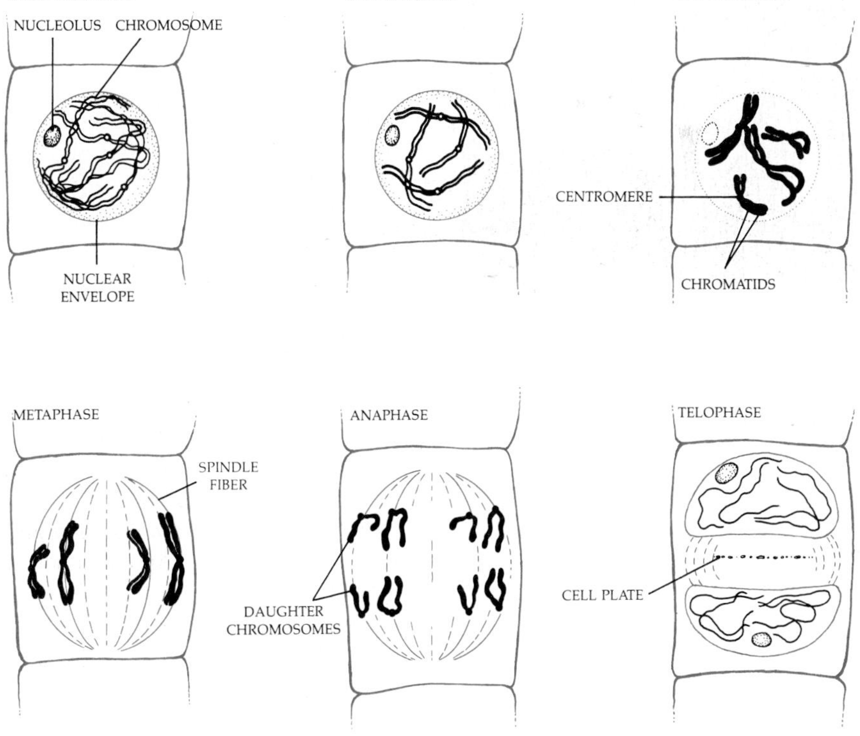

SUMMARY

Cell division in prokaryotes is a relatively simple process, in which two daughter chromosomes are attached to different spots on the interior of the cell membrane. As the membrane elongates, the chromosomes are separated. The cell membrane then pinches inward, and a new cell wall forms, completing the division of the daughter cells.

Cell division is a more complex process in eukaryotes, which contain a vast amount of genetic material, organized into a number of different chromosomes. Dividing cells pass through a regular sequence of cell growth and cell division known as the cell cycle. The cycle consists of a G_1 phase, during which the cytoplasmic material doubles; an S phase, during which the chromosomes are replicated; a G_2 phase, during which the tubulin of the spindle and other structures directly involved with mitosis are synthesized; mitosis, during which the replicated chromosomes are apportioned between the two daughter cells; and cytokinesis, during which the cytoplasm is divided, separating the two daughter cells. The first three phases of the cell cycle are known, collectively, as interphase.

When the cell is in interphase, the chromosomes are visible only as thin strands of threadlike material (chromatin). As mitosis begins, the chromatin—previously replicated during the S phase—condenses into pairs of identical chromosomal replicas held together at the centromere. At these early stages of mitosis, the chromosomal replicas are called chromatids. Simultaneously, the spindle is forming. In animal cells, it forms between the centrioles as they separate. In both animal and plant cells, some spindle fibers stretch from pole to pole and some are attached to the chromatids at their kinetochores, bodies adjacent to the centromeres. Prophase ends with the breakdown of the nuclear envelope and disappearance of the nucleoli. During metaphase, the chromatid pairs, apparently maneuvered by the spindle fibers, move toward the center of the cell. At the end of metaphase, the chromatid pairs are arranged on the equatorial plane. During anaphase, the sister chromatids separate, and each chromatid, now an independent chromosome, moves to an opposite pole. During telophase, a nuclear envelope forms around each group of chromosomes. The spindle breaks down, the chromosomes uncoil and once more become extended and diffuse, and the nucleoli reappear.

Cytokinesis in animal cells results from constrictions in the cell membrane between the two nuclei. In plant cells, the cytoplasm is divided by the coalescing of vesicles to form the cell plate, within which the cell wall is subsequently laid down. In both cases, the result is the production of two new, separate cells. As a result of mitosis, each has received an exact copy of the enormous skein of genetic material of the parent cell, and, as a result of cytokinesis, approximately half of the cytoplasm and organelles.

QUESTIONS

1. Distinguish among the following terms: cell cycle/cell division; mitosis/cytokinesis; chromatid/chromosome; centriole/centromere/kinetochore.

2. Describe the activities occurring during each phase of the cell cycle and the role of each phase in the overall process of cell division.

3. What is a chromosome? How is it related to chromatin?

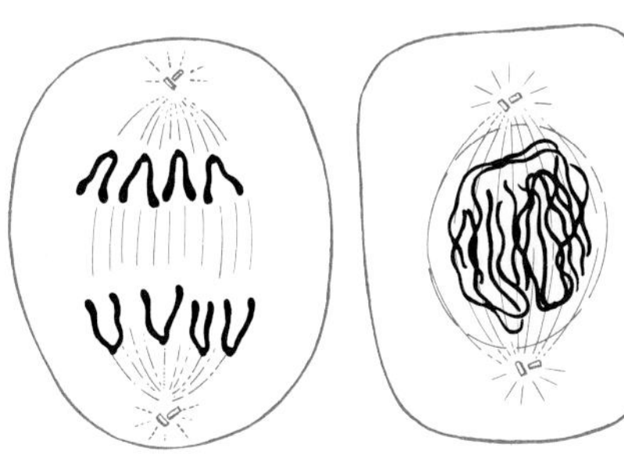

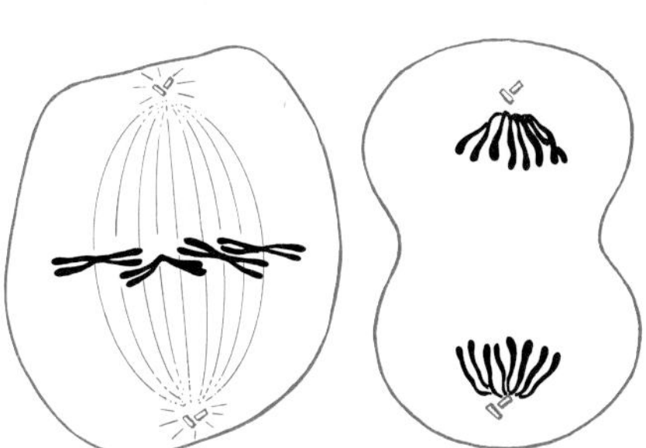

4. Why do we often refer to chromatids as sister chromatids? When are sister chromatids formed? How? When do they first become visible under the microscope?

5. Identify each of the stages of mitosis shown at the left and describe what is happening.

6. Look at Figure 7–5. Arrange the stages of mitosis you see there in the proper sequence. Are all the stages shown?

7. In what ways does cell division in plant cells differ from that in animal cells?

8. What is the function of cell division in the life of an organism? Suppose you, as an organism, were made up of a large single cell rather than trillions of small ones. How would you differ from your present self?

SUGGESTIONS FOR FURTHER READING

Books

ALBERTS, BRUCE, DENNIS BRAY, JULIAN LEWIS, MARTIN RAFF, KEITH ROBERTS, and JAMES D. WATSON: *Molecular Biology of the Cell,* Garland Publishing, Inc., New York, 1983.

Progressing from the molecules of which cells are composed, through an examination of cellular structure and function, to the interactions of cells within tissues, this outstanding text describes not only our current knowledge and how it was attained but also the many areas still to be explored. It is clearly written and filled with wonderful micrographs and explanatory diagrams. Highly recommended.

BAKER, J. J., and G. E. ALLEN: *Matter, Energy and Life: An Introduction to Chemical Concepts,* 4th ed., Addison-Wesley Publishing Co., Inc., Reading, Mass., 1981.*

A book for students who have had no previous chemistry or physics, which deals with such topics as the structure of matter, the formation of molecules, the course and mechanism of chemical reactions, as well as with the chemistry of living systems.

DE ROBERTIS, E. D. P., and E. M. F. DE ROBERTIS: *Cell and Molecular Biology,* 8th ed., Saunders College/Holt, Rinehart and Winston, Philadelphia, 1980.

An up-to-date, well-illustrated text integrating the fundamental roles of nucleic acids and proteins with cell structure and function.

DYSON, ROBERT D.: *Cell Biology: A Molecular Approach,* 2d ed., Allyn & Bacon, Inc., Boston, 1978.

Comprehensive, clear, up-to-date account of cell structure and how it relates to metabolism and transport functions.

LEDBETTER, M. C., and KEITH R. PORTER: *Introduction to the Fine Structures of Plant Cells,* Springer-Verlag, New York, 1970.

An excellent atlas of electron micrographs of plant cells, with detailed explanations.

LEHNINGER, ALBERT L.: *Principles of Biochemistry,* Worth Publishers, Inc., New York, 1982.

This introductory text is outstanding both for its clarity and for its consistent focus on the living cell.

* Available in paperback.

OPARIN, A. I.: *The Origin of Life,* Dover Publications, Inc., New York, 1938.*

Oparin, a Russian biochemist, was the first to argue that life arose spontaneously in the oceans of the primitive earth. Although his concepts have been somewhat modified in detail, they form the basis for the present scientific theories on the origin of living things.

PORTER, KEITH R., and MARY A. BONNEVILLE: *An Introduction to the Fine Structures of Cells and Tissues,* 4th ed., Lea & Febiger, Philadelphia, 1973.

An atlas of electron micrographs of animal cells; detailed commentaries accompany each. These are magnificent micrographs, and the commentaries describe not only what the pictures show but also the experimental foundations of our knowledge of cell ultrastructures.

SILK, JOSEPH: *The Big Bang: The Creation and Evolution of the Universe,* W. H. Freeman and Company, San Francisco, 1980.

A discussion of modern evidence concerning the formation of the solar system and the planet earth. An excellent, well-written introduction to cosmology.

STRYER, LUBERT: *Biochemistry,* 2d ed., W. H. Freeman and Company, San Francisco, 1981.

An introductory text, with many examples of medical applications of biochemistry. Handsomely illustrated.

THOMAS, LEWIS: *The Lives of a Cell: Notes of a Biology Watcher,* Viking Press, Inc., New York, 1974.*

THOMAS, LEWIS: *The Medusa and the Snail: More Notes of a Biology Watcher,* Viking Press, Inc., New York, 1979.*

Thomas, a physician and medical researcher, reveals the extent to which science can tune our intellectual antennae, broaden our perceptions, and extend our appreciation of ourselves and of the world around us. Anyone who wants to refute the contention that science destroys human values need look no further than these short, sensitive essays.

WEINBERG, STEVEN: *The First Three Minutes: A Modern View of the Origin of the Universe,* Basic Books, Inc., New York, 1977.*

A wonderful story, written for the intelligent nonscientist (characterized by the author as a smart old attorney who expects to hear some convincing arguments before he makes up his mind).

WOLFE, STEPHEN L.: *Biology of the Cell,* 2d ed., Wadsworth Publishing Co., Inc., Belmont, Calif., 1981.

An outstanding synthesis of cell structure and function.

Articles

ALBERSHEIM, PETER: "The Walls of Growing Plant Cells," *Scientific American,* April 1975, pages 81–95.

ALBRECHT-BUEHLER, GUENTER: "The Tracks of Moving Cells," *Scientific American,* April 1978, pages 69–76.

AUSTIN, P. R., C. J. BRINE, J. E. CASTLE, and J. P. ZIKAKIS: "Chitin: New Facets of Research," *Science,* vol. 212, pages 749–753, 1981.

CLARK, ROBERT L., and THEODORE L. STECK: "Morphogenesis in *Dictyostelium:* An Orbital Hypothesis," *Science,* vol. 204, pages 1163–1168, 1979.

DICKERSON, RICHARD E.: "Chemical Evolution and the Origin of Life," *Scientific American,* September 1978, pages 70–86.

* Available in paperback.

DUSTIN, PIERRE: "Microtubules," *Scientific American,* August 1980, pages 67–76.

EDMUNDS, LELAND N., JR., and KENNETH J. ADAMS: "Clocked Cell Cycle Clocks," *Science,* vol. 211, pages 1002–1013, 1981.

EYRE, DAVID R.: "Collagen: Molecular Diversity in the Body's Protein Scaffold," *Science,* vol. 207, pages 1315–1322, 1980.

FAUL, HENRY: "A History of Geologic Time," *American Scientist,* vol. 66, pages 159–165, 1978.

HADLEY, NEIL F.: "Surface Waxes and Integumentary Permeability," *American Scientist,* vol. 68, pages 546–553, 1980.

HOROWITZ, NORMAN H.: "The Search for Life on Mars," *Scientific American,* November 1977, pages 52–61.

LAZARIDES, ELIAS, and JEAN PAUL REVEL: "The Molecular Basis of Cell Movement," *Scientific American,* May 1979, pages 100–113.

LODISH, HARVEY F., and JAMES E. ROTHMAN: "The Assembly of Cell Membranes," *Scientific American,* January 1979, pages 48–63.

MAZIA, DANIEL: "The Cell Cycle," *Scientific American,* January 1974, pages 54–64.

MERTZ, WALTER: "The Essential Trace Elements," *Science,* vol. 213, pages 1332–1338, 1981.

MORELL, PIERRE, and WILLIAM T. NORTON: "Myelin," *Scientific American,* May 1980, pages 88–118.

PENZIAS, ARNO A.: "The Origin of the Elements," *Science,* vol. 205, pages 549–554, 1979.

PORTER, KEITH R., and JONATHAN B. TUCKER: "The Ground Substance of the Living Cell," *Scientific American,* March 1981, pages 56–67.

ROTHMAN, JAMES E.: "The Golgi Apparatus: Two Organelles in Tandem," *Science,* vol. 213, pages 1212–1219, 1981.

SATIR, BIRGIT: "The Final Steps in Secretion," *Scientific American,* October 1975, pages 28–37.

SATIR, PETER: "How Cilia Move," *Scientific American,* October 1974, pages 44–52.

SHARON, NATHAN: "Carbohydrates," *Scientific American,* November 1980, pages 90–116.

SLOBODA, ROGER D.: "The Role of Microtubules in Cell Structure and Cell Division," *American Scientist,* vol. 68, pages 290–298, 1980.

SNYDERMAN, RALPH, and EDWARD J. GOETZL: "Molecular and Cellular Mechanisms of Leukocyte Chemotaxis," *Science,* vol. 213, pages 830–837, 1981.

STAEHELIN, L. ANDREW, and BARBARA E. HULL: "Junctions between Living Cells," *Scientific American,* May 1978, pages 141–152.

STILLINGER, FRANK H.: "Water Revisited," *Science,* vol. 209, pages 451–457, 1980.

TRIMBLE, VIRGINIA: "Cosmology: Man's Place in the Universe," *American Scientist,* vol. 65, pages 75–86, 1977.

WICKNER, WILLIAM: "Assembly of Proteins into Membranes," *Science,* vol. 210, pages 861–868, 1980.

SECTION 2

Energetics

CHAPTER 8

The Flow of Energy

Life here on earth depends on the flow of energy from the thermonuclear reactions taking place at the heart of the sun. The amount of energy delivered to the earth by the sun is about 13×10^{23} (the number 13 followed by 23 zeros) calories per year. It is a difficult quantity to imagine. For example, the amount of solar energy striking the earth every day is about 1.5 billion times greater than the amount of electricity generated in the United States each year.

About one-third of this solar energy is reflected back into space as light (as it is from the moon). Much of the remaining two-thirds is absorbed by the earth and converted to heat. Some of this absorbed heat energy serves to evaporate the waters of the oceans, producing the clouds that, in turn, produce rain and snow. Solar energy, in combination with other factors, is also responsible for the movements of air and of water that help set patterns of climate over the surface of the earth.

A small fraction—less than 1 percent—of the solar energy reaching the earth becomes, through a series of operations performed by the cells of plants and other photosynthetic organisms, the energy that drives all the processes of life. Living systems change energy from one form to another, transforming the radiant energy from the sun into the chemical and mechanical energy used by everything that is alive (Figure 8–1).

8–1
The flow of energy in a biological system. Radiant energy from the sun is transformed to chemical energy by the grasses of the African plains. The zebras, eating the grasses, use this stored chemical energy for growth and convert some of it to kinetic energy, which may help to keep them from participating in another energy transfer involving the waiting lions. Of the radiant energy falling on the grass, less than 10 percent is converted to chemical energy. Less than 10 percent of the chemical energy stored in the grass is converted to chemical energy stored in the zebra, and less than 10 percent of the zebra's stored energy is converted to chemical energy stored in the body of its predator, the lion.

In this chapter, we shall look first at the general principles governing all energy transformations and then at the characteristic ways in which cells regulate the energy transformations that take place within living systems. In the chapters that follow, the principal and complementary processes of energy flow through the biosphere will be examined—glycolysis and respiration in Chapter 9 and photosynthesis in Chapter 10.

THE LAWS OF THERMODYNAMICS

Energy is such a common term today that it is surprising to learn that the word was coined less than 200 years ago, at the time of the development of the steam engine. It was only then that scientists and engineers began to understand that heat, motion, light, electricity, and the forces holding atoms together in molecules are all different forms of the same capacity to cause change, or, as it is often expressed, to do work. This new understanding led to the study of *thermodynamics*—the science of energy transformations—and to the formulation of its laws.

8–2
Electrical energy can be converted to light energy, as in these Las Vegas signs, for example. The energy emitted as an electron falls from one energy level to another is a discrete amount, characteristic for each atom. When electricity is passed through a tube of gas, electrons in the atoms of the gas are boosted to higher energy levels. As they fall back, light energy is emitted, producing, for example, the red glow characteristic of neon and the yellow glow characteristic of sodium vapor.

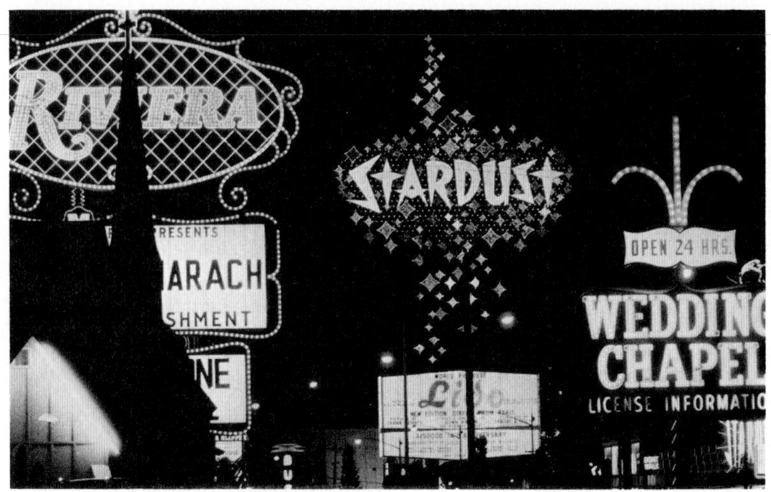

The First Law

The *first law of thermodynamics* states, quite simply: *Energy can be changed from one form to another, but it cannot be created or destroyed.* The total energy of any system plus its surroundings thus remains constant, despite any changes in form.

Electricity is a form of energy, as is light. Electrical energy can be changed to light energy (for example, by letting an electric current flow through the tungsten wire in a light bulb). Conversely, light energy can be changed to electrical energy, a transformation that is the essential first step of photosynthesis, as we shall see in Chapter 10.

Energy can be stored in various forms and then changed into other forms. In automobile engines, for example, the energy stored in the chemical bonds of gasoline is converted to heat, which is then partially converted to mechanical movements (kinetic energy). Some of the energy is converted back to heat by the friction of the moving engine parts, and some of it leaves the engine in the exhaust products. Similarly, when organisms oxidize carbohydrates, they convert the energy stored in chemical bonds to other forms. On a summer evening, for example, a firefly converts chemical energy to kinetic energy, to flashes of light, and to electrical impulses that travel along the nerves of its body. Birds and mammals convert chemical energy into the heat necessary to maintain their body temperature, as well as into kinetic energy, electrical energy, and other forms of chemical energy. According to the first law of thermodynamics, in these energy conversions, and in all others, energy is neither created nor destroyed.

In all energy conversions, however, some useful energy is converted to heat and dissipates. In an automobile engine, for example, the heat produced by friction and lost in the exhaust, unlike the heat confined in the engine itself, cannot produce work—that is, it cannot drive the pistons and turn the gears—because it is dissipated into the surroundings. But it is nevertheless part of the total equation. In a gasoline engine, about 75 percent of the energy originally present in the fuel is transferred to the surroundings in the form of heat—that is, it is converted to increased motion of atoms and molecules in the air. Similarly, the heat produced by "warm-blooded" animals, such as ourselves, is dissipated into the surrounding air or water.

$E = mc^2$

Protons and neutrons, as we noted in Chapter 1, are arbitrarily assigned an atomic weight of 1. One would therefore expect that an element with, for example, twice as many protons and neutrons as another element would weigh twice as much. This assumption is true—almost. If the weights of nuclei are measured with great accuracy, as they can be by instruments developed by modern physics, small nuclei always have proportionately slightly greater weights than larger nuclei. For example, the most common isotope of carbon, as you know, has a combined total of 12 protons and neutrons, and carbon, by convention, is assigned an atomic weight of 12. The hydrogen atom, however, has an atomic weight not exactly of 1, as would be expected, but of 1.008. Helium has two protons and two neutrons. It does not have a weight of exactly 4, however, or of 4.032 (four times the weight of hydrogen); it weighs, in relation to carbon, 4.0026. Similarly, oxygen, with a combined total of 16 protons and neutrons, has an atomic weight—in relation to that of carbon—of 15.995. In short, when protons and neutrons are assembled into an atomic nucleus, there are slight changes in weight, which reflect changes in mass.

One of the oldest and most fundamental concepts of chemistry is the law of conservation of mass—that mass is never created or destroyed. We assume the law of conservation of mass every time we write a chemical equation. Yet under conditions of extremely high temperature, atomic nuclei fuse to make new elements and there is a measurable decrease in mass. What happens to the mass "lost" in the course of this fusion? This is the question answered by Einstein's fateful equation $E = mc^2$. *E stands for energy, m stands for mass, and c is a constant equal to the speed of light. Einstein's equation means simply that under certain extreme and unusual conditions, mass is turned into energy.*

Albert Einstein in 1905, the year he published his paper on the theory of relativity. He was 26 years old and working at the Swiss Patent Office in Bern as a technical expert third class.

The sun consists largely of hydrogen nuclei. At the extremely high temperatures at the core of the sun, hydrogen nuclei strike each other with enough velocity to fuse. In a series of steps, four hydrogen nuclei fuse to form one helium nucleus. In the course of these reactions, energy is released, enough to keep the fusion reaction going and to emit tremendous amounts of radiant energy into space. Life on this planet depends on energy emitted by the sun in the course of this reaction. This same reaction provides, of course—as Einstein foresaw—the energy of the hydrogen bomb.

It was in the course of studies of engine efficiency that the notion of potential energy was first developed. A barrel of gasoline or a ton of coal could be assigned a certain amount of potential energy, measured in terms of the amount of heat it would liberate when burned. The efficiency of the conversion of the potential energy to "useful" energy depended on the design of the system.

Although these concepts were formulated in terms of engines running on heat energy, they apply to other systems as well. As we saw in Chapter 1, a boulder pushed to the top of a hill contains energy—potential energy. Given a little push, it rolls down the hill again, converting the potential energy to kinetic energy and the heat produced by friction. Water, as we saw in Chapter 6, may also possess potential energy. As it moves by bulk flow from the top of a waterfall or over a dam, it can turn waterwheels that turn gears and, for example, grind corn. Thus the potential energy of water, in this system, is converted to the kinetic energy of the wheels and gears and to heat, produced by the movement of the water itself and also by the turning wheels and gears. Molecules also contain potential energy, stored in the chemical bonds between their constituent atoms. When these bonds are broken in chemical reactions, the energy they contain can be used to form other chemical bonds or can be released as heat.

The first law of thermodynamics states that in energy exchanges and conversions, wherever they take place and whatever they involve, the total energy of the system and its surroundings after the conversion is equal to the total energy before the conversion. In the case of chemical reactions, this means that the energy of the products of the reaction plus the energy released in the reaction is equal to the initial energy of the reactants.

The Second Law

The energy that is released as heat in an energy conversion has not been destroyed—it is still present in the random motion of atoms and molecules—but it has been "lost" for all practical purposes. It is no longer available to do useful work. This brings us to the *second law of thermodynamics*, which is the more interesting one, biologically speaking. It predicts the direction of all events involving energy exchanges; thus it has been called "time's arrow." The second law states that *in all energy exchanges and conversions, if no energy leaves or enters the system under study, the potential energy of the final state will always be less than the potential energy of the initial state.* The second law is entirely in keeping with everyday experience. A boulder will roll downhill but never uphill. Heat will flow from a hot object to a cold one and never the other way. Our cells can process glucose to yield carbon dioxide and water but—unless we are plants and can capture the energy of the sun—our cells will not produce glucose from carbon dioxide and water.

A process in which the potential energy of the final state is less than that of the initial state is one that yields energy (otherwise it would be in violation of the first law). An energy-yielding process is called an *exergonic* ("energy-out") reaction. As the second law predicts, only exergonic reactions can take place spontaneously—that is, without an input of energy from outside the system. (Spontaneously, though the word has an explosive sound to it, says nothing about the rate of the reaction, just whether or not it can take place at all.) By contrast, a process in which the potential energy of the final state is greater than that of the initial state is one that requires energy. Such energy-requiring processes are known as *endergonic* ("energy-in") reactions, and in order for them to proceed, an input of energy is required that is greater than the difference in energy between the products and the reactants.

One important factor in determining whether or not a reaction is exergonic is already familiar to us: ΔH, the change in heat content of a system. As we noted in Chapter 3, the energy change that takes place when glucose, for instance, is oxidized can be measured in a calorimeter and expressed in terms of ΔH. The oxidation of a mole of glucose yields 673 kilocalories. Or,

$$C_6H_{12}O_6 + 6O_2 \longrightarrow 6CO_2 + 6H_2O$$
$$\Delta H = -673 \text{ kcal/mole}$$

Generally speaking, an exergonic chemical reaction is also an exothermic reaction—that is, it gives off heat and thus has a negative ΔH. However, there are exceptions. One of the most dramatic is found with a substance known as dinitrogen pentoxide, which decomposes spontaneously and with explosive force to nitrogen dioxide and oxygen, and in so doing, absorbs heat:

$$2N_2O_5 \longrightarrow 4NO_2 + O_2$$
$$\Delta H = +26.18 \text{ kcal/mole}$$

In short, another factor besides the gain or loss of heat affects the change in potential energy and thus the direction of a process. This factor is given the formal name of *entropy*, and it is a measurement of the disorder or randomness of a system.

Before we examine more closely why the decomposition of dinitrogen pentoxide is exergonic despite its positive ΔH, let us return to the more familiar example of water. The change from ice to liquid water and the change from liquid water to water vapor are both endothermic processes—a considerable amount of heat is removed from the surrounding air as they take place. Yet, under the appropriate conditions, they proceed spontaneously. The key factor in all three of these examples is the increase in entropy. In the case of the dinitrogen pentoxide, a solid is being changed into two gases, and two molecules are being converted into five. In the case of ice and liquid water, a solid is being turned into a liquid, and some of the bonds that hold the water molecules together in a crystal (ice) are being broken. As the liquid water turns to vapor, the rest of the hydrogen bonds are ruptured as the individual water molecules dance off, one by one. In every case, the disorder of the system has increased.

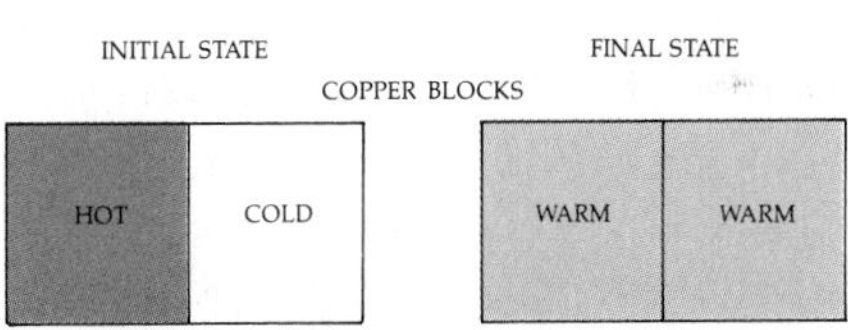

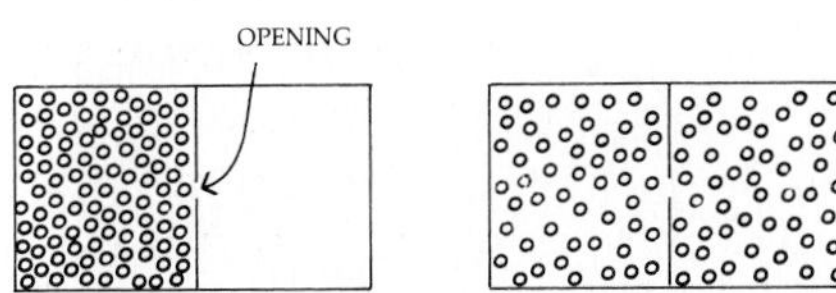

8–3
Some illustrations of the second law of thermodynamics. In each case, a concentration of energy—in the hot copper block, in the gas molecules under pressure, and in the neatly organized books—is dissipated. In nature, processes tend toward randomness, or disorder. Only an input of energy can reverse this tendency and reconstruct the initial state from the final state. Ultimately, however, disorder will prevail, since the total amount of energy in the universe is finite.

8-4
Organisms are experts in energy conversion. (a) *This hummingbird is drinking nectar, in which is stored energy from the sun. Hummingbirds convert large amounts of chemical energy to the kinetic energy of flight.* (b) *Heat energy is being generated from chemical energy by these rapidly growing skunk cabbages.* (c) *These bioluminescent mushrooms are transforming some of their chemical energy into light energy, thereby glowing in the dark.* (d) *A cheetah converting chemical energy to kinetic energy.*

(a)

The notion that there is more disorder associated with more numerous and smaller objects than with fewer, large ones is in keeping with our everyday experience. If I have 20 papers on my desk, the possibilities for disorder are greater than if I have 2 or even 10. If I cut each of the 20 in half, the entropy of the system—the capacity for randomness—increases. Also, the relationship between entropy and energy is a commonplace idea. If you were to find your room tidied up and your books in alphabetical order on the shelf, you would recognize that someone had been at work—that energy had been expended. For me to organize the papers on my desk similarly requires that I expend energy. Furthermore, it would be feasible to measure the energy expenditure in calories.

Now let us return to the question of the energy changes that determine the course of chemical reactions. Both the change in the heat content of the system (ΔH) and the change in entropy (which is symbolized as ΔS) contribute to the overall change in energy. This total change—the one that takes into account both heat and entropy—is called the *free energy change* and is symbolized as ΔG, after the American physicist Josiah Willard Gibbs (1839–1903), who was one of the first to put all of these ideas together.

The relationship between ΔG, ΔH, and entropy is given in the following equation:

$$\Delta G = \Delta H - T\,\Delta S$$

It states that the free energy change is equal to the change in heat (a negative value in exothermic reactions, remember) minus the change in entropy multiplied by the absolute temperature, T. In exergonic reactions, ΔH may be zero or may even be positive, but ΔG is always negative. As you can see in the equation, $T\,\Delta S$ is preceded by a minus sign. The greater the increase in entropy, the more negative ΔG will be; that is, the more exergonic the reaction will be.

With ΔG in mind, let us examine once more the combustion of glucose. The ΔH of that reaction is -673 kcal/mole. The free energy change, ΔG, is -686

(b)

(c)

kcal/mole. The increase in entropy has contributed 13 kcal/mole to the free energy change of the process. Thus both the change in heat and the change in entropy contribute to the lower energy state of the products of the reaction.

ΔG can also enable one to predict processes that occur when ΔH is zero or even positive, as with dinitrogen pentoxide. For instance, it confirms our earlier observations that heat will flow spontaneously from a hot object to a cold one, that dye molecules will diffuse spontaneously in a beaker of water, or that my desk will revert to disorder. In each of these processes, the final state has more entropy—and therefore less potential energy—than the initial state.

Previously we stated the second law in terms of the energy change between the initial and final states of a process. The second law can also be stated in another, simpler way: *All natural processes tend to proceed in such a direction that the disorder or randomness of the system increases.*

Living Systems and the Second Law

The universe, according to the present model, is a closed system. The matter and energy present in it 18 billion years ago, at the time of the primordial explosion, are all the matter and energy it will ever have. Moreover, after each and every energy exchange and transformation, the universe as a whole has less potential energy and more entropy than it did before. In this view, of course, the universe is running down. The stars will flicker out, one by one; life—any form of life on any planet—will come to an end. Finally, even the motion of individual molecules will cease. However, even the most pessimistic among us do not believe this will occur for another 20 billion years or so.

In the meantime, life can exist *because* the universe is running down. Although the universe as a whole is a closed system, the earth is not. As we noted at the beginning of this chapter, it is receiving an energy input of 13×10^{23} calories per year from the sun. Photosynthetic organisms are specialists at capturing the light energy released by the sun as it slowly burns itself out. They use this energy to

(d)

organize small, simple molecules (water and carbon dioxide) into larger, more complex molecules (sugars). In the process, the captured light energy is stored in the chemical bonds of sugars and other molecules. Living cells—including photosynthetic cells—can convert this stored energy into motion, electricity, light, and, by shifting the energy from one type of chemical bond to another, more convenient forms of chemical energy. At each transformation, of course, energy is lost to the surroundings as heat. But before the energy captured from the sun is completely dissipated, organisms use it to create and maintain the complex organization of structures and activities that we know as life.

OXIDATION-REDUCTION

You will recall from Chapter 1 that electrons possess differing amounts of potential energy depending on their distance from the atomic nucleus and the attraction of the nucleus for electrons. An input of energy will boost an electron to a higher energy level, but without added energy an electron will remain at the lowest energy level available to it.

Chemical reactions are essentially energy transformations in which energy stored in chemical bonds is transferred to other, newly formed chemical bonds. In such transfers, electrons shift from one energy level to another. In many reactions, electrons pass from one atom or molecule to another. These reactions, which are of great importance in living systems, are known as oxidation-reduction (or redox) reactions. The *loss* of an electron is known as *oxidation*, and the atom or molecule that loses the electron is said to be oxidized. The reason electron loss is called oxidation is that oxygen, which attracts electrons very strongly, is most often the electron acceptor.

Reduction is, conversely, the *gain* of an electron. Oxidation and reduction always take place simultaneously because an electron that is lost by the oxidized atom is accepted by another atom, which is reduced in the process.

Redox reactions may involve only a solitary electron, as when sodium loses an electron and becomes oxidized to Na^+, and chlorine gains an electron and is reduced to Cl^-. Often, however, the electron travels with a proton, that is, as a hydrogen atom. In such cases, oxidation involves the removal of hydrogen atoms, and reduction the gain of hydrogen atoms. For example, when glucose is oxidized, hydrogen atoms are lost by the glucose molecule and gained by oxygen:

$$C_6H_{12}O_6 + 6O_2 \longrightarrow 6CO_2 + 6H_2O + \text{energy}$$

The electrons are moving to a lower energy level, and energy is released.

Conversely, in the process of photosynthesis, hydrogen atoms are transferred from water to carbon dioxide, thereby reducing the carbon dioxide to form glucose:

$$6CO_2 + 6H_2O + \text{energy} \longrightarrow C_6H_{12}O_6 + 6O_2$$

In this case, the electrons are moving to a higher energy level, and an energy input is required to make the reaction occur.

In living systems, the energy-capturing reactions (photosynthesis) and energy-releasing reactions (glycolysis and respiration) are oxidation-reduction reactions. As we have seen, the complete oxidation of a mole of glucose releases 686 kilocalories of free energy (conversely, the reduction of carbon dioxide to form a mole of glucose stores 686 kilocalories of free energy in the chemical bonds of glucose).

8–5

The flow of biological energy. Chloroplasts, present in all photosynthetic eukaryotic cells, capture the radiant energy of sunlight and use it to convert water and carbon dioxide into carbohydrates, such as glucose, starch, and other foodstuff molecules. Oxygen is released as a product of the photosynthetic reactions.

Mitochondria, present in all eukaryotic cells, carry out the final steps in the breakdown of these carbohydrates and capture their stored energy in ATP molecules. This process, cellular respiration, consumes oxygen and produces carbon dioxide and water, completing the cycling of the molecules.

With each transformation, some energy is dissipated to the environment in the form of heat. Thus the flow of biological energy is one-way and can continue only so long as there is an input of energy from the sun.

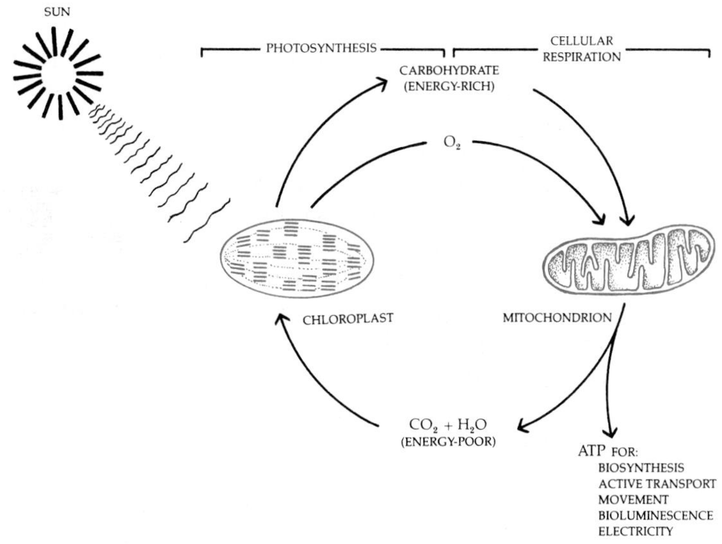

If this energy were to be released all at once, most of it would be dissipated as heat. Not only would it be of no use to the cell, but the resulting high temperature would be lethal. However, mechanisms have evolved in living systems that regulate these chemical reactions—and a multitude of others—in such a way that energy is stored in particular chemical bonds from which it can be released in small amounts as the cell needs it. These mechanisms generally involve sequences of reactions, some of which are oxidation-reduction reactions. Although each reaction in the sequence represents only a small change in the free energy, the overall free energy change for the sequence can be considerable.

METABOLISM

In any living system, energy exchanges occur through thousands of different chemical reactions, many of them taking place simultaneously. The sum of all these reactions is referred to as <u>*metabolism*</u> (from the Greek *metabole,* meaning "change"). If we were merely to list the individual chemical reactions, it would be difficult indeed to understand the flow of energy through a cell. Fortunately, there are some guiding principles that lead one through the maze of cell metabolism.

First, virtually all the chemical reactions that take place in a cell involve enzymes—large protein molecules that play very specific roles. Second, biochemists are able to group these reactions in an ordered series of steps, commonly called a pathway; a pathway may have a dozen or more sequential reactions or steps. Each pathway serves a function in the overall life of the cell or organism. Furthermore, certain pathways have many steps in common—for instance, those that are concerned with the synthesis of the different amino acids or the various nitrogenous bases. Some pathways converge; for example, the pathway by which fats are broken down to yield energy leads to the pathway by which glucose is broken down to yield energy.

Many living systems have pathways unique to them. Plant cells expend much of their energy building their cell walls, an activity not engaged in by animal cells. Red blood cells specialize in the synthesis of hemoglobin molecules, not made anywhere else in the animal body. It is not surprising that the distinctive differences in function among cells and organisms are correlated not only with their forms but also with their biochemistry. What is more surprising, however, is that much of the metabolism of even the most diverse of organisms is exceedingly similar; the differences in many of the metabolic pathways of humans, oak trees, mushrooms, and jellyfish are very slight. Some pathways—for example, those of glycolysis and respiration—are virtually universal, found in almost all living systems.

The magnitude of the chemical work carried out by a cell—and its consequent energy expenditure—can be imagined if one recognizes that, for the most part, the thousands of different molecules, large and small, found within a cell are synthesized there. The total of chemical reactions involved in synthesis is called *anabolism*. Cells also are constantly involved in the breakdown of larger molecules; these activities are known, collectively, as *catabolism*. Catabolism serves two purposes: (1) it releases the energy for anabolism and other work of the cell, and (2) it serves as a source of raw materials for anabolic processes.

Not only do living systems carry out this multitude of chemical activities, but they do so under what might seem, at first glance, extraordinarily difficult conditions. Most of their chemical reactions are carried out within living cells, and in cells, not two or three but thousands of different kinds of molecules are present. As we saw in Chapter 5, however, the cytoplasm of the living cell is highly structured, with the organelles, endoplasmic reticulum, vesicles, vacuoles, and the cytoskeleton itself effectively compartmentalizing the cell into different "work areas" (Figure 8–6). This segregates different reaction pathways from one another, enabling different chemical reactions to take place without mutual interference.

However, for any particular molecules to react with one another, it is not enough that they be in the same general region of the cell; they must be in extremely close proximity and, moreover, must collide with sufficient force to overcome the mutual repulsion of their electron clouds. The force required varies with the nature of the molecules; the more stable their initial state, the more forceful the collision must be. The force with which molecules collide depends on their kinetic energy, and the average kinetic energy of the molecules in a cell is quite moderate, as reflected in the moderate temperatures of living systems. In a group of molecules, it is likely that some proportion is moving with sufficient energy to cause a reaction to occur, but often this proportion is so small that the reaction, for all practical purposes, does not take place. How, then, can the complex chemical work of a cell be accomplished? The question can be answered in a single word: enzymes.

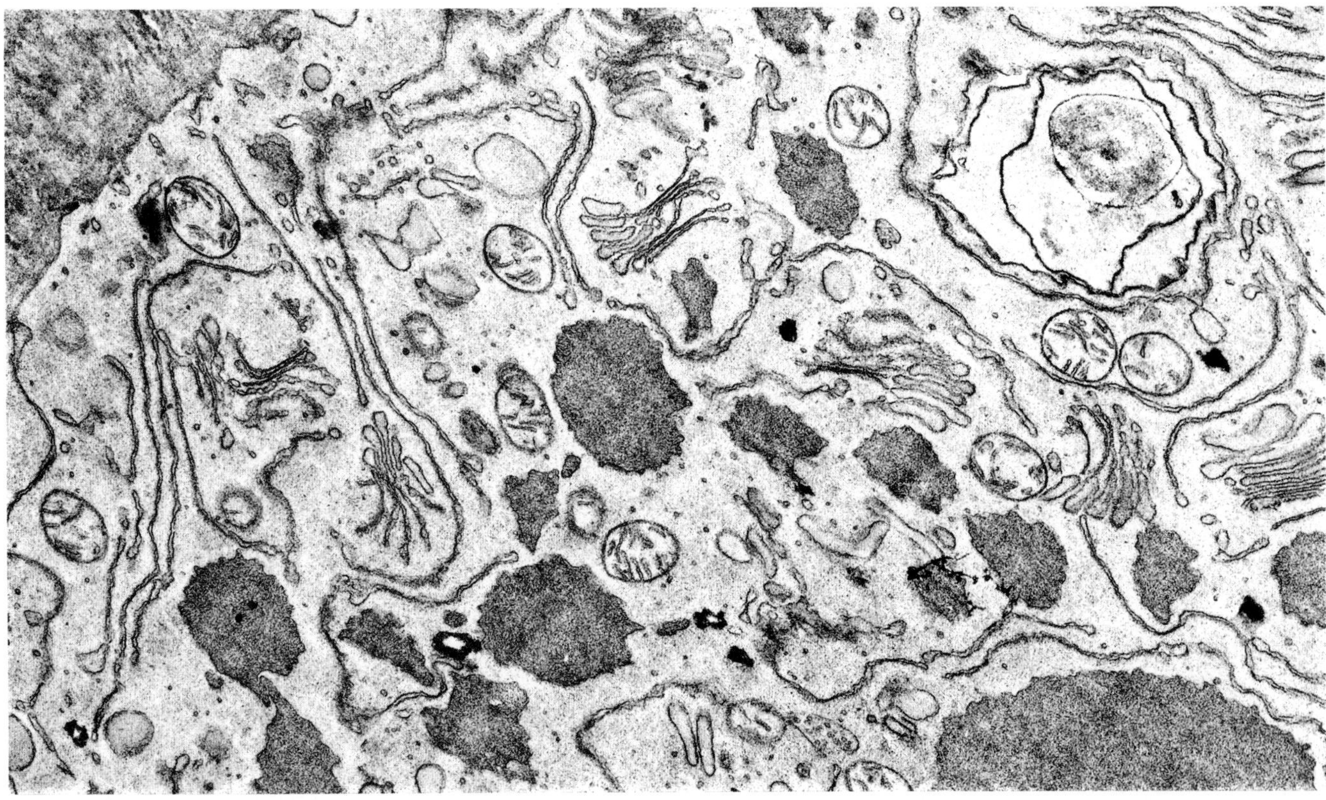

8–6
A chemical factory. Part of a cell from the root of a wheat plant. Cells such as this contain thousands of different organic molecules. These molecules form the many intricate structures of which the cell is composed and carry out the multitude of chemical reactions necessary for the life of the cell, its maintenance and growth, and its interactions with the cells around it. Many of the chemical activities of the cell are compartmentalized within the organelles, vesicles, and vacuoles, on the surfaces of the membranes, and in different regions of the cytoplasmic solution.

ENZYMES

To proceed at a reasonable rate, most chemical reactions require an initial input of energy. This is true even for exergonic reactions such as the oxidation of glucose or the burning of natural gas (methane). The added energy increases the kinetic energy of the molecules, enabling a greater number of them to collide with sufficient force not only to overcome their mutual repulsion but also to break existing chemical bonds within the molecules. The energy that must be possessed by the molecules in order to react is known as the *energy of activation*. Sometimes, as in the case of natural gas, a spark is all that is needed to supply enough energy. Once the reaction begins, it liberates energy that is transferred to the other methane molecules until all are moving rapidly enough to react almost simultaneously with explosive force.

In the laboratory, the energy of activation is usually supplied as heat. But, in a cell, many different reactions are going on at the same time, and heat would affect all of these reactions indiscriminately. Moreover, heat would break hydrogen bonds and would have other generally destructive effects on the cell. Cells get around this problem by the use of enzymes, globular proteins that are specialized to serve as catalysts. A *catalyst* is a substance that lowers the activation energy required for a reaction by forming a temporary association with the molecules that are reacting (Figure 8–7). This temporary association brings the reacting molecules close to one another and may also weaken the existing chemical bonds, making it easier for new ones to form. The lower activation energy in the presence of the

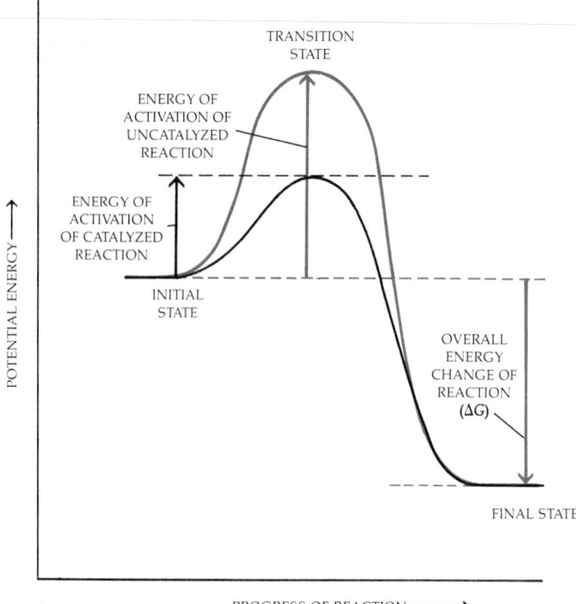

8–7
In order to react, molecules must possess enough energy—the energy of activation—to collide with sufficient force to overcome their mutual repulsion and to weaken existing chemical bonds. An uncatalyzed reaction requires more activation energy than a catalyzed one, such as an enzymatic reaction. The lower activation energy in the presence of the catalyst is often within the range of energy possessed by the molecules, and so the reaction can occur at a rapid rate with little or no added energy. Note, however, that the overall energy change (ΔG) from the initial state to the final state is the same with and without the catalyst.

catalyst is within the range of energy possessed by a greater proportion of the reacting molecules; as a result, the reaction goes more rapidly than it would in the absence of a catalyst. The catalyst itself is not permanently altered in the process, and so it can be used over and over again.

Because of enzymes, cells are able to carry out chemical reactions at great speed and at comparatively low temperatures. For instance, the combination of carbon dioxide with water,

$$CO_2 + H_2O \rightleftharpoons \underset{\text{CARBONIC ACID}}{H_2CO_3}$$

can take place spontaneously, as it does in the oceans. In the human body, however, this reaction is catalyzed by an enzyme, carbonic anhydrase, which is one of the fastest enzymes known, each enzyme molecule producing 10^5 (100,000) molecules of carbonic acid per second. The catalyzed reaction is 10^7 times faster than the uncatalyzed one. In animals, this reaction is essential in the transfer of carbon dioxide from the cells, where it is produced, to the bloodstream, which transports it to the lungs. As carbonic anhydrase illustrates, enzymes are typically effective in very small amounts. (Note also that enzyme names usually end in "-ase.")

Almost 2,000 different enzymes are now known, each of them capable of catalyzing a specific chemical reaction. However, different types of cells are able to manufacture different types of enzymes—no cell contains all the known enzymes. The particular enzymes that a cell can manufacture are a major factor in determining the biological activities and functions of that cell. A cell can carry out a given chemical reaction at a reasonable rate only if it has a specific enzyme that can catalyze that reaction. The molecule (or molecules) on which an enzyme acts is known as its *substrate*. For example, in the reaction diagrammed in Figure 8–8, sucrose is the substrate and sucrase is the enzyme.

8–8
A model of enzyme action. Sucrose, a disaccharide, is hydrolyzed to yield a molecule of glucose and a molecule of fructose. The enzyme involved in this reaction, sucrase, is specific for this process; as you can see, the active site of the enzyme fits the opposing surface of the sucrose molecule. The fit is so exact that a molecule composed, for example, of two subunits of glucose would not be affected by this enzyme.

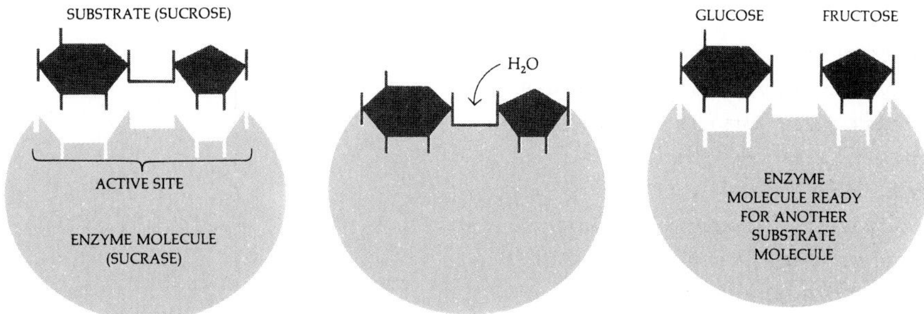

Enzyme Structure and Function

Enzymes are large to very large,* complex globular proteins consisting of one or more polypeptide chains (Figure 8–9). They are folded so as to form a groove or pocket into which the reacting molecule or molecules—the substrate—fit and where the reactions take place. This portion of the enzyme is known as the *active site*. Only a few amino acids of the enzyme are involved in any particular active site. Some of these may be adjacent to one another in the primary structure, but often the amino acids of the active site are brought close to one another by the intricate folding of the amino acid chain that produces the tertiary structure (Figure 8–10). In an enzyme with a quaternary structure, the amino acids of the active site may even be on different polypeptide chains.

* Different enzymes vary in their molecular weights from about 12,000 to more than 1 million. Amino acids have an average molecular weight of about 120, and this figure is used to estimate the number of amino acids in a polypeptide of known molecular weight.

8–9
Model of an enzyme. This enzyme (the digestive enzyme chymotrypsin) is composed of three polypeptide chains. The amino (NH_2) and carboxyl (COOH) ends of each are labeled. The numbers represent the positions of particular amino acids in the chains. Five disulfide bridges connect amino acids 1 and 122, 42 and 58, 136 and 201, 168 and 182, and 191 and 220. The three-dimensional shape of the molecule is a result of a combination of disulfide bridges and of interactions among the chains and between the chains and the surrounding water molecules, based on the positive or negative charges or the polarity of the various amino acids. As a result of this bending and twisting of the polypeptide chains, particular amino acids come together in a highly specific configuration to form the active site of the enzyme. Two amino acids known to be part of the active site are shown in color.

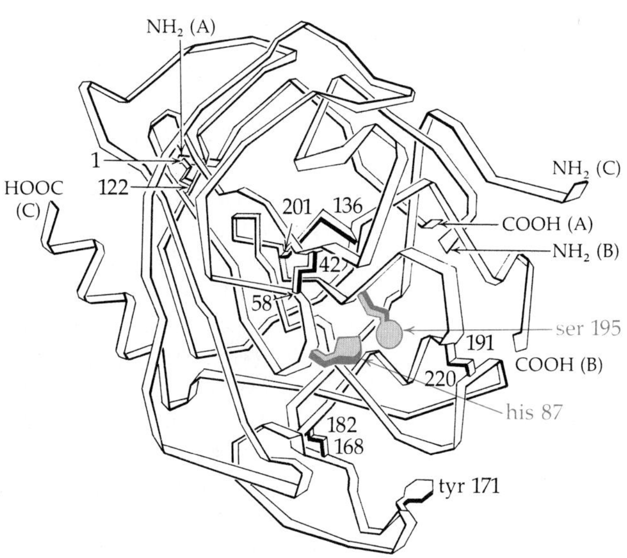

8-10

(a) *Primary structure of the enzyme lysozyme. This enzyme, which contains 129 amino acids, is found in many different types of animal cells. The sequence of the amino acids constitutes its primary structure. One consequential feature of the primary structure is the occurrence of the sulfur-containing amino acid cysteine. Covalent bonds form between the sulfur atoms, linking parts of the molecule together. Lysozyme breaks down the cell walls of many types of bacteria. It was first detected by Sir Alexander Fleming, better known for his discovery of penicillin, as a consequence of his having a bad cold. A droplet fell from his nose onto a culture dish containing bacteria, and he noticed that the bacteria in the vicinity of the droplet were destroyed. Lysozyme was subsequently discovered in tears, saliva, milk, and various other body fluids of many different animals. The molecule shown here was extracted from egg white. Lysozymes from other organisms might differ in some of their amino acids.*

(b) *Tertiary structure of the enzyme lysozyme. From this drawing you can gain a rough idea of the complex nature of the precise folding and coiling that make up the tertiary structure of an enzyme. The crevice that forms the active site runs horizontally across the molecule. The substrate is shown in a darker color.* (From R. E. Dickerson and I. Geis, The Structure and Action of Proteins, *W. A. Benjamin, Inc., Menlo Park, Calif., 1969. Copyright 1969 by Dickerson and Geis.)*

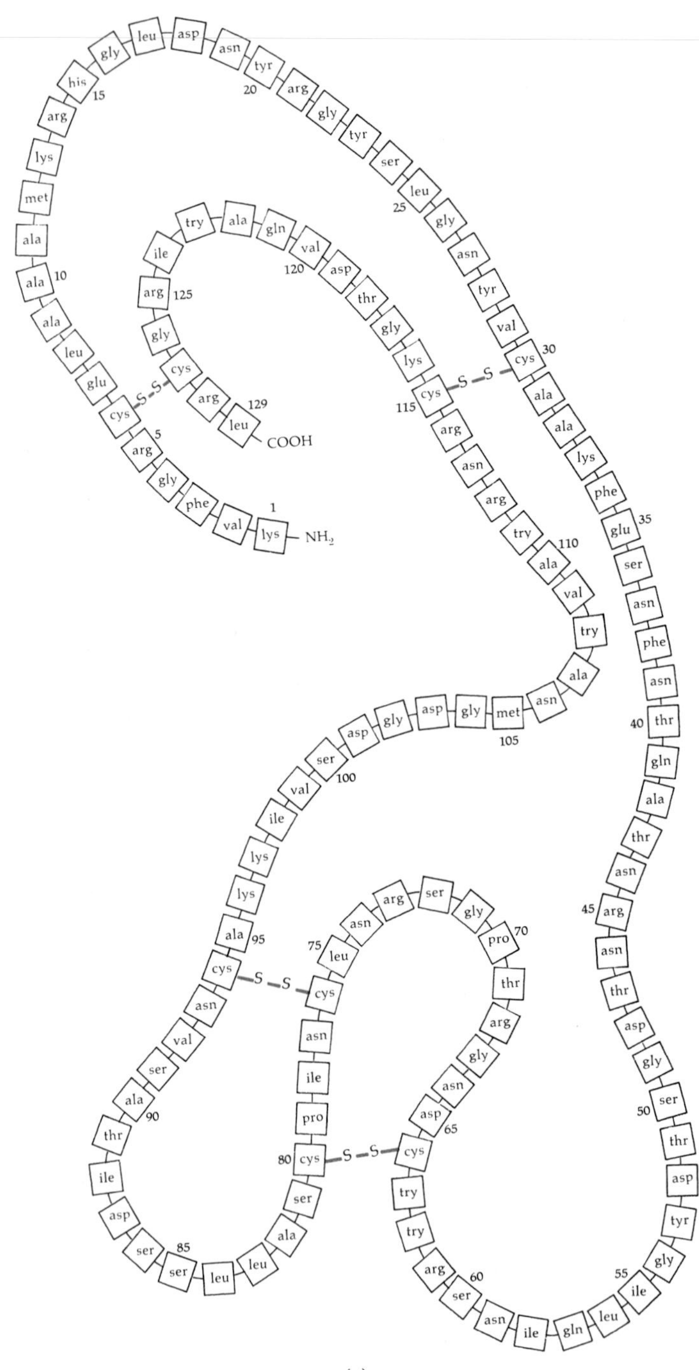

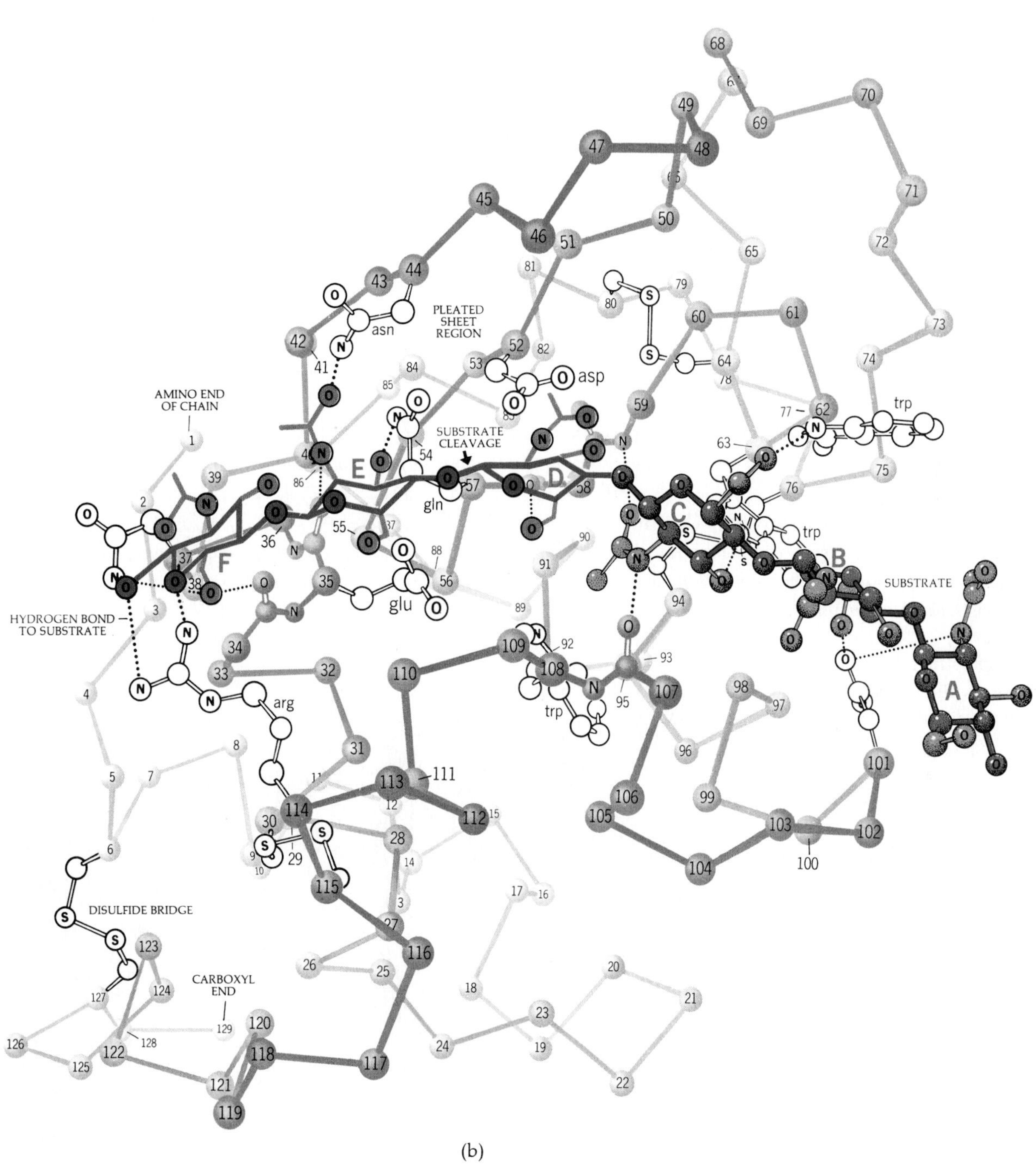
AMINO END
OF CHAIN
HYDROGEN BOND
TO SUBSTRATE
PLEATED
SHEET
REGION
SUBSTRATE
CLEAVAGE
SUBSTRATE
DISULFIDE BRIDGE
CARBOXYL
END
asn
asp
gln
glu
arg
trp
A
B
C
D
E
F

(b)

THE LINGERING DEATH OF VITALISM

The concept that living things are made up of the same chemical and physical components as nonliving things and that they follow the same chemical and physical laws is comparatively new. Less than 100 years ago, many biologists believed that living systems are qualitatively different from nonliving ones, containing within them a "vital spirit" that enables them to perform activities that cannot be carried on outside the living organism. This concept is known as vitalism and its proponents as vitalists. (Do not dismiss this "foolish" idea too rapidly; you may discover you too are a vitalist.)

In the seventeenth century, the vitalists were opposed by a group known as the mechanists. The French philosopher René Descartes (1596–1650) was a leading proponent of this group. The mechanists set about proving that the body worked essentially like a machine; the arms and legs move like levers, the heart like a pump, the lungs like a bellows, and the stomach like a mortar and pestle. By the nineteenth century, such simple mechanical models of living organisms had been abandoned, and the argument now centered on whether or not the chemistry of living organisms was governed by the same principles as the chemistry performed in the laboratory. The vitalists claimed that the chemical operations performed by living tissues could not be carried out experimentally in the laboratory, categorizing reactions as either "chemical" or "vital." The reductionists, as their opponents were now called (since they believed that the complex operations of living systems could be reduced to simpler and more readily understandable ones), achieved a partial victory when the German chemist Friedrich Wöhler (1800–1882) converted an "inorganic" substance (ammonium cyanate) into a familiar organic substance (urea). On the other hand, the claims of the vitalists were supported by the fact that, as chemical knowledge improved, many new compounds were found in living tissues that were never seen in the nonliving, or inorganic, world.

In the late 1800s, the leading vitalist was Louis Pasteur, who claimed that the marvelous changes that took place when fruit juice was transformed to wine were "vital" and could be carried out only by living cells—the cells of yeast. In spite of many advances in chemistry, this phase of the controversy lasted until almost the turn of the century.

However, in 1898, the German chemists Eduard and Hans Büchner showed that a substance extracted from the yeast cells could produce fermentation outside the living cell. This substance was given the name "enzyme," from zyme, *the Greek word meaning "yeast" or "ferment." A "vital" reaction was proved to be a chemical one, the theory of vitalism was thus disproved, and the subject was eventually laid to rest.*

The laboratory at Giessen, where Wöhler was a student.

The active site not only has a three-dimensional shape complementary to that of the substrate, but it also has a complementary array of charged or uncharged, hydrophilic or hydrophobic areas on the binding surface. If a particular portion of the substrate has a negative charge, any corresponding feature on the active site is likely to have a positive charge, and so on. Thus the active site not only recognizes and confines the substrate molecule but also orients it in a particular direction.

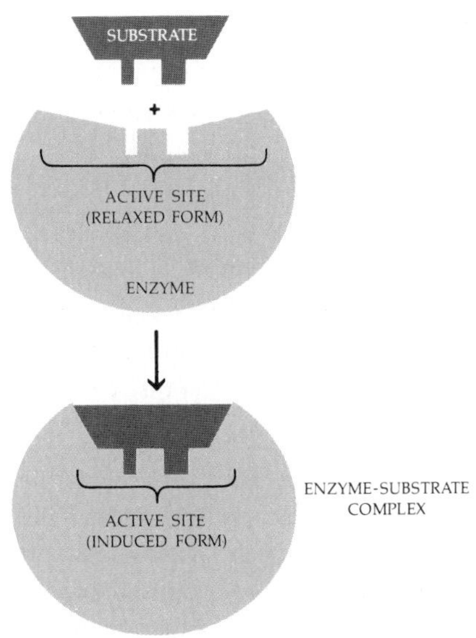

8–11
The induced-fit hypothesis. The active site is believed to be flexible and to adjust its conformation to that of the substrate molecule. This induces a close fit between the active site and the substrate.

The Induced-Fit Hypothesis

When the existence of active sites was first postulated by Emil Fischer in 1894, he compared the relationship between the active site and the substrate to that between a lock and a key. Within the last few years, however, studies of enzyme structure have suggested that the active site is considerably more flexible than a keyhole. The binding between enzyme and substrate appears to alter the conformation of the enzyme, thus inducing a close fit between the active site and the substrate. It is believed that this induced fit may put some strain on the reacting molecules and so further facilitate the reaction (Figure 8–11).

Cofactors in Enzyme Action

The catalytic activity of some enzymes appears to depend only upon the physical and chemical interactions between the amino acids of the active site and the substrate. Many enzymes, however, require additional nonprotein low-molecular-weight substances in order to function. Such nonproteins that are essential for enzyme function are known as *cofactors.*

Ions as Cofactors

Certain ions are cofactors for particular enzymes. For example, the magnesium ion (Mg^{2+}) is required in all enzymatic reactions involving the transfer of a phosphate group from one molecule to another. Its two positive charges hold the negatively charged phosphate group in position. K^+, Ca^{2+}, and other ions play similar roles in other reactions. In some cases, bonds between ions and the R groups of particular amino acids help to maintain certain folds in the tertiary structure or to hold a quaternary structure together.

Coenzymes and Vitamins

Nonprotein organic molecules may also play a crucial role in enzyme-catalyzed reactions. Such molecules are called *coenzymes.* For example, in some oxidation-reduction reactions, electrons—often traveling as a hydrogen ion with a pair of electrons—are passed to a coenzyme that serves as an electron acceptor. There are several different kinds of coenzymes in any given cell, each tailor-made to hold the electron at a slightly different energy level. As an example, let us look at just one, nicotinamide adenine dinucleotide (NAD^+), which is shown in Figure 8–12 on the next page. At first glance, NAD^+ looks unfamiliar, but if you look at it closely, you will find that you recognize most of its component parts. The two units labeled ribose are five-carbon sugars. They are linked by two phosphate groups. One of the sugars is attached to the nitrogenous base adenine. The other is attached to another nitrogenous base, nicotinamide. (A nitrogenous base plus a sugar plus a phosphate is called a *nucleotide,* and a molecule that contains two of them is called a dinucleotide. Watch out for nucleotides; they will return.)

8–12
Nicotinamide adenine dinucleotide (NAD) in its oxidized form, NAD^+, and its reduced form, NADH. Notice how the bonding within the nicotinamide ring shifts as the molecule changes from the oxidized to the reduced form, and vice versa.

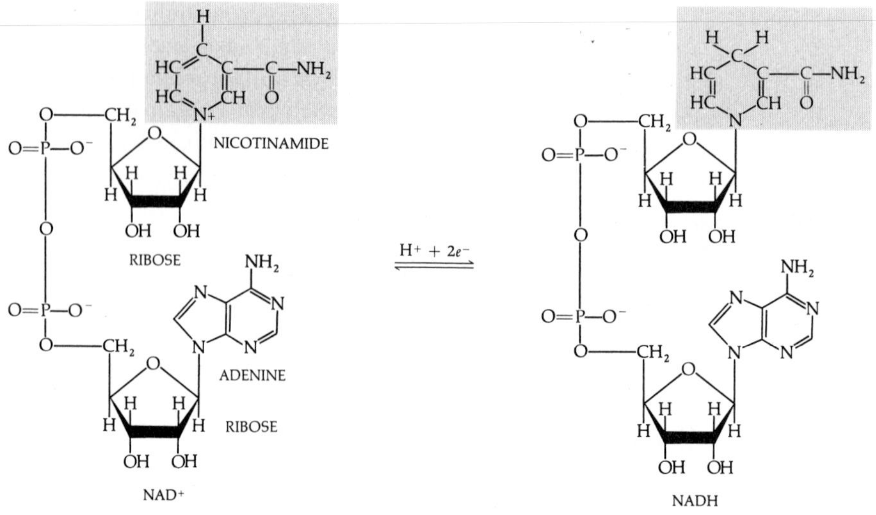

The nicotinamide ring is the business end of NAD^+, the part that accepts the electrons. Nicotinamide is a vitamin, niacin. Vitamins are compounds required in small quantities that humans and other animals cannot synthesize themselves and so must obtain in their diets (see Table 34–3, page 686). Thus we must eat foods containing niacin (which includes both nicotinamide and nicotinic acid but should not be confused with nicotine, found in tobacco). When nicotinamide is present, our cells can use it to make NAD^+. Many vitamins are coenzymes or parts of coenzymes.

Nicotinamide adenine dinucleotide, like other coenzymes, is recycled. That is, NAD^+ is regenerated when NADH passes its electrons on to another electron acceptor. Thus, although this coenzyme is involved in many cellular reactions, the actual number of NAD^+ molecules required is relatively small.

Effects of Temperature and pH

As we noted earlier, an increase in temperature increases the rate of uncatalyzed chemical reactions. This temperature effect also holds true for enzyme-catalyzed reactions—but only up to a point. As you can see in Figure 8–13, the rate of most enzymatic reactions approximately doubles for each 10°C rise in temperature and then drops off very quickly at about 40°C. The increase in reaction rate occurs because, at higher temperatures, more of the substrate molecules possess sufficient energy to react; the decrease in the reaction rate occurs as the enzyme molecule itself begins to move and vibrate, disrupting the hydrogen bonds and other relatively fragile forces that maintain its tertiary structure. A molecule that has lost its characteristic three-dimensional structure in this way is said to be *denatured*. Some denatured enzymes regain their activity on being cooled, indicating that their polypeptide chains have regained their necessary shape. Others, however, are permanently tangled and inactivated.

The pH of the surrounding solution also affects enzyme activity. The conformation of an enzyme depends, among other factors, on attractions and repulsions between negatively charged (acidic) and positively charged (basic) amino acids. As

8–13

The effect of temperature on the rate of an enzyme-controlled reaction. The concentrations of enzyme and reacting molecules (substrate) were kept constant. As you can see, the rate of the reaction, as in most chemical reactions, approximately doubles for every 10°C rise in temperature. In the enzymatic reactions of humans (body temperature 37°C) and other mammals, the maximum rate of reaction is attained at about 40°C. Above this temperature, the rate decreases, and at about 60°C the reaction stops altogether, presumably because the enzyme is denatured. Although the shape of the curve is similar for all enzymatic reactions, the temperature range through which an enzyme is active varies with the type of organism and the particular enzyme.

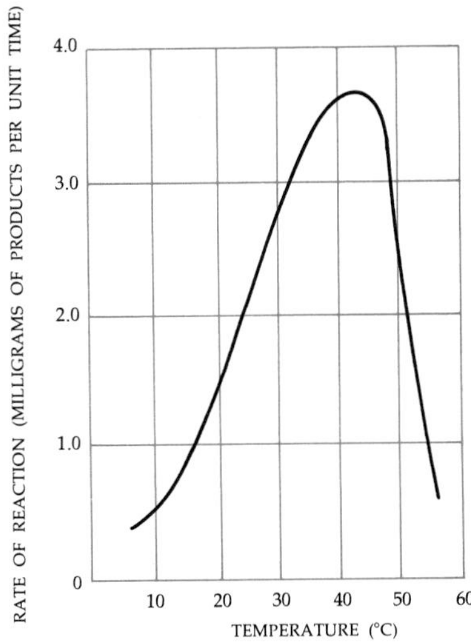

the pH changes, these charges change, and so the shape of the enzyme changes until it is so drastically altered that it is no longer functional. More important, probably, the charges of the active site and substrate are changed so that the binding capacity is affected. The optimum pH of one enzyme is not the same as that of another. The digestive enzyme pepsin, for example, works at the very low (highly acidic) pH of the stomach (page 43), in an environment where most other proteins would be permanently denatured.

Enzymatic Pathways

Enzymes typically work in series—the pathways we referred to earlier.

$$\text{product 1} \xrightarrow{\text{enzyme 1}} \text{product 2} \xrightarrow{\text{enzyme 2}} \text{product 3} \xrightarrow{\text{enzyme 3}} \text{product 4} \xrightarrow{\text{enzyme 4}} \text{product 5} \xrightarrow{\text{enzyme 5}} \text{end product}$$

Cells derive several advantages from this sort of arrangement. First, the groups of enzymes making up a common pathway can be segregated within the cell. Some are found in solution, as in the lysosomes, whereas others are embedded in the membranes of particular organelles. The enzymes located in membranes appear to be lined up in sequence, so the product of one reaction moves directly to the adjacent enzyme for the next reaction of the series. A second advantage is that there is little accumulation of intermediate products, since each product tends to be used up in the next reaction along the pathway. A third and most important advantage can be understood by considering the nature of chemical equilibrium.

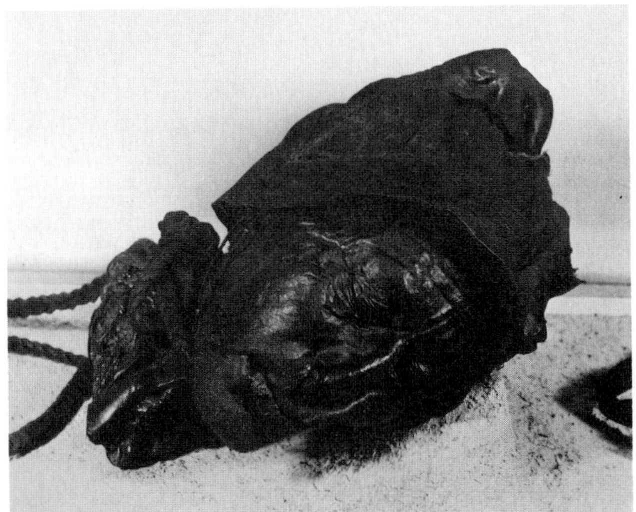

8–14

The body of a man with a noose around his neck was found in a Danish peat bog in 1950. He died some 2,000 years ago. The remarkable preservation of the body is the result of the extremely acidic pH of the peat bog, which almost completely inhibited the enzymatic activities of the microorganisms that customarily decompose organic molecules.

AUXOTROPHS

Sometimes, as a result of a genetic mutation, an organism is unable to synthesize a particular enzyme in an active form. When this occurs, the reactions of the pathway in which that enzyme participates cannot proceed to completion. The end product, which may be of critical importance to the organism, is not formed, and there may be, moreover, an accumulation of the substrate of the defective or missing enzyme. As we shall see in Chapter 18, such accumulations can lead to serious human disease or even death.

Although mutations resulting in defective or missing enzymes are no less serious for microorganisms, such as bacteria and fungi, they do provide biologists with a valuable tool for elucidating enzymatic pathways. Cells with a defect in a biosynthetic pathway are known as auxotrophs. They grow normally only if they are supplied with either the end product of the entire pathway or the product of the specific reaction normally catalyzed by the defective or missing enzyme. Auxotrophs can be produced by irradiating cells with x-rays; they are identified and isolated by their inability to grow on chemical media that will support the growth of normal cells. Studies of the exact chemical requirements of different auxotrophs have revealed the details of many biosynthetic pathways, particularly those involved in the synthesis of amino acids. As we shall see in Chapter 15, auxotrophs have also made a major contribution to our understanding of the mechanism by which hereditary information is converted into the structures and processes of the living cell.

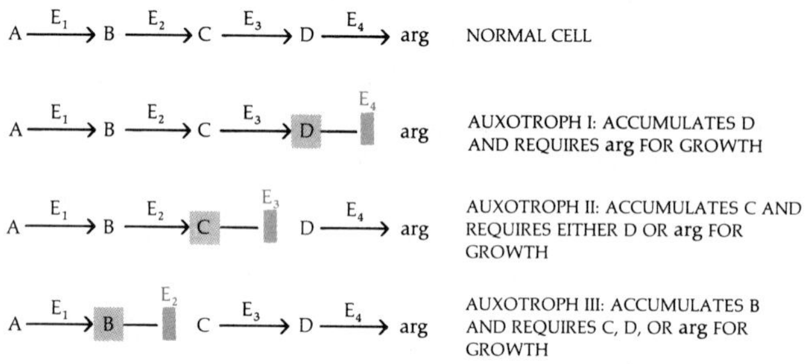

$A \xrightarrow{E_1} B \xrightarrow{E_2} C \xrightarrow{E_3} D \xrightarrow{E_4} arg$ NORMAL CELL

$A \xrightarrow{E_1} B \xrightarrow{E_2} C \xrightarrow{E_3} D \text{ — } E_4 \text{ | } arg$ AUXOTROPH I: ACCUMULATES D AND REQUIRES arg FOR GROWTH

$A \xrightarrow{E_1} B \xrightarrow{E_2} C \text{ — } E_3 \text{ | } D \xrightarrow{E_4} arg$ AUXOTROPH II: ACCUMULATES C AND REQUIRES EITHER D OR arg FOR GROWTH

$A \xrightarrow{E_1} B \text{ — } E_2 \text{ | } C \xrightarrow{E_3} D \xrightarrow{E_4} arg$ AUXOTROPH III: ACCUMULATES B AND REQUIRES C, D, OR arg FOR GROWTH

One group of auxotrophs of the red bread mold Neurospora crassa *have defective enzymes (indicated in color) at different points in the biosynthesis of the amino acid arginine (arg). These mutants, each deficient in one enzyme, retain the activity of the other enzymes of the arginine pathway. Four of the enzymes (E_1 through E_4) and four of the intermediate products (A through D) are shown in the diagram. The substance accumulated (indicated in gray) by each auxotroph and its requirements for growth reveal the sequence of the enzymes and the intermediates in the reaction pathway.*

As we have noted previously, chemical reactions can go in either direction. When net change ceases, the reaction is said to be at equilibrium. In the reaction

$$A + B \rightleftharpoons C + D$$

the point of equilibrium is reached when as many molecules of C and D are being converted to molecules of A and B as molecules of A and B are being converted to molecules of C and D.

The concentration of reactants does *not* have to equal the concentration of products in order for equilibrium to be established; only the *rates* of the forward and reverse reactions must be the same. Consider the reaction shown above. The

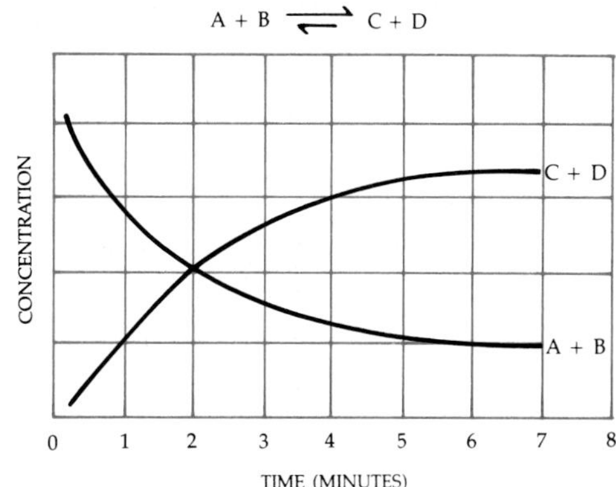

8–15
The changes in concentration of products and reactants in a reversible reaction. At first, only molecules of A and B are present. The reaction begins when A and B start to yield products C and D. At the end of two minutes, the concentrations of A + B and C + D are equal. As the reaction proceeds, the concentration of C + D will continue to increase to the point of chemical equilibrium (at about the sixth minute) and will thereafter remain greater than the concentration of A + B. This is the proportion at which the rates of forward and reverse reactions are the same.

different lengths of the arrows indicate that at equilibrium there is more C + D present than A + B. If only A and B molecules are present initially, the reaction occurs at first to the right, with A and B molecules converting into C and D molecules. Figure 8–15 shows the relative changes in concentration as the reaction continues. As C and D accumulate, the rate of the reverse reaction increases, and at the same time, the rate of the forward reaction decreases because of the decreasing concentrations of A and B. At about minute 6, the rates of the forward and reverse reactions equalize and no further changes in concentration take place. The proportions of A + B and C + D will remain the same. There will always be more C + D molecules in the system than A + B molecules, but the reaction will not go to completion—that is, not all of the A + B molecules will be converted to C + D molecules.

The relative proportions of A + B and C + D at equilibrium are determined by the free energy change (ΔG) of the reaction. Only if there were no net change in the free energy (that is, $\Delta G = 0$) would the concentrations of A + B and C + D be equal at equilibrium. The fact that the concentration of C + D molecules is greater than that of A + B molecules at equilibrium tells us that the potential energy of C + D is less than the potential energy of A + B. The reaction proceeding from A + B to C + D is exergonic (that is, it has a negative ΔG); conversely, the reaction from C + D to A + B is endergonic (positive ΔG). In any reversible reaction, the point of equilibrium will lie in the direction for which ΔG is negative. The larger the negative value of ΔG, the more strongly the reaction will be pulled in that direction.

This has important consequences in the sequential reactions that occur within living cells. If A + B and C + D molecules are in a closed system, equilibrium will be attained and there will be no subsequent change in their concentrations. If, however, the system is open and C + D molecules are continually removed from it, equilibrium will never be attained, and the conversion of A + B molecules into C + D molecules will continue. The sequential reactions of enzymatic pathways have the effect of removing the product of each reaction from the system, so that equilibrium is not attained. If, for example, product 2 in the enzymatic pathway shown on page 175 is used up (by being converted into product 3) almost as rapidly as it is formed, the reaction product 1 ⟶ product 2 can never reach equilibrium.

If the eventual end product is also used up, the whole series of reactions will move toward completion. Moreover, if any of the reactions along the pathway are highly exergonic (large negative ΔG), they will rapidly use up the products of the preceding reactions, pulling those reactions forward; similarly, the accumulation of the products from the exergonic reactions will push the subsequent reactions forward by increasing the concentrations of their reactants. The linking of reactions in enzymatic pathways, with exergonic reactions moving the entire series forward, is a key factor in the remarkable efficiency with which living organisms carry out their chemical activities.

Regulation of Enzyme Activity

Another remarkable feature of the metabolic activity of cells is the extent to which each cell regulates the synthesis of the products necessary to its well-being in the amounts and at the rates required, while avoiding overproduction, which would waste both energy and raw materials. Temperature, as we have noted, affects enzymatic reactions, as does pH; some enzymes are usually found at a pH that is not their optimum, suggesting that this discrepancy may not be an evolutionary oversight but a way of damping enzyme activity. The availability of reactant molecules or of cofactors is a principal factor in limiting enzyme action, and most enzymes probably work at a rate well below their maximum for this reason.

Living systems also have more precise ways of turning enzyme activity on and off. Some enzymes are produced only in an inactive form and, just when they are needed, are activated, usually by another enzyme. Pepsin, a digestive enzyme, is controlled in this way. It is synthesized by cells in the lining of the stomach in the form of pepsinogen, which contains 42 additional amino acids on the amino end of the molecule and is inactive. After pepsinogen is released into the stomach, a particular enzyme snips off these 42 amino acids, converting pepsinogen to pepsin, the active form. In this way, pepsin molecules (and other digestive enzymes) are prevented from digesting the proteins in the cells in which they are synthesized. Once activated, these enzymes cannot again be deactivated, however.

Allosteric Interactions

An ingenious mechanism by which an enzyme may be temporarily activated or inactivated is known as *allosteric interaction*. Allosteric interactions occur among enzymes that have at least two binding sites, one the active site and another, into which a second molecule, known as an allosteric effector, fits. The binding of the effector changes the shape of the enzyme molecule and either activates or inactivates it (Figure 8–16).

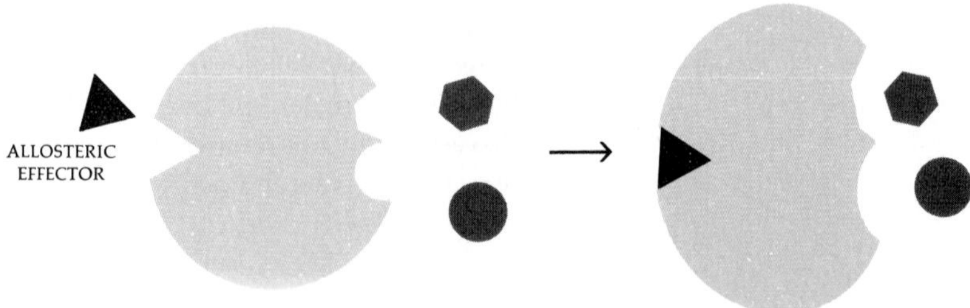

8–16
An allosteric ("other shape") effector can bind to an enzyme and, by disrupting the bonds determining its tertiary structure, alter the active site. As a consequence, the enzyme may be altered so that it cannot interact with its substrate, as shown here, or it may be activated.

SOME LIKE IT COLD

Enzyme action is exquisitely dependent on the tertiary and quaternary structure of the globular protein, particularly as it affects the active site. As a consequence, many enzymes are thermolabile—that is, they do not function at higher temperatures, even within the normal physiological range. One such enzyme is responsible for color in Siamese cats. It functions adequately in the cooler, peripheral areas of the body, such as the ears, nose, paws, and tip of the tail, but becomes inactive in the warmer areas of the body. For similar reasons, Himalayan rabbits are all black when raised at temperatures of about 5°C; white with black ears, forepaws, noses, and tails when raised at normal room temperatures; and all white when raised at temperatures above 35°C.

Thermolability has its advantages. In the northern seal, the newborns are white, as a result of their developing at a warm (internal) temperature. The newborns cannot swim and so are restricted to the ice floes, where their white coats provide them with color camouflage. By the time they are able to swim, their coats have turned brown and so blend with the dark Arctic waters. The thermolability of this same enzyme (tyrosinase) provides similar camouflage for the Arctic fox. During the summer it develops a white coat, which provides color protection during the winter months. During the winter, its dark coat develops, which is revealed when the white coat is shed in the spring. All of which illustrates that, as we shall see often, evolution is opportunistic.

(a)

(b)

(a) *The characteristic color pattern of the Siamese cat is a result of the thermolability of an enzyme affecting coat color. This enzyme, involved in controlling the synthesis of a dark pigment, is active only in the cooler, more peripheral areas of the body.*

(b) *Similarly, the enzyme controlling dark pigmentation in northern seals is active only at low temperatures. The newborns, which, like other mammals, develop within the warm bodies of their mothers, are white. New fur, growing in after the seals are exposed to the colder external environment, is brown. Although the white fur protects the newborn seals against most predators, it makes them highly desirable to human hunters, now their principal predators.*

8–17
Feedback inhibition. (a) *In this series of reactions, each step (the black arrows) is catalyzed by a specific enzyme. Enzyme* E_1*, which converts A to B, is allosteric, and the allosteric inhibitor is the product F. Thus, enzyme* E_1 *will be more active when amounts of F are low.* (b) *A branched metabolic pathway, A being converted to B, B being converted to C, and C being converted to both D and X. In this case, the enzyme at the branch,* E_3*, is allosteric. As the quantity of the allosteric inhibitor F increases, the upper branch of the pathway will become less active, and the pathway will be primarily shunted to the production of Z.*

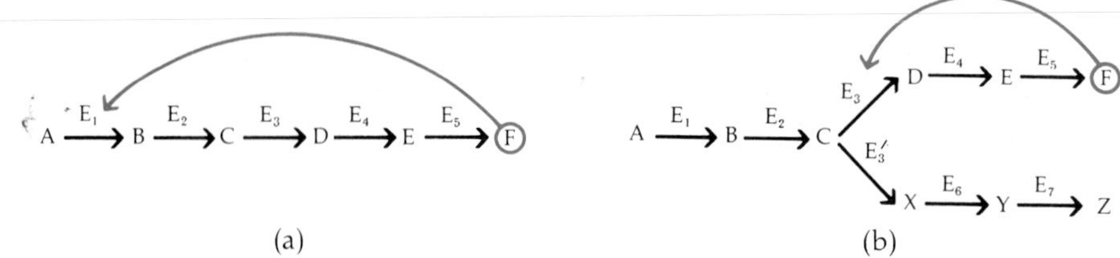

Allosteric interactions are often involved in *feedback inhibition*, which is a common means of biological control. A familiar nonbiological example of feedback inhibition is a thermostat that turns off the furnace when the room temperature reaches a desired level. In feedback inhibition of enzymatic reactions, one of the products, usually the last in the series, acts as an effector, inhibiting the function of one of the enzymes, often the first in the series (Figure 8–17a). Or, in a reaction that may take one of two directions, the effector may act to shunt the reactions along another pathway (Figure 8–17b).

Competitive Inhibition

Some compounds inhibit enzyme activity by temporarily occupying the active site of the enzyme; regulation in this way is known as competitive inhibition, because the regulatory compound and the substrate compete for the active site. The result of the competition depends on how many of each kind of molecule are present. For example, in the reaction series

$$A \xrightarrow{E_1} B \xrightarrow{E_2} C \xrightarrow{E_3} D \xrightarrow{E_4} E \xrightarrow{E_5} F$$

the final product F might be rather similar in structure to product D. F could occupy the active site of enzyme E_4, preventing D, the normal substrate, from binding to the enzyme. As F was used up by the cell, the active site of enzyme E_4 would once more become available to D.

Competitive inhibition is the mechanism of action of some drugs used to treat bacterial infections in animals. For instance, bacteria make the vitamin folic acid, which animal cells do not make (animals obtain folic acid from their food). One of the compounds in the metabolic pathway leading to folic acid is para-aminobenzoic acid (Figure 8–18). The drug sulfanilamide, as you can see, has a structure very similar to that of PABA. The two structures are so similar, in fact, that the enzyme involved in converting PABA to folic acid combines with the drug rather than with the PABA. Without folic acid, the bacterial cell dies, leaving the animal cell, which lacks this enzyme, unharmed.

8–18
para-Aminobenzoic acid (PABA) is one of the compounds in the metabolic pathway to folic acid in bacterial cells. Sulfanilamide, a drug, has a similar structure. It can combine with the enzyme that converts PABA to folic acid, thereby blocking the synthesis of folic acid, without which the bacterial cell cannot live.

H_2N–C_6H_4–C(=O)OH
PARA-AMINOBENZOIC ACID (PABA)

H_2N–C_6H_4–S(=O)(=O)–NH_2
SULFANILAMIDE

Noncompetitive Inhibition

In noncompetitive inhibition, the inhibitory chemical, which need not resemble the substrate, binds with the enzyme at a site on the molecule other than the active site. Lead, for instance, forms covalent bonds with sulfhydryl (SH) groups. Many enzymes contain cysteine, which has a sulfhydryl group. The binding of lead to such enzymes disrupts their tertiary structure and deactivates them, producing the symptoms associated with lead poisoning. Like competitive inhibition, noncompetitive inhibition is reversible, but such reversal is not accomplished by increased concentrations of substrate. In the case of lead, for example, the inhibition can be reversed by treatment with other sulfhydryl-containing compounds that bind the lead atoms more tightly than cysteine does.

8-19
The bacterial cell wall has polysaccharide backbones formed by alternating molecules of two sugars, NAG and NAM. The backbones (indicated by the thick black lines) are cross-linked by short peptide chains. One of the amino acids in these chains, lysine (shown in color), forms three peptide bonds—one with the amino acid above it, one with the amino acid below it, and another with the adjacent amino acid. These bonds are formed in an unusual way—by the enzymatic transfer of a peptide bond from one molecule to another. Penicillin blocks this transfer. Its structure mimics that of a dipeptide, and it binds to the active site of the enzyme, irreversibly inhibiting it. Without cross-links, the cell wall cannot hold the cell together. Thus penicillin acts specifically on growing bacterial cell walls.

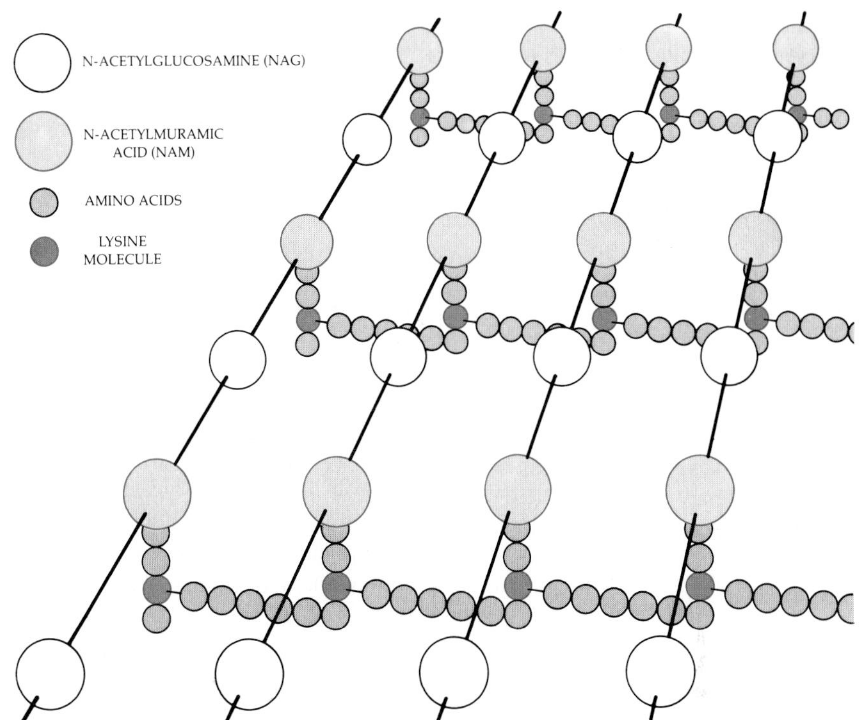

Irreversible Inhibition

Some substances inhibit enzymes irreversibly, either binding permanently with key functional groups of the active site or so thoroughly denaturing the protein that its tertiary structure cannot be restored. The nerve gases, which are among the most potent poisons known, irreversibly inhibit enzymes involved in the transmission of the nerve impulse, resulting in paralysis and death. Many useful drugs, including the antibiotic penicillin (Figure 8-19), are irreversible inhibitors of enzyme activity.

Regulation at the Source

Many enzymes are broken down rapidly, characteristically by other enzymes that hydrolyze peptide bonds. A highly efficient means of regulation of these quickly degraded enzymes is for the cell to produce them only when they are needed. The way in which bacterial cells turn on and off their enzyme production at the source is described in some detail in Chapter 15.

THE CELL'S ENERGY CURRENCY: ATP

All of the biosynthetic activities of the cell (and many other activities as well) require energy. A large proportion of this energy is supplied by a single molecule, *adenosine triphosphate*, or ATP. ATP is the cell's chief energy currency. Glucose and other carbohydrates are storage forms of energy and also forms in which energy is transferred from cell to cell and organism to organism. In a sense, they are like money in the bank. ATP, however, is like the change in your pocket or a subway token.

8–20
Adenosine triphosphate (ATP) is the cell's chief energy currency. The bonds between the three phosphate groups in the molecule are important in ATP function.

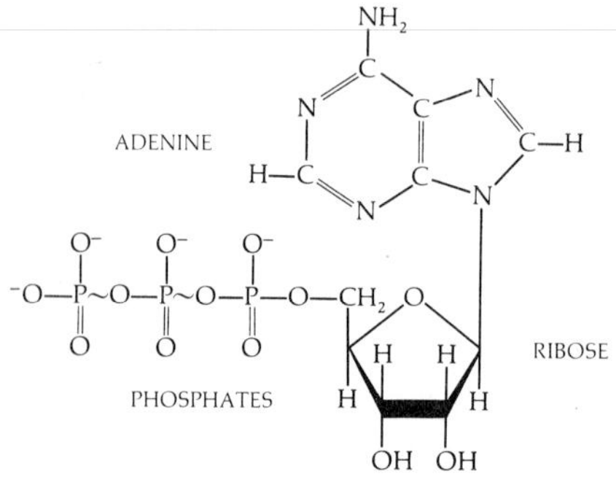

At first glance, ATP, as shown in Figure 8–20, appears to be a complex molecule. As with NAD^+, you will find, however, that its component parts are familiar to you. It is made up of adenine, the five-carbon sugar ribose, and three phosphate groups. These three phosphate groups with strong negative charges are covalently bonded to one another; this is an important feature in ATP function. Three linked phosphate groups are also characteristic of other molecules that play a role similar to that of ATP in certain cellular reactions; for example, guanosine triphosphate (GTP), another energy-carrying molecule, differs from ATP only in the substitution of the nitrogenous base guanine for adenine.

To understand the role of ATP and related triphosphate compounds, we must return briefly to the concept of the chemical bond. Because a chemical bond is a stable configuration of electrons, reacting molecules must possess a certain amount of energy in order to collide with sufficient force to overcome their mutual repulsion and to weaken existing chemical bonds, allowing the formation of new bonds. This energy is the energy of activation (Figure 8–7). Because of enzymes, which reduce the required energy of activation to a level already possessed by a significant proportion of the reacting molecules, the reactions essential to life are able to proceed at an adequate rate. As we have seen, however, the *direction* in which a reaction proceeds is determined by the free energy change, ΔG. Only if the reaction is exergonic (negative ΔG), will it proceed to any significant extent. Yet many cellular reactions, including synthetic reactions such as the formation of a disaccharide from two monosaccharide molecules, are endergonic (positive ΔG). In such a reaction, the electrons forming the chemical bonds of the product are at a higher energy level than the electrons in the bonds of the starting materials—that is, the potential energy of the product is greater than the potential energy of the reactants, an apparent violation of the second law of thermodynamics. Cells circumvent this difficulty by *coupled reactions* in which endergonic reactions are linked to exergonic reactions that provide a surplus of energy, making the entire process exergonic and thus able to proceed spontaneously. The molecule that most frequently supplies energy in such coupled reactions is ATP.

The internal structure of the ATP molecule makes it unusually suited to this role in living systems. In the laboratory, energy is released from the ATP molecule when the third phosphate is removed by hydrolysis, leaving ADP (adenosine diphosphate) and a phosphate:

$$ATP + H_2O \longrightarrow ADP + \text{phosphate}$$

In the course of this reaction, about 7 kilocalories of energy are released per mole of ATP. Removal of the second phosphate produces AMP (adenosine monophosphate) and releases an equivalent amount of energy.

$$\text{ADP} + \text{H}_2\text{O} \longrightarrow \text{AMP} + \text{phosphate}$$

The covalent bonds linking these two phosphates to the rest of the molecule are symbolized by a squiggle, ~, and were, for many years, called "high-energy" bonds—an incorrect and confusing term. These bonds are not strong bonds, like the covalent bonds between carbon and hydrogen, which have a bond energy of 98.8 kilocalories per mole. They are instead bonds that are easily broken, releasing an amount of energy—about 7 kilocalories per mole—adequate to drive many of the essential endergonic reactions of the cell. Moreover, the energy released does not arise entirely from the movement of the bonding electrons to lower energy levels. It is also a result of a rearrangement of the electrons in other orbitals of the ATP or ADP molecules. The phosphate groups each carry negative charges and so tend to repel each other. When a phosphate group is removed, the molecule as a whole undergoes a change in electron configuration that results in a structure with less energy.

ATP in Action

In living cells, ATP is sometimes directly hydrolyzed to ADP plus phosphate, releasing energy for a variety of activities. ATP hydrolysis provides, for example, a means for producing heat, as in an animal waking up from hibernation. Enzymes catalyzing the hydrolysis of ATP are known as ATPases; a variety of different ATPases have been identified. The protein "arms" in cilia and flagella (page 115), for example, are ATPase molecules, catalyzing the energy release that causes the microtubules to move past one another. The proteins that pump ions through cellular membranes against a concentration gradient (page 134) are not only carriers but also ATPases, releasing energy to power the pumping process.

Usually, however, the terminal phosphate group of ATP is not simply removed but is transferred to another molecule. This addition of a phosphate group is known as *phosphorylation*; enzymes that catalyze such transfers are known as kinases. Phosphorylation reactions transfer some of the energy of the phosphate group in the ATP molecule to the phosphorylated compound, which, thus energized, participates in a subsequent reaction.

For example, in the reaction

$$\text{W} + \text{X} \longrightarrow \text{Y} + \text{Z}$$

if the potential energy of W plus the potential energy of X were less than that of Y plus Z, the reaction would not take place to any significant extent. Chemists could drive the reaction forward by supplying outside energy, probably in the form of heat. The cell might handle it this way:

$$\text{W} + \text{ATP} \longrightarrow \text{W-P} + \text{ADP}$$

The potential energy of the products is less than that of the reactants, so the reaction will take place. However, much of the energy made available when the phosphate group was removed from ATP is conserved in the new compound W-phosphate, or W-P.

The next step in the reaction becomes

$$\text{W-P} + \text{X} \longrightarrow \text{Y} + \text{Z} + \text{P}$$

With the release of the phosphate from W, this second reaction also becomes one in which the potential energy of the products is less than the potential energy of the reactants and which, therefore, can take place.

Take, for instance, the formation of sucrose in sugarcane.

$$\text{glucose} + \text{fructose} \longrightarrow \text{sucrose} + H_2O$$

In this reaction, the potential energy of the products is 5.5 kilocalories per mole greater than the potential energy of the reactants. However, the sugarcane plant carries out this synthesis through a series of reactions, coupled to the breakdown of ATP and the accompanying phosphorylation of the glucose and fructose molecules. The overall reaction is

$$\text{glucose} + \text{fructose} + 2\text{ATP} \longrightarrow \text{sucrose} + 2\text{ADP} + 2\text{P}$$

Since the potential energy of 2 ADPs is about 14 kilocalories per mole less than the potential energy of 2 ATPs, the overall difference in products and reactants becomes 8.5 kilocalories per mole. The coupling of reactions permits sugarcane to form sucrose.

Where does the ATP come from? As we shall see in the next chapter, energy released in the cell's catabolic reactions, such as the breakdown of glucose, is used to "recharge" the ADP molecule to ATP. Thus the ATP/ADP system serves as a universal energy-exchange system, shuttling between energy-releasing reactions and energy-requiring ones.

One marvels at the process of evolution as exemplified by the intricate flower of an orchid, the shell of a chambered nautilus, or the opposable thumb and forefinger of the human hand. Remember that ATP, NAD, and indeed the place of each amino acid in the polypeptide chain of an enzyme are also the products of evolution and also, you must admit, quite marvelous. Perhaps even beautiful.

SUMMARY

Living systems convert energy from one form to another as they carry out essential functions of maintenance, growth, and reproduction. In these energy conversions, as in all others, some useful energy is lost to the surroundings at each step.

The laws of thermodynamics govern transformations of energy. The first law states that energy can be converted from one form to another but cannot be created or destroyed. The potential energy of the initial state (or reactants) is equal to the potential energy of the final state (or products) plus the energy released in the process or reaction. The second law of thermodynamics states that in the course of energy conversions, the potential energy of the final state will always be less than the potential energy of the initial state. The difference in potential energy between the initial and final states is known as the free energy change and is symbolized as ΔG. Exergonic (energy-yielding) reactions have a negative ΔG, and endergonic (energy-requiring) reactions have a positive ΔG. Factors that determine the ΔG

include ΔH, the change in heat content, and ΔS, the change in entropy, which, multiplied by the absolute temperature (T), is a measure of randomness or disorder:

$$\Delta G = \Delta H - T\,\Delta S$$

Another way of stating the second law of thermodynamics is that all natural processes tend to proceed in such a direction that the entropy of the system increases. To maintain the organization on which life depends, living systems must have a constant supply of energy to overcome the tendency toward increasing disorder. The sun is the original source of this energy.

The energy transformations in living cells involve the movement of electrons from one energy level to another and, often, from one atom or molecule to another. Reactions in which electrons move from one atom to another are known as oxidation-reduction reactions. An atom or molecule that loses electrons is oxidized; one that gains electrons is reduced.

Metabolism is the total of all the chemical reactions that take place in cells. Reactions resulting in the breakdown or degradation of molecules are known, collectively, as catabolism. Biosynthetic reactions are called anabolism. Metabolic reactions take place in series, called pathways, each of which serves a particular function in a cell. Each step in the pathway is controlled by a specific enzyme. The stepwise reactions of enzymatic pathways enable cells to carry out their chemical activities with remarkable efficiency in terms of both energy and materials.

Enzymes serve as catalysts, lowering the energy of activation and thus enormously increasing the rate at which reactions take place. The enzymes remain themselves unchanged in the process. They are large protein molecules folded in such a way that particular groups of amino acids form an active site. The reacting molecules, known as the substrate, fit precisely into this active site. Many enzymes require cofactors, which may be simple ions, such as Mg^{2+} or Ca^{2+}, or nonprotein organic molecules such as NAD. The latter are known as coenzymes, and they function as electron carriers. Different coenzymes hold electrons at slightly different energy levels. Many vitamins are parts of coenzymes.

Enzyme-catalyzed reactions are under tight cellular control. The rate of enzymatic reactions is affected by temperature and pH, which affect the attractions among the amino acids of the protein molecule and also between the active site and substrate.

A principal means for enzyme control is allosteric interaction. Allosteric interaction occurs when a molecule other than the substrate combines with an enzyme at a site other than the active site and in so doing alters the shape of the active site to render it either functional or nonfunctional. Feedback inhibition occurs when the product of an enzymatic reaction at the end or at a branch of a particular pathway acts as an allosteric effector, temporarily inhibiting the activity of an enzyme earlier in the pathway and thus temporarily stopping the series of chemical reactions.

Enzymes may also be regulated by competitive inhibition, in which another molecule, similar to the normal substrate, competes for the active site. Competitive inhibition can be reversed by increased concentrations of the substrate. Noncompetitive inhibitors bind elsewhere on the molecule, altering the tertiary structure so that the enzyme cannot function. Noncompetitive inhibition is also reversible, but not by the substrate. Irreversible inhibitors bind permanently to the active site or irreparably disrupt the tertiary structure. Another way in which cells regulate enzymes—and thus the energy transformations they control—is by producing them only when they are needed.

ATP supplies the energy that drives most of the endergonic processes that take place in living systems. The ATP molecule consists of a nitrogenous base, adenine; a five-carbon sugar, ribose; and three phosphate groups. The three phosphate groups are linked by two covalent bonds that are easily broken, each yielding about 7 kilocalories of energy per mole. ATP participates as an energy carrier in most series of reactions that take place in the cell.

QUESTIONS

1. Distinguish among the following terms: the first law of thermodynamics/the second law of thermodynamics; oxidation/reduction; metabolism/catabolism/anabolism; active site/substrate; competitive inhibition/noncompetitive inhibition/irreversible inhibition; ATP/ADP/AMP.

2. Why, in Figure 8–1, are there more plants than zebras and more zebras than lions? (Explain in terms of thermodynamics.)

3. At present, at least four types of energy conversions are going on in your body. Name them.

4. All natural processes proceed with an increase in entropy. How then do you explain the freezing of water?

5. The laws of thermodynamics apply only to closed systems, that is, to systems into which no energy is entering. Is an aquarium ordinarily a closed system? Could you convert it to one? A spaceship may or may not be a closed system, depending on certain features of its design. What would these features be? Is the earth a closed system?

6. Explain why it is that living systems, despite appearances, are not in violation of the second law of thermodynamics.

7. What is there about the orderliness of a living organism that most significantly distinguishes it from the orderliness of a machine, such as a computer or the telephone system?

8. What is the basis for the specificity of enzyme action? What is the advantage to the cell of such specificity? What might be its disadvantages?

9. Turn back to Figure 3–17, in which all the amino acids are shown, and try to make some educated guesses about which amino acids can substitute for one another in the structure of an enzyme, and what substitutions would have drastic effects.

10. When a plant does not have an adequate supply of an essential mineral, such as magnesium, it is likely to become sickly and may die. When an animal is deprived of a particular vitamin in its diet, it too is likely to become ill and may die. What is a likely explanation of such phenomena?

11. Most organisms cannot live at high temperatures. Explain at least one way in which high temperatures are harmful to organisms. Some bacteria and algae, however, live in hot springs, at temperatures far above those that can be tolerated by most organisms. How might such bacteria and algae differ from most other organisms?

12. In enzyme regulation by allosteric interaction, the inhibitor often works on the first enzyme of the series. In regulation by competitive inhibition, it often works on the last. Why this difference?

13. In a series of experiments with an enzyme that catalyzes a reaction involving substrate A, it was found that a particular substance X inhibited the enzyme. When the concentration of A was high and the concentration of X was low, the reaction proceeded rapidly; as the concentration of X was increased and that of A was decreased, the reaction slowed down; when the concentration of X was high and that of A was low, the reaction stopped. If the concentration of A was again increased, the reaction resumed. How can you explain these results?

14. When a sulfa drug, such as sulfanilamide, is prescribed for a bacterial infection, it is very important to remember to take the drug at the prescribed times and in the prescribed quantity. Why is this essential? Suppose you were instructed to take two tablets every three hours, and instead you took only one tablet every five hours. What do you think would happen?

15. Some human societies use the barter system for exchange of goods and services. However, all complex societies have some form of monetary exchange. What are the advantages of a monetary exchange? Relate your answer to the ADP/ATP system.

CHAPTER 9

How Cells Make ATP: Glycolysis and Respiration

ATP is the principal energy carrier in living systems. It participates in a great variety of cellular events, from chemical biosyntheses, to the flick of a cilium, the twitch of a muscle, or the active transport of a molecule across a cell membrane. It is involved in the propagation of an electric impulse along a nerve or, in some remarkable organisms, the electrocution of prey (Figure 9–1). In the following pages, we shall show in some detail how a cell breaks down carbohydrates and captures and stores the released energy in the terminal phosphate bonds of ATP. The oxidation of glucose (or other carbohydrates) is complicated in detail—so go slowly—but simple in its overall design.

AN OVERVIEW OF GLUCOSE OXIDATION

Oxidation, you will recall, is the loss of an electron. Reduction is the gain of an electron. Since, in spontaneous oxidation-reduction reactions, electrons go from higher to lower energy levels, a molecule usually releases energy as it is oxidized. In the oxidation of glucose, carbon-carbon bonds, carbon-hydrogen bonds, and oxygen-oxygen bonds are exchanged for carbon-oxygen and hydrogen-oxygen bonds, as oxygen atoms attract and hoard electrons. The summary equation for this process is

$$\text{glucose} + \text{oxygen} \longrightarrow \text{carbon dioxide} + \text{water} + \text{energy}$$

Or,

$$C_6H_{12}O_6 + 6O_2 \longrightarrow 6CO_2 + 6H_2O$$
$$\Delta G = -686 \text{ kcal/mole}$$

Living systems are experts at energy conversions. They are organized to trap this free energy so that it will not be dissipated randomly but can be used to do the work of the cell. About 40 percent of the free energy released by the oxidation of glucose is conserved in the conversion of ADP to ATP. As you will recall, about 75 percent of the energy in gasoline is "lost" as heat in an automobile engine, and only 25 percent is converted to useful forms of energy. The living cell is significantly more efficient.

In living systems, the breakdown of glucose takes place in two major stages. The first is known as *glycolysis*. The second is *respiration*, which, in turn, consists of two stages: the *Krebs cycle* and terminal *electron transport*. Glycolysis occurs in the cy-

9-1
(a) *The electric ray converts the chemical energy of ATP—provided by the oxidation of glucose and other organic compounds—to electrical energy, stunning and immobilizing its prey with electric discharges.* (b) *The prey shown here, a small reef fish, is then moved to the mouth by the pectoral fins and* (c) *is swallowed. The reef fish will be converted to chemical energy, which will, in turn, be converted to kinetic and electrical energy for the capture of additional prey. Only a fraction of the potential energy is transferred at each passage.*

Some other types of electric fish produce discharges of lower voltage that are used in establishing territories, in locating objects (including both potential prey and potential predators), and, perhaps, in communicating with members of the same species.

(a)

(b)

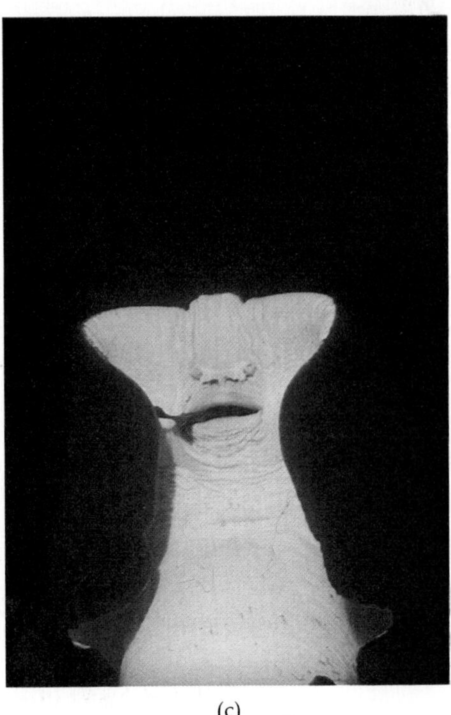

(c)

toplasm of the cell, and the two stages of respiration take place within the mitochondrion.

In glycolysis, the six-carbon glucose molecule is split into two molecules of a three-carbon compound, pyruvic acid (Figure 9-2). The electrons are accepted not by oxygen but by NAD^+ (page 174). The free energy change, ΔG, of this stage is -143 kcal/mole; this represents a relatively small proportion of the total energy of the glucose molecule. In respiration, the electrons and protons removed from the carbon atoms are accepted by oxygen. For this series of reactions, ΔG is -543 kcal/mole, a comparatively large energy yield.

In the course of glycolysis and respiration, about 38 molecules of ATP are regenerated from ADP in the breakdown of each molecule of glucose. As we shall see, the exact number of ATP molecules produced depends on at least two variables.

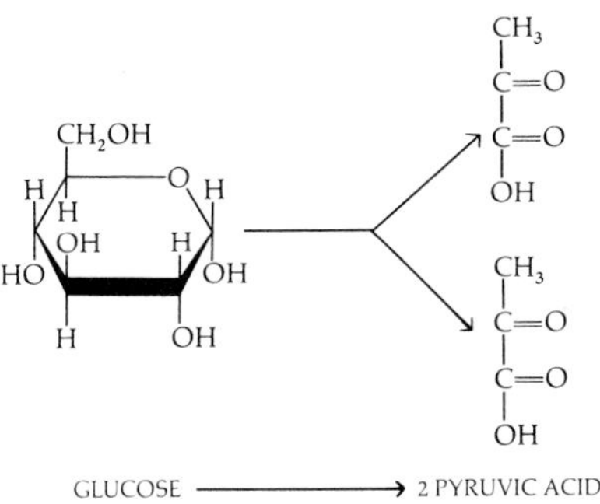

9-2
In glycolysis, the six-carbon glucose molecule is split into two three-carbon molecules of a compound known as pyruvic acid.

GLYCOLYSIS

Glycolysis—the lysis (splitting) of glucose—exemplifies the way the biochemical processes of a living cell proceed in small sequential steps. It takes place in a series of nine reactions, each catalyzed by a specific enzyme. Recent evidence suggests that these enzymes are sequentially arranged on portions of the microtrabecular network of the cytoskeleton.

As we examine the details of glycolysis, notice how the carbon skeleton is dismembered and its atoms rearranged step by step. Note especially the formation of ATP from ADP and of NADH from NAD^+. ATP and NADH represent the cell's net energy harvest from this reaction pathway.

9-3
The steps of glycolysis.

GLUCOSE

Step 1 HEXOKINASE ATP ADP

GLUCOSE 6-PHOSPHATE

Step 2 PHOSPHOGLUCO-ISOMERASE

FRUCTOSE 6-PHOSPHATE

Step 3 PHOSPHO-FRUCTOKINASE ATP ADP

FRUCTOSE 1,6-DIPHOSPHATE

Step 1. The first steps in glycolysis require an input of energy, which is supplied by coupling these steps to the ATP/ADP system. The terminal phosphate group is transferred from an ATP molecule to the carbon in the sixth position of the glucose molecule, to make glucose 6-phosphate. The reaction of ATP with glucose to yield glucose 6-phosphate and ADP is an exergonic reaction. Some of the energy released is conserved in the chemical bond linking the phosphate to the glucose molecule, which then becomes energized. This reaction is catalyzed by a specific enzyme (hexokinase), and each of the reactions that follows is similarly regulated by a specific enzyme.

Step 2. The molecule is reorganized, again with the help of a particular enzyme. The six-sided ring characteristic of glucose becomes the five-sided fructose ring. (As you know, glucose and fructose both have the same number of atoms—$C_6H_{12}O_6$—and differ only in the arrangements of these atoms.) This reaction can proceed approximately equally well in either direction; it is pushed forward by the accumulation of glucose 6-phosphate and the removal of fructose 6-phosphate as the latter enters Step 3.

Step 3. In this step, which is similar to Step 1, fructose 6-phosphate gains a second phosphate by the investment of another ATP. The added phosphate is bonded to the first carbon, producing fructose 1,6-diphosphate, that is, fructose with phosphates in the 1 and 6 positions. Note that in the course of the reactions thus far, two molecules of ATP have been converted to ADP and no energy has been recovered.

The enzyme catalyzing this step, phosphofructokinase, is an allosteric enzyme, and ATP is an allosteric effector inhibiting this enzyme. The allosteric interaction between them is the chief regulatory mechanism of glycolysis. If ATP is present in adequate quantities for other purposes of the cell, ATP inhibits the activity of the enzyme and so ATP production ceases and glucose is conserved. As the cell uses up its supply of ATP, the enzyme is released from inhibition and the breakdown of glucose resumes. This is one of the major control points of ATP production.

Step 4. The six-carbon sugar molecule is split into two three-carbon molecules, dihydroxyacetone phosphate and glyceraldehyde phosphate. The two molecules are interconvertible by the enzyme isomerase. However, because the glyceraldehyde phosphate is used up in subsequent reactions, all of the dihydroxyacetone phosphate is eventually converted to glyceraldehyde phosphate. Thus, all subse-

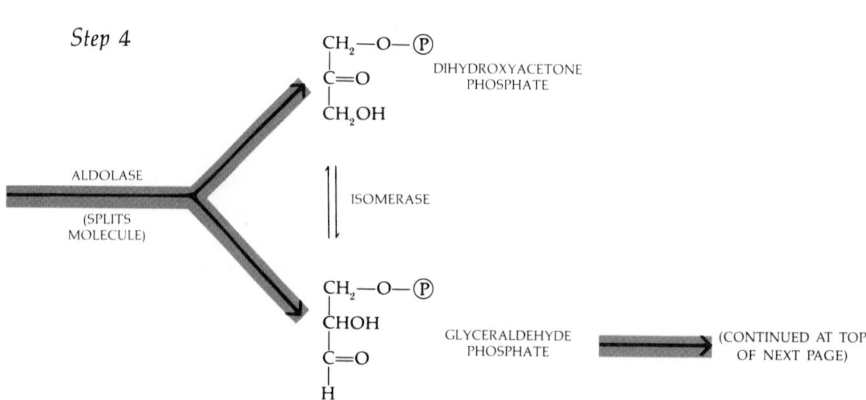

quent steps must be counted twice to account for the fate of one glucose molecule. With the completion of Step 4, the preparatory reactions are complete.

Step 5. Glyceraldehyde phosphate molecules are oxidized—that is, hydrogen atoms with their electrons are removed—and NAD^+ is reduced to NADH. This is the first reaction from which the cell harvests energy. Some of the energy from this oxidation reaction is also conserved in the attachment of a phosphate group to what is now the 1 position of the glyceraldehyde phosphate molecule. (The designation P_i indicates inorganic phosphate available as a phosphate ion in solution in the cytoplasm.) The properties of this bond are similar to those of the phosphate bonds of ATP, as indicated by the squiggle.

Step 6. This phosphate is released from the diphosphoglycerate molecule and used to recharge a molecule of ADP (a total of two molecules of ATP per molecule of glucose). This is a highly exergonic reaction (the ΔG is negative and large), and so it pulls all the preceding reactions forward.

Step 7. The remaining phosphate group is enzymatically transferred from the 3 position to the 2 position.

Step 8. In this step, a molecule of water is removed from the three-carbon compound. This internal rearrangement of the molecule concentrates energy in the vicinity of the phosphate group.

Step 9. The phosphate is transferred to a molecule of ADP, forming another molecule of ATP (again, a total of two molecules of ATP per molecule of glucose). This is also a highly exergonic reaction and thus pulls forward the preceding two reactions (Steps 7 and 8).

Summary of Glycolysis

The complete sequence begins with one molecule of glucose. Energy is invested at Steps 1 and 3 by the transfer of a phosphate group from an ATP molecule—one at each step—to the sugar molecule. The six-carbon molecule splits at Step 4, and from this point onward, the sequence yields energy. At Step 5, a molecule of NAD^+ is reduced to NADH, storing some of the energy from the oxidation of glyceraldehyde phosphate. At Steps 6 and 9, molecules of ADP take energy from the system, becoming phosphorylated to ATP.

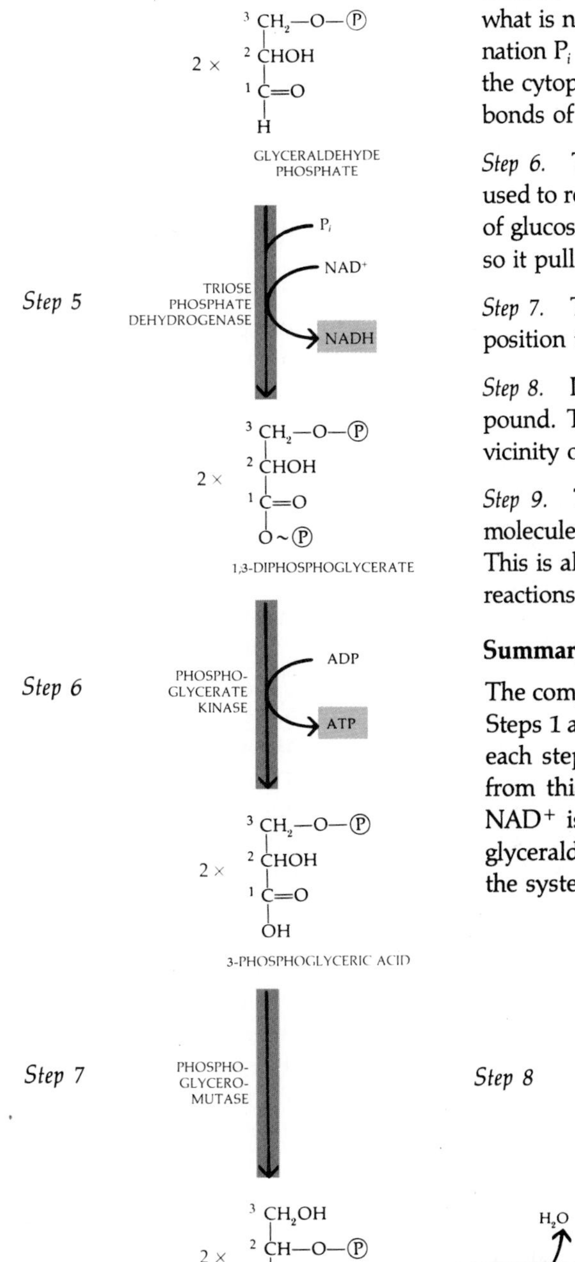

9-4
Summary of the two stages of glycolysis. The first stage utilizes 2ATP; the second stage yields 4ATP and 2NADH. Compounds other than glucose, such as the sugars galactose, mannose, and the pentoses, as well as glycogen and starch, can undergo glycolysis once they have been converted to glucose 6-phosphate.

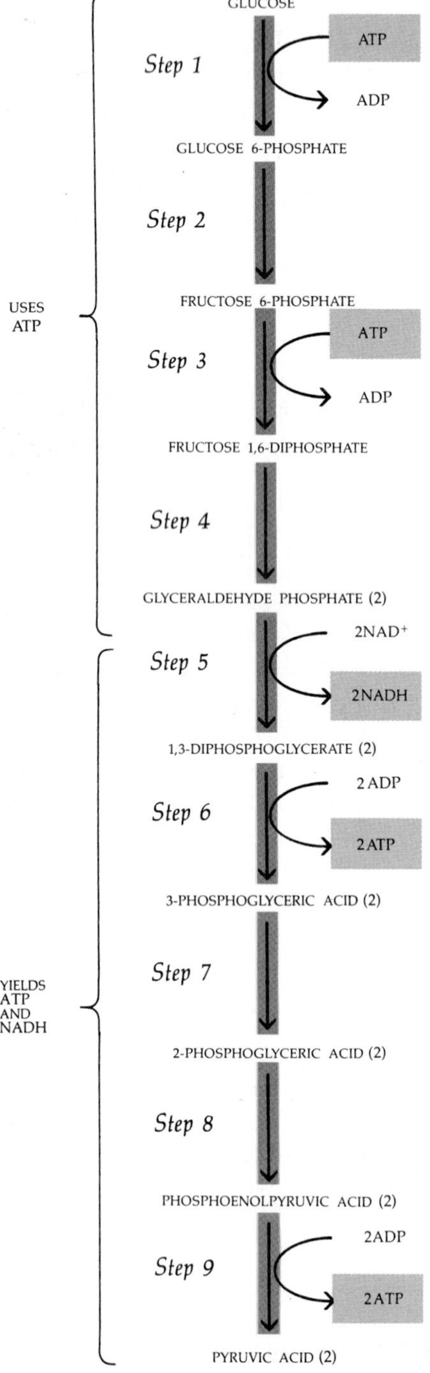

To sum up: The energy from the phosphate bonds of two ATP molecules is needed to initiate the glycolytic sequence. Subsequently, two NADH molecules are produced from two NAD^+ and four ATP molecules from four ADP:

$$\text{glucose} + 2\text{ATP} + 4\text{ADP} + 2\text{P}_i + 2\text{NAD}^+ \longrightarrow 2\text{ pyruvic acid} + 2\text{ADP} + 4\text{ATP} + 2\text{NADH} + 2\text{H}^+ + 2\text{H}_2\text{O}$$

Thus one glucose molecule has been converted to two molecules of pyruvic acid. The net harvest—the energy recovered—is two molecules of ATP and two molecules of NADH per molecule of glucose. The two molecules of pyruvic acid still contain a large amount of the potential energy that was stored in the original glucose molecule. This series of reactions is carried out by virtually all living cells—from prokaryotes to the eukaryotic cells of our own bodies.

ANAEROBIC PATHWAYS

Pyruvic acid can follow one of several pathways. Two pathways are anaerobic (without oxygen), and one is aerobic. We shall briefly discuss the anaerobic pathways and then follow the aerobic one, which is the principal pathway of energy metabolism for most cells in the presence of oxygen.

In the absence of oxygen, pyruvic acid can be converted either to ethanol (ethyl alcohol) or to lactic acid, depending on the type of cell. For example, yeast cells, present as a "bloom" on the skin of grapes, can grow either with or without oxygen. When the sugar-filled juices of grapes and other fruits are extracted and stored under anaerobic conditions, the yeast cells turn the fruit juice to wine by converting glucose into ethanol (Figure 9-5). When the sugar is exhausted, the yeast cells cease to function; at this point, the alcohol concentration is between 12 and 17 percent, depending on the variety of the grapes and the season at which they were harvested.

The formation of alcohol from sugar is called *fermentation.* Because of the economic importance of the wine industry, fermentation was the first enzymatic process to be intensively studied. (In fact, for many years enzymes were commonly referred to as "ferments." The scope of their activities is now known to be so broad that this word is no longer appropriate.) Louis Pasteur was the first to recognize the role of yeast cells in the process. The fermentation process was the setting for a phase of the vitalism-reductionism controversy (page 172) that came to an end around the beginning of the twentieth century.

Lactic acid is formed from pyruvic acid by a variety of microorganisms and also by some animal cells when O_2 is scarce or absent (Figure 9-6). It is produced, for example, in muscle cells during strenuous exercise, as by an athlete during a sprint. We breathe hard when we run fast, thereby increasing the supply of oxygen, but even this increase may not be enough to meet the immediate needs of the muscle cells. These cells, however, can continue to work by accumulating what is known as an oxygen debt. Glycolysis continues, using glucose released from glycogen stored in the muscle, but the resulting pyruvic acid does not enter the aerobic pathway of respiration. Instead, it is converted to lactic acid, which, as it accumulates, lowers the pH of muscle and reduces the capacity of the muscle fibers to contract, producing the sensations of muscle fatigue. The lactic acid diffuses into the blood and is carried to the liver. Later, when oxygen is more abundant (as a result of the deep breathing that follows strenuous exercise) and ATP demand is

$$\underset{\text{PYRUVIC ACID (FROM GLYCOLYSIS)}}{\begin{array}{c}CH_3\\ |\\ C{=}O\\ |\\ C{=}O\\ |\\ OH\end{array}} \xrightarrow{\ CO_2\ } \underset{\text{ACETALDEHYDE}}{\begin{array}{c}CH_3\\ |\\ C{=}O\\ |\\ H\end{array}} \xrightarrow{\ NADH \to NAD^+\ } \underset{\text{ETHANOL}}{\begin{array}{c}CH_3\\ |\\ H{-}C{-}OH\\ |\\ H\end{array}}$$

(a)

(b)

9–5

(a) The steps by which pyruvic acid, formed by glycolysis, is converted anaerobically to ethanol (ethyl alcohol). In the first step, carbon dioxide is released. In the second, NADH is oxidized, and acetaldehyde is reduced. Most of the energy of the glucose remains in the alcohol, which is the end product of the sequence. However, by regenerating NAD^+, these steps allow glycolysis to continue, with its small but sometimes vitally necessary yield of ATP.

(b) Consequences of anaerobic glycolysis. Yeast cells, visible on the grapes as dustlike "bloom," mix with the juice when the grapes are crushed. Storing the mixture under anaerobic conditions causes the yeast to break down the glucose in the grape juice to alcohol.

reduced, the lactic acid is resynthesized to pyruvic acid and back again to glucose or glycogen.

Why is pyruvic acid converted to lactic acid, only to be converted back again? The function of this reaction is simple: it regenerates the NAD^+ without which glycolysis cannot go forward (see Step 5, page 191). Even though this step seems to be wasteful in terms of energy consumption, it may be all-important in the economy of the organism, spelling the difference between life and death when an animal "out of breath" needs one last burst of ATP to escape from a predator or catch a prey.

An oxygen debt is usually accumulated during a short burst of intensive exercise, as during a rapid sprint. During sustained moderate exercise, the intake of oxygen by the lungs and the circulation of the blood supplying oxygen to the muscle tissues often catch up with the oxygen consumption in the muscle, thus producing the "second wind" phenomenon well known to runners.

The fact that glycolysis does not require oxygen suggests that the glycolytic sequence evolved early, before free oxygen was present in the atmosphere. Presumably, primitive one-celled organisms used glycolysis (or something very much like it) to extract energy from the organic compounds they absorbed from their watery surroundings.

$$\underset{\text{PYRUVIC ACID (FROM GLYCOLYSIS)}}{\begin{array}{c}CH_3\\ |\\ C{=}O\\ |\\ C{=}O\\ |\\ OH\end{array}} \xrightarrow{\ NADH \to NAD^+\ } \underset{\text{LACTIC ACID}}{\begin{array}{c}CH_3\\ |\\ H{-}C{-}OH\\ |\\ C{=}O\\ |\\ OH\end{array}}$$

9–6

The enzymatic reaction that produces lactic acid from pyruvic acid anaerobically in muscle cells. In the course of this reaction, NADH is oxidized and pyruvic acid is reduced. The NAD^+ molecules produced in this reaction and the one shown in Figure 9–5 are recycled in the glycolytic sequence. Without this recycling, glycolysis cannot proceed. Lactic acid accumulation results in muscle soreness and fatigue.

RESPIRATION

In the presence of oxygen, the next stage in the breakdown of glucose involves the stepwise oxidation of pyruvic acid to carbon dioxide and water—the process known as respiration. Respiration has two meanings in biology. One is the breathing in of oxygen and breathing out of carbon dioxide; this is also the ordinary, nontechnical meaning of the word. The second meaning of respiration is the oxidation of food molecules by cells. This process, sometimes qualified as cellular respiration, is what we are concerned with here.

9–7

Mitochondria are surrounded by two membranes. The inner membrane folds inward to make a series of shelves, or cristae. The enzymes and electron carriers involved in the final stage of cellular respiration are built into these internal membranes. The matrix is a dense solution containing enzymes involved in the earlier stages of cellular respiration, coenzymes, phosphates, and other solutes.

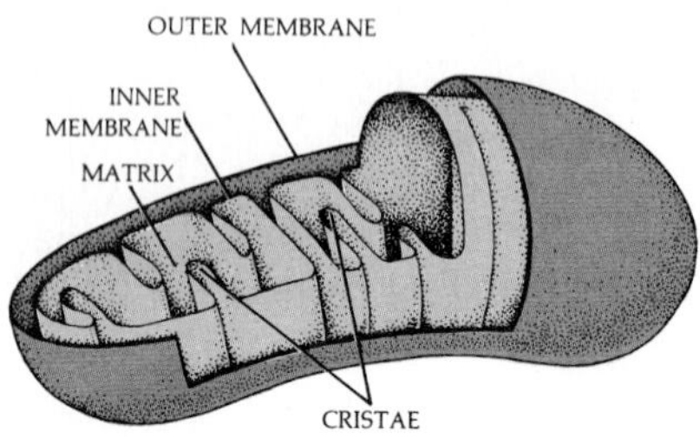

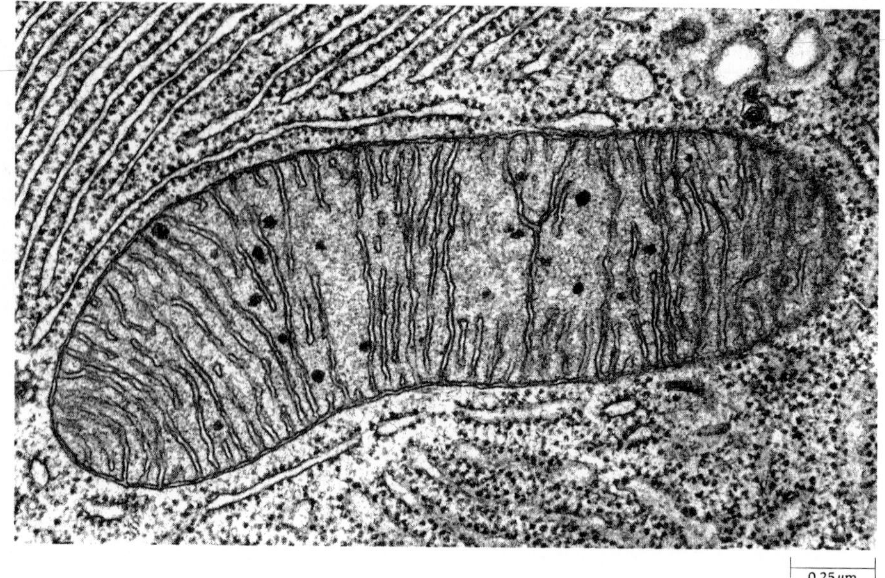

As we noted previously, cellular respiration takes place in two stages: the Krebs cycle and terminal electron transport. In eukaryotic cells, these reactions take place within the mitochondria. Mitochondria, as we saw in Chapter 5, are surrounded by two membranes. The outer one is smooth, and the inner one folds inward. The folds are called *cristae*. Within the inner compartment of the mitochondrion, surrounding the cristae, is a dense solution (the *matrix*) containing enzymes, coenzymes, water, phosphates, and other molecules involved in respiration. The outer membrane is permeable to most small molecules, but the inner one permits the passage of only certain molecules, such as pyruvic acid and ATP, and restrains the passage of others. As we shall see, this selective permeability of the inner membrane is critical to the ability of the mitochondria to harness the power of respiration to the production of ATP.

Some of the enzymes of the Krebs cycle are in solution in the inner compartment. Other Krebs cycle enzymes and the enzymes and other components of the electron transport chains are built into the membrane of the cristae. These inner membranes of the mitochondria are about 80 percent protein and 20 percent lipid. In the mitochondria, pyruvic acid from glycolysis is oxidized to carbon dioxide and water, completing the breakdown of the glucose molecule. Ninety-five percent of the ATP generated by heterotrophic cells is produced in the mitochondria.

9–8

(a) The three-carbon pyruvic acid molecule is oxidized to the two-carbon acetyl group, which is combined with coenzyme A to form acetyl CoA. The oxidation of the pyruvic acid molecule is coupled to the reduction of NAD^+. *Acetyl CoA enters the Krebs cycle. (b) Electron micrograph showing the enzymes involved in the oxidation of pyruvic acid to acetyl CoA. Each of the complexes visible here represents multiple copies of three different enzymes.*

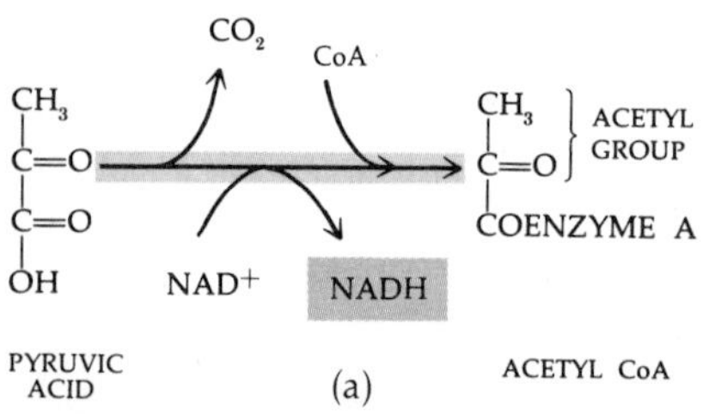

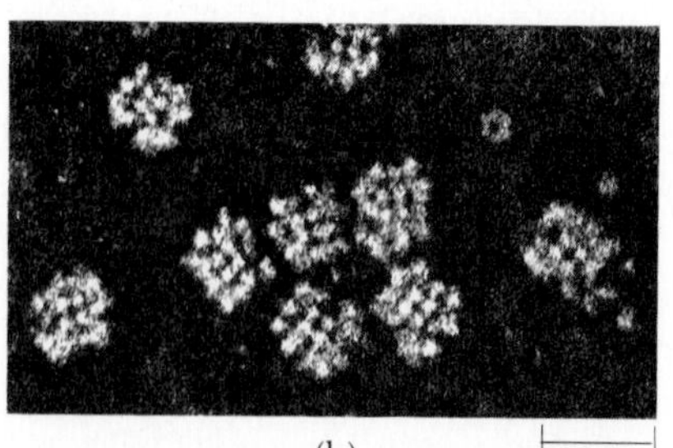

(b)

A Preliminary Step: The Oxidation of Pyruvic Acid

Pyruvic acid passes from the cytoplasm, where it is produced by glycolysis, and crosses the outer and inner membranes of the mitochondria. Before entering the Krebs cycle, the three-carbon pyruvic acid molecule is oxidized (Figure 9–8). The carbon and oxygen atoms of the carboxyl group are removed in the form of carbon dioxide, and a two-carbon acetyl group (CH_3CO) remains. In the course of this exergonic reaction, the hydrogen of the carboxyl group reduces a molecule of NAD^+ to NADH. The original glucose molecule has now been oxidized to two CO_2 molecules and two acetyl groups, and, in addition, four NADH molecules have been formed (two in glycolysis and two in the oxidation of pyruvic acid).

DISSECTING THE CELL

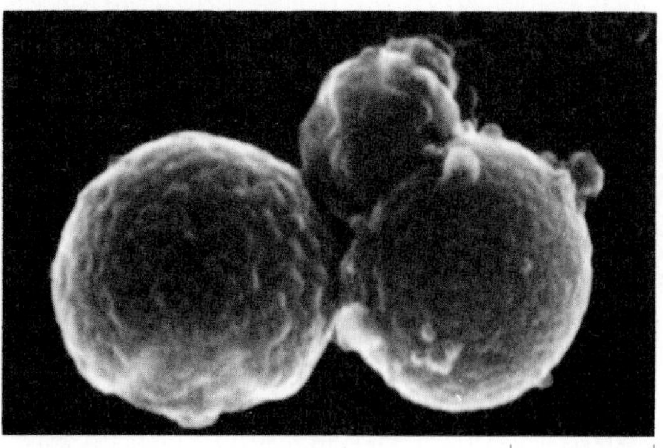

Scanning electron micrograph of isolated intact liver mitochondria.

In every living cell, many hundreds of chemical reactions proceed simultaneously. Molecules are continuously synthesized via certain enzymatic pathways, and molecules are broken down by other enzymatic pathways. Many of these reactions are mutually incompatible, as can be demonstrated by destroying the structure of cells and mixing their enzymes in a test tube. Chemical chaos results, and the enzymes are soon inactivated. In the cell, however, anabolic and catabolic pathways operate in harmony because biochemical reactions are spatially localized and compartmentalized within specific subcellular organelles. A living cell is the most intensely concentrated set of chemical reactions known. A cell carries out many more chemical reactions than any apparatus devised by chemical engineers, and all within the space of a few cubic micrometers. The cell's unique chemical versatility results from the compartmentalization of biochemical pathways within organelles.

In order to study the specific functions of any organelle type, the organelle must be dissected free from all other cell structures and collected in large quantities. Mitochondria, lysosomes, and other organelles are, of course, far too small for hand dissection, but cell biologists can prepare pure samples of any organelle type by the technique of preparative centrifugation.

Small particles, ranging in size from cells to macromolecules, can be separated by centrifugation if the particle types differ in size and density. Particles suspended in fluid and then subjected to strong gravitational force will move through the fluid at varying rates, the largest, densest particles settling most rapidly. Forces up to 400,000 times the force of gravity (400,000 g) can be generated in a test tube of suspended particles by rotating the tube at very high speeds in an ultracentrifuge. Thus, subcellular structures, such as mitochondria, nuclei, and intracellular membranes, can be separated into purified fractions by spinning fragmented cells at appropriate centrifugal forces.

For example, in order to determine which enzymatic pathways are present in mitochondria, a tissue, such as rat liver, is minced into small pieces and homogenized—that is, the cells are gently broken up by grinding the tissue in a glass tube fitted with a Teflon pestle. The tube contains a sucrose solution that is isotonic with intracellular fluid. The resulting suspension of cell organelles is then placed in an unbreakable test tube and spun in the centrifuge at low speed (700 g) for 10 minutes, so as to drive the bulkiest structures, the nuclei, to the bottom of the tube. All the lighter organelles remain suspended in the fluid, which is called the supernatant. The supernatant is transferred to another centrifuge tube and spun at a higher speed (10,000 g for 20 minutes), which sediments particles such as mitochondria and lysosomes into a pellet at the bottom of the tube. The supernatant, containing ribosomes and various membranes, is discarded, and the pellet is retained.

At this point, the mitochondria have been partially purified by means of differential centrifugation. This type of centrifugation separates particles of quite different size and density by "spinning down" the large particles to form a pellet. To separate particles of rather similar size, such as mitochondria and lysosomes, the more subtle technique of zonal centrifugation is used. To return to the experiment we have outlined, the pellet containing the mitochondria and lysosomes is resuspended and gently layered atop a sucrose density gradient in a centrifuge tube. A sucrose density gradient is prepared by layering sucrose solutions of differing densities one above the other so that the densest solution is at the bottom of the tube and the least dense solution is at the top of the tube. When organelles are centrifuged in a density gradient (120,000 g for 8 hours), each organelle type moves through the gradient at a different rate, depending on the organelle's density. Following centrifugation, mitochondria will occupy one zone in the gradient, lysosomes another, and other organelles will be found in other zones. By puncturing the bottom of the tube and removing the contents drop by drop, we can collect a pure sample of mitochondria. The purity can be verified by the electron microscope. If the procedure has been performed correctly, the isolated mitochondria will emerge with membranes intact and all enzymatic pathways functioning. It is then possible to test for the activities of specific enzymes, thus determining which biochemical functions are compartmentalized within mitochondria. Similarly, other cell constituents can be isolated and their biochemical activities determined.

Each acetyl group is momentarily accepted by a compound known as coenzyme A. Like the coenzymes we have examined previously, coenzyme A is a large molecule, a portion of which is a nucleotide and a portion of which is a vitamin (pantothenic acid, one of the B complex vitamins). The combination of the acetyl group and CoA is abbreviated as acetyl CoA. This reaction is the link between glycolysis and the Krebs cycle.

The Krebs Cycle

Upon entering the Krebs cycle (Figure 9–9), the two-carbon acetyl group is combined with a four-carbon compound (oxaloacetic acid) to produce a six-carbon compound (citric acid). In the course of the cycle, two of the six carbons are oxidized to CO_2, and oxaloacetic acid is regenerated—thus making this series literally a cycle. Each turn around the cycle uses up one acetyl group and regenerates a molecule of oxaloacetic acid, which is then ready to begin the sequence again.

In the course of these steps, some of the energy released by the oxidation of the carbon-hydrogen and carbon-carbon bonds is used to convert ADP to ATP (one

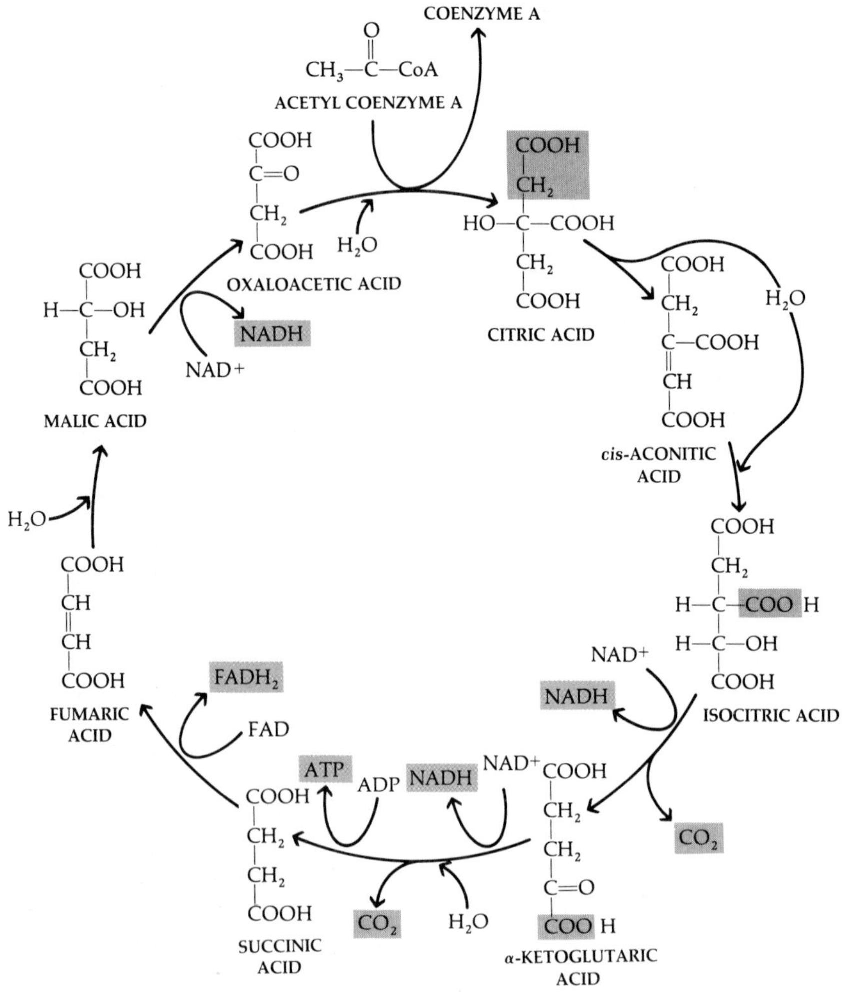

9–9
The Krebs cycle. In the course of the cycle, the carbons donated by the acetyl group are oxidized to carbon dioxide, and the hydrogen atoms are passed to electron carriers. As in glycolysis, a specific enzyme is involved at each step.

Coenzyme A shuttles back and forth between the oxidation of pyruvic acid and the Krebs cycle, linking these two stages of respiration.

9–10
Flavin adenine dinucleotide, an electron acceptor, in (a) *its oxidized form (FAD) and* (b) *its reduced form ($FADH_2$). Riboflavin is a vitamin made by all plants and many microorganisms. It is also known as vitamin B_2. It is a pigment; in its oxidized form it is a bright yellow.*

A related electron acceptor, flavin mononucleotide (FMN), consists of riboflavin and the first phosphate group shown here. It accepts electrons from NADH in the electron transport chain.

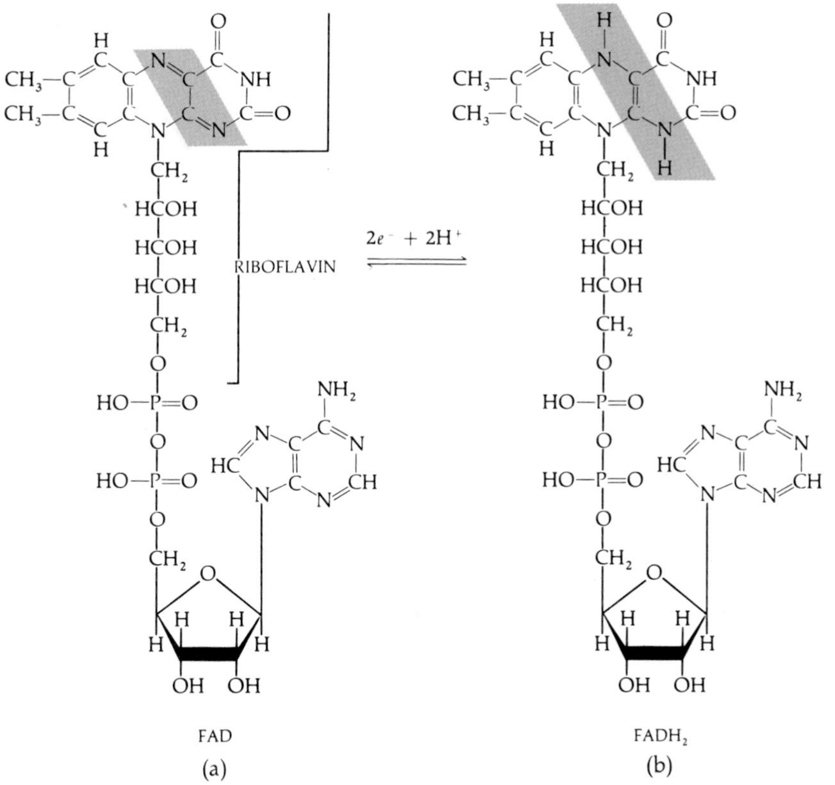

RIBOFLAVIN

$2e^- + 2H^+$

FAD
(a)

$FADH_2$
(b)

molecule per cycle), and some is used to produce NADH from NAD^+ (three molecules per cycle). In addition, some energy is used to reduce a second electron carrier, flavin adenine dinucleotide, abbreviated FAD (Figure 9–10). One molecule of $FADH_2$ is formed from FAD per turn of the cycle. No O_2 is required for the Krebs cycle; the electrons and protons removed in the oxidation of carbon are all accepted by NAD^+ and FAD.

$$\text{oxaloacetic acid} + \text{acetyl CoA} + \text{ADP} + P_i + 3NAD^+ + FAD \longrightarrow \text{oxaloacetic acid} + 2CO_2 + \text{CoA} + \text{ATP} + 3NADH + FADH_2 + 3H^+ + H_2O$$

Note that the oxaloacetic acid molecule the cycle ends with is not the same molecule with which the cycle began. If one begins with a glucose molecule in which the carbon atoms are radioactive, radioactive carbon atoms will appear among the four carbons of the oxaloacetic acid.

Electron Transport

The carbon atoms of the glucose molecule are now completely oxidized. Some of the energy of glucose has been used to produce ATP from ADP. Most of its energy, however, remains in electrons removed from the C—C and C—H bonds and passed to the electron carriers NAD^+ and FAD. These electrons are still at a high energy level.

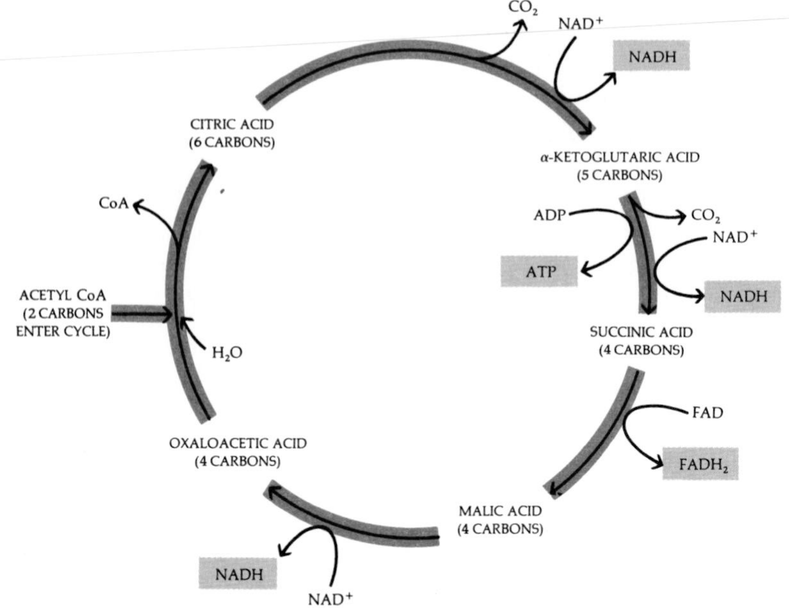

9–11
Summary of the Krebs cycle. One molecule of ATP, three molecules of NADH, and one molecule of $FADH_2$ represent the energy yield of the cycle. Two turns of the cycle are required to complete the oxidation of one molecule of glucose.

In the final stage of respiration, these high-energy-level electrons are passed step-by-step to the low energy level of oxygen. The energy they yield in the course of this passage is ultimately used to regenerate ATP from ADP. This step-by-step passage is made possible by a series of electron carriers, each of which holds the electrons at a slightly lower level.

These carriers make up what is known as an *electron transport chain*. At the top of the energy hill the electrons are held by NADH and $FADH_2$. Most of the energy of the glucose molecule now resides in these electron acceptors. The Krebs cycle yielded two molecules of $FADH_2$ and six molecules of NADH for each molecule of glucose. The oxidation of pyruvic acid to acetyl CoA yielded two molecules of NADH. Also, you will recall, two molecules of NADH were produced in glycolysis. In the presence of oxygen, the electrons held by these two NADH molecules are also transported into the mitochondrion where they are fed into the electron transport chain. In the process, NAD^+ is regenerated in the cytoplasm, allowing glycolysis to continue.

The principal components of the electron transport chain are molecules known as cytochromes (Figure 9–12). These molecules consist of a protein and a porphyrin ring, similar to that of heme, enclosing an atom of iron. Although similar, the protein structures of the individual cytochromes differ enough to enable them to hold electrons at different energy levels. The iron atom of each cytochrome alternately accepts and releases an electron, passing it along to the next cytochrome at a slightly lower energy level until the electrons, their energy spent, are accepted by oxygen (Figure 9–13). The energy released in this downhill passage of electrons is harnessed, as we shall see, to form ATP molecules from ADP. Such ATP formation is known as *oxidative phosphorylation*. At the end of the chain, the electrons are accepted by oxygen, which then combines with protons (hydrogen ions) from the solution to produce water.

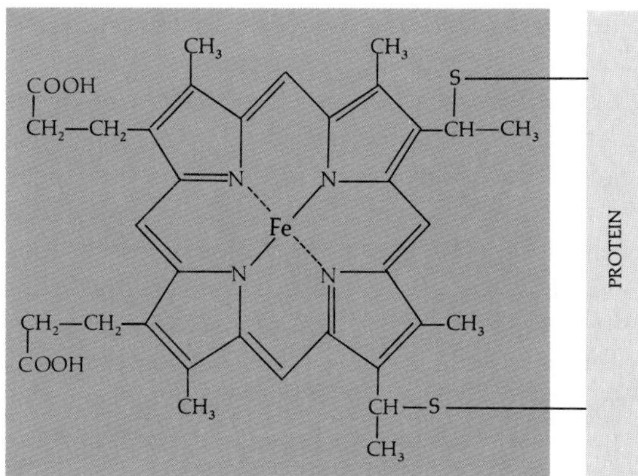

(a)

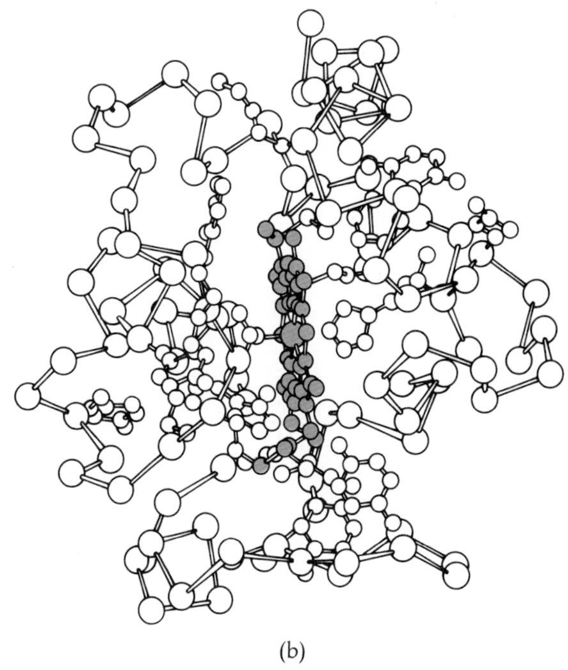

(b)

9–12
Cytochromes are molecules in which a heme group is held in an intricate protein structure. (a) *The heme group of cytochrome* c. *In the cytochromes, an atom of iron (Fe) is enclosed in a nitrogen-containing ring (a porphyrin ring). Cytochromes are involved in electron transfer. It is the iron within the molecule that actually combines with the electrons.* (b) *In the cytochrome* c *structure shown here, the heme group is in color.*

Quantitative measurements show that for every two electrons that pass from NADH to oxygen, three molecules of ATP are formed from ADP and phosphate. For every pair of electrons that passes from $FADH_2$, which holds them at a slightly lower energy level than NADH, two molecules of ATP are formed. In oxidative phosphorylation, the electron transfer potential of NADH and $FADH_2$ is converted to the phosphate transfer potential of ATP.

The Mechanism of Oxidative Phosphorylation: Chemiosmosis

It has long been known that as electrons move from high energy levels to a low energy level during their trip down the electron transport chain, the energy they release is harnessed to produce ATP from ADP and phosphate. How this is accomplished was, until quite recently, one of the most perplexing puzzles in biochemistry.

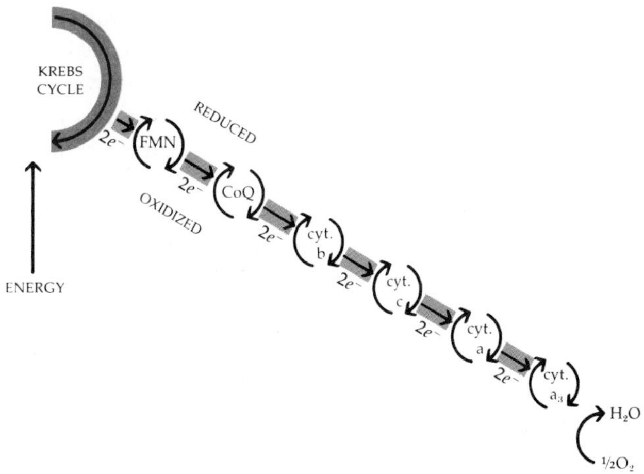

9–13
The principal electron-carrier molecules of the electron transport chain. At least nine other carrier molecules function as intermediates between the carriers shown here.

Flavin mononucleotide (FMN) and coenzyme Q (CoQ) transfer electrons and protons. The cytochromes transfer only electrons. The electrons carried by NADH enter the chain when they are transferred to FMN; those carried by $FADH_2$ *enter the chain farther down the line at CoQ. The electrons are ultimately accepted by oxygen, which combines with protons (hydrogen ions) in the solution to form water.*

At three transitions in the electron transport chain, significant drops occur in the amount of potential energy held by the electrons. As a consequence, a relatively large amount of free energy is released at each of these steps—as the electrons move from FMN to coenzyme Q, as they move from cytochrome *b* to cytochrome *c*, and as they move from cytochrome *a* to cytochrome a_3. In each case, the energy released by the passage of a pair of electrons is sufficient to phosphorylate one ADP to ATP. For many years, biochemists thought that the phosphorylations took place enzymatically and were chemically coupled to these specific points in the electron transport chain. They hypothesized that at each point an intermediate phosphorylated compound was formed and that, subsequently, the intermediate transferred its phosphate to ADP. Such a mechanism would be analogous to the phosphorylation steps of glycolysis (page 191). However, despite years of effort in many laboratories, no one was ever able to isolate or identify the postulated intermediate compounds or the enzymes catalyzing their formation. Hypothesis after hypothesis was proposed, experiment after experiment was designed and run, but the process occurring in the mitochondria remained baffling.

By the early 1960s, the perplexity and frustration among biochemists studying oxidative phosphorylation was great indeed. The answer to their difficulties came from Peter Mitchell, a British biochemist who had, for reasons of health, retired from academic life and was working in his private laboratory in Cornwall on the membrane transport processes of bacteria. All bacteria maintain a gradient of protons (H^+ ions) across the cell membrane, pumping protons out of the cell. As a result, the H^+ concentration is lower (and the pH therefore higher) inside the cell than outside. Mitchell had found that bacteria use the potential energy resulting from this gradient to power the active transport of needed substances into the cell. In 1961, he put forth the ingenious—and radical—proposal that the phosphorylation of ADP to ATP in the mitochondria is similarly powered by a proton gradient. This concept was at first received with considerable skepticism, but experimental work during the past 20 years by Mitchell and his coworker Jennifer Moyle, and by biochemists in many other laboratories, has revealed that oxidative phosphorylation is indeed driven by a proton gradient. Many of the details remain to be worked out, but the initial chemiosmotic hypothesis has been elevated to the status of the *chemiosmotic theory*, and, in recognition of his major conceptual breakthrough, Mitchell was awarded the 1978 Nobel Prize in chemistry.

How Chemiosmosis Works

The term "chemiosmosis" reflects the fact that the production of ATP in oxidative phosphorylation includes both chemical processes and transport processes across a selectively permeable membrane. Two distinct events take place in chemiosmosis: (1) the establishment of a proton gradient across the inner membrane of the mitochondrion, and (2) the release of the potential energy stored in the gradient and its capture in the formation of ATP from ADP and phosphate.

As electrons pass down the electron transport chain, the three relatively large releases of free energy power the pumping of protons from the mitochondrial matrix through the inner membrane to the intermembrane space—that is, the space between the inner and outer membranes of the mitochondrion. It has not been established whether the protons remain in the intermembrane space or pass through the outer membrane into the cytoplasm (the outer membrane, you will recall, is freely permeable to most ions and small molecules). Exactly how the

9–14
According to one hypothesis, the proton gradient of chemiosmosis is established as the electrons contributed by NADH follow a looping pattern across the inner mitochondrial membrane during their transit down the electron transport chain. With each passage of two electrons from the matrix side of the membrane to its outer side, two protons are transported by carrier molecules (loops indicated in color) and then discharged into the external medium. After discharge of the protons, the two electrons are transported back to the matrix side of the membrane by other carrier molecules (black loops). This hypothesis assumes that the carrier molecules (here designated as A through F) are fixed in the membrane in the necessary positions.

pumping of protons is accomplished is a matter of current investigation. One hypothesis proposes that the electron carriers in the chain are positioned so that the electrons travel a zigzag course from the inner to the outer surface of the inner membrane, traversing it three times in the course of their passage from NADH to oxygen. According to this hypothesis, each time two electrons travel from the inside of the membrane to the outside, they pick up two protons from the matrix and release them to the outside (Figure 9–14). Another hypothesis proposes that the energy released by the electrons triggers conformational changes in specific membrane transport proteins that enable them to carry protons against the concentration gradient from the matrix into the intermembrane space. The precise number of protons transported as each electron pair moves down the chain is also uncertain; it is thought to be at least six, but it may be more.

Like the boulder at the top of the hill, the water at the top of the falls, or the chemical energy in a stick of dynamite, the differential concentration of protons between the matrix and the outside represents potential energy. This potential energy results not only from the difference in pH (more H^+ ions outside than inside) but also from the difference in electric charge (Figure 9–15). Because the inner membrane is impermeable to virtually all charged particles, other positive ions cannot move into the matrix to neutralize the negative charge created when the protons are pumped out. The movement of negative ions out of the matrix, which would also neutralize the charge difference, is similarly blocked. The result is potential energy, available to power any process that would provide a channel allowing the protons to flow down the electrochemical gradient back into the matrix.

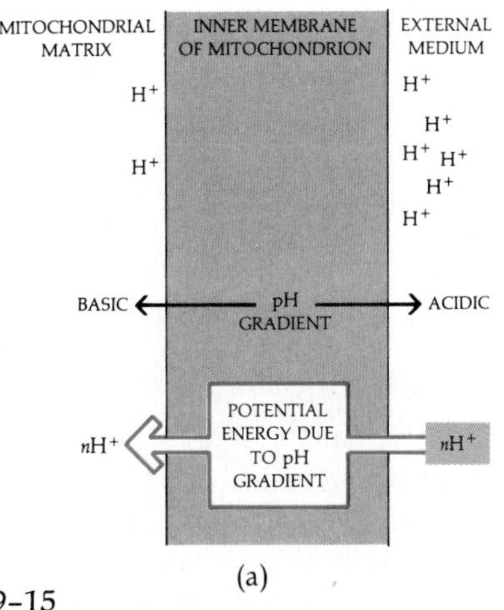

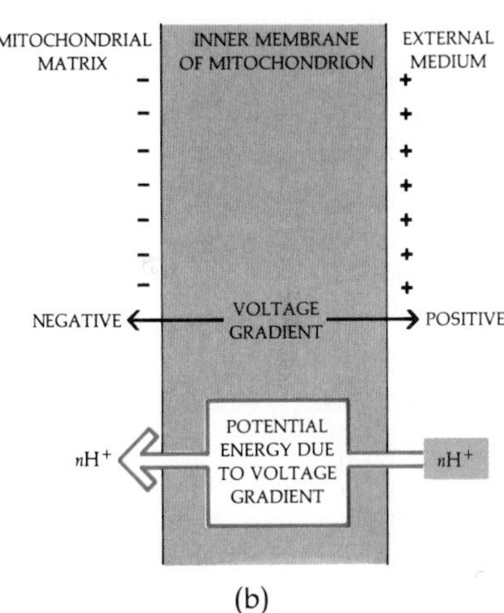

9–15
The potential energy stored by the pumping of protons from the mitochondrial matrix has two components: (a) a chemical gradient resulting from the difference in pH between the matrix and the external medium, and (b) a voltage gradient, resulting from the difference in electric charge on the inside and outside of the membrane. When a channel is provided that allows protons to reenter the matrix, they descend both gradients simultaneously. The contribution of the voltage gradient to the total potential energy is, however, greater than that of the chemical gradient.

Such a channel is provided by a large enzyme complex known as *ATP synthetase*. This complex consists of two major portions, or factors, known as F_O and F_1 (Figure 9–16a). F_O is embedded in the inner mitochondrial membrane, traversing the membrane from outside to inside. It is thought to have an inner channel, or pore, through which protons can pass. F_1 is a large globular structure, consisting of nine polypeptide subunits, that is attached to F_O on the matrix side of the membrane. In electron micrographs of the inner mitochondrial membrane, the F_1 units appear as protruding knobs (Figure 9–16b). With proper chemical treatment, the F_1 unit can be removed from the mitochondrial membrane and subjected to detailed study. It has been shown to have binding sites for ATP and ADP, and, in solution, it catalyzes the hydrolysis of ATP to ADP, thus functioning as an ATPase. Its usual function when attached to the F_O unit in the intact mitochondrion is, however, the reverse. As protons flow down the electrochemical gradient from the outside into the matrix, passing through the F_O unit and then the F_1 unit, the free energy released powers the synthesis of ATP from ADP and phosphate. It is not certain whether the phosphorylation of one molecule of ADP to ATP requires the passage of two, three, or four protons through the ATP synthetase complex.

Figure 9–17 summarizes chemiosmosis as it occurs in oxidative phosphorylation. Chemiosmotic power also has other uses in living systems. For example, it provides the power that drives the rotation of bacterial flagella (page 117). In photosynthetic cells, as we shall see in the next chapter, it is involved in the formation of ATP using energy supplied to electrons by the sun. And, it can be used to power other transport processes—the original object of Mitchell's research. In the mitochondrion, the energy stored in the proton gradient is used not only to drive the synthesis of ATP but also to carry other substances through the inner membrane. For example, both phosphate and pyruvic acid are carried into the mitochondrion by membrane proteins that simultaneously transport protons down the gradient.

9–16
(a) *Diagram of the ATP synthetase complex. The F_O portion is contained within the inner membrane of the mitochondrion, and the F_1 portion, which consists of nine subunits, extends into the mitochondrial matrix.*

(b) *The knobs protruding from the membrane of these vesicles are the F_1 portions of ATP synthetase complexes. The F_O portions to which they are attached are embedded in the membrane and are not visible in this electron micrograph. These vesicles were prepared by disrupting the inner mitochondrial membrane with ultrasonic waves. When the membrane is disrupted in this fashion, the membrane fragments immediately reseal, forming closed vesicles. These vesicles are, however, inside-out. The surface exposed to the external medium is the surface that faces the matrix in the intact mitochondrion. Such inside-out vesicles are an important tool in the continuing study of oxidative phosphorylation.*

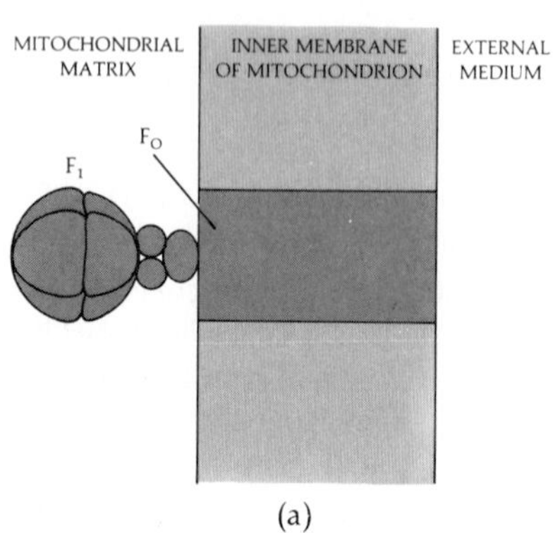

(a)

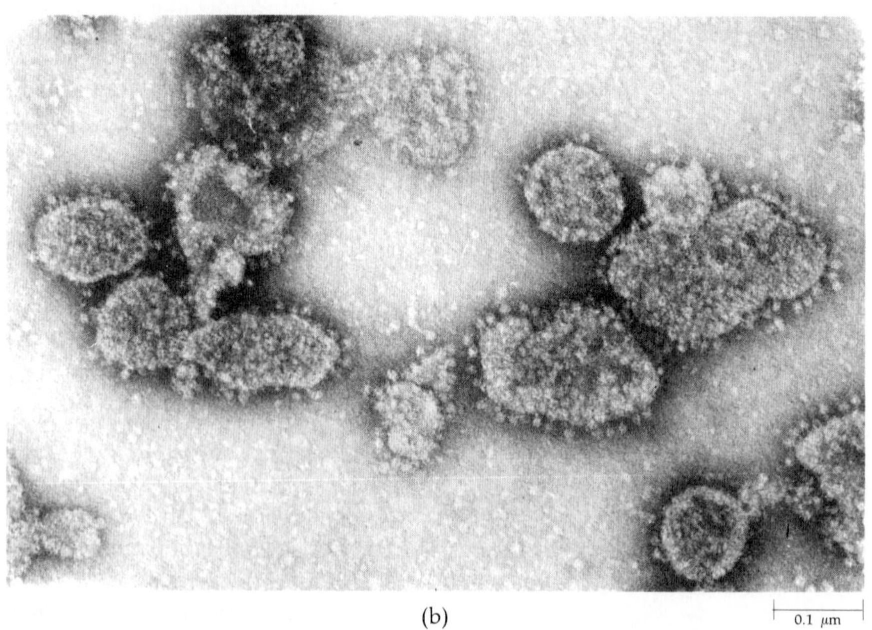

(b)

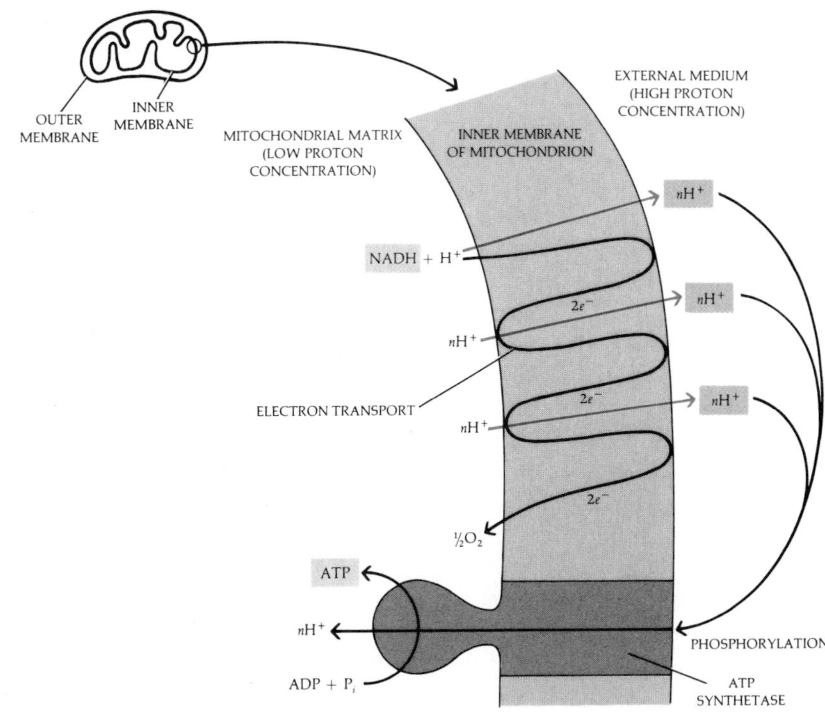

9–17
According to the chemiosmotic theory, protons are pumped out of the mitochondrial matrix as electrons are passed down the electron transport chain, which forms a part of the inner mitochondrial membrane. The movement of the protons down the electrochemical gradient as they pass through the ATP synthetase complex provides the energy by which ATP is regenerated from ADP and phosphate. The exact number of protons pumped out of the matrix as each electron pair moves down the chain is still to be determined, as is the number that must flow through ATP synthetase for each molecule of ATP formed.

You will recall that we mentioned earlier that "about" 38 molecules of ATP are formed for each molecule of glucose oxidized to carbon dioxide and water. One of the reasons for this vagueness is that the exact amount of ATP formed depends on how the cell apportions the energy made available by the proton gradient. The more of this energy that is used in other transport processes, the less of it that is available for ATP synthesis. The needs of the cell vary according to the circumstances, and so does the amount of ATP synthesized.

Control of Oxidative Phosphorylation

Electrons continue to flow along the electron transport chain, releasing energy to create and maintain the proton gradient, only if ADP is available to be converted to ATP. Thus oxidative phosphorylation is regulated by supply and demand. When the energy requirements of the cell decrease, fewer molecules of ATP are used, fewer molecules of ADP become available, and electron flow is decreased.

OVERALL ENERGY HARVEST

We are now in a position to see how—and how much of—the energy originally present in the glucose molecule has been recovered in the form of ATP. Bear in mind that because the proton gradient in the mitochondrion can be used for purposes other than ATP synthesis, the figures we give represent the maximum energy harvest.

Glycolysis, in the presence of oxygen, yields two molecules of ATP directly and two molecules of NADH. These NADH molecules, however, cannot cross the

9–18

Summary of glycolysis and respiration. Glucose is first broken down to pyruvic acid, with a yield of two ATP molecules and the reduction (dashed arrows) of two NAD^+ molecules to NADH. Pyruvic acid is oxidized to acetyl CoA, and one molecule of NAD^+ is reduced. (Note that this and subsequent reactions occur twice for each glucose molecule; this electron passage is indicated by solid arrows.) In the Krebs cycle, the acetyl group is oxidized and the electron acceptors NAD^+ and FAD are reduced. NADH and $FADH_2$ then transfer their electrons to the series of cytochromes and other electron carriers that make up the electron transport chain. As the electrons are passed "downhill," relatively large amounts of free energy are released during the passage from FMN to CoQ, from cytochrome b *to cytochrome* c, *and from cytochrome* a *to cytochrome* a_3. *These bursts of free energy transport protons through the inner mitochondrial membrane, establishing the proton gradient that powers the synthesis of ATP from ADP.*

inner membrane of the mitochondrion, and the electrons they carry must be "shuttled across" the membrane. In most cells, the energy cost of this process is quite low and each NADH formed in glycolysis ultimately results in the synthesis of three molecules of ATP. In these cells, the total gain from glycolysis is 8 ATP. In other cells, including those of brain, skeletal muscle, and insect flight muscle, the energy cost of the shuttle is higher; the electrons are at a lower energy level when they reach the electron transport chain, and they enter the chain at coenzyme Q, rather than at FMN. Thus, like the electrons carried by $FADH_2$ from the Krebs cycle, they yield only two molecules of ATP per electron pair. In such cells, the total gain from glycolysis is only 6 ATP. (This was the second factor in our earlier hedging about the number of ATPs formed.)

The conversion of pyruvic acid to acetyl CoA yields two molecules of NADH (inside the mitochondrion) for each molecule of glucose and so produces six molecules of ATP.

The Krebs cycle yields, for each molecule of glucose, two molecules of ATP, six of NADH, and two of $FADH_2$, or a total of 24 ATP.

As a balance sheet (Table 9–1) shows, the complete yield from a single molecule of glucose is a maximum of 38 molecules of ATP. Note that all but 2 of the 38 molecules of ATP have come from reactions taking place in the mitochondrion, and all but 4 result from the passage down the electron transport chain of electrons carried by NADH or $FADH_2$.

The free energy change (ΔG) that occurs during glycolysis and respiration is -686 kilocalories per mole. About 266 (7×38) kilocalories per mole have been captured in the phosphate bonds of the ATP molecules, an efficiency of almost 40 percent.

The ATP molecules, once formed, are exported across the membrane of the mitochondrion by a shuttle system that simultaneously brings in one molecule of ADP for each ATP exported.

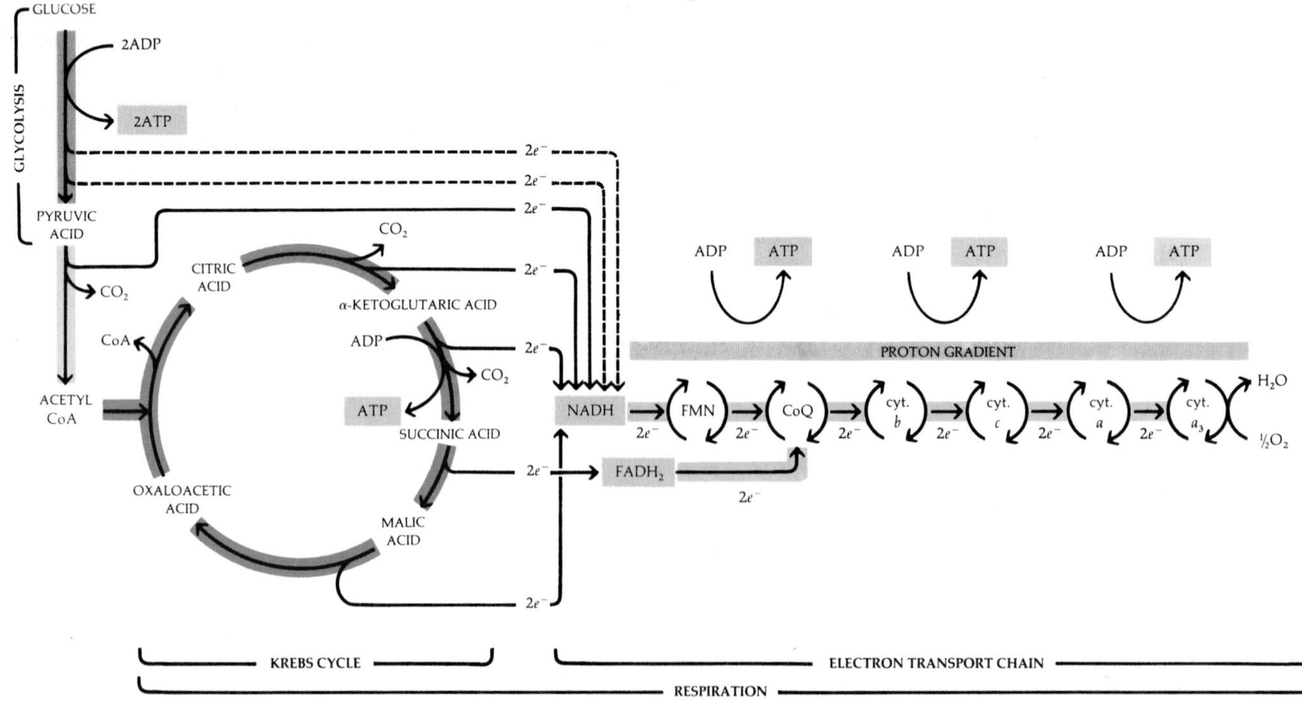

9-19
Energy changes in the oxidation of glucose. The complete respiratory sequence (glucose + $6O_2 \longrightarrow 6CO_2 + 6H_2O$) proceeds with an energy drop of 686 kcal/mole. Of this, almost 40 percent (266 kilocalories) is conserved in 38 ATP molecules. In anaerobic glycolysis (glucose $\longrightarrow$ lactic acid), by contrast, only 2 ATP molecules are produced, representing only about 2 percent of the available energy of glucose.

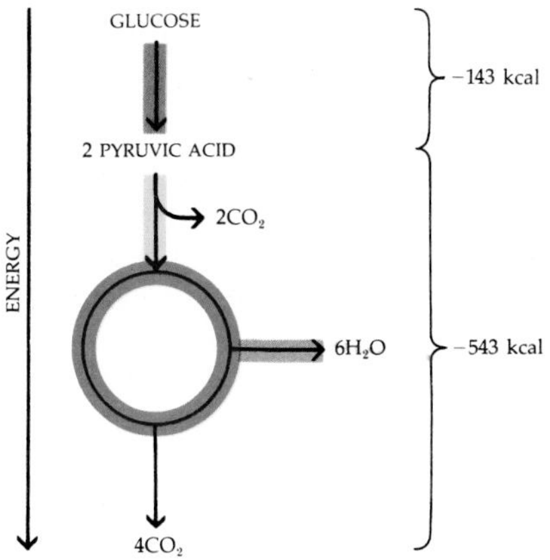

Table 9-1 Summary of Maximum Energy Yield from One Molecule of Glucose

In the cytoplasm				
Glycolysis:		2 ATP		$\longrightarrow$ 2 ATP
In the mitochondria				
From glycolysis:	2 NADH	$\longrightarrow$ 6 ATP		$\longrightarrow$ 6 ATP*
From respiration:				
Pyruvic acid $\longrightarrow$ acetyl CoA:	1 NADH	$\longrightarrow$ 3 ATP	(× 2)	$\longrightarrow$ 6 ATP
		1 ATP		
Krebs cycle:	3 NADH	$\longrightarrow$ 9 ATP	(× 2)	$\longrightarrow$ 24 ATP
	1 $FADH_2$	$\longrightarrow$ 2 ATP		

* In some cells, the energy cost of transporting the electrons from the NADH molecules formed in glycolysis across the inner mitochondrial membrane lowers the net yield from these 2 NADH to 4 ATP.

OTHER CATABOLIC PATHWAYS

Most organisms do not feed directly on glucose. How do they extract energy from, for example, fats or proteins? The answer lies in the fact that the Krebs cycle is a Grand Central Station for energy metabolism. Other foodstuffs are broken down and converted into molecules that can feed into this central pathway at some point or other.

Polysaccharides such as starch are broken down to their constituent monosaccharides and phosphorylated to glucose 6-phosphate and in this form enter the glycolytic pathway. Fats are first split into their glycerol and fatty acid components. These are then chopped up into two-carbon fragments and slipped into the Krebs cycle as acetyl CoA. Proteins are broken down into their constituent amino acids. The amino acids are deaminated (the amino groups removed), and the residual carbon skeleton is either converted to an acetyl group or to one of the larger carbon compounds of the glycolytic pathway or the Krebs cycle so that it can be processed at this stage of the central pathway. The amino groups, if not reutilized, are eventually excreted as urea or other nitrogen-containing wastes. These various degradative pathways are, collectively, catabolism.

9-20
The strategy of energy metabolism. Organisms extract energy from compounds by oxidizing them to carbon dioxide and water. NAD^+ is the major oxidizing agent. The NADH gives up its high-energy electrons and associated protons to electron acceptors in the electron transport chain. The electrons are passed down the chain ultimately to oxygen. The energy released in this process is used to phosphorylate ADP, converting it to ATP, which is used to drive the endergonic reactions of the organism. Note that both NADH and ATP are used in a cyclic fashion. Although they are critically important compounds, they are present in very small amounts, accomplishing their work as a result of constant and rapid turnover. The need for a constant supply of these molecules explains why a deprivation of oxygen brings about death in a very few minutes.

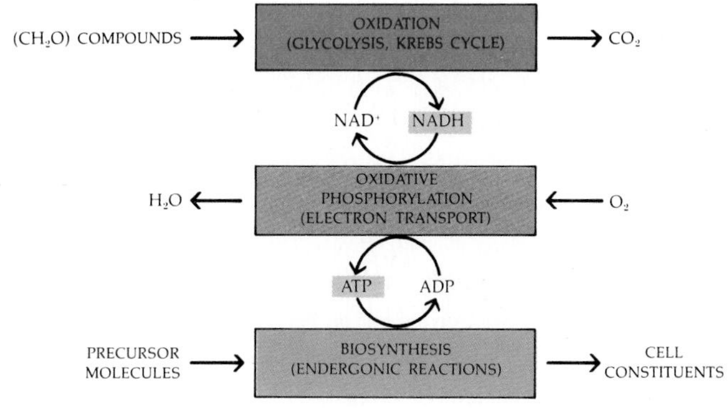

ETHANOL, NADH, AND THE LIVER

The human body can dispose fairly readily of most toxic products of its own manufacture, such as carbon dioxide and nitrogenous wastes. In contrast, most ingested toxic substances, such as ethanol (beverage alcohol), must first be broken down by the liver, which possesses special enzymes not present in other tissues.

It has been known for many years that heavy drinkers are at great risk for severe, and often fatal, liver disease. Studies conducted by Charles S. Lieber and his colleagues at the Bronx Veterans Administration Hospital and the Mount Sinai School of Medicine in New York City have demonstrated that the origin of the problem lies in the simple chemical steps involved in the breakdown of ethanol. Enzymes in the liver first oxidize ethanol (CH_3CH_2OH) to acetaldehyde (CH_3CHO), removing two hydrogen atoms and reducing a molecule of NAD^+; this is the reverse of the second reaction shown in Figure 9–5a on page 193. The acetaldehyde is then oxidized to acetic acid, which is, in turn, oxidized to carbon dioxide and water and eliminated from the body.

The chief culprits in the development of liver disease are the hydrogen atoms (electrons and protons) removed from ethanol. These "extra" hydrogens—carried by NADH—follow two principal pathways within the cell. Most are fed directly into the electron transport chain, producing water and ATP. Because of the high levels of NADH present in the cell from the oxidation of ethanol, the production of NADH by glycolysis and the Krebs cycle is reduced. As a result, sugars, amino acids, and fatty acids are not broken down but are instead converted to fats. The fats accumulate in the liver. The mitochondria also swell, presumably as a result of the distortion of their normal function—the electron transport chain is doing very heavy duty, while the Krebs cycle is effectively shut down.

Other hydrogen atoms are used in the synthesis of glycerophosphates and fatty acids from the carbohydrate skeletons that are not being processed in glycolysis and the Krebs cycle. More fats accumulate. It does not take long. In human volunteers fed a good high-protein, low-fat diet, six drinks (about 10 ounces) a day of 80 proof alcohol produced an eightfold increase in fat deposits in the liver in only 18 days. Fortunately, these early effects are completely reversible.

The liver cells work hard to get rid of the excess fats. The fats are not soluble in water (or in blood plasma). Before being released into the bloodstream, they are coated with a thin layer of protein. This coating and secretion process is carried out on the membranes of the endoplasmic reticulum. The liver cells of heavy drinkers show enormous proliferation of the endoplasmic reticulum.

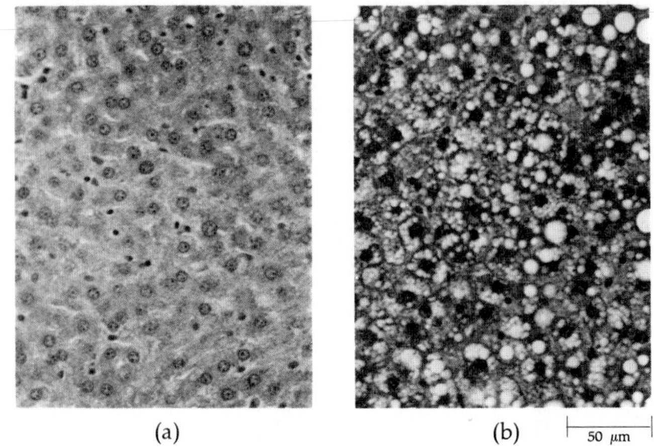

(a) Normal liver tissue from a rat fed a balanced liquid diet for 24 days. (b) In this liver tissue from another rat fed a liquid diet in which ethanol provided 36 percent of the total calories, many globular fat droplets have accumulated. This rat was also maintained on its special diet for 24 days.

As we noted, the fat deposits are initially reversible, but after a few years—depending on how much alcohol is consumed—liver cells, engorged with fat, begin to die, triggering the inflammatory process known as alcoholic hepatitis. Liver function becomes impaired. Cirrhosis is the next step; it is the formation of scar tissue, which interferes with the function of the individual cells and also with the supply of blood to the liver. This leads to the death of more cells. The liver can no longer carry out its normal activities—such as breaking down nitrogenous wastes—which is why cirrhosis is a cause of death. In fact, cirrhosis of the liver is the seventh leading cause of death in the nation, and the third leading cause of death between the ages of 25 and 65 in New York City.

Not so long ago, it was commonly believed that a good diet was all that was required to protect even a heavy drinker from the deleterious effects of alcohol. In fact, if one were just to add a few vitamins to the alcohol itself, some sophisticates maintained, most of the long-term physical damage of alcohol would disappear. This new evidence refutes these comforting notions, and it comes at a time when alcohol is enjoying a resurgence of popularity among persons of high school and college age. (In populations of postgraduate age, as in other human societies the world over, it never lost its status as the drug abuse of choice.)

9–21
Major pathways of catabolism and anabolism in the cell.

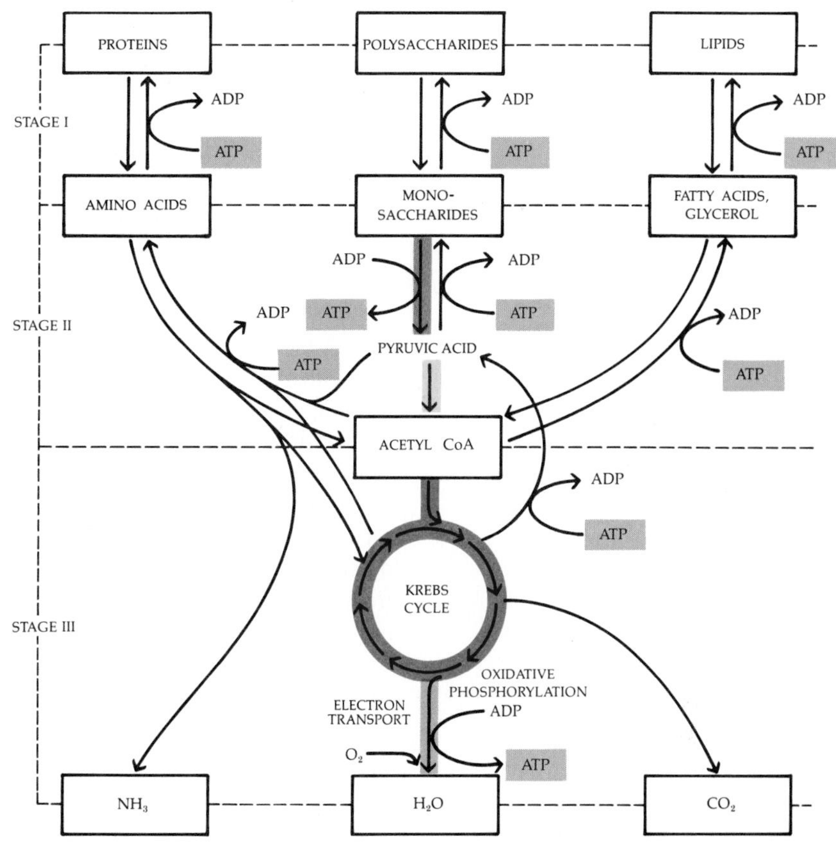

BIOSYNTHESIS

The pathways of glucose breakdown, central to catabolism, are also central to the biosynthetic, or anabolic, processes of life. These processes are the pathways of synthesis of the various molecules and macromolecules that make up an organism.

Since many of these substances, such as proteins and lipids, can be broken down and fed into the central pathway, you might guess that the reverse process can occur—namely, that the various intermediates of glycolysis and the Krebs cycle can serve as precursors for biosynthesis. This is in fact the case. However, the biosynthetic pathways, while similar to the catabolic ones, are distinctive. Different enzymes control the steps and the various critical steps of anabolism differ from the catabolic processes. These general pathways are sketched out in Figure 9–21.

SUMMARY

The oxidation of glucose is a chief source of energy in most cells. As the glucose is broken down in a series of small enzymatic steps, the energy in the molecule is repackaged in the phosphate bonds of ATP molecules.

The first phase in the breakdown of glucose is glycolysis, in which the six-carbon glucose molecule is split into two three-carbon molecules of pyruvic acid. A net yield of two molecules of ATP (from ADP) and two of NADH (from NAD^+) results from the process. Glycolysis takes place in the cytoplasm of the cell.

The second phase in the breakdown of glucose and other fuel molecules is respiration. It requires oxygen and, in eukaryotic cells, takes place in the mitochondria. It occurs in two stages: the Krebs cycle and terminal electron transport. (In the absence of oxygen, the pyruvic acid produced by glycolysis is converted to either ethanol or lactic acid by the process of anaerobic fermentation. NAD^+ is regenerated, allowing glycolysis to continue, producing a small but vital supply of ATP for the organism.)

In the course of respiration, the three-carbon pyruvic acid molecules are broken down to two-carbon acetyl groups, which then enter the Krebs cycle. In a series of reactions in the Krebs cycle, the two-carbon acetyl group is oxidized completely to carbon dioxide. In the course of the oxidation of each acetyl group, four electron acceptors (three NAD^+ and one FAD) are reduced, and another molecule of ATP is formed.

The final stage of respiration is terminal electron transport, which involves a chain of electron carriers and enzymes embedded in the inner membrane of the mitochondrion. Along this series of electron carriers, the high-energy electrons accepted by NADH in glycolysis and by NADH and $FADH_2$ in the Krebs cycle pass downhill to oxygen. At three points in their passage down the complete electron transport chain, large quantities of free energy are released that power the pumping of protons (H^+ ions) out of the mitochondrial matrix. This creates an electrochemical gradient of potential energy across the inner membrane of the mitochondrion. When protons pass through the ATP synthetase complex as they flow down the electrochemical gradient back into the matrix, the energy released is used to form ATP molecules from ADP and phosphate. This mechanism, by which oxidative phosphorylation is accomplished, is known as chemiosmosis.

In the course of the breakdown of the glucose molecule, a maximum of 38 molecules of ATP are formed. The exact number of ATP molecules formed depends on how much of the energy of the proton gradient is used to power other mitochondrial transport processes and on the shuttle mechanism by which the electrons from the NADH molecules formed in glycolysis are brought into the mitochondrion. Generally, almost 40 percent of the free energy released in glucose oxidation is retained in the form of newly synthesized ATP molecules.

Other food molecules, including fats, polysaccharides, and proteins, are utilized by being degraded to compounds that can slip into these central pathways at various steps. The biosynthesis of these substances also originates with precursor compounds derived from intermediates in the respiratory sequence and is driven by the energy derived from those processes.

QUESTIONS

1. Distinguish among the following: oxidation of glucose/glycolysis/respiration/fermentation; anaerobic pathways/aerobic pathways; FAD/$FADH_2$; Krebs cycle/electron transport.

2. Describe the process of fermentation. What conditions are essential if it is to occur? With some strains of yeast, fermentation stops before the sugar is exhausted, usually at an alcohol concentration in excess of 12 percent. What is a plausible explanation?

3. If oxygen-breathing organisms are so much more successful in converting energy than anaerobes, why are there any anaerobes left on this planet? Why didn't they all become extinct long ago?

4. Sketch the structure of a mitochondrion. Describe where the various stages in the breakdown of glucose take place in relation to mitochondrial structure. What molecules and ions cross the mitochondrial membranes during these processes?

5. Each $FADH_2$ molecule produced in the Krebs cycle results in the formation of only two molecules of ATP when its electrons pass down the electron transport chain. It was long thought that because these electrons enter the chain at coenzyme Q rather than at FMN they "missed" one of the sites of phosphorylation. What is a more accurate explanation?

6. Cyanide can combine with—and deactivate—cytochrome *a* and cytochrome a_3. In our bodies, however, cyanide tends to react first with hemoglobin and to make it impossible for oxygen to bind to the hemoglobin. Either way, cyanide poisoning has the same effect: it inhibits the synthesis of ATP. Explain how this is so.

7. When the F_1 portion of the ATP synthetase complex is removed from the mitochondrial membrane and studied in solution, it functions as an ATPase. Why does it not function as an ATP synthetase?

8. Certain chemicals function as "uncoupling" agents when they are added to respiring mitochondria. The passage of electrons down the chain to oxygen continues, but no ATP is formed. One of these agents, the antibiotic valinomycin, is known to transport K^+ ions through the inner membrane. Another, 2,4-dinitrophenol, transports H^+ ions through the membrane. How do these substances prevent the formation of ATP? Which would you expect to have the most profound effect on ATP formation? Why?

9. In the cells of a specialized tissue known as brown fat, the inner membrane of the mitochondrion is permeable to H^+ ions. These cells contain large stores of fat molecules, which are gradually broken down and the resulting acetyl groups are fed into the Krebs cycle. The electrons captured by NADH and $FADH_2$ are, in turn, fed into the electron transport chain and ultimately accepted by oxygen. No ATP is synthesized, however. Why not? Brown fat tissue is found in some hibernating animals and in mammalian infants that are born hairless, including human infants. What do you suppose the function of brown fat tissue is?

10. (a) As we have seen, a cell obtains a maximum of 38 molecules of ATP from each molecule of glucose that is completely oxidized. Account for the production of each molecule of ATP. (b) In the course of glycolysis, the Krebs cycle, and the electron transport chain, 40 molecules of ATP are actually formed. Why is the net yield for the cell only 38 molecules? (c) What other factors can reduce the yield of ATP?

11. Describe how the processes of the cell are adapted to the efficient use of a variety of foodstuffs, and to the efficient production of the variety of materials that the cell needs to manufacture for its own use.

12. In terms of the cell's economy, what do anabolic processes provide for the cell? What do catabolic processes provide? How are they dependent on each other?

CHAPTER 10

Photosynthesis, Light, and Life

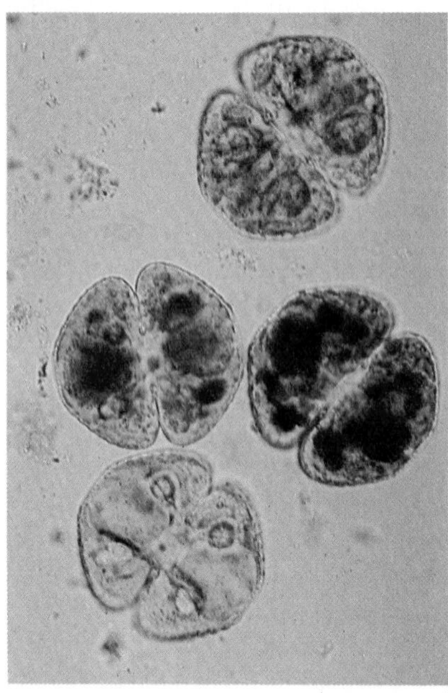

10–1
Cells of a green alga (Cosmarium botrytis) *that lives in fresh water. Each cell is an individual, self-sufficient photosynthetic organism. The green color is due to chlorophyll, in which the radiant energy of sunlight is converted to chemical energy.*

The first photosynthetic organism probably appeared 3 to 3.5 billion years ago. Before the evolution of photosynthesis, the physical characteristics of earth and its atmosphere were the most powerful forces in shaping the course of natural selection. With the evolution of photosynthesis, however, organisms began to change the face of our planet and, as a consequence, to exert strong influences on each other. Organisms have continued to change the environment, at an ever-increasing rate, up to the present day.

As we saw earlier, the atmosphere in which the first cells evolved lacked free oxygen. These earliest organisms were, of course, adapted to living in an environment without free oxygen, and, in fact, oxygen, with its powerful electron-attracting capacities, would have been poisonous to them. Their energy probably came from anaerobic glycolysis, using the organic molecules in the seas as fuel. This would have resulted not only in a decrease in the concentration of organic molecules but also in the accumulation of carbon dioxide in the atmosphere.

Then, it is hypothesized, there slowly evolved photosynthetic organisms that used carbon dioxide as their carbon source and released oxygen, as do most modern photosynthetic forms. As these photosynthetic organisms multiplied, they provided a new supply of organic molecules, and free oxygen began to accumulate. In response to these changing conditions, cell species arose for which oxygen was not a poison but a requirement for existence. (It has been proposed that the original function of the electron transport chain found in the cell membrane of aerobic bacteria—thought to be the forerunner of the mitochondrial electron transport chain—was to protect the cell from oxygen.)

As we saw in the last chapter, oxygen-consuming organisms have an advantage over those that do not use oxygen. A higher yield of energy can be extracted per molecule from the oxidation of carbon-containing compounds than from anaerobic processes, in which fuel molecules are not broken down completely. Energy released in cells by reactions using oxygen made possible the development of increasingly active, increasingly complex organisms. Without oxygen, the complex forms of life that now exist on earth could not have evolved.

Life on earth continues to be dependent on photosynthesis both for its oxygen and for its carbon-containing fuel molecules. Photosynthetic organisms capture

VAN NIEL'S HYPOTHESIS

For more than 100 years, it was generally assumed that in the equation

$$CO_2 + H_2O + light \longrightarrow (CH_2O) + O_2$$

the carbohydrate (CH_2O) resulted from the combination of the carbon and water molecules and that the oxygen was released from the carbon dioxide molecule. This entirely reasonable hypothesis was widely accepted. But, as it turned out, it was wrong.

The investigator who upset this long-held theory was C. B. van Niel of Stanford University. Van Niel, then a graduate student, was investigating photosynthesis in different types of photosynthetic bacteria. In their photosynthetic reactions, bacteria reduce carbon to carbohydrates, but they do not release oxygen. Among the types of bacteria van Niel was studying were the purple sulfur bacteria, which require hydrogen sulfide for photosynthesis. In the course of photosynthesis, globules of sulfur (S) are excreted or accumulated inside the bacterial cells. In these bacteria, van Niel found that this reaction takes place during photosynthesis:

$$CO_2 + 2H_2S \xrightarrow{light} (CH_2O) + H_2O + 2S$$

This finding was simple enough and did not attract much attention until van Niel made a bold extrapolation. He proposed that the generalized equation for photosynthesis is

$$CO_2 + 2H_2A \xrightarrow{light} (CH_2O) + H_2O + 2A$$

In this equation, H_2A stands for some oxidizable substance such as hydrogen sulfide, free hydrogen, or any one of several other compounds used by photosynthetic bacteria—or water. In the cyanobacteria, the eukaryotic algae, and the green plants, H_2A is water. In short, van Niel proposed that it was the water that was the source of oxygen in photosynthesis, not the carbon dioxide.

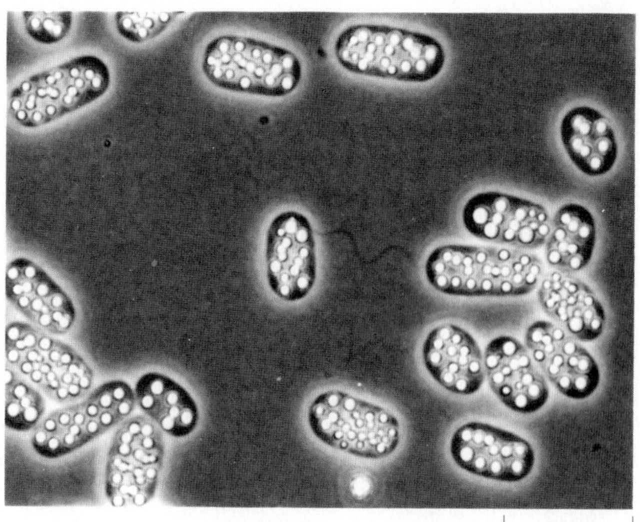

Purple sulfur bacteria. In these cells, hydrogen sulfide plays the same role as water does in the photosynthetic process of plants. The hydrogen sulfide (H_2S) is split, and the sulfur accumulates as globules, visible within the cells.

This brilliant speculation, first proposed in the early 1930s, was not proved until many years later. Eventually, investigators, using a heavy isotope of oxygen ($^{18}O_2$), traced the oxygen from water to oxygen gas:

$$CO_2 + 2H_2{}^{18}O \xrightarrow{light} (CH_2O) + H_2O + {}^{18}O_2$$

This confirmed van Niel's hypothesis. The overall concept of photosynthesis has remained unchanged from the time of van Niel's proposal. However, many of its details have subsequently been worked out, and more are still under active investigation.

light energy and use it to form carbohydrates and free oxygen from carbon dioxide and water, in a complex series of reactions. The overall equation for photosynthesis can be summarized as:

$$CO_2 + H_2O + \text{light energy} \longrightarrow \underset{\text{CARBOHYDRATE}}{(CH_2O)} + O_2$$

To understand how organisms are able to capture light energy and convert it into stored chemical energy, we must first look at the characteristics of light itself.

THE NATURE OF LIGHT

Over 300 years ago, the English physicist Sir Isaac Newton (1642–1727) separated visible light into a spectrum of colors by passing it through a prism (Figure 10–2). Then by passing the light through a second prism, he recombined the colors, producing white light once again. By this experiment, Newton showed that white light is actually made up of a number of different colors, ranging from violet at one end of the spectrum to red at the other. Their separation is possible because light of different colors is bent at different angles in passing through the prism. Newton believed that light was a stream of particles (or, as he termed them, "corpuscles"), in part, because of its tendency to travel in a straight line.

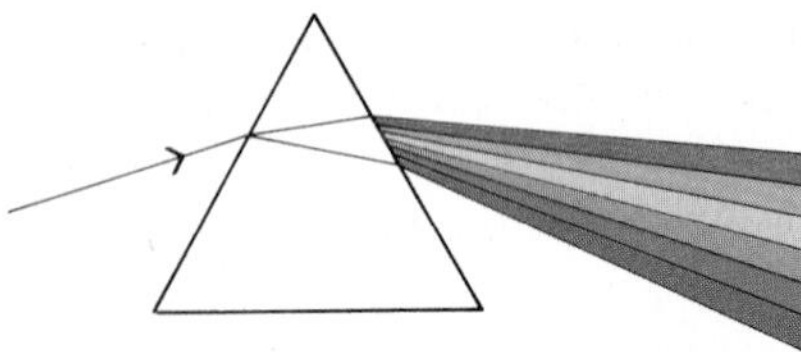

10–2
White light is actually a mixture of different colors, ranging from violet at one end of the spectrum to red at the other. It is separated into its component colors when it passes through a prism—"the celebrated phaenomena of colors," as Newton referred to it.

In the nineteenth century, through the genius of James Clerk Maxwell (1831–1879), it became known that what we experience as light is in truth a very small part of a vast continuous spectrum of radiation, the electromagnetic spectrum. As Maxwell showed, all the radiations included in this spectrum act as if they travel in waves. The wavelengths—that is, the distances from one wave peak to the next—range from those of gamma rays, which are measured in nanometers, to those of low-frequency radio waves, which are measured in kilometers. Within the spectrum of visible light, red light has the longest wavelength, violet the shortest. Another feature that these radiations have in common is that, in a vacuum, they all travel at the same speed—300,000 kilometers per second.

By 1900, it had become clear, however, that the wave model of light was not adequate. The key observation, a very simple one, was made in 1888: when a zinc plate is exposed to ultraviolet light, it acquires a positive charge. The metal, it was soon deduced, becomes positively charged because the light energy dislodges electrons, forcing them out of the metal atoms. Subsequently, it was discovered that this photoelectric effect, as it is called, can be produced in all metals. Every metal has a critical wavelength for the effect; the light (visible or invisible) must be of that wavelength or a shorter wavelength for the effect to occur.

With some metals, such as sodium, potassium, and selenium, the critical wavelength is within the spectrum of visible light, and as a consequence, visible light

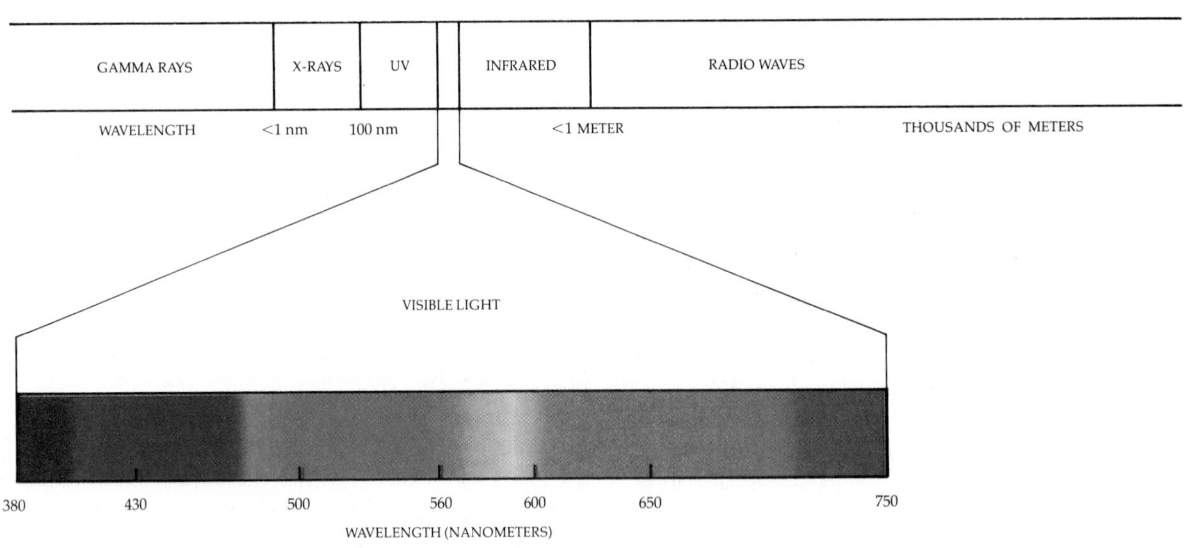

10–3
Visible light is only a small portion of the vast electromagnetic spectrum. For the human eye, the visible spectrum ranges from violet light, which is made up of comparatively short rays, to red light, the longest visible rays.

striking the metal can set up a moving stream of electrons (such a stream is an electric current). Burglar alarms, exposure meters, television cameras, and the electric eyes that open doors for you at supermarkets or airline terminals all operate on this principle of turning light energy into electrical energy.

The wave model of light would lead you to predict that the brighter the light—that is, the stronger, or more intense, the beam—the greater the force with which the electrons would be dislodged. But as we have already seen, whether or not light can eject the electrons of a particular metal depends not on the brightness of the light but on its wavelength. A very weak beam of the critical wavelength or a shorter wavelength is effective, while a stronger beam of a longer wavelength is not. Furthermore, as was shown in 1902, increasing the brightness of the light increases the number of electrons dislodged but not the velocity at which they are ejected from the metal. To increase the velocity, one must use a shorter wavelength of light. Nor is it necessary for energy to be accumulated in the metal. With even a dim beam of a critical wavelength, electrons may be emitted the instant the light hits the metal.

To explain such phenomena, the particle model of light was resurrected by Albert Einstein in 1905. According to this model, light is composed of particles of energy called *photons.* The energy of a photon is not the same for all kinds of light but is, in fact, inversely proportional to the wavelength—the longer the wavelength, the lower the energy. Photons of violet light, for example, have almost twice the energy of photons of red light, the longest visible wavelength.

The wave model of light permits physicists to describe certain aspects of its behavior mathematically, and the photon model permits another set of mathematical calculations and predictions. These two models are no longer regarded as opposed to one another; rather, they are complementary, in the sense that both—or a totally new model—are required for a complete description of the phenomenon we know as light.

The Fitness of Light

Light, as Maxwell showed, is only a tiny band in a continuous spectrum. From the physicist's point of view, the difference between radiations we can see and radiations we cannot see—so dramatic to the human eye—is only a few nanometers of wavelength, or, expressed differently, a small amount of energy. Why does this particular group of radiations, rather than some other, make the leaves grow and the flowers burst forth, cause the mating of fireflies and palolo worms, and, when reflecting off the surface of the moon, excite the imagination of poets and lovers? Why is it that this tiny portion of the electromagnetic spectrum is responsible for vision, for the rhythmic day-night regulation of many biological activities, for the bending of plants toward the light, and also for photosynthesis, on which all life depends? Is it an amazing coincidence that all these biological activities are dependent on these same wavelengths?

George Wald of Harvard, an expert on the subject of light and life, says no. He thinks that if life exists elsewhere in the universe, it is probably dependent on the same fragment of the vast spectrum. Wald bases this conjecture on two points. First, living things, as we have seen, are composed of large, complicated molecules held in special configurations and relationships to one another by hydrogen bonds and other weak bonds. Radiation of even slightly higher energies than the energy of violet light breaks these bonds and so disrupts the structure and function of the molecules. Radiations with wavelengths less than 200 nanometers—that is, with

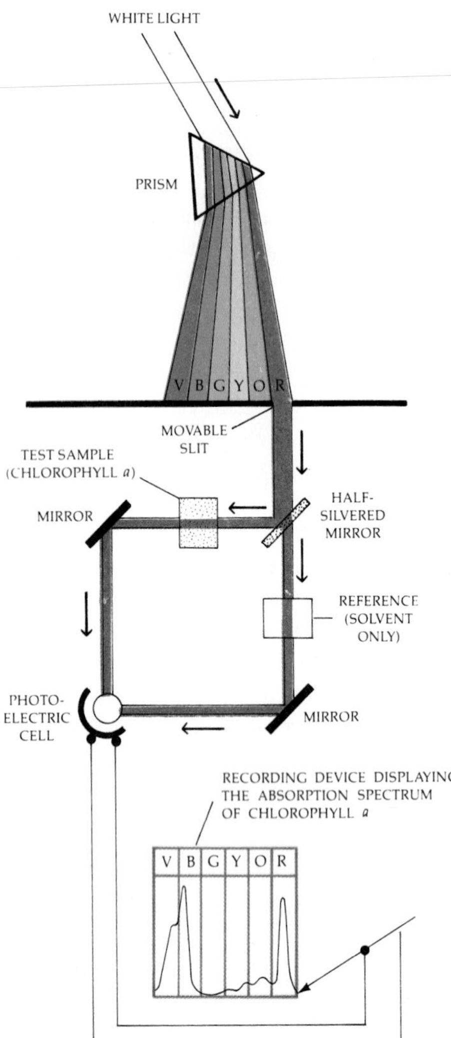

10–4
The absorption spectrum of a pigment is measured with a spectrophotometer. This device directs a beam of light of each wavelength at the object to be analyzed and records what percentage of light of each wavelength is absorbed by the pigment sample as compared to a reference sample. Because the mirror is lightly (half) silvered, half of the light is reflected and half is transmitted. The photoelectric cell is connected to an electronic device that automatically records the percentage absorption at each wavelength.

still higher energies—drive electrons out of atoms. On the other hand, light of wavelengths longer than those of the visible band—that is, with less energy than red light—is absorbed by water, which makes up the great bulk of all living things on earth. When this light reaches molecules, its lower energy causes them to increase their motion (increasing heat) but does not trigger changes in their electron configurations. Only those radiations within the range of visible light have the property of exciting molecules—that is, of moving electrons into higher energy levels—and so of producing chemical and, ultimately, biological changes.

The second reason that the visible band of the electromagnetic spectrum has been "chosen" by living things is that it, above all, is what is available. Most of the radiation reaching the earth from the sun is within this range. Higher-energy wavelengths are screened out by the oxygen and ozone high in the modern atmosphere. Much infrared radiation is screened out by water vapor and carbon dioxide before it reaches the earth's surface.

This is an example of what has been termed "the fitness of the environment"; the suitability of the environment for life and that of life for the physical world are exquisitely interrelated. If they were not, life could not, of course, exist.

CHLOROPHYLL AND OTHER PIGMENTS

In order for light energy to be used by living systems, it must first be absorbed. A pigment is any substance that absorbs light. Some pigments absorb all wavelengths of light and so appear black. Some absorb only certain wavelengths, transmitting or reflecting the wavelengths they do not absorb. Chlorophyll, the pigment that makes leaves green, absorbs light in the violet and blue wavelengths and also in the red; because it reflects green light, it appears green. Different pigments absorb light energy at different wavelengths. The absorption pattern of a pigment is known as the *absorption spectrum* of that substance (Figure 10–4).

Different groups of plants and algae use various pigments in photosynthesis. There are several different kinds of chlorophyll that vary slightly in their molecular structure (Figure 10–5). In plants, chlorophyll *a* is the pigment directly involved in the transformation of light energy to chemical energy. Most photosynthetic cells also contain a second type of chlorophyll—in plants, it is chlorophyll *b*—and a representative of another group of pigments called the carotenoids. One of the carotenoids found in plants is beta-carotene. The carotenoids are red, orange, or yellow pigments. In the green leaf, their color is masked by the chlorophylls, which are more abundant. In some tissues, however, such as those of a ripe tomato, the carotenoid colors predominate, as they do also when leaf cells stop synthesizing chlorophyll in the fall.

The other chlorophylls and the carotenoids are able to absorb light at wavelengths different from those absorbed by chlorophyll *a*. They apparently can pass the energy on to chlorophyll *a*, thus extending the range of light available for photosynthesis (Figure 10–6).

An *action spectrum* defines the relative effectiveness (per number of incident photons) of different wavelengths of light for light-requiring processes, such as photosynthesis, flowering, phototropism (the bending of a plant toward light), and vision. Similarity between the absorption spectrum of a pigment and the action spectrum of a process is considered evidence that that particular pigment is responsible for that particular process (Figures 10–7 and 10–8 on page 216).

(a)

10–5

(a) Chlorophyll a *is a large molecule with a central atom of magnesium held in a porphyrin ring. Attached to the ring is a long, hydrophobic carbon-hydrogen chain that may help to anchor the molecule in the internal membranes of the chloroplast. Chlorophyll* b *differs from chlorophyll* a *in having an aldehyde (CHO) group in place of the CH_3 group indicated by color. Alternating single and double bonds, such as those in the chlorophylls, are common in pigments.* (b) *The estimated absorption spectra of chlorophyll* a *and chlorophyll* b *within the chloroplast. (Prepared by Govindjee.)*

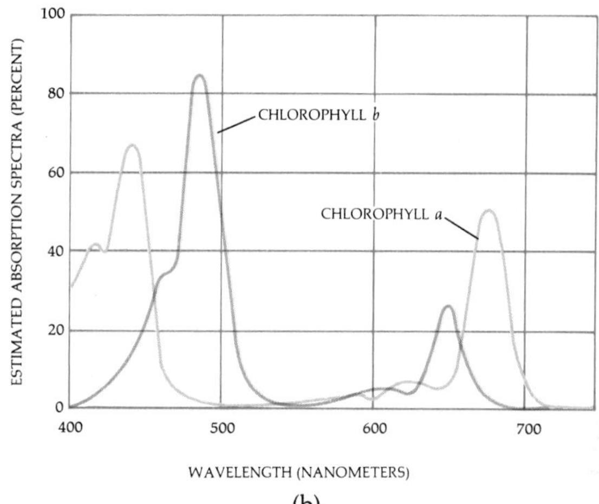

(b)

10–6

(a) Related carotenoids. Cleavage of the beta-carotene molecule at the point indicated by the arrow yields two molecules of vitamin A. Oxidation of vitamin A yields retinal, the pigment involved in vision. Absorption of light energy changes the electron configuration of retinal and triggers a nerve impulse in the retina. All of this explains why you were told to eat your carrots. (b) *The estimated absorption spectrum of carotenoids in the chloroplast. (Prepared by Govindjee.)*

BETA-CAROTENE

VITAMIN A

RETINAL

(a)

ESTIMATED ABSORPTION SPECTRUM (PERCENT)

WAVELENGTH (NANOMETERS)

(b)

10–7
Results of an experiment performed in 1882 by T. W. Englemann revealing the action spectrum of photosynthesis in a filamentous alga. Like more recent investigators, Englemann used the rate of oxygen production to measure the rate of photosynthesis. Unlike his successors, however, he lacked sensitive devices for detecting oxygen. As his oxygen indicator, he chose motile bacteria that are attracted by oxygen. In place of the mirror and diaphragm usually used to illuminate objects under view in his microscope, he substituted a "microspectral apparatus," which, as its name implies, produced a tiny spectrum of colors that it projected upon the slide under the microscope. Then he arranged a filament of algal cells parallel to the spread of the spectrum. The oxygen-seeking bacteria congregated mostly in the areas where the violet and red wavelengths fell upon the algal filament. As you can see, the action spectrum for photosynthesis Englemann revealed in this elegant experiment paralleled the absorption spectrum of chlorophyll. He therefore concluded that photosynthesis depends on the light absorbed by chlorophyll.

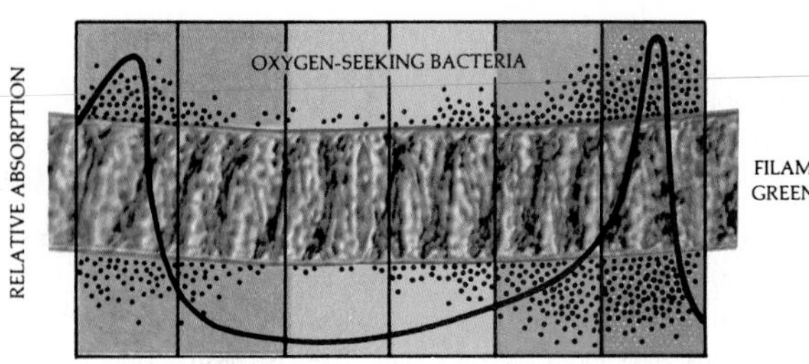

When pigments absorb light, electrons are boosted to a higher energy level. Three of the possible consequences are: (1) the energy may be dissipated as heat; (2) it may be reemitted immediately as light energy of a longer wavelength, a phenomenon known as fluorescence; (3) the energy may cause a chemical reaction, as happens in photosynthesis. Whether or not a particular pigment can cause a chemical reaction depends not only on its structure but also on its relationship with neighboring molecules. For example, if chlorophyll molecules are isolated in a test tube and light is permitted to strike them, they fluoresce. In other words, the molecules absorb light energy, and the electrons are momentarily raised to a higher energy level and then fall back again to a lower one. As they fall to a lower energy level, they release much of this energy as light. None of the light absorbed by isolated chlorophyll molecules is converted to any form of energy useful to living systems. Chlorophyll can convert light energy to chemical energy only when it is associated with certain proteins and embedded in a specialized membrane.

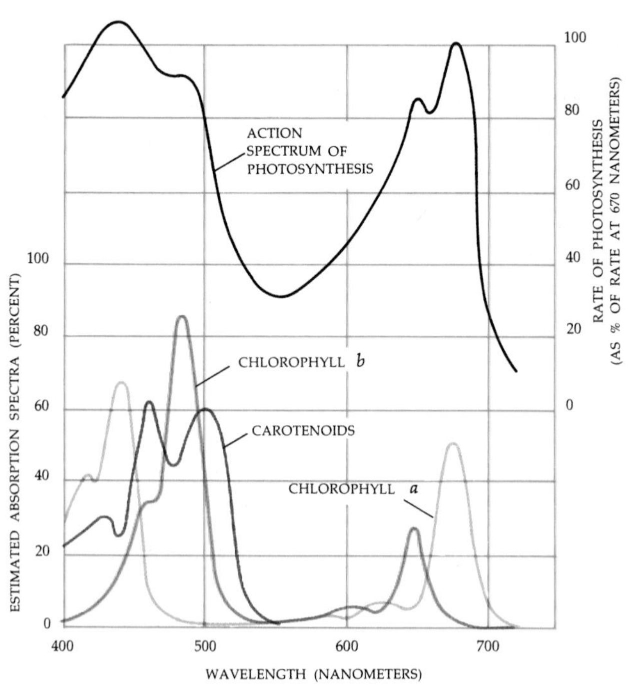

10–8
The upper curve shows the action spectrum for photosynthesis and the lower curves, absorption spectra for chlorophyll a, *chlorophyll* b, *and carotenoids in the chloroplast. Note that the action spectrum of photosynthesis indicates that chlorophyll* a, *chlorophyll* b, *and carotenoids all absorb light used in photosynthesis. (Prepared by Govindjee.)*

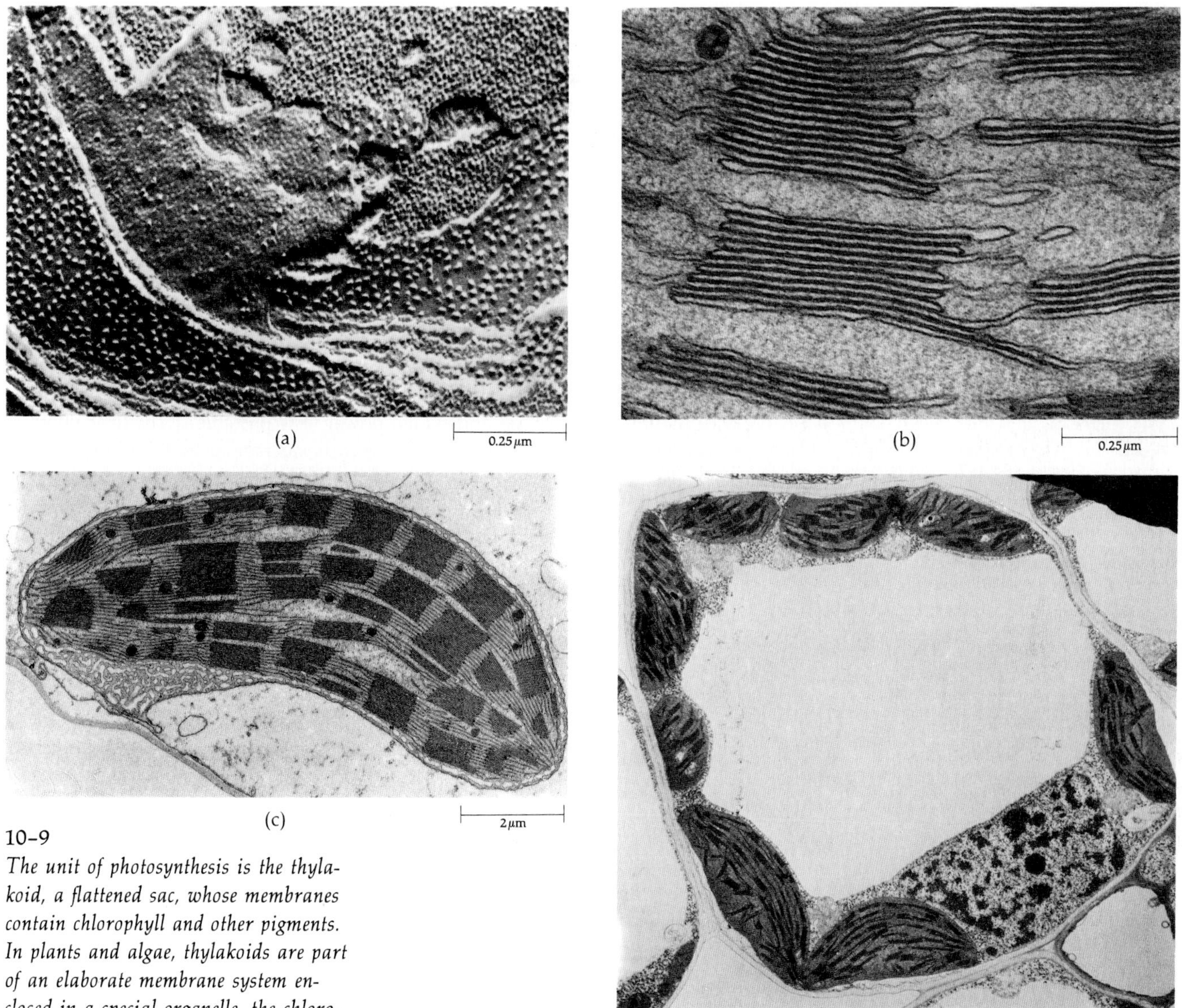

10–9
The unit of photosynthesis is the thylakoid, a flattened sac, whose membranes contain chlorophyll and other pigments. In plants and algae, thylakoids are part of an elaborate membrane system enclosed in a special organelle, the chloroplast. (a) *Internal membrane surface of a thylakoid prepared by the freeze-fracture technique (see page 133). The particles are believed to be enzymes involved in the light-capturing reactions of photosynthesis.* (b) *A stack of thylakoids from a plant cell. The inner compartments of the thylakoids are interconnected, forming the thylakoid space, which contains a solution whose composition differs from that of the stroma and the cytoplasm.* (c) *A chloroplast, showing the elaborate system of internal membranes comprising interconnected stacks of thylakoids.* (d) *A photosynthetic cell with eight chloroplasts visible. The center of the cell is filled with a large vacuole.*

PHOTOSYNTHETIC MEMBRANES: THE THYLAKOID

The structural unit of photosynthesis is the *thylakoid,* which usually takes the form of a flattened sac, or vesicle. In the photosynthetic prokaryotes, thylakoids may form a part of the cell membrane, or they may occur singly in the cytoplasm, or, as in the cyanobacteria, they may be part of an elaborate internal membrane structure (see Figure 4–8, page 85). In eukaryotes, the thylakoids form a part of the internal membrane structure of specialized organelles, the chloroplasts (Figure 10–9). The alga *Chlamydomonas,* for instance, has a single very large chloroplast; the cell of a leaf characteristically has 40 to 50 chloroplasts, and there are often 500,000 chloroplasts per square millimeter of leaf surface.

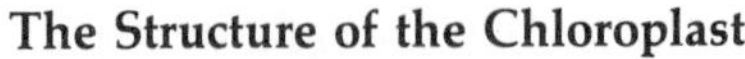

The Structure of the Chloroplast

Chloroplasts, like mitochondria, are surrounded by two membranes that are separated by an intermembrane space. The inner membrane, unlike that of the mitochondrion, is smooth. The thylakoids, in the interior of the chloroplast, constitute a third membrane system. Surrounding the thylakoids, and filling the interior of the chloroplast, is a dense solution, the stroma, which (like the matrix of the mitochondrion) is different in composition from the material surrounding the organelles in the cytoplasm. The thylakoids, as we noted earlier, are flattened sacs, or vesicles; they enclose an additional compartment, known as the thylakoid space, which contains a solution of still different composition. Thus, whereas the mitochondrion has two membrane systems (the outer and the inner) and two compartments (the intermembrane space and the matrix), the chloroplast has three membrane systems (outer, inner, and thylakoid) and three compartments (intermembrane space, stroma, and thylakoid space).

With the light microscope under high power, it is possible to see little spots of green within the chloroplasts of leaves. The early microscopists called these green specks grana ("grains"), and this term is still in use. Under the electron microscope, it can be seen that the grana are stacks of thylakoids. Some of the thylakoid membranes have extensions that interconnect the grana through the stroma that separates them.

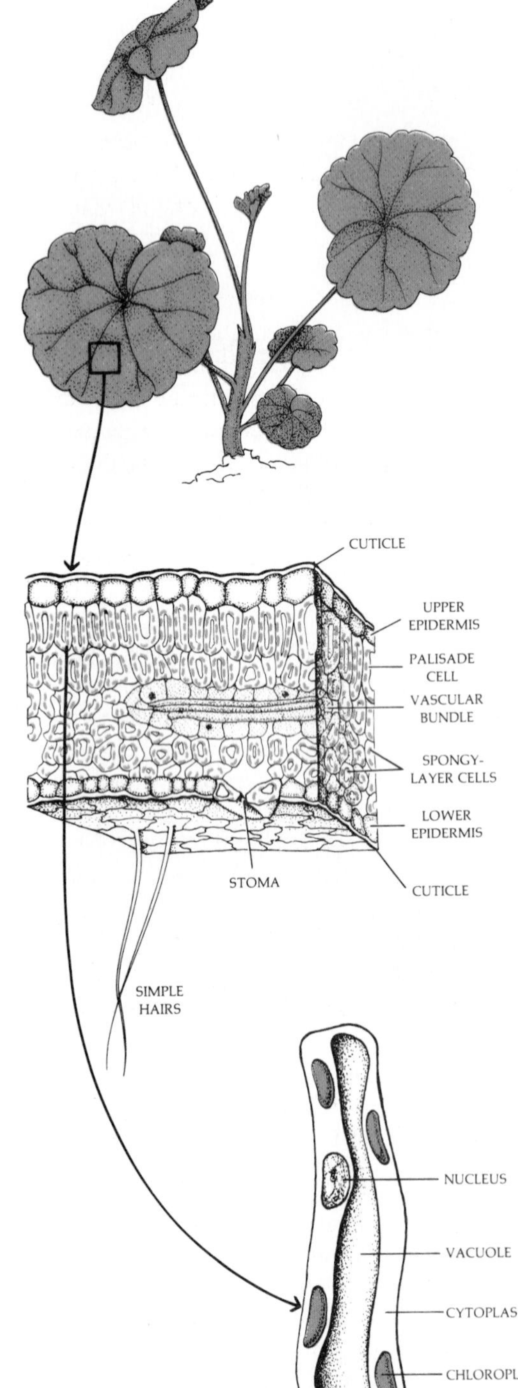

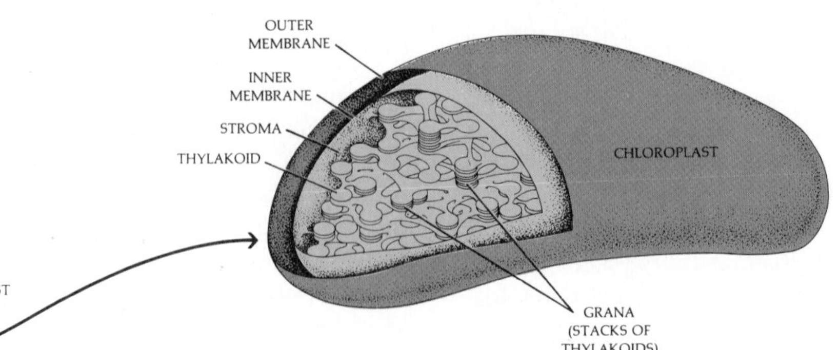

10–10

Journey into a chloroplast. The plant shown is a geranium, which you may recognize by the characteristic shape of its leaves. The inner tissues of the leaf are completely enclosed by transparent epidermal cells that are coated with a waxy layer, the cuticle. Oxygen, carbon dioxide, and other gases enter the leaf largely through special openings, the stomata (singular, stoma). These gases and water vapor fill the spaces between cells in the spongy layer, leaving and entering cells by diffusion. Water, taken up by the roots, enters the leaf by way of the vascular bundle, and sugars, the products of photosynthesis, leave the leaf by this route, traveling to nonphotosynthetic parts of the plant. Much of the photosynthesis takes place in the palisade cells, elongated cells directly beneath the upper epidermis. They have a large central vacuole and numerous chloroplasts that move within the cell, orienting themselves with respect to the light. Light is captured in the membranes of the disk-shaped thylakoids within the chloroplast.

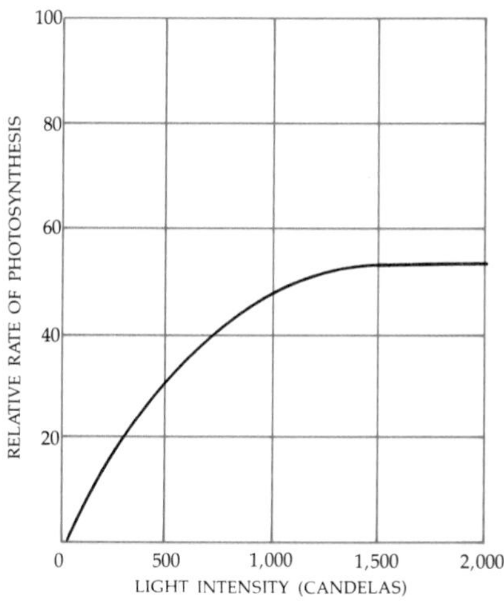

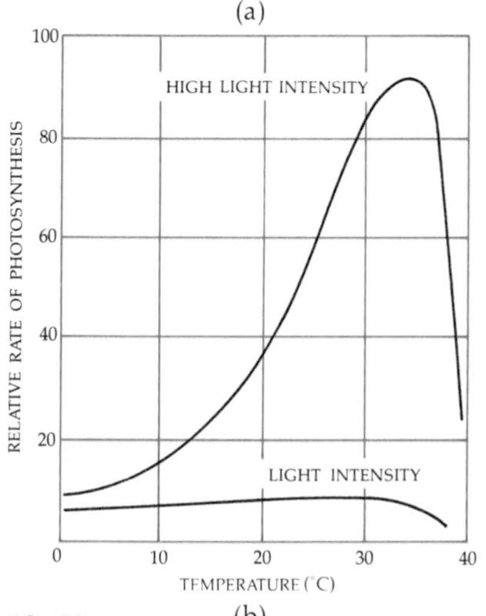

10–11

(a) *An increase in light intensity does not produce a corresponding increase in the rate of photosynthesis beyond about 1,200 candelas. A curve such as the one shown here indicates that some other factor—known as a rate-limiting factor—is involved in the process under study. Under field conditions,* CO_2 *concentration is commonly the rate-limiting factor.* (b) *At a low intensity of light, an increase in temperature does not increase the rate of photosynthesis. At a high intensity, however, an increase in temperature has a very marked effect. From these data, Blackman concluded that photosynthesis includes both light-dependent and light-independent reactions.*

All the thylakoids in a chloroplast are oriented parallel to each other. Thus, by swinging toward the light, the chloroplast can simultaneously aim all of its millions of pigment molecules for optimum reception, as if they were miniature electromagnetic antennae (which, of course, they are).

THE STAGES OF PHOTOSYNTHESIS

About 200 years ago, a Dutch physician, Jan Ingenhousz (1730–1799), demonstrated that light is required for the process we know as photosynthesis. It is now known that photosynthesis actually takes place in two stages, only one of which requires light. Evidence for this two-stage mechanism was first presented in 1905 by the English plant physiologist F. F. Blackman, as the result of experiments in which he measured the rate of photosynthesis under varying conditions.

Blackman first plotted the rate of photosynthesis at various light intensities. In dim to moderate light, increasing the light intensity increased the rate of photosynthesis, but at higher intensities, a further increase in light intensity had no effect. He then studied the combined effects of light and temperature on photosynthesis. In dim light, an increase in temperature had no effect. However, Blackman found that if he increased the light and also increased the temperature, the rate of photosynthesis was greatly accelerated (Figure 10–11). As the temperature increased above 30°C, the rate of photosynthesis slowed and finally the process ceased.

On the basis of these experiments, Blackman concluded that more than one set of reactions was involved in photosynthesis. First, there was a group of light-dependent reactions that were temperature-independent. The rate of these reactions could be accelerated in the dim-to-moderate light range by increasing the amount of light, but it was not accelerated by increases in temperature. Second, there was a group of reactions that were dependent not on light but rather on temperature. Both sets of reactions seemed to be required for the process of photosynthesis. Increasing the rate of only one set of reactions increased the rate of the entire process only to the point at which the second set of reactions began to hold back the first (that is, it became rate-limiting). Then it was necessary to increase the rate of the second set of reactions in order for the first to proceed unimpeded.

In Blackman's experiments, the temperature-dependent reactions increased in rate as the temperature was increased, but only up to about 30°C, after which the rate began to decrease. From this evidence it was concluded that these reactions were controlled by enzymes, since this is the way enzymes are expected to respond to temperature (see Figure 8–13). This conclusion has since proved to be correct.

Photosynthesis was thus shown to have both a light-dependent stage, the so-called "light" reactions, and an enzymatic, light-independent stage, the "dark" reactions. The terms "light" and "dark" reactions have created much confusion, for although the "dark" reactions do not require light as such—only the chemical products of the "light" reactions—they can occur in either light or darkness. Moreover, recent work has shown that the enzyme controlling one of the key "dark" reactions is indirectly stimulated by light. As a result, these terms are now falling into disfavor. They are being replaced by terms that more accurately describe the processes occurring during each stage of photosynthesis.

In the first stage of photosynthesis—the energy-capturing reactions—light energy trapped by the chloroplast is used to form ATP from ADP and to reduce electron-carrier molecules. In the second stage of photosynthesis, the chemical products of the first stage are used to reduce carbon from carbon dioxide to a simple sugar, thus converting the chemical energy of the carrier molecules to forms suitable for transport and storage in the algal cell or plant body. At the same time, a carbon skeleton is formed from which other organic molecules can be built. This incorporation of CO_2 into organic compounds is known as the *fixation of carbon*; the steps by which it is accomplished are termed the carbon-fixing reactions.

THE ENERGY-CAPTURING REACTIONS

The Photosystems

In the thylakoids, chlorophyll and other molecules are, according to the present model, packed into units called photosystems. Each unit contains from 250 to 400 molecules of pigment, which serve as light-trapping antennae. Once a photon of light energy is absorbed by one of the antenna pigments, it is bounced around (like a hot potato) among the other pigment molecules of the photosystem until it reaches a special form of chlorophyll *a*, which is the reaction center. When this particular chlorophyll molecule absorbs the energy, an electron is boosted to a higher energy level from which it is transferred to another molecule, an electron acceptor. The chlorophyll molecule is thus oxidized (minus an electron) and positively charged.

Present evidence indicates that there are two different photosystems. In Photosystem I, the reactive chlorophyll *a* molecule is known as P_{700} (P is for pigment) because one of the peaks of its absorption spectrum is at 700 nanometers, a slightly longer wavelength than the usual chlorophyll *a* peak. When P_{700} is oxidized, it bleaches, which is how it was detected. No one has managed to isolate pure P_{700}. Recent evidence indicates that P_{700} is not an unusual kind of chlorophyll but rather a dimer ("two-part") of two chlorophyll *a* molecules that has unusual properties because of its association with special proteins in the membrane and its position in relation to other molecules. Photosystem II also contains a specialized chlorophyll *a* molecule, which passes its electron on to a different electron acceptor. The reactive chlorophyll *a* molecule of Photosystem II is P_{680}.

The Light-Trapping Reactions

The two photosystems probably evolved separately, with Photosystem I coming first. As we shall see, Photosystem I can operate independently. In general, however, the two systems work together simultaneously and continuously, as shown in Figure 10–12. According to the current model, light energy enters Photosystem II, where it is trapped by the reactive chlorophyll *a* molecule P_{680}. An electron from the P_{680} molecule is boosted to a higher energy level from which it is transferred to an electron-acceptor molecule. The electron then passes downhill along an electron transport chain to Photosystem I. As electrons pass along this transport chain, a proton gradient is established; the potential energy of the gradient is used to form ATP from ADP in a chemiosmotic process similar to that in the mitochondrion. This process is known as *photophosphorylation*.

Three other events are taking place simultaneously:

1. The P_{680} chlorophyll molecule, having lost its electron, is avidly seeking a replacement. It finds it in the water molecule, which, while bound to a manganese-containing protein, is stripped of an electron and then broken into protons and oxygen gas.

10-12

Light energy trapped in the reactive chlorophyll a molecule of Photosystem II boosts electrons to a higher energy level. These electrons are replaced by electrons pulled away from water molecules, releasing protons (H^+) and oxygen gas. The electrons are passed from the electron acceptor along an electron transport chain to a lower energy level, the reaction center of Photosystem I. As they pass along this electron transport chain, some of their energy is packaged in the form of ATP. Light energy absorbed by Photosystem I boosts the electrons to another primary electron acceptor. From this acceptor, they are passed via other electron carriers to $NADP^+$ to form NADPH. The electrons removed from Photosystem I are replaced by those from Photosystem II. ATP and NADPH represent the net gain from the light-trapping reactions.

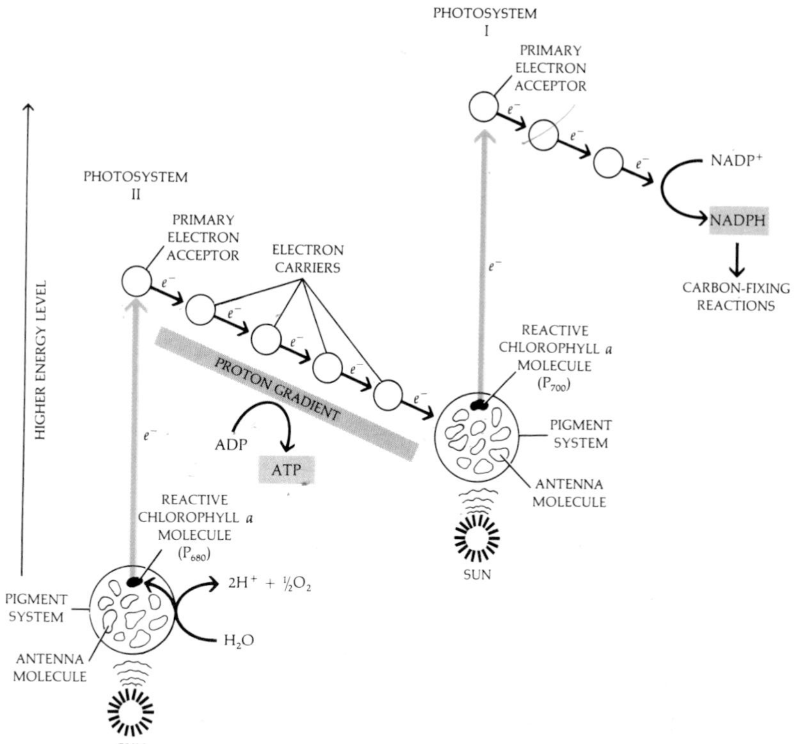

$NADP^+$

10-13

Although NAD^+ (Figure 8-12) and $NADP^+$ resemble one another very closely, their biological roles are distinctly different. NADH generally transfers its electrons to other electron carriers, which continue to pass them on down to successively lower electron levels in discrete steps. In the course of this electron transfer, ATP molecules are formed. NADPH provides energy directly to biosynthetic processes of the cell that require large energy inputs.

2. Additional light energy is trapped in the reactive chlorophyll molecule (P_{700}) of Photosystem I. The molecule is oxidized, and an electron is boosted to a primary electron acceptor from which it goes downhill to $NADP^+$ (Figure 10-13).
3. The electron removed from the P_{700} molecule is replaced by the electron that moved downhill from the primary electron acceptor of Photosystem II.

Thus in the light there is a continuous flow of electrons from water to Photosystem II to Photosystem I to $NADP^+$. In the words of Nobel laureate Albert Szent-Györgyi: "What drives life is . . . a little electric current, kept up by the sunshine."

The energy harvest from these steps is represented by an ATP molecule (whose formation releases a water molecule) and NADPH, which then become the chief sources of energy for the reduction of carbon. To generate one molecule of NADPH, four photons must be absorbed, two by Photosystem II and two by Photosystem I.

Cyclic Electron Flow

As we mentioned previously, there is also evidence that Photosystem I can work independently. When this occurs, no NADPH is formed. In this process, called *cyclic electron flow*, electrons are boosted from P_{700} to the primary electron acceptor of Photosystem I. They do not, however, pass down the series of electron carriers leading to $NADP^+$. Instead, they are shunted to the electron transport chain that connects Photosystems I and II and pass downhill through that chain back into the

10–14

In eukaryotic photosynthetic cells, cyclic electron flow bypasses Photosystem II and requires only Photosystem I. ATP is produced from ADP, but oxygen is not released and $NADP^+$ is not reduced. In some photosynthetic bacteria, which have only Photosystem I, cyclic electron flow is the principal photosynthetic mechanism.

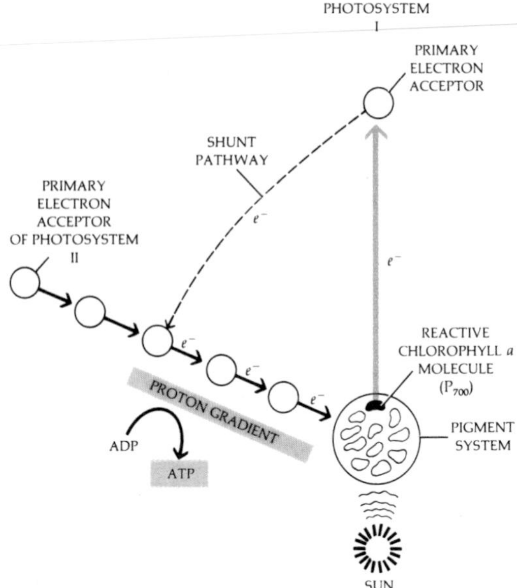

reactive P_{700} molecule (Figure 10–14). ATP is produced in the course of this passage. In the absence of $NADP^+$ (as, for example, when all of the available $NADP^+$ has been reduced to NADPH but has not yet been reoxidized in the carbon-fixing reactions), or when the cell needs additional supplies of ATP but not of NADPH, eukaryotic cells are able to synthesize ATP using the energy of sunlight to power cyclic electron flow. However, no oxygen is released and no carbon is reduced.

It is believed that the most primitive photosynthetic mechanisms worked by cyclic electron flow. This is also apparently the way in which some photosynthetic bacteria carry out photosynthesis. What is an alternative shunt pathway in eukaryotes is, in these bacteria, the principal pathway of electron flow in photosynthesis.

Photosynthetic Phosphorylation

The photophosphorylation of ADP to ATP as electrons pass down the electron transport chain from Photosystem II to Photosystem I (or, as they pass down a portion of that chain during cyclic electron flow) is a chemiosmotic process, similar in many ways to oxidative phosphorylation in the mitochondria. In both the mitochondria and the chloroplasts, the electron transport chains contain cytochromes, and the electron carriers and enzymes of these chains are embedded in membranes (the inner membrane of the mitochondrion and the thylakoid membrane of the chloroplast) that are impermeable to protons (H^+ ions). In both organelles, an electrochemical gradient of potential energy is established as protons are pumped through the membrane using energy released as electrons pass down the chain. Moreover, ADP is phosphorylated to ATP as protons flow down the potential energy gradient through ATP synthetase complexes consisting of two factors.

The topography of this chemiosmotic process is, however, slightly different in chloroplasts. In mitochondria, as we saw on page 201, protons are pumped out of the matrix into the external medium; they flow down the gradient from the exterior back into the matrix. Similarly, in chloroplasts the protons are pumped out of the stroma (analogous to the mitochondrial matrix); they are not, however, pumped out of the organelle itself but rather *into* the thylakoid space (Figure 10–15). The potential energy gradient is between this internal, third compartment and the stroma. When protons flow down the gradient, they move from the thylakoid space back into the stroma, where the ATP is synthesized.

Many of the details of photophosphorylation, like those of oxidative phosphorylation, remain to be worked out. For example, it is clear that protons are released into the thylakoid space when water is split into protons, free oxygen, and electrons at the reaction center of Photosystem II. Additional protons are pumped into the thylakoid space as electrons flow down the transport chain, but the exact number pumped and the mechanism are not yet certain. A similar uncertainty concerns the number of protons that must flow through the ATP synthetase complex to synthesize one molecule of ATP. The best estimate at the present time is three protons for each ATP synthesized.

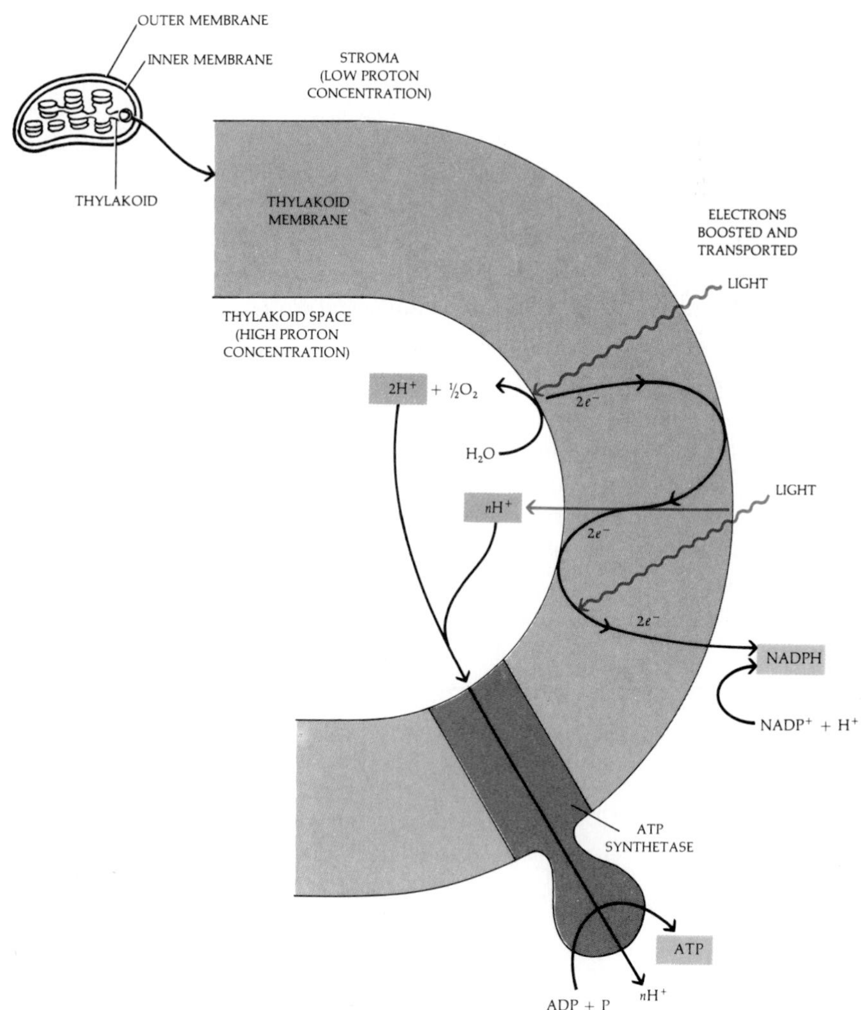

10–15
The chemiosmotic mechanism of photophosphorylation. In this process, electrons from chlorophyll a*—boosted to a high energy level by sunlight—flow down an electron transport chain in the thylakoid membrane. The energy they release as they move to a lower energy level is used to pump protons from the stroma into the thylakoid space, creating an electrochemical gradient of potential energy. As the protons flow down the gradient from the thylakoid space back into the stroma, ADP is phosphorylated to ATP by ATP synthetase. The chemical structures of the electron carriers and enzymes of the thylakoid membrane (including ATP synthetase) are only slightly different from those of the mitochondrial membrane.*

PHOTOSYNTHESIS WITHOUT CHLOROPHYLL

Halobacteria are rod-shaped cells, quite similar in appearance to Escherichia coli. *They grow best in very salty water, about seven times as salty as sea water. If the salt concentration is reduced to only about three times that of sea water, the cell wall falls apart, and as it is reduced still further, the cell membrane begins to break up. Walther Stoeckenius, then at Rockefeller University, separated the membrane fragments by centrifugation. One of the fractions, it turned out, was purple, and this, though the investigators did not know it at the time, was the first clue to a major energy source of the salt-loving bacteria.*

Halobacteria are aerobes, and when suitable substrates and adequate oxygen are available, they oxidize organic molecules, producing ATP by oxidative phosphorylation. Oxygen, however, is often unavailable in the salty waters in which halobacteria live. The secret of their success, it has now been shown, lies in the purple patches of the cell membrane, which provide an alternative, photosynthetic mechanism for producing ATP. The photosynthetic pigment of the halobacteria is not a form of chlorophyll, as in all other photosynthetic organisms, but retinal (Figure 10–6, page 215), which is also the visual pigment of the vertebrate eye. The membrane of the halobacteria contains molecules of retinal plus protein—the complex is called bacteriorhodopsin. When bacteriorhodopsin is excited by light and then returns to its original energy level, the energy released pumps protons across the membrane, out of the cell. This establishes a proton gradient that drives the phosphorylation of ADP to ATP, thus providing additional support that the chemiosmotic mechanism is indeed a universal one for the regeneration of ATP.

Darwin himself confessed a certain uneasiness when called upon to explain how an organ as complex as the eye might have arisen by the slow, cumulative steps of evolution. Was retinal "invented" twice? Or do these purple fragments of membrane hold clues both to the mechanisms of human vision and also to its origins?

A freeze-fracture preparation of the membrane of a halobacterium. The fine-grained regions with a hexagonal pattern are patches of purple membrane.

The proton gradient that drives photophosphorylation is established by the release of energy that originally entered the system from sunlight. Phosphorylation by isolated chloroplasts can also be driven by a proton gradient established artificially across the thylakoid membrane. In such experimental systems, phosphorylation proceeds in the dark, providing impressive evidence that the key factor in photophosphorylation, as in oxidative phosphorylation, is the establishment of a proton gradient.

10–16
Scanning electron micrograph of open stomata on the lower surface of a leaf. The carbon dioxide used in photosynthesis reaches the photosynthetic cells through these openings.

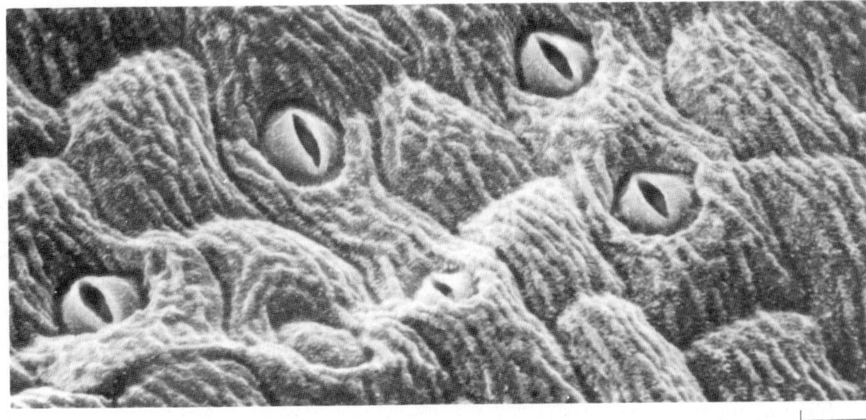

10–17
Calvin and his collaborators briefly exposed photosynthesizing algae to radioactive carbon dioxide ($^{14}CO_2$). After 5 seconds of illumination, the radioactive carbon atom was found almost entirely in the three-carbon compound phosphoglycerate. (The transient intermediate is inferred; it has not been isolated.) After 60 seconds, many compounds contained radioactive carbon.

O=C=O + $CH_2OPO_3^{2-}$ / C=O / H—C—OH / H—C—OH / $CH_2OPO_3^{2-}$

RIBULOSE
1,5-DIPHOSPHATE

↓

$CH_2OPO_3^{2-}$ / ⁻O(O=)C—C—OH / C=O / H—C—OH / $CH_2OPO_3^{2-}$

TRANSIENT
INTERMEDIATE

↓ H_2O

$CH_2OPO_3^{2-}$ / H—C—OH / COO^-

\+

COO^- / H—C—OH / $CH_2OPO_3^{2-}$

TWO MOLECULES
OF 3-PHOSPHOGLYCERATE

THE CARBON-FIXING REACTIONS

The reactions that we have just described are the energy-capturing reactions of photosynthesis. In the course of these reactions, as we saw, light energy is converted to electrical energy—the flow of electrons—and the electrical energy is converted to chemical energy stored in the bonds of NADPH and ATP. In the second stage of photosynthesis, this energy is used to reduce carbon. Carbon is available to photosynthetic cells in the form of carbon dioxide. In algae, such as those seen in Figure 10–1, the carbon dioxide is dissolved in the surrounding water. In plants, carbon dioxide reaches the photosynthetic cells through specialized openings in leaves and green stems, called stomata (Figure 10–16).

The Calvin Cycle: The Three-Carbon Pathway

The reduction of carbon takes place in the stroma in a cycle named after its discoverer, Melvin Calvin. The Calvin cycle is analogous to the Krebs cycle (page 198) in that, in each turn of the cycle, the starting compound is regenerated. The starting (and ending) compound is a five-carbon sugar with two phosphates attached, ribulose diphosphate (RuDP).

The cycle begins when carbon dioxide is bound to RuDP, which then splits to form two molecules of phosphoglycerate, or PGA (Figure 10–17). (Each PGA molecule contains three carbon atoms, hence the name, the three-carbon pathway.) The enzyme catalyzing this crucial reaction, RuDP carboxylase, is very abundant in chloroplasts, making up more than 15 percent of the total chloroplast protein. RuDP carboxylase, said to be the most abundant protein in the world, is located on the surface of the thylakoid membranes.

The complete cycle is diagrammed in Figure 10–18. As in the Krebs cycle, each step is regulated by a specific enzyme. At each full turn of the cycle, a molecule of carbon dioxide enters the cycle, is reduced, and a molecule of RuDP is regenerated. Three turns of the cycle introduce three molecules of carbon dioxide, the equivalent of one three-carbon sugar. Six revolutions of the cycle, with the introduction of six molecules of carbon dioxide, are necessary to produce the equivalent of a six-carbon sugar, such as glucose. The overall equation is

$$6RuDP + 6CO_2 + 18ATP + 12NADPH + 12H^+ + 12H_2O \longrightarrow 6RuDP + glucose + 18P_i + 18ADP + 12NADP^+$$

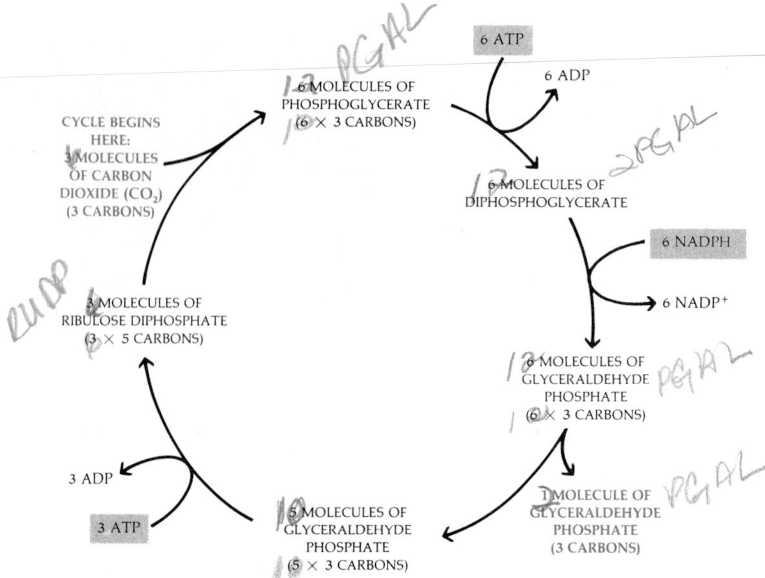

10–18

Summary of the Calvin cycle. At each full "turn" of the cycle, one molecule of carbon dioxide enters the cycle. Three turns are summarized here—the number required to make one molecule of glyceraldehyde phosphate. Three molecules of ribulose diphosphate (RuDP), a five-carbon compound, are combined with three molecules of carbon dioxide, yielding six molecules of phosphoglycerate, a three-carbon compound. These are reduced to six molecules of glyceraldehyde phosphate. Five of these three-carbon molecules are combined and rearranged to form three five-carbon molecules of RuDP. The "extra" molecule of glyceraldehyde phosphate represents the net gain from the Calvin cycle. The energy that drives the Calvin cycle is in the form of ATP and NADPH, produced by the light-trapping reactions.

The immediate product of the cycle itself is glyceraldehyde phosphate. This same three-carbon sugar-phosphate molecule is formed when the fructose diphosphate molecule is split at the fourth step in glycolysis (page 190).

The Four-Carbon Pathway

In most plants, the first step in the fixation of carbon is the binding of carbon dioxide to RuDP and its entrance into the Calvin cycle. Some plants, however, first bind carbon dioxide into the four-carbon compound oxaloacetic acid. (Oxaloacetic acid, you may recall, is also an intermediate in the Krebs cycle.) The carbon dioxide bound into oxaloacetic acid is ultimately bound to RuDP and enters the Calvin cycle, but not until it has passed through a series of reactions that transport it more deeply into the leaf. Plants that utilize this pathway, also known as the Hatch-Slack pathway, are commonly called C_4, or four-carbon, plants, as distinct from the C_3 plants in which carbon is bound first into the three-carbon compound phosphoglycerate (PGA).

In C_4 plants, the oxaloacetic acid is formed when carbon dioxide is bound to a compound known as phosphoenolpyruvate (PEP). This reaction is catalyzed by the enzyme PEP carboxylase (Figure 10–19). The oxaloacetic acid is then reduced to malic acid or converted (with the addition of an amino group) to aspartic acid.

COOH
|
$COPO_3^{2-}$
||
CH_2
PEP

$CO_2 + H_2O$ → PEP CARBOXYLASE

COOH
|
C=O
|
CH_2
|
COOH
OXALOACETIC ACID

→ MALIC ACID:
COOH
|
HOCH
|
CH_2
|
COOH

→ ASPARTIC ACID:
COOH
|
H_2NCH
|
CH_2
|
COOH

10–19

Carbon dioxide fixation by the C_4 pathway. Carbon dioxide is bound to phosphoenolpyruvate (PEP) by the enzyme PEP carboxylase. The resulting oxaloacetic acid is converted either to malic acid or aspartic acid. These steps later will be reversed, releasing carbon dioxide for use in the Calvin cycle.

10–20

In C_4 plants, the chloroplasts of mesophyll cells differ from those of bundle-sheath cells. As shown in this micrograph of portions of chloroplasts in two adjacent cells of a corn leaf, the mesophyll chloroplast (top) contains well-developed grana, while the grana of the bundle-sheath chloroplast are poorly developed. Notice the plasmodesmata that connect the two cells, providing channels through which substances can flow from one cell to the other.

These steps take place in mesophyll cells, whose chloroplasts are characterized by an extensive network of thylakoids, organized into well-developed grana. The next step is a surprise: the malic acid (or aspartic acid, depending on the species) is transported to bundle-sheath cells. The chloroplasts of these cells, which form tight sheaths around the vascular bundles of the leaf, have poorly developed grana and often contain large grains of starch. In the bundle-sheath cells, the malic (or aspartic) acid is decarboxylated to yield CO_2 and pyruvic acid. The CO_2 then enters the Calvin cycle. This process, summarized in Figure 10–21, physically separates the capture of CO_2 by the plant from the reactions of the Calvin cycle.

One might well ask why C_4 plants should have evolved such an energetically expensive and seemingly clumsy method of providing carbon dioxide to the Calvin cycle. To answer this question, we must consider both the function of the leaf as a whole and the properties of PEP carboxylase and RuDP carboxylase, the enzymes that catalyze the first step of the carbon-fixing reactions in C_4 and C_3 plants, respectively.

Carbon dioxide is not continuously available to the photosynthesizing cells. It enters the leaf by way of the stomata, specialized pores that open and close depending on, among other things, water stress. PEP carboxylase, the enzyme that catalyzes the formation of oxaloacetic acid in C_4 plants, has a higher affinity for carbon dioxide than does RuDP carboxylase. Even at low concentrations of carbon dioxide, the enzyme works rapidly to bind it to PEP. Compared with RuDP carboxylase, it fixes carbon dioxide faster, at lower levels, keeping the CO_2 concentration lower within the leaf. This maximizes the gradient of carbon dioxide between the cells and the outside air. Thus, when the stomata are open, carbon dioxide readily diffuses down the concentration gradient into the leaf. If the stomata must be closed much of the time—as they must be to conserve water in a hot, dry climate—the plant with C_4 metabolism will take up more carbon dioxide with each gasp (so to speak) than the plant that has only C_3 metabolism. Hence, the C_4 plant is at a distinct advantage in drought-ridden areas.

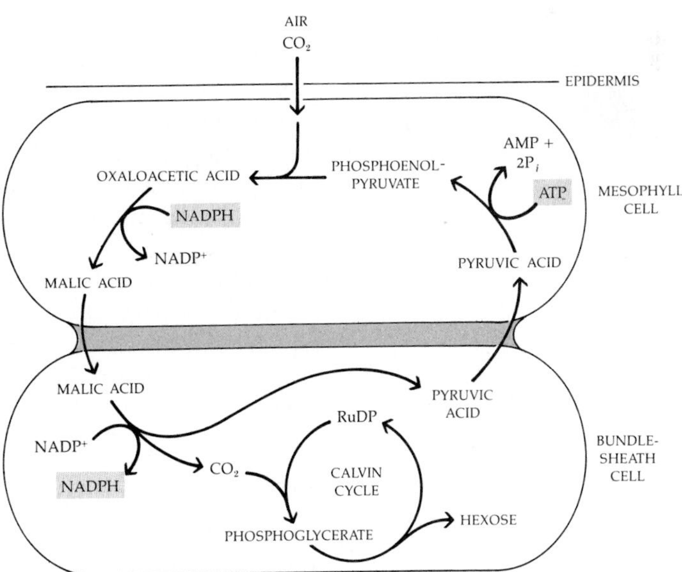

10–21

A pathway for carbon fixation in C_4 plants. CO_2 is first fixed in mesophyll cells as oxaloacetic acid. It is then transported to bundle-sheath cells, where the carbon dioxide is released. The CO_2 thus formed enters the Calvin cycle. Pyruvic acid returns to the mesophyll cell, where it is phosphorylated to PEP.

In the presence of ample carbon dioxide, RuDP carboxylase fixes carbon dioxide efficiently, feeding it into the Calvin cycle. However, when the carbon dioxide concentration in the leaf is low in relation to the oxygen concentration, this same enzyme catalyzes a reaction of RuDP with oxygen, rather than with carbon dioxide. This reaction leads to the formation of glycolic acid, the substrate for a process known as photorespiration. Photorespiration, which occurs in the peroxisomes (page 111) of photosynthetic cells, is the oxidation of carbohydrates in the presence of light and oxygen. Unlike mitochondrial respiration, however, it yields neither ATP nor NADH. Under normal atmospheric conditions, as much as 50 percent of the carbon fixed in photosynthesis by a C_3 plant may be reoxidized to CO_2 during photorespiration. Thus photorespiration greatly reduces the photosynthetic efficiency of C_3 plants.

High CO_2 and low O_2 concentrations limit photorespiration. Consequently, C_4 plants have another distinct advantage over C_3 plants. First, the RuDP carboxylase is sequestered in the bundle-sheath cells, in the interior of the leaf, where it is to some extent protected from atmospheric oxygen. Second, because the CO_2 fixed by the C_4 pathway is essentially "pumped" from the mesophyll cells into the bundle-sheath cells, it is delivered to RuDP carboxylase in a concentrated form. The ratio of CO_2 to O_2 is high enough that the enzyme catalyzes the reaction involving CO_2 and not the reaction leading to photorespiration. Any carbon dioxide that is released by photorespiration is immediately recaptured by PEP carboxylase in the mesophyll cells and fed back to RuDP and into the Calvin cycle. As a result of all of these factors, the net rates of photosynthesis in C_4 grasses, such as corn, sugarcane, and sorghum, can be two to three times the rates in C_3 grasses, such as wheat, oats, and rice.

C_4 plants evolved primarily in the tropics and are especially well adapted to high light intensities, high temperatures, and dryness. The optimal temperature range for C_4 photosynthesis is much higher than that for C_3 photosynthesis, and C_4 plants flourish even at temperatures that would eventually be lethal to many C_3 species. Because of their more efficient use of carbon dioxide, C_4 plants can attain the same photosynthetic rate as C_3 plants but with smaller stomatal openings and, hence, with considerably less water loss.

Perhaps the most familiar example of the competitive capacity of C_4 plants is seen in lawns in the summertime. In most parts of the United States, lawns consist mainly of C_3 grasses such as Kentucky bluegrass. As the summer days become hotter and drier, these dark green, fine-leaved grasses are often overwhelmed by rapidly growing crabgrass, which disfigures the lawn as its yellowish-green, broader-leaved plants slowly take over. Crabgrass, you will not be surprised to hear, is a C_4 plant.

The list of plants known to utilize the four-carbon pathway has grown to over 100 genera, at least a dozen of which include both C_3 and C_4 species. This pathway has undoubtedly arisen independently many times in the course of evolution and is another example of the exquisite adaptation of living systems to their environment.

THE PRODUCTS OF PHOTOSYNTHESIS

Glyceraldehyde phosphate, the three-carbon sugar produced by the Calvin cycle, may seem an insignificant reward, both for all the enzymatic activity on the part of the cell and for our own intellectual stress. However, this molecule and those

THE CARBON CYCLE

By photosynthesis, living systems incorporate carbon dioxide from the atmosphere into organic, carbon-containing compounds. In respiration, these compounds are broken down again into carbon dioxide and water. These processes, viewed on a worldwide scale, result in the carbon cycle. The principal photosynthesizers in this cycle are plants and the phytoplankton, the marine algae. They synthesize carbohydrates from carbon dioxide and water and release oxygen into the atmosphere. About 75 billion metric tons of carbon per year are bound into carbon compounds by photosynthesis.

Some of the carbohydrates are used by the photosynthesizers themselves. Plants release carbon dioxide from their roots and leaves, and marine algae release it into the water where it maintains an equilibrium with the carbon dioxide of the air. Some 500 billion metric tons of carbon are "stored" as dissolved carbon dioxide in the seas, and some 700 billion metric tons in the atmosphere. Some of the carbohydrates are used by animals that feed on the live plants, on algae, and on one another, releasing carbon dioxide. An enormous amount of carbon is contained in the dead bodies of plants and other organisms plus discarded leaves and shells, feces, and other waste materials that settle into the soil or sink to the ocean floors where they are consumed by decomposers—small invertebrates, bacteria, and fungi. Carbon dioxide is also released by these processes into the reservoir of the air and oceans. Another, even larger store of carbon lies below the surface of the earth in the form of coal and oil, deposited there some 300 million years ago.

The natural processes of photosynthesis and respiration balance one another out. For many millions of years, the carbon dioxide of the atmosphere, as far as we can tell, has remained constant. By volume, it is a very small proportion of the atmosphere, only about 0.03 percent. It is important, however, because carbon dioxide, unlike most other components of the atmosphere, absorbs heat from the sun's rays. Since 1850, carbon dioxide concentrations in the atmosphere have been increasing, owing in part to our use of fossil fuels, to our plowing of the soil, and to our destruction of forest land, particularly in the tropics. Some environmentalists predict that this increase in the carbon dioxide "blanket" will increase the temperature here on earth, with a consequential increase in the great deserts of the world. Others, looking on the sunnier side, foresee an increase in photosynthetic activity because of the increased carbon dioxide that will be available to plants and algae. Most, however, feel alarmed by the fact that although we do not know the consequences of what we are doing, we keep right on doing it.

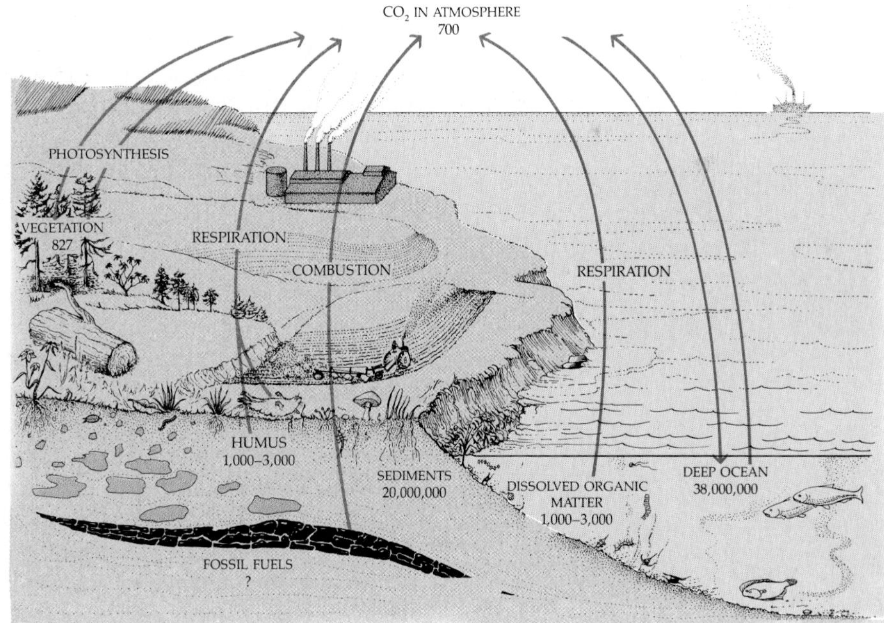

The carbon cycle. The arrows indicate the movement of carbon atoms. The numbers are all estimates of the amount of carbon stored, expressed in billions of metric tons. The amount of carbon released by respiration and combustion has, it is believed, begun to exceed the amount fixed by photosynthesis.

derived from it provide (1) the energy source for all living systems, and (2) the basic carbon skeleton for all organic molecules. Carbon has been fixed—that is, it has been brought from the inorganic world into the organic one.

Molecules of glyceraldehyde phosphate may flow into a variety of different metabolic pathways, depending on the activities and requirements of the cell. Often they are built up to glucose or fructose, following a sequence that is in many of its steps the reverse of the glycolysis sequence described in the previous chapter. (At some steps, the reactions are simply reversed and the enzymes are the same. Other steps—the highly exergonic ones of the downhill sequence—are bypassed.) Plant cells use these six-carbon sugars to make starch and cellulose for their own purposes and sucrose for export. Animal cells store them as glycogen. All cells use sugars, including glyceraldehyde phosphate and glucose, as the starting point for the manufacture of other carbohydrates, fats and other lipids, and, with the addition of nitrogen, amino acids and nitrogenous bases. Finally, as we saw in the preceding chapter, the carbon fixed in photosynthesis is the source of ATP energy for heterotrophic cells.

SUMMARY

In photosynthesis, light energy is converted to chemical energy, and carbon is fixed into organic compounds. The generalized equation for this reaction is

$$CO_2 + 2H_2A + \text{light energy} \longrightarrow (CH_2O) + H_2O + 2A$$

in which H_2A stands for water or some other substance from which electrons can be removed.

Light energy is captured by the living world by means of pigments. The pigments involved in photosynthesis in eukaryotes include the chlorophylls and the carotenoids. Light absorbed by the pigments boosts their electrons to higher energy levels. Because of the way the pigments are packed into membranes, they are able to transfer this energy to reactive molecules, probably chlorophyll *a* packed in a particular way.

Photosynthesis takes place within cellular organelles known as chloroplasts. These organelles are surrounded by two membranes. Contained within the membranes of the chloroplast are a solution of organic compounds and ions known as the stroma and a complex internal membrane system consisting of fused pairs of membranes that form sacs called thylakoids. The pigments and other molecules responsible for capturing light are located in and on these membranes.

Photosynthesis takes place in two stages, as summarized in Table 10–1. In the currently accepted model of the light-trapping reactions, light energy strikes antennae pigments of Photosystem II, which contains several hundred molecules of chlorophyll *a* and chlorophyll *b*. Electrons are boosted uphill from the reactive chlorophyll *a* molecule P_{680} to an electron acceptor. As the electrons are removed, they are replaced by electrons from water molecules, with the simultaneous production of free O_2. The electrons then pass downhill to Photosystem I along an electron transport chain; this passage generates a proton gradient that drives the synthesis of ATP from ADP (photophosphorylation). Light energy absorbed in antennae pigments of Photosystem I and passed to chlorophyll P_{700} results in the boosting of electrons to another electron acceptor. The electrons removed from P_{700} are replaced by the electrons from Photosystem II. The electrons are ultimately

Table 10-1 **Summary of the Stages of Photosynthesis**

	CONDITIONS	WHERE	WHAT APPEARS TO HAPPEN	RESULTS
Energy-capturing reactions	Light	Thylakoids	Light striking Photosystem II boosts electrons uphill. (These electrons are replaced by electrons from water molecules, which release O_2.) Electrons then pass downhill along an electron transport chain to Photosystem I; in the process, ATP is formed by chemiosmosis. Light hits Photosystem I, boosts electrons uphill to an electron acceptor from which they are passed to $NADP^+$ to form NADPH. Electrons are replaced in Photosystem I by electrons from Photosystem II.	Energy of light is converted to chemical energy stored in bonds of ATP and NADPH.
Carbon-fixing reactions	Do not require light, although some enzymes stimulated by light	Stroma	Calvin cycle. NADPH and ATP formed in energy-capturing reactions are used to reduce carbon dioxide. The cycle yields glyceraldehyde phosphate from which glucose or other organic compounds can be formed.	Chemical energy of ATP and NADPH is used to incorporate carbon into organic molecules.

accepted by the electron carrier molecule $NADP^+$. The energy yield from this sequence of reactions is contained in the molecules of NADPH and in the ATP formed by photophosphorylation.

Photophosphorylation also occurs in cyclic electron flow, a process that bypasses Photosystem II. In cyclic electron flow, the electrons boosted from P_{700} in Photosystem I do not pass to $NADP^+$ but are instead shunted to the electron transport chain that links Photosystem II to Photosystem I. As they flow down this chain, back into P_{700}, ADP is phosphorylated to ATP.

Like oxidative phosphorylation in the mitochondria, photophosphorylation in the chloroplasts is a chemiosmotic process. As electrons flow down the electron transport chain from Photosystem II to Photosystem I, protons are pumped from the stroma into the thylakoid space, creating a gradient of potential energy. As protons flow down this gradient from the thylakoid space back into the stroma, passing through an ATP synthetase, ATP is formed.

In the carbon-fixing reactions, which take place in the stroma, the NADPH and ATP produced in the energy-capturing reactions are used to reduce carbon dioxide to organic carbon. This is accomplished by means of the Calvin cycle. In the Calvin cycle, a molecule of carbon dioxide is combined with the starting material, a five-carbon sugar called ribulose diphosphate. At each turn of the cycle, one carbon atom enters the cycle. Three turns of the cycle produce a three-carbon molecule, glyceraldehyde phosphate. Two molecules of glyceraldehyde phosphate (six turns of the cycle) can combine to form a glucose molecule. At each turn of the cycle, RuDP is regenerated. The glyceraldehyde phosphate can also be used as a starting material for other organic compounds needed by the cell.

In C_4 plants, carbon dioxide is initially accepted by a compound known as PEP (phosphoenolpyruvate) to yield the four-carbon oxaloacetic acid. The oxaloacetic acid is then reduced to malic acid, which is, in turn, oxidized, and carbon dioxide is transferred to the RuDP of the Calvin cycle. Although C_4 plants use more energy to fix carbon, their net photosynthetic efficiency can be higher than that of C_3 plants.

The greater affinity of PEP carboxylase for carbon dioxide enables C_4 plants to capture ample CO_2 with minimal water loss. Also, photorespiration, a process in which fixed carbon is reoxidized to carbon dioxide and lost to the plant, is limited in C_4 plants. Under conditions of intense sunlight, high temperatures, or drought, C_4 plants are more efficient than C_3 plants.

QUESTIONS

1. Distinguish between the following: absorption spectrum/action spectrum; grana/thylakoid; stroma/thylakoid space; NAD^+/$NADP^+$; energy-capturing reactions/carbon-fixing reactions; Photosystem I/Photosystem II; C_3 photosynthesis/C_4 photosynthesis.

2. Why is it plausible to argue, as the Nobel laureate George Wald does, that wherever in the universe we find living organisms, we will find them (or at least some of them) to be colored?

3. Predict what colors of light might be most effective at stimulating plant growth. (Such is the principle of the special light bulbs used for plants.)

4. Sketch a chloroplast and label all membranes and compartments. In what ways does the structure of a chloroplast resemble that of a mitochondrion? In what ways is it different?

5. Describe in general terms the events of photosynthesis. Compare your description with Table 10–1.

6. In what ways are photophosphorylation and oxidative phosphorylation alike? In what ways are they different?

7. The experiment shown in Figure 4–4 on page 81 was attempted in Stockton, California, in 1973, but instead of making amino acids, the electric sparks caused the apparatus to explode. (No one was hurt, fortunately.) What is present in today's atmosphere that was not present in the primitive atmosphere and that would account for the explosion?

8. Given the scarcity of high-salt environments and the difficulties of surviving in them, explain in evolutionary terms why the halobacteria are found there.

9. Return to Figure 10–21 and describe the biochemical events taking place in each of the chloroplasts.

10. Consider the anatomy of the leaf of a C_4 plant. In which type of cell—mesophyll or bundle-sheath—would you expect the oxygen-releasing reaction of photosynthesis to occur? Why? Does this localization of oxygen release increase or decrease photorespiration?

11. In what ways is C_4 photosynthesis advantageous to those plants that have it? Many plants, however, have not evolved C_4 photosynthesis. Why is it advantageous to such plants *not* to have C_4 photosynthesis?

12. Trace a carbon atom through a series of biological events, such as those indicated in Figure 8–1 (page 157).

SUGGESTIONS FOR FURTHER READING

Books

ALBERTS, BRUCE, DENNIS BRAY, JULIAN LEWIS, MARTIN RAFF, KEITH ROBERTS, and JAMES D. WATSON: *Molecular Biology of the Cell,* Garland Publishing, Inc., New York, 1983.

This outstanding cell biology text includes a clear, up-to-date discussion of our current understanding of the processes that occur in mitochondria and chloroplasts. The authors' discussion of the experimental procedures used to study these processes is especially helpful.

CONANT, JAMES BRYANT (ed.): *Harvard Case Histories in Experimental Science,* vol. 2, Harvard University Press, Cambridge, Mass., 1964.

Case #5, Plants and the Atmosphere, *edited by Leonard K. Nash, describes the early work on photosynthesis, presented often in the words of the investigators themselves. The narrative illuminates the historical context in which the discoveries were made.*

LEHNINGER, ALBERT L.: *Bioenergetics: The Molecular Basis of Biological Energy Transformations,* 2d ed., The Benjamin/Cummings Publishing Company, Menlo Park, Calif., 1971.

A thorough account; Lehninger is one of the foremost experts on cellular energetics.

LEHNINGER, ALBERT L.: *Principles of Biochemistry,* Worth Publishers, Inc., New York, 1982.

This introductory text is outstanding both for its clarity and for its consistent focus on the living cell. There are numerous medical and practical applications throughout.

RABINOWITCH, EUGENE, and GOVINDJEE: *Photosynthesis,* John Wiley & Sons, Inc., New York, 1969.*

A lucid introduction, suitable for undergraduate students, to the processes of photosynthesis and to related physical and chemical concepts, such as entropy and free energy.

STRYER, LUBERT: *Biochemistry,* 2d ed., W. H. Freeman and Company, San Francisco, 1981.

A good introduction, handsomely illustrated, to cellular energetics.

Articles

BJORKMAN, O., and J. BERRY: "High-Efficiency Photosynthesis," *Scientific American,* October 1973, pages 80–93.

DICKERSON, RICHARD E.: "Cytochrome *c* and the Evolution of Energy Metabolism," *Scientific American,* March 1980, pages 136–153.

GOVINDJEE, and RAJNI GOVINDJEE: "The Primary Events of Photosynthesis," *Scientific American,* December 1974, pages 68–82.

HANSEN, J., et al.: "Climate Impact of Increasing Atmospheric Carbon Dioxide," *Science,* vol. 213, pages 957–966, 1981.

HINKLE, PETER C., and RICHARD E. McCARTY: "How Cells Make ATP," *Scientific American,* March 1978, pages 104–123.

MILLER, KENNETH R.: "The Photosynthetic Membrane," *Scientific American,* October 1979, pages 102–113.

NASSAU, KURT: "The Causes of Color," *Scientific American,* October 1980, pages 124–154.

SCHOPF, J. WILLIAM: "The Evolution of the Earliest Cells," *Scientific American,* September 1978, pages 110–138.

STOECKENIUS, WALTHER: "The Purple Membrane of Salt-loving Bacteria," *Scientific American,* June 1976, pages 38–46.

---

* Available in paperback.

SECTION 3

Genetics

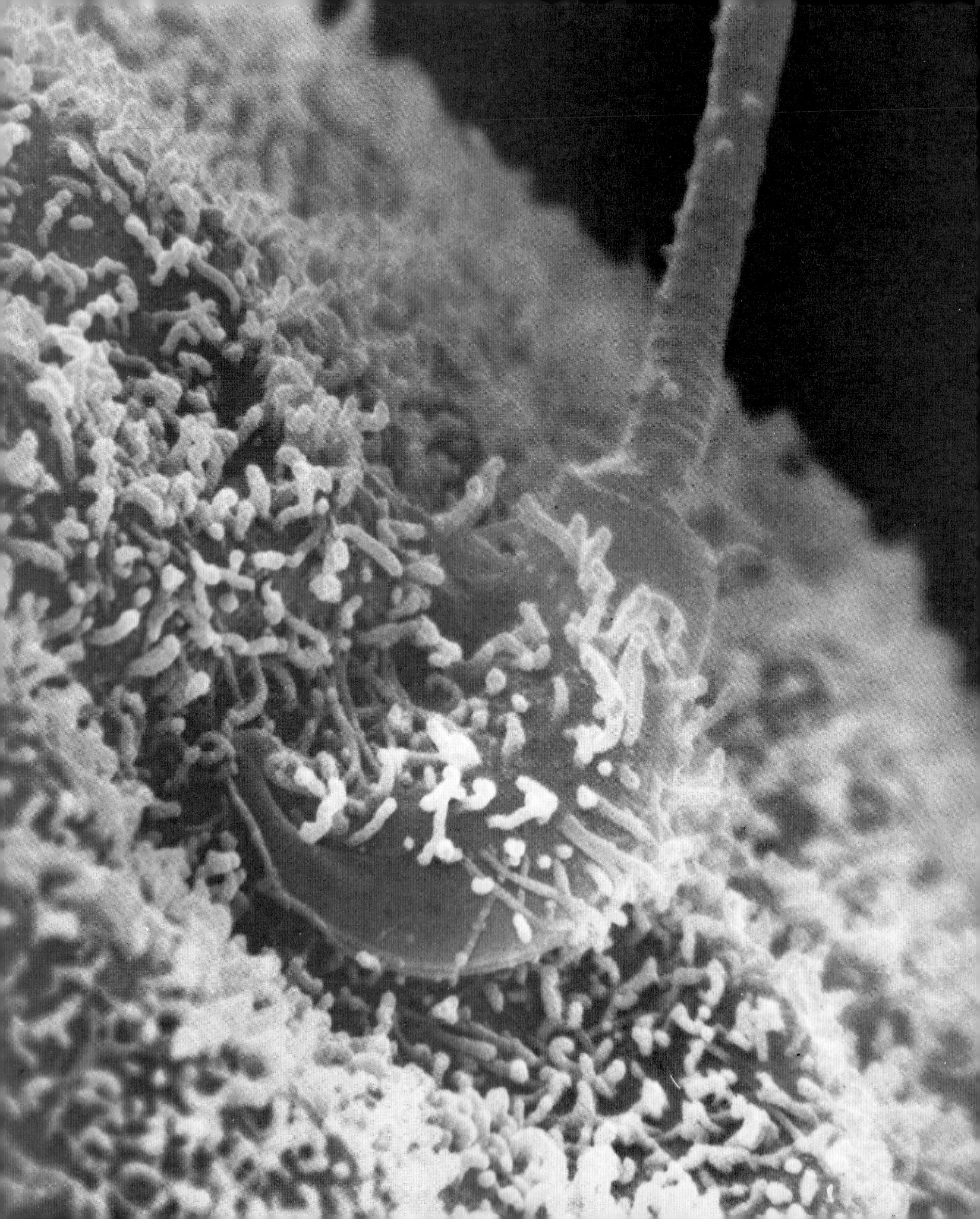

CHAPTER 11

From an Abbey Garden: The Beginning of Genetics

11–1
Sperm fertilizing a mammalian egg. All the information for the organism-to-be is carried in the genetic material of these two cells.

11–2
The protruding lip of the Hapsburgs is a famous example of an inherited trait. These portraits of members of the Hapsburg family encompass a period of about 300 years: (a) *Rudolph I (1218–1291), King of Germany,* (b) *Maximilian I (1460–1519), Holy Roman Emperor,* (c) *Charles V (1500–1558), Holy Roman Emperor,* (d) *Ferdinand I (1503–1564), Holy Roman Emperor.*

Among all the symbols in biology, perhaps the most widely used and most ancient are the hand mirror of Venus (♀) and the shield and spear of Mars (♂), the biologists' shorthand for female and male. Ideas about the nature of biological inheritance—the role of male and female—are even older than these famous symbols. Very early, it must have been noticed that certain characteristics—hair color, for example, a large nose, or a small chin—were common to parent and offspring. And throughout history, biological inheritance has been an important factor in the social organizations of mankind, determining the distribution of wealth, power, land, and royal privileges.

Sometimes a family trait is so distinctive that it can be traced through many generations. A famous example of such a characteristic is the Hapsburg lip (Figure 11–2), which has appeared in Hapsburg after Hapsburg, over and over again since at least the thirteenth century. Examples such as this have made it easy to accept the importance of inheritance in the formation of the individual, but it is only comparatively recently that we have begun to understand how this process works. In fact, the study of heredity as a science did not really begin until the second half of the nineteenth century. Yet the problems posed in this study are among the most fundamental in biology, since self-replication is the essence of the hereditary process and one of the principal properties of living systems.

EARLY IDEAS ABOUT HEREDITY

Far back in human history, people learned to improve domestic animals and crops by inbreeding and crossbreeding. In the case of date palms, male and female flowers are found on different trees, and artificial fertilization of the palm was well known to the ancient Babylonians and Egyptians. The nature of the difference between the two flowers was understood by Theophrastus (371–287 B.C.). "The males should be brought to the females," he wrote, "for the male makes them ripen and persist." That crossing a male donkey and a mare produced a mule was well known in the days of Homer. Both Plutarch and Lucretius noted in their writings that some children resemble their mothers, some resemble their fathers, and some even skip back a generation to resemble a grandparent; this fact, so easy to observe, continued to puzzle people for a very long time.

Many legends were based on bizarre possibilities of crossbreeding. The wife of

11–3
Only fairly recently has it been realized that living things come only from other living things of the same species. This picture from an old Turkish history of India shows a wakwak tree, which bears human fruit. According to the account, the tree is to be found on an island in the South Pacific.

Minos, according to Greek mythology, mated with a bull and produced the Minotaur. Folk heroes of Russia and Scandinavia were traditionally the sons of women who had been captured by bears, from which these men derived their great strength and so enriched the national stock. The camel and the leopard also crossbred from time to time, according to the early naturalists, who were otherwise unable—and it is hard to blame them—to explain an animal as improbable as the giraffe (the common giraffe still bears the species name of *camelopardalis*). Thus folklore reflected early and imperfect glimpses of the nature of hereditary relationships.

The first scientist known to have pondered the mechanism of heredity was Aristotle. He postulated that the male semen was made up of a number of imperfectly blended ingredients and that, at fertilization, it mixed with the "female semen," the menstrual fluid, giving form and power (*dynamis*) to this amorphous substance. No one had a better idea—or indeed many ideas at all—for two thousand years. Seventeenth-century medical texts show various stages in the coagulation of the embryo from the mixture of maternal and paternal semens. Indeed, many scientists as well as laymen did not believe that such mixtures were even always necessary; they held that life, at least the "simpler" forms of life, could arise by spontaneous generation. Worms, flies, and various crawling things, it was commonly believed, took shape from putrid substances, ooze, and mud, and a lady's hair dropped in a rain barrel could turn into a snake. Jan Baptista van Helmont, who also carried out experiments on the growth of plants, published his personal recipe for the production of mice: One need only place a dirty shirt in an open pot containing a few grains of wheat, and in 21 days mice would appear. He had performed the experiment himself, he said. The mice would be adults, both male and female, he added, and would be able to produce more mice by mating. As we saw in Chapter 4 (page 77), spontaneous generation did not lose its grip on the imagination until Pasteur's decisive disproof in 1864.

THE FIRST EXPERIMENTS

In 1677, the Dutch lens maker Anton van Leeuwenhoek discovered living sperm—"animalcules," he called them—in the seminal fluid of various animals, including man. Enthusiastic followers peered through Leeuwenhoek's "magic looking glass" (his homemade microscope) and believed they saw within each human sperm a tiny creature—the homunculus, or "little man" (Figure 11–4). This little man was the future human being in miniature. Once implanted in the female womb, the future human being was nurtured there; the only contribution that the mother made was to serve as an incubator for the growing fetus. Any resemblance a child might have to its mother, these theorists held, was because of "prenatal influences."

During the very same decade (the 1670s) that van Leeuwenhoek first saw human sperm cells, another Dutchman, Régnier de Graaf, described for the first time the ovarian follicle, the structure in which the human egg cell (the ovum) forms. Although the actual human egg was not seen for another 150 years, the existence of a human egg was rapidly accepted. In fact, de Graaf attracted a school of followers, the ovists, who were as convinced of their opinions as the animalculists, or spermists, were of theirs and who soon contended openly with them. It was the female egg, the ovists said, that contained the future human being in miniature; the animalcules in the male seminal fluid merely stimulated the egg to grow. Ovists and animalculists alike carried the argument one logical step further. Each homunculus had within it another perfectly formed homunculus, and in that was still another one, and so on—children, grandchildren, and great-grandchildren, all stored away for future use. Some ovists even went so far as to say that Eve had contained within her body all the unborn generations yet to come, each egg fitting closely inside another like a child's hollow blocks. Each female generation since Eve has contained one fewer than the previous generation, they explained, and after 200 million generations, all the eggs will be spent and human life will come to an end.

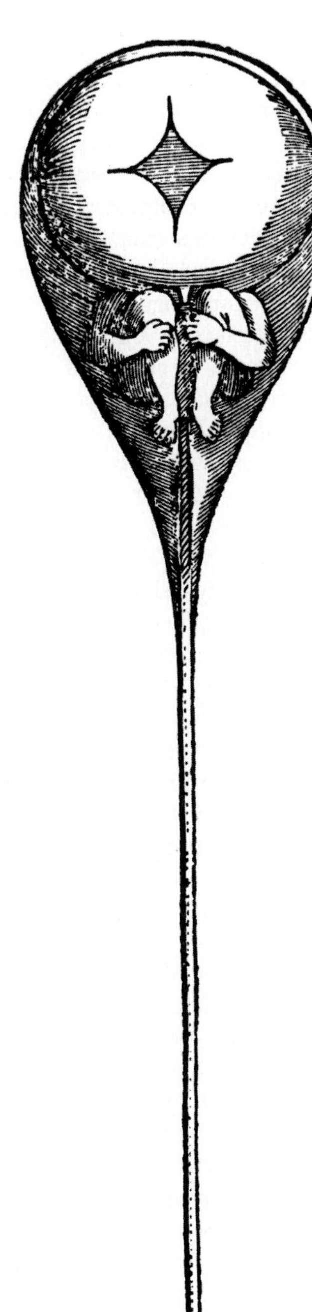

11–4
What the animalculists, or spermists, of the seventeenth and eighteenth centuries believed they saw when they looked through a microscope at sperm cells. This is a homunculus ("little man"), a future human being in miniature, in a sperm cell.

BLENDING INHERITANCE

By the middle of the nineteenth century, the concepts of the ovists and spermists began to yield to new data. The facts that challenged these earlier hypotheses came not so much from scientific experiments as from practical attempts by master gardeners to produce new ornamental plants. Artificial crossings of such plants showed that, in general, regardless of which plant supplied the pollen (which contains the male cells, or sperm) and which plant contributed the egg cells, or ova, both contributed to the characteristics of the new variety. But this conclusion raised even more puzzling questions: What exactly did each parent plant contribute? How did all the hundreds of characteristics of each plant get combined and packed into a single seed?

The most widely held hypothesis of the nineteenth century was blending inheritance. According to this concept, when the sex cells, or *gametes* (from the Greek word *gamos,* meaning "marriage"), combine, there is a mixing of hereditary material that results in a blend, analogous to a blend of two different-colored inks. On the basis of such a hypothesis, one would predict that the offspring of a black animal and a white animal would be gray, and their offspring would also be gray because the black and white hereditary material, once blended, could never be separated again.

You can see why this concept was unsatisfactory. It ignored the phenomenon of traits skipping a generation, or even several generations, and then reappearing. To Charles Darwin and other proponents of the theory of evolution, it presented particular difficulties. Evolution, according to Darwin, as we noted in the Introduction (page 9), takes place as natural selection acts on existing random variations. If the hypothesis of blending inheritance were valid, the small hereditary variations would disappear, like a single drop of ink in the many-colored mixture. Sexual reproduction would eventually result in complete uniformity, natural selection would have no raw material on which to act, and evolution would not occur.

THE CONTRIBUTIONS OF MENDEL

At about the same time that Darwin was writing *The Origin of Species,* an Austrian monk, Gregor Mendel, was beginning a series of experiments that would lead to a new understanding of the mechanism of inheritance. Mendel, who was born into a peasant family in 1822, entered a monastery in Brünn (now Brno, Czechoslovakia), where he was able to receive an education. He attended the University of Vienna for two years, pursuing studies in both mathematics and science. He failed his tests for the teaching certificate he was seeking and so retired to the monastery, of which he eventually became abbot. Mendel's work, carried on in a quiet monastery garden and ignored until after his death, marks the beginning of modern genetics.

11-5
In a flower, pollen develops in the anther and the egg cells in the ovule. Pollination occurs when pollen grains, trapped on the stigma, germinate and grow down to the ovule, where they release sperm cells. Egg and sperm unite, and the fertilized egg develops within the ovule. In the garden pea (Pisum sativum), *the ovule and embryo form the peas (the seeds).*

Pollination in most species involves the pollen from one plant (often carried by an insect) being caught on the stigma of another plant. This is called cross-pollination.

In the pea flower, however, the stigma and anthers are completely enclosed by petals, and the flower, unlike most, does not open until after fertilization has taken place. Thus the plant normally self-pollinates. In his crossbreeding experiments, Mendel pried open the bud before the pollen matured and removed the anthers with tweezers. Then he artificially pollinated the flower by dusting the stigma with pollen collected from other plants.

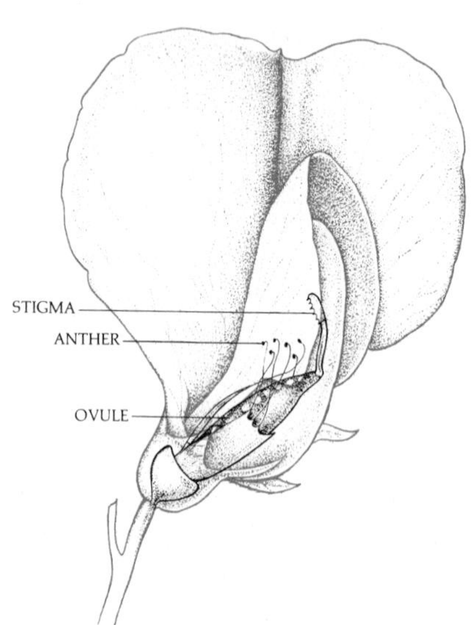

Mendel's great contribution was to demonstrate that inherited characteristics are carried by factors that are discrete units and that are parceled out in different ways—reassorted—in each generation. These discrete units eventually came to be known as *genes.*

For his experiments in heredity, Mendel chose the common garden pea. It was a good choice. The plants were commercially available, easy to cultivate, and grew rapidly. Different varieties had clearly different characteristics that "bred true," reappearing in crop after crop. Finally, the sexual structures of the pea flower are entirely enclosed by the petals, even when they are mature (Figure 11-5). Consequently, the flower normally self-pollinates, that is, sperm cells from the flower's own pollen fertilize its egg cells. Although the plants could be crossbred experimentally, accidental crossbreeding could not occur to confuse the experimental results. As Mendel said in his original paper, "The value and utility of any experiment are determined by the fitness of the material to the purpose for which it is used."

Mendel's choice of the pea plant for his experiments was not original. However, he was successful in formulating the basic principles of heredity—where others had failed—because of his approach to the problem. First, he tested a very specific model in a series of logical experiments. He planned his experiments carefully and imaginatively, choosing for study clear-cut and measurable hereditary differences. Second, he studied the offspring of not only the first generation but also the second generation. Third, and most important, he counted the offspring and then analyzed the results mathematically. Even though his mathematics was simple, the idea that a biological problem could be studied quantitatively was startlingly new. Finally, he organized his data in such a way that his results could be evaluated simply and objectively. The experiments themselves were described so clearly that other scientists could repeat and check them, as eventually they did.

The Principle of Segregation

Mendel began with 32 different types of pea plants, which he studied for two years to see which characteristics were clearly defined. As he said later in his report on this work, he did not want to experiment with traits in which the difference could be "of a 'more or less' nature, which is often difficult to define." As a result of these observations, he selected for study seven traits that appeared as conspicuously different characteristics in different varieties of plants. One variety of plant, for example, always produced yellow peas, or seeds, while another always produced green ones. In one variety, the seeds, when dried, had a wrinkled appearance, while in another variety they were smooth. The complete list of alternate traits is given in Table 11-1.

Mendel performed experimental crosses, removing the anthers from flowers and dusting the stigmas with pollen from a flower of another variety. He found that in every case in the first generation (now known as the F_1, for "first filial generation"), one of the alternate traits disappeared completely. All the plants produced as a result of a cross between yellow-seeded plants and green-seeded plants were as yellow-seeded as the yellow-seeded parent. All the flowers produced by plants resulting from a cross between a purple-flowered plant and a white-flowered plant were purple. Traits that appeared in the F_1 generation Mendel called *dominant.*

The interesting question was: What had happened to the alternative trait—the greenness of the seed or the whiteness of the flower—which had been passed on so

11–6
Garden pea plant with seeds (peas) in the approximate ratio of three yellow seeds for each green seed, as observed by Mendel.

Table 11-1 Results of Mendel's Experiments with Pea Plants

	ORIGINAL CROSSES			(F_2) SECOND GENERATION		
TRAIT	DOMINANT	×	RECESSIVE	DOMINANT	RECESSIVE	TOTAL
Seed form	Round	×	Wrinkled	5,474	1,850	7,324
Seed color	Yellow	×	Green	6,022	2,001	8,023
Flower position	Axial	×	Terminal	651	207	858
Flower color	Purple	×	White	705	224	929
Pod form	Inflated	×	Constricted	882	299	1,181
Pod color	Green	×	Yellow	428	152	580
Stem length	Tall	×	Dwarf	787	277	1,064

faithfully for generations by the parent stock? Mendel let the pea plant itself carry out the next stage of the experiment by permitting the F_1 plants to self-pollinate and produce a second generation, the F_2. The traits that had disappeared in the first generation reappeared in the F_2. In Table 11–1 are the results of Mendel's actual counts. These traits, which were present in the parent generation and reappeared in the F_2 generation, must also have been present somehow in the F_1 generation, although not apparent there. Mendel called these traits *recessive.*

If you analyze the results in Table 11–1 as Mendel did, you will notice that the dominant and recessive traits appear in the second, or F_2, generation in ratios of about 3:1. How do the recessives disappear so completely and then appear again, and always in such constant proportions? It was in answering this question that Mendel made his greatest contribution. He saw that the appearance and disappearance of traits and their constant proportions could be explained if hereditary characteristics are determined by discrete (separable) factors.* These factors, Mendel saw, must have occurred in the offspring as pairs, one factor inherited from each parent. These pairs of factors (genes) were separated again when the mature F_1 plants produced sex cells, resulting in two kinds of gametes, with one gene of the pair in each. Alternative forms of the same gene are now known as *alleles.*

The hypothesis that every individual carries pairs of factors for each trait and that the members of the pair segregate during the formation of gametes is known as Mendel's first law, or the *principle of segregation.*

Consequences of Segregation

According to Mendel's hypothesis, the two factors in a pair might be the same, in which case the self-pollinating plant would breed true. Yellow-seededness and green-seededness, for instance, are determined by alleles, different forms of the gene (factor) for seed color. When the alleles of a gene pair are the same, the organism is said to be *homozygous* for that particular trait; when the alleles of a gene pair are different, the organism is *heterozygous* for that trait.

* Mendel called these factors *Elemente.* As we noted previously, they are now known as genes, and we shall use this term to refer to them.

11-7

A pea plant homozygous for purple flowers is represented as WW in genetic shorthand. The gene for purple flowers is designated W because of a convention by which geneticists, in indicating a pair of alleles, use the first letter of the less common form (white). The capital indicates the dominant, the lowercase the recessive. A WW plant can produce gametes with only a purple-flower (W) allele. The female symbol ♀ indicates that this flower contributed the egg cells, or female gametes.

A white pea plant (ww) can produce gametes with only a white-flower (w) allele. The male symbol ♂ indicates that this flower contributed the sperm cells, or male gametes.

When a w sperm cell fertilizes a W egg cell, the result is a Ww pea plant, which, since the W allele is dominant, will produce purple flowers. However, this Ww plant can produce gametes with either a W or a w allele.

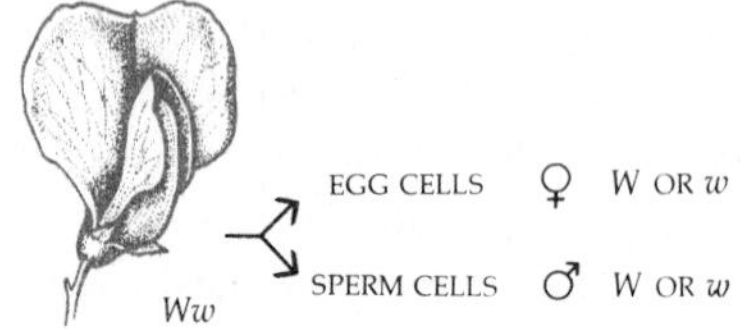

And so, if the plant self-pollinates, four possible crosses can occur:

♀ W × ♂ W → *purple flowers*
♀ W × ♂ w → *purple flowers*
♀ w × ♂ W → *purple flowers*
♀ w × ♂ w → *white flowers*

These results are summarized in Figure 11-8.

When gametes are formed, alleles are passed on to them, but each gamete contains only one of the possible alleles for any given trait. When two gametes combine in the fertilized egg, the alleles occur in matched pairs again. If the alleles are the same, both will be expressed. If the alleles are different, one may be dominant over the other; in this case, the organism will appear as if it had only the dominant allele. This outward appearance of a trait is known as the *phenotype.* However, in the genetic makeup, or *genotype,* each allele still exists independently and as a discrete unit, even though it may not be expressed in the phenotype. The recessive allele will separate from its dominant partner when gametes are again formed. Only if two recessive alleles come together—one from the female gamete and one from the male—will the phenotype then show the recessive trait.

When pea plants homozygous for purple flowers are crossed with pea plants having white flowers, only pea plants with purple flowers are produced, although each plant in the F_1 generation will carry an allele for purple and an allele for white. Figure 11-7 shows what happens in the F_2 generation if the F_1 generation self-pollinates. Notice that the result would be the same if an F_1 individual were cross-fertilized with another F_1 individual, which is the way these experiments are performed with animals and with plants that are not self-pollinating.

In order to test his hypothesis (diagrammed in Figure 11-8), Mendel performed two additional experiments. He crossed white-flowering plants with white-flowering plants and confirmed that they bred true—that only white-flowering plants were produced. Next he crossed one of his F_1 individuals, the result of a cross between purple- and white-flowering plants, with a true-breeding white-flowering plant. To the outside observer, it would appear as if Mendel were simply repeating his first experiment, crossing plants having purple flowers with plants having white flowers. But he knew that if his hypothesis were correct, his results would be different from those of his first experiment. In fact, he actually predicted the results of such a cross before he made it. Can you? Stop a moment and think about it.

The easiest way to analyze the possible result of such a cross is to diagram it, as in Figure 11-9. This experiment, which reveals the genotype of the parent with the dominant trait, is known as a testcross. A testcross is an experimental cross between an individual with the dominant phenotype for a given trait with another individual that is homozygous recessive for the trait. The resulting ratio of offspring will indicate whether the individual with the dominant phenotype was homozygous or heterozygous for the trait being studied. In the testcross shown in Figure 11-9, the cross revealed that the genotype of the plant being tested was *Ww* rather than *WW.*

The Principle of Independent Assortment

In a second series of experiments, Mendel studied crosses between pea plants that were different in two characteristics; for example, one parent plant had peas that were round and yellow, and the other had peas that were wrinkled and green. The round and yellow traits, you will recall (see Table 11-1), are dominant, and the wrinkled and green are recessive. As you would expect, all the seeds produced by a cross between the parental types were round and yellow. When these F_1 seeds were planted and the flowers allowed to self-pollinate, 556 seeds were produced. Of these, 315 showed the two dominant characteristics, round and yellow, but only 32 combined the recessive traits, green and wrinkled. All the rest of the seeds were unlike either parent; 101 were wrinkled and yellow, and 108 were round and green. Totally new combinations of characteristics had appeared.

11–8

A cross between a parent (P) pea plant with two dominant alleles for purple flowers (WW) and one with two recessive alleles for white flowers (ww). The phenotype of the offspring in the F_1 generation is purple, but note that the genotype is Ww. The F_1 heterozygote self-pollinates, producing four kinds of gametes, ♀ W, ♂ W, ♀ w, and ♂ w, in equal proportions. The W and w sperm cells and eggs combine randomly to form, on the average, $\frac{1}{4}$ WW (purple), $\frac{1}{2}$ Ww (purple), and $\frac{1}{4}$ ww (white) offspring. It is this underlying 1:2:1 genotypic ratio that accounts for the phenotypic ratio of 3 dominants (purple) to 1 recessive (white). The distribution of traits in the F_2 generation is shown by a Punnett square, named after the English geneticist who first used this sort of checkerboard diagram for the analysis of genetically determined traits.

11–9

A testcross. In order for a pea flower to be white, the plant must be homozygous for the recessive allele (ww). But a purple pea flower can come from a plant with either a Ww or a WW genotype. How could you tell such plants apart? Mendel solved this problem by breeding such plants with homozygous recessives. This sort of experiment is known as a testcross. As shown here, a phenotypic ratio in the F_1 generation of 2 purple to 2 white indicates a heterozygous purple-flowering parent. What would have been the result if the plant being tested had been homozygous for purple flowers?

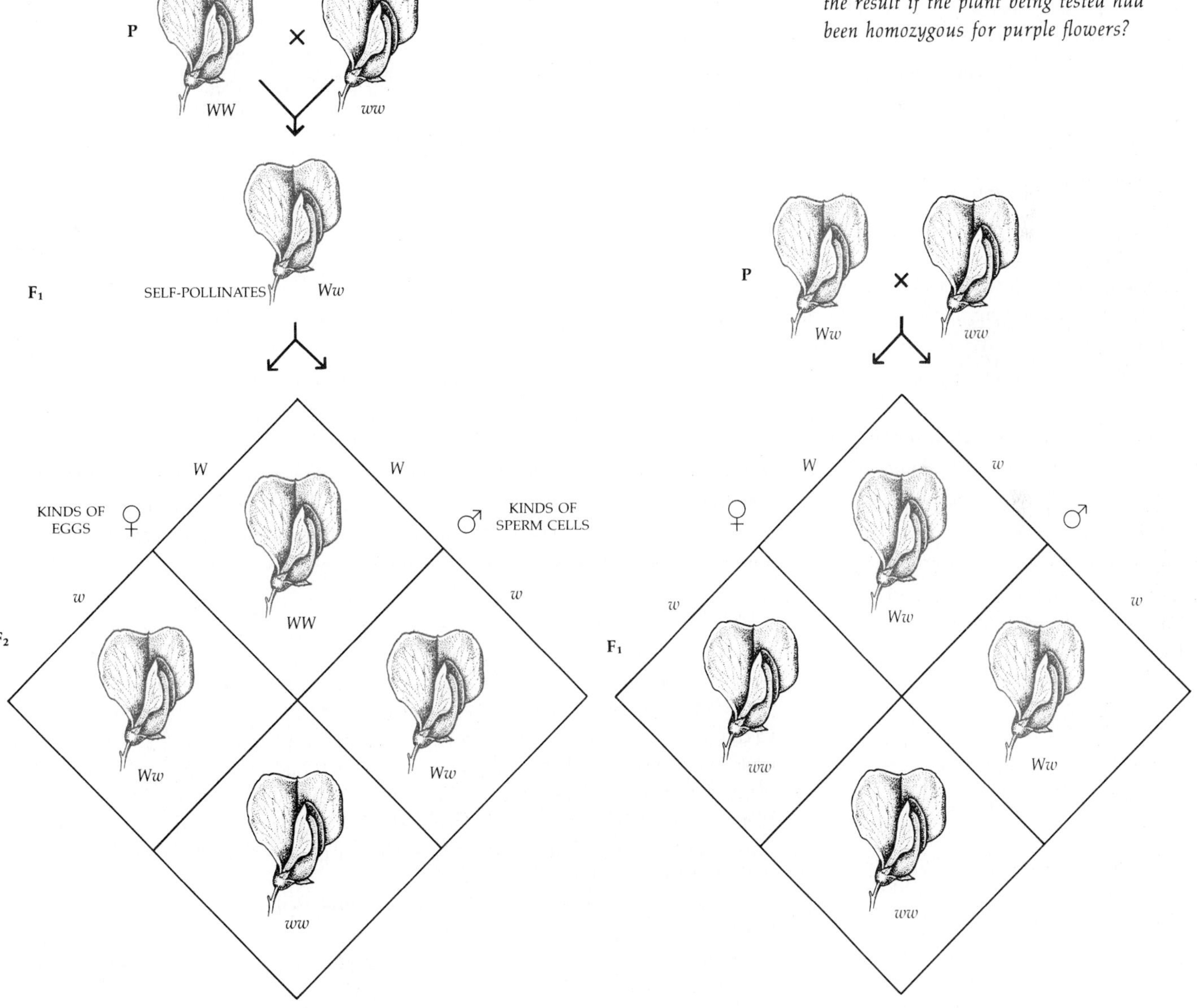

MENDEL AND THE LAWS OF CHANCE

In applying mathematics to the study of genetics, Mendel was stating that the laws of chance apply to biology as they do to the physical sciences. Toss a coin. The chance that it will turn up heads is fifty-fifty, or $\frac{1}{2}$. *The chance that it will turn up tails is also fifty-fifty, or* $\frac{1}{2}$. *The chance that it will turn up one or the other is certain, or one chance in one. Now toss two coins. The chance that one will turn up heads is again* $\frac{1}{2}$. *The chance that the second will turn up heads is also* $\frac{1}{2}$. *The chance that both will turn up heads is* $\frac{1}{2} \times \frac{1}{2}$, or $\frac{1}{4}$. *The probability of two independent events occurring together is simply the probability of one occurring alone multiplied by the probability of the other occurring alone. The chance of both turning up tails is similarly* $\frac{1}{2} \times \frac{1}{2}$. *The chance of the first turning up tails and the second turning up heads is* $\frac{1}{2} \times \frac{1}{2}$, *and the chance of the second turning up tails and the first heads is* $\frac{1}{2} \times \frac{1}{2}$.

We can diagram this in a Punnett square (see figure), which indicates that the combination in each square has an equal chance of occurring. It was undoubtedly the observation that one-fourth of the offspring in the F_2 *generation showed the recessive phenotype that indicated to Mendel that he was dealing with a simple case of the laws of probability.*

If there were three coins involved, the probability of any given combination would be simply the product of all three fifty-fifty possibilities: $\frac{1}{2} \times \frac{1}{2} \times \frac{1}{2}$, or $\frac{1}{8}$. *Similarly, with four coins, the probability of any given combination is* $\frac{1}{2} \times \frac{1}{2} \times \frac{1}{2} \times \frac{1}{2}$, or $\frac{1}{16}$. *The Punnett square on page 245 expresses the probability (or chance) of each of any one of four possible combinations.*

Notice that in planning his experiments, Mendel assumed that the gametes combined at random. Thus, the laws of probability could be employed.

If you toss two coins 4 times, it is unlikely that you will get the precise results diagrammed above. However, if you toss two coins 100 times, you will come close to the proportions predicted in the Punnett square, and if you toss two coins 1,000 times, you will be very close indeed. As Mendel knew, the relationship between dominants and recessives might well not have held true if he had been dealing with a small sample. The larger the sample, the more closely it will conform to results predicted by the laws of chance.

This experiment did not contradict Mendel's previous results. Round and wrinkled still appeared in the same 3:1 proportion (423 round to 133 wrinkled), and so did yellow and green (416 yellow to 140 green). But the round and the yellow traits and the wrinkled and the green ones, which had originally combined in one plant, sorted themselves out as if they were entirely independent of one another. From this, Mendel formulated his second law, the *principle of independent assortment.* This principle states that members of each pair of genes are distributed independently when the gametes are formed.

Figure 11–10 diagrams Mendel's interpretation of these results and shows why, in a cross involving two gene pairs, each pair with one dominant and one recessive allele, the phenotypes in the progeny will, on the average, be in the ratio of 9:3:3:1.

11–10

One of the experiments from which Mendel derived his principle of independent assortment. A plant homozygous for round (RR) *and yellow* (YY) *peas is crossed with a plant having wrinkled* (rr) *and green* (yy) *peas. The* F_1 *generation are all round and yellow, but notice how the traits will, on the average, appear in the* F_2 *generation. Of the 16 kinds of offspring, 9 show the two dominant traits* (RY), *3 show one combination of dominant and recessive* (Ry), *3 show the alternate combination* (rY), *and 1 shows the two recessives* (ry). *This 9:3:3:1 distribution is always the expected result from a cross involving two pairs of independent dominant-recessive alleles. (The letters* R *and* Y *are used because round and yellow are the less common forms in nature.)*

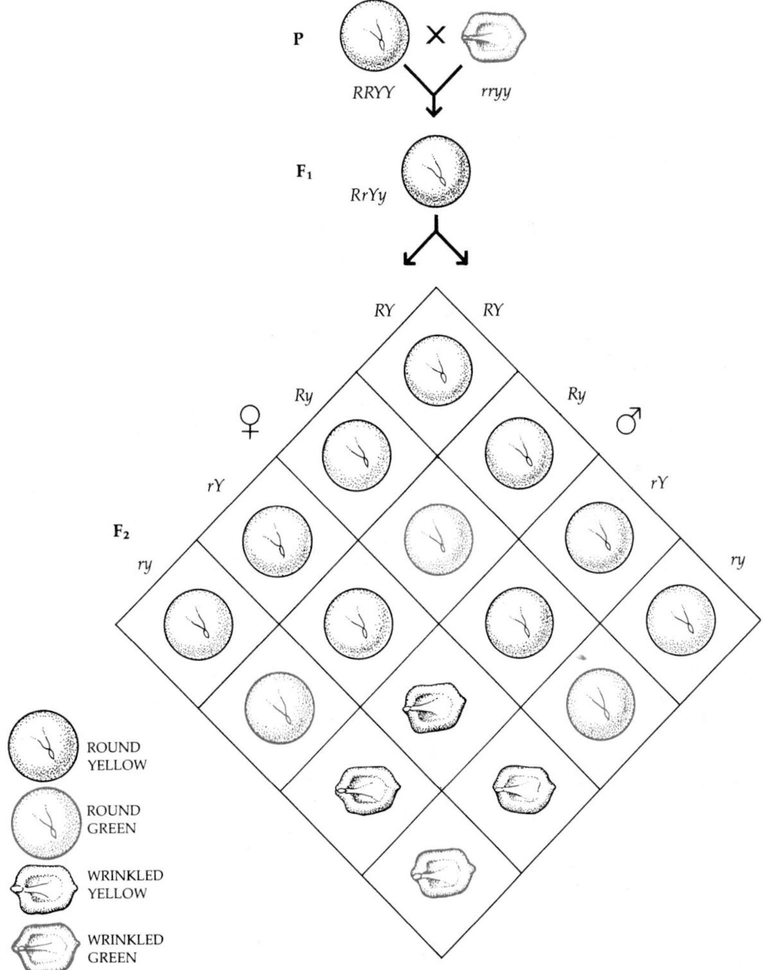

The 9 represents the proportion of F_2 progeny that will show the two dominant traits, 1 the proportion that will show the two recessive traits, and 3 and 3 the proportions that will show the two alternative combinations of dominants and recessives. This is true when one of the original parents is homozygous for both recessive traits and the other homozygous for both dominant ones, as in the experiment just described *(RRYY × rryy),* as well as when each original parent is homozygous for one recessive and one dominant trait *(rrYY × RRyy).*

A Testcross

Can you predict the outcome of a cross between a homozygous recessive for the two genes and a heterozygote? Such a cross is similar to the one analyzed in Figure 11–9 but involves two gene pairs instead of one. For simplicity, let us again study the distribution of round versus wrinkled *(R* versus *r)* and green versus yellow *(Y* versus *y)* in a cross between a heterozygote and a homozygous recessive. Draw a Punnett square with 16 squares (Figure 11–11a). Put the female symbol on one side and the male on the other. Assume that the heterozygote contributes the female

11-11
A cross involving two gene pairs.

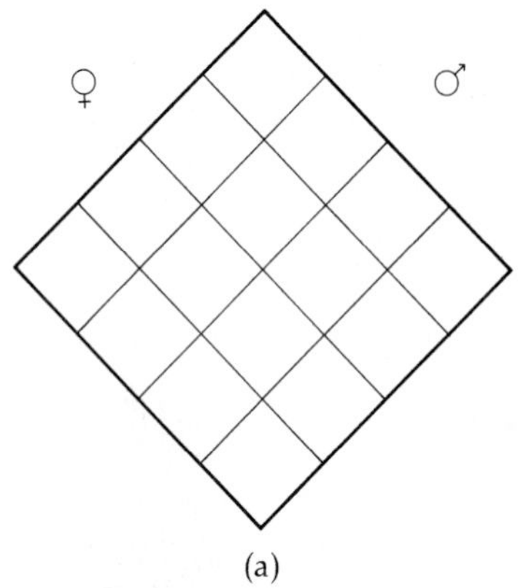

(a)

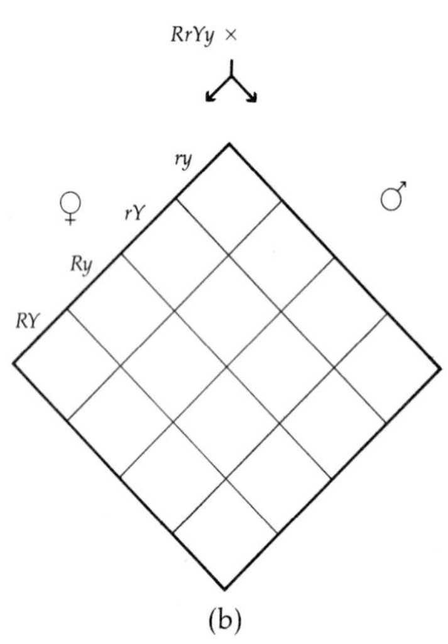

(b)

gametes. With a genotype of *RrYy*, the heterozygote can produce four kinds of gametes: *RY*, *Ry*, *rY*, and *ry*. At the head of each column at the left, where the female symbol is, put one of these possible combinations (Figure 11–11b). (Notice that in this step we are assuming, as Mendel did, that each of the possible kinds of gamete is produced in equal numbers.)

The homozygous recessive can produce only one type of gamete in terms of the traits being studied: *ry*. Put *ry* at the head of each column on the right (Figure 11–11c). The advantage of using a Punnett square is that it makes it impossible to overlook any combination of gametes.

Now, starting with the column on the far left, begin to fill in the squares. You are less likely to make a mistake if you fill in all the female gametes first (or all male gametes), a column at a time, than if you try to work with both male and female gametes at the same time. When you are halfway through, your square will look like the one in Figure 11–11d.

Next, fill in the symbols for the genes carried in the male gametes. Your square will then look like the one in Figure 11–11e.

Now count the phenotypes that this square predicts. Every capital *R* indicates a round seed (since *R* is dominant); there are eight capital *R*'s and, hence, eight round seeds. Conversely you can count eight wrinkled *(rr)* seeds. Similarly there are eight yellow *(Y)* seeds and eight green *(yy)* seeds.

Notice that each of the possible combinations of traits, round green, round yellow, wrinkled green, and wrinkled yellow, appears in equal proportions.

The Influence of Mendel

Mendel's experiments were first reported in 1865 before a small group of people at a meeting of the Brünn Natural History Society. None of them, apparently, understood what Mendel was talking about. But his paper was published the following year in the *Proceedings* of the Society, a journal that was circulated to libraries all over Europe. In spite of this, his work was ignored for 35 years, during most of which he devoted himself to the duties of an abbot, and he received no scientific recognition until after his death. He was, to use DuPraw's phrase (page 11), an odd traveler whose tale could not be made to fit.

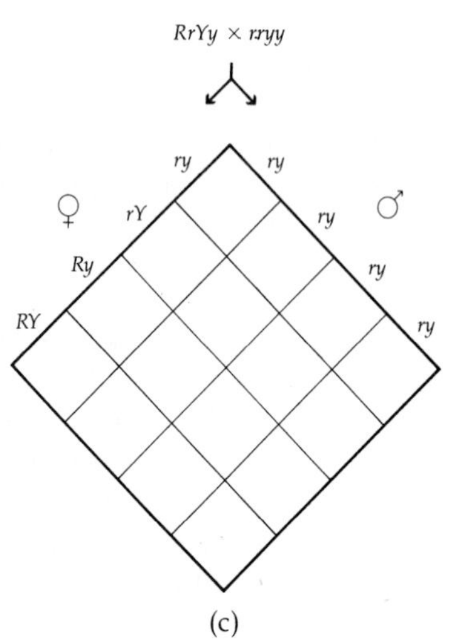

(c)

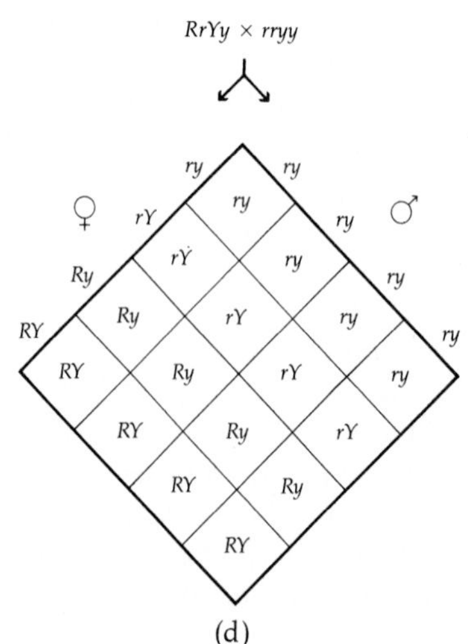

(d)

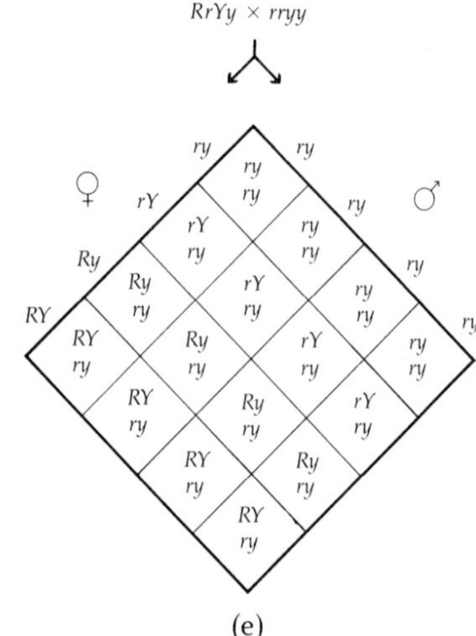

(e)

11–12
Gregor Mendel, holding a fuchsia, is third from the right in this photograph of members of the Augustinian monastery in Brünn in the 1860s. In his experiments carried out in the monastery garden, Mendel showed that hereditary determinants are carried as separate units from generation to generation. His discoveries explained how inherited variations can persist for generation after generation.

It was not until 1900, 35 years after Mendel first reported his work, that biologists were finally prepared to accept his findings. Within a single year, his paper was independently rediscovered by three scientists working in three different European countries. Each of them had done similar experiments and was searching the scientific literature for confirmation of his results. And each found, in Mendel's brilliant analysis, that much of his own work had been anticipated.

SUMMARY

In this chapter, we started with the earliest ideas about inheritance and traced the gradual development of these ideas into a science. The first question with which this new science was concerned was the mechanism of inheritance. How are hereditary characteristics passed from one generation to another?

By the middle of the nineteenth century, it was recognized that ova and sperm are specialized cells and that the ova and sperm both contribute to the hereditary characteristics of the new individual. But how were these special cells, called gametes, able to pass on the many hundreds of characteristics involved in inheritance? One answer to this question was the theory of blending inheritance, which held that the traits of the parents blended in the offspring, like a mixture of two fluids.

The revolution in genetics came when the blending theory was replaced by a unit theory. According to Mendel's principle of segregation, hereditary characteristics are determined by discrete factors (now called genes) that appear in pairs (alleles)—one of each pair inherited from each parent.

The genetic makeup of an organism is known as its genotype. Its appearance, or set of outward characteristics, is its phenotype. Both alleles in a pair may be alike (a homozygous condition), or they may be different (heterozygous). In the heterozygous pairs studied by Mendel, only one allele was detected in the phenotype. An allele that is expressed in the phenotype to the exclusion of the other is a dominant allele; one that is concealed in the phenotype is a recessive allele. When two such heterozygotes are crossed, the ratio of dominant to recessive in the phenotype of the F_2 generation is 3:1.

Mendel's other great principle, the principle of independent assortment, applies to the behavior of two or more gene pairs. This law states that the alleles of each gene segregate independently of the alleles of another gene. In crosses involving two independent gene pairs, the expected phenotypic ratio in the F_2 generation is 9:3:3:1. A testcross, in which an individual of the F_1 generation is crossed with a homozygous recessive, reveals the possible phenotypes in equal numbers.

QUESTIONS

1. Distinguish between the following terms: gene/allele; dominant/recessive; homozygous/heterozygous; genotype/phenotype; the F_1/the F_2.

2. In the experiments summarized in Table 11–1, which of the alternate traits appeared in the F_1 generation?

3. Why is a homozygous recessive always used in a testcross?

4. (a) In a pea plant that breeds true for tall, what possible gametes can be produced? (Use the symbol *D* for tall, *d* for dwarf.) (b) In a pea plant that breeds true for dwarf, what possible gametes can be produced? (c) What will be the genotype of the F_1 generation produced by a cross between these two types? (d) What will be the phenotype of the F_1 generation produced by a cross between these two types? (e) What will be the expected distribution of traits in the F_2 generation? Illustrate with a Punnett square.

5. In Figure 11–6, what are the possible parental types for the plant shown?

6. You have just flipped a coin five times and it has turned up heads every time. What are the chances that the next time you flip it, it will be tails?

7. The ability to taste a bitter chemical, phenylthiocarbamide (PTC), is due to a dominant allele. In terms of tasting ability, what are the possible phenotypes of a man both of whose parents are tasters? What are his possible genotypes?

8. If the man in Question 7 marries a woman who is a nontaster, what proportion of their children could be tasters? Suppose one of the children is a nontaster. What would you know about the father's genotype? Explain your results by drawing Punnett squares.

9. A taster and a nontaster have four children, all of whom can taste PTC. What is the probable genotype of the parent who is a taster? Is there another possibility?

CHAPTER 12

Meiosis and Sexual Reproduction

Mendelian genetics is concerned with the way in which hereditary traits are passed from parents to offspring. Most eukaryotic organisms reproduce sexually, a process that always involves two events: *fertilization* and *meiosis.* Fertilization is the means by which the different genetic contributions of the two parents are brought together to form the new genetic identity (genome) of the offspring. Meiosis is a special kind of nuclear division that is believed to have evolved from mitosis and uses much of the same cellular machinery. However, as you will see, it differs from it in some important respects.

HAPLOID AND DIPLOID

To understand meiosis, we must look again at the chromosomes. Every organism has a chromosome number characteristic of its particular species. A mosquito has 6 chromosomes per cell; a cabbage, 18; corn, 20; a sunflower, 34; a cat, 38; a human being, 46; a plum, 48; a dog, 78; and a goldfish, 94. However, the sex cells—or

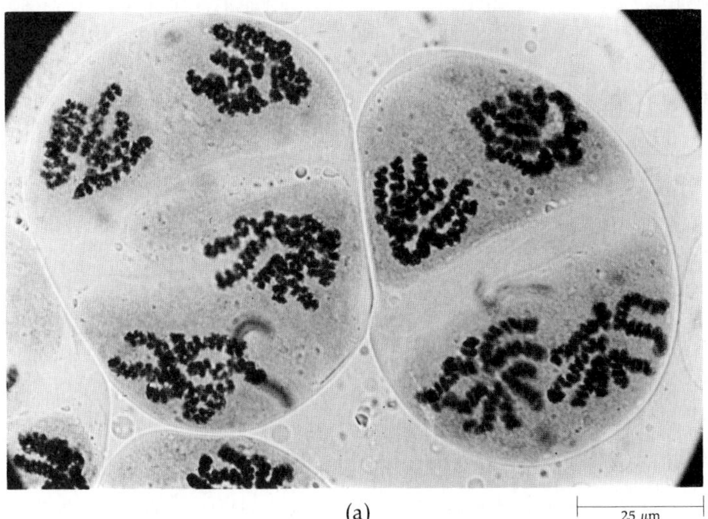

12–1
(a) *Meiosis in the formation of pollen in wake-robin* (Trillium erectum). *The clearly visible chromosomes are almost completely separated. Each diploid nucleus has divided to form four haploid sets of chromosomes.* (b) *Wake-robin flowers in the early spring.*

gametes—have exactly half the number of chromosomes that is characteristic of the somatic (body) cells of the organism. The number of chromosomes in the gametes is referred to as the *haploid* ("single") number, and the number in the somatic cells as the *diploid* ("double") number. Cells that have more than one double set of chromosomes are known as *polyploid*.

For brevity, the haploid number is designated n and the diploid number as $2n$. In humans, for example, $n = 23$ and therefore $2n = 46$. When a sperm fertilizes an egg, the two haploid nuclei fuse, $n + n = 2n$, and the diploid number is restored (Figure 12-2). The diploid cell produced by the fusion of two gametes is known as a *zygote*.

In every diploid cell, each chromosome has a partner. These pairs of chromosomes are known as homologous pairs, or *homologues*. The two resemble each other in size and shape and also, as we shall see, in the type of hereditary information each contains. One homologue is derived from one parent, and its partner is derived from the other parent.

In the special kind of nuclear division called meiosis, the diploid number of chromosomes is reduced to the haploid number, thus counterbalancing the effects of fertilization. Moreover, as we shall see, meiosis is in itself a source of new chromosome combinations.

12-2
Sexual reproduction is characterized by two events: the coming together of the gametes (fertilization) and meiosis. Following meiosis, the number of the chromosomes is single, or haploid (n). *Following fertilization, the number is double, or diploid* (2n).

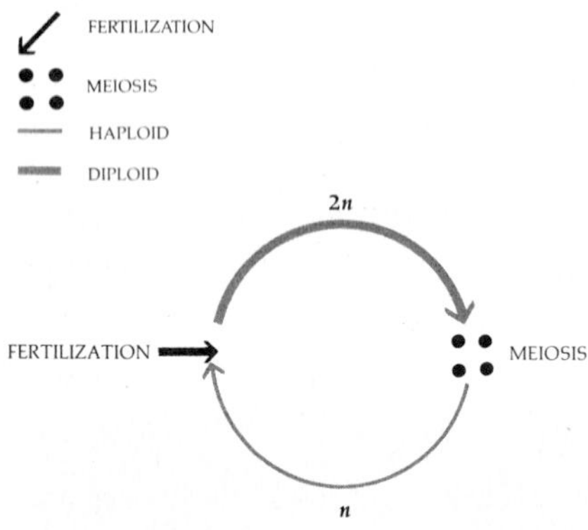

MEIOSIS AND THE LIFE CYCLE

Meiosis occurs at different times during the life cycle of different organisms (Figure 12-2). In the alga *Chlamydomonas* and the mold *Neurospora*, it occurs immediately after fusion of the mating cells (Figure 12-3); the cells are ordinarily haploid, and meiosis restores the haploid number.

In plants, such as ferns, a haploid phase typically alternates with a diploid phase (Figure 12-4). The common and conspicuous form of a fern is the sporophyte, the diploid organism. By meiosis, fern sporophytes produce spores, usually on the undersides of their fronds. These spores have only the haploid number of chromosomes. They germinate to form much smaller plants (gametophytes), typically only a few cell layers thick. In these plants all the cells are haploid. The small, haploid plants produce gametes, which fuse and then develop into a new, diploid sporophyte. This process, in which a haploid stage is followed by a diploid stage and again by a haploid stage, is known as *alternation of generations*. As we shall see, alternation of generations occurs, although not always in the same form, in all plants.

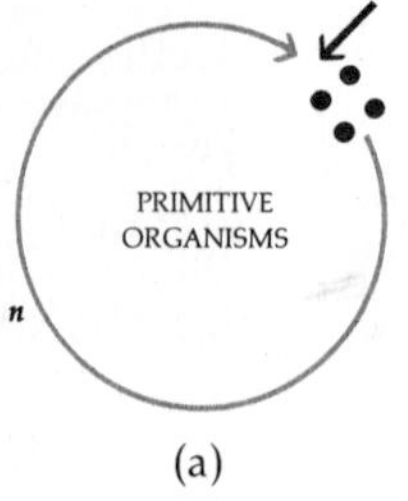

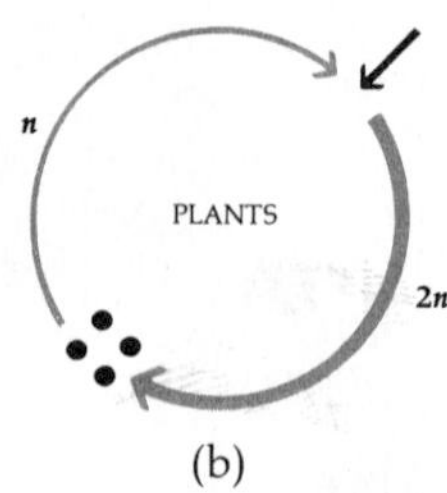

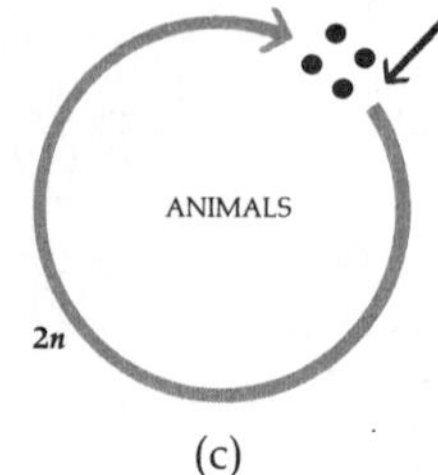

Fertilization and meiosis occur at different points in the life cycle of different organisms. (a) *In some organisms, meiosis occurs immediately after fertilization, and most of the life cycle is spent in the haploid state (signified by the thin line).* (b) *In plants, fertilization and meiosis are separated, and the organism characteristically has a diploid phase and a haploid phase.* (c) *In animals, meiosis is immediately followed by fertilization. As a consequence, during most of the life cycle the organism is diploid (signified by the thick line).*

12–3

The life cycle of Chlamydomonas *is of the type shown in Figure 12–2a. The organism is haploid for most of its life cycle (thin arrows). Fertilization, the fusing of cells of different mating strains (indicated here by + and −), temporarily produces the diploid state (thick arrows). The diploid zygote divides meiotically, forming four new haploid cells, which will probably divide repeatedly by mitosis (asexually) before entering another sexual cycle.*

12–4

The life cycle of a fern is of the type shown in Figure 12–2b. Following meiosis, spores, which are haploid, are produced in the sporangia and then are shed (extreme right). The spores develop into haploid gametophytes. In many species, the gametophytes are only one layer of cells thick and are somewhat heart-shaped, as shown here (bottom). From the lower surface of the gametophyte, filaments, the rhizoids, extend downward into the soil.

On the lower surface of the gametophyte are borne the flask-shaped archegonia, which enclose the egg cells, and the antheridia, which enclose the sperm. When the sperm are mature and there is an adequate supply of water, the antheridia burst, and the sperm cells, which have numerous flagella, swim to the archegonia and fertilize the eggs. From the fertilized (2n) *egg, the zygote, the* 2n *sporophyte grows out of the archegonium within the gametophyte. After the young sporophyte becomes rooted in the soil, the gametophyte disintegrates. When it becomes mature, the sporophyte develops sporangia, in which meiosis occurs, and the cycle begins again.*

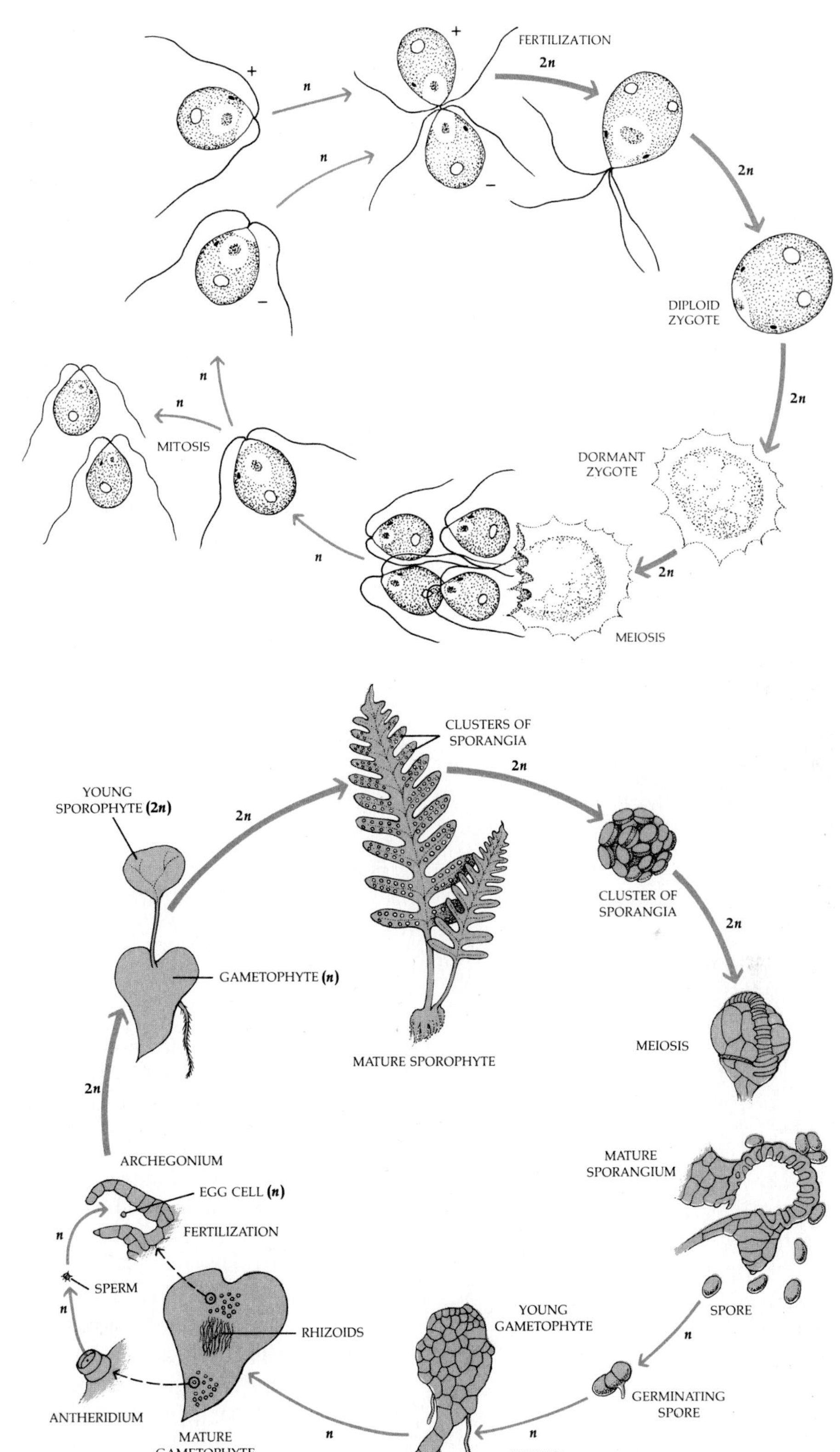

12–5
The life cycle of Homo sapiens. *Gametes, which are haploid, are produced by meiosis. At fertilization, they fuse, restoring the diploid number in the fertilized egg. The fertilized egg develops into a mature man or woman, who again produces haploid gametes. As is the case with most other animals, the life cycle is almost entirely diploid, the only exception being the gametes. This is the type of life cycle diagrammed in Figure 12–2c.*

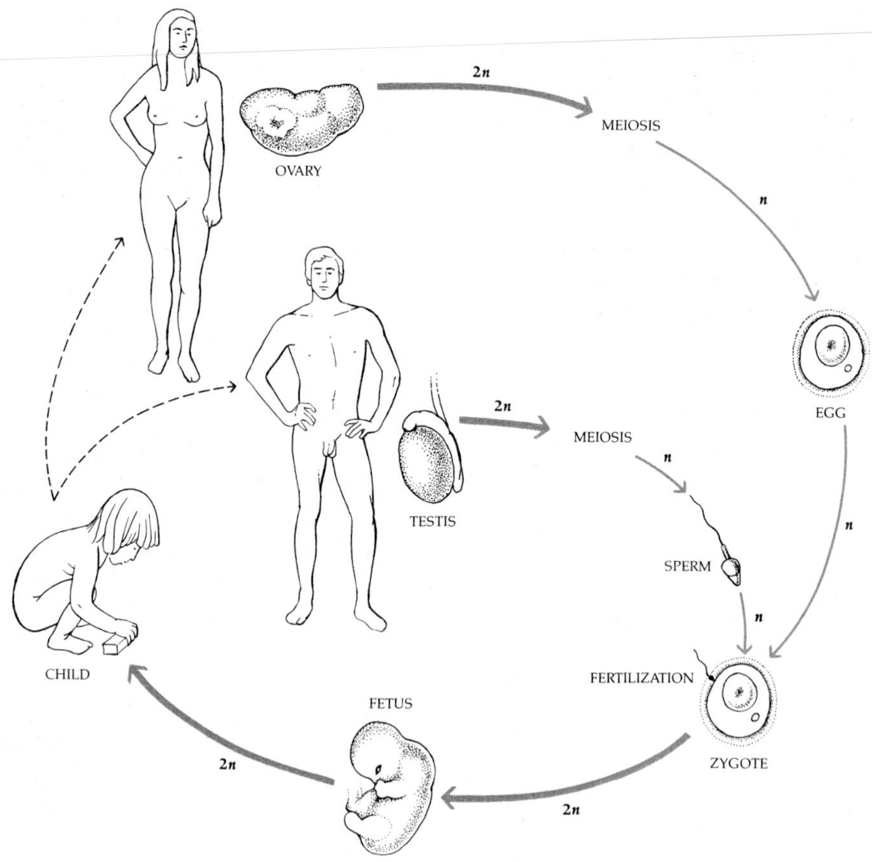

Human beings have the typical animal life cycle, in which meiosis immediately precedes fertilization and virtually all of the life cycle is spent in the diploid state (Figure 12–5).

MEIOSIS VS. MITOSIS

As we saw in Chapter 7, the events that occur in mitosis result in the formation of two daughter cells, each of which receives an exact copy of the parent cell's chromosomes. The events that take place during meiosis somewhat resemble those of mitosis, but there are some important differences:

1. Only diploid cells undergo meiosis.
2. Each diploid cell divides twice, producing a total of four cells.
3. Each of the four cells contains half the number of chromosomes present in the original cell.
4. The haploid cells produced by meiosis contain new combinations of chromosomes. That is, the homologous chromosomes, originally derived from the organism's parents, are assorted randomly among the four new haploid cells. (For example, whether a particular gamete produced by your body contains, for a particular pair of homologues, the chromosome provided by your mother or by your father is purely a matter of chance.)

12–6
(a) During prophase I of meiosis, chromosomes become arranged in homologous pairs. Each homologous pair consists of four chromatids and is therefore also known as a tetrad (from the Greek tetra, *meaning "four"). (b) Crossing over. During meiosis I, homologues, paired up as tetrads, connect at crossover points, where exchanges of segments of the chromosomes—crossing over—take place. Crossing over is depicted here, very diagrammatically, in three dimensions.*

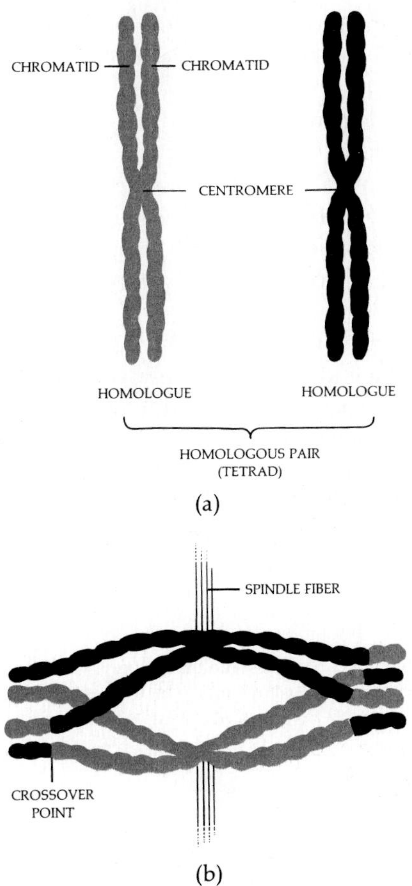

THE PHASES OF MEIOSIS

Meiosis consists of two successive nuclear divisions, conventionally designated meiosis I and meiosis II. In meiosis I, the homologues separate; in meiosis II, the chromatids separate. In the following discussion we shall describe meiosis in a plant cell in which the diploid number is 8 ($n = 4$). Four of the eight chromosomes were inherited from one parent and four from the other parent. For each chromosome from one parent there is a homologous chromosome, or homologue, from the other parent—a chromosome of similar size and shape.

During interphase, the chromosomes are replicated, so that by the beginning of meiosis I each chromosome consists of two identical sister chromatids held together at the centromere as shown in Figure 12–6a. In the first prophase of meiosis, *prophase I*, the chromatin condenses and the nuclear envelope begins to dissolve.

During this first prophase, an event occurs for which the mechanism is completely unknown. The homologous chromosomes come together in pairs. Once contact is made at any point between the two homologues, pairing extends, zipperlike, along the length of the chromatids. The alignment is very precise. Since each chromosome consists of two identical chromatids, the pairing of the homologous chromosomes actually involves four chromatids; this complex of paired homologous chromosomes is known as a *tetrad*. *Crossing over*, the interchange of segments of one chromosome with corresponding segments from its homologue, takes place at this time (Figure 12–6b). Again, the mechanism by which such precise transfers take place is unknown. As we shall see in the next chapter, such exchanges of chromosomal material are important sources of variation in the hereditary material.

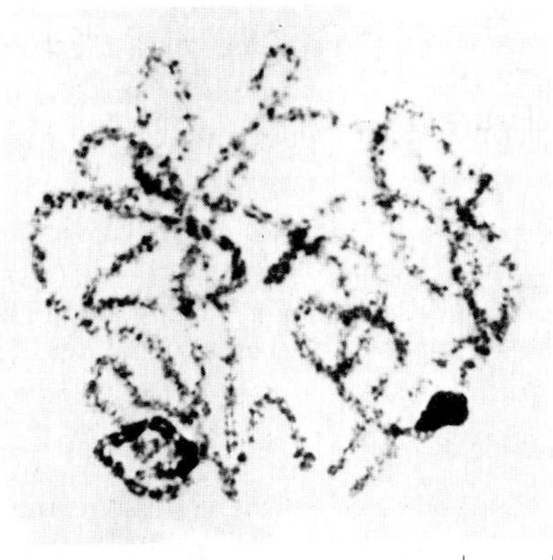

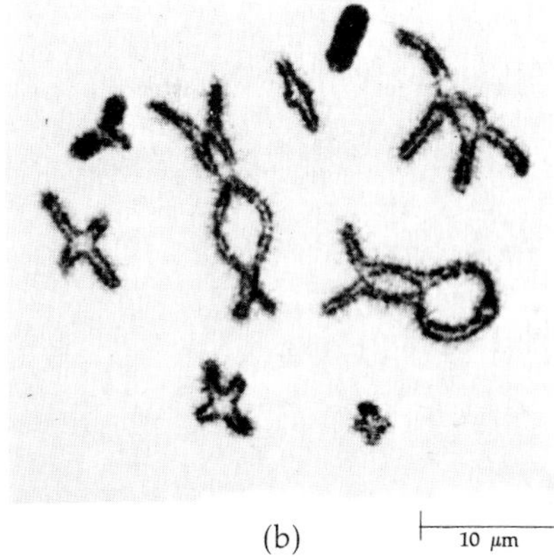

12–7
(a) Early prophase I in the formation of a sperm cell in a grasshopper. The homologous chromosomes are now paired; the individual chromatids are not visible, however, so each chromosome still appears as a single structure, and the tetrads appear double-stranded (rather than four-stranded). The dark area in the lower right is a sex chromosome, which is very prominent in the grasshopper. (b) Late prophase I. All four chromatids can be seen in most of the tetrads.

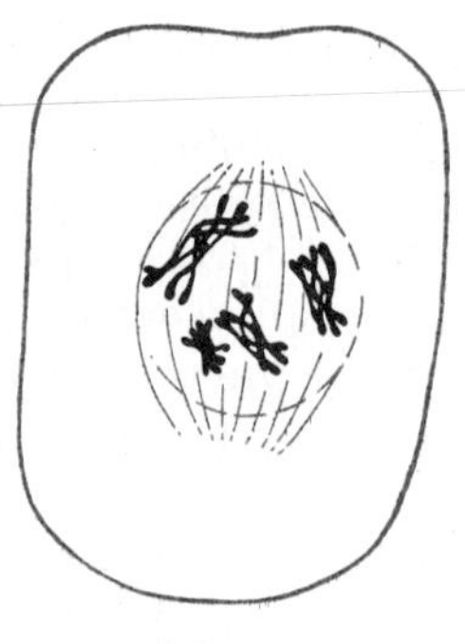

LATE PROPHASE I

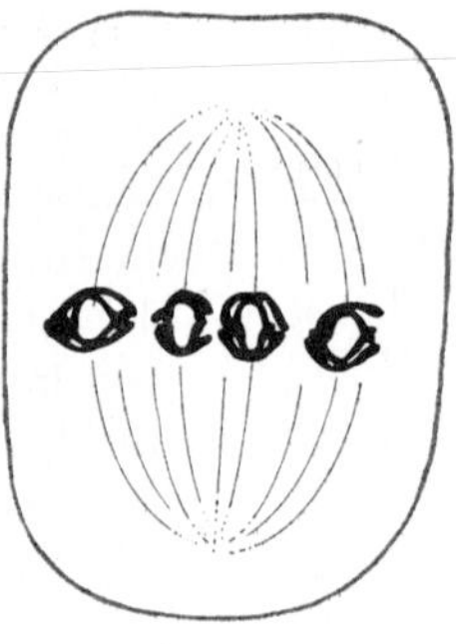
METAPHASE I

ANAPHASE I

If you remember the arrangement of the homologous chromosomes at this stage of meiosis, you will be able to remember all the subsequent events with little difficulty.

Toward the end of prophase I, the spindle apparatus is formed, the nucleolus disappears, and the nuclear envelope breaks down.

In *metaphase I*, the homologous pairs (four in this example) line up along the equatorial plane of the cell. Each pair consists of four chromatids (two homologous chromosomes). The spindle has formed, and spindle fibers attach to each homologue at its kinetochore. If this were an animal cell, centrioles and asters would be present.

At *anaphase I*, the homologues, each consisting of two chromatids, separate, as if pulled apart by the spindle fibers attached to their kinetochores. However, the two sister chromatids of each chromosome do not separate as they did in mitosis.

By the end of the first meiotic division, *telophase I*, the homologues have moved to the poles. Each group now contains only half the number of chromosomes as the original nucleus.* Moreover, these chromosomes may be different from any one that was present in the original cell because of exchanges taking place during crossing over. Depending on the species, new nuclear envelopes may or may not form, and cytokinesis may or may not take place.

* In counting, it is often difficult to know whether to count a chromosome that has replicated but has not divided as 1 or 2. It is customary to count such a chromosome as 1. The trick is to count centromeres.

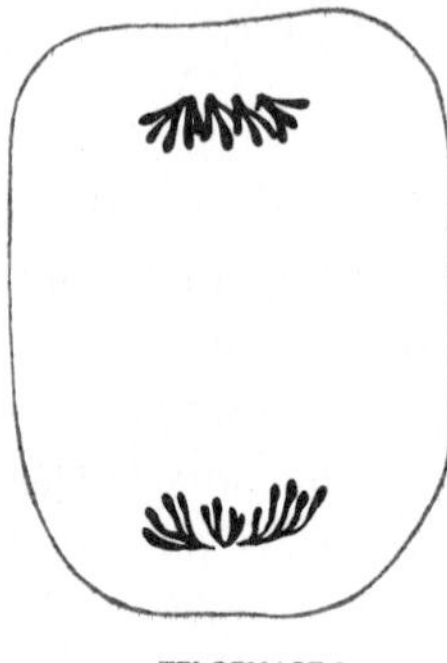

TELOPHASE I

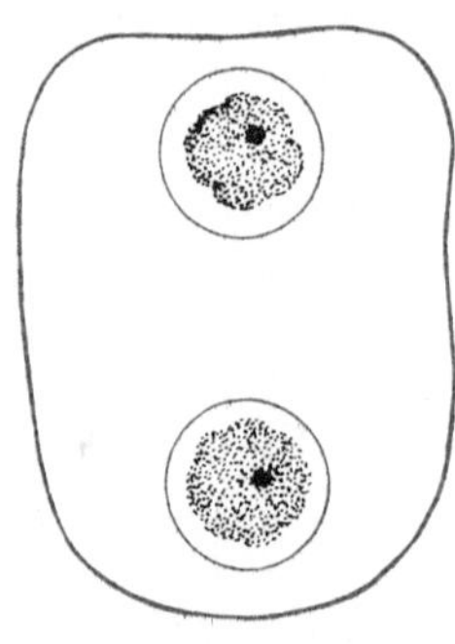
INTERPHASE II

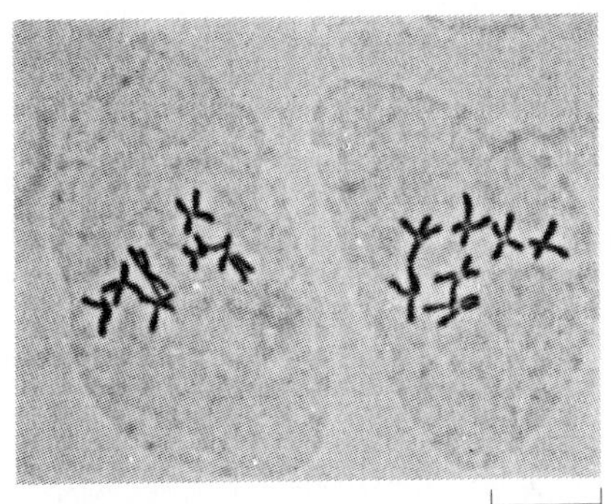

12–8
Metaphase II in crested wheat grass. The chromosomes (chromatid pairs) are approaching the equatorial plane.

Meiosis II resembles mitosis except that it is not preceded by replication of the chromosomal material. At the beginning of the second meiotic division, the chromosomes, which may have dispersed somewhat, condense fully again. There are four in each nucleus (the haploid number), and they are still in the form of two chromatids held together at the centromere.

During *prophase II*, the nuclear envelopes, if present, dissolve, and new spindle fibers begin to appear.

During *metaphase II*, the four chromatid pairs in each nucleus line up on the equatorial plane. At *anaphase II*, as in mitosis, the sister chromatids separate, and each resulting chromosome moves toward one of the poles.

During *telophase II*, the spindles disappear and a nuclear envelope forms around each set of chromosomes. There are now four nuclei in all, each containing the haploid number of chromosomes. Cell division (cytokinesis) proceeds as it does following mitosis. Cell walls form, dividing the cytoplasm, and the cells begin to differentiate into spores.

So, beginning with one cell containing eight chromosomes (four homologous pairs), we end with four cells, each with four chromosomes (no homologous pairs).

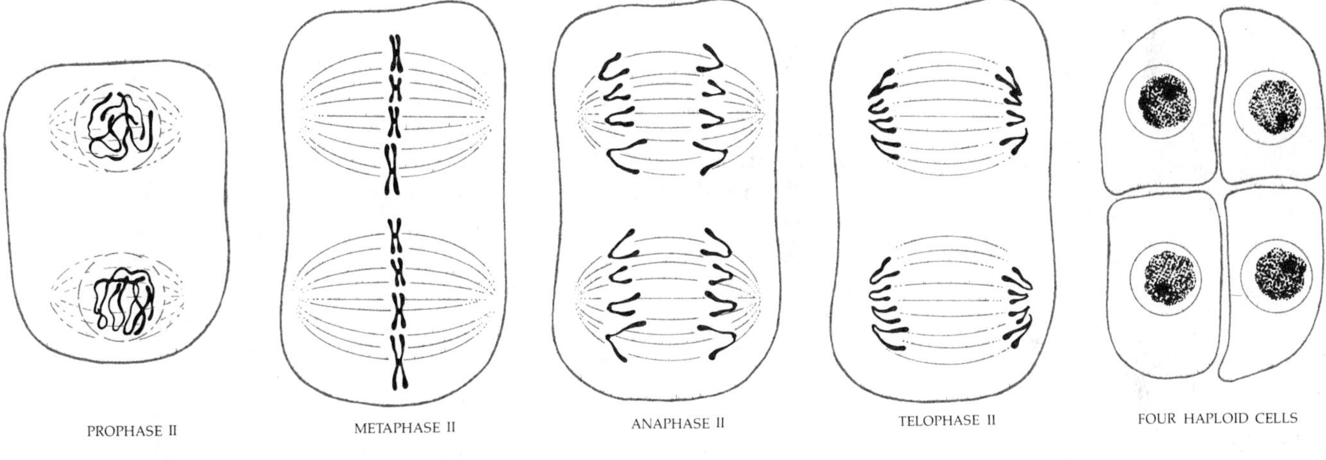

12–9
(a) *Anaphase II in the royal fern* Osmunda regalis. *The chromatids have separated, and the daughter chromosomes are moving to the opposite poles of the spindles.* (b) *The end of spore formation in* Osmunda regalis. *Each of these haploid cells can germinate to produce a gametophyte, the haploid phase in the life cycle.*

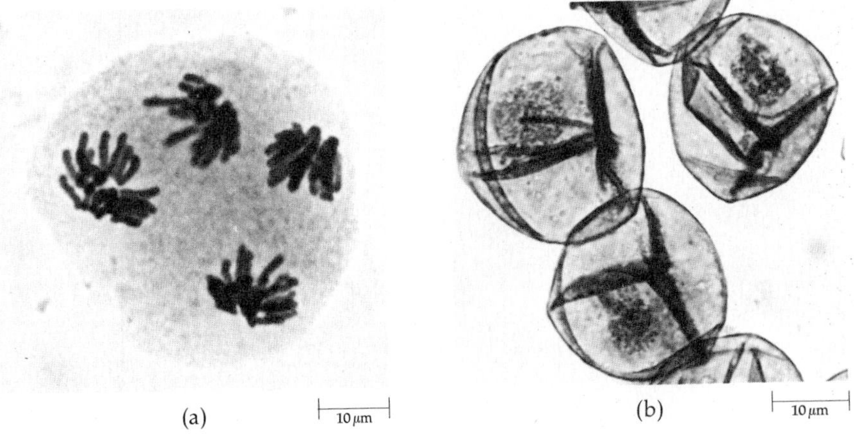

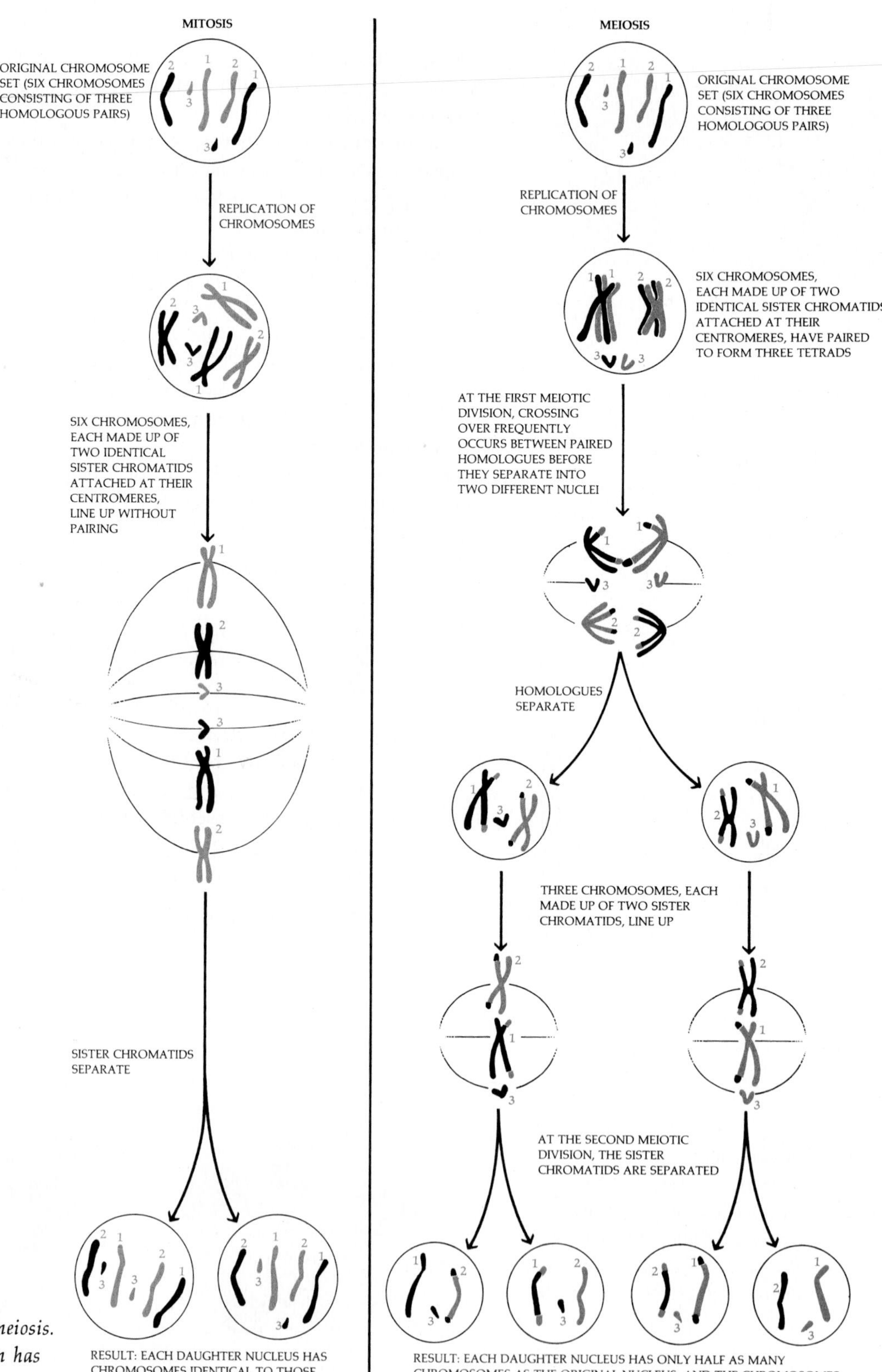
MITOSIS
ORIGINAL CHROMOSOME SET (SIX CHROMOSOMES CONSISTING OF THREE HOMOLOGOUS PAIRS)
REPLICATION OF CHROMOSOMES
SIX CHROMOSOMES, EACH MADE UP OF TWO IDENTICAL SISTER CHROMATIDS ATTACHED AT THEIR CENTROMERES, LINE UP WITHOUT PAIRING
SISTER CHROMATIDS SEPARATE
RESULT: EACH DAUGHTER NUCLEUS HAS CHROMOSOMES IDENTICAL TO THOSE IN THE ORIGINAL NUCLEUS
MEIOSIS
ORIGINAL CHROMOSOME SET (SIX CHROMOSOMES CONSISTING OF THREE HOMOLOGOUS PAIRS)
REPLICATION OF CHROMOSOMES
SIX CHROMOSOMES, EACH MADE UP OF TWO IDENTICAL SISTER CHROMATIDS ATTACHED AT THEIR CENTROMERES, HAVE PAIRED TO FORM THREE TETRADS
AT THE FIRST MEIOTIC DIVISION, CROSSING OVER FREQUENTLY OCCURS BETWEEN PAIRED HOMOLOGUES BEFORE THEY SEPARATE INTO TWO DIFFERENT NUCLEI
HOMOLOGUES SEPARATE
THREE CHROMOSOMES, EACH MADE UP OF TWO SISTER CHROMATIDS, LINE UP
AT THE SECOND MEIOTIC DIVISION, THE SISTER CHROMATIDS ARE SEPARATED
RESULT: EACH DAUGHTER NUCLEUS HAS ONLY HALF AS MANY CHROMOSOMES AS THE ORIGINAL NUCLEUS, AND THE CHROMOSOMES ARE NOT IDENTICAL

12–10
A comparison of mitosis and meiosis. In these examples, the organism has six chromosomes (2n = 6).

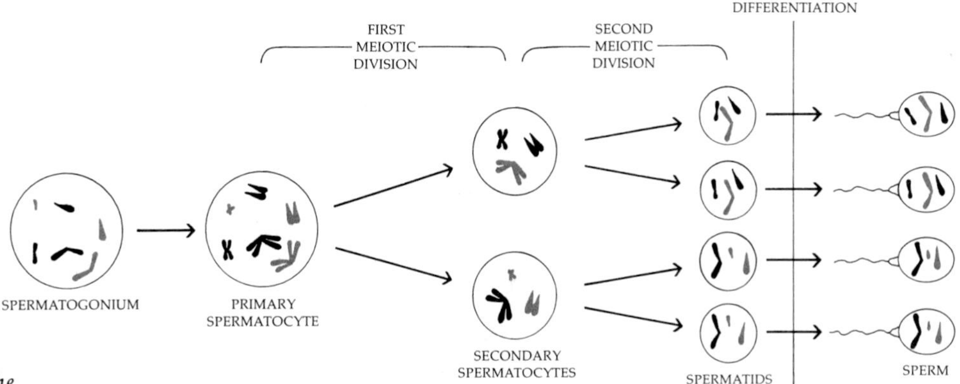

12–11
The series of changes resulting in the formation of sperm cells begins with the growth of spermatogonia into primary spermatocytes. At the first meiotic division, each primary spermatocyte divides into two haploid secondary spermatocytes. The second meiotic division results in the formation of four haploid spermatids. The spermatids differentiate into functional sperm. For simplicity, only six ($n = 3$) *chromosomes are shown.*

Meiosis in the Human Species

In all vertebrates, including humans, meiosis takes place in the reproductive organs, the testes of the male and the ovaries of the female. In the male, a cell known as a primary spermatocyte undergoes the two divisions of meiosis to produce four spermatids, each of which then differentiates into a sperm cell (Figure 12–11). The ejaculate of a normal human male contains 300 million sperm cells.

In females, the meiotic divisions are equal in terms of the chromosomes but unequal in the way the cytoplasm is apportioned (Figure 12–12). One egg cell (ovum) is produced and two or three polar bodies, which contain the other post-meiotic nuclei. These then disintegrate. As a result of this unequal division, the ovum is well supplied with essential materials for the developing embryo.

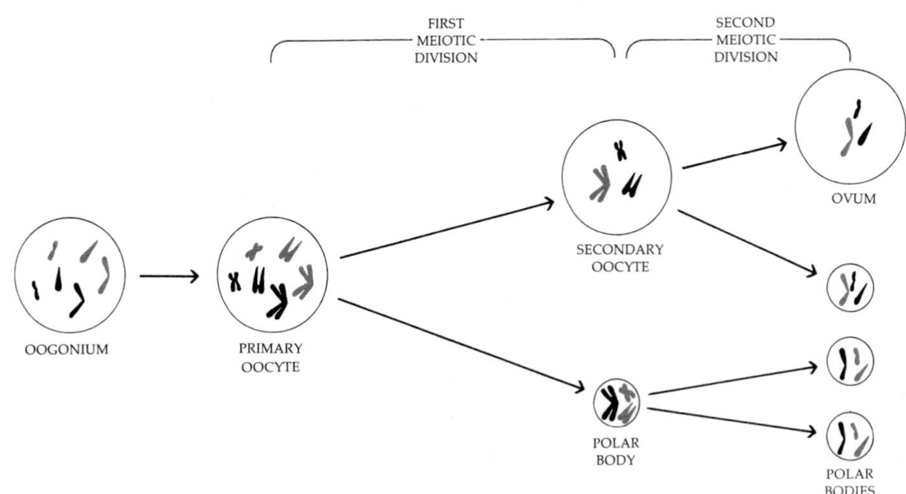

12–12
Formation of the ovum. A primary oocyte undergoes a meiotic division to produce a secondary oocyte and a polar body. This first meiotic division begins in the human female during the third month of fetal development and ends at ovulation, which may take place 50 years later. The second meiotic division, which produces the egg cell and a second polar body, does not take place until after fertilization. The first polar body may also divide.

THE CONSEQUENCES OF SEXUAL REPRODUCTION

Many organisms can reproduce both sexually and asexually (by mitosis and cytokinesis). Most one-celled organisms have a life cycle similar to that of Chlamydomonas *(Figure 12–3), in which either sexual or asexual reproduction may take place, often depending on environmental circumstances. Some one-celled organisms, such as amoebas, reproduce only asexually. Many plants can also reproduce asexually; many grasses, for instance, spread by means of underground stems (rhizomes). In animals, asexual reproduction can take place by budding, as in* Hydra, *or by the breaking off of a fragment of the parent animal, as occurs in sponges, sea anemones, and certain types of worms. In many organisms, including some fish, offspring are produced parthenogenetically (from unfertilized eggs). Because of the careful copying process of mitosis, asexually produced individuals are always genetically identical to their parents.*

As compared to asexual reproduction, sexual reproduction is expensive (in terms of energy expenditure) and inefficient (since it takes two cells to make a new organism instead of just one). Yet it is widely practiced throughout the multicellular world. The advantage of sexual reproduction seems to lie in its potential for variation in the offspring.

The diagram below shows the possible distributions of chromosomes at meiosis. The black chromosomes were originally of paternal origin, and the colored chromosomes of maternal origin. They were transmitted by replication and mitotic division to the cells from which the eggs and sperm were formed. In the course of meiosis, these chromosomes are distributed among the haploid cells. As you can see, chromosomes of maternal or paternal origin do not stay together but are assorted independently. (a) *If the original number of chromosomes is 4* ($n = 2$)*, the number of possible combinations of chromosomes is* 2^2*, or 4.* (b) *If the original number is 6, the number of possible combinations is* 2^3*, or 8, and* (c) *if there are 8 chromosomes, 16 different combinations* (2^4) *are possible.*

A human male with his 46 chromosomes is capable of producing 2^{23} *kinds of sperm cells—8,388,608 different combinations of chromosomes, equal in number to the population of New York City. Similarly, a human female is capable of producing* 2^{23} *kinds of egg cells—8,388,608 different combinations of chromosomes. And this does not take into account the additional variations introduced by recombinations, which we shall examine in Chapter 13.*

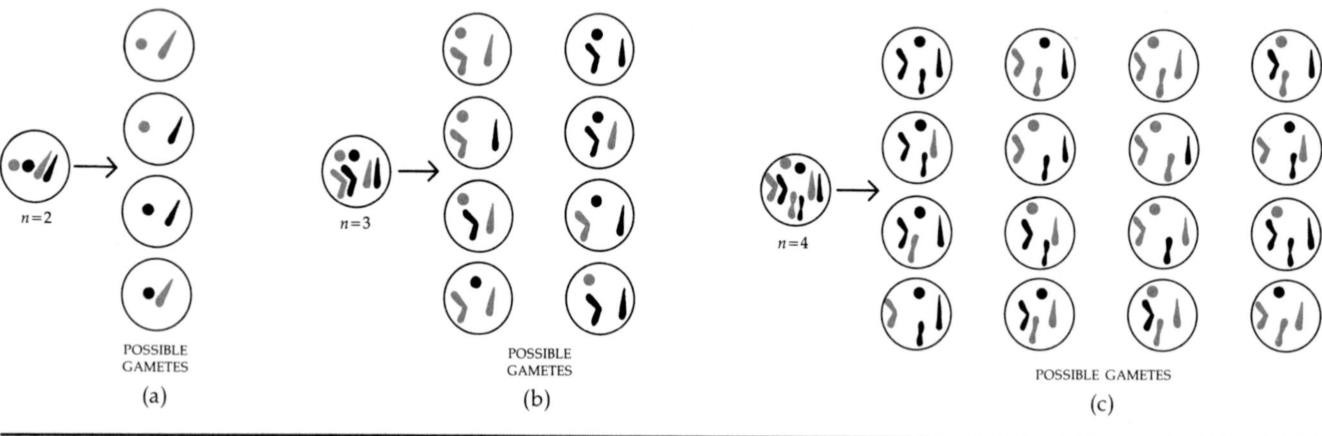

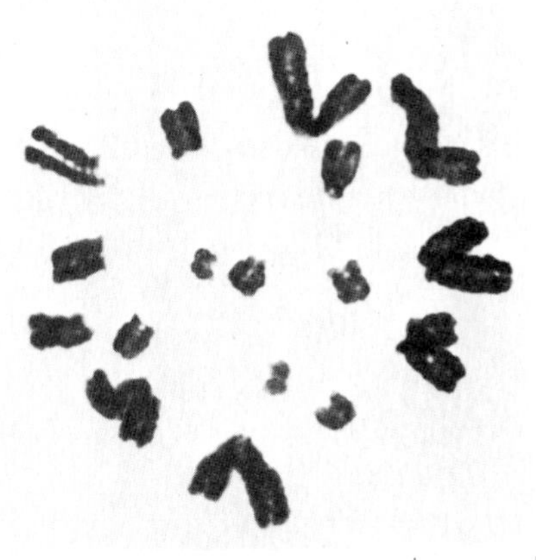

12–13
Chromosomes from a diploid cell of a grasshopper, during metaphase of a mitotic division. Each chromosome consists of two closely aligned chromatids. Note that even though these chromosomes are not paired, it is possible to pick out some of the homologues. The observation that chromosomes come in homologous pairs was one of Sutton's clues to the meaning of meiosis.

CYTOLOGY AND GENETICS MEET: SUTTON'S HYPOTHESIS

In 1902, shortly after the rediscovery of Mendel's work, William S. Sutton, a graduate student at Columbia University, was studying the formation of sperm cells in male grasshoppers. Observing the process of meiosis, Sutton noticed that the chromosomes that paired with one another at the beginning of the first meiotic division had physical resemblances to one another. In diploid cells, he noted, chromosomes apparently came in pairs. The pairing was obvious only at meiosis, but the discerning eye could find the homologues in the unpaired chromosomes when they became visible at the time of mitosis (Figure 12–13).

Sutton was struck by the parallels between what he was seeing and the first principle of Mendel. Suddenly the facts fell into place. Suppose chromosomes carried genes, the *Elemente* described by Mendel. This idea does not seem very startling to you now, with your superior knowledge, but remember that the gene was just an abstract idea or mathematical unit to the geneticist, and that the chromosome was just an unidentified colored body to the cytologist. Suppose, Sutton reasoned, alleles occurred on homologous chromosomes. Then the alleles could always remain independent and so could separate at meiosis, with new pairs of alleles forming when the gametes came together at fertilization. Mendel's principle of the segregation of alleles could be explained by the segregation of the homologous chromosomes at meiosis.

What about Mendel's second principle in relation to the movement of the chromosomes at meiosis? This principle, as you will recall, states that members of different pairs of genes assort independently. We can see that this can be true if—and this is an important point—the genes are on different pairs of chromosomes, as shown in Figure 12–14 on the next page.

As occurs often in the history of science, two other biologists recognized the correlation between the behavior of Mendel's *Elemente* and the observed movement of the chromosomes, but young Sutton's paper appeared first, and his presentation was by far the most convincing. Much more evidence was required, however, before, more than a decade later, most biologists were ready to concede that the little "colored bodies" performing their stereotyped, repetitive dance within the cell's nucleus actually held the secrets of the most ancient mysteries of heredity.

SUMMARY

Sexual reproduction involves a special kind of nuclear division called meiosis. Meiosis is the process by which the chromosomes are reassorted and cells are produced that have the haploid chromosome number (n). The other principal component of sexual reproduction is fertilization, the coming together of haploid cells, which restores the diploid number $(2n)$. There are characteristic differences among major groups of organisms as to where in the life cycle these events take place.

At the start of meiosis, the chromosomes arrange themselves in pairs, the members of which are known as homologues. One homologue of each pair is of maternal origin, and one is of paternal origin. Each homologue consists of two identical chromatids. Early in meiosis, crossing over occurs between homologues, resulting in exchanges of chromosomal material.

12–14

The chromosome distributions in Mendel's cross of round yellow and wrinkled green peas, according to Sutton's hypothesis. Although the pea has 14 chromosomes (n = 7), only 4 are shown here, the two carrying the genes for round or wrinkled and the two carrying the genes for yellow or green. (This is analogous to what Mendel did when he selected the two traits to study.) As you can see, one parent is homozygous for the recessives, one for the dominants. Therefore, the only gametes they can produce are RY *and* ry. *(Remember,* R *now stands not just for the allele but also for the chromosome carrying the allele, as do the other letters.) The* F_1 *generation, therefore, must be* Rr *and* Yy. *When a cell of this generation undergoes meiosis,* R *is separated from* r *and* Y *from* y *when the respective homologues separate at anaphase I. Four different types of haploid egg cells are possible, as the diagram reminds us, and also four different types of sperm nuclei. These can combine in 4 × 4, or 16, different ways, as illustrated in the Punnett square. (Yellow is indicated by the color, green by gray.)*

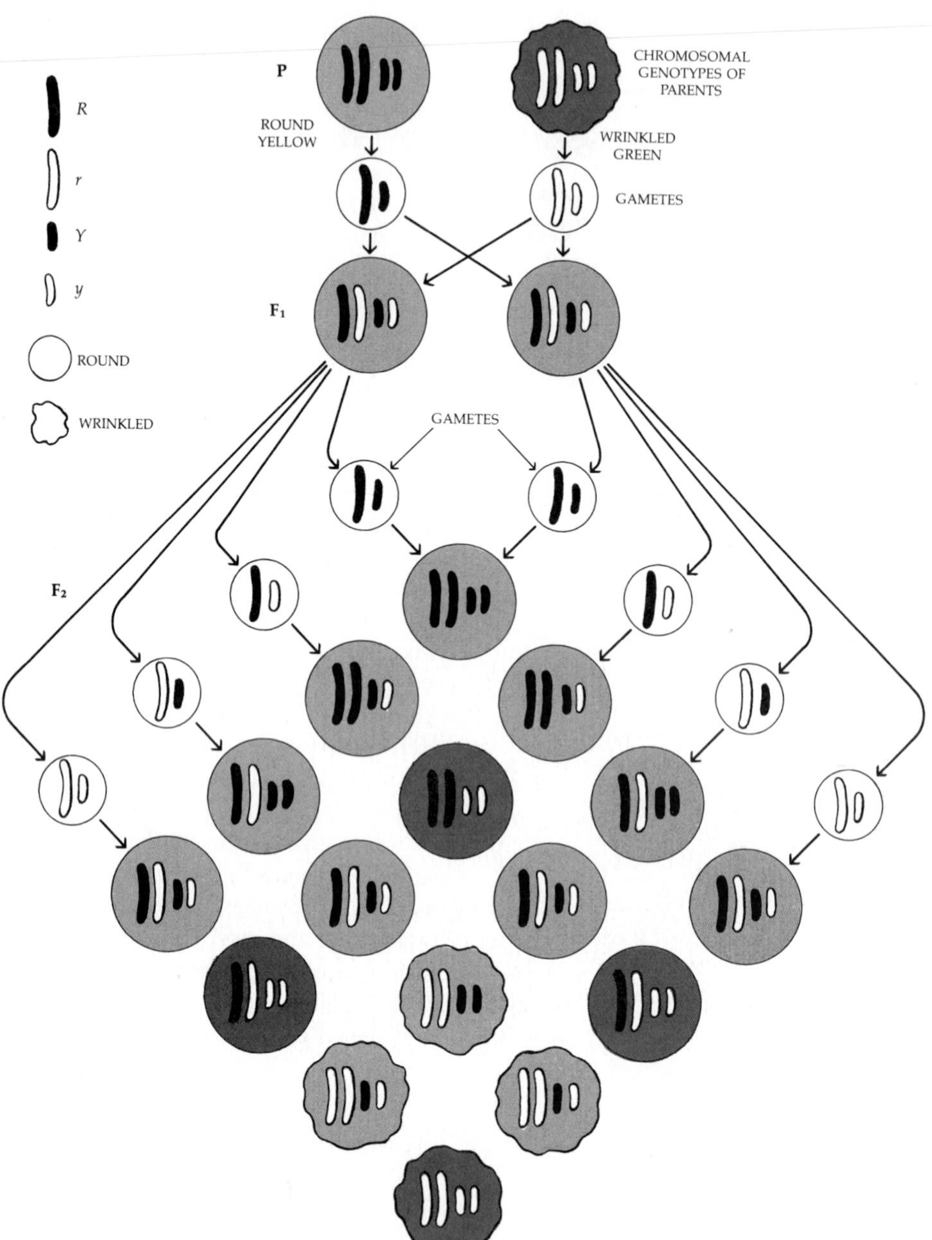

In the first stage of meiosis, the homologues are separated. Two nuclei are produced, each with a haploid number of chromosomes, which, in turn, consist of two chromatids each. The cell enters interphase, but the chromosomal material is not replicated. In the second stage of meiosis, the sister chromatids separate as in mitosis. When the two nuclei divide, four haploid cells result.

Each of the haploid cells produced by meiosis contains a unique assortment of chromosomes due to crossing over and random assortment of homologues. Thus meiosis is a source of the variation upon which evolution depends.

Sutton was among the first to notice the analogy between the behavior of the chromosomes at meiosis and the segregation and assortment of genes described by Mendel. On the basis of this observation, Sutton proposed that genes are carried on chromosomes.

QUESTIONS

1. Distinguish among the following: haploid/diploid/polyploid; sporophyte/gametophyte; gamete/zygote; meiosis I/meiosis II; homologue/tetrad.

2. Draw a diagram of a cell with six chromosomes ($n = 3$) at meiotic prophase I. Label each pair of chromosomes differently (for example, label one pair A^1 and A^2, and another B^1 and B^2, etc.).

3. Diagram the eight possible gametes resulting from meiosis in a plant cell with six chromosomes ($n = 3$). Label each chromosome differently, as in Question 2. Neglect crossing over.

4. Identify the stages of meiosis in crested wheat grass shown in the micrographs below. What stage of meiosis is visible in Figure 12–1?

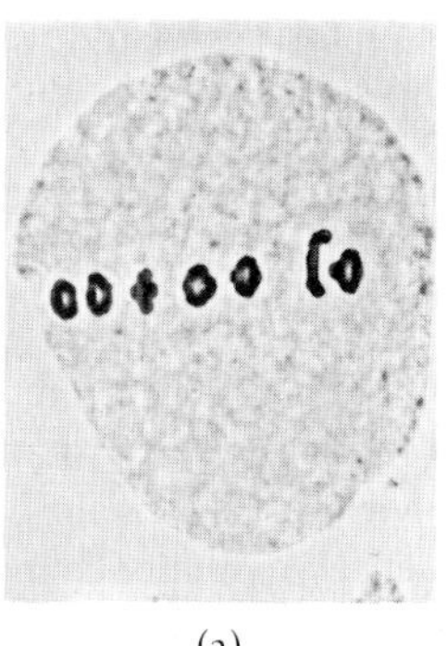
(a)

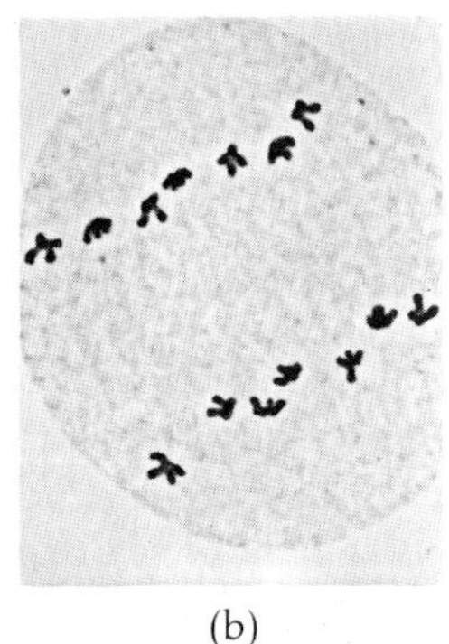
(b)

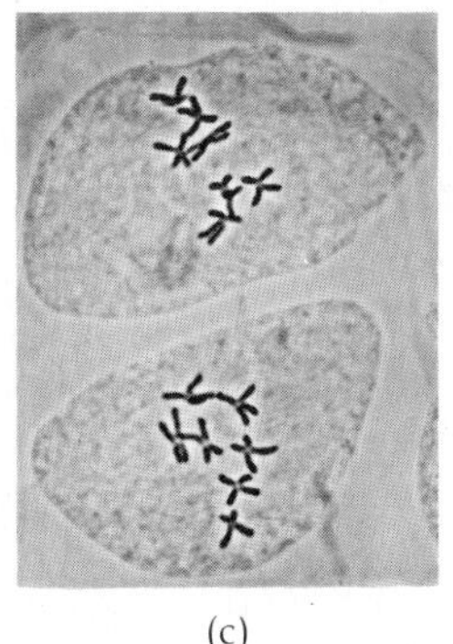
(c)

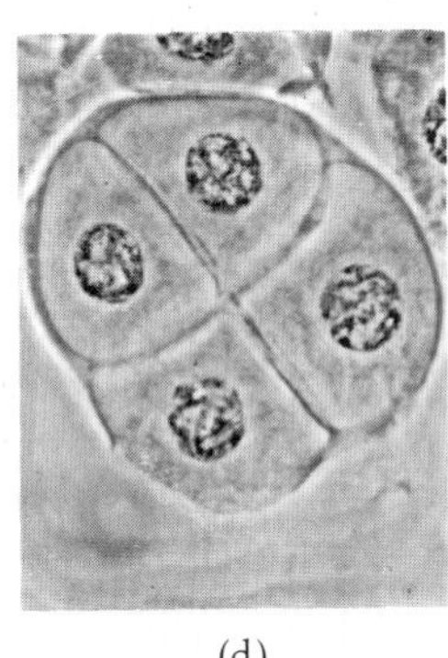
(d)

5. Compare and contrast the processes and the consequences of meiosis and mitosis.

6. In our bodies and in those of most other animals, both mitosis and meiosis occur. Describe the functions of these two processes in our bodies.

7. Is sexual reproduction—that is, fertilization and meiosis—possible with only one parent? Explain.

8. Mendel did not know of the existence of chromosomes. Had he known, what change might he have made in his second principle?

9. Compare metaphase of mitosis and metaphase II of meiosis both in terms of the positions of the chromosomes and the consequences.

CHAPTER 13

Genes and Chromosomes

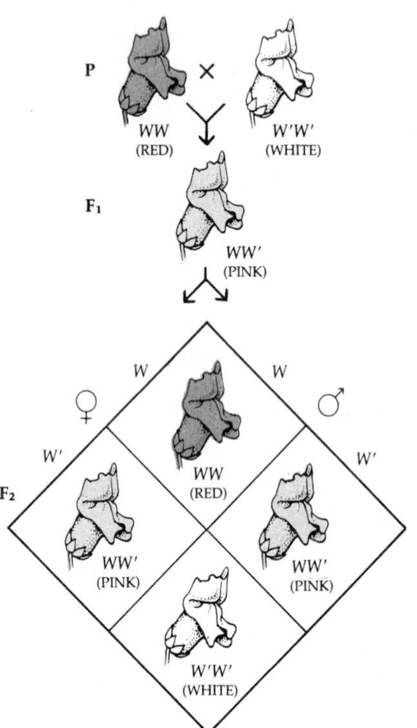

13–1
A cross between a red (WW) snapdragon and a white (W'W') snapdragon. This looks very much like the cross between a purple- and a white-flowering pea plant shown in Figure 11–8, but there is a significant difference because neither allele is dominant. The flower of the heterozygote is a blend of the two colors.

In the decades that followed the rediscovery of Mendel's experiments, some of the most brilliant work in the history of genetics was carried out. The first group of studies, based largely in England and Europe, were, like Mendel's, carried out mostly in plants. Then a group at Columbia University began to work with a tiny fly, *Drosophila melanogaster.* The wealth of data that emerged from these studies was so impressive that this period in genetic research, which lasted until World War II, has been characterized as the "golden age" of genetics (though some would argue that the golden age is now).

BROADENING THE CONCEPT OF THE GENE

The new studies in genetics, while confirming Mendel's work in principle, showed that the action of genes is not as simple and direct as Mendel's reported results had indicated. (This fact is not surprising when you recall that Mendel had carefully selected certain traits for study.) The effects of genes are influenced by their alleles, by nonallelic genes, and by their environment. In addition, genes change, or mutate. And many, indeed most, traits are influenced by more than one gene, just as most genes influence more than a single trait. Here are some examples.

Gene Interactions

Incomplete Dominance

Dominant and recessive traits are not always so clear-cut as in the pea plant. Some traits appear to blend. For instance, a cross between a red-flowering snapdragon and a white-flowering snapdragon produces heterozygotes that are pink (Figure 13–1). But when members of this generation self-pollinate, the traits begin to sort themselves out once again, showing that the alleles themselves, as Mendel had affirmed, remain discrete and unaltered. As we shall see, what occurs in the snapdragon and in similar crosses is actually a result of the combined effect of gene products. This phenomenon is known as *incomplete dominance,* or *codominance.*

Epistasis

Early in the present century, the British geneticist William Bateson and his group obtained some surprising results that at first seemed impossible to explain. They crossed two pure-breeding white-flowered varieties of sweet pea *(Lathyrus odoratus)* and found that the progeny all had purple petals. When these F_1 plants were allowed to self-pollinate, they found that of 651 plants that had flowered in the F_2 generation, 382 had purple petals and 269 were white. At first, these figures may seem meaningless, but if you examine them closely, you will see that they fit a 9:7 ratio, or in other words, they follow the 9:3:3:1 proportion for independent assortment of two nonallelic genes. Nine, you will remember, is the proportion of offspring that will show the effects of the two dominant alleles. So we can conclude that only a plant that has received at least one dominant allele from each gene pair can make the purple pigment (Figure 13-2).

A similar situation was seen in a cross between a pink-flowered and a white-flowered strain of salvia *(Salvia horminum)*. The F_1 plants all had purple flowers. In the F_2 generation, 255 had purple flowers, 92 pink, and 114 white. These results are harder to explain. Suppose, however, that you take the total, which is 461, and break it down into the expected results of the 9:3:3:1 ratio. The figures would be 259.3, 86.4, 86.4, and 28.9. Thus you can see that, as in the previous example, only

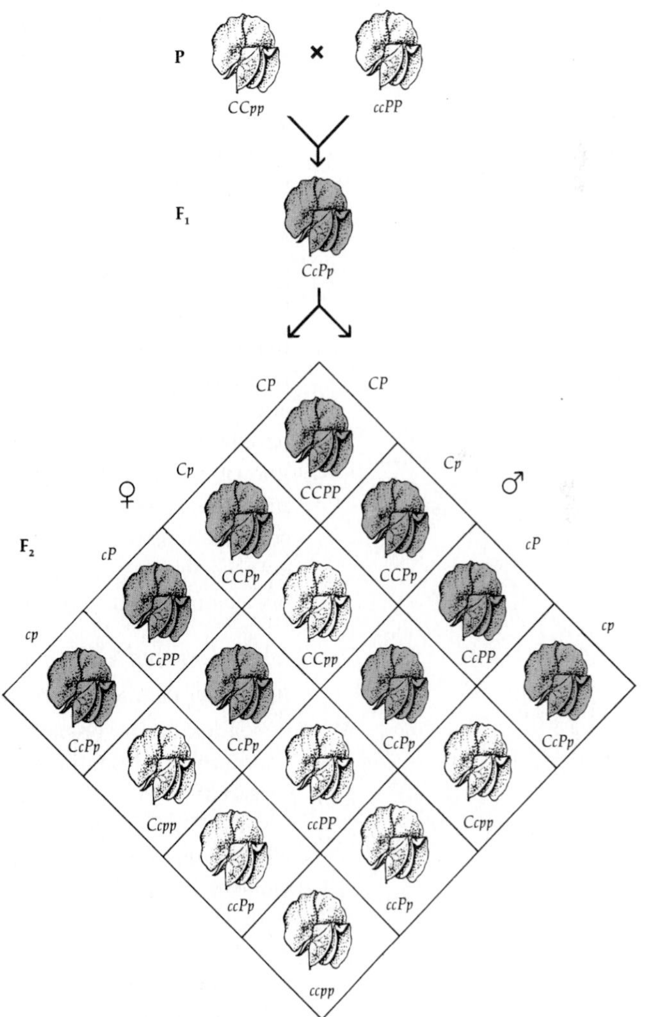

13-2
When two different varieties of white-flowered sweet pea plants are crossed, all of the F_1 *plants have purple flowers (indicated by gray). In the* F_2 *generation, the ratio of purple- to white-flowered plants is 9:7. The dominant allele* P *is responsible for producing the purple pigment, but this allele is expressed only if the dominant allele* (C) *of another gene pair is also present. These experiments with the sweet pea provided one of the first examples of epistasis, the modification of the expression of one gene by another, nonallelic gene.*

the plants with both dominants can make the purple pigment. If a plant has only one dominant, it either makes pink pigment or none, depending on which dominant is present. If both recessives are present, the plant cannot produce any pigment.

The modification of the expression of a gene by another, nonallelic gene is known as *epistasis.*

Genes and the Environment

Gene expression is always the result of the interaction of genetic potential with the environment. To take a common example, a seedling may have the genetic capacity to be green, to flower, and to fruit, but it will never turn green if it is kept in the dark, and it may not flower and fruit unless certain precise environmental requirements are met.

The water buttercup, *Ranunculus peltatus,* is a more striking example. It grows with half the plant body submerged in water. Although the leaves are genetically identical, the broad, floating leaves differ markedly in both form and physiology from the finely divided leaves that develop under water.

Temperature often affects gene expression. Primrose plants that are red-flowered at room temperature are white-flowered when raised at temperatures above 30°C (86°F). Similarly, as we noted in Chapter 8, Himalayan rabbits are white at high temperatures and black at low temperatures.

These are extreme examples of a universal verity: The phenotype of any organism is the result of interaction between genes and environment.

13–3
The water buttercup. Leaves growing above water are broad, flat, and lobed. The genetically identical underwater leaves are thin and finely divided.

Mutations

In 1902, a Dutch botanist named Hugo de Vries reported results of his studies on Mendelian inheritance in the evening primrose. Heredity in the primrose, he found, was generally orderly and predictable, as in the garden pea. Occasionally, however, a characteristic appeared that was not present in either parent or indeed anywhere in the lineage of that particular plant. De Vries called these abrupt hereditary changes *mutations,* and organisms exhibiting such changes came to be known as *mutants.* In fact, only about 2 of some 2,000 changes in the evening primrose observed by de Vries were actually mutations. The rest were due to new combinations of genes rather than to actual changes in any particular gene. However, de Vries's concept of mutation proved of great importance, although most of his examples were not valid.

Mutations and Genetic Research

Beginning with Mendel's experiments, the detection of the presence of a gene has depended on the phenotypic changes produced when one allele is different from another. Thus mutations have played, and continue to play, a major role in genetic research.

In 1927, H. J. Muller found that exposure to x-rays greatly increases the rate at which mutations occur. It was soon discovered that other radiations, such as ultraviolet light, and also some chemicals can act as *mutagens,* agents that produce mutations. The ability to increase the rate of mutation was an important research achievement. However, the discovery of the existence of a large number of mutagens in our ordinary environment was and is continuing to be a matter of considerable public concern.

13–4
Hugo de Vries is shown standing next to Amorphophallus titanum, *which has the largest flower cluster of any plant. De Vries, a Dutch botanist, was one of the rediscoverers of Mendel's work. He was also the first to recognize the nature of mutations and to assign them a role in evolutionary processes.*

Mutations and Evolutionary Theory

Mendel's work filled an important gap in Darwin's theory by explaining why small variations persist in populations. However, Mendelian principles presented new problems to the early evolutionists, because they appeared to offer no possibility for major changes in the genetic makeup of organisms. Segregation of traits explained how variation was maintained from generation to generation. Independent assortment explained how individuals could have characteristics in combinations not present in either parent and so perhaps be better adapted, in evolutionary terms, than either parent. But if all hereditary variations were to be explained by the reshuffling process proposed by Mendel, there would be little or no opportunity for major changes in organisms. Mutations are now recognized as the ultimate—and continual—source of the hereditary variations that make evolution possible.

Polygenic Inheritance and Pleiotropy

Continuous Variation

Mendel's experiments seemed to suggest that each gene affects a single characteristic in a one-to-one relationship. It was soon discovered, however, that a single trait is often affected by many genes (*polygenic inheritance*).

A trait affected by a number of genes does not show a clear-cut difference between groups of individuals—such as the differences tabulated by Mendel—but rather shows a gradation of small differences, which is known as *continuous variation*. If you make a chart of differences among individuals in any trait affected by a number of genes, you get a curve such as that shown in Figure 13–5.

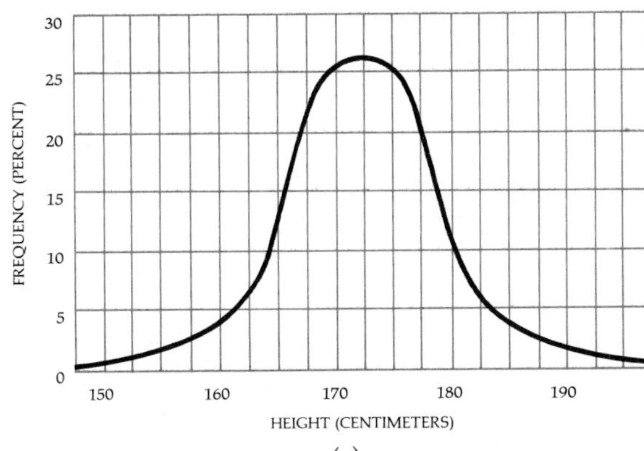

13–5
(a) Height distribution of males in the United States. Height is an example of polygenic inheritance; that is, it is affected by a number of genes. Such genetic traits are characterized by small gradations of difference. A graph of the distribution of such traits takes the form of a bell-shaped curve, as shown, with the mean, or average, usually falling in the center of the curve. The larger the number of genes involved, the smoother the curve.

(b) *A company of student recruits at the Connecticut Agricultural College about 70 years ago. The number of men in each group and their height (in feet and inches) are shown below the photograph.*

Fifty years ago, the average height of males in the United States was less than it is now but the shape of the curve was the same; in other words, the great majority fell within the middle range and the extremes in height were represented by only a few individuals. Some of these height variations are produced by environmental factors, such as diet, but even if all the men in a population were maintained from birth on the same type of diet, there would still be a continuous variation in height in the population. This is due to genetically determined differences in hormone production, bone formation, and numerous other factors.

Table 13–1 illustrates a simple example of polygenic inheritance, color in wheat kernels, which is controlled by two pairs of genes. Human skin color is under a similar kind of genetic control (although probably involving more than two genes), as are many other characteristics.

Pleiotropy

A single gene can affect more than one characteristic, a phenomenon known as *pleiotropy*. The frizzle trait in fowl (Figure 13–6) is an example of pleiotropy, as revealed by a single mutation—that is, a mutation involving only one gene. In these animals, a change in feather formation leads to drastic changes in many other aspects of their physiology.

Similarly, Hans Gruneberg showed that a single mutation affecting a gene involved in the formation of cartilage produces a whole complex of congenital deformities in rats. These include thickened ribs, a narrowing of the tracheal passage, blocked nostrils, a blunt snout, a loss of elasticity in the lungs, hypertrophy of the heart, and, needless to say, a greatly increased mortality. Since cartilage is one of the most common structural substances of the body, the widespread consequences of such a mutation are not difficult to understand. In fact, it is very likely that Mendel's allele for wrinkled peas, for example, affected other structural characteristics of the pea plant.

Table 13–1 The Genetic Control of Color in Wheat Kernels (Polygenic inheritance*)

Parents:	$R_1R_1R_2R_2 \times r_1r_1r_2r_2$ (Dark Red) (White)		
F_1:	$R_1r_1R_2r_2$ (Medium Red)		
F_2:	GENOTYPE	PHENOTYPE	
1	$R_1R_1R_2R_2$	Dark red	15 red to 1 white
2 } 4	$R_1R_1R_2r_2$	Medium-dark red	
2	$R_1r_1R_2R_2$	Medium-dark red	
4 } 6	$R_1r_1R_2r_2$	Medium red	
1	$R_1R_1r_2r_2$	Medium red	
1	$r_1r_1R_2R_2$	Medium red	
2 } 4	$R_1r_1r_2r_2$	Light red	
2	$r_1r_1R_2r_2$	Light red	
1	$r_1r_1r_2r_2$	White	

* Two genes are involved, each with two alleles: R_1 and r_1 for gene 1 and R_2 and r_2 for gene 2.

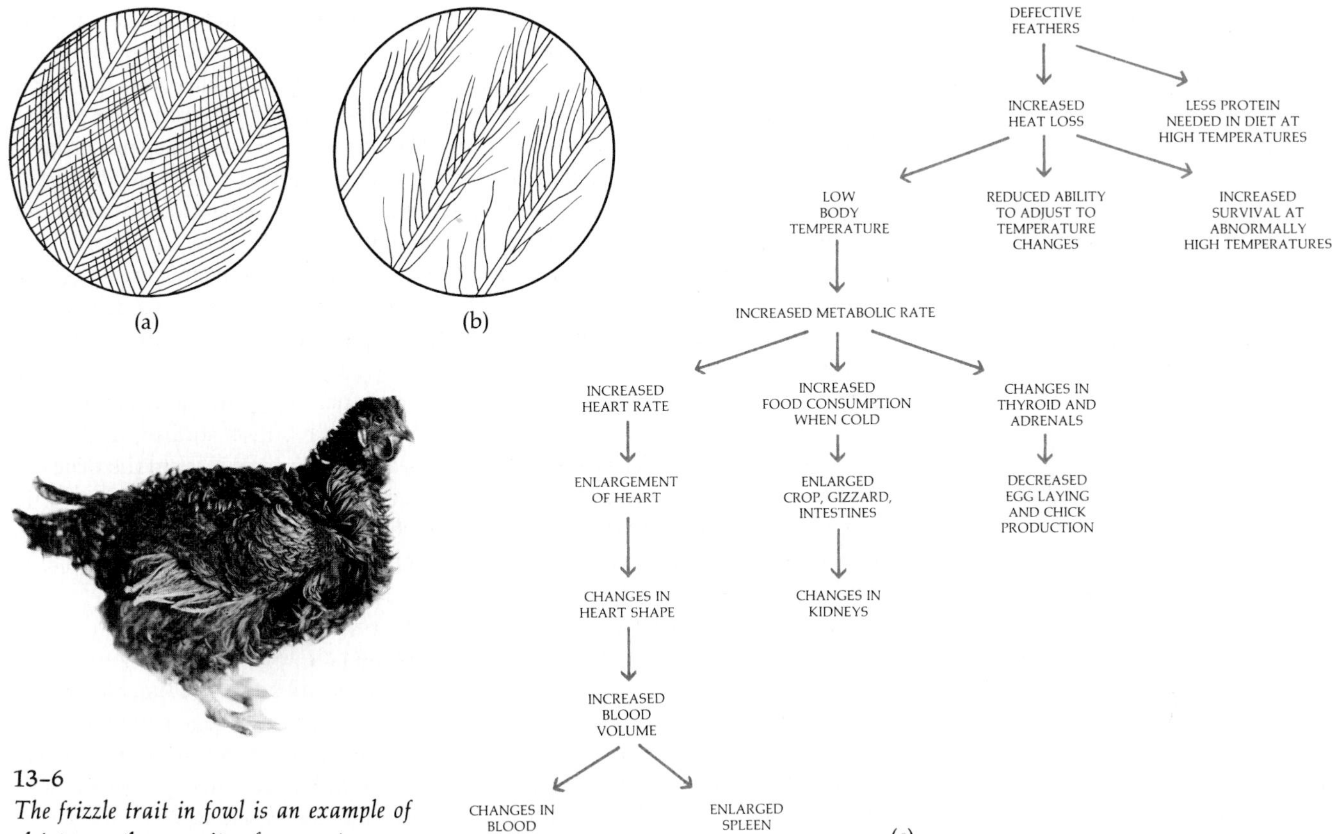

13–6
The frizzle trait in fowl is an example of pleiotropy, the capacity of a gene to produce a variety of phenotypic effects. "Frizzle" is manifested primarily by differences in the feathers. (a) *Under low magnification, feathers from normal birds show a closely interwebbed structure.* (b) *Feathers from birds homozygous for the frizzle trait are weak and stringy and provide poor insulation. Some of the consequences of the manifestation of this single gene are shown in* (c). *Notice that the frizzle trait is a liability at low temperatures but may increase survival at high temperatures.*

Multiple Alleles

Although any individual diploid organism can have only two alleles of any gene, multiple alleles can be present in a population. In humans, the three alleles of the ABO blood groups (to be discussed in Chapter 39) are probably the best known.

CHROMOSOMES AND GENES

One of the most remarkable features of this stage of genetic research is that the gene had no physical reality. It was a pure abstraction. The work of Sutton and other cytologists was known, but, to most researchers, it seemed in no way relevant to studies of segregation of characteristics and mutations and gene interactions. As late as 1916, Bateson wrote, "The supposition that particles of chromatin, indistinguishable from each other and indeed almost homogeneous under any known test, can by their material nature confer all the properties of life, surpasses the range of even the most convinced materialism."

Sex Determination

One line of cytological observations, however, was providing further evidence linking the "particles of chromatin" with heredity. As early as the 1890s, cytologists noticed that male and female organisms often show chromosomal differences, and they began to speculate that these differences are related to sex determination.

13–7
How the sperm cell determines sex in an organism (such as the human organism) in which the male is heterogametic. (a) *At meiosis, every egg cell receives an X chromosome from the mother.* (b) *A sperm cell may receive either an X chromosome or a Y chromosome.* (c) *If a sperm cell carrying an X chromosome fertilizes the egg, the offspring will be female* (XX). (d) *If a sperm cell carrying a Y chromosome fertilizes the egg, the offspring will be male* (XY).

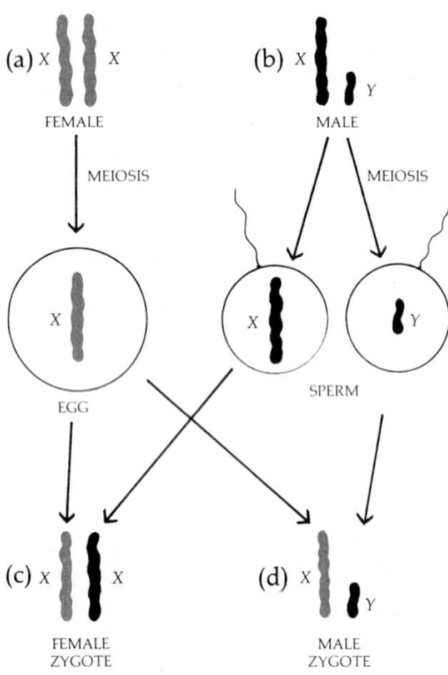

As Sutton had observed, the chromosomes of a diploid organism come in pairs. All of the pairs except one are the same in both males and females; these are called *autosomes*. In many organisms, one pair differs between males and females. These are known as the *sex chromosomes*. In many species, the two sex chromosomes are identical in the female but are dissimilar in the male, with one male sex chromosome resembling the female sex chromosomes and the other usually smaller and of a different shape. The sex chromosome that is similar in the cells of both males and females is called the X chromosome, and the unlike chromosome characteristic of the cells of males is called the *Y* chromosome. Thus we can characterize the two sexes as *XX* (female) and *XY* (male). In such species, the male is said to be *heterogametic*, since he can produce two types of gametes (Figure 13–7), and the female *homogametic*.

Not all organisms, however, have heterogametic males and homogametic females. In some insects, such as the grasshopper, which Sutton studied, there is no *Y* chromosome. In such cases, the females are characterized as *XX* and the males as *XO*. In birds, moths, and butterflies (and in occasional species in other groups), the chromosomes are reversed; the male has the two *X* chromosomes, and the female only one. The *Y* chromosome may or may not be present. In these organisms, it is the female that is heterogametic.

Human beings have 22 pairs of autosomes, which are structurally the same in both sexes. Females have a twenty-third matching pair, the sex chromosomes, *XX*. Human males, as their twenty-third pair, have one *X* and one *Y*. During meiosis, as each diploid spermatocyte undergoes meiotic division into four haploid sperm cells, two of the sperm cells receive *X* chromosomes and two receive *Y* chromosomes. The ovum always contains an *X* chromosome, since a human female does not normally possess the *Y* in any of her cells. Thus the zygote will become *XX* or *XY*, depending on whether an *X*-bearing sperm or a *Y*-bearing sperm fertilizes the egg. It is in this way that the sperm cell determines the sex of the future offspring, and it is the process of meiosis that governs the almost equal production of male and female babies. (In actual fact, the ratio of human male to human female births is about 106 to 100. The reason for this is not known, but it has been suggested that the male-determining sperm may have an advantage in getting to the egg.)

The correlation of the appearance of chromosomes with a particular characteristic—sex—lent strength, as you can see, to the gene-chromosome hypothesis. Stronger support was to come from a variety of studies carried out with the fruit fly, *Drosophila*.

THE GOLDEN AGE OF DROSOPHILA

Early in the 1900s, Thomas Hunt Morgan began a study of genetics at Columbia University, founding what was to be the most important laboratory in the field for several decades. By a remarkable combination of insight and good fortune, he selected *Drosophila melanogaster* as his experimental material.*

Drosophila means "lover of dew," although actually this useful little fly is not attracted by dew but feeds on the fermenting yeast that it finds in rotting fruit. The

* Biologists have often used for their experiments "insignificant" plants and animals—such as Mendel's pea plants, for instance, or Hertwig's sea urchins (page 103). Underlying this approach is the assumption that basic biological principles are universal, applying equally to all living things.

13–8
The fruit fly (Drosophila) *and its chromosomes. Fruit flies have only four pairs of chromosomes, a fact that simplified Morgan's experiments. Six of the chromosomes (three pairs) are autosomes (including the two small dot chromosomes in the center), and two are sex chromosomes.*

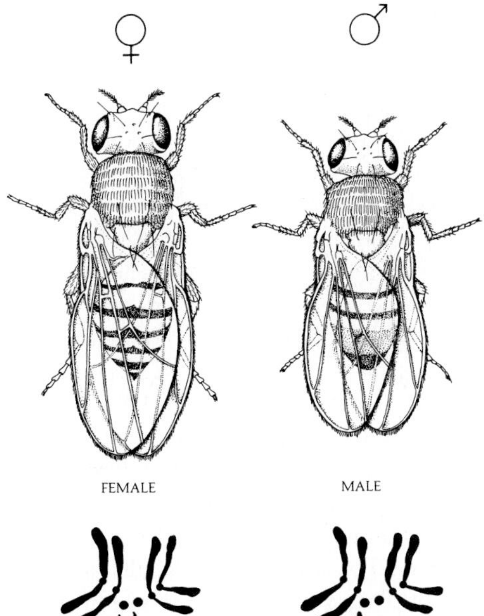

fruit fly was a likely choice for a geneticist since it is easy to breed and maintain. These tiny flies, each only 3 millimeters long, produce a new generation every two weeks. Each female lays hundreds of eggs at a time, and very large numbers can be kept in a half-pint bottle, as they were in Morgan's laboratory, known familiarly as the "Fly Room."

Also, *Drosophila* has only four pairs of chromosomes, a feature that turned out to be particularly useful, although Morgan could not have foreseen that. Three pairs of these are autosomes, and the fourth is an *XX* pair in the female and an *XY* pair in the male (Figure 13–8).

Sex-Linked Characteristics

The aim in Morgan's laboratory was, initially, to do breeding experiments similar to those Mendel had carried out with the pea plant. Such experiments involved examining under a magnifying lens hundreds—eventually tens of thousands—of individual fruit flies.

The investigators, at first, were looking for genetic differences among individual flies that they could study by interbreeding experiments. Shortly after Morgan established his colony, such a difference appeared. One of the prominent and readily visible characteristics of fruit flies is their brilliant red eyes. One day, a white-eyed fly, a mutant, appeared in the colony. The mutant, a male, was mated with a red-eyed female. All members of the first generation were red-eyed, indicating that the mutation was recessive. Morgan then crossbred the F_1 offspring, just as Mendel had done in his pea experiments. This is what resulted:

Red-eyed females	2,459
White-eyed females	0
Red-eyed males	1,011
White-eyed males	782

In short, the expected 3:1 ratio of dominant to recessive did not appear and, even more unexpected, all the white-eyed flies of the F_2 generation were males. Why were there no white-eyed females?

To explore the situation further, Morgan crossed the original white-eyed male with one of the F_1 females. The following results were obtained from this testcross:

Red-eyed females	129
White-eyed females	88
Red-eyed males	132
White-eyed males	86

In other words, females can be white-eyed; the trait behaves pretty much like a usual recessive one. So why were there no white-eyed females in the F_2 generation? Stop here a moment and see if you can answer this question. Morgan was able to.

Morgan and his coworkers examined these figures and, on the basis of them, formulated the following hypothesis: The gene for eye color is carried only on the *X* chromosome. (In fact, as it was later shown, the *Y* chromosome of *Drosophila* carries very little genetic information.) The allele for white eyes is a recessive. Thus a heterozygous female has red eyes—which is why there were no white-eyed females in the F_2 generation. However, a male that received an *X* chromosome carrying the allele for white eyes would always be white-eyed since no other allele would be present.

13-9
Offspring of a cross between a white-eyed female fruit fly and a red-eyed male fruit fly, illustrating what happens when a recessive allele, indicated by the white dot, is carried on an X chromosome. The F_1 females, with one X chromosome from the mother and one from the father, are heterozygous (Ww) and so will be red-eyed. But the F_1 males, with their single X chromosome received from the mother carrying the recessive (w) allele, will be white-eyed because the Y chromosome carries no gene for eye color. Thus the recessive allele on the X chromosome inherited from the mother will be expressed.

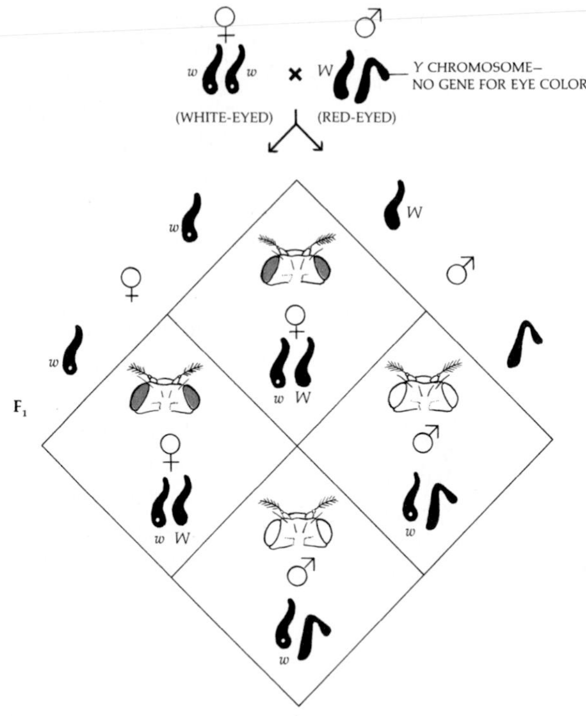

Further experimental crosses proved Morgan's hypothesis to be right (Figure 13-9). They also showed—in another example of pleiotropy—that white-eyed fruit flies are more likely to die before they reach adulthood than are red-eyed fruit flies, which explains their lower-than-expected numbers in the F_2 generation and the testcross.

These experiments introduced the concept of sex-linked traits, which are, as we shall see, important in the genetics of human beings as well as of fruit flies. More important, they convinced Morgan, and most other geneticists as well, that Sutton's hypothesis was right: Genes *are* on chromosomes.

Linkage

Mendel showed that certain pairs of alleles, such as those for round and wrinkled peas, assort independently of other pairs, such as those for yellow and green peas. However, as we noted previously, two genes (pairs of alleles) assort independently only if they are on different pairs of homologous chromosomes. (Note how fortunate Mendel was to have selected for study seven different traits, each affected by a single gene, with each of the seven genes located on a separate chromosome, especially given the fact that the garden pea has only seven different chromosomes, that is, $n = 7$. Modern geneticists suspect that Mendel may have had some results that he did not report because he did not know how to interpret them. However, his original papers were all lost.) Clearly, if two genes are on the same chromosome, then their alleles will not segregate independently. Such genes are said to be in the same linkage group.

13-10
Thomas Hunt Morgan at work in Columbia University's "Fly Room" in 1917. The camera-shy Morgan was photographed surreptitiously by a colleague who concealed a camera under a pile of milk bottles on his own desk. The half-pint milk bottles were used to house experimental Drosophila.

CALICO CATS, BARR BODIES, AND THE LYON HYPOTHESIS

A dark spot of chromatin—called a Barr body—can be seen at the outer edge of the nucleus of female mammalian cells in interphase. According to the Lyon hypothesis—named after Mary Lyon, who proposed it—this dark spot is an inactivated X chromosome. Early in embryonic life, according to Lyon, this inactivation occurs in one or the other X chromosome in each cell of the female mammal (except for those cells from which egg cells will form). Thus all the somatic cells of female mammals are not identical but are one of two types, depending on which of the X chromosomes is active.

Calico cats have coats that are both black and yellow. They are also almost always female. In cats, the alleles for black or yellow coat color are carried on the X chromosome, so calico cats neatly fit the Lyon hypothesis. There are, however, occasional male calicos. These are suspected of having an extra X chromosome, a supposition supported by the observation that they are almost always sterile.

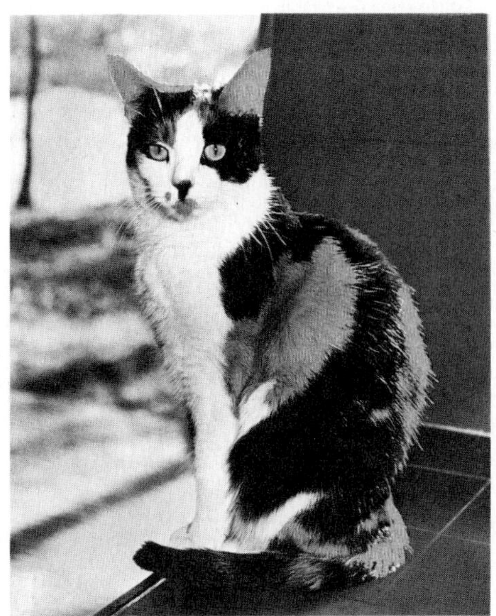

As increasing numbers of mutants were found in Columbia University's *Drosophila* collection, the mutations began to fall into four linkage groups, in accord with the four pairs of chromosomes visible in the cells. Moreover, one of the linkage groups was very small (see Figure 13–8).

Recombination

Large-scale studies of linkage groups soon revealed some unexpected difficulties. For instance, most fruit flies have light tan bodies and long wings, both of which are dominant traits. When these individuals were bred with mutant fruit flies having black bodies and short wings (both recessive traits), all the progeny had light tan bodies and long wings, as would be expected. Then the F_1 generation was inbred. Two outcomes seemed possible:

1. The genes for body color and wing length would be assorted independently, giving rise to Mendel's 9:3:3:1 ratios in the phenotypes and indicating that the genes for these traits were on different chromosomes.
2. The genes for the two traits would be linked. In this case, 75 percent of the flies would be tan with long wings and 25 percent, homozygous for both recessives, would be black with short wings.

In the case of these particular traits, the results closely resembled the second possibility, but they did not conform exactly. In a few of the offspring, the genes for these traits seemed to assort independently, not together; that is, some few flies appeared that were tan with short wings, and some that were black with long

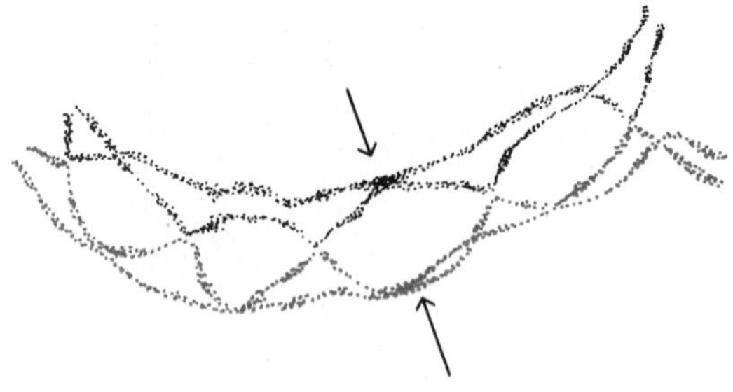

13–11
Homologous chromosomes of a grasshopper, as seen in prophase I. All four chromatids are visible. Crossing over—the exchange of genetic material—has probably occurred at the point at which these chromatids intersect. The arrows indicate the position of the centromeres.

wings. How could this be? Somehow genes that were presumed to be on the same chromosome had become separated.

To find out what was happening, Morgan tried a testcross, breeding a member of the F_1 generation with a homozygous recessive. If black and tan, long and short, assorted independently—that is, if they were on different chromosomes—25 percent of the offspring of this cross should be black with long wings, 25 percent tan with long wings, 25 percent black with short wings, and 25 percent tan with short wings. On the other hand, if the genes for color and wing size were on the same chromosome and so moved together, half of the testcross offspring should be tan with long wings and half should be black with short wings. But actually, as it turned out, over and over, in counts of hundreds of fruit flies resulting from such crosses, 41.5 percent were tan with long wings, 41.5 percent were black with short wings, 8.5 percent were tan with short wings, and another 8.5 percent were black with long wings.

Morgan was convinced by this time that genes are located on chromosomes. It now seemed clear that the genes for the two traits were located on the same chromosome since the traits did not show up in the 25:25:25:25 percentage ratios of independently assorted alleles. The only way in which the observed figures could be explained was if alleles could sometimes be exchanged between homologous chromosomes.

As we noted in Chapter 12, it now has been established that exchange of portions of homologous chromosomes—crossing over—takes place at the beginning of meiosis. Presumably, the actual exchange takes place when the homologues are closely grouped early in the first meiotic prophase. As the homologues begin to separate, chiasmata ("crosses") become visible where crossing over has occurred.

MAPPING THE CHROMOSOME

With the discovery of crossovers, it began to be clear not only that the genes are carried on the chromosomes, as Sutton had hypothesized, but also that they must be positioned at particular spots, or loci (singular, locus), on the chromosomes. Furthermore, the alleles of any given gene must occupy corresponding loci on homologous chromosomes. Otherwise, exchange of sections of chromosomes would result in genetic chaos rather than in an exact exchange of alleles.

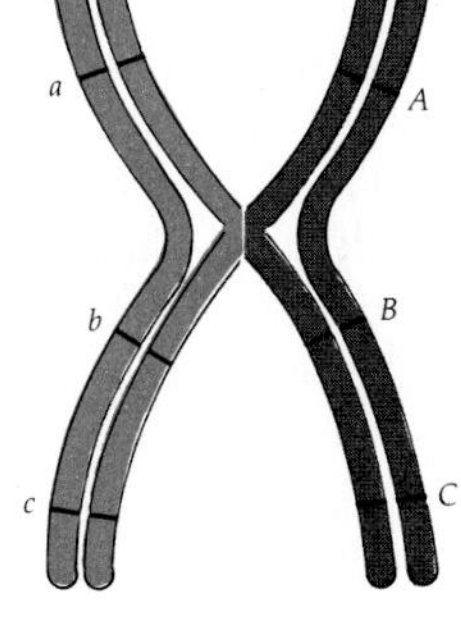

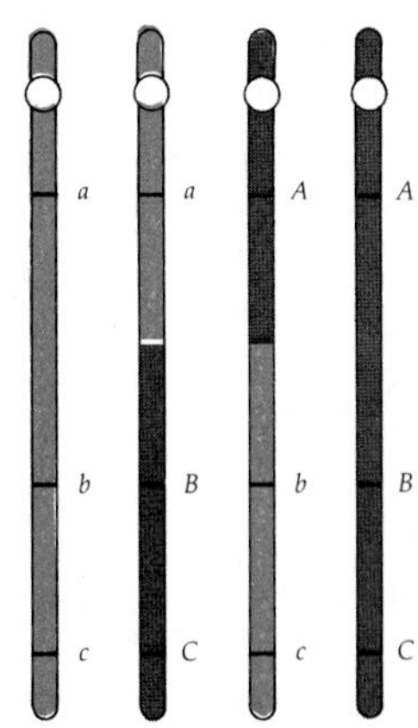

13–12
Crossing over takes place when breaks occur in the chromatids of homologous chromosomes during prophase I, when the chromosomes are paired. The broken end of each chromatid joins with the chromatid of a homologous chromosome. In this way, alleles are exchanged between chromosomes. The white circles symbolize centromeres.

As other traits were studied, it became clear that the percentage of recombinations between any two genes, such as those for tan body and long wing, was different from the percentage of recombinations between two other genes, such as those for tan body and long leg. In addition, as Morgan's experiments had shown, these percentages were fixed and predictable. It occurred to A. H. Sturtevant, who was an undergraduate working in Morgan's laboratory at this time, that the percentage of recombinations probably had something to do with the physical distances between the gene loci, or in other words, with their spacing along the chromosome. This concept opened the way to the "mapping" of chromosomes.

Sturtevant postulated (1) that genes are arranged in a linear series on chromosomes, like beads on a string; (2) that genes that are close together will be separated by crossing over less frequently than genes that are farther apart; and (3) that it should therefore be possible, by determining the frequencies of recombinations, to plot the sequence of the genes along the chromosome and the relative distances between them. In Figure 13–12, for example, you can see that in a crossover, the chances of a strand's breaking and rejoining with its homologous strand somewhere between *B* and *C* should be less likely than this happening somewhere between *A* and *C*. An analogy might be that of a traveler who maps landmarks along a road by recording the time it took to cover the distances between them.

CROSS	OFFSPRING		
AB × *ab*	*AB* + *ab*	(PARENTAL)	96%
	Ab + *aB*	(RECOMBINANT)	4%
AC × *ac*	*AC* + *ac*	(PARENTAL)	91%
	Ac + *aC*	(RECOMBINANT)	9%
BC × *bc*	*BC* + *bc*	(PARENTAL)	95%
	Bc + *bC*	(RECOMBINANT)	5%

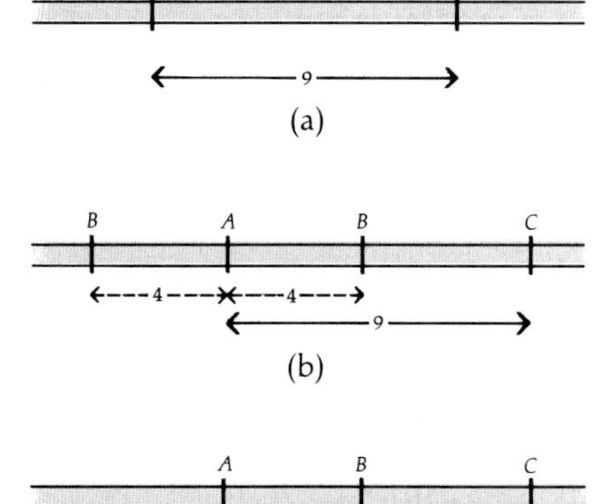

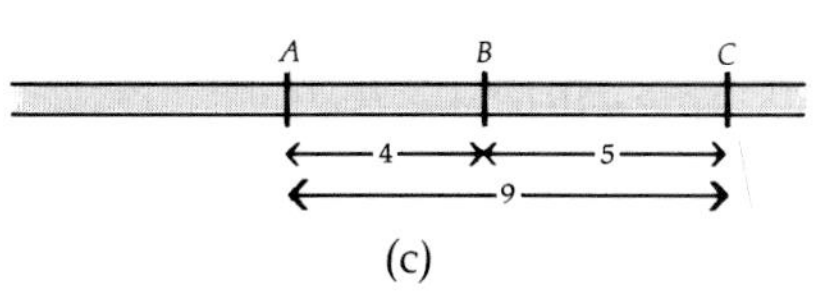

13–13
How to "map" a chromosome. Genes A *and* B *recombine with* a *and* b *in 4 percent of the offspring, and genes* A *and* C *with* a *and* c *in 9 percent of the offspring. Use as a unit of measure the distance that will give (on the average) one recombinant per 100 fertilized eggs.* (a) *Start with the largest number and establish the relative positions of* A *and* C *on the chromosome.* (b) A *and* B*, as you know from the data, are 4 units apart. Thus* B *could theoretically be either to the left of* A *or to the right.* (c) *However, if it were to the left,* B *and* C *would be 13 units apart, a distance that does not conform to the data.* B *must therefore be between* A *and* C.

13–14

A portion of the genetic map of Drosophila melanogaster, *showing relative positions of some of the genes on chromosome 2, as calculated by the frequency of recombinations. As you can see, more than one gene may affect a single characteristic, such as eye color.*

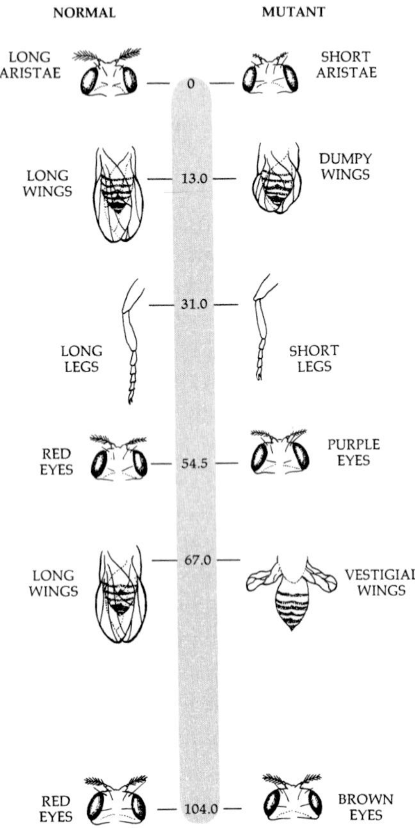

In 1913, Sturtevant began constructing chromosome maps using data from crossover studies in fruit flies. As a standard unit of measure, he arbitrarily took the map distance that would give (on the average) one recombination per 100 fertilized eggs. The genes with 10 percent recombination would be 10 units apart; those with 8 percent recombination would be 8 units apart. By this method, he and other geneticists constructed chromosome maps locating a variety of genes and their mutants in *Drosophila* (Figure 13–14). These map distances do not necessarily accurately reflect physical distances because breaking and rejoining of chromosome segments are more likely to occur at some sites on the chromosomes than at others. (To follow the analogy, the condition of the road would also influence our traveler.)

These important studies confirmed that not only are genes located on chromosomes, as Sutton hypothesized, but they also have fixed positions in a linear sequence.

GIANT CHROMOSOMES

In *Drosophila,* as in many other insects, certain cells do not divide during the larval (immature) stages of the insect. In such cells, however, the chromosomes continue to replicate, over and over again, but since daughter chromosomes do not separate from one another after replication, they simply become larger and larger. In 1933, such giant chromosomes were reported in the salivary glands of the larvae of *Drosophila.* As you can see in Figure 13–15, when stained these giant chromosomes are characterized by very distinctive dark and light bands.

These banding patterns became another useful tool for geneticists, enabling them to detect structural changes in the chromosomes themselves. For example, from time to time breaks occur in chromosomes, and portions of the chromosomes can become dislocated. Sometimes a whole segment is lost; usually the effect of such a loss—known as a *deletion*—is lethal. Sometimes a portion is transferred from one chromosome to another, which is known as *translocation.* (As we shall see in

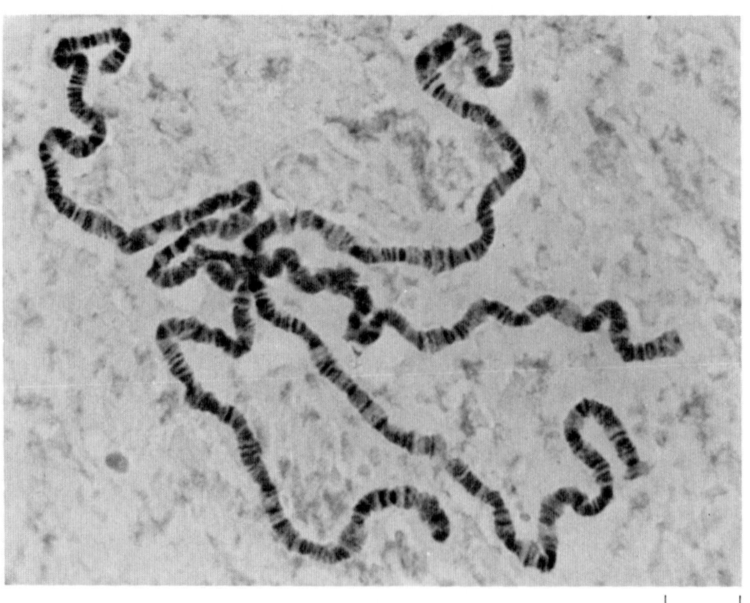

13–15

Chromosomes from the salivary gland of a Drosophila *larva, as shown by a photomicrograph. These chromosomes are 100 times larger than the chromosomes in ordinary somatic cells, and their details are therefore much easier to see. Because of the distinctive banding patterns, it is possible in some cases to associate genes with specific bands in particular chromosomes.*

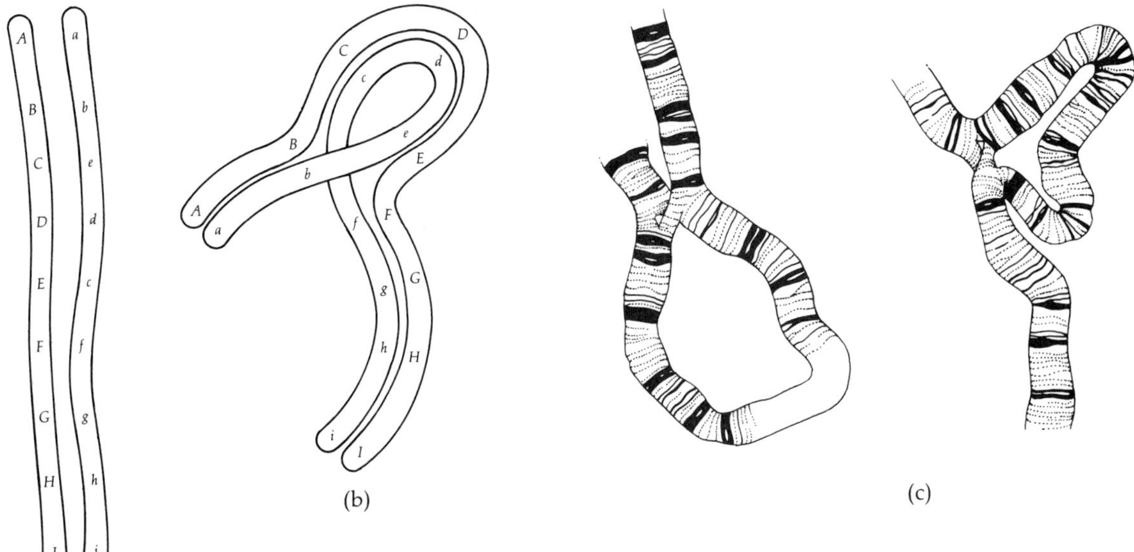

(a)

13–16
A chromosomal inversion occurs when a middle segment of the chromosome breaks off, is turned 180°, and rejoins by the "wrong" ends. (a) *An inversion results in a reversal of the sequence of alleles, as shown in the chromosome on the right.* (b) *When one member of a homologous pair contains a large inverted segment, that chromosome must loop inside the other for close pairing to occur.* (c) *In giant chromosomes, such loops are greatly magnified, making it possible to readily identify regions with inverted segments.*

Chapter 18, a specific translocation is associated with Down's syndrome in human beings.) In other cases, a double break in a chromosome occurs and a segment is turned 180° and then reincorporated in the chromosome; this phenomenon is known as <u>inversion</u>. The existence of such chromosomal aberrations had been hypothesized by Morgan's group on the basis of mapping studies. An inversion, for example, completely suppresses recombinations along the affected portion of the chromosome. Studies of the giant chromosomes of *Drosophila* confirmed the existence of these chromosomal aberrations (Figure 13–16), and it now became possible to assign genes to physical locations on the *Drosophila* chromosome.

Perhaps more important, almost a quarter of a century after the *Drosophila* work had begun, the fact that those little, seemingly homogeneous particles of chromatin are indeed the repositories of the mysteries of heredity was no longer only a brilliant hypothesis.

SUMMARY

The rediscovery of Mendel's work in 1900 was the catalyst for many new discoveries in genetics, leading to the modification and extension of some of Mendel's conclusions and, eventually, to the identification of chromosomes as the carriers of heredity.

Genes interact with one another and their environment. In incomplete dominance, the effects of both alleles of a single gene are apparent in the phenotype. Epistasis is an interaction between nonallelic genes. Gene effects may also be profoundly modified by the environment.

Mutations are abrupt changes in the genotype. They are the ultimate source of the genetic variations that Mendel studied and that provide the raw material for evolution.

Many characteristics are under the control of a number of separate genes. This phenomenon is known as polygenic inheritance. Such traits typically show continuous variation, as represented by a bell-shaped curve. Conversely, many genes affect two or more superficially unrelated characteristics; this property of a gene is known as pleiotropy. Some genes, such as those for blood type, have multiple alleles.

Strong support for the hypothesis that genes are on the chromosomes came from studies by Morgan and his group on the fruit fly *Drosophila. Drosophila,* because it is easy to breed and maintain, has been used in a wide variety of genetic studies. It has four pairs of chromosomes; three pairs are structurally the same in both sexes, but the fourth pair, the sex chromosomes, is different. In fruit flies, as in many other species (including humans), the two sex chromosomes are *XX* in females and *XY* in males.

At the time of meiosis, the sex chromosomes are segregated. Each egg cell receives an *X* chromosome, but half the sperm cells receive *X* chromosomes and half receive *Y* chromosomes. Thus in fruit flies, humans, and many (but not all) other organisms, it is the paternal gamete that determines the sex of the embryo.

In the early 1900s, experiments with mutations in the fruit fly showed that certain characteristics are sex-linked, that is, their genes are carried on the sex chromosomes. Recessive genes carried on the *X* chromosome appear in the phenotype far more often in males than in females (the *Y* chromosome carries less genetic information); a female heterozygous for a sex-linked characteristic will show the dominant trait, whereas a single recessive allele in the male, if carried on the *X* chromosome, will result in a recessive phenotype since no other allele is present.

Some genes assort independently in breeding experiments, and others tend to remain together (in violation of Mendel's principle of independent assortment). Genes that do not segregate at meiosis (because they are on the same chromosome) are said to be in the same linkage group.

Genes are sometimes exchanged between homologous chromosomes. Such recombinations can take place only if (1) the genes are arranged in a fixed linear array along the length of the chromosomes, and (2) the alleles of a given gene are at corresponding sites (loci) on homologous chromosomes. On the basis of these assumptions, chromosome maps, showing the relative positions of gene loci along *Drosophila* chromosomes, were developed from recombination data provided by breeding experiments.

Genetic studies indicated that chromosome breaks other than those resulting in crossovers may sometimes occur. A portion of a chromosome may be lost, or deleted, it may be translocated to another chromosome, or it may be inverted. Studies of the giant chromosomes of *Drosophila* larvae provided visual confirmation of these changes, as well as the final, conclusive evidence that the chromosomes are the carriers of the genetic information.

QUESTIONS

1. Distinguish among the following: pleiotropy/polygenic inheritance; incomplete dominance/epistasis; mutation/mutant/mutagen; heterogametic/homogametic; sex chromosome/autosome; inversion/deletion/translocation.

2. The so-called "blue" (really gray) Andalusian variety of chicken is produced by a cross between the black and white varieties. Only a single pair of alleles is involved. What color chickens (and in what proportions) would you expect if you crossed two blues? If you crossed a blue and a black? Explain.

3. In one strain of mice, skin color is determined by five different pairs of alleles. The colors range from almost white to dark brown. Would it be possible for any given pair of mice to produce offspring darker or lighter than either parent? Explain.

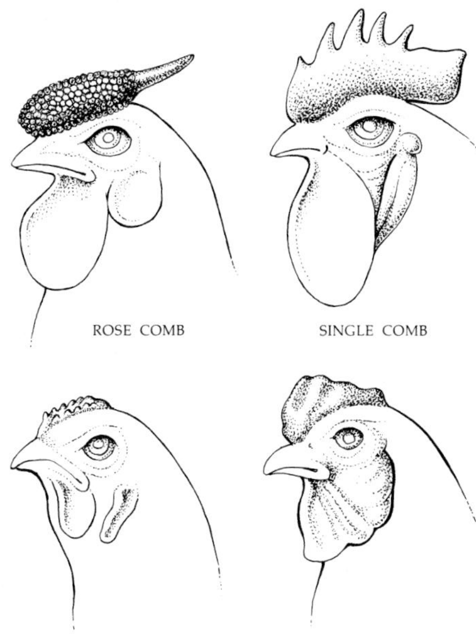

4. In chickens, two pairs of alleles determine comb shape, *R, r* and *P, p. RR* or *Rr* results in rose comb, whereas *rr* produces single comb. *PP* or *Pp* produces pea comb, and *pp* produces single comb. When *R* and *P* occur together, they produce a new type of comb: walnut. What would be the genotype of the F_1 cross resulting from *RRpp* × *rrPP*? The phenotype? If F_1 hybrids were crossbred, what would be the probable distribution of genotypes? Of phenotypes? (Illustrate this cross with a Punnett square.)

5. Draw a diagram similar to Figure 13–7 indicating sex determination in a robin.

6. A couple has three girls. What are the chances that the next child will be a boy?

7. Suppose you would like to have a family consisting of two girls and a boy. What are your chances, assuming you have no children now? If you already have one boy, what are your chances of completing your family as planned?

8. Segregation of alleles can occur at either of two stages of meiosis. Name the two stages and explain what happens in each of them.

9. Does crossing over necessarily result in a recombination of alleles? Explain.

10. Height and weight in animals follow a distribution similar to that shown in Figure 13–5. By inbreeding large animals, breeders are usually able to produce some increase in size among their stock. But after a few generations, increase in size characteristically stops. Why?

11. The genes for coat color in cats are carried on the X chromosome. Black *(b)* is the recessive and yellow *(B)* is the dominant. What coat colors would you expect in the offspring of a cross between a black female and a yellow male? What coat colors would you expect in the sons of a calico female, regardless of the coat color of the father?

12. Judging from the colors of the calico cat (page 271), at what stage in development does the X chromosome become inactivated?

13. In a series of breeding experiments, a linkage group composed of genes *A, B, C, D,* and *E* was found to show approximately the following recombination frequencies:

		GENE				
		A	*B*	*C*	*D*	*E*
GENE	*A*	—	8	12	4	1
	B	8	—	4	12	9
	C	12	4	—	16	13
	D	4	12	16	—	3
	E	1	9	13	3	—

RECOMBINATIONS PER 100 FERTILIZED EGGS

Using Sturtevant's standard unit of measure, "map" the chromosome.

14. You and a geneticist are looking at a mahogany-colored Ayrshire cow with a newly born red calf. You wonder if it is male or female, and the geneticist says it is obvious from the color which sex the calf is. He explains that in Ayrshires the genotype *AA* is mahogany and *aa* is red, but the genotype *Aa* is mahogany in males and red in females. What is he trying to tell you—that is, what sex is the calf? What are the possible phenotypes of the calf's father?

CHAPTER 14

The Path to the Double Helix

14–1
Electron micrograph of a bacterial virus (bacteriophage), showing its tadpole-shaped coat, composed entirely of protein, surrounded by its single molecule of DNA. The protein jacket was burst open and the DNA released by osmotic shock, produced by transferring the bacteriophage to distilled water. Studies with phages provided the essential confirmation that DNA carried the hereditary information.

By the early 1940s the existence of genes and the fact that they were in chromosomes were no longer in doubt. But what were these genes? What did they really do? A turning point in genetics came when scientists began to focus on the question of how it was possible for these little particles of chromatin to be the bearers of what they had come to realize must be an enormous amount of extremely complex information.

THE CHEMISTRY OF HEREDITY

The chromosomes, like all the other parts of a living cell, are composed of atoms arranged into molecules. As we noted previously, some scientists, a number of them eminent in the field of genetics, thought it would be impossible to understand the complexities of heredity in terms of the structure of "lifeless" chemicals. (Although they did not use the term to describe themselves, they were, in fact, the vitalists—page 172—of the twentieth century.) It was not until 1953 that all remaining support for these arguments was swept away.

The Language of Life

Chemical analysis revealed, as we mentioned previously, that the eukaryotic chromosome consists of both DNA and protein, about half and half. Thus, both were equal candidates for the role of the genetic material. Proteins seemed the more likely choice, because of their greater chemical complexity. Speculative thinkers in the field of biology were quick to point out that the amino acids, the number of which was so provocatively close to the number of letters in our own alphabet, could be arranged in a variety of different ways. They were seen as making up a sort of language—"the language of life"—that spelled out the directions for all the many activities of the cell. Many prominent investigators, particularly those who had been studying proteins, believed that the genes themselves were proteins, that the chromosomes contained master models of all the proteins that would be required by the cell, and that enzymes and other proteins active in cellular life were copied from these master models. This was a logical hypothesis, but, as it turned out, it was wrong.

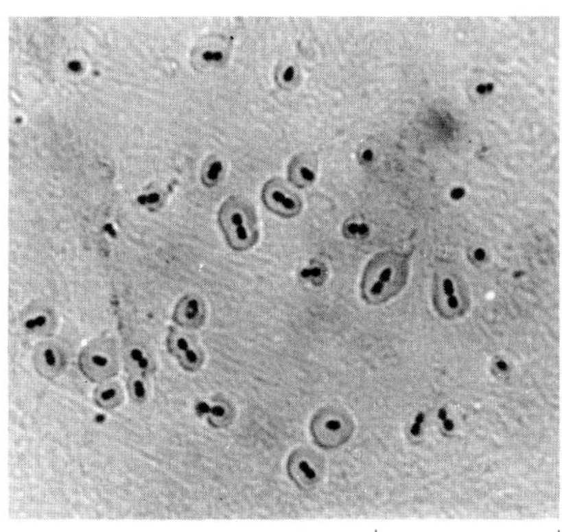

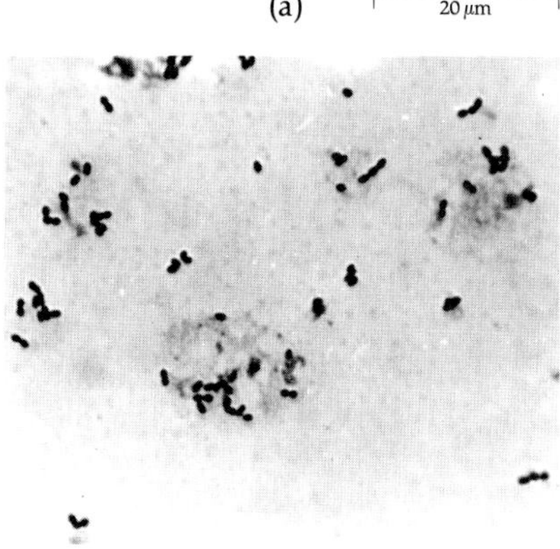

14–2
(a) *Encapsulated and* (b) *nonencapsulated forms of pneumococci. The capsule is made up of polysaccharides deposited outside the cell wall. The encapsulated form, which is resistant to phagocytosis by white blood cells, produces pneumonia; the mutant, nonencapsulated form is harmless.*

THE DNA TRAIL

The Transforming Factor

To trace the beginning of the hypothesis that proved right, it is necessary to go back to 1928 and pick up an important thread in modern biological history. In that year, an experiment was performed that seemed at the time very remote from either biochemistry or genetics. Frederick Griffith, a public health bacteriologist, was studying the possibility of developing vaccines against the bacterial cells, pneumococci, that cause one kind of pneumonia. In those days, before the development of modern antibiotics, bacterial pneumonia was a serious disease, the grim "captain of the men of death."

Pneumococci, as Griffith knew, come in either virulent (disease-causing) forms with capsules (polysaccharide coats) or nonvirulent (harmless) forms without capsules (Figure 14–2). (It is known now that the nonencapsulated pneumococcus is a mutant form, but in Griffith's time the term was not applied to bacteria.) Griffith was interested in finding out whether injections of heat-killed virulent pneumococci, which do not cause disease, could be used to immunize against pneumonia. In the course of various experiments, he performed one that gave him very puzzling results. He injected mice simultaneously with heat-killed virulent bacteria and with living but nonvirulent bacteria, each of which was harmless—but all the mice died. When Griffith performed autopsies on them, he found their bodies filled with living encapsulated (and therefore virulent) bacteria (Figure 14–3). Had the dead virulent cells come back to life or had something been passed from them to the living nonvirulent cells that endowed the living cells with the capacity to make capsules and therefore to be virulent?

Within the next few years it was shown that the same phenomenon could be reproduced in the test tube and these questions could be answered. It was found that extracts from the killed encapsulated bacteria, when added to the living harmless bacteria, could convert them to the virulent type with the ability to make capsules. Furthermore, once converted, they could transmit this characteristic to their progeny.

One of the laboratories that worked on the nature of this *transforming factor*, as it came to be called, was that of O. T. Avery at Rockefeller University. After almost a decade of patient chemical isolation and analysis, Avery and his coworkers were convinced that the chemical substance in the cellular extracts of killed bacteria that transmitted the new genetic quality was the molecule known as *deoxyribonucleic acid.* Subsequent experiments showed that a variety of genetic traits could be passed from one colony of bacterial cells to members of a similar colony by means of isolated deoxyribonucleic acid, which soon took on the now familiar abbreviation of DNA.

The Nature of DNA

At this point, we shall begin to examine the chemical structure of DNA. It was largely because of the seeming simplicity of its structure, as compared to that of proteins, that Avery's experiment, although beautifully designed and executed, did not receive full recognition for almost a decade.

DNA was first discovered in that same remarkable decade in which Darwin published *The Origin of Species* and Mendel presented his results to an audience of forty at the Natural History Society in Brünn. In 1869, a German chemist, Friedrich Miescher, had extracted a substance from the nuclei of cells that was white, slightly

14–3
Discovery of the transforming factor, a substance that can transmit genetic characteristics from one cell to another, resulted from studies of pneumococci, pneumonia-causing bacteria. One strain of these bacteria has capsules (protective outer layers); another does not. The capacity to make capsules and cause disease is an inherited characteristic, passed from one bacterial generation to another as the cells divide. (a) *Injection into mice of encapsulated pneumococci killed the mice.* (b) *The nonencapsulated strain produced no infection.* (c) *If the encapsulated strain was heat-killed before injection, it too produced no infection.* (d) *If, however, heat-killed encapsulated bacteria were mixed with live nonencapsulated bacteria and the mixture was injected into mice, the mice died.* (e) *Blood samples from the dead mice revealed live encapsulated pneumococci. Something had been transferred from the dead bacteria to the live ones that endowed them with the capacity to make capsules and cause pneumonia. This "something" was later isolated and found to be DNA.*

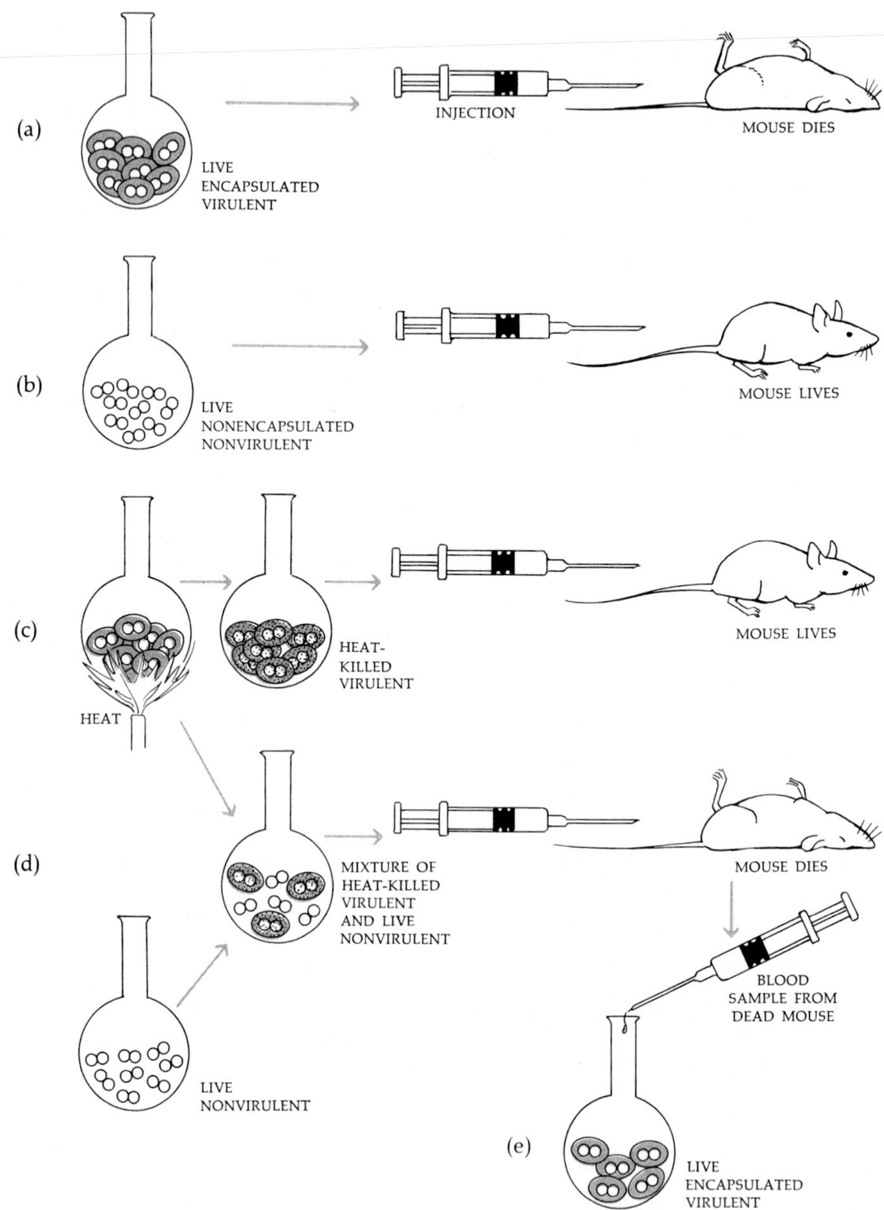

acidic, and contained phosphorus. He called it nucleic acid, which was later amended to deoxyribonucleic acid to distinguish it from a closely related chemical, ribonucleic acid (RNA), which subsequently was also isolated from cells.

In 1914, another German, Robert Feulgen, discovered that DNA had a remarkable attraction for a red dye called fuchsin, but he considered this finding so unimportant that he did not trouble to report it for a decade. Feulgen staining, as it was called when it finally made its way into use, revealed that DNA was present in all cells and was characteristically located in the chromosomes.

There was no particular interest in DNA for several decades since no role had been postulated for it in cellular metabolism. Most of the work on its chemistry was carried out by the great biochemist P. A. Levene in the 1920s. He showed that

14–4

(a) *A nucleotide is made up of three different parts: a nitrogenous base, a sugar, and a phosphate.* (b) *Each nucleotide in DNA contains one of the four possible nitrogenous bases, a deoxyribose sugar, and a phosphate.*

(a) PHOSPHATE — SUGAR — NITROGENOUS BASE

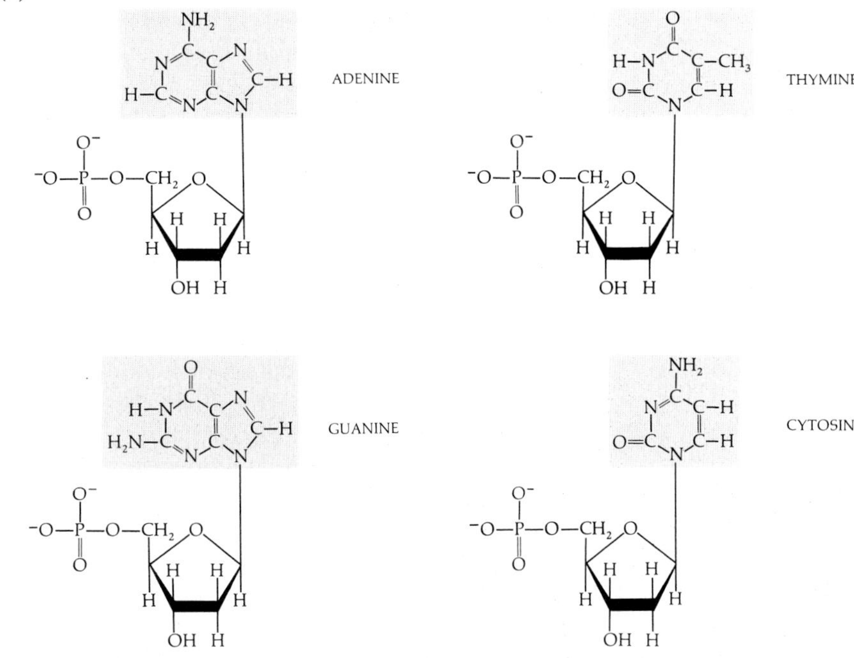

14–5

Max Delbrück and Salvador Luria at Cold Spring Harbor Laboratory of Quantitative Biology in 1953. They shared the Nobel Prize with A. D. Hershey in 1969 for "their discoveries concerning the replication mechanism and the genetic structure of viruses."

DNA could be broken down into four nitrogenous (nitrogen-containing) bases—adenine and guanine (the purines) and thymine and cytosine (the pyrimidines)—a five-carbon sugar, and a phosphate group. From the proportions of these components, he made two deductions, one correct and one incorrect:

1. Each nitrogenous base is attached to a molecule of sugar, which, in turn, is attached to a phosphate group to form a single molecule, a nucleotide (Figure 14–4). This deduction was right.
2. Since, in all the samples he measured, the proportions of the nitrogenous bases were approximately equal, he concluded that all four nitrogenous bases must be present in nucleic acid in equal quantity. Furthermore, he hypothesized that these molecules must be grouped in clusters of four—a tetranucleotide, he called it—that repeated over and over again along the length of the molecule. This deduction, which was wrong, dominated scientific thinking about the nature of DNA for more than a decade.

Because Levene's tetranucleotide theory was given great weight by his renown as a biochemist, biologists as a whole were slow to recognize the importance of Avery's experiment. Avery, like Mendel before him, was a traveler bearing an odd tale that did not fit.

The Bacteriophage Experiments

In 1940, Max Delbrück and Salvador Luria, both of whom had left Europe in the mass intellectual exodus of the 1930s, initiated a series of studies with another "fit material," destined to become as important to genetic research as the garden pea and the fruit fly. The fit material was a group of viruses that attacked bacterial cells and were therefore known as *bacteriophages*, "bacteria eaters." Every known type of bacterial cell is preyed upon by its own type of bacterial virus, and many bacteria are host to many different kinds of viruses. Delbrück, Luria, and the group that joined them in these studies agreed to concentrate on a series of seven related viruses that attacked *Escherichia coli,* a normal inhabitant of the human intestine. These viruses were numbered T1 through T7, with the T standing simply for "type." As it turned out, most of the early work was done on T2 and T4, which became known as the T-even bacteriophages.

These viruses were inexpensive to work with, easy to maintain in the laboratory, and demanded little space or equipment. Furthermore, they were phenomenal at reproducing themselves. Twenty-five minutes from the time a single virus infected a bacterial cell, that cell would burst open, releasing a hundred new viruses, all exact copies of the original virus. Another advantage (which was not discovered until after the research was begun) was that this group of bacteriophages has a highly distinctive shape (Figure 14–6a), and so can be readily identified with the electron microscope.

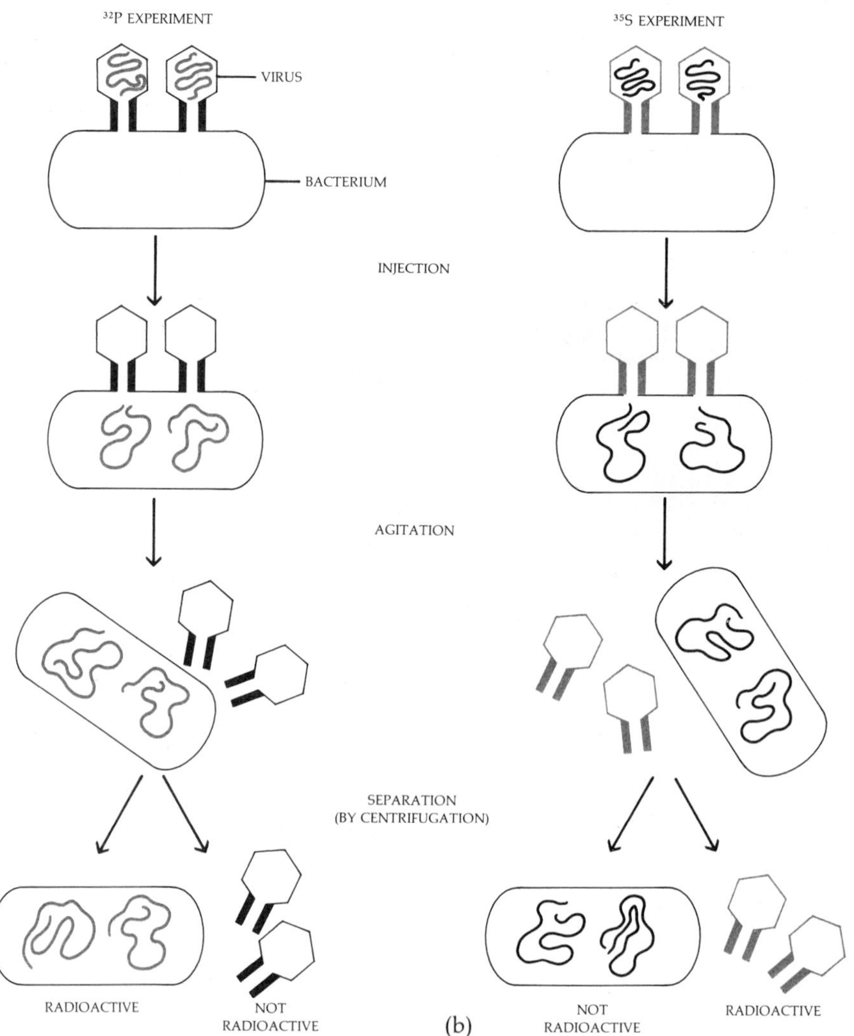

14–6
(a) *Electron micrograph of a T4 bacteriophage. Notice the highly distinctive "tadpole" shape. Each bacteriophage consists of a head, which appears hexagonal in electron micrographs, and a complex tail.* (b) *A summary of the Hershey-Chase experiments demonstrating that DNA is the hereditary material of a virus.*

According to electron-microscope studies of infected *E. coli* cells (broken open at regular intervals after infection), the bacteriophages do not multiply like bacteria. Except for a few fragments, they disappear the moment after infection, and for the first 10 to 11 minutes of the infectious cycle, not a single virus can be seen within the bacterial cell. Then, depending on when the cell is opened during the course of the infection, increasing numbers of completed bacteriophages can be seen and, mixed with them, odds and ends that appear to be bits of incomplete bacteriophages.

Chemical analysis of the bacteriophages revealed that they consist quite simply of DNA and of protein, the two leading contenders in the 1940s for the role of the genetic material. The chemical simplicity of the bacteriophage offered geneticists a remarkable opportunity. The viral genes—the hereditary material by which new viruses are made within the bacterial cell—had to be carried either on the protein or on the DNA. If it could be determined which of the two it was, then the gene would be chemically identified. In 1952, a simple but ingenious experiment was carried out.

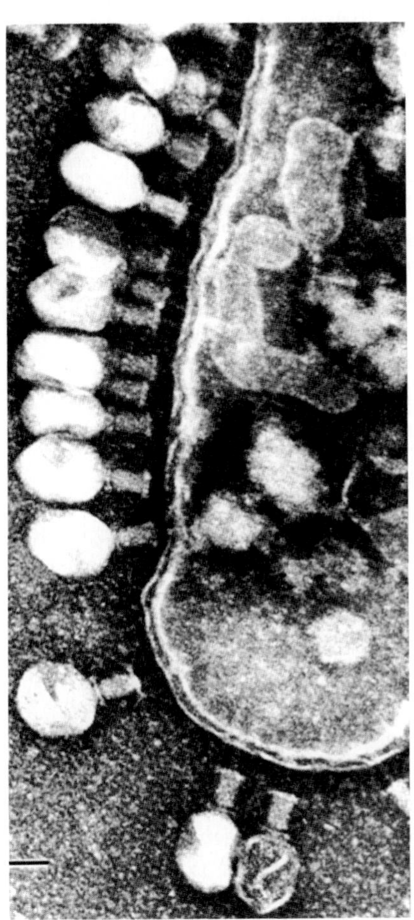

14–7
Electron micrograph of T4 bacteriophages attacking a cell of E. coli. *The viruses are attached to the bacterial cell by their tails. Some of the viruses have injected their DNA into the cell, as indicated by their empty heads. A complete cycle of virus infection takes only about 25 minutes. At the end of that period about a hundred new virus particles are released from the cell.*

Alfred D. Hershey and his laboratory assistant, Martha Chase,* prepared two separate samples of viruses, one in which the DNA was labeled with the radioactive isotope ^{32}P and the other in which the protein was labeled with the radioactive isotope ^{35}S. These were made by growing the *E. coli* host on a medium that contained the radioactive isotope. After a cycle of multiplication, the newly formed viruses all contained some of the radioactive isotope in place of the common nonradioactive isotope. If you recall the chemical structure of nucleic acids and proteins, you will note that DNA contains phosphorus but no sulfur, while the amino acid components of proteins contain no phosphorus but two of them (methionine and cysteine) contain sulfur. Thus ^{32}P and ^{35}S can serve as specific labels that distinguish DNA from protein.

One culture of bacteria was infected with ^{32}P-labeled phage and another with ^{35}S-labeled phage (Figure 14–6b). After the infectious cycle had begun, the cells were agitated in a blender and then spun down in a centrifuge to separate them from any viral material remaining outside the cells. The two samples—one containing extracellular material and the other intracellular material—were then tested for radioactivity. Hershey and Chase found that the ^{35}S had stayed outside the bacterial cells with the empty viral coats and the ^{32}P had entered the cells, infected them, and caused the production of new virus progeny. It was therefore concluded that the genetic material of the virus is DNA rather than protein.

Electron micrographs (Figure 14–7, for example) have now confirmed that the T4 bacteriophage attaches to the bacterial cell wall by its tail. They also indicate that the phage injects its DNA into the cell, leaving the empty protein coat on the outside. In short, the protein is just a container for the bacteriophage DNA. It is the DNA of the bacteriophage that enters the cell and carries the complete hereditary message of the virus particle, directing formation of new viral DNA and new viral protein.

* We are including the names of the scientists involved in these experiments, not only to give credit where it is due, but also because the names have become synonymous with the work. What we are describing now are the Hershey-Chase experiments.

Further Evidence for DNA

The role of DNA in transformation and in viral replication formed very convincing evidence that DNA was the chemical basis of the gene. Two other lines of experimental work also helped to lend weight to the argument. Alfred Mirsky, in a long series of careful studies conducted at Rockefeller University, showed that all the tissue cells of any given species contain equal amounts of DNA. The only exceptions are the gametes, which regularly contain just half as much DNA as the other cells of the same species, and certain unusual cells, such as polyploid cells and cells with giant chromosomes (see page 274).

Chargaff's Results

A second important series of contributions was made by Erwin Chargaff of Columbia University's College of Physicians and Surgeons. Chargaff analyzed the purine and pyrimidine content of the DNA of many different kinds of living things and found that, in contradiction to Levene's conclusions (see page 281), the nitrogenous bases do not always occur in equal proportions. The proportion of nitrogenous bases is the same in all cells of a given species but varies from one species to another. Therefore variations in base composition could very well provide a "language" in which the instructions controlling cell growth could be written. Some of Chargaff's results are reproduced in Table 14–1. Can you, by examining these figures, notice anything interesting about the proportions of purines and pyrimidines?

The Hypothesis Is Confirmed

The explanation of the way that the genetic information is contained in DNA was to be found in the structure of the DNA molecule. To fulfill its biological role, genetic material has to meet at least four requirements:

1. It must carry genetic information from cell to cell and from generation to generation. Further, it must carry a great deal of information. Consider how many instructions must be contained in the set of genes that directs, for example, the development of an elephant, or a tree, or even a *Paramecium.*
2. It must contain information for producing a copy of itself, for it is copied with every cell division and with great precision.
3. On the other hand, it must sometimes mutate. When a gene changes, that is, when a "mistake" is made, the "mistake" must be copied as faithfully as was the original. This is a most important property, for without the capacity to replicate "errors," there could be no evolution by natural selection.
4. There must be some mechanism for decoding the stored information and translating it into action in the living cell.

It was when the DNA molecule was found to have the size, the configuration, and the complexity required to code the tremendous store of information needed by living things and to make exact copies of this code that DNA became widely accepted as the genetic material.

The scientists primarily responsible for working out the structure of the DNA molecule were James Watson and Francis Crick, and their feat is one of the milestones in the history of science.

Table 14-1 Composition of DNA in Several Species

SOURCE	PURINES		PYRIMIDINES	
	ADENINE	GUANINE	CYTOSINE	THYMINE
Human being	30.4%	19.6%	19.9%	30.1%
Ox	29.0	21.2	21.2	28.7
Salmon sperm	29.7	20.8	20.4	29.1
Wheat germ	28.1	21.8	22.7	27.4
E. coli	24.7	26.0	25.7	23.6
Sea urchin	32.8	17.7	17.3	32.1

THE WATSON-CRICK MODEL

In the early 1950s, a young American scientist, James Watson, went to Cambridge, England, on a research fellowship to study problems of molecular structure. There, at the Cavendish Laboratory, he met physicist Francis Crick. Both were interested in DNA, and they soon began to work together to solve the problem of its molecular structure. They did not do experiments in the usual sense but rather undertook to examine all the data about DNA and to attempt to unify them into a meaningful whole.

The Known Data

By the time Watson and Crick began their studies, quite a lot of information on the subject had already accumulated:

1. The DNA molecule was known to be very large, and also very long and thin, and to be composed of nucleotides containing the nitrogenous bases adenine, guanine, thymine, and cytosine.
2. According to Levene's interpretation of his data, these nucleotides were assembled in repeating units of four.
3. Linus Pauling, in 1950, had shown that a protein's component chains of amino acids are often arranged in the shape of a helix and are held in that form by hydrogen bonds between successive turns of the helix. Pauling had suggested that the structure of DNA might be similar.
4. X-ray diffraction studies of DNA (Figure 14-8) from the laboratories of Maurice Wilkins and Rosalind Franklin at King's College, London, showed markings that almost certainly reflected the turns of a giant helix. (Neither Luria nor Pauling was permitted to visit England at that time and so did not see the x-ray diffraction photographs. The McCarthy era kept these two without passports. It has been suggested that the U.S. Passport Office may have determined the winners in the race for the double helix.)
5. Also crucial were the data of Chargaff indicating, as you perhaps noticed in Table 14-1, that (within experimental error) the amount of adenine is the same as the amount of thymine, and the amount of guanine is the same as the amount of cytosine: A = T and G = C.

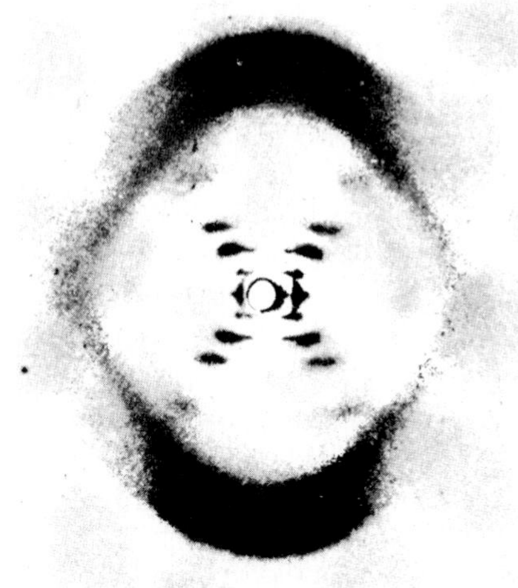

14-8
X-ray diffraction photograph of DNA taken by Rosalind Franklin. The reflections crossing in the middle indicate that the molecule is a helix. The heavy dark regions at the top and bottom are due to the closely stacked bases perpendicular to the axis of the helix.

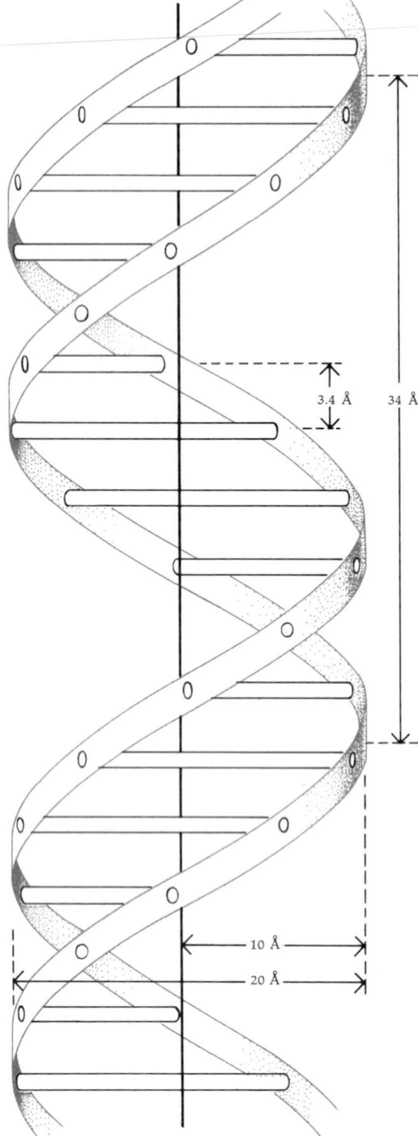

14–9
The double-stranded helical structure of DNA, as first presented in 1953 by Watson and Crick. The framework of the helix is composed of the sugar-phosphate units of the nucleotides. The rungs are formed by the four nitrogenous bases adenine and guanine (the purines) and thymine and cytosine (the pyrimidines). Each rung consists of two bases. Knowledge of the distances between the atoms, determined from x-ray diffraction pictures, was crucial in establishing the structure of the DNA molecule.

Building the Model

From these data, some of them contradictory, Watson and Crick attempted to construct a model of DNA that would fit the known facts and explain the biological role of DNA. In order to carry the vast amount of genetic information, the molecules should be heterogeneous and varied. Also, there must be some way for them to replicate readily and with great precision so that faithful copies can be passed from cell to cell and from parent to offspring, generation after generation.

On the other hand, Watson and Crick could not be sure that the chemical structure of DNA actually would reflect its biological function. After all, this idea had never really been tested rigorously. Perhaps DNA was merely some sort of biological clay on which some outside "vital force" operated. "In pessimistic moods," Watson has recalled, "we often worried that the correct structure might be dull—that is, that it would suggest absolutely nothing."

It turned out, in fact, to be unbelievably "interesting." By piecing together the various data, they were able to deduce that DNA does not have a single-stranded helix structure, as do proteins, but is an exceedingly long, entwined double helix.

The banister of a spiral staircase forms a single helix. If you take a ladder and twist it so the rails become spirals, keeping the rungs perpendicular, this would form a crude model of the molecule (Figure 14–9). The two rails, or sides, of the ladder are made up of alternating sugar and phosphate molecules. The perpendicular rungs of the ladder are formed by the nitrogenous bases—adenine (A), thymine (T), guanine (G), and cytosine (C)—one base for each sugar-phosphate, as Levene had shown, and two bases forming each rung. The nucleotides along any one chain of the double helix can occur in any order: ATGCGTACATTGCCA, and so on (Figure 14–10). Since a DNA molecule may be many thousands of nucleotides long, there is a possibility for great variety, one of the primary requirements for the molecule. Note also that each phosphate group is attached to one sugar at the 5′ position (the fifth carbon in the ring) and the other at the 3′ position. Thus the chain has a 5′ end and a 3′ end.

The paired bases meet across the helix and are joined together by hydrogen bonds, the relatively weak, omnipresent bonds that Pauling had demonstrated in his studies of the secondary structures of proteins. The distance between the two sides, or railings, according to Wilkins's measurements, was 20 angstroms. Two purines in combination would take up more than 20 angstroms (since they have two nitrogen rings), and two pyrimidines would not reach all the way across (since they have only one nitrogen ring). But if a purine paired in each case with a pyrimidine, there would be a perfect fit. The paired bases—the "rungs" of the ladder—would therefore always be purine-pyrimidine combinations.

The most exciting discovery came, however, when Watson and Crick set out to match the pairs of bases. They encountered another interesting and important restriction. Not only could purines not pair with purines and pyrimidines not pair with pyrimidines, but because of the configurations of the molecules, adenine could pair only with thymine, and guanine only with cytosine (Figure 14–10). Look at Chargaff's data (Table 14–1, page 285) again and see how well these chemical requirements confirm his data.

The double-stranded structure is shown in Figure 14–11. Note that the two strands run in opposite directions; that is, the direction from the 5′ end to the 3′ end of each strand is opposite. The strands are antiparallel.

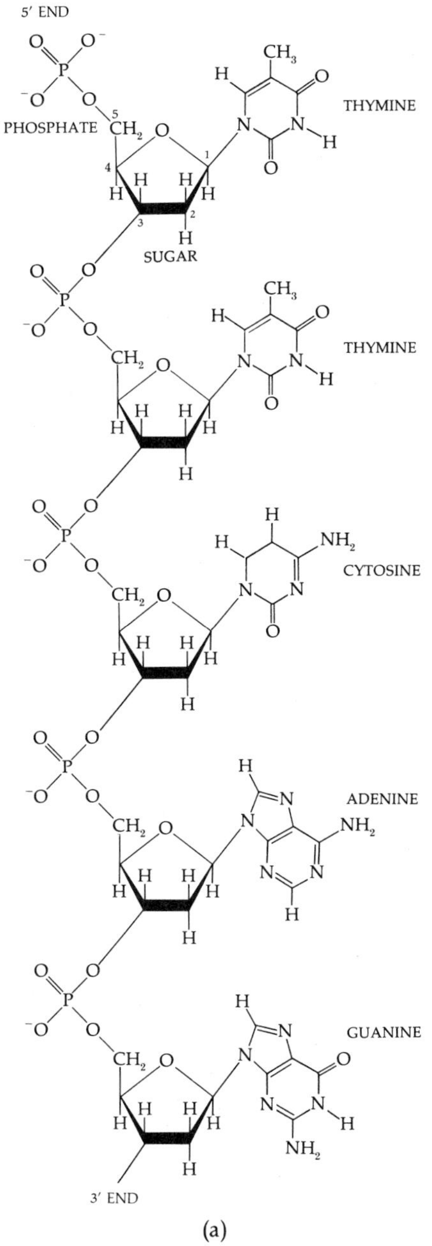

14–10
(a) *The structure of a portion of one strand of a DNA molecule. Each nucleotide consists of a sugar, a phosphate group, and a purine or pyrimidine base. Note the repetitive sugar-phosphate-sugar-phosphate sequence that forms the backbone of the molecule. Each phosphate group is attached to the 5′-carbon of one deoxyribose sugar and to the 3′-carbon of the deoxyribose sugar in the adjacent nucleotide. The sequence of bases varies from one DNA molecule to another. In the figure, the order of nucleotides (from the top) is TTCAG.* (b) *Diagram and space-filling model of the hydrogen-bonded base pair adenine-thymine. The two hydrogen bonds are indicated by colored dotted lines.* (c) *The guanine-cytosine base pair, showing its three hydrogen bonds. Guanine-cytosine pairs are slightly closer together, more compact, and thus slightly denser than adenine-thymine pairs.*

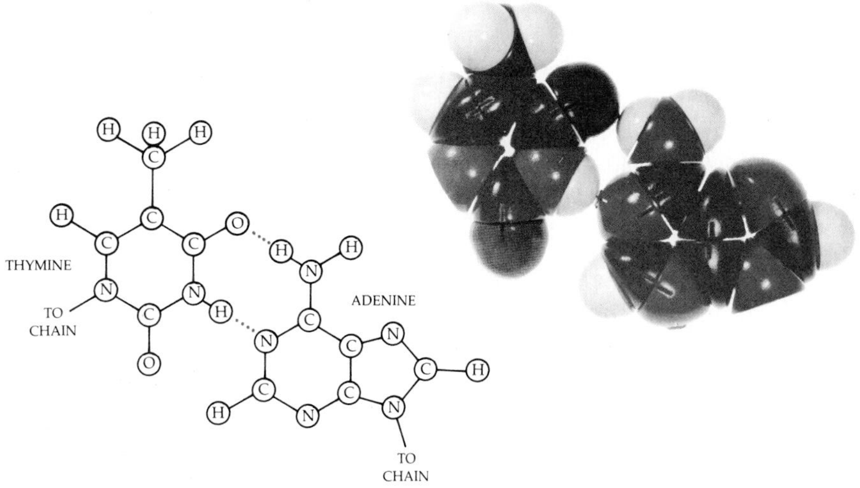

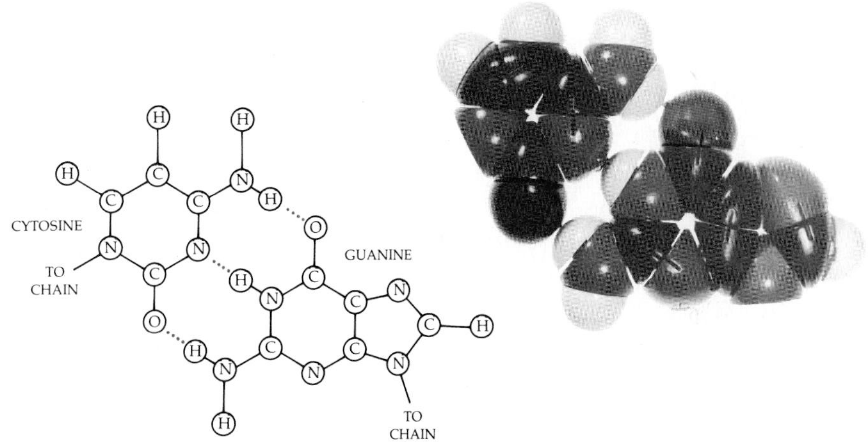

14–11

The double-stranded structure of a portion of the DNA molecule. Adenine can pair only with thymine, and guanine only with cytosine. Thus the order of bases along one strand determines the order of bases along the other. Note that the strands run in opposite directions.

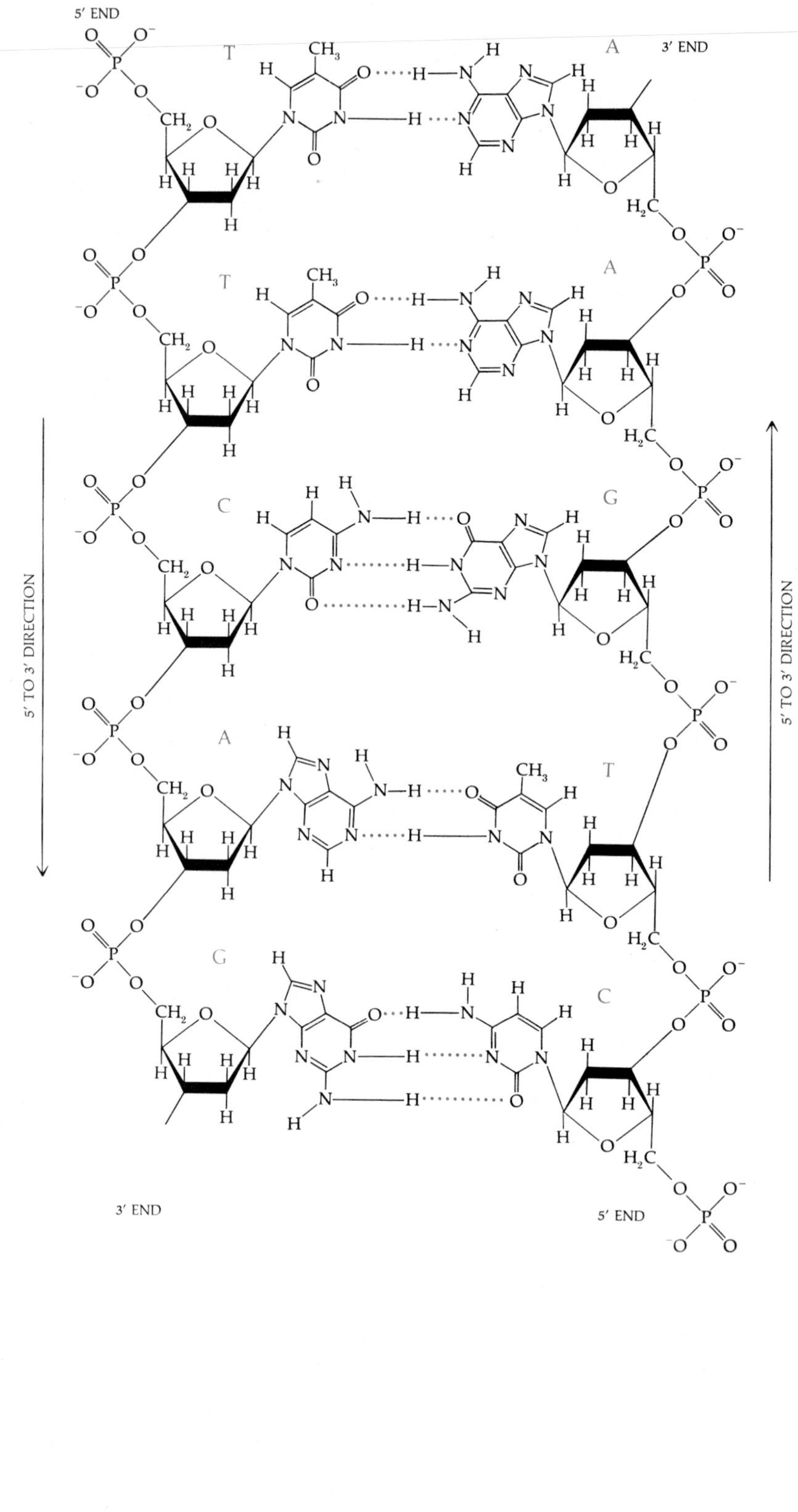

5′ END

3′ END

5′ TO 3′ DIRECTION

5′ TO 3′ DIRECTION

3′ END

5′ END

WHO MIGHT HAVE DISCOVERED IT?

Then there is the question, what would have happened if Watson and I had not put forward the DNA structure? This is "iffy" history which I am told is not in good repute with historians, though if a historian cannot give plausible answers to such questions I do not see what historical analysis is about. If Watson had been killed by a tennis ball I am reasonably sure I would not have solved the structure alone, but who would? Olby has recently addressed himself to this question. Watson and I always thought that Linus Pauling would be bound to have another shot at the structure once he had seen the King's College x-ray data, but he has recently stated that even though he immediately liked our structure it took him a little time to decide finally that his own was wrong. Without our model he might never have done so. Rosalind Franklin was only two steps away from the solution. She needed to realise that the two chains must run in opposite directions and that the bases, in their correct tautomeric forms, were paired together. She was, however, on the point of leaving King's College and DNA, to work instead on TMV [tobacco mosaic virus] with Bernal. Maurice Wilkins had announced to us, just before he knew of our structure, that he was going to work full time on the problem. Our persistent propaganda for model building had also had its effect (we had previously lent them our jigs to build models but they had not used them) and he proposed to give it a try. I doubt myself whether the discovery of the structure could have been delayed for more than two or three years.

There is a more general argument, however, recently proposed by Gunther Stent and supported by such a sophisticated thinker as Medawar. This is that if Watson and I had not discovered the structure, instead of being revealed with a flourish it would have trickled out and that its impact would have been far less. For this sort of reason Stent had argued that a scientific discovery is more akin to a work of art than is generally admitted. Style, he argues, is as important as content.

I am not completely convinced by this argument, at least in this case. Rather than believe that Watson and Crick made the DNA structure, I would rather stress that the structure made Watson and Crick. After all, I was almost totally unknown at the time and Watson was regarded, in most circles, as too bright to be really sound. But what I think is overlooked in such arguments is the intrinsic beauty of the DNA double helix. It is the molecule which has style, quite as much as scientists. The genetic code was not revealed all in one go but it did not lack for impact once it had been pieced together. I doubt if it made all that difference that it was Columbus who discovered America. What mattered much more was that people and money were available to exploit the discovery when it was made. It is this aspect of the history of the DNA structure which I think demands attention, rather than the personal elements in the act of discovery, however interesting they may be as an object lesson (good or bad) to other workers.

Francis Crick: "The Double Helix: A Personal View," Nature, *vol. 248, pages 766–769, 1974.*

Watson (left) and Crick in 1953 with one of their models of DNA. "DNA, you know, is Midas' gold," said Maurice Wilkins, with whom they shared the Nobel Prize. "Everyone who touches it goes mad."

DNA REPLICATION

An essential property of the genetic material is the ability to provide for exact copies of itself. Does the Watson-Crick model satisfy this requirement? In their published account, Watson and Crick wrote, "It has not escaped our notice that the specific pairing we have postulated immediately suggests a possible copying mechanism for the genetic material." Implicit in the double and complementary structure of the DNA helix is the method by which it reproduces itself. The molecule "unzips" down the middle, the paired bases separating at the hydrogen bonds. As the two strands separate, new strands form along each old one, using the raw materials in the cell. Each old strand forms a template, or guide, for the production of the new one. If a T is present on the old strand, only an A can fit into place in the new strand; a G will pair only with a C, and so on. In this way, each strand forms a copy of the original partner strand, and two exact replicas of the molecule are produced (Figure 14–12). The age-old question of how hereditary information is duplicated and passed on, duplicated and passed on, for generation after generation, had in principle been answered.

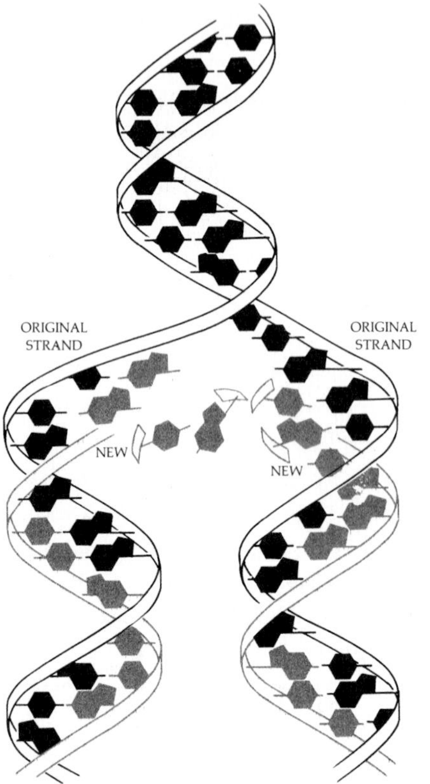

14–12
The DNA molecule shown here is in the process of replication, separating down the middle as the paired bases separate at the hydrogen bonds. (For clarity, the bases are shown out of plane.) Each of the original strands then serves as a template along which a new, complementary strand forms from nucleotides available in the cell.

A Confirmation

The Watson-Crick hypothesis of DNA replication is not the only possible one. Matthew Meselson and Franklin W. Stahl, working at the California Institute of Technology, devised an elegant experiment to choose among three possible models (Figure 14–13).

In designing their experiment, they took advantage of the availability of a heavy isotope of nitrogen (^{15}N) and an extremely sensitive method of separating macromolecules on the basis of weight. The method, which had been devised by Meselson while a graduate student, involves placing a solution of cesium chloride (CsCl) in a tube and spinning it in an ultracentrifuge. The small, dense CsCl molecules settle to form a continuous density gradient, less concentrated at the top of the tube and more concentrated at the bottom. If DNA molecules are present in this solution, they will settle at a level equivalent to their density (Figure 14–14a). (CsCl was selected for this procedure because the density gradient it forms overlaps that of DNA.)

Meselson and Stahl grew *E. coli* for several generations during which its sole nitrogen source contained ^{15}N, the heavy isotope of nitrogen. At the end of this period, the DNA of the bacterial cells contained a large proportion of heavy nitrogen. Although it was only about 1 percent denser than normal DNA, it formed a separate and distinct band in the cesium chloride gradient (Figure 14–14b and c).

They then placed a sample of cells containing heavy nitrogen in a medium containing ^{14}N and left them there long enough for the DNA to replicate once (as determined by a doubling of the number of cells). A sample of this DNA was spun in the ultracentrifuge (Figure 14–14d). A second generation was then grown in the ^{14}N medium. This DNA was also ultracentrifuged (Figure 14–14e).

Each sample of DNA contained more light DNA, as could be expected, because newly formed DNA had to incorporate the available ^{14}N. Moreover—and this was of crucial importance—the density of the first generation DNA was exactly halfway between that of heavy parent DNA and that of ordinary light DNA, as it should be if each molecule contained one old (heavy) strand and one new (light) strand, as predicted by Watson and Crick (Figure 14–13b). The second generation contained one-half half-heavy DNA and one-half light DNA, which again, exactly and ingeniously, confirmed the Watson-Crick hypothesis.

14–13
Three possible mechanisms of replication of DNA. Newly replicated strands are shown in color. (a) *Conservative replication. Each of the two strands of parent DNA is replicated, without strand separation. In the first generation, one daughter is all old DNA and one daughter is all new. The* F_2 *generation contains one helix composed of two old strands and three made up entirely of new strands.* (b) *Semiconservative replication of DNA. In the first generation, each daughter is half old and half new. The* F_2 *generation comprises two hybrid DNAs (half old, half new) and two DNAs made up entirely of new strands.* (c) *Dispersive replication. During replication, parent chains break at intervals and replicated segments are combined into strands with segments from parent chains. All daughter helices are part old, part new. The Meselson-Stahl experiment (Figure 14–14) was undertaken to determine which of these three possibilities was correct. Watson and Crick had predicted* (b).

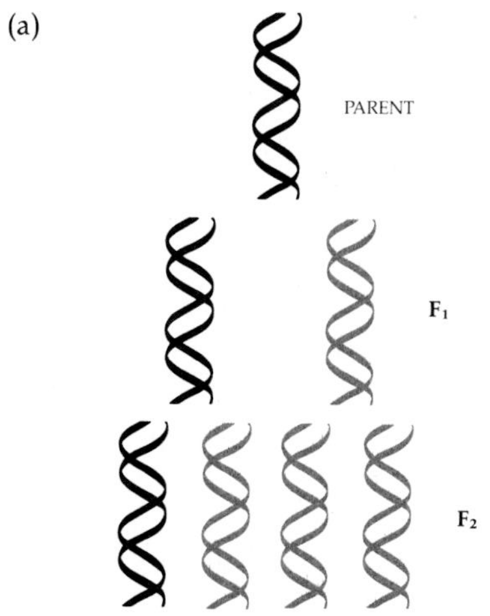

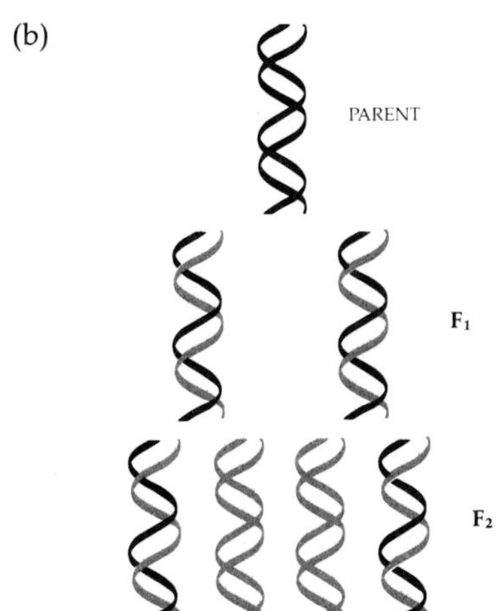

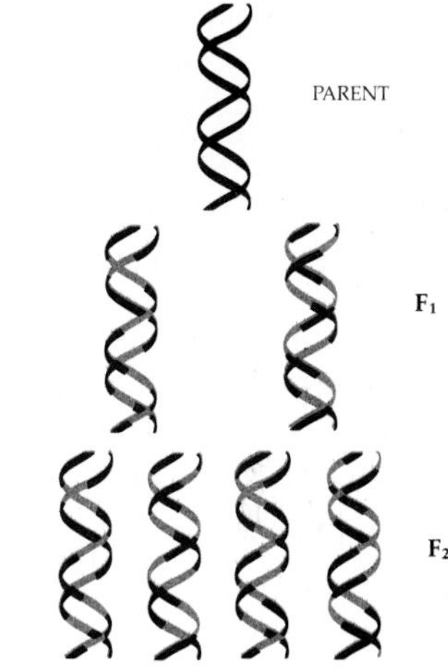

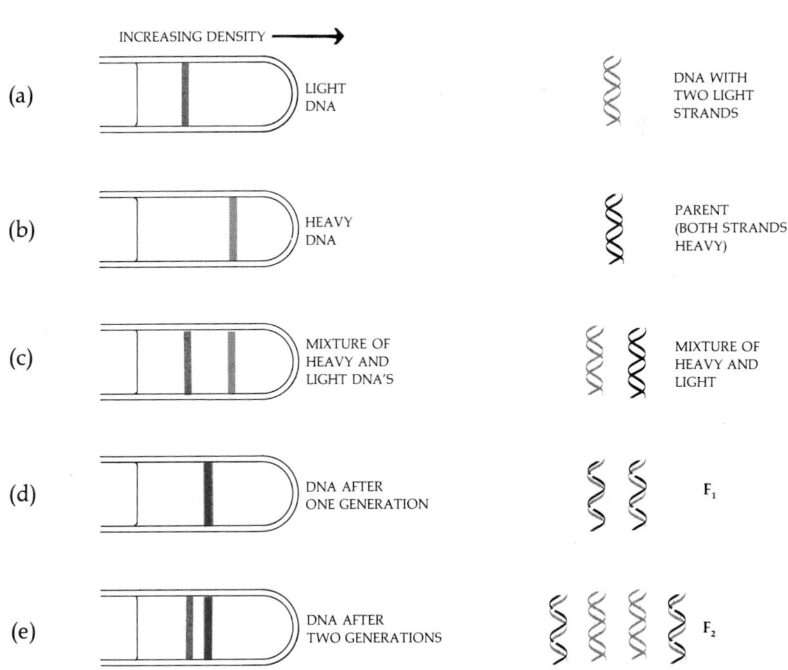

14–14
Meselson-Stahl experiment. (a) *Normal DNA settles in a band when ultracentrifuged in a density gradient of cesium chloride.* (b) E. coli *cells cultured in a medium containing heavy nitrogen (^{15}N) accumulate a "heavy" DNA, which forms a separate and distinct band.* (c) *For comparison, a mixture of heavy and light DNA separates into two distinct bands in a cesium chloride density gradient.* (d) *When cells grown in a "heavy"* ^{15}N *medium are permitted to multiply for one generation in a "light"* ^{14}N *medium, their DNA settles into a band midway in density between the heavy and light DNA.* (e) *When cells with heavy DNA are grown for two generations in the* ^{14}N *medium, their DNA settles into two bands—one of light DNA and one of intermediate-density DNA. The column on the right shows the investigators' interpretations. As you can see, the experiment confirmed the Watson-Crick hypothesis.*

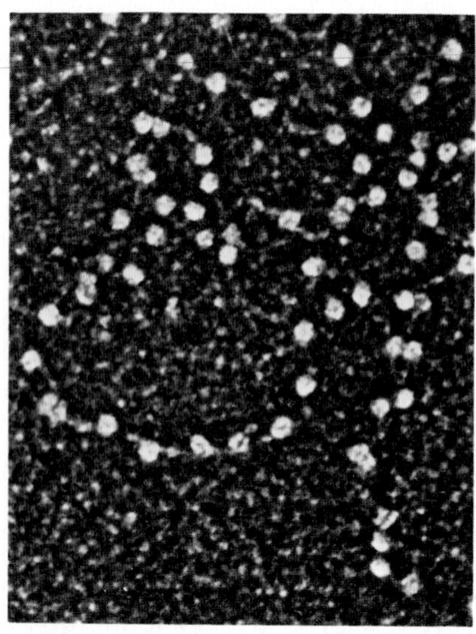

14-15
Electron micrograph of DNA polymerase molecules—the light-colored globules—bound to a DNA molecule, visible as a thin thread.

The Mechanics of DNA Replication

Although DNA is often referred to as a self-replicating molecule, this description is not precisely true. If DNA is placed in solution along with all the necessary components for new DNA, nothing happens. As in other biological reactions, special enzymes are needed. These enzymes, the DNA polymerases, link the nucleotides of the new DNA strand along the template of the old strand.

Along with the DNA polymerases, there are a dozen or more different enzymes that specialize in performing various operations on DNA. Some seem to help unwind the DNA for replication; others fill in gaps that arise in its synthesis; and a third group seals together the broken ends of DNA strands. Other enzymes are important in repairing damaged DNA. Such enzymes may also help to cause crossovers, thereby recombining DNA from different sources.

The Energetics of DNA Replication

The nucleotides required for DNA synthesis are assembled along biosynthetic pathways of the cell. They are not assembled in the form of the monophosphates shown in Figure 14-4, but rather as triphosphates—that is, adenine is provided as ATP, guanine as the analogous guanine triphosphate (GTP), and so forth. These "extra" P∼P groups provide the energy that powers the reactions mediated by DNA polymerase. As the nucleotide is attached to the growing DNA strand, the two phosphates are removed and released. Almost immediately, another enzyme breaks the bond between the two phosphates, releasing them as inorganic phosphates.

Why does the cell do it this way? The release of the energy-yielding P∼P group and the subsequent breaking of the bond between the two phosphates seem like a waste of carefully stored chemical energy. Is the cell really so profligate?

Measurement of the energy changes involved reveals an interesting point. The reaction in which the activated nucleotide is attached to the growing DNA strand is only slightly exergonic. Therefore, it could tend to go in either direction. Under certain equilibrium conditions, the DNA strand could come apart about as fast as it was synthesized, with the enzyme working both ways. However, with immediate degradation of the P∼P fragment, the reaction becomes highly exergonic. Thus the reverse reaction—which would necessitate reforging the P∼P group—becomes highly endergonic and so, for all practical purposes, does not occur. This is another example of the way in which cells exercise control over their biochemical activities.

DNA AS A CARRIER OF INFORMATION

You will recall that a necessary property of the genetic material is the ability to carry information. The Watson-Crick model showed that the DNA molecule is able to do this. The information is carried in the sequence of the bases, and *any* sequence of bases is possible. Since the number of paired bases ranges from about 5,000 for the simplest known virus up to an estimated 5 billion in the 46 human chromosomes, the possible variations are astronomical. The DNA from a single human cell—which if extended in a single thread would be almost 2 meters long—can contain information equivalent to some 600,000 printed pages of 500 words each, or a library of about a thousand books. Obviously, the DNA structure can well account for the endless diversity among living things. We shall consider the subject of the genetic code in the next chapter.

SUMMARY

Classical genetics had been concerned with the mechanics of inheritance—how the units of heredity are passed from one generation to the next and how changes in the hereditary material are expressed in individual organisms. In the 1930s, new questions arose and geneticists began to explore the nature of the gene—its structure, composition, and properties, and its role in the internal chemistry of living organisms.

During the 1940s, many investigators believed that genes were proteins, but others were convinced that the hereditary material was deoxyribonucleic acid (DNA). Important, although not widely accepted, evidence for the genetic role of DNA was presented by Avery in his experiments on the transforming factor of pneumococci. Confirmation of Avery's hypothesis came from studies with bacteriophages (bacterial viruses) showing that DNA and not protein is the genetic material of the virus.

Further support for the genetic role of DNA came from two more sets of data: (1) Almost all somatic cells of any given species contain equal amounts of DNA, and (2) the proportions of nitrogenous bases are the same in the DNA of all cells of a given species, but they vary in different species.

In 1953, Watson and Crick proposed a structure of DNA. The DNA molecule, according to their model, is a double-stranded helix, shaped like a twisted ladder. The two sides of the ladder are composed of repeating groups of a phosphate and a five-carbon sugar. The "rungs" are made up of paired bases, one purine base pairing with one pyrimidine base. There are four bases—adenine (A), guanine (G), thymine (T), and cytosine (C)—and A can pair only with T, and G only with C. The four bases are the four "letters" used to spell out the genetic message. The paired bases are joined by hydrogen bonds. On the basis of this structure, as revealed by Watson and Crick, the role of DNA as the carrier and transmitter of the genetic information became widely accepted.

When the DNA molecule replicates, the two strands come apart, breaking at the hydrogen bonds. Each strand forms a new complementary strand from nucleotides available in the cell. The semiconservative (one strand conserved) nature of this process was confirmed by studies using heavy isotopes.

Replication of DNA is enzymatically mediated, and energy is supplied by phosphate bonds.

QUESTIONS

1. One of the chief arguments for the erroneous theory that proteins constitute the genetic material is that proteins are heterogeneous. Explain why the genetic material must have this property. What feature of the Watson-Crick DNA model is important in this respect?

2. What are the steps by which Griffith demonstrated the existence of the transforming factor? Can you think of any implications of Griffith's discovery for modern medicine?

3. What characteristics of bacteriophages made them a useful experimental tool?

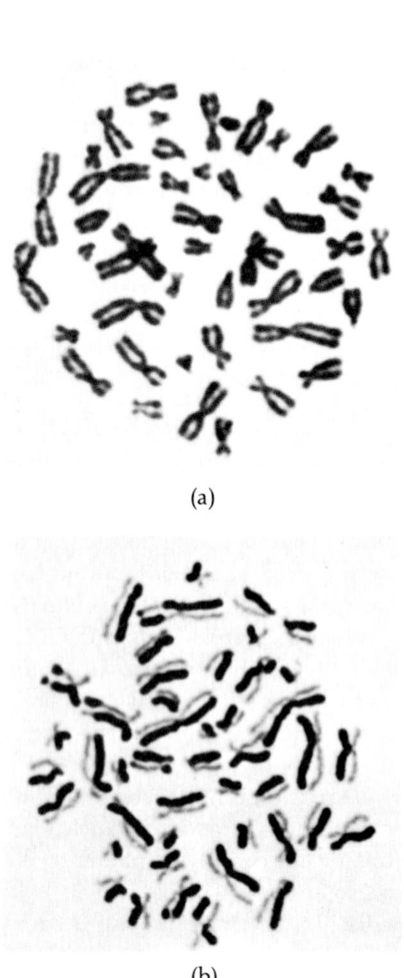
(a)

(b)

4. When the structure of DNA was being worked out, it became apparent that one purine base must be paired with a pyrimidine base, and that the other purine base must be paired with the other pyrimidine base. The evidence for this requirement came from two types of data. What were the data, and how did they indicate this structural requirement? Further consideration of the structures of the four nitrogenous bases indicated that adenine could pair only with thymine, and cytosine only with guanine. What features of the structures of the bases imposed this requirement on the structure of the DNA molecule?

5. Suppose you are talking to someone who has never heard of DNA. How would you support an argument that DNA is the genetic material? List at least five of the strong points in such an argument.

6. Suppose the Meselson-Stahl experiment was extended to the third generation. What would be the proportion of light to half-heavy DNA?

7. Eukaryotic cells are grown in a medium containing thymine labeled with ^3H. Then they are removed from the radioactive medium, placed in an ordinary medium, and allowed to divide. Studies of the distribution of the radioactive isotope are made after each generation to determine whether or not a chromatid contains radioactive material. Before they are placed in the nonradioactive medium, all the chromatids contain ^3H. After one generation in the nonradioactive medium, the ^3H is still divided evenly among the chromatids. Assuming each chromatid contains a single DNA molecule, explain the results. Is this consistent with the Watson-Crick hypothesis? What would be the distribution of the ^3H after two divisions in the nonradioactive medium? Why?

8. When 5-bromodeoxyuridine (BrdU) is added to a cell culture, it is used in place of thymidine (thymine plus deoxyribose sugar). All DNA synthesized thereafter will contain BrdU. After the cells replicate once in BrdU, their chromatids are not differentiable, as you can see in micrograph (a) in the margin. Both chromatids contain parent strands of original DNA combined with new strands of BrdU-substituted DNA. The original DNA stains darkly, so both sides are dark. Micrograph (b) shows the appearance of the chromosomes after the second cell division. As you will notice, one sister chromatid is dark, the other light. Does this confirm or refute the results of the Meselson-Stahl experiment? Explain your answer.

CHAPTER 15

The Code and Its Translation

The Watson-Crick model, which both established the structure of DNA and indicated the mechanism for its replication, led to the virtually universal acceptance of DNA as the repository of the hereditary information. The DNA molecule carries the instructions for the structure and function of the cell and also transmits these instructions to new cells and organisms. Yet a big question was left unanswered: How are the instructions carried out?

GENES AND PROTEINS

One Gene–One Enzyme

There were some major clues. In 1941, about the same time that the collaborative studies on bacteriophages began, George Beadle and Edward L. Tatum completed a series of important studies using the red bread mold *Neurospora.* Utilizing x-rays to increase the mutation rate, they then analyzed the resultant mutants by genetic mapping studies analogous to those performed with *Drosophila.* A number of the mutations proved to affect the activity of single enzymes, and Beadle and Tatum were able to show that a mutation in one gene could be correlated with the loss of function of one enzyme. Based on these experiments, for which they were awarded the Nobel Prize, they formulated the hypothesis that one gene is responsible for the production of one enzyme. This turned out to be an oversimplification, for although all enzymes are proteins, not all proteins are enzymes. Some, for instance, are hormones, like insulin; others are structural proteins, like collagen. These proteins, too, are under genetic control. This expansion of the original concept did not modify it in principle. "One gene–one enzyme," as it was first abbreviated, was simply amended to the less memorable but more precise "one gene–one polypeptide chain." (As we shall see, this, too, has now been amended.)

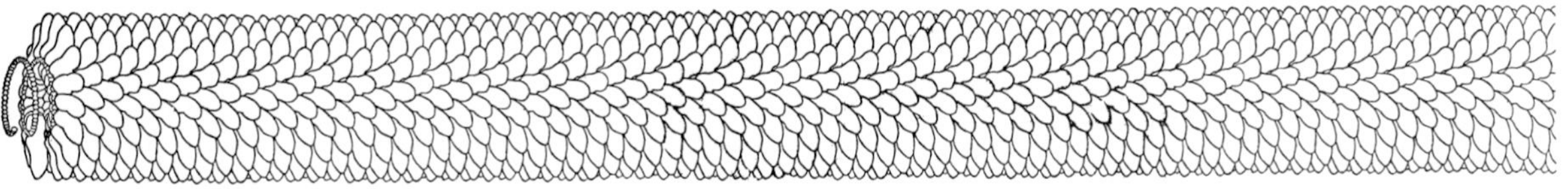

15–1
Diagram of tobacco mosaic virus (TMV), representing about half of its total length. This virus has a central core not of DNA but rather of RNA (ribonucleic acid). Its outer coat is composed of 2,200 identical protein molecules. If the RNA is removed from the protein coat and rubbed into scratches on a tobacco leaf, new TMV viruses are formed, complete with new protein coats. Studies with TMV were among the first to suggest that RNA might provide a link between DNA and proteins.

The Structure of Hemoglobin

Linus Pauling was one of the first to see some of the implications of the work of Beadle and Tatum. Perhaps, Pauling reasoned, human diseases involving hemoglobin, such as sickle cell anemia, could be traced to a variation from normal in the protein structure of the hemoglobin molecule. To test this hypothesis—which, at this stage, was pure speculation—he took samples of hemoglobin from people with sickle cell anemia, from others heterozygous for the gene, and from still others homozygous for the normal gene. To try to detect differences in these proteins, he used a process known as electrophoresis, in which organic molecules are dissolved in a solution and separated in a weak electric field.

As we noted in Chapter 3, individual amino acids may have a positive charge, a negative charge, or no charge at all. Therefore, a mutation that causes the substitution of one amino acid for another may change the total charge of the protein molecule, and the normal and the mutant protein molecules will, as a result, move differently in an electric field.

15–2
Results from electrophoresis of (a) *normal hemoglobin,* (b) *the hemoglobin of a person with sickle cell anemia, and* (c) *the hemoglobin of a person who is heterozygous for the sickle cell trait. Because of slight differences in electric charge, the normal and sickle cell hemoglobins move differently in an electric field. The normal hemoglobin is more negatively charged; hence, it is closer to the positive pole than the sickle cell hemoglobin. The hemoglobin of the heterozygote separates into the two different types.*

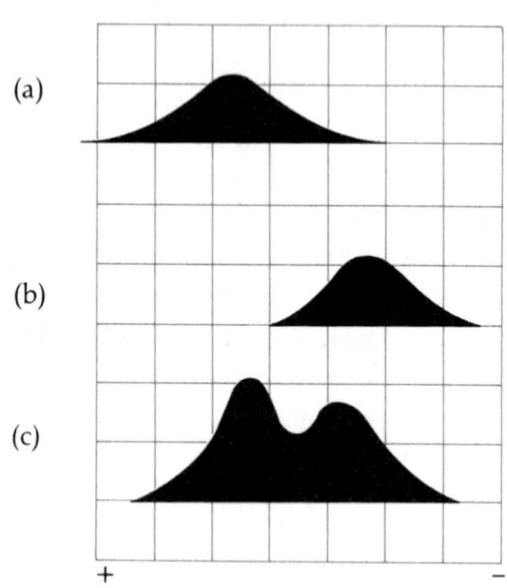

Figure 15–2 shows the results of Pauling's experiment. A person who has sickle cell anemia makes a different sort of hemoglobin than a person who does not have the disease. A person who is heterozygous (carrying one copy of the recessive allele for sickling and one copy of the allele for normal hemoglobin) makes both kinds of hemoglobin molecules; however, enough normal molecules are produced to prevent anemia. (Notice that the terms "dominant" and "recessive" are beginning to take on a different meaning.)

A few years later, Vernon Ingram, of Cambridge University in England, was able to show that the actual difference between the normal and the sickle cell hemoglobin molecules is one amino acid in 300, as we noted previously (page 73).

The Virus Coat

A third piece of evidence that DNA specifies the structure of proteins came from the studies of the bacteriophage group described in Chapter 14. You will recall that they showed that the introduction of viral DNA into the bacterial cell resulted in the production not only of more viral DNA, but also of the proteins of the virus coat.

THE GENETIC CODE

Given that genes are made of DNA and that specific proteins are the products of specific genes, what is the link between them? How does DNA influence protein synthesis? One early hypothesis was that the DNA somehow formed a template for protein production. But there was simply no way that a strand of protein could be matched up against a strand of DNA in a templatelike fashion. The relationship between DNA and protein had to be a more complicated one. If the proteins, with their 20 amino acids, were the "language of life," to extend the metaphor, the DNA molecule, with its four nitrogenous bases, could be envisioned as a code for this language. So the term *genetic code* came into being.

The Code Is a Triplet

As it turned out, the idea of a code was useful not only as a dramatic metaphor but also as a working analogy. Scientists, seeking to understand how the sequence of nucleotides stored in the double helix could order the quite dissimilar structures of protein molecules, approached the problem with methods used by cryptographers in deciphering codes. There are 20 biologically important amino acids, and there are four different nucleotides. As George Gamow, a Nobel Prize-winning physicist, pointed out, if a single nucleotide "coded" one amino acid, only four amino acids could be provided for. If two nucleotides specified one amino acid, there could be a maximum number, using all possible arrangements, of 4^2, or 16—still not quite enough to code all 20 amino acids. Therefore, following the code analogy, at least three nucleotides in sequence must specify each amino acid. This would provide for 4^3, or 64, possible combinations, or *codons*.

The three-nucleotide, or triplet, codon was widely and immediately adopted as a working hypothesis, although its existence was not actually demonstrated until a decade after the Watson-Crick discovery. Proof depended on answering yet another question: How is the information encoded in the DNA molecule translated into protein?

15-3
Chemically, RNA is very similar to DNA, but there are two differences in its nucleotides. One difference is in the sugar component; instead of deoxyribose, RNA contains ribose, which has an additional oxygen atom. The other difference is that instead of thymine, RNA contains the closely related pyrimidine uracil (U). A third, and very important, difference between the two nucleic acids is that most RNA does not possess a regular helical structure and is usually single-stranded.

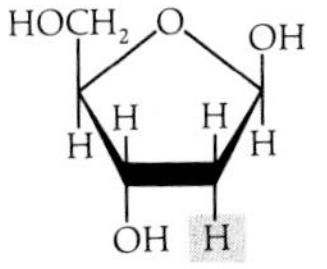

DEOXYRIBOSE 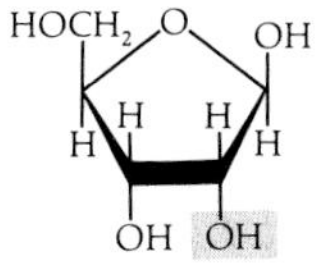 RIBOSE

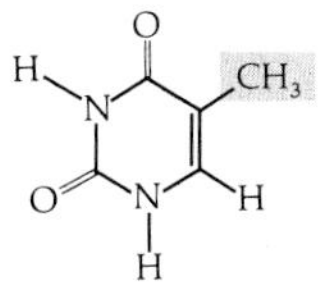

THYMINE 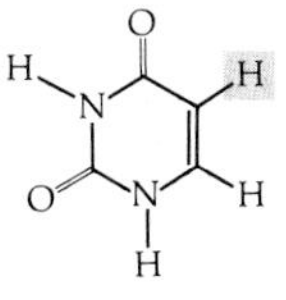 URACIL

THE RNAs

The search for the answer to this question led to the examination of another kind of molecule, *ribonucleic acid* (RNA), a close chemical relative of DNA (Figure 15-3). There were several clues indicating that RNA might play a role in the translation of genetic information into a sequence of amino acids. First, there was some circumstantial evidence. Unlike DNA, which is found primarily in the nucleus, RNA is found mostly in the cytoplasm, and it is there that most protein synthesis takes place. Embryologists noted that developing embryos of many different kinds contained high levels of RNA. Also, cells making large amounts of protein have numerous ribosomes. A rapidly growing *Escherichia coli* cell, for instance, contains about 15,000 ribosomes, constituting about one-half of the total mass of the cell. And ribosomes are two-thirds RNA.

Additional evidence came from experiments with viruses. When a bacterial cell is infected by a DNA- containing bacteriophage, RNA is synthesized from the virus DNA before virus protein synthesis begins. Also, some viruses contain no DNA—only RNA and protein; tobacco mosaic virus (Figure 15-1) is an example. When a tobacco leaf is infected with RNA removed from TMV, new viruses are produced, protein coats and all. In other words, RNA as well as DNA seemed to contain information about proteins.

Many of the experiments to define the role of RNA in protein biosynthesis were carried out using cell-free extracts of *E. coli* and of animal liver cells. ("Cell-free" in this context means simply that the cells have been broken apart.) A principal advantage of using such extracts is that they can be separated into various fractions and the fractions studied separately. The basic principles of protein synthesis are the same for both prokaryotic and eukaryotic cells, but there are some important differences, which will be described in Chapter 17. This chapter focuses on the process as it takes place in prokaryotes, particularly *E. coli.*

RNA Synthesis

In the 1960s, scientists established that the RNA in a cell and the RNA produced by a bacteriophage are copied, respectively, directly from the DNA of the cell or of the virus. A complex enzyme known as RNA polymerase catalyzes the copying reaction, and the process is known as *transcription*.

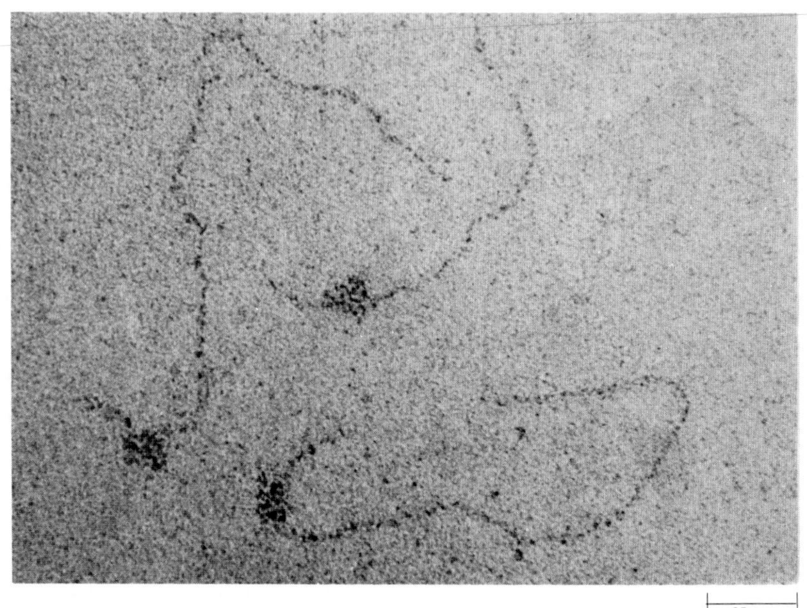

15–4
Molecules of the enzyme RNA polymerase bound to a DNA molecule. This enzyme catalyzes the transcription of RNA from DNA.

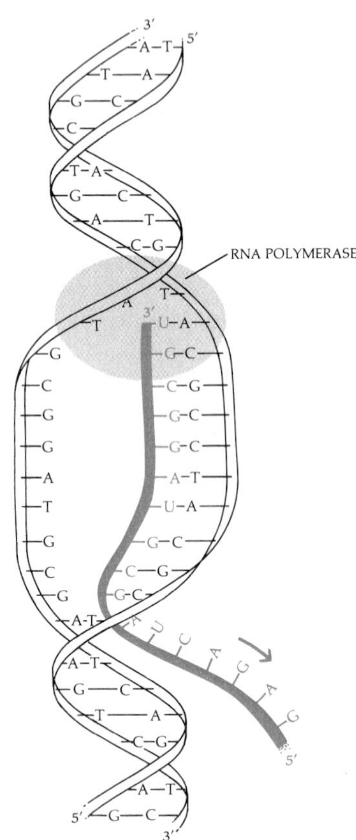

15–5
A schematic representation of RNA transcription. At the point of attachment of the enzyme RNA polymerase, the DNA opens up. Nucleotide building blocks are assembled into RNA using one strand of the DNA as a template. As the RNA polymerase moves along the DNA molecule, the hydrogen bonds between the two strands of the DNA re-form, squeezing out the newly formed, single-stranded RNA molecules.

Just as each new DNA strand is a negative copy of an existing strand, each new RNA molecule is copied from one of the two strands of DNA by the same base-pairing principle. The difference is that RNA contains uracil instead of thymine and that the sugar component is ribose instead of deoxyribose. Like a strand of DNA, each RNA molecule has a 5′ end and a 3′ end. The nucleotides, which are present in the cell as triphosphates, are added, one at a time, to the 3′ end of the growing RNA chain.

Three different types of RNA molecules are transcribed from the cell's DNA: messenger RNA (mRNA), transfer RNA (tRNA), and ribosomal RNA (rRNA). Messenger RNA molecules are working copies of the genetic DNA. Transfer RNA molecules are the dictionary for the translation of nucleotides to protein. The ribosomal RNA molecules are, along with proteins, assembled into ribosomes, the sites at which the translation takes place. Thus, each type plays its own particular role in the assembly of amino acids into proteins.

Messenger RNA

Messenger RNA molecules, the working copies, are long—1,000 to 10,000 nucleotides—and single-stranded. The instructions are carried by the molecule in the form of codons, the nucleotide triplets hypothesized by Gamow. The molecule begins with a short "leader" sequence that contains an initiation site, the point at which the mRNA attaches to the ribosome. The rest of the molecule is a linear sequence of nucleotides that precisely dictates the linear sequence of amino acids in a particular polypeptide chain. Accomplishing this biochemical feat involves transfer RNA, ribosomes, and a number of special enzymes.

Transfer RNA

Transfer RNA molecules are the adapters by which the language of nucleic acids is translated into the language of proteins. They are small molecules—from about 75 to 85 nucleotides long—and intricately folded, with several highly specific recogni-

15–6

(a) *The structure of a tRNA molecule. Such molecules consist of about 80 nucleotides linked together in a single chain. The chain always terminates in a CCA sequence. An amino acid can link to its specific tRNA at this end. Some nucleotides are the same in all tRNAs; these are shown in gray. The other nucleotides vary according to the particular tRNA. The symbols D, Y, ψ, and T represent unusual modified nucleotides characteristic of tRNA molecules.*

Some of the nucleotides are hydrogen-bonded to one another, as indicated by the colored lines. The unpaired nucleotides at the bottom of the diagram (indicated in color) are known as the anticodon. They serve to "plug in" the tRNA molecule to an mRNA codon.

(b) *The molecule folds over on itself, producing this three-dimensional structure. This is a photograph of a model.*

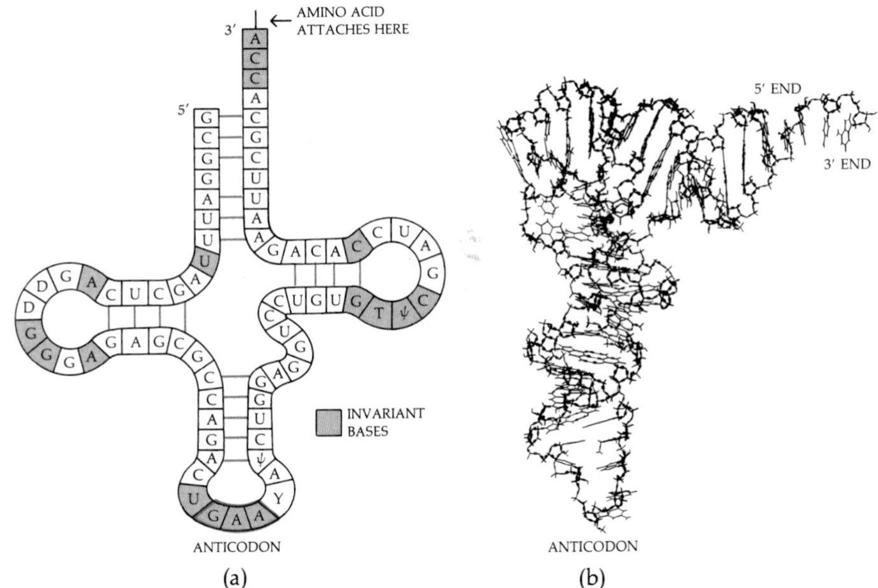

tion sites. One end of the molecule attaches to one particular amino acid; this attachment is catalyzed by an enzyme that recognizes the particular tRNA molecule and its specific amino acid. A loop within the molecule exposes a set of three nucleotides that form an anticodon, so called because it attaches to an appropriate codon in the mRNA molecule. Another loop attaches to the ribosome and another to a specific enzyme.

There are more than 20 kinds of tRNA molecules in every cell, at least one for every amino acid found in proteins. One end (the 3′ end) always terminates in a CCA sequence; this is the site of attachment of the amino acid. The other (5′) end often terminates in a guanine nucleotide. The other nucleotides vary according to the particular type of tRNA. All tRNA molecules can be represented in two dimensions by the cloverleaf shape shown in Figure 15–6a; the three-dimensional structure is shown in Figure 15–6b.

15–7

Diagram of a ribosome from an E. coli *cell. As you can see, it consists of two subunits, one larger than the other, and each composed of specific RNA and protein molecules. The terms 30S and 50S refer to relative densities of the two subunits, as indicated by sedimentation rates in the ultracentrifuge.*

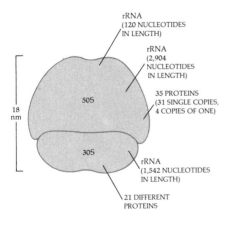

rRNA (120 NUCLEOTIDES IN LENGTH)
rRNA (2,904 NUCLEOTIDES IN LENGTH)
35 PROTEINS (31 SINGLE COPIES, 4 COPIES OF ONE)
50S
30S
18 nm
rRNA (1,542 NUCLEOTIDES IN LENGTH)
21 DIFFERENT PROTEINS

Ribosomal RNA

Ribosomes are about two-thirds RNA and one-third protein. Each is composed of two subunits, each with its characteristic RNAs and proteins (Figure 15–7). In the ribosomes of *E. coli,* which have been studied the most extensively, the smaller subunit has one type of rRNA, 1,542 nucleotides in length, and a single copy each of 21 different proteins. The larger subunit has two types of rRNA, one consisting of 120 nucleotides and the other of 2,904 nucleotides. It contains a single copy each of 31 different proteins and four copies of one protein. The way that the proteins and rRNAs fit together to form this complex and intriguing structure is not known, although the overall function of ribosomes in protein synthesis has now been described in some detail.

The smaller (30S) subunit has a binding site for messenger RNA. The larger (50S) subunit has two binding sites for transfer RNAs with their attached amino acids. When the two subunits are joined together, all the components needed for the assembly of amino acids into proteins are brought into the necessary relationship with one another.

15-8
The structures of methionine and N-formylmethionine. Note that the attachment of the formyl group prevents a peptide bond from forming on the "wrong" (amino) end of the amino acid.

$$\mathrm{H_2N{-}CH(CH_2CH_2SCH_3){-}COOH}$$

METHIONINE
(met)

$$\mathrm{OHC{-}NH{-}CH(CH_2CH_2SCH_3){-}COOH}$$

N-FORMYLMETHIONINE
(fMet)

PROTEIN SYNTHESIS

The synthesis of proteins is known as *translation*, since it is the transferring of information from one language to another. It takes place in three stages: initiation, elongation, and termination. The first stage begins when the smaller ribosomal subunit attaches to the strand of messenger RNA at its 5′ end, exposing its first, or initiator, codon. (Often the 3′ end is still bound to the DNA helix, with transcription continuing even as translation begins.) Next, the first tRNA comes into place. The first codon is always AUG (5′ to 3′), the first anticodon is always UAC (3′ to 5′),

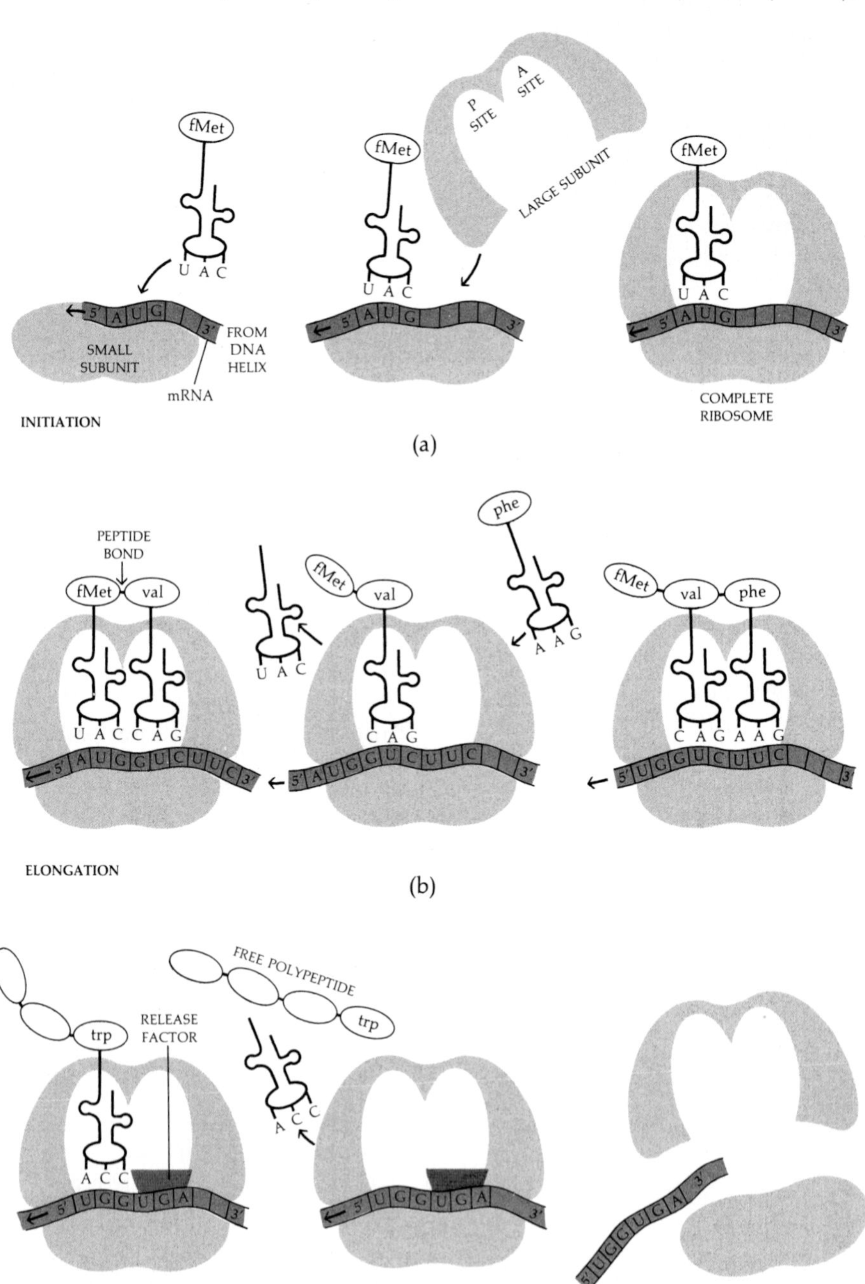

15-9
Three stages in protein synthesis. (a) Initiation. The smaller ribosomal subunit attaches to the 5′ end of the mRNA molecule. The first tRNA molecule plugs into the AUG codon on the mRNA molecule. The larger ribosomal subunit locks into place, with the tRNA occupying the P (peptide) site. The A (amino acid) site is vacant. The initiation complex is now complete.

(b) Elongation. A second tRNA with its attached amino acid moves into the A site, and its anticodon plugs into the mRNA. A peptide bond is formed between the two amino acids brought together at the ribosome. The second tRNA, with the dipeptide attached, moves to the P site from the A site as the first tRNA is released. A third tRNA moves into the A site and another peptide bond is formed. The growing peptide chain is always attached to the tRNA in the P site, and the new tRNA always occupies the A site. This step is repeated over and over until the polypeptide is complete.

(c) Termination. When the ribosome reaches a termination codon, the polypeptide is cleaved from the last tRNA and the tRNA is released from the P site. The A site is occupied by a release factor that triggers the dissociation of the two subunits of the ribosome.

15–10
Clusters of ribosomes reading the same mRNA strand. Such groups are called polyribosomes, or polysomes.

and the first amino acid is always a modified methionine, N-formylmethionine, or fMet (Figure 15–8), which is later removed. Then the larger subunit of the ribosome attaches to the smaller subunit, and the first tRNA attached to its N-formylmethionine is locked into the P (peptide) site on the larger subunit (Figure 15–9). The initiation stage is now complete.

At the beginning of the elongation stage, the second codon of the mRNA is positioned opposite the A (amino acid) site. A tRNA with an anticodon for that second mRNA codon plugs into the mRNA molecule and, with its amino acid, comes to occupy the A site on the ribosome. When both sites are occupied, an enzyme that is part of the larger ribosome unit forges a peptide bond between the two amino acids, attaching the first to the second. The first tRNA is released. The second tRNA, to which is now attached the N-formylmethionine and the second amino acid, moves to the P position. A third tRNA–amino acid moves into the A position opposite the third codon on the mRNA, and the step is repeated. The P position accepts the tRNA bearing the growing polypeptide chain; the A position accepts the tRNA bearing the new amino acid that will be added to the chain. As the ribosome moves along the mRNA chain, the ribosomal binding site of the mRNA molecule is freed, and another ribosome forms an initiation complex with it. A group of ribosomes reading the same mRNA molecule is known as a polysome (Figure 15–10).

Toward the end of the mRNA molecule is a codon that serves as a termination signal. When this codon is reached, translation stops and the two ribosomal subunits separate. The polypeptide chain is complete. As many as 3,000 proteins have been isolated from *E. coli*, each different and each assembled in this same way.

The working out of the details of this precise and elegant solution to the problem of translation was an awe-inspiring achievement. Even more awe-inspiring is the knowledge that at this very moment a similar process is taking place in virtually every cell of our own bodies.

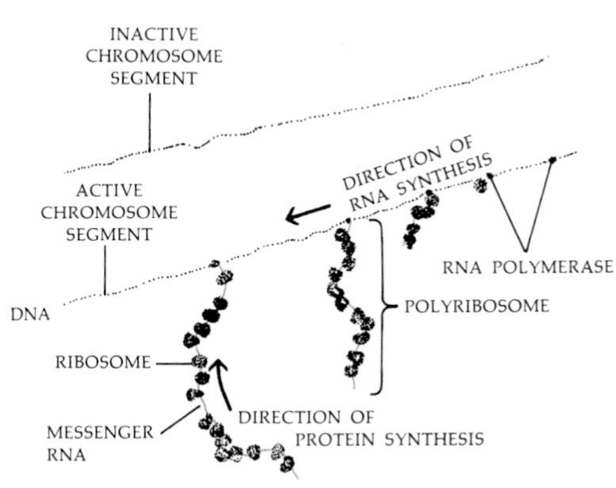

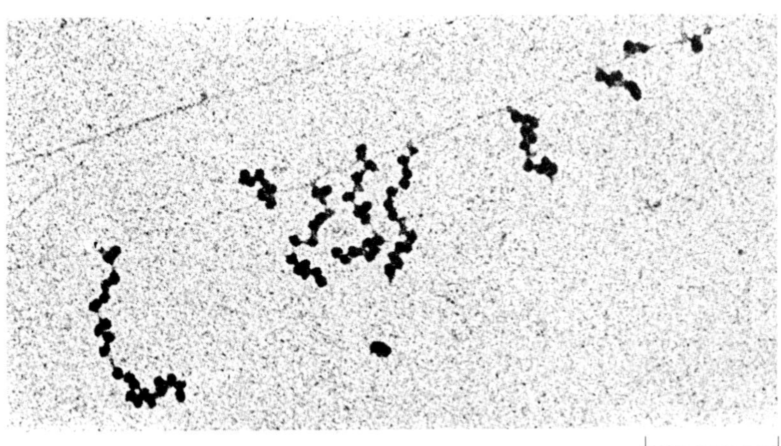

15–11
A bacterial gene in action. In the micrograph, you can see molecules of RNA polymerase, the enzyme that regulates the transcription of RNA from DNA. The one at the far right is approximately at the point where transcription begins. Several different mRNA strands (shown in color in the diagram) are being formed simultaneously. The longest one, at the left, was the first one synthesized. As each mRNA strand peels off the DNA molecule, ribosomes attach to the RNA, translating it into protein. The protein molecules are not visible in the micrograph.

BREAKING THE CODE

Long before all the details of transcription and translation became known, Marshall Nirenberg of the National Institutes of Health set out to test the messenger RNA hypothesis. He took cell-free extracts of *E. coli* and added to them radioactively labeled amino acids and crude samples of RNA from a variety of cell sources. All of the RNAs stimulated protein synthesis; the amounts of protein produced were small but measurable. In other words, the cellular material would start producing protein molecules even when the RNA "orders" it received were from a "complete stranger." Even the RNA from tobacco mosaic virus, which naturally multiplies only in cells of the leaves of tobacco plants, could be read as an mRNA by the ribosomes and tRNA molecules of the bacterial cell.

Nirenberg and Heinrich Matthaei next tried an artificial RNA. Perhaps if the cell extracts could read a foreign message and translate it into protein, they could read a totally synthetic message, one dictated by the scientists themselves. Severo Ochoa of New York University had developed an enzymatic process for linking ribonucleotides into a long strand of RNA. By this process, he had produced an RNA molecule that contained only one nitrogenous base, uracil, repeated over and over again. It was called "poly-U." What kind of protein would poly-U dictate?

Nirenberg and Matthaei prepared 20 different test tubes, each of which contained cell-free extracts of *E. coli* with ribosomes, tRNA, ATP, the necessary enzymes, and all the amino acids. In each test tube, one of the amino acids, and only one, carried a radioactive label. Synthetic poly-U was added to each test tube. In 19 of the test tubes, no radioactive polypeptides were produced, but in the twentieth tube, to which radioactive phenylalanine had been added, the investigators were able to detect newly formed, radioactive polypeptide chains. When the polypeptides were analyzed, they were found to consist only of phenylalanines, one after another. Nirenberg and Matthaei had dictated the message "uracil . . . uracil . . . uracil . . . uracil . . . uracil . . . uracil . . . ," and a clear answer had come back, "phenylalanine . . . phenylalanine. . . ." The experiment not only defined the first code word (UUU = phe) but also made available a method for defining the others.

H. G. Khorana at the University of Wisconsin succeeded in producing an artificial messenger in which two or three nucleotides were repeated over and over again in a known sequence: AGAGAGAGAG, UCUCUCUCUC, ACACACACAC, and UGUGUGUGUG. Each of these RNA chains, when used as a messenger in the cell-free system, produced polypeptide chains of alternating amino acids. Poly-AG produced arginine and glutamic acid over and over again; poly-UC, serine and leucine; poly-AC, threonine and histidine; and poly-UG, cysteine and valine (Figure 15–12). This is, of course, what you would expect from a triplet code, since the message would be read AGA . . . GAG . . . AGA. . . . These studies provided the

mRNA BASE SEQUENCE	READ AS	AMINO ACID SEQUENCE OBTAINED
$(AG)_n$	· · · AGA GAG AGA GAG · · ·	· · · arg-glu-arg-glu · · ·
$(AGC)_n$	· · · AGC AGC AGC · · ·	· · · ser-ser-ser · · ·
	· · · GCA GCA GCA · · ·	· · · ala-ala-ala · · ·
	· · · CAG CAG CAG · · ·	· · · gln-gln-gln · · ·

15–12
In Khorana's experiments, an artificial mRNA in which two nucleotides were repeated over and over produced a polypeptide chain of alternating amino acids. Depending on where the reading process began, an artificial mRNA with three different nucleotides could produce three different polypeptides, each consisting of only one type of amino acid.

15–13

The genetic code. The 20 amino acids are coded by 61 triplet codons, with three additional codons serving as stop signals. The codon triplets shown here are the ones that would appear in the mRNA molecule. Note that almost every amino acid is specified by more than one codon and some by as many as six. Often, the first two nucleotides are the same, with the codon differing only in the third. This suggests that the first two "letters" of the codon may be enough to hold the tRNA to the mRNA—a hypothesis formulated by Crick and known as "wobble."

ala	arg	asn	asp	cys	glu	gln	gly	his	ile	leu	lys	met	phe	pro	ser	thr	trp	tyr	val	stop
	AGA									UUA					AGC					
	AGG									UUG					AGU					
GCA	CGA						GGA			CUA				CCA	UCA	ACA			GUA	
GCC	CGC						GGC		AUA	CUC				CCC	UCC	ACC			GUC	UAA
GCG	CGG	AAC	GAC	UGC	GAA	CAA	GGG	CAC	AUC	CUG	AAA		UUC	CCG	UCG	ACG		UAC	GUG	UAG
GCU	CGU	AAU	GAU	UGU	GAG	CAG	GGU	CAU	AUU	CUU	AAG	AUG	UUU	CCU	UCU	ACU	UGG	UAU	GUU	UGA

15–14

Deletion or addition of nucleotides within a gene leads to changes in the protein produced. The original DNA molecule, the mRNA transcribed from it, and the resulting polypeptide are shown in (a). In (b) we see the effect of the deletion of a nucleotide pair (T-A), as indicated by the arrow. The reading frame for the gene is altered, and a different sequence of amino acids occurs in the polypeptide. A similar change results from the addition of a nucleotide pair (color), as shown in (c).

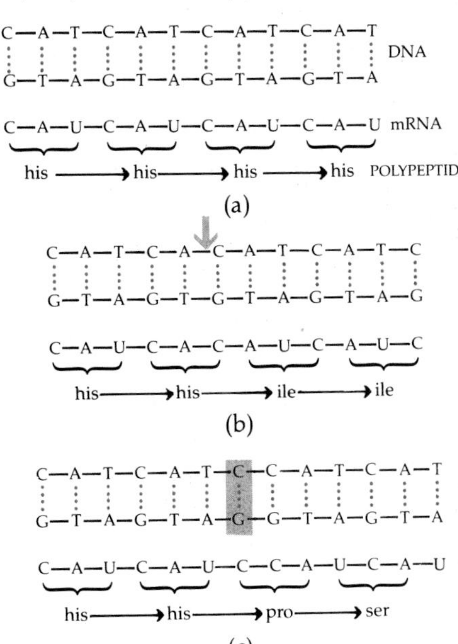

first clear demonstration (1) that mRNA is read sequentially (that is, one codon after another); (2) that how it is read depends on the reading frame—that is, the nucleotide at which translation starts; and (3) that the codon consists of an uneven number of nucleotides.

As a result of these kinds of experiments carried out by Nirenberg, Khorana, and others, the mRNA codons for all of the amino acids have been worked out. Of the 64 possible triplet combinations, 61 of them specify particular amino acids. With 61 combinations coding for 20 amino acids, you can see that there must be more than one codon for many of the amino acids. As shown in Figure 15–13, codons specifying the same amino acid often differ only in the third nucleotide, leading to the speculation that the first two may be sufficient to hold the tRNA in most instances.

Hemoglobin Reexamined

Some of the biological implications of these findings are strikingly clear. Let us take another look at sickle cell anemia, for example, in the light of Figure 15–13. Normal hemoglobin contains glutamic acid; sickle cell hemoglobin contains valine in the same position. In mRNA, GAA or GAG specifies glutamic acid (glu), and GUA, GUC, GUG, or GUU specifies valine (val). So the difference between the two is merely the replacement of one adenine by one uracil in a molecule that, since it dictates a polypeptide that contains more than 150 amino acids, must contain more than 450 bases. In other words, the tremendous functional difference—literally a matter of life and death—can be traced to a single "misprint" in over 450 nucleotides.

De Vries, some 80 years ago, defined mutation in terms of traits appearing in the phenotype. In the light of current knowledge, the definition is somewhat different: A mutation is a change in the sequence or number of nucleotides in a nucleic acid that can be transmitted from parent to offspring.

Most mutations involve only a single nucleotide substitution. As we saw in the case of sickle cell anemia, such a substitution can lead to changes in the protein produced by a gene. Other changes can result from deletion or addition of nucleotides within a gene (Figure 15–14). When this occurs, the reading frame of the gene may shift—that is, the way in which the nucleotides are grouped into triplets changes—resulting in the production of an entirely new protein.

BACTERIOPHAGE ϕX174 BREAKS THE RULES

In 1977, the complete nucleotide sequence of the DNA of a small bacteriophage known as ϕX174 was worked out by Frederick Sanger and his coworkers at the Medical Research Council in Cambridge, England. The DNA was known to code for nine proteins, whose sequences were also known. (Viruses borrow enzymes and other proteins from host cells, so they are able to get along with a relatively small molecular arsenal.)

Even before the last details were worked out, however, the investigators began to realize they had a serious and interesting problem: the DNA (which contained some 5,375 nucleotides) was insufficient to code for the known proteins by the triplet code hypothesis. It simply was not long enough. Were the concepts established by a quarter-century of intensive effort to be overthrown by a submicroscopic particle?

When the nucleotide sequence became known, its secret was revealed. The investigators had originally assumed that each gene was physically separate along the DNA molecule. However, it turns out that there are pairs of overlapping genes. In other words, different genes in ϕX174 are coded by the same regions of DNA using different reading frames. More recently, several additional examples of overlapping genes have been found in other viruses.

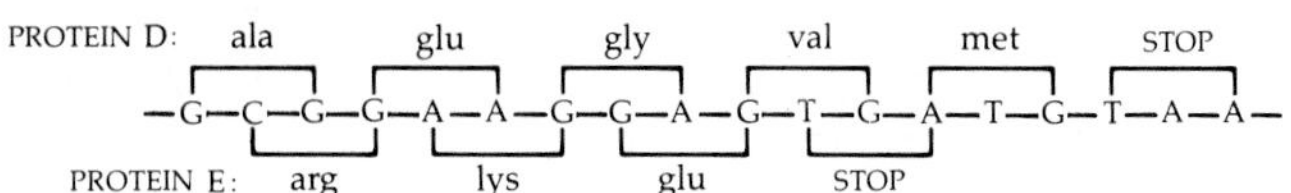

A segment of the DNA of ϕX174, showing a portion of the genes for proteins D and E. Both are coded by the same stretch of DNA; their reading frames, however, are different. The gene for protein E is completely contained within the gene for D, an entirely unrelated protein.

REGULATION OF PROTEIN SYNTHESIS

A cell does not need all of its enzymes and other proteins at the same time and in the same amounts. One way in which a bacterial cell regulates protein production is by the rapid degradation of mRNA. Another way is by allosteric interactions, described in Chapter 8 (page 178). A third strategy for protein regulation is utilization of a single mRNA molecule to transcribe a series of genes with related functions. For example, the genes for the seven different enzymes involved in the biosynthesis of the amino acid histidine are all transcribed onto a single mRNA molecule. Such a cluster of functionally related genes is called an *operon.*

Studies by François Jacob and Jacques Monod, working at the Pasteur Institute in Paris, showed that the expression of such operons is tightly controlled. The enzymes for the biosynthesis of the amino acid histidine, for example, are only synthesized when there is an insufficient level of histidine available to the cell in the growth medium. Similarly, the tryptophan operon—the cluster of genes for the enzymes involved in tryptophan synthesis—is transcribed only when there is an insufficient supply of tryptophan in the medium. Thus, the concept began to emerge that the DNA molecule consists not only of genes coding for proteins (structural genes, as these are called) but also of regulatory genes, switches that turn the structural genes on and off.

The *lac* Operon

To explore this concept further, Jacob and Monod selected for analysis a group of genes involved in the utilization of lactose as a carbon source—the *lac* operon, as it came to be called. They began by collecting hundreds of mutants of *E. coli*—auxotrophs (page 176)—in which the capacity to use lactose was affected. The *lac* operon,

15–15
The splitting of lactose (milk sugar) to galactose and glucose requires the enzyme beta-galactosidase. Normal E. coli *cells synthesize beta-galactosidase only when lactose is present.*

LACTOSE

H_2O

BETA-GALACTOSIDASE

GALACTOSE

GLUCOSE

they discovered by this method, codes for three proteins. One is beta-galactosidase, which splits the disaccharide lactose into glucose and galactose (Figure 15–15). When lactose is not present in the medium, normal *E. coli* cells have an average of only three molecules of beta-galactosidase per cell. In the presence of lactose, there are approximately 3,000 molecules of the enzyme in every normal *E. coli* cell, representing about 3 percent of all the cell's protein. The second protein is a permease (page 132) that transports lactose through the cell membrane. The third is transacetylase, which transfers an acetyl group from acetyl CoA to galactose. Mutants that lack the capacity to turn off production of these three proteins are at a considerable disadvantage since, by making these enzymes in the absence of a substrate, they squander their energies and resources.

As we mentioned previously, transcription of RNA from DNA depends on the activity of the enzyme RNA polymerase. RNA is transcribed along a DNA strand in the 3′ to 5′ direction. Just upstream of the operon—that is, on the 3′ side—is the site of attachment of the RNA polymerase molecule; transcription from the operon is regulated by preventing or facilitating attachment of the RNA polymerase to this site—the *promoter*—on the DNA strand.

In the absence of lactose, another protein, the *repressor*, prevents attachment of RNA polymerase to the promoter. The repressor binds to the DNA molecule in such a way that it blocks the site of attachment of the RNA polymerase (see Figure 15–16). The site of attachment of the repressor is called the *operator*. When the repressor is bound to the molecule, the RNA polymerase cannot attach itself, and transcription cannot proceed. In the presence of lactose, the repressor molecule is altered and can no longer attach to the operator sequence. Then the RNA polymerase attaches and moves along the DNA strand, transcribing the operon. A substrate, such as lactose, that inactivates the repressor and turns on the operon is known as an *inducer*. Thus, to sum up, beginning at the 3′ end of the DNA strand, there is a promoter site, where the RNA polymerase attaches, and, overlapping the promoter, an operator site, where the repressor attaches, and then the operon itself, consisting of the structural genes (see Figure 15–18).

More recently, another regulatory mechanism for the *lac* operon has been found. This mechanism, which further increases the efficiency of lactose utiliza-

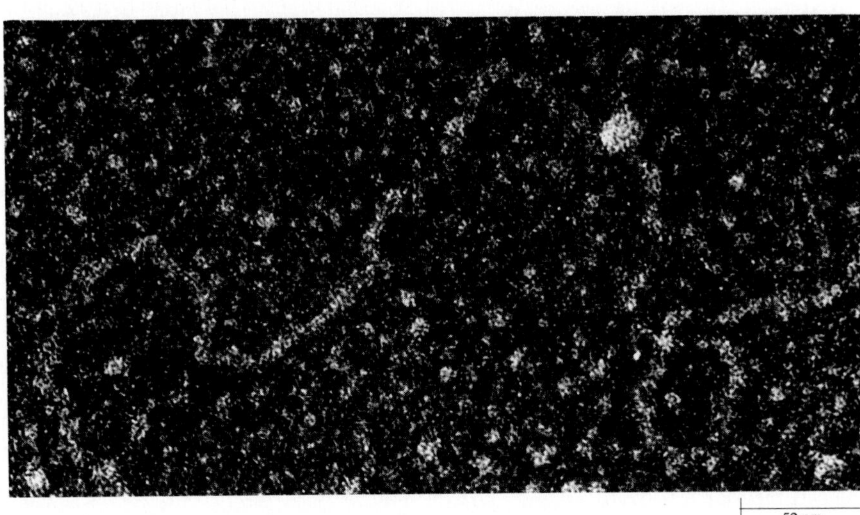

15–16
A repressor protein (the white spherical form at the upper right in the micrograph) attached to the operator of the lac *operon.*

15-17
In an operon system, the synthesis of proteins is regulated by interactions involving a repressor and an inducer. In the lac *operon, for instance, the repressor molecule prevents transcription from the operon unless lactose is present. In the presence of lactose, the repressor is inactivated, and beta-galactosidase and the other proteins are synthesized.* E. coli *cells that cannot produce the repressor are at a disadvantage because they use their resources uneconomically.*

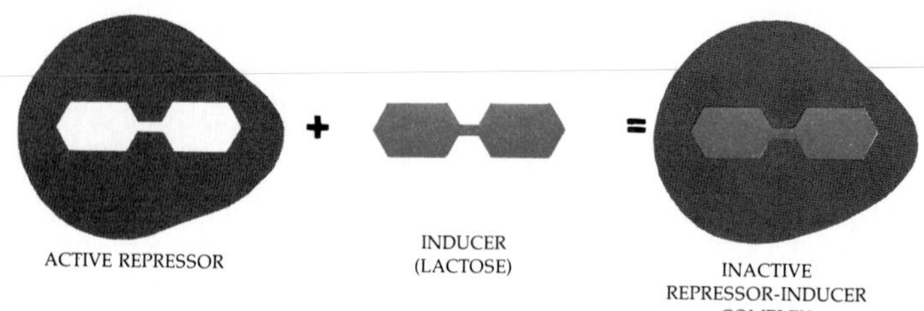

tion, involves a protein known as cyclic AMP binding protein, or CAP. The function of CAP is to inhibit *E. coli* from using lactose as an energy source, even when lactose is present, if some more efficient source of energy—such as glucose—is available. When the supply of glucose in the cell decreases, the level of cyclic AMP increases. (You encountered cyclic AMP previously in Chapter 5.) When cyclic AMP increases, it complexes with CAP, and then—and only then—CAP binds to the DNA strand. RNA polymerase can bind to the promoter site of the *lac* operon efficiently only if the CAP–cyclic AMP complex is already present. Thus two conditions must be met for the *lac* operon to be expressed: lactose must be present, and no carbon source that can be used more efficiently than lactose can be available.

The operon model has now been confirmed for a number of other enzyme systems, some containing as many as 10 structural genes.

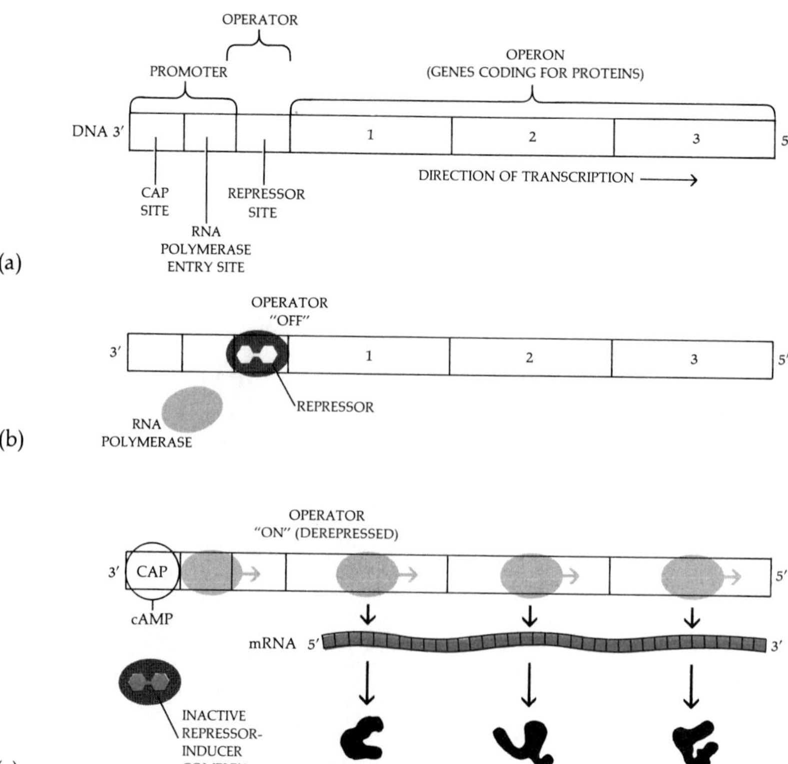

15-18
The lac *operon.* (a) *An operon is a group of structural genes (that is, genes that code for proteins, often enzymes that work sequentially in a particular reaction pathway). Adjacent to the operon on the bacterial chromosome is a segment of DNA that contains the promoter and the operator. The operator slightly overlaps the promoter. The promoter of the* lac *operon is the binding site for the CAP–cyclic AMP complex, as well as for RNA polymerase. RNA polymerase binds efficiently to the promoter only if the CAP complex is in place.*

(b) *In the absence of lactose, the repressor binds to DNA at the operator and so prevents the RNA polymerase from initiating transcription.*

(c) *In the presence of lactose, the repressor is inactivated and can no longer attach to the operator, and synthesis of mRNA proceeds. The genes of the operon are transcribed onto a single mRNA molecule.*

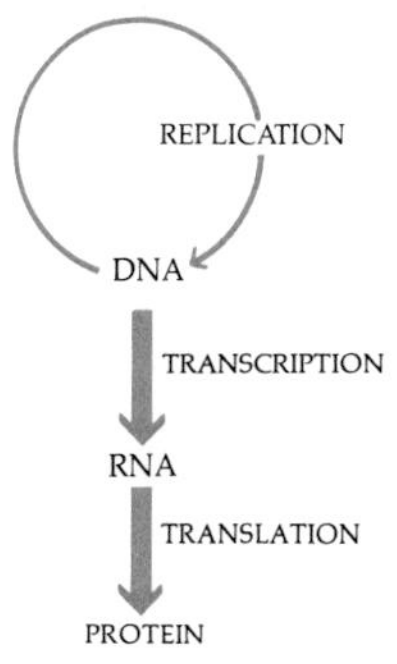

15–19
A summary of the information flow from DNA to protein.

SUMMARY

Genetic information is coded in the sequence of nucleotides in molecules of DNA, and these, in turn, determine the sequence of amino acids in molecules of protein.

The way in which DNA is translated into protein in bacterial cells has been worked out in considerable detail. Each series of three nucleotides along a DNA strand is the DNA code for a particular amino acid. The information is transcribed from the DNA to a long, single strand of RNA (ribonucleic acid). This type of RNA molecule is known as messenger RNA, or mRNA. The mRNA forms along the 3′ to 5′ strand of DNA, following the principles of base pairing first suggested by Watson and Crick. The mRNA therefore is complementary to the transcribed DNA strand.

In bacteria, even as the mRNA strand is being transcribed, its 5′ end becomes attached to a ribosome, a complex structure consisting of two subunits, each composed of specific RNA and protein molecules. At the point where the strand of mRNA is in contact with the ribosome, small molecules of another type of RNA, known as transfer RNA (tRNA), which serve as adapters between the mRNA and the amino acids, are bound temporarily to the mRNA strand. This bonding takes place by complementary base pairing between the mRNA codon and the tRNA anticodon. Each tRNA molecule carries the specific amino acid called for by the mRNA codon to which the tRNA attaches. Thus, following the sequence originally dictated by the DNA, the amino acid units are brought into line one by one and are formed into a polypeptide chain (see Figure 15–20 on the next page).

The genetic code has now been "broken," that is, it is now known which amino acid is called for by a given mRNA codon. Of the 64 possible triplet combinations of the four-letter nucleotide code, 61 combinations have been identified with one of the 20 amino acids that make up protein molecules. The other three triplets serve as "stop signals," terminating protein synthesis.

Mutations are now defined as changes in the sequence or number of nucleotides in the nucleic acid of a cell or organism. Most take the form of substitutions of one nucleotide for another, which may result in the change in an amino acid, as in the hemoglobin of sickle cell anemia.

A principal means of genetic regulation in bacteria is the operon system. An operon is a linear sequence of genes coding for a group of functionally related proteins. The operon is transcribed as a single mRNA molecule. Transcription from the operon is controlled by DNA sequences adjacent to the operon that bind specific proteins: the CAP–cyclic AMP complex, RNA polymerase, and a repressor. The binding site for the repressor is known as the operator. The binding site for RNA polymerase and the CAP complex is known as the promoter. The operator and repressor overlap; therefore, when the repressor is attached to the DNA molecule at the operator, RNA polymerase cannot attach to the promoter, and RNA transcription is blocked. When the repressor is inactivated by the presence of an inducer, RNA polymerase can attach to the DNA, permitting transcription and protein synthesis to take place.

Additional regulation of the operon is provided by the binding of the CAP complex. For example, when a better carbon source than lactose is present, cyclic AMP levels are low, and CAP does not bind to the promoter of the *lac* operon. As that carbon source is depleted, CAP complexes with cyclic AMP and then binds to the promoter. With lactose present and the CAP complex in place, RNA polymerase also binds to the promoter, and transcription from the operon takes place.

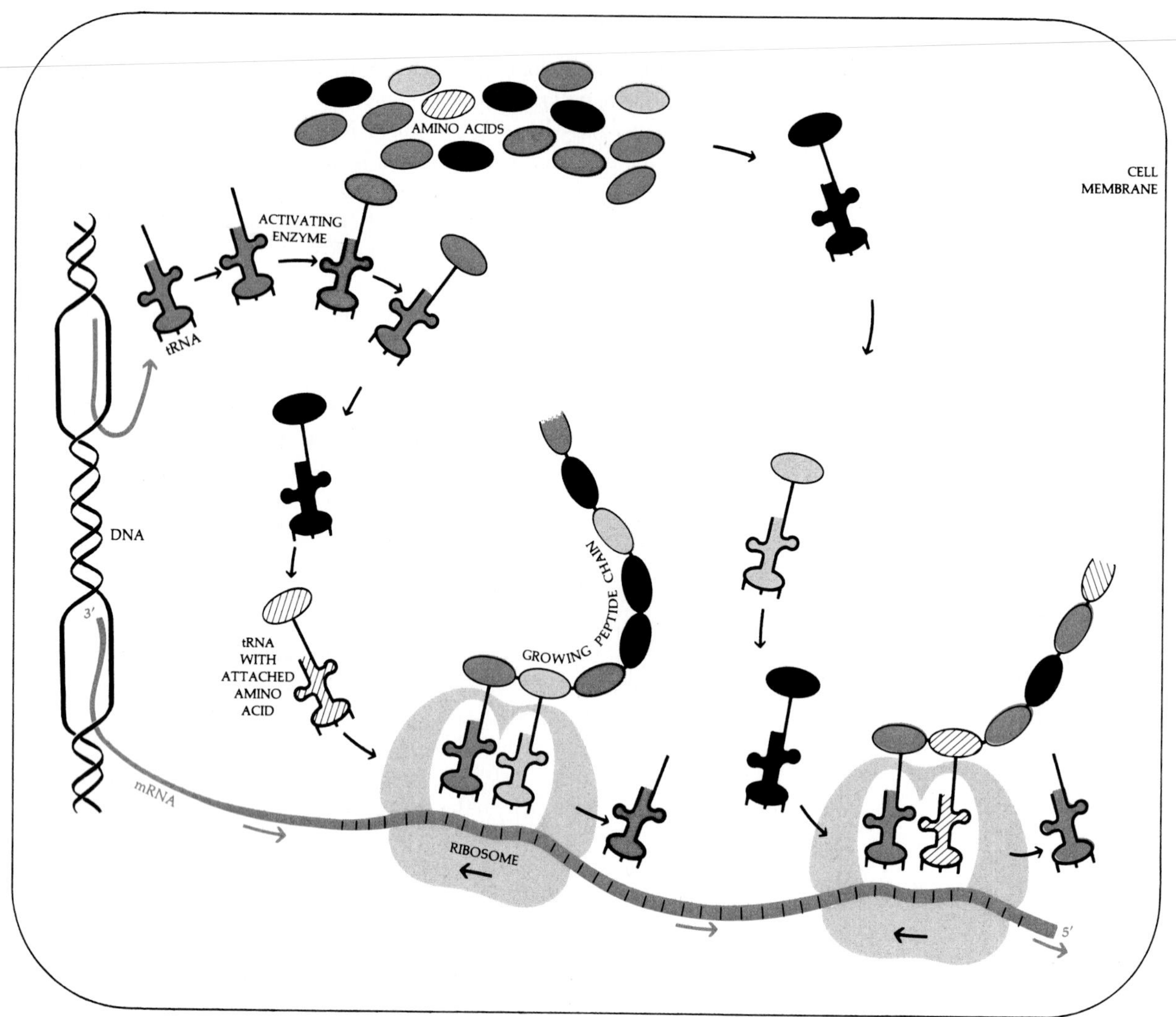

15–20

Summary of protein synthesis in a bacterial cell. At least 32 different kinds of tRNA molecules are transcribed from the DNA of the cell. These molecules are so structured that each can be attached at one end (by a special activating enzyme) to a specific amino acid. Each contains an anticodon that fits an mRNA codon for that particular amino acid.

The process begins when an mRNA strand is formed on the DNA template. At the point of attachment to the ribosome, the matching tRNA molecule, with its amino acid, plugs in momentarily to the codon in the mRNA. As the ribosome moves along the mRNA strand, a tRNA linked to its particular amino acid fits into place and the first tRNA molecule is released, leaving behind its amino acid, which is linked to the second amino acid by a peptide bond. As the process continues, the amino acids are brought into line one by one, following the exact order dictated by the DNA code.

QUESTIONS

1. Distinguish among the following: mRNA/tRNA/rRNA; code/codon/anticodon; transcription/translation; 30S subunit/50S subunit; P site/A site; operator/promoter; repressor/inducer.

2. In a hypothetical segment of one strand of a DNA molecule, the sequence of bases is (3′)-AAGTTTGGTTACTTG-(5′). What would be the sequence of bases in an mRNA strand transcribed from this DNA segment? What would be the sequence of amino acids coded by the mRNA? Does it matter where the transcription from DNA to mRNA begins? Show why.

3. Most of the bacterial DNA codes for messenger RNA, and most of the RNA produced by the cell is mRNA. Yet analysis of the RNA content of a cell reveals that, typically, rRNA is about 80 percent of the cellular RNA and that tRNA makes up most of the rest. Only about 2 percent is normally mRNA. How do you explain these findings? What is the functional explanation?

4. Deletion or addition of nucleotides within a gene leads to changes in the protein produced. The original DNA molecule, the mRNA transcribed from it, and the resulting polypeptide are:

```
T—C—G—T—C—G—T—C—G—T—C—G
:  :  :  :  :  :  :  :  :  :  :  :    DNA
A—G—C—A—G—C—A—G—C—A—G—C

U—C—G—U—C—G—U—C—G—U—C—G   mRNA
\_____/ \_____/ \_____/ \_____/
  ser  →  ser  →  ser  →  ser       polypeptide
```

Deleting the second T-A pair yields the following DNA molecule:

```
      ↓
T—C—G—C—G—T—C—G—T—C—G—T
:  :  :  :  :  :  :  :  :  :  :  :
A—G—C—G—C—A—G—C—A—G—C—A
```

How is the resulting amino acid sequence altered?
How does the addition of a C-G pair to the original molecule,

```
T—C—G—T—C—G—C—T—C—G—T—C
:  :  :  :  :  :  :  :  :  :  :  :
A—G—C—A—G—C—G—A—G—C—A—G
```

affect the amino acid sequence?

5. A culture of bacterial cells is grown in a medium in which both glucose and lactose are present in fixed amounts as the sole carbon sources. Describe the series of events that take place in the operon as the sugars are metabolized.

6. A person heterozygous for the gene for sickle cell anemia makes abnormal hemoglobin molecules but does not suffer from anemia. In what respect is this situation different from the definitions of "dominant" and "recessive" given in Chapter 11?

CHAPTER 16

Recombinant DNA

About a decade ago, just when it seemed as if the "mystery of life itself" had been so satisfactorily accounted for that all that was left was the tidying up of a few details, genetics underwent another revolution. Quite suddenly, molecular biologists became able to manipulate genes in ways never before even imagined—copying, dissecting, and modifying the genetic material and transferring genes between organisms eons apart in evolutionary history. This work has generated an avalanche of studies whose furthest implications have yet even to be imagined.

In the first part of this chapter, we are going to describe studies of ways in which DNA is replicated and recombined in nature. In the last part of the chapter, we are going to sketch the development of the revolutionary new technology by which researchers have themselves become able to replicate and recombine the genetic material. This new technology has come to be known as recombinant DNA, but note that the term applies equally to the processes that living systems accomplish so ably and in such a variety of ways without human intervention.

THE BACTERIAL CHROMOSOME

The knowledge and the impetus for this new work in genetics came largely from microbiology and, in particular, from studies of the familiar (but never boring) *E. coli.* The chromosome of *E. coli,* it is now known, is a continuous thread of double-stranded DNA almost a millimeter long but only 2 nanometers in diameter. It contains 2.2 million base pairs. The molecule must be folded tightly to fit into the cell, which is less than 2 micrometers long, about 1/500th the length of the chromosome. The packaging apparently involves both RNA and proteins, since treatment with specific RNA- and protein-digesting enzymes causes the DNA molecule to unfold. The packaging process must be quite precise because (1) genes located all along the molecule are continuously expressed and therefore must be available to RNA polymerase, and (2) despite the fact that the chromosome must be folded many thousands of times, it replicates in such a way that the two daughter molecules can separate without tangling.

The DNA molecule can be separated intact from the cell by using very gentle procedures. Electron microscopy has shown that the unfolded molecule forms a circle (which, in fact, was already known from some ingenious studies to be described later in this chapter) and that replication starts at a single initiation point and travels in both directions around the circle (Figure 16-1). This is known as theta replication since the replicating molecules resemble the Greek letter theta (Θ). As we saw in Chapter 7 (page 140), the replicated molecules are attached to the cell membrane, which ensures their separation at cell division.

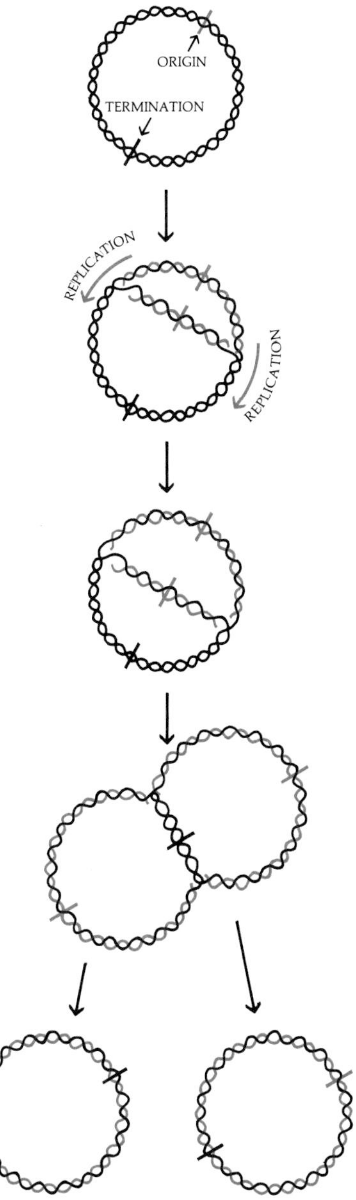

16-1
Diagram of bidirectional (theta) replication of the DNA of a circular chromosome, with points of origin and termination of replication indicated in color and black, respectively. The newly synthesized DNA strand is shown in color.

PLASMIDS AND CONJUGATION

Although the bacterial chromosome carries all the genes necessary for growth and reproduction of the cell, virtually all types of bacteria have been found to carry additional genes on much smaller DNA molecules called *plasmids*. Plasmids may carry as few as two genes or as many as thirty. Like the bacterial chromosome, they are circular and self-replicating. They usually replicate in synchrony with cell division by the theta mechanism (Figure 16-2) and are passed on to daughter cells. About a dozen different kinds of plasmids have been described in *E. coli* alone.

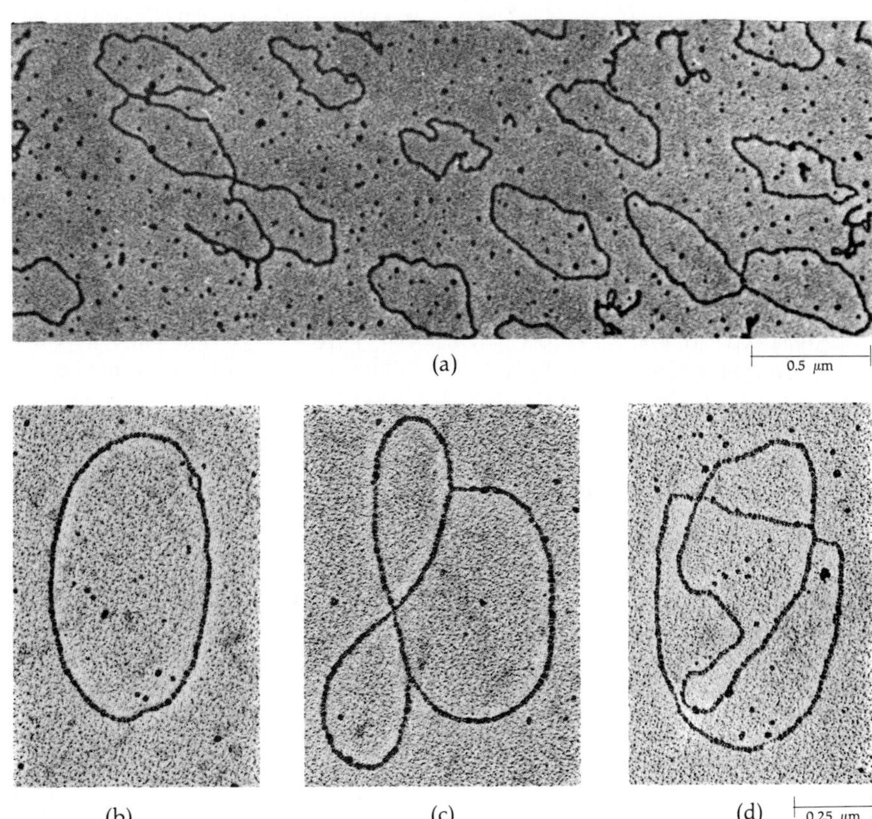

16-2
(a) *Plasmids from* Neisseria gonorrhoeae, *the bacterium that causes gonorrhea. The two pairs of connected plasmids visible in the micrograph are probably just completing replication.* (b) *A plasmid replicating. At almost 2:00 in the plasmid, you can see the two DNA double helixes beginning to separate (each consisting, as you will remember, of one old strand and one new one).* (c) *The replication process is more than half completed, and* (d) *the two plasmids are almost at the point of separation.*

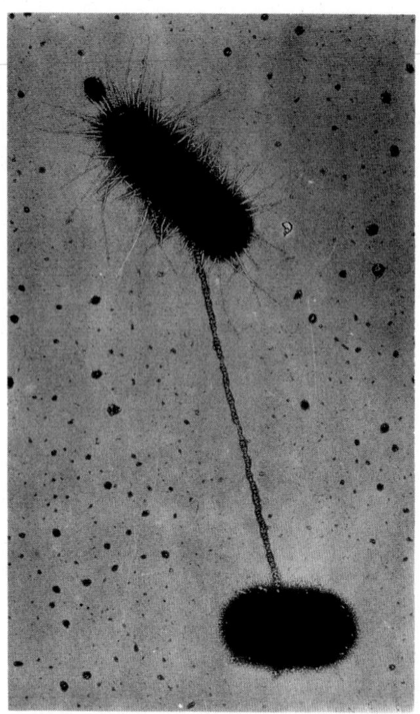

16–3
Electron micrograph of conjugating E. coli *cells. The F^+ cell at the top of the micrograph is connected to an F^- cell by a single, long pilus. Genes on the F plasmid are responsible for the production of these specialized pili, which are necessary for conjugation. Numerous shorter pili are visible on the F^+ cell.*

The F Plasmid

The first plasmid to be recognized as such was the F (for fertility) factor of *E. coli.* The F plasmid contains some 25 genes, many of which control the production of F pili, which are long appendages that extend from the surface of the cells containing the F plasmid. These cells, known as F^+ cells, can attach themselves to F^- cells (cells lacking the F plasmid) by the pili and transfer the F plasmid to the F^- cells, giving them the capacity to produce the special pili and to transfer the F plasmid. The transfer of DNA from one cell to another by cell-to-cell contact is known as *conjugation.*

The transfer of an F plasmid is diagrammed in Figure 16–4. Note that it involves a different mode of replication, best known as the rolling circle.

The F factor, like many other plasmids, can become spliced into the bacterial chromosome (Figure 16–5). Once integrated, it is no longer replicated independently but is passively copied along with the chromosome. A bacterial cell that contains the F plasmid as part of its chromosome is known as an Hfr (high frequency of recombination) cell. An Hfr cell has an astonishing property: when it attaches to an F^- cell, the bacterial chromosome itself (or a portion thereof) can be transferred from the Hfr cell to the F^- cell. In other words, the genes in the bacterial chromosome can be passed from one cell to another, resulting in a new gene combination in the recipient cell. Conjugation is thus, in effect, a form of mating.

In the Hfr cell at conjugation, one strand of the circular DNA molecule of the bacterial chromosome breaks and opens up within the F plasmid sequence, and rolling-circle replication begins. The 5′ end of a single strand of DNA passes from the Hfr cell to the F^- cell. Recombination may then occur between the chromosome of the recipient cell and portions of the chromosome of the donor cell, with new material replacing the old in the chromosome. By selecting suitable auxotrophic strains of bacteria, such recombinations can be detected.

Chromosome Mapping

Studies of bacterial conjugation revealed the fact that the genes of a bacterium are arranged in a regular linear order and that the bacterial chromosome is circular. (This surprising information was later confirmed by electron microscopy.) The essential laboratory tools in these conjugation experiments were a kitchen blender and a timer. Stop here a moment and see if you can figure out the role of these two comparatively humble instruments in mapping the bacterial genome. Here is a clue: transmission of the entire chromosome requires about an hour and a half at 37°C and proceeds at a constant rate.

During the approximately 90 minutes required for the transfer of the chromosome, a strand of DNA makes its way into the recipient cell. By separating the cells at various points in the process (which can be accomplished by whirling them at high speed in an electric blender) and then analyzing which genes have been transferred, it is possible to construct a map of the chromosome. The mapping process confirms that the genes are carried in a linear array on the chromosome: *A* is followed by *B*, *B* by *C*, and so on, down to *XYZ*. As different Hfr strains were studied, it was found that insertion sites for the F plasmid differed from strain to strain and that the chromosome could be transferred in either direction, depending on the orientation of insertion of the F plasmid. It was these studies that also first gave a clue that the *E. coli* chromosome is circular; that is, *Y* followed *X*, and *Z* followed *Y*, but next would come *A* and then *B*.

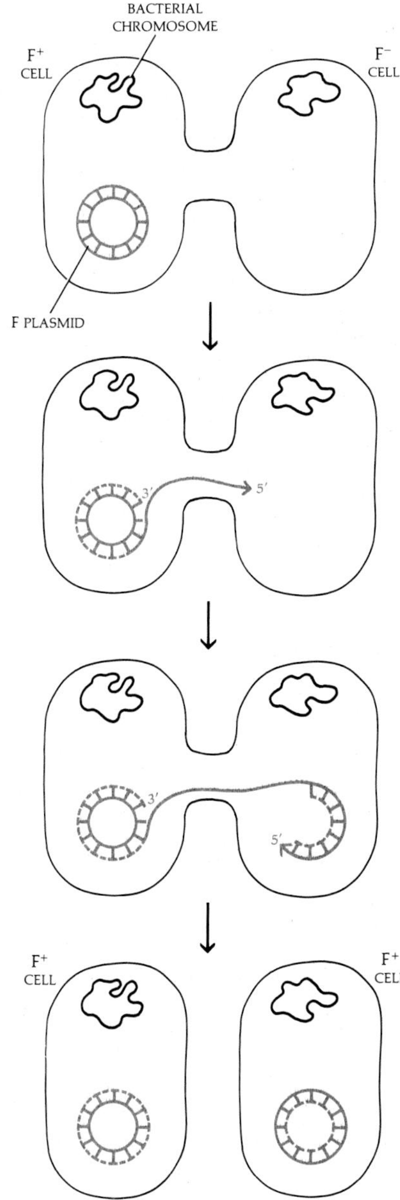

16–4
Transfer of an F plasmid from an F^+ cell to an F^- cell via rolling-circle replication. The plasmid converts the recipient cell to an F^+ cell. Transfer of DNA by cell-to-cell contact is known as conjugation.

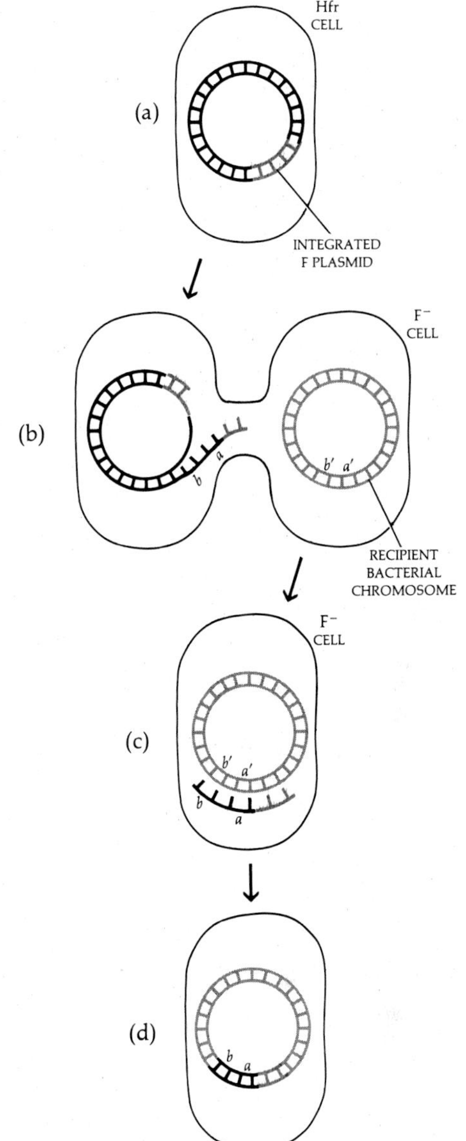

16–5
Bacterial conjugation. (a) *An F^+ cell is converted into an Hfr cell when an F plasmid becomes inserted into its chromosome.* (b) *A break occurs in the F plasmid sequence in the chromosome, and rolling-circle replication begins. Leading by its 5′ end, a single strand of the DNA containing a portion of the F sequence and the attached chromosomal DNA moves into the recipient F^- cell.* (c) *Donor DNA can recombine with recipient DNA, altering the genotype of the recipient bacterium* (d).

16-6
Known as R6, this plasmid confers resistance to six different drugs, including the antibiotics tetracycline, neomycin, and streptomycin.

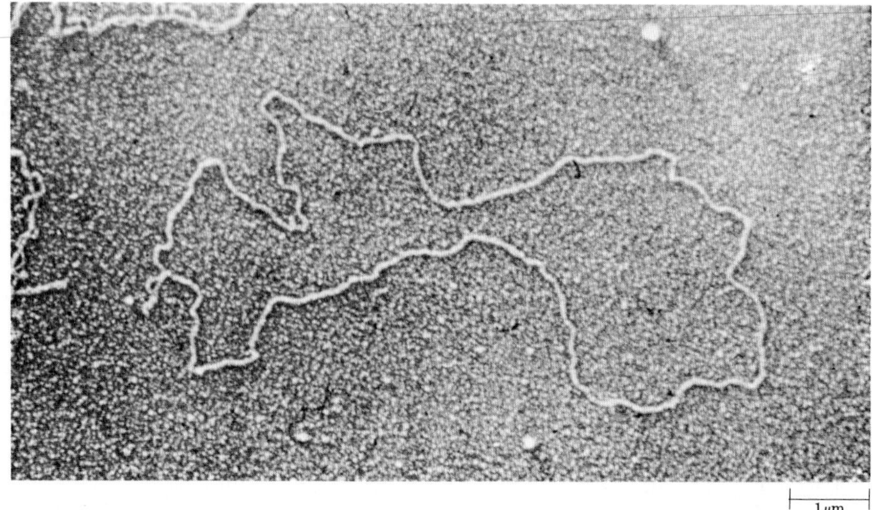

R Plasmids

In 1959, a group of Japanese scientists discovered that resistance to certain antibiotics and other antibacterial drugs can be readily transferred from cell to cell. It was subsequently found that the genes conveying drug resistance are often carried on plasmids (called R plasmids). Such plasmids can be transferred from cell to cell, like the F plasmid, and, under experimental conditions, 100 percent of a population of drug-sensitive cells can become resistant within an hour after being mixed with suitable drug-resistant bacteria.

Resistance genes can be transferred from one R plasmid to another, with a single plasmid collecting as many as 10 genes, making the cell it inhabits (and any it infects) resistant to 10 antibiotics. Resistance genes can also be transferred from plasmids to the bacterial chromosome, to viruses, and, most disturbing of all, to bacteria of other species. Thus the usually innocuous *E. coli* can pick up R plasmids by conjugation and transfer them to *Shigella,* a bacterium capable of causing a sometimes fatal form of dysentery. Infectious resistance has now been found among an increasing number of types of pathogens, including those responsible for typhoid, gastroenteritis, plague, undulant fever, meningitis, and gonorrhea.

Some R plasmids do not carry genes for the production of pili. As many as a hundred copies of such small plasmids may exist in a single cell and be transferred from mother to daughter cells at cell division. Also, if another plasmid is present in the cell that can initiate conjugation, these small R plasmids can be transferred at the same time.

VIRUSES AND TRANSDUCTION

Viruses, like bacteria, contain chromosomes that carry information for their own replication. The genetic material of a virus may be either DNA or RNA. As we saw in Figure 14-6, viruses multiply by inserting their nucleic acid into the host, where it is replicated. In the case of the DNA viruses, the DNA serves as the template for new DNA and for the transcription of messenger RNA. In the case of the RNA viruses, the RNA serves both as the template for more RNA and also as messenger

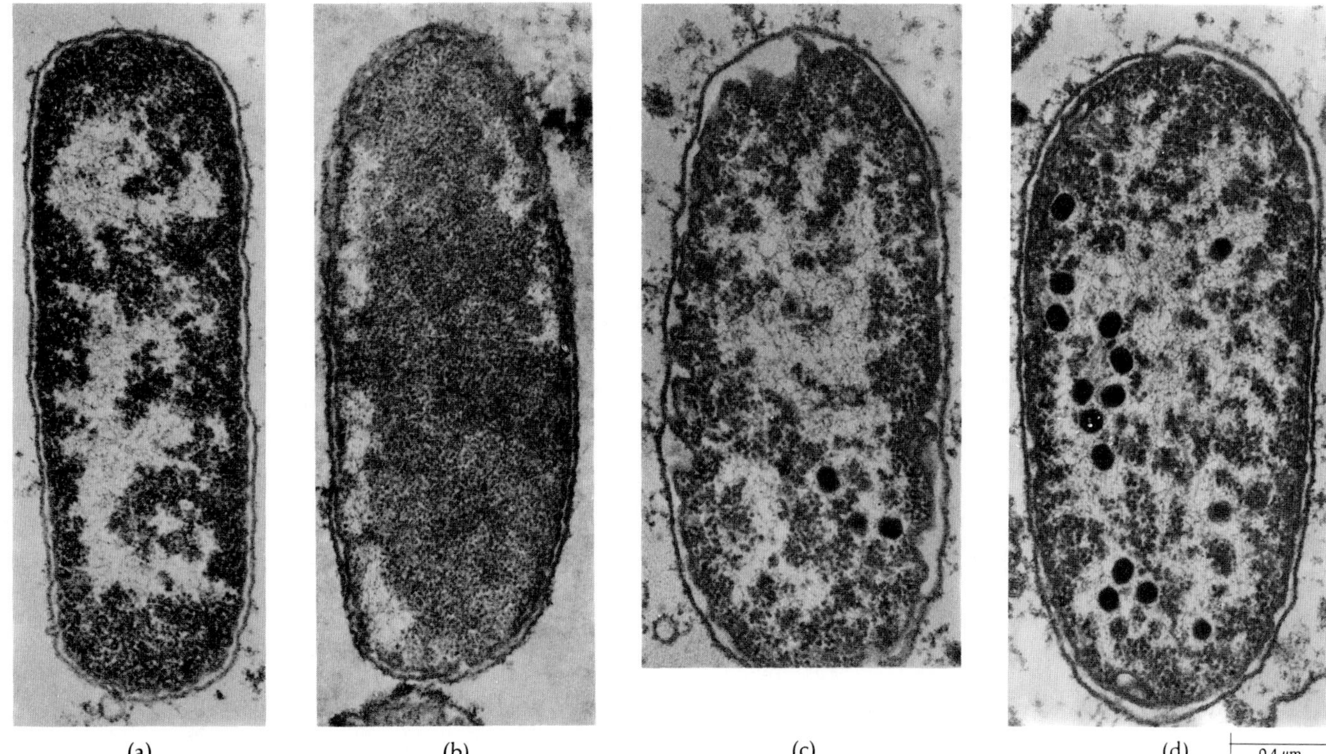

16–7
Stages in the replication of T4 bacteriophage in cells of E. coli. (a) *As the viral infection begins, the DNA of the bacterial cell is visible as electron-transparent areas.* (b) *After 5 minutes, the regions of DNA have changed form and moved toward the cell membrane.* (c) *After 15 minutes, the bacterial DNA has disappeared, replaced by vacuoles containing threads of replicating phage DNA; transcription from phage DNA is occurring simultaneously. The cell also contains phage "pre-heads," which are thought to contain little or no DNA. These "pre-heads" mature into complete, DNA-containing heads.* (d) *After 30 minutes, many phages, both complete and incomplete, are present.*

RNA. The messenger RNA codes for the protein coat of the virus, which may consist of only one type of protein repeated over and over, as in TMV (Figure 15–1), or of many different kinds, as in the T-even bacteriophages with their complex tails. It also codes for a few essential enzymes. Viruses use the raw materials of the cell—such as nucleotides and amino acids, the cell's ATP and other energy resources, and the cell's metabolic machinery. Hence they are obligate parasites; they cannot multiply outside the cell.

The number of genes carried by a viral chromosome varies widely from one virus type to another. The large bacteriophages of the T-even group carry approximately 100 genes, and their chromosomes are about 1/100th the size of *E. coli*'s. Small DNA bacteriophages such as ϕX174 contain only about 1/100th as much DNA as T4. There are RNA bacteriophages that are smaller still, and, by overlapping genes, they encode the information for four proteins in just 3,600 nucleotides of single-stranded RNA.

In some of the animal RNA viruses, as in the RNA bacteriophages, the RNA serves both as the template for more RNA and also as messenger RNA. In other RNA viruses (about which we shall have more to say in Chapter 17), the RNA codes for an enzyme known as *reverse transcriptase*, which transcribes the RNA to DNA. Reverse transcriptase is found in nature only in association with these *retroviruses* (from the Latin *retro*, "turning back").

With both RNA and DNA viruses, the infectious cycle is complete when the nucleic acid is packaged into the newly synthesized protein coat and the virus particle escapes from the host cell, ready to begin another cycle of infection.

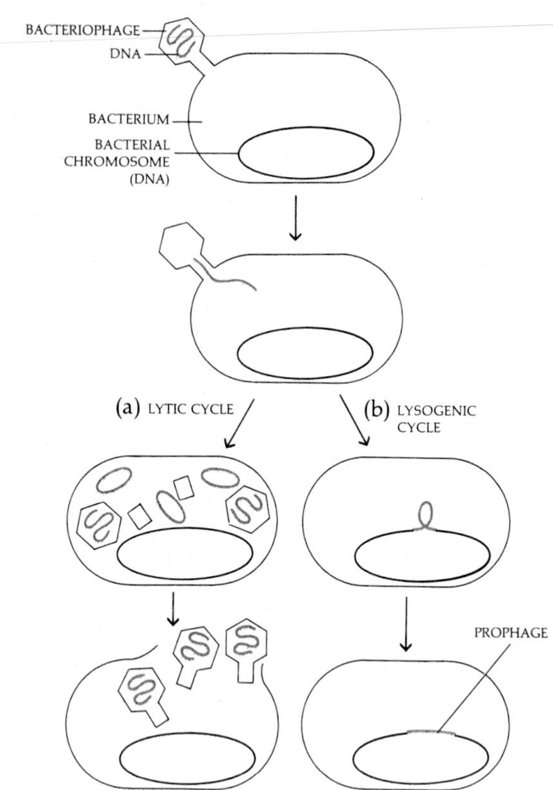

16-8
When certain types of bacterial viruses—temperate bacteriophages—infect bacterial cells, one of two events may occur. (a) *The viral DNA may enter the cell and set up an infectious, or lytic, cycle, such as we described in Chapter 14, or* (b) *it may become part of the bacterial chromosome, replicating with it so that it is passed on to daughter cells. Bacteria harboring such viruses are known as lysogenic. From time to time, such a virus, called a prophage, becomes activated and sets up a new infectious cycle.*

Temperance and Lysogeny

Early in the study of bacteriophages, it was noted that a virus infection could suddenly erupt in a colony of apparently uninfected bacterial cells. Such cells were termed lysogenic, because of their capacity to generate a cycle of cell lysis that spread through neighboring cells. The cause of lysogeny, it was discovered, was the capacity of certain viruses to set up a long-term relationship with their host cell, lying latent for many cellular generations before initiating an infectious cycle. Such viruses became known as *temperate bacteriophages*. These temperate phages, like the F plasmid, may become integrated into the host chromosome, replicating along with the chromosome. Such integrated bacteriophages are sometimes called *prophages*. Prophages break loose from the host chromosome spontaneously about once in every 10,000 cell divisions. Lysogeny may also be triggered in the laboratory by x-rays, ultraviolet light, or other agents that damage nucleic acids.

Transduction

As we noted, in the course of an infectious cycle, viruses exploit the resources of the host cell. The host DNA, in the course of such an infection, becomes fragmented, and when the viruses leave the cell, they may carry with them some of these fragments of the host chromosome in place of their own DNA.

Since the amount of DNA that can be packaged into the protein coat is limited, such viruses may lack some of their own necessary genetic information and so may not be able to complete an infectious cycle. However, the genes from their previous host may become incorporated into the chromosome of the new host. If the bacteria are auxotrophs, these recombinations can be detected. The transfer of genetic material from cell to cell by viruses is known as *transduction*, and the process just described is called *general transduction*.

When prophages break loose from the host chromosome to initiate a lytic cycle, they may, similarly, take a fragment of the host chromosome with them. In this case, the fragment is adjacent to the site of insertion of the prophage. When this occurs, the DNA fragments are not picked up at random but are, quite specifically, the portions of the host chromosome adjoining the insertion site. This transduction process is known as *restricted transduction*.

Transduction resembles conjugation in that it involves the transfer of bacterial genes from one bacterial cell to another. It differs from conjugation in that in transduction, the genes are carried by viruses.

Introducing Lambda

Lambda is the best studied of the temperate bacteriophages. There are several interesting and instructive features about the lambda chromosome. First, when the double-stranded DNA of the virus is packaged in the virus particle, it is linear—that is, it has two free ends. Once it is released in the cell, however, the two ends close because they are "sticky." On the 5′ end of each strand of DNA of the chromosome are protruding single strands of 12 nucleotides, which, as you can see in Figure 16–9, are exactly complementary to one another and so seal the molecule in a circle.

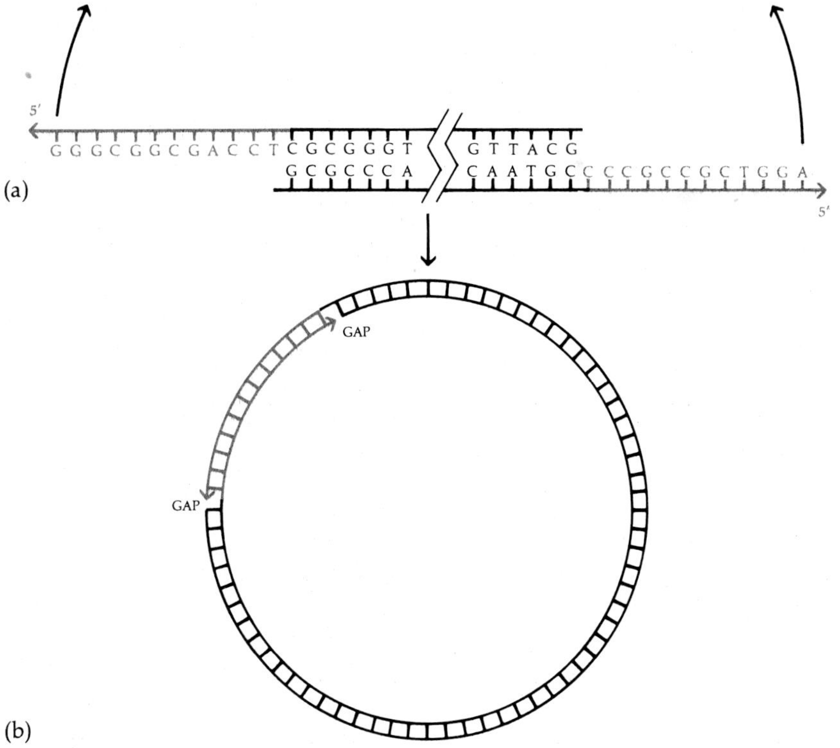

16–9
(a) *A portion of the lambda chromosome, showing the nucleotide sequences of the single-stranded "sticky" ends.* (b) *The DNA molecule forms a circle when these single strands come together and their complementary bases are paired. A special enzyme, DNA ligase, forms bonds between the phosphate groups, closing the "gaps."*

16–10
Integration of lambda DNA into the chromosome of E. coli. (a) *Lambda integrase recognizes the attachment sites on the DNA molecules of lambda and* E. coli *and brings them into close proximity.* (b) *The enzyme catalyzes the breakage of both circular DNA helixes and* (c) *their joining together,* (d) *resulting in the incorporation of the lambda prophage into the* E. coli *chromosome.*

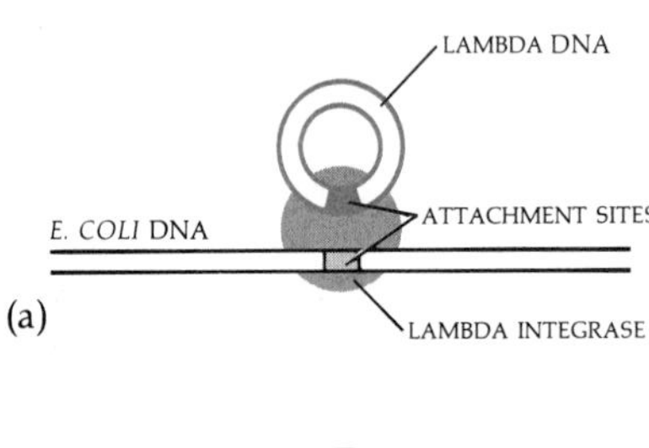

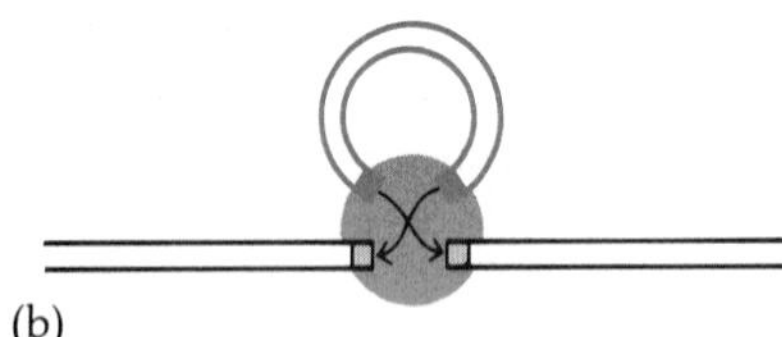

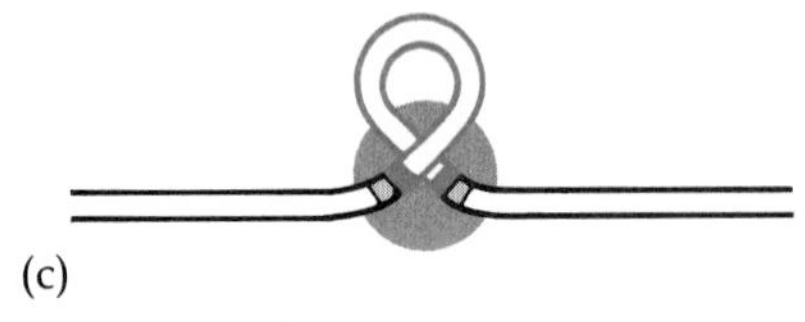

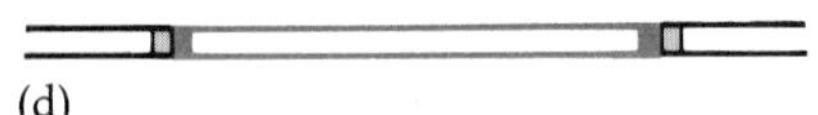

Second, lambda's integration into the *E. coli* chromosome depends on two short DNA sequences, or attachment sites, one on the bacterial chromosome and one on the bacteriophage DNA. Attachment is brought about by a special enzyme, lambda integrase, that recognizes both sequences, brings the two circular DNA helixes together, and initiates the cutting and sealing reactions (Figure 16–10). The enzyme is coded for by the virus.

Third, the lambda chromosome has been shown to consist of several operons that work sequentially. The operons are turned off when lambda is in its prophage state; turning on the first operon initiates the lytic cycle. Lambda codes for its own repressor protein when the virus is in its prophage state.

Thus temperate phages, as exemplified by lambda, resemble plasmids in that (1) they are self-replicating molecules of DNA, and (2) they may become integrated into the bacterial cell chromosome. They differ from plasmids in their capacity to manufacture a protein coat and thus to exist (though not to replicate) outside the cell, and also in their capacity to lyse the host cell.

MOVABLE GENETIC ELEMENTS

In the course of these studies, the picture began to emerge that genetic recombinations occur all the time in nature. They are of two basically different kinds. One kind involves exchanges between homologous segments of DNA. When two such segments of double-stranded DNA are aligned with one another, exchanges occur between the molecules in such a way that genes may be transferred from one molecule to the other. This phenomenon occurs as recombination in eukaryotic cells, as a result of crossing over; it also takes place during conjugation, transformation, and transduction in bacterial cells. (Transformation, you will recall—page 279—involves the transfer of free DNA into living cells.) Several models of how recombination takes place have been proposed; one of them, the "single-strand switch," is seen in Figure 16–11.

Studies of the bacterial chromosome, plasmids, and viruses led to the recognition of the second kind of gene transfer, the insertion of movable (and removable) genetic elements. The F plasmid and lambda phage were early examples. More recently two other kinds of movable genetic elements have been recognized: insertion sequences and transposons.

Insertion Sequences

Most mutations, as we noted in the previous chapter, are caused by the substitution of one nucleotide pair for another. As mapping analysis of *E. coli* became more and more detailed, however, another major source of mutations was uncovered: breaks in the continuity of genes resulting from the insertions of short sequences of DNA. The intriguing thing about these sequences is that the identical sequence often turns up in a number of different, independent mutants. The sequences range from 600 to 6,000 base pairs in length, hardly a gene's worth of DNA. These *insertion sequences* have no independent existence, as do viruses and plasmids. They are found only on the chromosome of a bacterium, a plasmid, or a virus. However, they have one extremely unusual property: while on the chromosome, they can generate copies of themselves that can then become inserted somewhere else in the same molecule or in another molecule—such as a plasmid or virus—inhabiting the same cell. Although they are commonly referred to as movable, they do not actu-

16–11
Diagram of genetic recombination between two homologous strands of DNA by the "single-strand switch" model. (a) *The homologous parental DNAs are indicated in color and black.* (b) *One strand of each DNA is broken,* (c,d) *switched over to the other molecule, and* (e) *joined to the opposite switched strand.* (f) *The exchange of strands between DNAs proceeds along the chromosome, and* (g) *at a specific point the switched strands are again broken and* (h) *resealed, completing the exchange and recombination of the genes.*

(a)

(b)

(c)

(d)

(e)

(f)

(g)

(h)

ally move in the sense of leaving the chromosome, as lambda does. Only a copy moves. Also unlike lambda, this copy apparently can become inserted anywhere in another DNA molecule.

Transposons

When two insertion sequences flank a DNA sequence, the entire sequence can move from place to place on a chromosome or from chromosome to chromosome. Such "jumping genes" are known as *transposons*. Transposons differ from insertion sequences in that (1) they are phenotypically detectable—that is, they function as genes, and (2) they require insertion sequences for their mobility. Resistance genes are transposons, it is now known, which is why they can jump so readily from plasmid to plasmid and from plasmid to bacterial chromosome to plasmid again.

Recently a transposon has been discovered that can serve as an on-off switch for protein production. The bacterium *Salmonella* can produce either one of two types of flagellar proteins. These proteins, which are responsible for the movement of the bacterial cells, trigger host resistance against the bacteria. Which gene is expressed is determined by a 950-base-pair sequence of DNA that regulates a small operon coding for one of the flagellar proteins and for a repressor of the gene for the other flagellar protein. If this 950-base-pair sequence is oriented in one direction, the first flagellar protein is synthesized and the second repressed. If the sequence is oriented in the other direction, the repressor is not produced and the second flagellar protein is synthesized. Thus, in effect, the regulatory segment is a transposon that serves as a flip-flop switch.

RESTRICTION ENZYMES

In the course of studies on lambda bacteriophage, a discovery was made that won the Nobel Prize and changed the course of DNA research. Lambda bacteriophages produced by one strain of *E. coli* were sometimes not able to infect another strain.

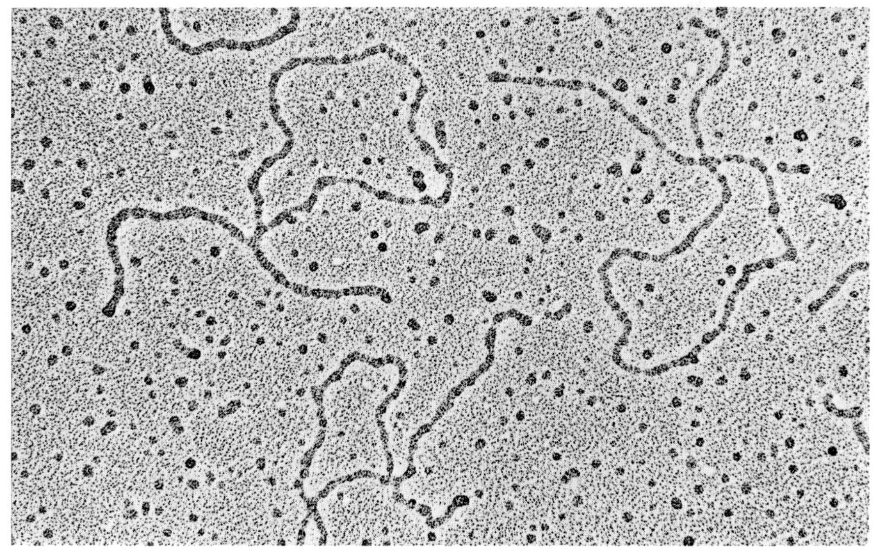

16–12
Transposons can carry genes conferring resistance to antibiotics from R plasmid to R plasmid and also into and out of the bacterial chromosome. This micrograph shows three transposons from a gonorrhea-causing bacterium that confers resistance to the antibiotic ampicillin.

16–13
(a) *Simian virus 40, or SV40, has an icosahedral (20-sided) protein coat enclosing its chromosome, a circular molecule of double-stranded DNA.* (b) *Because the SV40 chromosome has only one GAATTC sequence, Eco RI cleaves the SV40 DNA at only one point, which becomes the reference point for other manipulations.* (c) *The restriction enzyme Hin cleaves at 11 points, yielding a total of 12 fragments.* (d) *Hpa cleaves at only 4 points, yielding 5 fragments. The capacity to cleave the DNA molecule into uniform fragments led rapidly to the sequencing of its nucleotides.*

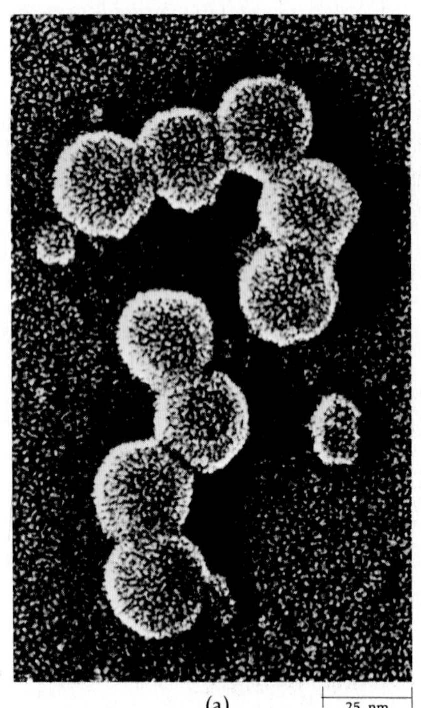

(a)

This restriction of growth occurs, it was later discovered, because of the presence in bacteria of special enzymes that came to be known as *restriction enzymes*. These enzymes cleave DNA molecules at specific sites, characterized by short sequences of nucleotides. The bacteria protect their own DNA from the restriction enzymes by adding a methyl group to one or more of the nucleotides in the recognition sequence. Methylation is also accomplished by special enzymes, which are commonly found paired with restriction enzymes. One restriction enzyme, for example, called Eco RI, cleaves DNA only at the sequence GAATTC. This enzyme is found in association with a methylating enzyme that methylates one of the adenines in the GAATTC sequence, thus protecting it from recognition. Bacteriophages, such as lambda, that are synthesized in a strain of *E. coli* with a particular methylating enzyme are similarly protected and so can multiply in those strains but not in others. More than 100 different restriction enzymes have now been isolated from different strains of bacteria.

Restriction enzymes have two extremely important features for researchers working in this field. The most important is that different enzymes cleave at different specific sites (Figure 16–13). As you may recall (page 72), the sequencing of amino acids in protein molecules was made possible by the discovery of chemical reagents and specific enzymes that cleave the macromolecules into short, uniform fragments susceptible to laboratory analysis. Similarly, discovery of restriction enzymes has made possible the sequencing of the giant molecules of DNA, now going forward in hundreds of laboratories around the world.

The second important feature of these enzymes is that not all of them make straight cuts; some cleave the two strands a few nucleotides apart. This can have interesting consequences. Look again at the sequence at which Eco RI cleaves the molecule: GAATTC. Now visualize the double-stranded molecule. It would be

5′ . . . GAATTC . . . 3′
3′ . . . CTTAAG . . . 5′

Reading in the 5′ to 3′ direction, Eco RI cleaves each strand between the guanine and the adenine:

↓
5′ . . . GAATTC . . . 3′
3′ . . . CTTAAG . . . 5′
↑

What is left, as you will readily see, is

. . . G   AATTC . . .
. . . CTTAA and   G . . .

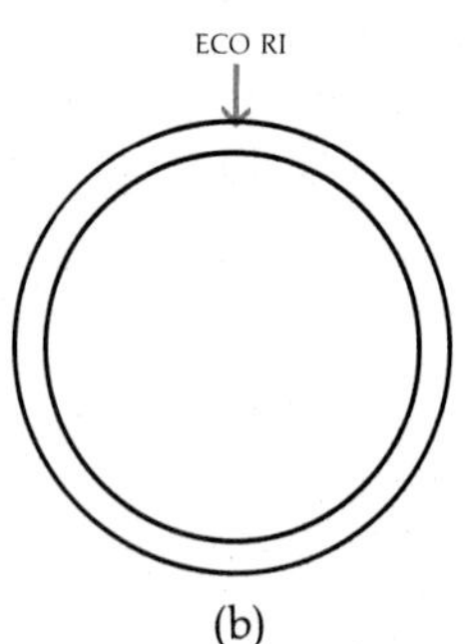

(b)

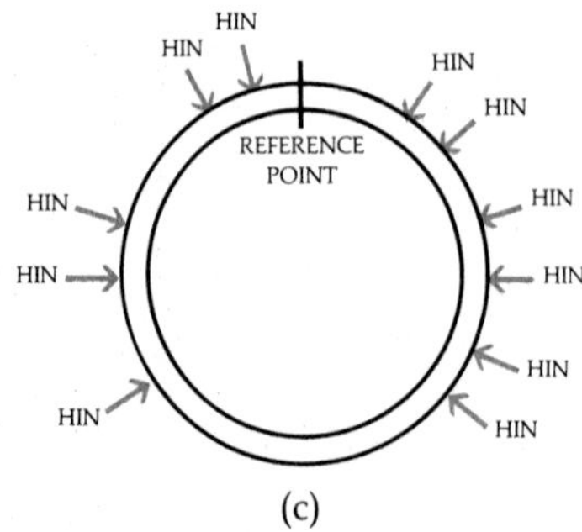

(c)

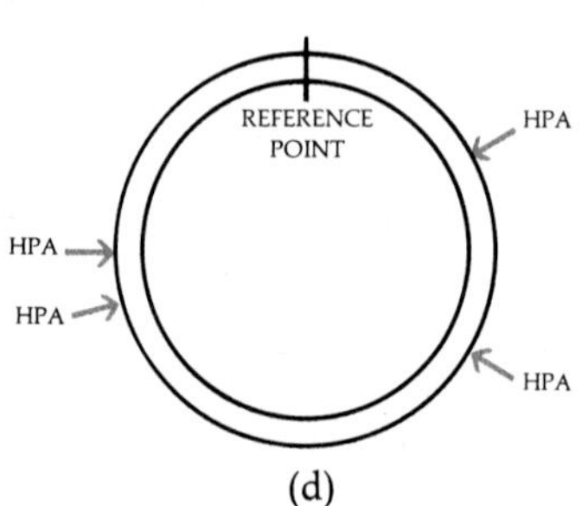

(d)

The "sticky" ends, . . . TTAA and AATT . . . , can join again with each other and also—and this is most important—with any other segment of DNA that has been cleaved by the same enzyme. Thus it is now possible to splice together segments of DNA from, apparently, a limitless variety of sources. (Such DNA molecules have been called chimeras after the mythological beast that was part lion, part goat, and part snake.)

SOME TRICKS OF THE TRADE

Work in microbial genetics gave molecular geneticists both the impetus and the tools to develop their own recombinant technology, which has now begun to rival the recombinatorial powers of the bacteria, viruses, and plasmids that were their mentors. In the next few pages we are going to review briefly some of the fundamental techniques of this fast-moving field. By oversimplifying, we make the procedures seem easier than they actually are, but the underlying ideas are so ingenious that they are what we would like to emphasize. In the next two chapters, we are going to give some examples of the research advances these new techniques have made possible—especially in the increased knowledge of the molecular genetics of eukaryotes, in general, and of humans, in particular.

DNA Cloning

One of the prerequisites of most of the current work in investigating and manipulating genes is the availability of multiple copies of the DNA segments being studied. These copies have been produced by harnessing the replicative powers of plasmids and viruses. Not long after the discovery of Eco RI, Stanley Cohen and coworkers from the Stanford University Medical School isolated a small plasmid of *E. coli,* designated pSC101 (note the initials of its discoverer), that makes the bacteria resistant to the antibiotic tetracycline. pSC101 has only one GAATTC sequence in its entire molecule, and as a consequence, it is cleaved at only one spot by Eco RI. These investigators have shown that the insertion of an additional segment of DNA into the plasmid does not affect either the uptake of the plasmid by *E. coli,* its capacity to make the recipient cells tetracycline-resistant, or its ability to replicate. Once replicated, the pSC101 chimera can be readily harvested and separated out on the basis of its density. The treatment of the isolated plasmids by Eco RI then releases multiple copies of the cloned DNA sequence that can be used in research.

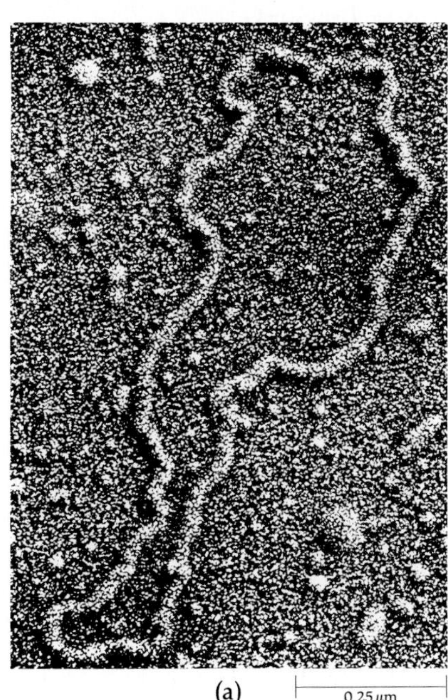

(a)

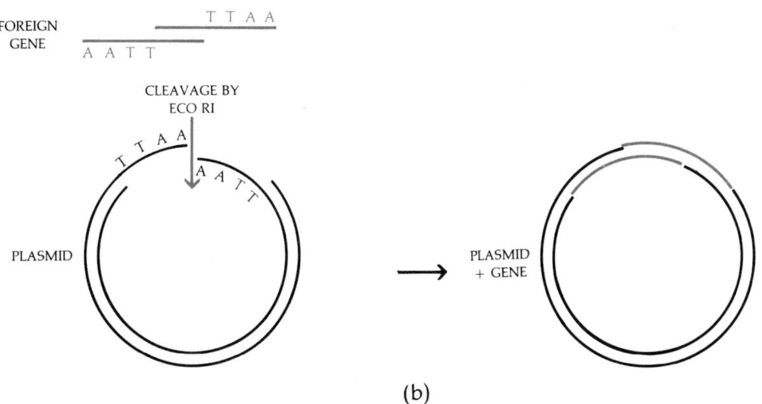

(b)

16–14
(a) *pSC101.* (b) *The enzyme Eco RI cleaves the plasmid pSC101 at the sequence GAATTC, leaving "sticky" ends exposed. These ends, consisting of TTAA and AATT sequences, can join with any other segment of DNA that has been cleaved by the same enzyme. Thus it is possible to splice a foreign gene into the plasmid. Once they are released into the medium in which bacteria are growing, such plasmids will be taken up by some of the cells and will multiply rapidly. Plasmids can then be separated from the bacterial chromosome and treated with Eco RI to release multiple copies of the cloned gene.*

Lambda phage, other plasmids, and the bacterial chromosome itself can be used similarly for making multiple copies. It is of interest that foreign genes, inserted into such carriers and introduced into *E. coli,* can produce mRNA and also proteins using the translation machinery of *E. coli.* This establishes once again the universality of the genetic code.

DNA Sequencing

In 1977, a complete genome was sequenced for the first time, that of the bacteriophage ϕX174 (see page 304). Frederick Sanger shared the Nobel Prize in 1980 for this accomplishment. It was his second. His first, received 22 years before, was for the first sequencing of a protein, insulin. A few months after the sequencing of ϕX174, investigators reported the complete nucleotide sequence for the DNA of simian virus 40 (SV40).

The development of techniques for cleaving DNA molecules into smaller pieces and cloning them into multiple copies now makes it possible, in principle, to determine the sequence of any isolated gene. One method is that worked out by Allan Maxam and Walter Gilbert (who shared the Nobel Prize with Sanger). Following this method, a collection of uniform DNA fragments is prepared by a restriction enzyme and is labeled on one end by removing a phosphate and substituting a radioactive one in its place. Then the mixture of identical labeled DNA molecules is separated into four separate portions. Each portion is treated chemically in such a way that one base—C, for example—and no other, is damaged and removed from the molecule and the strand broken at that site. Crucial to the method is the fact that the chemical treatment is regulated so that not all the Cs are damaged and those that are damaged are hit at random. Take, for example, a hypothetical 10-nucleotide sequence in which the base at the 5′ end bears a radioactive label, as indicated by the color:

5′ 3′

G—A—T—C—A—G—C—T—A—G

Excision of Cs at random results in the following mixture:

G—A—T—C—A—G

T—A—G

G—A—T

A—G—C—T—A—G

A—G

Two of these fragments, the two with which we are concerned, are labeled at the 5′ end.

The mixture of nucleotide fragments is then placed on a gel and subjected to electrophoresis, which separates them, quite simply, by their relative lengths. The shorter fragments move farther through the gel than do the longer fragments. The positions to which the two radioactive fragments move, which can be readily determined from the radioactive label, reveal the number of nucleotides the labeled fragments contain. This same procedure, repeated for all four bases, results in a sequence ladder from which the order of nucleotides can be read off directly (Figure 16–15).

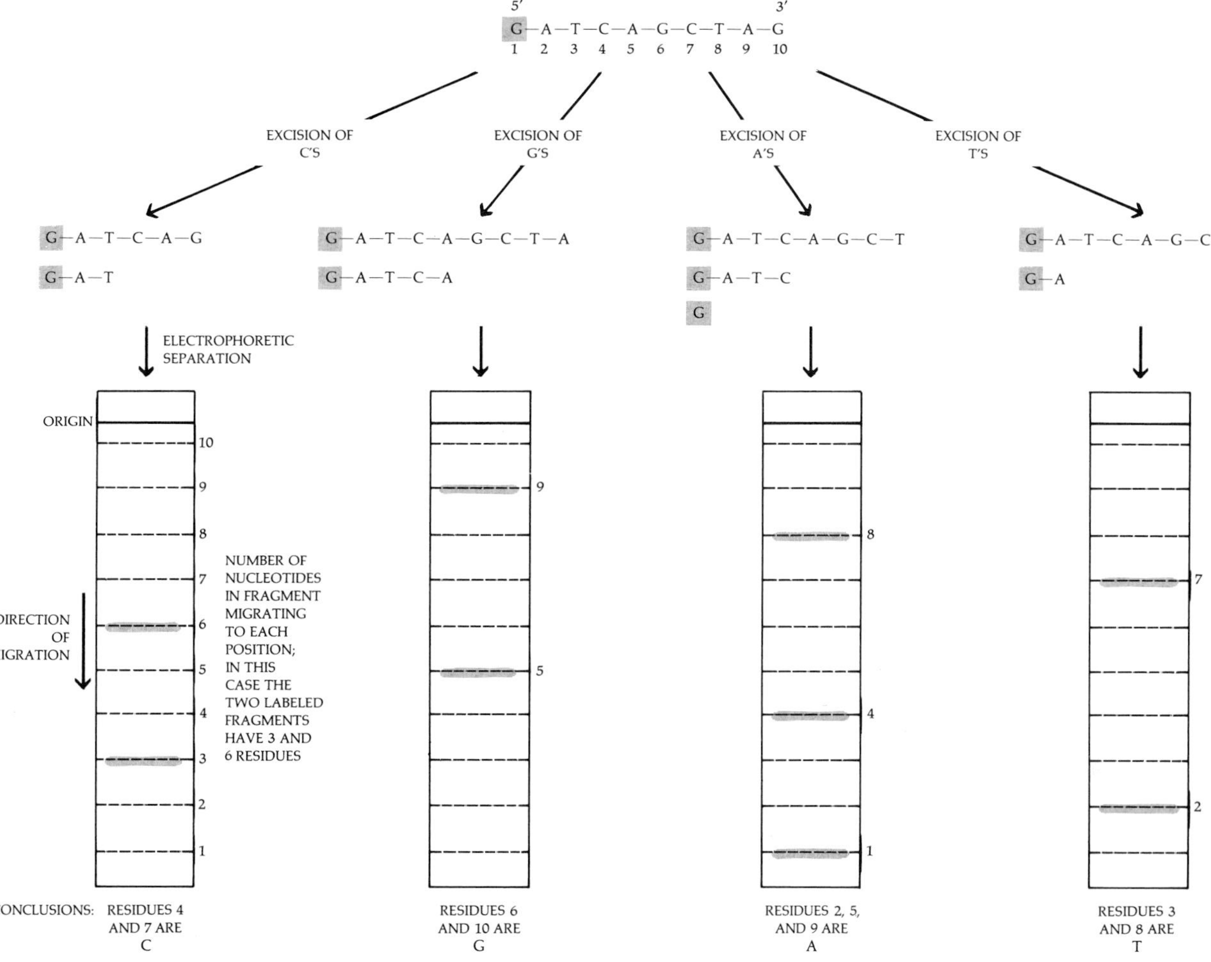

16–15

The sequencing of a segment of a DNA molecule, obtained by the action of restriction enzymes. The single-stranded segment (present in multiple copies) is radioactively labeled at one end. The solution containing the labeled DNA is divided into four portions, each of which is subjected to a different chemical treatment to break the molecule at only one of the four bases. The resulting fragments are then separated by electrophoresis on lanes that are calibrated to indicate the positions occupied by fragments of different lengths. Thus the location of the radioactivity reveals the number of nucleotides contained in the labeled fragments. By combining the information gained from each procedure, the sequence of the complete segment can be read directly off the lanes.

Hybridization Techniques

One of the earliest methods for studying large DNA and RNA molecules was hybridization. Its foundation is the base-pairing properties of the nucleic acids. If DNA is heated at 30°C, the hydrogen bonds holding the two strands together are broken and the strands separate. (This process is called denaturation.) When the solution is cooled, if the two strands are still in juxtaposition, the hydrogen bonds re-form, reconstituting the double helix.

When denatured DNAs from different sources are mixed together, they undergo random collisions. If two strands with nearly complementary sequences (the matching need not be exact) find each other, they will form a hybrid double helix. The extent to which the segments from two samples reassociate and the speed with which they do so provide an estimate of the similarity between nucleic acid sequences.

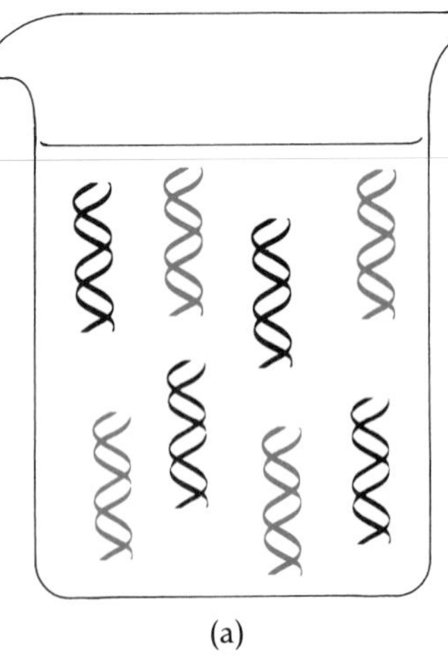

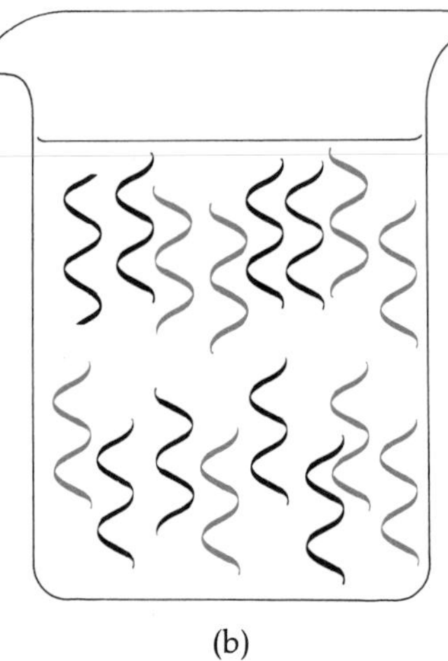

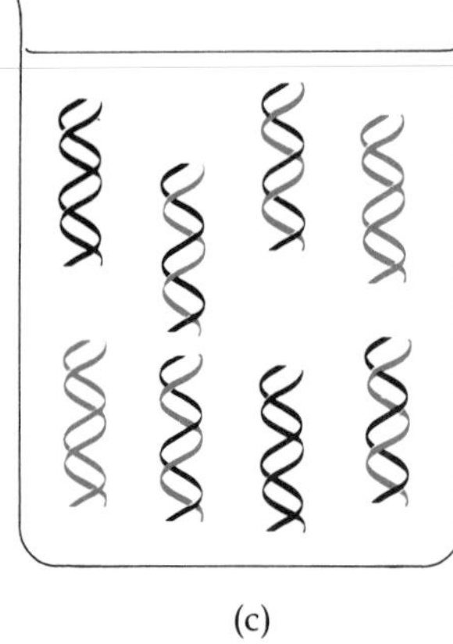

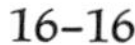
16–16
(a) *Two types of DNA are mixed in a beaker. One (indicated in color) is labeled with radioactive phosphorus.* (b) *When heated, the double-stranded DNA separates into two single strands.* (c) *When cooled, the DNA strands recombine in double helixes. The double-stranded molecules are then separated from any single-stranded molecules present, and the amount of radioactivity in the newly formed duplexes is analyzed. The extent to which radioactive and nonradioactive strands are able to form duplexes in a given length of time is an indicator of the similarities of the DNAs.*

The hybridization procedure has been elegantly adapted for detecting the presence of specific nucleic acids. According to this method, a highly purified sample of DNA or RNA is tagged with a radioactive label and is used as a probe to seek out a particular corresponding sequence. Thus, for example, messenger RNA could be used to seek out the chromosomal DNA from which it was transcribed. Some applications of this technique will be described in the chapters that follow.

SUMMARY

Studies of the replication of and recombinations among bacteria, plasmids, and viruses laid the foundations for current research in recombinant DNA.

The essential genetic information of bacteria, of which *E. coli* is the best-studied example, is coded in a circular double-stranded molecule of DNA. Other genes are found in plasmids, which are much smaller, also circular, DNA molecules. Plasmids often carry genes for drug resistance; as many as 10 such genes have been located on a single plasmid. Some plasmids can become reversibly integrated into the bacterial chromosome. Plasmids can be transferred from cell to cell. Such transfer of DNA by cell-to-cell contact is known as conjugation.

The F (fertility) factor of *E. coli* is a plasmid present in F^+ cells that can be transferred to recipient F^- cells; such cells then become F^+ and can transfer the F factor. When the F plasmid becomes integrated into the chromosome of an *E. coli* cell (making it an Hfr cell), part or (rarely) all of the chromosome can be transferred to another *E. coli* cell. At the time of transfer, the chromosome replicates by the rolling-circle mechanism, and a copy of the DNA enters the recipient cell linearly so that the bacterial genes enter the cell one after another in a fixed sequence. Because the rate at which they enter is constant at a given temperature, separation of conjugating cells at regular intervals provides a means for mapping the bacterial chromosome.

Viruses consist of either DNA or RNA wrapped in a protein coat. In the case of the DNA viruses, the DNA molecule is both replicated, to produce more viral DNA, and translated to produce the coat molecules and other proteins. In the case of the RNA viruses, the RNA genome can serve as the template for more viral RNA and also as messenger RNA. Alternatively, in some animal viruses, known as retroviruses, the viral RNA is transcribed into DNA molecules; this is brought about by the enzyme reverse transcriptase. On leaving the cell, newly formed virus particles may incorporate fragments of host DNA and transfer them to other cells. This form of genetic exchange is known as general transduction.

A temperate virus is a virus that can either set up an active infectious cycle or can, alternatively, become part of the host chromosome. When such a virus—known as a prophage—is released from the chromosome, it may carry adjacent portions of the chromosome with it, a process known as restricted transduction. Lambda is a temperate bacteriophage of *E. coli.* The lambda chromosome is linear when it is in the virus, but when it is released into the cytoplasm of *E. coli,* it forms a circle. The closing occurs because of the existence of "sticky" ends—single-stranded complementary DNA sequences on either end of the molecule. Lambda is integrated into the bacterial chromosome at a particular site.

Genetic recombinations can take place either because of exchanges between homologous sequences of DNA—substitution of one sequence for another—or by insertion of additional, new DNA into a host chromosome. Genetic exchanges in meiosis, transformation, conjugation, and transduction all take place by the former principle. The second type of gene transfer is characteristic of a plasmid, such as the F factor, which can be added to or subtracted from the chromosome.

More recently, insertion sequences have been discovered; these are short segments of DNA integrated into chromosomal DNA. Insertion sequences generate copies of themselves that may then become inserted, apparently at random, elsewhere in the bacterial chromosome or in plasmids. Transposons are genes flanked by insertion sequences and may similarly change location in a DNA molecule.

Restriction enzymes are bacterial enzymes that have the function of cleaving foreign DNA. Different restriction enzymes cleave the DNA at different sequences, usually four to six bases long. Bacteria protect themselves against their own restriction enzymes by methylating one or more nucleotides in the sequences bordering the cleavage sites of restriction enzymes.

Because a particular restriction enzyme always cleaves a molecule at the same DNA sequence, the use of restriction enzymes to fragment DNA yields uniform fragments for analysis. Instead of cleaving the molecule straight across, some restriction enzymes, including Eco RI, leave "sticky" ends. Thus any DNA cleaved by this enzyme can be joined readily to another molecule cleaved by the same enzyme.

The use of restriction enzymes has made it possible to insert DNA segments into plasmids and other self-replicating DNA molecules and so to produce multiple copies—clones—of the selected segments. Techniques have been developed for sequencing such DNA segments. Hybridization techniques make it possible to detect specific DNA sequences.

QUESTIONS

1. Distinguish among the following: theta replication/rolling-circle replication; transformation/conjugation/transduction; F^+ cell/F^- cell/Hfr cell; plasmid/virus; general transduction/restricted transduction; insertion sequence/transposon.

2. How did the mapping experiments with *E. coli* demonstrate that the bacterial chromosome was both linear and circular?

3. Suppose you treated a DNA molecule with a particular restriction enzyme and obtained five fragments, which you then sequenced. What would you do next in order to establish the order of the five fragments in the original molecule?

4. Describe the two possible outcomes of infection of a bacterial cell by a temperate virus.

5. *Escherichia coli* cells used in recombinant DNA studies are "disabled"; that is, they lack the capacity to synthesize their cell walls, for example, or to make a nitrogenous base, such as thymine. What are the consequences of such disablement?

CHAPTER 17

The Eukaryotic Chromosome

The discovery that the genetic code is universal tempted molecular biologists to believe that the eukaryotic chromosome would turn out to be simply a large-scale version of *E. coli*'s. This has not proved to be the case. As studies of eukaryotes have progressed, the gulf between eukaryotes and prokaryotes has widened rather than narrowed. It is now clear that the eukaryotic chromosome has many important differences from the bacterial chromosome—some expected and some very surprising.

CYTOLOGICAL OBSERVATIONS

As we noted in Chapter 7, the genetic material of eukaryotes is visible at mitosis and meiosis as condensed chromosomes. The DNA of chromosomes is categorized by its staining properties with dyes. The more open DNA (which stains weakly) is called *euchromatin*, and the more condensed DNA (which stains strongly) is called *heterochromatin*. During interphase the chromatin becomes dispersed.

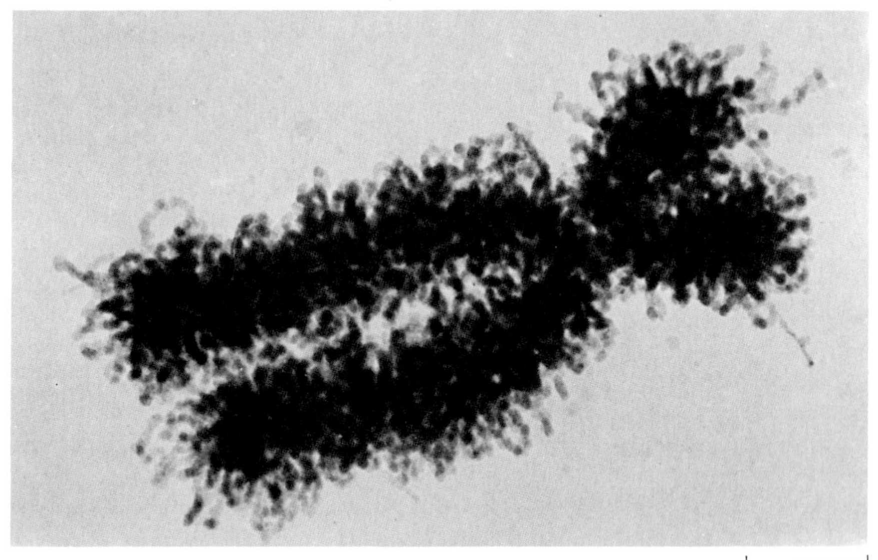

17–1
Human chromosome 12, shown at mitotic metaphase, before the chromatids have separated. The chromatin is highly condensed. Each chromatid, according to present evidence, contains a single circular double-stranded DNA molecule and is, by weight, about 60 percent protein.

17–2
Electron micrograph of the nucleus of the single-celled alga Chlamydomonas. *The dark body in the center is the nucleolus, where the RNA of the ribosomes is synthesized and the ribosomal subunits are assembled. Very fine rRNA strands are visible, around which ribosomal subunits are grouped. Notice also the nuclear envelope, with two clearly visible pores (arrows). There is a portion of a chloroplast to the left of the nucleus and a Golgi body below the nucleus to the left. Ribosomal proteins are assembled in the cytoplasm and migrate into the nucleus where the ribosomal subunits are assembled. The subunits then move into the cytoplasm where they combine to form the ribosomes.*

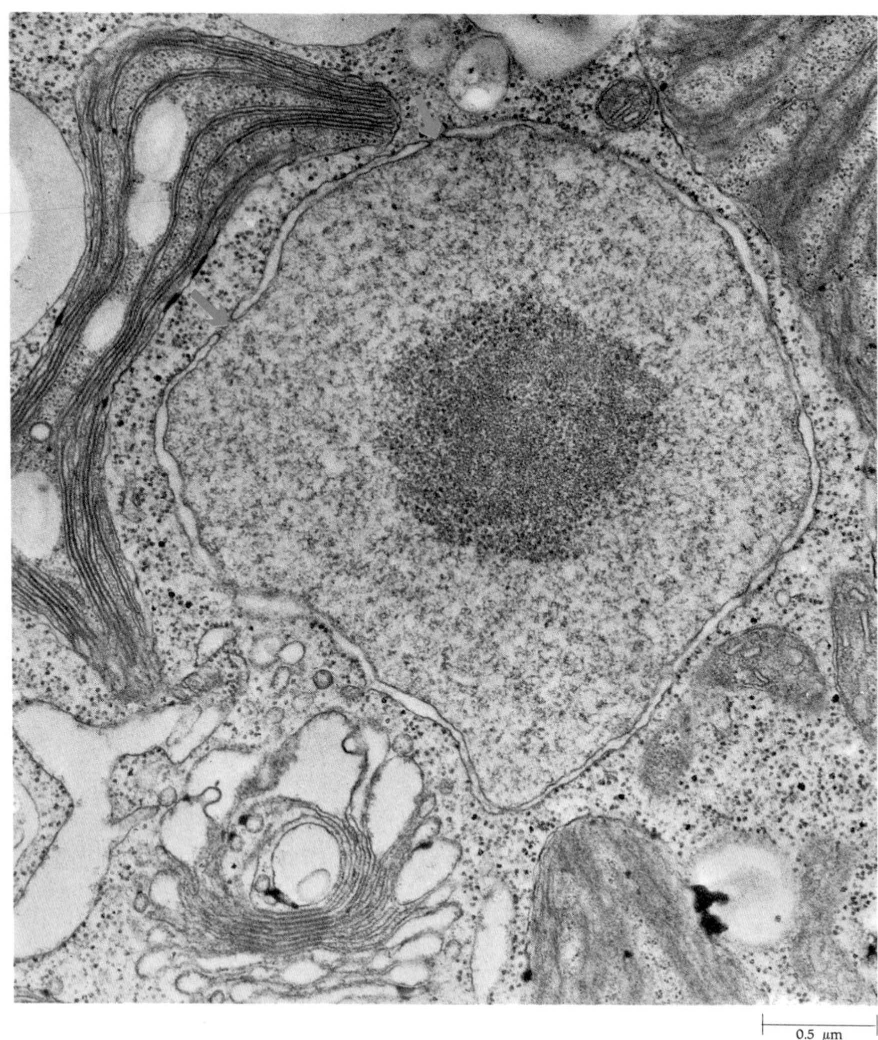

The Nucleolus

The most prominent body in the nucleus of the interphase cell is the nucleolus. The nucleolus is not actually a structural entity but is a cluster of loops of chromatin from different chromosomes. For example, 10 of the 46 human chromosomes contribute such chromatin loops to the nucleolus.

The nucleolus is the site of ribosome production. Ribosomes of eukaryotic cells resemble those of prokaryotes in that they consist of two subunits, each composed of RNA and proteins, and they are the sites of translation of mRNA into protein. They differ in that eukaryotic ribosomes are much larger than prokaryotic ribosomes, and their individual proteins and RNAs are different.

Three of the four rRNAs are transcribed on the loops of the nucleolus. The ribosomal proteins, which are translated in the cytoplasm from mRNA, are shipped back into the nucleus where they are combined with the rRNAs. The ribosome subunits are then exported to the cytoplasm.

THE CHROMOSOMAL PROTEINS

Chromatin is more than half protein. These proteins are of hundreds of different kinds. The predominant type, by weight, is a group of relatively small polypeptides, the *histones*. Histones are positively charged (basic) and so are attracted to the negatively charged (acidic) DNA. They are always present in chromatin, are very similar from species to species, and are synthesized in large numbers during the S phase of the cell cycle (see page 142), when DNA is also being synthesized. There are five distinct types of histones—H1, H2A, H2B, H3, and H4—with about 60 million copies of each type per cell.

The other proteins associated with the chromosome are the enzymes concerned with DNA and RNA synthesis, various promoters and regulators, plus a large number and variety of unidentified molecules believed to have regulatory or packaging functions.

The DNA Helix and Proteins

Recently, x-ray diffraction studies have shown that the DNA helix can assume several forms: A-DNA, which is partially unwound; B-DNA, the right-handed alpha helix, described by Watson and Crick; C-DNA, which is tightly coiled; and Z-DNA, which is a left-handed helix that results in zigs and zags of the phosphate backbone (hence, Z-DNA). It is hypothesized that changes in DNA configuration affect the binding of proteins to the molecule.

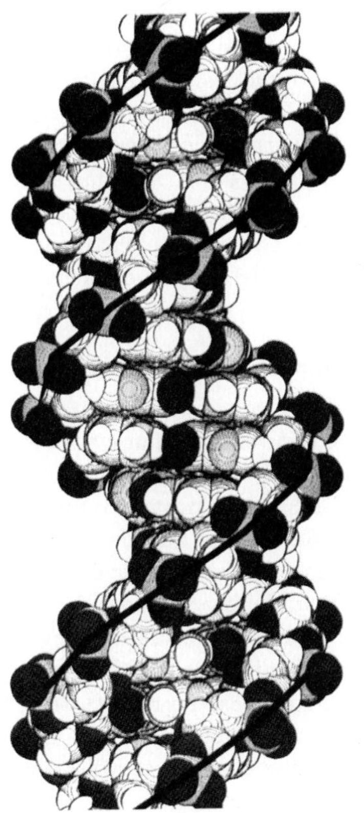

17–3
Computer-generated diagrams of B-DNA and Z-DNA. B-DNA is the form described by Watson and Crick. It consists of two strands of nucleotides twisted around each other, with the backbones forming a smooth, right-handed helix. In Z-DNA, the backbones of the nucleotide strands form a zig-zag, left-handed helix. The two forms are reversible, suggesting that the changes in structure might reflect some regulatory role.

17–4

(a) *Chromatin showing the beadlike nucleosomes. The distance between nucleosomes is about 10–11 nanometers, and the diameter of each bead is about 7 nanometers. Each nucleosome is composed of about 140 base pairs of DNA and an assembly of eight histone molecules.*

(b) *The structure of a nucleosome. The negatively charged DNA coils twice around a cluster of eight positively charged histone molecules, which form the core of the nucleosome. Outside the nucleosome, an H1 histone molecule (also positively charged) serves to clamp the DNA in place.*

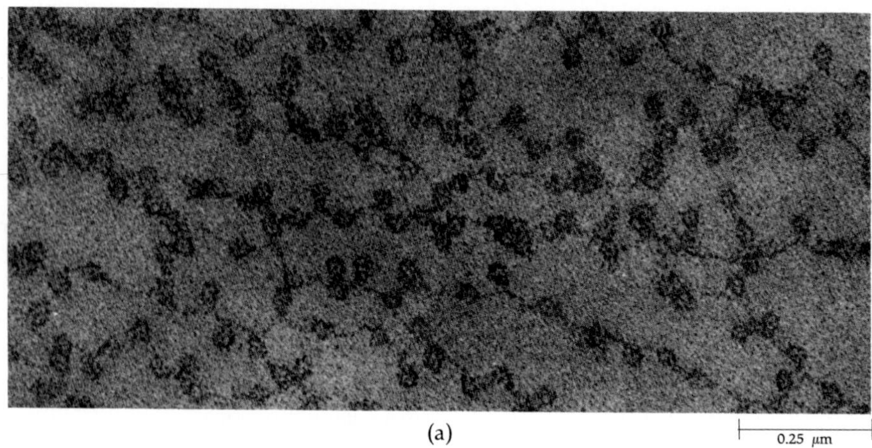

(a)

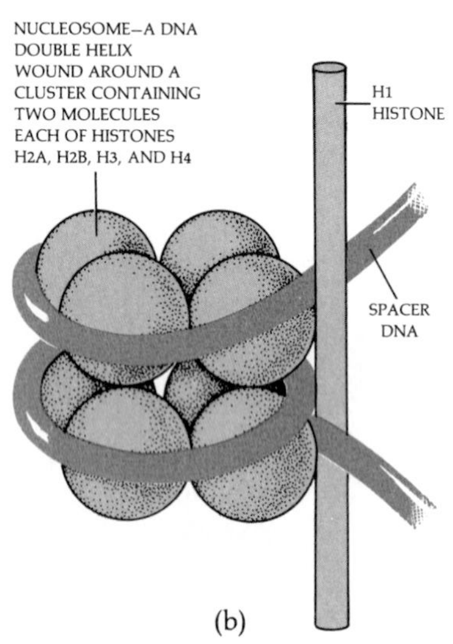

(b)

The Nucleosome

The nucleus of a human cell averages about 5 micrometers in diameter and contains about 2 meters of DNA. How so much DNA can be packaged in so small a space and still be available for replication and transcription has been a matter of considerable interest ever since DNA was identified as the carrier of the hereditary information. The problem is still not solved, but progress has been made.

The fundamental packing unit, it was discovered just about a decade ago, is the *nucleosome*, which is a complex of DNA and histone proteins. Its structure, as visualized in electron micrographs and deduced by biochemical evidence, resembles a bead on a string (Figure 17–4).

Each nucleosome consists of DNA and four of the five different types of histones. The fifth histone lies outside the nucleosome; the connecting "string" is DNA. Each nucleosome contains about 140 pairs of nucleotides, and the strand between the particles contains another 30 to 60 base pairs. Thus an average gene is made up of three or four nucleosomes. When a fragment of DNA is tied up in a bead, it is about one-sixth the length it would be if it were fully extended. The way in which the further condensing and packing of DNA comes about is not yet known. One of the proposed models is shown in Figure 17–5.

Chromosomal Proteins and Gene Regulation

The proteins of the chromatin are responsible for its condensation. There are several diverse lines of evidence relating the degree of condensation of the chromatin with the expression of its DNA.

First, heterochromatin, the more tightly packed form of chromatin, is characteristically found near the centromeres. The DNA of this region is never expressed, and this part of the chromosome is believed to play a structural role in the movement of the chromosomes in mitosis and meiosis (Figure 17–6).

Second, in general, the cells of multicellular plants and animals contain all the DNA that is present in the fertilized egg (see essay on page 333), but once a cell has differentiated to become a skin cell, a blood cell, or a liver cell, much of this DNA is never transcribed; skin cells never make hemoglobin, for instance. As a cell differentiates during embryological development, the proportion of heterochromatin to euchromatin increases.

17–5
Stages in the folding of a chromosome, according to one proposed model.

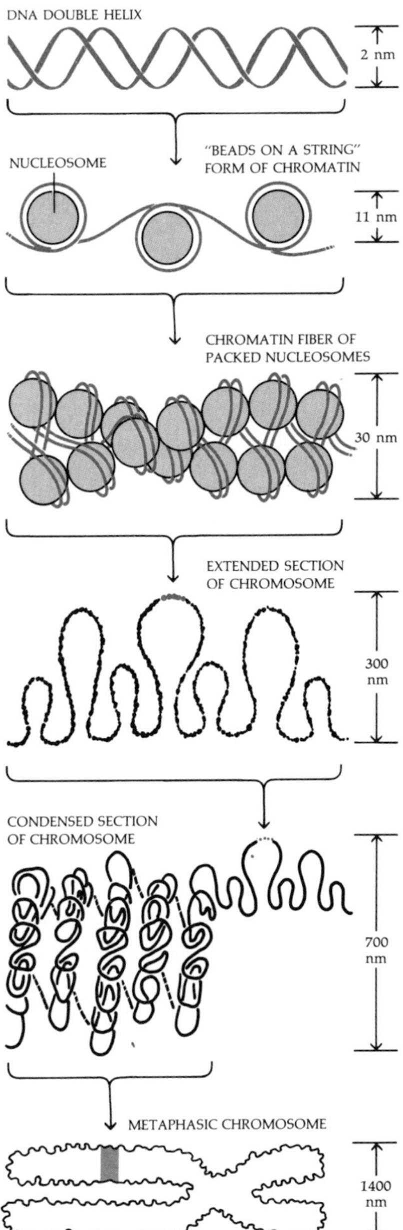

17–6
Stained human chromosomes, showing dark-staining heterochromatin (tightly condensed DNA) in the region of the centromeres.

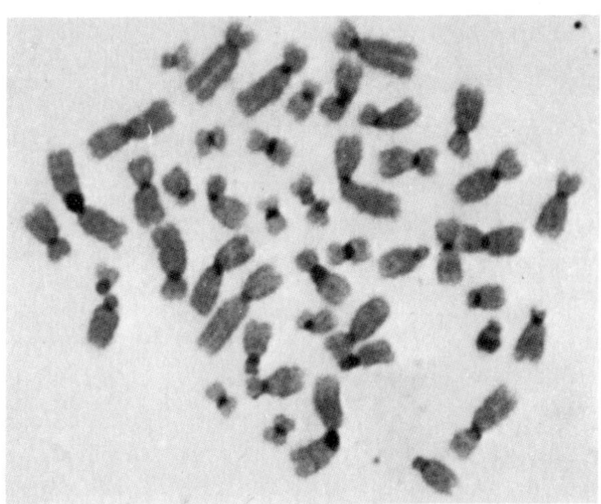

Third, no new RNA is produced during mitosis, when the chromosomal material is tightly condensed. Transcription takes place only during interphase. Fourth, no transcription takes place on the inactivated female chromosomes, the Barr bodies, which are tightly condensed (see essay, page 271).

A fifth type of evidence comes from studies of the giant chromosomes of insects (see page 274). At various stages of larval growth in insects, it is possible to observe diffuse thickenings, or "puffs," in various regions of these chromosomes (Figure 17–7). The puffs are open loops of DNA, and studies with radioactive isotopes indicate that these loops are sites of rapid RNA synthesis. When ecdysone, a hormone that produces molting in insects, is injected, the puffs occur in a definite sequence that can be related to the developmental stage of the animal. For example, in one species of *Drosophila,* ecdysone initiates three new puffs and causes increases in 18 other puffs within 20 minutes after it is injected; during this same period, 12 other puffs decrease in size. After 4 to 6 hours, five additional puffs can be seen. The looping out of the DNA occurs before RNA synthesis is initiated. The mechanism of this unraveling, or spinning out, of the chromosomal DNA is not known.

Somewhat similar observations have been made with another type of chromosomal material, lampbrush chromosomes, so called because they resemble the brushes used to clean kerosene lamps. They are found in the nuclei of vertebrate egg cells during the first prophase of meiosis. These cells are very actively engaged in the RNA and protein syntheses required for the rapid growth of the mature egg. Each lampbrush chromosome consists of two homologues held together by their centromeres and chiasmata (page 272), each homologue consisting of two sister chromatids. As you can see in Figure 17–8, each chromatid has a number of lateral loops, which branch out from the main axis. These loops appear to be reeled in and out from the body of the chromosome. Each of these loops consists of a thin fiber.

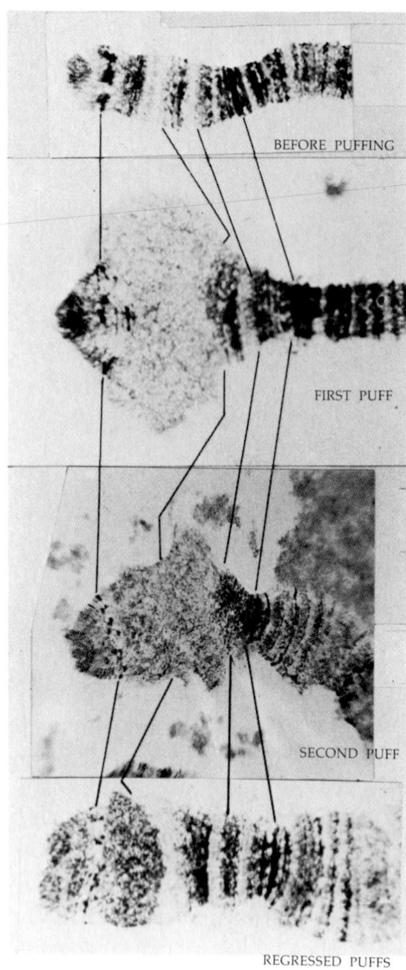

17–7
Observations of chromosome puffs support the concept that the DNA is somehow unwound to make it available for RNA transcription. These puffs were observed in chromosomes of the Brazilian gnat, which, like the fruit fly, has giant chromosomes in some of its cells. Puffs occur normally but can also be induced experimentally. These puffs occurred in response to a hormone that causes molting. As the micrographs indicate, the puffs occurred sequentially along one chromosome.

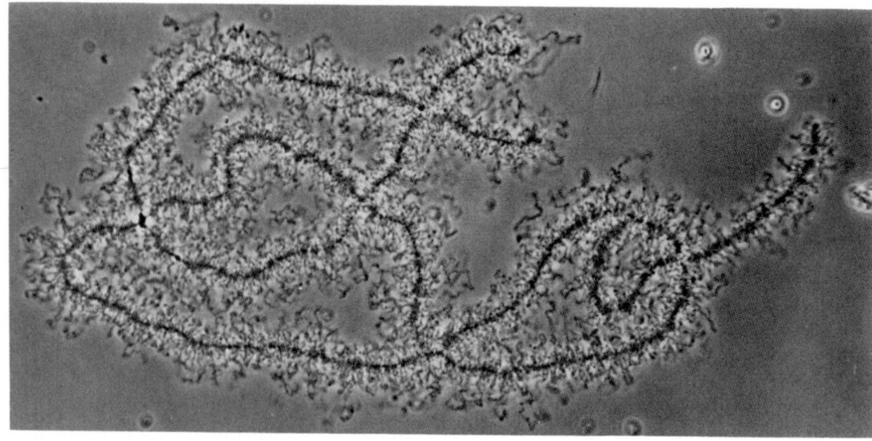

17–8
Two homologous lampbrush chromosomes. The individual chromatids are closely paired and so cannot be distinguished.

If this thin fiber is exposed to an enzyme that digests away the associated protein, all that remains is a thin filament about the diameter of a single DNA molecule. These loops, like the chromosome puffs in giant chromosomes, are also the sites of active RNA synthesis.

Finally, working with *Drosophila,* Muller showed some years ago that transposition of a gene from a euchromatic region of DNA to a heterochromatic region profoundly reduced its activity. (These transpositions were caused by breaking the chromosomes by exposure to x-rays and verified by studying banding patterns in giant chromosomes.)

Thus, information from various sources indicates that—although it may not be a cause-and-effect relationship—tightly coiled DNA does not transcribe RNA, and, conversely, the unfolding of chromosomes is correlated with RNA transcription.

DNA OF THE EUKARYOTIC CHROMOSOME

DNA is an "exquisitely thin filament," in the words of E. J. DuPraw, who calculates that a length sufficient to reach from the earth to the sun would weigh only half a gram. The DNA of each eukaryotic chromosome is believed to be in the form of a single molecule. In a human chromosome, each of these molecules is believed to be from 3 to 4 centimeters long. Thus the entire human body contains some 25 billion kilometers of DNA double helix.

Calculations of the amount of DNA present in eukaryotic cells have revealed two surprising facts. First, with some few exceptions, the amount of DNA per cell is the same for every diploid cell of any given species (which is not surprising), but the variations among different species are enormous. *Drosophila* has about 1.4×10^8 base pairs per haploid genome, only about 70 times more than *E. coli.* Humans

THE TOTIPOTENT CELL

Just over 20 years ago, R. W. Briggs and T. J. King were able to show that if nuclei were removed from embryonic frog cells and injected singly into a series of unfertilized eggs whose nuclei had been physically removed, some of the eggs would begin development and grow into normal adult frogs. This experiment showed that early development does not result in any permanent inactivation of genes or the loss of functional DNA.

Subsequently, J. B. Gurdon of the University of Oxford was able to do the same kind of experiment with nuclei from the intestinal cells of young tadpoles. Although very few of the recipient eggs developed normally, Gurdon's results showed that even nuclei of differentiated cells retain the capacity to function as egg nuclei—that is, that all of their genes are present and can be expressed.

At about the same time, F. C. Steward demonstrated that under certain conditions, a single differentiated cell of a carrot can be persuaded to reconstitute a whole carrot plant.

Thus we are led to the conclusion that differentiation between cells depends on inactivation of certain groups of genes and activation of others. The way in which this regulatory process is controlled is now a leading question in molecular genetics.

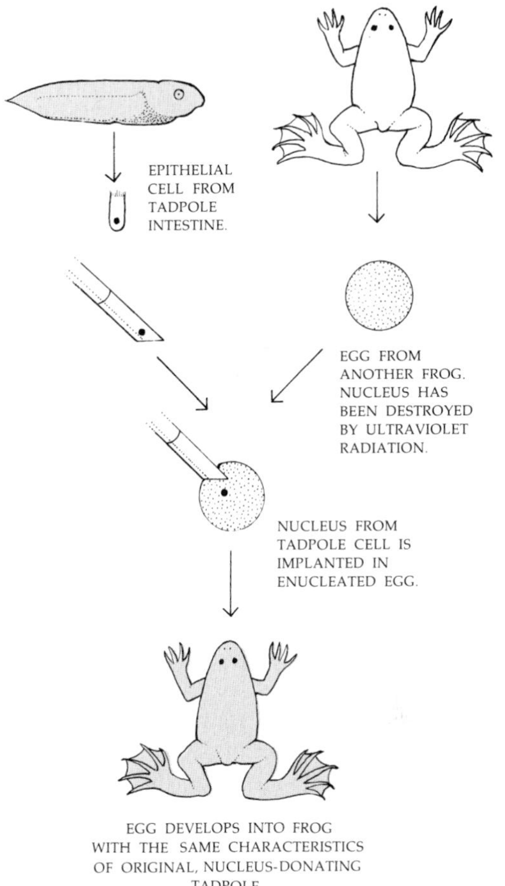
EPITHELIAL CELL FROM TADPOLE INTESTINE.
EGG FROM ANOTHER FROG. NUCLEUS HAS BEEN DESTROYED BY ULTRAVIOLET RADIATION.
NUCLEUS FROM TADPOLE CELL IS IMPLANTED IN ENUCLEATED EGG.
EGG DEVELOPS INTO FROG WITH THE SAME CHARACTERISTICS OF ORIGINAL, NUCLEUS-DONATING TADPOLE.

In experiments by J. B. Gurdon, the nucleus was removed from an intestinal cell of a tadpole and implanted into an egg cell in which the nucleus had been destroyed. In some cases, the egg developed normally, indicating that the tadpole intestinal cell nucleus contained all the information required for all the cells of the organism.

(with 2.87×10^9 base pairs) have 20 times as much as *Drosophila,* about as much as a mouse, but somewhat less than a toad (3.32×10^9 base pairs). The largest amount so far has been found in a salamander, with 8×10^{10} base pairs.

Second, in every eukaryotic cell, there is what appears to be a great excess of DNA, or at least of DNA whose functions are completely unknown. In humans, for example, only slightly more than 1 percent of all the DNA codes for proteins.

By contrast, prokaryotes have comparatively little DNA and they use it thriftily. In nearly all cases, each chromosomal DNA molecule contains only one copy of any given gene, and, except for regulatory and signaling sequences, virtually all of the DNA is expressed.

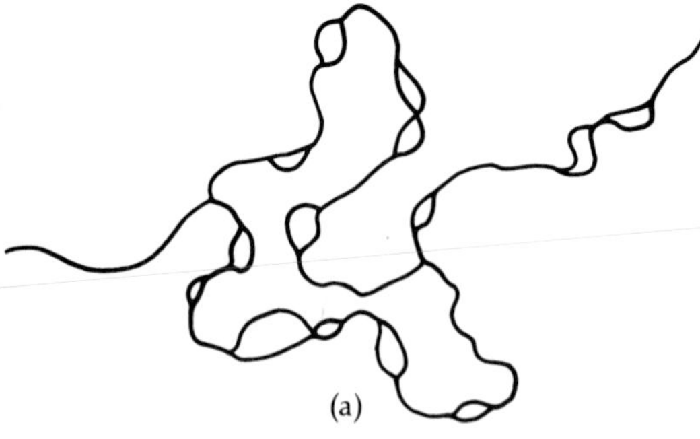

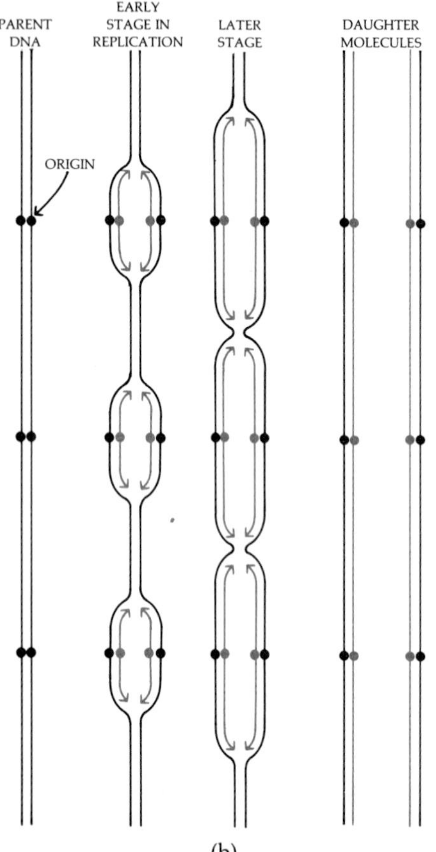

17–9
Replication of a eukaryotic chromosome. (a) This drawing of a replicating segment of DNA from a Drosophila *egg shows the "bubbles," or "eyes," formed by multiple points of bidirectional replication. (b) Bidirectional replication begins simultaneously at thousands of initiation sites, or origins. It proceeds until the daughter strands (color) are complete. The new double helixes then separate; each contains a parent strand (black) and a daughter strand (color).*

DNA Replication

DNA replication in eukaryotes occurs during the S phase of the cell cycle. It is bidirectional, as in bacteria (page 311), but it differs from *E. coli*'s in that it has multiple initiation sites—several thousand per chromosome. Replication at these sites, and indeed in all the chromosomes of the cell, proceeds synchronously. Each replication fork progresses at a relatively slow speed (0.5 micrometer per minute, which is 50 times slower than in *E. coli*), but each averages only about 30 micrometers, terminating when it meets another, oncoming replication fork (Figure 17–9). The newly synthesized DNA rapidly becomes associated with histones to form nucleosomes.

Classes of DNA

Hybridizing techniques, density gradient centrifugation, and, more recently, DNA sequencing, have revealed that the DNA of eukaryotic cells can be divided into three general classes of DNA. One is composed of short repetitive sequences, arranged in tandem (head to tail); it is sometimes known as satellite DNA because it separates from the other DNA during density gradient centrifugation. Half the DNA in one species of crab, for instance, is made up of ATATATAT and so on; *Drosophila virilis* has ACAAACT repeated 12 million times. About 10 percent of the DNA of the mouse, and about 20 to 30 percent of human DNA, is made up of short repeats. (Repetitive sequences, especially those in which purines—A or G—and pyrimidines—T or C—alternate, are especially likely to form Z-DNA.) Radioactive gene probes (page 324) have revealed that this satellite DNA is present in the tightly coiled heterochromatin (Figure 17–10). There is no evidence that it is ever transcribed.

The second class of DNA consists of moderate repeats (up to 100,000 copies), sometimes in tandem but more often dispersed throughout the DNA. Some of these sequences have been associated with particular RNAs and proteins, as we shall see; the role, or probably roles, of the others are not yet known.

The third class, which is about 70 percent of the DNA in both mice and humans, is made up of comparatively long nucleotide sequences present in only one or a few copies. Some of this DNA contains the structural genes, the genes that code for proteins.

Multiple Genes

Although most eukaryotic structural genes occur singly, several multiply occurring genes are known. The genes that code for the rRNAs are an example. There are four separate types of rRNA in the eukaryotic ribosome. Three of them (18S, 5.8S, and 28S rRNA) are transcribed onto a single mRNA, which is then processed, and the fourth (5S) is transcribed from a much-repeated single gene located elsewhere. The African toad *Xenopus*, for example, has about 300 copies (per haploid genome) of the gene that codes for the three rRNAs, all repeated in tandem on one chromosome that loops out to form the nucleolus (Figure 17–11). In between the genes are spacer sequences, which vary in length and composition. The 5S rRNA gene, of which there are an estimated 20,000 copies, is located on another chromosome. Human cells contain about 200 copies of the three-rRNA gene, spread out in small clusters on 10 different chromosomes. The genes that code for the many different tRNAs are also transcribed in tandem.

17–10
The use of a radioactive probe to determine the location of satellite DNA in the intact chromosome. The probe—RNA labeled with tritium—was prepared in a test tube, using as a template satellite DNA that had been separated from other chromosomal DNA in a density gradient. Salamander chromosomes were then exposed to the radioactively labeled RNA. As you can see, the RNA hybridized with the DNA located near the centromeres, revealing the location of the satellite DNA. The chromosomes are in meiotic metaphase I.

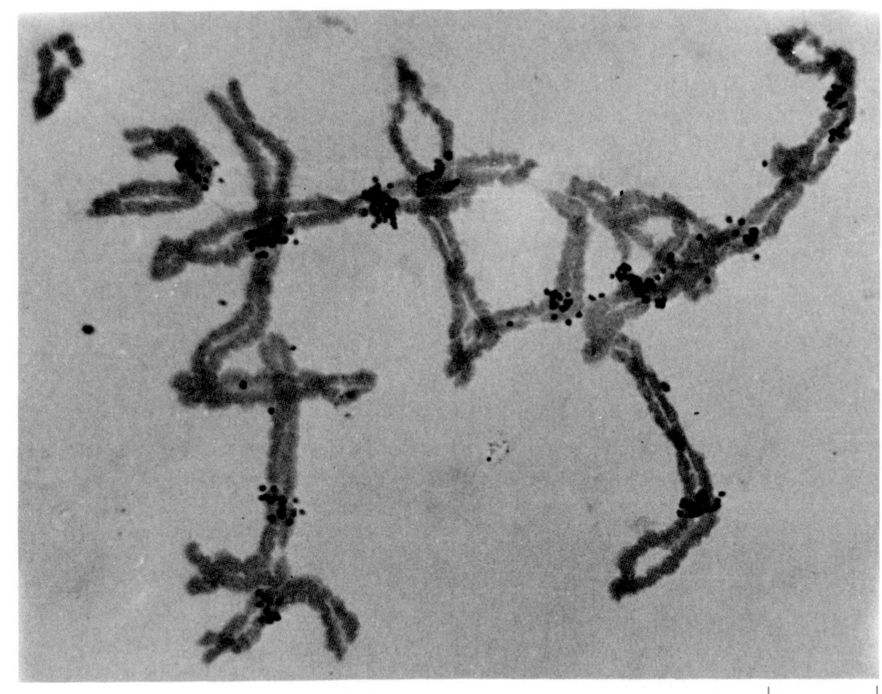

17–11
The gene coding for 18S, 5.8S, and 28S rRNA occurs repeatedly in tandem (head to tail). (a) *Electron micrograph of five such genes separated by nontranscribed spacer sequences.* (b) *An enlargement of one of the genes and its transcribed rRNA precursor molecules (the fine fibrils perpendicular to the DNA molecule). Enzymatic cleavage of the rRNA precursor molecules yields individual 18S, 5.8S, and 28S rRNA molecules.* (c) *Map of the rRNA gene.*

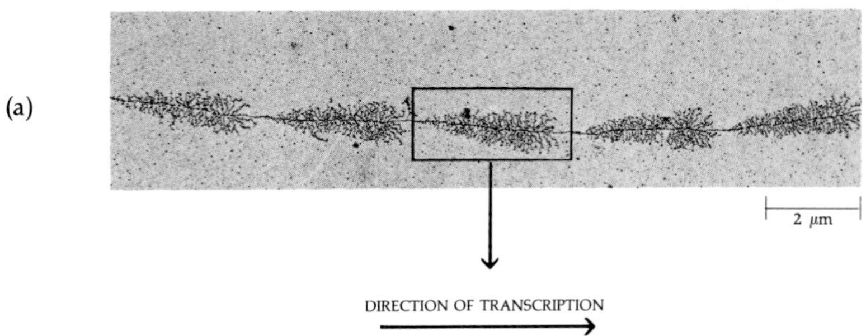

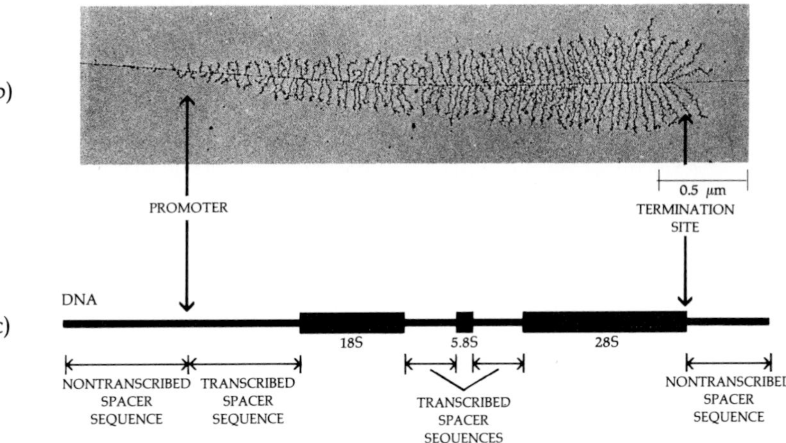

17–12

Electron micrograph of the five histone genes on a partially denatured DNA molecule. Spacer sequences between the genes are rich in adenine and thymine and can be denatured at a lower temperature than the histone-coding regions. In the eukaryotic chromosome, the five-gene units coding for histones are repeated in tandem; the DNA segment shown here was cloned in E. coli *and cut with a restriction enzyme.*

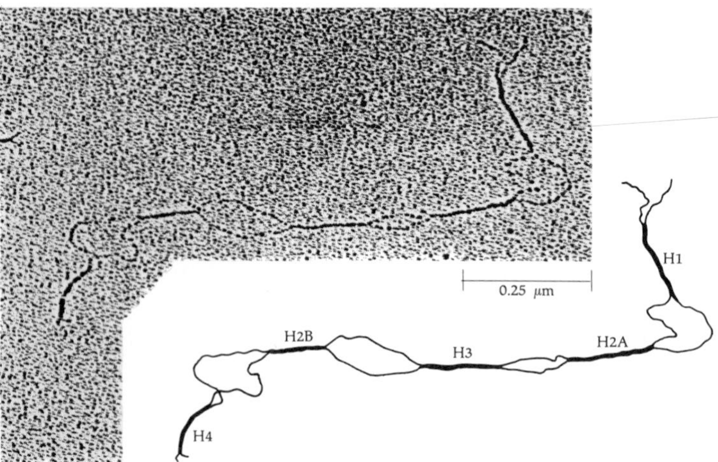

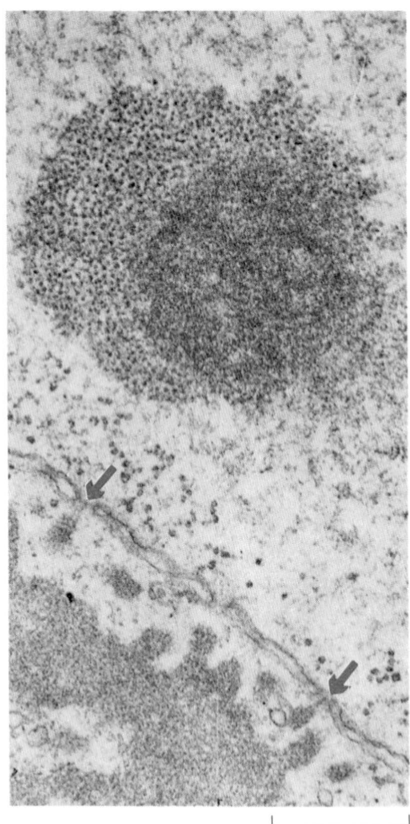

17–13

Electron micrograph of a frog oocyte, showing the nucleolus, which contains amplified genes coding for rRNAs. The arrows indicate materials—perhaps partially assembled ribosomal subunits—entering the cytoplasm through the nuclear pores.

Among the few multiply occurring genes coding for proteins are those that code for histones. The histone gene sequence, which codes for all five histone proteins (Figure 17–12), is repeated 10 to 20 times in *Xenopus,* 300 to 1,000 times in sea urchins.

Gene Amplification

Most multiply occurring genes are a stable part of the genome. However, in some instances, additional copies of genes are produced when needed. For example, in the oocytes of *Xenopus,* the rRNA genes are amplified about a thousandfold, producing many free rRNA-producing segments of DNA. The ribosomes formed on these segments are used in the manufacture of the many proteins needed for the development of the amphibian egg, which must be self-sufficient until the feeding (tadpole) stage is reached. This extra ribosomal DNA, formed at one stage in the early oocyte, is lost in subsequent cell divisions.

A similar phenomenon is seen in *Drosophila,* in which DNA sequences coding for the proteins of the hard outer coating of insect eggs are amplified about thirtyfold in the egg-forming cells of the female.

The Hemoglobin Family

Diploidy is a most important factor in the preservation of variability in eukaryotes. In a haploid organism, any genetic variations are immediately exposed to the selection process, whereas in the diploid organism, such variations may be stored as recessives, as with the allele for white flowers in pea plants. If one allele is doing the work, the other is free to experiment. Multiply occurring genes offer an even better shelter for variation. One of the duplicated genes may accumulate variations that prove useful to the organism, or alternatively, a recessive allele that has accumulated such variations may be duplicated.

Striking evidence that one or both of these events have indeed occurred is found in the mammalian hemoglobin genes.

Evolution of the Hemoglobins

Myoglobin is a relatively small (153 amino acids) oxygen-binding protein found in muscle cells. In some invertebrates and also in some primitive fish, oxygen is carried in the bloodstream in a myoglobinlike molecule. In more recently evolved

17-14
The proposed evolutionary relationships among some of the globin genes. The solid dots represent duplications of genes.

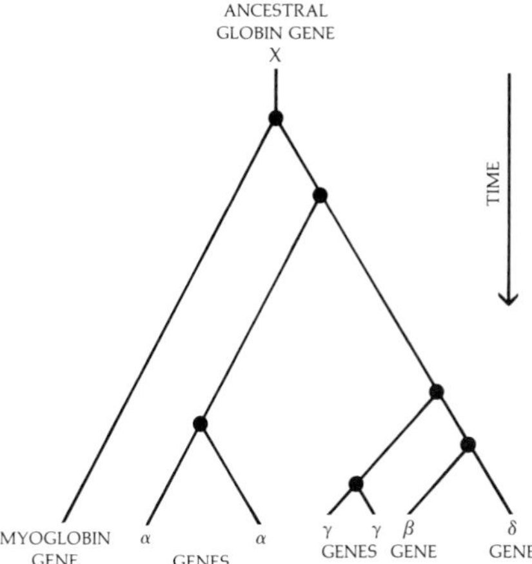

vertebrates, oxygen is transported in the bloodstream by hemoglobin, which is a complex of four polypeptide chains—two alpha (α) chains and two beta (β) chains. Each of these chains carries a heme group identical to that in myoglobin, and the four work together in such a way as to carry more oxygen than if they were separated.

In mammals, there are several additional hemoglobin polypeptides: an epsilon (ϵ), which appears early in embryonic life, and two gammas (γ), which appear later in the fetus. These substitute for the beta chain; thus the fetus has a hemoglobin made up of two alpha and two gamma chains. This combination has a higher affinity for oxygen than the alpha-beta hemoglobin, and so the fetus can pick up oxygen from the maternal bloodstream. Finally, primates have another type of hemoglobin, delta (δ), which can also substitute for beta. Alpha-delta hemoglobin makes up about 2 percent of adult human hemoglobin. Based on comparative studies between species, it appears that a series of gene duplications gave rise, step by step, to the array of hemoglobin polypeptides now present in humans. A reconstruction of these events is shown in Figure 17-14.

DNA sequencing has demonstrated that this beta family of hemoglobin genes occupies a single, long nucleotide sequence (Figure 17-15). Note also that the genes are arranged in the order in which they are expressed, with the embryonic epsilon first, then two genes coding for gamma, the fetal hemoglobin, and then the two adult globin genes, delta and beta. Whether this is a coincidence or reflects an underlying regulatory mechanism is not known.

DNA sequencing has also revealed the existence of duplicated globin sequences in this region that are similar to the structural genes but that are not expressed. These sequences, known as pseudogenes, are believed to have been disabled by mutations. It is interesting, however, to consider that further accumulation of mutations might turn the pseudogenes into functional genes and perhaps set them off on an evolutionary course of their own.

Introns and Exons

As we noted earlier, only a small percentage of eukaryotic DNA codes for protein. Some of the apparently excess DNA has been found in an unusual location—right in the middle of structural genes. Among the first to detect these extra DNA sequences—now known as introns—were Pierre Chambon and his colleagues at the University of Strasbourg in France, who were studying the genetic regulation of the production of ovalbumin. This protein—egg-white protein—is a chain of 386 amino acids produced in specialized cells of the oviducts of hens at egg-laying time. (Its regulation is of special interest because it is turned on and off by female hormones, but that is another story, which we shall touch upon in Chapter 41.)

17-15
Hemoglobin genes. (a) *The beta group of globin genes is located in a long DNA nucleotide sequence along chromosome 11. The epsilon gene is expressed early in embryonic life, followed by the two gamma genes (which code for the gamma chains that combine with alpha chains to produce fetal hemoglobin). The delta gene, found only in primates, and the beta gene are expressed in adult life. The group also includes two pseudogenes—DNA sequences similar to the structural genes, but which are not expressed.*

(b) *The alpha group, on chromosome 16, is less diversified. It includes two embryonic zeta genes, one pseudogene, and two very similar alpha genes, both of which are expressed in adults.*

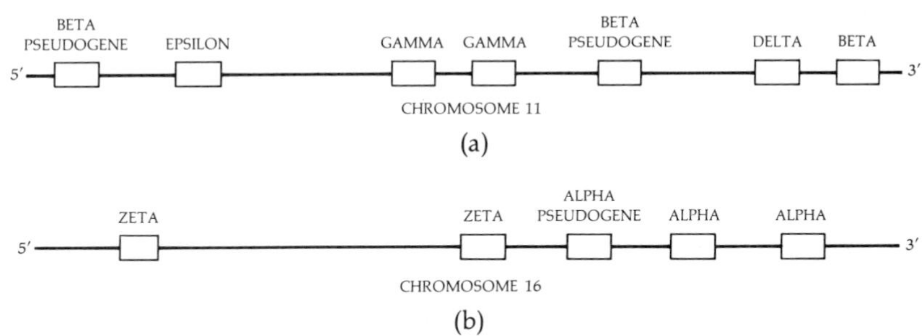

When recombinant DNA techniques became available, the investigators set out to apply them to the isolation of the ovalbumin gene. In order to do this, they isolated the messenger RNA for ovalbumin from the oviduct cells. (When hens are laying, ovalbumin mRNA makes up more than 50 percent of the RNA in these cells.) Using the reverse transcriptase enzyme described on page 315, the investigators made radioactive DNA copies of the messenger RNA with the intention of using them to isolate the ovalbumin gene. They selected a restriction enzyme that would not cleave the ovalbumin gene sequence (as determined from their copies) and used it to fragment the DNA of the cell. However, when the results were analyzed, it was found that the DNA probe had hybridized not with one fragment—the hoped-for gene—but with a number of different fragments of different lengths. Either the technique was faulty or the mRNA from the cytoplasm and the chromosomal DNA of the same gene were not colinear.

Although the former seemed more likely, the latter possibility was not entirely unexpected. Attempts to find mRNA molecules in the cell nucleus had revealed only much larger molecules, which were called, not very usefully, heterogeneous nuclear RNAs (hnRNAs). It was becoming clear that the original mRNA transcriptions—that is, the hnRNAs isolated from the nucleus—were heavily edited and processed into their final mRNA form before they were sent to the cytoplasm for translation.

In order to compare the DNA of the gene and the radioactive DNA produced from the cytoplasmic mRNA, Chambon began to sequence both types of molecules. In the course of doing so, he found that long nucleotide sequences present in the DNA of the original gene were not present in the mRNA that was isolated from the cytoplasm. In addition, and much more surprising, he found that these long stretches of untranslated DNA appear not only at the beginning or end of the gene, as might reasonably be expected, but also as *intervening sequences*, interrupting the sequences that code for proteins. These intervening sequences became known as *introns* and the protein-coding sequences as *exons* (signifying that they are expressed). Ovalbumin was found to have its coding sequence interrupted by seven introns (Figure 17–16).

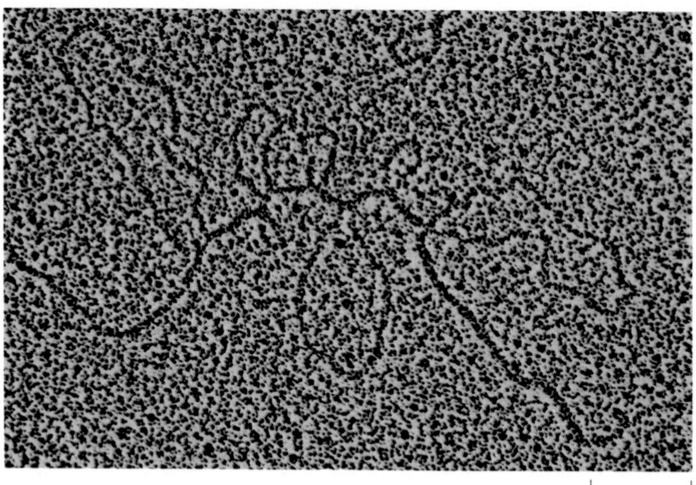

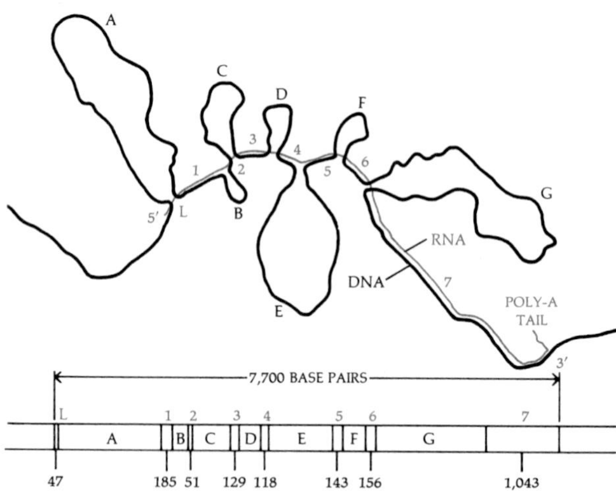

17–16
The results of an experiment in which a single strand of DNA containing the gene for ovalbumin was hybridized with the messenger RNA for ovalbumin. The complementary sequences of the DNA and mRNA are held together by hydrogen bonds; there are eight such sequences, the exons labeled L and 1 through 7 in the accompanying diagram. Some segments of the DNA do not have corresponding mRNA segments and so loop out from the hybrid; these are the seven introns, labeled A through G. Only the exons are translated into protein.

17–17
Small nuclear RNA molecules (snRNAs), present in snRNPs, excise introns and splice together exons, producing the finished mRNA molecule. The ends of the intron form base pairs with a segment of snRNA. Accomplishing excision and splicing in one step ensures that exons will not be assembled in the wrong order.

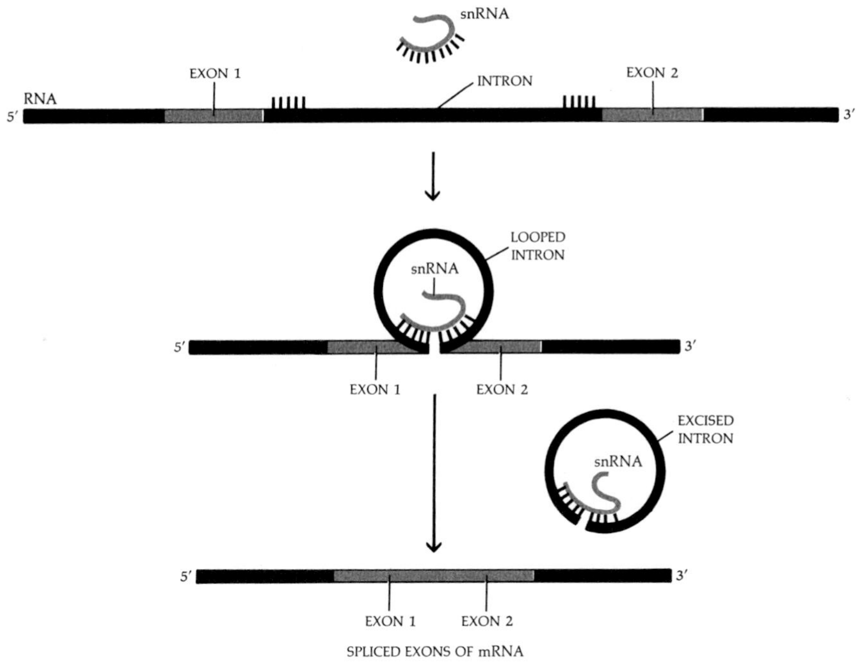

At just about the same time, in mid-1977, similar discoveries were made concerning the gene for beta globin. Since then just about all eukaryotic genes that have been analyzed have been found to contain intervening sequences, constituting, as they do in the case of ovalbumin, as much as 85 percent of the original sequence.

The original mRNA transcript—hnRNA—includes the introns, it has been found, but once produced, the introns are removed and the molecule spliced together. Introns must, of course, be excised very precisely; a shift of a single nucleotide could alter the reading frame and produce nonsense. Recent investigations have implicated some very small nuclear ribonucleoprotein particles, snRNPs (pronounced "snurps"), in this process. snRNP RNAs have been found to have sequences that are complementary to those at the ends of introns (Figure 17–17). There are about 10^6 snRNPs per cell.

17–18
An attractive hypothesis: Exons represent discrete functional regions in the protein for which the entire gene codes.

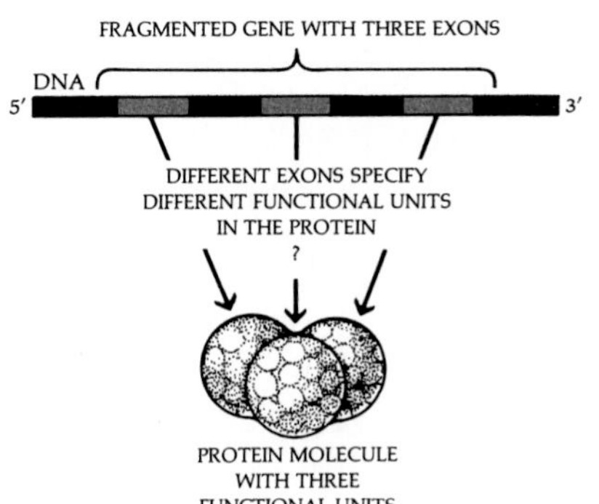

The Function of Introns

The way in which introns established themselves in the DNA sequence and their present function, if any, are not known. One suggestion is that they promote genetic recombination. Crossing over in intron sequences would be more likely to occur (just because of the distances involved). There are also suggestions that each exon codes for a different functional segment of the finished protein. In the globin genes, for example, which have three exons, the central one codes for the part of the molecule that holds the heme, and the other two code for parts of the molecule that fold around this central portion, stabilizing its configuration. According to this hypothesis, the introns serve to bring the DNA sequences for these functional modules, composed of three different polypeptides, together into a single gene.

17–19

Summary of stages in the expression of eukaryotic genes. The genetic information coded in the DNA is transcribed into an RNA copy. This copy is then edited, with the addition of the 5′ cap (7-methylguanosine) and a poly-A tail, the excision of the introns, and the splicing of the exons. The finished mRNA then goes to the cytoplasm, where it is translated into protein.

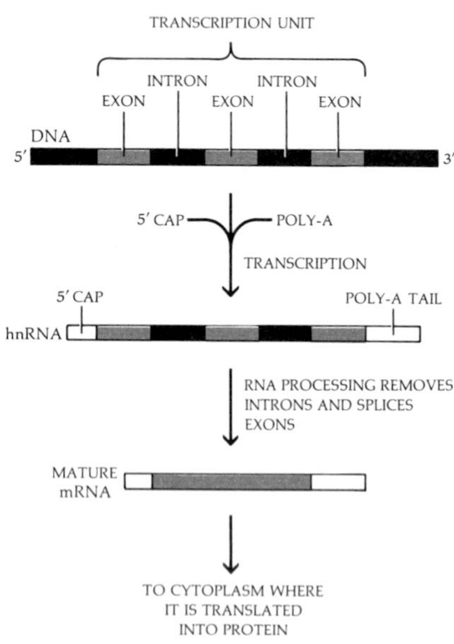

Other mRNA Editing

In addition to the excision of their introns, eukaryotic mRNA molecules undergo at least two other very specific alterations before leaving the nucleus. At the 3′ end of most mRNAs, a special enzyme adds a long sequence of adenine nucleotides, known as a poly-A tail. Its function is not known. At the other (5′) end, a molecule of an unusual base, 7-methylguanosine, is added; this is known as the 5′ cap. The 5′ cap is involved with the initiation of translation on the ribosome.

Jumping Genes

More than 30 years ago, Barbara McClintock, working in the Cold Spring Harbor Laboratories on Long Island, New York, discovered evidence of movable controlling elements—jumping genes—in eukaryotic cells. She was working with corn, performing genetic analyses and mapping studies somewhat similar to those that were going on at the same time in *Drosophila.* Each kernel on an ear of corn is, in essence, an embryo plant. On the basis of color changes and other variations in the embryos, she was able to deduce that these controlling elements changed position in the genome in the course of embryonic development and that these changes had profound effects on the expression of specific genes. Her work was, at that time, so far from the mainstream of genetics—an odd traveler's tale—that it was ignored. Only in the last few years, with the discovery of a host of movable genetic elements, first in bacteria and now in eukaryotes, has she received recognition for work stubbornly pursued over the intervening decades.

Sex-Determining Cassettes in Yeast

Yeast provided one of the first examples of a transposable genetic element in eukaryotes that was studied on a molecular scale. Yeasts are single-celled eukaryotes that mate by fusion of haploid cells in much the same way as *Chlamydomonas* (page 251). In order for two haploid cells to mate, they must differ in mating type, a restriction that ensures that two daughter cells will not mate with each other. In common bakers' yeast, *Saccharomyces cerevisiae,* there are two mating types, alpha and a; the type is determined by which of two genes, alpha or a, occupies a particular site on the yeast chromosome.

17–20

Yeast cells, as seen through a special microscope that polarizes the light. Yeasts are single-celled fungi that reproduce either asexually by budding or sexually by the fusion of haploid cells. The resultant zygote may form diploid buds or may undergo meiosis, producing four new haploid cells.

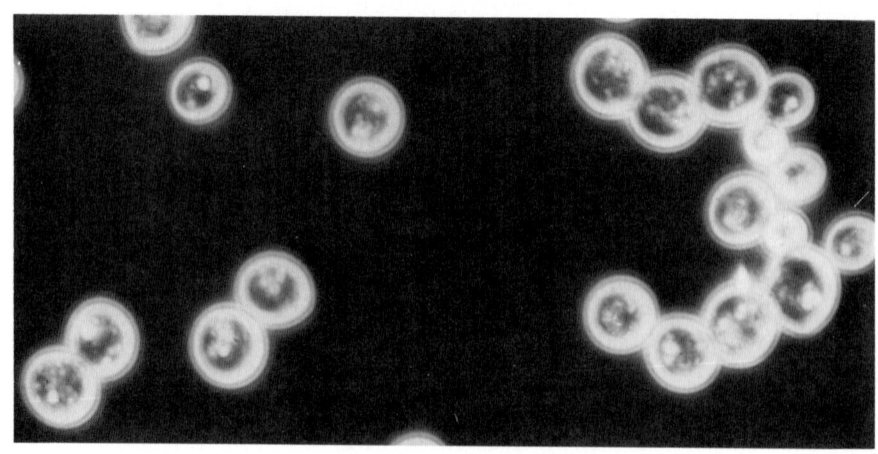

From time to time, certain strains of yeast cells switch mating types. This is accomplished by excision of gene alpha from the mating type locus and its replacement by gene a, or vice versa. The investigators who discovered this phenomenon call it a "cassette" mechanism, since the genes are inserted into and removed from a particular "slot." Gene probes have revealed that master cassettes are stored at the ends of the same chromosome that contains the "slot," and duplicates are copied from the masters and then inserted in the slot. Used cassettes are discarded.

17–21
Barbara McClintock. Her work was largely ignored for almost 40 years until 1981, when, at age 79, she received eight separate scientific awards, including a lifetime grant, and was hailed as a scientific prophet. James Watson said of her: "She is a very remarkable person, fiercely independent, beholden to no one."

Antibody Formation

Structural genes can also move about in the genome. The most striking case is seen in antibody formation. Antibodies, or immunoglobulins, are made by a particular group of lymphocytes (white blood cells). They are large protein molecules that recognize and combine with foreign macromolecules (antigens). The number and variety seem almost limitless; it has been estimated that a single individual is capable of making at least a million different kinds of antibodies. In the "one gene, one protein" days, this capacity for antibody formation posed a serious dilemma, since there are not a million genes coding for protein in the entire human genome.

Figure 17–22 shows the structure of an antibody molecule. As you can see, it is made up of two heavy (long) polypeptide chains and two light (short) chains. The heavy and light chains both have constant and variable regions. The constant regions of the chains are characteristic of the species of organism and the general class of antibody. The variable regions are responsible for the highly specific interaction with the antigens. It was suggested a decade or more ago that different types of antibody molecules might be made up of different combinations of light and heavy chains, but there was, at that time, no way to test the hypothesis. More recently, using recombinant DNA techniques, Susumu Tonegawa and his colleagues found that the genes for one special type of light chain are very close together in the DNA of lymphocytes making this particular chain, but far apart in lymphocytes that make another sort of light chain.

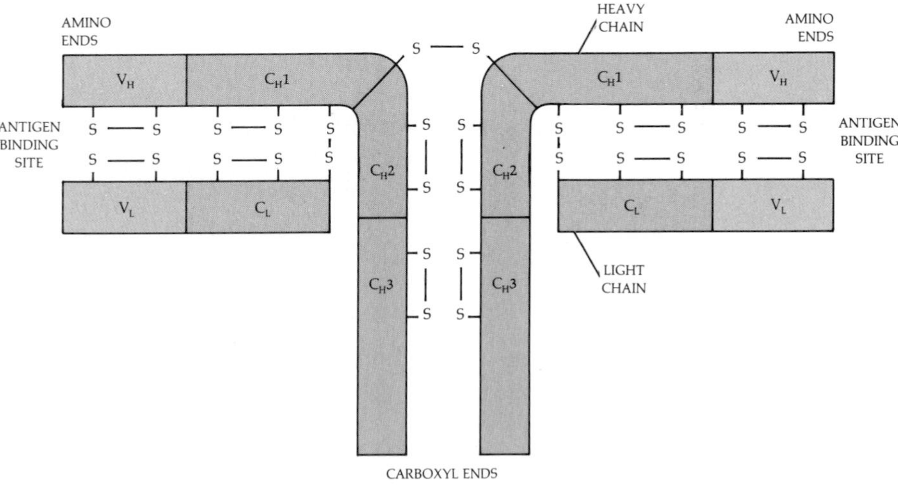

17–22
Structure of an antibody molecule. It has two light (L) and two heavy (H) chains, each of which has a variable (V) region and a constant (C) region. A single human being is capable of making at least a million different antibodies.

17–23
Diagram showing the assembly of an antibody heavy chain. Selected V, D, and J genes are transposed from different parts of the genome to form a complete variable gene that then joins to one of the constant (C) genes. The intervening sequence between the variable and constant sequences is removed from the RNA transcript to yield the finished mRNA.

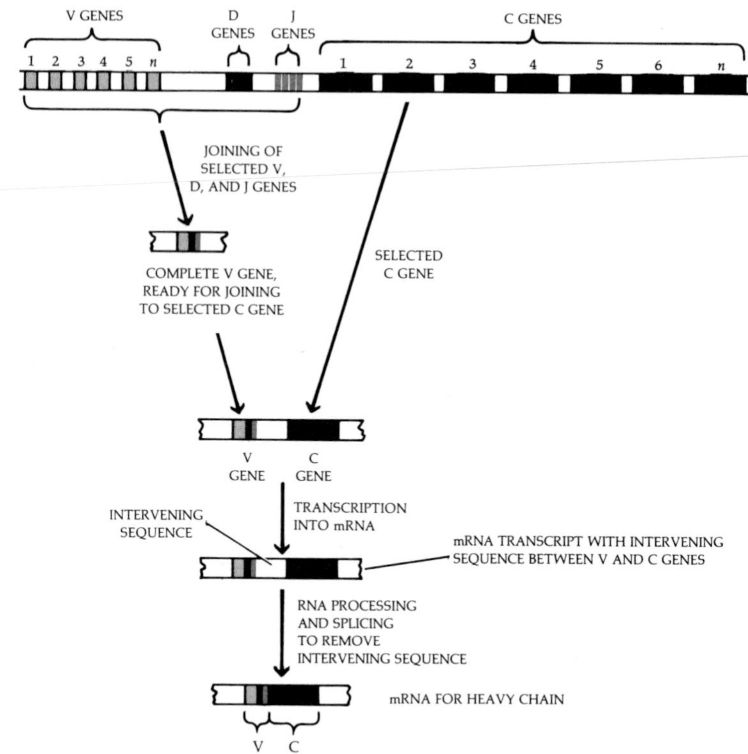

Tonegawa proposed that in the course of maturation of antibody-producing cells, the DNA undergoes rearrangements that result in a shuffling of gene sequences that brings the genes for different variable regions close to those for particular constant regions. It has been subsequently found that the DNA for the variable regions of heavy chains consists of about 400 different variable (V) genes, about 12 diversity (D) genes, and four joining (J) genes, which can be assembled in thousands of different ways.

Genes That Cause Cancer

As we noted previously, cancer is a group of diseases in which particular cells in the body cease to respond to whatever controls growth under normal conditions. They multiply autonomously, crowding out, invading, and destroying other tissues. Two lines of evidence have long linked the development of cancers with changes in the genetic material. First, translocations, deletions, and other chromosomal abnormalities are often visible in cancer cells. Second, most of the agents that cause cancer—including ultraviolet radiation, x-rays, and a variety of chemicals—are also mutagens.

With the discovery of a number of viruses that cause cancer in mice and other animals (though no human cancer viruses have been found), the mutation theory of cancer was briefly overshadowed. Most of the cancer viruses are RNA viruses, which also directed attention away from the DNA. Eventually, however, evidence began to emerge that the cancer viruses exert their effects by making changes in the cell's genetic material. In the case of some DNA viruses, it could be shown that the DNA introduced by the virus is actually incorporated into the host-cell DNA.

Especially interesting is the fact that all the cancer-causing RNA viruses were found to be retroviruses; that is, they all contain RNA sequences that code for reverse transcriptase, which can transcribe their RNA into DNA. Reverse transcripts of nucleic acids of RNA tumor viruses, such as the Rous sarcoma virus, which causes tumors in chickens, have now been located in the host chromosome.

Most recently, evidence is emerging that indicates the virus-borne DNA found in the chromosomes of cancerous cells is not of viral origin but instead came originally from a eukaryotic host chromosome. Thus it represents "extra" DNA transduced by the virus, just as lambda, for example, might carry around some of *E. coli*'s chromosome.

Returning to the basic concept that cancer comes about because of some revolutionary change in the normal control mechanism of a cell and adding to this the fact, revealed by McClintock more than 30 years ago, that transposition of genetic elements can bring about such changes, researchers are formulating a new picture of how cancer may originate. In 1982, molecular biologists were able to extract from human cancer cells segments of DNA that transform mouse fibroblasts growing in tissue culture to cancer cells. Such segments are called oncogenes. Some of these segments have been cloned, and hybridization techniques have shown that they are also present in the DNA of normal human cells. This finding suggests that cancers may result from changes in cell regulation brought about by the transposition of a normal gene from one site to another in the genome. Such transpositions could be brought about by external agents, such as radiation or chemical mutagens; by viruses, carrying genes from organism to organism; or by internal events in the cell, as with the yeast cassettes.

17–24
Scanning electron micrographs of (a) *normal cells growing in tissue culture and* (b) *the same type of cells after transformation with a cancer-causing virus. Observe that the cancer cells not only show striking surface changes but also have piled up on top of one another. The normal cells are inhibited by cell-to-cell contact and stop multiplying, whereas the cancer cells do not (see page 119).*

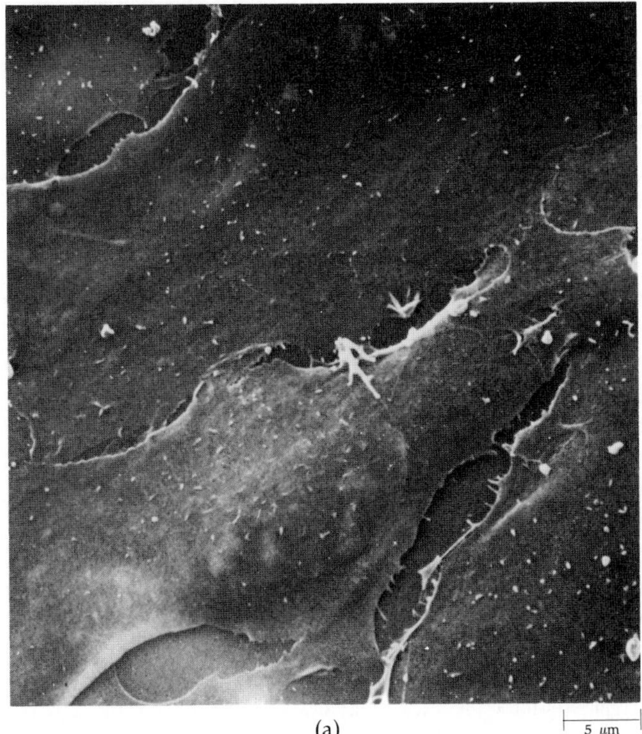

(a)

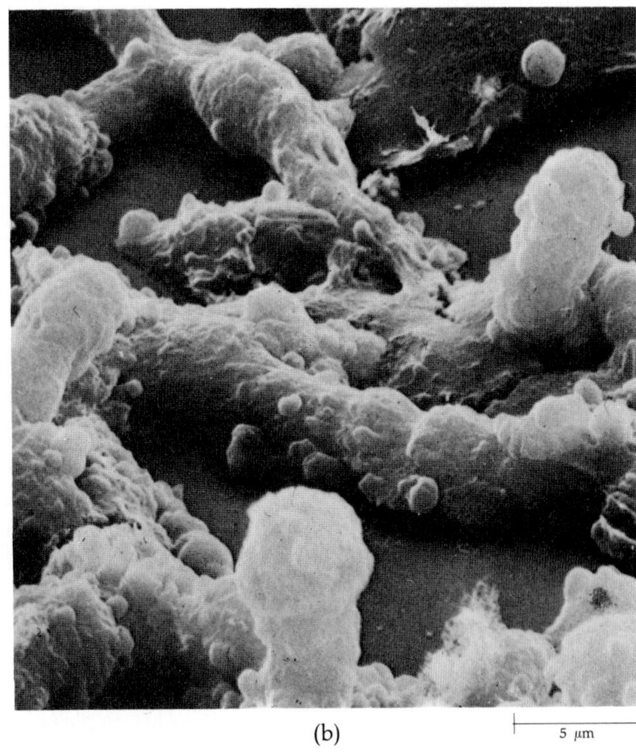

(b)

Viruses in Gene Transfer

The idea that a virus might be the vector for a cancer-causing gene receives some support from the recent discovery of two pseudogenes, each found at some distance from its gene family, that have unusual characteristics. One of them, a pseudogene for a mouse alpha globin, has no introns. (Introns, you will recall, are remarkably stable features of the hemoglobin family.) The second, an immunoglobulin pseudogene, not only has no introns but also has a poly-A tail. Moreover, it contains J (joining) and C (constant) sequences that are clearly connected. In short, both these pseudogenes have the features of processed mRNA rather than of DNA. It is hypothesized that mRNA was transcribed back into DNA and then found its way into chromosomal DNA, reaching cells of the germ line—that is, cells from which eggs or sperm were to be formed.

Reverse transcriptase, the enzyme that makes DNA copies from RNA, however, has only been found in association with RNA retroviruses. This leads to the intriguing speculation that these pseudogenes may have been picked up by a virus from a eukaryotic cell at some point in evolutionary history. They were then processed and carried to another host (transduction).

A gene that may have had a similar history is that for leghemoglobin, also, as its name implies, a member of the hemoglobin family. The unusual feature of leghemoglobin is that it is produced in plants—legumes—in which it is used in nitrogen fixation, the process by which atmospheric nitrogen is made available to living systems. The DNA coding for leghemoglobin has all the characteristics of a primitive globin gene. The best explanation, so far, of its presence in legumes is that it was transported there by a virus from an animal host.

Many years ago, Luria characterized viruses as "bits of heredity looking for a chromosome." More recently, Bernard D. Davis has stated, ". . . viruses may turn out to have evolved primarily to transfer blocks of nucleic acid between organisms," adding, ". . . it is not inconceivable that all the DNA in the living world may be part of an unbroken chain of low-frequency contacts."

17–25
The large tumor on this stem of a Begonia *plant is a crown gall—a swelling that contains cancerous, rapidly growing plant cells. Crown gall disease is caused by a kind of "genetic colonization" of the plant cells by plasmids from the common soil bacterium* Agrobacterium tumefaciens.

Crown Gall Disease

Among the most intriguing of the movable genetic elements are those that cause the tumors of crown gall disease in plants (Figure 17–25). Armin C. Braun of Rockefeller University demonstrated some years ago that the tumors are caused by infection with a common soil bacterium, *Agrobacterium tumefaciens,* and suggested that the tumor-causing bacterium exerts its effects by producing a permanent hereditary change in the infected cells. The fact that the plant cells (and their daughter cells) continue their cancerous behavior even when the bacteria are no longer present lent support to this idea.

Mary-Dell Chilton of Washington University and her coworkers have now discovered that the agents of crown gall disease are not the bacteria themselves but large plasmids carried by the bacteria. When the plant cell is infected, these plasmids (or portions of them) become incorporated into the DNA of the plant cell. The presence of the plasmid in the host chromosome not only results in the accelerated and uncontrolled pattern of growth characteristic of the disease but also in the production of some unusual proteins. These new proteins have two functions: first, they provide nutrients to the *Agrobacterium* cells, and second, they promote conjugation among them, thus encouraging the spread of the tumor-causing plasmids. In effect, the bacteria have used their plasmids to take over the host-cell chromosome and subvert the cell's entire economy to the benefit of *Agrobacterium.* Aside from those now achieved by human intervention, this is the only genetic recombination yet recorded between prokaryotes and eukaryotes.

THE DNA OF MITOCHONDRIA

We began this chapter with a statement of the one truism that seems to have emerged from the field of molecular genetics—that the genetic code is universal. Given the number of recent unexpected developments, it should not now be surprising to learn that this truism no longer holds. Mitochondria (page 194) have their own DNA, which, like that of *E. coli,* is not associated with histones and is circular. This relatively small molecule (16,569 base pairs in a human mitochondrion) has now been sequenced. Comparison of the DNA sequence with the RNAs and the few proteins made by the mitochondria has shown that there are differences in the code. For instance, UGA is tryptophan and not termination, and AUA is methionine, not isoleucine (see page 303). Also mitochondria have many fewer tRNAs than either *E. coli* or eukaryotes, not enough to translate all the possible codons by the conventional base pairing, even allowing for "wobble." These findings may shed some light on the question of the origin of mitochondria, a topic discussed in Chapter 21.

Another interesting feature of mitochondria that has been confirmed by DNA analysis is that all the mitochondria in an organism are inherited from the maternal side. It is possible that some diseases that have seemed hereditary in origin but whose genetic background could not be clarified may be traced to disturbances in the function of these important organelles.

SUMMARY

In eukaryotes, the genetic material is packaged into chromosomes, visible when condensed at mitosis or meiosis. In condensed chromosomes, the chromatin takes the form either of euchromatin, which is loosely packed, or heterochromatin, which is tightly packed. The nucleolus, visible in the interphase nucleus, is made up of loops of chromatin from different chromosomes. It is the site of synthesis of ribosomes.

The eukaryotic chromosome differs in many ways from the chromosome of prokaryotes. Its DNA is always associated with proteins, which constitute more than half of the weight of the chromosome. Most of these proteins are in the form of histones, which are relatively small, positively charged molecules. The DNA molecule wraps around histone cores to form nucleosomes, which are the basic packaging units of eukaryotic DNA. Much evidence links the transcription of DNA to the degree of condensation of the chromatin.

Eukaryotes have far more DNA than prokaryotes. The amount of DNA is constant in the cells of any given species but is not correlated with the size, complexity, or position on the evolutionary scale of the organism. DNA replication is bidirectional, as in prokaryotes; however, there are thousands of different replication initiation sites on each chromosome.

DNA hybridization and sequencing studies have revealed three classes of eukaryotic DNA: short multiple repeats, characteristically arranged in tandem; longer repeats, usually dispersed throughout the chromosomes; and single-copy DNA.

(The latter makes up about 70 percent of the chromosomal DNA in humans.) Only about 1 percent of the DNA actually codes for proteins.

Some genes come in families. Of these, some are uniform, such as those coding for the rRNAs and tRNAs and for the histones. Some are diverse, as in the myoglobin-hemoglobin family, coding for proteins with slightly different properties.

Most of the structural genes of eukaryotes contain one or more sequences that, although they are transcribed into RNA in the nucleus, are not present in the mRNA of the cytoplasm and thus are not translated into proteins. These segments are known as introns, whereas the expressed segments are known as exons.

DNA segments can move from place to place on the chromosome. Genes for mating type can change position in yeast, and regulatory elements have been detected in corn that, when they change position, activate other genes and so cause genetic changes. In the course of the maturation of antibody-producing cells, gene sequences coding for different parts of the antibody molecule are moved around in the chromosome, thus producing proteins with the vast diversity characteristic of antibodies. There is some evidence that some forms of cancer in animals may be a result of the insertion of normal genes in new locations. Such transpositions may be mediated by viruses. A tumor in plants has been shown to be the result of incorporation of a bacterial plasmid into the host chromosome. Thus, in the eukaryotic chromosome as well as in the prokaryotic one, there are many different mechanisms for transfer of genetic material.

An exception to the universality of the genetic code has been discovered: the DNA of mitochondria differs from the DNA of either eukaryotes or prokaryotes.

QUESTIONS

1. Distinguish among the following: nucleosomes/nucleolus; euchromatin/heterochromatin; chromosome puffs/lampbrush chromosomes; B-DNA/Z-DNA; myoglobin/hemoglobin/leghemoglobin; globin/globulin; gene/pseudogene; hnRNA/snRNP/snRNA; exon/intron; insertion sequence/intervening sequence.

2. In what ways are the chromosomes of prokaryotes and eukaryotes similar? In what ways are they different? Consider their composition, structure, and replication.

3. Describe the three classes of eukaryotic DNA. What is thought to be the function of each class?

4. What might be the advantage to an organism of having multiple copies of the genes for the rRNAs and tRNAs and for the histones? What might be the advantage of being able to make "extra" copies of particular genes at certain stages in its life?

5. The genes coding for histones are rich in guanine and cytosine compared to the spacers between them. How does this explain the appearance of the DNA molecule shown in Figure 17–12?

6. A hemoglobin molecule contains four polypeptide chains, two from the alpha family and two from the beta family. What benefit does a mammal derive from the capacity to make several different kinds of polypeptides in each family?

7. In what way does the control of mating in yeast resemble the switch from one flagellar protein to another in *Salmonella* (page 319)? How do they differ?

8. Although the human genome does not contain a million structural genes, a human individual is capable of making at least a million different kinds of antibodies. How is this possible?

9. What is the function of the intron in the assembly of a polypeptide chain for an antibody molecule?

10. Bacteriophages lack intervening sequences, but intervening sequences have been found in animal viruses. What do these findings suggest about the origin of viruses?

11. The existence of mating types is common among one-celled organisms. Can you explain their function?

CHAPTER 18

Human Genetics

The principles of genetics are, of course, the same for humans as they are for any other species. There are some important differences, however, particularly in methodology. Breeding experiments, so readily performed with fruit flies, bread molds, and pea plants, are not possible with humans, and even if they were, the long generation time would make them highly impractical. Most data on patterns of human heredity are based on family pedigrees. With the exception of those of us who belong to royal families, few have knowledge about their forebears that extends over more than four generations. Do you, for instance, know what your various great-great-grandparents looked like? More important, do you know what health problems they had and what caused their deaths?

18–1
Queen Victoria (seated center) and some of her immediate family. Seventeen of the people in this photograph, which was taken in 1894, are her direct descendants. These include Princess Irene of Prussia, standing to the right of Victoria and wearing a feather boa, and, to the left of Victoria, Alexandra (also wearing a boa), the future Tsarina of Russia. Nicholas II, to become the last Tsar of Russia, is standing beside Alexandra. Both Irene and Alexandra were carriers of hemophilia.

Most information about human genetics comes from medical sources, and it is here that human genetics, as a science, has a great advantage. Large numbers of people are kept under careful scrutiny by a variety of medical experts. As a result, a large amount of data has been accumulated on a number of disease-causing genetic abnormalities.

THE HUMAN KARYOTYPE

A *karyotype* is simply the graphic representation of the physical appearance of the chromosomes of a particular organism. The karyotype of *Drosophila* (page 269) has been known since the early 1920s. It was not until 1956, however, that it was established, after a flurry of controversy, that the diploid number of human chromosomes is 46: 44 autosomes and two sex chromosomes. Interest in karyotyping received great impetus three years later with the discovery that the presence of an extra chromosome could be associated with a major disease problem, Down's syndrome. Karyotypes are now prepared routinely for genetic counseling of couples at risk for Down's syndrome and other defects associated with chromosome abnormalities.

The chromosomes shown in a human karyotype are characteristically metaphase chromosomes, each consisting of two sister chromatids held together at their centromeres. To prepare a karyotype (Figure 18–2), white blood cells in the process of dividing are interrupted at metaphase by the addition of colchicine, which prevents the subsequent steps of mitosis from taking place. After treating and staining, the chromosomes are photographed, enlarged, cut out, and arranged according to size.

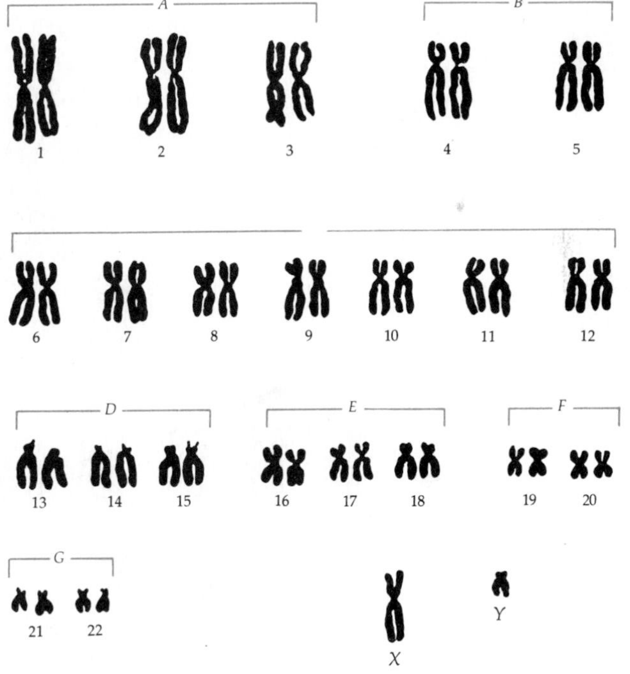

18–2
The normal diploid chromosome number of a human being is 46, 22 pairs of autosomes and the two sex chromosomes. The autosomes are grouped by size (A, B, C, etc.), and then the probable homologues are paired. A normal woman has two X chromosomes and a normal man, shown here, an X and a Y.

PREPARATION OF A KARYOTYPE

Chromosome typing for the identification of hereditary defects is being carried out at an increasing number of genetic counseling centers throughout the United States. The result of the procedure is a graphic display of the chromosome complement, known as a karyotype. The chromosomes shown in a karyotype are metaphase chromosomes, each consisting of two sister chromatids held together at their centromeres. White blood cells in the process of dividing have been interrupted at metaphase by the addition of colchicine, which prevents the subsequent steps of mitosis from taking place. After treating and staining, the chromosomes are photographed, enlarged, cut out, and arranged according to size. Certain abnormalities, such as an extra chromosome or piece of a chromosome, can be detected.

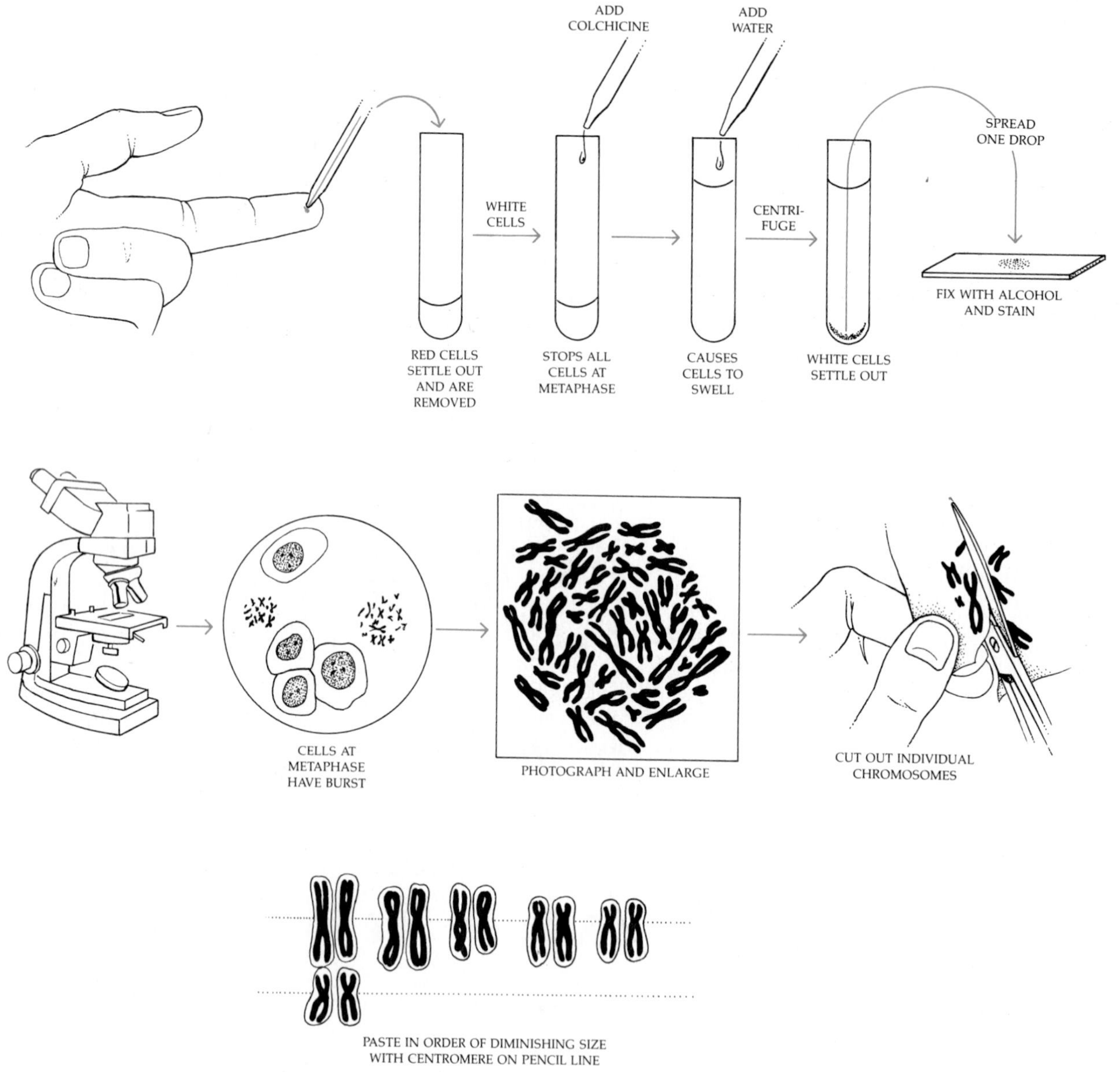

ADD COLCHICINE

ADD WATER

SPREAD ONE DROP

WHITE CELLS

CENTRIFUGE

FIX WITH ALCOHOL AND STAIN

RED CELLS SETTLE OUT AND ARE REMOVED

STOPS ALL CELLS AT METAPHASE

CAUSES CELLS TO SWELL

WHITE CELLS SETTLE OUT

CELLS AT METAPHASE HAVE BURST

PHOTOGRAPH AND ENLARGE

CUT OUT INDIVIDUAL CHROMOSOMES

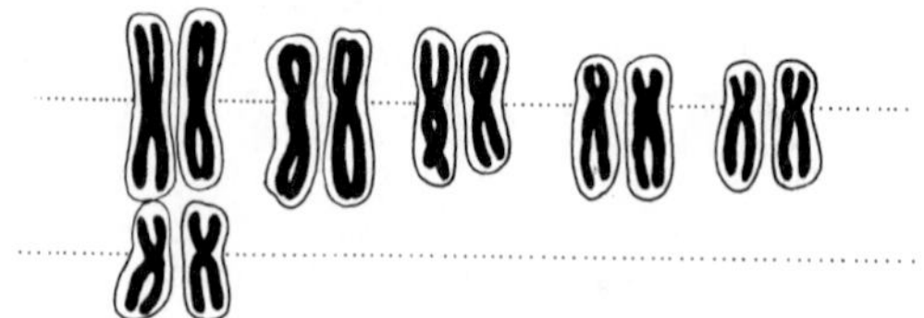

PASTE IN ORDER OF DIMINISHING SIZE WITH CENTROMERE ON PENCIL LINE

18–3
A standard map of the banding patterns of chromosomes 8 through 11 in the human karyotype as determined both at the metaphase stage (black bands) and at the early prophase stage of mitosis (colored bands). The early prophase chromosomes are much longer and thinner than the metaphase chromosomes, and many more bands can therefore be detected. All of the bands shown here are those that stain with reagents that appear to be specific for DNA sequences rich in adenine and thymine. Note how these chromosomes, which are similar in size and shape, can readily be distinguished by their banding patterns.

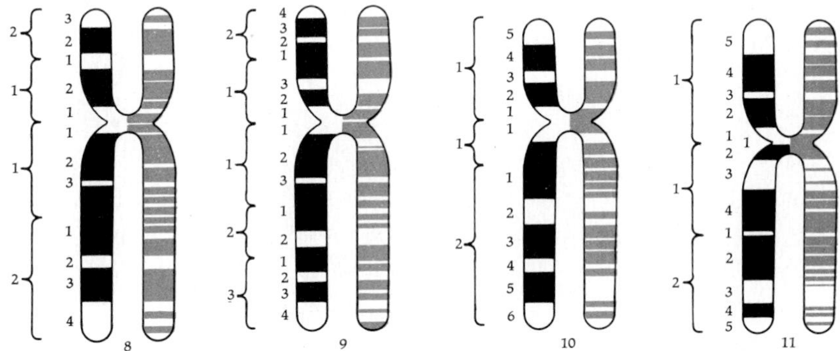

As you can see in Figure 18–2, many of the chromosomes within the various size groups closely resemble one another. Within the last decade, staining techniques have been developed that reveal characteristic banding patterns on the various chromosomes (Figure 18–3). The significance of these banding patterns is not understood, but, nevertheless, the bands form useful landmarks for the characterization of chromosomes.

CHROMOSOME ABNORMALITIES

Certain genetic diseases are caused by gross abnormalities in the number or structure of chromosomes. For example, from time to time, usually because of "mistakes" at the time of meiosis, homologues may not separate. In this case, one of the gametes has one chromosome too many, and the other, one too few. This phenomenon is known as *nondisjunction*. The cell with one too few (with the exception of an *XO*) cannot produce a viable embryo, but the cell with one too many sometimes can. The result is an individual with an extra chromosome in every cell of his or her body.

The presence of additional chromosomes often produces widespread abnormalities. Many infants with such abnormalities are stillborn. Among those who survive, many are mentally deficient. In fact, studies of abnormalities in human chromosome number among living subjects are often carried out on patients in mental hospitals. These patients frequently have abnormalities of the heart and other organs as well.

Down's Syndrome

One of the most familiar chromosomal abnormalities is the form of mental deficiency known as Down's syndrome, after the physician who first described it. (Because it usually involves more than one defect, it is referred to as a syndrome, a group of disorders that occur together.) Down's syndrome includes, in most cases, not only mental deficiency but also a short, stocky body type with a thick neck and, often, abnormalities of other organs, especially the heart.

Down's syndrome arises when a child inherits three, rather than two, copies of chromosome 21. In about 95 percent of the cases of Down's syndrome, the cause of the genetic abnormality is nondisjunction of chromosome 21. This results in an extra chromosome 21 (Figure 18–4b) in the cells of the child.

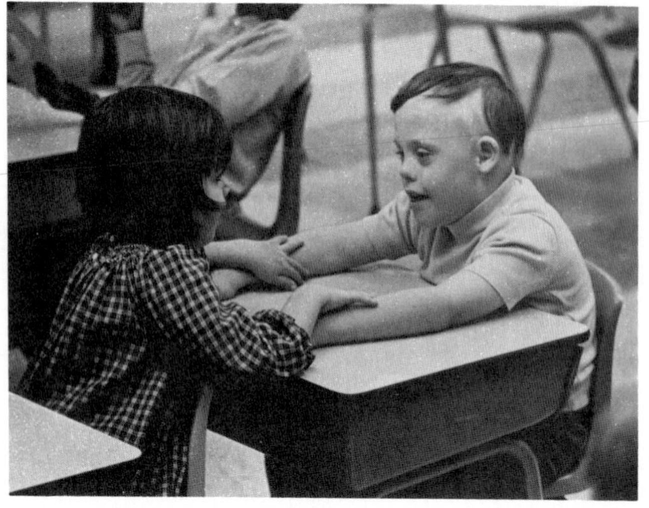

(a)

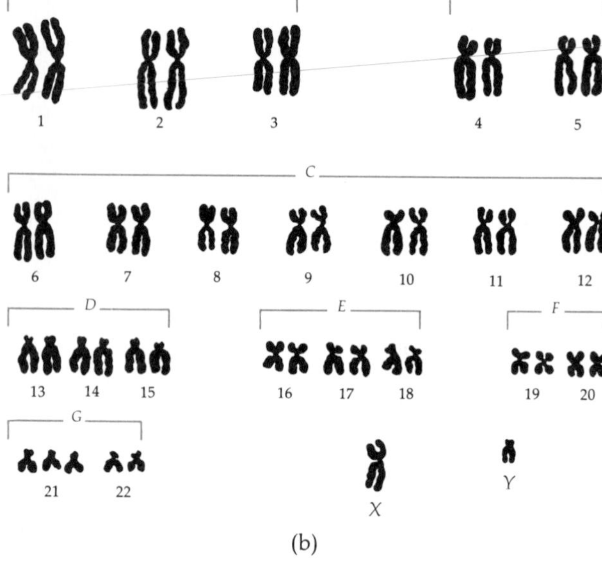

(b)

18–4
(a) *Although children with Down's syndrome share certain physical characteristics, there is a wide range of mental capacity among these individuals.* (b) *The karyotype of a male with Down's syndrome caused by nondisjunction. Note that there are three chromosomes 21.*

Down's syndrome may also be the result of an abnormality, known as translocation, in the chromosomes of one of the parents. As we saw in Chapter 13 (page 274), translocation occurs when a portion of a chromosome is broken off and becomes attached to another chromosome. The person with Down's syndrome caused by translocation usually has a third chromosome 21 (or, at least, most of it) attached to a larger chromosome, such as 15. Thus, in both cases, the individual has three chromosomes 21, or their equivalent.

When a child with Down's syndrome is found to carry a translocated chromosome 21, further investigation usually discloses that one parent, although phenotypically normal, has only 45 separate chromosomes—one chromosome being composed of most of chromosomes 15 and 21 joined together. The possible genetic makeups of the offspring of this parent are diagrammed in Figure 18–5. Three out

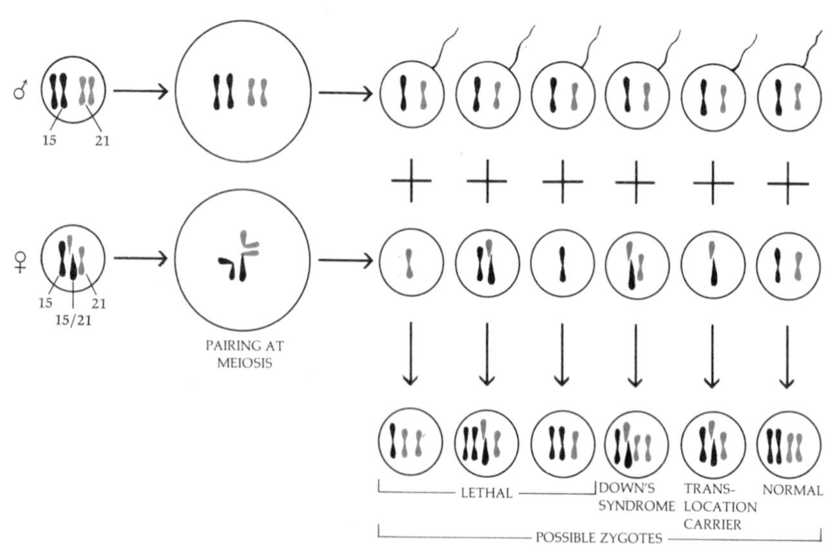

18–5
Transmission of translocation Down's syndrome. The father, top row, has normal pairs of chromosomes 21 and 15, and each of his sperm cells will contain a normal 21 and a normal 15. The mother (shown here as the translocation carrier) has one normal 15, one normal 21, and a translocation 15/21. She herself appears normal, but her chromosomes cannot pair normally at meiosis. There are six possibilities for the offspring of these parents; the infant will (1) die before birth (three of the six possibilities), (2) have Down's syndrome, (3) be a translocation carrier like the mother, or (4) be normal. Tests for the chromosomal abnormality can be made in prospective parents and in the fetus before birth.

18–6
The frequencies of births of infants with Down's syndrome in relation to the ages of the mothers. The number of cases shown for each age group represents the occurrence of Down's syndrome in every 1,000 births by mothers in that group. As you can see, the risk of having a child with Down's syndrome increases rapidly after the mother's age exceeds 40.

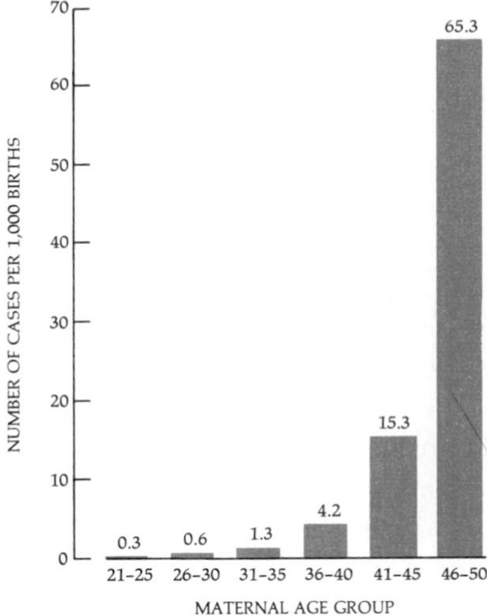

of the six possible combinations are lethal. One of the remaining three will produce Down's syndrome, one will be normal, and one will be a carrier.

Thus parents who have had a child with Down's syndrome are advised to have their karyotypes prepared. If either parent has the translocation, they are warned that they may transmit the condition. If the karyotypes of both parents are normal, they do not run a greater-than-average risk for the mother's age group of having another congenitally defective child.

It has been known for many years that Down's syndrome and a number of other defects involving nondisjunction are more likely to occur among infants born to older women (Figure 18–6). The reasons for this are not known, but the formation of egg cells is well under way in the human female before she is born, so the increasing incidence of abnormalities may be correlated in some way with the aging of the mother's reproductive cells. Recent studies, however, have also indicated that in about 20 percent of the cases of Down's syndrome due to nondisjunction, the extra chromosome comes from the father rather than the mother.

Abnormalities in the Sex Chromosomes

Nondisjunction may also produce individuals with extra sex chromosomes. An *XY* combination in the twenty-third pair, as you know, produces maleness, but so do combinations *XXY*, *XXXY*, and even *XXXXY*. These latter males are usually sexually underdeveloped and sterile, however. *XXX* combinations sometimes produce normal females, but many of the *XXX* women and all *XO* women (women with only one X chromosome) are sterile. It is not unusual for individuals with sex chromosome abnormalities to be mentally retarded.

Chromosome Deletions

Small chromosome deletions—detectable with chromosome banding techniques—can also result in congenital defects or illness. For example, a small deletion on the short arm of chromosome 11 is associated with Wilm's tumor, a cancer of the kidney found almost exclusively in infants and young children. Associated with this same deletion is a condition known as aniridia, which is a congenital absence of the iris (Figure 18–7). Not everyone with the deletion develops Wilm's tumor, but infants with aniridia and the chromosome deletion are at a high risk for this form of cancer. The testing of children with aniridia for the chromosome deletion has enabled medical researchers to detect Wilm's tumor before it has caused any symptoms and, in at least one case, to watch for the tumor before it developed. Wilm's tumor is curable when detected and treated early.

18–7
(a) *One of the few chromosomal abnormalities associated with cancer. The chromosome on the left is normal. The one on the right has a deletion, shown by the smaller size of the bracket. Such deletions have been found in children with Wilm's tumor.* (b) *The left eye of a 15-year-old boy who has this chromosomal deletion and who developed Wilm's tumor in infancy. Note the absence of an iris. An older half-brother and a maternal aunt also had aniridia and developed Wilm's tumor at an early age. Another brother and the boy's mother are phenotypically normal. Analysis of the mother's chromosomes revealed that although she carries the deletion in chromosome 11, the missing segment is present in her cells in chromosome 2.*

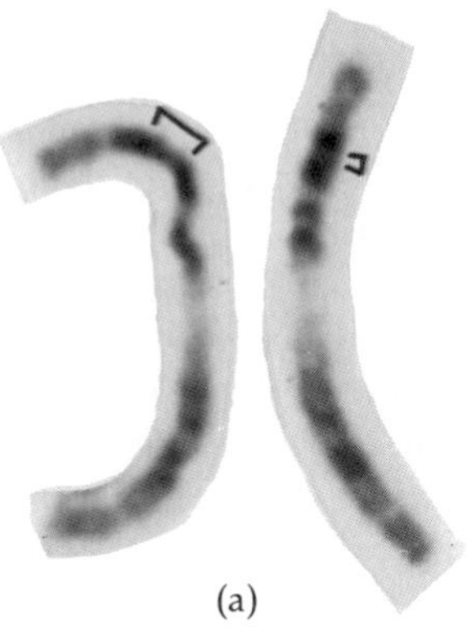
(a)

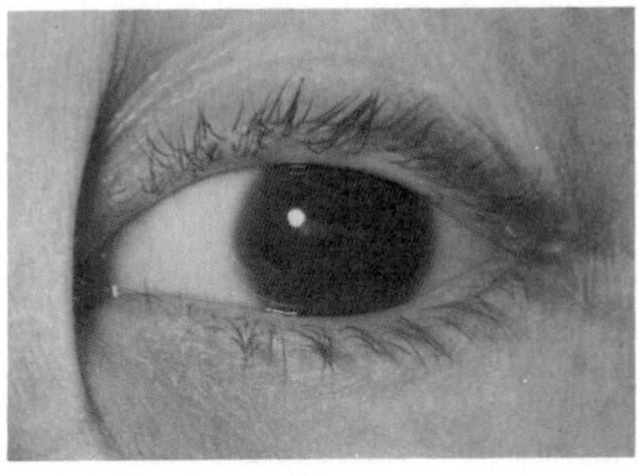
(b)

Amniocentesis

A procedure known as amniocentesis makes possible the prenatal detection of Down's syndrome, as well as certain other genetic conditions, in the fetus. A thin needle is inserted through the mother's abdominal wall and through the membranes that enclose the fetus, and a sample of the amniotic fluid surrounding the fetus is withdrawn (Figure 18–8). While the procedure must be done with great care, it is simple, quick, and relatively harmless. This fluid contains living cells sloughed off by the fetus. These cells, grown in tissue culture, can provide mitotic cells from which a karyotype can be made.

18–8
Amniocentesis. The position of the fetus is first determined by ultrasound. Then a needle is inserted into the amniotic cavity, and fluid containing fetal cells is withdrawn into a syringe. The cells can then be analyzed for genetic defects. The procedure is usually not performed until the sixteenth week of pregnancy, to ensure both that there are enough fetal cells in the amniotic cavity to make detection possible and that there is sufficient amniotic fluid so that removal of the small amount necessary for the test will not endanger the fetus.

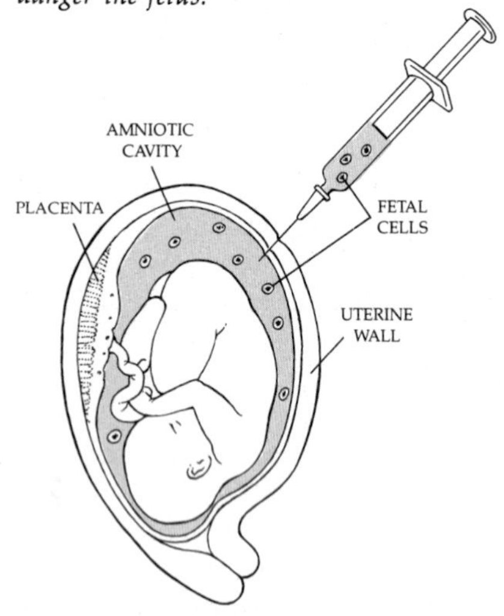

SEX-LINKED HEREDITARY TRAITS

Color Blindness

As in *Drosophila,* the Y chromosome of a human male carries less genetic information than the X chromosome. Genes for color vision, for example, are carried on the X chromosome in humans but not on the Y chromosome. Color blindness is produced by a recessive allele of the normal gene. The allele for complete color vision is dominant; a woman with one X chromosome with the normal allele and one X chromosome with the allele for color blindness will have normal color vision. If she transmits the X chromosome with the recessive allele to a daughter, the daughter will also have normal color vision if she receives an X chromosome with the normal allele from her father (that is, if he is not color-blind). If, however, the X chromosome with the recessive allele is transmitted from mother to son, he will be color-blind since, lacking a second X chromosome, he has only the recessive allele (Figure 18–9). As we noted earlier, characteristics such as eye color in *Drosophila* or human color blindness, which are controlled by alleles on the X chromosome, are said to be sex-linked.

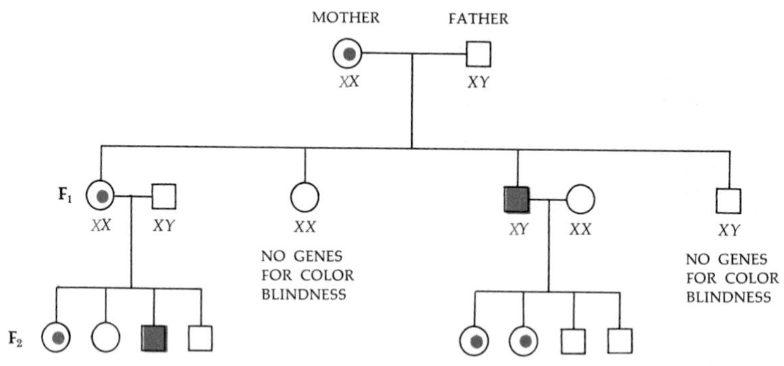

18–9
Color blindness in humans is determined by a recessive allele on the X chromosome. In the pedigree shown here, the mother has inherited one normal and one defective allele. The normal allele will be dominant, and she will have normal color vision. However, half her eggs (on the average) will carry the defective allele and half will carry the normal allele—and it is a matter of chance which kind is fertilized. Since her husband's Y chromosome, the one that determines a son rather than a daughter, carries no gene for color discrimination, the single gene the wife contributes (even though it is a recessive allele) will determine whether or not the son is color-blind. Therefore, half her sons (on the average) will be color-blind. Assuming that her children marry individuals with X chromosomes with the normal alleles, the expected distribution of the trait among her grandchildren will be as shown on the chart.

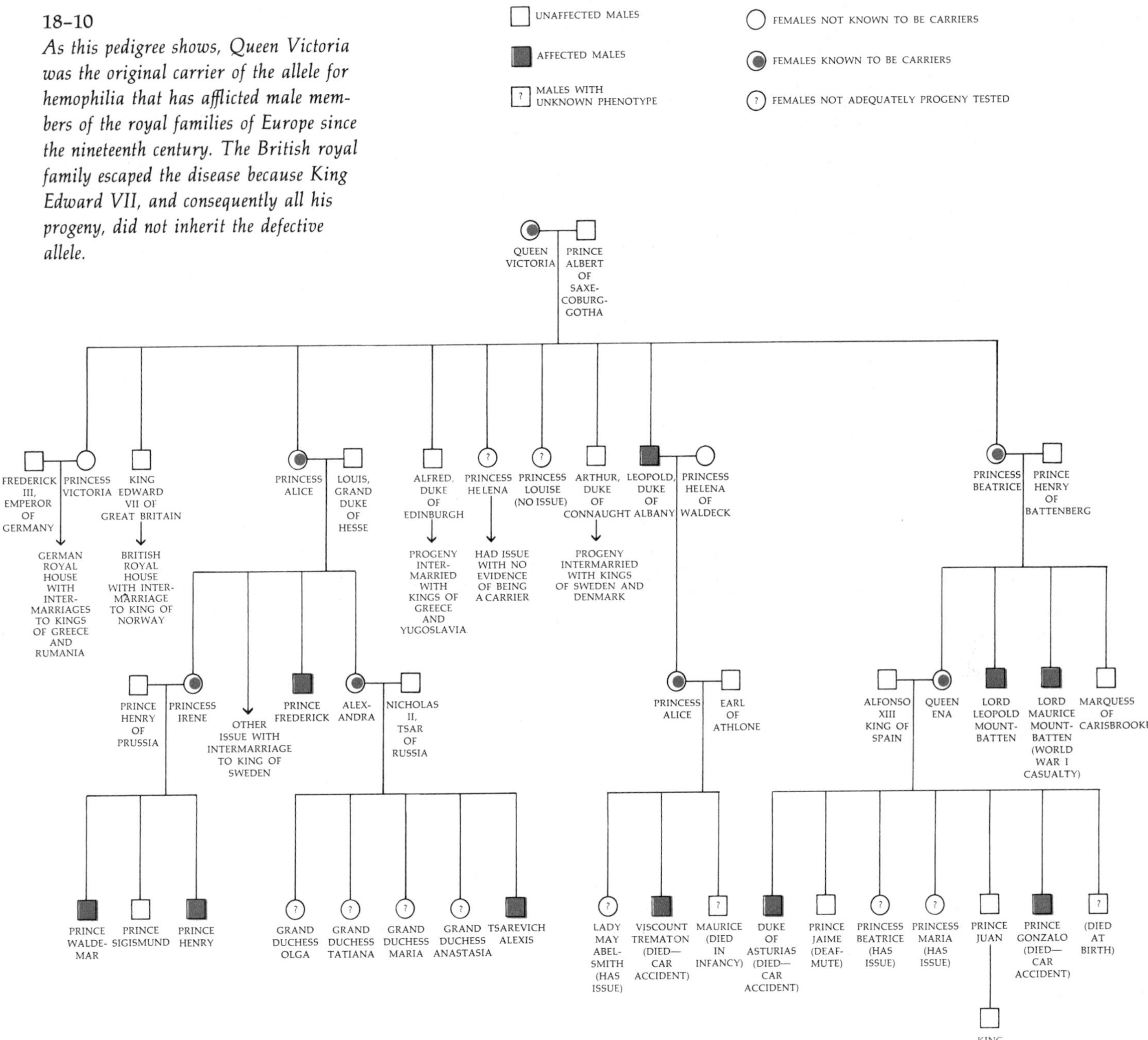

18–10
As this pedigree shows, Queen Victoria was the original carrier of the allele for hemophilia that has afflicted male members of the royal families of Europe since the nineteenth century. The British royal family escaped the disease because King Edward VII, and consequently all his progeny, did not inherit the defective allele.

Hemophilia

A classic example of a recessive allele transmitted on the X chromosome is the type of hemophilia that has afflicted some royal families of Europe since the nineteenth century (see Figure 18–1). Hemophilia is a group of diseases in which the blood does not clot normally. In some kinds of hemophilia, even minor injuries carry the risk of the patient's bleeding to death. Queen Victoria was probably the original carrier in her family (Figure 18–10). Because none of her forebears or collateral

18–11
Gregory Efimovitch Rasputin (1871–1916). Nicholas II and Alexandra of Russia fell under the sinister influence of Rasputin because of the hemophilia of their son Alexis, on whom Rasputin exerted apparently mystical healing powers. The allele for hemophilia had been inherited from Queen Victoria. Alexis did not die of his disease but was executed with other members of the royal family in 1918.

relatives were affected, we conclude that the mutation occurred on an X chromosome in one of her parents or in the cell line from which her own eggs were formed. One of her sons, Leopold, Duke of Albany, died of hemophilia at the age of 31. At least two of Victoria's daughters were carriers, since a number of *their* descendants were hemophiliacs. And so, through various intermarriages, the disease spread from throne to throne across Europe. In Tsarevitch Alexis, son of the last Tsar of Russia, the gene for hemophilia, inherited from Victoria, had considerable political consequences.

The genes for color blindness and hemophilia are in the distal portion of the long arm of the X chromosome.

MAPPING HUMAN CHROMOSOMES

In humans, as in *Drosophila,* the association of particular hereditary traits with the presence or absence of the second X chromosome makes it possible, of course, to assign a particular gene to the sex chromosome. In *Drosophila* and other organisms, the mapping of the genes on the autosomes was made possible by breeding experiments. Mapping human chromosomes has required the development of some wholly new techniques.

Mouse-Human Hybrid Cells

The most widely used method for correlating human genes with particular chromosomes is the production of cell hybrids. This involves the fusion of human cells grown in tissue culture with similarly cultivated cells of another species, usually a mouse. The cell that is produced has, at first, the full complements of both the human and mouse chromosomes. As the cells divide, however, for reasons that are not known, the human chromosomes are lost at random. In the simplest situation, a hybrid cell loses all but one of its human chromosomes and, by multiplying, produces a colony that represents a clone of these cells. Any human enzyme or other protein that can be detected in the cells must reflect the presence of a gene on the single remaining human chromosome, which is identifiable by its banding pattern.

Adaptations and refinements of this technique now make it possible to insert a single chromosome or chromosome fragments into cultured mouse cells, thus making location of genes more rapid and more precise.

Radioactive Probes

An obvious limitation of using cell hybrids is that this method can detect only those genes that produce proteins detectable in tissue culture. An alternative technique involves the use of radioactive nucleic acid probes. Such probes, radioactive samples of specific purified mRNAs or cloned DNA sequences, are introduced into cultured cells. If they hybridize with a specific segment of DNA on a human chromosome, their radioactive tags reveal their presence. Investigators at Yale University used this method to show that the alpha family of hemoglobin genes is located on human chromosome 16 and the beta family is on chromosome 11 (Figure 17–15, page 337).

More than 400 human genes have now been mapped using a combination of approaches.

INBORN ERRORS OF METABOLISM

In 1908, more than three decades before the "one gene–one enzyme" hypothesis of Beadle and Tatum (a turning point, you will recall, in the history of genetics), an English physician, Sir Archibald Garrod, presented a series of lectures in which he set forth a new concept of human diseases, one he called "inborn errors of metabolism." With a leap of the imagination that spanned almost half a century, Garrod postulated that certain diseases that are caused by the body's inability to perform particular chemical processes are hereditary in nature.

One such disease was described in 1649:

> The patient was a boy who passed black urine and who, at the age of fourteen years, was submitted to a drastic course of treatment which had for its aim the subduing of the fiery heat of his viscera, which was supposed to bring about the condition in question by charring and blackening his bile. Among the measures prescribed were bleedings, purgation, baths, a cold and watery diet, and drugs galore. None of these had any obvious effect, and eventually the patient, who tired of the futile and superfluous therapy, resolved to let things take their natural course. None of the predicted evils ensued, he married, begat a large family, and lived a long and healthy life, always passing urine black as ink.

Sir Archibald postulated that this disease, alcaptonuria, (1) is the result of an enzyme deficiency (Figure 18–12) and (2) is hereditary in nature. He was, though he would not have understood the term, the first molecular geneticist.

Alcaptonuria and a number of other "inborn errors of metabolism" are inherited as simple Mendelian recessives. For reasons that we now understand, heterozygous individuals are usually symptom-free. In the heterozygote, enough of the particular enzyme is produced from the "good" allele to make up for the nonfunctioning defective allele. The symptoms of the enzyme deficiency show up only in the recessive homozygote.

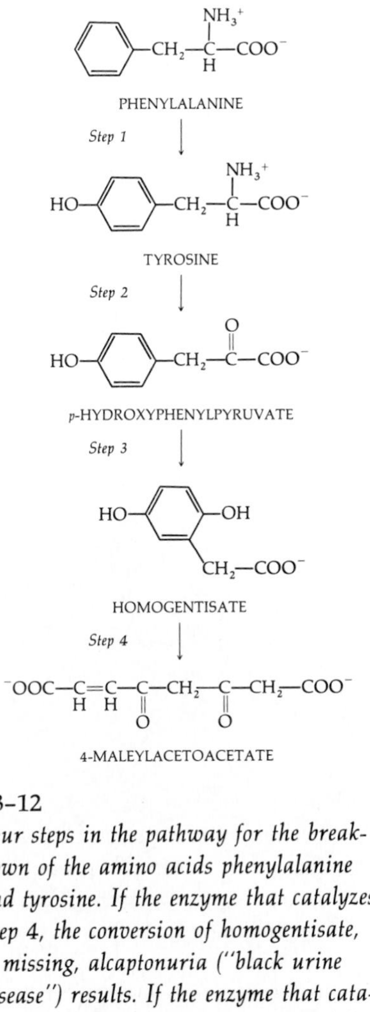

PHENYLALANINE

Step 1

TYROSINE

Step 2

p-HYDROXYPHENYLPYRUVATE

Step 3

HOMOGENTISATE

Step 4

4-MALEYLACETOACETATE

18–12

Four steps in the pathway for the breakdown of the amino acids phenylalanine and tyrosine. If the enzyme that catalyzes Step 4, the conversion of homogentisate, is missing, alcaptonuria ("black urine disease") results. If the enzyme that catalyzes Step 1, the conversion of phenylalanine to tyrosine, is defective, the result is an accumulation of phenylalanine and the disease known as phenylketonuria.

Phenylketonuria

One of the most familiar examples of such an "inborn error" inherited as a Mendelian recessive is phenylketonuria, or PKU, as it is commonly known. PKU is caused by a lack of the enzyme that normally breaks down the amino acid phenylalanine (Figure 18–12). When this enzyme is missing or deficient, phenylalanine and its abnormal breakdown products accumulate in the bloodstream. Thus it resembles alcaptonuria, with the important difference that in phenylketonuria the breakdown products that accumulate are harmful to the cells of the developing brain and can result in mental retardation. PKU is caused by a single recessive allele in the homozygous state. About 1 in every 15,000 infants born in the United States is homozygous for this allele.

Infants with PKU usually appear healthy and normal at birth, but after the first few months the symptoms of the disease set in, and without treatment severe mental retardation usually results. Many never learn to walk or talk and are subject to periodic convulsions and seizures. Most afflicted individuals must be hospitalized for their entire lives, which, in untreated persons, is seldom more than 30 years.

It is not yet known how the high levels of phenylalanine and derivative compounds bring about the tragic mental symptoms. However, the knowledge we do have is enough to effectively treat infants with PKU and prevent the symptoms from appearing. Many states now require routine tests of all newborn babies in order to detect PKU homozygotes. Those identified at birth are simply put on a special diet containing low amounts of phenylalanine—enough to supply dietary needs but not enough to permit toxic accumulations. Low phenylalanine diets have been developed that, on the basis of 20 years' experience, allow PKU homozygotes to develop normally.

Tay-Sachs Disease

Tay-Sachs disease is a recessive condition resulting in degeneration of the nervous system. As with PKU, Tay-Sachs homozygotes appear normal at birth and through the early months. However, by about eight months, symptoms of severe listlessness set in. Blindness usually occurs within the first year. Afflicted children rarely survive past their fifth year. The biochemical basis of this disease is now, at least in part, understood. Homozygous individuals lack an enzyme, ganglioside GM_2-hexosaminidase, which breaks down a special lipid found in brain cells. This enzyme is normally found in the lysosomes of brain cells and plays a crucial role in keeping the lipid from accumulating. In the child lacking the enzyme, the lysosomes of the brain cells fill with the lipid and swell, and the cells die. There is no therapy yet available for Tay-Sachs disease. While in the general population it is a rare disorder (1 in 300,000 births), it has a very high incidence among Jews of Central European extraction, Ashkenazic Jews, who make up more than 90 percent of the American Jewish population. The disease shows up once in every 3,600 births. It is estimated that approximately 1 in 28 Ashkenazic Jews is a carrier (a heterozygote).

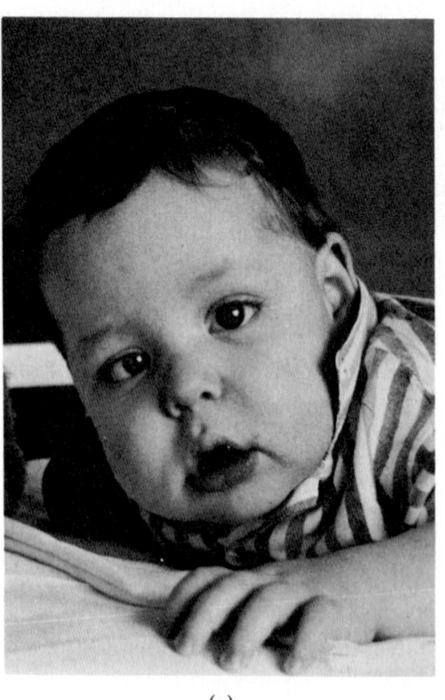

(a)

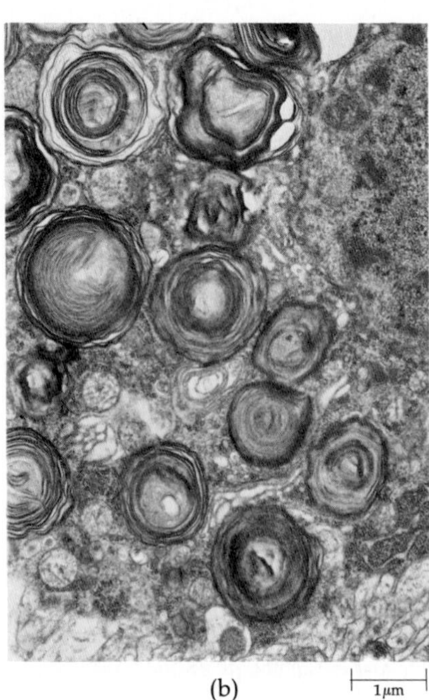

(b)

18–13
(a) *In this one-year-old infant with Tay-Sachs disease, deterioration of the brain, already begun, will progress rapidly. The child will probably die before he is six years old.* (b) *The disease is caused by the absence of an enzyme involved with lipid metabolism. Without the enzyme, harmful fatty deposits accumulate in the lysosomes of brain cells, as shown in this micrograph. A portion of the nucleus can be seen at the right of the micrograph.*

18–14

Scanning electron micrograph of deoxygenated blood from an individual heterozygous for the sickle cell allele. As you can see, some—but not all—of the red blood cells have sickled.

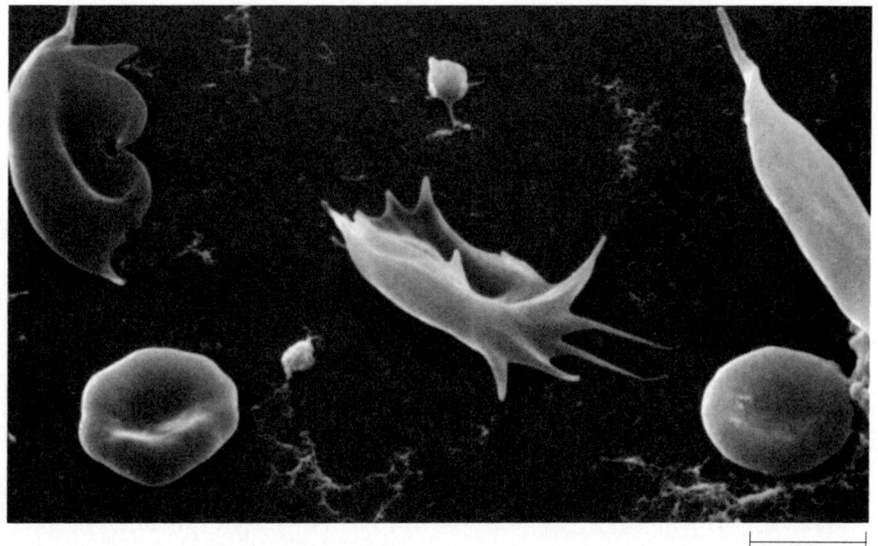

THE HEMOGLOBINS

As we have seen in the previous chapters, the globins of the hemoglobin molecule have played leading roles in the history of genetics. While many different hereditary variants of hemoglobin have been discovered, the most serious from a medical point of view is the one that causes sickle cell anemia. This hereditary disease occurs with a very high frequency among blacks. In the United States about 9 percent of blacks are heterozygous for the sickle cell gene, and about 0.2 percent are homozygous and therefore have the symptoms of sickle cell anemia.

Sickle cell anemia, you will recall, is due to a single amino acid substitution in the beta chain of the molecule. When sickle cell hemoglobin is deoxygenated, it becomes insoluble and forms bundles of stiff tubular fibers. These fibers distort the shape of the cell, making it more fragile and, more important, making it difficult for the red cells (which are normally very flexible) to make their way through small blood vessels. The symptoms of sickle cell disease and the hazards to life are caused by this occlusion of the blood vessels in the joints and in vital organs.

Heterozygotes for sickle cell anemia are symptomless. As with PKU, the "good" allele makes enough hemoglobin that the effects of the "bad" allele are not discernible. However, if blood samples are treated in ways that remove oxygen from all the hemoglobin molecules, some of the blood cells of a heterozygote will sickle (Figure 18–14). Thus it is possible to detect heterozygotes quite readily. If two heterozygotes marry, there is 1 chance in 4 of their having a child with sickle cell anemia, and a fifty-fifty chance of having a child who is, like themselves, a heterozygote and so a "carrier" of the disease.

Sickle cell hemoglobin is not the only known genetic alteration of the hemoglobin molecule. More than 100 hereditary variants that are due to amino acid substitution in the beta chains have now been found, and a sizable number of alpha-chain alterations have also been detected. Unlike sickle cell anemia, most of these hereditary conditions are rare, and often the symptoms of the condition are mild or nonexistent.

MEDICAL IMPLICATIONS OF RECOMBINANT DNA TECHNIQUES

When recombinant DNA studies were first begun, they were simultaneously hailed as a future panacea for humanity's major diseases and denounced as dangerous and irresponsible. Research using recombinant DNA was greatly curtailed for more than a year in the middle 1970s until safety guidelines were drawn up for the containment of infectious organisms, like *E. coli,* bearing foreign genes. The precautions now used are of two general types. First, the same general rules are applied as those governing the handling of infectious microorganisms, such as the viruses of rabies, influenza, or smallpox, and pneumococci, streptococci, or the bacteria causing syphilis. All of these microorganisms, with a few instructive exceptions, have been handled safely in the laboratory for decades. Second, the *E. coli* strains have been crippled in such a way that even were they to find their way into an animal host, they would be unable to survive because of nutritional requirements that can be met only in the laboratory.

Also, although the recombinatorial techniques have proved very successful, it has been difficult so far to get foreign genes to produce consequential amounts of protein in their bacterial hosts. These problems are due both to differences in regulatory mechanisms between prokaryotes and eukaryotes and also to differences in the processing of mRNA, such as removal of introns. Thus the spectre of runaway bacteria flooding a new host with foreign proteins has faded.

Finally, as more and more examples of recombinant DNA in nature are uncovered, the achievements of scientists in this regard are beginning to seem less revolutionary, in fact, almost conservative by comparison.

Although the fears have largely subsided, the hopes are still flourishing. It appears that the techniques may be useful in three different ways: first, in prenatal diagnosis (a hope already realized to some degree); second, in the production of proteins for medical use (a program that has met with some success but also with some difficulties); and third, in the treatment of hereditary diseases by the replacement of faulty genes with functional ones (which is still a glimmer on the horizon).

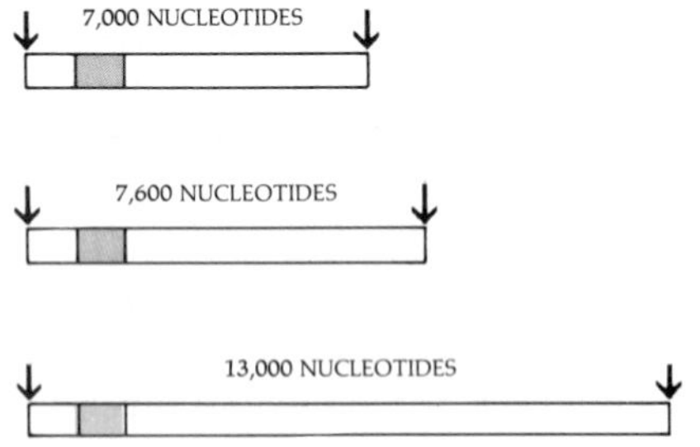

18–15
A test for sickle cell disease in the fetus. When human DNA is treated with the restriction enzyme Hpa I, three types of fragments containing the beta-globin gene are produced. In persons with the normal allele (gray) for beta globin, the fragments are either 7,000 or 7,600 nucleotides long. In persons with the sickle cell allele (color), the fragments are 13,000 nucleotides in length. A recognition site for the restriction enzyme Hpa I—present in persons with the normal beta allele—is missing in individuals carrying the sickle cell allele.

Prenatal Diagnosis of Genetic Disease

Sickle cell anemia can now be diagnosed prenatally using cells obtained by amniocentesis (page 354). The tools are a restriction enzyme to cleave the DNA of the fetal cells and a radioactive copy of the gene for beta globin. The diagnosis depends on the fact that the mutation in the globin gene that causes the abnormal protein to be produced is accompanied by another mutation, some 5,000 base pairs away, involving the cleavage site of that particular restriction enzyme. In persons with normal hemoglobin, the beta globin gene is found on a DNA fragment either 7,000 or 7,600 nucleotides in length, while in persons with sickle cell hemoglobin, the radioactive probe hybridizes with a longer (13,000-nucleotide) sequence (Figure 18–15).

In the first family studied with this technique, both parents were known carriers of the sickle cell allele and had previously produced an infant with sickle cell disease. The woman was pregnant again, and the parents wanted to know if the child would be healthy. DNA tests from both of the parents yielded both the short and the long fragments, as would be expected. Tests from their child with sickle cell anemia yielded only the long fragments. Tests of the fetal cells yielded both short and long fragments; the unborn child would be a carrier like the parents, but would not have the disease.

It seems likely that various versions of this technique can be extended to the diagnosis of other diseases.

Designer Genes

A number of medically useful proteins have, in the past, been available only in very small quantities from human sources or, in larger quantities, from other animals. Although such animal proteins (insulin from cows or pigs, for example) have made an invaluable contribution to human health, they are not as effective as the human proteins and they can trigger allergic reactions in some individuals. Early in the course of recombinant DNA research, biologists realized that if segments of human DNA that code for such proteins could be inserted into bacteria and made to function, the bacterial "factories" could provide a low-cost, virtually limitless source of the proteins.

The first synthesis of a mammalian protein in a bacterial cell was reported in 1977 by Keiichi Itakura and his associates at the City of Hope Medical Center. They selected the gene for the hormone somatostatin because (1) it was a small protein (only 14 amino acids), (2) it could be detected in very small amounts, and (3) it was a potentially useful compound.

The amino acid sequence of somatostatin is alanine-glycine-cysteine-lysine-asparagine-phenylalanine-phenylalanine-tryptophan-lysine-threonine-phenylalanine-threonine-serine-cysteine. Knowing this sequence, the investigators determined what one of the possible sequences of nucleotides in the DNA should be (you can do this, too) and then synthesized the gene, including an initiation codon, bonding together the nucleotides one by one. (The synthesis at that time was technically very difficult and required six months; now "gene machines" have been developed that are computer-controlled and can synthesize short nucleotide sequences automatically. These sequences can then be strung together to produce artificial genes of any length.)

Multiple copies were obtained by replicating the artificial gene in a system containing DNA polymerase and an adequate supply of nucleotides. These copies were then spliced into drug-resistant plasmids with the use of a restriction enzyme, and the plasmids were supplied to *E. coli* cells. Some cells took up the plasmids (as evidenced by the fact that they were now drug-resistant), but there was no evidence of somatostatin synthesis; Itakura's group needed some way to turn the gene on. They inserted the regulatory sequences for the *lac* operon into the plasmid upstream from the somatostatin gene. When they turned them on, they were able to detect somatostatin, but only in very small amounts. Cells degrade foreign proteins, and the hormone was being destroyed almost as fast as it was being produced. Finally, to protect the mammalian protein from the bacterial enzymes, they spliced the somatostatin gene onto the beta-galactosidase gene, the first structural gene of the *lac* operon. At the point of the splice, for reasons that we shall see shortly, they inserted the codon for methionine. The plasmids were reintroduced into host cells and at last clones of bacteria were obtained, all making the hybrid beta-galactosidase-somatostatin protein. The protein was then isolated and treated with the chemical cyanogen bromide, which cleaved it exactly at the methionine insert, releasing the somatostatin. The somatostatin, tested in laboratory animals, was found to have the biological activity of the natural hormone. In other words, it worked!

18–16

The somatostatin project. The gene for somatostatin, synthesized artificially, was fused to the beta-galactosidase gene in a bacterial plasmid. Introduced into E. coli, *the plasmid directs the synthesis of a protein that begins as beta-galactosidase but ends as somatostatin. Cyanogen bromide cleaves the protein at methionine, thus releasing the hormone intact (together with many fragments of beta-galactosidase, from which it can be separated).*

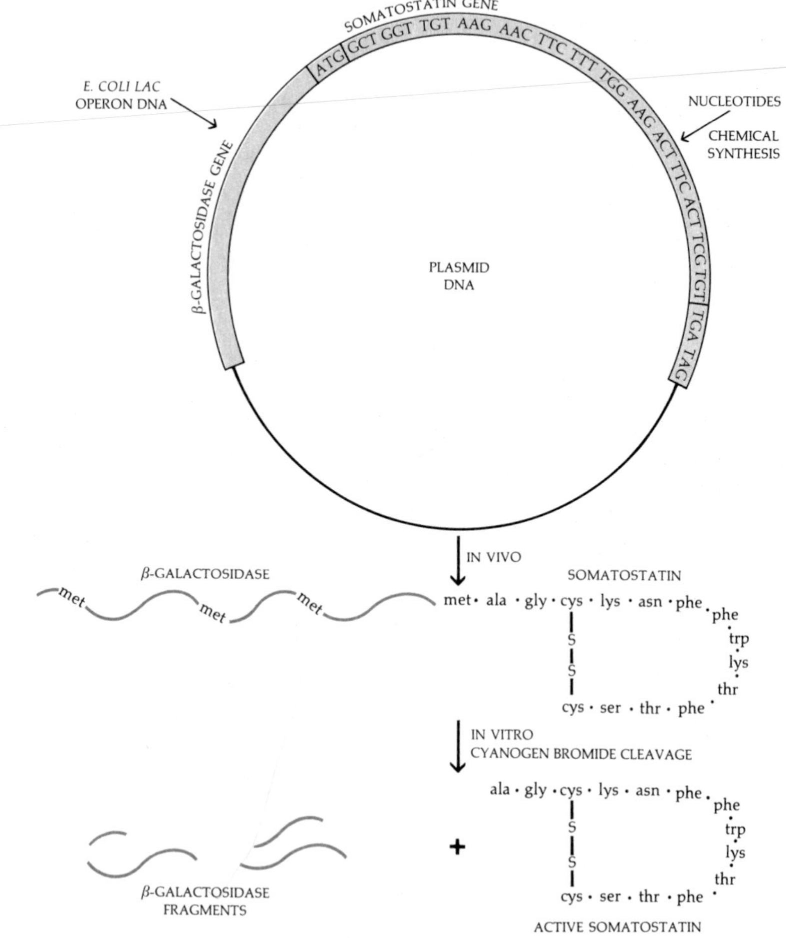

Thus, in the span of one scientific generation, molecular geneticists not only identified the genetic material in chemical terms and demonstrated how it is replicated, transcribed, and translated, but they also produced a gene of their own making and made it function. The late Philip Handler, then president of the National Academy of Sciences, described this "astonishing" accomplishment as "a scientific triumph of the first order." And so it is.

More recently, genes for ovalbumin, human insulin, growth hormone, and several enzymes have been successfully inserted into bacteria and have functioned in protein synthesis. Also of particular promise is the production of interferon and of viral proteins to be used for immunization procedures, topics that will be discussed further in Chapter 39.

Gene Therapy

Putting Genes into Mammalian Cells

The most extravagant hope in regard to the medical applications of recombinant DNA is that at some future time it may be possible to correct genetic defects by substituting "good" genes for "bad" ones. This is an enormously complex undertaking. It requires, first, the preparation of a gene that will be taken up by a

18–17

The mouse cell from which these chromosomes were taken was "cured" of thymidine kinase deficiency by incubation with the TK gene from herpes simplex virus. The chromosome that incorporated the TK gene (arrow) was identified with a radioactive probe.

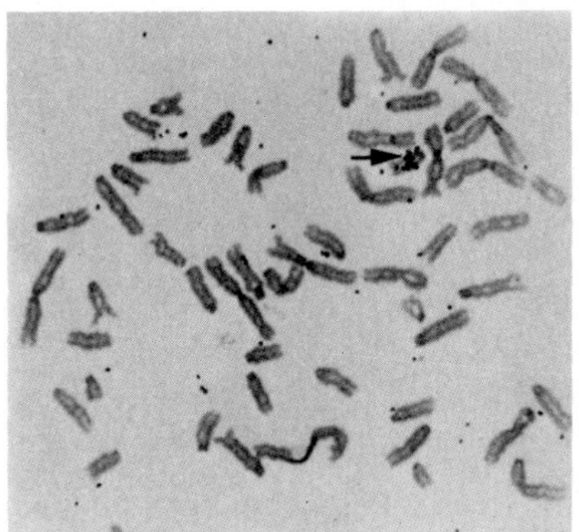

eukaryotic cell, become incorporated into a chromosome, and be expressed there—but this is only the beginning. The new gene must be established in a large number of cells of the appropriate type—blood cells should not produce somatostatin, for example—and be subject to the complicated (and as yet largely unknown) regulatory controls of the normal gene.

Actually, the first stage of the project has proved easier than expected; foreign genes will undergo recombination in the eukaryotic cell. The first successful transfer of a mammalian gene to a mammalian cell was achieved by inserting a rabbit beta-globin gene into the DNA of SV40 and using the virus to infect monkey cells in tissue culture. The recipient cells produced rabbit beta globin. (SV40 could not be used in humans, of course, because of its cancer-causing properties in other species.)

In another experiment, mouse cells that lacked the enzyme thymidine kinase (TK) were incubated with DNA containing a gene for TK from herpes simplex virus. About one in a million of the TK^- cells incorporated the TK gene. They were "cured" of their genetic defect and thus were able to survive in a medium in which the TK^- cells died (which, of course, is how the clones of "cured" cells were isolated).

Putting Genes into Mammalian Organisms

The pilot study for introducing genes into a whole organism was, once again, carried out using a virus. Rudolf Jaenisch infected early mouse embryos (4 to 16 cells) with mouse leukemia virus and then reimplanted the embryos into foster mothers to continue their development. In a substantial percentage of the resulting mouse pups, the virus had become integrated into the chromosomes—becoming a *provirus*, analogous to a prophage (page 316). Moreover, active virus production was detected in some tissues of the pups. When the animals carrying the provirus matured, they were mated and 50 percent of their offspring were born with the provirus, showing that it had become established in the chromosomes of the germ line. Of particular interest is the discovery that the provirus may occupy different sites in the chromosomes and that the time at which it is expressed depends on its chromosomal locus. For instance, virus production begins in the liver cells after 16 days gestation, in the spleen soon after birth, but not at all in some other somatic cells.

Jon W. Gordon and Frank H. Ruddle of Yale University have developed a technique for inserting a DNA sequence into the fertilized eggs of mice. When the eggs were injected with a rabbit beta-globin gene, the mice derived from the eggs were found to contain rabbit globin polypeptide in their red blood cells (Figure 18–18, on the next page). Of special interest is the fact that the gene was expressed only in the red blood cells and not in other tissues of the mouse, indicating that it had been incorporated in the "right" place and so had come under cellular control mechanisms. The gene was passed in a Mendelian distribution to subsequent generations.

Niels Bohr is said to have stated, "It is difficult to make predictions, especially about the future," and, in keeping with this admonition, all the investigators in this fast-moving field urge caution in assuming that gene therapy in humans is in any way close to being achieved. In view of recent events, however, it seems safe to predict that our knowledge of the genetics of the eukaryotic cell will continue to increase by quantum jumps, even if the direction in which the knowledge leads us is as yet unknown.

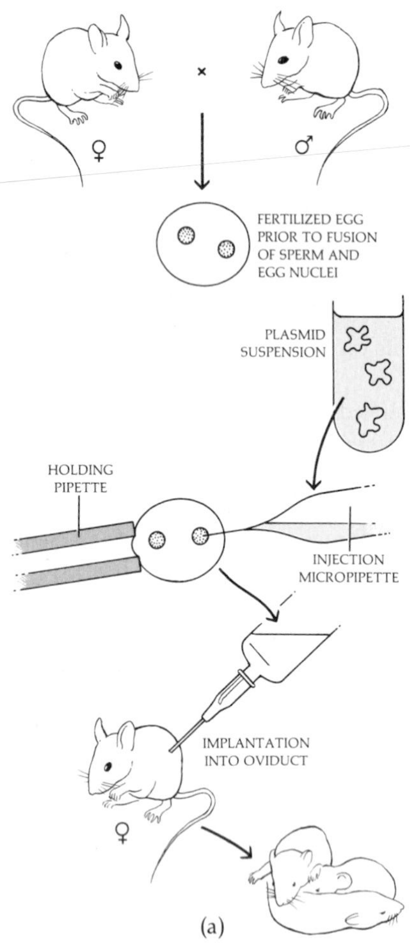

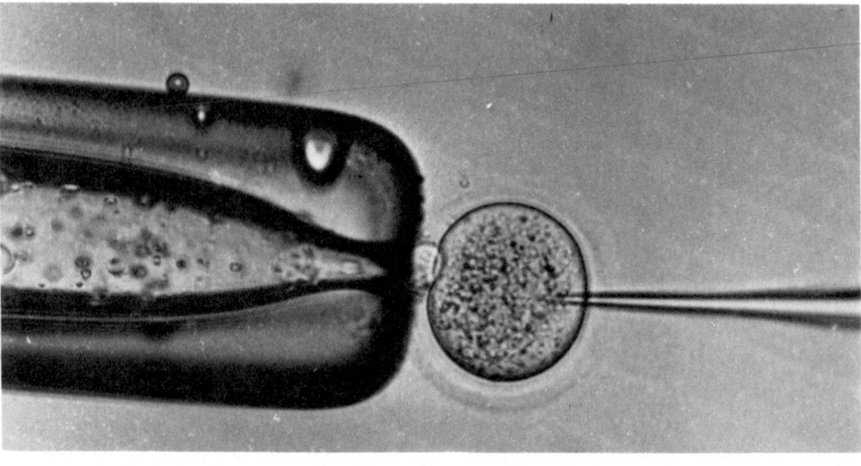

18–18
(a) *The procedure by which Gordon and Ruddle inserted the gene for rabbit beta globin into mice. The gene was spliced into plasmids, which were then injected into fertilized eggs before the egg and sperm nuclei fused. After injection, the fertilized eggs were implanted in a female mouse, who gave birth. Rabbit beta globin was present in the red blood cells of the offspring, and hybridization techniques revealed that the gene had been incorporated into their DNA.*

(b) *Injection of the plasmid suspension into a fertilized egg. The diameter of the micropipette tip is only about 0.5 micrometer, but Gordon says that the injection "is equivalent to your being speared by a telephone pole." Nevertheless, 75 percent of the eggs survive.*

SUMMARY

The normal diploid number of human chromosomes is 46: 44 autosomes and two sex chromosomes. The male is the heterogametic sex. A karyotype is a graphic representation of a set of chromosomes. In preparing a karyotype, metaphase chromosomes are paired in homologues and homologues are grouped by size. Nonhomologous chromosomes of similar size and shape can be distinguished by staining techniques that reveal chromosome banding. Chromosome mapping has been carried out using cell hybridization techniques and also with radioactive nucleic acid probes.

Visible chromosomal abnormalities include extra chromosomes (usually a result of nondisjunction—the failure of two homologues to separate at the time of meiosis), translocations, and deletions. Down's syndrome is among the disorders associated with an extra chromosome; it may be caused either by nondisjunction or, less commonly, by translocation. Extra sex chromosomes can also result from nondisjunction; this is often, but not always, associated with sterility and mental retardation. Aniridia and Wilm's tumor are defects associated with deletion of one small region of chromosome 11. Many genetic defects can now be detected in the fetus by the use of amniocentesis, the collection of fetal cells from the amniotic fluid.

Genetic defects expressed in males but carried in females are caused by mutant recessive alleles on the X chromosome. In humans, such sex-linked characteristics include color blindness and hemophilia.

Many genetic diseases are the result of deficiencies or defects in enzymes or other critical proteins. These are caused, in turn, by mutations in the alleles coding for the proteins. Such diseases are usually apparent only in the homozygote. They include phenylketonuria, Tay-Sachs disease, sickle cell anemia, and a number of other diseases related to variations in the hemoglobin molecules.

Recombinant DNA techniques are providing new means for early diagnosis of hereditary diseases. They are also being used in an attempt to produce biologically important proteins, and they may be of future importance in the treatment of human hereditary diseases.

QUESTIONS

1. Nondisjunction can occur at the first meiotic division or the second. How do the effects differ? Include diagrams with your answers.

2. Describe the two types of chromosomal abnormalities that can cause Down's syndrome. With which type is it possible to identify prospective parents who are at a higher-than-average risk? How?

3. Why is male-to-male inheritance of color blindness impossible? Under what conditions would color blindness be found in a woman? If she married a man who was not color-blind, what proportion of her sons would be color-blind? Of her daughters?

4. How do we know that Prince Albert was not responsible for the hemophilia in Queen Victoria's descendants?

5. A woman whose maternal grandfather was hemophiliac has parents who seem to be normal. She too seems normal, as does her husband. What are the chances that her first son will be normal? (*Hint:* Determine the genotype of the woman's mother and then the possible genotypes of the woman herself.)

6. Albinism results from the failure to make the pigment melanin, which is made from tyrosine. Like other enzyme deficiency diseases, it is transmitted as a recessive allele. In 1952 it was reported that two albinos, who had met at a school for the partially sighted, had married and had three children, all of whom had normal pigmentation. Assuming the children are not illegitimate, how do you explain that the children are normal?

7. If two healthy parents have a child with sickle cell anemia, what are their genotypes with respect to this allele? Having had one such child, what are their chances of having another child with the same disease?

8. What proportion of the children of the parents in Question 7 will be carriers of the sickle cell allele (that is, heterozygous)? What proportion of their children will not carry the allele? (Draw a Punnett square to diagram this problem.)

SUGGESTIONS FOR FURTHER READING

Books

BODMER, WALTER F., and L. L. CAVALLI-SFORZA: *Genetics, Evolution, and Man,* W. H. Freeman and Company, San Francisco, 1976.

An excellent undergraduate text with emphasis on human genetics.

DE ROBERTIS, E. D. P., and E. M. F. DE ROBERTIS, JR.: *Cell and Molecular Biology,* 7th ed., Saunders College/Holt, Rinehart and Winston, Philadelphia, 1980.

Almost half the book is concerned with the nucleus, gene expression, and gene regulation. Outstanding for its emphasis on structural features of the nucleus and chromosomes, and on human cytogenetics, and for its illustrations.

GOODENOUGH, URSULA: *Genetics,* 3d ed., Saunders College/Holt, Rinehart and Winston, New York, 1983.

An up-to-date, general introductory text with emphasis on molecular genetics. Outstanding for its clarity of explanation.

JACOB, FRANÇOIS: *The Logic of Life: A History of Heredity,* Pantheon Books, New York, 1973.

Jacob's principal theme concerns the changes in the way people have looked at the nature of living beings. These changes, which are part of our total intellectual history, determine both the pace and direction of scientific investigation. The opening chapters are particularly brilliant.

JUDSON, HORACE F.: *The Eighth Day of Creation: Makers of the Revolution in Biology,* Simon and Schuster, New York, 1979.*

A comprehensive study of the human and scientific aspects of molecular biology from the 1930s through the mid-1970s. As events unfold, the story is told from each participant's point of view. We are treated to an enlightening and personal glimpse of the development of scientific thought.

LEHNINGER, ALBERT L.: *Principles of Biochemistry,* Worth Publishers, Inc., New York, 1982.

This introductory text is outstanding both for its clarity and for its consistent focus on the living cell. There are numerous medical and practical applications throughout.

OLBY, ROBERT: *The Path to the Double Helix,* University of Washington Press, Seattle, Wash., 1975.

An account, written by a professional historian of science, of twentieth-century genetics. Olby is interested not only in the scientific concepts and experiments but also in the various personalities involved and their effects on one another and on the course of scientific discovery.

PETERS, JAMES A. (ed.): *Classic Papers in Genetics,* Prentice-Hall, Inc., Englewood Cliffs, N. J., 1959.*

Includes papers by most of the scientists responsible for the important developments in genetics: Mendel, Sutton, Morgan, Beadle and Tatum, Watson and Crick, and so on. You should find this book very interesting; the authors are surprisingly readable, and the papers give a feeling of immediacy that no modern account can achieve.

STRICKBERGER, MONROE W.: *Genetics,* 2d ed., The Macmillan Company, New York, 1976.

An up-to-date and cohesive account of the science, this book is much broader in coverage of classical genetics.

STRYER, LUBERT: *Biochemistry,* 2d ed., W. H. Freeman and Company, 1981.

An introductory text, with many examples of medical applications of biochemistry. Handsomely illustrated.

WATSON, J. D.: *The Double Helix,* Atheneum Publishers, New York, 1968.*

"Making out" in molecular biology. A brash and lively book about how to become a Nobel laureate.

WATSON, J. D.: *Molecular Biology of the Gene,* 3d ed., The Benjamin/Cummings Publishing Company, Menlo Park, Calif., 1976.*

For the student who wants to go more deeply into the questions of molecular biology, this is a classic account.

Articles

ABELSON, J., and E. BUTZ (eds.): "Recombinant DNA," *Science,* vol. 209, 1980. This entire issue is devoted to a series of important research papers and contains a valuable glossary of terms.

ANDERSON, W. F., and E. G. DIACUMAKOS: "Genetic Engineering in Mammalian Cells," *Scientific American,* July 1981, pages 106–121.

---

* Available in paperback.

BISHOP, J. MICHAEL: "Oncogenes," *Scientific American,* March 1982, pages 80–92.

BROWN, DONALD D.: "Gene Expression in Eukaryotes," *Science,* vol. 211, pages 667–674, 1981.

CAMPBELL, A. M.: "How Viruses Insert Their DNA into the DNA of the Host Cell," *Scientific American,* December 1976, pages 102–113.

CHAMBON, PIERRE: "Split Genes," *Scientific American,* May 1981, pages 60–71.

COHEN, S. N., and J. A. SHAPIRO: "Transposable Genetic Elements," *Scientific American,* February 1980, pages 40–49.

DAVIS, BERNARD D.: "Frontiers of the Biological Sciences," *Science,* vol. 209, pages 78–79, 1980.

DE ROBERTIS, E. M., and J. B. GURDON: "Gene Transplantation and the Analysis of Development," *Scientific American,* December 1979, pages 74–82.

FIDDES, J. C.: "The Nucleotide Sequence of a Viral DNA," *Scientific American,* December 1977, pages 54–67.

GILBERT, WALTER: "DNA Sequencing and Gene Structure," *Science,* vol. 214, pages 1305–1312, 1981.

GILBERT, WALTER, and LYDIA VILLA-KAMAROFF: "Useful Proteins from Recombinant Bacteria," *Scientific American,* April 1980, pages 74–94.

GORDON, JON W., and FRANK H. RUDDLE: "Integration and Stable Germ Line Transmission of Genes Injected into Mouse Pronuclei," *Science,* vol. 214, pages 1344–1345, 1981.

HALL, B. D.: "Mitochondria Spring Surprises," *Nature,* vol. 282, pages 129–130, 1979.

LEWIN, ROGER: "Biggest Challenge Since the Double Helix," *Science,* vol. 212, pages 28–32, 1981.

LEWIN, ROGER: "Do Jumping Genes Make Evolutionary Leaps?" *Science,* vol. 213, pages 634–636, 1981.

LEWIN, ROGER: "Evolutionary History Written in Globin Genes," *Science,* vol. 214, pages 426–429, 1981.

LEWIN, ROGER: "On the Origin of Introns," *Science,* vol. 217, pages 921–922, 1982.

MARX, JEAN L.: "A Movable Feast in the Eukaryotic Genome," *Science,* vol. 211, pages 153–155, 1981.

MARX, JEAN L.: "Restriction Enzymes: Prenatal Diagnosis of Genetic Disease," *Science,* vol. 202, pages 1068–1069, 1978.

NIRENBERG, M.: "The Genetic Code II," *Scientific American,* March 1963, pages 80–94.

RICH, A., and S. H. KIM: "The Three-Dimensional Structure of Transfer RNA," *Scientific American,* January 1978, pages 52–62.

SCHMID, CARL W., and WARREN R. JELINEK: "The Alu Family of Dispersed Repetition Sequences," *Science,* vol. 216, pages 1065–1070, 1982.

SMITH, MICHAEL: "The First Complete Nucleotide Sequencing of an Organism's DNA," *American Scientist,* vol. 67, pages 57–67, 1979.

PART II

Biology of Organisms

SECTION 4

The Diversity of Life

19–1
Representatives of two major groups of organisms. The plant is a euphorb, a succulent that resembles the cacti of the American deserts, and the insects are African grasshoppers. Euphorbs contain compounds toxic to most animals. These grasshoppers, which are unharmed by the toxins, concentrate them in their own tissues and so are themselves poisonous to other animals. The bright color of the grasshoppers is a warning signal to would-be predators. Red is a highly visible color to birds, the principal predators of grasshoppers.

CHAPTER 19

The Classification of Organisms

Part I of this book, the first three sections, dealt largely with cellular and subcellular aspects of living things. In Part II, we shall be concerned with the next level of organization, the whole organism. We shall begin, in Chapter 20, with the simplest of single-celled creatures, the bacteria (which, as we have seen, are astonishingly gifted and complex), and end with one of the most complicated of multicellular organisms, ourselves.

We newcomers on the planet earth share the biosphere with as many as 5 million other species of living things. One of the fundamental goals of observers of the natural world, recorded since before the time of Aristotle, is to perceive order in this diversity. *Taxonomy* is the classification of organisms (or, for that matter, of any other items, such as books in the library or groceries on the shelf). *Systematics* is the study of the relationships among organisms. Taxonomy may or may not reflect such relationships.

The very words "systematics" and "taxonomy" conjure up for some of us images of dusty archives, gray beards, and fossil bones. Thus, it may be surprising to learn that these subjects have become the focus of some of the most heated arguments now going on in modern biology.

WHAT IS A SPECIES?

One of the disagreements concerns the definition of species. *Species* in Latin simply means "kind," and so species, in the simplest sense, are different kinds of organisms. A more rigorous definition of species was set forth in 1940 by Ernst Mayr of Harvard University, who said that species are "groups of actually or potentially interbreeding natural populations which are reproductively isolated from other such groups." The phrase "actually or potentially" allows for the fact that although members of the human population of Greenland are not apt to interbreed with those of Patagonia, they are still members of the human species, or that transporting a group of insects to some remote island does not automatically make them members of another species. The words "groups" and "populations" are important in this definition also. The possibility that single individuals of different species may have occasional offspring—such as by the crossing of lions and tigers in a zoo—is unimportant in terms of the group. Mayr's definition conforms to common sense: if one species freely exchanged genes with another species, they could no longer retain those unique characteristics that identify them as different kinds of organisms.

(a)

(b)

(c)

19–2
These plants are all members of the genus Viola: (a) *the common blue violet,* Viola papilionacea, (b) *the long-spurred violet,* Viola rostrata, *and* (c) *the pansy,* Viola tricolor. *Although there is an overall similarity among all three species, there are clear-cut differences in leaf shape, flower color and size, and other characteristics.*

This definition works well for animal species and is generally accepted by zoologists. Many plants, however, as we shall see, can reproduce asexually and also can form fertile hybrids with other species. Bacteria, with their variety of forms of genetic exchange, do not fit this definition neatly, nor do the many unicellular eukaryotes that reproduce simply by cell division, forming clones of identical cells. Thus, although botanists and microbiologists use the term "species," they are more apt to consider it as a category of convenience, existing rather in the human mind than in the natural world.

A second problem with the concept of species is that it fulfills two not quite congruent needs. It is a category into which is placed an individual organism that conforms to certain fairly rigid definitions concerning its morphology (from the Greek word *morphe,* meaning "shape") and other characteristics. By another definition, it is a group or population of organisms, reproductively united but very probably changing as it moves through space and time, and occasionally producing splinter groups that may, in reproductive isolation, form new species. One usage emphasizes constancy, the other recognizes change.

Nevertheless, despite the inability of biologists to agree upon a definition of species, or the failure of some organisms to conform to one, species forms an indispensable biological grouping, and the species concept is the basis of systematics and taxonomy (and, as we shall see, of much of evolutionary theory as well).

The Naming of Species

A *genus* (plural, genera) is a group of closely related species. According to the binomial system of nomenclature devised by Linnaeus, a species name consists of two parts—the name of the genus plus the specific epithet (an adjective or modifier). The genus name is written first, as in *Drosophila melanogaster,* and it may be used alone when one is referring to members of the entire group of species making up that genus, such as *Drosophila* or *Paramecium.*

A specific epithet is meaningless when written alone, however, because many different species in different genera may have the same specific epithet. For example, *Drosophila melanogaster* is the fruit fly that played such an important role in the history of genetics; *Thamnophis melanogaster,* however, is a semiaquatic garter snake. Thus, by itself, the specific epithet would not identify any organism. For this reason, the specific epithet is always preceded by the genus name, or, in a context where no ambiguity is possible, the genus name may be abbreviated to its initial letter. Thus *Drosophila melanogaster* may be designated *D. melanogaster.*

Whoever describes a species first has the privilege of naming it. They may not name it after themselves; more often, it is named after a friend or colleague. *Escherichia,* for example, is named after Theodor Escherich, a German physician (*coli* simply means intestinal); and *Rhea darwinii,* an ostrichlike bird found in Patagonia, is named after Charles Darwin.

Names may be descriptive. The first discovered early fossil of a horse—which does not superficially resemble a modern horse at all—was named *Hyracotherium,* the "hyrax-like beast." (A hyrax looks like a big guinea pig and is said, by some, to be a distant relative of the elephant.) When O. C. Marsh of Yale University began, in the 1870s, to study fossil horses, he recognized that the little dog-sized *Hyracotherium* was an early equine and gave it the charming name Eohippus ("dawn horse"). However, *Hyracotherium* remains its official designation because it came first.

Some names are heartfelt. Thus, members of various mosquito genera have been given the specific epithets *punctor, tormentor, vexans, horrida, perfidiosus, abominator,* and *excrucians.* Others are frivolous. An English entomologist coined a whole series of generic names based on the pseudo-Greek ending *chisme,* pronounced "kiss me." Thus, there are squash bugs, stink bugs, and seed bugs known variously as *Polychisme, Peggichisme, Dolichisme,* and the promiscuous *Ochisme.* A new species of wasp in the genus *Lalapa* was recently given the species name *lusa;* there is a (presumably treacherous) beetle named *Ytu brutus,* and a horsefly called *Tabanus balazaphyre.* So, although species names may appear formidable and be unpronounceable, they are not necessarily as pompous (or even as informative) as they may seem.

These binomials are a necessary tool of science. Many species lack common names, and, even when common names do exist for a kind of organism, more than one name can be given to the same animal, such as groundhog and woodchuck, gnu and wildebeest, pill bug and sow bug and wood louse. Or names may vary from place to place. As shown in Figure 19–3, a robin in North America is distinctly

(a)

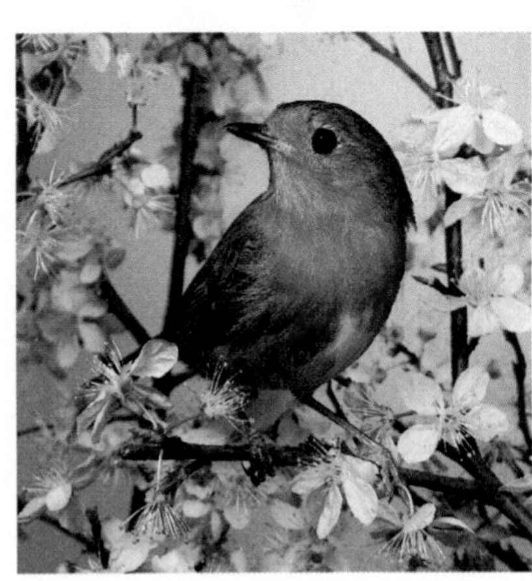

(b)

19–3
Though they are both called robins, (a) *the North American robin,* Turdus migratorius, *is a distinctly different bird from* (b) *the English robin,* Erithacus rubecula.

different from the English bird of the same name. "Mole" refers to a placental mammal in North America but to a marsupial in Australia. A yam in the southern United States is a totally different vegetable from a yam several hundred kilometers away in the West Indies. When different languages are involved, the problems of communication would be virtually insurmountable without a system of nomenclature universally recognized and agreed upon by biologists.

19–4
Species are named by a binomial system first devised by the eighteenth-century Swedish professor, physician, and naturalist Carolus Linnaeus (1707–1778). He is shown here wearing his collector's outfit. (Linnaeus was born Carl von Linné but latinized his name in the scholarly fashion of the time.)

TAXONOMY

The classification of organisms, taxonomy, is based on a hierarchical system—that is, it consists of groups within groups, with each group being ranked at a particular level. In such a system, a particular group is called a *taxon* (plural, taxa), and the level at which it is ranked is called a *category*. For example, in political geography, nation, state or province, and city are categories, whereas Canada, Ontario, and Toronto are taxa within those categories. Similarly, genus and species are categories, and *Homo* and *Homo sapiens* are taxa.

In the time of Linnaeus, three categories were in common use: the species; the genus, a group of species with similar appearance; and a category of much higher level, the *kingdom*. Naturalists recognized three kingdoms: plant, animal, and mineral. Kingdom is still the highest category used in biological classification. Between the level of genus and the level of kingdom, however, Linnaeus and subsequent taxonomists have added a number of other categories. Thus, genera are grouped into *families*, families into *orders*, orders into *classes*, and classes into *phyla* or *divisions*. (The categories of division and phylum are equivalent. The term "division" is generally used in the classification of prokaryotes, algae, fungi, and plants, whereas "phylum" is used in the classification of protozoa and animals.) These categories may be further subdivided or aggregated into a number of less frequently employed categories such as subphylum or superfamily. By convention, generic and specific names are written in italics, while the names of families, orders, classes, and other taxa whose categories rank above the genus level are not, although they are capitalized.

This system of classification makes it possible to generalize. Table 19–1 (page 376) shows the classification of two different organisms. Notice how much information is stored in the classification of an animal as a mammal, for example, or a plant as Anthophyta. Notice also that in progressing downward from kingdom to species, there is an increase in detail, proceeding from the general to the particular. In short, hierarchical classification is a highly useful means for storing and retrieving information.

As we noted previously, the species may be considered a biological reality, but the other categories exist only in the human mind. To take a familiar group as an example, some taxonomists, "lumpers," would combine all the cats except one into the single genus *Felis,* excluding only the cheetah, because of its nonretractable claws. Others, "splitters," would reserve the designation *Felis* for the smaller cats, such as the cougar, the ocelot, and the domestic cat, and divide the others into the larger cats *(Panthera),* including the lion, tiger, and leopard, and the bobtail cats *(Lynx).* More extreme splitters favor separate genera for the clouded leopard *(Neofelis)* and the snow leopard *(Uncia).* No one is arguing about the characteristics of the animals themselves but only about the weighing of similarities and differences. One taxonomist's family can easily be another's order.

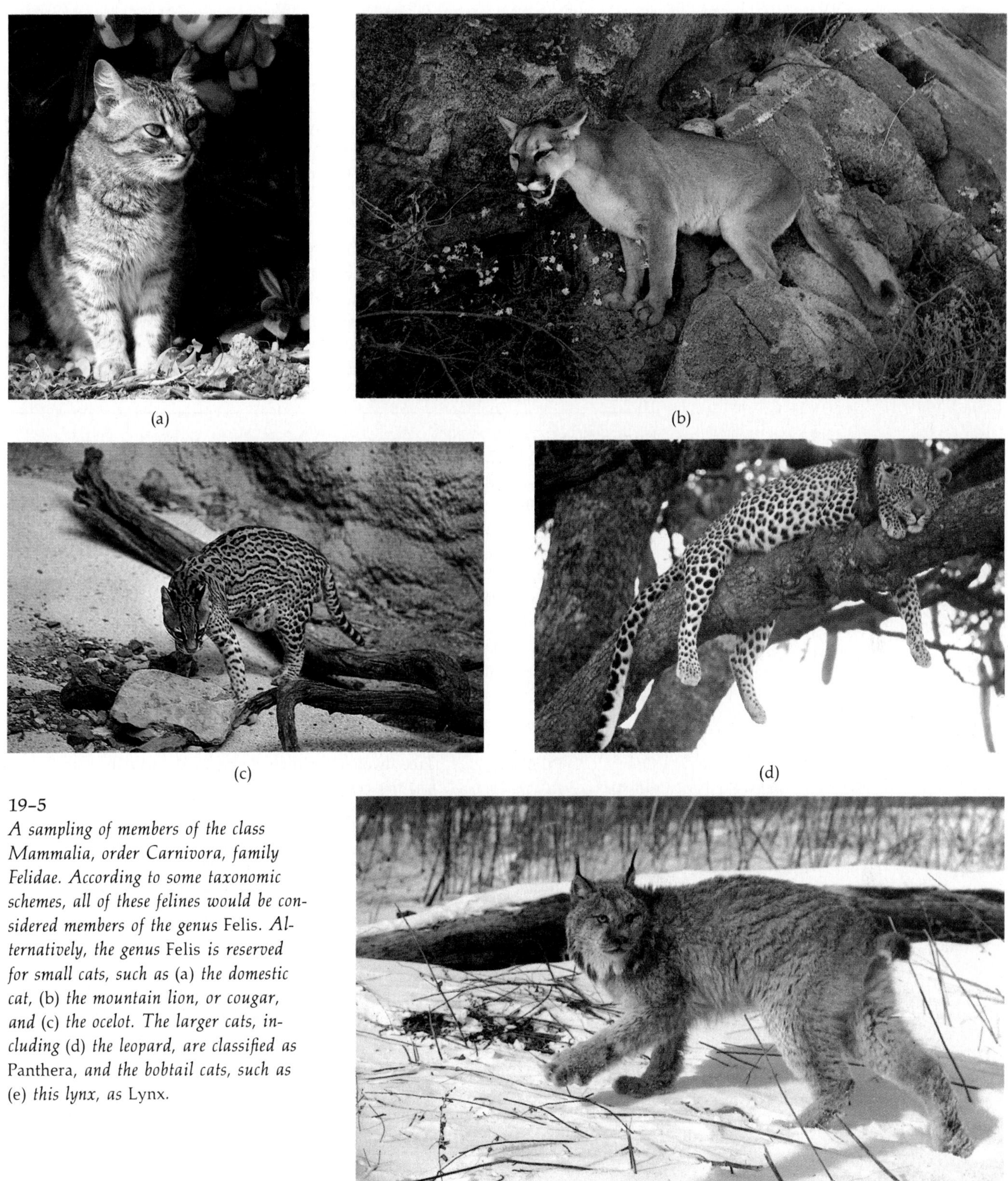

(a) (b) (c) (d) (e)

19–5

A sampling of members of the class Mammalia, order Carnivora, family Felidae. According to some taxonomic schemes, all of these felines would be considered members of the genus Felis. *Alternatively, the genus* Felis *is reserved for small cats, such as* (a) *the domestic cat,* (b) *the mountain lion, or cougar, and* (c) *the ocelot. The larger cats, including* (d) *the leopard, are classified as* Panthera, *and the bobtail cats, such as* (e) *this lynx, as* Lynx.

Table 19-1 Biological Classifications

Red Maple *(Acer rubrum)*

CATEGORY	TAXON	CHARACTERISTICS
Kingdom	Plantae	Multicellular terrestrial organisms that have chlorophylls *a* and *b* contained in chloroplasts, have rigid cell walls, and show structural differentiation
Division	Anthophyta	Vascular plants with seeds and flowers; ovules enclosed in an ovary; the angiosperms
Class	Dicotyledones	Embryo with two seed leaves (cotyledons)
Order	Sapindales	Soapberry order; usually woody plants
Family	Aceraceae	Maple family, characterized by watery, sugary sap; opposite leaves; winged fruit; chiefly trees of temperate regions
Genus	*Acer*	Maples and box elder
Species	*Acer rubrum*	Red maple

Human *(Homo sapiens)*

CATEGORY	TAXON	CHARACTERISTICS
Kingdom	Animalia	Multicellular organisms requiring complex organic substances for food
Phylum	Chordata	Animals with notochord, dorsal hollow nerve cord, gill pouches in pharynx at some stage of life cycle
Subphylum	Vertebrata	Spinal cord enclosed in a vertebral column, body basically segmented, skull enclosing brain
Superclass	Tetrapoda	Land vertebrates, four-limbed
Class	Mammalia	Young nourished by milk glands, skin with hair or fur, body cavity divided by a muscular diaphragm, red blood cells without nuclei, three ear bones (ossicles), high body temperature
Order	Primates	Tree dwellers or their descendants, usually with fingers and flat nails, sense of smell reduced
Family	Hominidae	Flat face; eyes forward; color vision; upright, bipedal locomotion
Genus	*Homo*	Large brain, speech, long childhood
Species	*Homo sapiens*	Prominent chin, high forehead, sparse body hair

SYSTEMATICS

As we noted previously, systematics is the study of the relationships among organisms. For Linnaeus and his immediate successors, the goal of classification was the revelation of the grand, unchanging design of special creation. After 1859, differences and similarities between organisms came to be seen as reflections of their evolutionary history (phylogeny). Thus genera came to be regarded as, ideally, groups of recently diverged sister species, and families as less recently divergent genera, and so on. Thus, classification began to attempt to fill two separate functions: providing useful methods of cataloging organisms and reflecting the sometimes erratic course of evolutionary change. Whether or not these functions are compatible is now under debate.

Homology and Phylogeny

Despite the change in perspective, actual changes in hierarchical classification have been few, so far. Organisms continue to be grouped on the basis of morphologic and other phenotypic similarities. From Aristotle on, biologists had recognized that superficial similarities were not useful taxonomic criteria; in other words, to use a simple example, birds and insects should not be grouped together simply because both have wings. A wingless insect is still an insect, and a flightless bird is still a bird on the basis of overall design. Linnaeus classified whales with mammals and not with fish, despite their external similarities.

However, post-Darwinian systematists are concerned, as their predecessors were not, with the origin of a similarity or difference. Does a similarity reflect inheritance from a common ancestor, or does it reflect adaptation to similar environments by organisms that do not share a common ancestor? A related question arises with differences between organisms: Does a difference reflect separate phylogenetic histories, or does it reflect instead the adaptations of closely related organisms to very different environments? A classic example is the vertebrate forelimb. The wing of a bird, the flipper of a whale, the foreleg of a horse, and the human hand have quite different functions and appearances. Detailed study of the structure of the underlying bones reveals, however, the presence of the same basic structure (Figure 19-6). Such structures, which have a common origin but not necessarily a common function, are said to be *homologous*. These are the features upon which post-Darwinian classification systems are ideally constructed.

19-6
The bones in these forelimbs are color-coded to indicate fundamental similarities of structure and organization. The crocodile, a reptile, is placed first because it is the closest to the ancestral type—the form from which all the others arose. (Note also the similarity between the forelimb of the crocodile and that of the human.) Structures that have a common origin but not necessarily a common function are known as homologues. Analogous structures, by contrast, are superficially similar but have an entirely different evolutionary background—for example, the spine of a cactus (a modified leaf) and the thorn of a rose (a modified branch).

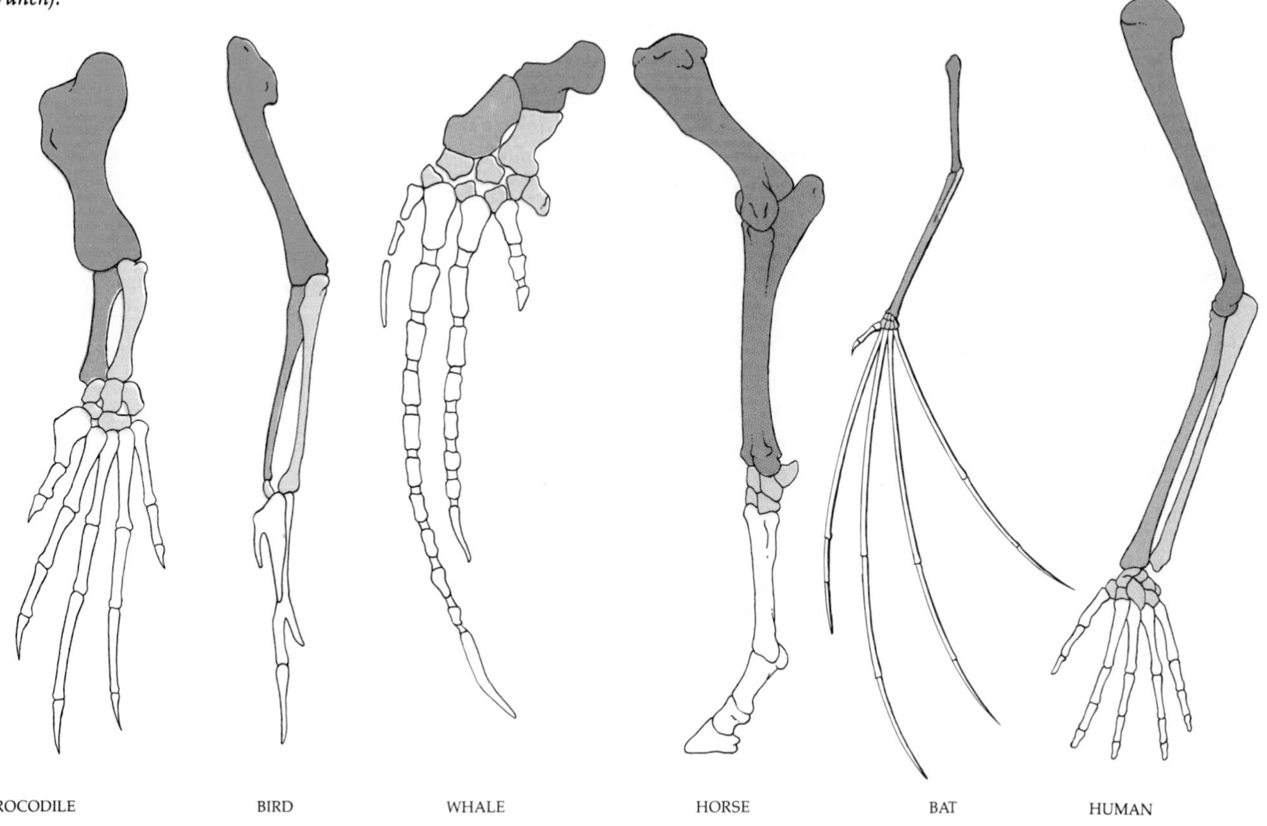

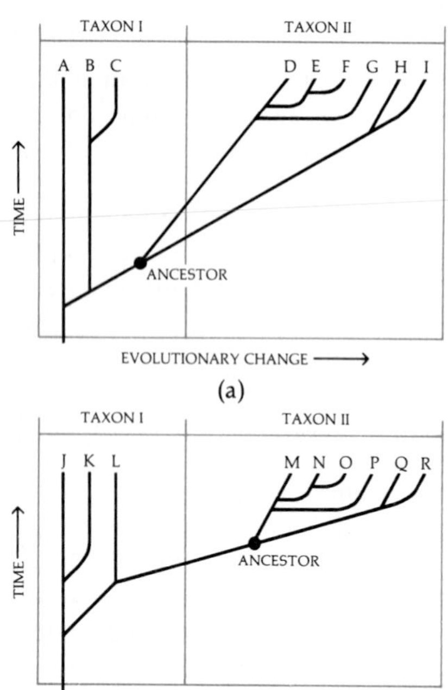

19–7

The evolutionary history of a group of related organisms can be represented by a phylogenetic tree. The vertical locations of the branching points indicate when particular taxa diverged from one another; the horizontal distances indicate how much the taxa have diverged, taking into account a number of different characteristics.

The two diagrams shown here represent the evolutionary histories of two different groups of taxa, labeled A through I in (a), *and J through R in* (b). *Each group has been classified using traditional classification methods. In* (a), *the ancestral taxon indicated is included in taxon I because of its close resemblance to taxa B and C. In* (b), *the ancestor is placed in taxon II, because of its close resemblance to taxon M. In each case, taxa I and II would themselves be members of a taxon at a higher categorical level, which would probably include other taxa as well.*

By contrast, other structures, which may have a similar function and superficial appearance, have an entirely different evolutionary background. Such structures are said to be *analogous*. Thus the wings of a bird and the wings of an insect would be said to be analogous, not homologous.

Decisions as to homology and analogy are seldom so simple. In general, the features most likely to be homologous—and thus useful in determining phylogenetic relationships—are those that are complex and detailed, consisting of a number of separate parts. This is true whether the similar feature is anatomical, as in the bones of the vertebrate forelimb, or is a biochemical pathway or a behavioral pattern. The more separate parts involved in a feature shared by several species, the less likely it is that the feature evolved independently in each.

The Monophyletic Ideal

In a classification scheme that accurately reflects evolutionary history, every taxon is, ideally, *monophyletic*. This means that every taxon, at whatever categorical level, should include a common ancestral species as well as the organisms descended from it. In other words, taxa should be real historical units. Thus a genus includes the most recent common ancestor and the species descended from that ancestor (and no other species). Similarly, a family should include a more distant common ancestor and the genera descended from it, and so on. A taxon that includes more than one ancestral line is said to be *polyphyletic*.

TAXONOMIC METHODS

The traditional means of determining the classification of a newly discovered specimen requires several different steps, involving different kinds of appraisals. First, it is tentatively assigned to a particular taxon on the basis of its overall outward similarities to other members of that taxon. Then these similarities are tested for homologies. Fossils are taken into account when possible. For instance, hares and rabbits (collectively known as lagomorphs) were long believed to be rodents, but the earliest fossil remains of lagomorphs and rodents show that the two groups had quite different origins. Conversely, bears, once considered a very distinct group of carnivores, are now known, on the basis of paleontology, to have diverged relatively recently from dogs. Various stages in the life cycle should also be examined; as you will see in the chapters that follow, some major decisions concerning relatedness of invertebrate phyla are based on resemblances among larval forms. In short, a judgment, when it is finally made, reflects the consideration and weighing of a large number of factors. Thus, it is not surprising that radically different classifications have sometimes been proposed for the same organism. For example, flamingos are classified with storks by some authorities, with ducks by others.

New Methodologies

Recently, two new methodologies—numerical phenetics and cladistics—have been proposed as replacements for this traditional "evolutionary" method. Pheneticists and cladists are fairly united in their reasons for seeking a change. Both criticize the traditional methodology as being based on subjective rather than objective criteria. Both claim that the traditional methodology is circular, using phylogeny to identify homology and then homology to establish phylogeny. Both claim that it is not really possible for a classification scheme to indicate both overall similarity and genealogy (branching patterns). They point out that some long-separated lineages have evolved in parallel and so continue to resemble one another more closely than organisms that have diverged rapidly from a recent common ancestor. Not only are the traditional methods suspect, but the goals are unattainable, according to this joint analysis. The remedies proposed by the two groups are, however, exactly opposite to one another.

Numerical Phenetics

Numerical phenetics is based only on a species' observable characteristics. To begin, the characteristics of the species being studied are divided into unit characters, that is, characters of two or more states that cannot logically be subdivided further. As shown in Table 19-2, these unit characters can be assigned numbers and coded for the computer as plus or minus or 0 (data not available). As many different characters as possible—at least 100—are taken into consideration. The data are then processed by computer, which scores the taxa according to the number of unit characters they have.

Table 19-2 Portion of a Coded Data Table for Phenetic Analysis of Four Taxa

UNIT CHARACTERS	TAXON A	TAXON B	TAXON C	TAXON D
1	+	+	−	0
2	+	+	+	+
3	+	+	+	−
4	−	+	0	0
5	+	+	+	+
6	+	+	−	+
7	+	+	−	0
8	0	−	+	+
9	+	+	+	+
10	+	+	+	−
11	+	0	−	0
12	+	+	+	−

Each character is given equal weight by this system with no subjective evaluation or prior knowledge taken into consideration. For instance, the possession of five fingers in a phenetic analysis would signify that lizards are more similar to humans than to snakes. The difference between homology and analogy is disregarded. Characters known to be more subject to ecological pressures—such as the shape of a leaf—are weighted equally with more constant characters—for example, the morphology of a flower. Pheneticists argue that such problems are resolved if enough characters are taken into consideration. Thus, for instance, despite the fact that one has five fingers and one does not, the relationship between lizard and snake would emerge when the other characters are taken into account.

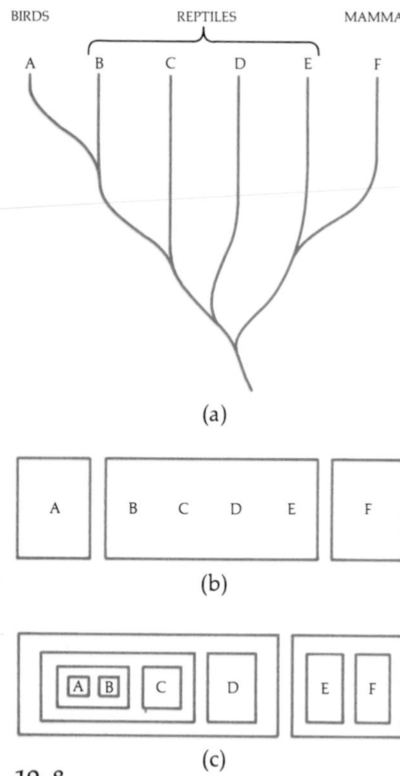

19–8

(a) *The phylogeny of some vertebrates.* (b) *According to traditional classification, taxa B through E are grouped together (class Reptilia). Taxa A and F, the birds and the mammals, are placed in class Aves and class Mammalia, because of their obvious biological distance from the other groups.* (c) *According to cladistic methodology, classification of taxa must be based solely on genealogy (branching patterns), with the more recent branches assigned lower categorical ranks in the hierarchical system.*

Cladistics

Cladistics is the most revolutionary of the three current methodologies. In contrast to numerical phenetics, which bases classification exclusively on degree of overall similarity, cladistics ignores overall similarity and is based exclusively on phylogeny. Cladists argue that the branching of one lineage from another in the course of evolution is the one event that can be measured objectively and that, therefore, taxonomy should be based solely on genealogy (see essay).

The goal of the cladists is the construction of monophyletic taxa, but their definition of monophyletic is more restrictive than that of the traditionalists. A monophyletic taxon, by cladistic standards, must include the ancestral species of the group and also all members of the group. None of the taxa in Figure 19–7, determined by traditional methods, would satisfy these requirements.

The substitution of cladistics for traditional methods of classification would result in some revolutions. Figure 19–8 shows a traditional phylogenetic tree of some vertebrates. Four of the modern groups (B through E), which include organisms such as crocodiles, turtles, lizards, and snakes, are placed in class Reptilia according to conventional schemes. Two of the groups, the birds and the mammals, are placed in separate classes because of their obvious biological differences. According to the cladist scheme, however, taxa must conform to branching patterns (genealogy), forming what the cladists refer to as "nested sets" within the hierarchy of the more inclusive taxa. Thus, crocodiles (B) end up with the birds (A), instead of in class Reptilia. Traditionalists protest that given the obvious biological differences between birds and crocodiles and the similarity, for instance, of crocodiles and lizards, a system that groups birds with crocodiles, instead of crocodiles with lizards, makes no biological sense at all. Cladists, however, reject the lack of consistency of the traditional system; class Reptilia, for the cladist, simply should not exist. It is analogous to a political taxon made up of the United States minus all states admitted since 1893.

Some Criteria

Systematics by any methodology is based largely on anatomy, and this will probably continue to be true in the future since a great body of data on comparative anatomy has been accumulated. However, new biochemical techniques are making some other methods available. As you will see in Chapters 20 and 21, biochemical criteria are helping to sort out the lineages of some of the one-celled organisms, groups difficult to distinguish on anatomy alone.

Amino Acid Sequences

One of the problems in comparing morphological similarities among organisms, as we have seen, is the difficulty of assigning appropriate weights to different kinds of similarities. Moreover, morphological differences are poor criteria in the study of structurally dissimilar organisms—fish and fungi, for instance. When techniques became available for the sequencing of amino acids in proteins, biologists realized that comparing the structures of homologous proteins might offer another technique for studying evolutionary relationships. It offered two advantages: it was objectively quantifiable, and it could be used to compare very diverse organisms.

One of the proteins first selected for analysis was cytochrome *c*, one of the carriers of the electron transport chain (page 199). Cytochrome *c* molecules from a

HOW TO CONSTRUCT A CLADOGRAM

A cladogram is a hypothesis of branching sequences. It looks, at first glance, like a phylogenetic tree, but it is not one. It contains no ancestors, only branching points. Branching points are determined by the appearance of evolutionary novelties. This concept is hard to understand without an example. Take the following organisms: lizard, mouse, trout, cow, shark. They have some characteristics—dorsal nerve cord, chambered heart, four appendages, jaws—that set them off, as a group, from other groups—clams or grasshoppers, for instance. Four of them have bony skeletons and one, the shark, does not. Three of them are characterized by an amniote egg (that is, an egg that contains its own water supply). Two of them have mammary glands. Thus, in terms of evolutionary novelties, the branching sequence shown in (a) *is constructed.*

The hypothesis can now be tested. For instance, cow and mouse not only both have mammary glands, but both also have hair, another evolutionary novelty. If the lizard and mouse had hair, but the cow lacked hair, the hypothesis would have to be reexamined. Or, if the mouse and trout had three ear bones, but the lizard did not, the cladogram would have to be reconstructed. Thus, every cladogram fulfills the scientific ideal of presenting a testable hypothesis.

From the cladogram, a hierarchical classification can be constructed:

Group: shark, trout, lizard, mouse, cow
Subgroup 1a: shark
Subgroup 1b: trout, lizard, mouse, cow
Subgroup 2a: trout
Subgroup 2b: lizard, mouse, cow
Subgroup 3a: lizard
Subgroup 3b: mouse, cow
Subgroup 4a: mouse
Subgroup 4b: cow

Cladists view such a hierarchical listing as representing "nested sets," that is, groups within groups, as shown in (b). *Examination of this figure reveals that, following the logical steps outlined, we have been led to put sharks in one high-ranking taxon and trout in another, along with the mammals. This sort of result has led traditionalists to cry out that cladists cannot tell a fish from a cow. In this we see the essence of the argument.*

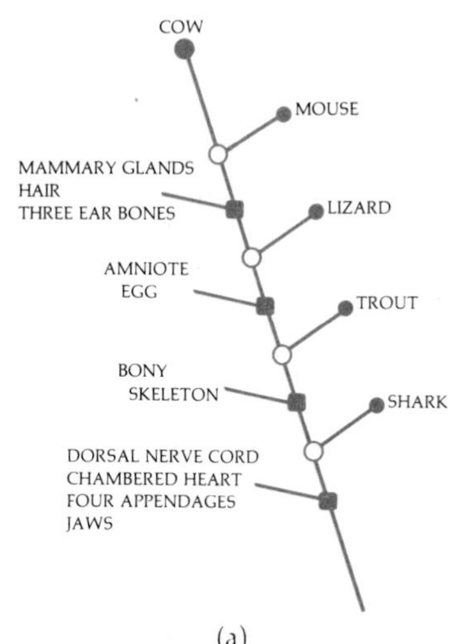

(a)

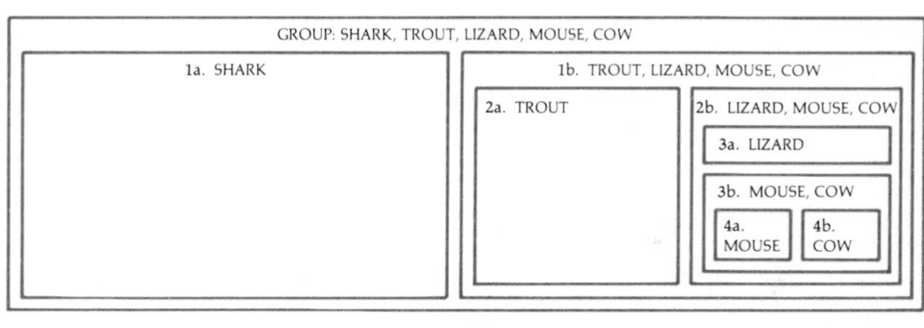

(b)

(a) *A cladogram for five types of vertebrates. Each branch point is defined by one or more characteristics interpreted as evolutionary novelties.* (b) *From the cladogram, the cladistic hierarchy, represented as "nested sets," can be constructed.*

great variety of organisms have now been sequenced, making it possible to determine the number of amino acids by which the molecules of the various organisms differ. From the amino acid data, it is possible to calculate the minimum number of nucleotide differences that would be necessary to account for the amino acid differences (Table 19–3). Presumably, the greater the number of amino acid (or nucleotide) differences between any two organisms, the more distant their evolutionary relationship; conversely, the smaller the number of differences, the closer their relationship. Figure 19–9 illustrates a phylogeny based on cytochrome *c* data. The results conform well, but not perfectly, with phylogenies constructed by more traditional methods. Clearly, protein structure is another useful parameter of evolutionary relationships.

Molecular Clocks

As the data on protein variations have accumulated, they too have become the source of controversy. One school of biologists maintains that the differences in protein structure represent functional differences among the molecules, just as differences in the structure of the beaks of birds represent adaptations to different food sources. An opposing school contends that these amino acid changes occur at random and that they do not represent the result of a selection process but merely mark off the passage of time, like grains of sand trickling through an hour glass or the decay of radioisotopes. From this viewpoint, the amino acid differences in the homologous proteins of different groups of organisms do not represent functional differences; instead, they indicate the time at which the groups diverged (Figure 19–10, page 384).

Table 19–3 The Minimum Number of Nucleotide Differences Required to Account for the Differences in the Cytochrome *c* Molecules of 20 Organisms

ORGANISM	1	2	3	4	5	6	7	8	9	10	11	12	13	14	15	16	17	18	19	20
1. Human	—	1	13	17	16	13	12	12	17	16	18	18	19	20	31	33	36	63	56	66
2. Monkey		—	12	16	15	12	11	13	16	15	17	17	18	21	32	32	35	62	57	65
3. Dog			—	10	8	4	6	7	12	12	14	14	13	30	29	24	28	64	61	66
4. Horse				—	1	5	11	11	16	16	16	17	16	32	27	24	33	64	60	68
5. Donkey					—	4	10	12	15	15	15	16	15	31	26	25	32	64	59	67
6. Pig						—	6	7	13	13	13	14	13	30	25	26	31	64	59	67
7. Rabbit							—	7	10	8	11	11	11	25	26	23	29	62	59	67
8. Kangaroo								—	14	14	15	13	14	30	27	26	31	66	58	68
9. Duck									—	3	3	3	7	24	26	25	29	61	62	66
10. Pigeon										—	4	4	8	24	27	26	30	59	62	66
11. Chicken											—	2	8	28	26	26	31	61	62	66
12. Penguin												—	8	28	27	28	30	62	61	65
13. Turtle													—	30	27	30	33	65	64	67
14. Rattlesnake														—	38	40	41	61	61	69
15. Tuna															—	34	41	72	66	69
16. Screwworm fly																—	16	58	63	65
17. Moth																	—	59	60	61
18. *Neurospora* (red bread mold)																		—	57	61
19. *Saccharomyces* (yeast)																			—	41
20. *Candida* (fungus)																				—

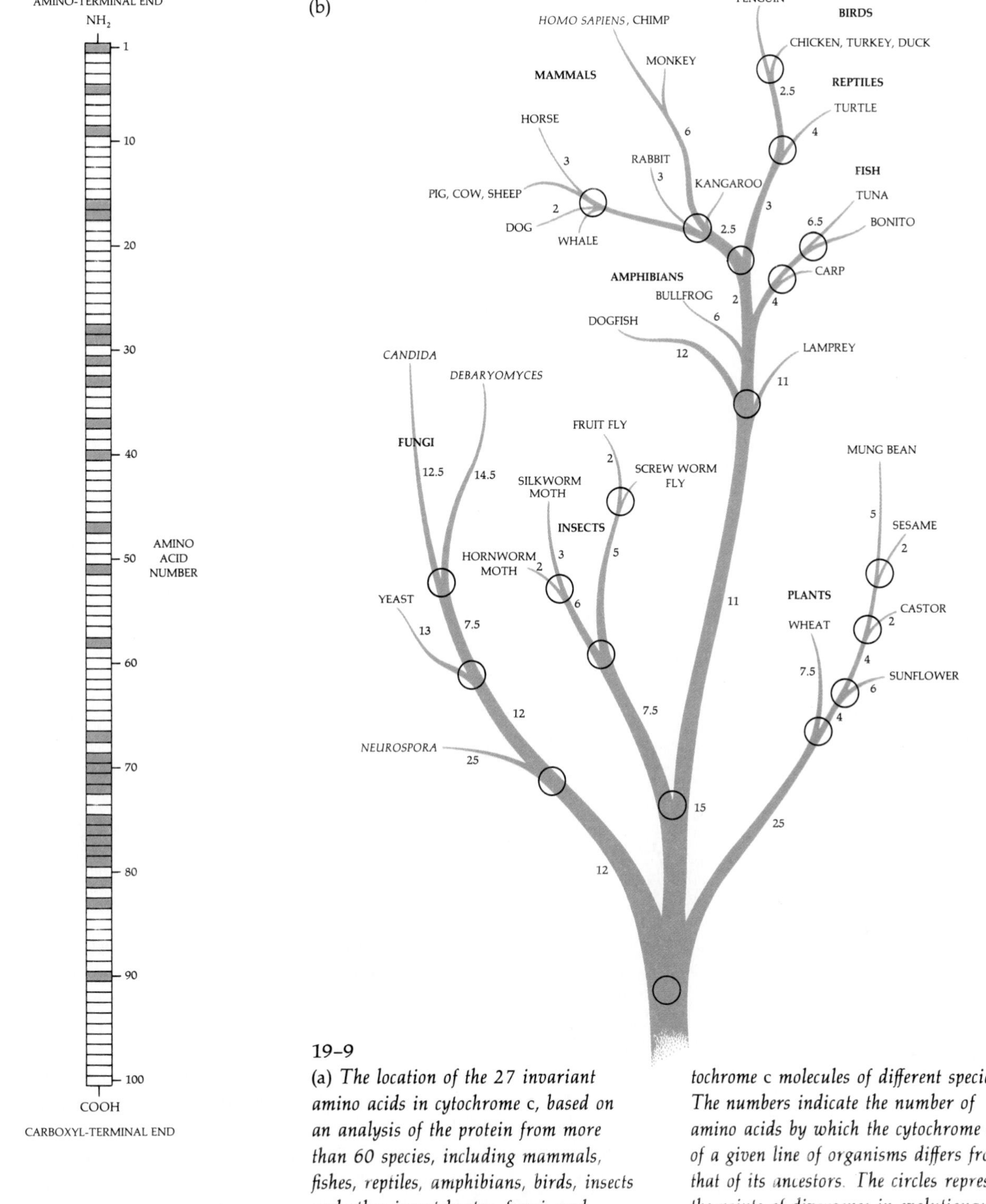

19–9

(a) The location of the 27 invariant amino acids in cytochrome c*, based on an analysis of the protein from more than 60 species, including mammals, fishes, reptiles, amphibians, birds, insects and other invertebrates, fungi, and plants.* (b) *The main branches of a phylogenetic tree based on comparisons of the amino acid sequences of the cytochrome* c *molecules of different species. The numbers indicate the number of amino acids by which the cytochrome* c *of a given line of organisms differs from that of its ancestors. The circles represent the points of divergence in evolutionary history. As you can see, this tree is in fairly good agreement with phylogenies constructed by more conventional means.*

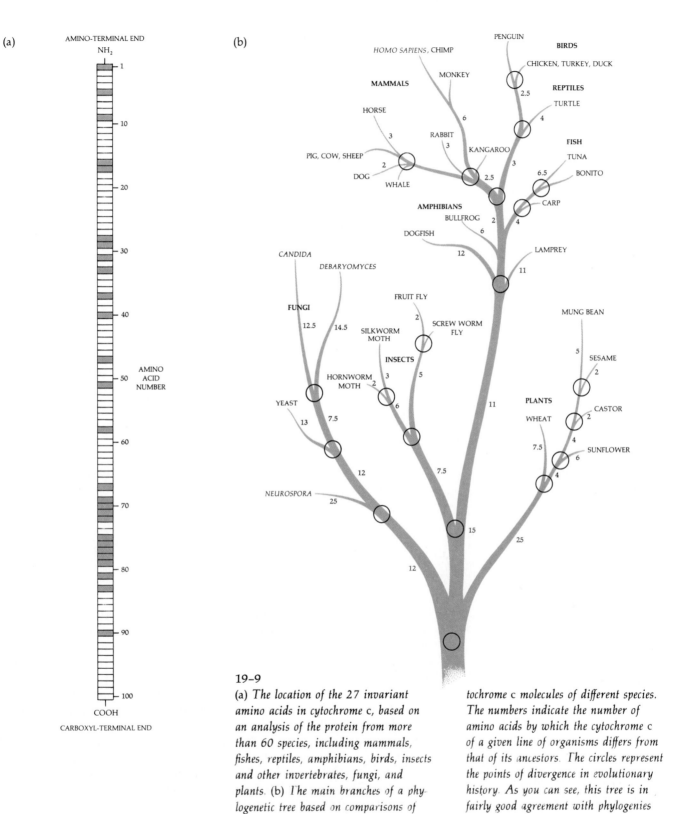

19–10
Different proteins accumulate changes at different rates. However, the relative constancy of the rates of change and consistent correlations among different proteins as to the extent of the change lend support to the concept that the changes are not functional but rather represent the ticking of a molecular clock.

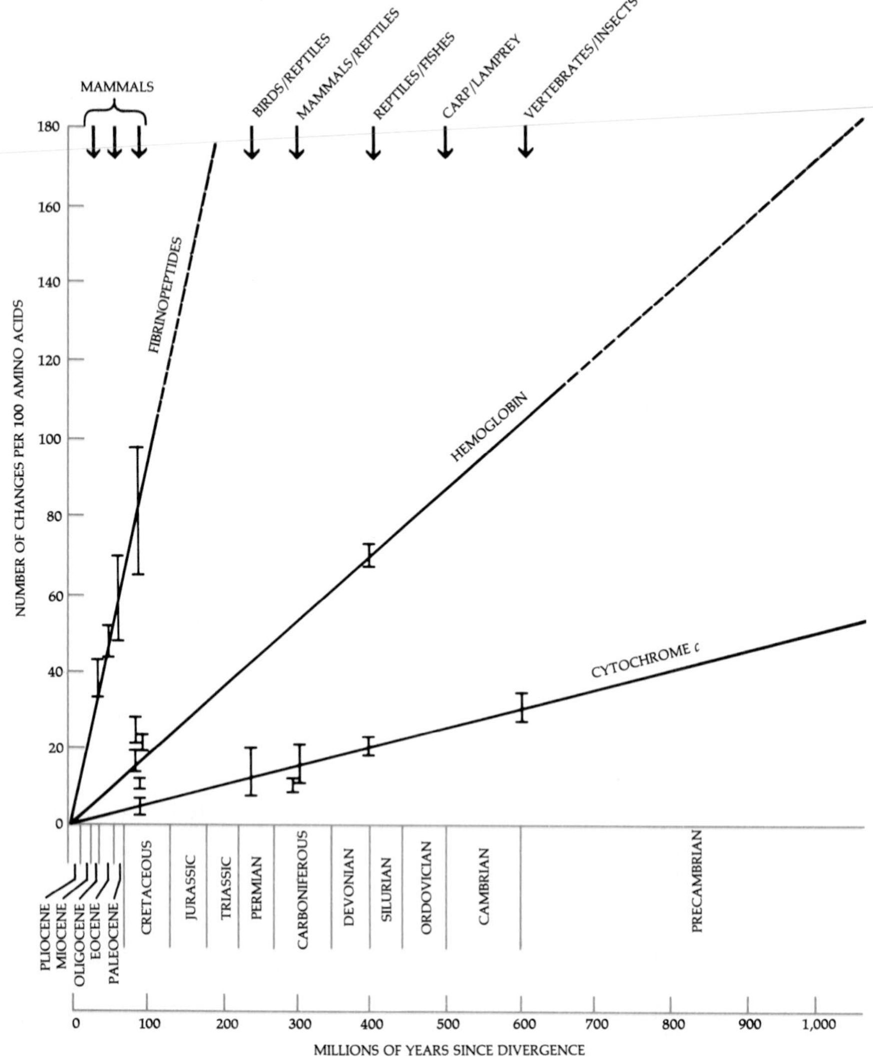

In support of the "random-tick" hypothesis, biological clock makers point out that two frog species that diverged millions of years ago, but remained similar enough in appearance to be included in a single genus, differ from one another in amino acid substitutions as much as a bat does from a whale. *Homo sapiens* and the chimpanzee, on the other hand, two species that differ anatomically and in a number of other characteristics but that diverged recently, according to paleontological evidence, have identical amino acid sequences in cytochrome *c* and some other proteins as well.

If cytochrome *c* and other ancient proteins are clocks, some apparently well-founded traditional concepts will be overthrown. For example, as we shall discuss in Chapter 56, a fossil primate, *Ramapithecus,* has long been thought to be an ancestor of the hominids (the lineage that branched off from the great apes and led to *Homo*). The problem is that *Ramapithecus* lived at least 14 million years ago, and, according to the molecular clocks, hominids diverged from the other primates only 5 to 6 million years ago.

Now that the discovery of restriction enzymes has made possible the sequencing of nucleic acids (see page 322), the use of homologous proteins to estimate evolutionary relationships has been largely abandoned. One reason is that nucleic acid sequencing is technically far easier than protein sequencing, dealing, as it does, with only four different nucleotides as compared to 20 amino acids. Also, it is more sensitive, since changes in nucleotides may not be reflected in changes in amino acids because of the many synonyms in the genetic code. Are changes from one synonym to another neutral? Some argue that they are, since they do not affect the structure or function of the proteins involved. Others maintain that not enough is known about transcription and translation and the roles of the various tRNAs to assume that selective forces are not at work. Finally, as we noted in Chapter 17, a large amount—most—of the DNA in the eukaryotic cell is never translated. Will some of this DNA prove to be truly neutral as to function, and will it also in some way reflect evolutionary divergence? We shall have to wait for an answer to these questions, but possibly not for very long.

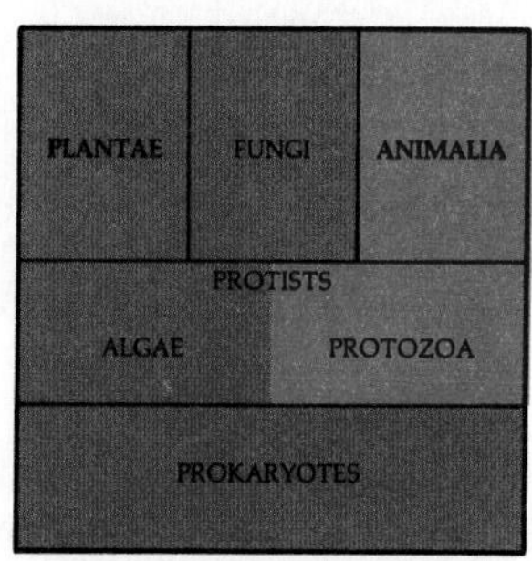

(a)

(b)

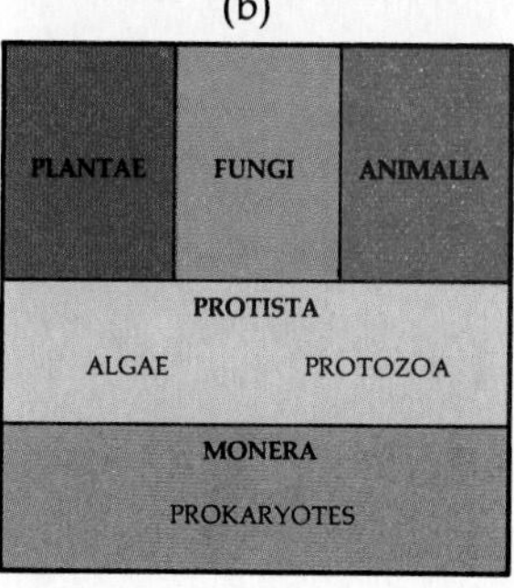

(c)

19–11
Three of the classification systems in common use. In (a), *all organisms are classified as either plants or animals. In* (b), *the prokaryotes are considered a third kingdom, Monera. In* (c), *five separate kingdoms are recognized.*

A QUESTION OF KINGDOMS

We shall close this chapter with a brief description of one of the current unresolved—and perhaps unresolvable—problems in classification, which has to do with the placement of taxa in the category of kingdom.

In Linnaeus's time, as we mentioned, three kingdoms were recognized—animals, plants, and minerals—and until very recently it was common to classify every living thing as either an animal or a plant. Kingdom Animalia included those organisms that moved and ate things, and whose limbs, organs, and bodies grew to a certain size and then stopped growing. Kingdom Plantae comprised all living things that did not move or eat and that grew indefinitely. Thus the fungi, algae, and bacteria were grouped with the plants, and the protozoa—the one-celled organisms that ate and moved—were classified with the animals (Figure 19–11a).

In the twentieth century, new data began to emerge. This was partly a result of improvements in the light microscope and, subsequently, the development of the electron microscope, and partly because of the application of biochemical techniques to studies of differences and similarities among organisms. As a result, the number of groups recognized as constituting different kingdoms has increased. The new techniques revealed, for example, the fundamental differences between prokaryotic and eukaryotic cells—differences sufficiently great to warrant placing the prokaryotes in a separate kingdom, Monera (Figure 19–11b).

Other studies have provided new information about the evolutionary history of the major types of organisms. As we shall see in Chapter 21, there is strong evidence that different lineages of eukaryotes arose independently from different prokaryotic ancestors. Moreover, different lineages of unicellular eukaryotes appear to have given rise to multicellular plants, fungi, and animals, and, at least in the case of plants and fungi, multicellularity arose several times. This history makes it impossible, on the basis of current knowledge, to establish monophyletic kingdoms and still have the kingdoms reflect similarities and differences among major groups of living organisms.

Most contemporary proposals concerning kingdoms are based not on evolutionary history, but rather on the cellular organization and the mode of nutrition of the organisms. The proposal we shall follow recommends five kingdoms: Monera, Protista, Fungi, Plantae, and Animalia (Figure 19–11c). In Appendix C, the major groups of organisms are classified within these five kingdoms. The members of the

19–12

Representatives of the five kingdoms. (a) *Prokaryotes. Cyanobacteria,* Nostoc, *chlorophyll-containing prokaryotes that form gelatinous colonies.* (b) *Protists.* Vorticella, *like most protists, is a single cell. It attaches itself to a substrate by a long stalk. A contractile fiber, visible in this micrograph, runs through the stalk.* (c) *Fungi. Inky cap mushrooms. Fungi are characterized by a multicellular underground network and also spores and sporangia, which, in mushrooms, are borne in these familiar structures.* (d) *Plants. Marsh marigolds. The flowers of angiosperms, attractive to pollinators, are among the principal reasons for their evolutionary success.* (e) *Animals. A luna moth. A nervous system with a variety of sense organs is a chief characteristic of the animal kingdom.*

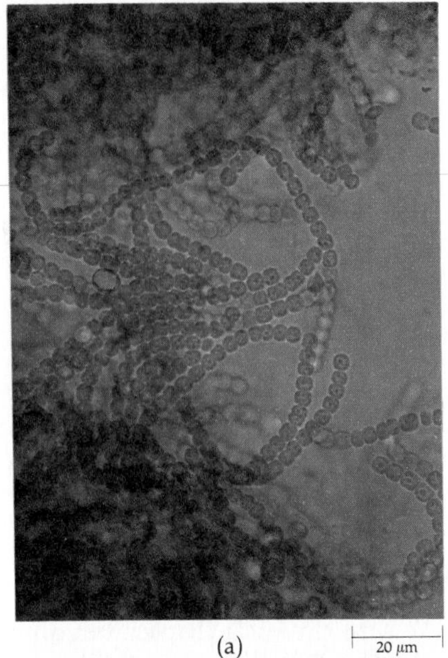

(a)

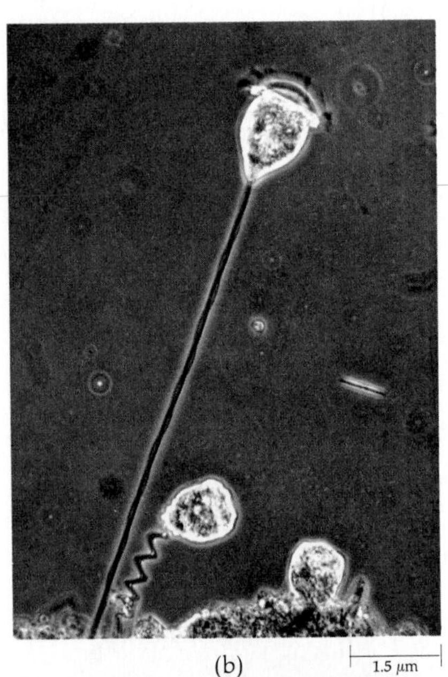

(b)

(c)

(d)

(e)

Table 19-4 Characteristics of Major Groups of Organisms

	PROKARYOTES	PROTISTS	FUNGI	PLANTS	ANIMALS
Cell type	Prokaryotic	Eukaryotic	Eukaryotic	Eukaryotic	Eukaryotic
Nuclear envelope	Absent	Present	Present	Present	Present
Mitochondria	Absent	Present	Present	Present	Present
Chloroplasts	None (photosynthetic membranes in some types)	Present (some forms)	Absent	Present	Absent
Cell wall	Noncellulose (polysaccharide plus amino acids)	Present in some forms, various types	Chitin and other noncellulose polysaccharides	Cellulose and other polysaccharides	Absent
Means of genetic recombination	Conjugation, transduction, transformation, or none	Fertilization (syngamy) and meiosis, conjugation, or none	Fertilization and meiosis, dikaryosis (page 441), or none	Fertilization and meiosis	Fertilization and meiosis
Mode of nutrition	Autotrophic (chemosynthetic and photosynthetic) and heterotrophic (saprobic and parasitic)	Photosynthetic and heterotrophic, or combination of these	Heterotrophic (saprobic and parasitic), by absorption	Photosynthetic	Heterotrophic, by ingestion
Motility	Bacterial flagella, gliding, or nonmotile	9 + 2 cilia and flagella, amoeboid, contractile fibrils	9 + 2 cilia and flagella in some forms, none in most forms	9 + 2 cilia and flagella in lower forms and in some gametes, none in most forms	9 + 2 cilia and flagella, contractile fibrils
Multicellularity	Absent	Absent in most forms	Present in most forms, but limited	Present in all forms	Present in all forms
Nervous system	None	Primitive mechanisms for conducting stimuli in some forms	None	None	Present, often complex

kingdom Monera, the prokaryotes, are, of course, identified on the basis of their unique cellular organization and biochemistry. Members of the kingdom Protista are unicellular eukaryotes, both autotrophic and heterotrophic. A few groups of relatively simple multicellular organisms are also included in this kingdom, because they are more similar to unicellular forms than to related fungi, plants, or animals. All the other multicellular eukaryotes are divided into three kingdoms, based primarily on their mode of nutrition: fungi absorb organic molecules from the surrounding medium, plants manufacture them by photosynthesis, and animals ingest them in the form of other organisms. These three groups of organisms have distinct ecological roles—plants are generally producers, animals are consumers, and fungi are decomposers. Table 19-4 summarizes some of the essential similarities and differences among these five kingdoms of organisms. We shall discuss each group in turn in the chapters that follow.

In truth, no system of kingdoms is really satisfactory. For instance, as we shall see in Chapter 23, there is a clear evolutionary sequence, with modern, living representatives, leading from certain one-celled algae to the flowering plants. So if we group all of the one-celled eukaryotes together, as we do in this text, we break up what would appear to be a clear evolutionary line to the plants. However, at the one-celled level of life, there are no useful criteria for separating plants from animals. Two species of single-celled, motile organisms may be almost identical in most respects, except that one has chloroplasts and the other does not. In some cases, the one that has chloroplasts can lose them from time to time and still continue to survive and reproduce indefinitely. Yet in a plant-animal division based on the capacity for photosynthesis, these two closely related forms are separated at the level of kingdom.

Whether organisms are classified in two, three, or five kingdoms, however, their genus and species designations are not affected, nor are most of the other categories in which they are now classified.

SUMMARY

As many as 5 million kinds of organisms are estimated to inhabit earth's biosphere. Scientists have sought order in this vast diversity of living things by classifying them, that is, by grouping them in meaningful ways. Taxonomy is the science of classifying groups (taxa) of organisms in formal categories; systematics is the study of relationships among the groups. Post-Darwinian systematics seeks to reflect phylogeny—the evolutionary history of organisms.

The basic unit of classification is the species, most easily defined as an interbreeding group of organisms that does not breed with other groups of organisms. Species are named by a binomial system that includes the name of the genus (written first) and the specific epithet. Genera are groups of similar species, presumably species that have recently diverged. Genera are grouped into families, families into orders, orders into classes, classes into phyla or divisions, and divisions or phyla into kingdoms. Depending on the classification system used, there may be as few as two kingdoms (plants and animals) or as many as five (Monera, Protista, Fungi, Plantae, and Animalia). The latter system is followed in this text.

Traditional taxonomic methodologies have sought to group organisms in ways that reflect similarities and differences based on evolutionary history. A major principle in this methodology is that of homology, similarity based on common ancestry rather than on adaptation to a similar environment (analogy). In a phylogenetic system, every taxon should be, ideally, monophyletic; that is, every taxon should consist of organisms descended from a common ancestor. This goal has proved difficult to achieve, especially at higher categorical levels.

Two new methodologies have been proposed, both designed to provide a more objective basis of classification. One, numerical phenetics, relies solely on the scoring of equally weighted, objectively observable similarities and differences among groups of organisms. It involves the accumulation of large amounts of data, the final analysis of which is done by computer. The other new methodology, cladistics, is based entirely on branching sequences (genealogy), determined by evolutionary novelties, and ignores overall similarities.

Differences in protein structure among groups of organisms provide criteria for comparisons that are objective and numerical and that can be used to compare organisms very diverse in morphology. The question of whether these differences are adaptive is a matter of current controversy; an alternative proposal is that proteins (and presumably the nucleic acids that dictate their structure) are molecular clocks, with changes in composition a direct reflection of elapsed time.

QUESTIONS

1. Distinguish between the following: systematics/taxonomy; category/taxon; genus/species; division/phylum; homology/analogy; monophyletic/polyphyletic; numerical phenetics/cladistics.

2. Identify which of the following are categories and which are taxa: undergraduates; the faculty of the University of Tennessee; the Cincinnati Bengals; major league baseball teams; the U.S. Marine Corps; Mozart's symphonies.

3. In Table 19–3, the calculated nucleotide differences are minima. What factors in calculating nucleotide changes on the basis of amino acid changes might lead to an underestimation of differences?

4. Based on your knowledge of protein structure, what sort of changes might be tolerated in a protein and what sort would not?

5. What are the major identifying characteristics of each of the five kingdoms?

CHAPTER 20

The Prokaryotes

In the next three chapters, we are going to describe the three kingdoms comprising the microorganisms: Monera (the prokaryotes), Protista (the algae, protozoa, and slime molds), and Fungi. Although they are dissimilar in many important ways, and many algae and fungi are quite large, the organisms in these three kingdoms are conventionally studied together in the science known as microbiology. Viruses and viroids, which are even smaller particles of naked RNA, will also be described in this chapter. Although viruses do not fit comfortably in any of the taxa of living systems, they too are traditionally a part of microbiology.

Monera is, in evolutionary terms, the oldest kingdom, and the contemporary prokaryotes are the most abundant organisms in the world. Although there are sometimes difficulties in defining prokaryotic species unambiguously, about 2,700 distinct species are currently recognized. Prokaryotes are the smallest cellular organisms; a single gram (about 1/28 of an ounce) of fertile soil can contain as many as 2.5 billion individuals.

The success of the prokaryotes, biologically speaking, is undoubtedly due to their rapid rate of cell division and their great metabolic diversity. Growing under optimum conditions, a population of *Escherichia coli,* probably the best-known prokaryote, can double in size every 20 minutes. Prokaryotes can survive in many environments that support no other form of life. They have been found in the icy wastes of Antarctica, the dark depths of the ocean, and even in the near-boiling waters of natural hot springs. Some prokaryotes are among the very few modern organisms that can survive without free oxygen, obtaining their energy by anaerobic processes (see page 192). Oxygen is lethal to some types (obligate anaerobes), whereas others can exist with or without oxygen (facultative anaerobes).

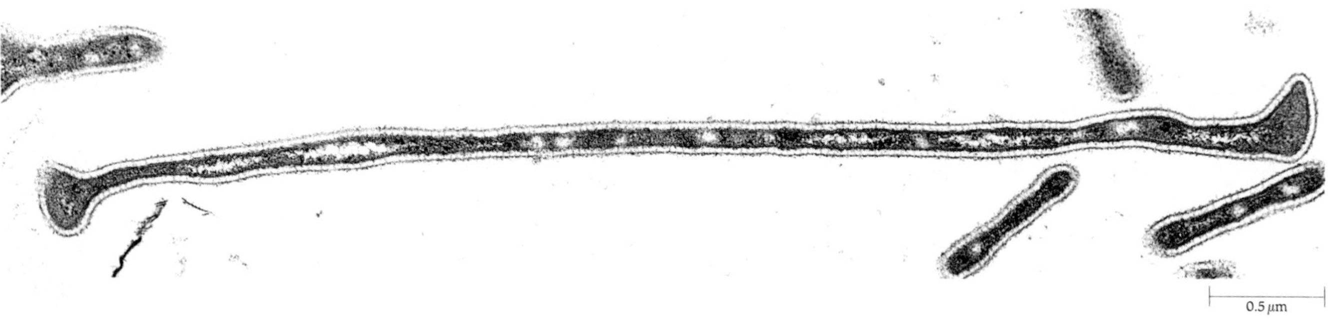

20–1
Thermophilic ("heat-loving") bacteria thrive at 92°C, a temperature close to the boiling point of water. This electron micrograph shows a single long filamentous type and several short rodlike types.

Some prokaryotes can form thick-walled spores. These spores are inactive, resistant forms that enable the cells to survive for long periods of time without water or nutrients or in conditions of extreme heat or cold. They may stay dormant for years, and some remain viable even when boiled in water for several hours.

From the ecological point of view, prokaryotes are most important as decomposers, breaking down organic material to forms in which it can be used by plants. They also play a major role in the process known as nitrogen fixation, by which nitrogen gas (N_2) is reduced to ammonia (NH_3) or ammonium ion (NH_4^+). Although nitrogen is abundant in the atmosphere, eukaryotes are not able to use atmospheric nitrogen, and so the crucial first step in the incorporation of nitrogen into organic compounds depends largely on certain species of prokaryotes; some of these species are free-living, others are found only in close association with plants. Some prokaryotes are photosynthetic, and a few species are both photosynthetic and nitrogen-fixing.

THE CLASSIFICATION OF PROKARYOTES

Until quite recently, the possibility of classifying the prokaryotes on an evolutionary basis seemed remote. Most of the characteristics used to determine phylogenetic relationships among eukaryotes—for example, intricate anatomical structures composed of interconnected parts and complex patterns of reproduction, development, and growth—simply do not exist in prokaryotes. Biologists studying the prokaryotes were forced to rely upon differences in the size and shape of individual cells, the general appearance of the colonies formed by some prokaryotes, type of movement, mode of nutrition, the presence or absence of spores, the presence or absence of a cell wall, and the ability of the cells to retain certain chemical dyes. Many of these characteristics do not reflect phylogenetic relationships. The prokaryotes have been in existence—and evolving in response to diverse selection pressures—for a very long time. Certain features, such as cell shape and colony form, have probably evolved over and over again. In contrast, other features, such as a cell wall, the capacity for photosynthesis, or the ability to form spores, have been lost independently in a number of lineages.

As a result of these difficulties, the classification scheme for the prokaryotes that has been in use for many years—and that is still the most widely used—is not hierarchical. The kingdom consists of two divisions: Cyanophyta, the cyanobacteria (known as the blue-green algae and classified with the eukaryotic algae until their prokaryotic structure was revealed), and Schizophyta, the bacteria. The cyanobacteria are generally classified into four groups, and the various genera of bacteria are classified into 19 groups. A condensed version of this scheme of classification, in which 12 of the 19 bacterial groups are placed together as the eubacteria, is given in Table 20-1 (page 393).

In recent years, studies of cell ultrastructure and biochemistry (particularly the details of metabolic pathways) have enabled biologists to begin establishing taxa that truly reflect evolutionary relationships. As we noted in the previous chapter, a crucial breakthrough has come with the development of techniques for determining the primary structure of proteins and for sequencing DNA and RNA molecules. The nucleotide sequences that are now being determined for many prokaryotes can reveal substitutions, additions, and deletions of nucleotides, as well as the translocation or inversion of larger segments of DNA. In general, as with eukaryotes, the smaller the number of differences in the nucleotide sequences of two organisms, the more recent their common ancestor.

20–2
Oscillatoria, *a filamentous cyanobacterium. The cyanobacteria were formerly regarded as algae, because of the similarity between their photosynthetic processes and those of the eukaryotic algae and the plants. Like the photosynthetic eukaryotes, they have chlorophyll* a *and produce oxygen during photosynthesis. Biochemical and electron microscope studies, however, have revealed their prokaryotic nature.*

These techniques are rapidly leading to a revolution in our understanding of prokaryotic phylogeny. One of the most striking discoveries thus far has been that a few genera of prokaryotes, traditionally classified with the eubacteria, differ very dramatically from all other organisms. These genera include the methanogens (bacteria that synthesize methane from carbon dioxide and hydrogen gas), the salt-loving halobacteria (see page 224), and the thermoacidophiles (bacteria that thrive in very acidic environments at high temperatures). They differ from all other prokaryotes in several significant ways. First, their cell walls have a unique composition. Second, they do not use the Calvin cycle for carbon reduction. Third, they require several unique coenzymes that serve the function of NAD^+ and FAD as electron acceptors. Fourth, and perhaps most important, the nucleotide sequences of their transfer RNAs and ribosomal RNAs are markedly different from those in all other organisms. A particularly interesting feature of the methanogens is that they would have been excellently suited to the conditions prevailing on earth during the earliest stages of biological evolution. It appears that the methanogens, extreme halophiles, and thermoacidophiles may be the few surviving representatives of a lineage that diverged very early from the common ancestor that gave rise to the other prokaryotes. Recently it has been proposed that they be placed in a new kingdom, Archaebacteria.

At the present time, the sequencing of the DNA and RNAs of many different prokaryotic species is proceeding rapidly in a number of laboratories. As the nucleotide sequences for an increasing number of prokaryotes are becoming available, the probable evolutionary relationships of different groups are being worked out. It is anticipated that in the relatively near future we will have an entirely new classification of the prokaryotes that provides a realistic picture of their evolutionary history. For now, however, we must work within the framework of the old system while pointing out, where possible, the outlines of the emerging new system.

Table 20-1 The Kingdom Monera

	FORM	MODE OF MOVEMENT	MODE OF REPRODUCTION	MODE OF NUTRITION	ECOLOGICAL ROLE
Division Cyanophyta *(Nostoc, Oscillatoria)*	Unicellular and filaments	Gliding or nonmotile	Binary fission	Photosynthetic	Carbon and nitrogen fixation
Division Schizophyta					
Eubacteria *(Escherichia, Streptococcus, Staphylococcus, Clostridium)*	Rod-shaped, spherical, spirillum, vibrio; rigid cell wall	Flagella or nonmotile	Binary fission	Chemoautotrophic, photosynthetic, heterotrophic	Decomposers, carbon and nitrogen fixation, pathogens (lockjaw, pneumonia)
Prosthecate and budding bacteria *(Hyphomicrobium, Caulobacter)*	Unusual shapes, appendages (prosthecae)	Flagella or nonmotile	Budding, binary fission (some members)	Most heterotrophic, some chemoautotrophic	Decomposers
Gliding bacteria *(Beggiatoa, Chondromyces)*	Filaments; large and short rods; flexible cell wall	Gliding (mechanism unknown)	Short filaments of cell break off; binary fission; some form fruiting bodies	Heterotrophic, chemoautotrophic	Decomposers, especially of macromolecules (myxobacteria)
Filamentous (sheathed) bacteria *(Sphaerotilus)*	Individual cells enclosed in common sheath	Flagella or nonmotile	Binary fission	Heterotrophic	Decomposers
Actinomycetes *(Streptomyces, Actinomyces, Mycobacterium, Corynebacterium)*	Branching multicellular filaments, many moldlike in appearance	Nonmotile	Spores, fragmentation, binary fission	Heterotrophic	Decomposers (degrading lipids and waxes), nitrogen fixation, pathogens (tuberculosis, leprosy, diphtheria)
Spirochetes *(Spirocheta, Treponema, Leptospira)*	Extremely long, helical, flexible	Twisting (axial filament)	Binary fission	Heterotrophic	Decomposers, pathogens (syphilis)
Rickettsiae *(Rickettsia, Chlamydia)*	Small (0.5 x 1.1 micrometers), rigid cell wall	Nonmotile	Binary fission	Heterotrophic	Pathogens (typhus, spotted fever, trachoma), all obligate intracellular parasites
Mycoplasmas *(Mycoplasma)*	Smallest free-living cells, no cell walls	Nonmotile	Binary fission	Heterotrophic	Pathogens (mycoplasmal pneumonia), many intracellular parasites

20-3
Cells of Escherichia coli, *a modern prokaryote that is a common inhabitant of the human digestive tract. The DNA is in the less dense (lighter-appearing) material in the center of each cell. The small, dense bodies in the cytoplasm are ribosomes. The two cells in the center have just finished dividing and have not yet separated completely.*

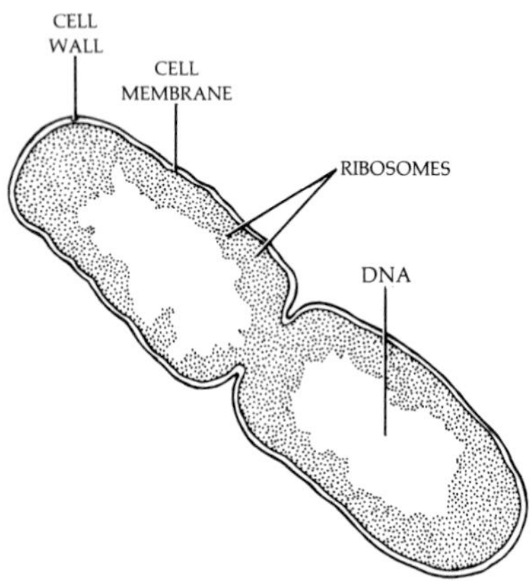

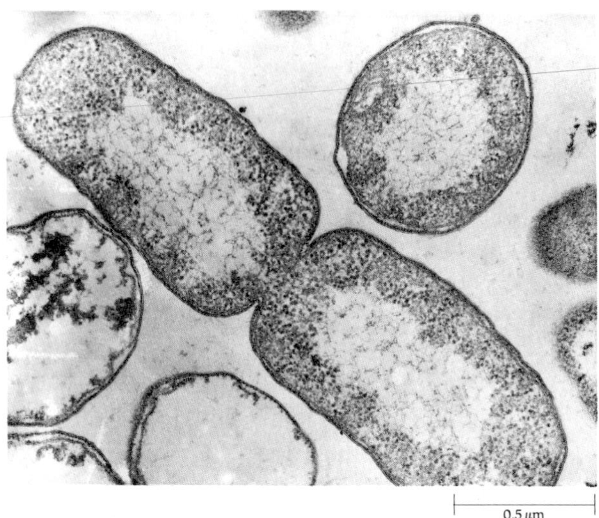

THE PROKARYOTIC CELL

The essential features of a prokaryotic cell were outlined in Chapter 4 and are reviewed in Figure 20-3. The cell shown here is the familiar *Escherichia coli*, one of the eubacteria. The eubacteria, in general, and *E. coli*, in particular, are the most thoroughly studied prokaryotes.

The Cytoplasm

The cytoplasm of prokaryotes is relatively unstructured. Except in the cyanobacteria, the cytoplasm is not divided or compartmentalized by membranes and does not contain a membrane-bound nucleus or any membrane-bound organelles. The most prominent feature within the cytoplasm is the chromosome. All prokaryotic chromosomes analyzed so far have proved to consist of one single, continuous ("circular") molecule of DNA associated with a small amount of RNA and nonhistone protein. A prokaryotic cell may also contain one or more plasmids (page 311).

The cytoplasm often has a fine granular appearance due to its many ribosomes. These are somewhat smaller than eukaryotic ribosomes but have the same general shape.

The Cell Membrane

The membrane surrounding a prokaryotic cell is similar in chemical composition to that of a eukaryotic cell (see page 100) but, except in the mycoplasmas (the smallest free-living cells known), the membranes of prokaryotes lack cholesterol or other steroids. In the aerobic prokaryotes, the cell membrane incorporates the electron transport system found in the mitochondrial membrane of eukaryotic cells. In the photosynthetic bacteria (but not in the cyanobacteria), the sites of photosynthesis are found in the cell membrane. In the photosynthetic purple bacteria and in aerobes with large energy requirements, the membrane is often extensively convoluted, with folds extending into the interior. These, of course, greatly increase the working surface of the membrane. Also, as we noted on page 140, the membrane appears to contain specific attachment sites (called mesosomes) for the DNA molecules; these sites are believed to play a role in ensuring the separation of the duplicate chromosomes at cell division.

20–4
This unusual micrograph, taken by scientists at the Bronx-Lebanon Hospital Center in New York, shows a bacterial cell exploding. This bacterium is a member of the species Staphylococcus aureus, *the cause of many human infections. Because water tends to move into the cell by osmosis, the contents of bacterial cells are under pressure. When this cell was treated with an antibiotic that damaged the cell wall, it exploded.*

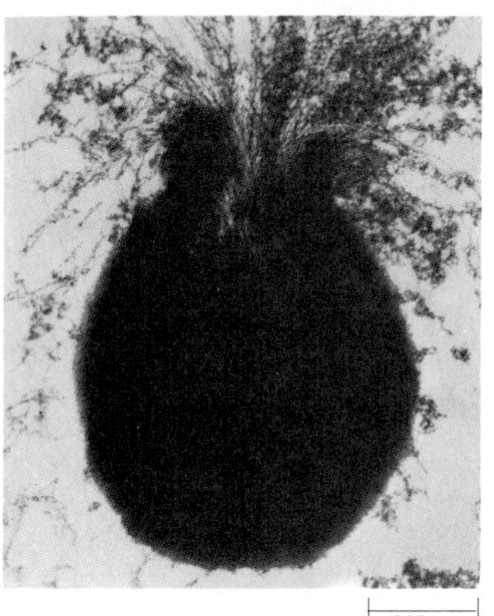

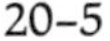

The Cell Wall

Almost all prokaryotes are surrounded by a cell wall, which ranges from 5 to 80 nanometers in thickness and gives the different types their characteristic shapes. Many prokaryotes have rigid walls, some have flexible walls, and only the mycoplasmas have no cell walls at all. Because most bacterial cells are hypertonic in relation to their environment, they would burst without their walls. (The mycoplasmas live as intracellular parasites in an isotonic environment.)

Chemical Structure of the Cell Wall

The cell walls of prokaryotes are complex and contain many kinds of molecules not present in eukaryotes. Except for the methanogens and their close relatives, the walls of prokaryotes contain complex polymers known as peptidoglycans, which are primarily responsible for the mechanical strength of the wall. In some prokaryotes, the wall consists of two layers, an inner peptidoglycan layer and an outer layer containing lipoproteins and lipopolysaccharides. Bacterial cell walls that lack the lipoprotein and lipopolysaccharide layer combine firmly with such dyes as gentian violet, and those in which it is present do not. Those that combine with the dyes are known as gram-positive, whereas the others are gram-negative, after Hans Christian Gram, the Danish microbiologist who discovered this distinction. Gram staining is widely used as a basis for classifying bacteria, since it reflects a fundamental difference in the architecture of the cell wall. It is thought that all gram-positive bacteria constitute a distinct phylogenetic lineage.

The architecture of the cell wall, in turn, affects various other characteristics of the bacteria, such as their patterns of susceptibility to antibiotics (see Figure 8–19, page 181). Gram-positive bacteria, such as *Staphylococcus,* are more susceptible to most antibiotics than are gram-negative bacteria, such as *E. coli.* They are also more susceptible to lysozyme (page 170), an enzyme found in nasal secretions, saliva, and other body fluids, which digests the cell walls of bacteria.

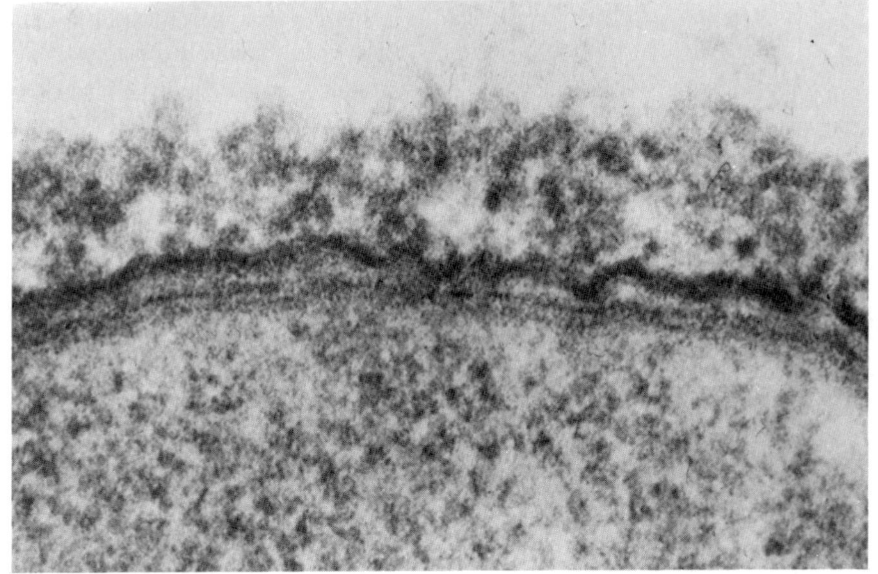

20–5
Most forms of meningitis, an infection of the membranes covering the brain and spinal cord, are caused by the bacterium Neisseria meningitidis. *In this electron micrograph of* N. meningitidis, *you can distinguish the cell membrane, the cell wall, and a fuzzy-appearing polysaccharide capsule outside the cell wall. Different strains of* N. meningitidis *are characterized by polysaccharide capsules of different composition. Scientists at the Walter Reed Army Institute of Research and Rockefeller University have used purified capsule polysaccharides to develop effective vaccines against two of the three major strains of this bacterium.*

20-6
(a) *A bacterial flagellum showing the basal end.* (b) *Diagram of a flagellum from* E. coli. *The basal body, which serves to anchor the flagellum to the cell wall and the cell membrane, consists of two pairs of rings surrounding a rod. The S and M rings are integrated into the cell membrane, the P ring is in the peptidoglycan layer of the cell wall, and the L ring is in the lipoprotein-lipopolysaccharide layer. The filament is made up of several protein chains that form a helix with a hollow core.*

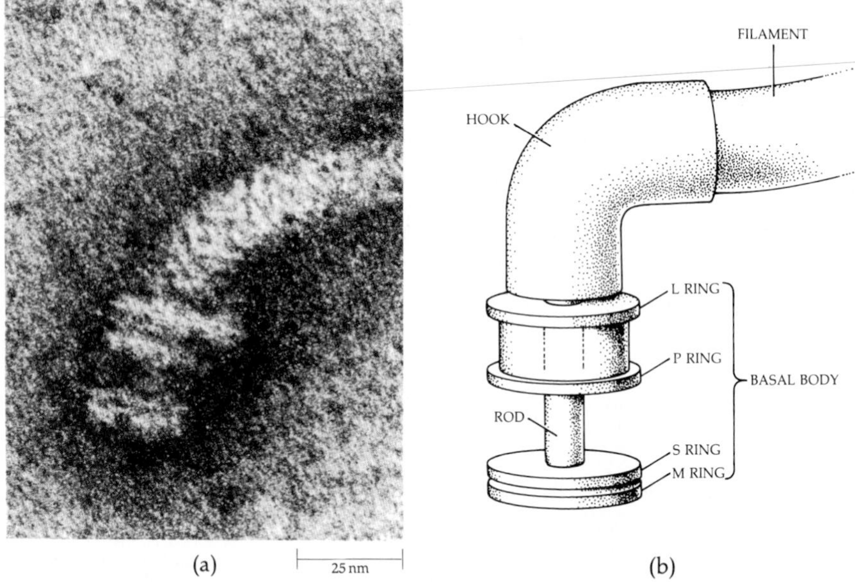

In certain bacteria, a gluey polysaccharide capsule, which is secreted by the bacterium, is present outside the cell wall (Figure 20–5). The function of the capsule is not entirely clear, but its presence is associated with pathogenic activity in certain organisms. For example, as shown in Figure 14–3, the encapsulated form of *Diplococcus pneumoniae* is virulent, whereas the nonencapsulated form is generally nonvirulent. It appears that the capsule may interfere with phagocytosis by host white blood cells.

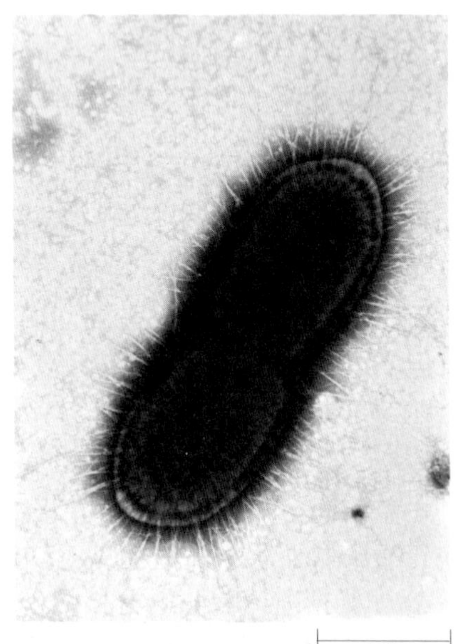

20-7
Pair of cells of Proteus mirabilis *stained to show the numerous pili.*

Flagella and Pili

Some types of bacteria have long, slender extensions, known as flagella and pili. Bacterial flagella, as we noted in Chapter 5, are made up of monomers of a protein known as flagellin, which are assembled in chains and wound around a hollow central core. The flagella of different species differ slightly in diameter (12 to 18 nanometers), probably due to slight differences in the composition of their flagellin.

Electron microscopy has recently revealed that bacterial flagella are anchored into the cell wall and membrane by a complicated assembly (Figure 20–6). The filament (the helix of flagellin monomers with its hollow core) terminates in a hook made up of a different protein. The hook is inserted into a basal body that consists of a rod and, in gram-negative bacteria, two pairs of rings, one pair of which is embedded in the cell wall and one pair in the cell membrane. In gram-positive bacteria, only the inner pair of rings is present.

Pili (singular, pilus) are assembled from protein monomers in much the same way that the filaments of flagella are. (You will not be surprised to learn that the protein is called pilin.) They are rigid, cylindrical rods that extend out from the cell, sometimes to a considerable distance. They are shorter and thinner than flagella and are often present in large numbers (hundreds on a single cell). They serve to attach bacteria to a food source, to the surface of a liquid (where oxygen is present), or, in the case of conjugating bacteria, to one another.

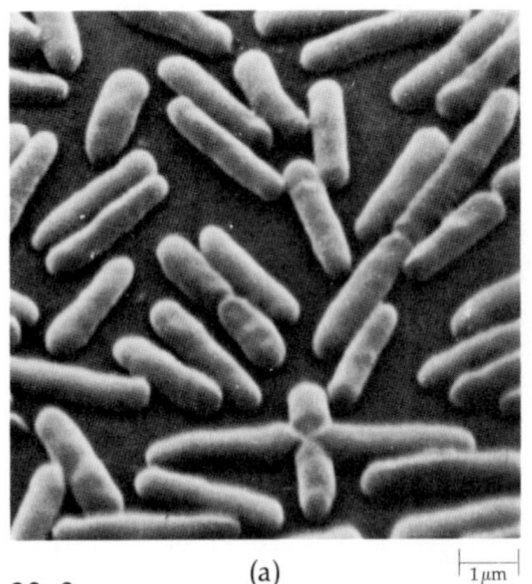

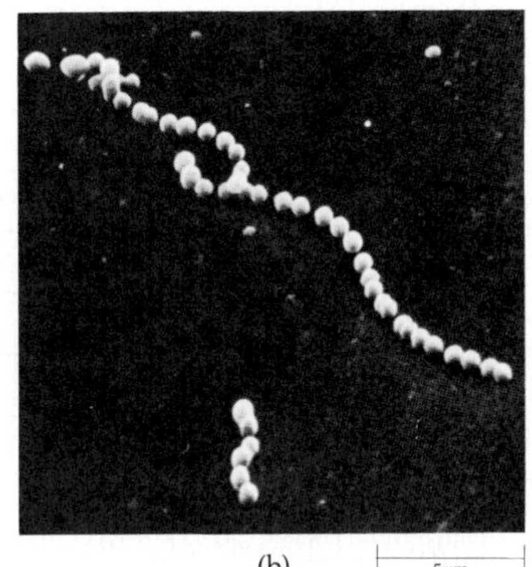

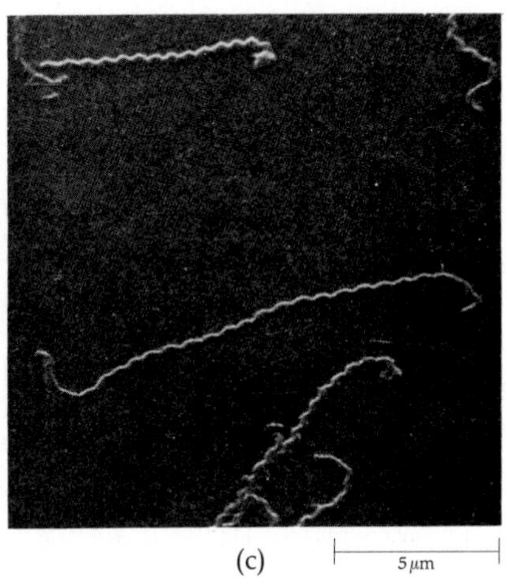

20–8
Three of the four major form-groups of bacteria: (a) *bacilli,* (b) *cocci, and* (c) *spirilla. The rod-shaped bacteria (bacilli) include those microorganisms that cause lockjaw* (Clostridium tetani), *diphtheria* (Corynebacterium diphtheriae), *and tuberculosis* (Mycobacterium tuberculosis), *as well as the familiar* E. coli. *Among the cocci are* Diplococcus pneumoniae, *the cause of bacterial pneumonia;* Streptococcus lactis, *which is used in the commercial production of cheese; and* Nitrosococcus, *soil bacteria that oxidize ammonia to nitrates. The spirilla, which are less common, are helically coiled bacteria. Cell shape is a relatively constant feature in most species of bacteria.*

DIVERSITY OF FORM

The oldest method of identifying microorganisms is by their physical appearance. The shape of individual prokaryotes is a result, as we noted previously, of their cell wall. In addition, different types of bacteria have characteristic patterns of growth, producing colonies that also have a distinctive shape, and many forms have sheaths or other coatings outside their cell walls. Bacteria exhibit considerable diversity of form, but many of the most familiar species fall into one of four form-groups (Figure 20–8). Straight, rod-shaped forms like *E. coli* are known as *bacilli*; spherical ones are called *cocci*; long, spiral rods are called *spirilla*; and short, curved rods, probably incomplete spirals, are called *vibrios*. Cocci may stick together in pairs after division (diplococci), they may occur in clusters (staphylococci), or they may form chains (streptococci). One bacterium that causes pneumonia is a diplococcus, while the staphylococci are responsible for many serious infections characterized by boils or abscesses.

Rod-shaped bacilli usually separate after cell division. When they do remain together, they spread out end to end in filaments, since they always divide in the same plane (transversely). In some genera, these filaments are funguslike in appearance and the combining form *myco-* (from the Greek word for "fungus") is part of the generic name. *Mycobacterium tuberculosis,* for example, the cause of tuberculosis, is a rod-shaped bacillus that forms a filamentous, funguslike growth.

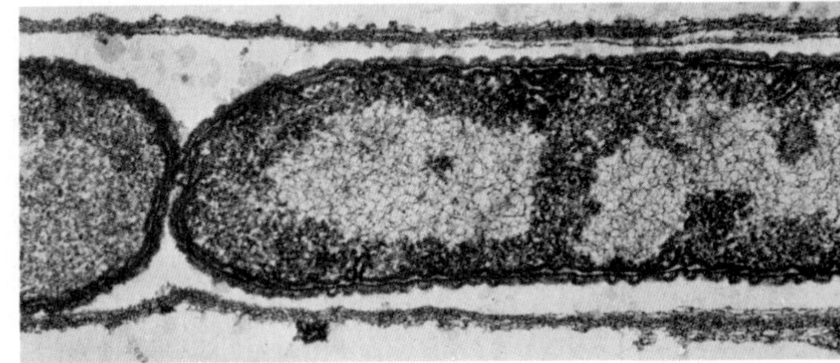

20–9
Sphaerotilus natans, *a filamentous bacterium enclosed in a sheath. The sheath is encrusted with particles of iron oxide, formed by the bacterium from soluble iron compounds. Filaments of* Sphaerotilus *are responsible for the brownish scum often seen on the surface of polluted streams.*

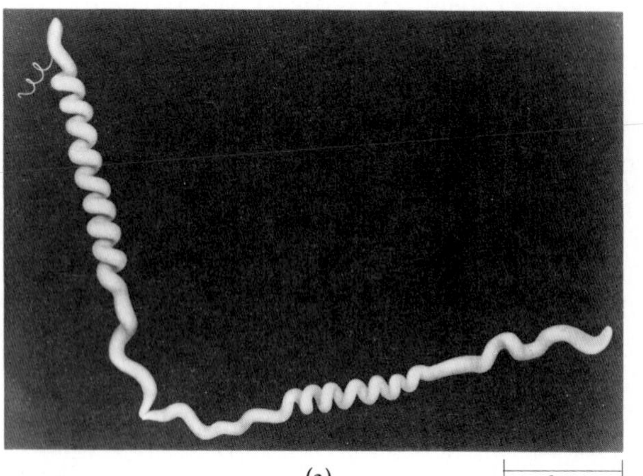

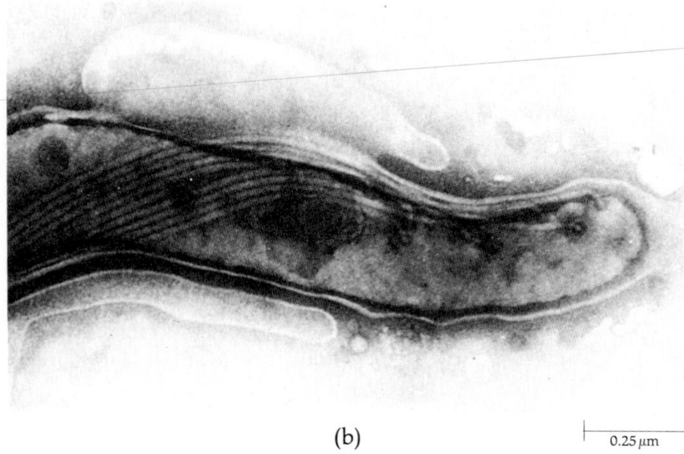

20–10

(a) *The spirochetes range in size up to 500 micrometers long, which is an enormous size for prokaryotes.* Treponema pallidum, *two of which are shown here, is the causative agent of syphilis.* (b) *The end of a spirochetal cell that has been stained to show the insertion points of two fibrils of the axial filament.* (c) *Cross section of a spirochete showing the fibrils of the axial filament between the cell membrane and the cell wall.*

Spirochetes are among the easiest microorganisms to identify (Figure 20–10). They are very long (5 to 500 micrometers) and slender (about 0.5 micrometer in diameter) and have an unusual structure, known as an axial filament, made up of two sets of fibrils attached at each end of the cell. The fibrils are identical to flagella in structure and so are recognized as modified flagella, or endoflagella. They are wrapped around the cell between the cell membrane and the delicate wall, with the fibrils of each set overlapping at the middle of the cell. All spirochetes are thought to belong to the same phylogenetic lineage.

Rickettsiae, another type of prokaryote distinguishable by their form, are shown in Figure 20–11.

TYPES OF MOTILITY

Prokaryotes move in a variety of ways, all of which are more or less mysterious. Flagella beat with a rotary movement. The flagella are so fine and the beat so fast, the motion of the flagellum itself cannot be seen. (The flagella of *Spirillum serpens*, for instance, have been clocked at 2,400 rpm.) Ingenious methods have been devised by which the cells can be tethered in place by the flagella. As a result, the cells rotate instead of the flagella, and the rotational movement can be observed. The basal end of the flagellum, with its complex structure, apparently acts as a motor, running on chemiosmotic power.

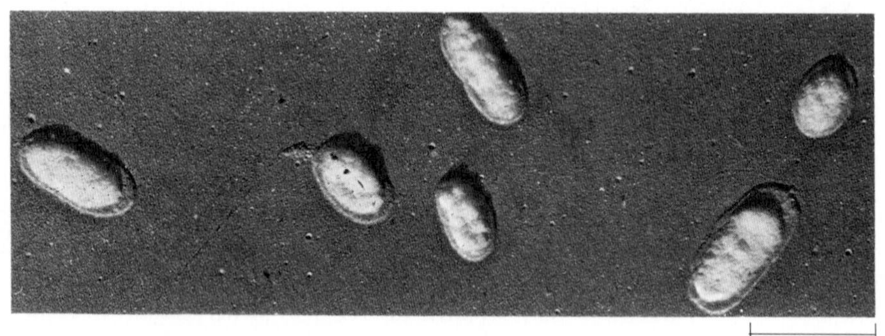

20–11

Rickettsiae are the smallest known cells. Typhus is caused by Rickettsia prowazekii, *shown here. It is transmitted by human body lice, and under crowded conditions, large numbers of people can be infected in a very short time. More human lives have been taken by rickettsial diseases than by any other infection except malaria. At the siege of Granada in 1489, 17,000 Spanish soldiers were killed by typhus, but only 3,000 in combat. In the Thirty Years War, the Napoleonic campaigns, and the Serbian Campaign during World War I, typhus was also the decisive factor.*

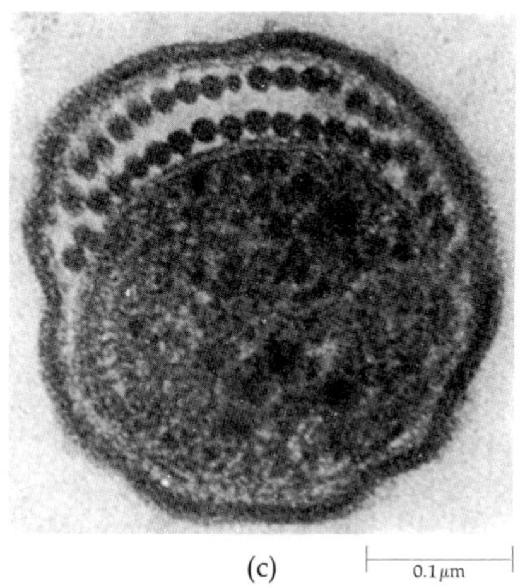

(c)

Typically, flagellated bacteria such as *E. coli* move rapidly in gently curved lines, or "runs," each lasting approximately a second. A series of runs is interrupted frequently by periods of tumbling, lasting about a tenth of a second. After each tumble, the bacterium starts off in a new direction on its next run. The frequency of tumbling controls the length of the runs and hence the overall direction of movement. Whether an *E. coli* cell moves in a run or tumbles is determined by the direction in which its flagella rotate. When the flagella, which originate at sites all over the cell surface, are rotating in a counterclockwise direction, they work together in an organized bundle at the rear of the organism. This counterclockwise rotation produces sustained swimming in one direction. But when the flagella reverse direction and rotate in a clockwise manner, the bundle falls apart, the flagella rotate independently, and the cell tumbles.

Another characteristic form of movement for some prokaryotes is gliding (Figure 20–12). Gliding requires attachment to a solid surface and the secretion of some sort of slime or mucus along which the cells slide, but so far the mechanism remains unknown.

Spirochetes move by a corkscrew motion caused, apparently, by rotation of the axial filament.

REPRODUCTION AND RESTING FORMS

Most prokaryotes reproduce by simple cell division (page 140), also called <u>*binary fission*</u>. In some forms, reproduction is by budding or by the fragmentation of filaments of cells. As they multiply, these prokaryotes, barring mutations, produce clones of genetically identical cells. Mutations do occur, however; it has been estimated that in a culture of *E. coli* that has divided 30 times, about 1.5 percent of the cells are mutants. Mutations, combined with the rapid generation time of prokaryotes, are responsible for their extraordinary adaptability. Further adaptability is provided by the genetic recombinations that take place as a result of conjugation, transformation, transduction, and exchanges of plasmids. It is not known how common such genetic recombinations are in nature or whether or not they occur in all types of prokaryotes.

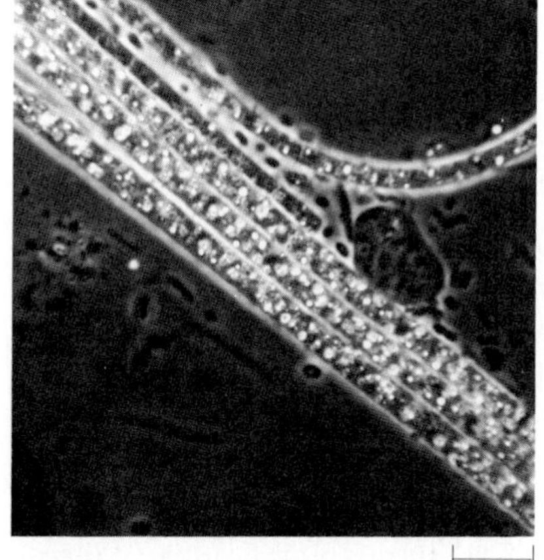

20–12
Beggiatoa, *gliding bacteria, are relatively large cells that form filaments and move in a "gliding" fashion similar to that of many cyanobacteria.* Beggiatoa *is a chemoautotroph. The conspicious granules in the cells are sulfur, produced by the oxidation of hydrogen sulfide—a process utilized by the bacteria for the production of energy.*

Many prokaryotes have the capacity to form spores, which are dormant, resting cells. Again, the process has been studied most extensively in the rod-shaped eubacteria. It occurs characteristically when a population of cells, growing very rapidly, has begun to use up its food supply. Each cell, at the beginning of sporulation (spore formation), contains two duplicate chromosomes. A cell membrane grows around one of the chromosomes, separating it from the rest of the cell, which then engulfs the newly formed cell. Thus the spore-to-be is now surrounded by two membranes, its own and that of the larger cell (these events are shown in Figure 20–13 on page 401). A spore coat, consisting of two layers, then forms around this smaller cell. The inner layer (the cortex) contains a peptidoglycan that is completely different from that present in the bacterial cell wall; the outer layer consists of proteins that are composed largely of hydrophobic amino acids. The mature spore is released from the cell in this protected state, and it remains dormant until appropriate events trigger its germination. Germination takes place rapidly, with the uptake of water, the dissolution of the spore coat, and the formation of a new cell wall. Genetic studies of the spore-making *Bacillus subtilis* indicate that some 50 genes, clustered in about five segments of the chromosome, are involved in spore production.

SENSORY RESPONSES IN BACTERIA

Even organisms as seemingly simple as prokaryotes respond to stimuli. Some photosynthetic bacteria, for example, respond to light, moving toward it. One such species is Halobacterium halobium *(see page 224), the salt-loving bacterium with the pigment rhodopsin in its cell membrane. It is tempting to speculate that rhodopsin is the light detector for* H. halobium, *just as it is for the light receptors of the vertebrate eye.*

Escherichia coli *and other flagellated bacteria show chemotactic responses, moving in "runs" toward chemical attractants such as sugars and tumbling away from noxious, repellent substances. As we have seen, the pattern of movement is determined by the direction of rotation of the flagella, which is controlled, in turn, by information the bacterium acquires from specialized chemoreceptors. Flagellated bacteria have chemoreceptors not only for nutritious substances but also for molecules that repel them—they do not simply sense the repellents by experiencing some harm they cause. Julius Adler and his colleagues at the University of Wisconsin have calculated that* E. coli *has some 20 different chemoreceptors, all of which are proteins. Twelve of these receptors are thought to detect attractants, and eight to detect repellents. The chemoreceptor molecules to which the sugars galactose, maltose, and ribose bind are located in the space between the cell membrane and the cell wall; other receptors are closely associated with the cell membrane. It is not known how information is transmitted from these protein molecules to the flagella, nor how information from the various chemoreceptors is integrated to produce counterclockwise or clockwise rotation of the flagella. There is evidence, however, that the bacteria are able to retain information about concentration levels for perhaps as long as 60 seconds—a primitive form of memory that enables them to compare concentrations at points many cell lengths apart.*

In 1975, a bacterial species from a Massachusetts swamp was shown to be magnetotactic—that is, the bacteria are able to orient themselves in a magnetic field and swim in a preferred direction. All of the bacteria observed were found to swim toward magnetic north. Detailed studies revealed that each bacterium accumulates about 20 opaque, roughly cubic granules of magnetite (Fe_3O_4), the iron ore from which magnets are made. The granules are arranged in a linear fashion, forming a single magnet with a north-seeking pole and a south-seeking pole. The interaction of the earth's magnetic field and the bacterium's internal magnet orients it toward the north; its flagella, which are at the end of the bacterium opposite the north-seeking pole of its magnet, then propel it in that direction. Because the earth's magnetic field has a vertical component as well as a horizontal component, this interaction also orients the bacterium downward—toward sediments rich in decaying organic matter.

After the discovery of these magnetotactic bacteria, scientists predicted that if such bacteria existed in the Southern Hemisphere, their polarity would be reversed. In 1980, magnetotactic bacteria were found in New Zealand and Tasmania, and, as predicted, their internal magnets are oriented so that the south-seeking poles are opposite the end at which the flagella are attached. Magnetotactic bacteria have also been isolated from sediments near the earth's magnetic equator. These bacteria are equally divided between north-seeking and south-seeking forms. There is virtually no vertical component to the earth's magnetism at the equator, so neither form has an advantage over the other. Their internal magnets, however, do prevent them from swimming upward, away from rich sources of food.

These examples illustrate a series of remarkably complex responses for such small and simply constructed organisms. They foreshadow the kinds of increasingly complex responses found among the eukaryotes.

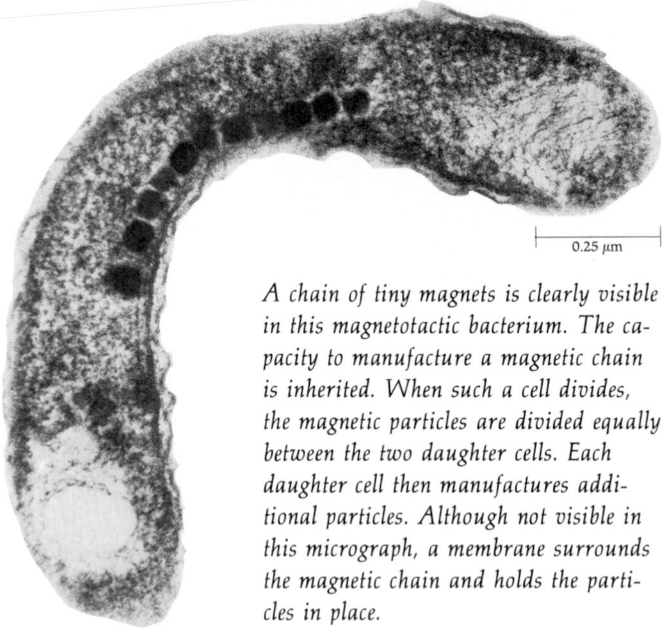

0.25 μm

A chain of tiny magnets is clearly visible in this magnetotactic bacterium. The capacity to manufacture a magnetic chain is inherited. When such a cell divides, the magnetic particles are divided equally between the two daughter cells. Each daughter cell then manufactures additional particles. Although not visible in this micrograph, a membrane surrounds the magnetic chain and holds the particles in place.

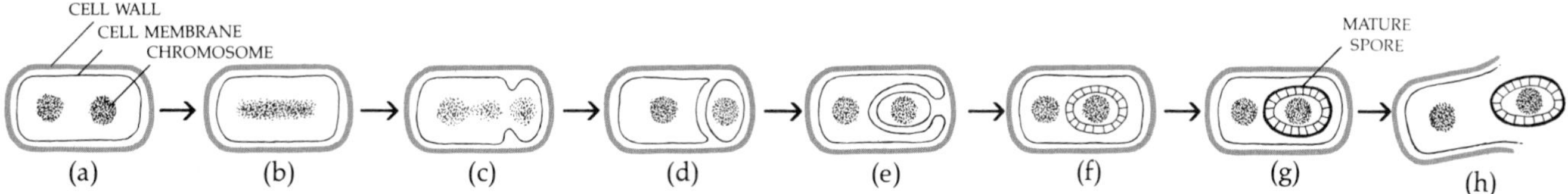

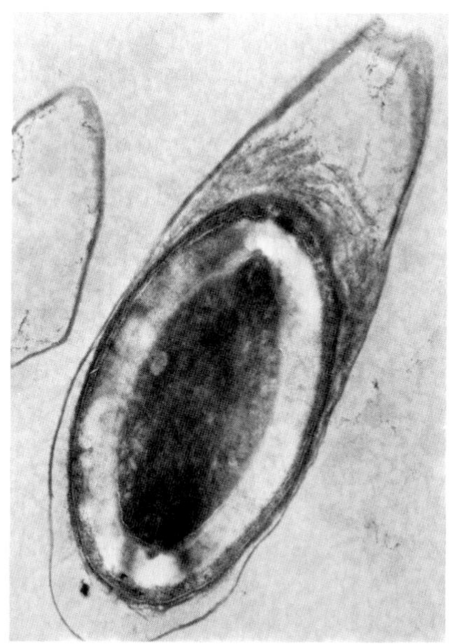

20–13
Endospore formation in Bacillus cereus. (a, b) *The two chromosomes within the vegetative cell have condensed into a rod-shaped form.* (c) *The transverse wall begins to form,* (d) *cutting off the spore material from the vegetative cell.* (e, f) *The vegetative cell grows around the spore, and the spore coat is formed.* (g, h) *The spore matures and is released from the cell. The micrograph shows an endospore of* Clostridium.

The myxobacteria, a type of gliding bacteria, form fruiting bodies, which are brightly colored collections of spores and slime large enough to be seen by the unaided eye (Figure 20–14).

Spore formation greatly increases the capacity of prokaryotic cells to survive. The spores of *Clostridium botulinum,* for instance, the bacterium that causes botulism, are not destroyed by boiling for several hours.

PROKARYOTIC NUTRITION

Heterotrophs

Most prokaryotes are heterotrophs; most types feed on dead organic matter. Bacteria and other microorganisms are responsible for the decay and recycling of organic material in the soil. Typically, different groups of bacteria play different, specific roles—such as the digestion of cellulose, starches, or other polysaccharides, or the hydrolysis of specific peptide bonds, or the breakdown of amino acids. Because of their high degree of nutritional specialization, prokaryotes are able to live in large numbers in the same small area with little competition and, indeed, with mutual assistance, as the activities of one group make food molecules available to another group. These combined activities release the nutrients and make them available to plants and, through plants, to animals. Thus, the bacteria are an essential part of ecological systems.

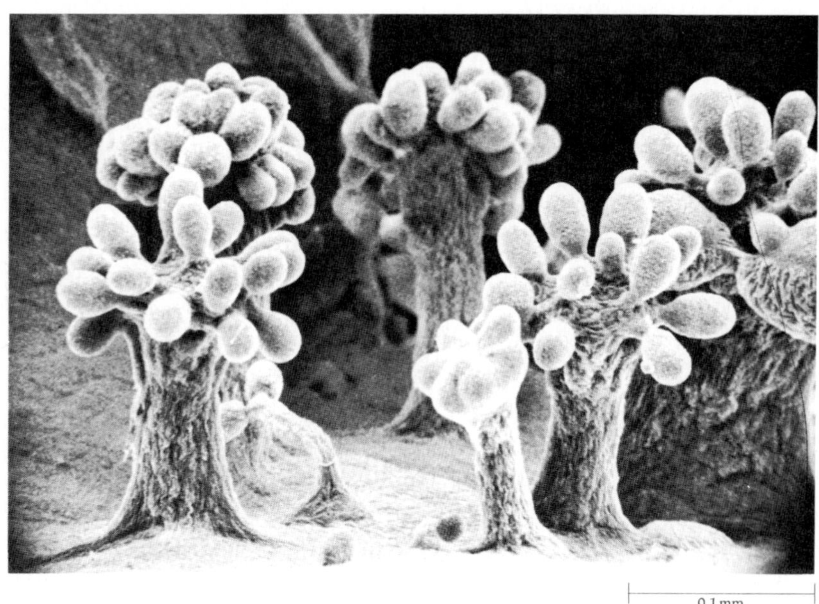

20–14
Scanning electron micrograph of the fruiting bodies of Chondromyces crocatus, *a myxobacterium. Each fruiting body may contain as many as 1 million cells.*

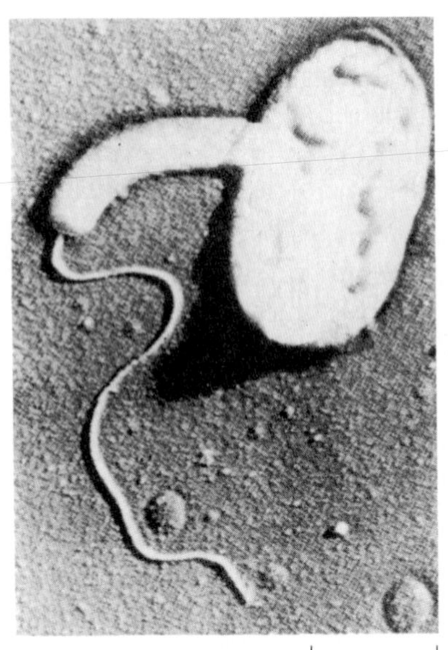

20–15
A cell of Bdellovibrio bacteriovirus *(left) is attacking a cell of* Erwinia amylovora *(upper right), the bacterium that causes fire blight of pears and apples.* Bdellovibrio bacteriovirus, *which is abundant in soil and sewage, moves by means of a single posterior flagellum.*

Some heterotrophic bacteria live in close association with other organisms. Some of these are parasites that break down organic material in the bodies of living organisms. The disease-causing (pathogenic) bacteria belong to this group, as do a number of other nonpathogenic forms. Some of these bacteria have little effect on their hosts, and some are actually beneficial. Cows and other ruminants can digest cellulose only because their stomachs contain bacteria and certain protozoans. Our own intestines contain a number of types of generally harmless bacteria (including *E. coli*). Some supply vitamin K, which is necessary for blood clotting. Others prevent us from developing serious infections. When the normal bacterial inhabitants of the human intestinal tract are destroyed—as can happen, for example, following prolonged antibiotic therapy—our tissues are much more vulnerable to disease-causing microorganisms. One group of very small bacteria, members of the genus *Bdellovibrio* (from *bdello,* the Greek word for "leech"), are parasites on other bacteria (Figure 20–15). They make a hole in the host cell wall and multiply between the wall and the cell membrane, digesting the host cell as they multiply.

Chemoautotrophs

Chemoautotrophic prokaryotes obtain their energy from the oxidation of inorganic compounds. Only prokaryotes are able to use inorganic compounds as an energy source.

An unusual group of chemoautotrophs are the methanogens (page 392). Although their existence was long known, they were not studied extensively until quite recently because they are obligate anaerobes—that is, they are poisoned by oxygen—and hence are difficult to isolate and impossible to grow under ordinary laboratory conditions. The methanogens are the final participants in decomposing processes involving organic matter in anaerobic environments, including marshes, lake sediments, and the digestive tracts of animals. They convert CO_2 and H_2 formed by the fermentation processes of other anaerobes to methane gas (CH_4). "Marsh gas," which gives some swamps their distinctive odor, is the methane produced by these organisms.

Certain chemoautotrophic bacteria are essential components of the nitrogen cycle, the process by which nitrogen compounds are cycled and recycled through ecosystems. One group oxidizes ammonia or ammonium (derived from the breakdown of organic materials, the activities of nitrogen-fixing prokaryotes, or, to a minor extent, from lightning or volcanic activities). The products of this reaction are nitrite (NO_2^-) and energy. Another group oxidizes nitrites, producing nitrate (NO_3^-) and energy. Nitrate is the form in which nitrogen moves from the soil into the roots of plants.

Sulfur is also required by plants for amino acid synthesis. Like nitrogen, it is converted to the form in which it is taken up by plant roots by the activities of chemoautotrophic bacteria that oxidize elemental sulfur to sulfate:

$$2S + 2H_2O + 3O_2 \longrightarrow 2H_2SO_4$$

Other sulfur bacteria, such as *Thiothrix* and *Beggiatoa,* obtain energy by oxidizing hydrogen sulfide.

These important biogeochemical cycles and the roles of microorganisms in them will be discussed further in Chapter 53.

20–16
A purple nonsulfur bacterium, Rhodospirillum rubrum. *The purple bacteria have bacteriochlorophyll rather than chlorophyll* a *as a photosynthetic pigment. They do not lyse water and do not release oxygen.*

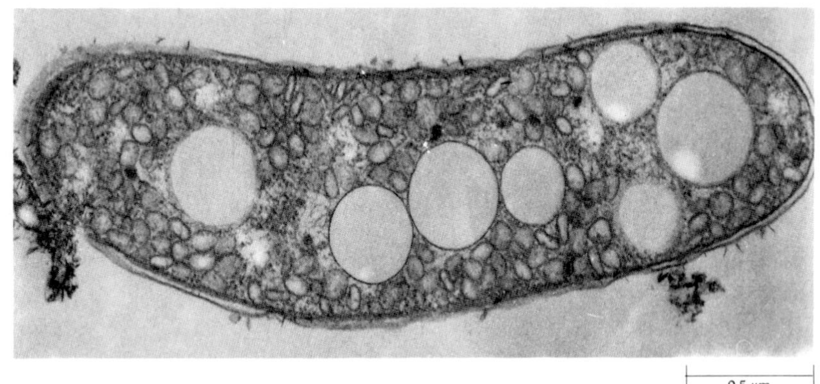

Photosynthetic Prokaryotes

Eubacteria

Among the eubacteria are three photosynthetic types, which are thought to represent two distinct phylogenetic lineages. One lineage includes the green sulfur bacteria, while the other includes both the purple sulfur bacteria and the purple nonsulfur bacteria. A number of nonphotosynthetic forms, including the familiar *Escherichia,* are believed to belong to the same lineage as the purple bacteria.

The distinctive colors of the photosynthetic bacteria result, of course, from the pigments they contain. The chlorophyll found in the green sulfur bacteria, chlorobium chlorophyll, is chemically similar to chlorophyll *a* (page 215). The chlorophyll found in the two groups of purple bacteria is bacteriochlorophyll, which differs chemically in several details from chlorophyll *a* and is a pale blue-gray. The colors of the purple bacteria are due to the presence of several different yellow and red carotenoids, which function as accessory pigments. In the purple nonsulfur bacteria, the colors may actually range from purple to red or brown.

In the photosynthetic sulfur bacteria, as we noted in Chapter 10, the sulfur compounds are the electron donors, playing the same role in bacterial photosynthesis that water does in photosynthesis in eukaryotes.

$$CO_2 + 2H_2S \xrightarrow{\text{light}} (CH_2O) + H_2O + 2S$$

In the photosynthetic nonsulfur bacteria, other compounds, including alcohols, fatty acids, and a variety of other organic substances, serve as electron donors for the photosynthetic reactions.

Photosynthesis by eubacteria is carried out anaerobically, and it never results in the production of molecular oxygen. However, in all species except the extreme halophiles (which, as we have seen, are members of a completely different phylogenetic lineage), carbon is fixed by means of the Calvin cycle.

Cyanobacteria

As we noted at the beginning of the chapter, the cyanobacteria constitute one of the two divisions of the kingdom Monera. Unlike other photosynthetic prokaryotes, but like all photosynthetic eukaryotes, the cyanobacteria contain chlorophyll *a* and lyse water during photosynthesis, producing molecular oxygen. They have several kinds of accessory pigments, including xanthophyll, which is a yellow

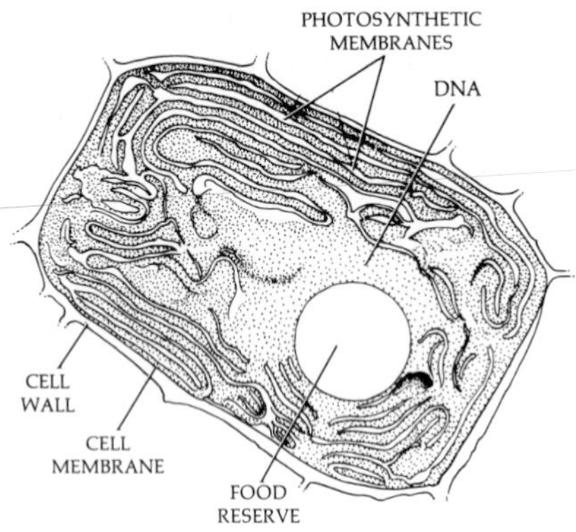

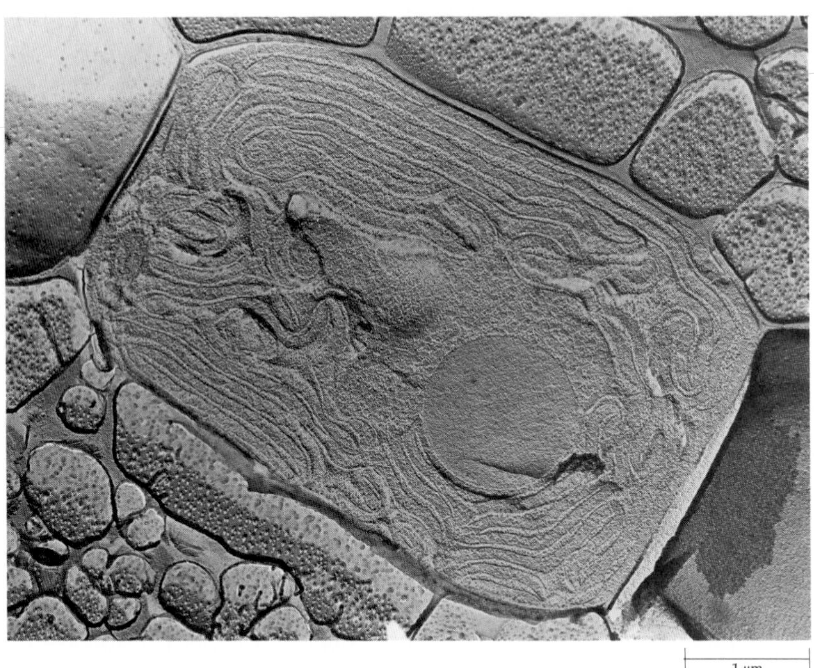

20–17
Electron micrograph of the cyanobacterium Anabaena cylindrica. *Cyanobacteria have no chloroplasts and no membrane-bound nucleus, such as are found in eukaryotic algae and plant cells. Photosynthesis takes place in chlorophyll-containing membranes within the cell, and the chromosome is a single molecule of DNA. The three-dimensional quality of this electron micrograph is due to freeze-fracturing (page 133).*

carotenoid, and several other carotenoids. (Carotenoids can also be found in photosynthetic eukaryotes and in some other prokaryotes.) The cells of cyanobacteria also contain one or two pigments known as phycobilins: phycocyanin, a blue pigment, which is always present, and phycoerythrin, a red one, which is often present. Chlorophyll and the accessory pigments are not enclosed in chloroplasts, as they are in plant cells, but are part of a membrane system distributed in the peripheral portion of the cell (Figure 20–17).

Cells of the cyanobacteria, as with many other prokaryotes, have an outer polysaccharide sheath, or coating. In some species, particularly those that spread up onto the land, this outer sheath is deeply pigmented; in most species, however, it is not pigmented, and the colors of the cells result from the carotenoids and phycobilins that they contain. Different species of cyanobacteria are golden yellow, brown, red, emerald green, blue, violet, or blue-black. The Red Sea was so named because of the dense concentrations, or "blooms," of red-pigmented cyanobacteria that float on its surface.

Some species of cyanobacteria are capable of nitrogen fixation. The photosynthetic, nitrogen-fixing cyanobacteria are among the most self-sufficient of organisms. They have the simplest nutritional requirements of any living thing, needing only nitrogen and carbon dioxide, which are always present in the atmosphere, a few minerals, and water.

The ecological importance of the cyanobacteria appears to be less than that of the nitrogen-fixing bacteria, at least for agriculture. However, in Southeast Asia, rice can be grown on the same land for years without addition of fertilizers because of the rich growth of nitrogen-fixing cyanobacteria in the rice paddies.

Because of their nutritional independence, the cyanobacteria are able to colonize bare areas of rock and soil. A dramatic example of such colonization was seen on the island of Krakatoa in Indonesia, which was denuded of all visible plant life by its cataclysmic volcanic explosion of 1883. Filamentous cyanobacteria were the first living things to appear on the pumice and volcanic ash; within a few years they had formed a dark-green gelatinous growth. The layer of cyanobacteria eventually became thick enough to provide a substrate for the growth of plants. It is very probable that the cyanobacteria were similarly the first colonizers of land in the course of biological evolution.

VIRUSES: A CASE APART

Viruses do not fit easily into any of the traditional kingdoms of living organisms, and the problem of categorizing them is made even more difficult by the fact that there is doubt about whether or not they should be considered living.

Structure of Viruses

In size, viruses range from about 17 nanometers (a hemoglobin molecule is 6.4 nanometers in diameter) to about 300 nanometers, larger than small bacteria. The larger ones are at the limits of resolution of the light microscope.

Viruses are made up of nucleic acid, either DNA or RNA, enclosed in a protein coat. The protein coat determines the specificity of the virus; a cell can be infected by a virus only if that type of cell has a receptor site for the viral protein. Thus cold viruses infect cells in the mucous membranes of the respiratory tract; measles and chicken pox infect skin cells; and polio infects the upper respiratory tract, the intestinal lining, and sometimes nerve cells. Even bacteria, as we noted in Section 3, have their own set of specific viruses, the bacteriophages.

In some virus infections, the protein coat is left outside the cell (see Figure 14–7, page 283); in others, the intact virus enters the cell, but once inside, the protein is destroyed by enzymes. The DNA of a DNA-containing virus serves as a template for more viral DNA and also for messenger RNA, which codes for viral enzymes, viral coat protein, and probably, at least in some cases, repressors and other regulatory chemicals. The virus uses the equipment of the host cell, including ribosomes, transfer RNA molecules, amino acids, and nucleotides. Many viruses use host enzymes as well as those coded for by their own nucleic acids, and some break up host DNA and recycle the nucleotides as viral DNA. In the case of the RNA viruses, the viral RNA not only serves as a template for more RNA but also acts directly as messenger RNA. As we noted in Chapter 16, viral RNA can also serve as a template for viral DNA. This phenomenon of reverse transcription is observed almost exclusively with cancer-causing viruses.

Virus particles are assembled within the host cell. They then leave the cell, often lysing the membrane. Some viruses, such as influenza virus, bud off from the host cell membrane and, in so doing, become wrapped in fragments of it. Each new viral particle is capable of setting up a new infectious cycle in an uninfected cell.

Origin of Viruses

Are viruses very simple forms of life? Can they give us clues to the nature of a precellular living system of which no fossil is likely to remain? As information increases on the nature of viruses and, in particular, of their relationships with host cells, this interpretation of viral origins seems less and less likely. Viruses are able to reproduce and to make their protein coats only because they are capable of commandeering the enzymes and other metabolic machinery of the host cell.

20–18

(a) *Adenovirus, one of the many viruses that cause colds in humans. This virus is an icosahedron. Each of its 20 sides is an equilateral triangle composed of identical protein subunits. Many viruses, and also Buckminster Fuller's geodesic domes, are constructed on this principle. There are 252 subunits in all. Within the icosahedron is a core of double-stranded DNA.* (b) *A model of the adenovirus, made up of 252 tennis balls.*

(c, d) *Electron micrograph and diagram of influenza virus. The virus is composed of a core of RNA surrounded by a lipoprotein envelope through which protrude stubby protein spikes. The influenza virus, for reasons that are not understood, mutates frequently. The changes in its nucleic acid alter its proteins and, hence, previously formed antibodies no longer "recognize" it. New strains of influenza virus are likely to arise more rapidly than new vaccines can be produced to combat them.*

(e, f) *Electron micrograph and diagram of a bacteriophage, showing the many different structural components of the protein coat. The DNA of the virus codes for all these structural proteins.*

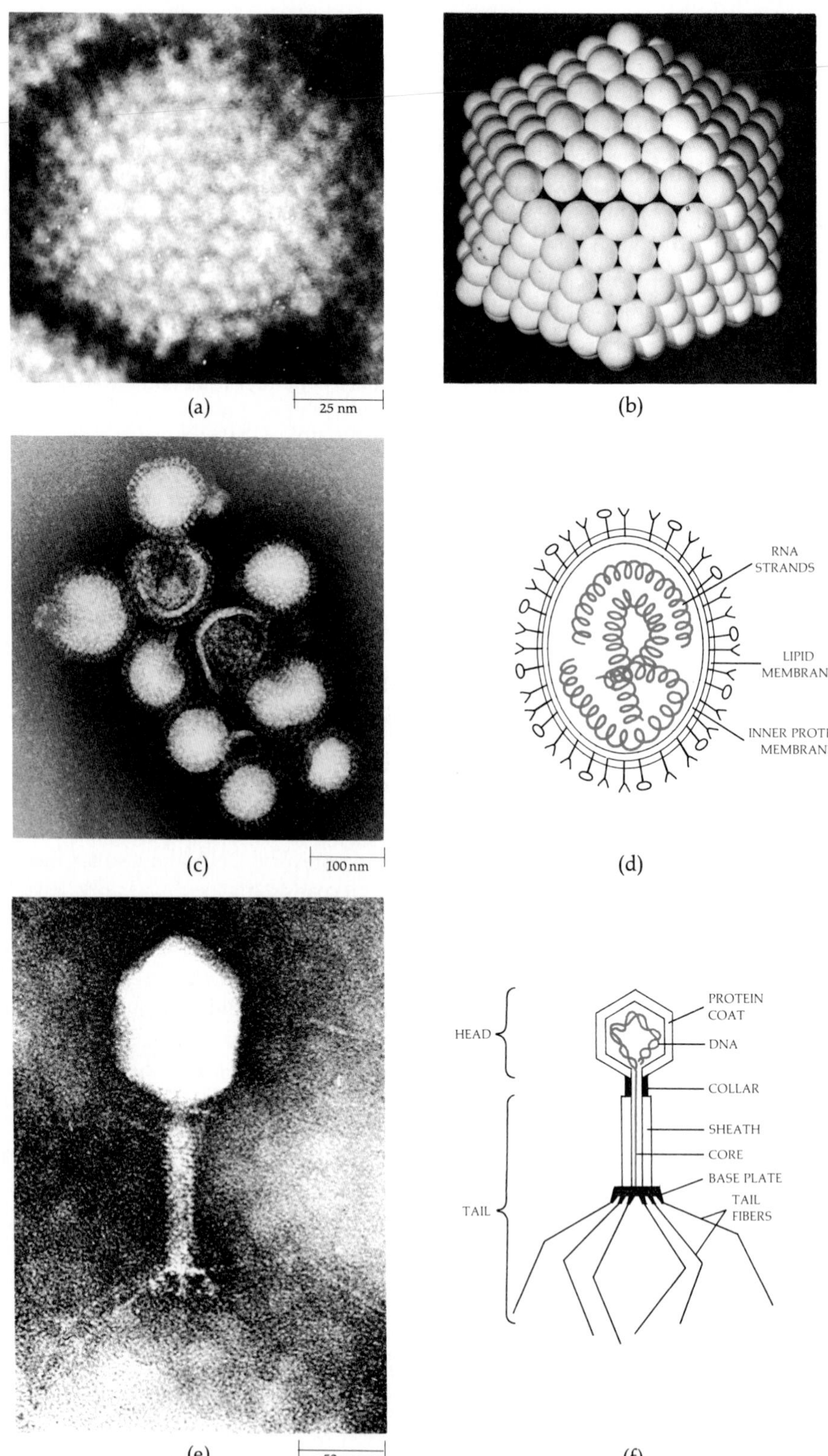

Without this machinery, they are as inert as any other macromolecule, lifeless by many criteria. It seems more likely that viruses are cellular fragments that have set up a partially independent existence.

VIROIDS: THE ULTIMATE IN SIMPLICITY

In the past few years, the agents that cause certain infectious diseases of plants have gained notoriety for their extreme simplicity and small size. These disease agents, known as viroids, consist of small molecules of RNA that replicate in susceptible cells. Viroids, unlike viruses, have no protein coat.

The first viroid to be characterized, known as PSTV, is the causative agent of potato spindle-tuber disease; in this disease, the plant produces elongated, gnarled potatoes (tubers) that sometimes have deep crevices in the surface. PSTV can also infect tomato plants—close relatives of potato plants—producing stunted growth and twisted leaves. PSTV is a single-stranded RNA molecule containing 359 nucleotides; the molecule can be either linear or a closed circle. The linear form folds back on itself in a hairpin-shaped structure, whereas the circular form is flattened. In each case, complementary base pairs are joined by hydrogen bonds, resulting in a double-stranded RNA structure similar to that of DNA. Under the electron microscope, both forms of PSTV appear as rods about 50 nanometers long (Figure 20–19).

The way in which viroids replicate is an enigma, as is the means by which they cause disease. Viroids are found almost exclusively in the nuclei of infected cells, and experiments have indicated that the RNA of viroids does not function as messenger RNA. Unlike the DNA or RNA of viruses, it is not translated into enzymes that participate in its own replication. Viroids are apparently dependent on the host cell not only for the raw materials for replication but also for the necessary enzymes.

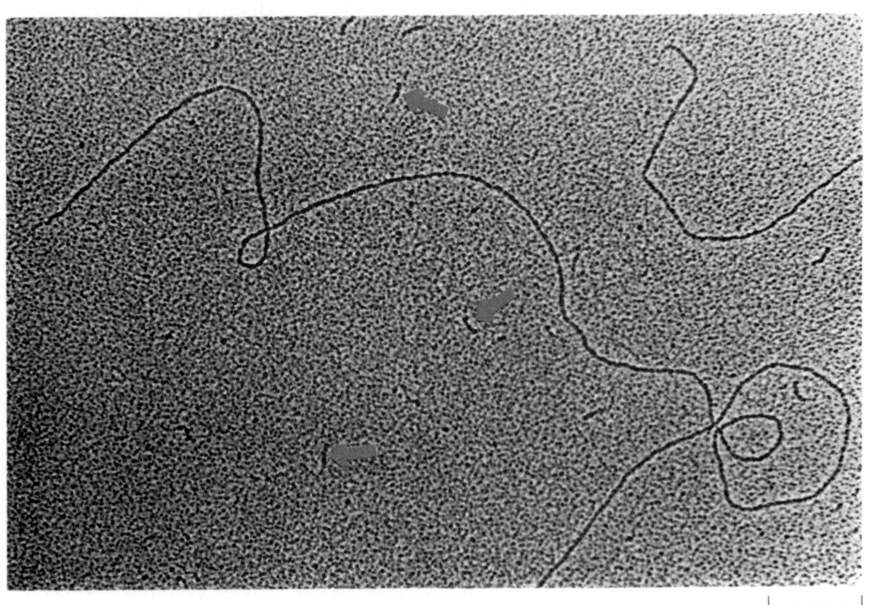

20–19
The short, rodlike structures indicated by arrows in this electron micrograph are potato spindle-tuber viroids. The much larger, looping structure is a portion of the DNA molecule from a bacteriophage, included to show the tremendous difference in size between a viroid and a virus.

Because of their location in the nucleus and their apparent inability to act as messenger RNAs, it has been suggested that viroids may cause their symptoms by interference with gene regulation in the infected host cells. Certain plant proteins, found in healthy cells, are present in significantly larger quantities in infected cells. Although nucleotide sequences complementary to PSTV have not been found in uninfected plants of host species, it has been postulated that PSTV may have originated from alterations of genes normally present in certain species of the potato family, which are its primary hosts.

Other viroids have been identified that cause diseases in citrus trees, cucumbers, and chrysanthemums. Although no viroids have been isolated from animal tissues so far, they are under suspicion as the cause of several infectious diseases of animals for which the pathogen has never been identified.

MICROORGANISMS AND HUMAN ECOLOGY

Symbiosis

Symbiosis ("living together") is a close and permanent association between organisms of different species. There are three types of symbiotic relationships. If the relationship is beneficial to both species, it is called *mutualism*. If one species benefits from the association while one is neither harmed nor benefited, it is called *commensalism*. If one species benefits and the other is harmed, the relationship is known as *parasitism*. Microorganisms and human beings are often symbionts. When the microorganisms are parasites, they are said to be pathogenic—disease-causing. The lines of demarcation between the different categories of symbiosis are not clear-cut. A bacterium such as *E. coli*, which lives in the lower intestine, may be commensal, depending on its host for food and shelter and neither helping nor harming. If it produces a needed vitamin or digestive enzyme, the relationship is then mutualistic. If it gets in the bloodstream and causes septicemia (blood poisoning), it is a parasite and a pathogen.

Evolutionary success is measured, as we noted, in terms of surviving progeny. A parasite that destroys its host is less likely to be successful by the evolutionary criterion than one that enjoys a long and comfortable relationship with its protector. Disease is likely to be the result of a sudden change in the parasite, in the host, or in their relationship. For instance, many persons harbor small numbers of *Mycobacterium tuberculosis* without any symptoms of disease; however, factors such as malnutrition, fatigue, or other diseases may weaken host defenses so that the signs of tuberculosis appear. Similarly, the herpes simplex viruses that cause "cold sores" or genital lesions, may remain latent for months or years at a time, with an outbreak occurring only in response to some change in the condition of the host.

How Microbes Cause Disease

The pathogenic effects of microbes are produced in a variety of ways. The viruses, as we have seen, enter particular types of cells and often destroy them. Bacteria may produce cell destruction also. Frequently, however, the effects we recognize as disease are caused not by the direct action of the pathogens but by toxins, or poisons, produced by them. For instance, diphtheria is caused by a bacillus, *Corynebacterium diphtheriae.* The organisms are inhaled and establish infection in the upper respiratory tract. They do not invade the bloodstream, but the powerful toxin they produce does. (This toxin is made only when the bacterium is harboring a particular prophage; see page 316.) The toxin is absorbed by the body cells.

Under the influence of the toxin, an enzyme involved in the assembly of activated amino acids into polypeptides becomes irreversibly bound to a fragment of NAD^+. The cells can no longer make proteins and so can no longer function.

Some diseases are the result of the body's reaction to the pathogen. In pneumonia caused by *Streptococcus pneumoniae,* the infection causes a tremendous outpouring of fluid and cells into the air sacs of the lungs, thus interfering with respiration. The symptoms caused by fungus infections of the skin similarly result from inflammatory responses.

A single disease agent can cause a variety of diseases. Skin infections of *Streptococcus pyogenes* cause the disease known as impetigo. Throat infections by the same bacteria are the familiar strep throat. Throat infections with strains of the bacteria that produce toxins (again as a result of a bacteriophage) are known as scarlet fever. Among the persons with untreated strep throat or scarlet fever, about 0.5 percent develop rheumatic fever, which is characterized by inflammatory changes in the joints, heart, and other tissues, apparently as a result of reactions involving the body's own immune system. Conversely, many agents may cause the same symptoms; the "common cold" can result from infection with any one of a large number of viruses.

20–20
Until the 1950s, poliomyelitis was one of the most feared of all childhood diseases. It was brought under control by immunization, initially with a killed-virus vaccine (Salk vaccine) that was subsequently replaced by a live-virus vaccine (Sabin vaccine), the one now in wide general use. The possible aftereffects of polio are depicted in this Egyptian hieroglyph, dated at approximately 1400 B.C.

Prevention and Control of Infectious Disease

Although microorganisms were first seen and depicted with remarkable accuracy by Antony van Leeuwenhoek in the late seventeenth century, they were not generally recognized as a cause of disease until 100 years ago. The relationship was made clear by the work of Louis Pasteur on disease in silkworms and by Robert Koch, who studied anthrax in sheep and tuberculosis in humans.

Recognition of microbes as disease agents opened the way to control measures, among the most important of which were the introduction of sterile procedures in hospitals. Among the agents spreading bacterial infection were physicians, surgeons, nurses, and hospital equipment. Childbed fever, for example, once a leading cause of sickness and death among young women, was transmitted almost exclusively by physicians who carried streptococci from patient to patient on their hands or unsterilized instruments.

Even more important than improvement in medical practices was the institution of public health measures, for example, the eradication of fleas, lice, mosquitoes, and other agents that carry disease; disposal of sewage and other wastes; protection of public water supplies; pasteurization of milk; and imposition of quarantines. Not long ago, for instance, infant mortality during the first years of life was often as high as 50 percent in some localities owing to infant diarrhea caused by contaminated milk and water; in some regions in the developing countries, it still remains quite high.

Many infectious diseases, both bacterial and viral, can be prevented by immunization (to be discussed in Chapter 39). Bacteria, in particular, are also susceptible to antimicrobial drugs, such as sulfa and penicillin, for which the battlefields of World War II were the proving grounds. In 1935, sulfanilamide, the first of the new "wonder drugs" for the control of bacterial infection, was discovered in Germany, and in 1940 the effects of penicillin (first noted by Alexander Fleming in 1929) were reported from England. Penicillin was the first-discovered antibiotic—by definition, a chemical that is produced by a living organism and is capable of inhibiting the growth of microorganisms. Many antibiotics are produced by bacteria, especially the actinomycetes; some are formed by fungi. Many, including penicillin, can now be synthesized in the laboratory. Ironically, with the advent of strains of bacteria resistant to these drugs and endemic in hospitals, the latter are once more becoming reservoirs of serious bacterial diseases.

Antimicrobial drugs made possible not only the treatment of battlefield wounds and of common infectious diseases but also the widespread and often life-saving use of major surgery as a treatment for cancer and other diseases. The tremendous decrease in deaths from infectious disease over the last decades is a chief cause of the present population explosion.

SUMMARY

The prokaryotes are currently classified in two divisions, Cyanophyta, the cyanobacteria (formerly known as the blue-green algae), and Schizophyta, the bacteria. The bacteria are further classified into 19 different groups, some of which reflect phylogenetic relationships and some of which are artificial. Several different major lineages of bacteria have been identified—photosynthetic green sulfur bacteria, photosynthetic purple bacteria and their nonphotosynthetic relatives, gram-positive bacteria, and spirochetes; there may well be others not yet identified. Recent studies have shown that a few genera of bacteria, including the methanogens, the extreme halophiles, and the thermoacidophiles, are so different from all other prokaryotes that they should perhaps be placed in a new, separate kingdom, Archaebacteria.

The prokaryotes are the only organisms with cells in which the DNA is not associated with histones and in which there is no membrane-bound nucleus or membrane-bound organelles. Prokaryotes are characteristically surrounded by a cell wall containing complex polymers known as peptidoglycans. Within the wall is a cell membrane that does not contain cholesterol and that often incorporates enzyme systems such as—in aerobic forms—the electron transport chain. Many species also have long, slender extensions, flagella and pili. The varied forms of prokaryotes are imparted by the cell walls, which may be rigid or flexible.

Prokaryotes move by means of their flagella, which have a complex structure and a rotary movement; by gliding, which involves the secretion of a mucus or slime; or not at all. Reproduction is usually by binary fission, but some forms reproduce by budding or fragmentation. Genetic variability is introduced by mutation and by genetic exchanges and recombinations. Many types of prokaryotes form tough, resistant spores.

Prokaryotes comprise both heterotrophs and autotrophs; the autotrophs include both chemoautotrophic and photosynthetic forms. Chemoautotrophs are found only among prokaryotes. The photosynthetic eubacteria use a variety of substances, including sulfur, as electron acceptors and do not produce oxygen gas. The cyanobacteria have a type of photosynthesis essentially like that in plants. They lack chloroplasts and their photosynthetic pigments are embedded in internal membranes.

Viruses are combinations of nucleic acids (DNA or RNA) and proteins. They are obligate intracellular parasites. The nucleic acid replicates within the host cell and directs the formation of new protein coats, utilizing the host cell's enzymes and other metabolic equipment. Viroids, which are the causative agents of certain plant diseases, consist of small molecules of RNA. Unlike viruses, they have no protein coat. Viroids are thought to interfere with gene regulation in infected host cells.

Microorganisms and human beings often live in symbiotic relationships. Symbiosis includes (1) commensalism, in which one member benefits, (2) mutualism, in which both benefit, and (3) parasitism, in which one is harmed. Pathogenic microorganisms cause disease by direct tissue destruction, by producing toxins, and by provoking host defenses. A symbiotic relationship may change from one type to another as a result of changes in the host or, sometimes, in the pathogen.

QUESTIONS

1. Distinguish between the following: cyanobacteria/bacteria; gram-positive/gram-negative; flagella/pili; axial filament/fibril; pathogen/toxin; virus/viroid.

2. Label the drawings below.

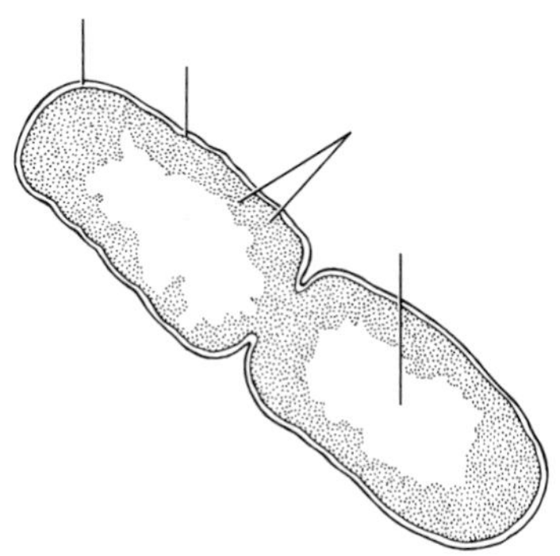

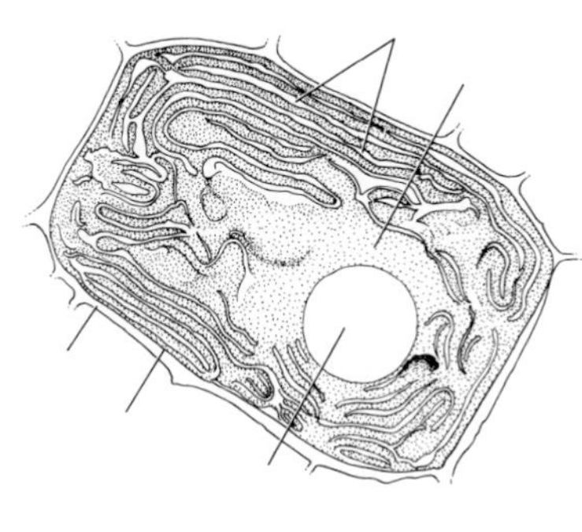

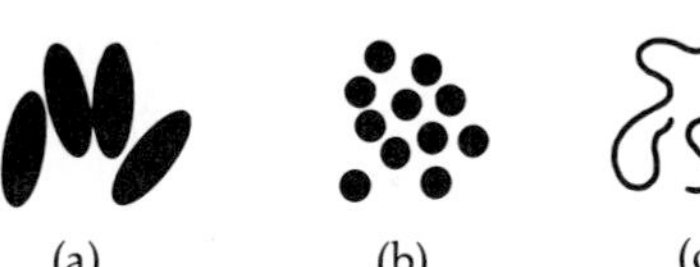

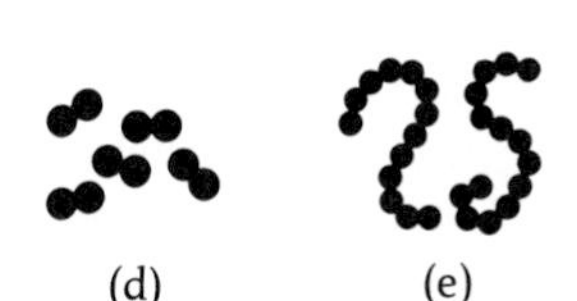

3. Identify the types of bacteria shown in the drawings on the left.

4. Many microorganisms produce antibiotics. What do you think their function might be for the organisms that produce them?

5. Four types of prokaryotes are photosynthetic: the green sulfur bacteria, the purple sulfur bacteria, the purple nonsulfur bacteria, and the cyanobacteria. Describe the similarities and differences between these groups of organisms and in their photosynthetic processes.

6. In addition to the four major types of photosynthetic prokaryotes, a few genera of salt-loving bacteria are photosynthetic. How do these cells differ from the other photosynthetic prokaryotes?

7. Before the structure of any virus was known, Crick and Watson predicted that the protein coats of viruses would prove to be made up of large numbers of identical subunits. Can you explain the basis of their prediction?

8. Some biologists consider viruses to be living organisms. By what criteria might viruses be considered alive?

9. Which of the three types of symbiotic association would, in your opinion, apply to the following relationships: dog and flea; human being and dog; maggot and wound; *E. coli* and human being; diphtherial microorganism and human being?

10. Prokaryotes are considered more primitive than eukaryotes. Does this mean they are all identical to forms of life that existed before eukaryotes arose? Explain your answer.

CHAPTER 21

The Protists

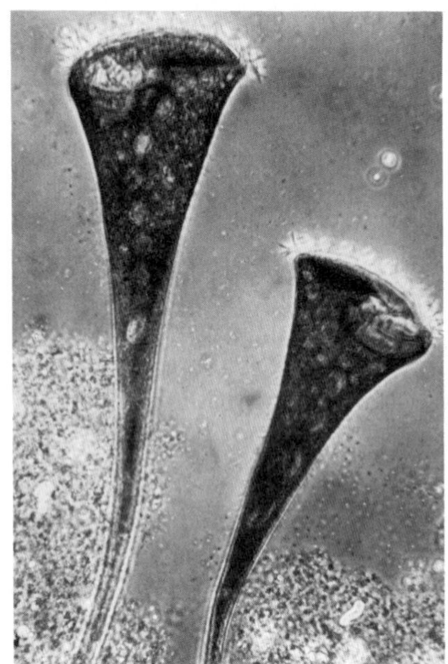

21–1
Stentor, a protozoan. Extended, it looks like a trumpet (this genus is named after "bronze-voiced" Stentor of the Iliad, *"who could cry out in as great a voice as 50 other men"). Stentor is crowned with a wreath of membranelles. These beat in rhythm, creating a powerful vortex that draws edible particles up to and into the funnel-like groove leading to the cytostome ("mouth"). Elastic protein threads, myonemes, run the length of the body. When contracted,* Stentor *is an almost perfect sphere.*

The protists are all eukaryotic. Most are unicellular, and those that are not have relatively simple multicellular structures. Apart from these two characteristics, they are an extremely diverse group of organisms. The kingdom Protista includes heterotrophs, among them a number of parasitic forms; photosynthetic autotrophs; and, as you will see, a few versatile organisms that are both heterotrophic and photosynthetic.

Protists are far from simple. Indeed, among one group, the protozoa ("first animals"), are found probably the most complex of all cells, with an astonishing variety of highly specialized structures. That single cells should be so complicated is actually not surprising. Among the protists, each cell is a self-sufficient organism, as capable of meeting all of the requirements of living as is any multicellular plant or animal.

THE EVOLUTION OF THE PROTISTS

The microfossil record indicates that the first eukaryotes (the protists) evolved at least 1.5 billion years ago. Eukaryotes, as we saw in Section 1, are distinguished from prokaryotes by their larger size, the separation of nucleus from cytoplasm by a nuclear envelope, the organization of their DNA, and their complex organelles, among which are chloroplasts and mitochondria.

The step from prokaryotes to the first eukaryotes was one of the big evolutionary transitions, second in importance only to the origin of life itself. The question of how it came about is a matter of current and lively discussion. One interesting hypothesis, gaining increasing acceptance, is that larger, more complex cells evolved partly as a result of certain prokaryotes' taking up residence inside other cells.

As we noted previously, oxygen began slowly to accumulate in the atmosphere about 2.5 billion years ago, as a result of the photosynthetic activity of the cyanobacteria. Those prokaryotes that were able to convert to the use of oxygen in ATP production gained a strong advantage, and so such forms began to prosper and increase. Some of these evolved into modern forms of aerobic bacteria. Others, according to this hypothesis, became symbionts within larger cells and evolved into mitochondria.

Several lines of evidence support the idea that mitochondria are descended from specialized bacteria. Mitochondria contain their own DNA, and this DNA is present in a single, continuous ("circular") molecule, like the DNA of bacteria. Many of the enzymes contained in the cell membranes of bacteria are found in mitochondrial membranes. Mitochondria contain ribosomes that resemble those of bacteria both in their small size and in details of their chemical composition, including some of the nucleotide sequences in the constituent rRNA molecules. Further, mitochondria appear to be produced only by other mitochondria, which divide within their host cell (Figure 21–2). (However, to make the situation more complex, there is not nearly enough DNA in mitochondria to code for all the mitochondrial proteins. Host-cell DNA is also required for the synthesis of mitochondrial enzymes and structural proteins.)

It is thought that the mitochondria are most likely derived from a lineage of purple nonsulfur bacteria in which the capacity for photosynthesis had been lost. Little is known, however, about the original cells in which these bacteria first set up housekeeping—or, indeed, if they actually existed. No contemporary prokaryotes provide clues about the origins of two key features of all eukaryotic cells—a nuclear envelope and chromosomes containing both DNA and protein. On the basis of studies of nucleotide sequences in transfer and ribosomal RNAs, it has been suggested that very early in the history of life a common ancestral lineage gave rise to three distinct lineages: the methanogens and their relatives (page 402), all of the other prokaryotes, and a group dubbed the "urkaryotes," which were the host cells in the evolution of the eukaryotes. If such cells existed, they probably had no means of using oxygen for cellular respiration and so were dependent entirely on anaerobic fermentation as an energy-releasing process. As we saw in Chapter 9, fermentation is relatively inefficient. Cells that acquired oxygen-utilizing respiratory assistants would have been more efficient than those lacking them and so would have outreproduced them.

In an analogous fashion, photosynthetic prokaryotes ingested by larger, nonphotosynthetic cells are believed to be the forerunners of chloroplasts. Such symbioses are thought to have occurred independently in a number of phylogenetic lineages, giving rise to the various groups of modern photosynthetic eukaryotes.

21–2
(a) *Mitochondria and* (b) *chloroplasts may have originated as symbiotic prokaryotes. This hypothesis is supported by the facts that both contain their own DNA and both, as shown here, replicate by division.*

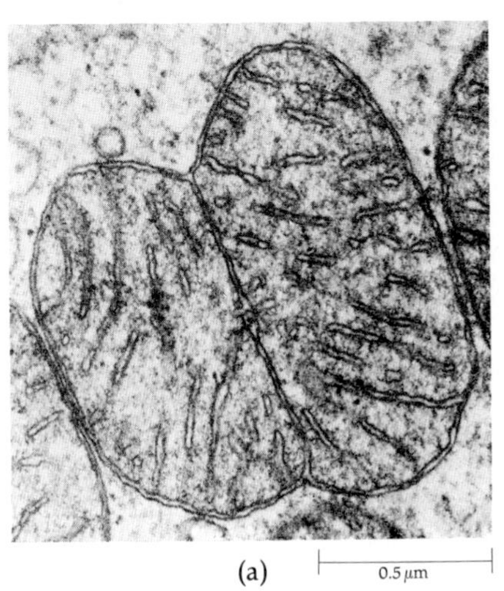

(a)

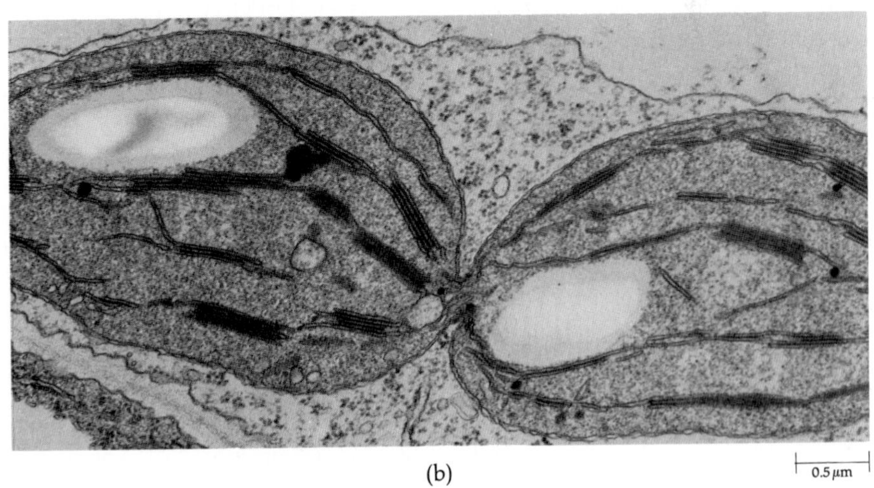

(b)

21–3
Plakobranchus, *a marine mollusk. The tissues of this animal contain chloroplasts, which it obtains by eating certain green algae.*

(a) *Ordinarily, structures called parapodia are folded over the animal's back, hiding the chloroplast-containing tissues.* (b) *When the parapodia are spread apart, however, the deep green tissues become visible. The chloroplasts carry on photosynthesis so efficiently that within a 24-hour cycle of light and darkness some individuals produce more oxygen than they consume.*

(a)

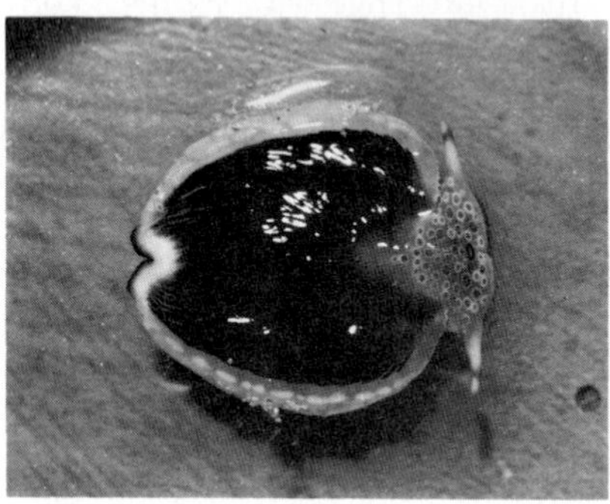

(b)

The endosymbiotic hypothesis accounts for the presence in eukaryotic cells of complex organelles not found in the far simpler prokaryotes. It gains support from the fact that many modern organisms contain intracellular symbiotic bacteria, cyanobacteria, or algae, indicating that such associations are not difficult to establish and maintain. Different species of cyanobacteria, for example, are found living within other cyanobacteria, some algae, amoebas, flagellated protozoans, water molds, plants, and sponges. These symbiotic cyanobacteria typically lack cell walls and are, functionally, chloroplasts. Moreover, they divide at the same time as the host cell, by a process similar to chloroplast division. A number of species of algae also form symbiotic associations with invertebrate animals (Figure 21–3). In such symbioses, whether ancient or modern, the smaller cells gain nutrients and protection, and the larger cells are given a new energy source.

Classification of Protists

Most of the modern unicellular eukaryotes—the protists—probably bear little or no resemblance to the first eukaryotes. As their complexity reminds us, they are not beginnings. They, like the modern prokaryotes, are the end products of millions of years of evolution, and in that time they have undergone tremendous diversification. (Those that have not changed much—amoebas might be an example—have survived over the millennia because they are exquisitely adapted to the environment in which they live.) Biologists generally agree (1) that the modern protists represent a number of quite different phylogenetic lineages, and (2) that all other eukaryotic organisms—fungi, plants, and animals—originated from primitive protists (Figure 21–4).

Table 21–1 The Kingdom Protista

Algae	
Division Euglenophyta	*Euglena* and related algae, all unicellular, mostly freshwater
Division Chrysophyta	Diatoms and related algae, all unicellular, mostly marine
Division Pyrrophyta	Dinoflagellates and related algae, all unicellular, mostly marine
Division Chlorophyta	Green algae, including unicellular, colonial, and multicellular forms; mostly freshwater
Division Phaeophyta	Brown algae, including the kelps; all multicellular; almost all marine
Division Rhodophyta	Red algae, all multicellular, mostly marine
Slime molds	
Division Myxomycota	Plasmodial slime molds
Division Acrasiomycota	Cellular slime molds
Protozoa	
Phylum Mastigophora	Flagellates, parasitic or free-living
Phylum Sarcodina	Amoeba-like organisms, free-living or parasitic
Phylum Ciliophora	Ciliates, free-living or parasitic
Phylum Opalinida	Opalinids, intestinal parasites of lower vertebrates
Phylum Sporozoa	Sporozoans, parasitic

21–4

Possible evolutionary relationships among the major groups of organisms. According to these hypotheses, the fungi, the various divisions of algae, and the modern protozoa evolved separately from single-celled eukaryotes. The origins of the plants have been traced to one group of algae. The animals are presumed to have arisen from a single-celled eukaryotic heterotroph, although direct evidence is lacking.

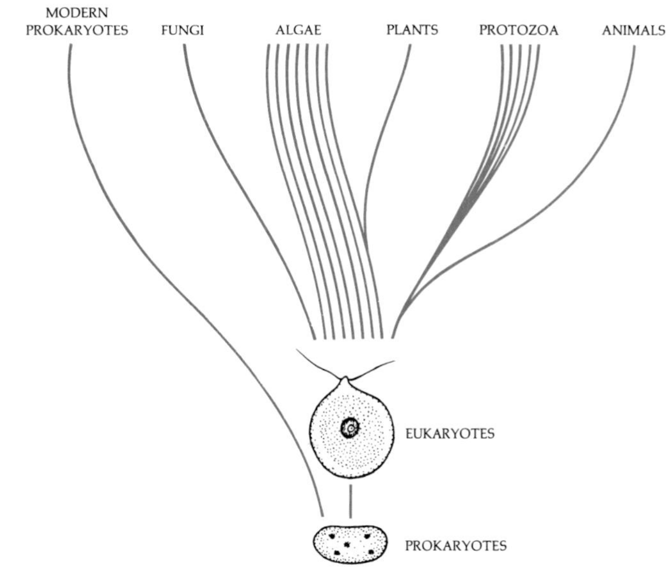

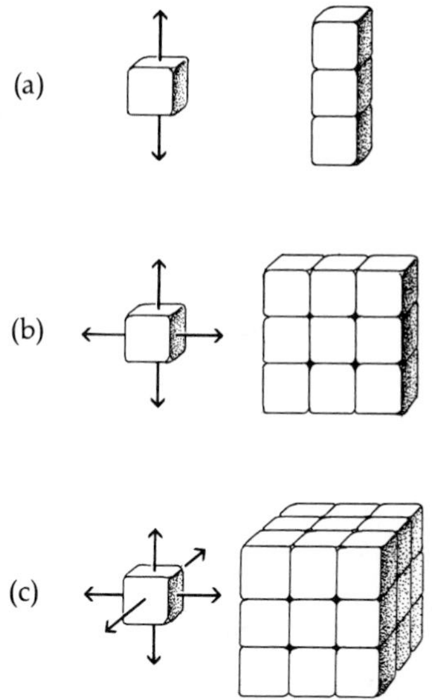

21–5

Different planes of cell division can produce different patterns of growth, all represented among the algae. (a) *Cell divisions in one plane produce a filament.* (b) *Divisions in two planes produce a one-celled layer.* (c) *Divisions in three planes produce a three-dimensional solid body. The arrows indicate the direction of growth.*

The classification of the protists, like that of the prokaryotes, is currently the subject of a major rethinking, as the electron microscope and modern biochemical techniques make available a vast amount of new information about these organisms. Such questions as the number of divisions or phyla constituting the kingdom, the relationships between the various divisions and phyla, the proper placement of particular species, and the criteria to be used in making these decisions are all topics of discussion and, in some cases, of heated controversy. For our purposes, the protists can be divided into three broad groups—the algae, in which the predominant mode of nutrition is photosynthesis; the slime molds, unusual heterotrophic organisms that resemble the fungi in some features but differ from them in others; and the protozoa, unicellular heterotrophs that are typically animal-like. Each of these informal groups encompasses two or more divisions or phyla (Table 21–1).

THE ALGAE

A number of distinct lines of algae had evolved by the early Paleozoic era, more than 450 million years ago (see page 457 for a geologic timetable). The approximately 26,000 described species now in existence can be grouped into six divisions. The term "algae" has been abandoned as a formal term in modern classification, because the various groups are not directly related to one another. Three of these divisions—the Euglenophyta, the Chrysophyta, and the Pyrrophyta—consist almost entirely of unicellular organisms. The other three divisions—Chlorophyta, Rhodophyta, and Phaeophyta—include groups that are multicellular.

Nearly all members of these divisions are photosynthetic. In general, they all have a relatively simple structure, which may be a single cell, a filament of cells, a plate of cells, or a solid body that may begin to approach the complexity of a plant body (Figure 21–5).

The unicellular algae are usually found floating near the surface of the oceans and inland waters. Together with small invertebrates and immature forms of larger animals they form the *plankton* (from *planktos*, the Greek word for "wanderer"). Each cell is a totally self-sufficient individual, dependent only on light from the sun and carbon dioxide and minerals from the water that surrounds it. These cells may live in colonies, and sometimes the cells of a single colony may be attached to one another as the gliding bacteria, for instance, are held together by sticky secretions; occasionally there is even a division of labor among the cells.

Most of the multicellular algae are adapted to living in shallow waters and along the shores. Here the waters are usually rich in nutrients, leached from the land or swept up in currents from the deeper waters. Living conditions are otherwise difficult, however. Along a rocky shore, for instance, multicellular algae—the seaweeds—are subject to great fluctuations of humidity, temperature, salinity, and light; to the pounding of the surf; and to the abrasive action of sand particles churned up by the waves. Life close to the shore is also crowded and competitive. Under these pressures, different groups of algae have become specialists at exploiting particular areas along tidal shores. The result is the zonation of life forms visible in many coastal areas (Figure 21–6).

Characteristics of the Algae

The divisions of algae vary widely in the number and position of their flagella (when present) and in their biochemical characteristics, especially pigmentation, nature of reserve food, and cell wall components (Table 21–2, page 421). In general, the cell walls of algae have a cellulose matrix, often with massive amounts of other polysaccharides that give certain algae a mucilaginous consistency.

21–6
Rocks along the coast of Cornwall, in England, showing the zonation of algal species from lower levels to higher levels. Zonation is based on a variety of factors, including light requirements, resistance to desiccation, predation, and competition.

21–7

A single cell of Prochloron, *a photosynthetic prokaryote discovered in 1975.* Prochloron *contains chlorophylls* a *and* b *and the same carotenoids found in green algae and plants. Biochemically, it has the characteristics that would be expected in a prokaryotic lineage that had given rise by symbiosis to the chloroplasts of the green algae.* Prochloron *is indeed symbiotic, but its hosts are sea squirts, soft-bodied animals that live along seashores and that belong to the lineage from which vertebrate animals arose.*

The names of some divisions are derived from the colors of the predominant accessory pigments, which mask the bright green of the chlorophylls. A wide variety of carotenoids is found in the chloroplasts of algae. The xanthophylls are yellowish-brown carotenoids; the xanthophyll fucoxanthin gives the brown algae their characteristic color and name. It is also found in the golden-brown algae and diatoms (division Chrysophyta). The red algae (Rhodophyta) owe their colors to several kinds of phycobilins, accessory pigments that are also characteristic of the cyanobacteria. In the green algae the color of the chlorophylls is usually not masked by accessory pigments. The wide variety of pigments present in the chloroplasts of the various divisions of algae suggests that (1) various types of oxygen-producing prokaryotes were in existence prior to the development of eukaryotic cells, and (2) the various divisions of algae may have evolved as a result of the establishment of symbiotic relationships with these different photosynthetic prokaryotes, which then evolved into modern chloroplasts (Figure 21–7).

A diversity of storage products is found among the algae, with most having distinctive carbohydrate food reserves, often in addition to lipids. The green algae and dinoflagellates store their carbohydrates as starch, as do the plants. In the brown algae, another glucose polymer, laminarin (Figure 21–8), takes the place of starch. The carbohydrate reserves of the red algae are biochemically similar to starch.

21–8

Laminarin, the principal storage product of brown algae. Like starch, it is made up of glucose residues, but unlike starch, there are only about 15 to 30 glucose units per molecule, and their linkage is different (*see page 58*).

CH₂OH CH₂OH CH₂OH

LAMINARIN

When algal cells divide, the cell membrane generally pinches inward from the margin of the cell (furrowing), just as in animals, fungi, and protozoa. Cell plates, like those of plants (see page 149), are formed during cell division only in one brown alga and a few green algae. Unlike plant cells, most algal cells—except those of the red algae—have centrioles.

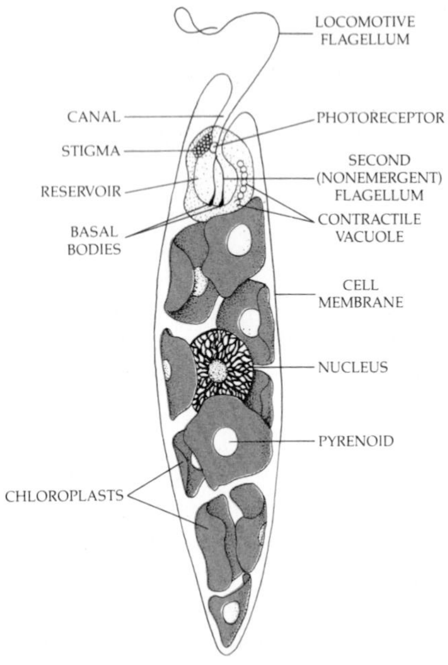

21-9
Euglena *is one of the most versatile of all one-celled organisms. Containing numerous chloroplasts, it is photosynthetic, but it can also absorb organic nutrients from the surrounding medium and can live without light. The pyrenoid is believed to be a food storage body.* Euglena *is pulled through the water by the whiplike motion of a single flagellum.*

Division Euglenophyta: Euglenoids

Euglenophyta are a small group of unicellular organisms (about 1,000 species), most of which are found in fresh water. They are named for the genus *Euglena*, the most common member of the group (Figure 21-9). About a third of the approximately 40 genera of euglenoids contain chloroplasts. These chloroplasts, which contain chlorophylls *a* and *b*, are very similar to those of the green algae but are enclosed by a triple—rather than a double—membrane. This suggests that euglenoids originally acquired chloroplasts by ingesting green algal cells; over long periods of time, most elements of the symbiotic green algae were lost, ultimately leaving only the chloroplast and the cell membrane. The nonphotosynthetic euglenoids are frequently able to absorb dissolved organic substances, and certain forms can ingest living prey, including other euglenoids. Euglenoids store their food as paramylum, an unusual polysaccharide that is found almost exclusively in this group.

Euglenoid cells are complex. *Euglena* is characteristically an elongated cell with a nucleus, two flagella (one of which is short and inactive), and numerous small emerald-green chloroplasts that give the cell its bright color. The flagella are attached at the base of a flask-shaped opening, the reservoir, at the anterior end of the cell. Emptying into the reservoir is the contractile vacuole, which collects excess water from all parts of the cell and discharges it into the reservoir. Only a few other photosynthetic organisms have contractile vacuoles, which are common features among the protozoans (see page 130). Euglenoid cells lack a cell wall but have a flexible series of protein strips, which make up the pellicle, inside the cell membrane (Figure 21-10). Unlike the stiff wall of plant cells, the flexible pellicle permits *Euglena* to change its shape, providing an alternative form of locomotion—wriggling—for mud-dwelling forms.

If one leaves a culture of *Euglena* near a sunny window, a clearly visible cloud will form in the water, and this will follow the light as it moves. If the light is too bright, however, the cells will swim away from it. *Euglena* is probably able to orient with respect to the light because of a pair of special structures: the stigma, or eyespot, which is a patch of pigment on the reservoir, and a photoreceptor on the locomotive flagellum. When the cell is in certain orientations with respect to light, the photoreceptor is shaded by the stigma. The amount of light striking the photoreceptor influences the action of the flagellum and therefore directs the movement of the cell.

Euglenoids reproduce asexually, dividing longitudinally to form two new cells that are mirror images of one another. If some strains of *Euglena* are kept at an appropriate temperature and in a rich medium, the cells may divide faster than the chloroplasts, producing nonphotosynthetic cells, which can survive indefinitely in a suitable medium containing a carbon source. It is tempting to speculate that some modern heterotrophs, including the nonphotosynthetic euglenoids, arose from autotrophs in an analogous manner.

21-10
A cell of Euglena, *broken open, showing the flexible protein strips that make up the pellicle. The large organelle at the upper right is a chloroplast. The spongy network below the chloroplast is endoplasmic reticulum. At the bottom left is a broken mitochondrion, identifiable by its characteristic cristae.*

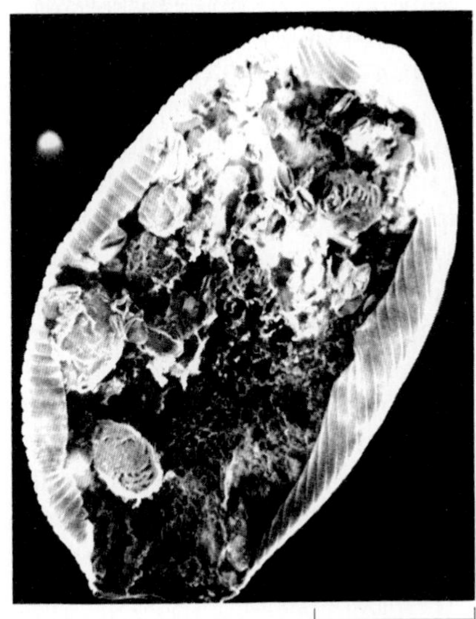

Division Chrysophyta: Diatoms and Golden-Brown Algae

This division includes two major classes, the diatoms (almost 10,000 species) and the golden-brown algae (about 1,500 species), and one minor class, the yellow-green algae (about 600 species). The chrysophytes have several identifying characteristics: (1) their photosynthetic pigments are chlorophyll *a* and chlorophyll *c*, which closely resembles the chlorophyll *b* of green algae and plants; (2) all species except the yellow-green algae contain a yellow-brown carotenoid, fucoxanthin, which functions as an accessory pigment and gives them their characteristic color; (3) their cell walls, which contain no cellulose, are often impregnated with silicon compounds and thus are very rigid; (4) they characteristically store food in the form of oils rather than starch. Because of their oil reserves, fresh water containing large numbers of these algae may have an unpleasant oily taste, as may fish caught in such waters. It has, in fact, been suggested that chrysophytes might be a potential source of oil for fuel.

Diatoms are a major component of the plankton and are an important source of food for small marine animals. Recently, the golden-brown algae have been found to be of extraordinary importance in the nannoplankton—components of the plankton so small that they pass through the fine mesh of an ordinary plankton net. It is now thought that the golden-brown algae may be the major food-producing organisms of the oceans.

Diatoms are enclosed in a fine double shell, the two halves of which fit together, one on top of the other, like a carved pillbox (Figure 21-11). Electron microscopy has shown that the fine tracings in diatom shells actually represent minute, intricately shaped pores or passageways connecting the interior of the cell to the exterior environment. The piled-up silicon shells of diatoms, which have collected over millions of years, form the fine, crumbly substance known as "diatomaceous earth," used as an abrasive in silver polish and toothpaste and for filtering and insulating materials.

Some other members of this division lack cell walls and are amoeboid. Except for the presence of chloroplasts, the amoeboid cells are indistinguishable from amoeboid protozoans (the sarcodines), and the two groups may be closely related.

21-11
Diatoms, chrysophytes. (a) *Side view showing the characteristic intricately marked shell.* (b) *Pillbox type seen from above and from the side. Notice that one cell is dividing. Each new cell will get half of the "pillbox" and then construct the other half. When the parent cell divides, one half is always smaller than the other. When this smaller cell divides, one of its daughter cells is, in turn, still smaller. When the individual cells are reduced in size to about 30 percent of the maximum diameter characteristic of the species, a cycle of sexual reproduction may be triggered. The offspring, which are full size, then undergo a new series of cell divisions.*

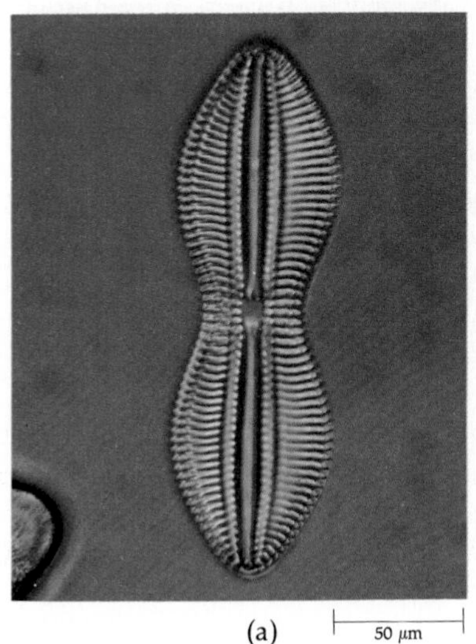

(a)

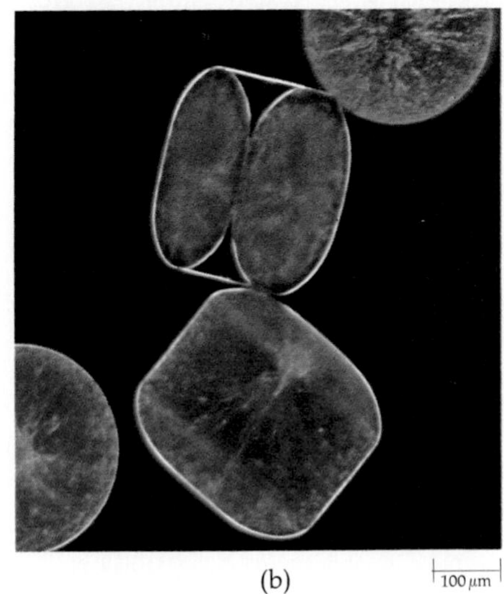

(b)

21-12
The silicon-containing walls of (a) *a diatom and* (b) *a golden-brown alga, as shown by the scanning electron microscope. Each diatom species has its characteristic pattern of perforations in the walls. The delicate markings of these shells, by which the species are identified, were traditionally used by microscopists to test the resolving power of their lenses.*

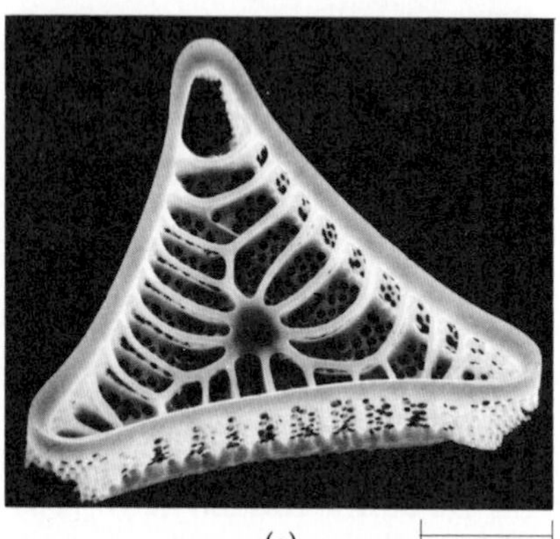

(a)

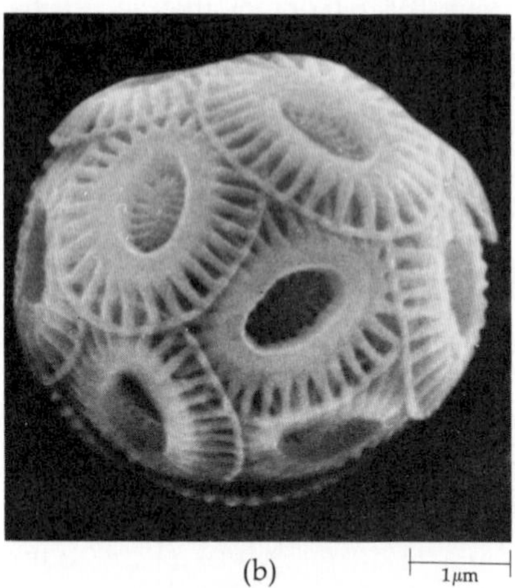

(b)

Reproduction in the chrysophytes usually takes place by cell division, an asexual process, but diatoms sometimes reproduce sexually. Gametes are produced by meiosis and then undergo fusion, or *syngamy*. (Syngamy in plants and animals is usually referred to as fertilization.) The resulting zygote expands to the full size characteristic of the species and undergoes cell division; the daughter cells then produce new silicon shells. Diatoms are diploid, except for the gametes.

Division Pyrrophyta: "Fire" Algae

The Pyrrophyta are also largely composed of single-celled algae, the dinoflagellates ("spinning flagellates"), of which about 1,100 different species are known, almost all of them marine forms. Like the chrysophytes, they are important components of the phytoplankton. Other members of the division are heterotrophs clearly related to the photosynthetic forms.

Many of the dinoflagellates are bizarre in appearance, with a stiff cellulose wall (theca) that often looks like a strange helmet or an ancient coat of armor. There are also "naked" forms, including some parasitic species. Dinoflagellates usually have two flagella that beat within grooves: one encircling the body like a belt and a second lying perpendicular to the first. The beating of the flagella in their respective grooves causes the cells to spin like tops as they move through the water. Many dinoflagellates are bioluminescent.

Dinoflagellates are often red in color (hence the name of the division), and the infamous red tides, in which thousands of fish die, are caused by great blooms of red dinoflagellates. The poison in these red tides has been traced to several species of dinoflagellates, including the armored *Gessnerium catenellum.* It is such an extraordinarily powerful nerve toxin that 1 gram of it would be enough to kill 5 million mice in 15 minutes. Blooms of *Gessnerium catenellum* appear regularly on the Pacific Coast and the Gulf of Mexico and have been reported off the New England coast. Mussels can also ingest the algae responsible for the red tides, concentrating the poison; these mollusks then become dangerous for consumption by vertebrates, including humans.

21-13
Dinoflagellates. (a) Ceratium tripos, *an armored dinoflagellate. You can see one flagellum in motion.* (b) Noctiluca scintillans, *a bioluminescent marine dinoflagellate. Yellow-brown diatoms that have been ingested can be seen inside the cell.*

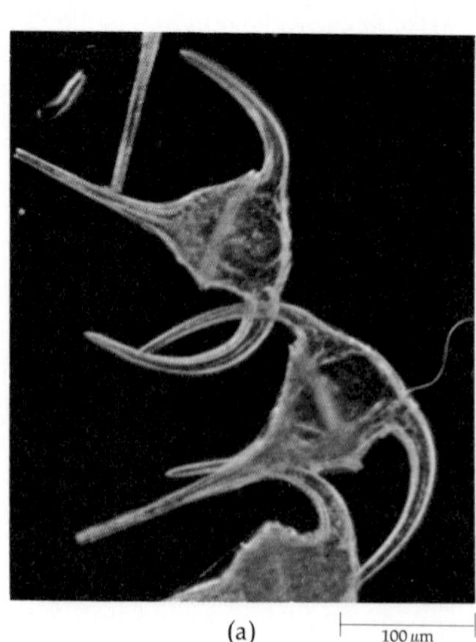

(a)

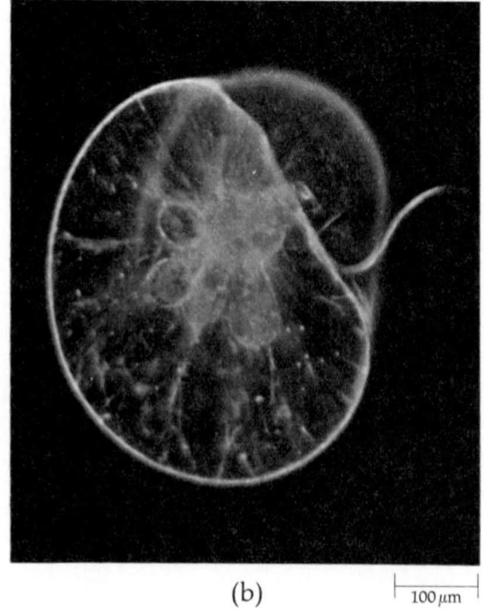

(b)

Table 21-2 Comparative Summary of Characteristics in the Six Major Divisions of Algae

DIVISION	NUMBER OF SPECIES	PHOTOSYNTHETIC PIGMENTS	FOOD RESERVE	FLAGELLA	CELL WALL COMPONENT	REMARKS
Euglenophyta (euglenoids)	1,000	Chlorophylls *a* and *b*, carotenoids	Paramylum and fats	1, 2, or 3 per cell, apical (at end of cell)	No cell wall; have a proteinaceous pellicle	Mostly freshwater; sexual reproduction unknown
Chrysophyta (diatoms, golden-brown algae, and yellow-green algae)	12,000	Chlorophylls *a* and *c*, carotenoids, including fucoxanthin	Leucosin and oils	None, 1, or 2, apical, equal or unequal	Pectin, frequently impregnated with silicon; cell wall lacking in some	Mostly marine
Pyrrophyta (dinoflagellates)	1,100	Chorophylls *a* and *c*, carotenoids	Starch and oils	None or 2, lateral	Cellulose	Mostly marine; sexual reproduction rare
Chlorophyta (green algae)	7,000	Chlorophylls *a* and *b*, carotenoids	Starch	Usually 2 per cell, identical	Cellulose and other polysaccharides	Mostly freshwater, but some marine
Phaeophyta (brown algae)	1,500	Chlorophylls *a* and *c*, carotenoids, including fucoxanthin	Laminarin and oils	2, lateral, in reproductive cells only	Cellulose and algin	Almost all marine, flourish in cold ocean waters
Rhodophyta (red algae)	4,000	Chlorophylls *a* and *d*, carotenoids, phycobilins	Floridean starch	None	Cellulose, pectin compounds, impregnated with calcium carbonate in some	Mostly marine; complex sexual cycle; many species tropical

Division Chlorophyta: Green Algae

The chlorophytes are the most diverse of all the algae, comprising at least 7,000 species. Although most green algae are aquatic, others occur in a wide variety of habitats, including the melting surface of snow, as green patches on tree trunks, and as symbionts in a variety of organisms. Of the aquatic species, a few groups are entirely marine, but the great majority are found in fresh water. Many green algae are unicellular and microscopic in size. *Chlamydomonas* (see Figure 21-14a), visible only as a fleck of green under the light microscope, is a chlorophyte. Some of the marine forms are large; *Codium magnum* in the Gulf of Mexico, for example, sometimes attains a breadth of 25 centimeters and a length of more than 8 meters.

The plants are believed to have originated among the Chlorophyta. Evidence for this theory includes the fact that both the plants and the chlorophytes—but no other group of algae—contain chlorophylls *a* and *b* and beta-carotene as their photosynthetic pigments and store their foods as starch. Most also have cellulose-containing cell walls.

21–14
The volvocine line. (a) Chlamydomonas, *a single-celled alga.* (b) Gonium *colonies. Each is composed of a shield-shaped mass of* Chlamydomonas-*like cells, held together in a gelatinous matrix.* (c) Pandorina, *in which the individual cells form an egg-shaped colony.* (d) Volvox, *a colonial alga. Each colony is made up of hundreds or thousands (depending on the species) of individual bright green biflagellate cells, attached to one another by fine threads of cytoplasm. Each colony forms a hollow sphere that spins through the water as a result of coordinated beating of the flagella. Daughter colonies form inside the mother colony. Many such mother-daughter combinations are visible in this micrograph.* Volvox *is the largest and most complex member of this evolutionary line.*

Paths to Multicellularity

The chlorophytes are of particular interest to students of evolution because, whereas the three divisions previously discussed contain only unicellular forms and the members of the two divisions that follow are almost entirely multicellular, division Chlorophyta has members of both types. Among the various green algae, it is possible to trace three different paths by which multicellular organisms have evolved from unicellular organisms.

The first pathway to multicellularity involves the association of individual cells in colonies. Colonies differ from true multicellular organisms in that, in colonies, the individual cells preserve a high degree of independent function. The cells are often connected by cytoplasmic strands, which integrate the colony sufficiently that it may be regarded as a single organism. The classic example of this pathway is called the volvocine line, after one of its more spectacular members, *Volvox.* It is based on the cohesion of *Chlamydomonas*-like cells in motile colonies. The simplest member of the volvocine line is *Gonium* (Figure 21–14b). The *Gonium* colony consists of 4 to 32 separate cells (depending on the species) arranged in a shield-shaped disk. The flagella of each cell beat separately, pulling the entire colony forward. Each cell in *Gonium* divides to produce an entire new colony.

A closely related colonial organism is *Pandorina* (Figure 21–14c), which consists of 16 or 32 cells in a tightly packed ovoid or ellipsoid shape. The colony is polar; that is, one end is different from the other: the eyespots are larger in the cells at one pole of the colony. Each cell has two flagella, and because all the flagella point outward, *Pandorina* rolls through the water like a ball. When the cells attain their maximum size, the colony sinks to the bottom, and each of the cells divides to form a daughter colony. The parent colony then breaks open like Pandora's box (which suggested its name), releasing new daughter colonies.

Volvox (Figure 21–14d), the culmination of the volvocine line, is a hollow sphere that is made up, according to species, of a single layer of 500 to 60,000 tiny biflagellate cells. The flagella of each cell beat in such a way as to turn the entire colony around its axis, spinning it majestically through the water. Like *Pandorina*, *Volvox* is polar; it orients its anterior end toward the light but moves away from

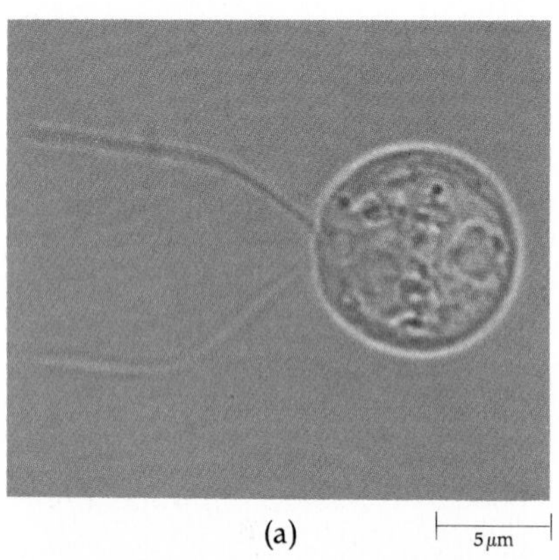
(a) 5 μm

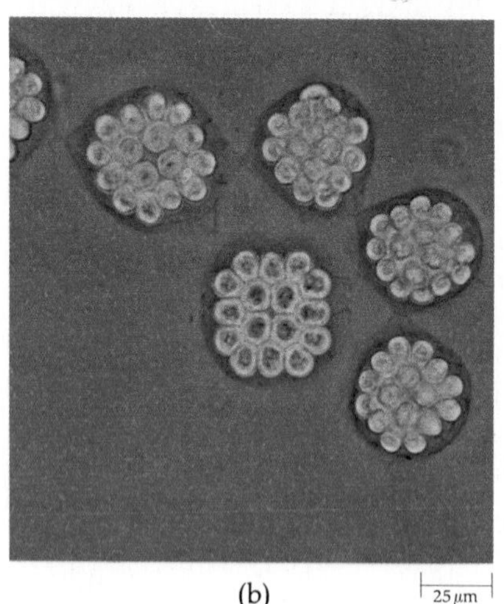
(b) 25 μm

(c) 5 μm

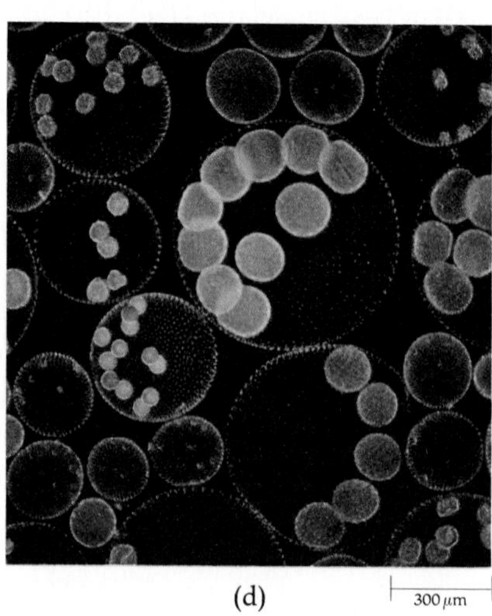
(d) 300 μm

(a)

(b)

(c)

21–15
The Chlorophyta include a variety of multicellular organisms in addition to those derived from colonial forms. (a) Valonia, *which is about the size of a hen's egg, is coenocytic, with many nuclei but no partitions separating them. It is common in tropical waters.* (b) Spirogyra *is a freshwater alga in which the cells all elongate and then are divided by transverse cross walls so that they are strung together in long, fine filaments. The chloroplasts form spirals that look like strips of green tape within each cell.* (c) Ulva, *or sea lettuce, is a marine alga in which the cells divide both longitudinally and laterally, with a single division in the third plane. This produces a broad thallus (vegetative body) two cells thick.*

light that is too strong. Most of the cells are vegetative and do not take part in reproduction. Only some of the cells of the lower hemisphere can form daughter colonies; in other words, there is specialization of function. These cells, which appear identical to others in the young colony, become larger, greener, and morphologically distinct as the colony matures. Then they enlarge and divide, forming new spheres of hundreds or thousands of cells. Daughter colonies remain inside the mother colony until the mother colony eventually breaks apart. In some species, the reproductive cells produce gametes and reproduction is sexual.

In these four extant genera of chlorophytes, *Chlamydomonas, Gonium, Pandorina,* and *Volvox,* there is a steady progression in size and complexity and also a trend toward specialization of function. In *Volvox,* this trend has led to organisms that are multicellular and differentiated in the same sense as the plants and animals. Nevertheless, this particular line represents an evolutionary "dead end," in that it has not given rise to a more complex group of organisms.

The second pathway to multicellularity results from repeated nuclear divisions that are not accompanied by a corresponding division of the cytoplasm and the formation of cell walls. An entire organism, such as *Valonia* (Figure 21–15a), may appear to be unicellular, but, in fact, it consists of many nuclei within a common cytoplasm. This type of organization is neither truly unicellular nor multicellular but is known as *coenocytic* (pronounced "see-no-sit-ik"; from the Greek *koinos,* "shared in common"). Some coenocytic green algae, such as *Cladophora,* are filamentous, while others, such as *Valonia* and *Codium magnum,* form more massive structures. A number of common seaweeds, especially tropical ones, are coenocytic. As we shall see in the next chapter, many of the fungi are also coenocytic.

A third pathway to multicellularity is exemplified by organisms in which nuclear division is followed by cytokinesis and the formation of cell walls. The cells, however, remain attached to one another. As we saw in Figure 21–5, such cell division can lead to the formation of filaments, sheets, or a three-dimensional body. Examples of green algae that have attained multicellularity by this pathway are *Spirogyra* and *Ulva* (Figure 21–15b and c).

Life Cycles in the Green Algae

A variety of different life cycles are observed among the green algae. In Figure 12–3 (page 251), we examined the life cycle of the unicellular green alga *Chlamydomonas.* The haploid cells of this alga usually reproduce asexually, with each cell undergoing two successive mitotic divisions, resulting in four new haploid cells. However, when essential nutrients, particularly nitrates, are in short supply, *Chlamydomonas* reproduces sexually. Haploid cells of different mating types function as gametes—they fuse, forming a diploid zygote around which a protective coat forms. The zygote, known as a zygospore, remains dormant until conditions are once more favorable for growth. It then undergoes meiosis, producing four haploid cells. The correlation between nutrient deficiency and sexual reproduction in *Chlamydomonas* and in many other organisms suggests that the rate of genetic recombination is subject to environmental influences.

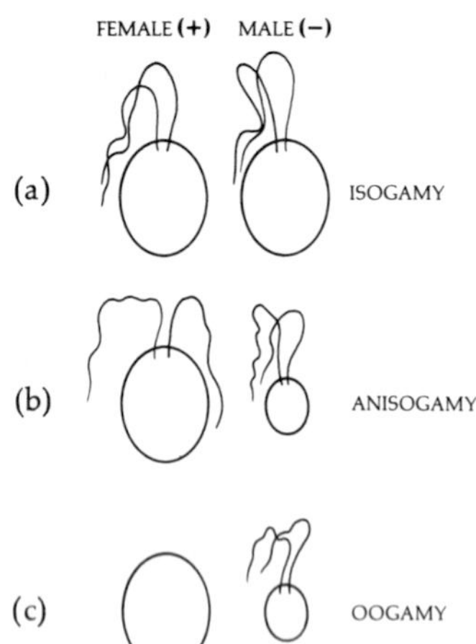

21–16
Evolution of sex. (a) *Isogamy. The gametes are similar in size and shape.* (b) *Anisogamy. One gamete, conventionally termed male, is smaller than the other.* (c) *Oogamy. One gamete, usually the larger is not motile. This is where sexual differentiation, with its mixed blessings and many complications, all began.*

In most species of *Chlamydomonas,* the cells of the two different mating types, conventionally designated as plus (+) and minus (−), are identical in size and structure; this condition is known as *isogamy* (Figure 21–16). In other species of *Chlamydomonas*, one gamete is larger than the other, but both are motile, a condition known as *anisogamy*. In still other species, one gamete, usually the larger, is not motile, a condition known as *oogamy*. The larger, nonmotile gametes are specialized for storing nutrients for the zygote, whereas the smaller gametes are specialized for seeking out and finding the first type of gamete. An individual that produces gametes that are nonmotile and (usually, though not always) larger is known as female. The entire range of differences between gametes that occurs among the algae (and among other types of organisms, as well) is exhibited in the various species of the single genus *Chlamydomonas.*

A more complex life cycle, characterized by alternation of generations, is found in some multicellular green algae, as well as in all plants. As we described in Chapter 12, in this type of life cycle, a diploid, spore-producing generation alternates with a haploid, gamete-producing generation. The gamete-producing form is known as the *gametophyte;* the individual cells of which it is composed are haploid (n) and so are the gametes it produces. The gametes fuse to form a diploid ($2n$) zygote. The zygote develops into the spore-producing form, the *sporophyte,* in which all the cells are diploid. Spores, which are produced in specialized structures of the sporophyte by meiosis, are always haploid. Spores differ from gametes in that when a spore germinates it forms a new organism, while a gamete must first unite with another gamete before further development occurs. The gametophyte and the sporophyte are always genetically different, since one is composed of haploid cells and the other of diploid cells. Although all gametes from a given gametophyte are genetically identical (since they are produced by mitosis from haploid cells), genetic recombination occurs both in the fusion of gametes of different types to form the zygote and in the meiotic divisions that produce spores. Moreover, the spores, with their new genetic combinations, may be dispersed quite widely from the parent sporophyte, enabling the organism to spread into new environments.

In some chlorophytes, such as the sea lettuce, *Ulva* (Figure 21–17), the two generations look alike and are said to be *isomorphic*. In other green algae, the sporophyte and the gametophyte do not resemble one another, and the generations are said to be *heteromorphic*. In fact, in some cases, the two generations of the same alga were once considered to be two entirely different species until the organism was studied in the laboratory (Figure 21–18).

21–17
In the sea lettuce, Ulva, *we can see the reproductive pattern known as alternation of generations, in which one generation produces spores, the other gametes. The haploid* (n) *gametophyte produces haploid isogametes, and the gametes fuse to form a diploid* (2n) *zygote. A sporophyte, a multicellular body in which all the cells are diploid, develops from the zygote. The sporophyte produces haploid spores by meiosis. The haploid spores develop into haploid gametophytes, and the cycle begins again. The micrograph shows isogametes of* Ulva *before cytoplasmic fusion.*

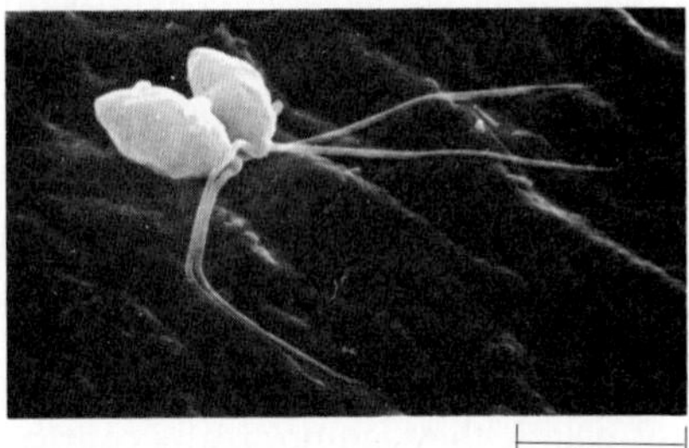

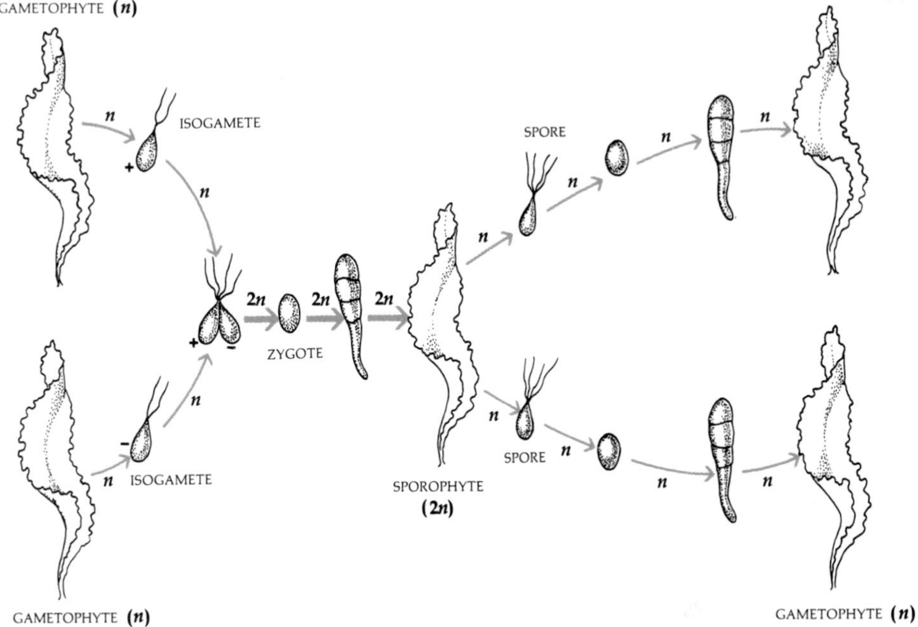

21–18
In some algae the alternating generations are so different that they once were believed to be different genera. At one time, the gametophyte of the Valonia*-like alga shown here was called* Halicystis *and the sporophyte* Derbesia. *As you can see, this alga exhibits anisogamy.*

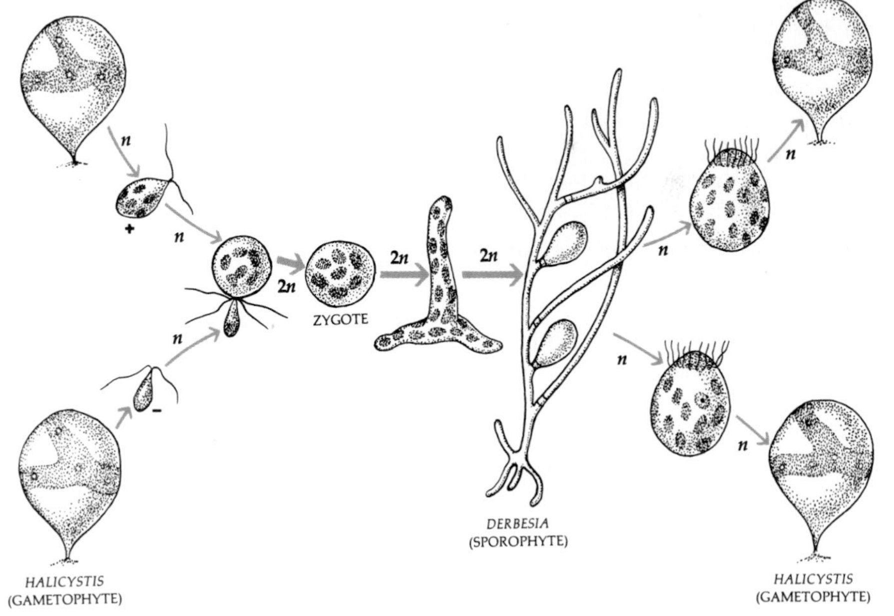

(a)

(b)

21–19
(a) *A brown alga,* Laminaria, *showing holdfasts, stipes, and blades.* (b) *Unlike the brown algae, most red algae are made up of filaments. The branched filaments of this red alga are hooked, enabling it to cling to other seaweeds.*

Division Phaeophyta: Brown Algae

The brown algae are the principal seaweeds of temperate and polar regions. An almost exclusively marine group, they dominate the rocky shores throughout the cooler regions of the world, and some, like the kelps, often form extensive beds offshore. The brown algae contain chlorophylls *a* and *c* and fucoxanthin, as do the Chrysophyta. They differ from most plants in that they store their food as an unusual polysaccharide (laminarin), or sometimes as oil, but never as starch, as do green algae and plants. Their cell walls contain cellulose.

The brown algae are often very large, and many have a variety of specialized tissues. Some of the giant kelps are nearly 60 meters long; these are annuals, reaching their full size in a single season. The body of an alga (or a plant) that is multicellular, but relatively unspecialized, is known as a *thallus*. In some kelps, the thallus is well differentiated into *holdfast* ("root"), *stipe* ("stalk"), and *blade* ("leaf"). The words are put into quotation marks because, although the corresponding parts of the alga superficially resemble the organs of plants, they are not really comparable in their internal organization. However, some kelps have strands of elongated conducting cells in the center of the stipe that are similar to the cells that conduct sugars in the vascular plants. Carbohydrates produced in the blades, which are exposed to sunlight, are thus transported to the stipe and holdfast, which may be far below the surface of the water.

There are no modern unicellular forms in this group, except for the gametes. The life cycles of most brown algae involve alternation of generations. In some species the two generations are isomorphic, whereas in others they are heteromorphic. In the larger kelps, the gametophyte is much smaller than the conspicuous, familiar sporophyte. In the rockweed *Fucus* and its relatives, the life cycle is superficially similar to that in higher animals (see Figure 12–2c, page 250). Meiosis occurs in specialized cells of the diploid organism, producing haploid cells that immediately undergo mitosis, giving rise to gametes.

Division Rhodophyta: Red Algae

Most of the seaweeds of the world are red algae, of which there are some 4,000 species. They are most commonly found in warm marine waters; less than 2 percent of the species are freshwater forms. Red algae usually grow attached to rocks or other algae. Their red color indicates that they absorb blue light, which is the color with the greatest penetration in water. As a result, red algae can grow at greater depths than other algae; some have been found attached 175 meters below the ocean surface in the clear water of the tropics. Although some are several meters long, red algae never attain the size of the largest of the brown algae.

The red algae contain chlorophylls *a* and *d*, carotenoids, and also certain phycobilins, which give them their distinctive colors. (Phycobilins and chlorophyll *a* are also found in the cyanobacteria, and it is believed that the chloroplasts of the red algae evolved from ancestors of the modern cyanobacteria.) The cell walls of most red algae include an inner layer composed of cellulose and an outer layer composed of mucilaginous carbohydrates, such as agar, which is used in culturing bacteria in the laboratory. In addition, many red algae have the capacity to deposit calcium carbonate in their cell walls. Such algae are called coralline algae, and they play an important role in building coral reefs.

The basic life cycle of the red algae involves alternation of generations. In most species, the gametophyte and sporophyte are isomorphic, but an increasing number of heteromorphic cycles are being discovered. The gametes are unusual in that neither type is motile; the male gamete is carried to the fixed female reproductive cell by the movement of the water. The red algae are one of the few groups of organisms that have no flagellated cells; they also lack centrioles.

THE SLIME MOLDS

The slime molds are a group of curious organisms, sometimes referred to as the "lower fungi." They are, however, usually classified with the protists because of their similarity to the protozoa, especially the amoebas. Two main groups are known, the plasmodial slime molds (division Myxomycota), with about 550 species, and the cellular slime molds (division Acrasiomycota), with 26 species.

Most slime molds live in cool, shady, moist places in the woods—on decaying logs, dead leaves, or other damp organic matter. One of the common plasmodial species *(Physarum cinereum),* however, is sometimes found creeping across city lawns. Plasmodial slime molds come in a variety of colors and can be spectacularly beautiful. The function of the pigments is not known with certainty, but they are probably photoreceptors because only pigmented species require light for spore production.

During their nonreproductive stages, the plasmodial slime molds are thin, streaming masses of protoplasm, which creep along in amoeboid fashion. As one of these plasmodia travels, it engulfs bacteria, yeast, fungal spores, and small particles of decayed plant and animal matter, which it digests. It may grow to weigh as much as 50 grams or more, and, since slime molds are spread thinly, this mass can cover an area more than a meter in diameter. The plasmodium is coenocytic; as it grows, the nuclei divide repeatedly and, in the early stages, synchronously.

Plasmodial growth continues as long as an adequate food supply and moisture are available. When either of these is in short supply, the plasmodium separates into many mounds of protoplasm, each of which develops into a mature sporangium (a structure in which spores develop) borne at the tip of a stalk. Meiosis takes place and cell walls form around the individual haploid nuclei to produce spores, which are resistant resting forms.

21–20
(a) *The plasmodium—a streaming mass of protoplasm—of a slime mold. Such a plasmodium, with its multiple nuclei, can pass through a piece of silk or filter paper and come out the other side apparently unchanged.* (b) *Sporangia of a plasmodial slime mold on a rotting log.*

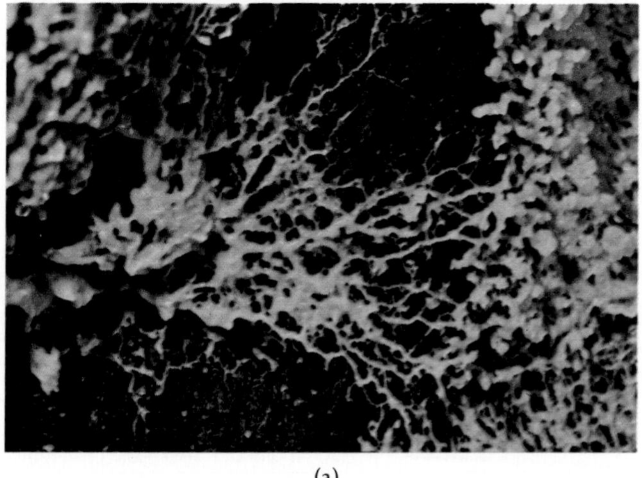

(a)

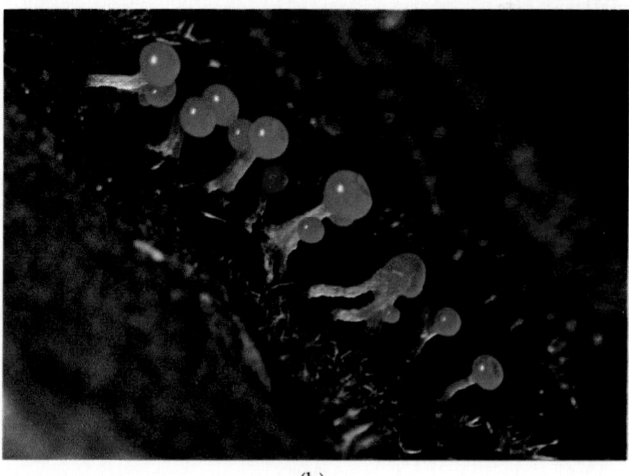

(b)

The spores germinate under favorable conditions, and each spore, depending on the species, produces one to four haploid, flagellated cells. Some of these cells fuse to form a zygote, from which a new plasmodium develops.

The life cycle of an acrasiomycete, or cellular slime mold, is shown on page 120. These slime molds also begin as amoeba-like organisms but differ from the plasmodial slime molds in that the amoebas, on swarming together, do not lose their cell membranes but retain their identity as individual cells.

THE PROTOZOA

The protozoa—"first animals"—are one-celled heterotrophs. Three phyla, which contain both free-living and parasitic members, can be distinguished on the basis of mode of locomotion: (1) by flagellar movement (phylum Mastigophora, the mastigophores), (2) by pseudopodia (phylum Sarcodina, the sarcodines), and (3) by ciliary movements (phylum Ciliophora, the ciliates). Two phyla are entirely parasitic: (1) Opalinida, the opalinids, which have ciliary movement, and (2) Sporozoa, the sporozoans, in which the motility of the cells is much reduced.

Protozoans usually reproduce asexually, by binary fission. Many also have sexual cycles, involving meiosis and the fusion of gametes, which results in a diploid ($2n$) zygote. The zygote is often a thick-walled, resistant resting cell, formed during periods of drought or cold. Some protozoans, notably the ciliates, undergo conjugation, in which nuclei are exchanged between cells.

Table 21-3 Comparative Summary of Characteristics in the Five Major Phyla of Protozoa

PHYLUM	NUMBER OF SPECIES	LOCOMOTOR STRUCTURES	MODE OF REPRODUCTION	REMARKS
Mastigophora	2,500	Flagella; some also form pseudopodia	Asexual (binary fission); sexual (meiosis and syngamy) in some species	Mostly parasitic; some free-living
Sarcodina	11,500	Pseudopodia; a few develop flagella at some stages of life cycle	Asexual or sexual	Mostly free-living; many have outer shells, or tests; about 33,000 fossil species known
Ciliophora	7,200	Cilia	Asexual; genetic exchanges through conjugation	Mostly free-living; have micronuclei and macronuclei
Opalinida	400	Cilia or flagella	Asexual or sexual, with flagellated gametes	All intestinal parasites of lower vertebrates
Sporozoa	6,000	None	Complex life cycle, involving both asexual and sexual reproduction	All parasitic; cause devastating diseases in humans and other animals, including malaria

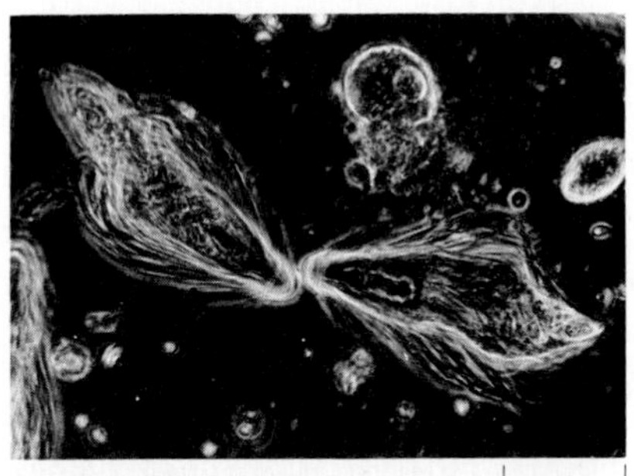

21-21
Two Trichonympha *cells. These flagellates, which break down cellulose, are responsible for the well-known proficiency of their termite hosts to digest wood.*

Phylum Mastigophora

The Mastigophora are regarded as the most primitive of the protozoans. Some authorities believe they are derived from photosynthetic flagellated cells, such as *Euglena,* that lost their chloroplasts; others think they may be the descendants of protists that never acquired photosynthetic symbionts. Almost all of the smaller cells have one or two flagella, and larger cells frequently have many. These flagella have the characteristic 9 + 2 structure (page 116). The mastigophores multiply asexually by binary fission (mitosis and cytokinesis), and, in some forms, sexually, by syngamy. They generally have no outer wall, and some are able to form pseudopodia, which are temporary extensions of the cell body that are used in locomotion and in engulfing food particles. Most are parasitic, but some are free-living. The former include *Trypanosoma gambiense* and *Trypanosoma rhodesiense,* flagellates that cause African sleeping sickness, and members of the genus *Trichonympha,* complex and beautiful flagellates that live as symbionts in the digestive tracts of termites, where they digest the wood ingested by the termite. A number of other mastigophores cause debilitating diseases in humans and their poultry and livestock.

Phylum Sarcodina

The sarcodines include amoebas and their relatives. They have no coat or wall outside their cell membrane and generally move and feed by the formation of pseudopodia. They take their name from the word "sarcode," coined in the early nineteenth century to describe the "simple, glutinous, and homogeneous jelly" of which, at one time, simple cells were thought to be composed. Despite their uncomplicated appearance, they are complex cells and, as we shall see (page 434), are even capable of complex behavior patterns—for example, when sensing and pursuing prey organisms.

Sarcodines are thought to have originated from mastigophores; some sarcodines may develop flagella during particular stages of their life cycle or under particular environmental conditions. They are found in both fresh and salt water. Some are parasites, such as those that cause amoebic dysentery in humans.

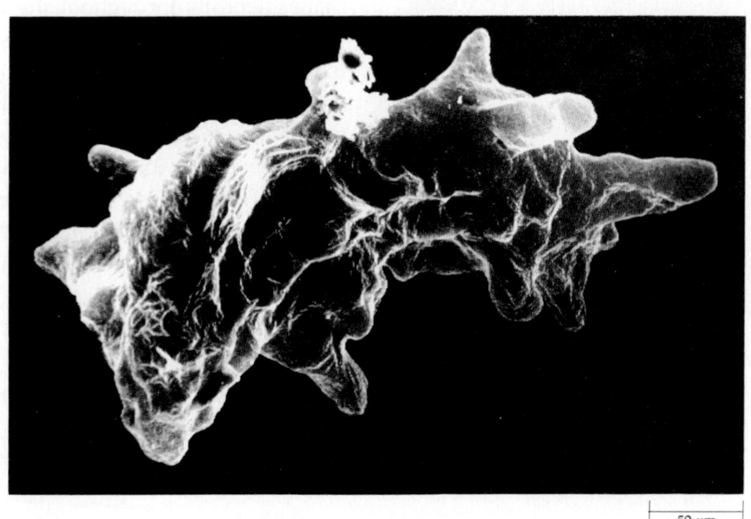

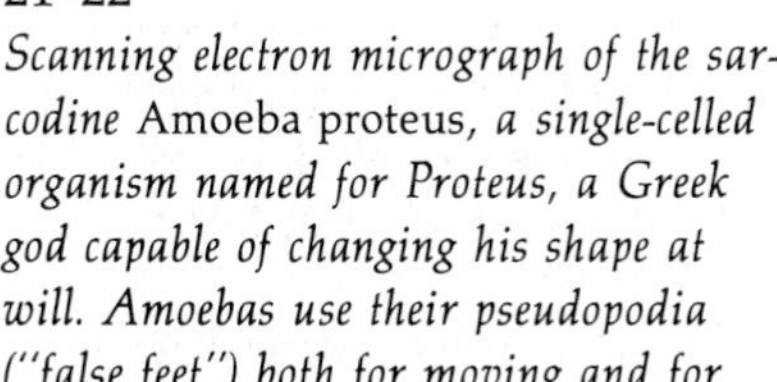
21-22
Scanning electron micrograph of the sarcodine Amoeba proteus, *a single-celled organism named for Proteus, a Greek god capable of changing his shape at will. Amoebas use their pseudopodia ("false feet") both for moving and for capturing prey.*

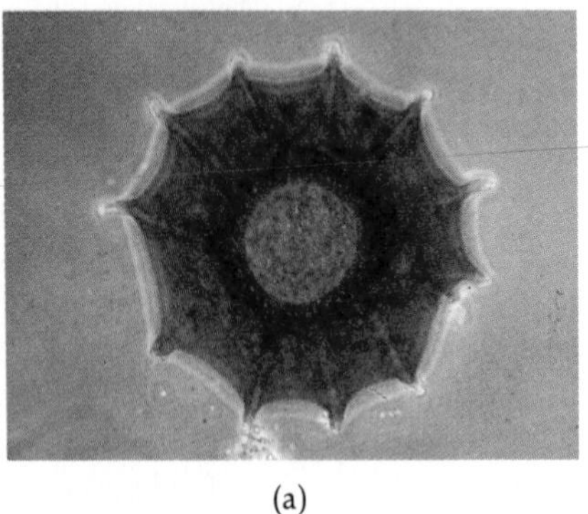

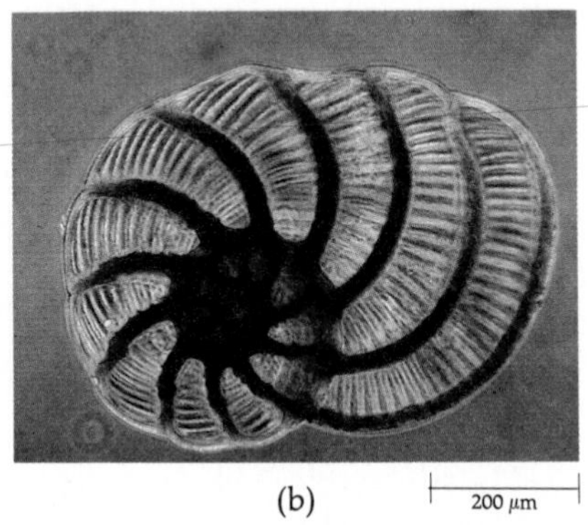

(a) (b)

21–23
Outer shells, or tests, are characteristic of certain groups of sarcodines. (a) *The brilliantly colored test of* Arcella dentata *consists of a proteinaceous material secreted by the organism.* (b) *The shell of a foraminiferan. In addition to as many as 7,000 living species of foraminiferans, there are about 30,000 extinct species, known only from their fossilized shells.*

Reproduction may be asexual or sexual. Asexual reproduction takes place by cell division accompanied by mitosis in which the nuclear envelope usually does not break down. In sexual reproduction, the cells, which are diploid, undergo meiosis, forming gametes, which then fuse to form zygotes.

Many of the sarcodines have outer shells, or tests, sometimes brightly colored. Some, like *Arcella,* secrete a proteinaceous material that hardens upon exposure. Others, like *Difflugia,* exude a sticky organic substance in which they deposit silicon-containing particles. These particles, which were previously ingested by the organism, are divided between the two daughter cells at cell division; each daughter cell then arranges them into an almost exact replica of the parental shell. Such shells, which are produced primarily by amoebas that live in sand or soil, on the mosses of bogs and forest floors, or in crevices in tree trunks, are thought to have evolved as a protection against abrasion or dehydration of the organism within. Some sarcodines, the "sun animals," resemble pincushions, with fine pseudopodia stiffened with microtubules (page 70) radiating from their bodies. *Actinosphaerium,* the protozoan shown in Figure 5–2, may reach as much as a millimeter in diameter and may sometimes be seen as a tiny white speck floating on the surface of a pond.

Other sarcodines, the Foraminifera (numerically the most common of all protozoa), have snail-like shells and live in the sea. Their shells are made of calcium carbonate, extracted from the sea water. The white cliffs of Dover and similar chalky deposits throughout the world are the result of the long accumulation of these discarded shells. The shells of the Foraminifera have been accumulating on the ocean bottom for millions of years, and in many areas, as a result of geologic changes, thick deposits of their skeletons (the "foraminiferan ooze") can be found on the surface of the land or under later rock formations. Since the skeletons have evolved over this long period of time, it is possible to date a particular stratum by the type of Foraminifera that it contains, a fact that has proved of immense practical value in locating oil-bearing strata.

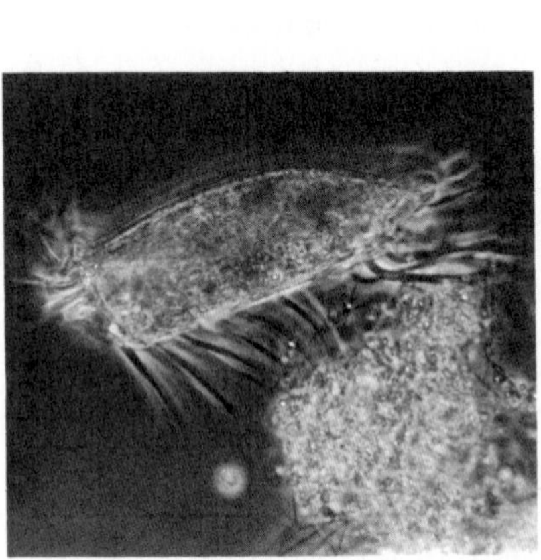

21–24
The cilia of some protozoans are clumped to form cirri. In Euplotes patella, *shown here, the cirri move individually, propelling the cell in a jerky motion.*

Phylum Ciliophora

The ciliophores, or ciliates, are the most highly specialized and complicated of the protozoans. About 7,200 species of ciliates are known, both freshwater and saltwater forms; most are free-living. They are believed to have been derived from primitive mastigophores, which, traveling in the opposite evolutionary direction from the sarcodines, developed elaborate ciliary systems (see Figure 5–23, page 115). They are characterized by 9 + 2 cilia. (Remember that cilia and flagella differ

21-25
Drawing of a Paramecium, *a ciliate. The body of this protist is completely covered by 9 + 2 cilia, although only a relative few are shown here. Like other ciliates, a* Paramecium *feeds largely on bacteria, smaller microorganisms, and other particulate matter. The beating of specialized cilia drives particles into the buccal cavity, where they are formed into food vacuoles that enter the cytoplasm. The food is digested in the vacuoles, and the undigested matter, still in vacuoles, is emptied out through the anal pore. The contractile vacuoles serve to eliminate excess water from the cell.*

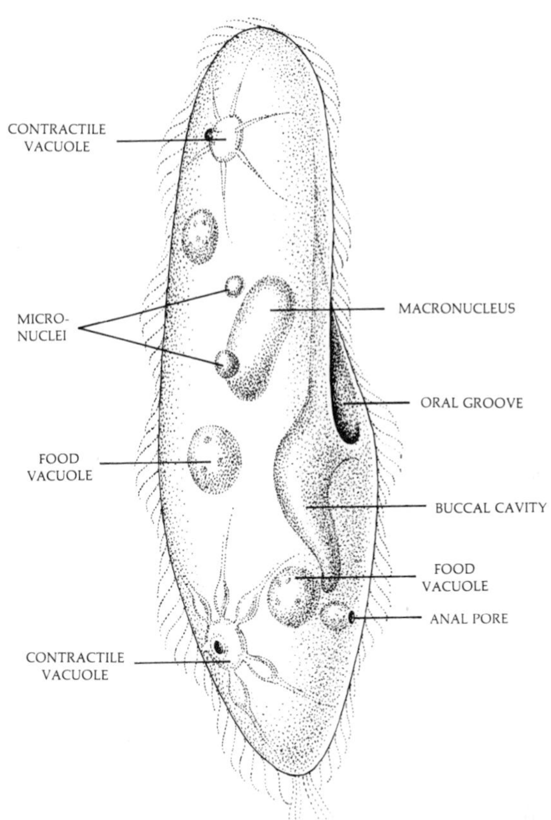

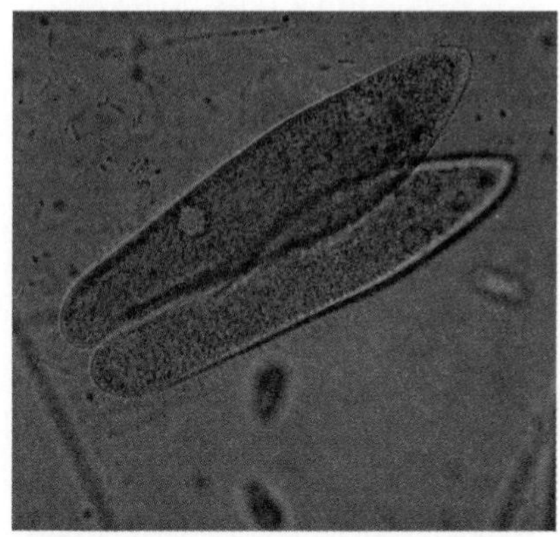

21-26
Two cells of Paramecium caudatum *in the process of conjugation. These cells, actually colorless, were alive when photographed through a special type of interference microscope. Its optical system adds different colors to regions of different opacity, producing the brilliant image seen here.*

only in length; cilia are shorter.) In some species, the cilia adhere to each other in rows, forming brushlike structures called membranelles or clumps of cilia called cirri, which can be used for walking or jumping. Cilia, membranelles, and cirri move in a coordinated fashion, although the way in which they are coordinated is not entirely understood. Some ciliates also have myonemes, contractile threads. All have a complex "skin," the cortex, which is bound on the outer side by the cell membrane. In a few groups, the cortex contains small barbs known as trichocysts, which are discharged when the cell is stimulated in certain ways.

The ciliates have another unusual feature: they have two kinds of nuclei, micronuclei and macronuclei. One or more of each kind is present in all cells. They also have a complex system for exchange of genetic information, in which cells conjugate and the micronuclei undergo meiosis. Cells then exchange haploid micronuclei that fuse, so that each cell has a new diploid micronucleus, which then divides. The old macronucleus dissolves, and a new macronucleus develops from one of the daughter micronuclei. The macronucleus in certain ciliates contains 50 to 100 times as much DNA as the micronucleus and so is believed to represent multiple copies of it. This view is supported by the fact that, in many ciliates, a cell can survive indefinitely without a micronucleus if even a small portion of a macronucleus is left, although the cell cannot conjugate. However, it cannot live without a macronucleus, even if it has a micronucleus. The macronucleus does not divide mitotically but is apportioned approximately equally between dividing cells as they constrict and separate.

THE EVOLUTION OF MITOSIS

Among modern prokaryotes and protists, it is possible to observe varying patterns of mitosis, which are believed to reflect evolutionary history. Among prokaryotes, you will recall, the replicated chromosomes (mostly DNA) attach to the cell membrane and so are separated as the newly formed cells divide. Among plant and animal cells, the nuclear envelope breaks down at mitosis, and the spindle fibers (composed of microtubules) are apparently involved in the mechanism of separation. Detailed analysis of some of the protists has revealed some intermediate stages between these two. In dinoflagellates, for instance, the chromosomes—which have no histones attached to the DNA and are always condensed—are anchored permanently to the nuclear envelope. The envelope remains intact during mitosis, and the chromosomes, as in bacteria, are separated as the enveloping membrane elongates and is pinched in two. Microtubules appear to be involved in the pinching-off process.

The macronucleus of ciliated protozoa appears to be apportioned in just this same way, but in the micronucleus of ciliates, both chromosomal and pole-to-pole spindle fibers are seen, as they are in slime molds. The spindle fibers appear to push the poles apart as the fibers grow, elongating the nucleus in preparation for its division.

As the history is reconstructed, in the earliest eukaryotes the nuclear envelope remained intact, and the chromosomes separated as it elongated. Next, microtubules, which first played an extranuclear role, entered the nuclear envelope, and, in an increasingly more organized way, aided in its elongation. Finally, the nuclear envelope, no longer useful in mitosis, began to fragment at prophase, and spindle fibers took over as a more efficient means for separating pairs of chromosomes in a diploid organism.

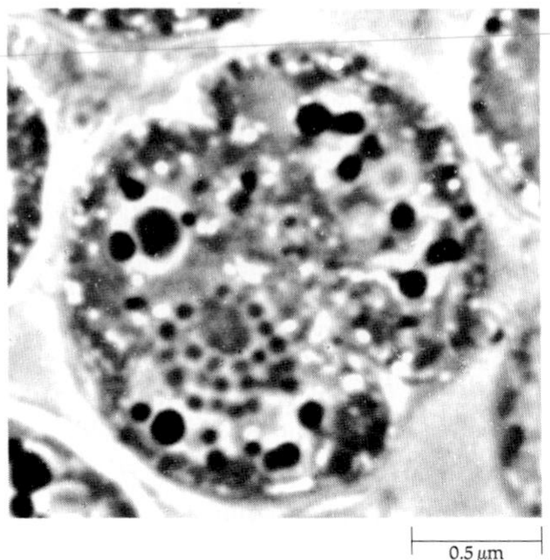

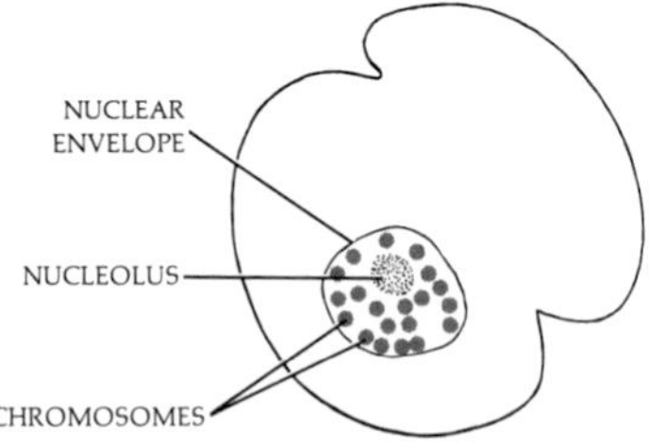

Interphase chromosomes in a dinoflagellate.

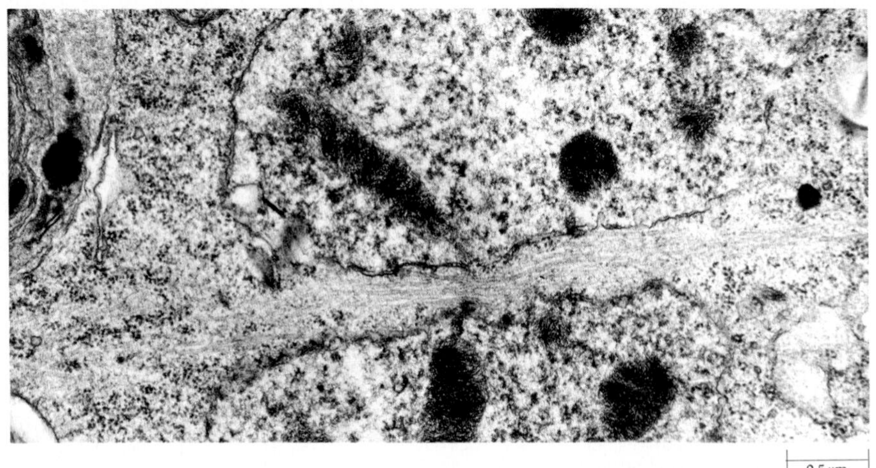

Microtubules in the cytoplasmic channel separating the two nuclei at mitosis. The dark areas are chromosomes.

The complexity of the conjugation process among ciliates reminds us again of the expenditure of energy and other resources involved in effecting exchanges of genetic information among individuals of the same species, from bacteria to *Homo sapiens.* The high biological cost of these activities is an indication of the great survival value, in evolutionary terms, of the genetic variability produced by such exchanges.

Phylum Opalinida

The opalinids have been found most often in the digestive tracts of frogs and toads, and, occasionally, in fishes and reptiles. They are structurally simple, with a uniform covering of cilia or flagella (Figure 21–27). At least two nuclei are present; in some species, the number of nuclei increases as the cell gets larger. The nuclei are not, however, differentiated into macronuclei and micronuclei. The opalinids may be distant relatives of the ciliates, or they may represent a distinct evolutionary lineage. Like certain groups of mastigophores, opalinids produce gametes that fuse to form a zygote; some authorities consider this an indication of a close relationship with those particular mastigophoran groups.

Phylum Sporozoa

All sporozoans are parasites. They are characterized by the lack of cilia or flagella and by complex life cycles. The best-known sporozoans are members of the genus *Plasmodium,* which cause malaria in many species of birds and mammals. A typical *Plasmodium* life cycle is shown in Figure 21–28. Malaria has generally been controlled by drugs that act on the parasite at various stages of its life cycle and by insecticides that kill the mosquitoes that transmit it. Recently, however, both the parasite and the mosquitoes have begun to develop resistance to the chemicals used to attack them, and malaria remains, on a worldwide basis, a principal cause of death among humans. With the advent of techniques for culturing the parasite in the laboratory, there is hope that vaccines against this devastating disease can be developed.

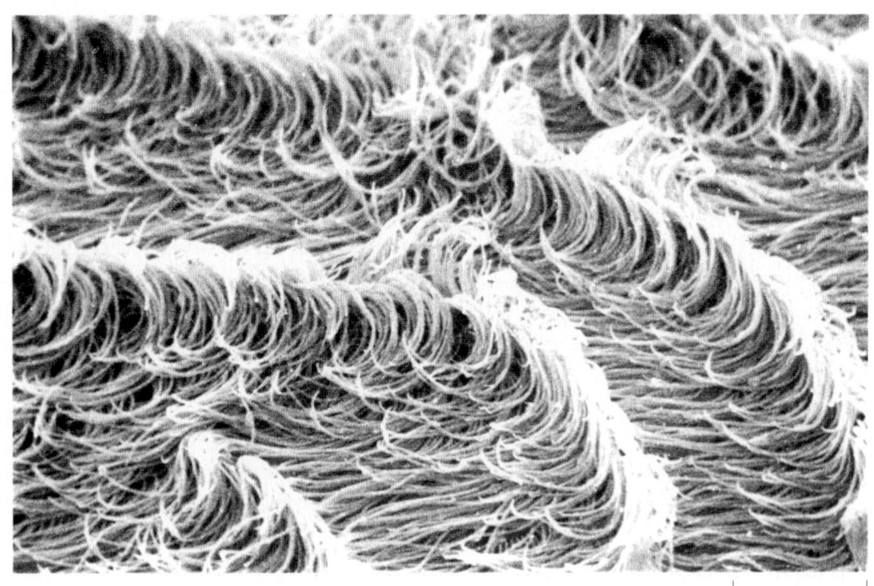

21–27
The surface of Opalina, *one of the parasitic opalinids, as seen through the scanning electron microscope. Its cilia beat continuously, about 40 to 60 times per second, in a coordinated beat that produces synchronous waves of motion.*

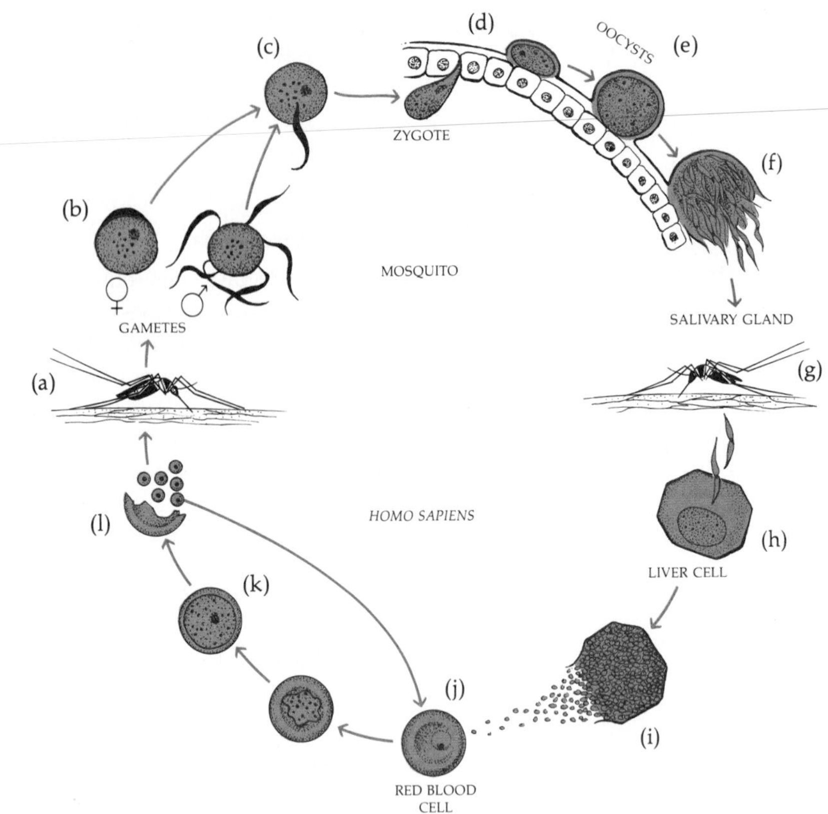

21–28
Life cycle of Plasmodium vivax, *one of the sporozoans that cause malaria in humans. The cycle begins* (a) *when a female* Anopheles *mosquito bites a person with malaria, and, along with the blood, sucks up gametes* (b) *of the protozoan. In the mosquito's digestive tract, the gametes unite* (c) *and form a zygote* (d). *From the zygotes, multinucleate structures called oocysts develop* (e), *which, within a few days, divide into thousands of very small, spindle-shaped cells, sporozoites* (f), *which then migrate to the mosquito's salivary glands. When the mosquito bites another victim* (g), *she infects the person with the sporozoites. These first enter liver cells* (h), *where they undergo multiple division* (i). *The products of these divisions (merozoites) enter the red blood cells* (j), *where again they divide repeatedly* (k). *They break out of the blood cells* (l) *at regular intervals of about 48 or 72 hours (depending on the species), producing the recurring episodes of fever characteristic of the disease. After a period of asexual reproduction, some of these merozoites become gametes, and, if they are ingested by a mosquito at this stage, the cycle begins anew.*

PATTERNS OF BEHAVIOR IN PROTISTS

In the last chapter, we saw that even at the prokaryotic level of organization, organisms are capable of behavioral responses to environmental stimuli, such as light, chemicals, and magnetic fields. Slightly more complex patterns of behavior are seen among the protists, at the eukaryotic level of organization. Photosynthetic protists, such as *Euglena,* are especially sensitive to light intensity, moving into areas with optimal levels of light but out of areas where the light is too bright. Nonphotosynthetic protists, such as amoebas, may also exhibit phototaxis. If a bright pinpoint of light is focused on the advancing pseudopodium of an amoeba, the pseudopodium withdraws. If the entire body of the amoeba is exposed to bright light, the cell contracts suddenly and will even extrude any half-digested food. However, if it cannot escape from the light, the amoeba, after a few moments' hesitation, will resume its normal activities. This type of response, by which the stimulus comes to be ignored and a previous behavior pattern restored, is known as *habituation*. It is an important component of the behavior of multicellular eukaryotes, especially animals. (If our own cells were not capable of habituation, we would be continuously responding to and distracted by, for example, the touch of our own clothing or background noises.)

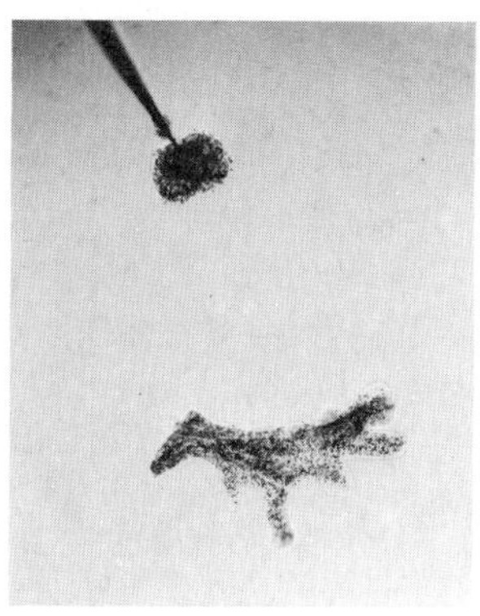
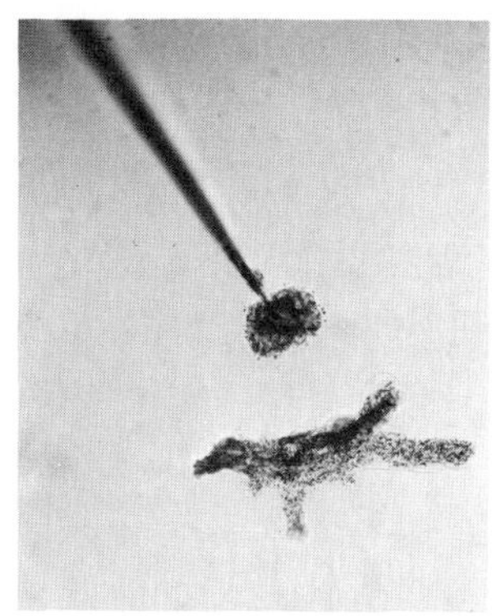
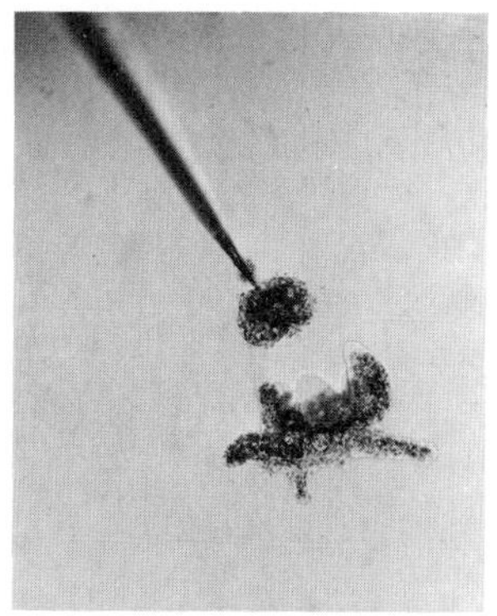
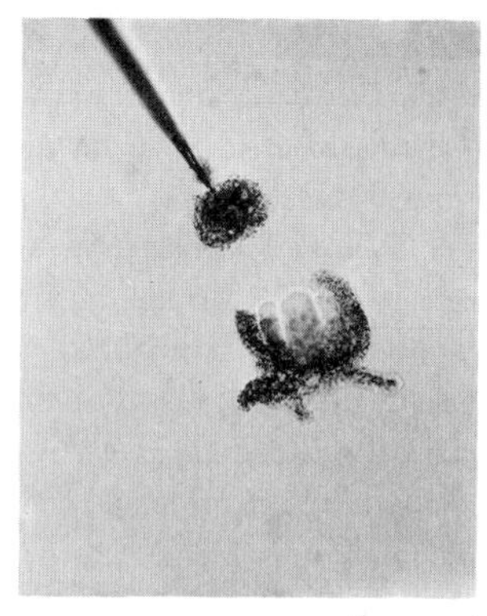

21–29
An amoeba pursues its prey. The initial stimulus produced by the "prey"—a fragment of Hydra *tentacle—induces pseudopodia from the amoeba and causes the amoeba to move toward the* Hydra *fragment. As the piece of tentacle is moved away, the amoeba moves after it and will remain in pursuit so long as the prey is close enough to continue to produce the stimulus.*

Amoebas are also chemotactic, that is, responsive to chemical stimuli. When something edible, an algal cell or a fellow protozoan, is in the vicinity of an amoeba, the amoeba can sense it at some distance, at least the length of its own body away. It then sends out a pseudopodium that is shaped quite specifically for its intended victim. A fine pincerlike projection will be formed for a small, quiet morsel; a much stronger and more massive pseudopodium will reach for a large ciliate or vigorously moving object. If the intended victim moves away, the amoeba will remain in pursuit so long as it is close enough to the prey to receive stimuli from it (Figure 21–29).

Avoidance in *Paramecium*

Other patterns of protist behavior have been studied extensively in ciliates. *Paramecium,* for example, shows behavior that appears more complex than that of amoebas but is just as simple in principle. These protists are responsive to a variety of stimuli, including very subtle changes in temperature and chemistry that we can detect only by finely calibrated instruments. But though the detection methods are highly sensitive, the responses are fixed (stereotyped).

As a *Paramecium* swims, the strongly beating cilia around the oral groove create currents that constantly bring a sample of water to the buccal cavity, which seems to be the testing area. In this way, the *Paramecium* continuously explores the environment that lies ahead, shifting and sampling (Figure 21–30).

21–30
Paramecium *samples a drop of India ink. Currents created by the strongly beating cilia draw particles of the ink toward the protist's buccal cavity.*

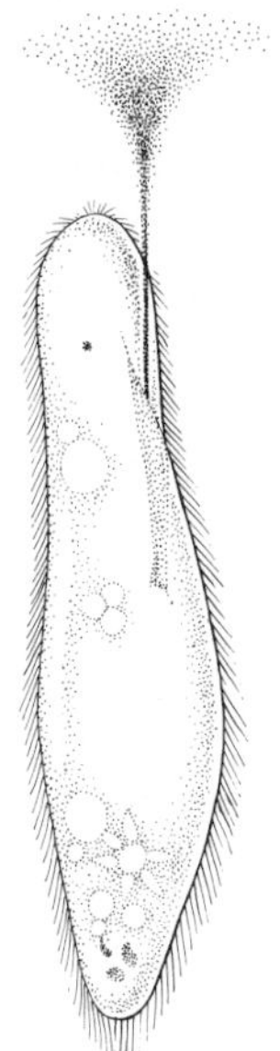

In general, *Paramecium* responds to changes in its environment by avoidance reactions. If it receives a negative stimulus, it turns away. If it turns, it always turns in the same direction, to its left, which is the aboral side—the side away from the oral groove; this occurs because of the relaxation of the beat of the body cilia. It does not matter which side the stimulus comes from. If the microscopist takes a blunt needle and jabs a *Paramecium* on the aboral side, it still turns toward that side; and once it has turned, it will continue in this new direction indefinitely.

As shown in Figure 21–31, if the negative stimulus is powerful, such as a poisonous chemical, the *Paramecium* will stop short, reverse its ciliary beat and back up, turn toward the aboral side about 30°, and then start forward again, testing. If necessary, it will repeat this performance. Under a strong negative stimulus, it will turn a full 360° and will continue to turn until an avenue of escape is found. It can find its way around solid objects in the same way.

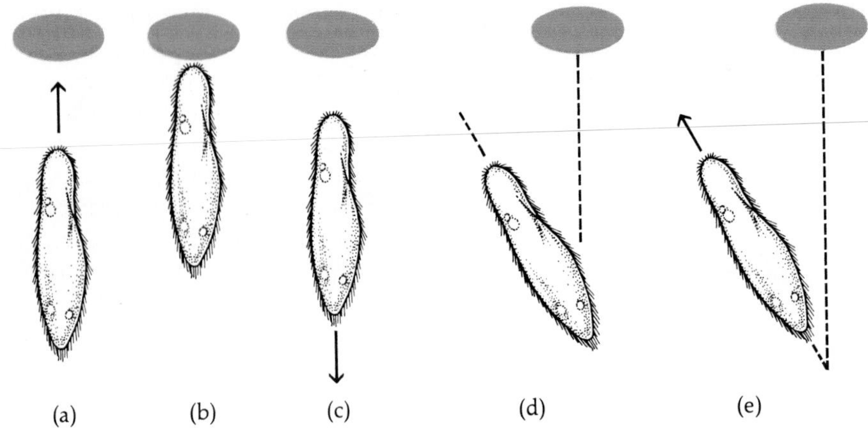

21–31
Avoidance behavior in Paramecium. *The colored area at the top of the figure represents a drop of a toxic substance, and the arrows indicate the direction of movement. The* Paramecium (a) *moves up to the substance,* (b) *tests it,* (c) *backs up,* (d) *turns 30°, and* (e) *moves forward in the new direction. Many protists are capable of this sort of simple behavior.*

The end result of this, of course, is about the same as if the organism were "attracted" to a favorable situation. For example, *Paramecium* is extremely temperature-sensitive. If you place the cells in a culture dish that can be made warmer at one end than another, you will find, by trying different combinations of temperatures, that they will tend to congregate in water that is about 27°C (80°F). If you place them on a slide with water that is below 27°C and then warm one little portion of it gently (you can do this by placing a hot needle on the cover slip), the cells will all gather in this warm spot. They reach it by chance as they move about through the water. But once they get into it, they literally cannot get out again, since as soon as they reach the edge of the warm water, they receive a sample of cooler water, which will turn them right back again. Similarly, on a slide that contains water that is too hot, they will trap themselves in an area that has been cooled slightly, even by only a few degrees.

They have also been shown to congregate in the same way in a spot that is slightly acidic, by avoiding the neutral or alkaline areas. Bacteria, which are a primary food of *Paramecium,* create a slightly acidic environment by their metabolism, and indeed, *Paramecium* itself creates a slightly acidic environment by giving off carbon dioxide; so their avoidance of neutral or alkaline areas enhances the tendency of these protists to group together and also to gather where there is food. This tendency is strengthened further by the fact that they are more likely to cling to some other object—such as the leaf or stalk of a plant, a pile of debris, or even a shred of filter paper—if they are in a slightly acidic medium.

As described here, the behavior of *Paramecium,* helpless to escape from a drop of warm water or forced to cling by the presence of carbon dioxide, seems very unlifelike. Yet, as one watches the cell negotiating the changes in its environment, its actions seem purposeful. Both of these impressions are true. The individual *Paramecium* has no choice as to its behavior, but the behavior of these protists, in general, is a result of millions of years of choice, the choice by natural selection.

SUMMARY

The kingdom Protista comprises a variety of eukaryotic organisms, mostly unicellular with some relatively simple multicellular forms. A major factor in the evolution of the eukaryotes may have been the establishment of symbiotic relationships with prokaryotic cells that, internalized, eventually became specialized as mitochondria and chloroplasts. It is thought that the protists represent a number of quite distinct phylogenetic lineages and, moreover, that all other eukaryotic organisms are derived from primitive protists. We have considered three major groups of protists: the algae (six divisions), the slime molds (two divisions), and the protozoa (five phyla).

Euglenophyta represent a small group of unicellular algae, mostly found in fresh water. They contain chlorophylls *a* and *b* and store carbohydrates in an unusual starchlike substance, paramylum. The cells lack a wall but have a flexible series of protein strips, which make up the pellicle, inside the cell membrane. The cells are highly differentiated, containing chloroplasts, a contractile vacuole, an eyespot, and flagella. No sexual cycle is known. This division also contains nonphotosynthetic forms.

The Chrysophyta—diatoms and golden-brown algae—are important components of freshwater and marine phytoplankton. They are unicellular. Diatoms are characterized by fine, double silicon-containing shells. They usually reproduce asexually, but syngamy also occurs.

The Pyrrophyta are unicellular biflagellates, many of which are marine. This division includes the dinoflagellates, which are characterized by two flagella that beat in different planes, causing the organism to spin; dinoflagellates often have stiff, bizarrely shaped cellulose walls.

The plants are thought to have originated from among the Chlorophyta, the green algae. The similarities between the two groups include chlorophylls *a* and *b* and beta-carotene as photosynthetic pigments, the storage of food reserves in the form of starch, and cellulose-containing cell walls. The widely diverse forms of green algae range from single-celled organisms such as *Chlamydomonas* to a variety of multicellular forms. Multicellularity has arisen in at least three different ways and in a number of different evolutionary lineages.

The reproductive cycles of green algae are often quite complex. In species with sexual cycles, the gametes of different mating types may be similar in size and structure (isogamy), different in size but both motile (anisogamy), or different in size, with one, usually the larger, not motile (oogamy). Some multicellular green algae have a life cycle known as alternation of generations, in which a haploid phase alternates with a diploid phase. The haploid (n) generation, known as the gametophyte, produces haploid gametes. The gametes fuse to form the zygote, which develops into a diploid $(2n)$ sporophyte. The sporophyte produces spores by meiotic division. A spore is a single cell that, unlike a gamete, can develop into an adult organism without combining with another cell. In organisms with alternation of generations, the spore, which is haploid, germinates to produce the haploid gametophyte.

The Phaeophyta (brown algae) and Rhodophyta (red algae) are the principal seaweeds. The brown algae, which include the kelps, are found more commonly in cooler water; the red algae, in the tropics. In some brown algae the thallus is differentiated into holdfast, stipe, and blade, analogous to the root, stem, and leaf of plants. Some kelps have tissues specialized for the conduction of sugar from the blades to nonphotosynthetic parts of the thallus.

The slime molds are heterotrophic, amoeba-like organisms that reproduce by the formation of spores. There are two principal divisions: Myxomycota (plasmodial slime molds), which are coenocytic, and Acrasiomycota (cellular slime molds).

Animal-like protists, the protozoans, are thought to have evolved from nonphotosynthetic, flagellated ancestors. Among the protozoans are some of the largest known cells and also the most complex. Three phyla—the Mastigophora (flagellates), the Sarcodina (amoebas and their relatives), and the Ciliophora (ciliates)—contain many free-living species, and their members may be identified on the basis of their locomotor structures. Two phyla—Opalinida and Sporozoa—contain only parasitic forms. The opalinids are ciliated, but the sporozoans have no locomotor organelles. In the course of their complex life cycles, the sporozoans, however, are transported quite efficiently from host to host.

Protists exhibit a variety of simple behavioral responses that prefigure the complex behaviors characteristic of multicellular eukaryotes. Among the protist responses are phototaxis (in both photosynthetic and nonphotosynthetic organisms), chemotaxis, avoidance, and habituation.

QUESTIONS

1. Describe the similarities and differences among the Euglenophyta, the Chrysophyta, and the Pyrrophyta.

2. In some classification schemes, the Chlorophyta, the Phaeophyta, and the Rhodophyta are placed in the plant kingdom. What similarities between the plants and these three divisions of algae might justify such a placement? What differences between the plants and these divisions justify their placement in the kingdom Protista? Which of these similarities and differences are most likely to be homologous and which analogous?

3. Describe the three different pathways to multicellularity that have been discussed.

4. Distinguish among the following: colonial organism/multicellular organism; syngamy/fertilization; isogamy/anisogamy/oogamy; isomorphic/heteromorphic; sporophyte/gametophyte; spore/gamete.

5. Explain alternation of generations, using as your example the sea lettuce, *Ulva*.

6. Name the five phyla of protozoans, and give the distinguishing characteristics of each.

7. Label the drawing at the left.

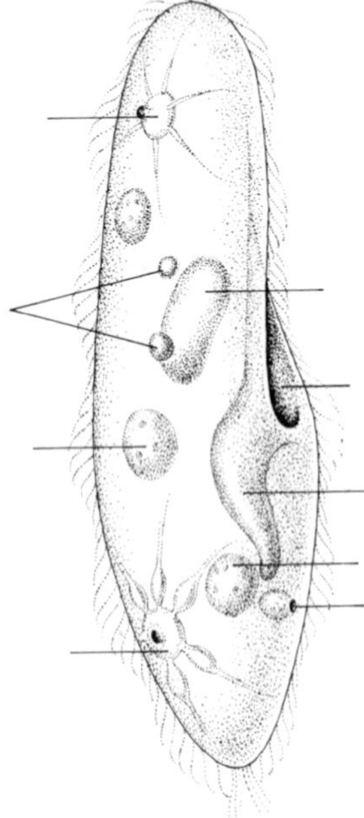

8. Consider the life cycle of *Plasmodium.* At what stages in the cycle do its numbers increase? Why might a parasite that requires several hosts find it advantageous to evolve a life cycle in which its numbers increase at several stages? Why might it be advantageous to a parasite to have a second host, such as the mosquito?

9. Given that light is essential for photosynthesis, it is easy to understand the adaptive significance of behavioral responses to light by photosynthetic protists. The significance of such responses for nonphotosynthetic organisms is not so obvious. What might be the advantages to an amoeba of a capacity to respond to light?

10. Arrange in order of evolutionary development: shell or skeleton, phagocytosis, active transport, multicellularity, symbiosis. (Like the rest of us, you will only be guessing, but be prepared to defend your guess.)

CHAPTER 22

The Fungi

The fungi are a group of organisms so unlike any others that, although they were long classified with the plants, it now seems appropriate to assign them to a separate kingdom. Although some fungi, including the yeasts, are unicellular, most species are coenocytic or multicellular organisms composed of masses of filaments. A fungal filament is called a *hypha*, and all the hyphae of a single organism are collectively called a *mycelium*. The complex, spore-producing structures of fungi, such as mushrooms, are tightly packed hyphae. In most groups of fungi, the cell walls are composed primarily of chitin (Figure 22–2), a polysaccharide that is never found in the kingdom Plantae. (It is, however, the principal component of the exoskeleton—the hard outer covering—of insects and other arthropods.)

A mycelium normally arises by the germination and outgrowth of a single cell, with growth taking place only at the tips of hyphae. Most fungi are nonmotile throughout their life cycle, although spores may be carried great distances by the wind. Growth of the mycelium substitutes for motility, bringing the organism into contact with new food sources and different mating strains. This growth can be quite rapid—a fungus may produce more than a kilometer of new mycelium within 24 hours.

22–1
Gill fungi on a dead quaking aspen. The only portions of these fungi visible are the spore-producing structures, composed of tightly packed hyphae; the bulk of the mycelium is below the surface of the dead trunk.

22–2
Fungal cell walls characteristically contain chitin (a) *rather than cellulose* (b), *the polysaccharide found in the cell walls of plants. Chitin resembles cellulose in that it is tough, inflexible, and insoluble in water. As you can see, they are also structurally very similar; in chitin, the hydroxyl (OH) group in the 2 position is replaced by a nitrogen-containing group.*

(a) CHITIN

H HNCOCH$_3$ CH$_2$OH H HNCOCH$_3$
O OH H H H H O O OH H H
H OH H H
H O O H H O
CH$_2$OH H HNCOCH$_3$ CH$_2$OH

(b) CELLULOSE

H OH CH$_2$OH H OH
O OH H H H H O O OH H H
H OH H H
H O O H H O
CH$_2$OH H OH CH$_2$OH

All fungi are heterotrophs with a highly characteristic means of nutrition. Because of their filamentous form, each fungal cell is no more than a few micrometers from the soil, water, or other substance in which the fungus lives, and is separated from it only by a thin cell wall. Because their cell walls are rigid, fungi are unable to engulf small microorganisms or other particles. They obtain food by absorbing dissolved inorganic or organic materials. Typically a fungus will secrete digestive enzymes onto a food source and then absorb the smaller molecules released. The mycelium may appear as a mass on the surface of the food source or may be hidden beneath the surface. Parasitic fungi often have specialized hyphae, called *haustoria* (singular, haustorium), that penetrate and absorb nourishment directly from the cells of the host organism. (As we shall see on page 651, some parasitic plants have analogous structures, also known as haustoria.)

The fungi, together with the bacteria, are the principal decomposers of the world. As we shall see in Section 8, their activities are as vital to the continued function of the earth's ecosystems as are those of the food producers. Some are also destructive; they may interfere with human activities by attacking our foodstuffs, our domestic plants and animals, our shelter, our clothing, and even our persons.

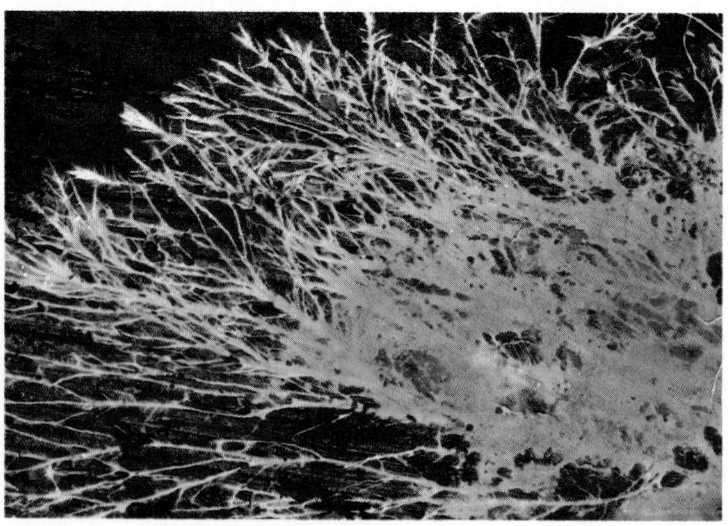

22–3
Mycelium of a fungus growing on the inner surface of the bark of a log.

22–4
A sporangium, an asexual reproductive structure of a fungus, from the black bread mold Rhizopus. *The contents of the sporangium are segregated from the rest of the mycelium by a cell membrane and a cell wall.*

REPRODUCTION IN THE FUNGI

Fungi reproduce both asexually and sexually. Asexual reproduction takes place either by the fragmentation of the mycelium (with each fragment becoming a new individual) or by the production of spores. In some of the fungi, spores are produced in reproductive structures called *sporangia*, and are borne on specialized hyphae called *sporangiophores*. Spores are often, but not necessarily, resting forms, surrounded by a tough, resistant wall. They are able to survive during periods of lack of water and extreme temperatures. Some airborne spores are very small and so can remain suspended in the air for long periods and be widely dispersed. Often the sporangia are raised above the mycelium by the sporangiophores; thus the spores are easily caught up and transported by air currents. The bright colors and powdery textures associated with many particular types of molds are produced by the spores.

Sexual reproduction typically involves the specialization of portions of the hyphae to form structures known as *gametangia*. The contents of a gametangium, like those of a sporangium, are separated from the hypha from which it is formed by a cell membrane and a complete cell wall, known as a *septum*. Sexual reproduction can occur in a variety of ways: (1) by fusion of gametes that have been released from the gametangia, (2) by penetration of a gamete into a gametangium, or (3) by fusion of gametangia.

Sometimes the fusion of fungal hyphae is not followed immediately by the fusion of nuclei. Thus strains of fungi may exist with two or more genetically distinct kinds of nuclei operating simultaneously. When such a combination contains two nuclei of complementary mating types, it is known as a *dikaryon*. Dikaryons are found uniquely among the higher fungi (divisions Ascomycota and Basidiomycota).

CLASSIFICATION OF THE FUNGI

The members of the kingdom Fungi are generally classified into five principal divisions: Chytridiomycota, commonly called the chytrids; Oomycota, the oomycetes; Zygomycota, the zygomycetes; Ascomycota, the ascomycetes; and Basidiomycota, the basidiomycetes. The criteria used in distinguishing these five divisions involve both features of basic structure and patterns of reproduction, particularly sexual reproduction (Table 22–1, on page 444). An additional taxon, Deuteromycota, or the Fungi Imperfecti, includes fungi in which sexual reproduction is unknown, either because it has been lost in the course of evolution or because it has not been observed. Also included in this taxon for convenience are certain other closely related fungi whose sexual stages are known. Taxonomists refer to groups like this as "waste baskets" because species are included here only because they do not fit in other taxa. Because of similarities between the patterns of asexual reproduction in many of the Fungi Imperfecti and the ascomycetes, this taxon is sometimes considered a class of division Ascomycota; most authorities, however, rank it at the level of division.

The first fungi were probably unicellular eukaryotic organisms that apparently have no living counterparts. These organisms are thought to have given rise to three distinct lineages, one leading to the modern chytrids, a second leading to the oomycetes, and a third leading to the zygomycetes. All three of these groups are characterized by a coenocytic organization, and with the exception of the repro-

22–5
The perforated cross wall of an ascomycete showing a nucleus squeezing through the perforation. The fungus is Neurospora crassa, *the red bread mold.*

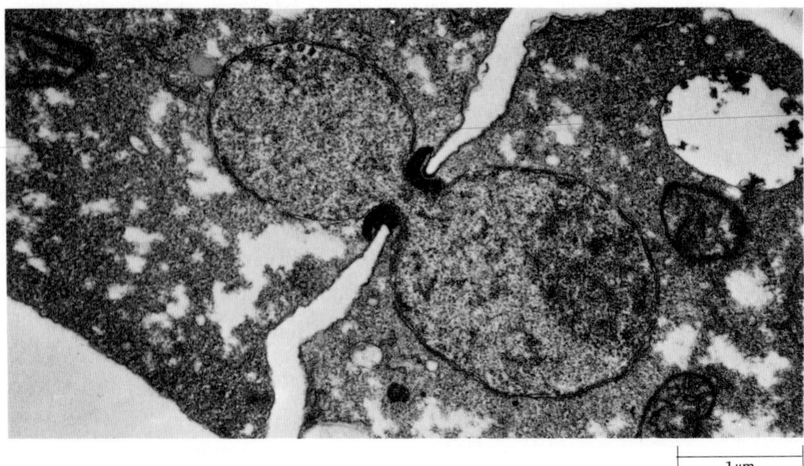

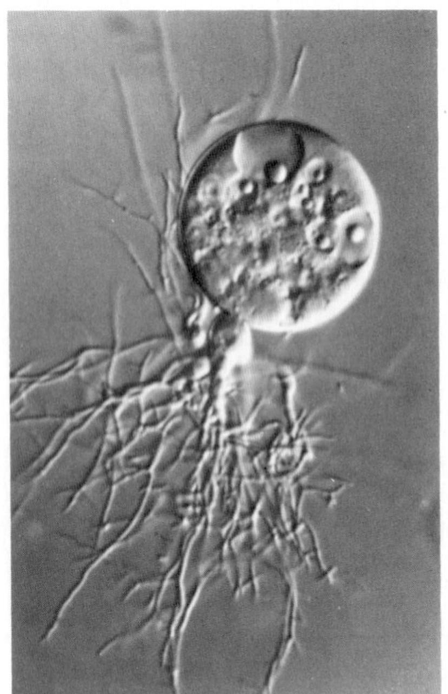

22–6
Chytridium confervae, *a common chytrid, as photographed through a differential-interference microscope. Note the slender, nutrient-absorbing rhizoids extending downward from the growing sporangium. The sporangium of this coenocytic organism contains numerous nuclei, vacuoles, and organelles, such as mitochondria. The rhizoids contain vacuoles and organelles but no nuclei.*

ductive structures, there is no compartmentalization of the mycelia. However, the chytrids and the oomycetes differ so significantly from all other fungi in a number of basic features that some authorities believe that they should not be classified with the fungi at all but should instead be placed in the kingdom Protista.

In both the ascomycetes and the basidiomycetes, the hyphae are septate—divided by transverse cell walls—but the walls are perforated, and the cytoplasm and even the nuclei (Figure 22–5) are able to flow through the septa. It is generally thought that the ascomycetes and the basidiomycetes are derived from a common ancestor, and that this ancestor and the zygomycetes evolved from an earlier common ancestor.

CHARACTERISTICS OF THE FUNGI

Division Chytridiomycota

The chytrids are thought to be the most ancient group of fungi. Although some species produce an extended mycelium, most consist of a small thallus, typically differentiated into a sporangium and rootlike anchoring structures, called *rhizoids* (Figure 22–6). The rhizoids, which absorb nutrients from the substrate, resemble hyphae but do not contain nuclei. When the chytrid is mature, the coenocytic sporangium cleaves into flagellated spores, each of which contains a single nucleus. Following free-swimming and encysted stages, each spore can germinate to form a new thallus.

Other chytrids are much more complex in their structure and reproduction. Chytrids are the only fungi with flagellated gametes, which may be similar or different in size and motility. In the genus *Allomyces,* the larger, less active gametes produce a hormone, appropriately named sirenin, that attracts the smaller, more active gametes. Like some algae and all plants, but alone among the fungi, *Allomyces* and one other closely related genus have a life cycle characterized by alternation of generations.

Chytrids are found in both fresh and salt water and in moist soil. They live as parasites on algae, plants, and other fungi, or as saprobes, feeding on dead algae, pollen grains, and other plant debris.

Division Oomycota

The oomycetes derive their name from *oion*, the Greek word for "egg." All oomycetes display oogamy—the differentiation of gametes into forms that are clearly male (sperm) and female (egg). However, the gametes of oomycetes, unlike those of chytrids, have no flagella and are nonmotile. In sexual reproduction, the sperm nucleus and the egg, each of which is borne in its own type of gametangium, fuse to produce a zygote (Figure 22–7). Oomycetes can also reproduce by forming asexual spores, each of which bears two flagella. Many oomycetes are aquatic (and thus they are also known as the water molds), but even the terrestrial forms can produce flagellated spores that require free water. Another distinctive characteristic of the oomycetes is the composition of their cell walls; the walls of oomycetes—unlike those of all other fungi—consist primarily of cellulose and only rarely contain chitin.

Most oomycetes are saprobes, living on dead organic matter. Some forms are parasitic and pathogenic, however, and as mycologist C. J. Alexopoulos has said, "At least two of them have had a hand—or should we say a hypha!—in shaping the economic history of an important portion of mankind." The first of these is *Phytophthora infestans (phytophthora* literally means "plant destroyer"), the cause of the "late blight" of potatoes, which produced the great potato famines in Ireland (Figure 22–8). The second economically important member of this group is *Plasmopara viticola*, the cause of downy mildew of grapes. This mildew threatened the entire French wine industry during the latter part of the nineteenth century.

22–7

Mating in the fungus Achyla ambisexualis, *an oomycete. The large spherical structure is the female gametangium. The dark bodies within it contain eggs. Encircling the female gametangium is the male gametangium. Fertilization tubes extending from the male into the female gametangium are barely visible. Sperm nuclei pass through these tubes to the egg nuclei. Development of the male gametangia and their attraction to the female are controlled by the production of a steroid hormone remarkably similar in structure to the human sex hormones.*

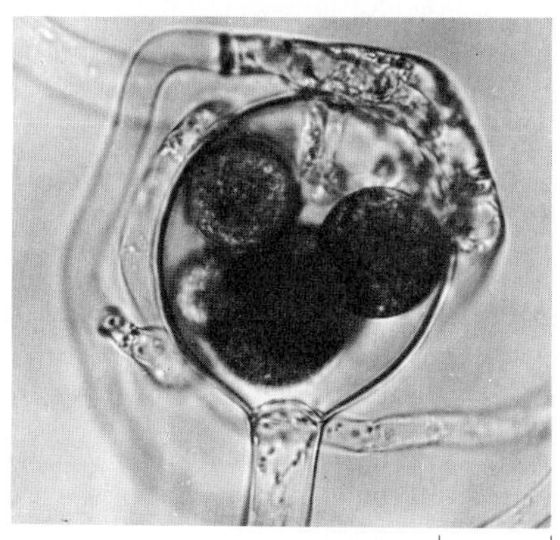

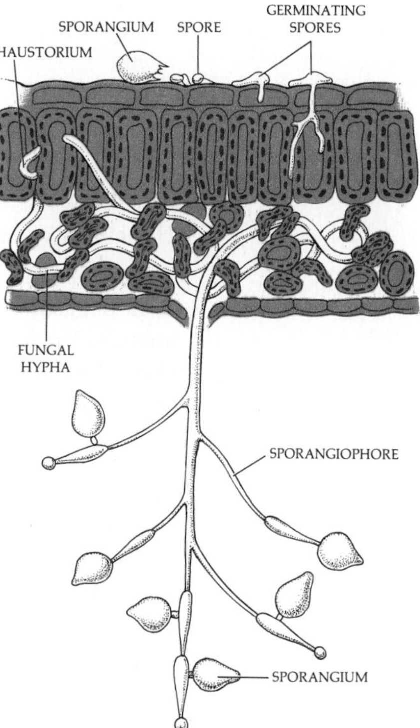

22–8

Phytophthora infestans, *cause of potato blight. Infection begins when an airborne sporangium alights on a leaf, releasing spores that move about in the film of water on the leaf's surface. These spores germinate, producing haustoria (specialized hyphae) that penetrate the epidermis and attack the photosynthetic cells in the interior of the leaf. Eventually aerial hyphae—sporangiophores—bear sporangia from which a new generation of asexual spores is released.*

Table 22-1 The Kingdom Fungi

DIVISION	NUMBER OF SPECIES	EXAMPLES	DISTINCTIVE CHARACTERISTICS	DISEASES	ECONOMIC USES
Chytridiomycota	About 650	*Chytridium*, *Allomyces*	Mostly aquatic; flagellated spores and gametes; cell walls contain chitin; vegetative body usually a thallus, rarely hyphal	Lettuce big vein; viruslike infection of corn	None
Oomycota	About 475	Potato blight fungus	Some aquatic; flagellated spores; formation of eggs and sperm in special gametangia; cell walls contain cellulose	Blights and mildews of plants; fish infections	None
Zygomycota	About 600	Black bread mold	Formation of zygospores (tough, resistant spores resulting from a fusion of gametangia); no flagellated cells	Few	None
Ascomycota	30,000	*Neurospora*, yeasts, morels, truffles	Formation of fine asexual spores (conidia); sexual spores in asci; hyphae divided by perforated septa; dikaryons; no flagellated cells	Powdery mildews of fruits, chestnut blight, Dutch elm disease, ergot	Food (morels, truffles), wine, beer, bread-making (yeasts)
Basidiomycota	25,000	Toadstools, mushrooms, rusts, smuts	Sexual spores in basidia; hyphae divided by perforated septa; dikaryons; no flagellated cells	Rusts, smuts	Food (mushrooms)
Deuteromycota (Fungi Imperfecti)	25,000	*Penicillium*	Fungi with no known sexual cycles; no flagellated cells	Ringworm, thrush	Cheeses, antibiotics

Division Zygomycota

The zygomycetes are terrestrial fungi, most of which live in the soil, feeding on dead plant or animal matter. Some are parasites of plants, insects, or small soil animals. Unlike chytrids and oomycetes, they produce no flagellated cells at any stage of the life cycle. In sexual reproduction, they produce zygospores, which are thick-walled, resistant spores that develop from a zygote.

One of the most common members of this division is *Rhizopus stolonifer*, the black bread mold. Infection begins when a spore germinates on the surface of bread, fruit, or some other organic matter and forms hyphae. Some of the hyphae extend rhizoids, which anchor the fungus to the substrate, secrete digestive enzymes, and absorb dissolved organic materials. Other specialized hyphae, the sporangiophores, push up into the air, and sporangia form at their tips. As the sporangia mature, they become black, giving the mold its characteristic color. They eventually break open, releasing numerous airborne spores, each of which can germinate to produce a new mycelium.

Sexual reproduction in *Rhizopus* occurs when the specialized hyphae of two different mating strains meet and fuse, attracted toward one another by hormones that diffuse in the form of gases. The two strains are designated + and −, since there are no morphological differences between them on which to base male and female designations. Septa, or cross walls, form behind the tips of the touching hyphae; the two tip cells thus formed are gametangia, one containing numerous + nuclei, the other containing numerous − nuclei. The two gametangia fuse, and the two types of nuclei then fuse, producing diploid nuclei. The resulting multinucleate cell, the zygote, forms a hard, warty wall and becomes a dormant zygospore. During this dormant stage, it can survive periods of extreme heat or cold or desiccation. At the end of dormancy, only one diploid nucleus remains, and this undergoes meiosis when the zygote germinates. Only one of the four nuclei produced by meiosis generally survives. It commonly gives rise to a sporangiophore, from which airborne spores are released.

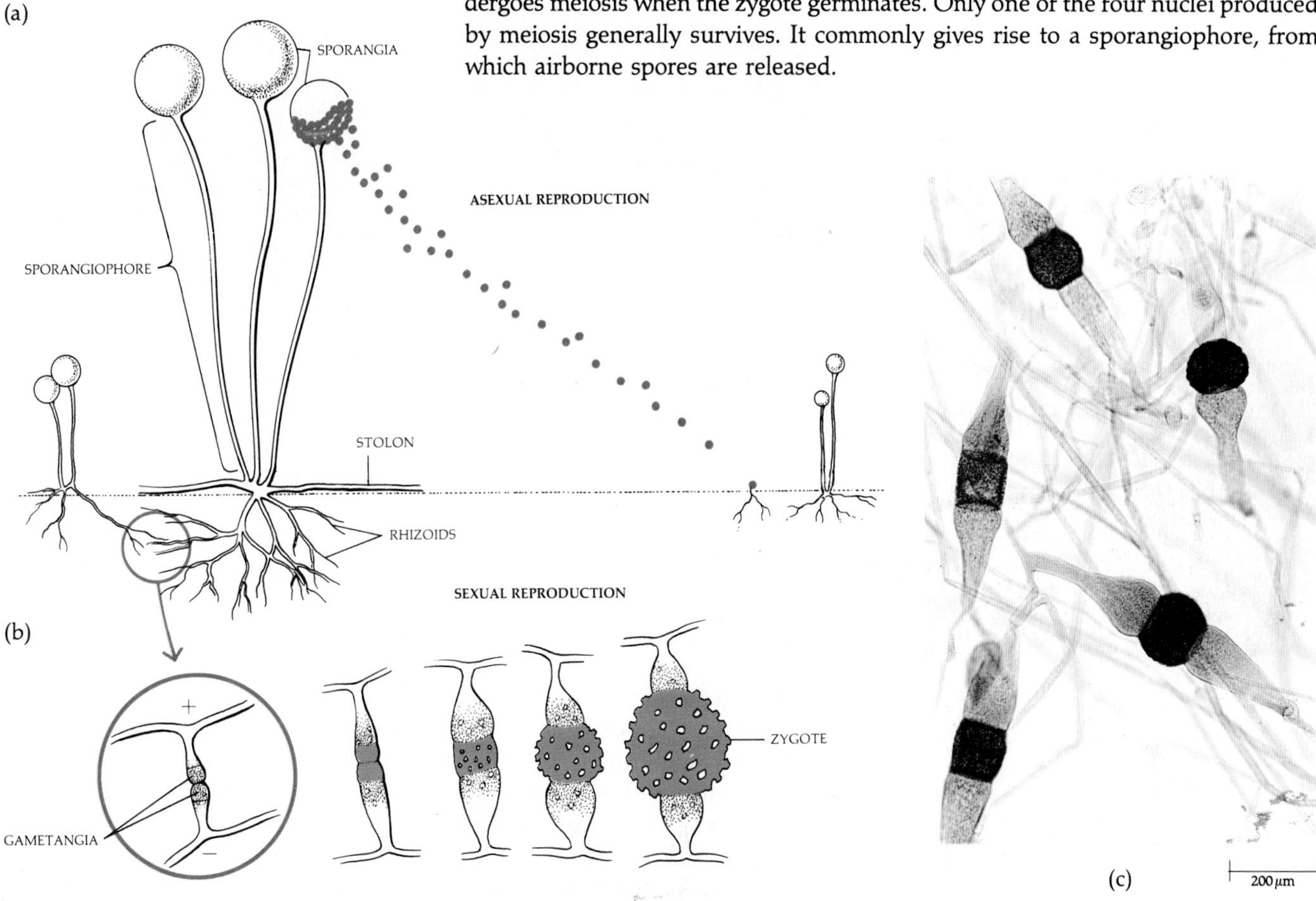

22–9
Asexual and sexual reproduction in Rhizopus. *The mold consists of branched hyphae, including rhizoids, which anchor the mycelium and absorb nutrients; stolons, which run above ground; and sporangiophores.* (a) *At maturity, the fragile wall of the sporangium disintegrates, releasing the asexual spores, which are carried away by air currents. Under suitable conditions of warmth and moisture, the spores germinate, giving rise to new masses of hyphae.* (b) *Sexual reproduction occurs when two hyphae from different mating strains come together, forming gametangia, which fuse to form a thick-walled, resistant zygote, commonly called a zygospore* (c). *After a period of dormancy, the zygote undergoes meiosis and germinates, producing a new sporangium.*

22–10
Two ascomycetes. (a) *A common morel. These (and truffles) are among the most prized of the edible fungi. The structure recognized as the morel is the ascocarp, in which asci and ascospores are produced.* (b) *Scarlet cup, a harbinger of spring in hardwood forests throughout the United States. It is usually found arising from a fallen branch.*

(a)

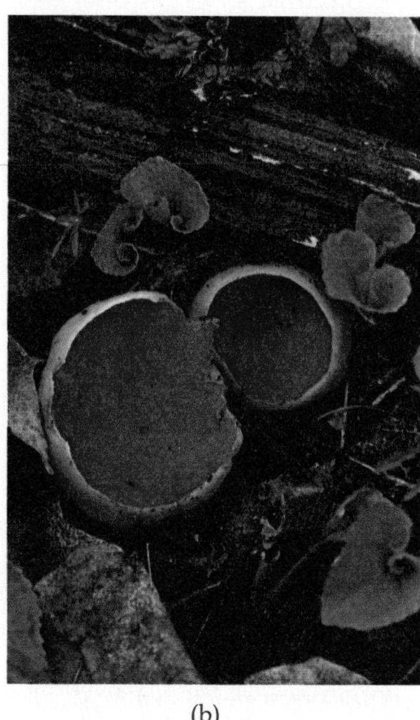

(b)

Division Ascomycota

The ascomycetes are the largest division of fungi, with some 30,000 species, plus an additional 25,000 species that are found only in lichens (page 449). Among the ascomycetes are the yeasts and powdery mildews, many of the common black and blue-green molds, and the morels and truffles prized by gourmets. Members of this group of fungi are the cause of many plant diseases, such as chestnut blight and Dutch elm disease, and are the source of many of the antibiotics. The red bread mold *Neurospora,* which played a major role in the history of modern genetics, is an ascomycete.

In ascomycetes, as we noted earlier, the hyphae are divided by cross walls, or septa. Each compartment generally contains a separate nucleus, but the septa have pores in them through which the cytoplasm and the nuclei can move (see Figure 22–5). Spores are formed sexually and asexually. Asexual spores are commonly formed either singly or in chains at the tip of a specialized hypha. They are characteristically very fine and so are often called conidia, from the Greek word for "dust."

Sexual reproduction always involves the formation of an ascus ("little sac"), a structure that is characteristic of this division (Figure 22–11a). Ascus formation is preceded by the fusion of gametes, gametangia, or unspecialized hyphae of different mating strains. The nuclei form pairs—dikaryons—that divide synchronously as the hypha grows. Eventually some of the nuclei fuse; this is the only truly diploid stage in the life cycle. The diploid nuclei immediately undergo meiosis, producing four haploid nuclei, which then usually divide mitotically, producing eight haploid nuclei (Figure 22–11b). Each of these nuclei becomes surrounded by a tough wall; a mature ascus contains eight of these spores (ascospores). In most ascomycetes, the asci are formed in complex structures called ascocarps. At maturity the asci become turgid and finally burst, releasing their ascospores explosively into the air.

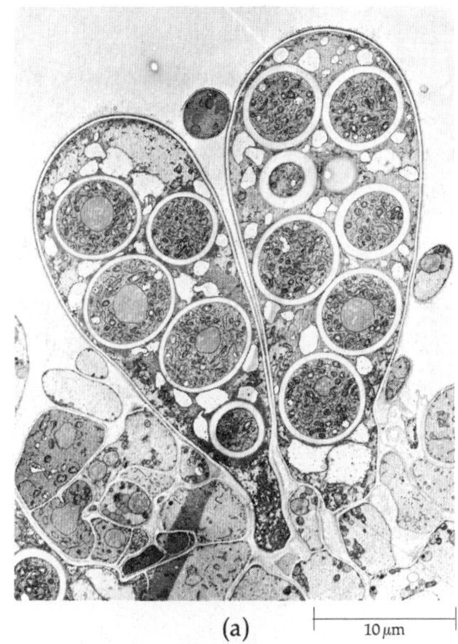

(a)

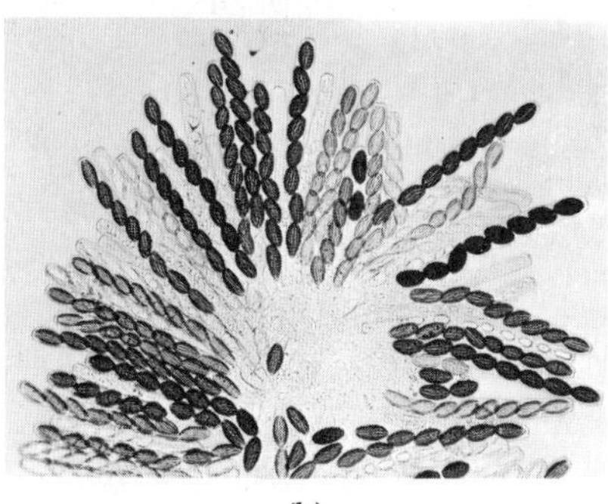

(b)

22–11
(a) *Electron micrograph of two asci in which ascospores are maturing. The closed ascus within which the sexually produced spores develop is the "trademark" of the ascomycete.* (b) *Asci of the red bread mold* Neurospora. *Each ascus contains eight haploid ascospores, lined up in the order in which they were produced by meiosis and the subsequent mitotic division.*

Single-celled ascomycetes are known as yeasts. Many yeasts are adapted to environments with high sugar content, such as the nectar of flowers or the surface of fruits and, as we noted in Chapter 9, they are responsible for the fermentation of fruit juice to wine. Yeasts are characteristically small, oval cells that reproduce asexually by budding. Sexual reproduction in yeasts occurs when two cells (or two ascospores) unite and form a zygote. The zygote may produce diploid buds or may undergo meiosis to produce four haploid nuclei. There may be a subsequent mitotic division. Within the zygote wall, which is now an ascus, walls are laid down around the haploid nuclei so that ascospores are formed, which are liberated when the ascus wall breaks down.

Many ascomycetes are plant parasites. Ergot, for instance, one of the most famous fungus-produced diseases, is caused by *Claviceps purpurea,* a parasite of rye. Although ergot seldom causes serious damage to a crop of rye, it is dangerous because a small amount mixed with rye grains is enough to cause severe illness among domestic animals or among the people who eat bread made with the flour. Ergotism is often accompanied by gangrene, nervous spasms, psychotic delusions, and convulsions. It occurred frequently during the Middle Ages, when it was known as St. Anthony's fire. In one epidemic in the year 994, more than 40,000 people died. Ergot, which causes muscles to contract and blood vessels to constrict, has various medical uses. It is also the initial source for the psychedelic drug lysergic acid diethylamide (LSD).

Division Basidiomycota

The most familiar basidiomycetes are mushrooms. The mushroom, or basidiocarp, which is the spore-producing body, is composed of masses of tightly packed hyphae. The mycelium from which the mushrooms are produced forms a diffuse mat, which may grow as large as 35 meters in diameter. Mushrooms usually form at the outer edges, where the mycelium grows most actively, since this is the area in which there is the most nutritive material. As a consequence, the mushrooms will appear in rings, and as the mycelium grows, the ring becomes larger and larger in diameter. Such circles of mushrooms, which might appear in a meadow overnight, were known in European folk legends as "fairy rings." They can make such a rapid appearance because most of the production of new protoplasm takes place underground, in the mycelium. The protoplasm then streams into the new hyphae of the fruiting body as it forms above ground.

The basidiomycetes, like the ascomycetes, have hyphae subdivided by perforated septa. Sexual reproduction is initiated by the fusion of haploid hyphae to form a dikaryotic mycelium (Figure 22–12). The dikaryotic mycelium may persist for years, forming an elaborate structure. Eventually, some of the nuclei fuse to form diploid nuclei that immediately undergo meiosis. Fusion and meiosis always take place in a specialized hypha called a *basidium* (from the Greek word for "club"). The spores (basidiospores) are formed externally on the basidium. Many of the larger basidiomycetes seem to have lost the capacity to produce asexual spores.

The best-known mushrooms belong to the group known as the gill fungi, which includes *Agaricus campestris,* the common field mushroom. Varieties of this species are among the most familiar of the mushrooms commercially cultivated in North America. The gill fungi also include most of the known poisonous mushrooms.

22–12
Life cycle of a basidiomycete. Basidiospores (upper left) germinate to produce primary monokaryotic mycelia. Secondary dikaryotic mycelia are formed by the fusion of hyphae from different mating types. The secondary mycelia grow and differentiate to form the reproductive structures (basidia). In mushrooms, the basidia form within the gills. After the basidium enlarges, the two nuclei, one from each mating strain, fuse. Meiosis follows almost immediately, resulting in the formation of four nuclei, from each of which a basidiospore develops. After the basidiospores are released, the basidiocarp disintegrates.

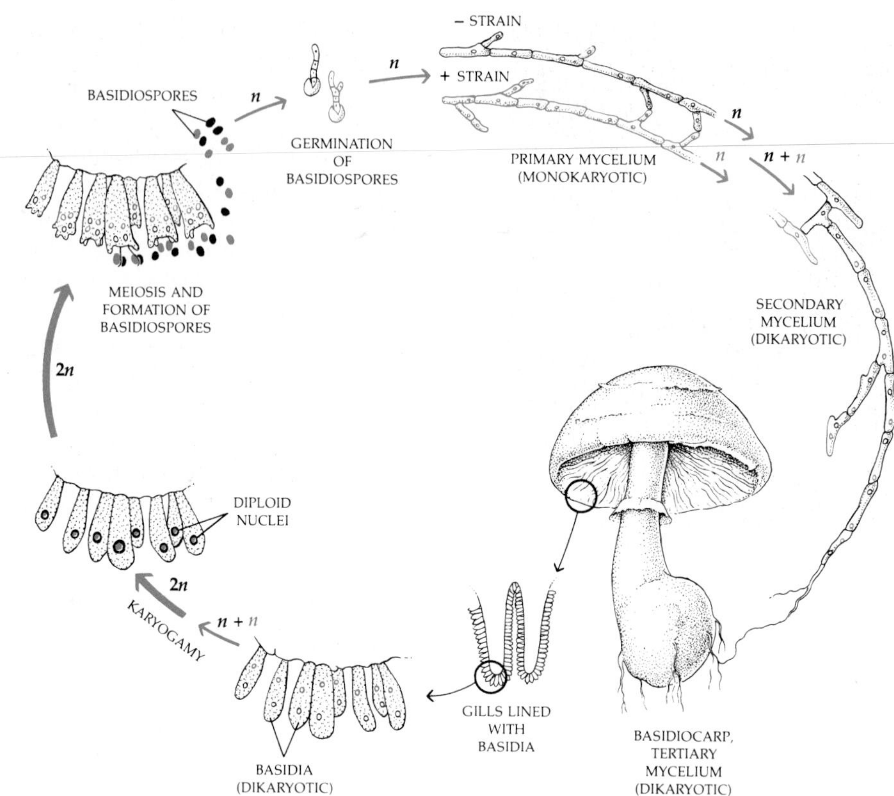

Mushrooms of the genus *Amanita* are the most highly poisonous of all mushrooms; even one bite of the white *Amanita bisporigera,* a "destroying angel," can be fatal. Some species of toxic mushrooms, such as *Psilocybe mexicana* (the source of psilocybin), are eaten for their hallucinogenic effects.

The spores of the gill mushrooms are found in the furrows, or "gills," under the cap. If you cut off the cap of a mature mushroom and place it on a piece of white paper, it will release fine spores that will trace out a negative copy of the gill structure. The spores, which come in a variety of colors, are a useful means of identifying various mushrooms. Other types of basidiomycetes include puffballs (a few of which are a meter in diameter), earthstars, stinkhorns, jelly fungi, and the parasitic rusts and smuts, some of which cause severe losses among cereal crops.

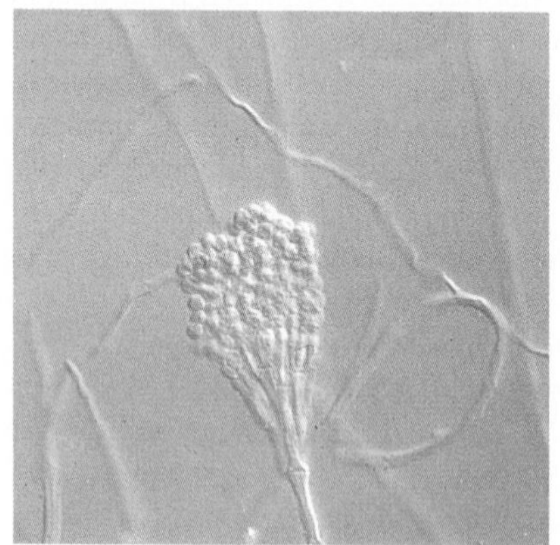

22–13
Penicillium, *an "imperfect fungus," showing conidiophores ("conidia-bearers"), which have formed at the tips of hyphae. Conidia are dust-fine asexual spores, characteristic of the ascomycetes.*

Division Deuteromycota

As we noted earlier, the deuteromycetes, or Fungi Imperfecti, are generally fungi in which sexual reproduction is unknown. About 25,000 species of deuteromycetes have been described, among them parasites that cause diseases of plants and animals. The most common human diseases caused by this group are infections of the skin and mucous membranes, such as ringworm (including "athlete's foot") and thrush (to which infants are particularly susceptible). A few deuteromycetes are of economic importance due to the part they play in the production of certain cheeses (Roquefort and Camembert, for example) and of antibiotics, including penicillin.

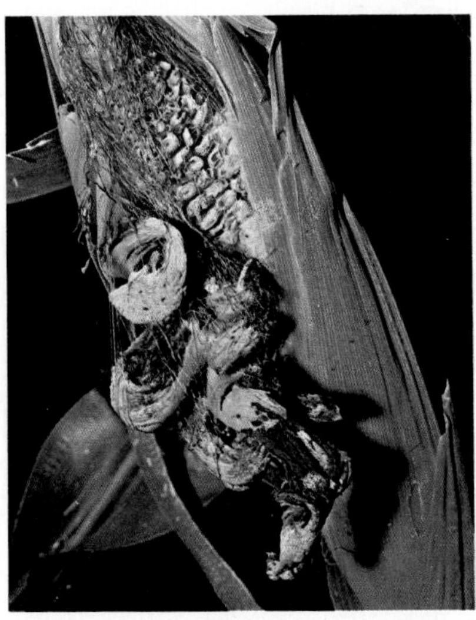

(a)

(b)

(c)

22–14
Three basidiomycetes. (a) *Corn smut, a common fungal disease of corn. The black, dusty-looking masses are spores.* (b) *A shelf fungus, which grows on decaying wood.* (c) Amanita bisporigera. *Members of this genus include the most beautiful and also the most poisonous of the mushrooms. The "skirt" near the top of the stalk is one of the identifying characteristics of this genus. One mushroom has been picked to show the gills, on which the sexual spores are formed. The toxin in* Amanita bisporigera *consists of two linked cyclopeptides, each containing eight amino acids. This protein binds to an RNA polymerase in liver cells and causes acute liver damage, resulting in death.*

SYMBIOTIC RELATIONSHIPS OF FUNGI

Although most fungi are saprobes, living on dead organic matter, a large number of fungi are parasitic on plants and animals, causing a variety of diseases. Fungi are also involved in other types of symbioses. Two of these—lichens and mycorrhizae—have been and are of extraordinary importance in enabling photosynthetic organisms to become established in previously barren terrestrial environments.

The Lichens

A lichen is a combination of a specific fungus and a green alga or a cyanobacterium. The product of such a combination is very different from either the photosynthetic organism or the fungus growing alone, as are the physiological conditions under which the lichen can survive. The lichens are widespread in nature. They occur from arid desert regions to the Arctic and grow on bare soil, tree trunks, sun-baked rocks, fence posts, and windswept alpine peaks all over the world; they have even been found growing in air spaces within Antarctic rocks, a few millimeters below the frigid rock surface. Lichens are often the first colonists of bare rocky areas; their activities begin the process of soil formation, gradually creating an environment in which mosses, ferns, and other plants can gain a foothold.

Lichens do not need an organic food source, as do their component fungi, and unlike many free-living algae and cyanobacteria, they can remain alive even when desiccated. They require only light, air, and a few minerals. They apparently

(a)

(b)

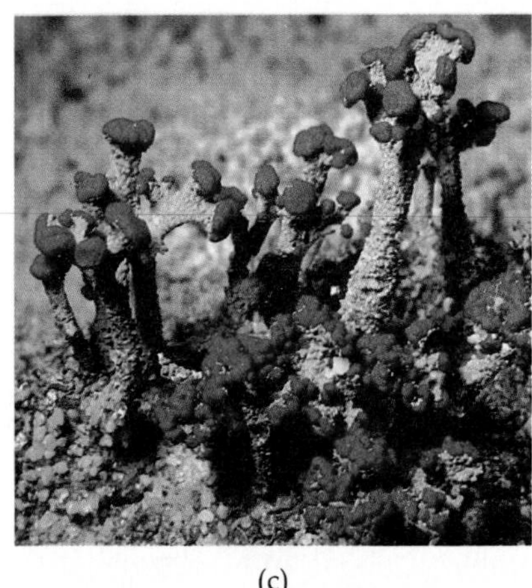

(c)

22–15
(a) Crustose ("encrusting") lichens on a rock. (b) Foliose ("leafy") lichen growing on a dead cedar in a North Carolina salt marsh. (c) British soldier lichen (Cladonia cristatella)*, a fruticose ("shrubby") lichen. Each soldier (so called because of the scarlet color) is 1 to 2 centimeters tall.*

absorb some minerals from their substrate (this is suggested by the fact that particular species are characteristically found on specific kinds of rocks or soil or tree trunks), but minerals reach the lichen primarily through the air and in rainfall. Because lichens rapidly absorb substances from rainwater, they are particularly susceptible to airborne toxic compounds. Thus, the presence or absence of lichens is a sensitive index of air pollution.

Many of the algae and cyanobacteria found in lichens are also commonly found as free-living species. Lichen fungi are thought to have free-living hyphal stages, but these fungi can generally be detected and identified only after they have encountered a suitable photosynthetic organism and formed a lichen. For these reasons, lichens are generally named and classified according to the species of the fungal complement. There are about 25,000 species of lichens, in almost all of which the fungus is an ascomycete. Photosynthetic organisms from some 26 different genera are found in symbiotic association with these fungi. The most frequent are the green algae *Trebouxia, Pseudotrebouxia,* and *Trentepohlia* and the cyanobacterium *Nostoc;* members of one of these four genera are found in about 90 percent of all lichens.

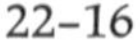

22–16
Scanning electron micrographs showing the establishment of the British soldier lichen (Cladonia cristatella) *in sterile laboratory culture.* (a) *An early interaction between fungal and algal components of the lichen.* (b) *Penetration of an algal cell by a fungal haustorium (arrow).*

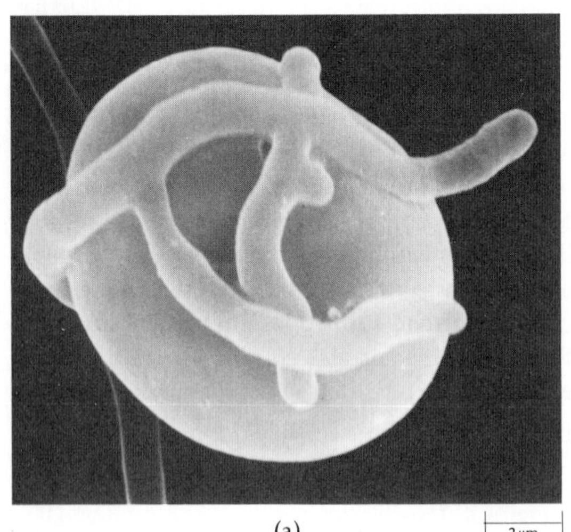

(a)

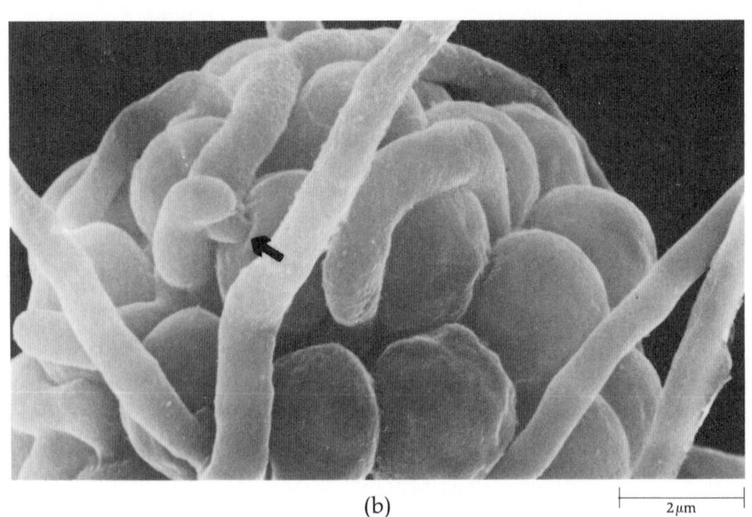

(b)

PREDACEOUS FUNGI

Among the most highly specialized of the fungi are the predaceous fungi that have developed a number of mechanisms for capturing small animals that they use as food. Some secrete a sticky substance on the surface of their hyphae in which passing protozoans, rotifers, small insects, or other animals become glued. More than 50 species of Fungi Imperfecti capture small roundworms (nematodes) that abound in the soil. In the presence of a population of roundworms (or even of water in which the worms have been growing), the hyphae of the fungi produce loops that swell rapidly, closing the opening like a noose when a nematode rubs against its inner surface. Presumably the stimulation of the cell wall increases the amount of osmotically active material in the cell, causing water to enter the cells and expand them rapidly.

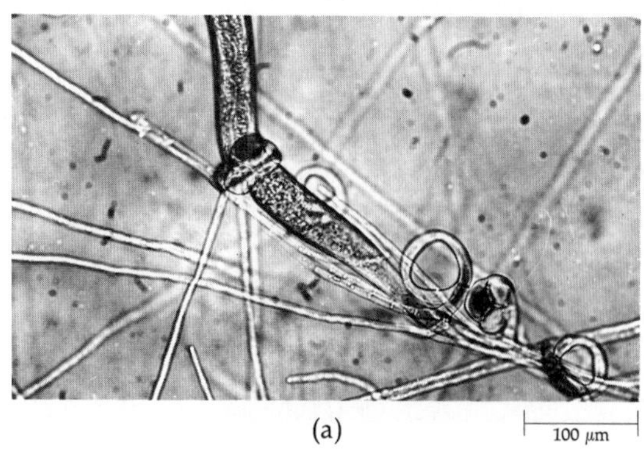

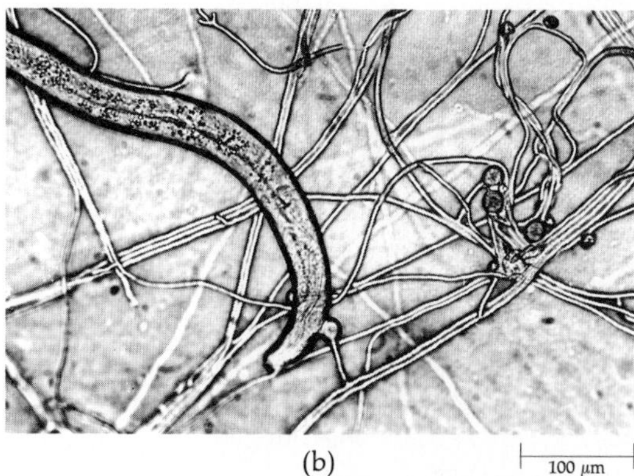

(a) *The predaceous imperfect fungus* Arthrobotrys dactyloides *has trapped a nematode. The traps consist of rings, each comprising three cells, which swell rapidly to about three times their original size and garrote the nematode. Once the worm has been trapped, fungal hyphae grow into its body and digest it. When triggered, the ring cells can expand completely in less than a tenth of a second. This species was appropriately called the "nefarious noose fungus" by the late W. H. Weston of Harvard University, who made vast contributions to our present knowledge of the fungi.* (b) *Another nematode-trapping fungus,* Dactylella drechsleri. *This species traps the worms with small adhesive knobs and was dubbed the "lethal lollipop fungus" by Weston.*

Lichens reproduce most commonly by the breaking off of fragments containing both fungal hyphae and photosynthetic cells. New individuals can also be formed by the capture of an appropriate alga or cyanobacterium by a lichen fungus in its free-living, hyphal stage. Sometimes the photosynthetic cells are destroyed by the fungus, in which case the fungus also dies. If the photosynthetic cells survive, a lichen is produced.

Mycorrhizae

Mycorrhizae ("fungus-roots") are symbiotic associations between fungi and the roots of vascular plants. The importance of mycorrhizae was first recognized in connection with efforts to grow orchids in greenhouses. Orchids have microscopic seeds that germinate to form a tiny pad of tissue called a protocorm. Cultivators of orchids found that the plants seldom developed beyond the protocorm stage unless they were infected with a particular kind of fungus. Subsequently it was found that if seedlings of many forest trees are grown in nutrient solutions and then transplanted to prairie or other grassland soils, they fail to grow. Eventually they may die from malnutrition, even if a soil analysis shows that there are abundant nutrients in the soil. If a small amount (0.1 percent by volume) of forest soil

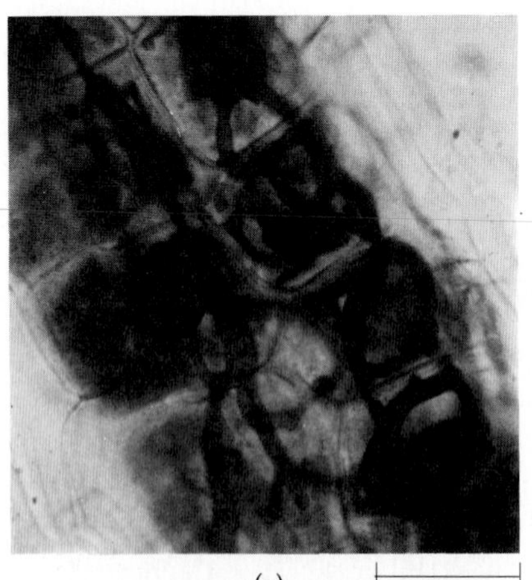

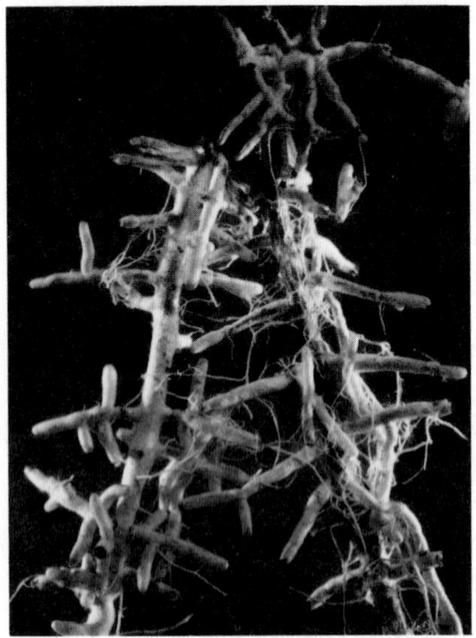

22–17
(a) *Surface view of an endomycorrhizal rootlet of fescue, a perennial grass, showing coiled hyphae in some of the cells. Mycorrhizal associations are especially important for grasses growing on nutrient-poor soils or at high elevations.* (b) *Ectomycorrhizal rootlets from a western hemlock. Hormones secreted by the fungus cause the root to branch in a special pattern. This growth pattern and the swollen hyphal sheath impart a characteristic appearance to the ectomycorrhizae. The narrow strands extending from the mycorrhizae are bundles of hyphae that function as extensions of the root system.*

containing fungi is added to the soil around the roots of the seedlings, however, they will grow promptly and normally. Mycorrhizae are now thought to occur in more than 90 percent of all families of plants.

In some mycorrhizal associations, known as endomycorrhizae, the fungal hyphae penetrate the cells of the root, forming coils, swellings, or branches (Figure 22–17a). The hyphae also extend out into the surrounding soil. Endomycorrhizae occur in about 80 percent of all vascular plants, and the fungal component is usually a zygomycete. In other associations, known as ectomycorrhizae, the hyphae form a sheath around the root but do not actually penetrate its cells (Figure 22–17b). Ectomycorrhizae are characteristic of certain groups of trees and shrubs, including pines, beeches, and willows; the fungus is usually a basidiomycete, but some associations involve ascomycetes, including truffles.

The exact relationship between roots and fungi is not known. Apparently the roots secrete sugars, amino acids, and possibly some other organic substances that are used by the fungi. Although the evidence is just now accumulating, it appears that the fungi convert minerals in the soil and decaying material into an available form and transport them into the root. It has been shown experimentally that mycorrhizae transfer phosphorus from the soil into roots, and there is evidence that water uptake is facilitated by the fungus.

A study of the fossils of early vascular plants has revealed that mycorrhizae occurred as frequently in them as they do in modern vascular plants. This has led to the interesting suggestion that the evolution of mycorrhizal associations may have been the critical step allowing plants to make the transition to the bare and relatively sterile soils of the then-unoccupied land.

SUMMARY

Fungi are typically composed of masses of filaments called hyphae; in most groups of fungi, the principal component of the hyphal walls is chitin. Fungi are heterotrophs, deriving their nutrition by absorption of organic compounds digested extracellularly by secreted enzymes.

Fungi form both asexual and sexual spores, although not all fungi form both kinds. The asexual spores may be formed in sporangia. The sexual cycle is initiated by the fusion of hyphae of different mating strains. In some groups of fungi, the nuclei in the fused hyphae immediately combine, and a zygote is formed. In others (the ascomycetes and basidiomycetes), the two genetically distinct nuclei usually remain separate, forming pairs—known as dikaryons—that divide synchronously, sometimes over prolonged periods. Once the nuclei fuse, meiosis always follows immediately.

The kingdom Fungi includes five principal divisions—Chytridiomycota, Oomycota, Zygomycota, Ascomycota, and Basidiomycota—as well as another major taxon, Deuteromycota, or Fungi Imperfecti, which is usually ranked at the level of division. The chytrids and the oomycetes are the only fungi with flagellated cells; both groups have flagellated asexual spores, and the chytrids also have flagellated gametes. The oomycetes, all of which exhibit oogamy, are the only fungi whose cell walls are composed principally of cellulose rather than chitin. The zygomycetes, which are thought to share a common ancestor with the lineage leading to the ascomycetes and basidiomycetes, are characterized by thick-walled zygospores. The hyphae of both ascomycetes and basidiomycetes are compartmentalized by perforated septa. In the ascomycetes, the sexual spores develop within an ascus (sac), and in the basidiomycetes, the sexual spores develop on a basidium (club). Most of the Deuteromycota have no known sexual cycle.

Fungi have an important ecological role as decomposers of organic material. They are also parasitic on many types of organisms, particularly plants, in which they often cause serious disease. Fungi participate in two additional types of symbioses that are of ecological significance—lichens and mycorrhizae.

Lichens are combinations of fungi and green algae or cyanobacteria that are morphologically and physiologically different from either organism as it exists separately. They are able to survive under adverse environmental conditions where neither partner could exist independently. The lichen represents a symbiotic relationship in which a fungus encloses photosynthetic cells and is dependent upon them for nourishment.

Mycorrhizae ("fungus-roots") are associations between soil-dwelling fungi and plant roots. There are two principal types: endomycorrhizae and ectomycorrhizae. Mycorrhizal associations facilitate the uptake of minerals by the roots of the plant and provide organic molecules for the fungus.

QUESTIONS

1. Distinguish between the following: hypha/mycelium; chitin/cellulose; sporangia/gametangia; conidia/ascospores; ascus/basidium; monokaryotic/dikaryotic; endomycorrhizae/ectomycorrhizae.

2. Give the distinguishing characteristics of the Zygomycota, Ascomycota, and Basidiomycota.

3. Some authorities believe that the Chytridiomycota and Oomycota should be placed in the kingdom Fungi, while others believe they are better placed in the kingdom Protista. What characteristics do the members of these two divisions share with the fungi of the three other major divisions? In what ways do the chytrids and oomycetes differ from those fungi?

4. As you can see from the last two chapters, our method of classification into kingdoms does not produce entirely satisfactory results. What are some of the difficulties unresolved (or even created) by this system? Can you suggest alternative solutions?

5. At the present time, most of the fungi that have no sexual reproduction cannot be classified with their sexually-reproducing relatives. Is this likely to change in the future? Why or why not?

6. Coenocytic organisms, such as some groups of fungi, show little differentiation. When differentiation does occur, as in gametangium formation, it is preceded by construction of a septum. In your opinion, why?

7. Most fungi can reproduce both asexually and sexually. What are the advantages and disadvantages of each type of reproduction?

8. What type of symbiotic relationship would you say exists between the fungus and the alga or cyanobacterium of a lichen? Between the fungus and the plant roots of a mycorrhizal association?

9. How have fungi aided photosynthetic organisms in the transition to land?

CHAPTER 23

The Plants

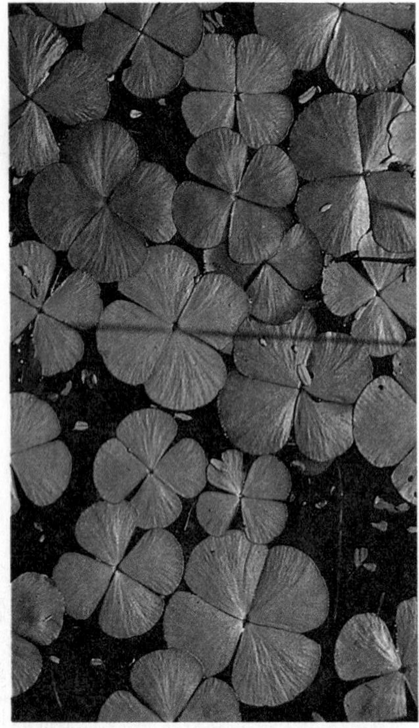

23–1
Although plants are primarily adapted for life on land, some, such as the water fern, Marsilea, *have returned to an aquatic existence. Like whales and dolphins,* Marsilea *retains the traces of its terrestrial sojourn. These include a water-resistant cuticle, stomata (openings through which gas exchange occurs), and a highly developed internal transport system.*

Plants are, quite simply, multicellular photosynthetic organisms primarily adapted for terrestrial life. Their characteristics are best understood in terms of their transition from water to land.

THE TRANSITION TO LAND

The story of plant evolution begins with one-celled algae floating on or near the surface of the water in the open seas. Such organisms have left little trace in the fossil record, but we assume that their way of life was much like that of modern members of the phytoplankton, although they may have been less complex and specialized. In the open water, light was abundant, and the oxygen, hydrogen, and carbon in the air and water were available to every floating cell, as they are today. Although groups of cells might form long filaments (as a consequence of not separating after cell division), each cell remained a separate functional entity and lived an independent existence.

As the cellular colonies multiplied, they probably started to exhaust the supplies of nitrogen, phosphorus, sulfur, and other minerals available in the open ocean. (In fact, this shortage of essential elements is the limiting factor in plans to farm the seas.) Life was probably more abundant near the shores, where the waters were rich in minerals washed from the land by rivers and streams, scraped from the coasts by the action of the waves, and brought up from the bottom by upwellings of coastal waters. Here, along the coasts, in a more variable environment than that of the open waters, the ancestor of the plants evolved.

The land offered a wealth of advantages to photosynthetic organisms. On land, light is abundant from daylight to dusk, unfiltered by the turbulent water. Carbon dioxide, needed for photosynthesis, is plentiful in the atmosphere and circulates more freely in air than in water. And, most important, the land was then unoccupied by competing forms of life.

The Ancestral Alga

As we noted in Chapter 21, all plants appear to have arisen from one group of green algae (division Chlorophyta). There are several lines of evidence leading to this conclusion. Some of the clues are biochemical: like the plants, the green algae contain chlorophylls *a* and *b* and beta-carotene as their photosynthetic pigments; they accumulate their food as starch; and their cell walls contain cellulose. Addi-

23–2
Cell division in Ulothrix, *a filamentous green alga. Formation of the cell plate is almost complete. Note the chloroplasts, with their deeply stained thylakoids, just outside the large nuclei.*

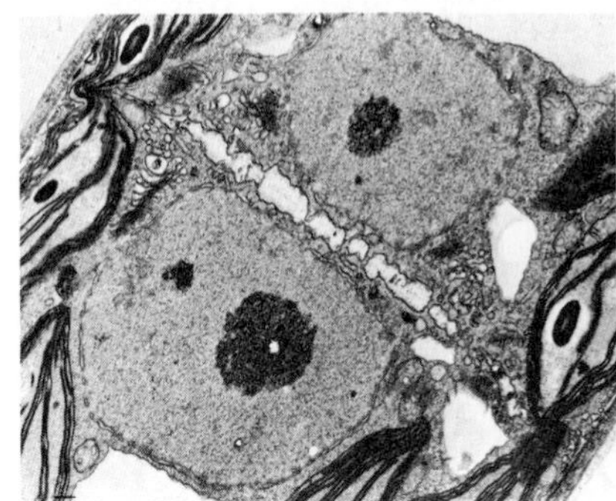

tional evidence is found in their pattern of cytokinesis. In almost all other organisms, division of the cytoplasm takes place by constriction and pinching off of the cell membrane. In plants and a few species of green algae (Figure 23–2), the cytoplasm is divided by the formation of a cell plate at the equator of the spindle. (Cell-plate formation is also seen in one brown alga, but this alga does not contain chlorophyll *b* or store its food as starch.)

The green alga ancestral to the plants was probably a relatively complex multicellular seaweed, perhaps differentiated—like many modern coastal seaweeds—into a holdfast and an upper photosynthetic area. It is thought to have been oogamous and to have had a life cycle involving heteromorphic alternation of generations—two reproductive characteristics shared by all plants. It invaded the land some 500 million years ago, perhaps accompanied and aided by a symbiotic fungus (page 449).

The Ancestral Plant

When a structural or biochemical characteristic is widespread among a group of modern organisms, it is presumed to have been present also in their common ancestor. One such characteristic of the plants, clearly associated with the transition to land, is the protective cuticle that covers the aboveground surfaces of plants and retards the loss of water from the plant body. The cuticle is formed of a waxy substance called cutin, secreted by the epidermal cells. (The likelihood that such a complex macromolecule was "invented" more than once is very slight.) Associated with the cuticle—in fact, made necessary by it—are pores through which the gas exchanges necessary for photosynthesis can take place (Figure 23–3).

Following the same reasoning, the ancestral plant is thought to have had a well-defined alternation of generations, a feature found also among the algae.

23–3
The aboveground surfaces of plants are characteristically covered with a waxy layer of cutin that retards water loss. Epidermal pores permit gas exchange. The pores shown in this scanning electron micrograph are those of Marchantia, *a liverwort. Unlike the stomata of vascular plants (Figure 10–16, page 225), they do not open and close in response to environmental cues.*

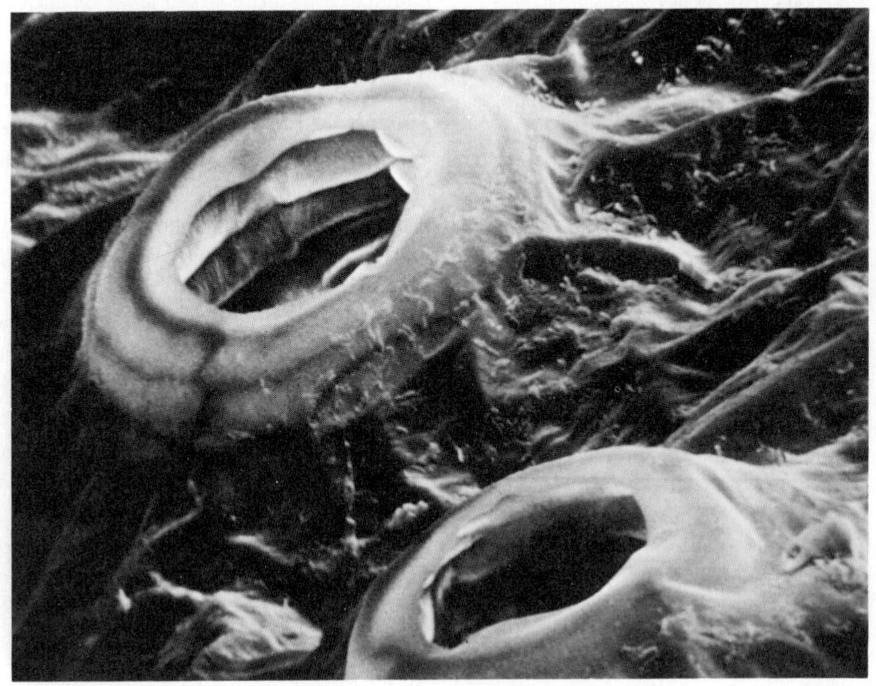

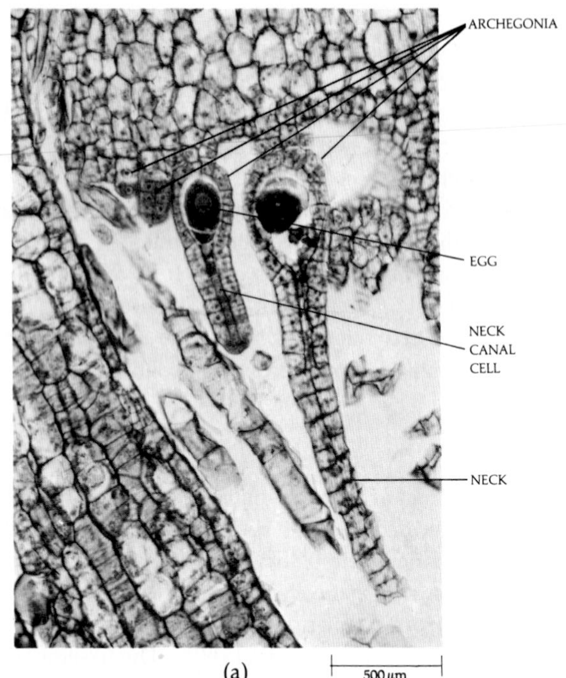

(a)

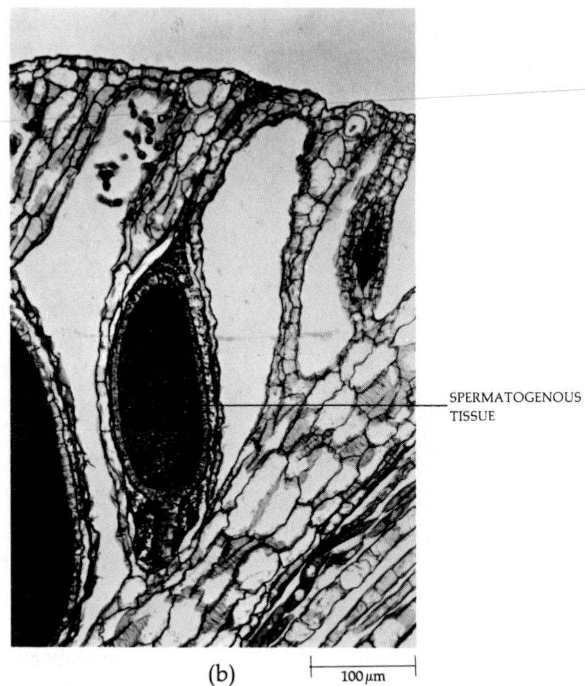

(b)

23–4
Multicellular gametangia of the liverwort Marchantia, *a member of division Bryophyta.* (a) *Female gametangia, or archegonia, in various stages of development. The archegonia are flask-shaped, with a single egg developing in the base of the flask.* (b) *Antheridium developing on the male gametophyte. The spermatogenous tissue will give rise to sperm cells, which, when mature, will swim to the egg through the neck canal of the archegonium.*

However, in contrast to the algae, the ancestral plant's gametophytes bore multicellular reproductive organs (gametangia) that were surrounded by a protective layer of sterile (nonreproductive) cells. These gametangia are called archegonia if they give rise to egg cells and antheridia if they give rise to sperm (Figure 23–4). Similarly, a sterile jacket layer developed around the spore-producing cells of the sporophyte, forming the sporangia.

A correlated adaptation was the retention of the fertilized egg (the zygote) within the female gametangium (the archegonium) and its development there into an embryo. Thus, during its critical early stages of development, the embryo, or young sporophyte, is protected by the female gametophyte. (In the algae, by contrast, the zygote leads an independent existence.) Eventually, antheridia and archegonia became obsolete in the most successful modern plants, being phased out as the result of a wholly new invention, the seed. But that is getting ahead of our story.

Not long after the transition to land, the plants diverged into two separate lineages. One gave rise to the bryophytes, a group that includes the modern mosses, hornworts, and liverworts, and the other to the vascular plants, a group that includes all of the dominant land plants. A principal difference between the bryophytes and the vascular plants is that the latter, as their name implies, have a well-developed vascular system that transports water, minerals, sugars, and other nutrients throughout the plant body. The bryophytes first appear in the fossil record in the Devonian period, more than 350 million years ago (Table 23–1). These ancient fossils are quite similar to bryophytes living today. The oldest fossils of vascular plants are from the Silurian period, which ended some 400 million years ago. The vascular plants, as we shall see, subsequently underwent a great diversification.

Table 23-1 Major Physical and Biological Events in Geologic Time

MILLIONS OF YEARS AGO	PERIOD	EPOCH	LIFE FORMS	CLIMATES AND MAJOR PHYSICAL EVENTS
Cenozoic Era				
	Quaternary	Recent Pleistocene	Planetary spread of *Homo sapiens;* extinction of many large mammals. Deserts on large scale.	Fluctuating cold to mild. Four glacial advances and retreats (Ice Age); uplift of Sierra Nevada.
$1\frac{1}{2}$–7	Tertiary	Pliocene	Large carnivores. First known appearance of hominids (humanlike primates).	Cooler. Continued uplift and mountain building, with widespread extinction of many species.
7–26		Miocene	Whales, apes, grazing mammals. Spread of grasslands as forests contract.	Moderate uplift of Rockies.
26–38		Oligocene	Large, browsing mammals. Apes appear.	Rise of Alps and Himalayas. Lands generally low. Volcanoes in Rockies.
38–53		Eocene	Primitive horses, tiny camels, modern and giant types of birds.	Mild to very tropical. Many lakes in western North America.
53–65		Paleocene	First known primitive primates and carnivores.	Mild to cool. Wide, shallow continental seas largely disappear.
Mesozoic Era				
65–136	Cretaceous		Extinction of dinosaurs at end of period. Marsupials, insectivores, and angiosperms become abundant.	Lands low and extensive. Elevation of Rockies at end of period.
136–195	Jurassic		Dinosaurs' zenith. Flying reptiles, small mammals. Birds appear. Gymnosperms, especially cycads, and ferns.	Mild. Continents low. Large areas in Europe covered by seas. Mountains rise from Alaska to Mexico.
195–225	Triassic		First dinosaurs. Primitive mammals appear. Forests of gymnosperms and ferns.	Continents mountainous. Large areas arid. Eruptions in eastern North America. Appalachians uplifted and broken into basins.
Paleozoic Era				
225–280	Permian		Reptiles evolve. Origin of conifers and possible origin of angiosperms; earlier forest types wane.	Extensive glaciation in Southern Hemisphere. Appalachians formed by end of Paleozoic; most of seas drain from North America.
280–345	Carboniferous Pennsylvanian Mississippian		Age of amphibians. First reptiles. Variety of insects. Sharks abundant. Great swamps, forests of ferns, gymnosperms, and horsetails.	Warm. Lands low, covered by shallow seas or great coal swamps. Mountain building in eastern U.S., Texas, Colorado. Moist, equable climate, conditions like those in temperate or subtropical zones, little seasonal variation, root patterns indicate water plentiful.
345–395	Devonian		Age of fish. Amphibians appear. Mollusks abundant. Lunged fish. Extinction of primitive vascular plants. Origin of modern groups of vascular plants.	Europe mountainous with arid basins. Mountains and volcanoes in eastern U.S. and Canada. Rest of North America low and flat. Sea covers most of land.
395–440	Silurian		Earliest vascular plants. Invasion of land by arthropods. Rise of fish and reef-building corals. Shell-forming sea animals abundant. Modern groups of algae and fungi.	Mild. Continents generally flat. Again flooded. Mountain building in Europe.
440–500	Ordovician		First primitive fish. Invertebrates dominant. First fungi. Invasion of land by plants.	Mild. Shallow seas, continents low; sea covers U.S. Limestone deposits; microscopic plant life thriving.
500–600	Cambrian		Age of marine invertebrates. Shelled animals.	Mild. Extensive seas. Seas spill over continents.
Precambrian Era				
Over 600			Earliest known fossils. Soft-bodied marine invertebrates.	Dry and cold to warm and moist. Planet cools. Formation of earth's crust. Extensive mountain building. Shallow seas. Accumulation of free oxygen.

CLASSIFICATION OF THE PLANTS

According to the classification scheme we follow, the modern plants are placed in 10 separate divisions (Table 23–2). Each of these is, by all available evidence, monophyletic—that is, all of its members are descended from a common ancestor. These divisions are also frequently grouped, for convenience, in ways that may or may not reflect phylogeny. For instance, the vascular plants, as a group, are often referred to as tracheophytes. They can be grouped into those without seeds (divisions Psilophyta, Lycophyta, Sphenophyta, and Pterophyta) and those with seeds. The seed plants also form two informal groups, the gymnosperms and the angiosperms. The gymnosperms are those with "naked," unprotected seeds (divisions Coniferophyta, Cycadophyta, Ginkgophyta, and Gnetophyta), and the angiosperms, with enclosed, protected seeds, are, formally speaking, the Anthophyta, the flowering plants.

23–5
A bryophyte, a haircap moss with spore capsules. The lower green structures are the gametophytes, which are haploid (n)*; the nonphotosynthetic stalks and capsules are the diploid* (2n) *sporophytes. The bryophytes are the only plants in which the gametophyte is the dominant, nutritionally independent generation.*

Table 23–2 A Classification of Living Plants

TAXON	COMMON NAME	NUMBER OF SPECIES
Division Bryophyta	Bryophytes	16,000
Class Hepaticopsida	Liverworts	6,000
Class Antherocerotopsida	Hornworts	100
Class Muscopsida	Mosses	9,500
Division Psilophyta	Whisk ferns	Several
Division Lycophyta	Club mosses	1,000
Division Sphenophyta	Horsetails	15
Division Pterophyta	Ferns	12,000
Division Coniferophyta	Conifers	550
Division Cycadophyta	Cycads	100
Division Ginkgophyta	Ginkgos	1
Division Gnetophyta	Gnetophytes	70
Division Anthophyta	Flowering plants (angiosperms)	235,000
Class Monocotyledones	Monocots	65,000
Class Dicotyledones	Dicots	170,000

DIVISION BRYOPHYTA: MOSSES, HORNWORTS, AND LIVERWORTS

Lacking water-gathering roots and specialized vascular tissues to transport water up the plant body, the bryophytes must absorb moisture through aboveground structures. As a consequence, they grow most successfully in moist, shady places and in bogs. Some of them, like sphagnum (peat moss), are able to absorb and hold large amounts of water; in effect, they maintain a watery existence even on land. Most bryophytes are tropical, but some species occur in temperate regions, and a few even reach the Arctic and Antarctic.

Most bryophytes are comparatively simple in their structure and are relatively small, usually less than 15 centimeters in height. A single moss plant may sprawl over a considerable area, but most liverworts are so small that they are noticeable only to a keen observer. In the damp environments frequented by the bryophytes, individual cells can absorb water and nutrients directly from the air or by diffusion from nearby cells. Like the lichens, they are sensitive indicators of air pollution.

Although the bryophytes may not be completely adapted to land, they do have some special structural adaptations to a terrestrial existence. As we noted earlier, bryophytes do not have roots. In the mosses, each individual plant is attached to the substrate by means of elongate single cells called rhizoids. Bryophytes also have small leaflike structures in which photosynthesis takes place. These structures lack the specialized tissues of the "true" leaves of the vascular plants and are only one or a few cell layers thick. For these reasons, the leaflike structures of the bryophytes and the leaves of the vascular plants are believed to have evolved separately. As in other plants, the body of a bryophyte is specialized for support and food storage. Thus the bryophytes resemble other plants far more than they do the algae, although they are not usually classified as "higher" plants and do not seem to have been their ancestors.

Bryophyte Reproduction

Bryophytes, like all plants, have a life cycle with alternation of generations (Figure 23–6). In contrast to the vascular plants, however, the bryophytes are characterized by a haploid gametophyte that is usually larger than the diploid sporophyte.

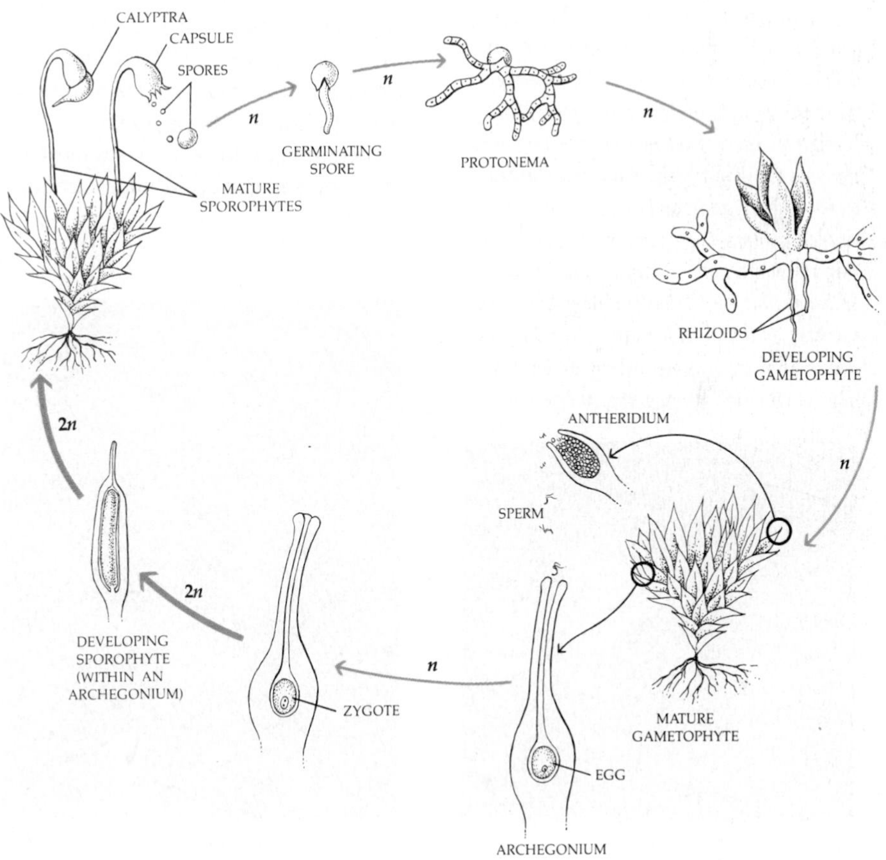

23–6
In a representative moss life cycle, spores are released from a capsule, which opens when a small lid bursts (upper left). The spore germinates to form a branched, filamentous protonema, from which a leafy gametophyte develops. Sperm cells, which are expelled from the mature antheridium, are attracted into the archegonium, where one fuses with the egg cell to produce the zygote. The zygote divides mitotically to form the sporophyte and, at the same time, the base of the archegonium divides to form the protective calyptra. The sporophyte consists of a capsule, which may be raised on a stalk—also part of the sporophyte—and a foot. Meiosis occurs within the capsule, resulting in the formation of haploid spores.

23–7
Protonema of a moss. Protonemata, from which the gametophytes develop, are characteristic of some bryophytes. They resemble filamentous green algae.

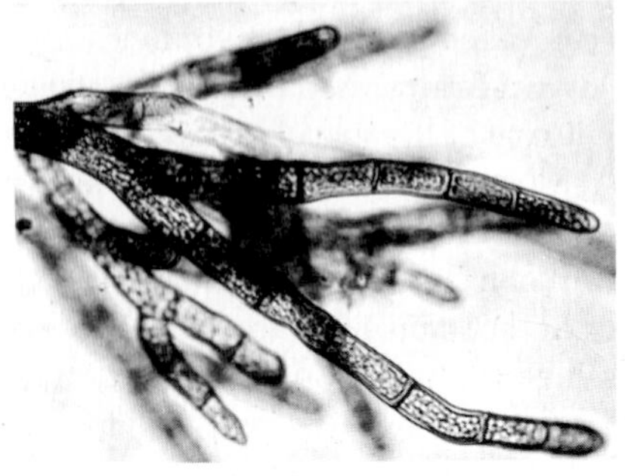

The life cycle of a moss begins when a haploid spore germinates to form a network of horizontal filaments known as protonemata (singular, protonema) (Figure 23–7). Individual plants (gametophytes) grow up like branches from this network. The multicellular antheridia and archegonia are borne on the gametophyte. When sufficient moisture is present, the sperm, which are biflagellate, are released from the antheridium and swim to the archegonium, to which they are attracted chemically. Without free water in which the sperm can swim, the life cycle cannot be completed.

Fusion of sperm and egg takes place within the archegonium. Inside the archegonium, the zygote develops into a sporophyte, which remains attached to the gametophyte and is often nutritionally dependent upon it. Typically the sporophyte consists of a foot, a stalk, and a single, large sporangium (or capsule), from which the spores are discharged.

Asexual reproduction, often by fragmentation, is also common. Many mosses and liverworts also produce minute bodies, known as gemmae, that can give rise to new plants (Figure 23–8).

THE VASCULAR PLANTS: AN INTRODUCTION

Evolutionary Developments in the Vascular Plants

Rhynia major, now extinct, is an example of the earliest known vascular plants, dating back some 400 million years. As you can see in Figure 23–9, it does not look much like any modern vascular plant and, indeed, is much more primitive in appearance than a bryophyte. However, it differs from all bryophytes in one important respect: within its stem is a central cylinder of vascular tissue, specialized for conducting water and dissolved substances up the plant body and products of photosynthesis down.

Beginning with a very simple vascular plant, such as *Rhynia,* it is possible to trace some major evolutionary trends, as well as several key innovations. One early

23–8
(a) *Young liverwort growing on a rock. In the liverwort* Marchantia, *the archegonia and the antheridia are formed on different plants, the female* (b) *and the male* (c) *gametophytes. The zygote, formed in the archegonium, develops into the sporophyte, which remains attached to the female gametophyte. The bowl-shaped gemma cups visible in* (c) *contain minute bodies, the gemmae, which are splashed out by the rain and grow in the vicinity of the parent plant. Asexual reproduction by fragmentation or by gemmae is common among the liverworts.*

(a)

(b)

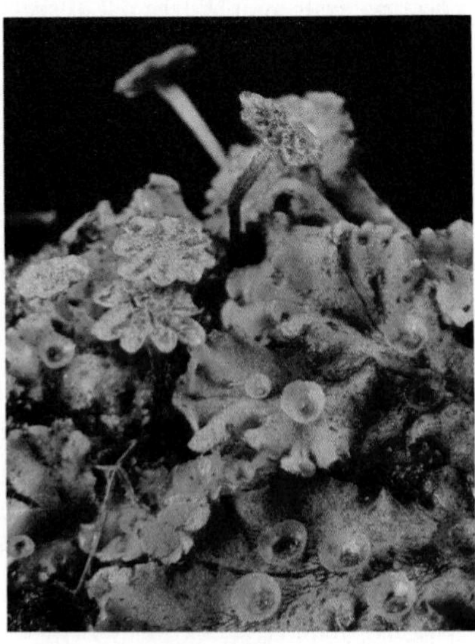

(c)

innovation was the root, a structure specialized for the absorption of water. Another was the leaf. Two distinct types of leaves evolved, the microphyll and the megaphyll. The microphyll contains only a single strand of vascular tissue, whereas the megaphyll typically contains a complex system of veins. These two types of leaves seem clearly to have originated in different ways (Figure 23–10). Of the modern plants, only the Psilophyta, the Lycophyta, and the Sphenophyta have microphylls.

23–9
Rhynia major, *one of the earliest known vascular plants. It lacked leaves and roots. Its aerial stems, which were photosynthetic, were attached to an underground stem, or rhizome. The aerial stems were covered with a cuticle and contained stomata. The dark structures at the tips of the stems are sporangia, which released the spores by splitting longitudinally.*

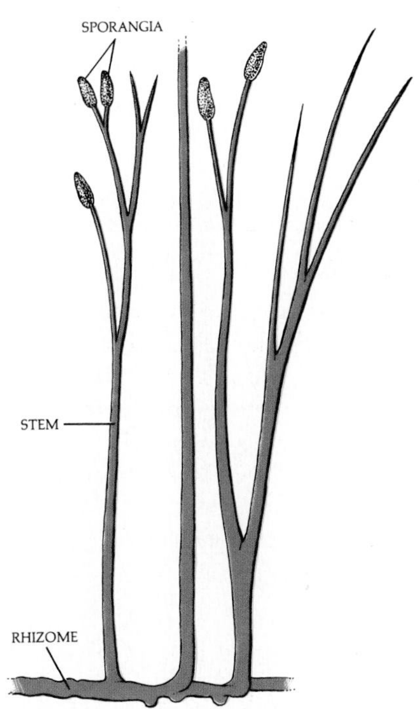

An important evolutionary trend has been the development of increasingly complex and efficient vascular systems. The conducting system in modern vascular plants consists of two distinct tissues: the *phloem* (pronounced "flow-em"), in which sugar and other soluble organic materials are transported, and the *xylem*, in which water and minerals are transported. The conducting elements of the phloem are *sieve cells* or *sieve-tube members*, and the conducting elements of the xylem are *tracheids* and *vessel members*. In stems, the longitudinal strands of xylem and phloem are side by side, either in vascular bundles or arranged in two concentric layers (cylinders), in which one tissue (typically the phloem) occurs outside the other.

Another major trend is the reduction in size of the gametophyte. In all vascular plants, the gametophyte is smaller than the sporophyte. However, in the more primitive vascular plants, the gametophyte is separate and nutritionally independent of the sporophyte. In the more recently evolved groups—the gymnosperms and the angiosperms—the gametophyte has been reduced to microscopic size and dependency status.

Also related to the reproductive cycle is the trend toward heterospory. The earliest vascular plants produced only one kind of spore (homospory) in one kind of sporangium. Upon germination, such spores typically produce gametophytes on which both antheridia and archegonia form. In plants that are heterosporous, two different kinds of gametophytes develop: one bearing archegonia and the other, antheridia. (As the gametophytes become reduced, archegonia and antheridia decrease in size until, in the angiosperms, they have disappeared.)

The final innovation among the vascular plants, and perhaps the most important to survival on land, was the seed. The seed is a complex structure in which the young sporophyte, or embryo, is contained within a protective outer covering, the seed coat. The seed coat, which is derived from tissues of the parent sporophyte, protects the embryo from drying out while it remains dormant, sometimes for years, until conditions are favorable for its germination. The earliest known seeds were fossilized in late Devonian deposits some 350 million years ago.

All of these developments will be discussed in greater detail in the pages that follow.

23–10
According to one widely accepted theory, (a) *microphylls evolved as lateral outgrowths of the stem. Representatives of the three divisions of modern plants with microphylls are shown in Figure 23–11.*

(b) *Megaphylls evolved by fusion of branch systems and thus have a complex vascular network. The great majority of vascular plants have megaphylls.*

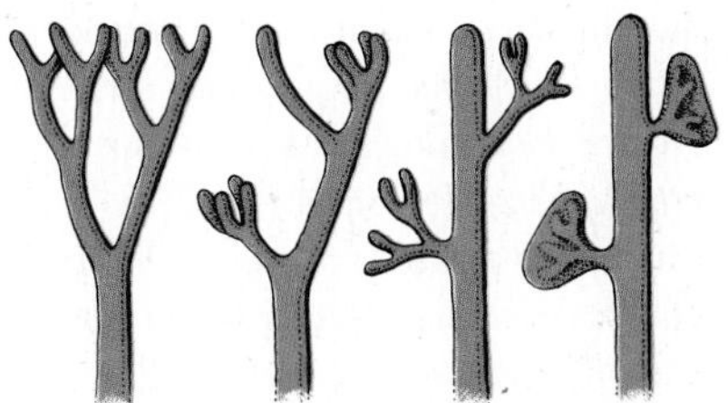

THE SEEDLESS VASCULAR PLANTS

Four divisions of seedless vascular plants have living representatives: Psilophyta (whisk ferns), Lycophyta (club mosses), Sphenophyta (horsetails), and Pterophyta (ferns), the largest group.

Division Pterophyta: The Ferns

Ferns are vascular plants that can usually be distinguished from most other plants by their large feathery leaves, which, in most species, unroll from base to tip as they develop. According to the fossil record, ferns first appeared almost 400 million years ago, and they are still relatively abundant. Most of the 12,000 living species are found in the tropics, but many also occur in temperate and even arid regions. Because they have flagellated sperm and need free water for fertilization, those species growing in arid regions exploit the seasonal occurrence of water for sexual reproduction.

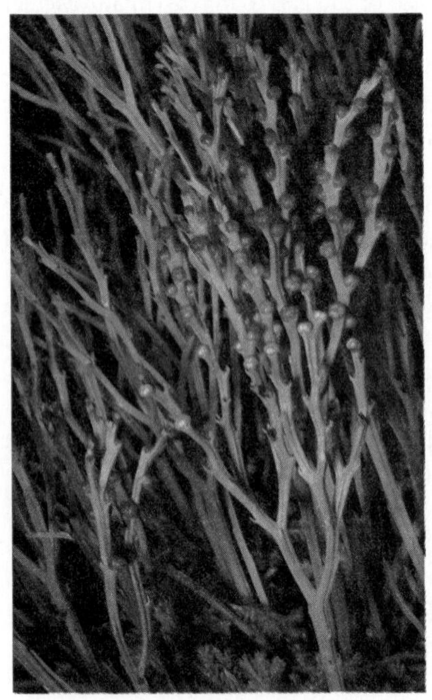

(a)

(b)

(c)

23–11
Representatives of three divisions of seedless vascular plants. (a) *The whisk fern,* Psilotum, *one of the two living genera of division Psilophyta. The bulbous structures are the sporangia, which occur in groups of three.* Psilotum *is unique among living vascular plants in that it lacks both roots and leaves. If you look closely, however, you can see small, scalelike outgrowths below the sporangia.*

(b) *The club mosses of the genus* Lycopodium *are the most familiar members of division Lycophyta. In this genus, the sporangia are borne on specialized leaves, sporophylls, which are aggregated into a cone at the apex (top) of the branches, as shown in this running ground pine,* Lycopodium complanatum. *The airborne, waxy spores give rise to small, independent subterranean gametophytes. The sperm, which are biflagellate, swim to the archegonium, where the young sporophyte, or embryo, develops.*

(c) *The horsetails, division Sphenophyta, of which there is only one living genus* (Equisetum), *are easily recognized by their jointed, finely ribbed stems, which contain silicon. At each node, there is a circle of small, scalelike leaves. Spore-bearing structures are clustered into a cone at the apex of the stem. The gametophytes are independent, and the sperm are coiled, with numerous flagella.*

(a)

(b)

23–12
Ferns. (a) *The immature sporophyte of many common ferns develops as a "fiddle head," which uncoils and spreads as it elongates.* (b) *A cinnamon fern. The sporangia are borne on separate stalks, visible in the center of the photograph.*

23–13
Fern spores develop on the sporophyte in sporangia, which are usually found in clusters (sori) on the underside of a leaf (sporophyll), as shown here. Fern spores give rise to tiny gametophytes that, although photosynthetic, are barely visible to the naked eye.

The stems of ferns are usually not as complex as those of gymnosperms and angiosperms, and are often reduced to a creeping underground stem (rhizome). Although ferns do not exhibit secondary growth—the type of growth that results in increase of girth and formation of bark and woody tissue—some grow very tall. For instance, *Cyathea australis,* a tree fern found on Norfolk Island in the South Pacific, sometimes reaches 28 meters in height.

The leaves (fronds) of ferns are often finely divided into pinnae; these divided leaves spread widely and so collect more light, and apparently they are thus adapted to growing on the forest floor in diffused light. The sporangia commonly are on the undersurface of the leaves or, sometimes, on specialized leaves. Spore-bearing leaves are called *sporophylls.* (As we shall see, the reproductive structures of flowers are also sporophylls.) The sporophylls may resemble the other green leaves on the plant or may be nonphotosynthetic stalks (modified leaves). The sporangia of ferns commonly occur in small clusters known as sori (singular, sorus).

In the ferns, as in all the vascular plants, the dominant generation is the sporophyte. The gametophyte of the homosporous ferns begins development as a small algalike filament of cells, each filled with chloroplasts, and then develops into a flat structure, often only one layer of cells in thickness. Although this gametophyte is small, it is nutritionally independent, as is the sporophyte. All but a few genera of ferns are homosporous, and a single gametophyte usually produces both antheridia and archegonia. The sperm are coiled and multiflagellate. The life cycle of a fern is shown in Figure 12–4 on page 251.

THE SEED PLANTS

The humid Carboniferous period, which ended some 280 million years ago, was the age when most of the earth's coal deposits were formed from lush vegetation that sank so swiftly into the warm, marshy soil that there was no chance for much of it to decompose. Seed plants were in existence by the close of this period; according to the fossil record, some of the fernlike plants and even some of the club mosses had seedlike structures.

In the Permian period (225 to 280 million years ago), there were worldwide changes of climate, with the advent of widespread glaciers and drought. Terrestrial plants and animals were under strong selection pressures to evolve specialized structures that would enable them to survive during periods when no water was available. Those organisms in which water-conserving, protective structures had previously evolved were at a great advantage. The amphibians gave way to the scaly-skinned reptiles, which may have been better suited to the harsh climate. Similarly, by the close of this period, which ended the Paleozoic era, plants that had seeds gained a major evolutionary advantage and came to be the dominant plants of the land.

The two major modern groups of plants are both seed plants: the *gymnosperms,* which have naked seeds, and the *angiosperms* (from the Greek word *angio,* meaning "vessel"—literally, a seed borne in a vessel), which have protected seeds.

Gymnosperms

It was during the Permian period that the gymnosperms evolved. There are four groups of gymnosperms with living representatives—three small divisions (Cycadophyta, Ginkgophyta, and Gnetophyta) and one large and familiar division (Coniferophyta).

Formation of the Seed

The seed is a protective structure in which the embryonic plant can lie dormant until conditions become favorable for its survival. Thus, in its function, it parallels the spores of bacteria or the resistant zygotes of the freshwater algae. In structure it is far more elaborate, however. A seed includes the embryo (the young, dormant sporophyte), a store of nutritive tissue, and an outer protective coat.

To understand the structure of the seed, it is necessary to return for a moment to the alternation of generations found in all plants and in some of the green algae (see pages 424–425). In the green algae, the two generations, sporophyte and gametophyte, are independent and are generally about the same size. In the seedless vascular plants, including the ferns, the gametophytes, although still independent, are smaller than the sporophytes. In the seed plants, the gametophytic generation is reduced still further and is totally dependent on the sporophyte.

All gymnosperms are heterosporous, producing two different types of spores in two different types of sporangia. Spores that give rise to male gametophytes are known as *microspores,* and they are formed in structures known as microsporangia. Spores from which female gametophytes develop are *megaspores,* and they are formed in megasporangia. A megasporangium contains a megaspore mother cell, which gives rise by meiosis to a megaspore, and it is surrounded by one or two layers of tissue, the integument. The entire structure—the megasporangium, its protective integument, and its contents—is known as the *ovule.*

(a)

(b)

(c)

(d)

23–14

Representative gymnosperms. (a) *Male and female plants of* Zamia pumila, *the only species of cycad native to the United States. It is common in the sandy woods of Florida. The stems are mostly or entirely underground and, along with the rootstalks, were used by the Seminole Indians as food. The two large gray cones in the foreground are female cones; the smaller brown cones are male cones.*

(b) *Leaves and fleshy seeds of* Ginkgo biloba, *the only surviving species of the Ginkgophyta, a lineage that extends back to the late Paleozoic era.* Ginkgo *is especially resistant to air pollution and is commonly cultivated in urban parks and along city streets.*

(c) *A large seed-producing plant of* Welwitschia mirabilis, *a gnetophyte, growing in the Namib Desert of southern Africa.* Welwitschia *produces only two adult leaves, which continue to grow for the life of the plant. As growth continues, the leaves break off at the tips and split lengthwise; hence, older plants appear to have numerous leaves.*

(d) *A branch of a conifer, the Ponderosa pine (also called western yellow pine), bearing a female cone. When the cone was mature, it opened and released its winged seeds, two of which have become caught between the scales. Like other pines, the Ponderosa has flexible, needlelike leaves, which are held together in a bundle (fasicle). Ponderosa pine, one of the principal forest trees of the Rockies from Canada to Mexico, is a staple of the Northwest lumber industry.*

COAL AGE PLANTS

About 100 billion metric tons of carbon dioxide are fixed in photosynthesis every year. Almost the same amount is released into the atmosphere by living organisms in the course of respiration—less only 1 part in 10,000. This very slight imbalance is caused by the burying of organisms in sediment or mud under conditions in which oxygen is excluded and decay is only partial. This accumulation of partially decayed material is known as peat. The peat may eventually become covered with sedimentary rock and so be placed under pressure. Depending on time, temperature, and other factors, peat may become compressed into coal, petroleum, or natural gas—the so-called fossil fuels.

During certain periods in the earth's history, the rate of fossil

(a)

fuel formation was greater than at other times. One such period, aptly called the Carboniferous, occurred some 300 million years ago. What are now the temperate regions of Europe and North America were then tropical to subtropical, supporting year-round growth. The lands were low, covered by shallow seas or swamps. The dominant plants of the Carboniferous period were lycopod trees and giant horsetails; these disappeared during the Permian period, a time of world-wide drought and extensive glaciation. Other plants with surviving descendants include several families of ferns and one group of gymnosperms, the conifers. The flowering plants had not yet put in an appearance.

(b)

(a) *A reconstruction of a Carboniferous swamp forest. The tall trees (left to middle) and the sparsley branched trees with terminal tufts in the center of the drawing are both lycopods. Tree ferns are visible in the left to middle foreground and giant horsetails at the right. The tall trees at the far right are primitive conifers. As you can see, these plants all had shallow root systems, as do modern swamp plants, and so were easily toppled by the winds.* (b) *A fossil of a branch of a giant horsetail. As in modern horsetails, a circle of leaves is located at each node.*

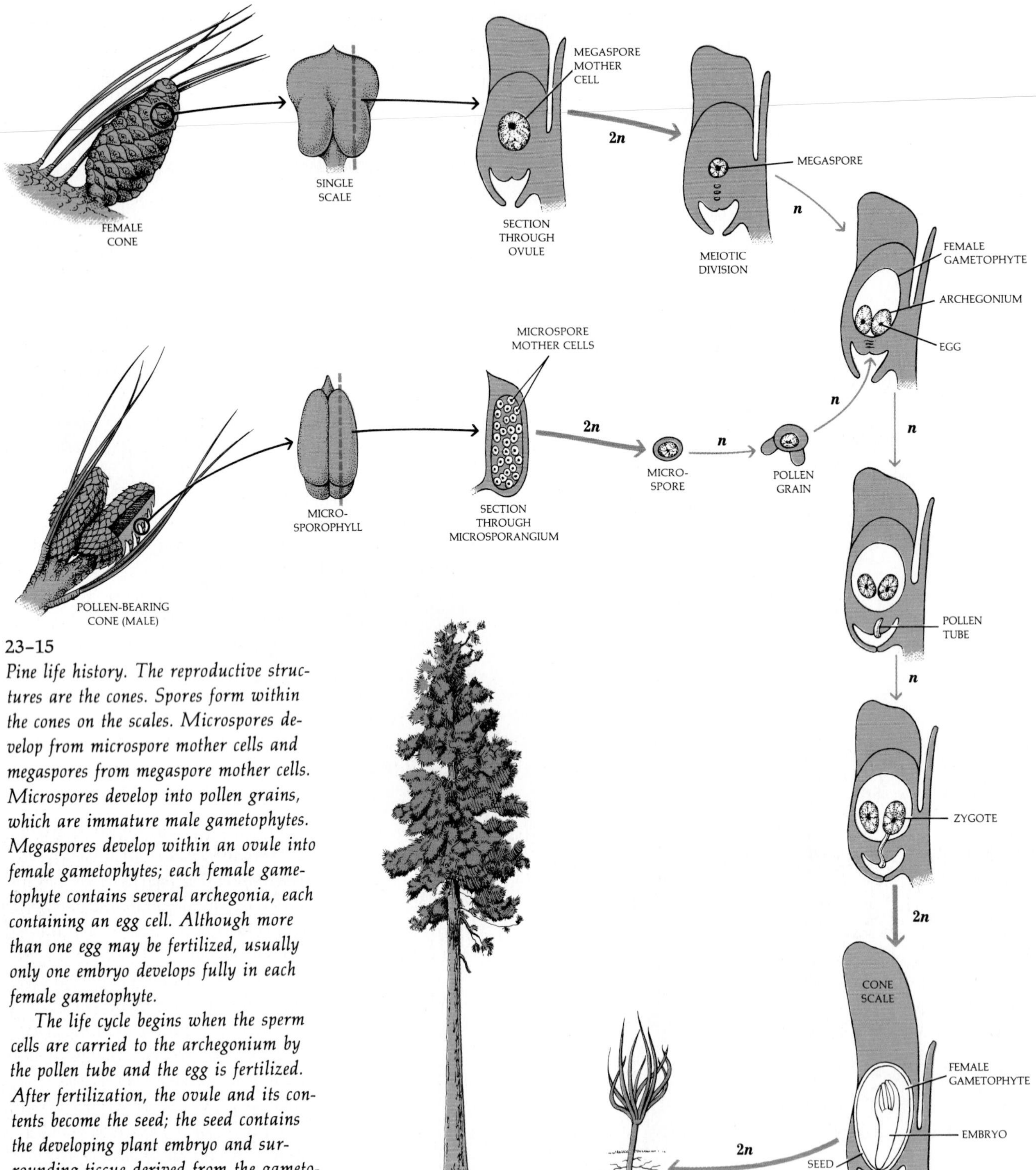

23–15

Pine life history. The reproductive structures are the cones. Spores form within the cones on the scales. Microspores develop from microspore mother cells and megaspores from megaspore mother cells. Microspores develop into pollen grains, which are immature male gametophytes. Megaspores develop within an ovule into female gametophytes; each female gametophyte contains several archegonia, each containing an egg cell. Although more than one egg may be fertilized, usually only one embryo develops fully in each female gametophyte.

The life cycle begins when the sperm cells are carried to the archegonium by the pollen tube and the egg is fertilized. After fertilization, the ovule and its contents become the seed; the seed contains the developing plant embryo and surrounding tissue derived from the gametophyte, which nourishes the embryo. As the seed matures, the cone opens, releasing the winged seeds, which germinate, producing the seedling (sporophyte). Both types of cones develop on the same sporophyte.

With this information in mind, let us look at a specific example, the formation of a pine seed (Figure 23–15). A pine tree—the mature sporophyte—has two types of cones, which produce the two types of spores. The small, male cones superficially resemble those of the club mosses, but on the large, female cones, the scales that bear the ovules are much thicker and tougher than the sporophylls of the male cones.

In the male cones, specialized microspore mother cells inside the microsporangia undergo meiosis to produce haploid microspores. Each microspore differentiates into a microscopic, windborne pollen grain, an immature male gametophyte. The wind is an unreliable messenger, disseminating the pollen grains at random, and wind-pollinated plants characteristically produce pollen in great quantities (Figure 23–16).

Within the ovules of the female cones, the megaspores are formed by meiosis. Of the four cells produced within the ovule by each meiotic sequence, three disintegrate and the remaining one—the megaspore—develops into a tiny female gametophyte. This haploid gametophyte grows within the ovule and develops two or more archegonia, each of which contains a single egg cell. The development from the megaspore into the gametophyte with its egg cells may take many months—slightly more than a year, for example, in some common pines.

As the ovule ripens, it secretes a sticky liquid. When the female cone becomes dusted with pollen, some of the pollen sifts down between the cone scales and comes into contact with the liquid. Pollen grains, caught in the sticky liquid, are drawn to the ovule as the liquid dries. Here the pollen grain develops into a mature male gametophyte. This gametophyte produces two nonmotile cells, the male gametes, or sperm. These are carried toward the egg within the pollen tube, which is produced by the male gametophyte and which grows through the tissues of the ovule.

Because the drought-resistant pollen is blown to the female cones by the wind, and the sperm are carried to the egg by the pollen tube, the pines and other conifers are not dependent on free water for fertilization. Thus they are able to reproduce sexually when (and where) ferns and bryophytes cannot.

23–16

(a) *Male cones of jack pine* (Pinus banksiana) *shedding pollen. The pollen grains are immature male gametophytes, which complete their maturation when they reach the ovules, embedded in the female cones. There they release sperm cells that fuse with the egg cells.*

(b) *Female cone. The female gametophytes develop in ovules on the base of a scale of the cone, and the eggs are fertilized there. Each scale contains two ovules. When the seeds are mature, they drop from the cone.*

23–17

Pine seed. The outer layers of the ovule have hardened into a seed coat, enclosing the female gametophyte and the embryo, which now consists of an embryonic root and a number of embryonic leaves, the cotyledons.

When the seed germinates, the root will emerge from the seed coat and penetrate the soil. When the root absorbs water, the tightly packed cotyledons will elongate and swell with the moisture, rising above ground on the lengthening stem and forcing off the seed coat. During this period, the cotyledons absorb nutrients that are stored in the gametophyte tissue and are essential for the growth of the embryo into a seedling.

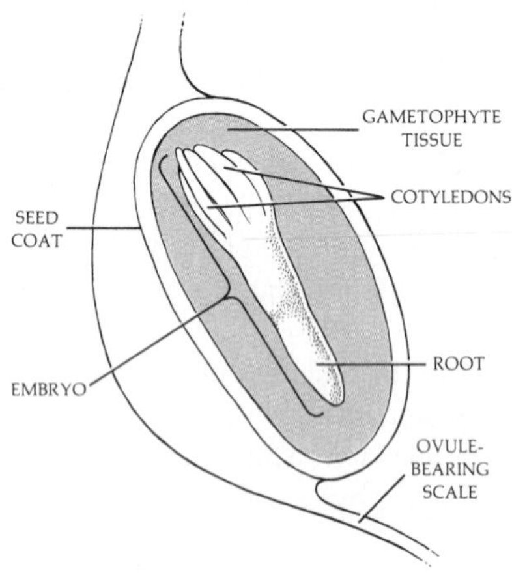

Following fertilization, the zygote begins to divide and forms the embryo, the young sporophyte. As the ovule matures, its outer walls harden into a seed coat, enclosing both the embryo and the female gametophyte (the latter provides food for the embryo when the seed germinates). After the cone matures, it opens, releasing its seeds. In most conifers the seeds are winged and are spread by the wind.

Figure 23–17 shows a cross section of a pine seed, which has been described as "three generations under one roof." The seed coat and the wing on which the seed is carried arise from the hardened outer layers of the ovule, derived from the mother sporophyte. The next layer represents the body of the female gametophyte; swollen and packed with stored food reserves, it grows and displaces the original sporophyte tissue. The inner core is the embryo with its many *cotyledons,* the embryonic leaves, which will appear as the first leaves of the shoot of the new sporophyte when the seed germinates. The lower part of the embryo will develop into the first root.

Under favorable conditions, the seed—the mature ovule and its contents—will germinate and give rise to a seedling. If conditions remain favorable, the seedling will ultimately become a mature pine tree, producing spores that give rise to gametophytes that, in turn, produce gametes.

The Conifer Leaf

Another feature commonly associated with conifers, although not with all gymnosperms, is their needlelike leaf. Figure 23–18 shows a cross section of a pine leaf, which may be 10 or more centimeters long but only 1 to 3 millimeters in diameter. In the center you can see veins, the vascular transport system, which carries water in one set of conducting cells (the tracheids) and sugars in another (the sieve cells). Outside the veins are the cells in which photosynthesis takes place. The ducts on the flat sides of the needles carry resin, a substance that is released if the plant is wounded and may serve to close the break. The outside layer of cells (epidermis) is porous but very hard.

The needlelike leaf, despite its slender form, is a megaphyll (page 461). It is well adapted to long periods of low humidity, such as in northern winters, and to moisture-losing, sandy soils. One or both of those are characteristic of many regions in which modern conifers are abundant.

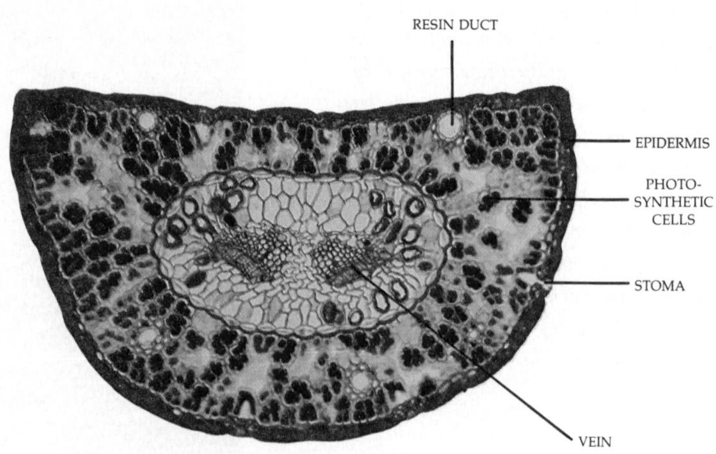

23–18

Cross section of a pine leaf stained to show the tissues and cell types. The hard outer covering and compact shape protect the leaf from water loss, an essential factor in the survival of these trees in areas in which there is little rainfall, or in which the water is locked in the ground as ice during many months of the year.

Angiosperms: The Flowering Plants

It is believed that the angiosperms evolved from a now-extinct group of gymnosperms. They appear in the fossil record in abundance during the Cretaceous period, about 120 million years ago, as the dinosaurs were vanishing. Of the numerous angiosperm genera that appeared suddenly at that time, many seem to have been very similar to our modern genera.

Daniel Axelrod, a paleobotanist at the University of California, believes that angiosperms probably arose long before the Cretaceous, during the Permian period. They are likely to have originated on the less fertile hills and uplands of tropical areas, the richer lowlands being crowded with club mosses, ferns, and gymnosperms. Once established, they spread into the lowlands, where they became the dominant plant forms and were deposited as fossils. During this mid-Cretaceous period, the climate of the earth was warmer and more uniform than it is at present, and by the end of the Cretaceous period, much of the land was covered with a rich forest of angiosperms, reaching almost as far north as the Arctic Circle.

Angiosperms, like other vascular plants, contain chlorophylls *a* and *b* and beta-carotene, and have megaphylls, stomata, and a cuticle impervious to water. The modern forms have a more highly evolved vascular system than is found in other groups (Figure 23–19). They also have two new, interrelated structures that distinguish them from all other plants: the flower and the fruit. Both are devices by which plants induce, reward, trick, and even seduce animals into carrying out the plants' reproductive strategies.

About 235,000 different species of angiosperms are known. Just listing them, one after another, would require a book twice the length of this one. They dominate the tropical and temperate regions of the world, occupying well over 90 percent of the earth's vegetative surface. With only very minor exceptions, all of our agricultural species are angiosperms.

The angiosperms include not only the plants with conspicuous flowers but also most of the great trees, the oak, the willow, the elm, the maple, and the birch; all the fruits, vegetables, nuts, and herbs; the cactus and the coconut; and all the corn, wheat, rice, and other grains and grasses that are the staples of the human diet and the basis of agricultural economy all over the world. These tremendously diverse plants are classified in two large groups: class Monocotyledones (the monocots), with about 65,000 species, and class Dicotyledones (the dicots), with about 170,000 species. Among the monocots are such familiar plants as the grasses, lilies, irises, orchids, cattails, and palms. The dicots include many of the herbs, almost all the shrubs and trees (other than conifers), plus many other plants. The major differences between these two classes are summarized in Table 23–3 and will be discussed further in Section 5.

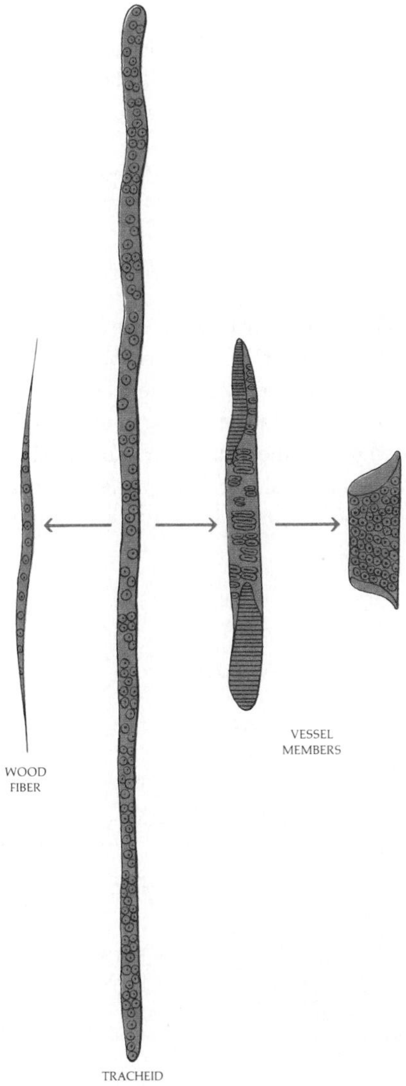

23–19

Evolutionary relationships among some cells of the xylem of vascular plants. Tracheids, which are the only water-conducting cells of the conifers, are believed to resemble more primitive (that is, earlier evolved) cells. Tracheids are elongated cells with thin areas (pits) in their lateral walls through which water moves from one tracheid to another up the trunk from the roots. Tracheids also provide mechanical support. Wood fibers, specialized for support, and vessel members, specialized for conducting water, are presumed to have evolved from the primitive water-conducting and supporting tracheids of early vascular plants. In the most highly evolved vessels, the end walls of the individual cells (vessel members) disintegrate during development, and the members are stacked on top of one another, leaving a continuous tube.

Table 23-3 Principal Differences between Monocots and Dicots

CHARACTERISTIC	MONOCOTS	DICOTS
Flower parts	Usually in threes	Usually in fours or fives
Pollen grains	Have one furrow or pore	Have three furrows or pores
Cotyledons ("seed leaves")	One	Two
Leaf venation	Major veins usually parallel	Major veins usually netlike
Vascular bundles in young stem	Scattered	In a ring
Secondary (woody) growth	Usually absent	Usually present

The Flower

Figure 23-20 is a diagram of a simple flower. The central structure is the *carpel*, the female reproductive structure. (A single carpel or a group of fused carpels is also known as a pistil, because of its resemblance to an apothecary's pestle.) The carpel is believed to be a sporophyll (a spore-bearing leaf) that, in the course of evolution, infolded in such a way that the ovule is attached to its inner surface (Figure 23-21). A single carpel may contain one or more ovules, and a single flower one or more carpels. The carpels may be separate (blackberry) or fused (tomato and apple).

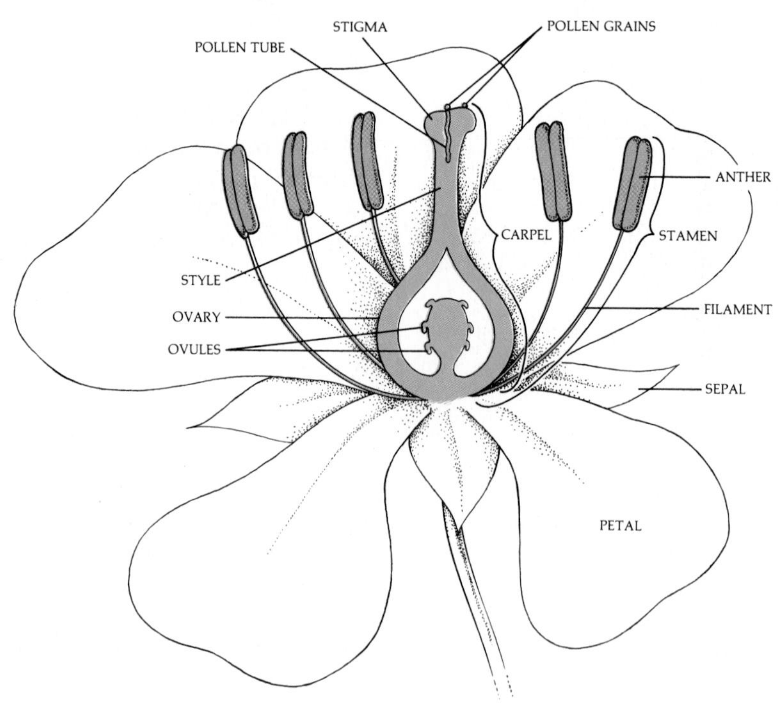

23-20
Structure of a flower. The reproductive structures are shown in color. Some flowers have only the male structures, some only the female; such flowers are said to be imperfect. A flower that possesses both stamens and carpel, such as this flower, is known as a perfect flower.

The petals and sepals are, like the stamens and carpel, modified leaves.

23–21
Presumed evolutionary development of simple and compound ovaries. A leaf-shaped carpel with ovules along its edges (left) folded in on itself, and the edges fused to form a simple ovary. Compound ovaries were formed by the fusing of separate infolded carpels.

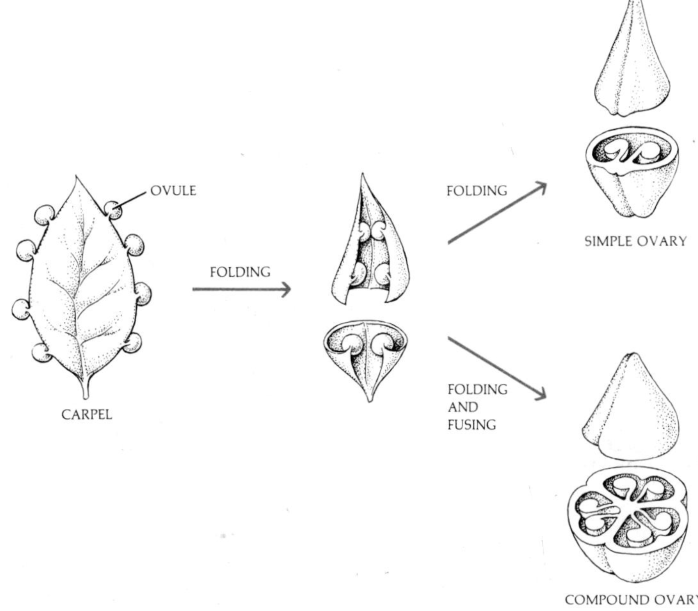

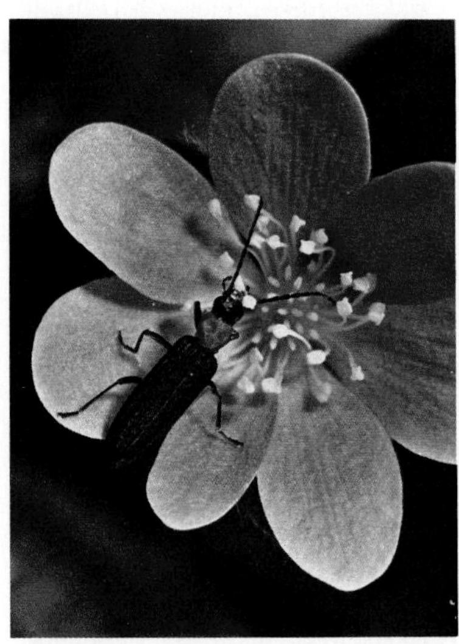

23–22
Flower of a round-leaved hepatica (Hepatica americana). *Notice the open, bowl shape and the numerous and separate floral parts. The insect is a pollen-eating beetle.*

The swollen base of the carpel or carpels is the *ovary*, within which is the ovule, or ovules, in which the female gametophyte develops from a megaspore. The tip of the carpel or carpels has become specialized as a *stigma*, a sticky surface to which pollen grains adhere. The stigma and ovary are connected by a slender column of tissue, the *style*.

The pollen grains (immature male gametophytes) develop in the *stamen*, which, like the carpel, is a sporophyll. It consists of the *anther*, which contains microsporangia in which the pollen grains develop, and a supporting *filament*.

The other parts of the flower are often specialized to attract insects and other pollinators. Pollen grains produced in the anthers are carried to the stigma of (usually) another flower, where they germinate, developing pollen tubes that grow down through the style toward the ovule. The sperm cells are carried by the pollen tubes to the female gametophyte, which typically consists of only seven cells. The extraordinary events of angiosperm fertilization, which give rise not only to an embryo sporophyte but also to a special nutritive tissue, will be discussed in detail in Chapter 29 (page 593).

Evolution of the Flower

Plants are, generally speaking, immobile. Thus, uniting the sperm of one individual with the egg of another presents a problem, to which the flower—that apt symbol of spring and romance—is a solution. The evolution of the flower is the elaboration of devices to lure animal pollinators and increase the efficiency of their activities. The primitive flower is believed to have resembled that of the modern hepatica, shown in Figure 23–22, which has numerous floral parts, each clearly separate from the other. By comparing this type of flower with some of the more specialized ones, such as an orchid (Figure 23–23) or a composite (Figure 23–24), it is possible to see four main trends in flower evolution:

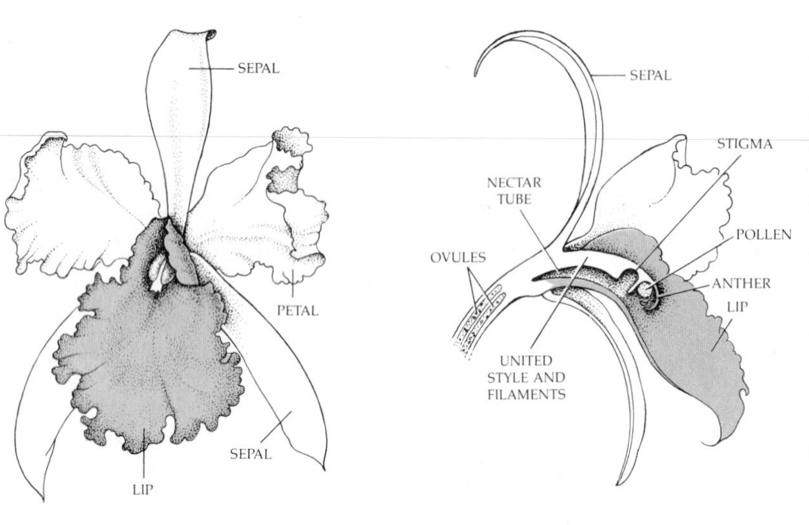

(a)

(b)

23–23
Orchidaceae, with about 20,000 species, is the largest family of flowering plants. Its flowers are highly specialized. (a) *The parts of an orchid flower. The lip is a modified petal that can serve as a landing platform for insects.* (b) Cymbidium *orchids. This genus of Asian orchids contains about 30 species, most of which grow in tropical or subtropical regions.*

1. Reduction in number of floral parts. Most specialized flowers have few stamens and few carpels.
2. Fusion of floral parts. Carpels and petals, in particular, have become fused, sometimes elaborately so.
3. Elevation of free floral parts above the ovary. In the primitive flower, the floral parts arise at the base of the ovary (Figure 23–25a). Such ovaries are said to be superior. In the more advanced flowers, the free portions of the floral parts are above the ovary (Figure 23–25b); this is an important adaptation by which the ovules are believed to be protected from foraging insects. Such ovaries are said to be inferior.
4. Changes in symmetry. The radial symmetry of the primitive flower has given way, in more advanced forms, to bilaterally symmetrical forms.

The Agents of Evolution

The early gymnosperms from which the angiosperms evolved were probably wind-pollinated, as are modern gymnosperms. And, as in the modern gymnosperms, the ovule probably exuded droplets of sticky sap in which pollen grains were caught and drawn toward the female gametophytes. Insects, probably beetles, feeding on plants must have come across the protein-rich pollen grains and the sticky, sugary droplets. As they began to depend on these new-found food supplies, they inadvertently carried pollen from plant to plant.

Beetle pollination must have been more efficient than wind pollination for some plant species because, clearly, selection began to favor those plants that had insect pollinators. The more attractive the plants were to the beetles, the more frequently they would be visited and the more seeds they would produce. Any chance variations that made the visits more frequent or that made pollination more efficient thus offered immediate advantages; more seeds would be formed, and more offspring would be likely to survive. Nectaries (nectar-secreting structures) evolved, which lured the pollinators. Plants developed white or brightly colored flowers that

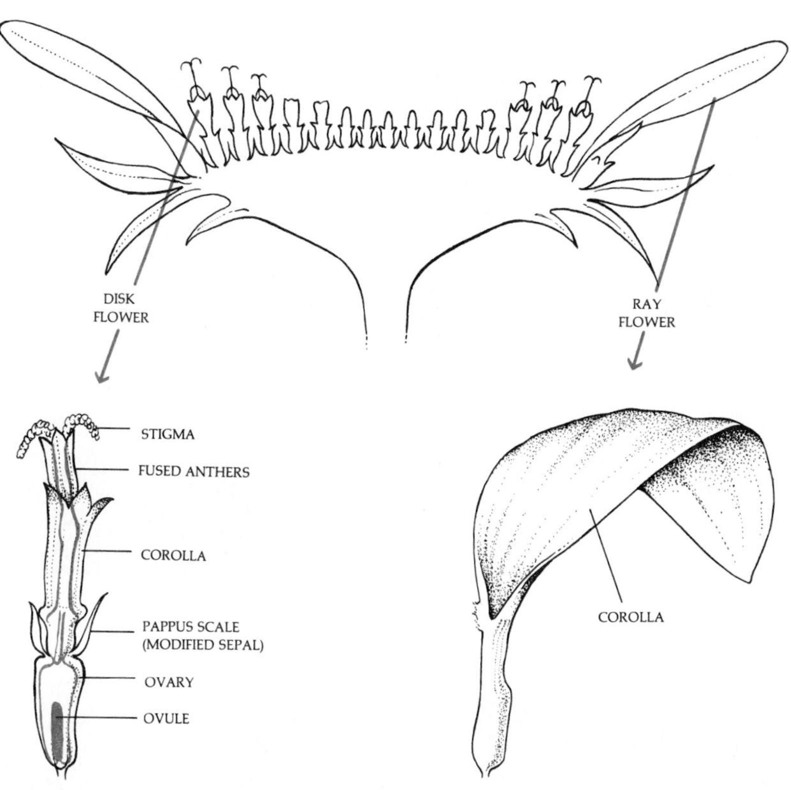

23–24

Composites (family Asteraceae), with some 13,000 species, are the second largest family of flowering plants. Their flowers are also highly specialized.

(a) *The organization of the head of a composite. The individual flowers are subordinated to the overall effect of the head, which acts as a large, single flower in attracting insects. The internal structure of the disk flower is indicated in color.*

(b) *The head of the sunflower,* Helianthus annuus *("annual flower of the sun"), is composed of numerous separate florets, each comprising a pair of fused carpels forming a single ovary and fused anthers enclosed in a small corolla of fused petals. The ray flowers (with yellow petals) are often sterile.*

23–25

(a) *Superior and* (b) *inferior ovaries. In flowers with superior ovaries, the floral parts are attached below the ovaries. In flowers with inferior ovaries, the floral parts are attached above, and the ovaries are protected.*

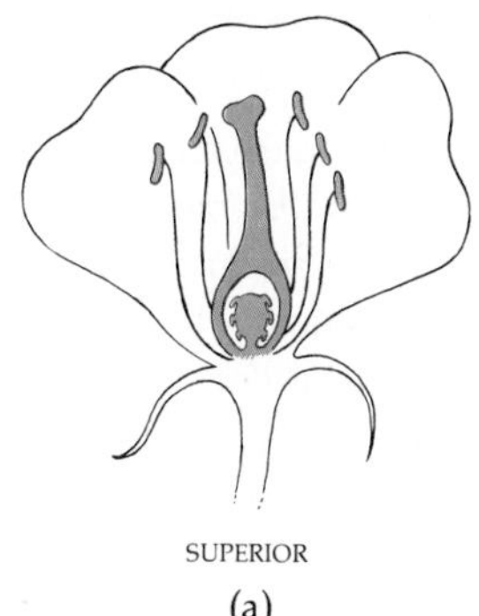

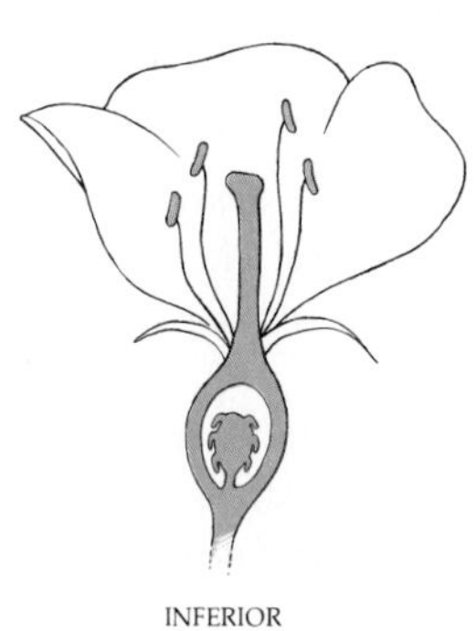

(a)

(b)

(c)

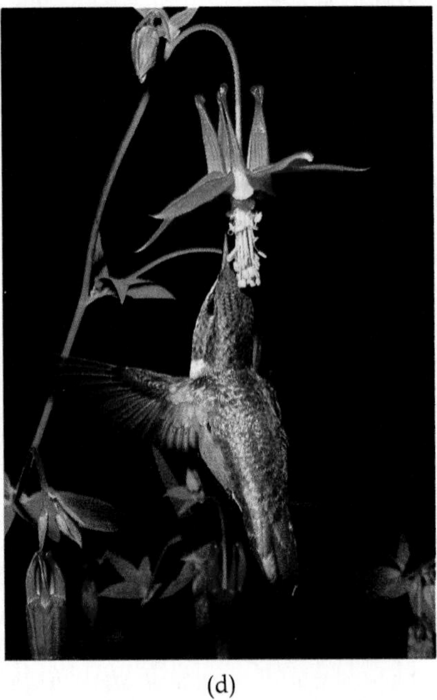

(d)

(e)

23–26
Pollinators. (a) *A honey bee foraging in a flower of* Salvia. *Notice that the anthers are depositing pollen grains on the bee's thorax.*

(b) *A longhorn beetle pollinating a lily. The pollen-covered head of the beetle is brushing against a stigma of the flower.*

(c) *A gossamer-winged butterfly sipping nectar in a daisy. Notice the long, sucking tongue of the butterfly.*

(d) *Female rufous hummingbird probing for nectar in a columbine. Her head is collecting pollen, which she will carry to another flower. Flowers pollinated by birds are scentless, bright red or orange, and have copious nectar that makes the visit worthwhile.*

(e) *A bush baby, a primate of the Old World, licking the flower of a baobab tree. The baobab, which provides food and shelter for a variety of small animals of the African plains, has flowers that open only at night, when the bush babies are active. Other mammalian pollinators include many species of bats and a few rodents, marsupials, and other primates.*

called attention to the nectar and other food supplies. The carpel, originally a leaf-shaped structure, became folded on itself, enclosing and protecting the ovule from hungry pollinators. By the beginning of the Cenozoic era, some 65 million years ago, the first bees, wasps, butterflies, and moths had appeared. These are insects for which flowers are often the only source of nutrition for the adult forms. From this time on, flowers and certain insect groups have had a profound influence on one another's history, each shaping the other as they evolved together.

A flower that attracts only a few kinds of animal visitors and attracts them regularly has an advantage over flowers visited by more promiscuous pollinators: its pollen is less likely to be wasted on a plant of another species. In turn, it is an advantage for the insect to have a "private" food supply that is relatively inaccessible to competing insects. Many of the distinctive features of modern flowers are special adaptations that encourage regular visits (constancy) by particular pollinators. The varied shapes, colors, and odors allow sensory recognition by pollinators. The diverse, sometimes bizarre, structures such as deep nectaries and complex landing platforms that are found, for example, in orchids, snapdragons, and irises, represent ways of excluding indiscriminate pollinators and of increasing the precision of pollen deposits on the animal messengers.

Evolution of the Fruit

A fruit is the mature, ripened ovary of an angiosperm and contains the seeds. Like the flower, the fruit evolved as a payment to an animal visitor for transportation services. In most cases, the chief requirement is that the seed be transported some distance from the parent plant, where it is more likely to find open ground and sunlight. A great variety of fruits, adapted for many different dispersal mechanisms, have evolved in the course of angiosperm history (Figure 23–27). A familiar example is provided by the many edible fleshy fruits that become sweet and brightly colored as they ripen, attracting the attention of birds and mammals,

23–27
Angiosperms are characterized by fruits. A fruit is a mature ovary, enclosing the seed or seeds. It often also includes accessory parts of the flower. The fruit aids in dispersal of the seed. Some fruits are borne on the wind, some are carried from one place to another by animals, some float on water, and some are even forcibly ejected by the parent plant.

(a) *In milkweed, the fruit bursts open when it is ripe, releasing seeds with tufts of silky hair that aid in their dispersal.*

(b) *The tough seeds of these blackberries will pass unharmed through the digestive tract of the dormouse. If they are deposited in a suitable environment, they will germinate, forming new blackberry plants.*

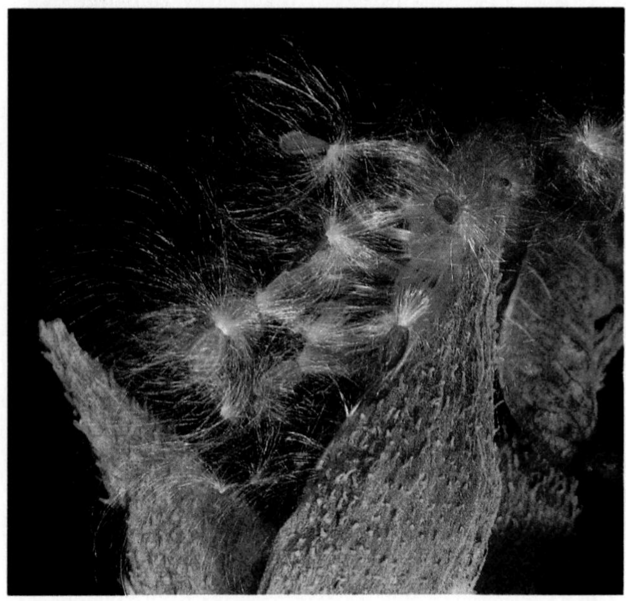
(a)

(b)

including ourselves. The seeds within the fruits pass through the digestive tract hours later and are often deposited some distance away. In some species the seed coat's exposure to digestive juices or bile is a prerequisite for germination. The seeds themselves may be bitter or toxic, as in apples, discouraging animals from grinding up and digesting the seeds.

Biochemical Evolution

Different groups of angiosperms produce characteristic bad-tasting or toxic compounds that, at one time, were thought of as by-products of plant metabolism and termed "secondary plant substances." The mustard family (Brassicaceae), for example, is characterized by the pungent taste and odor associated with cabbage, horseradish, and mustard. Milkweeds (family Asclepiadaceae) contain cardiac glycosides, heart poisons, that have potent effects in vertebrates, potential predators of this group of plants. Bitter-tasting quinine is derived from tropical trees and shrubs of the genus *Cinchona.* Nicotine and caffeine are plant products. Mescaline comes from the peyote cactus; tetrahydrocannabinol from *Cannabis sativa;* opium from a poppy; cocaine from the coca leaf. These chemicals, which are powerful defenses against animal predators, are now recognized as other products of angiosperm evolution that have contributed in a major way to the dominance of the flowering plants.

Angiosperms represent the most successful of all plants in terms of numbers of individuals, numbers of species, and their effects on the existence of other organisms. The anatomy, physiology, and reproductive processes of this dominant group of plants will be explored further in Section 5.

SUMMARY

Plants are multicellular photosynthetic organisms adapted for life on land. Among their adaptations are a waxy cuticle, pores through which gases are exchanged, protective layers of cells surrounding the reproductive cells, and retention of the young sporophyte within the female gametophyte during its embryonic development.

On the basis of similarities between the two groups, the ancestors of plants are believed to have been green algae (division Chlorophyta). These similarities include chlorophylls *a* and *b* and beta-carotene as their photosynthetic pigments, food reserves stored in the form of starch, and cellulose cell walls. All plants are oogamous and have a life cycle characterized by heteromorphic alternation of generations, two features observed in some multicellular green algae.

The bryophytes and the vascular plants diverged early in the history of plant evolution. In the classification system followed in this text, the bryophytes are placed in one division (Bryophyta) and the living vascular plants in nine separate divisions. The vascular plants, which are characterized by specialized conducting tissues, can be informally grouped into the seedless vascular plants (divisions Psilophyta, Lycophyta, Sphenophyta, and Pterophyta) and the seed plants. The seed plants can be grouped into the gymnosperms, or naked-seed plants (divisions Coniferophyta, Cycadophyta, Gnetophyta, and Ginkgophyta), and the angiosperms, or flowering plants (division Anthophyta).

The mosses and other modern bryophytes are small plants found usually in moist locations. Most lack specialized vascular tissues and all lack true leaves, although the plant body is differentiated into photosynthetic, food-storing, and anchoring tissues, as in higher plants. In sexual reproduction, the female gamete (egg) develops in the archegonium; the male gametes (sperm) develop in the antheridium. The sperm, which are flagellated, swim to the archegonium and fertilize the egg cell. The resultant zygote develops into an embryo within the archegonium. In the bryophytes—and in no other members of the plant kingdom—the gametophyte (n) is dominant and the sporophyte ($2n$) is smaller, attached, and often nutritionally dependent.

Among living seedless vascular plants, the ferns (division Pterophyta) are the most numerous. They are characterized by large, often finely dissected leaves called fronds. These leaves are, like the leaves of the seed plants, megaphylls. The sporophyte is the dominant generation, but in most ferns the gametophytes are independent. The sperm are flagellated, and free water is needed for fertilization. Sporangia are typically formed on the underside of specialized leaves (sporophylls) of the sporophyte.

Gymnosperms and angiosperms are seed plants. The largest group of naked seed plants (gymnosperms) are the conifers, which are characterized by cones. On the scales of the smaller, male cone, microspores differentiate into male gametophytes, which are released in the form of windblown pollen. The female gametophyte develops from a megaspore on a scale of a separate, larger cone within an ovule composed of the tissue of the parent sporophyte. Within the female gametophyte, archegonia form. The male gametophyte germinates and produces a pollen tube through which male reproductive cells (the sperm) enter the archegonium and fertilize the egg cell. The seed consists of the mature ovule and its contents, including the young embryo and the female gametophyte, which serves as nutritive tissue. The seed, which is shed from the female cone, can remain dormant for long periods of time and hence is adapted to withstanding cold and drought.

The angiosperms, of which there are about 235,000 species, are the dominant modern plants. Angiosperms are characterized by the flower and the fruit. Flowers attract pollinators, and fruits enhance the dispersal of seeds. The reproductive structures of the flower are the stamens, composed of filament and anther, and the carpels, composed of ovary, style, and stigma. The stamens and the carpels are highly specialized sporophylls. The ovule or ovules are enclosed in the ovary, which is the base of the carpel or group of fused carpels. Pollen grains, the immature male gametophytes, are formed in the anthers and germinate on the sticky surface of the stigma, the tip of the carpel.

The various shapes and colors of flowers evolved under selection pressures for more efficient pollinating mechanisms. Major trends in flower evolution include reduction and fusion of floral parts, a change in the position of the ovary relative to the other flower parts to a more protected (inferior) position, and a shift from radial to bilateral symmetry.

In addition to the flower and the fruit, a third factor in the success of the angiosperms has been the evolution of bad-tasting or toxic chemicals that discourage the predations of foraging animals.

QUESTIONS

1. Distinguish among the following: moss/club moss; spore/sporangium/sporophyte/sporophyll; homosporous/heterosporous; gametophyte/antheridium/archegonium; xylem/phloem; microphyll/megaphyll; conifer/gymnosperm; microspore/megaspore; carpel/stamen.

2. Bryophytes, among the plants, and amphibia, among the animals, often live in habitats intermediate between fresh water and dry land, rather than between salt water and land. Propose a physiological argument (referring back to Chapter 6) to explain why invasions of the land were more likely by organisms previously adapted for life in fresh water.

3. Where did the liverworts get their name? (The answer is not in the text, but a little research should tell you.) What does this name tell you about the human perspective of the rest of the biosphere?

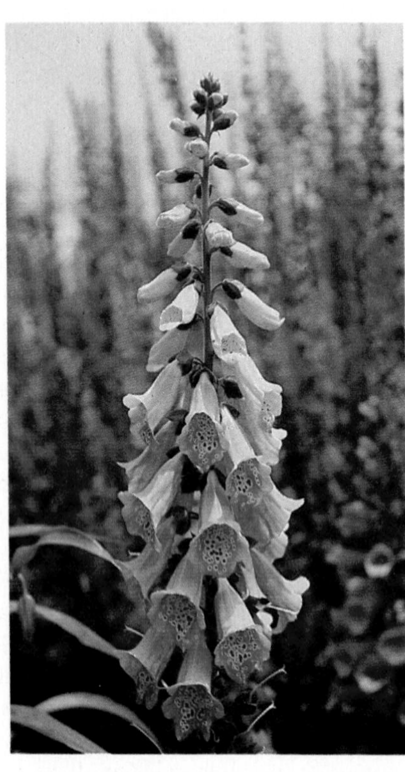

4. Vascular plants are characterized by the presence of lignin, which stiffens and supports the plant body as it grows upward toward the light. What is the primary source of support for the photosynthetic portions of the multicellular seaweeds?

5. Tall trees today are not appreciably taller than plants of the Devonian forests. What factors select for tallness in trees? What factors, by contrast, select against tallness? What kind of evolutionary innovation might be required to alter the optimal balance among these various factors?

6. In some areas on earth, large gymnosperms are either dominant or manage to coexist with angiosperms. List some advantages for a big plant's being a gymnosperm.

7. Describe the sporophylls of a club moss, a horsetail, a fern, and an angiosperm.

8. Sketch a pine seed, label it, and indicate the origin and ploidy of its components. How many generations are represented in the tissues of the seed?

9. Sketch a flower and label the structures. What is the function of each floral part?

10. Compare and contrast the structure of a "primitive" flower, such as an hepatica, with that of a "specialized" one, such as a sunflower or an iris.

11. A major trend in the evolution of angiosperms is a reduction in the number of floral parts. Why should reduction, rather than greater elaboration, be so common?

12. The dark spots on the foxglove flowers at the left are called nectar guides. What do you suppose their function is?

CHAPTER 24

The Animal Kingdom I: Introducing the Invertebrates

24–1
Animals are characterized by their mobility, usually a result of the contraction of assemblies of protein fibers within specialized (muscle) cells. Gonionemus murbachii, *a jellyfish, is shown here swimming actively, its bell contracted by muscles around its margin.*

Animals are many-celled heterotrophs, and their principal mode of nutrition is ingestion. They depend directly or indirectly for their nourishment on photosynthetic autotrophs—algae or plants. Typically they digest their food in an internal cavity and store food reserves as glycogen or fat. Their cells, unlike those of most other eukaryotes, do not have walls. Generally, animals move by means of contractile cells (muscle cells) containing characteristic proteins. Reproduction is usually sexual, with gametes being the only haploid stages. As adults, most animals are fixed in size and shape, in contrast to plants, in which growth often continues for the lifetime of the organism. The most complex animals—the octopuses, the insects, and the vertebrates—have many kinds of specialized tissues, including elaborate sensory and neuromotor mechanisms not found in any of the other kingdoms.

For most of us, animal means mammal, and mammals are, in fact, the chief focus of attention in Section 6. However, the mammals, or even the vertebrates as a whole, represent only a small fraction of the animal kingdom. More than 90 percent of the different species of animals are invertebrates—that is, animals without backbones—and most of these species are insects. Indeed, the enormous variety displayed by the invertebrates is partly why they are so endlessly fascinating to study. There are also two practical reasons for the study of invertebrates. First, by virtue of their variety and their sheer numbers, they are of great ecological importance. Second, the invertebrates, confronted with the same biological problems we face, demonstrate a spectrum of ingenious solutions. In this way, they illuminate the fundamental nature of these problems and so lead us to a deeper understanding of the physiology of the so-called "higher" animals, including ourselves.

THE DIVERSITY OF ANIMALS

The unifying characteristic among all the animals is their mode of nutrition. Unlike plants, which are passive recipients of energy from the sun, animals must seek out food sources or, alternatively, devise strategies for ensuring that the food comes to them. Thus mobility—of the entire organism, or its parts, or both—is a requirement for animal survival. Mobility in general and the hunting of prey organisms in particular require efficient systems of integration and control. In short, muscle and nerve, distinguishing features of the animal kingdom, have their origin in heterotrophy.

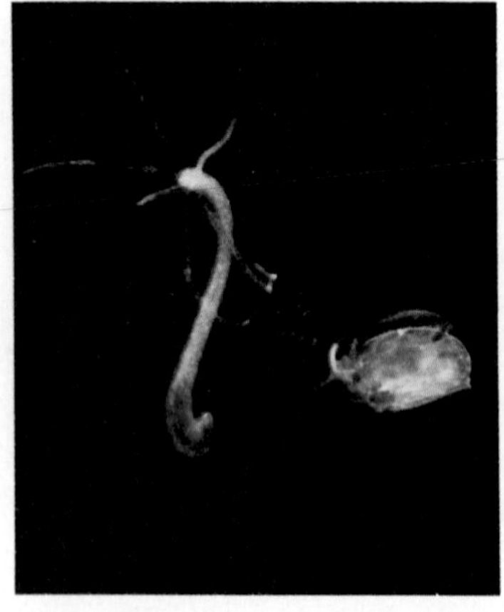
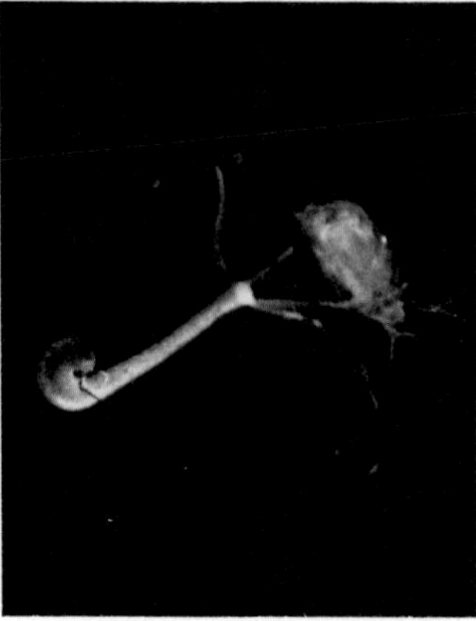
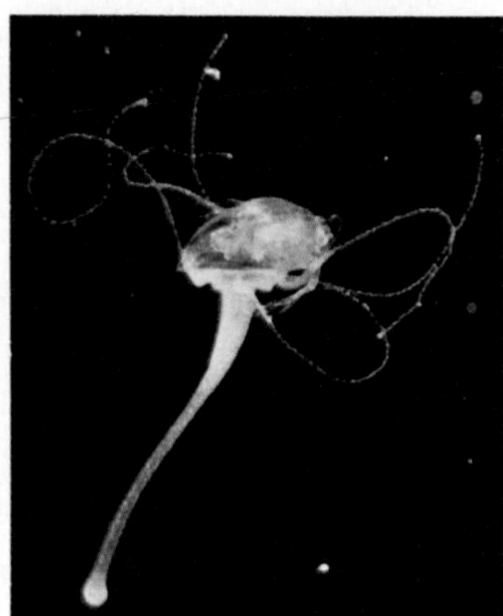

24–2
Some animals actively seek their food, whereas others arrange to have it delivered (see Figure 24–3). Here a Hydra *encounters a small crustacean* (Daphnia), *grasps it with its tentacles, engulfs it, and digests it. These activities are coordinated by the animal's simple but effective nervous system.*

Paradoxically, these unifying characteristics of the animal kingdom are also the key to its diversity. Given the properties of living tissue and the characteristics of our planet—particularly the force of gravity and the physical properties of water and air—there are only a few basic ways that locomotion, the capture of food, self-defense, and coordination can be accomplished. In the course of evolution, however, as animals have adapted to new or changing environments, those few basic ways have been "reinvented," refined, and elaborated upon, resulting in the great diversity of structural and functional detail that we see today.

In Chapter 20, we commented upon the tremendous versatility of the prokaryotes, as exemplified by the wide range of environments they inhabit and the many ways in which they satisfy their energy requirements. Among the more than 1 million species of animals that have been described, we see a similar pattern of adaptation to many different ways of life. Thus, for instance, on and around a single coral head, only a meter or two in diameter, one finds a dazzling array of different forms—reef fish, crustaceans, sponges, jellyfish, starfish, sea urchins, anemones, and the coral animals themselves. Similarly, to take a terrestrial example, a single spadeful of soil turns up earthworms, pillbugs, spiders, nematodes, and various other tiny animals. The branch of a single tree may harbor a dozen different kinds of insects, differing from one another in their nutritional specializations; in many instances, the adults even differ from the immature forms of the same species.

24–3
In the last century, Louis Agassiz, America's leading naturalist, characterized a barnacle as "nothing more than a little shrimp-like animal, standing on its head in a limestone house and kicking food into its mouth." Although the adult barnacle is immobile, its food-getting strategy, employing six pairs of bristly appendages, is quite efficient.

THE ORIGIN AND CLASSIFICATION OF ANIMALS

Animals, like plants, presumably had their origins among the protists. In the case of animals, however, we have fewer clues about which protists most closely resemble the ancestral ones, and multicellularity may well have arisen independently in more than one lineage. The fossil record indicates that by the Cambrian period, which ended some 500 million years ago, most, if not all, of the invertebrate phyla were already in existence. Although numerous efforts have been made to reconstruct the evolutionary history of the invertebrates, the Precambrian fossil record has so far proved inadequate to verify any hypotheses about the chronological order of their appearance. Most of the evidence for charts such as Figure 24–4 comes from studies of modern forms.

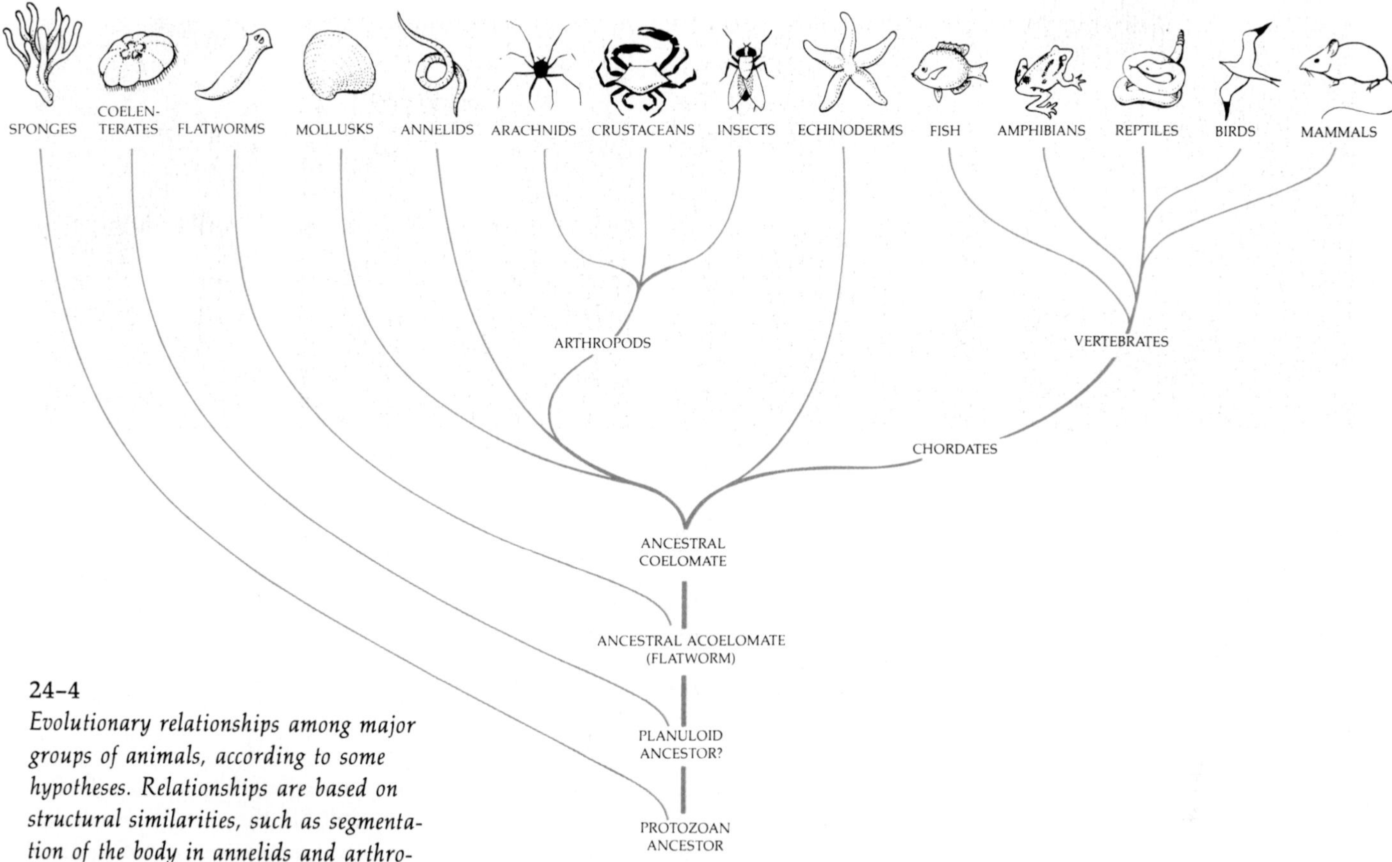

24–4
Evolutionary relationships among major groups of animals, according to some hypotheses. Relationships are based on structural similarities, such as segmentation of the body in annelids and arthropods, and also on resemblances among larval forms and in developmental patterns. The earliest invertebrate fossils available in quantity are from the Cambrian period, and by this time, the major groups of invertebrates had already diverged.

Modern animals are classified in about 30 phyla, each of which is thought to be monophyletic. A number of criteria are used in determining the classification of an animal and in attempting to elucidate the evolutionary relationships among the different phyla. Among the most important are the number of tissue layers into which the cells are organized, the basic plan of the body and the arrangement of its parts, the presence or absence of body cavities and the manner in which they form, and the pattern of development from fertilized egg to adult animal. Table 24–1 on the next page provides a short outline of animal classification and indicates the key features that are used in grouping the various phyla. As we examine the diversity of animals and their most significant adaptations, we shall devote most of our attention to the phyla indicated in boldface type.

In the remainder of this chapter, we shall look at the so-called lower invertebrates, characterized by relatively simple body plans. In the three subsequent chapters, we shall consider the more complex animals, all of which possess the type of body cavity known as a coelom.

PHYLUM PORIFERA: SPONGES

Sponges seem to have had a different origin from other members of the animal kingdom and to have traveled a solitary evolutionary route. For this reason, they are generally placed in a subkingdom of their own, the Parazoa ("beside the

Table 24-1 An Outline of Animal Classification*

I. Subkingdom Parazoa: phylum **Porifera**
II. Subkingdom Mesozoa: phylum Mesozoa
III. Subkingdom Eumetazoa
  A. Radially symmetrical animals: phyla **Cnidaria** and Ctenophora
  B. Bilaterally symmetrical animals
    1. Acoelomates (animals that lack a body cavity): phyla **Platyhelminthes,** Gnathostomulida, and Rhynchocoela
    2. Pseudocoelomates (animals with the type of body cavity known as a pseudocoelom): phyla **Nematoda,** Nematomorpha, Acanthocephala, Kinorhyncha, Gastrotricha, Rotifera, and Entoprocta
    3. Coelomates (animals with the type of body cavity known as a coelom)
      a. Protostomes (animals in which the mouth appears before the anus during embryonic development): phyla **Mollusca, Annelida,** Sipuncula, Echiura, Priapulida, Pogonophora, Pentastomida, Tardigrada, Onychophora, and **Arthropoda**
      b. Lophophorates (protostomes, but with some deuterostome characteristics): phyla Phoronida, Bryozoa, and Brachiopoda
      c. Deuterostomes (animals in which the anus appears before the mouth during embryonic development): phyla **Echinodermata,** Chaetognatha, Hemichordata, and **Chordata**

* A summary of the distinctive characteristics of each phylum can be found in Appendix C.

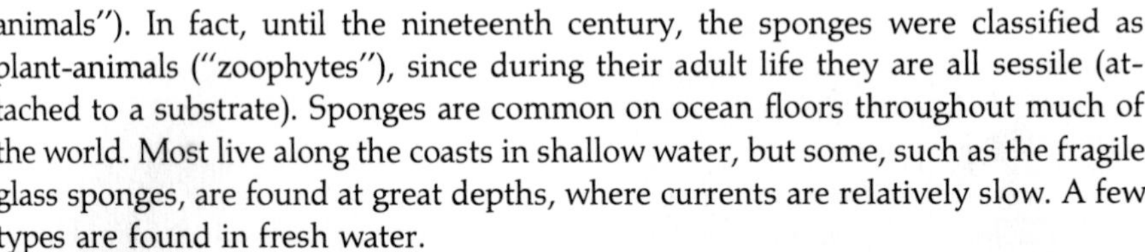

animals"). In fact, until the nineteenth century, the sponges were classified as plant-animals ("zoophytes"), since during their adult life they are all sessile (attached to a substrate). Sponges are common on ocean floors throughout much of the world. Most live along the coasts in shallow water, but some, such as the fragile glass sponges, are found at great depths, where currents are relatively slow. A few types are found in fresh water.

24-5
A purple sponge, photographed at a depth of about 10 meters in the waters off the Bahamas. Pigments within its cells give this sponge its brilliant color, which may also be enhanced by the refraction, or bending, of light as it passes through the water.

A sponge is essentially a water-filtering system, made up of one or more chambers through which water is driven by the action of numerous flagellated cells. Sponges are made up of a relatively few cell types, the most characteristic of which are the *choanocytes*, or collar cells, the flagellated cells that line the interior cavity of the sponge (Figure 24-6). The protozoan choanoflagellates have similar cells, and it is possible that sponges arose from colonial forms of such organisms.

Sponges are somewhere between a colony of cells and a true multicellular organism. The cells are not organized into tissues or organs, yet there is a form of recognition among the cells that holds them together and organizes them. If the body of a living sponge is squeezed through a fine sieve or cheesecloth, it is separated into individual cells and small clumps of cells. Within an hour, the isolated sponge cells begin to reaggregate, and as these aggregations get larger, canals, flagellated chambers, and other features of the body organization of the sponge begin to appear. This phenomenon has been used as a model for the analysis of cell adhesion, recognition, and differentiation, all of which are basic biological features of development in higher organisms.

The outer surface of a sponge is covered with epithelial cells, some of which contract in response to touch or to irritating chemicals, and in so doing, close up pores and channels. Each cell acts as an individual, however; there is little coordination among them. Between the epithelial cells and the choanocytes is a middle,

24–6
The body of a simple sponge is dotted with tiny pores, from which the phylum derives its name (Porifera, or "pore bearers"). Water containing food particles is drawn into the internal cavity of the sponge through these pores and is forced out the osculum. The water is moved by the sucking effect of flow of the local currents across the osculum and by the beating of the flagella protruding from the collars of choanocytes. The collar of each choanocyte is made up of about 20 retractile filaments and surrounds a single flagellum, the lashing of which directs a current of water through the filaments. Minute particles are filtered out and cling to one or more filaments and are then drawn into the cell. A sponge 10 centimeters high filters more than 20 liters of water a day.

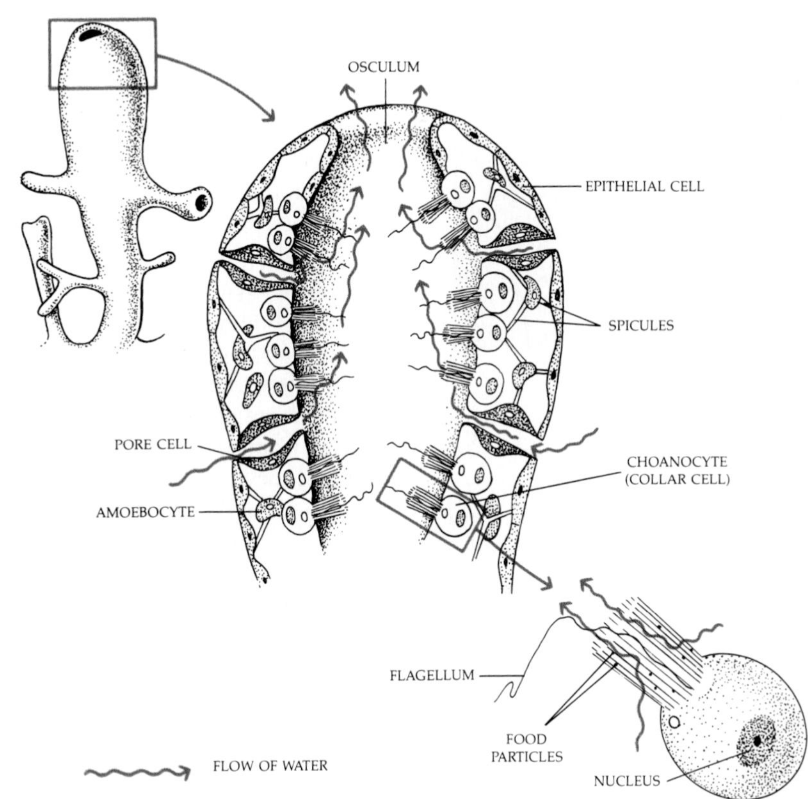

24–7
The skeleton of Euplectella speciosissima, *a species of glass sponge picturesquely known as Venus's flower basket. These fragile sponges, with their delicate silica-containing skeletons, are usually found at great depths. According to the fossil record, glass sponges were present in the Ordovician period, which began 500 million years ago.*

jellylike layer, and in this layer are amoebocytes, amoeba-like cells that carry out various functions. Amoebocytes play several roles in reproduction, secrete skeletal materials, and, most important, carry food particles from the choanocytes to the epithelial and other nonfeeding cells. Because all the digestive processes of sponges are carried out within single cells, even a giant sponge—and some stand taller than a man—can consume nothing larger than microscopic particles.

The sponge shown in Figure 24–6 is a small and simple one. In larger sponges, the body plan, although it is essentially the same, looks far more complex. These sponges, which need correspondingly more food, have highly folded body walls that greatly increase the filtering and feeding surfaces. We have already encountered this evolutionary stratagem for increasing biological work surfaces at the cellular level—as in the inner membrane of the mitochondrion—and we shall be encountering it repeatedly in different animal structures.

The approximately 5,000 species of sponges are grouped into four classes, according to their skeletal structure, which serves for protection, stiffening, and support. In the class Calcarea, the skeleton consists of individual spicules of calcium carbonate. Members of class Hexactinellida, the glass sponges, have spicules of silica fused in a continuous and often very beautiful latticework (Figure 24–7). The largest class, Demospongiae, has unfused silica spicules, or a tough, keratinlike protein called spongin, or a combination of the two. The cleaned and dried proteinaceous skeletons of this group are the "natural" sponges available commercially. Members of the fourth and smallest class, Sclerospongiae, have skeletons that contain all three kinds of material—calcium carbonate, silica, and spongin.

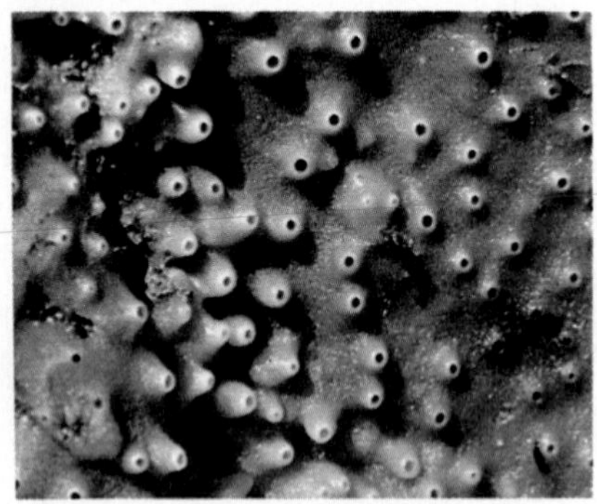

24–8
A breadcrumb sponge showing some of the many openings (oscula) through which water leaves the animal. The oscula usually protrude above the rest of the animal. This arrangement allows the natural water flow in the habitat to draw water through the sponge.

Reproduction in Sponges

The reproduction of sponges exhibits many of the features characteristic of sessile or slow-moving animals. Asexual reproduction is quite common, either by fragments that break off from the parent animal, or by gemmules, aggregations of amoebocytes within a hard, protective outer layer. Production of such resistant forms is found, in general, only among freshwater organisms. In the ocean, conditions are relatively unchanging, but the freshwater environment is much harsher. Invertebrates that live in fresh water are more likely to have protected embryonic forms than even closely related marine species.

Sexual reproduction in sponges is highly specialized. The simplest and most primitive form of fertilization is external, with sperm and egg cells shed into the water, where they unite. This primitive method of fertilization is retained in many phyla that are otherwise highly evolved. In most sponges, however, fertilization is internal. Gametes appear to arise from enlarged amoebocytes, but there are reports that choanocytes can also form gametes. The sperm cells, which resemble those of other animals, are carried by the water currents out of the osculum of one sponge and into the interior cavity of another sponge. There they are captured by choanocytes and transferred to amoebocytes, which then transfer them to ripe eggs—a method of fertilization unique to the sponges. Most sponges even provide a certain amount of maternal care, retaining the young during the early stages of development. The embryonic sponge develops into a flagellated, free-swimming larva that, after a short life in the plankton, locates an appropriate site, settles, and develops into an adult sponge.

Most kinds of sponges are hermaphrodites (from Hermes and Aphrodite); that is, they have male and female reproductive structures in the same individual. This is a great advantage for animals with little or no mobility. In animals with separate sexes, a given individual can mate only with members of the opposite sex, but for a hermaphrodite any partner—or the gametes of any partner—will suffice.

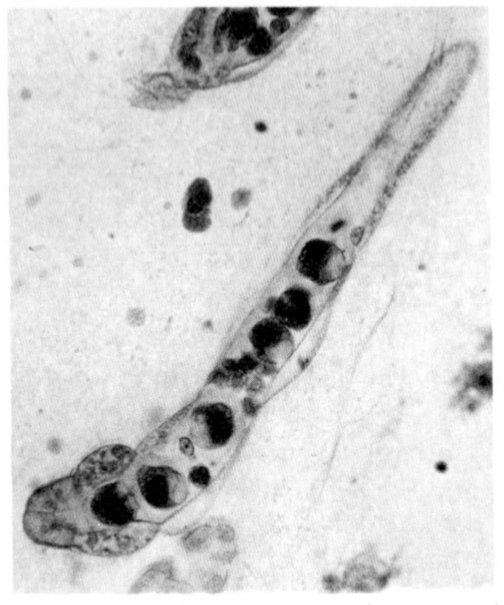

24–9
Dicyema typoides *is a mesozoan found in the kidneys of octopuses living in the Gulf of Mexico. Mesozoans typically reproduce asexually during one phase of the life cycle and sexually during another phase. This individual is a sexually reproducing adult, forming both sperm and egg cells.*

PHYLUM MESOZOA: MESOZOANS

Like the sponges, the mesozoans are so different from all other animals that they are placed in a subkingdom of their own. About 50 species of mesozoans are known; they are extremely simple wormlike animals that live as parasites inside a variety of marine invertebrates. The 20 to 30 cells that make up the body of a mesozoan are organized in two layers—a mass of reproductive cells surrounded by a single layer of ciliated cells (Figure 24–9). There are no organs or body cavities of any kind. As their name ("middle animals") suggests, the mesozoans may be primitive forms, transitional between protozoans and multicellular animals. Some authorities believe, however, that they are flatworms (phylum Platyhelminthes) that have become simplified as an adaptation to a parasitic way of life.

RADIALLY SYMMETRICAL ANIMALS

Two phyla, Cnidaria (jellyfish, sea anemones, and corals) and Ctenophora (comb jellies and sea walnuts), consist of gelatinous animals in which the adult form is generally radially symmetrical. In radial symmetry, the body parts are arranged around a central axis, like spokes around a hub (Figure 24–10).

The animals in these two phyla are also characterized by the *coelenteron* ("hollow gut"), a digestive cavity with only one opening. Within this cavity, enzymes are

24–10
In organisms with radial symmetry, any plane through the animal that passes through the central axis divides the body into halves that are mirror images of one another.

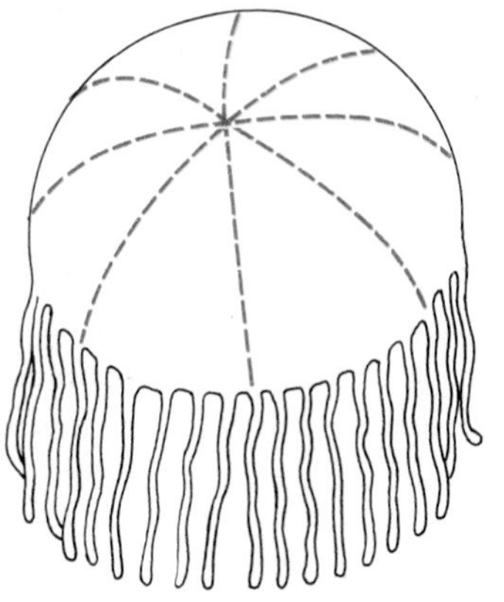

released that break down food, partially digesting it extracellularly, as our own food is digested within the stomach and intestinal tract. The food particles are then taken up by the cells lining the cavity; they complete the digestive process and pass the products on to the other cells of the animal. Inedible remains are ejected from the single opening.

Phylum Cnidaria

The cnidarians are a large and often strikingly beautiful group of aquatic organisms. The cells of cnidarians, unlike those of sponges and mesozoans, are organized into distinct tissues, and their activities are coordinated by a nervous system.

As you can see in Figure 24–11a, the basic body plan is a simple one: the animal is essentially a hollow container, which may be either vase-shaped, the *polyp*, or bowl-shaped, the *medusa*. The polyp is usually sessile; the medusa, motile. Both consist of two layers of tissue: *ectoderm* and *endoderm* (from the Greek *ektos,* "outer," and *endon,* "inner," plus *derma,* "skin"). Because they have two tissue layers, cnidarians are said to be diploblastic. Between the two layers is a gelatinous filling, the *mesoglea* ("middle jelly"), which is made of a collagenlike material. In the polyp form, the mesoglea is sometimes very thin, but in the medusa, it often accounts for the major portion of the body substance.

A distinctive feature of these animals, the *cnidocyte* (Figure 24–11b), gives the phylum its name. Cnidarians are carnivores. They capture their prey by means of tentacles that form a circle around the mouth. These tentacles are armed with

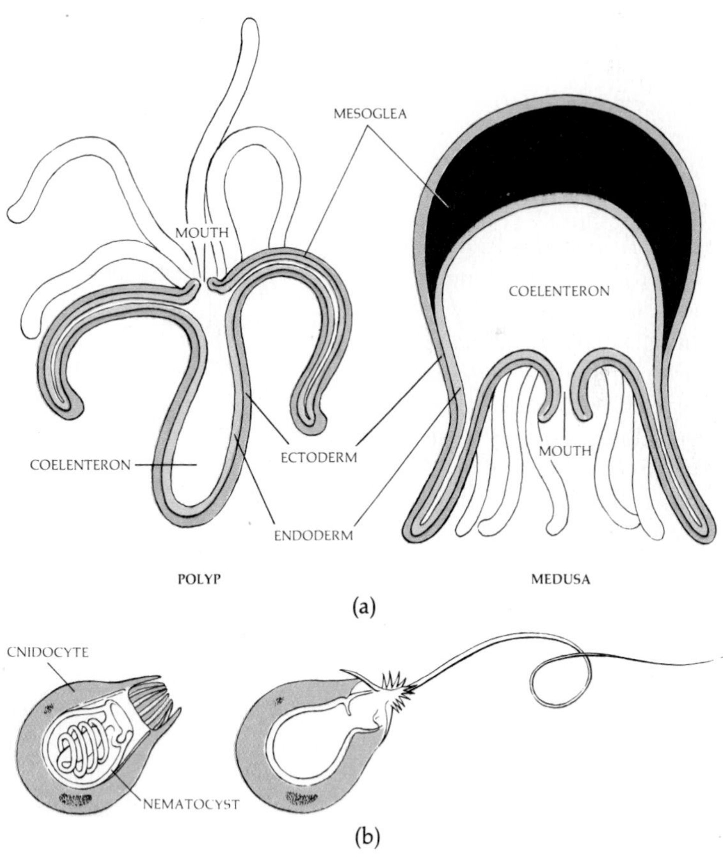

24–11
(a) *Among cnidarians, there are two basic body plans: the vase-shaped polyp (left) and the bowl-shaped medusa (right). The coelenteron is a digestive cavity with a single opening. The cnidarian body has two tissue layers, ectoderm and endoderm, with gelatinous mesoglea between them.*

(b) *Cnidocytes, specialized cells located in the tentacles and body wall, are a distinguishing feature of cnidarians. The interior of the cnidocyte is filled by a nematocyst, which consists of a capsule containing a coiled tube, as shown on the left. A trigger on the cnidocyte, responding to chemical or mechanical stimuli, causes the tube to shoot out, as shown on the right. The capsule is forced open and the tube turns inside out, exploding to the outside. The cnidocyte cannot be "reloaded"; it is absorbed and a new cell grows to take its place.*

cnidocytes, special cells that contain *nematocysts* (thread capsules). Nematocysts are discharged in response to a chemical stimulus or touch. The tubular nematocyst threads, which are often poisonous and may be sticky or barbed, can lasso prey, harpoon it, or paralyze it—or some useful combination of all three. The toxin apparently produces paralysis by attacking the lipoproteins of the nerve cell membrane of the prey.

Nematocysts occur only in this phylum, with some interesting exceptions. Certain other invertebrates, including nudibranchs (a kind of mollusk) and flatworms, can eat cnidarians without triggering the nematocysts. The nematocysts are then moved to the surface of the predator and can be fired in their new host's defense.

The cnidarian life cycle is characterized by an immature larval form, known as the *planula*, which is a small, free-swimming ciliated organism. Following this larval stage, some cnidarians go through both a polyp and a medusa stage in their life cycles. In such species, polyps reproduce asexually and medusas sexually. This sort of life cycle, in which the sexually reproductive form is distinctly different from the asexual form, superficially resembles alternation of generations in plants. There is, however, no alternation between haploid and diploid forms as there is in plants; the only haploid forms are the gametes. The cnidarian life cycle allows for rapid asexual reproduction (by the polyp), dispersal and genetic recombination (by the medusa), and habitat selection (by the planula larva).

The approximately 9,000 species of cnidarians are grouped in three major classes: Hydrozoa, in which the polyp is usually the dominant form; Scyphozoa, predominantly medusoid, exemplified by the common jellyfish; and Anthozoa, which includes the sea anemones and the reef-building corals, and has only the polyp.

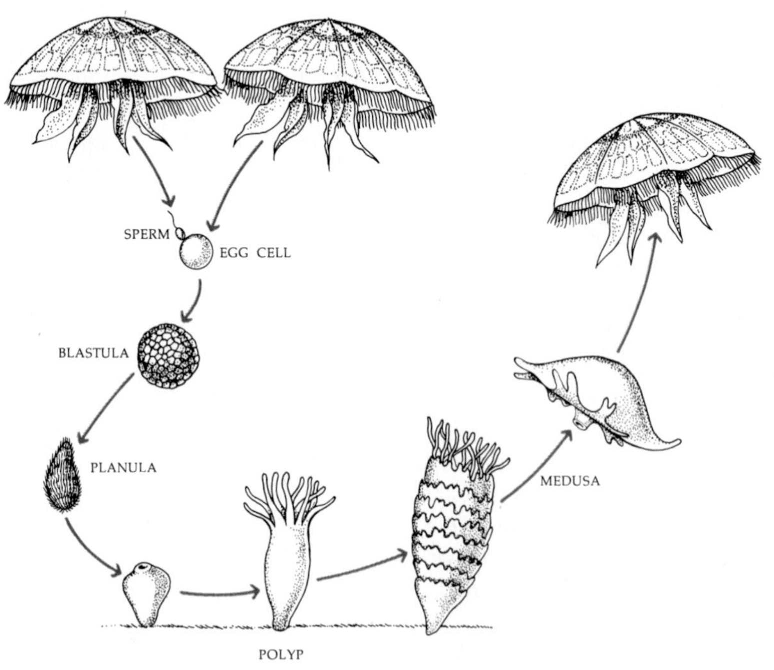

24–12
The life cycle of the cnidarian Aurelia. *Sperm and egg cells are released from adult medusas into the surrounding water. Fertilization takes place, and the resulting zygote develops first into a hollow sphere of cells, the blastula. It then elongates and becomes a ciliated larva called a planula. After dispersal, the planula settles to the bottom, attaches by one end to some object, and develops a mouth and tentacles at the other end, thus transforming into the polyp stage. The body of the polyp grows and, as it grows, begins to form medusas, stacked upside down like saucers. In a phase of rapid asexual reproduction, these bud off, one by one, and grow into full-sized jellyfish.*

24–13
The structure of the body wall of Hydra. *The outer layer of cells, the ectoderm, is primarily for protection, while the inner layer, the endoderm, performs the digestive function. One type of digestive cell, the gland cell, secretes the digestive enzymes that are released into the coelenteron. Gland cells can also produce a gas bubble that enables the animal to float to the surface. The other type of digestive cell, the nutritive cell, using its flagella, mixes the food as it is being processed and then extends pseudopodia that collect the food particles for further digestion.*

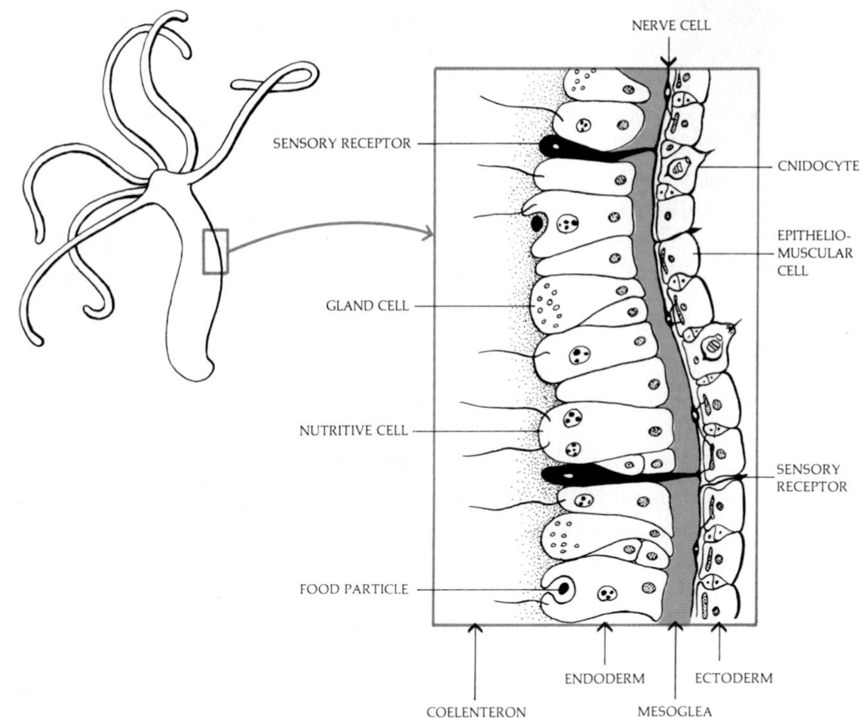

Class Hydrozoa

Among the most thoroughly studied of the cnidarians are species of *Hydra* and related genera, which are small, common freshwater forms, convenient to keep in the laboratory. Figure 24–13 shows a small section of the body wall of *Hydra.* The ectoderm is composed largely of epitheliomuscular cells, which perform a covering, protective function and also serve as muscle tissue. Each cell has contractile fibers, myonemes, at its base and so can contract individually, like the contractile epithelial cells of the sponge. The endoderm is mostly made up of cells concerned with digestion; these cells also contain contractile fibers. In *Hydra,* as in other polyps, the contractile fibrils of the ectoderm run lengthwise in the animal and the fibrils of the endoderm cells run circularly, so the body walls can stretch or bulge, depending on which group contracts.

In addition to cnidocytes and epitheliomuscular cells, which are independent effectors—cells that both receive and respond to stimuli—*Hydra* contains two other types of nerve cells: sensory receptor cells and cells connected into a network, the nerve net. Sensory receptor cells are more sensitive than other epithelial cells to chemical and mechanical stimuli, and when stimulated they transmit their impulses to an adjacent cell or cells. The adjacent cell may be simply an epitheliomuscular cell, an effector, which then responds. Note that this system is one step more complicated than the epitheliomuscular cell or cnidocyte, which acts as both receptor and effector. The nerve net, a loose connection of nerve cells lying at the base of the epithelial layer, is the simplest example of a nervous system that links an entire organism into a functional whole. It coordinates the muscular contractions of *Hydra,* making possible a wide variety of activities (Figures 24–2 and 24–14). However, there is no center of operations for the nervous system. This type of conducting system occurs in *Hydra* and certain other cnidarians.

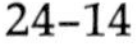
24–14
Hydra *has a nervous system that integrates the body into a functional whole, making possible a range of fairly complex activities. For instance,* Hydra *may float, glide on its base, or, as shown here, it may travel by a somersaulting motion.*

24–15
Colonial hydrozoans. (a) Obelia *is made up of two types of polyp, a* Hydra*-like form—shown here with tentacles extended—which is the feeding polyp, and a reproductive form, which lacks tentacles. You can see two of these reproductive forms in the axils of the branches. At the left is a newly released free-swimming medusa.*

(b) *Cnidarians of the order Siphonophora are large, floating colonies made up of both polyps and medusas. The polyps are feeding forms and the medusas are reproductive forms; in some species, nonreproductive medusas are swimming bells. The colony produces a gas-filled float, or pontoon. In the Portuguese man-of-war shown here, the large float serves also as a sail. The blue strands are composed of reproductive and feeding individuals; the purple strands, which may grow as long as 15 meters, are made up of stinging, food-gathering polyps armed with nematocysts. A large colony can kill a human being.*

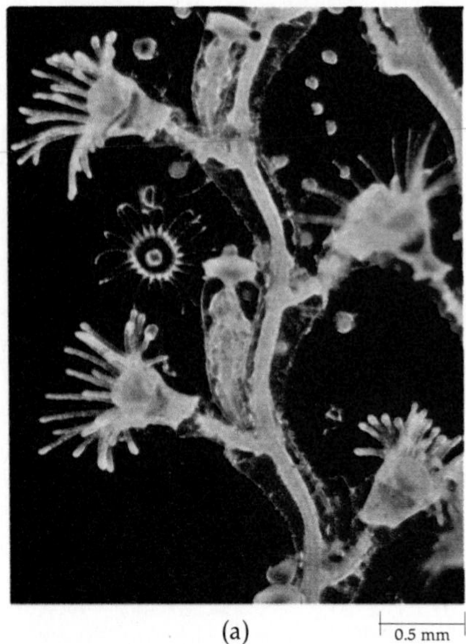

(a)

(b)

Hydra, a relatively simple hydrozoan that lives as a solitary polyp and has no medusa stage, is thought to be descended from more complex ancestors. Most hydrozoans are colonial marine animals and have both hydroid (polyp) and medusoid forms at different times in their life cycles. Members of the genus *Obelia,* for example, spend most of their lives as colonial polyps (Figure 24–15a). The colony arises from a single polyp, which multiplies by budding. The new polyps do not separate but remain interconnected so that their body cavities form a continuous channel, through which food particles are circulated. Within the colony are two types of polyps: feeding polyps with tentacles and cnidocytes, and reproductive polyps from which tiny medusas bud off. These medusas produce eggs or sperm that are released into the water and fuse to form zygotes.

Many colonial hydrozoans, with their division of labor between feeding and reproductive forms, are very like a single organism (Figure 24–15b). Such a high degree of specialization of structure and function among social organisms is seen in other phyla only among the social insects.

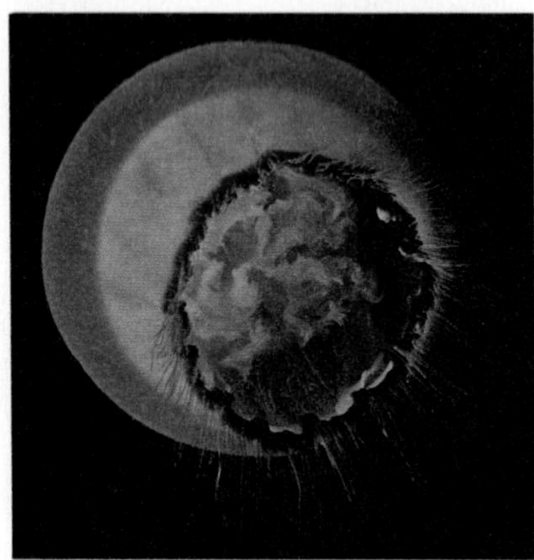

24–16
The moon jellyfish (Aurelia limbata) *has four frilly mouth lobes that gather minute organisms into the coelenteron. It has short tentacles and usually grows to only about 30 centimeters in diameter. This specimen was photographed in the cold waters of the Atlantic.*

Class Scyphozoa

A second major class of cnidarians is Scyphozoa, or "cup animals," in which the medusa form is dominant. Although the fairly simple polyp form reproduces itself asexually in addition to budding off medusas, elaborate colonies are not formed. Scyphozoan medusas, known more commonly as jellyfish, range in size from less than 2 centimeters in diameter up to animals 4 meters across and trailing tentacles 10 meters long. In the adult animal, the mesoglea is so firm that a large, freshly beached jellyfish can easily support the weight of a man. The mesoglea of some jellyfish contains wandering, amoeba-like cells, which serve to transport food from the nutritive cells of the endoderm.

In scyphozoans, the epitheliomuscular cells, which are more specialized than those of *Hydra,* underlie the ectoderm, contracting rhythmically to propel the medusa through the water (see Figure 24–1). The contractions are coordinated by concentrations of nerve cells in the margin of the bell. These nerve cells connect

24–17

(a) The statocyst is a specialized receptor organ that orients the jellyfish with respect to gravity. When the bell tilts, gravity pulls the statolith, a grain of hardened calcium salts, down against the hair cells. This stimulates the nerve fibers and signals the animal to right itself. (b) Eyespot (ocellus), the simplest type of photoreceptor organ. Ocelli of this sort are found among the cnidarians.

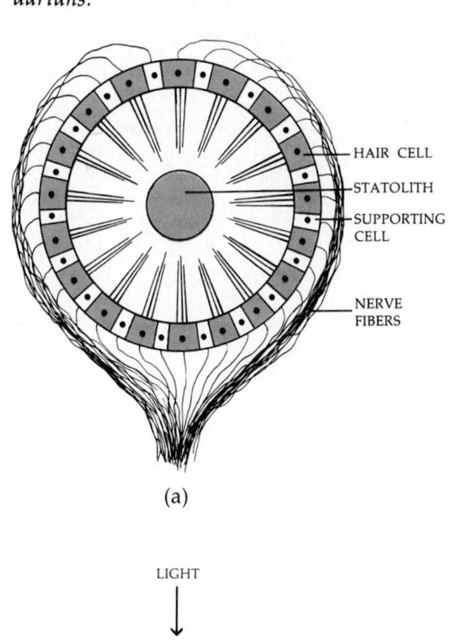

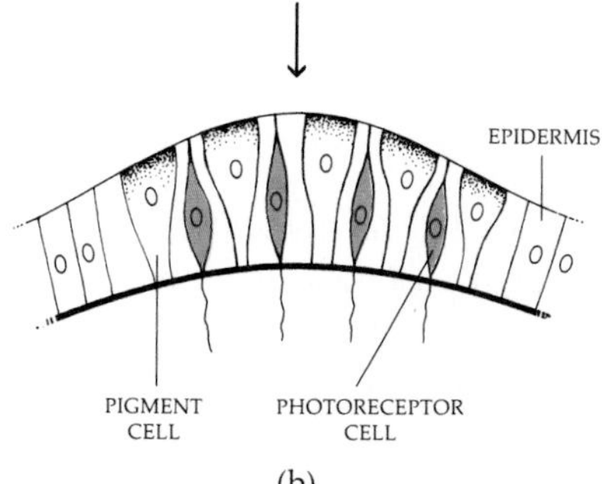

with fibers innervating (providing the nerve supply for) not only the musculature but also the tentacles and the sense organs.

The bell margin is liberally supplied with sensory receptor cells sensitive to mechanical and chemical stimuli. In addition, the jellyfish has two types of multicellular sense organs: statocysts and light-sensitive ocelli. *Statocysts* are specialized receptor organs that provide information by which an animal can orient itself with respect to gravity (Figure 24–17a). The statocyst, which seems to have been one of the first types of special sensory organ to have appeared in the course of evolution, has persisted apparently unchanged to the present day, appearing in many animal phyla. (As we shall see in Chapter 32, similar structures also play a role in orienting the growth of plant roots with respect to gravity.) *Ocelli*, which may have evolved even earlier, are groups of pigment cells and photoreceptor cells (Figure 24–17b). They are typically located at the bases of the tentacles.

Class Anthozoa

Anthozoans ("flower animals")—the sea anemones and corals—are cnidarians that, like *Hydra*, have no medusa stage. These polyps reproduce both asexually, by budding, division, or fragmentation, and sexually, by the production of gametes. The zygote develops into a planula larva that may travel some distance from the parents before attaching itself to a suitable substrate and developing into an adult animal.

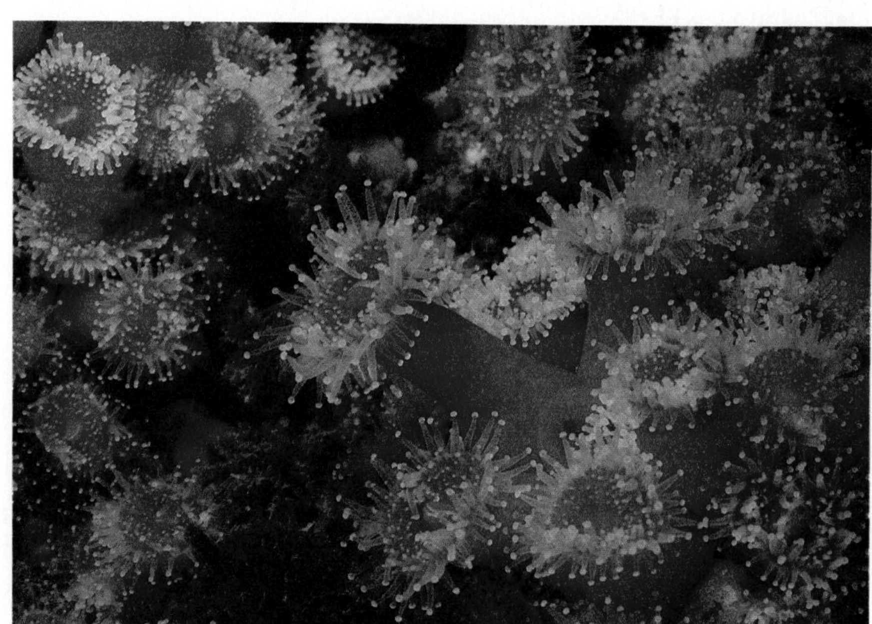

24–18

Pink sea anemones, photographed near Point Loma, California. The flowerlike appearance is deceptive; anemones are carnivorous animals belonging to a class of cnidarians that, like Hydra, *have dropped the medusa stage. In common with other cnidarians, their tentacles are equipped with stinging nematocysts. The tentacles move food into the coelenteron, which is divided longitudinally by partitions.*

(a)

(b)

(a) *A coral reef in the New Hebrides, islands in the Pacific, east of Australia. Staghorn coral is in the foreground.* (b) *Living gorgonian coral, photographed in the Caribbean. Notice the individual polyps, extended from their limestone skeletons.*

THE CORAL REEF

The coral reef is the most diverse of all marine communities. The reef structure itself is formed by colonial anthozoans. Each polyp in the colony secretes its own calcium-containing skeleton, which then becomes part of the reef. The photosynthetic activity of the reef is carried out almost entirely by symbiotic algae living within the tissues of the corals; in fact, as much as half of the mass of a coral reef may consist of green algae. Carbon, oxygen, and dissolved minerals flow over the reef as a result of the movement of waves and the ocean currents. The reef furnishes both food and shelter for other sea animals, including numerous species of reef fishes and a tremendous variety of invertebrates, such as sponges, sea urchins, polychaetes, and crustaceans. The coral polyps and algae that form the reef can grow only in warm, well-lighted surface water, where the temperature seldom falls below 21°C.

The largest coral-created land masses in the world are the 2,000-kilometer-long Great Barrier Reef, off the northeast shores of Australia, and the Marshall Islands in the Pacific. Other reefs are found throughout tropical waters and as far north as Bermuda, which is warmed by the Gulf Stream.

Unlike hydrozoan polyps, anthozoans have a gullet lined with epidermis and a coelenteron divided by vertical partitions. In most corals, which are colonies of anthozoans, the epidermal cells secrete protective outer walls, usually of calcium carbonate (limestone), into which each delicate polyp can retreat. The limestone-forming polyps are the most ecologically important of the cnidarians. A coral reef is composed primarily of the accumulated limestone skeletons of the corals, covered by a thin crust occupied by the living colonial animals. A reef is both the structural and nutritional basis of the complex coral reef community (see essay).

24-19
Mnemiopsis leidyi, *a ctenophore that is common along the Atlantic and Gulf coasts of the United States.*

Phylum Ctenophora

About 90 species, commonly known as comb jellies and sea walnuts, make up the phylum Ctenophora ("comb bearers"). These small animals, which are abundant in the open ocean, are readily identified by comblike plates of fused cilia that are arranged in eight longitudinal bands on the body surface (Figure 24-19). The coordinated beating of the cilia is their primary means of locomotion, although a few species also use muscular movements of the body wall for more rapid swimming. Comb jellies and sea walnuts are bioluminescent, giving off bright flashes of light that are particularly visible at night.

Ctenophores, like cnidarians, are characterized by a coelenteron and two tissue layers, ectoderm and endoderm, with a gelatinous mesoglea between. Amoeboid cells and muscle fibers are scattered through the mesoglea, which makes up most of the body substance. Two long tentacles, containing specialized cells that secrete a sticky substance, are used to capture food.

Reproduction is sexual, and all individuals are hermaphrodites. The fertilized eggs develop into free-swimming larvae that gradually change into the adult form.

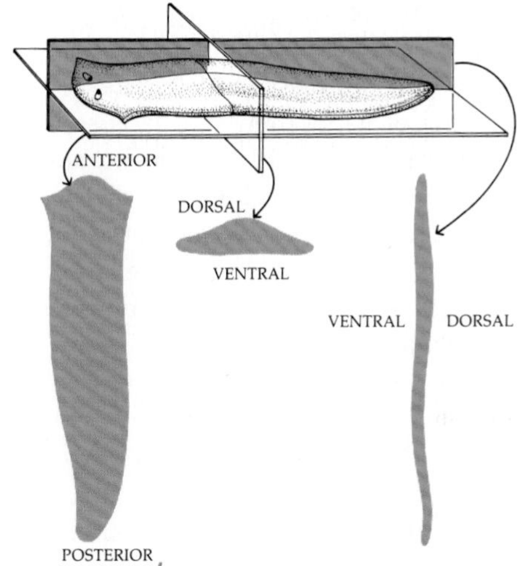

24-20
In a bilaterally symmetrical organism, such as the planarian shown here, the right and left halves of the body are mirror images of one another. The upper and lower (or back and front) surfaces are known as dorsal and ventral. With some exceptions, the end that goes first is termed anterior and the rear, posterior.

BILATERALLY SYMMETRICAL ANIMALS: AN INTRODUCTION

All of the remaining animals, including ourselves, are bilaterally symmetrical. The apparent exceptions, such as the echinoderms, which are radially symmetrical as adults, pass through a bilateral stage during their development. In bilateral symmetry, the body plan is organized along a longitudinal axis, with the right half an approximate mirror image of the left half (Figure 24-20). A bilaterally symmetrical animal can move more efficiently than a radially symmetrical one (which is better adapted to a sedentary existence). It also has a top and a bottom or, in more precise terms, a *dorsal* and a *ventral* surface. These terms are applicable even when the organism is turned upside down, or, as with humans, stands upright, in which case dorsal means back and ventral means front. Most bilateral organisms also have a distinct "headness" and "tailness," *anterior* and *posterior*. Having one end that goes first—cephalization—is characteristic of actively moving animals. In such animals, many of the sensory cells are collected into the anterior end. With the aggregation of sensory cells, there came a concomitant gathering of nerve cells; this gathering is the forerunner of the brain. Structures useful in capturing and consuming prey are also generally located in the anterior region of the animal, whereas digestive, excretory, and reproductive structures tend to be located toward the posterior.

The bilateral animals are all triploblastic; that is, they have three tissue layers. The third layer is the *mesoderm*, located between the ectoderm and the endoderm. These three layers can be detected very early in the embryological development of bilateral animals and give rise to the various specialized tissues of the adult animal. Thus they are known as the "germ layers." Generally speaking, covering and lining tissues, as well as nerve tissues, are derived from ectoderm; digestive structures from endoderm; and muscles and most other parts of the body from mesoderm. This general pattern makes functional sense. When sensory and nerve cells developed, they did so as specializations of the outer layer, the ectoderm, where sensation was most important. Similarly, digestive structures developed in the inner layer, the endoderm, which surrounds the food-containing cavity. As animals became more complex and additional structures developed for locomotion, internal transport, excretion, and reproduction, they did so from the "new," middle layer, the mesoderm.

Among the triploblastic animals, there are three basic body plans, determined by the presence or absence of a body cavity—a *coelom* (pronounced "see-loam")—in addition to the digestive cavity. In the simplest type of plan (Figure 24–21b), the three tissue layers are packed together and there is no body cavity other than the digestive cavity. Animals with this type of body plan are known as *acoelomates*. A more complex body plan is found in the *pseudocoelomates*. These animals have an additional cavity located between the endoderm and the mesoderm (Figure 24–21c); this cavity is known as a *pseudocoelom* because of its location and because it lacks the epithelial lining characteristic of a coelom. The *coelomates* (mollusks, annelids, and all of the more complex animals) have a true coelom, which is a fluid-filled cavity *within* the mesoderm (Figure 24–21d). Within the coelom, the gut and other internal organs—lined with epithelium—are suspended by double layers of mesoderm known as mesenteries. The advantages of these increasingly complex body plans will become apparent as we examine the animals themselves.

(a) TWO-LAYERED, NO COELOM (CNIDARIANS, CTENOPHORES)

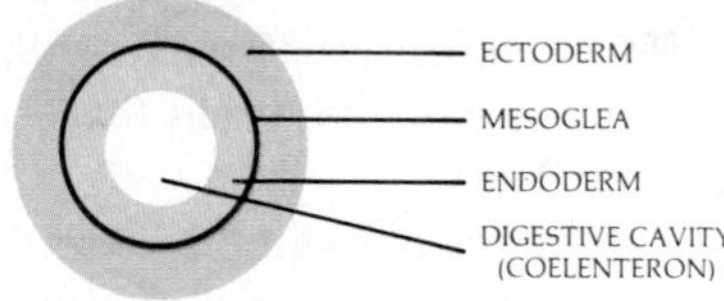

(b) THREE-LAYERED, NO COELOM (FLATWORMS, ETC.)

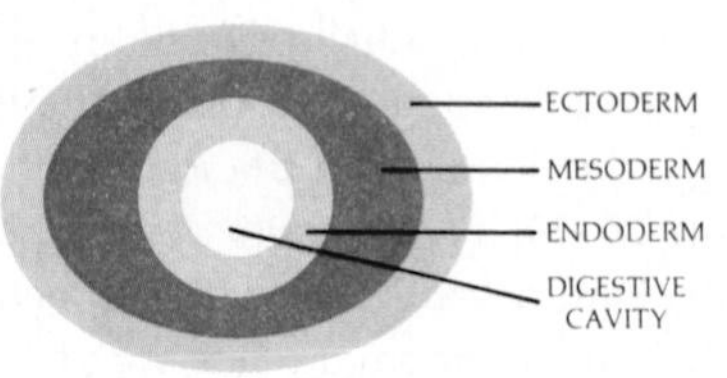

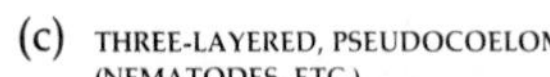

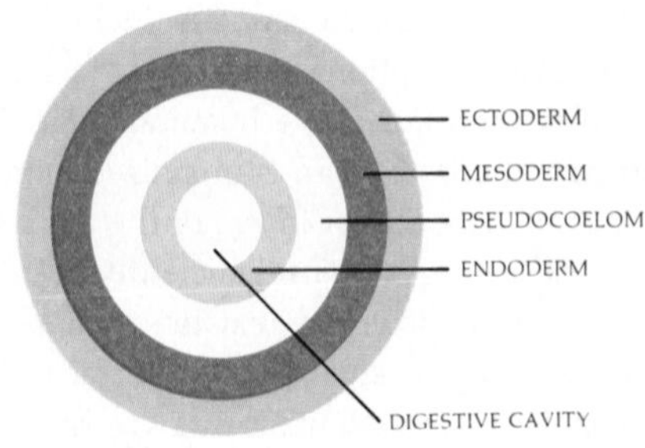

(d) THREE-LAYERED, COELOM (ANNELIDS, MOLLUSKS, ARTHROPODS, CHORDATES, ETC.)

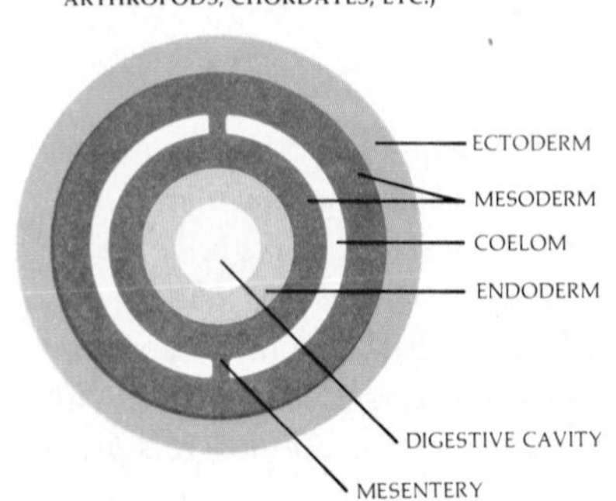

24–21
Basic body plans of the animal phyla, as shown in cross section. (a) *A body that consists of only two tissue layers is characteristic of cnidarians and ctenophores.*

(b) *Acoelomates, such as flatworms and ribbon worms, have three-layered bodies, with the layers closely packed on one another.*

(c) *Pseudocoelomates, such as nematodes, have three-layered bodies with a pseudocoelom between the endoderm and mesoderm.*

(d) *Coelomates—mollusks, annelids, and most other animals, including vertebrates—have bodies that are three-layered with a cavity, the coelom, within the middle layer (mesoderm). The mesodermal mesenteries suspend the gut within the body wall.*

Note that the term coelom, although it sounds similar to coelenteron and comes from the same Greek root, meaning "cavity," refers quite specifically to a cavity within the mesoderm, whereas the coelenteron is a digestive cavity lined by endoderm.

24-22

(a) *Example of a flatworm: a freshwater planarian.* (b) *The nervous system is indicated in color. Note that some of the fibers are aggregated into two cords, one on each side of the body. Clusters of nerve cells in the head, known as ganglia, are the beginnings of a brain. The ocelli are light-sensitive areas.* (c) *Planarians, like other flatworms, but unlike cnidarians, have three layers of body tissues. The only body cavity is the digestive cavity; thus flatworms are acoelomates.* (d) *A carnivore, the planarian feeds by means of its extensible pharynx.*

PHYLUM PLATYHELMINTHES: FLATWORMS

The flatworms are the simplest animals with bilateral symmetry and three distinct tissue layers. Moreover, not only are their tissues specialized for various functions, but also two or more types of tissue cells may combine to form organs. Thus, while sponges are made up of aggregations of cells and cnidarians are largely limited to the tissue level of organization, flatworms can be said to exemplify the organ level of complexity.

Like the cnidarians, but unlike most other bilaterally symmetrical animals, the flatworms have a digestive cavity with only one opening. Since the animal cannot feed, digest, and eliminate undigested residues simultaneously, food cannot be processed continuously.

Flatworms are acoelomates with solid bodies, and they have no circulatory system for the transport of oxygen and food molecules. Thus all cells must be within diffusion distance of sources of oxygen and of food. Flatworms have solved this problem in two ways. First, the body is flattened, which keeps the cells close to the external oxygen supply, and second, the digestive cavity is branched, carrying food particles to all regions of the body.

The flatworms are believed by some zoologists to have evolved from the cnidarians (or, perhaps, vice versa), not by way of either adult form, however, but from the ciliated planula larva. Others are persuaded that the flatworms had independent origins among the ciliates. Still another group maintains that they are degenerate annelids. The new molecular techniques for determining phylogenetic relationships (page 382) may ultimately resolve this question.

About 13,000 species of flatworms have been described, and they are placed in three classes. Class Turbellaria contains mostly free-living forms, whereas the members of classes Trematoda (flukes) and Cestoda (tapeworms) are parasitic.

Class Turbellaria

The free-living flatworms form a large and varied group, and we shall single out just one type for examination, the freshwater planarian (Figure 24-22). The ectoderm of a planarian is made up of cuboid epithelial cells, many of which are ciliated, particularly those on the ventral surface. Ventral ectodermal cells secrete mucus, which provides traction for the planarian as it moves by means of its cilia along its own slime trail. Planarians are among the largest animals that can use cilia for locomotion. The cilia in larger species are usually employed for moving water or other substances along the surface of the animal, as in the human respiratory tract, rather than for propelling the animal.

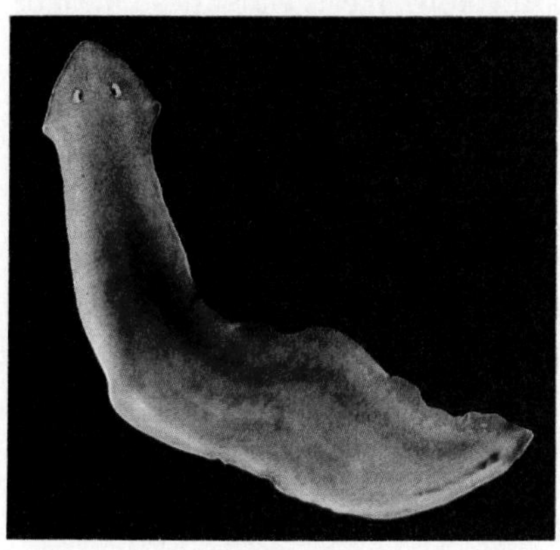

(a)

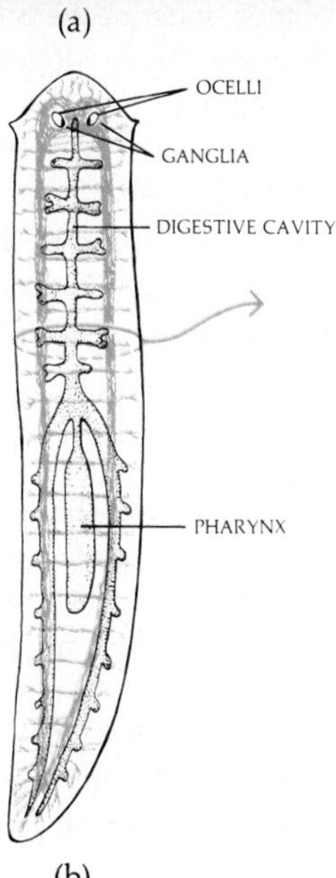

(b)

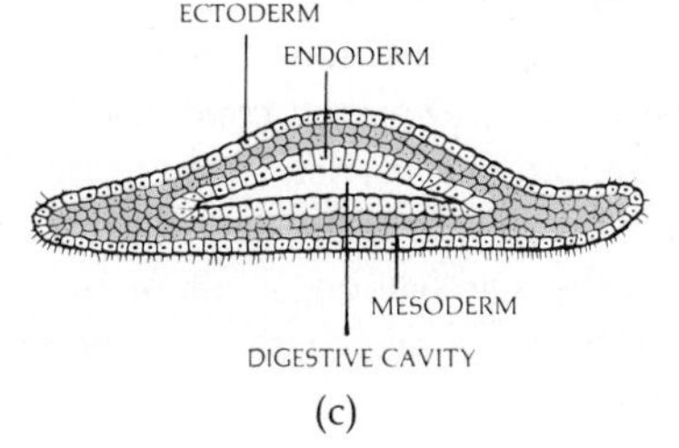

(c)

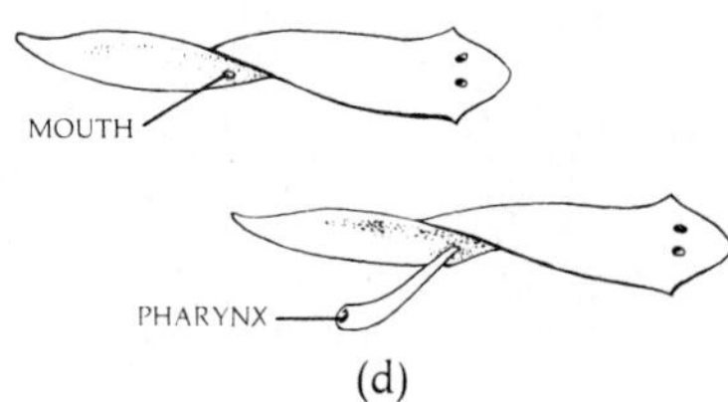

(d)

24–23
A free-living marine flatworm. The many species belonging to phylum Platyhelminthes, though widely varied in color and shape, are nearly all small and flat-bodied. This flatworm, a member of the order Polycladida, has a highly branched digestive cavity, extending into all parts of its body.

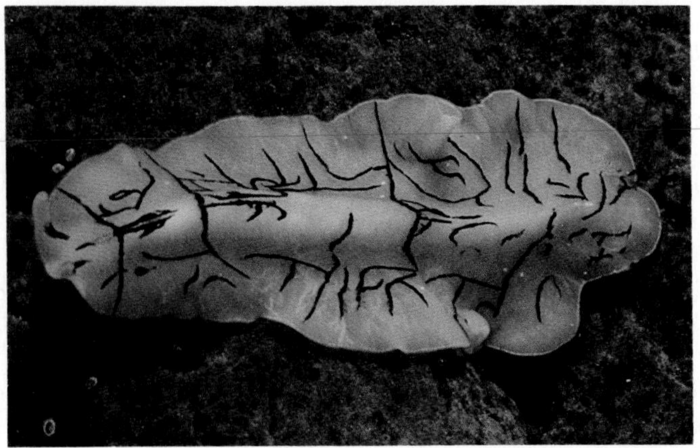

24–24
Flatworms have a tubular excretory system. The system usually consists of two or more branching tubules running the length of the body. In the planarian and its relatives, the tubules open to the body surface through a number of tiny pores. At the ends of the side branches are small bulblike structures known as flame cells. Within each of these cells, a tuft of cilia in constant motion resembles the flickering of a flame. Water and some waste materials from the tissue fluids are moved by the cilia through tubules to the excretory pores, where the collected liquid leaves the body.

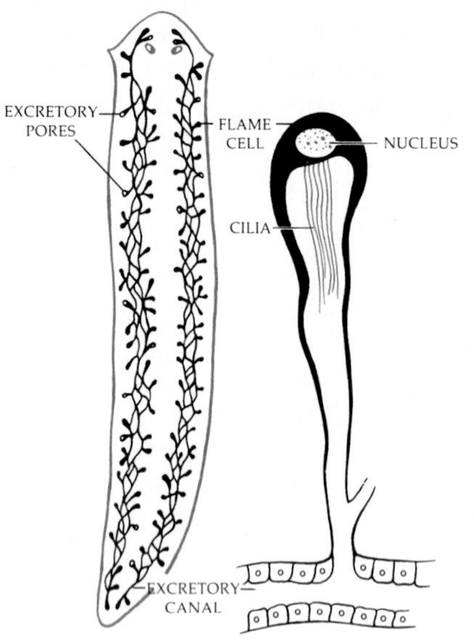

Like other turbellarians, the planarian is carnivorous. It eats either dead meat or other slow-moving animals, including smaller planarians. It feeds by means of a muscular organ, the pharynx, which is free at one end. The free end can be stretched out through the mouth opening. Muscular contractions in the tubular pharynx cause strong sucking movements, which tear the meat into microscopic bits and draw them into the internal cavity, where they are phagocytized by the cells of the endoderm. Because this digestive cavity has three main branches, planarians are placed in the order Tricladida.

Unlike the sponges or cnidarians, most flatworms have an excretory system (Figure 24–24). In the planarian, the system is a network of fine tubules that runs the length of the animal's body. Side branches of the tubules contain flame cells, each of which has a hollow center in which a tuft of cilia beats, moving water along the tubules to the exit pores between the epidermal cells. The flame-cell system appears to function largely to regulate water balance; most of the metabolic waste products probably diffuse out through the ectoderm or the endoderm.

In most respects, the free-living flatworms are relatively simple, and the rest of the bilaterally symmetrical animals may have evolved from something quite like them. However, by contrast, they have complicated reproductive systems with internal fertilization. Planarians, for example, are hermaphrodites, like most other flatworms, and when they mate, each partner deposits sperm in the copulatory sac of the other (Figure 24–25). These sperm then travel along special tubes, the *oviducts*, to fertilize the eggs as they become ripe.

The Planarian Nervous System

The evolution of bilateral symmetry brought with it marked changes in the organization of the nervous system as well as of other systems. Even among primitive flatworms, the neurons (nerve cells) are not dispersed in a loose network, as in *Hydra*, but are instead condensed into longitudinal cords. In the planarians, this condensation is carried further, and there are only two main conducting channels, one on each side of the flat, ribbonlike body. These channels carry impulses to and from the aggregation of nerve cells in the anterior end of the body (see Figure 24–22b). Such aggregations of nerve cell bodies are known as *ganglia* (singular, ganglion).

24–25

(a) Planarians have both male and female reproductive structures, and mating involves a mutual exchange of sperm. The erect penis of each partner is inserted into the copulatory sac of the other. The vas deferens is a duct leading from the testes, where sperm are produced by meiosis, to the penis. (b) Two planarians mating.

Planarians may also reproduce asexually, by fragmentation or by fission. Fragments of the tail, for example, may regenerate new heads.

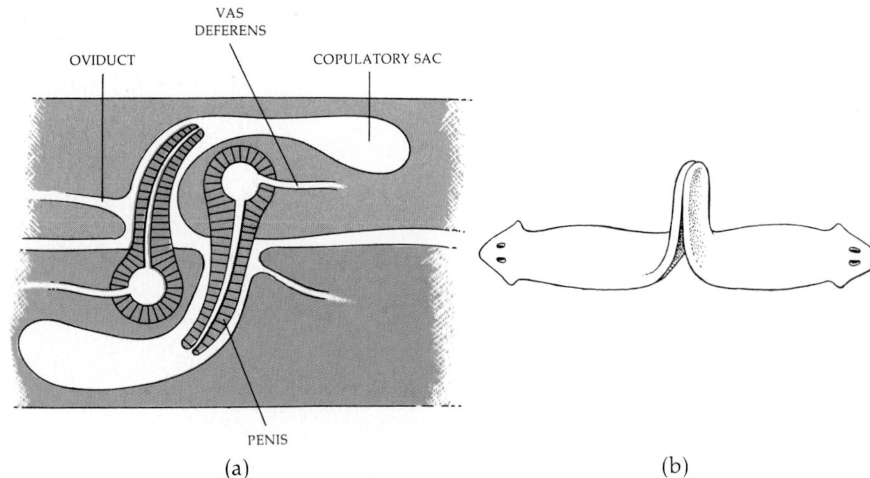

The ocelli of the planarian are usually inverted pigment cups (Figure 24–26). They have no lenses, and they cannot form an image. However, they can distinguish light from dark and can tell the direction from which the light is coming. Planarians are photonegative; if you shine a light on a dish of planarians from the side, they will move steadily away from the source of light.

Among the epithelial cells are receptor cells sensitive to certain chemicals and to touch. The head region in particular is rich in chemoreceptors. If you place a small piece of fresh liver in the culture water so that its juices diffuse through the medium, the planarians will raise their heads off the bottom and, if they have not eaten recently, will lope directly and rapidly (on a planarian scale) toward the meat, to which they then attach themselves to feed. The animal locates the food source by repeatedly turning toward the side on which it receives the stimulus more strongly until the stimulus is equal on both sides of its head. If the chemoreceptor cells are removed from one side of the head, the animal will turn constantly toward the intact side.

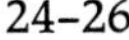

24–26

Ocellus of the planarian. The light stimulus is received by the ends of the photoreceptor cells adjacent to the pigment cup. Thus the light travels first through the fibers carrying the signals to the cerebral ganglia. In this inside-out organization, the planarian ocellus resembles the vertebrate eye.

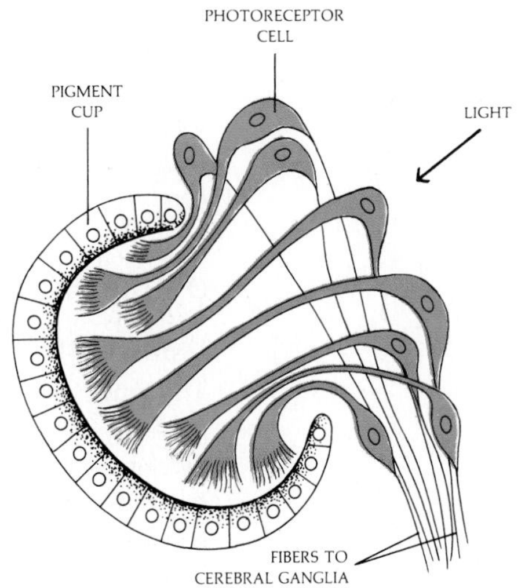

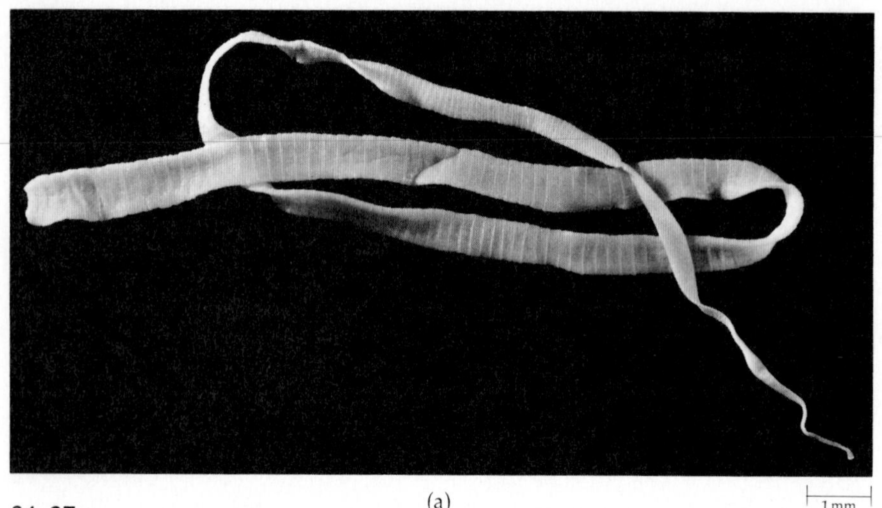

(a)

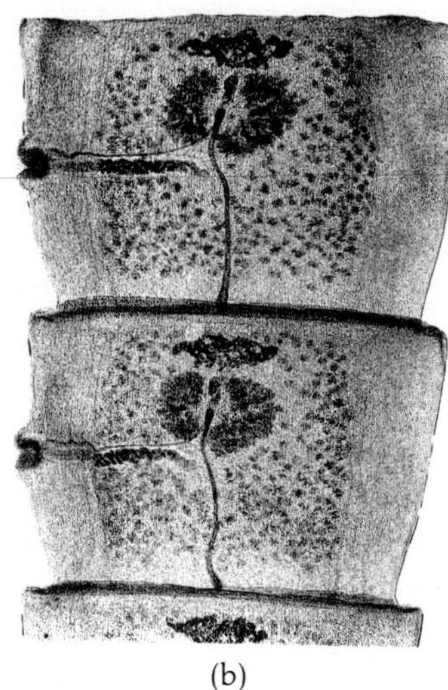

(b)

24–27
(a) *Human tapeworm,* Taenia solium. *Tapeworms are intestinal parasites that lack any digestive system of their own. They cling by their heads, which are equipped with hooks and suckers, and absorb the food molecules digested by their hosts through their body walls.*

(b) *Their bodies, posterior to the heads, are divided into segments known as proglottids. Each proglottid is a sexually complete hermaphroditic unit in which sperm and ova are produced. Proglottids break off when the ova are mature, and new ones are formed. Characteristically, the ova are eaten and develop to a larval form in one species, and the adult worm parasitizes members of another species. The dark areas are the reproductive structures; the opening is the genital pore.*

Classes Trematoda and Cestoda

Phylum Platyhelminthes includes also the trematodes, or flukes (class Trematoda), and the tapeworms (class Cestoda), parasitic forms that can cause serious and sometimes fatal diseases among vertebrates. Members of both of these parasitic classes have a tough outer layer of cells that is resistant to digestive fluids and, usually, suckers or hooks on their anterior ends by which they fasten to their victims. Trematodes feed through a mouth, but the tapeworms—which have no mouths, digestive cavities, or digestive enzymes—merely hang on and absorb predigested food molecules through their skin (Figure 24–27). Tapeworms are found in the intestines of many vertebrates, including humans, and may grow as long as 5 or 6 meters. They cause illness not only by encroaching on the food supply but also by producing wastes and by obstructing the intestinal tract. The most common human tapeworm, the beef tapeworm, infects people who eat the undercooked flesh of cattle that have grazed on land contaminated by human feces containing tapeworm segments.

All parasites, including parasitic flatworms, are believed to have originated as free-living forms and to have lost certain tissues and organs (such as the digestive tract) as a secondary effect of their parasitic existence, while developing adaptations of advantage to the parasitic way of life. Such adaptations also often include a complex life cycle involving two or more hosts (see essay).

OTHER ACOELOMATES

Phylum Gnathostomulida

This small phylum contains about 80 species of tiny marine worms. Their distinguishing characteristic is a unique pair of hard jaws, from which the phylum derives its name (from the Greek *gnathos,* "jaw," plus *stoma,* "mouth"). Gnathostomulids are abundant along coastal shorelines, where they live in the spaces between particles of sand and silt, using their hard jaws to scrape bacteria and fungi from the particles.

THE POLITICS OF SCHISTOSOMIASIS

Many of the parasitic platyhelminths have a complex life cycle involving two different hosts. In the case of three species of the trematode genus Schistosoma, *the hosts are freshwater snails and humans. The life cycle begins when small, swimming larvae are released from a snail; a single infected snail can shed some 100,000 larvae in its six-month lifetime. The larvae, if successful, attach to human skin and penetrate it. They feed, mature, and mate in the bloodstream. The female lays her eggs in the capillaries of the bladder wall or the intestine (depending on the species). The symptoms of schistosomiasis (also sometimes called bilharzia, or snail fever) are caused by the eggs' lodging in the liver and spleen, blocking blood vessels and causing hemorrhages.*

Eggs leave the human host by passing through the blood vessel wall into the bladder or intestinal tract where they are flushed out with the urine or feces. If the eggs are deposited in fresh water, they hatch immediately into ciliated larvae that seek out the particular species of snail in which they multiply asexually.

Schistosomiasis now affects some 200 to 300 million people in 71 countries in Asia, Africa, and South America. Ironically, western-style progress has contributed to its spread. Snails thrive in the still waters of irrigation canals and artificial lakes. The disease spread rapidly through Upper Egypt following the construction of the Aswan High Dam and the digging of permanent irrigation canals. Before the creation of the huge manmade Lake Volta in Ghana, the schoolchildren of the villages along the riverbank had an incidence of schistosomiasis of 1 percent. Now the prevalence in some lakeside villages is 100 percent.

Medical progress has not followed technology into the tropics. When the colonial period ended, the major research programs into tropical diseases ended too. Western medical research is geared overwhelmingly to the conquest of diseases affecting mainly rich, urban, older men and women. For instance, more than $800 million was spent on cancer research by the U.S. government alone in 1980, as compared to $3 million worldwide on schistosomiasis and $5 million on malaria. Cancer affects some 10 million people, schistosomiasis 200 million, malaria 300 million. The challenge of schistosomiasis and the other tropical parasitic diseases lies not so much in the disease process itself as in providing incentives (and therefore funding) to cure the diseases of the politically powerless rural poor.

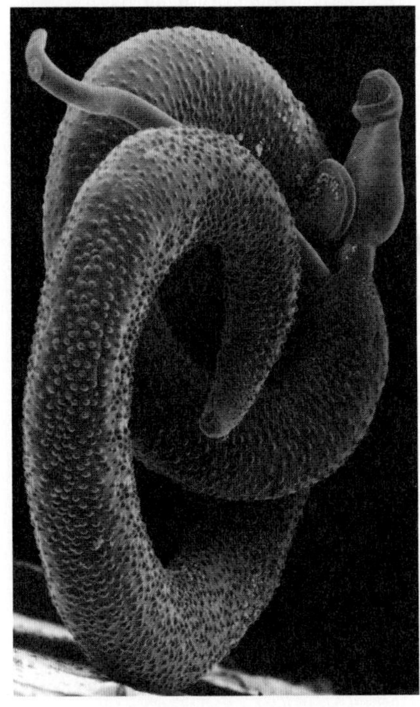

(a)

(b)

(a) *A scanning electron micrograph of male and female worms of* Schistosoma mansoni, *one of the three species that cause schistosomiasis in humans. The slender female is enveloped by the much larger body of the male. The two worms remain intertwined except when the female moves away to lay her eggs—about 300 per day. Adult worms live from 3 to 30 years, during which egg production continues unabated.* (b) *In the snail-infested waters near Aswan, Egypt, bathing children are rapidly infected by* Schistosoma *larvae. During the years that pass before the symptoms of disease become apparent, large quantities of eggs are shed by the children, continually infecting new generations of snails.*

24-28
(a) *A living gnathostomulid,* Problognathia minima, *as seen through the phase-contrast microscope. This species lives in the intertidal sand flats along the coast of Bermuda. It moves by slowly gliding between the sand grains, moving its head rhythmically from side to side.*

(b) *A ribbon worm in an embarrassing situation. Ribbon worms range in length from less than 2 centimeters to 30 meters and are virtually all the colors of the spectrum. All members of the phylum, however, have thin bodies (seldom more than 0.5 centimeter thick), a mouth-to-anus digestive tract, a circulatory system, and a long, muscular tube that can be thrust out to grasp prey.*

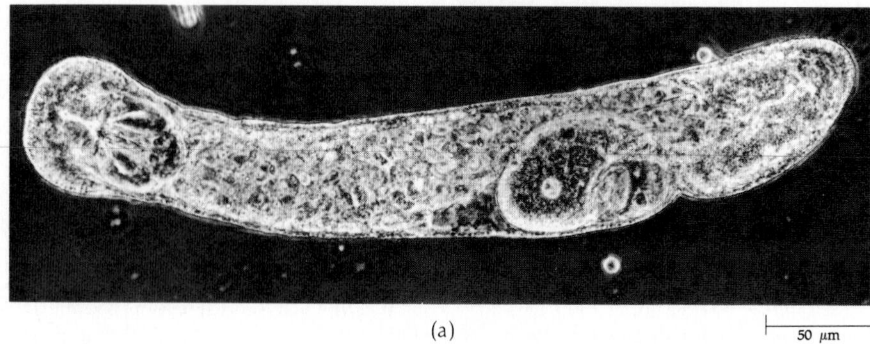

(a)

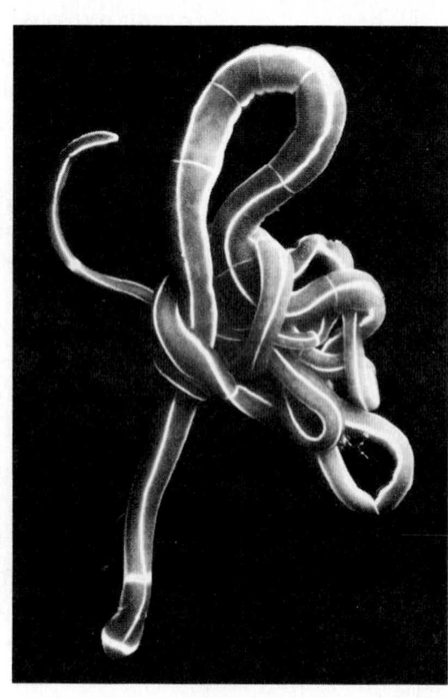
(b)

These tiny worms have no coelom or pseudocoelom, and the digestive cavity has only one opening. Some zoologists believe that they are closely related to the free-living flatworms, whereas others think they are degenerate forms, derived from pseudocoelomates.

Phylum Rhynchocoela

Phylum Rhynchocoela ("snout" plus "hollow") consists of about 650 species of acoelomate worms, commonly called ribbon worms or nemertines. They are characterized by a long, retractile, slime-covered hollow tube (proboscis). The proboscis, sometimes armed with a barb, seizes prey and draws it to the mouth, where it is engulfed. Some inject a paralyzing poison into their prey.

The ribbon worms are of special interest to biologists attempting to reconstruct the evolution of the invertebrates. They appear to be closely related to the flatworms, but they exhibit two significant new features. Ribbon worms have a one-way digestive tract beginning with a mouth and ending with an anus. This is a far more efficient arrangement than the one-opening digestive system of the cnidarians and flatworms. In the one-way tract, food moves assembly-line fashion, with the consequent possibilities (1) that eating can be continuous and (2) that various segments of the tract can become specialized for different stages of digestion. These worms also have a circulatory system, usually consisting of one dorsal and two lateral blood vessels that carry the colorless blood.

Reproduction in the ribbon worms, however, is more primitive than in most flatworms. Ordinarily the sexes are separate, and fertilization is external. Asexual reproduction by fragmentation of the body and regeneration of whole worms from the parts is also fairly common.

PSEUDOCOELOMATES

One major phylum (Nematoda) and six minor phyla are characterized by a pseudocoelom, a body cavity between the endoderm and the mesoderm (see Figure 24-21c). The pseudocoelom, which is essentially a sealed, fluid-filled tube, significantly increases the effectiveness of the animal's muscular contractions. In addition to working against the water or a substrate, muscles must also have something to work against in an animal's body—otherwise the body just bends in the direction of contraction and a floppy, uncoordinated motion results. Because it resists bending, the pseudocoelom functions as a hydrostatic skeleton within the animal's body, causing the body to return to its original shape after the muscles have contracted. It makes possible a great advance over the simple, rather flaccid movements of the acoelomate worms and most cnidarians.

All of the pseudocoelomates have a one-way digestive tract, but they lack a circulatory system. The movement of fluids within the pseudocoelom, however, compensates for this.

Phylum Nematoda

About 12,000 species of nematodes (roundworms) have been described and named, but some authorities think that there may be as many as 400,000 to 500,000. Most are free-living, microscopic forms. It has been estimated that a spadeful of good garden soil usually contains about a million nematodes. Some are parasites; most species of plants and animals are parasitized by at least one species of nematodes.

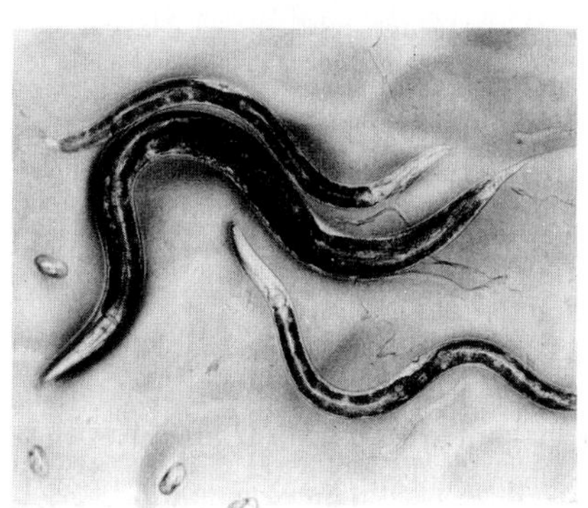

24–29
An adult and two juvenile nematodes, Caenorhabditis elegans. *Strands of shed cuticle can be seen extending from the right-hand side of the adult nematode. Four eggs are visible at the lower left of the micrograph.*

Nematodes are cylindrical, unsegmented worms and are covered by a thick, continuous cuticle, which is molted periodically as they grow (Figure 24–29). An interesting, and unique, feature of nematode construction is the absence of circular muscles. The contraction of the longitudinal muscles acting against both the tough, elastic cuticle and the internal hydrostatic skeleton gives the worm its characteristic whipping motion in water. The mouth of a nematode has a muscular pharynx and often is equipped with piercing stylets. Reproduction is sexual, and the sexes are usually separate.

Humans are hosts to about 50 species of parasitic nematodes, which are a major cause of death and disability throughout the world. In North America, the most common parasitic nematodes are pinworm *(Enterobius),* whipworm *(Trichuris),* hookworm *(Ancylostoma),* intestinal roundworm *(Ascaris),* and *Trichinella.* The latter causes trichinosis, which is transmitted by eating uncooked or undercooked pork, a single gram of which may contain 3,000 cysts (resting forms) of *Trichinella.* Ingestion of only a few hundred of these cysts can be fatal. On a worldwide basis, it is estimated that 650 million people are infected by *Ascaris,* 450 million by *Ancylostoma,* 250 million by *Filaria,* the cause of filariasis, and 50 million by *Onchocerca,* the cause of "river blindness" in fertile regions of west Africa and in the coffee-growing highlands of Mexico and Guatemala. Research is revealing that many of these worms have unique enzymes and metabolic pathways directly related to such features of the parasitic life style as migration within the body of the host, attachment to host tissues, and the massive production of eggs.

Minor Pseudocoelomate Phyla

Six other phyla, quite diverse, share a body plan based upon the pseudocoelom. Phylum Nematomorpha consists of about 230 species of horsehair worms that resemble nematodes. The adults, which do not feed, are free-living and reproduce in water (Figure 24–30a); the juveniles are parasites of arthropods. The 500 species of spiny-headed worms, phylum Acanthocephala, are parasitic throughout their life cycle. The larval forms develop in the tissues of arthropods, and the adults live and reproduce in the intestines of vertebrates. The adult worms, which superficially resemble tapeworms, lack a digestive system and have a proboscis bearing hooks that attach to the intestinal wall of the host (Figure 24–30b and c).

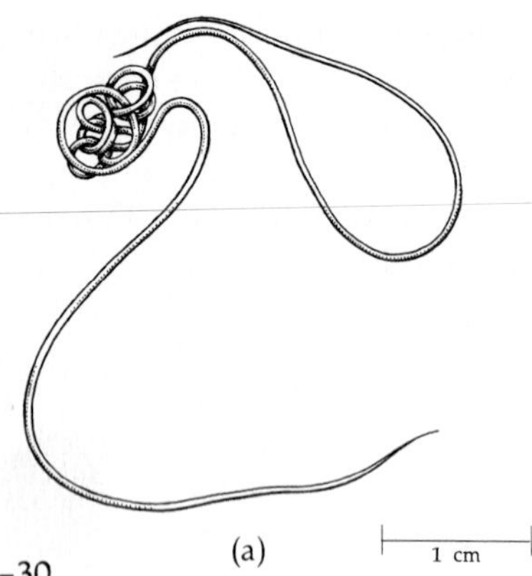

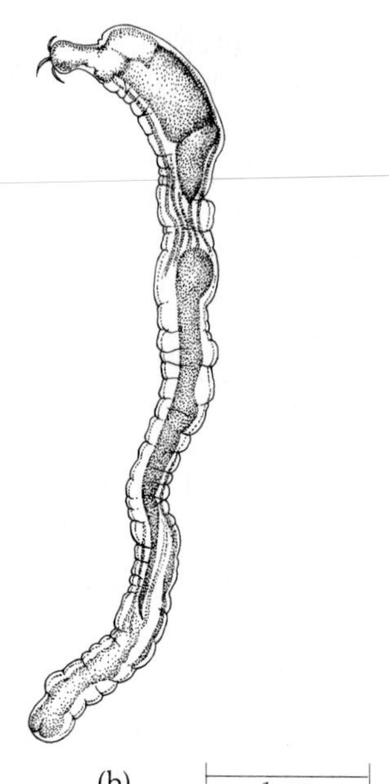

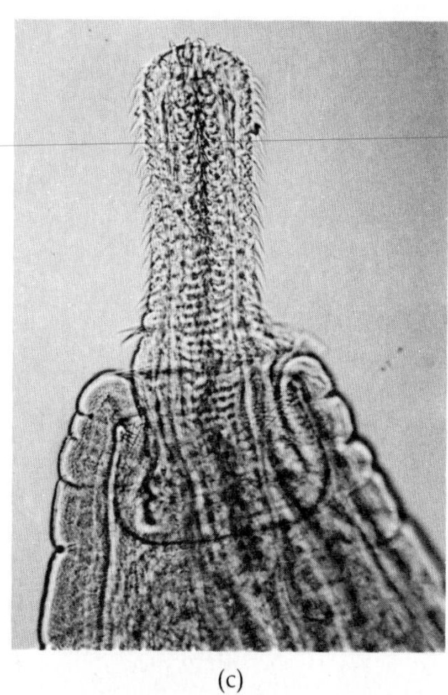

24–30
Two phyla of pseudocoelomates consist of worms that are parasitic during at least one stage of the life cycle. (a) *A horsehair worm, phylum Nematomorpha. The specimen from which this drawing was prepared was found in a garden hose on Long Island, New York. It was probably carried into the hose by a cricket or a grasshopper that it parasitized during its juvenile stage. Because of the intricate knots into which they tie themselves, nematomorphs are also known as gordian worms, after Gordius, King of Phrygia, who tied the gordian knot severed by Alexander the Great.*

(b) *An adult male of phylum Acanthocephala. Known commonly as spiny-headed worms, acanthocephalans are characterized by a hook-bearing proboscis* (c).

The members of three pseudocoelomate phyla are found, along with the acoelomate gnathostomulids, in the spaces between sand and silt particles along shorelines (Figure 24–31). Phylum Kinorhyncha contains about 100 species of tiny, burrowing marine worms that feed on diatoms in muddy ocean shores. Short-bodied, these worms are covered with spines and have a spiny, retractile proboscis. The sexes are separate. Most of the approximately 400 species in phylum Gastrotricha, however, are hermaphrodites. They live in sandy shores of both fresh and salt water, where they feed on algae, protozoans, and dead organic matter. Many of the 1,500 to 2,000 species of phylum Rotifera are found in the same environment in fresh water, as well as around the vegetation in ponds and along lake shores. Other members of the phylum are marine, and some are terrestrial. Rotifers are sometimes called "wheel animalcules" because the beating of a crown of cilia around the mouth causes them to spin through the water like tiny wheels. Rotifers have a muscular pharynx with hard jaws; they feed on algae, bits of vegetation, protozoans, and other animals even smaller than themselves. The sexes are separate in rotifers, but in some groups the females produce eggs that can develop without fertilization, a phenomenon known as *parthenogenesis*.

Phylum Entoprocta contains about 75 species of stalked, usually sessile animals that superficially resemble both hydrozoans and a group of coelomate animals known as bryozoans, or ectoprocts (see page 524). Their internal structure, however, identifies them as pseudocoelomates. They have a U-shaped digestive tract, and both the mouth and anus are located within a circle of tentacles (Figure 24–32). (In the bryozoans, by contrast, the anus is located outside a circle of tentacles, and the tentacles are structurally different from those of the entoprocts.) Reproductive patterns vary among the entoprocts. Some species have separate sexes, some are simultaneous hermaphrodites, and in others a single reproductive organ produces sperm at one stage in the life cycle and eggs at a later stage. As we shall see in the next chapter, this phenomenon, known as sequential hermaphroditism, is not limited to the entoprocts.

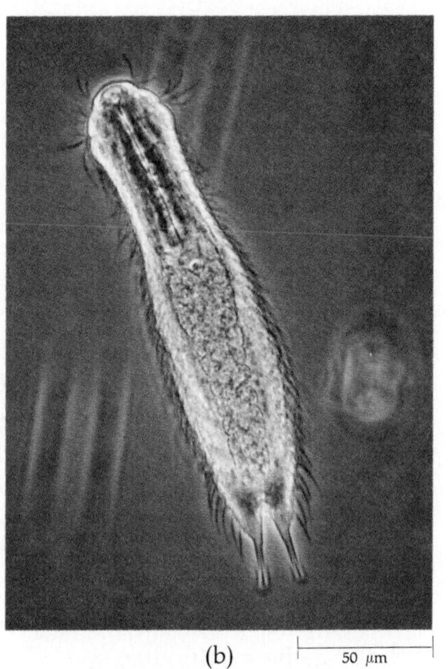

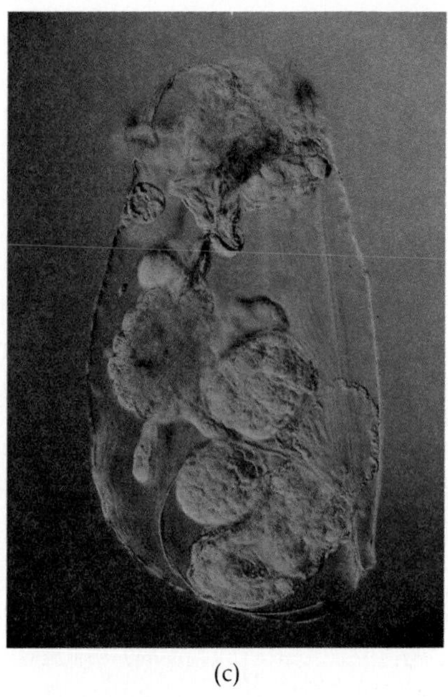

24–31
Among the residents of the sand and silt of shorelines are members of three pseudocoelomate phyla—Kinorhyncha, Gastrotricha, and Rotifera. (a) Centroderes spinosus, *a kinorhynch. Unable to swim, a kinorhynch burrows by forcing fluid into its head; when the head is anchored in the mud by its spines, the animal can pull the rest of its body forward.* (b) Chaetonotus, *a common gastrotrich. Gastrotrichs can both swim and crawl, clinging to surfaces by means of adhesive tubes that project from the sides of their bodies.* (c) *A female rotifer of the genus* Asplanchna, *commonly found in freshwater plankton. The cilia that propel the rotifer through the water are barely visible at the anterior end of the organism. The large yellow-green structure in the center of the rotifer is the stomach; beside and below the stomach are three developing embryos.*

24–32
The most common freshwater entoproct in the United States is Urnatella gracilis, *which forms branching colonies, securely attached to a substrate. Most entoprocts are colonial, but members of a few species live as solitary individuals.*

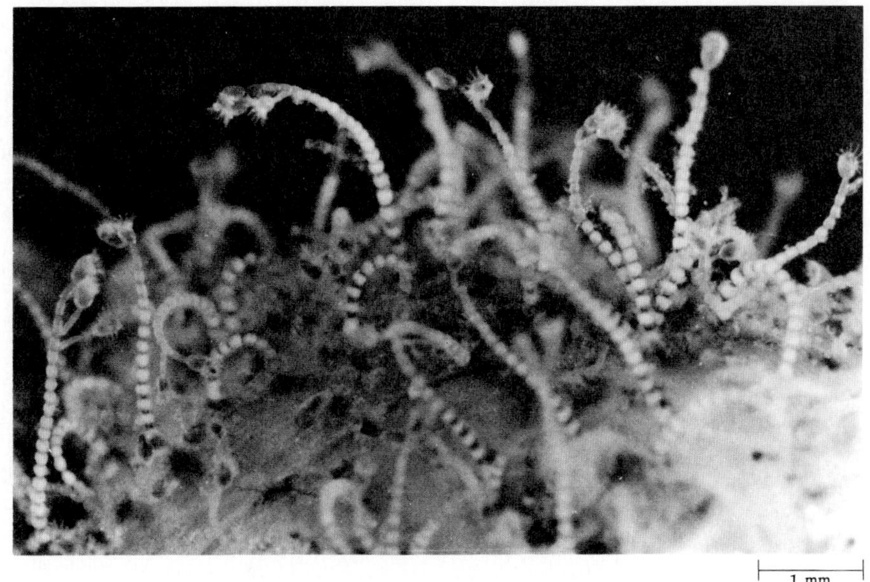

SUMMARY

Animals are multicellular heterotrophs that depend directly or indirectly on plants or algae as their source of food energy. Almost all digest their food in an internal cavity. Most are motile. Reproduction is usually sexual. More than 90 percent of the animal species are invertebrates, animals without backbones.

Modern animals are classified into about 30 phyla. Main criteria for classification include the number of tissue layers, the basic body plan and the arrangement of body parts, the presence or absence of body cavities, and the pattern of development from fertilized egg to adult.

Both the sponges (phylum Porifera) and the extremely simple, parasitic mesozoans (phylum Mesozoa) are so different from all other animals that they are placed in separate subkingdoms. Sponges are composed of a number of different cell types, including choanocytes, which are the feeding cells; epithelial cells, some of which are contractile; and amoebocytes, which perform a variety of functions in reproduction, the secretion of skeletal materials, and the transport of food particles within the animal. In the sponge, there is little coordination among the various cells. Sponges reproduce both asexually and sexually, with a highly specialized form of internal fertilization. Most sponges, like many other sessile or slow-moving animals, are hermaphroditic.

The animals in phylum Cnidaria and phylum Ctenophora share three major characteristics: (1) radial symmetry; (2) the coelenteron, a cavity in which food is partially digested extracellularly; and (3) a two-layered body plan, in which the two tissue layers, the ectoderm and the endoderm, are divided by a jellylike substance, the mesoglea. The jellyfish, sea anemones, and corals of phylum Cnidaria are distinguished from all other animals by their special stinging cells, the cnidocytes; the sea walnuts and comb jellies of phylum Ctenophora are identified by comblike plates of fused cilia, arranged in eight rows on the body surface. Adult cnidarians may take the form of either the polyp or the medusa; in many species, the life cycle includes an asexually reproducing polyp form, a sexually reproducing medusa form, and a ciliated planula larva. Cnidarians have a simple nervous system that coordinates the movements of the animal, a variety of sensory cells, and two types of specialized sensory organs, statocysts and ocelli.

The animals in all of the remaining phyla have primary bilateral symmetry, with a distinct "headness" and "tailness" and a concomitant clustering of nerve cells in the anterior region. They also have three distinct tissue layers—ectoderm, mesoderm, and endoderm. The simplest of the bilateral animals are the acoelomates (phyla Platyhelminthes, Gnathostomulida, and Rhynchocoela), which have no body cavity other than the digestive cavity.

The flatworms, phylum Platyhelminthes, are characterized by a flattened body, a branched digestive system with only one opening, and an excretory system, involving flame cells, that serves largely to maintain water balance. Ocelli are present, as are sensory cells responsive to touch and to various chemicals; the nerve cells are organized into longitudinal cords. Flatworms may be free-living (class Turbellaria) or parasitic (classes Trematoda and Cestoda).

Other acoelomates include the tiny marine worms of phylum Gnathostomulida and the ribbon worms of phylum Rhynchocoela. The ribbon worms, which are characterized by a retractile, prey-seizing proboscis, are the most primitive animals with a one-way (mouth-to-anus) digestive tract and a circulatory system.

Seven phyla of animals have a body plan based on the pseudocoelom, a fluid-filled cavity between the endoderm and the mesoderm. The pseudocoelom functions as a firm hydrostatic skeleton, enabling these animals to move more efficiently than the cnidarians or the acoelomate worms. The pseudocoelomates all have a one-way digestive tract but lack a circulatory system.

The roundworms (phylum Nematoda) are the largest and most important group of pseudocoelomates. Most are free-living, but many are parasitic, causing a variety of serious diseases in plants and animals, including humans. Unlike other types of worms, roundworms have only longitudinal muscles and move in a characteristic whipping manner.

Other pseudocoelomates include the horsehair worms (phylum Nematomorpha) and the spiny-headed worms (phylum Acanthocephala), which are both parasitic; the tiny marine and freshwater animals of phyla Kinorhyncha, Gastrotricha, and Rotifera, most of which are bottom-dwellers; and the Entoprocta, which superficially resemble the hydrozoans of phylum Cnidaria. Among the various pseudocoelomates, the entire range of sexual reproductive patterns in animals is seen: separate sexes, simultaneous hermaphrodites, sequential hermaphrodites, external fertilization, internal fertilization, and parthenogenesis.

QUESTIONS

1. Distinguish among the following: cnidocyte/nematocyst; polyp/medusa; diploblastic/triploblastic; endoderm/mesoderm/ectoderm; radial symmetry/bilateral symmetry; acoelomate/pseudocoelomate/coelomate.

2. In which phylum or phyla do you find each of the following and what is its function: choanocyte; coelenteron; cnidocyte; statocyst; planula larva; flame cell?

3. Describe the similarities and differences among the three classes of phylum Cnidaria.

4. How are the ocelli of jellyfish and planarians similar? How are they different?

5. A pseudocoelom (or, as we shall see in the next chapter, a coelom) provides hydrostatic support for an animal. Why is a coelenteron less likely to fill this same function?

6. How do the parasitic flatworms of phylum Platyhelminthes differ from the free-living flatworms? What general features are characteristic of adaptation to a parasitic way of life?

7. Virtually all of the animals in phyla Gnathostomulida, Kinorhyncha, and Gastrotricha and many of the animals in phylum Rotifera (as well as some animals we have yet to meet) live in the spaces between sand and silt grains along shorelines. What specializations enable such a large number of different animals to occupy the same microenvironment?

8. On the basis of your knowledge of sexual reproduction, what do you think might be the advantages to an organism of functioning as a male in the earlier stages of its life and then as a female in the later stages, or vice versa?

9. What might be the advantages of parthenogenetic reproduction?

CHAPTER 25

The Animal Kingdom II: The Protostome Coelomates

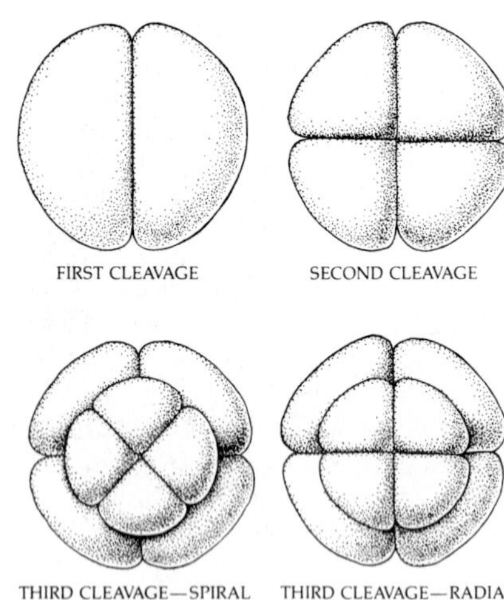

25–1
Egg cleavages, showing spiral and radial patterns at the third cleavage. Spiral cleavage is characteristic of protostomes (acoelomates, pseudocoelomates, mollusks, annelids, and arthropods); radial cleavage is characteristic of deuterostomes (echinoderms and chordates).

The coelom, a fluid-filled cavity within the mesoderm, characterizes all the remaining phyla in the animal kingdom. Although it may seem less dramatic than other evolutionary innovations, the coelom is extremely important. Within this cavity, organ systems—suspended by the mesenteries—can bend, twist, and fold back on themselves, increasing their functional surface areas and filling, emptying, and sliding past one another, surrounded by lubricating coelomic fluid. Consider the human lung, constantly expanding and contracting in the chest cavity, or the 6 or 7 meters of coiled human intestine; neither of these could have evolved before the coelom. The coelom, like the pseudocoelom, also constitutes a hydrostatic skeleton, stiffening the body in somewhat the same way water pressure stiffens and distends a fire hose.

The various phyla of coelomate animals fall into two broad groups that roughly correspond to major branches of the phylogenetic tree. These groups are based on characteristic features of embryonic development. When a fertilized egg—the zygote—begins to divide, the early cell divisions usually follow one of two patterns. In mollusks, annelids, arthropods, and a number of minor coelomate phyla (as well as in acoelomates and pseudocoelomates), the early divisions are spiral, occurring in a plane oblique to the long axis of the egg. In echinoderms, chordates, and a few other coelomate phyla, the cleavage pattern is radial, parallel to and at right angles to the axis of the egg (Figure 25–1). With both types of cleavage, the embryo gradually develops to a stage known as the blastula, which is a hollow ball of cells. Next, an opening, the blastopore, appears. Among the animals with spiral cleavage, the mouth (stoma) develops at or near the blastopore, and this group is called the *protostomes*—"first the mouth." In the animals with radial cleavage, the anus forms at or near the blastopore and the mouth forms secondarily elsewhere; thus these animals are known as *deuterostomes*—"second the mouth." These differences are so fundamental that they are believed to have originated very early in animal evolution, before the branchings that gave rise to the modern coelomate phyla.

Another characteristic difference between the two groups of animals is the manner in which the coelom forms (Figure 25–2). In the protostomes, the coelom usually forms by a splitting of the mesoderm, and the process is said to be *schizocoelous* (from the Greek *schizo,* "split"). In the deuterostomes, however, the coelom is usually formed by an outpouching of the cavity of the embryonic gut, a process said to be *enterocoelous* (from the Greek *enteron,* "gut"). Enterocoelous formation of

25–2

Key features of embryonic development in coelomate animals. (a) *In protostomes, the mesoderm takes shape between the endoderm and the ectoderm, in the region around the blastopore. The coelom arises from a splitting of the solid mesoderm. As the cells continue to multiply, the blastocoel—the original embryonic cavity of the blastula—is obliterated. The blastopore of protostomes becomes the mouth.*

(b) *In deuterostomes, the mesoderm originates from outpocketings of the embryonic gut. These outpocketings create cavities within the mesoderm that become the coelom. The blastopore becomes the anus, and the mouth develops elsewhere.*

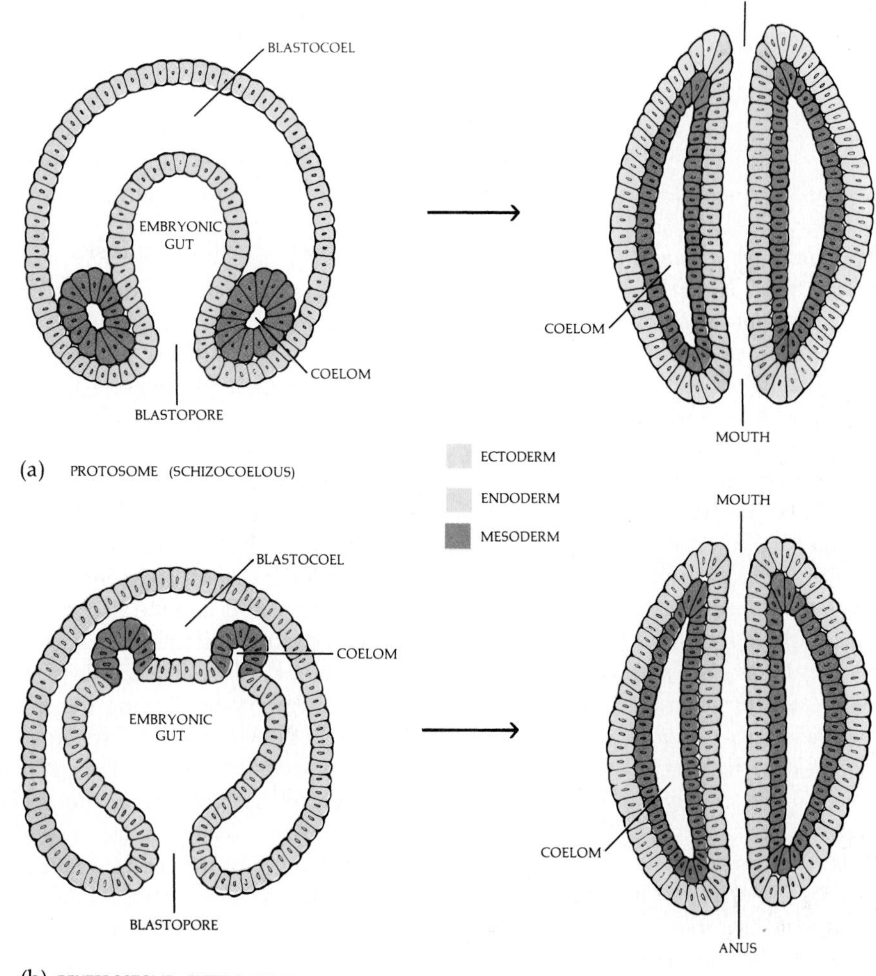

the coelom is thought to have arisen from the schizocoelous process several times in the course of animal evolution.

In the next two chapters, we shall focus on the protostome coelomates—the mollusks, annelids, and other protostomes in this chapter, and the arthropods in the next. In the final chapter of this section, our attention will be devoted to the deuterostomes.

PHYLUM MOLLUSCA: MOLLUSKS

The mollusks constitute one of the largest phyla of animals, both in numbers of living species (at least 47,000, and perhaps many more) and in numbers of individuals. Their name is derived from the Latin word *mollus,* meaning "soft," and refers to their soft bodies, which are generally protected by a hard, calcium-containing shell. In some forms, however, the shell has been lost in the course of evolution, as in slugs and octopuses, or greatly reduced in size and internalized, as in squids.

Mollusks exhibit a tremendous diversity of form and behavior. The three major classes range from largely sedentary or sessile filter-feeding animals, such as clams and oysters (class Bivalvia), through aquatic and terrestrial snails and slugs (class Gastropoda), to the predatory cuttlefish, squids, and octopuses (class Cephalopoda), which are not only the most active of the mollusks but also the most intelligent of all invertebrates.

Characteristics of the Mollusks

Structurally, mollusks are quite distinct from all other animals. The basic body plan is shown in Figure 25–3a. As you can see, the hypothetical ancestor displayed a clear bilateral symmetry. Among modern mollusks, only the polyplacophorans (chitons) and monoplacophorans, both relatively small classes, bear any obvious resemblance to the archetypal model. Modern mollusks, however, all have the same fundamental body plan. There are three distinct body zones: a *head-foot*, which contains both the sensory and motor organs; a *visceral mass*, which contains the well-developed organs of digestion, excretion, and reproduction; and a *mantle*, a specialized tissue formed from folds of the dorsal body wall, which hangs over and enfolds the visceral mass and which secretes the shell. The *mantle cavity*, a space between the mantle and the visceral mass, houses the gills; the digestive, excretory, and reproductive systems discharge into it. Water sweeps into the mantle cavity (usually propelled by cilia on the gills), passing across the surface of the gills and aerating them. Water leaving the mantle cavity carries excreta and, in season, gametes, both of which are discharged downstream from the gills.

A characteristic organ of the mollusk, found only in this phylum, and in all classes except the bivalves, is the *radula* (Figure 25–4), a movable, tooth-bearing strap of chitinous material suggestive of a tongue. The radular apparatus, which can be projected out of the mouth and then drawn back in with a licking movement, serves both to scrape off algae and other food materials and also to convey them backward to the digestive tract. In some species, it is also used in combat.

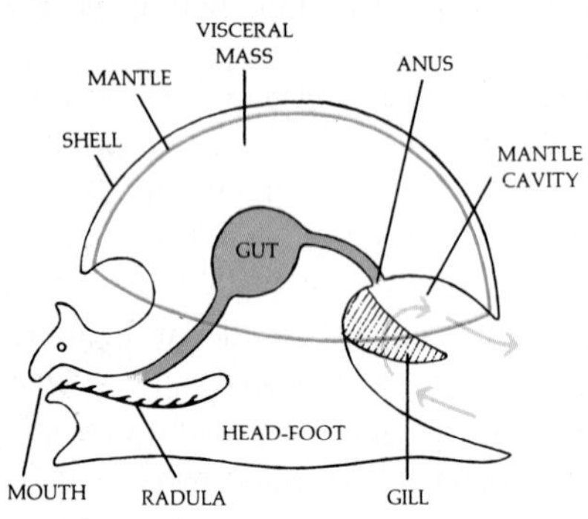

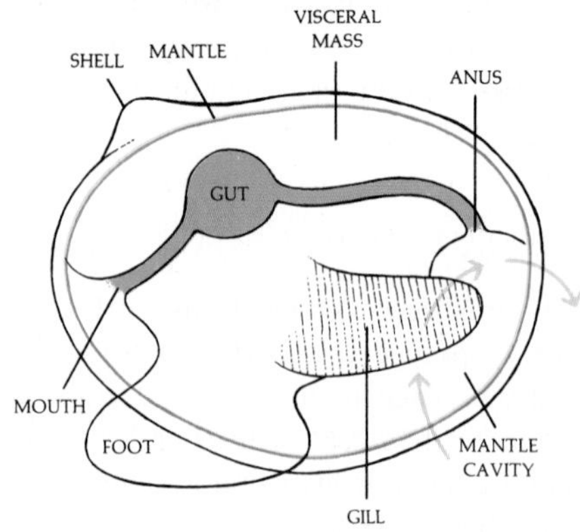

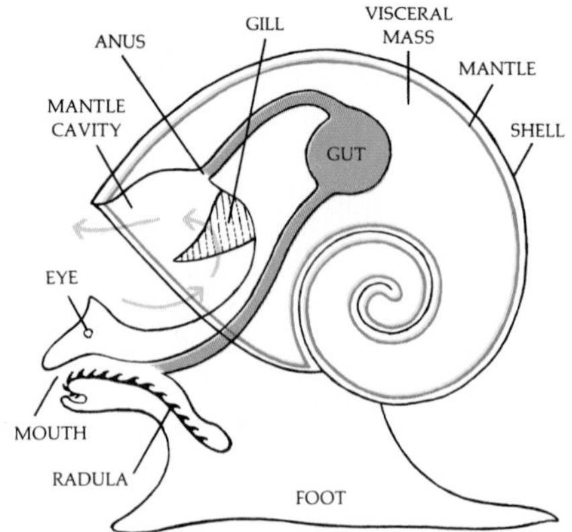

25–3
Mollusks are characterized by soft bodies composed of a head-foot, a visceral mass, and a mantle, which can secrete a shell. They exchange gases with the surrounding water through gills, except for the land snails, in which the mantle cavity has been modified for air breathing. A hypothetical primitive mollusk is shown in (a).

The three major modern classes are the bivalves, the gastropods, and the cephalopods. (b) *The bivalves, such as the clam shown here, are generally sedentary and feed by filtering water currents, created by beating cilia, through large gills.* (c) *In the gastropods, exemplified by the snail, the visceral mass has become coiled and rotated through 180° so that mouth, anus, and gills all face forward and the head can be withdrawn into the mantle cavity.* (d) *In the cephalopods, such as the squid, the head is modified into a circle of arms, and part of the head-foot forms a tubelike siphon through which water can be forcibly expelled, providing for locomotion by jet propulsion. The arrows indicate the direction of water movement.*

25-4
(a) *A vertical section through the head of a land snail to show the radula. Teeth on the radula rasp and tear food materials and then convey them to the esophagus, a narrow tube leading to the stomach.* (b) *Radular teeth from a minute African land snail that feeds on bits of dead leaves. The large teeth (upper left) cut and tear the leaf; the smaller teeth (lower right) pull the pieces into the snail's mouth.*

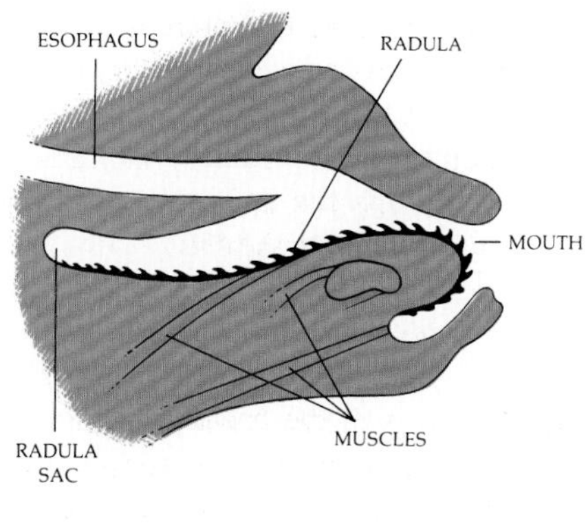

(a)

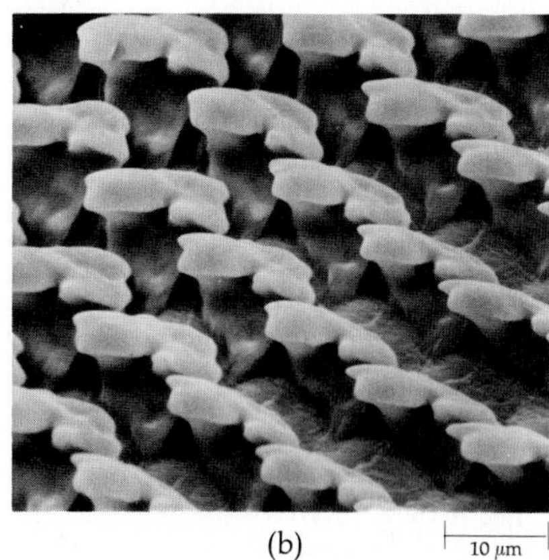

(b)

Supply Systems

In the protists and in the smaller and simpler animals, oxygen and food molecules are supplied to cells—and waste products are removed from them—largely by diffusion, aided by the movement of external fluids. Internal transport may also be assisted, as we have seen, by wandering, amoeboid cells, as in the cnidarians, or by the movement of fluids in a body cavity, as in the pseudocoelomates. For larger, thicker animals, a more effective method of providing each cell with a direct and rapid line of supply is a circulatory system that propels extracellular fluid—blood—around the body in a systematic fashion.

The molluscan circulatory system consists of a muscular pumping organ, the heart, and vessels that carry the blood to and from the heart. The heart usually has three chambers: two of them (atria) collect blood from the gills, and the third (the ventricle) pumps it to the other body tissues. Except for the cephalopods, mollusks have what is known as an open circulation; that is, the blood does not circulate entirely within vessels—as it does in the annelids, for example—but is collected from the gills, pumped through the heart, and released directly into spaces in the tissues from which it returns to the gills and then to the heart. Such a blood-filled space is known as a *hemocoel* ("blood cavity"). In the mollusks, the hemocoel has largely replaced the coelom, which is reduced to a small area around the heart and to the cavities of the organs of reproduction and excretion. Cephalopods, which are extremely active animals, have a closed circulatory system of continous vessels and accessory hearts that propel blood into the gills.

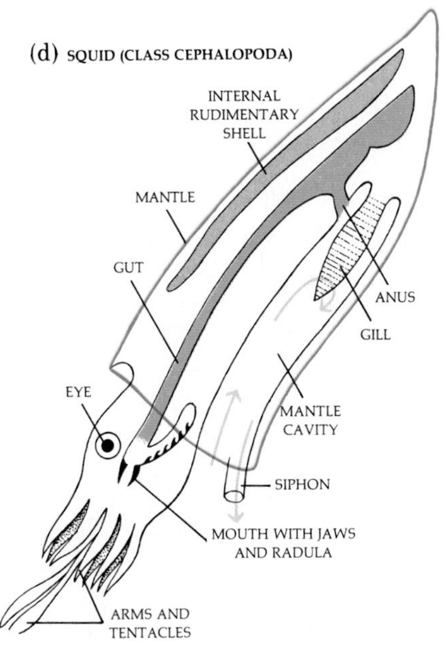

Oxygen enters the body of a mollusk through the moist epidermis of the mantle and the gills. A gill is a structure with an increased amount of surface area, through which gases can diffuse from the surrounding water, and a rich blood supply for the transport of these gases. Oxygen travels inward by diffusion, down the concentration gradient. The gradient exists because the surface film, exposed to the dissolved oxygen in the passing water, contains more oxygen than does the blood within the gills, which was depleted of oxygen as it passed through the body tissues. Carbon dioxide, produced by cellular respiration, moves out to the surface film and then into the surrounding water by the same principle. In fact, all gas exchange in animals, whether water-dwelling or land-dwelling, takes place across moist membranes.

The digestive tract of all mollusks is extensively ciliated and has many different working areas. Food is taken up by the cells lining the digestive glands arising from the stomach and the anterior intestine, and then is passed into the blood. Undigested materials are compressed into mucus-coated fecal pellets, which are discharged through the anus into the mantle cavity and are carried away from the animal in the water currents. This packaging of digestive wastes in solid form prevents fouling of the water passing over the gills.

Nitrogenous wastes produced by the metabolic activities of the cells are removed by one or two tubular structures known as *nephridia*. One opening of each nephridium is in the coelom surrounding the heart, and the other opening discharges into the mantle cavity. Coelomic fluid is forced, under pressure, into the nephridium; as it passes through the tubule, water, sugar, salts, and other needed materials are returned to the tissues through the walls of the nephridium, while other materials are absorbed into it for excretion. Thus the excretory system is concerned not only with the problem of water balance, as are the contractile vacuoles of *Paramecium* and the flame cells of planarians, but also with the homeostatic regulation of the chemical composition of body fluids.

Minor Classes of Mollusks

Four of the seven classes of mollusks with living representatives are fairly small. About 250 species of wormlike marine animals, known as solenogasters, constitute the class Aplacophora ("no plates"). Although they have no shell and the foot is greatly reduced, the presence of a radula identifies them unequivocally as mollusks.

The approximately 600 species of chitons are placed in class Polyplacophora, and, as we noted previously, they bear some resemblance to the hypothetical ancestral mollusk (Figure 25–3a). Common as grazers on surf-swept ocean shores,

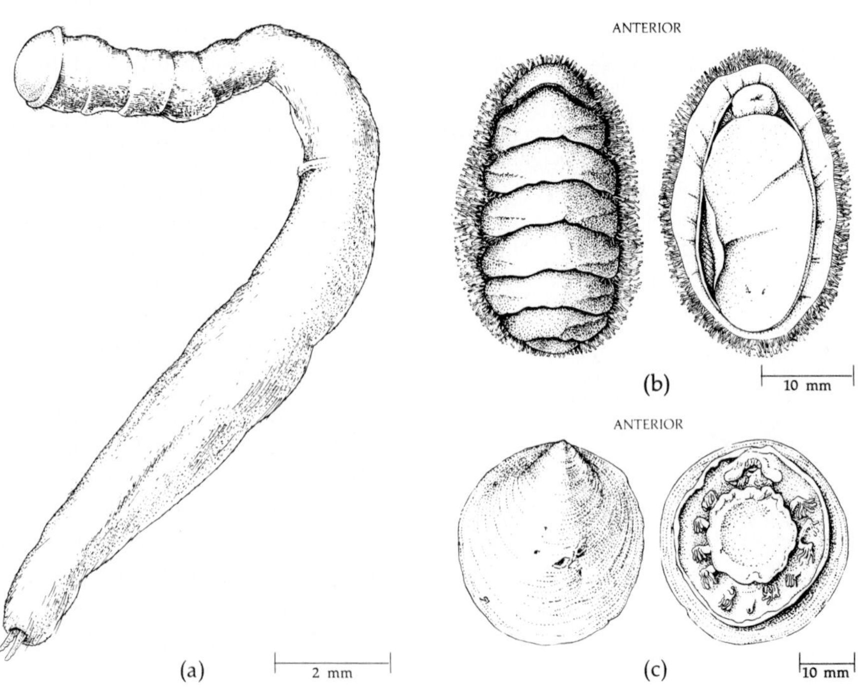

25–5
Representatives of the four minor classes of mollusks. (a) *A solenogaster, class Aplacophora. Its wormlike body is covered with fine setae. The pronglike structures at the posterior of the animal are the gills. This specimen, a member of the species* Chaetoderma nitidulum, *was found in St. Margaret's Bay, Nova Scotia.*

(b) *Dorsal (left) and ventral (right) views of a chiton, class Polyplacophora. The dorsal shell of eight plates is fringed by a girdle of hard spicules. As seen from the ventral surface, the bulk of the chiton's body is its foot, which has been pulled aside at the left to reveal the gills within the mantle cavity. At the anterior of the animal is the mouth, through which it extends the radula when grazing.*

(c) *Dorsal and ventral views of* Neopilina galatheae, *the first known living representative of class Monoplacophora. The large central structure of the ventral surface is the foot, which is surrounded by five pairs of gills.*

(d) *A tusk shell, class Scaphopoda. This mollusk burrows in sand or mud with the large foot that extends from the wide end of the shell. The specialized tentacles bring food particles, such as diatoms, to the mouth hidden within the shell.*

chitons have a somewhat flattened body, covered by a dorsal shell formed from a series of eight plates. On either side of the body, a series of gills are suspended between the mantle and the foot and are continuously aerated by water flowing from the sides toward the posterior of the animal.

Class Monoplacophora contains just eight living species. The first species, *Neopilina galatheae,* was discovered in the 1950s in a deep ocean trench off the coast of Costa Rica. The only previously known representatives of the class were fossils from the Cambrian period, which ended 500 million years ago. *Neopilina,* which is little more than 2.5 centimeters long, also has an organization suggestive of the hypothetical ancestor. It has a large, single dorsal shell, but is unusual in having five pairs of gills, six pairs of nephridia, and eight pairs of retractor muscles, with which it can pull the shell down securely over its soft body.

About 350 species of tusk or tooth shells make up class Scaphopoda. These familiar residents of the seashore have long, tubular shells, open at both ends. They live a sedentary life, with the wide end of the shell, containing the head (which bears structures used in feeding) and the foot, buried in the sand or mud. Water, carrying dissolved gases, enters and leaves the mantle cavity through the exposed, narrow end of the shell.

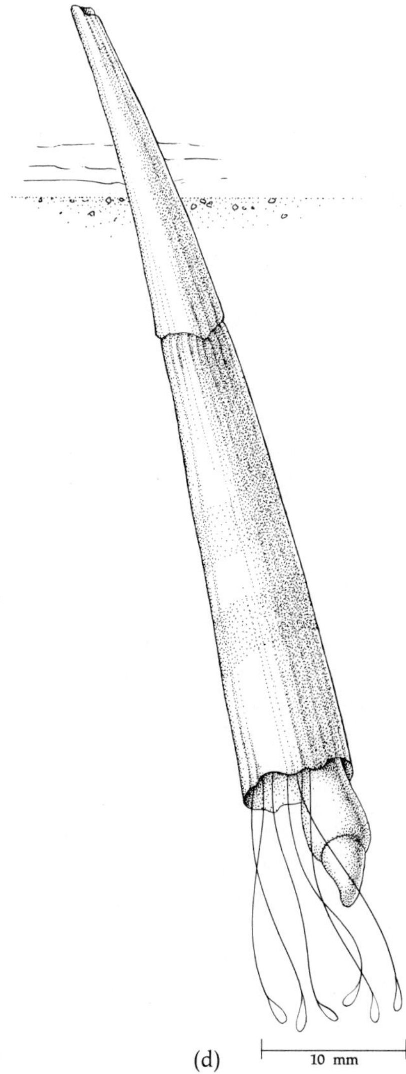

(d)

Class Bivalvia

The approximately 7,500 species of bivalves include such common animals as clams, oysters, scallops, and mussels. They derive their name from the two parts, or valves, into which the shell is divided. The left and right valves are connected dorsally by a hinge with a flexible ligament. One or two large *adductor muscles,* familiar as the delectable portion of the scallop, are used to close the shell tightly in times of danger. These muscles are so strong in scallops that the animals can move swiftly through the water by clapping their shells together, thereby eluding predators such as starfish.

In this class of mollusks, the body has become flattened between the valves, and "headness" has generally disappeared (see Figure 25-3b). The bivalves are sometimes called Pelecypoda—"hatchet foot"—because the muscular foot is often highly developed in this group. A clam, using its "hatchet foot," can dig itself into sand or mud with remarkable speed. However, many bivalves are sessile, and some of them secrete strong strands of protein by which they anchor themselves to rocks.

Abundant in both salt and fresh water, most bivalves are filter-feeding herbivores; they live largely on microscopic algae. Their gills, which are large and elaborate, collect food particles. Water is circulated through the sievelike gills by the beating of gill cilia. Small organisms and particles of food are trapped in mucus on the gill surface and swept toward the mouth by the cilia; the gills also sort particles by size, rejecting sand and other larger particles.

Throughout the molluscan phylum, there is a wide range of development of the nervous system. The bivalves have three pairs of ganglia of approximately equal size—cerebral, visceral, and pedal (supplying the foot)—and two long pairs of nerve cords interconnecting them. They have statocysts, usually located near the pedal ganglia, and sensory cells for discrimination of touch, chemical changes, and light. The scallop has quite complex eyes; a single individual may have a hundred or more eyes located among the tentacles on the fringe of the mantle. The lens of this eye cannot focus images, however, so it does not appear to serve for more than the detection of light and dark, including passing shadows cast by other moving organisms.

25-6
A bivalve. The blue eyes of this scallop are visible among its tentacles.

In the earliest mollusks, the sexes were separate and fertilization was external. This primitive condition is retained in most bivalves, but internal fertilization has evolved in a number of different bivalve lineages. Specialized "brood pouches," in which the young are protected during their early development, are found in hermaphroditic species as well as in the females of some species with separate sexes.

Class Gastropoda

The gastropods, which include the snails, whelks, periwinkles, abalones, and slugs, are the largest group of mollusks (at least 37,500 species). They have either a single shell or, as a secondary evolutionary development, no shell. Gastropods are common in both salt and fresh water and on land. The group is unusual in that some members are able to digest cellulose and other structural carbohydrates. In addition to herbivores, class Gastropoda includes omnivores, a wide variety of specialized carnivores, scavengers, and even some parasites.

Gastropods have lost the bilateral symmetry characteristic of other mollusks and have become asymmetrical through a curious anatomical rearrangement called *torsion*. Torsion, which is a separate phenomenon from the coiling of the shell, is a twisting through 180° of the rest of the body relative to the head-foot. It occurs as one side of the larval gastropod's body grows much more rapidly than the other side. As a result of torsion, the shell, mantle cavity, and visceral mass are moved so that parts that once were located at the rear of the body now lie over the head (see Figure 25-3c). Other events in gastropod development usually result in a spiral coiling of the visceral mass and the shell. In response to the displacement and consequent crowding of the internal organs, the gill and nephridium of the right side have been lost in many species. In cases where the shell has subsequently been lost and snails have evolved into slugs, the mantle cavity has generally moved back toward its original position—but the missing gill and other organs have not been regained.

Land-dwelling snails do not have gills, but the area in their mantle cavities once occupied by gills is rich in blood vessels, and the snail's blood is oxygenated there. Thus the mantle cavity has, in effect, become a lung. Moreover, as with all lungs, the opening is reduced to retard evaporation. Some snails that were probably once land dwellers have returned to the water, but they have not regained gills. Instead, they bob up to the surface at intervals to entrap a fresh bubble of air in their mantle cavities.

25-7
A land-dwelling gastropod. The shell, which is secreted by the mantle and grows as the soft body grows, covers and protects the visceral mass. The head contains sensory organs, including two eyes at the tips of the longer tentacles.

Gastropods, which lead a more mobile, active existence than bivalves, have a ganglionated nervous system with as many as six pairs of ganglia connected by nerve cords (Figure 25-8). There is a concentration of nerve cells at the anterior end of the animal, where the tentacles, which have chemoreceptors and touch receptors, are located. In some of the animals, the eyes are quite highly developed in structure; they appear, however, to function largely in the detection of changes in light intensity, like the eyes of the scallop.

In some gastropods, the primitive form of reproduction—separate sexes with external fertilization—is retained. In most gastropods, however, fertilization is internal, and hermaphroditism has evolved repeatedly in this group. The simultaneous hermaphroditism found in many snails and slugs probably resulted from the difficulties of such slow-moving animals in finding mates. In some species, the animals are sequential hermaphrodites: they are males when they are younger, then females when they are older and larger.

25–8
The ganglionated nervous system of a gastropod. The cerebral ganglia supply the tentacles and eyes; the pleural ganglia, the mantle; the pedal ganglia, the foot muscles; and the subesophageal, supraesophageal, and visceral ganglia supply the visceral mass. Counterclockwise torsion of the visceral mass during development results in the figure-eight pattern seen here, in which one major nerve passes over and another under the esophagus, a tubular structure leading from the mouth to the stomach.

The operculum, a hardened plate attached to the foot, functions as a trapdoor when the snail withdraws into its shell.

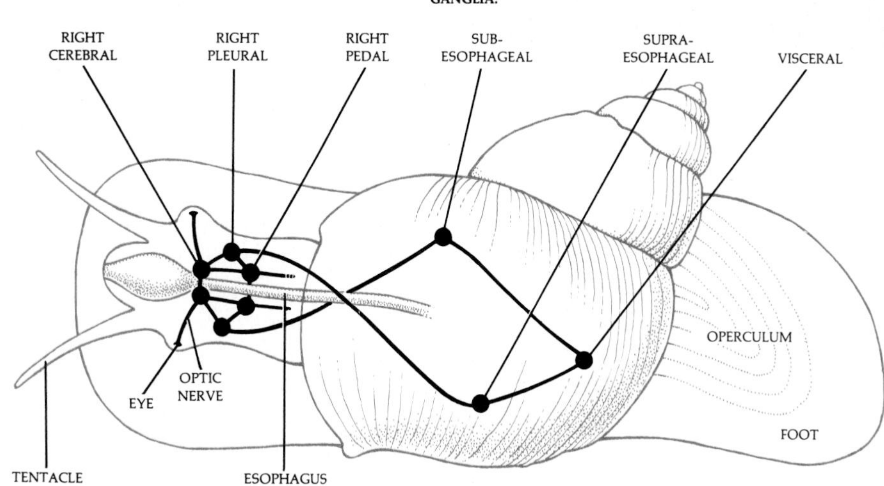

25–9
A squid. Jet-propelled, the squid is moving to the left. If you look closely, you can see its siphon just below and to the right of its eye.

Class Cephalopoda

The cephalopods (the "head-foots") are, in many respects, the most evolutionarily advanced animals to be found among the invertebrates. The 600 living species in this strictly marine class rival the vertebrates in complexity and, in some cases, in intelligence. Active predators, they compete quite successfully with fish.

Although obviously mollusks, the cephalopods have become greatly modified (see Figure 25–3d). The large head has conspicuous eyes and a central mouth surrounded by arms, some 90 in the chambered *Nautilus,* 10 in the squid, and 8 in the octopus. The octopus body seldom reaches more than 30 centimeters in diameter (except on the late late show), but giant squids sometimes attain sea-monster proportions. One caught in the Atlantic some hundred years ago was about 6 meters long, not counting the tentacles.

Nautilus, as the only modern shelled cephalopod, offers an indication of some of the steps by which this class disposed of the shell entirely. The animal occupies only the outermost portion of its elaborate and beautiful shell, the rest of which serves as a flotation chamber. In the squid and its relative, the cuttlefish, the shell has become an internal stiffening support, and in the octopus, it is lacking entirely.

Freedom from the external shell has given the mantle more flexibility. The most obvious effect of this is the jet propulsion by which cephalopods dart through the water. Usually, water taken into the mantle cavity bathes the gills and is then expelled slowly through a tube-shaped structure, the siphon; but when the cephalopod is hunting or being hunted, it can contract the mantle cavity forcibly and suddenly, thereby squirting out a sudden jet of water. Contraction of the mantle-cavity muscles usually shoots the animal backward, head last, but the squid and the octopus can turn the siphon in almost any direction they choose. Cephalopods also have sacs from which they can release a dark fluid that forms a cloud, concealing their retreat and confusing their enemies. These colored fluids were at one time a chief source of commercial inks. *Sepia* is the name of the genus of cuttlefish from which a brown ink used to be obtained.

BEHAVIOR IN THE OCTOPUS

Behaviorally, the octopus is the best studied of the cephalopods. The octopus is a sea dweller that creeps about actively on its arms or swims rapidly through the water by strong rhythmic muscular contractions that expel water from the mantle cavity. The octopus, like the other cephalopods, is carnivorous, living on smaller sea animals, usually crabs. It bites the crab, or other prey, with its parrotlike beak, injecting a toxin from its salivary glands. It then bundles up the paralyzed animal in its web and carries it home to eat. When it is not actively in pursuit of food, the octopus, lacking any protective shell, lives in small caves behind rocks or in reefs or wreckage.

Curious and able to use its arms with great dexterity, the octopus makes an extremely apt experimental subject. In a seawater tank, it will gather together bricks, shells, and any movable debris into a crude sort of house, and there it sits and watches, often bobbing its head up and down. Although the eye is remarkably similar to our own and equally acute, apparently the octopus does not have stereoscopic vision, and this head bobbing seems to be the way it estimates distance, fixing on an object from two points, just as surveyors triangulate a distant landmark.

When confronted by an object larger than itself, an octopus literally pales. It flattens and turns almost white except for a dark area around the eyes and a dark trim along the edges of its web. This makes the octopus seem larger than it is and probably serves to deter a would-be predator. By contrast, the sight of a crab can so excite an octopus that its arms weave about and its skin color darkens, breaking out in patches of bright blue, pink, or purple, depending on the species. These color changes are brought about by the contraction of muscles that draw out small sacs of pigment, the chromatophores, to form flat plates of color. When the muscles relax, the chromatophores return to their original size and the patches of color disappear.

By using a system of rewards (crabs, for example) and punishments (mild electric shock), the investigator can readily teach the octopus to seize certain objects and not to seize others. Such experiments approximate what an octopus must learn in nature—that some objects are edible, some are not, and some may bite or sting. A small crab, for example, is a meal, but if it is carrying a sea anemone on its shell, it had better be left alone. Because of the comparative ease with which such tests can be performed, owing to the octopus's natural curiosity and appetite, the animal has been the subject of a great many experiments on vision, touch, and learning.

The manipulatory powers of the octopus are great, and the arms are very sensitive to texture and are rich in chemoreceptors. However, the animal seems to have difficulty processing and coordinating sensory data. As Martin Wells, of Churchill College, Cambridge, who has carried out many studies on octopus behavior, points out, the problems of handling data received from eight extensible arms, each of which has several hundred suckers and can move separately and in any direction, are probably insurmountable. As we shall see in the next chapter, the severe limitations of movement provided by an articulated skeleton prove, in contrast, to be a great advantage for many types of specialized activities.

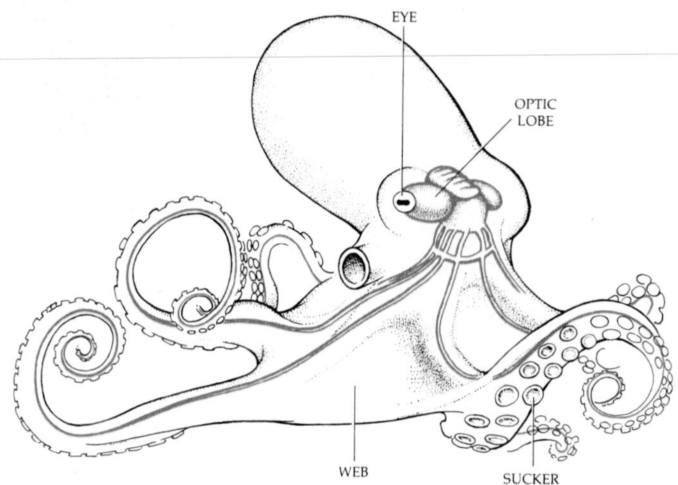

A simplified view of the brain and nervous system of the octopus. The large lobes behind the eyes are the optic lobes, concerned with collecting and analyzing data from the retina. The octopus often finds its prey by reaching its arms into a crevice into which it cannot see. The suckers on the arms of an octopus are softer and more flexible than our fingertips and so can make fine distinctions among textures. They also contain chemoreceptors that can detect sugars, salts, and other chemicals in dilutions well below the range of discrimination of the human tongue.

25–10
Octopus macropus. *Note its well-developed eyes and the suckers on the undersurface of the arms.*

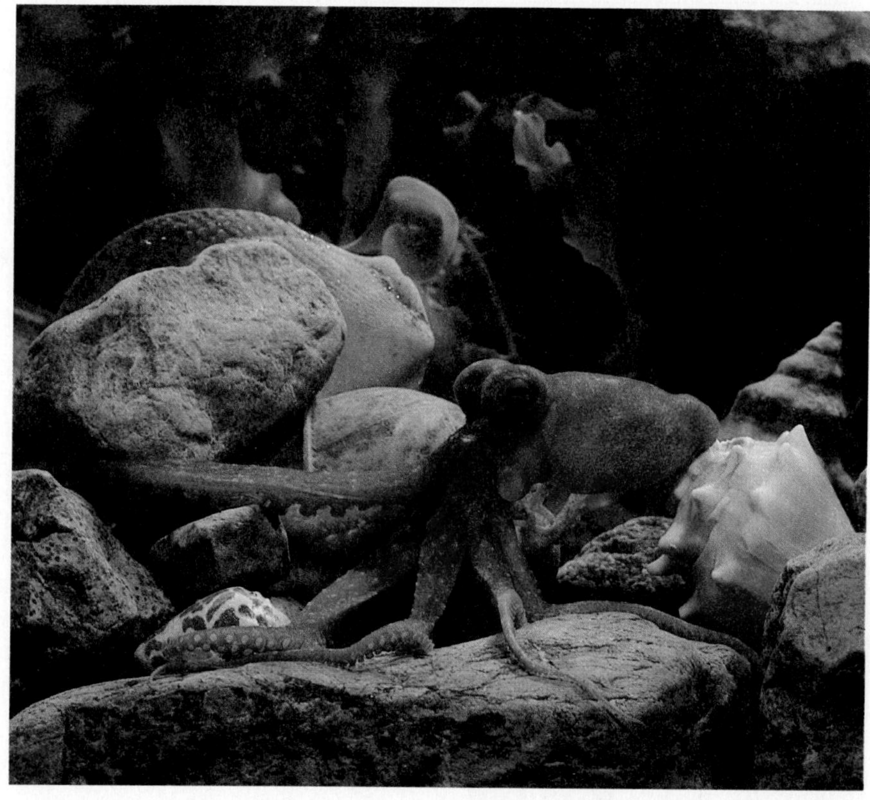

The cephalopods have well-developed brains, composed of many groups of ganglia, in keeping with their highly developed sensory systems and their lively, predatory behavior. These large brains are covered with cartilaginous cases. The rapid responses of the cephalopods are made possible by a bundle of giant nerve fibers that control the muscles of the mantle. Many of the studies on conduction of the nerve impulse are made with the giant axon of the squid, which is large enough to permit the insertion of an electrode.

In cephalopods, the sexes are always separate, and fertilization is internal. Courtship and mating behavior are complex, and males often fight for access to females. Fertilization occurs when the male uses one of his arms to transfer packets of sperm, called spermatophores, from his mantle cavity to the mantle cavity of the female. The female produces an intricate, gelatinous egg mass in which the developing embryos are protected until they hatch as miniature adults. In *Octopus* and some other cephalopod genera, the mother guards the egg masses, cleaning and aerating them.

Evolutionary Affinities of the Mollusks

Although the mollusks and the annelids (segmented worms) are, as we are about to see, quite different in their basic body plans, they have an important similarity that seems to suggest an evolutionary link. This is the trochophore larva (Figure 25–11).

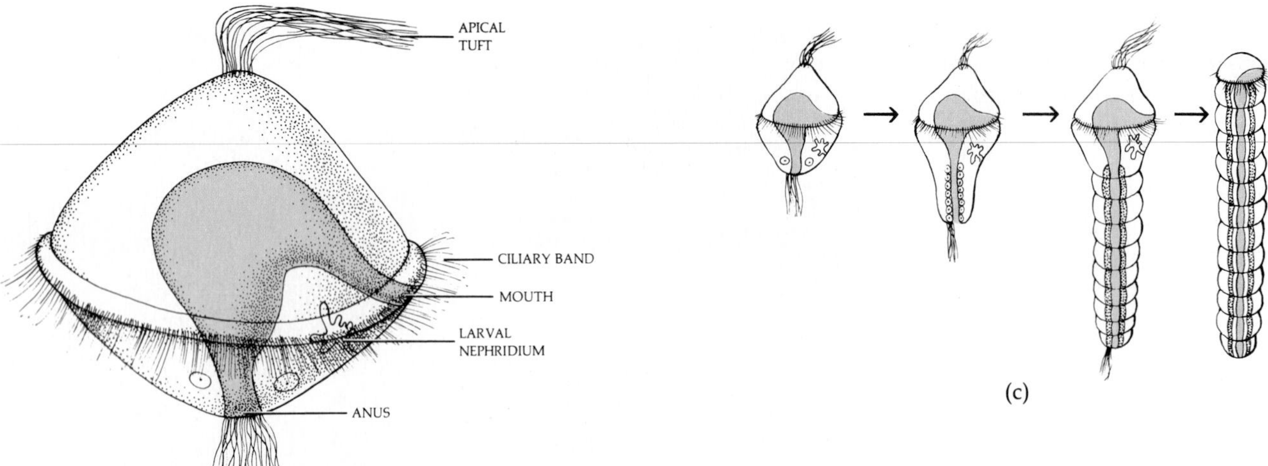

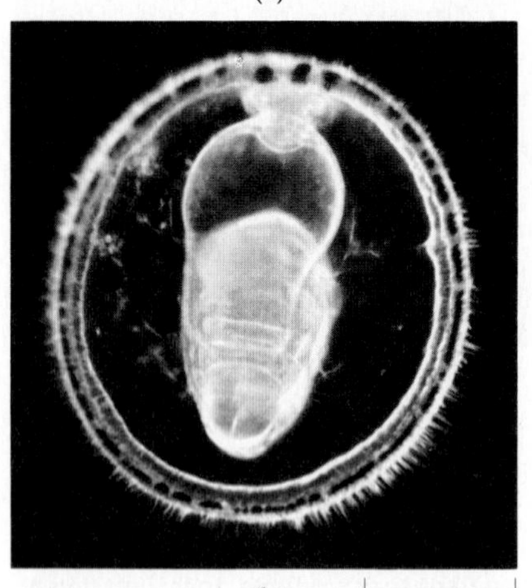

25-11
(a) *The trochophore larva. Although their adult forms are very different, certain annelids and mollusks have larvae of this type.* (b) *This particular larva will develop into a marine annelid—a polychaete worm.*

(c) *Development of the annelid trochophore into a segmented worm. The process begins with the elongation of the lower part of the trochophore. The elongated region then becomes constricted into segments, which soon develop bristles. The apical tuft disappears, and the upper part of the trochophore becomes the head. The worm will continue to grow throughout its lifetime by adding new segments just in front of its rear segment.*

Most marine mollusks (except the cephalopods) and marine annelids pass through this very distinct larval form during their development. With the exception of *Neopilina* and *Nautilus,* however, adult mollusks do not exhibit signs of the segmentation so characteristic of the annelids, and no traces of segmentation are seen in the larval development of any known mollusk. Most, but not all, authorities think that the segmental patterns seen in the gills, nephridia, and some other structures of *Neopilina, Nautilus,* and some fossil bivalves were late evolutionary developments, unrelated to the development of segmentation in the annelids. If so, the lineages giving rise to the mollusks and to the annelids diverged from an unsegmented coelomate ancestor that was characterized by the trochophore larva.

PHYLUM ANNELIDA: SEGMENTED WORMS

This phylum includes almost 9,000 species of marine, freshwater, and terrestrial worms, including the familiar earthworms. The term annelid is from the Latin for "ringed," and refers to the most distinctive feature of this group, which is the division of the body into segments, or *metameres*. The metameres are visible as rings on the outside and are separated by partitions (septa) on the inside. This segmented pattern is found in a modified form in arthropods, too, such as dragonflies, millipedes, and lobsters, which are thought to have evolved from the same ancestors that gave rise to modern annelids.

The annelids have a segmented coelom, a tubular gut, and a closed circulatory system that transports oxygen (diffused through the skin or through fleshy extensions of the skin) and food molecules (from the gut) to all parts of the body. The excretory system is made up of paired nephridia, which typically occur in each segment of the body except the head. Annelids have a centralized nervous system and a number of special sensory cells, including touch cells, taste receptors, light-sensitive cells, and cells concerned with the detection of moisture. Some also have well-developed eyes and sensory antennae.

The three classes of annelids are Oligochaeta (terrestrial worms, with some freshwater and marine relatives), Polychaeta (mostly marine worms), and Hirudinea (leeches). It is generally agreed that the hirudineans are derived from oligochaetes, but the question of whether the polychaetes or the oligochaetes came first remains unresolved. However, we shall begin with class Oligochaeta, since its most familiar member—the earthworm—is such a clear exemplar of annelid structure.

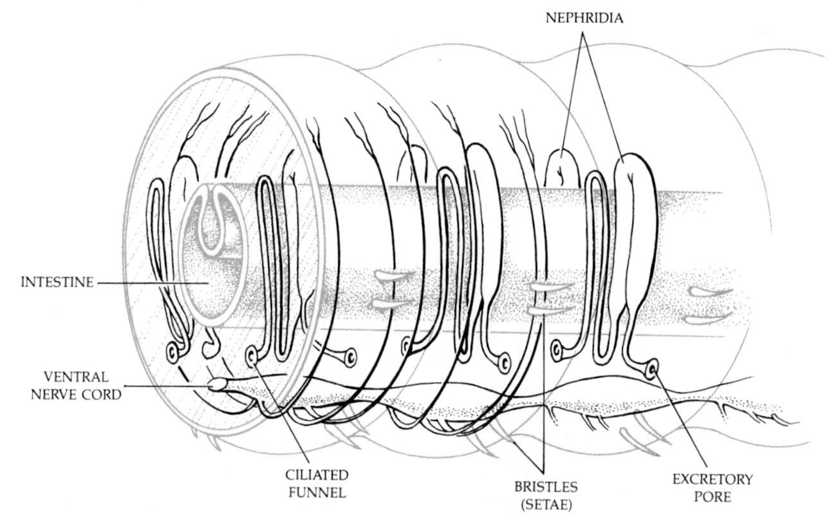

25–12
Three segments of the earthworm, an annelid. On each segment are four pairs of bristles, which are extended and retracted by special muscles. These are used by the worm to anchor one part of its body while it moves another part forward. Two excretory tubes, or nephridia, are in each segment (except the first three and the last). Each nephridium really occupies two segments, since it opens externally by a pore in one segment and internally by a ciliated funnel in the segment immediately in front of it. The intestine, nephridia, and other internal organs are suspended in the large fluid-filled coelom, which also serves as a hydrostatic skeleton.

Class Oligochaeta: The Earthworm

Figure 25–12 shows a portion of the body of an earthworm, a representative oligochaete. Note how the body is compartmentalized into regular segments. Most of these segments, particularly the central and posterior ones, are identical, each exactly like the one before and the one after. Each identical segment contains four pairs of bristles, or *setae*; two nephridia; three pairs of nerves branching off from the central nerve cord running along the ventral surface; a portion of the digestive tract; and a left and right coelomic cavity. The chief exceptions to this rule of segmented structure are found in the most forward segments. In these, specialized areas of the nervous, digestive, circulatory, and reproductive systems are found.

The tubelike body is wrapped in two sets of segmented muscles, one set running longitudinally and the other encircling the segments. When the earthworm moves, it anchors some of its segments by its setae, and the circular muscles of the segments anterior to the anchored segments contract, thus extending the body forward. Then its forward setae take hold, and the longitudinal muscles contract while the posterior anchor is released, drawing the posterior segments forward. The ultimate in a hydrostatic skeleton is provided by the coelom of the earthworm. Not only is the coelom partitioned by the septa between segments, but it is also divided into left and right compartments within each segment. This arrangement, in combination with the two sets of segmented muscles, allows exquisite control over movements of small parts of the body.

Digestion in Earthworms

The digestive tract of the earthworm (Figure 25–13) is a long, straight tube. The mouth leads into a strong, muscular *pharynx*, which acts like a suction pump, helping the mouth to draw in decaying leaves and other organic matter, as well as dirt, from which organic materials are extracted. The earthworm makes burrows in the earth by passing such material through its digestive tract and depositing it outside in the form of castings, a ceaseless activity that serves to break up, enrich, and aerate the soil. Darwin, in his study of earthworms *(The Formation of Vegetable Mould through the Action of Worms)*, estimated 15 tons as the weight of castings thrown up annually on an acre of land—perhaps 20 ounces per worm per annum, he calculated.

25–13

The digestive tract of an earthworm. The mouth leads into a muscular pharynx, which sucks in decaying vegetation and other material. These are stored in the crop and ground up in the gizzard with the help of soil particles. The rest of the tract is a long intestine (gut) in which food is digested and absorbed.

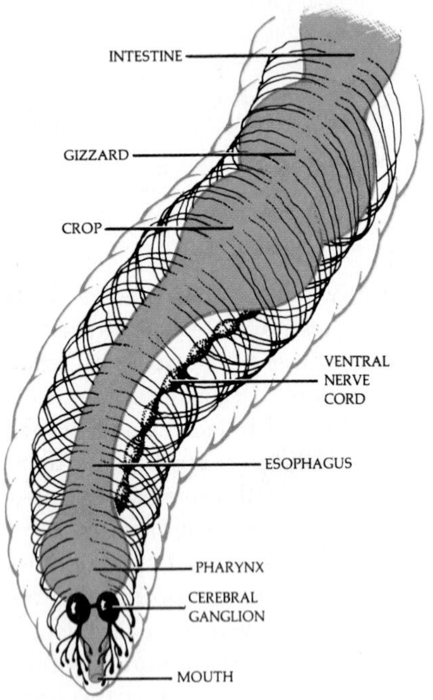

25–14

The circulatory system of the earthworm is made up of longitudinal vessels running the entire length of the animal, one dorsal and several ventral. Smaller vessels (the parietal vessels) in each segment collect the blood from the tissues and from the subneural vessel and feed it into the muscular dorsal vessel, through which it is pumped forward. In the anterior segments are five pairs of hearts—muscular pumping areas in the blood vessels—whose irregular contractions force the blood downward to the ventral vessel, from which it returns to the posterior segments. The arrows indicate the direction of blood flow.

The narrow section of digestive tract posterior to the pharynx, the *esophagus,* leads to the *crop,* where food is stored. In the *gizzard,* which has thick, muscular walls lined with protective cuticle, the food is ground up with the help of the ever-present soil particles. The rest of the digestive tract is made up of a long intestine, which has a large fold along its upper surface that increases its surface area. The intestinal epithelium consists of enzyme-secreting cells and ciliated absorptive cells.

Circulation in Earthworms

The circulatory system of the earthworm (Figure 25–14) is composed of longitudinal vessels running the entire length of the worm, one dorsal and several ventral. The largest ventral vessel underlies the intestinal tract, supplying blood to it, to the smaller ventral vessels that surround the nerve cord, and, by means of many small branches, to all the tissues of the body. Numerous small capillaries in each segment carry blood from the tissues to the dorsal vessel. Also in each segment are larger parietal ("along the wall") vessels transporting blood from the subneural vessel to the dorsal vessel, which also collects nutrients from the intestinal tract. The muscular dorsal vessel propels forward the fluids collected in this way from all over the animal's body.

Connecting the dorsal and ventral vessels, and so completing the circuit, are five pairs of hearts, muscular pumping areas in the blood vessels. Their irregular contractions force the blood down to the ventral vessels and also forward to the vessels that supply the more anterior segments. Both the hearts and the dorsal vessel have valves that prevent backflow. Note that annelids have a closed circulatory system in which the blood flows entirely through vessels. Evolution of a closed circulatory system made possible a degree of control, not previously feasible, over the composition of the circulating body fluid.

Respiration in Earthworms

The earthworm, unlike the mollusks, has no special respiratory organs; respiration takes place by simple diffusion through the body surface. The gases of the atmosphere dissolve in the liquid film on the earthworm's body, which is kept moist by secreted mucus and by coelomic fluid released through dorsal pores. Oxygen diffuses inward to the network of capillaries just underlying the body surface and is consumed by body cells as the blood circulates. Carbon dioxide, which is picked up by the blood, diffuses out into the air through the body surface.

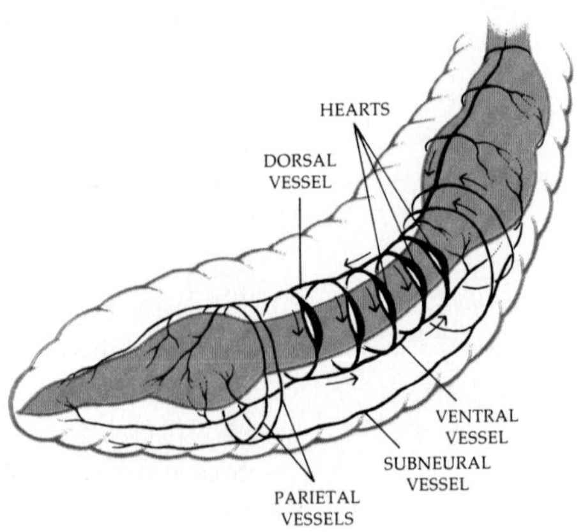

Excretion in Earthworms

In annelids, as in mollusks, the nephridia remove nitrogenous wastes from the coelomic fluid and regulate its chemical composition. The excretory system of the earthworm consists of one pair of nephridia for each segment (see Figure 25–12). Each nephridium consists of a long, convoluted tubule that begins with a ciliated funnel opening into the coelomic cavity of the anteriorly adjacent segment. Coelomic fluid is carried into the funnel by the beating of the cilia and is excreted through an outer pore. During its passage through the nephridium, needed water, other molecules, and ions are reabsorbed and waste products are secreted into the fluid.

The Nervous System of Earthworms

The earthworm has a variety of sensory cells. It has touch cells, or *mechanoreceptors*. These contain tactile hairs, which, when stimulated, trigger a nerve impulse. Patches of these hair cells are found on each segment of the earthworm. The hairs probably also respond to vibrations in the ground, to which the earthworm is very sensitive. The earthworm does not have ocelli or eyes—as one might expect, since it rarely emerges from underground—but it does have light-sensitive cells. Such cells are more abundant in its anterior and posterior segments, the parts of its body most likely to be outside of the burrow. These cells are not responsive to light in the red portion of the spectrum, a fact exploited by anglers who search for worms in the dark using red-lensed flashlights.

25–15
Earthworms mating. The worms' heads are facing in opposite directions, and their ventral surfaces are in contact. The clitellum, a thickened band that surrounds the body of each, secretes mucus. The mucus holds the worms together during copulation, which may take as long as two hours. Sperm cells are released through pores in specialized segments of one worm into the sperm receptacles of its partner. After the partners separate, the clitellum secretes a cocoon into which first the eggs and then the sperm are released. The eggs are fertilized within this cocoon.

Among the earthworm's most sensitive cells are those that detect moisture. The cells are located on its first few segments. If an earthworm emerging from its burrow encounters a dry spot, it swings from side to side until it finds dampness; failing that, it retreats. However, when the anterior segments are anesthetized, the earthworm will crawl over dry ground. The animal also appears to have taste cells. In the laboratory, worms can be shown to select, for example, celery in preference to cabbage leaves and cabbage leaves in preference to carrots.

Each segment of the worm is supplied by nerves that receive impulses from sensory cells and by nerves that cause muscles to contract. The cell bodies for these nerves are grouped together in clusters (ganglia). The movements of each segment are directed by a pair of ganglia. Movement in each segment is triggered by movement in the adjacent anterior segment; thus a headless earthworm can move in a coordinated manner. However, an earthworm without its cerebral ganglia moves ceaselessly; in other words, the cerebral ganglia inhibit and modulate activity.

There are also, as in planarians, conducting channels made up of nerve fibers bound together in bundles, like cables, which run lengthwise through the body. These nerve fibers are gathered together in a fused, double nerve cord that runs along the ventral surface of the body. The nerve cords contain fast-conducting fibers that make it possible for the earthworm to contract its entire body very quickly, withdrawing into its burrow when disturbed.

Reproduction in Earthworms

Earthworms are hermaphrodites, and some can reproduce parthenogenetically. In most species, however, two earthworms, held together by mucous secretions from the clitellum (a special collection of glandular cells), exchange sperm and separate (Figure 25–15). Two or three days later, the clitellum forms a second mucous sheath

25–16
This annelid, a polychaete worm, unlike the more familiar earthworm, has a well-differentiated head with sensory appendages and lateral parapodia ("side feet") with many setae.

25–17
Leeches are primarily bloodsuckers with digestive tracts specially adapted for storage of blood. They range in size from 1 to 30 centimeters and are found mostly in inland waters or in damp places on land, where they parasitize fish, turtles, and other vertebrates. A few species are predatory, feeding on worms and insects that they swallow whole.

Shown here is a medicinal leech, of the type commonly used for bloodletting. Bloodletting, one of the most common forms of medicinal treatment as late as the 1800s, was used for patients suffering from whooping cough, gout, drunkenness, rheumatism, sore throat, and asthma, among other ailments. Patients were commonly "bled to faintness," losing as much as 1.5 liters of blood.

surrounded by an outer, tougher protective layer of chitin. This sheath is pushed forward along the animal by muscular movements of its body. As it passes over the female gonopores, it picks up a collection of mature eggs, and then, continuing forward, it picks up the sperm deposited in the spermathecas. Once the mucous band is slipped over the head of the worm, its sides pinch together, enclosing the now fertilized eggs in a small capsule from which the infant worms hatch.

Class Polychaeta

The polychaetes, which are almost all marine, differ from the earthworms and other oligochaetes in a number of ways (Figure 25–16). The most striking difference is that they typically have a variety of appendages, including tentacles, antennae, and specialized mouthparts. Each segment contains two fleshy extensions, *parapodia*, which function in locomotion and also, because they contain many blood vessels, are important in gas exchange. In polychaetes, as in other segmented animals, there is a tendency for the division of labor between segments to lead to *tagmosis*—the formation of groups of segments into body regions with functional differences. This tendency is more pronounced in the polychaetes than in the oligochaetes and often results in distinct head, trunk, and tail regions.

Polychaetes have diverse life styles. Some are motile predators, using their strong jaws to feed on small animals. Others are more sedentary, feeding on materials suspended in the water or deposited in bottom sediments. Many polychaetes live in elaborately fashioned tubes constructed in the mud or sand of the ocean bottom. Usually the sexes are separate, fertilization is external, and there is a free-swimming trochophore larva.

Class Hirudinea

Hirudineans are the leeches, which have flattened, often tapered, bodies with a sucker at each end (Figure 25–17). In most species, the setae have been lost, and the animals either creep along with a loping movement or swim with undulating motions of the body. Like the earthworm and other oligochaetes, leeches are hermaphrodites.

Bloodsucking leeches attach themselves to their hosts by their posterior sucker, and then, with their anterior sucker, either slit the host's skin with their sharp jaws or digest an opening through the skin by means of enzymes. Finally, they secrete a special chemical (hirudin) into the host's blood to prevent it from coagulating.

MINOR PROTOSTOME PHYLA

In addition to the three major phyla of protostome coelomates—Mollusca, Annelida, and Arthropods—there are seven minor phyla with living representatives. The first four of these phyla contain bottom-dwelling marine worms that show varying degrees of similarity to the annelids.

The peanut worms, phylum Sipuncula, range from 1 to more than 60 centimeters in length and are characterized by a long proboscis that can be retracted into the stout, bulblike body (Figure 25–18a). Although the approximately 300 species of this phylum have neither segmentation nor setae, their trochophore larvae are very similar to those of the polychaete annelids.

Phylum Echiura contains about 100 species that are sometimes called spoon worms. Most have a long proboscis that, unlike the proboscis of the peanut worms, cannot be retracted into the stout body; it can contract, however, forming a structure that resembles the bowl of a spoon (Figure 25–18b). As in the peanut worms, the sexes are separate, fertilization is usually external, and there is a trochophore larva. The adult worms have setae but show no traces of segmentation.

Nine species of burrowing worms, ranging up to 20 centimeters in length, make up phylum Priapulida. They are characterized by a retractile proboscis bearing spines that are used to capture small, soft-bodied prey (Figure 25–18c). Priapulids are not segmented internally, but the external surface of the body is marked by anywhere from 30 to 100 superficial rings; setae are found only on the males of one species. These worms most closely resemble the pseudocoelomate kinorhynchs

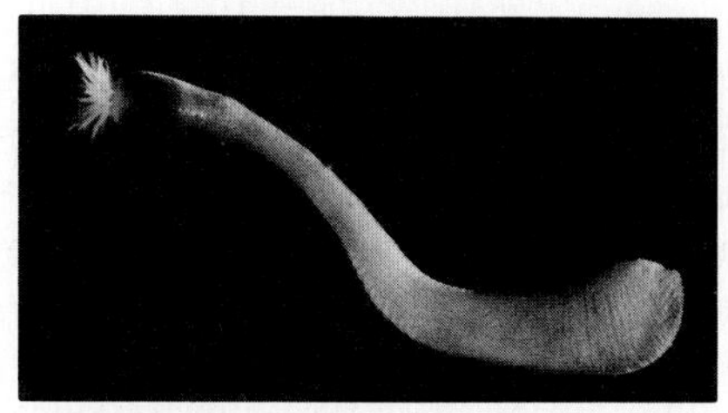

(a)

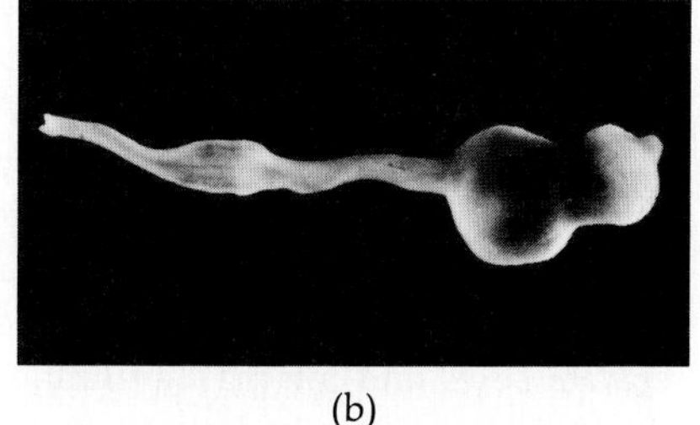

(b)

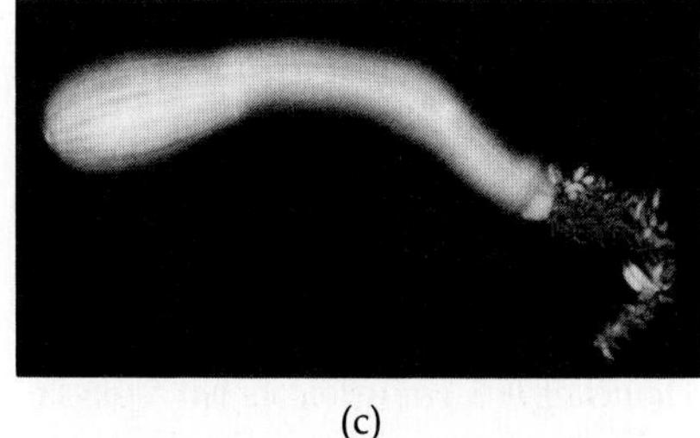

(c)

25–18

Representatives of the smaller phyla of protostome worms. (a) Thimeste lageniformis, *a sipunculan. The retractile proboscis, here everted from the body, represents about one-third of the worm's total length. Food particles, trapped by the mucus-covered tentacles, are moved to the mouth by cilia. When disturbed, these worms contract into the shape of a peanut.* (b) *An echiuran,* Lissomyema mellita. *Members of this genus live in empty gastropod shells, actively pumping water through the shell. Passing food particles are caught in a ciliated groove in the long proboscis and are carried to the mouth.* (c) Priapulis caudatus, *the first priapulid described, with its anterior proboscis everted. The bushy tail at the posterior of the animal consists of tubular structures that are thought to have a respiratory function. In some species the tail is muscular and bears hooks; it presumably anchors the worm when it is burrowing.*

25–19
The most stunning members of phylum Pogonophora are the giant tube worms that live near fissures in the earth's crust, deep in the Pacific Ocean. These worms, first discovered in 1977, sometimes reach 1.5 meters in length. Their bodies harbor large quantities of symbiotic chemoautotrophic bacteria that are thought to supply them with organic molecules.

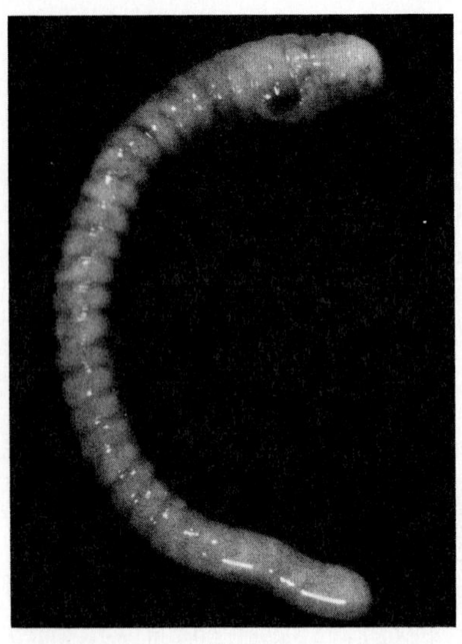

25–20
Porocephalus crotali, *a pentastomid. Adults of this genus live as parasites in the lungs of snakes. Their life cycle includes two larval stages, each parasitic in a different host, which may be a snake or a mammal. Although not visible in this photograph, the mouth of a pentastomid is surrounded by four projections. At one time, these projections were erroneously thought to be additional mouths; hence the name "five mouths."*

(see page 502) but may have a true coelom. Very little is known about either their embryonic development or their phylogenetic relationships.

The 100 species of beard worms, phylum Pogonophora, have a segmented posterior end with setae but have no mouth or digestive tract. These very slender worms live in long tubes buried in deep-sea sediments (Figure 25–19). The anterior region of the body bears a crown of tentacles that are thought to provide surfaces for the uptake of nutrients.

The remaining three phyla—Pentastomida, Tardigrada, and Onychophora—have varying combinations of annelid and arthropod characteristics. These animals are distinguished from all others by unjointed appendages bearing claws; arthropod appendages, by contrast, are jointed. Like the arthropods, however, they have an external cuticle that is molted periodically.

The 70 species of wormlike pentastomids (Figure 25–20) are all parasites of vertebrate respiratory systems. Pentastomids are so highly specialized for their parasitic existence—lacking, for example, circulatory, respiratory, and excretory organs—that it has been difficult to determine the primitive form from which they may have evolved. Although their larvae resemble young tardigrades, recent studies indicate that they are most closely related to the arthropods. Pentastomids have separate sexes and internal fertilization and, like many other parasites, produce tremendous numbers of eggs.

Phylum Tardigrada contains about 350 species of tiny segmented animals often called "water bears" (Figure 25–21). Common in fresh water and in the film of moisture on mosses, they lumber along on their four pairs of stubby legs. Their protective cuticle is thin, but they are able to survive drying out; when conditions are unfavorable, they enter a state of suspended animation and remain dormant until moisture is once more available. The sexes are separate, but in some species males are unknown and the females produce eggs that develop parthenogenetically.

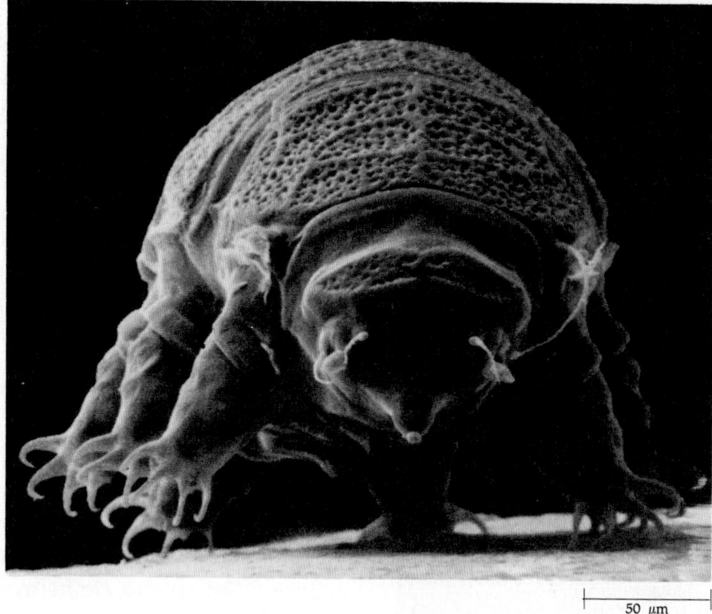

25–21
A "water bear," phylum Tardigrada. These remarkable animals are able to survive extremes of temperature as well as extreme desiccation. They are found not only in temperate climates, moving in their slow, deliberate way over the surfaces of mosses and lichens, but also in the Arctic, in the tropics, and even in hot springs. Their mouths are equipped with piercing stylets, through which they suck the juices of plant cells or of such small animals as rotifers and nematodes.

25–22
Peripatus, *an onychophoran, found living under a log in a rain forest in South America. Its short, unjointed legs are the only external sign of segmentation. Its body is covered with an unjointed chitinous cuticle, which is, however, soft enough that it bends as the animal moves.*

The caterpillarlike animals of phylum Onychophora (Figure 25–22) have a particularly striking combination of annelid and arthropod characteristics. The annelid characteristics shared by the 70 species in this phylum include relatively soft bodies, segmentally arranged nephridia, muscular body walls, and ciliated reproductive tracts. On the other hand, their jaws (derived from appendages), protective cuticle, relatively large brains, and circulatory and respiratory systems resemble those of arthropods. Their antennae and eyes are similar to those of both the polychaete annelids and the arthropods. Although it is reasonable to consider onychophorans primitive animals, suggestive of a stage in the early evolution of the arthropods, their reproduction is very advanced. Although some onychophorans lay eggs, most give birth to live young. Moreover, in some species the embryos develop within a uterus and are nourished through a placenta-like structure, analogous to that of the mammals. Modern onychophorans are all terrestrial, living inconspicuously in moist habitats, mainly in the Southern Hemisphere. Fossil evidence, however, indicates that the earliest species were marine. As onychophorans made the transition to land, they, like the arthropods and the vertebrates, successfully solved the problem of protecting and nourishing the developing young.

THE LOPHOPHORATES

The animals in three additional phyla, known collectively as the lophophorates, are, strictly speaking, protostomes (that is, the mouth forms before the anus during embryonic development). Nevertheless, they display some deuterostome characteristics, leading some authorities to regard them as primitive deuterostomes.

The lophophorates, which are all aquatic, include the phoronid worms (phylum Phoronida), the bryozoans, or "moss animals" (phylum Bryozoa, or Ectoprocta), and the lamp shells (phylum Brachiopoda). Although the three phyla have di-

25–23
A phoronid worm, Phoronopsis harmeri. *The extensive lophophore is in the shape of a horseshoe, with each of the ends forming a spiral coil. Cilia on the tentacles of the lophophore direct water currents through the groove between the two coils. Food particles are trapped in mucus on the tentacles and carried by the cilia to the mouth, which is at the base of the groove. Although phoronids are not attached to their tubes, which may be chitinous or leathery, they never leave them. When disturbed, the animal can completely withdraw into the safety of the tube.*

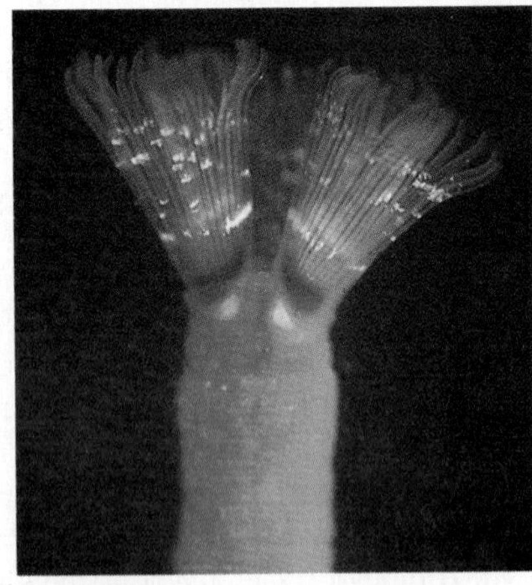

verged considerably, all of these animals have a characteristic food-gathering organ known as the lophophore. Located at the anterior end of the animal, the lophophore consists of a crown of hollow, ciliated tentacles used in filter-feeding (Figure 25–23). The cavity within the tentacles is an extension of the coelom, and gases are exchanged between the coelomic fluid and the surrounding water through the thin walls of the tentacles.

Lophophorates do not display the segmentation characteristic of annelids and arthropods, but the coelom of the adult forms of some species and of the embryos of most species is partitioned into several major compartments. A similar sort of partitioning is characteristic of many deuterostomes. Other deuterostome characteristics shared by the lophophorates include radial cleavage in the early stages of embryonic development and, in some species, enterocoelous formation of the coelom.

The simplest of the lophophorates are the 18 species of phylum Phoronida. These worms, which range from 4 to 25 centimeters in length, live in tubes in or on soft ocean bottoms in shallow waters. The lophophore projects above the top of the tube but can be withdrawn into the tube when the animal is disturbed. Although phoronids are often found clustered together, each animal is independent of the others. Most species are hermaphroditic, and one species is known to reproduce asexually.

In contrast to the phoronids, the tiny bryozoans, or ectoprocts, are colonial animals, often with a considerable division of labor among the members of the colony. Their colonies can be found on virtually any type of firm surface in salt water and, less frequently, in fresh water. About 4,000 living species of bryozoans are known, and most are hermaphrodites; the freshwater forms, however, reproduce asexually as well as sexually. Bryozoans secrete a hard, protective covering around themselves, from which only the individual lophophores extend (Figure 25–24). Their name derives from the superficial resemblance of their colonies to patches of moss. As we noted earlier (page 502), the pseudocoelomate Entoprocta have a similar appearance, and before the differences in internal structure were fully understood, the entoprocts were also included in the phylum Bryozoa.

25–24
A colony of bryozoans, Flustrellidra hispida. *Note the extended lophophores with which the members of the colony catch passing algal cells. Because of the protective covering that bryozoans secrete around themselves, their colonies are also known as "sea mat."*

25-25
The brachiopod Terebratulina septen, *with its shell closed and open. When the shell is open, the only structure visible is the large, spiral lophophore, which constitutes about two-thirds of the animal's body. Water is circulated through the shell by the moving tentacles of the lophophore. Its cilia sweep small organisms in the water toward the mouth, which is under the lophophore, toward the left. The brachiopod is anchored to the substrate by a stalk, or pedicel.*

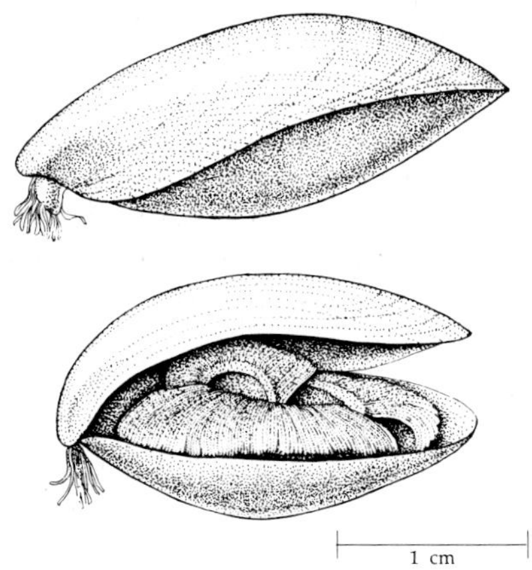

Another case of mistaken classification on the basis of inadequate information involved the lamp shells, phylum Brachiopoda. Until well into the nineteenth century, these animals, which resemble a Greek or Roman oil-burning lamp (Figure 25-25), were classified with the bivalve mollusks. The lophophore, however, clearly distinguishes the brachiopods from the bivalves. Moreover, the two valves of the brachiopod shell are dorsal and ventral, rather than left and right as in the bivalves. There are about 250 species of brachiopods living today, but the fossil record reveals an additional 30,000 extinct species.

SUMMARY

One of the most significant innovations in the course of animal evolution was the coelom, a fluid-filled cavity within the mesoderm. The coelom not only functions as a hydrostatic skeleton but also provides space within which the internal organs can be suspended by the mesenteries.

The coelomate animals are divided into two broad groups on the basis of their embryonic development. In the protostomes, the cleavage of the fertilized egg is spiral, the mouth develops at or near the blastopore, and the coelom is formed by a splitting of the solid mesoderm. In the deuterostomes, the cleavage pattern is radial, the anus develops at or near the blastopore, and the coelom is formed by outpocketings of the primitive gut.

The soft-bodied animals of phylum Mollusca are classified into four relatively small classes and three major classes: Bivalvia (oysters and clams), Gastropoda (snails), and Cephalopoda (octopuses and their relatives). The basic body plan of all mollusks is the same—a head-foot, a visceral mass, and a shell-secreting mantle—but it has been modified in the course of adaptation to different ways of life. Shells are often present but may be reduced or absent. Mollusks are also characterized by a toothed tongue, the radula, and an open circulatory system with an efficient three-chambered heart. In most mollusks, respiration is carried out by means of gills, thin-walled structures that are an extension of the epidermis and are located in the mantle cavity. Excretion is accomplished by special organs, the nephridia, tubular structures that collect fluids from the coelom and exchange salts and other substances with body fluids as the urine passes along the tubules for excretion. Nervous systems and behavior vary among the species, reaching a zenith of complexity in the brainy octopus.

On the basis of the similarities of their trochophore larvae, the mollusks and the segmented worms of phylum Annelida are thought to have diverged from a common coelomate ancestor. Annelids are characterized by bodies that are conspicuously segmented, both internally and externally; well-developed coeloms; tubular digestive tracts; paired nephridia in each segment; and closed circulatory systems, often with contractile vessels. The phylum includes terrestrial worms, such as the earthworms (class Oligochaeta), marine worms (class Polychaeta), and leeches (class Hirudinea). Unlike the oligochaetes, the polychaetes have a variety of appendages and exhibit tagmosis—the formation of groups of segments into body regions with functional differences. Annelids have relatively complex nervous systems, consisting of a ventral nerve cord, which divides to encircle the pharynx at the anterior end of the animal, and pairs of ganglia (clusters of nerve cells), one pair to each segment, which receive impulses from a variety of sensory receptors and trigger motor activities in each segment.

Of the seven minor phyla of protostome coelomates, four consist of bottom-dwelling marine worms that show varying similarities to the annelids. These groups include the peanut worms (phylum Sipuncula), the spoon worms (phylum Echiura), the priapulids (phylum Priapulida), and the beard worms (phylum Pogonophora). Three other phyla—Pentastomida, Tardigrada, and Onychophora—all characterized by unjointed, clawed appendages and an external cuticle, have combinations of annelid and arthropod features. The onychophorans are of particular interest, for they suggest a stage in the evolution of arthropods from a segmented, coelomate ancestor common to both the annelids and the arthropods.

Lophophorates, although protostomes, resemble deuterostomes in the radial cleavage of the fertilized egg and in the compartmentalization and, in some species, the enterocoelous formation of the coelom. The marine worms of phylum Phoronida, the colonial Bryozoa, and the lamp shells (phylum Brachiopoda) are all lophophorates. They are characterized by the lophophore, a crown of hollow tentacles bearing cilia that sweep food particles into the mouth.

QUESTIONS

1. Distinguish between the following terms: protostome/deuterostome; schizocoelous/enterocoelous; coelom/hemocoel; planula/trochophore.

2. What modifications of the basic molluscan body plan have occurred in bivalves, gastropods, and cephalopods?

3. Many mollusks have lost or may be in the evolutionary process of losing their shells. What are the advantages to an organism of having a shell? Of losing one?

4. Smallness may also be an advantage to an organism. What are some of the advantages of smallness?

5. Why does any heart truly worthy of the name have at least two chambers?

6. Label the drawing below.

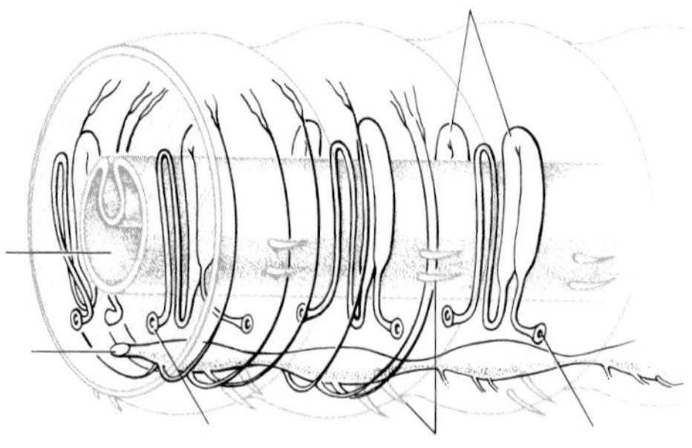

7. Nematodes have only longitudinal muscles in their body walls but have very high internal fluid pressures; earthworms, with both longitudinal and circular muscles, have low fluid pressures. Can you suggest a mechanism by which internal fluid at high pressure can circumvent the need for certain muscles?

8. Describe the structure and function of the lophophore. How does it differ from the tentacles of a cnidarian, such as *Hydra?*

CHAPTER 26

The Animal Kingdom III: The Arthropods

Phylum Arthropoda, the "joint-footed" animals, is by far the largest of the animal phyla. About 850,000 species of insects and other arthropods have been classified to date, and estimates of the total number are as high as 10 million. As to the number of individuals, it has been calculated that, of insects alone, as many as 10^{18}—a billion billion—are alive at any one time. Arthropods are abundant in virtually all habitats. It has been estimated that over every square kilometer in the temperate zone there are, at certain seasons, some 20 million individual arthropods, layered in the atmosphere, like plankton.

CHARACTERISTICS OF THE ARTHROPODS

Arthropods are characterized by their jointed appendages. In the more highly evolved members of the phylum, the number of appendages is reduced, and they are more specialized and efficient. Especially among the insects and crustaceans, these appendages include not only walking legs but also a kit of marvelously adapted tools—jaws, gills, tongs, egg depositors, sucking tubes, claws, antennae, paddles, and pincers.

26–1
A harlequin beetle in flight. Arthropods are characterized by a variety of specialized appendages. Notice the three pairs of jointed legs, the jointed antennae, the intricately colored wings, and the segmented body.

26–2
Arthropods, like the annelids, are segmented. In the more primitive arthropods, such as this millipede, the segmented pattern remains clearly visible in the adult animal. However, unlike annelids, adult arthropods have rigid, jointed exoskeletons and appendages. In millipedes, which are members of class Diplopoda, most of the body segments are fused into "diplosegments," each of which has two pairs of legs.

All arthropods are segmented, a characteristic that strongly suggests a common ancestry with the annelids (Figure 26–2). In the course of arthropod evolution, however, the body has become shorter, with fewer segments, which have become fixed in number and more specialized. In many arthropods, tagmosis (page 520) has been carried much farther than in the polychaete annelids, with the fusion of segments to form distinct body regions—a head, a thorax (sometimes fused with the head to form a cephalothorax), and an abdomen. But the basic segmented pattern is often still clearly evident in the immature stages (witness the caterpillar) and can be discerned in the adult by examination of the appendages, the musculature, and the nervous system.

At some point well after the lineage leading to the arthropods diverged from that leading to the annelids, further major branchings occurred (see Figure 24–4, page 483). These branchings gave rise to three principal types of arthropods: chelicerates, aquatic mandibulates, and terrestrial mandibulates. The conspicuous differences in the appendages of these three types can be clearly distinguished by even an inexperienced eye. In both the aquatic mandibulates (class Crustacea) and the terrestrial mandibulates (class Insecta and four smaller classes), the most anterior appendages are one or two pairs of *antennae*, and the next are *mandibles* (jaws). Differences in the development and structure of the mandibles suggest that they evolved independently in the two groups and do not reflect a common mandibulate ancestor. The chelicerates, which include class Merostomata (horseshoe crabs), class Pycnogonida (sea spiders), and class Arachnida (spiders, mites, scorpions, and their relatives), have no antennae and no mandibles. Their first pair of appendages consist of *chelicerae* (singular, chelicera), which take the form of pincers or fangs. Chelicerates may also have *book lungs* or *book gills*; these structures, which are not present in mandibulates, derive their name from their resemblance to the leaves of a partially opened book.

Despite the huge number of arthropods and the richness of their diversity, there are a number of features shared by all members of this phylum.

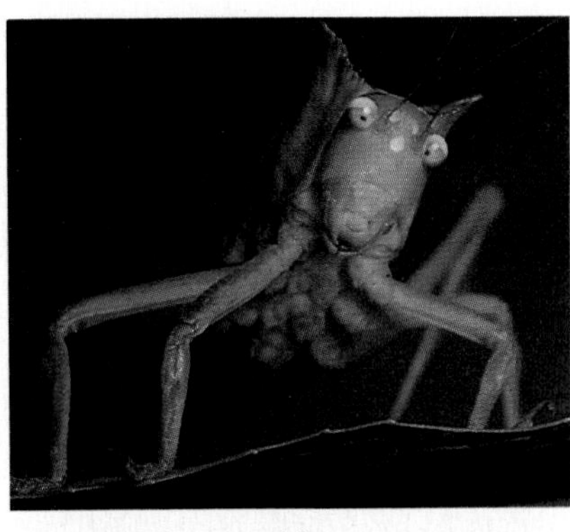

(a)

(b)

26–3
(a) *A many-jointed exoskeleton, as in this South American katydid, is characteristic of the arthropods. The slits in each foreleg are the insect's ears.*

(b) *Molting. An exoskeleton, once formed, does not grow and so must be periodically discarded. The old cuticle is at the top, and below is the newly emerged "soft-shelled" blue crab. The new exoskeleton, already formed, has begun to expand and harden.*

The Exoskeleton

The most striking characteristic of all arthropods is their articulated (jointed) exoskeleton. This exoskeleton, or cuticle, is secreted by the underlying epidermis and is attached to it; it is made up of an outer, often waxy layer, composed of lipoprotein, a hardened middle layer, and an inner flexible one, both composed principally of chitin and proteins. The exoskeleton not only covers the surface of the animal but also extends inward at both ends of the digestive tract and, in insects, lines the tracheae (breathing tubes) as well. The cuticle serves as protection against predators, and it is often waterproof, keeping exterior water out and interior water in. It is used for food grinders in the foregut, for wings, and for tactile hairs. Cuticle even forms the lens of the arthropod eye.

The exoskeleton may form a veritable coat of armor, as it does in beetles and in some of the crustaceans (in which it is often infiltrated with calcium salts), but at the joints it is flexible and thin, permitting free movement. Muscles are attached to the various portions of the exoskeleton, just as they are attached to the various bones of the endoskeleton in vertebrates. When the muscles contract, the exoskeleton moves at its joints; such movements can be exquisitely precise because the force is brought to bear on very small areas, for example, on the different sections of an insect's leg.

The exoskeleton has certain disadvantages. It does not grow (as the bony vertebrate endoskeleton does), and so it must be discarded and re-formed many times as the animal grows and develops. Molting is dangerous; the newly molted animal is soft and hence particularly vulnerable to predators and, in the case of terrestrial forms, subject to water loss. Many arthropods go into hiding until their new cuticle has hardened. Molting is also costly in terms of metabolic expenditures (although a number of insects and some freshwater crustaceans limit their losses by thriftily eating the old exoskeleton).

The fact that the exoskeleton can be waterproofed made possible the evolution of terrestrial forms among the arthropods. Arthropoda is the only invertebrate phylum with many species that are well adapted to withstand the drying action of the air.

Internal Features

Arthropods, like annelids, have a tubular mouth-to-anus gut. They also have a coelom, but the arthropod coelom—like that of the mollusks—is markedly reduced, consisting only of the cavities of the reproductive and excretory organs. (As you will recall, the coelom serves as a hydrostatic skeleton in annelids, but the arthropods, with their exoskeleton, require less internal stiffening.) Like most mollusks, arthropods also have an open circulatory system in which blood flows through free spaces in the tissues—the hemocoel—as well as through vessels. Blood returns from the hemocoel to the tubular heart through special valved openings.

The insects and some other terrestrial forms have an unusual means of respiration, consisting of a system of cuticle-lined air ducts, known as *tracheae*, that pipe air directly into the various parts of the body (Figure 26–4a). Gases must pass through the tracheae largely by diffusion, thus placing a limit on the size of insects. Some terrestrial arthropods, such as spiders, have book lungs, either in addition to or instead of tracheae. Excretion in terrestrial forms is by means of *Malpighian tubules* (Figure 26–4b), which are attached to and empty into the midgut or hindgut.

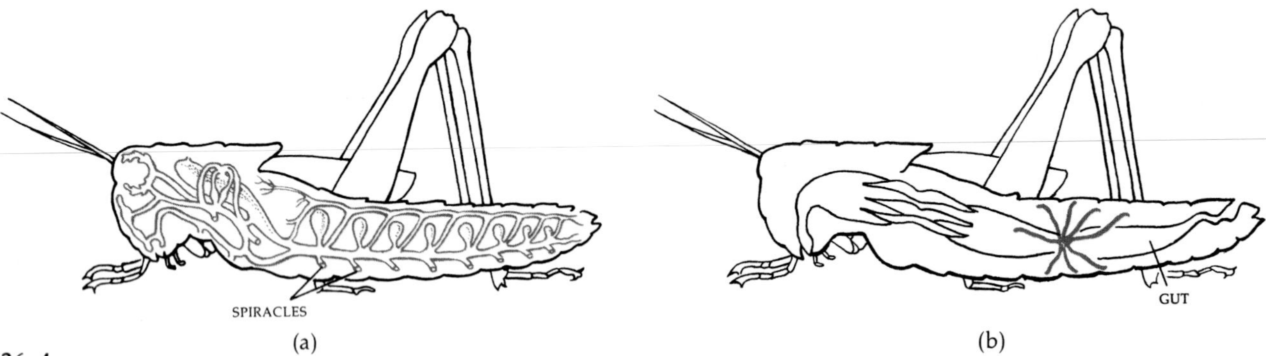

26–4
(a) Respiration by means of a system of internal tubes (called tracheae) is found almost exclusively among terrestrial arthropods. Tracheae are usually branched, like those shown here, and open to the outside by spiracles that may be closed to conserve water. The tubes are lined with spiral rings and cuticle, which keeps them open. A tracheal system is one of the most efficient respiratory systems in the animal kingdom because it delivers oxygen directly to the cells.

(b) Malpighian tubules represent another exclusively arthropod characteristic, although like tracheae, they are not found in all classes. These tubules collect water and nitrogenous wastes from the hemocoel and empty them into the gut. The wastes (in the form of uric acid or guanine) are excreted with the feces.

These tubules absorb wastes from the body cavities. Respiration by tracheae, book lungs, or book gills and excretion by Malpighian tubules are found almost exclusively in the arthropods (although not all arthropods have these features).

The Arthropod Nervous System

The bulk of the arthropod nervous system consists of a double chain of segmental ganglia running along the ventral surface (Figure 26–5). The double chains part to encircle the esophagus, ending in three fused pairs of dorsal ganglia. These fused dorsal ganglia, which perform a number of specialized functions, constitute a brain. However, many arthropod activities are controlled at the segmental level, as in annelids. For example, members of a number of species can move, eat, and carry on other functions normally, even after the brain is removed. In fact, in the arthropods generally, the brain appears to act not so much as a stimulator of the action of the animal but as an inhibitor, as in the earthworm. The grasshopper, for example, can walk, jump, or fly with its brain removed; indeed, the brainless grasshopper responds to the slightest stimulus by jumping or flying.

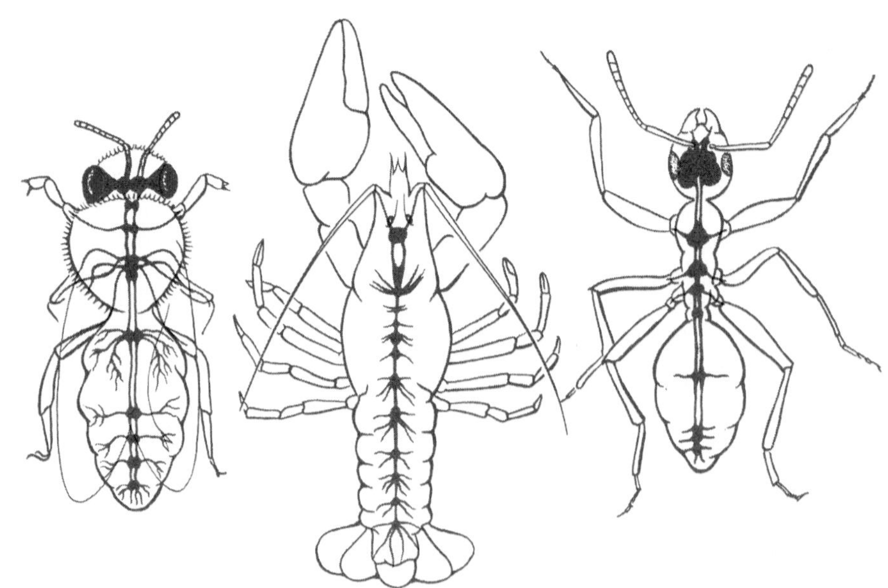

26–5
Arthropod nervous systems: a bee, a crayfish, and an ant. The brain consists of three fused pairs of dorsal ganglia at the anterior end of a double chain of ganglia. The ganglia are interconnected by two bundles of nerve fibers running along the ventral surface. Because of the arthropod's segmental type of nervous system, many arthropod activities are controlled at a local level, and a number of species can carry on some of their normal functions after the brain has been removed.

The extreme consequences of this releasing of inhibition can be seen in the praying mantis. Mantises are carnivorous and cannibalistic, and the female, being larger than the male, frequently captures her mate and, grasping him with her forelegs, begins to eat him, head first. This decapitation results in the release of strong motor activities by which the headless male struggles loose from the grasp of the female, mounts her, and mates with her. The headless male is more likely to copulate than an intact male; investigators seeking to breed mantises have found that males of strains that do not mate readily in captivity will often do so after decapitation.

SUBDIVISIONS OF THE PHYLUM

The Chelicerates

As we noted previously, chelicerates have neither antennae nor mandibles. The first pair of appendages, the chelicerae, which bear pincers or are sharp and fang-like, are used for biting prey. The second pair are the *pedipalps,* which may bear pincers, be modified as walking legs, or serve as sense organs. Posterior to the pedipalps are a series of jointed walking legs. The body segments have become fused into two tagmata—an anterior cephalothorax and a posterior abdomen, which is unsegmented in most chelicerates but conspicuously segmented in scorpions. Except for some mites, all chelicerates are carnivorous. The sexes are almost always separate.

Two classes of chelicerates are relatively small. Class Merostomata consists of only four species of horseshoe crabs (Figure 26–6). In viewing a horseshoe crab from below, one can see the chelicerae, pedipalps, and four pairs of walking legs. Posterior to the walking legs are a series of flaplike book gills, derived from modified limbs. *Limulus,* a bottom dweller that feeds upon such small animals as annelids and clams, is common in the shallow waters along the East Coast of the United States.

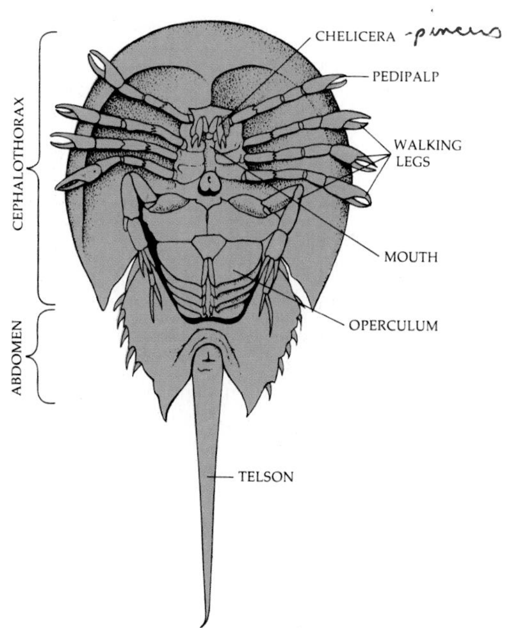

26–6
Because the body of the horseshoe crab is largely covered with a heavy shield, or carapace, its segmented body plan is only evident from the underside. The operculum is a flat, movable plate that covers and protects the book gills. Horseshoe crabs are often called living fossils because their remarkably similar ancestors date back to the Silurian period. There are now only four living species.

26–7
A male sea spider, Nymphon gracile, *carrying two large masses of developing eggs. Among pycnogonids, the eggs are always carried by the males on a subsidiary pair of specialized legs. Sea spiders have neither respiratory nor excretory systems; gases and waste products are thought to diffuse through the large surface area provided by the long, thin legs.*

The approximately 500 species of sea spiders (class Pycnogonida) have slender bodies and four (or, rarely, five) pairs of legs, which are often very long (Figure 26–7). They feed on soft-bodied invertebrates—particularly cnidarians—by sucking juices from their bodies. Although sea spiders are common in coastal waters, most species are quite small and inconspicuous.

All of the remaining chelicerates belong to class Arachnida, with about 57,000 species.

Class Arachnida

The arachnids, which include spiders, ticks, mites, scorpions, and daddy longlegs, are almost all terrestrial, except for a few species that have returned to the water. Arachnids have four pairs of walking legs, and their chelicerae and pedipalps are often highly specialized. In spiders, ducts from a pair of poison glands lead through the chelicerae, which are sharp and pointed and are used for biting and paralyzing prey (Figure 26–8). Scorpions use their pedipalps to handle and to tear food. Male spiders also use the pedipalps to transfer semen to the female.

Spiders, like most arachnids, live on a completely liquid diet. All are predatory. The prey is bitten and often paralyzed by the chelicerae, and then enzymes from the midgut are poured out over the torn tissues to produce a partially digested broth. The liquefied tissues of the prey are pumped into the stomach and then to the intestine, where digestion is completed and the juices absorbed. Arachnids respire by means of tracheae or book lungs, or both. Book lungs are a series of leaflike plates within a chitin-lined chamber into which air is drawn and expelled by muscular action.

On the posterior portion of the spider's abdominal surface is a cluster of spinnerets, modified appendages from which a fluid protein exudes that polymerizes into silk as it is exposed to air. Silk is used not only for the variety of prey-snaring webs made by the different species but for a number of other purposes as well—such as a drop line, on which the spider can make a defensive dive, draglines for marking a course, gossamer threads for ballooning, hinges for trap doors, an egg case, lining for a burrow, the shroud of a victim, or a wrapping for an edible offering presented to the female of certain species by the courting male. Most spiders can spin several kinds and thicknesses of silk. Webs are species-specific, and web-building is a genetically programmed behavior.

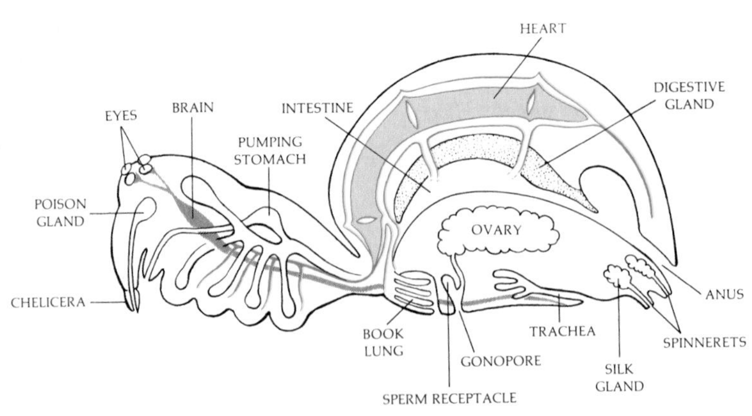

26–8
A spider, an arachnid. Ducts from the poison gland open at or near the tips of the chelicerae. The flow of poison is voluntarily controlled by the spider. Only a few spiders are dangerous to human beings; perhaps the most dangerous are members of the species shown here, the black widow.

(a)

(b)

(c)

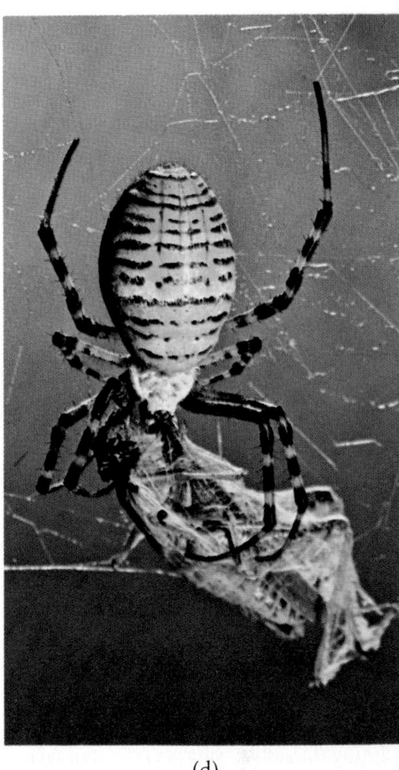

(d)

26–9
Some arachnids. (a) *A scorpion belonging to the genus* Centruroides, *a highly toxic genus. Scorpions sting in self-defense, releasing a venom that attacks the nervous systems of their victims.* (b) *A wolf spider. These common spiders have eight eyes, four small ones in a row beneath four large ones. Note that the segments of the body have become fused into two regions—cephalothorax and abdomen.* (c) *An adult red velvet mite of the genus* Trombidium. (d) *A banded garden spider* (Argiope trifasciata) *has caught a grasshopper in its web. The prey, immobilized by venom, is being wrapped in silk. Webs, which may be as much as a meter in diameter, are reconstructed daily.*

The Aquatic Mandibulates: Class Crustacea

The 25,000 species of crustaceans include crabs, crayfish, lobsters, shrimp, prawns, barnacles, *Daphnia* (water fleas), and a number of smaller forms. Some crustaceans, such as the familiar pillbugs, or sowbugs, are adapted to life in moist land environments. Crustaceans differ from the terrestrial mandibulates, such as insects, in that they have legs or leglike appendages on the abdomen as well as the thorax and have two pairs of antennae as compared to the insects' one pair.

Among the crustaceans, the sexes are usually separate, but there are exceptions. Barnacles, which are sessile as adults, are simultaneous hermaphrodites. Some species of shrimp are sequential hermaphrodites; when the animal is small, it is male, but when it reaches a size that is effective for carrying eggs, it becomes female. Most marine crustaceans have larval stages that swim about before developing into the adult animals.

The Lobster

Figure 26–10 shows the structure of a lobster, a representative crustacean. A crayfish, which is a freshwater form, differs morphologically from the lobster only in minor respects.

The lobster has 19 segments, the first 13 of which are united on the dorsal side in a combined head and thorax, or cephalothorax. A heavy shield, or carapace, arises from the head and covers the thorax. Carapaces are common among crustaceans. The abdomen consists of six distinct segments. (To lobster eaters, the abdomen is the "tail," but to biologists, the designation "tail" is usually reserved for areas of the body posterior to the anus.)

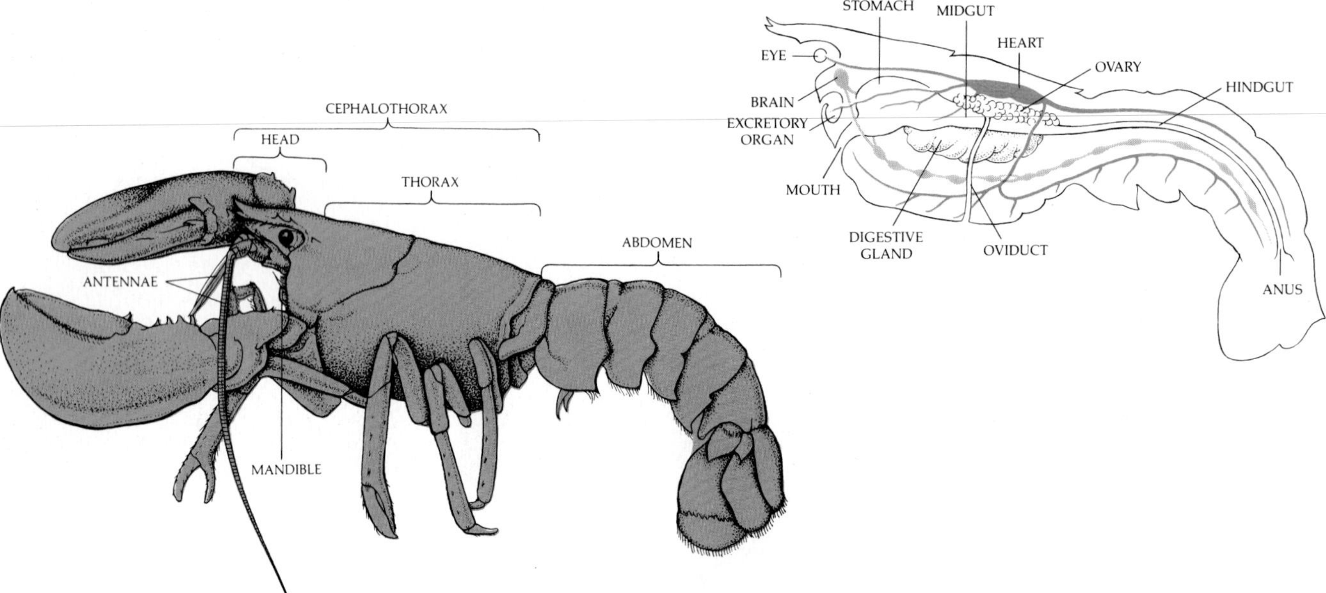

26–10
A representative crustacean, the American lobster, Homarus americanus. *The lobster has 19 pairs of appendages, including antennae, mouthparts, and legs specialized for feeding, walking, and swimming.*

The various appendages have special functions. The antennae, of which there are two pairs, are sensory. The mandibles, or jaws, are used for crushing food; like all arthropod jaws, they move laterally, opening and closing from side to side like a pair of ice tongs rather than up and down like the jaws of vertebrates. Crustacean mandibles are thought to have arisen from a pair of limbs, the bases of which took on a chewing function. The maxillae and maxillipeds are also modified limbs and serve chiefly to collect food, mince it, and pass it on to the mouth. The thorax of the lobster bears five pairs of walking legs, the first pair of which are modified as claws. The claws are unequal in size in the full-grown lobster; the larger claw is used for defense and for crushing food, and the smaller claw, which has the sharper teeth, seizes and tears the prey. The next two pairs of walking legs have small pincers, which can also seize prey. The last pair of walking legs also serve to clean the appendages of the abdomen. These flattened appendages, or swimmerets, are used, like flippers, for swimming. The claws and abdomen are filled almost entirely with the large striated muscles many of us enjoy eating. These muscles are extremely powerful. A lobster can snap its "tail" ventrally with enough force to shoot backward through the water, and a large lobster can shatter the shell of a clam or oyster with its crushing claw. Any one of its appendages can be regenerated over a series of molts if lost, and a lobster will often drop off (autotomize) a claw or leg that is held by a predator, in order to escape. Severely damaged legs are also autotomized, reducing blood loss.

The anterior and posterior regions of the digestive tract, the foregut and hindgut, are lined wth cuticle. As a consequence, most of the food is absorbed through the midgut and through the cells of the large digestive gland, an appendage of the midgut. Respiration is accomplished by the flow of water over 20 pairs of feathery gills attached on or near the bases of the legs. A pair of excretory organs is located in the head; wastes extracted from the blood are collected into a bladder and excreted from a pore at the base of each of the second antennae.

When lobsters mate, the sperm are deposited in or near the female gonopore; the eggs are then fertilized as they are laid. The fertilized eggs cling by means of a sticky secretion to the swimmerets of the female until they hatch.

(a)

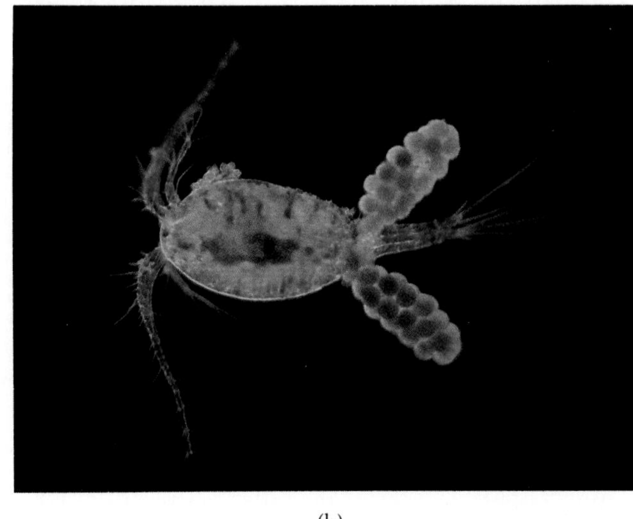

(b)

(c)

26–11

Some crustaceans. (a) *A cleaner shrimp of the Indian and Pacific Oceans, photographed near the Philippine Islands.* (b) *A copepod, an inhabitant of the plankton. Copepods are the most numerous animals not only in the plankton but also in the world; the individuals of a single genus* (Calanus) *are thought to outnumber all other animals put together.* (c) *Pillbugs (also known as sowbugs or wood lice) are terrestrial crustaceans found in damp places.*

Other crustaceans include the barnacles (Figure 24–3, page 482) and the crabs (Figure 26–3b, page 529). Crabs resemble lobsters, but the relative proportions of the body have changed; the cephalothorax is broader and larger than the abdomen, which is tucked under the rest of the body.

Terrestrial Crustaceans

Unlike other arthropod groups, almost all crustaceans are aquatic; some crabs, however, are amphibious or terrestrial. Amphibious crabs continue to respire with gills, carrying around water in their thoracic cavities with which to keep the gills wet. The true land crabs have lost some of the gill structures but have an area of highly vascularized epithelial tissue through which oxygen is exchanged. The land snails, you will recall, solved the respiratory problems involved in the transition from water to land in an analogous way.

The Terrestrial Mandibulates: Myriapods

Terrestrial mandibulates are identified by their single pair of antennae and by their mandibles, which, as we noted earlier, differ from those of the crustaceans. They respire through tracheae, and excretion is by means of Malpighian tubules. In addition to the insects, there are four smaller classes of relatively unspecialized terrestrial mandibulates. There has been little tagmosis in these arthropods, and their bodies consist of a head region followed by an elongated trunk with many distinct segments, all more or less alike. With a few exceptions, all segments have paired appendages, and the members of these four classes are known collectively as myriapods ("many-footed").

The most familiar myriapods are the centipedes (class Chilopoda) and the millipedes (class Diplopoda). The approximately 3,000 species of centipedes prefer damp places—under logs or rocks or in basements. They are all carnivorous, feeding on cockroaches and other insects, as well as on soft-bodied annelids. A centipede (Figure 26–12) has one pair of appendages on each body segment, unlike a millipede (see Figure 26–2, page 528), which appears to have two pairs of appendages per segment. Each body ring of the millipede, however, actually represents two segments fused into a double unit. About 7,500 species of millipedes have been described, and they too prefer damp environments. Unlike centipedes, they are herbivorous, usually feeding on bits of decaying vegetation.

26–12
A centipede of the genus Scolopendra, *photographed in southern Africa. In centipedes, the appendages of the first segment are modified as poison claws. Prey is killed with the claws and then chewed with the mandibles.*

Less familiar are the approximately 300 species of class Pauropoda and the 130 species of class Symphyla. Unlike other arthropods, they are soft-bodied, living in moist soil, leaf litter, and other decaying matter. Although both pauropods and symphylans are abundant, they are quite small and, with the exception of one species of symphylan that is a common pest in greenhouses, they usually go about their business unnoticed by human eyes.

The Terrestrial Mandibulates: Class Insecta

The insects constitute the largest class, by far, of the arthropods. In fact, more than 70 percent of all animal species on earth are insects.

Insects are the only invertebrates capable of flight. When insects began to fly—more than 240 million years ago—they were able to move into and exploit a life zone almost totally unoccupied by any other form of animal life. In terms of both numbers of species and numbers of individuals, they are the dominant terrestrial organisms of this planet.

(a)

(b)

26–13
The late J. B. S. Haldane, who was noted for his crusty disposition, as well as for his scientific achievements, was once asked what his study of biology had revealed to him about the mind of God. "Madame," he replied, "only that He had an inordinate fondness for beetles."

Shown here is everybody's favorite beetle, a ladybird ("ladybug") (a), *accompanied by a stag beetle* (b), *so called because of the antlerlike mandibles characteristic of the pugnacious males.*

There are about 30 orders of insects, of which we shall briefly describe the four largest: Diptera, Lepidoptera, Hymenoptera, and Coleoptera. The Diptera ("two-winged") include the familiar flies, gnats, and mosquitoes. The Lepidoptera ("scale-winged") are the moths and the butterflies. Hymenoptera ("membrane-winged") include ants, wasps, and bees, many species of which are social. The Coleoptera ("shield-winged") are the beetles, most of which have a pair of hard protective forewings, which pivot forward out of the way during flight (see Figure 26–1, page 527), and a pair of membranous hindwings used for flying. Of the more than 750,000 classified species of insects, at least 275,000 are beetles.

Insect Characteristics

Figure 26–14 shows a grasshopper. Here you can see many of the characteristic features of insects: three main divisions—the head, the thorax, and the abdomen; three pairs of legs; one pair of antennae; and a set of complex mouthparts. In the less specialized insects, such as the grasshopper, the mouthparts are used for handling and masticating food (Figure 26–15), but in the more highly specialized groups, the mouthparts are often modified into sucking, piercing, slicing, or sponging organs. Some are exquisitely adapted to draw in nectar from the deep, tubular nectaries of specialized flowers.

Most adult insects have two pairs of wings made up of light, strong sheets of chitin; the veins in the wings are chitinous tubules that serve primarily for strengthening. In some primitive orders, wings never evolved, and in other orders, such as fleas and lice, wings were lost secondarily, returning the insect to the condition of its wingless ancestors. Other species may have short, nonfunctional wings in one or both sexes.

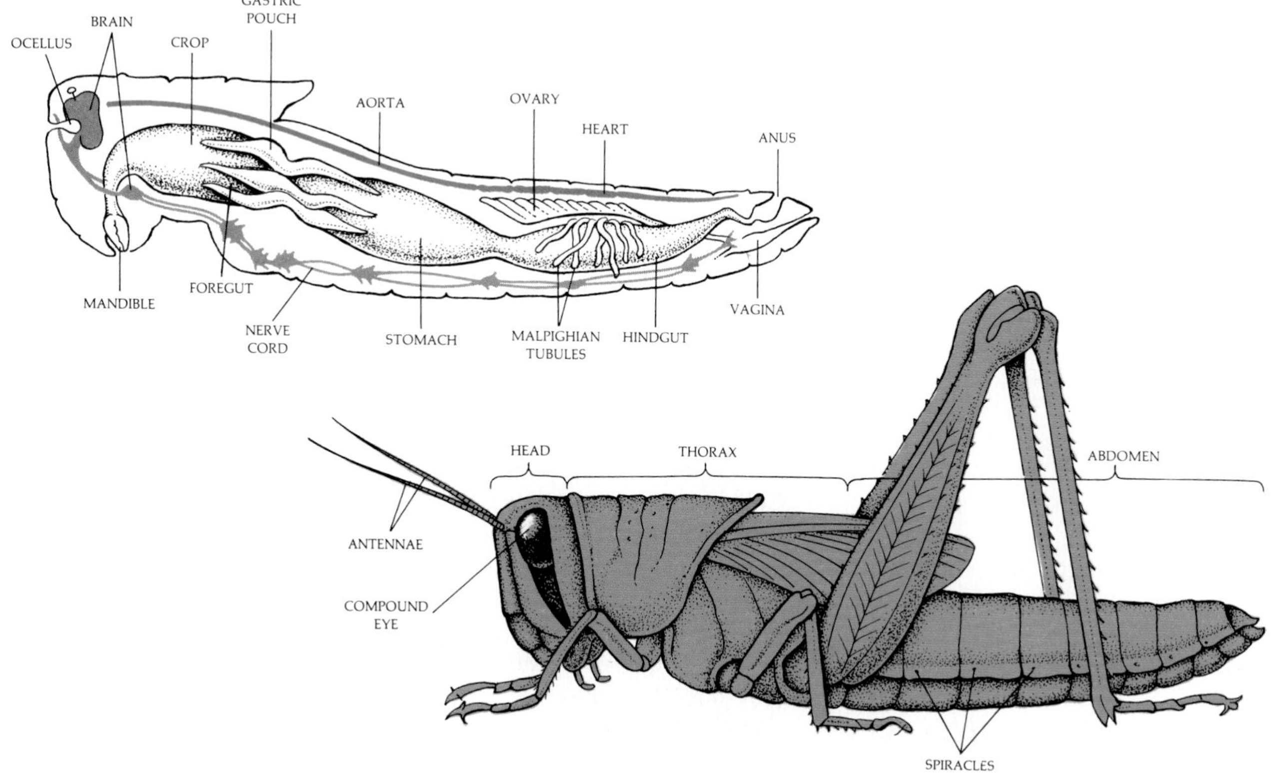

26–14
In the grasshopper, an insect, the head consists of six fused segments that have appendages specialized for tasting and biting. Each of the three segments of the thorax carries a pair of legs (three pairs in all), and two of them carry wings (in the grasshopper, the forewings are hardened as protective covers). The spiracles in the abdomen open into a network of chitin-lined tubules through which air circulates to various tissues of the body. This sort of tubular breathing system is found in the terrestrial mandibulates and some arachnids. Excretion takes place by Malpighian tubules that empty into the hindgut.

26–15
Mouthparts of a grasshopper. The mandibles are crushing jaws. The labium and the labrum are the lower and upper lip. The maxillae move food into the mouth, and the palpi assist in tasting.

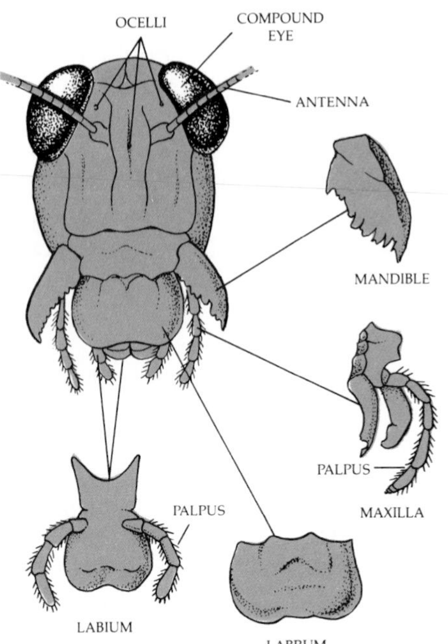

Digestive, Excretory, and Respiratory Systems

The foregut and hindgut of the insect digestive tract are lined with cuticle. Salivary gland fluids are carried with food into the crop, where digestion begins. The stomach, or midgut, which lies mainly in the abdomen, is the chief organ of absorption. Insects have digestive enzymes as specialized as their mouthparts; the structure of the enzymes depends on whether the insect dines on blood, seeds, other insects, eggs, flour, cereal, glue, wood, paper, or your woolen clothes.

Excretion is carried out through Malpighian tubules. In the grasshopper and many other insects, the nitrogenous wastes are eliminated in the form of nearly dry crystals of uric acid, an adaptation that promotes water conservation.

The respiratory system consists of a network of cuticle-lined tubules through which air circulates to the various tissues of the body, supplying each cell directly. Muscular movements of the animal's body improve the internal circulation of air. The amount of incoming air and also the degree of water loss is regulated by the opening and closing of the spiracles.

Insect Life Histories

The life histories of insects are fundamentally different from those of marine invertebrates, particularly forms that are sessile or slow-moving as adults. Although many marine invertebrate larvae feed, their primary biological function seems to be the invasion and selection of new habitats at some distance from the parental habitat. Such larvae may migrate great distances, resulting in the dispersal of the species over a wide area. The immature forms of most insects, by contrast, have limited mobility and usually pass through the various stages of their development very close to the location where the adult female originally laid the eggs—

26–16
Some immature insect forms. (a) *Scarab beetle (white grub) larva in soil.* (b) *Black swallowtail larva (caterpillar).* (c) *Mosquito larvae and a pupa (on right). Mosquito larvae are aquatic, hanging to the undersurface of the water with respiratory tubes.* (d) *Tent caterpillars on their protective web.*

(a)

(b)

sometimes in the exact spot. Insect young are voracious feeders, acquiring the resources needed for their own growth and development and, in some species, storing up reserves for an adult stage in which they will not feed. The capacity to fly has, of course, given most adult insects an extraordinary mobility, and they have little difficulty finding mates or locating new habitats in which to lay their eggs—thereby ensuring dispersal of the species. One corollary of their mobility is that the sexes are separate in all insects. In many insects, some degree of parental care has evolved, with one or both parents protecting or feeding the young, or both. Among the termites, ants, wasps, and bees, such care has led ultimately to complex societies with a substantial division of labor, as we shall see in Chapter 55.

Young, growing insects change not only in size but often in form, a phenomenon known as metamorphosis. The extent of change varies. In some species, the young insect, although sexually immature, looks like a small adult; it grows larger by a series of molts until it reaches full size. In others, like the grasshopper, the newly hatched young are wingless but may gain wingpads in later immature stages; otherwise they are similar to the adult. These immature, nonreproductive forms are known as *nymphs*. Almost 90 percent of the insects, however, undergo a complete metamorphosis, in which adults are drastically different from their immature forms. These immature feeding forms are all correctly referred to as *larvae*, although they are also commonly known as caterpillars, grubs, or maggots, depending on the different species. Following the larval period, the insect undergoing complete metamorphosis enters a pupal stage within which extensive remodeling of the organism occurs. The adult (sexually mature) insect emerges from the pupa. Both eggs and pupae (which are nonfeeding) can endure lengthy cold or dry seasons, a critical adaptation for terrestrial organisms.

An insect that undergoes complete metamorphosis exists in four different forms in the course of its life history. The first form is the egg and the embryo. The second form is the larva, the animal that hatches from the egg; larvae eat and grow.

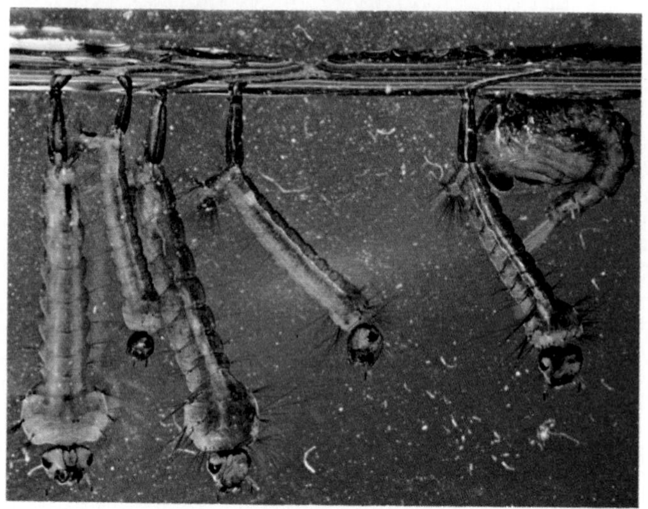

(c)

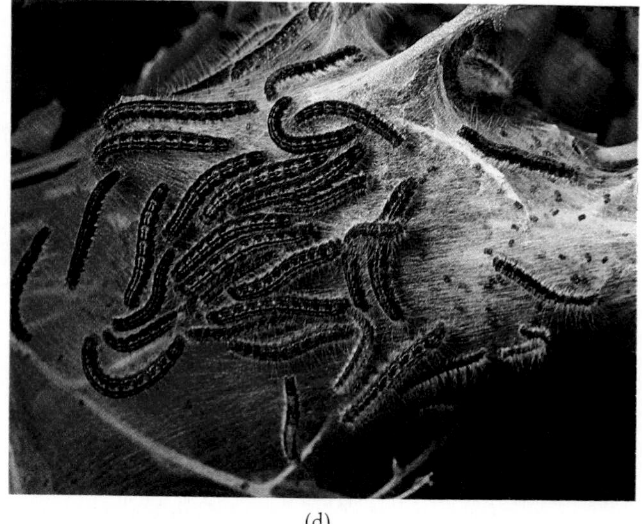

(d)

(a)  (b) 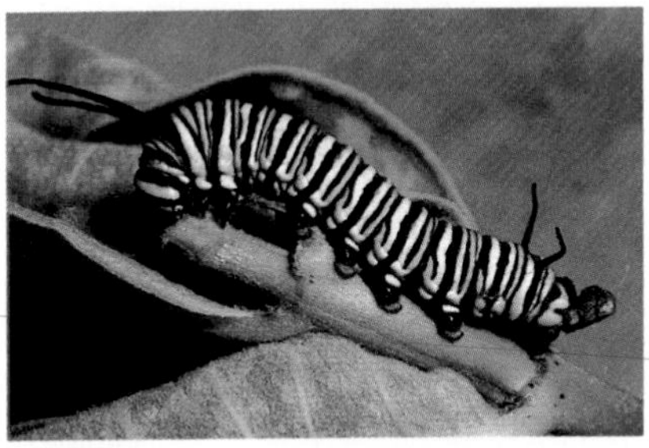 (c)

(d) (e) (f)

(g) (h) (i)

26–17

Developmental stages of a monarch butterfly. (a) Larva hatching and (b) eating its rubbery egg shell. (c) A caterpillar at the fifth molt. (d) When ready to pupate, the caterpillar attaches itself to a leaf with a patch of silk. Its skin splits, revealing the pupa. (e) Inside the pupa, development of the butterfly begins by recycling the tissues of the caterpillar. (f) The pupating butterfly remains inside the gold-studded chrysalis for two to three weeks. (g) The butterfly forces its swelling thorax out of the pupa and (h) pulls free, forcing blood into its wings. (i) The butterfly rests until its wings harden and it can fly.

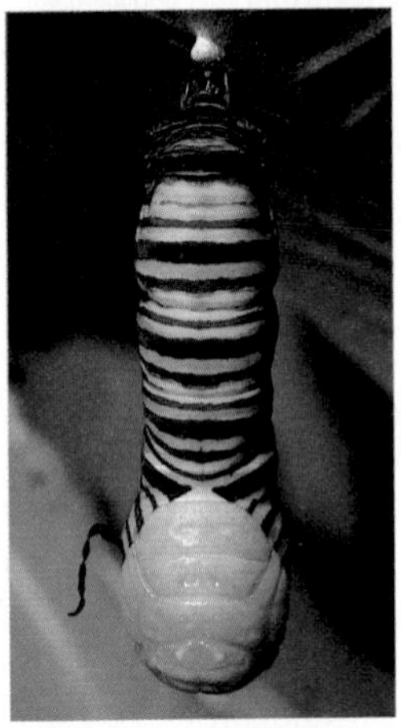

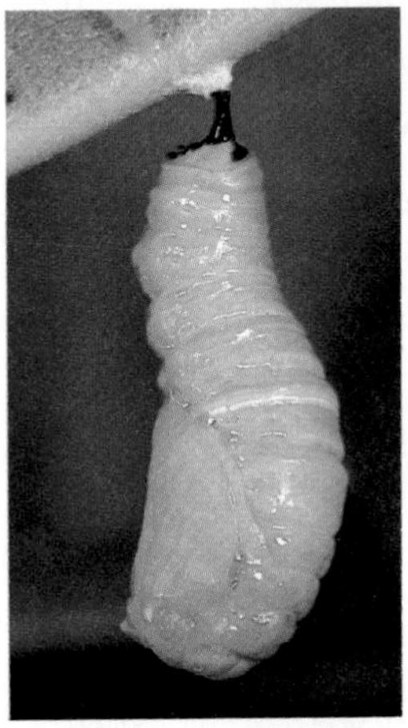

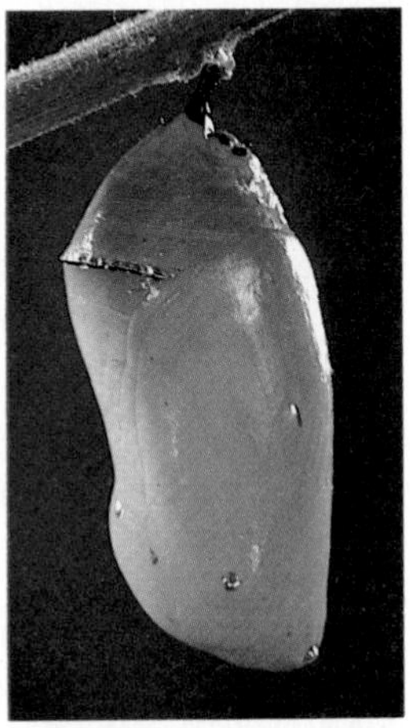

In many larvae, such as those of flies, growth takes place not by an increase in the number of cells, as in most animals, but by an increase in the size of the cells, in somewhat the same way that growth takes place in certain plant tissues. During the course of its growth, the larva molts a characteristic number of times—twice in the fruit fly, for example. The stages between molts are known as *instars*. Then, when the larva is full-grown, it molts to form the pupa. During this outwardly lifeless pupal stage, many of the larval cells break down, and entirely new groups of cells, set aside in the embryo, begin to proliferate, using the degenerating larval tissue as a culture medium. These groups of cells are known as imaginal discs since they form the imago, the adult insect, which, according to Aristotle, is the perfect form or ideal image that the immature form is "seeking to express." These imaginal discs develop into the complicated structures of the adult.

Molting and metamorphosis are under hormonal control (Figure 26-18). Like many hormone-controlled processes, they are the end result of an interplay of several hormones: brain hormone, molting hormone (ecdysone), and juvenile hormone. At intervals during larval growth, brain hormone, produced by neurosecretory cells in the brain, is released into the blood. It stimulates the release, in turn, of molting hormone from a gland in the thorax. The molting hormone stimulates not only molting but also the formation of a pupa and the development of adult structures. The latter are held in check, however, by a third hormone, the juvenile hormone. Only when the production of juvenile hormone declines, in later larval life, can metamorphosis to the adult form take place.

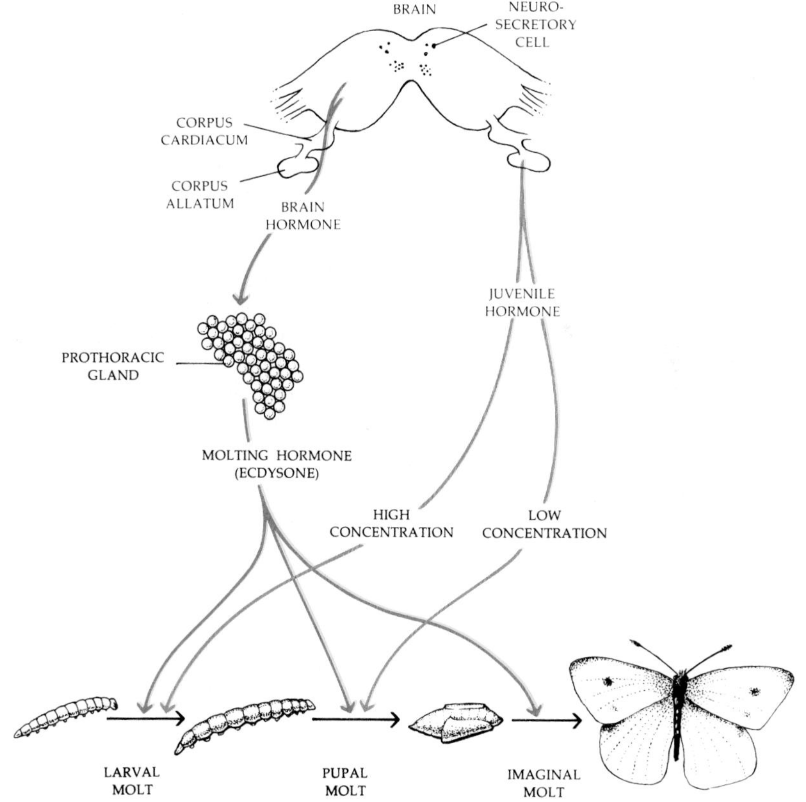

26-18
Molting is under hormonal control. In insects, a hormone (brain hormone) produced by neurosecretory cells in the brain and released from corpora cardiaca stimulates the prothoracic gland, which, in turn, produces molting hormone (ecdysone). Although all molts require ecdysone, whether or not metamorphosis occurs depends on a third hormone, juvenile hormone, produced by the corpora allata. Continued presence of juvenile hormone at high concentrations ensures that larval molts occur during the first portion of the life history. In later larval life, production of juvenile hormone declines, permitting adult structures to develop.

REASONS FOR ARTHROPOD SUCCESS

Among all the invertebrates, why are the arthropods, in general, and the insects, in particular, so spectacularly successful? One important reason is undoubtedly the nature of the exoskeleton, which waterproofs, provides protection, and makes possible the evolution of the many finely articulated appendages characteristic of this phylum.

Another reason, which applies especially to insects, is their small size and the high specificity of diet and other requirements of each species. As a consequence, many different species can live in a single small environment—in a few cubic centimeters of soil, on a small plant, or on or within the egg or body of a single animal—without competing with one another. The varied and highly specialized mouthparts are a reflection of this specificity of diet.

The diversity of insects and the specificity of their requirements may be, in part, an evolutionary response to the great diversity of microenvironments provided by the vascular plants. Not only are there 235,000 species of angiosperms alone, but also the structure of each individual plant is so complex that it provides a variety of resources that can be exploited by insects with different requirements and adaptations. This is a much richer environment than open waters, the seashore, or the soil. And, virtually every cubic centimeter of it is available to insects because of the extraordinary mobility that results from their small size and their capacity for flight. The resources provided by vascular plants were, in effect, an evolutionary laboratory in which great numbers of variations could be tested, leading to the diversity of adaptations we see today. This process, of course, continues.

Another factor in the success of insects is the complete metamorphosis that occurs in the vast majority of insect species. In these insects, the adaptations for feeding and growth that are found in the larvae are separated from the adaptations for dispersal and reproduction found in the adults. The adaptations for each function can be more finely honed because compromises between conflicting needs are minimal. Another consequence of complete metamorphosis is that the dietary and other requirements of the larvae and the adults of the same species are so different that they do not compete with each other. In effect, they occupy different environments.

A final reason for success is undoubtedly the arthropod nervous system, with its fine control over the various appendages and the many extraordinarily sensitive sensory organs found in great diversity throughout the phylum. Sensory perception among arthropods, especially insects, is so important that we shall devote the remainder of this chapter to that topic and to several examples of arthropod behavior.

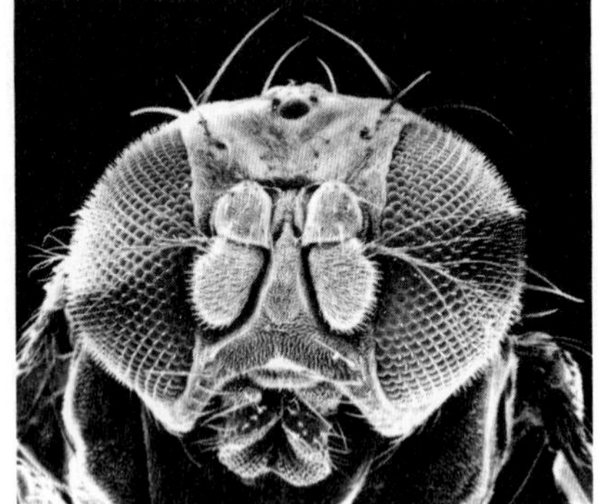

26–19
The head of Drosophila, *as shown by the scanning electron microscope. Note the large compound eyes on either side of the head. Although insect eyes cannot change focus, they can define objects only a millimeter from the lens, a useful adaptation for an insect.*

Arthropod Senses and Behavior

Vision: The Compound Eye

The most conspicuous sensory organ of the arthropods is the compound eye, which is an evolutionary development characteristic of this one phylum. The basic structural unit of this eye is the *ommatidium* (Figure 26–20). A dragonfly has some 30,000 ommatidia. Each ommatidium is covered by a cornea, usually with a round or hexagonal surface; these are visible under low-power magnification as individual facets of the eye. Underlying the cornea is a group of eight retinular cells surrounded by pigment cells. The light-sensitive portion of the ommatidium is the

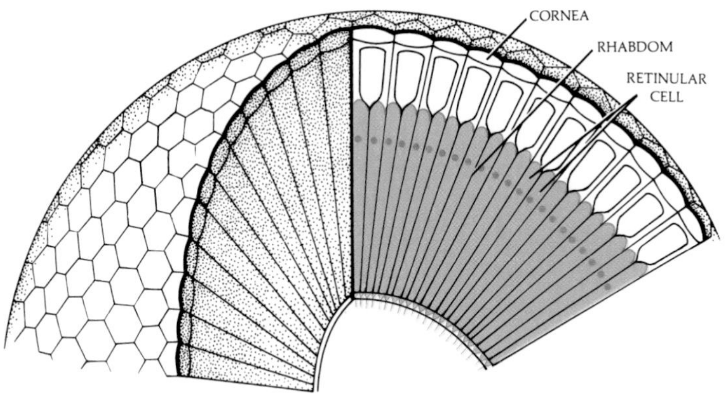

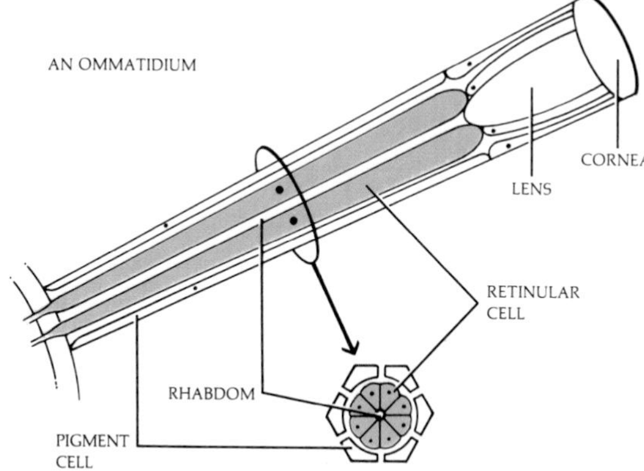

26–20
Structure of the compound eye. The eye is composed of a large number of structural and functional units called ommatidia. Each ommatidium has its own cornea, which forms one of the facets of the compound eye, and its own light-focusing lens. The light-sensitive part is the rhabdom, which is surrounded by the retinular cells, which transmit the stimulus. The ommatidium is surrounded by pigment cells that prevent light from traveling from one ommatidium to another.

rhabdom, which is the central core of the ommatidium. Nerve fibers carry the stimulus from each ommatidium to the brain. The pigment cells prevent light from traveling from one ommatidium to another. An ommatidium is much larger than a vertebrate photoreceptor, and so there are far fewer in an equivalent space. Hence, the image has less resolution, like a newspaper picture under high magnification.

Although the compound eye is deficient in acuity, offering less detail than the vertebrate eye, it is better for detecting motion because each ommatidium is stimulated separately and so has a separate visual field. Also, each ommatidium responds to stimuli more rapidly than does a vertebrate photoreceptor. Ability to detect motion can be measured accurately in the laboratory by testing a phenomenon known as flicker fusion. In this test, a light is flicked on and off with increasing rapidity until the observer sees the flicker as a continuous beam. The beam is perceived as continuous because stimulation of any retinal cell persists for a brief period even after the stimulus disappears. So, in effect, the flicker-fusion test is a measurement of how quickly the photoreceptor cell recovers from one stimulus and becomes sensitive to another. It is possible to test flicker-fusion rates in animals by training experiments in which the animal learns to associate a flickering light with a reward (usually food) and a steady beam with no reward, or vice versa. Such tests have proved that the compound eye greatly exceeds the camera eye of vertebrates in this respect. A bee would see in clear outline a moving figure that we would see as blurred, and if the bee went to the movies, the film seen by us as a continuous picture would jerk along from frame to frame for the bee. The ability to perceive motion is extremely important for an insect since it must be able to make out objects when it is flying at high speed (which, as far as the visual apparatus is concerned, presents the same problems as following a moving object).

In addition to, or instead of, compound eyes, many of the arthropods possess simple eyes, or ocelli, which seem generally to serve only for light detection. Most insects have two or three ocelli, and spiders, which do not have compound eyes, may have as many as eight ocelli, depending on the species.

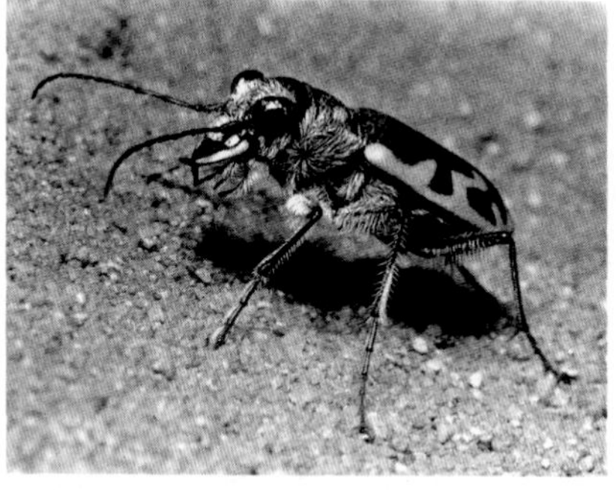

26–21
The hairs on the legs of this tiger beetle are touch receptors, or sensilla. At the base of the hairs are sensory cells. When a hair is touched or bent, nerve impulses are initiated.

Touch Receptors

The body surfaces of terrestrial arthropods are often covered with sensory-receptor units known as sensilla (singular, sensillum), or "little sense organs." Most of the sensilla take the form of fine spines, or setae, composed of hollow shafts of cuticle.

FIREFLY LIGHT: AN ADVERTISEMENT, A WARNING, A SNARE

(a) *Two fireflies of the species* Photuris hebes *mating as they hang beneath a goldenrod leaf.* (b) *A female of the genus* Photuris *devours a male of the species* Photinus tanytoxus.

Fireflies, like many other insects, are beetles (of the family Lampyridae). As is characteristic of other higher insects, they undergo a complete metamorphosis. The eggs are laid in moist soil, hatching into larvae in about three weeks. (Even the embryos glow a little.) They spend up to two years as larvae; the larvae are carnivorous, feeding on annelids, insects, and mollusks, which they subdue with poison from their mandibles. The larvae are also luminescent, producing a glow that waxes and wanes over a period of seconds.

After molting through a number of instars, the larvae pupate over a period of two to three weeks. During this time the adult luminescent organs—lanterns—develop on the ventral surface of the terminal abdominal segments. The adult firefly that emerges from the pupa is, of course, the reproductive form. It lives only a relatively short time—one to four weeks—and its function during this period is to find a mate and reproduce.

The survival value of the larval luminescence is not clear. The best hypothesis is that the larvae taste bad and that the glow is a warning signal to predators. Compounds related to the skin poisons of toads have been found in some species. Experiments reported by Albert Carlson of the State University of New York at Stony Brook have shown that mice that bite into larvae reject them violently, flinging the carcass away and scrubbing their mouths with their forefeet. It has also been shown that mice can be readily trained to avoid glowing larvae.

In the adult firefly, the luminescent organs are under neural control and are used to solve the all-important (for the firefly) problem of finding a mate. Each species of firefly has its own signal pattern composed of light flashes lasting only a fraction of a second; the male signals of 130 species have been recorded. Males emit their signals in flight; the females are stationary, near or on the ground. Females flash in response to the signals from males of their own species, aiming their lantern toward the flashing male. When the male sees the answering flash, he alights and walks toward the female, still flashing. In some species, this courtship flashing is different from the previous seeking pattern of flashing. Copulation lasts minutes to hours, after which the female lays her eggs. The male immediately resumes his signaling and searching activities, looking for another mate. Thus, on any given summer evening, there are many more males than females seeking mates—perhaps a ratio of 50 to 1—and competition among males is very intense.

Unlike the larvae, most adult flies are not carnivorous. Some species do not feed at all. Females of the genus Photuris, *however, are exceptions. They are predaceous carnivores, and their prey is the male firefly of other species. They attract them by mimicking the signals of the females of the males' own species.*

At the bases of these shafts are sensory cells. In their simplest form, the sensilla are touch receptors. In these, when the hair is touched or bent, the sensory cell responds and initiates nerve impulses. Such receptors are found, in particular, on the antennae and the legs. In addition to being stimulated by direct contact, they can also be stimulated by vibrations and air currents. A spider monitors what is going on in its web by sensing vibrations transmitted through the threads when the web is touched. Soldier termites of certain species strike the ground or the walls of their nest with their heads when threatened or disturbed; the vibrations they produce warn their colony mates. A fly perceives the air currents from the movement of a hand or fly swatter and so escapes; a fly in a glass jar is much less likely to be disturbed by such movements.

(a)

(b)

Photuris versicolor is one of the most extensively studied. Within three days after mating, the female undergoes a behavioral change. Instead of responding only to the flashing pattern of a male of her own species, she becomes an active hunter, flying to an area where the prey species is active. When a male firefly flashes nearby, she responds—and not with her own signal but with a signal characteristic of the female of the male's own species. A Photuris versicolor female can respond appropriately to the male signals of at least five different firefly species. The male's dilemma is acute. If he hesitates, he loses his chance to mate, given the 50 to 1 competition. If he rushes in, he risks being devoured by the aggressive female.

Males of the Photinus macdermotti species have responded to this dilemma by finding a means of stalling for time while holding off the competition. These tiny males are preyed upon by the females of at least three different Photuris species, so their incentives are particularly strong. Their weapons are also false flashes. As they approach a female, they emit flashes that mimic the female predator of the other species, thereby deterring the rival males of their own species. Also, by injecting their flashes into the flash patterns of rival males, they can disrupt their signals and get the females to respond to them rather than their rivals.

James Lloyd, of the University of Florida, who has been a leader in these studies for almost 20 years, reminds us that on the same summer evenings that the fireflies are engaged in these rituals of courtship, mating, deceit, treachery, and death, thousands of other species of insects are also occupied in similar activities. Are fireflies more complex than other insects, he asks, or are comparable but unilluminated dramas taking place throughout the insect world?

Proprioceptors

Proprioceptors are sensory receptors that provide information about the position of various parts of the body and the stresses and strains on them. A type common in the arthropods is the campaniform sensillum (Figure 26–22a). Campaniform sensilla are located in thin, stretchable areas of the cuticle. When the cells are twisted or stretched, a nerve fiber signals the central nervous system.

Touch receptors can also serve as proprioceptors. The praying mantis, for example, is capable of making a lightning-swift strike at a moving object. When it sights a potential victim, the insect moves its entire head to bring it into binocular range, since the eyes themselves do not move (Figure 26–22b). Movement of the head results in the stimulation of proprioceptive hairs on the head and thorax of

26–22
(a) Campaniform sensillum, a type of proprioceptor common in arthropods.

(b) Since its eyes do not move, the praying mantis must move its entire head to bring its victim into binocular range. The movement of its head sends impulses through proprioceptive hairs on its head and thorax to its legs. The position of its prey and the movement of its legs are thus automatically coordinated, giving the mantis the ability to strike at a victim swiftly and at the proper range.

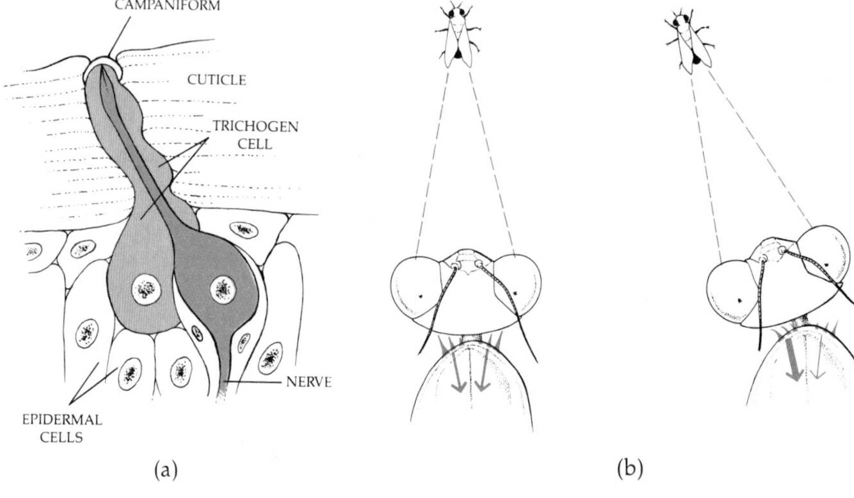

the insect. On the basis of the impulses received from these hairs, the position of the prey and the movement of its own legs are automatically coordinated by the mantis. If these hairs are removed, the mantis can strike a moving object only if the object is directly in front of it.

Communication by Sound

Arthropods, particularly insects, have developed complex forms of sensory communication. A number of species, such as the locusts, grasshoppers, and crickets, call to one another by sounds made by rubbing their legs or wings together or against their bodies (Figure 26–23). Five distinct types of calls are known: (1) calling by males and (2) calling by females, both of which are long-range sounds; (3) courtship sounds by males and (4) aggressive sounds by males, both of which are short-range; and (5) alarm sounds, which may be given either by males or by females. Recognition of and response to the sound may be based on its pattern, rhythm, or frequency (pitch). Some, but not all, insects are unable to distinguish frequencies, or the differences between high and low notes, and so are essentially "tone deaf." The effectiveness of calling songs is often increased by group singing, such as the famous chorus of male seventeen-year cicadas, which can attract females from distances far greater than an individual "voice" would reach. Insects produce songs and respond to appropriate songs without ever having heard a song before.

26–23
The katydid produces its characteristic loud, shrill sounds by rubbing the scraper at the base of its right wing against the file at the base of its left wing.

Arthropods have a variety of sensory receptors that enable them to detect the calls produced by other members of their species or the sounds of approaching predators. Many moths, in particular, have a remarkable ability to detect the sounds of predatory bats. Sound waves are set in motion by the vibration of some object, for example, a beating wing, a flexible area of an arthropod's skeleton, a human larynx, or a violin string. The vibrations not only cause the molecules of the surrounding medium (air or water) to travel away from the object in small bursts but also produce a series of changes in air (or water) pressure. The sound receptors of arthropods may be sensitive either to the impact of the moving molecules or to the pressure changes.

26–24
The most elaborate of the insect sound receptors is the tympanic organ. The tympanic air sacs are covered by a membranous drum, and the sensory cells are so arranged in the organ that they are stimulated by movements of the drum or the air-sac walls. Tympanic organs respond to pressure changes in the air or in water. This is a cross section of the tympanic organ of a noctuid moth.

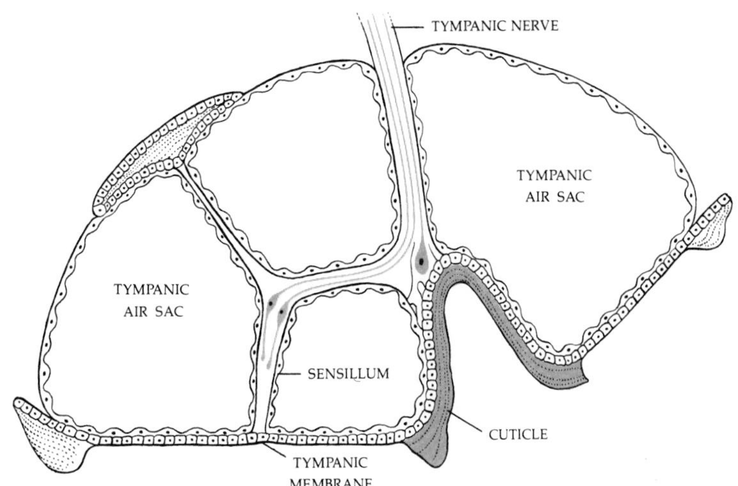

The simplest of the arthropod sound receptors are sensilla with tactile hairs that vibrate when they are struck by moving molecules in the air; the vibrations of these hairs, in turn, trigger nerve impulses in the animal. The antennae of the male mosquito, for example, contain thousands of such hairs, which are sensitive to the sounds produced by the vibrating wings of the female mosquito in flight. The response of the male to these sounds serves to bring the sexes together. When the male mosquito first emerges from its pupal shell, it is sexually immature and also deaf, with its antennal hairs lying flat along the shafts. When the male matures sexually, some 24 hours later, the hairs almost simultaneously become erect and are now free to vibrate when a female approaches.

Other insects have developed special structures that respond to the pressure changes in sound waves; these structures, which may be located on the legs (see Figure 26–3a), the thorax, or the abdomen, are known as tympanic organs. In these organs, a fine membrane, the tympanum (or eardrum), is stretched across one or more closed air sacs (Figure 26–24). The tympanic membrane vibrates in response to the pressure differences in sound waves of certain frequencies, and this vibration is transmitted to underlying receptor cells. In many insects, the tympanic air sacs abut portions of the tracheal system and, in some species, are adjacent to the air sacs of the tympanic organ on the opposite side of the body. As a result, the pressure changes of the sound waves are received by the tympanic membrane not only on its outer surface but also on its inner surface, after transmission from the outside through the tracheae or the opposite tympanic organ. The differences in the pressures on the two sides of the membrane enable the insect to detect the direction from which the sound is coming.

26–25
The two large appendages extending from either side of the head of this cricket are palpi, which contain specialized chemoreceptors (taste organs).

Communication by Pheromones

The use of chemicals for communication is common among organisms, and the substances employed range from the sex attractants of the little algal cell *Chlamydomonas* to Chanel No. 5. Many insects communicate by chemicals known as *pheromones*. These chemical messengers, usually produced in special glands, are discharged into the environment, where they act on other members of the same species.

26–26
The lavishly plumed antennae of the male Cecropia moth are receptors for the alluring pheromones released by the females.

Among the best studied of the pheromones are the mating substances of moths. Female gypsy moths, by the emission of minute amounts of a pheromone commonly known as disparlure, can attract male moths that are several kilometers downwind. One female produces about one-millionth (10^{-6}) of a gram of disparlure, enough to attract more than a billion males if it were distributed with maximum efficiency. The male moth characteristically flies upwind, and the pheromone, of course, disperses downwind. When a male moth detects the pheromone of a female of the species, he will fly toward the source. Since the male can detect as little as a few hundred molecules per milliliter of the attractant, disparlure is still potent even when it has become widely diffused. If the male loses the scent, he flies about at random until he either picks it up again or abandons the search. It is not until he is quite close to the female that he can fly "up the gradient" and use the intensity of the odor as a locating device. Figure 26–26 shows the antennae of a male Cecropia moth by which the pheromone emitted by the female is detected.

Programmed Behavior

Complex patterns of unlearned, genetically transmitted behavior—such as web-building in spiders—are another arthropod characteristic. This rigid programming of behavior may be a necessary correlate to the shortness of the life span of the smaller arthropods. It provides an interesting contrast to the more flexible behavior patterns of the mammals, with their comparatively long life spans, long periods of learning, and more complex brains.

SUMMARY

Arthropoda constitutes the largest animal phylum in both number of species and of organisms. Arthropods are segmented animals with jointed chitinous exoskeletons and a variety of highly specialized appendages and sensory organs. In most groups, the segments are combined, forming a head, a thorax (sometimes combined with the head as a cephalothorax), and an abdomen. The arthropods are also characterized by an open circulatory system and a nervous system consisting of a series of ganglia, a pair per segment, interconnected by a double ventral nerve cord. Tracheae (cuticle-lined breathing tubes), book gills, book lungs, and Malpighian tubules (excretory ducts leading into the hindgut) are found almost exclusively among arthropods.

There are three major groups of arthropods: the chelicerates, characterized by chelicerae (fangs or pincers) and pedipalps; the aquatic mandibulates, with two pairs of antennae and a pair of mandibles (jaws); and the terrestrial mandibulates, with one pair of antennae and a pair of mandibles that differ from those of the aquatic mandibulates. The chelicerates include the horseshoe crabs (class Merostomata), the sea spiders (class Pycnogonida), and the spiders, scorpions, mites, and ticks (class Arachnida). The aquatic mandibulates (some of which actually live in moist environments on land) all belong to class Crustacea and include such familiar animals as lobsters, crabs, shrimp, and barnacles. Terrestrial mandibulates include four relatively small classes (Chilopoda, Diplopoda, Pauropoda, and Symphyla) and the largest class in the animal kingdom, Insecta, with more than 750,000 species. The insects are the only invertebrates capable of flight.

In the life histories of most insects, dispersal and habitat selection are carried out by the highly mobile adult forms, whereas the immature forms have limited mobility and feed voraciously in a fairly restricted area. In the course of their development, most insects pass through a complete metamorphosis that is controlled by the interaction of at least three hormones. The stages in the life history are egg, larva (caterpillar, maggot, grub, etc.), pupa (sometimes with a cocoon or some other protective covering), and adult. In a minority of species, including the grasshopper, the hatchling looks much like a miniature adult; in these, the immature form is known as a nymph.

Among the factors contributing to the extraordinary success of the arthropods are their exoskeleton, their generally small size, and their great specialization in both diet and habitat. Additional factors in the success of the insects are the capacity for flight and complete metamorphosis, which allows for greater refinement of adaptations for feeding and for reproduction and dispersal, as well as reducing competition between adults and immature forms.

The finely tuned arthropod nervous system, with its diverse sensory organs, has also been important in arthropod success. Among the most important sensory receptors are the compound eye, touch receptors, proprioceptors, and tympanic organs. Arthropods communicate with members of the same species by sound and also by pheromones—chemicals released by one individual that affect the behavior or physiology of another. Arthropod behavior is notable not only for its complexity and diversity but also for the extent to which it is programmed in the nervous system, that is, not learned.

QUESTIONS

1. Distinguish among the following terms: chelicerae/mandibles; nephridia/Malpighian tubules; nymph/larva/pupa; hormone/pheromone; brain hormone/molting hormone/juvenile hormone; ommatidium/sensillum.

2. Describe each of the following arthropod structures and explain its function: book gills, tracheae, spiracle, spinnerets, compound eye, tympanic organ.

3. Note that both the shelled mollusks and the arthropods have an open blood system. Name an obvious feature shared by the members of these two groups. What is the correlation between these two structural similarities?

4. A larval or medusoid stage occurs in the life histories of many marine invertebrates. Yet in group after group of freshwater and terrestrial invertebrates, each supposedly separately evolved from marine ancestors, the free-living larva or medusa has been lost. What might be the selective factor underlying the loss?

5. How would you distinguish an insect from an arachnid? From a crustacean?

6. Describe respiration in a clam, a terrestrial snail, an earthworm, a lobster, an arachnid, and an insect. How does respiration in these animals differ from that in a cnidarian? How is it similar?

7. Compare the nervous systems of a clam, an octopus, an annelid, and an arthropod. How do they differ from those of a *Hydra* and a planarian?

8. The arthropods, which include some of the most active animals, have open circulatory systems, often considered inefficient. Annelids, from which arthropods may have evolved, have closed circulatory systems. Presumably, half of the annelid system must have been lost. How could the acquisition of a relatively rigid exoskeleton make superfluous the vessels returning blood to the heart?

CHAPTER 27

The Animal Kingdom IV: The Deuterostomes

Based on characteristic features of embryonic development, the coelomate animals fall into two broad groups. In the protostomes—the subject of the two preceding chapters—the cleavage pattern in the early cell divisions of the embryo is spiral, the mouth develops at or near the blastopore, and the coelom results from a splitting of the mesoderm (schizocoelous formation). By contrast, in the deuterostomes, the early cleavage pattern is radial, the anus develops at or near the blastopore, the mouth forms secondarily elsewhere, and the coelom is formed by outpocketings of the embryonic gut (enterocoelous formation).

These features of embryonic development are shared by four strikingly different phyla of animals: Echinodermata (starfish, sea urchins, and related forms), Chaetognatha (arrow worms), Hemichordata (acorn worms and their relatives), and Chordata. The largest subphylum of Chordata includes all the vertebrate animals—fish, amphibians, reptiles, birds, and mammals, among them *Homo sapiens*.

PHYLUM ECHINODERMATA: THE "SPINY-SKINNED" ANIMALS

27–1
An echinoderm, a feather star. According to available evidence, echinoderms evolved from bilaterally symmetrical, mobile animals into radially symmetrical, sessile forms, such as this.

The echinoderms include the sea lilies and feather stars (class Crinoidea), starfish (class Asteroidea), brittle stars (class Ophiuroidea), sea urchins and sand dollars (class Echinoidea), and sea cucumbers (class Holothuroidea). The majority of the echinoderm species are known only through fossils, but the 6,000 living species are abundant throughout the oceans of the world. Particularly in deep waters, echinoderms often make up the bulk of living tissue.

Most adult echinoderms are radially symmetrical, like most cnidarians, but the symmetry is imperfect with some traces of bilaterality. The larvae, however, are bilaterally symmetrical; their metamorphosis into radially symmetrical adults occurs as different parts of the body grow at different rates. The echinoderms are believed to have evolved from an ancestral, bilateral, mobile form that settled down to a sessile life and then became radially symmetrical. The feather stars (Figure 27–1) represent this second hypothetical stage. In the third evolutionary stage, some of the animals, as represented by the starfish and sea urchins, became mobile again. Following this line of reasoning, one might expect an eventual return to bilateral symmetry in this group, and, in fact, this is seen to some extent in the soft, elongated bodies of sea cucumbers.

27-2
Echinoderms are characterized by a calcium-containing skeleton of rods, plates, or spicules embedded just below the skin, a water vascular system of canals, and tube feet. The five-part body plan, clearly visible in the starfish shown here, is also characteristic of all echinoderms.

The characteristic features of this phylum are clearly visible in the most familiar of the echinoderms, the starfish.

Class Asteroidea: Starfish

The body of a starfish consists of a central disk from which radiate a number of arms. Most starfish have five arms, which was the ancestral number, but some have more. Like all echinoderms, a starfish has an interior skeleton that typically bears projecting spines, the characteristic from which the phylum derives its name. The skeleton is made up of tiny, separate calcium-containing plates held together by the skin tissues and by muscles.

The central disk of a starfish contains a mouth on the lower surface, above which is the stomach. A starfish has no head or brain, but rings of nerves around the mouth provide coordination of the arms, any one of which may lead the animal in its sluggish, creeping movements along the sea bottom. Each arm contains a pair of digestive glands and also a nerve cord, with an eyespot at the end. These latter are the only sensory organs, strictly speaking, of the starfish, but the epidermis contains thousands of neurosensory cells (as many as 70,000 per square millimeter) concerned with touch, photoreception, and chemoreception. The sexes are separate in most starfish, and each arm also has its own pair of sperm-producing or egg-producing organs, which open directly to the exterior through small pores. Although reproduction is usually sexual, with external fertilization, some starfish multiply asexually, regenerating whole animals after division of the central disk. A few species can reproduce asexually by regeneration of shed arms.

The coelom forms a complicated system of cavities and tubes and helps provide for circulation. Respiration is accomplished by many small, fingerlike projections, the skin gills, which are protected by spines. Waste removal is carried out by amoeboid cells that circulate in the coelomic fluid, picking up the wastes and then escaping through the thin walls of the skin gills, where they are ejected.

The *water vascular system* (Figure 27-3) is a unique feature of this phylum. This system, which is a modified coelomic cavity, creates a hydrostatic skeleton for the *tube feet*, unusual locomotor structures found only in the echinoderms. Each arm of a starfish contains two or more rows of fluid-filled tube feet, which are interconnected by a central ring and radial canals. Water filling the soft, hollow tubes makes them rigid enough to walk on. At one end of each tube foot is a sucker, and at the other end, a rounded muscular sac, the ampulla. When the ampulla con-

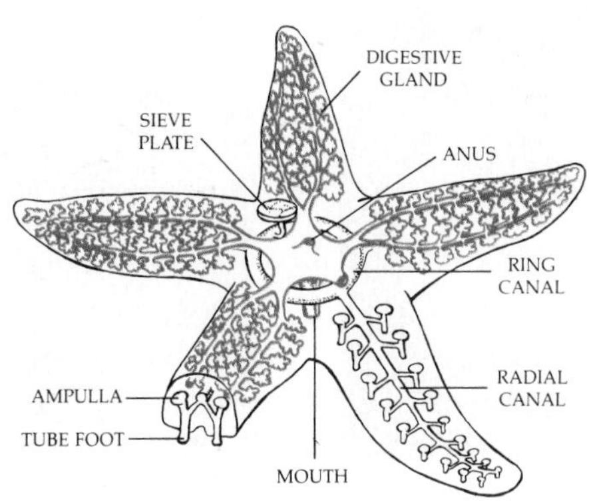

27-3
The water vascular system of the starfish supports its means of locomotion. Five radial canals, one for each arm, connect the ring canal with many pairs of tube feet, which are hollow, thin-walled cylinders ending in suckers. At the other end of each tube foot is a rounded muscular sac, the ampulla. When the ampulla contracts, the water in it, prevented by a valve from flowing back into the radial canal, is forced under pressure into the tube foot. This stiffens the tube, making it rigid enough to walk on, and extends the foot until it attaches to the substrate by its sucker. The muscles of the foot then contract, forcing the water back into the sac and creating the suction that holds the foot to the surface. If the tube feet are planted on a hard surface, such as a rock or a clamshell, their combined force will be great enough to pull the starfish forward or to open the clam.

27–4
Brittle stars, Ophiothrix fragilis.

tracts, the water is forced under pressure through a valve into the tube foot; this extends the foot, which attaches to the substrate by its sucker. When the muscles at the base of the tube feet contract, the animal is pulled forward.

If the tube feet are planted on a hard surface, such as a rock or a clam shell, the collection of tubes will exert enough force to pull the starfish forward or to pull apart a bivalve mollusk, a feat that will be appreciated by anyone who has ever tried to open an oyster or a clam. When attacking bivalves, which are its staple diet, the starfish everts its stomach through its mouth opening and then squeezes the stomach tissue through the opening that it has made between the bivalve shells. The stomach tissues can insinuate themselves through a slit as narrow as 0.1 millimeter to digest the soft tissue of the prey.

Other Echinoderms

The members of the four additional classes of echinoderms show a number of variations on the body plan of the starfish. The sea lilies and feather stars (Figure 27–1), which are thought to be the most ancient of the echinoderms, spend most of their lives either attached or clinging to the substrate. As we have seen with so many sessile animals, the food-gathering structures (in this case, the arms, tube feet, and mouth) are directed upward, enabling the animal to gather food items from the surrounding water.

Brittle stars (Figure 27–4), which are also known as serpent stars, look like starfish with particularly skinny arms. The arms bend from side to side in a snakelike motion, enabling the animal to crawl about on the ocean bottom. Their tube feet, which lack suckers, are used in gathering and handling food, rather than for locomotion. As you might suspect from their name, the arms of brittle stars break easily. These echinoderms, like many crustaceans (see page 533), can autotomize body parts seized by a predator, thereby making good their escape.

In the sea urchins and sand dollars (Figure 27–5a and b), arms are not present, but the five-part body plan is clearly visible in five paired rows of tube feet that extend through the strong skeleton. As in the starfish, the tube feet are used primarily for locomotion. These animals are formidably armed with movable spines and, in some species, with poison glands hidden among the spines. Typically they are also equipped with complex and powerful jaws that are used to graze on algae and other organic materials adhering to the substrate.

27–5
More echinoderms. (a) *A sea urchin;* (b) *a sand dollar, or sea biscuit; and* (c) *a sea cucumber.*

(a)

(b)

(c)

Sea cucumbers (Figure 27–5c), with their relatively soft bodies and partial return to bilateral symmetry, bear little superficial resemblance to the other echinoderms, but they too have five rows of tube feet on the body surface. The calcium-containing skeleton is greatly reduced, usually consisting of ossicles in the skin. The body wall is often tough and leathery. Modified tube feet around the mouth look like tentacles and are used in gathering food. Some sea cucumbers feed on plankton in the surrounding water, while others feed primarily on organic matter in the bottom deposits. Some simply ingest the bottom sediment, extracting whatever nutrients it may contain as it passes through the digestive tract.

PHYLUM CHAETOGNATHA: ARROW WORMS

Although there are only about 60 species in phylum Chaetognatha ("bristle jaws"), these arrow-shaped animals are among the most abundant predators in marine plankton. They range from 1 to 10 centimeters in length and, with a flick of the tail, can shoot forward to capture other small planktonic animals (Figure 27–6). Arrow worms are unsegmented, but their bodies have three distinct regions—head, trunk, and tail. Their embryonic development identifies them as deuterostomes, but they do not seem to be closely related to any other deuterostome phylum. All arrow worms are hermaphroditic, with self-fertilization occurring in some species. The newly hatched young resemble miniature adults and, without passing through a dramatically different larval stage, soon begin a life of active predation.

PHYLUM HEMICHORDATA: ACORN WORMS

Most of the 80 species in phylum Hemichordata are acorn worms (Figure 27–7). As you can see, the body is divided into three regions—a proboscis, with which the animal burrows in ocean sediments in shallow waters, a short collar, and a long trunk. A few species of sessile hemichordates, known as pterobranchs, look more like bryozoans than acorn worms, but they have the same three body regions and share other hemichordate characteristics.

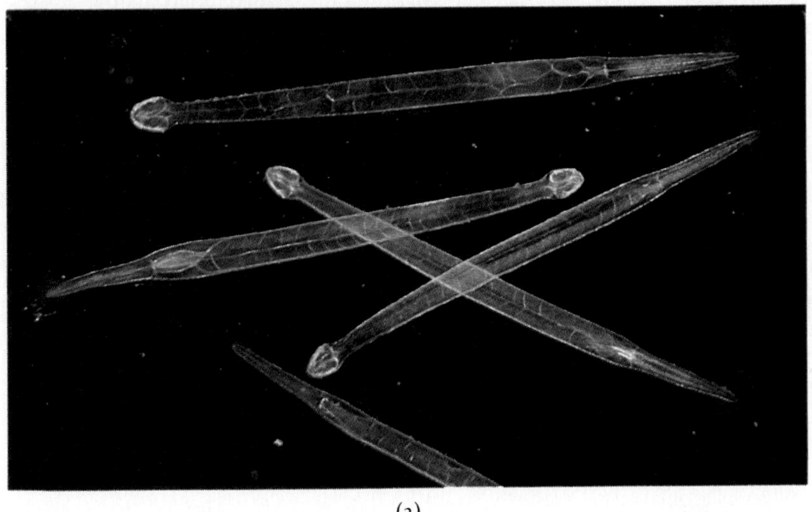

(a)

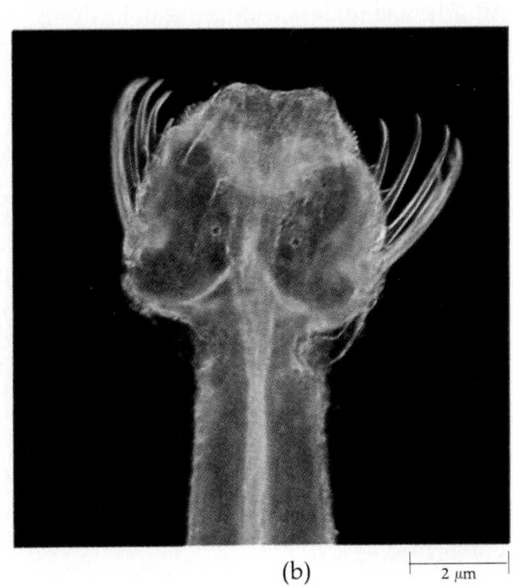

(b)

27–6
(a) *The predatory arrow worms are an important component of marine plankton, feeding on copepods and, occasionally, small fish. They usually feed near the surface at night, descending to deeper waters during the day.*

(b) *An arrow worm is marvelously equipped for its predaceous activities. The head bears two large eyes on the dorsal surface and numerous sharp spines, used to spear prey. Within the chamber leading to the mouth there are strong teeth. This particular arrow worm, however, has been captured by a medusa of the hydrozoan* Obelia. *When photographed, it was dead, its tail well within the coelenteron of its captor.*

27–7
An acorn worm, Glossobalanus sarniensis, *burrowing in shell gravel. Its body consists of three regions—a proboscis, a short collar, and a long trunk, only a portion of which is visible here. Notice the gill slits in the anterior portion of the trunk. Such slits are one of the identifying characteristics of the chordates.*

The hemichordates are of particular interest to evolutionary biologists because they have features characteristic of both the echinoderms and the chordates. Some of the hemichordates have ciliated larvae that are almost identical to the larvae of starfish. Moreover, the coelomic cavities of the tentacles of the pterobranchs provide a hydrostatic skeleton similar to the water vascular system of echinoderms, although there are no tube feet. The hemichordate nervous system is, in some respects, similar to that of the echinoderms, but it includes both ventral and dorsal nerve cords that are joined by a ring at the posterior limit of the collar. Elsewhere in the animal kingdom, a principal nerve cord on the dorsal side of the body is found only among the chordates. The dorsal nerve cord of chordates is hollow, unlike the ventral nerve cords in other animals, which are solid; in some hemichordates, the anterior portion of the dorsal cord is also hollow. The strongest evidence of a close relationship between the hemichordates and the chordates is provided by the pharynx, a structure in the anterior portion of the trunk, which is perforated by holes known as pharyngeal gill slits. As we are about to see, such a pharynx with gill slits is one of the identifying characteristics of the chordates.

PHYLUM CHORDATA: THE CEPHALOCHORDATES AND UROCHORDATES

The phylum Chordata includes some 43,000 species, grouped in three subphyla: the Cephalochordata, or lancelets, which include *Branchiostoma* (formerly called *Amphioxus*); the Urochordata, or tunicates, of which the most familiar are the sea squirts; and the Vertebrata, or vertebrates. Thus the term "invertebrate" refers to all animals except the members of one subphylum of the Chordata.

Subphylum Cephalochordata includes only 28 species. The best-known cephalochordate is *Branchiostoma* (Figure 27–8), a small, blade-shaped, semitransparent animal found in shallow marine waters all over the warmer parts of the world. Although it can swim very efficiently, it spends most of its time buried in the sandy bottom, with only its mouth protruding above the surface. This animal exemplifies all four of the salient features of the chordates. The first is the *notochord*, a rod that extends the length of the body and serves as a firm but flexible axis. The notochord is a structural support. Because of it, *Branchiostoma* can swim with strong undulatory motions that move it through the water with a speed unattainable by the flatworms or aquatic annelids.

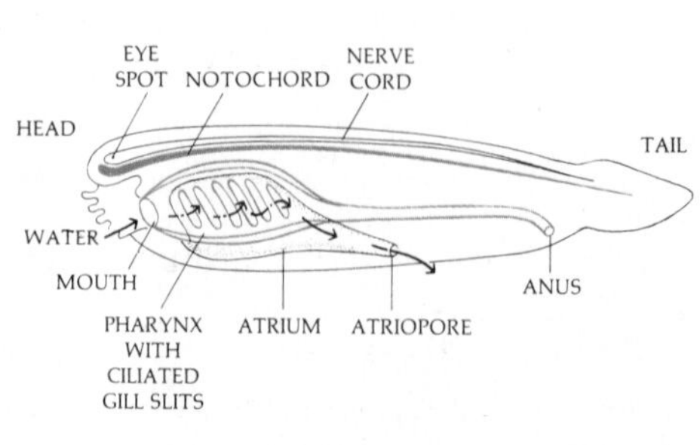

(a)

(b)

27–8
(a) Branchiostoma, *a lancelet, exemplifies four distinctive chordate characteristics: (1) a notochord, the dorsal rod that extends the length of the body; (2) a dorsal, tubular nerve cord; (3) pharyngeal gill slits; and (4) a tail.* Branchiostoma *retains these four characteristics throughout its life; many chordates, however, have a notochord, pharyngeal gill slits, and a tail only during their immature stages.*

(b) *Lancelets, with their anterior ends protruding from the substrate. Much of the body tissue consists of segmental blocks of muscle. The square, yellowish structures along the side of the body are gonads. The sexes are separate in lancelets, and sperm or egg cells are released into the atrium. The gametes pass out through the atriopore, with fertilization occurring externally.*

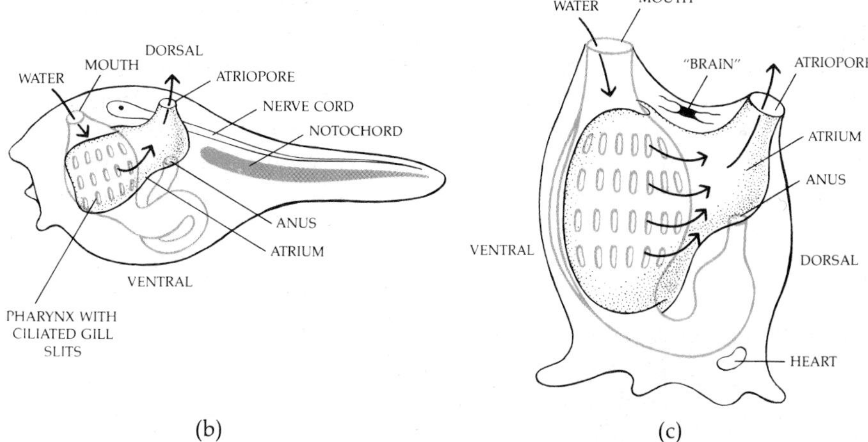

27–9 (a)
(a) *Living tunicates, or sea squirts,* Ciona intestinalis.

Two stages in the life of a tunicate: (b) *the larva, and* (c) *the adult form. After a brief free-swimming existence, the larva settles to the bottom and attaches at the anterior end. Metamorphosis then begins. The larval tail, with the notochord and dorsal nerve cord, disappears, and the animal's entire body is turned 180°. The mouth is carried backward to open at the end opposite that of attachment, and all the other internal organs are also rotated back. As some biologists reconstruct the past, tunicate larvae wriggled up the rivers, where they gave rise, after many generations, to the ancestral vertebrates.*

The second chordate characteristic is the *dorsal, hollow nerve cord,* a tube that runs beneath the dorsal surface of the animal, above the notochord. (The principal nerve cords in other phyla, as we have noted, are solid and are almost always near the ventral surface.)

The third characteristic is a *pharynx with gill slits.* The pharyngeal apparatus becomes highly developed in fishes, in which it serves a respiratory function, and traces of the gill slits remain even in the human embryo. In *Branchiostoma,* the pharynx serves primarily for collecting food. The cilia on the sides of the pharyngeal gill slits pull in a steady current of water, which passes through the slits into a chamber known as the atrium and then exits through the atriopore. Food particles are collected in the sievelike pharynx, mixed with mucus, and channeled along ciliated grooves to the intestine.

The fourth characteristic is a *tail,* posterior to the anus, consisting of blocks of muscle around an axial skeleton. Most of the body tissue of *Branchiostoma* is made up of blocks of muscles, the myotomes.

Although *Branchiostoma* usefully exemplifies the chordates, many biologists believe it is more likely to be a degenerate form of fish rather than a truly primitive member of the phylum. Other possible candidates for the ancestral form are found among the tunicates of subphylum Urochordata. Although adult tunicates do not have all of the typical chordate features, the larvae, which resemble *Branchiostoma,* are clearly chordates possessing the four chordate characteristics (Figure 27–9). About 1,300 urochordate species are known, and they are found in the plankton and on ocean bottoms throughout the world in both shallow and deep waters. They are commonly known as tunicates because the body of the adult is covered by a firm, protective tunic that contains cellulose.

PHYLUM CHORDATA: THE VERTEBRATES

The vertebrates are a large (about 41,700 species) and familiar subphylum of the chordates. All vertebrates have a vertebral column, or backbone, as their structural axis. This is a flexible, usually bony support that develops around the notochord, supplanting it entirely in most species. Dorsal projections of the vertebrae encircle the nerve cord along the length of the spine. The brain is similarly enclosed and protected by the bony skull plates. Between the vertebrae are cartilaginous disks,

27-10
The fetus of a long-legged bat. The bones have been stained red and the cartilage blue, so that you can see the extent to which the skeleton is still cartilaginous. Notice the legs, for example. Only the red areas are bone; these will gradually grow and replace the cartilage as the animal matures.

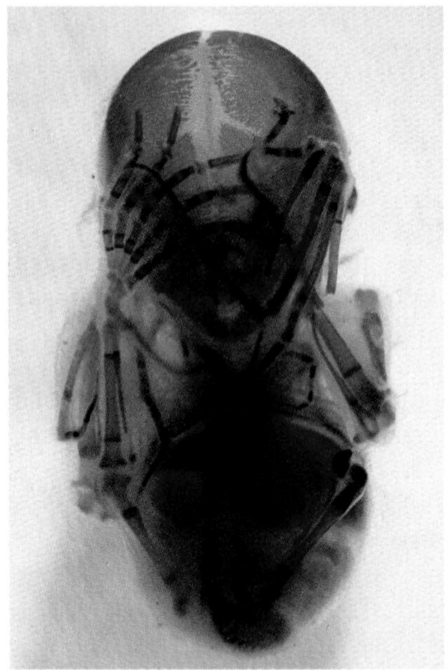

which give the vertebral column its flexibility. Associated with the vertebrae are a series of muscle segments by which sections of the vertebral column can be moved separately. This segmented pattern persists in the embryonic forms of higher vertebrates but is largely lost in the course of development.

One of the great advantages of a bony endoskeleton is that it is composed of living tissue that can grow with the animal. In the developing vertebrate embryo, the skeleton is largely cartilaginous, with bone gradually replacing cartilage in the course of maturation (Figure 27-10). The growing portions of the bones characteristically remain cartilaginous until the animal reaches its full size.

There are seven living classes of vertebrates: the fish (comprising three classes), the amphibians, the reptiles, the birds, and the mammals. Their evolution is clearly documented in the fossil record.

Classes Agnatha, Chondrichthyes, and Osteichthyes: Fish

The first fish were jawless and had a strong notochord running the length of their bodies. Today these jawless fish (class Agnatha), once a large and diverse group, are represented only by the hagfish and the lampreys. They have a notochord throughout their lives, like *Branchiostoma.* Although their ancestors had bony skeletons, modern agnaths have a cartilaginous skeleton. Lacking true bones, they are very flexible; a hagfish can actually tie itself in a knot. Many cyclostomes ("round mouths"), as they are called, are highly predatory, attaching to other fish by their suckerlike mouths and rasping through the skin into the viscera of their hosts. The juvenile lamprey, which resembles *Branchiostoma,* however, feeds by sucking up mud containing microorganisms and organic debris—as, most probably, did the primitive Agnatha.

The sharks (including the dogfish) and skates, the Chondrichthyes, the second major class of fish, also have a completely cartilaginous skeleton. Like agnaths, their ancestors were also bony animals. Their skin is covered with small, pointed teeth (denticles), which resemble vertebrate teeth structurally and give the skin the texture and abrasive quality of coarse sandpaper.

The third major class of fish includes those with bony skeletons, the Osteichthyes. This group includes the trout, bass, salmon, perch, and many others—most of the familiar freshwater and saltwater fish.

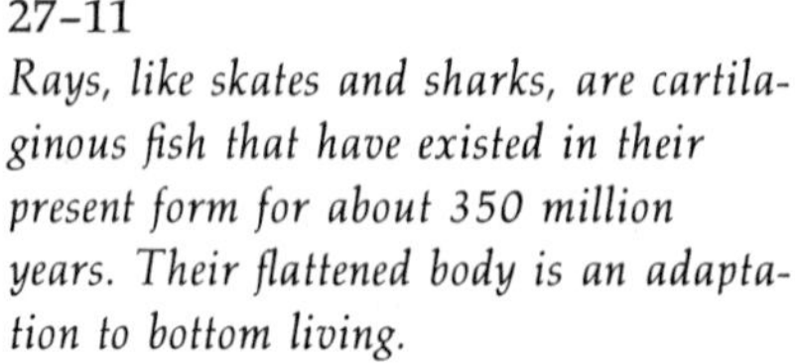

27-11
Rays, like skates and sharks, are cartilaginous fish that have existed in their present form for about 350 million years. Their flattened body is an adaptation to bottom living.

27-12
The heavily armored placoderm is the ancestor of two major classes of present-day fish—the Chondrichthyes (cartilaginous fish) and the Osteichthyes (bony fish).

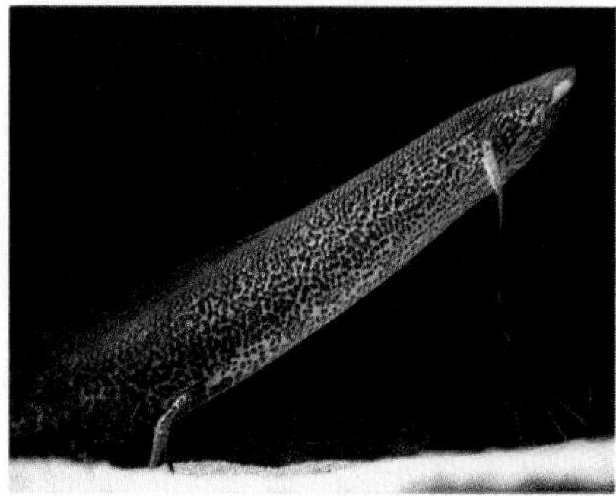

27-13
A modern lungfish. When the dry seasons come, members of this African species wriggle downward into the mud, which eventually hardens around them. Mucus glands under the skin secrete a watertight film around the body, preventing evaporation. Only the mouth is left exposed. During this period, they take a breath only about once every two hours.

According to present evidence, fish evolved in fresh water. The chondrichthyans moved to the sea early in their evolution, while the bony fish went through most of their evolution in fresh water and spread to the seas at a much later period. Some still make this difficult physiological transition in each lifetime. Salmon, for example, hatch in fresh water, spend most of their lives in salt water, and return to fresh water to spawn. At breeding time, eels travel from the fresh waters of Europe and North America to the Sargasso Sea (an area of the south Atlantic), from which distant point the young begin the long, difficult journey, often lasting many years, back to the rivers and lakes.

The Transition to Land

Another characteristic of the bony fishes is that the early forms seem to have had lungs or lunglike structures, as well as gills. These lungs, however, were not efficient enough to serve as more than accessory structures to the gills. They were a special adaptation to fresh water, which unlike ocean water, may become stagnant (depleted of oxygen) because of decay of organic matter or of algal bloom. Lunged fish apparently evolved independently several times, and they were the most common fish in the later Devonian period, a time of recurring drought. In most of them, the lung evolved into an air bladder, or swim bladder; many modern osteichthyans have gas-filled swim bladders that serve as flotation chambers or organs of sound production. A fish raises or lowers itself in the water by adding gases to or removing them from the air bladder via the bloodstream. Still other primitive fish evolved into the modern lungfish (Figure 27-13). These fish can live in water that does not have sufficient oxygen to support other fish life. Lungfish surface and gulp air into their lungs in much the same way that certain aquatic but air-breathing snails bob to the surface to fill their mantle cavities.

In yet others, skeletal supports evolved that served to prop up the thorax of the fish. These fish could gulp air even when their bodies were not supported by water. It is believed that these osteichthyans could waddle, dragging their bellies on the ground, along the muddy bottom of a drying stream bed to seek deeper water or perhaps even make their way from one water source to another one nearby. Thus the transition to land may have begun as an attempt to remain in the water.

Class Amphibia

Amphibians descended from air-breathing lunged fish. Modern amphibians include frogs and toads (which lack tails as adults) and salamanders (which have tails throughout their lives). They can readily be distinguished from the reptiles by their thin, usually scaleless skins, which serve as respiratory organs. Adult frogs also have lungs, into which they gulp air, but some salamanders respire entirely through their skins and the mucous membranes of their throats. Because water evaporates rapidly through their skins, amphibians can die of desiccation in a dry environment. Those found in deserts spend the drier times of the day far below the surface of the sand.

Most frogs in cold climates have two life stages, one in water and the other on land (hence their name, from *amphi* and *bios*, meaning "two lives"). The eggs are laid in water and are fertilized externally. They hatch into gilled larvae (tadpoles). The tadpoles later develop into adults that lose their gills and develop lungs. The adults may live out of the water, at least in the summer. However, there are many

27–14
Many frogs are clearly fishlike in their larval (tadpole) stages. As adults, they require water to reproduce, and their moist skins are an important accessory respiratory organ. Frogs, like all adult amphibians, are carnivores. This green frog, Rana clamitans, *has caught a grasshopper with a flick of its long tongue, which is attached at the front of the mouth and which has a sticky, flypaper-like surface.*

variations on this theme. Some of the American salamanders fertilize their eggs on land; the males deposit sperm packets that are picked up by the females. Many modern amphibians skip the free-living larval stage. The eggs, which may be laid on land, in a hollow log or cupped leaf, or may even be carried by the parent, hatch into miniature versions of the adult. Some salamanders, such as the mud puppy and the axolotl, never complete their metamorphosis, remaining essentially aquatic larvalike forms (see Figure 35–4, page 691), even as sexually mature adults. In some species, these larvalike forms can be induced to metamorphose into "adult" forms by administration of hormones, indicating that the genetic capacity for this later developmental stage has not been lost.

27–15
A midland painted turtle. The dorsal carapace of turtles and tortoises is partly fused to the vertebral column and the ribs, with a mosaic of horny plates on the surface. Unlike other reptiles, most tortoises do not molt but add new epidermal scales on the undersurface. This results in the addition of a growth ring each year.

Class Reptilia

As you will recall, the vascular plants were freed from the water by the evolution of the seed. Analogously, the vertebrates became truly terrestrial with the evolution in the reptiles of the amniote egg (Figure 27–16), an egg that retains its own water supply and so can survive on land. The reptilian egg, which is much like the familiar hen's egg in basic design, contains a large yolk, the primary food supply for the developing embryo, and abundant albumen (egg white), which supplies additional nutrients and water. A membrane, the amnion, surrounds the developing embryo with a liquid-filled space that substitutes for the ancestral pond. The gilled stage is passed in a shelled egg or in the maternal oviduct or uterus. In mammals also, although their eggs typically develop internally, the embryos are enclosed in water within the amnion and pass through a gilled stage (although the gills are never functional) before birth.

Reptiles are characteristically four-legged, although the legs are absent in snakes and some lizards. In keeping with their terrestrial existence, reptiles have a dry skin, usually covered with protective scales. Modern reptiles, of which there are about 6,000 species, include lizards, snakes, turtles, and crocodiles.

27-16

Amniote egg. The membranes, which are produced as outgrowths from the embryo as it develops, surround and protect the embryo and the yolk (its food supply). The egg shell and egg membrane, which are waterproof but permeable to gases, are added as the early embryo passes down the maternal reproductive tract.

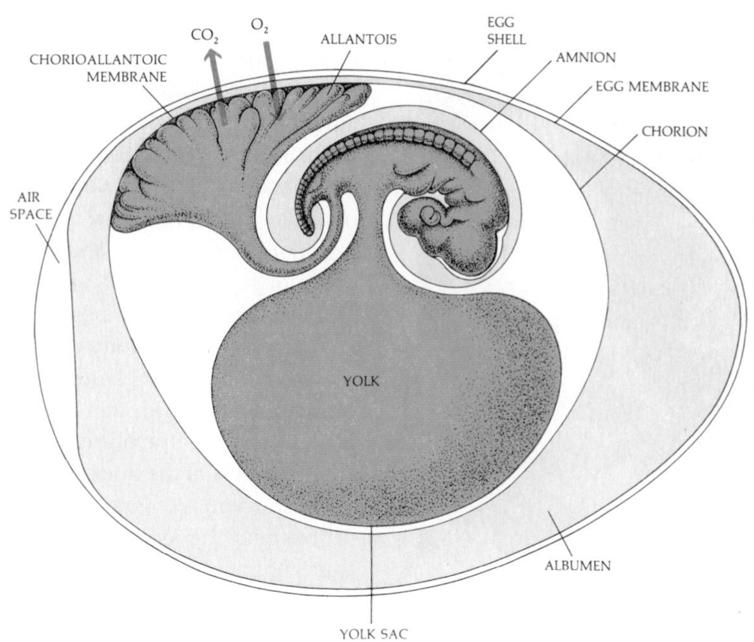

Evolution of the Reptiles

By late in the Carboniferous period (see Table 23-1, page 457), the first reptiles had begun to evolve from closely similar amphibian ancestors. During the succeeding Permian period, there was an explosive increase in the number of reptilian species. (During this same period, conifers began to replace the ferns and other "amphibious" plants, suggesting that a drier climate may have been a primary selective force in both of these events.) During the Permian and much of the Triassic, the dominant land vertebrates were mammal-like reptiles, an abundant and diverse group, members of which were later to give rise to the mammals. Around the beginning of the Triassic period, several of the more specialized reptile groups arose, including turtles, lizards, and the thecodonts, ancestors of the Archosauria, or ruling reptiles, perhaps the most spectacular of all the land's inhabitants so far.

27-17

Alligators and crocodiles, the largest modern reptiles, lay their eggs on land, and their skins are reinforced with epidermal horny scales. Crocodiles, which are essentially tropical animals, have more slender snouts than alligators. Alligators have jaws that are broader and more rounded anteriorly; they are also reported to be less aggressive. The animal shown here is a female American alligator; if you look closely, you can see her young, hitching a ride on mother's tail.

27-18
These newly hatched black snakes are sunning themselves, a behavior characteristic of ectotherms—animals that take in heat from the environment. Black snakes, like all other snakes, are carnivorous, devouring their prey whole.

The origin and rise of the Archosauria were associated with improvements in reptilian locomotion. Vertebrates first came to the land with all four limbs sprawled far out to the side; turtles have retained this sprawling gait. Among the early Archosauria, there was a progressive tendency toward bipedalism, with the concomitant freeing of the front legs for other purposes—flight, for example. There were three major groups of archosaurs: the pterosaurs, or flying reptiles; the crocodilians, which, reverting to (or perhaps retaining) four-legged posture, became our modern crocodiles and alligators; and the dinosaurs, a varied and splendid group of reptiles.

Until recently, the dinosaurs were assumed to have been *ectothermic*—that is, to have regulated their body temperatures within broad limits by taking in heat from the environment or giving it off to the environment. Some biologists are now contending, however, that at least some groups of dinosaurs were *endothermic*—that is, their body temperatures were regulated internally.

The largest of the dinosaurs, *Brachiosaurus,* was 25 meters in length and weighed, it is estimated, 50 tons, far larger than any land animal that has succeeded it. Throughout the long Mesozoic era, the dinosaurs dominated the life of the land, rulers of the earth for 150 million years. Then, they vanished; the reasons for their extinction are not known, but it did occur during a period of climatic change and may have been related to problems of temperature regulation or changes in the vegetation or the disappearance of swampy habitats. They left only a single line of descendants, the birds.

Class Aves: Birds

Birds are essentially reptiles specialized for flight (Figure 27-19). Their bodies contain air sacs, and their bones are hollow. The frigate bird, a large seagoing bird with a wingspread of more than 2 meters, has a skeleton that weighs only 110 grams (about 4 ounces). The most massive bone in the bird skeleton is the keel, or breastbone, to which are attached the huge muscles that operate the wings. Flying birds have jettisoned all extra weight; the female's reproductive system has been trimmed down to a single ovary, and even this becomes large enough to be functional only in the mating season.

Birds have feathers, which is their outstanding, unique physical characteristic. They maintain a high and constant body temperature, which distinguishes them from most of the modern reptiles (although a few, such as leatherback turtles, show some degree of endothermy). In modern birds, feathers make flight possible and also serve as insulation. (Only animals that are endothermic require insulation; insulation would be a disadvantage for animals that warm their bodies by exposure to the environment.) Birds also have scales, a reminder of their reptilian ancestry. Many birds are born at a very immature stage, and virtually all birds require a long period of parental care.

Evolution of Flight

How did flight evolve? Biologists agree that evolution occurs by a series of changes, each of which, to be conserved by natural selection, must be of survival value. Being able to fly not very well is a dubious advantage. Until recently, the most popular theory for the origin of flight has been that the ancestors of the birds were tree-dwelling reptiles and that flight evolved as a way to extend or brake jumps from branch to branch.

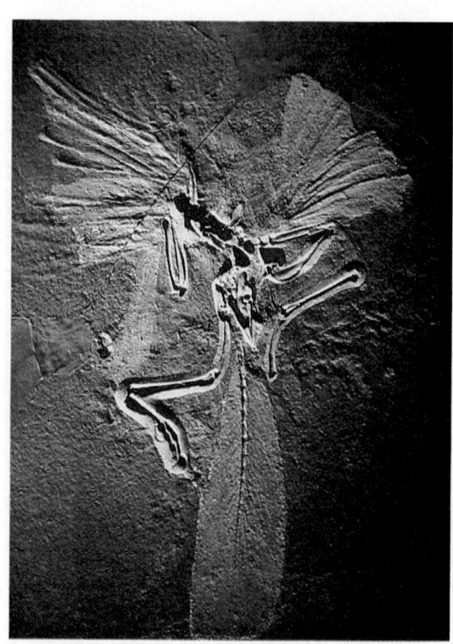

27–19
The oldest known fossil bird, Archaeopteryx, *dates from the middle Jurassic period, about 150 million years ago. It still had many reptilian characteristics. The teeth and the long, jointed tail are not found in modern birds. The clearly evident feathers may have been related as much to endothermy as to flight.*

However, John Ostrom of Yale, having studied the anatomy of the five known specimens of *Archaeopteryx* and many related forms as well, has come to support a second theory: that the ancestors of birds were ground-dwelling reptiles. *Archaeopteryx,* according to the fossil evidence, is a close relative of small, bipedal, carnivorous dinosaurs known as theropods. The only major distinctions are that *Archaeopteryx* has feathers and fused collar bones ("wishbone"), like modern birds. But why should feathers evolve in a ground-dwelling organism? According to Ostrom, feathers were originally an adaptation not for flight but for insulation. This raises an interesting question, because for ectothermic animals, such as most modern reptiles, insulation is a disadvantage. Only the endothermic animals, such as birds and mammals, require insulation as a help in conserving heat produced by a high metabolic rate. In other words, the theropods, according to this hypothesis, were endothermic.

Evidence that the original function of feathers was not flight is provided by anatomical studies showing that the wing feathers apparently were not attached to the bones of the "hand" in *Archaeopteryx,* as they are in modern birds, but were merely embedded in the skin. Also the breastbone and its keel are lacking, indicating that the wing muscles were not well developed. *Archaeopteryx* was clearly not a good flyer, if indeed it could fly at all.

As Ostrom reconstructs it, the early stages in flight began with a feathered, endothermic, carnivorous dinosaur running after its prey, flapping its long feathered arms, and leaping. The fact that the feathered forelimbs of *Archaeopteryx* end in claws, as did the elongated arms of the theropods, lends support to this image. The long "wing" feathers may have originated to serve as cagelike traps—natural nets—for capturing prey; some modern predatory birds use their wings in this way (Figure 27–20). Thus flight is seen as the culmination of a long, successful predatory leap.

27–20
A red-shouldered hawk capturing a mouse.

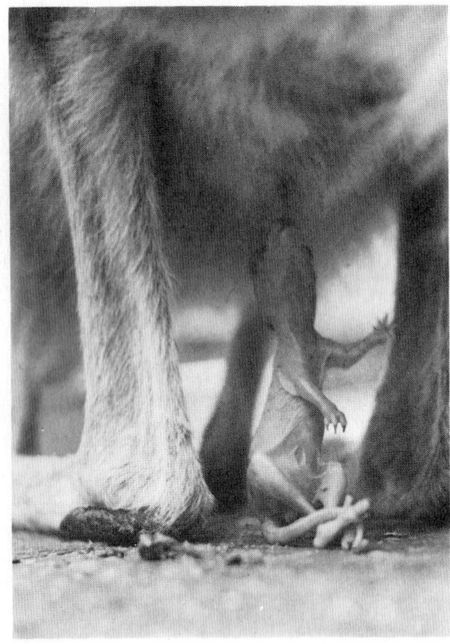

(a)

(b)

27-21
Marsupial infants are born at an immature stage and continue their development attached to a nipple in a special protective pouch of the mother. (a) *This tiny kangaroo accidentally became dislodged from its mother's pouch. As you can see, it is still attached to the nipple. After the picture was taken, the baby was restored to the pouch, with no apparent ill effects from its premature introduction to the outside world.* (b) *Opossum infants spend about two weeks in the womb and about three months in their mother's pouch. A newborn opossum is much smaller than a honey bee.*

Class Mammalia

Mammals also descended from the reptiles. Characteristics distinguishing mammals from other vertebrates are that mammals (1) have hair, (2) provide milk for their young from specialized glands (mammae), and (3) like birds, but unlike other vertebrates, maintain a high body temperature by generating heat metabolically.

In nearly all mammalian species, the young are born alive, as they are in some fish and reptiles, which retain the eggs in their bodies until they hatch. Some very primitive mammals, however, the *monotremes*, such as the duckbilled platypus, lay eggs with shells but nurse their young after hatching. The *marsupials*, which include the opossum and the kangaroo, also bear their young alive. They differ from the largest group of mammals, however, in that the infants are born at a tiny and extremely immature stage and are often kept in a special protective pouch in which they suckle and continue their development (Figure 27-21). Most of the familiar mammals are *placentals*, so called because they utilize their efficient nutritive connection, the placenta, between the uterus and the embryo for a relatively long period of time. As a result, the young develop to a much more advanced stage before birth. Thus the young are afforded protection during their most vulnerable period. The earliest placentals were small, shy, and probably nocturnal, thus avoiding the carnivorous dinosaurs. They undoubtedly lived mostly on insects, grubs, worms, and eggs. Shrews, which are believed to closely resemble these primitive mammals, have retained their elusive habits.

Mammals have fewer, but larger, skull bones than the fish and reptiles, an example of the fact that "simpler" and "more primitive" may have quite opposite meanings. In the mammals, as in some other vertebrates, a bony platform or partition has developed that separates nasal and food passages far back in the throat, making it possible for an animal to breathe while eating. The lower mammalian jaw, unlike that of reptiles, consists of a single bone. Mammals, unlike snakes or lizards, cannot move the upper jaw in relation to the brain case, nor can they unhinge their jaws—an ability that makes it possible for a large anaconda, for example, to swallow a pig whole.

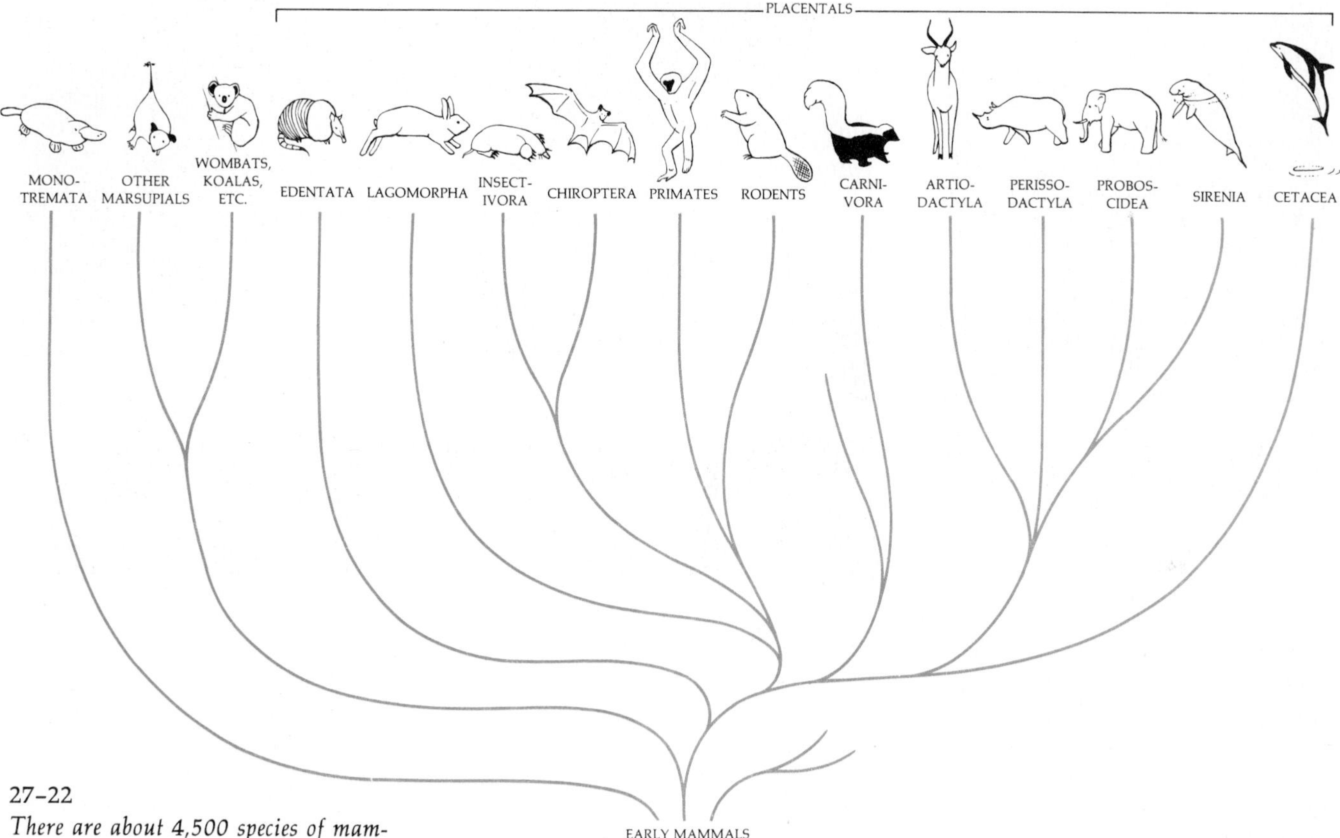

27–22
There are about 4,500 species of mammals, divided into three subclasses: the monotremes, the marsupials, and the placentals. Twelve of the sixteen orders of placentals are shown here.

The major evolutionary lines of mammals are summarized in Figure 27–22. The primates, the order to which we belong, are placental mammals that retain all four kinds of teeth (canines, incisors, premolars, and molars) and have opposable first digits (thumbs and usually big toes), two pectoral mammae, frontally directed eyes, and a relatively large brain with a convoluted cerebral cortex. We are distinguished from the other primates by our upright posture, long legs and short arms, high forehead and small jaw, and sparse body hair.

Of all the mammals, humans are among the least specialized. We are omnivores. Unlike the carnivores, which are meat eaters, and the several orders of herbivores, we eat a wide variety of fruits, vegetables, and other animals. Our hands closely resemble those of a primitive reptile (see Figure 19–7, page 378), in contrast to the highly specialized forelimbs developed by, for example, whales, bats, and horses. We cannot see as well as an eagle. Our sense of smell is much less keen than a dog's, and our sense of taste far less sensitive than a housefly's. Many animals can run faster, swim more powerfully, and climb trees with more agility (though few can do all three). Humans have, however, one area of extreme specialization: the brain. Because of our brain, we are unique among all the other animals in our capacity to reason, to speak, to plan, to learn, and so, to some extent, to control our own future and that of the other organisms with which we share this planet.

(a) (b) (c) (d) (e) (f)

27–23

An assortment of mammals. (a) Lagomorphs, such as the jackrabbit shown here, have two pairs of upper incisors, whereas rodents, such as beavers (b), have only one. In both lagomorphs and rodents, the teeth grow continually.

Carnivores, such as this nursing harbor seal (c) and lion (d), are adapted to hunt and kill for food. The lion's prey, a zebra, is a perissodactyl (an odd-toed ungulate). (e) Hippopotamuses, which are artiodactyls (two-toed ungulates), graze by night and spend most of the daylight hours resting in the water.

(f) Elephants, the Proboscidea, are the largest land mammals living today; some reach a weight of 7.5 metric tons.

SUMMARY

The four deuterostome phyla are Echinodermata, Chaetognatha, Hemichordata, and Chordata. Echinoderms include the sea lilies and feather stars, starfish, brittle stars, sea urchins and sand dollars, and sea cucumbers. Although echinoderms have bilaterally symmetrical larvae, the adult forms of most species are radially symmetrical, with a five-part body plan. Echinoderms have an internal, calcium-containing skeleton that typically bears spines. Their most unusual characteristic is the water vascular system, a modified coelomic cavity that provides a hydrostatic skeleton for the tube feet and creates suction for the clinging and pulling activities associated with locomotion and feeding.

The arrow worms, phylum Chaetognatha, have three distinct body regions and are among the most active predators in the marine plankton. Hemichordates, of which the majority are acorn worms, have a mixture of echinoderm and chordate characteristics. Their larvae, the coelomic cavities in some members of the phylum, and portions of the nervous system resemble those of echinoderms. In the hemichordates, however, the nerve cord is dorsal (rather than ventral), a feature found elsewhere only among the chordates. Hemichordates also have a pharynx with gill slits, another of the identifying features of the chordates.

The phylum Chordata comprises three subphyla: the Cephalochordata (lancelets), the Urochordata (tunicates), and the Vertebrata. The primary characteristics of the chordates are the notochord, a flexible longitudinal rod running just ventral to the nerve cord and serving as the structural axis of the body (present only in embryonic life in most vertebrates); the nerve cord, which is a hollow tube located dorsally; a pharynx with gill slits; and a tail. *Branchiostoma* illustrates the basic chordate body plan. It is hypothesized that the tunicate larva, which also shows these characteristics, resembles the primitive chordate from which the vertebrates evolved.

The vertebrates, the largest subphylum of the chordates, all have a vertebral column, a flexible and usually bony support that develops around and supplants the notochord and encloses the nerve cord, and a cranium enclosing the brain. The vertebrates include the fish (three living classes), the amphibians, the reptiles, the birds, and the mammals.

Some primitive bony fish, forerunners of the amphibians, were aided in the transition to land by the development of lungs or lunglike structures and by strong skeletal supports that could support the body of the fish out of water. Most modern amphibians remain incompletely adapted to life on land and must spend part of the life cycle in water. Vertebrates became truly terrestrial with the development, in the reptiles, of the amniote egg. The diversification of the reptiles that followed their conquest of the land ultimately gave rise not only to a great variety of reptiles (most of which have been extinct for about 65 million years) but also to the birds and the mammals.

QUESTIONS

1. Distinguish among the following terms: exoskeleton/endoskeleton; tube foot/ampulla; ectotherm/endotherm; monotreme/marsupial/placental.

2. Describe the water vascular system of a starfish. What are its similarities to and differences from the hydrostatic skeleton of an earthworm?

3. As we have seen, the larvae of gastropods, starfish, and tunicates all have bilateral symmetry, but the symmetry of the adults is dramatically different. How is the change in symmetry accomplished?

4. Describe the identifying characteristics of the phylum Chordata. What is the functional significance of each?

5. How do the three living classes of fish differ from one another?

6. Consider the graceful swimming of a fish or squid. What is the role of the semirigid beam running the length of the animal, whether the "pen" of a squid, the notochord of a *Branchiostoma,* or the vertebral column of a fish?

7. What anatomical features make possible the flight of birds?

8. Among fish and reptiles, some species are oviparous (that is, they lay eggs from which the young hatch) and some are viviparous (giving birth to live young). Name some of the advantages of each alternative. Why are all birds oviparous?

SUGGESTIONS FOR FURTHER READING

Classification

Books

AYALA, FRANCISCO J., and JOHN A. KIGER, JR.: *Modern Genetics,* The Benjamin/Cummings Publishing Company, Menlo Park, Calif., 1980.

Contains a good chapter on phylogeny and the comparison of proteins and nucleotides.

DOBZHANSKY, THEODOSIUS, et al.: *Evolution,* W. H. Freeman and Company, San Francisco, 1977.

A good general introduction to evolution with chapters on taxonomy and phylogeny.

ELDREDGE, NILES, and JOEL CRACRAFT: *Phylogenetic Patterns and the Evolutionary Process,* Columbia University Press, New York, 1980.

An introduction to the interplay between taxonomy, systematics, and evolutionary theory.

KESSEL, R. G., and C. Y. SHIH: *Scanning Electron Microscopy in Biology: A Students' Atlas on Biological Organization,* Springer-Verlag, New York, 1976.

An atlas filled with marvelous scanning electron micrographs of specialized structures of prokaryotes, protists, fungi, plants, and animals, as well as of whole organisms.

Articles

CONNIFF, RICHARD: "The Name Game," *Science 82,* June 1982, pages 66–67.

EIGEN, MANFRED, et al.: "The Origin of Genetic Information," *Scientific American,* April 1981, pages 88–118.

MAYR, ERNST: "Biological Classification: Toward a Synthesis of Opposing Methodologies," *Science,* vol. 214, pages 510–516, 1981.

PATRUSKY, BEN: "Molecular Evolution: A Quantifiable Contribution," *Mosaic,* March/April 1979, pages 12–22.

Prokaryotes

Books

BURNET, MACFARLANE, and DAVID O. WHITE: *Natural History of Infectious Disease,* 4th ed., Cambridge University Press, New York, 1972.

A general introduction to the ecology of infectious diseases and their influence on human activities.

STANIER, R. Y., E. A. ADELBERG, J. L. INGRAHAM, and M. L. WHEELIS: *The Microbial World,* 4th ed., Prentice-Hall, Inc., Englewood Cliffs, N.J., 1976.

An introduction to the biology of microorganisms, with special emphasis on the properties of bacteria. It is widely considered one of the most authoritative accounts.

ZINSSER, HANS: *Rats, Lice, and History,* The Atlantic Monthly Press/Little, Brown and Company, Boston, 1935.*

A classic popular account of the influence of infectious diseases and their animal vectors on the course of human history.

Articles

BLAKEMORE, RICHARD P., and RICHARD B. FRANKEL: "Magnetic Navigation in Bacteria," *Scientific American,* December 1981, pages 58–65.

BUTLER, P. JONATHAN G., and AARON KLUG: "The Assembly of a Virus," *Scientific American,* November 1978, pages 62–69.

DIENER, T. O.: "Viroids," *Scientific American,* January 1981, pages 66–73.

FOX, G. E., et al.: "The Phylogeny of Prokaryotes," *Science,* vol. 209, pages 457–463, 1980.

SIMONS, KAI, HENRIK GAROFF, and ARI HELENIUS: "How an Animal Virus Gets into and out of Its Host Cell," *Scientific American,* February 1982, pages 58–66.

WOESE, CARL R.: "Archaebacteria," *Scientific American,* June 1981, pages 98–122.

Protists

Books

BONNER, J. T.: *The Cellular Slime Molds,* 2d ed., Princeton University Press, Princeton, N.J., 1968.

A record of experimental work with a small but fascinating group of organisms.

CURTIS, HELENA: *The Marvelous Animals,* Natural History Press, Garden City, N.Y., 1968.

An informal introduction to one-celled organisms.

JURAND, A., and G. C. SELMAN: *The Anatomy of* Paramecium aurelia, The Macmillan Company, New York, 1964.

An exploration, mainly by electron microscopy, of the astonishing complexity of a single-celled organism.

MARGULIS, LYNN: *Symbiosis in Cell Evolution: Life and Its Environment on the Early Earth,* W. H. Freeman and Company, San Francisco, 1981.*

A fascinating discourse proposing the origin of eukaryotic cells by serial symbiotic events.

* Available in paperback.

PICKETT-HEAPS, J. D.: *Green Algae: Structure, Reproduction and Evolution in Selected Genera,* Sinauer Associates, Inc., Sunderland, Mass., 1975.

A beautifully illustrated book providing much insight into the variety of form and function in the cells of the green algae.

SLEIGH, M. A.: *The Biology of Protozoa,* American Elsevier Publishing Company, New York, 1973.*

A general biology of the protozoa, including chapters on structure, metabolism, reproduction, and ecology, with many excellent illustrations.

Articles

SCHOPF, J. WILLIAM: "The Evolution of the Earliest Cells," *Scientific American,* September 1978, pages 110–138.

SCHWARTZ, ROBERT M., and MARGARET O. DAYHOFF: "Origins of Prokaryotes, Eukaryotes, Mitochondria, and Chloroplasts," *Science,* vol. 199, pages 395–403, 1978.

Fungi

Books

AHMADJIAN, V.: *The Lichen Symbiosis,* Blaisdell Publishing Company, Waltham, Mass., 1967.

In this small but fascinating book, Ahmadjian describes the nature of the relationship between the fungal and algal components of a lichen.

ALEXOPOULOS, C. J., and C. W. MIMS: *Introductory Mycology,* 3d ed., John Wiley & Sons, Inc., New York, 1979.

A thorough introduction to the fungi and related groups of heterotrophic protista.

LARGE, E. C.: *The Advance of the Fungi,* Dover Publications, Inc., New York, 1962.*

A fascinating popular account of the closely interwoven histories of fungi and humans, first published in 1940.

SMITH, A. H.: *The Mushroom Hunter's Field Guide,* The University of Michigan Press, Ann Arbor, Mich., 1980.

A clear, concise, well-illustrated guide to edible mushrooms, enlivened with good advice and pertinent anecdotes.

Articles

AHMADJIAN, VERNON: "The Nature of Lichens," *Natural History,* March 1982, pages 30–37.

FRIEDMANN, E. IMRE: "Endolithic Microorganisms in the Antarctic Cold Desert," *Science,* vol. 215, pages 1045–1053, 1982.

MATOSSIAN, MARY K.: "Ergot and the Salem Witchcraft Affair," *American Scientist,* vol. 70, pages 355–357, 1982.

RUEHLE, JOHN L., and DONALD H. MARX: "Fiber, Food, Fuel, and Fungal Symbionts," *Science,* vol. 206, pages 419–422, 1979.

Plants

Books

DAWSON, E. Y.: *Marine Botany: An Introduction,* Holt, Rinehart and Winston, Inc., New York, 1966.

A short, lively text that covers seaweeds, marine bacteria, fungi, phytoplankton, and sea grasses.

---

* Available in paperback.

RAVEN, PETER H., RAY F. EVERT, and HELENA CURTIS: *Biology of Plants,* 3d ed., Worth Publishers, Inc., New York, 1981.

This general botany text contains an excellent presentation of the evolution of plants and related organisms, as well as a wealth of information on prokaryotes, protists, and fungi.

Article

MULCAHY, DAVID L.: "Rise of the Angiosperms," *Natural History,* September 1981, pages 30–35.

Animals

Books

BARNES, ROBERT D.: *Invertebrate Zoology,* 4th ed., Saunders College/Holt, Rinehart and Winston, Philadelphia, 1980.

One of the best general introductions to protozoans and invertebrates.

BUCHSBAUM, RALPH, and LORUS J. MILNE: *The Lower Animals: Living Invertebrates of the World,* Doubleday & Company, Inc., Garden City, N.Y., 1962.

A collection of handsome photos of the invertebrates accompanied by a text prepared by two noted zoologists but directed toward the general reader.

DESMOND, ADRIAN J.: *The Hot-Blooded Dinosaurs: A Revolution in Paleontology,* The Dial Press, Inc., New York, 1976.*

Desmond, a historian of science, describes the development of evolutionary theories, particularly as they were influenced by the discovery of dinosaur fossils, and presents a lively review of the evidence that is leading an increasing number of paleontologists to the conclusion that dinosaurs were endothermic. (If you are not interested in the history of paleontology, you may want to begin in the middle.) The text is well written and the illustrations are wonderful.

EVANS, HOWARD E.: *Life on a Little-Known Planet,* E. P. Dutton & Co., Inc., New York, 1978.*

Professor Evans is the author of many popular articles and books on insects. This book profits from his wide knowledge, clarity, and humor.

FEDUCCIA, ALAN: *The Age of Birds,* Harvard University Press, Cambridge, Mass., 1980.

A history of the birds, tracing their evolution from reptilian ancestors through the diversifications that gave rise to the major groups of modern birds. Well-illustrated with many photographs and drawings.

HANSON, EARL D.: *The Origin and Early Evolution of Animals,* Wesleyan University Press, Middletown, Conn., 1977.

Recommended to the serious student who is interested in how the experts make decisions concerning evolutionary relationships.

HICKMAN, CLEVELAND P.: *Biology of the Invertebrates,* 2d ed., The C. V. Mosby Company, St. Louis, 1973.

Probably the best general text on the invertebrates.

KLOTS, ALEXANDER B., and ELSIE B. KLOTS: *Living Insects of the World,* Doubleday & Company, Inc., Garden City, N.Y., 1975.

A spectacular gallery of insect photos. The text is informal but informative, written by experts for the general public.

ROMER, ALFRED: *The Procession of Life,* World Publishing Co., Cleveland, 1968.*

A history of evolution, written by an expert but as readable as a novel.

---

* Available in paperback.

RUSSELL-HUNTER, W. D.: *A Biology of Higher Invertebrates,* The Macmillan Company, New York, 1968.*

RUSSELL-HUNTER, W. D.: *A Biology of Lower Invertebrates,* The Macmillan Company, New York, 1968.*

Short, authoritative accounts, with emphasis on function.

WELLS, M. J.: *Brain and Behavior in Cephalopods,* Stanford University Press, Stanford, Calif., 1962.

Experimental analyses of behavior in the octopus and squid.

Articles

BUFFETAUT, ERIC: "The Evolution of the Crocodilians," *Scientific American,* October 1979, pages 130–144.

CARLSON, ALBERT D., and JONATHAN COPELAND: "Behavioral Plasticity in the Flash Communication Systems of Fireflies," *American Scientist,* vol. 66, pages 340–346, 1978.

EBERHARD, WILLIAM G.: "Horned Beetles," *Scientific American,* March 1980, pages 166–182.

HORRIDGE, G. ADRIAN: "The Compound Eye of Insects," *Scientific American,* July 1977, pages 108–122.

KIRSCH, JOHN A. W.: "The Six-Percent Solution: Second Thoughts on the Adaptedness of the Marsupialia," *American Scientist,* vol. 65, pages 276–288, 1977.

LA BARBERA, MICHAEL, and STEVEN VOGEL: "The Design of Fluid Transport Systems in Organisms," *American Scientist,* vol. 70, pages 54–60, 1982.

LLOYD, JAMES E.: "Mimicry in the Sexual Signals of Fireflies," *Scientific American,* July 1981, pages 138–145.

MANSOUR, TAG E.: "Chemotherapy of Parasitic Worms: New Biochemical Strategies," *Science,* vol. 205, pages 462–469, 1979.

MICHELSEN, AXEL: "Insect Ears as Mechanical Systems," *American Scientist,* vol. 67, pages 696–706, 1979.

OSTROM, JOHN H.: "Bird Flight: How Did It Begin?" *American Scientist,* vol. 67, pages 46–56, 1979.

RUSSELL, DALE A.: "The Mass Extinctions of the Late Mesozoic," *Scientific American,* January 1982, pages 58–65.

SILVERSTEIN, ROBERT M.: "Pheromones: Background and Potential for Use in Insect Pest Control," *Science,* vol. 213, pages 1326–1332, 1981.

VALENTINE, JAMES W.: "The Evolution of Multicellular Plants and Animals," *Scientific American,* September 1978, pages 140–158.

VOGEL, STEVEN: "Organisms That Capture Currents," *Scientific American,* August 1978, pages 128–139.

---

* Available in paperback.

SECTION 5

Biology of Plants

CHAPTER 28

The Plant: An Introduction

For most of earth's history, the land was bare. A billion years ago, seaweeds clung to the shores at low tide and perhaps some gray-green lichens patched a few inland rocks. But, had anyone been there to observe it, the earth's surface would generally have appeared as barren and forbidding as the bleak Martian landscape. According to the fossil record, plants first began to invade the land a mere half billion years ago. Not until then did the earth's surface truly come to life. As a film of green spread from the edges of the waters, other forms of life—the heterotrophs—were able to follow. The shapes of these new forms and the ways in which they lived were determined by the plant life that preceded them. Plants supplied not only their food—their chemical energy—but also their nesting, hiding, stalking, and breeding places.

And so it is today. In all terrestrial communities except those created by human activities, the character of the plants still determines the character of the animals and other forms of life that inhabit a particular area. Even we members of the human species, who have seemingly freed ourselves from the life of the land and even, on occasion, from the surface of the earth, are still dependent on the photosynthetic events that take place in the green leaves of plants.

In Chapter 23, we traced the evolution of the plants; if you have not already read that chapter, we recommend that you do so now. In this chapter, we shall focus on the group of plants that evolved most recently, the angiosperms (division Anthophyta). These plants are characterized by specialized reproductive structures—flowers—in which sexual reproduction takes place, in which the seed is formed, and from which the fruit develops.

28–1
One of the most successful angiosperms, showing two characteristics of this large group. One is the flower, in which the seed develops. The other is the fruit, which aids in dispersal of the seed. The fruit, in this case, bears a plumelike structure that is caught by the wind.

The angiosperms are by far the most abundant plants, with about 235,000 species. The group is divided into two large classes, the dicots (170,000 species) and the monocots (65,000 species). The differences between the two groups are summarized in Figure 28–2 on the next page.

THE PLANT BODY

As we discussed in Chapter 23, the ancestor of the plants was probably a multicellular green alga that floated on or just below the water's surface. Like modern plants, its photosynthetic pigments were chlorophylls *a* and *b* and carotenoids, especially beta-carotene, all of which were contained in chloroplasts. It had a membrane-bound nucleus and mitochondria and other cellular organelles, as well as an external cell wall containing cellulose. Its energy source was sunlight, and it obtained oxygen, carbon dioxide, and minerals from the water in which it lived.

Plants have the same few and relatively simple requirements: light, water, oxygen, carbon dioxide, and certain minerals. From these simple materials they, like their ancestors, make the sugars, amino acids, nucleotides, and other organic substances on which all plant and animal life depends. But there is an important difference. In the simplest photosynthesizing organism—the single algal cell or filament of cells—each of the needed materials is immediately available to every cell. In a plant, however, the individual cells can no longer function autonomously but can survive only through cooperation and division of labor among the many cells and tissues forming the plant body. The reasons for this division of labor are easy to understand. The water and minerals needed by plants are found mostly in the soil; thus plants came under strong selection pressures for the development of complex and extensive root systems. Sunlight cannot reach these belowground structures, however, and photosynthesis is relegated to another part of the plant body. As plants began to crowd each other and compete for light, selection pressures favored those with more extensive and efficient light-collecting surfaces—leaves.

The stem raises these photosynthesizing surfaces into the sunlight. Through the specialized vascular tissues of the stem, water and minerals from the soil are transported to the leaves, and the products of photosynthesis formed in the leaves are conducted to flowers, roots, and other nonphotosynthetic parts of the plant.

Three *tissue systems* are continuous throughout the plant body. The *vascular tissue* system, composed of xylem and phloem, is embedded in the *ground tissue* system, and, as we shall see in this chapter, the principal differences in the structure of leaves, stems, and roots lie in the relative distribution of the vascular and ground tissue systems. The third tissue system, the *dermal tissue* system, provides an outer protective covering for the entire plant body.

Figure 28-3 diagrams the external structure of a familiar plant, the garden geranium *(Pelargonium)*, indicating some of the terms used to describe parts of the plant body. In the rest of this chapter, we shall describe the three plant organs: leaves, stems, and roots.

28-2
The angiosperms are divided into two broad groups: the dicots and the monocots. The names refer to the fact that the embryo in the dicots has two cotyledons ("seed leaves") and in the monocots has one. In the dicots, the vascular tissues are arranged around a central core in the stem; in monocots, they are scattered. The veins of dicot leaves are usually netted; those of monocot leaves are usually parallel. Dicots characteristically have taproots, and monocot roots are often fibrous. As we noted previously (page 473), there are also characteristic differences in the number of floral parts and in the structure of the pollen grains.

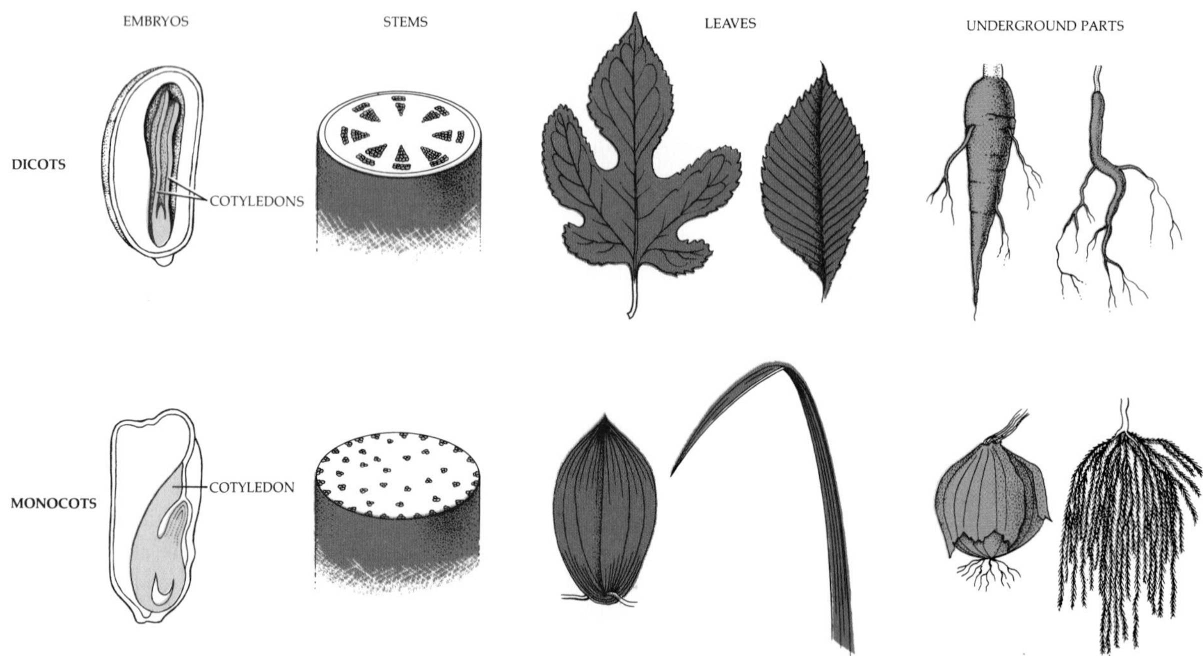

28–3
The body plan of a flowering plant. The aboveground shoot system consists of the stem, the leaves, whose primary function is photosynthesis, and the flowers, the reproductive structures. Leaves appear at regions on the stem known as nodes. The portions of the stem between nodes are called internodes. The belowground structures, the roots, supply water and minerals to the stem, leaves, and flowers.

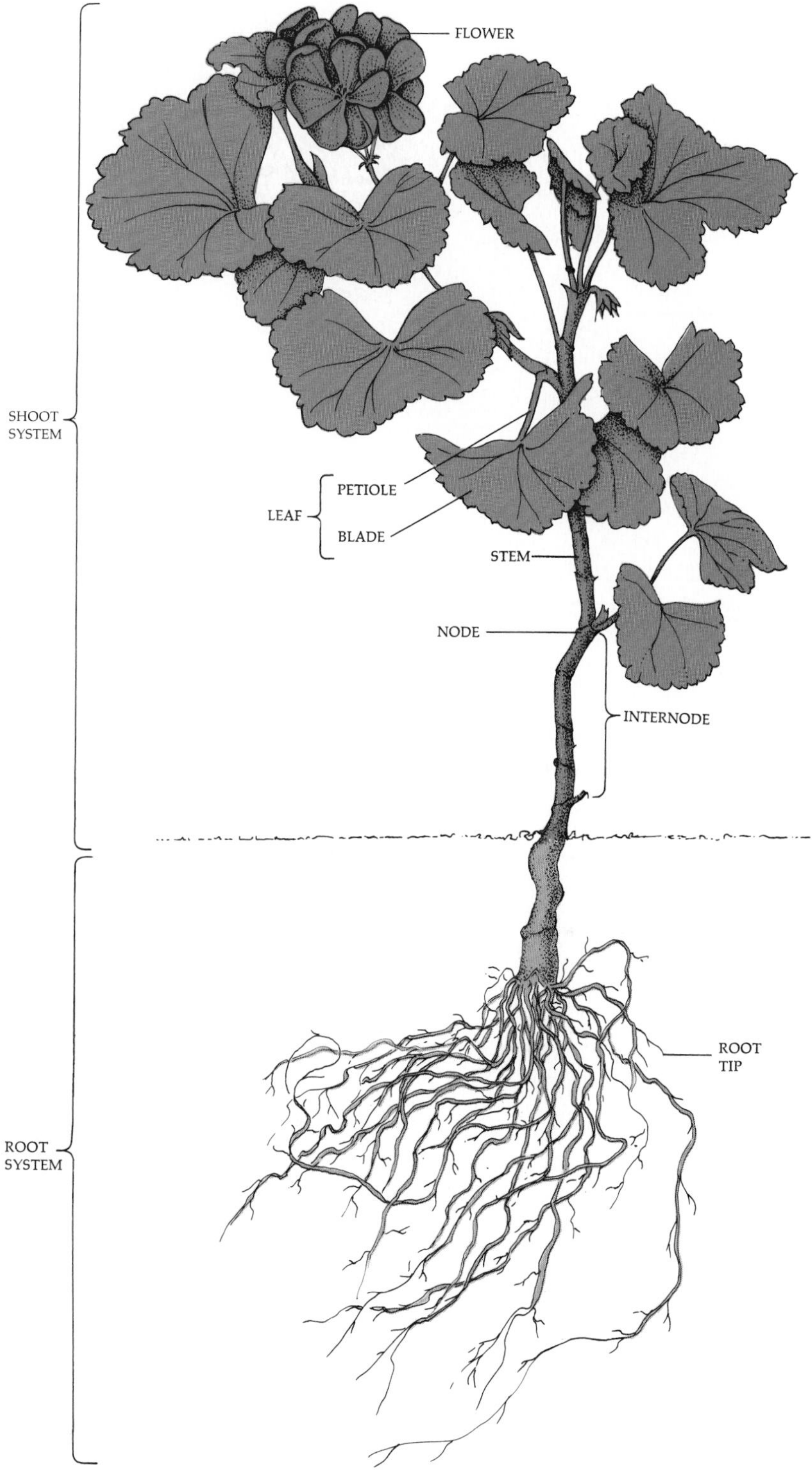

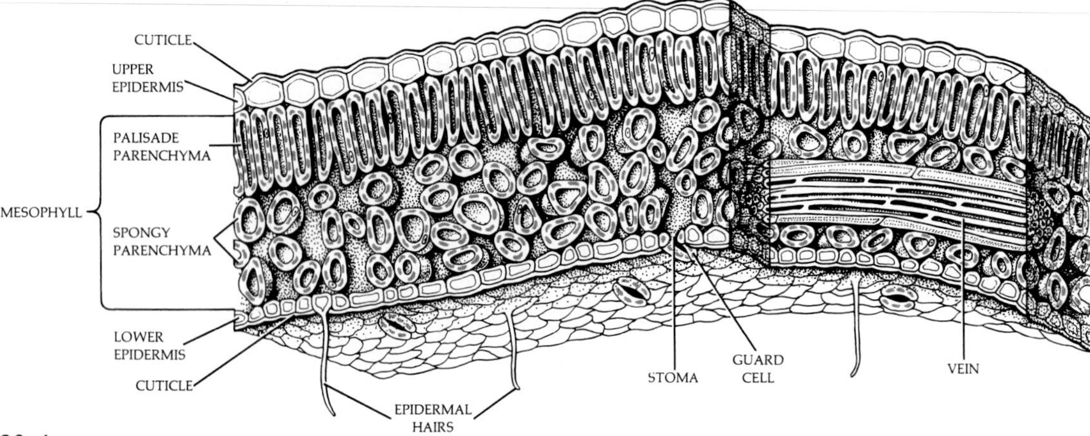

28–4

Diagram of the interior of a leaf. Photosynthesis takes place in the palisade cells and, to a lesser extent, in the spongy parenchyma. Chloroplasts are indicated in green. Note that the chloroplasts of the parenchyma cells are all near the cell surface, with the centers of the cells filled with large vacuoles. The chloroplasts move within the cytoplasm, orienting themselves to the sun. The veins carry water and solutes to and from these mesophyll cells. The interior of the leaf is entirely enclosed by epidermal cells covered with a waxy layer, the cuticle. Openings in the epidermis, the stomata, permit the exchange of gases. The guard cells surrounding the stomata also have chloroplasts, which may (or may not) play a role in opening and closing the stomata (see pages 616–618).

LEAVES

Leaf Structure

A leaf is a compromise between two conflicting evolutionary pressures. The first is to expose a maximum photosynthetic surface to sunlight; the second is to conserve water while, at the same time, providing for the exchange of gases necessary for photosynthesis.

The photosynthetic cells of leaves are of a general type known as *parenchyma*. These cells, which constitute the ground tissue of the leaf, are many-sided with thin, flexible cell walls. There are two types of photosynthetic parenchyma cells: *palisade parenchyma*, consisting of long columnar cells in which most photosynthesis takes place, and *spongy parenchyma*, which consists of irregularly shaped cells with large air spaces surrounding them. These spaces are filled with gases, including water vapor.

Palisade and spongy parenchyma make up the *mesophyll*, or "middle leaf." The mesophyll is completely enclosed in an almost airtight wrapping made up of epidermal cells, which secrete a waxy substance called *cutin*. The cutin forms a coating, the *cuticle*, over the outer surface of the epidermis. The epidermal cells and the cuticle are transparent, permitting light to penetrate to the photosynthetic cells of the mesophyll.

Substances move into and out of the leaf by two quite different routes: vascular bundles and stomata. Water and minerals are supplied to the leaf cells by the vascular bundles, which are known in the leaf (and only in the leaf) as *veins*. The veins pass through the petioles (leaf stalks) and are continuous with the vascular tissues of the stem and root. Xylem is the vascular tissue concerned primarily with water and mineral transport. Other groups of cells, the phloem, transport sugars and other products carried in water from the photosynthetic (autotrophic) cells to the other (heterotrophic) cells of the plant.

Veins form distinctive patterns in leaf blades. There is a conspicuous difference between the arrangement of the veins of monocots and dicots: in monocots, the major veins are usually all parallel; in dicots, the venation (vein pattern) is netted and either palmate (fanlike) or pinnate (featherlike), as shown in Figure 28–2.

Gases—oxygen and carbon dioxide—move into and out of the leaf by diffusion through stomata. A stoma consists of a small opening, or pore; it is surrounded by two specialized cells in the leaf epidermis, called guard cells (Figure 28-5). The exchange of gases is, as we saw in Chapter 10, necessary for photosynthesis. However, as gases are exchanged, water escapes from the leaf; about 90 percent of the water that leaves the plant body is lost through the stomata.

Stomata are commonly most abundant on the undersurface of leaves. They may be very numerous. For example, on the lower surface of tobacco leaves there are about 19,000 stomata per square centimeter. On the upper surface there are about 5,000 stomata per square centimeter.

28-5
Scanning electron micrograph of open and closed stomata on the surface of a sundew leaf. The stomata lead into air spaces within the leaf that surround the thin-walled spongy parenchyma cells. The air in these spaces, which make up 15 to 40 percent of the total volume of the leaf, is saturated with water vapor that has evaporated from the photosynthetic cells.

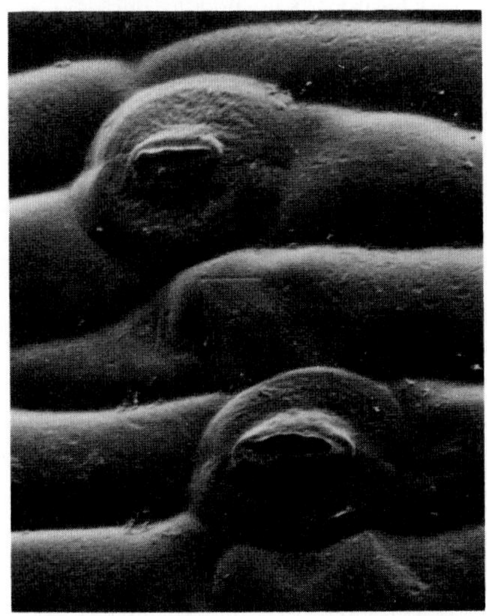

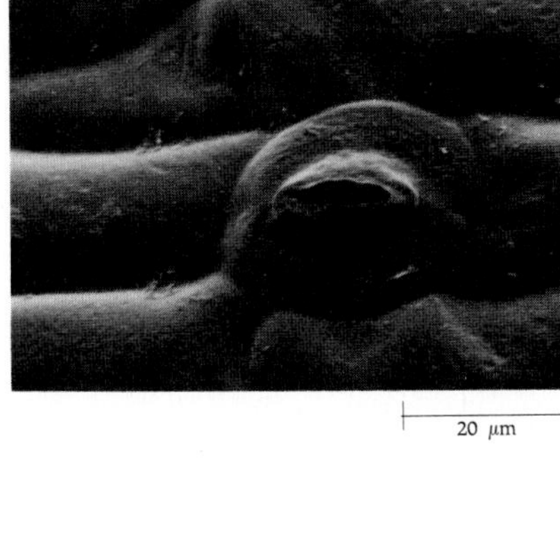

Leaf Adaptations and Modifications

Leaves come in a variety of shapes and sizes, ranging from broad fronds to tiny scales. Some of these differences can be correlated with the environments in which the plants live. Large leaves with broad surfaces are often found in plants that grow under the canopy in a tropical rain forest, where water is plentiful but where there is intense competition for light. Leaves of such plants sometimes have "drip tips," which facilitate the runoff of rainwater. Leaves with small surfaces are associated with dry climates. In conifers, for example, the photosynthetic surfaces are greatly reduced and there is an extra-thick layer of epidermis and cuticle (see page 470). In such plants, photosynthesis is reduced but so is water loss, so that the trees are able to survive long periods of water deprivation—as when water is locked up in ice. Similarly, angiosperms found in dry habitats often have small, leathery leaves. This reduction in leaf surface reaches its extreme in plants such as the desert cacti in which the leaves are modified as spines—hard, dry, and nonphotosynthetic structures. (The terms "spine" and "thorn" are often used interchangeably; however, thorns are technically modified branches.) In these plants, photosynthesis takes place in the fleshy stems, which are also water-storage organs.

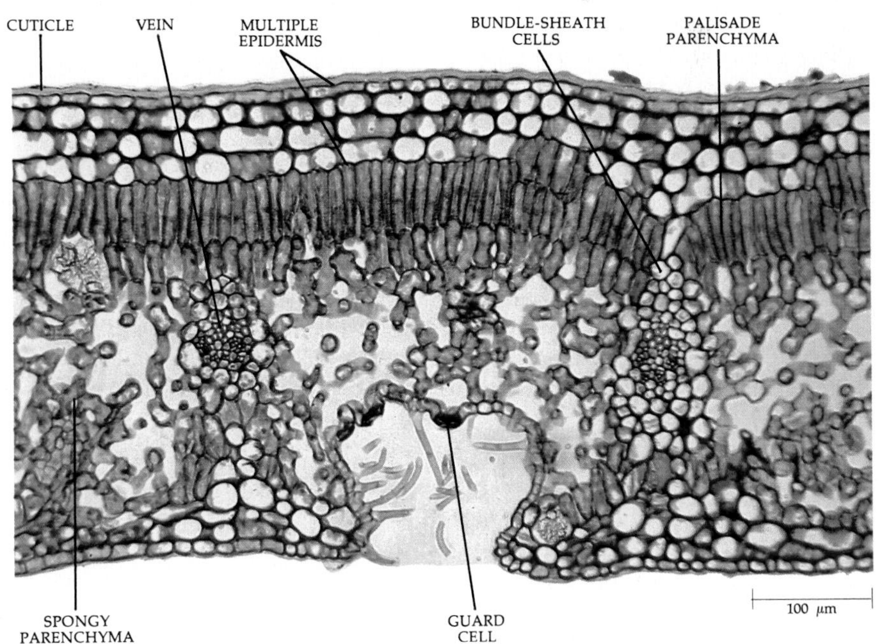

28-6
Cross section of the leaf of an oleander, a plant adapted to a dry climate. Note the very thick single-layered cuticle at the top covering the multiple epidermis, so called because it consists of four layers of cells. The vein is seen in cross section, which is why it looks different from the vein in Figure 28-4. The stoma is contained within a stomatal crypt, which is lined with epidermal hairs. Note that this leaf contains bundle-sheath cells.

CARNIVOROUS PLANTS

Among the most spectacular modified leaves are those of the carnivorous plants. These plants, of which there are over 350 species, are meat-eaters; their diets include insects, other invertebrates, and even some vertebrates, such as small birds and frogs. Unlike carnivorous animals, they do not utilize their prey for energy but rather as a source of mineral elements, particularly nitrogen and phosphorus. As Darwin noted in his book on this subject, published in 1875, these plants are usually found in swamps, bogs, and peat marshes where acids leach the soil of nutrients.

(a) *The sundew is a tiny plant, often only 2 to 5 centimeters across, with club-shaped tentacles on the upper surface of its leaf. These tentacles secrete a clear, sticky liquid that attracts insects. When an insect is caught by one tentacle, the others bend inward toward it until the insect drowns, its air passages filled with a mucilage-like fluid. The tentacles also secrete digestive enzymes.*

(b) *Pitcher plants attract insects into their flowerlike tubular leaves by means of nectar. Following the nectar over the rim, the insect finds itself on a carpet of fine, transparent hairs. When the nectar trail ends, and the insect turns to exit, it encounters the points of these fine hairs, pointing downward and blocking its way. If it moves inward, it encounters a slick waxed surface from which it slides or drops into a foul-smelling broth of rainwater, digestive enzymes, and decomposing bacteria at the pitcher's base.*

(c, d) *The Venus flytrap has leaves that close around any insect moving on the surface of the leaf. The closing of the leaf is triggered by touching one or two of the three trigger hairs in the middle of each leaf lobe. Venus flytraps are found in nature only on the coastal plain of North and South Carolina, usually on the edges of wet depressions and pools. It has long been believed that the Venus flytrap lures its victims by exuding nectar, but recent studies indicate that ants and other insects visit the leaves by chance. (Despite its name, its usual diet in the wild consists of crawling invertebrates.)*

(a)

(b)

(c)

(d)

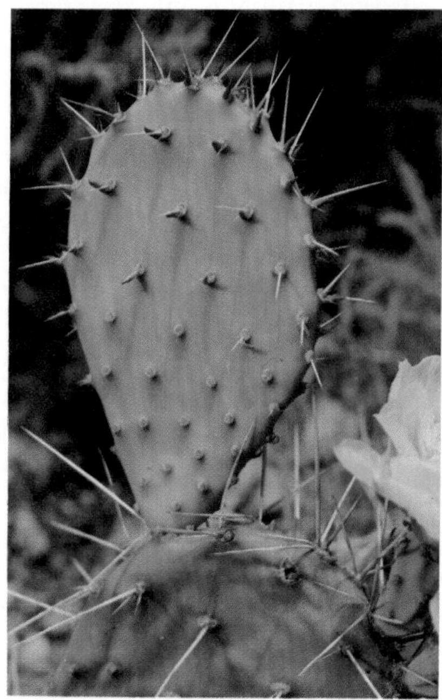

(a)

(b)

(c)

28–7
Modified leaves. (a) *Spines on a prickly-pear cactus.* (b) *Succulent leaves, adapted for water storage* (Sedum). (c) *Tendrils of a pea plant.*

In many plants, leaves are succulent; that is, they are adapted for water storage. Other leaves are specialized for food storage. A bulb, such as the onion, is a large bud consisting of a short stem with many leaves modified for food storage. The "head" of a cabbage also consists of a compressed stem bearing numerous thick, overlapping leaves. In some plants, the petioles become thick and fleshy: celery and rhubarb are two familiar examples. The tendrils of some plant species—the garden pea, for instance—are modified leaves.

THE STEM

The Structure of the Stem

The stem holds the leaves up to the light and provides for the transportation of substances to and from the leaves. The outer surface (dermal tissue) of a young green stem, like that of the leaf and the root, is made of epidermal cells. Like the leaf, the green stem is covered with a waxy cuticle, contains stomata, and is photosynthetic.

Ground Tissue

The bulk of the tissue of a young stem is ground tissue. As in the leaf, the ground tissue is composed mostly of parenchyma cells. The turgor (page 131) of these cells provides the chief support for young green stems.

The ground tissue of stems also may contain specialized supporting tissues formed from *collenchyma* or *sclerenchyma* cells. Collenchyma cells (Figure 28–8a) differ from the thin-walled parenchyma cells of the stem in having primary walls (see page 100) that are thickened at the corners or in some other uneven fashion. Their name derives from the Greek word *colla,* meaning "glue," which refers to their characteristic thick, glistening walls. Collenchyma cells are often located just inside the epidermis, forming either a continuous cylinder or distinct vertical strips of supporting tissue.

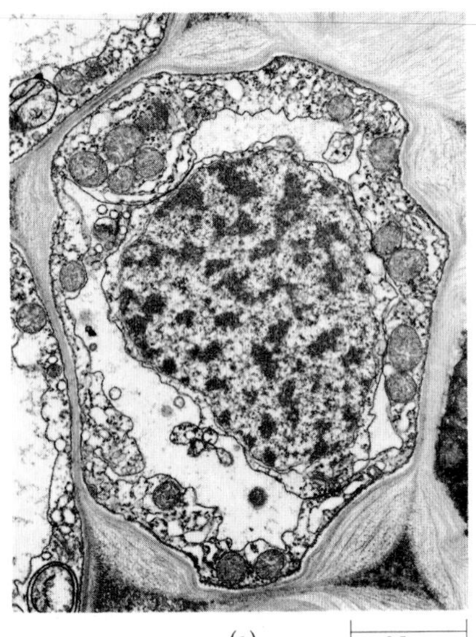

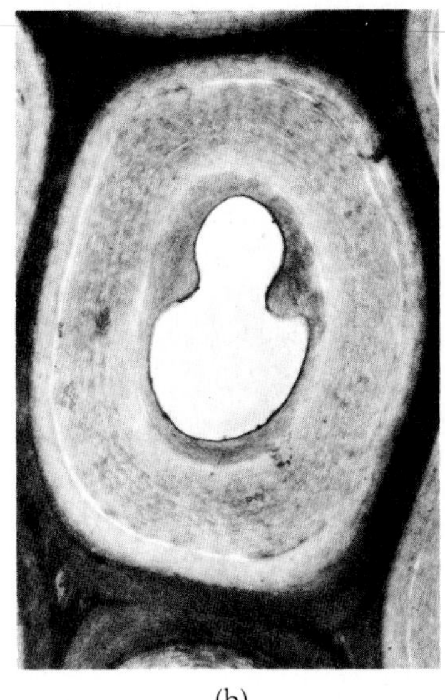

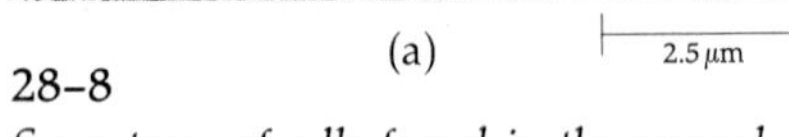

(a) 2.5 μm (b) (c) 25 μm

28–8

Some types of cells found in the ground tissue of stems. (a) *Collenchyma cell. Its irregularly thickened cellulose walls contain pectin and store water and minerals. They are plastic, and so the cell can continue to grow.* (b) *Cross section of a fiber, a sclerenchyma cell. Its specialization is thickened, often lignified cell walls that give it strength and rigidity. Many fibers, but not all, are dead at maturity, like the cell shown here.* (c) *Sclereids, another type of sclerenchyma cell, have very thick lignified walls. Sclereids, or stone cells, are often found also in seeds and fruits. These sclereids are from a pear; they give the fruit its characteristic gritty texture.*

Sclerenchyma cells are of two types: fibers and sclereids. The name is derived from the Greek *skleros,* meaning "hard." Fibers, which are elongated, somewhat elastic cells (Figure 28–8b), typically occur in strands or bundles arranged in patterns characteristic of the plant. These supporting cells are often associated with the vascular tissues. Plant fibers such as flax, hemp, jute, and sisal have long been used in human artifacts, including baskets, rope, and cloth. Sclereids (Figure 28–8c), which are variable in form, are also common in stems. Layers of sclereids are found in seeds, nuts, and fruit stones as well, where they form the hard outer coverings.

Sclerenchyma cells differ from collenchyma in three respects: (1) they have secondary walls in addition to primary walls (see page 101); (2) the secondary walls often contain lignin, a complex macromolecule that impregnates the cellulose and toughens and hardens it; and (3) the cells are often dead at maturity, with only their cell walls remaining.

Vascular Tissues

The vascular tissues, phloem and xylem, consist of specialized conducting cells, supporting fibers, and parenchyma cells, which store food and water.

The conducting cells of the phloem, as we noted previously, transport the products of photosynthesis, chiefly in the form of sucrose, from the leaves to the nonphotosynthetic cells of the plant. In angiosperms, these conducting cells are called sieve-tube members (Figure 28–9a). A *sieve tube* is a vertical column of sieve-tube members joined by their end walls. These end walls, called *sieve plates,* have openings leading from one sieve-tube member to the next (Figure 28–9b).

The sieve-tube members, which are alive at maturity but lack a nucleus, are filled largely with a sucrose-containing watery fluid, called sieve-tube sap. In dicots and some monocots, they are characterized by the presence of a protein-containing substance called slime, or P-protein (the "P" stands for phloem). Its function is unknown, although some botanists believe that P-protein, along with the polysaccharide callose, serves to seal sieve-plate pores in response to injury. In mature sieve-tube members, P-protein lies along the inner surface of the cell wall and is

28–9

In angiosperms, the conducting elements of the phloem are sieve tubes, made up of individual cells, the sieve-tube members. These cells, which lack nuclei at maturity, are usually found in close association with companion cells, which do have nuclei. Sieve-tube members are joined to other sieve-tube members at their ends by sieve plates. (a) *Longitudinal view of mature sieve-tube members and a sieve plate in the stem of the squash* (Cucurbita maxima). *P-protein lines the inner surface of the cell walls of the sieve-tube members.* (b) *Face view of a sieve plate between two mature sieve-tube members. In this electron micrograph, the pores of the sieve plate are open. In most cut sections of phloem tissue, however, the pores are plugged with callose, a polysaccharide deposited by the sieve-tube members in response to injury. Callose, which prevents leakage from the sieve tube, is also deposited as part of the normal aging process.* (c) *Longitudinal section, showing mature and immature sieve-tube members. The arrows point to P-protein bodies in immature cells.*

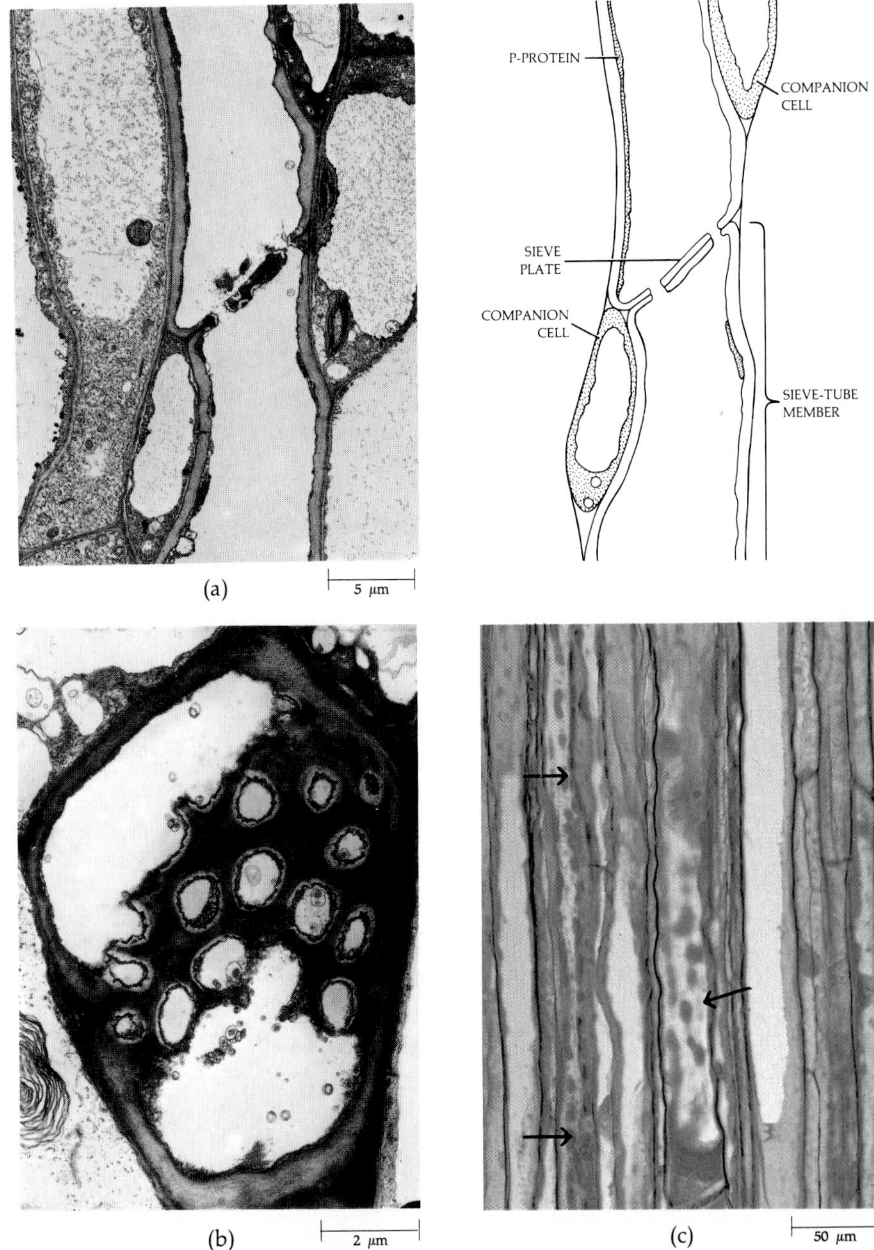

continuous from one cell to the next through the sieve plates. The nucleus of a sieve-tube member disintegrates as the cell matures, as do many of the organelles.

Sieve-tube members are characteristically associated with specialized parenchyma cells called *companion cells,* which contain all of the components commonly found in living plant cells, including a nucleus. Companion cells are responsible for the active secretion of substances into and out of the sieve-tube members and are thought also to provide nuclear functions for the sieve tubes and fulfill their energy requirements. A sieve-tube member and its companion cell arise from the same mother cell.

28–10

Tracheids and vessel members are the conducting cells of the xylem in angiosperms. (a) Tracheids are a more primitive and less efficient type of conducting cell. Water moving from one tracheid to another passes through pits. Pits are not perforations but areas in which there is no secondary cell wall. Water moving from one tracheid to another passes through two primary cell walls and the middle lamella (see page 100).

Vessel members differ from tracheids in that the primary walls and middle lamellae of vessel members are perforated at the ends where they are joined with other vessel members. (b) There may be numerous perforations in adjoining walls of vessel members, or (c) the adjoining walls may break down completely as the cells mature, forming a single opening. Vessel members are also characteristically shorter and wider than tracheids, and their adjoining walls are less oblique. Vessel members are connected with other vessel members and also with other cells by pits in the side walls.

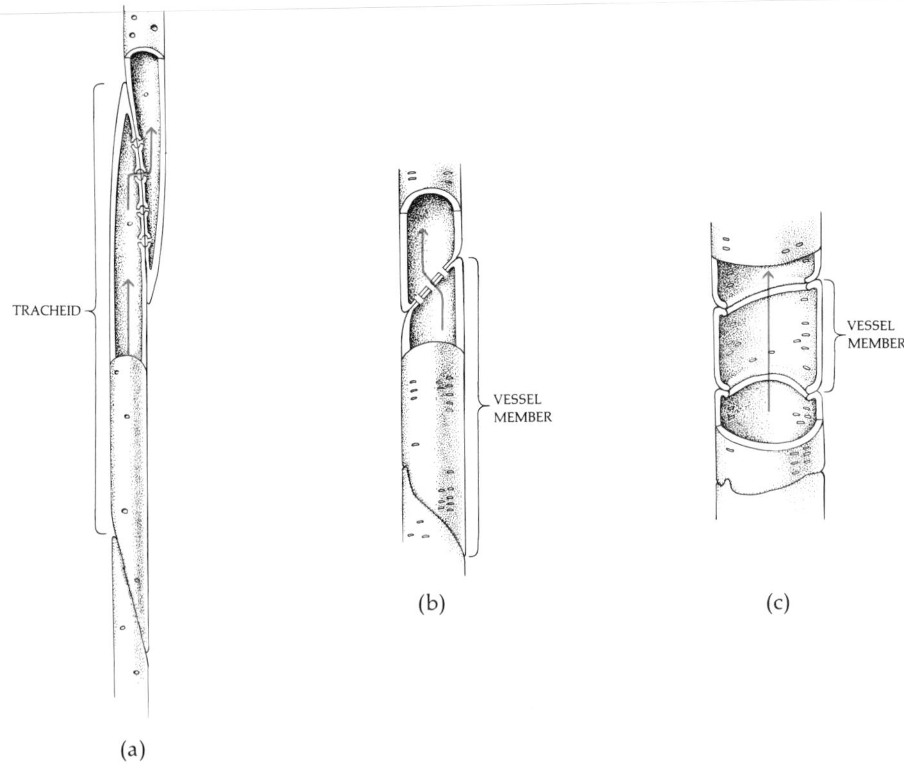

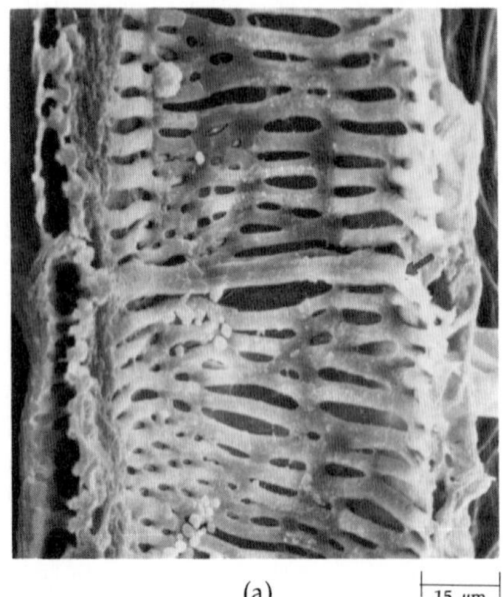

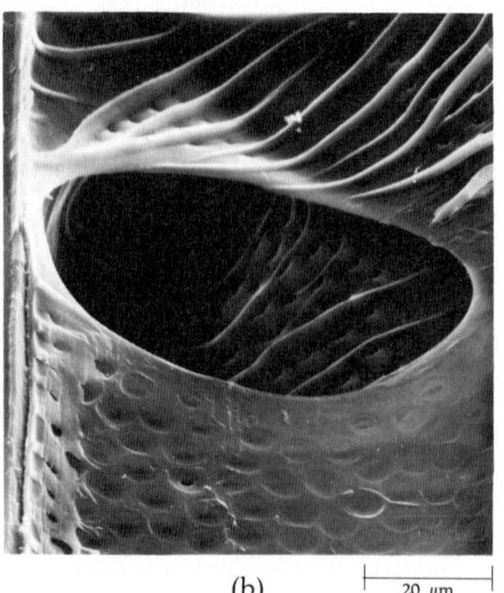

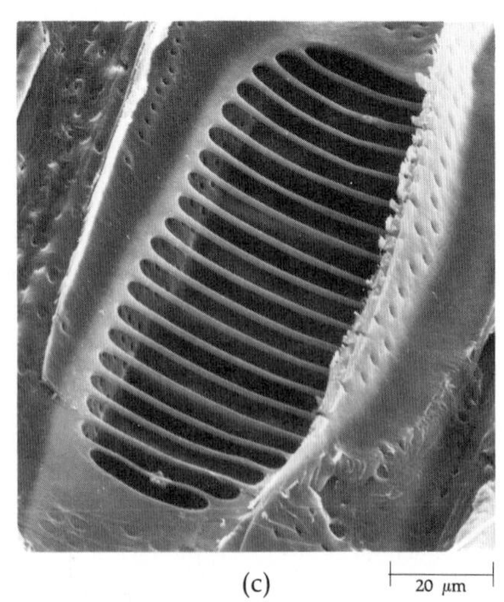

28–11

Scanning electron micrographs of xylem vessels. (a) Parts of two vessel members of squash, cut lengthwise so that half of the cylinder is seen from the inside. The arrow points to the boundary between the two members. Notice also the numerous pits in the walls of the two vessel members.

The end walls between adjacent vessel members may break down entirely, as in linden (b), or only partially, but in a definite pattern, as in red alder (c).

28-12
Cross sections of two dicot stems and one monocot stem. (a) *In alfalfa, also a dicot, the cylinder is made up of separate vascular bundles.* (b) *In this young stem of the linden, a dicot, the vascular tissue forms a continuous cylinder. The stem contains mucilage ducts, which stain red.* (c) *In corn, a monocot, numerous vascular bundles are scattered throughout the ground tissue.*

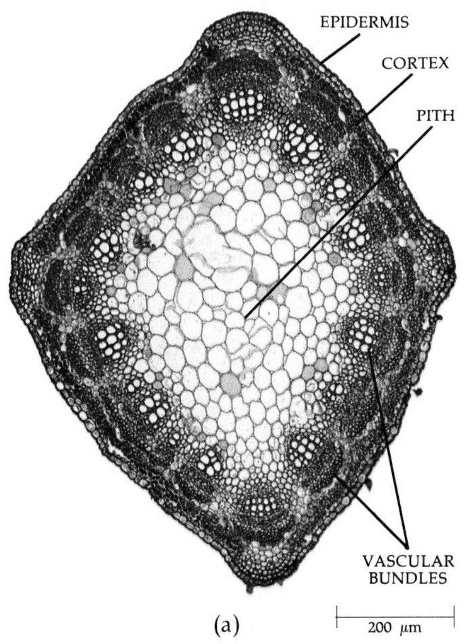

Specialized cells in the xylem conduct water and minerals from the roots to other parts of the plant body. It is customary to think of xylem as transporting water up and phloem as transporting sugars down, but if you think of the various shapes of plants you can see that water must also often be transported laterally, as along a tendril, or even down, as to the branches of a weeping willow. Conversely, sugars must often go upward, as into a flower or fruit.

In angiosperms, the conducting cells of the xylem are tracheids and vessel members. Both of these cell types have thick secondary walls impregnated with lignin, and both are dead at maturity. Tracheids are long, thin cells that overlap one another on their tapered ends (Figure 28–10a). These overlapping surfaces contain thin areas, pits, where no secondary wall has been deposited. Water passes from one tracheid to the next through the pits. Vessel members, which are much larger, also differ from tracheids in that their end walls contain one or more perforations or are broken down entirely (Figure 28–10b and c). Thus, the vessel members form a continuous *vessel*, which is a more effective conduit than a series of tracheids. Lower vascular plants and most gymnosperms have only tracheids. Most angiosperms have both tracheids and vessels.

Stem Patterns

In monocot stems and young stems of dicots, the xylem and the phloem are usually arranged in longitudinal parallel strands, the vascular bundles, which are embedded in the ground tissue. In young dicot stems, the vascular bundles are arranged in a ring, the vascular cylinder, around a central area of ground tissue called the *pith* (Figure 28–12a and b). The cylinder of ground tissue outside the vascular bundles is called the *cortex*. Within each bundle, the xylem is characteristically on the inside, adjacent to the pith, and the phloem is on the outside, adjacent to the cortex. In monocots, the vascular bundles are scattered throughout the ground tissue (Figure 28–12c).

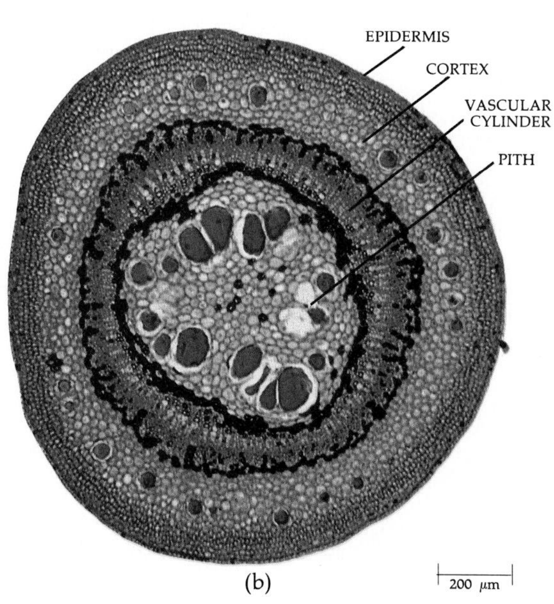

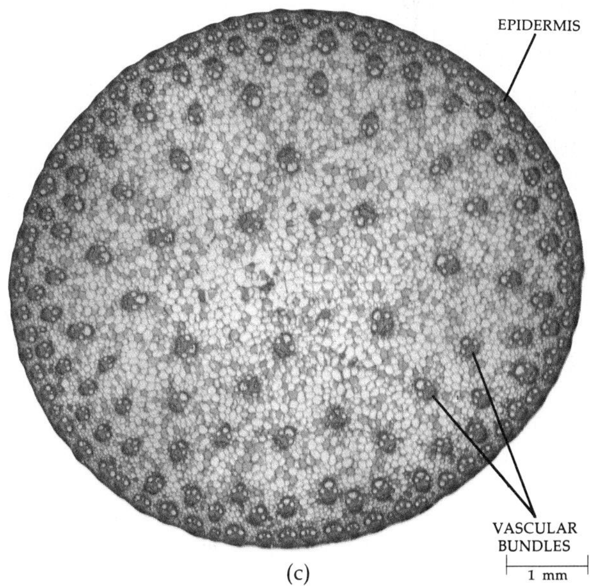

Special Adaptations of Stems

The stems of some climbing plants coil themselves around the structures on which they are growing. Others produce modified branches in the form of tendrils. (As we saw previously, tendrils may also be modified leaves.) The tendrils of English ivy, grape, and Virginia creeper are all modified stems.

Runners, such as those found in most varieties of strawberry, are long, slender stems that grow along the surface of the soil. Rhizomes are underground stems that often, as in the grasses, form buds and from these buds produce upright stems bearing leaves and flowers.

Stems also may be adapted for food storage. White potatoes are enlarged stems known as tubers. In some species of plants that grow in arid environments, stems are adapted for water storage. The water-storing tissues consist of large parenchyma cells that lack chloroplasts; the stem of a cactus may be 98 percent water by weight.

ROOTS

Roots are specialized structures that anchor the plant and, as we noted at the beginning of the chapter, take up water and essential minerals. In an older plant, the root system may make up more than half of the plant body. The lateral spread of tree roots is usually greater than the spread of the crown of the tree. In a study made on a four-month-old rye plant, the total surface area of the fibrous root system was calculated to be 639 square meters, 130 times the surface area of the leaves and stem. Root growth is affected by soil conditions and availability of water. The deepest known root was that of a pine growing on sandy, porous soil. It had penetrated to a depth of about 6.5 meters.

The Structure of the Root

The internal structure of the angiosperm root is comparatively simple. In dicots and most monocots, the three tissue systems (dermal, ground, and vascular) are arranged in three concentric layers: the *epidermis*, the *cortex*, and the *vascular cylinder* (Figure 28–13).

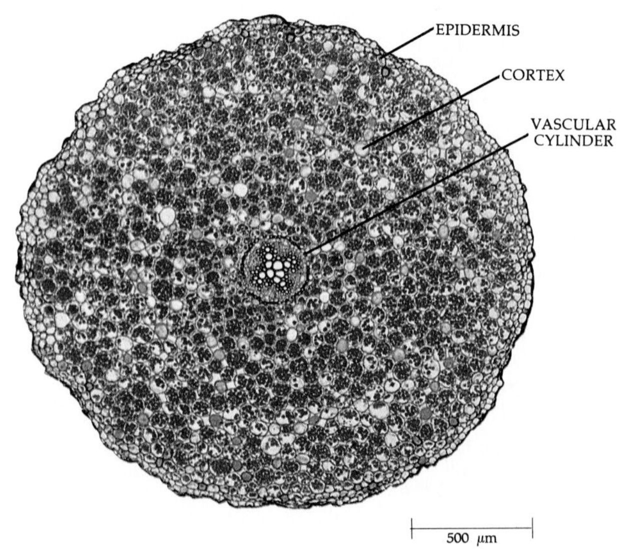

28–13
Root of a buttercup, a dicot, in cross section. Most of the cortical cells contain starch grains, stained purple in this preparation. The vascular cylinder is shown in more detail in Figure 28–16.

28–14
A radish seedling. Note the discarded seed coat, the cotyledons, and the primary root with its numerous root hairs. Most of the uptake of water and minerals takes place through the root hairs, which begin to form just behind the growing tip of the root.

The Epidermis

The epidermis, which encloses the entire surface of the young root, absorbs water and minerals from the soil and protects the internal tissues. Its cuticle is very thin compared with that found on the surface of the leaf epidermis.

The epidermal cells of the root are characterized by fine, tubular outgrowths, known as root hairs (Figure 28–14). Root hairs are slender extensions of the epidermal cells; in fact, the nucleus of the epidermal cell is often found within the root hair. In the study of the rye plant previously mentioned, the roots were estimated to have some 14 billion root hairs. Placed end to end, they would have extended more than 10,000 kilometers. Much of the water and minerals that enter the root is taken up through these delicate outgrowths of the epidermis. In many species, however, mycorrhizal associations (page 451) seem to substitute for root hairs.

The Cortex

As you can see in Figure 28–13, the cortex occupies by far the greatest volume of the young root. The cells of the cortex are parenchyma cells, as in the ground tissue of the leaf and young stem; however, root parenchyma lacks mature chloroplasts. The cells contain stored starch and other organic substances. There are many air spaces in the cortex, and oxygen-containing air from the soil enters these spaces through the epidermal cells and is used by the cortical cells in respiration.

Unlike the rest of the cortex, the cells of the innermost layer, the *endodermis*, are compact and have no spaces between them. Each endodermal cell is encircled by a *Casparian strip*, a waxy band within the cell wall (Figure 28–15). The strip is continuous and is not permeable to water. Therefore, water and dissolved substances, which can move freely around the other cortical cells and through their cell walls, must pass through the cell membranes of endodermal cells. As you will recall (page 132), water, oxygen, and carbon dioxide pass easily through cell membranes, but many ions and other substances do not. Thus, the membranes of the endodermal cells regulate the passage of such substances into the vascular tissues of the root and so determine what enters the plant body.

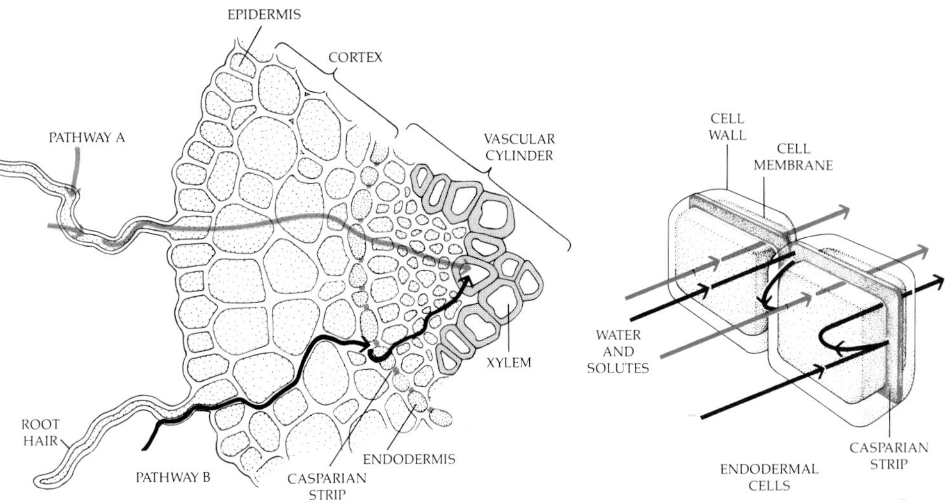

28–15
Diagrammatic cross section of a root, showing the two pathways of uptake of water and dissolved substances. Most of the solutes and some of the water entering the root follow pathway A, indicated in color; the solutes move by active transport and diffusion and the water by osmosis through the cell membranes and plasmodesmata (page 118) of a series of living cells. Most of the water and some of the solutes entering the root follow pathway B, indicated in black, flowing through the cell walls and along their surfaces. Notice, however, the location of the Casparian strip and how it blocks off pathway B all around the vascular cylinder of the root. In order to pass the Casparian strip, both water and solutes must be transported through the cell membranes of the endodermal cells, as in pathway A. After the water and solutes have crossed the endodermis, most of the solutes continue along pathway A, and most of the water returns to pathway B for the remaining distance to the xylem.

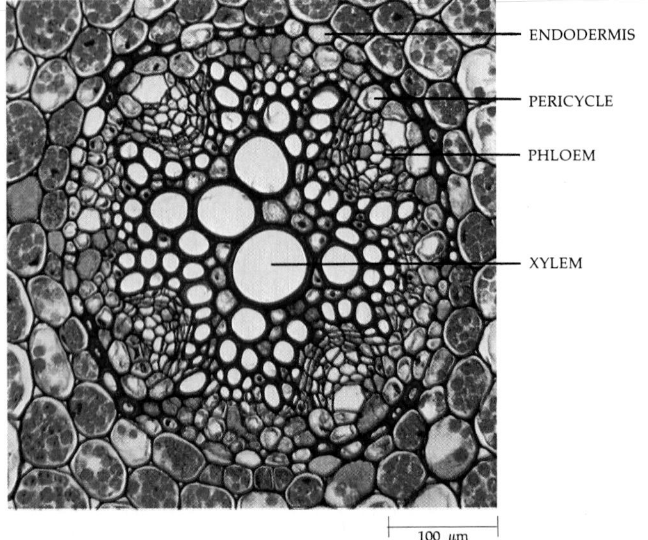

28–16
Details of the vascular cylinder of the buttercup root shown in Figure 28–13. The endodermis, which is outside the pericycle, is considered part of the cortex. The endodermis contains the Casparian strip.

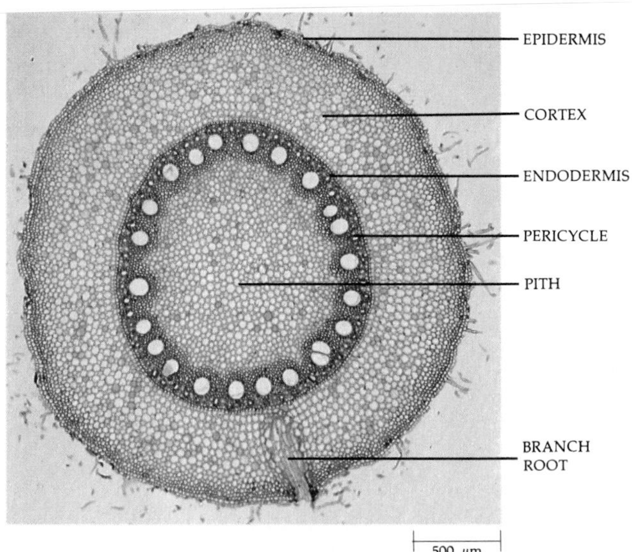

28–17
Cross section of the root of a corn plant (a monocot), showing the vascular cylinder enclosing the pith. Part of a branch root, which emerges from the pericycle, can be seen in the lower portion of the micrograph.

The Vascular Cylinder

The vascular cylinder of the root consists of the vascular tissues (xylem and phloem) completely surrounded by one or more layers of cells, the *pericycle*. Branch roots arise from the pericycle. In most species, the vascular tissues of the root are grouped in a solid cylinder, as shown in Figure 28–13. Figure 28–16 shows the details of the vascular cylinder of a buttercup root (a dicot). In some monocots, however, the vascular tissues form a hollow cylinder around a pith, a central core of ground tissue (Figure 28–17).

The vascular tissues of the root are continuous with those of the stem, and yet, as we have seen, their arrangement in the root is quite different from that in the stem. The change from the patterns observed in the root to those found in the stem is a gradual one. The region of the plant axis between stem and root in which the change occurs is known as the transition region.

Patterns of Root Growth

The first root of a plant, which originates in the embryo, is called the primary root. In many dicots, this root develops into a *taproot*, which, in turn, gives rise to lateral or branch roots (Figure 28–18a). In monocots, the primary root is usually short-lived and the final root system develops from the base of the stem; roots of this kind are called *adventitious roots*. ("Adventitious" describes any structure growing from other than the "usual" place.) These adventitious roots and their branches develop into a fibrous root system (Figure 28–18b).

28–18

Types of root systems. (a) Taproot system of a carrot (a dicot). (b) Fibrous root system characteristic of grasses (monocots).

Special Adaptations of Roots

Aerial roots are adventitious roots produced from aboveground structures. Some aerial roots, such as those of English ivy, cling to vertical surfaces and thus provide support for the climbing stem. In some plants, such as corn, aerial roots serve as prop roots (Figure 28–19a). Trees that grow in swamps, such as the red mangrove and the bald cypress, often have prop roots.

In swampy areas the soil is low in oxygen. Some trees that grow in swampy habitats develop roots that grow out of the water and serve not only to anchor but also to supply the root cells with the oxygen needed for respiration. Black mangroves, for example, have roots whose tips grow upward out of the mud and serve this aerating function (Figure 28–19b).

Most roots are storage organs, and in some species the roots are highly specialized for this function. Beets and carrots are examples of roots with an abundance of storage parenchyma.

ADAPTATIONS TO CLIMATE CHANGE

As we noted in Chapter 23, the angiosperms evolved during a relatively mild period in the earth's history. As the climate grew colder, some became extinct, and others were able to survive only near the equator. Some, presumably because of adaptations to drought (perhaps reflecting their highland origins), were able to survive in the cold, when water is locked in ice. Chief among such adaptations is the capacity to remain dormant.

Depending on their characteristic patterns of active growth, dormancy, and death, modern plants are classified as annuals, biennials, and perennials.

(a)

(b)

28–19

(a) Prop roots of corn. These are adventitious roots, arising from the stem. (b) Aerial roots (pneumatophores) of a black mangrove. The root tips push up out of the mud in which these coastal trees grow and take up oxygen needed by the roots for respiration.

28-20
Annual, biennial, and perennial plants. (a) *A desert annual, the Mexican gold-poppy. Many desert plants are annuals, racing through the entire life cycle from seed to flower to seed during a brief period following the seasonal rains.*

(b) *A biennial, the wallflower,* Erysimum. *Like other members of the mustard family, plants of this genus are found mostly in cooler regions of the Northern Hemisphere. This plant was photographed in Alpine County, California.*

(c) *An ancient perennial, the dawn redwood,* Metasequoia. *Twenty-five million years ago, the temperate zone of the Northern Hemisphere was covered with great forests of* Metasequoia. *Long thought to be extinct, living trees were discovered in western China in 1944. This tree in Philadelphia grew from seeds planted in 1947. Like the bald cypress and the larch, but unlike other conifers, the dawn redwood is deciduous, shedding all its leaves in the fall.*

(a)

(b)

Annuals, Biennials, and Perennials

Among *annual* plants, the entire cycle from seed to vegetative plant to flower to seed again takes place within a single growing season. Annuals include many of our weeds, wild flowers, garden flowers, and vegetables. All vegetative organs (roots, stems, and leaves) die, and only the dormant seed, which is characteristically highly resistant to cold, desiccation, and other environmental hazards, bridges the gap between one generation and the next. Annual plants are usually soft-stemmed (nonwoody), or *herbaceous*. (An herb, botanically speaking, is a nonwoody seed plant, a definition that differs from the more common, culinary designation.)

In *biennial* plants, the period from seed germination to seed formation spans two growing seasons. The first season of growth often results in a short stem, a rosette of leaves near the soil surface, and a root. The root is often a storage root, as in a sugar beet or a carrot. In the second growing season, the accumulated food reserves are mobilized for flowering, fruiting, and seed formation, after which the plant dies. Species that are characteristically biennials may, in some locations, complete their cycle in a single season or may require three or more years under adverse conditions.

Perennials are plants in which the vegetative structures persist year after year. The herbaceous perennials pass unfavorable seasons as dormant, underground roots, rhizomes, bulbs, or tubers. The woody perennials, which include vines, shrubs, and trees, survive above ground. Woody perennials flower only when they become adult plants; a horse chestnut, for instance, may not flower until it is 25 years old.

The great advantage to being woody and perennial is that height can be added each growing season. Thus leaves can overtop and shade neighbors, flowers can be more obviously displayed, and seeds can be dispersed over a wider area. The ancestral angiosperm is believed to have been a woody perennial. Modern plants in favorable climates, such as the tropical rain forest, may live year after year with little change during the annual cycle, just as did the ancestral plants. Perennials that live in areas where part of the year is unfavorable to growth show a variety of

(c)

adaptations. Some, such as cacti and most conifers, undergo little apparent change, although their rates of metabolism, and therefore of growth, change with the seasons. Among dicots, many of the common vines, shrubs, and trees are deciduous. Deciduous plants—plants that drop their leaves annually—are found in regions where there is a marked seasonal variation in available water. The broad leaf of a deciduous perennial presents a much more efficient light-collecting surface than the needlelike leaf of a conifer (see page 470). The annual replacement of leaves is expensive, however, in terms of energy and other resources and can only take place in areas where the soil is fertile enough to supply the necessary nutrients and where the growing season is long enough to show an energy profit.

SUMMARY

Plants are multicellular photosynthetic organisms adapted to life on land. The largest and most recently evolved division of plants is the Anthophyta, or angiosperms, the flowering plants. Angiosperms are divided into two classes: monocots and dicots. Characteristic differences include patterns of leaf venation, the arrangement of xylem and phloem in stems and roots, the number of seed leaves (cotyledons), and patterns of root growth.

The plant body has specialized photosynthetic areas (leaves), conducting and supporting structures (stems), and organs that anchor the plant in the soil and absorb water and minerals from it (roots). Three tissue systems—the dermal, the ground, and the vascular—are continuous throughout the plant body.

The leaf blade is attached to the stem by the petiole. The ground tissue of the leaf is composed principally of photosynthetic parenchyma cells. Parenchyma cells, which are thin-walled and many-sided, are the most common cells of the plant body. In the leaf, they include palisade cells and spongy parenchyma. Palisade cells, in which most of the photosynthesis takes place, are elongated cells with large central vacuoles. Spongy parenchyma cells, also photosynthetic, are surrounded by large air spaces. The palisade and spongy parenchyma cells make up the mesophyll, or "middle leaf."

The upper and lower surfaces of the leaf are one or more layers of transparent epidermal cells covered with a waxy layer, the cuticle. Specialized pores, the stomata, open and close, regulating the exchange of gases and the release of water vapor. Veins, the vascular bundles of the leaf, conduct water and minerals to the mesophyll cells of the leaf (through the xylem) and transport sugars away from them (through the phloem). The vascular tissues of the leaf are continuous with those of the stem.

Green stems, like leaves, have an outer layer of epidermal cells covered with a cuticle. The bulk of the young stem is ground tissue, made up largely of parenchyma cells, and may be divided into an outer cylinder (the cortex) and an inner core (the pith). Collenchyma cells and sclerenchyma cells (fibers and sclereids) may also be present.

The vascular tissues consist of phloem and xylem. In angiosperms, the conducting cells of the phloem are the sieve-tube members, living cells with perforated end walls that form continuous sieve tubes. Associated closely with each sieve-tube member is a companion cell. The conducting tissues of the xylem are made up of a series of tracheids or vessel members. Tracheids and vessel members characteristically have thick secondary walls and are dead at maturity. Phloem and xylem also contain parenchyma cells and fibers.

Young roots have an outer layer of epidermis and a very thin cuticle. Extensions of the epidermal cells form root hairs, which greatly increase the absorptive surface of the root. Beneath the epidermis is the ground tissue of the root, the cortex, composed mostly of parenchyma cells, often with large storage capacities. The innermost layer of the cortex is the endodermis, a single layer of specialized cells whose walls contain a waterproof zone, the Casparian strip. Just inside the endodermis is another layer of cells, the pericycle, from which branch roots arise. Within the pericycle are the xylem and phloem.

Angiosperms are classified as annuals, biennials, or perennials, depending on whether the plant body dies at the end of one growing season (annual) or after two seasons (biennial) or whether vegetative portions of the plant body persist (perennial). In areas where there is marked seasonal variation in the availability of water, perennials are often deciduous.

QUESTIONS

1. Distinguish among the following: monocot/dicot; dermal tissue system/ground tissue system/vascular tissue system; parenchyma/collenchyma/sclerenchyma; sieve-tube member/tracheid/vessel member; annual/biennial/perennial.

2. Sketch the interior of a leaf, labeling the principal cells and tissues. Describe the function of each of the labeled parts.

3. Some leaves are specialized for functions other than photosynthesis. List three other functions for which leaves may be specialized, and give an example of each specialization.

4. Sketch cross sections of (a) a dicot stem, (b) a monocot stem, and (c) a root. Label the principal cells and tissues in each sketch, and describe the function of each of the labeled parts.

5. What two cell layers are present in roots but absent in stems? What functions do the cells of these layers perform in the roots, and why are they necessary for the survival of the plant?

6. We have seen that the tissue of the root cortex contains many air spaces. Also, we noted that some plants growing in swampy areas have aerial roots. You may have discovered that overwatering can kill house plants. What do these data indicate about roots?

7. It has often been said that the principal differences among root, stem, and leaf are quantitative in nature rather than qualitative. Explain.

8. What are the major adaptations that enable plants to survive the periodic drought (winter) of temperate regions?

9. Where might you expect to find a leaf with stomata only on the bottom? Only on the top?

10. Reexamine the oleander leaf in Figure 28-6, relating its various structural features to the ecology of the oleander.

11. Label the micrograph at the left.

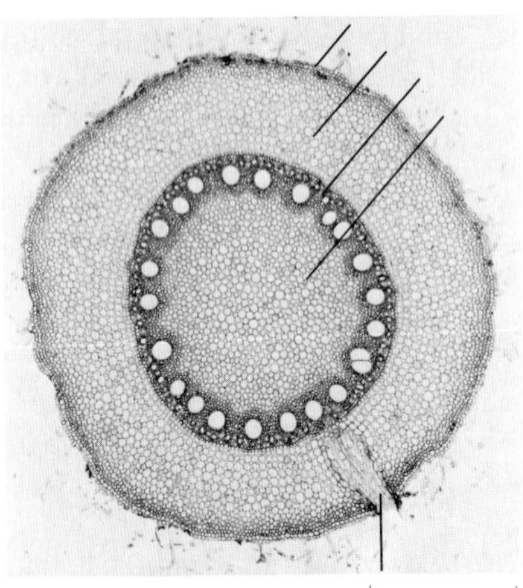

CHAPTER 29

Plant Reproduction, Development, and Growth

For flowering plants, a new cycle of life begins when a pollen grain—often brushed from the body of a foraging insect—comes into contact with the stigma of a flower of the same species. As we saw in Chapter 23, a flower is exquisitely designed to achieve this act of pollination. Unlike the gonads of animals, which are permanent organs that develop in the embryo, flowers, the reproductive structures of plants, are transitory. After pollination, some parts of the flower become the fruit; the others die and are discarded.

THE FLOWER

Most flowers consist of four sets of floral appendages (Figure 29-1); each floral part is thought to be, evolutionarily speaking, a modified leaf (see page 472). The floral parts may be arranged spirally on a more or less elongated stalk, or similar parts—such as the petals—may be located at one level in a whorl.

The outermost parts of the flower are the *sepals*, which are commonly green and obviously leaflike in structure. The sepals, collectively known as the *calyx*, enclose and protect the flower bud. Next are the *petals*, collectively called the *corolla*; these may also be leaf-shaped but are often brightly colored. The petals advertise the presence of the flower among the green leaves, attracting insects or other animals that visit flowers for their nectar or for other edible substances and so carry pollen from flower to flower. Calyx and corolla together are known as the *perianth*.

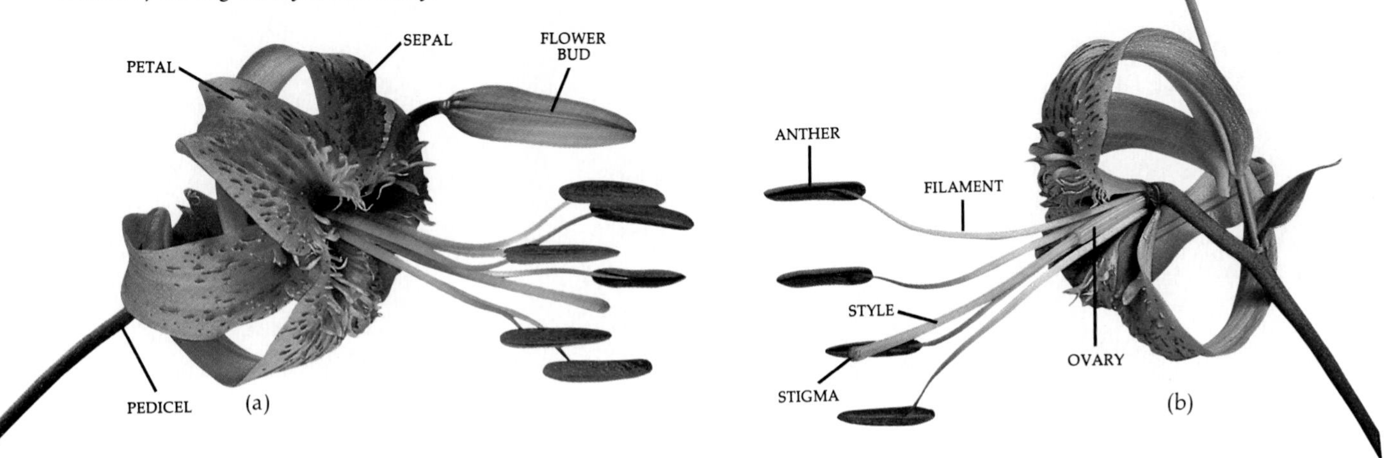

29-1
Parts of a lily flower. (a) *A complete flower in which both the sepals and the petals are brightly colored.* (b) *A lily partially dissected to reveal the ovary. The pollen-bearing anther and its filament make up the stamen. The carpel consists of the stigma, style, and ovary.*

29-2
Pollen grains. The walls of the pollen grain protect the male gametophyte on the journey between the anther and the stigma. These outer surfaces, which are remarkably tough and resistant, are often elaborately sculptured. As you can see, the pollen grains of different species are distinctly different: (a) *common ragweed (spiny pollen grains such as these, which cause hay fever, are common among the composites);* (b) *sand verbena, a dicot;* (c) *corn, a monocot (smooth pollen grains are found in most wind-pollinated plants).*

(d) *Pollen grains of cotton adhering to a stigma.*

Within the corolla are the stamens. Each stamen consists of a single elongated stalk, the filament, and at the end of the filament, the anther. The pollen grains, formed within the anther, contain immature male gametophytes. (For a review of the concept of alternation of generations, see page 250.) When ripe, the pollen grains are released, often in large numbers, through narrow slits or pores in the anther.

The centermost appendages of the flower are the carpels, which contain the female gametophytes. A single flower may have one carpel or several carpels, which may be separate or fused together. Typically a single carpel or fused carpels consist of a stigma, which is a sticky surface specialized to receive the pollen; a slender stalk, the style, down which the pollen tube grows; and a base, the ovary. Within the ovary are one to many ovules, each of which encloses a female gametophyte with a single egg cell. After the egg cell is fertilized, the ovule develops into a seed.

A flower that contains all four floral parts—sepals, petals, stamen, and carpel—is known as a complete flower.

In some species, flowers are either male (staminate) or female (carpellate). Male and female flowers may be present on the same plant, as in corn, squash, oaks, and birches; such plants are said to be monoecious ("in one house"). Species in which the male and female flowers are on separate plants, such as the tree of heaven *(Ailanthus)*, the date palm, and the American mistletoe, are known as dioecious ("in two houses").

THE POLLEN GRAIN

By the time a pollen grain is released from its parent flower, it usually consists of three haploid cells—two sperm cells contained within a larger cell (known as a tube cell), each of which has a haploid nucleus (three haploid nuclei in all). The tube cell, in turn, is enclosed by the thickened outer wall of the pollen grain. The pollen grain contains its own nutrients and has so tough an outer coating that intact grains thousands of years old have been found in peat bogs.

As you will recall, in many multicellular algae and more primitive plants, such as ferns, there is a distinct cycle of alternation of generations in which the sporo-

(a)

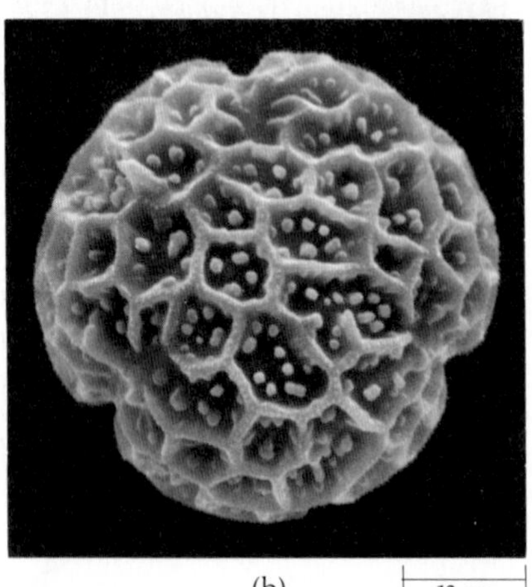

(b)

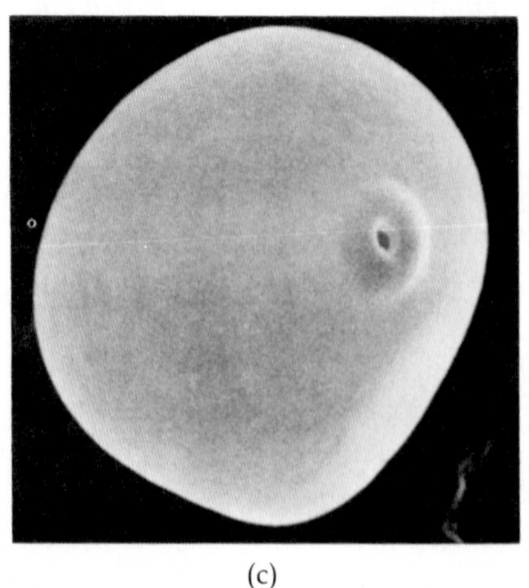
(c)

(d)

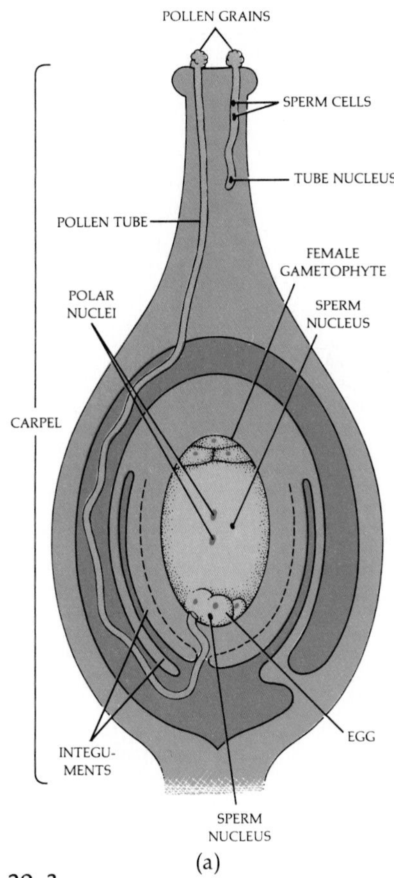

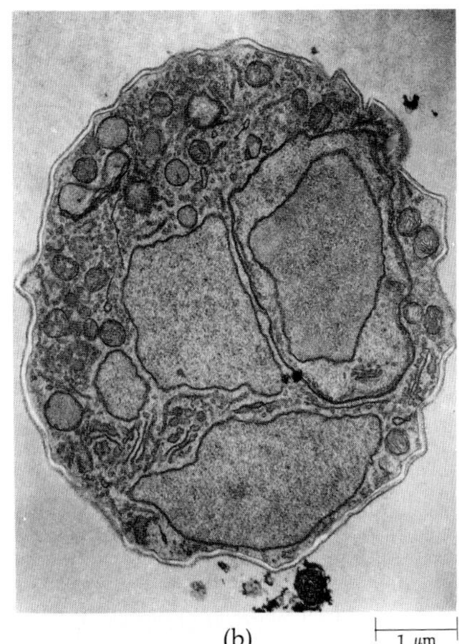

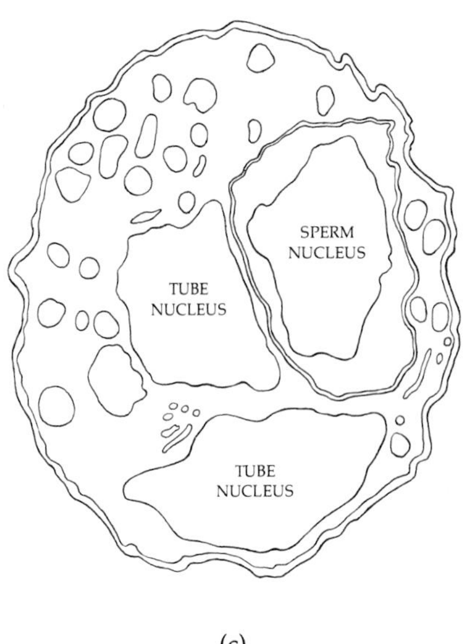

29–3

(a) *Fertilization in angiosperms. The pollen tube of the male gametophyte, or pollen grain, grows down through the style and enters the ovule, in which the female gametophyte has developed to a seven-cell stage. One of the sperm nuclei unites with the egg nucleus, and the zygote is formed. The other sperm nucleus fuses with the two polar nuclei that are present in a single large cell (which in the drawing fills most of the ovule). From the resulting triploid* (3n) *cell, the endosperm will develop. The carpel shown here contains a single ovule.*

(b, c) *Electron micrograph and diagram of a cross section of a pollen tube. Two lobes of the single tube nucleus are visible. This nucleus directs the formation of the pollen tube and eventually disintegrates. Numerous mitochondria are visible in the tube cell, as are several plastids. At the right is a sperm cell, which is actually a cell within a cell.*

phyte produces spores that produce gametophytes that produce gametes, with gametophyte and sporophyte having separate existences. In the course of plant evolution, the gametophyte stage has been steadily reduced in size. In the angiosperms, all that remains of the male gametophyte is the tough, tiny pollen grain and the pollen tube that grows from it. The sperm cells are the gametes.

FERTILIZATION

Once on the stigma, the pollen grain germinates, and, under the influence of the tube nucleus, a pollen tube grows down through the style into an ovule (Figure 29–3). This may be a long distance; in the case of *Petunia,* for instance, the length of the pollen tube is 900 times the diameter of the pollen grain. The number of grains reaching the stigma is often greater than the number of ovules available for fertilization, creating intense competition among pollen tubes, the race going to the swift.

The ovule contains the female gametophyte, which has also become reduced in size in the course of evolution. In many species it consists of seven cells, with a total of eight haploid nuclei. One of the seven cells is the egg, containing a single haploid nucleus. On either side of the egg are two small cells known as synergids. At the opposite end of the gametophyte are three small cells, the antipodal cells, whose function, if any, is unknown. The large central cell contains two haploid nuclei, called the polar nuclei because they move to the center from each end, or pole, of the gametophyte.

The nucleus of one of the two sperm cells carried by the pollen tube unites with the egg nucleus. The fertilized egg cell, or zygote, develops into the embryo—the young sporophyte. The nucleus of the second sperm cell unites with the two nuclei of the central cell. From the resultant $3n$ cell, a specialized tissue called the *endosperm* develops. It surrounds and nourishes the embryo. These extraordinary phenomena of fertilization to form a zygote and triple fusion to form an endosperm—together called "double fertilization"—take place, in all the natural world, only among the flowering plants.

(a)

(b)

(c)

(d)

(e) (f) (g) (h)

29–4

Flowers—forms and variations. (a) *Stamens and stigma of a white crocus.* (b) *In* Hibiscus, *a column of stamens is fused around the style.* (c) *Corn* (Zea mays), *a monoecious flower. The tassels, at the top of the stem, are the male (pollen-producing) flowers. Each thread of "silk," seen emerging from the ear of corn, is the combined stigma and style of a female flower.* (d) *The passion flower* (Passiflora), *with five stamens and three styles borne on a stalk in the center of the elaborate flower.* (e) *In the southern magnolia, the carpels form a cone-shaped receptacle from which curved styles emerge. The yellow stamens, which surround the carpels, have dropped off. The magnolia represents a very primitive form of flower with numerous separate floral parts arranged in a clearly evident spiral pattern.* (f) *Water hemlock, a member of the carrot family,* (g) *lupine* (Lupinus polyphyllus), *and* (h) *blueberry* (Vaccinium) *provide examples of inflorescences, or flower clusters.*

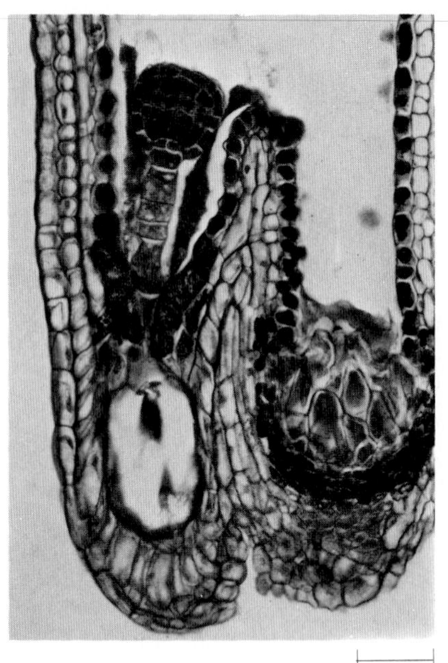

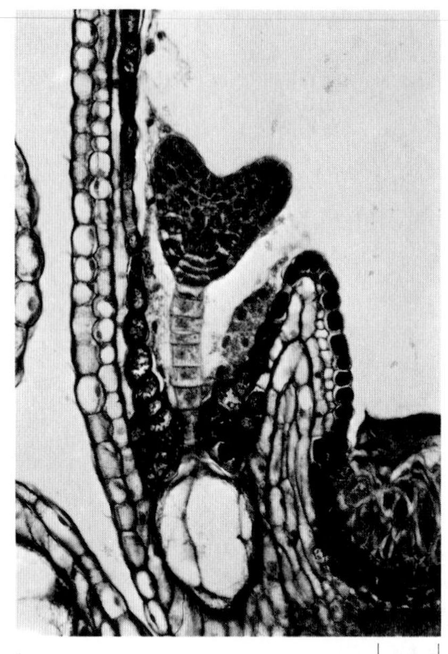

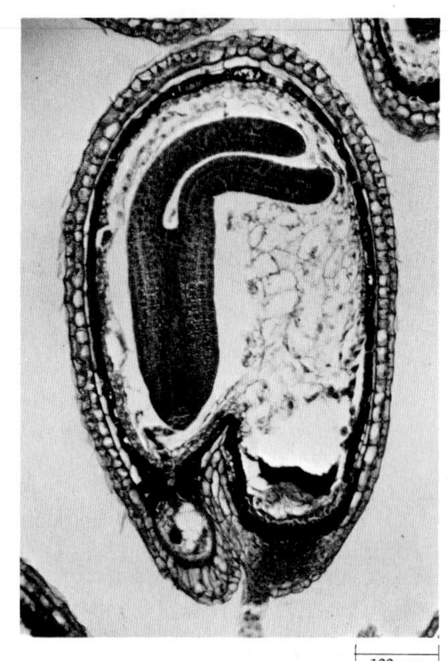

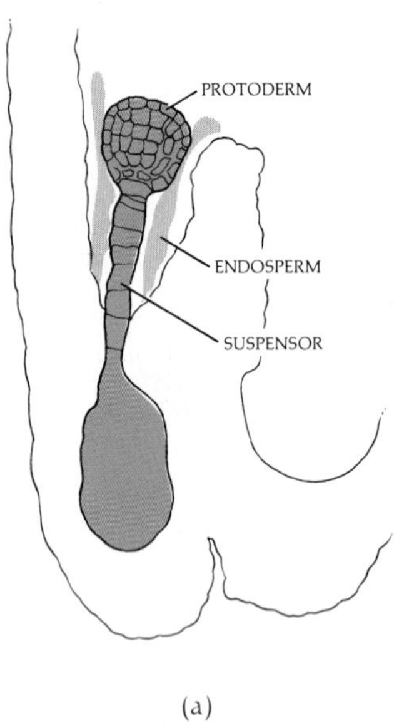

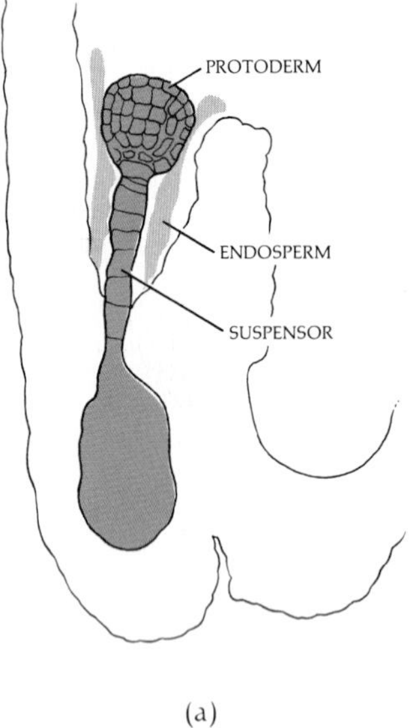

(a)

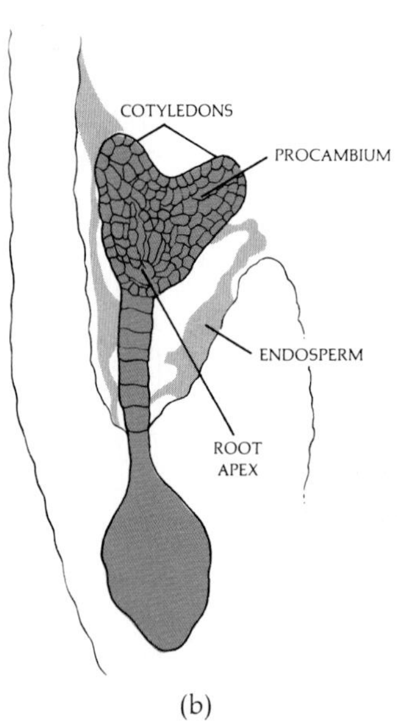

(b)

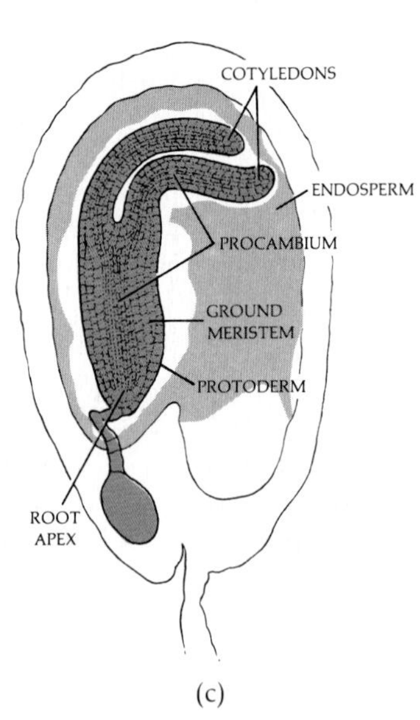

(c)

29–5
Some stages in the development of the embryo (shown in color) of shepherd's purse (Capsella bursa-pastoris), *a dicot.* (a) *The embryo proper is globular. The protoderm, the first embryonic tissue to differentiate, is the future epidermis. The large cell at the bottom is the basal cell of the suspensor.* (b) *The cotyledons are beginning to emerge. The procambium, a second embryonic tissue, will later give rise to the vascular tissues of the plant.* (c) *The cotyledons have developed further. Additional tissue differentiation has produced the ground meristem, the embryonic tissue from which the ground tissues of leaves, stems, and roots are derived. The three embryonic tissues, known as the primary meristems, are continuous between the cotyledons and the axis of the embryo.* (d) *The mature embryo. The apical meristems of the root and the shoot—the root apex and the shoot apex—are clearly differentiated.*

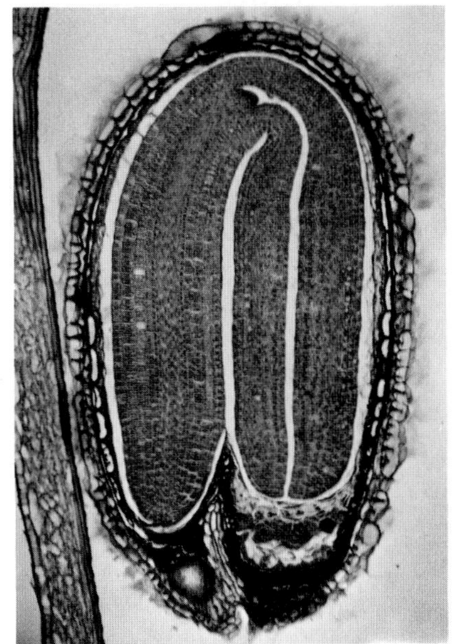

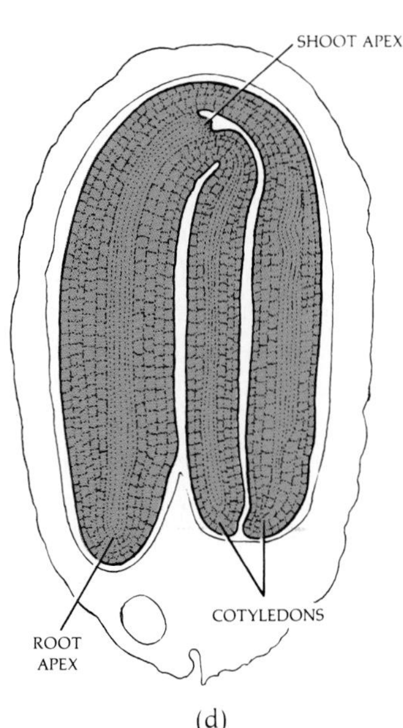

THE EMBRYO

Following "double fertilization," the 3*n* cell divides mitotically to produce endosperm. The zygote also divides mitotically, and as the embryo grows, its cells begin to *differentiate*—become different from one another. The embryo begins to take on a characteristic form, a process known as *morphogenesis*.

In its earliest stages, the embryo proper consists of a globular mass of cells. Suspensor cells, also formed from divisions of the fertilized egg cell, are believed to be actively involved in the nutrition of the embryo. As the development of the embryo proceeds, changes in its internal structure result in the beginnings of the tissue systems of the plant. At the same time, or slightly later, emergence of the one cotyledon in monocots, or the two in dicots, occurs. The stages in the development of a dicot embryo are shown in Figure 29–5.

In the earliest stages of embryonic growth, cell division takes place throughout the body of the young plant. As the embryo grows older, however, the addition of new cells becomes gradually restricted to certain parts of the plant body: the *apical meristems* of the root and the shoot. During the rest of the life of the plant, all the primary growth—which chiefly involves the elongation of the plant body—originates in these meristems.

THE SEED AND THE FRUIT

The seed consists of the embryo, which develops from the fertilized egg; the stored food, which consists of or derives from the endosperm; and the seed coat, which develops from the outermost layer or layers (integuments) of the ovule (Figure 29–6). At the same time, the fruit develops from the wall of the ovary. As the ovary

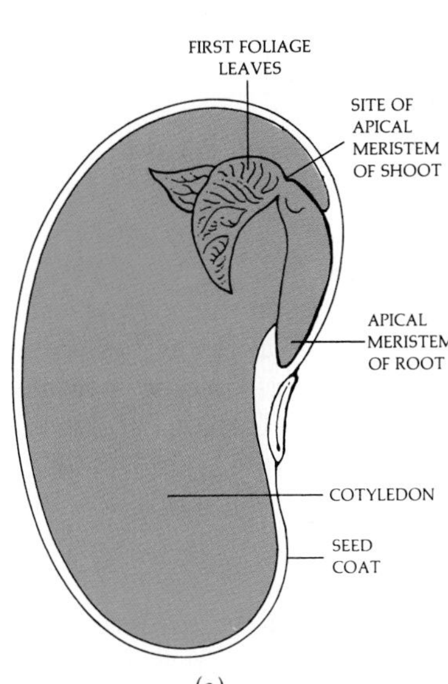

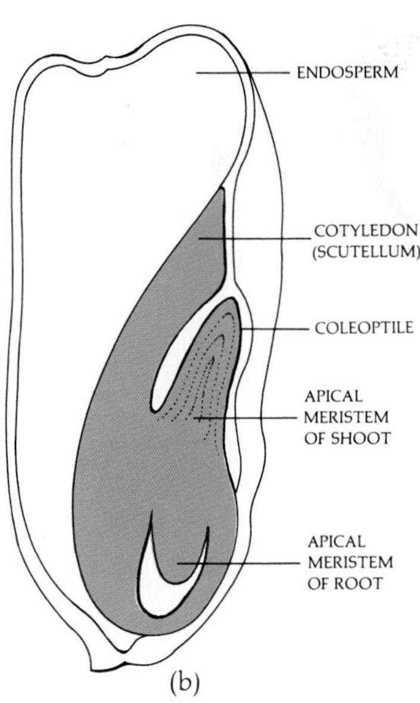

29–6
Seeds. (a) *In dicots such as the common bean, the endosperm is digested as the embryo (shown in color) grows, and the food reserve is stored in the fleshy cotyledons.* (b) *In corn and other monocots, the single cotyledon, known as the scutellum in corn and other grains, absorbs food reserves from the endosperm. The coleoptile is a sheath that encloses the apical meristem of the shoot.*

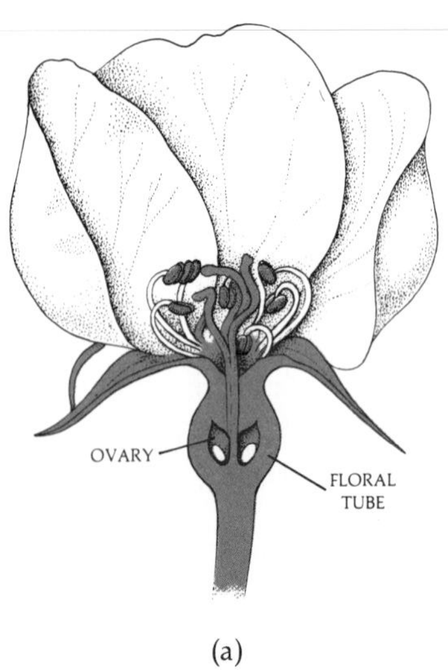

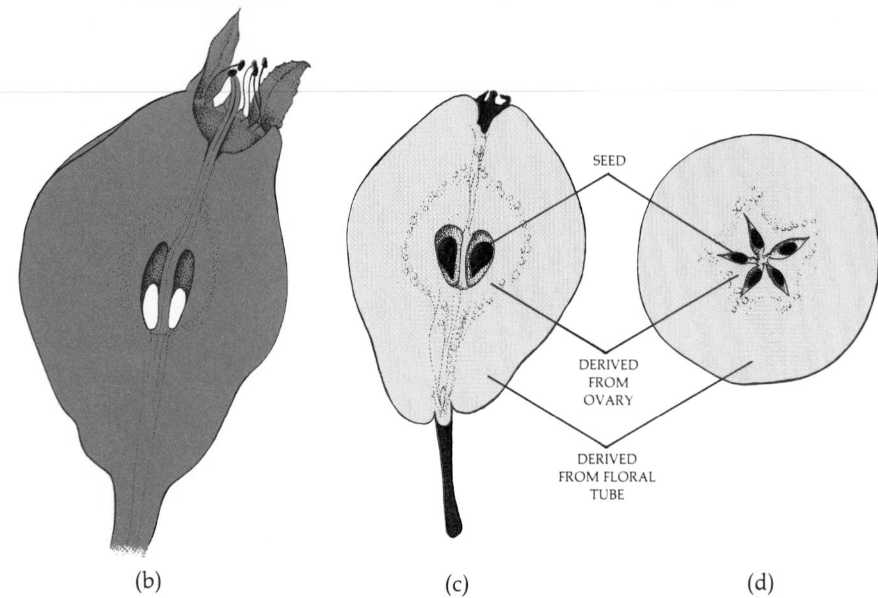

29–7
Development and structure of the pear. (a) *Flower of the pear. The ovary is the basal portion of the carpel.* (b) *Older flower, after petals have fallen.* (c) *Longitudinal section and* (d) *cross section of the mature fruit. The core of the pear is the ripened ovary wall. The fleshy, edible part develops from the floral tube.*

ripens into a fruit and the seeds form, the petals, stamens, and other floral parts may fall away (Figure 29–7).

Types of Fruits

As you know from your own observations, fruits have many forms. They are generally classified as simple, aggregate, or multiple (Figure 29–8), depending on the arrangement of the carpels from which they develop. Simple fruits develop from one carpel or the fused carpels of a single flower. Aggregate fruits, such as magnolia, raspberry, and strawberry, are formed from a number of separate carpels of one flower. Multiple fruits consist of the carpels of more than one flower. A pineapple, for example, is a multiple fruit formed from a flower cluster of many separate flowers. The ovaries of these flowers become fused as they mature.

Simple fruits are by far the most diverse of the three groups. When ripe, they may be soft and fleshy or dry. There are three main types of fleshy fruit—the berry, the drupe, and the pome. In the berry, examples of which are tomatoes, dates, and grapes, there are one to several carpels, each of which may have one or many seeds. The inner layer of the fruit wall is usually fleshy.

In the drupe, there are also one to several carpels, but each usually contains only a single seed. The inner wall of the fruit is stony and usually tightly adherent to the seed. Some familiar drupes are the peach, cherry, olive, and plum. The peach is a typical drupe; the skin, the succulent, edible portion of the fruit, and the stone are three layers of the mature ovary wall. The almond-shaped structure within the stone is the seed.

A highly specialized sort of fleshy fruit is the pome, which is characteristic of the subfamily of roses that produces rose hips. The pome is derived from an inferior ovary (page 475) in which the fleshy portion comes largely from the perianth, or floral tube. Apples and pears are pomes.

Dry fruits are classified into two groups, dehiscent and indehiscent. In dehiscent fruits, the tissues of the mature ovary wall break open, releasing the seeds. In indehiscent fruits, the seeds remain in the fruit after the fruit is shed from the parent plant.

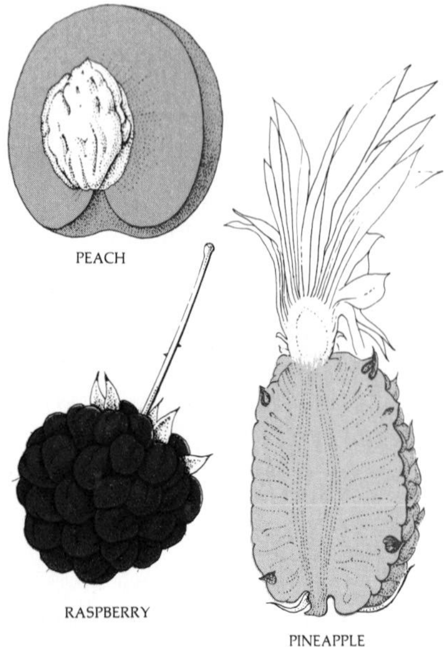

29–8
A simple fruit, an aggregate fruit, and, on the right, a multiple fruit.

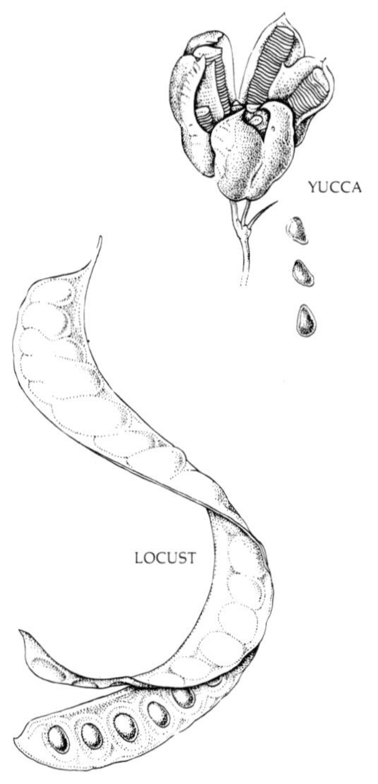

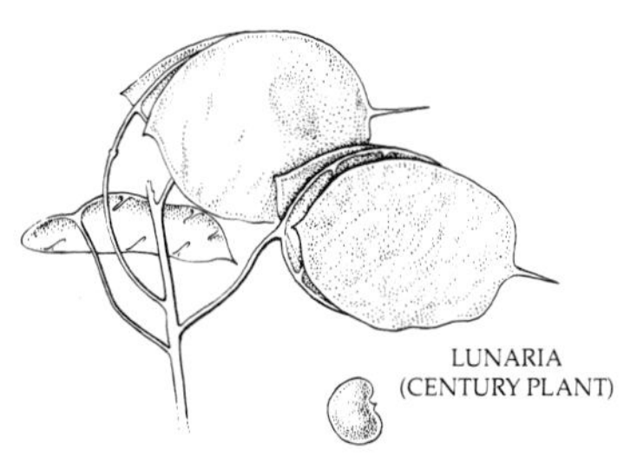

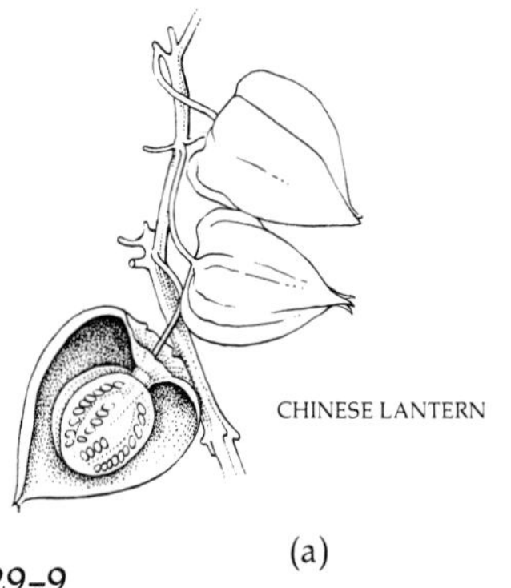

(a)

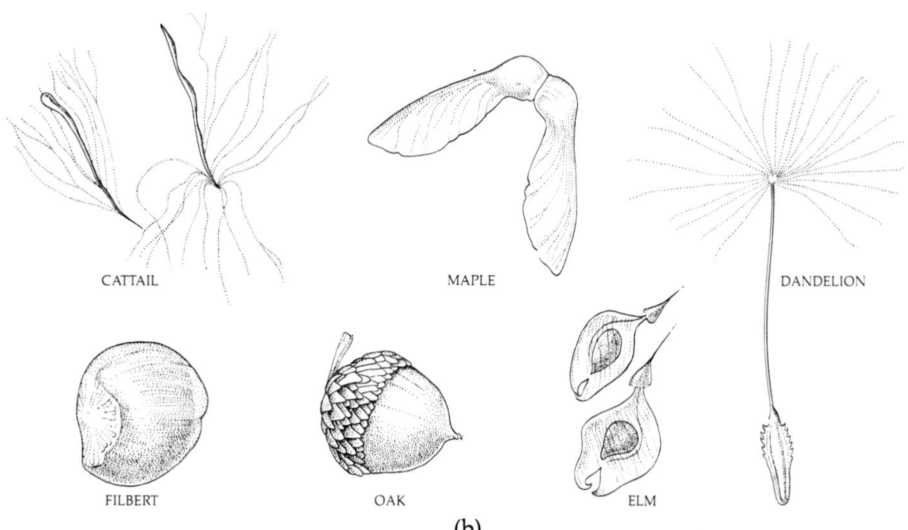

(b)

29–9
Examples of (a) *dehiscent and* (b) *indehiscent fruits.*

There are several sorts of dehiscent fruits. A fruit derived from a single carpel in which the ovary wall splits down one side is known as a follicle; the fruits of columbines and milkweeds are examples. In the pea family (legumes), the ovary splits down two sides; the pod is the mature ovary wall and the peas are the seeds. Other examples of dehiscent fruits are shown in Figure 29–9a.

Indehiscent fruits (Figure 29–9b) occur in many plant families. The most common is the achene, a small, single-seeded fruit. Winged achenes, such as those found in the elm and the ash, are called samaras. The most familiar kind of indehiscent fruit is the nut, which resembles the achene but has a stony coat and is derived from a compound ovary (an ovary formed from fused carpels). Examples of nuts are acorns and hazelnuts. Note that the word nut is used very indiscriminately in common speech: peanuts are legumes; pine nuts are seeds; almonds and coconuts are drupes.

Seed Dormancy

The seeds of most wild plants require a period of dormancy before they will germinate. This genetic requirement ensures that the seed will "wait" at least until the next favorable growth period. Some seeds can remain dormant and yet viable—with the embryo in a state of suspended animation—for hundreds of years. The record for dormancy, as far as is known, has been set by some seeds of East Indian lotus from a deposit at Pulantien, Manchuria. As determined by radioisotope dating, their age at the time of germination, in 1982, was 466 years.

The seed coat apparently plays a major role in maintaining dormancy. In some species, the seed coat seems to act primarily as a mechanical barrier, preventing the entry of water and gases, without which growth is not possible. In these cases, growth is initiated by the seed coat's being worn away in various ways—such as being abraded by sand or soil, burned away in a forest fire, or partially digested as it passes through the digestive tract of a bird or other animal. In other species, dormancy seems to be maintained chiefly by chemical inhibitors in the seed coat. These inhibitors undergo chemical changes in response to various environmental factors, such as light or prolonged cold or a sudden rise in temperature, which neutralize their effects, or they may be eroded or washed away by rainfall. Eventually, the embryo resumes growth.

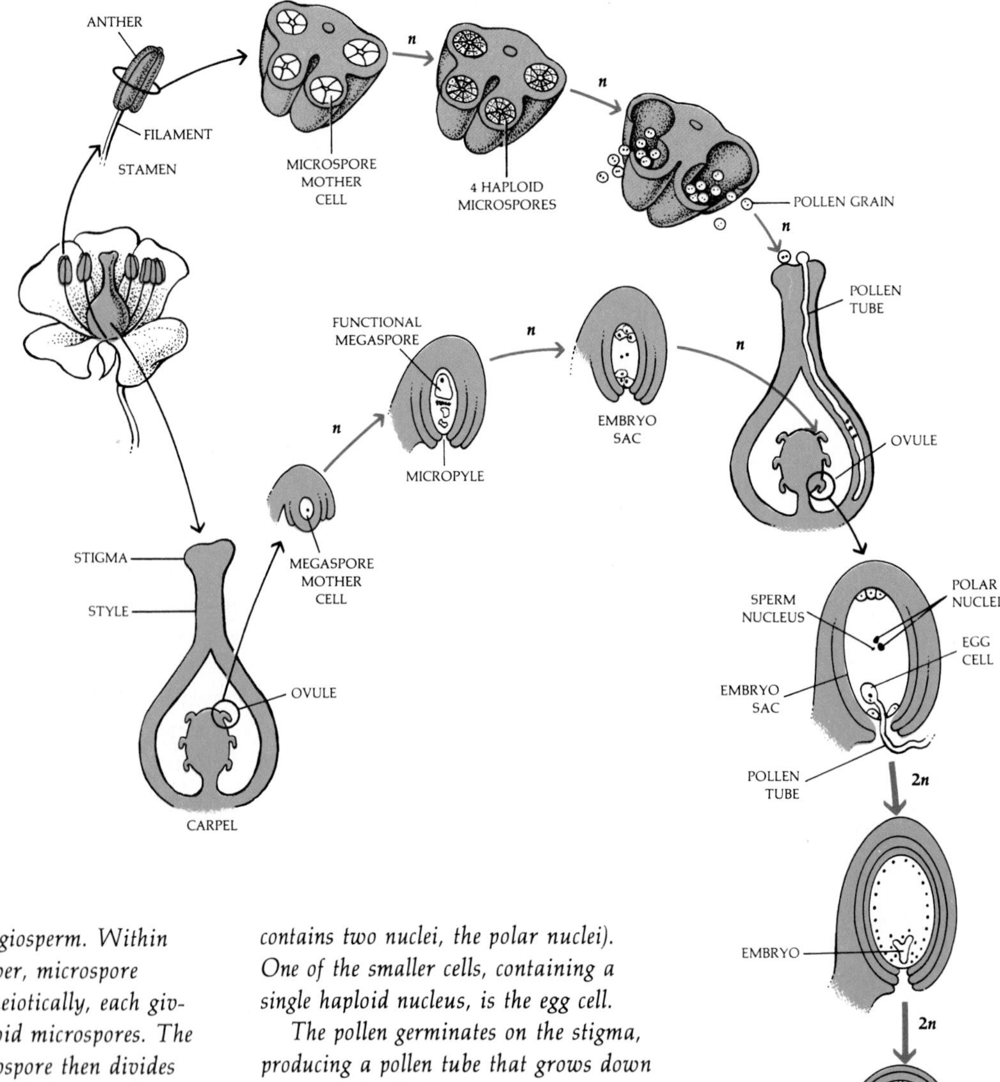

29–10
Life history of an angiosperm. Within the anther of the flower, microspore mother cells divide meiotically, each giving rise to four haploid microspores. The nucleus in each microspore then divides mitotically, and the microspore develops into a two-celled pollen grain, which is an immature male gametophyte. One of the cells subsequently divides again, usually upon germination, resulting in three cells per pollen grain: two sperm cells and the tube (pollen tube) cell.

Within the ovule, a megaspore mother cell divides meiotically to produce four haploid megaspores. Three of the megaspores disintegrate; the fourth divides mititically, developing into an embryo sac—the female gametophyte—consisting of seven cells with a total of eight haploid nuclei (the large central cell contains two nuclei, the polar nuclei). One of the smaller cells, containing a single haploid nucleus, is the egg cell.

The pollen germinates on the stigma, producing a pollen tube that grows down the style into the ovary. The growing tube enters the ovule through a small opening known as the micropyle. The two sperm cells pass through the tube into the embryo sac; one sperm nucleus fertilizes the egg cell, the other merges with the polar nuclei, forming the triploid (3n) endosperm. The embryo undergoes its first stages of development while still within the ovary of the flower, and the ovary itself matures to become a fruit. The seed, released from the mother sporophyte in a dormant form, eventually germinates, forming a seedling.

THE STAFF OF LIFE

Grains are the small, one-seeded fruits of grasses. Because they are dry, they can be stored as food. The collecting and storing of grains from wild grasses is believed to have been an important impetus to the agricultural revolution of some 11,000 years ago (see Chapter 56). Today we are heavily dependent on cultivated wheat, rice, corn, rye, and other grains. In many countries, they constitute the principal component of the human diet.

The fruit of wheat, sometimes known as the kernel, is made up of the embryo, the endosperm, and the surrounding seed coat. More than 80 percent of the bulk of the wheat kernel and 70 to 75 percent of its protein are in the endosperm. White flour is made from the endosperm. Wheat germ, the embryo, forms about 3 percent of the kernel. It is usually removed as wheat is processed because it contains oil, which makes the grain more likely to spoil. Bran is the seed coat plus the aleurone layer (outer part of endosperm); it constitutes about 14 percent of the kernel. The bran is also removed when wheat is milled to make white flour. Actually, the presence of the bran somewhat decreases the caloric value of the wheat kernel. We are unable to digest bran because it is mostly cellulose; its bulk therefore tends to speed the passage of food through our intestinal tracts, resulting in decreased absorption. The wheat germ and the bran, which contain most of the vitamins, are sometimes used for human consumption but more often are fed to livestock.

Wheat is about 9 to 14 percent protein, most of which, as we noted, is contained in the endosperm. Its protein value is diminished, however, by its deficiency in certain essential amino acids, notably lysine.

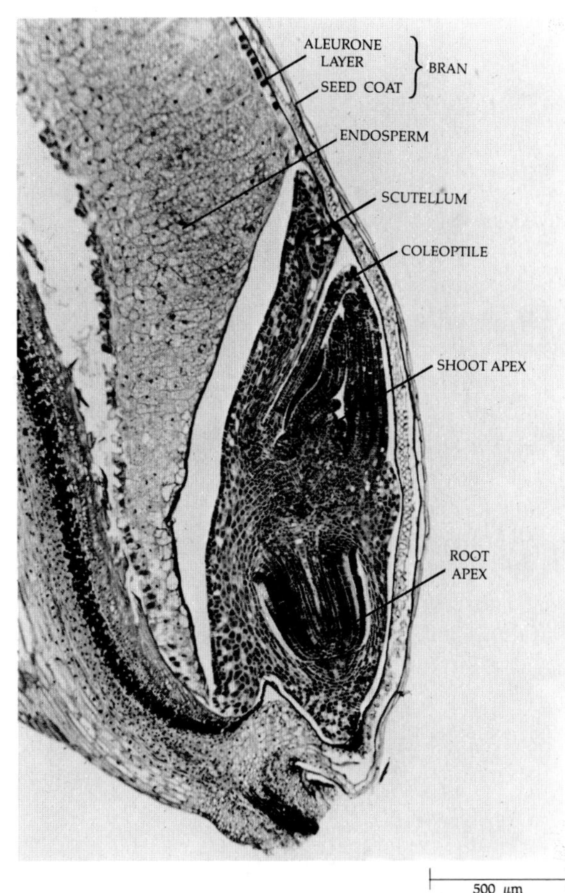

Longitudinal section of a wheat kernel.

The dormancy requirement in seeds apparently evolved only recently, geologically speaking, among groups of plants subjected to the environmental stress of increasing winter cold characteristic of the most recent Ice Age. By this time—only 1.5 to 2 million years ago—the angiosperms were already a highly diversified group, and different populations responded to these pressures in different ways. This explains why even closely related plants have different mechanisms for maintaining and breaking dormancy.

PRIMARY GROWTH

A dormant seed characteristically contains very little moisture (only about 10 to 15 percent of its total weight). Germination begins with a massive entry of water into the seed. The seed coat ruptures and the young sporophyte emerges.

Primary growth begins immediately. It involves the elongation of stems and roots, the formation of branches, and the differentiation of the three tissue systems of the plant. As we noted earlier, all primary growth originates in the apical meristems of the shoot and root.

The existence of such meristematic areas, which add to the plant body throughout the life of the plant, is one of the principal differences between plants and animals. Higher animals stop growing when they reach maturity, although the cells of certain "turnover" tissues, such as the skin or the lining of the intestine, continue to divide. Plants, however, continue to grow during their entire life span.

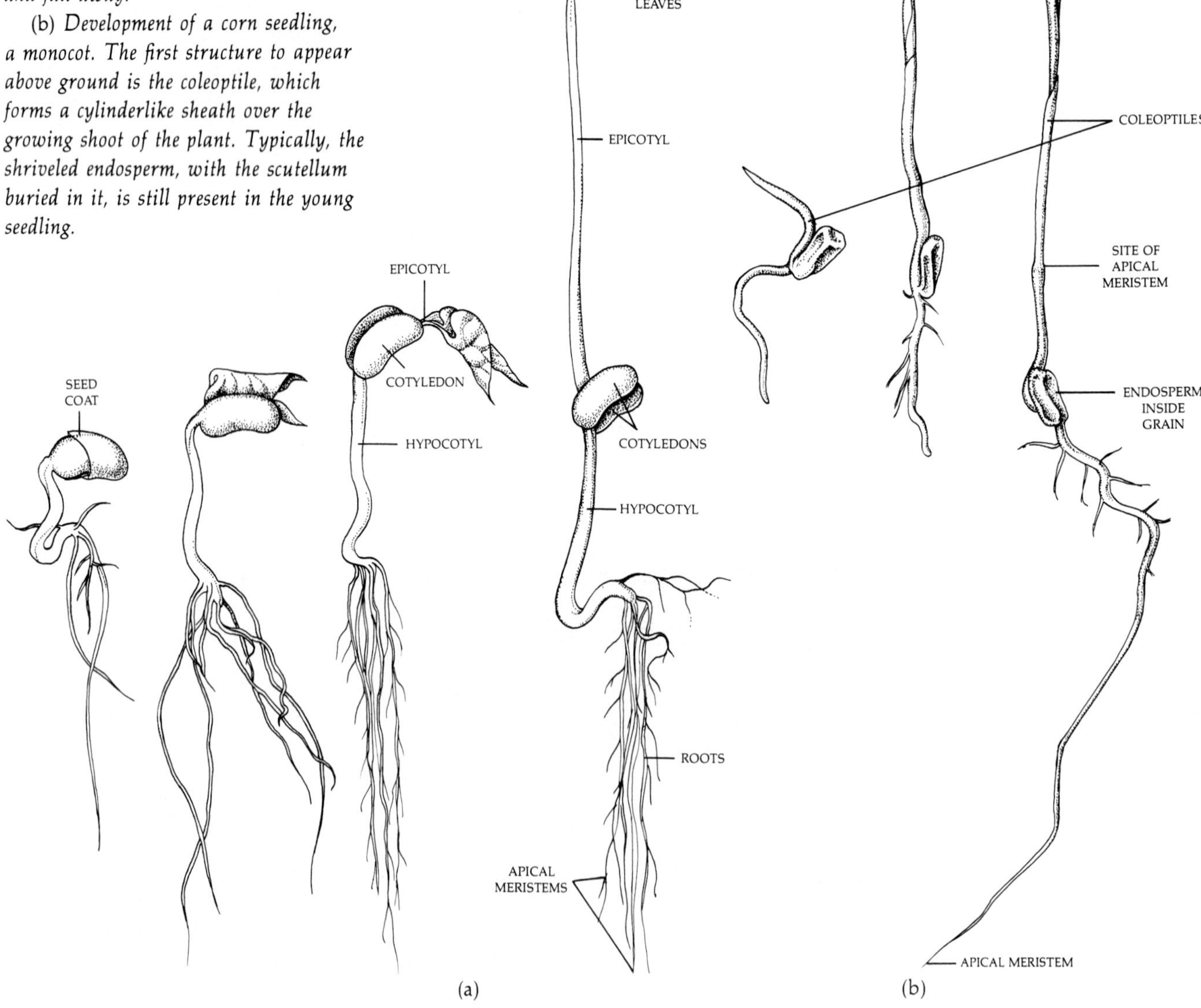

29–11
(a) *Development of a bean seedling, a dicot. Prior to germination, the seed imbibes water and swells, bursting the seed coat. The young root emerges first, then the hypocotyl ("below the cotyledons"). The cotyledons eventually will shrivel and fall away.*

(b) *Development of a corn seedling, a monocot. The first structure to appear above ground is the coleoptile, which forms a cylinderlike sheath over the growing shoot of the plant. Typically, the shriveled endosperm, with the scutellum buried in it, is still present in the young seedling.*

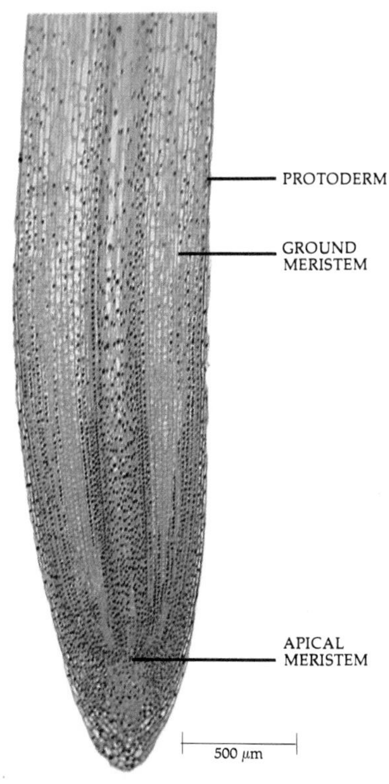

29–12
Longitudinal section of the root tip of an onion, a monocot. The primary meristematic tissues—protoderm, ground meristem, and procambium—can be distinguished close to the apical meristem.

Growth in plants is the counterpart, to some extent, of mobility in animals. Plants "move" by extending their roots and shoots, both of which involve changes in size and form. By growth, a plant modifies its relationship with the environment, turning toward the light and extending its roots. The sequence of growth stages in plants thus corresponds to a whole series of motor acts in animals, especially those concerned with the search for food and water. In fact, growth in plants serves many of the functions that we group under the term "behavior" in animals.

Primary Growth of the Root

The first part of the embryo to break through the seed coat, in nearly all seed plants, is the embryonic root, or radicle. At the very tip is the root cap, which protects the apical meristem as the root is pushed through the soil. The cells of the root cap wear away and are constantly replaced by new cells from the meristem.

Certain cells in the meristem retain the capacity to produce new cells and thus perpetuate the meristem. All the other cells in the root—which are the progeny of these relatively few meristematic cells—eventually differentiate, some becoming cells of the root cap and others forming the complex tissues of the root. The maximum rate of cell division occurs at a point well above the tip of the meristem. Then, just above the point where cell division is greatly reduced, the cells gradually elongate, growing to 10 or more times their previous length, often within the span of a few hours. This process of elongation is the principal cause of root growth, although, of course, growth is ultimately dependent on the production of new cells that become part of the zone of elongation.

As the cells elongate, they begin to differentiate. The first cells to mature are the sieve-tube members (the conducting cells of the phloem), followed by the vessel members (the conducting cells of the xylem). In the region of the root where the xylem first forms, the endodermis also takes shape. To the inside of the endodermis, the pericycle forms. This tissue gives rise to branch roots. In this same general region of the root, the epidermal cells differentiate and begin to extend root hairs into the crevices between the soil grains.

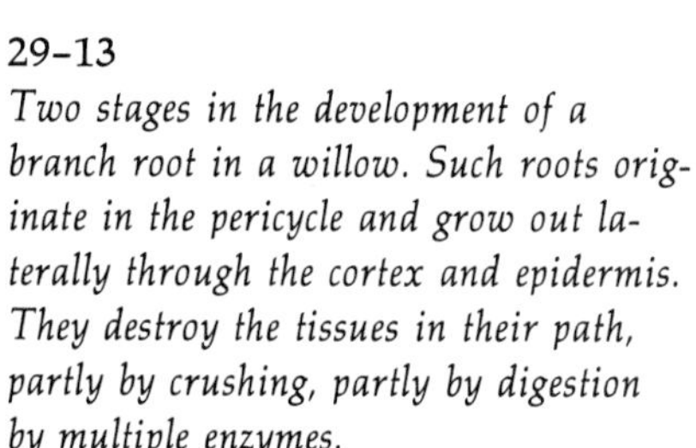

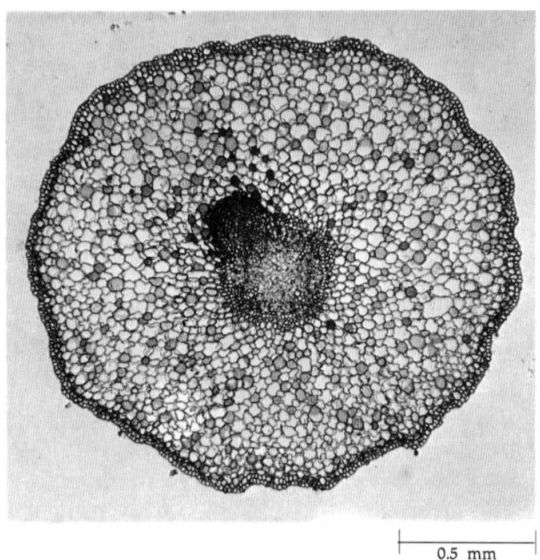

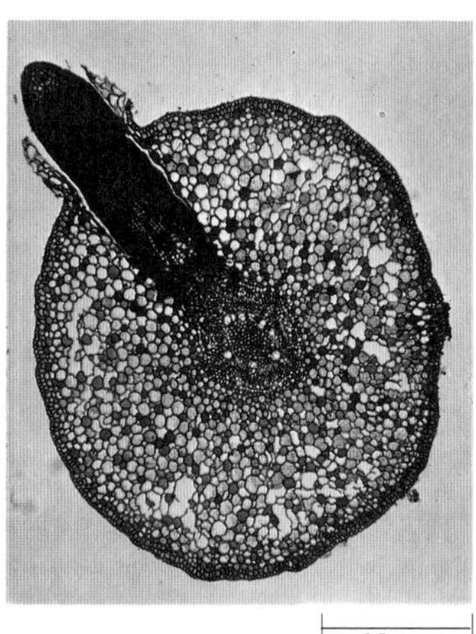

29–13
Two stages in the development of a branch root in a willow. Such roots originate in the pericycle and grow out laterally through the cortex and epidermis. They destroy the tissues in their path, partly by crushing, partly by digestion by multiple enzymes.

29–14
The growth regions of a dicot root. New cells are produced by the division of cells within the apical meristem. The cells above the meristem undergo a characteristic series of changes as the distance increases between them and the root tip. First, there is a maximum rate of cell division, followed by cell elongation with relatively little further division. As the cells elongate, they differentiate into various specialized tissues of the root. The protoderm becomes the epidermis, the ground meristem becomes the cortex, and the procambium becomes the xylem and phloem. Some of the cells produced in the apical meristem form the root cap.

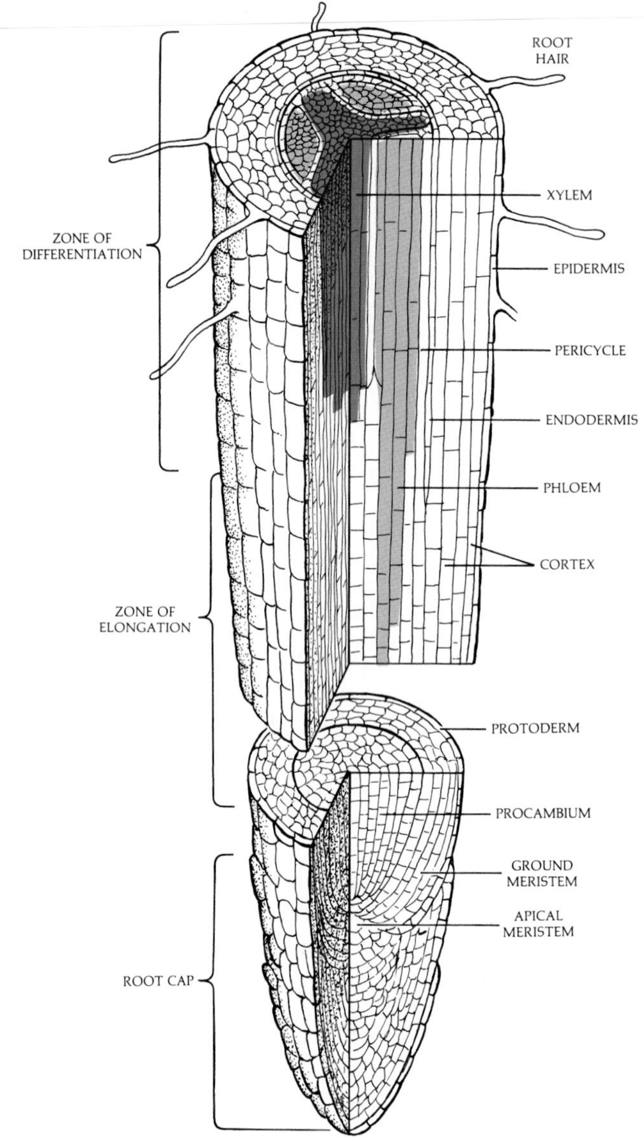

This same basic pattern of growth, as shown in Figure 29–14, is seen in the first root tip of a seedling and is repeated over and over again in the growing root tips of a tree 50 meters tall.

Primary Growth of the Shoot

The shoot includes all the aboveground parts of the plant. The pattern of growth of the developing shoot tip is similar to that seen in the root: first, cell division takes place; next, cell elongation; and finally, differentiation. However, because of the regular occurrence of nodes and their appendages—the leaves and buds—the growth zones are not as distinct in the shoot as they are in the root. Also, no covering analogous to the root cap is produced over the shoot tip.

As in the root, the outermost layer of cells develops into the epidermis. In the shoot, these cells have a relatively conspicuous waxy cuticle. Other cells differentiate to form ground tissue, which in dicots includes the cortex and the pith, and the primary vascular tissues—the primary xylem and the primary phloem. The pattern of development is more complicated, however, than in the root tip since the apical meristem of the shoot is the source of tissues that give rise to new leaves, branches, and flowers. (At the time of flowering, the apical meristem forms the floral parts and ceases to exist.)

Figure 29-15a shows the shoot tip of a lilac. Here you can see the apical meristem, which is very small, and the beginnings—primordia—of leaves. As you can see, leaves are formed in an orderly sequence at the shoot tip. The leaf originates by the division of cells in a localized area along the side of the shoot apex. The vascular tissue differentiates upward, becoming part of the general vascular system that connects the plant from root to leaf tip. In some species, leaves arise simultaneously in pairs opposite one another, as in the lilac. In other species, the leaves occur spirally or in circles (whorls) at the nodes. As the internodes elongate, the young leaves become separated so that the leaves clustered so tightly together around the apex in Figure 29-15a will eventually be spaced out along the stem of the plant.

Buds, Branches, and Flowers

As the growing tip of the shoot elongates, small masses of meristematic tissue are left just above the point at which the leaf joins the stem (the leaf axil). These new meristematic regions, the lateral buds, remain dormant until after the growth of the adjacent leaf and internode is complete. In many species, the buds do not develop at all unless the apical meristem of the shoot is damaged or removed. (This phenomenon, known as apical dominance, will be discussed further on page 633.)

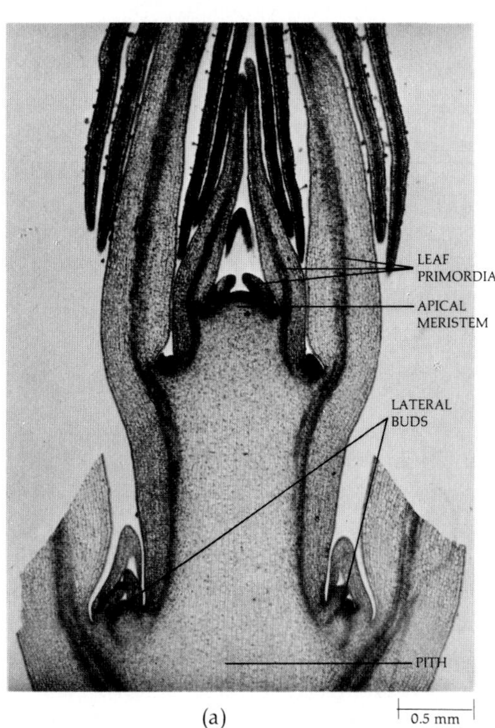

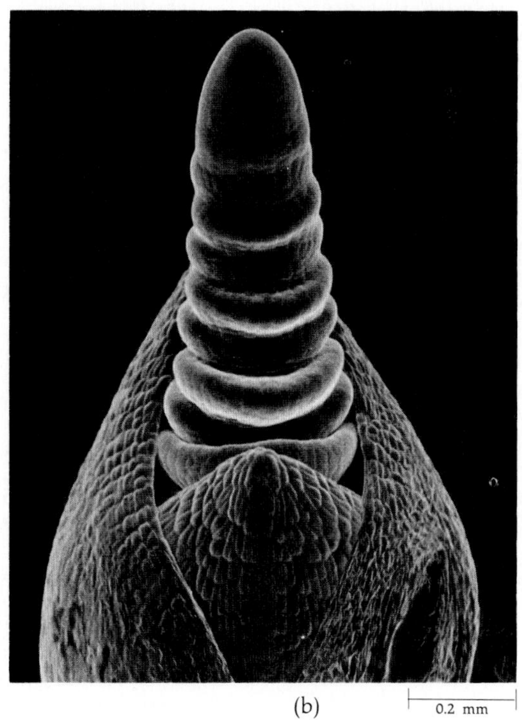

29-15
(a) *Longitudinal section of the shoot tip of a lilac* (Syringa vulgaris), *a dicot. At the tip of the stem is the apical meristem, the central zone of cell division. The leaf primordia, which are young leaves, originate along the sides of the shoot apex. In this picture, two leaf primordia can be seen arising from the apical meristem. Successively older leaf primordia had formed on both sides. The developing stem below the apical meristem and the lateral buds on either side are also regions of active cell division.* (b) *Scanning electron micrograph of a shoot tip of a monocot (wheat). The leaves are first formed as ridges, visible below the apical meristem.*

WHY THE GRASS GROWS

In most plants, as we have emphasized, growth of the shoot proceeds from its tip, the apical meristem. Grass plants, by contrast, have growth zones at the base of the leaf. The grass leaf consists of a leaf blade, the broad part, and a leaf sheath, which fits around the plant stem. When the blade is cut off, cells in the sheath, in response to some unknown signal, are activated, producing more blade. This useful adaptation to herbivore grazing has resulted, over the eons, in the piteous spectacle of thousands of hominids vibrating behind their lawnmowers from spring to fall.

Another strong contender in this evolutionary race is the dandelion, whose familiar flower and fruit are shown in Figure 28–1. If a dandelion plant is lopped off down to the root by either hominid or herbivore, three more grow in its place.

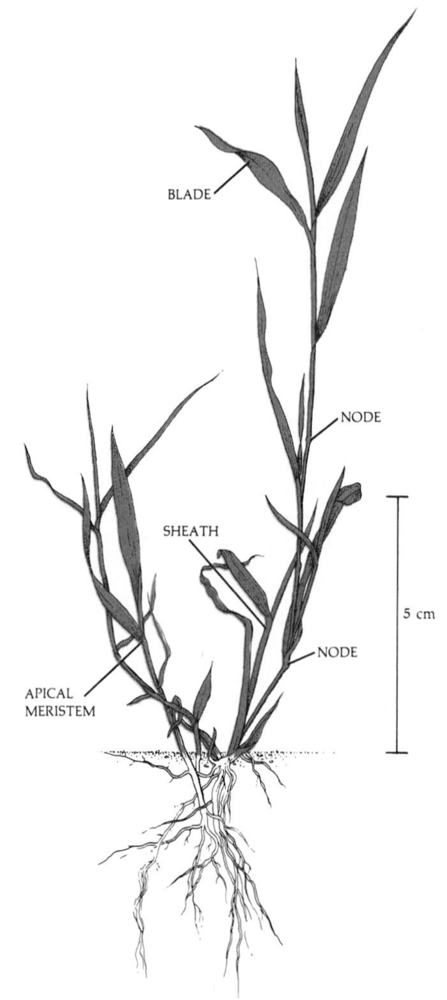

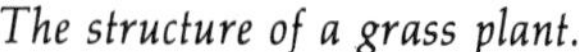

The structure of a grass plant.

In some species, some buds are destined to become lateral branches or specialized shoots, such as rhizomes, tubers, or flowers. In other species, whether or not a particular bud will become a flower or a branch is determined by environmental conditions, particularly day length. (The regulation of flowering is discussed in Chapter 32.)

SECONDARY GROWTH

Secondary growth is the process by which woody plants increase the thickness of trunks, stems, branches, and roots (Figure 29-16). The so-called "secondary tissues" are not derived from the apical meristems; they are the result of the production of new cells by *lateral meristems* that are known as the vascular cambium and the cork cambium.

The *vascular cambium* is a thin, cylindrical sheath of tissue between the xylem and the phloem. In plants with secondary growth, the cambium cells divide continually during the growing season, adding new xylem cells—that is, secondary xylem—toward the inside of the cambium, and secondary phloem toward the outside.

29–16
(a) *Stem of a dicot before the onset of secondary growth.*

(b) *Beginnings of secondary growth. Secondary xylem and secondary phloem are produced by the vascular cambium, a meristematic tissue formed late in primary growth. As the trunk increases in diameter, the epidermis is stretched and torn. This apparently triggers the formation of the cork cambium, from which cork is formed, replacing the epidermis.*

(c) *Cross section of a three-year-old stem, showing annual growth layers. Rays are rows of living cells that transport nutrients and water laterally (across the trunk). On the perimeter of the outermost growth layer of xylem is the vascular cambium, encircled by a band of secondary phloem. The primary phloem and also the cortex will eventually be sloughed off. In an older stem, the thin cylinder of active secondary phloem is immediately adjacent to the vascular cambium. The tissues outside the vascular cambium, including the phloem, constitute the bark.*

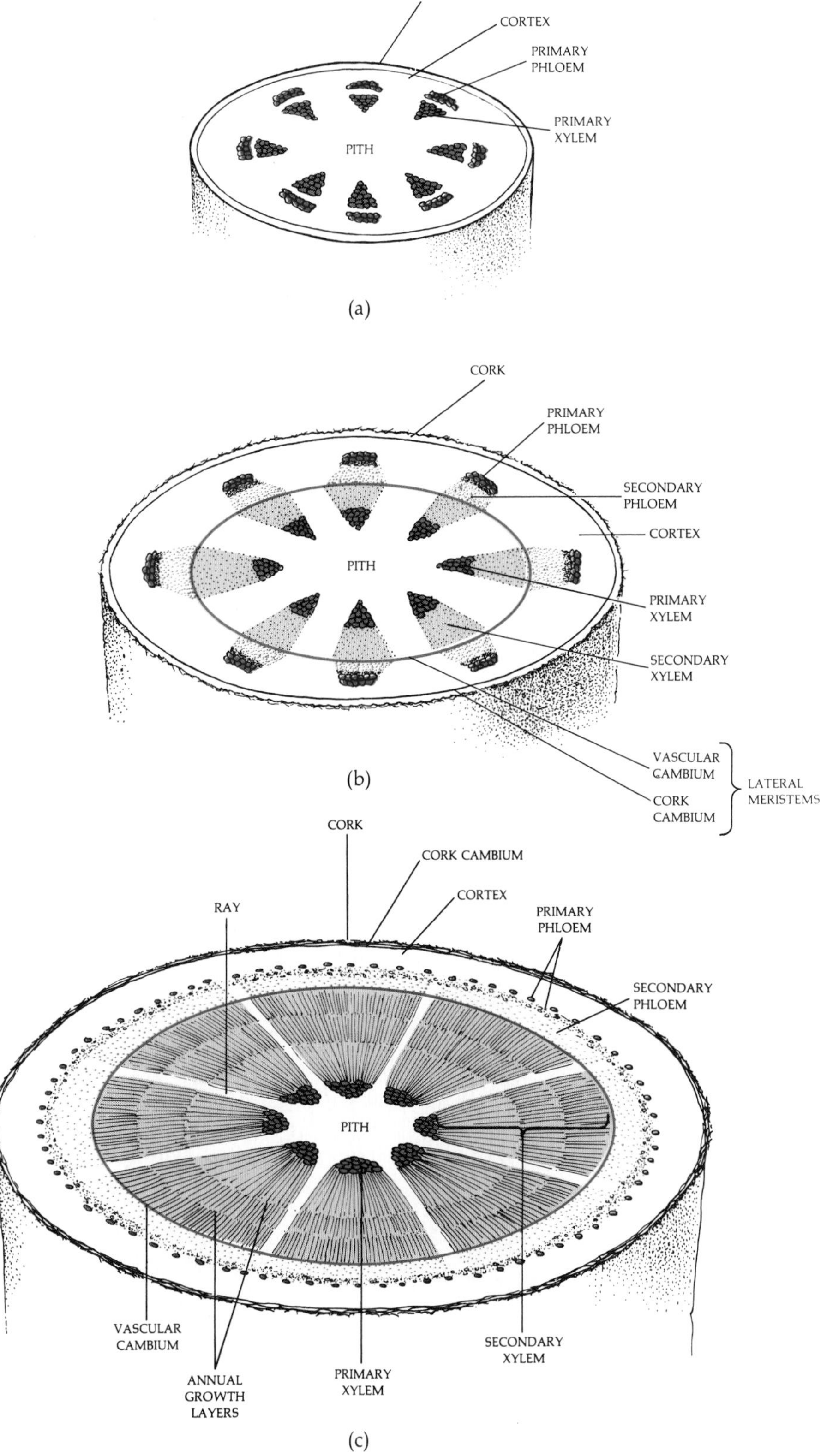

Some daughter cells remain as a cylinder of undifferentiated cambium. As the tree grows older, the living parenchyma cells of the xylem in the center of the trunk die, and their neighboring vessels cease to function. This nonfunctioning xylem is called heartwood, as distinct from sapwood, which consists of living parenchyma cells and functional vessels.

As the girth of stems and roots increases by secondary growth, the epidermis becomes stretched and torn. In response to this tearing process, a new type of cambium, the *cork cambium*, forms from the cortex. Cork (phellem), which is a dead tissue at maturity, is produced from the cork cambium and replaces the epidermis as the protective covering of woody stems and roots. The cork cambium forms anew each year, moving further inward until, finally, there is no cortex left. The tissues outside the vascular cambium, including the phloem, constitute the bark.

By the continuous formation of secondary layers of xylem and (to a lesser extent) phloem, tree trunks increase their diameter as primary growth increases their height. Season after season the new xylem forms visible growth layers, or rings. Each growing season leaves its trace, so that the age of a tree can be estimated by counting the number of growth rings in a section near its base. Since the rate of growth of a tree depends on climatic conditions, it is also possible to estimate from the annual growth layers fluctuations in temperature and rainfall that occurred many years ago (see essay).

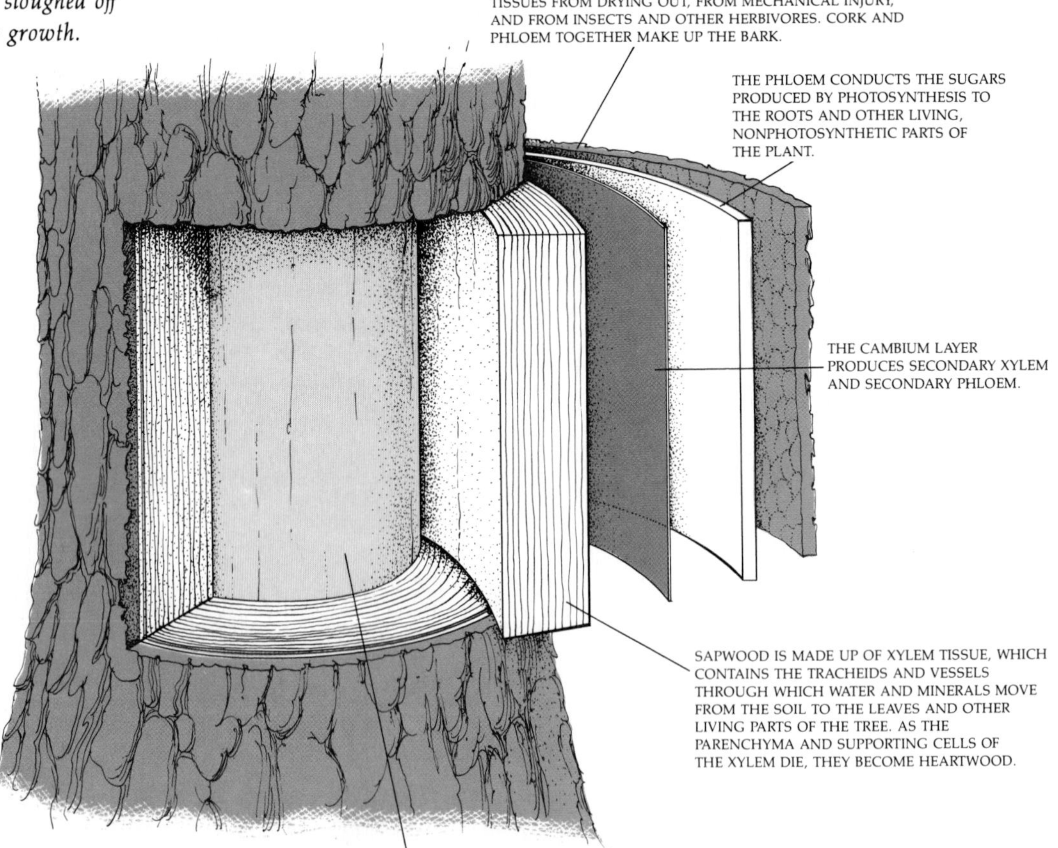

CORK, WHICH IS A DEAD TISSUE, PROTECTS THE INNER TISSUES FROM DRYING OUT, FROM MECHANICAL INJURY, AND FROM INSECTS AND OTHER HERBIVORES. CORK AND PHLOEM TOGETHER MAKE UP THE BARK.
THE PHLOEM CONDUCTS THE SUGARS PRODUCED BY PHOTOSYNTHESIS TO THE ROOTS AND OTHER LIVING, NONPHOTOSYNTHETIC PARTS OF THE PLANT.
THE CAMBIUM LAYER PRODUCES SECONDARY XYLEM AND SECONDARY PHLOEM.
SAPWOOD IS MADE UP OF XYLEM TISSUE, WHICH CONTAINS THE TRACHEIDS AND VESSELS THROUGH WHICH WATER AND MINERALS MOVE FROM THE SOIL TO THE LEAVES AND OTHER LIVING PARTS OF THE TREE. AS THE PARENCHYMA AND SUPPORTING CELLS OF THE XYLEM DIE, THEY BECOME HEARTWOOD.
HEARTWOOD, COMPOSED ENTIRELY OF DEAD CELLS, IS THE CENTRAL SUPPORTING COLUMN OF THE MATURE TREE.

29–17
A tree trunk showing the relationships of the successive concentric layers. The heartwood is composed entirely of dead cells. The cortex, which is outside the phloem in a green stem, is sloughed off in the process of secondary growth.

THE RECORD IN THE RINGS

As we have seen, each growing season in the life of a tree leaves its record in the ring of secondary xylem formed in the trunk. The growth rings of trees are visible because of differences in the density of wood produced early in the growing season and that produced late in the season. The early wood has large cells with thin walls, whereas the late wood has smaller cells with proportionately thicker walls. Within a given growth layer, the change from early wood to late wood is quite gradual, but a distinct change is visible where the small, thick-walled cells of the late wood of one growing season abut the larger, thin-walled cells of the next spring.

The width of the individual growth layers may vary greatly from year to year, depending on such environmental factors as light, temperature, rainfall, available soil water, and the length of the growing season. In particular, the width of a growth layer is a fairly accurate indicator of the amount of rainfall in a particular year. Under favorable conditions—that is, during periods of adequate or abundant rainfall—the growth rings are wide; under unfavorable conditions, they are narrow.

In semiarid regions, where there is very little rain, the tree is a sensitive rain gauge. An excellent example is the bristlecone pine (Pinus longaeva) *of the western Great Basin. Each growth ring is different, and a study of the rings tells a story that dates back thousands of years.*

The oldest known living specimen of bristlecone pine is 4,900 years old. Dendrochronologists—scientists who conduct historical research using the growth rings of trees—have been able to match samples of wood from living and dead trees. In this way, they have built up a library containing a continuous series of rings dating back more than 8,200 years. This record of average ring width has provided a valuable guide to past conditions of precipitation and temperature. The widths of the growth rings of bristlecone pines at the higher elevations (the upper tree line) have been found to be closely related to temperature changes. The rings of these trees have revealed that the summers in the White Mountains of California were relatively warm from 3500 B.C. *to 1300* B.C.*, and the tree line was about 150 meters higher than it is at present. Summers were cool from 1300* B.C. *to 200* B.C.

Information provided by the growth rings of both conifers and angiosperms is being used not only to reconstruct past climatic conditions but also to help predict future conditions. With more accurate knowledge of past climates than is provided by human records—even those of the past few centuries—it is possible to determine cyclic patterns of temperature change and of rainfall and drought. Such information is of considerable importance, for example, in planning the wise management and allocation of our finite resources of fresh water.

(a)

(b)

250 mm

(a) *A bristlecone pine. These pines, which grow near the timberline, are the oldest living trees.* (b) *A portion of a cross section of wood from a bristlecone pine, showing the variation in width of the annual rings. This section begins approximately 6,260 years ago; the band of rings in color represents the years from 4240* B.C. *to 4210* B.C.

29–18

Wild strawberry plants reproduce asexually by means of aboveground stems, or runners. Roots and leaves develop at every second node along the modified stems. These plants also form flowers and reproduce sexually.

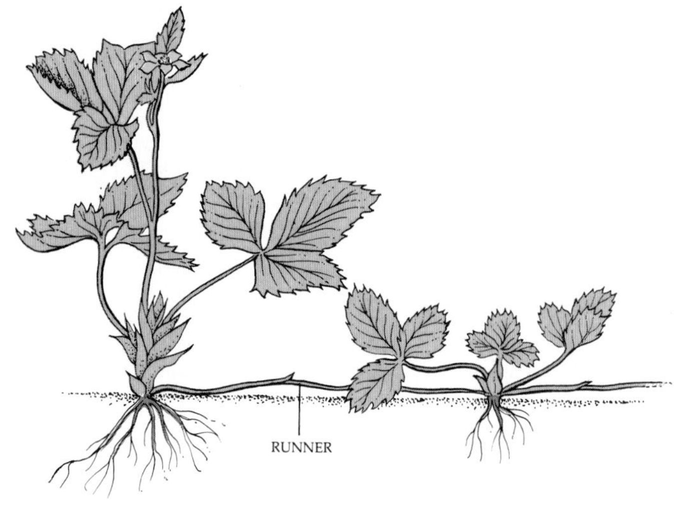

29–19

Common reed grass (Phragmites communis) *propagates asexually by means of rhizomes. Each stand of reed grass is, in effect, a single individual. The tassels at the tops of the stalks are flowers; the seeds produced through sexual reproduction give rise to new stands of reed grass, genetically different from the parent plants.*

ASEXUAL REPRODUCTION

Asexual reproduction is common among plants. It takes several different forms, but always with the same result: the offspring are identical to the parent plant. One of the most familiar types of asexual reproduction occurs by means of horizontal stems growing either above ground (runners) or below ground (rhizomes). Strawberries are a familiar example of plants that propagate by runners, as are spider plants, commonly grown as hanging plants. Plants that reproduce by rhizomes include potatoes, many flowering garden perennials, such as lilies-of-the-valley and irises, the sod-forming grasses of lawns and pastures, and bamboos and phragmites. Both runners and rhizomes develop adventitious roots.

The production of clones from runners and rhizomes is a very efficient way for a plant to spread and capture territory. The little plants that develop in this way have a continuous source of nourishment from the parent plant and a lower mortality than seedlings.

Plants may also develop from unfertilized seeds. The dandelion is an example; natural populations of dandelions consist of mixtures of groups of genetically identical individuals produced by parthenogenesis. Apparently dandelions are such successful plants that their best strategy is to stick with a particular winning combination, genetically speaking, rather than to experiment with new genotypes.

The capacity of plants to reproduce asexually has been exploited in the development of cultivated varieties of plants for food or ornamental use. Such plants are, of course, genetically identical to the parent stock, and so vegetative reproduction is a way of preserving uniformity. Many plants are reproduced by stem cuttings, which simply involves placing young stems in soil or some other planting medium and protecting them from drying out until adventitious roots appear. Rooting can often be facilitated by hormone treatment (see page 634). Many stem cuttings will also form roots when placed in water. Another artificial form of plant propagation is grafting, in which a stem cutting is inserted in a slit in the bark of a rooted woody plant. The wound is sealed with tape, and if the stem cutting and the rooted plant have not been damaged too much by the surgery, the grafted shoot will "take" and begin growing. Most fruit trees and roses are propagated in this way.

Table 29-1 Summary of Main Cell Types in Angiosperms

CELL TYPE	ORIGIN	LOCATION	CHARACTERISTICS	FUNCTION
Apical meristem	Embryonic cells	Apices of shoots and roots	Many-sided, small, thin-walled cells; vacuoles usually small	Origin of primary meristematic tissues
Epidermis	Protoderm	Surface of entire primary plant body	Flattened, variable in shape, overlaid by cuticle; specialized guard cells	Protective covering
Parenchyma	Protoderm, ground meristem, procambium, vascular cambium, cork cambium, wound tissues	Everywhere, usually dominant in pith, cortex, mesophyll	Many-sided, usually thin-walled; abundant air spaces between cells	Photosynthesis, storage, wound healing, among others
Collenchyma	Ground meristem of leaf and stem	Peripheral in cortex of stem and in leaves	Elongate, with irregularly thickened primary walls	Support for young stems and leaves
Sclereid	Protoderm, ground meristem, procambium, vascular cambium, and cork cambium	Anywhere; in pith and cortex of stems; in leaves and flesh of fruits; seed coats	Irregular; massive secondary wall; alive or dead at maturity	Produces hard texture, mechanical support
Fiber	Procambium or vascular cambium; ground meristem	Primary and secondary xylem and phloem; cortex	Very long, narrow cell, with secondary cell wall; alive or dead at maturity	Support
Tracheid	Procambium or vascular cambium	Primary or secondary xylem	Elongate, with pits in walls; dead at maturity	Conduction of water and solutes
Vessel member	Procambium or vascular cambium	Interconnected series (= vessels) in primary or secondary xylem	Elongate, with pits in walls and end walls perforated; dead at maturity	Conduction of water and solutes
Sieve-tube member	Procambium or vascular cambium	Primary or secondary phloem, usually with companion cells; form interconnected series (= sieve tubes)	Elongate, with specialized sieve plates; nucleus lacking at maturity	Conduction of organic solutes
Vascular cambium	Procambium and parenchyma of pith rays in stems; procambium and pericycle in roots	Lateral, between secondary phloem and xylem tissues	Elongate, spindle-shaped	Produces secondary xylem and phloem
Cork (phellem)	Cork cambium	Surface of stems and roots with secondary growth	Flattened cells, compactly arranged; dead at maturity, the cells often air-filled	Restricts gas exchange and water loss

Many economically important plants are sterile and can only be propagated vegetatively; these include pineapples, bananas, seedless grapes, navel oranges, and numerous ornamental plants.

SUMMARY

The flower is the structure of sexual reproduction in the angiosperms. The pollen grains, which contain the male gametophytes, are produced by the anthers. Anther plus filament is known as the stamen. The carpel typically consists of the stigma—an area on which the pollen grains germinate—a slender style, and, at its base, the ovary. The ovary contains one or more ovules, and within each ovule is a female gametophyte that contains an egg cell. One of the sperm nuclei of the pollen grain fertilizes the egg cell; the other sperm nucleus unites with the two polar nuclei of the female gametophyte. The nutritive endosperm, a triploid ($3n$) tissue, results from the second fertilization. The phenomena of fertilization and triple fusion, found only among the flowering plants, are known as "double fertilization." The ovule becomes the seed, and the ovary becomes the fruit.

The angiosperm seed, or mature ovule, consists of the embryo, the seed coat, and stored food. The petals, stamens, and other floral parts of the parent plant may fall away as the ovary ripens into fruit and the seeds form. The seed commonly enters a dormant period.

When the seed germinates, growth of the root and shoot proceeds from the apical meristems of the embryo. Certain cells within the meristem divide continuously. Others elongate and then differentiate, forming, according to their position, the various specialized cells of the plant body (see Table 29–1). In the root, cells produced by the apical meristem form a root cap, which protects the root tip as it is pushed through the soil.

Leaf primordia arise from the apical meristem of the shoot. As the nodes are separated by elongation of the internodes, small masses of meristem (buds) are formed in the axils of the leaves. These buds may remain dormant or may give rise to branches or specialized shoots.

Secondary growth is the process by which the woody plants increase their girth. Such growth arises primarily from the vascular cambium, a sheath of meristematic tissue completely surrounding the xylem and completely surrounded by phloem. The cambium cells divide during the growing season, adding new xylem cells (secondary xylem) on their inner surfaces and new phloem cells (secondary phloem) on their outer surfaces. As the trunk increases in girth, the epidermis is eventually ruptured and destroyed and is replaced by cork.

Many species of plants have the capacity to reproduce asexually, producing populations of genetically identical individuals.

QUESTIONS

1. Distinguish among the following: ovary/ovule/egg cell/seed; pollination/fertilization; epidermis/endodermis/endosperm; apical meristem/lateral meristem; protoderm/procambium/ground meristem; primary growth/secondary growth; runner/rhizome.

2. Identify the floral parts visible in the photograph at the left. Can you name the flower?

3. In many plants, pollen production in the anthers occurs prior to or after the full development of the carpel of the same flower. What are the consequences of this shift in time frames?

4. J. B. S. Haldane said, "A higher plant is at the mercy of its pollen grain." Explain.

5. Which of the following is a berry: blackberry, strawberry, mulberry, grape, pumpkin?

6. The jack pine produces seeds with very tough seed coats. These seed coats rupture, and the seeds germinate, only after exposure to intense heat, as in a forest fire. What might be the advantages for the jack pine of such seeds?

7. Sketch a dicot embryo at the time of seed release. Identify each part in terms of the future development of the plant body.

8. Suppose you carve your initials 1 meter above the ground on the trunk of a mature tree that is growing at an average of 15 centimeters a year. How high will your initials be at the end of 2 years? At the end of 20 years?

9. As a result of secondary growth, most of our common trees increase in girth as they increase in height. What limitations would a lack of secondary growth impose on the form and mechanical support of a tree?

10. Sketch a tree trunk with secondary growth. Compare your sketch with Figure 29–16c.

11. Porcupines often eat all of the bark off young trees, at the height they can reach. When they do this, the tree dies. Why?

12. Consider a mature tree. Which parts of the tree are composed of living cells? Why might it be advantageous for living cells to make up such a small fraction of the mass of the tree? Similarly, why might it be advantageous for a deciduous tree to invest as little material as possible in making leaves?

13. Many features deliberately bred into cultivated seeds and plants would be detrimental or fatal to wild varieties. Give four examples of such features, and explain why each would be disadvantageous to the plant in the wild.

CHAPTER 30

Transport Systems in Plants

As we noted in Chapter 23, the plants that have come to dominate the modern landscape are those that have found the best solution (so far) to the problems of transporting needed materials to the various cells and tissues of the plant body. In this chapter, we shall discuss first the movement of water and minerals in the xylem; second, the transport of sugars and other organic substances in the phloem; and, at the end of the chapter, a related subject, factors affecting the availability of nutrients to the plant. You may find it helpful to review the properties of water (especially pages 34 to 37) and the ways in which it moves (pages 126 to 131) before proceeding. The transport processes of plants depend on the extraordinary properties of this very common liquid.

THE MOVEMENT OF WATER AND MINERALS

Transpiration

A plant needs far more water than an animal of comparable weight. In an animal, most of its water remains in its body and recirculates. By contrast, in plants, more than 90 percent of the water that enters the roots is given off into the air as water vapor. Thus, for example, a single corn plant needs 160 to 200 liters of water from seed to harvest, and 1 acre of corn requires almost 2 million liters of water a season. British ecologist John L. Harper describes the terrestrial plant as "a wick connecting the water reservoir of the soil with the atmosphere." The loss of water vapor from the plant body is known as *transpiration*.

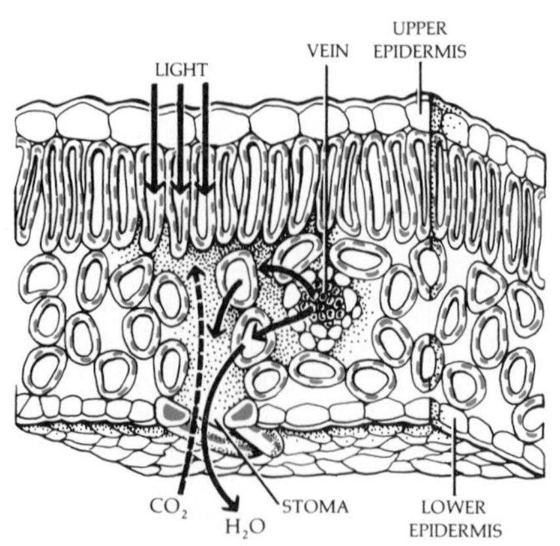

30–1
As carbon dioxide, essential for photosynthesis, enters the leaf through the stomata, water vapor is lost.

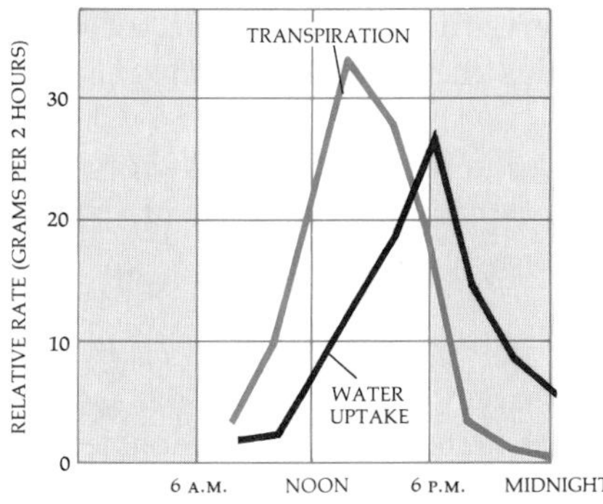

30–2

Measurements in ash trees show that a rise in water uptake follows a rise in transpiration. These data suggest that the loss of water generates forces for its uptake.

30–3

Guttation droplets on the edge of a rose leaf. Guttation, the loss of liquid water, is a result of root pressure. The water escapes through specialized pores located near the ends of the principal veins of the leaf. Guttation, which is restricted to relatively small plants, usually occurs at night when the air is moist.

The Uptake of Water

As we noted earlier, water enters the body of most plants almost entirely through the roots. During periods of rapid transpiration, water may be removed from around the roots so quickly that the water in the soil in the vicinity of the roots becomes depleted. Water will then move slowly by diffusion and capillary action through the soil toward the depleted region near the roots. The roots also obtain additional water by growing beyond the depleted region; the main roots of corn plants, for example, grow an average of 52 to 63 millimeters a day.

Root cells, like other living parts of the plant, contain a higher concentration of solutes (both organic and inorganic) than does soil water. As a result, water from the soil enters the roots by osmosis. The osmotic potential is sufficient to move water a short distance up the stem, a phenomenon known as root pressure. Guttation (Figure 30–3) is a visible consequence of root pressure. But how can water reach 20 meters high to the top of an oak tree, travel three stories up the stem of a vine, or move 125 meters up in a tall redwood?

One important clue is the observation that during times when the most rapid transpiration is taking place—which is, of course, when the flow of water up the stem must be the greatest—xylem pressures are characteristically negative (less than atmospheric pressure). The existence of negative pressure can be demonstrated readily. If you peel a piece of bark from a transpiring tree and make a cut in the xylem, no sap runs out. In fact, if you place a drop of water on the cut, the drop will be drawn in.

What is the pulling force? It is not simple suction, as the negative pressure might indicate. Suction simply removes air from a system so that the water (or other liquid) is pushed up by atmospheric pressure. But atmospheric pressure is only enough to raise water (against no resistance) about 11 meters at sea level, and many trees are much taller than 11 meters.

The Cohesion-Tension Theory

According to the now generally accepted theory, the explanation is to be found not only in the properties of the plant but also in the properties of water, to which the plant has become exquisitely adapted. As we pointed out in Chapter 2, in every water molecule, two hydrogen atoms are covalently bonded to a single oxygen atom. Each hydrogen atom is also held to the oxygen atom of a neighboring water molecule by a hydrogen bond. The cohesion resulting from this secondary attraction is so great that the tensile strength in a thin column of water can be as much as 140 kilograms per square centimeter (2,000 pounds per square inch). This means that it requires a negative pressure of more than 140 kilograms per square centimeter to pull the column of water apart.

In the leaf, water evaporates, molecule by molecule, from the cells into the intercellular spaces within the leaves. The water potential of the leaf cell falls, and the water from the vessels or tracheids moves, molecule by molecule, into the leaf cell. But each molecule in the xylem vessel is linked to other molecules in the vessel. They, in turn, are linked to others, forming one long, narrow, continuous thread of water reaching right down to a root tip. As the molecule of water moves through the stem and into the leaf, it tugs the next molecule along behind it.

Because the diameter of the vessels is very small and because the water molecules adhere to the walls, even as they are cohering to one another, gas bubbles, which could rupture the column, do not usually form. The pulling action, molecule

30–4
(a) *A simple model that illustrates the cohesion-tension theory. A piece of porous clay pipe, closed at both ends, is filled with water and attached to the end of a long, narrow glass tube also filled with water. The water-filled tube is placed with its lower end below the surface of a volume of mercury contained in a beaker. As water molecules evaporate from the pores in the pipe, they are replaced by water "pulled up" through the narrow glass tube in a continuous column. As the water evaporates, mercury rises in the tube to replace it.* (b) *Transpiration from plant leaves results in sufficient water loss to create a similar negative pressure.*

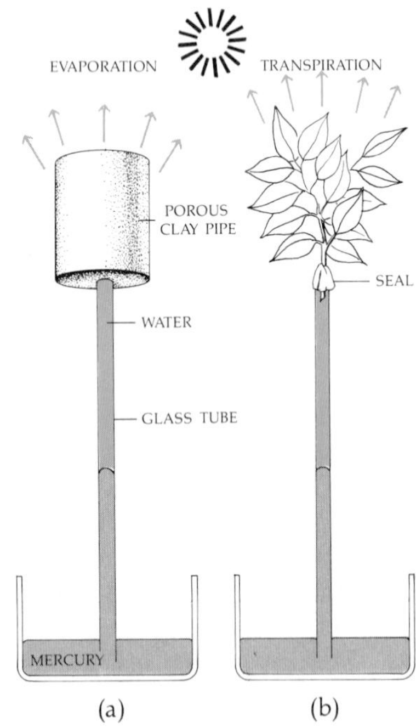

by molecule, causes the negative pressure observed in the xylem. The technical term for a negative pressure is tension, and this theory of water movement is known as the *cohesion-tension theory*.

The principle of cohesion-tension is illustrated in Figure 30–4. As indicated in the diagram, the power for this process comes not from the plant, which plays only a passive role in transpiration, but from the energy of the sun.

Factors Influencing Transpiration

Transpiration is costly, especially when water supply is limited. Various factors affect the rate of water loss. One is temperature; the rate of evaporation doubles for every temperature rise of about 10°C. Humidity is also important. Water is lost much more slowly into air already laden with water vapor. Air currents also affect transpiration. A breeze cools your skin on a hot day because it blows away the water vapor that has accumulated near the skin surface and so facilitates evaporation. Similarly, wind blows away the water vapor from leaf surfaces, which makes the concentration gradient steeper and so hastens the diffusion of water molecules. Leaves of plants that grow in exposed, windy areas are often hairy; these hairs are believed to protect the leaf surface from the wind and so retard transpiration (Figure 30–5). By far the most important factor affecting transpiration, however, is the regulatory effect exercised by the opening and closing of the stomata.

Mechanism of Stomatal Movements

As shown in Figure 30–6, each stoma has two surrounding guard cells. Stomatal movements are caused by changes in the turgor of these cells. When the guard cells are turgid, they bow out, opening the stoma; when they lose water, they collapse, and the stoma closes.

Turgor, as we saw on page 131, is maintained or lost due to the osmotic movement of water into or out of cells. As you will recall, water moves across the cell membrane from a solution of low solute concentration into a solution of high solute concentration. The active accumulation of solutes in the guard cells causes water to move into them by osmosis; conversely, a decrease in the solute concentration of guard cells results in the osmotic movement of water out of the cells.

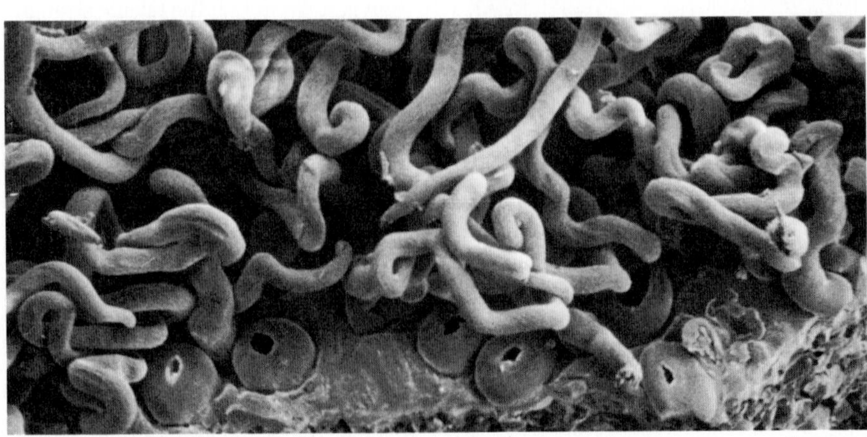

30–5
Leaf hairs create a layer of still air over the epidermis, restricting gas exchange and water loss from the stomata (visible in the lower portion of the micrograph).

30-6
Mechanism of stomatal movement. A surface view is shown in (a) *and a cross section in* (b). *Each stoma has two guard cells that open the stoma when they are turgid and close it when they lose turgor. In many species, the guard cells have thickened walls adjacent to the stoma. As turgor increases, the thinner parts of the cell wall are stretched more than the thicker parts, causing the cells to bow out and the stoma to open.*

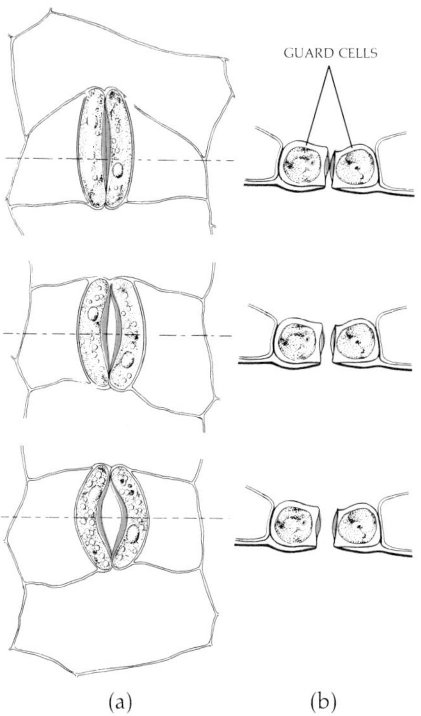

Factors Affecting Stomatal Movements

A number of environmental factors affect the opening and closing of the stomata. The principal one is the availability of water. When the water available to a leaf drops below a certain critical point (which varies from species to species), the stomata close, thus limiting the evaporation of the remaining water. This is not a direct effect; closing of the stomata occurs before the leaf as a whole loses turgor and wilts.

The plant's capacity to anticipate water stress depends on the action of the hormone abscisic acid, about which we shall have more to say in the chapter that follows. Several lines of evidence link the hormone with stomatal movements. First, during the initial stage of water stress, levels of the hormone have been shown to increase markedly in many species of plants. Second, application of abscisic acid to a leaf causes stomatal closing within a few minutes. Third, studies of a mutant strain of tomato plant that fails to close its stomata as the water supply decreases showed the mutant to be deficient in abscisic acid production. Replacement therapy with the hormone enables the tomato plant to regulate its stomata normally. (It was studies of this "wilty mutant," as it is known, that provided the first clues to the role of the hormone in preventing water stress. The mutant was produced by x-irradiation of normal seeds.)

Stomatal movement also occurs quite independently of the loss or gain of water. Among the other factors that affect stomatal movement are the carbon dioxide concentration, light, and temperature. In most species, an increase in carbon dioxide concentration in the intercellular spaces of the leaf causes the stomata to close. The magnitude of this response varies greatly from species to species and with the degree of water stress that a particular plant has undergone. In corn, the stomata may respond to changes in the internal carbon dioxide concentration within seconds. The site for sensing the level of carbon dioxide is inside the guard cells.

In most species, the stomata close regularly in the evening, when photosynthesis is no longer possible, and open in the morning. These movements in response to light occur even when there are no changes in the water available to the plant. Opening in the light and closing in the dark may be a result of changes in the carbon dioxide concentration—photosynthesis uses carbon dioxide and thus reduces the internal concentration, whereas respiration causes an increase in the carbon dioxide concentration. Light may, however, have a more direct effect. Blue light has long been known to stimulate stomatal opening independently of the carbon dioxide concentration. When isolated guard cells from onion are illuminated with blue light in the presence of potassium ions, they swell. The blue-absorbing pigment (which is thought to be located in either the cell membrane or the membrane of the vacuole) promotes the uptake of potassium (K^+) ions by the guard cells. The blue-absorbing pigment, which would be yellow in color, is thought to be a flavin.

Within normal ranges, temperature changes have little effect on the stomata, but temperatures higher than 35°C can cause stomatal closing in some species. The closing can be prevented, however, by holding the plant in air that contains no carbon dioxide. This suggests that temperature changes work primarily by affecting the carbon dioxide concentration within the leaf. An increase in temperature results in an increase in respiration and thus in the concentration of intercellular carbon dioxide. Many plants in hot climates close their stomata regularly at midday, apparently because of both the effect of temperature on carbon dioxide accumulation and increased water stress.

The Role of Potassium

Within the last decade, it has been shown that the common denominator in all these factors affecting the stomata is the potassium ion, K^+. Techniques that make it possible to measure the concentration of potassium ion within a single guard cell show that potassium levels rise when the stomata open and drop when they close. The evidence indicates that potassium ions are actively transported between the guard cells and the reservoir provided by the surrounding epidermal cells. Water follows by osmosis, causing the observed changes in turgor.

In many, but not all, species, chloride ions, Cl^-, accompany the K^+ ions across the membrane, thus maintaining electrical neutrality. In other, perhaps all, species, H^+ ions move in the opposite direction, producing a decreased hydrogen ion concentration within the guard cells of open stomata. (It was this variety of secondary effects that obscured for many years the primary effect of the movement of potassium.)

Crassulacean Acid Metabolism

As we have noted, the stomata of most plants are open during the day and closed at night. In some species, however, the stomata close in the daytime and open at night. Such plants include a variety of succulents, including cacti, pineapples, and members of the stonecrop family (Crassulaceae), plants adapted to hot, dry climates. They take in carbon dioxide at night, converting it to malic and isocitric acids. During the day, when the stomata are closed, the carbon dioxide is released from these organic acids and used immediately in photosynthesis. This process is known as Crassulacean acid metabolism, and plants that employ it are known as CAM plants. It is analogous to the C_4 pathway in photosynthesis described on page 226, although it apparently evolved independently.

Mineral Requirements of Plants

In addition to water and the energy-rich sugars and other compounds produced by photosynthesis, plant cells require a number of different chemical elements, which are found in the earth in the form of minerals and which are transported in water solution through the xylem. A *mineral* is a naturally occurring inorganic substance, usually solid, with a definite chemical composition. Some minerals, such as diamond, sulfur, and copper, consist of single elements. Others, such as quartz (SiO_2) and calcite ($CaCO_3$), are compounds. Of the 92 different elements found in the rocks and soil, oxygen and silicon are by far the most abundant, followed by aluminum and iron.

The mineral elements required by plants are usually classified either as macronutrients, which are generally expressed as a percentage of dry weight of the plant and which range from about 0.5 to 3 or 4 percent, or as micronutrients, of which as little as a few parts per million may be present. Determination of the amounts present of many of the elements is made by burning the plant completely, which permits the carbon, hydrogen, oxygen, nitrogen, and sulfur to escape as gases; the ash is then analyzed. Proportions of each element vary in different species and in the same species grown under different conditions. Also, the ash will often contain elements, such as silicon, which are present in the soil and are taken up by plants but which are not generally thought to be required for growth.

Mineral requirements can also be identified by studying the capacity of plants to grow in distilled water to which small amounts of various minerals are added. This sounds easier than it is. Sometimes it has been found that a substance—boron, for example—is needed in such small amounts that it is almost impossible to set up experimental conditions that exclude it. Thus, it is difficult to prove that its absence is lethal.

(a)

(b)

30-7
Some plants have special mineral requirements. (a) *Plants of the mustard family, such as the wintercress shown here, use sulfur in the synthesis of the mustard oils that give the plants their characteristic sharp taste.* (b) *Horsetails incorporate silicon into their cell walls, making them indigestible to most herbivores but useful, at least in colonial America, for scouring pots and pans.*

Uptake of Minerals

Mineral ions are taken up in solution by the roots and travel through plants with the transpiration stream. As you saw in Figure 28-15 (page 585), the cells of the endodermis play a major role in determining which ions enter the xylem.

The ionic composition of plant cells is far different from the ionic composition of the medium in which the plant grows. For example, in one study, cells of pea roots were found to have a concentration of potassium ions 75 times greater than that of the nutrient solution. Similarly, in another study, the vacuoles of rutabaga cells were shown to contain 10,000 times more K^+ than the external solution.

Thus it seems evident that mineral ions are brought into plant cells by active transport. Support for this hypothesis comes from observations indicating that the uptake of ions is an energy-requiring process. For instance, if roots are deprived of oxygen or poisoned so that respiration is curtailed, ion uptake is drastically decreased. Also, if a plant is deprived of light, it will cease to absorb ions and eventually will release them back into the soil.

One way that roots move ions such as potassium against a concentration gradient is by the action of carrier proteins in the cell membrane (see page 133). Such a protein combines with the ion and then, utilizing the energy of ATP, carries it to the other side of the membrane and releases it. Because the membrane is only slowly permeable to ions such as potassium, the continued action of the "pump proteins" results in differences in the concentration of the transported ions on one side of the membrane as compared to the other.

As a result of these differences in ion concentrations, there is often a difference of electric charge between the inside and the outside of the membrane. In the nerve cells of animals, stimulation of the cell membrane results in an abrupt change in its permeability to ions. The rapid movement of ions across the membrane is the basis for transmission of the nerve impulses. Similar events apparently involved in certain plant responses will be described in Chapter 32.

Functions of Minerals

Plants require mineral elements for many different functions. For example, mineral ions affect osmosis (page 129) and so regulate water balance. Because in many plants several mineral ions can serve interchangeably in this role, this requirement is described as nonspecific. On the other hand, a mineral element may be a functional part of an essential biological molecule, and in this case the requirement is highly specific. An example is the magnesium atom in the chlorophyll molecule. Some minerals—phosphorus and calcium, for example—are required constituents of cell membranes, and some control membrane permeability. Others are indispensable components of a variety of enzyme systems that catalyze biological reactions in the cell. Still others provide a proper ionic environment in which biological reactions can occur. Because mineral elements are involved in many fundamental processes, the effects of mineral deficiencies are typically very wide-ranging, affecting a number of structures and functions in the plant body.

HALOPHYTES: A FUTURE RESOURCE?

Unlike most animals, most plants do not require sodium and, moreover, cannot survive in brackish waters or saline soils. In such environments, the solution surrounding the roots often has a higher solute concentration than the cells of the plant and so water tends to move out of the roots by osmosis. Even if the plant is able to absorb water, it faces additional problems from the high level of sodium ions. If the plant takes up water and excludes sodium ions, the solution surrounding the roots becomes even saltier, increasing the likelihood of water loss through the roots. The salt may even become so concentrated that it forms a crust around the roots, effectively blocking the supply of water to the roots. Another problem is that sodium ions may enter the plant in preference to potassium ions, depriving the plant of an essential nutrient as well as inhibiting some enzyme systems.

Some plants, however—known as halophytes—can grow in saline environments such as deserts, salt marshes, and coastal areas. All of these plants have evolved mechanisms for dealing with high sodium concentrations, and for some of them sodium appears to be a required nutrient. The adaptations of halophytes vary. In many halophytes, the sodium-potassium pump (page 134) seems to play a major role in maintaining a low sodium concentration within the cells while simultaneously ensuring that a sufficient supply of potassium ions enters the plant. In some species, the pump operates primarily in the root cells, pumping sodium back to the environment and potassium into the root. The presence of calcium ions (Ca^{2+}) in the soil solution is thought to be essential for the effective functioning of this mechanism.

Other halophytes take in sodium through the roots but then either secrete it or isolate it from the living cytoplasm of the plant body. In Salicornia *(pickleweed), a sodium-potassium pump (or a variant of it) operates in the membranes of the vacuoles of leaf cells. Sodium ions enter the cells but are immediately pumped into vacuoles and isolated from the cytoplasm. In such plants, the solute concentration of the vacuoles is higher than that of the environment, establishing the necessary osmotic potential for the movement of water into the roots. In other genera, the salt is pumped into the intercellular spaces of the leaves and then secreted from the plant. In* Distichlis palmeri *(Palmer's grass), the salt exudes through specialized cells (not the stomata) onto the surface of the leaf. In* Atriplex *(saltbush), it is concentrated by special salt glands and pumped into bladders. The bladders expand as the salt accumulates, finally bursting. Rain or the passing tide washes the salt away.*

Halophytes are of current interest not only because of the light they may shed on the osmoregulatory mechanisms of plants, but also because of their potential as crop plants. In a world with an ever-increasing need for food, vast areas are unsuitable for agricultural purposes because of the salinity of the soil. For example, there are over 30,000 kilometers of desert coastline and about 400 billion hectares of desert with potential water supplies that are, however, too salty for crop plants. Moreover, each year about 200,000 hectares of irrigated crop land become so salty that further agriculture is impossible. When arid land is heavily irrigated, as in large areas of the western United States, salts from the irrigation waters accumulate in the soil. This occurs because in both evaporation from soil and transpiration from plants, essentially pure water is given off, leaving all solutes behind. Over many years, the salt concentration of the soil increases, eventually reaching levels that cannot be tolerated by most plants. It has been suggested that the ancient civilizations of the Near East ultimately fell because their heavily irrigated land became so salty that food could no longer be grown on it.

One way of extending the life of irrigated crop lands and of bringing presently barren areas into agricultural use would be to breed salt-tolerance into conventional crop plants. Thus far, however, such efforts have met with little success. Scientists at the University of Arizona's Environmental Research Laboratory are taking what appears to be a more promising approach. They have gathered halophytes from all over the world and are engaged in an extensive research program to determine optimal growing conditions, potential yields, and the nutritional value and palatability of the seeds and vegetative parts of the various species. Their results suggest that a number of halophyte species have great potential for use in livestock feed and, quite possibly, for human consumption as well.

(a)

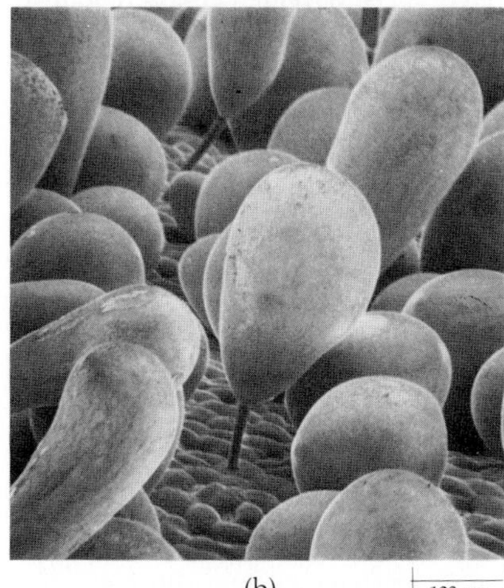

(b)

(a) Atriplex *(saltbush), a plant native to Australia, is one of several halophytes being evaluated as potential crop plants.* (b) *The surface of a leaf of* Atriplex. *Salt is pumped from the leaf tissues through narrow stalk cells into the large, expandable bladder cells.*

Table 30–1 lists the mineral elements required by plants, the form in which they usually are absorbed, and some of the uses plants make of them. You might anticipate that organisms make use of what is most readily available; indeed, they seem to have done this when life originated from elements in the gases of the primitive atmosphere. But the table reveals some findings that you might not expect. Sodium, for instance, which is one of the most abundant of the elements, is apparently not required at all by most plants. The fact that plants evolved with no functions requiring sodium is even more striking when you consider that sodium is vital to animals. (Sodium is the principal osmoregulator in animals and, as we shall see in Chapter 40, is necessary for transmission of the nerve impulse.) In the seas, where both plant and animal life seem to have originated, sodium is the most abundant mineral element and is far more readily available than potassium, which it closely resembles in its essential properties. Potassium, however, is the principal osmoregulator in plants. Similarly, although silicon and aluminum are almost always present in large amounts in soils, few plants require silicon and apparently none requires aluminum. On the other hand, most plants need molybdenum, which is relatively rare.

THE MOVEMENT OF SUGARS: TRANSLOCATION

As we have seen, the photosynthetic cells of the plant, which are most abundant in the leaves, provide the food energy for all of the other cells of the plant. The process by which the products of photosynthesis are transported to other tissues is known as *translocation*.

Evidence for the Phloem

The fact that water is transported by vessels in the xylem was recognized by botanists 300 years ago. Early evidence supporting the role of the phloem in sugar transport came from observations of trees that had a complete ring of bark removed from them. As we saw in the last chapter, bark contains the phloem but not the xylem. When a photosynthesizing tree is "girdled" in this manner, the tissue above the ring becomes swollen, indicating the accumulation of fluid moving downward in the phloem from the photosynthesizing leaves.

The role of the phloem in the movement of sugar was not generally agreed upon, however, until well into the twentieth century. A principal reason for this long delay is that the sieve-tube members are so delicate. When they are cut with even the thinnest dissecting blade, callose and P-protein plug the pores in the sieve plates, impeding or preventing the movement of substances through them (see Figure 28–9, page 581).

Convincing evidence for the role of the phloem was obtained when radioactive tracers became available. If plants carry out photosynthesis in air containing carbon dioxide made with carbon 14 ($^{14}CO_2$), the sugars produced will carry a radioactive tag. Studies of the passage of the radioactivity through the plant have shown conclusively that sugars are transported in the sieve tubes (see essay on page 623.)

Table 30-1 A Summary of Mineral Elements Required by Plants

ELEMENT	FORM IN WHICH ABSORBED	APPROXIMATE CONCENTRATION IN WHOLE PLANT (AS % OF DRY WEIGHT)	SOME FUNCTIONS
Macronutrients			
Nitrogen	NO_3^- (or NH_4^+)	1–4%	Component of amino acids, proteins, nucleotides, nucleic acids, chlorophyll, and coenzymes.
Potassium	K^+	0.5–6%	Involved in osmosis and ionic balance and in opening and closing of stomata. Activator of many enzymes.
Calcium	Ca^{2+}	0.2–3.5%	Component of cell walls. Enzyme cofactor. Involved in cell membrane permeability.
Phosphorus	$H_2PO_4^-$ or HPO_4^{2-}	0.1–0.8%	Formation of energy-carrying phosphate compounds (ATP and ADP). Component of phospholipids, nucleic acids, and several essential coenzymes.
Magnesium	Mg^{2+}	0.1–0.8%	Part of the chlorophyll molecule. Activator of many enzymes.
Sulfur	SO_4^{2-}	0.05–1%	Component of some amino acids, proteins, and coenzyme A.
Micronutrients			
Iron	Fe^{2+}, Fe^{3+}	25–300 parts per million (ppm)*	Required for chloroplast development. Component of cytochromes and nitrogenase.
Chlorine	Cl^-	100–10,000 ppm	Involved in osmosis and ionic balance. Probably essential in photosynthesis in the reactions in which oxygen is produced.
Copper	Cu^{2+}	4–30 ppm	Activator of some enzymes.
Manganese	Mn^{2+}	15–800 ppm	Activator of some enzymes. Required for oxygen release in photosynthesis.
Zinc	Zn^{2+}	15–100 ppm	Activator of many enzymes.
Molybdenum	MoO_4^{2-}	0.1–5.0 ppm	Required for nitrogen metabolism.
Boron	BO^{3-} or $B_4O_7^{2-}$(borate or tetraborate)	5–75 ppm	Influences Ca^{2+} utilization. Other functions unknown.
Elements Essential to Some Plants or Organisms			
Cobalt	Co^{2+}	Trace	Required by nitrogen-fixing microorganisms.
Sodium	Na^+	Trace	Involved in osmosis and ionic balance, probably for many plants not essential. Required by some desert and salt-marsh species and may be required by all plants that utilize C_4 photosynthesis.

* Parts per million (ppm) equal units of an element by weight per million units of oven-dried plant material; 1% equals 10,000 ppm.

RADIOACTIVE TRACERS IN PLANT RESEARCH

Radioactive tracers can be used in a number of ways to study the synthesis, transport, and use of materials within plants. Initially, in any tracer study, the radioactive isotope must be incorporated into the plant. Radioactive carbon, for example, will be taken up by a plant if its leaves are exposed to carbon dioxide containing carbon 14. Or, radioactive phosphorus will be taken up if the roots are exposed to a solution containing phosphorus-32 ions.

The length of time the plant is exposed to the radioactive material is determined by the information the investigators hope to obtain. For example, in studies to determine the time required for carbon dioxide to be incorporated into the various products of photosynthesis, a sequence of exposure times would be used. In studies focusing on the location of a particular product formed from the radioactive substance, the length of exposure would depend on the time required for the chemical reactions under study.

After exposure to the radioactive substance, the plant is quick-frozen and freeze-dried. In whole-plant autoradiography, the plant is flattened and then pressed against a sheet of x-ray film. Radiation given off by the isotope exposes the film adjacent to the portions of the plant in which the tracer is located. By comparing the flattened plant with the developed film, investigators can determine the location of the radioactive substance within the plant.

In tissue autoradiography (histoautoradiography), the freeze-dried plant tissues are embedded in paraffin, resin, or a similar material. Next, they are sliced into very thin sections, which are mounted on microscope slides. The tissue sections are then placed in contact with a photographic emulsion or film. As in whole-plant autoradiography, the radiation from the isotope exposes the film in contact with the portions of the tissue section containing the radioactive material. After an appropriate interval of time, the film is developed. Comparison, under the microscope, of the developed film and the underlying tissue section reveals the exact location of the radioactive substance in the plant tissues.

In the study illustrated here, three leaflets of a broad bean plant were enclosed in a flask and were exposed to $^{14}CO_2$ *and light for 35 minutes* (a). *During that time, the radioactive carbon dioxide was incorporated into sugars, which were then being transported to other parts of the plant. A cross section* (b) *and a longitudinal section* (c) *from the stem were placed in contact with autoradiographic film for 32 days. When the film was developed and compared with the underlying tissue sections, it was apparent that the radioactivity (visible as dark specks on the film) was confined almost entirely to the sieve tubes.*

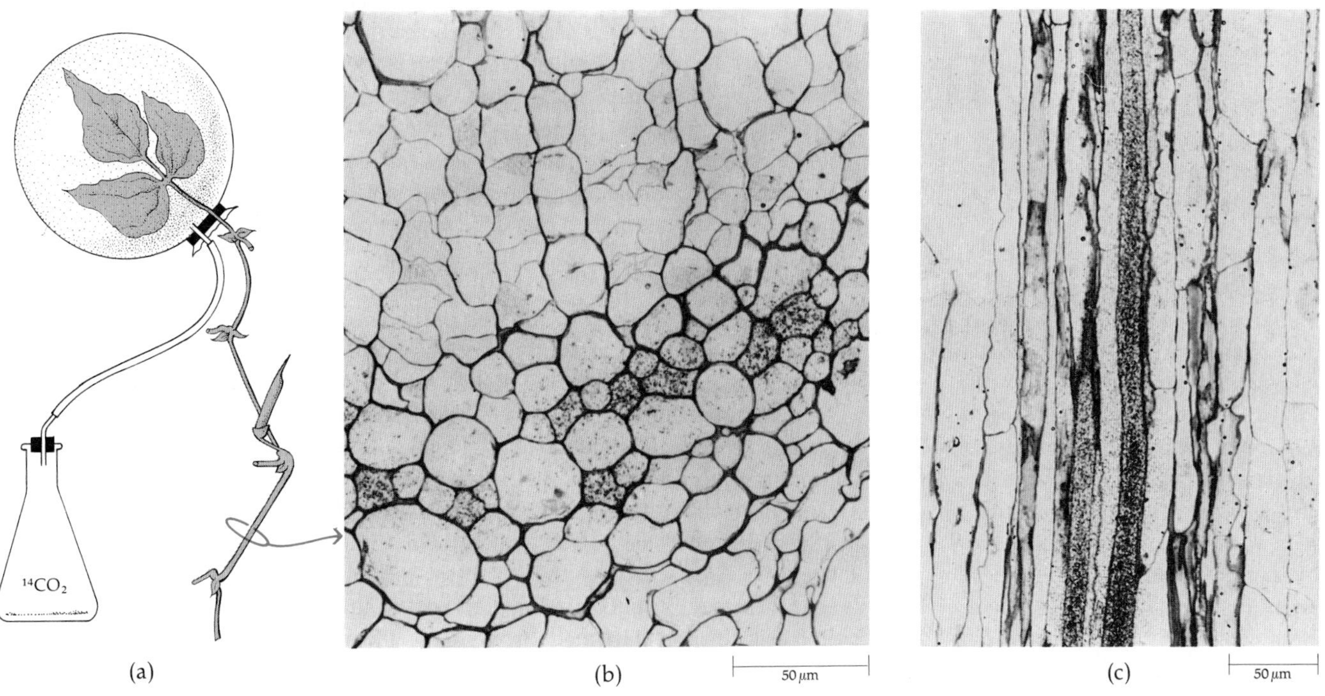

(a)

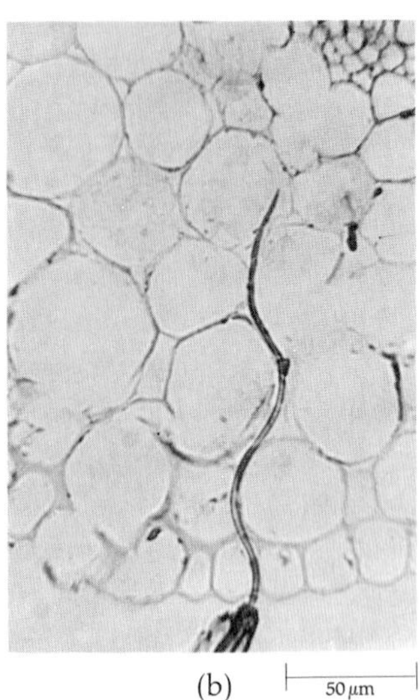

(b)

30–8
Assistance by aphids. (a) *Aphids are very small insects that feed on plant sap; you probably have seen them on rose bushes.* (b) *The aphid, as the micrograph reveals, drives its sharp mouthparts, or stylets, like a hypodermic needle between the epidermal cells and then taps the contents of a single sieve-tube member. If the aphid is anesthetized, it is possible to sever its body from the stylets and leave them undisturbed in the cell. The fluid often continues to exude through the stylets from the sieve tube for many hours, and pure samples of the fluid flowing through the sieve tube can be collected for analysis without damaging the sieve tube or interfering with its function.*

Assistance by Aphids

The technical problem of sampling the phloem without plugging up the sieve tubes has been cleverly circumvented with the help of aphids, which are very small sap-sucking insects. The aphid drives its sharp mouthparts, or stylets, through the epidermis of a plant, traversing cortical or mesophyll cells, and then taps the contents of a single sieve-tube member (Figure 30–8). If the aphid is anesthetized, it is possible to sever the aphid's body from the stylets, leaving the latter in place in the cell. The solution exudes through the stylets from the sieve tube, and pure samples of sieve-tube sap can be collected for analysis.

Data gathered by this technique indicate that sieve-tube sap contains (by weight) 10 to 25 percent solutes, more than 90 percent of which are sugars, mostly sucrose. Low concentrations of amino acids and other nitrogen-containing substances are also present. Tracer stains indicate that the rate of movement of these substances along the sieve tube is remarkably fast: in one set of experiments, for example, it was estimated that the sap was moving at a rate of about 100 centimeters per hour, far faster than could be accounted for by diffusion.

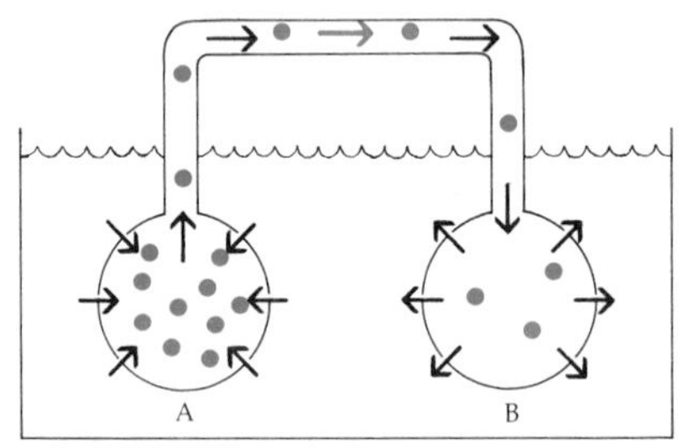

30–9
Model of pressure-flow hypothesis. Bulbs A and B, which are interconnected and are permeable to water, are placed in a bath of distilled water. Bulb A contains a higher concentration of sucrose than bulb B. Water enters bulb A from the medium, increasing the hydrostatic pressure and pushing the solution to bulb B. If B were connected to a third bulb, C, with a still lower concentration of sucrose, as sieve-tube members are connected in a series, hydrostatic pressure building up in B would push the solution on to C, and so on.

The Pressure-Flow Hypothesis

The movement of sugars and other organic solutes in translocation follows what is known as a *source-to-sink pattern*. The principal sources of the solutes are the photosynthesizing leaves, but storage tissues may also serve as important sources. All plant parts unable to meet their own nutritional needs may act as sinks, that is, as importers of organic solutes. Thus, storage tissues act as sinks when they are importing solutes and as sources when they are exporting solutes.

The most widely accepted current explanation for the source-to-sink movement in translocation is the *pressure-flow hypothesis*. According to this hypothesis, the solutes move in solutions that, in turn, move because of differences in water potential caused by concentration gradients of sugar. The principle underlying this hypothesis can be illustrated by a simple physical model consisting of porous bulbs permeable only to water and connected by glass tubes (Figure 30–9). The first bulb contains a solution in which there is a dissolved material, such as sucrose, and the second, to make the example as simple as possible, contains only water. When these interconnected bulbs are placed in distilled water, water will enter the first by osmosis. The entry of water will increase the hydrostatic pressure within this bulb and cause the water and the solutes in it to move along the tube to the second bulb, where the pressure again builds up. If the second bulb is connected with a third one containing water or a sucrose concentration lower than that now in the second one, the solution will flow from the second to the third by the same process, and so on indefinitely down the sucrose gradient. This hypothesis is supported by the fact that distinct gradients in the concentrations of sucrose and other sugars have been demonstrated along the phloem tissues during the summer months. Moreover, it can account for the known rates of movement in the phloem.

In leaves, sucrose has to be moved into sieve tubes against a concentration gradient. In sugar beets, it has been shown that sucrose molecules are moved from the mesophyll cells and surrounding spaces directly into the sieve tubes of the leaf veins by active transport. It appears to involve a cotransport of sucrose molecules and hydrogen ions by way of a specific permease on the sieve-tube membrane. The added sugar decreases the water potential in the sieve tube and causes water to move into the sieve tube from the xylem by osmosis. At a sink—for example, a root tip—sugar molecules are removed from the sieve tube, again as the result of active transport. The water molecules follow the sugar molecules out, again by osmosis. Thus the water flows in at one end of the sieve tube and out at the other. Between these two points, the water and its solutes, including sugar, move passively by bulk flow. At the sink, sugars may be either utilized or stored, but most of the water returns to the xylem and is recirculated in the transpiration stream.

Companion cells, because of their dense appearance and many mitochondria, are believed to be very active metabolically, and it is hypothesized that one of their functions is to meet the energy requirements of the sieve-tube members with which they are associated. Note, however, that the actual flow of the sugar solution is a passive process, requiring no metabolic activity on the part of the phloem cells. The speed of transport depends on the differences in concentration between source and sink.

30–10

Diagrams of soil layers of three major soil types. (a) *The litter of the northern coniferous forest is acidic and slow to decay, and the soil has little accumulation of humus, is very acidic, and is leached of minerals.* (b) *In the cool, temperate deciduous forest, decay is somewhat more rapid, leaching less extensive, and the soil more fertile. Such soils have been used extensively for agriculture, but they need to be prepared by adding lime (to reduce acidity) and fertilizer.* (c) *In the grasslands, almost all of the plant material above the ground dies each year, as do many of the roots, and so much organic matter is constantly returned to the soil. In addition, the finely divided roots penetrate the soil extensively. The result is highly fertile soil, often black in color, with a topsoil sometimes more than a meter in depth.*

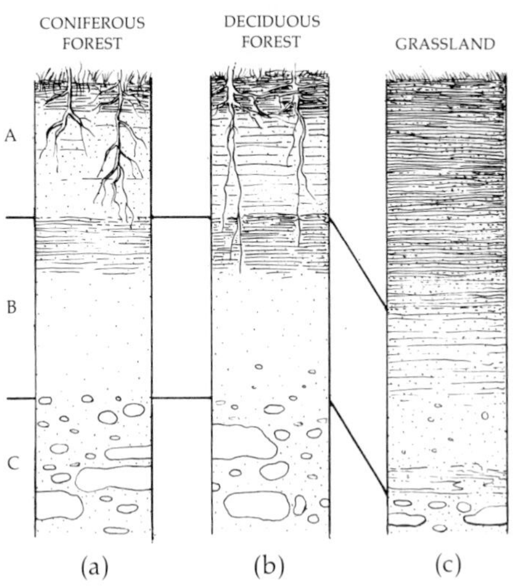

SOILS AND PLANT NUTRITION

The composition of the soil influences the availability of minerals for plant nutrition. Soil, the uppermost layer of the earth's crust, is composed of weathered rock associated with organic material, both living and in various stages of decomposition. It typically has three layers: the A horizon, the B horizon, and the C horizon. The A horizon, or topsoil, is the zone of maximum organic accumulation (humus). The B horizon, or subsoil, consists of inorganic particles in combination with mineral nutrients that have leached down from the A horizon. The C horizon is made up of loose rock that extends down to the bedrock beneath it. Figure 30–10 shows profiles of three common soil types.

The mineral content of the soil depends in part on the parent rock from which the soil is formed. These differences in mineral content can be extremely localized, with sharp lines of demarcation. Geologists sometimes use types of vegetation or changes in color or growth patterns of plants as indicators of mineral deposits.

Table 30-2 Soil Classification

	DIAMETER OF PARTICLES (MICROMETERS)
Coarse sand	200–2,000 (0.2–2 millimeters)
Sand	20–200
Silt	2–20
Clay	Less than 2

In most soils, however, the mineral content is more dependent on biological factors. In an undisturbed environment, most of the mineral nutrients stay within the system—the soil itself and the plants, microorganisms, and small soil animals that it supports and contains. If, however, the vegetation is repeatedly removed, as when crops are harvested or grasslands are overgrazed, or the top, humus-rich layer of the A horizon is eroded, the soil rapidly becomes depleted and can be used for agricultural purposes only if it is heavily fertilized.

Another factor influencing the mineral content of soils is the soil composition. The smaller fragments of rock are classified as sand, silt, or clay, according to size (Table 30-2). Water and minerals drain rapidly through soil composed of large particles (sandy soil). Soil composed of small particles (clay) holds the water against gravity. Moreover, the small clay particles are negatively charged and so hold positively charged ions, such as calcium (Ca^{2+}), potassium (K^+), and magnesium (Mg^{2+}). However, a pure clay soil is not suitable for plant growth because it is usually too tightly packed to let in enough oxygen for the respiration of plant roots, soil animals, and most soil microorganisms. Clay soils that contain enough large particles to keep the soil from packing are known as loams, and these are generally the best soils for plant growth.

The pH of the soil also affects its capacity to retain minerals. In an acidic soil, hydrogen ions replace other positively charged ions clinging to clay particles, and these nutrient ions easily leach out of the soil. Soil pH also affects the solubility of certain nutrient elements. Calcium, for example, increases in solubility (and therefore in its availability to plant roots) as pH increases, whereas iron decreases in availability as pH increases. Crops such as alfalfa, sweet clover, and other legumes have high calcium requirements and so grow well only in alkaline soil. Rhododendrons and azaleas, on the other hand, have high requirements for iron, which is abundant only when the soil is acidic.

Soils and plant life interact. Plants secrete hydrogen ions, which help to degrade rock surfaces and also to release positively charged ions. As they decay, plant parts constantly add to the humus, thereby changing not only the content of the soil but also its texture and its capacity to hold minerals and water. In turn, the plants are dependent upon the mineral content of the soil and its holding capacity. As these improve, plants increase in both number and size and also often change in kind, thereby producing further changes in the soil. Thus, under natural conditions, the soil is constantly changing its composition.

30-11
Indian pipe. It was once believed that this nonphotosynthetic angiosperm, with its strange white waxy flowers, lived on decaying matter in the soil. It is now known, however, that it is dependent on mycorrhizal fungi that transfer nutrients from other plants to the parasitic Indian pipe.

SYMBIOSES AND PLANT NUTRITION

Mycorrhizae

Two types of symbiotic relationships play important roles in plant nutrition. One of these is mycorrhizae ("fungus-roots"), the associations between roots and fungi that we described in Chapter 22. In these associations, the fungi apparently serve to extract nutrients from the soil and make them available to plants, thus enabling plants to prosper on nutrient-poor soils. Recent studies also indicate that the fungi may screen out chemicals, making it possible for plants to live in soils that would otherwise be toxic.

30–12
Effects of mycorrhizae on tree nutrition. Nine-month-old seedlings of white pine were grown for two months in a sterile nutrient solution and then transplanted to prairie soil. The seedlings on the left were transplanted directly. The seedlings on the right were placed in forest soil containing fungi for two weeks before being transplanted to the prairie.

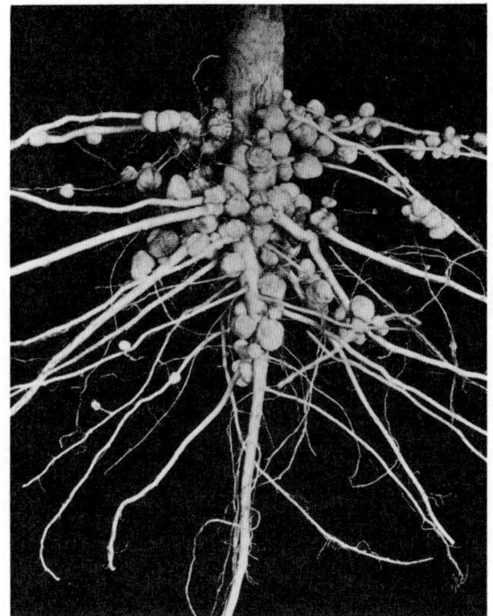

30–13
Nitrogen-fixing nodules on the roots of a soybean plant, a legume. These nodules are the result of a symbiotic relationship between a soil bacterium (Rhizobium) *and root cells.*

Rhizobia and Nitrogen Fixation

All but one of the elements listed in Table 30–1 are derived from the weathering of rocks. The exception is nitrogen, which enters the soil by way of the atmosphere.

Although nitrogen gas constitutes about 78 percent of the air, it is not available to plants. Molecular nitrogen—the nitrogen in the atmosphere—consists of two atoms of nitrogen held together by a triple bond, an exceptionally strong bond. A large quantity of energy is required to break it. The process by which atmospheric nitrogen is incorporated into nitrogen compounds that can be utilized by plants is known as *nitrogen fixation* (the nitrogen cycle is discussed in Chapter 53).

On a worldwide basis, most nitrogen fixation is carried out by a few types of microorganisms, including both free-living and symbiotic forms of cyanobacteria and heterotrophic bacteria. Of the various classes of nitrogen-fixing organisms, the symbiotic bacteria are by far the most important in terms of total amounts of nitrogen fixed. The most common of the nitrogen-fixing symbiotic bacteria is *Rhizobium,* which invades the roots of leguminous plants, such as clover, peas, beans, vetches, and alfalfa (Figure 30–13).

The beneficial effects to the soil of growing leguminous plants are so obvious that they have been recognized for hundreds of years. Where leguminous plants are grown, some of the "extra" nitrogen may be released into the soil; it then becomes available to other plants. In modern agriculture, it is common practice to rotate a nonleguminous crop, such as corn, with a leguminous one, such as alfalfa. The leguminous plants are then either harvested, leaving behind the nitrogen-rich roots, or, better still, plowed back into the field. A crop of alfalfa that is plowed back into the soil may add as much as 350 kilograms of nitrogen to the soil per hectare, frequently enough to grow a crop of a nonleguminous plant.

Nitrogen fixation is also carried out commercially, with the required energy supplied by fossil fuels. As the cost of these fuels increases, so does the price of the nitrogen-containing fertilizer produced. At present, the equivalent of 2 million barrels of oil a day is required for the production of nitrogen-containing fertilizers. In this perspective, investigation of the mechanism of nitrogen fixation by microorganisms takes on great practical significance.

The Symbiotic Relationship

The symbiosis between a species of *Rhizobium* and a legume is quite specific; for example, the species of bacteria that invade and induce nodule formation in clover roots will not induce nodules on the roots of soybeans. Rhizobia enter the root-hair tips of legumes while the plants are still seedlings (Figure 30–14a). The bacteria induce these epidermal cells to produce cellulose tubes, or infection threads, through which the bacteria move to the cortical cells of the root (Figure 30–14b). Within the cortex, the infection threads branch, becoming populated with multiplying bacteria. Soon there is a proliferation of the surrounding cortical cells, presumably the result of hormones released during the growth of the rhizobia. The infection threads typically pass close to the cell nuclei and seem to cause the nuclei to degenerate. Near the cell nuclei, vesicles form in the threads and then break open, releasing the bacteria into the cytoplasm of the host cells (Figure 30–14c).

Soon after their release, the bacteria begin to grow, increasing in size some tenfold. Ammonium (NH_4^+) produced by the bacteria is combined with carbon compounds synthesized by the photosynthetic cells of the plant to produce amino acids. A key molecule in the nitrogen-fixation process is an enzyme called nitrogenase, which is composed of two different polypeptides, one containing iron and the other molybdenum.

30–14
(a) *Tip of a root hair of a clover seedling, with several rhizobia and some soil particles.* (b) *Two branches of an infection thread (which, as you can see, is passing near the cell nucleus, at the top of the micrograph).* (c) *Cross section of an infected nodule, showing bacteria.*

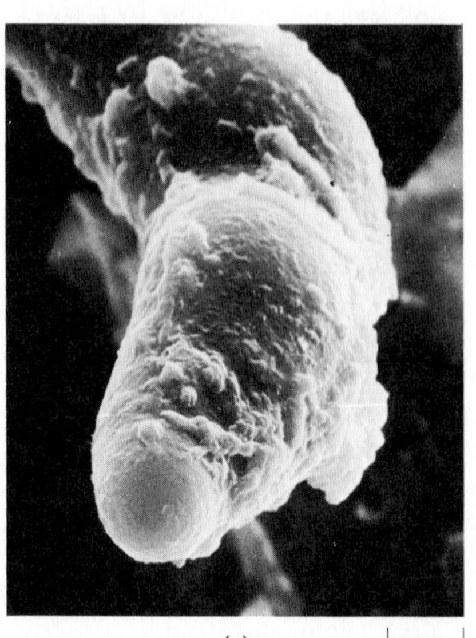

(a)

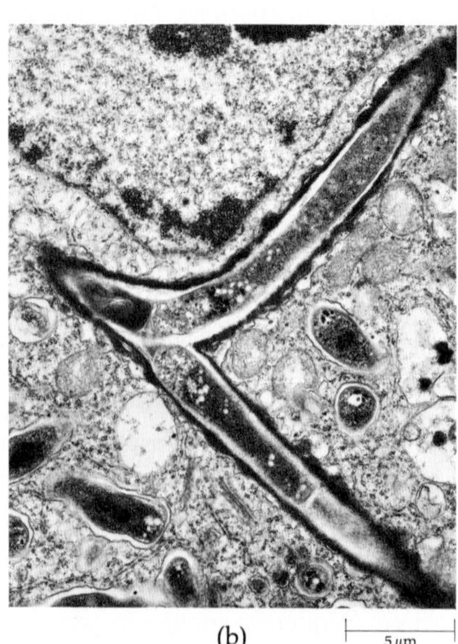

(b)

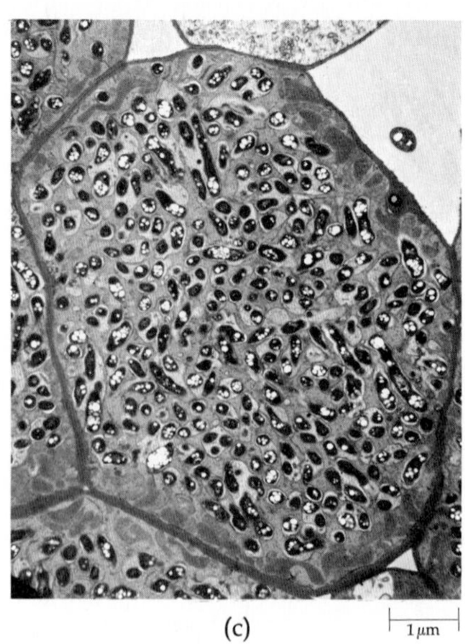

(c)

Recombinant DNA and Nitrogen Fixation

Genetic mapping of a free-living nitrogen-fixing bacterium has shown that the 17 genes known to be involved in nitrogen fixation are clustered on one portion of the chromosome. In experiments at the University of Sussex, biologists have succeeded in transferring this gene cluster to a plasmid and then introducing the plasmid into *Escherichia coli* cells. The *E. coli* cells were then found to synthesize nitrogenase and to fix nitrogen.

In other experiments at Cornell University, the gene cluster has been transferred to yeast cells (eukaryotes) and has remained intact during repeated cell divisions. These experiments raise the hope, of course, that nitrogen-fixing genes can be transferred to the cells of plants such as corn, which could then, in effect, make their own nitrogen-containing compounds. Thus far, however, the yeast cells have not demonstrated the capacity to fix nitrogen, perhaps because of differences in start/stop signals or in the cytoplasmic makeup of prokaryotic versus eukaryotic cells. An alternative way of conferring nitrogen-fixation capability on plants would be to transfer to nonlegumes the genes involved in the leguminous plant's contribution to the symbiotic association. This might be accomplished either by plasmids or by somatic cell hybridization.

The legumes are by far the largest group of plants that enter into a nitrogen-fixing partnership with symbiotic bacteria. There are, however, numerous nitrogen-fixing symbioses that involve plants other than legumes. Sweet fern (an angiosperm), for example, forms nodules that are induced by and contain nitrogen-fixing actinomycetes (moldlike bacteria), rather than *Rhizobium.* Like rhizobia, the actinomycetes enter the host plant by way of a root hair infection. The resulting symbiosis allows the plants to carry out a pioneering role in the revegetation of barren land. Sweet fern, for instance, is commonly planted along highways in Massachusetts for this purpose.

SUMMARY

Transpiration is the loss of water vapor by plants. As a consequence of transpiration, plants require large amounts of water. Water enters the plant from the soil through the roots and travels through the plant body by means of the conducting cells of the xylem—vessel members and tracheids. The current and widely accepted theory of how water moves through the xylem is the cohesion-tension theory. According to this theory, water moves through the tracheids and vessels under negative pressure (tension). Because the molecules of water cling together (cohesion), a continuous column of water molecules is pulled from the root, molecule by molecule, by the evaporation of water above.

Diffusion of gases, including water vapor, into and out of the leaf is regulated by the stomata. The stomata are opened and closed by the guard cells as a consequence of changes in turgor. Turgor is increased or decreased by the movement of potassium ions into or out of the guard cells, followed by the osmotic movement of water. The active transport of potassium ions is regulated, in turn, by a variety of factors, including water stress, abscisic acid, carbon dioxide concentration, light, and temperature.

Minerals, which are naturally occurring inorganic substances, are brought into the plant from the soil and are carried with the transpiration stream in the xylem.

They fulfill a variety of functions in plants. Some of the functions are nonspecific, such as the effects on osmosis. Others are specific, such as the presence of magnesium in the chlorophyll molecule. A number of minerals are essential components of enzyme systems.

The movement of organic compounds from the photosynthetic parts of the plant is known as translocation. It takes place in the phloem and follows a source-to-sink pattern. According to the pressure-flow hypothesis, sugars are pumped into the sieve tubes in the leaf by active transport and are pumped out of them in various parts of the plant body where they are needed for energy. Water moves into and out of the sieve tubes by osmosis, following the sugar molecules. These processes create a difference in water potential along the sieve tube, and thus water and the sugars dissolved in it move by bulk flow along the sieve tube.

Characteristics of the soil affect the availability of minerals to plants. These characteristics include the rock from which the soil was formed, the presence of humus on the soil surface, and soil composition and pH.

Two types of symbiotic relationships are important in plant nutrition: mycorrhizae and relationships involving nitrogen-fixing bacteria. Mycorrhizae, "fungus-roots," are associations between roots and soil-dwelling fungi that facilitate the uptake of minerals by the roots. The associations between nitrogen-fixing bacteria, such as rhizobia, and the roots of certain plants, particularly legumes, make it possible for some plants to utilize atmospheric nitrogen directly.

QUESTIONS

1. Distinguish among the following: transpiration/translocation; cohesion-tension theory/pressure-flow hypothesis; source/sink; A horizon/B horizon/C horizon; rhizobia/mycorrhizae.

2. What properties of water discussed in Section 1 are important to the movement of water and solutes through plants?

3. Transpiration has often been described as a "necessary evil" to the plant. Why is it necessary? How is it evil?

4. Gardeners advise removing many leaves of a plant after transplanting. How does this help the plant to survive?

5. Consider a tree transpiring most rapidly at midday and an investigator with a sensitive instrument for measuring changes in the diameter of the trunk. If water is pulled up from the top (cohesion-tension theory), what changes in diameter should be observed from night to day? (The change was, in fact, one of the early bits of evidence for the theory.)

6. In Figure 30–4, why doesn't air enter the tube through the top of the enclosed porous pipe or the leaf, even though water vapor can easily escape? How can the porous pipe (and, by analogy, the leaf) be permeable to air or water but effectively impermeable to an air-water interface?

7. Identify the cells and tissues through which a molecule of water travels from the time it enters the root until it is used in photosynthesis.

8. How is it physically possible for an increase in the turgor of the guard cells to open the stomata? Consider what would happen if you partially inflated a cylindrical balloon, applied a strip of adhesive tape along its length, and then inflated it further. What does the experiment suggest about the role of wall thickenings in the guard cells?

9. When K^+ ions move out of the guard cells, they move into adjacent epidermal cells. How would the influx of K^+ ions affect these cells? What effect would this have on the stomata?

10. Using the techniques described in this chapter for the analysis of sieve-tube sap, how would you measure the rate of movement? (*Hint:* Aphids are available.)

11. A particular sugar molecule was produced by photosynthesis in a leaf of a perennial plant late one August. Throughout the following winter, it was stored in a root of the plant. The next spring, it was oxidized in the process of respiration, providing energy for the growth of a new shoot tip. Identify the cells and tissues through which this sugar molecule traveled between its synthesis and its ultimate use, indicating the mode of transport during each stage of its journey.

12. As you will recall from Chapter 10, the immediate product of the Calvin cycle is glyceraldehyde phosphate, a 3-carbon sugar (page 226). What chemical reactions did the sugar molecule in Question 11 probably undergo between its synthesis and its use?

13. (a) Experts in flower arranging advise recutting the stems of flowers while holding them under water. Explain why. (b) Some florists advise adding ordinary table sugar to the water in which cut flowers are placed. When this is done, some types of flowers will remain fresh for several weeks. What is the explanation?

14. (a) Nitrogen-fixation is an energy-requiring process. What is the source of the energy used by *Rhizobium?* (b) If new symbiotic associations are developed in which the nitrogen-fixing bacteria produce greater yields of nitrogen-containing compounds, what price is the host plant likely to pay? Why?

CHAPTER 31

Hormones and the Regulation of Plant Growth

31–1
This young bean seedling has just broken through the soil. Its growth into a mature plant will depend on light, water, temperature, and minerals from the soil and also on the interaction of many internal factors, among which are plant hormones.

We saw in Chapter 29 that as a plant grows it does far more than increase its mass and volume. It differentiates, forming a variety of cells, tissues, and organs, and undergoes morphogenesis, taking on the shape characteristic of the adult sporophyte. Moreover, many of its activities are finely tuned to its environment and to the changing patterns of the seasons. Many of the details of how these processes are regulated are not known, but it is clear that plant development and growth depend on the interplay of a number of internal and external factors. Chief among the internal factors are the plant hormones.

A *hormone*, by definition, is a substance that is produced in one tissue and transported to another, where it exerts one or more highly specific effects. Hormones help the plant integrate the growth, development, and metabolic activities of its various tissues. Typically they are active in very small quantities. In the shoot of a pineapple plant, for example, only 6 micrograms of auxin, a common growth hormone, are found per kilogram of plant material. One enterprising plant physiologist calculated that the weight of the hormone in relation to that of the shoot is comparable to the weight of a needle in 20 metric tons of hay.

The term "hormone" comes from the Greek word meaning "to excite." It is now clear, however, that many hormones also have inhibitory influences. So, rather than thinking of hormones as stimulators, it is perhaps more useful to consider them as chemical messengers. But this term also needs qualification. As we shall see, the response to the particular "message" depends not only on its content but also upon how it is "read" by its recipient. Moreover, a response to any particular hormone is influenced by a variety of other factors in the internal environment of the plant, chief among which are likely to be other hormones. In this chapter we shall, for clarity, discuss the major groups of hormones one at a time, but try to keep in mind that we are always, in fact, dealing with interactions among them.

AUXINS

The first recognized plant hormone to be isolated was auxin, known also as indoleacetic acid, or IAA. Several different substances with activity similar to that of auxin have now been isolated from plant tissues, and others have been synthesized in the laboratory. These substances are known collectively as auxins (Figure 31–2).

IAA is synthesized in the plant by enzymatic conversion of the amino acid

CH_2COOH (structure) IAA

CH_2COOH (structure) NAPHTHALENACETIC ACID

$O—CH_2COOH$, Cl, Cl (structure) 2,4-D

31–2
Auxins. IAA (indoleacetic acid), isolated from plant tissues, is the most common natural auxin. Naphthalenacetic acid, a synthetic auxin, is commonly used to induce the formation of adventitious roots in cuttings and to reduce fruit drop in orchard crops. 2,4-D, also a synthetic auxin, is used as an herbicide.

tryptophan. It is produced principally by the apical meristems of shoots and is transported to other parts of the plant, moving in one direction only. (If a cut portion of a stem is turned upside down, IAA moves from bottom to top.) Auxin causes cells in the growing region of the shoot to elongate; if the shoot apex is removed, growth stops. If the hormone is applied to the cut surface, growth resumes. If the auxin is applied to the intact plant, no growth effect on the stem occurs until relatively high concentrations are reached. Then growth is inhibited. In short, it appears as if the apex produces the optimal concentration of auxin to which the shoot can respond positively.

The root, which also receives its auxin from the shoot apex, is more sensitive to the hormone. Auxin in very small amounts is required for root growth; however, even a slight increase in concentration can inhibit root growth.

31–3
Auxin, apparently produced by developing seeds, promotes the growth of fruit. (a) *Normal strawberry,* (b) *strawberry from which all seeds have been removed, and* (c) *strawberry in which three horizontal rows of seeds were left. If a paste containing auxin is applied to* (b), *the strawberry grows normally.*

Mechanism of Action of Auxin

The growth effects of auxin on shoots and roots are the result of rapid elongation of cells. Under the influence of auxin, the plasticity of the cellulose-containing cell wall increases and the cell expands in response to the turgor exerted by the fluid contents of the cell vacuole (see page 131). According to one current theory, IAA binds with and activates a proton pump on the plant cell membrane. Thus activated, the pump brings hydrogen ions out of the cell into the cell wall. The hydrogen ions, in turn, activate a pH-dependent enzyme already present in the cell wall that breaks the cross-links between the cellulose molecules. The molecules slide past one another as the pressure is exerted against the cell wall, and then the cross-links re-form. This is a rapid response, taking place in 10 or 15 minutes and reaching a maximum in 30 minutes.

This is only part of the story, however. Auxin also triggers RNA transcription within the cell, followed by an increase in the biosynthesis of proteins. These proteins have not been identified, but it is reasonable to believe, as a working hypothesis, that some of them are enzymes involved in the construction of new cell wall materials necessary for the continued growth of the cell.

Apical Dominance and Other Auxin Effects

In most dicot species, the growth of lateral buds is inhibited by auxin. If you pinch off the growing tip (apical meristem) of the stem of the house plant *Coleus,* for example, the lateral buds begin to grow vigorously, producing a plant with a bushier, more compact body; these derepressed buds can be repressed again by auxin application. Similarly, if you treat the "eyes" (actually lateral buds) of a potato with auxin, they will be inhibited from sprouting. As a result, the tubers can be stored longer. This phenomenon is known as *apical dominance*.

Auxin also plays a role in the seasonal initiation of activity in the vascular cambium (page 606) and the development of secondary xylem. It is also involved in the development of fruits (Figure 31–3).

31-4
The upper row of holly cuttings was treated with a synthetic auxin 21 days before the picture was taken. The cuttings in the lower row were not. Note the growth of adventitious roots on the plants in the upper row.

Synthetic Auxins

Synthetic auxins, unlike IAA, are not readily broken down by natural plant enzymes or by bacteria. Their long-lasting effects make them better suited for commercial purposes. The rooting preparations commonly used by gardeners to stimulate the growth of branch roots and adventitious roots in stem cuttings contain a synthetic auxin. Another synthetic auxin, 2,4-D, is commonly used as an herbicide.

For reasons not known, 2,4-D and related compounds are toxic to dicots at concentrations not harmful to monocots and so are commonly used on lawns to control broad-leaved weeds. During United States participation in the Vietnam War, an herbicide known as Agent Orange (a mixture of 2,4-D and a related synthetic auxin, 2,4,5-T) was used in massive quantities for defoliation of forests. There is growing concern that some of the health problems experienced by many Vietnam veterans and Vietnamese people may be a result of their exposure to Agent Orange. The toxic effects are attributed not to the herbicides themselves but to a trace contaminant of 2,4,5-T known as dioxin.

CYTOKININS

The isolation of auxin spurred the search for other growth-promoting factors in plants, particularly for a hormone that would stimulate cell division. Such a factor was first detected in coconut "milk," which is a liquid endosperm. It or related compounds have now been found in all plants examined, especially in actively dividing tissues, such as germinating seeds, fruits, and roots. They are called cytokinins; the synthetic cytokinin most commonly used in research is known as kinetin. Cytokinins resemble the purine adenine (Figure 31-5).

Responses to Cytokinin and Auxin Combinations

Studies of responses to combinations of auxin and cytokinins are helping physiologists glimpse how plant hormones work together to produce the total growth pattern of the plant. Apparently, an undifferentiated plant cell—such as the meristematic cell—has two courses open to it: either it can enlarge, divide, enlarge, and

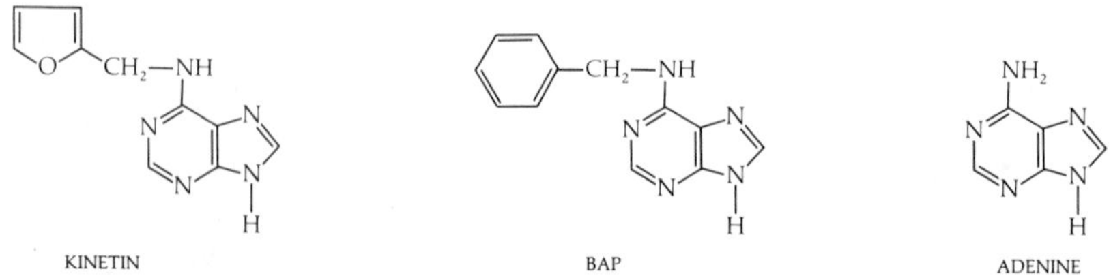

31-5
Cytokinins. (a) *Zeatin and isopentenyl adenine have been isolated from plant material.* (b) *Kinetin and 6-benzylamino purine (BAP) are commonly used synthetic cytokinins. Note the resemblances between the purine adenine and a portion of the molecule of each of these cytokinins. The significance of the resemblance between adenine and the hormones of this group is not known. It may merely be another example of biological economy: the use of a single major biosynthetic pathway to produce a number of functionally different products.*

(a)

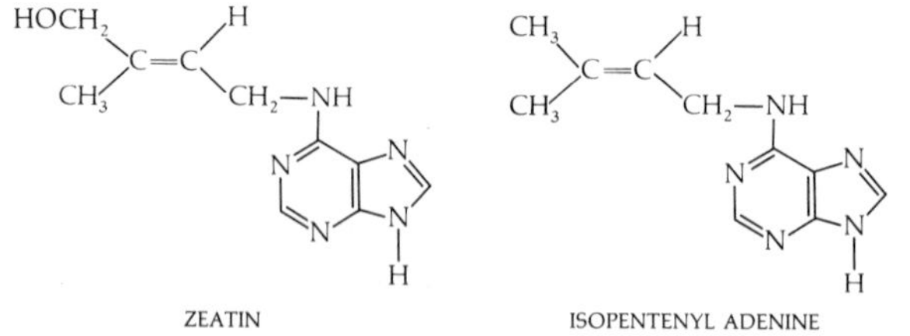

ZEATIN

ISOPENTENYL ADENINE

(b)

KINETIN

BAP

ADENINE

31-6
Two buds forming on undifferentiated tissue (callus) from a geranium following treatment with both auxin and a cytokinin. Callus from some types of plants will continue to grow as undifferentiated tissue, or roots, or buds, depending on the relative proportions of auxin and cytokinins.

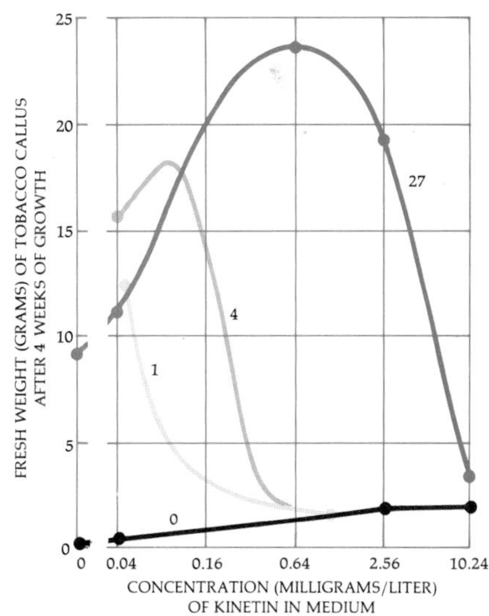

31-7
The response of tobacco cells in tissue culture to combinations of auxin (IAA) and a cytokinin. The concentrations of IAA (mg per liter) are indicated on the curves. Kinetin alone has little effect on the growth of tobacco callus. Auxin alone causes the culture to grow to a weight of about 10 grams, regardless of the concentration used. When both hormones are present, growth is greatly increased. Notice, however, that when optimum concentrations are exceeded, the growth rate declines.

divide again, or it can elongate without cell division. The cell that divides repeatedly remains essentially undifferentiated, whereas the elongating cell tends to differentiate and become specialized. In studies of tobacco stem tissue, the addition of auxin to the culture medium in which the tissue was growing produced rapid cell expansion, so that giant cells were formed. Kinetin alone had little or no effect. IAA plus kinetin resulted in rapid cell division, so that large numbers of relatively small cells were formed.

By slight alterations in the relative concentrations of IAA and kinetin, investigators have been able to affect the development of undifferentiated cells growing in tissue culture. At roughly equal concentrations of the two hormones, the cells remain undifferentiated, forming a callus. When a higher concentration of IAA is present, undifferentiated tissue gives rise to organized roots. With higher concentrations of kinetin than auxin, buds appear (Figure 31-6).

However, lest you think this is simple, we shall describe another tissue culture study, in which tuber tissue of the Jerusalem artichoke was used. In this study, it was shown that a third factor, the calcium ion (Ca^{2+}), can modify the action of the auxin-cytokinin combination. Auxin plus low concentrations of kinetin was shown to favor cell enlargement, but when calcium ions were added to the culture, there was a steady shift in the growth pattern from cell enlargement to cell division. High concentrations of calcium ions apparently prevent the cell wall from expanding, and at such concentrations the cell switches course and divides. Thus, not only do hormones modify the effects of hormones, but these combined effects may, in turn, be modified by nonhormonal factors, such as calcium ion and, undoubtedly, many others.

Other Cytokinin Effects

Cytokinins have been shown to be involved with the previously described phenomenon of apical dominance. Local application of the hormone to the repressed buds releases them from inhibition. In intact plants, cytokinins are synthesized in the roots and travel upward through the xylem, reaching the lower buds first and in highest concentration. As the dominating apex of the plant grows and moves away from these lower buds, the influence of the cytokinins overcomes that of the auxin.

31–8
Gibberellin has caused bolting and flowering in these cabbage plants. The plants on the left were not treated.

Another, apparently unrelated function of cytokinins is preventing the senescence of leaves. In many plants, for example, the lower leaves turn yellow and drop off as the upper, new leaves develop; if cytokinin is applied to the lower leaves, they remain green. Similarly, excised leaves remain green when maintained in a nutrient solution containing kinetin.

Studies on the way in which cytokinins promote cell division have shown that the hormone is required for some process that takes place after DNA replication is complete but before mitosis begins. The process does not involve the formation of new messenger RNA. The relationship, if any, between the mechanism of action of the cytokinins and their resemblance to adenine is not known.

THE GIBBERELLINS

Unlike the auxins and cytokinins, the gibberellins were first discovered serendipitously. A Japanese scientist was studying a disease of rice plants called "foolish seedling disease." The diseased plants grew rapidly but were spindly and tended to fall over under the weight of the developing grains. The cause of the symptoms, he found, was a chemical produced by a fungus, *Gibberella fujikuroi,* which infected the seedlings. The substance, which was named gibberellin, and many closely related substances were subsequently isolated not only from the fungus but also from many species of plants. More than 50 are now known (Figure 31–9).

Gibberellins characteristically produce hyperelongation of the stem, such as that seen in the foolish seedlings, in many species of plants. Particularly striking effects are seen in some plants that are genetic dwarfs. In some of these, application of gibberellin makes them indistinguishable from normal plants, suggesting that these dwarfs lack a gene needed for gibberellin production.

Gibberellins can also produce bolting, a phenomenon observed in many plants, especially biennials, that first grow as rosettes. Just before flowering, the flower stem elongates rapidly ("bolts"). Bolting and flowering, which normally occur only after an environmental cue (such as a cold season), occur in some plants following gibberellin treatments.

Gibberellins can also produce cellular differentiation. In woody plants, gibberellins stimulate the vascular cambium to produce secondary phloem. Both phloem and xylem development occur in the presence of both gibberellins and auxins. In the intact plant, interactions between gibberellins and auxins determine the relative rates of production of secondary phloem and secondary xylem.

GIBBERELLIC ACID (GA_3)

GA_7

GA_4

31–9
Three of the more than 57 gibberellins that have been isolated from natural sources. Gibberellic acid (GA_3) is the most abundant in fungi and the most biologically active in many tests. The minor structural differences that distinguish the other two gibberellins are indicated by arrows.

31–10
An experiment investigating the action of gibberellin in barley seeds. Forty-eight hours before the picture was taken, each of these seeds was cut in half and the embryo removed. The seed at the bottom was treated with plain water, the seed in the center was treated with a solution of 1 part per billion of gibberellin, and the seed at the top was treated with 100 parts per billion of gibberellin. As you can see, digestion of the starchy storage tissue has begun to take place in the seeds treated with gibberellin.

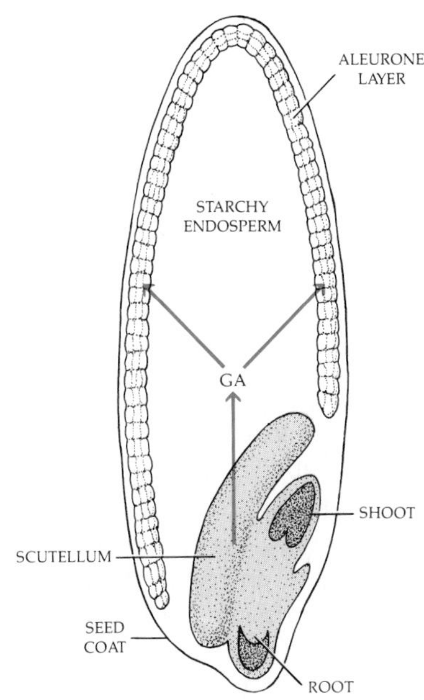

31–11
The action of gibberellin in a barley seed. The embryo, shown in color, releases gibberellic acid (GA), which diffuses to the aleurone layer. The gibberellin induces the aleurone cells to synthesize enzymes that digest the food reserves of the endosperm into smaller molecules. These molecules are absorbed by the scutellum and transported to the growing regions of the embryo.

Gibberellins and Seed Germination

The highest concentrations of gibberellins have been found in immature seeds, although they are present in varying amounts in all parts of the plants.

In grass seeds, there is a specialized layer of cells, the aleurone layer, just inside the seed coat (see page 601). These cells are rich in protein. During the early stages of germination, the embryo produces gibberellin, which diffuses to the aleurone layer. In response to the gibberellin, the aleurone cells produce enzymes that hydrolyze the starch and proteins in the endosperm, converting them to sugars and amino acids that the embryo and then the seedling can use (Figure 31–11). In this way, the embryo itself calls forth the substances needed for its survival and growth at the time they are required.

The production of the hydrolytic enzymes is the result of the synthesis of specific new messenger RNA molecules on the DNA template. Thus it is hypothesized that gibberellins, in this instance at least, are acting as derepressors, turning on certain genes.

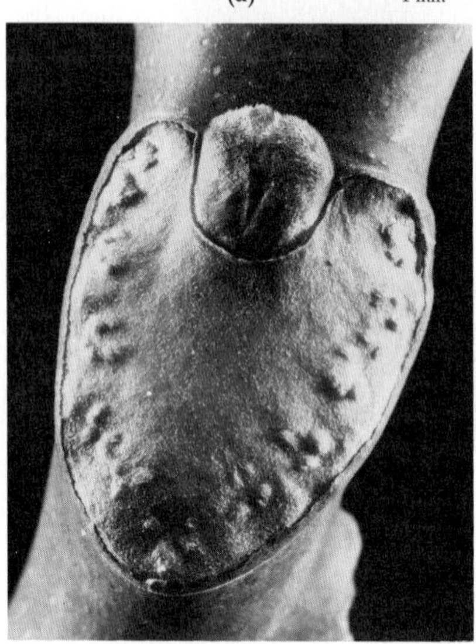

31–12
(a) *Abscission zone in a maple leaf, as seen in a longitudinal section. The abscission zone is a layer of structurally weak cells that forms across the base of the petiole. Under the influence of ethylene, enzymes are produced that cause cell wall dissolution in these cells.* (b) *After the leaf drops off, the protective layer forms a covering, the leaf scar, on the stem. This leaf scar is on a branch of* Ailanthus, *the tree of heaven.*

ETHYLENE

Ethylene is an unusual hormone in that it is a gas, a simple hydrocarbon, $H_2C{=}CH_2$. Its effects have been known for a long time, although it has only recently been considered a hormone. In the early 1900s, many fruit growers made a practice of improving the color and flavor of citrus fruits by "curing" them in a room with a kerosene stove. (Long before this, the Chinese used to ripen fruits in rooms where incense was being burned.) It was long believed that it was the heat that ripened the fruits. Ambitious fruit growers, who went to the expense of installing more modern heating equipment, found to their sorrow that this was not the case. As experiments showed, the incomplete combustion products of kerosene were actually responsible for ripening the fruits. The most active gas was identified as ethylene. As little as 1 part per million of ethylene in the air will speed the ripening process. Subsequently, it was found that ethylene is produced by plants and fungi, as well as by kerosene stoves, that it appears just before and also during fruit ripening, and that it is responsible for a number of changes in color, texture, and chemical composition that take place as fruits mature. It also is involved in the senescence of floral parts that follows pollination and precedes fruit development. The ethylene-synthesizing system is apparently located on the cell membrane, from which the hormone is released.

Auxin at certain concentrations causes a burst of ethylene production in some plants, and it is now believed that some of the effects on fruits and flowers generally attributed to auxin are related to the release of ethylene. Also, ethylene has been shown to be the intermediary in apical dominance. Auxin induces ethylene production in lateral buds, whereas the cytokinins inhibit it.

Ethylene and Leaf Abscission

Plants drop their leaves at regular intervals, either as a result of normal aging of the leaf or, in the case of deciduous trees, in response to an environmental cue. This process of *abscission* ("cutting off") is preceded by changes in the *abscission zone*, at the base of the petiole. In woody dicots, the abscission zone consists of two cell layers: (1) a structurally weak layer in which the actual abscission occurs and (2) a protective layer that forms a leaf scar on the stem (Figure 31–12).

Once leaf senescence has begun, ethylene is the principal regulator of leaf drop. It acts by promoting the synthesis and release of cellulase, an enzyme that breaks down plant cell walls (and which is also involved in the ripening process). Auxin inhibits abscission if applied to the leaf before senescence begins; once the abscission layer is formed, auxin promotes abscission by stimulating ethylene production.

ABSCISIC ACID

Soon after the discovery of the growth-promoting hormones, plant physiologists began to speculate that growth-inhibiting hormones would be found, since it is clearly advantageous to the plant not to grow at certain times and in certain seasons. Not long afterward, an inhibitory hormone, which was called dormin, was isolated from dormant buds. Subsequently, the same hormone was discovered in cotton bolls (the fruit of the cotton plant), where it was believed to promote abscission, and so was called abscisic acid, or ABA (Figure 31–13). The name given

31-13
Abscisic acid, an inhibitor that blocks the action of the growth-promoting hormones, producing dormancy in buds and leaves. It also plays a major role in stomatal closing and, as we shall see in the next chapter, in geotropism.

CH_3 CH_3 CH_3 OH CH_3 COOH O

ABSCISIC ACID

by the scientists studying cotton boll abscission proved dominant, which is somewhat unfortunate since it is now known that, in most plants at least, ABA has little to do with abscission.

ABA does, however, bring about dormancy. Application of abscisic acid to vegetative buds changes them to winter buds, converting the outermost leaf primordia to bud scales. ABA is also present in the seeds of many plant species, where it is a major factor in maintaining seed dormancy. Moreover, as we noted earlier, ABA brings about the closing of stomata under conditions of impending water shortage. Thus, ABA has come to be known as the stress hormone in recognition of its role as a protector of the plant against unfavorable environmental conditions. Abscisic acid also inhibits the effects of gibberellin in promoting bolting; this inhibition is, in turn, reversed by cytokinins.

31-14
Winter bud of a pignut hickory. These buds form in one growing season and develop in the next. The buds contain small amounts of water and high concentrations of proteins and lipids, which prevent the formation of ice crystals that could kill the cells. Bud scales, which are small, tough modified leaves, protect the delicate tissues of the bud from mechanical injury and desiccation. As the first shoots emerge in the spring, the dead bud scales are forced off. Note the leaf scar at the base of each of the two lower buds.

HORMONAL CONTROL OF FLOWERING

Flowering also appears to be under the control of a hormone or hormones, although to date no specific one has been isolated and identified. Some of the earliest experiments on this hypothetical flowering hormone were carried out by a Russian scientist, M. Kh. Chailakhyan, in the 1930s. Working with a species of chrysanthemum, Chailakhyan found that if the upper portion of the stem was stripped of its leaves and the leaves on the lower stem were exposed to an appropriate light cycle, the plant would flower, a phenomenon known as *photoinduction.* (We shall have more to say about light cycles and flowering in the next chapter.) If, however, only the upper, leafless stem and its buds were exposed, no flowering occurred. He interpreted these results as indicating that the leaves form a hormone that moves to the apical meristem of the plant and initiates flowering. He named this hypothetical hormone florigen, the "flower maker."

Subsequent experiments showed that the flowering response does not take place if the leaf is removed immediately after photoinduction. But if the leaf is left on the plant for a few hours after the induction cycle is complete, it can then be removed without stopping flowering. The flowering hormone can pass through a graft from a photoinduced plant to a noninduced plant. If a branch is girdled, florigen movement ceases. This led to the conclusion that florigen moves by way of the phloem, the system by which most organic substances are transported. However, despite this strong evidence for the existence of florigen, a specific hormone has never been isolated.

Both auxin and gibberellin have been shown to induce flowering in some plants under some circumstances, suggesting that flowering may involve not a single hormone but a combination of hormones, both stimulatory and inhibitory, in combination with other factors.

31–15
Experiments that indicate the existence of a flowering hormone. (a) *When certain plants, such as the cocklebur shown here, are exposed to an appropriate light cycle, those with leaves flower and those without leaves do not. When even one-eighth of a leaf remains on a plant, flowering occurs, and the illumination of a single leaf—not necessarily the whole plant—suffices. These experiments indicate that a chemical originating in the leaves causes the plant to flower.* (b) *This conclusion is supported by experiments on branched plants. Exposure of one branch to the light induces flowering on the other branch as well, even when only a portion of a leaf is present on the lighted branch.* (c) *When two plants are grafted together, exposure of one of the plants to the light cycle induces flowering in both the lighted plant and the grafted one.*

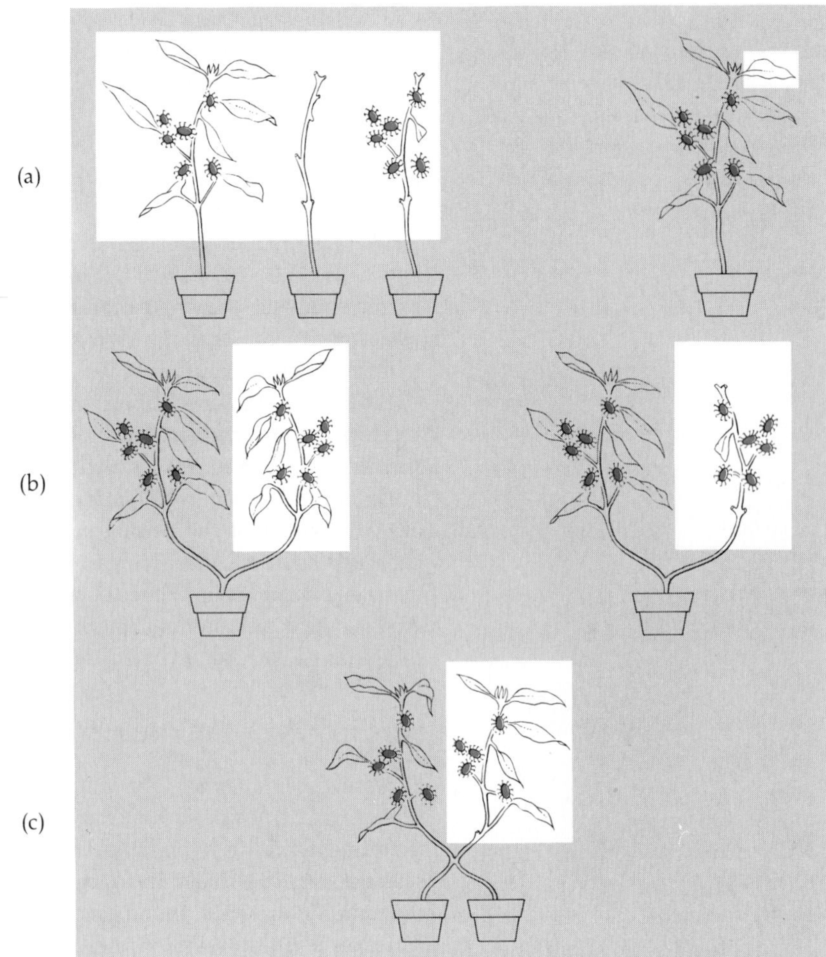

SUMMARY

A hormone is a chemical that is produced in particular tissues of an organism and carried to other tissues of the organism, where it exerts one or more specific influences. Characteristically, it is active in extremely small amounts. Hormones are important factors in plant development and growth. Five major groups of hormones have been isolated from plants: auxins, cytokinins, gibberellins, ethylene, and abscisic acid. Other classes of hormones and other growth-regulating substances may also be present.

Auxin is produced principally in rapidly dividing tissues, such as apical meristems. It causes lengthening of the shoot, chiefly by promoting cell elongation. It is involved in the phenomenon of apical dominance, in which lateral buds are inhibited, thus restricting growth principally to the apex of the plant. In low concentrations, auxin promotes the initiation of branch roots and adventitious roots; in higher concentrations, it inhibits the growth of the main root system. In developing fruits, auxin produced by seeds stimulates growth of the ovary wall; diminished production of auxin is correlated with abscission of fruits and leaves. The capacity of auxin to produce such varied effects appears to result from different responses of the various target tissues and the presence of other factors, including other hormones.

The cytokinins promote cell division. It is possible, by altering the concentrations of auxin and cytokinins, to alter patterns of growth in undifferentiated plant tissue in tissue culture.

The gibberellins were first isolated from a parasitic fungus that causes abnormal growth in rice seedlings. They were subsequently found to be natural growth hormones also present in plants. Gibberellins induce bolting and flowering in many plants. In some dwarf plants, application of gibberellins restores normal growth. Gibberellins are also involved in embryo and seedling growth. In grasses, they stimulate the production of hydrolyzing enzymes that act on the stored starch and proteins of the endosperm, converting them to sugar and amino acids, which nourish the seedling.

Ethylene is a gas that is produced by fruits during the ripening process. Ethylene promotes the ripening of fruits and is now considered a natural growth regulator. It is primarily responsible for leaf abscission and, in conjunction with auxin, for apical dominance. Abscisic acid, a growth-inhibiting hormone, induces dormancy in vegetative buds, maintains dormancy in seeds, and also effects stomatal closing.

There is evidence for the existence of a flowering hormone (or hormones), but it has not yet been isolated. A combination of hormones (and other factors) may be responsible for flowering.

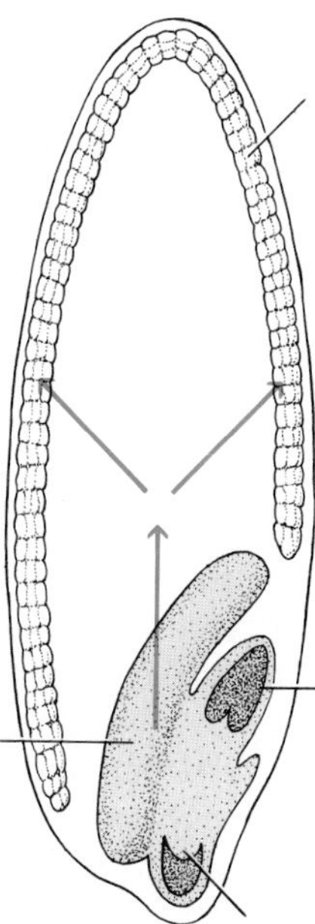

QUESTIONS

1. How would you detect the presence of each of the following hormones: auxin, cytokinin, gibberellin, ethylene, and abscisic acid?

2. Describe the phenomenon of apical dominance and the role of each of the three hormones involved.

3. The distinctions we make among the plant hormones refer to general classes of their effects rather than to the precise chemical names of the substances actually produced by particular plants. Why might it be relatively easy to test chemicals for hormonal activity but be particularly difficult to determine just which chemical a plant actually produces and uses?

4. Label the drawing in the margin. In terms of the labeled structures, describe the action of gibberellin in initiating growth of the embryo.

5. One bad apple can spoil the whole barrel. Explain.

CHAPTER 32

Plant Responses to Stimuli

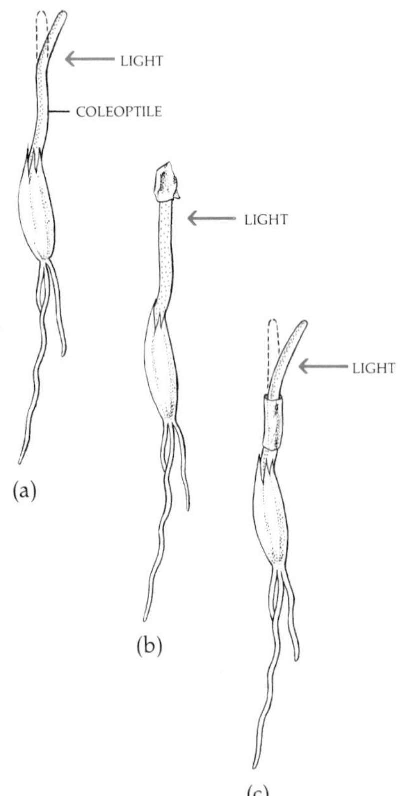

32–1
The Darwins' experiment. (a) *Light striking a growing coleoptile (such as the tip of this oat seedling) causes it to bend toward the light.* (b) *Placing an opaque cover over the tip of the seedling inhibits this bending response, but* (c) *an opaque collar placed below the tip does not. These experiments indicate that something produced in the tip of the seedling and transmitted down the stem causes the bending.*

The capacity to perceive stimuli and respond to them is commonly associated with organisms with nervous systems—that is to say, with animals. However, as we shall see, plants are able to sense and react to a range of environmental factors and, perhaps more important, to anticipate environmental changes and prepare for them. In this chapter we shall be concerned particularly with some plant responses to light, gravity, changing seasons, diurnal cycles, and touch.

PHOTOTROPISM

One of the most obvious and useful responses of plants is their bending toward light, or *phototropism*. Charles Darwin and his son Francis performed some of the first experiments on phototropism. Working with grass seedlings, they noted that the bending takes place just below the tip, in the lower part of the coleoptile (the sheath that surrounds the shoot tip). Then they showed that if they covered the terminal portion of the coleoptile with a cylinder of metal foil or a hollow tube of glass blackened with India ink and exposed the plant to a light coming from the side, the characteristic bending of the seedling did not occur. If, however, light was permitted to penetrate the cylinder, bending occurred normally. Bending also occurred normally when the lightproof cylinder was placed below the tip. "We must therefore conclude," they stated, "that when seedlings are freely exposed to a lateral light some influence is transmitted from the upper to the lower part, causing the material to bend."

In 1926, the Dutch plant physiologist Frits W. Went succeeded in separating this "influence" from the plants that produced it. Went cut off the coleoptile tips from a number of oat seedlings. He placed the tips on a slice of agar (a gelatinlike substance), with their cut surfaces in contact with the agar, and left them there for about an hour. He then cut the agar into small blocks and placed a block off-center on each stump of the decapitated plants, which were kept in the dark during the entire experiment. Within one hour, he observed a distinct bending *away* from the side on which the agar block was placed (Figure 32–2).

Agar blocks that had not been in contact with a coleoptile tip produced either *no* bending or only a slight bending *toward* the side on which the block had been placed. Agar blocks that had been exposed to a section of coleoptile lower on the shoot produced no physiological effect.

Went interpreted these experiments as showing that the coleoptile tip exerted its effects by means of a chemical stimulus (in short, a hormone) rather than a physical stimulus, such as an electrical impulse. This chemical stimulus came to be known as auxin, a term coined by Went from the Greek word *auxein,* "to increase." Auxin was the first plant hormone to be discovered.

The phototropism observed by the Darwins, it is now known, results from the fact that, under the influence of light, auxin migrates from the light side to the dark side of the tip. The cells on the dark side, which contain more auxin, elongate more rapidly than those on the light side, causing the plant to bend toward the light—a response with high survival value for young plants. (As with the effect of light on stomatal opening, only light in the blue region of the spectrum—less than 500 nanometers in wavelength—produces the phototropic response.)

GEOTROPISM

Another response with high survival value is the capacity of a young plant to respond to gravity, righting itself so that the shoot grows up and the root grows down; this response is known as *geotropism*, or gravitropism. It has been hypothesized that this upward growth of the shoot is also due to auxin. When a plant is lying on its side, according to this argument, auxin migrates to the lower side of the stem, producing differential growth that results in a bending upward of the shoot. However, evidence from recent studies indicates that the accumulation of IAA on the lower surface takes place too slowly to account for the observed curvature, which can begin as rapidly as 10 minutes after a seedling is placed on its side.

The downward growth of the root was similarly attributed to the inhibitory effect of accumulated auxin on the lower surface of the root. This hypothesis has been considerably weakened by a failure to consistently find more auxin on one side of the root than the other. Now it has been shown that, when a seedling needs righting, an inhibitor produced in the root cap slows the growth of cells on the underside of the root. This inhibitor appears to be abscisic acid, the "stress hormone."

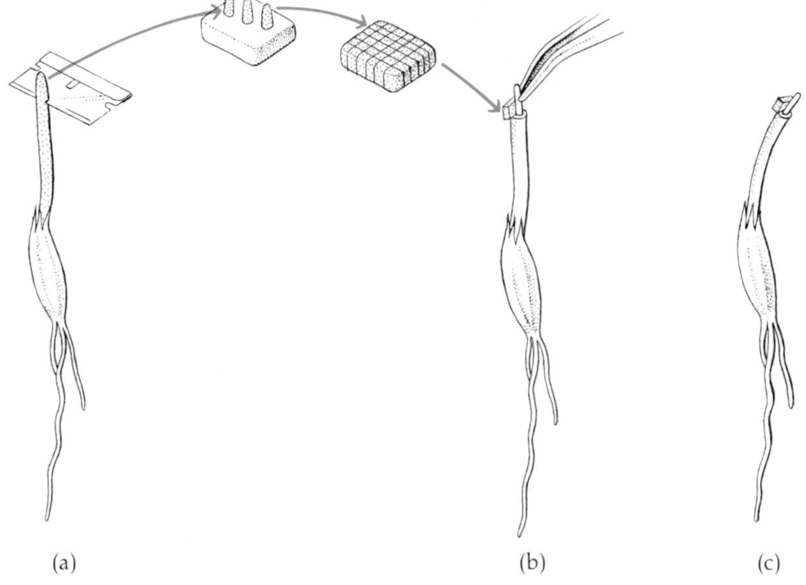

32–2
Went's experiment. (a) *He cut the coleoptile tips from oat seedlings and placed them on a slice of agar. After about an hour, he cut the agar in small blocks and* (b) *placed each block, off-center, on a decapitated seedling (the leaf is pulled up).* (c) *The seedlings bent away from the side on which the block was placed.*

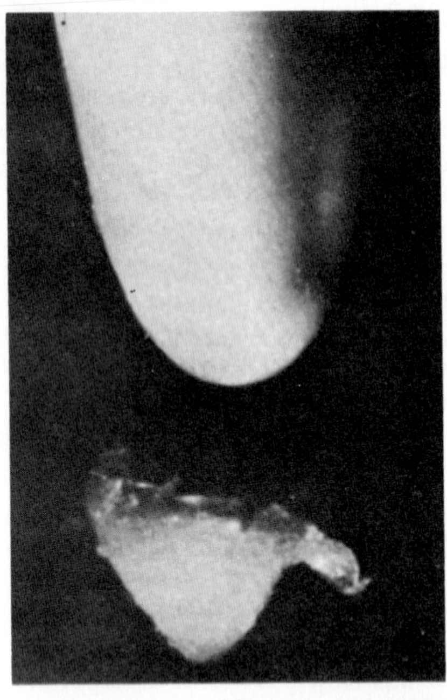

(a)

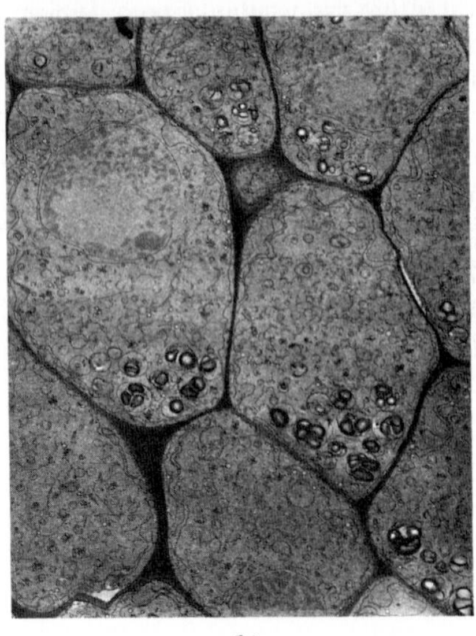

(b)

32–3

(a) *In addition to protecting the dividing cells of the apical meristem, the cells of the root cap (shown here pulled off the root) function as statocysts and produce the abscisic acid involved in root geotropism.* (b) *In this electron micrograph, starch-containing plastids—the statoliths—are visible in the lower portions of the root cap cells.*

A main question remains: How does a seedling "know" it is on its side? Hormones are soluble, and so gravity itself will not have any effect on their distribution. The answer to this question seems to lie in the inner cells of the root cap. These cells are analogous to the statocysts found in many animals (see page 491). Like statocysts, they contain statoliths—particles that move in response to gravity. In jellyfish, the statoliths are grains of hardened calcium salts; in the core cells of the root cap, they are starch-containing plastids. When a root is growing vertically, the plastids collect near the lower walls of the core cells. If the root is placed in a horizontal position, however, the plastids slide downward and come to rest near what were previously vertically oriented walls. After several hours, the root curves downward and the plastids return to their original position. Exactly how the movement of the plastids is translated into chemical gradients is not known, but such changes may be detected by the cell membranes, which, in turn, activate the chemical signals.

PHOTOPERIODISM

In many regions of the biosphere, the most important environmental changes affecting plants (and, indeed, land organisms in general) are those that result from the changing seasons. Plants are able to accommodate to these changes because of their capacity to anticipate the yearly calendar of events: the first frost, the spring rains, long dry periods, long growing spells, and even the time that nearby plants of the same species will be in flower. For many plants, all of these determinations are made in the same way: by measuring the relative periods of light and darkness. This phenomenon is known as *photoperiodism*.

Photoperiodism and Flowering

The effects of photoperiodism on flowering are particularly striking. Plants are of three general types: day-neutral, short-day, and long-day. Day-neutral plants flower without regard to day length. Short-day plants flower in early spring or fall; they must have a light period *shorter* than a critical length—for instance, the cocklebur flowers when exposed to 16 hours or less of light. Other short-day plants are poinsettias, strawberries, primroses, ragweed, and some chrysanthemums.

Long-day plants, which flower chiefly in the summer, will flower only if the light periods are *longer* than a critical length. Spinach, potatoes, clover, henbane, and lettuce are examples of long-day plants.

Ragweed and spinach, for example, will both bloom if exposed to 14 hours of daylight, yet one is designated as short-day and one as long-day. The important factor is not the absolute length of the photoperiod but rather whether it is longer or shorter than a particular critical interval for that variety. And in some varieties 5 or 10 minutes' difference in exposure can determine whether or not a plant will flower. Note that the photoperiodic initiation of flowering can take place only if the plant has passed from its juvenile stage into a phase designated as "ripeness to flower." In woody perennials, it may take decades for a plant to reach this stage.

Measuring the Dark

Following the early studies by the Beltsville group (see essay on page 646), other investigators, Karl C. Hamner and James Bonner, began a laboratory study of photoperiodism. They used the cocklebur as the experimental organism. As we

32–4
The cocklebur, a short-day plant that can stand a lot of abuse, has been important in experimental studies of photoperiodism. Each "burr" is an inflorescence with two flowers.

mentioned previously, the cocklebur is a short-day plant, requiring 16 hours or less of light per 24-hour cycle to flower. It is particularly useful for experimental purposes because a single exposure under laboratory conditions to a short-day cycle will induce flowering two weeks later, even if the plant is immediately returned to long-day conditions. Also, as we saw in Figure 31–15 (page 640), the cocklebur can withstand a good deal of rough treatment.

In the course of these studies, in which they tested a variety of experimental conditions, the investigators made a crucial and totally unexpected discovery. If the period of darkness is interrupted by as little as a 1-minute exposure to the light of a 25-watt bulb, flowering does not occur. Interruption of the light period by darkness has no effect whatsoever on flowering. Subsequent experiments with other short-day plants showed that they, too, required periods not of uninterrupted light but of uninterrupted darkness (Figure 32–5).

What about long-day plants? They also measure darkness. A long-day plant that will flower if it is kept in a laboratory in which there is light for 16 hours and dark for 8 hours will also flower on 8 hours of light and 16 hours of dark if the dark is interrupted by even a brief exposure to light.

Photoperiodism and Phytochrome

Following up on the clues from the Hamner and Bonner experiments, the Beltsville group was able to detect and eventually to isolate the pigment involved in photoperiodism. This pigment, which they called *phytochrome*, exists in two different forms (Figure 32–6). One form, known as P_r, absorbs red light with a wavelength of 660 nanometers; phytochrome is synthesized in the P_r form. The other form, P_{fr}, absorbs far-red light with a wavelength of 730 nanometers. P_{fr}, which is the active form of the pigment (that is, it can trigger a biological response), promotes flowering in long-day plants and inhibits flowering in short-day plants.

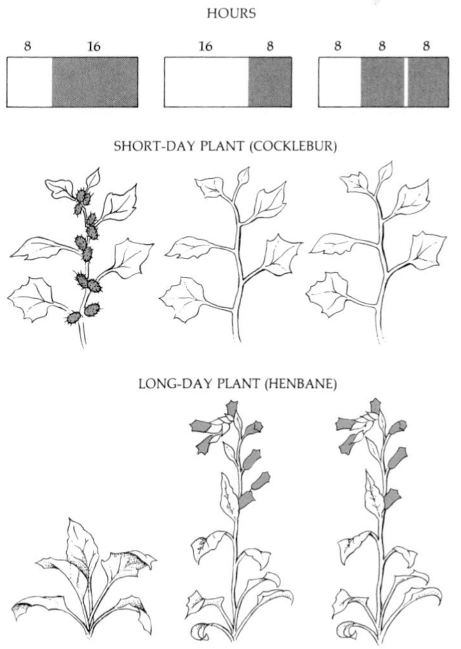

32–5
Experiments on photoperiodism showed that plants measure the darkness rather than the light. Short-day plants flower only when the darkness exceeds some critical value. Thus, the cocklebur, for instance, will flower on 8 hours of light and 16 hours of darkness. If the 16-hour period of darkness is interrupted even very briefly, as shown on the right, the plant will not flower.

The long-day plant, on the other hand, which will not flower on 16 hours of darkness, will flower if the darkness is interrupted. Long-day plants flower only when the darkness is less than some critical value.

THE DISCOVERY OF PHOTOPERIODISM

Fifty years ago a mutant appeared in a field of tobacco plants growing near Washington, D.C. The new variety had unusually large leaves and stood over 3 meters tall. As the season progressed, the regular plants flowered, but Maryland Mammoth, as it came to be called, merely grew bigger and bigger. Two scientists from the Department of Agriculture, W. W. Garner and H. A. Allard, took cuttings from the Mammoth and put them in the greenhouse, where they would be safe from frost. These cuttings flowered in December, although by then they were only 1.5 meters tall, half the size of their parent. New Maryland Mammoths grew from their seed, and these, too, did not flower until December.

Coincidentally, these same researchers were carrying out experiments with the Biloxi variety of soybean. Agriculturalists were interested in spacing out the soybean harvest by making successive sowings of seeds at two-week intervals from early May through June. But spacing out the planting had no effect; all the plants, no matter when the seeds were sown, came into flower at the same time—in September.

The investigators started growing these two kinds of plants—Maryland Mammoth tobacco and Biloxi soybeans—under a wide variety of controlled conditions of temperature, moisture, nutrition, and light. They eventually found that the critical factor in both species was the length of daylight during the 24-hour cycle. Neither plant would flower unless the day length was shorter than a critical number of hours. Consequently, Biloxi soybeans, no matter when they were planted, all flowered after the days became short enough, which was in September, and the Maryland Mammoth, no matter how tall it grew, would not flower until December, after the days became even shorter. (Note that these mutant strains, like many mutant forms, could not have survived in nature, at least not at the latitudes at which they appeared. They would have died without reproducing.)

Garner, Allard, and their coworkers (who have since become known as the Beltsville group, for the small town in Maryland where they carried out their studies) went on to test and confirm this discovery with many other species of plants. Following this single lead, they were able to answer a host of questions that had long troubled both professional botanists and amateur gardeners. Why, for example, is there no ragweed in northern Maine? The answer, they found, is that ragweed starts producing flowers when the day is less than 14.5 hours long. The long summer days do not shorten to 14.5 hours in northern Maine until August, and then there is not enough time for ragweed seed to mature before the frost. Why doesn't spinach reseed itself in the tropics? Because spinach needs 14 hours of light a day for a period of at least two weeks in order to flower and such long days never occur in the tropics. As you can see, the discovery of the photoperiodic control of flowering not only provided an explanation of plant distribution but was of great practical importance.

Photoperiodism has now been demonstrated in many species of insects, fish, birds, and mammals, influencing such diverse phenomena as the metamorphosis from caterpillar to butterfly, sexual behavior, migration, molting, and seasonal changes in coat or plumage.

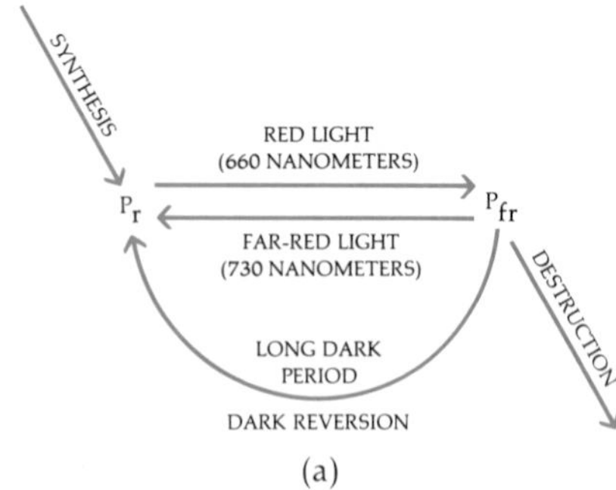

(a)

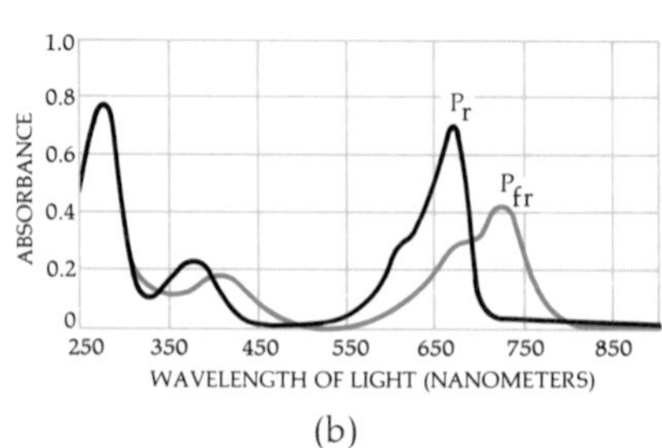

(b)

32–6
(a) Phytochrome is synthesized (from amino acids) in the P_r form. P_r changes to P_{fr} when exposed to red light, which is present in sunlight. P_{fr} is the active form, the one that induces the biological response. P_{fr} reverts to P_r when exposed to far-red light. In darkness, P_{fr} slowly reverts to P_r or is degraded. (b) Absorption spectra of the two forms of phytochrome. The similarities between the action spectra of the biological responses and the absorption spectra of the pigment provided important evidence that phytochrome was the pigment responsible for the responses. The reversible absorbance changes provided the necessary clues both for detecting the pigment in plant extracts by spectrophotometry (page 214) and for isolating it.

(a) *The relative length of day and night determines when plants flower. The four curves depict the annual change in day length in four North American cities at four different latitudes. The black lines indicate the effective photoperiod of three different short-day plants. The cocklebur, for instance, requires 16 hours or less of light. In Chicago, it can flower as soon as it matures, but in Winnipeg the buds do not appear until early in August, so late that frost usually kills the plants before the seed is mature.*

(b) *Relationship between day length and the developmental cycle of plants in the temperate zone.*

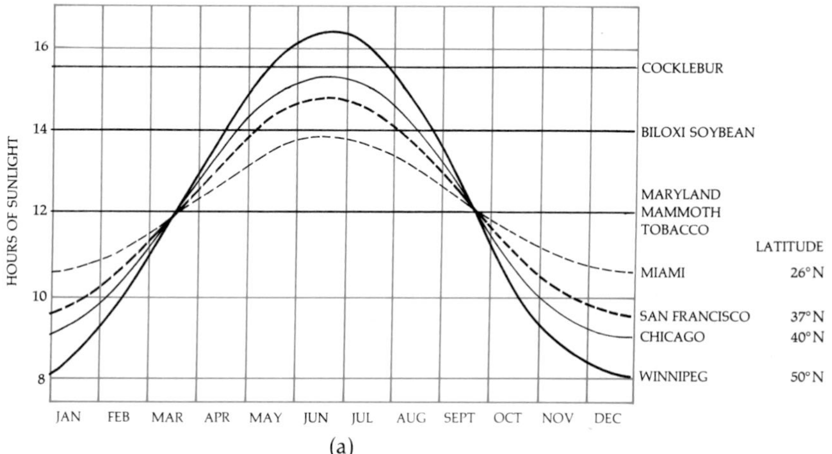

(a)

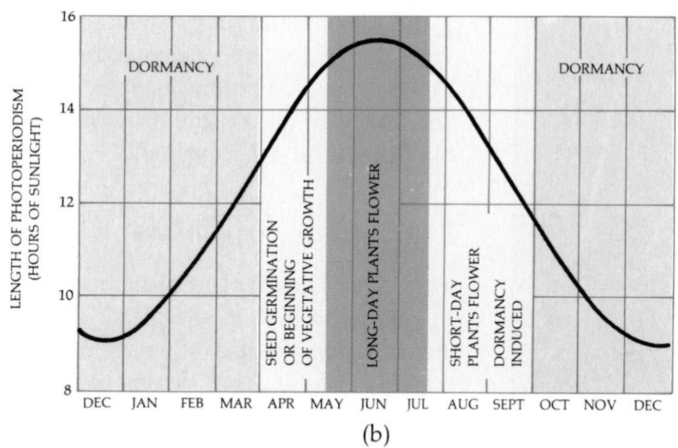

(b)

When P_r absorbs red light, it is converted to P_{fr}. This conversion takes place in daylight or in incandescent light; in both of these types of light, red wavelengths predominate over far-red. The persistence of the effect of even a very brief red-light exposure is due to the fact that P_{fr}, the active phytochrome, is relatively stable in the dark. When P_{fr} absorbs far-red light, it is converted back to P_r. In the dark, P_{fr} is either converted slowly to P_r or the phytochrome molecule is degraded and replaced by newly synthesized P_r.

Other Phytochrome Responses

The $P_r \rightleftharpoons P_{fr}$ conversion, it is now known, acts as an off-on switch in a number of plant responses to light. Many types of small seeds, such as lettuce, germinate only when they are in loose soil, near the surface. This ensures that the seedling will reach the light before it runs out of stored food. Red light, a sign that sunlight is present, stimulates seed germination by converting phytochrome to the active form. As with flower induction, exposure of the seeds to far-red light prevents the

32–7
Dark-grown seedlings, such as the ones on the right, are thin and pale with longer internodes and smaller leaves than the normal seedling on the left. This group of characteristics, known as etiolation, has survival value for the seedling because it increases its chances of reaching light before its stored energy supplies are used up.

effect. The exposures can be alternated repeatedly and the seeds respond only to the final stimulus of the series. The inhibition of seed growth by far-red light is interpreted as an adaptation that prevents seeds from germinating in the soil under a leaf canopy. As light filters through green leaves, the red is absorbed, leaving far-red predominant. Light-requiring plants have a better chance of survival if the seeds remain dormant, for years if necessary, until the required light is available.

Phytochrome is also involved in the early development of seedlings. When a seedling develops in the dark, as it normally does underground, the stem elongates rapidly, pushing the shoot up through the soil layers. A seedling grown in the dark will be elongated and spindly with small leaves (Figure 32–7). It will also be almost colorless, because the chloroplasts do not synthesize chlorophyll until they are exposed to light. Such a seedling is said to be etiolated. When the seedling tip reaches the light, normal growth begins. Phytochrome is involved in the switching from etiolated to normal growth. If a dark-grown bean seedling is exposed to only 1 minute of red (660 nanometers) light, it will respond with normal growth. If, however, the exposure to red light is followed by a 1-minute exposure to far-red light, thus negating the original exposure, etiolated growth continues.

The way in which phytochrome acts is not known. One recent suggestion is that it alters the permeability of the cell membrane, permitting particular substances to enter the cell, or, perhaps, inhibiting their entry, and that these substances, which probably include hormones, regulate the cell's activities. An increase in the concentrations of all the growth-promoting hormones can be detected immediately after phytochrome activation.

CIRCADIAN RHYTHMS

How can a spinach plant distinguish a 14-hour day from a 13.5-hour day? Seeking an answer to this question leads us to another group of readily observable phenomena. Some species of plants, for instance, have flowers that open in the morning and close at dusk. Others spread their leaves in the sunlight and fold them toward the stem at night (Figure 32–8). As long ago as 1729, the French scientist

(a)

(b)

32–8
Leaves of the wood sorrel, day (a) *and night* (b). *Darwin believed that the folding of the leaves conserved heat energy. A more recent but also unproved hypothesis is that the "sleep movements" protect the leaves from absorbing moonlight on bright nights, thus protecting photoperiodic phenomena.*

32–9
"Sleep movement" rhythms in the bean plant. Many legumes, such as the bean, orient their leaves perpendicular to the rays of the sun during the day and fold them up at night. These movements can easily be transcribed on a rotating chart by a delicately balanced pen-and-lever system attached to the leaf by a fine thread (a). *The rhythm will persist for several days in continuous dim illumination. A representative recording is seen in* (b).

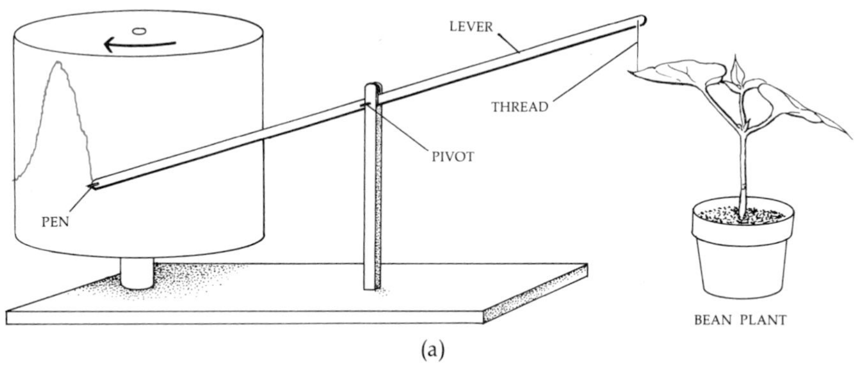

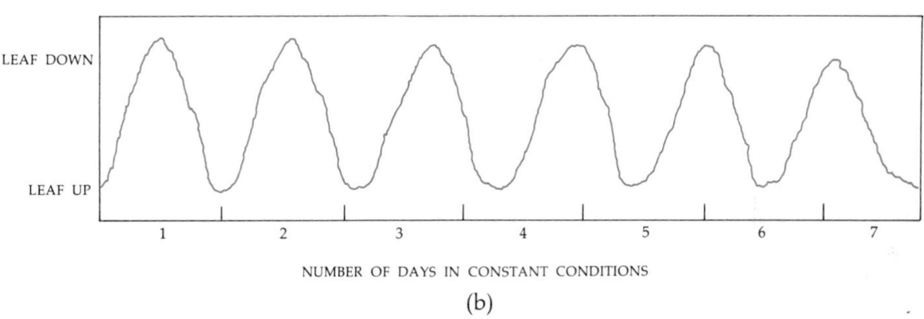

Jean-Jacques de Mairan noticed that these diurnal (daily) movements continue even when the plants are kept in constant dim light. More recent studies have shown that less evident activities, such as photosynthesis, auxin production, and the rate of cell division, also have daily rhythms. The rhythms continue even when all environmental conditions are kept constant. These regular day-night cycles are called *circadian rhythms*, from the Latin words *circa*, meaning "about," and *dies*, "day." They have been found throughout the plant and animal kingdoms.

Biological Clocks

Are these rhythms internal—that is, caused by factors within the plant or animal itself—or is the organism keeping itself in tune with some external factor? For a number of years, biologists debated whether it might not be some environmental force, such as cosmic rays, the magnetic field of the earth, or the earth's rotation, that was setting the rhythms. Strong evidence in support of an internal biological clock is that circadian rhythms are not exact. Different species and different individuals of the same species often have slightly different, but consistent, rhythms, often as much as an hour or two longer or shorter than 24 hours. Attempts to settle this recurrent controversy have led to numerous expeditions under an extraordinary variety of conditions. Organisms have been taken down into salt mines, shipped to the South Pole, flown halfway around the world in airplanes, and, most recently, orbited in satellites. As a result, it is now generally agreed that the rhythms are endogenous—that is, they originate within the organism itself. The mechanism by which they are controlled is known as a *biological clock*.

Setting the Clock

Although circadian rhythms probably originate within the organisms themselves, they can be modified by external conditions—which is, of course, important in keeping the organisms in tune with their environment. For instance, a plant whose natural daily rhythm shows a peak every 26 hours when grown under continuous dim light can adjust its rhythm to 14 hours of light and 10 hours of darkness. It can also adjust to 11 hours of light and 11 hours of dark (or 22 hours). Such adjustment to an externally imposed rhythm is known as entrainment. If the new rhythm is too far removed from the original one, however, the organism will "escape" the entrained rhythm and revert to its natural one. A plant that has been kept on an artificial or forced rhythm, even for a long period of time, will revert to its normal internal rhythm when returned to continuous dim light.

Clock Functions

Biological clocks play a role in many aspects of plant and animal physiology, synchronizing internal and external events. For example, some flowers secrete nectar or perfume at certain specific times of the day or night. As a result, pollinators—which have their own biological clocks—have become programmed to visit these flowers at these times, thereby ensuring maximum rewards for both the pollinators and the flowers. However, for most organisms, the "use" of the clock for such purposes is a secondary development. The primary function of the clock is to permit organisms to recognize the changing seasons of the year by "comparing" external rhythms of the environment, such as changes in day length, to their own relatively constant internal rhythms.

The Nature of the Clockwork

When the on-off switch of phytochrome was first discovered, it was hypothesized that the reversion of P_{fr} to P_r might be the time-measuring system of the plant, functioning by the same principle as an hourglass. Experiments have disproved this hypothesis, however; for instance, in many plants, P_{fr} disappears within three or four hours of darkness. However, in some plants, the circadian rhythm of sleep movements can be rephased by conversion of P_r to P_{fr}. The best evidence to date indicates that the timing mechanism involves rhythmic changes in the cell membrane, either in the protein components, in the fatty acids, or both. Thus, phytochrome conversion might reset the clock by producing changes in membrane structure.

TOUCH RESPONSES

Twining and Coiling

Many plants support themselves and climb by the use of modified stems or leaves that form tendrils. Tendrils often move in a spiral as they grow; this process, called circumnutation, increases the tendrils' chances of finding a support. When the apex of a tendril touches any object, it responds to the touch by forming a tight coil, which is the result of the outer surface of the tendril growing more rapidly than the inner surface. In the garden pea, for example, stroking the tendril with a glass rod for 2 minutes can induce a curling response that lasts more than 48 hours. Curling requires light, apparently for ATP production, since ATP will substitute for light in

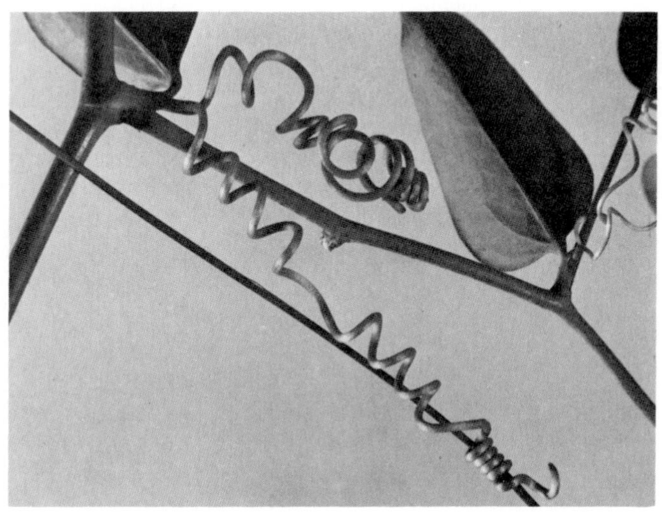

32–10
Tendrils of Smilax *(greenbrier). Twisting is caused by varying growth rates on different sides of the tendril.*

32–11
Dodder (Cuscuta) *parasitizing roadside weeds. The flowers are tiny, waxy, and absolutely colorless.*

excised tendrils. Also, like many growth responses, it involves auxin and ethylene; excised tendrils coil in the presence of auxin, even in the absence of touch.

Recent studies by M. J. Jaffe of Ohio University have shown that pea tendrils can store the "memory" of tactile stimulation. For example, if the tendrils are kept in the dark for three days and then stroked, they will not coil, perhaps because of the requirement for ATP. If, however, they are illuminated beginning as long as two hours after being stroked, they will show the coiling response.

A complex repertoire of coiling behavior is seen in the parasitic plant called dodder *(Cuscuta).* A mature dodder plant looks like a tangle of cooked spaghetti. The plants, which have no leaves and no chlorophyll, range in color from almost white to light yellow to pale orange and have a tiny waxy flower. When a dodder seed germinates, the seedling anchors itself with a small temporary root and begins to circumnutate. It can sense a host plant up to 8 centimeters away, presumably by chemical stimulus, and moves toward it. When it touches its new host (or any other object), it coils itself rapidly around it, pulling up its roots behind it. It then develops haustoria (see page 440) that it sinks into the vascular tissues of the host plant, tapping its food and water supply. Once its feeding network is established, the tip begins to circumnutate again, seeking a new host. Thus alternately winding and looping, a single dodder plant can crochet itself into a mat that covers as much as a kilometer.

Rapid Movements in the Sensitive Plant

A more rapid response to touch is seen in the sensitive plant, *Mimosa pudica.* A few seconds after a leaf is touched, the petiole drops and the leaflets fold (Figure 32–12). This response is a result of a sudden change in turgor in the so-called motor cells in the jointlike thickenings called pulvini (singular, pulvinus) at the base of leaflets and leaves. These same cells are involved with sleep movements, which are also seen in *Mimosa.* Depending on the variety and the stimulus, a single leaf or all the leaves on the plant may be affected. A series of reactions appears to be involved in the spread of the stimulus. The sensory stimulus is apparently translated into an

(a)

(b)

(c)

32–12
Sensitive plant (Mimosa pudica). *(a) Normal position of leaves and leaflets.* (b, c) *Successive responses to touch. It has been hypothesized that these reactions may prevent wilting (when they occur in response to strong winds), startle insects, or dismay larger herbivores. Collapse of the leaflets, which is caused by rapid turgor changes in cells of the petioles, is accompanied by release of tannins by the cells. These have an astringent taste and so are repellent to herbivores. In the evolution of the touch response in* Mimosa, *the release of tannins may have been more important than the rapid movements.*

electrical signal, similar to a nerve impulse in animals, that passes along the petiole and can be detected by microelectrodes placed in the plant. The electrical signal, in turn, triggers a chemical signal that acts on the motor cells. The loss of volume of the motor cells is accompanied by movement of potassium ions out of the cells, presumably as a result of changes in the membrane. Tannins are also excreted by the motor cells into extracellular spaces and may protect the plant against predation.

The Effects of Touch on Plant Growth

In addition to the specialized touch responses of some plants, touch and other mechanical stimuli have widespread effects on patterns of growth, according to recent evidence produced by Jaffe. Although botanists have long known that plants grown in a greenhouse tend to be taller and more spindly than plants of the same species grown outside, It was not until the 1970s that his systematic studies revealed that regular rubbing or bending of stems inhibits their elongation and results in shorter, stockier plants. Plants in a natural environment are, of course, subjected to similar stimuli, in the form of wind, raindrops, and the movements of passing animals. This response also involves both electrical signals and a change in cell membrane permeability to solutes. The change in membrane permeability is believed to affect the relative proportions of hormones available to the plant cells.

Responses in Carnivorous Plants

Turgor changes in key motor cells are also involved in the capture of prey by the carnivorous Venus flytrap (see essay, page 578). The leaves of the Venus flytrap have two lobes, and each leaf half is equipped with three sensitive hairs. When an insect alights on one of these leaves, it brushes against the hairs, setting off an electrical impulse that triggers the closing of the leaf. The toothed edges mesh like a bear trap, snapping shut in half a second. Once the insect is trapped, the leaf halves gradually squeeze closed, and the captive animal is pressed against digestive glands located on the inner surface of the trap. The movement of the insect stimu-

32–13
(a) *Diagram of a sundew tentacle with electrodes in place.* (b) *A record of electrical impulses produced by positioning a fruit fly so that its feet stroked the head of a tentacle. The recording ended when the contact was broken as the tentacle bent.*

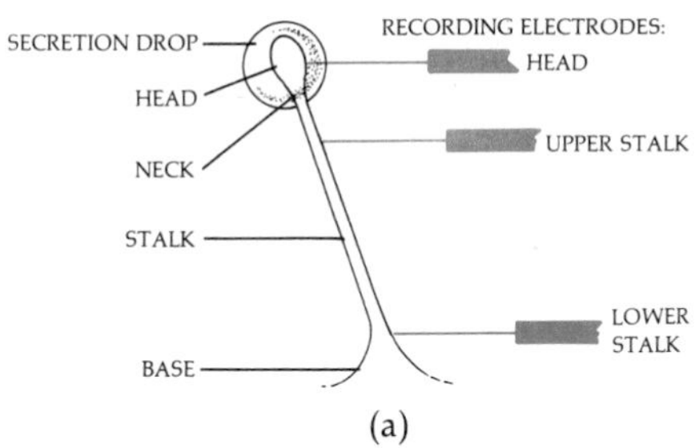

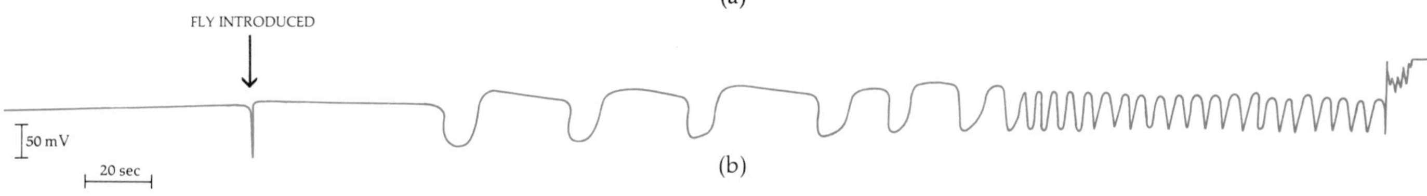

lates the trap to close more tightly and glands on the upper leaf surface to secrete digestive enzymes.

Studies by investigators at Washington University in St. Louis have shed new light on how the sundew traps insects. As you saw on page 578, the club-shaped leaves of the plant are covered with tiny tentacles. A sticky droplet surrounding the tip of each tentacle attracts insects. When an insect is caught on the tip of a tentacle, the surrounding tentacles bend in rapidly, carrying the prey to the center of the leaf, where it is digested. Microelectrodes placed in the tentacles (Figure 32–13) have revealed that this response, like the triggering of the Venus flytrap, is accompanied by an electrical impulse that moves down the tentacle.

The electrical impulses that have been observed in plants are the same, in principle, as the nerve impulses of animals (see page 769). Many botanists expect that electrical signals will be found to coordinate a variety of activities involving cells in different parts of the plant body.

SUMMARY

Plants keep in tune with their environment, responding to external stimuli and anticipating environmental changes.

Two responses with high survival value for young plants are phototropism, the bending of a plant toward the light, and geotropism, the capacity of the shoot to grow up and the root to grow down in response to gravity. Phototropism results from the migration of auxin from the lighted side to the dark side of the shoot tip. As a consequence of the increased levels of auxin, the cells on the dark side elongate more rapidly, bending the plant toward the light. The downward bending of the root, however, results from inhibition of growth in the cells on its lower surface. This inhibition is caused by abscisic acid, produced in the root cap in response to movements of starch-containing plastids in the root cap core cells.

Photoperiodism is the response of organisms to changing periods of light and darkness in the 24-hour day. Such a response controls the onset of flowering in many plants. Some plants will flower only when the periods of light exceed a critical length. Such plants are known as long-day plants. Other plants, short-day plants, flower only when the periods of light are less than some critical period.

Day-neutral plants flower regardless of photoperiod. Interruption of the dark phase of the photoperiod, by even a brief exposure to light, can serve to reverse the photoperiod effects, indicating that the dark period rather than the light period is the critical factor.

Phytochrome, a pigment commonly present in small amounts in the tissues of plants, is the receptor molecule for transitions between light and darkness. The pigment exists in two forms, P_r and P_{fr}. P_r absorbs red light of a wavelength of 660 nanometers, whereas P_{fr} absorbs far-red light of a 730-nanometer wavelength. P_{fr} is the active form of the pigment; among its many known effects, it promotes flowering in long-day plants, inhibits flowering in short-day plants, promotes germination in lettuce seeds, and promotes normal growth in seedlings. Its mechanism of action appears to involve changes in the cell membrane.

Circadian rhythms are regular cycles of growth and activity occurring approximately on a 24-hour basis. Many of these rhythms have been shown to be independent of the organism's environment and to be controlled by some endogenous regulator—a biological clock. The principal function of the biological clock is to provide the timing mechanism necessary for photoperiodism. Its chemical nature is not known.

Some species of plants exhibit specific, rapid movements in response to touch. Examples include the winding of tendrils, the collapse of the leaves of the sensitive plant *(Mimosa)*, the triggering of the carnivorous Venus flytrap, and the bending of the tentacles of the sundew. Recent studies indicate that all vascular plants respond to touch and other mechanical stimuli with altered growth patterns, resulting in shorter, stockier plants. Plant responses to touch involve various combinations of electrical impulses, chemical changes, and changes in turgor.

QUESTIONS

1. Distinguish among the following: phototropism/photoperiodism/photoinduction; circadian rhythm/biological clock.

2. Why is a biological clock necessary for photoperiodism?

3. Traveling from north to south, one can find varieties of the same species with different photoperiodic requirements. How would you expect them to differ?

4. Photoperiodic systems are not nearly as sensitive to low levels of illumination as are many visual systems. Why might extreme sensitivity be a disadvantage in a photoperiodic system? What might be the effect of the widespread use of bright lights for street illumination?

5. Suppose you were given a chrysanthemum plant, in bloom, one autumn, and you decided to keep it indoors as a house plant. What precautions would you need to take the following autumn to ensure that it would bloom again?

6. Carefully examine the graph in Figure 32–9. What is happening to the rhythm of the plant's "sleep movements" toward the end of the week? Speculate as to why this might be happening. It has also been observed that as the plant is maintained over time under the experimental conditions, it becomes rather sickly. What factors do you suppose might be involved in this change?

SUGGESTIONS FOR FURTHER READING

Books

BRADY, JOHN: *Biological Clocks, Studies in Biology, No. 104,* University Park Press, Baltimore, Maryland, 1979.*

An interesting and well-written introduction to the subject of biological clocks and their experimental study.

CUTLER, DAVID F.: *Applied Plant Anatomy,* Longman, Inc., New York, 1978.*

An interestingly written textbook on the fundamentals of plant anatomy showing some of the ways in which plant anatomy can be applied to solve many important everyday problems.

GALSTON, A. W., P. J. DAVIES, and R. L. SATTER: *The Life of the Green Plant,* 3d ed., Prentice-Hall, Inc., Englewood Cliffs, N.J., 1980.*

A comprehensive and up-to-date description of the functioning of the green plant, especially strong in plant physiology. Written for students without an advanced background in biology or chemistry.

GREULACH, VICTOR: *Plant Structure and Function,* 2d ed., The Macmillan Company, New York, 1983.

A short, lucid introduction to botany, with emphasis on physiology.

HEYWOOD, VERNON H. (ed.): *Flowering Plants of the World,* Mayflower Books, Inc., New York, 1978.

The best available guide for students to the families of flowering plants.

LEDBETTER, M. C., and KEITH PORTER: *Introduction to the Fine Structure of Plant Cells,* Springer-Verlag, New York, 1970.

An excellent atlas of electron micrographs of plant cells, with a detailed explanation of each.

RAVEN, PETER H., RAY F. EVERT, and HELENA CURTIS: *Biology of Plants,* 3d ed., Worth Publishers, Inc., New York, 1981.

An up-to-date and handsomely illustrated general botany text, especially strong in evolution and ecology.

RAY, PETER M.: *The Living Plant,* 2d ed., Holt, Rinehart and Winston, Inc., New York, 1972.*

An outstanding, short text.

SALISBURY, FRANK B., and CLEON W. ROSS: *Plant Physiology,* 2d ed., Wadsworth Publishing Co., Inc., Belmont, Calif., 1978.

A good, modern plant physiology text for more advanced students.

SCHNELL, DONALD E.: *Carnivorous Plants of the United States and Canada,* John F. Blair, Winston-Salem, N.C., 1976.

Descriptions of forty-five species of carnivorous plants with many color photographs. This book is not only interesting reading but also useful as a field guide, and it contains a chapter on growing techniques.

TROUGHTON, JOHN H., and F. B. SAMPSON: *Plants: A Scanning Electron Microscope Survey,* John Wiley & Sons, Inc., New York, 1973.*

A scanning electron microscope study of some anatomical features of plants and the relationship of these features to physiological processes.

---

* Available in paperback.

ZIMMERMANN, MARTIN H., and CLAUDE L. BROWN: *Trees: Structure and Function,* Springer-Verlag, New York, 1975.

An up-to-date discussion of how trees work, with emphasis on structure as it relates to function.

Articles

BRILL, WINSTON J.: "Biological Nitrogen Fixation," *Scientific American,* March 1977, pages 68–81.

BRILL, WINSTON J.: "Nitrogen Fixation: Basic to Applied," *American Scientist,* vol. 67, pages 458–466, 1979.

HITCH, CHARLES J.: "Dendrochronology and Serendipity," *American Scientist,* vol. 70, pages 300–305, 1982.

JUNIPER, BARRIE E.: "Geotropism," *Annual Review of Plant Physiology,* vol. 27, pages 385–406, 1976.

PETTITT, JOHN, SOPHIE DUCKER, and BRUCE KNOX: "Submarine Pollination," *Scientific American,* March 1981, pages 134–143.

RICK, CHARLES M.: "The Tomato," *Scientific American,* August 1978, pages 76–87.

RUEHLE, JOHN L., and DONALD H. MARX: "Fiber, Food, Fuel, and Fungal Symbionts," *Science,* vol. 206, pages 419–422, 1979.

SECTION 6

Biology of Animals

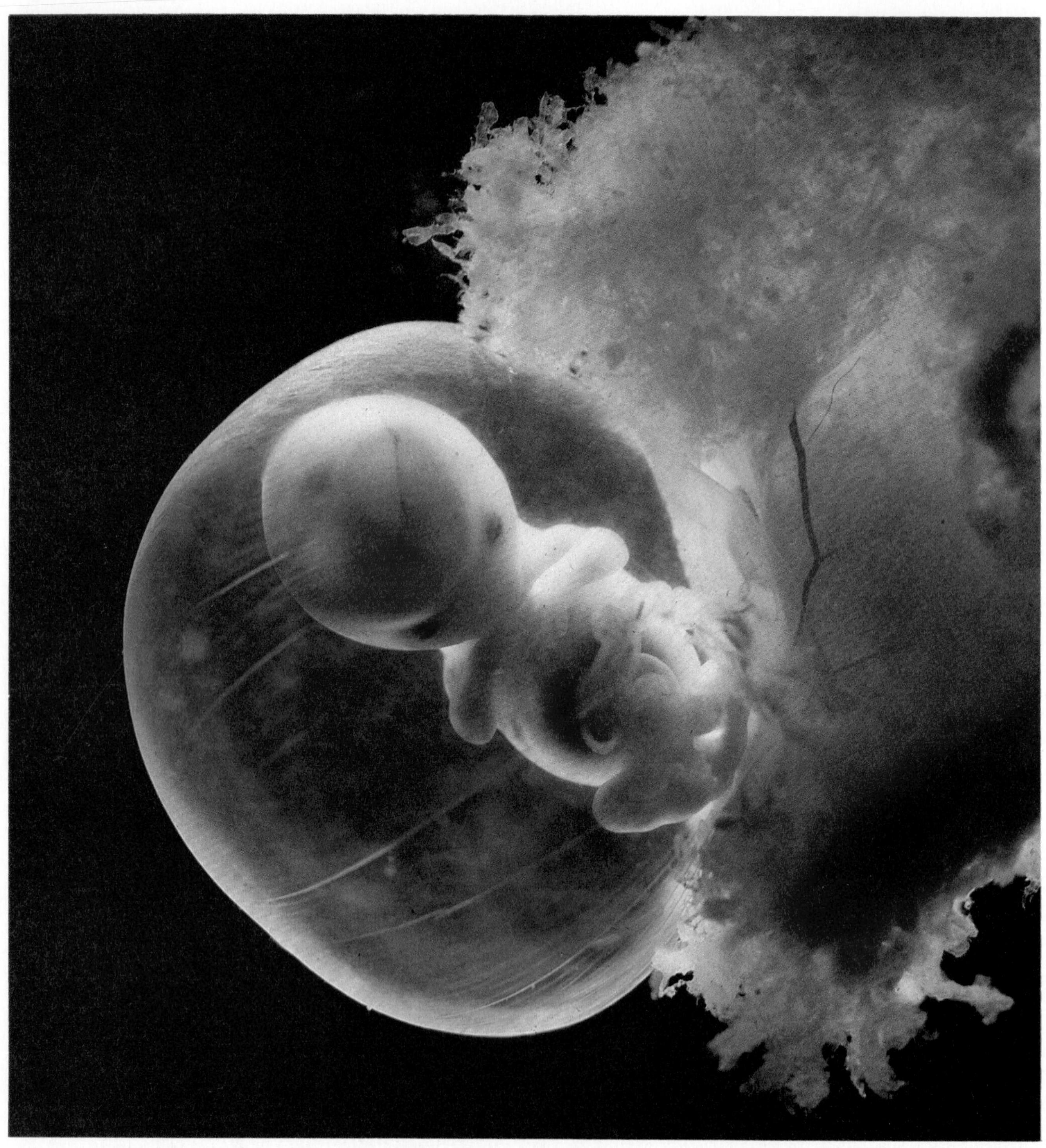

33–1
Genus, Homo; *species,* sapiens. *Embryo at 7 weeks.*

CHAPTER 33

The Human Animal: An Introduction

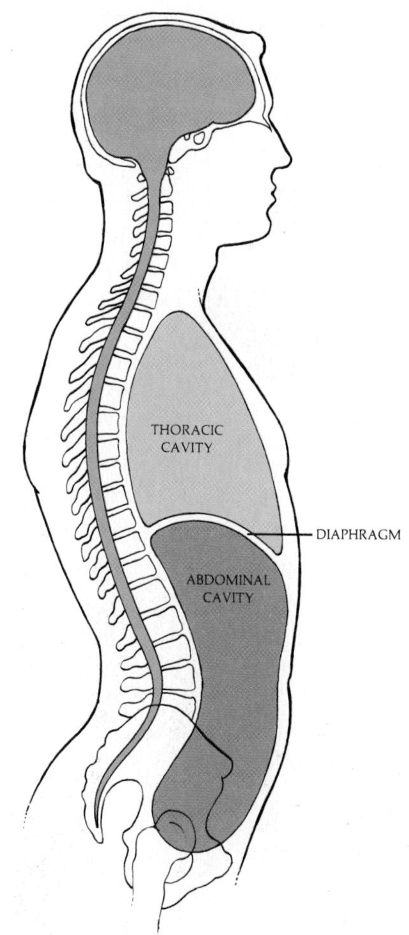

33–2
Humans, like other vertebrates, are characterized by a dorsal central nervous system (brain and spinal cord) enclosed in vertebrae and the skull. As in other mammals, a muscular diaphragm divides the coelom into the thoracic cavity and the abdominal cavity.

Even those of us who have most marveled at the exquisite architecture of an orchid flower, or who are as content as Leeuwenhoek to watch the intricate and varied movements of a *Paramecium* and its neighbors, approach the subject of the biology of our own systems with a quickened interest. Indeed for some, student and scientist alike, the principal and perhaps the only reason for studying "lower forms" is the extent to which such studies bear directly on human welfare. But as *Escherichia coli*, T2 and T4 bacteriophages, and *Drosophila* remind us, there is no way to study only the human species, any more than it is possible to study only a fruit fly.

The central theme in this section of the book is vertebrate physiology. The major focus will be on *Homo sapiens*; however, we shall rely on numerous examples from other organisms to gain insight into important physiological principles and adaptations. Because we are part of a continuum of nature, it is logical to study other animals in order to understand the human animal; it is equally logical to study the human animal (now one of the best understood) in order to gain a deeper understanding of animal life in general.

The human being is a vertebrate and as such has a bony, articulated (jointed) endoskeleton that supports the body and grows as it grows. The spinal cord, which is dorsal, is surrounded by bony segments, the vertebrae, and the brain is enclosed in a protective casing, the skull.

As in other vertebrates, and most invertebrates as well, the human body contains a cavity, or coelom. In humans, the coelom is divided into two parts, the thoracic cavity and the abdominal cavity (Figure 33–2). These are separated by a dome-shaped muscle, the diaphragm. The thoracic cavity contains the heart, the lungs, and the upper portion of the digestive tract. The abdominal cavity contains a large number of organs, including the stomach, intestines, and liver.

Human beings are also mammals. One of the most important characteristics of mammals is that they are warm-blooded. More precisely, they are homeotherms; that is, they maintain a high and relatively constant body temperature. As a consequence, mammals (and birds, which are also homeotherms) are able to achieve and sustain levels of physical activity and mental alertness generally far greater than those of animals whose temperatures rise and fall with that of their external environment. A concomitant of homeothermy is a high metabolic rate, which requires relatively large and constant supplies of food (fuel) molecules and oxygen.

33–3

(a) *The skeleton of a human adult contains 206 bones. Twenty-nine are in the skull, including 14 face bones and 6 small bones (ossicles) of the ears. There are 27 bones in each hand and 26 in each foot. Bones are living organs, made up not only of connective tissue but of the other tissue types as well. They are surrounded by a fibrous sheath, which contains blood vessels that nourish the bone tissues. Calcium compounds, secreted by osteoblasts (immature bone cells), give bone its characteristic hardness.* (b) *The structures of the right knee are shown from the front. A knee is a hinge joint between the femur and the tibia, covering which is the "knee cap," or patella (removed in this drawing). The patella acts as a stop so that the hinge can open to 180° but no further. As with all bones, bones of the knee joint are bound together by ligaments. The collateral ligaments bind the joint externally; they are slack when the knee is flexed and taut when the knee is extended. The cruciate ligaments cross within the joint, stabilizing it. The menisci are C-shaped wedges of cartilage, resting on the upper surface of the tibia and cushioning the joint.*

Knees are particularly susceptible to injury, as Joe Namath, Bobby Orr, Wilt Chamberlain, and others will sadly attest. The reason for the knee's vulnerability becomes clear when you contemplate the analogous problem of fastening together two match sticks end-to-end with several rubber bands, to produce a hinge that is both strong and flexible and can not only swing back and forth but also, to some extent, twist, bend, and rotate. Common knee injuries are torn ligaments, especially the collateral ligaments (as a result of the knee's being hit from the side), and crushed menisci. When the shock-absorbing wedges of menisci are lost, bone chips begin to accumulate in the joint.

(a)

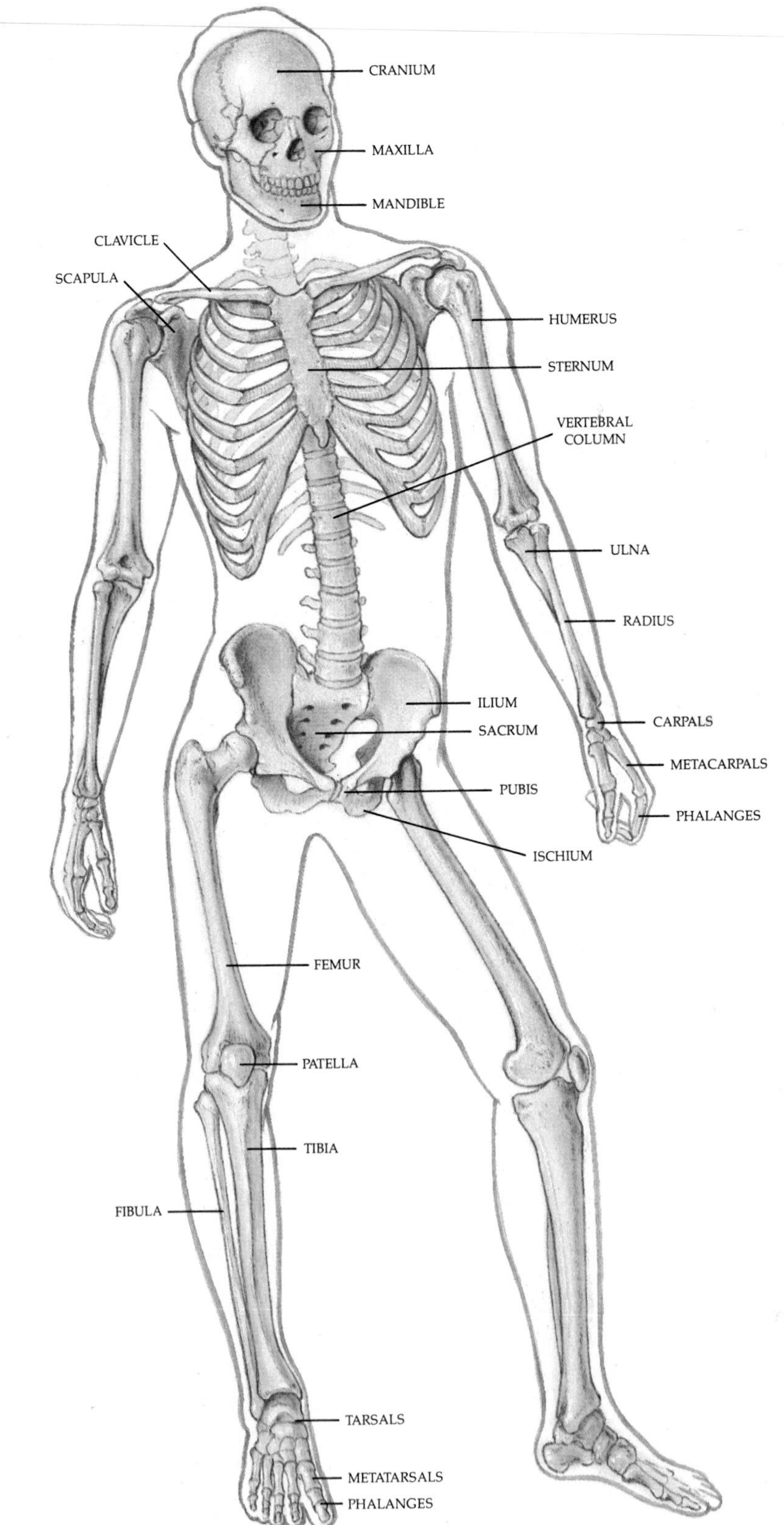

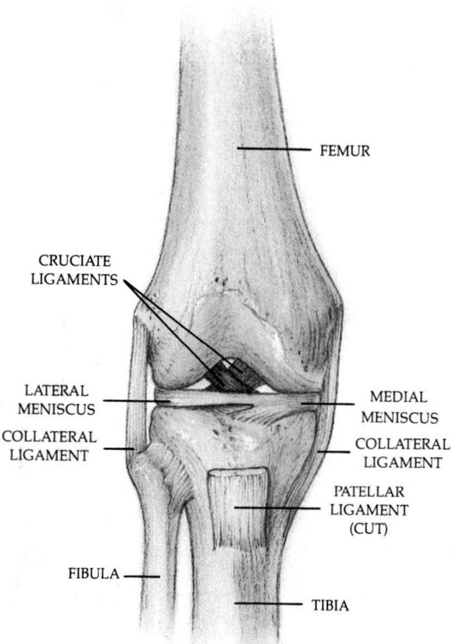

Mammals have other important characteristics. They have hair or fur rather than scales or feathers. In keeping with their high levels of activity and mental alertness, they have the most highly developed systems for receiving, processing, and correlating information from the environment (although, as we noted in Chapter 26, they are surpassed in complexity and variety of sense organs by some of the invertebrates). All mammals (except the monotremes) give birth to live young, as distinct from laying eggs, which all birds and most fish, amphibians, and reptiles do. Mammals nurse their offspring, which involves a relatively long period of parental care and makes possible a long learning period. Contrast this, for example, with most insects and nearly all species of fish, amphibians, and reptiles, in which the young are independent from the moment they hatch from the egg. There is a tendency among the large mammals, in particular, toward fewer offspring per litter and prolongation of parental care. Humans, for instance, rarely have more than two surviving young per birth, only two mammary glands with which to nurse them, and an extraordinarily long period of infancy and childhood, with dependency on parents often lasting well past physical maturity. Finally, for better or worse, *Homo sapiens* is by far the most intelligent of all mammals.

ORGANIZATION OF THE HUMAN BODY

Although they greatly resemble one-celled organisms in their requirements, the cells of multicellular organisms differ from one-celled organisms in that they develop and function as part of an organized whole. Cells are organized into tissues, groups of cells similar in structure and function. Different kinds of tissues, united structurally and coordinated in their activities, form organs, such as the skin. Organs that function together in an integrated and organized way make up *organ systems*. The digestive system, for example, is composed of a number of different organs, each of which carries out a specific activity that contributes toward the overall process. The focus of this section is on the various functions of the body, and so in the chapters that follow we shall be examining it principally in terms of organ systems.

CELLS AND TISSUES

Although there are about 200 different cell types in the human body, they are generally classified into four types of tissue: (1) epithelial, (2) connective, (3) muscle, and (4) nerve.

Epithelial Tissue

Epithelial tissues consist of continuous sheets of cells that provide a protective covering over the whole body and contain various sensory nerve endings. They also form the lining membranes of internal organs, cavities, and passageways and cover internal organs. Hence, as a moment's reflection will reveal, everything that goes into and out of the body must pass through epithelial cells. One surface of the epithelial sheet is always attached to an underlying layer, called the basement membrane, composed of a fibrous polysaccharide material produced by the epithelial cells themselves.

SQUAMOUS

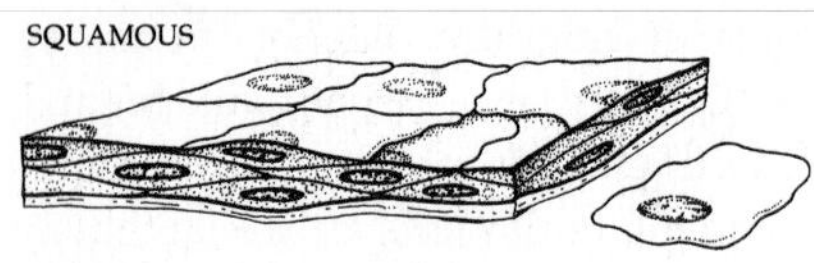

CUBOIDAL

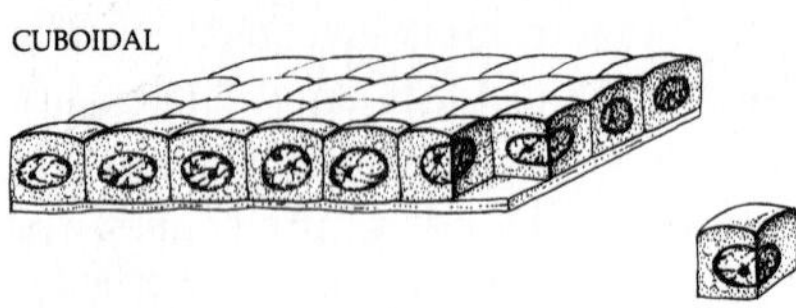

COLUMNAR

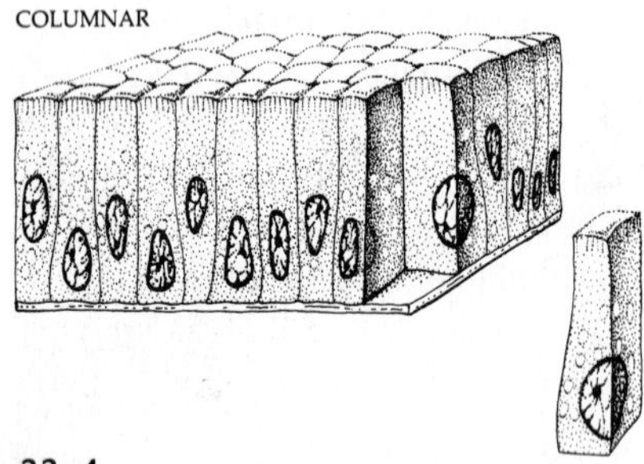

33–4
The three types of epithelial cells that cover the inner and outer surfaces of the body. Squamous cells, which usually perform a protective function, make up the outer layers of the skin and the lining of the mouth and other mucous membranes. There are usually several layers of these flat cells piled on top of one another. Cuboidal and columnar cells, in addition to lining various passageways, are often involved in active transport and other energy-requiring activities.

Epithelial cells are often specialized for synthesis and secretion of products for export. For example, *glands* are made up of modified epithelial cells that produce specific substances, such as perspiration, saliva, milk, hormones, or digestive enzymes. Gland cells may be isolated, such as the goblet cells that secrete the mucus that lubricates the linings of body cavities and passages, or clustered together in organs.

Epithelial tissues are classified according to the shape of the individual cells as squamous, cuboidal, or columnar (Figure 33–4). They may consist of only a single layer of cells (simple epithelium), as found in the inner lining of the circulatory system, or several layers (stratified epithelium), as found in the outer layer (epidermis) of the skin.

Connective Tissue

Connective tissue binds together, supports, and protects the other three kinds of tissue. Unlike epithelial tissue cells, the cells of connective tissue are widely separated from one another by large amounts of intercellular material, the matrix, which anchors and supports the tissue. The matrix comprises a ground substance, which is more or less fluid and amorphous (formless), and fibers synthesized by the tissue cells. There are several different types of fibers depending on the tissue: (1) connecting and supporting fibers, such as collagen, which is a major component of skin, tendons, ligaments, and bones; (2) elastic fibers, which are found, for example, in the walls of large blood vessels; and (3) reticular fibers, which form networks inside solid organs, such as the liver. While epithelial tissues are classified according to cell shape, connective tissues are grouped by the characteristics of their intercellular matrix.

Bone, like other connective tissues, consists of cells, fibers, and ground substance (Figure 33–5). It is distinctive in that the intercellular matrix is impregnated with hard crystals. Bone tissue, despite its strength, is amazingly light; our bones make up only about 18 percent of our weight.

Blood and lymph are connective tissues in which the matrix is plasma. The characteristic cells of blood and lymph are described in Chapters 36 and 39.

Muscle Tissue

Muscle cells are specialized for contraction. Every function of muscle—from running, jumping, smiling, and breathing to propelling the blood through the arteries and ejecting the fetus from the uterus—is carried out by the contraction of muscle cells in concert.

Muscle tissue can be classified in various ways. Classifying by appearance yields two types, as shown in Figure 33–6: striated muscle, which has stripes under the microscope, and smooth muscle (no stripes). Classifying by location, there are three categories: skeletal muscle, cardiac muscle, and, again, smooth muscle. The latter surrounds the walls of internal organs, such as the digestive organs, uterus, bladder, and blood vessels. In a third type of classification, muscle tissue is categorized as voluntary or involuntary. Skeletal muscle is voluntary. Smooth muscle and cardiac muscle are not, except in rare individuals, under direct voluntary control.

Contraction of each of these types of muscle cell depends on the interaction of two proteins: actin and myosin. In skeletal and cardiac muscle, these proteins are arranged in regular, repeating assemblies, resulting in the characteristic striations.

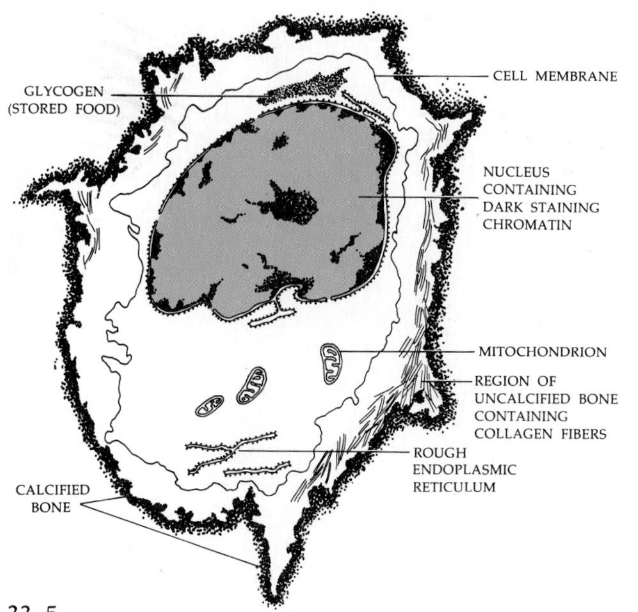

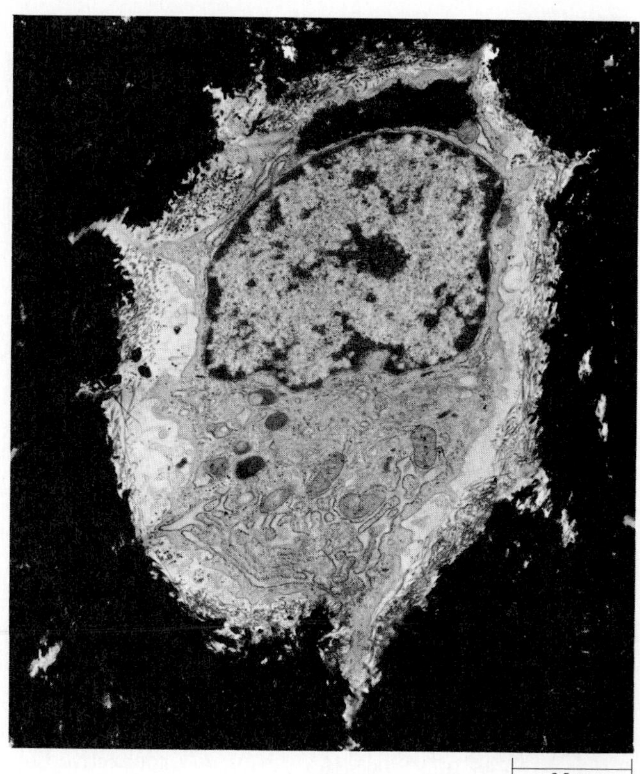

33–5
Electron micrograph and diagram of a bone cell (osteocyte). Young bone cells (osteoblasts) produce the intercellular bone matrix, an organic material consisting of collagen fibers and ground substance. This matrix gradually calcifies and hardens.

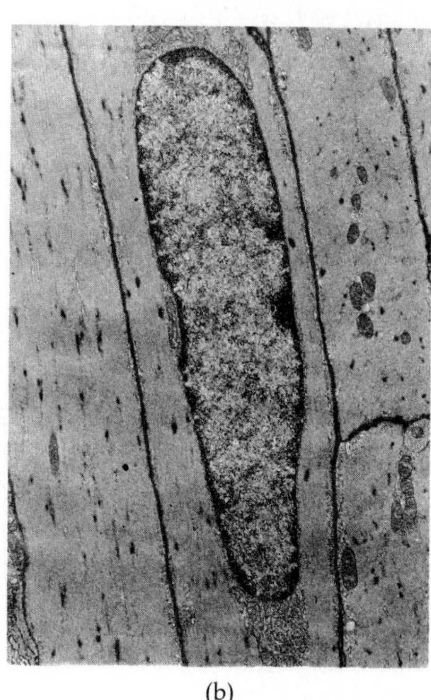

33–6
(a) *Photomicrograph of skeletal muscle fibers, showing striated pattern. The arrow indicates a small blood vessel, a capillary. Within it, you can see red blood cells, which carry oxygen to the muscle fibers, each of which is a single cell.*

(b) *Electron micrograph of smooth muscle from the spermatic duct of a mouse. Smooth muscle is made up of long, spindle-shaped cells containing contractile proteins. Portions of a number of smooth muscle cells can be seen in the picture. The nucleus of one cell, distinguishable by its granular texture, can be seen in the middle of the micrograph. Very thin fibrils run lengthwise through the cells and make up the bulk of the cytoplasm. A number of mitochondria are visible. Rhythmic movements of the spermatic duct help move the sperm from the testes to the urethra.*

33–7
Muscles attached to bone move the vertebrate endoskeleton. They often work in antagonistic pairs, with one relaxing as the other contracts. For example, when you move your hand toward your shoulder, as shown here, the biceps contracts and the triceps relaxes. When you move your hand down again, the triceps contracts, while the biceps relaxes. The muscles that move the skeleton, such as those diagrammed here, are known as skeletal muscles. They are striated, as shown in Figure 33–6a.

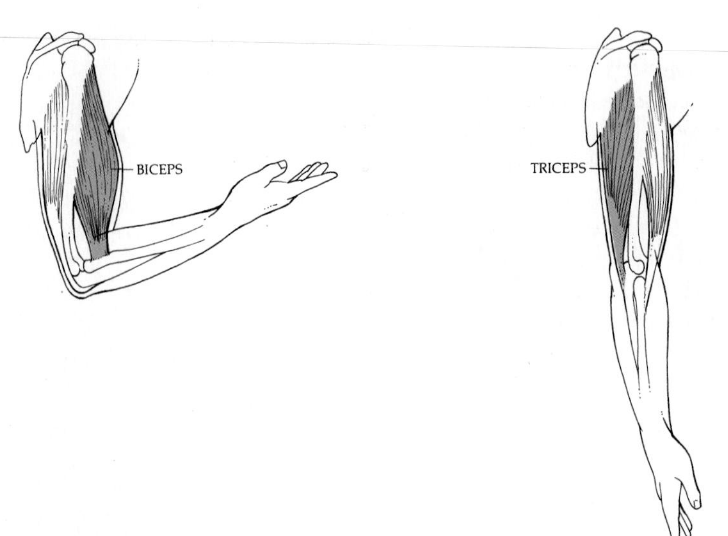

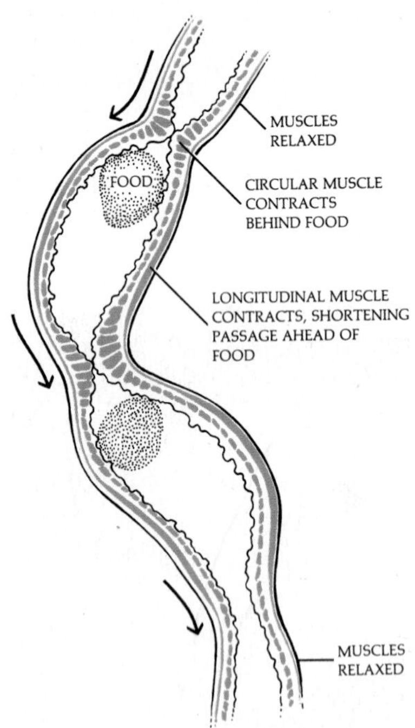

33–8
Alternating contractions of circular and longitudinal muscles move food along the digestive tract, a process known as peristalsis. Similar arrangements of muscles are involved in many other animal functions, including, for example, the locomotion of the earthworm (page 517).

Striated Muscle

Striated muscle, as mentioned previously, includes both skeletal and cardiac muscle. Some 40 percent of a man's body, by weight, is skeletal muscle; women characteristically have less, around 20 percent. A skeletal muscle is typically attached to two or more bones, either directly or by means of the tough strands of connective tissue known as tendons. Some of these tendons, such as those that connect the finger bones and their muscles in the forearm, are very long. When the muscle contracts, the bones move around a joint, which is held together by ligaments and generally contains a lubricating fluid. Most of the skeletal muscles of the body work in antagonistic groups, one flexing, or bending, the joint, and the other extending, or straightening, it (Figure 33-7). Also, two antagonistic groups may contract together to stabilize a joint. Such muscle action makes it possible for us (and others) to stand upright.

A skeletal muscle, such as the biceps, consists of bundles of muscle fibers held together by connective tissue. Each fiber is a single cell, cylindrical or spindle-shaped (tapering at both ends), with many nuclei, formed by the fusion of a large number of small, mononucleate cells during embryonic development. These fibers are very large cells—50 to 100 micrometers in diameter and, often, several centimeters long. Their internal structure is described in Chapter 42.

Cardiac muscle resembles skeletal muscle in its assemblies of actin and myosin and, thus, in its striated appearance. It differs from skeletal muscle in that its cells are usually mononucleate.

Smooth Muscle

Although smooth muscle cells also contain actin and myosin, the assemblies do not form a striated pattern. Smooth muscle cells are spindle-shaped and mononucleate. Functionally, smooth muscle contracts much less rapidly than skeletal striated muscle, its contractions are more prolonged, and it is under involuntary control. In most hollow organs, such as the intestines, smooth muscle fibers are organized into sheets arranged in two layers—an outer, longitudinal layer and an inner, circular layer—which can contract alternately, thus, for example, moving food along the intestinal tract (Figure 33-8). In blood vessels, bundles of smooth

33–9
Neurons. (a) *A motor neuron, characterized by an axon, that, in a large vertebrate, may be a meter or more in length.* (b) *An interneuron, showing the complex system of dendrites and an axon with many branches. Such a neuron forms connections—synapses—with many other nerve cells.* (c) *A sensory neuron, which transmits impulses from sensory receptors at the ends of the dendrite branches. In this type of neuron, the cell body is off to one side and the dendrite and axon form a single long fiber.*

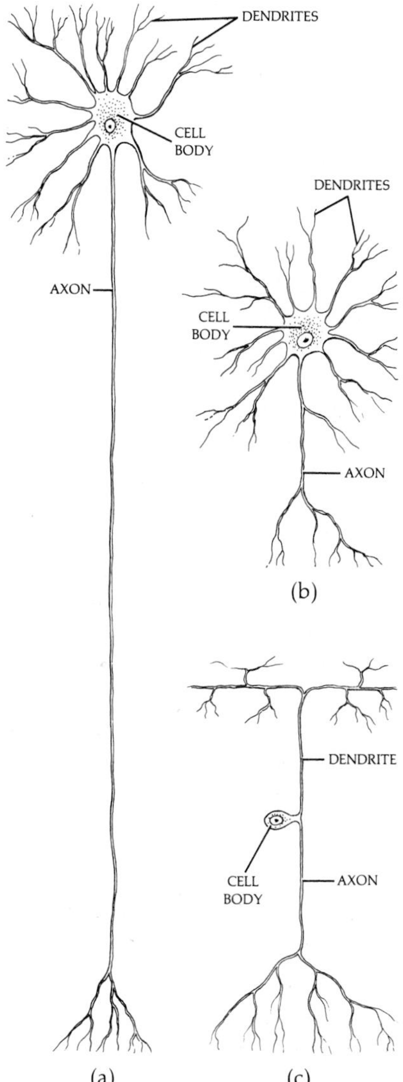

muscle fibers encircle the arterial walls, constricting the vessels as the fibers contract.

Nerve Tissue

The fourth major tissue type is nerve tissue. The major functional units of nerve tissue are the neurons, which transmit nerve impulses. Nerve tissue also comprises a second type of cell that supports and insulates neurons. Such cells are called neuroglia when they are in the central nervous system (brain and spinal cord) and Schwann cells in the peripheral nervous system (see page 764). These supporting cells are believed to supply the neurons with nutrients and other molecules and play an important part in maintaining the ionic composition of nerve tissue, which is, as we shall see, central to its function.

Neurons are specialized to receive signals, from either the external or the internal environment, and to transmit them in the form of electrical impulses to other neurons, muscles, or glands. Functionally, there are three types of neurons: *sensory neurons*, which receive information and relay it to the central nervous system; *motor neurons*, which relay signals from the central nervous system to effector organs (muscles or glands); and *interneurons*, which transmit signals within the central nervous system.

A neuron consists of the cell body, which contains the nucleus and much of the metabolic machinery of the cell; the *dendrites*, usually numerous, short, threadlike cytoplasmic extensions—processes—that receive stimuli from other cells; and the *axon*, a single long process that carries the nerve impulse away from the cell body to other cells or organs. Both dendrites and axons are also called nerve fibers.

Neurons may reach astonishing lengths. For example, the axon of a single motor neuron may extend from the spinal cord down the whole length of the leg to the toe. Or a sensory neuron may send a dendrite down to the toe and an axon up the

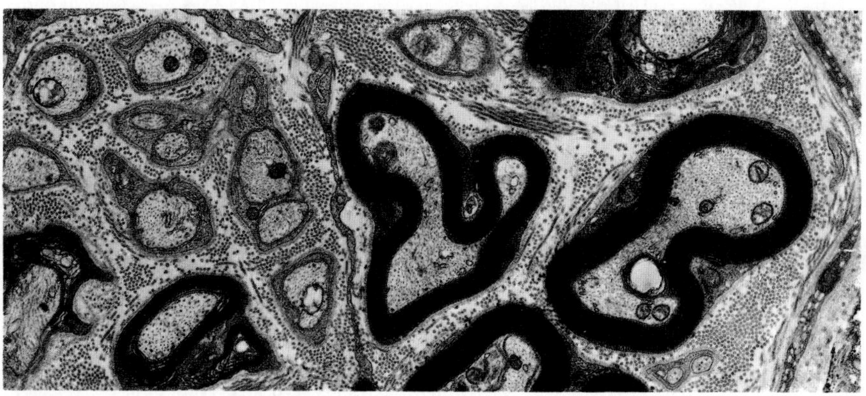

33–10
Cross section of a nerve showing the separate fibers. These fibers, which may be axons or long dendrites, are insulated from one another by Schwann cells. In some cases, the Schwann cells have wrapped themselves around the nerve fibers, forming myelin sheaths, the dark borders. These fatty layers are opaque to electrons, for reasons explained on page 92, and so they appear black in electron micrographs. Under the light microscope (or to the unaided eye) they are a glistening white.

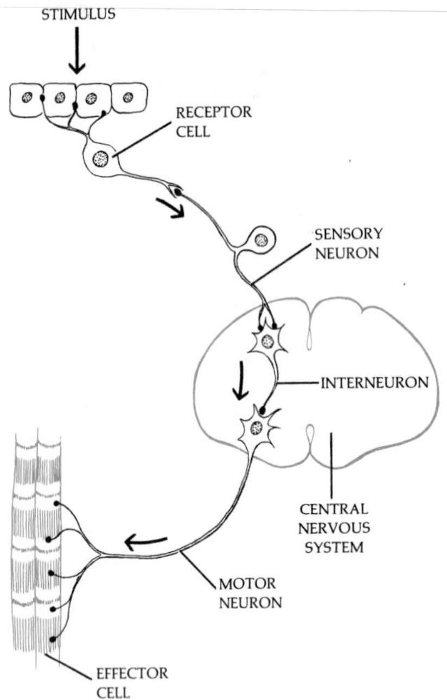

33–11

A reflex arc, showing the function of the three types of neurons. The sensory neuron is stimulated by a sensory receptor. It relays the signal to an interneuron, located entirely within the central nervous system. From the interneuron, the signal is transmitted to a motor neuron that stimulates an effector, shown here as a muscle cell. These basic components of the reflex arc are found in all vertebrates, from the simplest to the most complex.

entire length of the spinal cord to terminate in the lower part of the brain. In an adult human, such a cell might be close to 2 meters long (5 meters in a giraffe).

Nerves are bundles of many nerve fibers from many neurons—usually hundreds and sometimes thousands. Each fiber is capable of transmitting a separate message, like the wires in a telephone cable.

Every organ in the human body contains these four tissue types, as shown, for example, in Figure 33–12.

FUNCTIONS OF THE ORGANISM

In Chapter 28, we examined the interdependent tissues and organs of the plant body in terms of the challenges imposed by the transition from life in the water to life on the land. Similarly, the organ systems of the animal body "make sense" only when seen as solutions to particular problems presented by the relationship between the organism and its environment.

Before we further employ this metaphor of problem and solution, however, we should stop for a moment to see what we really mean by biological problem solving. An organism confronts its "problems" with a set of genetic instructions. If they work, the organism lives and passes on these instructions to its offspring. If its instructions are better than those carried by its neighbor, its offspring will probably be more numerous—and so "problems" get "solved."

Energy and Metabolism

A major problem for any living system—indeed, *the* major problem—is that posed by the second law of thermodynamics: to maintain the high level of organization characteristic of such systems in the face of the universal trend toward disorder. To do this, organisms need sources of energy and raw materials to maintain and operate their energy-extracting machinery. In broad outline, the complexities of the human digestive, respiratory, and circulatory systems can be viewed as particular ways of meeting these requirements that evolved over the long period of animal evolution.

Homeostasis

A second problem is also imposed by the laws of physics and chemistry. The molecular structures and chemical reactions characteristic of living systems can take place only within certain quite stringent limits of temperature and pH. Elements needed in micro amounts are disruptive—lethal—except within a very narrow range of concentration. Living systems are characterized by a capacity for controlling their internal environment. This characteristic, as we noted in Chapter 1, is called *homeostasis*.

Even a single cell, as we have seen, is capable of maintaining an internal composition distinctly different from its external environment. However, it is extremely vulnerable to changes in temperature or chemical composition of the medium in which it lives and is, of course, unable to function unless it is surrounded by liquid water. It is likely that the strongest evolutionary pressures toward larger size and multicellularity were related to the maintenance of homeostasis.

As noted in Chapter 5 (page 97), the larger the organism, the smaller the surface-to-volume ratio. Thus larger animals can resist external change better than smaller ones. A multicellular animal maintains a tightly controlled inner environment protected from drastic changes in temperature and chemical balance and

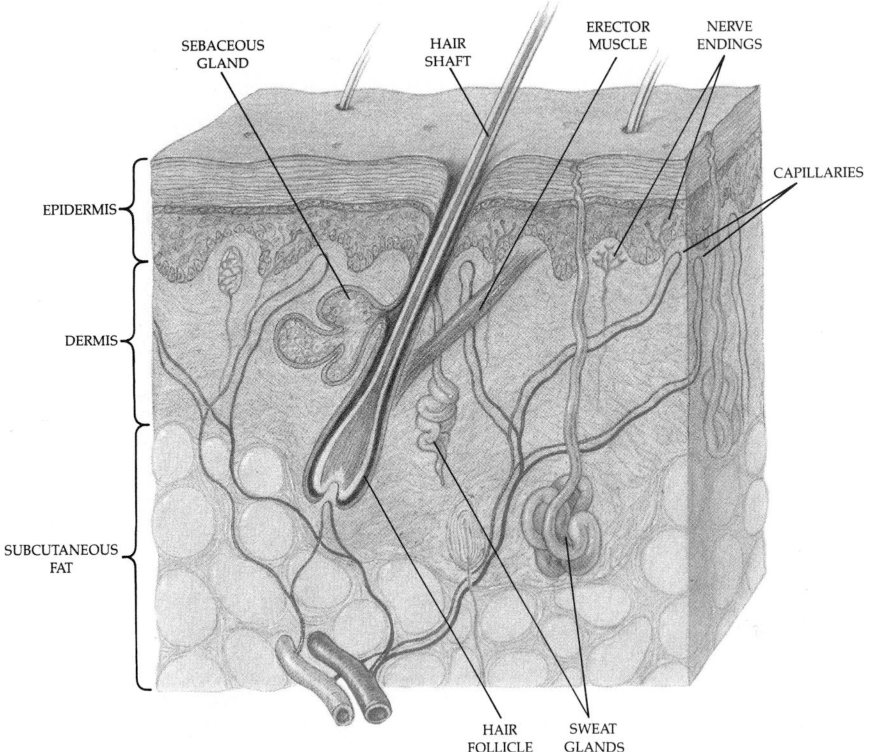

33–12
A section of human skin, the body's largest organ, showing the epidermis and the dermis, which together make up the skin, and the underlying subcutaneous fat. The epidermis is made up mostly of epithelial cells and consists of two layers: an inner layer of living cells, and an outer layer of dead cells filled with keratin. The dermis, consisting mostly of connective tissue, contains sensory nerve endings, erector muscles, which raise the hair when contracted, and sweat and sebaceous glands, which are modified epithelial cells. The latter produce a fatty substance that lubricates the skin surface. Fatty tissue, which makes up the insulating layer below the dermis, is also a form of connective tissue, as is the blood found in the capillaries.

shielded from invaders. The world in which your individual cells function and flourish is decidedly different from the world around you (but not so very different from the warm soup in which we all began).

Integration and Control

A third problem faced by the multicellular organism is posed by multicellularity itself: coordinating the activities of each of the enormous complex of living cells, making tissues and organs responsive to the overall needs of the organism, needs that change with fluctuations in the external environment. There are two major control systems in the animals, the endocrine system (the hormone-secreting glands and their products) and the nervous system. Speaking very generally, the endocrine system is responsible for changes that take place over a relatively long time period—minutes to months—whereas the nervous system is involved with more rapid responses—milliseconds to minutes. However, as you will see in Chapters 40 to 43, the more we learn about these systems, the more they are seen to be closely interrelated. For instance, the production of sex and other hormones was once believed to be under the control of the "master" pituitary gland; more recently it has been discovered that the pituitary, though perhaps a master in some aspects of the endocrine world, is actually an executive secretary of the hypothalamus, a major brain center. And within little more than the last decade, it has been discovered that the embryonic development of certain major brain centers is profoundly influenced by hormones regulated by the pituitary.

Even anatomically, the endocrine and nervous systems are not distinct. One of the body's most important glands, the adrenal medulla, the source of adrenaline (also called epinephrine), is not, strictly speaking, a gland. That is, it is not modified epithelial tissue but is rather a large ganglion—a collection of nerve cell bodies—whose nerve endings secrete the hormone.

In addition to the interplay between the endocrine and nervous systems, the nervous system itself has subdivisions that interact with one another in a finely tuned system of checks and balances. One subdivision, the somatic system, innervates skeletal muscle. Another, the autonomic ("involuntary") system, innervates smooth muscle, cardiac muscle, and glands. The autonomic system is further subdivided into the sympathetic and the parasympathetic divisions. The sympathetic division is most active in times of stress or danger. (To remember which is which, it is helpful to associate sympathy with emotions.) Among the major effectors of the sympathetic division is the adrenal medulla, and the overall effects of general sympathetic stimulation are those we associate with a "rush of adrenaline." The parasympathetic division plays its major role in supporting everyday activities such as digestion and excretion. The functions of the endocrine and nervous systems will be described in greater detail in Chapters 40 to 43. We mention them now because every organ system to be discussed is under the influence of both the endocrine system and the sympathetic and parasympathetic divisions of the autonomic nervous system.

Feedback Control

The body's integration and control systems characteristically act through negative feedback loops. The simplest example from everyday life of such a system is the thermostat that regulates your furnace. When the temperature in your house drops below the preset thermostat level, the thermostat turns the furnace on. When the temperature rises above the preset level, the thermostat turns the heat off. In a living organism, systems are seldom completely on or off, and homeostatic control is much more finely modulated. The principle, however, is the same. For instance, endocrine cells in the pancreas produce two hormones vital in carbohydrate metabolism: the alpha cells produce glucagon, and the beta cells produce insulin. When the glucose concentration of the blood falls, glucagon is produced, which causes release of glucose from the liver. When glucose levels in the blood rise, the beta cells release insulin, which causes uptake of glucose by body cells. Thus, in effect, glucose concentration is controlled by two separate negative feedback systems, one stimulatory and one inhibitory.

Negative feedback loops may involve both the nervous and endocrine systems. One important homeostatic function of the body, for example, is keeping the blood volume constant. A hormone known as antidiuretic hormone (ADH), which is produced by the pituitary gland, alters the permeability of tubules of the kidney to water and so decreases water excretion. The production of ADH is controlled by sensory receptors in the circulatory system, particularly in the heart, that measure blood pressure. When blood pressure goes up, firing of these receptors inhibits the release of ADH; as blood pressure goes down, the stimulus from the receptors decreases, ADH production increases, water is retained by the kidney, and blood pressure rises.

Some negative feedback systems involve additional relay loops. The thyroid gland, for instance, produces the thyroid hormone, thyroxine, which, among other effects, steps up cellular metabolism. The amount of thyroxine produced depends on the production of another hormone, thyroid-stimulating hormone (TSH) from the pituitary gland. A major factor regulating the production of TSH is the concentration of thyroid hormone in the circulating blood. Thus, though one more step is involved, the principle is the same: the concentration of the hormone itself or the response to the hormone by a target tissue inhibits the synthesis or secretion of the hormone in question.

Many functions are controlled by a number of separate feedback control pathways. An example is shown in Figure 33-13, which illustrates the pathways important in temperature regulation. Core body temperature is controlled by a thermostat located in the hypothalamus, an area of the brain. The thermostat collects information from a number of thermoreceptors, integrates it, compares the results with its thermostat setting, and marshalls appropriate responses. The subject of temperature control will be explored in more detail in Chapter 38.

Continuity of Life

The fourth problem an organism faces, following the dictates of its genes, is to multiply. Reproduction, as we have seen, may be carried out in a variety of ways, but in mammals it is always sexual and always involves formation of gametes and their union to form a zygote, or fertilized egg, and the development of the zygote into an adult. The last chapter of this section will describe the almost miraculous, though oft-repeated, phenomena by which this single cell develops into a human being.

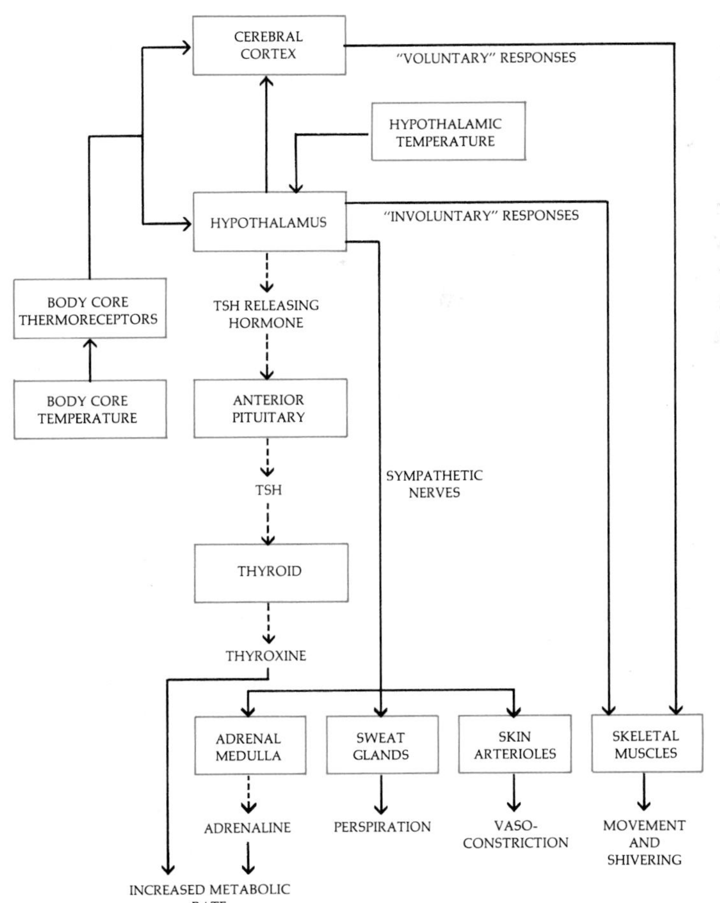

33-13
Body temperature in mammals is regulated by a complex network of activities involving both the nervous and endocrine systems. The chief controlling center is the hypothalamus, a region in the brain. In some animals (although not, apparently, in humans), the hormonal pathway—indicated by dashed lines—is of major importance. TSH is thyroid-stimulating hormone, which is produced by the pituitary and which stimulates the production of thyroid hormone. Thyroid hormone increases cellular metabolism, apparently by acting directly on the mitochondria. Many behavioral responses, such as seeking sun or shelter, are also involved.

SUMMARY

We are vertebrates. As such, we have a bony, jointed supporting endoskeleton that includes a skull and a vertebral column enclosing the central nervous system. Our bodies contain a coelom that is divided by a muscle, the diaphragm, into two major compartments, the abdominal cavity and the thoracic cavity.

Our bodies are organized into tissues and organs. Tissues are groups of cells that are structurally, functionally, and developmentally similar. Various types of tissues are grouped in different ways to form organs, and organs are grouped to form organ systems. The four principal tissue types of which our bodies are made are epithelial tissue, connective tissue, muscle, and nerve. Epithelium serves as a covering or lining for the body and its cavities. Glands are composed of specialized epithelial cells. Their secretions include mucus, perspiration, milk, saliva, hormones, and digestive enzymes. Connective tissues are characterized by their capacity to secrete substances, such as collagen and other fibers, that make up the intercellular matrix. They serve to support, strengthen, and protect the other tissues of the body. Muscle cells are specialized for contraction. Muscle is categorized as striated muscle or smooth muscle. In striated muscle, which includes cardiac and skeletal muscle, the striped pattern is due to regular assemblies of specialized proteins, actin and myosin. Smooth muscle, or involuntary muscle, is under the control of the autonomic nervous system, whereas striated muscle (with the exception of cardiac muscle) is under somatic control. Nerve cells, or neurons, are specialized for the conduction of an electrical impulse. Neurons consist of dendrites, a cell body, and an axon. Neurons are surrounded and supported by neuroglia or Schwann cells.

A multicellular animal processes nutrients to yield energy and structural materials. It regulates its internal environment, a process known as homeostasis. It coordinates the activities of its many tissues and organ systems in response to changes in both the internal and external environments. These integration and control systems characteristically function through negative feedback loops. Finally, as dictated by its genes, the organism reproduces itself.

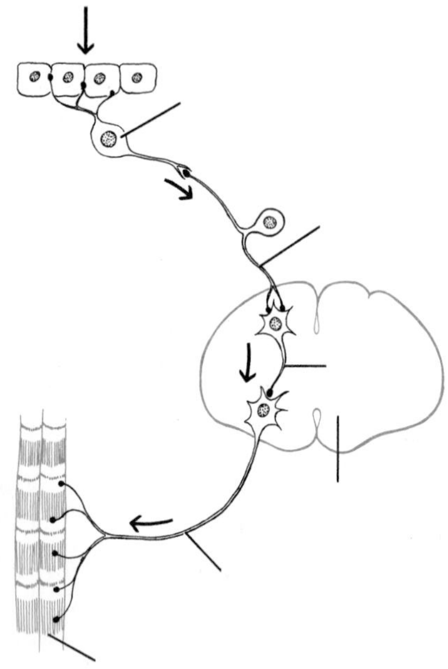

QUESTIONS

1. Distinguish among the following terms: animal/vertebrate/mammal/*Homo sapiens;* diaphragm/coelom; endoskeleton/exoskeleton; muscle cell/muscle fiber; neuron/neuroglia; somatic/autonomic; sympathetic/parasympathetic.

2. What is the functional significance of each of the four tissue types? Give an example of each type.

3. What are the three types of epithelial tissue? What is the basis for classification of epithelial cells?

4. What is the major structural difference between connective tissue and epithelial tissue? What functions of connective tissue account for this structural difference?

5. Name the types of muscle cells in an animal body.

6. Label the drawing in the margin.

7. What are the major survival problems of an organism? Compare the solutions to these problems achieved by plants with those achieved by vertebrates.

CHAPTER 34

Energy and Metabolism I: The Digestive System

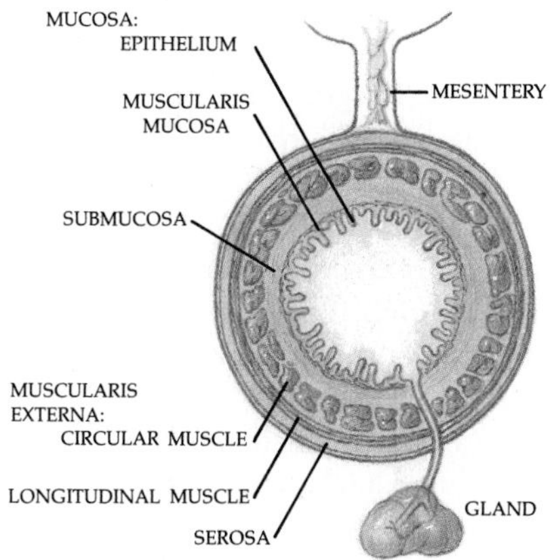

34–1
The layers of the digestive tract include (1) the mucosa, (2) the submucosa, which contains nerves and blood and lymph vessels, (3) the muscularis externa, and (4) the serosa (also known as the visceral peritoneum), which covers the outer surfaces of the organs in the abdominal cavity. Glands outside the digestive tract, principally the pancreas and liver, discharge digestive enzymes and bile into the tract through various ducts. A mesentery is a fold of the peritoneum that connects the digestive tract with the posterior abdominal wall.

Digestion is the breakdown of food materials into molecules that can be delivered to and utilized by the individual cells of the body. These molecules may serve as energy sources; they may provide essential chemical elements, such as calcium, nitrogen, or iron; or they may be complete molecules—certain amino acids, fatty acids, and vitamins—that cells need but cannot synthesize for themselves.

In Section 4, we traced the evolution of digestive systems from the phagocytic vacuole of protists to the coelenteron of jellyfish, the digestive cavity of flatworms, and, finally, the major evolutionary breakthrough, exemplified by ribbon worms, the invention of the one-way digestive tract. As we noted in Chapter 24, the great diversification of the Animalia was determined, in large part, by their differing adaptations for obtaining food.

DIGESTIVE TRACT IN VERTEBRATES

In vertebrates, digestion takes place in a long convoluted tube, extending from mouth to anus, known as the digestive tract or, rather less elegantly but just as accurately, as the gut. The inner surface of the gut is continuous with the outer surface of the body, and so, technically, the cavity of the gut is outside the body. Nutrient molecules actually enter the body only when they pass through the walls of the digestive tract. Thus, the process of digestion involves two components: breakdown of food molecules and their absorption into the body.

The digestive tube begins with the oral cavity and includes the mouth, pharynx, esophagus, stomach, small intestine, large intestine, and anus. Each of these areas is specialized for a particular phase in the overall process of digestion, but the fundamental structure of each is similar. The tube, from beginning to end, has four layers: (1) the innermost layer, the *mucosa*, which is made up of epithelial tissue, an underlying basement membrane, and connective tissue, with a thin outer coating of smooth muscle in some places; (2) the *submucosa*, which is made up of connective tissue and contains nerve fibers and blood and lymph vessels; (3) the *muscularis externa*, muscle tissue; and (4) the *serosa*, an outer coating of connective tissue (Figure 34–1).

The intestinal epithelium contains many mucus-secreting goblet cells and, in some parts of the gut, glands that secrete digestive enzymes. Along most of the digestive tract, the muscularis is made up of two layers of smooth muscle, the inner layer, in which the orientation is circular, and the outer layer, in which the cells are longitudinally arranged. Coordinated contractions of these muscles produce ring-like constrictions that mix the food, as well as the wavelike motions, known as *peristalsis*, that move food along the digestive tract (see Figure 33–8, page 664). At several points the circular layer of muscle thickens into heavy bands, called *sphincters*. These sphincters, by relaxing or contracting, act as valves to control passage of food from one area of the digestive tract to another.

The Oral Cavity

The mechanical breakdown of food begins in the mouth. In humans, as in most other mammals, the first tearing and grinding of food is done by the teeth. (The gizzards of birds and earthworms serve the same function.) Children have 20 teeth, which are gradually lost and replaced by a second set of 32 teeth as the jaw grows larger. Of the 16 adult teeth in each jaw, four are incisors, flat chisel-like structures specialized for cutting; two are canines, used by carnivores for stabbing and tearing; four are premolars ("in front of the molars"), each of which has two cusps, or protuberances, and so are called bicuspids; and six are molars, each of which has four or five cusps. The premolars and molars are used for grinding. Molars do not replace temporary teeth but are added as the jaw grows. The crown of the tooth—the visible part—is covered with enamel, the hardest substance in the body (mostly calcium phosphate); the root, within the gum, is covered with cement, a substance similar to bone. The bulk of the tooth is composed of dentine, another bonelike material, which forms slowly during the life of the tooth. The pulp cavity within the tooth contains the cells that produce dentine and also the nerve endings and blood vessels.

In mammals, the tongue serves largely to move and manipulate food and to mix it with saliva. Some vertebrates, though, such as the hagfish and lampreys, have tongues equipped with horny "teeth." The tongues of frogs and toads flip out (they are attached at the front, not the back) to catch insects. Mammalian tongues carry taste buds (Figure 34–3). In humans, the tongue has developed a secondary function of formulating sounds for communication.

As the food is being chewed, it is moistened by saliva, a watery secretion produced by three pairs of large salivary glands plus numerous minute glands, the buccal glands, which are located in the mucosal lining of the mouth. On the average, we produce 1 to 1.5 liters of saliva every 24 hours. The saliva, which contains mucus, lubricates the food so that it can be swallowed easily. Saliva is slightly alkaline, owing to the presence of sodium bicarbonate. In humans, saliva also contains a digestive enzyme, amylase, which begins the breakdown of starches. (Carnivores, such as dogs, which characteristically tear and gulp their

34–2
The basic pattern of dentition in mammals includes, in each quadrant, two incisors, one canine ("eye tooth"), two premolars, and three molars. The various species of mammals have modifications of this pattern that reflect their dietary habits. Rodents (whose name comes from the Latin rodere, *"to gnaw") are characterized by their sharp, chisel-shaped incisors that grow throughout their lives. Most rodents have no canines. Large predatory carnivores, such as the lion, have canines adapted for stabbing and slicing, and large molars, which can easily crush the bones of large herbivores. In elephants, tusks are modified incisors; they are used for attack and defense and for rooting food from the ground or breaking branches. (The tusks of other mammals, such as walruses, are modified canines.) The modern horse is a grazing animal; its incisors clip off grasses, which are then ground by the large, flat molars. Humans, with a relatively unspecialized diet, have a correspondingly unspecialized dentition. Similarly, though less spectacularly, differences in anatomy and physiology of other components of the digestive system reflect their evolutionary history.*

GENERALIZED MAMMAL

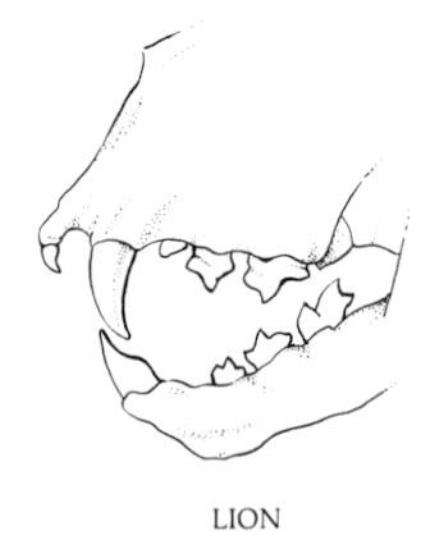

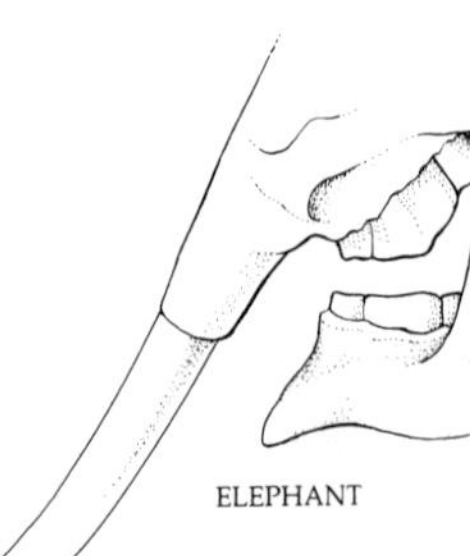

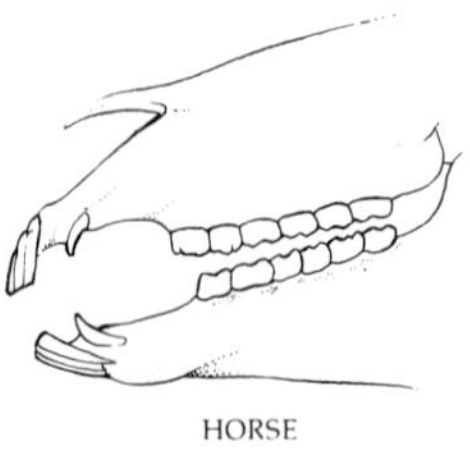

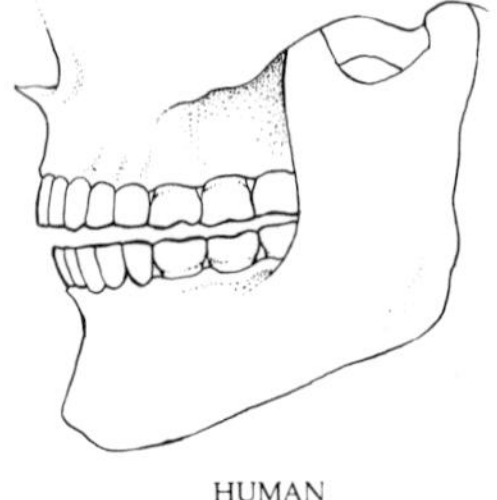

34–3

(a) The outer pore of a taste bud. The receptor cells of the taste bud are just visible within the pore. This scanning electron micrograph shows a portion of the surface of the human tongue. Other taste buds are found on the roof of the mouth, the pharynx, and the larynx. (b) Longitudinal section of a taste receptor.

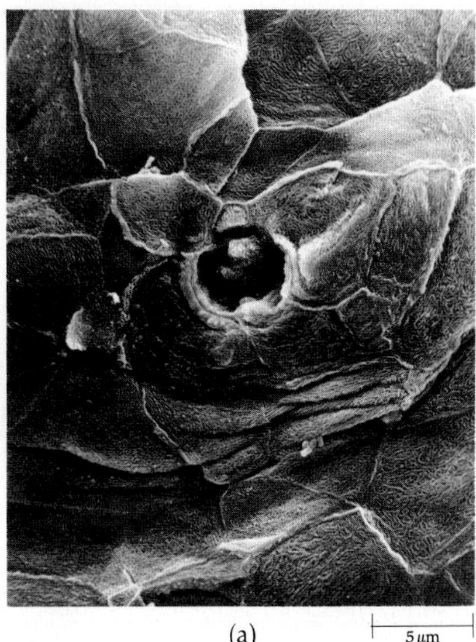

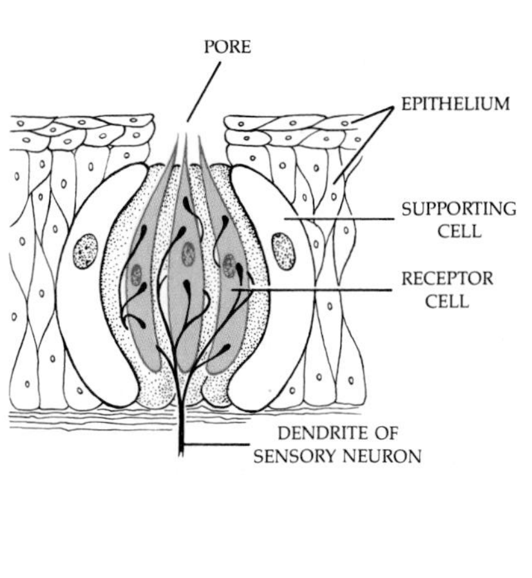

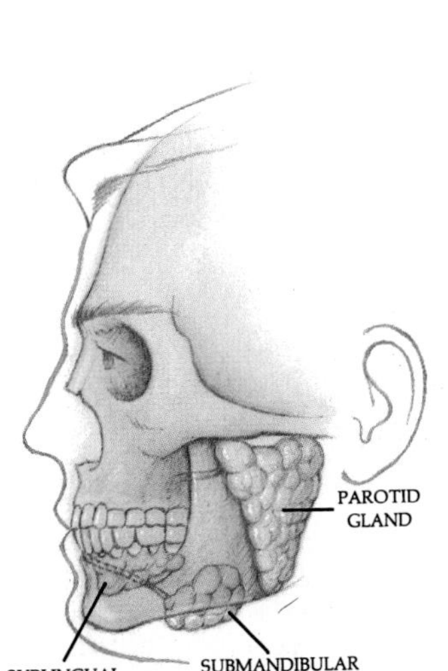

34–4

The bulk of the saliva is produced by three pairs of salivary glands. Additional amounts are supplied by minute glands, the buccal glands, in the mucous membrane lining the mouth. The parotid glands are the sites of infection by mumps virus.

food, have no digestive enzymes in their saliva.) Like all digestive enzymes, amylase works by hydrolysis; that is, the breaking of each bond involves addition of a molecule of water.

The secretion of saliva is controlled by the autonomic nervous system. It can be initiated by reflexes originating in taste buds and in the walls of the mouth, and also by the mere smell or anticipation of food. (Think hard, for a moment, about eating a lemon.) Fear inhibits salivation; at times of great danger or stress, the mouth may become so dry that speech is difficult. On the other hand, the presence of a noxious substance in the mouth or stomach stimulates the copious production of a watery saliva as a protective reaction.

The Pharynx and Esophagus: Swallowing

From the mouth, food is propelled backward toward the esophagus, a muscular tube (about 25 centimeters long in adults). Swallowing, the passing of food to the esophagus and through the esophagus on to the stomach, begins as a voluntary action. However, once under way in humans, it continues involuntarily. In humans, the upper part of the esophagus is striated muscle, but the lower part is smooth muscle. (Dogs, cats, and other animals that gulp their food have striated muscle along the whole length of the esophagus.) Both liquids and solids are propelled along the esophagus by peristalsis. This process is so efficient that we can swallow water while standing on our heads.

The esophagus passes through the diaphragm, which separates the thoracic and abdominal cavities, and opens into the stomach, which, with the remaining digestive organs, lies in the abdomen. The abdominal cavity is completely lined by the *peritoneum*, a thin layer of connective tissue covered by moist epithelium. (The serosa is that portion of the peritoneum that covers the outer surface of the stomach and intestinal walls.) The stomach, intestines, and other organs in the abdominal cavity are suspended by folds of peritoneum known as *mesenteries*, as shown in Figure 34–1. Mesenteries consist of double layers of tissue, with blood vessels, lymph vessels, and nerves lying between the two layers.

34–5
The human digestive tract.

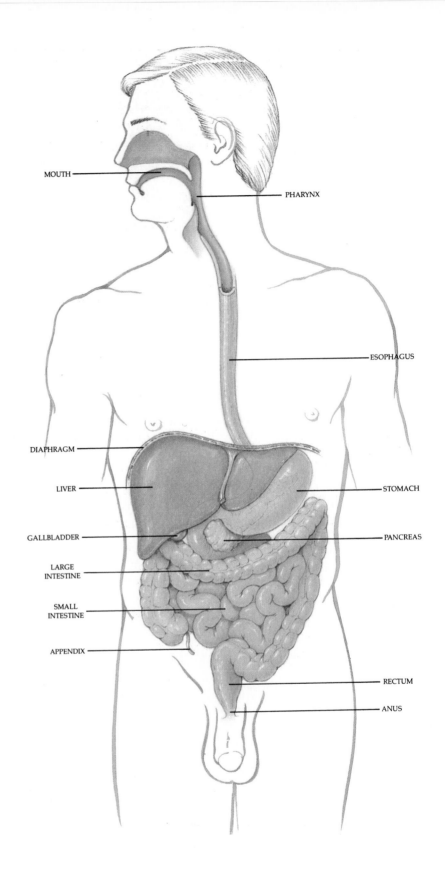

MOUTH
PHARYNX
ESOPHAGUS
DIAPHRAGM
LIVER
STOMACH
GALLBLADDER
PANCREAS
LARGE INTESTINE
SMALL INTESTINE
APPENDIX
RECTUM
ANUS

34–6
Separation of the digestive and respiratory systems in mammals makes it possible for them to breathe while eating. The pharynx, the common passageway of the two systems, is at the back of the mouth and connects with the trachea and the esophagus. (a) *A diagrammatic view of the upper portion of the respiratory and digestive systems.* (b) *Swallowing. As the food mass descends, the epiglottis tips downward, and the food mass passes into the esophagus.*

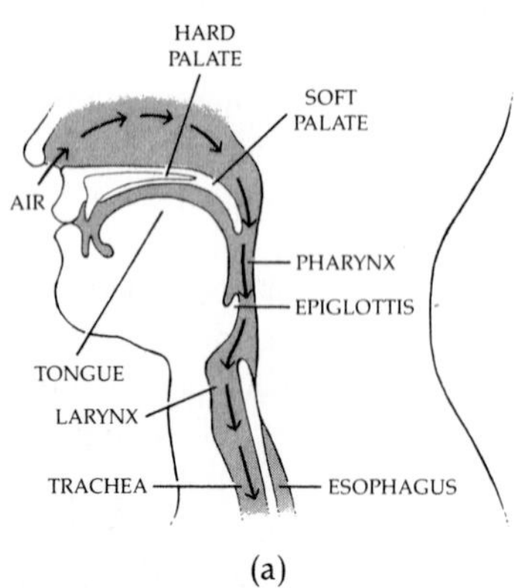

(a)

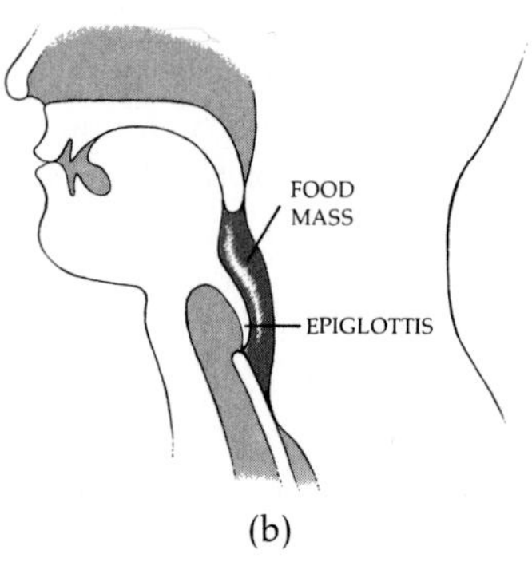

(b)

THE HEIMLICH MANEUVER

Food strangulation claims the lives of almost 3,000 persons a year in this country alone, more than accidents involving firearms or airplanes. It occurs when a food mass enters the trachea rather than the esophagus (Figure 34–6). If the food becomes lodged, the victim cannot speak or breathe and, if the airway is completely blocked, will die in four or five minutes. (The fact that the victim cannot speak helps onlookers to distinguish it from a heart attack; although the symptoms are similar, heart attack sufferers can talk.)

The food can nearly always be dislodged by the Heimlich maneuver, a procedure so simple that it has been carried out successfully in at least two instances by eight-year-olds. There are three steps: (1) Stand behind the victim and wrap your arms around his or her waist. (2) Make a fist with one hand, grasp it with your other hand, and then place the fist against the victim's abdomen, slightly above the navel and below the rib cage. (3) Press your fist into the victim's abdomen with a quick upward thrust. The sudden elevation of the diaphragm compresses the lungs and forces air up the trachea, pushing the food out. Repeat several times if necessary.

If the victim is sitting, the procedure can be carried out in the same way. If the victim is lying on his back, face him, and kneel astride the hips. Put the heel of one hand on the abdomen above the navel and below the rib cage, put the other hand on top of the first hand, and press with a quick upward thrust. You can even do it on yourself: place your hands at your waist, make a fist, and press quickly upward.

The victim should be seen by a physician immediately after the emergency treatment because it is possible to break a rib or cause other internal injuries, especially if the movements are performed incorrectly. But, considering the alternative, it is well worth the risk.

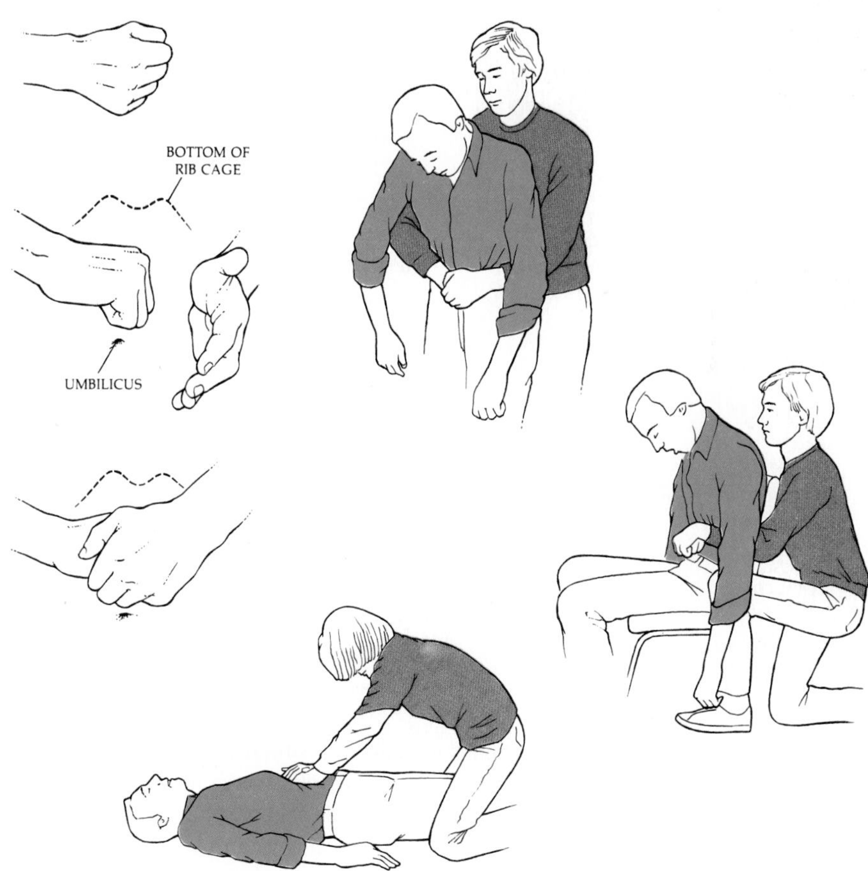

The maneuver can be carried out with the victim in a standing, sitting, or lying position. If the person is larger than you are, it is preferable to have him sitting or lying down. Note how the fist is made (you might want to practice this) and the position of the hands. These are both very important.

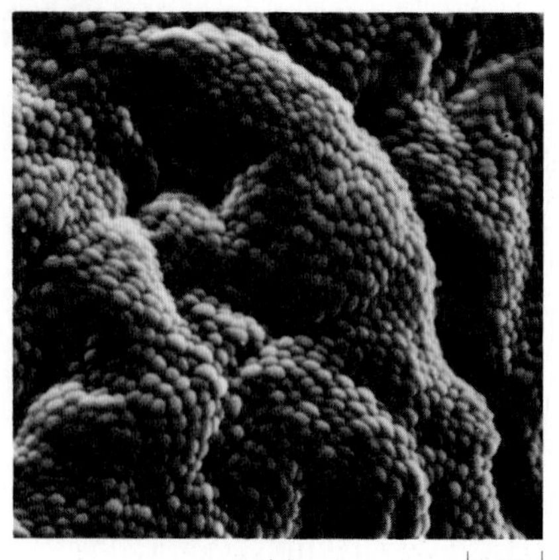

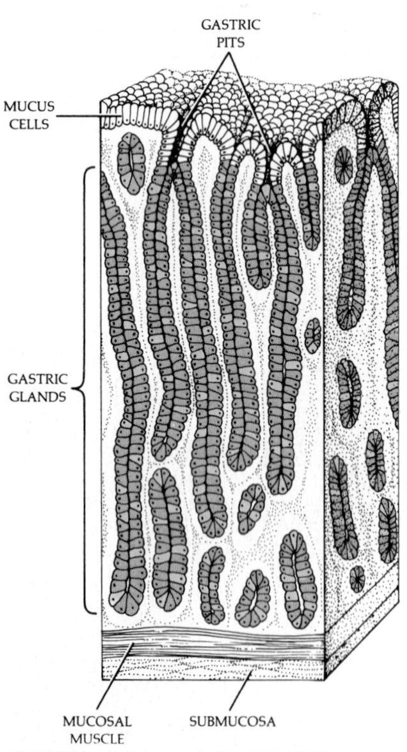

34–7
(a) *Surface of the stomach, as shown in a scanning electron micrograph. The numerous indentations are gastric pits.* (b) *A cross section of stomach mucosa. The parietal cells secrete hydrochloric acid and the chief cells produce pepsinogen. Mucus, secreted by the goblet cells of the epithelium, coats the surface of the stomach and lines the gastric pits, protecting the stomach surface.*

The Stomach: Storage and Liquefaction

The stomach is essentially a collapsible elastic bag, which, unless it is fully distended, lies in folds. Stomachs vary widely in storage capacity. A hyena's stomach can hold an amount equivalent to one-third of the animal's own body weight, for instance, whereas a mammal that eats small, regularly available, and nutritious food items—such as seeds or insects—needs only a small storage organ. The human stomach, distended, holds 2 to 4 liters of food. The mucosal layer of the stomach is very thick and contains numerous gastric pits (Figure 34–7). Mucus-secreting epithelial cells cover the surface of the stomach and line the gastric pits. Opening into the bottoms of the gastric pits are the gastric glands, whose walls contain the parietal cells, which produce hydrochloric acid (HCl), and the chief cells, which produce pepsinogen (see page 178).

As a consequence of the HCl secretion, the pH of gastric juice is normally between 1.5 and 2.5, far more acidic than any other body fluid. The burning sensation you feel if you vomit is caused by the acidity of gastric juice acting on unprotected membranes. Normally, the mucus in the stomach forms a barrier between the epithelium and the gastric juices and so prevents the stomach from digesting itself. The HCl kills most bacteria and other living cells in the ingested food. It also loosens the tough, fibrous components of tissues and erodes the cementing substances between cells. HCl initiates the conversion of pepsinogen to its active form, pepsin, by splitting off a small portion of the molecule. Once pepsin is formed, it acts on other molecules of pepsinogen to form more pepsin. Pepsin, which breaks proteins down into peptides, is active only at the low pH of the normal stomach.

The stomach is influenced by both the nervous and endocrine systems. Anticipation of food and the presence of food in the mouth stimulate churning movements of the stomach and the production of gastric juices. Fear and anger decrease the stomach's motility. When food reaches the stomach, its presence causes the release of the hormone gastrin from gastric cells into the bloodstream. Gastrin acts on the cells of the stomach to increase their secretion of gastric juices.

In the stomach, food is converted into a semiliquid mass as a result of the action of the pepsin, the hydrochloric acid, and the churning motions. The stomach empties gradually through the pyloric sphincter, which separates the stomach and small intestine; the food mass is forced out, little by little, by peristalsis. The stomach is usually empty four hours after ingestion of a meal.

The Small Intestine: Final Digestion and Absorption

In the small intestine, the breakdown of food begun in the mouth and stomach is completed. The resulting nutrient molecules are then absorbed from the digestive tract into the circulatory system of the body, from which they are delivered to the individual cells.

Anatomically, the small intestine is characterized by circular folds in the submucosa; numerous microscopic fingerlike projections, *villi*, on the mucosa; and tiny cytoplasmic projections, *microvilli*, on the surface of the individual epithelial cells. All of these structural features increase the surface area of the small intestine. The small intestine is, on the average, 6 meters long in the living adult; the total surface area of the human small intestine is approximately 300 square meters, about the size of a doubles tennis court. The upper 25 centimeters, known as the duodenum, is the most active in the digestive process; the rest is principally concerned with absorption of nutrients.

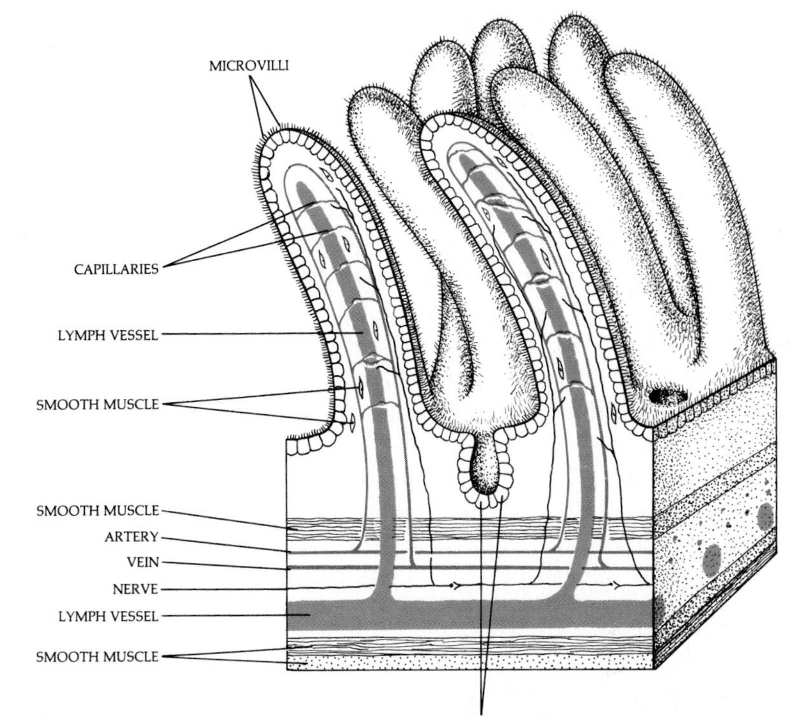

34–8
Longitudinal section of villi of small intestine. Nutrient molecules are absorbed through the walls of the villi and, with the exception of fat molecules, enter the bloodstream by means of the capillaries. Fats—hydrolyzed to fatty acids and glycerol, resynthesized into new fats, and packaged into chylomicrons—are taken up by the lymphatic system. The villi can move independently of one another; their motion increases after a meal.

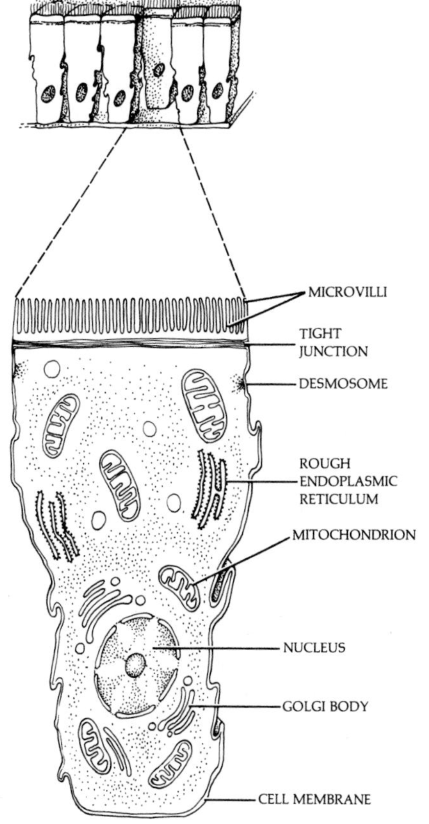

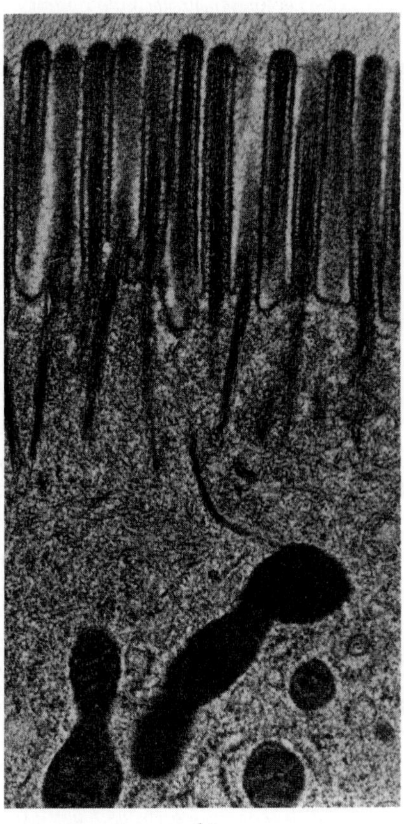

34–9
(a) *A columnar epithelial cell from the lining of the intestine. Notice, in particular, the tight junctions and desmosomes that bind these cells together in a continuous sheet.* (b) *Microvilli on the surface of an intestinal epithelial cell. These cellular extensions greatly increase the absorptive surface of the intestine. They also contain digestive enzymes, which are part of the cell membrane, and carrier molecules involved in active transport. Several mitochondria are visible at the bottom of this electron micrograph. Their abundance indicates that these cells have high energy requirements, probably due, in large part, to active transport of nutrient molecules from the contents of the intestine.*

MOTHER'S MILK—IT'S THE REAL THING

One special adaptation of the Mammalia—the one from which the class derives its name—is the modification of sweat glands to produce the complex, highly nutritive fluid known as milk. In the course of evolution, the various species of mammals have developed milks especially suited to the nourishment of their own young. Human milk, for instance, contains about the same amount of fat (3.7 percent) as cow's milk, but it has less protein and more carbohydrate (in the form of lactose, milk sugar). The milk of a harp seal, on the other hand, is twice as concentrated as either cow's milk or human milk (45 percent water as compared to 87 percent water), has no carbohydrate, is 43 percent fat, and so fulfills the higher energy requirements of an animal that maintains a high body temperature in water. A kangaroo can provide an even more specialized diet; one mother carrying two joeys of different ages in her pouch supplies different milks (through different teats) to each.

In addition, each species produces milk with particular vitamins and minerals and, perhaps most important, a number of antibodies, other immunologically active proteins, and various white blood cells concerned with protecting the infant of that species against infection. For example, human breast milk contains antibodies that protect against Staphylococcus, E. coli, *and polio, in particular. This immunological protection is especially important in the first few months of life, before the infant's own immune system is functioning fully. Another important feature of mother's milk is that it comes in a container that keeps it fresh, temperature-controlled, and free from bacterial contamination.*

Various commercially produced formulas exist that, mixed with water, offer a substitute for human milk. Where families have access to clean water, sterilized bottles, good pediatric care, and money for an adequate amount of formula, milk substitutes can be used successfully by women who cannot or choose not to nurse their babies. However, at a World Health Organization assembly in May 1981, 118 countries voted to regulate the promotion of infant formula (the United States cast the only vote against regulation). These regulations were aimed particularly at the promotion of infant formula in the Third World, where access to clean water or to the facilities for boiling water is often simply not available. In fact, bottles are often not even washed between feedings. Another reason for the regulations is that the formula may be overdiluted to save money. Adding this to the absence in formulas of antibodies and other protective substances, it can be seen that formula-fed infants run a much higher risk of infectious disease than breast-fed babies. Many die of "bottle-baby disease," severe diarrhea, and some starve due to watered-down formula. Ironically, women in Third World countries often use milk substitutes not as a matter of necessity but because bottle feeding is associated with progress, social status, and the adoption of Western conventions.

The length of time mammalian infants depend on milk for nourishment also varies with the species. Rabbits nurse for only a few days, cats for six to ten weeks, hyenas and humans for a year, elephants for four or five. However, with the exception of Western humans, once infancy is over, not only does milk consumption stop but there is also a decline in the level of lactase, the enzyme that breaks down lactose (milk sugar). This enzyme, present in most human infants, is absent in adults of most animal species and is much diminished in most human beings after the age of four. Lactose ingestion by individuals lacking the enzyme may cause intestinal cramps and diarrhea, as the lactose is broken down by bacterial fermentation. Natural selection for lactose digestion in adult humans is thought to have begun some 10,000 years ago when certain groups began to domesticate and milk mammals. Individuals of northern European descent are much more likely to have this enzyme as adults than are those of African descent. This distribution of the allele (or of the gene that turns it on or off) is apparently correlated with the fact that northerners were more likely to drink milk (including reindeer milk) as adults, whereas people from hot climates, if they used milk at all, were likely to use fermented milk products, such as cheese or yogurt, in which lactose is missing. In any case, this finding throws considerable doubt on the policy of shipping dried cow's milk to famine-ridden countries of Africa and Asia.

The small intestine contains a variety of digestive secretions, some produced by the intestinal cells and some by the pancreas and liver (see Table 34–1). Other epithelial cells, the goblet cells of the mucosa, secrete mucus. The pancreatic enzymes enter the small intestine through the pancreatic duct, about 10 centimeters below the pyloric sphincter.

Table 34-1 Principal Digestive Enzymes

ENZYME	SOURCE	SUBSTRATE	SITE OF ACTION
Salivary amylase	Salivary glands	Starches	Mouth
Pepsin	Stomach mucosa	Proteins, pepsinogen	Stomach
Pancreatic amylase	Pancreas	Starches	Small intestine
Lipase	Pancreas	Fats	Small intestine
Trypsin	Pancreas	Polypeptides, chymotrypsinogen	Small intestine
Chymotrypsin	Pancreas	Polypeptides	Small intestine
Carboxypeptidase	Pancreas	Polypeptides	Small intestine
Deoxyribonuclease	Pancreas	DNA	Small intestine
Aminopeptidase	Small intestine	Polypeptides	Small intestine
Dipeptidase	Small intestine	Dipeptides	Small intestine
Maltase	Small intestine	Maltose	Small intestine
Lactase*	Small intestine	Lactose	Small intestine
Sucrase	Small intestine	Sucrose	Small intestine
Enterokinase	Small intestine	Trypsinogen	Small intestine
Phosphatases	Small intestine	Nucleotides	Small intestine

* Often absent in adults, especially those of recent African origin.

In the small intestine, pancreatic amylases continue the breakdown of starch begun in the mouth. Lipases hydrolyze fats into glycerol and fatty acids. Other specialized enzymes—lactase, sucrase, and maltase—cleave disaccharides to monosaccharides. Three types of enzymes break down proteins. One group breaks apart the long protein chains. Each enzyme in this group acts only on the bonds linking particular amino acids, so that several enzymes are required to break a single large protein into shorter peptide fragments. (These enzymes, as we saw in Chapter 3, were among the tools that made possible the amino acid sequencing of proteins.) A second type of enzyme acts only on the end of a chain, some on the amino end and some on the carboxyl end, splitting off dipeptides (pairs of amino acids). A third group of enzymes then comes into action, breaking the remaining dipeptides into single amino acids. The amino acids and dipeptides are absorbed through the epithelial cells of the intestine and finally enter the bloodstream.

In addition to these many different enzymes, the small intestine receives an alkaline fluid from the pancreas, which neutralizes the stomach acid, and bile, which is produced in the liver and stored in the gallbladder. Bile contains a mixture of salts that, like laundry detergents, emulsify fats, breaking them apart into droplets. In this form they can be attacked by the lipases. Bile contains some sodium bicarbonate, which, with the pancreatic fluid, neutralizes the stomach acid. Neutralizing the acidity is essential because the intestinal enzymes are optimally active at a pH of 7 to 8 and would be denatured by the acidic pH of the gastric juices that enter the intestine.

Table 34-2 Gastrointestinal Hormones

HORMONE	SOURCE	MAJOR STIMULUS FOR PRODUCTION	MAJOR ACTIONS
Gastrin	Stomach	Protein-containing food in stomach, also parasympathetic nerves to stomach	Stimulates secretion of gastric juices
Secretin	Duodenum	HCl in duodenum	Stimulates secretion of alkaline bile and pancreatic fluids
Cholecystokinin	Duodenum	Fats and amino acids in duodenum	Stimulates release of pancreatic enzymes and stimulates release of bile from gallbladder

The digestive activities of the small intestine are coordinated and regulated by hormones (Table 34-2). In the presence of food, the duodenum releases secretin, a hormone that stimulates the pancreas and liver to secrete alkaline fluid, and cholecystokinin (formerly called pancreozymin), a hormone that acts on the pancreas to stimulate the secretion of enzymes and that triggers the emptying of the gallbladder. At least eight other substances are suspected of being gastrointestinal hormones.

In addition to hormonal influences, the intestinal tract is also regulated by the autonomic nervous system. Stimulating the vagus nerve, which carries parasympathetic nerve fibers, causes the pancreas to secrete digestive enzymes and the intestine to contract. Thus, a complex interplay of stimuli and checks and balances serves to activate digestive enzymes, to adjust the chemical environment, and to regulate movements of the intestines.

Absorption of Nutrients

Most of the absorption of food molecules takes place through the walls of the small intestine. Monosaccharides are rapidly absorbed by active transport and facilitated diffusion (page 133) from the small intestine into the blood vessels of the villi. Amino acids and dipeptides are absorbed by active transport.

Small fatty acids also enter the blood vessels of the intestines directly, but large fatty acids and cholesterol enter mucosal cells by passive diffusion and are incorporated into droplets, known as chylomicrons, that are secreted into the lymph and then enter the bloodstream. In the bloodstream, these chylomicrons are gradually broken apart. Fats are delivered in the form of fatty acids to cells such as muscle cells, where they are oxidized for energy, or to fat cells, where they are stored. (One gram of fat, as we noted previously, contains more than twice as many calories as 1 gram of protein or glycogen; hence it is a very useful storage form for mobile organisms.) Cholesterol is picked up by cells in the liver, where it is stored, secreted in the bile, or packaged for delivery to other body cells. It is used by all cells in the synthesis of cell membranes and, by specialized cells, for the production of steroid hormones.

AIDS TO DIGESTION

Compared to the microscopic photosynthetic cells of the open waters, plant life is large and tough. By logic, one would expect terrestrial herbivores to have sets of enzymes capable of breaking down cellulose and other structural polysaccharides found in plants. Not so. Terrestrial animals have evolved another solution: symbiotic relationships with bacteria and protozoa that perform these activities in their behalf, receiving, in return, a free food supply, protection (except from the host itself), and good housing.

In the warm, wet, airless interior of a mammalian digestive tract, symbionts break down plant products anaerobically. The glucose they use themselves; the host's share is mainly fatty acids, which are absorbed through the host's intestine. In addition, the microorganisms synthesize many vitamins, especially those of the B group, which are utilized by the host.

The fermentation process, the anaerobic oxidation of organic molecules, takes place in different areas in the intestinal tracts of different species. In the horse family, it occurs in the colon, which is enlarged, and in the cecum, a blind sac at the junction of the small and large intestines. (The cecum persists in humans in a vestigial form as the trouble-making appendix.) Rabbits and other lagomorphs have an enlarged cecum in which bacterial fermentation takes place and also a curious adaptation, peculiar to this group. At night, in their burrows, rabbits produce, from the cecum, a special type of feces, different from those produced in the daytime, that consist almost entirely of bacteria. These feces they eat, thus digesting and absorbing additional nutrients, obtaining vitamins produced by the bacteria, and recycling their beneficial symbionts.

The most elaborate specialization, anatomically speaking, is found among the ruminants. It is probably not a coincidence that the ruminants are among the most successful terrestrial herbivores, with their rise in evolutionary prominence taking place at the time the grasslands were undergoing rapid expansion. Ruminants hastily shred grass blades and other plant parts and store them, regurgitating them as cud and completing the chewing process at leisure. The actual fermentation takes place in the rumen, a large esophageal pouch. The contents of the rumen are then passed to the stomach where absorption begins. Bacteria and ciliates in the rumen not only break down the cellulose and other polysaccharides but also synthesize proteins, using ammonia and urea as nitrogen sources. Some of these microorganisms are also digested by the host, thus providing it with protein.

Ruminant digestion is a major activity. In cattle, the rumen-stomach complex makes up about 15 percent of the weight of the animal. To keep the mixture moist, a cow secretes 60 liters of saliva a day. She also produces about 2 liters of gas a minute, most of which escapes in sweet, chlorophyll-scented belches.

Omnivorous humans have no such symbionts and so are totally unable to utilize cellulose, which is eliminated as roughage. However, like other mammals, we are dependent on the bacteria of our digestive tracts for the synthesis of vitamins, and, as in the rabbit, a major component of our feces is symbiotic bacteria.

The Large Intestine: Absorption of Water and Elimination

The chief function of the large intestine is the absorption of water, sodium, and other minerals. Some of its epithelial cells secrete mucus, which lubricates the drying mass of undigested food residue. In the course of digestion, large amounts of water—approximately 7 liters per day—enter the stomach and small intestine by osmosis from the body fluids or as secretions of the glands lining the digestive tract, and also in saliva, bile, and pancreatic secretions. When the absorption process is interfered with, as in diarrhea, severe dehydration can result. Mortality from infant diarrhea, still the chief cause of infant deaths in many countries, is principally a consequence of water loss.

The large intestine also harbors a considerable population of symbiotic bacteria (including the familiar *E. coli*), which break down additional food substances. Living on these food substances, largely materials we lack the enzymes to digest, they synthesize amino acids and vitamins, some of which are absorbed into the bloodstream. These bacteria are our chief source of vitamin K.

The appendix is a blind pouch off the large intestine, which may become irritated, inflamed, and then infected. If it ruptures, as a consequence of inflammation and swelling, it spills its bacterial contents into the abdominal cavity. The appendix is an important site of interaction among cells involved with the production of antibodies.

The bulk of fecal matter consists of bacteria (mostly dead cells) and cellulose fibers, along with other indigestible substances. It is stored briefly in the rectum and eliminated through the anus as feces.

MAJOR ACCESSORY GLANDS

The Pancreas

The pancreas is a specialized digestive organ, developing both in the embryo and in the course of evolutionary history from the upper portion of the small intestine. The bulk of the tissue of the pancreas resembles salivary gland tissue and, like the salivary glands, secretes an amylase that plays a major role in the breakdown of starch, as well as other digestive enzymes (see Table 34–1). The pancreas retains its connections with the intestinal tract, and these enzymes travel through the pancreatic duct directly into the upper part of the small intestine.

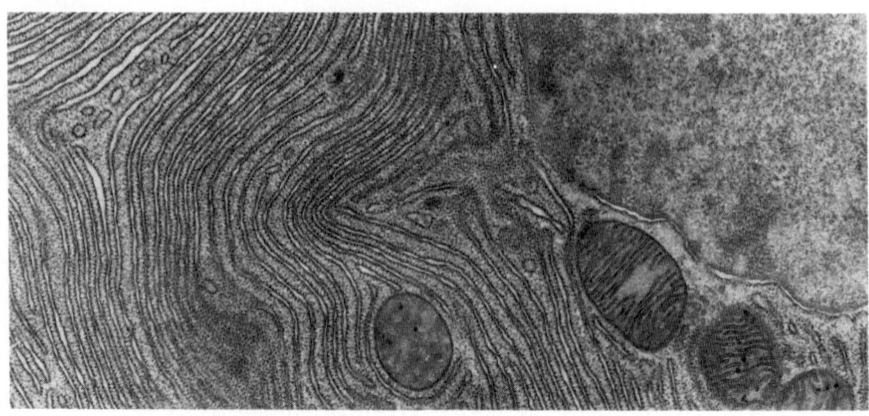

34–10

Portion of pancreatic cell involved in the production of digestive enzymes. Notice the extensive rough endoplasmic reticulum characteristic of cells that synthesize proteins for export. The cell nucleus can be seen on the right-hand side of the micrograph, and, just outside the nuclear envelope, there are several mitochondria.

The pancreas is also an endocrine gland. Clusters of pancreatic cells, known as the islets of Langerhans (named after their discoverer, not some romantic Pacific atoll) secrete insulin, glucagon, and somatostatin, which are released into the bloodstream.

The Liver

The liver, the largest internal organ of the body, is a three-pound chemical factory with an extraordinary variety of processes and products. It stores and releases carbohydrates, playing a central role in the regulation of blood glucose. It packages fats for circulation to other organs and storage sites. It processes amino acids, converting them to carbohydrates, channeling them to other tissues of the body, and synthesizing essential proteins such as enzymes and clotting factors from them. It manufactures the plasma proteins that make the blood hypertonic in relation to the interstitial ("between-the-tissues") fluids and so prevents the movement by osmosis of water from bloodstream to tissues. It is the major source of plasma lipoproteins, the molecules that transport cholesterol, fats, and other water-insoluble substances in the bloodstream. It regulates blood cholesterol by picking up protein-coated cholesterol particles from the blood, storing the cholesterol, releasing it as needed (packaged in new lipoprotein carriers), synthesizing it if necessary, and secreting some in the bile. (Gallstones are precipitates of cholesterol.) It stores fat-soluble vitamins, such as A, D, and E. It produces bile (which is then stored in the gallbladder). It breaks down hemoglobin to bilirubin, a yellow pigment that is released into the bile and excreted through the intestinal tract. It inactivates a number of hormones, thus playing an important role in hormone regulation. It breaks down a variety of foreign substances, some of which—alcohol, for instance—may form metabolic products that damage liver cells and interfere with their functions (see page 206).

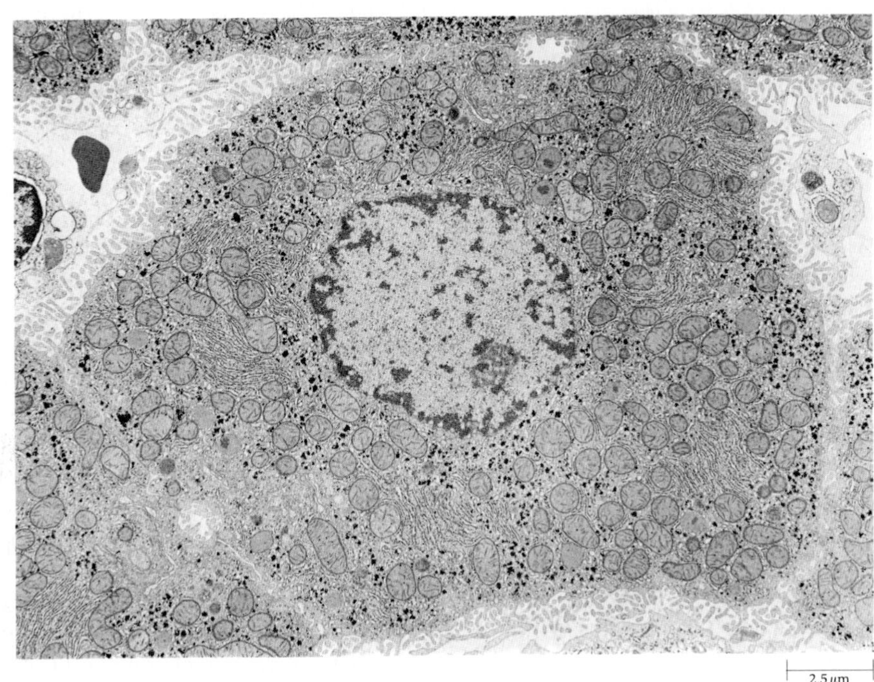

34–11
Electron micrograph of a liver cell. The dark granules are glycogen. Despite the variety of activities carried out by the liver, the cells of the liver all resemble one another in appearance and in function, with no division of labor among cell groups, as there is in the pancreas, for example (page 668). Liver cells, as you would expect, have many mitochondria and abundant rough endoplasmic reticulum.

H_3C CH_3
H_3C
CH_3
H_2C
HO
VITAMIN D

(a)

(b)

34–12
(a) *Vitamin D is a steroid-like substance that is produced in the skin by the action of ultraviolet rays from the sun on cholesterol (the absorbed energy opens the second ring of the molecule). According to one current hypothesis, the ancestors of modern* Homo sapiens *originated in the tropics and were all dark-skinned. In those populations that moved northward, however, the screening effects of the darker pigment, which inhibits the production of vitamin D, caused selection for lighter skin color, whereas no such selection pressures occurred in the sun-drenched areas nearer the equator. In fact, in the tropics, selection would favor dark skin, which prevents carcinogenic ultraviolet rays from reaching the dermis. Nor did such selection occur among the Eskimos* (b), *who eat a diet rich in fish oils, a major source of vitamin D.*

REGULATION OF BLOOD GLUCOSE

As we noted at the beginning of this chapter, a major function of digestion is to provide an energy source for each cell of the body. Although vertebrates rarely eat 24 hours a day, their blood sugar—the major cellular energy supply—remains extraordinarily constant. The liver plays a central role in this critical process. Glucose and other monosaccharides are absorbed into the blood from the intestinal tract and are passed directly to the liver by way of the portal vein. The liver converts some of these monosaccharides to glycogen and fat. (The liver stores enough glycogen to satisfy the body's needs for about four hours.) The fat is stored in the fat cells, which can also form fat from glucose. The liver similarly breaks down excess amino acids (which are not stored) and converts them to glucose. The nitrogen from the amino acids is excreted in the form of urea, and the glucose is stored as glycogen.

Whether the liver takes up or releases glucose and the amount it takes up or releases are determined primarily by the concentration of glucose in the blood. The concentration of glucose is, in turn, regulated by a number of hormones and is influenced by the autonomic nervous system.

Insulin, as we have mentioned previously, stimulates the uptake of glucose by cells, thus decreasing blood glucose. Glucagon, a hormone, also produced by the pancreas, promotes the breakdown of glycogen, thus increasing blood glucose. Somatostatin, the third hormone synthesized by the pancreas, lowers the production of both insulin and glucagon.

SOME NUTRITIONAL REQUIREMENTS

The energy requirements of the body can be met by carbohydrates, proteins, or fats—the three principal types of food molecules. Carbohydrates and proteins supply about the same number of calories per gram of dry weight, and fats supply about twice as many as either of them. (Because carbohydrates and proteins are always combined with water in the form in which they are eaten, the actual bulk of ingested material is usually nine or ten times greater.)

In addition to the need for calories, the cells of the body need the 20 different amino acids required for assembling proteins. When any one of the amino acids necessary for the synthesis of a particular protein is unavailable, the protein cannot be made and the other amino acids are converted to carbohydrates and oxidized or stored. Vertebrates are not able to synthesize all 20 amino acids. Humans (and albino rats) can synthesize 12, either from a simple carbon skeleton or from another amino acid. The other eight, which must be obtained in the diet, are known as essential amino acids.

Mammals also require but cannot synthesize certain polyunsaturated fats that provide fatty acids needed for fat synthesis. These can be obtained by eating plants or insects (or eating other animals that have eaten plants or insects).

Vitamins are an additional group of molecules required by living cells that cannot be synthesized by animals. They are characteristically required only in small amounts. Many of them function as coenzymes. Table 34–3 (page 686) indicates some of the vitamins required in the human diet. There is no clear evidence that the ingestion of amounts of any particular vitamin in excess of the amounts available in well-balanced diets has any beneficial effects on a healthy individual, and some, including vitamins A, D, and K, are toxic in large dosages.

AMINO ACIDS AND NITROGEN

Like fats, amino acids are formed within living cells using sugars as starting materials. But while fats are made up only of carbon, hydrogen, and oxygen atoms, all available in the sugar and water of the cell, amino acids also contain nitrogen. Most of the earth's supply of nitrogen exists in the form of gas in the atmosphere. Only a few organisms, all microscopic, are able to incorporate nitrogen from the air into compounds—ammonia, nitrites, and nitrates—that can be used by living systems. Hence the proportion of the earth's nitrogen supply available to the living world is very small.

Plants incorporate the nitrogen in ammonia, nitrites, and nitrates into carbon-hydrogen compounds to form amino acids. Animals are able to synthesize some of their amino acids, using ammonia as a nitrogen source. The amino acids they cannot synthesize, the so-called essential amino acids, must be obtained either directly or indirectly from plants. For adult human beings, the essential amino acids are lysine, tryptophan, threonine, methionine, phenylalanine, leucine, valine, and isoleucine.

People who eat meat usually get enough protein and the correct balance of amino acids. People who are vegetarians, whether for philosophical, esthetic, or economic reasons, have to be careful that they get enough protein and, in particular, the essential amino acids.

Until recently, agricultural scientists concerned with the world's hungry people concentrated on developing plants with a high caloric yield. Increasing recognition of the role of plants as a major source of amino acids for human populations has led to emphasis on the development of high-protein strains of food plants and of plants with essential amino acids, such as "high-lysine" corn.

Another approach to the right balance of amino acids is to combine certain foods. Beans, for instance, are likely to be deficient in tryptophan and in the sulfur-containing amino acids, but they are a good-to-excellent source of isoleucine and lysine. Rice is deficient in isoleucine and lysine but provides an adequate amount of the other essential amino acids. Thus rice and beans in combination make just about as perfect a protein menu as eggs or steak, as some nonscientists seem to have known for quite a long time.

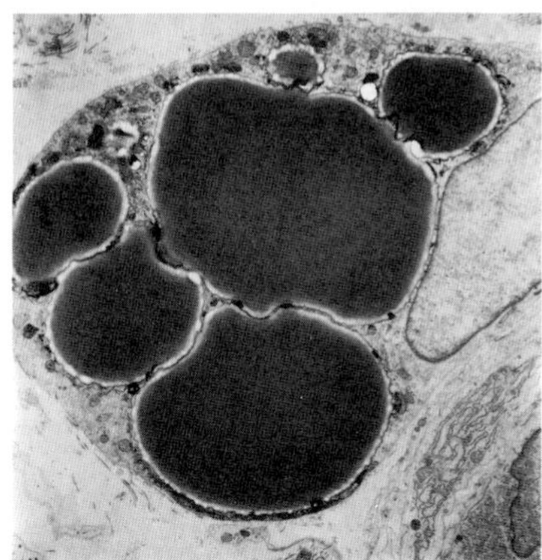

34–13
A fat cell. The large stored droplets of fat that nearly fill the cell have been fixed with osmium for electron microscopy. When excess calories are taken in in the diet, fat accumulates in these specialized cells, and when caloric intake is less than sufficient, fat is mobilized, broken down to glycerol and fatty acids, and released into the bloodstream. The cell nucleus is to the right.

The Price of Affluence

The major nutritional problem among North Americans (and also Europeans) is obesity. In the United States, 30 percent of middle-aged women and 15 percent of middle-aged men are obese—that is, they weigh more than 120 percent of their appropriate weight. Obesity is correlated with a significant increase in coronary artery disease, diabetes, and other disorders.

In addition to excess calories, our diets appear to contain additional health hazards as compared to those of other cultures. We consume, on the average, about twenty times as much salt as our bodies need. Excess salt has been correlated with hypertension (high blood pressure). Another hazard is animal fat, such as that present in beef and pork. It has been known for some time that diets high in animal fats somehow interfere with the regulation of blood cholesterol (page 680), implicated in atherosclerosis and heart attacks. More recently, attention has been focused on the role of dietary fibers in decreasing colon cancer. As a result of a diet high in fat and protein and low in carbohydrate, the feces of North Americans have only about half the bulk of their Far Eastern or African counterparts, and are eliminated from the intestinal tract much more slowly. Also, the rate of colon cancer is far higher among North Americans, suggesting that the retention of a fat-laden, compact fecal mass in the lower bowel may contribute to cancer development.

A final nutritional hazard, also related to affluence, is our willingness to experiment with our own bodies and to adopt extreme diets and follow nutritional fads—fasting, liquid protein, "macrobiotic," high protein, low protein, high fiber, megavitamin—without consideration of our requirements as biological organisms and the dictates of our long, omnivorous evolutionary history.

Table 34-3 Vitamins

DESIGNATION: LETTER AND NAME	MAJOR SOURCES	FUNCTION	DEFICIENCY SYMPTOMS
A, carotene	Egg yolk, green or yellow vegetables, fruits, liver, butter	Formation of visual pigments, maintenance of normal epithelial structure	Night blindness; dry, flaky skin
B-complex vitamins:			
B_1, thiamine	Brain, liver, kidney, heart, whole grains	Formation of cofactor involved in Krebs cycle	Beri-beri, neuritis, heart failure
B_2, riboflavin	Milk, eggs, liver, whole grains	Part of electron carrier FAD	Photophobia, fissuring of skin
B_6, pyridoxine	Whole grains, liver, kidney, fish, yeast	Coenzyme for amino acid metabolism and fatty acid metabolism	Dermatitis, nervous disorder
B_{12}, cyanocobalamin	Liver, kidney, brain	Maturation of red blood cells, coenzyme in amino acid metabolism	Anemia, malformed red blood cells
Biotin	Egg yolk, synthesis by intestinal bacteria	Concerned with fatty acid synthesis, CO_2 fixation	Scaly dermatitis, muscle pains, weakness
Folic acid	Liver, leafy vegetables	Nucleic acid synthesis, formation of red blood cells	Failure of red blood cells to mature, anemia
Niacin (nicotinic acid)	Whole grains, liver and other meats, yeast	Part of electron carriers NAD, NADP, and of CoA	Pellagra, skin lesions, digestive disturbances
Pantothenic acid	Present in most foods, especially eggs, liver, yeast	Forms part of CoA	Neuromotor disorders, cardiovascular disorders, GI distress
C, ascorbic acid	Citrus fruits, tomatoes, green leafy vegetables	Vital to collagen and ground (intercellular) substance	Scurvy, failure to form connective tissue fibers
D_3, calciferol	Fish oils, liver, fortified milk and other dairy products, action of sunlight on lipids in the skin	Increases Ca^{2+} absorption from gut, important in bone and tooth formation	Rickets (defective bone formation)
E, tocopherol	Green leafy vegetables, milk, eggs, meat	Maintains resistance of red cells to hemolysis, cofactor in electron transport chain	Increased red blood cell fragility
K, naphthoquinone	Synthesis by intestinal bacteria, leafy vegetables	Enables synthesis of clotting factors by liver	Failure of blood coagulation

SUMMARY

Digestion is the process by which food is broken down into molecules that can be taken up by the bloodstream or lymph system and so distributed to the individual cells of the body. It occurs in successive stages, regulated by an interplay of hormones and nervous stimuli. In mammals, food is processed initially in the mouth, where the breakdown of starch begins in humans. It moves through the esophagus to the stomach, where gastric juices destroy bacteria and begin to break down proteins.

Most of the digestion occurs in the upper portion of the small intestine; here, digestive activity, which is performed by enzymes, is almost completely under hormonal regulation. The breakdown of starch by amylases continues, fats are hydrolyzed by lipases, and proteins are reduced to single amino acids or dipeptides. Monosaccharides, amino acids, and dipeptides are absorbed into the blood vessels of the villi; fats are absorbed into the lymphatic vessels. Hormones secreted by duodenal cells stimulate the functions of the pancreas and the liver. The pancreas releases an alkaline fluid containing digestive enzymes; the liver produces bile, which is also alkaline and emulsifies fats.

Water is reabsorbed from the residue of the food mass as it passes through the large intestine. The large intestine contains symbiotic bacteria, which are the source of certain vitamins. Undigested residues are eliminated from the large intestine.

The chief energy source for cells in the mammalian body is glucose circulating in the blood. The organ principally responsible for maintaining a steady supply of glucose is the liver, which stores glucose (in the form of glycogen) when glucose levels in the blood are high and breaks down glycogen, releasing glucose, when the levels drop. These activities of the liver are regulated by a number of different hormones.

Requirements for good nutrition include calories (which can be obtained from carbohydrates, fats, or proteins), essential amino acids, certain essential fatty acids, vitamins, and minerals.

QUESTIONS

1. Distinguish, in terms of function, between: gastrointestinal hormones/gastrointestinal enzymes; gastric juice/bile; vitamins/essential amino acids; digestion/absorption; villi/microvilli.

2. Make a flow chart of the digestive system, listing tubes, chambers, valves, and so on, and specifying the functions of each.

3. If you oxidize a pound of fat, whether butter or the fat in your own body cells, about 4,200 kilocalories are released. Suppose that you go on a weight-reducing diet, limiting yourself to 1,000 kilocalories a day. You spend most of your time sitting in a well-heated library (studying, of course) and so you expend only about 3,000 kilocalories a day. How many pounds will you lose each week?

4. Four principal symptoms of cirrhosis of the liver and other liver diseases are jaundice (a yellowing of the eyes and skin), uncontrolled bleeding, increased sensitivity to drugs, and swelling of parts of the body, such as the legs and abdominal cavity. Explain each of these symptoms in terms of liver function.

5. Most of the protein-digesting enzymes are secreted in an inactive form and are themselves activated by special enzymes secreted into the digestive tract. Explain the adaptive value of this two-step process.

6. Name at least three factors affecting concentrations of cholesterol in the bloodstream.

7. Humans are omnivores rather than strict herbivores or carnivores or specialists that use a single organism as a food source. Humans have an unusually large number of substances specifically required in their diets. How might these two phenomena be related?

CHAPTER 35

Energy and Metabolism II: Respiration

Heterotrophs obtain energy by the oxidation of organic molecules. This process usually (although not always) requires oxygen, and when it does, it is called respiration. Actually, respiration has two meanings in biology. At the biochemical level, it refers to the oxygen-requiring chemical reactions that take place in the mitochondria and are the chief source of energy for eukaryotic cells. At the level of a whole organism, it designates the process of taking in oxygen from the environment and returning carbon dioxide to it. The latter process—which is, of course, essential for the former—is the subject of this chapter.

Oxygen consumption is directly related to energy expenditure, and, in fact, energy requirements are usually calculated by measuring oxygen intake or the release of carbon dioxide. The energy expenditure at rest is known as basal metabolism. Metabolic rates increase sharply with exercise. A person exercising consumes 15 to 20 times the amount of oxygen that he or she consumes sitting still, with the oxygen consumption increasing in proportion to the energy expenditure.

35–1
Oxygen is required for the energy-yielding reactions that take place in the mitochondria. The oxygen consumption of running animals—such as these gazelles—increases linearly with their speed.

35-2
Atmospheric pressure is usually measured by means of a mercury barometer. (a) *To make a simple mercury barometer, fill a long glass tube, open at one end only, with mercury. Closing the tube with your finger,* (b) *invert it into a dish of mercury. Remove your finger. The mercury level will drop until the pressure of its weight inside the tube is equal to the atmospheric pressure outside. At sea level, the height of the column will be about 760 millimeters.*

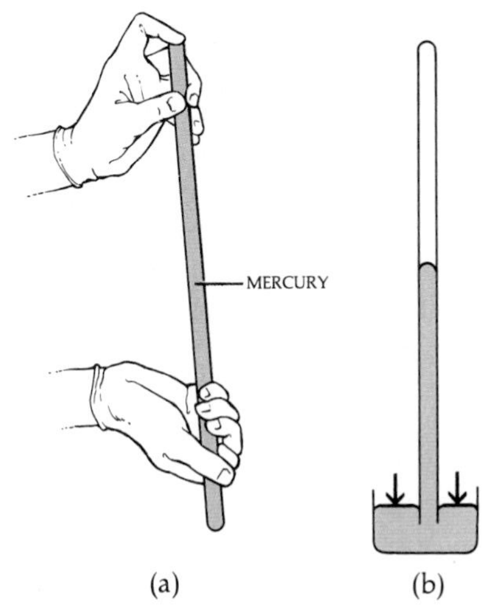

Table 35-1 **Composition of Dry Air**

GAS	% OF VOLUME
Oxygen	21
Nitrogen	77
Argon	1
Carbon dioxide	0.03
Other gases*	0.97

* Includes hydrogen, neon, krypton, helium, ozone, xenon, and now, unfortunately, in some environments, radon.

DIFFUSION AND AIR PRESSURE

In every organism from an amoeba to an elephant, gas exchange—the exchange of oxygen and carbon dioxide—takes place by diffusion. This chapter begins, therefore, with a discussion of the principles of diffusion, particularly as they refer to the movement of gases between air and a liquid or between two liquids.

Diffusion, you will recall (page 127), is the net movement of particles from a region of higher concentration to a region of lower concentration as a result of their random motion. In describing gases, scientists speak of the pressure of a gas rather than its concentration. At sea level, the air around us exerts a pressure on our skin of 1 atmosphere (about 15 pounds per square inch). This pressure is enough to raise a column of water about 10 meters (394 inches) high or a column of mercury 760 millimeters (29.91 inches). Atmospheric pressure is generally measured in terms of mercury simply because mercury is relatively heavy—so the column will not be inconveniently tall (Figure 35-2).

The total pressure of a mixture of gases, such as air, is the sum of the pressures of the separate gases in the mixture. The pressure of each gas is proportional to its concentration. Oxygen, for instance, makes up about 21 percent by volume of dry air (Table 35-1). Thus the pressure of the oxygen component of dry air is 21 percent of 760 millimeters of mercury, or about 159 millimeters of mercury (mm Hg). This is known as the partial pressure of oxygen and is abbreviated P_{O_2}. In air that contains water vapor (also a gas), the volume of O_2 is proportionately less and so the P_{O_2} is less—around 155 mm Hg.

When one speaks of the P_{O_2} of the blood, one means the pressure of dry gas with which the dissolved O_2 in the blood is in equilibrium. Thus, venous blood with a P_{O_2} of 40 mm Hg would be in equilibrium with air in which the partial pressure of oxygen was 40 mm Hg.

If a liquid containing no dissolved gases is exposed to air at atmospheric pressure, each of the gases in the air diffuses into the liquid until the partial pressure of each gas in the liquid is equal to the partial pressure of the gas in the air. So if the venous blood with a P_{O_2} of 40 mm Hg were exposed to the usual mixture of air, oxygen would move from the air into the venous blood until the blood P_{O_2} equaled 155 mm Hg and equilibrium was reached. Conversely, if a liquid containing a gas is exposed to air containing a lower concentration of that gas (in terms of partial pressure), the gas will leave the liquid until the partial pressures of the air and the liquid are equal. In short, gases move from a region of higher partial pressure to a region of lower partial pressure.

We are so accustomed to the pressure of the air around us that we are unaware of its presence or of its effects on us. However, if you visit a place—such as Mexico City—which is at a comparatively high altitude and therefore has a lower atmospheric pressure (and of course, a lower P_{O_2}), you will feel lightheaded at first and will tire easily. Regular inhabitants of such places breathe more deeply and have more red blood cells for oxygen transport. If they have lived at high altitudes since childhood, they are likely to have enlarged hearts that circulate oxygen-carrying blood more rapidly.

The consequences of increased gas pressures are seen in deep-sea divers. Early in the history of deep-sea diving, it was found that when divers come up from the bottom too quickly, they get the "bends," which are always painful and sometimes fatal. The bends develop as a result of breathing compressed air. High pressures force more nitrogen from the air in the lungs into solution in the blood and tissues.

If the body is rapidly decompressed, the gas expands, and nitrogen bubbles form in the blood, like the carbon dioxide bubbles that appear in a bottle of soda when you first remove the top. The nitrogen bubbles lodge in the capillaries, stopping blood flow, or invade nerves or other tissues.

EVOLUTION OF RESPIRATORY SYSTEMS

The biochemical means by which organisms obtain energy in the absence of oxygen are not very efficient, as we saw in Chapter 9. Before oxygen was present in the atmosphere, the only forms of life were one-celled organisms; the only present-day forms that can carry out all life processes totally without oxygen are a few types of bacteria and yeasts and some parasites, such as intestinal worms. The transition to an atmosphere containing oxygen was responsible for one of the giant steps in evolution (page 210).

Oxygen enters and moves within cells by diffusion. Within the cell, it takes part in the oxidation of organic compounds that serve as cellular energy sources. In this process, carbon dioxide is produced, which then diffuses out of the cell along the concentration (partial pressure) gradient. This is true of all cells, whether an amoeba, a *Paramecium,* a liver cell, or a brain cell. But, substances can move effectively by diffusion only for very short distances, less than 1 millimeter. These limits pose no problem for very small animals, in which each cell is quite close to the surface, or for animals in which much of the body mass is not metabolically active—like the "jelly" (mesoglea) of jellyfish. (Many eggs and embryos also respire in this simple way, particularly early in development.) However, diffusion cannot possibly meet the needs of large organisms in which cells in the animal's interior may be many centimeters from the air or water serving as the oxygen source. As organisms increased in size in the course of evolution, there also evolved circulatory and respiratory systems that transport large numbers of gas molecules by bulk flow. (Remember that whereas diffusion is the result of the random movement of individual molecules, bulk flow is the overall movement of molecules in response to pressure or gravity.)

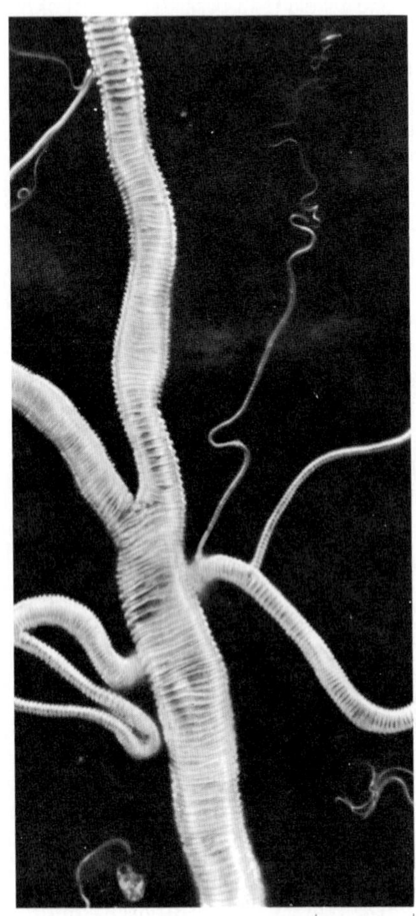

35–3
Insects and some other terrestrial arthropods breathe by means of tracheae. These are inward tubular extensions of the exoskeleton, which are thickened in places, forming spiral supports that hold the tubes open. The tracheae are lined with chitin, which is molted when a new exoskeleton is formed. Gas exchange takes place at the thin, moist terminal ends of the tubules. This micrograph shows part of the tracheal system of a cockroach.

An early stage in the evolution of gas-transport systems is exemplified by the earthworm (page 518). As is the case with most other kinds of worms, earthworms have a network of capillaries just one cell layer beneath the surface of their bodies. Oxygen and carbon dioxide diffuse directly through the body surface into and out of the blood as it travels through these capillaries. The blood picks up oxygen by diffusion as it travels near the surface of the animal and releases oxygen by diffusion as it travels past the oxygen-poor cells in the interior of the earthworm's body. Conversely, blood picks up carbon dioxide from the cells and releases it by diffusion as it travels near the surface of the animal. Thus the gases move into and out of the earthworm by diffusion but are transported within the animal by bulk flow.

This system is particularly suitable for worms because their tube shape exposes a proportionately large surface area. Some worms can adjust their surface area in relation to oxygen supply. If you have an aquarium at home, you may be familiar with tubifex worms, which are frequently sold as fish food. When these worms are placed in water low in oxygen, such as a poorly aerated aquarium, they stretch out as much as 10 times their normal length and thus increase the surface area through which oxygen diffusion occurs and decrease the diffusion distance.

Insects and some other arthropods, as we have seen (page 538), have evolved a different strategy. Air is piped directly into the tissues by a network of chitin-lined tubules (Figure 35–3). In large insects, in particular, diffusion is assisted by body movements, which propel the air into and out of the spiracles. This system is fine for small organisms but is a major limitation on the size that can be obtained by insects and other tracheal-system breathers. The planet may eventually be taken over by cockroaches or ants, but it is safe to bet they will not be the giant forms of science fiction.

Evolution of Gills

Gills and lungs are other ways of increasing the respiratory surface. Gills are usually outgrowths (Figure 35–4), whereas lungs are ingrowths, or cavities. The respiratory surface of the gill, like that of the earthworm, is a layer of cells, one cell thick, exposed to the environment on one side and to circulatory vessels on the other. The layers of gill tissue may be spread out flat, stacked, or convoluted in various ways. The gill of a clam, for instance, is shaped like a steam-heat radiator (which is also designed to provide a maximum surface-to-volume ratio).

The vertebrate gill is believed to have originated primarily as a feeding device. Primitive vertebrates respired mostly through their skin. They filtered water into their mouths and out of what we now call their gill slits, extracting bits of organic matter from the water as it went through. (*Branchiostoma*—page 554—which is believed to resemble closely the ancestral vertebrate, feeds in this way.)

In the course of time, numerous selection pressures, chiefly involved with predation, came into operation. As one consequence, there was a trend toward an increasingly thick skin, even one armored or covered with scales. Such a skin is not, of course, useful for respiratory purposes. At the same time, related forces were operating to produce animals that were larger and swifter and so more efficient at capturing prey and escaping predators. Such animals had larger energy requirements—and consequently larger oxygen requirements. These problems were solved by the "capture" of the gill for a new purpose: respiratory exchange.

35–4
Gills are essentially outpocketings of the epithelium that increase the surface area exposed to water. Often gills are covered by an exoskeleton, as in crustaceans, or a flaplike shield (the gill cover), as in fish. In the axolotl, the amphibian shown here, the external nature of the gills is clearly evident.

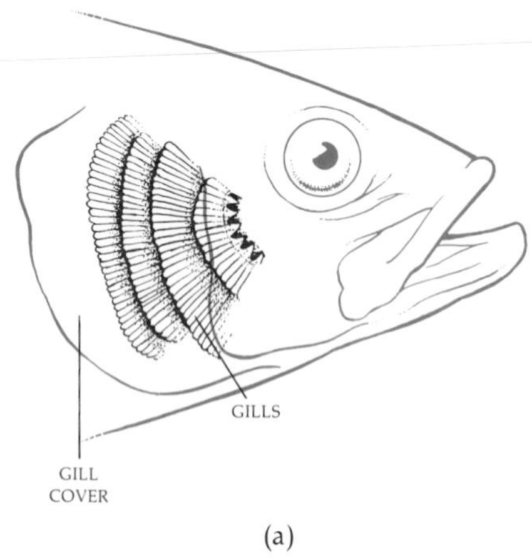

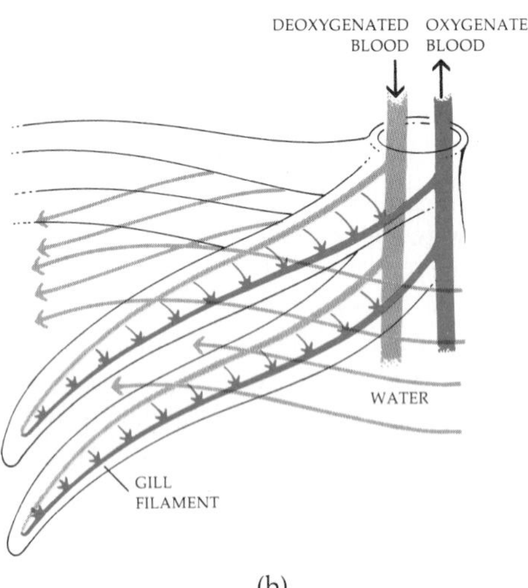

35–5
In fish, oxygen enters the blood by diffusion from water flowing through the gills (a). *The anatomical structure of the gills maximizes the rate of diffusion, which is proportional not only to the surface areas exposed but also to differences in concentration of the diffusing molecules. The greater the difference in concentration of a molecule, the more rapid its diffusion.* (b) *The circulatory vessels are arranged so that the blood is pumped through them in a direction opposite to that of the oxygen-bearing water. This countercurrent arrangement results in a far more complete transfer of oxygen to the blood than if the blood flowed in the same direction as the water.*

The surface area and blood supply of the gill epithelium have slowly increased over the millennia. The modern gill is the result of this evolutionary process.

In most fish, the water (in which oxygen is dissolved) is pumped in at the mouth by oscillations of the bony gill cover and flows out across the gills (Figure 35–5). In the gills of fish, the circulatory vessels are arranged so that the blood is pumped through them in a direction opposite to that of the oxygen-bearing water. This countercurrent arrangement (see page 129) results in a far more efficient transfer of oxygen to the blood than if the blood flowed in the same direction as the water. Also the fish can regulate the rate of water flow, and sometimes assist it, by opening and closing its mouth. Fast swimmers, such as mackerel, obtain enough oxygen to meet their energy needs by keeping their mouths open as they swim. As a result, water moves rapidly over the gills. Such fish have become so dependent on this method of respiration that if they are kept in an aquarium or any other space where their motion is limited, they will suffocate.

Evolution of Lungs

Lungs are internal cavities into which oxygen-containing air is taken. They have a disadvantage as compared with gills; it is more efficient from the point of view of diffusion to have a continuous flow across the respiratory surface. Air, however, is a far better source of oxygen than is water; 20 percent of the air of the modern atmosphere is oxygen, as compared to 0.5 percent by volume in water (at 15°C). Not only must more water be processed to obtain a given amount of oxygen, but water also weighs a great deal more. A fish spends up to 20 percent of its energy in the muscular work associated with respiration, whereas an air breather expends only 1 or 2 percent of its energy in respiration. Also, oxygen diffuses about 300,000 times more rapidly through air than through water, and so can be replenished much more quickly in air as it is used up by respiring organisms.

Lungs are not essential for air breathing. As we saw, earthworms are air breathers. The overwhelming advantage of gas exchange across an internal surface, however, is that the respiratory areas can be kept moist without a large loss of water by evaporation. Although lungs are largely a vertebrate "invention," they are found in some invertebrates. Land-dwelling snails, for example, have independently evolved lungs that are remarkably similar to the lungs of some amphibians.

Some primitive fish had lungs as well as gills, although the lungs were not efficient enough to serve as more than accessory respiratory structures. These lungs were probably a special adaptation to life in fresh water, which, unlike ocean water, may stagnate (become depleted of oxygen). A few species of lungfish still exist, which, by coming to the surface and gulping air into their lungs, can live in water that does not have sufficient oxygen to support other fish life.

Amphibians and reptiles have relatively simple lungs, with small internal surface areas, although their lungs are far larger and more complex than those of the lungfish. The lungs of lungfish developed directly from the pharynx, the posterior portion of the mouth cavity, which leads to the digestive tract. In amphibians, reptiles, and other air-breathing vertebrates, we see the evolution of the windpipe, or *trachea,* guarded by a valve mechanism, the glottis, and nostrils, which make it possible for the animal to breathe with its mouth closed. Amphibians still rely largely on their skin for gas exchange (Figure 35–6), but reptiles breathe almost entirely through their lungs.

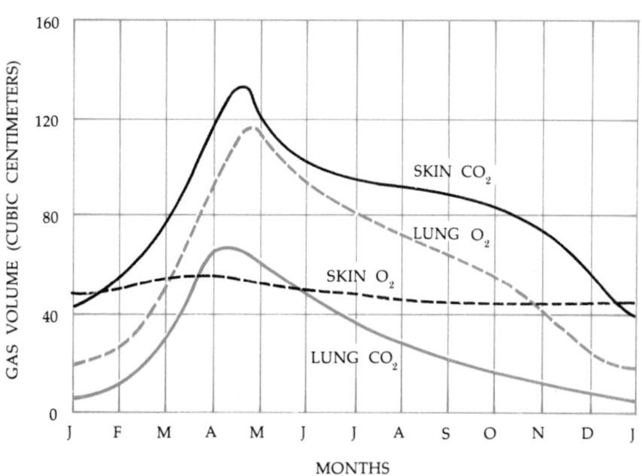

35–6
Frogs have relatively small and simple lungs, and a major part of their respiration takes place through the skin. The chart shows the results of a study of a frog in which oxygen and carbon dioxide exchanges through the skin and lungs were measured simultaneously for one year. Can you explain why the respiratory activity peaks in April and May and is lowest in December and January?

An important feature of all vertebrate lungs is that the exchange of air with the atmosphere takes place as a result of changes in lung volume. Such lungs are known as ventilation lungs. Frogs gulp air, force it into their lungs in a swallowing motion, and then open the glottis and let the air out again. In reptiles, birds, and mammals, air is moved in and out of the lungs as a consequence of changes in the size of the lung cavity, brought about by muscular contractions and relaxations. In short, frogs ventilate their lungs by positive pressure; mammals, birds, and reptiles by negative pressure. The various types of respiratory systems are summarized in Figure 35–8.

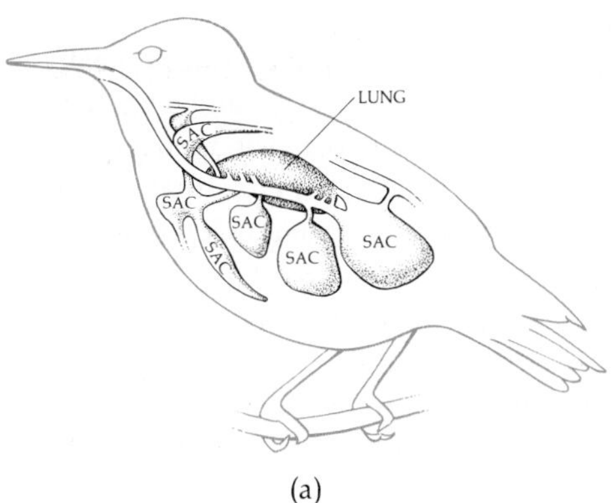

(a)

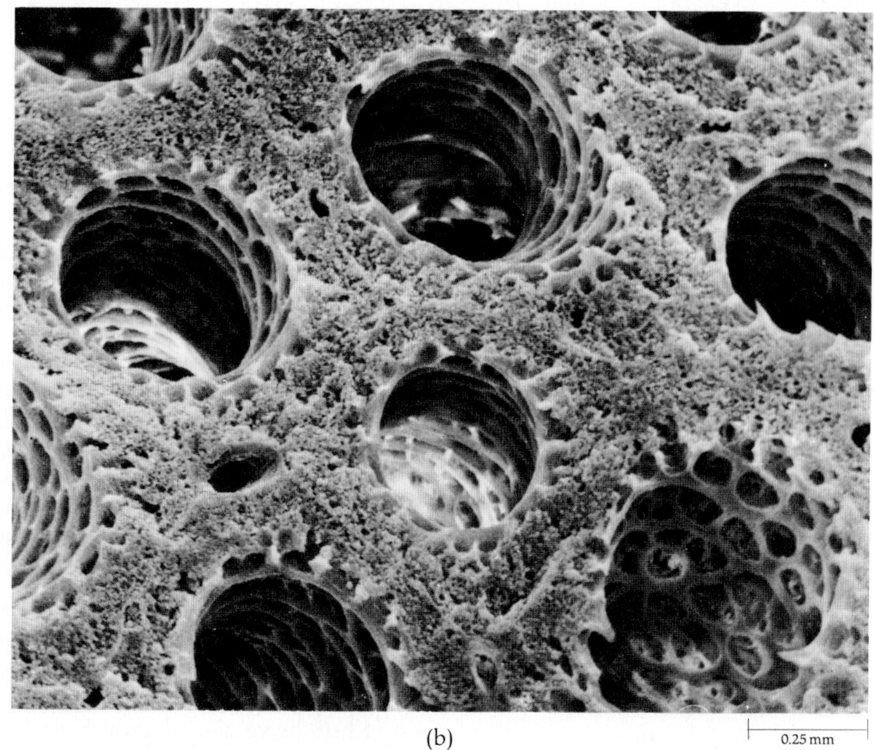

(b)

35–7
(a) *The lungs of birds are extraordinarily efficient. They are small and are expanded and compressed by movements of the body wall. Each lung has several air sacs attached to it, which empty and fill like balloons at each breath. No gas exchange takes place in the sacs. They appear rather to act as bellows. Thus, the lung is almost completely flushed with fresh air at every breath; there is little residual "dead" air left in the lungs, as there is in mammals.* (b) *Scanning electron micrograph of lung tissue from a domestic fowl. The tubes visible here are ventilated by air sacs, into which fresh air is drawn. Gas exchange takes place in the broad meshwork of air capillaries and blood capillaries, which make up the spongelike respiratory tissue seen here surrounding the tubes.*

35-8
Respiratory systems. (a) *Gas exchange across the entire surface of the body is found in a wide range of small organisms from protozoans to earthworms.* (b) *Gas exchange across the surface of a flattened body is seen, for example, in flatworms. Flattening increases the surface-to-volume ratio and also decreases the distance over which diffusion has to occur within the body.* (c) *External gills increase the surface area but are unprotected and therefore easily damaged. External gills are found in polychaete worms and some amphibians. Gas exchange usually takes place across the rest of the body surface as well as the gills.* (d) *With internal gills, a ventilation mechanism draws water over the highly vascularized gill surfaces, as in fish.* (e) *Gas exchange at the terminal ends of fine tracheal tubes that branch through the body and penetrate all the tissues is characteristic of insects and some other arthropods.* (f) *Lungs are highly vascularized sacs into which air is drawn by a ventilation mechanism. Lungs are found in all air-breathing vertebrates and some invertebrates, such as terrestrial snails.*

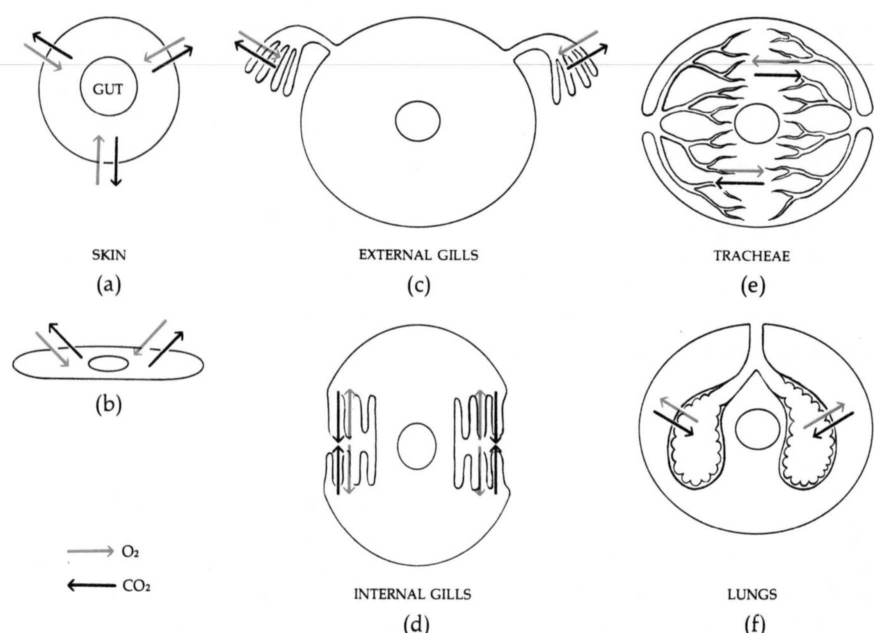

RESPIRATION IN LARGE ANIMALS: SOME PRINCIPLES

Let us stop for a moment to sum up this discussion. In large animals, both diffusion and bulk flow move oxygen molecules between the external environment and actively metabolizing tissues. This movement occurs in four stages:

1. Movement by bulk flow of the oxygen-containing external medium (air or water) to a thin membrane close to small blood vessels in lungs or gills.
2. Diffusion of the oxygen across this membrane into the blood.
3. Movement by bulk flow of the oxygen with the circulating blood to the tissues where it will be used.
4. Diffusion of the oxygen from the blood into the tissue cells.

Carbon dioxide, which is produced in the tissue cells, follows the reverse path as it is eliminated from the body.

HUMAN RESPIRATORY SYSTEM

In *Homo sapiens,* inspiration (breathing in) and expiration (breathing out) usually take place through the nose. The nasal cavities are lined with hairs and cilia, both of which trap dust and other foreign particles. The epithelial cells that line the cavities secrete mucus, which humidifies the air and collects debris that can be removed by swallowing, sneezing, or spitting. The cavities have a rich blood supply, which keeps their temperature high, warming the air before it reaches the lungs.

From the nasal passages, the air goes to the pharynx and from there to the larynx, located in the upper front part of the neck. An adult human larynx is shaped somewhat like a triangular box, with its point downward. Across it are stretched the vocal cords, which are two ligaments drawn taut across the lumen of the respiratory tract. Vibrations of these cords, by expired air, cause the sounds made in speech. The vocal cords are influenced by the male hormone. At puberty, the cords in males become longer and thicker, sometimes so rapidly that the adolescent male temporarily loses control over them, occasionally emitting embarrassing squeaks. Laryngitis, which is simply an inflammation of the vocal cords, interferes with their vibration, and so you "lose your voice."

From the larynx, inspired air travels through the trachea, which is a long membranous tube, also lined with ciliated epithelial cells. The walls of the trachea are strengthened by rings of cartilage that prevent it from collapsing during inspiration or when food from the adjacent esophagus presses on it. The trachea leads into the *bronchi* (singular, bronchus), which subdivide into smaller and smaller passageways, the *bronchioles*. The bronchi and bronchioles are surrounded by thin layers of smooth muscle. Contraction and relaxation of this muscle alters resistance to air flow.

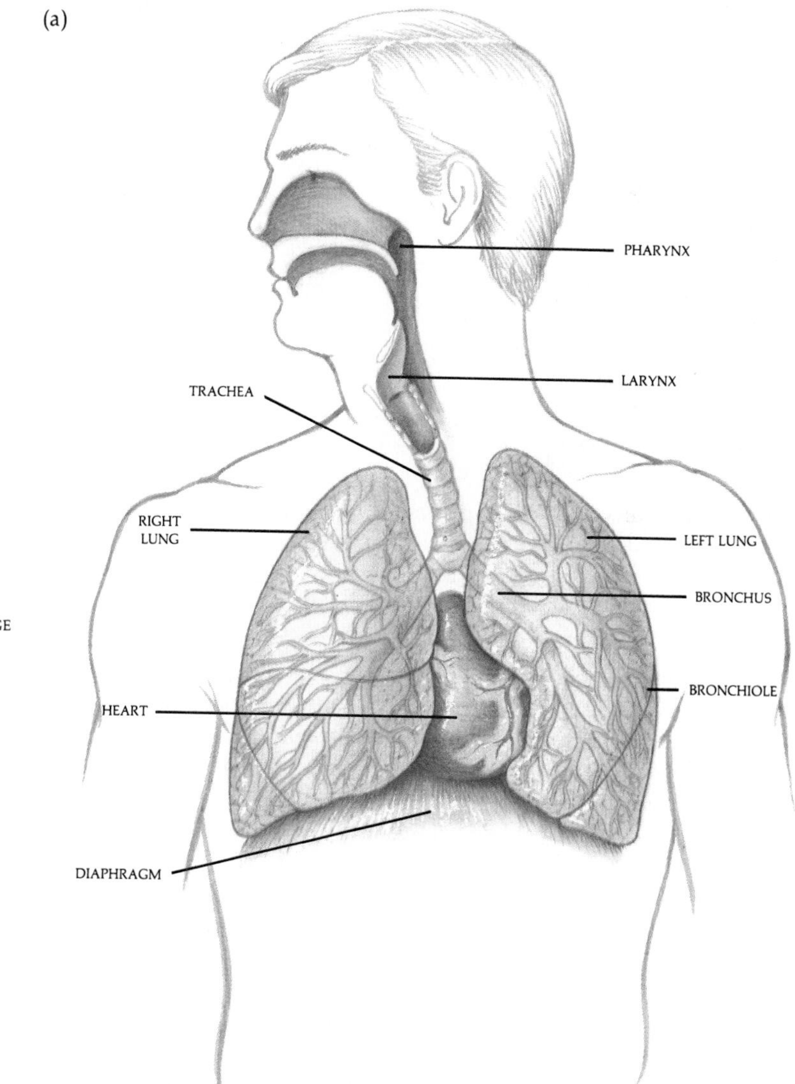

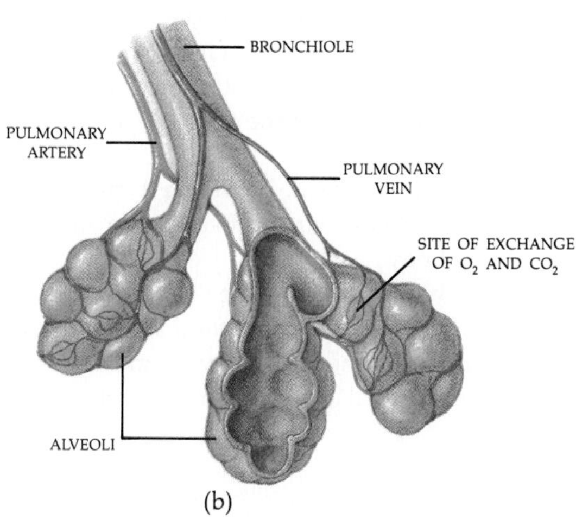

35–9
The human respiratory system. (a) *Air enters through the nose or mouth and passes into the pharynx and down the trachea, bronchi, and bronchioles to the alveoli* (b) *in the lungs. The alveoli, of which there are some 300 million in a pair of lungs, are the sites of gas exchange. Oxygen and carbon dioxide diffuse into and out of the bloodstream through the walls of the alveoli into and out of the capillaries.*

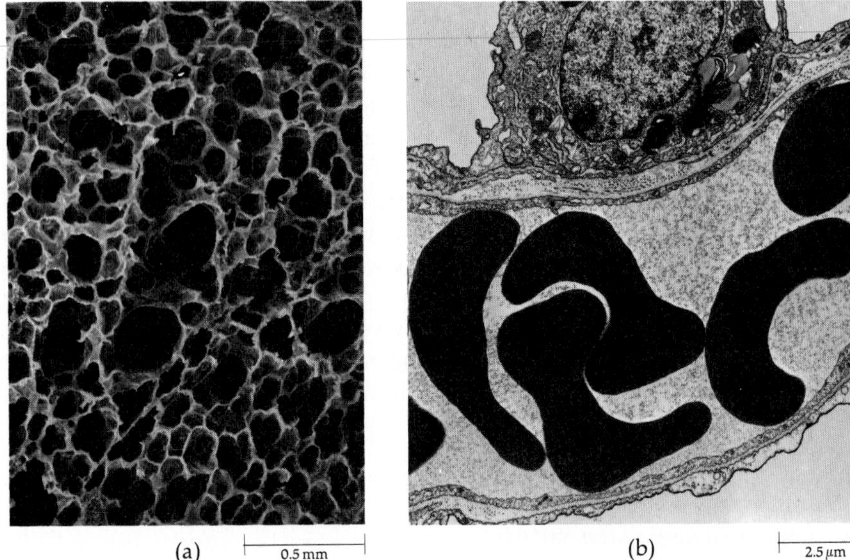

35–10
(a) *Scanning electron micrograph of lung tissue, showing numerous alveoli.* (b) *Longitudinal section of an alveolar capillary. The large, dark, irregularly shaped structures are red blood cells. A portion of an alveolar cell is visible at the top.*

Cilia along the trachea, bronchi, and bronchioles beat continuously, pushing mucus and foreign particles embedded in mucus up toward the pharynx, from which it is generally swallowed. We are usually aware of this production of mucus only when it is increased above normal as a result of an irritation of the membranes.

The actual exchange of gases takes place in small air sacs, the *alveoli,* which are clustered in bunches like grapes around the ends of the smallest bronchioles. Each alveolus is about 0.1 or 0.2 millimeter in diameter, and each is surrounded by capillaries. The walls of the capillaries and of the alveoli each consist of only a single layer of flattened epithelial cells separated from one another only by a thin (basement) membrane; thus the barrier between an alveolus and its capillaries is only about 0.3 micrometer. Gases are exchanged between the air in the alveoli and the blood in the capillaries by diffusion.

A pair of human lungs has about 300 million alveoli, providing a respiratory surface of some 70 square meters, or about 750 square feet—approximately 40 times the surface area of the entire human body.

The lungs are surrounded by a thin membrane known as the pleura, which also lines the thoracic cavity. The pleura secretes a small amount of fluid that lubricates the surfaces so that they slide past one another as the lungs expand and contract. Pleurisy is an inflammation of these membranes that causes them to secrete excess fluid that collects in the thoracic cavity.

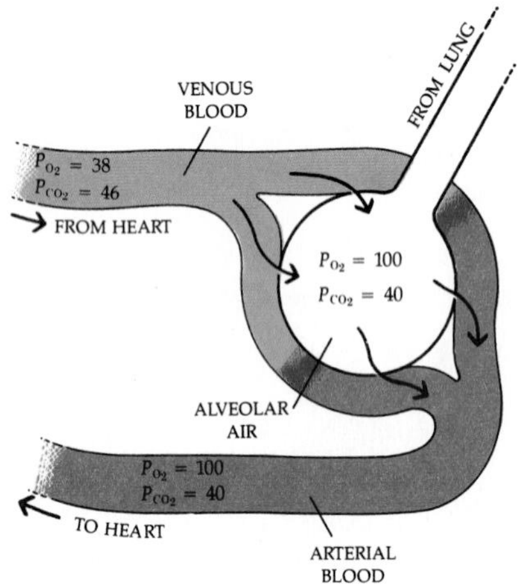

35–11
Gases are exchanged by diffusion as a consequence of the different partial pressures of oxygen and carbon dioxide in the alveolus and the alveolar capillary. The figures indicate millimeters of mercury.

MECHANICS OF RESPIRATION

Air flows into or out of the lungs when the air pressure within the alveoli differs from the pressure of the external air (atmospheric pressure). When alveolar pressure is greater than atmospheric pressure, air flows out of the lungs, and expiration occurs. When alveolar pressure is less than atmospheric pressure, air flows into the lungs, and inspiration occurs.

CANCER OF THE LUNG

Lung cancer is the most common cause of death from cancer among men and is the most rapidly increasing form of cancer in the United States. In fact, between 1950 and 1982, the overall cancer death rate increased in the United States due entirely to an increase in the number of deaths from lung cancer.

About 129,000 men and women will develop lung cancer in the United States this year, and more than 90 percent of them will die in less than three years. Most of these patients will be cigarette smokers. The death rate from lung cancer among U.S. males and females more than tripled during the period from 1947 to 1982.

The middle and lower lobes of a cancerous lung are shown at the right. The cancer is the solid grayish-white mass. Its rapid growth has replaced most of the normal lung tissue in the middle lobe. It has also probably begun to spread to other parts of the body. Notice the bronchial branch that leads into the cancer and is destroyed by it. The lung tissue remaining around the cancer is compressed, airless, and dark red. Away from the cancer, in the lower lobe, the lung is filled with air and is light pink.

The lung is normally protected by ciliated cells lining the trachea and bronchi. The cilia catch and sweep out particulates in the respired air. Cancer cells arising in the bronchi generally do not have cilia.

The pressure in the lungs is varied by changes in the volume of the thoracic cavity. These changes are brought about by the contraction and relaxation of the muscular diaphragm and of the intercostal ("between-the-ribs") muscles. We inhale by contracting the dome-shaped diaphragm, which flattens it and lengthens the thoracic cavity, and by contracting those intercostal muscles that pull the rib cage up and out. These movements enlarge the thoracic cavity; the pressure within it falls, and air moves into the lungs. Air is forced out of the lungs as the muscles relax, reducing the volume of the thoracic cavity. Usually, only about 10 percent of the air in the lung cavity is exchanged at every breath, but as much as 80 percent can be exchanged by deliberate deep breathing.

Whales and other large aquatic mammals suffocate on land because of the inability of their intercostal muscles to expand their massive chests to inhale when compressed under the weight of their bodies.

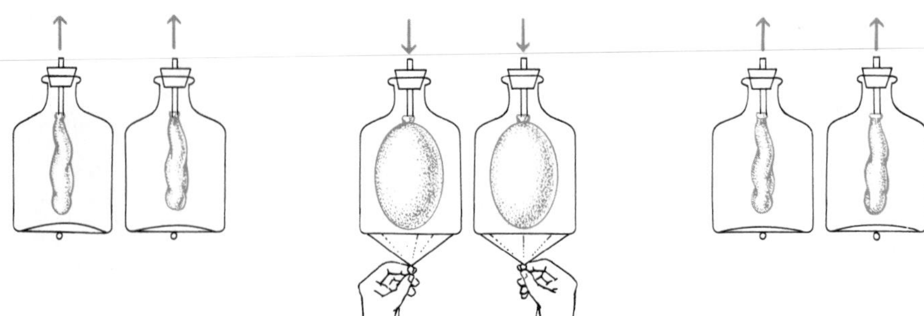

35-12

A model illustrating the way air is taken into and expelled from the lungs, showing the action of the diaphragm.

TRANSPORT AND EXCHANGE OF GASES

Hemoglobin and Its Function

Oxygen is relatively insoluble in blood plasma (the liquid part of blood)—only about 0.3 milliliter of oxygen will dissolve in 100 milliliters of plasma at normal atmospheric pressure. All active animals—and this includes even the earthworm—have in their blood special oxygen-carrying protein molecules, known as respiratory pigments, that raise the oxygen-transporting capacity of blood as much as seventyfold. (The only exception is some Antarctic fishes that have no respiratory pigments at all.) Hemoglobin is the respiratory pigment found among vertebrates and in a wide variety of invertebrate species representing many different phyla. A form of oxygen carrier (hemocyanin), which contains copper rather than iron, is the most common respiratory pigment of mollusks and arthropods. (Unlike hemoglobin, which is red, hemocyanin is blue when combined with oxygen.) Other respiratory pigments are known, all of which are a combination of a metal-containing unit and a protein chain. In most invertebrates, respiratory pigments are simply dissolved in the blood plasma; in vertebrates and echinoderms, the pigments are carried in red blood cells.

Hemoglobin, as you will recall, is made up of four subunits, each of which comprises a heme unit and a polypeptide chain (Figure 3-27). The heme unit consists of a porphyrin ring with one atom of iron in the center. The iron in each heme unit can combine with one molecule of oxygen; thus each hemoglobin molecule can carry four molecules of oxygen. The oxygen molecules (O_2) are added one at a time:

$$Hb_4 + O_2 \rightleftharpoons Hb_4O_2$$

$$Hb_4O_2 + O_2 \rightleftharpoons Hb_4O_4$$

$$Hb_4O_4 + O_2 \rightleftharpoons Hb_4O_6$$

$$Hb_4O_6 + O_2 \rightleftharpoons Hb_4O_8$$

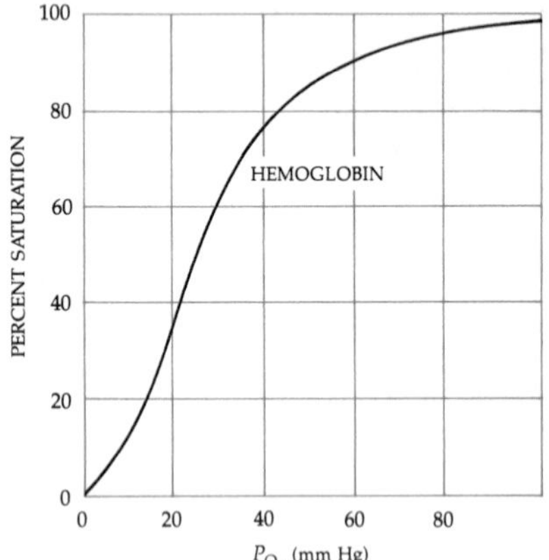

35-13

Oxygen-hemoglobin association-dissociation curve. This curve represents figures for normal adult human hemoglobin at 38°C and at a normal pH. As the partial pressure of oxygen rises, hemoglobin picks up oxygen. As the P_{O_2} drops, the oxygen and the hemoglobin dissociate.

Combination of the first subunit (Hb) with O_2 increases the affinity of the second for O_2, and oxygenation of the second increases the affinity of the third, and so on. (As O_2 is taken up, the two beta polypeptide chains of hemoglobin move closer together, and this movement is apparently the reason the shift in affinity takes place.) As a consequence of the change in affinity of the hemoglobin molecule for oxygen, the curve relating the uptake of oxygen to P_{O_2} is not a straight line, but has a characteristic sigmoid shape (Figure 35-13). When hemoglobin is fully oxygenated, it enables our bloodstreams to carry about 60 times as much oxygen as could be transported by an equal volume of plasma alone.

Whether oxygen combines with hemoglobin or is released from it depends on the partial pressure of oxygen (P_{O_2}) in the surrounding blood plasma. Oxygen diffuses from the air into the alveolar capillaries. In these capillaries, where the P_{O_2} is high, most of the hemoglobin is combined with oxygen. In the tissues, however, where the P_{O_2} is lower, oxygen is released from the hemoglobin molecules into the plasma and diffuses into the tissues. This system compensates automatically for the oxygen requirements of the tissues. For example, in adult humans, the P_{O_2} as the blood leaves the lungs is about 100 mm Hg; at this pressure, the hemoglobin is saturated with oxygen. As the hemoglobin molecules travel through the tissue capillaries, the P_{O_2} drops, and as it drops, the oxygen bound to the hemoglobin molecules is given up. Little oxygen is yielded as the P_{O_2} drops from 100 mm Hg to 60 mm Hg. (This built-in safety factor ensures that oxygen is delivered to the tissues, particularly in individuals who live at high altitudes or who have heart and lung diseases that result in lower blood P_{O_2}.) However, as the P_{O_2} drops below 60 mm Hg, oxygen is given up much more readily (Figure 35–14).

The P_{O_2} of the blood in the tissue capillaries is normally about 40 mm Hg. As a consequence, when the blood leaves the capillaries, its hemoglobin is still usually 70 percent saturated. The oxygen still carried by hemoglobin represents a reserve supply of this gas should the demand increase—as a result, for example, of exercise.

Carbon dioxide is more soluble than oxygen in the blood, and a small amount of it is simply dissolved in the plasma. About another 25 percent is bound to the amino groups of the hemoglobin molecule. However, most of the carbon dioxide (about 65 percent) is carried in the blood as bicarbonate ion (HCO_3^-). Bicarbonate ion is produced in a two-stage reaction. First, carbon dioxide combines with water to form carbonic acid. This reaction is catalyzed by the enzyme carbonic anhydrase found in red blood cells. Then carbonic acid, a weak acid, dissociates to yield bicarbonate and hydrogen ions:

$$CO_2 + H_2O \xrightleftharpoons{\text{carbonic anhydrase}} \underset{\text{CARBONIC ACID}}{H_2CO_3} \rightleftharpoons \underset{\text{BICARBONATE ION}}{HCO_3^-} + \underset{\text{HYDROGEN ION}}{H^+}$$

35–14

These curves show how the amount of oxygen carried by the hemoglobin is related to oxygen pressure. When oxygen pressure reaches 100 mm Hg—the pressure usually present in the human lung—the hemoglobin becomes totally saturated with oxygen. As the pressure drops, the oxygen bound to the hemoglobin molecule is given up. Therefore, when blood carrying oxygen reaches the capillaries, where pressure is only about 40 mm Hg or less, it gives up some of its oxygen to the tissues. A curve located to the right signifies that the oxygen is given up more readily at a given pressure. (a) *Small animals have higher metabolic rates and so need more oxygen per gram of tissue than larger animals. Therefore, they have blood that gives up oxygen more readily.* (b) *The llama, which lives in the high Andes of South America, has a hemoglobin that enables its blood to take up oxygen more readily at the low atmospheric pressures.* (c) *The fetus must take up all its oxygen from the maternal blood. The hemoglobin of mammalian fetuses has a greater affinity for oxygen than does the hemoglobin of adult mammals, and so the oxygen tends to leave the maternal blood and enter the fetal blood.*

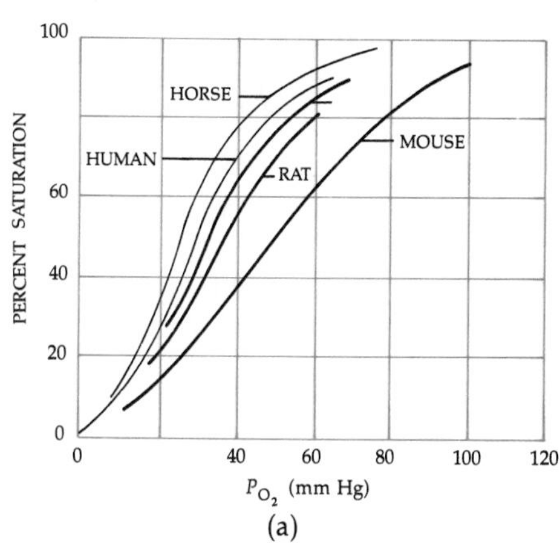

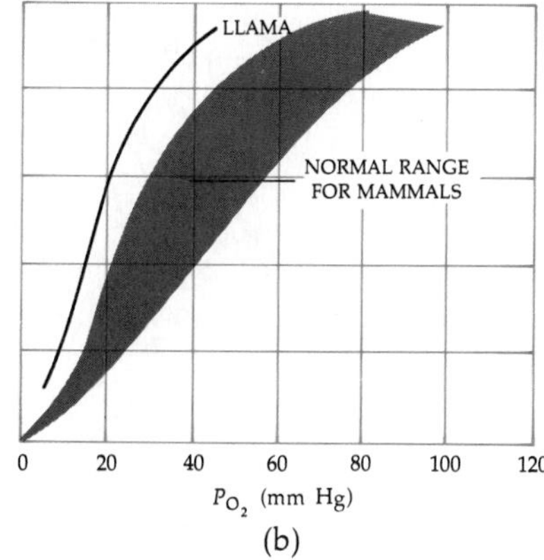

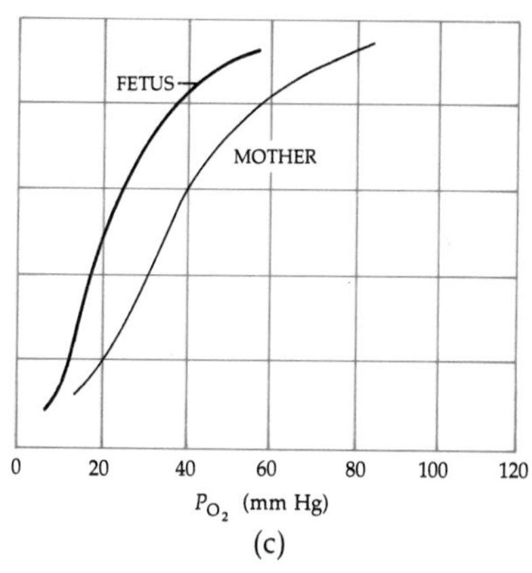

As you can see by the arrows, this reaction can go in either direction. The direction it actually takes depends on the partial pressure of carbon dioxide in the blood. In the tissues, where the partial pressure of carbon dioxide in the blood is low, bicarbonate dissociates to form carbon dioxide and water. The carbon dioxide diffuses from the plasma into the alveoli and flows out of the lung with the expired air.

The partial pressure of carbon dioxide in the blood also affects hemoglobin's affinity for oxygen. In the tissues, where more carbon dioxide is taken up by the blood, the blood becomes increasingly acidic. As the acidity increases, hemoglobin's affinity for oxygen decreases and so it gives up its oxygen more readily.

Table 35-2 illustrates the gas exchange occurring in the lung during the respiratory process.

Table 35-2 Composition of Respiratory Gas at Standard Atmospheric Pressure

	INSPIRED AIR		EXPIRED AIR		ALVEOLAR AIR	
GAS	% OF VOLUME	PARTIAL PRESSURE (mm Hg)	% OF VOLUME	PARTIAL PRESSURE (mm Hg)	% OF VOLUME	PARTIAL PRESSURE (mm Hg)
O_2	20.71	157	14.6	111	13.2	100
CO_2	0.04	0.3	4.0	30	5.3	40
H_2O	1.25	9.5	5.9	45	5.9	45
N_2	78.00	593	75.5	574	75.6	574

Myoglobin and Its Function

Myoglobin is a respiratory pigment found in skeletal muscle. Structurally, it resembles a single subunit of the hemoglobin molecule (Figure 35-15). Myoglobin's affinity for oxygen is greater than hemoglobin's and so it picks up oxygen from

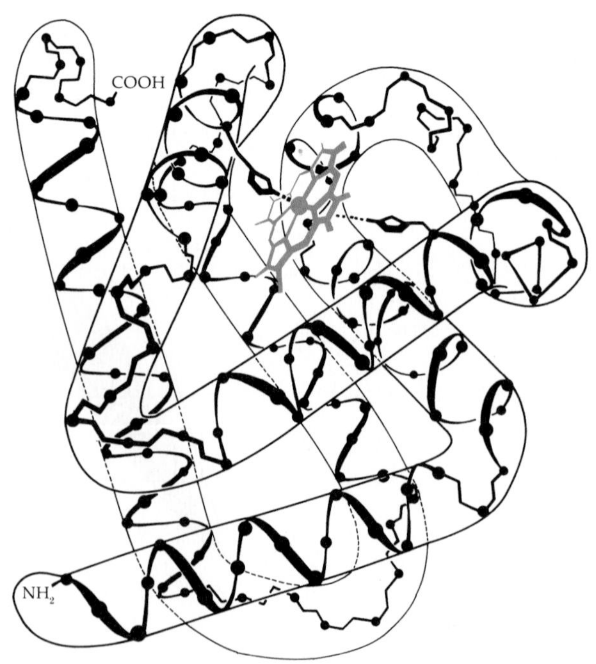

35-15
Tertiary structure of myoglobin, as deduced from x-ray diffraction analyses. Myoglobin closely resembles a single subunit of the four-subunit hemoglobin molecule. The heme group is shown in color.

35–16
Comparison of the oxygen association-dissociation curves of myoglobin and hemoglobin. Note that myoglobin remains 80 percent saturated with oxygen until the partial pressure of oxygen falls below 20 mm Hg. Therefore, myoglobin retains its oxygen in the resting cell and relinquishes it only when strenuous muscle activity uses up the available oxygen provided by hemoglobin.

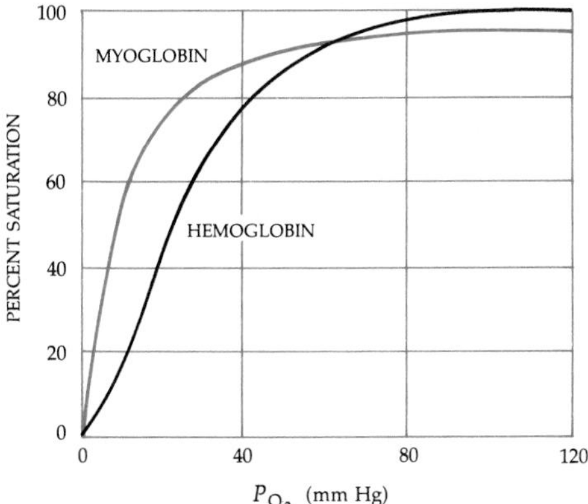

hemoglobin. It begins to release significant amounts of this oxygen only when the P_{O_2} in skeletal muscle falls below 20 mm Hg (Figure 35–16). Thus, when the muscle is at rest or engaged in only moderate activity, the myoglobin holds on to its oxygen. During strenuous exercise, however, when muscle cells are using oxygen rapidly and the partial pressure of oxygen in the muscle cells drops to zero, myoglobin gives up its oxygen. Thus myoglobin provides an additional reserve of oxygen for active muscles.

CONTROL OF RESPIRATION

The rate and depth of respiration are controlled by respiratory neurons in the brainstem. These neurons are responsible for normal breathing, which is rhythmic and automatic, like the beating of the heart. Unlike the beating of the heart, however, which few of us can control voluntarily, breathing may be brought under voluntary control within certain limits.

The respiratory neurons in the brain become active spontaneously. They activate the motor neurons in the spinal cord that cause the diaphragm and intercostal muscles to contract. Periodically the respiratory neurons are inhibited to allow expiration. In addition to their own spontaneous activity, the respiratory neurons receive signals from receptors sensitive to carbon dioxide, oxygen, and hydrogen ions as well as to the degree of stretch of the lungs. Chemoreceptor cells located in the carotid arteries, which supply oxygen to the brain, signal the respiratory neurons when the concentration of oxygen in the blood decreases. These cells also monitor the concentration of dissolved carbon dioxide or hydrogen ion, which is simultaneously monitored by centers in the brain. Thus, a number of regulatory mechanisms are at work.

Control of P_{CO_2} is of overriding importance. If the P_{CO_2} increases only slightly, breathing immediately becomes deeper and faster, permitting more carbon dioxide to leave the blood until the carbon dioxide level has returned to normal. If you deliberately hyperventilate (breathe deeply and rapidly) for a few moments, you will feel faint and dizzy because of the blood's (and, therefore, the brain's) increased alkalinity. You can, as we noted, deliberately increase your breathing rate by contracting and relaxing your chest muscles, but breathing is normally under automatic control. It is impossible to commit suicide by deliberately holding your breath; as soon as you lose consciousness and the CO_2 levels rise, the automatic controls take over once more.

35–17
Location of the carotid body, one of the receptors that monitors the concentration of dissolved oxygen (P_{O_2}) in the blood and also, to a lesser extent, monitors P_{CO_2} and H^+. Carbon dioxide and hydrogen ion concentrations are also measured directly by neurons in the brain.

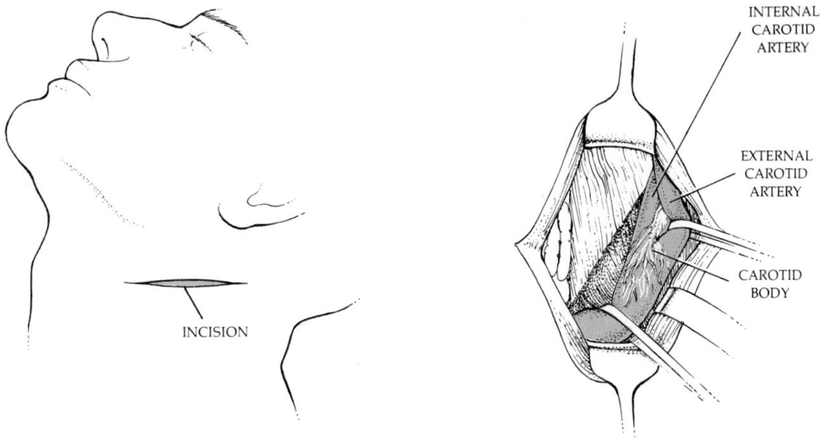

DIVING MAMMALS

Most mammals have about the same oxygen requirements as humans. Structurally, their respiratory organs are very similar; yet marine mammals can dive to great depths and stay submerged for long periods of time. A sperm whale, for example, has been found at a depth of more than 1,000 meters and has been known to stay submerged for 75 minutes. A Weddell seal, a much smaller animal, can dive to 600 meters and stay submerged for 70 minutes, or can swim submerged for between 2,000 and 4,400 meters—1 to 2 miles. Are their lungs different from ours? Or their blood? Do they have special oxygen reserves, like a submarine?

Studies in a variety of diving mammals have shown that none has lungs significantly larger, in proportion, than ours. In fact, seals (and perhaps other diving mammals also) exhale before diving or early in the dive. Blood volume is greater, however. In humans, blood is about 7 percent of the body weight, whereas in diving marine mammals, it is 10 to 15 percent. The blood vessels are proportionately enlarged, and they appear to serve as a reservoir of oxygenated blood. The proportion of red blood cells is higher and myoglobin is more concentrated, giving the muscles a very dark color. Still, physiologists calculate, these adaptations would not provide enough oxygen for long dives.

A major survival factor, studies in diving mammals have shown, is a group of automatic reactions known, collectively, as the diving reflex. During a dive, the heart rate slows and blood supply is reduced to one-tenth or one-twentieth of normal to tissues that are tolerant of oxygen deprivation, such as the digestive organs, skin, and muscles. Muscles obtain energy by anaerobic glycolysis, producing large amounts of lactic acid and building up an oxygen debt (see page 192). Most of the oxygen is shunted to the heart and brain, whose cells would begin to die after about four minutes without oxygen. Blood supply to the adrenal glands is reduced only slightly. In a pregnant female, the fetus similarly is given high priority for the available oxygen.

Recent studies by Martin Nemiroff of the University of Michigan Medical School indicate that the diving reflex is also present in human beings. In humans, its primary survival value is during birth, when the infant may be cut off from an oxygen supply during the later stages of labor. Operation of the reflex can be demonstrated in a very simple way: immersing the face in cold water causes the heart rate to slow down.

These observations by Nemiroff came in the course of investigating about 60 near-drownings. It has been generally assumed that a person who is submerged for four minutes or more will die, or, at best, suffer irreversible brain damage. In the group of 60, 15 had been rescued after four minutes or more in the chilly waters of Michigan lakes. Of these 15, 11 survived without brain damage. One of these, a college student who crashed through the ice in an automobile accident, was under water for 38 minutes; he finished the college semester with a 3.2 average. Another survivor resumed his career as a practicing physician. All the survivors had been submerged in cold (below 21° C) water, which slowed the metabolic demands of their cells and therefore reduced their need for oxygen. Some of those rescued required assisted breathing for as long as 13 hours. Moral: Even if a drowning victim is cold, blue, with no pulse, no heartbeat, and the eyes fixed in a glassy stare—as was the case with the student—if he or she has been in cold water, efforts at resuscitation should be started immediately and continued indefinitely in a hospital. We are not Weddell seals, but owing to an ancient survival strategy, we have unexpected powers of underwater survival.

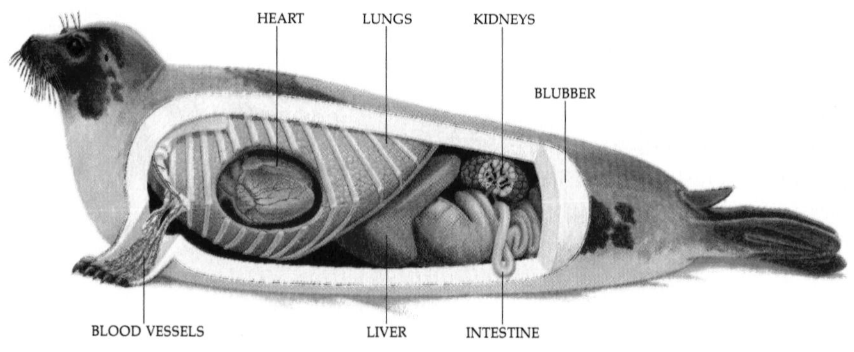

A seal exhales as it dives, reducing the likelihood of decompression sickness (the bends). It has 1.5 times as much blood as a land mammal of the same size and, when it dives, can reduce its heartbeat from 120 to 30 beats per minute.

The receptor cells sensitive to oxygen concentration provide a kind of back-up system for the carbon dioxide sensors. In some cases of drug poisoning—for example, by morphine or barbiturates—the brainstem cells sensitive to carbon dioxide become depressed. This causes a decrease in the breathing rate, leading ultimately to a reduction in the oxygen concentration in the blood. The oxygen sensors are then stimulated, and they maintain breathing. Massive overdoses of these drugs, however, depress the activity of the respiratory neurons themselves.

SUMMARY

Most organisms obtain energy from oxidation of carbon-containing compounds. This process requires oxygen and releases carbon dioxide. Respiration is the means by which an organism obtains the oxygen required by its cells and rids itself of carbon dioxide.

Oxygen is available in both water and air. It enters into cells and body tissues by diffusion, moving from regions of higher partial pressure to regions of lower partial pressure. However, movement of oxygen by diffusion requires a relatively large surface area exposed to the source of oxygen and a short distance over which the oxygen has to diffuse. Selection pressures for increasingly efficient means of gas exchange led to the evolution in vertebrates of gills and lungs. Both gills and lungs present enormously increased surface areas for the exchange of gases. They also have a rich blood supply for transporting these gases to and from other parts of the animal's body. Respiration in large animals involves both diffusion and bulk flow. Bulk flow brings air or water to the lungs or gills and circulates oxygen and carbon dioxide in the bloodstream. Gases are exchanged between the blood and the air in the lungs or the water around the gills and between the blood and the tissues by diffusion.

In humans, air enters the lungs through the trachea, or windpipe, and goes from there into a network of increasingly smaller tubules, the bronchi and bronchioles, which terminate in small air sacs, the alveoli. Gas exchange actually takes place across the alveolar walls. Air moves in and out of the lungs as a result of changes in the pressure within the lungs, which, in turn, result from changes in the size of the thoracic cavity. The muscles that bring about respiration are under the control of neurons that originate in the spinal cord.

Respiratory pigments increase the oxygen-carrying capacity of the blood. In vertebrates, the respiratory pigment is hemoglobin. Each hemoglobin molecule has four subunits, each of which can combine with one molecule of oxygen. The addition of each molecule of oxygen increases the affinity of the molecule for each subsequent molecule of oxygen. Conversely, the loss of each molecule of oxygen facilitates the loss of the subsequent one.

The rate and depth of respiration are controlled by respiratory neurons in the brainstem that respond to signals caused by very slight changes in the carbon dioxide or hydrogen ion concentration of the blood. They are responsive to larger changes in oxygen concentration.

QUESTIONS

1. Distinguish among the following: gills/lungs; hemoglobin/myoglobin/hemocyanin; trachea/pharynx/larynx; bronchi/bronchioles/alveoli.

2. Why are calculations of partial pressure used in determining the movement of gases between liquids and the atmosphere rather than concentrations? When does a gas move from an area of lower concentration to one of higher concentration?

3. Sketch and label a diagram of the human respiratory system. When you have finished, compare your drawing to Figure 35–9.

4. What are the advantages and disadvantages of obtaining oxygen from air rather than from water? You might be able to think of several of each besides those mentioned in the text.

5. Carbon monoxide (CO) attaches to hemoglobin; the resulting compound does not readily dissociate, can no longer combine with oxygen, and is a brighter red than normal hemoglobin. From these facts, suggest how you might recognize and give assistance to a victim of carbon monoxide inhalation.

6. One of the results of long-term smoking is the loss of bronchial cilia. What effects would you expect this to have on normal lung function?

CHAPTER 36

Energy and Metabolism III: Circulation of the Blood

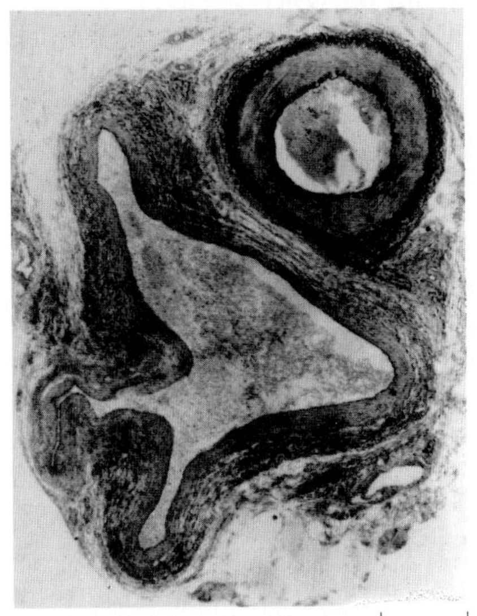

36–1
A large vein and a small artery. The artery can be identified by its thick, muscular wall.

The blood is the medium in which the nutrient molecules processed by digestion and the oxygen molecules taken in by way of the lungs (or gills) are delivered to the individual cells. The blood also carries a number of other important substances, such as hormones, enzymes, and antibodies, and waste materials, including urea and carbon dioxide.

In vertebrates, the blood is propelled by a specialized muscle, the heart, through a closed system of continuous vessels. The heart and vessels together are known as the *cardiovascular system* (from *cardio,* meaning "heart," and *vascular,* meaning "vessel"). In this system, the heart pumps the blood into the large arteries, from which it travels to branching, smaller arteries (the *arterioles*) and then into very small vessels, the *capillaries.* Through the thin walls of the capillaries, nutrients, oxygen, carbon dioxide, and other molecules are exchanged between the blood and the fluids surrounding the body cells (the interstitial fluids). From the capillaries, the blood passes into small veins, the *venules,* then into larger veins, and through them, back to the heart. Thus the heart, arteries, and veins are, in essence, the means for getting the blood to and from the capillaries, where the actual function of the circulatory system is carried out.

COMPOSITION OF THE BLOOD

Blood Plasma

An individual weighing 75 kilograms (165 pounds) has about 6 liters of blood. (Blood is about 8 percent of body weight.) About 60 percent of it is a straw-colored liquid called *plasma,* which is 90 percent water. With the exception of the oxygen and carbon dioxide carried by hemoglobin, most of the molecules needed by the individual cells, as well as waste products from the cells, are carried along in the heavy traffic of the bloodstream dissolved in the plasma. In addition, the plasma contains plasma proteins; these differ from the other molecules carried in the plasma in that they are not nutrients for or waste products of tissue cells, but function in the bloodstream itself.

Because of the dissolved plasma proteins, the osmotic potential of the blood is greater than that of the interstitial fluid. These proteins act to prevent excessive loss of fluid from the bloodstream to the tissues. They also serve to bind certain ions and small molecules, thus preventing them from leaving the bloodstream, and to transport fats, cholesterol, and otherwise insoluble molecules in the bloodstream.

Plasma proteins are of three major types: albumin, whose chief function is to maintain the plasma hyperosmotic to the interstitial fluid; globulins, a key group of which act to defend the body against foreign invaders; and fibrinogen, which is responsible for blood clotting. (Plasma from which fibrinogen has been removed as a result of clotting is blood serum.)

Blood Cells

The other 40 percent of the blood that is not plasma is made up of the red blood cells (erythrocytes), white blood cells (leukocytes), and platelets.

Red Blood Cells (Erythrocytes)

Red blood cells are specialized for the transport of oxygen. As a mammalian red blood cell matures, it extrudes its nucleus and mitochondria, and its other cellular structures dissolve. Almost the entire volume of a mature red blood cell is filled with hemoglobin, about 300 million molecules per cell. There are about 5 million red blood cells per cubic millimeter of blood—some 25 trillion (25×10^{12}) in the whole body. Because red blood cells, lacking a nucleus, cannot repair themselves, their life span is comparatively short, some 120 to 130 days. At this moment, in your body, red blood cells are dying at a rate of about 2 million per second and new ones are being formed to replace them at the same rate.

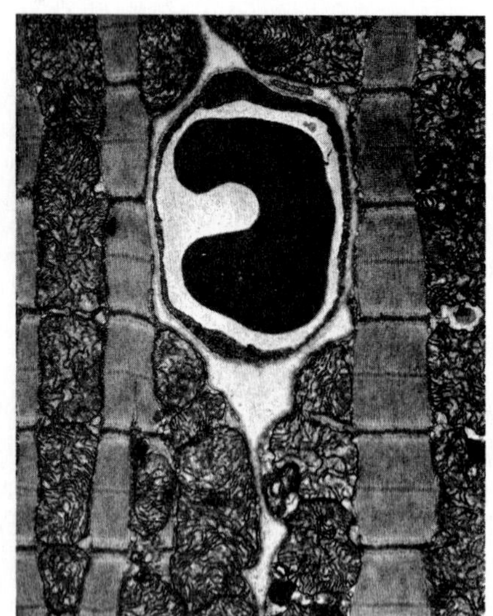

36–2
Electron micrograph of cardiac muscle, showing a red blood cell within a capillary in the muscle. The striated pattern of the cardiac muscle is clearly visible. Note also the large and numerous mitochondria in the muscle and the extremely thin wall of the capillary. The heart, like all the other tissues of the body, is dependent on a fine capillary network for its supply of oxygen and other blood-borne materials.

White Blood Cells (Leukocytes)

There are about 6,000 to 9,000 white blood cells per cubic millimeter of blood—1 or 2 white blood cells for every 1,000 red blood cells. These cells are nearly colorless, are larger than red blood cells, contain no hemoglobin, and have a nucleus. Unlike red blood cells, white blood cells are not confined to the vascular system but can migrate out into the interstitial fluid. As we shall see in Chapter 39, they play a major role in defending the body against viruses, bacteria, and other invading pathogens.

Platelets

Platelets, so called because they look like little plates, are colorless, round or biconcave disks smaller than red blood cells (about 3 micrometers in diameter). Platelets are cytoplasmic fragments of unusually large cells, megakaryocytes, found in the bone marrow. They are, in effect, little bags of chemicals that initiate the clotting of blood and the aggregation of other platelets.

Blood Clotting

Blood clotting is a complex phenomenon; at least 15 factors involved in the process have been identified. It is not initiated when blood encounters air, as was reasonably assumed for a long period of time. The sequence of events begins when plasma encounters a rough surface, such as a torn tissue. This stimulates a chain of chemical reactions resulting in the activation of a substance called thromboplastin. Thromboplastin acts to convert the enzyme prothrombin, a plasma protein produced in the liver, to its active form, thrombin:

$$\text{prothrombin} \xrightarrow{\text{thromboplastin}} \text{thrombin}$$

Platelets as well as several other factors normally present in the bloodstream are required for this series of reactions, which involves several enzymatic steps.

36–3
Beginning of the formation of a clot. Red blood cell enmeshed in fibrin.

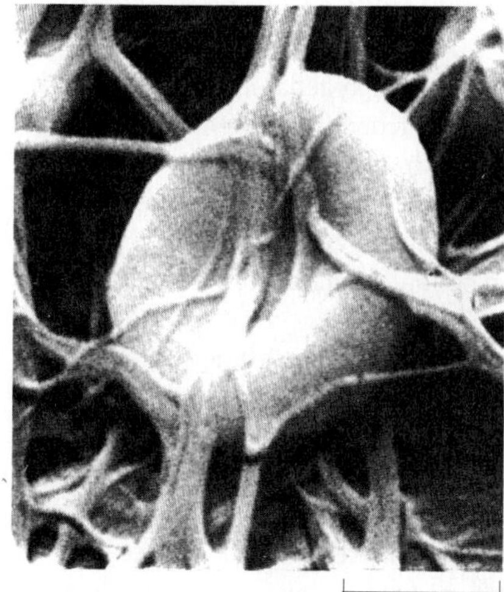

Thrombin converts fibrinogen, a soluble plasma protein, to fibrin:

$$\text{fibrinogen} \xrightarrow{\text{thrombin}} \text{fibrin}$$

The fibrin molecules clump together, forming an insoluble network that enmeshes red blood cells and platelets to form a clot (Figure 36–3). The clot contracts, pulling together the edges of the wound. When blood is removed from the body and placed in a glass test tube, it congeals to form a clot. The clot eventually contracts, leaving a clear fluid, the serum.

Hemophilia is a genetically determined disease that affects blood clotting (see page 355). In the most common type of hemophilia, one of the factors involved in the chain of reactions required to activate prothrombin, Factor VIII, occurs in a defective form. As a result, the blood does not clot easily. These hemophiliacs can be treated by administration of Factor VIII extracted from normal blood.

THE HEART

Evolution of the Heart

In its simplest form—as in the earthworm—the heart is a muscular contractile part of a circulatory vessel. In the course of vertebrate evolution, the heart has undergone some structural adaptations, as shown in Figure 36–4.

Fish have a single heart divided into an atrium, the receiving area for the blood, and a ventricle, the pumping area from which the blood is expelled into the vessels.

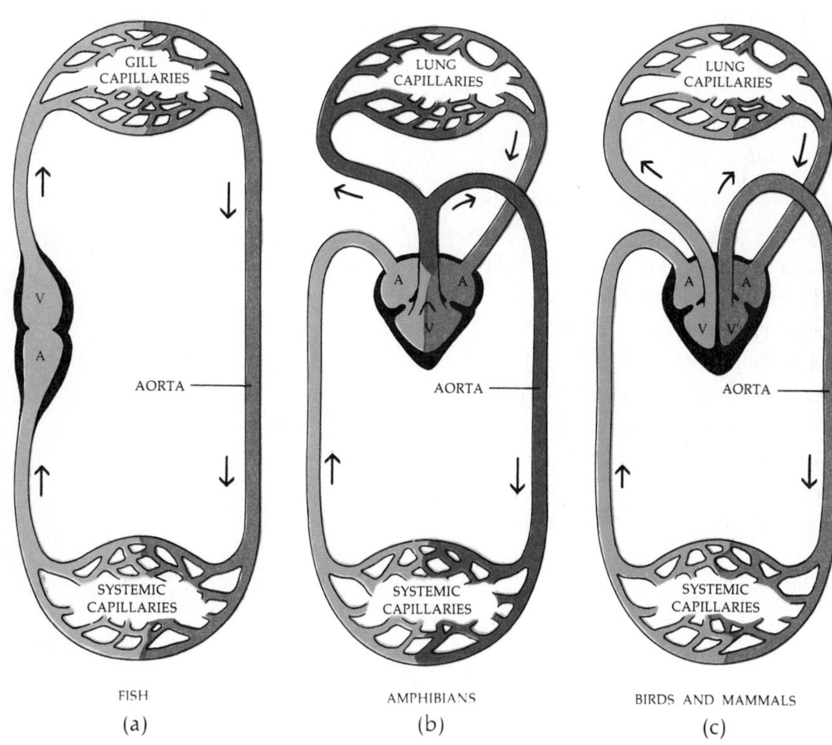

36–4
Vertebrate circulatory systems. Oxygenated blood is indicated in color. (a) *In the fish, the heart has only one atrium (A) and one ventricle (V). Blood oxygenated in the gill capillaries goes straight to the systemic capillaries without first returning to the heart.* (b) *In amphibians, the single primitive atrium has been divided into two separate chambers. Oxygen-rich blood from the lungs enters one atrium, and oxygen-poor blood from the tissues enters the other. Little mixing of the blood occurs in the ventricle, despite its lack of a structural division. From the ventricle, oxygen-rich blood is pumped to the body tissues at the same time oxygen-poor blood is pumped to the lungs. Branches from the pulmonary circulation carry oxygen-poor blood to the skin, a major respiratory organ in amphibians.* (c) *In birds and mammals, the atrium and the ventricle are divided into two separate chambers, so that there are, in effect, two hearts—one for pumping oxygen-poor blood through the lungs and one for pumping oxygen-rich blood through the body tissues.*

The ventricle of the fish heart pumps blood directly into the capillaries of the gills, where it picks up oxygen and releases carbon dioxide. From the gills, oxygenated blood is carried to the tissues. By this time, however, most of the propulsive force of the heartbeat has been dissipated by the resistance of the capillaries in the gills, so that the blood flow through the rest of the tissues (the systemic circulation) is relatively sluggish.

In amphibians, there are two atria; one receives oxygenated blood from the lungs, and the other receives deoxygenated blood from the systemic circulation. Both atria empty into a single ventricle; the two kinds of blood remain relatively unmixed, although the ventricle is not divided. The oxygenated blood is pumped into the systemic circulation at the same time the deoxygenated blood is pumped through the lungs. The pulmonary circulation sends branches to the skin, a major site of oxygen uptake in amphibians.

In birds and mammals, the heart is functionally separated longitudinally into two organs, the right heart and the left heart, each with an atrium and a ventricle. The right heart receives blood from the tissues and pumps it into the lungs, where it becomes oxygenated. From the lungs it returns to the left heart, from which it is pumped into the body tissues. This efficient, high-pressure circulation system, with its full separation of oxygenated and deoxygenated blood, is necessary to maintain the high metabolic rate of both birds and mammals, with their constant body temperature and their generally high level of physical and mental activity.

The Human Heart

Figure 36–5 shows a diagram of the human heart. Its walls are made up predominantly of a specialized type of muscle—cardiac muscle. Blood returning from the body tissues enters the right atrium through two large veins, the superior and inferior *venae cavae*. Blood returning from the lungs enters the left atrium through the pulmonary veins. The atria, which are thin-walled compared to the ventricles, expand as they receive the blood. Both atria then contract simultaneously, assisting

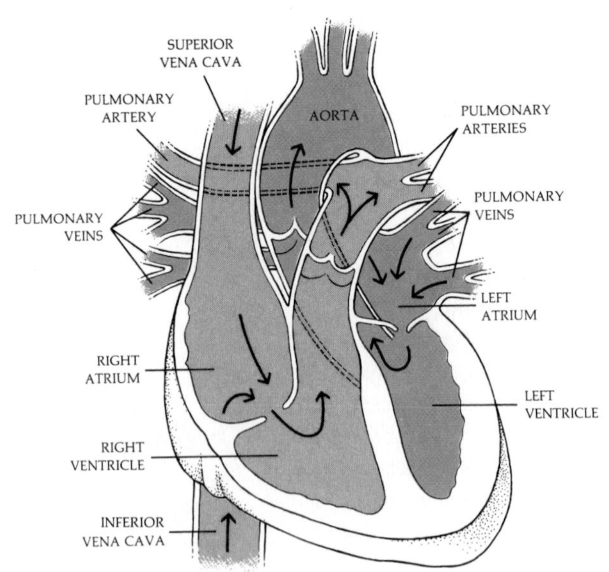

36–5
The human heart. Blood returning from the systemic circulation through the superior and inferior venae cavae enters the right atrium and from there passes to the right ventricle, which propels it through the pulmonary arteries to the lungs, where it is oxygenated. Blood from the lungs enters the left atrium through the pulmonary veins and from there passes to the left ventricle and then through the aorta to the body tissues.

the flow of blood through the open valves into the ventricles. Then the ventricles contract simultaneously; the valves between the atria and ventricles are closed by the pressure of the blood in the ventricles. The right ventricle propels the deoxygenated blood into the lungs through the pulmonary arteries; the left ventricle propels oxygenated blood into the *aorta,* from which it travels to the other body tissues. Valves between the ventricles and the pulmonary artery and the aorta close after the ventricles contract, thus preventing backflow of blood.

A heart has an inherent capacity to contract without external stimulation. A vertebrate heart will continue to contract even after it is removed from the body if it is kept in an oxygenated nutrient solution. In vertebrate embryos, the heart begins to beat very early in development, before the appearance of any nerve supply. In fact, embryonic heart cells isolated in a test tube will beat.

Contraction of the heart is initiated by a special area of the heart, the *sinoatrial node,* which is located in the right atrium and functions as the pacemaker (Figure 36–6). This pacemaker is composed of specialized cardiac muscle cells that can spontaneously initiate their own impulse and contract. From the pacemaker the impulse spreads throughout the right and left atria. As it passes along the surface of the individual heart muscle cells, it activates their contractile machinery, and they contract. The impulse travels very quickly, so that the cells of both atria are activated almost simultaneously.

About 100 milliseconds after the pacemaker fires, the impulse reaches and stimulates a second area of nodal tissue, the *atrioventricular node.* The atrioventricular node is the only electrical bridge between the atria and the ventricles. It consists of slow-conducting fibers. Thus the atrioventricular node imposes a delay between the atrial and ventricular contractions, so that the atrial contraction is completed before the beat of the ventricles begins. From the atrioventricular node, impulses are carried rapidly by special muscle fibers, the *bundle of His* (named after its discoverer), to the walls of the right and left ventricles, which then contract almost simultaneously.

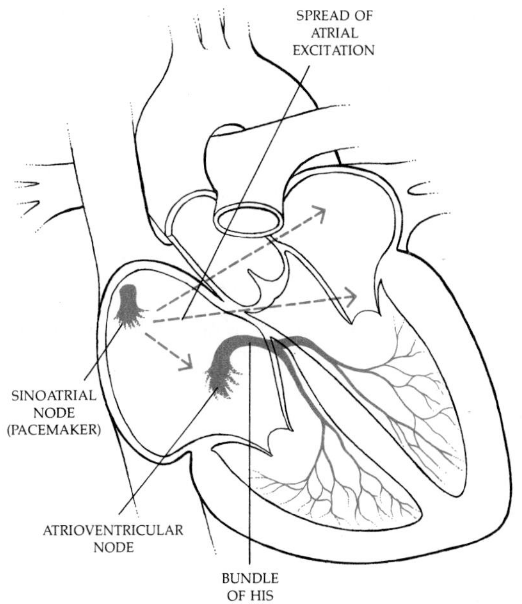

36–6
The conducting system of the heart. The beat of the mammalian heart is controlled by a region of specialized muscle tissue in the right atrium, the sinoatrial node, which functions as the heart's pacemaker. Some of the nerves regulating the heart have their endings in this region. Excitation spreads from the pacemaker throughout the atria, causing both atria to contract almost simultaneously. When the wave of excitation reaches the atrioventricular node, its conducting fibers pass the stimulation to the bundle of His, which triggers almost simultaneous contraction of the ventricles. Because the fibers of the atrioventricular node conduct relatively slowly, the ventricles do not contract until after the atrial contraction has been completed.

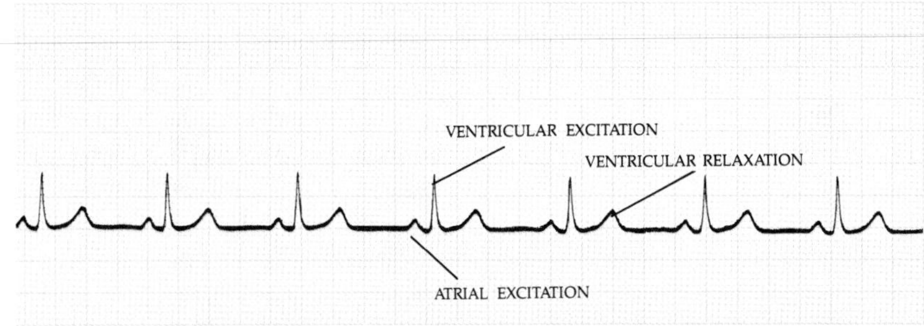

36–7
An electrocardiogram showing seven normal heartbeats. Each beat is denoted by a series of waves that record the electrical activity of the heart during contraction. Analysis of the rate and character of these waves can reveal abnormalities in heart function. Atrial relaxation occurs during ventricular excitation; therefore its record is obscured.

When the impulse from the conducting system travels across the heart, electric current generated on the heart's surface is transmitted to the body fluids, and from there some of it reaches the body surface. Appropriately placed electrodes on the surface connected to a recording instrument can measure this current. The output, an electrocardiogram, is important in assessing the heart's capability to initiate and transmit the impulse. In a normal electrocardiogram (Figure 36–7), the first small hump represents the current generated by the passage of the impulse through the atria, the spike by its passage through the ventricles, and the final hump by the return of the ventricles to their resting state. Atrial relaxation occurs during ventricular excitation; therefore, the record of its electrical activity is obscured.

If you listen to a heartbeat, you hear "lubb-dup, lubb-dup." The deeper, first sound ("lubb") is the closing of the valves between the atria and the ventricles; the second sound ("dup") is the closing of the valves leading from the ventricles to the arteries. If any one of the four valves is damaged, as from rheumatic fever, blood may leak back through the valve, producing the sloshy noise characterized as a "heart murmur."

Although the autonomic nervous system does not initiate the vertebrate heartbeat, it does control its rate. Fibers from the parasympathetic division travel in the vagus nerve (a large nerve that runs through the neck) to the pacemaker. Parasympathetic stimulation has a slowing effect on the pacemaker and thus decreases the rate of heartbeat. Sympathetic nerves stimulate the pacemaker, increasing the rate of heartbeat. Adrenaline from the adrenal medulla affects the heart in the same way that the sympathetic nerves do.

In a healthy adult at rest, the heart beats about 70 times a minute. This rate can more than double during strenuous exercise.

THE BLOOD VESSELS

Structure and Blood Flow

In humans, the diameter of the opening of the largest artery, the aorta, is about 2.5 centimeters, that of the smallest capillary only 8 to 10 micrometers, and that of the largest vein, the vena cava, about 3 centimeters. Arteries, veins, and capillaries differ in the structure of their walls (Figure 36–8). Of the three types of vessels, the arteries have the thickest, strongest walls, made up of three layers. The inner layer, or endothelium (a type of epithelial tissue), forms the lining of the vessels; the middle layer contains smooth muscle and elastic tissues; the outer layer, also elastic, is made of collagen and other supporting tissues. Because of their elasticity,

36–8
Structure of blood vessels. Arteries have thick, tough, elastic walls that can withstand the pressure of the blood as it leaves the heart. Capillaries have walls only one cell thick. Exchange of gases, nutrients, and wastes between the blood and the cells of the body takes place through these thin capillary walls. Veins usually have larger lumens (passageways) and always have thinner, more readily extensible walls that minimize resistance to the flow of blood on its return to the heart.

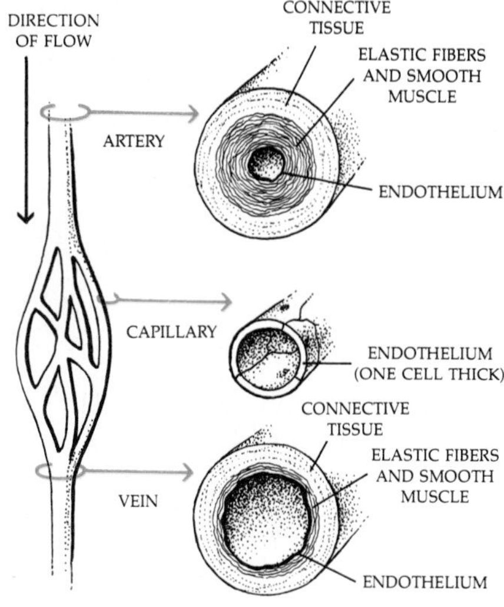

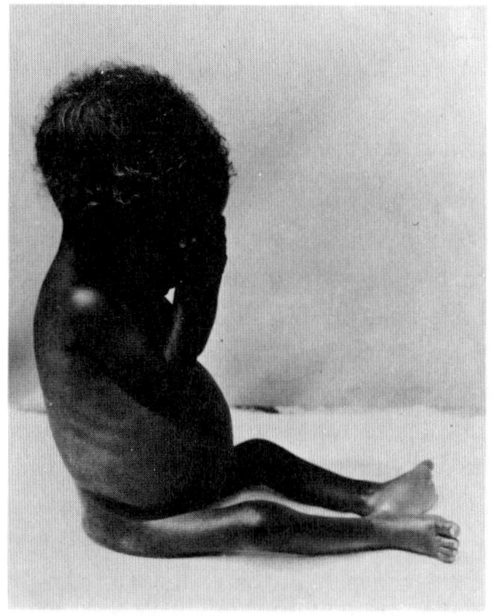

36–9
A child with kwashiorkor, a West African word that means "the sickness a child develops when another child is born." The swelling characteristic of this disease is caused by a deficit in plasma proteins; because of the deficit, fluids are not drawn back into the blood by osmotic forces to the extent that they are in well-nourished persons. Children with kwashiorkor are usually also weak and apathetic and have skin discoloration. The syndrome, which is the result of a deficiency in protein but not in total calories, usually develops after a child is weaned; hence its name.

the arteries are stretched when the blood is pumped into them, and then recoil slowly. Consequently, by the time the blood leaves the arteries, it is flowing smoothly through the vessels, rather than in spurts, as it does when it leaves the heart. The pulsation felt when the fingertips are placed over an artery close to the body surface—as in the wrist—represents the alternating expansion and recoil of an elastic arterial wall.

As we mentioned earlier, the heart pumps blood into the arteries, which branch to form arterioles and then capillaries. At the arteriolar end of the capillaries, the pressure of the circulating blood—the hydrostatic pressure—is greater than the osmotic potential, forcing some of the watery component of blood to leave the capillaries along with nutrient molecules and oxygen. By the time the blood reaches the venous end of the capillaries, the hydrostatic pressure has dropped, due to resistance in the capillary bed, to a point where it is now less than the osmotic potential of the blood. Fluids move back into the capillaries. Consequently, there is little total fluid loss from the blood as it moves through the capillaries. (The relatively small amount that is not returned to the blood by osmosis is channeled back into the bloodstream by the lymphatic system, as we shall see later in this chapter.) When loss of fluid into tissues does occur, as, for example, when the endothelium is damaged by a blow, swelling (edema) is produced. The edema associated with certain forms of malnutrition is also due to the movement of fluids out of the capillaries (Figure 36–9).

The walls of the capillaries consist of only one layer of cells (Figure 36–10), the endothelium, and the lumen (passageway) of the smallest capillary is about 8 to 10 micrometers in diameter, just wide enough for red blood cells to move in single

36–10
Electron micrograph of a capillary from cardiac muscle. Portions of two endothelial cells can be seen, fitting together to enclose the capillary lumen.

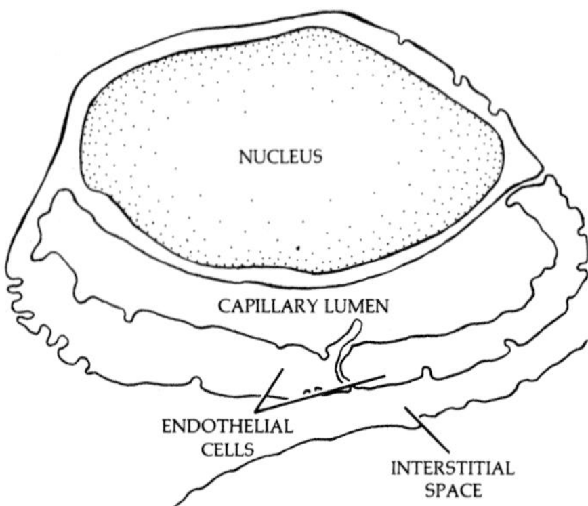

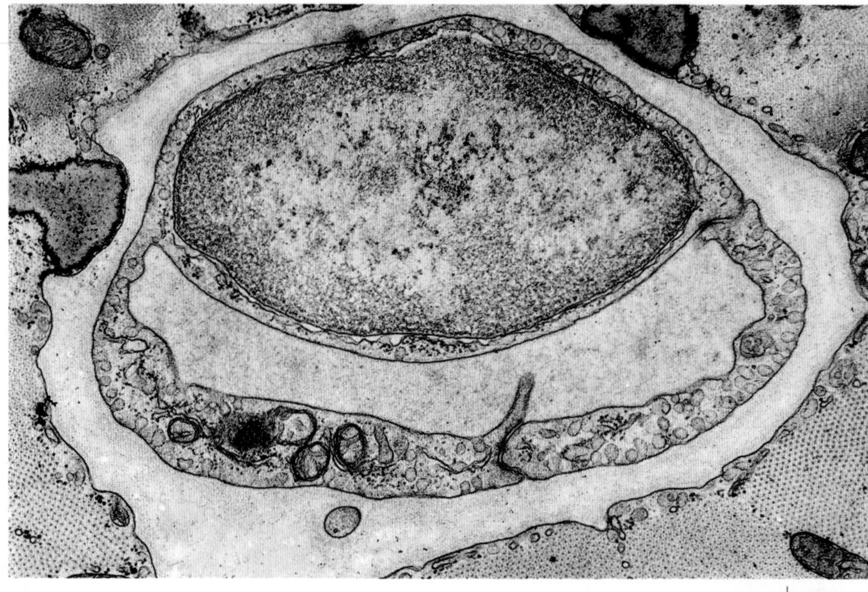

file. The total length of the capillaries in a human adult is more than 80,000 kilometers. Because of both of these facts, blood moves slowly through the capillary system. As it moves, gases (oxygen and carbon dioxide), hormones, and other materials are exchanged with the surrounding tissues by diffusion through junctions between the endothelial cells and through the cytoplasm of the endothelial cells. Organic molecules, such as glucose, are probably moved by transport systems of the endothelial cells. Pinocytotic vesicles (page 135) have been observed in the endothelial cells; they presumably serve to ferry dissolved material into or out of the capillaries.

No cell in the human body is farther than 130 micrometers—a distance short enough for rapid diffusion—from a capillary. Even the cells in the walls of the large veins and arteries depend on this capillary system for their blood supply, as does the heart itself, like all the other organs of the body.

At the venous end of the capillary beds, blood passes into the venules, smallest of the veins, then into larger veins, and finally back into the heart, either through the superior (anterior) or inferior (posterior) vena cava. Like those of the arteries, the walls of the veins are three-layered, but they are less elastic and more pliable. An empty vein collapses, whereas an empty artery remains open. The veins, with their thin walls and relatively large diameters, offer little resistance to flow, facilitating the return of the blood to the heart.

36–11
Valves in the veins open to permit movement of blood toward the heart but close to prevent backflow.

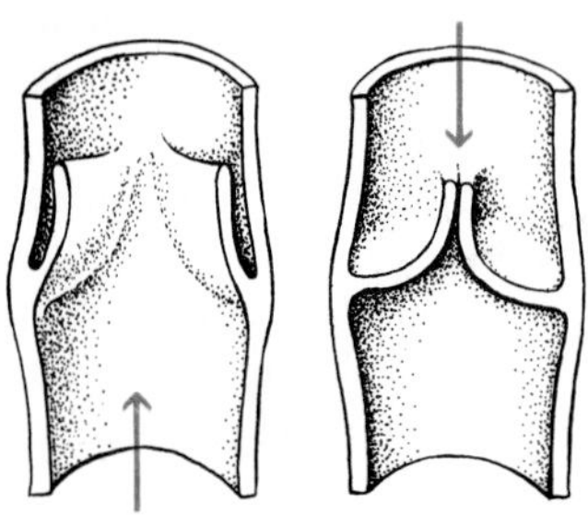

There is a large drop in hydrostatic pressure as the blood goes through the capillaries. The return of blood to the heart through the veins is enhanced by body movements, which squeeze the veins between contracting muscles, raise the pressure within the veins, and increase the blood flow back to the heart. (If you have to stand still for long periods of time, try contracting your leg muscles periodically to move blood back toward the heart and prevent blood pooling.) Valves in the veins prevent backflow (Figure 36–11). Also, as the thorax expands on inspiration, the elastic walls of the veins in the thorax dilate, venous pressure in the region of the heart decreases, and return of blood to the heart increases.

The Vascular Circuits

There are two principal circuits in the vascular system of air-breathing vertebrates: the pulmonary circuit and the systemic circuit (Figure 36–12). In the pulmonary circuit of mammals, deoxygenated blood leaves the right ventricle of the heart through the pulmonary artery. This artery divides into right and left branches, which carry the blood to the right and left lungs, respectively. Within the lungs, the arteries divide into smaller and smaller vessels that finally become capillaries through which oxygen and carbon dioxide are exchanged. Blood flows from the capillaries into small venules and then into larger and larger veins, finally draining into the four pulmonary veins that carry the blood, now oxygenated, to the left atrium of the heart. The pulmonary arteries are the only arteries that carry fully deoxygenated blood, and the pulmonary veins are the only veins that carry fully oxygenated blood.

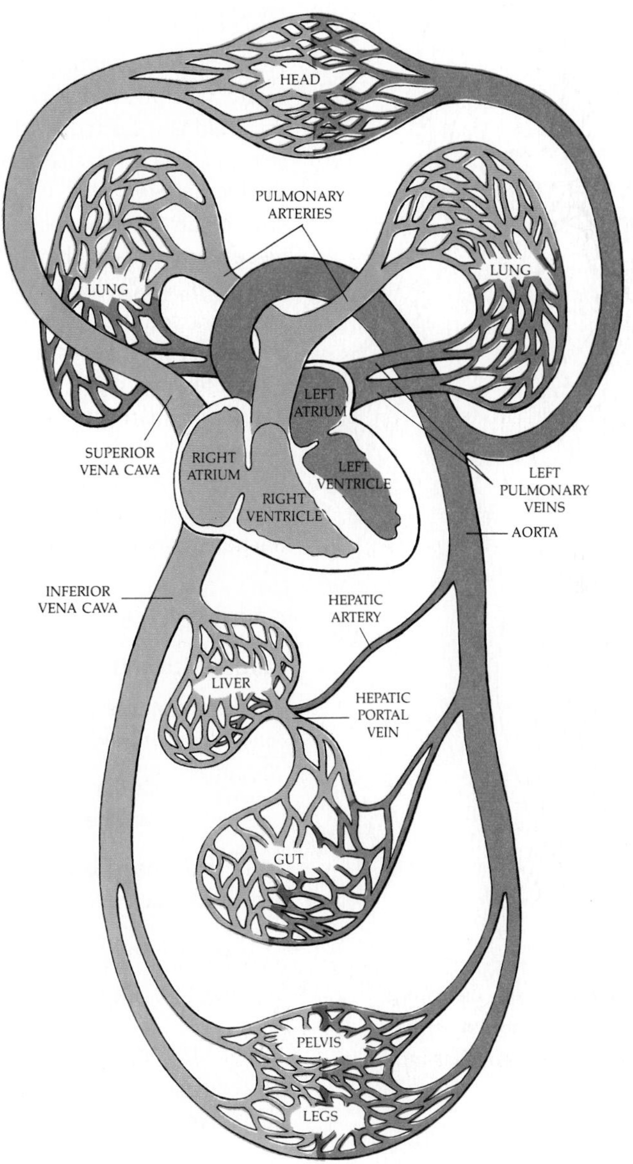

36–12
Diagram of the human circulatory system. Oxygenated blood is shown in color. Every tissue of the body contains capillaries that supply oxygen and nutrients to every cell of these tissues and carry off carbon dioxide and other wastes.

The systemic circuit is much larger. Many arteries branch off the aorta after it leaves the left ventricle. The first two branches are the right and left coronary arteries, which bring oxygenated blood to the heart muscle itself. Another major subdivision of the systemic circulation supplies the brain. If the circulation of freshly oxygenated blood to the brain is cut off for even five seconds, unconsciousness results; after four to six minutes, brain cells are damaged irreversibly.

The hepatic portal system is a special subdivision of the systemic circulation. Venous blood collected from the digestive organs is shunted via the hepatic portal vein through the liver. There it goes through a second capillary system before it is emptied into the inferior vena cava. (This passage from capillaries to veins to capillaries is called a portal system.) In this way, the products of digestion can be directly processed by the liver. The liver also receives freshly oxygenated blood directly from a major artery, the hepatic artery.

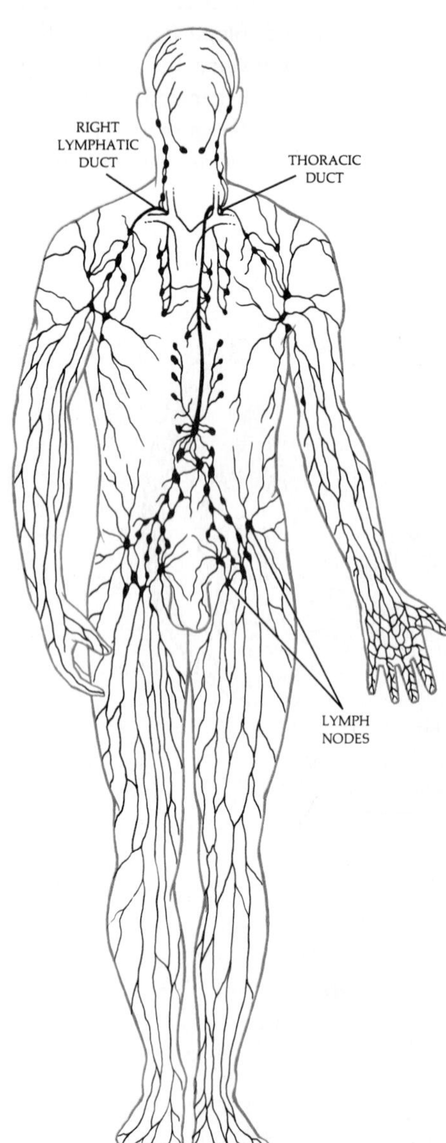

36–13
Human lymphatic system. Lymphatic fluid reenters the bloodstream through the thoracic duct and right lymphatic duct.

LYMPHATIC SYSTEM

As we noted previously, not quite all of the fluid forced out of the capillaries by the hydrostatic pressure is returned to the capillaries by osmosis. In higher vertebrates, this fluid is collected by the lymphatic system, which routes it back to the bloodstream. The lymph also serves to transport fats absorbed from the digestive tract to the bloodstream. The lymphatic system is like the vascular system in that it consists of an interconnecting network of progressively larger vessels. The larger vessels are, in fact, similar to veins in their structure. The small vessels are much like the blood capillaries; the most important difference is that, rather than forming part of a continuous circuit, the lymph capillaries begin blindly in the tissues. Interstitial fluid seeps into the lymph capillaries, where it travels in the form of lymph to large ducts, from which it is emptied into the superior vena cava. Some nonmammalian vertebrates have lymph "hearts," which help to move the fluid. In mammals, lymph is moved by contractions of the body muscles, with the valves in the lymph vessels preventing backflow, as in the venous system of the blood. Also, recent studies have shown that lymph vessels contract rhythmically; these contractions may be the principal factor propelling the lymph.

CARDIOVASCULAR DYNAMICS

Cardiac Output

The total volume of blood pumped by the heart per minute is called the *cardiac output*. It is defined as:

cardiac output (liters per minute)	=	heart rate (beats per minute)	×	stroke volume (liters per beat)

Thus, if the heart beats 72 times per minute and ejects 0.07 liter of blood into the aorta with each beat, then

cardiac output = 72 beats per minute × 0.07 liter of blood per beat

cardiac output = 5 liters per minute

Blood Pressure

The contractions of the ventricles propel the blood into the arteries with considerable force. Blood pressure is a measure of the force per unit area with which blood

36–14

In mammals, blood goes from the right heart to the lungs, from the lungs to the left heart, and, from the left heart, it enters the systemic circulation, moving from arteries to arterioles to capillaries to veins. Blood pressure varies in the different areas of the cardiovascular system. The fluctuations in blood pressure produced by three heartbeats are shown in each section of the diagram. Note the fall in pressure as blood traverses the arterioles of the systemic circulation.

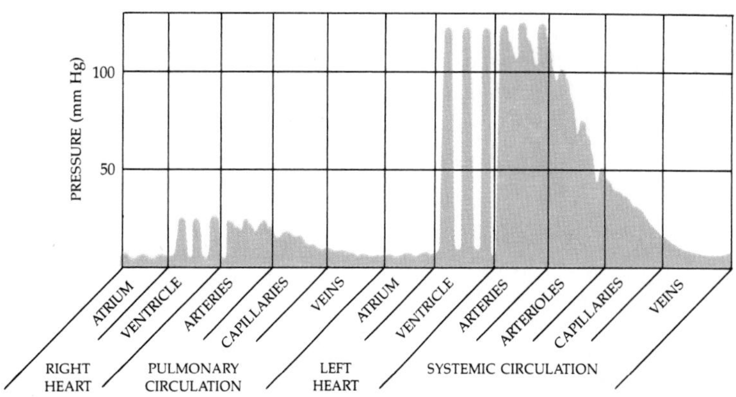

pushes against the walls of the blood vessels. It is conventionally described in terms of how high it can push a column of mercury. For medical purposes, it is usually measured at the artery of the upper arm. Normal blood pressure in a young adult is generally about 120 millimeters of mercury (120 mm Hg) when the ventricles are contracting (the systolic blood pressure) and 80 mm Hg when the ventricles relax (diastolic pressure); this is stated as a blood pressure of 120/80.

The pressure is generated by the pumping action of the heart and changes with the rate at which it contracts. Blood flow is directly proportional to blood pressure; the greater the pressure, the greater the flow. The regulation of blood flow depends on a very simple physical principle: Fluid flow through a tube is proportional to the fourth power of the radius of the tube (r^4). The diameter of the arterioles, which directly supply the capillaries, can be altered by rings of smooth muscle in the vessel walls. As the smooth muscle contracts, the opening of the arteriole gets smaller, and blood flow through the arteriole (and the capillary bed it feeds) decreases (Figure 36–15). Conversely, when the smooth muscle relaxes, the arteriole opens wider, and blood flow into the capillaries increases. These smooth muscles are controlled by autonomic nerves (chiefly sympathetic nerves), the hormones adrenaline and noradrenaline (norepinephrine), and the levels of other chemicals that are produced locally in the tissues themselves.

36–15

Diagram of a capillary bed. The muscular wall of the arteriole controls the blood flow through the capillaries. The arteriolar muscles, which are innervated by the sympathetic division of the autonomic nervous system, make it possible to regulate the blood supply to different regions of the body depending on the physiological state of the organism at different times.

The oxygen supply can be distributed according to the requirements of different parts of the body at different times, with skeletal muscles getting more during exercise, for example, or the intestinal tract more during digestion. Of particular importance is a constant flow of blood to the brain. For example, falling during fainting actually prevents serious damage to the brain cells as a result of inadequate blood supply. (These responses are often thwarted by well-meaning bystanders anxious to get the affected individual "back on his feet." In fact, holding a fainting person upright can lead to severe shock and even death.) Psychological events also influence blood flow distribution. Familiar examples are blushing, turning pale with fear, angina pectoris (pain in the chest and left arm) precipitated by emotion, and erection of the penis or clitoris as a result of erotic stimulation.

Cardiovascular Regulating Center

Activity of the nerves controlling the smooth muscle of the blood vessels is coordinated with the activity of nerves regulating heart rate and strength of heartbeat by the *cardiovascular regulating center*. This center is located in the medulla, a small

36–16
Autonomic regulation of the rate of heartbeat. Sympathetic fibers stimulate the sinoatrial node, whereas parasympathetic fibers, which are contained in the vagus nerve, inhibit it.

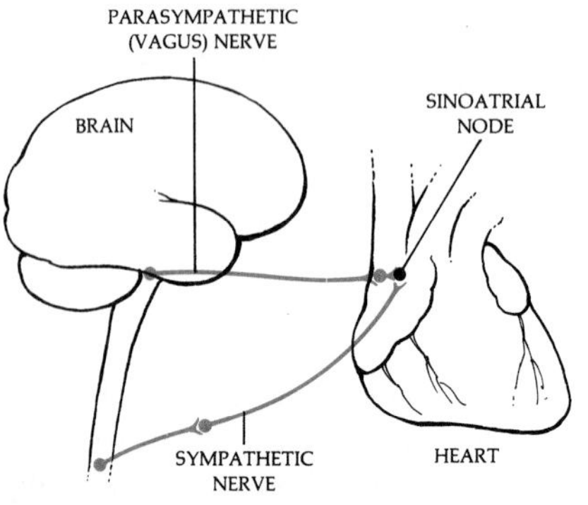

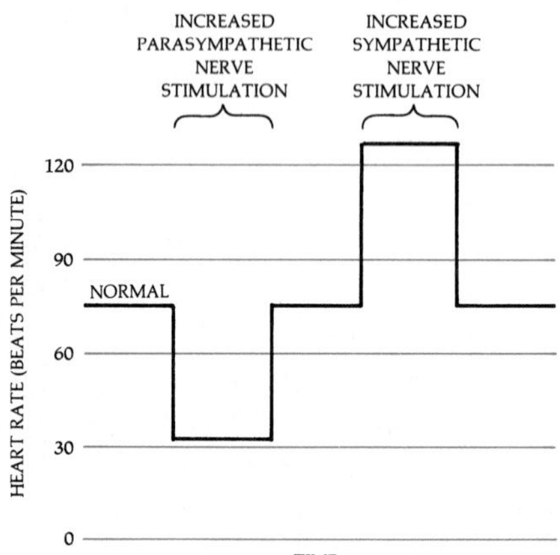

part of the brain continuous with the spinal cord. It controls the sympathetic and parasympathetic nerves to the heart as well as the nerves to the smooth muscle in the arterioles. Thus, if blood flow through a particular body area is increasing due to dilation of blood vessels, the heart is simultaneously stimulated to develop greater pressure to support the greater flow.

The cardiovascular regulating center integrates the reflexes that control blood pressure. It receives information about existing blood pressure from specialized stretch receptors in the carotid arteries, the venae cavae, the aorta, and the heart. The effector organs of the reflex are, as we have indicated, the heart and blood vessels.

The blood-pressure reflex is another example of negative-feedback control. When pressure falls, the activity of the heart is increased and the blood vessels are constricted, which raises the pressure again. Conversely, heart activity is decreased and the blood vessels are dilated in response to high pressure.

DISEASES OF THE HEART AND BLOOD VESSELS

Cardiovascular diseases cause more deaths in the United States than accidents and all other diseases combined. According to recent estimates, more than 42 million people in this country—18 percent of the total population—have some form of cardiovascular disease.

About 56 percent of the cardiovascular deaths in the United States are caused by heart attacks. A heart attack is the result of an insufficient supply of blood—ischemia (local anemia)—to an area of heart muscle; with their oxygen supply cut off, the cardiac muscle cells may die. A heart attack can be caused by a blood clot—a thrombus—that forms in the blood vessels of the heart itself or by a clot that forms elsewhere in the body and travels to the heart and lodges in a vessel there. (A wandering clot such as this is known as an embolus.) A heart attack can also result from a blocking of a blood vessel due to atherosclerosis. Recovery from a heart attack depends on how much of the heart tissue is damaged and whether or not other blood vessels in the heart can enlarge their capacity and supply these tissues, which then may recover to some extent. One-fifth of all deaths in the United States are caused by ischemic heart disease.

Angina pectoris is a related condition and refers to the pain occurring when the heart muscle receives an insufficient blood supply (but not so little that muscle dies)—often a result of a narrowing of the vessels. The symptoms, like those of a heart attack, are a pain in the center of the chest and, often, in the left arm and shoulder.

A stroke is caused by interference with the blood supply to the brain. This may be the result of a thrombus, an embolus, or the bursting of a blood vessel in the brain. Its effects depend on the severity of the damage and where it occurs in the brain.

Atherosclerosis contributes to both heart attacks and strokes. In this disease, the linings of the arteries thicken in places due to the accumulation of abnormal smooth muscle cells and to deposits of cholesterol, fibrin (clotting material), and cellular debris, and the inner surfaces become roughened. The arteries, becoming inelastic, no longer expand and contract. Blood moves with increasing difficulty through the narrowed vessels, and thrombi and emboli thus form more easily and are more likely to block the vessel.

36–17

A heart attack. When a clot forms in a blood vessel, the cells in the area supplied by this vessel are deprived of oxygen and may die. The severity of the heart attack depends, in part, on the extent of damage to the heart muscle.

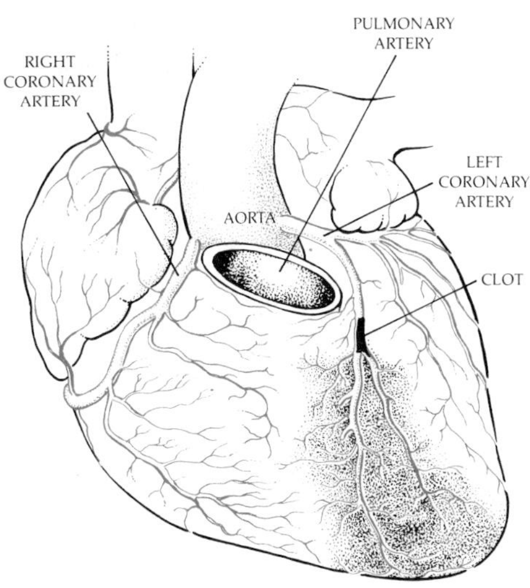

The causes are not clear-cut. Atherosclerosis is associated with high levels of cholesterol in the blood, high blood pressure, cigarette smoking, and lack of physical activity. Female hormones seem to protect against atherosclerosis; the condition is much less common in premenopausal women than it is among men of the same age. This difference in susceptibility to atherosclerosis is a major reason that, in the United States, women now live almost 10 years longer than men.

In the United States, the mortality from ischemic heart disease reached a peak in the mid-1960s and then, quite unexpectedly, began a decline that is continuing until this day; the total decrease has, astonishingly, exceeded 25 percent. The increase, when it occurred, was attributed to increased cigarette smoking, increased stress, and an increasingly sedentary existence. The decreases, however, which are occurring at all age levels and in both sexes, do not correlate well with any known environmental factor. Perhaps if we could find out what it is that we are doing right, we could do it even better.

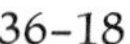

36–18

Atherosclerosis is the accumulation of fatty substances (mostly cholesterol) and the growth of abnormal muscle cells and fibrous tissue in the wall of the artery. Arteriosclerosis—hardening of the arteries—may subsequently occur if calcium becomes deposited in the arterial walls. (a) *A normal artery in cross section.* (b) *Fatty deposits have begun to form, and the space left for blood flow has decreased.* (c) *Plaques composed of abnormal muscle cells engorged with cholesterol begin to fill the lumen. The artery wall is becoming rigid and its inner surface roughened. The roughening promotes clotting.* (d) *A clot has formed, blocking the vessel completely. If this were a coronary artery, the result would be a heart attack.*

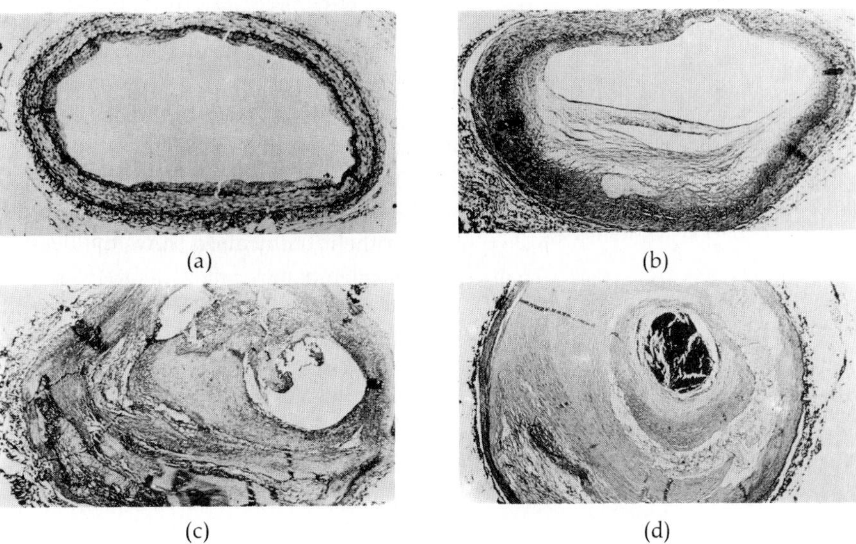

Hypertension—chronically increased arterial blood pressure—affects some 6 million people in the United States. In the great majority of cases, its cause is unknown; it is associated with narrowing of the arterioles. Chronic high blood pressure places additional strain on the walls of the arteries and increases the chances of emboli. In the United States, hypertension is twice as common among blacks as among whites. Even though the cause of hypertension is generally unknown, the disease is treated by drugs that act upon the autonomic nervous system to produce arteriolar dilation, and by measures that decrease blood volume.

SUMMARY

Oxygen, nutrients, and other molecules essential for each individual cell, as well as waste products, are carried in the blood. The blood is composed of plasma, red blood cells (erythrocytes), white blood cells (leukocytes), and platelets. Plasma, the fluid part of the blood is chiefly composed of water, in which are dissolved or suspended the nutrients and other molecules, such as hormones, enzymes, waste substances, and other specialized compounds necessary to the life functions. Red blood cells contain oxygen-bearing hemoglobin. Platelets serve in the formation of clots.

In vertebrates the blood is pumped by the muscular contractions of the heart into a circuit of arteries, arterioles, capillaries, venules, and veins. This network ultimately services every cell in the body. The essential function of the circulatory system is performed by the capillaries, through which substances are exchanged with the interstitial fluid surrounding the individual cells.

Heart structure in vertebrates varies with the demands of their differing metabolic rates. The fish has a two-chambered heart; the atrium receives deoxygenated blood, and the ventricle expels blood through the oxygenating gill capillaries into the systemic circulation. In amphibians there are two separate atria and only one ventricle. Oxygen-rich blood is received in one atrium, and oxygen-poor blood in the other. Despite the lack of an anatomical division in the ventricle, the two types of blood mingle little in the ventricle; the oxygenated blood is pumped through the systemic circulation at the same time the deoxygenated blood is pumped through the lungs. Birds and mammals have a double circulation system, made possible by a four-chambered heart that functions as two separate pumping organs. One side pumps deoxygenated blood to the lungs, and the other pumps oxygenated blood to the body tissues.

Synchronization of the heartbeat is controlled by the sinoatrial node (the pacemaker), located in the right atrium, and by the atrioventricular node, which delays the stimulation of ventricular contraction until the atrial contraction is completed. The rate of heartbeat is secondarily under neural and hormonal control. Parasympathetic stimulation slows the heartbeat; sympathetic stimulation and adrenaline accelerate it.

Although deaths from cardiovascular disease are decreasing, it remains, by far, the major cause of death in affluent countries.

QUESTIONS

1. Distinguish between the following: blood/plasma; systolic/diastolic; aorta/vena cava; atrium/ventricle; right heart/left heart; sinoatrial node/atrioventricular node.

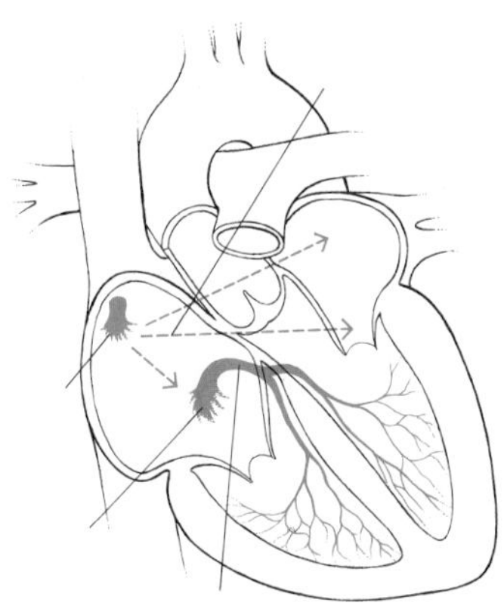

2. Label the diagram at the left.

3. The valves of the heart are not directly controlled by nerves. Yet, in most individuals, they open and shut at precisely the right points in the cardiac cycle for efficient heart operation. How is this precise timing possible? What does determine just when the valves will open and shut?

4. What is the advantage of the fibers of the atrioventricular node being slow-conducting?

5. Trace the course of a single red blood cell from the right ventricle to the right atrium in a mammal.

6. Blood serum is the portion of the plasma remaining after a clot is formed. Name some of the components of the plasma that would not be present or would be present in a lesser amount in serum.

7. How does the radius of a blood vessel affect the blood flow through it?

8. If a vertical tube full of water is connected to one of your arteries, how high will the water go?

9. Explain the reasons for the changes in blood pressure shown in Figure 36–14.

10. When you are very frightened, you turn quite pale. What is occurring to cause this change? Why is such an adaptation useful?

11. When an accident victim suffers blood loss, he or she is transfused with plasma rather than with whole blood. Why is plasma effective in meeting the immediate threat to life?

12. Give two reasons why atherosclerotic arteries are much more susceptible to clot formation and blockage than are normal, healthy arteries.

13. What is probably the immediate cause of death by crucifixion? (If you need a clue, see page 715.)

CHAPTER 37

Homeostasis I: Excretion and Water Balance

As we noted in Chapter 33, one of the advantages of multicellularity is the capacity to create a controlled internal environment in which the component cells live and function. Many factors contribute to the control of the internal environment: regulation of blood sugar (page 684) is, for example, of major importance, as is elimination of carbon dioxide (page 699).

As we noted in the previous chapter, the blood plays a major role in determining the internal chemical environment of the body. It is the supply line for chemicals taken up by the individual cells, and it carries away the wastes released by these cells. The blood can function as an efficient supply and sanitation system only because cellular wastes are constantly removed from the bloodstream. Notice that this sort of excretory mechanism is quite different in principle from the elimination of feces from the intestinal tract. In the latter case, the bulk of what is eliminated is material, such as cellulose, that was never actually in the body in the first place, in that it never passed through the walls of the digestive tract. Removal of substances from the blood, by contrast, is a very selective process of monitoring, analysis, selection, and rejection.

In this chapter, we shall look first at the factors involved in the control of the internal chemical environment of an organism. Second, we shall examine the evolution, structure, and function of the kidney—the chief organ involved in solving this problem in land-dwelling vertebrates. Finally, we shall see what is known of the control mechanisms that regulate kidney function.

REGULATING THE CHEMICAL ENVIRONMENT

Regulation of the internal chemical environment of an organism involves finding solutions to three different—yet interwoven—problems: (1) excreting metabolic wastes, (2) regulating concentrations of ions and other chemicals, and (3) maintaining water balance.

Excretion of Metabolic Wastes

The two chief metabolic wastes that cells release into the bloodstream are carbon dioxide and nitrogen compounds, mostly ammonia, produced by the breakdown of amino acids. In animals, carbon dioxide is eliminated from the lungs or other respiratory organs, or it diffuses out into water through the skin. Aquatic animals often excrete nitrogenous wastes in the form of ammonia (NH_3). In land-dwelling organisms, which do not have an unlimited water supply, ammonia must be converted to some other form since it is highly toxic, even in low concentrations, if not immediately excreted.

All birds, terrestrial reptiles, and insects eliminate nitrogenous wastes in the form of uric acid or uric acid salts, which can be excreted as crystals. In birds, the uric acid and uric acid salts are mixed with the fecal wastes in the cloaca (the common exit chamber for the digestive, urinary, and reproductive tracts), and the combination is dropped as a semisolid paste, familiar to frequenters of public parks and admirers of outdoor statuary. This nitrogen-laden substance forms a rich natural fertilizer; guano, the excreta of seabirds, accumulates in such quantities on islands where these birds gather in great numbers that at one time it was harvested commercially.

Mammals excrete nitrogenous waste products largely as urea, which like ammonia or uric acid comes principally from the breakdown of amino acids, with the removal of the amino group and the formation of ammonia. The ammonia is quickly converted to urea, which is relatively nontoxic. It then diffuses into the bloodstream. Urea, unlike uric acid, must be dissolved in water for excretion.

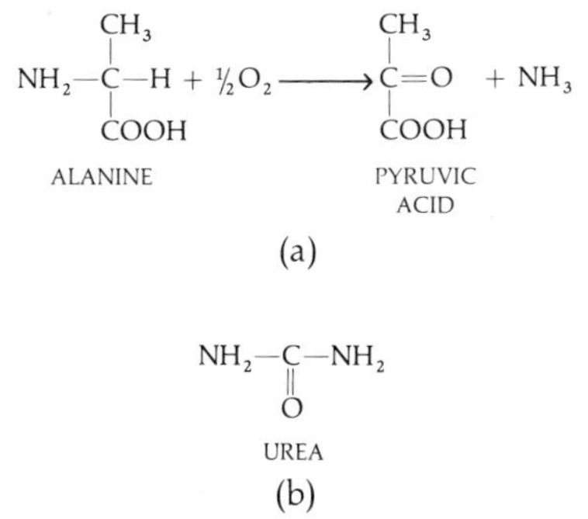

$$NH_2{-}\underset{\substack{|\\ COOH}}{\overset{\substack{CH_3\\ |}}{C}}{-}H + \tfrac{1}{2}O_2 \longrightarrow \underset{\substack{|\\ COOH}}{\overset{\substack{CH_3\\ |}}{C}}{=}O + NH_3$$

ALANINE PYRUVIC ACID

(a)

$$NH_2{-}\underset{\substack{\|\\ O}}{C}{-}NH_2$$

UREA

(b)

37-1
(a) *Deamination—the removal of the amino group—is the first step in the breakdown of amino acids. The products of the reaction are ammonia and a carbon skeleton, which can be broken down to yield energy or converted to sugar or fat.* (b) *Urea, the principal form in which nitrogen is excreted in most mammals. It is formed in the liver by the combination of two molecules of ammonia with one of carbon dioxide through a complex series of energy-requiring reactions. What would be the other product of this reaction?*

Controlling Concentrations of Chemicals

Kidneys are more accurately regarded as regulatory rather than excretory organs. Excretion by the kidneys is highly selective. For instance, although about half of the urea in the blood that enters a normal mammalian kidney is eliminated, almost all the amino acids are retained. Glucose (blood sugar) is not excreted unless it is present in high amounts, as in diabetes mellitus (sugar diabetes). The presence of glucose in the urine is, in fact, a basis for the diagnosis of this form of diabetes and the evaluation of its treatment. Chemical regulation also involves maintaining closely controlled concentrations of ions such as Na^+, K^+, H^+, Mg^{2+}, Ca^{2+}, Cl^-, and HCO_3^-. These ions are important for their specific roles in various chemical processes, such as maintenance of protein structure, membrane permeability, propagation of the nerve impulse, and muscle contraction. Some are important in the regulation of the pH of the blood, which is also regulated, in part, by the kidneys.

Maintaining Water Balance

A primary function of the excretory system is the regulation of water loss. The concentration of a particular substance in the body depends not only on the absolute amount of the substance but also on the amount of water in which it is dissolved. Thus, although the fundamental problem is always the same—the chemical regulation of the internal environment—the solution to the problem varies widely. A major factor is the availability of water to the organism.

The problem of water balance is such a universal one, biologically speaking, and is so important to the survival of the organism that we shall digress for a moment to discuss it in more detail before examining the anatomy and physiology of the kidney.

37–2
Because of their high concentration of blood urea, cartilaginous fish, such as this blacktip reef shark, have body fluids isotonic with sea water.

37–3
Some marine animals, such as the turtle, have special glands in their heads that can excrete sodium chloride at a concentration about twice that of sea water. Since ancient times, turtle watchers have reported that these great armored reptiles come ashore, with tears in their eyes, to lay their eggs, but it is only recently that biologists have learned that this is not caused by an excess of sentiment—as is the case with Lewis Carroll's mock turtle—but is, rather, a useful solution to the problem of excess salt from ingestion of sea water. Marine birds similarly secrete a salty fluid through their nostrils.

WATER BALANCE

An Evolutionary Perspective

The earliest organisms probably had a salt and mineral composition much like that of the environment in which they lived. The early organisms and their surroundings were probably also isotonic; that is, each had the same total effective concentration of dissolved substances, so water did not tend to move either into or out of these organisms by osmosis. When organisms moved to fresh water (a hypotonic—less concentrated—environment), they had to develop systems for "bailing themselves out," since fresh water tended to move into their bodies; the contractile vacuole of *Paramecium* is an example of such a bailing device.

If the first vertebrates—the fish—evolved in fresh water, as is generally believed, the first function of the kidneys, phylogenetically speaking, was probably to pump water out and to conserve salt and other desirable solutes, such as glucose. In freshwater fish today, the kidney works in just this way, primarily as a filter and reabsorber of solutes. The urine of these fish is hypotonic—that is, it has a concentration of solutes lower than that of body fluids.

Saltwater fish, to avoid loss of water by osmosis to their environment, either maintain body fluids that are isotonic with sea water or have systems that pump water in and salts out. The hagfish, a group of cartilaginous fish, maintain body fluids about as salty as the salt waters of the surrounding ocean and so do not tend to lose water by osmosis. The body fluids of another group of cartilaginous fish, the sharks, are also isotonic with sea water. They achieve their isotonicity in a different way. Because they have an unusual tolerance for urea, instead of constantly pumping this waste out, as do all other fish, sharks retain a higher concentration of it in their blood, which would otherwise be hypotonic to salt water. (This is a striking example of what has been called the "opportunism of evolution.")

In the bony fish, which spread to the sea much later than the cartilaginous fish, the body fluids are hypotonic to the marine environment, having a solute concentration only about one-third that of sea water. Thus they are constantly in danger of losing so much water to their environment that the solutes in their body fluids become so concentrated that their cells die. To compensate for their osmotic water loss, they drink sea water. This restores their water content but leads to a new problem—how to eliminate the excess salt ingested. This problem has been solved by the evolution of special gland cells in the gills that excrete excess salt. Hence these fish can take in salt water and still remain hypotonic. (Freshwater fish, conversely, have salt-absorbing cells in their gills.)

Sources of Water Gain and Loss in Terrestrial Animals

Since terrestrial animals do not always have automatic access to either fresh or salt water, they must regulate water content in other ways, balancing off gains and losses. Terrestrial animals gain water by drinking fluids, by eating water-containing foods, and as an end product of the oxidative processes that take place in the mitochondria, as we saw in Chapter 9. When 1 gram of glucose is oxidized, 0.6 gram of water is formed. When 1 gram of protein is oxidized, only about 0.3 gram of water is produced. Oxidation of 1 gram of fat, however, produces 1.1 grams of water because of the high hydrogen content of fat (the extra oxygen comes from the air).

Some animals can derive all their water from food and oxidation of nutrient molecules and do not therefore require fluids. The kangaroo rat of the American desert, for example, can live its entire existence without drinking water if it eats the right type of food. It is not surprising that it prefers a diet of fatty seeds, which yield a large amount of water on oxidation. If it is fed high-protein seeds, such as soybeans—the oxidation of which produces a large amount of nitrogenous waste and a relatively small amount of water—it will die of dehydration unless some other source of water is available.

37–4
The kangaroo rat, a common inhabitant of the American desert, may spend its entire life without drinking water. It lives on seeds and other dry plant materials, and in the laboratory it can be kept indefinitely on a diet of only fatty seeds such as barley or rolled oats. Analysis shows that the kangaroo rat is highly conservative in its water expenditures. It has no sweat glands, and being nocturnal, it searches for food only when the external temperature is relatively cool. Its feces have a very low water content, and its urine is highly concentrated. Its major water loss is through respiration, and even this loss is reduced by the animal's long nose in which some cooling of the expired air takes place, with condensation of water from it.

On the average, a human takes in about 2,300 milliliters of water (about 2.5 quarts) a day in food and drink, and gains an additional 200 milliliters a day by oxidation of nutrient molecules. Water is lost from the lungs in the form of moist exhaled air, is eliminated in the feces, is lost by evaporation from the skin, and is removed from the blood and excreted as urine. The latter is usually the major route of water loss. In a normal human adult, the rate of water excretion in urine averages 1,500 milliliters a day, but the actual amount produced may vary from 500 to 2,300 milliliters. (A minimum output of about 500 milliliters of water is necessary for health since this much water is needed to remove potentially toxic waste products.) However, there will be a variation of less than 1 percent in the fluid content of the body.

Water Compartments

The body has three principal water compartments: (1) the plasma (7 percent of body fluid), (2) the interstitial fluid and the lymph (28 percent of body fluid), and (3) the intracellular fluid, the fluid within the cells (65 percent of body fluid). Water absorbed from the digestive tract, the major source of water gain, passes largely into the intestinal capillaries and enters the plasma. This passage is mainly a result of osmosis. Because of the active transport of food molecules and salts into the capillaries from the intestinal tract, the plasma becomes hypertonic in relation to the intestinal contents, and so water tends to follow the dissolved molecules into the plasma. (The loss or gain of water from the plasma is of extreme importance for maintaining a stable blood pressure and for normal cardiac function.) Hydrostatic pressure forces watery fluid through the capillary walls into the interstitial fluid, as we mentioned on page 711. However, most of this fluid reenters the plasma osmotically or via the lymphatic vessels. Water remaining in the interstitial compartment comes in contact with the tissue cells. Since their membranes are permeable to water, water moves freely into the cells. Thus water is constantly

37–5
The fluid compartments of the human body, showing the main routes of exchange among them. The interstitial fluid forms the environment in which the cells of the body live and multiply. The arrows indicate exchanges between various compartments. An animal's body is about two-thirds water.

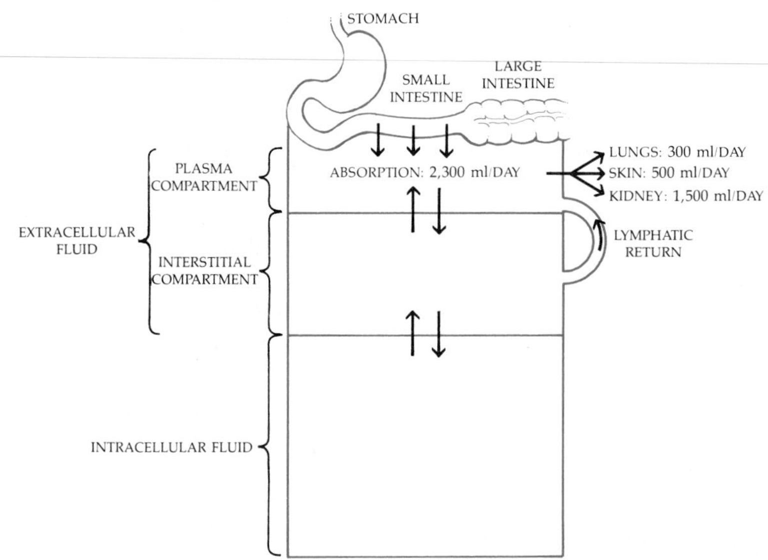

moving from compartment to compartment. (Rates of exchange between compartments are measured by administering a traceable substance, such as a harmless dye molecule, that passes readily through cell membranes, and then analyzing the concentration of the tracer material that turns up in each compartment.)

A number of factors affect the movement of water from one compartment to another. Dehydration (water loss greater than water intake) increases the solute concentration of the extracellular fluid; water therefore moves out of the cells, including the cells of the mucous membranes of the oral cavity, producing the sensation of dryness that we associate with thirst.

Human sweat, unlike that of most other mammals, contains salt. (The reason for salt excretion by the skin is not known, but it has been suggested that the little crumbs of salt so produced clung to the fur of our primate ancestors and that these delicacies served as rewards for their companions who groomed them.) In cases of profuse sweating, if the water is replaced without the salt, water will move into body cells, diluting their contents. The effects of such dilution are particularly severe on the central nervous system, and water intoxication may produce disoriented behavior, convulsions, coma, and even death before the excess water can be excreted.

A number of physiological malfunctions, such as the retention of salts, which may occur as a consequence of kidney disease, or the loss of plasma proteins as a result of starvation, lead to the accumulation of interstitial fluids (edema).

37–6
A mammal is in water balance when the total amount of water lost in expired air, evaporation from skin, and in urine and feces equals the total amount of water gained by the intake of food and fluids and by the oxidation of food molecules.

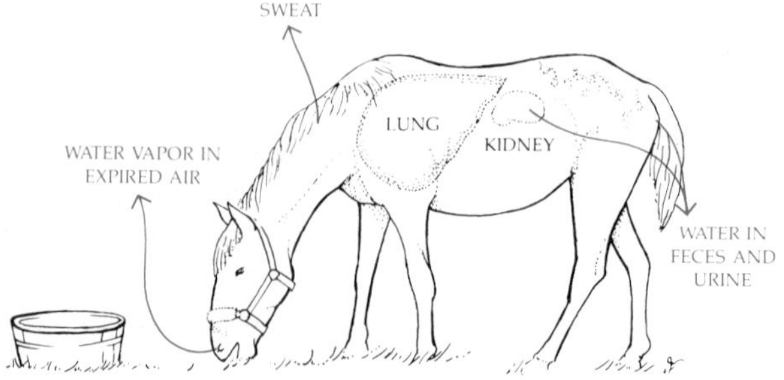

37–7

(a) *In longitudinal section, the human kidney is seen to be made up of two regions. The outer region, the cortex, contains the fluid-filtering mechanisms. The loops of Henle traverse the inner region, the medulla, as do the collecting ducts carrying the urine, which merge and empty into the funnel-shaped renal pelvis, which, in turn, empties into the ureter.*

(b) *The nephron is the functional unit of the kidney. Blood enters the nephron through the afferent arteriole leading into the glomerulus. Fluid is forced out by the pressure of the blood through the thin capillary walls of the glomerulus into Bowman's capsule. The capsule connects with the long renal tubule, which has three regions: the proximal tubule; the loop of Henle, which extends into the medulla; and the distal tubule. As the fluid travels through the tubule, almost all the water, ions, and other useful substances are reabsorbed into the bloodstream through the peritubular capillaries. Other substances are secreted from the capillaries into the tubules. Under the influence of antidiuretic hormone, the fluid is concentrated in the collecting duct by diffusion of water out of the duct. The remaining waste materials and water are excreted from the body as urine.*

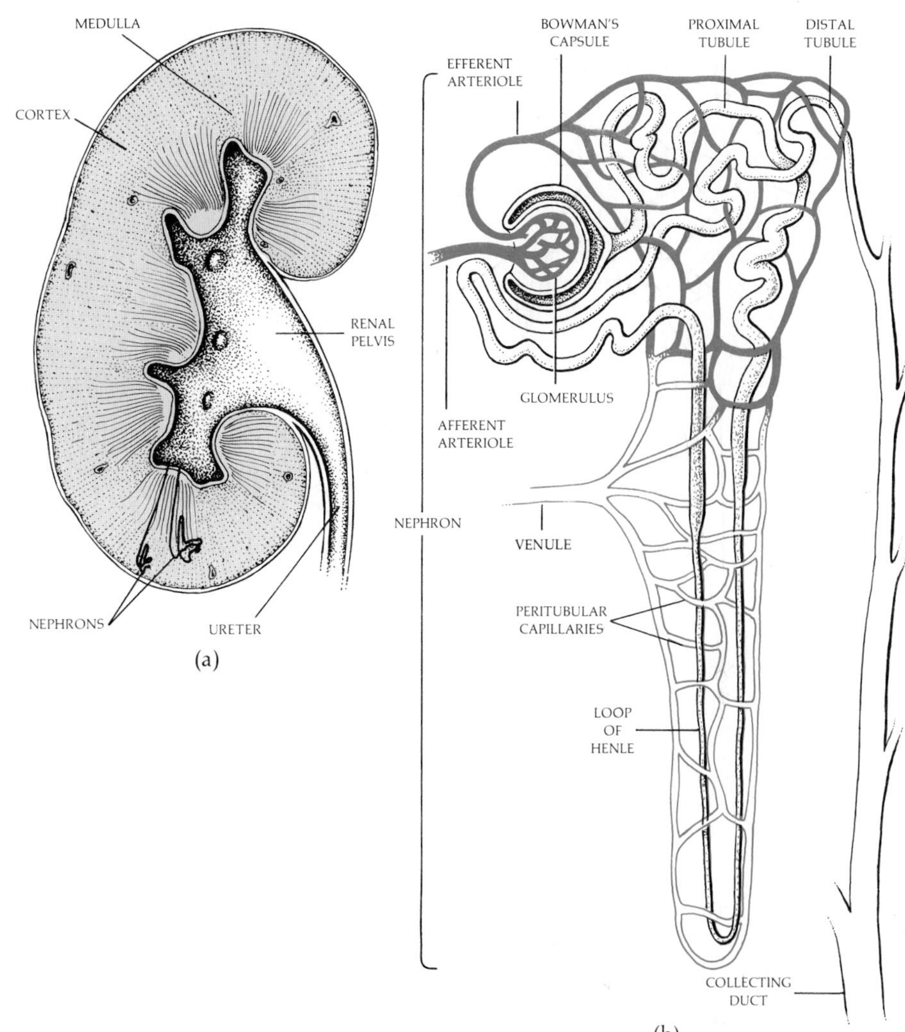

THE KIDNEY

In land-dwelling vertebrates, the organ chiefly responsible for the complex functions involved in regulating the internal chemical environment of the body is the kidney. The two human kidneys are dark red, bean-shaped structures about 10 centimeters long that lie at the back of the body, behind the stomach and liver.

In humans and other mammals, the functional unit of the kidney is the *nephron*. It consists of a cluster of capillaries known as the *glomerulus* and a long narrow tube, the *renal tubule* (Figure 37–7). The renal tubule originates as a bulb called *Bowman's capsule;* it is then made up of the *proximal* (near) and *distal* (far) *tubules,* which are separated by the *loop of Henle*. It ends as the straight *collecting duct.* Each of the two human kidneys contains about a million nephrons with a total length of some 80 kilometers (50 miles) in an adult human.

37–8

The section of kidney tissue shown in this scanning electron micrograph was prepared by the freeze-fracture technique. A glomerulus occupies the upper center of the micrograph, enclosed at the bottom and sides by Bowman's capsule. The space within the capsule is continuous with the lumen of the proximal tubule, the beginning of which you can see at the lower right.

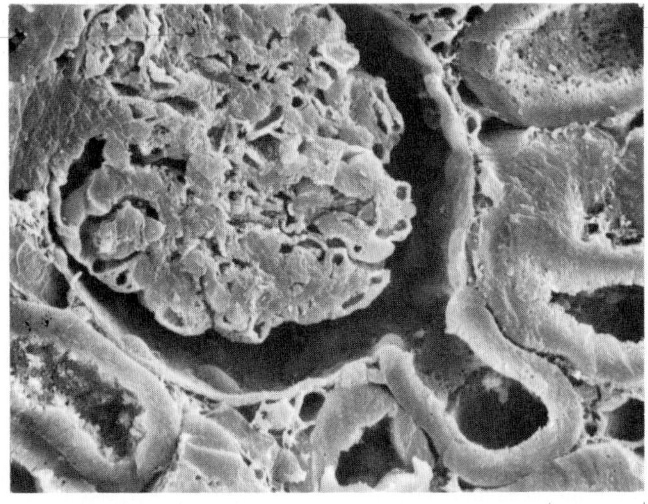

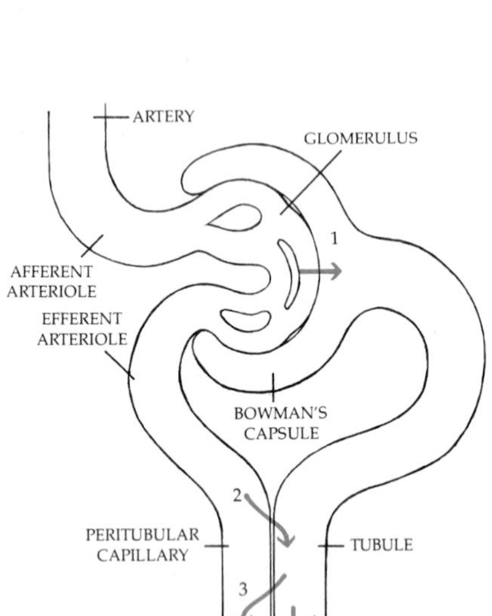

37–9

The four basic components of renal function: (1) filtration through the glomerular capillaries into Bowman's capsule; (2) secretion from the peritubular capillary into the tubule; (3) reabsorption from the tubule to the peritubular capillary; (4) excretion. The pressure in the glomerular capillaries is regulated by constriction of the afferent and efferent arterioles.

Function of the Kidney

Blood enters the kidney through the renal artery, which divides into progressively smaller branches, the arterioles, which each lead, in turn, to a glomerulus. Unlike the other capillary beds, a glomerulus lies between two arterioles—the one leading in is the afferent arteriole, and the one leading out is the efferent arteriole. The efferent arteriole then divides again into capillaries—the peritubular ("around-the-tubes") capillaries—which surround the renal tubule and then merge to form a venule that empties into the renal vein.

The constriction of the afferent and efferent arterioles keeps the blood pressure within the glomerulus at a level about twice that in other capillaries. As a consequence, about one-fifth of the blood plasma that enters the kidney is forced into Bowman's capsule through the capillary walls of the glomerulus. This process, the first in the formation of urine by the kidney, is called *filtration*, and the fluid entering the capsule is referred to as the *filtrate*. Except for the absence of large molecules, such as proteins, the filtrate has the same chemical composition as the plasma.

The second process in the formation of urine is *secretion*, in which molecules left in the plasma after filtration are selectively removed by the cells of the tubular walls and are secreted by them into the filtrate. Penicillin, for example, is removed from the circulation in this way.

The third major process in urine formation is *reabsorption*. During the passage through the renal tubule, most of the water and solutes that entered the tubule in the first place via filtration are transported back into the bloodstream by the cells of the tubular walls. Secretion and reabsorption of many solutes, including glucose, amino acids, and salt ions, take place by active transport, mainly in the proximal convoluted tubule. As a consequence, the kidney has a high energy requirement, higher on a per-gram basis than even the heart.

Finally, the remaining fluid—now the urine—leaves the nephron and passes into the renal pelvis, which is, in essence, a funnel. From this funnel, the urine trickles continuously through the ureter, to the bladder, which stores the urine until it is excreted through the urethra (Figure 37–11).

37–10
An electron micrograph of the cells of the proximal tubule of the kidney, where secretion and reabsorption of many solutes take place by active transport. Notice the many mitochondria and their position close to the brush border on the surface of the cells bordering the lumen of the tubule (the light area at the upper right of the micrograph). This border is made up of many very fine, hairlike extensions that, like microvilli, greatly increase the absorptive capacity of the cell. It is assumed that the mitochondria provide a sufficient nearby energy source to power active transport mechanisms residing in the convoluted cell surface. A red blood cell can be seen within the peritubular capillary, in the lower left of the micrograph.

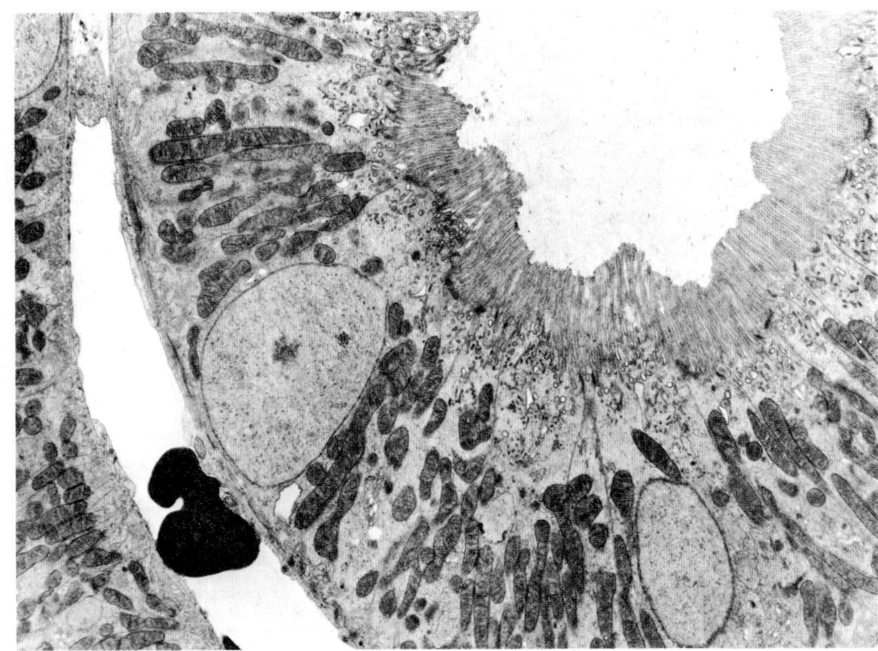

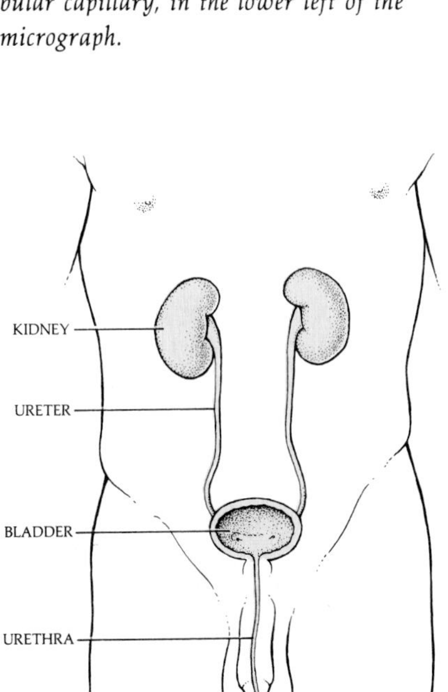

37–11
Urinary system in the human male. Urine formed in the kidneys travels through the ureters to the bladder where it is stored. Urine leaves the body by way of the urethra, which, in the male mammal, also serves as the passageway for semen.

Water Conservation: The Loop of Henle

The control of urinary water loss is a major mechanism by which body water is regulated. Animals with free access to fresh water typically excrete the wastes from their bloodstream in a copious urine that is hypotonic in relation to their blood. The daily urinary output of a frog, for instance, totals 25 percent of its body weight. Most terrestrial animals cannot afford to be so profligate with water, however. In response to evolutionary pressures, birds and mammals have developed the ability to excrete hypertonic urines—that is, urines that are more concentrated than their body fluids. In mammals, this ability is associated with the hairpin-shaped section of the renal tubule known as the loop of Henle. By sampling the fluids in and around different portions of the nephron, including different regions of the loop, and then analyzing the ions present in each sample, physiologists have been able to determine how such a deceptively simple-looking structure makes this function possible.

As is the case with the glomerulus, structure is the key to function. The first factor of importance is the pathway formed by the basic structure of the nephron (Figure 37–12). From Bowman's capsule, the filtrate first enters the proximal tubule, descends the loop of Henle, ascends it, and then passes through the distal tubule into the collecting duct. The second important factor is the permeability to water and salt of the walls in different parts of the nephron and the presence within them of carriers for the active transport of salt. The walls of the proximal tubule are freely permeable to water and contain carrier proteins that pump sodium ions (Na^+) out of the tubule by active transport, with chloride ions (Cl^-) following passively. The walls of the descending branch of the loop of Henle are also freely permeable to water but are relatively impermeable to salt. In the ascending branch of the loop the situation is quite different. The walls of the ascending branch are

37–12
The formation of hypertonic urine in the human nephron. The fluid entering the proximal convoluted tubule is isotonic with the blood plasma. Although sodium chloride is pumped from the tubule here, the fluid remains isotonic because water also moves out by osmosis. As the fluid descends the loop of Henle, it becomes increasingly concentrated as water moves by osmosis into the surrounding zone of high salt concentration. As the fluid ascends the loop of Henle, however, it becomes less and less concentrated as sodium chloride is pumped out. By the time it reaches the distal convoluted tubule, it is hypotonic in relation to the blood plasma, and it remains hypotonic throughout the distal tubule. The fluid then passes down the collecting duct, once more traversing the zone of high salt concentration.

From this point onward, the urine concentration depends on ADH (antidiuretic hormone). If ADH is absent, the walls of the collecting duct are not permeable to water, no additional water is removed, and a less concentrated urine is excreted. If ADH is present, the cells of the collecting duct are permeable to water, which moves by osmosis into the surrounding salty fluid, as shown in the diagram. In this case a highly concentrated (hypertonic) urine is passed down the duct to the renal pelvis, the ureter, the bladder, and finally out the urethra.

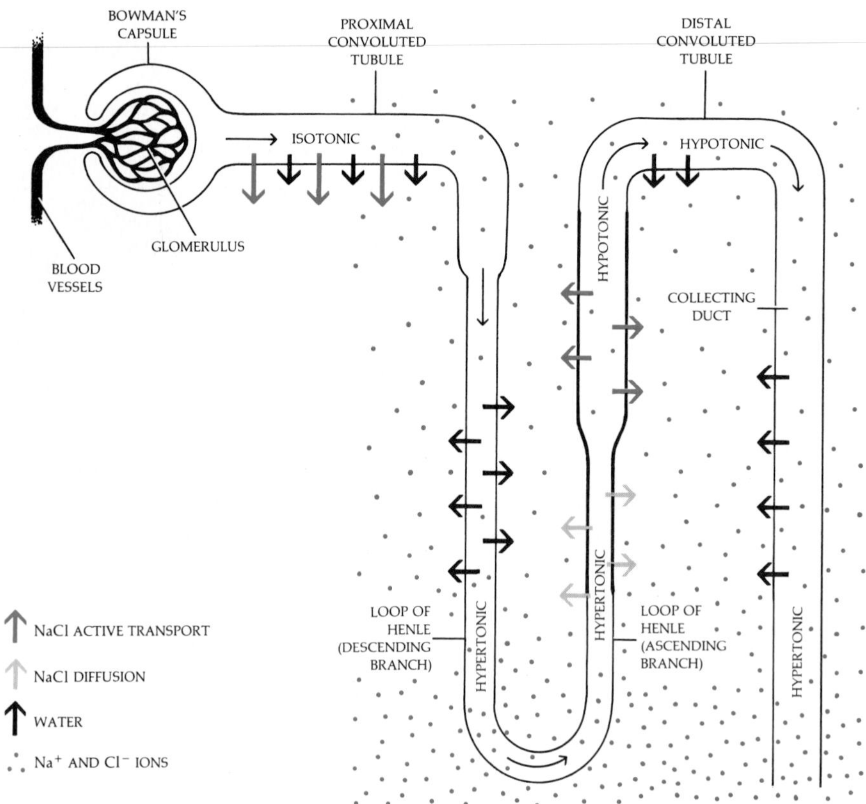

impermeable to water but permit some movement of salt: in the lower portion, limited diffusion of salt can occur, whereas in the upper portion, carrier proteins actively transport chloride ions out of the tubule, with sodium ions following by diffusion. In the distal tubule, the walls are once more permeable to water and relatively permeable to salt. And, except under certain circumstances, so are the walls of the collecting duct.

To understand the formation of a hypertonic urine, we must first consider the consequences of the structure of the ascending branch of the loop of Henle. In its upper portion, negatively charged chloride ions are actively pumped from the tubule into the surrounding interstitial fluid; positively charged sodium ions follow passively by diffusion, maintaining electrical balance. Water, however, cannot move through these walls. As a result, large quantities of sodium and chloride ions accumulate in the interstitial fluid bathing both branches of the loop of Henle and the collecting duct. With that in mind, let us now follow the filtrate as it moves through the nephron.

The fluid entering the proximal tubule from Bowman's capsule is isotonic with the blood plasma; that is, it has the same solute concentration as does the blood plasma. In the proximal tubule, sodium ions are pumped out, with chloride ions following passively. Water also moves out of the tubule by osmosis, following after these ions. Thus, as the fluid enters the descending branch of the loop, it is still isotonic with the blood plasma. The pathway of the loop, however, carries the fluid through a zone of high salt concentration—a result of the properties of the ascending branch. As we have noted, the walls of the descending loop are freely permeable to water, so large quantities of water move by osmosis from the tubule into the interstitial fluid (and from there into the peritubular capillaries) with only very small amounts of salt ions diffusing into the tubule. The fluid within the tubule is now hypertonic in relation to the blood plasma. As it travels through the ascending branch, large quantities of salt are removed by the action of the carrier proteins, making the tubular fluid hypotonic in relation to the blood plasma. In humans it remains hypotonic during its passage through the distal convoluted tubule. However, the fluid must again pass through the zone of high salt concentration as it flows down the collecting duct. Whether or not the walls of the collecting duct are permeable to water depends on the presence or absence of a hormone, antidiuretic hormone (ADH). If ADH is present, the walls of the collecting duct permit the passage of water. Water moves by osmosis through the walls, leaving within the duct a urine that is isotonic with the surrounding briny interstitial fluid but hypertonic in relation to the body fluids as a whole. In this way, mammals that need to conserve water are able to excrete a fluid, the urine, far more concentrated than the plasma from which it is derived. The water extracted from the urine is returned to the bloodstream via the peritubular capillaries. If ADH is absent, the urine remains hypotonic.

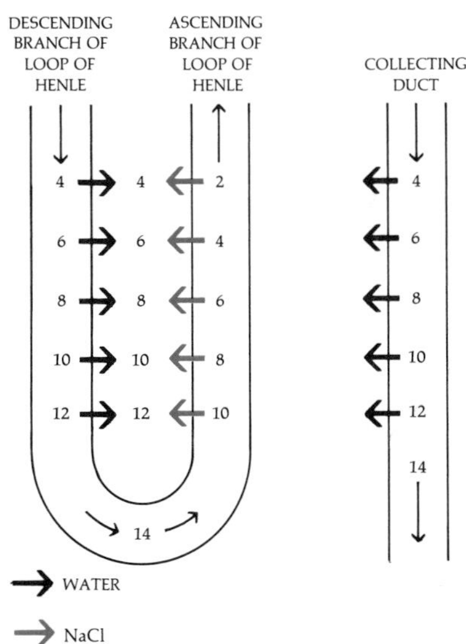

37–13

The effect of countercurrent flow in the loop of Henle. At any given horizontal level, the difference in relative salt concentration between the two branches of the loop is moderate. However, the longer the loop, the greater the concentration difference between top and bottom and the higher the concentration of salt in the interstitial fluid bathing the loop and the collecting duct.

Although the ion pumps in the ascending branch of the loop of Henle make possible the concentration differences on which this system depends, a key role is played by the hairpin structure of the loop as a whole. The flow of fluid through the loop provides another example of a countercurrent system. As we noted earlier (see page 129), a countercurrent flow can help to maintain differences in solute concentration from one end to the other of a system. When it is coupled to an active transport system, as in the loop of Henle, it can actually multiply the differences (Figure 37–13) and is known as a countercurrent multiplier. The longer the loop, the greater the concentration differences that can be established. Since the primary factor limiting the concentration of the urine is the salt concentration surrounding the collecting duct, it should come as no surprise to learn that those mammals that excrete the most hypertonic urine also have the longest loops of Henle.

Control of Kidney Function: Aldosterone and ADH

In mammals, at least two hormones act on the nephron to affect the composition of the urine. One of these is aldosterone, which is produced by the adrenal cortex. Aldosterone stimulates reabsorption of sodium ions from the distal tubule and secretion of potassium ions into it. When the adrenal glands are removed, or when they function poorly (as in Addison's disease), sodium chloride and water are lost in the urine, and the tissues of the body become depleted of them. Generalized weakness results, and if a patient with Addison's disease is not given hormone replacement therapy, the fluid loss can eventually be fatal. Aldosterone production is controlled by a complex feedback circuit involving, as you would expect, potassium ion levels in the bloodstream and reflexes in the kidneys themselves.

We have already mentioned the second hormone, ADH. This hormone is formed in a portion of the brain, the hypothalamus, and is stored and released from the pituitary gland. As noted earlier, it affects the amount of water excreted in the urine. ADH acts on the membranes of the collecting ducts of the nephrons and increases their permeability to water, so more water moves, by diffusion, back into the blood from the nephron. The amount of ADH released depends on the osmotic concentration of the blood and also on the blood pressure. Osmotic receptors that monitor the solute content of the blood are located in the hypothalamus and thus influence ADH release directly. Pressure receptors that detect changes in blood volume are found in the walls of the heart, in the aorta, and in the carotid arteries. Stimuli received by these receptors are transmitted to the hypothalamus. Factors that increase the concentration of solutes in the blood or decrease blood pressure, or both, increase the production of ADH and conservation of water in the body. Such factors include dehydration and hemorrhage. Factors that decrease blood concentration, such as the ingestion of large amounts of water, or that increase blood pressure—adrenaline, for instance—signal the hypothalamus to decrease production and release of ADH, and so more water is excreted. Cold stress inhibits ADH secretion and so increases urinary flow. Alcohol also suppresses ADH secretion and increases urinary flow, a phenomenon familiar to imbibers of beer and other alcoholic beverages. Pain and emotional stress trigger ADH secretion and thus decrease urinary flow.

SUMMARY

The excretory system helps maintain homeostasis of the internal chemical environment. This regulatory process involves: (1) eliminating the by-products of metabolism, especially the nitrogenous compounds produced in the breakdown of amino acids, (2) controlling the ion content of the body fluids and helping to regulate the pH of the blood, and (3) maintaining water balance.

Mammals excrete nitrogenous wastes chiefly in the form of urea, which is formed in the liver from ammonia (produced by deamination of amino acids) and carbon dioxide.

Water regulation involves equalizing water gain and loss. The principal source of water gain in most mammals is in the diet; water is also formed as a result of oxidation of nutrient molecules. Water is lost in the feces and the urine, by respiration, and from the skin. Although the amount of water taken in and given off may vary widely from animal to animal and also from time to time in the same animal, depending largely on environmental circumstances, the volume of water in the body remains very nearly constant. The principal water compartments of the body are the plasma, the interstitial fluids, including the lymph, and intracellular fluids. The major factor determining exchange of water among compartments of the body is osmotic potential.

The kidney is the organ chiefly responsible for the regulation of the chemical composition of the blood. The functional unit of the kidney is the nephron, which consists of a long tubule with a closed front end (Bowman's capsule) and, enclosed in the capsule, a cluster of capillaries, the glomerulus. Blood entering the glomerulus is under sufficient pressure to force plasma (minus the larger proteins) through the capillary walls into Bowman's capsule. Cells of the renal tubule regulate the chemical composition of the body fluids by selectively reabsorbing molecules from the glomerular filtrate or secreting molecules into it. As the filtrate passes through the tubule, most of the water and solutes are reabsorbed from it and are returned to the blood. Other molecules are secreted from the peritubular capillaries into the tubule. Some of the water, along with most of the urea and other substances that are not reabsorbed, is excreted as urine.

An important means of water conservation in mammals is the capacity for excreting a urine that is hypertonic in relation to the blood. The loop of Henle is the portion of the mammalian nephron that makes possible the production of hypertonic urine.

The function of the nephron is influenced by hormones, chiefly antidiuretic hormone (ADH), which is released from the pituitary gland, and aldosterone, an adrenal cortical hormone. ADH increases the return of water to the blood and so decreases water excretion. Aldosterone increases the reabsorption of sodium ions, which are followed by chloride ions and water, and the secretion of potassium ions.

QUESTIONS

1. Distinguish between the following: afferent arteriole/efferent arteriole; Bowman's capsule/glomerulus; extracellular fluid/intracellular fluid; aldosterone/ADH.

2. Explain the following terms in relation to kidney function: filtration, secretion, reabsorption, and excretion.

3. Sketch a nephron, indicating the pathways of the blood and the glomerular filtrate.

4. Diagram the paths of (a) glucose and (b) excreted urea through the human kidney.

5. Reexamine the micrographs in Figures 37–8 and 37–10. Describe the function of each structure visible in them in terms of the overall function of the nephron.

6. Why does a high-protein diet require an increased intake of water? (You should be able to think of two different reasons.) Why does a person lose some weight after shifting to a low-salt diet, even without reducing the caloric intake? Given the fact that amino acids in excess of the body's requirements are broken down by the liver, not stored, what is the advantage of a high-protein diet? What might be a disadvantage?

7. Sodium and chloride ions are actively pumped from the renal tubule in both the proximal and distal convoluted tubules and the upper portion of the ascending branch of the loop of Henle. How does the mechanism of active transport in the convoluted tubules differ from the mechanism in the ascending branch of the loop of Henle?

8. The text asserts that the primary factor limiting the concentration of the urine is the salt concentration surrounding the collecting duct. Why is this so?

9. In Figure 37–12, the labels within the tubule indicate the concentration of solutes in the fluid in relation to their concentration in the blood plasma. From the information given about the movement of ions and of water, can you determine the concentration of the fluid in relation to the concentration of the surrounding interstitial fluid?

CHAPTER 38

Homeostasis II: Temperature Regulation

For life forms, the seeming wastelands of the deserts and the poles represent great extremes of hot and cold. Yet, in fact, measured on a cosmic scale, the temperature differences between them are very slight. Life exists only within a very narrow temperature range. The upper and lower limits of this range are dictated by the nature of biochemical reactions, which sustain life and which are all extremely sensitive to temperature change.

Biochemical reactions take place almost entirely in water, the principal constituent of living things. The slightly salty water characteristic of living tissues freezes at −1 or −2°C. Molecules that are not immobilized in the ice crystals are left in such a highly concentrated form that their normal interactions are completely disrupted. In Chapter 8, we saw that, as temperature rises, the movement of molecules increases and the rate of biochemical reactions goes up rapidly. In fact, it is a convenient generalization that the rates of most biochemical reactions about double for every 10°C increase in temperature. The upper temperature limit for life is apparently set by the point at which proteins begin to lose their functional three-dimensional conformation (a process known as denaturation). Once denaturation occurs, enzymes and other proteins whose function depends on a specific shape are inactivated. As a consequence of this restriction to a very narrow tem-

38–1
Temperature regulation involves behavioral responses, as well as physiological and anatomical adaptations. Here a jackrabbit seeks shelter from the Arizona sun in the shade of a mesquite tree. Note the large ears, which are highly vascularized.

perature range, living organisms must either find external environments that range from just below freezing to between 45 and 50°C, or they must create suitable internal environments. (However, there are exceptions: cyanobacteria have been found in hot springs at 85°C, and recently bacteria have been discovered in the superheated waters of submarine vents, where heat escapes through fissures in the earth's crust. In the laboratory, these bacteria metabolize and multiply at temperatures of about 100°C. The special adaptations that make this possible are not yet known.)

The ways in which temperature requirements are met—and, in particular, the internal regulation of temperature, another example of homeostasis—are the subject of this chapter. Before going into the details of these processes, however, it is worth taking a moment to consider what excellent temperature regulators most mammals are. One of the simplest and most dramatic demonstrations of this capacity was given some 200 years ago by Dr. Charles Blagden, then secretary of the Royal Society of London. Dr. Blagden, taking with him a few friends, a small dog in a basket, and a steak, went into a room that had a temperature of 126°C (260°F).* The entire group remained there for 45 minutes. Dr. Blagden and his friends emerged unaffected. So did the dog. (The basket had kept its feet from being burned by the floor.) But the steak was cooked.

38-2
Life processes can take place only within a very narrow range of temperature. The temperature scale shown here is the Kelvin, or absolute, scale. Absolute zero (0 K) is equivalent to −273.1°C, or −459°F, and is the temperature at which all molecular motion ceases.

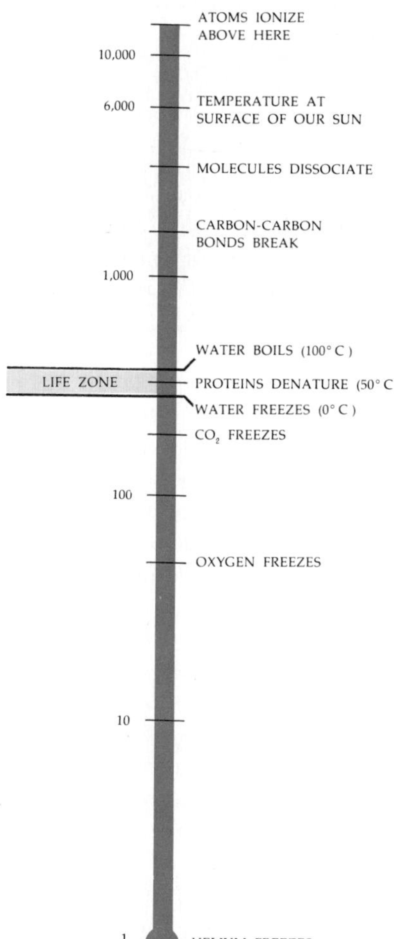

PRINCIPLES OF HEAT BALANCE

Water balance, as we saw in the preceding chapter, requires that the loss of water through urine, sweat, and respiration equal the water ingested. Similarly, maintaining a constant temperature depends on heat gains equaling heat losses. For living organisms there are two primary sources of heat gain: one is the radiant energy of the sun; the other is the heat generated by exothermic chemical reactions in an organism (Figure 38-3).

Heat Transfer

Heat is lost by transfer to a cooler body. If the two bodies are in direct contact, the movement of heat is called *conduction*. Conduction of heat consists of the direct transfer of the kinetic energy of molecular motion, and it always occurs from a region of higher temperature to a region of lower temperature. Some materials are better heat conductors than others. When you step out of bed barefoot on a cool morning, you probably prefer to step on a wool rug than on the bare floor. Although both are at the same temperature, the rug feels warmer. If you touch something metal, such as a brass doorknob, it will feel even cooler than the wood floor. These apparent differences in temperature are actually differences in the speed at which these different types of materials conduct heat away from your body. The doorknob, like all metals, is an excellent conductor, and wood is a better heat conductor than wool.

Water is a better conductor than air. You are quite comfortable in air at 21°C (70°F) but may be uncomfortable in water at the same temperature. Fat and air are poor conductors and so can serve as insulators. Animals that need to conserve heat are typically insulated with either fur or feathers, which trap air, or with fat or blubber.

* A Celsius to Fahrenheit temperature conversion scale can be found in Appendix B at the back of the book.

(a)

(b)

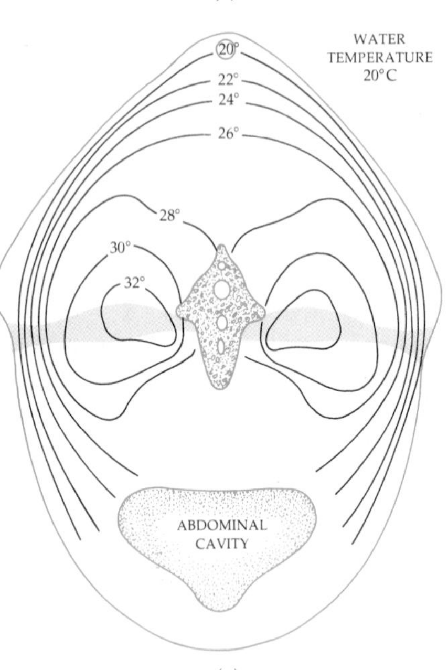

(c)

38–3

Heat sources. (a) *Within the wintering hive, bees maintain their temperature by clustering together in a dense ball; the lower the temperature, the denser the cluster. The clustered bees produce heat by constant muscular movements of their wings, legs, and abdomens. In very cold weather, the bees on the outside of the cluster keep moving toward the center, while those in the center move to the colder outside periphery. The entire cluster moves slowly about on the combs, eating the stored honey, which is their energy source. The photograph shows the upper half of an opened wintering hive.* (b) *Mound-building birds incubate their eggs in large compost heaps. The parent birds start by digging a pit, 3 meters wide and 1 meter deep, and raking plant litter into it. Following the spring rains, when the litter begins to decompose, the birds cover the fermenting heap with a layer of sand up to a meter in depth. The female lays her eggs in this mound, which is heated from beneath by the chemical reactions in the litter. The male regulates the temperature of the eggs by scraping away the sand around them to expose them to air or sun or piling up warm sand around them at night. The entire cycle from the beginning of mound building until the last egg hatches takes about a year.* (c) *Muscular activity and chemical ractions are also sources of internal heat. This diagram shows the temperatures in a cross section of a 70-kilogram big-eye tuna. The heat is produced by metabolism, particularly in the hard-working, red, myoglobin-containing muscle tissue, indicated in light color.*

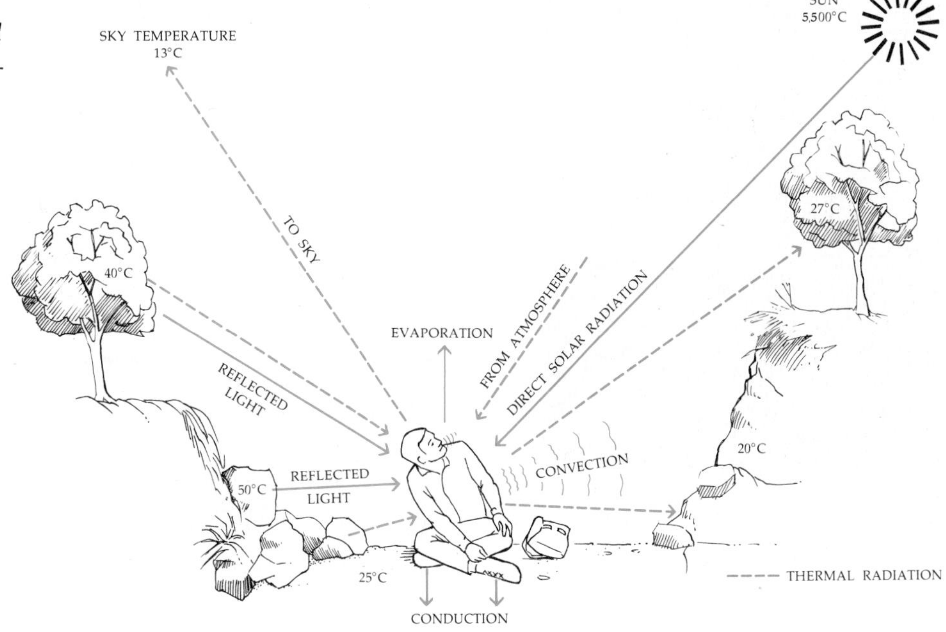

38–4
Heat exchanges between a mammal and its environment. The core body temperature of the man is 37°C. The air temperature is 30°C and there is no wind movement.

Conduction in fluids (air or water) is always influenced by *convection*, the movement of air or water in currents. Because both air and water become lighter as they get warmer, they move away from a heat source and are replaced by colder air or water, which again moves away as it warms.

Radiation is the transfer of energy by electromagnetic waves in the absence of direct contact, as between the sun and an organism. The energy may be transferred as light or heat, depending on the wavelength of the radiation (see Figure 10–3 on page 212). Light energy falling on an object is either absorbed as heat or reflected. Dark objects absorb more than light-colored ones.

Another route of heat exchange is by evaporation. As we saw in Chapter 2, every time a gram of water changes from a liquid to a gas, it takes more than 500 calories away with it. Many organisms, including ourselves, have exploited this property of water as a means for rapid adjustment of the heat balance. These routes of heat transfer are summarized in Figure 38–4.

Size and Temperature

Heat is transferred into or out of any object, animate or inanimate, across the body surface, and transfer of heat, like diffusion of gases, is proportional to the surface area exposed. The smaller the object, as we saw in Chapter 5, the larger its surface-to-volume ratio. Ten pounds of ice, separated into individual ice cubes, will melt far faster than the equivalent volume in a solid block. For the same reason, it is much more difficult for a small animal to maintain a constant body temperature than it is for a large one.

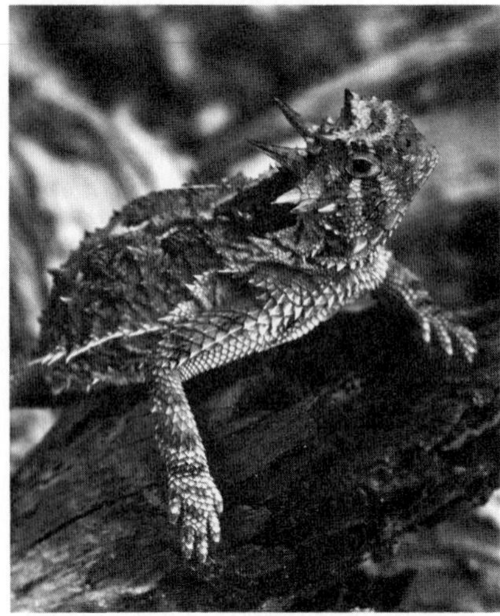

(a)

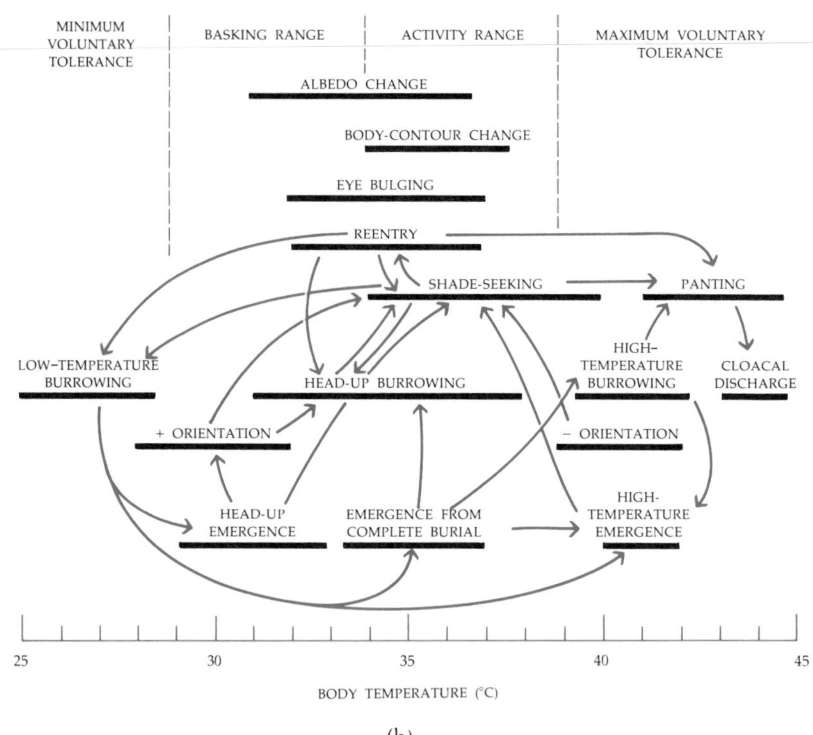

(b)

38–5

In laboratory studies, the internal temperature of reptiles was shown to be almost the same as the temperature of the surrounding air. It was not until observations were made of these animals in their own environment that it was found that they have behavioral means for temperature regulation. By absorbing solar energy, reptiles can raise their temperature well above that of the air around them. (a) *Shown here is a horned lizard (often but not accurately known as a horned toad) that, having been overheated by the sun, has raised its body to allow cooling air currents to circulate across its belly.* (b) *This chart is based on field studies of the behavior of a horned lizard in response to temperature fluctuations. Changes in albedo (whiteness) are produced by pigment changes in the epithelial cells. Albedo changes affect the reflection versus absorption of light rays.*

"Cold-blooded" and "Warm-blooded"

In common parlance, animals are often characterized as "cold-blooded" and "warm-blooded." This fits in with everyday experience: a snake is usually cold to the touch, and a living bird feels warm. Actually, however, a "cold-blooded" animal may create an internal temperature for itself that is warmer than that of a "warm-blooded" one. Another approach is to classify animals as ectotherms and endotherms. Ectotherms are warmed from the outside in, and endotherms are warmed from the inside out. These categories correspond approximately but not completely with "cold-blooded" and "warm-blooded." As we saw in Figure 38–3c, a large fish, such as a tuna, which is considered "cold-blooded," generates a considerable amount of metabolic heat. Finally, animals may be characterized as poikilotherms and homeotherms. A *poikilotherm* (from the Greek word *poikilos,* meaning "changeable") has a variable temperature, and a *homeotherm,* a constant one. These terms are generally used as synonymous with "cold-blooded" and "warm-blooded," though again the correspondence is not perfect. Fish that live deep in the sea, where the temperature of the water remains constant, have a body temperature far more constant than a bat or a hummingbird, whose temperatures, as we shall see, fluctuate widely depending on their state of activity. Nevertheless, poikilotherm and homeotherm are the most widely accepted terms and are those that we shall use in the rest of this discussion.

POIKILOTHERMS

Most aquatic animals are poikilotherms, with their temperatures varying with that of the surrounding water. Although the metabolic processes of such animals generate heat, it is usually quickly dissipated, even in large animals. In most fish, for example, heat is rapidly carried from the core of the body by the bloodstream and is lost by conduction into the water. A large proportion of this body heat is lost from the gills. Exposure of a large, well-vascularized surface to the water is necessary in order to acquire enough oxygen, as you will recall from Chapter 35. This same process rapidly dissipates heat, so fish cannot maintain a body temperature significantly higher than that of the water. They also cannot maintain a temperature lower than that of the water, since they have no means of unloading heat.

In general, large bodies of water, for the reasons we discussed in Chapter 2, maintain a very stable temperature. At no place in the open ocean does the temperature vary more than 10°C in a year. (By contrast, temperatures on land may vary annually in a given area by as much as 60 to 70°C.) Because water expands as it freezes, ice floats on the surface of the water, insulating the water so that life continues beneath the surface of the ice. In shallower water, where greater temperature changes occur, fish seek an optimal temperature, presumably the one to which their metabolic processes are adapted. However, because they can do almost nothing to make their own temperature different from that of the surrounding water, they can be quickly victimized by any rapid, drastic changes in water temperature.

Terrestrial reptiles—snakes, lizards, and tortoises—are also poikilotherms, but they can often maintain remarkably stable body temperatures during their active hours, even though ambient temperatures vary. By careful selection of suitable sites, such as the slope of a hill facing the sun, and by orienting their bodies with a maximum surface exposed to sunlight, they can heat themselves rapidly (as rapidly as 1°C per minute), even on mornings when the air temperature, as on deserts or in the high mountains, may be close to 0°C. As soon as their body temperatures go above the preferred level, lizards move to face the sun, presenting less exposed surface, or they seek shade. By such behavioral responses, lizards are able to keep

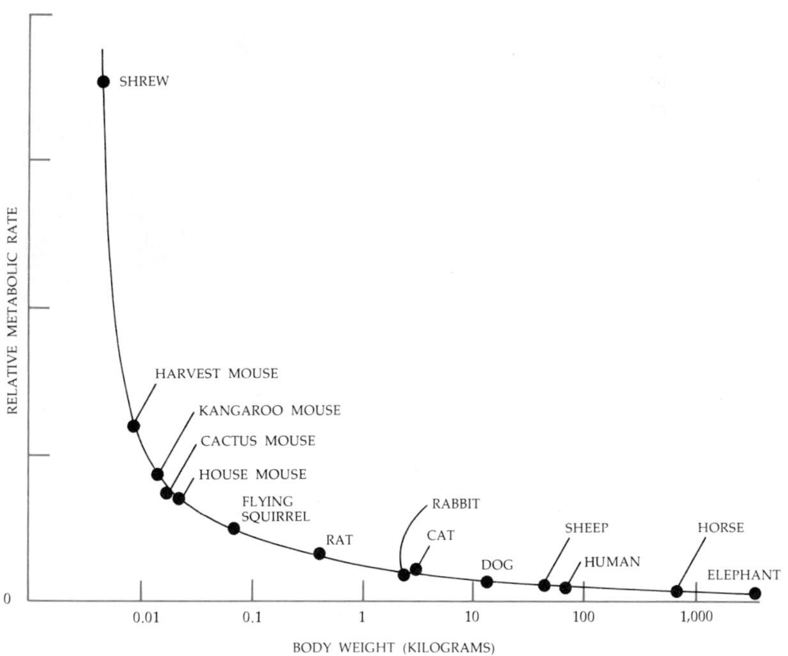

38–6
Observed metabolic rates of mammals. Each division on the abscissa represents a tenfold increase in weight. The metabolic rate of very small mammals is much higher than that of larger mammals, owing principally to their greater surface-to-volume ratios.

their temperatures oscillating within a quite narrow range. At night, when their body temperatures drop and they become sluggish, these animals seek the safety of their shelters. There they will not become immobilized in an exposed position, where they would be vulnerable to predators.

HOMEOTHERMS

Homeotherms are animals that maintain a constant body temperature despite fluctuations in their environment, and most maintain a body temperature well above that of their surroundings. Among modern organisms, only birds and mammals are true homeotherms. All homeotherms are endotherms, with the oxidation of glucose and other energy-yielding molecules within body cells as their primary source of heat. In terms of energy requirements, their cost of living is high: the metabolic rate of a homeotherm is about twice that of a poikilotherm of similar size at similar temperature. In short, homeotherms pay a high price for their independence. Also, the smaller the size, the higher the price. Small homeotherms have a proportionally larger heat budget than large ones, because of surface-to-volume ratios (see Figure 38–6).

Because it is heated from within, a warm-blooded mammal is warmer at the core of its body than at the periphery. (Our temperature usually does not reach 37°C, or 98.6°F, until some distance below the skin surface.) Heat is transported from the core to the periphery largely by the bloodstream. At the surface of the body, the heat is transferred to the air, as long as air temperature is less than body temperature. Temperature regulation involves increasing or decreasing heat production and increasing or decreasing heat loss at the body surface.

(a)

(b)

(c)

38–7
The size of the extremities in a particular type of animal can often be correlated with the climate in which it lives. (a) *The fennec fox of the North African desert has large ears that help it to dissipate body heat.* (b) *The red fox of the eastern United States has ears of intermediate size, and* (c) *the Arctic fox has relatively small ears. Like all mammals, these foxes are homeotherms, animals that maintain a constant internal temperature despite variations in external temperature.*

The Mammalian Thermostat

The remarkable constancy of temperature characteristic of humans, and many other animals as well, is maintained by an automatic system—a thermostat—in the hypothalamus. Like the thermostat that regulates your furnace, the thermostat in the hypothalamus receives information about the temperature, compares it to the set point of the thermostat, and, on the basis of this comparison, initiates appropriate responses. Unlike your furnace thermostat, however, the hypothalamic thermostat receives and integrates information from widely scattered temperature receptors. Also, rather than just controlling an on-off switch, the hypothalamic thermostat has a variety of responses at its command, as summarized in Figure 33–13 on page 669.

Under ordinary conditions, the skin receptors for hot and cold are probably the most important sources of information about temperature change. However, the hypothalamus itself contains receptor cells that monitor the temperature of the blood flowing through it. As some interesting experiments have demonstrated, information received by these hypothalamic receptors overrides that from other sources. For example, in a room in which the air is warmer than body temperature, if the blood circulating through a person's hypothalamus is cooled, he or she will stop perspiring, even though the skin temperature continues to rise.

The elevation of body temperature known as fever is due not to a malfunction of the hypothalamic thermostat but to its resetting. Thus, at the onset of fever, an individual typically feels cold and often has chills; although the body temperature is rising, it is still lower than the new thermostat setting. The substance primarily responsible for the resetting of the thermostat is a protein released by the white blood cells in response to a pathogen. The adaptive value of fever, if any—and we can only presume that there is one—is a subject of current research. One possibility, supported by some evidence, is that it creates an environment inhospitable to pathogens. Another is that it stimulates the release of a substance important in immunological reactions.

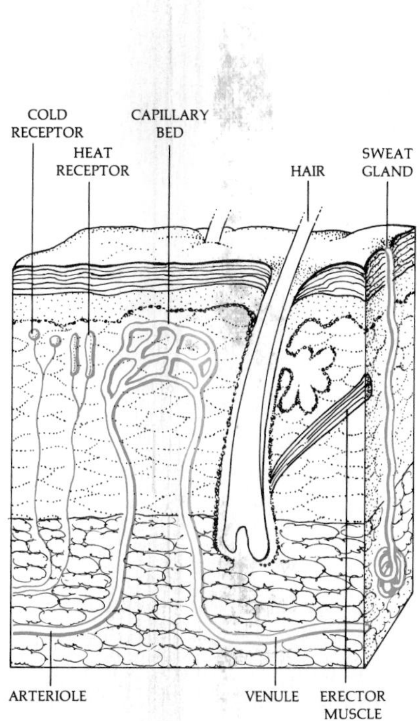

38–8

Cross section of human skin showing structures involved in temperature regulation. In cold, the arterioles constrict, reducing the flow of blood through the capillaries, and the hairs, each of which has a small erector muscle under nervous control, stand upright. Beneath the skin is a layer of subcutaneous fat that serves as insulation, retaining the heat in the underlying body tissues. With rising temperatures, the arterioles dilate and the sweat glands secrete a salty liquid. Evaporation of this liquid cools the skin surface, dissipating heat (approximately 540 calories for every gram of H_2O*).*

Regulating as Body Temperature Rises

In mammals, as the body temperature rises above its thermostat setting, the blood vessels near the skin surface are dilated, and the supply of blood to the skin increases. If the air is cooler than the body surface, heat can be transferred from the skin to the air. Heat can also be lost from the surface by the evaporation of saliva or perspiration. Some animals—dogs, for instance—pant, so that air passes rapidly over their large, moist tongues, evaporating saliva. Cats lick themselves all over as the temperature rises, and evaporation of their saliva cools their body surface. Horses and human beings sweat from all over their body surfaces. For animals that dissipate heat by water evaporation, temperature regulation at high temperatures necessarily involves water loss, which, in turn, stimulates thirst and water conservation by the kidneys. Dr. Blagden and his friends were probably very thirsty.

Regulating as Body Temperature Falls

When the temperature of the circulating blood begins to fall below the thermostat setting, blood vessels near the skin surface are constricted, limiting heat loss from the skin. Metabolic processes increase. Part of this increase is due to increased muscular activity, either voluntary (shifting from foot to foot) or involuntary (shivering). Part is due to direct stimulation of metabolism by the endocrine and

38–9
(a) *Dogs unload heat by panting, which involves short, shallow breaths. When it is hot, a dog pants at a rate of about 300 to 400 times a minute, compared with a respiration rate of from 10 to 40 times a minute in cool surroundings. Evaporation of water from the tongue, unlike evaporation from the skin, does not result in the loss of salt or other important ions.* (b) *Elephants, lacking sweat glands, wet down their thick, dry skins with mud or water. They also unload heat by flapping their ears, which are highly vascularized.*

nervous systems. Adrenaline stimulates the release and oxidation of glucose. Autonomic nerves to fat increase its metabolic breakdown. In some mammals, though apparently not in humans, the thyroid gland increases its release of thyroxine, the thyroid hormone. Thyroxine appears to exert its effects directly on the mitochondria.

Most mammals have a layer of subcutaneous fat that serves as insulation. Homeotherms also characteristically have hair or feathers that, as temperature falls, are pulled upright by erector muscles under the skin, trapping air, which insulates the surface. All we get, as our evolutionary legacy, are goose pimples.

Meeting Energy Costs

As we noted previously, maintaining a constant temperature adds significantly to an animal's energy budget. Birds, in particular, have high metabolic requirements. Although comparatively small, they maintain a higher body temperature—40 to 42°C—than most mammals. Also, unlike many other small homeotherms, they spend much of their time exposed. Flight compounds their problems. To fly, birds must keep their weight down, and so they cannot store large amounts of fuel. Because of this and the high energy requirements of their way of life, birds need to eat constantly. A bird eating high-protein foods, such as seeds and insects, commonly consumes as much as 30 percent of its body weight per day. Bird migrations are dictated not so much by a need to seek warmer weather as by a need for longer days with more daylight hours for feeding themselves and their young.

Cutting Energy Costs

Less energy is consumed by a sleeping animal than by an active one. Bears, for example, nap for most of the winter, living on fat reserves, permitting their temperature to drop several degrees, and keeping their energy requirements down.

Turning down the thermostat saves more fuel. Some small animals have different day and night settings. A hummingbird, for instance, has an extremely high rate of fuel consumption, even for a bird. Its body temperature drops every evening when it is resting, thereby decreasing its metabolic requirements and its fuel consumption.

38–10
Heat conservation measures. (a) *Hibernation saves energy by turning down the thermostat. This dormouse has prepared for hibernation by storing food reserves in body fat.* (b) *Note the heat-conserving reduction in surface-to-volume ratio in this hibernating chipmunk.* (c) *Many small animals, such as these penguin chicks, huddle together for warmth. Huddling decreases the effective surface-to-volume ratio. A few of the chicks have lifted their heads from the warmth of the huddle to observe the photographer at work.*

(a)

(b)

Hibernation, which comes from *hiber*, the Latin word for "winter," is another means of adjusting energy expenditures to food supplies. Hibernating animals do not stop regulating their temperatures altogether—like the hummingbird, they turn down their thermostats. Hibernators are mostly small animals, including a few insectivores, hamsters, ground squirrels, some South American opossums, some bats, and a few birds, such as the poor-will of the southwestern United States. In these animals, the thermostat is set very low, often close to that of the surrounding air (if the air is above 0°C). The heartbeat slows; the heart of an active ground squirrel, for instance, beats 200 to 400 times a minute, whereas that of a hibernating one beats 7 to 10 times a minute. The metabolism, as measured by oxygen consumption, is reduced 20 to 100 times. Apparently, even aging stops; hibernators have much longer life spans than similar nonhibernators. However, despite these profound physiological changes, hibernators do not cease to monitor the external environment. If the animals are exposed to carbon dioxide, for example, they breathe more rapidly, and those of certain species wake up. Similarly, animals of many species wake up if the temperature in their burrows drops to a life-threatening low of 0°C. Hibernators of some species can be awakened by sound or by touch.

Arousal from hibernation can be rapid. In one experiment, bats kept in a refrigerator for 144 days without food were capable of sustained flight after 15 minutes at room temperature. As indicated, however, by the arousals at 0°C, the arousal is a process of self-warming rather than of collecting environmental heat. Breathing becomes more regular and then more rapid, increasing the amount of oxygen available for consumption; subsequently, the animal "burns" its stored food supplies as it returns to its normal temperature and the fast pace of a homeothermic existence.

(c)

ADAPTATIONS TO EXTREME TEMPERATURES

We rely mainly on our technology, rather than our physiology, to allow us to live in extreme climates, but many animals are comfortable in climates that we consider inhospitable or, indeed, uninhabitable.

Adaptations to Extreme Cold

Animals adapt to extreme cold largely by increased amounts of insulation. Fur and feathers, both of which trap air, provide insulation for Arctic land animals. Fur and feathers are usually shed to some extent in the spring and regrow in the fall.

Aquatic homeotherms, such as whales, walruses, seals, and penguins, are insulated with fat (neither fur nor feathers serve as effective insulators when wet). In general, the rate of heat loss depends both on the amount of surface area exposed and on temperature differences between the body surface and the surroundings. These marine animals, which survive in extremely cold water, can tolerate a very great drop in their skin surface temperature; measurements of skin temperature have shown that it is only a degree or so above that of the surrounding water. By permitting the skin temperature to drop, these animals, which maintain an internal temperature as warm as that of a person, expend very little heat outside of their fat layer and so keep warm—like a diver in a wet suit.

Numerous animals similarly permit temperatures in the extremities to drop. By so doing, they conserve heat. The extremities of such animals are actually adapted to live at a different internal temperature from that of the rest of the animal. For example, the fat in the foot of an Arctic fox has a different thermal behavior from

38–11
Aquatic mammals, such as this bull fur seal, have a heavy layer of fat beneath the skin that acts, in effect, like a diver's wet suit, insulating them against heat loss. On land, they have a problem unloading heat. One way they do it is by keeping their fur moist; another is by sleeping, which, in some species, reduces heat production by almost 25 percent.

that of the fat in the rest of its body, so that its footpads are soft and resilient even at temperatures of −50°C. Also, for many of these animals, particularly those that stand on the ice, this capacity is essential for another reason. If, for example, the feet of an Arctic seabird were as warm as its body, they would melt the ice, which might then freeze over them, trapping the animal until the spring thaws set it free.

Countercurrent Exchange

In many Arctic animals, the arteries and veins leading to and from the extremities are juxtaposed in such a way that the chilled blood returning from the legs (or fins or tail) through the veins picks up heat from the blood entering the extremities through the arteries. The veins and arteries are closely apposed to give maximum surface for heat transfer. Thus the body heat carried by the blood is not wasted by

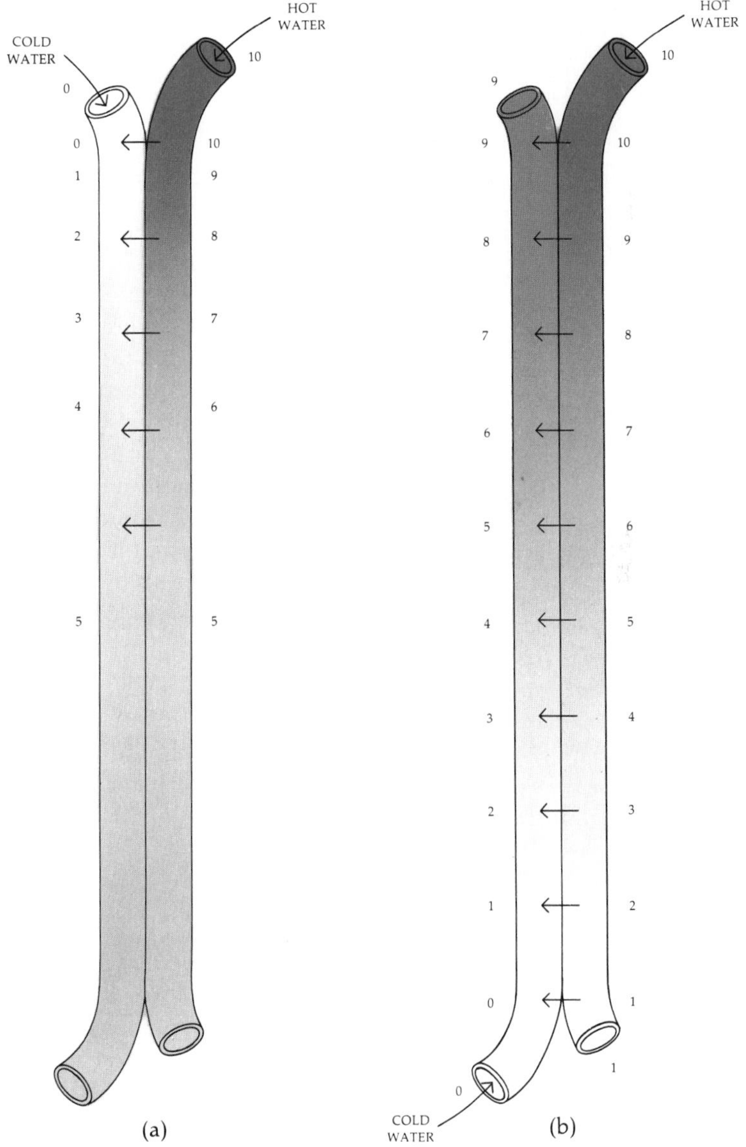

38–12
The principle of countercurrent heat exchange is illustrated by a hot-water pipe and a cold-water pipe placed side by side. In (a), *the hot water and cold water flow in the same direction. Heat from the hot water warms the cold water until both temperatures equalize at 5. Thereafter, no further exchange takes place, and the outflowing water in both pipes is lukewarm. In* (b), *the flow is in opposite directions (that is, it is countercurrent), so that heat transfer continues for the length of the pipes. The result is that the hot water transfers most of its heat as it travels through the pipe, and the cold water is warmed to almost the initial temperature of the hot water at its source.*

being dissipated to the cold air at the extremities. Instead it serves the useful function of warming the chilled blood that would otherwise put a thermal burden on the body. This arrangement, which serves to keep heat in the body and away from the extremities, is, of course, another example of countercurrent exchange.

38–13
By facing the sun, a camel exposes as small an area of body surface as possible to the sun's radiation. Its body is insulated by fat on top, which minimizes heat gain by radiation. The underpart of its body, which has much less insulation, radiates heat out to the ground cooled by the animal's shadow. Note also the loose-fitting garments worn by the camel driver. When such garments are worn, sweat evaporates off the skin surface. With tight-fitting garments, the sweat evaporates from the surface of the garment instead, and much of the cooling effect is lost. The loose robes, which trap air, also serve to keep desert dwellers warm during the cold nights. Other adaptations of the camel to desert life include long eyelashes, which protect its eyes from the stinging sand, and flattened nostrils, which retard water loss.

Adaptations to Extreme Heat

We are efficient homeotherms at high external temperatures, as Dr. Blagden's experiment showed. Our chief limitation in this regard is that we must evaporate a great deal of water in order to unload body heat, and so our water consumption is high. The camel, the philosophical-looking "ship of the desert," has several advantages over human desert dwellers. For one thing, the camel excretes a much more concentrated urine; in other words, it does not need to use so much water to dissolve its waste products. In fact, we are very uneconomical with our water supply; even dogs and cats excrete a urine twice as concentrated as ours.

Also, the camel can lose more water proportionally than a person and still continue to function. If a human loses 10 percent of body weight in water, he or she becomes delirious, deaf, and insensitive to pain. If the loss is as much as 12 percent, the person will be unable to swallow and so cannot recover without assistance. Laboratory rats and many other common animals can tolerate dehydration of up to 12 to 14 percent of body weight. Camels can tolerate the loss of more than 25 percent of their body weight in water, going without drinking for as long as an entire week in the summer months, three weeks in the winter.

Finally, and probably most important, the camel can tolerate a fluctuation in internal temperature of 5 to 6°C. This tolerance means that it can store heat by letting its temperature rise during the daytime (which the human thermostat would never permit) and then release heat during the night. The camel begins the next day at below its normal temperature—storing up coolness, in effect. It is estimated that the camel saves as much as 5 liters of water a day as a result of these internal temperature fluctuations.

Camels' humps were once thought to be water-storage tanks, but actually they are localized fat deposits. Physiologists have suggested that the camel carries its fat in a dorsal hump, instead of distributed all over the body, because the hump, acting as an insulator, impedes heat flow into the body core. An all-over fat distribution, which is decidedly useful in Arctic animals, would not be in inhabitants of hot climates.

Small desert animals usually do not unload heat by sweating or panting; because of their relatively large surface areas, such mechanisms would be extravagant in terms of water loss. Rather, they regulate their temperature by avoiding direct heat. Most small desert animals are nocturnal, like the earliest mammals, from which they are descended.

SUMMARY

Life can exist only within a very narrow temperature range, from about 0°C to about 50°C, with few exceptions. Animals must either seek out environments with suitable temperatures or create suitable internal environments. Heat balance requires that the net heat loss from an organism equal the heat gain. The two primary sources of heat gain are the radiant energy of the sun and cellular metabolism. Heat is lost by conduction, the transfer of thermal energy from one object to another; by radiation via electromagnetic waves; and by evaporation.

Animals that take in energy primarily from the outside are known as ectotherms; animals that rely on internal energy production are known as endotherms. Most ectotherms are poikilotherms, animals whose internal temperature fluctuates with that of the external environment. Most endotherms are homeotherms, animals that maintain a constant internal temperature.

In mammals, temperature is regulated by a thermostat in the hypothalamus that senses temperature changes in the circulating blood and triggers appropriate responses. As body temperature rises, blood vessels in the skin dilate, increasing the blood flow to the surface of the body. Evaporation of water from any body surface increases heat loss.

As body temperature falls, energy production is increased by increased muscular activity (including shivering) and by nervous and hormonal stimulation of metabolism. Homeotherms in cold climates are usually insulated by subcutaneous fat and fur or feathers.

Energy can be conserved by setting the thermostat lower. Some small homeotherms, such as hummingbirds, have different day and night settings. Others, hibernators, make seasonal adjustments.

Adaptations to extreme cold principally involve (1) insulation by heavy layers of fat and fur or feathers, and (2) countercurrent mechanisms for conserving heat within the body. Adaptations to extreme heat are largely water conservation measures and also include many behavioral responses.

QUESTIONS

1. Distinguish among the following terms: endotherm, ectotherm, poikilotherm, and homeotherm.

2. Compare the surface-to-volume ratio of an Eskimo igloo with that of a California ranch house. In what way is the igloo well suited to the environment in which it is found?

3. Compare hibernation in animals with dormancy in plants. In what ways are they alike? In what ways are they different?

4. Plants, as well as animals, are warmed above air temperature by sunlight, and plants cannot actively seek shade. Why might the leaves in the sunlight at the top of an oak tree be smaller and more extensively lobed (more fingerlike) than the leaves in the shady areas lower on the tree?

5. In terms of temperature regulation, what is the advantage of having temperature monitors that are located externally, as on the outside of a building or in the skin? Why is it also important to monitor the internal temperature? Which should be the principal source of information for your thermostat?

6. Diagram a heat-conserving countercurrent mechanism in the leg of an Arctic animal.

CHAPTER 39

Homeostasis III: The Immune Response and Other Defenses

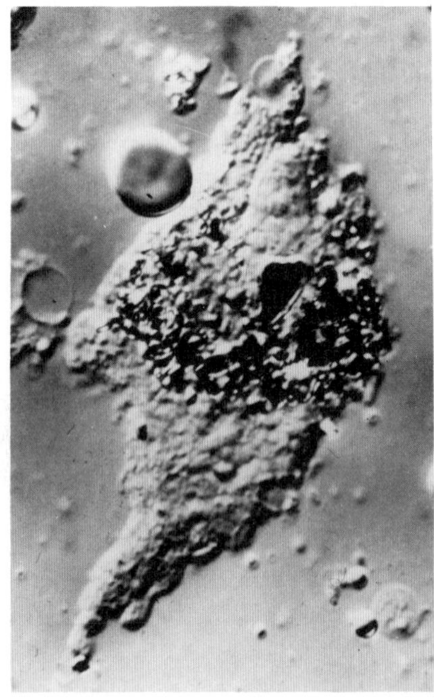

39–1
A macrophage, a phagocytic white blood cell, in action in a polluted lung. The macrophage has ingested something angular, probably a stone chip. White blood cells are important both in nonspecific and specific responses to microorganisms and other possible pathogens.

Every living thing, including yourself at this moment, is surrounded by potentially harmful microorganisms. In the course of evolution, organisms have developed an elaborate network of defenses designed to exclude these would-be invaders or to overcome them should they gain entry. In vertebrates, there are three general categories of defenses: (1) nonspecific defenses, in particular the inflammatory response; (2) interferon, which might be described as a semispecific response; and (3) the immune response, which is highly specific.

NONSPECIFIC DEFENSES

Anatomic Barriers

The body's first line of defense is its outer wrapping of skin and mucous membranes. The skin, with its tough layer of keratin, is an impregnable barrier as long as it is intact. Mucous membranes, though more fragile, are protected, as their name indicates, by a layer of mucus. Membranes in the respiratory tract are carpeted with cilia (page 115), which sweep microorganisms, dirt, and debris away from the epidermal surface. The lower intestinal tract and also the skin harbor a resident population of bacteria that defend their home territory against other microorganisms. The extremely acidic contents of the stomach provide an inhospitable environment for potential immigrants. Some membranes are constantly flushed with fluids, such as saliva, tears, and nasal secretions, that contain antimicrobial substances, including lysozyme (page 170). Despite these defenses, the membranes are the most common sites of entry of microorganisms or their toxins.

The Inflammatory Response

If a microorganism penetrates the outer barrier, it encounters a second line of defense, consisting of a variety of agents carried by the circulating blood and lymph. Suppose, for example, you nick your skin. Some of the injured cells immediately release histamine and other chemicals that lead to distension of the nearby capillaries. The consequent increase in local blood flow is primarily responsible for producing the outward manifestation of the *inflammatory response*, making the area surrounding the wound look red and feel hot. Granulocytes, which are circulating white blood cells, push their way through the distended

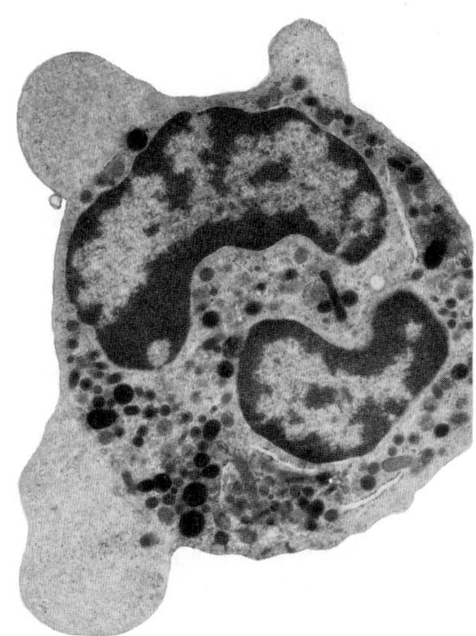

39–2
White blood cells move through the bloodstream and lymphatic tissues by amoeboid motion, forming and retracting pseudopodia. Many of these cells are also, like amoebas, phagocytic, engulfing bacteria and other small particles. This cell is a granulocyte; note the many-lobed nucleus and the numerous small granules in the cytoplasm.

capillary walls, crowding into the site of the injury, attracted by chemicals released by the cellular response to injury.

Granulocytes are classified by their staining abilities as neutrophils, eosinophils, and basophils. Neutrophils are by far the most numerous, making up 50 to 70 percent of all white blood cells. When the first signs of inflammation appear, neutrophils, which are usually round and move freely through the bloodstream, begin to stick to the inner surface of the endothelium (the lining of the blood vessels). They then develop amoeboid projections that enable them to push their way between the endothelial cells into the infected tissues. Here they phagocytize microorganisms and other foreign particles, just as an amoeba engulfs its prey (page 435). The cytoplasmic granules from which granulocytes derive their name are actually lysosomes, and once the offending microparticle is within the neutrophil, lysosomes fuse with the phagocytic vacuole. Most phagocytized microbes are killed and digested within the vacuole by microbicidal proteins and other lytic enzymes from the lysosomes. Some microbes—such as the encapsulated pneumococcus that played such a significant role in the earliest days of molecular genetics (page 279)—have evolved their defenses against the defenses of their hosts and so remain virulent.

Basophils and eosinophils are also phagocytic. Basophils contain granules that rupture readily, releasing chemicals, such as histamine, that enhance the inflammatory response. Basophils, as well as mast cells, which are noncirculating cells found in connective tissue, are important components of allergic reactions. The role of eosinophils is not clear; they are found in increased numbers during infections involving internal parasites, such as worms.

Another type of circulating white blood cell that plays a major role in the inflammatory response is the monocyte. Monocytes, like neutrophils, are attracted to the site of an infection by chemicals released by both bacterial and host cells, generally arriving later than the neutrophils. Once on location, they are transformed into macrophages, becoming enlarged, ameoba-like, and phagocytic. Macrophages also lodge in the lymph nodes, spleen, liver, lungs, and connective tissues, where they entrap microbes that may have penetrated the initial defenses. They are also important, as we shall see, in the immune response.

Inflammation can produce systemic as well as local symptoms. One common response to bacterial infection is an increase (up to fivefold) in the synthesis and release of neutrophils. Another common response is fever, caused by a protein released by bacteria or host cells in the course of the inflammatory response.

Table 39-1 White Blood Cells

	PERCENT OF TOTAL
Granulocytes	
Neutrophils	50–70
Eosinophils	1–4
Basophils	0.1
Lymphocytes	20–40
Monocytes	2.8

INTERFERON

Interferon differs from the other defense mechanisms of the body in two ways: (1) it acts not to suppress or destroy the invading microorganism directly but rather to stimulate the body's own cells to resist the invader, and (2) it is active against only one class of pathogens, viruses.

The clue that led to the discovery of interferon's existence was the observation that an animal—human or mouse—infected with one virus was not usually susceptible to an infection with another virus. The source of this resistance was traced, with some difficulty, to a protein that was given the name of interferon. Actually, it is now known that several slightly different proteins are involved, so "interferons" is more precise but seldom used. Interferons are very small proteins that bind readily with other molecules and are active in very small amounts. As a result, the isolation of an interferon took almost two decades.

When a cell is invaded by a virus, it releases interferon, which then interacts with receptor sites on the membranes of surrounding cells. Thus stimulated, these cells produce antiviral enzymes that block the translation of viral messenger RNA to protein. Only a very few molecules of interferon seem to be required to protect the surrounding cells from viral infection.

Until recently, the only interferon available for research purposes was the miniscule amount collected from mammalian cells exposed to virus in tissue culture. One-millionth of an ounce cost $1,500 in 1978. Using these small, precious quantities, medical investigators were able to establish that interferon could be used clinically in the control of some virus infections, such as herpes infections of the cornea. Moreover, there was suggestive evidence that interferon was effective against certain forms of cancer. Now, as a result of recombinant DNA techniques, interferon is being produced in much larger quantities, and the hopes raised by interferon's discovery a quarter-century ago will be fulfilled or put to rest.

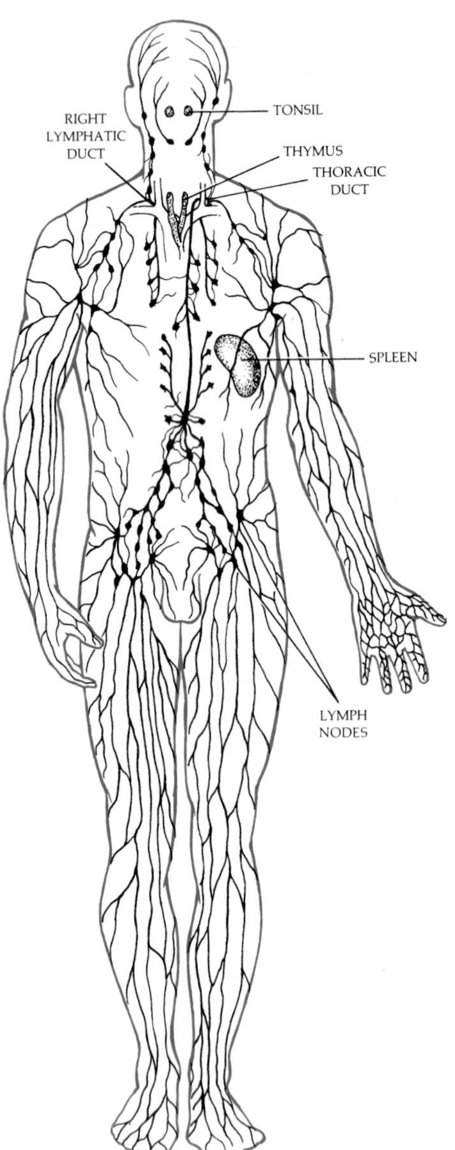

39–3
The immune system includes the lymph vessels, the numerous lymph nodes, the thymus, spleen, and tonsils, and also a complex interacting society of white blood cells.

THE IMMUNE RESPONSE

The immune response differs from the previously described defensive strategies in its high degree of specificity. This specificity derives from the actions and interactions of two remarkable groups of cells, known simply as B lymphocytes and T lymphocytes (or sometimes, even more simply, as B cells and T cells). Most of the rest of this chapter will be taken up with descriptions of their activities. First, however, we shall describe the arena in which these cells operate—the immune system.

The Immune System

As an organ system, the immune system is more diffuse than the gastrointestinal system, for instance, or the excretory system, but it resembles these in being a functional, integrated unit. It includes the lymph vessels, which, as you will recall, are the route for the return of interstitial fluids to the circulatory system, and also, in particular, the lymph nodes. A *lymph node* is a mass of spongy tissues separated into compartments by connective tissue. Microorganisms, other foreign particles, and tissue debris entering the intercellular spaces of any tissue are caught up in the interstitial fluid, swept into the channels of the lymphatic system, and trapped in the lymph nodes. Single lymph nodes are distributed throughout the body, but most are found clustered in particular areas, such as the neck, armpits, and groin. Lymph nodes serve as filters, removing microbes, foreign particles, tissue debris, and dead cells from the circulation. Those near the respiratory system, for example, are often filled with particles of soot or tobacco smoke. Lymph nodes near a cancer site may contain cancerous cells that have broken loose from the primary growth; in cancer operations, such lymph nodes are often excised in order to remove any still viable cancer cells. Lymph nodes also filter out bacterial cells and other microorganisms that have managed to make their way past the first line of defense, the skin and mucous membranes. Lymph nodes are also, as you would expect, densely populated by lymphocytes and macrophages.

The structure of a lymph node is shown in Figure 39–4. Lymph enters the node by the afferent lymphatic vessels, filters through the peripheral lymphatic sinuses into the central cistern, and trickles out through the efferent lymphatic vessel. The

39-4
Diagram of a lymph node. The cortical areas contain follicles, made up of reticulum cells, that are the germinating centers for lymphocytes. Thin, branched tubules, the medullary cords, traverse the node, adjacent to and within the central cistern. One cord is shown in longitudinal section, and the others are shown in cross section. Each cord consists of a small, central blood vessel, or venule, and an outer network of cells, the mantle. The mantle, which is exposed to the lymphatic fluid on its outer surface, is typically distended with lymphocytes and other immunologically active cells that enter the bloodstream as the blood travels through the lymph nodes. Lymph nodes, which range in size from 1 millimeter in diameter to as large as a grape, may increase in size during periods of exceptional activity.

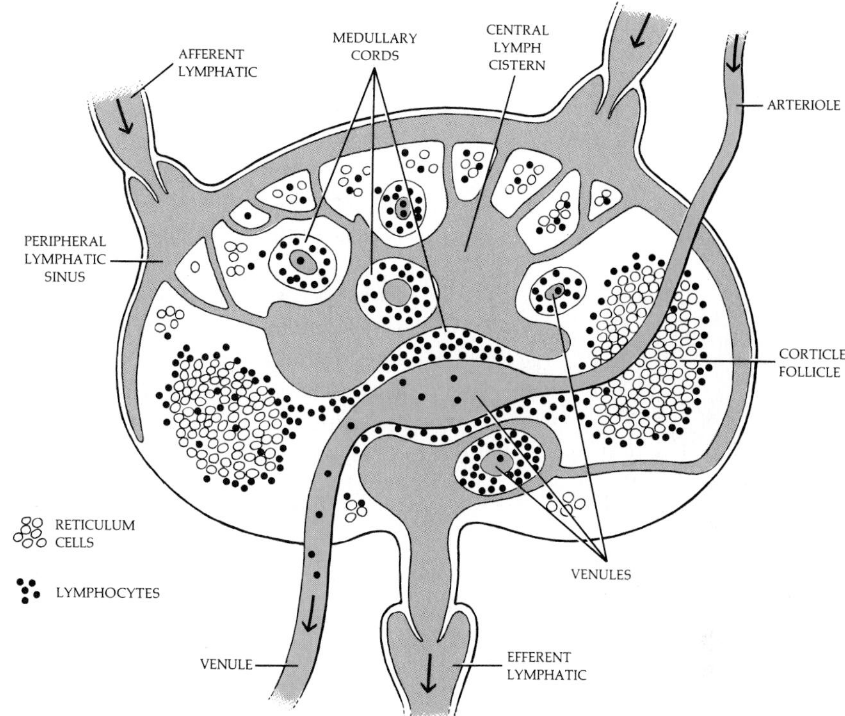

valves within the vessels prevent backflow. The peripheral cortical areas of the node contain follicles where lymphocytes proliferate after exposure to a foreign substance. The incoming lymph carries bacteria, viruses, and other potential pathogens into the node, exposes them to the lymphocytes in the follicles, and collects lymphocytes and other immunologically active cells, which are then circulated throughout the lymphatic system.

The spleen and the tonsils are also rich in lymphocytes and particle-trapping cells. In the spleen, foreign materials enter by way of the blood rather than the lymph; hence this organ is most important in the case of blood-borne infections. The tonsils trap airborne particles. Patches of lymphoid tissue with large lymphoid follicles—known as Peyer's patches—are actually embedded in the wall of the intestine, lying between the inner lining of mucous membrane and the outer muscular coat and defending the body against the billions of microorganisms that inhabit the normal intestinal tract.

The bone marrow and the thymus gland, also part of the immune system, are the primary production sites for the B lymphocytes and T lymphocytes.

B Lymphocyte Responses and the Formation of Antibodies

B lymphocytes are the major protagonists in one type of immune response: the formation of *antibodies*. Antibodies are large globular proteins, also called immunoglobulins, as we noted on page 341, where their structure was described. Each antibody makes a precise three-dimensional combination with a particular molecule or part of a molecule that the body recognizes as foreign. Any such molecular configuration that can trigger the formation of antibodies is known as an *antigen*.

Virtually all foreign proteins and most polysaccharides can act as antigens. The surface of a single cell, such as a bacterial cell, may have a number of different antigens, each of which can elicit formation of a specific antibody. An antigen and its antibody fit together according to the same general principles as an enzyme and its substrate.

Biochemists have identified five classes of antibodies: IgG, IgM, IgA, IgD, and IgE. (The Ig stands for immunoglobulin.) IgG, gamma globulin, constitutes the principal type of circulating antibodies and is the type with which the great majority of the studies on antibody structure and function have been done. The IgM molecules are the first to be produced in the course of infection and last a relatively short time. IgA antibodies are present principally in external secretions, such as tears, saliva, and milk. IgD can be found on the surfaces of antibody-forming cells; its function is not known. IgE, according to present evidence, plays a major role in the expulsion of parasites, such as worms, from the intestinal tract. It also is involved in allergic reactions.

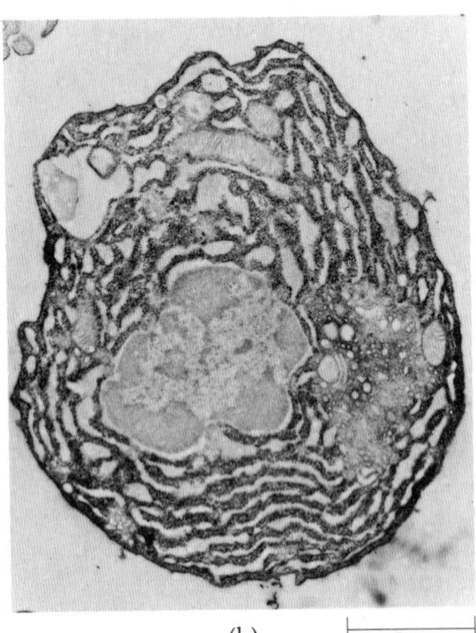

39–5
As a B lymphocyte (a) *differentiates into a plasma cell* (b), *it grows larger, its nucleus becomes relatively smaller and less dense, and there is a large increase in endoplasmic reticulum and ribosomes.*

The B Lymphocyte: A Life History

At any given time, about 2 trillion (2×10^{12}) lymphocytes are on patrol in the human body, moving through the bloodstream, squeezing out between the endothelial cells that form the walls of the blood vessels, migrating through the lymph system, and clustering in the lymph nodes and other lymphoid tissues. These prowling lymphocytes, which comprise 98 percent of the 2×10^{12} present, are small, round, nondividing, metabolically inactive cells. When a particular B lymphocyte meets its destiny in the form of a particular antigen, the B lymphocyte enlarges, its nucleolus swells, polysomes form, and the production of macromolecules begins. At the same time, microtubules form and the lymphocyte begins to divide. B lymphocyte proliferation often takes place in the follicles of the lymph nodes, and therefore these germinal centers enlarge during an infection.

There are two types of daughter cells formed, of which one is the *plasma cell*. Plasma cells continue to divide and to produce antibodies; mature plasma cells can make 3,000 to 30,000 antibody molecules per cell per second. It takes about five days and eight cell generations to produce fully mature antibody-synthesizing cells working at this maximum capacity. Thus, if the microorganism is multiplying also, it may take about this long for the immune system to catch up. In the days before antibiotic therapy, about all the family doctor could do was to await the "break in the fever" that signaled this catching-up process and, often, forecast the patient's eventual recovery. Antibiotics, by suppressing the rate of multiplication of bacteria, enable antibody production to overtake the infection more rapidly.

The second type of cell produced by the antigen-stimulated lymphocyte is the *memory cell*. Memory cells are also antibody-producing cells. They differ from plasma cells in their longevity. Following an infection, memory cells continue to circulate for long periods of time—up to a lifetime. Thus, the second time a particular pathogen gains entry to the body, large-scale production of antibodies to the invader begins immediately, often preventing any significant multiplication of the pathogen.

This rapid response by the memory cells is the source of immunity to many infectious diseases—such as smallpox, measles, mumps, and polio—following an infection. It is also the basis for vaccination against a number of diseases (see essay on page 754). Vaccines are prepared using a closely related pathogen, as in the smallpox vaccine; a killed pathogen, as in the Salk vaccine against polio; or one that has lost some of its pathogenic properties, usually as a result of growing in tissue culture, as in the widely used oral (Sabin) vaccine against polio. Most recently, a vaccine has been prepared using recombinant DNA technology to produce purified components of the protein coat of the virus causing hoof-and-mouth disease, a serious problem in the livestock industry. This method of preparing vaccines has the advantage of eliminating any possibility of disease-causing infection or of introducing contaminants that might produce adverse side reactions.

The Action of Antibodies

Antibodies act against invaders in one of three ways: (1) they may coat the foreign particle so that it can be taken up by phagocytic cells; (2) they may combine with it in such a way that they interfere with one of its vital activities—for example, preventing attachment of a virus to a host cell by covering the protein coat of the virus at the site where it attaches to the host cell membrane; or (3) they may themselves, in combination with another blood component known as *complement,* actually lyse and destroy foreign cells. Complement is a group of at least 11 different proteins found in blood. In particular combinations, they function as lytic enzymes, acting at the point on the bacterial cell wall where antigen and antibody combine. By digesting holes in the foreign cells, complement causes them to burst. In addition to this lytic function, complement may itself coat the foreign cell and promote phagocytosis by other cells, and it mediates and enhances the inflammatory response.

The Structure of Antibodies

Determining the structure of antibodies was made possible by the peculiar properties of cancer cells. Virtually any type of cell in the body can turn cancerous. When it does, it produces a clone of rapidly multiplying malignant cells, all with similar properties. In the cancer known as multiple myeloma, the cancer cell is a plasma cell—in other words, a mature antibody-producing lymphocyte. As the cell multiplies, its descendants continue to turn out its one particular antibody, which, as the cells multiply, is produced in greater and greater quantities.

Using the antibodies produced by a single patient with multiple myeloma, Gerald Edelman and his colleagues at Rockefeller University were able to determine the amino acid sequence of an IgG molecule. Each antibody, as we noted in Chapter 17 (page 341), is a complex protein consisting of four subunits, two identical light chains and two identical heavy chains. The light chains have about 214 amino acids each, and the heavy chains have about twice as many. Each of the four chains has a region where the sequence of amino acids is apparently common to all IgG antibodies; this is known as the constant (C) region. (More recently, it has been demonstrated that each of the five types of antibodies described on the facing page has a different characteristic constant region.) Each chain also has a variable (V) region, some 107 amino acids in length. Within this variable region, some of the sequences of amino acids are the same, but at approximately 40 positions the amino acids vary from one type of IgG to another. It is these variable amino acid sequences that, as the chain folds, come together to form the active region of the molecule, the region that recognizes and binds a specific antigen.

39–6
This drawing of an antibody molecule, based on a bead model made by Gerald Edelman and his colleagues, suggests how the four chains may be folded. Each bead represents an amino acid, of which there are more than 1,200. The variable regions are shown in color.

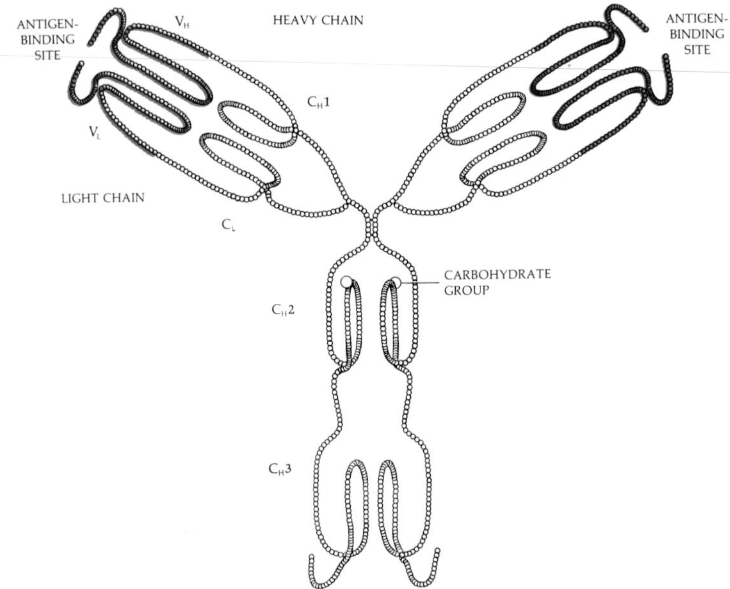

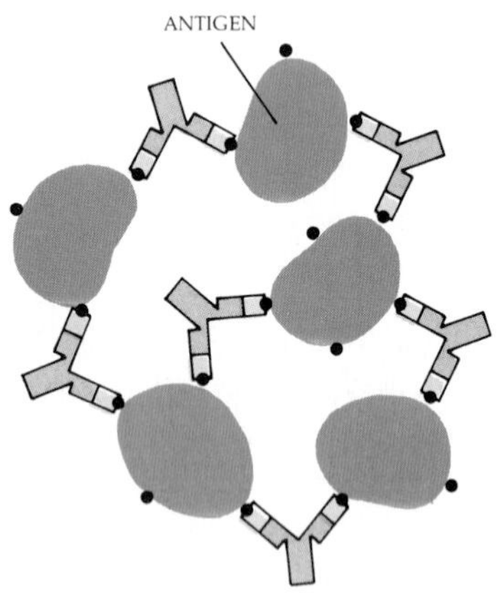

39–7
Because each antibody molecule has two antigen-binding sites, one antibody can bind to antigens on different cells or particles, thus causing them to stick together (agglutination).

Monoclonal Antibodies

In 1975, using the cell fusion techniques employed in chromosome mapping studies, mouse hybrid cells were produced that synthesized pure (monoclonal) antibodies. One cell type of the hybrid was a cancer cell. Noncancer cells die out quickly in tissue culture; the use of cancer cells provided the hybrids with immortality. The other half of the hybrid was a B lymphocyte that had been stimulated to make a particular antibody. Some six years and many experiments later, a similar hybrid was produced between a human cancer cell and a human B lymphocyte. This technique is beginning to make available relatively large quantities of purified antibody that can be used for research purposes, in medical diagnosis, and also for therapeutic trials against various human diseases.

The Clonal Theory of Antibody Formation

Perhaps the most intriguing fact about the immune response is the great variety of antigens against which a single individual can produce antibodies. It is estimated that a mouse, for instance, can form antibodies against at least a million different antigens. Moreover, antibodies can be formed not only against the natural, common invaders that an individual organism might reasonably expect to encounter in the course of its own life and those that its ancestors might have encountered, but also against synthetic antigens that are chemically unlike any substance found in nature.

The most widely accepted current model of antibody formation, proposed more than 20 years ago by Australia's Macfarlane Burnet, is that of clonal selection. According to this model, each individual has a vast variety of different lymphocytes, each genetically fitted with the capacity to synthesize one type of antibody. An antigenic stimulus would not affect the great bulk of lymphocytes, but only those genetically able to produce an antibody to that antigen. The result of the antigen-antibody interaction would be the proliferation of those particular lymphocytes, producing clones of plasma cells, all producing that same antibody. The clonal selection model predicted that (1) only a very small number of lymphocytes

39–8
The clonal selection model of antibody formation. (a) *An immature B lymphocyte, with antibodies on its surface, encounters the antigen molecules for which the antibodies are specific, and* (b) *antibodies and antigens bind together.* (c) *The B lymphocyte then begins to divide and differentiate* (d), *forming plasma cells and memory cells. Plasma cells secrete circulating antibodies. Memory cells persist in the circulation, secreting antibodies only following a subsequent encounter with the specific antigen.*

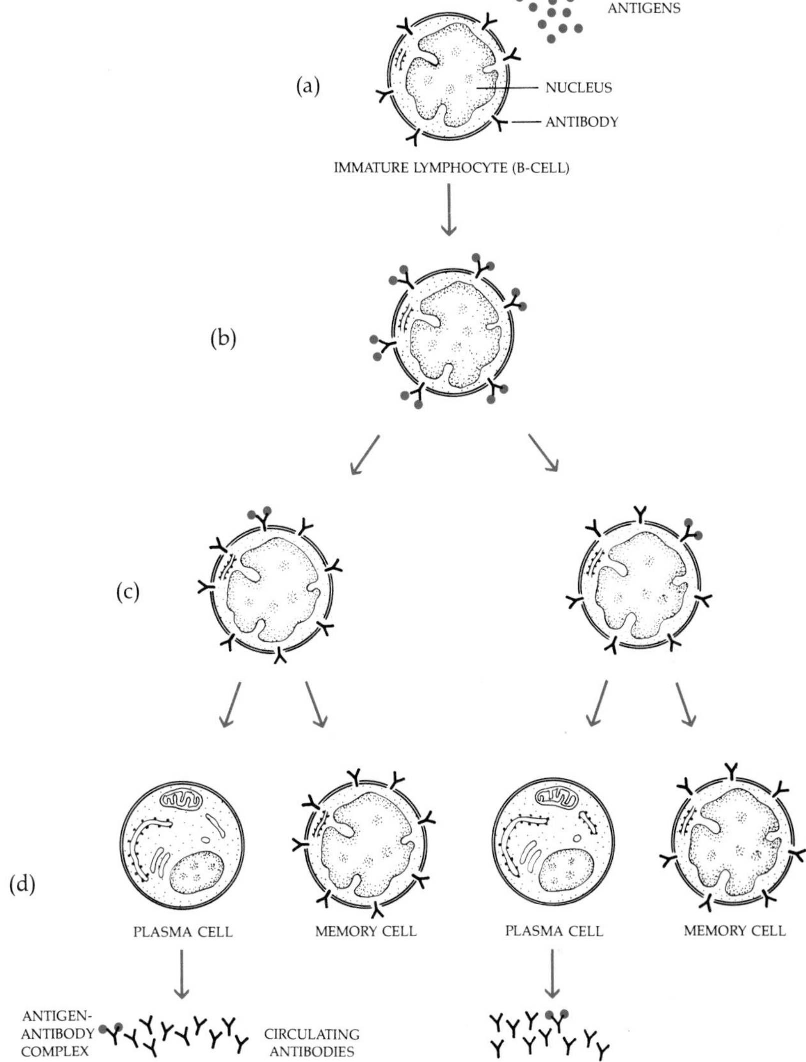

would respond to a given antigen, and (2) any one plasma cell would always form only one antibody. These predictions have now been confirmed. (Note that the studies involving multiple myeloma and the production of monoclonal antibodies constitute part of this confirmation.) In short, to borrow an analogy from Gerald Edelman, antibodies are not "tailor-made." Rather, the antigen shops around in a huge "ready-to-wear" department store until it finds antibodies that fit. Once the ready-made antibody is selected, the manufacturer begins mass production of that particular size and style.

The Genetics of Antibody Formation

In terms of genetics, this model would seem to predict that all the genes for making all the antibodies exist in each lymphocyte (indeed, if you follow the argument further, in every cell in the body), and that all but those coding for a single antibody are repressed in each lymphocyte's clone of plasma cells.

DEATH CERTIFICATE FOR SMALLPOX

In the early days of medicine, the human body was the laboratory and the disease itself the instructor. Smallpox was one of the great teachers. Unlike other infections, which might go unreported or misdiagnosed, the pox left its unmistakable trace in history. It originated in the Far East and seems to have first been introduced into Europe by the returning Crusaders. Here it flourished and spread, until by the eighteenth century one in ten persons died of the pox, and 95 percent of those who survived their childhood had had it. About half had permanent scars and many were blinded. Young women studied their reflections in the mirror, waiting their almost inevitable turn. They learned to scratch their legs and feet at the first sign of the disease in the knowledge that the ugly lesions would then localize in these decorously concealed locations and so, perhaps, spare their faces and bosoms. No one could miss the fact that a person who had suffered an attack was thereby protected from a future one. Pox scars were required of domestic servants, particularly nursemaids, as a prime certificate of employability.

It came to be recognized that some outbreaks of pox were more severe than others (owing, we know now, to mutations of the virus). Since one had to have the disease some time, it was reasoned, it was advantageous to choose which pox and when. Intentional infection of children with material preserved from a mild attack was first practiced in the Far East. The Chinese did it in the form of powdered scabs, "heavenly flowers," used as snuff. The Arabs carried matter from pox pustules around in nutshells and injected it under the skin on the point of a needle. Smallpox inoculation, known as variolation, was introduced into England in 1717 by Lady Mary Wortley Montagu, wife of the British Ambassador to Turkey. In 1746, a Hospital for the Inoculation against Smallpox had been established for the poor of London, where they could be confined during the course of the deliberate infection. Although the induced disease was usually mild, it produced serious illness in some persons. Moreover, since the pox caused by variolation was as contagious as normally contracted smallpox, it appears to have been responsible for some epidemics.

Edward Jenner was an English country doctor who, despite the derision of his colleagues, listened to the tales of country folk. They told him that smallpox never infected milkmaids or other persons who had had cowpox, a mild disease of farm animals that sometimes was transmitted to humans. Jenner tried variolation on several persons who had had cowpox and was unable to induce the customary infection. Then, in 1796, he performed his classic experiment on the farm boy Jamie Phipps. First he inoculated the eight-year-old with fluid taken from a pustule of a milkmaid with cowpox. Subsequently he inoculated the boy with material from a smallpox lesion. Fortunately— for himself, for Jamie, and for us all—the child did not contract smallpox. (The success of the cowpox inoculation depended, of course, on the antigenic similarity between the two naturally occurring pathogens.) Jenner called the process vaccination, from vacca, *the Latin word for cow. By 1800, at least 100,000 persons had been vaccinated and smallpox began to lose its hold on the Western world. It was not until almost 100 years later that Louis Pasteur discovered serendipitously that a virus or bacterium grown in tissues other than those of its normal host would often lose its virulence while retaining its immunogenicity. (He retained the term vaccination, in recognition of the earlier work of Jenner.) Pasteur's discovery provided the basis for most of our modern vaccines, such as those against poliomyelitis, diphtheria, and measles.*

Despite the availability of an effective vaccine for smallpox, by the 1950s there were still an estimated 2 million new cases of smallpox every year. However, these were largely confined to the Far East, mostly among the urban poor of India, whose crowded living conditions provided as fertile a soil as the European cities had two centuries earlier. In 1973, the World Health Organization (WHO) declared war on smallpox, which involved the mass production of vaccine and, more important, search-and-destroy missions charged with finding each new case of the disease and vaccinating every susceptible person who might be exposed to it; 150,000 persons were in the task force. On April 23, 1977, an international commission declared India free of smallpox and so it has remained.

In that same year, WHO recommended that smallpox research laboratories be closed and their supply of virus destroyed. One such laboratory was at the University of Birmingham in England. On July 25, 1978, during the last six months of the laboratory's existence, a woman medical photographer working on the floor below contracted smallpox; the virus, as it turned out, had made its way out of the laboratory through a duct system and the photographer was unvaccinated. The disease was diagnosed by the head of the laboratory, a prominent virologist, who was responsible for the woefully inadequate safety precautions. He committed suicide by cutting his throat. The photographer died five days later. Those appear to have been the last two deaths from smallpox on this planet.

Thus, until recently, the clonal selection theory was in some difficulty because of simple mathematics. A single individual is apparently capable of making antibodies for a million different antigens. However, a human cell does not contain this many structural genes in its entire genome.

Recently, as we saw in Chapter 17, the seeming paradox was resolved. The variable regions of both the light and heavy chains in mouse antibodies are coded for by some 300 DNA sequences scattered through the genome, which are transposed and assembled into a variety of different arrangements as the lymphocyte matures. After transposition, the gene segments are ordered on the chromosome in the sequence of their appearance (Figure 17-23), separated by noncoding sequences (introns). They are then transcribed into RNA, and the introns are excised before the RNA is translated into the protein of the final antibody molecule. The number of possible combinations of these gene sequences has been calculated at 18 billion, enough to account for the enormous diversity of antibody molecules.

A child with smallpox, showing the pustules characteristic of the disease.

T Lymphocytes and Their Functions

Until relatively recently in the 200-year history of immunology, circulating antibodies were believed to be the sole effectors of immunity. It is now known, however, that there is another category of highly specific immune response that is effected not by antibodies circulating in the blood and other fluids, but by cell-to-cell interactions involving the other class of lymphocytes, the T lymphocytes. This is sometimes known as the *cell-mediated response.*

Discovery of the T Lymphocytes

The T of T lymphocyte stands for thymus, and the discovery of these cells and their activities came about as a result of research directed at uncovering the function of the thymus gland. In humans, the thymus gland is located in the chest just behind the breastbone. The gland is large in infants and atrophies after puberty. In adults, surgical removal of the thymus gland has virtually no physiological effect. The first clue to the function of the thymus came when a British investigator succeeded in removing the thymus gland in newborn mice—quite a feat when you consider that these infants are only 2 centimeters long and weigh about a gram. The animals survived, but they were sickly and runty due, further study showed, to their decreased resistance to infections. Following this lead, it was found that as early as the eleventh day of fetal life in the mouse (and the eighth week in the human) primitive blood cells creep into the embryonic thymus gland. These are the future T lymphocytes, or T cells. (With the discovery of T cells, the other, antibody-producing lymphocytes, which develop in the bone marrow and fetal liver, came to be known as B cells.)

Whereas the circulating antibodies produced by B cells are primarily active against viruses and bacteria and the toxins they may produce, T cells act against other eukaryotic cells, particularly the body's own cells that are infected by a virus or other microorganism. When a virus, for example, is multiplying within a cell, it is protected from the action of antibodies; however, its presence is reflected by the appearance of new antigens on the surface of the infected cell, which make it possible for the T lymphocyte to find the infected cell and lyse it, exposing the viruses to antibody action. T lymphocytes are also the principal agents in the immune response to tissue transplants. (The thymectomized mice were also deficient in their capacity to reject grafts of foreign tissue.)

The Functions of T Lymphocytes

On the basis of present knowledge, three functional types of T lymphocytes can be distinguished. One group has the cytolytic functions just described. These cells resemble B lymphocytes in that, upon exposure to its specific antigen, the T lymphocyte divides and its daughter cells differentiate into either active cells or memory cells. The active cells have receptors on their surfaces that combine with the antigens on the surfaces of the target cells. (The nature of these receptors is not known.) The combination triggers the release of powerful chemicals, known as lymphokines, that attract macrophages and stimulate phagocytosis. Some of the activated T lymphocytes, known as "killer cells," secrete cytotoxins that destroy target cells directly, and some secrete interferon. Thus, a whole battery of defenses is mobilized, including the inflammatory response and antibodies, in addition to the cytolytic effects of the T lymphocytes.

The other two types of T lymphocytes have only recently been discovered. Members of one group, known as the helper T lymphocytes, are necessary for the activation of both B lymphocytes and the cytolytic T lymphocytes. Members of the other group, suppressor T lymphocytes, moderate the activities of B lymphocytes and other T lymphocytes, inhibiting their differentiation or multiplication, or both.

Elucidation of this complex network of specific immune responses is one of the most intriguing areas of modern medical research, both because of the basic biological knowledge that is being uncovered and also because of its tremendous promise for the control of many medical problems.

DISORDERS ASSOCIATED WITH THE IMMUNE SYSTEM

Allergies

When certain individuals are exposed to particular environmental antigens—such as pollen, dusts, and certain foods—the production of IgE antibodies by specific plasma cells is stimulated, as is the formation of memory cells. Upon reexposure to the same antigen, more IgE antibodies are formed. These antibodies circulate and attach themselves to basophils and mast cells. When the antigen combines with the attached antibody, the basophil or mast cell releases histamine and other chemicals that induce the inflammatory response. This reaction typically occurs on an epithelial cell surface, producing either increased mucus secretion, as in "hay fever"; hives or dermatitis; or cramps and diarrhea, as in the case of food allergies. Systemic reactions may result if the basophils and mast cells release their chemicals into the circulation causing dilation of the blood vessels, leading to a potentially dangerous fall in blood pressure, and constriction of the bronchioles (a syndrome known as anaphylactic shock). The normal function of IgE antibodies and the evolutionary background of this decidedly maladaptive response—in which the response is far more dangerous than the antigen—remains a mystery.

Antihistamines, which counteract the histamine effect, suppress some of the symptoms of an allergic reaction. (Most decongestants, on the other hand, promote the release of histamines, raising some questions about taking "allergy pills" that offer a combination of antihistamines and decongestants.) Steroid hormones related to cortisone are used in more severe cases. These drugs act by suppressing the production of white blood cells and so of inflammation and the immune response in general.

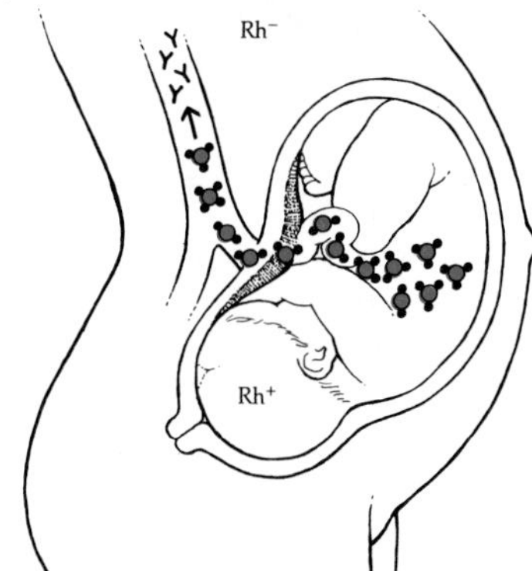

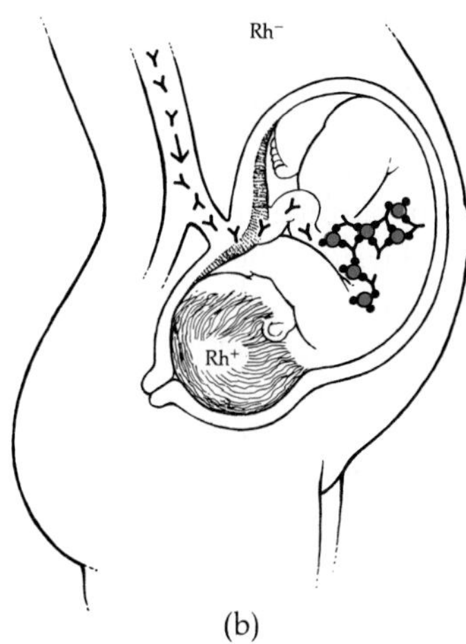

39–9

The events leading to Rh disease. (a) *Late in the first pregnancy—or during the birth of the child—fetal blood cells spill across the barrier that separates the maternal and fetal circulations. Antigens on the red blood cells of the Rh-positive fetus stimulate the production of antibodies by the mother's immune system. These antibodies remain in the mother's bloodstream indefinitely.* (b) *Late in the second pregnancy, the antibodies pass through the barrier from the maternal blood to the blood of the fetus. If the fetus is Rh positive, the antibodies react with the antigens on its red blood cells, destroying them.*

Autoimmune Diseases

The immune system can ordinarily distinguish between "self" and "not-self." Substances that are present during embryonic life, when the immune system is developing, will not be antigenic in later life. This recognition occasionally breaks down, however, and the immune system attacks body cells. Myasthenia gravis, lupus erythematosus, several types of anemia, and certain other disorders have been identified as autoimmune diseases—that is, diseases in which an individual makes antibodies against his or her own cells. It is possible that other disorders, such as rheumatoid arthritis, multiple sclerosis, and juvenile-onset diabetes, the causes of which are not yet known, may prove to be autoimmune diseases also.

The Rh Factor

A once common medical problem caused by the immune system is hemolytic anemia of the newborn, which is due to a blood factor, the Rh factor (named after the rhesus monkeys in which the research leading to its discovery was carried out). During the last month in the uterus, the human baby usually acquires antibodies from its mother. Most of these antibodies are beneficial. An important exception, however, is found in the antibodies formed against the Rh factor, which is a genetically determined substance found on the surface of red blood cells. If a woman who lacks the Rh factor (that is, an Rh-negative woman) has children fathered by a man homozygous for the Rh factor, all the children will be Rh positive; if he is a heterozygote, about half of the children will be Rh positive. When she is first pregnant with an Rh-positive child, an Rh-negative woman will generally form antibodies to the Rh factor at the time of delivery, when blood from the infant enters the mother's bloodstream. These antibodies persist and in subsequent pregnancies can be passed into the fetal bloodstream during the last month in the uterus, causing destruction of the red blood cells in an Rh-positive fetus. Now that its causes are recognized, Rh disease can be prevented by injecting the Rh-negative mother, within 72 hours of her delivery, with antibodies against the fetal Rh red blood cells in her system, thus destroying the cells and preventing them from triggering antibody production.

TISSUE AND ORGAN TRANSPLANTS

Blood Transfusion

Blood transfusions are such a commonplace component of modern medical practice that it is difficult to realize that it was only as recently as 1900 that such transfusions were made safe enough to be practicable. At about the turn of the century, Karl Landsteiner set out to discover why blood transfusions between humans were sometimes safe and effective but, more often, led to serious complications and sometimes death. Mixing samples of blood taken from members of his laboratory staff, Landsteiner found that sometimes the red blood cells would clump together, or agglutinate, and sometimes they would not. From these experiments, he came to realize that there were different categories of blood and that agglutination was caused by mixing blood of different categories. Soon after, the four major blood groups were determined: A, B, AB, and O. Persons who belong to blood group A, it was found, can receive blood from donors of blood group A or blood group O. Persons who belong to blood group B can receive blood from

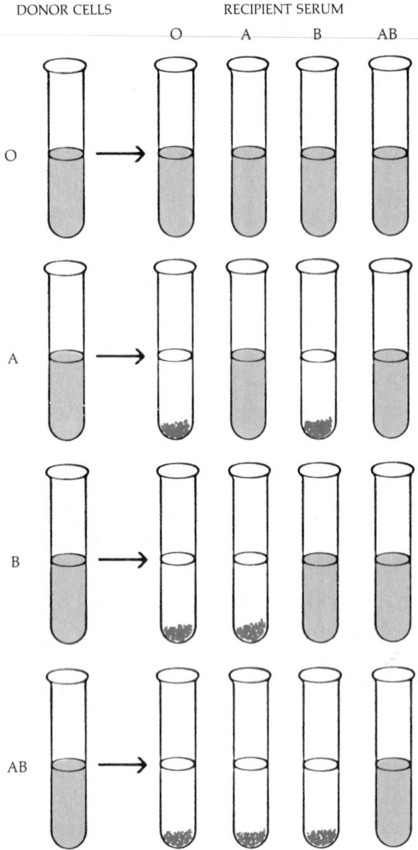

39-10
Severe and sometimes fatal reactions can occur following transfusions of blood of a different type from the recipient's. These reactions are the result of agglutination of the blood cells caused by antibodies present in the recipient's blood. Antibodies in the donor's blood are generally of little consequence because they are so diluted in the recipient's blood. Blood-group reactions to transfusions can be demonstrated equally well in test tubes, as shown here. The blood that is shown agglutinating has natural antibodies against the donor blood. Persons with type O blood used to be called universal donors and those with type AB blood, universal recipients. Now other factors are checked as well.

Table 39-2 Blood Groups

GROUP	GENOTYPE	REACTION WITH ANTIBODIES		ANTIBODIES IN BLOOD PLASMA
		ANTIBODY A	ANTIBODY B	
O	O/O	−	−	Antibody A, antibody B
A	A/A, O/A	+	−	Antibody B
B	B/B, O/B	−	+	Antibody A
AB	A/B	+	+	None

persons with blood groups B or O. Persons with AB blood can receive blood from anyone (groups A, B, AB, or O). Persons with type O blood can receive only type O blood. (These data are summarized in Figure 39-10.)

This incompatibility of different blood groups is due to the presence of antigens on the surface of the red blood cells and the presence of antibodies to these antigens in the bloodstream of the recipient. (For reasons that are still not clear, these antibodies are present even though an individual has not been previously exposed to the antigens.) Individuals with type A blood have the A antigen on the surface of their blood cells. They do not have antibodies against the A antigen but do against the B, and therefore can accept transfusions from other individuals with type A blood, but not from those with type B. Individuals with type B blood have the B antigen. They do not have antibodies against the B antigen but do against the A. They can accept transfusions from type B donors but not type A. AB individuals have both antigens and so do not have antibodies to either. They can receive blood from all groups. O individuals have neither antigen but both antibodies. They can receive blood only from other type O individuals; however, they can provide transfusions for any group.

Since Landsteiner's initial discoveries, additional antigens—such as the Rh factor—have been discovered on the surface of red blood cells, and modern blood typing takes these into account also. But it was his observation—relatively simple in retrospect—that opened the door to this great medical advance.

Inheritance of Blood Types

Blood types are inherited. The A, B, AB, and O blood groups are examples of multiple alleles (A, B, and O). The A and B alleles are codominant, while the O allele is recessive. If a person has type AB blood, it means that at least one of his or her parents had an A allele and the other, a B allele. If a person has type A blood, it means that an A allele was inherited from one parent and either an A or O allele from the other. A person with type O blood must have had parents who each carried one O allele, although they may have been, phenotypically, either A or B. Blood types have been used to settle questions of paternity in legal actions, although, as you can readily see, they can only prove that someone is not the father of a particular child, not that he is.

Organ Transplants

People with extensive burns die because of the loss of body fluids from exposed areas and from infection. If skin is taken from one part of the patient's body and transplanted to the burned area, the new tissue anneals to the exposed area, vascularization takes place, and the tissue grows and spreads out. If a skin transplant is taken from another individual (except the patient's identical twin), the initial stages of healing and vascularization take place, but then, on about the fifth to the seventh day, large numbers of host cells infiltrate the graft, and by the tenth to twelfth day it has died. The cells that infiltrate the graft are mainly lymphocytes and macrophages, and present evidence indicates that T lymphocytes are responsible for the graft rejection.

In an effort to reduce transplant failures due to graft rejection, patients are often given drugs to suppress the immune response. However, as we noted previously, infection is a major complication among patients requiring skin transplants, and it is the leading cause of death among kidney transplant recipients; so general suppression of the immune response is obviously not the ideal solution. It would be far more valuable to be able to selectively suppress the proliferation of only those T lymphocytes active against the foreign cells or to alter the graft in some way to make it more acceptable to the host. (The likelihood of rejection of a transplanted organ decreases greatly after several months have passed. Moreover, a second graft from the same donor is not more readily tolerated than the first. Thus, rejection seems to involve a change in the graft rather than in the host. This observation may lead to improved procedures for suppressing transplant rejection.)

Because of the great medical potential of organ grafts, this field of immunological research is extremely active. The 1980 Nobel Prize in physiology or medicine was awarded to three investigators—Baruj Benacerraf of Harvard Medical School, Jean Dausset of the University of Paris, and George Snell of the Jackson Laboratory in Bar Harbor, Maine, for their work on the antigens triggering transplant rejection.

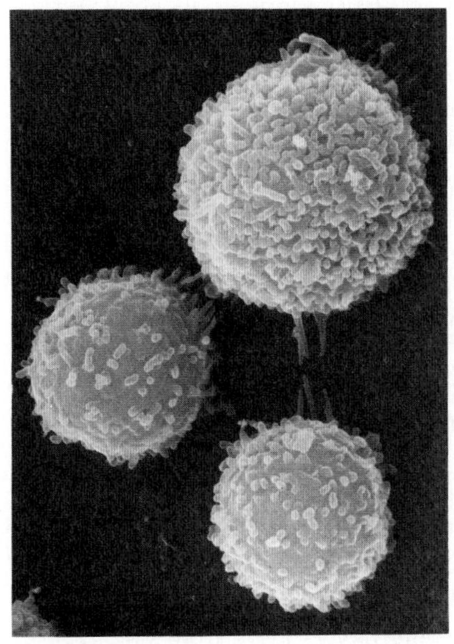

39–11
A macrophage and two lymphocytes. Note the numerous folds and microvilli that greatly increase the surface area of these cells. According to present evidence, lymphocytes, macrophages, and other cells involved in the immune response recognize one another by the pattern of antigenic glycoproteins embedded in the cell membranes. They can sense the presence of another cell and respond by extending microvilli toward it.

The Major Histocompatibility Complex

Studies on T lymphocyte functions and on tissue transplant rejections are now converging on a group of antigens known as the major histocompatibility complex, or MHC. (*Histo* is a Greek word signifying "tissue." The complex was first studied in the context of a search for compatible tissue grafts—that is, grafts that would not be rejected by the host.) These antigens, which are glycoproteins found on the cell surface, are coded for by at least 20 different genes, and each of these genes has as many as 8 to 10 alleles; so the total number of different combinations is virtually astronomical. In fact, it is predicted that no two persons will ever be found to have the same major histocompatibility complex. Each is as individual as a fingerprint. Thus, these antigens provide for extremely accurate discrimination between self and "not-self." (They are also, incidentally, far more accurate than blood typing in resolving questions of paternity.)

According to the picture that is just now emerging, there are two different sets of MHC cell-surface molecules, known simply as Class I and Class II. Class I molecules are necessary for the action of cytolytic T lymphocytes. These T cells can only recognize a foreign antigen—a viral antigen, for instance—if it is presented along with Class I molecules. This is the mechanism by which the action of these lymphocytes is directed against the body's own infected cells.

Class II molecules are present only on cells of the immune system and identify such cells to each other. They are involved in the interactions among the B lymphocytes, T lymphocytes, and macrophages that result in the highly organized cooperative effort of the immune response. Thus, this work in the very practical science of immunology has led medical researchers to what is perhaps only the first glimpse of a cellular communication network, probably involving far more than immune responses.

IMMUNITY AND CANCER

Cancer cells resemble the host's own cells in many ways; yet, within the host, they act like foreign organisms, invading and "choking off" or competing with normal tissues. Moreover, virtually all cancer cells have antigens on their cell surfaces that differ from the antigens of the normal host cells and can be recognized as foreign. Does this mean that the body can mount an immune response against its own cancers? A growing number of cancer researchers believe not only that cancer can induce an immune response but also that it usually does so. In fact, according to this hypothesis, it usually does so successfully, overwhelming the cancer before it is ever detected by the patient or the physician; the cancers that are discovered represent occasional failures of the immune system. This conclusion suggests that bolstering the patient's immune response may provide a means for cancer prevention or control. It also suggests that the cell-mediated immune response may have had its evolutionary origins as a defense not only against invaders but also against the treacherous, malignant cells of one's own body.

SUMMARY

Organisms have evolved a number of responses that exclude or destroy microorganisms, other foreign invaders, and cells not typically "self." The inflammatory response, which is nonspecific, involves the release of histamine and other chemicals, causing capillary distension, a local increase in temperature, cell death, and the mobilization of phagocytic white cells at the site of infection.

Interferon provides another and very different type of defense, which is directed against viruses. Interferons are small proteins produced by virus-infected cells that protect nearby cells from viral infection.

The immune response is highly specific and involves two types of white blood cells: B lymphocytes and T lymphocytes. B lymphocytes are major protagonists in the formation of antibodies, large protein molecules whose active regions are complementary to foreign molecules called antigens. The combination of antigen and antibody immobilizes the invader, destroying it or rendering it susceptible to phagocytosis.

There are five classes of antibodies (immunoglobulins), of which IgG (gamma globulin) is the most common and the most intensively studied. Antibodies consist of four subunits: two identical light chains and two identical heavy chains. Each of the four chains has a constant (C) region—a region common to all antibodies of its class—and a variable (V) region, which differs from one type of antibody to another. The antigen-binding sites—of which there are two on each antibody molecule—are formed by foldings of the variable regions of the light and heavy chains.

The most widely accepted model of antibody formation is the clonal selection theory. According to this model, each individual is endowed with a vast variety of B lymphocytes, each genetically capable of synthesizing antibodies against one particular antigen. Upon encountering that particular antigen, the B lymphocyte matures and divides, resulting in a clone of plasma cells all synthesizing antibodies against that particular antigen. Memory cells are also produced, which persist in the bloodstream following infection and produce antibodies immediately upon subsequent exposure to the same antigen. This memory-cell response is the cause of the rapid and enhanced immunity following vaccination or many viral infections. The capacity of the B lymphocytes to produce a tremendous variety of antibodies is accounted for by the large number of gene sequences (at least 300) coding for the variable regions of antibodies and by the transposition of these gene sequences in the course of lymphocyte development, thus creating an enormous variety of structural genes coding for antibodies.

T lymphocytes are responsible for cell-mediated immunity. They destroy cells by cell-to-cell contact. These cytolytic activities are directed principally against the body's own cells that are harboring intracellular viruses or other parasites. Two other types of T lymphocytes are also known: T helper cells, which promote immune reactions involving B lymphocytes and other T lymphocytes, and T suppressor cells.

Disorders associated with the immune system include allergies, Rh disease, and autoimmune diseases caused by an individual's immune reactions to his or her own tissues. Rejection of organ transplants is due principally to immune reactions between T lymphocytes and the transplanted tissues.

The major histocompatibility complex (MHC) consists of a group of genetically determined antigenic glycoproteins found on virtually all cell surfaces. There are two classes of MHC molecules. Class I molecules are involved in the recognition of self and "not-self" and are essential in the identification by T cells of the body cells that have been altered by intracellular parasites. Class II molecules are present on cells of the immune system and are responsible for recognition among these cell types.

QUESTIONS

1. Distinguish between the following: antibody/antigen; B lymphocyte/T lymphocyte; plasma cell/memory cell; constant region/variable region; light chain/heavy chain.

2. An early theory of antibody formation, the instructive theory, hypothesized that the antibody was molded into its special configuration by its encounter with the antigen. Since then, new data on the structure of proteins have invalidated this model. What are the new data?

3. It can never be proved that someone is the father of a particular child, but, as you can see when you fill in the final two columns of the table at the left, it is possible to prove that someone could *not* be the father. In the famous Charlie Chaplin paternity case in the 1940s, the baby's blood was B, the mother's A, and Chaplin's O. If you had been the judge, how would you have decided the case?*

PHENOTYPES OF PARENTS		CHILDREN POSSIBLE	CHILDREN NOT POSSIBLE
A	A		
A	B		
A	AB		
A	O		
B	B		
B	AB		
B	O		
AB	AB		
AB	O		
O	O		

4. Predict the results of the following experiment: A mouse of strain A receives a skin graft from another strain A mouse and one from a strain B mouse. Two weeks later, the same mouse is grafted with skin from strain B and strain C. What are the fates of these grafts?

5. What prominent feature in the structure of a virus makes it more susceptible than a bacterial cell to control by vaccination?

* As a matter of fact, Chaplin was judged guilty. Blood-group data are not admitted as evidence by some states in cases of disputed parentage.

CHAPTER 40

Integration and Control I: The Nervous System

The principal regulatory systems of the human body are the nervous system and the endocrine system. As we have glimpsed in the previous chapters, these two systems operate continuously to integrate and control the body's activities. Functionally, a nervous system differs from an endocrine system chiefly in its capacity for rapid response—a nerve impulse can travel through an organism in a matter of milliseconds. Hormones, however, move at a somewhat slower rate and, characteristically, elicit a slower response. Plants, which (with some exceptions) are not noted for their liveliness, rely largely on an elaborate interplay of hormones to coordinate their activities. But animals, while also relying on hormones, are generally characterized by nervous systems—networks of specialized nerve cells.

Invertebrate nervous systems range from the very simple to the complex (Figure 40–1). In the cnidarian *Hydra*, the neurons form a diffuse network. They receive information from sensory receptor cells, which can be found among the epithelial cells on the inner (feeding) and outer surfaces of the animal. The neurons stimulate epitheliomuscular cells that cause movements in the body wall.

40–1
In Hydra (a), *a cnidarian, the nerve impulse, which can travel in either direction, spreads out diffusely along the nerve net from the area of stimulation. In a planarian* (b), *there are two longitudinal nerve cords, with some aggregation of ganglia and sense organs at the anterior end. In annelids, such as the earthworm* (c), *the longitudinal nerve cords are fused together in a double ventral nerve cord. In the crayfish* (d), *an arthropod, the nerve cord is also double and ventral, with a series of ganglia, almost as large as the brain, that control particular segments of the body.*

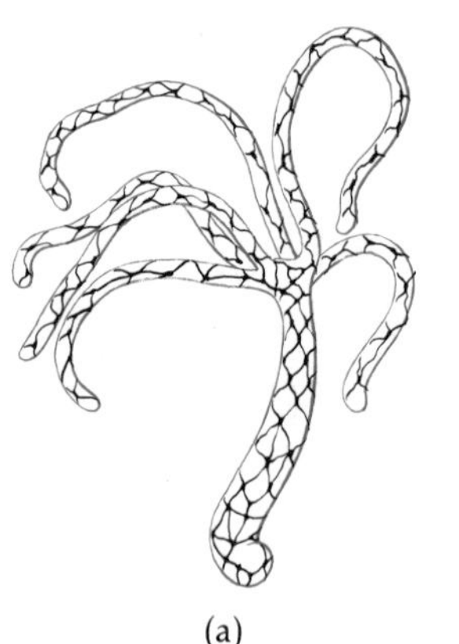

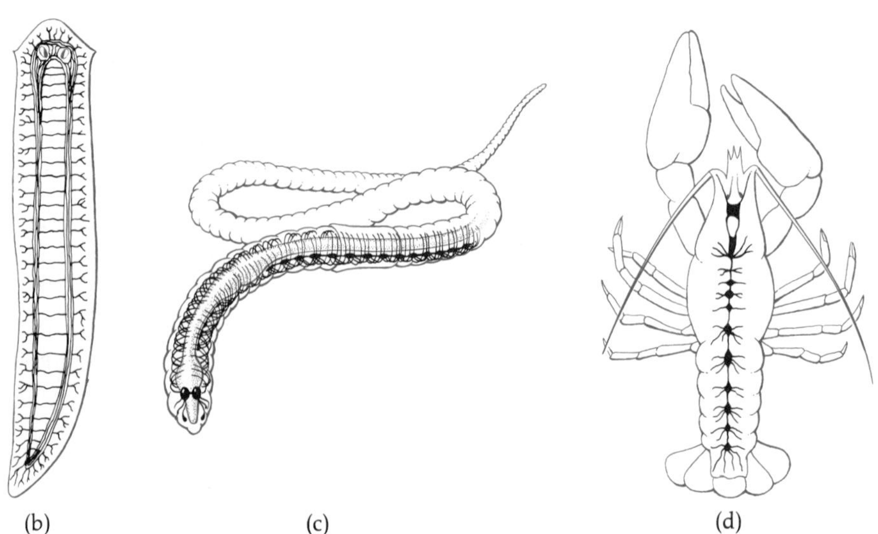

The nervous system of the planarian, a flatworm, is more highly organized than that of *Hydra.* Some of the nerve net is condensed into two cords, and there are two clusters of nerve cell bodies, known as ganglia (singular, ganglion), at the anterior end of the body.

In the earthworm, the two cords have come together in a fused, double nerve cord that runs along the ventral surface of the body. Along the nerve cord are ganglia, one for each segment of the earthworm's body. The nerve cord forks just below the pharynx, and the two forks meet again in the head, terminating in two large ganglia.

Arthropods, such as the crayfish, also have a double ventral nerve cord and, in addition, may have sizable clusters of nerve cell bodies in the head region. Collectively, these ganglia are large enough to be called a brain. The nervous system also contains many other ganglia interconnected by nerve fibers that run along the ventral surface. In animals with this ladder type of nervous system, many quite complicated activities—for example, the complex movements of the finely articulated appendages—are coordinated by the nearest ganglion.

In vertebrates, the nervous system is dorsal rather than ventral, and the central nervous system is enclosed and protected by the bones of the vertebral column and the skull. The trend in vertebrate evolution has been toward increased centralization of control in the brain—cephalization—a trend that may be still continuing.

ORGANIZATION OF THE VERTEBRATE NERVOUS SYSTEM

The nervous system of vertebrates has many subdivisions that can be distinguished by anatomical, physiological, and functional criteria. The primary and most obvious is the subdivision of the system into the central nervous system and the peripheral nervous system. The central nervous system includes the spinal cord and the brain. The peripheral subdivision includes the sensory (afferent) and motor (efferent) pathways that carry information to and from the central nervous system. The motor system has two further subdivisions: the autonomic ("involuntary") system, which relays signals to smooth muscle, to cardiac muscle, and to glands, and the somatic ("voluntary") system, which stimulates skeletal muscle (Figure 40–2).

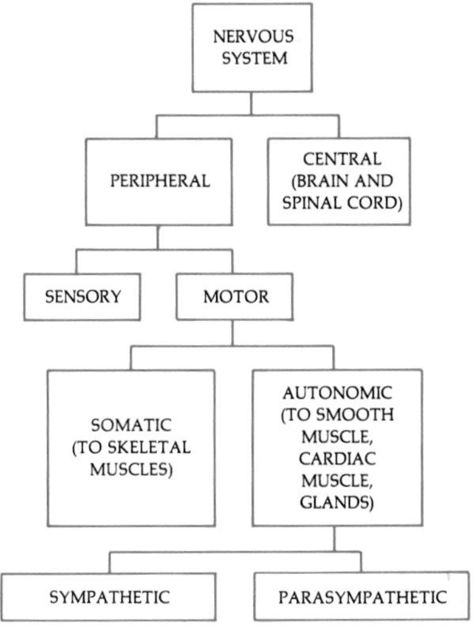

40–2
Subdivisions of the vertebrate nervous system.

The functional unit of the vertebrate nervous system, as we noted in Chapter 33, is the neuron, a nerve cell characterized by a cell body, an axon, and, often, many dendrites. Neurons are surrounded and insulated by other nerve cells, called neuroglia in the central nervous system and Schwann cells in the peripheral nervous system. In vertebrates, as in other animals, nerve cell bodies are often found in clusters; such clusters are called ganglia when they are outside the central nervous system, and *nuclei* when inside the central nervous system. Axons and long dendrites are referred to as nerve fibers. Bundles of nerve fibers are called *tracts* when they are in the central nervous system, and *nerves* when they are in the peripheral nervous system.

Neurons communicate with one another across junctions known as *synapses,* which may be electrical or chemical in nature. Synapses will be described in greater detail later in the chapter.

With these troublesome definitions behind us, we shall first describe the various subdivisions of the vertebrate nervous system in terms of their function. Then we shall discuss the ways in which signals travel throughout this vast and complex communications network.

The Central Nervous System

The central nervous system is made up of the spinal cord and brain. Your spinal cord, which is a slim cylinder about as big around as your little finger, can be seen in cross section to be divided into a central butterfly-shaped area of gray matter and an outer area of white matter. The gray matter is mostly interneurons (which, as you will recall, transmit signals within the central nervous system), the cell bodies of motor neurons, and neuroglia. The white matter consists of bundles of nerve fibers—tracts—running longitudinally along the spinal cord.

The spinal cord is continuous with the brainstem, which is the stalk of the brain. The brainstem contains nerve tracts transmitting signals to and from the spinal cord and also the cell bodies of the neurons whose axons innervate the muscles and glands of the head. In addition, within the brainstem are centers for some of the important automatic regulatory functions, such as respiration and blood pressure. (These functions can also be influenced by other parts of the brain.)

The structure of the brain will be described in Chapter 43.

The Peripheral Nervous System

The peripheral nervous system is made up of neurons whose fibers extend out of the central nervous system into the tissues and organs of the body. These include both motor (efferent) neurons, which carry signals out, and sensory (afferent) neurons, which carry signals into the central nervous system. The fibers of motor and sensory neurons are bundled together into nerves, which are classified as *cranial nerves,* nerves connecting with the brain (such as the optic nerve), and *spinal nerves,* those that connect with the spinal cord.

Figure 40–3 shows a segment of the human spinal cord, with pairs of spinal nerves entering and emerging from the cord through spaces between the vertebrae. The motor fibers of each pair innervate the muscles of a different area of the body, and the sensory fibers receive impulses from sensory receptors in the same area. In humans there are 31 such pairs.

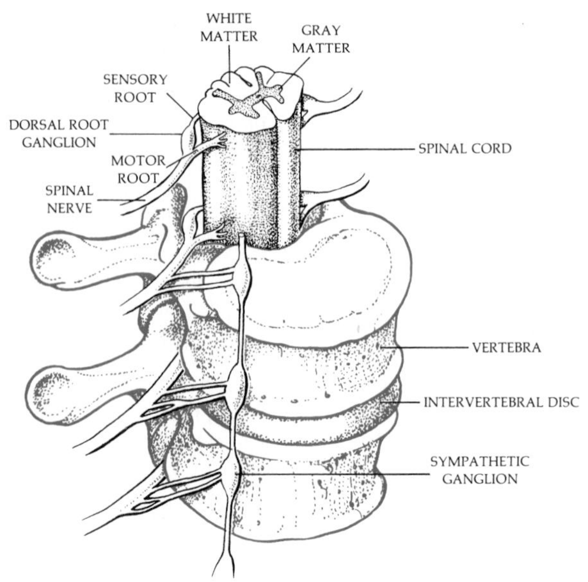

40–3
A segment of the human spinal cord. Each spinal nerve divides into two fiber bundles, the sensory root and the motor root, at the vertebral column. The sensory root connects with the cord dorsally; the cell bodies of the sensory neurons are in the dorsal root ganglia. The motor root connects ventrally with the spinal cord. The sympathetic ganglia, which form a chain, are part of the autonomic nervous system. The butterfly-shaped gray matter within the spinal cord is composed mostly of interneurons, cell bodies of motor neurons, and glial cells. The surrounding white matter consists of ascending and descending nerve tracts.

40–4

A polysynaptic reflex arc. In this example, nerve endings in the skin, when appropriately stimulated, transmit signals along the sensory neuron to an interneuron in the spinal cord. The interneuron transmits the impulse to a motor neuron. As a result of the stimulation of the motor neuron, muscle fibers contract. Other interneurons are also stimulated by the sensory neuron and carry the sensory information to the brain.

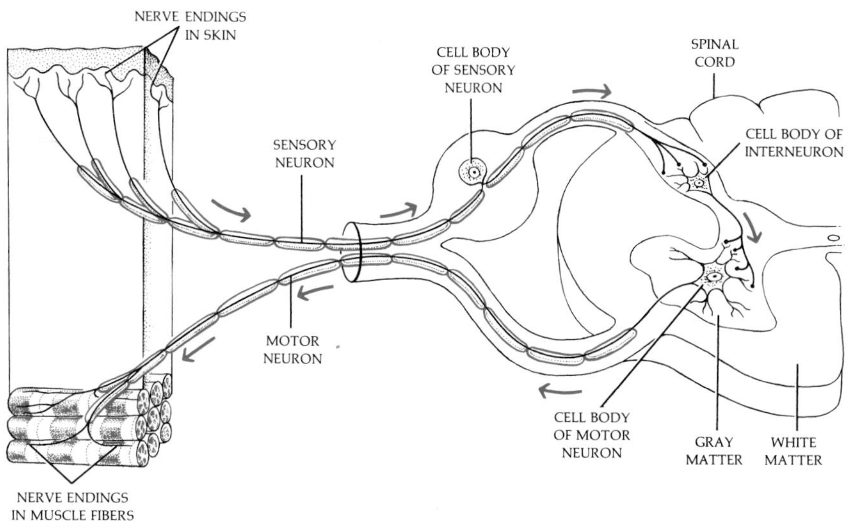

As you can see in Figure 40–3, as the spinal nerves reach the central nervous system, the motor and sensory fibers separate from each other. The cell bodies of the sensory neurons are in the dorsal root ganglia outside the spinal cord, and the sensory fibers feed into the dorsal side of the spinal cord. Here they may synapse immediately with interneurons or motor neurons as in a simple reflex, they may turn and ascend toward the brain, or they may do both. Fibers from the motor neurons emerge from the spinal cord on the ventral side. The cell bodies of motor neurons are found in the spinal cord, where they synapse with other neurons. The three types of neurons—sensory, motor, and interneurons—are interconnected in reflex arcs (Figure 40–4). A stimulus received by a sensory neuron is transmitted to the central nervous system, where the neuron synapses with either a motor neuron (a monosynaptic reflex) or one or more interneurons (a polysynaptic reflex) that, in turn, synapse with a motor neuron that completes the arc and carries out the reflex action. Other interneurons may transmit information concerning the event to other parts of the central nervous system. Thus, for instance, if your fingers touch a hot stove, your hand will automatically withdraw from the stove. Almost simultaneously, your brain will become aware of what has happened and you will take further action or make an appropriate comment.

Divisions of the Peripheral Nervous System: Somatic and Autonomic

As we noted earlier, there are two important subdivisions of the peripheral nervous system: the *somatic* and the *autonomic*. The autonomic nervous system consists of the nerves that control cardiac muscle, glands, and smooth muscle (the type of muscle found in the walls of blood vessels and in the digestive, respiratory, excretory, and reproductive tracts). The somatic system controls the skeletal muscles—that is, the muscles that can move at will.

You will readily recognize that the distinction here between "voluntary" and "involuntary" is not clear-cut. Skeletal muscles—part of the somatic or "voluntary" system—often move involuntarily, as in a reflex action. On the other hand, it has been reported that some individuals, such as practitioners of yoga or those who have had biofeedback training, can control the rate of contraction of cardiac muscle and the contractions of some smooth muscle, both of which are part of the autonomic, or "involuntary," system.

Anatomically, the motor neurons of the somatic system are distinct and separate from those of the autonomic nervous system, although fibers of both types may be carried in the same nerve. The cell bodies of the motor neurons of the somatic system are located within the central nervous system, with long fibers running without interruption all the way to the skeletal muscles, which they innervate. In the autonomic nervous system, the fibers of motor neurons also leave the central nervous system; however, they do not usually travel all the way to their target organ but instead synapse outside the central nervous system with a second motor neuron, which then innervates the effector (Figure 40-5). The neurons whose fibers emerge from the central nervous system are known as *preganglionic,* while those

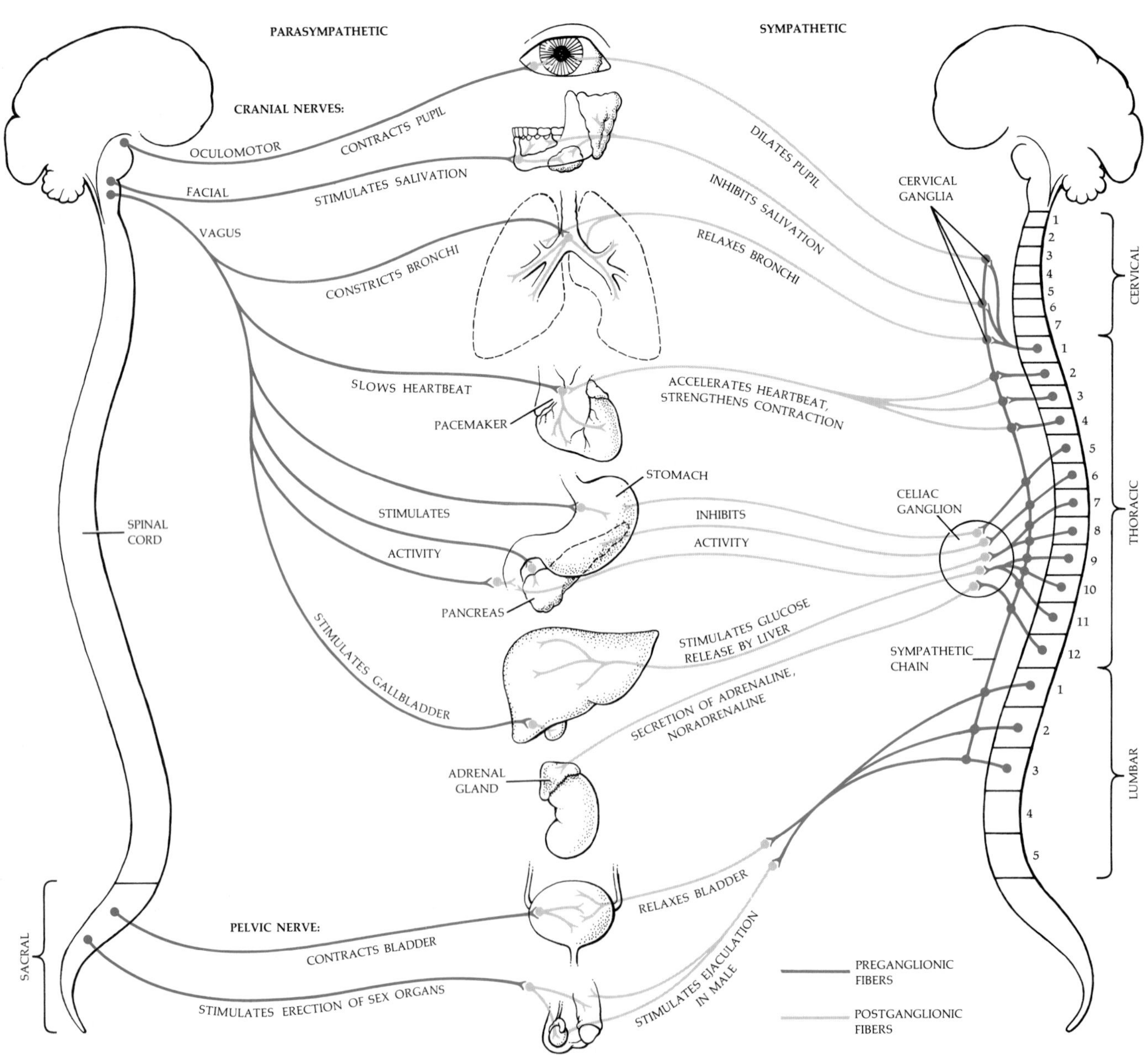

whose fibers terminate in the effectors are known as *postganglionic*. This two-neuron pathway constitutes a characteristic difference between the autonomic and somatic systems.

Another major difference between these systems is that, in vertebrates, the somatic system cannot inhibit its effector; it can only stimulate or not stimulate it. The autonomic system, by contrast, can stimulate or inhibit the activity of its effector. In the somatic system, sensory input comes from neurons monitoring environmental changes. The autonomic nervous system receives its sensory input from some of the same neurons as the somatic nervous system and also from sensory neurons monitoring changes in the interior of the body, such as the group signaling changes in blood pressure. These neurons are involved in reflexes similar to the one described earlier. One difference, however, is that, in reflex arcs involving the autonomic system, you are usually not conscious that the reflex action has taken place.

40–5
The autonomic nervous system. It differs from the somatic system anatomically in that the motor fibers emerging from the central nervous system do not travel without interruption to their effectors. Instead, they synapse outside the central nervous system with a second motor neuron, which then innervates the effector. The fibers emerging from the central nervous system are known as preganglionic fibers, and those terminating in the effectors are known as postganglionic fibers.

The autonomic nervous system consists of the sympathetic and the parasympathetic divisions. The preganglionic fibers of the parasympathetic division exit from the base of the brain and from the sacral region of the spinal cord and synapse with the postganglionic neurons at or near the target organs. The sympathetic division originates in the cervical, thoracic, and lumbar regions. Preganglionic fibers of the sympathetic division synapse with postganglionic neurons in the chain of sympathetic ganglia or in other ganglia, such as the celiac ganglion, which is part of the solar plexus. The chemical transmitter at the effector is usually noradrenaline in the sympathetic division and acetylcholine in the parasympathetic division.

Most, but not all, internal organs are innervated by both divisions, which usually function in opposition to each other. In general, the sympathetic division stimulates functions involved in "fight-or-flight" reactions, and the parasympathetic division stimulates more tranquil functions, such as digestion.

Divisions of the Autonomic Nervous System: Sympathetic and Parasympathetic

Neuroanatomists divide the autonomic nervous system into two divisons: the *sympathetic division* and the *parasympathetic division*. These two divisions are distinct anatomically, physiologically, and functionally. In fact, generally speaking, the effects of the two divisions are antagonistic. Most organs are under the control of both sympathetic and parasympathetic nerves, and these work in close cooperation for the ultimate homeostatic regulation of the body's functions.

There are three major physical differences between the sympathetic and parasympathetic divisions:

1. They differ in the sites of emergence of their nerves from the central nervous system. The sympathetic division originates in the cervical, thoracic (chest), and lumbar (lower back) regions of the spinal cord. The parasympathetic division emerges through the cranial (brain) region and the sacral ("tail") region of the spinal cord.
2. As previously stated, in the autonomic nervous system, there is always a relay system of two neurons connecting the central nervous system and the effector organ. These neurons synapse at a ganglion. In the sympathetic division, this synapse is usually close to the central nervous system. Thus, characteristically, the preganglionic axon is short, and the postganglionic axon is long. In the parasympathetic division, the opposite is true: the synapse is close to or embedded in the target organ. Thus the preganglionic axon is long, and the postganglionic axon is short.
3. Most postganglionic sympathetic nerve endings release noradrenaline. All postganglionic parasympathetic endings release acetylcholine. (Acetylcholine is also the transmitter substance at the preganglionic endings for both divisions.)

The parasympathetic division is generally concerned with rest and rumination; it is particularly active, for example, after a heavy meal or following orgasm. Parasympathetic stimulation also slows down the heartbeat, increases the movements of the smooth muscles in the walls of the intestine, and stimulates secretions of the salivary glands.

The sympathetic system prepares the body for action. The physical characteristics of rage, for example, result from a simultaneous discharge of many sym-

Table 40-1 Comparison of Parasympathetic and Sympathetic Divisions

	PARASYMPATHETIC DIVISION	SYMPATHETIC DIVISION
Preganglionic fibers	Long; emerge from cranial and sacral regions of spinal cord	Short; emerge from cervical, thoracic, and lumbar regions of spinal cord
Location of synapse	In small local ganglia near or in innervated organs	In a series of ganglia close to the spinal cord or in ganglia halfway between spinal cord and innervated organs
Postganglionic fibers	Short	Long
Neurotransmitter	Preganglionic and postganglionic: acetylcholine	Preganglionic: acetylcholine; postganglionic: usually noradrenaline
General effects	Promotes "vegetative" and restorative functions, such as digestion	Promotes "fight-or-flight" responses; inhibits "vegetative" functions

pathetic nerve fibers. Certain of these cause the blood vessels in the skin and intestinal tract to contract; this contraction increases the return of the blood to the heart, raising the blood pressure and allowing more blood to be sent to the muscles. The heart beats both faster and stronger. The pupils dilate. The muscles underlying the hair follicles in the skin contract; this is probably a legacy from our furry forebears, which looked larger and more ferocious with their hair standing on end. The rhythmic movement of the intestine stops, and the sphincters, muscles at the end of the intestines and the opening of the bladder, relax. These reactions inhibit digestive operations, but the relaxing of the sphincters may also, in extreme

40–6
A cat prepared for "fight or flight," showing the effects of stimulation by the sympathetic nervous system. This drawing is from Darwin's The Expression of the Emotions in Man and Animals, *published in 1872.*

cases, have the disconcerting consequence of allowing involuntary defecation or urination. The adrenal glands pour out adrenaline. This hormone causes the release of large quantities of glucose by the liver, skeletal muscle, and adipose tissues. This glucose is the extra energy source for the muscles. As a consequence, the body as a whole is prepared for "fight or flight"—or, at least, action that would have been appropriate at some earlier stage in our cultural evolution.

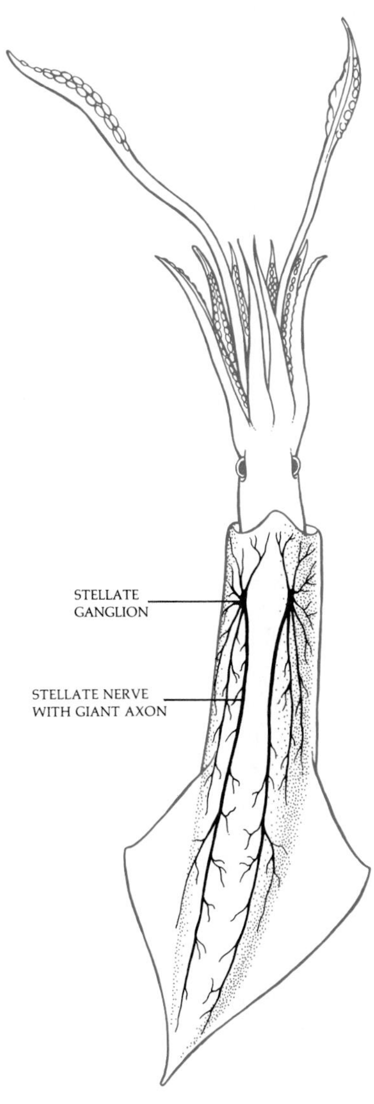

40–7
The squid, one of the chief protagonists in the research leading to the elucidation of the nature of the nerve impulse. The stellate nerves contain the giant axons used in all the early studies of the action potential.

THE NERVE IMPULSE

Some 200 years ago, Luigi Galvani observed that the passage of an electric current along the nerve of a frog's leg made the muscle twitch. Since that time, it has been known that nerve conduction is associated with electrical phenomena, and so, before progressing further, it might be useful to review some simple principles of electricity. One basic concept is that there are two types of electric charge, positive and negative; like charges repel one another, and unlike charges attract one another. Thus, negatively charged particles tend to move toward an area of positive charge, and vice versa. (In a metal, such as copper, only negative charges move, but the principle is the same.) Materials that permit the movement of charged particles, such as a copper wire or a solution containing ions, are known as conductors. Materials that do not permit the movement of charged particles, such as fat and rubber, are insulators.

The difference in the amount of electric charge between a region of positive charge and a region of negative charge is called the *electric potential.* As we saw in Chapter 9 (page 201), an electric potential is a form of potential energy, like a boulder at the top of a hill or water behind a dam. This potential energy is converted to electrical energy (electricity) when charged particles are allowed to move through a solution or along a wire between two regions of different charge. The difference in potential energy between the two regions is measured in volts or, if it is very small, in millivolts.

For a time, it was assumed that a nerve impulse was an electric current traveling along a nerve axon by a flow of electrons, in the same way electricity travels along a wire. However, this model did not hold up under critical scrutiny. First, it was shown that the axon is a poor conductor of electricity—meaning that the electrons flow through it for only a short distance before the current dies out. By contrast, the nerve impulse is undiminished from one end of the axon to the other. Second, although the impulse is fast by biological criteria, it is much slower than an electric current. Finally, unlike an electric current, the strength of the impulse is always the same. There is either no response to the stimulation of a nerve fiber, or there is a maximum response—all or nothing.

Major advances in understanding the nature of the nerve impulse occurred when it became possible to monitor changes in an individual neuron. The organism that first made this possible was the squid, which has large, long motor nerve fibers. (These innervate the muscles responsible for the rapid expulsion of water from the mantle cavity that produces the escape response described on page 513.) Thus, the squid, with its giant axons, took its place in the history of biology, along with the garden pea, the fruit fly, *Escherichia coli,* and T4.

Measurements within an axon are made with microelectrodes, the tips of which are fine enough to penetrate a living cell without interfering with its activities. The microelectrodes are connected with a very sensitive voltmeter called an oscilloscope, which measures voltage (in millivolts) in relation to time (in milliseconds).

40–8
The electric potential of the membrane of the axon is measured by microelectrodes connected to an oscilloscope. (a) *When both electrodes are outside the membrane, no potential is recorded.* (b) *When one electrode penetrates the membrane, the oscilloscope shows that the interior is negative with respect to the exterior and that the difference between the two is about 70 millivolts. This is the resting potential.* (c) *When the axon is stimulated and a nerve impulse passes along it, the oscilloscope shows a brief reversal of polarity—that is, the interior becomes positive in relation to the exterior. This brief reversal in polarity is the action potential.*

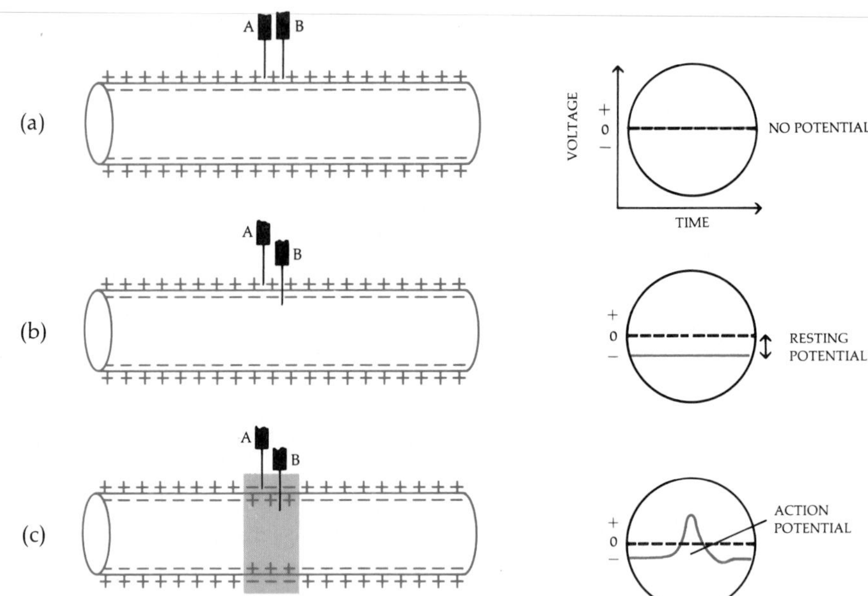

When both electrodes are outside the neuron, no voltage difference is recorded (Figure 40–8a). When one electrode penetrates the axon, a voltage difference of about 70 millivolts between the outside and the inside of the axon can be detected. The interior of the membrane is negatively charged with respect to the exterior (Figure 40–8b). This is the *resting potential* of the membrane. When the axon is stimulated, the oscilloscope records a very brief reversal of polarity—that is, the interior becomes positively charged in relation to the exterior (Figure 40–8c). This reversal in polarity is called the *action potential.* The action potential traveling along the membrane is the *nerve impulse.* It travels more slowly than an electric current, because it is actually a series of miniature electric circuits, but it is undiminished in intensity during the course of its transmission.

The action potentials recorded from any one neuron are always the same. Figure 40–9 shows, for example, the action potentials produced by a single sensory neuron in the skin of a cat in response to pressure. All the impulses are the same size (all-or-nothing response). The only variation is the frequency with which impulses are produced.

The Ionic Basis of the Action Potential

To understand the action potential, we must reexamine the chemistry of the cell and, in particular, the nature of the cell membrane. A difference in total electric charge between the inside and the outside of the membrane—an electric potential—is a characteristic common to all cells. It is produced by differences in the distribution of ions on either side of the membrane.

Let us look first at the axon of a motor neuron in its resting state. In this resting state, the concentration of potassium ions (K^+) in the cytoplasm of the axon is 30 times greater than the concentration outside, and the concentration of sodium ions (Na^+) is about 10 times greater outside than inside. The cytoplasm of the axon also contains large, negatively charged organic ions.

40–9
(a) Nerve impulses can be monitored by electronic recording instruments. The impulses from any one neuron are all the same; that is, each impulse has the same duration and voltage change as any other. (b) *Nerve impulses from a sensory neuron (touch receptor) in cat skin. The skin was touched and pressed in at varying depths, as indicated by the figures at the left. The more deeply the skin was pressed in, the more rapidly nerve impulses were produced. The vertical lines represent individual action potentials on a compressed time scale. As you can see, all the action potentials are the same size, but their frequency increases with the intensity of the stimulus.*

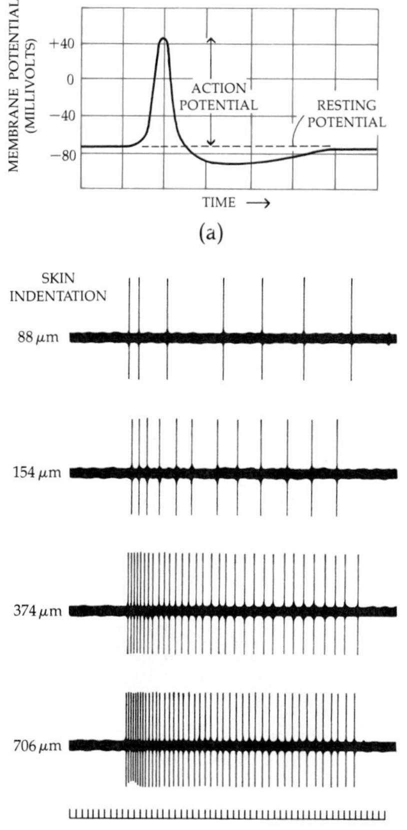

The distribution of ions on either side of the membrane is governed by three factors: (1) the diffusion of particles down a concentration gradient (page 128), (2) the attraction of particles with opposite charges and the repulsion of particles with like charges, and (3) the properties of the membrane itself. The membrane is not permeable to the large, negatively charged organic ions (anions) nor, in its resting state, is it permeable to Na^+ ions. It is, however, somewhat permeable to K^+ ions. Moreover, it contains a sodium-potassium pump (page 134), which pumps Na^+ ions out and K^+ ions in.

Now let us examine the forces at work on the K^+ ions, which, according to current hypotheses, are responsible for the electric potential of the membrane at rest (the resting potential). Because of the concentration gradient, K^+ ions tend to move out of the cell; in fact, if no other influence were at work, they would, of course, move down the concentration gradient until their distribution was equal on either side of the membrane. However, because of the impermeability of the membrane, anions cannot follow the K^+ ions out of the cell. Thus, as K^+ ions leave, an excess negative charge builds up inside the cell. This excess of negative charge impedes the further outward movement of the positive K^+ ions. As a result, an equilibrium is reached at which there is no net movement of K^+ ions across the membrane. At the point of equilibrium, there is a slight excess of negative charge inside the cell. This is the resting potential (Figure 40–10).

When the membrane is stimulated, it suddenly becomes highly permeable to Na^+ at the site of the stimulation. The Na^+ ions rush in, moving down their concentration gradient and attracted, at first, by the negative charge inside the cell. This influx of positively charged ions momentarily reverses the polarity of the membrane so that it becomes more positive on the inside than on the outside, producing the action potential (Figure 40–11). The change in Na^+ permeability lasts for only about half a millisecond; then the membrane regains its previous impermeability to Na^+. During this time, the permeability to K^+ increases, and there is an outward flow of K^+ ions due to the concentration gradient and also to the positive charge inside the cell at the peak of the action potential. This outward flow of positive K^+ ions counteracts the previous inward flow of positive Na^+ ions, and the resting potential is quickly restored. (The actual number of ions involved is very small. Only a very few Na^+ ions need enter to cause the reversal of polarity of the membrane, and only the same small number of K^+ ions need move out of the

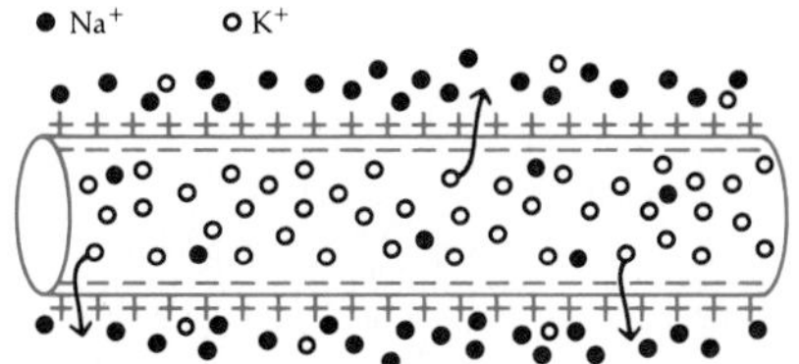

40–10
Resting potential. The membrane of the axon is polarized, with a slight excess of negative charge inside the membrane caused by the outward diffusion of K^+ *ions.*

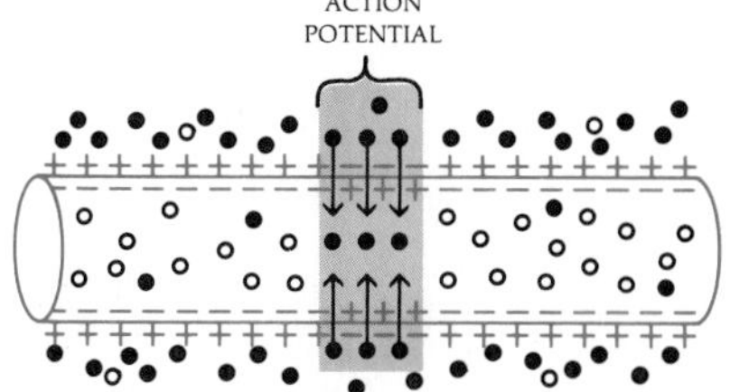

40–11
The action potential. A portion of the membrane becomes momentarily permeable to Na^+ *ions. The* Na^+ *ions rush in, and the polarity of the membrane is reversed.*

40–12
Propagation of the nerve impulse. In advance of the action potential, a small segment of the membrane becomes slightly depolarized, owing to the flow of charged ions along the inside of the membrane. When the membrane becomes depolarized in this way, its permeability to Na^+ increases, and Na^+ ions move in, creating a new action potential and depolarizing another segment of membrane. The nerve impulse is the action potential traveling along the membrane.

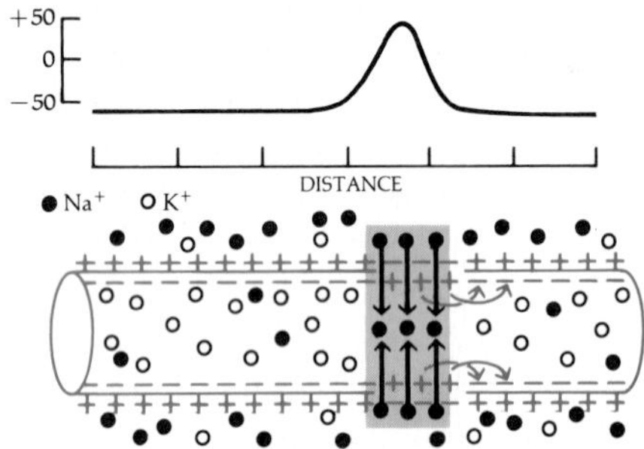

cell to restore the resting potential.) Subsequently, the sodium-potassium pump restores the Na^+ and K^+ concentrations to their original levels, restoring the resting potential and thus making it possible for the neuron to fire again.

Propagation of the Impulse

An important feature of the nerve impulse is that this transient reversal in polarity continues to move along the axon, renewing itself indefinitely, just as a flame traveling along a fuse ignites the fuse as it travels. The action potential is self-propagating because at the peak of the action potential, when the inside of the membrane at the active region is comparatively positive, positively charged ions move from this region to the adjacent area inside the axon, which is still comparatively negative. As a consequence, the adjacent area, in turn, becomes depolarized—that is, less negative (Figure 40–12). When it becomes slightly depolarized, its permeability to Na^+ increases. Na^+ ions rush in and create a new action potential, which, in turn, depolarizes the next adjacent area of the membrane. Thus the nerve impulse travels along the axon. (The segment of the axon behind the nerve impulse has a brief refractory period during which it cannot be reexcited and which therefore keeps the action potential from going backward.) Because of this renewal process, repeating itself along the length of the membrane, an axon, which, as we noted previously, is a very poor conductor of an ordinary electric current, is capable of transmitting a nerve impulse over a considerable distance with absolutely undiminished strength.

The Myelin Sheath

As we noted in Chapter 33, about 90 percent of nerve tissue cells are not neurons but are cells that support the neurons physically and also nourish, protect, and insulate them. The best understood of these cells are the Schwann cells, which form myelin sheaths around the long fibers of neurons in the peripheral nervous system of vertebrates (Figure 40–13).

The Function of the Myelin Sheath

The myelin sheath, according to Nobel laureate John Eccles, is "a brilliant innovation." Because of it, the propagation of the nerve impulse is much more rapid in vertebrates than in invertebrates. The sheath is not just an insulator. Its most important feaure is that it is interrupted at regular intervals by openings, or nodes.

40–13
(a) *Formation of a myelin sheath by a Schwann cell. As the Schwann cell grows, it extends itself around and around the axon and gradually extrudes its cytoplasm from between the layers.*
(b) *Electron micrograph of a cross section of a mature myelin sheath.*

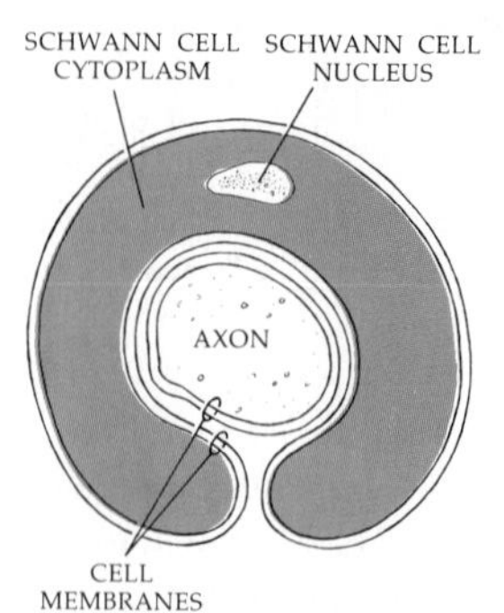

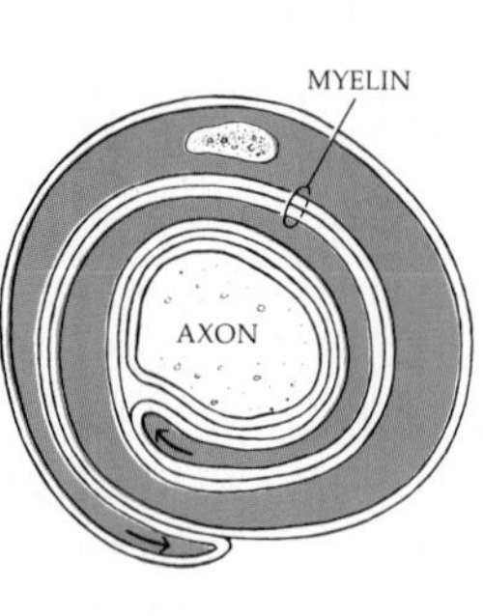

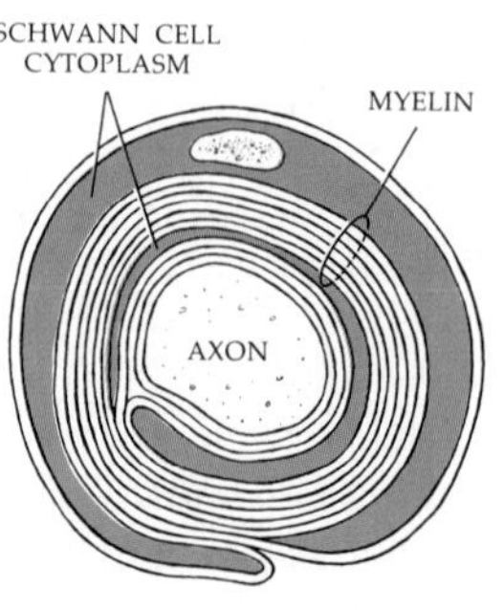

(a)

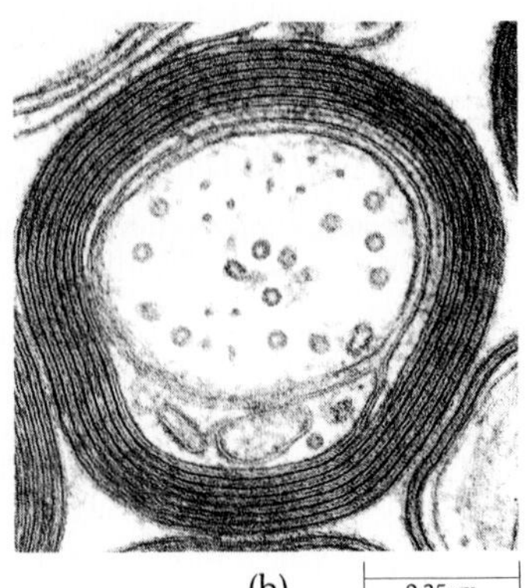

(b)

40–14
In an unmyelinated fiber (a), *the action potential travels continuously along the axon, whereas in a myelinated fiber* (b), *the impulse jumps from node to node, greatly accelerating the transmission.*

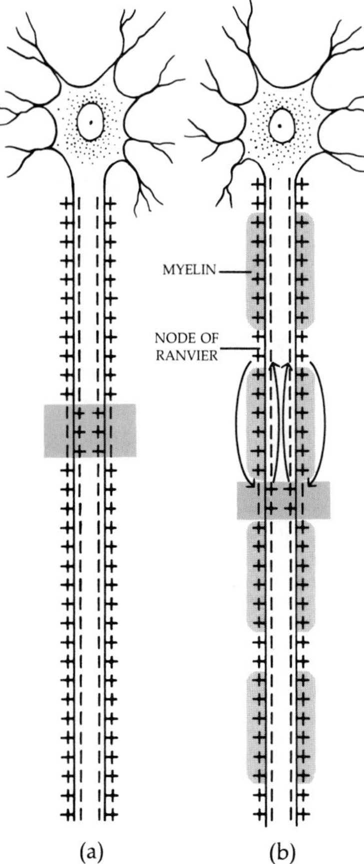

In myelinated fibers—which include all the large nerve fibers of vertebrates—the impulse, rather than creeping along the membrane like water along a wick, jumps from node to node (Figure 40–14). This saltatory (leaping) transmission increases the velocity up to 50 times. Some large, myelinated nerve fibers, for example, conduct impulses as rapidly as 200 meters per second, compared with velocities of only a few millimeters per second in small, unmyelinated fibers. Also, because only a small portion of the axon's membrane must be repolarized, there is an enormous saving in expenditure of energy by the sodium-potassium pump.

THE SYNAPSE

Signals are transmitted from one neuron to another across the specialized junction known as the synapse. As we noted before, synapses may be electrical or chemical. In the former, the electrical signal is passed directly through the membranes of closely juxtaposed neurons. Such electrical signals, which are common in lower invertebrates, have been identified at some sites in the mammalian brain. In chemical synapses, which make up the vast majority of connections between neurons in the mammalian nervous system, the two neurons never touch; there is a space of about 20 nanometers between them, known as the synaptic cleft (Figure 40–15). Signals travel across the synaptic cleft by means of chemical substances

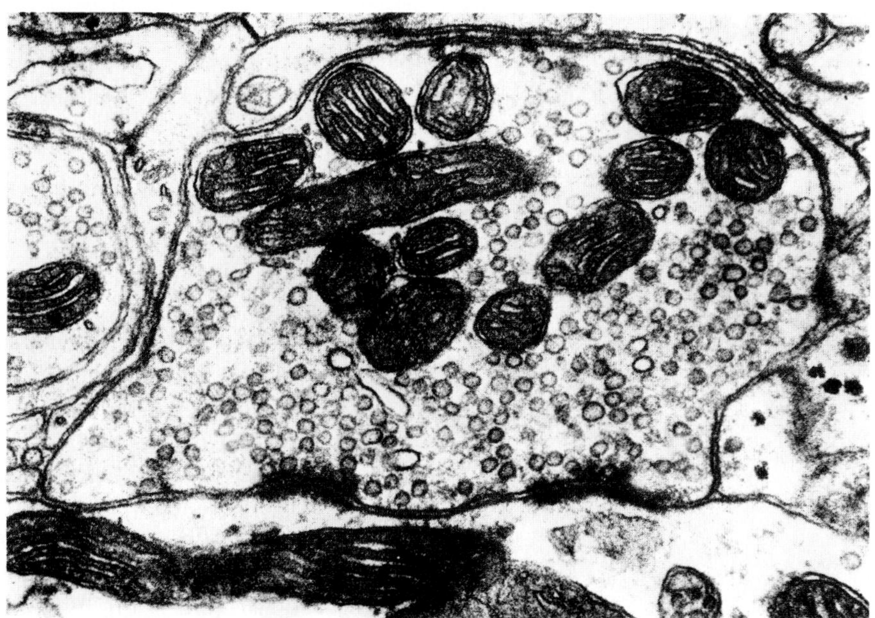

40–15
A synapse between the axon of one neuron, which fills most of the micrograph, and the dendrite of another, a small portion of which is visible at the bottom of the micrograph. Note the many small vesicles; these contain neurotransmitter, which is released into the synaptic cleft, the space between the two neurons. Note also the two fuzzy patches on the membrane of the axon terminus at the synaptic cleft. These are believed to be the active zone of the synapse, where the vesicles release the neurotransmitter.

40-16

Acetylcholine (a) *is the principal neurotransmitter in the somatic system and in the parasympathetic division of the autonomic system. Noradrenaline* (b) *is usually the neurotransmitter for the postganglionic endings of the sympathetic division of the autonomic nervous system. Noradrenaline and adrenaline* (c) *are also secreted by the adrenal medulla, the central portion of the adrenal gland.*

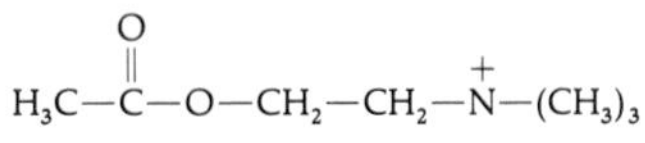

$$H_3C-\overset{\overset{\displaystyle O}{\|}}{C}-O-CH_2-CH_2-\overset{+}{N}-(CH_3)_3$$

ACETYLCHOLINE

(a)

HO, HO — OH H H — C—C—N—H — H H

NORADRENALINE

(b)

HO, HO — OH H H — C—C—N—CH_3 — H H

ADRENALINE

(c)

known as *neurotransmitters.* These neurotransmitters are contained in numerous small vesicles present in synaptic knobs at the ends of the axon. Arrival of the action potential at the axon terminus causes these vesicles to empty their contents into the synaptic cleft. The transmitter substance diffuses across the cleft and combines with receptor molecules on the membrane of the postsynaptic cell, changing the permeability of the membrane. Two principal neurotransmitters are acetylcholine and noradrenaline (Figure 40-16).

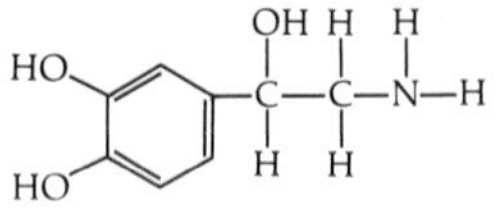

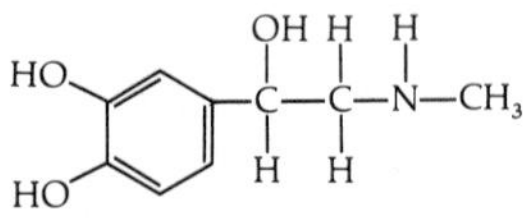

After their release, neurotransmitters are either rapidly destroyed by specific enzymes, diffuse away, or are taken up again by the synaptic knob. Such removal or destruction of the transmitter puts a halt to its effect. This in itself is an essential feature in the control of the activities of the nervous system.

Unlike the nerve impulse along the fiber—an all-or-nothing proposition—signals transmitted by chemicals across a synaptic cleft can modulate one another. That is, some may excite and some may inhibit. Each excitatory signal received by a neuron can cause a very small decrease in membrane voltage. If many such reactions occur at about the same time, they can cause a large enough depolarization of the axon to initiate a nerve impulse. The neurotransmitters released at inhibitory synapses, however, have a reverse effect. They stabilize the membrane of the postsynaptic neuron near its resting potential or even increase the membrane voltage, thereby decreasing the likelihood of an impulse's being generated.

A single neuron may receive signals from hundreds, even thousands, of synapses, and, based on the summation of excitatory and inhibitory signals, it will or will not be excited and initiate an impulse. Synapses are therefore relay and control points that are extremely important in the functioning of the nervous system.

SUMMARY

The nervous system, along with the endocrine system, integrates and controls the numerous functions that enable an animal to regulate its internal environment and to react to or deal with its external one.

The functional unit of the nervous system is the neuron, consisting of dendrites, which receive impulses, a cell body, which contains the nucleus and metabolic machinery, and an axon, or nerve fiber, which relays the impulses to other cells.

The central nervous system consists of the brain and the spinal cord, which are encased, in vertebrates, in the skull and vertebral column. That part of the nervous system outside the central nervous system constitutes the peripheral nervous system. Cranial nerves enter and emerge from the brain. Spinal nerves enter and emerge from the vertebral column in pairs, each pair consisting of both motor fibers and sensory fibers. Each of these pairs innervates the skeletal muscles and receives signals from sensory receptors of a different and distinct area of the body.

In vertebrates, the peripheral nervous system has two major divisions: (1) the somatic nervous system, and (2) the autonomic nervous system, which controls cardiac muscle and the smooth muscles and glands involved in the digestive, circulatory, urinary, and reproductive functions. In the autonomic system, the motor neurons arising from the spinal cord synapse outside the central nervous system with second neurons. These postganglionic neurons stimulate or inhibit the effectors. The autonomic system has two divisions: (1) the sympathetic division, which is largely responsible for reactions to the external environment; and (2) the parasympathetic division, which controls restorative activities, such as digestion and rest. The two divisions are anatomically distinct. The postganglionic neurotransmitter in the sympathetic division is usually noradrenaline; in the parasympathetic division, it is acetylcholine.

At rest, there is a difference in electric charge between the inside and outside of the nerve fiber membrane—the resting potential. When appropriately stimulated, an action potential, a transient reversal in membrane polarity, occurs. This action potential traveling along the fiber membrane is the nerve impulse. Because all action potentials are the same size, the message carried by a particular fiber can be varied only by a change in the frequency or pattern of stimulation of action potentials. In myelinated cells, the nerve impulse leaps from node to node of the myelin sheath, thereby speeding conduction.

Nerve cells transmit signals to other neurons across a junction called a synapse. In most synapses, the signal crosses the synaptic cleft in the form of a chemical, a neurotransmitter, that stimulates or inhibits the production of an action potential in the target cell.

QUESTIONS

1. Distinguish among the following: neuron/nerve/tract; ganglia/nuclei; afferent/efferent; preganglionic/postganglionic; axon/dendrite; resting potential/action potential; adrenaline/noradrenaline/acetylcholine; gray matter/white matter.

2. What are four significant differences between the somatic and autonomic nervous systems?

3. What are three significant differences between the sympathetic and parasympathetic divisions?

4. If a neuron is placed in a medium where ionic concentrations are the same as those of its own cytoplasm, what effect will this have on the resting potential?

5. Describe the way in which the nerve impulse propagates itself. Draw a diagram of this process.

6. Assume that three preganglionic neurons, A, B, and C, make adjacent synapses on the same postganglionic neuron, D. No postganglionic impulse is initiated as a result of single preganglionic impulses in A, B, or C, nor is it initiated if impulses arrive simultaneously in all three, in A and B, or in A and C. Only if impulses arrive together at the synapses of B and C will D fire an impulse. Explain these results in terms of excitatory and inhibitory neurotransmitters and their effects on the postganglionic membrane voltage.

CHAPTER 41

Integration and Control II: The Endocrine System

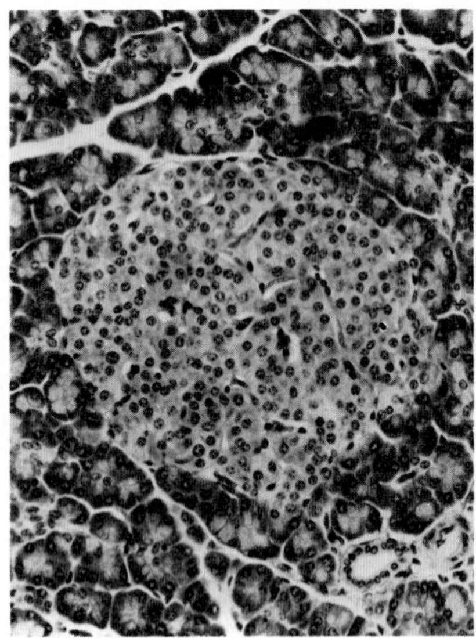

41–1

A cross section of pancreatic tissue. The pancreas is both an endocrine and an exocrine gland. The group of small cells in the center of the micrograph are islet cells, which are the endocrine cells of the pancreas. Different types of islet cells secrete the hormones insulin, glucagon, and somatostatin. The surrounding exocrine cells produce digestive enzymes, which are carried through the pancreatic duct to the small intestine.

Since the beginnings of agriculture, domestic animals have been castrated in order to make their flesh fatter and more tender, and to make their behavior more manageable. Eunuchs (human castrates) were traditionally, and for obvious reasons, used as harem guards, and as recently as the eighteenth century, selected boys were castrated before puberty in order to retain the purity of their soprano voices for church and opera choirs. As a consequence of these early practices, the effects of the male sex hormone—or more precisely, of its absence—were the first endocrine influences to be recognized and studied.

In 1849, A. A. Berthold, a physician and professor in Göttingen, carried out the first experiment to determine the way in which the testes, the pair of male sex glands, function. Using six young cockerels, he set up three experimental groups: (1) Two birds were castrated; these became typical capons in which the combs, wattles, plumage, aggressiveness, crow, and sexual urge characteristic of the mature cock failed to develop. (2) Two additional birds were castrated, and the removed testes were reimplanted in the same animal but at a new site, distant from ducts, nerves, or any possible previous connections except with the circulatory system. These birds developed into normal cocks. (3) Two birds were subjected to "sham" surgery; that is, surgical incisions were made, but no tissues were removed or transplanted. They, too, developed normally. (These last animals served as controls; without them, Berthold could not be sure that the results obtained in other groups were not related to surgical stress alone.)

This experiment initiated a wholly new concept of controlling mechanisms: substances produced by particular tissues of the body—the testes, in this case—could be carried by the bloodstream and could exert specific effects on distant tissues. Such substances came to be known as *hormones,* from the Greek word meaning "excite." It is now known, however, that many hormones act as inhibitors rather than exciters. (The juvenile hormone of insects, page 541, is an example.) Thus, the hormones are better described as chemical messengers, establishing communications between the various parts of the organism. Some of the communications are concerned with homeostatic regulation, the perpetual adjustment of the internal environment. Others are involved with change—the changes that occur with sexual maturation or responses to external events.

Hormones are secreted by glands—epithelial tissues specialized for secretion. Glands are classified as exocrine or endocrine. *Exocrine glands* secrete their products into ducts; examples are the sweat glands of the human skin and digestive glands.

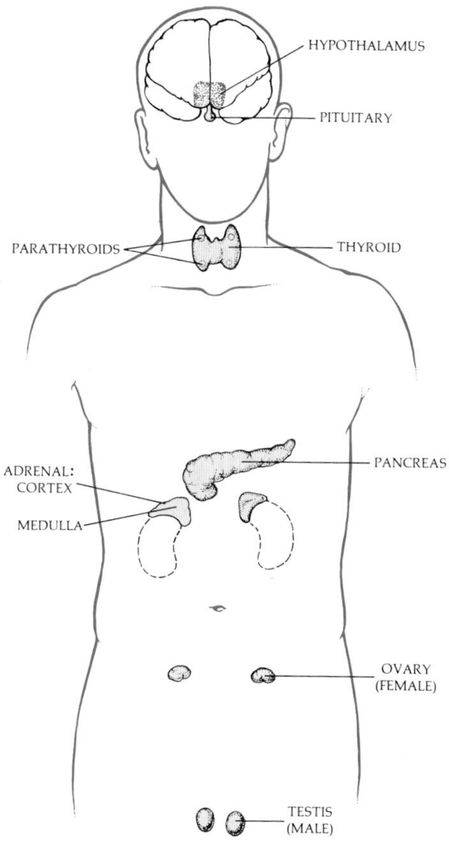

41–2
Some of the hormone-producing (endocrine) organs. The pituitary releases hormones that, in turn, regulate the hormone secretions of the sex glands, or gonads, the thyroid, and the adrenal cortex (the outer layer of the adrenal gland). The pituitary is itself under the regulatory control of the area in the brain known as the hypothalamus, indicated in color. The hypothalamus thus is the major link between the nervous and endocrine systems.

Endocrine glands are glands that secrete hormones. They secrete them into the bloodstream (or, more precisely, into the extracellular space, from which the substances diffuse into the bloodstream), so they are sometimes referred to as "ductless" glands. (A summary table listing the major endocrine glands and the hormones they secrete can be found on page 792).

As we saw in the last chapter, neurons make close, anatomical connections with the cells or organs that they influence, like talking on the telephone. Hormones, by contrast, broadcast their messages. Whether or not these messages are received and acted upon depends upon the receptivity of the target tissue as much as on the chemical characteristics of the hormone. Moreover, target tissues may be receptive under some circumstances and not under others. For example, prolactin causes the production of milk—but only by the mammary gland and only when it is acting in concert with estrogen, progesterone, thyroxine, adrenal steroids, and growth hormone.

SOME CHARACTERISTICS OF HORMONES

A hormone is an organic molecule secreted in one part of an organism that, in the case of vertebrates, diffuses or is transported by the bloodstream to another part of the organism, where it has a specific effect on some target organ or tissue. Hormones (with the exception of the prostaglandins, to be discussed later) are of three general chemical types: steroids (see page 63), peptides or proteins, and amino acid derivatives. They are characteristically active in very small amounts. It has been calculated, for example, that the concentration of adrenaline normally present in your bloodstream can be approximated by the concentration of one teaspoonful in a lake 2 meters deep and 100 meters in diameter. As befits such potent chemicals, hormones are under tight control. One aspect of their control is the regulation of their production. With very few exceptions, hormones are under negative feedback control, as described in Chapter 33. Another, equally important aspect is that they are rapidly degraded in the body; the steroids, peptides, and proteins are broken down by the liver, and the amines (amino acid derivatives) by enzymes in the blood.

In the following pages, we shall discuss some of the principal endocrine glands of mammals, the hormones they secrete, and the regulation of their secretion.

ANDROGENS: THE MALE SEX HORMONES

The principal male sex hormone is testosterone, a steroid hormone produced primarily by the interstitial cells of the testes, or testicles (Figure 41–3). Male sex hormones, including testosterone, are known collectively as *androgens*. Androgens are first produced in early embryonic development, causing the male fetus to develop as a male rather than a female. After birth, androgen production continues at a very low level until the child is about 10 years old. Then there is a surge in testosterone, resulting in the onset of sperm production (which marks the beginning of puberty) accompanied by enlargement of the penis and testes and also of the prostate and other accessory organs. In the healthy human male, a high level of testosterone production continues into the fourth decade of life, when it begins gradually to decline.

As can be readily observed, testosterone also has effects on other parts of the body not directly involved in the production and deposition of sperm. In the

41–3
The testis is composed of tightly packed seminiferous tubules, one of which is shown in cross section in this drawing. Within the tubules are sperm cells in various stages of development, surrounded, supported, and nourished by Sertoli cells. The interstitial cells, which are found in connective tissues between the tubules, are the source of testosterone.

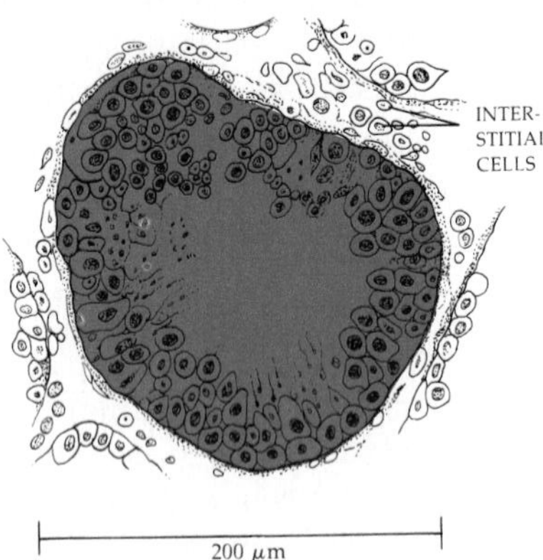

human male, these effects include growth of the larynx and an accompanying deepening of the voice, increase in skeletal size, and a characteristic distribution of body hair. Androgens stimulate the biosynthesis of proteins and so of muscle tissue. They stimulate the apocrine sweat glands, whose secretions attract bacteria and so produce the body odors associated with sweat after puberty. And they may cause the sebaceous glands of the skin to become overactive, resulting in acne. Such characteristics, associated with sex hormones but not directly involved in reproduction, are known as *secondary sex characteristics.*

In other animals, testosterone is responsible for the lion's mane, the powerful musculature and fiery disposition of the stallion, the cock's comb and spurs, and the bright plumage of many adult male birds. It is also responsible for a variety of behavior patterns such as the scent marking of dogs, the courting behavior of sage grouse, and various forms of aggression toward other males found in a great many vertebrate species.

Regulation of Testosterone Production

The production of testosterone is regulated by a negative feedback system involving, among other components, a gonadotropic (gonad-stimulating) hormone called luteinizing hormone (LH) that is produced by the pituitary gland.

The pituitary, which is about the size and shape of a kidney bean, is located at the geometric center of the human skull. It is lodged just beneath the area of the brain known as the hypothalamus, and the pituitary's production of gonadotropic hormones is controlled by this brain center. LH is carried by the blood to the interstitial tissues of the testes, where it stimulates the output of testosterone. As the blood level of testosterone increases, it, in turn, slows the release of LH from the pituitary (Figure 41–4).

The testes are also under the influence of a second pituitary hormone, follicle-stimulating hormone (FSH), which acts upon the Sertoli cells of the testes and, through them, on the developing sperm. It is believed that the Sertoli cells release a protein hormone that, in turn, inhibits FSH production. This hormone, tentatively called inhibin, appears to be a small protein, but it has not yet been isolated and identified. Its discovery is of more than academic interest; it would appear a likely candidate for the long-sought male contraceptive.

41–4
Hormonal control of testicular function. The hypothalamus stimulates the anterior pituitary gland to produce LH (luteinizing hormone), which, in turn, stimulates production and release of testosterone from the interstitial cells of the testes. The production of LH by the pituitary and gonadotropin-releasing hormone (GnRH) by the hypothalamus are inhibited by the resulting increased concentration of testosterone.

Similarly, in another negative feedback system, FSH (follicle-stimulating hormone) acts on the Sertoli cells, which, in turn, produce the hormone inhibin, which inhibits FSH production. The combined action of testosterone and FSH promotes sperm formation.

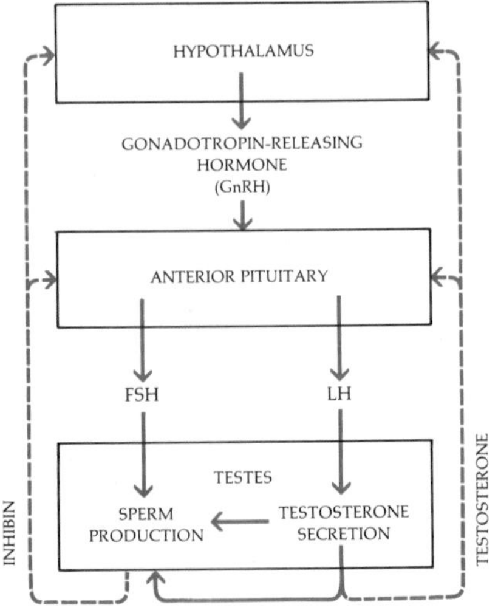

41–5

Secondary sex characteristics among vertebrate males. Many of the conspicuous characteristics of male animals, such as the peacock's tail, are associated with the competition of males for females.

In the human male, the rates of testosterone release are fairly constant. In many animals, however, male hormone production is triggered by changes in temperature, daylight, or other environmental cues and changes seasonally. Testosterone production may also be affected by social circumstances. Recent studies on bulls, for example, showed that after a bull sees a cow, his blood level of LH rises as much as seventeenfold; within about half an hour, the blood level of testosterone reaches its peak. In many animal societies, including wolves and cape hunting dogs (the wild dogs of the African plains), socially inferior males may never become sexually mature, presumably because of depressed male hormone production. Human testosterone production also may vary according to the emotional climate. In a study made among Army GIs during the Vietnam War, testosterone levels in recruits in basic training and among combat troops were markedly lower than the normal levels of testosterone found in men in behind-the-lines assignments.

ESTROGENS: THE FEMALE SEX HORMONES

Estrogens are the female sex hormones. These hormones, of which estradiol is the most important, are produced by the ovaries under the stimulation of FSH from the pituitary. Estrogens stimulate the development of the breasts and the external genitalia and the distribution of body fat. In conjunction with progesterone, which is also produced by the ovaries, estrogens help to prepare the endometrium—the lining of the uterus—for the implantation of the embryo. Estrogens, progesterone, FSH, and LH all interact to produce the menstrual cycle, which is timed and controlled by the hypothalamus. (The menstrual cycle is described in Chapter 44.)

41-6

The chemical structures of testosterone, estradiol, and progesterone. Note that the hormones differ only very slightly chemically, in contrast to the great differences in their physiological effects—another example of the extreme specificity of biochemical actions. All of these belong to a group of chemicals known as steroids, characterized by the four-ring structure shown here.

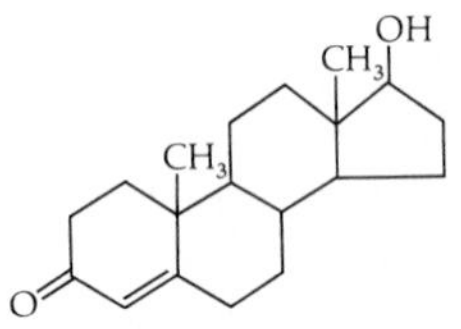

OH CH_3 CH_3 O

TESTOSTERONE

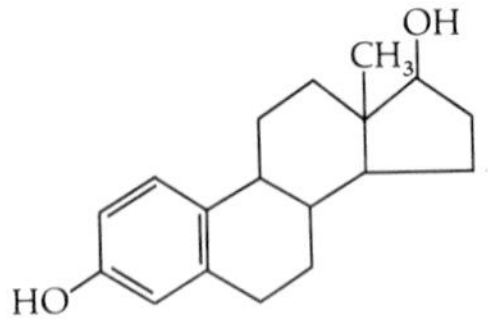

OH CH_3 HO

ESTRADIOL

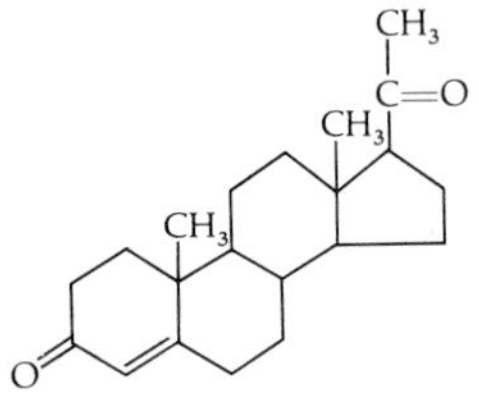

CH_3 C=O CH_3 CH_3 O

PROGESTERONE

HORMONES OF THE ADRENAL CORTEX

The adrenal cortex—the outer layer of the adrenal gland—is the other major source of steroid hormones. The adrenal (ad-renal) glands, as their name implies, are on top of the kidneys. About 50 different steroids, all very similar in structure, have been isolated from the adrenal cortex of various mammals. Some of these corticosteroids, as they are known, undoubtedly represent steps in the synthesis of various hormones, though most of them have some hormonal activity. In humans, there are two major groups of adrenocortical steroids, the glucocorticoids and the mineralocorticoids.

Cortisol is thought to be the most important glucocorticoid in humans. Cortisol and the other glucocorticoids are concerned primarily with the formation of glucose from protein and fat. At the same time, they inhibit the uptake of glucose by most cells, with the notable exceptions of cells of the brain and the heart, thus favoring the activities of these organs at the expense of other body functions. Their release increases during periods of stress such as facing new situations, engaging in athletic competitions, and taking final exams. Thus they work in concert with the sympathetic nervous system.

The glucocorticoids also act to suppress inflammation and are sometimes used medically as anti-inflammatory agents in the treatment of diseases such as arthritis. However, their serious side effects limit their usefulness. (Side effects of these drugs include reduced ability to combat infection; redistribution of body fat, resulting in "buffalo hump" and "moon face"; and mental disturbances.)

41-7
The chemical structures of two steroid hormones secreted by the adrenal cortex. Cortisol is a glucocorticoid, and aldosterone is a mineralocorticoid. Note, however, the very minor structural differences between the two.

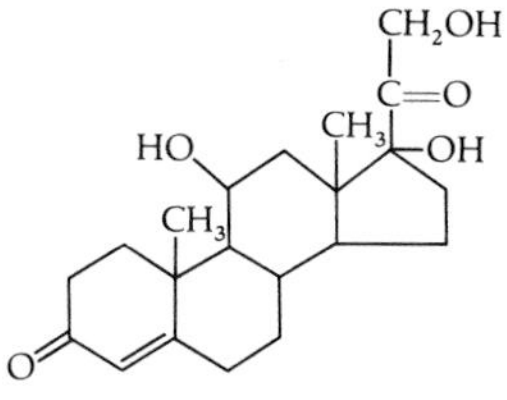

CORTISOL

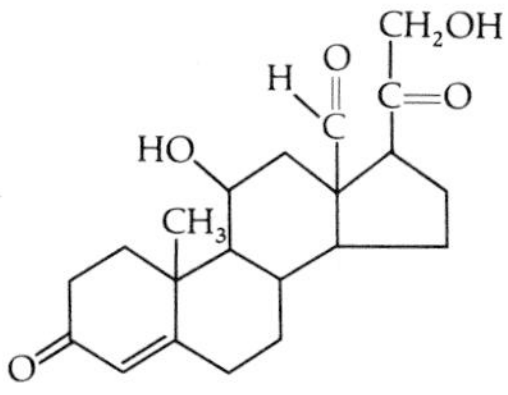

ALDOSTERONE

The cortisol group of hormones are secreted in response to adrenocorticotropic hormone (ACTH). ACTH is secreted by the pituitary gland when it is stimulated by a releasing hormone from the hypothalamus. As with testosterone, the secretion of these steroids is inhibited by negative feedback exerted on the pituitary and the hypothalamus.

A second group of hormones secreted by the adrenal cortex comprises the mineralocorticoids, of which aldosterone is the primary example. These corticosteroids are concerned with the regulation of ions, particularly sodium and potassium ions. The mineralocorticoids affect the transport of ions across the cell membranes of the nephrons and, as a consequence, have major effects both on ion concentrations in the blood and on water retention and loss. An increase in aldosterone secretion, as we noted in Chapter 37, results in greater reabsorption of sodium ions in the distal tubule of the nephron and increases the secretion of potassium ions into it. A deficiency in mineralocorticoids precipitates a critical loss of sodium ions from the body in the urine and, with it, a loss of water, leading, in turn, to a reduction in blood pressure.

The adrenal cortex is also a source of male sex hormones, which is why an adrenal tumor may result in increased production of these hormones and the production of facial hair and other masculine characteristics in a woman. Bearded ladies in the circus are often victims of such tumors.

THYROID HORMONE

Thyroid hormone (thyroxine) is also controlled by the hypothalamus-pituitary circuit, as we noted in Chapter 33. Its release from the thyroid gland is stimulated by thyroid-stimulating hormone (TSH), also known as thyrotropin, from the pituitary gland. Thyroxine differs from androgens, estrogens, and corticosteroids in its chemical structure: thyroxine is an amino acid combined with four atoms of iodine (Figure 41-8).

Thyroxine (or its metabolic product, triiodothyronine) accelerates the rate of cellular respiration. Hyperthyroidism, the overproduction of thyroxine, results in nervousness, insomnia, and excessive excitability; heat intolerance and excessive sweating; increased heart rate and blood pressure; and weight loss. Hypothyroidism in infancy affects development, particularly of the brain cells, and if not treated in time, can lead to permanent mental deficiency and dwarfism. In adults, hypothyroidism is associated with dry skin, intolerance to cold, and lack of energy.

Hypothyroidism may be caused by insufficient iodine, which is needed to make thyroxine, and, in these cases, it is often associated with goiter, an enlargement of the thyroid gland.

The thyroid gland is also the source of the hormone calcitonin, whose major action is to inhibit the release of calcium ions from bone. Its secretion is controlled directly by the calcium concentration of the fluid surrounding the thyroid cells.

41-8
Thyroxine, the principal hormone produced by the thyroid gland. Note the four iodine atoms in its structure. Triiodothyronine differs from thyroxine by having one less iodine atom on the OH-bearing ring. Because we need iodine for thyroxine, it is an essential component of the human diet. Where iodine is present in the soil, it is available in minute quantities in drinking water and in plants. In the United States, table salt is ordinarily iodized or must be specifically labeled as being uniodized.

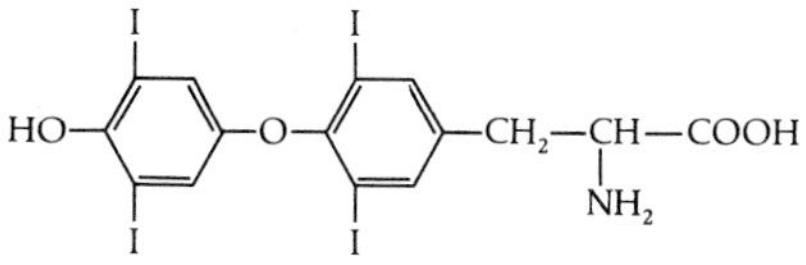

THYROXINE

THE PITUITARY HORMONES

The pituitary gland was once considered the master gland of the body, as it is the source of hormones stimulating the gonads, the adrenal cortex, and the thyroid. However, it is now known that this "master" gland is itself regulated by hormones from the hypothalamus that stimulate or, in some cases, inhibit the production of pituitary hormones.

The pituitary, about the size of a kidney bean, has three lobes: the anterior, the intermediate, and the posterior. The anterior lobe is the source of at least six different hormones. Four of these are the tropic hormones just described: FSH, LH, ACTH, and TSH, all regulated by hormones from the hypothalamus and by negative feedback systems.

Somatotropin, also called growth hormone, is produced as well by the anterior pituitary. It stimulates protein synthesis and promotes the growth of muscle and bone. As is the case with most of the hormones, growth hormone is best known by the effects caused by too much or too little. If there is a deficit in growth hormone production in childhood, a midget results, the so-called "pituitary dwarf." Excessive growth hormone production during childhood results in a giant; most circus giants are the result of this excess. Excessive growth hormone production in the adult does not lead to giantism, since growth of the long bones no longer occurs, but to acromegaly, an increase in the size of the jaw and the hands and feet, adult tissues that are still sensitive to the effects of growth hormone. Growth hormone also affects glucose metabolism. It inhibits the uptake and oxidation of glucose by many types of cells, and it stimulates the breakdown of fatty acids for energy, thus conserving glucose. This hormone is now being produced by recombinant DNA techniques, which will extend its use in medical treatment and facilitate the study of its role in glucose metabolism.

Prolactin, also produced by the anterior pituitary, stimulates the secretion of milk in mammals. Production of prolactin is controlled by an inhibitory hormone produced by the hypothalamus. As long as the infant continues to nurse, the nerve impulses produced by the suckling of the breast are transmitted to the hypothalamus, which then decreases production of prolactin-inhibiting hormone. The pituitary then releases prolactin, which, in turn, acts upon the breast to maintain the production of milk. Once suckling ceases, the synthesis and release of prolactin decreases and so milk production stops. Thus supply is regulated by demand.

In many vertebrates, the intermediate lobe of the pituitary gland is the source of melanocyte-stimulating hormone. In reptiles and amphibians, this hormone stimulates color changes associated with camouflage or with behavior patterns such as aggression or courtship. In humans, in which its secretion is greatly reduced, it has no known function.

The posterior lobe of the pituitary gland stores the hormones produced by the hypothalamus.

THE PITUITARY-HYPOTHALAMUS AXIS

The pituitary gland lies beneath the hypothalamus and is directly under its influence. The pituitary is also under the influence, by way of the hypothalamus, of other parts of the brain. The hypothalamus is the source of at least nine hormones that act either to stimulate or inhibit the secretion of hormones by the anterior

41–9
Relationships between the pituitary and the hypothalamus. The anterior lobe is connected with the hypothalamus by a network of capillaries. Neurosecretory cells of the hypothalamus secrete releasing hormones directly into these capillaries. They are carried in the blood to the anterior pituitary, where they affect the production of other hormones. Nerve fibers (in color) connecting the hypothalamus and the posterior lobe of the pituitary transmit oxytocin and ADH, which are produced in the hypothalamus and released from the posterior pituitary.

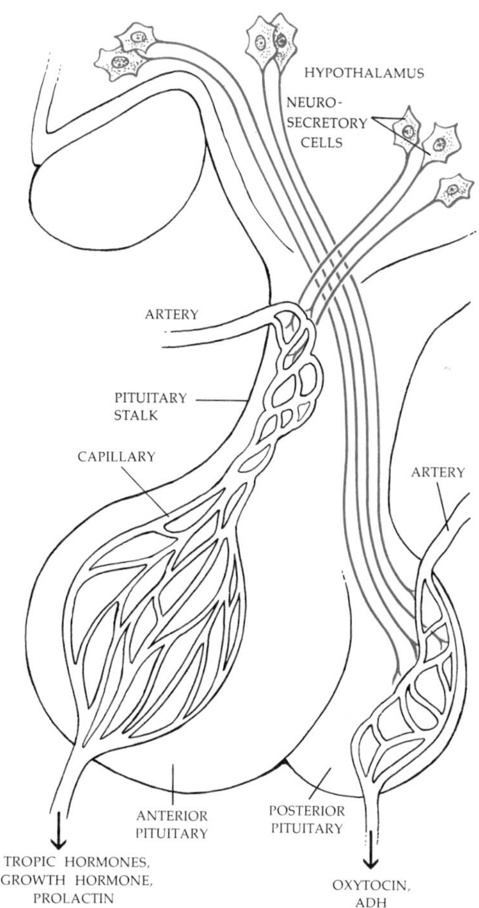

pituitary. These hormones are small peptides, one only three amino acids in length. They are unusual not only for their small size but also for the way in which they reach their target gland. Secreted from the nerve endings of hypothalamic nuclei, they travel only a few millimeters to the pituitary, apparently never entering the general circulation. However, they make this brief passage by way of the circulatory system (Figure 41–9), and so they technically meet this criterion of hormones.

The first of these hormones to be discovered was TRH, thyrotropin-releasing hormone. As its name implies, it stimulates the release of thyrotropin (TSH) from the pituitary gland. The second was GnRH, which controls the release of the gonadotropic hormones LH and FSH. These discoveries were made by two groups of scientists, one headed by Roger Guillemin at the Salk Institute (requiring the brains of 5 million sheep) and the other by Andrew Schally in New Orleans (using a comparable supply of pork). The third hormone found was not a releaser but, surprisingly, an inhibitor. Because it inhibited the growth hormone somatotropin, it was given the name of somatostatin. Several more releasing hormones and one other inhibitory hormone (for prolactin) have now been isolated, and the search is not over.

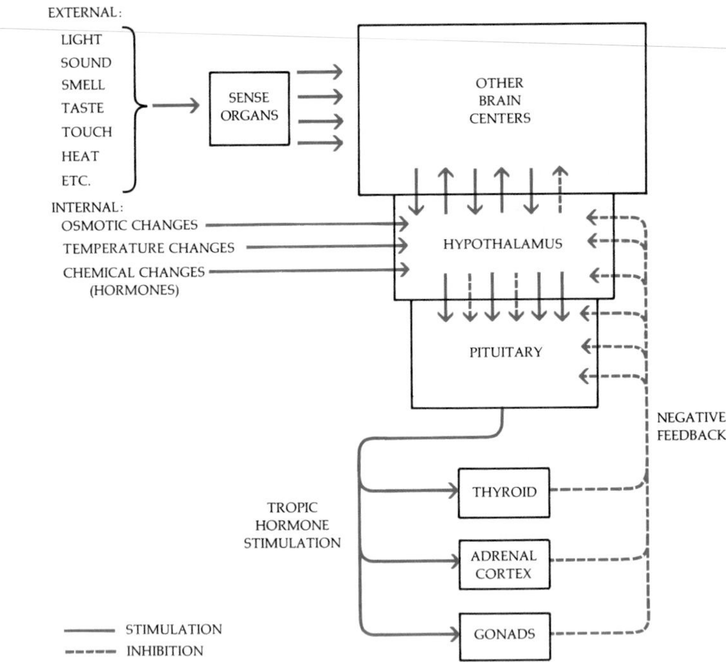

41–10
The production of many hormones is regulated by a complex negative feedback system involving the pituitary and the hypothalamus. The hypothalamus controls the pituitary's secretion of tropic hormones, and these, in turn, stimulate the secretion of hormones from the thyroid, adrenal cortex, and gonads (the testes or the ovaries). As the concentration of the hormones produced by these target glands rises in the blood, the hypothalamus decreases its production of releasing hormones, the pituitary decreases its hormone production, and production of hormones by the target glands also slows. By way of the hypothalamus, which receives information from many other parts of the brain, hormone production is also regulated in response to changes in the external and internal environments.

As we noted earlier, integration and control of an organism's functions involves both homeostasis and the capacity to respond to change. This dual feedback control system involving both the pituitary and the hypothalamus admirably serves these two goals. The pituitary feedback system provides for constancy. However, it can be overridden by the hypothalamic system, which takes into account not only the balance between output and input but also the changes elsewhere in the body and in the external environment, thus enabling the system to respond appropriately.

The hypothalamus is also the source of two hormones stored in and released from the posterior pituitary: oxytocin and antidiuretic hormone (ADH). Oxytocin is responsible for the "letting down" of milk that occurs when the infant begins to suckle. The hormone promotes contraction of the muscle fibrils around the milk-secreting cells of the mammary glands. It also accelerates childbirth by increasing uterine contractions during labor. These contractions also cause the uterus to regain its normal shape after delivery. The release of oxytocin is under the control of the nervous system and may be triggered by increasing pressure within the uterine wall or by movements of the fetus. In experimental animals, labor can be induced by mechanical stimulation of the uterus or electrical stimulation of the hypothalamus. The hormone is also present in males, but its function in the male, if any, is unknown.

ADH, as we noted in Chapter 37, decreases the excretion of water by the kidneys. It achieves this effect by increasing the permeability of the membranes of cells in the collecting ducts of the nephrons so that more water passes through them and back into the blood from the urine. (ADH is sometimes called vasopressin because it increases blood pressure in many vertebrates; it has no such effect in

41–11
Primary structures of antidiuretic hormone (ADH) and oxytocin, hormones produced in the hypothalamus and released from the posterior lobe of the pituitary. Note that each consists of only nine amino acids and that they differ by only two. This similarity of structure gives rise to some cross action by these hormones.

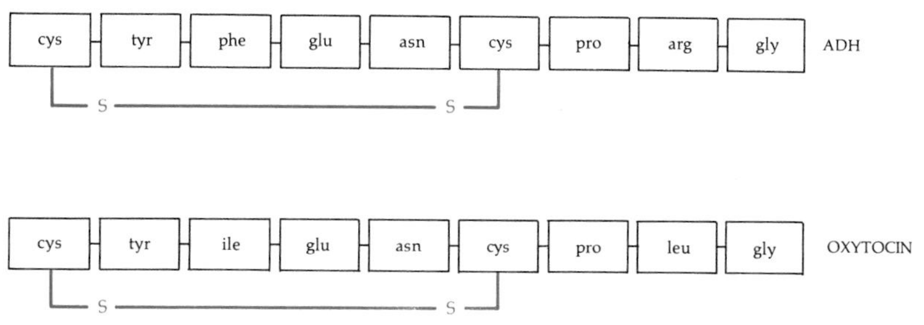

humans, however, except in very high concentrations.) Oxytocin has some ADH effect and ADH some oxytocin effect. This cross action is not surprising because each one of these hormones consists of only nine amino acids, and differences in the two hormones involve differences of only two amino acids among the nine (see Figure 41–11).

ADRENALINE AND NORADRENALINE

The adrenal medulla, the central portion of the adrenal gland, actually does not conform to the definition of a gland in that it is not made up of glandular epithelium but rather of nervous tissue. It is, in essence, a large ganglion whose nerve endings secrete adrenaline and noradrenaline into the bloodstream. These hormones increase the heart rate, raise blood pressure, stimulate respiration, dilate the respiratory passages, and promote the activity of the enzyme that breaks down glycogen to glucose 6-phosphate, thus increasing concentrations of glucose in the bloodstream. The adrenal medulla is stimulated by nerve fibers of the sympathetic division and so acts as an enforcer of sympathetic activity.

PANCREATIC HORMONES

As we mentioned previously, the islet cells of the pancreas are the source of insulin and glucagon, two hormones concerned with the metabolism of glucose. (These two hormones are synthesized and secreted by distinctly different islet cells.) Insulin is secreted in response to a rise in blood sugar or amino acid concentration (as after a meal). It lowers the blood sugar by stimulating cellular uptake of glucose and by stimulating the conversion of glucose to glycogen.

When there is an insulin deficiency, as in persons with diabetes mellitus, the concentration of blood sugar rises so high that not all the glucose entering the kidney can be reabsorbed; the presence of glucose in the urine forms the basis of simple tests for diabetes. The loss of glucose is accompanied by loss of water. The resulting dehydration, which can lead to collapse of circulation, is one of the causes of death in an untreated diabetic.

Glucagon increases blood sugar by stimulating the breakdown of glycogen to glucose in the liver and by stimulating the breakdown of fats and proteins, which decreases glucose utilization.

Somatostatin, originally found in the hypothalamus, has now also been isolated from a third class of islet cells in the pancreas and is believed to participate in controlling the synthesis of insulin.

Thus, as we have seen, at least five different hormones (and probably six) are involved in regulating blood sugar: growth hormone, cortisol, adrenaline, insulin, and glucagon. This tight control over blood glucose ensures that glucose is always available for brain cells. Unlike other cells in the body, which can derive energy from the breakdown of amino acids and fats, brain cells can utilize only glucose and so are immediately affected by low blood sugar. After several days of fasting, however, the brain begins to use fatty acids as an energy source. An insulin overdose, which causes a rapid drop in blood sugar, can result in coma and death.

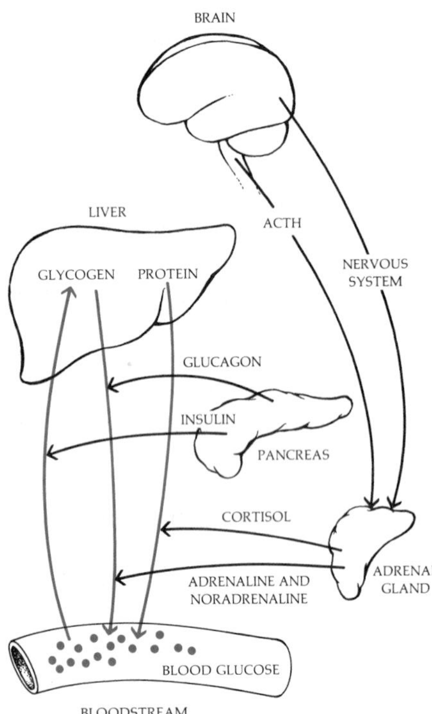

41–12
Hormonal regulation of blood glucose. When blood sugar concentrations are low, the pancreas releases glucagon, which stimulates the breakdown of glycogen and the release of glucose from the liver. When blood sugar concentrations are high, the pancreas releases insulin, which removes glucose from the bloodstream by increasing its uptake by cells and promoting its conversion into glycogen, the storage form. Under conditions of stress, ACTH, produced by the pituitary, stimulates the adrenal cortex to produce cortisol and related hormones, which increase the breakdown of protein and its conversion to glucose in the liver. At the same time, the adrenal medulla releases adrenaline and noradrenaline, which also raise blood sugar. The more recently discovered hormone somatostatin also plays a role, as yet not clearly defined, in the regulation of glucose metabolism.

PARATHYROID HORMONE

The pea-sized parathyroid glands, the smallest of the known endocrine glands, are located behind or within the thyroid gland. They produce parathyroid hormone (parathormone), which plays an essential role in mineral metabolism, specifically in the regulation of calcium and phosphate ions, which exist in a reciprocal relationship in the blood. Calcium is normally present in mammalian blood in concentrations of about 6 milligrams per 100 milliliters of whole blood. A rise or fall of more than 2 or 3 milligrams per 100 milliliters can lead to such severe disturbances in blood coagulation, muscle contraction, and nerve function that death may follow within hours. Parathyroid hormone maintains blood calcium levels by reducing the excretion of calcium ions by the kidneys.

Parathyroid hormone also stimulates the release into the bloodstream of calcium from bone, which contains 99 percent of the body's total calcium. Thus, parathyroid hormone and calcitonin from the thyroid work as a fine-tuning mechanism, regulating blood calcium. The production of both hormones is regulated directly by the concentration of calcium ions in the blood.

Hyperparathyroidism, caused by tumors of the parathyroids, occasionally occurs in humans. When there is too much parathyroid hormone, the bones lose too much calcium, becoming soft and fragile, and the vertebrae may shrink, producing a loss of height. Removal of the parathyroid glands without hormone replacement therapy results in violent muscular contractions and spasms, leading to death.

MELATONIN: THE PINEAL HORMONE

The pineal gland is a small lobe in the forebrain, lying near the center of the brain in humans. In lower vertebrates, the pineal gland contains light-sensitive cells and so is sometimes called the third eye. It is involved with photoperiodism; for instance, shading the pineal can prevent the seasonal enlargement of the ovaries in some species.

The pineal gland secretes the hormone melatonin. The production of melatonin rises sharply at night and falls rapidly in the daytime in many species, including rats, chickens, and humans. Exposure to light during the dark cycle interrupts production of melatonin, recalling the studies on photoperiodism in plants (page 645).

In some species of vertebrates—particularly among fish and reptiles—melatonin causes the melanophores, pigment-containing cells, to contract, thus causing the blanching of the skin associated with darkness. In sparrows, injections of melatonin induce roosting and the lowering of body temperature, both of which are

41–13

The pea-sized parathyroid glands, the smallest of the known endocrine glands, are located behind or within the thyroid gland. They produce parathyroid hormone (parathormone), which increases concentrations of blood calcium. Calcitonin, a hormone produced by the thyroid gland, decreases blood calcium.

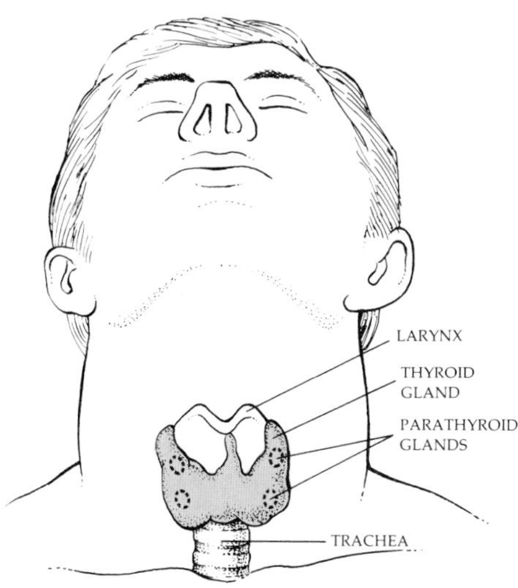

characteristic nighttime events in these birds. Melatonin also inhibits the development of the gonads in species as disparate as chickens and hamsters; hence, the decreased production of melatonin during long-day periods is believed to be associated with the preparation of the gonads for seasonal mating. In humans, the pineal gland may be involved in sexual maturation; tumors of the pineal have been associated with precocious puberty.

Thus there are a few tempting clues suggesting that the pineal gland may function as a biological timekeeper, but the way in which light affects the gland and the way in which melatonin alters physiological responses—if it does—are yet to be discovered.

PROSTAGLANDINS

The prostaglandins are a group of recently discovered hormone-like chemicals. They were given this name because they were first detected in semen and were thought to be produced by the prostate gland. (Actually, the seminal vesicles are the chief source of prostaglandins in semen.) Now a large series of these hormones is known, all related structurally but with a variety of different, and sometimes directly opposite, effects.

Prostaglandins differ from other hormones in a number of ways: (1) They are fatty acids. (All are formed by the oxygenation of arachidonic acid, a 20-carbon polyunsaturated fatty acid.) (2) They are produced by cell membranes in most—if not all—organs of the body, as opposed to other hormones, which are produced by particular glands. (3) They often exert their effects in the tissue that produces them. (4) Their target tissues may be those of another individual (if their role in fertilization has been correctly interpreted). (5) They are among the most powerful of all known biological molecules, even more potent than most hormones, producing marked effects in very small doses. A concomitant of their extraordinary potency is the fact that they are produced in very small amounts and are rapidly broken down by enzyme systems in the body. If it were not for their presence in unusually large amounts in semen, they might never have been discovered.

Prostaglandins were first studied for their possible role in reproductive physiology. Among their many effects one of the most striking is their capacity to induce contractions in smooth muscle. This is believed to play several important roles in reproduction. The walls of the uterus, which are composed of smooth muscle, normally contract in continuous waves. After sexual intercourse, prostaglandins from semen are found in the female reproductive tract, where they increase the rhythmic contractions of the uterine wall and oviducts. It is believed that this action assists both the sperm on its journey to the oviduct and the egg cell as it travels from the oviduct to the uterus. The semen of some infertile males has been found to be poor in prostaglandins, and the uterus of infertile females is often unresponsive to prostaglandins. Another major source of prostaglandins is menstrual fluid, and the contractions of the uterus increase when the endometrium is shed during a menstrual period.

Between 30 and 50 percent of all women of child-bearing age experience painful abdominal cramps during the first day or two of each menstrual period, a condition known as dysmenorrhea. Recent research has revealed that the menstrual fluid of such women contains concentrations of prostaglandins two to three times higher than the levels found in the menstrual fluid of women without dysmenorrhea.

CIRCADIAN RHYTHMS

Virtually every living organism, from one-celled algae to bean plants (page 649) to Homo sapiens, *exhibits circadian rhythms in many biological functions. These rhythms are in synchrony with the solar day but are not direct responses to the light-dark cycle. They characteristically persist in continuous darkness, though they then begin to deviate from a rigid 24-hour cycle (hence the name circadian, meaning "about a day"). If an organism exhibiting circadian rhythms is placed in an artificial, abnormal day-night environment—one of 26 hours, for instance, the circadian rhythms become adjusted to this artificial day, a process known as entrainment. The phenomenon of entrainment demonstrates that the rhythms are responsive to, though not dependent on, the environmental day-night cycle. If the day-night pattern is changed radically, as when a traveler passes through several time zones or a worker changes from a day shift to a night shift, the normal rhythms are disrupted, falling out of synchrony with each other and with the environment. Gradually, they become reset in the new cycle.*

Students of human physiology and behavior have now discovered many such rhythms. It has long been known, for instance, that we are more likely to be born between 3 and 4 A.M. and also to die in these same early morning hours. Body temperature fluctuates as much as 1°C (about 2°F) during the course of 24 hours, usually reaching a peak at about 4 P.M. and a low about 4 A.M. Alcohol tolerance is greatest, fortunately, at 5 P.M., while tolerance to pain is lowest at 6 P.M. in most persons. Respiration, heart rate, and urinary excretion of potassium, calcium, and sodium, all vary according to the time of day (heart rate by as much as 20 beats per minute). Synthesis of various hormones follows circadian rhythms: serum levels of corticosteroids peak between 4 A.M. and 8 A.M. in people on a normal sleep-wake cycle; growth hormone levels rise about an hour after falling asleep; prolactin peaks around 3 A.M.; testosterone about 9 A.M.

Knowledge about circadian rhythms may have important practical consequences. A study by the Federal Aviation Agency showed that pilots flying from one time zone to another—from New York to Europe, for instance—exhibit "jet lag," a general decrease in mental alertness, an inability to concentrate, and an increase in decision time and physiological reaction time. Measurements of various bodily functions show that the body may be "out of synch" for as much as a week after such a flight. This brings into question not only schedules for airline personnel but also present policies of speeding diplomats to foreign capitals at times of international crisis or of air transport of troops into combat. It may be significant that the Three Mile Island accident occurred at 4 A.M. and that the workers on duty at that time had been alternating between the day shift and the night shift every week.

Medical researchers are finding that failure to take circadian fluctuations into account can lead to mistakes in diagnosis and treatment. For example, blood-pressure readings may vary as much as 20 percent in the course of a day, so that a person might be found to be within the normal range at one time of day and diagnosed as hypertensive at another. Numbers of white blood cells may vary as much as 50 percent in a 24-hour period, rendering the same immunosuppressive therapy ineffectual at one time of day and life-threatening at another. Studies of cancer chemotherapy in mice indicate that proper timing of drug doses can mean a doubling of survival rate. Determination of a patient's chronobiology may someday become as routine in therapy as a blood-typing or taking a medical history.

Present evidence indicates that the increased prostaglandin levels not only cause stronger, more rapid contractions of the uterine walls but also reduce the blood supply to the tissue. As a result, less oxygen is available to the actively contracting muscles, creating an oxygen debt (page 192) and its accompanying pain.

Clinical studies in this country and in Europe have demonstrated that several compounds that inhibit prostaglandin synthesis are highly effective in reducing or, in some cases, completely eliminating the painful cramping associated with dysmenorrhea.

As we noted previously, prostaglandins that are closely related chemically may differ widely in their physiological effects. Although many of them stimulate contractions of smooth muscle, as previously described, others inhibit smooth muscle contraction. Furthermore, one may affect the smooth muscle of bronchioles, another the smooth muscle of blood vessels. One, produced by platelets, is a potent stimulus for platelet aggregation (see page 706) and vasoconstriction; another, produced by endothelial cells, is a potent inhibitor of platelet aggregation and a vasodilator. The way in which they exert these multitudinous effects is not known.

Prostaglandins have also been implicated in inflammatory and immune responses and, by extension, in diseases such as rheumatoid arthritis and asthma. Discovery of the prostaglandins has also provided a solution to another long-standing physiological puzzle. Aspirin is one of the oldest and most effective medications known, but how it acts has been a mystery. Now it has been discovered that aspirin exerts its effects, at least in part, by inhibiting the synthesis of prostaglandins, thus producing its well-known soothing effects on inflammation and fever.

The Nobel Prize in Physiology or Medicine for 1982 was awarded to Bengt Samuelsson and Sune Bergström of Sweden and John Vane of England for their work in isolating and characterizing the prostaglandins.

CYCLIC AMP

(a)

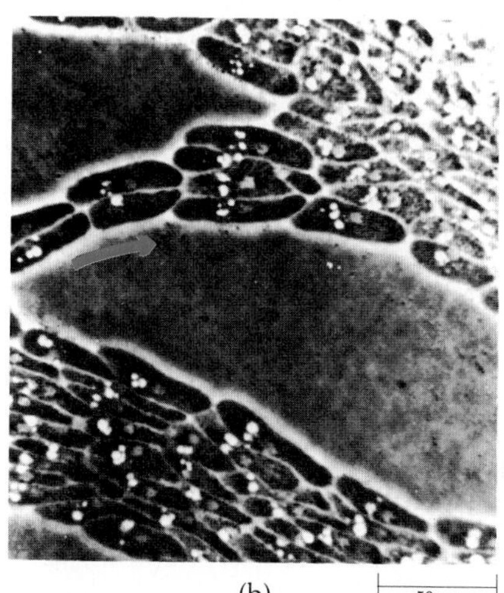

(b)

41–14
(a) Cyclic AMP (adenosine monophosphate) acts as a "second messenger" within the cells of vertebrates. Following stimulation by various hormones—the "first messengers"—cyclic AMP is formed from ATP. "Cyclic" refers to the fact that the atoms of the phosphate group form a ring. (b) *Cyclic AMP is also the chemical that attracts the amoebas of the cellular slime molds, causing them to aggregate into a slug-like body, which then behaves like a multicellular organism. The arrow shows the direction in which the cells are moving.*

MECHANISMS OF ACTION OF HORMONES

How do hormones exert their highly specific effects on their target tissues? Recent research has revealed two quite different mechanisms underlying hormone specificity. Both involve receptor molecules: in one, the receptor molecules are intracellular, in the nucleus or cytoplasm; in the other, the receptor molecules are in the cell membrane. The steroid hormones and thyroid hormone are members of the first group; the amine, peptide, and protein hormones appear to be members of the second group.

Intracellular Receptors

The steroid hormones and thyroid hormone are soluble in lipids and pass freely through cell membranes. Their specificity of action depends on whether or not, once within the cell, they encounter an appropriate receptor molecule. Estradiol (Figure 41–6) is the best-studied example of this group of hormones. The current working hypothesis is that the estradiol molecule enters the target cell by diffusion. Once within the cytoplasm of the target cell, it combines with a specific protein molecule, the receptor. The hormone-receptor complex moves to the nucleus, where it binds to the chromosome, initiating RNA transcription leading to protein synthesis.

These findings explain how the very slight differences in configuration among steroid molecules can be correlated with such drastically different effects. The protein receptors, like enzymes, are highly specific in their combining properties.

Membrane Receptors

The second group of hormones acts by combining with receptors on (actually, probably in) the membranes of target cells, setting in motion a "second messenger" that is responsible for the sequence of events inside the cell. The second messenger in many cases is a chemical known as cyclic AMP (Figure 41–14). These findings have important medical implications. For example, it has long been known that juvenile diabetes—diabetes in young persons—is caused by a deficiency of the hormone insulin, which, as we noted, promotes the uptake of glucose by cells. It was assumed by analogy that the diabetes found commonly in older persons and associated with obesity had the same cause. It has now been found, however, that adult diabetes often results from a decrease in the number of receptor sites on target cell membranes rather than from a shortage of insulin. Such patients are treated most effectively by diet.

Another example of this type of hormone is adrenaline. Figure 41-15 summarizes the steps in the release of glucose from a liver cell stimulated by adrenaline. Adrenaline molecules bind to a receptor on the outer surface of the cell membrane. This event activates an enzyme, adenylate cyclase, that is bound on the inner surface of the membrane. Adenylate cyclase converts ATP to cyclic AMP (cAMP). The cyclic AMP binds to another enzyme, protein kinase, and activates it. This enzyme activates another enzyme, which, in turn, activates yet another. The final enzyme, phosphorylase *a*, then breaks down glycogen at a high rate, producing glucose 1-phosphate, which is further broken down to glucose and released from the cell. Since each enzyme greatly increases the rate of the particular reaction it catalyzes (page 168), the number of molecules involved in these reactions is am-

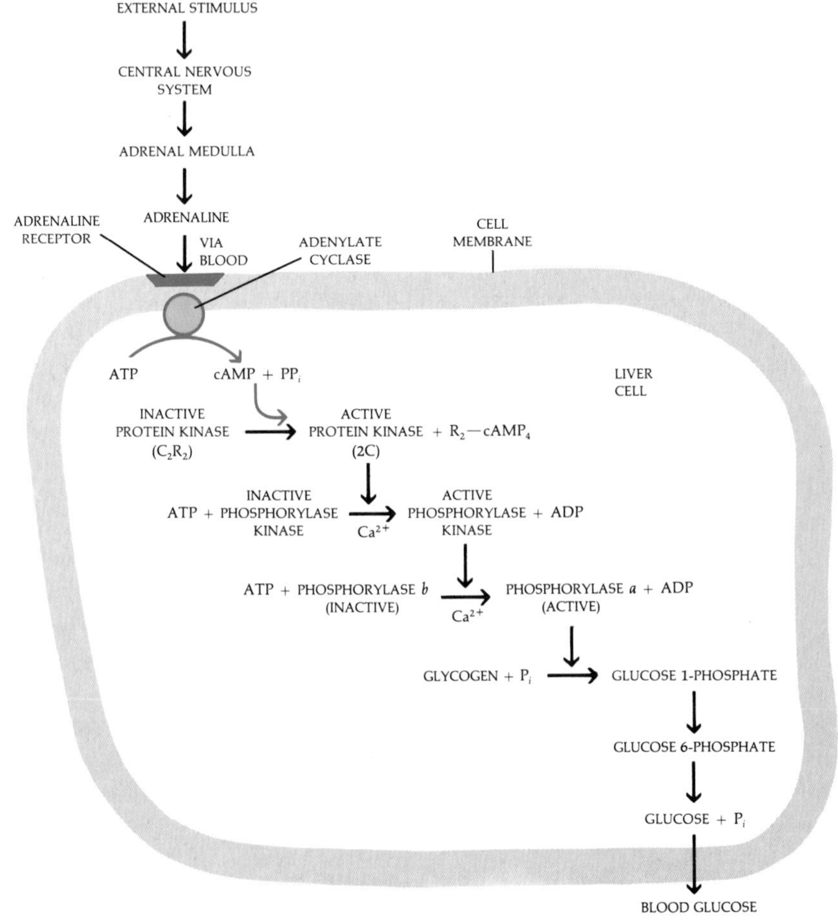

41-15
Adrenaline triggers an amplification cascade in liver cells. Binding of a few molecules of adrenaline to its specific receptors on the cell surface initiates a series of enzymatic reactions that result in the release of a very large amount of glucose into the blood.

plified at each step. Thus, the binding of a few molecules of adrenaline to cells in the liver leads to the activation of an estimated 25 million molecules of phosphorylase *a* and the consequent release of many grams of glucose into the blood.

Just about the same time that the role of cyclic AMP in mammalian cells was established—for which Earl W. Sutherland was awarded the Nobel Prize—biologists studying that peculiar group of organisms known as the cellular slime molds (see page 120) isolated a chemical of great importance in this biological system. As you will recall, the cells of the cellular slime mold begin as individual amoebas and then come together to form a single organism. The chemical that calls them together, which was named acrasin, was identified as cyclic AMP. More recently, it has been found that the hormone insulin is present in fruit flies, earthworms, protozoa, fungi, and even *Escherichia coli,* and there is preliminary evidence for other "mammalian" hormones, including ACTH, glucagon, and somatostatin in unicellular organisms. Their function in these organisms is not known, but they may also, like acrasin, serve in cell-to-cell communication. These new discoveries of the universality of hormones provide other examples of the long thread of evolutionary history linking all organisms.

SUMMARY

Hormones are organic molecules secreted in one part of an organism that, in the case of vertebrates, are carried by the bloodstream to other tissues and organs, where they exert specific effects. The effects of a hormone on its target tissues depend on the presence in the target cells of receptors specific for that hormone.

The production of many hormones is regulated by negative feedback systems involving the anterior lobe of the pituitary gland and an area of the brain, the hypothalamus. Under the influence of hormones secreted by the hypothalamus, the pituitary produces tropic hormones that, in turn, stimulate the target glands to produce hormones. These hormones then act upon the pituitary or the hypothalamus (or both) to inhibit the production of the tropic hormones. Production of the steroid hormones of the testes, ovaries, and adrenal cortex is regulated by the hypothalamus-pituitary regulatory system. The production of thyroxine, the hormone of the thyroid gland, is also under hypothalamus-pituitary control.

Besides producing the tropic hormones, the anterior lobe of the pituitary also secretes somatotropin (growth hormone) and prolactin.

The hypothalamus is the source of at least nine peptide hormones (sometimes called releasing hormones) that act upon the anterior lobe of the pituitary gland. It also produces the hormones ADH and oxytocin, which are stored in and released from the posterior lobe of the pituitary.

The adrenal medulla is a source of the hormones adrenaline and noradrenaline, which are released in times of stress and have rapid effects of short duration.

The islet cells of the pancreas are the source of three hormones involved in the regulation of blood glucose: insulin, which lowers blood sugar by stimulating cellular uptake of glucose; glucagon, which raises blood sugar by stimulating the breakdown of storage forms of glucose; and somatostatin. Blood sugar is also under the influence of adrenaline, cortisol, and somatotropin.

The major function of the hormone released by the parathyroid glands is control of the levels of calcium and phosphate ions in blood. Parathyroid hormone maintains blood calcium levels by stimulating release of calcium from bone and

Table 41-1 Some of the Principal Endocrine Glands of Vertebrates and the Hormones They Produce

GLAND	HORMONE	PRINCIPAL ACTION	MECHANISM CONTROLLING SECRETION	CHEMICAL COMPOSITION
Pituitary, anterior lobe	Growth hormone (somatotropin)	Stimulates bone and muscle growth, inhibits oxidation of glucose, promotes breakdown of fatty acids	Hypothalamic hormone(s)	Protein
	Prolactin	Stimulates milk production and secretion in "prepared" gland	Hypothalamic hormone(s)	Protein
	Thyroid-stimulating hormone (TSH) (thyrotropin)	Stimulates thyroid	Thyroxine in blood; hypothalamic hormone(s)	Glycoprotein
	Adrenocorticotropic hormone (ACTH)	Stimulates adrenal cortex	Cortisol in blood; hypothalamic hormone(s)	Polypeptide (39 amino acids)
	Follicle-stimulating hormone (FSH)	Stimulates ovarian follicle, spermatogenesis	Estrogen in blood; hypothalamic hormone(s)	Glycoprotein
	Luteinizing hormone (LH)	Stimulates corpus luteum and ovulation in female, interstitial cells in male	Progesterone or testosterone in blood; hypothalamic hormone(s)	Glycoprotein
Hypothalamus (via posterior pituitary)	Oxytocin	Stimulates uterine contractions, milk ejection	Nervous system	Peptide (9 amino acids)
	Antidiuretic hormone (ADH, vasopressin)	Controls water excretion	Osmotic concentration of blood; plasma-volume receptors	Peptide (9 amino acids)
Thyroid	Thyroxine, other thyroxinelike hormones	Stimulate and maintain metabolic activities	TSH	Iodinated amino acids
	Calcitonin	Inhibits release of calcium from bone	Concentration of Ca^{2+} ions in blood	Polypeptide (32 amino acids)
Parathyroid	Parathyroid hormone (parathormone)	Stimulates release of calcium from bone, promotes calcium uptake from gastrointestinal tract, inhibits calcium excretion	Concentration of Ca^{2+} ions in blood	Polypeptide (34 amino acids)
Adrenal cortex	Cortisol, other glucocorticoids	Affect carbohydrate, protein, and lipid metabolism	ACTH	Steroids
	Aldosterone	Affects salt and water balance	Reflexes in the kidney, K^+ ions in blood	Steroid
Adrenal medulla	Adrenaline and noradrenaline	Increase blood sugar, dilate some blood vessels, increase rate of heartbeat	Nervous system	Catecholamines (amino acid derivatives)
Pancreas	Insulin	Lowers blood sugar, increases storage of glycogen	Concentration of glucose in blood; somatostatin	Polypeptide (51 amino acids)
	Glucagon	Stimulates breakdown of glycogen to glucose in the liver	Concentration of glucose and amino acids in blood	Polypeptide (29 amino acids)
Ovary, follicle	Estrogens	Develop and maintain sex characteristics in females, initiate buildup of uterine lining	FSH	Steroids
Ovary, corpus luteum	Progesterone and estrogens	Promote continued growth of uterine lining	LH	Steroids
Testis	Testosterone	Supports spermatogenesis, develops and maintains sex characteristics of males	LH	Steroid
Pineal	Melatonin	Involved in regulation of circadian rhythms	Light-dark cycles	Catecholamine

reducing excretion of calcium. Calcitonin, secreted by the thyroid gland, inhibits release of calcium from bone.

The pineal gland is the source of melatonin and is believed to be involved with regulation of circadian and seasonal physiological changes. Its function in humans is not known.

The prostaglandins are a group of fatty acids that resemble hormones in exerting effects on specific target tissues but that are unlike hormones in that they often act directly upon the tissues that produce them or, in some cases, on the tissues of another individual. Prostaglandins are formed in most, if not all, tissues of the body and affect such diverse systems as the cardiovascular and reproductive systems.

Hormones act by at least two different mechanisms. Steroid hormones and thyroid hormone, which are lipid-soluble molecules, enter the cell and, after combining with an intracellular receptor, exert a direct influence on the transcription of RNA. Other hormones, such as adrenaline, insulin, and glucagon, combine with receptor molecules on the surface of the target-cell membranes. This combination triggers the release of a "second messenger" and sets off a series of events within the cell that is responsible for the end results of hormone activity. Cyclic AMP has been identified as the second messenger in many of these interactions.

QUESTIONS

1. Distinguish among the following: endocrine/exocrine; androgen/testosterone; estrogen/estradiol; anterior pituitary/posterior pituitary; adrenal cortex/adrenal medulla; thyroxine/thyrotropin/triiodothyronine; thyroid-stimulating hormone/thyrotropin-releasing hormone; ADH/aldosterone; insulin/glucagon.

2. Diagram the feedback system regulating the production and release of thyroid hormone.

3. In terms of the feedback system diagrammed in Question 2, explain why a shortage of iodine produces goiter.

4. Describe how the concentration of calcium ion in the bloodstream is regulated by the thyroid and parathyroid glands.

5. Which hormones act to increase the level of blood glucose? To decrease it? How does each hormone exert its effects?

6. How does a steroid hormone exert its specific effects on a target cell? In what way is the action of a protein hormone different?

7. What are the classical experimental steps by which an unknown organ or tissue can be diagnosed as endocrine in function? How might this procedure prove deceptive?

8. In the study of circadian rhythms, what is the difference between entrainment and resetting? What is the relationship between circadian rhythms and photoperiodism?

CHAPTER 42

Integration and Control III: Sensory Receptors and Skeletal Muscle

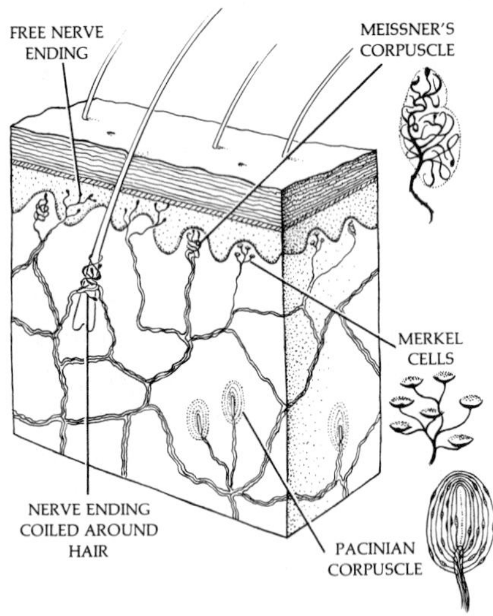

42–1
Some sensory receptors present in human skin. Merkel cells and Meissner's corpuscles respond to touch, as do the nerve endings surrounding the hair follicles. (Actually, when many hairs are present, the other touch receptors are sparse or absent.)

In the Pacinian corpuscle, the best studied of the four, the specialized ending of a single myelinated nerve fiber is encapsulated by a corpuscle, which is composed of many concentric layers of connective tissue. Pressure on these outer layers stimulates the ending. The free nerve endings are largely pain and temperature receptors.

Our sensory equipment is the means by which we know the world around us. It provides the mechanisms by which we and other animals are informed of predator and prey, friend and foe, whether something is good to eat or not, and changes in the weather or the seasons. It identifies infant to mother, mother to infant, and mate to mate. It is also our source of pleasure and of esthetic enjoyment and provides our tools for learning. Our capacity to act on the information provided by our sensory equipment—to move toward desirable objects, to move away from harmful or potentially harmful objects, and to perform the multitudinous activities of our daily lives—depends on our skeletal muscles. In this chapter, we shall first examine our sensory equipment and then consider the marvelous molecular machinery that enables us to smile, to frown, to perform precise manipulative activities, or to run the 100-meter dash.

SENSORY RECEPTORS

Our sense organs, like the other organs of the body of any animal, are the products of adaptation and evolution, and by these long processes, they have been tailored to our specific requirements. For example, we think we see what is visible and hear what is audible, but visible and audible are actually not properties of the objects themselves but of our particular sensory equipment. What we see is very different from what an insect sees, and a bat or a fish, acoustically speaking, might as well be living on another planet. Moreover, it is likely that our insensitivity to a great number of the potential stimuli that surround us is also as useful as our sensitivity to others.

Types of Sensory Receptors

Our sensory receptors are many and varied. Like most animals, we have mechanoreceptors (touch, hearing, position), chemoreceptors (taste and smell), photoreceptors (vision), temperature receptors, and pain receptors. (We do not have electroreceptors or magnetoreceptors, but some animals do.)

Some sensory receptors are small and relatively simple in structure. Look again at the human skin (Figure 42–1) as an example. The simplest receptors are the free nerve endings, which are receptors for pain and temperature and perhaps for other sensations as well. Slightly more complex are the combinations of free nerve

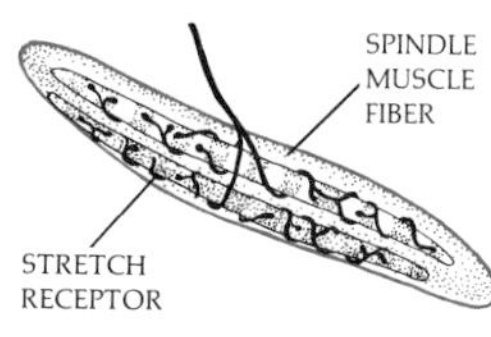

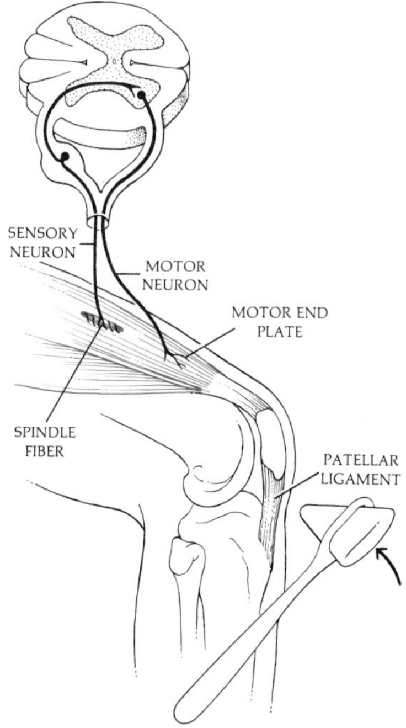

42–2
A light tap on the knee, as with a rubber-headed hammer, causes the lower leg to fly forward in the knee-jerk reflex. This reaction occurs because the tap pulls the patellar ligament, which, in turn, gives a sudden tug to muscles of the upper leg. The stretching of muscles is sensed by the stretch receptors surrounding the muscle spindle, which send a message, encoded in action potentials, to the spinal cord. There the sensory neurons synapse directly with motor neurons that stimulate the muscles to contract. This reflex helps to maintain balance and posture.

endings with a hair and its follicle. Each of these little organs is an exquisitely sensitive mechanoreceptor. When the hair is touched or bent, it sets off an action potential in the nerve endings of a sensory neuron that is carried directly to the central nervous system. Three other types of mechanoreceptors are also shown, each a combination of one or more free nerve endings and an outer layer or layers of connective tissue. Meissner's corpuscles and Merkel cells are both involved with touch. They are found in particularly sensitive areas of the skin, such as the fingertips, palms, lips, and nipples, and are especially abundant where hairs are not present. They are responsible for the extraordinary cutaneous sensitivity of these parts of the body and are associated with the ability, for example, to read Braille, do certain magic tricks, crack a safe, or enjoy a kiss. The Pacinian corpuscles, lying deeper within the tissues, respond to pressure and vibrations. The free nerve ending of the corpuscle is surrounded by layers of connective tissue and fluid. This layered structure is easily deformed, so it responds to even very slight pressure changes, but it also, because of its structure, quickly adjusts so that the nerve ending stops firing when pressure is sustained. This is one reason that we can adapt rapidly to continuous pressure, such as that produced by sitting in a chair or leaning against a solid object. Of all the sensory receptors shown here, the simplest in structure—the free nerve endings associated with pain—are the least understood in terms of how they function. It seems likely that some of them are not mechanoreceptors, as are the others, but rather chemoreceptors, responding to very small amounts of some chemical released by cells when they are injured.

Sensory receptors can also be categorized as interoceptors, exteroceptors, and proprioceptors. *Interoceptors* include the mechanoreceptors and chemoreceptors that are sensitive to blood pressure and CO_2 concentration in the carotid arteries. The temperature sensors of the hypothalamus are also interoceptors. We are usually not conscious of signals from these receptors, although sometimes the signals result in perceptions of, for example, pain, hunger, thirst, nausea, or the sensations, produced by stretch receptors, of having a full bladder or bowel.

Exteroceptors are the most familiar, including those in the ear, the eye, the olfactory epithelium, and the touch receptors and temperature receptors of the skin.

Proprioceptors (from the Latin *proprius,* meaning "self") inform us about the orientation of our body in space and the position of our arms, legs, and other body parts. Because of proprioceptors you can, with your eyes shut, touch your nose with your fingers or tie your shoelace in the dark. The semicircular canals of the ear are major proprioceptive organs in many animals. Among the most common and most active proprioceptors are specialized muscle fibers, known as spindle fibers, which act as gauges monitoring how far and how fast the muscle is being stretched. The actual sensory receptors are nerve endings that are wrapped around the muscle spindle fibers located deep within the muscle. Muscle spindle fibers are responsible for the well-known knee-jerk reflex (Figure 42–2). A normal knee-jerk reflex indicates normal muscle activity and the absence of degenerative disease of the peripheral nerves, the myelin sheath, or the spinal cord. (Note that this reflex arc is unusual in that it is monosynaptic; that is, the sensory neuron synapses directly with the motor neuron.)

All sense organs, as different as they may be in structure and function, are essentially similar. They are all transducers. That is, they convert energy from one form to another. Also, they all use the same code to transmit information to the nervous system; this code is the action potential. The differences among the senses lie not in the signals as they are transmitted but in their pattern and its reception and interpretation in the brain. We shall discuss this further in the following chapter.

Light Reception: Vision

Eyes have evolved independently several times in the course of evolutionary history. Among the most highly developed of modern light-receptor systems are the compound eye of the arthropods (see page 542), the eye of the octopus (see page 515), and the vertebrate eye, of which the human eye, shown in Figure 42–3, is an example.

The vertebrate type of eye is often called a camera eye. In fact, it has a number of features in common with an ordinary camera equipped with several expensive accessories, such as a built-in cleaning and lubricating system, an exposure meter, and an automatic field finder. Light from the object being viewed passes through the transparent cornea and lens, which focus an inverted image of the object on the light-sensitive retina in the back of the eyeball. (Though the eye may be like a camera, the retina is not like a piece of film. As we shall see, much processing of information takes place in the retina itself.)

In mammals, fine focusing of the image on the retina is brought about by contracting or relaxing the ciliary muscles, thus changing the curvature, and hence the focal length, of the lens (Figure 42–4). (Fish and amphibians, which lack ciliary muscles, focus their eyes the same way one focuses a camera with a lens of fixed focal length. Muscles within the eye change the position of the lens, drawing it

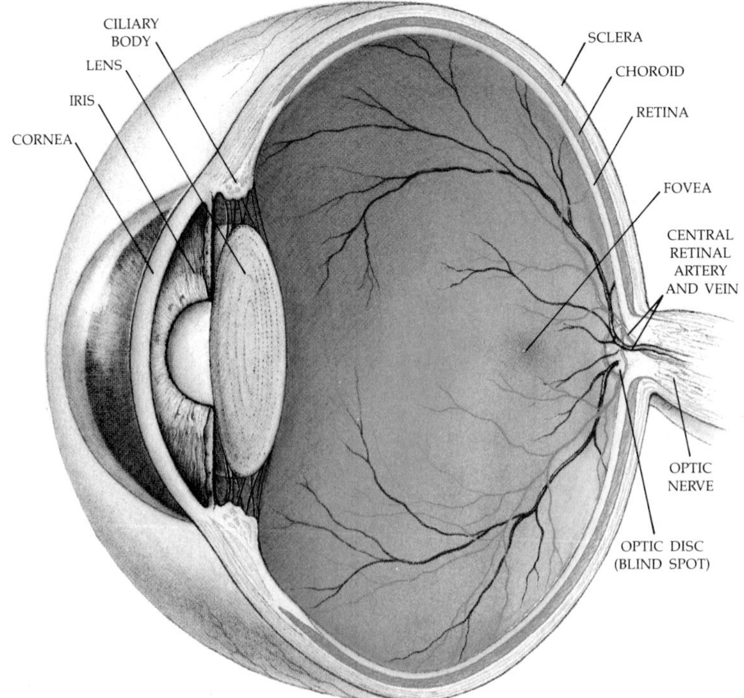

42–3
The eye is a complex organ composed of three layers of tissue, which form a fluid-filled sphere. The outer layer, the sclera, is white fibrous connective tissue that serves a protective function. The anterior portion of the sclera, the cornea, is transparent. The middle layer, the choroid, contains blood vessels. Its anterior portion is modified into the ciliary body, the suspensory ligament, and the iris. The ciliary body is a circle of smooth muscle from which extend the suspensory ligaments that hold the lens in place. The colored part of the eye, the iris, is a circular structure attached to the ciliary body. The pupil is a hole in the center of the iris. The innermost layer, the retina, contains the light-sensitive cells, the rods and cones. Most of the eyeball is recessed in the orbit, protected by the bony socket of the skull.

42–4
Focusing the eye. Preliminary focusing is done by the cornea, and fine focusing by the lens, which is like a jelly-filled, transparent rubber balloon. The lens is held in place by suspensory ligaments attached to the ciliary body, which encircles the lens. The ciliary body contains the ciliary muscle, which is the muscle of accommodation. When it contracts, the lens is released, and, as it returns to its more nearly spherical shape, its curvature is increased. (a) *The normal eye views distant objects with the lens stretched and flattened;* (b) *for viewing near objects, the lens is relaxed and more convex, bending the light rays more sharply.* (c) *Nearsightedness occurs when the eyeball is too long for the lens to focus a distant image on the retina.* (d) *It is corrected by a concave lens that bends the light rays out enough so that the image is focused.* (e) *Farsightedness is the result of an eyeball too short for the lens to focus a near object.* (f) *It is corrected by a convex lens that bends the light rays in before they reach the lens of the eye.*

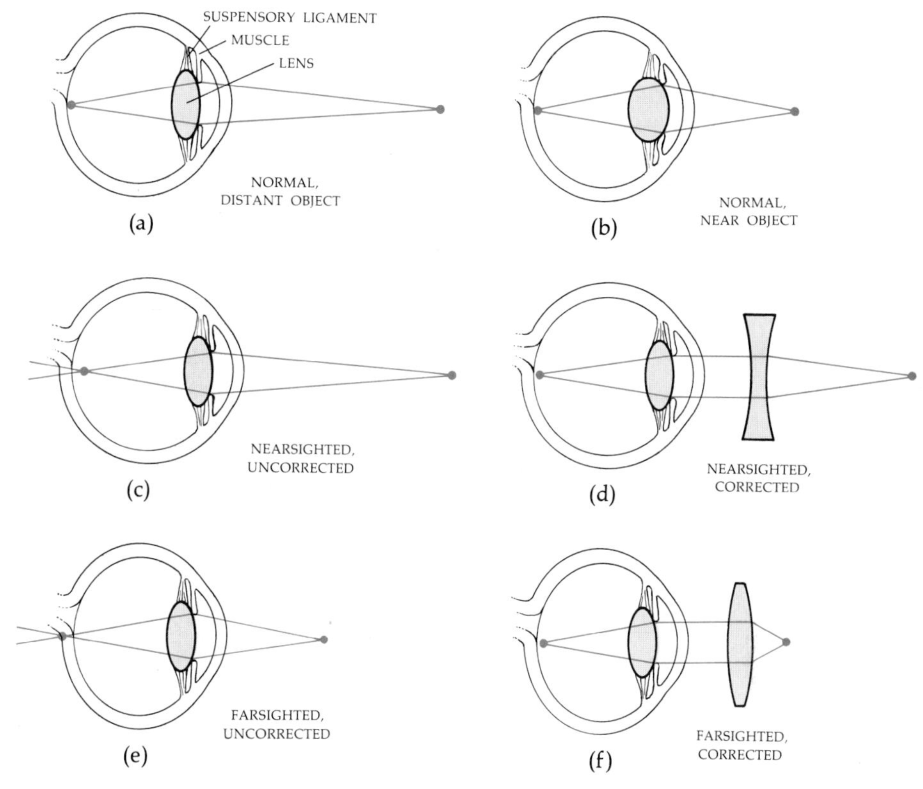

back toward the retina in order to focus more distant objects.) The iris controls the amount of light entering the eye by regulating the size of the pupil.

Stereoscopic vision—that is, vision in three dimensions—depends on viewing the same visual field with both eyes simultaneously. When both eyes are trained on a distant object, both eyes see almost the same image. A nearby object, however, will present a slightly different image to each eye (parallax). We learn to relate the disparity in the two images of the nearby object to the distance between us and the object. When the eyes are trained on a very distant object, however, the disparity is too small; hence, we are not able to make such judgments for faraway objects. Instead, we estimate their distance on the basis of known sizes of the objects or such visual clues as the size and relative position of houses, people, or cars in the same field of vision. A young child viewing a distant object—an airplane, for example—may see it as little rather than far away.

Tree-dwelling animals, such as the probable forerunners of *Homo sapiens,* usually have overlapping visual fields in the front that provide them with the stereoscopic vision essential for the distance judgments involved in moving from branch to branch. Predatory animals also tend to have stereoscopic vision, but animals that are more likely to be hunted than to hunt usually have an eye placed on each side of the head, giving a wide total visual field. Some birds with laterally placed eyes—the woodcock, the cuckoo, and certain species of crow—have binocular fields of vision both in front of them and behind them.

42–5
Some vertebrate adaptations. (a) *Mud skipper eyes are directed upward like periscopes.* (b) *Eagles and other predatory birds often have eyes as large as ours (although the skull and brain are much smaller), two foveas (see page 799) in each retina, and strong powers of near and far accommodation.* (c) *Rabbits have eyes set on each side of their skulls, an adaptation that allows them to watch on both sides for predators while they are feeding.* (d) *This ring-tailed cat is nocturnal. Like many other nocturnal animals, it has large eyes. Nocturnal animals characteristically have mainly rod cells, which are more light-sensitive than cone cells and are involved in night vision. Naturally, without cones they lack any color vision.*

(a)

(b)

(c)

(d)

The Retina

The retina of the vertebrate eye is anatomically inside out—that is, the photoreceptors of the eye are pointed toward the back of the eyeball—and light must reach them by passing through all the other nerve cell layers in the retina (Figure 42–6). Only about 10 percent of the light falling on the cornea reaches the retina. Of this light, that which is not captured by the photoreceptors is absorbed by pigmented epithelium that lines the back of the eyeball, just behind the photoreceptors. Some nocturnal vertebrates have a reflecting layer (tapetum) behind the photoreceptors that increases the likelihood of dim light stimulating the photoreceptors. This reflecting layer makes the animal's eyes seem to shine at night when light is directed into them.

The photoreceptor cells are connected to the ganglion cells (whose axons form the optic nerve, which leads into the brain) by an intervening group of cells called bipolar cells. When light is captured by the photoreceptor cells, it causes a change in their membrane polarity; this, in turn, causes a change in the membrane polarity of the bipolar cells, which finally influence the firing rate of the ganglion cells.

There are about 125 million photoreceptors in the retina and about 1 million ganglion cell axons in the optic nerve, a reduction of 125:1. Thus, many receptors converge on many fewer bipolar cells and, these, in turn, feed into far fewer ganglion cells. Also, other neurons in the retina—horizontal and amacrine cells—participate in these connections and interactions. Therefore, a point-to-point representation is not transmitted—as with a television camera, for instance—but there is processing of the information before it even leaves the retina (see essay on page 802).

Ganglion cell axons from all over the retina converge at the rear of the eyeball and bundle together like a cable to form the optic nerve, which connects the retina

42–6

The retina of the vertebrate eye. Light (shown here as entering from the right) must pass through a layer of cells to reach the photoreceptors (the rods and cones) at the back of the eye. Signals from the photoreceptor cells are then transmitted through the neurons known as bipolar cells to the ganglion cells, whose axons converge to become the optic nerve. Horizontal and amacrine cells, other neurons in the retina, also participate in the elaborate transmission paths. Before leaving the retina, some processing of information occurs in these pathways.

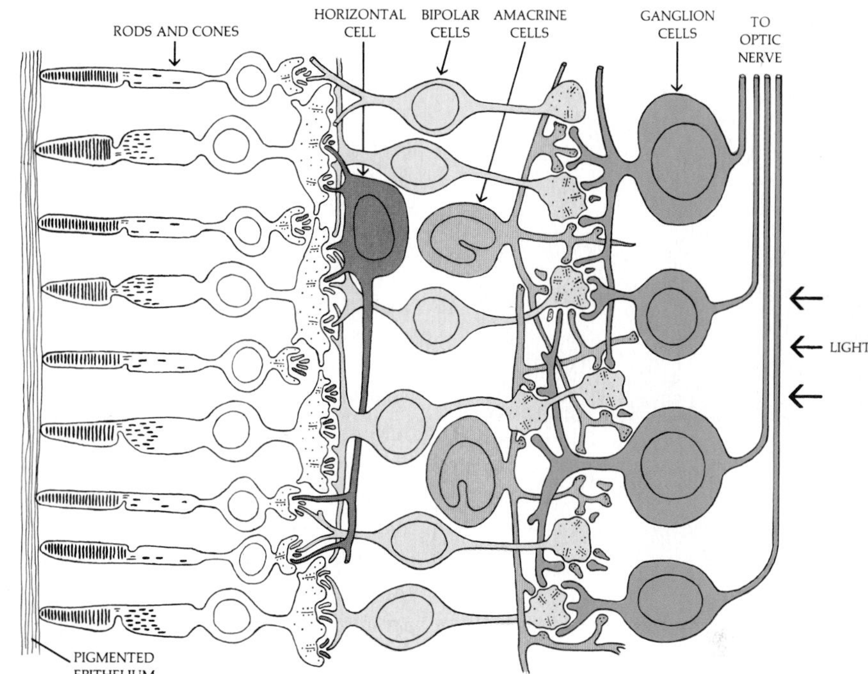

with the rest of the brain. The point at which the axons pass out of the retina is a blind spot, since photoreceptor cells are absent there. We generally are not aware of the existence of the blind spot since we usually see the same object with both eyes and the "missing piece" is always supplied by the other eye. You can demonstrate the existence of the blind spot with the help of Figure 42–7.

Rods and Cones

The photoreceptor cells are of two types, named, because of their shapes, rods and cones. Rods are responsible for black-and-white vision; cones, for color vision.

Rod vision does not provide as great a degree of resolution as cone vision, but the rod system is more light sensitive than the cone system. Dim light does not stimulate the cone system, which is why the world becomes colorless to us at night. Some nocturnal animals, such as toads, mice, rats, and bats, have retinas made up almost entirely of rods, and some diurnal animals, such as some reptiles and squirrels, have almost entirely cones.

As shown in Figure 42–3, there is an area in the human retina known as the *fovea*. This is the only area in the retina at which the imperfect lens of our eye forms a really sharp image. In this area, the photoreceptor cells consist entirely of closely packed cones, and these cones, instead of having the 125:1 relationship of the average photoreceptor cells and ganglion cells, make one-to-one connections with the bipolar and ganglion cells. The one-to-one connections and the close packing of the cones provide greater resolution, giving a crisper picture.

42–7

With this diagram, you can prove for yourself that you have a blind spot in each eye. Hold the book about 30 centimeters (12 inches) from your face, cover your left eye, and gaze steadily at the X while slowly moving the book toward your face. Note that at a certain distance, the image of the dot becomes invisible. Then cover your right eye and gaze steadily at the dot while you move the book toward your face. What happens to the X?

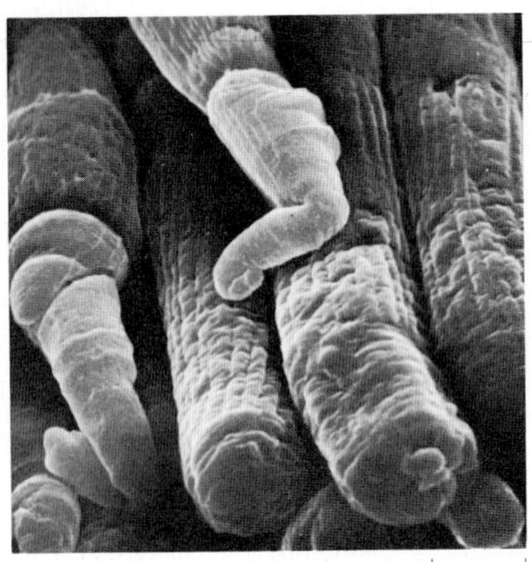

42–8
Rods and cones as shown by the scanning electron microscope. A single photon is sufficient to cause a response in a photoreceptor; as few as five to eight photons in the blue-green frequency range can result in conscious perception.

42–9
Eye-hand coordination is a characteristic of the higher primates, such as the female olive baboon shown here with her daughter.

Birds, which rely on vision above all other senses, may have two or three foveas. Also, the photoreceptors of birds tend to be more tightly packed. We have about 160,000 cones per square millimeter of eye, or some 5 to 10 million cones per eye. A hawk, with an eye of about the same size, has some 1 million cones per square millimeter and therefore a visual acuity about eight times that of a human being.

Among the mammals, only primates have a central fovea for sharp vision. In *Homo sapiens,* highly developed vision is undoubtedly closely correlated with the development of the use of the hands for fine manipulative movements. This reasonable hypothesis is supported by the fact that the human eye is so constructed that the sharpest images—images within the fovea—can be made of objects that are within the reach of the hand (Figure 42–9).

Visual Pigments

Both rods and cones contain light-sensitive compounds—visual pigments—consisting of a protein, called an opsin, and a carotenoid, called retinal, derived from vitamin A. The opsin part of the molecule differs in different types of photoreceptors; the retinal is the same. As we noted previously, the eyes of arthropods, mollusks, and vertebrates all evolved separately, and yet each of the three types contains almost identical visual pigments. Since no animal can synthesize carotenoids, all vision depends on substances derived in the diet directly or indirectly from plants. (The relationships among beta-carotene, vitamin A, and retinal are summarized in Figure 10–6, page 215.)

In rods, the retinal-opsin combination is rhodopsin, sometimes called visual purple. When rhodopsin absorbs light, the retinal changes shape (Figure 42–11), causing it to dissociate from its opsin. The opsin then also changes shape. This breakdown of rhodopsin results in a change in the resting potential of the membrane of the rod cell, which causes a change in the rod cell's release of transmitter to bipolar cells. An enzymatic reaction converts retinal back to its original shape, and then it recombines with opsin, restoring rhodopsin and so completing the cycle.

In humans, there are three different types of cones, each containing one of three different visual pigments. Each of these pigments is made up of retinal and a slightly different opsin. As long ago as 1802, a three-receptor hypothesis for color vision was proposed; this hypothesis, supported by a number of psychological studies, has been greatly bolstered by the identification of the three separate cone types and their pigments. Each pigment is most sensitive to the wavelength of one of the three fundamental colors—blue, green, or red. Different shades of color stimulate different combinations of these cones, and the signals, after being processed in the retina, are further processed by the brain and so become what we perceive as color.

Visible light is only a small portion of the vast electromagnetic spectrum (see Figure 10–3, page 212). For the human eye, the visible spectrum ranges from violet light, which is made up of comparatively short light waves, to red light, the longest visible to us. Fortunately, we cannot see in the infrared (heat-wave) spectrum. Otherwise, we would see everything through an infrared glow emitted by our own bodies. Our photoreceptors are, however, sensitive to ultraviolet. Ordinarily, these waves are filtered out by yellow pigment in the lens, but persons who have had cataracts removed can read by ultraviolet light.

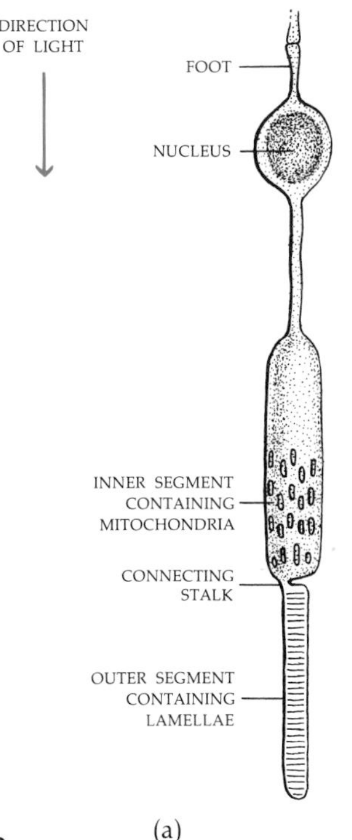

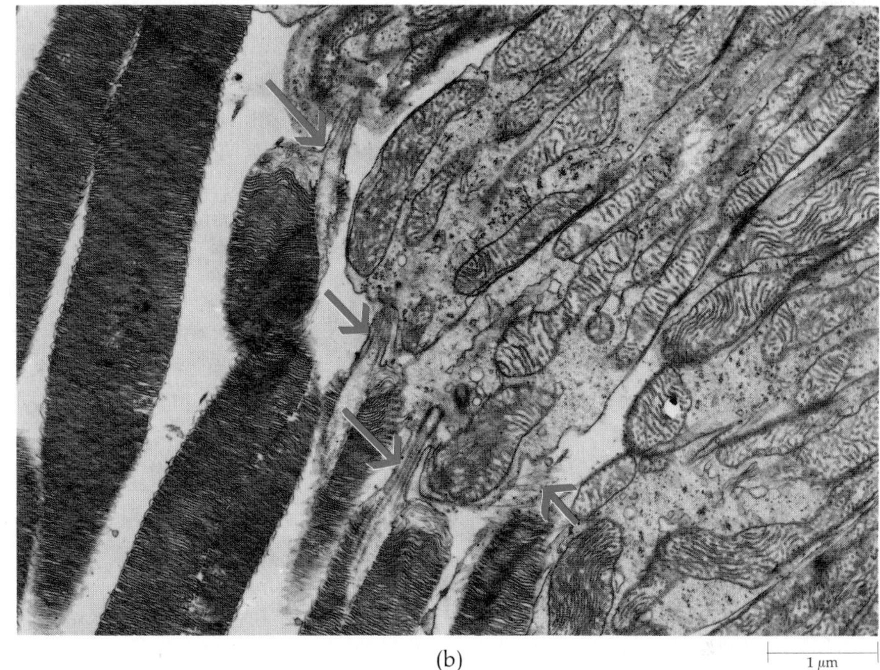

42–10

(a) *Diagram of a rod from the human retina. The molecules of light-sensitive pigment, rhodopsin, are located on the lamellae. The foot makes synaptic contact with bipolar and horizontal cells.* (b) *An electron micrograph of photoreceptor cells. The ends pointing toward the back of the eyeball consist of a stack of membranes, piled up one on top of the other like poker chips. The light-sensitive pigment is built into these membranes. In the adjoining part of the cell, new pigment molecules and other substances required by the light-receptor area are synthesized and transported through a narrow connecting stalk (indicated by the arrows), separating the two portions of the cell, to the membrane-packed portion, where the actual work of this highly specialized cell is done. A cross section of the stalk reveals, surprisingly, that it has the internal structure of a cilium, lacking only the two central microtubules. The cells shown here are rods, but cones are built along the same general principles.*

42–11

When the retinal portion of the rhodopsin molecule is excited by light, it changes shape, setting in motion a series of events that result in the sensation of light. Retinal is changed back to its original conformation by an enzyme, completing the visual cycle. The arrow indicates the bond around which the side chain of the "resting" retinal rotates to form the "excited" retinal.

CH_3 CH_3 CH C(CH_3) CH CH CH CH C CH_3 CH $HC{=}O$ CH_3

RESTING

LIGHT

ENZYME

CH_3 CH_3 CH C(CH_3) CH C(CH_3) $HC{=}O$ CH CH CH CH CH_3

EXCITED

WHAT THE FROG'S EYE TELLS THE FROG'S BRAIN

The frog's eye differs from the human eye in that it has no central fovea and the rods and cones are distributed uniformly over the surface of the retina. It resembles the human eye, however, in that these photoreceptor cells transmit their signals to a far fewer number of ganglion cells, whose axons make up the optic nerve leading from the eye to the brain.

By inserting microelectrodes into the axons while exposing the frog's eye to various kinds of stimuli, Jerome Lettvin and coworkers at the Massachusetts Institute of Technology found that different ganglion cells responded to different stimuli—light on, light off, or a big moving shadow, for instance. Most interesting, one type of ganglion cell responded only to a small moving object; in other words, it was a bug detector. An object bigger than a bug would not stimulate these particular ganglion cells even if it was in motion, and a bug-sized object would not stimulate them if it was motionless. The existence of the bug detector corresponds nicely with certain well-known features of the animal's behavior: a frog will strike only—and virtually always—at a small moving object and will literally starve surrounded by dead insects. Thus, the information about an extremely important aspect of the frog's world is processed right in the retina itself.

Vision and Behavior

Colors often play an important role in eliciting behavior in vertebrates. Studies of the three-spined stickleback, a small freshwater fish, have shown that fighting behavior among the males is elicited by the red color of their underbellies, which develops during the mating season. Niko Tinbergen, who kept a laboratory in England full of aquariums containing sticklebacks, found that the male sticklebacks would rush to the sides of their tanks and assume threatening postures every time a red mailtruck passed on the street outside. Similarly, it has been found that fighting behavior among English robins can be evoked by a small tuft of red feathers. Among birds, bright plumage, especially of the males, plays an important part in attracting the opposite sex and in courtship ceremonies. Most mammals are nocturnal and do not have color vision. The chief exceptions are found among the higher primates, including humans.

Chemoreception: Taste and Smell

If we were to lose our sense of taste or smell, our lives would lose many of their pleasures but we would not be greatly handicapped in our essential activities. Blindness or deafness is a far more serious threat. For many animals, however, the sense of smell is their window on the world, much as vision is ours.

Many of the smaller mammals are nocturnal, and most of them live close to the ground in habitats in which their highly developed sense of smell provides the greatest amount of information about their environment. Presumably, it was the tree-dwelling habits of our immediate ancestors that made the sense of smell less useful for the primates.

Although the organs of chemoreception seem much simpler than those of vision or hearing, we know less about how they work. We do not know exactly how the various different chemicals identify themselves to sensory receptors in such a way to permit recognition and discrimination among them.

42–12
Amphibians and reptiles have, in addition to nostrils, a special olfactory organ, the vomeronasal organ, in the roof of the mouth. Food substances are apparently tested in this organ. When a snake lashes its forked tongue in the air, it is collecting samples for testing in its vomeronasal cavities. The snake shown here is an Eastern hognose.

Taste

Fish, particularly bottom feeders, have taste cells scattered over the surface of their bodies, and these cells play a major role in determining their behavior. In the carp, for example, the part of the brain that receives information from these taste cells is larger than all the other sensory centers combined. The catfish, also a bottom feeder, has taste cells on its body and, in addition, trails barbels, or "whiskers," along the bottom. These are richly supplied with gustatory nerve endings, and when they are stimulated, the fish turns and snaps at the source of the stimulus.

In terrestrial vertebrates, taste cells are located inside the mouth, where they act as sentinels, providing a final judgment on what is and what is not to be swallowed. The taste receptors and the supporting cells around them form the lemon-shaped clusters known as taste buds (see Figure 34–3, page 673). We are able to distinguish four primary tastes: sweet, sour, salty, and bitter. While each primary taste appears to stimulate a different type of taste bud, a single taste bud may respond to more than one category of substance.

Most animals seem to have the same general range of taste discrimination, although there is some difference in what animals "like" and "don't like." Birds, for example, will readily eat seeds and insects with a bitter taste, whereas for most other vertebrates, bitterness serves as a warning signal. Cats are among the few animals that do not prefer substances with a sweet taste, and very few taste receptors that respond to sugar can be found in the cat's tongue.

Smell

In fish, taste and smell are operationally very similar since both involve the detection of substances dissolved in the surrounding water. Taste receptors and smell receptors differ anatomically in fish as in higher vertebrates, however, and the centers of taste and of smell within the brain are entirely distinct.

In terrestrial animals, smell can be defined as the chemoreception of airborne substances. To be detected, however, these substances must first be dissolved in the watery layer of mucus overlying the olfactory epithelium. In humans, this epithelium, which is located high within the nasal passages, is comparatively small. The part within each nasal passage is only about as large as a postage stamp. Each of these areas contains some 50 million receptor cells (Figure 42–13). Even with our relatively insensitive olfactory equipment, we are able to discriminate some 10,000 different odors. Much of what we call flavor in food is actually a result of volatile substances reaching the olfactory epithelium.

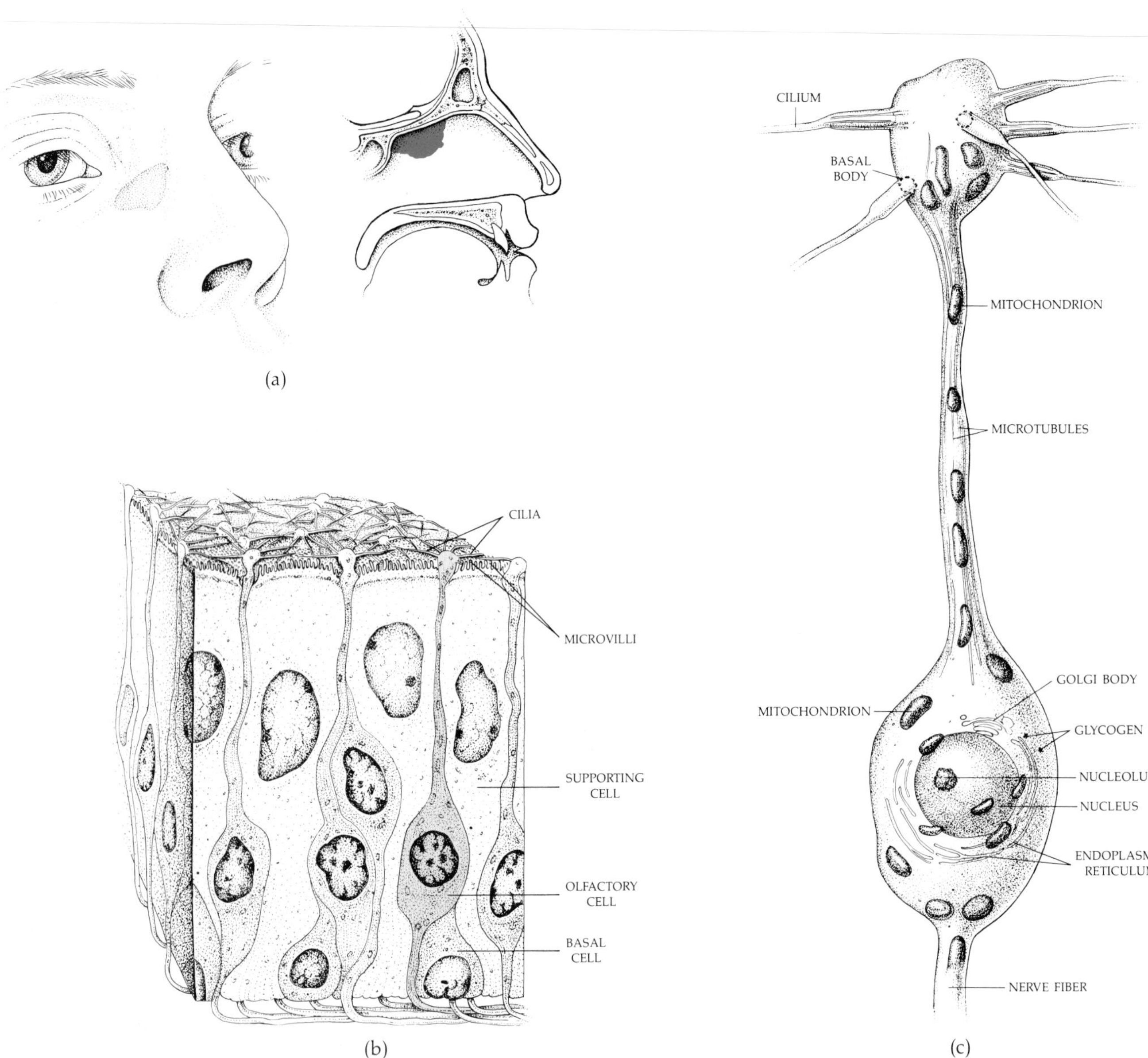

42–13

(a) *A patch of special tissue, the olfactory epithelium, arching over the roof of each nasal cavity, is responsible for our sense of smell.* (b) *The olfactory epithelium is composed of three types of cells: supporting cells, basal cells, and olfactory cells. Supporting cells are tall and columnar, wider near the surface than they are deep within the tissue, and their outer surfaces are covered with microvilli, similar to the microvilli found on the surface of intestinal cells. The basal cells, triangular in shape, are found along the deepest layer of the epithelium. Their function is unknown. They may give rise to new supporting cells when these are needed. The olfactory cells are the sensory receptors.* (c) *The olfactory cell is long and narrow. The cilia protruding from its upper surface are believed to be the odor receptors, although the way in which they function is not known. This outermost part of the cell is connected with the cell body by a long stalk containing microtubules arising from deep inside the cell. From the deepest portion of the cell body, a nerve fiber extends through the underlying tissue, transporting the sensory message to the brain.*

One of the current theories of odor discrimination was developed in the 1960s by John Amoore while he was still an undergraduate at Oxford University in England. According to the present version of Amoore's theory, all scents are made up of combinations of "primary" odors: the seven proposed by Amoore are camphoric, musky, floral, pepperminty, etherlike, pungent, and putrid. There are also, according to this hypothesis, different types of olfactory receptors corresponding to the primary odors. Molecules of a certain shape fit into correspondingly shaped receptors, just as a piece of a jigsaw puzzle fits into its own particular space. All substances that have a pungent odor, for example, have the same general shape and fit into the same type of receptor. Some molecules, depending on which way they are oriented, can fit into more than one receptor and so can evoke in the brain two different types of signals. From the signals received from the various different types of cells, the brain constructs a "picture" of an odor.

Chemical Communication in Mammals

Chemorecption plays a role in the behavior of most mammals. The males of many species—including domestic dogs and cats—scent-mark their territories with urine as a warning signal to other males. Males are attracted to females by special odors associated with estrus (being "in heat"). Such chemical messages among members of the same species are known as pheromones. (Insect pheromones are discussed on page 547.)

Among mice, juvenile females reach sexual maturity earlier when exposed to the odor of an adult male. Pregnant females may resorb their fetuses in the presence of the odor of a strange male. Both of these effects are mediated by pheromones in the urine of the male.

In female rhesus monkeys, vaginal secretions of a volatile fatty acid have been shown to increase near the midpoint of the estrus cycle and to be attractive to male monkeys. Secretions of a similar compound also increase at the midpoint in the human menstrual cycle, but its attractiveness to human males has not been demonstrated. It has also been hypothesized that the chemicals (also fatty acids), which are the sources of "body odor" and are produced, beginning at puberty, by the apocrine sweat glands, originally played such a role. The pubic and axillary hair that develops at about the same time would appear to have the function of retaining and amplifying these odors.

The most convincing evidence to date of pheromonal interactions in humans was assembled by Martha McClintock while still an undergraduate at Radcliffe College. McClintock found a significant tendency toward synchronization of menstrual cycles over the school year among women who were roommates and close friends. The effect appeared to correlate directly with the time these women spent together and not with other external factors, such as food habits or photoperiodism. This finding was later extended by demonstrating that the odor from one female's underarm secretions affected the timing of another female's cycle, independent of any other contact.

Sound Reception: Hearing

Figure 42–14 diagrams the structure of the human ear. Sound travels through the ear canal to the tympanic membrane (the eardrum), in which it sets up a vibration. This vibration is transferred to a series of three very small and delicate bones in the middle ear, which are called, because of their shapes, hammer (malleus), anvil

42–14
The structure of the human ear. Sound waves enter the outer ear and pass to the tympanic membrane, which they cause to vibrate. These vibrations are transmitted through the hammer and anvil to the stirrup. The space containing these three small bones (and which connects with the auditory tube) is the middle ear. The stirrup is attached to another membrane, which covers the oval window and leads to the inner ear. As the stirrup vibrates, it pushes against the oval window membrane, setting up vibrations in the fluid in the inner ear. The three semicircular canals are additional fluid-filled chambers within the bony labyrinth of the inner ear. Each one is in a plane perpendicular to the other two. Their function is to monitor the position of the body in space and to maintain equilibrium. Movements of the head set the fluid in these canals in motion, activating their sensitive hair cells and sending action potentials to the brain.

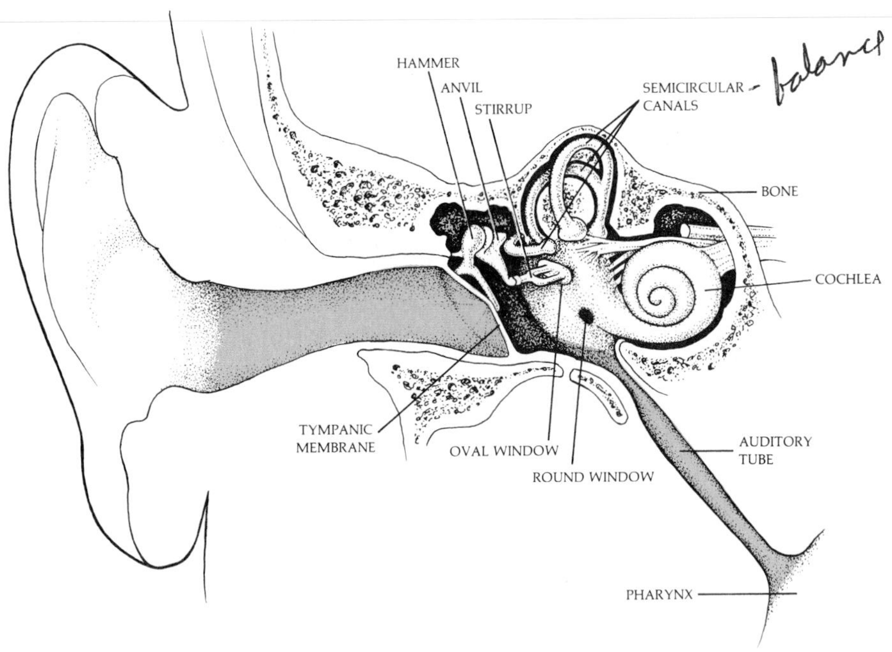

(incus), and stirrup (stapes). Vibrations in the tympanic membrane cause the stirrup to push gently and rapidly against the membrane covering the oval window, which leads to the fluid-filled inner ear.

The pressure must be amplified as it passes from the tympanic membrane to the inner ear because fluid is more difficult to move than air. Amplification is achieved in two ways. First, the oval window membrane is smaller than the tympanic membrane, resulting in more force per unit area—that is, more pressure—as the total force is transferred. Second, the three bones of the middle ear act as a lever system to increase the force of the vibration.

The middle ear is connected with the upper pharynx by the auditory tube (also known as the eustachian tube). This connection makes it possible to equalize the air pressure in the middle ear with atmospheric pressure; it also unfortunately makes the middle ear a fertile breeding ground for infectious microorganisms that enter the body through the nose or mouth.

The hair cells, which are the sensory receptors, are in the inner ear. They lie along a long membrane, the basilar membrane. In mammals, the membrane is coiled in a bony, spiral passageway, the *cochlea* (snail). Figure 42–15 shows a diagram of the cochlea partially uncoiled. It consists essentially of three fluid-filled canals separated by membranes. The upper and lower canals are connected with one another at the far end of the spiral. At the near end of each of these two canals are movable membranes, the oval window membrane and the round window membrane. The stirrup vibrates against the oval window membrane at the near end of the upper canal, causing pressure waves in the fluid contained in these canals. The waves travel the length of the cochlea, around the far end, and back again to the round window membrane at the near end of the lower canal. As the oval window membrane moves in, the round window membrane moves out, keeping the pressure equalized. You will notice that, although hearing is generally

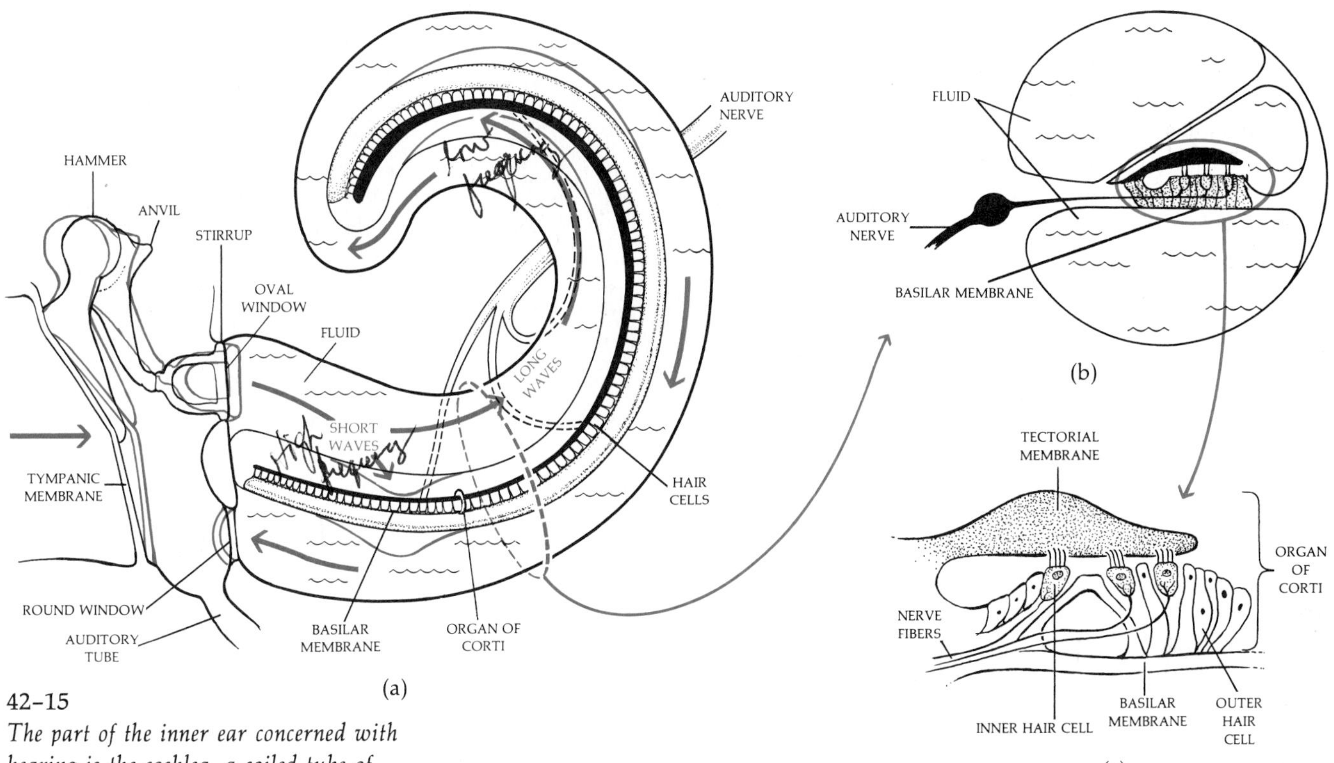

42–15

The part of the inner ear concerned with hearing is the cochlea, a coiled tube of 2.75 turns—shown here (a) *as if it were partially uncoiled. Vibrations transmitted from the tympanic membrane to the stirrup cause the stirrup to push against the oval window membrane, resulting in pressure waves in the fluid that fills the cochlear canals. Pressure waves in the fluid set up vibrations in the basilar membrane, stimulating the sensory cells in the organ of Corti, which rests on the basilar membrane. Sounds at different frequencies (or pitch) have their maximum effect on different areas of the membrane. The round window prevents the pressure from building up in the cochlea. A cross section of the cochlea is shown in* (b), *and a close-up of the organ of Corti in* (c).

considered to be the detection of airborne sounds, our hearing cells are ultimately stimulated by movements in fluid.

The central canal contains the organ of Corti, which rests on the basilar membrane of the canal. The movement of the fluid waves along the outer surface of the central canal causes vibrations in the basilar membrane, which, in turn, stimulate individual sensory cells, the hair cells, within the organ of Corti.

The basilar membrane is narrower and less plastic at the end nearer the middle ear. Hence it does not vibrate uniformly along its length; instead different areas of the membrane respond to different frequencies of sound. Thus, different hair cells are stimulated by different frequencies. Human beings are capable of very fine discriminations of sound; such discriminations are made by the brain on the basis of the signals received from the various hair cells.

In general, the human ear can detect sounds ranging from 16 to 20,000 cycles per second, although children can hear up to 25,000 cycles per second. (Middle C is 256 cycles per second.) From middle age onward, there is a progressive loss in ability to hear the higher frequencies.

Our inability to hear very low-pitched sounds (below 125 cycles per second at normal levels of intensity) is a useful adaptation since if we could hear at such low pitches, we would be barraged with sounds caused by the movements of our own bodies, our moving blood, and those conducted through our bones. We do hear some sounds conducted directly through the skull—for example, the noise we make when we chew celery. We also hear our own voices, mainly through the skull. This is why a recording of one's own voice sounds so startlingly unfamiliar.

42–16
Scanning electron micrographs of hair cells from the organ of Corti of a guinea pig before (a) and after (b) exposure to 24 hours of loud noise, comparable to that of a loud rock concert. On some of the hair cells, the orderly arrangement of cilia has been disrupted. Other hair cells have degenerated, losing their cilia.

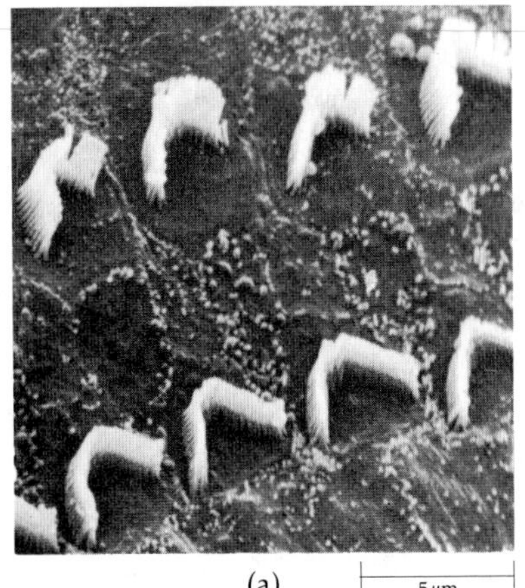

(a)

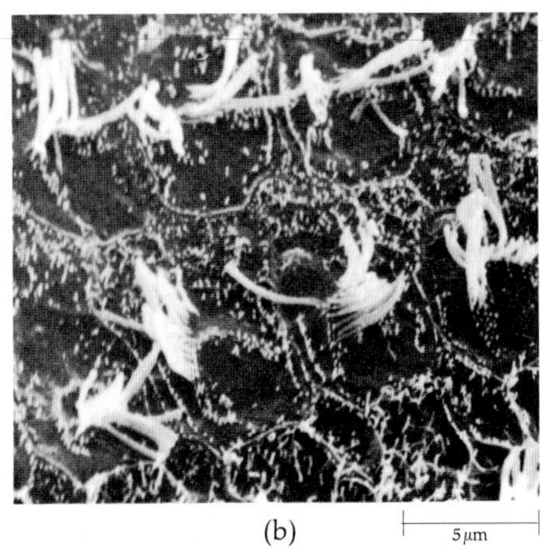

(b)

Dogs, as all dog trainers know, can hear very high sounds (40,000 cycles per second) and mice squeak to one another at 80,000 cycles per second. An elephant named Lois, tested by investigators at the University of Kansas and the Ralph Mitchell Zoo in Independence, Kansas, could not hear sounds higher than 12,000 cycles per second but could hear sounds at 16 cycles per second at levels of intensity inaudible to human beings. (When Lois heard a sound, she pressed a lever with her trunk and was rewarded with half a cup of Kool-Aid.)

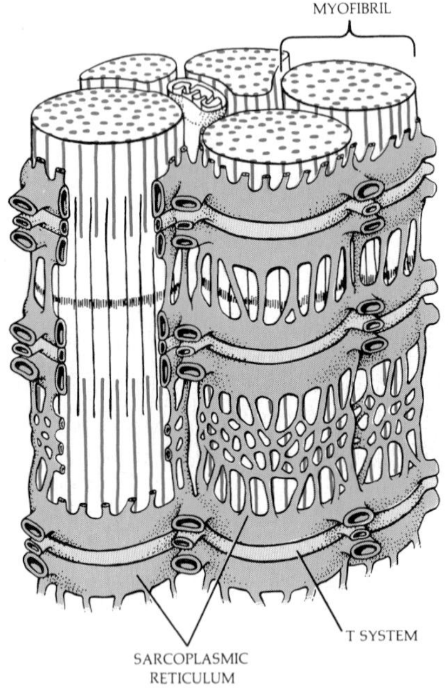

42–17
Six myofibrils, each of which is surrounded by the specialized endoplasmic reticulum of muscle cells, the sarcoplasmic reticulum. The sacs of the sarcoplasmic reticulum contain calcium ions, which, when released, trigger muscle contraction. Traversing the myofibrils, perpendicular to the sarcoplasmic reticulum, is the T system, which is continuous with the cell membrane.

SKELETAL MUSCLE

The activities an animal undertakes as a result of the information received by its sensory receptors and processed by its brain depend upon skeletal muscle, the effector of the somatic nervous system. Muscle, as we noted in Chapter 33, is the principal tissue, by weight, in the vertebrate body. However, even if there were only a single muscle cell per organism, it would be worthy of our attention and admiration because skeletal muscle is the one tissue of the biological world whose function on a macroscopic scale has been traced down to, and actually visualized in terms of, the very molecules that make it possible, a wonderful correspondence between structure and function.

Structure and Function of Skeletal Muscle

As we noted earlier, a skeletal muscle consists of bundles of muscle fibers—often hundreds of thousands of fibers—held together by connective tissue. Each fiber is a single, multinucleated cell 10 to 100 micrometers in diameter and, often, several centimeters long. Each muscle fiber is surrounded by an outer cell membrane called the *sarcolemma*. Like the membrane of the nerve fiber, the sarcolemma can propagate an action potential.

Embedded in the cytoplasm of each skeletal muscle fiber (cell) are some 1,000 to 2,000 smaller structural units; they appear ribbonlike in electron micrographs but they are actually cylindrical strands (see Figure 33–6, on page 663). These strands, which are called *myofibrils* (from *myo*, the prefix for "muscle"), run parallel for the length of the cell. The nuclei are crowded to the periphery of the cytoplasm by the myofibrils and can typically be found at the cell surface, just beneath the sarcolemma.

Each myofibril is encased in a sleevelike membrane structure, the *sarcoplasmic reticulum*, which is a specialized endoplasmic reticulum. The sacs of the sarcoplasmic reticulum contain calcium ions (Ca^{2+}), which, as we shall see, play an essential role in muscle contraction. Running perpendicular to the myofibrils is a system of transverse tubules, the T system; the tubules are continuous with the sarcolemma.

Each myofibril is, in turn, composed of units called *sarcomeres*. The repetition of these units gives the muscle its characteristic striated pattern.

Figure 42–18 shows a sarcomere as seen in a longitudinal section of muscle. Each sarcomere is about 2 or 3 micrometers in length. The Z line is the dense black line seen in the electron micrograph; the I band is the relatively clear, broad stripe that the Z line bisects; and the A band is the large, dense stripe in the center of the sarcomere, bisected by the central H zone. As the diagram shows, each sarcomere is composed of two types of filaments running parallel to one another. The thicker filaments in the central portion of the sarcomere are composed of the protein known as myosin; the thinner filaments are made almost entirely of actin, also a protein. The Z line is where the thin (actin) filaments from adjacent sarcomeres interweave.

The sarcomere is the functional unit of skeletal muscle, the contractile machinery. When the muscle is stimulated, the thin (actin) filaments of the sarcomere slide past the thick (myosin) filaments. Since the thin filaments are anchored into a Z line, this causes each sarcomere to shorten, and thus the myofibril as a whole contracts. According to the current hypothesis, known as the sliding filament model, cross bridges between the thick and thin filaments form, break, and re-form rapidly, as one filament "walks" along the other (Figure 42–19).

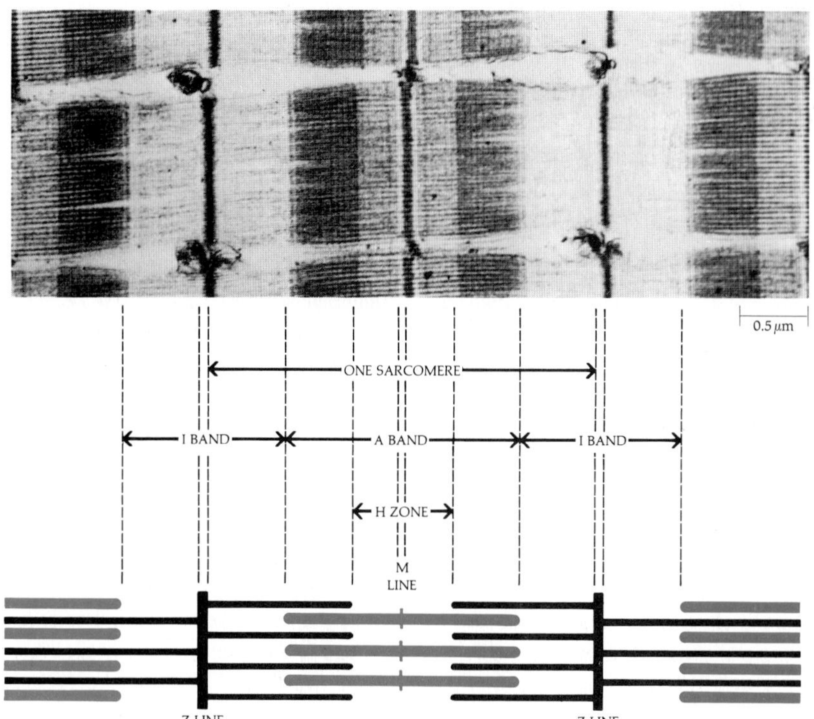

42–18
Electron micrograph and diagram of a sarcomere, the contractile unit of muscle. Each sarcomere is composed of an array of thick and thin protein filaments arranged longitudinally. The way the thin and thick filaments overlap makes the muscle fiber appear striated (banded). The Z line is where thin filaments from adjoining sarcomeres interweave. The I band is a region that contains only thin fibers. The A band marks the extent of the thick fibers. The part of the A band where there are no thin filaments is called the H zone. The thick filaments are interconnected and held in place at the M line. Muscle contraction is caused by the sliding of the thin filaments between the thick ones.

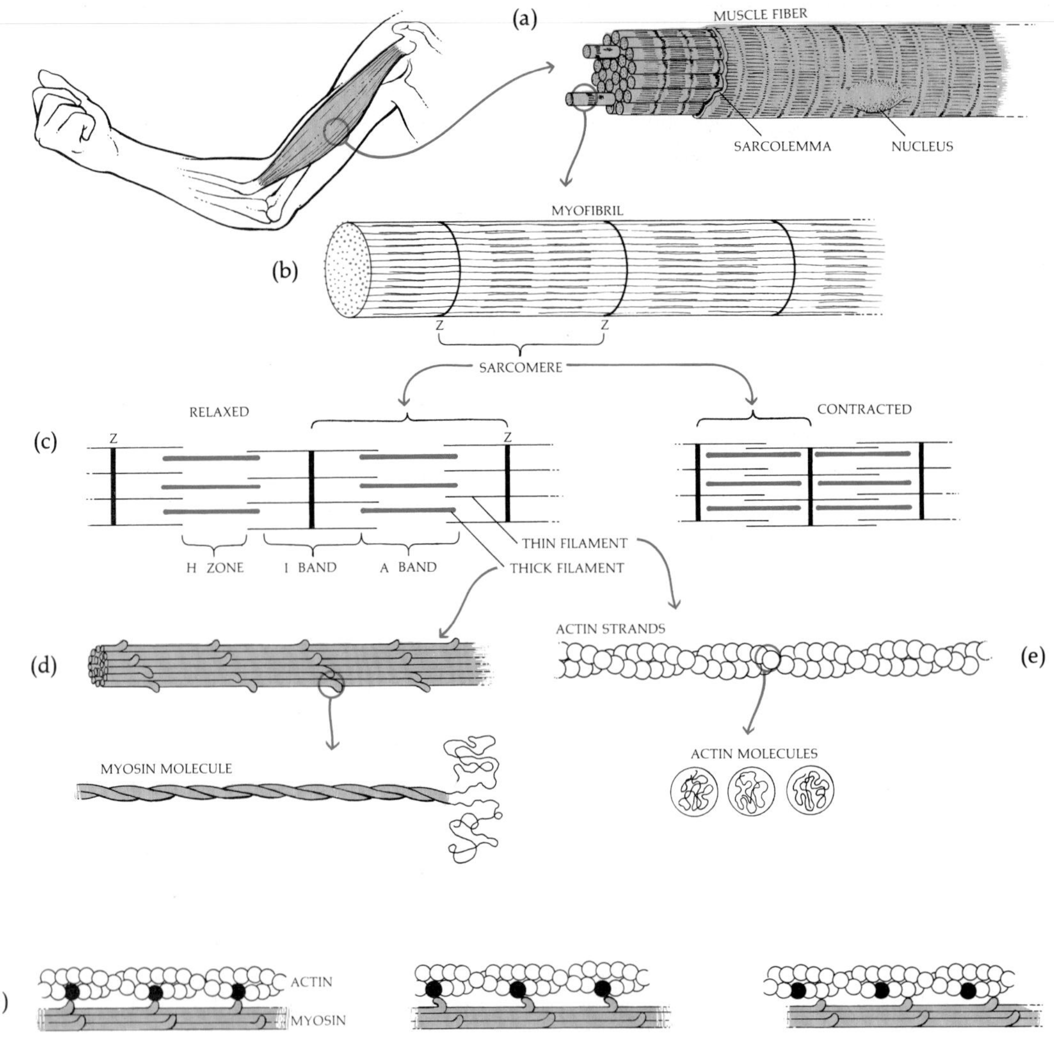

42–19
(a) *Skeletal muscle is composed of individual muscle cells, the muscle fibers. These are cylindrical cells, often many centimeters long, with numerous nuclei.* (b) *Each muscle fiber is made up of many cylindrical subunits, the myofibrils. These fibrils, which contain contractile proteins, run from one end of the cell to the other.* (c) *The myofibril is divided into segments, sarcomeres, by thin, dark partitions, the Z lines. The Z lines appear to run from myofibril to myofibril across the fiber. The sarcomeres of adjacent myofibrils are in line with each other, giving the muscle cell its striated appearance.* (d) *Each sarcomere is made up of thick and thin filaments. Chemical analysis shows that the thick filaments consist of bundles of a protein called myosin. Each individual myosin molecule is composed of two protein chains wound in a helix; the end of each chain is folded into a globular "head" structure.* (e) *Each thin filament consists primarily of two actin strands coiled about one another in a helical chain. Each strand is composed of globular actin molecules.* (f) *The globular myosin heads protruding from the thick filaments serve as hooks or levers, attaching to the actin molecules of the thin filament and pulling them toward the center of the H zone, shortening the sarcomere and contracting the myofibril. When the myofibrils contract, the fiber shortens, and when enough fibers shorten, the entire muscle shortens, producing skeletal movements.*

TWITCH NOW, PAY LATER

Vertebrate muscle fibers differ in their speed of contraction, which depends on the ATPase activity of their myosin heads, and in their energy metabolism. There are two general categories, red fibers and white fibers, most familiar to poultry fans as dark meat and light meat.

Red fibers are slow-twitch, pay-as-you-go fibers. They are relatively small, with large surface-to-volume ratios, a good blood supply, numerous mitochondria, and an abundance of myoglobin, which is mainly responsible for their color. As these characteristics would indicate, they have an abundant supply of oxygen, and oxidative phosphorylation provides most of their energy. In general, they use fatty acids as their fuel. (These fatty acids are stored in red fibers, which is why drumsticks tend to be greasy.) They are designed for relatively continuous use that requires endurance. Thus in the domestic chicken, which walks some 14 hours a day, the leg muscles are mostly red fibers and the wing muscles—the "breast"—are white; whereas in a migratory bird—a wild duck, for instance—the "breast" is mostly red fibers.

White fibers, on the other hand, are fast-twitch, high-powered, twitch-now-pay-later muscles. They have fewer capillaries, fewer mitochondria, and less myoglobin. Their fuel supply is glucose or glycogen, which they break down by anaerobic glycolysis, resulting in lactic acid accumulation and an oxygen debt. They are designed for quick spurts of power. The strong, fast muscles of the legs of a rabbit or a frog are mostly white fibers.

The average human being has a fifty-fifty distribution of fast-twitch (white) and slow-twitch (red) fibers in the major skeletal muscles. However, recent studies have shown that the muscles of long-distance runners and swimmers average about 80 percent slow-twitch fibers, whereas sprinters tend to have about 75 percent fast-twitch fibers.

The proportion of slow-twitch fibers to fast-twitch fibers is genetically determined. (Identical twins have identical fiber patterns in their muscles, regardless of the type of muscular activity they engage in.) Training can cause fast-twitch fibers to behave more like slow-twitch fibers, but the reverse is not true. Some trainers recommend the use of fiber typing as part of a total body profile to determine what type of sport a would-be athlete should pursue. Russians and East Germans, who take their sports very seriously, are doing so already.

Actin and Myosin

The actin strands of the thin filament are composed, it has been found, of many globular actin molecules assembled in a long chain. As shown in Figure 42–19e, each thin filament is composed primarily of two such actin chains wound around each other.

The thick filaments are composed of bundles of myosin molecules. A myosin molecule is composed of two long protein chains, each consisting of some 1,800 amino acids and each with a globular "head" at one end. In the molecule these two chains are wound around each other, with the globular "heads" free. These globular heads have two crucial functions: they are the binding sites that link the thin and thick filaments, and they also act as enzymes that split ATP to ADP, thus providing the energy for muscle contraction.

When a muscle fiber is stimulated, the heads of the myosin molecules move away from the thick (myosin) filament toward the thin (actin) filament, to which they attach themselves. The heads move with a swiveling, oarlike motion, pulling the thick filament, pushing the thin one. Thus, a repeated cycle of attachment, breaking away, and reattachment moves the two filaments, ratchetlike, past one another. The thin filaments on opposite sides of the H zone move toward one another, so the Z lines bordering the sarcomere are pulled together.

The contraction of the sarcomeres is dependent on ATP in two ways: hydrolysis of ATP by the myosin molecule provides the energy for the cycle, and combination

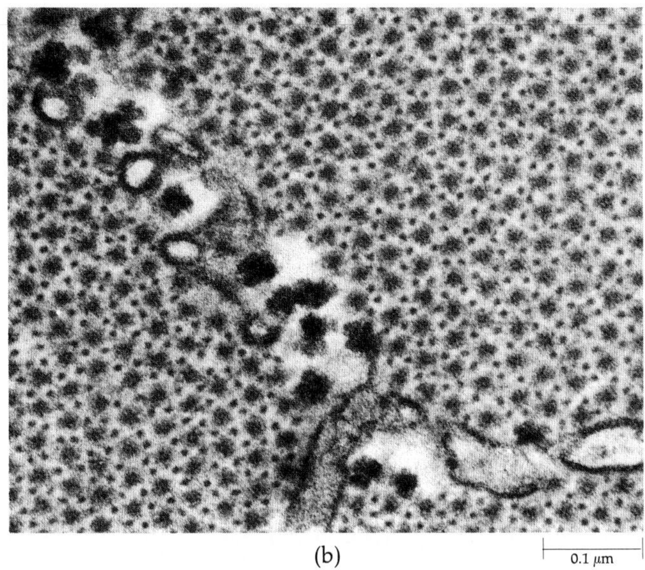

42–20
(a) *Electron micrograph of sarcomeres, contracted. You should be able to identify the Z lines, the A bands, the H zones, the I bands, and the thin (actin) and thick (myosin) filaments.* (b) *Transverse section of myofibrils, showing the hexagonal array of filaments. Each thick filament is surrounded by six thin filaments.*

of a new ATP molecule with the myosin molecule releases the myosin head from the binding site on the actin molecule. The stiffened muscles of a corpse—rigor mortis—are due to the absence of ATP and the subsequent locking of all the actin-myosin crossbridges.

Regulation of Muscle Contraction

Regulation of muscle contraction depends upon two other groups of organic molecules, troponin and tropomyosin, plus calcium (Ca^{2+}) ions. As shown in Figure 42–21, tropomyosin molecules are long, thin double cables that lie along the actin molecules of the thin filament, blocking the cross-bridge binding sites on those molecules. The troponin molecules are complexes of globular proteins that are located at regular intervals on the tropomyosin chains. When Ca^{2+} combines with the troponin molecules, they undergo conformational changes that result in shifting of the tropomyosin chains and exposure of the cross-bridge binding sites.

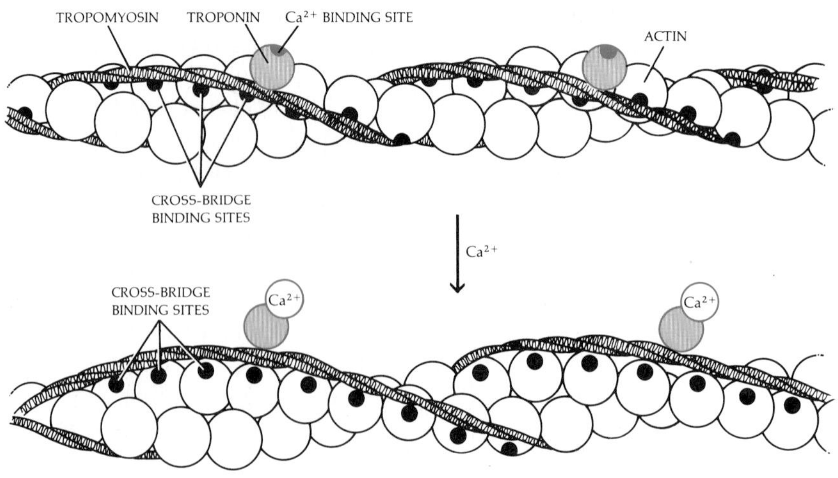

42–21
Tropomyosin and troponin molecules, both proteins, have a regulatory role in muscle contraction. When calcium ions are not present, tropomyosin molecules, long, thin double cables, block the cross-bridge binding sites on the thin (actin) filaments. Troponin molecules, which are globular proteins, are situated at regular intevals on the long tropomyosin chain. When calcium ion binds to troponin, the tropomyosin molecule shifts position, exposing the binding sites and permitting the cross bridges to form.

The source of the Ca^{2+} ions, as we mentioned earlier, is the sarcoplasmic reticulum. When the muscle fiber is stimulated, the electrical impulse passes along the sarcolemma through the T system and alters the sarcoplasmic reticulum, which then releases Ca^{2+} ions. These ions continue to be released only as long as the fiber is stimulated; once stimulation stops, the ions are pumped back into the sacs of the sarcoplasmic reticulum by active transport. Thus it is the Ca^{2+} ions that turn the contractile machinery on and off.

Initiation of Muscle Contraction: The Neuromuscular Junction

Skeletal muscle is stimulated to contract by a signal received from a motor neuron (Figure 42–22a). A motor neuron typically has a single long axon that branches as it reaches the muscle (Figure 42–22b). At the end of each branch, the nerve fiber emerges from the myelin sheath and becomes embedded in a groove on the surface of a muscle fiber, forming the *neuromuscular junction* (Figure 42–22c). As is the case with most synapses between neurons, the signal travels across the neuromuscular junction by means of a chemical transmitter—in this case, acetylcholine. However, unlike synaptic transmission between neurons, this is a direct, one-to-one excitatory response. The acetylcholine combines with receptors on the sarcolemma, depolarizing the muscle cell membrane and initiating an action potential.

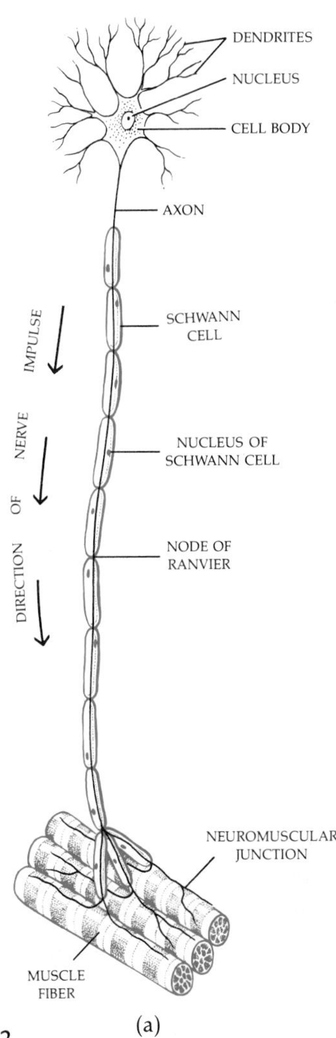

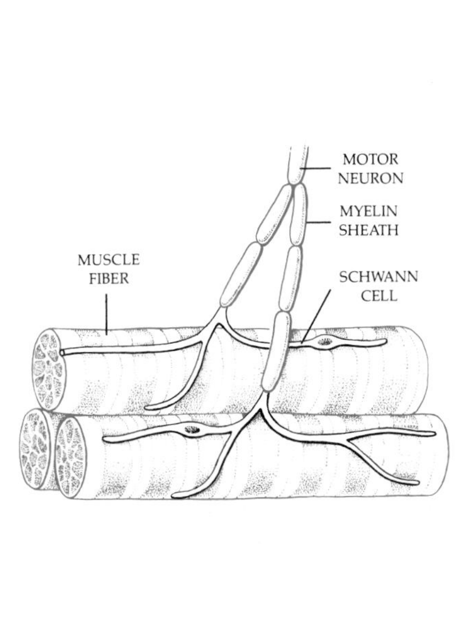

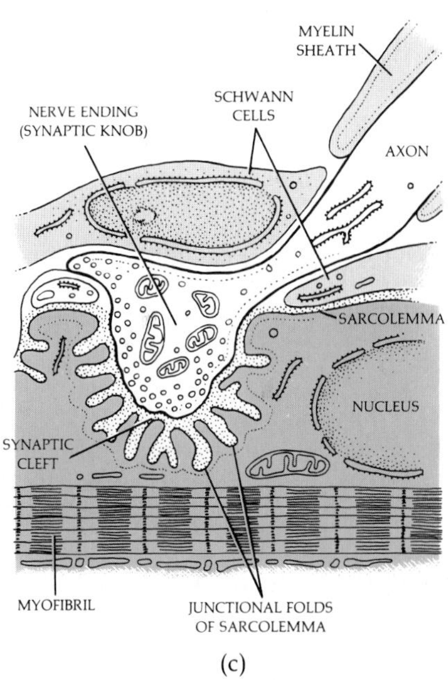

42–22

(a) *A motor neuron. The stimulus is received by the dendrites, which transmit the signal to the cell body and to the axon. The action potential travels along the axon, which is insulated by a myelin sheath composed of Schwann cell membranes.* (b) *The axon of each motor neuron divides into branches, each forming a neuromuscular junction with a different muscle fiber (cell). The motor neuron and the numerous muscle fibers that it innervates are known as a motor unit. Stimulation of a motor axon stimulates all of the fibers in that motor unit. Within a given muscle, fibers of different motor units are intermingled.* (c) *A neuromuscular junction. An action potential conducted along the axon of a motor neuron releases acetylcholine from synaptic vesicles into the synaptic cleft. This transmitter agent combines with receptor sites on the sarcolemma, altering the membrane permeability and initiating an action potential in the muscle fiber.*

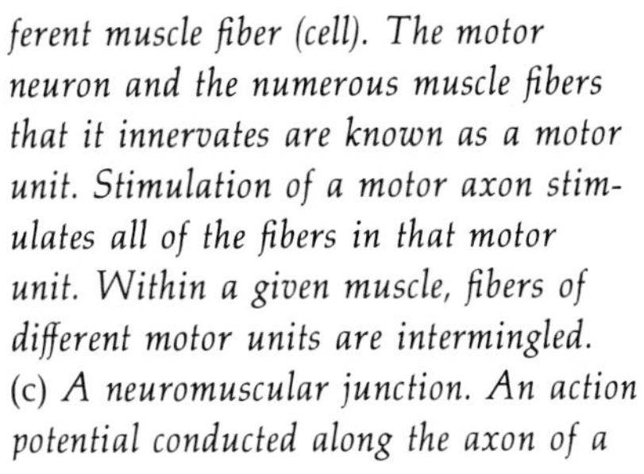

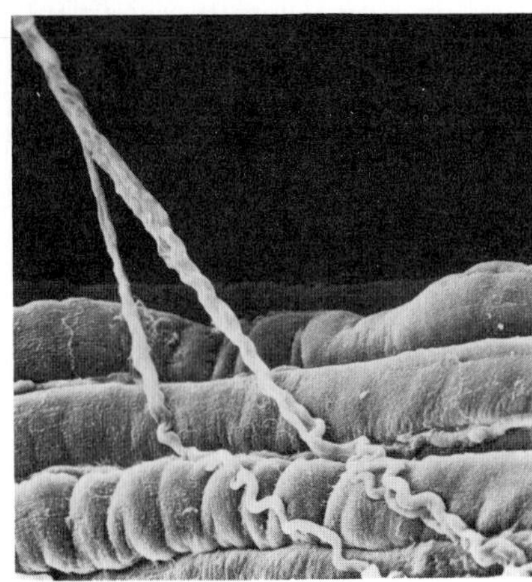

42–23
A scanning electron micrograph showing the axon of a motor neuron and three skeletal muscle fibers.

Many drugs act specifically on the neuromuscular junction. Curare, for instance, a plant extract that Indians of South America used to poison their arrow tips, blocks excitation of the muscle by binding to acetylcholine receptors on the sarcolemma and produces paralysis. It is used medically as a muscle relaxant during surgery. The bacterial toxin of botulism—the most poisonous substance known—prevents nerve endings from liberating acetylcholine and so kills by paralysis of the muscles controlling breathing.

The Motor Unit

The axon of a single motor neuron and all the muscle fibers it innervates are known as a motor unit. The number of muscle fibers in a motor unit is determined by the fineness of control. In a muscle that moves the eyeball, for instance, a motor unit may contain as few as two or three muscle fibers, whereas in the biceps, each motor unit contains more than a thousand. Within a given muscle, fibers of different motor units are intermingled. A slight movement may involve the contraction of only a few motor units. The strength of contraction of a muscle as a whole depends on the number of motor units that have been activated and the frequency with which they are stimulated. Smoothly coordinated activities require the asynchronous contractions of different groups of fibers in antagonistic groups of muscles.

THE KNEE-JERK REVISITED

Look again at Figure 42–2 on page 795. The rubber hammer taps the ligament. The muscle spindles are stretched. The stretching stimulates the sensory nerve endings, generating action potentials. The nerve impulses travel along the myelinated sensory fibers, whose cell bodies are in a dorsal root ganglion, and enter the spinal cord along the dorsal root, which passes between two vertebrae. The fibers synapse with the dendrites of a motor neuron. The electrical signal travels toward the cell

42–24
(a) *A motor neuron terminal. The larger, darker bodies are mitochondria. Below them are the spherical synaptic vesicles. At the bottom of the micrograph is a portion of a muscle cell. The light area immediately below the neuron terminal is the synaptic cleft, and below that is a junctional fold in the sarcolemma. The muscle fiber underlying the sarcolemma is shown in a transverse section, as in Figure 42–20b.* (b) *Synaptic vesicles discharging into the synaptic cleft.*

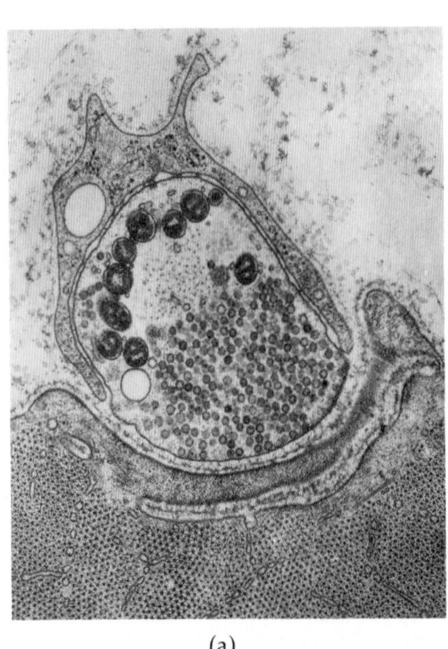

(a)

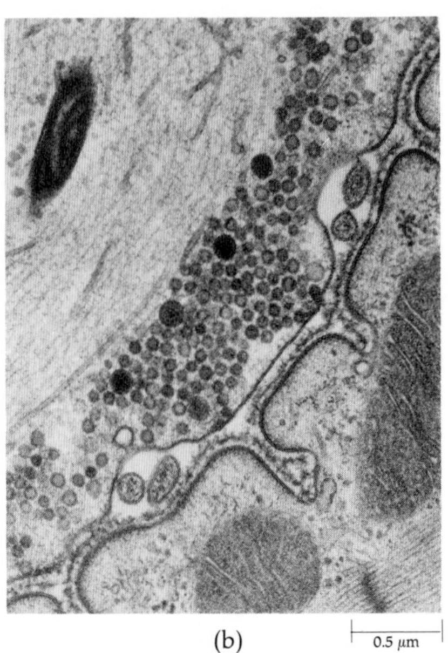

(b)

body of the motor neuron, which is in the spinal cord, and from the cell body to the axon, which emerges from the spinal cord. When the nerve impulse arrives at the axon terminal, it causes the release of packets of acetylcholine into the neuromuscular junction. The acetylcholine diffuses across to the muscle and causes the depolarization of the cell membrane (sarcolemma) of the muscle fiber, producing an action potential. The action potential travels through the T system, which is continuous with the sarcolemma, and stimulates the sarcoplasmic reticulum to release calcium ions. The calcium ions attach to the troponin complexes, and the long, cable-like tropomyosin molecules shift position on the thin (actin) filaments, exposing the binding sites. The globular heads of the myosin molecules attach themselves to the binding sites and, with the hydrolysis of ATP, release and reattach, ratchetlike. The sarcomeres contract, the myofibrils contract, the muscle fibers contract, the muscle contracts, and the lower part of the leg jerks forward—in much less time, one would fervently hope, than it takes to read about it.

SUMMARY

Sensory receptors are cells or groups of cells particularly adapted to detecting changes in the environment and so allowing the animal to initiate appropriate responses. Receptors may be classified according to the type of stimuli received as mechanoreceptors, chemoreceptors, photoreceptors, temperature receptors, and pain receptors. They can also be classified as interoceptors, which monitor internal conditions; proprioceptors, which respond to joint position or stretch within the body and provide our positional sense; and exteroceptors, which respond to conditions of the external environment, such as eyes, ears, and receptors for touch, pressure, heat, and cold in the skin.

The major human sense organ is the eye, which provides much of our information about our environment. The outer, transparent layer of the eye is the cornea, behind which lies the lens. The image is focused on the light-sensitive retina by the cornea and by changes in the shape of the lens produced by the ciliary muscles. Light passes through the eyeball to the retina where it is received by photoreceptor cells, the rods and the cones. Rods, which are more light-sensitive than cones, are responsible for night vision. Cones provide greater resolution than rods and are responsible for color vision. The photoreceptor cells transmit signals to the bipolar cells, from which they are relayed to a network of ganglion cells, whose axons form the optic nerve. Some vertebrates, notably birds and humans, have areas of the retina specialized for acute vision; such an area, where cones are concentrated, is known as a fovea.

Chemoreception involves the detection of specific molecules. In humans, taste buds in the oral cavity are specialized for the discrimination of four primary tastes: sweet, sour, salty, and bitter. Detection of airborne molecules takes place in the olfactory epithelium. The mechanism of chemoreception is unknown. According to current hypotheses, chemoreceptors respond to the three-dimensional, hand-in-glove fit of particular molecules.

Hearing is a form of mechanoreception. In humans, the outer ear functions as a funnel for capturing and routing sound waves to the tympanic membrane (eardrum). Vibrations in the tympanic membrane are transmitted via the bones of the middle ear (the hammer, anvil, and stirrup). The motion of the stirrup against the oval window membrane, which leads to the cochlea, causes vibrations in the fluid of the cochlea in the inner ear. The resulting movements of the basilar membrane stimulate the hair cells of the organ of Corti. The pattern of stimulation is then relayed to the brain.

Skeletal muscles are the effectors of the somatic nervous system. Each muscle is made up of muscle fibers—long, multinucleated cells bound together by connective tissues. Every fiber is surrounded by an outer membrane, the sarcolemma.

Each muscle cell contains 1,000 to 2,000 small strands, the myofibrils, running parallel to the length of the cell. Each myofibril is surrounded by specialized endoplasmic reticulum, the sarcoplasmic reticulum, and traversed by transverse tubules, the T system. Myofibrils are made up of units called sarcomeres. Sarcomeres consist of alternating thin and thick filaments, and contraction comes about as a result of the filaments' sliding past one another. The thin filaments are composed of actin, troponin, and tropomyosin; and the thick filaments, of myosin. The globular heads of the myosin molecules serve as binding sites to link thick and thin filaments and as enzymes for the splitting of ATP, which provides the power for the muscle contraction.

Muscle fibers contract in response to stimulation by a motor neuron, whose axon terminals release the transmitter acetylcholine, which causes depolarization of the fiber membranes. The resulting action potential travels along and into the muscle fiber by way of the membrane of the transverse tubules. This causes the sarcoplasmic reticulum to release stored calcium ions (Ca^{2+}) into the fiber cytoplasm. The sudden increase in the intracellular Ca^{2+} concentration permits the interaction between actin and myosin that brings about contraction of the sarcomeres.

QUESTIONS

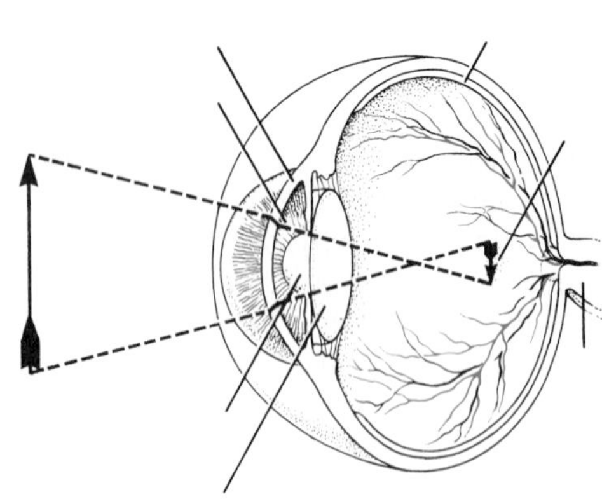

1. Distinguish among the following terms: exteroceptor/interoceptor/proprioceptor; cochlea/organ of Corti; taste/smell; rod/cone; muscle fiber/muscle cell/myofibril; thick filament/thin filament; troponin/tropomyosin; sarcolemma/sarcomere/sarcoplasmic reticulum.

2. What are the five classes of stimuli to which our sense organs respond?

3. Label the drawing at the left.

4. Why don't we see objects upside down?

5. Why do you lose your sense of taste when you have a cold?

6. Consider the process that occurs when another person speaks a word and you then hear the word. Describe the steps involved from the time the air leaves the lungs of the speaker until you are aware of the word spoken.

7. Why is hearing considered a form of mechanoreception?

8. What advantage do mammals gain by having such a long basilar membrane in the organ of Corti?

9. Draw a diagram of the sarcomere relaxed and contracted. What happens to the size of the I and A bands and the H zone as a sarcomere contracts? Explain how this provides the essential clue to the contractile mechanism.

10. Explain the events whereby an action potential in a motor neuron causes a muscle to contract.

CHAPTER 43

Integration and Control IV: The Brain

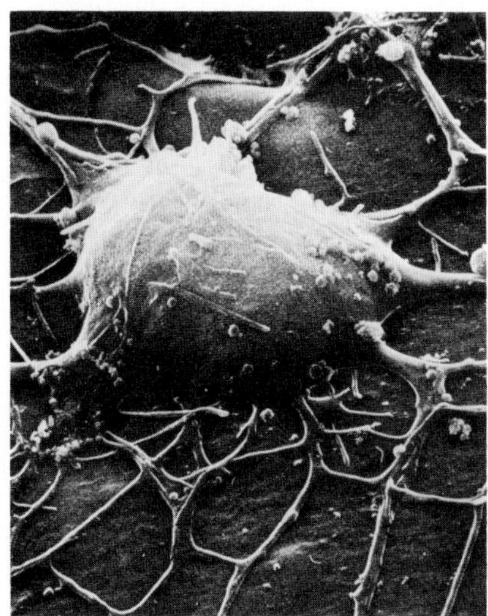

43–1
A neuron from a human brain, growing in tissue culture.

The human brain, which weighs about 1,400 grams (3 pounds), has the consistency of semisoft cheese. Like the spinal cord, the brain is made up of both white matter—the fiber tracts, made white by their fatty myelin sheaths—and gray matter. The gray matter consists of nerve cell bodies and glial ("glue") cells. The glial cells apparently nourish and support the neurons and may also play a role in ionic balance and thus in electric potentials. In some areas of the brain, neurons and glial cells are so densely packed that a single cubic centimeter of gray matter contains some 6 million cell bodies, with each neuron connected to as many as 80,000 others. The human brain is, by far, the most complex and highly organized structure on this planet.

THE STRUCTURAL ORGANIZATION OF THE BRAIN

The vertebrate brain has its beginnings, both in evolutionary history and embryological development, as a series of three hollow bulges at the anterior end of the spinal cord. The bulges, called ventricles, are filled with the same cerebrospinal fluid that fills the spinal canal. These three areas of the brain are known as the hindbrain, the midbrain, and the forebrain, and even though they become folded onto one another in the course of human development, they can still be identified and these terms are used to describe the principal regions of the human brain.

Hindbrain and Midbrain

The hindbrain and midbrain in mammals can be seen as a knobby extension of the spinal cord, the *brainstem*, and a convoluted structure known as the *cerebellum*. The brainstem, like the spinal cord, contains nuclei (clusters of nerve cell bodies) involved with certain reflexes; the centers of the brainstem control heartbeat and respiration, among other functions (which is why a blow to the base of the skull is so dangerous). It contains sensory and motor neurons that serve the skin, muscles, and other structures of the head. The brainstem also contains all the nerve fibers that pass between the spinal cord and higher brain centers. Many of these fiber tracts cross over in the brainstem, so that the right side of the brain receives messages from and sends signals to the left side of the body, and vice versa.

In the lower vertebrates, a major part of the midbrain is made up of the optic lobes, which receive fibers from the optic nerves. In mammals, the analysis of

Table 43-1 Some Major Structures of the Human Brain

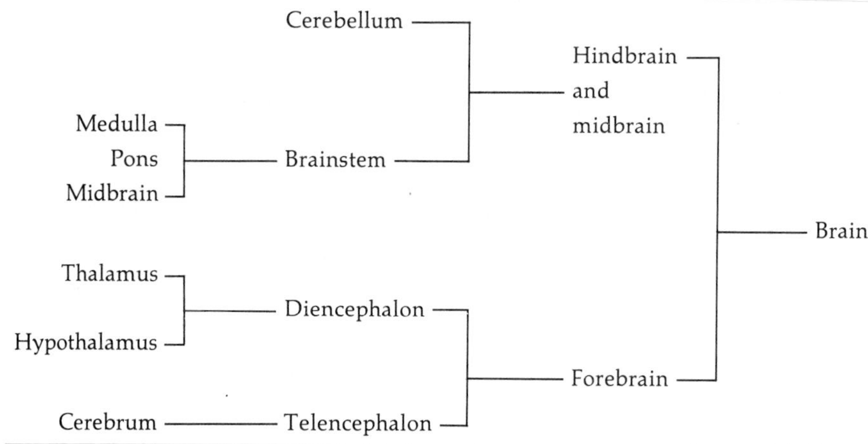

Structure			
	Cerebellum	Hindbrain and midbrain	Brain
Medulla Pons Midbrain	Brainstem		
Thalamus Hypothalamus	Diencephalon	Forebrain	
Cerebrum	Telencephalon		

visual information has become a function of the forebrain, and the midbrain serves primarily as a relay-and-reflex center.

The cerebellum, an outgrowth of the primitive hindbrain, is concerned with the execution and fine-tuning of complex patterns of muscular movements. It is much larger in homeotherms than in the more slow-moving fish and reptiles and reaches its greatest relative size in birds, in which it is associated with the exquisite coordination necessary for flight.

Forebrain

As you can see in Figure 43-2, the primitive forebrain is divided into two major parts; these are called the *diencephalon* and the *telencephalon*. The diencephalon, which contains the thalamus and the hypothalamus, is a major coordinating center of the brain. The thalamus, two egg-shaped masses of gray matter, constitutes the main relay center between the brainstem and the higher brain centers. Its nuclei process and sort sensory information. The hypothalamus, lying just below the

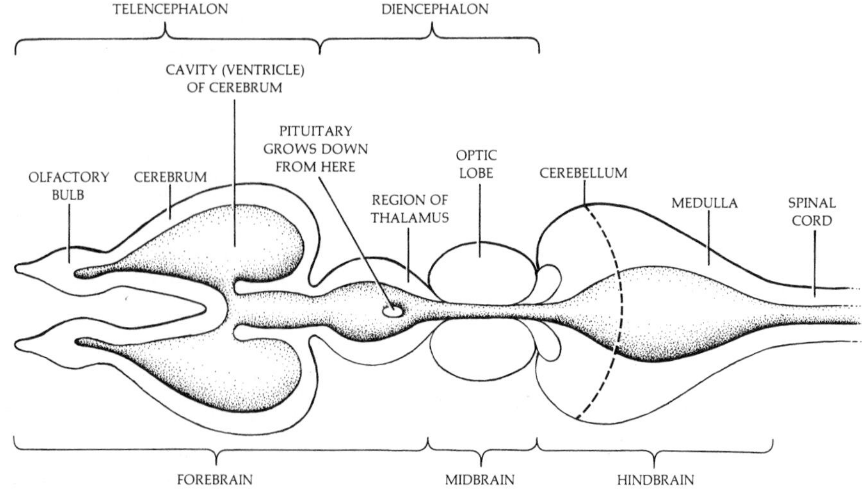

43-2
A dorsal view showing the general pattern of organization of the vertebrate brain. The brain has been cut horizontally to show the ventricles (cavities within the brain). The brain ventricles are continuous with the spinal canal.

43–3
The brains of various vertebrates. The brainstems (indicated in color) include the pons, medulla, and midbrain. They are approximately the same in the various species. However, the cerebral hemispheres have become progressively larger in the course of evolution. The cerebral cortex, the outer surface of the cerebral hemispheres, reaches its greatest development in the primates, particularly Homo sapiens. *The olfactory bulb, present as a stalked knob, is hidden in the human brain by the much more developed cerebrum.*

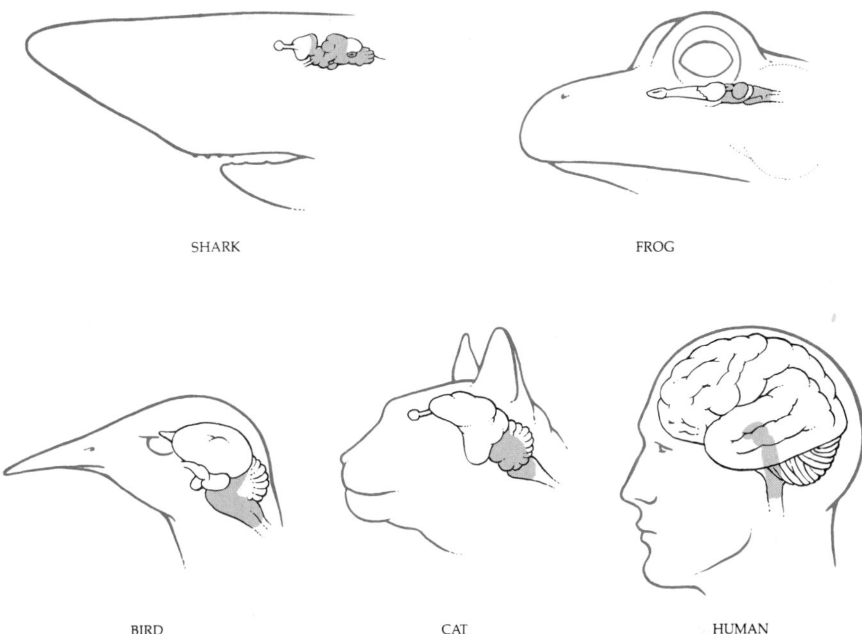

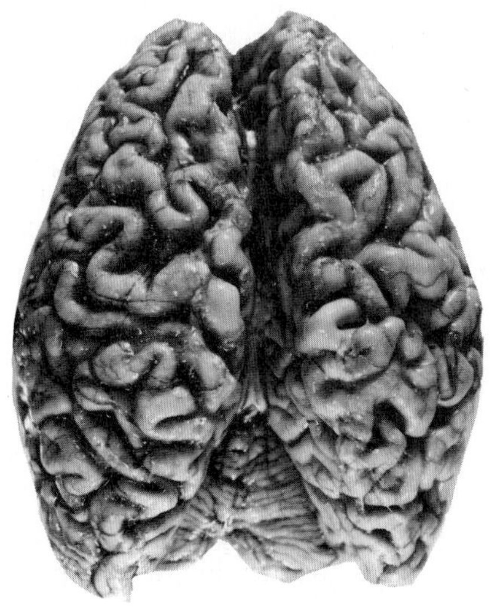

43–4
A brain, viewed from above. The corpus callosum is visible within the deep groove separating the two cerebral hemispheres, and the cerebellum can be seen below the hemispheres at the base of the brain. This specimen is somewhat desiccated, so that the many convolutions of the cerebral cortex are clearly visible. By these, you can immediately distinguish this brain as human.

thalamus, contains nuclei responsible for the activities associated with sex, hunger, thirst, pleasure, pain, and anger. As we have seen, it is the major center for integration of the nervous and endocrine systems. It contains the thermostat of mammals (page 739). It is the source of ADH and oxytocin, which are stored in and released from the posterior lobe of the pituitary (page 783) and of releasing hormones, small peptides that control the secretion of hormones from the anterior pituitary.

The telencephalon ("end brain") is the most anterior portion of the brain and the structure that has changed the most in the course of vertebrate evolution. In the most primitive vertebrates, the fishes, it is concerned almost entirely with olfactory information and is known as the rhinencephalon, or "smell brain." In reptiles, and especially birds, the most prominent structure of the telencephalon is the corpus striatum, which is involved in the control of complicated stereotyped behavior. In mammals, the *cerebrum*, the most anterior part of the telencephalon, is greatly increased in size in relation to other parts of the brain. This increase reaches its greatest extent (so far) in the human brain, in which the many folds and convolutions of the surface of the cerebrum, the cerebral cortex, greatly increase its surface area. In humans, the cerebrum occupies 80 percent of the total brain volume. The wrinkling and folding of the cortex allows its enormous area of 2,500 square centimeters to fit within the confines of the skull.

The cerebral hemispheres are connected at their base by a tightly packed, relatively large mass of fibers called the corpus callosum.

Brain Circuits

Obviously, each of these parts of the brain does not work separately, but rather, information is relayed by complex networks of neurons and synapses from one area of the brain to another. Two examples of such integrating circuits are the reticular activating system and the limbic system.

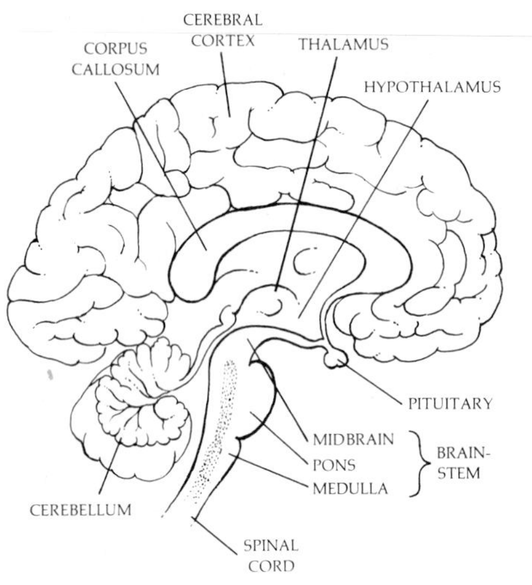

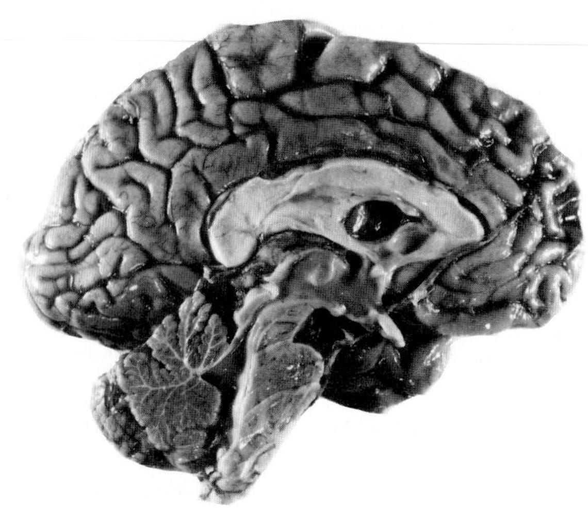

43–5
Longitudinal section of a human brain. The gray matter, characteristic of the cortex, consists of nerve cell bodies, glial cells, and unmyelinated fibers. The white matter is made up predominantly of myelinated fibers.

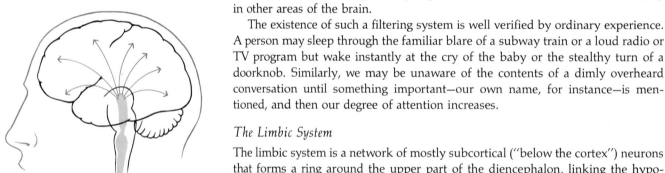

The Reticular Activating System

The reticular activating system is made up of the reticular formation, a core of tissue running through the entire brainstem, and neurons in the thalamus that function as an extension of this system. It is of particular interest to physiological psychologists because it is involved with arousal and with that hard-to-define state we know as consciousness. All the sensory systems have fibers that feed into the reticular activating system, which apparently filters incoming stimuli and discriminates the important from the unimportant. Stimulation of this system, either artificially or by incoming sensory impulses, results in increased electrical activity in other areas of the brain.

The existence of such a filtering system is well verified by ordinary experience. A person may sleep through the familiar blare of a subway train or a loud radio or TV program but wake instantly at the cry of the baby or the stealthy turn of a doorknob. Similarly, we may be unaware of the contents of a dimly overheard conversation until something important—our own name, for instance—is mentioned, and then our degree of attention increases.

43–6
The reticular formation is a diffuse network of neurons in the brainstem. Here, incoming stimuli are monitored, analyzed, and relayed to other areas of the brain. The reticular activating system includes the reticular formation and its thalamic extension. It is concerned with general alertness and the direction of attention.

The Limbic System

The limbic system is a network of mostly subcortical ("below the cortex") neurons that forms a ring around the upper part of the diencephalon, linking the hypothalamus to the cerebral cortex and other structures. It is thought to be the circuit by which drives and emotions, such as anger, thirst, and desire for pleasure, are translated into complex actions, such as seeking food, drinking water, or courting a mate. The brain structures of the limbic system include the hippocampus and other primitive regions of the brain. It is phylogenetically primitive, corresponding to the telencephalon (rhinencephalon) of reptiles.

The Cortex

Of the approximately 100 billion nerve cells in the human brain, about 10 billion are in the cerebral cortex. The cortex, which literally means bark, is a thin layer of gray matter about 1.5 to 4 millimeters thick, covering the surface of the cerebrum.

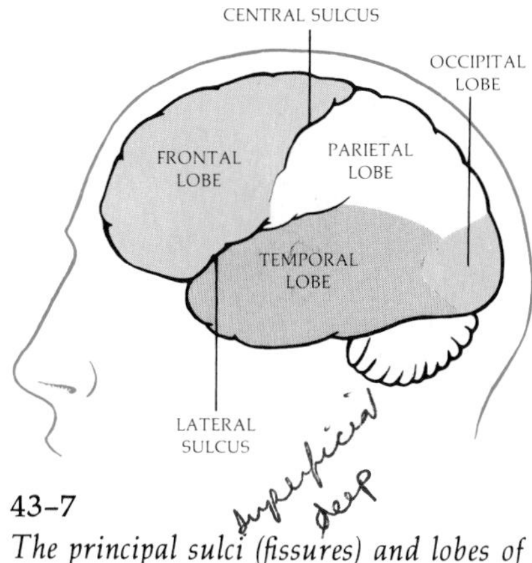

43–7
The principal sulci (fissures) and lobes of the human cerebral cortex.

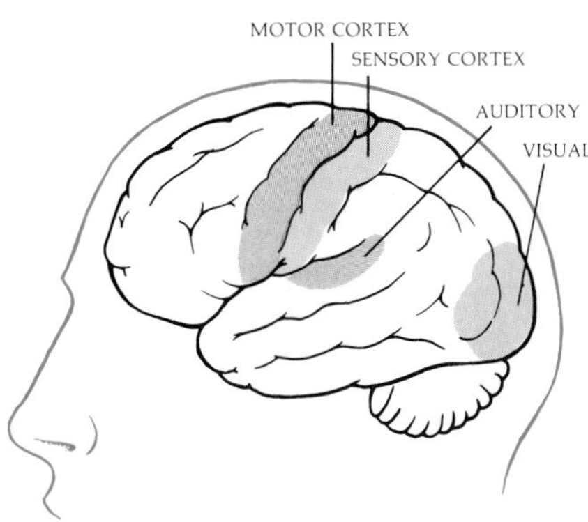

43–8
The human cerebral cortex, showing the location of the motor and sensory areas, on either side of the central sulcus, and the auditory and visual zones. The motor and sensory cortices span the brain like earphones. Functionally, the left and right motor and sensory cortices are mirror images, with the left cortex receiving signals from and sending signals to the right side of the body, and vice versa.

In *Homo sapiens,* as in other primates, each of the cerebral hemispheres is divided into lobes by two deep fissures, or grooves, in the surface. The principal fissures are the central sulcus, which runs down the side of each hemisphere, and the lateral sulcus (Figure 43–7). There are four lobes on each side, or hemisphere, of the cerebrum: frontal, parietal, temporal, and occipital.

Besides these anatomical designations, certain areas of the cortex have been mapped functionally (Figure 43–8). Some of the information is based upon observation of human patients and experimental animals that have had particular areas of the cortex destroyed. Other studies involve stimulating particular areas of the cortex and observing what takes place in various parts of the body, or stimulating various sensory receptors and noting electrical discharges in parts of the cortex. Such investigations have been carried out in experimental animals as well as in humans undergoing brain surgery. (Touching the brain itself produces no sensation of pain because it has no sensory receptors.)

These studies have shown that the area just anterior to the central sulcus, in the frontal lobe, contains neurons concerned with integration of muscular activities. Each point on the motor cortex, as it is called, is involved in the movement of a different part of the body. The relative amount of cortex allocated to a particular group of muscles varies from animal to animal. For instance, in humans, the areas controlling the hand and fingers are very large (Figure 43–9, on page 822), whereas the area controlling a cat's paw is somewhat smaller and that of a horse's hoof very small.

Posterior to the central sulcus, in the parietal lobe, is the sensory cortex, which is involved with the reception of tactile stimuli and taste. Its greatest representation is for those parts of the body that are most richly endowed with sensory receptors: fingertips, tongue, lips, face, genitalia.

The auditory cortex is in the temporal lobe, partially buried within the lateral sulcus (see Figure 43–8). Different regions of the auditory cortex respond best to different frequencies of sound.

The visual cortex occupies the occipital lobe. By using a tiny point of light to stimulate very small regions of the retina, one after the other, investigators have been able to show that each region of the retina is represented by a corresponding but larger region of the visual cortex. This cortical region contains a variety of cells, different groups of which respond to different types of visual stimuli.

The unmapped areas of the cortex were once known as the association, or "silent," cortex. The term "association" was introduced when this part of the cortex was thought to function as a sort of giant switchboard, interconnecting the motor and sensory cortices. More recent studies suggest, however, that the organization of the brain is more vertical than horizontal; for example, severing the motor from the sensory cortex by deep vertical cuts in the association cortex appears to have little, if any, effect on an animal's behavior. Anatomical studies support the interpretation that communication between the sensory and motor areas of the cortex takes place primarily via lower brain centers, particularly the thalamus. In fact, much of the association cortex can be destroyed (as in the once popular frontal lobotomy) without destroying basic mental activities.

But these association areas are not actually "silent." The neurosurgeon Wilder Penfield, who was largely responsible for mapping the sensory and motor areas of the cortex, found that in a small percentage of patients stimulation of certain association areas of the cortex—especially in the temporal lobe—evoked particular

43–9
A combined cross section of one hemisphere of the human cerebrum, indicating the functional areas of the motor and sensory cortices. The motor and sensory cortices are located on either side of the central sulcus in each hemisphere. The motor cortex is indicated in black; stimulation of these areas causes responses in corresponding parts of the body. The sensory cortex is in color; stimulation of various parts of the body produces electrical activity in corresponding parts of this cortex. Notice the relatively huge motor and sensory areas associated with the hand and the mouth. This map is based largely on studies done by neurosurgeon Wilder Penfield on patients undergoing surgical treatment for epilepsy. Because nerve fibers usually cross over in the brainstem, the motor and sensory cortices of the right hemisphere are associated with the left side of the body, and vice versa.

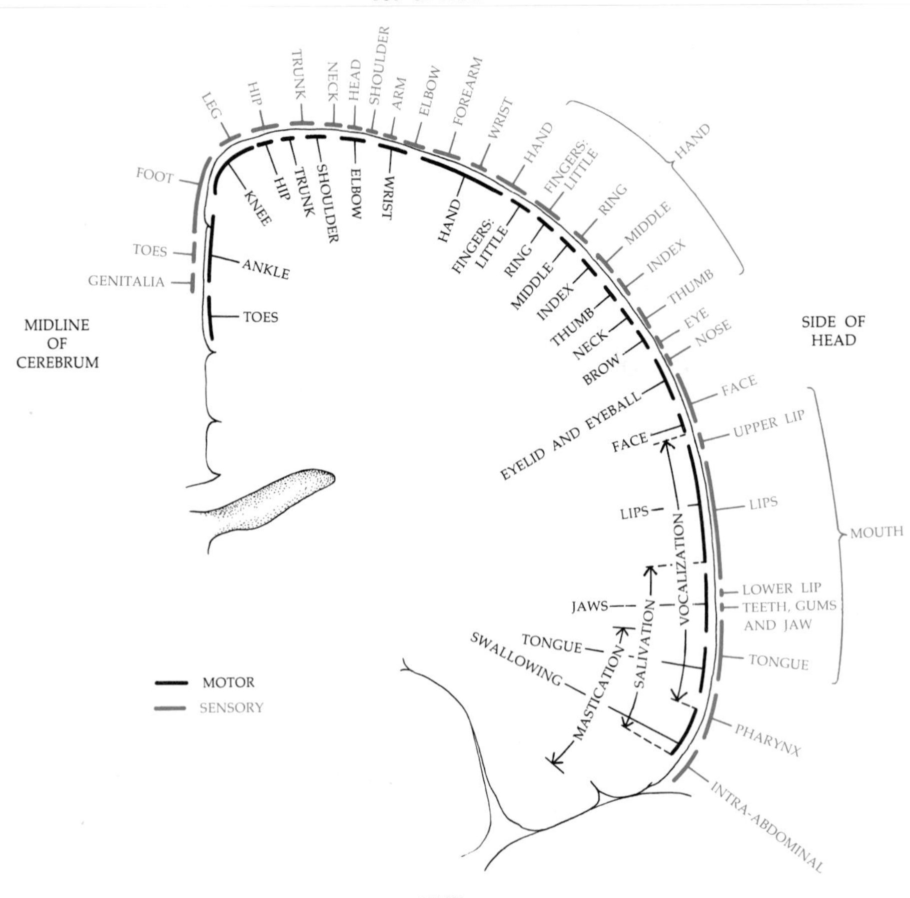

memories so vivid that patients reported that they felt they were actually reliving the past events, including episodes from childhood, fragments of a song, images of persons known long ago. But in general, these "silent" areas do not respond in any specific way to stimulation, although the association cortex shows steady background activity, apparent (but nonspecific) responses to stimuli, increased activity before motor movements (in the planning stage), and increased activity following electrical stimulation of other areas of the brain.

The proportion of association area to motor and sensory cortices is much higher in primates than in other mammals (even mammals such as cats) and is very large in humans. This suggests that these areas have something to do with what is special about the human mind. Also, about half of the association area of the cortex is in the frontal lobes, the part of the brain that has developed most rapidly during the recent evolution of *Homo sapiens.* The frontal lobes are responsible for our high forehead, as compared with the beetle brow of our most immediate ancestors. Public appraisal of their function is reflected in the terms "high brow" and "low brow." The present scientific consensus is that the so-called association cortex is concerned with long-range planning and the organization of ideas.

Left Brain/Right Brain

For more than 100 years it has been known that injury to the left side of the brain often resulted in impairment or loss of speech (aphasia), whereas a corresponding injury to the right side of the brain usually did not. Two areas in the left hemisphere concerned with speech have now been mapped; most of this work has been carried out with patients who have had a stroke (an occlusion of the blood supply to a particular portion of the brain). These are known as Broca's area and Wernicke's area, each named for the nineteenth-century neurologist who first identified it. Ninety percent of all right-handed people and sixty-five percent of all left-handed people have these speech areas in the left hemisphere of the brain.

43–10
The human cerebral cortex, showing the areas associated with speech. Damage to Broca's area—the more anterior area—affects motor control of speech. Damage to Wernicke's area affects conceptual aspects of language.

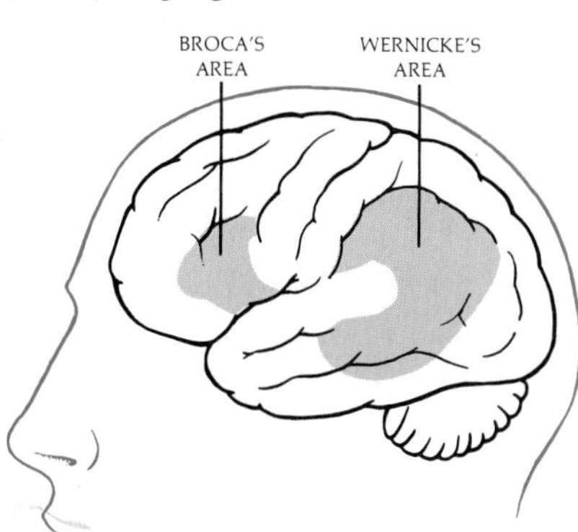

Broca's area is located just anterior to the segment of the motor cortex that controls muscles of the lips, tongue, jaw, and vocal cords. Damage to this area results in slow and labored speech—if speech is possible at all—but does not affect comprehension. Wernicke's area is just adjacent to the area of the cortex where auditory signals are received. Localized damage in Wernicke's area results in speech that is fluent, grammatically correct, and articulate, but the content of the speech is meaningless and the patient's comprehension is poor.

Because the left hemisphere in most people controls both language and the more capable right hand, it has traditionally been regarded as dominant, while the right hemisphere has been described as mute, minor, or passive. However, severe perceptual and spatial disorders are now known to result from injury to the right hemisphere. For example, a person who has sustained such an injury may have difficulty orienting himself in space, may get lost easily and have difficulty finding his way in new or unfamiliar areas, and may have trouble recognizing familiar faces or voices. Musical talent also appears to reside in the right hemisphere: damage to the right hemisphere may result in loss of musical ability, leaving speech unimpaired. Alexander Luria, the eminent Russian neuropsychologist, described a composer whose best work was done after he became aphasic following a massive left hemisphere stroke.

The acquisition of different functions by the two cerebral hemispheres is seen as yet another way of increasing the functional capacity of the brain without increasing the size of the skull (which is calculated to be as large as possible in relation to the optimum size of the birth canal). There is some evidence that this lateralization of function—that is, differentiation of function between the two cerebral hemispheres—is part of the developmental process. It is well known, for instance, that in a young child areas of the right hemisphere can take over following damage to the left hemisphere, and completely normal speech may develop. The possibility of the right hemisphere assuming this left-hemisphere function is closely correlated with the age at which the injury occurs.

It appears that songbirds also have some lateralization of function in their brains. Studies of songbirds indicate that their musical ability resides in the left hemisphere. Fernando Nottebohm at Rockefeller University has shown, for instance, that canaries with lesions in the left hemisphere sing much less than normal birds, and when they do sing, their song is grossly distorted. Canaries with right hemisphere lesions show only minor changes in song production. Also, as in children with left-hemisphere damage, the right hemisphere of canaries can, with time, take over the functions of the damaged left hemisphere.

The song nuclei—the regions of the brain controlling song—can be readily identified anatomically. Nottebohm has found that the size of these regions of the brain controlling song varies from bird to bird and from season to season. Males with

43–11
The optic chiasm viewed from below. The chiasm is the structure formed by the crossing over of fibers traveling from the retina to the visual cortex. Because of this crossing over, the right and left cortical areas each "see" only the opposite half of the visual field.

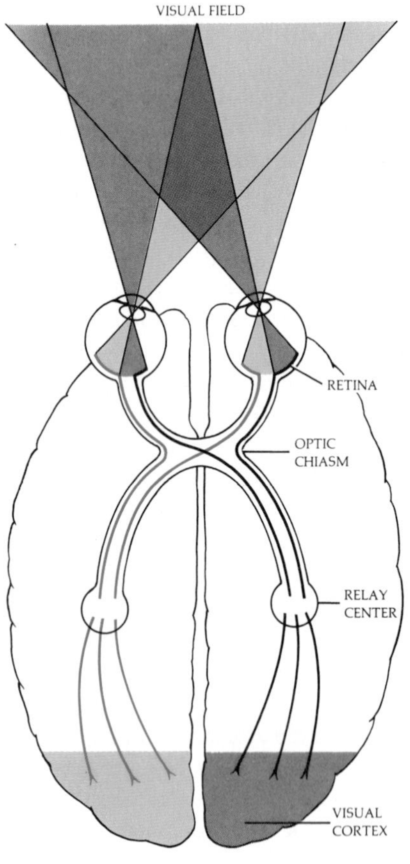

small song centers generally have small repertoires, and the song centers of females, which sing very little, are one-fourth the size of those of males. The size of the nuclei increases in the early spring and then decreases in the summer and fall.

Split Brain

As we mentioned previously, the two cerebral hemispheres are connected by the corpus callosum. In some cases of epilepsy (which is an electrical storm in the brain), severing the corpus callosum lessens the severity of the epileptic attacks. In the 1960s, Roger Sperry and his coworkers at the California Institute of Technology launched a series of studies on such patients, who came to be known as split-brain patients. The name arose because, as Sperry and his associates showed, once the corpus callosum was severed, the two hemispheres of the brain in these patients were functionally separated. In general, it was found that split-brain patients are able to carry out their normal activities, but under controlled experimental conditions, they behave as if they had two separate brains. If such patients are asked to identify by touch objects that they cannot see, for instance, they can name those they can feel with the right hand but not those they can feel only with the left hand. This is apparently because information from the right hand goes to the left cerebral hemisphere, where the speech centers of the brain are located, and information from the left hand goes to the right brain, which is mute. If, for instance, a patient's left hand is given a plastic object shaped like the number 2, which the patient cannot see, he or she is unable to identify the object verbally but can readily tell the experimenter what it is by extending two fingers.

In split-brain patients, if a picture is briefly flashed before them while their eyes are held in a fixed position, the right hemisphere sees only the left side of the picture and the left hemisphere sees only the right side. If a word is flashed to the left hemisphere, the patient is able to identify and write the word correctly. If the word is flashed to the right hemisphere, the patient is unable to speak or write the word. However, such a person is quite able to select an object corresponding to the word with his left hand from a group of objects hidden from the left hemisphere by a screen bisecting the visual field.

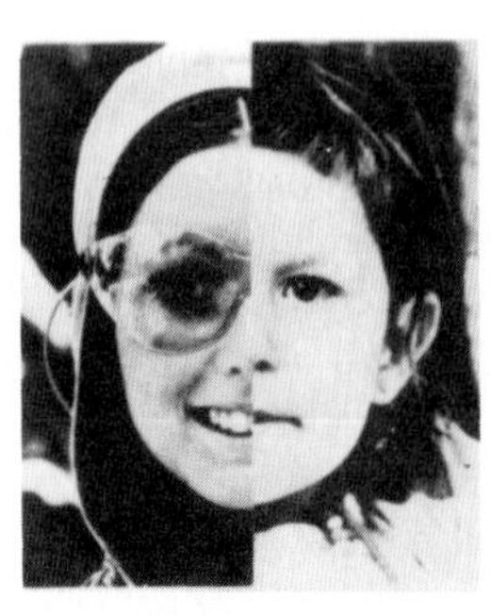

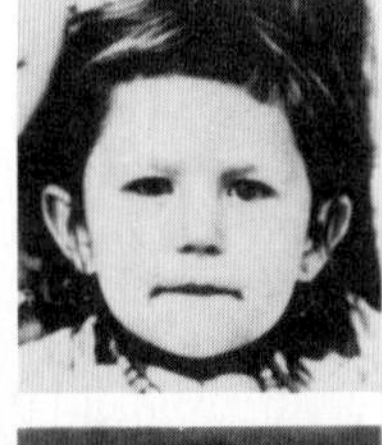

(a) (b)

43–12
A split face (a) *used in tests of patients whose brains have been surgically divided by severing the corpus callosum is made up from two of the faces shown in the selection* (b). *The patient, wearing headgear that restrains eye movements, sees the picture projected briefly on a screen. The left side of the brain recognizes the child; the right side sees the woman with glasses.*

Sperry's findings created a flurry of interest in the psychological consequences of these "two realms of consciousness," and some authorities now believe that the popular speculations about them have far outstripped the hard evidence.

CHEMICAL ACTIVITY IN THE BRAIN

Neurotransmitters

As we noted in Chapter 40, transmission of nerve impulses across synapses in the human nervous system generally involves the release of chemicals into the synaptic cleft. In the peripheral nervous system, you will recall, the major neurotransmitters are acetylcholine and noradrenaline. Acetylcholine is also found in the brain, although at a relatively few synapses. Noradrenaline is a major neurotransmitter in the reticular formation and thus may play a role in arousal and attention. There is some evidence that severe depression may be related to an abnormally low level of noradrenaline at particular synapses; the two major types of antidepressants in clinical use apparently act by increasing the amount of noradrenaline at such synapses. Noradrenaline is also concentrated in synapses in the hypothalamus and other parts of the limbic system.

Many other neurotransmitters have been found in the central nervous system, including dopamine, serotonin (5-hydroxytryptamine), and gamma-aminobutyric acid (GABA), all of which, like noradrenaline, are amino acid derivatives. Dopamine is a transmitter for a relatively small group of neurons involved with muscular activity. Parkinson's disease, which is characterized by muscular tremors and weakness, is associated with a decrease in the number of dopamine-producing neurons and thus with the level of dopamine in certain areas of the brain. Serotonin is found in the reticular formation; increasing concentrations of serotonin are correlated with inhibition of the reticular activating system and with sleep. GABA is a major inhibitory transmitter.

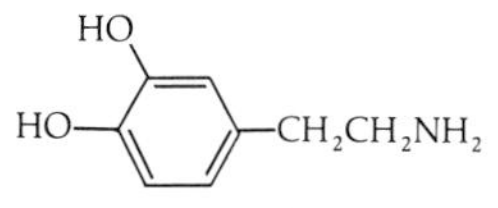

43–13
The structures of some biogenic amines that function as neurotransmitters in the brain. Biogenic amines are produced by slight modifications of an amino acid by a nerve cell.

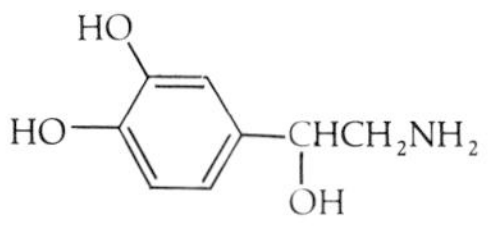

HO
HO— —$CHCH_2NH_2$
OH

NORADRENALINE

HO
HO— —$CH_2CH_2NH_2$

DOPAMINE

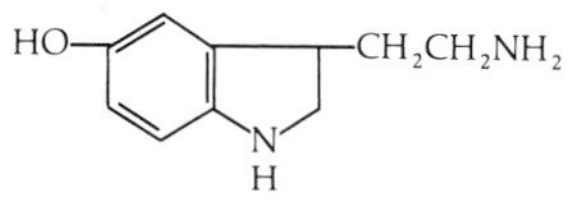

HO— —$CH_2CH_2NH_2$
N
H

SEROTONIN (5-HYDROXYTRYPTAMINE)

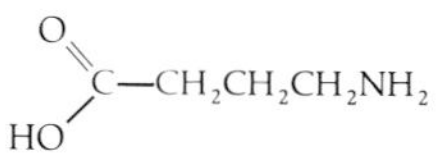

O
C—$CH_2CH_2CH_2NH_2$
HO

GAMMA-AMINOBUTYRIC ACID

Almost all drugs that act in the brain to alter mood or behavior do so by enhancing or inhibiting the activity of the neurotransmitter systems. Caffeine, nicotine, and the amphetamines, for example, stimulate brain activity by substituting for the biogenic amines at chemical synapses. Chlorpromazine and other phenothiazines, which act as tranquilizers, depress reticular system activity by blocking dopamine receptors. LSD inhibits brain serotonin.

The Endorphins

The word "opium" comes from the Greek *opion,* meaning "poppy juice." Since the time of the Greeks, poppy juice and its derivatives have been used for control of pain. They are the most potent painkillers known, and their physiological effects are greatly enhanced by the fact that they produce euphoria. They also are addictive, and although the pharmaceutical industry, urged on by the possibility of great profits, has made repeated attempts to develop an opium derivative that is nonaddictive, their efforts have been uniformly unsuccessful. (Demerol was a recent failure.)

All substances with opiate action are related chemically and have similarities in their three-dimensional structures. Thus, it has long been suspected that the opiates act upon the brain by binding to specific receptors. Using opium derivatives labeled with radioactive isotopes, investigators were able to show that the

43–14

Opiate receptors in the brain of a rat. A thin slice of tissue was placed on a slide and incubated with a radioactively labeled opiate. The opiate molecules bound to their receptors, like a lock to a key. After drying, the slide was pressed against photographic film for a month. Radiation given off by the labeled molecules exposed the film adjacent to the regions of the tissue in which the opiates were localized. The resulting image shows the relative density and distribution of the opiate receptors.

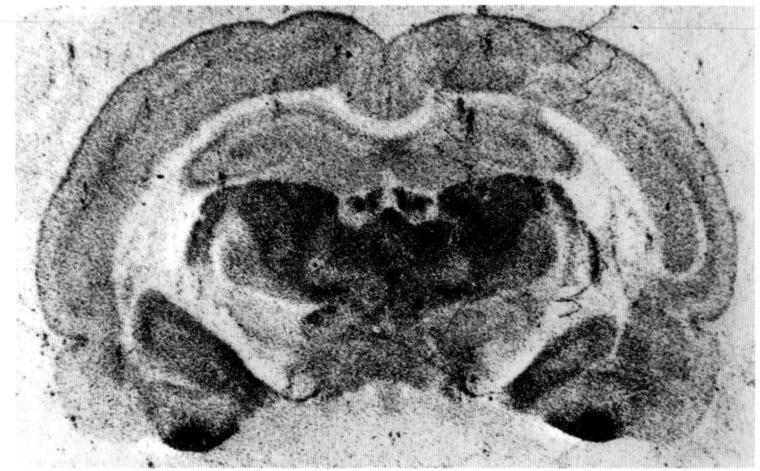

central nervous system does, indeed, have receptors for opiates. These receptors are located primarily in the limbic system, brainstem, and spinal cord. When opiates bind to them, the production of nerve impulses by neurons in this circuit decreases. Such receptors have been found not just in humans but in all other vertebrates tested.

Why would vertebrate brains have opiate receptors? Only one answer seemed logical: Because vertebrate brains themselves must produce opiates. This rather startling conclusion has now been found to be true. Nine naturally occurring substances have now been isolated that have opiate properties. They have been given the name of *endorphins,* for endogenous morphine-like substances.

Two types of endorphins are now recognized. One group, known as the enkephalins, is widespread in the central nervous system, including the spinal cord, the brainstem, and the limbic system, and is also abundant in the adrenal medulla. The enkephalins that have been identified are two pentapeptides (peptides containing five amino acids) that differ by only one amino acid (Figure 43–15). According to recent evidence, the two enkephalins are produced in multiple copies on a single polypeptide chain.

The other endorphins are produced primarily by the pituitary gland and perhaps by other tissues as well. Recently it has been found that the most common of these, beta-endorphin, is synthesized as part of a long peptide chain that also contains ACTH, the hormone that is released by the anterior pituitary and stimulates the adrenal cortex. As you can see in Figure 43–15, there is a great deal of overlap in the primary structures of the various opioid peptides. The functional relationships among the various endorphins are not known.

43–15

The amino acid sequences of three endorphins: (a) *methionine-enkephalin,* (b) *leucine-enkephalin, and* (c) *beta-endorphin. The enkephalins are found in the brain tissue and adrenal medulla of vertebrates, and beta-endorphin is isolated from the pituitary gland. The first four amino acids of each sequence are identical.*

(a) tyr — gly — gly — phe — met

METHIONINE-ENKEPHALIN

(b) tyr — gly — gly — phe — leu

LEUCINE-ENKEPHALIN

(c) tyr — gly — gly — phe — met — thr — ser — glu — lys — ser — gln — thr — pro — leu — val — thr

gln — gly — lys — lys — his — ala — asn — lys — val — ile — ala — asn — lys — phe — leu

BETA-ENDORPHIN

43–16

Is exercise addictive? Concentrations of beta-endorphin in the blood increase substantially during exercise. These recent findings may explain the state of euphoria commonly known as "runner's high." Also, the depression reported by regular joggers on days when they are not able to exercise may be, in fact, a withdrawal symptom.

The endorphins are of great interest to medical researchers because of the insight they may afford into the relief of two extremely serious (and related) medical problems, opiate addiction and pain. The endorphins are believed to function as natural analgesics (pain relievers). Individuals in stressful situations—soldiers in battle, athletes at crucial moments in a contest—have often reported being unaware of what later proved to be an extremely painful injury and so being able to continue to function in a life-threatening (or victory-threatening) situation. This interpretation of the role of the endorphins is supported by the fact that adrenaline and ACTH are also released during stress.

Morphine, heroin, and other exogenous opiates combine with the endorphin receptors, relieving stress, elevating mood, and soothing pain. However, it is hypothesized that these external opiates, acting by negative feedback, reduce the normal production of endorphins, resulting in ever-increasing dependence on the artificial source—in other words, addiction.

ELECTRICAL ACTIVITY OF THE BRAIN

The Electroencephalogram

Another way in which the brain is studied is by electroencephalography. An electroencephalogram (EEG) is a record of continuous electrical activity of the brain as measured by the difference in electric potential between an electrode placed on a specific area of the scalp and a "neutral" electrode placed elsewhere on the body, or the difference in electric potential between pairs of electrodes on the head. The voltages that arise are very weak—about 300 microvolts is the maximum in a normal adult—and extremely sensitive recording equipment is required. (A microvolt is a millionth of a volt, or a thousandth of a millivolt.)

EEG recordings represent the combined activity of large numbers of neurons located in the vicinity of the recording electrode. Electroencephalography is an extremely useful tool for the diagnosis and monitoring of epilepsy, cerebral tumors, and brain damage because there are characteristic EEG patterns that can be correlated with certain levels and types of brain activity.

43–17
The wave patterns of alpha and beta waves. Note that the frequency is much greater and the amplitude (voltage) is much lower for the beta waves.

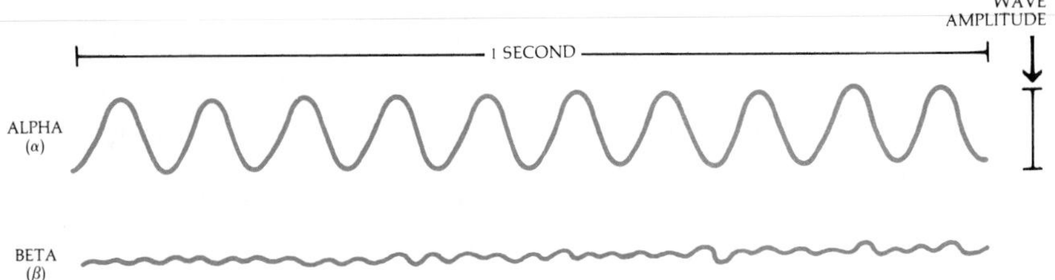

Alpha, Beta, and Delta Waves

One characteristic EEG pattern is the alpha wave, a slow, fairly irregular wave with a frequency of about 8 to 12 cycles per second, recorded at the back of the head. Alpha waves are usually produced during periods of relaxation; in most persons they are conspicuous only when the eyes are shut. About two-thirds of the general population have alpha waves that are disrupted by attention. Of the remaining third, about one-half have almost no alpha waves at all and about one-half have persistent alpha waves that are not easily disrupted by attention.

Another EEG pattern is the beta wave. Beta waves are lower amplitude (lower voltage) than alpha waves, but their frequency is greater, from 18 to 32 cycles per second. Beta waves occur in bursts and are associated with mental activity and excitement.

Waves of lower frequencies than alpha waves, called delta waves, are normally seen in infants or in adults during sleep. Delta waves in adults who are awake are signs of mental disturbance or brain injury.

43–18
A continuous discharge of alpha waves seems to be associated with tranquility and well-being. By concentrating on a musical tone that sounds when alpha waves are produced, it is possible to learn to turn on those waves at will. This technique is known as biofeedback training.

Sleep, Dreams, and the EEG

Sleep is characterized by repeated cycles in which the sleeping person passes through distinct stages during which the EEG waves get larger and slower. In the course of eight hours of sleep, a young adult typically passes through five such cycles. During each cycle there is usually a period of rapid, low-amplitude waves similar to those seen in alert persons. This stage, called paradoxical sleep, is associated with rapid eye movements (REMs), in which the eyes make rapid, coordinated movements as if the sleeper were watching some scene of intense activity.

REM sleep differs from the other sleep stages. For example, although there is electroencephalographic evidence of alertness, the muscles are more relaxed than in ordinary deep sleep. Further, although the sleeper is harder to awaken from REM than from deep sleep, once awakened, he or she is more alert. Fluctuations in heart rate, blood pressure, and respiration are also seen during REM sleep. Many male subjects experience erections of the penis during REM sleep, and women have similar erections of the clitoris with secretions of vaginal fluid. A person awakened during a period of REM sleep nearly always reports that he or she has been dreaming.

The significance of REM sleep is not known. At one time, REM sleep was thought to be the period when all dreaming occurred. On the basis of more recent work, it has been suggested that dreaming can occur at any time during sleep but that conditions for recalling dreams are most favorable when subjects are awakened during REM sleep. Another proposal is that REM sleep serves as an information-processing period during which data from the previous waking cycle are sorted, processed, and stored. Supporting the hypothesis that this stage of sleep is

43–19
(a) *Brain wave recordings of a student with conspicuous alpha rhythm. Electrodes were placed at four positions on the head. When the student was asked to multiply 18 × 14, the alpha rhythm was suppressed and then resumed after the problem was solved (10 seconds of the recording are omitted in order to keep it on the page).* (b) *Brain waves from a student with a complete absence of alpha rhythms. According to a recent study, people who have almost no alpha rhythms under any conditions think almost exclusively by visual imagery, whereas people with persistent alpha rhythm tend to be abstract thinkers.*

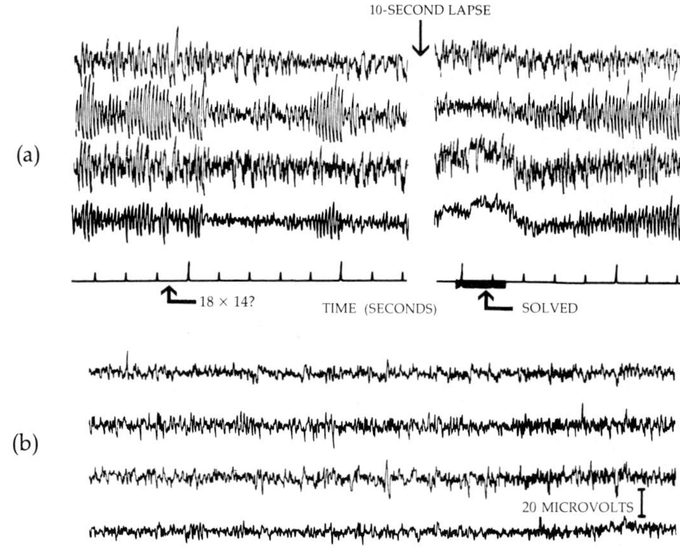

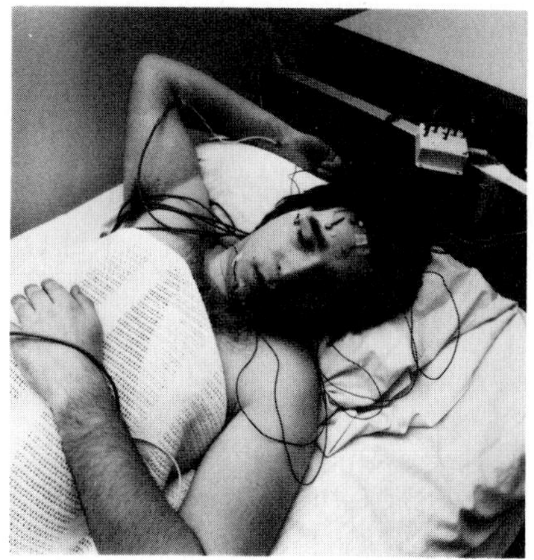

43–20
Volunteer wired to record brain waves, eye movements, breathing rate, and other physical functions.

essential to the memory process are studies showing that laboratory rats forget tasks they have learned if they are deprived of REM sleep. Similarly, evidence indicates that the student who stays up all night cramming for an exam will not have as good a recollection of the material studied as the student who studies and then sleeps.

Neuromagnetism

The EEG measures the sum total of electrical activity of the cerebral cortex, as recorded at the surface of the scalp. Within the last decade, technological advances have made possible the detection and localization of pinpoints of electrical activity deep within the human brain. The principle involved is well known to students of physics: any flow of electricity is accompanied by a magnetic field. The instrument used is called the SQUID (for Superconducting Quantum Interference Device). Brain tissue is essentially transparent to magnetic fields, so that by aiming two or more magnetic detectors at a source within the brain, it is possible to record and plot the source of activity from subcortical structures—the hippocampus, for example, a structure associated with memory (see page 833). The technique is being used to correlate discrete areas of activity in the brain with particular mental processes. In addition, the SQUID is being used by clinical neurologists to localize the small areas of damage within the brains of epileptics that are responsible for their seizures. Precise location of such areas should make it possible to destroy them nonsurgically with microelectrodes.

Single-Cell Recordings

Another method for studying the electrical activity of the brain is by the use of microelectrodes to record the electrical responses of single cells. Many groups of investigators are using the microelectrode technique in studies using invertebrates as models in the long, fruitful tradition that began with the squid axon. Invertebrates are useful because, first of all, their nerve cells are large and unmyelinated.

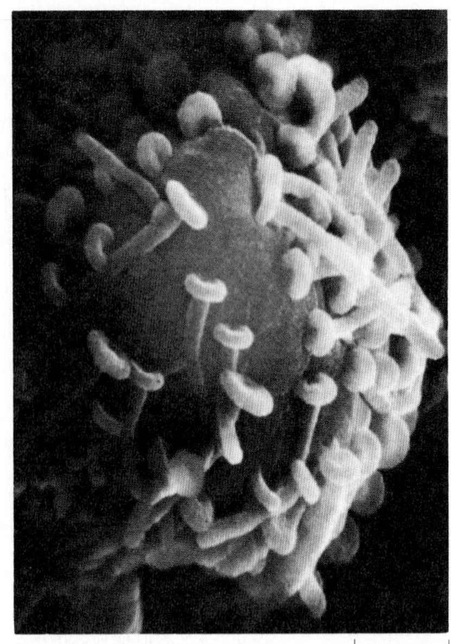

43–21
Synaptic knobs of the sea hare Aplysia *as shown by the scanning electron microscope. Much current research in neurobiology is being carried out in invertebrate systems.*

Moreover, their nervous systems have far fewer neurons, numbering in the thousands rather than in the billions. Thus, invertebrate behavior circuits are relatively easy to trace—easy, that is, in contrast to the bewildering complexity of the circuits in the vertebrate nervous system and of the behaviors associated with their functioning.

A notable example of such work is that of Eric Kandel and his associates at Columbia University on the gill withdrawal reflex of the sea hare *(Aplysia).* If the underside of the animal is touched gently, it quickly withdraws its siphon and delicate gills in a protective reaction. If it is touched repeatedly, it becomes habituated to the stimulus and ceases to withdraw. Habituation is regarded as a very simple form of learning.

The stimulation, it was found, causes 24 sensory neurons in the area to fire. These, in turn, activate interneurons and six motor neurons that control the movement of the gill and cause it to contract. (Kandel is able to stimulate the sensory neurons and record from the motor neurons simultaneously.) Habituation is associated with a gradual decrease in the amount of neurotransmitter released by the repeatedly stimulated sensory neurons. This decrease is reflected, in turn, by a decline in the response of the motor neurons controlling the gill.

David H. Hubel and Torsten Wiesel of Harvard University have used microelectrodes to record the responses of individual cells in the visual cortex of the cat to different visual stimuli. They have obtained some very surprising results. One class of cortical neurons, for example, responds only to a horizontal bar; as the bar is tipped away from the horizontal, discharges from the cells slow and cease. Another class responds best to a vertical bar, ceasing to fire as the bar is turned toward the horizontal. A third class fires only when the bar is moved from left to right; another only to a right-to-left movement. Another class of cells turns out to be a

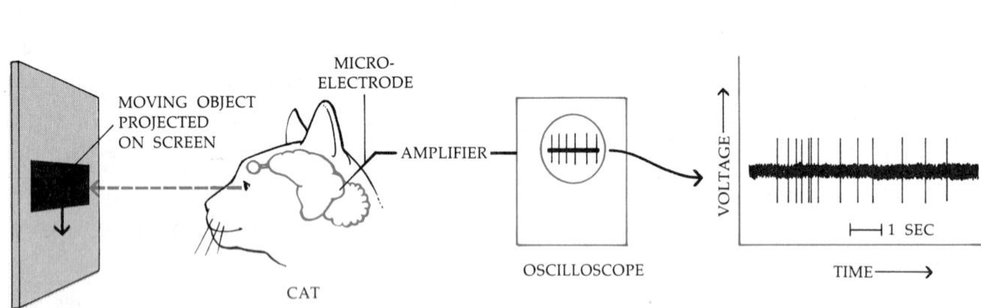

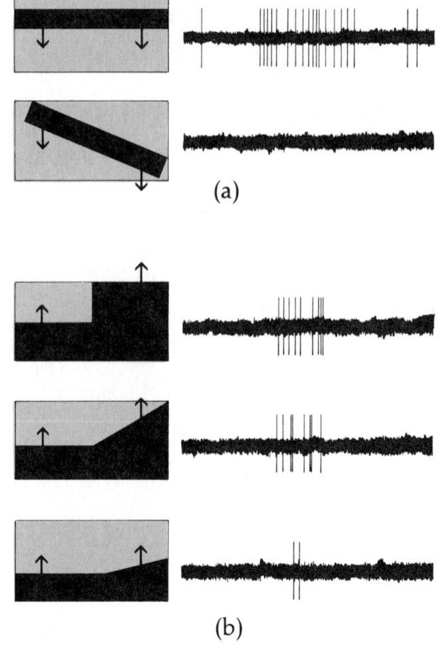

43–22
The experimental method of Hubel and Wiesel. A cat with a small electrode recording from a single cell in its visual cortex observes a screen on which a moving shape is projected. (a) *and* (b) *represent tracings from two different neurons. The first neuron* (a) *responds to a horizontal bar; as the bar is turned toward the vertical, the action potential ceases. The second neuron* (b) *responds to a right angle. Note the reduction in response as the angle increases. The arrows indicate the direction of movement of the shape on the screen.*

right-angle detector; such cells respond only to an angle moving across the visual field and are most excited when the angle is a right angle. Here again is further evidence (see page 802) that what the eye transmits to the brain is not a picture but rather coded and processed information. An important component of this visual information-processing system appears to be a set of very specifically tuned neurons, each responsive to one aspect of the shape of an object in the visual field.

SEX AND THE BRAIN

As we noted earlier, male songbirds have a song center that is located in the left cerebral hemisphere. Enlargement of the song center, which is associated with the establishment of territories and the initiation of courtship behavior, is produced by an increase in circulating testosterone. If a male songbird is castrated, the song center does not enlarge and the bird never sings. If the castrated bird is injected with testosterone, the song center enlarges and he begins to sing. Female birds normally do not sing and both hemispheres are the same size. However, if adult female songbirds are given testosterone, their song centers enlarge and they begin to sing, although their musical repertoires never become as large or varied as those of the males. Thus, in songbirds, as exemplified by canaries, there are differences between the brains of male and female animals, and these differences can be traced, in part, to the influence of testosterone on the brain itself.

Rats don't sing but they do have certain patterns of clearly defined behavior that differ between the sexes. The most obvious—and highly necessary—differences are in mating behavior. When a female rat in estrus is exposed to a sexually competent male, she crouches and arches her back, exposing her genitals. (This female mating posture is called lordosis.) The male mounts her, grabs her flanks, and makes pelvic thrusts that result in intromission and ejaculation. Lordosis is eliminated by removal of the ovaries and is restored by estrogen administration. In fact, genital presentation by a female rat depends on the presence of estrogen in cells of the hypothalamus.

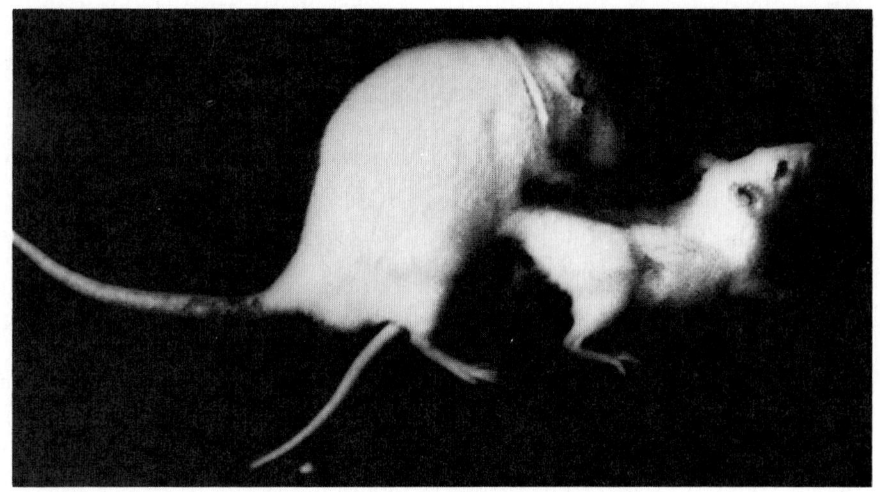

43–23
Lordosis, the characteristic arched posture of the female rat in estrus, exposes the genital area to allow fertilization by the male (left), who has just finished mounting. In experimental animals, the frequency of genital presentation can be correlated with the amount of estrogen administered.

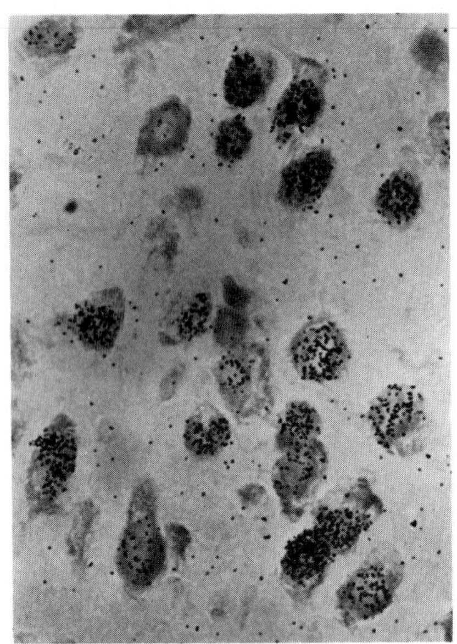

43–24
This autoradiograph of a section of brain tissue from an adult female rat shows the localization of tritium-labeled estradiol in cells of the hypothalamus. The black dots represent the presence of estradiol bound to an estrogen receptor. In rats and many other species, the hypothalamus has been shown to play a key role in the hormonal regulation of sexual behaviors.

In adult male rats, the normal repertoire of mating behavior ceases if the rat is castrated; the behavior is restored by testosterone injections. Studies using radioactive testosterone have shown that this hormone can also be localized in neurons in certain areas of the brain, particularly in a group of neurons in the hypothalamus. (This same area of the hypothalamus is the source of gonadotropin-releasing hormone.) Injection of testosterone to this area produces an increase in electrical activity of these neurons and an increase in sexual activity. Electrical stimulation of this area also produces an increase in sexual activity. However, administration of estrogen to these same castrated male rats does not produce lordosis or any other sex-specific female behavior. In short, there is a difference between the male and female brain in its response to hormones.

This differentiation between the male and female brain occurs during a critical period in early development, the researchers have found. Rats have a 21-day gestation period. About eight days prior to birth, the embryonic male testes begin to secrete testosterone, and this secretion continues until the tenth day after birth. If male rats are castrated on the day of birth (thus depriving them of testosterone for about half the normal period), they never develop male sexual behavior. If they are given testosterone as adults, they do not respond to it. However, if these same adult rats are given estrogen, they show the lordosis behavior in the presence of normal males. By contrast, if male rats are castrated on the tenth day after birth or thereafter, they will not respond to the administration of estrogens. Thus, it is concluded that in rats sex hormones affect the brain in two stages: (1) an early period, during which differentiation of brain cells occurs, and (2) a sexually mature period, in which the differentiated brain cells respond to the hormones by evoking particular patterns of behavior.

Scientists are interested in what other types of behavior might be influenced by sex differences in the cellular organization of the brain. In rhesus monkeys, for instance, the critical burst of testosterone production occurs entirely in utero, around the middle of the 168-day gestation period. Female rhesus monkeys that are masculinized by high doses of testosterone given to their mothers during the critical period show an increase in rough-and-tumble play, an increase in aggressive behavior toward other individuals, and a decrease in maternal imitative behavior. The critical period in humans—the period of the surge in testosterone production—occurs about the second to third month in embryonic development.

The question of what aspects of human behavior—if any—may be influenced by sex differences in the brain is a fascinating one to scientists and nonscientists alike. It is, however, generally agreed that even if such differences exist, they are largely overcome by the powerful forces of environment and culture.

LEARNING AND MEMORY

For scientists interested in brain research, perhaps the greatest challenge is to understand the mechanisms of learning and memory. If we define learning as a change in behavior based on experience, and if the functions of the brain—the mind—are to be explained in terms of the atoms and molecules and the structures composed of them, then learning must involve changes in these atoms or molecules or structures. But what is this change, and where and how does it take place?

Almost 60 years ago, the late Karl Lashley set out to locate this physical change, the trace of memory, which he called the engram. He taught rats and other animals to solve particular problems and then performed operations to see whether he could remove the portion of the cortex containing that particular engram. But he never found the engram. As long as he left enough brain tissue to enable the animal to respond to the test procedures at all, he left memory as well. And the amount of memory that remained was generally proportional to the amount of remaining brain tissue. Lashley concluded that memory is "nowhere and everywhere present." Although much more sophisticated techniques are now available for brain research, modern neurobiologists have had little more success than Lashley in finding memory's hiding place.

There are two distinct types of memory, short-term and long-term. A simple example of short-term memory is looking up an unfamiliar number in a phone book; you usually remember it just long enough to dial it. (In fact, the capacity of short-term memory is about seven items, the number of digits necessary for a local call.) If you call the number enough times, it is transferred to long-term memory. This laying down of long-term memory is analogous to the establishment of a footpath. The more frequently the path is traveled, the better established it becomes. The analogy is strengthened by the familiar experience of consciously retrieving a name, for instance, by seeking out related information that puts one "on the right track." In light of this concept, it is hypothesized that Penfield evoked specific memories in his stimulation of the temporal lobe (page 821) by making contact with such a memory pathway. It is generally believed that the establishment of such a pathway involves alterations in the synapses by which cells communicate with one another, but there is no clear-cut evidence of such alterations.

The concept of two kinds of memory is supported by experience with patients with memory deficits; it is possible to lose one kind of memory and not the other. A blow to the head can result, for instance, in the soap-opera kind of amnesia for prior events while not interfering with short-term memory or the establishment of new long-term memories. (Typically, "lost" memories return, indicating that what has been lost is not the memory itself but rather the capacity to retrieve it.) Conversely, injury to the hippocampus, which is part of the limbic system, does not affect already established long-term memories but does interfere with the transfer of short-term memories to long-term memory. A patient with bilateral destruction of the hippocampus can remember where he lived as a child but not where he lives now, for example. He can carry on an apparently normal conversation, but if the person he was talking to leaves the room and returns a minute or two later, he will not recognize him.

The problem of memory and learning appears at present as bewildering (and as fascinating) as did the problem of human heredity a century ago. Some scientists—Kandel is one of them—believe that a simple model is needed, the equivalent of *Drosophila* and T4. Others contend that the enormous complexities of the vertebrate brain will never be understood in terms of simple invertebrate models and single cells but rather that the secrets lie in the vast communication network itself. Stay tuned.

SUMMARY

The human brain is the most complex structure known. Like the primitive vertebrate brain, it is made up of the hindbrain, midbrain, and forebrain. The hindbrain consists of the medulla, pons, and cerebellum, the brain structure associated with coordination of fine-tuned movements. The medulla, pons, and midbrain make up the brainstem, which controls vital functions such as heartbeat and respiration and serves as the passageway for ascending and descending fiber tracts. The reticular formation within the brainstem is concerned with attention, arousal, and consciousness.

The posterior part of the forebrain, the diencephalon, comprises the thalamus, the hypothalamus, and parts of the limbic system. It contains nuclei associated with basic drives and emotions such as hunger, thirst, anger, fear, and sex and is the center for the integration of the nervous and endocrine systems.

The anterior part of the forebrain, the telencephalon, consists, in humans, of the two cerebral hemispheres connected by the corpus callosum and covered by the much convoluted cerebral cortex. In both experimental animals and humans, it has been possible to correlate particular areas of the cerebral cortex with particular functions. These areas include the motor cortex, sensory cortex, and parts of the cortex concerned with vision, hearing, and speech. Most of the human cortex is "silent"; that is, it has no direct sensory or motor function. This so-called association cortex, about half of which is in the frontal lobes, appears to be involved in long-term planning and the organization of ideas. It is the part of the brain that has developed most rapidly in human evolution.

In the motor and sensory cortices, the two cerebral hemispheres are mirror images of one another, with the right hemisphere controlling the left side of the body, and vice versa. However, the speech centers are found only in one hemisphere, nearly always in the left one, and other faculties, such as spatial orientation and musical ability, appear to be associated with the right hemisphere. Usually the functions of the two hemispheres are integrated, but studies of patients whose corpus callosum has been severed indicate that the two hemispheres can function independently and confirm that they differ in their capacities.

Neurotransmitters in the brain include acetylcholine and noradrenaline, which are also found in the peripheral nervous system, as well as serotonin, dopamine, gamma-aminobutyric acid, and many others. Brain neurons also, it has been discovered, contain receptors for opiates, and this discovery has led to the detection of a group of chemical substances known as the endorphins, which resemble opiates in their structure and activities.

Wavelike variations in electric potential can be recorded from the brain surface by electroencephalography. Three principal types of brain waves can be detected in normal subjects: beta waves, of relatively high frequency, associated with alertness and attention; alpha waves, of lower frequency, associated with relaxation; and delta waves, of still lower frequency, which normally appear only during sleep.

The slow, rhythmic waves characteristic of deep sleep are interrupted by periods of rapid, high-frequency waves. These periods, which are known as REM sleep or paradoxical sleep, are typically accompanied by rapid eye movements and associated with dreaming.

Electrical activity in the brain is also studied through the detection of neuromagnetism and by the use of microelectrodes capable of detecting changes in the electric potential of individual cells. Such studies are being performed with both invertebrate nervous systems and more complex vertebrate brains.

Sex hormones can act directly on brain nuclei to influence behavior. In birds, song centers enlarge and song repertoires develop under the influence of testosterone. In rats, sex hormones affect the brain in two stages: (1) an early period, during which differentiation of brain cells occurs, and (2) a sexually mature period, in which the differentiated brain cells respond to the hormones by evoking particular patterns of behavior.

Two types of memory have been identified: short-term memory and long-term memory. Studies on the mechanisms of memory and learning involve the search for changes in the physical or chemical structure of brain cells or the connections among them.

QUESTIONS

1. Distinguish among the following: reticular system/limbic system; cerebellum/cerebrum; Broca's area/Wernicke's area; alpha waves/beta waves/delta waves; motor cortex/sensory cortex.

2. Sketch the human cerebral cortex and indicate on it the areas that have been mapped.

3. What functions might be affected by damage (from stroke, accident, or disease) to the cerebellum? To the reticular system? To the dorsal portion of the cerebral cortex anterior to the central sulcus?

4. Relate the relative sizes of various portions of the brains of the animals in Figure 43-3 with the habitats in which they live and the ways in which they live, obtain food, and escape predators.

5. Diagram the major structures of the vertebrate brain.

CHAPTER 44

The Continuity of Life I: Reproduction

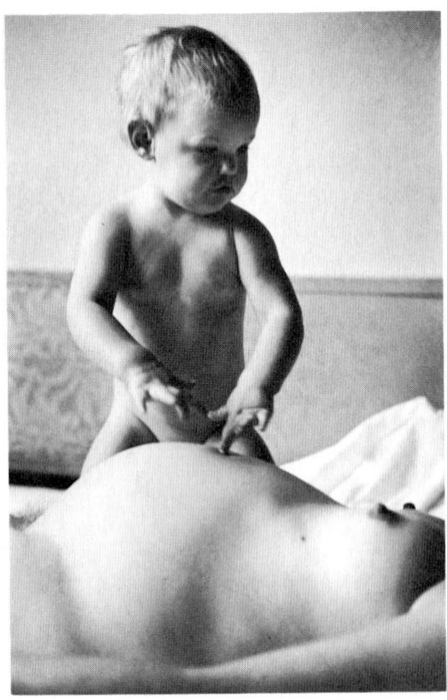

44–1
Mother and child.

Most vertebrates—and all mammals—reproduce sexually. As you will recall,* sexual reproduction involves two events: meiosis and fertilization. In vertebrates, which are almost always diploid, meiosis produces gametes, the only haploid forms in the life cycle. The gametes are specialized for motility (sperm) or for production and storage of nutrients (eggs) and are produced by separate individuals. In many lower organisms (insects, in particular), the generations are nonoverlapping, as is the case in annual plants; but in vertebrates, parents not only survive after their young are produced but often are essential to the rearing of the young. This evolutionary trend toward increasing parental care becomes pronounced among birds and reaches its fullest expression among certain of the mammals, ourselves included.

In most species of fish and in amphibians, as in many invertebrates, fertilization is external. Among organisms that produce amniote eggs (reptiles, birds, and monotreme mammals), fertilization is internal. The outer protective shell, produced as the egg moves down through the female reproductive tract, is laid down after the egg cell is fertilized, enclosing the embryo and its membranes. Fertilization is also internal among marsupial and placental mammals, in which the embryo develops within the mother and is nourished by her.

In the following pages, we are going to describe sexual reproduction in humans. We shall first trace the development of the male gametes, then the development of the female gametes, and, finally, we shall describe the special structures and activities that provide for fertilization and the subsequent implantation of the developing embryo.

THE MALE REPRODUCTIVE SYSTEM

The primary sexual organs are the gonads, the organs in which the gametes are formed. The male gonads are the testes (singular, testis). The testes develop in the abdominal cavity of the male embryo and, in the human male, descend into an external sac, the scrotum. This descent usually occurs before birth. The function of the scrotum, apparently, is to keep the testes in an environment cooler than the abdominal cavity. A temperature 3°C lower than that of the body is necessary for

* Or turn back to page 250 for a reminder.

the sperm to develop. Sperm are not produced in an undescended testis, and even temporary immersion of the testes in warm water—as in a hot bath—has been known to produce temporary sterility. (This effect is not reliable enough, however, to recommend its use as a birth control measure.) When the temperature outside the scrotal sac is warm, the sac is thin and hangs loose in multiple folds. When the outside temperature is cold, the muscles under the skin of the scrotum contract, drawing the testes close to the body. In this way, a fairly constant testicular temperature is maintained.

Spermatogenesis

Each testis is subdivided into about 250 compartments (lobules), and each of these is packed with tightly coiled seminiferous ("seed-bearing") tubules. These are the sperm-producing regions of the testes. Between the tubules are the interstitial cells, the sources of testosterone (page 778). Each seminiferous tubule is about 80 centimeters long, and the two testes together contain a total of about 500 meters of tubules. The sperm are produced continuously within the tubules (Figure 44–3).

The tubules contain two types of cells: spermatogenic (sperm-producing) cells and Sertoli cells. The spermatogenic cells pass through several stages of differentiation. Once production of sperm begins at puberty in the human male, it goes on continuously and so, in a single seminiferous tubule, it is possible to find cells in all the different stages of spermatogenesis. It is unusual, however, to find all of the stages in a single cross section, because spermatogenesis characteristically occurs in waves that travel down the tubules.

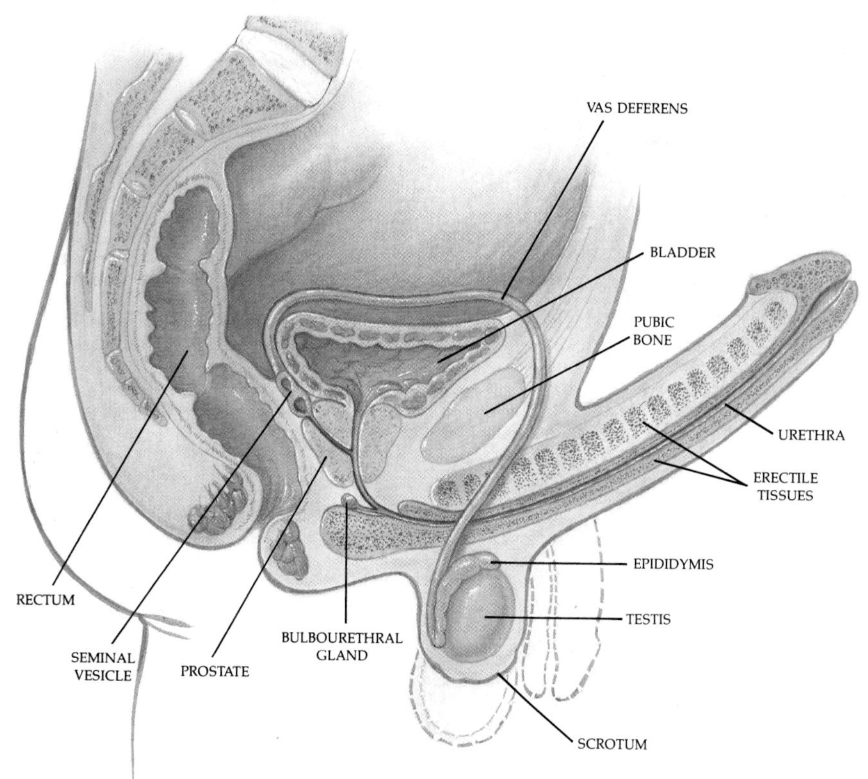

44–2
Diagram of the human male reproductive tract, showing the penis and scrotum before (dashed lines) and during erection. Sperm cells formed in the seminiferous tubules of the testis enter the epididymis. From there they move to the vas deferens, where most of them are stored. The vas deferens merges with a duct from the seminal vesicle and then, within the prostate gland, joins the urethra. The sperm cells are mixed with fluids, mostly from the seminal vesicles and prostate gland. The resulting mixture, the semen, is released through the urethra of the penis. The urethra is also the passageway for urine, which is stored in the bladder.

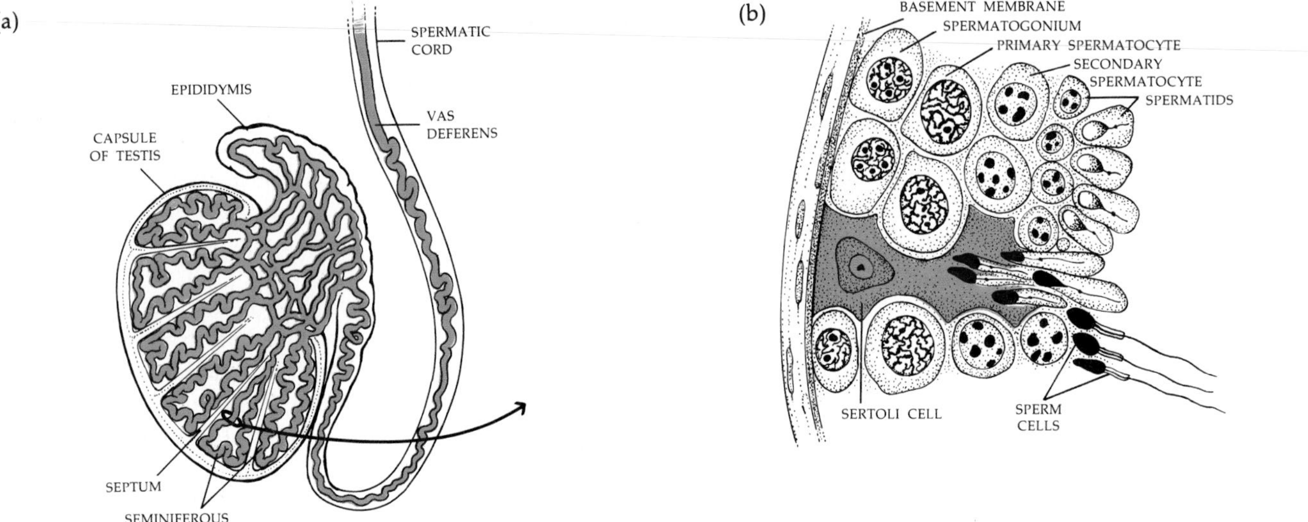

44–3
The testis (a) *is made of tightly packed coils of seminiferous tubules, containing sperm cells in various stages of development. The entire developmental sequence takes eight to nine weeks. As shown in the idealized cross section* (b), *spermatogonia develop into cells known as primary spermatocytes. In the first meiotic division, these divide into two equal-sized cells, the secondary spermatocytes. In the second meiotic division, four equal-sized spermatids are formed. These differentiate into sperm cells. (It is unusual to find all these stages in a single cross section.) The Sertoli cells support and nourish the developing sperm. The sperm cells leave the testis through the epididymis and the vas deferens.*

Cells in the first stage of spermatogenesis, the spermatogonia, line the basement membrane of the tubules. Spermatogonia are diploid and have, in the human, 44 autosomes and 2 sex chromosomes, an *X* and a *Y*. Spermatogonia divide continuously. Some of the cells produced by these mitotic divisions remain undifferentiated, whereas others, in the course of their successive mitotic divisions, move away from the basement membrane and begin to differentiate, giving rise to primary spermatocytes. Primary spermatocytes undergo the first meiotic division* to produce two secondary spermatocytes, each of which contains 22 autosomes and either an *X* chromosome or a *Y* chromosome; each of the 23 chromosomes consists of two chromatids. The secondary spermatocytes undergo the second meiotic division to produce spermatids, each of which contains the haploid number of single chromosomes. Spermatids develop without further division into sperm cells, or spermatozoa. It takes eight to nine weeks for a spermatogonium to differentiate into four sperm cells. During this time the developing cells receive nutrients from adjacent Sertoli cells.

Differentiation of Spermatids

A spermatid is a small spherical or polygonal ("many-sided") cell that develops into a sperm cell. The sequence of changes by which the quite unremarkable spermatid becomes the highly specialized, very extraordinary sperm cell is an excellent example of cell differentiation. Differentiation is a process that, as we shall see in the next chapter, is an essential component of embryonic development.

The first visible sign of differentiation of a spermatid is the appearance of vesicles containing small, dark granules within the Golgi body, which, as in most cells, lies close to the nuclear envelope. These vesicles enlarge and coalesce into a single vesicle, the acrosome. The acrosome contains enzymes that will help the sperm to penetrate the protective layer surrounding the egg. The position of the acrosomal vesicle determines the polarity of the sperm; that is, it establishes where the anterior end, or "head," is going to be.

---

* For a review of meiosis, see pages 253 to 255.

During the early stages of acrosome formation, the cell's pair of centrioles moves to the outer cell membrane at the end of the cell opposite the acrosome. From one of these centrioles a fine thread grows out, the beginning of the sperm flagellum. The centrioles then move back to the nucleus, carrying the cell membrane inward with them. One centriole eventually lodges in a notch formed at the posterior end of the nucleus, while the other, the one associated with the flagellum, lies at right angles to the first (Figure 44–4).

As the flagellum, or tail, grows, it becomes apparent that its axial filament has the characteristic $9 + 2$ structure found in cilia and flagella generally. Mitochondria aggregate about its basal end, forming a continuous spiral, providing a ready energy source (ATP) for the flagellar movement. The rest of the axial filament, almost to its tip, is surrounded by nine additional protein fibers tightly coiled in a helix that forms a fibrous sheath. These fibers, which are somewhat thicker than the microtubules within the flagellum, presumably play some role in sperm motility.

During this period, the nucleus condenses, apparently by eliminating water. Once the tail has formed, the cell rapidly elongates. Longitudinal bundles of microtubules can be seen in the cell at this time, and they may play a role in changing the shape of the cell. As the cell lengthens, the bulk of the cytoplasm, together with the Golgi body, is sloughed away.

In its final form, the sperm cell consists of the acrosome, the tightly condensed nucleus, the mitochondria, a pair of centrioles, one of which is now serving as the basal body for the flagellum, and the long, powerful flagellum itself, all bounded by the cell membrane (Figure 44–5). In the fully differentiated sperm, all other functions have been subordinated to the task of providing motility for delivery of the "payload," the DNA and its associated protein, condensed and coiled in the sperm head. A young adult human male may produce several hundred million sperm per day; a ram may produce several billion.

44–4
Mammalian spermatids in the process of differentiation. (a) *The nucleus almost fills the right half of the electron micrograph. The two centrioles have moved to a position just behind the nucleus (see arrow), and the flagellum has begun to form from one of them. To the left of the nucleus are several mitochondria.*
(b) *The nucleus occupies the center of this micrograph. At the upper right, the flagellum is forming and, at the left, you can see the beginning of the acrosome.*

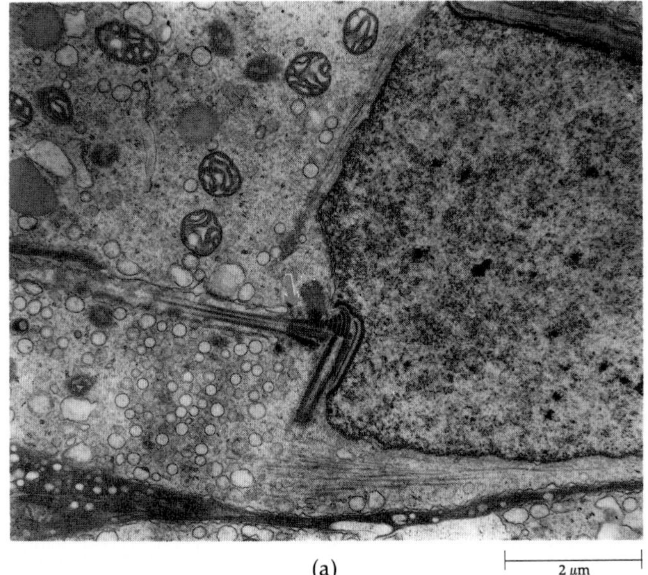

(a)

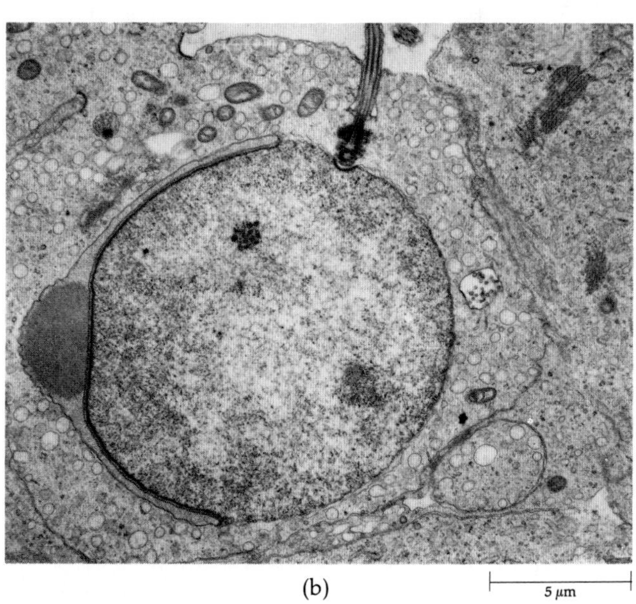

(b)

44-5
Diagram of a human sperm cell. The mature cell consists primarily of the nucleus, carrying the "payload" of tightly condensed DNA and associated protein, the very powerful tail (flagellum), and mitochondria, which provide the power for sperm movement. The acrosomal vesicle is a specialized lysosome (see page 110) containing enzymes that help the sperm penetrate the protective layer of an unfertilized egg cell.

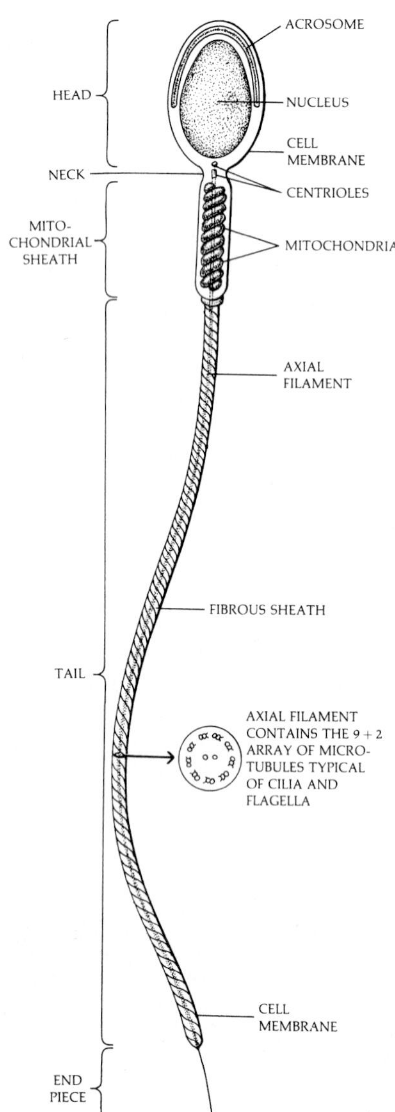

Pathway of the Sperm

From the testis, the sperm are carried to the epididymis, which consists of a coiled tube 7 meters long, overlying the testis. It is surrounded by a thin, circular layer of smooth muscle fibers. The sperm are nonmotile when they enter the epididymis and gain motility only after some 18 hours there. (Full motility occurs only after they have entered the female reproductive tract.) From the epididymis, the sperm pass to the *vas deferens*, where most of them are stored. The vas deferens (plural, vasa deferentia), an extension of the tightly coiled tubules of the epididymis, leads from the testis into the abdominal cavity. The vas deferens and its accompanying nerves, arteries, and connective-tissue wrapping constitute the spermatic cord. The spermatic cords, one from each testis, are found along the same path the testes took during their descent in the embryo.

Within the posterior wall of the abdominal cavity, the vasa deferentia lead around the bladder where they merge with the ducts of the seminal vesicles and then enter the urethra. Each vas deferens is covered with a heavy, three-layered coat of smooth muscle whose contractions propel the sperm along it. Severing of the vasa deferentia (vasectomy) offers a relatively safe and almost painless means of birth control (Figure 44-6). The nerves and blood vessels of the spermatic cords are left intact.

The urethra, which runs through the prostate gland and terminates at the tip of the penis, serves both for the excretion of urine and the ejaculation of sperm.

The Penis and Orgasm in the Male

The function of the penis is to deposit sperm cells within the reproductive tract of the female. The penis, in various forms, has evolved independently in a number of species of insects and in other invertebrates. It is found among some reptiles and birds; all flightless birds have penes and so do all ducks, flightless or not. In most reptiles and birds, however, one opening, the cloaca, serves as the passage for eggs or sperm and also for the elimination of wastes. These animals mate by juxtaposition of their cloacae. Only among mammals is the penis found in all species.

The human penis is formed of three cylindrical masses of spongy erectile tissue, each of which contains a large number of small spaces, each about the size of a

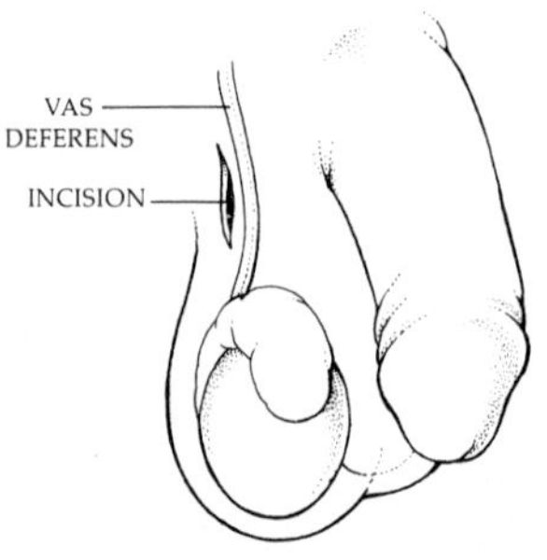

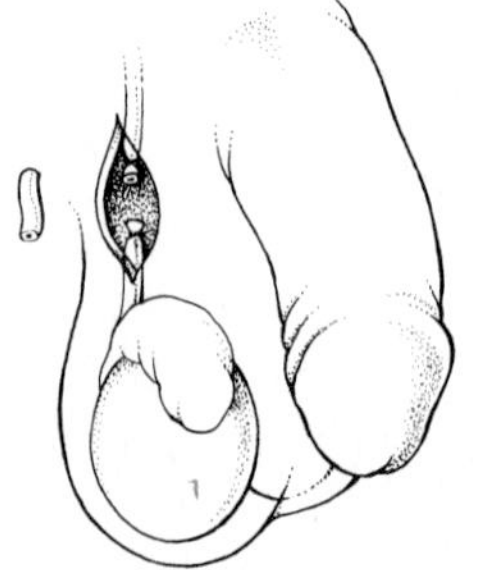

44-6
During a vasectomy, the vas deferens on each side is severed, and the cut ends are folded back and tied off, preventing the release of sperm from the testis. The sperm cells are reabsorbed by the body, and the semen is normal except for the absence of sperm. The procedure seems to have no effect at all on testosterone levels, sexual potency, or performance.

44–7
A cross section of a human penis. The penis is formed of three cylindrical masses of spongy erectile tissue that contain a large number of small spaces, each about the size of a pinhead. Erection of the penis is caused by dilation of the blood vessels carrying blood to the spongy tissues.

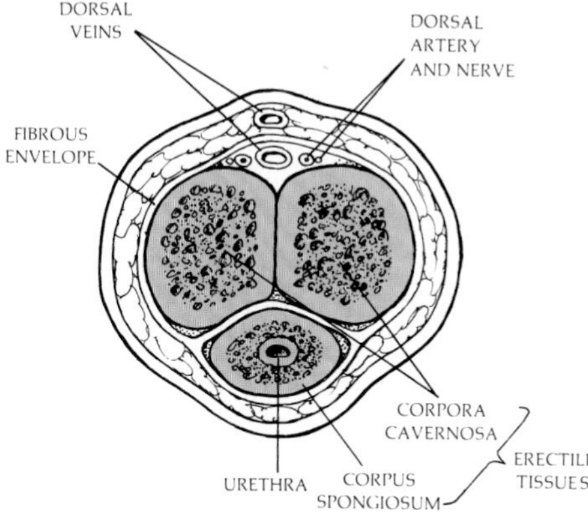

pinhead. Two of these masses are in the upper portion of the penis, and the third lies beneath them, surrounding the urethra (Figure 44–7). This third mass is enlarged at the distal (far) end to form the glans penis, which is a smooth protective cap over the spongy tissues. At the basal end, it is enlarged to form the bulb of the penis, which is embedded below the pelvic cavity and is surrounded by muscles that participate in orgasm. The exterior portion of the penis is covered by a loose, thin layer of skin, which at the end forms an encircling fold over the glans. This fold, the foreskin, is sometimes surgically removed (circumcision). The urethra terminates in a slitlike opening in the glans.

Erection of the penis, which can be elicited by a variety of stimuli, is caused by dilation of the blood vessels carrying blood to the spongy tissues, resulting in the collection of blood within the spaces. As the tissues become distended, they compress the veins and so inhibit the flow of blood out of the tissues. With continued stimulation, the penis and the underlying bulb become hard and enlarged.

Erection is accompanied by discharge of a small amount of fluid from the bulbourethral glands (pea-shaped organs at the base of the penis). This serves as a lubricant to facilitate the movement of spermatozoa along the male urethra and to aid penetration of the penis into the female. Continued stimulation, such as may be produced by repeated thrusting of the penis in the vagina, characteristically leads to contraction of the muscles in the scrotum (raising the testes close to the body) and of the muscles encircling the epididymis and vas deferens. These contractions move the spermatozoa toward and into the urethra. As this occurs, the seminal vesicles secrete a fructose-rich fluid that nourishes the sperm cells. This fluid contains a high concentration of prostaglandins, which cause contractions in the musculature of the uterus and oviducts and so may assist the sperm in reaching the egg. The prostate gland adds a thin, milky, alkaline fluid that helps neutralize the normally acidic pH of the female reproductive tract. Finally, the muscles surrounding the bulb are stimulated. These muscle contractions propel the sperm out through the urethra (ejaculation) and produce some of the sensations associated with orgasm.

The sperm, along with the secretions from the seminal vesicles, the prostate gland, and the bulbourethral glands, constitute semen. The volume of semen measures from 3 to 6 milliliters per ejaculation. Even though sperm constitute less than 10 percent of the semen, about 300 to 400 million sperm cells are present in each ejaculate of a normal adult male. Of these, only one can fertilize each egg cell. The other sperm cells apparently play an accessory role, however, perhaps by bringing about chemical changes necessary for fertilization. Males that produce fewer than 20 million sperm per milliliter of fluid are generally sterile.

44–8
Human sperm. About 300 to 400 million sperm cells are present in the ejaculate of an average, healthy adult male.

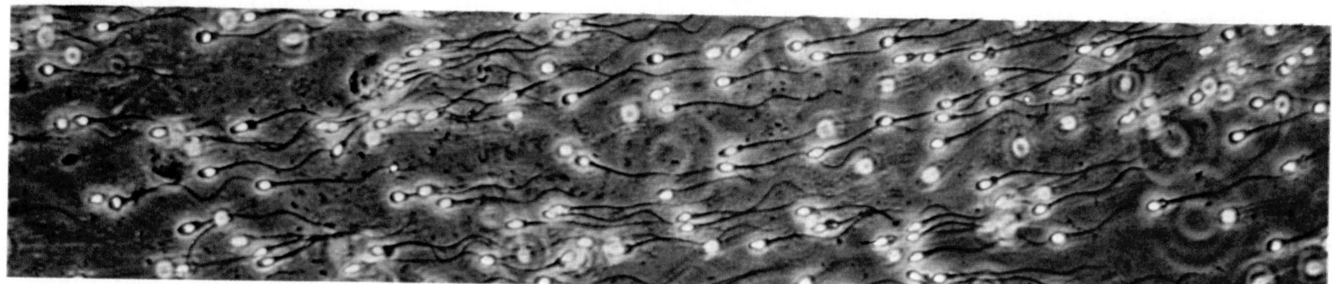

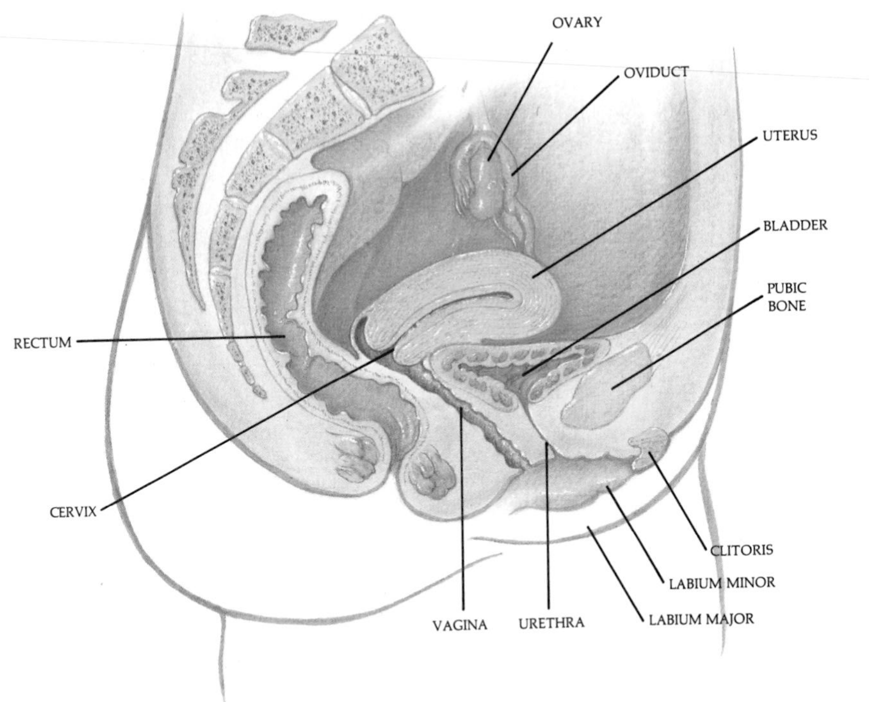

44–9
The female reproductive organs. Notice that the uterus lies at right angles to the vagina. This is one of the consequences of the bipedalism and upright posture of Homo sapiens *and one of the reasons that childbirth is more difficult for the human female than for other mammals.*

THE FEMALE REPRODUCTIVE SYSTEM

The uterus is a hollow, muscular, pear-shaped organ slightly smaller in size than a clenched fist (about 7.5 centimeters long and 5 centimeters wide) in the nonpregnant female. It lies almost horizontally in the abdominal cavity and is on top of the bladder (Figure 44–9). The uterus is lined by the endometrium, which has two principal layers, one of which is shed at menstruation and another from which the shed layer is regenerated. The smooth muscles in the walls of the uterus move in continuous waves. This motion possibly increases the motility of both the sperm on its journey to the oviduct (sometimes called the uterine tube) and the oocyte, from which the egg develops, as it passes from the oviduct to the uterus. These contractions increase when the endometrium is shed during a menstrual period, and they are greatest when a woman is in labor. The muscular sphincter guarding the opening of the uterus is the cervix. The sperm pass through this opening on their way toward the oocyte. The cervix dilates to allow the fetus to emerge at the time of birth.

The vagina is a muscular tube about 7.5 centimeters long that leads from the cervix of the uterus to the outside of the body. It is the receptive organ for the penis and also the birth canal. Its exterior opening is between the urethra, the tube leading from the bladder, and the anus. The lining of the vagina is rich in glycogen, which bacteria normally present in the vagina convert to lactic acid. As a consequence, the vaginal tract is mildly acidic, with a pH between 4 and 5.

The external genital organs of the female are collectively known as the vulva. The clitoris, which corresponds to the penis in the male,* is about 2 centimeters long and, like the penis, is composed chiefly of erectile tissue. The clitoris has two

* In the early embryo, the structures are identical.

bulbs (analogous to the penile bulb of the male) that lie on either side of the opening of the vagina. The labia (singular, labium) are folds of skin. The labia majora are fleshy and, in the adult, covered with pubic hair. They enclose and protect the underlying, more delicate structures, the labia minora. (Embryologically, they are homologous with the scrotum in the male.) The labia minora are thin and membranous.

Oogenesis

The gamete-producing organs in the female are the ovaries, each a solid mass of cells about 3 centimeters long. They are suspended in the abdominal cavity by ligaments (bands of connective tissue) and mesenteries. The oocytes, from which the eggs develop, are in the outer layer of the ovary.

In human females, the primary oocytes begin to form about the third month of fetal development. By the time of birth, the two ovaries contain some 2 million primary oocytes, which have reached prophase of the first meiotic division. These primary oocytes remain in prophase until the female matures sexually. Then, under the influence of hormones, the first meiotic division resumes and is completed at about the time of the release of the oocyte from the ovary (ovulation). Of these 2 million primary oocytes, about 300 to 400 reach maturity, usually one at a time, about every 28 days after puberty, to become secondary oocytes, which develop into mature egg cells (ova). Given this timetable, a moment's reflection will reveal that as many as 50 years may elapse between the beginning and the end of the first meiotic division in a particular oocyte.

Maturation of the primary oocyte involves both meiosis and a great increase in size. This size increase reflects the accumulation of stored food reserves and metabolic machinery, such as messenger RNA and enzymes, required for the early stages of development. At meiosis an oocyte does not divide to form four ova.

44–10
(a) *Stages in the development of an ovum. Oocytes develop near the surface of the ovary within follicles. Some of these stages take place before birth in the human female. After an oocyte is discharged from a follicle (ovulation), the remaining cells of the ruptured follicle give rise to the corpus luteum, which secretes estrogens and progesterone. If the ovum is not fertilized, the corpus luteum is reabsorbed in two to three weeks. If the ovum is fertilized, the corpus luteum persists, sustaining the production of estrogens and progesterone, which maintain the uterus during pregnancy.* (b) *Ovulation. The oocyte is being discharged from the ovarian follicle.*

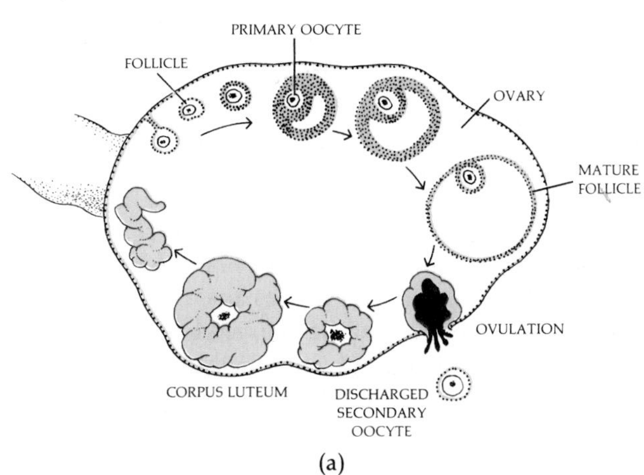

(a)

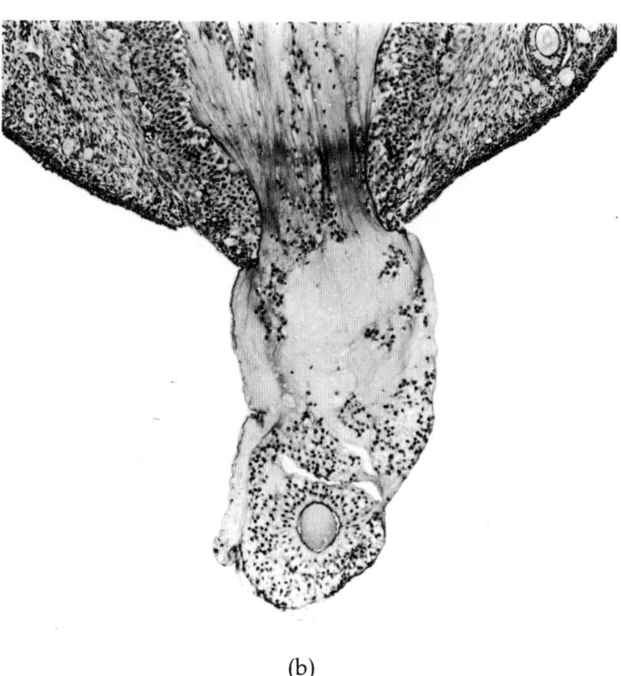

(b)

Instead a single ovum and one to three polar bodies are formed (see Figure 12–12, page 257). When the oocyte is ready to complete meiosis, the nuclear envelope fragments, and the chromosomes move to the surface of the cell. As the nucleus divides, the cytoplasm of the oocyte bulges out. One set of chromosomes moves into the bulge, which then pinches off into a small cell, the first polar body. The rest of the cellular material forms the large secondary oocyte. The first meiotic division is completed a few hours before ovulation. The second meiotic division does not take place until after fertilization. This division produces the ovum and another small polar body.

As a consequence of these unequal cell divisions, most of the accumulated food reserves of the oocyte are passed on to a single ovum. The first polar body may also divide, although there is no functional reason for it to do so. All the polar bodies eventually die.

Oocytes develop near the surface of the ovary. An oocyte and the specialized cells surrounding it are known as an ovarian follicle. The cells of the follicle supply nutrients to the growing oocyte and also secrete estrogens, the hormones that initiate the buildup of the endometrium. During the final stages of its growth, the follicle moves to the surface and produces a thin, blisterlike elevation that eventually bursts, releasing the oocyte.

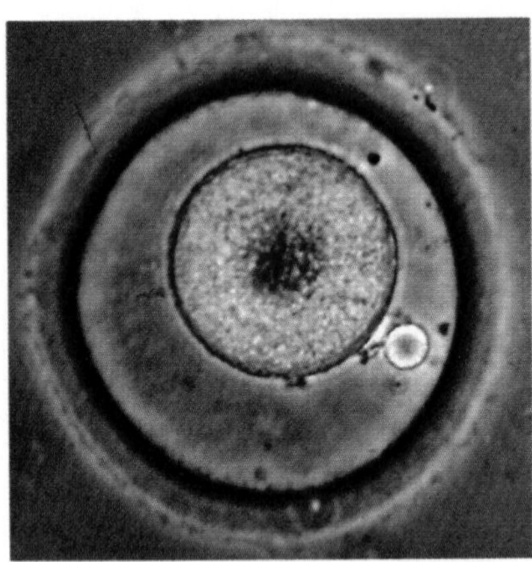

44–11
An unfertilized human egg. The dark strands within the nucleus are chromosomes. A polar body is at the lower right. (Roberts Rugh and Landrum B. Shettles, M.D., From Conception to Birth: The Drama of Life's Beginnings, *Harper & Row, Publishers, Inc., New York, 1971.)*

A human oocyte is about 100 micrometers in diameter, which is very large for a cell. It contains an unusually large supply of ribosomes, enzymes, amino acids, and all the other cellular machinery that will be used in the early, rapid stages of biosynthesis characteristic of embryonic cells.

The Menstrual Cycle

Although the menstrual cycle does not require an environmental cue, as do reproductive cycles in many other vertebrates, it is clearly under the influence of external factors to some extent. For example, some women find that emotional upset delays a menstrual period or eliminates it completely.

The cycle is the result of a complex feedback system involving the sex hormones estrogen and progesterone, the pituitary gonadotropins LH and FSH, and gonadotropin-releasing hormone (GnRH) from the hypothalamus. Estrogen in low concentrations inhibits the production of FSH and GnRH (and so of LH). In high concentrations, estrogen stimulates the production of LH and GnRH (and so of FSH). Estrogen and progesterone together inhibit the production of GnRH.

At the beginning of the cycle, during menstruation, hormone levels are low (Figure 44–12). After the menstrual flow ceases and under the influence of the gonadotropic hormones FSH and LH, an egg cell and its follicle begin to mature. The follicle, as it enlarges, secretes increased amounts of estrogens. (Usually a number of follicles begin to enlarge simultaneously, but only one becomes mature enough to release its egg cell, and the others regress.) The estrogens stimulate the regrowth of the endometrium in preparation for implantation of a fertilized egg cell. The rapid rise in estrogen levels near the midpoint of the cycle triggers a sharp increase in production of LH by the pituitary gland. The spurt of high LH stimulates the follicle to release the egg cell, which begins its passage to the uterus. Under the continued stimulus of LH, the cells of the emptied follicle grow larger and fill the cavity, producing the corpus luteum ("yellow body"). The cells of the corpus luteum, as they increase in size, begin to synthesize significant amounts of

44-12
Diagram of events taking place during the menstrual cycle. The cycle begins with the first day of menstrual flow, which is caused by the shedding of the endometrium, the lining of the uterine wall. The increase of FSH and LH at the beginning of the cycle promotes the growth of the ovarian follicle and its secretion of estrogen. Under the influence of estrogen, the endometrium regrows. The sudden rise in estrogen just before midcycle triggers a sharp increase of LH from the pituitary, which stimulates the release of the egg cell (ovulation). (It is not known what role, if any, is played by the simultaneous increase in FSH.) Following ovulation, LH and FSH levels drop. The follicle is converted to the corpus luteum, which secretes estrogen and also progesterone. Progesterone further stimulates the endometrium, preparing it for implantation. If pregnancy does not occur, the corpus luteum degenerates, the production of progesterone and estrogen falls, the endometrium begins to slough off, FSH and LH concentrations increase once more, and the cycle begins anew.

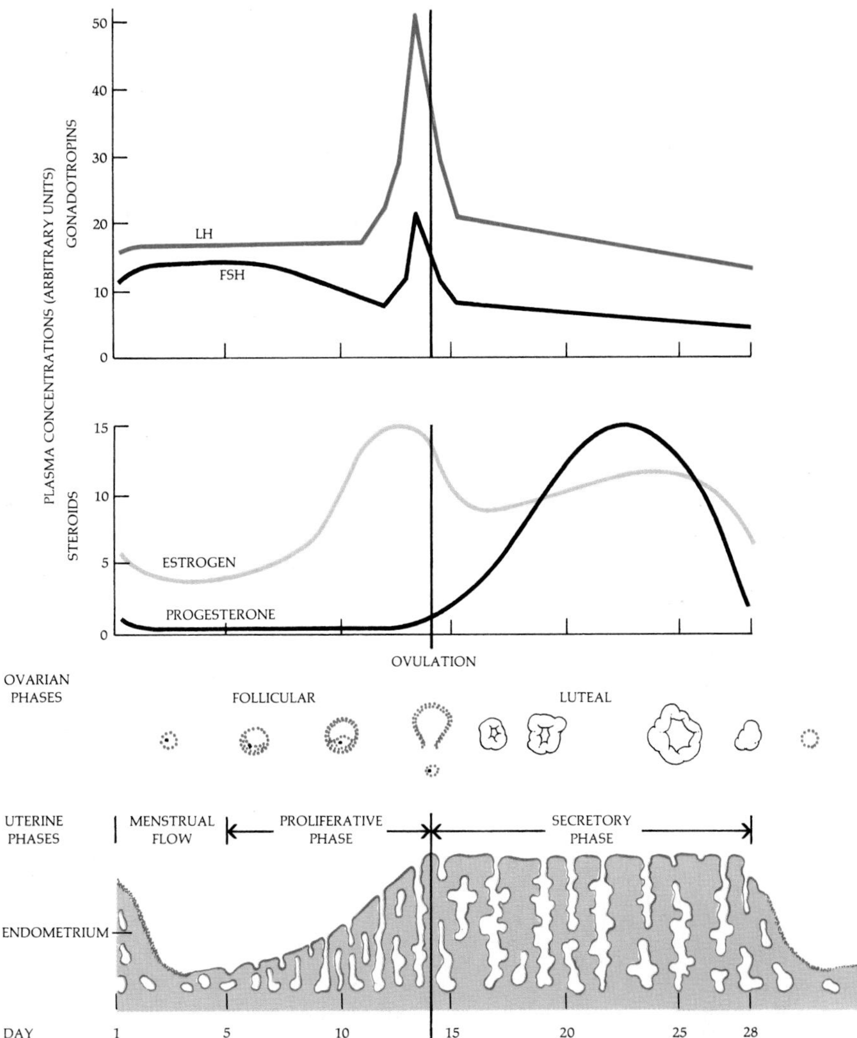

progesterone as well as estrogens. As the progesterone levels increase, estrogen and progesterone together inhibit the production of GnRH and so of the gonadotropic hormones (LH and FSH) from the pituitary. As a result of this drop in level of the gonadotropic hormones, production of ovarian hormones drops. Without hormonal support, the endometrium can no longer sustain itself, and a portion of it is sloughed off in the menstrual fluid. Then, in response to the now low level of ovarian hormones, the level of pituitary gonadotropic hormones begins to rise again, followed by development of a new follicle and a rise in estrogens as the next monthly cycle begins.

The cycle usually lasts about 28 days, but individual variation is common. Moreover, even in women with cycles of average length, ovulation does not always occur at the same time in the cycle (which is the reason the "rhythm method" is an unreliable means of birth control).

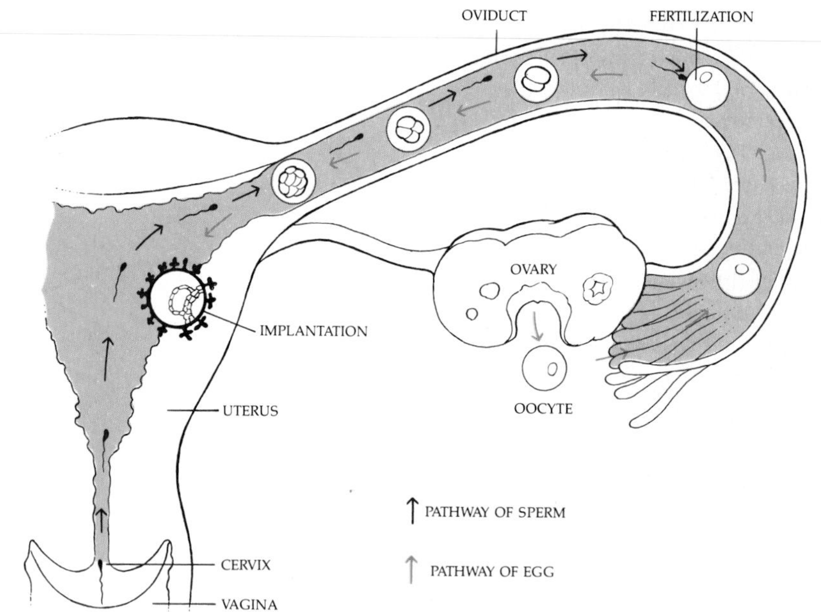

44–13

Fertilization of the egg by sperm. About once a month in the nonpregnant female of reproductive age, an oocyte is ejected from the ovary and is swept into one of the oviducts. Fertilization, when it occurs, normally takes place within an oviduct, after which the fertilized egg passes down the oviduct and becomes implanted in the lining of the uterus. Muscular movements of the oviduct, plus the beating of the cilia that line it, propel the egg cell down the oviduct toward the uterus. If the egg cell is not fertilized, it dies, usually within 12 to 24 hours. A sperm cell has an average life of 48 hours within the female reproductive tract.

Pathway of the Egg

When the oocyte is released from the follicle at ovulation, it is swept into the adjacent oviduct by the movement of the funnel-shaped opening of the oviduct over the surface of the ovary and by the beating of cilia that line the fingerlike projections surrounding this opening.

The smooth muscles in the walls of the oviducts contract in continuous peristaltic waves that move the oocyte from the ovary to the uterus in about three days. An unfertilized egg, however, lives only about 24 hours after it is ejected from the follicle. So, fertilization, if it is to occur, must occur in an oviduct. If the egg cell is fertilized, it becomes implanted in the endometrium three to four days after the young embryo reaches the uterus, six or seven days after the egg cell was fertilized. If the egg cell is not fertilized, it dies, and the endometrial lining of the uterus is shed at menstruation. Fertilized eggs implanted in the endometrium are sometimes lost in abnormal menstrual flow, but it is difficult to estimate the number of such very short-lived pregnancies.

Statistics show that 25 percent of women who conceive become pregnant within one month of trying to do so, 63 percent within six months, 75 percent within nine months, and 90 percent within 18 months.

Orgasm in the Female

Under the influence of a variety of stimuli, the clitoris and its bulbs become engorged and distended with blood, as does the penis of the male. The distension of the tissues is accompanied by the secretion into the vagina of a fluid that both lubricates the walls of the vagina and neutralizes its acidic, and therefore spermicidal, environment.

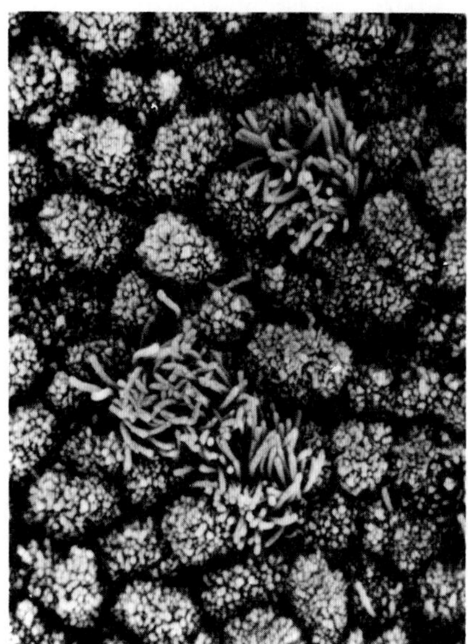

44–14
Scanning electron micrograph of the inner lining of a human oviduct. The cells of the lining of the oviduct have numerous microvilli. Note the cilia on several of the cells shown here. The beating of these cilia help propel the egg cell along the oviduct.

At orgasm, the cervix drops down into the upper portion of the vagina, where the semen tends to form a pool. The female orgasm also may produce contractions in the oviducts that propel the sperm upward. It has been calculated that it would take a sperm cell at least two hours to make its way up the oviducts under its own power, but sperm have been found in the oviducts as soon as five minutes after intercourse. Orgasm in the female, however, is not necessary for conception.

CONTRACEPTIVE TECHNIQUES

A variety of contraceptive techniques is now available for couples who wish to prevent or defer pregnancy. In Table 44–1 on the next page, they are rated in order of effectiveness, in terms of the average number of pregnancies per year among the women of child-bearing age using the technique. In most cases, two figures are given for effectiveness. The first, lower, figure is an "ideal" figure, obtainable when the method is used consistently and correctly. The second figure is an average figure, reflecting actual experience.

"The pill" consists of a combination of estrogen and progesterone. When taken daily, it keeps the level of these ovarian hormones in the blood high enough to shut off production of the pituitary hormones FSH and LH (see pages 782–784). Without FSH the ovarian follicles do not ripen, and in the absence of LH no ovulation occurs, so pregnancy is not possible.

ESTRUS

Females of almost all species except *Homo sapiens* will mate only during estrus, their fertile period. Estrus may occur only once a year (as in wolves and deer), about once a month (as in cows and horses), or every few days (as in rats and mice).

The periods of estrus may last from only a few hours to three or four weeks. In animals, such as dogs, that produce eggs continuously during estrus, the eggs may be fertilized at different times and by different males, which explains in part why a mixed-breed litter can contain such an astonishing variety of siblings. Also, there may be a great size variation even among purebred pups, because many of them are actually different ages at birth. In some mammals, such as cats, rabbits, and minks, although the egg is mature and the female receptive during estrus, ovulation occurs only under the stimulation of copulation, obviously a very efficient system, ensuring maximum economy in the utilization of gametes. There is suggestive evidence that in some women, also, ovulation may be triggered by sexual intercourse.

The human female appears to be one of the few female animals receptive to mating during infertile periods. Some anthropologists speculate that this receptivity coevolved with the establishment of strong pair-bond relationships between human or prehuman males and females. A consequence of this pair-bond relationship is a society based on a family unit, in contrast to many other primate groups, in which the social and breeding unit is a troop or band. The establishment of such family units is seen, in turn, as the basis for the traditional division of labor between the sexes, with the female concentrating on childbearing and the home and the male on hunting, protecting the family unit, and defending territory. If the anthropologists are right, this behavioral adaptation on the part of the human female has profoundly influenced the shape of human civilization.

Table 44-1 Methods of Birth Control Currently Available

METHOD	MODE OF ACTION	EFFECTIVENESS (PREGNANCIES PER 100 WOMEN PER YEAR)	ACTION NEEDED AT TIME OF INTERCOURSE	REQUIRES INSTRUCTION IN USE	POSSIBLE UNDESIRABLE EFFECTS
Vasectomy	Prevents release of sperm	0	None	No	Usually produces irreversible sterility
Tubal ligation	Prevents passage of egg cell to uterus	0	None	No	Usually produces irreversible sterility
"The pill" (estrogen and progesterone)	Prevents follicle maturation and ovulation	0–10	None	Yes, timing	Early—some water retention, breast tenderness, nausea; late—increased risk of cardiovascular disease
Intrauterine device (coil, loop, IUD)	Possibly prevents implantation	1–5	None	No	Menstrual discomfort, displacement or loss of device, uterine infection
"Minipill" (progesterone alone)	Probably prevents sperm from entering uterus	1–10	None	Yes, timing	?
"Morning-after pill" (50 × normal dose of estrogen)	Arrests pregnancy, probably by preventing implantation	?	None	Yes, timing	Breast swelling, nausea, water retention, cancer (?)
Condom (worn by male)	Prevents sperm from entering vagina	3–10	Yes, male must put on after erection	Not usually	Some loss of sensation in male
Diaphragm with spermicidal jelly	Prevents sperm from entering uterus, jelly kills sperm	3–17	Yes, insertion before intercourse	Yes, must be inserted correctly each time	None known
Vaginal foam, jelly alone	Spermicidal, mechanical barrier to sperm	3–22	Yes, requires application before intercourse	Yes, must use within 30 minutes of intercourse; leave in at least 6 hours after	None usually, may cause irritation
Withdrawal	Removes penis from vagina before ejaculation	9–25	Yes, withdrawal	No	Frustration in some
Rhythm	Abstinence during probable time of ovulation	13–21	None	Yes, must know when to abstain	Requires abstinence during part of cycle
Douche	Washes out sperm that are still in the vagina	?–40	Yes, immediately after	No	None

SUMMARY

In vertebrates, reproduction is characteristically sexual and involves two parents, one of which produces sperm and the other eggs. Sperm and eggs are formed by meiosis in the gonads (the testes and the ovaries). The male gametes are produced by meiosis in the seminiferous tubules of the testes. The spermatogonia become primary spermatocytes; then, after the first meiotic division, secondary spermatocytes; and, following the second meiotic division, spermatids, which then differentiate into sperm cells. These sperm cells enter the epididymis, a tightly coiled tubule overlying the testis, where they are partially mobilized. The epididymis is continuous with the vas deferens, which carries sperm through the spermatic cord, along the posterior wall of the abdominal cavity and around the bladder. Below the bladder in the region of the prostate gland, the two vasa deferentia merge with the ducts of the seminal vesicles and then with the urethra, which leads out through the penis.

The penis is composed largely of spongy erectile tissue that can become engorged with blood, enlarging and hardening it. At the time of ejaculation, sperm are propelled along the vasa deferentia by contractions of a surrounding coat of smooth muscle. Secretions from the seminal vesicles, the prostate, and the bulbourethral glands are added to the sperm as they move toward the urethra. The resulting mixture, the semen, is expelled from the urethra by muscular contractions involving, among other structures, the base of the penis. These muscular contractions also contribute to the sensations of orgasm.

The female gamete-producing organs are the ovaries. The primary oocytes develop within nests of cells called follicles. The first meiotic division begins in the female fetus and is completed at ovulation. The second meiotic division is completed at fertilization. On the average, one potential ovum is produced every 28 days; it travels down the oviduct to the uterus. If it is fertilized (which usually takes place in an oviduct), it becomes implanted in the lining of the uterus (the endometrium). If it is not fertilized, it degenerates and the endometrial lining is shed at menstruation.

The production of oocytes and the menstrual cycle are regulated by a complex feedback system involving gonadotropin-releasing hormone (GnRH), the gonadotropic hormones FSH and LH, and the sex hormones estrogen and progesterone. Before ovulation, FSH and LH stimulate the ripening of the follicle and the secretion of estrogen. After ovulation, the corpus luteum, which forms from the emptied follicle, produces both estrogen and progesterone. Progesterone and estrogen both stimulate the growth of the endometrium.

In many vertebrates, gamete production and mating are cyclic. The part of the cycle during which mating occurs in most female animals is known as estrus. Although ovulation is cyclic in human females, they, unlike most other animals, are continuously receptive to sexual intercourse, a behavior pattern that may have coevolved with strengthened pair bonding.

QUESTIONS

1. Consider the kinds of organisms that generally have external fertilization and those that generally have internal fertilization. What differences in their life styles require these differing mechanisms?

2. How is the shelled egg of a hen fertilized?

3. Would a vasectomy affect the structures associated with orgasm?

4. Explain the changes in the plasma concentrations of the gonadotropins and sex hormones during the menstrual cycle as shown on the graphs below.

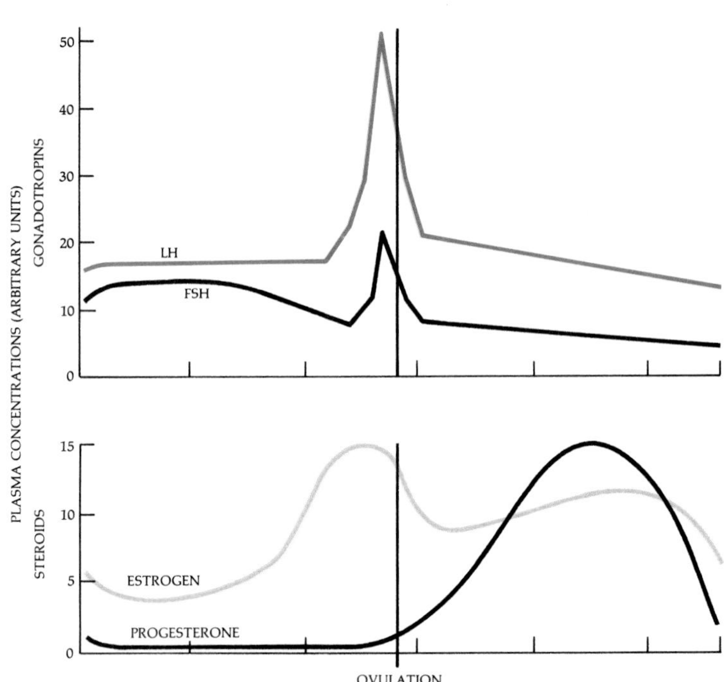

5. As you will recall from Chapter 18, the frequency of Down's syndrome increases with maternal age. On the basis of your knowledge of oogenesis, suggest a factor that may contribute to this increased frequency and relate it to the chromosomal abnormalities of Down's syndrome.

6. What would be an appropriate method of contraception for a couple who have infrequent intercourse (once a month)? For a couple who have frequent intercourse (three times a week), with plans to have children? For a couple who have frequent intercourse but do not wish to have children? Under what circumstances, if any, would you elect to have a vasectomy or a tubal ligation?

7. During which days in the menstrual cycle is a woman most likely to become pregnant? (Include data on the longevity of eggs and sperm in making this calculation.) Why is the use of the calendar method of birth control much less effective than other methods?

CHAPTER 45

The Continuity of Life II: Development

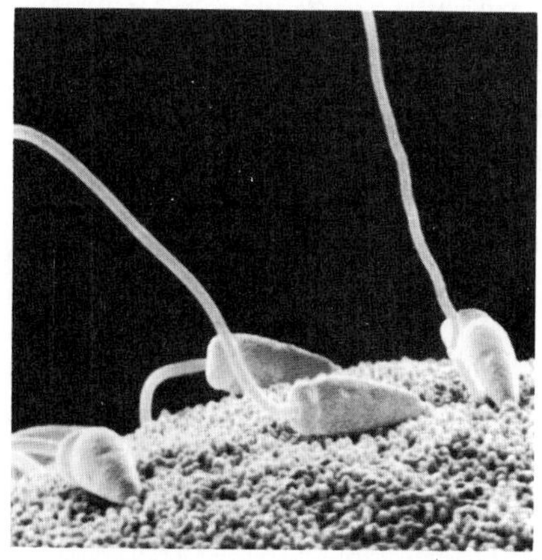

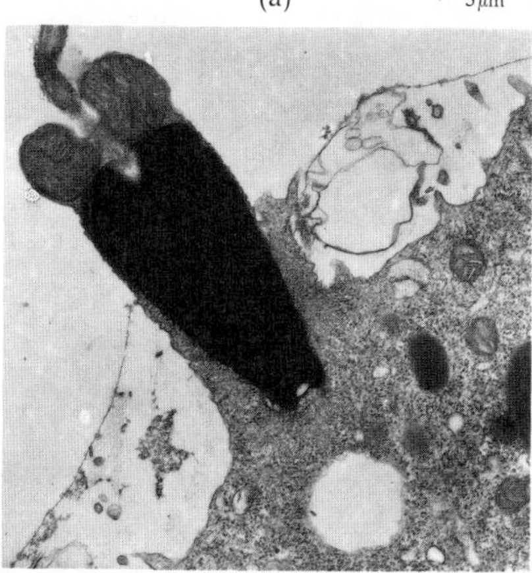

45–1
Egg and sperm cells from a sea urchin: (a) sperm cells on the egg surface, and (b) a sperm penetrating the egg. The dark, torpedo-shaped body is the sperm nucleus.

In this chapter, we shall describe the process by which a single cell—the fertilized egg—becomes a complete organism, consisting of billions of cells and closely resembling its parent organisms. This process of development, as it is known, involves growth, differentiation, and morphogenesis—that is, increase in size, specialization of cells, tissues, and organs, and shaping of the adult body form. The course of development in many species has been minutely observed and meticulously documented. Indeed, embryology, the study of early development, first commanded the attention of biologists more than 100 years ago—long before genetics was an established scientific discipline. Yet the underlying processes of development are almost as mysterious as they were a century ago. As a biological challenge, development ranks second only to the functioning of the brain.

Most of the experimental work on development has been carried out on animals other than humans. (Even though abortion is, at this writing, legal in the United States, experimentation on human fetuses is everywhere strictly prohibited.) However, the basic patterns of development are remarkably the same throughout the animal kingdom and so, as in genetics, for example, or cytology, we understand our own species by observations of others.

DEVELOPMENT OF THE SEA URCHIN

Let us begin by following the course of development in a sea urchin. The sea urchin has long been a favorite of embryologists because (1) the eggs, which are large and produced in great numbers, are fertilized and develop externally, making it possible to study their development under relatively simple laboratory conditions; (2) both egg and developing embryo are almost transparent, so that it is possible to observe many of the early events without disrupting them; and (3) the process is rapid. In about 48 hours, the zygote develops into a larval form, known as the pluteus. Moreover, sea urchins are abundant in pleasant places such as Woods Hole, Massachusetts, where biologists like to spend the summer.

Development begins with fertilization of the egg by the sperm (Figure 45–1). Like the mammalian sperm cell, the sperm cell of the sea urchin consists of a highly condensed, tightly packed nucleus, a small amount of cytoplasm, and a long flagellum, all surrounded by a cell membrane. The egg cell is much larger than the sperm cell (Figure 45–2a). Fertilization—the fusion of sperm and egg—has at least four consequences: (1) Changes take place in or on the outer membrane of the fertilized egg that prevent entry of other sperm (Figure 45–2b and c). (2) The

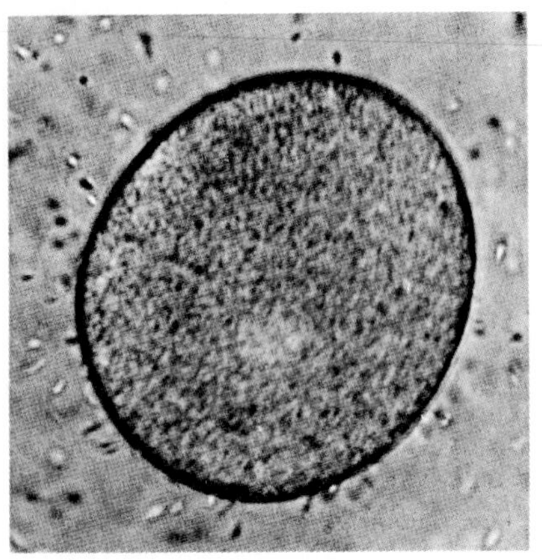

(a) *Numerous spermatozoa can be seen surrounding an unfertilized egg.*

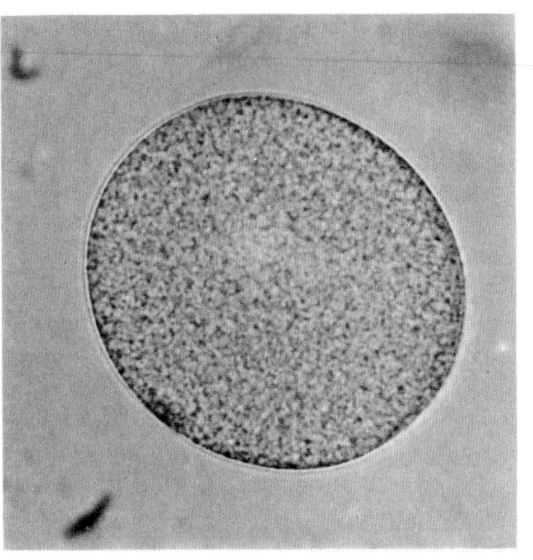

(b) *Fertilized egg; the fertilization membrane has just begun to form. The light area slightly above the center is the diploid nucleus.*

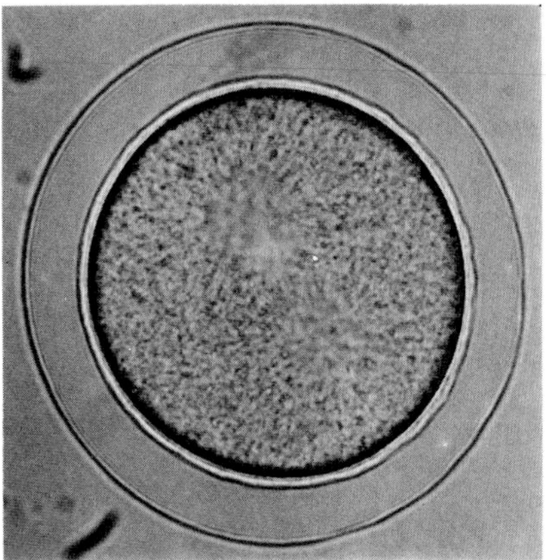

(c) *The fertilization membrane is fully formed. The egg has begun to divide; if you look closely, you can see that there are two nuclei.*

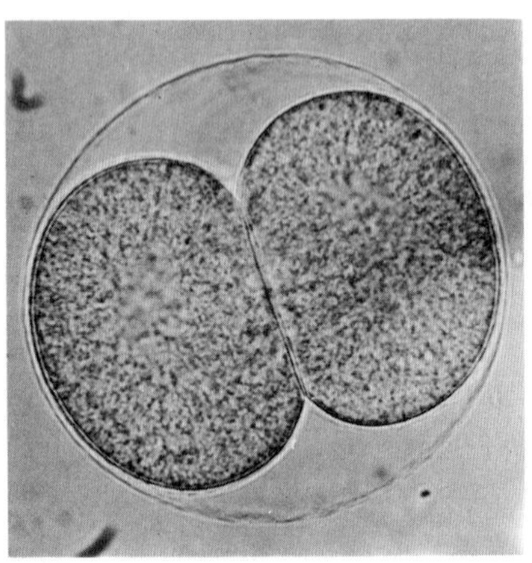

(d) *The first division.*

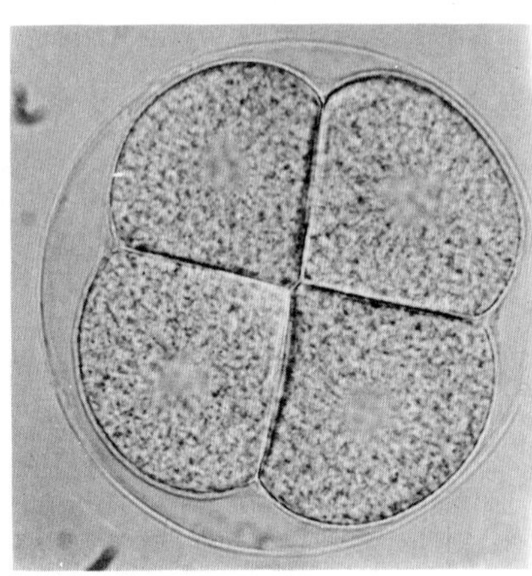

(e) *Four-cell stage.*

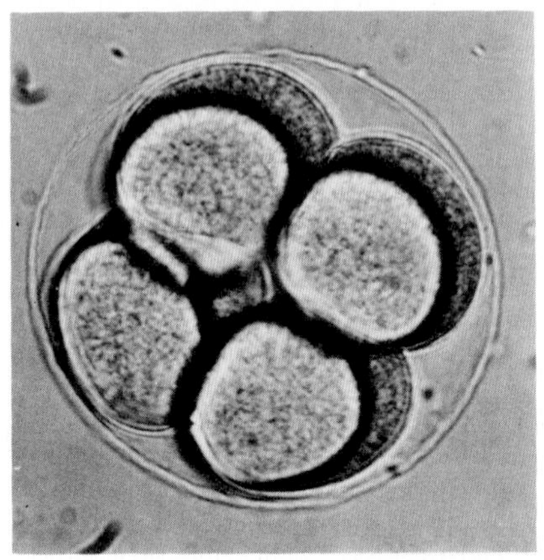

(f) *Eight-cell stage; the upper four blastomeres (blastula cells) are smaller than the lower four.*

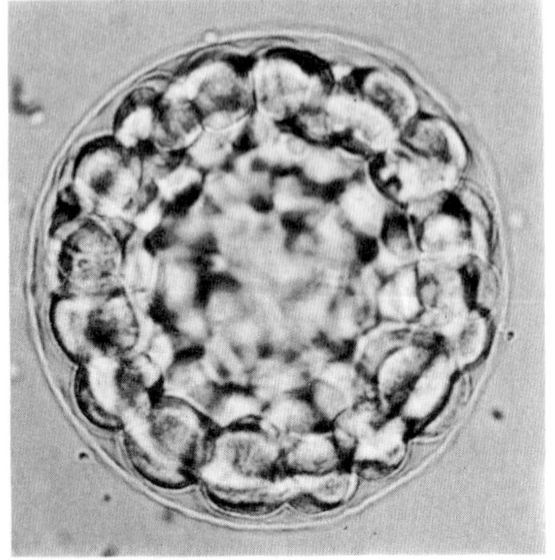

(g) *The blastocoel forms.*

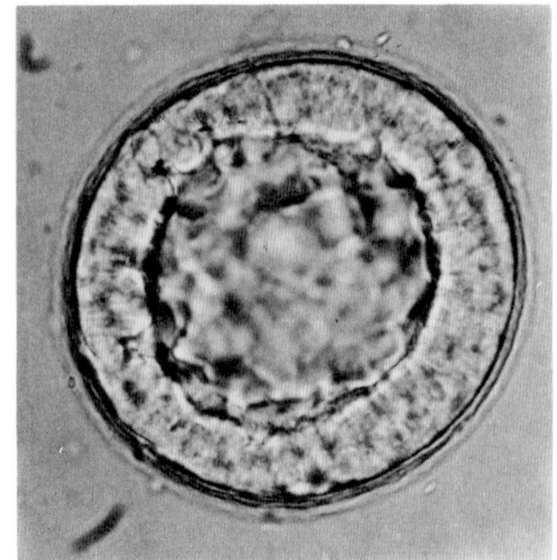

(h) *The mature blastula.*

0.1 mm

45–2
The early development of the sea urchin. Notice that as the egg divides, the cells become progressively smaller, so that in the blastula stage, at the time of hatching from the egg membrane, they are barely distinguishable, although the magnification has not changed.

genetic material of the male is introduced into the female gamete, and the two nuclei fuse, producing a diploid nucleus. The genotype of the new individual is thus established. (3) The egg is activated metabolically, as evidenced by a dramatic increase in protein synthesis. (4) The egg begins to divide by mitosis, and the developmental chain of events is set in motion.

In many species, although activation and mitosis follow fertilization, they can also proceed without it. In the sea urchin egg, for instance, even exposure to a hypertonic solution can start the developmental chain of events; frogs' eggs can be activated and will divide after being pricked with a glass needle or given a mild electric shock.

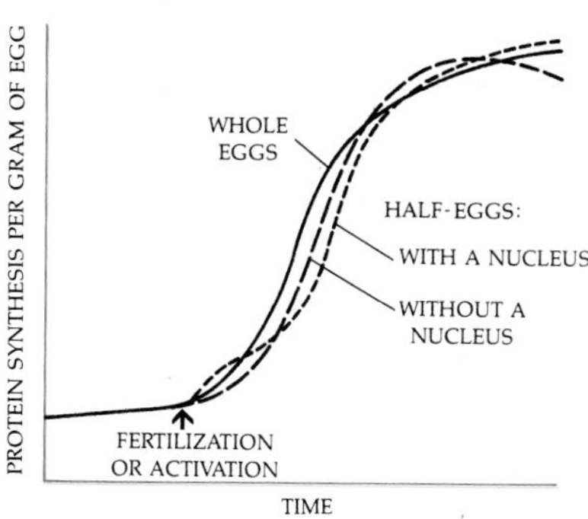

45–3
Rates of protein synthesis during the first few hours after activation in whole eggs, half-eggs with a nucleus, and half-eggs without a nucleus. The initial pattern of protein synthesis is the same whether the activated egg contains a nucleus or not, indicating that the mRNA for early protein synthesis is present in the egg before fertilization.

In fact, in some species it has been possible to demonstrate that activation of the egg does not require any nucleus at all, male or female. Sea urchin eggs can be divided in two, half with a nucleus, half without. Each half can then be activated artificially. Protein production in the half without a nucleus is as great initially, on a gram-for-gram basis, as in the half with a nucleus or in the whole egg (Figure 45–3), although, of course, the enucleated half does not long survive. In other words, the egg in the course of its differentiation transcribes all the information needed for the early rounds of biosynthetic activity, and the transcribed mRNA is inactive in the egg until fertilization initiates translation.

The egg divides about once an hour for 10 hours. This process is known as *cleavage*, and the embryo at this stage is called a *morula*. As the cells continue to divide, a fluid-filled cavity forms in the center. This cavity is the blastocoel, and the entire group of cells, the developing organism, is now called the *blastula* (see Figure 45–2h). The sea urchin blastula is about the same size as the egg cell from which it developed. Cleavage has not changed the total volume but has greatly altered the surface/volume ratio (page 98) and also the ratio of volume of the nucleus to that of the cytoplasm of the individual cells, called blastomeres.

The formation of the blastula is followed by a process known as *gastrulation* (Figure 45–4). In the sea urchin, gastrulation begins with the formation of the blastopore, an opening into the blastula. Cells on the inner surface of the blastula near the blastopore break loose and move toward the opposite pole. These cells are the primary mesenchyme. Next, the entire cell layer at one pole turns inward at the blastopore, moving through the blastocoel to the opposite pole, forming a new cavity, the gastrocoel, or *archenteron*. The archenteron will develop into the primitive gut, and the blastopore will become the anus. (Sea urchins, you will recall, are deuterostomes—page 506—like other echinoderms and like vertebrates.)

As a result of the movements that take place at gastrulation, three layers have been formed: an outside layer, the ectoderm; a middle layer, the mesoderm, which formed from the primary mesenchyme cells; and an inner layer, the endoderm (Figure 45–5, on page 855). Also, the axis of the embryo has been established.

Once gastrulation is completed, evidence of cell differentiation can be observed. Cells in the mesoderm begin to secrete calcium-containing granules that develop into tiny three-armed spicules. These become the supporting skeleton for the pluteus. At the point at which the endoderm touches the opposite surface of the blastocoel, the ectodermal cells curve inward to form the mouth of the larva. The archenteron subdivides into mouth, stomach, intestine, and anus, and a ring of long cilia forms around the mouth region. The single fertilized egg cell has become a number of specialized, differentiated cells, performing specific functions, such as digestion; producing new organelles, such as cilia; and secreting new products, such as the calcium-containing skeleton.

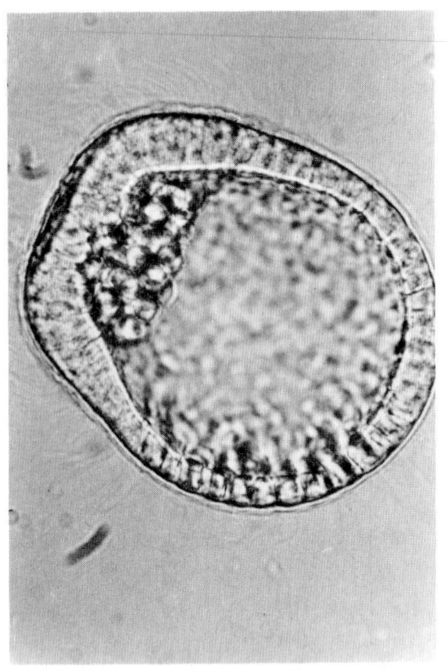

(a) *The beginning of gastrulation; the blastopore has begun to form at the upper left, and cells near the blastopore have begun to migrate across the blastocoel.*

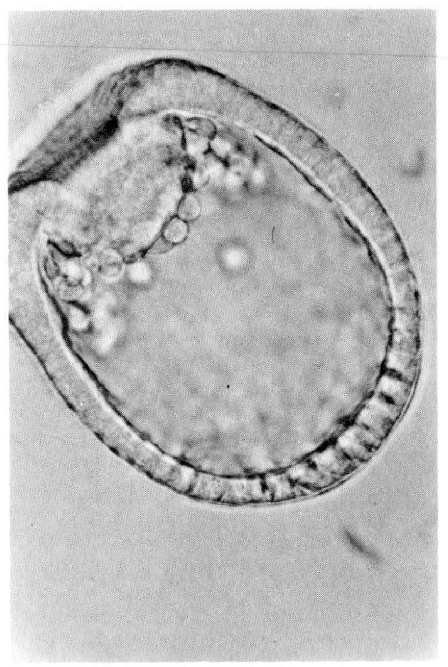

(b) *The outer cell layer begins to fold inward at the blastopore, forming the archenteron.*

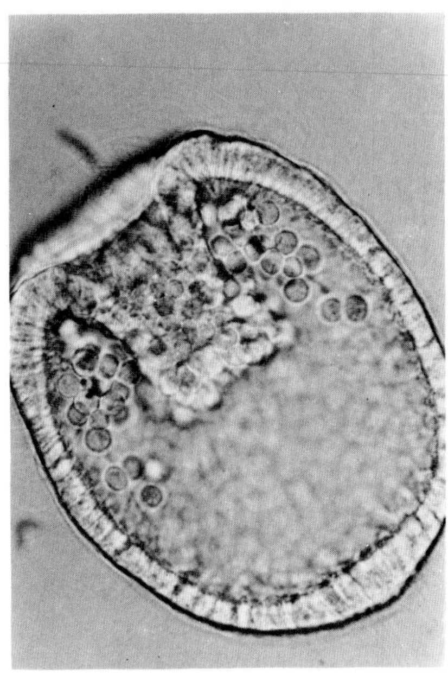

(c) *The outer layer of cells continues to move across the blastocoel.*

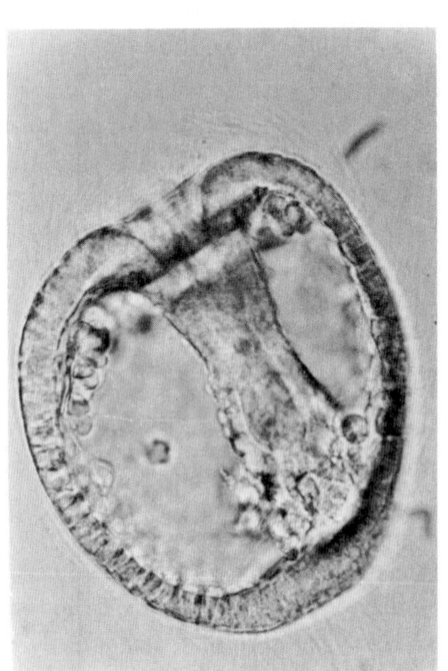

(d) *The mature gastrula.*

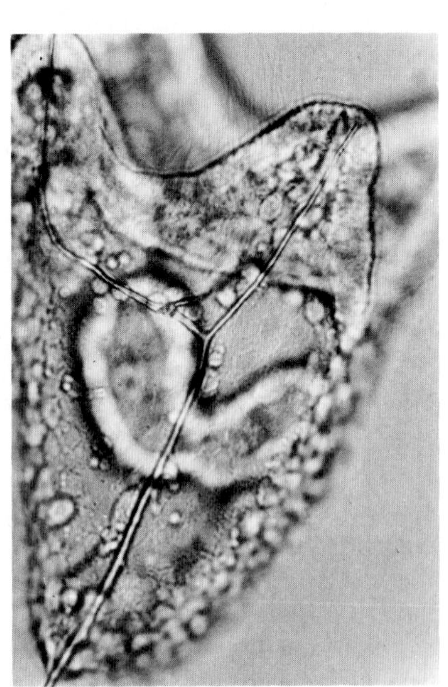

(e) *Gastrula cells differentiate and organize to form the pluteus larva.*

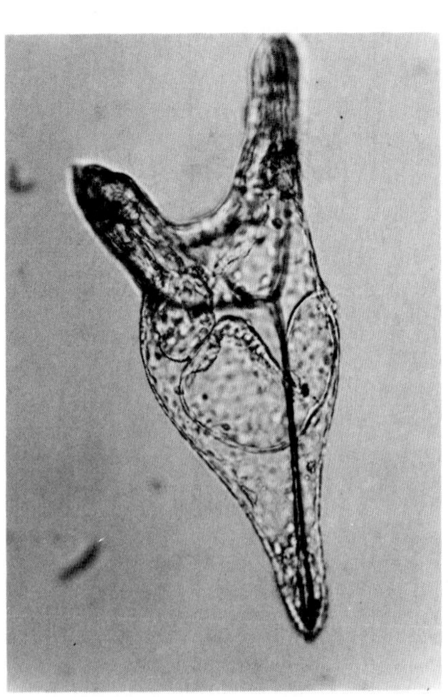

(f) *Within 48 hours after fertilization, the egg has developed into this multicellular organism, the pluteus.*

45–4
Gastrulation in the sea urchin (shown at left).

45–5
The sea urchin gastrula. Gastrulation produces a three-layered embryo. The archenteron becomes the gastrointestinal tract, and the blastopore becomes the anus. The blastocoel is almost entirely obliterated.

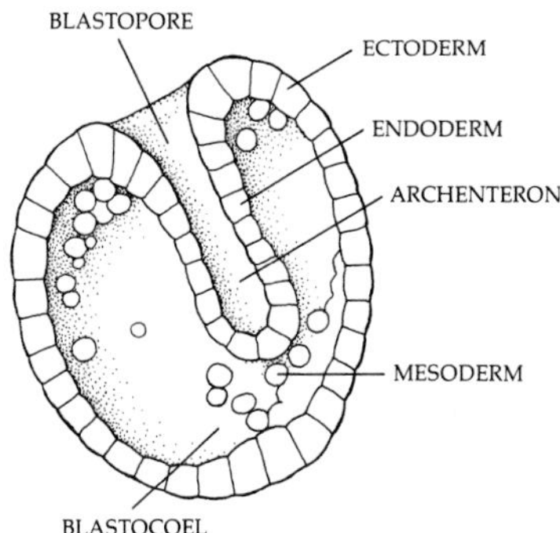

The Influence of the Cytoplasm

The sea urchin egg contains a relatively small amount of stored food (yolk), which, because of its weight, settles in the lower half of the egg. This half is called the vegetal half; the upper half is called the animal half. At cleavage, the first two cell divisions run from the pole of the animal half to the pole of the vegetal half, perpendicular to one another—the way you would quarter an apple. If these four cells are separated, each can develop into a normal pluteus. If, however, the embryo is divided in half along the third cleavage line, which splits the embryo across the equator, the two halves develop abnormally. The top (animal) half has large tufts of cilia but no gut. The bottom (vegetal) half is almost all gut, with no mouth, no arms, no cilia, and only a few, if any, spicules. In other words, it is not just the nucleus that controls the differentiation of cells. The cytoplasm surrounding the nucleus also has a profound effect.

Let us describe another experiment. If you put the early blastula of a sea urchin in calcium-free water and agitate the water, the cells will separate. The cells still look the same, and as we shall see, neither their genetic material nor their cytoplasm has been altered in any crucial way. But they simply stop dividing and very shortly they deteriorate and die. Returning the cells to normal sea water does not restore their developmental potential. Suppose, however, the cells are returned to normal sea water and, by gently stirring, are brought into contact with one another. Under these conditions the cells reaggregate, arrange themselves in their former pattern, and once more begin to divide.

In short, early development is determined by the nucleus, by the cytoplasm, and by interactions among cells and between cells with their environment.

DEVELOPMENT OF THE AMPHIBIAN

The eggs of frogs and many other amphibians are laid in shallow water and, like sea urchin eggs, are fertilized externally; hence, they can be readily observed. The amphibian egg, however, contains a much larger amount of yolk. The animal half and the vegetal half of the egg differ markedly in appearance. For example, in frogs of the genus *Rana*, the yolk is massed in the lower hemisphere of the unfertilized egg, and the upper two-thirds of the egg is covered by a heavily pigmented layer (Figure 45–6a). Following fertilization, there is a massive reorganization of the cytoplasm. When the sperm penetrates the egg, the pigment cap rotates toward the point of sperm penetration, and a gray crescent appears on the side of the egg, opposite the point of sperm entry (Figure 45–6b).

45–6
The formation of the gray crescent in a frog's egg. (a) *Before fertilization, the upper two-thirds of the egg is covered by a heavily pigmented layer, and the greater part of the yolk is massed in the lower hemisphere. The nucleus of the egg is near the pole of the upper hemisphere.* (b) *The egg has been fertilized; the sperm nucleus has entered at the right and is moving toward the center of the egg. The trail it leaves behind it is caused by the disruption of pigment granules clustered near the egg surface. The whole pigment cap has rotated toward the point of sperm penetration. The cytoplasmic region on the other side of the egg, from which the pigment layer has moved away, becomes the gray crescent.*

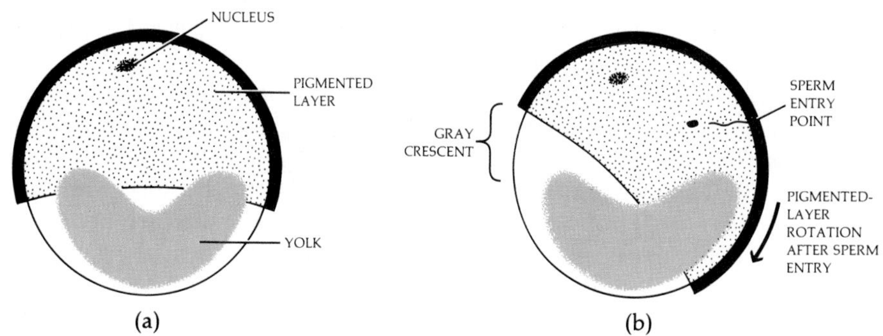

45–7
The importance of the gray crescent in development was demonstrated by separating the two cells formed by the first division of the egg. When the egg on the left divided, half of the gray crescent passed into each of the two new cells. When these cells were separated from each other, each formed a complete embryo. The first division of the egg on the right resulted in all the gray crescent going to one cell and none to the other. When these cells were separated, the one without the crescent did not develop.

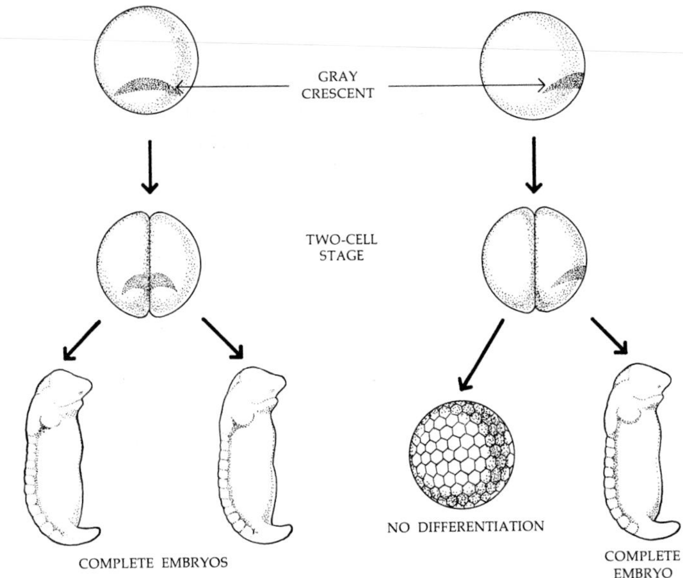

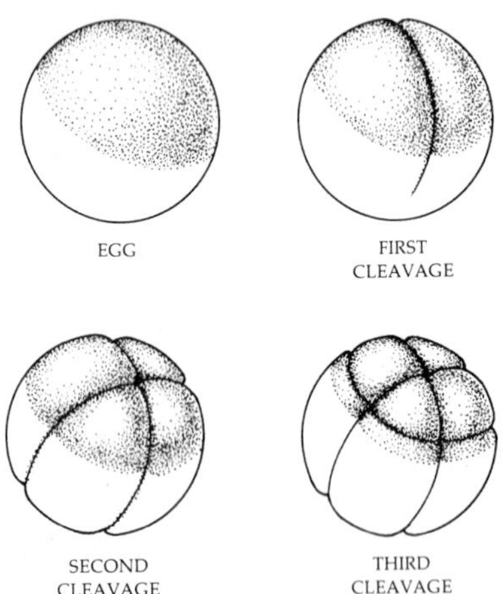

45–8
Eggs that contain a large amount of yolk concentrated in one hemisphere, such as those of the frog, cleave unequally. The first two cleavages split the frog's egg through the poles to produce four cells shaped like the segments of an orange. The third cleavage separates the lower, yolkier (vegetal) part from the upper, less yolky (animal) part. As you can see, the four cells in the animal hemisphere are much smaller than the four in the yolky vegetal hemisphere. Subsequently, the yolkier cells cleave much more slowly than the less yolky ones.

In a series of brilliant experiments by Hans Spemann in the early 1900s, it was shown that the cytoplasm associated with the gray crescent is of critical importance in the later development of the embryo. For instance, if cells of an amphibian egg are divided at the two-cell stage, each blastomere may develop into a normal embryo. Whether or not both cells develop normally depends on whether or not each has received some of the cytoplasm containing the gray crescent (Figure 45–7).

These experiments add further support to the hypotheses that in the early stages of development, the nuclei are equivalent in their potential for directing future development and that differences in the cytoplasm, rather than in the nuclei, determine the course of early development.

The Amphibian Egg: Cleavage and Gastrulation

Cleavage in the amphibian egg differs from that in the sea urchin egg chiefly because of differences in the yolk content. When yolk is absent or present only in small amounts, as in the sea urchin, cleavage involves the whole egg. When larger amounts of yolk are present, as in frogs, the egg divides unevenly (Figure 45–8), with fewer cell divisions in the vegetal hemisphere. As with the sea urchin egg, cleavage results in the formation of a blastula and the subsequent appearance of a blastopore. In amphibians, the blastopore is a crescent-shaped slit; it always forms at the boundary between the gray crescent and the vegetal hemisphere (Figure 45–9a).

Gastrulation in the two types of egg differs in detail but not in principle. In the sea urchin egg, one wall of the gastrula moves inward forming a tubular extension toward the other wall. In the yolky frog's egg, the cells also migrate inward, moving over the lip of the blastopore as if they were being hauled over a pulley (Figure 45–9b and c). The direction in which they move is the future axis of the animal. These mesodermal cells along the axis are called the *chordamesoderm*.

At the end of gastrulation in the amphibian egg, as in the sea urchin egg, there are three tissue layers: endoderm, mesoderm, and ectoderm.

45-9

Development of the frog gastrula.
(a) *The blastopore forms in the blastula.* (b) *Cells from the outer surface move into the blastopore, obliterating the blastocoel and creating the archenteron.* (c) *Three layers of tissues have been produced: ectoderm, endoderm, and mesoderm.* (d) *The ectoderm on the dorsal surface, overlying the notochord, thickens and flattens.* (e) *The neural plate and neural folds begin to form.*

In this diagram and those in Figures 45-10, 45-16, 45-17, and 45-18, ectoderm is blue, neural ectoderm is green, mesoderm is red, and endoderm is yellow.

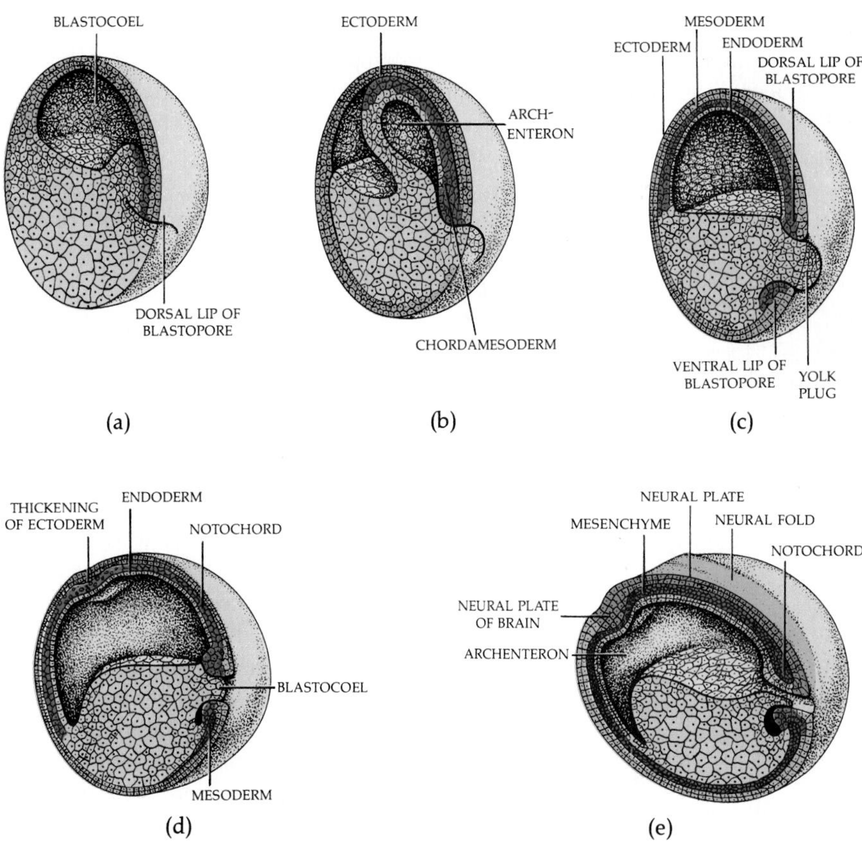

45-10

The formation of the neural tube from the neural plate in the frog. The thickened elevations of ectoderm on the right and left sides of the neural plate curve inward, forming the neural groove. The ridges bordering the neural groove then meet and fuse. Finally, the resulting neural tube pinches off from the rest of the ectodermal tissue.

The Amphibian: Neural Tube Formation

At the end of gastrulation, the first visible signs of differentiation appear. The chordamesoderm becomes the notochord (see page 554). The ectoderm above the chordamesoderm begins to thicken, forming the neural plate (Figure 45-10a). The ridges of the neural plate meet and fuse to form the *neural tube*, which pinches off from the rest of the ectodermal tissue (Figure 45-10b and c). Thus the principal features of the vertebrate have been established.

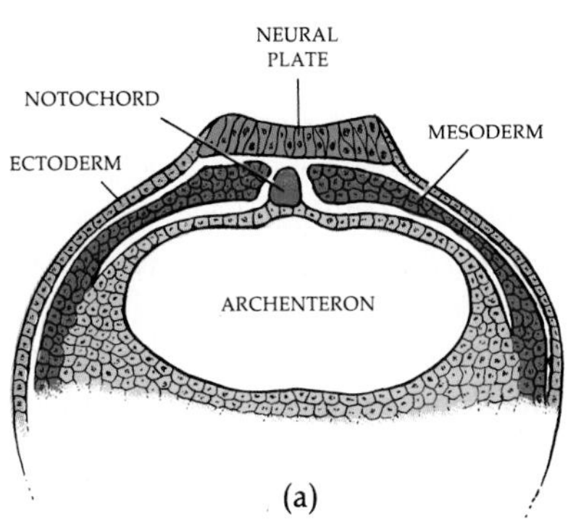

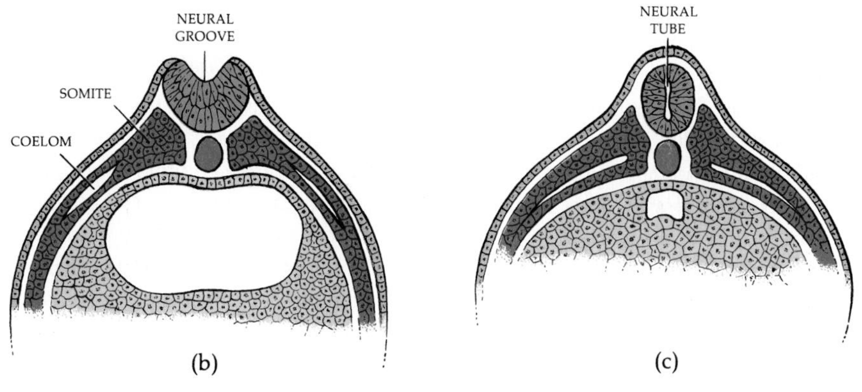

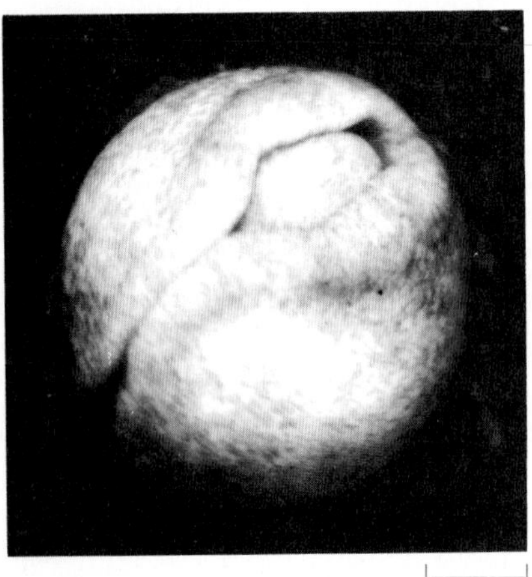

45–11
An amphibian embryo, showing the neural folds and neural groove.

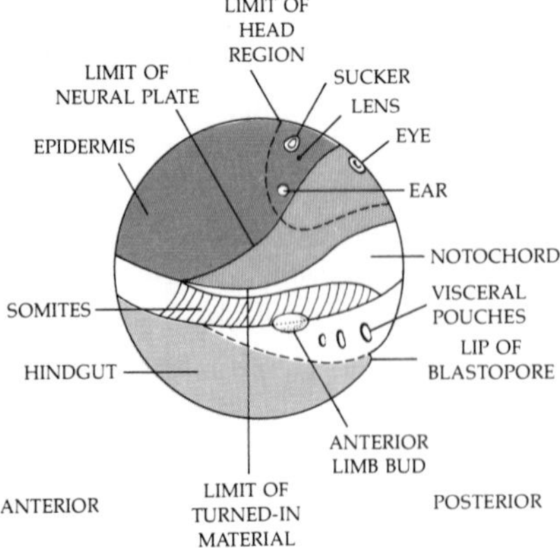

45–12
By the late blastula or early gastrula stage, the developmental fate of groups of cells in the amphibian embryo has already been determined. By applying small amounts of a harmless dye to cells on the surface of the blastula, it is possible to trace their morphogenetic movements and construct a "fate map," such as the one shown here.

The mesodermal tissue on either side of the notochord breaks up into two longitudinal masses that segment, forming blocks of tissue called *somites*. In the adjacent mesoderm, a cavity, the coelom, forms between two layers of mesoderm. The endoderm, which contains the yolk, continues to take up much of the body mass until the stored nutrients are finally absorbed by the larva, the tadpole. The movements of the cells are so regular and predictable that it is possible to chart the fate of various cells of the late blastula (Figure 45–12).

The Organizer

The tissue that lies at the dorsal lip of the blastopore of the amphibian is one of the most thoroughly investigated in all of embryology. Spemann, in a continuation of the studies described earlier (see Figure 45–7), discovered that he could divide the developing blastula in two and sometimes obtain two normal embryos. Once formation of the blastula was complete and gastrulation had begun, however, he found that he could obtain only one normal embryo from the operation. The embryo that was normal always developed from the half that contained the dorsal lip of the blastopore. (This was actually an extension of his previous experiment of separating cells in the first cleavage stage. The dorsal lip forms in the blastula at the position once occupied in the egg by the gray crescent.)

In 1924, Hilde Mangold, a student of Spemann, excised the dorsal lip from one salamander embryo and grafted it into the belly region of another embryo. When the embryo that had received the graft developed, it was found that a second embryo had formed within its tissues, a sort of Siamese twin to the first. Because the two salamanders differed in color, it was possible to distinguish the tissues of the host embryo from those of the transplant. The transplanted dorsal lip had developed into a notochord, as it would have had it not been removed in the first place. This in itself was noteworthy because previous experiments had shown that other portions of an embryo of similar age, if transplanted, would develop strictly according to the site to which they were transplanted. Even more remarkable, however, was the fact that the rest of the second embryo was made up of tissues of the host. As countless repetitions of this experiment have shown, the transplanted dorsal lip has the effect of organizing tissues of the host into a second embryo (Figure 45–13). This is how the dorsal lip of the blastopore (the prospective chordamesoderm) came to be called the *organizer*.

We do not know the nature of the organizer or how it exerts its effect. However, it is known that in chordate embryos of every species, including *Homo sapiens*, the chordamesoderm serves this same function of primary organizer, inducing differentiation of the overlying ectoderm and setting in motion the other events of development.

Induction and Inducers

It is now known that the action of Spemann's organizer is just one of many tissue interactions that take place in development. This general process is known as *embryonic induction*. Embryonic induction occurs when two different types of tissue come in contact with one another in the course of development and one tissue induces the other to differentiate.

Numerous experiments have shown that induction usually takes place, even when direct contact is prevented between the two tissues, if chemical exchanges are allowed, as by placing a porous filter between them, but not if chemical exchanges

45–13

Experiment showing the activity of the organizer. (a) *The dorsal lip of the blastopore from one amphibian embryo is grafted onto another embryo. At gastrulation the grafted tissue moves into the interior of the embryo* (b, c) *through a second blastopore created in the host. In* (b) *and* (c) *the embryo is viewed from a different angle so the host blastopore is not visible. Two neural plates are formed* (d), *and a double embryo develops* (e). *In* (f) *you see the structure of the secondary embryo under the yolk of the host embryo. The colored cells are derived from the graft, and the light cells are host material that has been induced to undergo these differentiations.*

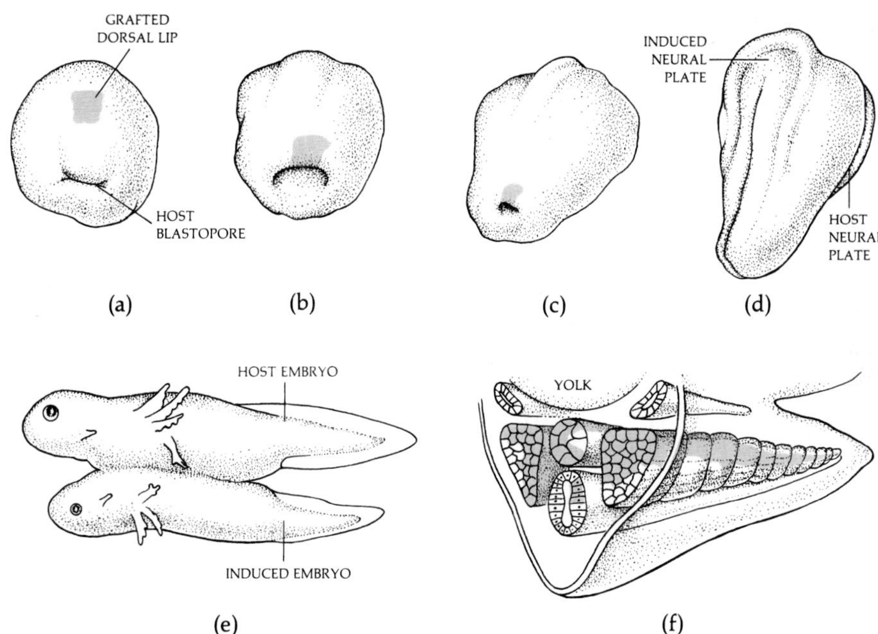

are prevented, as by a piece of metal foil. The chemical substance or substances exchanged must be of some very general nature. For example, the capacity of the organizer to induce neural plate formation is not species-specific; thus one can provoke the formation of a secondary neural tube in a chick embryo by means of a graft of tissue from a rabbit embryo. Moreover, experiments have shown that a number of chemicals, some of which never occur in embryos, will induce neural plate formation in embryonic ectoderm. It is not, then, that the inducing substance endows the ectoderm with the capacity to form nerve tissue; rather, it merely evokes a potential that is already present in the responding tissue and initiates a new pattern of gene activity.

DEVELOPMENT OF THE CHICK

The chick egg, familiar to us all, is different in many respects from the eggs previously examined. First, and most obvious, it is surrounded by a shell, which permits it to develop on the land. (Such eggs, as noted on page 559, are called amniote eggs.) Second, related to its terrestrial development, it is surrounded by a system of membranes. Third, it contains a large amount of yolk. In accord with this large amount of stored food, development in the chick can take much longer than in the sea urchin or the frog and does not stop with the emergence of a still immature, larval form, such as the pluteus or the tadpole.

The yolk of the chick egg, like that of other bird and also reptile eggs, is so large and dense that cleavage does not involve most of the egg mass. The only part of the egg that cleaves is a thin layer on top. These cleaving cells sit like a cap on the top of the yolk mass. A blastula like this is known as a *blastodisc*. Birds, reptiles, and mammals develop from a blastodisc.

45–14
Blastodisc of early chick.

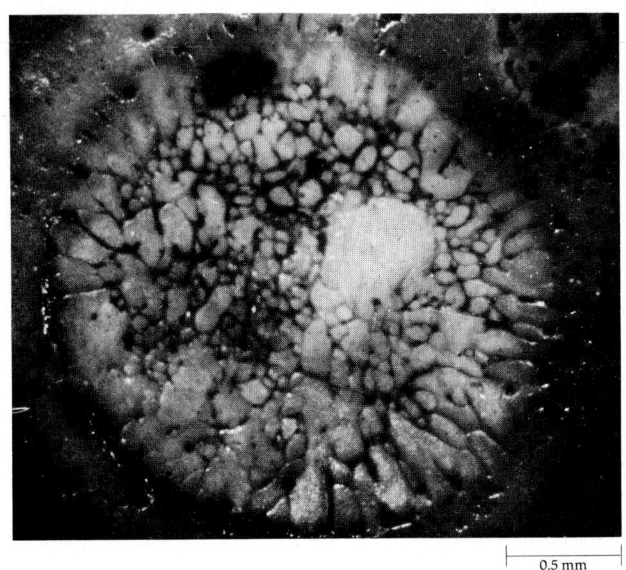

45–15
Blastulas of (a) *a sea urchin,* (b) *a frog, and* (c) *a bird. In the sea urchin and the frog, the blastula is a hollow sphere of cells. In the bird, it forms a blastodisc that sits on the surface of the yolk.*

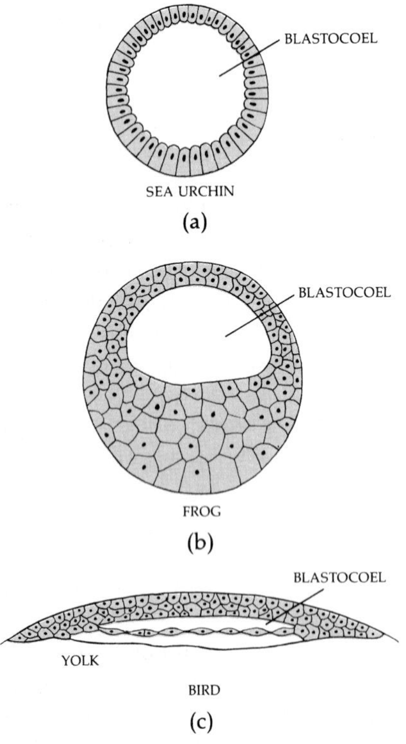

If you break open a fertilized chick egg when it is first laid, you can see the blastodisc as a white mass about 2 millimeters in diameter on top of the yolk (Figure 45–14). Microscopic examination will show that this mass is made up of many cells (close to 100,000) and that the cells are in two layers: the upper layer will give rise to the ectoderm and the lower layer to the endoderm. The space between them is the blastocoel (Figure 45–15). If the egg is left intact and development continues, within a short period of time, a visible line, known as the *primitive streak,* appears on the surface of the blastodisc; the primitive streak is, in effect, an elongated blastopore. Cells migrate through the primitive streak to form the mesoderm and some endoderm, thereby establishing the three characteristic tissue layers: ectoderm, mesoderm, and endoderm (Figure 45–16). On the completion of gastrulation, the neural plate forms, and subsequently the neural tube, as in the frog.

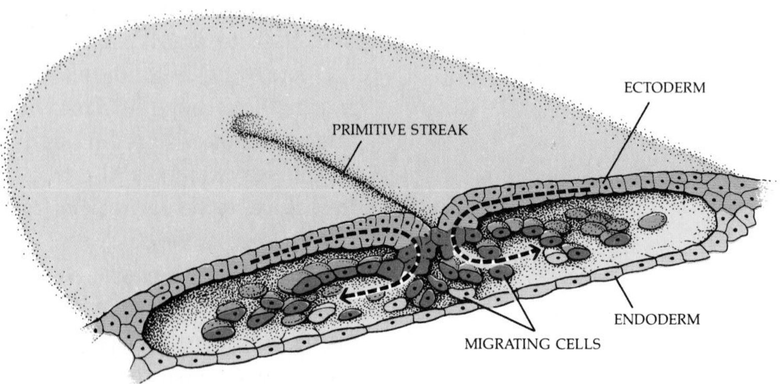

45–16
Gastrulation of the chick blastodisc. Cells migrate into and through the primitive streak to form the mesoderm and some endoderm. The result is a thin embryo with the three characteristic tissue layers.

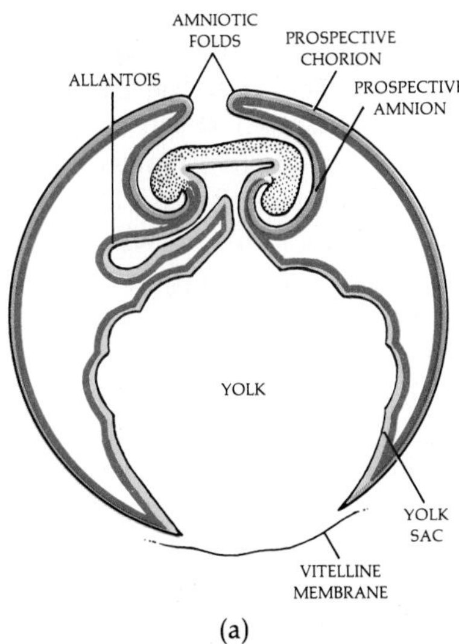

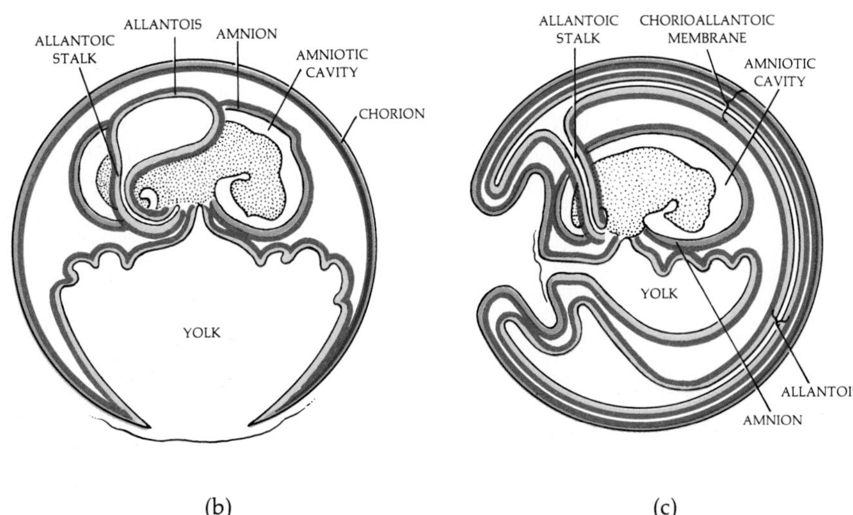

THE FOUR EXTRAEMBRYONIC MEMBRANES:
YOLK SAC (ENDODERM AND MESODERM)
ALLANTOIS (ENDODERM AND MESODERM)
AMNION (MESODERM AND ECTODERM)
CHORION (MESODERM AND ECTODERM)

45–17
Development of the extraembryonic membranes of the chick. In early development, body folds of the embryo begin to separate from the underlying yolk. One membrane, the yolk sac, grows around and almost completely encloses the yolk. A second, the allantois, arises as an outgrowth of the rear of the gut. The third and fourth are elevated over the embryo by a folding process during which the membrane is doubled. When the folds fuse, two separate membranes are formed. The inner one is the amnion, and the outer one is the chorion. The chorion eventually fuses with the allantois to form the chorioallantoic membrane, which, in the later stages of development, encloses embryo, yolk, and all the other structures. Note that each membrane is composed of two types of primary tissue.

Extraembryonic Membranes of the Chick

Eggs develop in water. In the amniote egg, this water is contained within a set of membranes formed from the tissues of the embryo, the *extraembryonic membranes*. These membranes begin as extensions of the blastodisc. Each is formed from a combination of two of the basic tissue types.

As Figure 45–17 shows, the yolk gradually becomes surrounded by a membrane of mesoderm and endoderm, the *yolk sac membrane*. The function of this membrane is nutritive; the endodermal cells digest the yolk, and the blood vessels formed in the mesodermal component carry food molecules from the yolk into the embryo proper.

The extraembryonic ectoderm and mesoderm on the outside of the coelomic space fold upward around and over the embryonic region. These so-called amniotic folds eventually fuse with each other over the embryo, enclosing it inside a fluid-filled cavity, the *amniotic cavity*. The saline fluid in the cavity serves to buffer the embryo against mechanical and thermal shocks. As Figure 45–17b shows, the fusion of the amniotic folds creates two membranes, both composed of a layer of ectoderm and mesoderm, separated by coelomic space, which is continuous with the extraembryonic coelom. The inner membrane is the *amnion*, and the outer membrane is the *chorion*.

As development proceeds and the embryo becomes increasingly demarcated from the yolk mass (although always attached to it by the yolk sac stalk), a ventral and posterior pouch is formed from the primitive hindgut. This is the *allantois*, and, as you can see in Figure 45–17b, its wall is composed of endoderm and mesoderm. At first, the function of the allantois is excretion. Since bird and reptile eggs are closed off from their environment by the shell, there is no way to dispose of the products of nitrogen metabolism, most of which are toxic. Evolution provided a dual answer to this problem. First, the nitrogenous wastes are converted to uric acid, which is insoluble and of low toxicity. The uric acid is then voided from the embryo into the allantois, which essentially serves as an embryonic garbage bag. (When a chick hatches, the accumulated uric acid can be found adhering to the abandoned shell.)

As development proceeds further, the allantois expands in volume and pushes out into the extraembryonic coelom, gradually enveloping the embryonic region as it does so. Eventually, as Figure 45–17c shows, the allantois effectively obliterates the extraembryonic coelom, and its walls fuse with the chorionic membrane lying under the eggshell. Since the allantois is plentifully supplied with blood vessels, the *chorioallantoic membrane*, formed by the fusion of the allantoic wall with the chorion, acts as an efficient respiratory membrane for the embryo during its later development.

Organogenesis

While the extraembryonic membranes are forming, the development of the embryo itself continues. These later stages of development, following cleavage and gastrulation, are generally termed *organogenesis*, the formation of the organ systems. Organogenesis begins with the inductive interaction between ectoderm and underlying chordamesoderm. Each of the three primary tissues formed during gastrulation now proceeds to undergo growth, differentiation, and morphogenesis. This process is essentially the same in all vertebrates.

Ectoderm

The neural tube stretches out along the dorsal surface, growing and becoming longer and thinner as the embryo increases in size. The cells in the neural tube at first appear to be all alike, but some of them, the future motor neurons, begin to extend long processes that grow out beyond the neural tube and invade the peripheral organs and tissues.

When the neural tube is formed, some of the ectodermal cells at the crests of the neural folds are left behind in the surrounding tissue, as shown in Figure 45–18. Some of these cells from the neural crest now migrate into positions around the neural tube and form connections between the dorsal part of the tube and the surrounding tissues. Meanwhile, mesodermal cells have begun to migrate toward the neural tube and underlying notochord and to aggregate around them. Here they differentiate into cartilage and give rise to the hollow vertebral column. If a portion of neural tube is transplanted to another region of the mesoderm, mesodermal cells will similarly aggregate around it as a result of tissue interaction.

The brain begins as three bulges in the foremost part of the neural tube (Figure 45–19). Almost immediately, two saclike protrusions, the optic vesicles, appear on both sides of the forming brain. These spherical vesicles enlarge and come into contact with the epidermis (outer skin), while still remaining connected to the neural tube by the narrowing optic stalk. Within this stalk, the optic nerve will develop.

45–18
During the formation of the neural tube, some of the ectodermal cells at the crests of the neural folds are pinched out as the folds come together. Eventually, some of these neural crest cells will migrate down toward the notochord and aggregate into ganglia. These cells become sensory neurons that send extensions up to connect with the dorsal part of the spinal cord and nerve fibers into the surrounding tissues. Others become Schwann cells, which surround and insulate the nerve fibers, some become the pigment cells (melanocytes) found at the base of the epidermis, and some form the adrenal medulla.

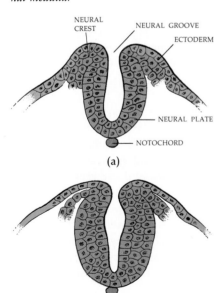

NEURAL CREST
NEURAL GROOVE
ECTODERM
NEURAL PLATE
NOTOCHORD
(a)

(b)

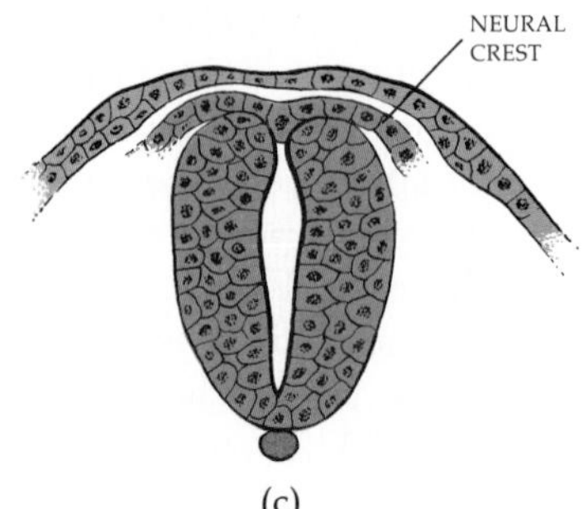

(c)

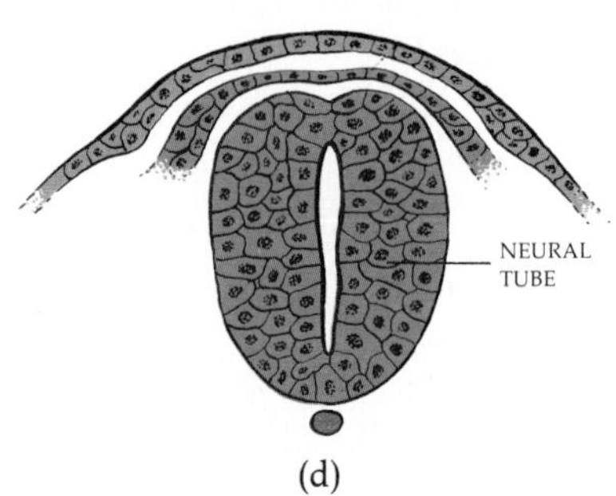

(d)

45–19

The development of the brain and associated sense organs. At the front end of the neural tube, local swellings produce three distinct bulges—the forebrain, the midbrain, and the hindbrain. The forebrain then bulges laterally, and two optic vesicles appear and develop a cuplike shape. At the same time, the surface epidermis folds inward to meet the optic vesicles. The earliest stages—primordia—of ears and nostrils also appear at first as epidermal infoldings.

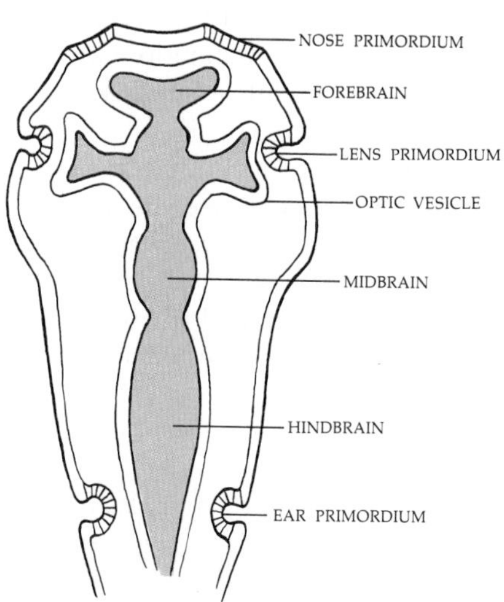

When the optic vesicle comes into contact with the inner surface of the epidermis, the external surface of the vesicle flattens out and pushes inward. The vesicle thus becomes a double-walled cup. The rim of the cup becomes the edge of the pupil. The opening of the cup is large at first, but the rims bend inward and converge, so that the opening of the pupil becomes smaller. At the point where the optic vesicle touches the epidermis, the epidermis begins to differentiate into a lens, becoming transparent (Figure 45–20).

The differentiation of the eye lens is an example of a *secondary induction*. The inducing tissue (the optic vesicle) is itself the result of the primary induction of dorsal ectoderm by dorsal mesoderm. In fact, the formation of the complete eye involves at least six separate inductions, occurring in an orderly sequence, each one linked to the one before, with primary induction by the organizer as the starting point.

Other infoldings and differentiations of head ectoderm, associated with the presence of the embryonic brain, lead to the elaboration of the organs of smell and hearing.

Mesoderm

The series of somites that came to lie on each side of the notochord shortly after gastrulation now differentiate into three kinds of cells: (1) sclerotome cells, which later form skeletal elements; (2) dermatome cells, which become part of the developing skin; and (3) myotome cells, which form most of the musculature. In the lower vertebrates, this segmented pattern, with its simple double series of back muscles, is retained in adulthood; it is well suited to the side-to-side motion of the body in swimming and crawling. In the higher vertebrates, however, this initial segmented arrangement is almost obliterated by modifications and additions of the muscles associated with terrestrial locomotion.

Mesoderm lateral to the somites forms the kidneys, the gonads, and the ducts of the excretory and reproductive systems. The ventral mesoderm splits into two

45–20

The eye develops as a result of interactions between the epidermis and a lateral outgrowth of the forebrain, the optic vesicle. When the optic vesicle reaches the epidermis, the overlying epidermis begins to thicken and differentiate. The optic vesicle flattens out and invaginates, becoming the double-walled optic cup. The invaginated wall will become the retina of the eye; the outer, thinner wall develops into the pigment layer. The rim of the optic cup will later form the edge of the pupil. The thickened epidermal layer pinches off to become the transparent lens, and the overlying epidermis, which also becomes transparent, forms the cornea. The connection with the brain remains as the optic stalk, within which is the optic nerve. Note that the retina is thus a differentiated extension of the brain.

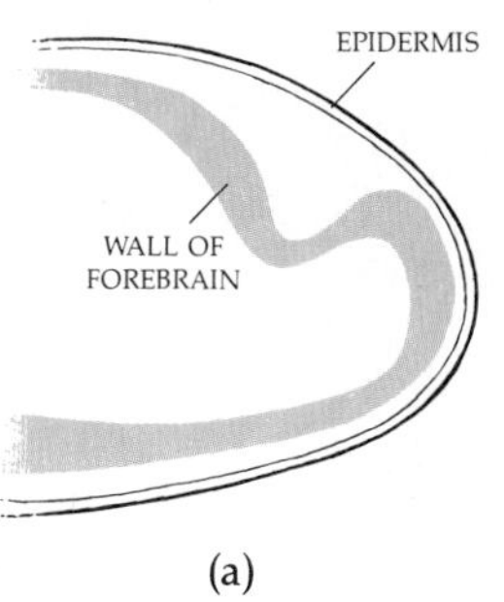

(a)

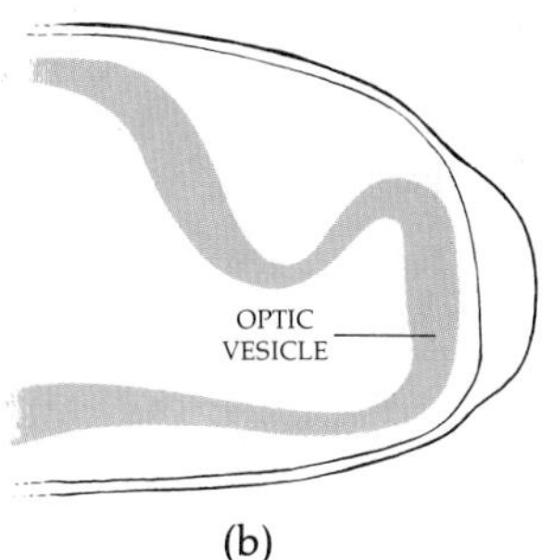

(b)

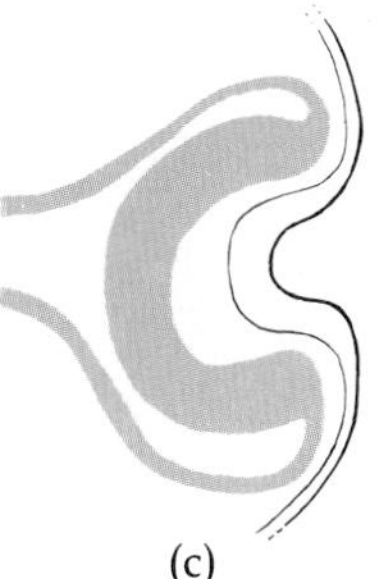
(c)

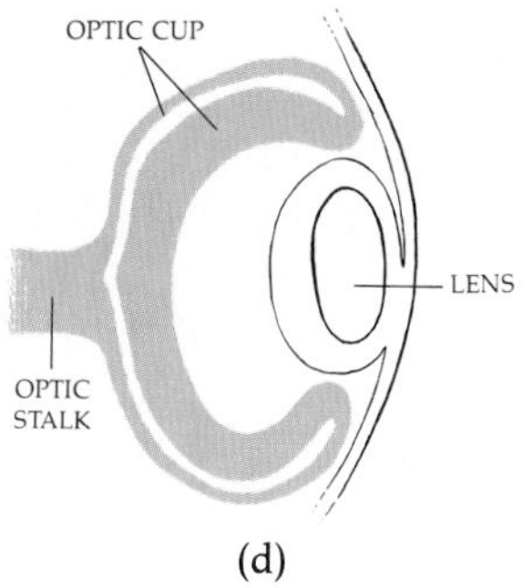

(d)

layers. One sheet of mesoderm lines the thoracic and abdominal cavities, and another becomes the outer layer of the internal organs. The space between the two sheets is the coelom.

Endoderm

The endoderm differentiates into tissues of the respiratory and digestive tracts and a number of related organs. Early in development, endodermal pouches develop at the anterior end of the endodermal tube (the gut). They push laterally until they meet the ectoderm, which folds inward to meet the pouches, producing a series of grooves on the surface of the embryo. In lower vertebrates, endoderm and ectoderm fuse and a perforation forms, around which gill filaments develop. In terrestrial vertebrates, including humans, portions of these endodermal pouches develop into auditory tubes (which connect the pharynx and the middle ear), tonsils, parathyroid glands, and the thymus gland. Posterior to the gill region, the lungs develop as similar outpocketings, branching into two sacs; more posterior still, outpocketings from the primitive gut begin to differentiate into liver, gallbladder, and pancreas.

No organ system is derived from only one type of tissue. For example, the lining of the intestine is of endodermal origin; these lining cells secrete the digestive juices and absorb the digested materials. (Secretion and absorption are the principal functions of the intestine.) However, the functional structure of the intestine also includes muscles, connective tissue, blood vessels, nerves, and an outer wrapping, which are made up of tissues derived from mesoderm and ectoderm.

45–21
Development of the chick embryo. (a) *At 4 days, large vessels have grown out from the embryo into the yolk sac. Note the heart, above the junction of the blood vessels with the embryo.* (b) *At 5 days, the yolk sac has grown further around the yolk. Its endodermal cells secrete lipase, an enzyme that digests fat in the yolk. Nutrient molecules are carried to the embryo via the mesodermal blood vessels.* (c) *By 8 days, the large, pigmented eyes are clearly visible.* (d) *At 11 days, the limbs are developing. The brain is visible through the transparent skull.* (e) *By 14 days, feathers have appeared.* (f) *By 21 days, just prior to hatching, a white knob is visible on the end of the beak; this is the egg tooth, which is used in pecking open the shell.*

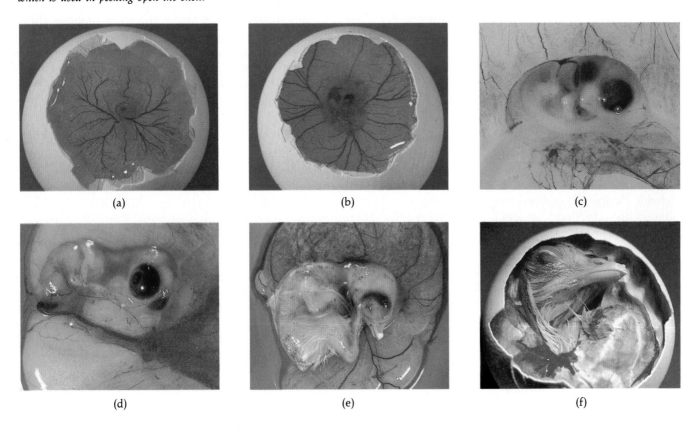

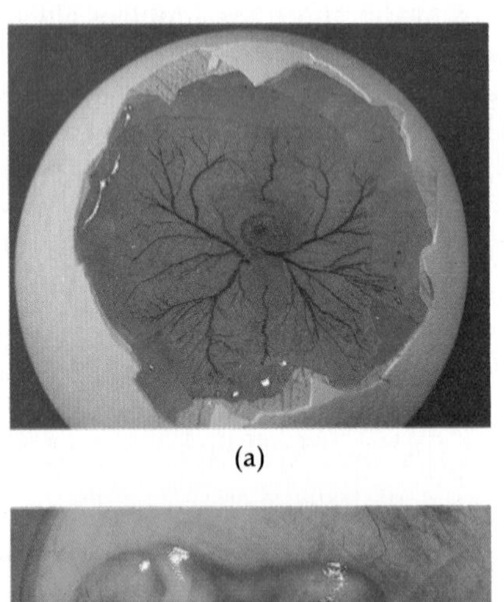

(a)

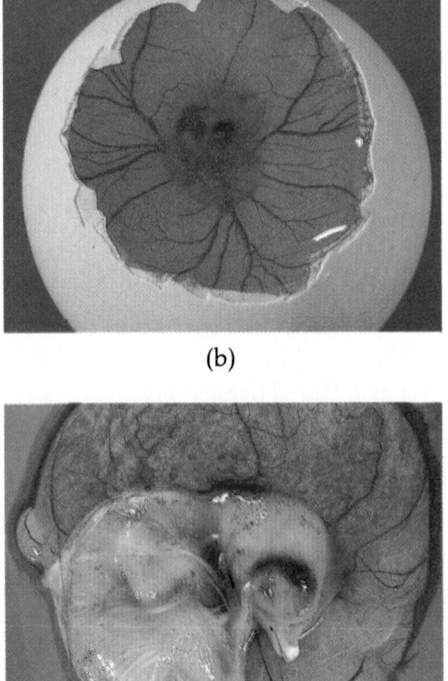

(b)

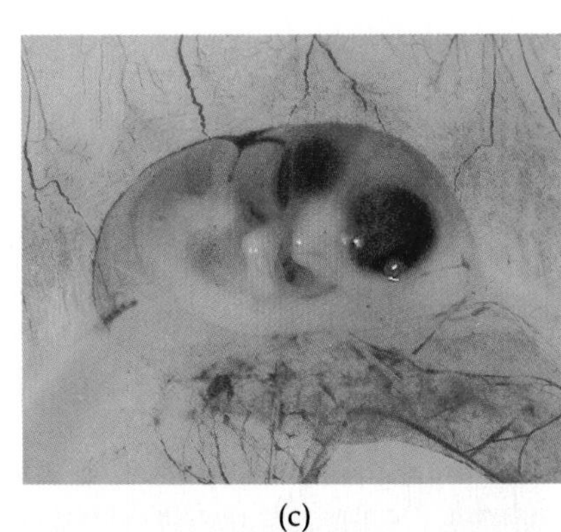

(c)

(d)

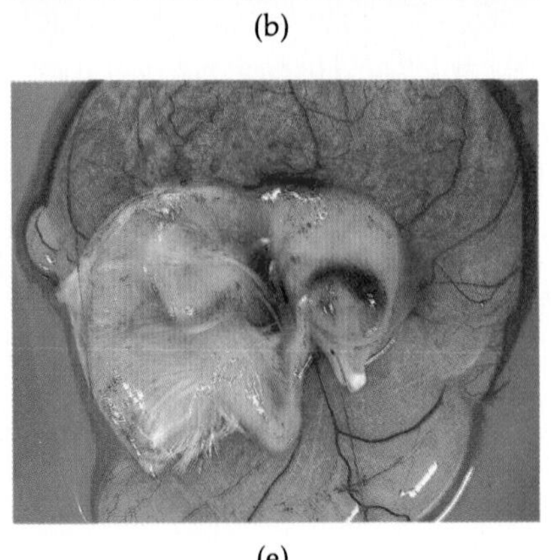

(e)

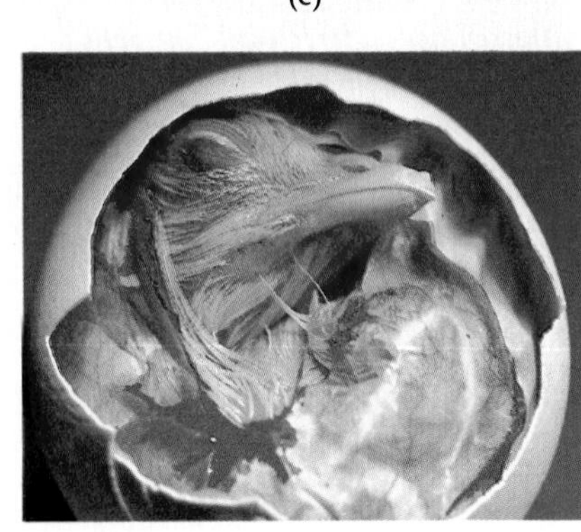

(f)

45-22
The principal channels of vertebrate development.

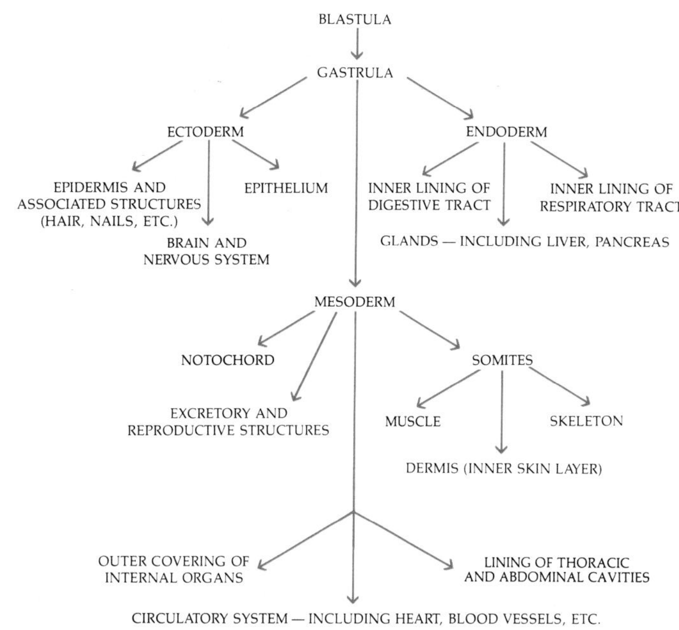

DEVELOPMENT OF THE HUMAN EMBRYO

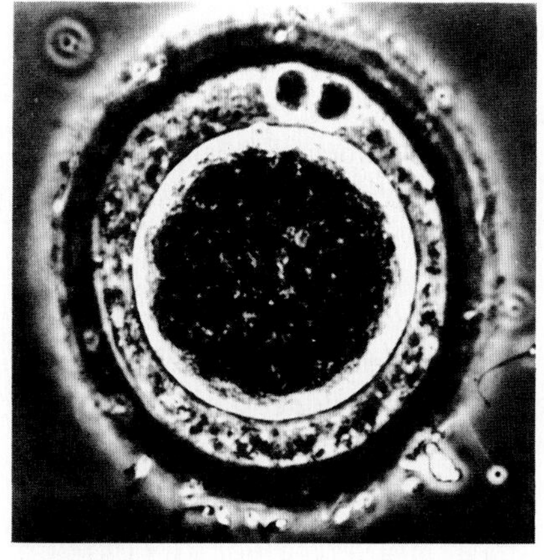

45-23
A human egg surrounded by sperm cells. At the top are two polar bodies. Surrounding the ovum is a layer of mucoprotein, the zona pellucida, which separates the egg from a layer of follicle cells that provide nourishment to the egg.

Human egg cells, as we noted in the preceding chapter, mature in the ovary and are released at approximately 28-day intervals. Fertilization usually takes place in the oviduct. As in other species, fertilization results in introduction of the genetic material of the father, changes in the outer surface of the egg that prevent entry of other sperm cells, metabolic activation of the egg, and cleavage. After fertilization, the egg continues its passage down the oviduct, where the first cell divisions take place. At about 36 hours after fertilization, the fertilized egg divides to form two cells (Figure 45-24a); at 60 hours, the two cells divide to form four cells. At three days, the four cells divide to form eight. In this early stage (Figure 45-24b), all of the cells are of equal size, as they are in the sea urchin. During this period, the embryo is entirely on its own, following its own genetic program and supplying its own metabolic requirements. All it needs is a suitable fluid environment containing progesterone.

By about five days after fertilization, the blastula consists of some 120 cells, and a fluid-filled cavity has begun to form within it; because of this cavity, it is now known as a *blastocyst* (Figure 45-24c). A cross section of the blastocyst looks somewhat like a ring with the stone on the inside. The inner cell mass (the stone) is a ball of cells at one pole of the blastula that will develop into the embryo proper. The ring is called the *trophoblast* (from the Greek word *trophe*, "to nourish"). It is composed of a double layer of cells and completely encloses the developing embryo. The trophoblast is the precursor of the placenta and will give rise to the chorion. (In the chick, the surrounding membranes, you will recall, grew out from cells at the edge of the embryo early in development. It is as if the human embryo has skipped this step entirely, arriving, as it were, prepackaged.)

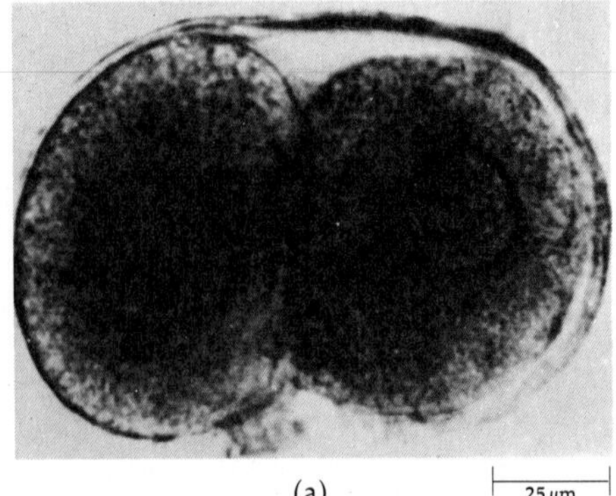

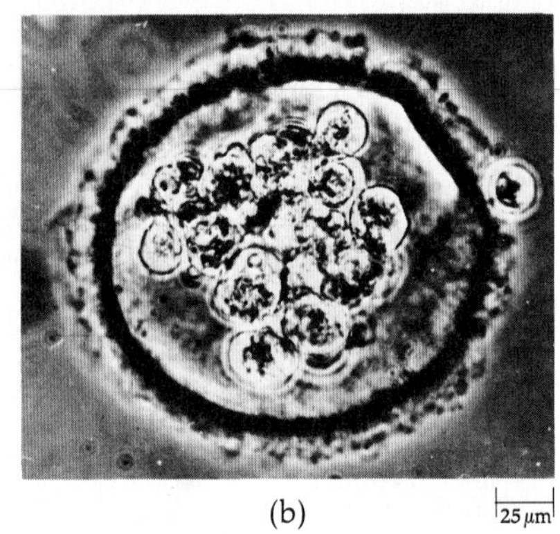

45–24
(a) *A human embryo at the two-cell stage. The cells are still surrounded by an outer membrane. Sometimes they separate at this stage, resulting in identical twins.* (b) *The cells continue to divide, but since the volume of the embryo does not increase, they are still easily contained within the same membrane.* (c) *By about five days after fertilization, the blastocyst has formed. The embryonic cell mass is at the right; the remainder of the cells make up the trophoblast.*

At about the sixth day, the trophoblast makes contact with the tissues of the uterus and releases a protein hormone, chorionic gonadotropin. This hormone prevents menstruation by stimulating the corpus luteum to continue its production of estrogens and progesterone in large amounts. Pregnancy tests involve the detection of chorionic gonadotropin in blood or urine. The trophoblast cells, multiplying rapidly by now, chemically induce changes in the endometrium and invade it. As the embryo penetrates the endometrial tissues (*implantation*), it becomes surrounded by ruptured blood vessels and the nutrient-filled blood escaping from them. The trophoblast thickens and develops fingerlike projections that invade the uterine lining to develop into extraembryonic membranes.

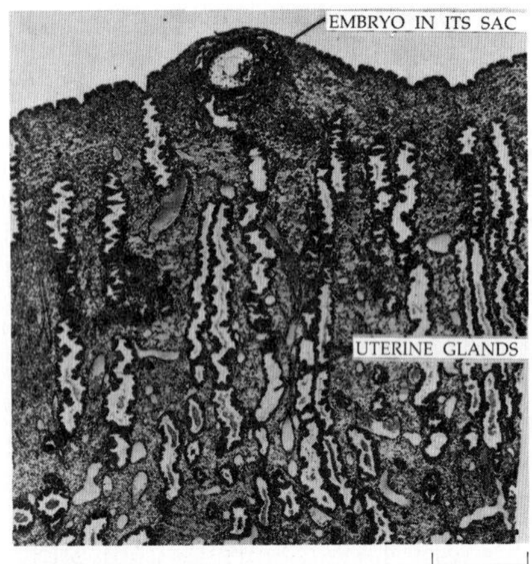

45–25
Implantation. On the sixth or seventh day after fertilization, the embryo invades the lining of the uterus. Subsequently, the placenta begins to form. This organ is the site at which oxygen, nutrients, and other materials are exchanged between the mother and fetus; it is also the source of hormones that help to maintain pregnancy. Implantation usually occurs three to four days after the young embryo reaches the uterus. Once implantation has been achieved, the embryo is no longer independent but is parasitic on the mother.

Human Embryonic Membranes

As the embryo becomes implanted, the embryonic membranes begin to develop. These have interesting similarities to and differences from the avian-reptilian membranes discussed previously. In the first place, the yolk sac, which develops first, has no yolk. This lack of yolk is a secondary evolutionary development. The monotremes (egg-laying mammals) produce eggs with yolks. The marsupials (pouched mammals) produce eggs with yolks, but the yolks are discarded at the first cleavage. The placental mammals produce eggs with no yolks; instead there is a prominent cavity where the yolk "used to be." Cleavage and the cellular migrations at gastrulation proceed as if yolk were still there. Germ cells (the cells that will later give rise to sperm and eggs) are segregated in the endoderm of the yolk sac during the first month of fetal life.

In humans, the allantois connects with the hindgut and eventually becomes a primitive urinary bladder. Wastes are not segregated as uric acid and stored, as they are in the chicken, but are transported as urea and ammonia to the maternal bloodstream. In the course of development, the endodermal component of the allantois (the actual sac) becomes greatly reduced, but the allantoic mesoderm spreads out; it is the source of the major blood vessels on the embryonic side of the placenta. The allantoic stalk eventually becomes the umbilical cord, the embryo's principal connection with the uterine tissues. The blood vessels of the allantois transport oxygen and nutrients that have diffused in from the mother's circulatory system and carry off carbon dioxide and other wastes.

The third membrane is the amnion. The space between the amnion and the embryo is the amniotic cavity. This cavity fills with a saline fluid. For the genetic

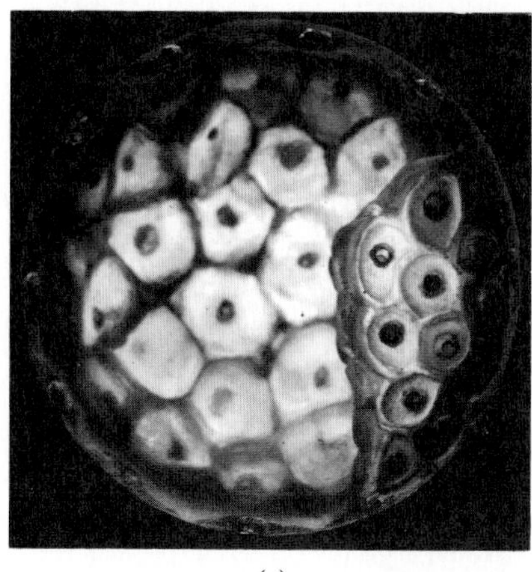

(c)

testing procedures described in Chapter 18, fluid samples are taken from the amniotic cavity. The samples obtained by amniocentesis contain cells that have been sloughed off by the embryo (see page 354).

The fourth membrane is the chorion, which is a combination of ectoderm from the trophoblast and mesoderm grown out from the embryo itself. By about the fourteenth day, chorionic villi begin to form: the formation of the villi represents the beginning of the mature placenta.

The Placenta

By the end of the third week after conception, the placenta covers 20 percent of the uterus. It is a disc-shaped mass of spongy tissue through which all exchanges between mother and fetus take place. The placenta is formed from a maternal tissue, the endometrium, as well as from the chorion of the fetus, and has a rich blood supply from both. However, the fetal and maternal circulatory systems are not directly connected, so maternal and fetal blood cells do not mix. Molecules, including food and oxygen, diffuse from the maternal bloodstream through the placental tissue and into the blood vessels that carry them into the fetus. Similarly, carbon dioxide and other waste products from the fetus are picked up from the placenta by the maternal bloodstream and carried away for disposal through the mother's lungs and kidneys. The mother's blood volume begins to increase in response to these extra demands. Her appetite increases and the absorption of certain nutrients, such as calcium and iron, is increased.

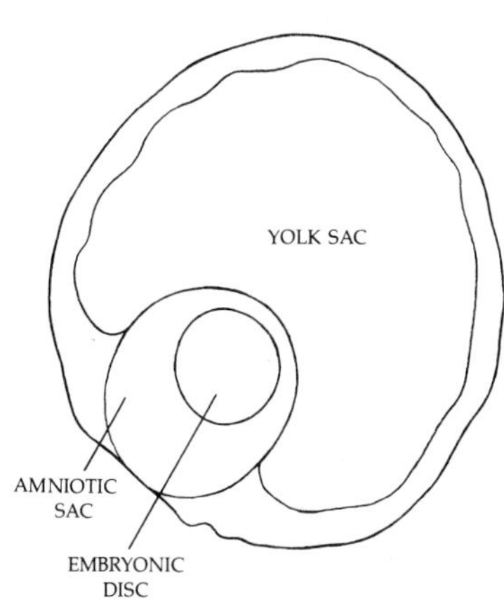

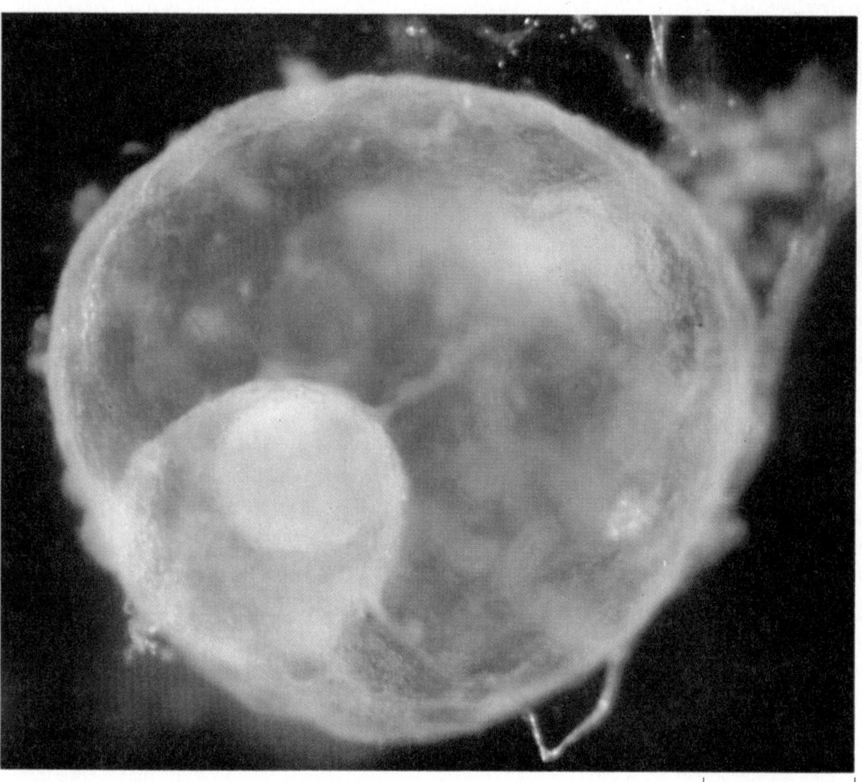

45–26
Human embryo at 13 days, showing the amniotic sac and the yolk sac.

45–27
From the placenta, numerous fingerlike chorionic villi project into the maternal blood space in the wall of the uterus. This space is kept charged with blood from branches of the uterine artery. Across the thin barrier separating fetal from maternal blood, exchange of materials takes place: soluble food substances, oxygen, water, and salts pass into the umbilical vein from the mother's blood; carbon dioxide and nitrogenous waste, brought to the placenta in the umbilical artery, pass into the mother's blood. The placenta is thus the excretory organ of the fetus as well as its respiratory surface and its source of nourishment.

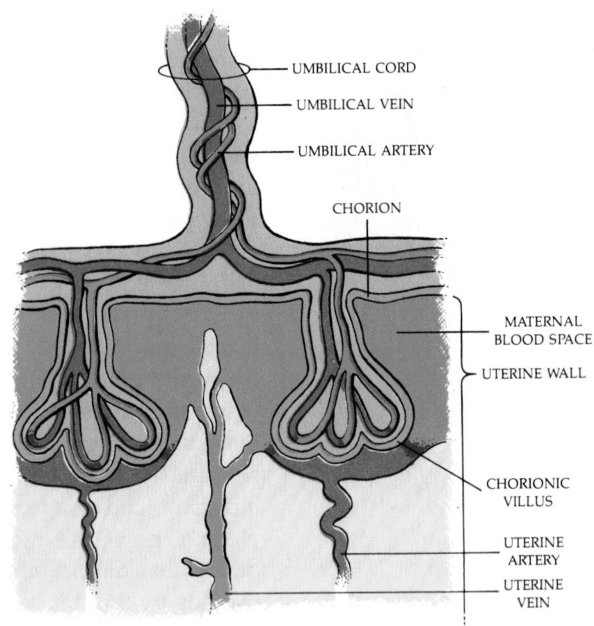

As the embryo grows larger, it remains attached to the placenta by the umbilical cord, which permits it to float freely in its sac of amniotic fluid.

Although the placenta is a foreign tissue, it does not evoke an immunological response, for reasons that are not understood.

Hormones and Pregnancy

The chorion (the outer membrane of the trophoblast) is a source of the hormone chorionic gonadotropin. By the time of the first missed menstrual period, if pregnancy has occurred—that is, within a week after implantation—a pregnancy test based on detection of this hormone will probably be positive.

As the placenta matures, it also begins its own production of estrogens and progesterone. By the end of the third month of pregnancy, the placenta has taken over completely the production of estrogens and progesterone from the corpus luteum; chorionic gonadotropin is no longer produced and the corpus luteum degenerates (Figure 45–28).

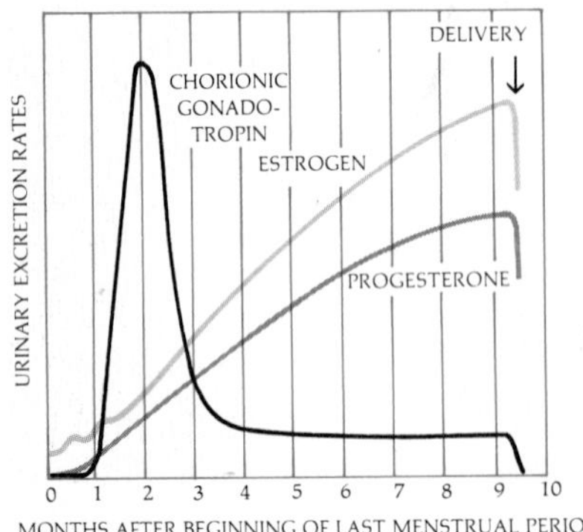

45–28
Levels of estrogen, progesterone, and chorionic gonadotropin excreted in the urine during pregnancy. Urinary excretion rates indicate the concentrations of these hormones in the blood.

The First Trimester

During the second week of its life, the embryo grows to 1.5 millimeters in length, and its major body axis begins to develop. (In this and subsequent measurements, the embryo is measured from crown to rump.) As it elongates, a primitive streak forms, very similar in appearance to the primitive streak of the chick. Cells migrating through the primitive streak form the mesoderm, establishing the three-layered embryo. The dorsal part of the body can be seen to be divided into somites.

During the third week, the embryo grows to 2.3 millimeters long, and most of its major organ systems begin to form. The neural groove (Figure 45–29) is the beginning of the central nervous system (spinal cord and brain), which is the first organ system to develop. By 22 days, the very rudimentary heart, still only a tube, begins to flutter and then to pulsate. From this time on, the heart will not stop its 100,000 beats per day until the death of the individual. Soon after, the eyes begin to form.

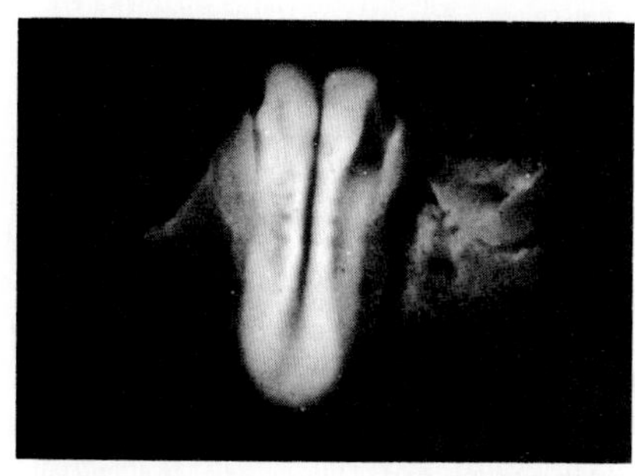

45–29
As the embryo develops, two ridges form and enlarge. The groove between them is called the neural groove. From the neural groove, the dorsal nerve cord—one of the "trademarks" of the chordates—develops. These two folds soon close. This embryo is 18 days old.

Also, by this time, about 100 cells have been set aside (in the yolk sac) as germ cells, from which the egg or sperm cells of the individual will eventually develop. These cells begin to crawl, amoeba-like, toward the site at which the reproductive organs will develop. It has been hypothesized that the germ cells are set aside early in order to minimize the possibility of copying errors that might occur during the many early divisions other cells undergo. Such "errors" would probably be unimportant in somatic cells, which express only a small amount of their genetic information, but could be of major importance in cells that carry half of the genetic instructions for organisms yet to be conceived.

By the end of the first month, the embryo is 5 millimeters in length and has increased its mass 7,000 times. The neural groove has closed, and the embryo is now C-shaped. At this stage, it can be clearly seen that the tissues lateral to the notochord are arranged in paired somites. Each embryo has 40 pairs of somites, from which muscles, bones, and connective tissues will develop. The heart, even as it beats, develops from a simple set of paired, contracting tubes to a four-chambered vessel.

By 38 days, the germ cells reach their destination, the developing gonads. Although the rudimentary gonads have begun to form by this time, male and female embryos are still morphologically identical. Whether or not the infant develops as a male is determined by the influence of some as yet unknown chemical whose production is directed by a gene or genes situated on the *Y* chromosome. Under the direction of this substance, the primitive gonad differentiates as a testis and the germ cells become spermatogonia. In the absence of this substance, the gonad develops as an ovary and the germ cells become oogonia. Apparently the interac-

(a)

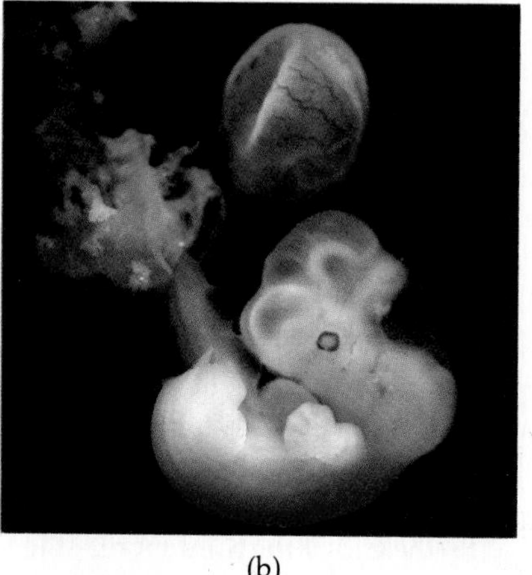

(b)

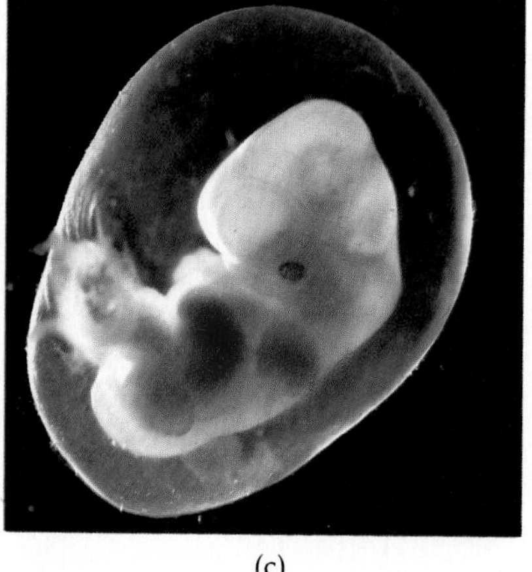

(c)

45–30
Human embryos, early in their development. (a) *In this four-week-old embryo, the neural tube is still open. The protuberances near the top are the future arms; the beginnings of the legs, which develop more slowly, are visible at the bottom. This embryo is 5 millimeters long.* (b) *By five weeks, the embryo is 10 millimeters in length. The major divisions of its brain, an eye, hands, and a long tail can be seen. The yolk sac is at the left.* (c) *At six weeks, the embryo has grown to 15 millimeters in length. As it floats securely in the amniotic cavity, its heart beats rapidly. The brain continues to grow, and the eyes are more developed. The dark red object in the region of the stomach is the liver. Skin folds have formed that will give rise to external ears.*

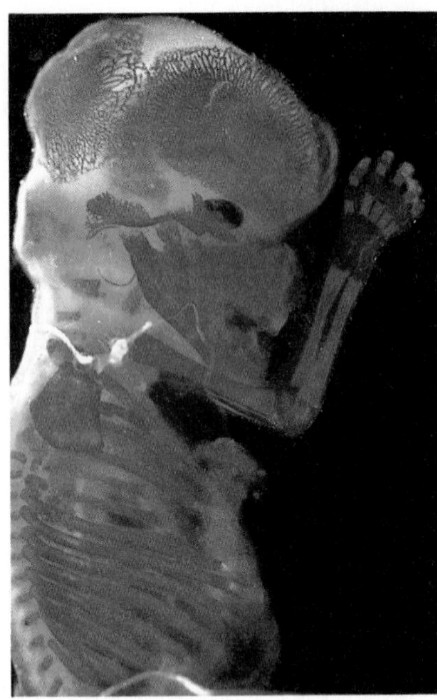

45–31
By the seventh week of embryonic life, the skeleton begins to turn from cartilage to bone, as revealed by differential staining. A gradual layering of membranes, one on top of another, will give rise to the skull, which protects the developing brain.

tions between the germ cells and the primitive gonad are necessary for organogenesis to proceed. As the testis forms, it begins its secretion of androgens, and under the influence of androgens, the external genitalia and other structures become masculinized. If a female embryo is exposed to androgens, it will become similarly masculinized, although, gonadally speaking, it will be female.

Gallbladder and pancreas are present at this stage, and there is clear differentiation of the divisions of the intestinal tract. The liver now constitutes about 10 percent of the body of the embryo and is its main blood-forming organ. Arms, legs, elbows, knees, fingers, and toes are all forming during this time (Figure 45–31). As another reminder of our ancestry, there is a temporary tail.

By the end of the second month, the major steps in organ development are more or less complete, and the embryo, despite its very small size (3 centimeters), is almost human-looking, and from this time on it is generally referred to as a fetus. Its head is still relatively large, because of the early and rapid development of the brain, but the size of the head in proportion to the body will continue to decrease throughout gestation (and throughout childhood as well).

The first two months are the most sensitive period in human development as far as the possible influence of external factors is concerned. For example, when the arms and legs are mere rudiments (fourth and fifth weeks), a number of substances can upset the normal course of events and result in limb abnormalities. The experience with the presumably safe tranquilizer thalidomide in the 1960s is a tragic and familiar example. Because of the widespread publicity about the "thalidomide babies," now becoming thalidomide adults, there was an increase in research into teratogens (substances that cause fetal deformities). A large number of drugs have been found to cause birth defects, and pregnant women and their doctors have become far more cautious about the use of any drug during these critical first months. Heavy alcohol consumption also affects normal organogenesis; frequently, alcoholic mothers have miscarriages during the first trimester.

Infections may also affect the development of the embryo. Rubella (German measles) is a very mild disease in children and adults. Yet when contracted by the mother during the fourth through the twelfth weeks of pregnancy, it can have damaging effects on the formation of the heart, the lens of the eye, the inner ear, and the brain, depending on exactly when the infection occurs in relation to embryonic development. Similarly, exposure to x-rays at doses that would not affect an adult or even an older fetus may produce permanent abnormalities.

During the third month, the fetus begins to move its arms and kick its legs, and the mother may become aware of its movements. Reflexes, such as the startle reflex and (by the end of the third month) sucking, first appear at this time. Its face becomes expressive; the fetus can squint, frown, or look surprised. Its respiratory organs are fairly well formed by this time but, of course, are not yet functional. The external sexual organs begin to develop.

By the end of the third month, the fetus is about 9 centimeters long from the top of its head to its buttocks and weighs about 15 grams (0.5 ounce). It can suck and swallow, and occasionally does swallow some of the fluid that surrounds it in the amniotic sac. The finger, palm, and toe prints are now so well developed that they can be clearly distinguished by ordinary fingerprinting methods. The kidneys and other structures of the urinary system develop rapidly during this period, although waste products are still disposed of through the placenta. By the end of this period—the first trimester of development—all the major organ systems have been laid down.

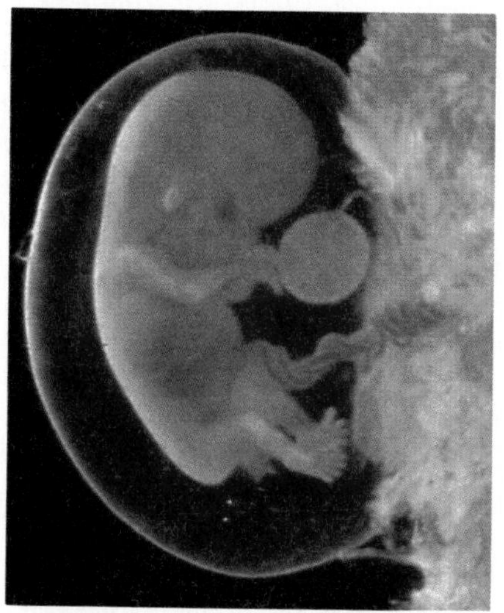

45–32
Embryo at 2 months. The yolk sac is now smaller in comparison to the embryo but still persists. The umbilical cord, connecting the embryo to the placenta, contains both veins and arteries. Portions of the skeleton are visible through the skin. (Roberts Rugh and Landrum B. Shettles, M.D., From Conception to Birth: The Drama of Life's Beginnings, *Harper & Row, Publishers, Inc., New York, 1971.)*

The Second Trimester

During the fourth month, movements of the fetus become obvious to the mother. Its bony skeleton is forming and can be seen with x-rays. The body is becoming covered with a protective cheesy coating. The four-month-old fetus is about 14 centimeters long and weighs about 115 grams (4 ounces).

By the end of the fifth month, the placenta covers about 50 percent of the uterus. The fetus has grown to almost 20 centimeters and now weighs 250 grams. It has acquired hair on its head, and its body is covered with a fuzzy, soft hair called the lanugo, from the Latin word for "down." Its heart, which beats between 120 and 160 times per minute, can be heard with a stethoscope. The five-month-old fetus is already discarding some of its cells and replacing them with new ones, a process that will continue throughout its lifetime. Nevertheless, a five-month-old fetus cannot yet survive outside of the uterus. The youngest fetus on record to survive was about 23 weeks old and required continuous assistance in breathing, taking food, and maintaining its body temperature.

During the sixth month, the fetus has a sitting height of 30 to 36 centimeters and weighs about 680 grams. By the end of the sixth month, it could survive outside the mother's body, although probably only with respiratory assistance in an incubator. Its skin is red and wrinkled, and although teeth are only rarely visible at birth, they are already forming dentine. The cheesy body covering, which helps protect the fetus against abrasions, is now abundant. Reflexes are more vigorous. In the intestines is a pasty green mass of dead cells and bile, known as meconium, which will remain there until after birth.

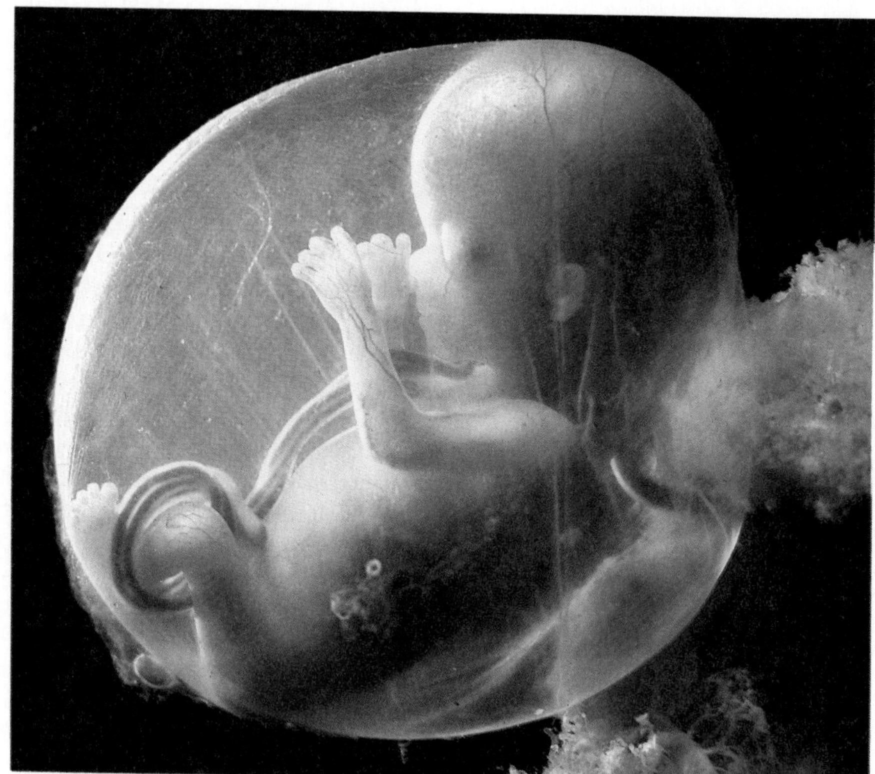

45–33
Although the head of a 12-week-old fetus is disproportionately large, its features are clearly human. Its fingers and toes are fully developed, and external ears and eyelids have formed. The lids are fused and will remain closed for the next three months.

45–34
A fetus at 18 weeks of age. When the thumb comes close to the mouth, the head turns, and the lips and tongue begin sucking motions—an important reflex for survival after birth.

The Final Trimester

During the final trimester, the fetus increases greatly in size and weight. In fact, the fetus normally doubles in size just during the last two months. During this period, many nerve tracts are forming, and the number of brain cells is increasing at a very rapid rate. By the seventh month, brain waves can be recorded, through the abdomen of the mother, from the cerebral cortex of the fetus. Some recent research indicates that the protein intake of the mother is important during this period if the child is to have full development of its brain. Infants weighing less than 2,000 grams (4 pounds, 6 ounces) at birth are at high risk of death or severe brain damage.

As the fetal period progresses, the physiology of the fetus becomes increasingly like the physiology of the adult, and so agents that affect the mother also threaten the late fetus. A depressingly familiar example is found in the infants born with heroin or methadone addiction.

During the last month of pregnancy, the baby usually acquires antibodies from its mother. Antibodies, which are far too large to diffuse across the placenta, are conveyed by highly selective active transport. The immunity they confer is only temporary. Within one to two months after birth, the maternal antibodies will be gradually replaced by antibodies manufactured by the baby's own immune system.

Weight at birth is the major factor in infant mortality. Low-weight infants (under 2,500 grams, or 5.5 pounds) are 40 times as likely to die within a month of birth, and two-thirds of all babies who die are low weight. A black child in the United States is more than twice as likely to be born underweight as a white child and twice as likely to die within four weeks of birth as a white child.

During this last month, the growth rate of the baby begins to slow down. (If it continued at the same rate, the child would weigh 90 kilograms—about 200 pounds—by its first birthday.) The placenta begins to regress and becomes tough and fibrous.

Birth

The date of birth is calculated as about 266 days after conception or 280 days after the beginning of the last regular menstrual period, but only some 75 percent of babies are born within two weeks of the scheduled time. Labor is divided into three stages: dilation, expulsion, and placental stages. Dilation, which usually lasts from 2 to 16 hours (it is longer with the first baby than with subsequent births), begins with the onset of contractions of the uterus and ends with the full dilation, or opening, of the cervix. At the beginning of this stage, uterine contractions occur at intervals of about 15 to 20 minutes and are relatively mild. By the end of the dilation stage, contractions are stronger and occur about every 1 to 2 minutes. At this point, the opening of the cervix is about 10 centimeters. Rupture of the amniotic sac, with the expulsion of fluids, usually occurs during this stage.

The second, or expulsion, stage lasts 2 to 60 minutes. It begins with the full dilation of the cervix and the appearance of the head in the cervix, called crowning. Contractions at this stage last from 50 to 90 seconds and are 1 or 2 minutes apart.

The third, or placental, stage begins immediately after the baby is born. It involves contractions of the uterus and the expelling of small amounts of fluid and blood and the placenta with the umbilical cord attached (also called the afterbirth).

45–35
A human fetus, shortly before birth, showing the protective membranes surrounding it and the uterine tissues. The cervical plug is composed largely of mucus. It develops under the influence of progesterone and serves to exclude bacteria and other infectious agents from the uterus. In 95 percent of all births, the fetus is in this head-down position.

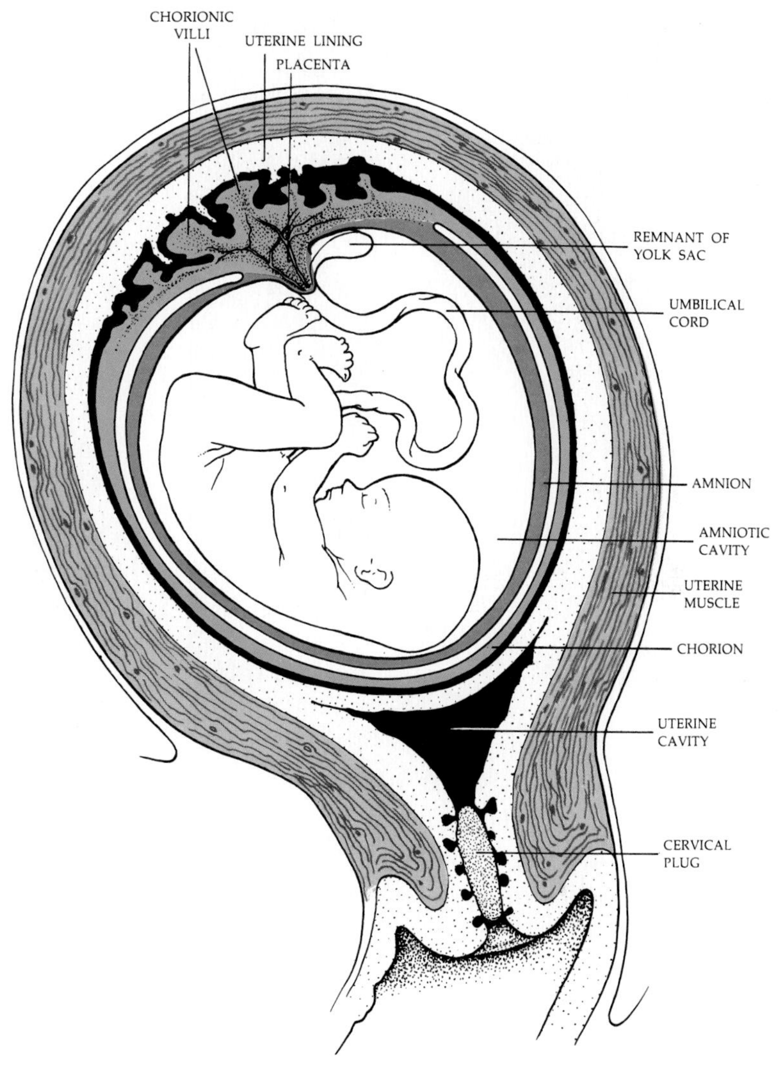

The placenta now weighs about 500 grams, about one-sixth of the weight of the infant. Minor uterine contractions continue; they help to stop the flow of blood and to return the uterus to its prepregnancy size and condition.

The baby emerges from the warm, protective enclosure where it has been nourished and permitted to grow for nine months. The umbilical cord—until that moment, its lifeline—is severed. The baby cries as it takes its first breath, starts to breathe regularly, and so begins its independent existence.

EPILOGUE

When did this particular human life begin? When the sperm encountered the egg? When the embryo became a fetus, becoming obviously human? At the first heartbeat? The first brainwave? When the infant becomes viable as an independent entity? In the past, these matters were discussed by philosophers and theologians

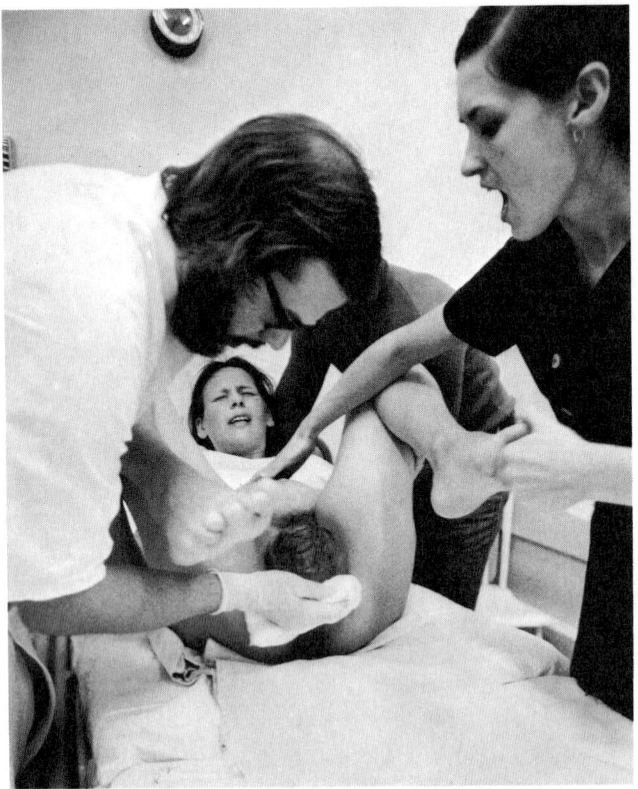

(a)

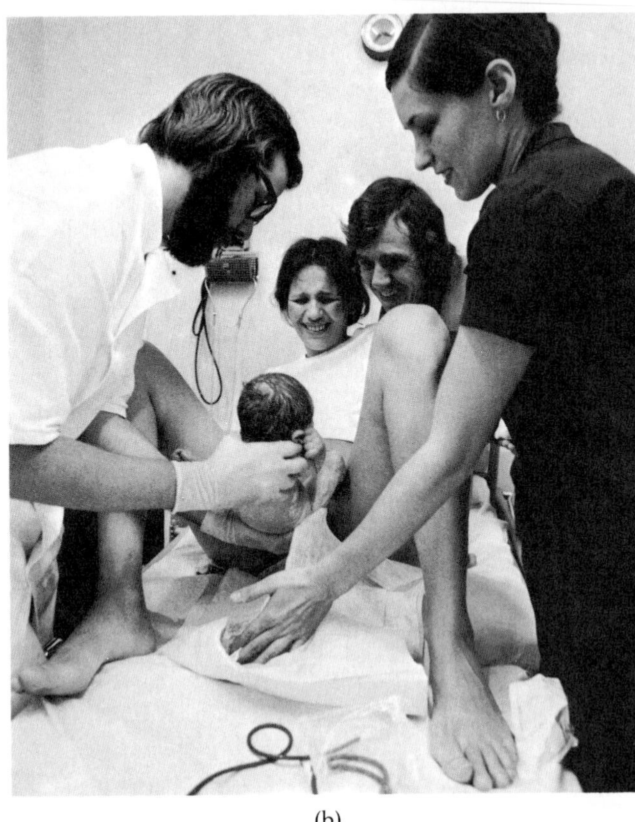

(b)

45–36
The birth of a baby. (a) *The baby's head appears.* (b) *The parents see their newborn infant for the first time. When the umbilical cord—the baby's lifeline throughout its development—is severed, the infant will begin its separate existence.*

who were concerned with the question of when the soul enters the body. These issues have been revived in the ethical and legal controversies concerning abortion. In the evolutionary sense, however, none of these events marks the beginning of life. Life began more than 3 billion years ago and has been passed on since that time from organism to organism, generation after generation, to the present, and stretches on into the future, farther than the mind's eye can see. Each new organism is thus merely a temporary participant in the continuum of life. So is each sperm, each egg, indeed, in a sense, each living cell. Every individual is a unique blend of heredity and experience, never to be duplicated and therefore irreplaceable; but from the perspective of the biological continuum, a human life lasts no longer than the blink of an eye.

SUMMARY

By the process of development, a fertilized egg becomes a complete organism, consisting of billions of cells and closely resembling its parent organisms. This process involves growth, differentiation, and morphogenesis.

Development in most species of animals begins with fertilization—the fusion of egg and sperm. Fertilization results in: (1) changes in the egg cell membrane that prevent fertilization by another sperm cell, (2) introduction of the genetic material of the male parent into the egg, (3) metabolic activation of the egg cell, and (4) cell division.

Development takes place in three stages: cleavage, gastrulation, and organogenesis. During cleavage, the original egg cell divides, with very slight, if any, change in overall volume. In eggs containing little or no yolk, such as those of the sea urchin, cleavage results in cells of approximately equal size. In yolkier eggs, such as those of the frog, cleavage is unequal, with fewer and larger cells in the yolky (vegetal) hemisphere. In eggs with a great deal of yolk, such as the hen's egg, cleavage is limited to a small, nonyolky disc at the top of the egg. In all cases, when cleavage is complete, the embryo consists of a sphere of cells, the blastula, with a central cavity, the blastocoel.

Gastrulation involves the movement of cells into new relative positions and results in the establishment of three tissue layers: endoderm, mesoderm, and ectoderm. In the course of gastrulation, an opening forms, the blastopore. Cells from the outer surface of the embryo move through the blastopore into the blastocoel and form a new cavity, the archenteron, which will be the primitive gut. In the chick and the human, the homologue of the blastopore is the primitive streak. When the movement of cells is complete, the central part of the dorsal surface of the gastrula flattens, becoming the neural plate, which folds to form the neural tube. The cells underlying the neural plate become notochord (a mesodermal tissue), and the somites become embryonic muscle tissue. The endodermal cells, on the interior of the embryo on the ventral side, become gut.

The small area of tissue on the dorsal lip of the blastopore prior to gastrulation is known as the organizer because, as shown by experiments on amphibians, it induces the cells overlying it to form the neural tube. (It itself forms the notochord.) If it is implanted in another embryo, it will induce the formation of a second neural tube in that embryo.

Other organ systems also begin to develop by this same process of embryonic induction, in which one tissue induces changes in the growth or migration of an adjacent tissue with which it comes in contact. The nature of the inducing substances is not known.

Each of the three primary tissue layers established during gastrulation gives rise to particular tissues and organs. The epidermis and the nervous system arise from ectoderm. The notochord, muscles, bones and cartilage, heart and blood vessels, excretory organs and gonads, the inner lining of the skin, and the outer linings of the digestive and respiratory tracts arise from mesoderm. The lungs, salivary glands, pancreas, liver, and inner lining of the digestive tract develop from endoderm. The body cavity, or coelom, is a space, or discontinuity, between two layers of mesoderm.

In amniote eggs, such as those of reptiles, birds, and mammals, the embryo, as it develops, forms four extraembryonic membranes: the yolk sac, amnion, chorion, and allantois.

The human embryo develops as a cell mass, a blastodisc, and an extraembryonic layer of cells, the trophoblast. During these early stages, there is a large increase in the number of cells but little or no increase in the total size of the embryo. When the embryo descends into the uterus, at about its sixth day of development, the trophoblast develops rapidly and invades the maternal tissues. The trophoblast becomes the chorion and eventually the fetal part of the placenta. Like the chick embryo, the human embryo is surrounded by an amnion filled with amniotic fluid. Also present are a yolk sac with no yolk and an allantoic membrane, which develops to become part of the umbilical cord, linking the embryo to the placenta.

When the embryo is about 2 weeks old, a primitive streak forms, followed by the development of a neural plate and neural ridges, which fold to form a neural tube. Although the embryo is still very small (about 2.5 millimeters long), most of the major organs have begun to form in these very early weeks, which is why damage caused to the embryo by viral infection, x-rays, or drugs during this period can be widespread. By the end of the second month, the embryo, now called a fetus, is almost human-looking, although it only weighs about 1 gram. By the end of the third month, all of the organ systems have been laid down. Birth occurs, on the average, 266 days after fertilization of the ovum.

QUESTIONS

1. Distinguish among the following: blastula/gastrula; blastocoel/archenteron; ectoderm/mesoderm/endoderm; fetus/embryo.

2. Describe, in general terms, the end results of each of the following events: fertilization, cleavage, gastrulation, organogenesis.

3. Follow the course of a single cell from its place of origin in the fertilized egg of a frog to its position in the eye cup of an early frog embryo. List, for each stage, all of the influences on this cell that might affect its history.

4. In birds, the fetus is basically masculine and fetal estrogens are necessary to produce female characteristics. Why would this not be a workable arrangement in mammals?

SUGGESTIONS FOR FURTHER READING

Books

BRECHER, EDWARD M., et al.: *Licit and Illicit Drugs: The Consumers Union Report on Narcotics, Stimulants, Depressants, Inhalants, Hallucinogens, and Marijuana—Including Caffeine, Nicotine, and Alcohol,* Little, Brown and Company, Boston, 1972.

A well-balanced, straightforward, and nonsensational review.

CALDER, NIGEL: *The Mind of Man,* 2d ed., Viking Press, Inc., New York, 1973.

A well-written, fast-moving, journalistic report on the "drama of brain research."

ECCLES, JOHN C.: *The Understanding of the Brain,* McGraw-Hill Book Company, New York, 1977.

A general survey, written for the layman, by one of the great leaders in the field of neurophysiology.

ECKERT, ROGER, and DAVID RANDALL: *Animal Physiology,* W. H. Freeman and Company, San Francisco, 1978.

The emphasis is on basic principles. Well-written and handsomely illustrated.

GANONG, W. T.: *Review of Medical Physiology,* 10th ed., Lange Medical Publications, Los Angeles, 1981.*

A concise, accurate, up-to-date summary of human physiology. An excellent reference book.

---

* Available in paperback.

GORDON, MALCOLM S., et al.: *Animal Physiology: Principles and Adaptations*, 4th ed., The Macmillan Company, New York, 1982.

A good animal physiology textbook, intended for an advanced course. The authors emphasize function as it relates to the survival of organisms in their natural environments.

KUFFLER, STEPHEN W., and JOHN G. NICHOLLS: *From Neuron to Brain*, Sinauer Associates, Inc., Sunderland, Mass., 1976.

A lively account of the cellular basis of the workings of the nervous system, based largely on work with which these two distinguished neurobiologists have had first-hand experience.

OATLEY, KEITH: *Brain Mechanisms and Mind*, E. P. Dutton & Co., Inc., New York, 1972.

A readable, well-illustrated survey written by an experimental psychologist for the general reader.

ROMER, ALFRED: *The Vertebrate Story*, 4th ed., The University of Chicago Press, Chicago, 1971.*

The history of vertebrate evolution, written by an expert but as readable as a novel.

ROMER, ALFRED, and THOMAS S. PARSONS: *The Vertebrate Body*, 5th ed., W. B. Saunders Company, Philadelphia, 1977.

A thorough account of comparative vertebrate anatomy. The discussion of embryological development is outstanding.

RUGH, ROBERTS, and LANDRUM B. SHETTLES: *From Conception to Birth: The Drama of Life's Beginnings*, Harper & Row, Publishers, Inc., New York, 1971.

This is an account of the history of life before birth. The book describes in detail the development of the unborn child from the moment of fertilization and also the changes in the mother during pregnancy. It discusses such related topics as birth control, congenital malformation, labor, and delivery of the baby. There are numerous illustrations, including a group of magnificent color photographs of the developing fetus.

SCHMIDT-NIELSEN, KNUT: *Animal Physiology: Adaptation and Environment*, 2d ed., Cambridge University Press, New York, 1979.

Schmidt-Nielsen is concerned with underlying principles of animal physiology—the problems animals have to solve in order to survive. The emphasis is on comparative physiology, and the lucid exposition is illuminated by many interesting examples.

SCHMIDT-NIELSEN, KNUT: *Desert Animals*, Oxford University Press, New York, 1964.

Although considered the definitive work on the physiological problems relating to heat and water, this readable book also contains numerous anecdotes—such as that about Dr. Blagden—and many fascinating personal observations.

SMITH, HOMER W.: *From Fish to Philosopher*, Doubleday & Company, Inc., Garden City, N.Y., 1959.*

Smith was an eminent specialist in the physiology of the kidney. Writing for the general public, he explains the role of this remarkable organ in the story of how, in the course of evolution, organisms have increasingly freed themselves from their environments.

THOMPSON, R. F.: *Introduction to Physiological Psychology*, Harper & Row, Publishers, Inc., New York, 1975.

Intended for the undergraduate student, this text presents an up-to-date survey of the biological foundations of psychology.

---

* Available in paperback.

VANDER, A. J., J. H. SHERMAN, and DOROTHY S. LUCIANO: *Human Physiology,* 3d ed., McGraw-Hill Book Company, New York, 1980.

Most highly recommended. The text is a model of clarity, and the diagrams, many of which we have borrowed, are splendid.

Articles

BARCHAS, JACK D., et al.: "Behavioral Neurochemistry: Neuroregulators and Behavioral States," *Science,* vol. 200, pages 964–973, 1978.

BEACONSFIELD, PETER, GEORGE BIRDWOOD, and REBECCA BEACONSFIELD: "The Placenta," *Scientific American,* August 1980, pages 95–102.

BLUM, FLOYD E.: "Neuropeptides," *Scientific American,* October 1981, pages 148–168.

EPEL, DAVID: "The Program of Fertilization," *Scientific American,* November 1977, pages 128–138.

GUILLEMIN, ROGER: "Peptides in the Brain: The New Endocrinology of the Neuron," *Science,* vol. 202, pages 390–402, 1978.

HELLER, H. CRAIG, LARRY I. CRAWSHAW, and HAROLD T. HAMMEL: "The Thermostat of Vertebrate Animals," *Scientific American,* August 1978, pages 102–113.

JERNE, NIELS KAI: "The Immune System," *Scientific American,* July 1973, pages 52–60.

LESTER, HENRY A.: "The Response to Acetylcholine," *Scientific American,* February 1977, pages 106–118.

LLINÁS, RODOLFO R.: "Calcium in Synaptic Transmission," *Scientific American,* October 1982, pages 56–66.

MARX, JEAN L.: "How the Brain Controls Birdsong," *Science,* vol. 217, pages 1125–1126, 1982.

McEWEN, BRUCE S.: "Interactions between Hormones and Nerve Tissue," *Scientific American,* July 1976, pages 48–58.

MOOG, FLORENCE: "The Lining of the Small Intestine," *Scientific American,* November 1981, pages 154–176.

MORELL, PIERRE, and WILLIAM T. NORTON: "Myelin," *Scientific American,* May 1980, pages 89–116.

PERUTZ, M. F.: "Hemoglobin Structure and Respiratory Transport," *Scientific American,* December 1978, pages 92–125.

ROSE, NOEL R.: "Autoimmune Diseases," *Scientific American,* February 1981, pages 80–103.

ROUTTENBERG, AREH: "The Reward System of the Brain," *Scientific American,* November 1978, pages 154–164.

SNYDER, SOLOMON H.: "Opiate Receptors and Internal Opiates," *Scientific American,* March 1977, pages 44–56.

STALLONES, RUEL A.: "The Rise and Fall of Ischemic Heart Disease," *Scientific American,* November 1980, pages 53–59.

PART III

Biology of Populations

SECTION 7

Evolution

CHAPTER 46

The Evidence for Evolution

As we noted in the Introduction,* the concept of evolution brought about profound changes in the science of biology (and in almost every other aspect of human concern, but that is another story). This intellectual revolution came about for two reasons. First, evidence was accumulating that brought the doctrine of special creation into question; what was missing was a new paradigm with which to replace it. Second, Darwin proposed a theory of how "descent with modification," as he called it, came about, and thus he made evolution scientifically intelligible. It was like dropping a grain of sand in a supersaturated solution; everything crystallized.

In these chapters, we shall look briefly at the accumulating evidence and then at Darwin's theory and its subsequent modifications.

LINES OF EVIDENCE

The Number of Species

One line of evidence that had been building up for more than a century was the existence of an enormous number of species. It was easier to believe that each kind of organism was created separately if one confined oneself to observations of a limited area in the temperate zone, although even there the existence of a score or more of very similar beetles (none of which seemed to serve a higher purpose) was something of an embarrassment. It is no coincidence that Darwin and Wallace had come to doubt the doctrine of special creation while traveling in the tropics, where the widest variety of species is found and where, even today, many species new to science are encountered.

The creationists sought to accommodate doubters of the fixity of species by tinkering with the definition of species. Some had remained the same, they said; other species represented altered forms of the originally created species and so were not really species at all in the special creation sense. Some creationists said it was the genera that were specially created and not the species at all, and so on. The species problem was the first serious fissure in the monolith of special creation. (And, indeed, as we shall see in Section 8, the question of why there are so many species is still a controversial question among many biologists, although not for the same reasons.)

46–1
Two peppered moths (Biston betularia) *on a lichen-encrusted tree trunk.*

* If you have not read the Introduction, we suggest you do so now.

46–2
In 1847, Henry Walter Bates, at the age of 22, formerly an apprentice to a hosiery manufacturer in Leicester, embarked on a trip up the River Amazon with his friend Alfred Russel Wallace. Their purpose: to gather facts "towards solving the problem of the origin of species." In his 11 years in Brazil, Bates collected 14,712 species of insects and other animals, of which some 8,000 were previously undescribed. Bates, like Wallace, was one of many scientists, naturalists, and explorers with whom Darwin corresponded. This drawing of Bates, reproduced from his book The Naturalist on the River Amazon, *shows him collecting a curl-crested toucan.*

Biogeography

A second, somewhat related puzzle that confronted the naturalist explorers—Darwin and Wallace among them—was the distribution of plants and animals in the various regions of the world, or biogeography, as it is now known. According to creationist doctrine, each species was created specially for a particular way of life and placed in the locality for which it was suited—hence, no polar bears in the tropics, for instance. Darwin began his voyage with this point of view. Why then, he wondered, are places so geographically similar populated by very different organisms? Why, for example, did remote oceanic islands have no terrestrial mammals but only peculiar species of bats? Why did England and Europe have rabbits galore, whereas similar areas in South America had only the Patagonian hare—taxonomically not a rabbit or hare at all but a rodent—and Australia a marsupial that resembled a hare?

Or, speaking of Australia, why did this island continent lack placentals and contain a large array of marsupials, all clearly related to one another and found only rarely elsewhere on the planet? For example, was each of the 57 separate species of kangaroos created separately and dropped off in Australia? And why only in Australia? Or was there perhaps an ancestral marsupial that gave rise to all these clearly related forms? (The fossils of extinct marsupials had been described many years before Darwin visited Australia on the *Beagle;* he remarks on "this wonderful relationship on the same continent between the living and the dead.")

Concerning the Galapagos Islands, the question became still more finely focused. The 13 species of finches on the Galapagos resembled South American species and also resembled one another. Had each been created separately and distributed among these volcanic islands (which were generally considered to have been formed much more recently than the mainland)?

Darwin wrote in his notebook: "Seeing this gradation and diversity of structure in one small, intimately related group of birds, one might really fancy that from an original paucity of birds in this archipelago, one species had been taken and modified for different ends."

These observations about geographic distribution did not, of course, disprove the possibility of special creation. (In any case, because of its reliance on a supernatural agency, it cannot be disproved.) What the evidence did, however, was suggest that living things are what and where they are today because of events that happened in the past. Biogeography introduced the notion that life has a history and that changes occur in the course of that history.

The Fossil Record

Geologic studies, as well as the collecting of specimens of plants and animals, were among Darwin's duties as a naturalist aboard the *Beagle.* The coasts of South America were of particular interest because they showed evidence of widespread upheaval, with many geologic strata exposed. The strata, like those studied by William Smith on the British Isles, contained successive deposits of marine shells, some of them found high above sea level. Because of the gradually increasing percentage of modern species in the more recently deposited strata, Darwin was able to estimate their relative ages and to correlate strata from different localities, as Smith had before him (page 3).

In the course of his geologic studies, Darwin came across many fossils of extinct mammals. Among the most interesting to him were those of giant armadillos.

(a)

(b)

(c)

46–3
The rhea of South America (a), *the ostrich of Africa* (b), *and the emu of Australia* (c) *are all very large, flightless birds found in similar habitats on the different continents. Was each a product of a separate act of special creation? For Darwin, biogeographic differences such as these cast doubts on the doctrine of special creation.*

Nowhere else in the world were surviving species of these strange armored mammals found. Was it only a coincidence that extinct armadillos were also found buried in these same South American plains? Here again Darwin encountered tangible evidence of change and history.

Nowhere in the geologic record—either in his own observations or in the reports of others—did Darwin find exactly what he was seeking: evidence of a gradual transition between one species and another. (We shall return to this problem at the end of the section.) Thus, as Darwin reiterates in *The Origin of Species,* the fossil record provided little evidence for him of how evolution came about. For Darwin's contemporaries, however, as for modern observers, the fossil record provided the overwhelming evidence that evolution had indeed occurred. During the decades immediately following the first publication of *The Origin of Species,* many new fossil

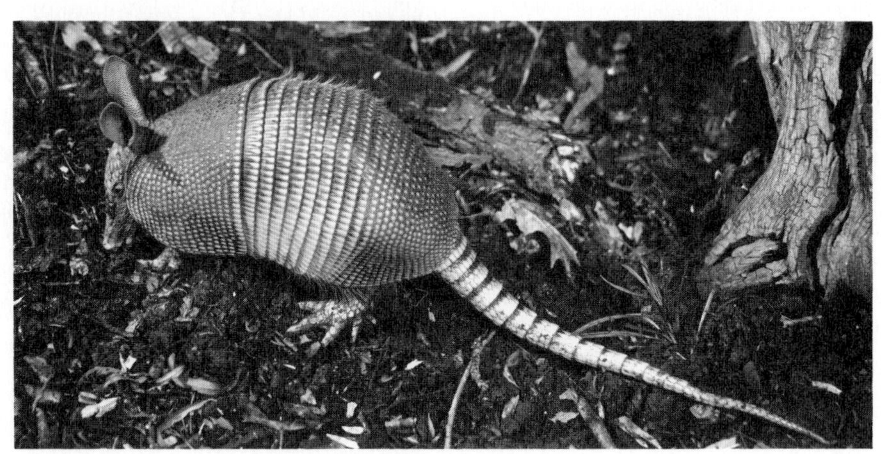

46–4
Armadillos, mammals related to sloths and anteaters, were among the wholly new families of animals Darwin encountered in South America. He was particularly interested in the clear resemblance between the living and fossil forms he found there and later came to regard such resemblances as evidence of evolution. The nine-banded armadillo, shown here, is the one species that has recently begun to spread northward into the southwestern United States.

finds were made. One of these was of *Archaeopteryx,* in 1862; Darwin describes this find in his later editions, noting its combination of avian and reptilian features (see Figure 27–19, page 561).

The most impressive early success in correlating Darwin's theory of evolution with fossil evidence stemmed from the discovery of a long series of fossil horses. In 1879, Othniel C. Marsh of Yale University published a description of horse genealogy, leading from the tiny dawn horse, Eohippus (Figure 46–5), through a number of stages of successively larger animals. These changes in size were accompanied by concomitant major evolutionary changes in teeth, legs, and feet.* Thomas H. Huxley, who characterized himself as Darwin's bulldog, contending with Darwin's critics and arguing the case for evolution before the public, leaned heavily on horse genealogy in presenting the evidence for evolution. Describing Marsh's six stages of horses, spanning some 60 million years, before a meeting of The Zoological Society in 1880, Huxley stated:

> With respect to the interpretations of these facts, two, and only two, appear to be imaginable. The one assumes that these successive forms of equine animals have come into existence independently of one another. The other assumes that they are the result of the gradual modification undergone by the successive members of a continuous line of ancestry. As I am not aware that any zoologist maintains the first hypothesis, I do not feel called upon to discuss it. The adoption of the second, however, is equivalent to the acceptance of the doctrine of evolution so far as horses are concerned, and in the absence of evidence to the contrary, I shall suppose that it is accepted.

Many similar examples have now been uncovered in which lineages can be traced through numerous layers of successive strata, with the higher strata containing those more closely resembling the modern form.

---

* A more complete discussion of the evolutionary history of the horse follows, on page 940.

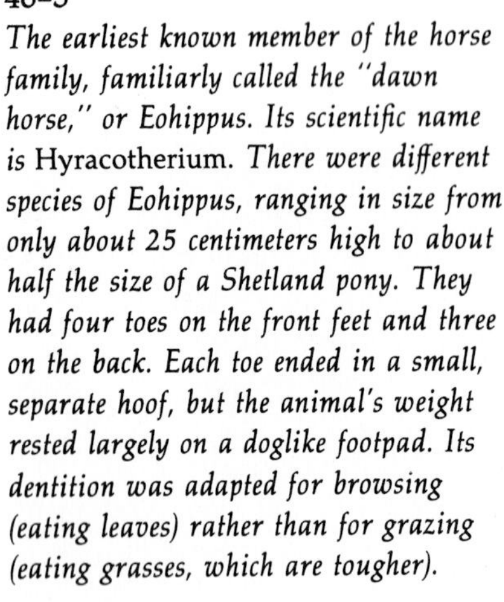

46–5
The earliest known member of the horse family, familiarly called the "dawn horse," or Eohippus. Its scientific name is Hyracotherium. *There were different species of Eohippus, ranging in size from only about 25 centimeters high to about half the size of a Shetland pony. They had four toes on the front feet and three on the back. Each toe ended in a small, separate hoof, but the animal's weight rested largely on a doglike footpad. Its dentition was adapted for browsing (eating leaves) rather than for grazing (eating grasses, which are tougher).*

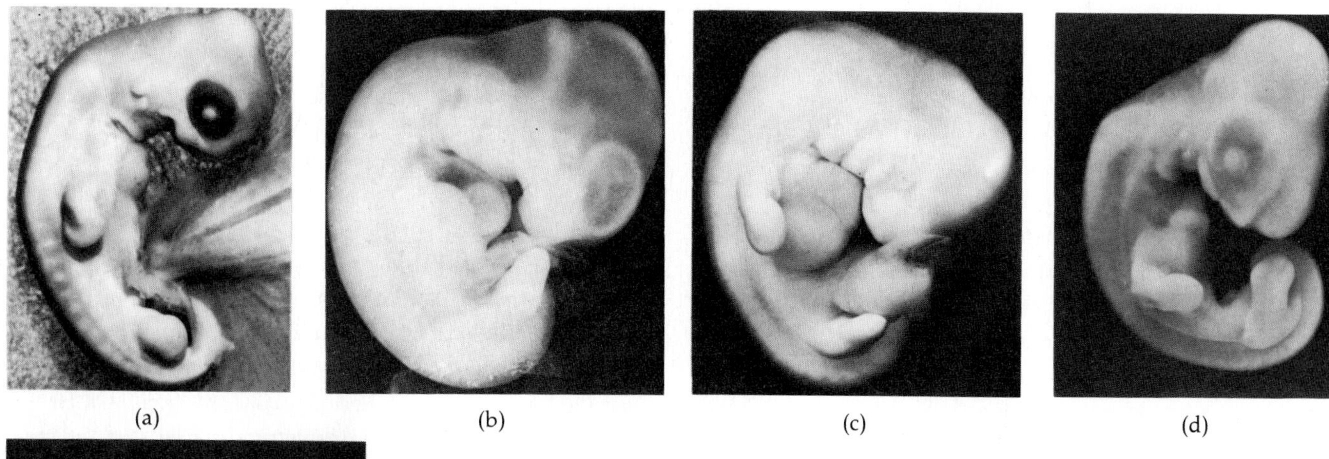

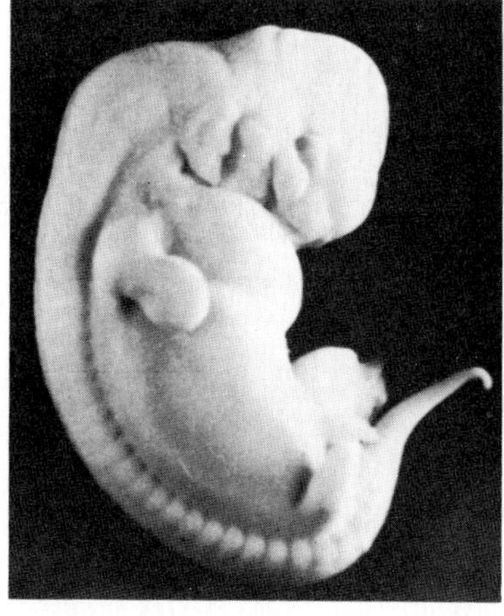

46–6

One of these embryos is human; can you tell which? Homologies among vertebrates are clearly evident early in development, as the photos reveal. Embryo (a) *is a turtle,* (b) *is a mouse,* (c) *is human,* (d) *is a chick, and* (e) *is a pig.*

Homology

A fourth major line of evidence was that of homology, which we discussed to some extent in Chapter 19. All mammals, whether a giraffe or a mouse, have seven cervical vertebrae; if one were starting from scratch, one might choose somewhat different body plans for a giraffe, for instance, than for a meadow mouse. The forelimbs of animals as diverse as humans, dogs, whales, chickens, lizards, and frogs are all constructed of bones arranged in the same pattern. All vertebrates have four limbs, never six, or eight, or a hundred, and all have gill pouches, at least at some stage of their development (Figure 46–6). Whales and even some snakes retain the vestiges of pelvic and leg bones, for which they have no use (Figure 46–7).

Why should an organism created specifically for a particular place in the grand design be modeled on an obsolete pattern and be built of hand-me-down materials? (These were real questions to Darwin, not rhetorical ones; by all evidence, he left England as a creationist.)

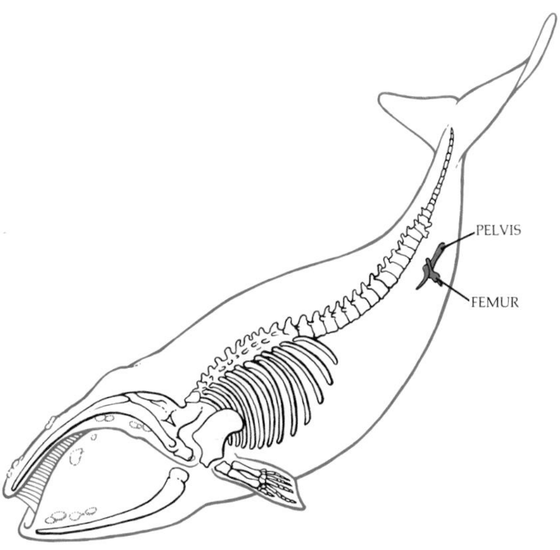

46–7

Many organisms retain traces of their evolutionary history. For example, the whale retains pelvic and leg bones as useless vestiges.

THE RECORD IN THE ROCKS

The earth's long history is recorded in the rocks that lie at or near its surface, layer piled upon layer, like the chapters in a book. These layers, or strata, are formed as rocks in upland areas, are broken down to pebbles, sand, and clay, and are carried to the lowlands and the seas. Once deposited, they slowly become compacted and cemented into a solid form as new material is deposited above them. As continents and ocean basins change shape, some strata sink below the surface of an ocean or a lake, others are forced upward into mountain ranges, and some are worn away, in turn, by water, wind, or ice or are deformed by heat and pressure.

Individual strata may be paper-thin or many meters thick. They can be distinguished from one another by the types of parent material from which they were laid down, the way the material was transported, and the environmental conditions under which the strata were formed, all of which leave their traces in the rock. They can be distinguished, moreover, by the types of fossils they contain. Small marine fossils, in particular, can be associated with specific periods in the earth's history. The fossil record is seldom complete in any one place, but because of the identifying characteristics of the strata, it is possible to piece together the evidence from many different sources. It is somewhat like having many copies of the same book, all with chapters missing—but different chapters, so it is possible to reconstruct the whole.

The geologic eras—Precambrian, Paleozoic, Mesozoic, and Cenozoic—which are the major volumes of the geologic record, were identified and named in the early nineteenth century. These eras were subdivided into periods, many of which are named, quite simply, for the areas in which the particular strata were first studied or studied most completely: the Devonian for Devonshire in southern England, the Permian for the province of Perm in Russia, the Jurassic for the Jura Mountains between France and Switzerland, and so on.

Early attempts to date the various eras and periods were based simply on their relative ages compared to the age of the earth; obviously, a stratum occurring regularly above another was younger than the one below it. The first scientific estimate was made in the mid-1800s by the famous British physicist Lord Kelvin. On the basis of his calculations of the time necessary for the earth to have cooled from its original molten state, Kelvin maintained that the planet was about 100 million years old, a calculation that posed considerable difficulties for Darwin. (Kelvin was not aware of the existence under the earth's surface of radioactive materials that heat the planet from within.) In the last 40 years, however, new methods for determining the ages of strata have been developed involving measurements of the decay of radioactive isotopes. As a result, the estimated age of the earth has increased, in little more than a century, from 100 million years to 4.6 billion.

Geologic strata are now dated by analysis of radioactive isotopes contained in crystals of igneous rock (rock formed from molten material) associated with particular strata. As we noted in Chapter 1, many naturally occurring isotopes are radioactive. All the heavier elements—atoms that have 84 or more protons in the nuclei—are unstable and, therefore, radioactive. All radioactive elements emit energy (as particles or rays) at a fixed rate; this process is known as

Adaptation

Finally, there was the question of adaptation. Adaptation is a word with several meanings in biology. First, it can mean a state of being adjusted to the environment; every living organism is adapted in this sense, just as Abraham Lincoln's legs were, as he remarked, "just long enough to reach the ground." Second, adaptation can mean a process, which may take place either within the lifetime of an individual (physiological adaptation)—such as producing more red blood cells in response to dwelling at high altitudes—or in a population over the course of the passage of generations (evolutionary adaptation). Third, adaptation is commonly used to refer to a particular characteristic, that which is adapted, such as an eye or a hand. In this discussion, we shall concentrate on the last definition because it has been the center of the evolutionary controversy. How is it possible for an organ of such perfection as the eye, for example, to come into existence from nothingness, the creationists argued (as some argue today). What use would half an eye be?

In the badlands of Montana, geologic strata have been carved out by the Missouri River and its tributaries. The lower strata are from the late Cretaceous period and the upper ones are from the Paleocene (see page 457). These formations are of particular interest because they may hold clues to one of the major mysteries in evolutionary history—the worldwide extinction of the dinosaurs.

radioactive "decay." The rate of decay is measured in terms of half-life: the half-life of a radioactive element is defined as the time in which half the atoms in a sample of the element lose their radioactivity and become stable. Since the half-life of an element is constant, it is possible to calculate the fraction of decay that will take place for a given isotope in a given period of time. The radiometric clock starts to tick when the crystal is formed.

Half-lives vary widely, depending on the isotope. The radioactive nitrogen isotope ^{13}N has a half-life of 10 minutes, and tritium has a half-life of 12.25 years. The most common isotope of uranium (^{238}U) has a half-life of 4.5 billion years. The uranium atom undergoes a series of decays, eventually being transformed to an isotope of lead (^{206}Pb). Thus the proportion of ^{238}U to ^{206}Pb in a given rock sample, for example, is a good indication of how long ago that rock was formed. Five different isotopes are now commonly employed as radioactive clocks, and in thousands of instances, rocks have been dated by three or more independent clocks.

Any theory of evolution requires, as Darwin knew well, that the earth have a long history. Thus, these radioactive clocks are doubly important to modern students of evolution. First, they demonstrate conclusively that the earth's age is close to 5 billion years; in other words, the earth is indeed old enough. Second, they provide the tools for estimating the relative ages of various fossils and so, of unraveling the details of the earth's biological past.

In the course of his career as a naturalist, Darwin amassed an enormous amount of information about living organisms; assembled in someone else's brain, much of it would have been charming, unrelated trivia. On the basis of this vast knowledge, Darwin knew that not all adaptations—"contrivances," he called them—were perfect, and this was his strongest point in dealing with the creationist's objections, at least in his arguments with himself. Adaptations were simply as good as they had to be, long enough to reach the ground.

For instance, he uses a page to describe the highly ingenious workings of an orchid flower by which the flower "arranges" for a bee to fall into a pool of water, held in the highly modified petal that forms the lip, so it cannot fly out of the flower. Instead, the bee must crawl through narrow passageways and so deposit and pick up pollen in the course of its exit. However, Darwin also describes orchids and other types of plants with much less elaborate flowers, offering insects only bright colors and a few drops of nectar, and finally plants with simple, inconspic-

46–8
An American bittern hiding among the reeds. As Darwin noted, in cases of mimicry or camouflage, each chance "improvement" in the disguise would be conserved by natural selection, with the end result representing the accumulation of such changes. Note that the behavior of the bittern—standing motionless with its neck elongated—is as important as its anatomical adaptations. Similarly, the orientation of the moths shown in Figure 46–1 plays an important role in concealing them.

uous flowers and windblown pollen. In short, there are gradations and varieties of adaptations, not merely an array of perfect solutions.

From his correspondence with Wallace and Bates, Darwin knew of remarkable examples of defensive mimicry among insects. He mentions a walking stick insect that not only is shaped like a stick but also is overgrown with material that exactly resembles native mosses. Darwin used this as a near-perfect example of how gradual "improvements," each carefully conserved, could be continuously advantageous to the individual organisms (as, of course, could a gradual improvement in the capacity to detect light and focus an image, leading ultimately to the eye).

DARWIN'S THEORY

All of Darwin's contemporaries—whether amateurs, as was Darwin himself, strictly speaking, or academics—had access to much of this same information. The revolution would have eventually taken place even if Darwin had never existed, just as the structure of DNA would someday have been discovered in the absence of Watson and Crick. It is doubtful that Wallace alone could have persuaded the scientific world to accept his theory of evolution. He might well have been just another traveler with an odd tale that could not yet be made to fit. It was Darwin's dogged intellectual pursuit of the subject and his vast body of personal knowledge that made it come about at the time and in the way that it did—as an intellectual avalanche.

Although it is now some 125 years since the first publication of *The Origin of Species,* Darwin's and Wallace's original concept of how evolution comes about still provides the basic framework for our understanding of the process. Their concept rested on four premises:

1. Organisms beget like organisms—in other words, there is stability in the process of reproduction.
2. In any given population, there are chance variations among individual organisms—that is, variations that are not produced by the environment—and some of the variations are inheritable.
3. In most species, the number of individuals that survive and reproduce in each generation is small compared with the number produced.
4. Which individuals will survive and reproduce and which will not are determined to a significant degree by the interaction between these chance variations and the environment. Some variations enable individuals to produce more offspring than other individuals. Darwin termed these "favorable" variations and argued that inherited favorable variations tend to become more and more common from one generation to the next. This is the process that Darwin called *natural selection*.

EVOLUTION IN ACTION

Darwin believed evolution to be such a slow process that it could never be observed directly. However, the effects of human civilization have produced such extremely strong selection pressures on some organisms that it has been possible to observe not only the results but also the actual process of evolution by natural selection.

(a)

(b)

46–9

The two varieties of Biston betularia, *the peppered moth, resting on* (a) *a lichen-covered tree trunk in an unpolluted English countryside and* (b) *a dark tree trunk, near Manchester. If you look carefully, you will notice that there are two moths in each photograph.*

Kettlewell demonstrated that on lichen-covered trees the black moths are more likely to be eaten by birds (which are visually oriented), whereas on polluted trees the light-colored moths are more likely to be eaten.

The Peppered Moth

One of the best-studied examples of natural selection in action is that of *Biston betularia,* the peppered moth. These moths were well known to British naturalists of the nineteenth century, who remarked that they were usually found on lichen-covered trees and rocks. Against this background, the light coloring of these moths made them practically invisible, concealing them from predatory birds. Until 1845, all reported specimens of *Biston betularia* had been light-colored, but in that year one black moth of this species was captured in the growing industrial center of Manchester.

With the increasing industrialization of England, smoke particles began to pollute the foliage in the vicinity of industrial towns, killing the lichens and leaving the tree trunks bare. In heavily polluted districts, the trunks and even the rocks and ground became black. During this period, more and more black *Biston betularia* were found. This black form spread through the population until black moths made up 98 percent of the Manchester population. Where did the black moths come from? Eventually, it was demonstrated that the black color was the result of a rare, recurring mutation. The moth had always been there, in very small numbers.

Why were they increasing so rapidly? H. B. D. Kettlewell was among those who hypothesized that the color of the moths protected them from predators, notably birds. In the face of strong opposition from entomologists—all of whom claimed they had never seen a bird eat a *Biston betularia* of any color—he set out to test his hypothesis. He marked a sample of moths of each color by carefully putting a spot on the underside of the wings, where it could not be seen by a predator. Then he released known numbers of marked individuals into a bird reserve near Birmingham, an industrial area where 90 percent of the local population consisted of black moths. Another sample was released into an unpolluted Dorset countryside, where no black moths ordinarily occurred. He returned at night with light traps to recapture his marked moths. From the area around Birmingham, he recovered 40 percent of the black moths, but only 19 percent of the light ones. Within the county of Dorset, 6 percent of the black moths and 12.5 percent of the light moths were retaken. Clearly, if you are a moth, it is advantageous to be black near Birmingham but light around Dorset.

To clinch the argument, Kettlewell placed moths on tree trunks in both locations, focused hidden movie cameras on them, and was able to record birds actually selecting and eating the moths, which they do so rapidly that it is not surprising it was not previously observed. When equal numbers of black and light moths were available near Birmingham, the birds seized 43 light moths and only 15 black ones; near Dorset, they took 164 black moths but only 26 light forms. (Industrial melanism, the tendency for dark-colored forms to replace light-colored forms in polluted areas, has been found among some 70 other moth species in England and some 100 species of moths in the Pittsburgh area of Pennsylvania. It has also been demonstrated in many species of butterflies.)

By the late 1890s, black moths formed more than 95 percent of the population in English industrial areas. Then, in the mid-twentieth century, strong pollution controls were instituted in England, and the heavy soot accumulation has begun to decrease. The light moths are now increasing in proportion to the black forms.

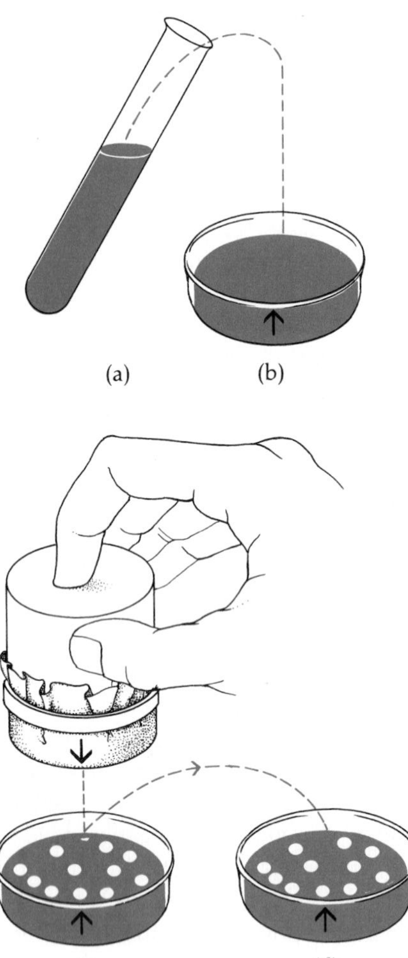

46–10
The Lederbergs' replica-plating method for detecting and isolating drug-resistant bacteria. (a) *The bacteria are cultured in a broth containing nutrient molecules.* (b) *A sample of the cell suspension is spread over the surface of a petri dish containing broth solidified with agar.* (c) *The plate is incubated until colonies appear. A piece of velveteen held by a ring that fits snugly around a cylindrical block is used to transfer a sample of each colony to a replicate plate, another petri dish containing a solidified medium* (d).

There is a moral to this story. Note that the black moth is not absolutely superior to the light one, or vice versa. It is, as Darwin realized, entirely a matter of time and place.

Drug Resistance in Bacteria

A more recent example concerns the development of resistance in bacteria. Following the rapid development of antibiotics after World War II, bacteria soon began to appear that were resistant to these agents, so widely hailed as "miracle drugs." What caused this resistance? Changes in the metabolism of individual bacteria or changes in the bacterial population—that is, evolution?

The question, important for medical as well as scientific reasons, was answered by a beautifully simple experiment performed by Joshua and Esther Lederberg. A set of petri dishes were prepared, each containing agar (a jellylike substance) and each marked on the side with an arrow. Bacterial cells grown in a nutrient broth were spread in a thin layer on one such dish. Within 24 hours, visible colonies began to appear, clones of the original scattered cells. The experimenters prepared a cylindrical block the same diameter as the inside of the petri dish, covered one end of it with a piece of velveteen and marked it on the side with an arrow. Then the arrow was lined up with the arrow on the petri dish, and the velveteen surface was pressed against the agar, picking up bacteria. The velveteen surface was then pressed against agar containing penicillin in another petri dish. Again, the position was marked with an arrow. Each colony was thus placed in the new dish in exactly the same position it had occupied originally. Usually, the bacterial colonies failed to survive in the penicillin medium, but eventually the Lederbergs found what they were looking for: bacteria growing on the penicillin plate.

In the next step of the experiment, they located the colony on the original, untreated petri dish that corresponded with the colony that had grown on the penicillin-laced agar. They transferred a sample of cells from this untreated colony to another petri dish containing no penicillin. The cells were allowed to grow, and when samples of these cells were tested in a medium containing penicillin, it was found that they too were penicillin-resistant, even though they had never been exposed to penicillin. Like the black moths, they were simply variants produced by chance in the original population and selected by their environment.

THE THEORY TODAY

Since Darwin's time, massive additional evidence has accumulated supporting the fact of evolution—that all living organisms present on earth today have arisen from simpler forms in the past over the course of the earth's long history. Indeed, all of modern biology is an affirmation of this relatedness of the many species of living things and of their gradual divergence from one another over the course of time. Cytology and biochemistry have revealed innumerable new homologies at the submicroscopic and molecular levels of organization. Many new examples of the effects of geography on variation within and among species have been discovered. Even more compelling, perhaps, are the additional examples yielded by the fossil record of lineages changing through time. (The most dramatic of these, the evolution of hominids, will be discussed in detail in Chapter 56.)

Since the publication of *The Origin of Species,* the important question, scientifically speaking, about evolution has not been whether or not evolution has taken place. That is no longer an issue among the vast majority of modern biologists. Today, the central and still fascinating questions for biologists concern the details of how evolution takes place. This is the subject of the four chapters that follow.

SUMMARY

There were five principal categories of evidence available to Darwin at the time he formulated his theory of evolution: (1) the large number and diversity of species; (2) the fact that particular kinds of organisms were confined to particular geographic areas, apparently as a consequence of natural barriers; (3) the fossil record, which showed that organisms had a long history and had changed in the course of time; (4) homologies, or similarities, among structures of diverse organisms, implying a common ancestry; and (5) adaptations, the evidence that gradual changes occur in populations over time in response to selective forces in the environment.

The strong new element in Darwin's theory of evolution was that he saw it as a two-stage process involving, first, inheritable variations produced by chance and, second, natural selection, the operation of the environment upon these variations.

Two modern examples demonstrating natural selection acting on chance variations are the increase in number of the black form of *Biston betularia* in industrial areas and the development of drug resistance in bacteria.

QUESTIONS

1. Since Darwin's time, biochemical analyses and electron microscopy have uncovered many new homologies. Name five.

2. Could the fossil record disprove that evolution has occurred? How?

3. Could the doctrine of special creation be disproved? How?

4. Many of Darwin's arguments concerning evolution focused on imperfections rather than perfections. Can you explain why?

5. Darwin emphasized that when mammals (cats, goats, sheep, rodents, and so on) were introduced onto remote islands they often thrived there, yet there were no indigenous mammals on the islands except bats. Can you explain why he considered this important?

6. What would have been the Lamarckian explanation (see page 5) for the development of drug resistance in bacteria? How did the Lederbergs' experiment disprove Lamarck's proposal as to how evolution takes place?

CHAPTER 47

The Genetic Basis of Evolution

47–1
According to Theodosius Dobzhansky (1900–1975), "Nothing in biology makes sense except in the light of evolution." The publication in 1937 of his Genetics and the Origin of Species *is considered to mark the birth of the modern synthetic theory.*

As we noted in Chapter 11, Mendel's principles were rediscovered early in the twentieth century, initiating the emergence of genetics as a major field of biological study. These developments in genetics had great consequences for the study of evolution. A major weakness in Darwin's theory was the lack of a valid mechanism of heredity. Mendelian genetics explained, as Darwin could not, why genetic traits are not "blended out" but can disappear and reappear (like whiteness in pea flowers).

It is instructive, however, for anyone interested in the course of science, that the early geneticists, including Bateson, de Vries, and even T. H. Morgan, did not accept Darwin's theory. Working with selected strains of organisms under laboratory conditions, they did not see the great range of variation in natural populations that Darwin had seen as a naturalist. The chief agent of evolution, in their eyes, had to be mutation. Most mutations they knew to be harmful; these were eliminated by natural selection. Occasionally, a useful mutation would come along and thus, suddenly and fortuitously, the species would be set on a new evolutionary course.

Darwinian evolution and Mendelian genetics became reconciled when biologists stopped thinking in terms of individual organisms and genotypes and began thinking in terms of populations, genes, and the frequencies of alleles. As we shall see, the resulting revised theory of evolution is the prevailing one today, although it too is now being questioned and modified. This theory is known as neo-Darwinian, because it explains evolution in terms of the Darwinian two-step model of variation and selection. It is also known as the *synthetic theory*, because it combines Darwin's theory of evolution with the principles of Mendelian genetics. (Here "synthetic" does not mean "artificial," which is the connotation it has for us in these days of manmade fibers and artificial flavorings, but has its original meaning of "the putting together of two or more different elements.")

The branch of biology that emerged from this synthesis of Darwinian evolution and Mendelian principles is known as *population genetics*. A *population* is defined as an interbreeding group of organisms. All of the fish of one particular species in a pond constitute a population, as do all the fruit flies in a bottle. A population is unified and defined by its *gene pool*, which is simply the sum total of all the alleles of all the genes of all the individuals in the population. From the viewpoint of the population geneticist, the individual organism is only a temporary vessel, holding a small sampling of the gene pool for a moment of time.

47–2
J. B. S. Haldane (1892–1964), whose comment on the plenitude of beetles has already been noted (page 536), was one of several mathematicians responsible for shaping the synthetic theory. Haldane once remarked that he would lay down his life for two brothers or four first cousins, thereby anticipating the basic tenet of sociobiology by some 40 years, as we shall see in Chapter 55.

In natural populations, some alleles increase in frequency from generation to generation, and others decrease. If an individual has a favorable combination of alleles, these alleles are more likely to be present in an increased proportion in the next generation. If the combination is not favorable, its representation in the next generation will be reduced or perhaps eliminated.

Fitness, in the context of population genetics, does not mean physical well-being or optimal adaptation to the environment. The only criterion—the measurement—of an individual's fitness is the relative number of surviving offspring, that is, the extent to which its genotype is present in succeeding generations. Evolution is the result of such accumulated changes in the gene pool. Thus, the gene pool changes slowly over time.

THE CONCEPT OF THE GENE POOL

For purposes of this discussion, imagine, if you will, a population of sexually reproducing diploid organisms in which a particular gene has only two alleles, *A* and *a*, and the frequency of allele *A* is 0.8 and the frequency of allele *a* is 0.2. (Frequencies are equivalent to percentages; if the frequency of allele *A* is 0.8, the percentage of allele *A* is 80 percent.) At every new generation, after these alleles segregate and recombine, according to Mendelian principles, we sample the population again. Will we expect to find the same frequency of *A* and *a*? If not, why not?

What can cause a change in the relative frequency of alleles? Mutation can. If, for example, there is a tendency for *A* to mutate to *a* (and it is stronger than the tendency for *a* to mutate to *A*), the frequency of *a* will increase. Migration into and out of a population can cause a change, if the migrants do not have the same relative proportions of *A* and *a* as the parent population. Chance can change the proportion of alleles. If we take too small a sample, it may not be representative of the parent population. Similarly, if the population itself becomes too small, the next generation may not be a representative sample of the alleles present in the previous one. Finally, differences in survival rates or reproduction rates (natural selection) can change the proportion of alleles, as in the case of the peppered moth discussed in the last chapter. In fact, natural selection is the principal agent producing changes in the frequency of alleles in a gene pool.

Alleles and Genotypes: The Hardy-Weinberg Principle

A gene pool is, of course, an abstraction. Genes exist only in genotypes, and genotypes exist only in organisms. What is the relationship between the gene pool and the organisms in the population? Or, to put it in more practical terms, how do we, given a population of individual organisms, make a correlation between them and the change in the frequency of an allele? The answer to this question was formulated independently, in 1908, by G. H. Hardy, an English mathematician, and G. Weinberg, a German physician, and is known as the Hardy-Weinberg principle. It is almost as simple as Mendel's principles, and it is as fundamental to population genetics as Mendel's principles were to classical genetics.

To demonstrate the Hardy-Weinberg principle, let us return to our hypothetical population with its alleles *A* and *a* and engage in some very simple algebra. Let the symbol p represent the frequency of allele *A* and the symbol q represent the frequency of allele *a*. Then, $p + q = 1$, 1 being 100 percent, or all the alleles of that

SURVIVAL OF THE FITTEST

The phrase "survival of the fittest" is often used in describing the Darwinian theory. In the early twentieth century, the doctrine of survival of the fittest in natural populations was used by some individuals to defend gross social inequalities and ruthless competitive tactics in industry on the grounds that they were merely in accord with the "laws of nature." This philosophy was referred to by some as social Darwinism. However, in fact, very little evolutionary change fits the concept of "nature red in tooth and claw." One fuchsia flowering a little brighter than its neighbors and better able to catch the attention of a passing hummingbird is a more pertinent model of the struggle for survival. Fitness, as measured by population geneticists, is determined solely by the relative number of descendants of an individual in a future population.

Natural selection in operation. A bee is making her choice.

47–3
The relationship between the frequency of allele a in the population and the frequency of the genotypes AA, Aa, and aa. Naturally, the more AAs there are, the lower the frequency of a. Because of the interrelationship of AA, Aa, and aa, a change in the frequency of either allele results in a corresponding and symmetrical change in the frequencies of the other allele and of the genotypes.

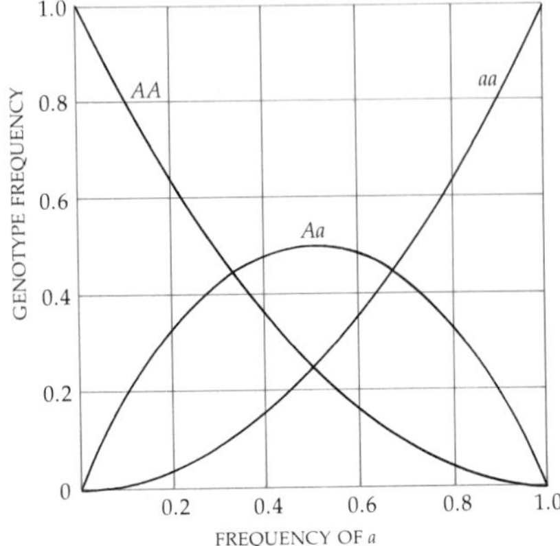

particular gene. Let us assume that the relative frequencies of *A* and *a* are the same in both male and female (as they are for most alleles in natural populations). If males and females mate at random with respect to the alleles *A* and *a*, the frequency with which male gametes carrying allele *A* will combine with female gametes with allele *A* will be $p \times p$, or p^2 (0.64 in our particular example). Similarly, the frequency with which male gametes carrying allele *a* will combine with female gametes with allele *a* will be $q \times q$, or q^2 (0.04). The probability of a male gamete carrying allele *A* combining with a female gamete carrying allele *a* is $p \times q$, or, in our example, 0.8×0.2, or 0.16; the probability of a male gamete carrying allele *a* combining with a female gamete carrying allele *A* is the same. Thus, the total probability of *Aa* combinations is $2pq$ (0.32). So, all of the possible combinations of two alleles in a population can be represented by the expression $p^2 + 2pq + q^2$, in which p equals the frequency of one allele and q the frequency of another. The expression $p^2 + 2pq + q^2 = 1$ is equivalent to $(p + q)^2$. As a moment's reflection will convince you, the relative frequencies of the genotypes will remain constant generation after generation if the allele frequencies do not change.

Multiple Alleles

In the previous discussion, we talked about only two alleles, but actually there may be many alleles of the same gene. For instance, in the case of the gene for red eyes carried on the female chromosome in *Drosophila* (see Chapter 13), there are a variety of alleles, producing intermediate colors known by names such as apricot, buff, eosin, and honey, as well as the mutant allele that results in white eyes. More than 15 alleles have been recognized, and probably more could be detected if there were ways of distinguishing finer gradations of color.

The Hardy-Weinberg formula applies equally well when there are more than two alleles of the same gene, although the calculations then become more complex. For example, the genotype frequencies of three alleles are expressed by the algebraic expansion of $(p + q + r)^2$, with r representing the frequency in the gene pool of the third allele.

THE AGENTS OF CHANGE

The Hardy-Weinberg principle states that in the absence of mutation, migration, and natural selection, and in a population that is sufficiently large (that is, no sampling error), the relative frequencies of alleles will remain the same, and, if mating is random with respect to the alleles in question, the proportions of each genotype will remain the same.

Since no population is free of all of these agents of change, it is natural to question the usefulness of the Hardy-Weinberg principle. However, to borrow an analogy, Newton's first law says a body remains at rest or maintains a constant velocity when not acted upon by an external force. Bodies are always acted upon by external forces, but this first law is an essential premise for examining the nature of such forces. Similarly, the frequencies of alleles in populations are always changing, but in the absence of the Hardy-Weinberg principle, we would not know how to detect change or to uncover the forces responsible for it. According to the Darwinian (and neo-Darwinian) theory, natural selection is the major force in changing allele frequencies. We shall examine the workings of natural selection and some of the current controversies about it in Chapter 49. Here, let us look at some of the other agents that change the frequencies of alleles in a population. There are four: mutations, migration, sampling errors, and, in some instances, nonrandom mating.

Mutations

Mutations, from the point of view of population genetics, are inheritable changes in the genotype. They occur by chance, in the sense that the type of mutation produced cannot be influenced by the environment. The rate of mutation can be influenced by environmental factors (such as radiation), however, and also different genes have different rates of mutation for reasons that probably have to do with the chemical composition of the gene in question and its arrangement in the chromosome. The average rate of mutations detectable in the phenotype varies between 1 in 1,000 to 1 in 1,000,000 gametes per generation, depending on the allele involved. If you assume that there are about 100,000 genes (pairs of alleles) per human, you can assume that each new human being will carry two new mutations. Thus, although the incidence of mutation in any given gene is low, the number of new mutations per population generation is very high. Mutations are generally

47–4
An example of the sometimes dramatic effects of mutation. The ewe in the middle is an Ancon, an unusually short-legged strain of sheep. The first Ancon on record was born in the late nineteenth century into the flock of a New England farmer. By inbreeding (the trait is transmitted as a recessive), it was possible to produce a strain of animals with legs too short to jump the low stone walls that traditionally enclosed New England sheep pastures. A similar strain was produced in northern Europe as a result of an independent mutation. At one time, it was proposed that evolution takes place in sudden, large jumps such as this—a concept sometimes referred to as the "hopeful monster" theory. One reason this concept has been abandoned is that nearly all mutations producing dramatic changes in the phenotype are harmful, as this one would be in a wild population.

regarded as the raw material for evolutionary change; they provide variation for other evolutionary forces to act on, but they do not determine the direction of evolutionary change.

Migration: Gene Flow

Gene flow is the movement of alleles into or out of a population as a result of the immigration or emigration of breeding individuals or, as in the case of plants, the introduction of gametes (in the form of pollen) from other populations. Immigration can introduce new alleles into a population or can change allele frequencies. Its overall effect is to decrease the difference between populations, whereas the effect of natural selection is likely to increase differences, producing populations more suited for different local conditions. Thus, gene flow often counteracts natural selection and, also, as we shall see, can inhibit speciation.

Sampling Errors: Genetic Drift

As we stated previously, the Hardy-Weinberg principle holds true only if the population is large. This qualification is necessary because the Hardy-Weinberg equilibrium depends on the laws of probability. These laws—the laws of chance—apply equally well to flipping coins, rolling dice, or betting at roulette. In flipping coins, it is possible for heads to show up five times in a row, but, on the average, heads will show up half the time and tails half the time. The more times the coin is flipped, the more closely the expected frequencies of half (0.50) and half (0.50) are approached.

Consider, for example, an allele, say *a*, that has a frequency of 1 percent. In a population of 1 million, 20,000 *a* alleles would be present in the gene pool (remember that each diploid individual carries two alleles for any given gene). But in a population of 50, only one copy of allele *a* would be present. If the individual carrying this allele failed to mate or were destroyed by chance before leaving offspring, allele *a* would be completely lost. Similarly, if 10 of the 49 individuals without allele *a* were lost, the frequency of *a* would jump from 1 in 100 to 1 in 80. This phenomenon, a change in the gene pool that takes place as a result of chance, is *genetic drift*. The role of genetic drift in determining the course of evolution is a matter of controversy. There are at least two situations in which it has been shown to be important.

The Founder Effect

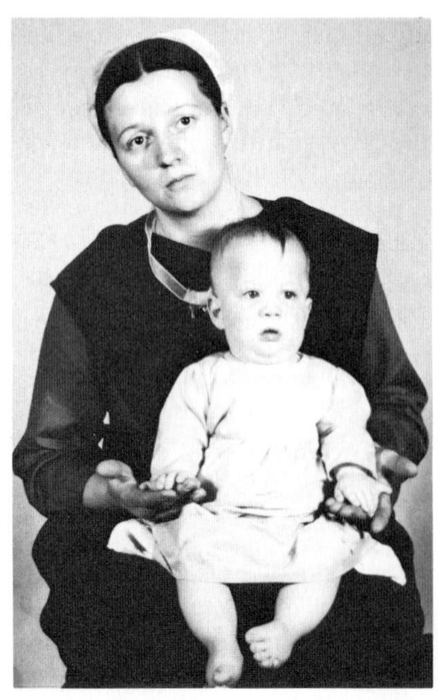

47–5
Among the Old Order Amish, a group founded by only a few couples some 200 years ago, there is an unusually high frequency of a rare allele. In its homozygous state, the allele results in extra fingers and dwarfism. This Amish child is a six-fingered dwarf.

A small population that branches off from a larger one may or may not be genetically representative of the larger population from which it was derived. Some rare alleles may be overrepresented or may be lost completely. As a consequence, even when and if the small population increases in size, it will have a different genetic composition—a different gene pool—from that of the parent group. This phenomenon, a type of genetic drift, is known as the *founder effect*.

An example of the founder effect is found in the Old Order Amish of Lancaster, Pennsylvania (Figure 47–5). Among these people, there is an unprecedented frequency of a recessive allele that, in the homozygous state, causes a combination of dwarfism and polydactylism (extra fingers). Since the group was founded in the early 1770s, some 61 cases of this rare congenital deformity have been reported, about as many as in all the rest of the world's population. Approximately 13 percent of the persons in the group, which numbers some 17,000, are estimated to carry this rare allele. The entire colony, which has kept itself virtually isolated from the rest of the world, is descended from only a few individuals. By chance, one of those must have been a carrier of the allele.

The Bottleneck

Population bottleneck, another type of genetic drift, is a situation in which a population is drastically reduced in numbers by an event having little or nothing to do with the usual forces of natural selection. For example, from the 1820s to the 1880s, the northern elephant seal was hunted so heavily along the coast of California and Baja California that it was rendered almost extinct, with perhaps as few as 20 individuals remaining. Since 1884, when the seal was placed under the protection of the United States and Mexican governments, the population has increased to more than 30,000, all presumably descendants of this small group. Studies of blood samples taken from 124 seal pups showed them to be homozygous for some 21 gene loci, indicating (by comparison with other mammalian groups) a drastic loss of genetic variability. A population bottleneck is likely not only to eliminate some alleles entirely but also to cause others to become overrepresented in the gene pool. The high rate of Tay-Sachs disease among Ashkenazic Jews (page 358) is attributed, for example, to a population bottleneck experienced by these people in the Middle Ages.

All population biologists and other students of evolution agree that genetic drift and related sampling errors play a role in determining the evolutionary course of populations. Its relative importance, as compared to natural selection, is a matter of current debate, a subject to be discussed further in the following chapters.

47–6
At one time, the lesser white snow goose (Chen hyperborea, *"the goose from beyond the north wind") and the blue goose* (Chen caerulescens) *were believed to represent distinct species. More recently, it has been found that they actually represent one species, with individuals of two colors. The white is recessive; heterozygotes and homozygotes for blue are both the same dark blue color. However, birds mate preferentially with animals of their own color. As a result, there are more homozygotes than if mating were random.*

Nonrandom Mating

Changes in the Hardy-Weinberg equilibrium can also be produced by nonrandom mating. A form of nonrandom mating particularly important in plants is self-pollination (as in the pea plants Mendel studied). In animals, nonrandom mating is often behavioral. For example, snow geese may be either white or blue; white snow geese tend to mate preferentially with other white geese and blue with blue. Thus, assuming only two alleles are involved, there will be a decrease in the heterozygote (represented as $2pq$ in the Hardy-Weinberg expression), with concomitant increases in the two homozygotes (p^2 and q^2).

Note that nonrandom mating can cause changes in the genotype frequencies without necessarily producing any changes in the frequencies of the alleles in question. If, of course, all female snow geese, white or blue, preferred to mate only with blue snow geese, and, as a consequence, some of the blue males mated with more than one female, and some of the white males with none, changes in the frequencies of the two alleles and also of the phenotypes would occur. In fact, as we shall see, such nonrandom mating produced by female choice is an important agent of natural selection in many species.

SUMMARY

Population genetics is a synthesis of the Darwinian theory of evolution with the Mendelian principles of genetics. A population, for the population geneticist, is an interbreeding group of organisms sharing a common gene pool (the sum of all the alleles of all the genes of all the individuals in the population). Evolution is, by definition, a change in the composition of the gene pool.

The Hardy-Weinberg principle is the application of Mendelian genetics to population biology. It states that in a large population of diploid sexually reproducing individuals in which random mating occurs, genetic segregation alone does not change the proportions of genotypes or the frequencies of alleles in the gene pool. For a gene with two alleles, the mathematical expression of the Hardy-Weinberg equilibrium is $p^2 + 2pq + q^2 = 1$, where p equals the frequency of one allele and q equals the frequency of the other allele. Thus, in any given population, the homozygotes for each allele are represented by p^2 and q^2, and the heterozygotes by $2pq$.

Natural selection—the differential reproduction of genotypes—is the principal cause of change in the Hardy-Weinberg equilibrium. The relative frequency of alleles in the gene pool can also be changed by mutation, gene flow (migration), genetic drift (sampling errors), or, in some instances, nonrandom mating.

Mutations provide the raw material for evolution but mutation rates are usually so low that mutations, in themselves, do not change allele frequencies or affect the Hardy-Weinberg equilibrium.

Gene flow, the movement of alleles into or out of the gene pool, may introduce new alleles or alter the proportions of alleles already present.

The Hardy-Weinberg equilibrium is maintained only if the population is large enough. In a small population, chance may cause certain alleles to increase or to decrease in frequency and perhaps even to disappear; this phenomenon is known as genetic drift. The founder effect and population bottleneck are types of genetic drift.

Nonrandom mating causes changes in the proportions of genotypes but not necessarily in allele frequencies.

QUESTIONS

1. The Hardy-Weinberg principle was first formulated, by both Hardy and Weinberg, in response to a question raised by students of Mendelian genetics: Why, if some alleles are dominant and some are recessive, don't the dominants drive out the recessives? What is the fallacy in the reasoning underlying that question? How does the Hardy-Weinberg expression answer that question?

2. Suppose, in a breeding experiment, you combined 7,000 *AA*s and 3,000 *aa*s. In the second generation, what would be the frequencies of the three genotypes? What would be the values in the third generation, assuming Hardy-Weinberg assumptions hold?

3. Among black Americans, the frequency of sickle cell anemia is about 0.0025. What is the frequency of heterozygotes? When one black American marries another, what is the probability that both will be heterozygotes? If both are heterozygotes, what is the probability that one of their children will have sickle cell anemia?

4. How would the following affect the Hardy-Weinberg equilibrium and the phenotypes in the population: Increased mutation of A to a (assuming A is a dominant)? Increased mutation of a to A (assuming a is a recessive)?

5. How does selection for the heterozygote affect the relative frequency of the two alleles in question? How does selection against the heterozygote affect this frequency?

6. How does self-fertilization, as in flowers, affect the Hardy-Weinberg equilibrium?

7. What is the difference between gene flow and genetic drift? How does each affect the gene pool of a population?

CHAPTER 48

Variability: Its Extent, Preservation, and Promotion

Like begets like, we know now, because of the remarkable precision with which the DNA is replicated and because a complete complement is transferred from every cell to its daughter cells at cell division. The DNA in the sperm or eggs of any individual is, except for occasional mutations, a true copy of one-half of the DNA that individual received from its father and mother. In fact, these mechanisms of replication and transmission serve not only to link us to our immediate ancestors but also to link all living things to one another.

This fidelity of duplication is, of course, essential to the survival of the individual organisms of which a population is composed. However, if evolution is to occur, there must be variations among individuals. Such variations make it possible for populations to change as conditions do; they are the raw material on which evolutionary forces act. A major aspect of research in modern population genetics is concerned with the extent of such variability (far greater even than Darwin could have realized), the way variations are preserved and fostered in gene pools, and the extent to which random factors determine the course of evolution and the maintenance of variation.

48–1
As Darwin observed, variations exist among the members of any given species. This photo was taken at the Woodstock music festival held in Bethel, New York, in 1969.

THE EXTENT AND ORIGINS OF VARIATIONS

To Linnaeus and other creationists, each species was represented by a perfect type, and any variations from this type were considered as imperfections. For Darwin, and for modern biologists as well, variation is a characteristic of the population; there is no ideal type, only a range of variants, changing in time and space. It was this concept, this view of nature, that made possible the development of Darwin's theory. The extent of variation has been revealed in a number of different ways.

Artificial Selection

Experiments in evolution were carried out by animal breeders and horticulturists for centuries before any concept of evolution was formulated. This process of choosing which individuals should be represented in the next generation and which should not was called artificial selection by Darwin, and he saw it as a direct analogy to the process of natural selection. First-time readers of *The Origin of Species* are sometimes surprised to find themselves, in the very first chapter, caught up in a treatise on pigeon breeding, one of the many subjects on which Darwin made himself an expert. By selecting for breeding birds with particular traits—such as a tail with more feathers or a larger beak—pigeon fanciers had been able, over the years, to produce a number of exotic breeds: the short-faced tumbler, the pouter, the barb, the trumpeter, and the fan tail. These breeds of birds, all developed from the same wild species and still able to interbreed, differed widely in appearance, more widely in fact than many animals of different species. What the pigeon breeders showed, to put it in modern terms, is that there is a large amount of variability hidden in the gene pool and that this latent variability can be expressed under selection pressures.

The breeds of dogs (Figure 48–3) provide another example of the variants present in a single species, an example more familiar to most modern readers and also well known to Darwin.

48–2
Even in populations in which the individuals appear almost identical, such as the members of this group of gannets, we know that variations exist because individuals have no difficulty in identifying their own mates or offspring. To gannets, all people probably look alike.

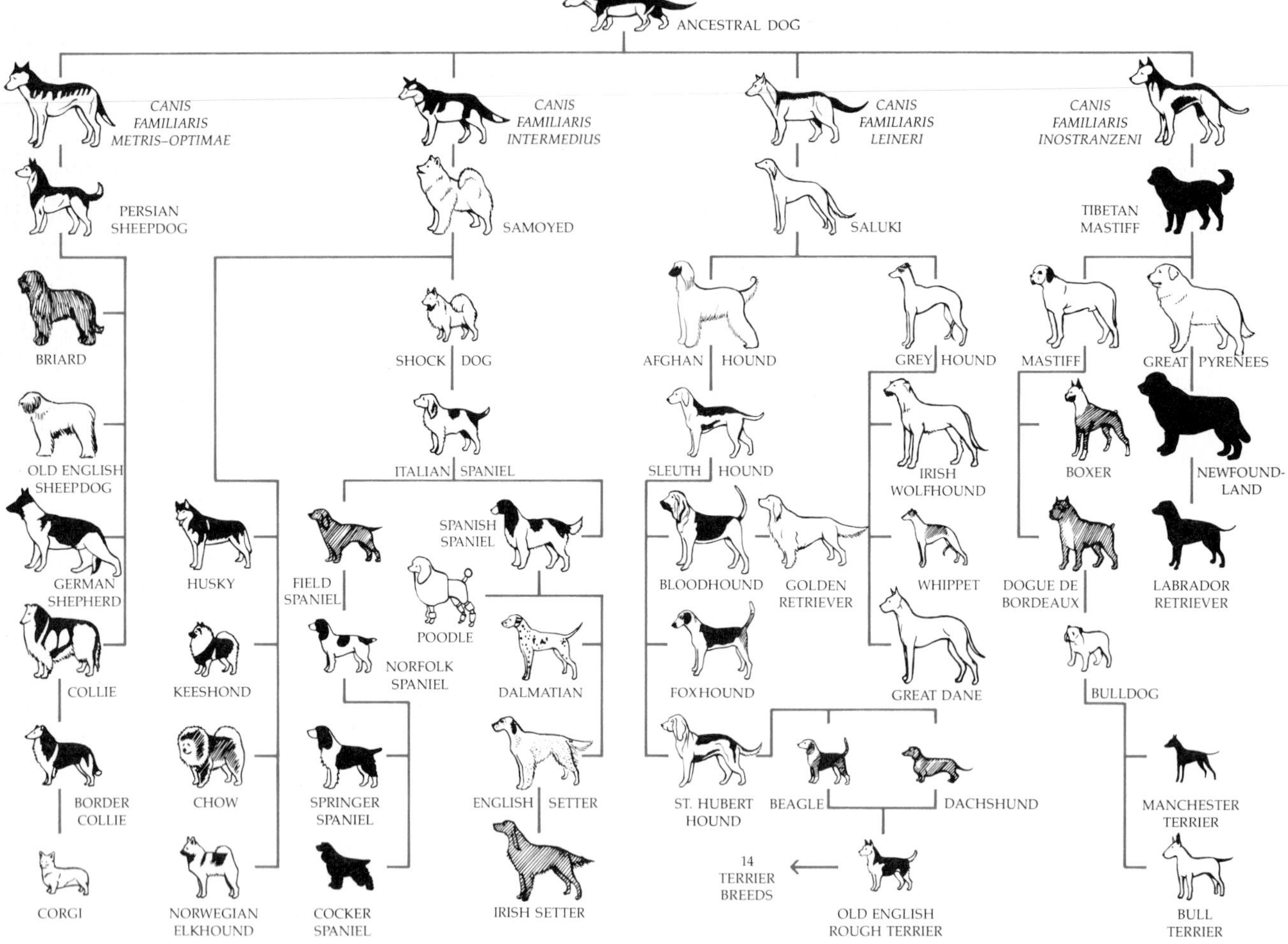

48–3
All breeds of dogs have been produced by artificial selection, a demonstration of the tremendous potential for variability in a single species.

Bristle Number in Drosophila

Artificial selection has also been carried out in the laboratory. In studies with *Drosophila melanogaster,* for example, an easily observable hereditary trait, the number of bristles on the ventral surface of the fourth and fifth abdominal segments (Figure 48–4), was chosen for selection. In the starting stock, the average number of bristles was 36. Two separate groups were interbred, one selected for increase of bristles and one for decrease. In every generation, individuals with the fewest bristles were selected and crossbred with each other, and so were individuals with the highest number of bristles. Selection for low bristle number resulted in a drop over 30 generations from an average of 36 to an average of about 30 bristles. In the high-bristle-number line, progress was at first rapid and steady. In 21 generations, bristle number rose steadily from 36 to an average of about 56 (Figure 48–5). No new genetic material had been introduced; within the single population the potential for a wide range of bristle numbers already existed. Subsequent experiments in *Drosophila* and other organisms have shown that the choice of bristle number was not merely a fortunate accident. Many characteristics studied in breeding experiments have revealed a comparable range of natural variability.

48–4
Scanning electron micrographs of (a) *a fruit fly* (Drosophila melanogaster) *and* (b) *the ventral surface of its posterior abdominal segments. The results of selection experiments to increase and decrease the number of bristles on the ventral surface of the fourth and fifth abdominal segments are shown in Figure 48–5.*

There is a second part to the bristle-number story. The low-bristle-number line soon died out because it became sterile. Presumably, changes in factors affecting fertility had also taken place during selection. When sterility became severe in the high-bristle line, a mass culture was started; members of the high-bristle line were permitted to interbreed without selection. The average number of bristles fell sharply, and in five generations dropped from 56 to 40. Thereafter, as this line continued to breed without selection, the bristle number fluctuated up and down, usually between 40 and 45, which still was higher than the original 36. At generation 24, selection for high bristle number was begun again for a portion of this line. The previous high bristle number of 56 was regained, and this time there was no loss in reproductive capacity. Apparently, the genotype had become reorganized in such a way that alleles controlling bristle number were present in more favorable combinations with alleles affecting fertility.

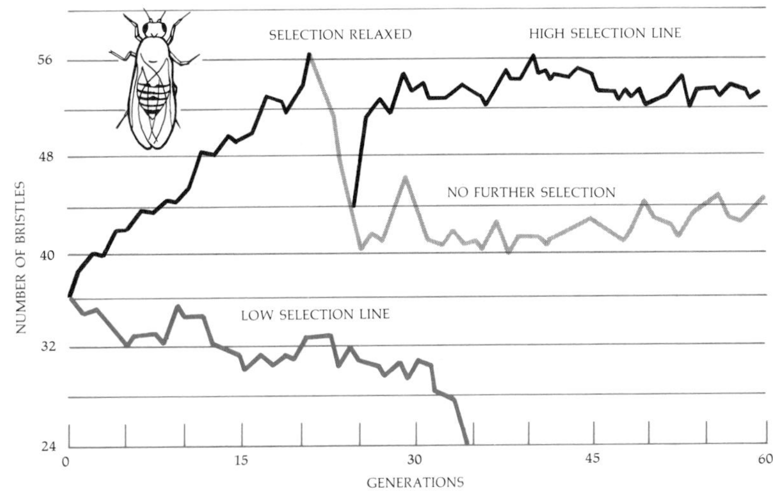

48–5
The results of an experiment with Drosophila melanogaster, *demonstrating the extent of latent variability in a natural population. From a single parental stock, one group was selected for an increase in the number of bristles on the ventral surface (high selection line) and one for a decrease in the bristle number (low selection line). As you can see, the high selection line rapidly reached a peak of 56, but then the stock began to become sterile. Selection was abandoned at generation 21 and begun again at generation 24. This time, the previous high bristle number was regained, and there was no apparent loss in reproductive capacity. Note that after generation 24 the stock interbreeding without selection was also continued, as indicated by the line of gray. After 60 generations the freely breeding group from the high selection line had 45 bristles. The low selection line died out owing to sterility.*

Mapping studies have shown that bristle number is controlled by a large number of genes, at least one on every chromosome and sometimes several at different sites on the same chromosome. Therefore, although we do not know how important bristle number is to the survival of the animal, selection for this trait in some way disrupted the entire genotype. Livestock breeders are well aware of this consequence of artificial selection. Loss of fertility is a major problem in virtually all circumstances in which animals have been purposely inbred for particular traits. This result emphasizes the fact that in natural selection it is the entire phenotype that is selected, rather than certain isolated traits, as is usually the case in artificial selection.

Quantifying Variability

Analysis at the molecular level provides a newer method for assessing variability. Some genes, as we saw in Section 3, code for proteins. Proteins therefore reflect the structure of the genes coding for them. J. L. Hubby and R. C. Lewontin ground up fruit flies from a natural population and extracted proteins from them. From this homogenate, they were able to isolate 18 functionally different enzymes. Then, by electrophoresis—the same method used by Pauling to separate the variants of hemoglobin—they analyzed each of the 18 enzymes separately, looking for variations in amino acid structure. (Such enzymes that are the same functionally but differ structurally are known as isozymes.)

Of the 18 enzymes studied in this way, 9 were found to be structurally indistinguishable by this method; in other words, the genes that had produced these proteins were the same throughout the entire population of fruit flies studied. However, 9 of the 18 enzymes were found to be made up of two or more isozymes. Thus, without any direct analysis of the genes themselves, the investigators were able to conclude that among the fruit flies studied there were two or more alleles of the gene responsible for each of these 9 enzymes. One enzyme had as many as six slightly different structural forms; that is, at least six alleles for the gene coding for that enzyme were present in the gene pool.

Each fruit-fly population examined was heterozygous for almost half of the genes tested. Each individual, it was estimated, was probably heterozygous for about 12 percent of its genes. Similar studies on humans, using accessible tissues such as blood or placenta, indicate that at least 25 percent of the genes in any given population are represented by two or more alleles, and individuals are heterozygous for at least 7 percent of their genes, on the average.

Now that methods for sequencing nucleotides in DNA have been developed (page 322), it will be possible to make direct comparisons of the genetic material. Since all changes in nucleotides do not result in changes in amino acid sequences, and all changes in amino acid sequences are not detectable by electrophoresis, DNA sequencing is expected to reveal a greater extent of genetic variation.

Explaining the Extent of Variation

The findings of Hubby and Lewontin and others after them have, like most important scientific discoveries, raised major new questions. Many geneticists had previously thought that the individuals of a population should be close to genetic uniformity, as a result of a long history of selection for "optimal" genes. Yet, as these studies revealed, natural populations are far from uniform.

One school of geneticists, the selectionists, claim that even such small variations as those in enzyme structure are maintained by various forms of natural selection that favor some genotypes at some times and in some areas, and others at other times or in other localities. In the selectionist's view, all variation directly or indirectly affects fitness. An opposing school, the neutralists, claim that the observed variations in the protein molecules are so slight that they do not make any difference in the function of the organism and so are not affected by natural selection. (Such neutralism is, of course, supported by the evidence of the molecular clocks, described on page 382.) Neutral alleles, according to the neutralist argument, accumulate as a result of random mutations and may increase in frequency by genetic drift. This difference of opinion is not currently resolvable.

PRESERVATION AND PROMOTION OF VARIABILITY

Sexual Reproduction

By far the most important method by which eukaryotic organisms promote variation in their offspring is sexual reproduction. Sexual reproduction produces new genetic combinations in three ways: (1) by independent assortment at the time of meiosis—as diagrammed on page 258, (2) by crossing over with genetic recombination, and (3) by the combination of two different parental genomes. By the process of sexual recombination, alleles are assorted in new combinations in every generation. In contrast, consider, for a moment, organisms that reproduce only asexually, with every reproductive cell giving rise to an entire new organism. Except when a mutation has occurred in the duplication process, the new organism will exactly resemble its only parent. In the course of time, various clones may form, each carrying one or more mutations, but unless the same mutations occur in the same clones, potentially favorable combinations are never going to accumulate in one genotype.

On the debit side, organisms that reproduce sexually must produce twice as many reproductive cells, since two cells combine to make the new individual. So the cost of each new individual is twice as great, not counting the expenditure of time and effort and the possible dangers involved in seeking a mate, courtship, and copulation. The only advantage to the organism of sexual reproduction, speaking strictly scientifically, is the promotion of variation, the production of new combinations of alleles among the offspring.

On the assumption that organisms operate on a cost-efficient basis, biologists therefore hypothesize that, because the cost is twice as great, sexual reproduction must confer at least a twofold advantage. The demonstration of such an advantage to the individual organism is a matter of current controversy.

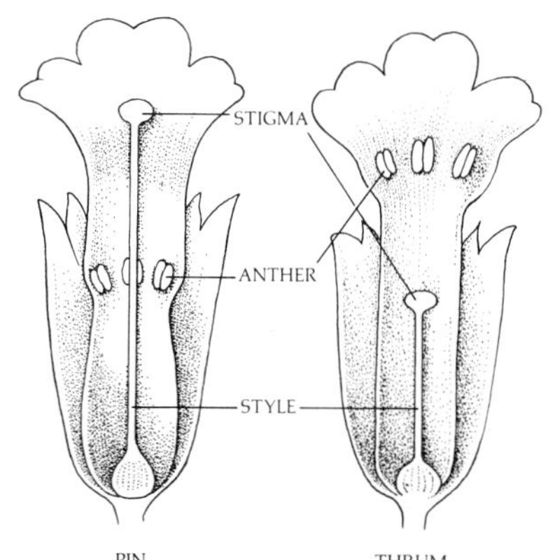

48–6
Diagrams of a pin type and a thrum type of a primrose. Notice that the pollen-bearing anthers of the pin flower and the pollen-receiving stigma of the thrum flower are both situated about halfway up the length of the flower and that the pin stigma is level with the thrum anthers. An insect foraging for nectar in these plants would collect pollen in different areas of its body, so that thrum pollen would be deposited on pin stigmas, and vice versa.

Mechanisms that Promote Outbreeding

Many ways have evolved by which new genetic combinations are promoted in sexually reproducing populations. Among plants, a variety of mechanisms ensure that the sperm-bearing pollen is from a different individual than the stigma it lights upon. Some plants, such as the holly and the date palm, have male flowers on one tree and female flowers on another. In others, such as the avocado, the pollen of a particular plant matures at a time when its own stigma is not receptive. In some species, anatomical arrangements inhibit self-pollination (Figure 48–6).

Some plants have genes for self-sterility. Typically, such a gene has multiple alleles—s^1, s^2, s^3, and so on. A plant carrying the allele s^1 cannot pollinate a plant with an s^1 allele; one with an s^1/s^2 genotype cannot pollinate any plant with either of those alleles; and so forth. In one population of about 500 evening primrose plants, 37 different self-sterility alleles were found, and it has been estimated that there are more than 200 alleles for self-sterility in red clover. A plant with a rare self-sterility allele is more likely to be able to pollinate another plant than is a plant with a common self-sterility allele. As a consequence, the self-sterility system strongly encourages variability in a population; selection for the rare allele makes it more common, whereas more common alleles become rarer.

Among animals, even in those invertebrates that are hermaphrodites, such as earthworms, slugs, and snails, an individual seldom fertilizes its own eggs. Among mammals, in particular, behavioral strategies promote outbreeding (mating with unrelated individuals). Often, for example, young males leave the family group as they reach reproductive age; this occurs among lions, gorillas, and baboons, to mention only a few. Among hunting dogs of the African plains, it is the young females who leave at reproductive age.

Most human cultures discourage inbreeding (breeding between closely related individuals). Many pretechnological societies, for example, demand that a young man choose a wife from another village rather than from his own community, and virtually all cultures have strong prohibitions against incest. Such prohibitions are particularly interesting when you consider that intermarriages of brother and sister or father and daughter would tend to keep property or power within the family and so could be socially and economically advantageous. Followers of Freud maintain that incest is forbidden by strong psychological taboos, as reflected in the Oedipus myth. Some biologists hold that the prohibition stems from the observed ill effects of incestuous matings (through greater chances of deleterious alleles becoming homozygous) and has been merely reinforced by cultural restraints. Others maintain that it is a genetically determined behavioral mechanism, analogous to those observed in other mammals.

48–7
Male lions leave the pride in which they are born at the age of about 3 years. These young nomadic males are often found in small groups, such as the one shown here. They are characteristically of the same age and are probably brothers or, at least, from the same pride. They will probably eventually take over a pride of their own. This social system ensures outbreeding.

Diploidy

Another factor in the preservation of variability in eukaryotes is diploidy. In a haploid organism, any genetic variations are immediately expressed in the phenotype and are therefore exposed to the selection process. On the other hand, in the diploid organism, such variations may be stored as recessives, as with the allele for white flowers in Mendel's pea plants. The extent to which a rare allele is protected is revealed by the following table:

FREQUENCY OF ALLELE *a* IN GENE POOL	GENOTYPE FREQUENCIES			PERCENTAGE OF ALLELE *a* IN HETEROZYGOTES
	AA	*Aa*	*aa*	
0.9	0.01	0.18	0.81	10
0.1	0.81	0.18	0.01	90
0.01	0.9801	0.0198	0.0001	99

As you can see, the lower the frequency of allele *a*, the smaller the proportion of it exposed in the *aa* homozygote becomes, and the removal of the allele by natural selection slows down accordingly (see Figure 47-3, page 894). This result should be of special interest to proponents of eugenics, the science of improving the human gene pool through controlled breeding. For instance, consider a recessive genetic disorder with an allele frequency of about 0.01 in the human population. Individuals with the *aa* genotype make up 0.0001 of the population (1 child for every 10,000 born). It would take 100 generations, roughly 2,500 years, of a program of sterilization of defective homozygous individuals to halve the allele frequency (to 0.005) and reduce the number born with this genetic disorder to 1 in 40,000, which should be enough to discourage even the most zealous eugenicist.

NATURAL SELECTION AND VARIABILITY

In the course of the controversy that led to the formulation of the synthetic theory, anti-Darwinists argued that natural selection would serve only to eliminate the "less fit," and so would tend to reduce variation in a population, acting in effect as an anti-evolutionary force. Modern population genetics has shown this not to be true: natural selection often acts to promote variability.

Balanced Polymorphism

Polymorphism is the coexistence within a population of two or more phenotypically distinct forms. A population may be polymorphic because one phenotype is replacing the other, as in the case of the peppered moths (transient polymorphism), or it may be polymorphic because two or more types are maintained in fairly stable proportions by natural selection (balanced polymorphism). Many examples of balanced polymorphism are known, of which we shall mention two.

Shell Color in Snails

Among snails of the genus *Cepaea,* individuals with banded and unbanded shells have coexisted for at least 10,000 years, according to fossil evidence. These snails occupy a variety of habitats, including rocks, bogs, and woodlands, and are preyed upon by birds, among which are song thrushes. Thrushes seize the snails and take

(a)

(b)

48–8
(a) *Polymorphism in land snails: a banded and an unbanded snail.* (b) *An "anvil," where song thrushes break land snails open in order to obtain the soft, edible parts. From the evidence left by the empty shells, investigators have been able to show that in areas where the background is fairly uniform, unbanded snails have a survival advantage over the banded type. Conversely, in colonies of snails living on dark, mottled backgrounds (such as woodland floors), banded snails are preyed upon less frequently.*

them to nearby rocks—"anvils"—where they break them open and eat the soft bodies, leaving the shells. Analysis of the shells around the thrush anvils reveals that in habitats where the background is uniform (rocks, for instance) banded snails are more likely to be captured (Table 48–1), whereas in habitats where the backgrounds are mottled (bogs and woodlands), unbanded snails are more likely to be the victims. Thus, polymorphism in this case is promoted by the heterogeneous environment of the mollusks.

Table 48–1 A Comparison of Numbers of Individuals with Banded Shells Collected Alive in a Uniform Habitat and Killed by Thrushes at Nearby Anvils

	NUMBER OF SHELLS			PERCENT
	BANDED	UNBANDED	TOTAL	BANDED
Living	264	296	560	47.1
Killed by thrushes	486	377	863	56.3

Human Blood Groups

The human species is polymorphic for blood types A, B, AB, and O. Apparently, the three alleles associated with these blood types are part of our ancestral legacy, since the same blood types are also found in other primates. A great deal is known about the chemistry of the different blood groups and about the allele frequencies in different populations. Yet we know very little about how this balanced polymorphism has been maintained.

Some population biologists regard the blood types as probably neutral in selective value. Others maintain that polymorphism in human blood groups is a result of selection. For example, among Caucasian males, the life expectancy is greatest for those with group O and least with group B; exactly the opposite is true for Caucasian females with these blood types. People with type A blood run a relatively higher risk of cancer of the stomach and of pernicious anemia. People with type O blood have a higher risk of duodenal ulcers and are more likely to contract Asian flu. Most of these conditions, however, would not act as selective forces since they generally occur in individuals who are past reproductive age.

It has been suggested that there are correlations between the different blood types and susceptibility to such diseases as plague, leprosy, tuberculosis, syphilis, and smallpox—all diseases that, in the past, could have been powerful selective forces. However, so far, such correlations have not been substantiated.

The geographic distributions of the A, B, AB, and O groups are irregular. For example, there is a large predominance of O in the Western Hemisphere and an increase in the B allele as one moves from Europe toward Central Asia. The B allele is totally absent in American Indians and Australian aborigines who have not mixed with Europeans. Although most American Indians are type O, the Blackfoot tribe has the highest frequency of type A blood found anywhere in the world (55 percent). Even within an area as small as the British Isles, there are significant variations in allele frequency, going both from north to south and from east to west. These differences may reflect some differences in the selective forces favoring particular blood groups under particular conditions, they may be the result of population migrations and genetic drift, or some combination of both.

Heterozygote Superiority

Heterozygote superiority is the favoring by natural selection of an individual heterozygous for a particular gene over either homozygote. In other words, an individual with an *Aa* genotype is selected in favor of *AA* or *aa*.

Sickle Cell Anemia

The best-studied example of heterozygote superiority involving a single gene locus is found in association with sickle cell anemia. Individuals homozygous for sickling almost never become parents. Therefore, almost every time one sickling allele encounters another in a homozygous individual, two sickling alleles are removed from the gene pool. At one time, it was thought that the sickling allele was maintained in the population by a steady influx of new mutations. Yet, in some African tribes, as much as 45 percent of the population is heterozygous for sickling, and to replace the loss of sickle alleles by mutations alone would require a rate of mutation about 1,000 times greater than any other known human mutation rate.

In the search for an alternative explanation, it was discovered that the sickling allele is maintained at high frequencies because the heterozygote has a selective advantage. In many regions of Africa, malaria is one of the leading causes of illness and death, especially among young children. Studies of the incidence of malaria among young children showed that susceptibility to malaria is significantly lower in individuals heterozygous for sickling than in normal homozygotes. Moreover, for reasons that are not known, women who carry the sickling allele are more fertile than normal heterozygotes.

Thus, although two alleles for sickle cell hemoglobin are eliminated virtually every time they appear in the homozygous state, selection for the heterozygote maintains the allele in the gene pool.

Heterosis or Hybrid Vigor

When two different inbred strains of a crop plant are crossed with one another, the resultant strain is often found to be superior to either of the parent strains. Hybrid corn is a striking example. The development of hybrid corn caused a revolutionary improvement in the corn crop of the United States in the 1930s, because of the increased size and hardiness of the plants derived by crossing two different varieties to produce the seeds for each planting (Figure 48–9). This phenomenon is known as *heterosis*. Heterosis is believed to be one of the most widespread genetic mechanisms by which polymorphism is maintained in a population.

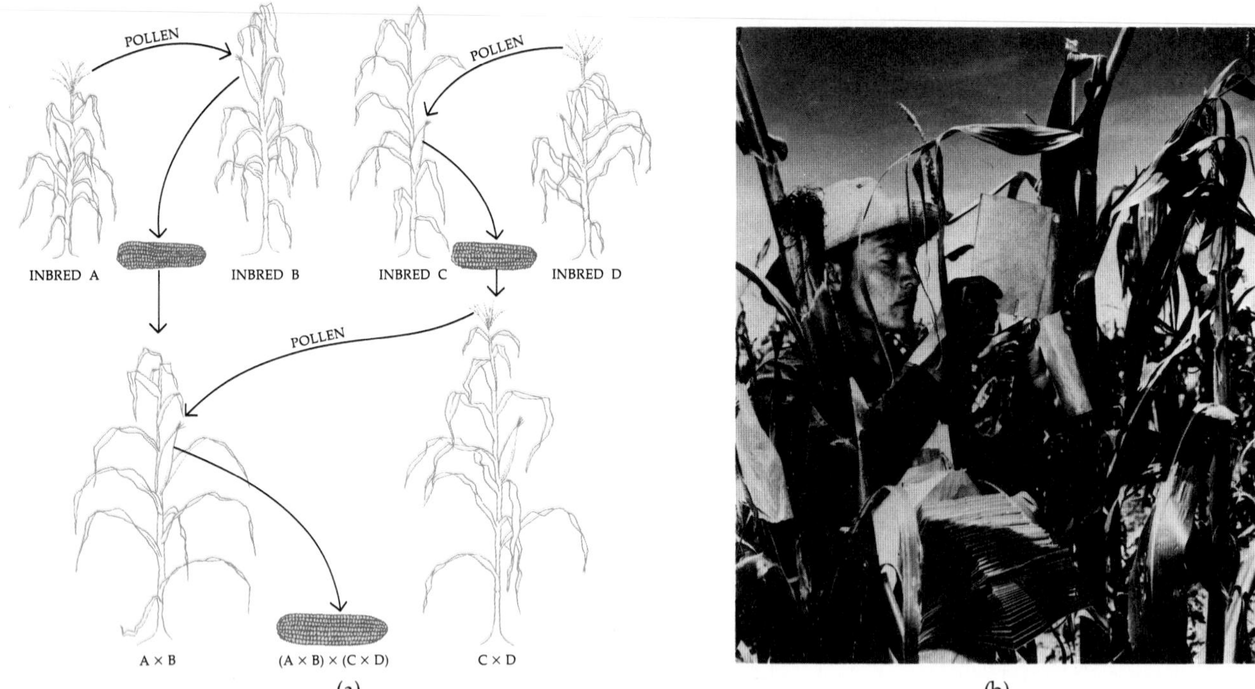

48–9
(a) *Hybrid corn is derived by first crossing strains A with B and C with D and then crossing the resulting single-cross plants to produce double-cross seed for planting. The increased size and hardiness of the hybrid are probably due to its increased heterozygosity.* (b) *Collecting pollen for hybridization.*

Heterosis, also known as hybrid vigor, is apparently the result of the fact that hybrids are, by definition, heterozygous at far more loci than most natural varieties. The superiority of the hybrid may be because hybrids are less likely to be homozygous for deleterious alleles, or because of heterozygote superiority.

Geographic Variations: Clines and Ecotypes

Sometimes variations within the same species follow a geographic distribution and are correlated with temperature, humidity, or some other environmental condition. Such a graded variation in a trait or a complex of traits is known as a *cline*.

Many species exhibit north-south clines of various characteristics. House sparrows, for example, tend to have a smaller body size in the warmer parts of the range of the species, and a larger body size in the cooler parts (Figure 48–10). In cooler regions, a larger body size is advantageous for heat conservation. Conversely, among mammals, ears, tails, and other extremities are relatively longer in the warmer areas of a species' range; such adaptations allow for heat radiation from the animal. Plants growing in the south often have slightly different requirements for flowering or for ending dormancy than the same plants growing in the north, although they all may belong to the same species.

A species that occupies many different habitats may appear to be slightly different in each one. Each group of distinct phenotypes is known as an *ecotype*. Are the differences among ecotypes determined entirely by the environment or are there genetic differences too? Figure 48–11 illustrates a series of experiments carried out by Jens Clausen, David Keck, and William Hiesey, under the auspices of the Carnegie Institution of Washington. One of the plants studied was the perennial *Potentilla glandulosa,* a relative of the strawberry. Experimental gardens were established at various altitudes, and wild plants collected from near each of the

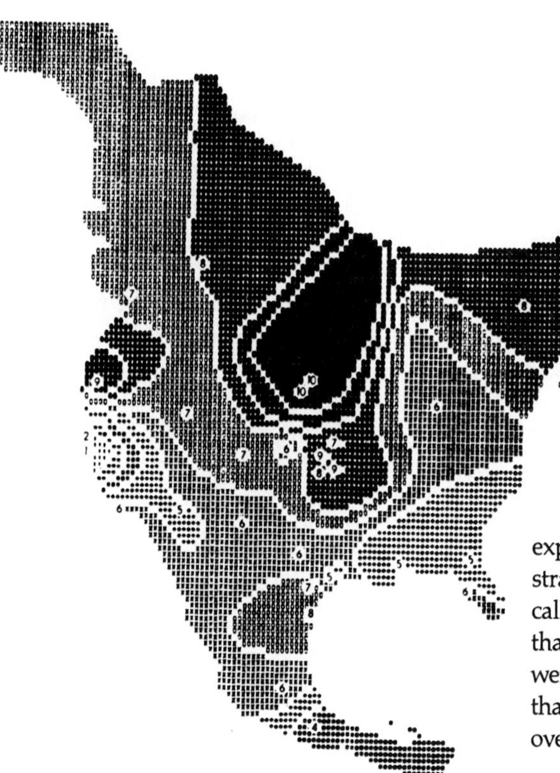

48–10
The distribution by size of male house sparrows in North America. The higher numbers indicate larger sizes, based on a composite of 16 different measurements of the birds' skeletons. This map was generated and drawn by a computer.

experimental sites were grown in all the gardens. *(Potentilla glandulosa,* like the strawberry, reproduces asexually by runners, making it possible to study genetically uniform replicates.) Under these conditions it was possible to demonstrate that many of the phenotypic differences among the ecotypes of *Potentilla glandulosa* were due to genetic variations produced by natural selection. It is not surprising that in their very different environments, different traits were selected for and so, over time, genetic differences arose.

Frequency-Dependent Selection

Polymorphism may be maintained in a population in situations in which the more common allele is selected against. For instance, students of animal behavior report that predators form a "search image" that enables them to hunt a particular kind of prey more efficiently. If the prey individuals differ, for example, in color, the most common color will be preyed upon disproportionately. If, as a result, individuals with that color should become less common, selection pressure on them would be relaxed. Self-sterility alleles (page 906) are also an example of a very different kind of frequency-dependent selection.

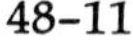

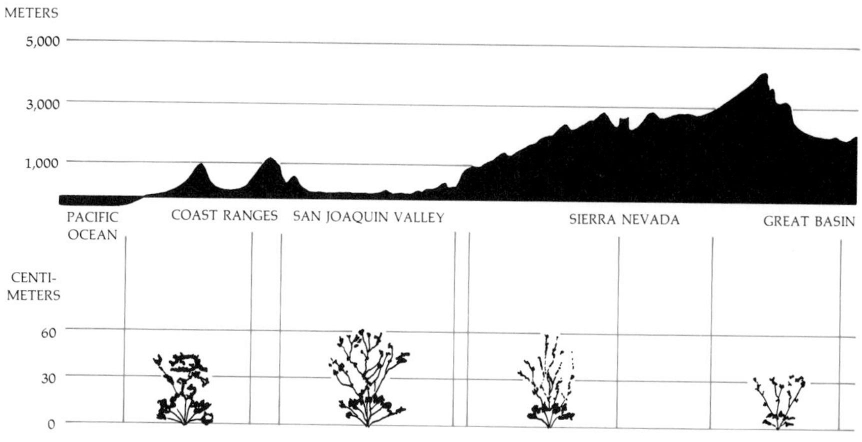

48–11
Ecotypes of Potentilla glandulosa, *a relative of the strawberry. Notice the correlation between the height of the plant and the altitude at which it grows; there are other phenotypic differences among the plants as well. When plants of the four geographic races are grown under identical conditions, many of these phenotypic differences persist and are passed on to the next generation, indicating that these plants are genotypically, as well as phenotypically, different.*

VARIATION AND THE EUKARYOTIC CHROMOSOME

The newly won capacity to analyze chromosomal DNA of eukaryotic organisms has, as we noted in Chapter 17, turned up a number of surprises. Large segments of DNA, it has been shown, have the capacity to produce duplicates of themselves and to disperse these duplicates to other locations in the chromosome and even to other chromosomes. These duplicate genes are then free to travel their own evolutionary course, leaving their functions to be carried out by the parent, original genes; the duplicate genes are therefore free from constraints.

Evolutionary biologists are speculating that all present-day structural genes had their beginnings in a very few protogenes that were then duplicated and modified over the last 4 billion years. Most important, there is clear evidence that this duplication and modification process is continuing at this very moment. This process of gene amplification has undoubtedly played a major role in evolution, and as new information unfolds, major changes may take place in evolutionary theory.

SUMMARY

The extent of genetic variability in a population is a major determinant of its capacity for evolutionary change. Natural populations can be shown by breeding experiments—artificial selection—to harbor a wide spectrum of genetic variations. Such variability is preserved in populations by sexual reproduction and by diploidy (which shelters the recessive allele). Outbreeding is promoted by mechanisms such as self-sterility alleles in plants, anatomic adaptations that inhibit self-fertilization, and in animals, behavioral strategies.

Natural selection may promote variability. Balanced polymorphism is the continued existence in stable proportions in a population of two or more distinct phenotypes, owing to the action of natural selection. In cases of heterozygote superiority, the heterozygote is selected over either homozygote, thus maintaining both alleles in the population. Heterosis, or hybrid vigor, is the result either of heterozygote superiority or of the masking of the effects of recessive alleles. In either case, it maintains two or more alleles in the population. Continuous variations that follow a geographic distribution are known as clines. Distinct groups of phenotypes occupying different habitats are called ecotypes. In these cases, the variability of the species is maintained by the heterogeneity of the environment.

QUESTIONS

1. Distinguish between genotypic variability and phenotypic variability. Which is acted on by natural selection? Which is necessary for evolution?

2. What is the difference between artificial selection and natural selection? Under what circumstances might they become indistinguishable?

3. Under what circumstances would two alleles be maintained in a population in exactly the same numbers?

4. Why is a large body size often associated with increasing cold in a cline, as in male sparrows?

CHAPTER 49

Natural Selection

According to Darwin's own account, the concept of natural selection came to him in the fall of 1838 upon his reading of Malthus "for amusement." Malthus's gloomy essay emphasized that, given the fact that populations increase geometrically and resources linearly, the human population is doomed to exceed its resources. And so, indeed, are all populations. Only a small fraction of individuals that might exist are born and survive. According to Darwin, those that do survive are those that are "favoured," to use his own term, by reason of slight, advantageous variations. This process of survival of the "favoured" he termed natural selection, by analogy with the artificial selection practiced by breeders of domestic animals and plants.

In terms of population genetics, natural selection is now defined more rigorously as the differential net reproduction of genotypes. This differential reproductive success may result in changes in the frequencies of alleles in the population —that is, in evolution. According to neo-Darwinian theory, natural selection is the major force in evolution.

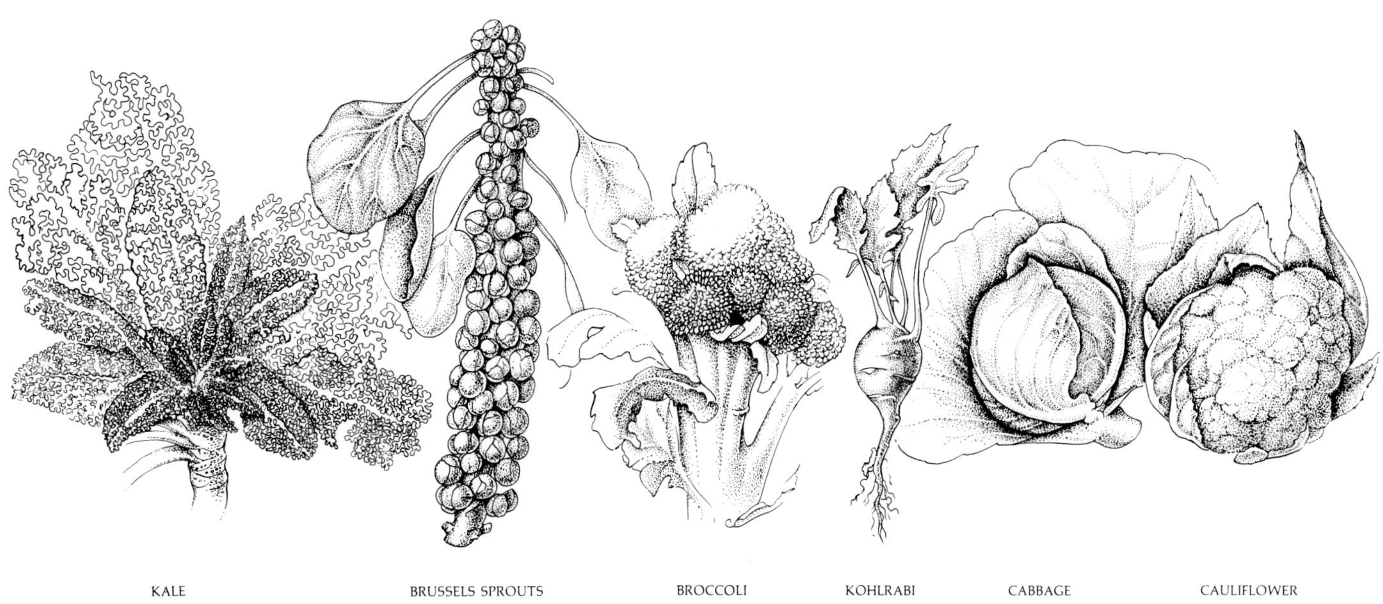

49–1
Six vegetables produced from a single species of plant (Brassica oleracea, *a member of the mustard family). They are the result of selection for leaves (kale), lateral buds (brussels sprouts), flowers and stem (broccoli), stem (kohlrabi), enlarged terminal buds (cabbage), and flower clusters (cauliflower). Kale most resembles the wild plant. Artificial selection, as practiced by animal breeders and agriculturalists, gave Darwin the clue to the concept of natural selection.*

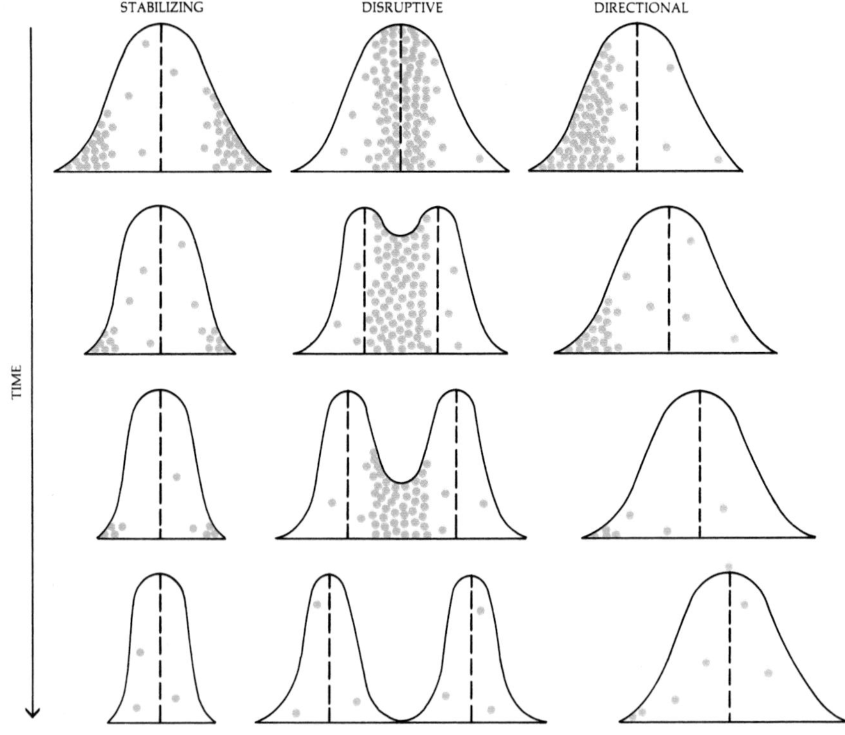

49-2
The three different types of natural selection. The dots represent individuals that failed to reproduce or that left less than the average number of offspring. Stabilizing selection involves the elimination of extremes. In disruptive selection, intermediate forms are eliminated, producing two divergent gene pools. Directional selection, which is the gradual elimination of one phenotype in favor of another, produces adaptive change, as may disruptive selection. In these graphs, the vertical axes denote the proportion of individuals in a population with a particular characteristic, and the horizontal axes, the varying dimensions of whatever characteristic is being considered.

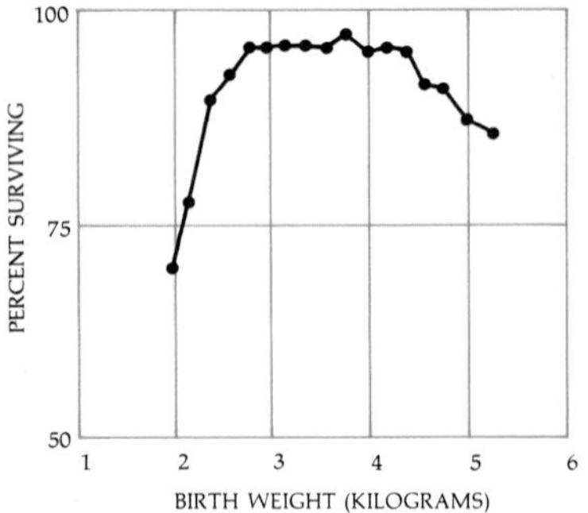

49-3
Relationship between weight at birth and survival in a particular group of human babies. The death rate is higher for babies weighing either more or less than about 3.8 kilograms.

TYPES OF SELECTION

Three general types of natural selection operate within populations: stabilizing, disruptive, and directional (Figure 49-2).

Stabilizing selection, a process that is always in operation in all populations, is the elimination of extreme individuals. Many mutant forms are probably immediately weeded out in this way, often in the zygote or the embryo. Clutch size in birds, for instance, is a result of stabilizing selection. Clutch size (the number of eggs a bird lays) is determined genetically. As you can see in Table 49-1, it is a disadvantage, in terms of surviving young, for the Swiss starling to have a clutch size of less than 4 or more than 5. Females whose genotypes dictate a clutch size of 4 or 5 will have more surviving young, on the average, than members of the same species that lay more or fewer eggs.

A somewhat similar example is seen in a study of the correlation between birth weight and survival for 13,730 babies born in London over a ten-year period (1935–1945). As you can see in Figure 49-3, the optimum weight for this population of babies was about 3.8 kilograms. At this birth weight, 98.2 percent of the babies

Table 49-1 Survival in Relation to Number of Young in Swiss Starling

Number of young in brood	1	2	3	4	5	6	7	8
Number of young marked	65	328	1,278	3,956	6,175	3,156	651	120
Percentage of marked birds recaptured after 3 months	–	1.8	2.0	2.1	2.1	1.7	1.5	0.8

survived. The overall survival rate was 95.5 percent. To the extent that birth weight of offspring is determined by heredity, stabilizing selection maintained this birth weight in this population.

The second type of selection, *disruptive selection*, increases the two extreme types in a population at the expense of intermediate forms. Disruptive selection in the laboratory gave rise to the high- and low-bristle-number lines of *Drosophila* described on page 902.

A particularly clear instance of disruptive selection in action has been demonstrated in studies of plants growing on soils contaminated by salts of heavy metals, such as lead and zinc, because of mining operations. The boundaries between contaminated and uncontaminated areas are often very sharp. Plants growing on uncontaminated soil are unable to survive on contaminated soil; plants of the same species growing on contaminated soil are able to survive in the uncontaminated areas but cannot compete with those already growing there. Thus, the two extreme phenotypes have been "favoured" at the expense of the intermediate form, resulting in the development of very marked differences between the two groups in the 50 years or so since the mining operations were discontinued and the plants began to colonize the area.

Disruptive selection of this sort may result in the formation of two new species, a subject we shall discuss in the next chapter.

The third type, the one with which we shall be most concerned, is *directional selection*. Directional selection acts to increase the proportion of individuals with an extreme phenotypic characteristic and so is likely to result in the gradual replacement of one allele or group of alleles by another in the gene pool.

Natural selection results in adaptation, with its several meanings and multiple manifestations. Consider, for example, a squirrel. Note how its tail serves as a

(a)

(b)

49–4
(a) *The woodpecker has a number of adaptations that enable it to obtain food. These include two toes pointing backward with which the woodpecker clings to the tree bark, strong tail feathers that prop it up, a strong beak that can chisel holes in the bark, strong neck muscles that make the beak work as a hammer, air spaces in the skull that cushion the brain during hammering, and a very long tongue that can reach insects under the bark.*

(b) *The woodpecker finch* (Camarhynchus pallidus) *is a rare phenomenon in the bird world because it is a tool user. Like the true woodpecker, it feeds on grubs, which it digs out of trees using its beak for a chisel. Lacking the woodpecker's long, barbed tongue, it resorts to an artificial probe, a twig or cactus spine, to dislodge the grub. The woodpecker finch shown has selected a cactus spine, which it inserts into the grub hole. The bird has succeeded in prying out the grub, which it eats. If the pick selected turns out to be an efficient tool, the bird will carry it from tree to tree in its search for grubs.*

counterbalance as the animal leaps and turns; in addition, the same marvelous structure serves as a parasol, a blanket, an aerial rudder, and, in case of mishap, a parachute. Pluck a burr from your clothing and consider the artful contrivances by which it clings there. Consider the love and devotion characteristic of the domesticated dog; these are adaptations related to procurement of food and shelter as stringently selected for as the beak of a hawk. Thread a needle; your capacity to do so represents the cumulative effect of millions of years of selective pressures for digital dexterity and eye-hand coordination. (The needle itself made its appearance a mere 10,000 years ago.)

WHAT IS SELECTED?

Natural selection acts on the phenotype. In the days of fruit-fly genetics, when populations were being scanned for white eyes and dumpy wings, phenotype became synonymous, operationally speaking, with physical appearance. In terms of evolutionary theory, however, it must be regarded as including such important characteristics as the optimum temperature at which a particular enzyme works, or the speed of response to a stimulus. In short, the phenotype includes all observable attributes of an organism.

A phenotype characteristically is the expression of many different genes and so, as a corollary, any particular phenotypic trait may be arrived at by a number of genotypic routes. For example, selection for high and low bristle number led, in the first series of experiments, to sterility. The second attempt to select for high bristle number led to the same number of bristles (that is, the same phenotypic trait) but without a reduction in fertility. Obviously, different genotypes were responsible.

49–5
Owing to a rapid rise in IQ since World War II, Japanese schoolchildren now have an average IQ of 111 as compared to an average of 100 in the United States. This difference, seen among children as young as six, is most marked in tests of block designs, mazes, picture arrangement, and object assembly.

A particularly complex relationship between genotype and phenotype is illustrated by the IQ score. First, the IQ test itself is made up of a number of subtests of verbal comprehension, mathematical abilities, spatial concepts, and even value judgments (such as what do you do if you are the first to see a fire in a crowded theater). Scores achieved in the subtests by any individual may vary widely. Second, many variables are at work. Among those that can be measured are prenatal and perinatal nutrition, socioeconomic background, weight at birth, parental IQs, inborn errors of metabolism (PKU is one, but there are many others), chromosomal aberrations (as in Down's syndrome), oxygen deprivation at birth, and motivation and attitude, to name a few. The IQ has a reality: the fact that an individual has an IQ of 95 or 125, or 145, or 165 predicts fairly accurately his or her performance in certain social situations. It also has genetic components, and some of these components, at least in past human history, probably had selective value. However, what was being selected, genotypically speaking, is virtually impossible to define.

The long and generally fruitless debate about the relative importance of heredity and environment in determining IQ has been enlivened in recent months by the discovery that in Japan, the average IQ of schoolchildren has increased seven points in a single generation. About 10 percent of the Japanese population has an IQ of more than 130, as compared to only 2 percent of Americans and Western Europeans.

Conversely, the same genotype may produce quite different phenotypes (Figure 49–6). For example, among human identical twins (that is, twins produced from a single fertilized ovum and so having the same genotype), noticeable differences—such as body weight—between the two are often apparent even at birth, owing to differences in the intrauterine environment.

49–6
A pinyon pine growing out of a crack in a rock in California. Usually, trees of this species are tall and straight; however, the constant, strong winds in which this one has grown have produced this phenotype.

(a)

(b)

49–7
Coordinated anatomical and behavioral characteristics protect the eyed hawkmoth from predators. (a) *At rest, the moth positions itself on the bark of a tree in a way that enhances the camouflaging effect of its cryptically patterned wings.* (b) *Disturbed, the moth parts its wings, revealing a pair of "eyes," reported to be sufficiently menacing to scare away a bird.*

Any particular phenotype lasts, in terms of evolutionary history, for as long as the blink of an eye. In the case of sexually reproducing organisms, the genotype is as unique and transient as the phenotype, shuffled and recombined at every generation. Only the individual genes survive. The great men and women of history have long since vanished, but each of us may carry one or two of their alleles as part of our human legacy.

However, to retrace our steps, it is clearly very rare that a single allele can determine a winning phenotype. In *Biston betularia,* black wing color is determined by a single allele. However, the peppered coloring may be the result of several different genes, and, along with the mottled appearance, there are other phenotypic traits, such as selecting the right background, lying very still while on the tree trunk, and positioning oneself correctly so as to enhance the camouflage effect, all under genetic control (Figure 49–7).

Similarly, the number of young produced in a clutch (page 914) can be described as a simple trait, something like pushing the right number for copies on the Xerox machine. However, it must involve a large number of coadapted physiological (how much calcium for the shells) and behavioral (when to stop mating) traits. Groups of genes that work together are known as *coadaptive gene complexes,* or, when they are linked together on one chromosome, supergenes. Some of the chromosomal inversions seen in the giant chromosomes of *Drosophila* (page 275) may increase fitness by protecting supergenes from disassembly by recombination.

EVOLUTION AND THE IDEA OF PROGRESS

The concept of evolution has always carried with it some notion of progress. This vision is reinforced by the evidence that evolution has, over time, produced organisms increasingly larger, more complex in structure, and more sophisticated in design, culminating—as if it were preordained—in something so marvelous as ourselves. Natural selection does tend to push populations toward better solutions to environmental problems, as we shall see in the pages that follow. However, evolutionary change does not necessarily mean improvement by anthropomorphic criteria, nor does it necessarily even result in organisms that are better adapted to their immediate environment.

Fecundity and Longevity

Imagine a rapidly expanding population in which the environment allows every organism that is born to survive and reproduce to its maximum extent. Does natural selection act on such a population? Stop a moment to answer this question before you read on. If your answer is yes, you are correct. Selection also acts on fecundity, the total number of eggs produced by a female or the total number fertilized by a male. In an expanding population, genotypes that can churn out copies of themselves faster than other genotypes will increase in the population.

Longevity, on the other hand, highly prized by most people over 30, is virtually valueless from an evolutionary standpoint. Survival beyond reproductive age is of no value and may actually be detrimental, as when a postreproductive adult competes with its own young for food or some other resource. The only exceptions are those rare cases in which older animals make social contributions to the survival of related members of the younger generation, as in the case of matriarchal elephants or an occasional human.

49–8
A cave-dwelling crayfish, like other animals adapted for life in the lightless interiors of caves, is sightless and unpigmented. Its antennae are longer than those of an ordinary crayfish, and, moving slowly and silently, it is able to detect even slight disturbances in the water.

Developmental and Structural Constraints

Evolution has to work with what is available, not only in the chance variations in the genetic material but also in the three-dimensional phenotypic expression of these genes. To take a simple example, all of us vertebrates are limited structurally to a body with a long vertebral column and four legs, one on each corner, with which our ancestors waddled up on shore. A structural engineer setting out to build a flying machine or a submarine would scrap these blueprints and start from scratch, but evolution can only build on past history. Many of our human ailments, such as unusual difficulty in childbirth and a propensity to low back pain, can be traced directly to the conversion of this basically quadrupedal form to an upright stance. In *The Panda's Thumb,* Stephen Jay Gould points out that such absence of perfection and jerry-built contraptions (the panda's thumb, for example) are more convincing evidence that evolution has occurred than are the more usually cited examples of exquisite adaptation.

The conservatism of evolution is nowhere more evident than in embryological development. Here the complexity of building an organism from a single cell and a meter or two of encoded instructions is apparently so great that, as we saw in Figure 46–6 (page 885), it is only possible to proceed along certain well-established ancient pathways.

Eyeless Arthropods and Other Degenerates

Natural selection may lead to the loss of capacities rather than the increase of complexity. For example, cave animals that live in perpetual darkness often lack eyes; for an organism living in the dark, eyes are vulnerable and easily injured and there is no selective pressure to maintain them. Birds on islands where there are no predators may lose the ability to fly. In more extreme cases, as we saw in Chapter 24, parasitism can lead to degeneration of digestive organs and respiratory systems. Such organisms are highly evolved, in the sense of being extremely specialized—finely tuned to a particular way of life—although they may appear primitive and simple.

49–9
The Red Queen from Through the Looking Glass: *"You have to run faster than that to stay in the same place."*

The Red Queen Effect

Natural selection acts on the here and now; thus it is always a generation behind. If the selective force is steady and unidirectional—a gradual warming trend, for instance, or an increase in the speed of a predator—a population may manage to track the changes, always remaining a little behind. This is known as the Red Queen effect (Figure 49–9). A population can continue on such a course only if it possesses sufficient genetic variability. One of the many reasons given for the demise of the dinosaurs is that in their long span of ascendancy, they had become so well adjusted to their environment, which was at that time relatively unchanging, that when environmental changes began to occur, they lacked the genetic resources to keep up.

In an environment that fluctuates from generation to generation, a population that adapts—except by phenotypic flexibility—is always going to be slightly out of tune.

Sexual Selection

As Darwin recognized, many of the conspicuous adaptations of animals have to do not with survival on a day-to-day basis but with sexual selection, the "struggle between the members of one sex, generally the males, for the possession of the other sex." Males produce many more gametes than females, and so the female has a larger investment, in terms of time, energy, and resources, in each fertilized egg. When parental care is involved, females almost always are the ones caring for the young. Hence, males, in general, seeking to inseminate as many females as possible, are the competitors, and the females, with a higher stake in each mating, seek the best possible genetic partner and so are the choosers. The competition may either be direct—with other males for territories, harems, or privileges of consort—or indirect, as with nest-building and displays.

Sexual selection is the chief cause of sexual dimorphism, those differences between males and females that do not have to do directly with reproduction, such as the extravagant plumage of birds, the oversized antlers of deer and elk, and the larger size, mantle, and huge canines of the male baboon (Figure 49–10). Sexual dimorphism is most marked among animal species in which only a few of the males breed. Among birds, for example, monogamous pairs, such as swans or geese, look very much alike. Among polygamous groups, the males are more brightly colored, culminating in birds such as pheasants, in which a single male may have a large number of females. Among those few birds in which the females do the courting and the males do the choosing, such as the phalarope, it is the female that is the gaudier, with the stay-at-home male the drab partner. For those of us weary of comparisons between the resplendent peacock and the dowdy peahen, it is comforting to realize that it is the female whose size, color, and general behavior are considered to be optimal for the environment, while the dimorphic characteristics of the male are only useful for threat, display, and other bids for attention. In fact, these characteristics may well be maladaptive with respect to other traits, such as conspicuousness to predators.

Darwin, recognizing that such males were not "fitter to survive in the struggle for existence," categorized sexual selection as a force separate from that of natural selection. However, with fitness stringently redefined in terms of relative numbers of surviving offspring, such a distinction is now invalid.

(a)

(b)

(c)

(d)

(e)

49–10
Sexual selection is caused by competition for mates, usually among males, and results in differences between the two sexes. (a) *In many species of birds, such as the Chinese golden pheasants shown here, the females are colored in a way that blends with their surroundings, thus protecting them and their young, and the males have bright, conspicuous, and sometimes ornate plumage.* (b) *In elk, large antlers command respect from other males and are attractive to females.* (c) *Even though male and female baboons eat the same foods, the males have large canines, as shown here, and are almost twice as large as the females, features of importance in male-male competition.* (d) *Male elephant seals are also much larger than females and are made even more formidable by an inflatable proboscis. Dominant males command large harems, while inferior males go mateless.* (e) *In species in which mating couples form monogamous pairs, such as these albatrosses, males and females look alike.*

PATTERNS OF EVOLUTION

Natural selection produces different patterns of evolution. It may produce widely different phenotypes in closely related organisms, remarkably similar phenotypes in distantly related organisms, or the organisms themselves may become the forces of selection through their interactions with other species.

Divergent Evolution

Divergent evolution occurs when a population becomes isolated from the rest of the species and, owing to particular selection pressures, begins to follow a different evolutionary course. For example, *Ursus arctos,* the brown bear, is distributed throughout the Northern Hemisphere, ranging through the deciduous forests up through the taiga and into the tundra, as it was some 1.5 million years ago. As is characteristic of such widespread species, there are many local ecotypes or races. During one of the massive glaciations of the Pleistocene, a population of *Ursus arctos* was split off from the main group, and, according to fossil evidence, this group, under selection pressure from the harsh environment, evolved into the polar bear, *Ursus maritimus* (Figure 49–11). Brown bears, although they are members of the order of carnivores and closely related to dogs, are mostly vegetarians, supplementing their diet only occasionally with fish and game. The polar bear, however, is almost entirely carnivorous, with seals being its staple diet. The polar bear differs physically from the other bears in a number of ways, including, besides its white color, its carnivore-type teeth, its streamlined head and shoulders, and the stiff bristles that cover the soles of its feet, providing insulation and traction on the slippery ice.

Convergent Evolution

Organisms that occupy similar environments often come to resemble one another even though they may be only very distantly related phylogenetically. When they are subjected to similar selection pressures, they show similar adaptations. The whales, a group that includes the dolphins and porpoises, are similar in many exterior features to sharks and other large fish, but the fins of whales conceal the remnants of a tetrapod hand. Whales are warm-blooded, like their land-dwelling ancestors, and they have lungs rather than gills. Similarly, two families of plants invaded the deserts, giving rise to the cacti and the euphorbs. Both evolved large fleshy stems with water-storage tissues and protective spines, and they appear superficially similar. However, their quite different flowers reveal their widely separate evolutionary origins.

49–11
The polar bear is one example of divergent evolution. The white coloring of the polar bear is probably related to its need for camouflage while hunting rather than to a need to hide from predators, since it is one of the world's largest carnivores. (Its greatest enemies of all time are commercial fishermen and hunters who prize the skins for trophies.) Unlike their southern cousins, the brown bears, polar bears do not hibernate (except for females with young) but roam the ice and icy waters all winter long in search of fish, seals, young walruses, and other prey.

49–12
An example of convergent evolution is provided by (a) *the bull fur seal and* (b) *the king penguin shown here. Although one is a mammal and the other a bird, both have streamlined, fishlike bodies and a layer of insulating fat below the skin. The penguin swims by means of its flipperlike wings, using its webbed feet as rudders. The fur seal also swims primarily with its webbed forelimbs, using them like oars.*

(a)

(b)

Biogeography offers the most massive evidence of convergent evolution. In the biomes—the world's major groupings of organisms—unrelated plants and animals widely separated in time and space have come to resemble one another in their physical characteristics as a consequence of having to solve similar environmental problems.

Parallel Evolution

Parallel evolution is used to describe a situation in which lineages have changed in similar ways so that the evolved descendants are as similar to each other as their ancestors were. The most striking example is the parallel between the evolution of marsupials in Australia and the evolution of placental mammals in other parts of the world (Figure 49–14).

49–13
Members of (a) *the euphorb family and* (b) *the cactus family have been separated for millennia of evolutionary history, with the cacti evolving in the deserts of the New World, and the euphorbs in the desert regions of Asia and Africa. Members of both families have fleshy stems adapted for water storage, protective spines, and greatly reduced leaves.*

(a)

(b)

49–14
Some of the more striking examples of the parallel evolution of placental mammals and Australian marsupials.

COEVOLUTION

When two or more populations interact so closely that each is a strong selective force on the other, simultaneous adjustments occur that result in *coevolution*.

We have previously mentioned several examples of coevolution. One of the most important, in terms of sheer numbers of species and individuals involved, is the coevolution of flowers and their pollinators, described in Chapter 23. Here is one more example, also involving plants and insects, those two ancient allies and enemies.

Milkweed, Monarchs, and Mimics

As we noted in Chapter 23, various families of plants have evolved chemical defenses, so-called "secondary substances," that, because they are toxic, bad-tasting, or both, deter predation by herbivores. The bitter white sap of plants of the milkweed family contains a cardiac glycoside, similar to digitalis, that acts as a deterrent to most herbivores. In the course of the evolutionary race to stay in the same place, some species of insects, including monarch butterflies, have evolved enzymes that enable the caterpillars to feed on the milkweeds without being poisoned.

Monarch caterpillars not only utilize the plant tissues for food, they ingest and store up the toxic glycoside, which is then present in their adult forms, the butterflies. The butterflies, in turn, are distasteful and poisonous to their predators, which are insectivorous birds.

Like many small animals that taste bad, monarchs, as part of this same defensive strategy, came to have conspicuous warning colors that deter predators. Such warning colors are common in the animal world. They are found not only among insects but also among poisonous reptiles and amphibians. The skunk is a familiar mammalian example of an animal whose distinctive markings remind us to keep our distance. As a result of convergent evolution, especially among insects, unrelated species often come to resemble one another in their warning characteristics—bees, wasps, and hornets are examples. This type of mimicry is known as Müllerian, after F. Müller, who first described the phenomenon. Other insects that feed on milkweeds have the bright warning colors characteristic of the monarch.

49–15

(a) *A monarch butterfly egg (center of photograph) clings to the bud of a milkweed.* (b) *When the egg hatches, two to four days after it is laid, a caterpillar emerges that feeds on the milkweed, ingesting and storing the toxic compounds produced by the plant. The monarch caterpillar and also the monarch butterfly* (c) *thus become unpalatable and poisonous. The conspicuous coloration of caterpillar and butterfly warns would-be predators. The viceroy butterfly* (d) *does not feed on milkweeds and is not poisonous, but it is protected by its similarity to the monarch.*

(a)

(b)

(c)

(d)

49–16
A blue jay that has never before tasted a monarch butterfly is fed one in a controlled experiment. Shortly after ingesting the butterfly, the blue jay begins to show signs of being uncomfortable and then vomits. Upon subsequently being offered a monarch, the bird refused it.

Next to appear on the evolutionary scene, as events are reconstructed, were several species of butterflies that do not taste bad and that are not poisonous but that have coloration similar to that of the monarchs, and so escape predation. Such deceptive mimicry is known as Batesian because it was first described by Wallace's friend and traveling companion H. W. Bates (page 882) in 1862.

Laboratory experiments confirm the selective value of Batesian mimicry. Jane Brower, working at Oxford University, made artificial models by dipping mealworms in a solution of quinine, to give them a bitter taste, and then marking each one with a band of green cellulose paint. Other mealworms, which had first been dipped in distilled water, were painted green like the models, so as to produce mimics, and still others were painted orange to indicate another species. These colors were chosen deliberately: orange is a warning color, since it is clearly distinguishable, and green is usually found in species that are not repellent and that therefore benefit from being inconspicuous.

(a)

(b) (c) (d) (e)

49–17
Model and mimics. (a) *A yellowjacket* (Vespula) *and some of its mimics:* (b) *a sand wasp,* (c) *a masarid wasp,* (d) *an anthidiine bee, and* (e) *a syrphid fly. All except the fly are Müllerian mimics; that is, they all sting. The fly is Batesian. A Batesian mimic has been compared to an unscrupulous retailer who copies the advertisement of a successful firm. Müllerian mimics, by contrast, are reputable tradesmen who share a common advertisement and divide its costs.*

The painted mealworms were fed to caged starlings, which ordinarily eat mealworms voraciously. Each of the nine birds tested received models and mimics in varying proportions. After initial tasting and violent rejection, the models were generally recognized by their appearance and avoided. In consequence, their mimics were protected also.

Batesian mimicry obviously works to the advantage only of the mimic. The model, on the other hand, suffers from attacks not only by inexperienced predators but by predators who have had their first experience with mimic rather than with model. The mimetic pattern will be at its greatest advantage if the mimic is rare—that is, less likely to be encountered than the model. However, even when as many as 60 percent of the green-banded worms were mimics, 80 percent of the mimics escaped predation. A mimic also has a better chance if it times its emergence to appear after the model, thus reducing its chances of being encountered first. You will not be surprised to learn that in any given area, mimics do, indeed, generally emerge from the chrysalis after the model.

SUMMARY

Natural selection is the differential reproduction of genotypes resulting from interactions between individual organisms and their environment. According to neo-Darwinian concepts, it is the major force in evolution.

The major types of selection are stabilizing, in which extreme types are eliminated from the population; disruptive, in which the extreme types are selected at the expense of the intermediate form; and directional, in which one of the extremes is favored, pushing the population along a particular evolutionary pathway.

Natural selection can act only on characteristics expressed in the phenotype. The unit of selection, according to neo-Darwinian theory, is the entire phenotype—the organism. In extreme cases, a single allele may be decisive. Groups of coadapted genes operate together in gene complexes.

Evolution by natural selection does not necessarily produce a population with the best possible relationship to its environment. In a changing environment, many of its members will be at least one generation behind. The potential for evolutionary change is constrained by the extent of variability in the gene pool and the possibility of variation in structure and development of the individual organisms. Sexual selection, the result of competition for mates, can greatly increase differential reproduction without improving solutions to other environmental problems.

Observed patterns of evolution include divergent evolution, convergent evolution, and parallel evolution. In the first, similar, related species become more dissimilar; in the second, dissimilar species, only distantly related, come to resemble one another as a result of similar selection pressures; in the third, species neither diverge nor converge but evolve in parallel due to similar environmental pressures.

Coevolution is the result of strong directional selection upon two or more groups of organisms as a result of their interactions. An example is the relationship between milkweeds, monarchs, and monarch mimics.

QUESTIONS

1. Distinguish among the following: disruptive/directional/stabilizing selection; parallel evolution/convergent evolution/divergent evolution/coevolution; Batesian/Müllerian mimicry.

2. If you were a poisonous butterfly, would you rather have a Batesian mimic or a Müllerian one? Why?

3. What tool-using animals other than the woodpecker finch can you think of?

4. What characteristics in the contemporary human population are probably being maintained by stabilizing selection? Which might be subject to directional selection?

CHAPTER 50

On the Origin of Species

The central issue in the study of evolution concerns, of course, the source of the major changes that have taken place in the history of living organisms. Can the gradual accumulation of small changes, so well documented by the population geneticists, account for the great panorama of large-scale evolution—the transition to land, the rise of the dinosaurs, the appearance of the hominids? Or, put more succinctly, does microevolution explain macroevolution?

Macroevolution, by most definitions, is evolution above the species level, the emergence of new, higher taxa of organisms. Traditionally, macroevolution has been studied by paleobiologists, investigators of the fossil record. Microevolution, change within populations, has been the focus of attention of population geneticists. The bridge across this gulf, where population biologists and paleobiologists may someday meet, is the process of speciation, a subject that Darwin, despite the title of his great work, did not really address. The origin of species, however, is of great current interest.

MODES OF SPECIATION

In Chapter 19, we offered the following definition of species: "Species are groups of actually or potentially interbreeding natural populations which are reproductively isolated from other such groups." The essential feature of this definition is that of reproductive isolation. One knows that two separate species exist when the two can occupy the same space without interbreeding. In terms of population genetics, members of a species share a common gene pool effectively separated from the gene pools of other species. A central question, then, is how one pool of genes splits off from another to begin a separate evolutionary journey. A subsidiary question is how two species, often very similar to one another, inhabit the same place at the same time and yet remain reproductively isolated.

According to current perspectives, speciation is most commonly the result of the geographic separation of a population (interbreeding group) of organisms: this process is known as *allopatric* ("other country") *speciation*. Under certain circumstances, speciation may also occur without geographic isolation, in which case it is known as *sympatric speciation*.

50–1
Two subspecies that have formed on either side of a natural geographic barrier, the Tana River in Kenya: (a) *the common giraffe,* Giraffa camelopardalis rothschildi, *and* (b) *the reticulated giraffe,* Giraffa camelopardalis reticulata. *The two are considered subspecies because they are phenotypically distinct, but they interbreed readily in zoos, producing fertile hybrids, and hybrids are also found in nature.*

(a)

(b)

Allopatric Speciation

Geographic Races

Every widespread species that has been carefully studied has been found to contain geographically representative populations that differ from each other to a greater or lesser extent, as with the ecotypes of *Potentilla glandulosa* (page 910). Such populations that differ somewhat genetically but still form part of the same species—that is, they can still interbreed—are known as races. (A subspecies is a race that is sufficiently different morphologically to have been formally recognized by a Latin name; an example is given in Figure 50–1.) A species composed of geographic races of this sort is obviously particularly susceptible to speciation if geographic barriers arise.

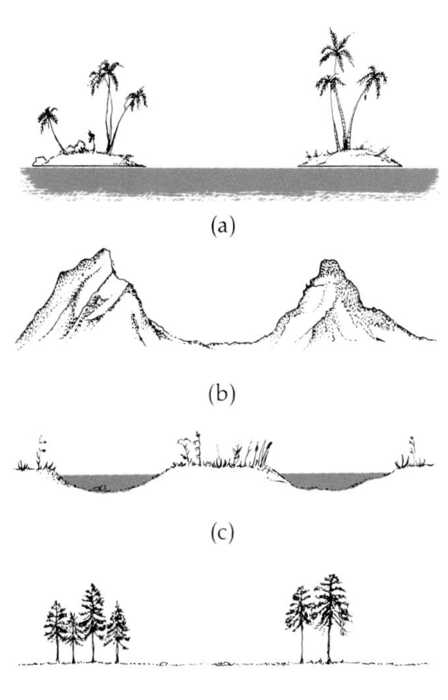

50–2
For different species, islands, genetically speaking, may be (a) *true islands,* (b) *mountaintops,* (c) *ponds, lakes, or even oceans, or* (d) *isolated clumps of vegetation.*

Geographic barriers are of many different types. Islands are frequent sites for the development of unusual species. The breakup of Pangaea (see essay, pages 930–931) profoundly altered the course of evolution, setting the island continent of Australia adrift, for example, as a veritable Noah's ark of marsupials. Populations of many organisms can become cut off from one another by barriers less obvious than oceans. For a plant, an island may be a mist-veiled mountaintop, and for a fish, a freshwater lake. A forest grove may be an island for a small mammal. A few meters of dry ground can isolate two populations of snails. Islands may form by the creation of barriers between formerly contiguous geographic zones. The Isthmus of Panama, for instance, has repeatedly submerged and reemerged in the course of geologic time. With each new emergence, the Atlantic and Pacific oceans became "islands," populations of marine organisms were isolated, and some new species formed. Then, when the oceans joined again (with the submergence of the Isthmus), the continents separated and became, in turn, the "islands."

Often the isolated population is a small, peripheral one and so is more likely to differ from the parent population (because of nonrandom sampling) and less likely to be subjected to gene flow. Once separated, the isolated population may begin to diverge genetically until interbreeding under natural conditions would no longer

THE BREAKUP OF PANGAEA

In the early 1900s, German geologist Alfred Wegener proposed that continents had migrated in the course of the earth's history and that this migration was the cause of mountain building and other geologic phenomena. Wegener's theory was rejected because of the discovery of a thick crust below the ocean floor that seemed to render the movement of continents impossible, and for decades the theory of continental drift was viewed with as much suspicion as reports of flying saucers and extrasensory perception (despite the fact that numerous students of geography had noted how neatly the coastlines of South America and Africa mirror one another).

In the past 20 years, continental drift has become respectable, having become a part of the new theory of plate tectonics. According to this theory, the outermost layer of the earth is divided into a number of segments, or plates. These plates, on which the continents rest, slide around on the surface of the earth, moving in relation to one another. The geologic expression of this relative motion occurs mainly at the boundaries between the adjacent plates. Where plates collide, volcanic islands such as the Aleutians may be formed or mountain belts such as the Andes or Himalayas may be uplifted. At the boundaries where plates are separating, volcanic material wells up to fill the void. It is here that ocean basins are created. Plates may also move parallel to the boundary that joins them but in opposite directions, or in the same direction but at different speeds, as with the San Andreas fault.

About 200 million years ago, all the major continents were locked together in a supercontinent, Pangaea. Several reconstructions of Pangaea have been proposed, one of which is shown here. It is generally agreed that Pangaea began to break up about 190 million years ago, about the time the dinosaurs approached their zenith and the first mammals began to appear. First, the northern group of continents (Laurasia) split apart from the southern group (Gondwana). Subsequently, Gondwana broke into three parts: Africa–South America, Australia–Antarctica, and India. India drifted northward and collided with Asia about 50 million years ago. This collision initiated the uplift of the Himalayas, which continue to rise today as India still pushes northward into Asia.

By the end of the Cretaceous period, about 65 million years ago, South America and Africa had separated sufficiently to have formed half the South Atlantic, and Europe, North America, and Greenland had begun to drift apart; however, final separation between Europe and North America–Greenland did not occur until the Eocene (43 million years ago). During the Cenozoic era, Australia finally split from Antarctica and moved northward to its present position, and the two Americas were joined by the Isthmus of Panama, which was created by volcanic action.

From the outset, of course, the theory of continental drift has been closely interwoven with that of evolution. One of the earliest and most impressive pieces of evidence in favor of the new theory

occur. At this point, speciation is said to have taken place. Every isolated population does not, of course, become a species. It may rejoin the parent group or it may perish, which is probably the fate of most small isolated populations.

Sympatric Speciation

Hybridization

A hybrid is the offspring of parents of different species. Hybrids can occur in animals (such as the mule), but they are far more common in plants. Kentucky bluegrass, for instance, is a promiscuous hybridizer and has crossbred with many related species, producing hundreds of hybrid races, each well adapted to the ecological area in which it grows. Such hybrids, spreading by means of rhizomes, are sometimes able to outcompete both parents. Mountain lilacs offer another example (Figure 50–3, on page 932). (As we noted in Chapter 19, such hybrids may

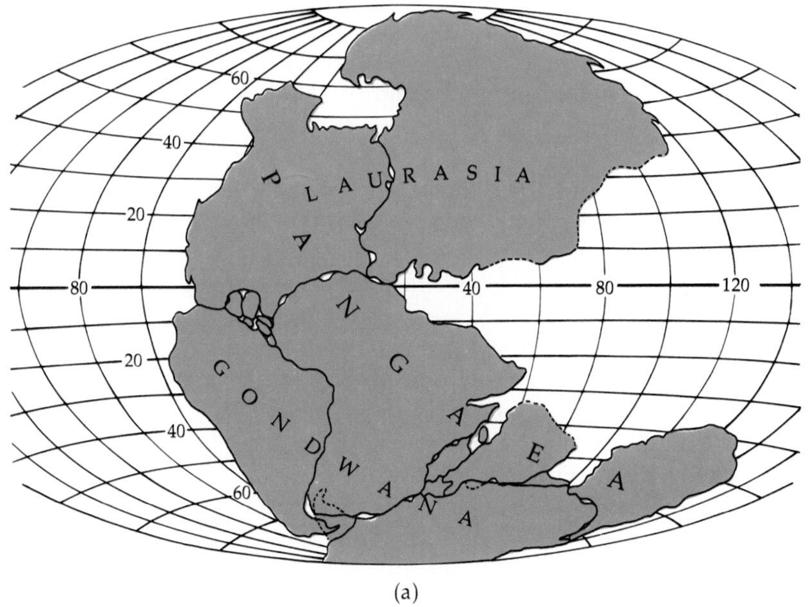

(a)

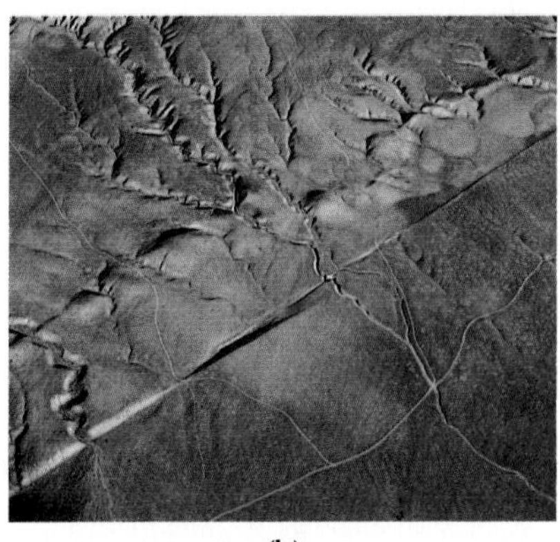

(b)

(a) *Pangaea.* (b) *The San Andreas fault is a boundary between two giant moving plates. The fault, running through San Francisco and continuing southeast of Los Angeles, is responsible for California's notorious earthquakes.*

was the discovery of fossil remains of a small, snaggle-toothed reptile, Mesosaurus, *found in coastal regions of Brazil and South Africa but nowhere else. Early in 1982, a team of American scientists returned from Antarctica with the first fossil of a land animal ever found there—a marsupial. This find supports the theory that marsupials migrated by land from South America (where only the opossum remains) across Antarctica to Australia before the two separated, some 55 million years ago.*

be considered species in that they are genetically isolated from their parents and from each other, but they do not conform to the working definition because they do not interbreed.)

Hybrids in both plants and animals are often sterile because the chromosomes cannot pair at meiosis (having no homologues), a necessary step for producing viable gametes (Figure 50-4a). In plants, however, fertile forms may arise from these infertile hybrids by the process of *polyploidy*, in which the cells acquire more than two sets of chromosomes. Polyploid cells arise at a low frequency as the result of a "mistake" in mitosis, in which the chromosomes divide but the cell does not. If such cells divide by further mitosis so that they eventually produce a new individual asexually, that individual will have twice the number of chromosomes as its parent. Polyploid individuals can be produced deliberately in the laboratory by the use of the drug colchicine, which prevents separation of chromosomes during mitosis.

50–3

The establishment of hybrid populations is an important evolutionary mechanism in many groups of plants, such as the mountain lilac. The map shows a portion of northern California (note San Francisco Bay at the bottom), a geologically complex region. Two relatively widespread and distinct species of mountain lilac, the coastal Ceanothus gloriosus *and the interior* Ceanothus cuneatus *(which ranges far to the east out of the area depicted), have produced three series of hybrid populations. Each hybrid is variable but stabilized and is able to grow better than either parent in the areas where it occurs. Leaves from representative populations are shown on the map. Clusters of species such as these tend to be common in geologically diverse areas.*

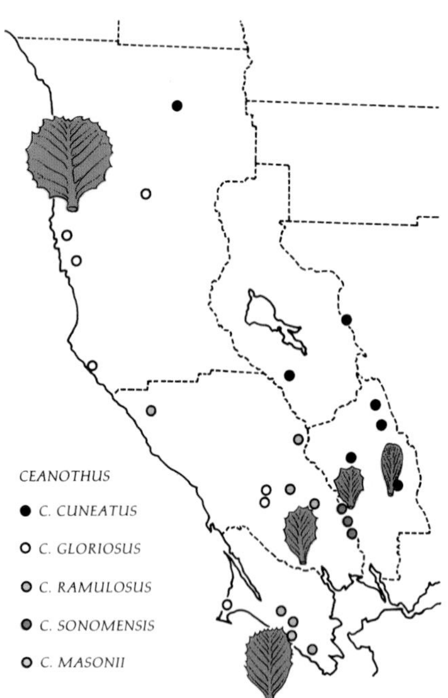

If polyploidy occurs in a sterile hybrid, each chromosome from each parent will be present in duplicate. The duplicate chromosomes can then pair, meiosis will be normal, and fertility restored (Figure 50–4b). Approximately half of the 235,000 kinds of flowering plants have had a polyploid origin, and many important agricultural species, including wheat, are hybrid polyploids. Sympatric speciation by hybridization and polyploidy is an important, well-established phenomenon in plants.

Disruptive Selection

Sympatric speciation as a result of disruptive selection (page 915) is a more controversial issue. Cases of sympatric speciation are difficult to prove because of the time factor: either (1) the two forms are in the process of diverging, as in the plants growing on the mine tailings, in which case it is not possible to prove that complete separation will ever occur, or (2) the two forms have separated completely, and it is not possible to prove that they were not formed allopatrically and have since been reunited.

One often-cited example of sympatric speciation produced by natural selection has been reported by Guy Bush of the University of Texas. The genus *Rhagoletis* comprises a group of species of small, brightly colored flies whose larvae feed on developing fruit. Each species of *Rhagoletis* feeds on fruits of only one plant family; the host fruits of the family serve not only as food but also as the rendezvous for courtship and mating followed by the deposition of eggs. *Rhagoletis pomonella,* for example, is a species that feeds on hawthorns. In 1865, farmers in the Hudson River Valley began to report that these flies had begun attacking their apples. The infestation then spread rapidly to orchards in adjacent areas of Massachusetts and Connecticut. There are now two distinct sympatric races of *R. pomonella,* with a variety of different genetic traits, that are isolated by their reproductive behavior and so may be regarded as being on the path to speciation.

Maintaining Genetic Isolation

Once speciation has occurred, the now-separate species can live together without interbreeding, despite the fact that some are so similar phenotypically that only an expert with a microscope can tell them apart—*Drosophila* offers several examples. What factors operate to maintain genetic isolation of closely related species? Isolating mechanisms may be conveniently divided into two categories: premating mechanisms, which prevent mating between members of different species, and postmating mechanisms, which prevent the production of fertile offspring from such matings as do occur. One of the most significant things about the postmating isolating mechanisms is that, in nature, they are rarely tested. The premating mechanisms alone usually prevent interbreeding.

Premating Isolating Mechanisms

Premating isolating mechanisms in vertebrates often involve elaborate behavioral rituals, visual signals, or frequently, a combination of the two (Figure 50–5). Visual behavioral signals may also be important in arthropods, as, for example, in the flashings of fireflies (see page 544).

Bird songs, frog calls, and the strident love notes of cicadas and crickets all serve to identify members of a species to one another. For example, studies of leopard frogs have recently revealed that some of the populations that can crossbreed

50–4

(a) *An organism, such as a mule, that is a hybrid between two different species and is produced from two haploid* (n) *gametes, can grow normally because mitosis is normal. It cannot reproduce, however, because the chromosomes cannot pair at meiosis.* (b) *If polyploidy occurs and the chromosome number doubles, the hybrid can produce viable gametes. Since each chromosome will have a partner, the chromosomes can pair at meiosis. The resultant gametes will be diploid* (2n).

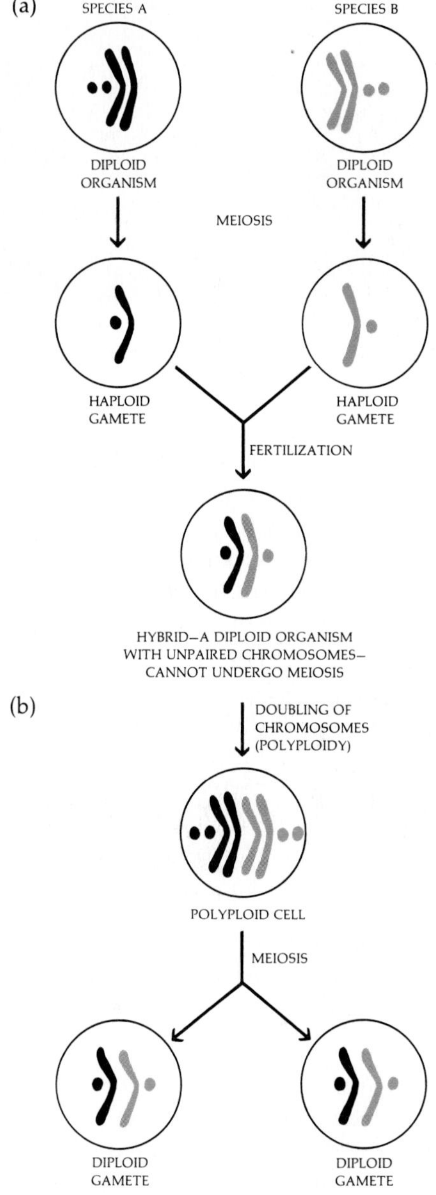

50–5

Sticklebacks, small freshwater fish, have elaborate mating behavior. The male at breeding time, in response to increasing periods of sunlight, changes from dull brown to the radiant colors shown here. He builds a nest and begins to court females, zigging toward them and zagging away from them. A female ready to lay eggs responds by displaying her swollen belly. The male leads her down to the tunnel-like nest. He prods her tail and, in response, she lays her eggs and swims off. He follows her through, fertilizes the eggs, and stays to tend the brood. If either partner fails in any step of this quite elaborate ritual, no young are produced.

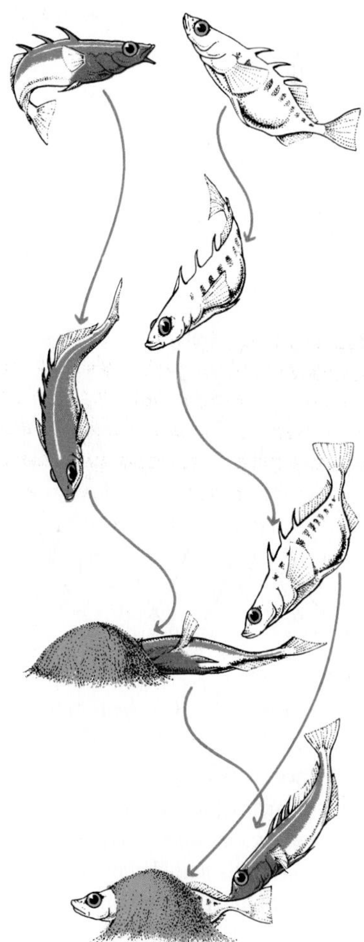

50–6
A male frigate bird displays his crimson pouch. Throughout the courtship period, the pouch remains bright and inflated, even when the bird is flying or sleeping. The female will not mate with a male that does not have these species-specific characteristics.

under laboratory conditions do not do so in the wild. They are effectively isolated from one another by differences in their mating calls. (For this reason, artificial breeding experiments are not always a good indication of whether individuals belong to the same or different species.)

In many species of invertebrates, pheromones (page 547) serve to bring the two sexes together. They may serve as signals to attract the male, as in the case of the Cecropia moth, or to trigger the release of gametes by the female, as with oysters. Because such pheromones are species-specific, they also act, in effect, as isolating mechanisms.

Temporal differences also play an important role in sexual isolation. Species differences in flowering times are important isolating mechanisms in plants. Most mammals—we are a notable exception—have seasons for mating, often controlled by temperature or by day length. Figure 50–7 shows the mating calendar of species of frogs near Ithaca, New York.

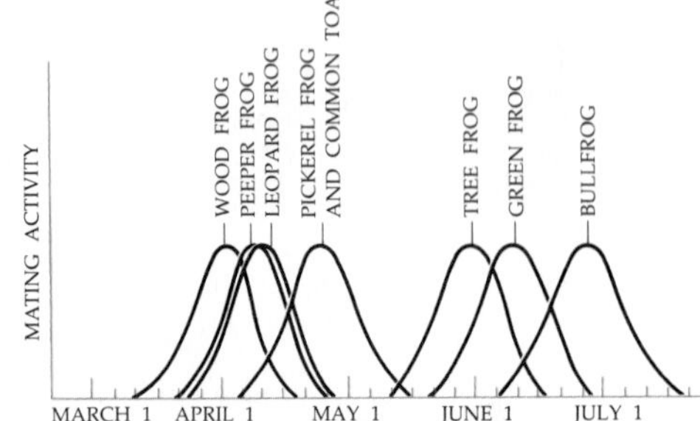

50–7
Mating timetable for various frogs and toads that live near Ithaca, New York. In the two cases where two different species have mating seasons that coincide, the breeding sites differ. Peepers prefer woodland ponds and shallow water; leopard frogs breed in swamps; pickerel frogs mate in upland streams and ponds; and common toads use any ditch or puddle.

Postmating Isolating Mechanisms

On the rare occasions when members of two different species attempt to mate or succeed in doing so, anatomical or physiological incompatibilities often maintain genetic isolation. These postmating isolating mechanisms are of several different types. For instance:

1. Differences in the shape of the genitalia may prevent insemination, or differences in flower shape may prevent pollination.
2. The sperm may not be able to survive in the reproductive tract of the female, or the pollen tube may not be able to grow on the stigma.
3. The sperm cell may not be able to fuse with the ovum.
4. The ovum, once fertilized, may not develop.
5. The young may survive but may not become reproductively mature.
6. The offspring may be hardy but sterile—the mule, for example.

CREATING SEXUAL CHAOS

Among the most destructive of the agricultural pests plaguing U.S. farmers are cotton bollworms (Heliothis zea) *and tobacco budworms* (Heliothis virescens). *Their combined efforts cause nearly $1 billion in cotton, corn, and soybean losses each year. They are currently controlled almost entirely by insecticides, all of which cause the death of other insect species, including those that prey on members of the genus* Heliothis.

An alternative method for controlling harmful insects is the use of sex pheromones to lure males to their death. In 1982, scientists from the U.S. Department of Agriculture (USDA) were experimenting with synthetic sex pheromones, looking for compounds that would either attract male budworms to traps or block their receptors so they could not find receptive females. While they were testing one of these in the field, they discovered that it was particularly effective in an unexpected way: it lured male bollworms as well as male budworms, and even more surprising, it caused the male bollworms to cross the species barrier and mate with the female budworms.

The first isolating mechanism having failed owing to the duplicity and ingenuity of the scientists, a second was encountered. Members of the two species differ in the shapes of their genitalia. These minute differences do not prevent copulation, but once the mismatched pair copulate, they cannot separate again and so are locked permanently in a deadly embrace.

USDA entomologists are cautious but hopeful that this "sexual chaos," as they call it, could prove a more effective means of insect control. One major problem is the existence of other isolating mechanisms—such as differences in times of mating—that may keep the two species apart despite the best scientific efforts.

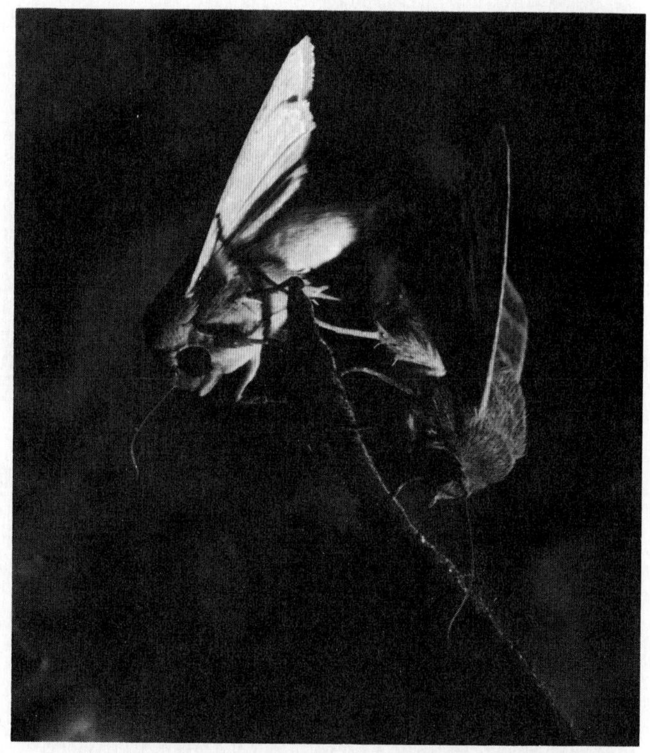

Tobacco budworms mating. One of the factors in maintaining genetic isolation among insect species is this lock-and-key fit of the genitalia. When bollworms attempt to mate with budworms, the key jams in the lock.

Postmating mechanisms reinforce premating ones. A female cricket that answers to the wrong song or a frog whose individual calendar is not synchronized with that of the rest of the species will contribute less (or nothing) to the gene pool. As a consequence, there is a steady selection for premating isolating mechanisms.

An Example: Darwin's Finches

Admirers of Charles Darwin find it particularly appropriate that one of the best examples of speciation is provided by the finches observed by Darwin on his voyage to the Galapagos Islands. All the Galapagos finches are believed to have arisen from one common ancestral group—perhaps either a single pair or even a single female bearing a fertilized egg—transported from the South American mainland, some 950 kilometers away. How she or they got there is, of course, not known, but it may have been the result of some particularly severe storm. (Peri-

(a)

(b)

(c)

(d)

(e)

(f)

50–8
Six of the 13 different species of Darwin's finches. Except for the warbler finch (a), *which resembles a warbler more than a finch, the species look very much alike; the birds are all small and dusky-brown or blackish, with stubby tails. The most obvious differences among them are in their bills, which vary from small, thin beaks to large, thick ones.*

(b) *The small ground finch* (Geospiza fuliginosa) *and* (c) *the large ground finch* (Geospiza magnirostris) *are both seed eaters.* Geospiza magnirostris, *with a somewhat larger beak than* G. fuliginosa, *is able to crack larger seeds.*

(d) *The cactus ground finch* (Geospiza conirostris) *lives on cactus blooms and fruit. Notice that its beak is larger and more pointed than those of the other two ground finch species.*

(e) *The small tree finch* (Camarhynchus parvulus) *and* (f) *the large tree finch* (Camarhynchus psittacula), *both insectivorous, take prey of different sizes.*

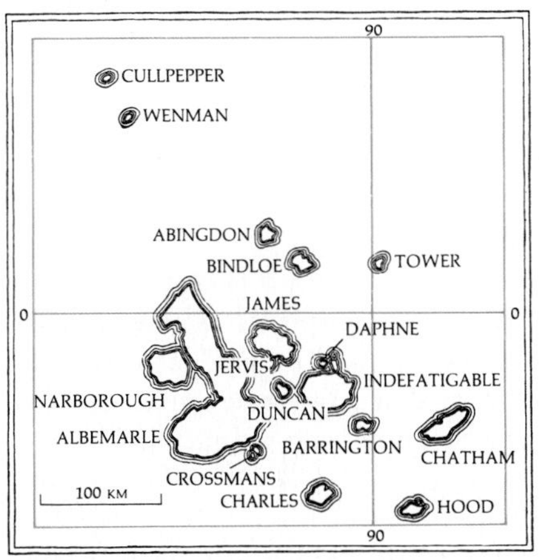

50–9
The Galapagos Islands, some 950 kilometers west of the coast of Ecuador, have been called "a living laboratory of evolution." Species and subspecies of plants and animals that have been found nowhere else in the world inhabit these islands. "One is astonished," wrote Charles Darwin in 1837, "at the amount of creative force . . . displayed on these small, barren, and rocky islands. . . ."

odically, for example, some American birds and insects appear on the coasts of Ireland and England after having been blown across the North Atlantic.) It is very likely that finches were the first land birds to colonize the islands.

We do have an idea of what greeted these unwilling adventurers. The Galapagos archipelago consists of 13 main volcanic islands, with many smaller islets and rocks. On some of the islands, craters rise to heights of more than a kilometer. The islands were pushed up from the sea more than a million years ago, and most of them are still covered with black basaltic lava.

The major vegetation is a dreary grayish-brown thornbush—making up vast areas of dense, leafless thicket—and a few tall tree cactuses. Inland and high up on the larger islands, the air is more humid, and there one can find rich, black soil and tall trees covered with ferns, orchids, lichens, and mosses, kept damp by a mist that forms around the volcanic peaks. During the rainy season, the area is dotted by sparkling shallow crater lakes. Thus the habitats available to the finches were highly diversified.

From the small ancestral group, 13 different species arose (Figure 50–8), plus one to the northeast on Cocos Island, 1,000 kilometers away. Apparently the various islands were near enough to one another that, over the years, small founding groups could emigrate successfully. The islands were far enough apart, however, that once a group was established, there would be little or no gene flow between it and the parent group for a period of time long enough for genetic barriers to develop. Cocos is so far away that it was successfully invaded only once. (Finches are not very good long-distance fliers; if they were, this natural experiment in evolution would have been a failure.) Development of a new species in this way requires, it is estimated, at least 10,000 years of geographic isolation, but since there are many islands, several species could have been evolving at the same time.

The ancestral type was a finch, a smallish bird with a short, stout, conical bill especially adapted for seed-crushing (Figure 50–10). The ancestor is believed to have been a ground-feeding finch, and six of the Galapagos finches are ground finches. Four species of ground finches live together on most of the islands. Three of them eat seeds and differ from one another mainly in the size of their beaks, which, in turn, of course, influences the size of seeds they eat. The fourth lives largely on the prickly pear and has a much longer and more pointed beak. The two other species of ground finch are usually found only on outlying islands, where some supplement their diet with cactus.

In addition to the ground finches, there are six species of tree finches, also differing from one another mainly in beak size and shape. One has a parrotlike beak, suited to its diet of buds and fruit. Four of these tree finches have insect-eating beaks, each adapted to a different size range of insects. The sixth, and most remarkable of the insect eaters, is the woodpecker finch, which has a beak like a woodpecker's that it uses as a chisel. Lacking the woodpecker's long, prying tongue, it carries about a twig or cactus spine to dislodge insects from crevices in the bark, as we saw in Figure 49–4.

By all ordinary standards of external appearance and behavior, the thirteenth species of Galapagos finch would be classified as a warbler. Its beak is thin and pointed like a warbler's, it even has a warblerlike habit of flicking its wings partly open, and, warblerlike, it searches leaves, twigs, and ground vegetation for small insects. However, its internal anatomy and other characteristics clearly place it among the finches, and there is general agreement that it, too, is a descendant of the common ancestor or ancestors.

50–10

Evolutionary tree of the Galapagos finches. According to this hypothesis, all are derived from a single ancestral species. There are six ground finches, including the two cactus-feeding finches (Geospiza *species); six tree finches* (Camarhynchus *species), including the woodpecker finch; and one warbler finch* (Certhidea). *Another warblerlike finch is found 1,000 kilometers away on Cocos Island.*

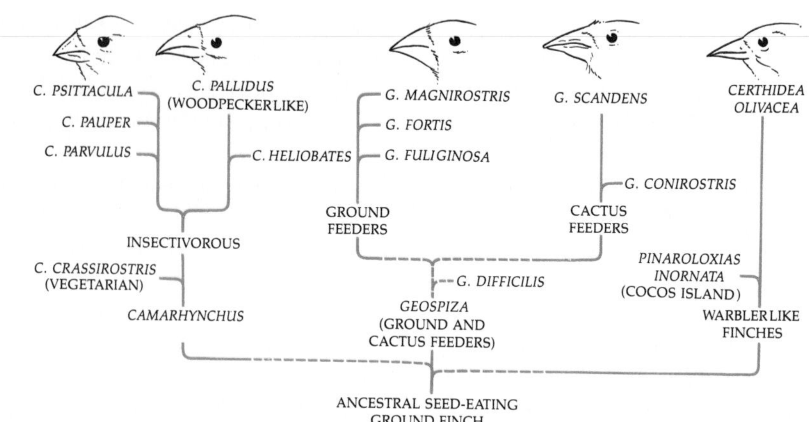

Galapagos finches recognize each other by their beaks. A male finch may mistakenly pursue a finch of another species—the finches often look very much alike from the rear—only to lose interest as soon as he sees the beak. He will not court a female of another species. The beak is a conspicuous feature in courtship, during which food is passed from the beak of the male to that of the female.

THE EVIDENCE OF THE FOSSIL RECORD

Although we may make inferences about macroevolution from the observation of living organisms, the only real proving ground of macroevolutionary theory is the fossil record itself, and any conclusions about evolution on a large scale must be tested against this record.

Phyletic Change

According to paleobiologists, the fossil record reveals three broad patterns of evolutionary change. One, known as *phyletic change*, or *anagenesis*, is change within a single lineage. According to this model, under the pressures of directional selection, a species gradually accumulates changes until eventually it is so distinct from its antecedents that it may properly be considered a new species—a new kind—of organism. Darwin's concept of evolution emphasized the slow, gradual accumulation of change, and phyletic change is, of course, a large-scale version of the microevolutionary events so ably documented by population geneticists.

Cladogenesis

The second pattern of evolutionary change is the splitting of lineages, or *cladogenesis* (the forming of branches). Species formed by cladogenesis are the contemporaneous descendants of a common ancestor—like Darwin's finches.

Paleobiologists have tended to give more weight to the role of cladogenesis in evolution than to phyletic change. Ernst Mayr, a systematist and a leader in the formulation of modern evolutionary theory, agrees. He proposes that the formation of new species by the splitting off of small populations from the parent stock (the founder effect) is responsible for almost all major evolutionary change. In such

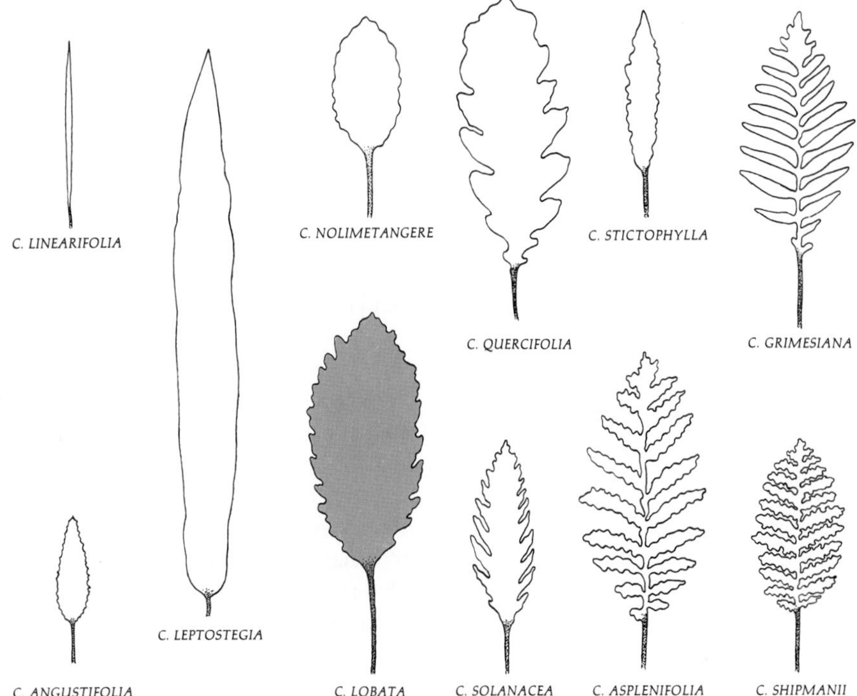

50–11
Another example of adaptive radiation. The species of Cyanea *(a genus found only in the Hawaiian Islands) differ widely in leaf size and leaf shape. All of these species probably evolved from ones like* Cyanea lobata *(bottom center).* Cyanea linearifolia *(upper left) is usually found in dry, sunny locations. Plants with fernlike leaves, like those on the right, are found in shady locations, where the divided wide-bladed leaves are more efficient in light-gathering.*

small populations, favorable genetic combinations, if present, can increase rapidly in number and frequency without being diluted out by gene flow. Hence, evolution probably does not occur steadily and gradually but in spurts, accounting for the sudden appearance of new species in the fossil record. These two models—phyletic change and cladogenesis—have not been viewed as alternatives, however, but rather as complementary to one another. The question, until recently, was simply a matter of degree.

Adaptive Radiation

Another pattern of evolution seen in the fossil record is that of adaptive radiation. George Simpson, a leading paleobiologist, emphasizes that adaptive radiation is the major pattern of macroevolution. In the fossil record, it is disclosed as the sudden (in geologic time) diversification of a group of organisms that share a common ancestor, often itself newly evolved. It is associated with the opening up of a new biological frontier that may be as vast as the land or the air, or, as in the case of the Galapagos finches, as small as an archipelago. Adaptive radiation is generally considered as combining both cladogenesis and anagenesis.

The fossil record contains many examples of adaptive radiation. For example, some 300 million years ago, the reptiles, liberated from an amphibian existence by the "invention" of the amniote egg (page 559), diversified rapidly into terrestrial environments. A similar, even more rapid burst of evolution later gave rise to the birds (page 560). As the dinosaurs became extinct, the mammals similarly burst forth on the evolutionary scene, with many different kinds appearing simultaneously in the fossil record.

Extinction

A phenomenon particularly well documented by the fossil record is that of extinction. Only a small fraction of the species that have ever lived are presently in existence—certainly less than 1/10 of 1 percent, perhaps less than 1/1,000 of 1 percent. Certain times in the earth's history have seen the elimination of major life forms: the trilobites that dominated the early Paleozoic seas; the dinosaurs that disappeared at the end of the Mesozoic era; and the great land mammals that became extinct during the Pleistocene, victims of its harsh climatic changes or, perhaps, of the predatory activities of a new life form, ourselves. Extinction is seen as playing a major role in evolution by clearing the way for new species.

Equus: A Case Study

As a case study in evolution, let us return to horses, whose history is particularly well documented (Figure 50–12). At the beginning of the Tertiary period, some 65 million years ago (Table 23–1, page 457), the horse lineage was represented by *Hyracotherium,* a small herbivore (25 to 50 centimeters high at the shoulders) with three toes on its hind feet, four toes on its front feet, and doglike footpads on which its weight was carried. Its eyes were halfway between the top of its head and the tip of its nose, and its teeth had small grinding surfaces and low crowns, which probably could not have stood much wear. Its teeth indicate that it did not eat grass; in fact, there was probably not much grass to eat. It probably lived in forests and browsed on succulent leaves.

Slightly higher in the fossil strata, one finds larger horses, all three-toed, still browsing. Some of these clearly had molar teeth with large crowns that continued to grow as they were worn away (as do those of modern horses).

The strata of the Oligocene yield fossils of still larger horses. In these the middle digit of each foot had expanded and was now the weight-bearing surface. In some, the two non-weight-bearing digits had become greatly reduced. These horses were the direct ancestors of modern horses.

These changes can be correlated with changes in the environment. In the time of little *Hyracotherium,* the land was marshy and the chief vegetation was leaves; the teeth of *Hyracotherium* were adapted for browsing. By the Miocene, the grasslands began to spread; horses whose teeth became adapted to grinding grasses (which have much coarser blades than most dicots do) survived, whereas those who remained browsers did not. The placement of the eye higher in the head may have facilitated watching for predators while grazing. The climate became drier, and the ground became harder; reduction of the number of toes, with the development of the spring-footed gait characteristic of modern horses was an adaptation to harder ground and to larger size. An animal twice as high as another tends to weigh about eight times as much. Little *Hyracotherium* was probably as fast as the modern horse, but a larger, heavier horse with the foot and leg structure of *Hyracotherium* would have been too slow to escape from predators. (During this same period, predators were developing adaptations that rendered them better able to catch large herbivores, including horses.)

Thus the evolution of *Equus,* viewed from the long retrospective of the geologic record, represents a fairly straightforward accumulation of adaptive changes related to pressure for increased size and for grazing, and so might be considered—as it was by Professors Marsh and Huxley (page 884)—as an example of phyletic change. As the number of fossil specimens increased, it became increasingly evi-

50–12

The modern horse and some of its ancestors. Only one of the several branches represented in the fossil record is shown. Over the past 60 million years, small several-toed browsers, such as Hyracotherium, *were replaced in gradual stages by members of the genus* Equus, *characterized by, among other features, a larger size; broad molars adapted to grinding coarse grass blades; a single toe surrounded by a tough, protective keratin hoof; and a leg in which the bones of the lower leg had fused, with joints becoming more pulleylike and motion restricted to a single plane.*

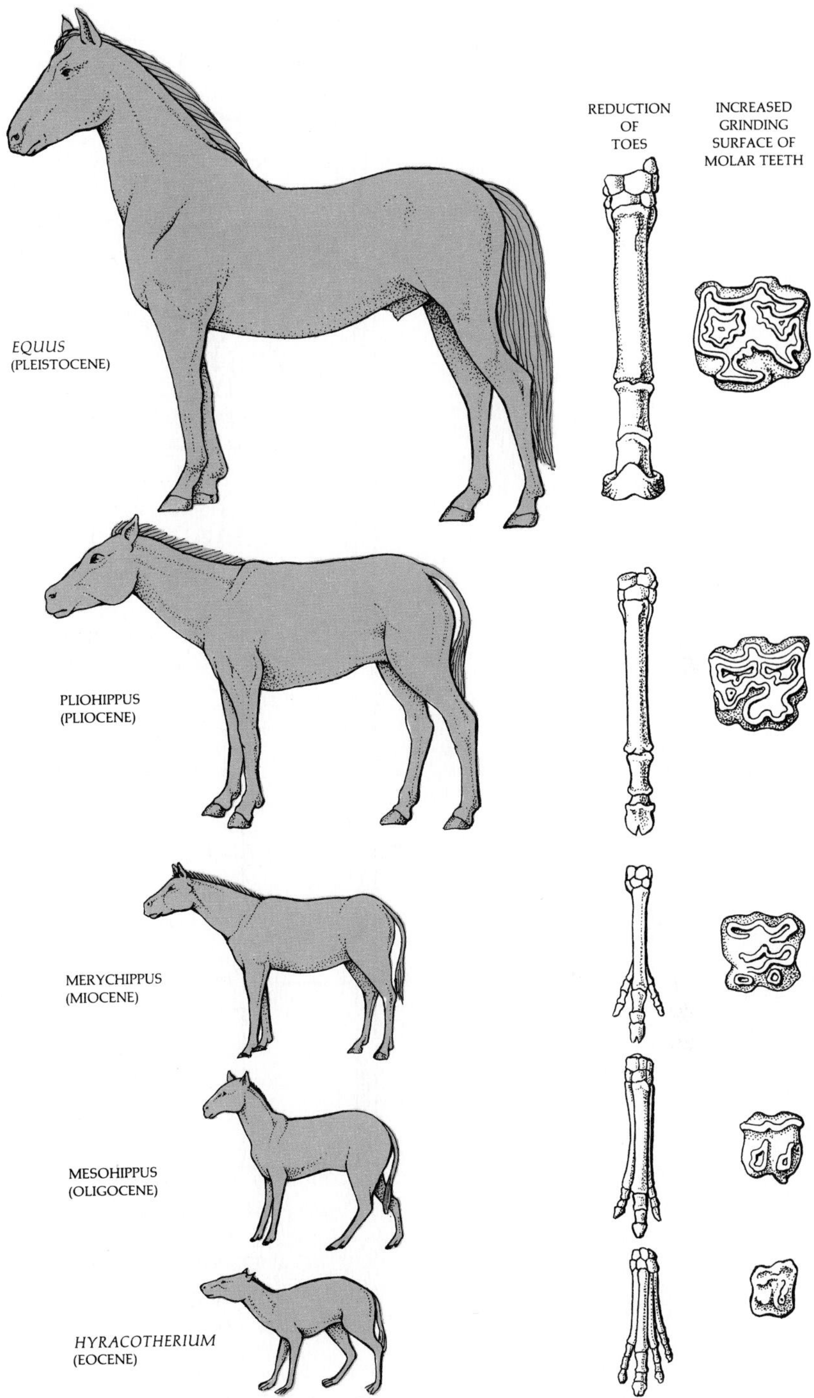

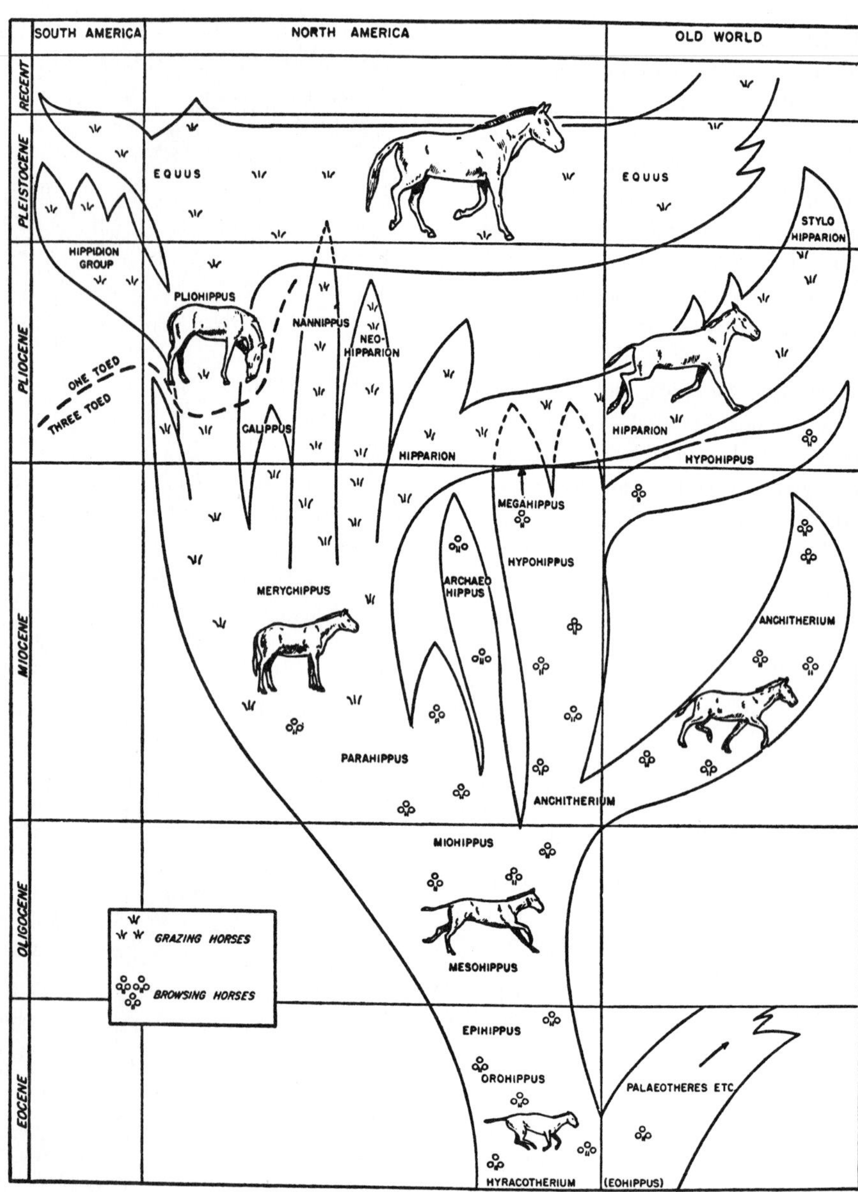

50–13
The lineages of the horse family, according to George Gaylord Simpson.

dent that at any given time many different species of horses coexisted, only some of which survived (Figure 50–13). Thus both anagenesis and cladogenesis are seen as contributing toward the evolution of *Equus.*

ON THE IMPERFECTION OF THE GEOLOGICAL RECORD

In Chapter X of *The Origin of Species,* Darwin discusses the problem of the lack of intermediate forms in the fossil record: ". . . the distinctness of specific forms, and their not being blended together by innumerable transitional links, is a very obvious difficulty." Darwin blamed this lack on "The Imperfection of the Geological Record" (the title of Chapter X), closing the chapter with his well-known description of the geological record as ". . . a history of the world imperfectly kept, and written in a changing dialect; of this history, we possess the last volume alone, relating only to two or three countries. Of this volume, only here and there a short chapter has been preserved; and of each page, only here and there a few lines."

Many fossils have, of course, been discovered in the 100 years since Darwin's death, and some of the volumes are now far more complete. Nevertheless, few examples of gradual change within forms have been found. Until recently, this discrepancy between the neo-Darwinian model of slow phyletic change, and the poor documentation of such change in the fossil record has been ascribed, in the Darwinian tradition, to the imperfection of the record itself. Reconciliation between theory and evidence has been forestalled by the lack of communication between paleobiologists and population geneticists; they attend separate scientific gatherings, publish in different scientific journals, and speak in a different dialect. (The situation is somewhat analogous to the lack of communication between geneticists and cytologists half a century ago.)

PUNCTUATED EQUILIBRIA

About a decade ago, two young scientists, Niles Eldredge of the American Museum of Natural History and Stephen Jay Gould of Harvard University, ventured the radical proposal that perhaps the fossil record was not so imperfect after all. Both Eldredge and Gould have backgrounds in geology and invertebrate paleontology, and both were impressed with the fact that there was very little evidence of phyletic change in the fossil species they studied. Typically, a species would appear abruptly in the fossil strata, last 5 million to 10 million years, and disappear, not much different morphologically than when it first appeared. Another species, related but distinctly different—"fully formed"—would take its place, persist with little change, and disappear equally abruptly. Suppose, Eldredge and Gould argued, that these long periods of no change ("stasis" is the word they use) punctuated by gaps, are not flaws in the record but *are* the record, the evidence of what really happens.

How could it be that a new species would make such a sudden appearance? They found their answer in the model of allopatric speciation. If new species formed principally in small populations on the geographic periphery of the range of the species, if speciation occurred rapidly (by rapidly, paleobiologists mean in thousands rather than millions of years), and if the new species then outcompeted the old one, taking over its geographic range, the resulting fossil pattern would be the one observed. In the words of Gould: "Thus the fossil record is a faithful rendering of what evolutionary theory predicts, not a pitiful vestige of a once bountiful tale."

As first proposed by Eldredge and Gould in 1972, punctuated equilibria seemed to refer principally to the tempo of evolution. The population geneticists and Darwin before them emphasized gradual change. There was clearly room, however, for the idea that populations would change more rapidly at some times than at others, particularly in periods of environmental stress, as in the case of the peppered moth (page 889).

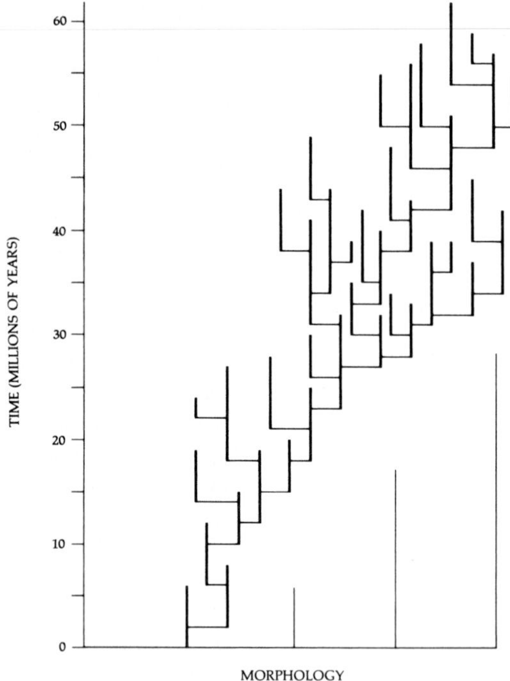

50–14
Model of species selection. The direction of speciation plays no role in the direction of evolutionary change; speciation events moving to the left are equal in number to events moving to the right. However, the average rate of speciation increases to the right and the average rate of extinction decreases in the same direction, so the direction of speciation plays no role in the direction of evolution.

As the new model has become more fully developed, particularly by Steven M. Stanley of Johns Hopkins (also a paleobiologist), it has become more radical. Its proponents now argue that not only is speciation the principal mode of evolutionary change (as Mayr stated some 30 years ago) but that natural selection occurs among species as well as among individuals. Thus, at any one time in evolutionary history it may be possible to find a number of related species coexisting, each departing in a different way from the ancestral type. For example, as the horses evolved, some species were larger, others remained small. Some became grazers, others remained browsers. The overall trends in horse evolution resulted from the differential survival of species, according to this model, rather than from the phyletic change that occurred within species. Thus, in this new formulation, species take the place of individuals, and speciation and extinction substitute for birth and death. In short, there are two mechanisms of evolution, according to this proposal: in one, natural selection acts on the individual, and in the other, it acts on the species.

Will the punctuated equilibrium model be assimilated into neo-Darwinian theory? Or will some radical new evolutionary concept spread through the scientific strata, outcompeting the old ideas? At this writing, it is too early to tell. All that is clear is that this challenge to orthodoxy has stimulated a vigorous debate, a reexamination of evolutionary theory, and a reappraisal of the evidence. Although it is referred to in some circles as anti-Darwinian, we are willing to wager that Darwin would have been delighted.

SUMMARY

A major question in evolutionary theory is whether or not microevolution (the gradual changes that take place below the species level) can account for macroevolution (the diversity among families, orders, classes, and phyla). The process of speciation is currently considered of central importance by those seeking an answer to this question.

Two principal modes of speciation are recognized, allopatric and sympatric. Allopatric speciation occurs in geographically isolated populations. Sympatric speciation occurs in plants by the formation of hybrids accompanied by polyploidy; it can also take place by disruptive selection. The key event in speciation is genetic isolation. Once species have become genetically isolated, they can once again inhabit the same geographic area without interbreeding because of numerous elaborate behavioral, anatomical, and physiological mechanisms.

According to paleobiologists, the fossil record discloses four components of evolutionary change: anagenesis, cladogenesis, adaptive radiation, and extinction. Phyletic change, or anagenesis, is gradual change within a single lineage over time. Cladogenesis, or splitting evolution, is evolutionary change produced by the branching off of one population from another to form a new species. Adaptive radiation is the rapid formation of new species from a single ancestral group, characteristically to fill a new ecological zone. Extinction is the disappearance of a species from the earth. Traditionally, macroevolution has been regarded as the product of a combination of these patterns.

Paleobiologists have recently presented evidence for an additional pattern of macroevolution known as punctuated equilibrium. They propose that major changes in evolution take place in small peripheral populations that have undergone speciation. Speciation, according to this model, is analogous to the role of mutation in population genetics; it provides variations on which natural selection acts. Major changes in evolution take place as a result of natural selection among species, according to the punctuationalists, as well as within populations.

QUESTIONS

1. Distinguish between the following: anagenesis/cladogenesis; allopatric/sympatric; hybrid/polyploid; premating isolating mechanisms/postmating isolating mechanisms.

2. Define genetic isolation. Why is it so important a factor in the species concept?

3. Give three possible reasons why polyploidy is more common in species of plants than of animals.

4. Describe the separate steps involved in the formation of the distinct species of Galapagos finches. What physical features of the Galapagos made possible the evolution of Darwin's finches?

5. What sort of data will be required to resolve the controversy between gradualism and punctuated equilibria?

SUGGESTIONS FOR FURTHER READING

Books

BATES, MARSTON, and PHILIP S. HUMPHREY, eds.: *The Darwin Reader,* Charles Scribner's Sons, New York, 1956.*

A collection of Darwin's writings, including The Autobiography *and excerpts from* The Voyage of the Beagle, The Origin of Species, The Descent of Man, *and* The Expression of the Emotions. *Darwin was a fine writer, and you can discover here the wide range of his interests and concerns at different periods of his life.*

CALDER, NIGEL: *The Restless Earth,* Viking Press, Inc., New York, 1972.

A handsomely illustrated report on the new geology, written for the general reader.

CARLQUIST, SHERWIN: *Island Life: A Natural History of the Islands of the World,* Natural History Press, Garden City, N.Y., 1965.

An exploration of the nature of island life and the intricate and unexpected evolutionary patterns found in island plants and animals.

DARWIN, CHARLES: *The Origin of Species by Means of Natural Selection,* or *The Preservation of Favored Races in the Struggle for Life,* Doubleday & Company, Inc., Garden City, N.Y., 1960.*

Darwin's "long argument." Every student of biology should, at the very least, browse through this book to catch its special flavor and to begin to understand its extraordinary force.

DARWIN, CHARLES: *The Voyage of the Beagle,* Natural History Press, Garden City, N.Y., 1962.*

Darwin's own chronicle of the expedition on which he made the discoveries and observations that eventually led him to his theory of evolution. The sensitive, eager young Darwin that emerges from these pages is very unlike the image many of us have formed of him from his later portraits.

DOBZHANSKY, THEODOSIUS, FRANCISCO J. AYALA, G. LEDYARD STEBBINS, and JAMES W. VALENTINE: *Evolution,* W. H. Freeman and Company, San Francisco, 1977.

The most complete general textbook on evolution for advanced undergraduate and graduate students.

ELDREDGE, NILES: *The Monkey Business: A Scientist Looks at Creationism,* Washington Square Press, New York, 1982.*

Eldredge presents a clear, straightforward analysis of the "scientific creationism" controversy and also provides interesting background information on the broader issue of the history of evolutionary biology.

FUTUYMA, DOUGLAS J.: *Evolutionary Biology,* Sinauer Associates, Inc., Sunderland, Mass., 1979.

A textbook for an advanced course in evolution, focusing on controversial issues and problems. More up-to-date and more difficult than Dobzhansky, et al.

GOULD, STEPHEN JAY: *Ever Since Darwin: Reflections in Natural History,* W. W. Norton & Company, New York, 1977.*

GOULD, STEPHEN JAY: *The Panda's Thumb: More Reflections in Natural History,* W. W. Norton & Company, New York, 1980.*

Collections of thoughtful, witty, and well-written essays from Natural History. *Evolution is the central theme.*

---

* Available in paperback.

LACK, DAVID: *Darwin's Finches,* Harper & Row, Publishers, Inc., New York, 1961.*

This short, readable book, first published in 1947, gives a marvelous account of the Galapagos, its finches and other inhabitants, and of the general process of evolution.

LEWONTIN, R. C.: *The Genetic Basis of Evolutionary Change,* Columbia University Press, New York, 1974.*

This book is intended for the advanced student, but it is also highly recommended for any reader interested in the problems and perspectives of population genetics.

LEWONTIN, RICHARD: *Human Diversity,* W. H. Freeman and Company, New York, 1982.

A leading population geneticist, Lewontin analyzes the biological bases of human variation and also explores their often controversial social ramifications—such as IQ, skin color, and sex-linked differences. This is a wise and beautiful book that, transcending politics and biology, becomes a celebration of human potential.

MAYR, E.: *Populations, Species, and Evolution: An Abridgement of Animal Species and Evolution,* Harvard University Press, Cambridge, Mass., 1970.*

A masterly, authoritative, and illuminating statement of contemporary thinking about species—how they arise and their role as units of evolution.

MOOREHEAD, ALAN: *Darwin and the Beagle,* Harper & Row, Publishers, Inc., New York, 1969.*

A delightful narrative of Darwin's journey, beautifully illustrated with contemporary or near contemporary drawings, paintings, and lithographs.

SCIENTIFIC AMERICAN: *Evolution,* W. H. Freeman and Company, San Francisco, 1978.*

This reprint of the September 1978 issue of Scientific American *provides an overview of the history of life as understood by neo-Darwinian theory. Such topics as the evolution of ecological systems and of behavior are discussed, in addition to the evolution of the major types of organisms that have inhabited our planet.*

SIMPSON, GEORGE GAYLORD: *Horses,* Oxford University Press, New York, 1951.

The story of the horse family in the modern world and through 60 million years of history, as recorded by a leading paleontologist.

SIMPSON, GEORGE GAYLORD: *Penguins: Past and Present, Here and There,* Yale University Press, New Haven, Conn., 1976.

As Simpson says, "Penguins are beautiful, interesting, inspiring, and funny." They are also excellent examples of evolution and adaptation, as this informal account demonstrates.

SIMPSON, GEORGE GAYLORD: *Splendid Isolation: The Curious History of South American Mammals,* Yale University Press, New Haven, Conn., 1980.

Charles Darwin found in the South American fauna some of the major clues that led to the formulation of The Origin of Species. *Today, their history can be seen as an almost ideal natural experiment in evolution, told in Simpson's informal and engaging style.*

STANLEY, STEVEN M.: *Macroevolution: Pattern and Process,* W. H. Freeman and Company, San Francisco, 1979.

Stanley, a paleontologist, is an advocate of the punctuated equilibrium model of evolution, and this text provides the clearest complete statement of macroevolution from this perspective.

---

* Available in paperback.

STANLEY, STEVEN M.: *The New Evolutionary Timetable,* Basic Books, Inc., New York, 1982.

This book, written for a general audience, is an excellent introduction to evolutionary theory, in general, and to the current controversies surrounding it.

WICKLER, WOLFGANG: *Mimicry in Plants and Animals,* World University Library, London, 1968.*

Many examples and illustrations of a delightful subject.

Articles

ABELSON, PHILIP H.: "Creationism and the Age of the Earth," *Science,* vol. 215, page 119, 1982.

CARSON, HAMPTON L., *et al.*: "Hawaii: Showcase of Evolution," *Natural History,* December 1982, pages 16–72. A series of 10 articles on the evolution of the flora and fauna of the Hawaiian Islands.

ELDREDGE, NILES, and STEPHEN JAY GOULD: "Punctuated Equilibria: An Alternative to Phyletic Gradualism," in T. J. M. Schopf (ed.), *Models in Paleobiology,* Freeman Cooper & Company, San Francisco, 1972, pages 82–115.

FAUL, HENRY: "A History of Geologic Time," *American Scientist,* vol. 66, pages 159–165, 1978.

STEBBINS, G. LEDYARD, and FRANCISCO J. AYALA: "Is a New Evolutionary Synthesis Necessary?" *Science,* vol. 213, pages 967–971, 1981.

WILLIAMSON, PETER G.: "Morphological Stasis and Developmental Constraint: Real Problems for Neo-Darwinism," *Nature,* vol. 294, pages 214–215, 1981.

---

* Available in paperback.

SECTION 8

Ecology

CHAPTER 51

Population Dynamics and Life-History Strategies

Ecology is the study of the interactions of organisms with their physical environment and with each other. As a science, it seeks to discover how an organism affects, and is affected by, its living and nonliving environment and to define how these interactions determine the kinds and numbers of organisms found in a particular place at a particular time.

Ecology is both the oldest and youngest of the major subdivisions of biology. It is at least as old as the inquiring human mind—clearly not a subject any australopithecine could safely ignore. However, as a "hard" science it is young because it is only very recently that biologists have been able to devise ways to analyze the multitude of variables that affect a population of organisms in their natural environment, to study them quantitatively, and to construct models, pose hypotheses, and formulate and test predictions. The great ecologists of 30 years ago were sensitive observers of nature; those of today are wizards at calculus and statistics, although undoubtedly equally sensitive.

We are going to begin these chapters on ecology by looking at the smallest unit, the population. (A population, you will recall, is a group of interbreeding organisms, members of the same species found in the same locality.) Later chapters will be concerned with interactions between and among the populations of different species (the community) and with the interactions between the community and its physical environment (the ecosystem), ending with a planetary overview (the biosphere).

PROPERTIES OF POPULATIONS

As we noted in Chapter 1, different properties emerge at different levels of organization. A population has very different properties from the individuals it comprises. As a simple illustration, picture a patch of woods, or a field, or a lake shore to which you return from time to time, which perhaps your parents and grandparents visited and where, with any luck, your children may walk someday. If you go there in the spring, you will see, for the sake of the argument, squirrels, birds, wildflowers, and beetles, and generally, the same species in about the same numbers. Yet, year after year, it is unlikely that you see the same individual

51–1
An ecologist at work in a tropical rain forest. In the rain forest, many populations of both animals and plants spend their entire lives in the treetops.

squirrels and birds, very unlikely that you encounter the same beetles, and seldom the same individual wildflowers. The individual is transitory, but the population somehow endures, in the same place and in approximately the same numbers, year after year.

Among the properties of populations that are not properties of individual organisms are growth patterns, carrying capacity, mortality patterns, and age structure.

Table 51-1 Reproductive Capacity of the Housefly *(Musca domestica)**

GENERATION	NUMBERS IF ALL SURVIVE
1	120
2	7,200
3	432,000
4	25,920,000
5	1,555,200,000
6	93,312,000,000
7	5,598,720,000,000

* In one year, about seven generations are produced. The numbers are based on each female laying 120 eggs per generation, each fly surviving just one generation, and half of these being females. Adapted from E. J. Kormondy, *Concepts of Ecology,* Prentice-Hall, Inc., Englewood Cliffs, N.J., 1969.

Patterns of Growth

As Darwin noted almost 150 years ago, the reproductive potential of almost all species is very high. It can be calculated that a single female housefly producing 120 eggs (with half developing into females) can, within seven generations, give rise to almost 6×10^{12} houseflies (Table 51-1). With bacteria that have a generation time of 20 minutes, a single bacterium can give rise to 8 bacteria within one hour, 512 in three hours, and 262,144 in six hours. Fortunately for us, few organisms ever attain their maximum rate of growth.

The rate of increase of a population is the increase in the number of individuals in a given unit of time per individual present. Thus, the annual rate of growth in the United States at present is about 9.6 per thousand, and in the world it now approaches 17 per thousand. In the absence of net emigration or immigration, the increase is equal to the birth rate minus the death rate, and thus it can be equal to zero or it can be a minus figure, as it is now in some countries. This property of a population is called its *intrinsic rate of growth.*

The type of growth in which the number of individuals increases by a constant rate is known as exponential growth. It is described by the equation $dN/dt = rN$. dN/dt equals the growth rate of the population, the change in number over time. This equation states that the growth rate is equal to r, the rate of increase, times N, the number of individuals *already present.* (Compound interest is calculated in the same way; the more you have, the more you get.) A graph of the equation is shown in Figure 51-2. As you can see, although the rate of increase remains the same, the growth of the population—the increase in the number of individuals—changes markedly over time. Moreover, the slope of the growth curve is slight when the population is small, and then increases as the population gets larger. A curve such as this is most closely approximated by microorganisms cultivated in the laboratory, where resources are constantly renewed, or by the initial stages of seasonal "blooms" of algae, or by the recent growth of the human population (Figure 51-3).

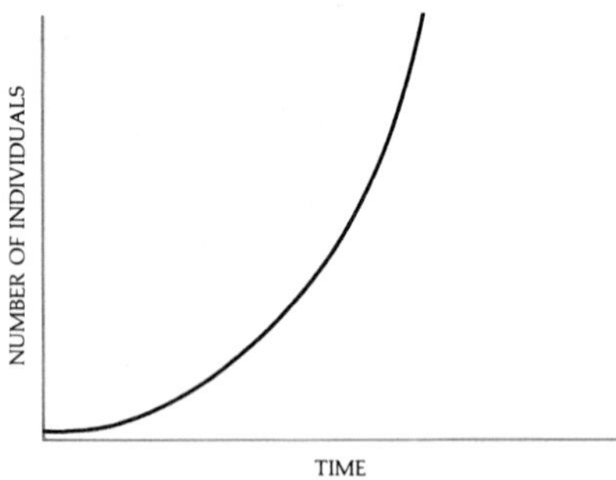

51-2
Exponential growth curve. The rate of increase remains constant, but the growth in the number of individuals increases rapidly as the size of the population increases.

Obviously, a population cannot long continue to increase exponentially without reaching some environmental limits imposed by shortages of food, space, oxygen, nesting or hiding places, or accumulation of its own waste products. In nature, exponential growth is characteristic of so-called fugitive or opportunistic species that rapidly use up a local resource and then either enter a phase of dormancy or move on. Weeds and insects are typical opportunists.

Sometimes, a population may hit a limit prematurely and so "crash," and die. An example can be found in the gypsy moth infestations that recurrently plague the northeastern United States. If a population grows rapidly enough, it can wipe out its food supply before the caterpillars complete metamorphosis and reach the reproductive stages. Or, alternatively, it may reach such a density that it is highly

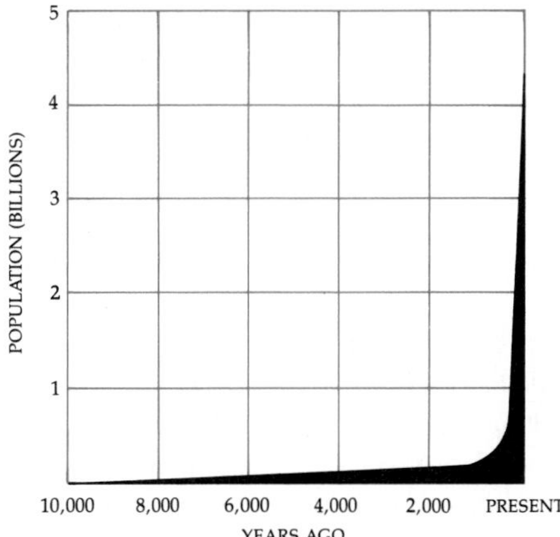

51–3
Growth of the human population. Note the resemblance to an exponential growth curve.

susceptible to infectious disease (a virus, in the case of the gypsy moth). Controlling such populations by the use of insecticides may act to prevent such crashes and to maintain the population at a lower but constant level.

Carrying Capacity

Most populations, however, as we suggested at the beginning of this chapter, are fairly stable in size. For such populations, the number of individuals is usually determined not by reproductive potential but by the environment. The number at which the population stabilizes is known as the *carrying capacity*. For animal species, the carrying capacity is often determined by food supply. For plants, the determining factor may be access to sunlight or the availability of water.

Limiting factors may vary seasonally. When the wildebeests migrate through the Serengeti, all the lions are well fed, but when the big herd animals are gone, the lion cubs often starve to death, and some of the less capable adults as well. The carrying capacity of a herd of deer is not the number counted in the springtime when the new fawns appear, but the number that can survive over several winters.

When environmental factors limit population growth, the growth pattern is more closely approximated by the equation:

$$\frac{dN}{dt} = rN\left(\frac{K-N}{K}\right)$$

In this equation, r is the rate of increase, as in the previous exponential equation, and r is again multiplied by the number (N) present at any one time. K stands for the carrying capacity, the number of organisms the environment can support over a period of time. Note that when N is very small, $(K - N)/K$ is close to 1, and the curve approximates the exponential growth curve (Figure 51–4). As N increases, $K - N$ decreases, and growth slows down. This slowing down represents a decline in the rate of increase due, typically, to competition among the individuals in the population for some limited resource. The curve that results is S-shaped, or sigmoid (from the Greek letter sigma).

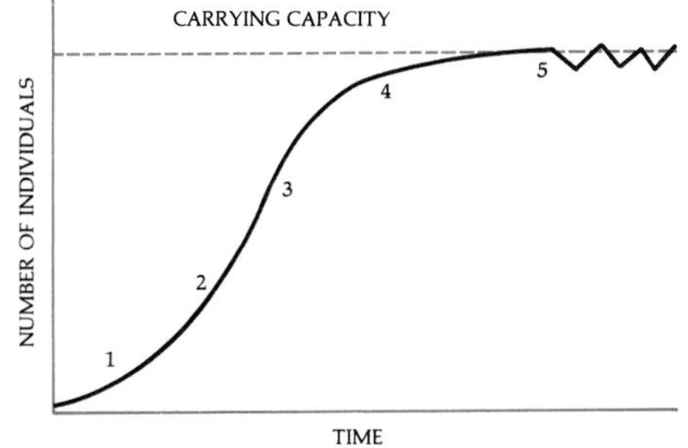

51–4
Biological growth curves are usually sigmoid, or S-shaped. As with exponential growth, there is an establishment phase (1) and a phase of rapid acceleration (2). Then, as the population approaches environmental limits, the growth rate slows down (3 and 4), and finally stabilizes (5), though minor fluctuations around K may continue.

51–5
Representative survivorship curves. In the oyster, mortality is extremely high during the free-swimming larval stage, but once the individual attaches itself to a favorable substrate, life expectancy levels off. In Hydra, *the mortality rate is the same at all ages.*

The curve at the top of the graph is for a hypothetical population in which all individuals live out the average life span of the species—a population, in other words, in which all individuals die at about the same age. The fact that the curve for humans approaches this hypothetical curve suggests that the human population as a whole is reaching some genetically programmed uniform age of mortality.

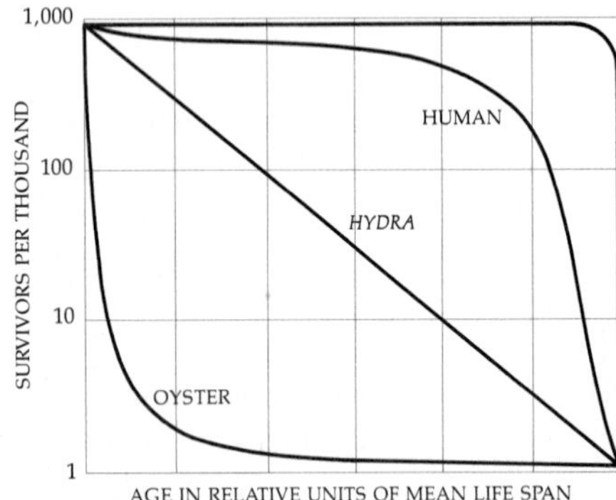

This model has practical applications. If, for instance, one wants to control a population of rats, killing half of them may merely reduce the population to the point at which it increases most rapidly. A more effective approach would be to reduce the carrying capacity, which, in the case of rats, usually means tighter control of garbage disposal. Similarly, if one wants to achieve maximum long-term productivity in the harvesting of a particular type of economically valuable fish, the population should not be harvested below the level of rapid growth unless one is willing to wait a long time for the population to recover.

Mortality Patterns

Another important property affecting the composition of a population is its mortality pattern. Figure 51–5 shows three different patterns of mortality. Often natural populations show a combination of patterns. For instance, in a study made of the saltmarsh song sparrows of San Francisco Bay, it was estimated that of every 100 eggs laid, 26 are lost before hatching. Of the 74 live nestlings, only 52 leave the nest, and of these 52, 80 percent die the first year. The remaining 10 breed the following season, but during the next year, 43 percent of these die, leaving only 6 out of the original 100. Each subsequent year, mortality among the survivors amounts to 43 percent, apparently regardless of age. In the case of birds, and many other animals, maximum longevity of the individual does not affect mortality patterns of the population. Thus, the early survivorship curve for the sparrows resembles that of the oyster, while the later curve is more *Hydra*-like.

Age Structure

Another property of populations is age structure, the proportions of different individuals of different ages in the population. A population that is not increasing in size will attain a stable age structure. Knowledge of the age structure is useful in

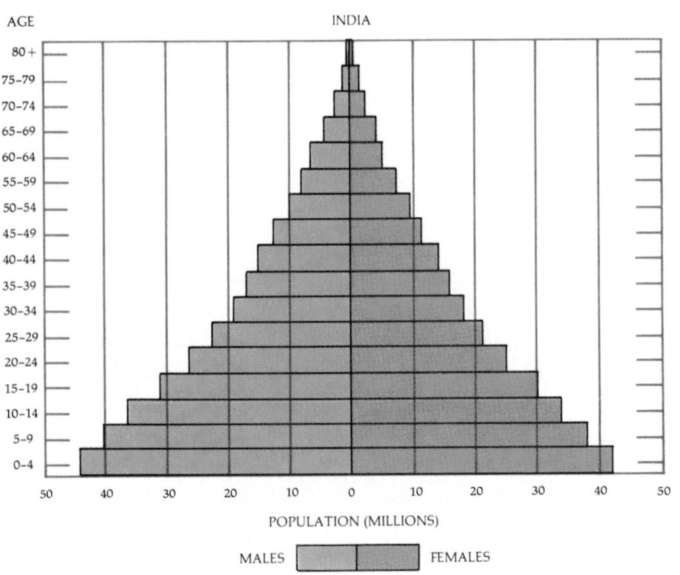

51–6
Graph showing the age structure of the human population of India. This pyramidal shape is characteristic of so-called "underdeveloped," or "developing," nations, with half the population under 20 years of age. The population can stay the same size only if death rates are as high as birth rates. Even if members of the present generation in India limit their family size to only two children per couple (which means cutting the current birth rate in half), population growth will not level off until about the year 2040—and at a level of well over a billion.

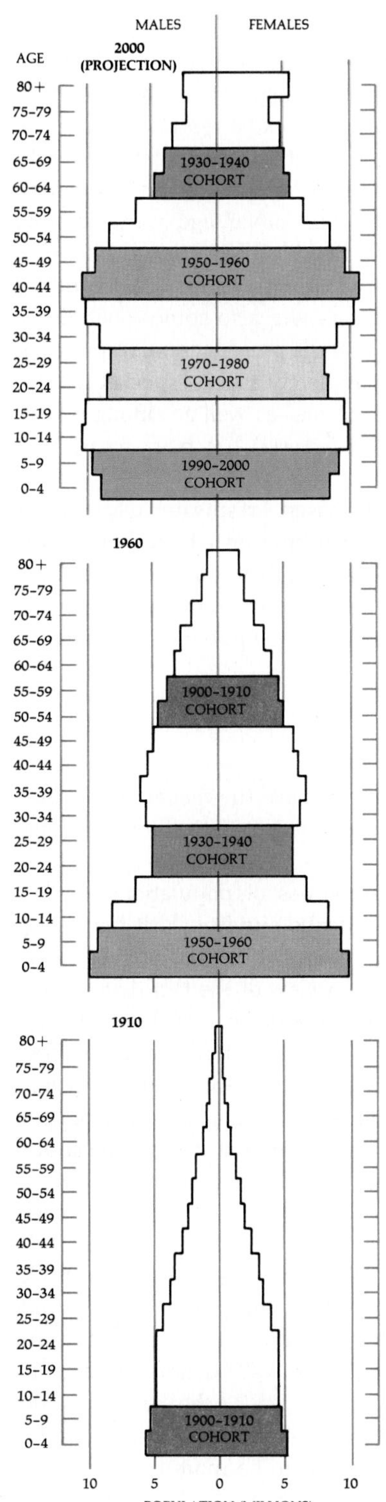

51-7

Age structure of the United States population. In 1910, the graph of age structure was shaped like a pyramid, although its base—that is, the number of persons in the youngest age groups—was not as large as that of India's. In subsequent years (actual and projected), the proportion of the population that is more than 40 years of age has steadily increased. Note the decrease in the growth of the population during the depression years of 1930 to 1940 and the bulge produced by the "baby boom" of the 1950s. Although birth rates are now low in the United States, as they were through the 1970s, the increase in the number of individuals during the 1950s is reflected in an increase in population growth in the 1980s as the boom babies reach reproductive age. The term "cohort" refers to the group of individuals born during the decade indicated.

predicting future changes in population size. If a large proportion of a population is young, as in India, population growth may continue at a high rate even though reproductive individuals only replace themselves. This age structure is a matter of considerable importance in predicting the growth of human populations. It is one of the reasons why the population of the United States has continued to increase despite the fact that, on the average, young couples are having slightly fewer than two children. (Another reason for the increase is continued immigration.)

The Asexual Advantage

Vegetative Reproduction

As we noted in Chapter 29, many plants reproduce by means of runners, or stolons, and by so doing, may be able to grow to a very large size. All of the plants produced represent a single genotype. A young plant that develops in this way has a continued supply of resources from the parent plant and thus a much higher probability of survival. Figure 51-8 shows the differences in survivorship in a species of buttercup among plants growing from seeds as compared to plants growing from stolons. Note the very high mortality in the early stages of growth from seed—an oysterlike curve—compared to the uniform risk of death of the plantlet grown from a stolon.

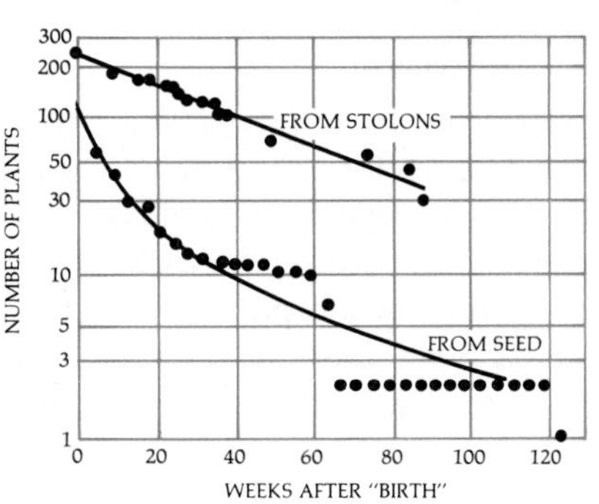

51-8

Survivorship curves of populations of Ranunculus repens, *a species of buttercup. Of 100 plants that started from seed (lower curve), only two (2 percent) were still alive 20 months later. Of 225 plants originating from stolons, 30 (more than 15 percent) were alive after 20 months (upper curve). Such plants receive support from the parent plant during early growth.*

Parthenogenesis

Parthenogenesis is reproduction from an unfertilized egg. In species in which the male gamete determines the sex of the offspring, parthenogenesis always results in all female offspring. Hence, it is far more efficient than sexual reproduction. If the houseflies of Table 51-1 had reproduced parthenogenetically, each female would have had twice as many female young in every generation, and the population would have been 358×10^{12}. Parthenogenesis lacks the advantage of the parental support system supplied by vegetative growth, trading it off for the possibilities of larger numbers and, usually, wider dispersal of the young. Dandelions reproduce parthenogenetically. They form conspicuous flowers and also some functionless pollen grains, which may be taken as indications that the present asexual species of dandelions evolved recently from sexual ones. Completely asexual species are also found among small invertebrates—rotifers, for example—as well as among plants. Recently, a species of lizard (of the genus *Cnemodophorus*) has been found that apparently only reproduces parthenogenetically.

Some organisms alternate sexual and asexual phases. Freshwater *Daphnia,* for instance, multiply by parthenogenesis when the plankton on which they feed is abundant. Then, in response to some environmental cue, they start producing both males and females. Typically, the asexual phase occurs when conditions are favorable for rapid local growth, and the sexual phase when the population is facing a less certain future and less homogeneous conditions.

LIFE-HISTORY STRATEGIES

The term "strategy," as used by biologists, does not have the meaning of a purposeful plan. It is used to refer to a group of coadapted, genetically determined traits.

Among the most interesting and variable properties of populations are *life-history strategies*, coadapted traits affecting reproductive survival. Is it better, for example, to have two million almost microscopic young at one spawning, like an oyster, or only one large infant, like an elephant? In view of the fact that physiological constraints might prevent an oyster from producing a 10-kilogram offspring, this question may sound silly. However, as a result of studies such as those on clutch size in birds (page 914), biologists have become aware that life-history strategies vary from individual to individual within a population and also from one population to another population of related organisms. In other words, the strategies are genetically determined variations subject to natural selection.

The Alternatives

Alternative life-history strategies have been called by a variety of names. G. E. Hutchinson of Yale University has called them prodigal and prudent, noting that despite the apparent value judgments of these terms, prodigal may be successful under some circumstances where prudent is not. Others have defined prodigal as opportunistic, and prudent as an equilibrium strategy, since the prodigal traits would appear more adaptive for a "weedy" species, a colonizer of open fields, for instance, whereas the prudent strategies would appear to be more beneficial to a population at carrying capacity. For this same reason, Robert MacArthur and E. O. Wilson proposed that the strategies be classified as r-selected or K-selected; these

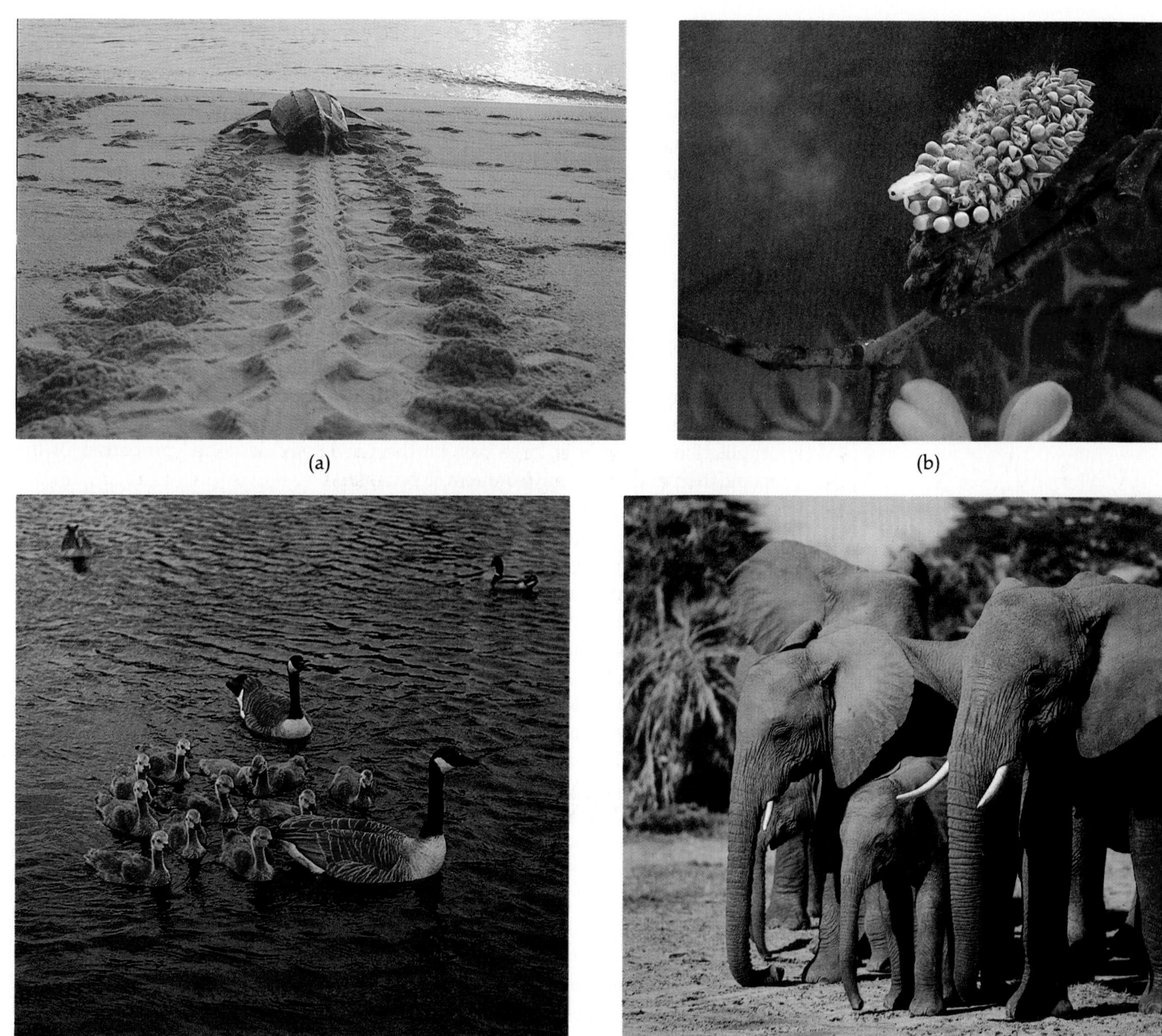

51–9

Life-history strategies. (a) *A leatherback turtle returning to the sea after laying her eggs in a hole dug in the sand. The eggs, which she has covered over with sand, will be incubated by the warmth of the sun. Once the eggs are laid, the mother turtle does not participate further in raising the young.* (b) *A female water bug lays her eggs on the back of the male. The male then carries the eggs, aerating them, until they hatch. In the photograph, one egg can be seen hatching at the left.* (c) *Baby Canada geese remain with their parents through the summer after they are hatched and migrate with them to the winter feeding grounds. In proportion to life span, the time alloted to parental care by birds is longer than for any other animal group.* (d) *Elephant calves are suckled by their mothers for at least two years, and the young are zealously guarded by the mother and the sisters and the aunts.*

Table 51-2 Alternative Life-History Strategies

PRODIGAL (r-SELECTED)	PRUDENT (K-SELECTED)
Many young	Few young
Small young	Large young
Rapid maturation	Slow maturation
Little or no parental care	Intensive parental care
Reproduction once ("Big Bang")	Reproduction many times

latter categories, although criticized in some quarters as oversimplifications, are the ones most in use in ecological parlance. The alternatives listed in Table 51-2 obviously represent the extremes in a continuum of possibilities. The best choice in the continuum—the choice made by natural selection operating on a particular population—depends in large part on the previously described properties of the population and of the environment it occupies.

Early or Late

Suppose you are a plant with a very short life expectancy—a flowering annual, for example. The choice made for you, by natural selection, is to put as much energy as possible into one reproductive effort, a big burst of seeds, since these have a much better chance than you of survival. On the other hand, if you are an oak tree, once you have established a place for yourself, your life span is likely to be a very long one, more than a hundred years. However, the likelihood of your progeny finding space on the forest floor is very small. You grow as tall as possible, for 20 years or more, since height aids in seed dispersal, and put out relatively small numbers of acorns per year, every year, over your entire life span.

Mackerel are like oak trees. Once a mackerel reaches a large size, its chances of continued survival are extremely good. However, the mortality among eggs and larvae is always very high, and, in years in which the plankton growth is low, there are no survivors at all. So it is better for a mackerel to concentrate its efforts first on reaching a safe size and thereafter to expend a relatively small amount of reproductive effort each season over a long period of time.

Early breeding, however, can greatly influence the rate of population growth. Reconsider the housefly of Table 51-1. If each generation time were shortened by only two days (from a minimum of 13 to 11), there would be time for one more generation, and the total number of possible progeny per season would increase by 330×10^{12}. Bacteria and other microorganisms can multiply so rapidly because of their very short generation time.

The prize for precocious development among multicellular animals should perhaps be awarded to a fungus-eating gall midge,* a small insect found on mushroom beds. These flies can reproduce sexually, but when abundant food is available, a female can produce parthenogenic eggs, which are retained inside her body. The larvae develop inside the mother, devour her tissues, and, completely skipping the usual metamorphosis, emerge with eggs inside their own tissues.

* This gall midge was borrowed from Stephen Jay Gould, *Ever Since Darwin.*

Within two days, larvae emerge from these eggs, devour their own mother, and are soon devoured in turn. Eventually, in response to some environmental signal, the parthenogenic cycle is broken, and the females produce normal males and females that fly off in search of new mushrooms.

Precocity does not always pay, however. Among many of the larger mammals, the rate of juvenile survival can be correlated with the experience of the mother, her size, or her social position, which are often determined by her age. Similarly, infant mortality is higher among teenage mothers in the United States than among women over 20, and higher still if the teenagers are from poor families, as they often are.

Small or Large

Among the factors particularly susceptible to natural selection is seed size in plants. Table 51–3 shows the average seed weights of populations of various species of goldenrod *(Solidago)* growing in a field (a new environment open to exploitation), and a prairie (an environment where populations are closer to carrying capacity). Five of the six species examined were found in both habitats. As you can see from the table, there were remarkable differences both among species and among populations. When seed size was plotted against seed number, it was found, as expected, that the smaller the seeds on any given plant, the more numerous they were. Thus, in this situation, the predictions of MacArthur and Wilson regarding patterns of life-history traits (see Table 51–2) were fulfilled.

Because dandelions reproduce asexually, dandelions growing in a single locality often consist of several different populations, each composed of genetically identical individuals. Otto Solbrig compared two such populations of dandelions growing together in various localities near Ann Arbor, Michigan. One genotype, genotype D, outperformed the other under all environmental conditions, both in the number of plants that survived and in their total dry weight. On the other hand, genotype A always produced more seeds and produced them earlier, so it always got a head start whenever a newly disturbed area became available for occupation.

Some Consequences

Populations following *r*-strategies would appear to lead riskier lives, both as individuals and as species. Typically, they rapidly exploit an environment and move on, and there may not, for example, always be another mushroom. However, such

Table 51–3 Mean Weight (in micrograms) per Seed of *Solidago* Species in Old Field and Prairie Communities

	OLD FIELD	PRAIRIE
S. nemoralis	26.7 ± 2.1	104.0 ± 8.3
S. missouriensis	17.6 ± 0.6	39.3 ± 3.2
S. speciosa	19.5 ± 1.6	146.3 ± 11.7
S. canadensis	27.3 ± 2.3	58.3 ± 11.1
S. gigantea	— —	50.8 ± 4.2
S. graminifolia	24.5 ± 2.7	10.6 ± 1.6

Adapted from P. A. Werner and W. J. Platt, "Ecological Relationships of Co-occurring Goldenrods *(Solidago:* Compositae)," *American Naturalist,* vol. 110, pages 959–971, 1976.

populations possess remarkable recuperative powers, because a population can be built up quickly from only a few individuals. Among *K*-strategy populations, by contrast, the individual has a much greater chance of survival, especially when maturity is reached, but a *K*-strategy population may not be able to recover from population densities much below their equilibrium level. The California condor and the whooping crane are *K*-strategists. Each does not begin to reproduce until it is about four years old, each raises only a single chick per season, and each is threatened with extinction.

SUMMARY

Ecology is the study of the interactions of organisms with their physical environment and with each other. The primary unit for ecological study is the population, a group of organisms in the same locality sharing the same gene pool. All populations have various properties that are determined by the sum total of the properties of the individuals in the population. One of these is the intrinsic rate of increase, which is equal to the birth rate minus the death rate, plus or minus the effects of immigration and emigration. The growth of a population at a constant rate is known as exponential growth and is represented by the equation $dN/dt = rN$. In this equation, r stands for the rate of increase (per individual per unit of time), and N stands for the number of individuals in the population at any given time (t). A key aspect of exponential growth is that, although the rate of increase remains the same, the growth rate (dN/dt) increases as the size of the population increases. Exponential growth is approached very seldom in natural populations.

Another characteristic of a population is its carrying capacity, the number of individuals of that particular species that can be supported by local resources. The model that more nearly represents natural growth is given by the equation:

$$\frac{dN}{dt} = rN\left(\frac{K - N}{K}\right)$$

In this equation, K represents the carrying capacity. The graph of this equation at first resembles the exponential growth curve, rising slowly while N is still small and then shooting up rapidly as N increases. However, unlike the exponential curve, it levels off as the carrying capacity is approached, producing an S-shaped curve.

Populations also have characteristic mortality patterns, with varying risk of death at varying ages. A related feature is the age structure of the population, the proportion of individuals of different ages. Age structure is an important factor in predicting future growth.

Asexual reproduction can increase total production of young at the cost of loss of variability among the progeny.

Populations are also characterized by their life-history strategies, groups of coadapted, genetically programmed traits subject to natural selection. The strategic options include many young versus few young, small young versus large young, rapid maturation versus slow maturation, little or no parental care versus intensive parental care, and reproducing once versus reproducing many times. Some combinations of these traits are favorable for organisms of opportunistic species undergoing exponential growth, whereas others will be selected for in more stable populations living at or near the carrying capacity.

QUESTIONS

1. Distinguish between the following terms: exponential growth/sigmoid growth; opportunistic species/equilibrium species; *r*-strategies/*K*-strategies.

2. An old French riddle: "The pond lilies in a certain pond grow at a rate such that each day they cover twice as much of the pond as they did the day before. The pond is of a size that it will be completely covered at the end of 30 days. On what day is the pond half-covered? One-tenth covered? One-hundredth covered?"
What is the relevance of this riddle to human ecology?

3. What would be the appearance of the survivorship curve of a population of annual plants? Of automobiles? Of salmon? Of butterflies? Of dishes in a dishwasher?

4. Explain how a change in each of the following factors would affect the growth rate of a population: age at first reproduction; time between generations; pre-reproductive mortality; post-reproductive mortality; length of period of parental care.

5. Suppose that you have a "farm" on which you grow, harvest, and sell edible freshwater fish. The growth of the fish population is sigmoid. You, of course, wish to obtain maximum yields from your "farm" over a number of years. To ensure this, how large should you allow the population to become before you begin harvesting? Identify the point on the sigmoid growth curve (Figure 51–4) at which you should begin harvesting.
How large a population should you leave unharvested? Identify the point on the curve at which you should take no more fish from the population.
Factors in addition to the pattern of harvesting will affect the yields of fish obtained. What are some of these factors, and how might you adjust them to further increase the yields?

6. Note that in the three graphs in Figure 51–7 there is a marked difference in the proportion of individuals more than 80 years old, but the vertical axis has not become appreciably longer. What do these data suggest? Does the age-structure graph of India (Figure 51–6) support your conclusion?

CHAPTER 52

Interactions in Communities: Competition, Predation, and Symbioses

Populations are made up of organisms. Communities are made up of populations. Ecologically speaking, a community comprises all the populations of organisms inhabiting a common environment and interacting with one another. These interactions affect the numbers of individuals in each population and determine the numbers of species in the community. They are also major forces of natural selection. Thus, as we shall see, ecology and evolution are intertwined, in what G. E. Hutchinson aptly called "the ecological theater and the evolutionary play."

We shall discuss three principal categories of interactions: competition, predation, and symbioses.

52–1
Some members of a coral reef community on the coast of Palau, one of the Caroline Islands. The crown-of-thorns starfish (Acanthaster) *is devouring the coral; the white areas are exposed coral skeleton. As a result of such predation, the total amount of coral is reduced, but the number of different species of coral increases as competition between them decreases.*

COMPETITION

Competition is the interaction between organisms using the same resource present in limited supply. The resource may be food, water, light, nesting sites, or burrows, to name a few. It takes place among organisms of the same species (where it is most intense) and also among those of different species. Interspecific ("between species") competition is usually greatest among organisms at the same trophic (feeding) level, with plants competing with other plants for sunlight and water, herbivores with other herbivores, carnivores with other carnivores. Moreover, within these categories, the more alike two species are, the more intense the competition between them. As we have noted previously, a key element in the scientific process is the asking of good questions. The question of how similar two or more species can be and still coexist in the same community has been, and continues to be, one of the most productive in the field of ecology.

The Principle of Competitive Exclusion

The principle of competitive exclusion was formulated by the Russian biologist G. F. Gause. Gause's principle states that if two species are in competition for the same limited resource, one or the other will be more efficient at utilizing or controlling access to this resource and will eventually eliminate the other in situations in which they occur together. Gause's principle has been valuable in ecological practice because it has been amenable to testing. Gause's own, now classic, ex-

52–2
Results of Gause's experiment with two species of Paramecium *demonstrate Gause's principle that if two species are in competition for the same limited resource—in this case, food—one eliminates the other.* Paramecium caudatum *and* Paramecium aurelia *were first grown separately under controlled conditions and with a constant food supply. As you can see,* P. aurelia *grew much more rapidly than* P. caudatum, *indicating that* P. aurelia *uses available food supplies more efficiently. When the two protozoans were grown together, the more rapidly growing species outmultiplied and eliminated the slower-growing species.*

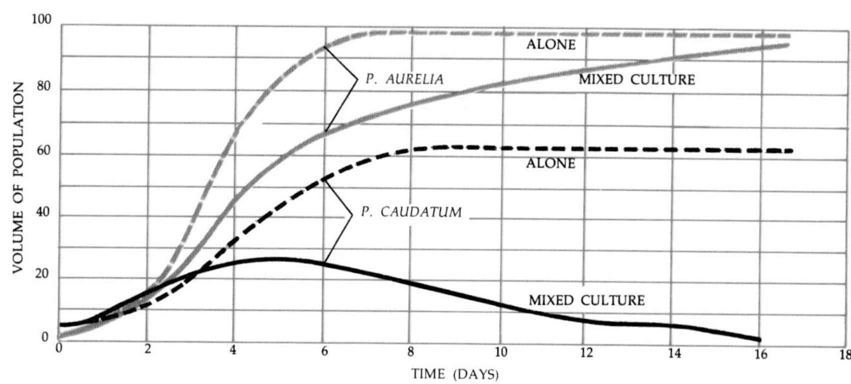

periment involved laboratory cultures of two species of *Paramecium: Paramecium aurelia* and *Paramecium caudatum.* When the two species were grown under identical conditions in separate containers, both grew well, but *P. aurelia* multiplied much more rapidly than *P. caudatum,* indicating that the former used the available food supply more efficiently than the latter. When the two were grown together, the former rapidly outmultiplied the latter, which soon died out (Figure 52–2).

The fastest growing species is not always the winner, however, as observed with two species of duckweed, *Lemna gibba* and *Lemna polyrrhiza. Lemna gibba* grows more slowly in pure culture than *L. polyrrhiza,* yet *L. gibba* always replaces *L. polyrrhiza* when they are grown together. The plant bodies of *L. gibba* have air-filled sacs that serve as little pontoons, so that these plants form a mass over the other species, cutting off the light. As a consequence, the shaded *L. polyrrhiza* dies out (Figure 52–3).

It is possible to change the culture conditions so that the outcomes of both the *Paramecium* and *Lemna* experiments are reversed. However, regardless of conditions, one species is always eventually eliminated.

52–3
An experiment with two species of floating duckweed, tiny angiosperms found in ponds and lakes. One species, Lemna polyrrhiza, *grows more rapidly in pure culture than the other species,* Lemna gibba. *But* L. gibba, *which has tiny air-filled sacs, floats on the surface and so shades the other species, making it the victor in the competition for light.*

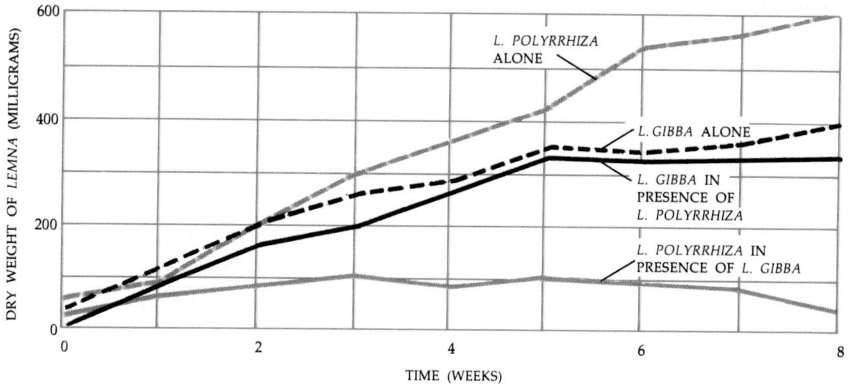

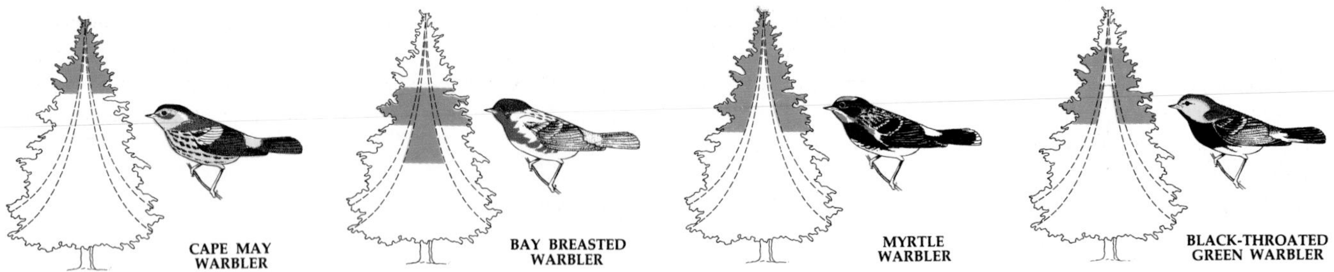

52–4
A demonstration of Gause's principle: the feeding zones in a spruce tree of five species of North American warblers. The colored areas in the tree indicate where each species spends at least half its feeding time. In this way, all five species feed in the same trees with reduced competition.

Testing Gause's Principle

In nature, similar species are often found in the same community. For example, some New England forests are inhabited by five closely related species of warbler, all about the same size and all insect eaters. Why do these birds not eliminate one another by competitive exclusions?

The late Robert MacArthur, an extraordinarily brilliant and innovative young ecologist, took this question for his doctoral thesis. He showed that these birds—now known as MacArthur's warblers—have different feeding zones in the trees (Figure 52–4). Because they exploit slightly different resources—that is, insects in different parts of the trees—the competition is reduced to a point where the species can coexist.

Similar accommodations are found among plants. In a bog, for example, mosses of the genus *Sphagnum* often appear to form a continuous cover, and several species are usually involved. How can these species continue to coexist? When the situation is examined in more detail, it is found that there are semiaquatic species growing along the bottoms of the wettest hollows; other species grow in drier places on the sides of the hummocks, which they help to form; and still other species grow only in the driest conditions on the tops of the hummocks, where they are eventually succeeded by one or more species of flowering plants. Therefore, although all the species of *Sphagnum* coexist, in the sense that they are all present in the same bog, they actually occupy different microhabitats and continually replace one another as the characteristics of each microhabitat change.

The Ecological Niche

Gause's competitive exclusion principle gave rise to the concept of the *ecological niche.* This term is a somewhat misleading one because the word "niche" has a connotation of space, whereas an ecological niche is not the space occupied by an organism but rather the role that it plays in its ecosystem. The simplest definition is that the niche is an organism's profession, as distinct from its habitat, which is its address.

A working definition of niche is more complex, however, and includes many more factors than the way in which an organism makes its living. A niche is, in fact, the total environment and way of life of all the members of a particular species of organism in the ecosystem. Its description includes physical factors, such as the temperature limits within which the organisms can survive and their requirements for moisture. It includes biological factors, such as the nature and amount of food

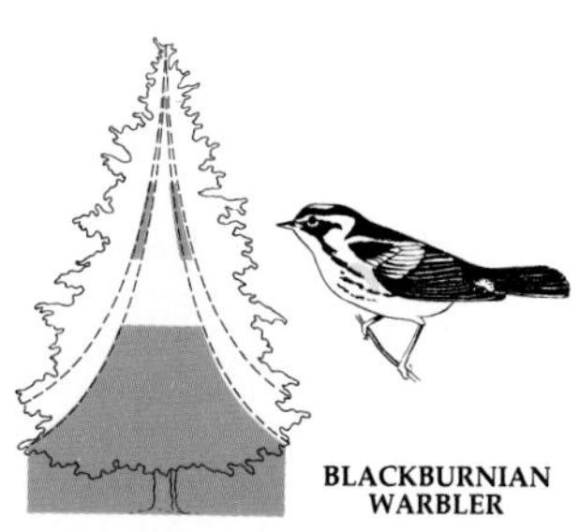

sources. It includes aspects of the behavior of the organism, such as patterns of movement and daily and seasonal activity cycles. Obviously, only one or two of these factors can be studied at any one time, but all are likely to influence the interactions of the members of a species with the members of other species in the community.

Fundamental and Realized Niche

As the studies on warblers suggest, competing species may limit one another's range of utilization of resources. A study by J. H. Connell of competition between two species of barnacles in Scotland offers a good example. Barnacles, as you will recall, are crustaceans. When they change from their free-swimming, larval form into their adult, sedentary form, they cement themselves to rocks and secrete shells. Once attached, barnacles remain fixed, so that by making careful records, one can determine the history of a particular population. One can identify exactly which barnacles have died and which new ones have arrived between visits to the study site.

One of the barnacle species studied, *Chthamalus stellatus,* is found in the high part of the intertidal seashore. As the tides go in and out, the barnacles are exposed to wide fluctuations of temperature and salinity and to the hazards of desiccation. The other species, *Balanus balanoides,* occurs lower down, where conditions are more constant. Although *Chthamalus* larvae, after their period of drifting in the plankton, often attach to rocks in the lower, *Balanus*-occupied zone, adults are rarely found there. Connell was able to show that in the lower zone, *Balanus,* which grows faster, ousts *Chthamalus* by crowding it off the rocks and growing over it or undercutting it. Isolated from contact with *Balanus, Chthamalus* lived with no difficulty in the lower zone. Connell removed *Chthamalus* from the higher zone and showed that *Balanus* was unable to move up to take its place. Although *Chthamalus* is driven from the lower zone by competition from the faster-growing *Balanus,* it survives in the intertidal community by special physiological adaptations that enable it to inhabit areas where *Balanus* cannot live.

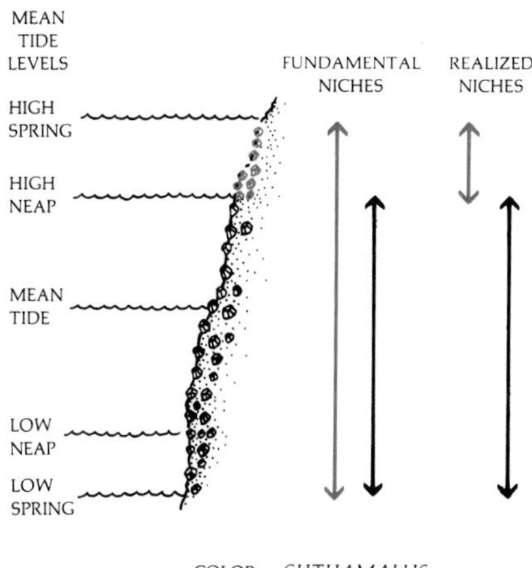

52–5
Interspecific competition in Balanus *and* Chthamalus *barnacles. The larvae of both species settle over a wide range, but the adults live in quite precisely restricted areas. The upward limits of the* Balanus *area are determined by physical factors such as desiccation. The* Chthamalus *barnacles, however, are prevented from living in the* Balanus *area not by physical factors—they would probably thrive there since the area is less physically limiting—but by the* Balanus *barnacles. The* Balanus *grows faster, and whenever it comes upon* Chthamalus *in the* Balanus *area, it either pries it off the rocks or grows right over it.*

Studies such as these generated the concepts of the fundamental niche, niche overlap, and the realized niche (Figure 52–6). The fundamental niche describes the physical limits of tolerance of the organism. The realized niche is that portion of the fundamental niche actually utilized; it is determined by physical factors and also by competition with other organisms. In its fundamental niche, *Chthamalus,* for example, can occupy both the intertidal zone and the lower zone, but because of niche overlap with *Balanus, Chthamalus* actually occupies a smaller area, its realized niche.

Niche Shift

As shown in Figure 52–6, when members of two species are competing for the same resource, there are members of the populations whose niches overlap. Organisms with overlapping characteristics will do less well in evolutionary terms; that is, there is selection pressure against them. As a result of this selection pressure, the two species tend to diverge in those characteristics that overlap. This phenomenon of niche shift is known as *character displacement.*

Darwin's finches provide a good example of character displacement. As we noted in Chapter 50, the large, medium, and small ground finch species are very similar except for differences in overall body size and in the sizes of their beaks.

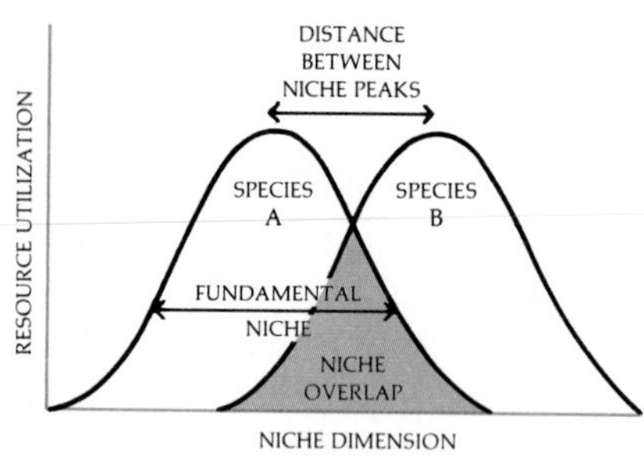

52–6
One dimension of an ecological niche. The two bell-shaped curves represent resource utilization by two species in a community. The niche dimensions might represent living space, as in the case of barnacles; foraging space, as in the case of warblers; size of seeds eaten, as in the case of finches; and so on. Competition is most intense in areas of niche overlap, leading to restriction of one or both species in living space, foraging space, or the size of seeds eaten, and so on. The narrowing of the niche (from the fundamental to the realized niche) takes place because of natural selection against individuals with overlapping characteristics.

These differences in beak size are correlated with the fact that they eat seeds of different sizes. Because of these different eating habits, interspecific competition for food is reduced. On islands such as Abingdon and Bindloe, where all three species of ground finch exist together, there are clear-cut differences in beak size. On Charles and Chatham Islands (see page 937), the large species is not found, and the beak sizes of the medium ground finches found on these islands are larger and overlap the beak sizes of the large finches found on Abingdon and Bindloe. Daphne and Crossmans, which are very small islands, each have only one species; Daphne has the medium-sized finch and Crossmans, the small finch. These two populations have similar beak sizes, which are intermediate between those of the medium-sized and small finches on the larger islands (Figure 52–7)

Family Rivalries

Most of the definitive observations on competition have been carried out on pairs of very similar organisms, such as those previously described. One notable exception is a study carried out by Henry Horn of Princeton University. Horn calculated that most leaves are able to photosynthesize at 90 percent of their maximal rate with as little as 25 percent of full sunlight. However, a leaf must receive at least 2 percent of full sunlight to pay for itself metabolically. In some trees, most of the leaves grow at the tips of well-lighted branches. Such monolayered trees, as Horn calls them, include the sugar maple, the beech, and the hemlock. Other trees are characteristically multilayered; these include the silver maple, the gray birch, the big-tooth aspen, and the white pine. In an open area, as in a cleared field, multilayered trees, with their higher photosynthetic capacity, grow more rapidly and can overtop and suppress monolayered trees. However, a multilayered sapling cannot grow beneath an adult tree, either multilayered or monolayered, because not enough light can reach the sapling's inner leaf layers to make them self-supporting. However, a monolayered sapling can grow beneath an adult tree. So the field, if left undisturbed, will gradually be taken over by species of monolayered trees, and the multilayered species continue to survive only by invading new, disturbed areas where they have a competitive advantage.

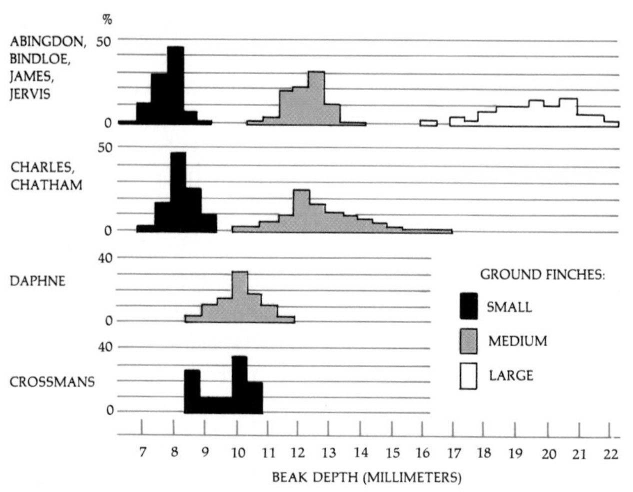

52–7
Beak sizes in three species of ground finch found on the Galapagos Islands. Beak measurements are plotted horizontally, and the percentage of specimens of each species is shown vertically. Daphne and Crossmans, which are very small islands, each have only one species of ground finch. These species have beak sizes halfway between those of the medium-sized and small finches on the larger islands.

52–8
The silver maple is a multilayered tree and the sugar maple a monolayered tree, as depicted in these diagrams. In open spaces, where the intensity of sunlight is high, the multilayered tree is competitively superior to the monolayered tree, but in the shaded understory of a forest, the monolayered tree has the competitive advantage, and it will eventually replace the multilayered species.

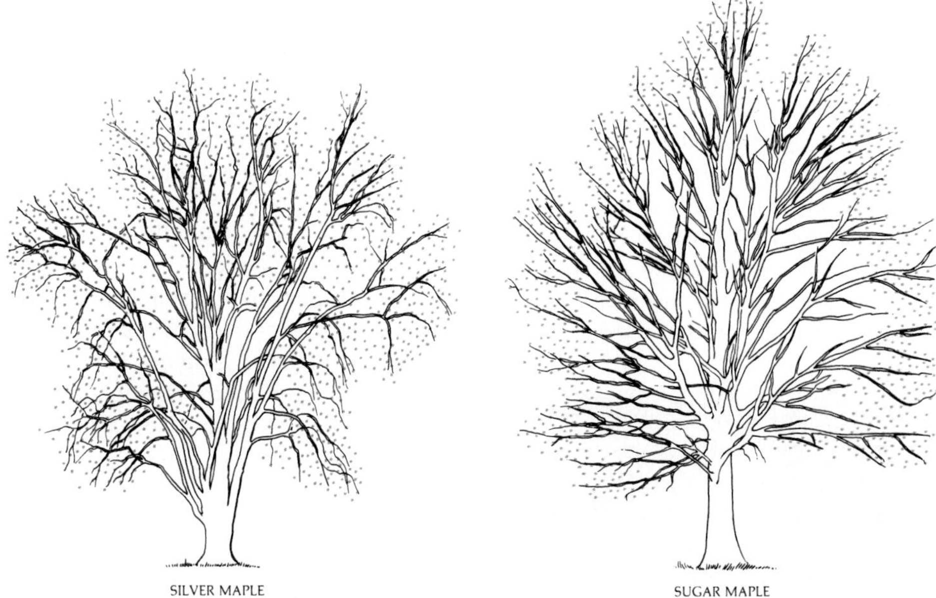

The African Ungulates

East Africa is notable for the numbers and diversity of the large herbivores—both browsers (leaf eaters) and grazers (grass eaters)—that occupy the same ecosystems and, often, even the same herds. The vegetation that supports these animals looks far less luxuriant and much more homogeneous than the animals that live upon it. Closer inspection reveals, however, that the animals' niches do not overlap. Like MacArthur's warblers, the East African ungulates survive by partitioning the resources.

The subdividing of resources by the browsers is the more obvious process. Only giraffes reach the taller branches, 3 or 4 meters off the ground. Gerenuks, long-legged, long-necked antelopes, stand on their hind legs to nibble delicately in the middle branches. The lower branches, a meter or so from ground level, are browsed upon by the rhinos; and the smaller antelopes, such as the dik-dik, hide in the thickets and browse on buds, new twigs, and seed pods.

The adaptations by which the grazers apportion the resources are less obvious, though equally effective. Zebras, which, unlike the browsers, have front teeth in both their upper and lower jaws, eat the older, tougher grass stems. Wildebeests specialize in younger stems and blades nearer the ground level. Gazelles, with their tiny muzzles, nip off the smallest green shoots of grass and also crop the little dicots that grow among the grasses and are almost entirely ignored by the larger herbivores. Thus, not only do the zebras inadvertently leave enough for the wildebeests and the wildebeests for the gazelles, but the tramping and cropping of the larger animals actually encourage the growth of the younger vegetation on which the smaller grazers depend.

Winner Takes All

These reported studies of competitive interactions have, of course, emphasized adaptations that make possible the coexistence of species rather than the elimination of one by another. This is clearly a biased view, it being difficult to study

52–9
Like many other plants of arid regions, this species of acacia has thorns that protect it against herbivores. The graceful gerenuk, however, which also evolved in the savanna, has special adaptations that enable it to circumvent these defenses. These giraffe-necked gazelles obtain water as well as food from the plants they eat, and never need to drink.

interspecies dynamics after one of the protagonists has left. Just as competition within species leads to the elimination of the great majority of individual organisms, as Darwin observed, competition between species may lead to elimination of a species from the community. One example is the disappearance from many localities of bluebirds, owing, in large part, to the usurpation of their nesting sites by starlings. Starlings were introduced into Central Park, in New York City, in 1891, and are now found throughout the United States, whereas some of us have never seen a bluebird.

A similar example, also involving birds and human interference, is seen among the gull species of the northeastern United States. Herring gulls are such successful exploiters of garbage dumps that they have greatly increased in numbers at the expense of other species, notably the ring-billed gull and the laughing gull. These once plentiful birds have now also virtually disappeared, again owing to competition for nesting sites.

PREDATION

Every organism within a community is part of a *food web* (Figure 52–10). Such a web may be very complex, comprising more than 100 different types of organisms. The relation of any species to others in its food web is an important dimension of its ecological niche.

Predation affects the number of organisms in a population, the diversity of species, and their evolution.

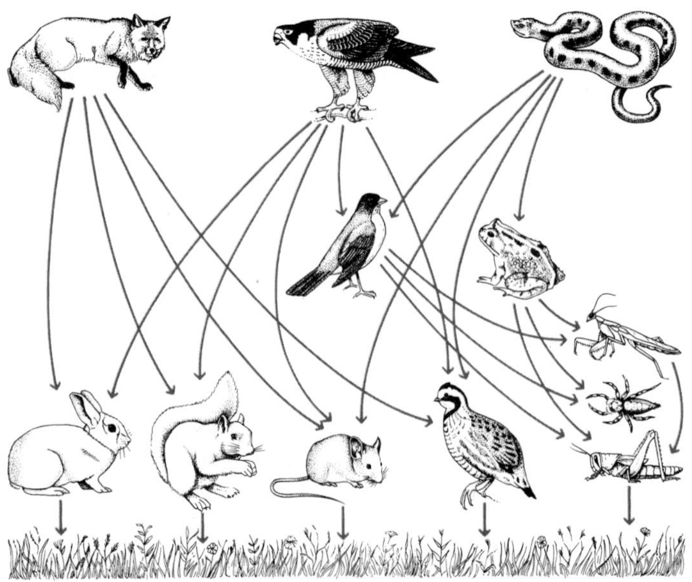

52–10
Diagram of a food web. The arrows point from each animal to its energy sources. This food web is much simplified; in reality, many more animals and plants would be involved.

Predation and Numbers

For many populations, predation is the major cause of death, yet, paradoxically, it is not at all clear that predation necessarily reduces the numbers of an equilibrium population below carrying capacity. However, when predation is heavier on certain age groups—juveniles versus adults, for instance—or certain life stages—such as caterpillars versus butterflies—it can alter the structure of a population and promote adjustments in life-history strategies.

Predation, especially of large herbivores, tends to cull animals in poor physical condition. Wolves, for instance, have great difficulty overtaking healthy adult moose or even healthy calves. A study of Isle Royale, an island in Lake Superior, showed that in some seasons more than 50 percent of the animals the wolves killed had lung disease, although the incidence of such individuals in the population was less than 2 percent. Thus, many of the animals killed by predators, according to this study, are animals that would have died anyway. (Modern human hunters, however, with their superior weapons and their desire for a "prize" specimen, are more likely to injure or destroy strong, healthy animals.)

Although predators may not always limit prey populations, the availability of prey constitutes a major component of the carrying capacity of predator populations, often stringently affecting their size. This is evident in relatively simple situations, such as when a bloom of phytoplankton results from an upwelling of nutrients due to ocean currents and then is followed by a corresponding increase in zooplankton. A more complex example is that of the lynx and the snowshoe hare; the data (Figure 52–11) are based on pelts received yearly by the Hudson's Bay Company over a period of more than 100 years. As you can see, there are oscillations in population density that occur about every 10 years. Generally speaking, a rise in the hare population is followed by a rise in the lynx population; the hare population then plummets, and the lynx population follows.

This example, which has been studied by ecologists over the last 40 years, can be interpreted in a variety of ways. The traditional explanation is that overpredation by the lynx reduces the snowshoe hare population. The lynx population, heavily dependent on the snowshoe hare as prey, is reduced in turn. The reduction in predation then permits the snowshoe hares to increase in number, followed by an increase in the number of lynx, and so on.

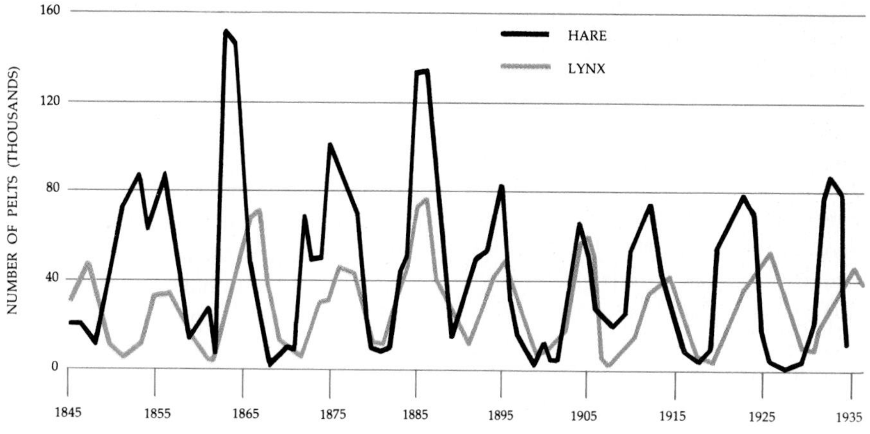

52–11
The number of lynx and snowshoe hare pelts received yearly by the Hudson's Bay Company over a period of almost 100 years, indicating a pattern of 10-year oscillations in population density. The lynx reaches a population peak every 9 or 10 years, and these peaks are followed in each case by several years of sharp decline. The showshoe hare follows a similar cycle, with a peak abundance generally preceding that of the lynx by a year or more.

(a)

(b)

52–12
(a) Prickly-pear cactus growing on a pasture in Queensland, Australia. Such rapid and environmentally destructive spread often occurs when alien organisms are introduced into a region where they have no competitors or predators. (b) *The same pasture three years after the introduction of the cactus moth.*

A second explanation—and perhaps the most popular one at the moment—is that the hare population undergoes a regular 10-year cycle caused, perhaps, by diseases associated with crowding or the effects of its own predatory activities on the vegetation it consumes. This latter hypothesis is supported by two recent discoveries: (1) when overbrowsed, certain types of plants put out new shoots and leaves that contain chemicals toxic to hares, and (2) on an island where there are no lynx, the hare population undergoes similar cycles.

A third possibility is that the lynx undergo a regular cycle, independent of the hares, perhaps associated with some other factor, such as changes in the habits of their own predators, the human hunters. A decrease in the lynx population may permit growth of the hares, or the two may oscillate independently. Thus, whereas the lynx and the hare used to serve as a simple model of predator-prey relationships, it is now perhaps even more instructive as an example of the difficulties in dealing with ecological variables.

The limiting of a prey species by a predator has been clearly demonstrated by the introduction of alien species. When prickly-pear cactus, for instance, was brought to Australia from South America, it escaped from the garden of the gentleman who imported it and spread into fields and pastureland until more than 12 million hectares were so densely covered with prickly pears that they could support almost no other vegetation (Figure 52–12). The cactus then began to take over the rest of Australia at the rate of about 400,000 hectares a year. It was not brought under control until a natural predator was imported—a South American moth, whose caterpillars feed only on the cactus. Now only an occasional cactus and a few moths can be found. (Note, however, that introduction of the moth was a risky experiment.)

Few predator-prey relationships are this simple. Most predators have more than one prey species, as indicated by the food web shown on page 968. Characteristically, when one prey species becomes less abundant, predation on other species increases so the proportions of each in the predator's diet fluctuate.

Escape from Predation

Prey organisms elude their predators in a variety of ways. Members of some species, masters of camouflage, hide in plain sight (Figure 52–13). Physical hiding places, such as burrows, are important for many species, especially of small animals. Some animals use temporal strategies. Newborn wildebeest calves, for example, are extremely vulnerable to predation by hyenas, lions, and other large carnivores, and there are no hiding places for them on the African plains. Wildebeest herds often number in the thousands, and if wildebeest calves were produced at a steady rate throughout the year, as is the case with other species of antelopes that are able to conceal their young, most of the calves would die. However, the births of the calves are synchronized so almost all appear within the same short interval of time, with 80 percent being born within the same three-week period. Though many calves are lost to predation as the carnivores feast to satiety, the chance of survival of each individual is greatly increased, and the carrying capacity of the predators is minimized.

Periodic cicadas provide an even better example of escape in time. These insects emerge only at thirteen- or seventeen-year periods, thereby outwaiting their predators. Thus, at each appearance, they are, in effect, an alien species with no natural enemies in the ecosystem.

(a)

(b)

(c)

(d)

(e)

(f)

52–13

Concealment and camouflage. (a) Bark katydid; (b) stone grasshopper in the desert; (c) horned lizard; (d) a francolin, an East African bird; (e) viper buried in sand; (f) a leaflike insect (Anaea). *The physical adaptation of these animals is dependent, in large part, on a crucial behavioral adaptation. In times of danger, all of these animals remain absolutely still.*

(a)

(b)

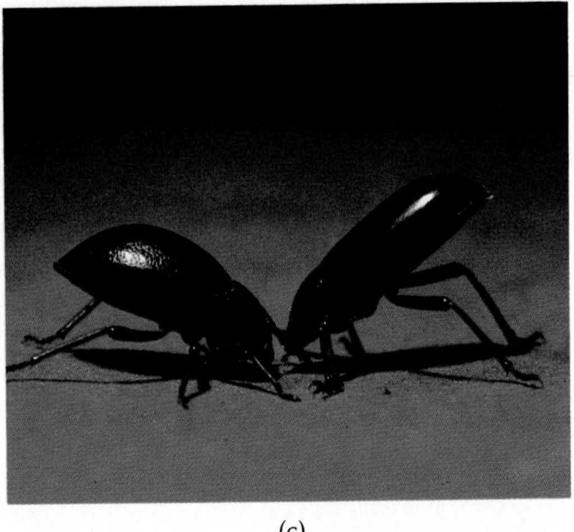

(c)

52–14
Effects of predator-prey interactions.
(a) *The beetle* Eleodes longicollis *has glands in its abdomen that secrete a foul-smelling liquid. When disturbed,* Eleodes *stands on its head and sprays the liquid at the potential predator.*
(b) *A grasshopper mouse that has solved the problem of how to eat* Eleodes. *The mouse drives the posterior of the beetle into the ground and eats it head first.*
(c) Megasida obliterata *is a Batesian mimic of* Eleodes longicollis. *It resembles the latter physically and, as you can see, emphasizes this resemblance by also standing on its head. However,* Megasida *has no similar glands, no noxious secretion, and no spray.*

Predation and Species Diversity

Predation may eliminate a prey species. On the other hand, many studies have shown that predation is often an important factor in maintaining species diversity in a community. R. T. Paine, for example, studied a community on the rocky coast of Washington. In this community, the principal predator was the starfish *Pisaster.* At the beginning of the experiment, there were 15 prey species, including several species of barnacles and several kinds of mollusks, including mussels, limpets, and chitons. Paine systematically removed the starfish from one area of the community, 8 meters by 2 meters in size. At the end of the experiment, the number of prey species had been reduced to eight. Populations of mussels and barnacles had taken over.

In England, on chalky soils, grasses grow more vigorously than the annual flowering plants, but the annuals are able to maintain their populations because rabbits crop the grasslands. When a virus epidemic severely reduced the rabbit population, the grasses took over, and the wildflowers disappeared.

Similarly, Jane Lubchenco, now of Oregon State University, was able to show in a series of experiments that the herbivorous marine snail *Littorina littorea* controls the abundance and type of algae in higher intertidal pools on the New England coast. In such pools, the snails' preferred food (the green alga *Enteromorpha)* is competitively superior, and its removal by the snails permits the growth of other algal species. However, in exposed areas, where *Enteromorpha* is competitively inferior, its removal by the snails facilitates the growth of the dominant species and decreases the number of species in the area.

The Arms Race

Predators and prey are engaged in an ever-escalating arms race, the evidence of which is present in every community. For example, the beetle *Eleodes longicollis* has a gland located in the tip of its abdomen that contains a noxious secretion. When threatened, the beetle does a quick headstand and sprays the chemical. Grasshopper mice, which feed on insects, recognize *Eleodes longicollis.* They utilize them as prey by jamming them butt-end into the earth and eating them head first.

The seeds of legumes are the frequent prey of bruchid weevils, a kind of beetle. The adult lays its eggs on the developing pods, and the larvae, when they hatch, burrow into the seeds. Each larva occupies one seed, which it consumes as it grows. The legumes have countered with a variety of defenses. Some have evolved seeds so tiny that one will not sufficiently nourish a larva to pupation and so they are not preyed upon. Others produce a chemical that inhibits the protein-digesting enzymes. These defenses are still successful against most insects, but bruchids have evolved metabolic pathways that bypass the enzyme block. Soybeans appear to be a momentary victor: bruchid larvae laid in soybean seeds die. Chemicals have been isolated from the seeds that inhibit bruchid development.

Bats, Moths, and Biosonar

In 1793, Italian scientist Lazzaro Spallanzani made a study of how animals find their way around in the darkness. He captured bats in the bell tower of the Cathedral of Pavia, blinded them, and turned them loose. Several days later, he captured the same bats again; not only had the blind animals found their way back to the bell tower but also their stomachs were full of the freshly caught flying insects that are their normal diet. Guessing that the bats might "hear" their way through the darkness, Spallanzani plugged the ear canals of some captive bats and found that the deafened animals were wholly disoriented and bumped into obstacles at random.

In the 1930s, Donald Griffin, then an undergraduate at Harvard, repeated some of Spallanzani's experiments while monitoring bats with microphones sensitive to ultrasound. He found that bats emit cries above the frequency range of human hearing. When these sound waves hit a solid object, they echo back to the sensitive ears of the bat. The time required for the sound waves to return to the bat and the direction from which they return enable the bat to determine the distance and direction of the object. Moreover, the bat can tell if the object is moving toward it, moving away from it, or is stationary. If the frequency (that is, the pitch) of the returning sound is higher than the emitted sound, the distance between the bat and the object is decreasing; if the frequency is lower, the distance between bat and object is increasing; if the frequency is the same, the distance between the two is unchanging. (This is an example of the Doppler effect, which we experience whenever the whistle of a train or the wailing siren of an emergency vehicle approaches and then recedes from us.) Using the information provided by the echoes, bats can navigate skillfully through a dark room strung with wires little thicker than a human hair or can catch an insect as small as a mosquito.

Few other animals can hear the shrill cries of bats. Exceptions are the moths that are the normal prey of insectivorous bats. Recent studies by Canadian biologists M. Brock Fenton and James H. Fullard indicate that more than 95 percent of the moth species in Ontario have functioning ears, which may be located on the thorax, abdomen, or mouthparts. The ears are not sensitive to all frequencies of sound; instead, they are "tuned" to the frequencies most often emitted by the bats that are their predators. Moreover, the moths can detect the bats before the bats can detect them. Studies by the late Kenneth Roeder of Tufts University showed that some moths can detect loud bat cries at distances of up to 40 meters. By contrast, bats apparently cannot detect moths any farther than 20 meters away, and generally bats can echolocate only at distances ranging from 1 to 10 meters. On hearing the bat, the moth takes evasive action. If the bat is still at some distance, the

(a)

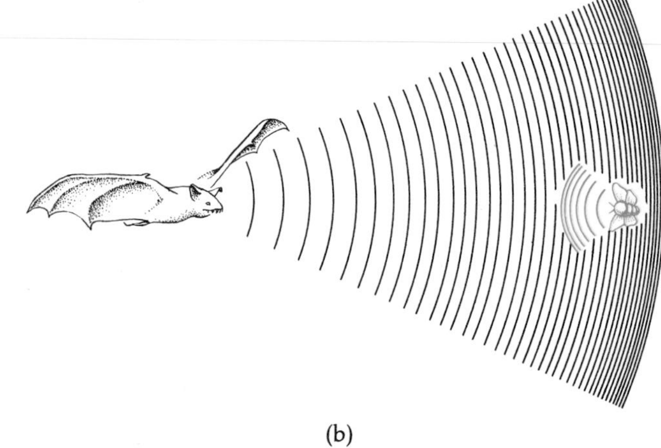

(b)

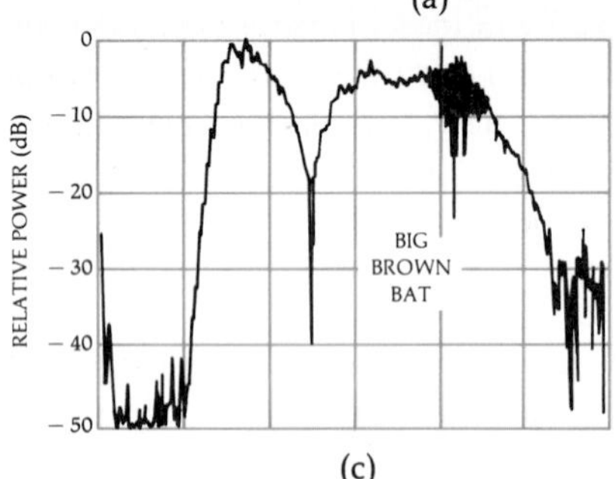

(c)

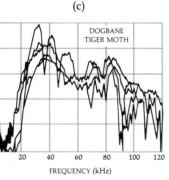

(d)

52–15
(a) *A big brown bat,* Eptesicus fuscus. (b) *As a bat flies, it emits shrill cries that are reflected back to it by any objects in the vicinity—for instance, as shown here, a moth.*

Graphs of the intensity (expressed in decibels) and frequency (expressed in kilohertz; 1 kHz = 1,000 cycles per second) of (c) *the echolocation calls of a big brown bat and* (d) *the clicks of its prey, a dogbane tiger moth. By mimicking the echolocation calls of the bat, the moth is able to jam the bat's information-processing system, allowing it to escape.*

moth flies in the opposite direction, often changing direction frequently. If the bat is close, the moth may suddenly collapse its wings and dive to the ground.

More recent studies have demonstrated that some moths have evolved an even more sophisticated defense. As the bat closes in on the moth, the moth emits a series of high-frequency clicks that mimic the echolocation cries of the bat, jamming the bat's information-processing system and so enabling the moth to make its escape (Figure 52–15).

SYMBIOSIS

As we saw in Chapter 20, symbiosis ("living together") is a close and long-term association between organisms of two different species. If one species benefits and the other is harmed, the relationship is known as parasitism. If the relationship is beneficial to both species, it is called mutualism. If the relationship is beneficial to one species, and the other receives neither benefit nor harm, it is known as commensalism. Long-continued symbiotic relationships can result in profound evolutionary changes in the organisms involved, as in the case of lichens (page 449), one of the oldest and most ecologically successful mutualistic relationships.

Parasitism

Parasitism may be considered as a special form of predation in which the predator is considerably smaller than the prey. The plants and animals in a natural community support hundreds of parasitic species—in fact, certainly thousands and perhaps millions if one were to count viruses.

As with other forms of predation, infectious diseases are most likely to wipe out the very young, the very old, and the disabled—either directly or, often, indirectly, by making them more susceptible to other predators or to the effects of climate or food shortages. It is predictable that a parasite-caused disease should not be too virulent or too efficient. If a parasite were to kill all the hosts for which it is adapted, it too would perish.

This principle is particularly well illustrated by a series of misadventures on the continent of Australia. There were no rabbits in Australia until 1859, when an English gentleman imported a dozen from Europe to grace his estate. Six years later he had killed a total of 20,000 on his own property and estimated he had 10,000 remaining. In 1887, in New South Wales alone, Australians killed 20 million rabbits. By 1950, Australia was being stripped of its vegetation by the rabbit hordes. In that year, rabbits infected with myxoma virus were released on the continent. (Myxoma virus causes only a mild disease in South American rabbits, its normal hosts, but is usually fatal to the European rabbit.) At first, the effects were spectacular, and the rabbit population steadily declined, yielding a share of pastureland once more to the sheep herds, on which much of the economy of the country depends. But then occasional rabbits began to survive the disease, and their litters also showed resistance to the myxoma virus.

A double process of selection had taken place. The original virus was so rapidly fatal that often a rabbit died before there was time for it to be bitten by a mosquito and thereby infect another rabbit; the virus strain then died with the rabbit. Strains less drastic in their effects, on the other hand, had a better chance of survival since they had a greater opportunity to spread to a new host. (A rabbit that survives an initial infection is immune to reinfection by that virus.) So selection began to work in favor of a less virulent strain of myxoma virus. Simultaneously, rabbits that were resistant to the original virus began to proliferate (as in the Lederbergs' experiment; see page 890). Now, as a result of coevolution, the two are reaching an equilibrium, like most hosts and parasites.

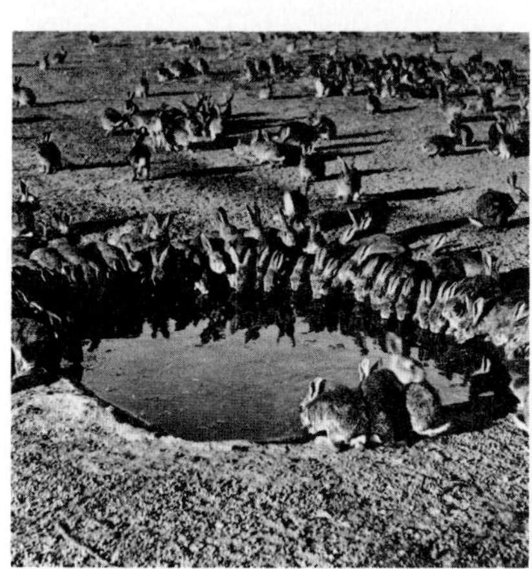

52–16
Rabbits crowd a water hole in Australia. Imported from Europe as potentially valuable herbivores, they soon overran the countryside. The myxoma virus was introduced to control them, but now host and parasite are coexisting.

Mutualism

If, indeed, current prevailing theories as to the origins of eukaryotic cells (page 412) are right, we owe our very origins to mutualism. Examples of present-day mutualistic associations are so abundant that we must forcibly limit ourselves to only a few. Perhaps among the most significant are those that take place underground, between roots and nitrogen-fixing bacteria (page 627), and in mycorrhizae (page 451). Studies of underground root systems have revealed that intertwining roots often actually form grafts so that water, minerals, and organic materials are exchanged among plants of different species. As a result of such grafts, root stumps may be kept alive indefinitely. Some particularly colorful symbiotic relationships are illustrated in Figure 52–17.

Ants and Acacias

Trees and shrubs of the genus *Acacia* grow throughout the tropical and subtropical regions of the world. In Africa and tropical America, where plants are preyed upon by large herbivores, acacia species are protected by thorns. (Acacias that evolved where there are no large browsers—Australia, for instance—lack thorns.)

On one of the African species of *Acacia,* ants of the genus *Crematogaster* gnaw entrance holes in the walls of the thorns and live permanently inside them. Each colony of ants inhabits the thorns on one or more trees. The ants obtain food from nectar-secreting glands on the leaves of the acacias and eat caterpillars and other herbivores that they find on the trees. Both the ants and the acacias appear to benefit from this mutualistic association.

In the lowlands of Mexico and Central America, the ant-acacia relationship has been extended to even greater lengths. The bull's-horn acacia is found frequently in cutover or disturbed areas, where it grows extremely rapidly. This species of acacia has a pair of greatly swollen thorns, several centimeters in length, at the base of most leaves. The petioles bear nectaries, and at the very tip of each leaflet is a small structure, rich in oils and proteins, known as a Beltian body. Thomas Belt, the naturalist who first described these bodies, noted that their only apparent

(a)

(b)

(c)

(d)

52–17
Symbioses. (a) *Sea anemones on the back of a snail shell occupied by a hermit crab. The anemones protect and camouflage the crab and, in turn, gain mobility—and so a wider feeding range—from their association with the crab. Hermit crabs, which periodically move into new, larger shells, will coax their anemones to move with them.* (b) *Cleaner fish are permitted to approach larger fish with impunity because they feed off the algae, fungi, and other microorganisms on the fish's body. The fish recognize the cleaners by their distinctive markings. Other species of fish, by closely resembling the cleaners, are able to get close enough to the large fish to remove large bites of flesh. What would probably happen if the number of cleaner mimics began to approach the number of true cleaners?* (c) *Aphids suck sap from the phloem, removing certain amino acids, sugars, and other nutrients from it and excreting most of it as "honeydew," or "sugar-lerp," as it is called in Australia, where it is harvested as food by the aborigines. Some species of aphids have been domesticated by some species of ants. These aphids do not excrete their honeydew at random, but only in response to caressing movements of the ant's antennae and forelimbs. The aphids involved in this symbiotic association have lost all their own natural defenses, including even their hard outer skeletons, relying upon their hosts for protection.* (d) *Oxpeckers live on the ticks they remove from their hosts. An oxpecker forms an association with one particular animal, such as the wart hog shown here, conducting most of its activities, including courtship and mating, on the back of its host.*

(a)

(b)

(c)

(d)

52–18

Ants and acacias. (a) *The beginning. A queen ant cuts an entrance into a thorn on a seedling bull's-horn acacia. She will hollow out the thorn and raise her first brood inside it.* (b) *The tip of an acacia leaf. The orange structures at the tips of the leaves are Beltian bodies. They are a source of protein-rich food for the ants.* (c) *A worker ant drinking from the nectary.* (d) *Warriors in a battle for possession of an acacia. Such battles occur when the branches of acacias grow to touch each other. The largest colony usually wins.*

function was to nourish the ants. Ants live in the thorns, obtain sugars from the nectaries, eat the Beltian bodies, and feed them to their larvae.

Worker ants, which swarm over the surface of the plant, are very aggressive toward other insects, and indeed toward animals of all sizes. They become alert at the mere rustle of a passing mammal, and when their tree is brushed by an animal, they swarm out and attack at once, inflicting painful, burning stings. The effect has been described as similar to walking into a large nettle. Moreover, and even more surprisingly, other plants sprouting within as much as a meter of occupied acacias are chewed and mauled and their bark girdled, and twigs and branches of other trees that touch an occupied acacia are similarly destroyed. Not surprisingly, acacias inhabited by these ants grow very rapidly, soon overtopping other vegetation.

Daniel Janzen, who first analyzed the ant-acacia relationship in detail, removed ants from acacias either by insecticides or by removing thorns or entire occupied branches. Acacias without their ants grew slowly and usually suffered severe damage from insect herbivores. Their stunted bodies were soon overshadowed by competing species of plants and vines. As for the ants, according to Janzen, these particular species can live only on acacias.

There is an epilogue to the ant-acacia story. Three new species—a fly, a weevil, and a spider—have recently been discovered that mimic the ants that inhabit the acacias. So expert is their mimicry (probably involving chemical recognition signals as well as appearance) that the patrolling ants do not recognize them as interlopers, and so they enjoy the hospitality and protection of the ant-acacia complex.

THE NUMBER OF SPECIES

What determines the number of species in a community? This, too, has been a very productive ecological question, and one that is far from answered. MacArthur and E. O. Wilson of Harvard, framing this question in terms of islands, formulated the hypothesis—known as the island biogeography model—that for an island of any given size there is a "right" number of species. Furthermore, this number will be stable over a long period of time, with new arrivals and extinctions balancing one another out.

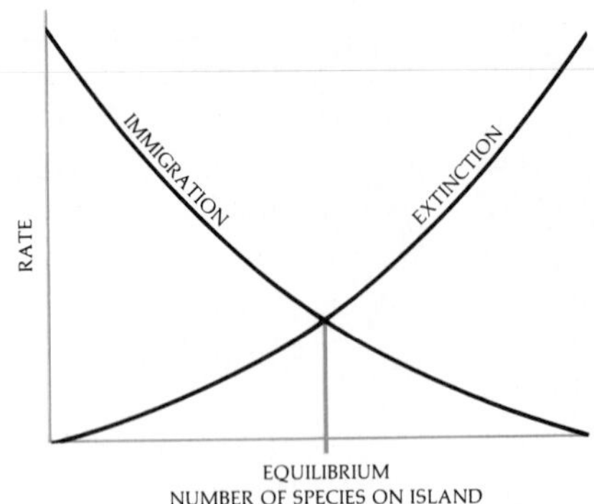

52–19
Equilibrium model of diversity of species on an island. The immigration rate declines as more species reach the island because existing species will become better established and, also, fewer migrants will belong to new species. The extinction rate increases more rapidly at high species number because of increased competition between species. The equilibrium number of species (colored line) is determined by the intersection of the immigration and extinction curves.

This model was tested in a very ingenious way. Wilson and Daniel Simberloff selected a number of little mangrove islands off the Florida Keys and counted the number of species of arthropods on each. They then destroyed all the animal life by the simple expedient of covering the small islands with plastic tents and fumigating them. Their plant life intact, the islands were soon colonized again from the mainland, and the recolonization process was monitored. As predicted by the model, the total number of species present on an island after recolonization tended to be the same as the total number before the island was disturbed. However, and this is an important point, the species composition was often quite different from what it had been previously. Moreover, once the equilibrium number had been reached, the species composition continued to change, with extinction and immigration balancing one another out (Figure 52–19).

The equilibrium concept was tested on a considerably broader scale some 2 million years ago when the Isthmus of Panama formed, bridging the American continents. They had been separated for some 50 or 60 million years, since the beginning of the age of mammals, and so each had evolved phylogenetically distinct lines of land mammals, with no families common to both continents. North America had 25 families of mammals and South America, 30. With the formation of the land bridge, there was a great interchange of mammals. During this interchange, the number in North America rose to 35 families and the number in South America, to 40. After the interchange, each continent contained a mixture of species from both continents and some species had become extinct, but once more there were about 25 families of mammals in North America and about 30 in South America.

However, lest you think that such a complex question can be disposed of so easily, we shall present evidence against the equilibrium model at the close of the next chapter.

SUMMARY

A community is a group of locally interacting populations. Three major types of interspecific interaction in communities are competition, predation, and symbiosis.

The more similar organisms are, the more intense competition is between them. The competitive exclusion principle, formulated by Gause, states that if two species are in competition for the same limited resource, one or the other will be eliminated. Each species in a community is considered as occupying a particular ecological niche, which is defined by the way in which members of the species utilize the resources of the ecosystem. Gause's principle predicts only one species per ecological niche per community. Analyses of situations in which similar species coexist have demonstrated that their niches are not identical, but that the resources are subdivided in such a way that coexistence is possible. A fundamental niche represents the resources that would be utilized by a species if competitors were not present; a realized niche describes the resources actually utilized.

Niche overlap describes the situation in which members of more than one species utilize the same limited resource. In communities in which niche overlap occurs, natural selection may result in an increase in the differences between the competing species, or niche shift. This phenomenon, also known as character displacement, minimizes competition between the competing species.

Every organism in a community is part of a food web. Such webs, which may be very complex, represent feeding relationships among the various organisms. The size of a predator population is often limited primarily by the availability of prey. However, in equilibrium communities, predation is not necessarily the major factor in regulating the population size of prey organisms, which are influenced more by their own food supply. Predator-prey interactions may produce cycles in population numbers, may increase species diversity, and may often have a strong effect on the evolution of both predator and prey populations.

Symbiosis may take the form of parasitism, mutualism, or commensalism. In parasitism, one species benefits and the other is harmed. It is to the advantage of the parasite, however, not to destroy its host. In mutualism, both species benefit. In commensalism, one species benefits, and the other is not affected.

According to the model of island biogeography, the number of species in an ecosystem reaches an equilibrium between immigration and extinction. Species composition may vary widely but the number remains approximately the same. For new species to gain a foothold, established species must become extinct, leading to a continual turnover in species composition.

QUESTIONS

1. Distinguish among the following: niche/habitat; fundamental niche/realized niche; multilayered tree/monolayered tree; character displacement/competitive exclusion; symbiosis/parasitism/mutualism/commensalism.

2. Compare the effects of interspecific (between species) and intraspecific (within species) competition. What is the principal reason for the difference?

3. Consider the following chain of eater and eaten: plants–herbivores–carnivores. What are likely to be the critical resources for which organisms compete at each level? Are the levels different with respect to the mixture of resources, competition, and predation that might be involved in population regulation?

4. Contrast the likely population dynamics of a predator-prey versus a parasite-host system.

5. In the long and ruthless war between coyotes and sheep herders, studies have shown that (a) coyotes kill sheep, and (b) the percentage of sheep lost from herds in areas where coyotes have been exterminated is about the same as the percentage lost in areas where coyotes are still present. How could you explain this finding?

6. Plagues (population explosions) of locusts and cicadas are associated with lag periods of 7, 13, and 17 years. What is significant about these numbers?

7. In the opinion of some ecologists, animals that eat seeds, such as Galapagos ground finches, should be regarded as predators, while animals that eat leaves, such as deer, should be regarded as parasites. Justify this classification of herbivores as either predators or parasites.

8. A national park or game reserve might be considered an island, from the point of view of the species living there. What lesson in management of such a park or reserve might be learned from the experiment carried out on the mangrove islands by Simberloff and Wilson?

CHAPTER 53

Ecosystems

53–1
The energy on which life depends enters the living world in the form of light. It is converted to chemical energy by photosynthetic organisms, such as the wild barley shown here. Such photosynthetic organisms are, in turn, the energy source for all heterotrophs, including ourselves.

The interactions of the various organisms in a natural setting have two consequences: (1) a flow of energy through autotrophs (usually photosynthetic organisms) to heterotrophs, which eat either autotrophs or other heterotrophs; and (2) a cycling of materials, which move from the abiotic (nonliving) environment through the bodies of living organisms and back to the abiotic environment. The cycling of materials is dependent upon a third category of organisms, the decomposers, that break down dead organic materials into a form in which they can be used by autotrophs.

Such a combination of biotic and abiotic components through which energy flows and materials cycle is known as an ecological system, or *ecosystem*. Taking a large, astronautical view, the entire surface of the earth can be seen as a single ecosystem. This view is useful when studying materials that are circulated on a worldwide basis, such as carbon dioxide, oxygen, and water. A suitably stocked aquarium or terrarium is also an ecosystem, and such models may be useful in studying certain ecological problems, such as the details of transfer of a particular mineral element. Most studies of ecosystems and their component communities have been made, however, on more or less self-contained natural units—for example, a pond, a swamp, or a meadow.

THE FLOW OF ENERGY

Productivity

In an ecosystem, the flow of energy begins with photosynthesis and the fixation of carbon into organic molecules. Photosynthetic efficiency is low. Generally, only about 1 percent of the light that falls on a plant—the annual incident radiation—is converted to chemical energy. Even the most productive stands of vegetation utilize no more than 3 percent. Yet the total amount of energy available in the light that reaches the surface of this planet is so enormous that some 120 billion metric tons of new organic material are produced every year, despite the low conversion efficiency.

One of the properties of ecosystems is *productivity*, which is the total amount of energy converted to organic compounds in a given length of time. It might be considered as analogous to the rate of gross income of a business. *Net productivity* is productivity less the cost of all the metabolic activities of the producer; this cost might be considered as equivalent to the cost of doing business. Net productivity is thus comparable to the rate of net profit. It is usually measured in terms of the rate of increase of energy (calculated in calories) or biomass (measured in grams or metric tons). (*Biomass*, a convenient shorthand term, means the total dry weight of all the organisms being measured. It is equivalent to the standing crop and, unlike productivity, does not involve any time factor.) Productivity is not necessarily dependent on biomass. If you cut your lawn, the grass grows faster; you decrease biomass, but you increase productivity. Similarly, a mature forest, with a huge biomass, may have lower productivity than an open field.

Trophic Levels

Energy passes from one organism to another along a particular food chain. A *food chain* is a sequence of organisms related to one another as prey and predator. The first is eaten by the second, the second by the third, and so on, in a series of *trophic levels*, or feeding levels.

In most ecosystems, food chains are linked together in complex food webs, with many branches and lateral connections (see Figure 52–10 on page 968). Such a web may involve more than a hundred different species, with predators characteristically taking more than one type of prey, and each type of prey being exploited by several different species of predators. However, any particular chain in any web has only a few links. A recent study of 102 top predators (animals themselves free of predation) showed that there are usually only three links (four levels) in a food chain—plant to herbivore to carnivore to carnivore. For only one top predator were there more than five links (six species) involved.

Producers

A primary producer is always at the first trophic level. On land, the primary producer is usually a plant; in aquatic ecosystems, it is usually a photosynthetic alga. These photosynthetic organisms use light energy to make carbohydrates and other compounds, which then become sources of chemical energy. Producers far outweigh consumers; 99 percent of all the organic matter in the biosphere is made up of plants and algae. All heterotrophs combined account for only 1 percent.

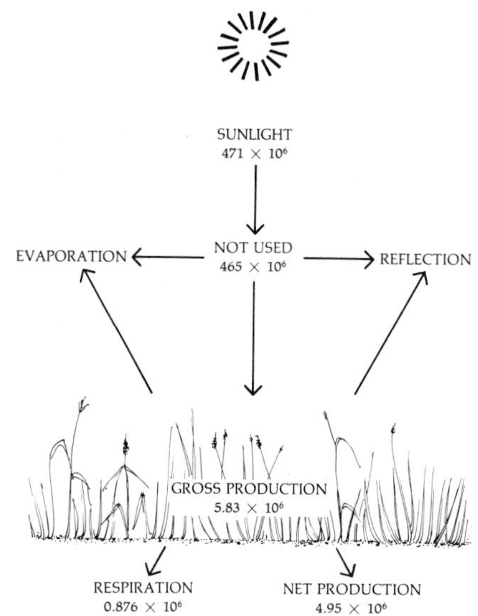

53–2
Calculation of the productivity of a field in Michigan in which the vegetation was mostly perennial grasses and herbs. Measurements are in terms of calories per square meter per year. In this field, the net production—the amount of chemical energy stored in plant material—was 4,950,000 calories per square meter per year. Thus, slightly more than 1 percent of the 471,000,000 calories per square meter per year of sunlight reaching the field was converted to chemical energy and stored in the plant tissues.

A CHEMOSYNTHETIC ECOSYSTEM

Throughout this text, we have been informing you that life on earth depends on radiant energy from the sun, which is converted by plants or algae to chemical energy by the process of photosynthesis. Not so! Now, a new type of ecosystem has been discovered, based on chemosynthesis.

Its discoverer was Alvin, *the little research submarine from Woods Hole Oceanographic Institution, and its crew of scientific investigators, led by John B. Corliss of Oregon State University. The place was the Galapagos Rift, a boundary between tectonic plates. Sea water seeps into the fissures in the volcanic rocks erupting along this boundary, becomes heated, and rises again. As a consequence, oases of warmth are created in the normally near-freezing waters 2.5 kilometers below the surface. More important, the water reacts with the rocks deep within the crust. Here, under extreme heat and pressure (300°C, 280 kilograms per square centimeter), chemical reactions take place. The crucial one, according to current evidence, is the reduction of sulfate in the sea water to hydrogen sulfide, with heat supplying the necessary energy for the reaction. Chemosynthetic bacteria oxidize the sulfide, obtaining the energy needed to extract carbon from carbon dioxide in the sea water and fix it in organic molecules, thus becoming the primary producers in the ecosystem.*

About a dozen such oases have now been located, each with a somewhat different animal community. (According to the investigators, the types of animals that are present are probably the result of whatever floating larvae happened to colonize the area in its early stages of development.) Among the most spectacular inhabitants are clams, measuring some 20 centimeters in diameter, which cover the lava floor. Other permanent inhabitants include crabs, mussels, and octopods. The bivalves feed on the bacteria, and the crustaceans and octopods on the bivalves.

One oasis, known as the Garden of Eden, is dominated by huge tube worms. These tube worms, which lack mouths and digestive tracts, apparently utilize organic carbon synthesized by symbiotic chemosynthetic bacteria. Thus, although most life on earth is, indeed, based on radiant energy from the sun, there are clearly highly viable alternatives not only on this planet but perhaps on others as well.

(a)

(b)

(a) *Among those responsible for the discovery of the rift communities is the 16-ton submersible* Alvin, *which has a depth capacity of 4,000 meters.* Alvin *is 7.6 meters long and can accommodate a crew of three, one pilot and two scientific observers. Based in Woods Hole, Massachusetts,* Alvin *is transported to its research sites by its mother vessel* Lulu, *a large catamaran that carries a laboratory and living quarters. In transit,* Alvin *is cradled beneath* Lulu, *between her two pontoons.*

(b) *Giant tube worms, 1.5 meters tall, growing in the depths of the Pacific Ocean at the Garden of Eden oasis. These tube worms are dependent on chemosynthetic bacteria as their energy source. Also visible in the photograph are eyeless, colorless crabs, also part of the food web for which the bacteria are the primary producers. The fish is probably an occasional visitor.*

Primary productivity is the total amount of energy converted to organic compounds by a primary producer. *Net primary productivity* is primary productivity minus the energy costs involved—that is, the energy lost in respiration.

Consumers

Energy enters the animal world through the activities of the herbivores, animals that eat plants and algae. A herbivore may be a caterpillar, an elephant, a sea urchin, a snail, or a field mouse; each type of ecosystem has its characteristic complement of herbivores. Of the organic material consumed by herbivores, much is eliminated undigested. Some of the chemical energy from digested food is transformed to other types of energy—heat or motion—or is used in the digestive process itself. A fraction of the material ingested is converted to new (animal) biomass. Increase in animal biomass is the sum of the increase in weight of individual animals plus the weight of new offspring.

The next level in the food chain, the secondary consumer level, is made up of carnivores, meat-eating animals that devour herbivores. The carnivore may be a lion, a minnow, a starfish, a robin, or a spider, but in each of these cases, only a small part of the organic substance present in the body of the herbivore becomes incorporated into the body of the carnivore. Some food chains have third and fourth consumer levels, but as we mentioned previously, five links are usually the limit, regardless of the ecosystem.

Detritivores

Detritivores are organisms that live on the refuse, or detritus, of a community—dead leaves, branches, and tree trunks, the roots of annual plants, feces, carcasses, even the discarded exoskeletons of insects. They include large scavengers, such as vultures, jackals, and crabs, as well as smaller organisms, such as earthworms, fungi, and bacteria. Detritivores can be regarded as consumers that utilize dead prey rather than living prey. The smallest detritivores—fungi and bacteria—are often referred to as decomposers. Many of them have evolved specializations that enable them to exploit sources of chemical energy, such as cellulose and nitrogenous waste products, that cannot be used by animals.

In a forest community, more than 90 percent of the plant biomass is eventually consumed by detritivores rather than by herbivores. Some of this energy flows through the food web by way of consumers that feed on detritivores, while the rest of it is used in the metabolic processes of the detritivores themselves. As a result,

53–3

Consumers. (a) *Primary consumers: two harvest mice eating wheat kernels.* (b) *A secondary consumer: a garter snake eating a toad, also a carnivore.* (c) *A top-level consumer: a sea otter, photographed in a kelp bed off the coast of California. Sea otters eat sea urchins, mollusks, and crabs, which they break open using rocks.*

(a)

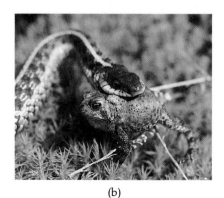

(b)

(c)

(a)

(b)

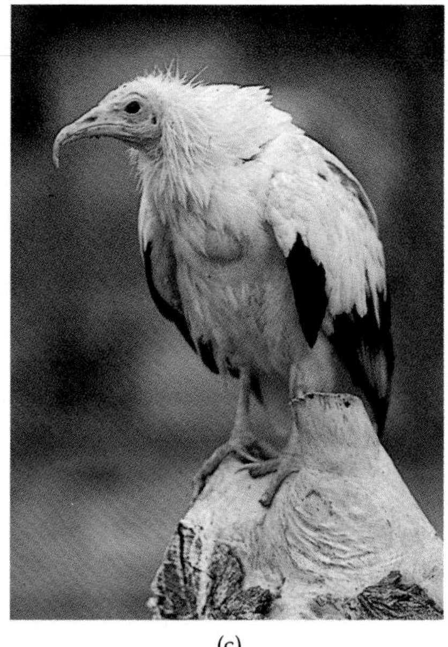

(c)

53–4
Many birds, such as these vultures, are scavengers, living on dead animals. In East Africa, several species of vulture are often seen sharing the same carcass. (a) *The lappet-faced vulture—the largest and most powerful, with a huge, heavy beak—is the one that can first break into a carcass. Other vultures are often seen standing by a carcass awaiting its arrival.* (b) *Ruppell's griffon, also a vulture, specializes in reaching its long, snaky, featherless neck far inside the carcass to feed on the intestines and other internal organs.* (c) *The smaller Egyptian vulture, with its fine beak, can tear off scraps of flesh, such as from the skull, where the coarser beaks cannot reach.*

essentially all of the energy stored by plants (and, in aquatic communities, by algae) is ultimately used to support life. Energy stored in organic matter goes unused only when it is trapped in an environment in which detritivores cannot live, such as in a peat bog.

Efficiency of Energy Transfer

The shortness of food chains is generally attributed to the inefficiency involved in the transfer of energy from one trophic level to another. In general, only about 10 percent of the energy stored in a plant is converted to the biomass of the herbivore that eats the plant. Much is expended metabolically, as the cost of doing business; much is lost as heat during energy transfer; and much plant material is eliminated undigested. Thus, the net profit is only about 10 percent. A similar profit-to-loss relationship is found at each succeeding level. Thus, if an average of 1,500 calories of light energy per square meter of land surface is utilized by plants per day, about 15 calories are converted to plant material. Of these, about 1.5 calories are incorporated into the bodies of the herbivores that eat the plants, and about 0.15 calorie is incorporated into the bodies of the carnivores that prey on the herbivores. Although meat is a more efficient source of calories and nutrients than vegetation, carnivores must usually expend more energy in procuring prey than herbivores do, and so net productivity may be about equivalent.

To give a concrete example, Lamont Cole of Cornell University, in his studies of Cayuga Lake, calculated that for every 1,000 calories of light energy utilized by algae in the lake, about 150 calories are reconstituted as small aquatic animals. Of these 150 calories, 30 calories are converted to smelt. If trout eat the smelt, about 6 calories are transferred, and if we eat the trout, we gain about 1.2 calories from the original 1,000. (Note that if we ate the smelt instead of the trout, we would derive about five times as much food energy from the original 1,000 calories. This, in short, is the argument for living lower on the food chain.)

ENERGY COSTS OF FOOD GATHERING

How much does a calorie cost, in terms of calories? For the expenditure of 1 calorie, an organism in most natural populations obtains from 2 to 20 calories in food energy. This is true for organisms whose expenditures are very high—such as the hummingbird, which spends up to 330 calories a minute—as well as for organisms whose expenditures are very low—such as the damselfly, which uses less than a calorie a day.

In simple human societies, in which individuals obtain their food without fossil-fuel energy subsidies, the ratio of food calories gained to calories invested is similar to that which prevails for the rest of the animal kingdom. Hunter-gatherers average 5 to 10 calories for each calorie spent; shifting agriculture (which requires no fertilizer) yields about 20 calories per calorie spent.

As is true in most societies (the social insects and civilized man are among the exceptions), almost the entire adult population has a share in the business of getting food. In the United States, about 20 percent of the population is involved in the food supply system. (Only about 2 percent are actually farmers; the rest are involved in food processing, transportation, and marketing.) Thus 80 percent are, for better or worse, free for other pursuits.

On the surface, it would seem that we expend less energy than most animals on the mundane work of food supply. Not so. At the turn of the century, for each calorie expended, including human labor, fuel for farm machinery and food transport, and the energy cost of fertilizer, we received about a calorie in return. Today, in the United States, as in other "advanced" technological societies, for every calorie invested we get a return of 0.1 calorie. This cost figure does not include the energy used for heating or lighting or running private automobiles (even those that bring the food home from the market) or electric can openers.

Obviously, other animals cannot live so profligately; their energy income must exceed their expenditures. Like these other organisms, we also are dependent almost exclusively on solar energy. There is an important difference, however; because of our technology, we have been able to draw upon energy stored millions of years ago. It is only in the last decade that we have come to realize that not only are these resources finite but also they may soon be expended.

From Robert M. May, "Energy Costs of Food Gathering," Nature, *vol. 225, page 669, 1975.*

A Bushman, a hunter-gatherer, returns triumphantly with a porcupine he has killed.

53–5
Pyramid of energy flow for a river system in Florida. A relatively small proportion of the energy in the system is transferred at each trophic level. Much of the energy is used metabolically and is measured as calories lost in respiration.

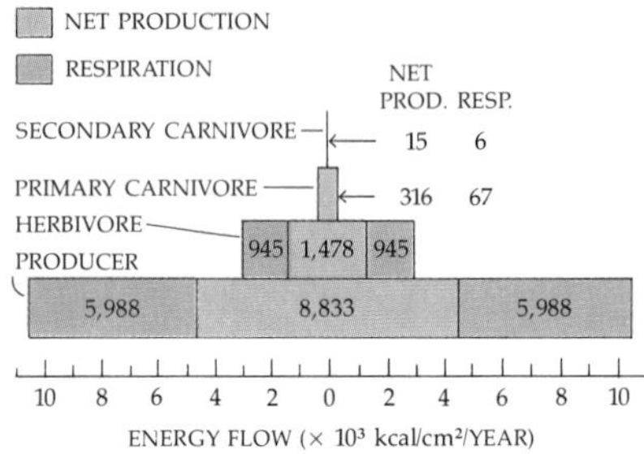

As these figures suggest, the 10 percent "rule of thumb" is not very exact. Actual measurements show more than a tenfold difference in transfer efficiencies, from less than 1 percent to well over 10 percent, depending on the species involved. The flow of energy with large losses at each successive level can be depicted as a pyramid such as that shown in Figure 53–5.

The energy relationships between the trophic levels determine the structure of an ecosystem in terms both of numbers of organisms and also of biomass. Figure 53–6a, for example, shows a pyramid of numbers for a grassland ecosystem. In this type of ecosystem, the primary producers (the individual grass plants) are small, and so a large number of them is required to support the primary consumers (the herbivores). In an ecosystem in which the primary producers are large (for instance, trees), one primary producer may support many herbivores, as indicated in Figure 53–6b.

Most pyramids of biomass take the form of an upright pyramid, as shown in Figure 53–7a, whether the producers are large or small. Pyramids of biomass are inverted only when the producers have very high reproduction rates. For example, in the ocean, the standing crop of phytoplankton may be smaller than the biomass of the zooplankton that feeds upon it (Figure 53–7b). Because the growth rate of the phytoplankton is much more rapid than that of the zooplankton, a small biomass of phytoplankton can supply food for a larger biomass of zooplankton. Like pyramids of numbers, pyramids of biomass indicate only the quantity of organic material present at one time; they do not give the total amount of material produced or, as do pyramids of energy, the rate at which it is produced.

53–6
Pyramids of numbers for (a) *a grassland ecosystem, in which the number of primary producers (grass plants) is large, and* (b) *a temperate forest, in which a single primary producer, a tree, can support a large number of herbivores.*

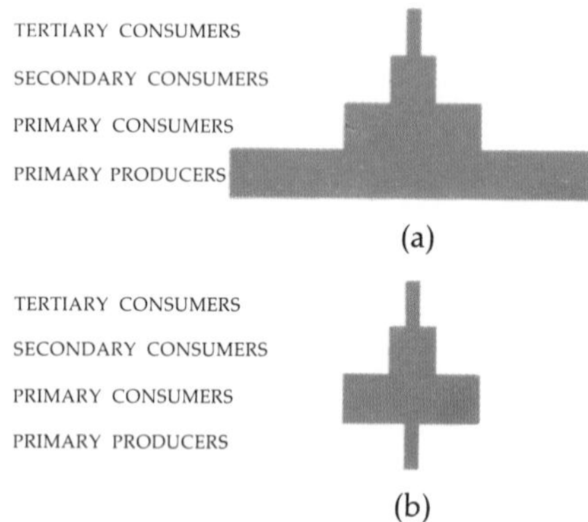

BIOGEOCHEMICAL CYCLES

Energy takes a one-way course through an ecosystem, but many substances cycle through the system. Such substances include water, nitrogen, carbon, phosphorus, potassium, sulfur, magnesium, calcium, sodium, chlorine, and also a number of minerals, such as iron and cobalt, that are required by living systems in only very small amounts. The water cycle is shown on page 46, and the carbon cycle is on page 229. You may want to look at them again at this time.

Movements of inorganic substances are referred to as *biogeochemical cycles* because they involve geological as well as biological components of the ecosystem. The geological components are the atmosphere, which is made up largely of gases, including water vapor; the lithosphere, the solid crust of the earth; and the hydrosphere, comprising the oceans, lakes, and rivers, which cover three-fourths of the earth's surface.

The biological components of biogeochemical cycles are producers, consumers, and detritivores. The detritivores, primarily bacteria and fungi, break down dead

53–7
Pyramids of biomass for (a) *a field in Georgia and* (b) *the English Channel. Such pyramids reflect the mass present at any one time; hence the seemingly paradoxical relationship between phytoplankton and zooplankton.*

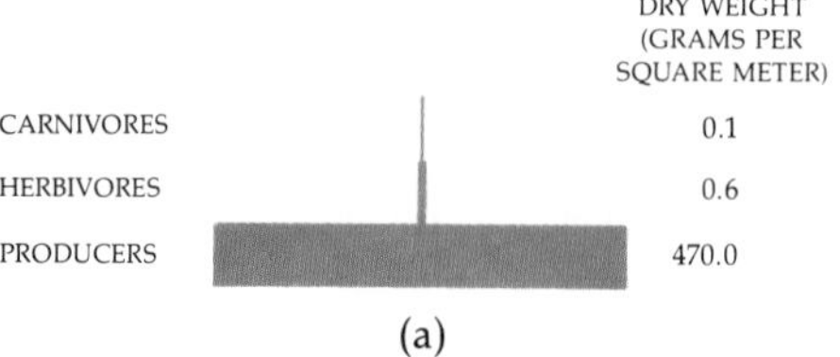

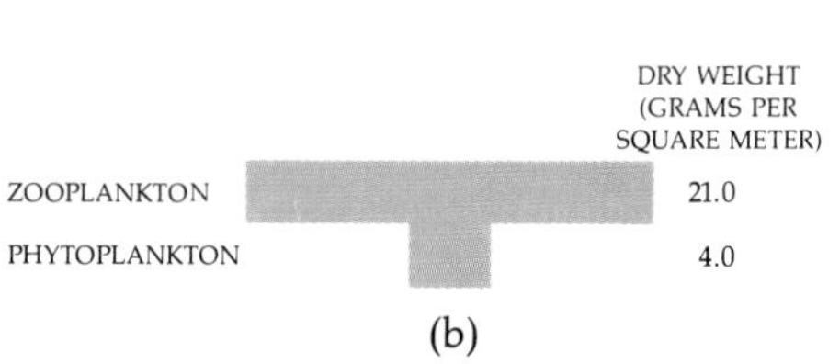

and discarded organic matter, completing the oxidation of the energy-rich compounds formed by photosynthesis. As a result of the metabolic work of the detritivores, waste products are broken down to inorganic substances that are returned to the soil or water. As we shall see in the case of the nitrogen cycle, the role of each detritivore may be very specialized—the performance of a single energy-yielding metabolic step. It has been calculated that for almost every reaction in the biosphere in which the conversion of one compound to another yields at least 15 kilocalories per mole, some organism has evolved that can exploit this energy to survive. From the soil or water, inorganic substances are taken back into the tissues of the primary producers, passed along to consumers, and then released to the detritivores, from which they enter the plants again, repeating the cycle.

The cycling through the ecosystem of one important mineral, phosphorus, is shown in Figure 53–8. Other minerals in the ecosystem undergo similar cycles, differing only in details.

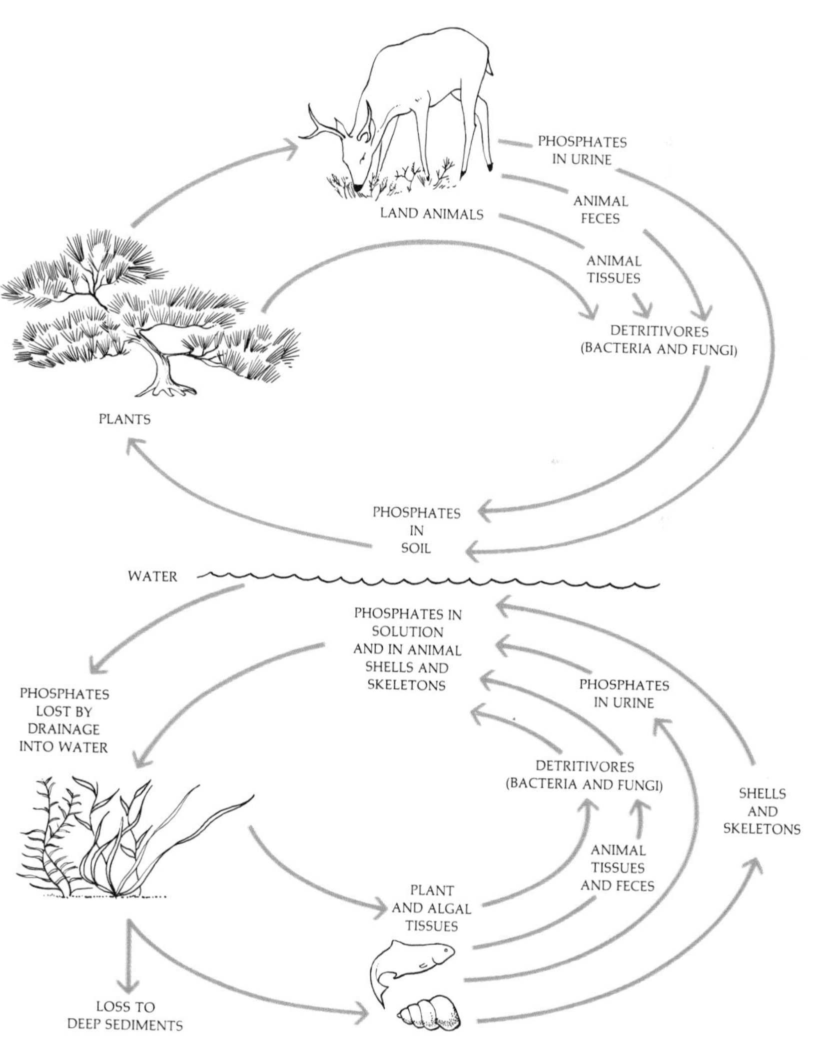

53–8
The phosphorus cycle. Phosphorus is essential to all living systems as a component of energy-carrier molecules, such as ATP, and also of the nucleotides of DNA and RNA. Like other minerals, it is released from dead tissues by the activities of detritivores, taken up from soil and water by plants and algae, and cycled through the ecosystem.

The Nitrogen Cycle

The chief reservoir of nitrogen is the atmosphere; in fact, nitrogen makes up 78 percent of the gases in the atmosphere. Most living things, however, cannot use elemental atmospheric nitrogen to make amino acids and other nitrogen-containing compounds; they are therefore dependent on nitrogen present in soil minerals. So, despite the abundance of nitrogen in the biosphere, a shortage of nitrogen in the soil is often the major limiting factor in plant growth. The process by which this limited amount of nitrogen is circulated and recirculated throughout the world of living organisms is known as the *nitrogen cycle*. The three principal stages of this cycle are (1) ammonification, (2) nitrification, and (3) assimilation.

Much of the nitrogen found in the soil is the result of the decomposition of organic materials and so is in the form of complex organic compounds, such as proteins, amino acids, nucleic acids, and nucleotides. These nitrogenous compounds are usually rapidly decomposed into simple compounds by soil-dwelling organisms. Certain soil bacteria and fungi are mainly responsible for the decomposition of dead organic materials. These microorganisms use the proteins and

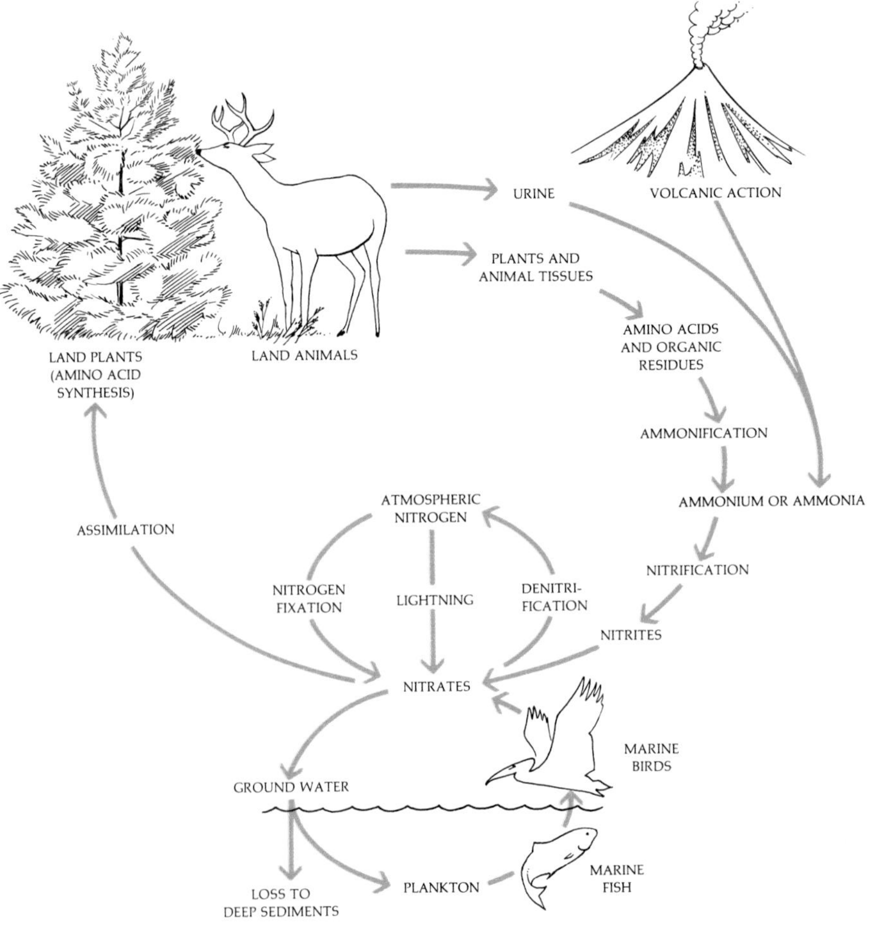

53–9
The nitrogen cycle. Although the reservoir of nitrogen is in the atmosphere, where it makes up 78 percent of dry air, the movement of nitrogen in the ecosystem is more like that of a mineral than of a gas. Only a few microorganisms are capable of nitrogen fixation.

amino acids to build their own needed proteins and release the excess nitrogen in the form of ammonia (NH_3) or ammonium (NH_4^+). This process is known as *ammonification.*

Several species of bacteria common in soils are able to oxidize ammonia or ammonium. This oxidation of ammonia or ammonium, known as *nitrification,* is an energy-yielding process, and the energy released is used by these bacteria as their primary energy source. One group of bacteria oxidizes ammonia (or ammonium) to nitrite (NO_2^-):

$$2NH_3 + 3O_2 \longrightarrow 2NO_2^- + 2H^+ + 2H_2O$$

Nitrite is toxic to plants, but it rarely accumulates. Members of another genus of bacteria oxidize the nitrite to nitrate, again with a release of energy:

$$2NO_2^- + O_2 \longrightarrow 2NO_3^-$$

Although plants can utilize ammonium directly, nitrate is the form in which most nitrogen moves from the soil into the roots, and hence these bacteria play a crucial role in the recycling of nitrogen.

Once the nitrate is within the plant cell, it is reduced back to ammonium. In contrast to nitrification, this process, called *assimilation,* requires energy. The ammonium ions thus formed are transferred to carbon-containing compounds to produce amino acids and other nitrogenous organic compounds needed by the plant.

The nitrogen-containing compounds of land plants are returned to the soil with the death of the plants (or of the animals that have eaten the plants) and are reprocessed by soil organisms and microorganisms, taken up by the plant roots in the form of nitrate dissolved in the soil water, and reconverted to organic compounds. In the course of this cycle, a certain amount of nitrogen is always "lost," in the sense that it becomes unavailable to plants.

The main source of nitrogen loss is the removal of plants from the soil. Soils under cultivation often show a steady decline of nitrogen content. Nitrogen may also be lost when topsoil is carried off by soil erosion or when ground cover is destroyed by fire. Nitrogen is also leached away by water percolating down through the soil to the groundwater. In addition, when oxygen is not present, numerous types of bacteria in the soil can break down nitrates, releasing nitrogen into the air and using the oxygen for the oxidation of carbon compounds (respiration). This process, known as *denitrification,* takes place in poorly drained (hence, poorly aerated) soils.

As you can see, if the nitrogen lost from the soil were not continuously replaced, virtually all life on this planet would finally flicker out. The "lost" nitrogen is returned to the soil by nitrogen fixation, the process by which gaseous nitrogen from the air is incorporated into organic nitrogen-containing compounds and thereby brought into the nitrogen cycle. Some small amount of nitrogen is fixed by abiotic processes; most nitrogen fixation, however, is carried out by a few types of microorganisms, the most important of which are species of bacteria that live in symbiosis with plants (see pages 627–629). Fifty million metric tons of nitrogen are added to the soil each year, of which 45 million tons are biological in origin. (The other 10 percent is largely in the form of chemical fertilizers.) Just as all organisms are ultimately dependent on photosynthesis for energy, they all depend on nitrogen fixation for their nitrogen.

53–10
Weir at Hubbard Brook Experimental Forest in New Hampshire. Water from each of six experimental ecosystems was channeled through a weir, such as this one built where the water leaves the watershed, and was analyzed for chemical elements. The watershed behind the weir in this photograph has been stripped of vegetation. The experiments showed that such deforestation greatly increased the loss of nutrient elements from the system.

Recycling in a Forest Ecosystem

Continuing studies of a deciduous forest ecosystem have shown that the plant life of a community plays a major role in its retention of nutrient elements. The studies have been carried out in the Hubbard Brook Experimental Forest in the White Mountain National Forest in New Hampshire. The investigators first established a procedure for determining the mineral budget—input and output, "profit" and "loss"—of areas in the forest. By analyzing the content of rain and snow, they were able to estimate input, and by constructing concrete weirs that channeled the water flowing out of selected areas, they were able to calculate output. (A particular advantage of the site is that bedrock is present just below the soil surface, so that little material leaches downward.) They discovered, first, that the natural forest was extremely efficient in conserving its mineral elements. For example, the annual net loss of calcium from the ecosystem was 9.2 kilograms per hectare (a hectare is about 2.5 acres). This represents only about 0.3 percent of the calcium in the system. In the case of nitrogen, the ecosystem was actually accumulating this element at a rate of about 2 kilograms per hectare per year. There was a similar, though somewhat smaller, net gain of potassium.

In the winter of 1965–1966, all of the trees, saplings, and shrubs in one 15.6-hectare area of the forest were cut down completely. No organic materials were removed, however, and the soil was undisturbed. During the following summer, the area was sprayed with a herbicide to inhibit regrowth. During the four months from June through September, 1966, the runoff of water from the area was 4 times higher than in previous years. Net losses of calcium were 10 times higher than in the undisturbed forest and of potassium, 21 times higher. The most severe disturbance was seen in the nitrogen cycle. Plant and animal tissues continued to be decomposed to ammonia or ammonium, which then were acted upon by nitrifying bacteria to produce nitrates, the form in which nitrogen is usually taken up by plants. However, no plants were present, and the nitrate, a negatively charged ion, was not held in the soil. Net losses of nitrate nitrogen averaged 120 kilograms per hectare per year from 1966 to 1968. As a side effect, the stream draining the area became polluted with algal blooms, and its nitrate concentration exceeded the levels established by the U.S. Public Health Service for drinking water.

Trees are now beginning to grow again on the devastated site, and the runoff of nutrients has dramatically decreased.

CONCENTRATION OF ELEMENTS

The elements needed by living organisms often are present in their tissues in higher concentrations than in the surrounding air, soil, or water. This concentration of elements comes about as a result of the selective uptake of such compounds by living cells, amplified by the channeling effects of the ecological pyramids of biomass and energy. Under natural circumstances, this concentration effect is usually valuable; animals generally have a greater requirement for minerals than do plants because so much of the biomass of plants is cellulose. In some special cases, the effect of the accumulation may be quite dramatic. As we noted previously, foraminiferans built the white cliffs of Dover, and the thousands of kilometers of coral reefs in the world's warmer waters are composed of calcium extracted from sea water over the millennia by one tiny polyp after another.

Foreign substances can also get caught up in biogeochemical cycles and, as they are passed from one living organism to another, reach high concentrations as they approach the top of the food chain.

DDT is probably the best known of the toxic substances whose effects were amplified in this way (Figure 53–11). Another was strontium-90, a radioactive element produced by nuclear testing in the 1950s. It is closely related to calcium and can take its place in many biochemical reactions. In dairy lands, strontium-90 made its way through grasses into dairy cows and into milk; from there it became concentrated in the bones and teeth of children.

By 1959 the bones of children in North America and Europe averaged an estimated 2.6 picocuries of strontium-90 per gram of bone calcium, compared with 0.4 picocurie per gram of calcium in the bones of adults. This amount of radioactivity has not been proved dangerous, but exposure to radioactive elements is known to cause leukemia, bone cancers, and genetic abnormalities and generally to shorten the life span. And, most important, the minimum exposure that can produce such effects is not established. Since the half-life of strontium-90 is 28 years (in other words, it takes 28 years for half of the element to lose its radioactivity), this exposure still continues.

The channeling of strontium-90 along the food chain could have been anticipated. Other results were less predictable. For example, in the Arctic only slight radiation exposure from the atomic tests was expected because the amount of fallout that reaches the ground at the poles is much less than it is in the temperate zones. However, Eskimos in the Arctic were discovered to have concentrations of radioactivity in their bodies that were much higher than those found in the inhabitants of the temperate regions. The key link in the chain was the lichens. Lichens, which obtain their minerals largely from the rain, had absorbed a large amount of fallout material, little of which had time to decay and none of which was dissipated by absorption into the soil. In the winter, caribou live almost exclusively on lichens, and, at the top of the food chain, Eskimos live largely on caribou.

53–11
(a) *Concentration of DDT residues being passed along a simple food chain. As organic matter is transferred from one level to the next in the chain, about nine-tenths of it is usually respired or excreted; only the remaining 10 percent forms new biomass. The losses of DDT residue, however, are small in proportion to the loss of other organic matter through respiration and excretion. Consequently, the concentration of DDT increases as the material passes along the chain, and high concentrations occur in the carnivores.* (b) *An eaglet and an egg that will never hatch, photographed in a nest near the Muskegon River in Michigan. DDT causes a bird's liver to break down the hormones that mobilize calcium at the time of egg production, resulting in thin-shelled, fragile eggs, like the one shown here. Birds at the top of food chains, such as the osprey, peregrine falcon, and bald eagle, were principal victims. With the banning of DDT in many countries, dramatic recoveries have been observed in populations of all three species. For example, more than 13,000 bald eagles were counted in 1981, as compared with about 3,000 in 1976.*

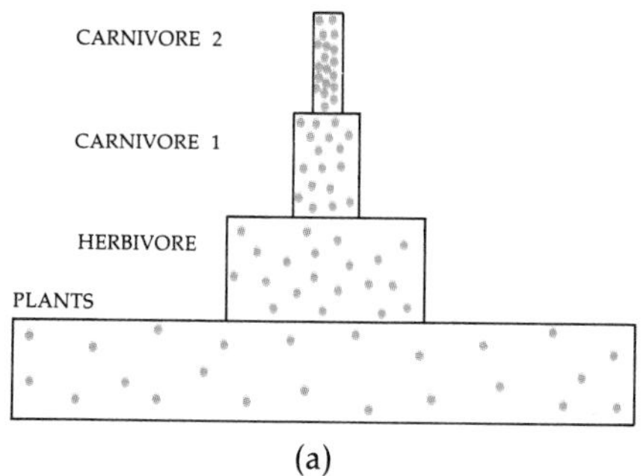

(a)

(b)

(a)

(b)

53–12
(a) *An early stage of succession on a rocky slope. Lichens have begun to accumulate soil, and a bladder fern has sprung up in a small crevice.* (b) *A young stand of loblolly pine is taking over an abandoned field in Wake County, North Carolina. The plow furrows are still clearly visible. The field was abandoned perhaps 20 years ago.*

ECOLOGICAL SUCCESSION

Land laid bare by a lava flow, a landslide, or a plow is slowly repopulated, being captured first by opportunistic, fugitive species (*r*-strategists), which are gradually replaced over time by species better adapted for equilibrium high-density communities (*K*-strategists), and these in turn compete for a place in the next community. The succession of plant forms (and the animals they bring with them) is so regular and predictable that ecological succession was at one time viewed as analogous to the developmental processes of a single organism, with each stage making way for the next by "preparing" the soil and other environmental factors. Now ecological succession is viewed rather as the outcome of a series of contests, such as that between monolayered and multilayered trees described on page 967.

Results of Succession

As ecosystems pass through the various stages of succession, one type of community is replaced by another, distinctly different type. Although the changes that take place differ as to the exact species involved, they appear to have certain features in common.

First, there is an increase in total biomass. Compare, for example, a recently abandoned field, which is at an early successional stage, with a deciduous forest, which is at a late one.

Second, there is a decrease in net productivity in relation to biomass. In other words, the biomass of mature ecosystems does not tend to increase with time, as

53–13
When Mt. St. Helens exploded on May 18, 1980, its shock waves leveled all of the fir trees in an area of about 18,000 hectares. Subsequent mudflows, moving at speeds of up to 80 kilometers an hour, buried the remaining vegetation. As life gradually returns to the devastated areas, one of the earliest colonizers of the mudflows has been Lupinus lepidus, *a nitrogen-fixing plant. This individual, photographed in August of 1982, probably sprouted from roots that survived beneath the mud.*

does that of systems in earlier stages of succession, since mature systems are dominated by slower-growing species. (Agriculture achieves high net productivity by exploiting the features of the immature ecosystem.)

Third, as we would expect from the Hubbard Brook experiments, mature systems have a greater capacity to entrap and hold nutrients within the system.

Fourth, the number of species increases. There is also a general increase in the size of organisms, the length of their lives, and the complexity of their life cycles.

The increase in the number of species as the ecosystem matures raises two interesting questions. First, does the increased complexity of the ecosystem, resulting from its increased diversity, make the system more stable? Not long ago ecologists were more or less agreed that complexity lends stability, pointing to the "fragility" of ecosystems such as the Arctic tundra, where there are relatively few species, and the susceptibility to disease of agricultural crops consisting of vast stands of a single species. Now, ecologists have come to recognize that it is often, in fact, the younger systems, with their readily available store of *r*-strategists, that are able to make the most rapid recovery from injury. The more complex systems, such as the tropical rain forest, are often the most fragile, taking the longest to restore their original state.

Second, as the system matures, does the number of species reach an equilibrium? Do arrivals balance out departures? This outcome would be predicted by the model of island biogeography. An opposing hypothesis, however, holds that the numbers and kinds of species in a community are the product of the history of the community, and that as long as an ecosystem exists, the number of species should continue to increase. Immigration and extinction may explain what happens on a tiny islet near the mainland, but on larger islands—Jamaica, Great Britain, New Guinea—where geographic isolation is possible, speciation occurs. Niches shift, making room for more species by subdividing the resources more and more finely. According to this point of view, the ecosystem becomes more complex and diverse, opening new opportunities.

SUMMARY

An ecosystem is a unit of biological organization made up of all the organisms in a given area and the environment in which they live. It is characterized by interactions between the living (biotic) and nonliving (abiotic) components that result in (1) a flow of energy from the sun through autotrophs to heterotrophs, and (2) a cycling of minerals and other inorganic materials.

Within an ecosystem, there are trophic (feeding) levels. All ecosystems have at least two such levels: autotrophs, which are plants or algae, and herbivores, which are usually animals. The autotrophs, the primary producers in the ecosystem, convert a small proportion (about 1 percent) of the sun's energy into chemical energy. The herbivores, which eat the autotrophs, are the primary consumers. A carnivore that eats a herbivore is a secondary consumer, and so on. About 10 percent of the energy transferred at each trophic level is stored in body tissue; of the remaining 90 percent, part is used in the metabolism of the organism and part is unassimilated.

The movements of water, carbon, nitrogen, and minerals through ecosystems are known as biogeochemical cycles. In such cycles, inorganic materials from the air, water, or soil are taken up by primary producers, passed on to consumers, and eventually transferred to detritivores, chiefly bacteria and fungi. The detritivores break down dead and discarded organic material and return it to the soil or water in a form in which it can be used again by the primary producers.

Ecological succession is the sequence of changes in the type of vegetation and other organisms on a particular site, resulting from competitive interaction among the various organisms involved, with opportunistic (*r*-strategy) species eventually giving way to equilibrium (*K*-strategy) species.

Maturation of ecosystems is accompanied by increases in biomass and in the number of species and by decreases in net productivity.

QUESTIONS

1. Distinguish among the following: biotic/abiotic; food chain/food web; producer/consumer/detritivore; net productivity/biomass; pyramid of numbers/pyramid of energy; ammonification/nitrification/assimilation.

2. Describe what happens to the light energy striking a temperate forest ecosystem. What happens when it strikes a cornfield? A pond? A field on which cattle are grazing?

3. Describe what happens to a mineral nutrient in each of the environments listed in Question 2.

4. In what ways does a farm crop, such as corn, resemble an immature ecosystem? In what ways does a farm crop differ from an immature ecosystem? In what ways might it be advantageous to make farm crops more like ecosystems?

5. Consider each of the organisms below and list the effects of each on its ecosystem. Consider how the organism receives its inputs of energy and nutrients, where its outputs (metabolic wastes, offspring, dead carcasses) go, and its effects on other organisms.
   a. Nitrogen-fixing bacterium
   b. Earthworm
   c. Oak tree or grass plant
   d. Deer or grasshopper
   e. Lion or wolf

6. How would you apply the ideas of this chapter to arguments about the preservation of an area of "wilderness"?

7. What are the implications of mineral cycling for human practices of fertilization of land and harvesting of crops? How are these implications different for nutrients whose major inorganic reservoir is the atmosphere rather than the soil?

CHAPTER 54

The Biosphere

54-1
Photograph of earth taken from the Apollo 17 spacecraft during the final lunar landing mission in 1972. Virtually the entire continent of Africa is visible, as is the Arabian Peninsula. The Mediterranean Sea is at the top left of the photograph. At the bottom, Antarctica is blanketed under a heavy cloud cover.

The biosphere, that part of the earth in which life exists, is only a thin film on the surface of this little planet. It extends about 8 to 10 kilometers above sea level and a few meters down into the soil, as far as roots penetrate and microorganisms are found. It includes the surface waters and the ocean depths. It is patchy, differing in both depth and density.

THE SUN

The sun powers the biosphere. It is responsible for the wind and the weather, as well as for the energy flow that characterizes life.

Every day, year in and year out, energy from the sun arrives at the upper surface of the earth's atmosphere at an average rate of 1.94 calories per square centimeter per minute, a total of about 10^{24} calories per year. This is known as the solar constant, and, although it is only a tiny fraction of the total energy radiated by the sun, it is a tremendous quantity of energy. Of this total amount of energy, about 30 percent is reflected back into space from the clouds and dust, held closely to the earth's surface, that make up the earth's atmosphere. Because of these reflections, earth, as seen from outer space, is a shining planet, brighter than Venus. About 20 percent of the energy is absorbed in various layers of the earth's atmosphere. Of this 20 percent, about 1 to 3 percent is absorbed by oxygen and ozone (O_3) in the outermost layers of the atmosphere. This percentage, though small, is very important because it represents most of the ultraviolet radiation. Ultraviolet and other high-energy radiations damage organic molecules; until photosynthesis produced this protective oxygen-ozone layer, living systems were probably able to survive only below the surface of the waters. Of the 20 percent of incoming radiation absorbed in the atmosphere, 17 percent is taken up in the lower layers of the atmosphere, mostly by water vapor, dust, and water droplets. This absorption of radiation warms the atmosphere slightly, though much of the energy is stored as latent heat in water vapor.

The remaining 50 percent of the incoming radiation reaches the surface. A small amount of this is reflected from bright areas of the earth's surface, but most is absorbed. Energy absorbed by the oceans warms the surface of the water, and water molecules evaporate. Water, as a gas in the atmosphere, moves with the winds. When it condenses, clouds form, and the resulting rain is a source of water

THE ATMOSPHERE

The atmosphere that lies between the earth and outer space forms four concentric shells, or layers, distinguished by temperature differences. The lowest layer, closest to the earth, is called the troposphere. It extends out about 10 kilometers and so covers all points on the earth's surface, although the tallest mountain peaks extend almost to its limits. About 75 percent of all the molecules in the atmosphere are contained in this layer. At its outer boundary, the temperature is below −50°C. It derives its name from tropos, *"to turn"; it is the layer that turns over (see Figure 54–4). Almost all of the phenomena included under the broad general heading of weather take place within the troposphere.*

The layer above the troposphere is the stratosphere, which extends to an altitude of about 50 kilometers. In the stratosphere, the temperature increases with altitude; the temperature at the outer boundary of the stratosphere is only slightly lower than the temperature at the earth's surface. The major reason for the warming of the stratosphere is its layer of ozone, which is proportionately densest near the outer boundary. Ozone (O_3) is formed when molecules of oxygen gas (O_2) are broken apart by radiant energy and recombine. Ozone molecules absorb most of the ultraviolet rays of the sun, rays that would be lethal to most forms of life should they reach the earth's surface. The solar energy absorbed by the ozone is changed to heat and so the temperature rises. About 99 percent of the atmosphere lies below the outer boundary of the stratosphere.

In the mesosphere, the third layer, there is a gradual decrease in temperature as the concentration of ozone decreases in the steadily thinning atmosphere. Then, above the mesosphere, we encounter another of those surprising temperature reversals: the temperature increases after we enter the thermosphere. This increase is, in fact, easy to understand if you recall that temperature and heat are not the same; temperature is a measure of the average kinetic energy of molecules, whereas heat also depends on the number of molecules in a given volume (page 36). The individual molecules of the mesosphere, unshielded from the sun's rays, move at great speed, but there are few molecules, so there is little heat. At the outer boundary of the thermosphere, however, the thin atmosphere blends with the hydrogen and helium atoms in the space between the stars.

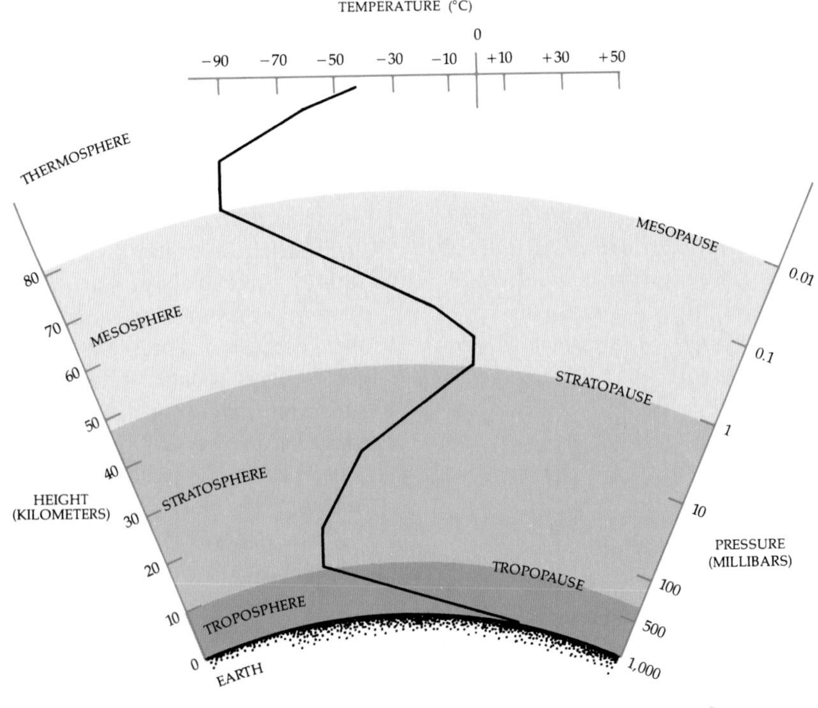

There are four main divisions of the atmosphere: troposphere, stratosphere, mesosphere, and thermosphere. The boundaries between them are determined by abrupt changes in average temperature.

for the dry land areas. The sun powers this water cycle (page 46). The unequal heating of the waters of the oceans, combined with the rotation of the earth, causes the movements of the ocean currents.

Light energy absorbed by the earth is radiated from the earth's surface in longer (infrared) wavelengths—that is, as heat. The gases of the atmosphere are transparent to visible light, but carbon dioxide and water, in particular, are not transparent to infrared rays, and so the heat is held in the atmosphere, warming the surface of the earth. This is known as the greenhouse effect; the floor of the greenhouse (or the surface of the earth) absorbs incoming light and reradiates it as heat, and the glass of the greenhouse (carbon dioxide and water vapor) prevents the heat from escaping, so that the air inside a greenhouse on a sunny day is warmer than the air outside (Figure 54–2).

All of these factors balance out; heat loss and heat gain are in equilibrium. If they were not, life would be in trouble. Yet the balance is a delicate one. Increase the earth's reflectivity, thicken its cloud cover, increase or decrease the CO_2 content of its atmosphere, or decrease its ozone layer, and the entire system would change in response. The nature and outcome of such a change is a matter of current intense interest and concern.

CLIMATE, WIND, AND WEATHER

The amount of energy received by various parts of the earth's surface is not uniform. At the equator, the sun's rays are almost perpendicular to the earth's surface, and this region receives more energy per unit area than the regions to the north and south, with the polar regions receiving the least (Figure 54–3a). Moreover, because the earth, which is tilted on its axis, rotates once every 24 hours and completes an orbit around the sun about once every 365 days, the angle of the sun's radiation, and so the amount of energy reaching different parts of the surface, changes hour by hour and season by season (Figure 54–3b).

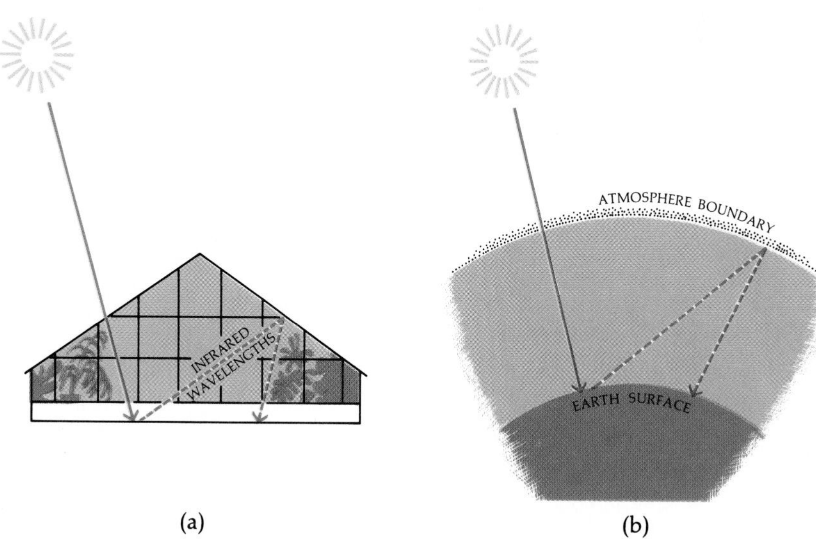

54–2
The greenhouse effect. (a) *Light rays penetrate the glass of the greenhouse, are absorbed by the plants and soil, and are then reradiated as longer-waved infrared radiation (heat). The glass does not permit these rays to escape, and so the heat remains within the greenhouse.* (b) *Carbon dioxide, like glass, is transparent to light but absorbs the infrared rays, preventing their escape. An increase in atmospheric carbon dioxide, such as may be caused by the burning of fossil fuels, could lead to an increase in the temperature of the biosphere as a whole.*

54–3
(a) *A beam of solar energy striking the earth near one of the poles is spread over a wider area of the earth's surface than is a similar beam striking the earth near the equator.* (b) *In the Northern and Southern Hemispheres, temperatures change in an annual cycle because the earth is slightly tilted on its axis in relation to its pathway around the sun. In winter, the Northern Hemisphere tilts away from the sun, which decreases the angle at which the sun's rays strike the surface and also decreases the duration of daylight, both of which result in lower temperatures. In the summer, the Northern Hemisphere tilts toward the sun. Note that the polar region of the Northern Hemisphere is continuously dark during the winter and continuously light during the summer.*

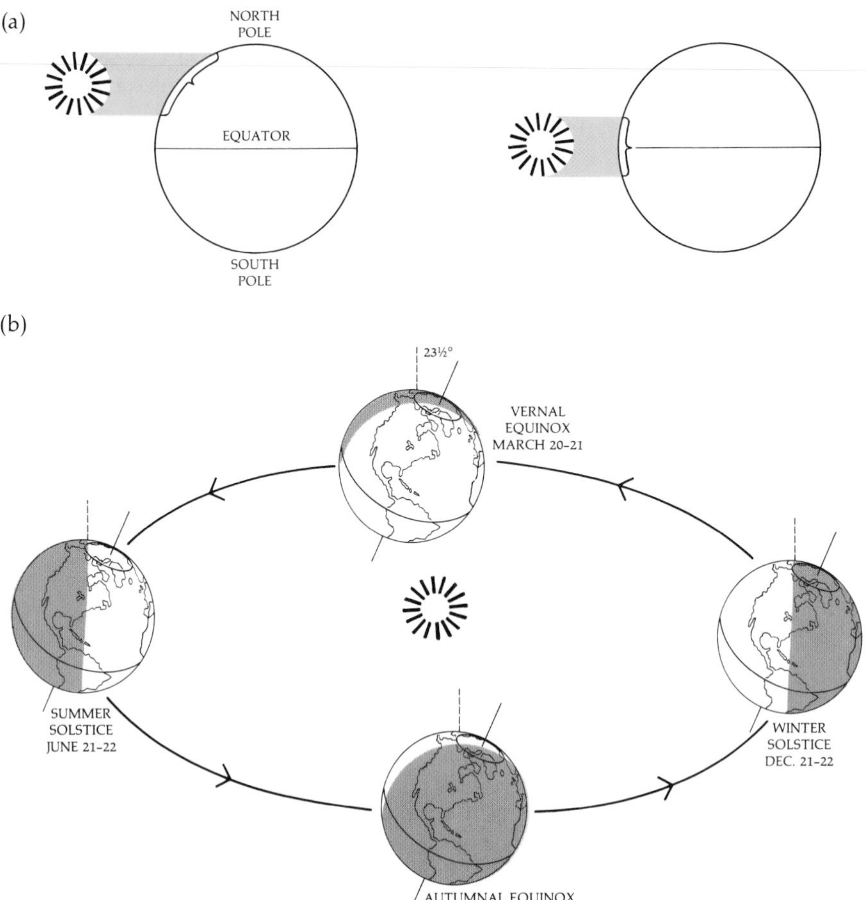

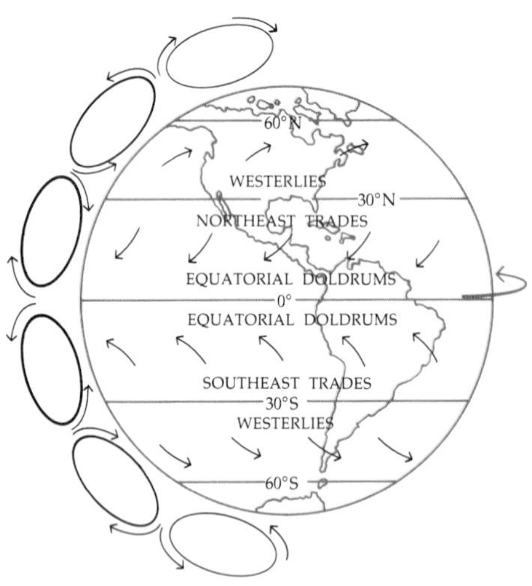

54–4
The earth's surface is covered by belts of air currents, which determine the major patterns of wind and rainfall. Air rising at the equator loses moisture in the form of rain, and falling air at latitudes of 30° north and south is responsible for the great deserts found at these latitudes.

Temperature variations over the surface of the earth and the earth's rotation establish major patterns of air circulation and rainfall. These patterns depend, to a large extent, on the fact that cold air is denser than warm air. As a consequence, hot air rises and cold air falls. As air rises, it encounters lower pressure and consequently expands, and as a gas expands, it cools. Cooler air holds less moisture, so rising, cooling air tends to lose its moisture in the form of rain or snow.

The air is warmest along the equator, the region heated most intensely by the sun. This air rises, creating a low-pressure area (the doldrums) that draws air from north and south of the equator. As equatorial air rises, it cools, loses most of its moisture, and then falls again at latitudes of about 30° north and south, the regions where most of the great deserts of the world are found. This air warms, picks up moisture, and rises again at about 60° latitude (north and south); this is the polar front, another low-pressure area. A third, weaker belt rising at the polar front descends again at the poles, producing a region in which, as in other areas of descending air, there is virtually no rainfall. The winds caused by this heat transfer from equator to poles are twisted by the spinning motion of the earth's rotation, thus creating the major wind patterns (Figure 54–4).

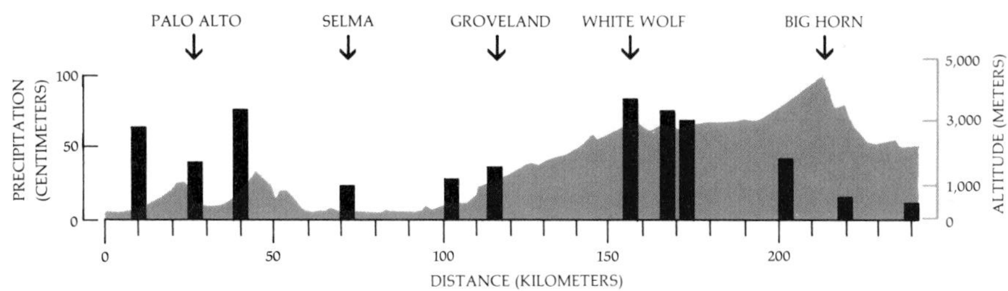

54–5
The mean annual rainfall (vertical columns) in relation to altitude at a series of stations from Palo Alto on the Pacific Coast across the Coast Ranges and the Sierra Nevada. The prevailing winds are from the west, and there are rain shadows on the eastern slopes of both mountain ranges.

The worldwide patterns are modified locally by a variety of factors. For example, along our own West Coast, where the winds are prevailing westerlies, the western slopes of the Sierra Nevada have abundant rainfall, while the eastern slopes are dry and desertlike (Figure 54–5). As the air from the ocean hits the western slope, it rises, is cooled, and releases the water. Then, after passing the crest of the mountain range, the air descends again, becomes warmer, and its water-holding capacity increases, resulting in a so-called "rain shadow" on the eastern slope.

Large bodies of water—lakes and oceans—moderate temperature variations and provide moisture. Thus, for example, areas near the Great Lakes have milder winters and cooler summers than areas at the same latitude but farther inland. The major currents of the oceans (Figure 54–6) affect land temperatures along the coasts.

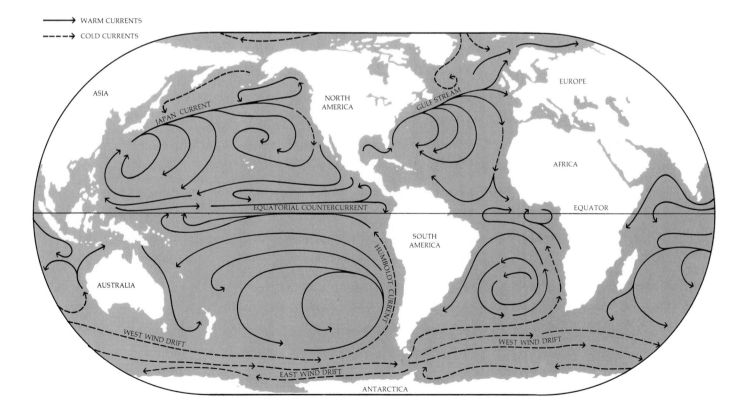

54–6
The major currents of the oceans have profound effects on climate. Because of the warming effects of the Gulf Stream, Europe is milder in temperature than is North America at similar latitudes. The eastern coast of South America is warmed by water from the equator, and the Humboldt Current brings cooler weather to the western coast of South America.

THE EARTH'S SURFACE

Much of the iron and other heavier materials of which the earth is composed is collected in a dense core in the center. Surrounding the core is a lighter layer of solid and molten rock. The outermost layer is made up of overlapping, mobile plates on which the continents rest.

The continents themselves are composed largely of relatively light igneous rocks (from the Latin word *ignis,* for "fire"); igneous rock is formed directly from molten material. The surfaces of the continents change constantly. They are crumpled by contractions and collisions as the continents rise, sink, and collide because of the motion of the plates on which they are carried (page 930). As a consequence, the earth's surface is not at all uniform but varies widely from place to place in its composition and in its height above sea level. Both the mineral content of the earth's surface and the altitude affect the growth of plants and other living organisms.

BIOMES

The land surface of the earth can be seen as being divided into a number of geographic areas distinguished by particular types of dominant plants. Thus each major continent has, for example, deserts, grasslands, and deciduous forests. These categories of characteristic plant life are called *biomes.*

The distribution of biomes (Figure 54–7) results from three kinds of physical factors: (1) the distribution of heat from the sun and the relative seasonality of different portions of the earth; (2) global patterns of air circulation, particularly the directions in which the prevailing moisture-bearing winds blow; and (3) such geologic factors as the distribution of mountains and their height and orientation.

The communities of plants and associated animal life that make up a biome are discontinuous, but a community may closely resemble another community on the other side of the planet. For instance, the deserts of the world look remarkably similar. However, when one looks more closely, one will see that although the physical features of the environment—temperature and rainfall—are the same, the organisms are not. But they look and act alike. The same is true of the chaparral of California and the Mediterranean, the grasslands of North and South America, and so on.

Thus, to repeat, a biome is a class, or category, not a place. When we speak of the tropical forest biome, we are not speaking of a particular geographic region, but rather of all the tropical forests on the planet. As with most abstractions, important details are omitted. For example, the boundaries are not so sharp as shown on maps nor are all areas of the world easy to categorize. However, the biome concept emphasizes one important truth: Where the climate is the same, the organisms are also very similar, even though these organisms are not related genetically and are far apart in their evolutionary history. The organisms of the biomes provide many examples of convergent evolution.

Temperate Deciduous Forest

Deciduous forests occupy areas where there is a warm, mild growing season with moderate precipitation, followed by a colder period less suited to plant growth. Leaf-shedding in the deciduous forest evolved as a protection against water loss.

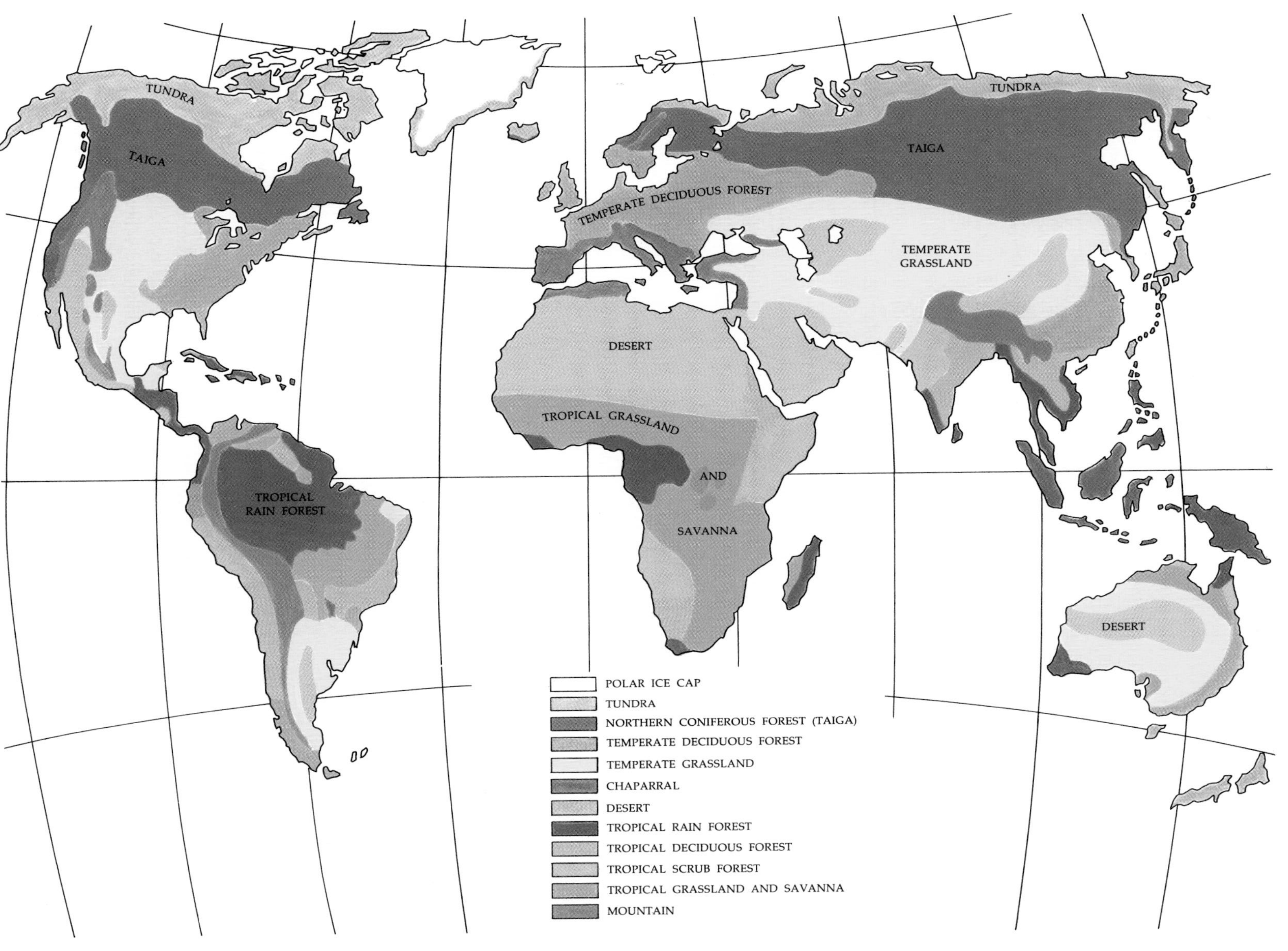
TUNDRA
TAIGA
TUNDRA
TAIGA
TEMPERATE DECIDUOUS FOREST
TEMPERATE GRASSLAND
DESERT
TROPICAL GRASSLAND
AND
SAVANNA
TROPICAL RAIN FOREST
DESERT
POLAR ICE CAP
TUNDRA
NORTHERN CONIFEROUS FOREST (TAIGA)
TEMPERATE DECIDUOUS FOREST
TEMPERATE GRASSLAND
CHAPARRAL
DESERT
TROPICAL RAIN FOREST
TROPICAL DECIDUOUS FOREST
TROPICAL SCRUB FOREST
TROPICAL GRASSLAND AND SAVANNA
MOUNTAIN

54–7
Biomes of the world.

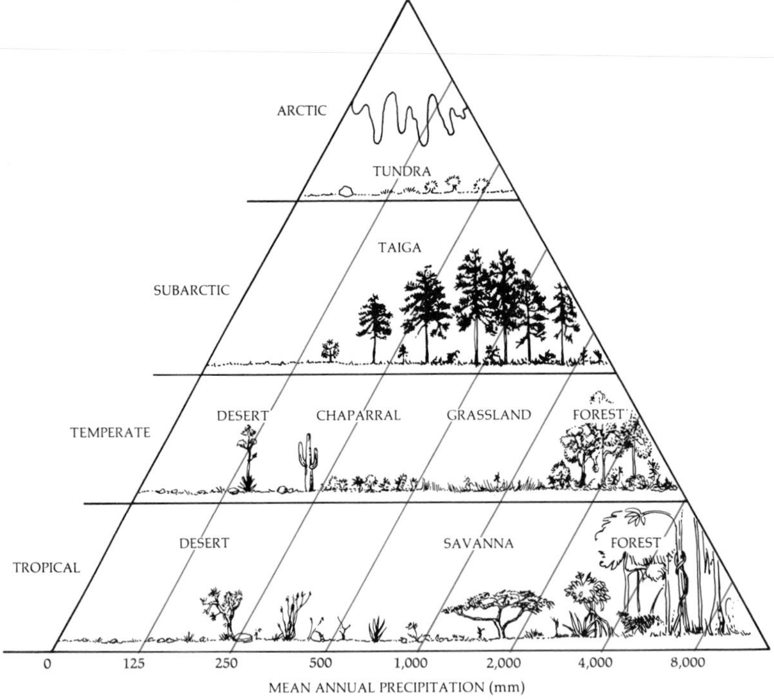

54–8
The principal determinants of the structure of plant communities are temperature, which decreases with increasing latitude, and precipitation.

(As you will recall, most of the water lost in transpiration escapes through the stomata of the leaves.) However, discarding leaves is expensive, both in terms of energy utilization (for the production of new organic materials) and of minerals. For the water-saving benefits of defoliation to exceed the costs, in terms of nutrients and energy, deciduous trees must have a growing season of at least four months. Also, soils that are poor or leached of minerals—such as those of pine barrens—are unable to support a deciduous forest.

In deciduous woodlands there are up to four layers of plant growth.

1. The tree layer, in which the crowns form a continuous canopy. The canopy is usually between 10 and 35 meters high. The tree layer usually has only one or two dominant species of trees, the final victors in the competition for light (page 966). The presence of other species of trees in the canopy layer is likely to be due to some disturbance, such as fire, a wind storm, or cutting, that has permitted a fugitive species to gain a transient roothold.
2. The shrub layer, which grows to a height of about 5 meters. Shrubs and bushes resemble trees in that they are woody and deciduous, but they branch at or close to the ground.
3. The field layer, made up of grasses and other herbaceous (nonwoody) plants, including annual flowering plants, that typically bloom in the spring before the trees regain their leaves. Bracken and other ferns, whose large leaf areas make them efficient interceptors of light trickling through the canopy, are often conspicuous members of the field layer.
4. The ground layer, which consists of mosses and liverworts. The ground is also often covered with leaf litter.

54–9
The deciduous forest of North America. (a) *A birch forest, showing layers of plant growth. The shrub layer is made up of thimbleberry and mountain maple, a favorite browse of deer.* (b) *A raccoon fishing and* (c) *a red fox killing a cottontail rabbit.* (d) *Large-flowered trillium* (Trillium grandiflorum). (e) *Caesar's mushrooms, the reproductive structures of the fungus* Amanita caesarea, *appear in scattered groups on the forest floor during the summer. All of these organisms are common inhabitants of the deciduous forests.*

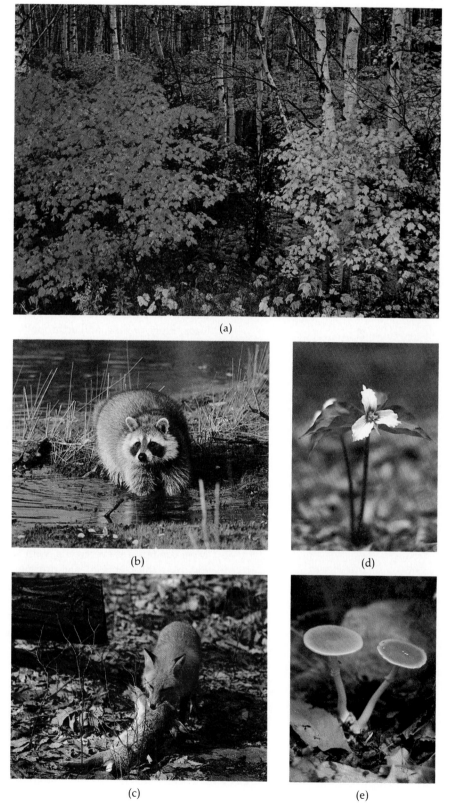

(a)

(b)

(d)

(c)

(e)

The dominant trees of deciduous forests vary from region to region, depending largely on the local rainfall. In the northern and upland regions of North America, oak, birch, beech, and maple are the most prominent trees. Before the chestnut blight struck North America, an oak-chestnut forest extended from Cape Ann, Massachusetts, and the Mohawk River valley of New York to the southern end of the Appalachian highland. Maple and basswood predominate in Wisconsin and Minnesota, and maple and beech in southern Michigan, becoming mixed with hemlock and white pine as the forest moves northward. The southern and lowland regions have forests of oak and hickory. Along the southeastern coast of the United States, the wet, warm climate supports an evergreen forest of live (nondeciduous) oak and magnolia.

The deciduous forest supports an abundance of animal life. Smaller mammals, such as chipmunks, voles, squirrels, raccoons, opossums, and white-footed mice, live mainly on nuts and other fruits, mushrooms, and insects. The wolves, bobcats, gray foxes, and mountain lions, in the areas where they have not been driven out by the encroachments of civilization, feed on these smaller mammals. Deer live mainly on the forest borders, where they browse on shrubs and seedlings.

Beneath the ground layer is the soil of the forest, often a rich, gray-brown topsoil. Such soil is composed largely of organic material—decomposing leaves and other plant parts and decaying insects and other animals—and the bacteria, protozoa, fungi, worms, and arthropods that live on this organic matter. The roots of plants penetrate the soil to depths measurable in meters and add organic matter to the soil when they die. Carnivorous arthropods carry fragments of their prey to considerable depths in the soil. The myriad passageways left by dead roots and fungi and by the earthworms and other small animals that inhabit the forest make the soil a sponge that holds water and minerals. Thus, land where deciduous forests have stood often makes good farmland.

Temperate deciduous forests once covered most of eastern North America, as well as most of Europe, part of Japan and Australia, and the tip of South America. In the United States, only scattered patches of the original forest still remain.

Coniferous Forests

The Taiga

Most conifers are evergreens, with small, compact leaves protected against water loss by a thick cuticle. The dropping of leaves is a more efficient adaptation to changing seasons, and conifers generally cannot compete with deciduous trees in temperate zones with adequate summer rainfall and rich soil. The boundary between the deciduous forest and the northern coniferous forest occurs where summers become too short and winters too long for deciduous trees to grow well.

The northern coniferous forest, also called *taiga,* is characterized by long, severe winters and a constant cover of winter snow. It is composed chiefly of evergreen needle-leaved trees such as pine, fir, spruce, and hemlock. A thick layer of needles and dead twigs, matted together by fungal mycelia, covers the ground. Along the stream banks grow deciduous trees, such as tamarack, willow, birch, alder, and poplar. There are very few annual plants.

The principal large animals of the North American coniferous forest are elk, moose, mule deer, black bears, and grizzlies. Among the smaller animals are

(a)

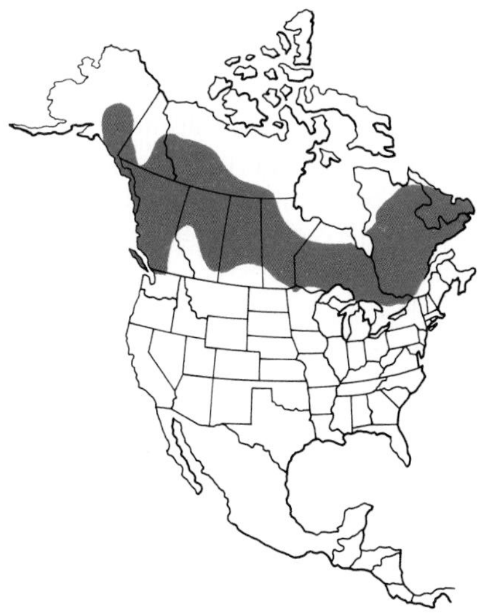

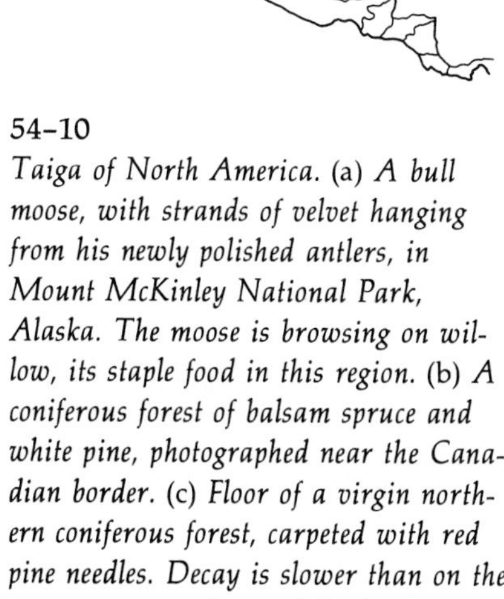

(b)

(c)

54–10

Taiga of North America. (a) *A bull moose, with strands of velvet hanging from his newly polished antlers, in Mount McKinley National Park, Alaska. The moose is browsing on willow, its staple food in this region.* (b) *A coniferous forest of balsam spruce and white pine, photographed near the Canadian border.* (c) *Floor of a virgin northern coniferous forest, carpeted with red pine needles. Decay is slower than on the warmer, wetter floors of the deciduous forest, and there is no undergrowth due to the shade cast by the mature trees.*

porcupines, red-backed mice, snowshoe hares, shrews, wolverines, lynxes, warblers, and grouse. The small animals use the dense growths of the evergreens for breeding and for shelter. Wolves feed upon these mammals, particularly the larger ones. The black bear and grizzly bear are omnivores—devouring leaves, buds, fruits, nuts, berries, fish, the supplies of campers, and occasionally the flesh of other mammals. Porcupines are bark eaters and may seriously damage trees by girdling them. Moose and mule deer are largely browsers. The ground layer of the coniferous forest is less richly populated by invertebrates than that of the deciduous forest because the accumulated litter is slower to decompose.

The Pacific Northwest

The massive evergreen coniferous forests of the Pacific Northwest are adaptations to the winter-wet/summer-dry environment of that region. Because photosynthesis is limited by lack of moisture during the warm season, deciduous trees are at a disadvantage. The evergreen conifers, however, can synthesize carbohydrates all year round. Also, these trees, because of their massive size, can store water and nutrients for use during the dry season, and their thick barks and high crowns protect them from the fires characteristic of the region. These forests, which escaped the Pleistocene glaciation, are among the most ancient in North America, both in terms of individuals and populations.

The Tundra

Where the climate is too cold and the winters too long even for conifers, taiga grades into *tundra*. The tundra is a form of grassland that occupies one-tenth of the earth's land surface, forming a continuous belt across northern North America, Europe, and Asia. Its most characteristic feature is permafrost, a layer of permanently frozen subsoil. During the summer the ground thaws to a depth of a few centimeters and becomes wet and soggy; in winter it freezes again. Because of this freeze-thaw process, which tears and crushes roots, roots are small and shallow. In this environment of drying winter winds and abrasive, wind-driven snow, only small, stunted plants survive.

The virtually treeless vegetation of the tundra is dominated by herbaceous plants, such as grasses, sedges, and rushes, and woody shrubs like heather. Beneath these is a well-developed ground layer of mosses and lichens, particularly the lichen known as reindeer moss. All the flowering plants are perennials.

The largest animals of the Arctic tundra are the musk oxen and caribou of North America and the reindeer of the Old World. Lemmings (small rodents with short tails) and ptarmigans (pigeon-sized grouse) are tundra herbivores. The white fox and the snowy owl of the Arctic are among the principal predators, feeding largely on lemmings.

During the brief Arctic summer, flies, mosquitoes, and other insects emerge in great numbers, and migratory birds visit, taking advantage of the insect hordes and the long periods of daylight to feed their young. The growing season in many areas of the tundra is less than two months.

54–11
(a) *Tundra of North America on a long Arctic day. Cotton grass surrounds a kettle hole formed by a chunk of glacial ice.* (b) *The Arctic tern is one of a number of bird species that breed in the tundra, taking advantage of the long summer days to gather food for their nestlings. The terns winter in the Antarctic, following migration routes of 13,000 to 18,000 kilometers. Within three months after hatching, the young must be ready to migrate.* (c) *A snowshoe hare in its winter coat. Hair on its feet facilitates its traveling over the snow.*

(a)

(b)

(c)

(a)

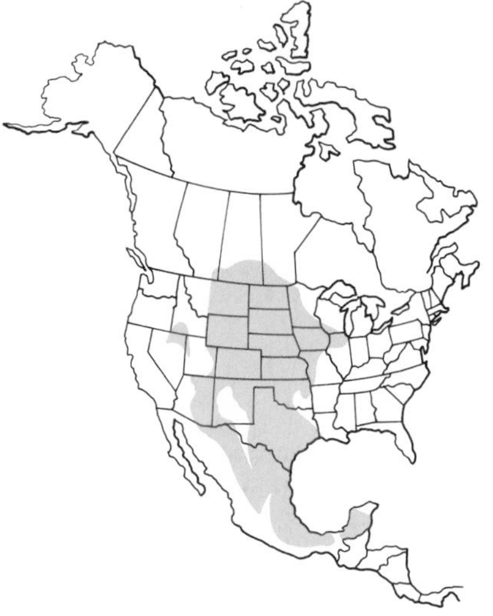

(b)

54–12
The grasslands of North America. (a) *Short-grass prairie with two familiar inhabitants, a black-tailed prairie dog peering out of its burrow and a bull snake.* (b) *A June day on a tall-grass prairie in North Dakota. The cottonwood grove by the prairie creek is characteristic of this biome. A thunderstorm is gathering on the horizon.*

Temperate Grasslands

Grasslands, which are transitional areas between deserts and temperate forests, are usually found in the interior areas of continents. They are characterized also by periodic droughts, rolling to flat terrain, hot-cold seasons (rather than the wet-dry seasons of the tropical grasslands), and fires. The great grasslands of the world include the plains and prairies of North America, the steppes of Russia, the veld of South Africa, and the pampas of Argentina.

The vegetation is largely bunch or sod-forming grasses, often mixed with legumes (clovers and wild indigo) and a variety of annuals. In North America, there is a transition from the more desertlike, western short-grass prairie (the Great Plains), through the richer, more moist, tall-grass prairie (the Corn Belt), to the eastern temperate deciduous woodland. Grasslands are drier to the west, where they are in the rain shadow of the Rocky Mountains.

The grasslands of the world support small, seed-eating rodents and also large herbivores, such as the bison of early America, the gazelles and zebras of the African veld, the wild horses, wild sheep, and ibex of the Asiatic steppes, and now the domestic herbivores. These large, grass-eating mammals, in turn, support carnivores, such as lions, tigers, and wolves, as well as omnivorous humans. The herds of grazing animals and periodic fires serve to maintain the nature of the grassland, destroying tree seedlings and preventing their encroachment.

Tropical Grasslands: Savannas

Savannas are tropical grasslands with scattered clumps of trees (Figure 54–13). The transition from open forest with grassy undergrowth to savanna is gradual and is determined by the duration and severity of the dry season and, often, by fire and grazing and browsing animals.

In the savanna, the critical competition is for water, and grasses are the victors. Grasses are well suited to a fine, sandy soil with seasonal rain because their roots

54–13
A savanna in East Africa, with a giraffe, zebras, and an impala. The trees in the background are acacias.

form a dense network capable of extracting the maximum amount of water during the rainy period. During dry seasons, the aboveground portions of the plants die, but the deep roots are able to survive even many months of drought. The balance between woody plants and grasses is a delicate one. If rainfall decreases, the trees die. If rainfall increases, the trees increase in number until they shade the grasses, which, in turn, die. If the grasses are overgrazed (which often happens when people begin to use the savanna for agricultural purposes and introduce livestock), enough water is left in the soil so that the woody plants can increase in number, and the grassland is eventually destroyed.

The best-known savannas are those of Africa, which are inhabited by the most abundant and diverse group of large herbivores in the world, including the gazelle, the impala, the eland, the buffalo, the giraffe, the zebra, and the wildebeest.

Chaparral

Regions with mild, rainy winters and long, hot, dry summers, such as the southern coast of California, are dominated by small trees or, often, by spiny shrubs with broad, thick evergreen leaves. In the United States, such areas are known as *chaparral*. Similar communities are found in areas of the Mediterranean (where they are called the *maquis*), in Chile (where they are the *matorral*), in southern Africa, and along portions of the coast of Australia. Although the plants of these various areas are phylogenetically unrelated, they closely resemble one another in their growth patterns and characteristic appearance.

Mule deer live in the North American chaparral during the spring growing season, moving out to cooler regions during the summer. The resident vertebrates—brush rabbits, lizards, wren-tits, and brown towhees—are generally small and dull-colored, matching the dull-colored vegetation.

(a)

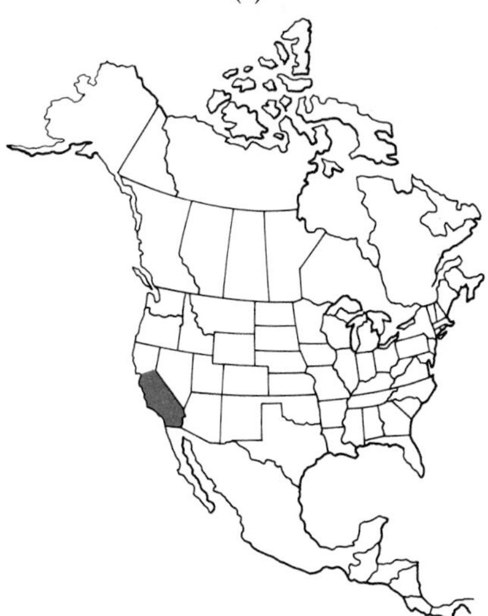

(b)

(c)

54–14
The North American chaparral. (a) *A cacomistle, or ring-tailed cat, a common inhabitant of the chaparral.* (b) *The bushy vegetation that characterizes the chaparral grows as dense as a mat on the foothills of southern California. It is the result of long, dry summers, during which much of the plant life is semidormant, followed by a brief, cool rainy season. The name comes from* chaparro, *the Indian word for the scrub oak that is one of the prominent components of the chaparral. Chaps, the leather leggings worn by the cowboys making their way through this dense, dry growth, has the same derivation.*

(c) *Mediterranean chaparral, or* maquis. *This photograph was taken on the island of Sardinia, off the western coast of Italy. Note the limestone on which the plants are growing.*

The Desert

The great deserts of the world are located at latitudes of about 30°, both north and south, and extend poleward in the interiors of the continents. These are areas of falling, warming air and, consequently, little rainfall. Only about 5 percent of North America is desert. The Sahara Desert, which stretches all the way from the Atlantic coast of Africa to Arabia, is the largest in the world (almost equal to the size of the United States) and is increasing in size, spreading along its southern boundaries. This spread is due in large part to the growth of the human population, resulting, in turn, in intensified grazing by domestic animals along these margins of the desert.

(a)

(b)

(c)

(d)

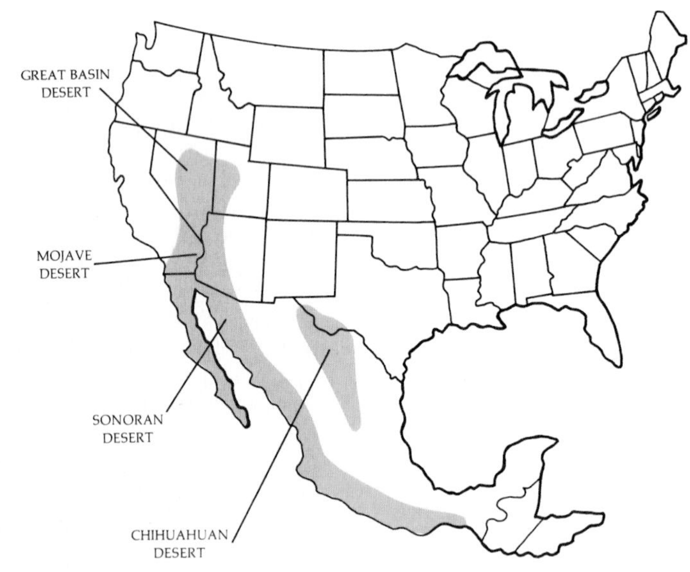

54–15

The North American deserts. The Sonoran stretches from southern California to western Arizona and down into Mexico. A dominant plant, the giant saguaro cactus (a) *is often as much as 15 meters high, with a widespreading network of shallow roots. Water is stored in a thickened stem, which expands, accordionlike, after a rainfall.*

To the southeast is the Chihuahuan desert, one of whose principal plants is the agave (b)*, or century plant, a monocot.*

North of the Sonoran is the Mojave, whose characteristic plant is the grotesque Joshua tree (c)*. This plant was named by early Mormon colonists who thought that its strange, awkward form resembled a bearded patriarch gesticulating in prayer. The Mojave contains Death Valley, the lowest point on the continent (90 meters below sea level), only 130 kilometers from Mt. Whitney, whose elevation is more than 4,000 meters.*

The Mojave blends into the Great Basin, a cold desert bounded by the Sierra Nevada to the west and the Rockies to the east. It is the largest and bleakest of the American deserts. The dominant plant form is sagebrush (d)*, shown here in the background. The large green plant in the foreground is shadscale, and the yellow-flowered plant to the left is rabbitbrush.*

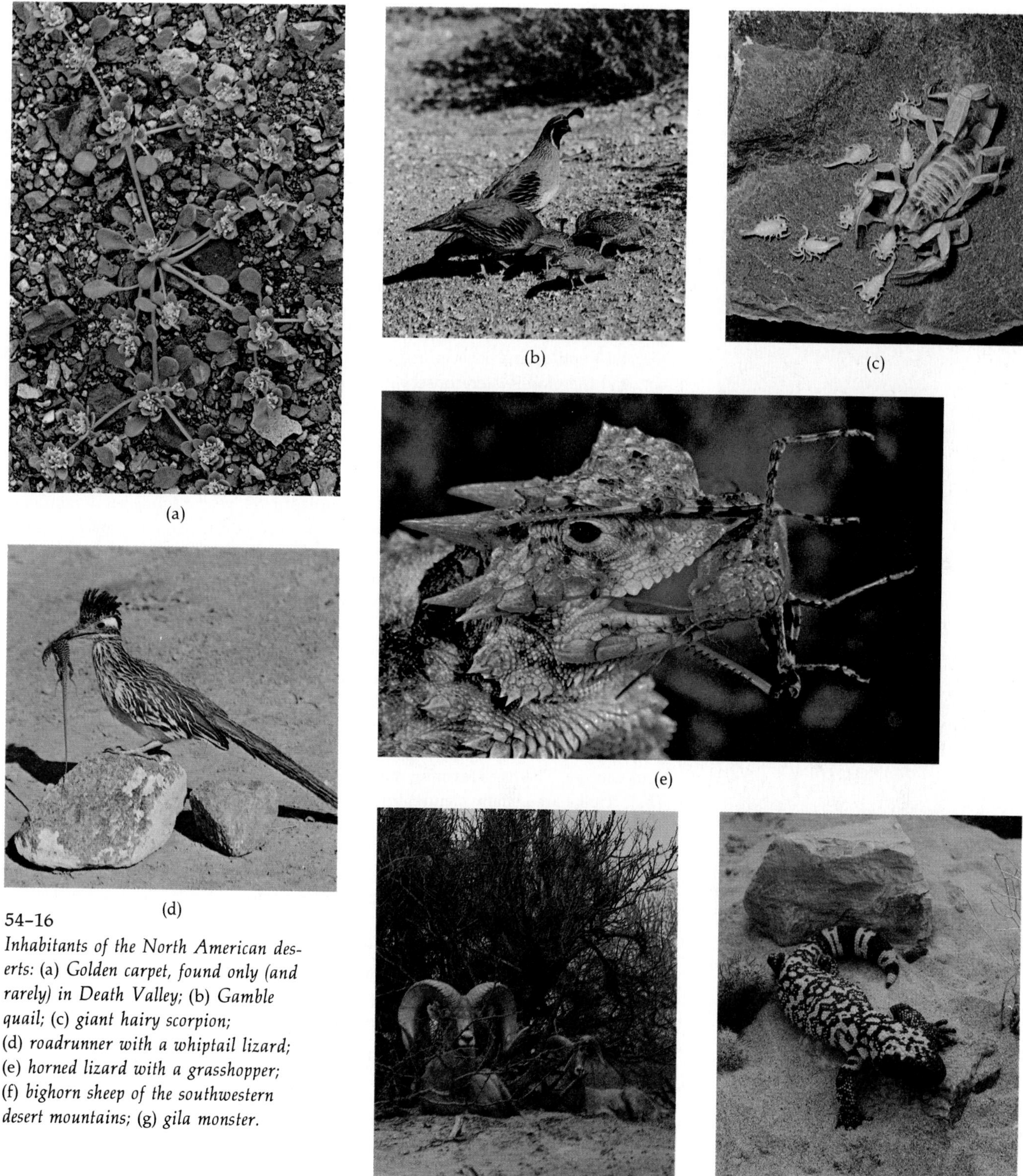

54–16
Inhabitants of the North American deserts: (a) *Golden carpet, found only (and rarely) in Death Valley;* (b) *Gamble quail;* (c) *giant hairy scorpion;* (d) *roadrunner with a whiptail lizard;* (e) *horned lizard with a grasshopper;* (f) *bighorn sheep of the southwestern desert mountains;* (g) *gila monster.*

Desert regions are characterized by less than 25 centimeters of rain a year. Because there is little water vapor in the air to moderate the temperature, the nights are often extremely cold. The temperature may drop as much as 30°C at night, in comparison with the humid tropics, where day and night temperatures vary by only a few degrees.

Many desert plants are annuals that race from seed to flower to seed during periods when water is available, and during the brief growing seasons, the desert may be carpeted with flowers. Many of the perennials are succulents (adapted for water storage). Some, like cacti, have no leaves, some are drought-deciduous (dropping their leaves in dry seasons), and some have small, leathery water-conserving leaves. C_4 and CAM photosynthesis (pages 226 and 618), both of which conserve water, are common among desert plants, as are extensive root systems that are able to trap large amounts of water during the brief periods it is available.

The animals that live in the desert are also specially adapted to this extreme climate. Reptiles and insects, for example, have waterproof outer coverings and dry, and therefore water-conserving, excretions. The few mammals of the desert are small and nocturnal and obtain what little water they require from the plants they eat.

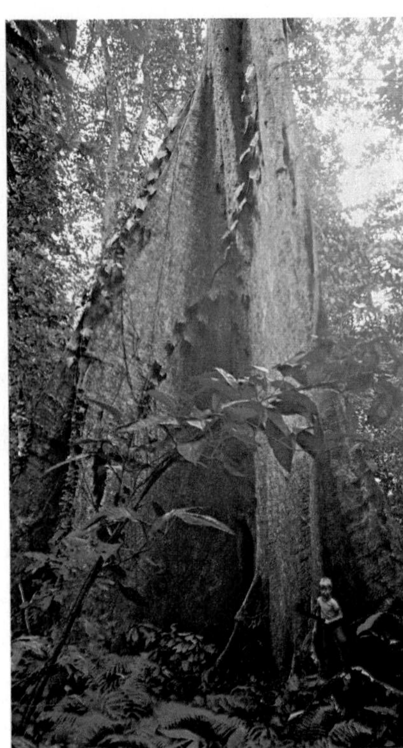

54–17
A tropical rain forest, Rancho Grande National Park, in Venezuela. Notice the height of the trees and the huge, buttressed trunk of the tree in the center. The vine is a philodendron.

Tropical Rain Forest

In the tropical rain forest, rainfall is abundant all year round; total rainfall is between 200 and 400 centimeters per year, and a month with less than 10 centimeters of rain is considered relatively dry. There are more species of plants and animals in the tropical rain forest than in all the rest of the biomes of the world combined. As many as 100 species of trees can be counted on 1 hectare, but each species may be represented by only one tree. By contrast, a comparable area in a deciduous forest in the northeastern United States typically contains only a few tree species, but each species is represented by many individual trees.

The critical competition among plants of the tropical forest is for light. About 70 percent of all species of plants are trees. The upper tree story consists of solitary giants 50 to 60 meters tall. A lower story of trees characteristically forms a continuous canopy. The trees forming the canopy are remarkably similar in appearance. Their trunks are usually slender and branch only near the crown. The crowns are high up and relatively small as a result of crowding. Because the soil is perpetually wet, their roots do not reach deep into it, and trunks often end in thick buttresses that provide broader anchorage. Their leaves are large, leathery, and dark green; their bark is thin and smooth; and their flowers are generally inconspicuous and greenish or whitish in color.

Woody vines, or lianas, are abundant, especially where an opening has appeared in the forest, as a result, for example, of a tree's falling; vines as long as 240 meters have been measured. There are also many epiphytes, which are plants that grow on other plants, often high above the forest floor (Figure 54–18). The epiphytes of the tropical rain forest germinate in the branches of trees and obtain water from the humid air of the canopy. Unlike the plants that have contact with the moist floor, epiphytes need to conserve water between rainfalls. Some epiphytes resemble desert succulents, having fleshy water-storing leaves and stems. Others have spongy roots or cup-shaped leaves that capture moisture and organic debris; many of these epiphytes can take up nutrients from decaying organisms in these storage tanks. A variety of plants, including ferns, orchids, mosses, and bromeliads, have exploited this life style.

(a)

(b)

54–18
*Inhabitants of the tropical rain forest. (a) Epiphytes, such as this bromeliad, grow in the canopy of the tropical rain forest, obtaining water and minerals from the moist air. Bromeliads are members of the pineapple family that are especially adapted to the tropical forest biome. (b) A splendid parakeet (*Neophema splendida*), also a denizen of the canopy.*

An extraordinary variety of insects, birds, and other animals, including mammals, have moved into the treetops along with the vines and epiphytes to make it the most abundantly and diversely populated area of the tropical rain forest.

Little light reaches the forest floor (from 0.1 to 1 percent of the total), and the few plants that are found there are adapted to growing at low light intensities. Many of these, such as the African violet, are familiar to us as house plants.

Plants also compete for nutrients. The nutrient cycles are tight, and turnover is rapid. There is almost no accumulation of leaf litter on the forest floor, such as we find in our northern forests; decomposition is too rapid. Everything that touches the ground disappears almost immediately—carried off, consumed, or rapidly decomposed. In many places the ground is bare.

The soils of tropical rain forests are relatively infertile. Many are chiefly composed of a red clay; these red soils are known as laterites, from the Latin word *later*, or "brick." When laterite soils are cleared, in many cases they either erode rapidly or form thick, impenetrable crusts that cannot be cultivated after a season or two. Tropical soils are generally deficient in minerals, and what minerals are found there are likely to be leached out by heavy rainfall. Most of the nitrogen, phosphorus, calcium, and other nutrients are found in the plants rather than the soil, and trees that most effectively store these nutrients may be the ones that win the competition for light.

As a consequence of this intense competition for nutrients, soils where tropical forests have stood are very poor agriculturally and will support crops for only a few years. In the tropics, the traditional agricultural practices have been those of clearing and short-term cultivation. With the rapidly expanding human population, these traditional practices have become immensely destructive because they are now carried out on such a wide scale. Although the tropical rain forest now forms about half of the forested area of the earth, some ecologists predict that, at the present rate of destruction, it may almost all have disappeared by the year 2000, and with it thousands of species of plants and animals found nowhere else in the world.

The tropical rain forest, with its vast diversity of plants and animals, presents the ultimate challenge to ecologists seeking to answer the central ecological question of what determines the number of species. One hypothesis holds that the rain forest is so diverse because of its age, which has brought about the finer and finer subdivision of resources and the opening of new opportunities for speciation. Another, more recent hypothesis is that the diversity is the result of instability or nonequilibrium, with disruptions and changes in habitat preventing the long-term competition that would ultimately eliminate species and decrease diversity.

Other Tropical Forests

Where rainfall is more seasonal, tropical rain forests give way to tropical deciduous forests, which are dominated by trees that lose their leaves during the dry seasons. Tropical scrub forests, which have trees with small, water-conserving leaves, are found where rainfall is limited all year.

EFFECTS OF ALTITUDE

The mean atmospheric temperature decreases about 0.5°C for each degree of increase in latitude. Increases in elevation produce a similar effect, and in general, a change in mean atmospheric temperature corresponding to a 1° increase in latitude occurs with each rise of about 100 meters in elevation. This relationship has important consequences for the distribution of land organisms. For example, plants and animals characteristic of Arctic regions may approach or even reach the equator in mountain ranges running from north to south (Figure 54–19).

There are, however, important differences between high-latitude habitats and high-altitude habitats. In mountains the air is clearer and the solar radiation is more intense. Most of the water vapor in the atmosphere—which plays a major role in preventing heat from radiating away from the earth at night—occurs below 2,000 meters. Consequently, nights in mountains are often much cooler than they are in regions with the same mean temperature but at higher latitudes and lower elevations. Moreover, there is pronounced seasonal variation both in day length and in temperature near the poles and relatively little variation near the equator. Those "Arctic" organisms that do range toward the equator in the mountains must make physiological adjustments for such differences between high-latitude and high-altitude environments.

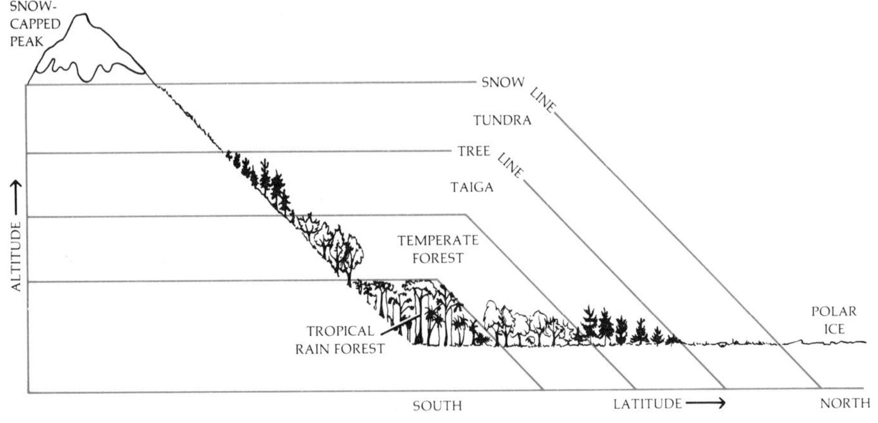

54–19
We can experience a similar series of plant communities whether we travel north for hundreds of kilometers or ascend a mountain.

SUMMARY

The biosphere is the part of the earth that contains living organisms. It is a thin film on the surface of the planet, irregular in its thickness and its density. The earth's atmosphere is a blanket of gas and dust molecules held close to the surface by gravity. The atmosphere filters out harmful ultraviolet rays from the sun and impedes the escape of heat from the earth's surface, resulting in the greenhouse effect. The biosphere is affected by the position and movements of the earth in relation to the sun and the movements of air and water. These conditions cause wide differences in temperature and rainfall from place to place and season to season on the earth's surface. These differences are reflected in differences in the kinds of plant and animal life.

Biomes are groupings of communities of organisms with common patterns of climate and distinctive vegetation, distributed over a wide area. The temperate deciduous forest is an important biome of eastern North America and Eurasia, where temperatures are warm in the summer and cold in the winter. In the deciduous forest, the trees lose their leaves in the fall, which reduces water loss during the winter months, when the water is locked in ice.

North of the deciduous forest is the taiga, the worldwide subarctic coniferous forest. The trees of this forest, the conifers, retain their small needlelike leaves, which are especially adapted for conservation of water and protection against extreme cold. Other needle-leafed evergreen forests are found along the Pacific Coast and in the southeastern United States and also in Southeast Asia.

Between the taiga and the northern polar region is the tundra, which is characterized by permafrost, a layer of permanently frozen subsoil. The dominant vegetation consists of low-growing perennials.

The temperate grasslands lie between the deciduous forests and the deserts and have a rainfall intermediate between the two. The growth of trees is prevented not only by a shortage of rain but also by grazing animals, which eat the young shoots, and by recurrent prairie fires. Savannas are tropical grasslands intermediate between tropical rain forests and deserts.

Chaparral, a shrubland biome found on the southern California coast and in the Mediterranean area, is characterized by dry summers but abundant winter rainfall.

Desert regions have very little rainfall and high daytime temperatures. The vegetation is characterized by annual plants with extremely short growing seasons. Cacti and other perennial desert plants are highly adapted for the conservation of water.

The richest biome, in terms of species diversity, is the tropical rain forest, where neither water shortages nor extreme temperatures limit plant growth. Here the trees are broad-leaved evergreens, which are characteristically covered with vines and epiphytes. Decomposition is so rapid that there is almost no accumulation of organic material or humus in the soil. Tropical soils are often red clays (called laterites), which erode or solidify when the forest is cleared. Tropical deciduous forests occur where rainfall is highly seasonal.

In general, increases in latitude and altitude result in similar temperature effects and, consequently, in similar distributions of land organisms.

QUESTIONS

1. Distinguish between the following: biosphere/atmosphere; tundra/taiga; solstice/equinox; troposphere/stratosphere.

2. What is the position of the earth in relation to the sun at the vernal equinox? At the winter solstice? What is the day-night cycle at the South Pole on these dates?

3. What are the eight major biomes? Describe the principal abiotic features of each.

4. Name a plant and an animal associated with each biome, and describe the special adaptations of each.

5. Although each of the biomes we have considered is sufficiently different from all the others to warrant its identification as a distinct biome, there are important similarities among some of the biomes. Consider the following groups of biomes: tropical rain forest/tropical deciduous forest; tropical deciduous forest/temperate deciduous forest/taiga; savanna/temperate grasslands/tundra. Describe the essential similarities and the most significant differences in the environmental factors affecting the members of each group. How do these factors affect the types of plants characterizing each biome?

6. The rate of decomposition of plant litter, animal wastes, and dead plants and animals varies from biome to biome. Describe the differences in decomposition rates in the following biomes: tropical rain forest, temperate deciduous forest, taiga. What factors in each biome are important in causing these differences? What are the consequences of these differing decomposition rates for mineral recycling, soil quality, and the size and diversity of detritivore populations?

CHAPTER 55

The Evolution of Social Behavior

As long ago as the turn of the century, evolutionary systematists recognized that patterns of behavior could be used to construct phylogenies in the same way as similarities and differences in morphology (see Figure 55–2 on the next page). In other words, behavioral characteristics are as much the products of natural selection as the shape of a tooth or the curve of a beak.

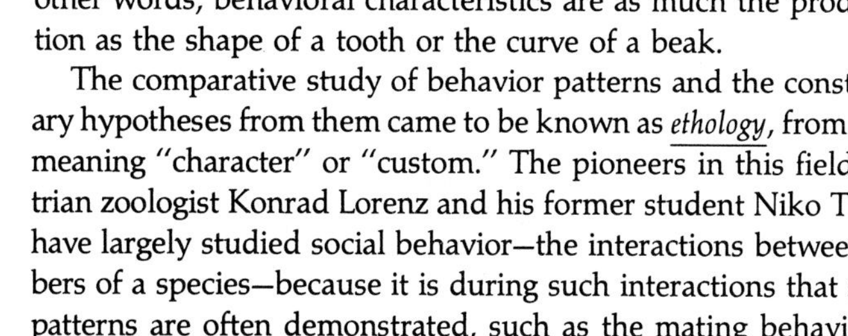

The comparative study of behavior patterns and the construction of evolutionary hypotheses from them came to be known as *ethology,* from the Greek word *ethos,* meaning "character" or "custom." The pioneers in this field have been the Austrian zoologist Konrad Lorenz and his former student Niko Tinbergen. Ethologists have largely studied social behavior—the interactions between two or more members of a species—because it is during such interactions that stereotyped behavior patterns are often demonstrated, such as the mating behavior of the stickleback (page 933). Such patterns are often very elaborate and, the more elaborate they are, the more species-specific they are likely to be.

Nina Leen, Time-Life Picture Agency

55–1
Lorenz and followers. Many species of precocial birds (birds that are able to walk and feed as soon as they are born) follow the first moving object they see after hatching. The object is usually mother goose (for goslings, that is) but can be a matchbox on a string or a member of another species. In many species, this phenomenon, called "imprinting," determines mate selection in the adult.

The factors governing the evolution of behavioral characteristics are the same as those that apply to any other trait. First, there are variations among individual organisms in behavioral traits. Second, some of these variations are genetically determined; that is, they are influenced by the presence of particular alleles. Third, individuals with favorable variations will tend to have greater relative reproductive success. This is the process called natural selection. Fourth, as a result of natural selection, certain alleles tend to increase generation after generation; in other words, evolution occurs.

When biologists concerned with evolutionary theory began to analyze social behavior, some disturbing questions began to emerge. How can you explain, in a system based on comparative reproductive success, the evolution of sterile castes—worker bees, for instance—in insects? How is it that in many vertebrate societies only a few of the males breed, with their right to breed apparently unchallenged by other males? Why, among animals in groups, do certain ones utter warning cries or exhibit other forms of behavior that attract attention to the individual issuing the warning, thus threatening the warner's life? Altruism is, by definition, behavior that benefits others and is performed at some risk or cost to the doer. How then can acts of altruism be explained in terms of natural selection acting on the individual organism? Ponder these questions a little, while we describe some of the observations of social behavior that gave rise to them.

55–2

Phylogeny of the Pelecaniformes, as determined by (a) *traditional morphological methods and* (b) *analysis of behavior. This order of birds includes not only pelicans but also tropic birds, frigate birds, boobies, cormorants, and anhingas. Note that a division occurs between the tropic birds (genus* Phaëthon) *and the frigate birds* (Fregata), *on the one hand, and the other members of the order in that the latter bow and the former do not. The pelicans* (Pelecanus) *are separated from the remaining members of the order by hopping, sky pointing, and wing waving in the latter and not the former. Gannets* (Morus) *and boobies* (Sula) *are the only members of the order to engage in head wagging, while the anhingas and cormorants* (Phalacrocorax) *share the behaviors of pointing and kink-throating. Morphology and behavior may be closely related in some instances; for example, anhingas and cormorants both have extremely long snakelike necks, well suited to pointing and kink-throating.*

(a)

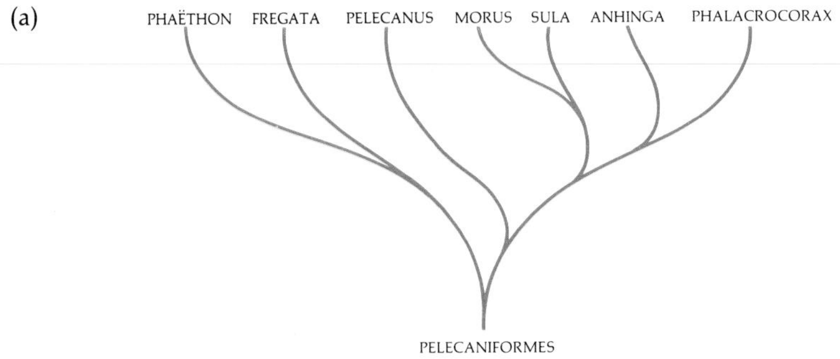

(b)

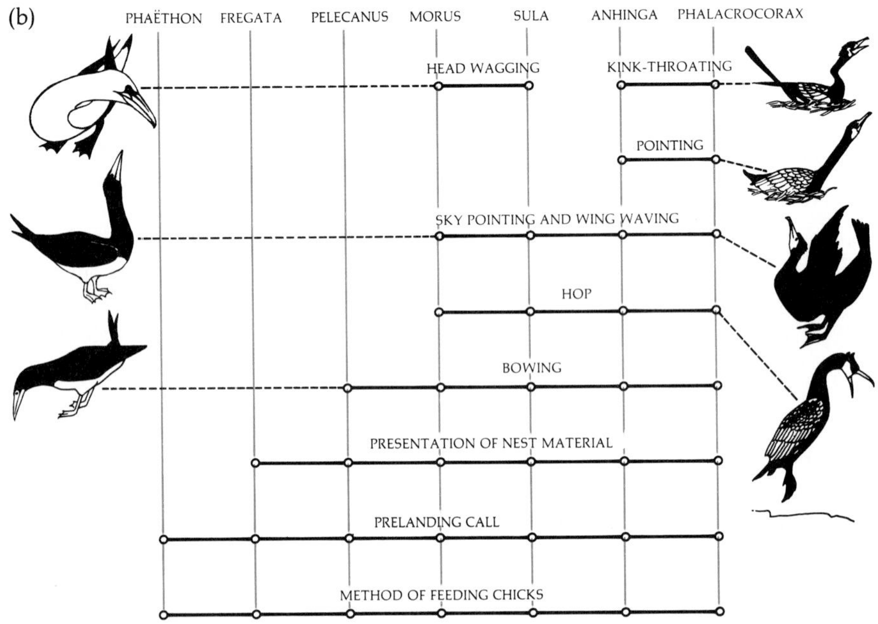

55–3

The gazelle in the center is stotting, performing a series of stiff-legged, high bounces. Typically, stotting occurs when an animal is alarmed, and it alerts other members of the herd to danger. Because it attracts the attention of a predator to the stotter, it appears to be an altruistic act.

INSECT SOCIETIES

Insect societies are by far the most ancient of all societies and, with the possible exception of modern human societies, by far the most complex. Also, of all animal organizations, insect societies are probably the best understood, in terms of both their evolution and the interplay of forces that keep them together. Social insects include termites and hymenopterans (ants, wasps, and bees).

Stages of Socialization

As with other animals, the social insects evolved from forms that were originally solitary. Among bees, for example, true sociality appears to have evolved on at least eight separate occasions, and among wasps four times.

Most species of bees and wasps are solitary and others show varying degrees of sociality; thus it is possible to reconstruct the various stages of social evolution by the analysis of the various present-day species. Among the solitary species, the female builds a small nest, lays her eggs in it, stocks it with a food supply, seals it off, and leaves it forever. She usually dies soon after, so there is no overlap between generations.

Among subsocial or presocial species, the mother returns to feed the larvae for some period of time, and the emerging young may subsequently lay their eggs in the same nest or comb. However, the colony is not permanent (usually being destroyed over the winter), there is no division of labor, and all females are fertile.

Eusocial, or "truly social," insects are characterized by cooperation in caring for the young and a division of labor, with sterile individuals working on behalf of reproductive ones. All ants and termites and some species of wasps and bees—for example, honey bees—are eusocial.

Honey Bees

A honey-bee society usually has a population of 30,000 to 40,000 workers and one adult queen. Each worker, always a female, begins life as a fertilized egg deposited by the queen in a separate wax cell. (Drones, or male bees, develop from unfertilized eggs.) The fertilized egg hatches to produce a white, grublike larva that is fed

55–4
Wasps, most of which are not truly social insects, do not tend their young but often provide for them. (a) *Wasps of the* Apanteles, *a solitary and parasitic genus, inject their eggs under the skins of caterpillars. The resultant larvae eat the internal tissues of the caterpillar, chew their way to the surface, and spin the cocoons shown here. Adults emerge from these cocoons.* (b) *The potter wasp builds a graceful clay vase, lays the eggs inside, and stocks it with caterpillars for the larvae. A caterpillar, paralyzed but not dead, provides up to 40 days' food.*

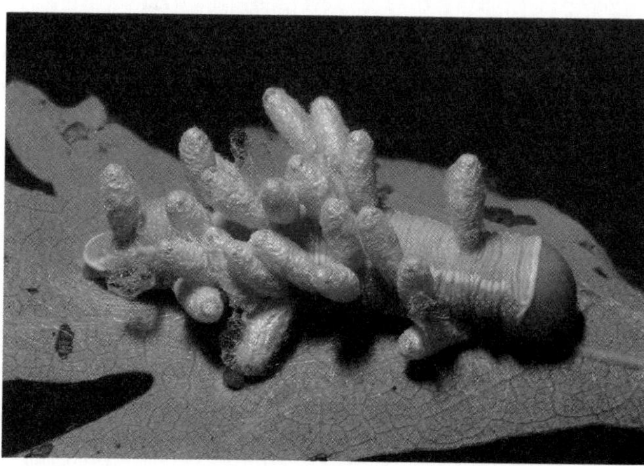

(a)

(b)

(a)

(b)

55–5
Honey-bee workers. (a) *The first segment of each of the three pairs of legs has a patch of bristles on its inner surface. Those of the first and second pairs are pollen brushes, which gather the pollen that sticks to the bee's hairy body. On the third pair of legs, the bristles form a pollen comb that collects pollen from the brushes and the abdomen. From the comb, the pollen is forced up into the pollen basket, a concave surface fringed with hairs on the upper segment of the third pair of legs. Transfer of pollen to the pollen basket occurs in midflight. The sting is at the tip of the abdomen.*

(b) *The mouthparts are fused into a sucking tube containing a tongue with which the bee obtains nectar. The antennae, attached to the head by a ball-and-socket joint, contain receptors for touch and for odors. The large compound eyes cannot see red (which is black, or colorless, to them) but can see ultraviolet, which is invisible to human eyes.*

almost continuously by the nurse workers; each larval bee eats about 1,300 meals a day. After the larva has grown until it fills the cell, a matter of about six days, the nurses cover the cell with a wax lid, sealing it in. It pupates for about 12 days, after which the adult bee emerges.

The newly emerged adult worker rests for a day or two and then begins successive phases of employment. She is first a nurse, bringing honey and pollen from storage cells to the queen, drones, and larvae. This occupation usually lasts about a week, but it may be extended or shortened, depending on the conditions of the colony. Then she begins to produce wax, which is exuded from the abdomen, passed forward by the hind legs to the front legs, chewed thoroughly, and then used to enlarge the comb. During this stage of employment as a houseworking bee, she may also remove sick or dead comrades from the hive, clean emptied cells for reuse, or serve as a guard at the hive entrance. During this period, she begins to make brief trips outside, seemingly to become familiar with the immediate neighborhood of the hive. It is only in the third and final phase of her existence that the worker honey bee forages for nectar and pollen. At about 6 weeks of age, she dies.

The Queen

Queenship is not genetically determined. The differences between a queen and a worker depend on the substance fed the queen-to-be in the larval stage and on the pheromonal influence she, in turn, exerts upon her subjects.

Queens are raised in special cells that are larger than the ordinary cells and shaped somewhat like a peanut shell. At one time, it was believed that these larvae became queens because they were fed special glandular secretions, known as royal jelly. Many attempts were made to identify the substance in royal jelly that conferred queenhood, but without success. According to the best recent evidence, queens become queens because of a generally more nutritious diet, especially rich in protein, as compared to the mostly carbohydrate (honey) diet fed the workers.

The queen exerts influences on her subjects by means of pheromones (page 547), of which there appear to be several. The influence of one of the pheromones, queen substance, inhibits ovarian development in the worker bees and prevents them from becoming queens or from producing rival queens.

If a hive loses its queen, workers will notice her absence very quickly and will

(a)

(b)

(c)

55-6
Honey bees. (a) *Workers tending honey and pollen storage cells. The honey is made from nectar processed by special enzymes in the workers' bodies.* (b) *Workers feed the queen (top center) and lick queen substance from her body. Queen substance, a pheromone, prevents sexual maturation in the workers.* (c) *Four pupae in different stages of development. A fully developed worker at the far left has shed her pupal skin and is ready to emerge, which she will do by gnawing through the wax cap.*

become quite agitated. Very shortly, they begin enlarging worker cells to form emergency queen cells, and the larvae in the enlarged cells are then fed the special diet. Any diploid larva so treated will become a queen.

The Annual Cycle

One of the important differences between subsocial and eusocial bees is that colonies of eusocial bees survive the winter. Within the wintering hive, as we saw on page 734, bees maintain their temperature by clustering together in a dense ball; the lower the temperature, the denser the cluster.

In the spring, when the nectar supplies are at their peak, so many young may be raised that the group separates into two colonies. The new colony is always founded by the old queen, who leaves the hive, taking about half of the workers with her. The group stays together in a swarm for a few days, gathered around the queen, after which the swarm will settle in some suitable hollow tree or other shelter found by its scouts.

As the old queen is preparing to leave the hive, the new queens are getting ready to emerge. These two events are synchronized by sound signals transmitted through the comb. As these signals are exchanged, the workers remain motionless. During this period, ovarian development begins in some of the workers, a few of which lay eggs. The unfertilized eggs develop into males, or drones. After the old queen leaves the hive, a new young queen emerges, and any other developing queens are destroyed. The young queen then goes on her nuptial flight, exuding a pheromone (apparently also the queen substance) that entices the drones of neighboring colonies. She mates only on this one occasion (although she may mate with more than one male) and then returns to the hive to settle down to a life devoted to egg production.

During her nuptial flight, the queen receives enough sperm to last her entire life, which may be some five to seven years. These are stored in a special organ in her reproductive tract and are released, one at a time, to fertilize each egg as it is being laid. The queen usually lays unfertilized eggs only in the spring, at the time males are required to inseminate the new queens.

The drones' only contribution to the life of the hive is their participation in the nuptial flight. Since they are unable to feed themselves, they become an increasing liability to the social group. As nectar supplies decrease in the fall, they are stung to death by their sisters or are driven out.

VERTEBRATE SOCIETIES

No vertebrate society has as rigid a caste system as the societies of the truly social insects. Studies have revealed, however, that social interactions of vertebrates are also governed by species-specific behavior, which results in dominance hierarchies and strongly affects comparative reproductive success of the individuals involved.

Pecking Orders

One type of social dominance among vertebrates that has been studied in some detail is the pecking order in chickens. A pecking order is established whenever a flock of hens is kept together over any period of time. In any one flock, one hen usually dominates all the others; she can peck any other hen without being pecked in return. A second hen can peck all hens but the first one; a third, all hens but the first two; and so on through the flock, down to the unfortunate pullet that is pecked by all and can peck none in return.

Hens that rank high in the pecking order have privileges such as first chance at the food trough, the roost, and the nest boxes. As a consequence, they can usually be recognized on sight by their sleek appearance and confident demeanor. Low-ranking hens tend to look dowdy and unpreened and to hover timidly on the fringes of the group.

During the period when a pecking order is being established, frequent and sometimes bloody battles may ensue, but once rank is fixed in the group, a mere raising or lowering of the head is sufficient to acknowledge the dominance or submission of one hen in relation to another. Life then proceeds in harmony. If new members are added to a flock, the entire pecking order must be reestablished, and the subsequent disorganization results in more fighting, less eating, and less

55–7
Wolves have dominance hierarchies of both males and females. (a) *Here a subordinate female is licking the muzzle of the dominant female. This same muzzle-nuzzle gesture is used by pups begging for food. Usually only the dominant male and dominant female breed, and the rest of the pack cooperate in caring for the young. Caring includes guarding the den and providing food, which is swallowed at the kill and then regurgitated for the pups. (Some domestic dogs regularly vomit at the sight of a puppy, which should be recognized not as a sign of disapproval but as a social reflex.)* (b) *Dominant male (center) surrounded by pack in greeting ceremony.*

(a)

(b)

55–8
A subordinate baboon turns his buttocks toward a superior male. This gesture, known as presenting, is used by females to indicate their readiness to mate. Presenting is also used between males and between females to signify submission or conciliation or to beg for special favors. The superior is reassuring the subordinate with a pat on the back.

tending to the essential business, from the poultry farmer's point of view, of growth and egg laying.

Pecking orders reduce the breeding population. Cocks and hens low in the pecking order copulate much less frequently than socially dominant chickens. Thus the final outcome is the same as if the social structure did not exist: the stronger and otherwise dominant animals eat better and leave the most offspring. However, because of the social hierarchy, this comes about with a minimum expenditure of lives and energy.

Territories and Territoriality

Most vertebrates stay close to their birthplaces, occupying a home range that is likely to be the same home range as that occupied by their parents. Even migratory birds that travel great distances are likely to return year after year to the same areas. Often these home ranges are defended, either by individuals (or more likely, mating pairs) or by groups against other individuals or groups of the same species or closely related species that use the same resources. Areas so defended are known as *territories*, and the behavior of defending an area against rivals is known as *territoriality*.

Territoriality in Birds

Territoriality was first recognized by an English amateur naturalist and bird watcher, Eliot Howard, who observed that the spring songs of male birds served not only to court the females but also to warn other males of the same species to stay away from the terrain that the prospective father had selected for his own. In general, a breeding territory is established by a male. Courtship of the female, nest building, raising of the young, and often feeding are carried out within this territory. Frequently the female also participates in territory defense.

By virtue of territoriality, a mating pair is assured of a monopoly of food and nesting material in the area and of a safe place to carry on all the activities associated with reproduction and care of the young. Some pairs carry out all their domestic activities within the territory. Others perform mating and nesting activities in the territories, which are defended vigorously, but gather food on a nearby communal feeding ground, where the birds congregate amicably together. A third type of territory functions only for courtship and mating, as in the bower of the bowerbird or the arena of the prairie chicken. In these territories, the males prance, strut, and posture—but very rarely fight—while the females look on and eventually indicate their choice of a mate by entering his territory. Males that have not been able to secure a territory for themselves are not able to reproduce; in fact, there is evidence from studies of some territorial species, such as the Australian magpie, that adults that do not secure territories do not mature sexually.

Territorial Defense

Even though territorial boundaries may be invisible, they are clearly defined and recognized by the territory owner. With birds, for example, it is not the mere proximity of another bird of the same species that elicits aggression, but its presence within a particular area. The territory owner patrols his territory by flying from tree to tree. He will ignore a nearby rival outside his territory, but he will fly off to attack a more distant one that has crossed the border. Animals of other species are generally ignored unless they are prey or predators or are in competition for some limited resource.

55–9

Territories come in many shapes and sizes. (a) *The male Uganda kob displays on his stamping ground, which is about 15 meters in diameter and is surrounded by similar stamping grounds on which other males display. A female signifies her choice by entering one of the stamping grounds and grazing there. Only a small proportion of males possess stamping grounds, and those that do are the only ones that breed.*

(b) *A fiddler crab's territory is a burrow, from which he signals with his large claw, beckoning females and warning off other males.*

(c) *Howler monkeys shift their territories as they move through the jungle canopy, but maintain spacing between groups by chorusing.*

(d) *Territoriality is common among reef fish. For many species, a territory is a crevice in the coral, but for others, such as the skunk clown fish shown here, the territory is a sea anemone. The fish is covered by thick slime that partially protects it from the poison of the tentacles, but its acceptance by the anemone is chiefly a consequence of behavioral adaptation of the fish, which even mates and raises its brood among the tentacles.*

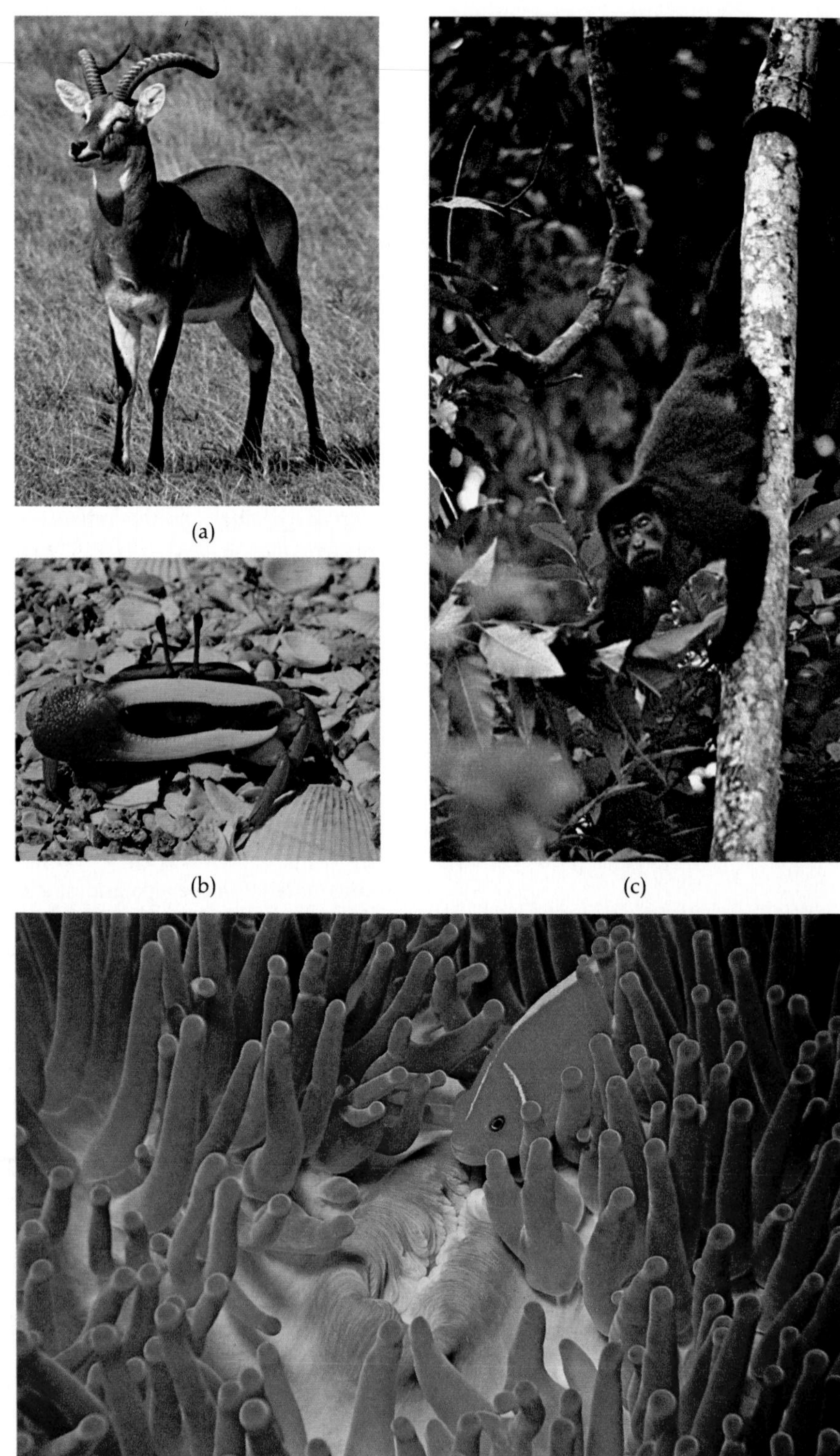

Once an animal has taken possession of a territory, he or she is virtually undefeatable on it. Among territory owners, prancing, posturing, scent marking, and singing and other types of calls usually suffice to repel intruders.

Similarly, the expulsion from communal territories is typically accomplished by ritual rather than by force. For example, among the red grouse of Scotland, the males crow and threaten only very early in the morning, and then only when the weather is good. This ceremony may become so threatening that weaker members of the group leave the moor. Those that leave often starve or are killed by predators. Once the early-morning contest is over, the remaining birds flock together and feed side by side for the rest of the day.

KIN SELECTION

In 1962, V. C. Wynne-Edwards caused a commotion in biological circles by proposing that individuals that failed to reproduce were doing so for the benefit of the society to which they belonged. In this way, a society could maintain its population at a level always slightly below its resources, he argued, and so the entire group would benefit. Such behavior was perpetuated by the increased survival of groups whose members behaved with such altruistic self-restraint. Wynne-Edwards' hypothesis of group selection proved extremely important. Although his ideas concerning altruistic behavior have now been rejected by almost all biologists (including Wynne-Edwards), his proposal served to galvanize a whole series of extremely fertile speculations and studies that have revolutionized the way modern biologists view and study social behavior.

Group selection was rejected on fairly simple grounds. If there were alleles for breeding and alleles for restraint-of-breeding, clearly the alleles for breeding would soon wipe out the other alleles in the population. In other words, to use a term that did not make its appearance until later, it would not be an evolutionarily stable strategy.

Rejection of Wynne-Edwards' hypothesis left a vacuum, which was filled, in part, by a proposal put forward by W. D. Hamilton, based largely on studies of social insects. As Darwin himself realized, the evolution of sterile castes of insects posed a special problem for evolutionary theory. If evolution is based on the number of surviving offspring, how can natural selection acting on the individual organism result in the sterility of a large proportion of a population? Darwin concluded that, in some cases, natural selection might act not only on individuals but on families. Conceptually the idea is a simple one: members of families share inheritable traits; thus there are variations among families as well as among individuals, and families that have favorable variations are likely to leave more survivors than other families.

Hamilton elaborated Darwin's suggestion, modifying it to the concept of the gene pool. According to population genetics, the measure of fitness becomes not the number of surviving offspring, but rather the increase or decrease in particular alleles in the gene pool. A mother takes risks and makes sacrifices for her offspring in order to increase the representation of her alleles in the gene pool. Or, to put the same thing another way, she is programmed genetically to take certain risks or to make certain sacrifices for her offspring. To the extent that taking these risks or making these sacrifices increases her contribution to the gene pool, the alleles dictating this risk-taking program will increase in the next generation, and so on.

However, it is not just mothers and offspring that share alleles. For example, consider brothers—that is, sons of the same mother. Brothers with the same father share, on an average, half of their alleles with one another. Therefore, any allele that favorably influenced altruistic behavior among brothers could similarly increase its representation in the gene pool.

Hamilton's hypothesis, based on this principle, is called *kin selection*. Kin selection is differential reproduction among lineages of closely related individuals. Tests of the hypothesis are based on degrees of relatedness of individuals. As a start, calculate your degree of relatedness to yourself, your parents, and your siblings. Half your alleles came from your mother and half from your father; thus the probability that you will share a particular allele with a particular parent is 0.5 (this is often abbreviated as $r = 0.5$, with r standing for relatedness). You and your sibling (ignoring the sex chromosome) each received half of your total alleles from your mother. Therefore, you each have a 50 percent chance of getting the same particular allele, so your total chance is 0.5×0.5, or 0.25 ($r = 0.25$). Your genetic relationship with your father is the same, so the total relatedness between two human siblings is $0.25 + 0.25$, or 0.5. Note that human siblings, like most other siblings with the same father, have the same degree of relatedness to each other as to their parents and to their own offspring.

Now consider the hymenopterans. As we noted earlier, in this order of insects, a fertilized (and therefore diploid) egg develops into a female, and an unfertilized, haploid egg develops into a male—a phenomenon known as haplodiploidy. Thus, for a female hymenopteran, her genetic relationship with her mother is the same as that of the usual vertebrate ($r = 0.5$), but on the paternal side, since the father is haploid, each sister will get an exact replica of the father's alleles. As a consequence, the average coefficient of relationship between sister hymenopterans is not 0.5, but 0.75 ($0.25 + 0.5$). Thus, if you are a female bee, more of your alleles will be represented in a reproductive sister (that is, a new queen) than in a daughter (Table 55–1).

Hence, from this perspective, the hive's workers are not laboring for the queen; they are working for one another, and the queen thus becomes merely the machine by which they produce more of their own genotype (0.75 at a time).

The workers' relationship with their brothers is quite different, however. The brother gets no alleles from his father, since he has no father. He gets half his

55–10
In a haplodiploid species, daughters have a 50 percent chance of sharing an allele from the mother (A or a) and a 100 percent chance of sharing an allele from the father (Z).

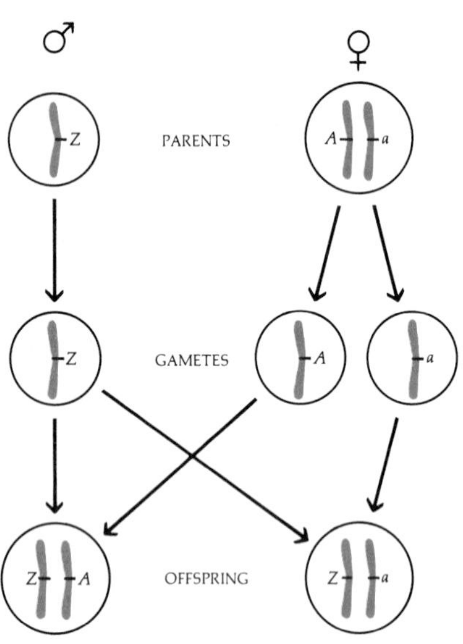

Table 55–1 Proportion of Alleles Shared by Relatives

	Normal diploid sex determination			Haplodiploid sex determination	
Alleles shared with his or her:	♂	♀	Alleles shared with his or her:	♂	♀
Mother	0.5	0.5	Mother	1	0.5
Father	0.5	0.5	Father	0	0.5
Son	0.5	0.5	Son	0	0.5
Daughter	0.5	0.5	Daughter	1	0.5
Sister	0.5	0.5	Sister	0.5	0.75
Brother	0.5	0.5	Brother	0.5	0.25

(a)

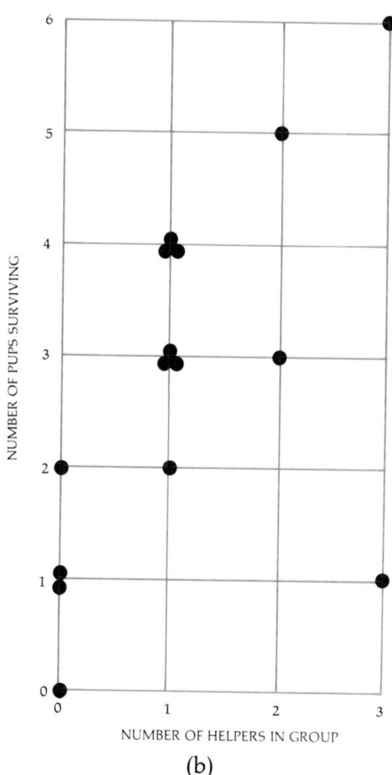

(b)

(c)

55-11

(a) *Parents groom a helper. Helpers are also provided food by parents who, as the chart* (b) *clearly shows, have reasons to encourage their assistance.* (c) *This young silver-backed jackal, by helping raise his younger brother, probably increases his inclusive fitness more than if he raised his own family.*

mother's alleles, and that is all. His chances of getting the same half as any given sister are 0.5 × 0.5, or 0.25. However, if he mates, his daughters inherit all of his alleles. Moreover, such a mating would probably be with the new queen of a neighboring colony—not with one of his sisters. On this basis one would predict that the male bee would be "selfish," performing little in the way of service to the hive, and that the females would invest little energy in caring for their brothers, both of which predictions turn out to be true.

Hamilton's example of kin selection among bees came under some criticism because all workers do not have the same father, since the queen may be inseminated by more than one male. Also, obviously eusociality does not depend exclusively on haplodiploidy, since the highly social termites, which are not haplodiploid, also have sterile workers (both male and female). Nor were his ideas entirely new; as we noted earlier (page 893), J. B. S. Haldane had remarked some decades previously that he would lay down his life for two brothers or four first cousins, thus anticipating Hamilton's thesis. However, Hamilton's exposition of these ideas served to establish a new evolutionary perspective, that of *inclusive fitness*. The single criterion of Darwinian fitness was the relative number of an individual's offspring that survive to reproduce. The criterion of inclusive fitness is the relative number of an individual's alleles that are passed on from generation to generation, either as a result of his or her own reproductive success, or that of closely related individuals.

Hamilton's hypothesis is useful because it is testable. For example, the family unit of the silver-backed jackal consists of a monogamous pair, young pups, and, often, one to three older siblings from a previous litter that serve as helpers.

Patricia Moehlman, who carried out her studies near Ndutu Lodge in Tanzania, studied 17 litters of jackal pups. In litters with no helpers, an average of only one pup was raised. With a single helper, three pups survived, and in one family with three helpers, six pups survived. Thus, by being helpers, young jackals (with a relatedness of 0.5 to their siblings, just as they would have to their own offspring) produced more copies of their own alleles than if they had reproduced themselves.

Among Florida scrub jays, the situation is somewhat different. Here again, breeding success is positively correlated with the number of helpers, which are usually close relatives of the breeding pair (Table 55-2). However, the success is somewhat less than would be achieved by the young jay's setting up its own nest and rearing its own young (Table 55-3). The reasons that the young jays stay on as helpers is hypothesized therefore to be the shortage of available territories for young males. This hypothesis is supported by the fact that as soon as a territory becomes available, the male helper leaves. One of the most common ways to get a new territory is to inherit part of the parents' territory. Thus, one is led to predict that females will help less than males and that older, more dominant males will help more than the younger males, both predictions that have been shown to be true.

Table 55-2 Benefit of Helpers to Breeding Pairs of Florida Scrub Jays

	NO HELPERS	WITH HELPERS
Inexperienced pairs	1.03	2.06
Experienced pairs	1.62	2.20

From data of G. E. Woolfenden, analyzed by S. T. Emlen, in J. R. Krebs and N. B. Davies (eds.), *Behavioural Ecology: An Evolutionary Approach*, Blackwell Scientific Publications, Oxford, 1978.

Table 55-3 A Scrub Jay Helper Would Do Better by Setting up Its Own Breeding Territory, If It Could Find One, Than by Helping Its Parents at Home (results are calculated on the basis of the helper's being a full sibling to the young—that is, sharing one-half of its alleles)

OPTION		RESULT
1. Stay at home and help	Young produced by experienced parents with no help	1.62
	Young produced by pair with one helper	1.94
	Extra young due to presence of helper	0.32
	Genetic equivalents to helper ($r = 0.5$)	0.16
2. Go off and rear own young	Young reared by first-time breeders	1.36
	Genetic equivalents ($r = 0.5$)	0.68

From S. T. Emlen, in J. R. Krebs and N. B. Davies (eds.), *Behavioural Ecology: An Evolutionary Approach*, Blackwell Scientific Publications, Oxford, 1978.

THE SELFISH GENE

Many years ago Samuel Butler remarked that a chicken was just an egg's way of making another egg. Updating this concept, some members of the new generation of ethologists see an organism as just a gene's way of making more genes. The argument is simple: The individual organism is transient. The genome is fragmented at every generation. All that can survive from generation to generation is the gene. The way it survives is in the form of replicas. The more replicas, the better the chance of survival. The organism is the gene's survival machine, and it so programs the machine that it will turn out gene copies at a maximum rate, regardless of personal cost. Thus, for example, there is a mite in which the female has a brood of one son and twenty daughters. The son mates with his sisters while still within the mother and dies before he is born. Not much of a life for the male mite but a great low-risk, high-return strategy for his genes and a wise maternal investment.

THE KINSHIP OF LIONS

Among the lions of the Serengeti, the females are the stable occupants of the territories. They hunt together, and cooperative hunting is undoubtedly the ecological basis for social living. They are always related—usually sisters or half-sisters—and this relatedness is reflected in the unusual degree of cooperation among them. All the adult females in a pride come into estrus at about the same time, a phenomenon apparently synchronized by pheromones. Thus the young are all born at about the same time, and the females nurse each other's cubs.

The males of the pride are the transients, holding the territory and the females for only a few months or, at most, a few years. The males are also often (but not always) related. Young males are driven from the pride by the adult males as they begin to mature, and they tend to stay together as nomads until they can find a pride to take over. Cooperation among the males is important for hunting and especially for taking over a pride, which may involve defeating the residents. There is little competition for females among the males. At the time of takeover, the intruding males often kill the young cubs. Following the death of the cubs, the females come into estrus again and so are available to perpetuate the genes of the newcomers, at least during the brief period of their reign.

Two lionesses, probably sisters, with their cubs.

The selfish-gene concept is admitted, even by its highly articulate sponsor, Richard Dawkins, to be an oversimplification. It has served, however, to bring into sharp focus questions on the evolution of survival strategies and behavior.

CONFLICTS OF INTEREST

In directing their attention to the survival of genes rather than of individuals, biologists began to realize that there were opportunities for serious conflicts of interest. These conflicts could occur not only between those individuals that were in obvious direct competition for some limited resource—such as territory, a food supply, or a mate—but also among individuals previously envisaged as working together for some common good—such as members of the nuclear family.

For example, a mammalian female invests in her offspring by providing nourishment in the form of milk. Nursing in mammals delays the onset of estrus and so keeps females from becoming pregnant. It is to the offspring's benefit to monopolize this supply for as long as possible; the mother, however, can maximize her fitness by weaning the offspring at an appropriate time and beginning a new investment.

Another example of intergenerational conflict occurs when a new male takes over a female with young. The new male's evolutionary goal, as programmed by his genes, is to produce his own young as rapidly as possible. The existing offspring (fathered by another male) may interfere with this objective, either by

55–12
Mother langur holding an infant that has been fatally wounded by an infanticidal male at the time of take-over of her troop.

competing with their half-siblings, or by delaying estrus in the female. Sarah Blaffer Hrdy of Harvard University was the first to report the practice of infanticide by males at the time of take-over. Her observations, which were made on Hanuman langurs in India, were greeted with considerable skepticism when first reported in 1971, but similar acts have now been observed in other primates and also in lions (see essay on page 1029). Infanticide can clearly represent a successful evolutionary strategy.

Male vs. Female

Among the most closely fought conflicts of interest are those that occur between the partners of a reproductive pair. A female, by definition, is the individual that produces the larger gametes. Thus, given the same investment of resources as the male, the female of any species virtually always produces far fewer gametes than the male. Male gametes are cheap; a single insemination costs a male almost nothing. Female gametes are relatively expensive, metabolically speaking, and if the female also must carry the embryo in her body and care for the young after the birth, her investment, in terms of her total reproductive potential, may be very large indeed.

In most animal species, all that the male contributes to the next generation is his genes. In such situations, it is to the interest of the female's selfish genes to find themselves in the best possible company, thus promoting their survival in the next generation. Hence, since all the females can be inseminated by relatively few males, intense competition may arise among the males to prove to the females that they are the best endowed. Such proof may take the form of elaborate courtship displays, in which strength, vigor, stamina, and other desirable traits may be displayed, or it may take the form of dominance over other males. Since the same qualities are likely to be involved in both cases, the distinction is not always clear.

Males also attract females by offers of food or other resources. This offer may be made in the form of a territory or a nesting site. In some species of birds and insects, males actually bring food offerings—a fly gift-wrapped in silk, for example.

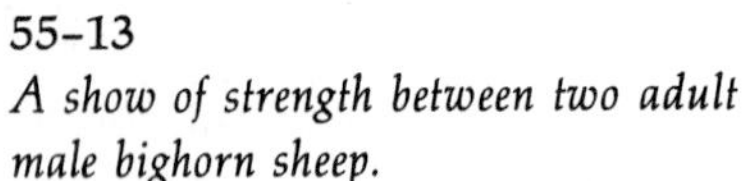

55–13
A show of strength between two adult male bighorn sheep.

ARTS AND CRAFTS OF BOWERBIRDS

The bowerbirds are a family of fruit-eating songbirds found only in New Guinea and Australia. They are named for their elaborate nuptial structures. Their bowers include huts 1 to 2 meters in diameter, stick towers several meters high, walled avenues, decorated lawns, and moss platforms with parapets. The buildings are decorated with colored objects, including fruits, shells and rocks, flowers, and, if the birds live near human settlements, coins, marbles, false teeth, bottle tops, eyeglasses, and toothbrushes. At least six species use a tool, such as a wad of bark, to paint the bower with crushed plant matter, charcoal, or stolen blue laundry powder.

The male birds, as you can well imagine, spend much of their time building, adorning, and perfecting their bowers, and protecting them from raids by rival males. Young males watch adults. Adults steal ornaments from other adult males and damage their bowers. Some of the males are brightly colored with elaborate crests and plumes, and others are drab-colored and lack any special plumage; there is an inverse relationship between the splendor of the bower and the color and elaborateness of plumage of the birds themselves.

Male birds call from their nests and display to females if they approach. The male may carry one of the brightly colored objects during his display, and extend it toward the female, though she has never been seen to take it; observers interpret this part of the display as having originated in courtship feeding, just as the bower itself probably was once, in evolutionary history, a nest. Often several different adult males of the same species are competing within half a kilometer or less of one another. The female selects one, enters his bower, copulates briefly, and departs. Then she builds her own utilitarian nest some distance from the bower of her mate, lays her eggs, and cares for the young entirely on her own.

What are the rewards to the female of this activity? If she raises her brood, she will probably have sons who have inherited their father's artistic skills and so will be able to attract females whose activities will perpetuate her genes.

(a)

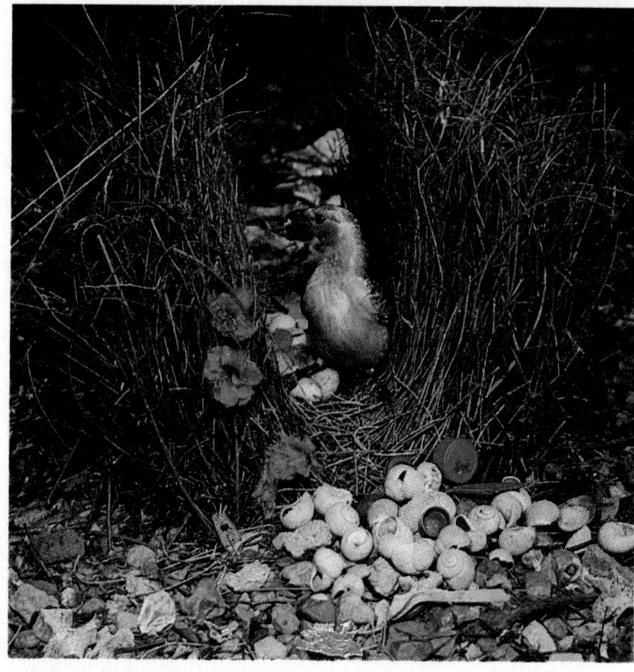

(b)

(a) *A pair of satin bowerbirds at a nest.*
(b) *Great bowerbird, decorating with hibiscus.*

Among praying mantids the male himself may be the offering, programmed by his genes to make this final sacrifice, if necessary, for their survival. Because the male that commands the most resources is likely also to be the male that is superior genetically, the offering or the territory may be less important as a food resource than as a symbol of the male's strength and desirability (see essay).

In some few species, males make an equal contribution to the care of the young. Selection for such a reproductive strategy would depend on at least two factors:

(a)

(b)

55–14
(a) *Courtship feeding in terns. The female (on the left) is soliciting the fish from the male. If he is ready to mate, he will give it to her (although she may have to ask several times). If she is ready, she will eat it; otherwise, she returns it. Such rituals are common among monogamous pairs of birds.* (b) *Copulating praying mantids. This male mantid is lucky—so far. Female mantids usually eat their mates, often decapitating them before copulation. Decapitation of the male mantid releases inhibition and results in his copulating even more vigorously, thus helping to ensure that his sacrifice will not have been in vain.*

first, that the additional parental investment would result in a significant increase in surviving young and, second, that the male has some way of being sure that the offspring he cares for are his own. Females in such situations give up the opportunity of mating with the highest-ranking male in return for a male that will provide care for the young. Selection for such a reproductive strategy would include a means for the female to be sure that the male she has selected will remain with her after her eggs have been fertilized.

A long courtship, such as is seen among some species of birds, serves the purposes of both male and female. It gives each of them a chance to assess the desirable qualities of the other; being choosey becomes equally important for a male under a monogamous system. For her, it symbolizes commitment and keeps him away from other females. It may also bring material benefits; for example, many female birds refuse to copulate until the nest is built. For him, the long courtship provides assurance that she has not already been inseminated. She loses nothing by cheating, whereas he loses his entire investment if she does.

Studies at Duke University have shown that the togetherness of turtle doves—the traditional symbol of matrimonial bliss—may be more pragmatically interpreted as the male's ensuring that the female does not bear eggs fertilized by another male. The male keeps the female in sight all during nest-building activities. But once the eggs appear, he spends hours away from the nest.

Such suspiciousness on the part of the male may be warranted. Many birds—such as members of the heron family—share communal nesting areas, with nests within a meter or so of one another. Males and females take turns at the nest. Bird watchers have reported observing matings between the female on one nest and the male on another. To test whether or not such matings were actually successful, an ethologist captured some of the monogamously paired males, vasectomized them, and released them. The females paired with the sterile males were all able to lay fertile eggs and raise normal-sized clutches, with the help, of course, of their duped partners.

DOMINANCE HIERARCHIES AND THE SELFISH GENE

Territories, dominance hierarchies, and pecking orders can also be viewed from the selfish-gene perspective. An animal that is not breeding because it is low in the dominance hierarchy is not sacrificing itself for the good of the group or of the species, as group selection hypotheses would maintain; it is waiting for its chance. If the individual is very low in the hierarchy, such an opportunity may never come, but still its chances of perpetuating its genotype are better than if it challenges a superior and is killed or badly injured.

Studies of bird territories confirm the hypothesis that individuals that lack territories are waiting for their turn. An empty territory is immediately filled, as in the previously cited example of Florida scrub jays. Some years ago a group of ecologists accidentally tested the waiting hypothesis in a more rigorous way. They undertook a study in Maine to determine whether or not the warblers nesting in the spruce trees exerted a controlling effect on the caterpillars of the spruce bud worm. They mapped the positions of the singing males within an area of 16 hectares (40 acres), found 148 mating pairs, and began to shoot them all. After 3 weeks, they had shot 302 male warblers (and a lesser number of females), and there were still male warblers singing throughout the 16 hectares. The moral of this story—among others—is that it sometimes pays to wait your turn.

REPRODUCTIVE STRATEGIES IN MALE BABOONS

Baboons are among the best studied of monkeys not only because they are interesting but also because they are terrestrial, a great advantage to the earthbound field worker. They are large for monkeys, and the adult male, which weighs about 55 kilograms, is about twice the size of the female and has a mantle and prominent canine teeth. Since both males and females eat the same diet, the large canines are reasonably assumed to function chiefly in fighting and display.

There are four species of baboons, all of which are organized into bands ranging from 8 to almost 100 individuals. There are many males in each band and, in all species except the hamadryas, the bands have social dominance hierarchies, one for males and one for females. In the hamadryas, which is a desert species, the bands are subdivided into family groups with a single adult male, one or more females, and their young, totaling up to 7 or 8 members.

Hamadryas males have a strong inhibition against taking each other's females. In an experiment, two adult males, who knew each other, and a strange adult female were trapped. One male was enclosed with the female, while the other was allowed to watch from a cage 10 meters away. The male caged with the female immediately made grooming, mounting, and herding advances toward the female, as he would toward a member of his harem. Fifteen minutes later, the second male was put in the same enclosure with the pair. He not only refrained from fighting over the female but avoided even looking at the pair. The animals were separated, and two days later the experiment was repeated again, this time switching the males. Once again, the male that had watched the activities between the male-female pair was strongly inhibited in his behavior.

The only problem with this evolutionarily stable bourgeois strategy is that it provides little opportunity for a young male to set up housekeeping. A further feature of the program is that this inhibition applies only to adult males and females. Young males are attracted to juvenile females, which they "kidnap" from their family groups before they are sexually mature and train to follow them. Thus the younger male has a detour around the genetic inhibition barrier, and new family units can come into existence.

(a)

(b)

(a) *A hamadryas band. Note the grouping by one-male units.* (b) *A young male hamadryas with a juvenile female that he has stolen from another one-male unit. New family groups originate in this way.*

EVOLUTIONARILY STABLE STRATEGIES

Sometimes the best behavioral strategy depends not on the rigid dictates of one's selfish genes but on what the other fellow is doing. For instance, you may have played a game in which you and your opponent each put your right hand behind your back and arrange your fingers to symbolize a rock, a piece of paper, or a pair of scissors. Then—one, two, three—you bring your hands forward. Scissors cuts paper, rock breaks scissors, paper wraps rock, and so on; in other words, whether or not you win depends on what your opponent does. If values are given to the various results of the encounters, the best strategy to use in situations of this sort can be computerized and calculated—an undertaking that goes under the general heading of "game theory." The models produced can be used in a variety of ways, such as planning battle maneuvers.

A strategy is evolutionarily stable, according to John Maynard Smith, the principal developer of this idea, if a population of individuals using this strategy cannot be invaded by a mutant adopting a different strategy.

Maynard Smith has applied game theory to show how natural selection would maintain polymorphisms in behavior. His simplest example is that of hawk and dove, terms used not to signify real birds but to denote two extremes of behavior (actually doves can be pretty aggressive). Hawks always fight, to the death, if need be. Doves display but never fight. The behavior is genetically programmed and passed on to the offspring.

Arbitrarily assigned pay-offs for winners and losers are summarized in Table 55-4. The game assumes that doves and hawks reproduce their own kind in proportion to the pay-offs.

In a population in which all organisms were doves, every encounter would pay off an average of +15. Obviously group selection would favor such a behavioral strategy. However, suppose one mutant with hawk behavior found its way into the group. It would always win and with a +50 for a win, hawks would soon spread through the population. Hawk strategy would never take over completely, how-

Table 55-4 The Game between Hawk and Dove (pay-offs: winner +50, loser 0, injury −100, display −10; pay-off matrix: average pay-offs in a fight to the attacker)

	OPPONENT	
ATTACKER	HAWK	DOVE
Hawk	(a) $\frac{1}{2}(50) + \frac{1}{2}(-100) = -25$	(b) $+50$
Dove	(c) 0	(d) $\frac{1}{2}(50 - 10) + \frac{1}{2}(-10) = +15$

Rules of the game:

(a) When a hawk meets a hawk, we assume that on half of the occasions it wins, and on half the occasions it suffers injury.

(b) Hawks always beat doves.

(c) Doves always immediately retreat against hawks.

(d) When a dove meets a dove, we assume that there is always a display, and it wins on half of the occasions.

From J. Maynard Smith, "Evolution and the Theory of Games," *American Scientist*, vol. 64, pages 41–45, 1976.

ever, because every hawk-hawk encounter pays off an average of −25, and so hawks would begin to disappear from the population. At some point, a balance would be reached at which the advantages of being a hawk would be just equal to those of being a dove. For the pay-offs listed in Table 55-4, the stable equilibrium occurs when there are 7 hawks for every 5 doves. If this balance should be upset by an invasion of either hawks or doves, natural selection would act until stable equilibrium was reached.

Another, slightly more complex way to reach a stable equilibrium is if individuals play a mixed strategy, playing hawk 7 out of 12 times and dove 5 out of 12. This also offers a stable equilibrium for the values assigned.

Maynard Smith introduced a third option into his hawk-dove game—a strategy he termed bourgeois. The bourgeois individual is the owner of a territory. A bourgeois always fights when another individual, hawk or dove, invades its territory and never fights when it is the intruder (Table 55-5). Using the same system of pay-offs and losses, Maynard Smith showed that given these three possible strategies, bourgeois is the only evolutionarily stable strategy (sometimes referred to as ESS).

No one imagines that real life is so simple or that exact numbers can be so assigned to complex situations. However, bourgeois, for example, is not an imaginary strategy. Territorial animals always have an advantage on their own territory. As Tinbergen showed some years ago, if a male stickleback is placed in a test tube and moved into the territory of a rival male, his posture becomes less and less aggressive the farther within the territory he is transported. Similarly, a male cichlid will dart toward a rival male intruder, but as the defender chases the intruder back into his own territory, he will begin to swim more and more slowly, his caudal fins seemingly working harder and harder, as if he were making his way against a current that increases in strength the farther he pushes into the other male's home ground. The fish know just where the boundary lies and, after chasing each other back and forth across it, will usually end up each one trembling and victorious on his own side.

Table 55-5 The Hawk, Dove, Bourgeois Game (pay-offs: winner +50, loser 0, injury −100, display −10; pay-off matrix: average pay-offs in a fight to the attacker)

	OPPONENT		
ATTACKER	HAWK	DOVE	BOURGEOIS
Hawk	−25	+50	+12.5
Dove	0	+15	+7.5
Bourgeois	−12.5	+32.5	+25

Rules of the game:

1. The pay-offs for encounters between hawks and doves are the same as in Table 55-4.
2. When bourgeois meets either hawk or dove, we assume it is owner half the time, and therefore plays hawk, and intruder half the time, and therefore plays dove. Its pay-offs are therefore the average of the two numbers above it in the matrix.
3. When bourgeois meets bourgeois, on half the occasions it is owner and wins, while on half the occasions it is intruder and retreats. There is never any cost of display or injury.

From J. Maynard Smith, "Evolution and the Theory of Games," *American Scientist*, vol. 64, pages 41–45, 1976.

55–15
Food sharing among nonrelated individuals is an example of reciprocal altruism. Here chimpanzees share a delicacy—the body of a red colobus monkey that one of the chimps has chanced upon. Meat is not actively sought as food by chimps but is highly prized when found.

RECIPROCAL ALTRUISM

Seemingly altruistic acts may pay off, as we have seen earlier, because they increase the probability of survival of genes shared by the performer of the act and the beneficiary. In 1971, Robert Trivers proposed another model for seemingly self-sacrificing behavior, which he termed *reciprocal altruism.* According to this model, an altruistic act is performed with the expectation that the favor will be returned. It might involve warning of danger, removing a tick from an inaccessible portion of the anatomy, or sharing food, to cite just a few possible examples. It also might involve acts of cooperation between organisms of different species, as well as between members of the same species.

In terms of game theory, the possibilities can be cast as cooperator or cheater. Any numerical figure can be assigned as the reward, provided that (1) in any encounter in which one individual cooperates and one cheats, the cheater always wins, and (2) if both individuals cooperate, the pay-off is higher than if both cheat.

In terms of both game theory and biological evolution, the only solution is to cheat, even though, paradoxically, both players would do better if they cooperated. If all players were programmed to cooperate, one mutant cheat could sweep through the population, winning all the prizes. Thus, all cooperating is not an evolutionarily stable strategy.

The solution is to add two more conditions. The first is that the players meet more than once and are able to recognize each other. The second is that each player cooperates on the first move and, on the second move, does whatever the other player did on the preceding move (Tit for Tat). Under these conditions, cooperation pays off and cheating is punished, so cheaters cannot invade and cooperation becomes an evolutionarily stable strategy.

One problem with Tit for Tat is how to get it started; a sufficient proportion of cooperators must be present in the original population for the reciprocal behavior to become established. A possible, and reasonably probable, solution is that cooperation originates in a group of related individuals—that is, as kin selection—and spreads from the original cluster.

Another problem is that individual players must have a way of recognizing one another. Thus, reciprocal altruism would probably not work in a large impersonal society, such as that of honey bees. It might well have become established in early human societies, which is why the model is of special interest. Psychologists have performed studies, some involving very young human infants, that have demonstrated that the ability to recognize and remember the faces of others is highly developed among humans, which may be a related phenomenon.

THE BIOLOGY OF HUMAN BEHAVIOR

It is tempting—indeed almost irresistible—to draw parallels between human behavior and that observed in other species. The extent to which these concepts concerning the evolution of behavior can be extrapolated to the human species is a matter of current controversy. One group of biologists maintains that the human species is basically no different from any other species, that our genes are as selfish as any, and that if we seek to modify human behavior for the common good, we should understand its roots.

An opposing group maintains that while early human ancestors may, in the past, have been governed by their genes, modern humans are so much a product of their culture and of their individual experience that such analyses are no longer valid. Moreover, they may be dangerous. The concept that biology determines human behavior lies at the roots of all notions of racial superiority, they point out. Thus it has provided the rationale for slavery, exploitation, and genocide. More commonly, the notion that our behavior is, to some extent, biologically determined allows us to forgive ourselves for, and even to justify, violence, aggressiveness, docility, and greed.

SUMMARY

Ethology is the study of the evolution of behavior. It emphasizes the ecological and genetic bases of behavior, contrasting behavior patterns among various species.

Insect societies are the largest and most complex of animal societies, possibly excluding only modern human social organizations. In the honey-bee colony (and among other social hymenopterans and termites as well), the queen is the only female reproductive form. She and her brood are tended by sterile workers. The behavior and physiology of the members of the hive are controlled by the exchange of chemical signals in the form of pheromones.

Animal societies are often organized in terms of social dominance. Social hierarchies may take the form of pecking orders, in which higher-ranking animals have privileged access to food and other resources and reproduce more. Territoriality is a system of social dominance in which only animals with territories reproduce and animals with superior territories reproduce more.

A central question in the study of social behavior is the mechanism by which natural selection can result in behavior that limits the reproductive potential of individuals in the society. An early explanation was that of group selection, which hypothesized that animals were genetically programmed to refrain from breeding for the good of the society as a whole. The group-selection concept was rejected because there was no demonstrable way that group selection could be maintained against individual selection for increased reproduction.

This hypothesis has been largely supplanted by that of kin selection, which introduced the concept of inclusive fitness. The kin-selection hypothesis was first formulated in terms of the social organization of hymenopterans. In these insects, females are produced from fertilized, diploid eggs and males by unfertilized, haploid eggs. Therefore, as in diploid species, females share half their alleles with their mother. However, if they have the same (haploid) father, they share three-quarters of their alleles with each other (instead of one-half). Thus by tending the queen, their sister, the workers produce more copies of their own alleles than if they reproduced themselves. Previous concepts of evolutionary fitness had focused on the relative number of surviving offspring in future generations; inclusive fitness focuses on the proportion of specific alleles introduced as replicas into future generations.

Kin selection has been demonstrated in other species. Among jackals, for example, young adults can contribute more of their own alleles to the future gene pool by helping their parents raise the next litter of young than by breeding themselves. From this viewpoint, social behavior is regulated by "the selfish gene," which programs the individual not necessarily for the individual's own well-being, or even survival, but only for the perpetuation of the allele by any means.

Competition for representation in the future gene pool can occur not only between individuals obviously competing for the same resources—such as for food or mates—but also between parents and offspring and between members of a reproductive pair.

The selfish-gene concept can also be invoked to explain the establishment of pecking orders and dominance hierarchies. The individual that is low in a social hierarchy is programmed by its genes to accept such an inferior status because the individual's chances of reproducing are better if it waits than if it embarks on a probably futile challenge.

An evolutionarily stable strategy is one that is determined by what other individuals in the society are doing. It can take the form of genetic polymorphism, with certain proportions of individuals always behaving in particular ways, or of individuals behaving in one way a fixed proportion of the time and in another way the rest of the time. An evolutionarily stable strategy, by definition, is resistant to upset by an invader utilizing an alternative strategy.

Altruism is behavior that carries a cost to the individual that performs it and benefits some other individual or individuals. Some acts of altruism are probably based on inclusive fitness. Others may be based on reciprocal altruism, the performing of an unselfish act with the expectation that it will be returned.

QUESTIONS

1. Distinguish between the following: group selection/kin selection; altruism/reciprocal altruism; reciprocal altruism/evolutionarily stable strategy.

2. Among organisms in which both sexes are diploid, the relatedness between parents and children is always 0.5, and the relatedness between siblings has a theoretically possible range from 0 to 1 and averages 0.5. Explain why.

3. Haldane was being too generous in his offer (see page 1027). Explain why.

4. In order to protect their young against infanticidal males, females might refuse to mate with murderers. Would this be a successful strategy?

5. Some recent studies suggest that females in good health are more likely to produce males and those in poor condition are more likely to produce females. How can this be explained as an evolutionary strategy?

CHAPTER 56

Human Evolution and Ecology

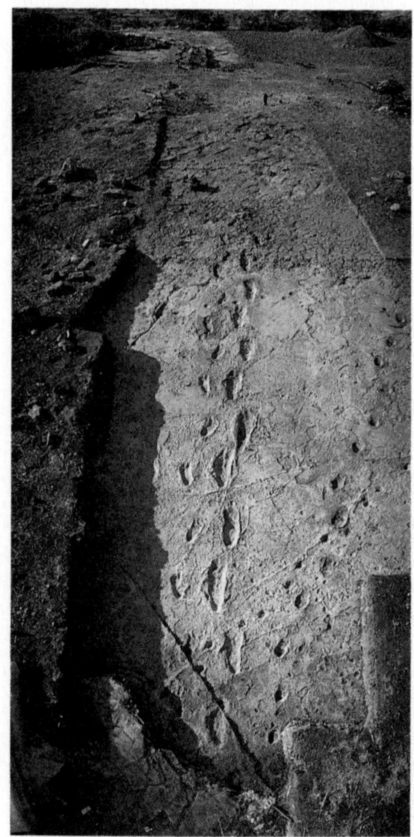

56–1
Fossil prints of the ancestors of Homo sapiens, *left in volcanic ash that fell 3.8 million years ago. Those at the left seem to be those of a child. At the right of the hominid trail are the tracks of an extinct three-toed horse, Hipparion.*

Where does the story of human evolution begin? We might start with a chance combination of chemicals in some warm Precambrian sea. Or perhaps even with the formation of a small planet 150 million kilometers from a star. Or it might begin more than 4.5 billion years later, when some little tribe of hominids found they could sharpen a digging stick or hone the flat edge of a stone. In any case, it is a very long story, measured in human terms, and many of its details are lost to us, probably forever.

For present purposes, let us start the story about 200 million years ago, in the early Mesozoic era, at about the time of the first dinosaurs. In this same period of time—give or take a few million years—the first mammals appeared, arising from a primitive reptilian stock. Our information about these mammals is very slight. The entire length of the Jurassic and Cretaceous periods has left us with only a few fragments of skulls and some occasional teeth and jaws. From these scraps of evidence, we know that the first mammals were about the size of a mouse. They had sharp teeth, indicating that they were basically carnivorous. Since they were too small to attack most other vertebrates, however, they are assumed to have lived on insects and worms, supplementing their diet with tender buds, fruits, and perhaps eggs. These first mouse-sized mammals were probably nocturnal, judging by the large size of their eye sockets, and they were almost certainly warm-blooded. If such an animal were alive today, it would be classified as an insectivore, something like a ground shrew.

For about 130 million years, these small mammals led furtive existences in a land dominated by reptiles. Then suddenly, as geologic time is measured, the giant reptiles, the dinosaurs, disappeared. The cause of the dinosaur extinction is one of the great biological mysteries. It occurred at a time when, geologists believe, there was a drop in the average temperature and, perhaps more important, a marked increase in seasonal temperature fluctuations. In any case, by the end of the Cretaceous period all of the dinosaurs had disappeared forever, and about 65 million years ago an explosive radiation of the mammals began.

The early mammals immediately diverged into the two dozen or so different lines that included (1) the monotremes, or egg-laying mammals, of which the duckbilled platypus is one of the few remaining examples; (2) the marsupials, such as the kangaroos, opossums, koala bears, and others, whose young are born in embryonic form and continue their development in pouches; and (3) the placen-

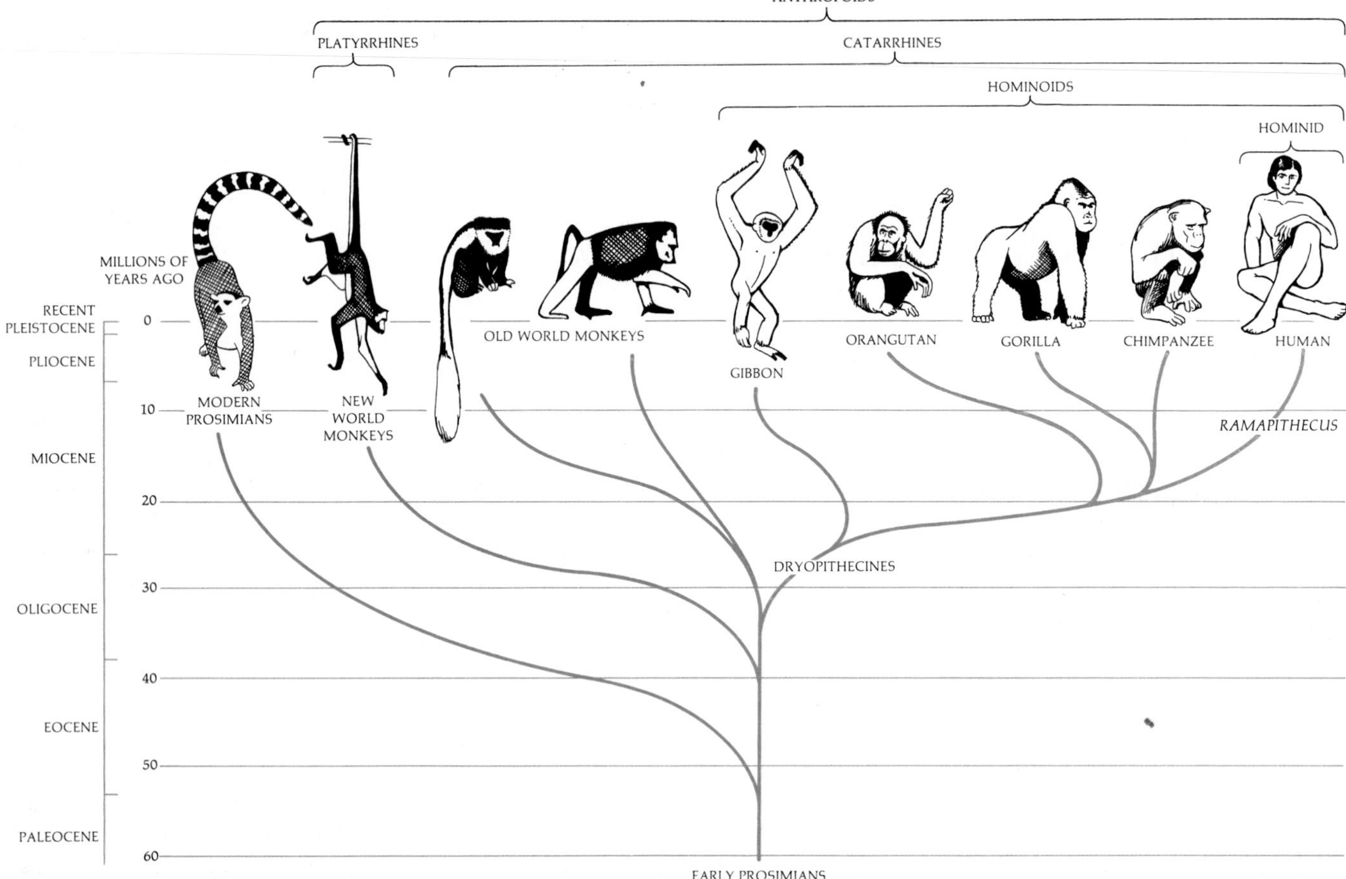

56–2

A tentative phylogenetic tree of the primates, based on fossil evidence. Biochemical evidence—molecular clocks—indicate a more recent divergence between hominids (members of the human family Hominidae) and their closest living relatives, the gorilla and the chimpanzee.

tals, by far the largest group. Among the placentals are carnivores, ranging in size from the saber-toothed tiger down to small, weasel-like creatures; herbivores, which include not only the many wild grazing animals but also most of our domesticated farm animals; the omnipresent rodents; and such odd groups as the whales and dolphins, the bats, the modern insectivores, and the primates. We are placental mammals and members of the primate order, as are tarsiers, lemurs, monkeys, and apes, among others.

TRENDS IN PRIMATE EVOLUTION

Primate evolution began when a group of the small, shrewlike mammals took to the trees. Most trends in primate evolution seem to be related to various adaptations to arboreal life.

The Primate Hand and Arm

With a few exceptions, primates have five digits with a divergent thumb. The divergent thumb, which can be brought into opposition to the forefinger, greatly increases gripping powers and dexterity. There is an evolutionary trend among the primates toward finer manipulative ability that reaches its culmination in humans (Figure 56–3).

Compared to other mammals, however, primates are relatively unspecialized. Their extremities resemble those of the primitive mammals—indeed, of the reptiles—more closely than do the extremities of mammals of most of the other major orders (see page 377). The first four-legged mammals all had five separate digits on each hand and foot, and each digit except the thumb and the first toe had three separate segments that made it flexible and capable of independent movement. In the course of evolution, most mammals developed hooves and paws more suited for running, seizing prey, and digging; other mammals developed flippers for swimming. The primates retained and elaborated on the primitive five-digited pattern.

In the basic quadrupedal structure of the early mammals and reptiles, the forearm has two long bones (the radius and the ulna), a pattern that provides for flexibility. Among mammals, it is the primates, in particular, that can twist the radius, the bone on the thumb side, over the ulna so that the hand can be rotated through a full semicircle without moving the elbow or the upper arm. Similarly, only a few mammals have the ability to move the upper arm freely in the shoulder socket. A dog or horse, for instance, usually moves its legs in only one plane, forward and backward; some lemurs, South American monkeys, apes, and humans are among the few higher mammals that can rotate the arm widely in the socket.

Primates also have nails rather than claws. Nails leave the tactile surface of the digit free and so greatly increase the sensitivity of the digits for exploration and manipulation.

Visual Acuity

Another result of the move to the trees is the high premium placed on visual acuity, with a decreasing emphasis on the role of olfaction, the most important of the senses among many of the other mammalian orders. (Flying produced similar evolutionary pressures among the birds, which were also evolving rapidly during this same period.) This shift from dependence on smell to dependence on sight has anatomical consequences. Among the primates there can be traced a steady evolutionary trend toward frontally directed eyes and stereoscopic vision.

Almost all primate retinas have cones as well as rods; cones, as we discussed on page 799, are concerned with color vision and with fine visual discrimination. All primate retinas have foveas, areas of closely packed cones that produce sharp visual images.

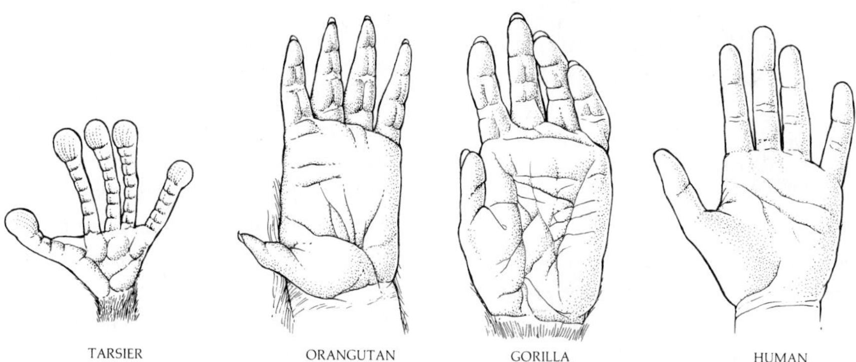

56–3
Some primate hands. The hand of the tarsier has enlarged skin pads for grasping branches. In the orangutan, the fingers are lengthened and the thumb reduced, which provide for efficient brachiating. The gorilla's hand, which is used in walking as well as handling, has shortened fingers. The human thumb is larger proportionately than that of any other primate, and opposition of thumb and fingers, on which the handling ability depends, is greatest in humans.

56-4
Life in the treetops made maternal care a major factor in infant survival. Also the necessity for carrying the young for long periods resulted in strong selection pressures for reduced numbers of offspring. (a) *Anthropoids, such as this vervet monkey, usually have single births.* (b) *A mother chimpanzee with her infant. Field studies suggest that bonds between mother and offspring and perhaps also among siblings last well into adulthood, perhaps for a lifetime.*

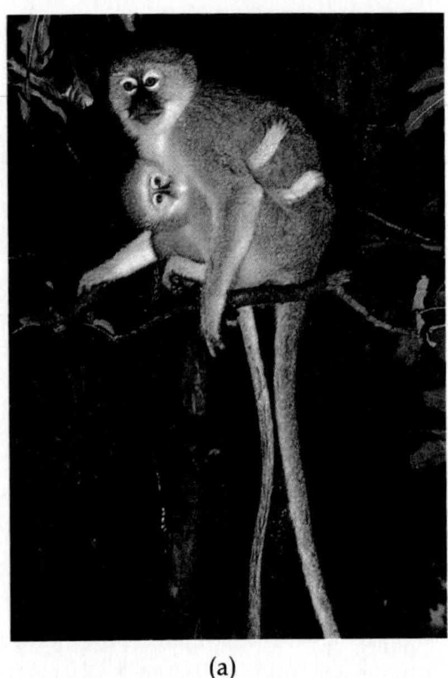
(a)

(b)

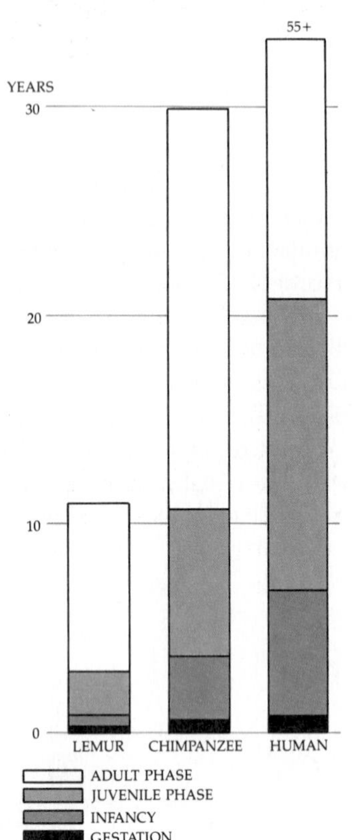

56-5
In the course of primate evolution there has been a trend toward longer periods of juvenile dependency and thus longer periods of learning.

Care of the Young

Another principal trend in primate evolution is toward increased care of the young. Because mammals, by definition, nurse their young, they tend to have longer, stronger mother-child relationships than other vertebrates (with the exception, in some cases, of birds). In the larger primates, the young mature slowly and have long periods of dependency and learning (Figure 56-5)

Uprightness

Another adaptation to arboreal life is an upright posture. Even quadrupedal primates, such as monkeys, sit upright. One consequence of this posture is a change in the orientation of the head, allowing the animal to look straight ahead while in a vertical position; it is this characteristic, above all others, that makes primates look so "human" to us. Vertical posture was an important preadaptation for the upright stance of modern humans.

MAJOR LINES OF PRIMATE EVOLUTION

Prosimians

Primates are divided into two major groups: the *prosimians* (lorises, bush babies, tarsiers, and lemurs) and the higher primates, the *anthropoids* (monkeys, apes, and humans). During the Paleocene and the Eocene (about 65 to 38 million years ago), a great abundance and variety of prosimians inhabited the tropical and subtropical forests that spread much farther north and south of the equator than they do today. Modern prosimians (Figure 56-6) are mostly small, arboreal, furry, and vegetarian or insectivorous, and many are nocturnal.

Monkeys

Monkeys, along with the apes and humans, make up the higher primates, the anthropoids. The monkeys are generally larger than prosimians, their skulls are

56–6
Two prosimians. (a) *A ring-tailed lemur, combing his tail. The second digit of each foot is a special grooming claw.* (b) *A native of Indonesia, the little tarsier (about the size of a kitten) has existed relatively unchanged for some 50 million years. It has stereoscopic vision and a larger brain than the lemur's. Living entirely in trees, it has hands and feet with enlarged skin pads for grasping branches. As you may have guessed from the owl-like eyes, tarsiers are nocturnal.*

(a)

(b)

more rounded, and they are generally more intelligent. They have full stereoscopic vision and also color discrimination. They are all diurnal.

Monkeys generally move in bands composed of adult males and females, infants, and juveniles. The females protect and otherwise care for the young, and the males often serve protective functions for the group, ranging from observing and warning to attacking predators outright. In some species, males join together to mob a predator, harassing it by hooting and calling, jumping on branches until they fall on enemy heads, sometimes throwing sticks and branches, and, in some species, by group defecation and urination.

The monkeys arose from prosimian stock apparently during the Eocene period; related fossil forms are found in the Oligocene. There are two principal groups, the New World monkeys, also known as platyrrhines (meaning flat-nosed), and the Old World monkeys, the catarrhines (downward-nosed). The separation of these groups took place very early in their history, with the platyrrhines evolving in South America and the catarrhines in Africa, quite possibly during the Oligocene, some 26 to 38 million years ago.

56–7
Early in their evolution, the anthropoids split into two main lines, the platyrrhine, or flat-nosed (left), and the catarrhine, or downward-nosed (right). New World monkeys are platyrrhines; Old World monkeys and the hominoids are catarrhines. There are other characteristic anatomical differences between the two groups.

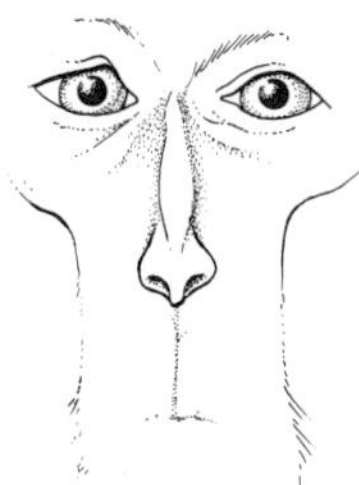

CULTURAL EVOLUTION AMONG MACAQUES

In insects, such as the honey bee, behavior is largely determined by what Ernst Mayr terms a "closed program"—that is, a program that is largely genetically determined, with little possibility of modification. Such programs, obviously, are useful for organisms with a short life span and little opportunity for learning. Among mammals, the evolutionary trend has been toward "open programs," which allow adaptive changes in behavior. This trend is particularly in evidence among the primates.

The macaques of the Japanese island of Koshima offer an example. These primates (also known as rhesus monkeys) lived and fed in the inland forest until about 30 years ago, when a group of Japanese researchers began throwing sweet potatoes on the beach for them. The group quickly got used to venturing onto the beach, brushing sand off the potatoes, and eating them. One year after the feeding started, a two-year-old female the scientists had named Imo was observed carrying a sweet potato to the water, dipping it in with one hand, and brushing off the sand with the other. Soon, other macaques began to wash their potatoes, too. Only macaques that had close associations with a potato washer took up the practice themselves. Thus it spread among close companions, siblings, and their mothers, but adult males, which were rarely part of these intimate groups, did not acquire the habit. However, when the young females that learned potato washing matured and had offspring of their own, all of them learned potato washing from their mothers. Today all of the macaques of Koshima dip their potatoes in the salt water to rinse them off, and many of them, having acquired a taste for salt, dip them between bites.

This was only the beginning. Later, the scientists began scattering wheat kernels on the beach. Like the others, Imo, then four years old, had been picking the grains one by one out of the sand. One day she began carrying handfuls of sand and wheat to the shore and throwing them into the water. The sand sank, the wheat kernels floated to the top, and Imo collected the wheat and ate it. The researchers were particularly intrigued by this new behavior since it involved throwing away food once collected, much less a part of the macaques' normal behavioral repertory than holding on to food and cleaning it off. Washing wheat spread through the group in much the same way that washing potatoes had. Now the macaques, which had never even been seen on the beaches before the feeding program began, have taken up swimming. The youngsters splash in the water on hot days. Some of them dive and bring up seaweed, and at least one has left Koshima and swum to a neighboring island, perhaps as a cultural missionary.

56–8
Only New World monkeys, such as the spider monkey, can hang by their tails. The platyrrhine monkeys are the only primates native to the Americas.

Despite their long chronological separation, members of the two groups closely resemble each other, an example of parallel evolution.

The New World monkeys, from South and Central America, are all strictly arboreal in their habits, and many of them use their tails as a fifth prehensile limb, which none of the Old World monkeys can do. Any monkey you see hanging by its tail is definitely a New World monkey. The New World monkeys include marmosets, howler monkeys, the spider monkey, and the capuchin—the familiar monkey of the organ-grinder.

The Old World monkeys include both arboreal and terrestrial species. Among the tree dwellers are the colobus monkeys, langurs, the mangabeys, and the guenons. Their tails are used for balance rather than as prehensile organs. The ground dwellers, which came from tree-dwelling ancestors, include the macaques, or rhesus monkeys, and some baboons. These ground dwellers walk on all fours.

Apes

Hominoids, now represented by the apes and us, first appeared, according to the fossil record, some 30 million years ago, and fossil apes are known in large numbers from deposits in Kenya and Uganda, ranging in age from 25 million to some 18 million years ago. These Miocene fossils have been given the genus name of *Dryopithecus,* a combination of the Greek for "oak" and "ape," reflecting the belief that these primates were forest dwellers. Dryopithecines have now been found throughout Europe and Asia, as well as Africa. They flourished for some 20 million years, eventually giving rise to the modern hominoids (see Figure 56–2).

The modern apes comprise four principal genera: *Hylobates* (gibbons), *Pongo* (orangutans), *Pan* (chimpanzees), and *Gorilla* (gorillas). Apes, with the exception of the gibbons, are larger than monkeys. Their brain case is larger in proportion to their size. Like the Old World monkeys, they are catarrhines. They are all capable of brachiation—that is, swinging from one arm and then the other with their bodies upright. Although, among modern genera, only the gibbons move primarily in this way, brachiation may have played a role in the transition from the body structures associated with the crouching position characteristic of the lower primates to the body structure that makes possible our erect posture. Apes have relatively long arms and short legs. As a result, even when they are on all fours, their bodies are partially erect.

The social structures of the genera vary. Gibbons form permanent pairs, each couple living by itself with its offspring. Male orangutans are solitary; adult females are usually accompanied by one or two youngsters. Gorillas and chimpanzees are highly social.

Gorillas, the largest of living primates, are among the shyest and the gentlest, according to recent field studies. Gorillas have almost completely abandoned trees, with the exception of some of the smaller animals that sleep in nests in the lower branches. The larger gorillas sleep on the ground, and all feed on ground plants. They live in groups ranging from 8 to 24 individuals, with about twice as many females as males, and a number of juveniles and infants. Each gorilla troop has a large, mature (silver-backed) male as a leader.

Chimpanzees, the primates generally considered to be the closest to humans, usually move in groups, feeding on wild fruits, seeds, and pods of a large variety, and occasionally killing smaller animals for meat. Within a group there is a male dominance hierarchy; females are usually subordinate to adult males, but there is also a female dominance hierarchy. Temporary (two- or three-day) bonds may form between a male and a female in estrus, but otherwise there are no pair bonds, and receptive females typically mate with a series of males. However, there are often friendships among members of the group, and friends may spend as much as two hours a day in mutual grooming. Males sometimes cooperate in capturing prey, which consists most often of other primates, such as infant baboons and colobus monkeys. Prey may be shared among the group. Offspring maintain strong bonds with their siblings and their mothers, apparently throughout their lifetimes. Chimps are gregarious, curious, boisterous, and extroverted. Although groups move through home ranges, there is not a strong sense of territoriality, and the bands are not well knit and exclusive, as they are in some other primate groups. The composition of the group changes as newcomers are taken in and other members wander off.

Thus, as you can see, it is difficult to make correlations between human behavior and the diverse behavior of other higher primates.

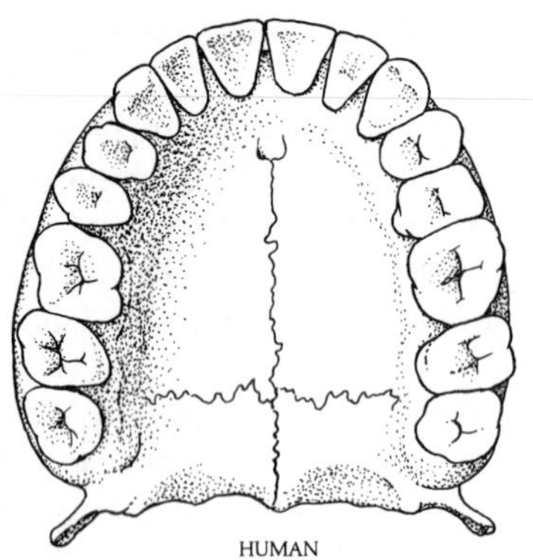

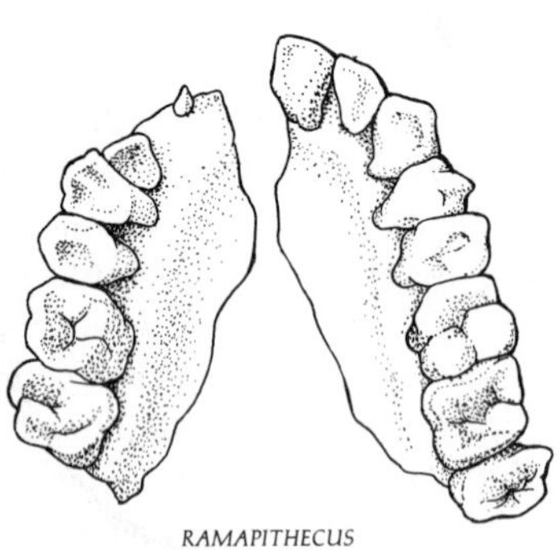

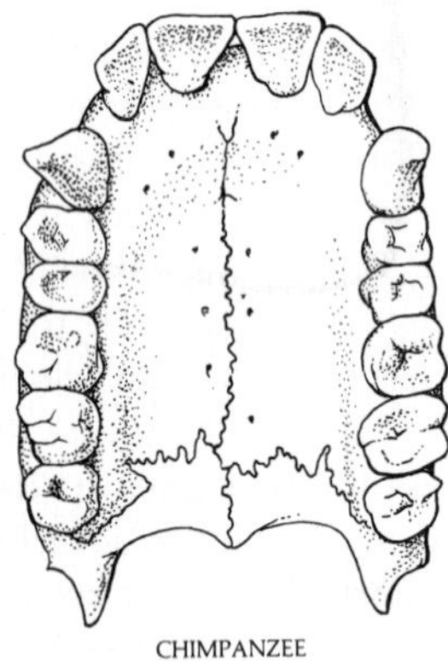

56–9
A comparison of the upper jaws of a human, Ramapithecus, *and a chimpanzee. The jaw of* Ramapithecus *is more rounded and thus more like the human jaw than that of the ape.*

Ramapithecus

Between 10 and 15 million years ago, *Dryopithecus* diverged into at least three different genera: *Sivapithecus, Gigantopithecus,* and *Ramapithecus.* Of these, *Ramapithecus* is generally (but not unanimously) believed to be the first known hominid. *Ramapithecus* is known primarily from fossil fragments of upper and lower jaws dated about 12 to 14 million years ago (Figure 56–9). When the pieces of jawbone are pieced together, it is clear that *Ramapithecus* had a smaller and broader dental arch, as compared to other large contemporary primates or, indeed, to modern apes. The jaw fragments also indicate that *Ramapithecus* was comparatively small, smaller than a chimpanzee. Second, adjacent molars on the jaw fragments of *Ramapithecus* show marked differences in wear in comparison to contemporary primitive primates, in which no sharp variations in wear are evident. This suggests that the molar eruption sequence of *Ramapithecus* was much slower; as a consequence, there is reason to believe that *Ramapithecus* matured more slowly than its hominoid contemporaries—as do modern hominids.

Finally, the teeth and their condition indicate that tasks such as biting off and tearing up vegetation, for which apes use their front teeth, were not carried out to the same extent by the teeth of *Ramapithecus.* Physical anthropologists hypothesize that ramapithecines used their forelimbs for these purposes, and, on the basis of this hypothesis, they further hypothesize a trend toward bipedalism. Additional support for bipedalism is found in the fact that the back teeth are specialized for more powerful chewing, which may indicate that *Ramapithecus* had abandoned the fruits and other moist vegetation of the forest for the harder, drier food (such as tough rhizomes and dried seeds) of the bush and savanna.

Evidence based on molecular clocks, however, indicates that the gibbons diverged from the other apes about 12 million years ago, the orangutans about 10 million years ago, and that the split between humans and the African apes (chimps and gorillas) occurred only about 6 million years ago. If the molecular evidence is valid, *Ramapithecus,* which existed 4 to 6 million years before this, is clearly not a member of the human family tree. Recently, David Pilbeam of Harvard pieced together an unusually complete skull of *Sivapithecus.* The teeth resemble those of *Ramapithecus,* but the rest of the skull is much like that of an orangutan. In other words, the hominidlike aspects of the jaw of *Ramapithecus* may reflect parallel evolution with hominids and not a common ancestry. The outcome of this dispute is awaited with particular interest because it is considered a test of the accuracy of molecular clocks.

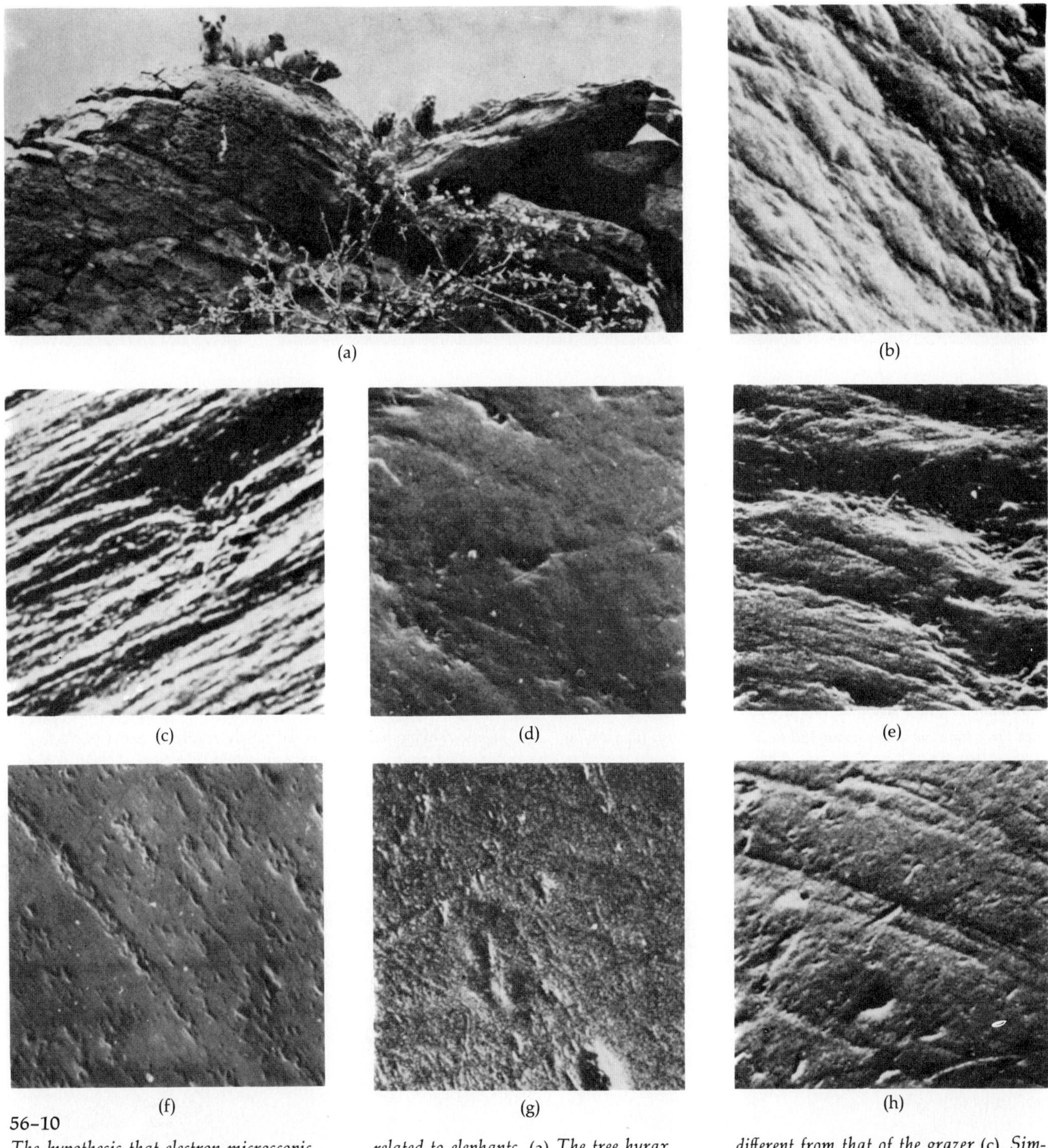

56–10
The hypothesis that electron microscopic studies of tooth surfaces can reveal diet was tested in two species of hyrax. These little animals, the conies of the Bible, look like large guinea pigs but are actually related to elephants. (a) *The tree hyrax and the rock hyrax look alike and share the same burrows, but the former is a browser and the latter a grazer. The wear pattern on the browser* (b) *is distinctly different from that of the grazer* (c). *Similar wear patterns distinguish the diets of a chimpanzee* (d), *an omnivorous bush pig* (e), *a cheetah* (f), Ramapithecus (g), *and* Australopithecus robustus (h).

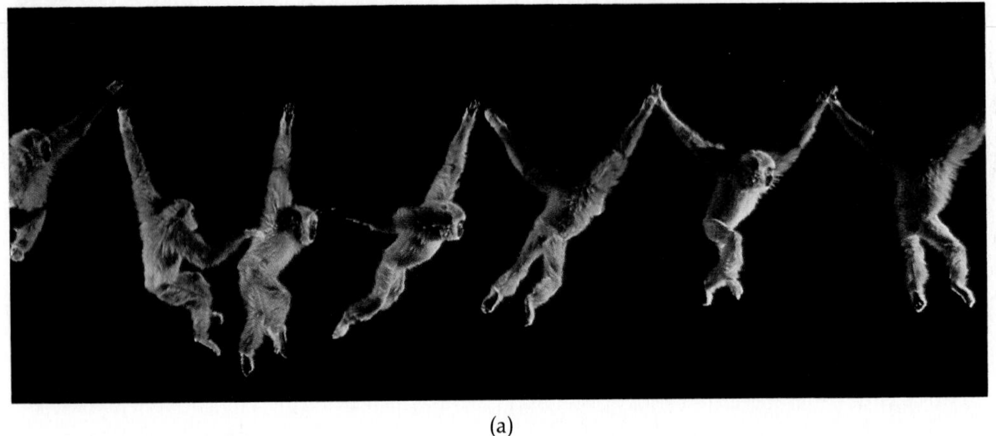

(a)

(b)

ANTHROPOID APES

The anthropoid apes include gibbons, orangutans, gorillas, and our closest living relatives, chimpanzees. Gibbons are the smallest of the apes. They live in pairs or in small family groups, and, along with some human beings, they are the only monogamous anthropoids. Gibbons can stand and walk upright, but usually they move through the trees, swinging arm over arm (brachiating) as shown above (a). *The rarest and least studied of the large primates, the modern orangutan* (b), *is found only in the tropical rain forests of Borneo and Sumatra. Males, which can weigh as much as 100 kilograms (220 pounds), are usually found alone.*

A male gorilla is about as tall as a tall man but weighs more than three times as much (some 180 to 275 kilograms). (c) *Gorillas have a beetling brow, a strong, heavy jaw, and, on top of the skulls of adult males, a bony crest to which the strong jaw muscles are attached. Its legs are shorter in proportion to its body than are ours, but its arms are much longer—with a span of about 3 meters—and extremely powerful.*

Gorillas usually walk on all fours, but they can stand erect, as they do when challenging an enemy, and they also walk erect for short distances. They live principally on the ground and feed on ground plants. In (d), *a mature male is feeding on a bamboo shoot.*

Each gorilla troop, consisting of from 8 to 24 individuals, is led by a mature male. Although reportedly placid by nature, when threatened, males will give fearsome exhibitions of power, thrashing with broken branches, standing erect, beating their chests, barking and roaring, even hitting themselves under the chin to make their teeth rattle.

A female gorilla of reproductive age mates only with the troop's highest-ranking male that is not closely related to her. Once a female gives birth, she goes through a period of celibacy, lasting 2 to 4 years, during which lactation occurs. In (e), *an infant suckles as its mother rests.*

(c)

(d)

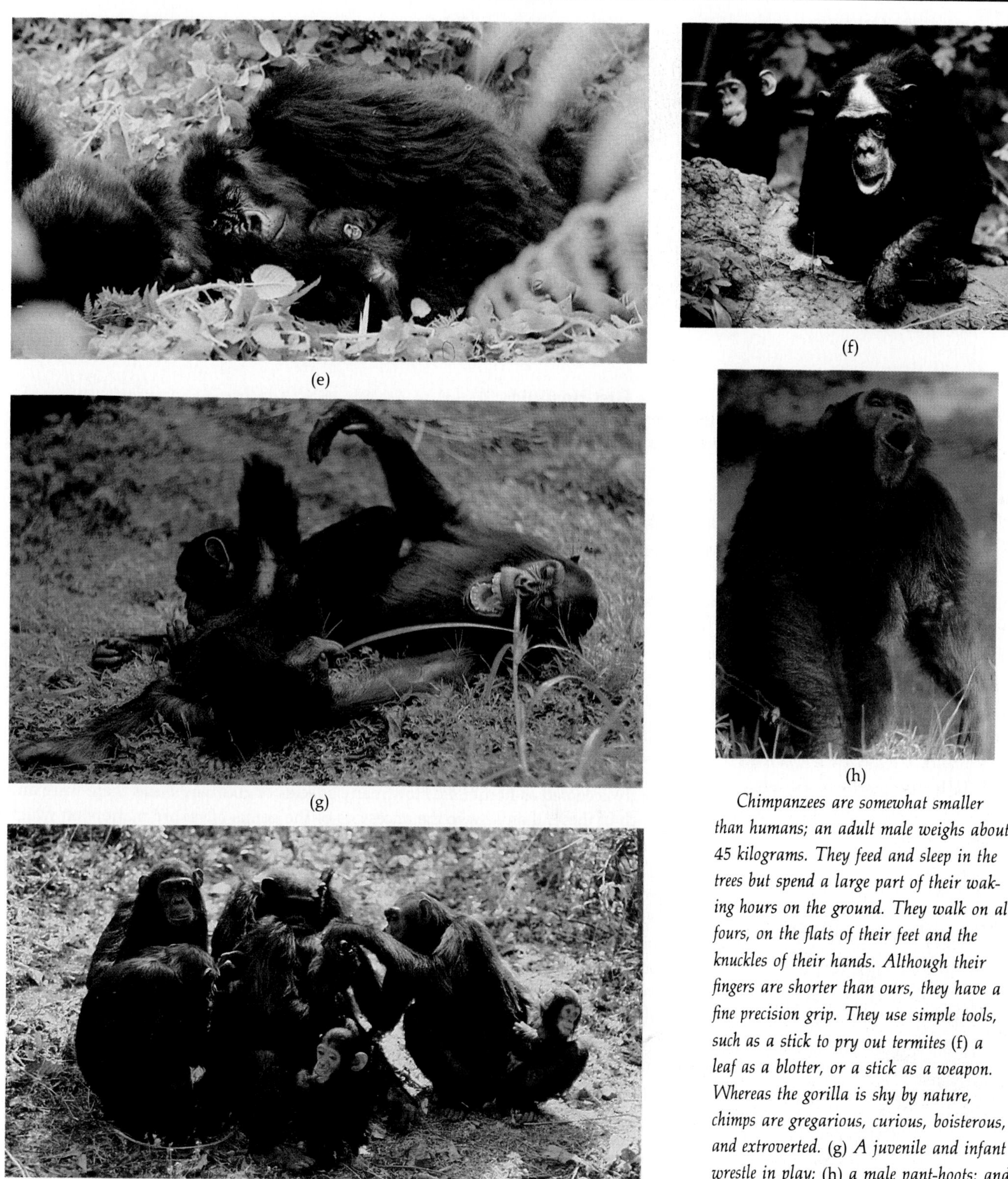

(e) (f) (g) (h) (i)

Chimpanzees are somewhat smaller than humans; an adult male weighs about 45 kilograms. They feed and sleep in the trees but spend a large part of their waking hours on the ground. They walk on all fours, on the flats of their feet and the knuckles of their hands. Although their fingers are shorter than ours, they have a fine precision grip. They use simple tools, such as a stick to pry out termites (f) *a leaf as a blotter, or a stick as a weapon. Whereas the gorilla is shy by nature, chimps are gregarious, curious, boisterous, and extroverted.* (g) *A juvenile and infant wrestle in play;* (h) *a male pant-hoots; and* (i) *a chimpanzee group sits and grooms.*

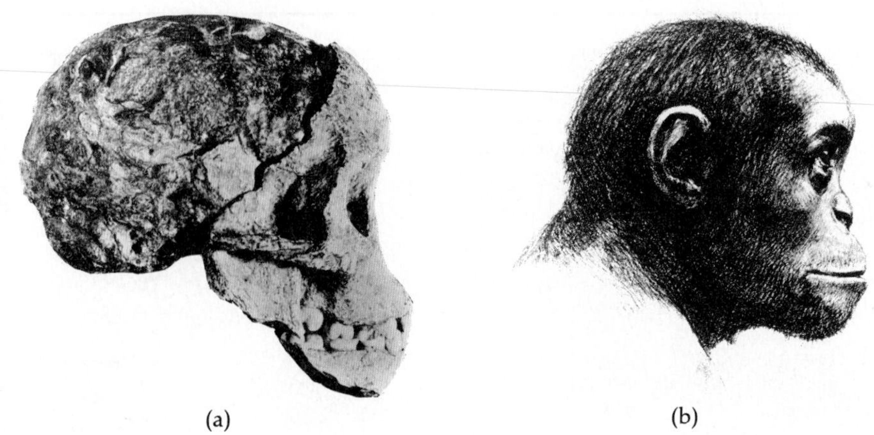

56–11
The skull of a child, found in a limestone quarry in Taung, South Africa, in 1924, was the first specimen discovered of the earliest known hominid, an australopithecine. Shown here are (a) *the skull and* (b) *a reconstruction drawing of the head of the child, which was about five or six years old when it died. The australopithecines were erect-walking creatures with brains somewhat larger proportionately than the gorilla's and with teeth that are more similar to ours than to an ape's.*

The First Hominids

In 1924 an explosion in a quarry that was being mined in South Africa loosened a piece of rock containing the skull of a child. This specimen, along with many others, was crated up and sent to anatomist Raymond Dart in Johannesburg. Dart was quick to see that the little creature had many humanlike features that distinguished it from both modern apes and their ancestors. These included the rounded appearance of the skull, the size and shape of the brain case, and the roundness of the jaw. Also, the point of attachment of the vertebral column to the skull indicated that the young animal was a biped.

Dart reported his find in the British journal *Nature,* naming the new fossil *Australopithecus* ("southern ape"). His announcement was largely ignored. The scientific community had not yet recovered from the effects of the Piltdown forgery.* Besides, it was generally agreed at that time that Europe, with all its cultural advantages, was the cradle of humankind, not darkest Africa.

Subsequent fossil discoveries, however, have confirmed Dart's findings and their interpretation. The australopithecines were bipedal ground walkers and are clearly accepted as hominids. How many species of *Australopithecus* there were and which of these, if any, were the ancestors of the genus *Homo* are matters on which hardly anyone agrees at the present moment.

The oldest australopithecine fossils have been dated at 3.2 to 3.6 million years ago. They belong to the species known as *Australopithecus afarensis,* if you accept Donald Johanson's Lucy (Figure 56–12a) as representing a distinct species, and to *Australopithecus africanus* if you do not. These earliest hominids were small (under four feet tall) and slight (less than 50 pounds). It is clear from various skeletal features, particularly the pelvis, that they walked upright. Their discovery has settled an important issue in human evolution: upright posture evolved long before brain size began any dramatic increase (Figure 56–12b). These early australopithecines seem to have lived in the open, on grasslands and savannas, and to have been generalized feeders. It is not known whether or not they were hunters of big game.

---

* In the Piltdown forgery, the skull of a man and the jaw of an ape, carefully doctored to fit prevailing ideas concerning early hominids, were planted as fossils and, embarrassingly enough, widely accepted by anthropologists.

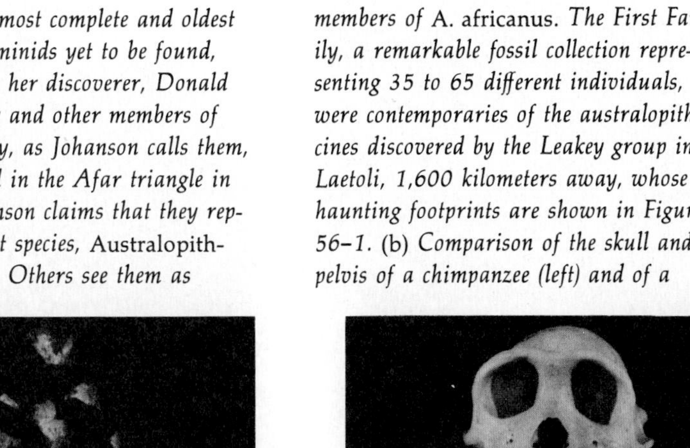

56-12
(a) *One of the most complete and oldest of the early hominids yet to be found, named Lucy by her discoverer, Donald Johanson. Lucy and other members of the First Family, as Johanson calls them, were discovered in the Afar triangle in Ethiopia. Johanson claims that they represent a distinct species,* Australopithecus afarensis. *Others see them as members of* A. africanus. *The First Family, a remarkable fossil collection representing 35 to 65 different individuals, were contemporaries of the australopithecines discovered by the Leakey group in Laetoli, 1,600 kilometers away, whose haunting footprints are shown in Figure 56–1.* (b) *Comparison of the skull and pelvis of a chimpanzee (left) and of a member of the First Family (right). Note that the skulls are very similar but the pelvises are totally different. The pelvis on the right is almost the same as the modern human pelvis. The conclusion: hominids walked fully upright before there was any significant increase in brain size.*

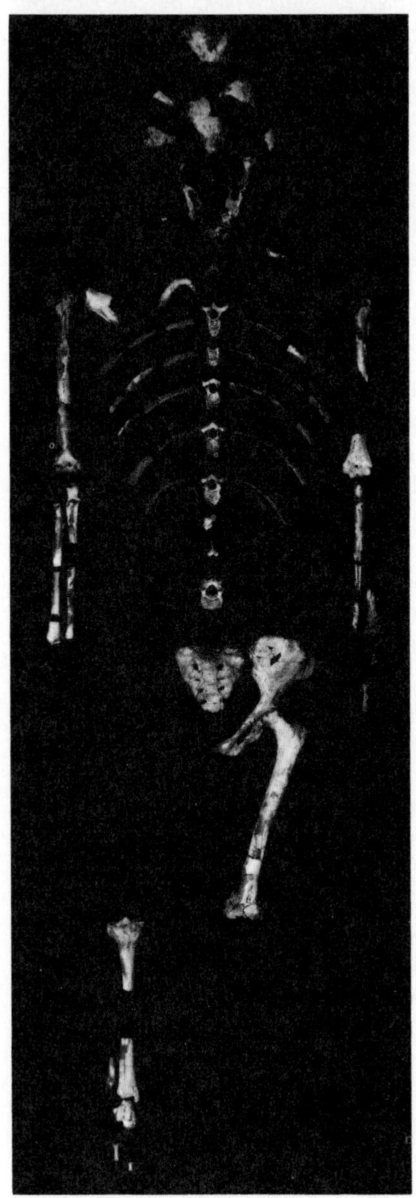

(a)

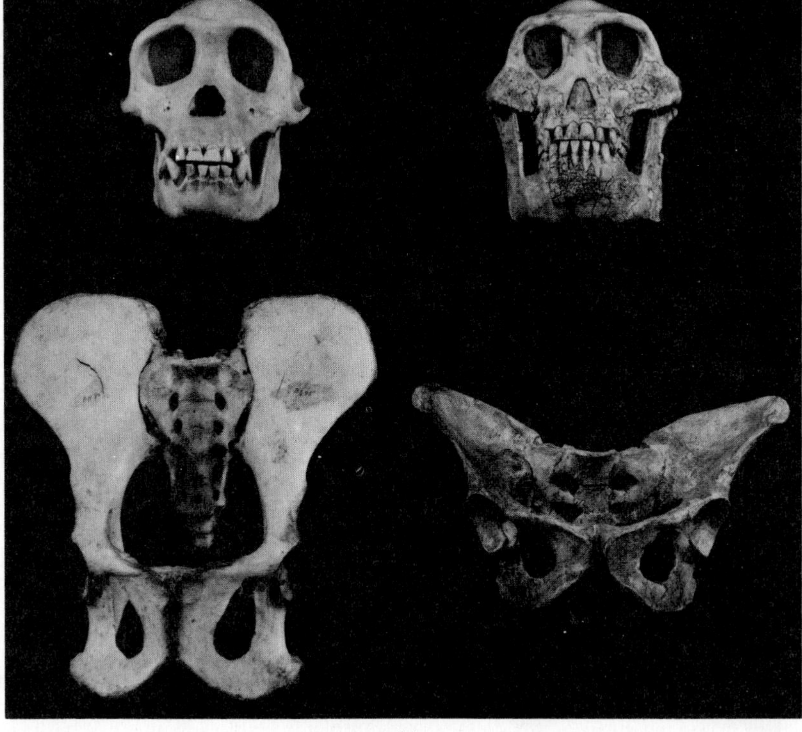

(b)

Australopithecus robustus is distinguished from the *A. africanus* (and/or *A. afarensis*) group by the fact that it appeared later in time (from perhaps 2.3 to 1.3 million years ago), was much more heavily built, and seems to have eaten a vegetarian diet. Almost everyone now agrees that *A. robustus* was not the ancestor of modern humans but became extinct some time ago. *Zinjanthropus,* the so-called "nutcracker man," discovered by the Leakeys in 1959 in Olduvai Gorge, is now generally assigned to this species, although some experts claim that there are two heavily built australopithecines, *A. robustus* and *Australopithecus boisei* (which is even more robust), to which Leakey's nutcracker man belongs. The discovery of *Zinjanthropus,* the culmination of 20 years of exploration and preparation by Louis Leakey, marked the beginning of the great surge of interest in and support for research into human origins.

56–13
Louis and Mary Leakey, the British anthropologists, with one of their sons at Olduvai Gorge in the Serengeti plain of Tanzania. Olduvai, from which many important hominid fossils have been recovered, represents a geologic sequence that is almost continuous from the present to nearly 2 million years ago. Like our Grand Canyon, it was cut by river action. Because it was by the side of a lake, it was visited by many animals and was apparently a popular hominid camping ground. Louis Leakey died in 1975, but Mary Leakey and their son Richard are still leading figures in the search for hominid fossils.

In 1962, Leakey reported an even more stunning discovery. Also from Olduvai and contemporaneous with Zinj, as he was affectionately known, was another hominid. This animal was more lightly built than the robust australopithecines and had a larger brain case. Also, it appeared to have been a tool user. Leakey named it *Homo habilis,* the earliest known member of the genus *Homo.*

Homo habilis is the most controversial of the hominid species. Some paleoanthropologists do not admit the existence of the taxon, saying the fossils belong to the *A. africanus* group. Others place it between *A. africanus* and *Homo,* whereas the *A. afarensis* contingent sees it as the link between that species and modern humans (Figure 56–15).

To sum this all up, there now seems to be no doubt that the earliest hominids, on the basis of the fossil evidence, were of two general types: the heavier built, robust forms and the lighter, gracile forms. Each of these may be represented by more than one species, for a minimum of two and a maximum of five. The robust form was an evolutionary dead end. A gracile form was probably our ancestor.

56–14
Pebble tools such as these have been found in fossil strata at Olduvai and other sites in eastern Africa. From pebbles of lava and quartz, flakes were struck off in two directions at one end, making a somewhat pointed implement, or in a row on one side, making a chopper. The tools measure up to 10 centimeters in length and were probably used to prepare plant food and to butcher game. The flakes were also apparently used as scrapers. The oldest have been dated at between 2.5 and 2.7 million years ago.

How Did It Happen?

What prompted one group of primitive apes to make the first fateful step to bipedalism? We know from pollen specimens that grasslands were spreading and forests contracting during the Miocene. The grasslands may have represented a new adaptive zone. The horse and other grazing animals evolved rapidly during this time, filling the new niches available to herbivores. Baboons also seem to have become terrestrial in this period. We do not know why one group of apes and not another moved into the grasslands. One possible explanation is that this group was driven out of the woodland, then contracting in size, by competition with other primates. Perhaps they possessed some important characteristic that we do not know about.

56–15

The many scenarios of human origins. (a) Until little more than a decade ago, the hominid line was generally considered as a single lineage evolving gradually from Australopithecus *through* Homo erectus *to* Homo sapiens. *(b) Next it became generally accepted that there were two types of australopithecines, a robust and a gracile (slightly built) form, and that the robust form represented an evolutionary dead-end. (c–e) The questions now remain of the status of* H. habilis *and of* A. afarensis. *(f) Punctuated equilibriumists are pleased to point out that the more fossils that are discovered, the more different contemporaneous hominid species are found to have existed. Selection among these species fits the evidence better, they point out, than a gradual phyletic change from one to another.*

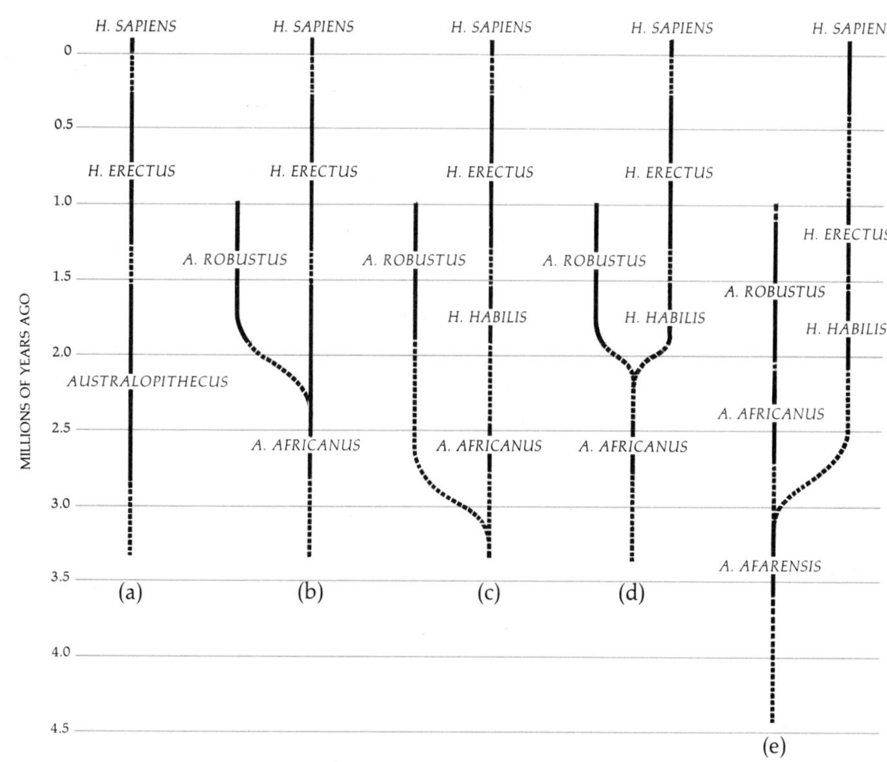

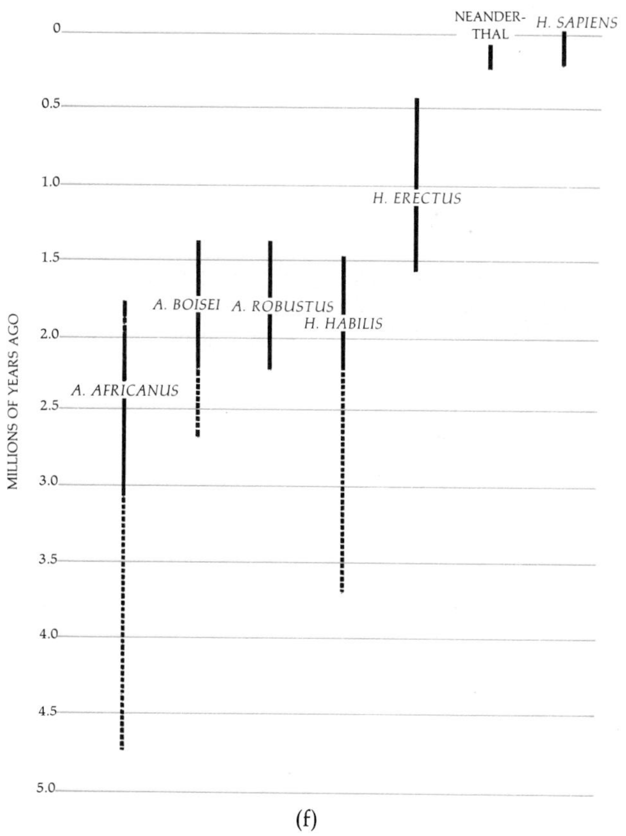

Standing upright on the ground would have greatly increased the range of vision of a ground-dwelling species with highly developed eyesight; other mammals, including bears, squirrels, and prairie dogs, often raise themselves on their two hind legs to look around. Standing upright also, of course, frees the hands, which then can be employed for picking protein-rich grains from wheat and other grasses, for tool making and using, for carrying food, and for brandishing weapons. Use of weapons for defense correlates well with the reduction in the size of the canines among fossil hominids—as compared with baboons, for example, or gorillas, among which the males have large canines that are used in aggressive display. Owen Lovejoy of Kent State University speculates that bipedalism evolved in association with pair-bonding; because the hands were freed for carrying, males could carry food back to females, who were then able to remain with their young in the comparative safety of the home camp.

The advantages of bipedalism must have been very great for natural selection to have operated in its favor, because the first bipedal primates must have moved slowly and inefficiently compared with the swift movement of other large animals, either on the ground or in the trees.

THE EMERGENCE OF *HOMO SAPIENS*

By about 1.5 million years ago, a group of hominids had evolved that are placed, without dispute, in the genus *Homo.* Within this genus are two species, the now-extinct *Homo erectus* and our own species, *Homo sapiens.*

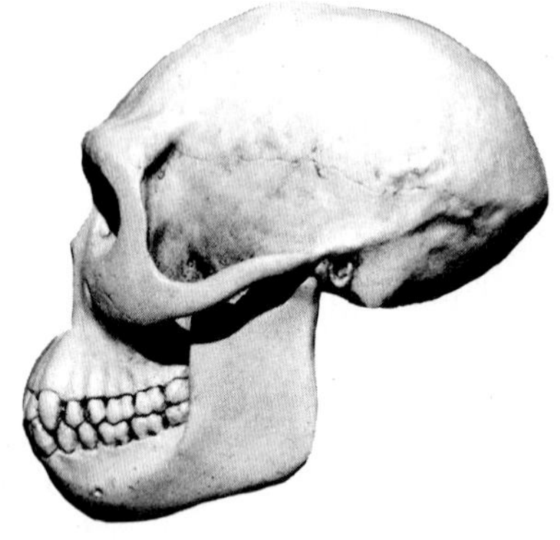

56–16
A reconstruction of the skull of H. erectus *from the Solo River, Java. Much larger than earlier hominids,* H. erectus *had a much larger brain. The skull walls are thick and heavy, and the brow ridges are prominent. The protruding occipital crest (clearly visible in this photograph) was the point of attachment of strong, heavy neck muscles. The jaw is prognathous (protruding) and chinless.*

Homo erectus

Homo erectus was distinctly different from even the most advanced australopithecine. Members of this species had body skeletons much like our own and were about the same size as we are. The bones of their legs indicate that they had a stride similar to our own. The chief differences between *H. erectus* and *H. sapiens* are in the skull. Skull specimens of *H. erectus* are thick and massive, with low foreheads. Their jaws are large and chinless, with large teeth. Brain capacity ranges from 700 to 1,100 cubic centimeters, overlapping that of modern humans.

Also, unlike the australopithecines, which have been found only in Africa, members of *H. erectus* seem to have occupied a much larger range. Fossil specimens have also been found in Java (Java man) and China (Peking man). There is also a jaw from Heidelberg and some skull fragments from Vértesszöllös, Hungary, that are believed to represent this species. The fossils cover a period of about a million years, from about 1.5 million to 400,000 to 300,000 years ago. Along with the skeletal fragments have been uncovered a number of clues about the social life of *H. erectus.*

Hunting

Probably the most significant factor in the evolution of *H. erectus* was that they were hunters. Field studies have shown that baboons and chimpanzees eat meat occasionally and obviously enjoy it (see Figure 55–15 on page 1036), and the gracile australopithecines also probably had a diet that included meat. Members of the species we call *H. erectus* became hunters on a different scale, however. During various stages of the Pleistocene epoch there seem to have been far more grasslands than now exist. Some areas that are now deserts were savannas. In addition, to the north, much land that is now forest was prairie or steppe, which was able to support large numbers of grazing and browsing animals. Judging by the fossil record, the hominids were greatly outnumbered by these large herbivores and so would have had little difficulty finding game. And indeed, from sites as far apart as eastern Africa and China and dating back from 2 million to 350,000 years ago, the bones of very large animals—elephants, rhinoceroses, antelopes, bears, hippopotamuses, and giant baboons—are found in the debris of human campsites.

We do not know how these early hunters killed animals so much larger and more ferocious than themselves. Stone implements found at the sites were apparently made to be held in the hand and were used for chopping, cutting, or pounding. These tools therefore appear to have been used for butchering or preparing food rather than for killing game. There is some evidence that these early hunters killed animals by stampeding them into marshes or over cliffs, perhaps with the help of fire to frighten them. American Indians slaughtered game in this way until very recent times.

Certainly hundreds of thousands of years passed before humans developed weapons as sophisticated as spears with stone or metal tips or bows and arrows. Anthropologists who have studied primitive tribes have been impressed by the fact that, in general, their weapons are not efficient and that hunters take little interest in perfecting them. These modern hunters emphasize patience, endurance, skill in stalking, and highly specialized knowledge of animals and terrain, and perhaps prehistoric hunters did also. Clearly the capacity to cooperate with other members of the hunting party would have been of extreme importance. It is not known whether or not *H. erectus* was capable of speech, but it would have been of great advantage to him ("Head off the hippopotamus!").

Tools

In the hands of *H. erectus,* the pebble tool (see Figure 56–14) became a new and highly distinctive implement, the hand ax (Figure 56–17). Hand axes were used extensively throughout Africa, India, and much of the Near East and Europe; tens of thousands have been found in the valleys of the Somme and Thames alone. All these hand axes closely resemble one another. Unlike the pebble tools, they were clearly fashioned according to a formal pattern. Thus we see the emergence of a clear cultural tradition, with skills and learning passed from one generation to another. Also, the wide distribution of the hand ax indicates communication and exchange among the groups of humans roaming over these vast territories at that time.

56–17
The hand ax is a stone that has been worked on all its surfaces to provide what appears to be a gripping surface and various combinations of cutting edges, sometimes with a more or less sharp point. Making such an implement requires both skill and time, as reported by anthropologists who have tried. Hand axes came into use about 1 million years ago and are associated with H. erectus.

Fire

The hominids of Laetoli and the Afar triangle would have been familiar with fire. The seasonally dry grasslands in which they lived resembled, by all fossil evidence, the African savanna of today, in which brush fires caused by lightning are common. Also these sites of early hominid activity (and Olduvai as well) are all within the Rift Valley, a long geologic fault that runs from the Afar triangle southward through Tanzania. The valley at this time was undergoing violent volcanic activity (hence the preservation of the Laetoli footprints), which would have been another major and dangerous source of fire. Probably fire was first captured from sources such as these. The earliest evidence of the use of fire has been reported from a Rift Valley site in Kenya and dated at 1.4 million years ago. Late robust australopithecine fossils and early *H. erectus* fossils have been found nearby. It is assumed that if the fire was under hominid control, *H. erectus* was the fire user.

No other evidence of the human use of fire older than about 500,000 years has been found. One of the oldest sites is in the Choukoutien caves, near Peking, China, which were inhabited by *H. erectus* (Peking man) 350,000 years ago.

The possession of fire would have greatly changed the life of early hominids. It would have greatly extended the range of their diet, not only making meat easier to chew but also, and perhaps more important, making it possible to eat plant parts that, uncooked, would have been too tough, too bitter, or too toxic. The smaller teeth and jaws of *H. erectus,* as compared to the australopithecines, may be associated with the use of fire.

Fire also would have greatly improved our ancestors' social life, providing a center for daily activities and a reason for bringing back game and other food to a home base.

Dwelling Places

All higher primates, some modern humans excepted, are continuously on the move, and though they range back and forth over the same territory, they may sleep in a different place every night. Any member of the group that cannot keep up may be left behind. Early hominids were probably also nomads, as are most modern hunter-gatherers, but there is evidence that at least seasonally they maintained base camps. Some of these camps appear to have been fairly permanent, in the sense that they were returned to year after year.

The australopithecines and the earliest humans probably camped mostly in the open. The campsites found in the Rift Valley are near streams or lakes, which would not only have provided water for the campers but would also have attracted game.

Once hominids controlled fire, they were able to occupy caves that otherwise, occupied by cave bears and saber-toothed tigers, would have been too dangerous for habitation. The Choukoutien caves are among the earliest caves known to have been used for human habitation. The caves near Shanidar (Figure 56–18) in the remote mountains of Iraq have been occupied continuously—to the present day—for more than 100,000 years.

56–18
This large cave near the little village of Shanidar in northern Iraq has been continuously inhabited for more than 100,000 years. Nine Neanderthal skeletons have been found in the cave, including one who was, according to analysis of fossil pollen, buried on a bed of woody branches and June flowers gathered from the hillside.

56–19

Increase in brain volume in the course of hominid evolution. Some of the increase can be correlated with the increase in body size that was also taking place during this period, but most of it is believed to represent the results of strong selection pressures for intelligence.

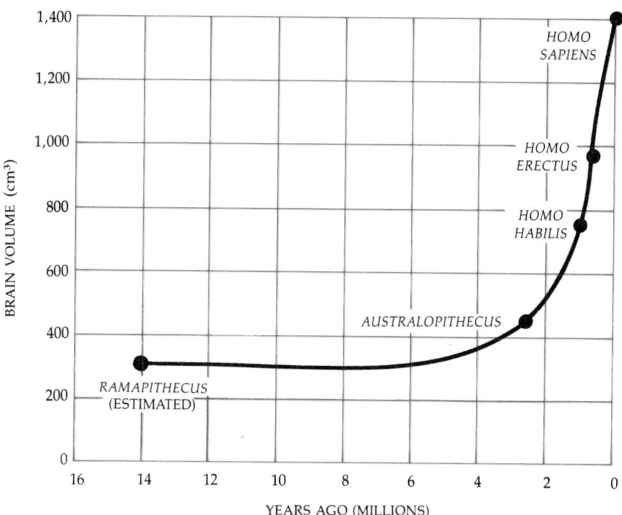

Increase in Intelligence

In mammals, the ratio of brain size to body size can be correlated with the intelligence of the species under consideration. It is therefore reasonable to assume that the enormous increase in cranial capacity that took place in the course of human evolution signifies strong selection pressures for intelligence. In fact, insofar as one can compare the evolution of the human brain with the evolution, for instance, of the foreleg of the horse (which is considered very rapid), the human brain evolved 100 times more quickly. The increase in cranial capacity (Figure 56–19) took place well after bipedalism and the early use of tools and well before the development of sophisticated tools or weapons, which did not take place until a mere 30,000 to 40,000 years ago.

Homo sapiens

The earliest fossils that are classified as *H. sapiens* come from Swanscombe in England and Steinheim in Germany. They consist only of some skull fragments, but these clearly indicate that the brains of these hominids were larger than those of *H. erectus* and that their skulls were less massive. These fossils are dated at about 400,000 to 200,000 years ago.

The Neanderthals

The period from about 150,000 years ago to about 35,000 years ago abounds with specimens of what we have come to call the Neanderthals. They have been found largely in Europe but also in the Near East and Central Asia. The Neanderthals stood as erect as we do but were more heavily built and more muscular. They had a brain capacity as large as ours, a long, low, massive skull, a prognathous (protruding) face, a low forehead, and heavy brow ridges. They are now usually classified as a variety of *H. sapiens.*

Neanderthals used hand-held stone tools that were somewhat more sophisticated than those of *H. erectus*. Some of the stone tools appear to have been used for scraping hides, suggesting that Neanderthals wore clothing made of animal skins, which would certainly have been in keeping with the climate in which they lived.

Neanderthals buried their dead, often with food and weapons and, in at least one instance, with spring flowers. Formal burials such as these suggest a belief in life after death. Whether or not Neanderthals were able to speak is a matter of current debate, but it is difficult to understand how a society could hold such abstract concepts without some means of exchanging ideas among its members.

The Cro-Magnons

About 34,000 years ago, specimens of the Neanderthals disappear abruptly from the fossil record and are replaced by what are known as the Cro-Magnons, who are physically indistinguishable from us. We do not know what became of the Neanderthals. Perhaps they were exterminated in warfare, although there is no evidence of this. Perhaps they were simply unable to compete for food and living space with the better-equipped Cro-Magnon type. Perhaps they interbred, although there are no clear traces of intermediate forms. Some have suggested that modern people brought with them some disease to which they themselves were resistant and the Neanderthals were not. In any case, soon after the appearance of the Cro-Magnons, there were no other hominids in Europe, and within the course of 10,000 to 20,000 years, this new variety of primate had spread over the face of the planet.

The Cro-Magnons, when they first appeared in Europe, came bearing a new, quite different, and far better tool kit (Figure 56–20). Their stone tools were essentially flakes—which, of course, had been in use for more than 2.5 million years—but they were struck from a carefully prepared core with the aid of a punch (a tool made to make another tool). These flakes, usually referred to as blades, were smaller, flatter, and narrower, and, most important, they could be and were shaped in a large variety of ways. They included, from the beginning, various scraping and piercing tools, flat-backed knives, awls, chisels, and a number of different engraving tools. Using these tools to work other materials, especially bone and ivory, Cro-Magnons made a variety of projectile points, barbed points for spears and harpoons, fishing hooks, and needles. Thus, although the Cro-Magnons lived much the same sort of existence as that of their forebears, they apparently lived it with more possessions, more comfort, and more style.

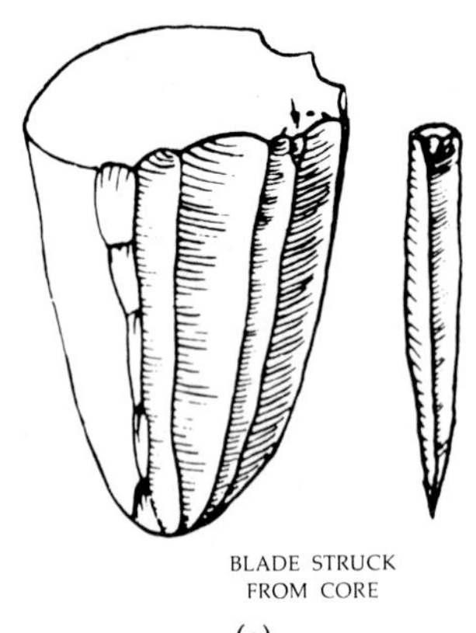

(a)

(b)

56–20
(a) *Cro-Magnon culture was characterized by a great increase in the types of specialized tools made from long and relatively thin flakes with parallel sides, called blades.* (b) *These beautifully worked "laurel leaf" blades are often more than 30 centimeters long and only 0.5 centimeter thick.*

Perhaps our closest emotional links to this most immediate ancestor are the cave paintings of western Spain and southern France (Figure 56–21). The examples that remain to us, many surprisingly untouched by time, clearly form part of a rich artistic tradition that endured for at least 20,000 years. The cave drawings are almost entirely of animals, nearly all game animals, and they are deep within the caves, so they must have been viewed (as they must have been painted) by the light of crude lamps or torches.

The meaning of these drawings and paintings has long been a matter of debate. Some of the animals are marked with darts or wounds (although very few appear to be seriously injured or dying). Such markings have led to the suggestion that the figures are examples of sympathetic magic, in which there is the notion that one can exert control over another creature by taking symbolic action against its image. The fact that many of the animals appear to be pregnant suggests that they may symbolize fertility. Many appear also to be in motion. Perhaps these animals, so vital to the hunters' welfare, were migratory in these areas, and they may have seemed to vanish at certain times in the year, mysteriously returning, heavy with young, in the springtime. This return of the animals might have been an event to be solicited or celebrated in much the same spirit as the rites of spring or Easter are celebrated by more recent peoples.

56–21
In almost every cave, a small number of the animals show wounds, as seen in this steppe wisent from the Cave of Niaux in France. Although the animals are shown in realistic detail, the weapons and the hunters, if shown at all, are abstractions, perhaps to keep their identity a secret. The illusion of movement is greatly enhanced by the patterns of light and shadow in the dark, narrow recesses of the cave.

Paleolithic cave art came to an end perhaps 8,000 or 10,000 years ago. Not only were the tools and pigments laid aside but the sacred places—for such they seem to have been—were no longer visited. New forces were at work to mold the course of human existence, and the end of this first great era in human art is our clearest line of demarcation between the old life of *H. sapiens* the hunter and the new life that was to come.

The Agricultural Revolution

The most important event in the cultural evolution of the human species was the change to an agricultural way of life. The reasons for such a transition are not clear, but one factor seems to have been a change in the climate. The most recent of the

56–22
A rock painting from Spain. Such paintings date from postglacial times, about 8000 to 3000 B.C. In contrast to the earlier cave paintings, paintings such as this were executed on sheltered rock surfaces, frequently show human figures, and are small in scale. These paintings typically show forest animals rather than the great plains animals that stalk the caves of Lascaux, France, and Altamira, Spain. The deer are apparently being driven toward the archers by beaters. The hunting of small, nonmigratory animals such as these is believed to mark the transition between the nomadic, hunting-gathering existence of Homo sapiens *and the agricultural revolution.*

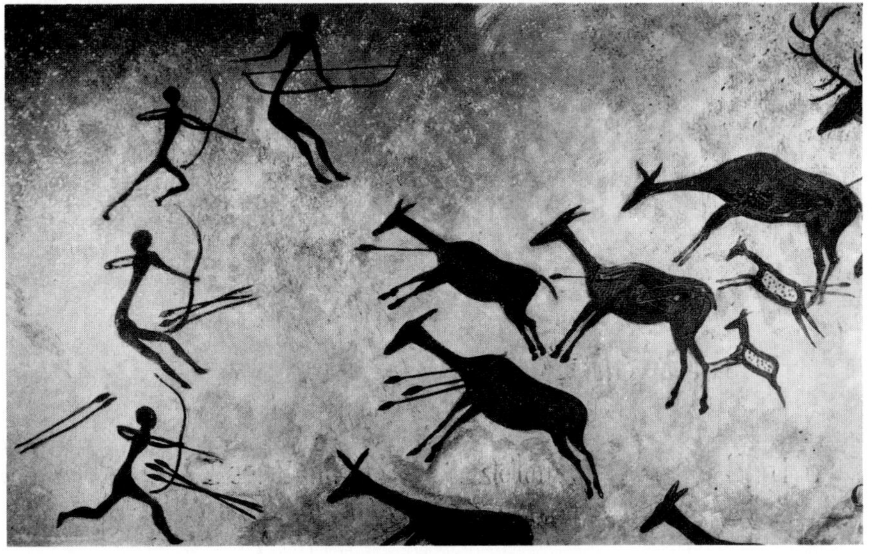

glaciations began to retreat about 18,000 years ago, withdrawing slowly for about 10,000 years. As the glaciers retreated, the plains of northern Europe and of North America, once cold grasslands, or steppes, gave way to forest. Many of the great herbivores that roamed these steppes retreated northward and eventually vanished; the woolly mammoth was last seen in Siberia more than 10,000 years ago. While some animals became extinct, humans adapted, as they had during the entire period of violent climatic changes that have marked their evolutionary history. With the migratory animals gone, human hunters shifted their attention to smaller game. The hunting and gathering of small animals—instead of large migratory herbivores—undoubtedly resulted in a less nomadic existence for the hunters and formed a prelude to the agricultural revolution. (However, it should be noted that similar fluctuations had occurred previously in hominid history without leading to a revolution in the way of life.)

The earliest traces of agriculture, dating back about 11,000 years, are found in areas of the Near East, which are now parts of Iran, Iraq, and Turkey. Here were the raw materials required by an agricultural economy: cereals, which are grasses with seeds capable of being stored for long periods without serious deterioration, and herbivorous herd animals, which can be readily domesticated. The grasses in these areas were wild wheats and barley, which still grow wild in the foothills. The animals were wild sheep and goats.

By 8,100 years ago, agricultural communities were established in eastern Europe. By 7,000 years ago (about 5000 B.C.), agriculture had spread to the western Mediterranean and up the Danube River into central Europe and, by 4000 B.C., to Britain. During this same period, agriculture originated separately in Central and South America, and perhaps slightly later in the Far East.

THE POPULATION EXPLOSION

One immediate and direct consequence of the agricultural revolution was an increase in population. It is estimated that about 25,000 years ago there may have been as many as 3 million people. By the close of the Pleistocene epoch, some 10,000 years ago, the human population probably numbered a little more than 5 million, spread over the entire world. By 4000 B.C., about 6,000 years ago, the population had increased enormously, to about 90 million, and by the time of Christ, it is estimated, there were 130 million people. In other words, the population increased more than 25 times between 10,000 and 2,000 years ago.

By 1650, the world population had reached 500 million, many people lived in urban centers, and the development of science, technology, and industrialization had begun, bringing about further profound changes in human life and its relationship to nature.

In 1982, there were more than 4.5 billion people on our planet, and the number is increasing at a rapid rate. About 141 people are added to the world population every minute, about 200,000 each day, and 74 million every year. China, after conducting the largest census in history, announced that it had a population of more than 1 billion, almost one-fourth of the world's total. Its population had increased 313 million since 1964. China has made progress with birth-control programs, according to this report, and its population is growing by 1.4 percent a year. The birth rate in the United States plunged dramatically from 18.4 babies born per 1,000 people in the population in 1970, to 14.8 in 1975, about the same as

THE ICE AGES

During most of our planet's history, its climate appears to have been warmer than it is at the present time. However, these long periods of milder temperatures have been interrupted periodically by Ice Ages, so called because they are characterized by glaciations, or persistent accumulations of ice and snow. Such glaciations occur whenever the summers are not hot enough and long enough to melt ice that has accumulated during the winter. In many parts of the world, an alteration of only a few degrees in temperature is enough to begin or end a glaciation.

An early Ice Age appears to have occurred at the beginning of the Paleozoic era, some 600 million years ago. Another, marked by extensive glaciations in the Southern Hemisphere, closed the Paleozoic, some 250 million years ago. The conifers evolved during this period and possibly the angiosperms, as older forest types disappeared. A more recent, less severe cold, dry period occurred at the end of the Mesozoic, about 65 million years ago, and was perhaps a principal cause of the extinction of the dinosaurs.

The most recent Ice Age began during the Pleistocene epoch, about 1.5 million years ago. The Pleistocene has been marked by four extensive glaciations that have covered large areas of North America, England, and northern Europe. Between the glaciations there have been intervals, called interglacials, during which the climate has become warmer. In each of these four Pleistocene glaciations, sheets of ice, thicker than 3 kilometers in some regions, spread out locally from the poles, scraped their way over much of the continents—reaching as far south as southern Illinois in North America and covering Scandinavia, most of Great Britain, northern Germany, and northern Russia—and then receded again. We are living at the end of the fourth glaciation, which began its retreat only some 10,000 years ago.

Glaciers are accumulations of snow and ice that flow across a land surface as a result of their own weight. This glacier is flowing into the sea. As the glacier moves forward, its edges melt; the rocks carried within it have become concentrated in the dark band at the right. This is part of the Greenland ice sheet, which has an area of about 1,726,000 square kilometers and at some places is more than 3 kilometers thick. Glaciers covered much of North America and Europe during the period of the evolution of the genus Homo.

During these periods of violent climatic changes, the fossil record shows that the populations of these regions were under extraordinary evolutionary pressures. Plant and animal populations moved, changed, or became extinct. In the interglacial periods, during which the average temperatures were at times warmer than those of today, the tropical forests and their inhabitants spread up through today's temperate zones. During the periods of glaciations, only animals of the northern tundra could survive in these same locations. Rhinoceroses, great herds of horses, large bears, and lions roamed Europe in the interglacial periods. In North America, as the fossil record shows, there were camels and horses, saber-toothed cats, and great ground sloths, one species as large as an elephant. In the colder periods, reindeer ranged as far south as southern France, while during the warmer periods, the hippopotamus reached England. It was during this most recent period of glaciations and interglacials that hominids of the genus Homo *evolved.*

The reason for these large changes in temperature is one of the most controversial issues in modern science. They have been variously ascribed to changes in the earth's orbit, variations in the earth's angle of inclination toward the sun, migration of the magnetic poles, fluctuations in the solar energy, migration of the continents, higher elevations of the continental masses, continental drift, and combinations of these and other causes. Also under debate is the question of whether this period of glaciations is over—perhaps for another 200 million years—or whether we are merely enjoying a brief interglacial before the ice sheets begin to creep toward the equator once more.

China's reported rate, and has currently risen to about 16 births per 1,000. The world's population is growing rapidly—at about 1.7 percent per year. If this rate of increase is sustained, there will be close to 6 billion people on earth by the year 2000.

Life vs. Death

The size of a population is, of course, the result of both its birth rate and its death rate, and the enormous recent increase in the world's population is actually more a product of the decline in the death rates, especially among the young, than an increase in births. For example, in Pakistan, the population was 20 million in 1911, 34 million in 1950, and 77 million in 1978. During this period, birth rates seem to have remained the same—about 45 per 1,000—but the death rate is two-thirds of what it was only 20 years ago.

Population growth also depends on the age of the population. In Pakistan, again, almost half of the population is under 15, so if the ideal of a two-child family were to be obtained in the near future, Pakistan's population could still double in the next half century. (The effects of population structure on population growth are discussed on pages 954–955.)

The Rich and the Poor

The unequal distribution of growth makes the problem more difficult. First, births are occurring at the greatest rates in precisely those areas where the new arrivals have the least chance of an adequate diet, good housing, schools, medical care, or future occupations. Moreover, because the affluent citizens of developed countries are not constantly reminded of the soaring population rate by the problems of hunger and crowding, so evident to the poor of other countries, they may feel that it is not their problem. Yet a child born to a middle-class American will consume, in his or her lifetime, a far greater amount of the limited resources of the world—more than twice the amount of food, for instance—than a child born in a less developed country. In the words of Jean Mayer, chairman of the National Council on Hunger and Malnutrition in the United States:

> Rich people occupy much more space, consume more of each natural resource, disturb the ecology more, and create more land, air, water, chemical, thermal and radioactive pollution than poor people. So it can be argued that from many viewpoints it is even more urgent to control the number of the rich than it is to control the numbers of the poor.

Birth Rates, Death Rates, and Social Security

Despite the fact that the most recent phase of the population increase is related to a reduction in death rates, particularly from infectious diseases, some experts believe that further reductions in death rates and a general increase in the standard of living will reduce the rate of population growth. There seems to be a correlation between economic deprivation and high birth rates. It is in the underdeveloped countries that the greatest rates of population growth are found. In India, for example, where large numbers of people are now starving, the annual rate of population increase is 2 percent. Yet, most Indian women do not seek help in birth control until they have three or four children. The desire for large families is deeply rooted in the Indian culture. Moreover, in Indian tradition, children provide security for their parents in old age. Two recent studies of life expectancy in India show that, with the high death rate among children, it is necessary for a mother to bear five children if the couple is to be 95 percent certain that one son will survive the father's sixty-fifth birthday.

Barry Commoner, of Washington University in St. Louis, is of the opinion that population pressures can be significantly reduced by reducing death rates, particularly among infants. Wherever income rises and death rates drop below a critical point, birth rates begin to fall too, he points out. This was true historically in industrialized nations, and it is becoming true in some countries where living standards are just beginning to rise. However, for many countries, a significant rise in the standard of living seems an almost unattainable goal and, in some cases, a rapidly receding one.

It is clear that a reduction in birth rate is not directly correlated with the availability of birth-control measures. United States birth rates are comparatively low, 16 per 1,000, but they were equally low in the 1930s during the Depression. The Caribbean island of Jamaica has a Family Planning Center in almost every small town, dispensing pills and contraceptive devices at very low cost, yet the current birth rate is 27 per 1,000 population.

56–23
Field workers weeding a rice plot at the International Rice Research Institute in the Philippines. The new strains of rice are dwarf varieties with short, stiff stalks that respond well to heavy fertilization. Older, nondwarf strains become too tall when heavily fertilized and tend to fall over, spilling the grain.

OUR HUNGRY PLANET

The consequences of the increase in the human population include the problems of pollution, depletion of fossil fuels, destruction of natural resources, and extinction of other species. By far the most difficult problem and also the most urgent is hunger and starvation. Of the world's 4.5 billion people, at least 1 billion are inadequately nourished. It is estimated that about one-third of the deaths that now occur worldwide are due directly or indirectly to malnutrition. Perhaps even more important, in terms of the world's future, are the effects, both physical and psychological, of the prolonged chronic hunger of so large a proportion of our population.

The effort to increase agriculture by the development of new crop plants—especially grains—is called the Green Revolution. Enormous progress is being made. From 1950 to 1970, the production of wheat in Mexico increased from 270,000 metric tons per year to 2.35 million; the corn harvest increased a more modest 250 percent, with yields per hectare almost doubling. Between 1950 and the present, India has increased its production of food grains about 2.8 percent a year. (Its population during this period has increased about 2.1 percent a year, which is significantly less.) Since 1971, China, the most populous nation in the world, has become agriculturally self-sustaining. Most of this success has come about as a result of improved varieties of crop plants, combined with better techniques of irrigation and fertilization. Moreover, and perhaps most important, the full potential of the Green Revolution has not yet been realized.

Despite its acknowledged success, this massive effort has come under criticism in recent years. One reason is the increasing cost of fertilizer; these new grains require intensive cultivation. Because the large landowners are able to afford the investment in fertilizer and farming equipment that the small-scale farmers cannot, these new agricultural developments are seen as accelerating the consolidation of farm lands into a few large holdings by the very wealthy. Most serious of all, although food production is still outstripping population growth, the Green Revolution, at its most productive, will not be able to keep pace indefinitely with the rapid growth of the world's population. (If the population had remained at its 1950 levels, there would be enough food to feed everyone on the planet adequately today.)

Finally, there is a more fundamental though more elusive reason for the dissatisfaction with the Green Revolution. When it was first introduced, it appeared to many to be an almost magical solution to problems so enormous and distressing that they had seemed insoluble. It is now clear, however, that poverty and famine and the unrest and violence they may bring will not be solved by a "technological fix." The Green Revolution must, of course, go forward. At the same time, we must recognize that the broader solutions are social, political, and ethical, involving not only the growth of crops but their distribution, not only the limiting of populations but the raising of living standards of these populations to tolerable levels.

SCIENCE AND HUMAN VALUES

In the Introduction, we drew an analogy between the goals of science and those of art, religion, or philosophy. There is an important difference, however. The raw materials of science are our observations of the phenomena of the natural universe. Science is limited to what is observable and measurable and, in this sense, is rightly categorized as materialistic. Hunches are abandoned, hypotheses superseded, theories shattered, but the observations endure, and, moreover, they are used over and over again, sometimes in wholly new ways. It is for this reason that scientists stress and seek objectivity. (In the arts, by contrast, the emphasis is on subjectivity—experience as filtered through the individual consciousness.)

Because of this emphasis on objectivity, value judgments cannot be made in science in the way that such judgments are made in philosophy, religion, and the arts, and indeed in our daily lives. Whether or not something is good or beautiful or right in a moral sense, for example, cannot be determined by the scientific method. Such judgments, even though they may be supported by a broad consensus, are not subject to testing by the scientific method.

At one time, sciences, like the arts, were pursued for their own sake, for pleasure and excitement and satisfaction of the insatiable curiosity with which we are both cursed and blessed. In the twentieth century, however, the sciences have spawned a host of giant technological achievements—the H-bomb, the Salk vaccine, DDT, indestructible plastics, nuclear energy plants, perhaps even ways to manipulate our genetic heritage—but have not given us any clues about how to use them wisely. Moreover, science, as a result of these very achievements, appears enormously powerful. It is thus little wonder that among members of the present generation, as perhaps not in any previous one, there are many individuals who are angry at science, as one would be angry at an omnipotent authority who apparently has the power to grant one's wishes but who refuses to do so.

The reason that science cannot and does not solve the problems we want it to is inherent in its nature. Most of the problems we now confront can be solved only by value judgments. For example, science gave us nuclear weapons and also some predictions as to the extent of the biological damage that their use might cause. Yet it could not help us, as citizens, in weighing the risk of damage from fallout against the desire for a strong national defense. In fact, in the 1950s, the most outspoken advocate of atomic testing was one of the country's most prominent physicists, as was the most ardent opponent.

Science has produced the information for the construction of the artificial kidney and now of the artificial heart and can explain how they work. Yet the application of scientific methods cannot help us decide who, given limited availability of such complex and expensive machines, should be treated by them and who should be abandoned. Similarly, science can predict the possible extent of damage to a particular area from the use of pesticides, but it cannot weigh this damage against the reduction of food crops or the increase in malaria that would occur were they prohibited.

At one time, it was traditional for scientists to abstain from such political and ethical controversies, maintaining quite rightly that scientists have no special qualifications in such matters. "Living in an ivory tower" became a well-worn phrase. Now there is an increasing tendency for scientists as citizens to take public stands, particularly on matters involving the applications of science. For example, a large number of distinguished scientists, including several Nobel laureates, recently issued a public statement questioning the safety of nuclear reactors. Also, in recent years, a small group of molecular geneticists took the unprecedented move of calling for a general moratorium on the research in which they themselves were involved.

It is one of the ironies of this so-called "age of science and materialism" that probably never before have ordinary individual men and women, including scientists, been confronted with so many moral and ethical problems.

In the age of the dinosaurs, the earliest primates survived, it would appear, largely by their wits; now, if we are to survive the monsters of our own creating, we will have to do it, again, by the contents of our own skulls. For within the human mind—that complex collection of neurons and synapses—resides the uniquely human capacity to accumulate knowledge, to plan with foresight, and so to act with enlightened self-interest and even, on occasion, with compassion.

SUMMARY

The first mammals arose from primitive reptilian stock about 200 million years ago and coexisted with the dinosaurs for 130 million years. The extinction of the dinosaurs was followed by a rapid adaptive radiation of the mammals. The primates are an order of mammals that became adapted to arboreal life. Primates are characterized by five-digited extremities adapted for grasping, nails rather than claws, and freely movable limbs. They are dependent more upon vision than upon smell, and the higher primates all have stereoscopic vision with foveas for fine focus and cones for color vision.

The two principal groups of living primates are the prosimians and the anthropoids. Prosimians were widespread and abundant during the Paleocene and Eocene epochs, some 65 to 38 million years ago. Modern prosimians include lorises, bush babies, lemurs, and tarsiers.

The anthropoids include the New World monkeys, the Old World monkeys, and the hominoids (apes and humans). Early in hominoid evolution, in the Miocene epoch, the hominoids diverged. The ramapithecines, known only from some jaw fragments dated 12 to 14 million years ago, are believed by some paleoanthropologists to be ancestral to humans. They have a broader dental arch and teeth adapted for powerful chewing, and evidence suggests that *Ramapithecus* was bipedal. However, the biochemical evidence of the molecular clocks indicates a more recent divergence between hominids and other hominoids.

From 3.8 million to 1 million years ago, groups of hominids lived that walked erect and, some of whom at least, used simple tools. At least two species are now generally recognized: *Australopithecus africanus* and *Australopithecus robustus.* Three more are proposed: *Australopithecus afarensis, Australopithecus boisei,* and *Homo habilis.*

Homo erectus lived about 1.5 million to 300,000 years ago. Individuals of the species were successful hunters of very large animals, indicating that they lived in fairly large groups, shared food, and worked together cooperatively. The hand ax is associated with *H. erectus,* and its widespread distribution indicates cultural exchanges between groups. Some groups lived in caves and had fire, two developments that are probably related.

The Neanderthals are generally considered a variety of *Homo sapiens.* Neanderthal fossils date from about 150,000 to 35,000 years ago, the period of the last glacial advance. The majority of specimens have been found in Europe. Neanderthals had fire, inhabited caves, hunted large animals, and probably wore clothing. They used stone tools of a characteristic type. They buried their dead, sometimes with food and weapons. Neanderthals disappeared about 34,000 years ago.

The Cro-Magnons, who replaced the Neanderthals, were nomads and hunters, like their predecessors. They had tools (blade tools) distinctly different, smaller, and far more varied than had been seen previously. Their art, which covers a period of some 20,000 years, is evidence of a culture rich not only in beauty but also in mystery, magic, and tradition.

The most important event in the cultural evolution of the human species was the advent of agriculture, more than 10,000 years ago.

About 8000 B.C., when agriculture had its beginning, there were probably about 5 million people in the world. By 4000 B.C., the population had increased to about 90 million; by the time of Christ, there were probably around 130 million people in the world; and by 1650, there were about 500 million. In 1980, there were 4.5 billion, and the population continues to increase, especially in the underdeveloped countries.

QUESTIONS

1. Distinguish among the following: primate/prosimian; monkey/ape; hominid/hominoid/anthropoid.

2. If you were to meet a gathering of the following, how would you distinguish one from the other: *Homo erectus?* Cro-Magnon? Peking Man? Neanderthal? *Ramapithecus?* Lucy? *Zinjanthropus? Australopithecus robustus? Australopithecus africanus?*

3. Name five evolutionary trends among primates, and discuss the probable selective value for each.

4. What song was playing on the tape deck in the camp at Afar when the First Lady of the First Family was discovered?

5. In an editorial entitled "Food, Overpopulation, and Irresponsibility" that appeared recently in a scientific journal, the authors concluded: "Because it creates a vicious cycle that compounds human suffering at a high rate, the provision of food to the malnourished nations of the world that cannot, or will not, take very substantial measures to control their own reproductive rates is inhuman, immoral, and irresponsible." What is your opinion?

SUGGESTIONS FOR FURTHER READING

Ecology

Books

CODY, MARTIN L., and JARED M. DIAMOND (eds.): *Ecology and Evolution of Communities,* Belknap Press, Cambridge, Mass., 1975.

A collection of papers, for the more serious, mathematically inclined student.

COLINVAUX, PAUL: *Introduction to Ecology,* John Wiley & Sons, Inc., New York, 1973.

A good, general text by a teacher who clearly enjoys his subject, designed primarily for second-year students and science majors. It is notable for its stress on and criticism of the concepts of ecology.

COLINVAUX, PAUL: *Why Big Fierce Animals Are Rare,* Princeton University Press, Princeton, N.J., 1978.*

A series of essays for the general reader.

EHRLICH, PAUL R., ANNE H. EHRLICH, and JOHN P. HOLDREN: *Ecoscience: Population, Resources, Environment,* W. H. Freeman and Company, San Francisco, 1977.*

Required, although not cheerful, reading for all concerned with the biological, economic, and social consequences of human population growth. Topics covered include basic geology and ecology, human population dynamics, food supplies, energy and material resources, environmental disruption, and approaches to solutions for the many problems confronting us. The authors present a wealth of useful data, even though you may not agree with all their conclusions.

ELTRINGHAM, S. K.: *The Ecology and Conservation of Large African Mammals,* Macmillan Press Ltd., London, 1979.

This book is highly recommended not only as an ecology text for undergraduates—the purpose for which it was written—but also for anyone who has visited East Africa or dreams of going there.

EMMEL, THOMAS C.: *An Introduction to Ecology and Populations,* W. W. Norton & Company, New York, 1973.*

A short, sound text for the beginner, logically presented and sensibly organized.

HARPER, JOHN L.: *Population Biology of Plants,* Academic Press, Inc., New York, 1977.

The world from a plant's perspective. A wonderful introduction to the ecology of populations and a must for any student of botany.

HUTCHINSON, G. EVELYN: *The Ecological Theater and the Evolutionary Play,* Yale University Press, New Haven, Conn., 1965.

By one of the great modern experts on freshwater ecology, this is a charming and sophisticated collection of essays on the influence of environment in evolution—and also on an astonishing variety of other subjects.

HUTCHINSON, G. EVELYN: *An Introduction to Population Ecology,* Yale University Press, New Haven, Conn., 1978.

A beautifully written and handsomely produced collection of lectures by one of the founders of the field, who also happens to be a marvelous humanistic scholar and raconteur.

---

* Available in paperback.

KREBS, CHARLES J.: *Ecology: The Experimental Analysis of Distribution and Abundance,* Harper & Row, Publishers, Inc., New York, 1972.

This excellent, modern text focuses on populations and on interactions among them. Requires some knowledge of mathematics.

MAY, ROBERT M. (ed.): *Theoretical Ecology: Principles and Applications,* 2d ed., Sinauer Associates, Inc., Sunderland, Mass., 1980.*

A multiauthored volume for upper level undergraduates or graduate students.

RICKLEFS, ROBERT E.: *Ecology,* 2d ed., Chiron Press, Inc., New York, 1979.

This outstanding textbook provides a comprehensive introduction to the basic concepts of modern ecology. Beautifully written and rich with examples, it is highly recommended.

SMITH, ROBERT L.: *Ecology and Field Biology,* 2d ed., Harper & Row, Publishers, Inc., New York, 1974.

The outstanding sections of this text are those that deal with descriptions of biomes and communities and the plants and animals within them. Although meant to accompany a course in field biology, this book would be an asset and a pleasure to the amateur naturalist.

STORER, JOHN H.: *The Web of Life,* New American Library, Inc., New York, 1966.*

One of the first books ever written on ecology for the layman. In its simple presentation of the interdependence of living things, it remains a classic.

Articles

CONNELL, JOSEPH H.: "Diversity in Tropical Rain Forests and Coral Reefs," *Science,* vol. 199, pages 1302–1309, 1979.

DIAMOND, JARED M.: "Niche Shifts and the Rediscovery of Interspecific Competition," *American Scientist,* vol. 66, pages 322–331, 1978.

HANSEN, J., et al.: "Climate Impact of Increasing Atmospheric Carbon Dioxide," *Science,* vol. 213, pages 957–966, 1981.

HORN, HENRY S.: "Forest Succession," *Scientific American,* May 1975, pages 90–98.

LUBCHENCO, J.: "Plant Species Diversity in a Marine Intertidal Community: Importance of Herbivore Food Preference and Algal Competitive Abilities," *American Naturalist,* vol. 112, pages 23–29, 1978.

MAY, ROBERT M.: "The Evolution of Ecological Systems," *Scientific American,* September 1978, pages 160–175.

STEARNS, C.: "Life-History Tactics: A Review of the Ideas," *Quarterly Review of Biology,* vol. 51, pages 3–47, 1976.

WARING, RICHARD H.: "Land of the Giant Conifers," *Natural History,* October 1982, pages 54–63.

Social Behavior

Books

ALCOCK, JOHN: *Animal Behavior: An Evolutionary Approach,* Sinauer Associates, Inc., Sunderland, Mass., 1975.

A general text on animal behavior.

---

* Available in paperback.

ALTMANN, JEANNE: *Baboon Mother and Infants,* Harvard University Press, Cambridge, Mass., 1980.

This book, the result of a nine-year study in Kenya, has been called the most significant contribution to the understanding of primate behavior to date. It is outstanding both for its quantitative data and for the author's ability to write clearly and interestingly.

BROWN, JERRAM: *The Evolution of Behavior,* W. W. Norton & Company, New York, 1975.

A text for advanced undergraduates and graduates, with emphasis on ethology and evolution.

DAWKINS, RICHARD: *The Selfish Gene,* Oxford University Press, New York, 1976.*

Dawkins argues that we are "survival machines" programmed to preserve the "selfish molecules" known as genes. A wit-sharpening account and also an entertaining one.

HRDY, SARAH BLAFFER: *The Langurs of Abu: Female and Male Strategies of Reproduction,* Harvard University Press, Cambridge, Mass., 1981.

This book, which has been described as "spell-binding," is notable not only for its original observations but also for its contributions to the theoretical structure of sexual combat.

HRDY, SARAH BLAFFER: *The Woman that Never Evolved,* Harvard University Press, Cambridge, Mass., 1981.

Both a sociobiologist and a feminist, Hrdy examines what it means to be female. Males are almost universally dominant over females in primate species, and Homo sapiens *is no exception. Yet in studying our own remote ancestors, the living primates, she discovers that the female is competitive, independent, sexually assertive, and has just as much at stake in the evolutionary game as her male counterpart.*

KREBS, J. R., and N. B. DAVIES: *An Introduction to Behavioral Ecology,* Sinauer Associates, Inc., Sunderland, Mass., 1981.

The best introduction to the general subject of social behavior. Well written with many good illustrations and an excellent balance between examples and theory.

KUMMER, HANS: *Primate Societies: Group Techniques of Ecological Adaptation,* Aldine, Atherton, Inc., Chicago, 1971.

Most of this book is concerned with behavior in baboons. This book is greatly enriched both by the author's personal experience in observing baboons and other primates and by his interpretation of primate behavior patterns in terms of evolutionary and ecological principles. Some wonderful and unusual photographs, also by the author.

MECH, L. DAVID: *The Wolf: The Ecology and Behavior of an Endangered Species,* Natural History Press, Garden City, N.Y., 1970.

The definitive book on the social history of wolves.

MOSS, CYNTHIA: *Portraits in the Wild: Animal Behaviour in East Africa,* Hamish Hamiton, London, 1976.

A description of some of the studies carried out on various wild species. Full of fascinating bits of information and firsthand observations.

SCHALLER, GEORGE: *The Serengeti Lion: A Study of Predator-Prey Relations,* University of Chicago Press, Chicago, 1972.*

One of the first, best, and most comprehensive of the modern field studies in animal ecology and behavior. Winner of the National Book Award.

---

* Available in paperback.

TINBERGEN, NIKO: *Curious Naturalists,* Natural History Library, Doubleday & Company, Inc., Garden City, N.Y., 1968.*

Some charming descriptions of the activities and discoveries of scientists studying the behavior of animals in their natural environment.

VAN LAWICK-GOODALL, JANE: *In the Shadow of Man,* Houghton Mifflin Company, Boston, 1971.

A personal account of 11 years spent observing the complex social organization of a single chimpanzee community in Tanzania.

WILSON, E. O.: *The Insect Societies,* Harvard University Press, Cambridge, Mass., 1971.

A comprehensive and fascinating account of the social insects. An unusual nominee for the National Book Award.

WILSON, E. O.: *Sociobiology: The New Synthesis,* Harvard University Press, Cambridge, Mass., 1975.*

In this extremely interesting, beautifully written, and beautifully illustrated book, Wilson undertakes to set forth the biological principles that govern social behavior in all kinds of animals. The last chapter, which concerns the sociobiology of man, has become a focus of the current controversy about the inheritance of behavioral traits in Homo sapiens.

Articles

DIAMOND, JARED M.: "Evolution of Bowerbirds' Bowers: Animal Origins of the Aesthetic Sense," *Nature,* vol. 297, pages 99–102, 1982.

DIAMOND, JARED M.: "Rediscovery of the Yellow-Fronted Gardener Bowerbird," *Science,* vol. 216, pages 431–434, 1982.

EMLEN, STEPHEN T., and LEWIS W. ORING: "Ecology, Sexual Selection, and the Evolution of Mating Systems," *Science,* vol. 197, pages 215–223, 1977.

HAMILTON, W. D., and ROBERT AXELROD: "The Evolution of Cooperation," *Science,* vol. 211, pages 1390–1396, 1981.

HORN, HENRY S.: "Sociobiology," in Robert M. May (ed.): *Theoretical Ecology,* Sinauer Associates, Sunderland, Mass., 1981.

LIGON, J. DAVID, and SANDRA H. LIGON: "The Cooperative Breeding Behavior of the Green Woodhoopoe," *Scientific American,* July 1982, pages 126–134.

MAYNARD SMITH, JOHN: "The Evolution of Behavior," *Scientific American,* September 1978, pages 176–192.

MAYNARD SMITH, JOHN: "Evolution and the Theory of Games," *American Scientist,* vol. 64, pages 41–45, 1976.

MAYNARD SMITH, JOHN, and G. R. PRICE: "The Logic of Animal Conflict," *Nature,* vol. 246, pages 15–18, 1973.

MAYR, ERNST: "Behavior Programs and Evolutionary Strategies," *American Scientist,* vol. 62, pages 650–659, 1974.

MOEHLMAN, PATRICIA: "Jackals of the Serengeti," *National Geographic,* vol. 158, pages 840–850, 1980.

WOOLFENDEN, G. E., and J. W. FITZPATRICK: "The Inheritance of Territory in Group Breeding Birds," *BioScience,* vol. 28, pages 104–108, 1978.

---

* Available in paperback.

Human Ecology and Evolution

Books

JOHANSON, DONALD, and MAITLAND EDEY: *Lucy: The Beginnings of Humankind,* Simon and Schuster, New York, 1981.*

A brash, exciting, firsthand account of the discovery of Lucy and of the conflicts and controversies among those seeking to uncover human origins.

LEAKEY, RICHARD E.: *The Making of Mankind,* E. P. Dutton & Co., Inc., New York, 1981.

LEAKEY, RICHARD E., and ROGER LEWIN: *Origins,* E. P. Dutton & Co., Inc., New York, 1977.*

Beautifully illustrated, thoughtful, and imaginative books about 3 million years of human history, and how we know what we do about it, with some speculations on human nature.

SANDERS, N. K.: *Prehistoric Art in Europe,* Penguin Books, Ltd., Harmondsworth, England, 1968.

An illustrated history of art's first 30,000 years.

UCKO, PETER J., and ANDRÉE ROSENFELD: *Paleolithic Cave Art,* McGraw-Hill Book Company, New York, 1967.*

This book not only presents the cave art, in photos and drawings, but also examines its contents and context and discusses the various interpretations of it.

Articles

HAY, RICHARD L., and MARY D. LEAKEY: "The Fossil Footprints of Laetoli," *Scientific American,* February 1982, pages 50–57.

HOLDEN, CONSTANCE: "The Politics of Paleoanthropology," *Science,* vol. 312, pages 737–740, 1981.

ISAAC, GLYNN: "The Food-Sharing Behavior of Protohuman Hominids," *Scientific American,* April 1978, pages 90–119.

JOHANSON, D. C., and T. D. WHITE: "A Systematic Assessment of Early African Hominids," *Science,* vol. 203, pages 321–330, 1979.

LEWIN, ROGER: "Ethiopian Stone Tools Are World's Oldest," *Science,* vol. 211, pages 806–807, 1981.

SIMONS, ELWYN L.: "*Ramapithecus,*" *Scientific American,* May 1977, pages 28–35.

TRINKAUS, ERIK, and WILLIAM W. HOWELLS: "The Neanderthals," *Scientific American,* December 1979, pages 118–133.

WALKER, ALAN, and RICHARD E. LEAKEY: "The Hominids of East Turkana," *Scientific American,* August 1978, pages 54–66.

WASHBURN, SHERWOOD L.: "The Evolution of Man," *Scientific American,* September 1978, pages 194–208.

---

* Available in paperback.

Appendixes

APPENDIX A

Metric Table

	QUANTITY	NUMERICAL VALUE	ENGLISH EQUIVALENT	CONVERTING ENGLISH TO METRIC
Length	kilometer (km)	1,000 (10^3) meters	1 km = 0.62 mile	1 mile = 1.609 km
	meter (m)	100 centimeters	1 m = 1.09 yards	1 yard = 0.914 m
			= 3.28 feet	1 foot = 0.305 m
	centimeter (cm)	0.01 (10^{-2}) meter	1 cm = 0.394 inch	= 30.5 cm
	millimeter (mm)	0.001 (10^{-3}) meter	1 mm = 0.039 inch	1 inch = 2.54 cm
	micrometer (μm)	0.000001 (10^{-6}) meter		
	nanometer (nm)	0.000000001 (10^{-9}) meter		
	angstrom (Å)	0.0000000001 (10^{-10}) meter		
Area	square kilometer (km^2)	100 hectares	1 km^2 = 0.3861 square mile	1 square mile = 2.590 km^2
	hectare (ha)	10,000 square meters	1 ha = 2.471 acres	1 acre = 0.4047 ha
	square meter (m^2)	10,000 square centimeters	1 m^2 = 1.1960 square yards	1 square yard = 0.8361 m^2
			= 10.764 square feet	1 square foot = 0.0929 m^2
	square centimeter (cm^2)	100 square millimeters	1 cm^2 = 0.155 square inch	1 square inch = 6.4516 cm^2
Mass	metric ton (t)	1,000 kilograms =	1 t = 1.103 ton	1 ton = 0.907 t
		1,000,000 grams		1 pound = 0.4536 kg
	kilogram (kg)	1,000 grams	1 kg = 2.205 pounds	1 ounce = 28.35 g
	gram (g)	1,000 milligrams	1 g = 0.0353 ounce	
	milligram (mg)	0.001 gram		
	microgram (μg)	0.000001 gram		
Time	second (sec)	1,000 milliseconds		
	millisecond	0.001 second		
	microsecond	0.000001 second		
Volume (solids)	1 cubic meter (m^3)	1,000,000 cubic centimeters	1 m^3 = 1.3080 cubic yards	1 cubic yard = 0.7646 m^3
			= 35.315 cubic feet	1 cubic foot = 0.0283 m^3
	1 cubic centimeter (cm^3)	1,000 cubic millimeters	1 cm^3 = 0.0610 cubic inch	1 cubic inch = 16.387 cm^3
Volume (liquids)	kiloliter (kl)	1,000 liters	1 kl = 264.17 gallons	1 gal = 3.785 l
	liter (l)	1,000 milliliters	1 l = 1.06 quarts	1 qt = 0.94 l
	milliliter (ml)	0.001 liter	1 ml = 0.034 fluid ounce	1 pt = 0.47 l
	microliter (μl)	0.000001 liter		1 fluid ounce = 29.57 ml

Temperature Conversion Scale

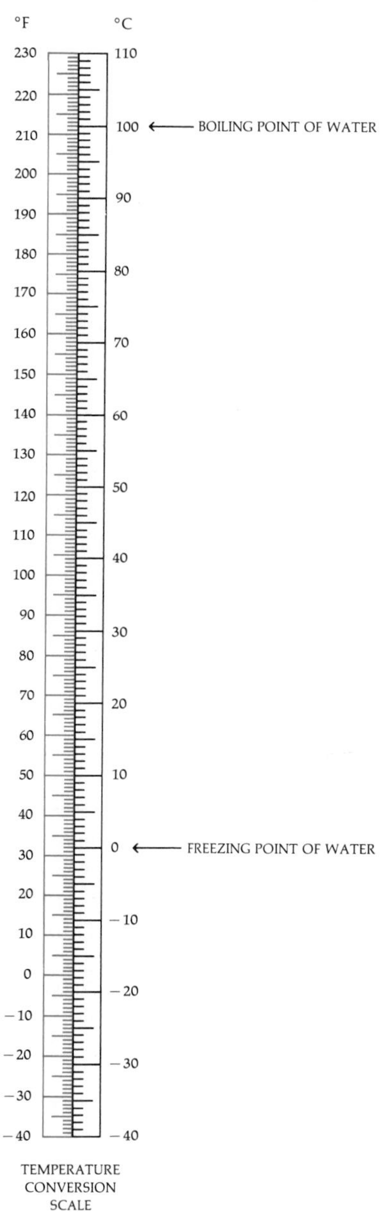

TEMPERATURE
CONVERSION
SCALE

FOR CONVERSION OF FAHRENHEIT TO CELSIUS,
THE FOLLOWING FORMULA CAN BE USED:

$$°C = \frac{5}{9}(°F - 32)$$

FOR CONVERSION OF CELSIUS TO FAHRENHEIT,
THE FOLLOWING FORMULA CAN BE USED:

$$°F = \frac{9}{5}°C + 32$$

Classification of Organisms

There are several ways to classify organisms. The one presented here follows the overall scheme described at the end of Chapter 19. Organisms are divided into five major groups, or kingdoms: Monera, Protista, Fungi, Plantae, and Animalia.

The chief taxonomic categories are kingdom, division or phylum, class, order, family, genus, species. (The taxonomic categories of division and phylum are equivalent. The term "division" is generally used in the classification of prokaryotes, algae, slime molds, fungi, and plants, whereas "phylum" is used in the classification of protozoa and animals.) The following classification includes all of the generally accepted divisions and phyla. Certain classes and orders, particularly those mentioned in this book, are also included, but the listing is far from complete. The number of species given for each group is an estimated number of living (that is, contemporary) species described and named to date.

KINGDOM MONERA

Monerans (prokaryotes) are cells that lack a nuclear envelope, chloroplasts and other plastids, mitochondria, and 9 + 2 flagella. Prokaryotes are unicellular but sometimes occur as filaments or other superficially multicellular bodies. Their predominant mode of nutrition is absorption, but some groups are photosynthetic or chemoautotrophic. Reproduction is primarily asexual, by fission or budding, but conjugation occurs in some species.

DIVISION

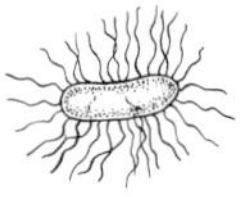

DIVISION SCHIZOPHYTA: bacteria. Unicellular prokaryotes; reproduction usually asexual by binary fission; nutrition usually heterotrophic. About 2,500 species.

DIVISION CYANOPHYTA: cyanobacteria, or blue-green algae. Unicellular or colonial; prokaryotic; chlorophyll, but no chloroplasts; nutrition usually autotrophic; reproduction by fission. Common on damp soil and rocks and in fresh and salt water. Although some 7,500 species of cyanobacteria have been described, a more reasonable estimate puts the number at about 200 distinct, nonsymbiotic species.

KINGDOM PROTISTA

Eukaryotic organisms, including unicellular and multicellular algae, unicellular or multinucleate heterotrophs (slime molds), and unicellular or simple colonial heterotrophs (protozoa). Their modes of nutrition include photosynthesis, absorption, and ingestion. Reproduction is both sexual (in some forms) and asexual. They move by 9 + 2 flagella (or cilia) or pseudopodia or are nonmotile. Nearly 55,000 living species and about 35,000 extinct species known only from their fossils.

DIVISION

DIVISION EUGLENOPHYTA: euglenoids. Unicellular photosynthetic (or sometimes secondarily heterotrophic) organisms with chlorophylls *a* and *b*. They store food as paramylum, an unusual carbohydrate. Euglenoids usually have a single apical flagellum and a contractile vacuole. Sexual reproduction is unknown. Euglenoids occur mostly in fresh water. There are some 1,000 species.

DIVISION CHRYSOPHYTA: diatoms, golden-brown algae, and yellow-green algae. Unicellular photosynthetic organisms with chlorophylls *a* and *c* and the accessory pigment fucoxanthin. Food stored as the carbohydrate leucosin or as large oil droplets. Cell walls consisting mainly of pectic compounds, sometimes heavily impregnated with siliceous materials. There may be as many as 12,000 living species.

CLASS

Class Bacillariophyceae: diatoms. Chrysophyta with double siliceous shells, the two halves of which fit together like a pillbox. They are sometimes motile by the secretion of mucilage fibrils along a specialized groove, the raphe. There are many extinct species and nearly 10,000 living species.

Class Chrysophyceae: golden-brown algae. A diverse group of organisms, including flagellated, amoeboid, and nonmotile forms, some naked and others with a cell wall that may be ornamented with siliceous scales. About 1,500 species.

DIVISION

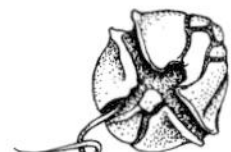

DIVISION PYRROPHYTA: "fire" algae. Unicellular photosynthetic organisms with chlorophylls *a* and *c*. Food is stored as starch. Cell walls contain cellulose. The division contains some 1,100 species, mostly biflagellated organisms, of which the great majority belong to the following class:

CLASS

Class Dinophyceae: dinoflagellates. Pyrrophyta with lateral flagella, one of which beats in a groove that encircles the organism. They probably have no form of sexual reproduction, and their mitosis is unique. More than 1,000 species.

DIVISION

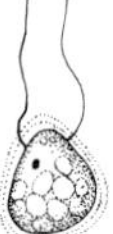

DIVISION CHLOROPHYTA: green algae. Unicellular or multicellular, characterized by chlorophylls *a* and *b* and various carotenoids. The carbohydrate food reserve is starch. Motile cells have two lateral or apical flagella. True multicellular genera do not exhibit complex patterns of differentiation. Multicellularity has arisen at least three times, and quite possibly more often. There are about 7,000 known species and possibly many more.

DIVISION PHAEOPHYTA: brown algae. Multicellular marine organisms characterized by the presence of chlorophylls *a* and *c* and the pigment fucoxanthin. Their food reserve is a carbohydrate called laminarin. Motile cells are biflagellate, with one forward flagellum and one trailing flagellum. A considerable amount of differentiation is found in some of the kelps, with specialized conducting cells for transporting photosynthate to dimly lighted regions of the alga present in some genera. There is, however, no differentiation into leaves, roots, and stem, as in the plants. About 1,500 species.

DIVISION RHODOPHYTA: red algae. Primarily marine organisms characterized by the presence of chlorophyll *a* and pigments known as phycobilins. Their carbohydrate reserve is a special type of starch (floridean). No motile cells are present at any stage in the complex life cycle. The algal body is built up of closely packed filaments in a gelatinous matrix and is not differentiated into leaves, roots, and stem. It lacks specialized conducting cells. There are some 4,000 species.

DIVISION

DIVISION MYXOMYCOTA: plasmodial slime molds. Heterotrophic amoeboid organisms that form a multinucleate plasmodium that creeps along as a mass and eventually differentiates into sporangia, each of which is multinucleate and eventually gives rise to many spores. Predominant mode of nutrition is by ingestion. About 550 species.

DIVISION ACRASIOMYCOTA: cellular slime molds. Heterotrophic organisms in which there are separate amoebas that eventually swarm together to form a mass but retain their identity within this mass, which eventually differentiates into a compound sporangium. Principal mode of nutrition is by ingestion. Seven genera and about 26 species.

PHYLUM

PHYLUM MASTIGOPHORA: mastigophores (flagellates). Protozoa with flagella, most of which are symbiotic forms such as *Trichonympha* and *Trypanosoma,* the cause of sleeping sickness. Some free-living species. Reproduction usually asexual by binary fission. About 2,500 species.

PHYLUM SARCODINA: sarcodines. Protozoa with pseudopodia, such as amoebas. No stiffening pellicle; some produce shells. Reproduction may be asexual or sexual. About 11,500 living species and some 33,000 fossil species.

PHYLUM CILIOPHORA: ciliates. Protozoa with cilia, including *Paramecium* and *Stentor.* Reproduction is asexual, but genetic exchange through the phenomenon of conjugation is also common. About 7,200 species.

PHYLUM OPALINIDA: opalinids. Parasitic protozoa found mainly in the digestive tracts of frogs and toads. Covered uniformly with cilia or flagella. Reproduction asexual by fission or sexual, with flagellated gametes. About 400 species.

PHYLUM SPOROZOA: sporozoa. Parasitic protozoa; usually without locomotor organelles during a major part of their complex life cycle. Includes *Plasmodium,* several species of which cause malaria. About 6,000 species.

KINGDOM FUNGI

Eukaryotic filamentous or, rarely, unicellular organisms. The filamentous forms consist basically of a continuous mycelium; this mycelium becomes septate (partitioned off) in certain groups and at certain stages of the life cycle. Chitin is predominant in the cell walls. Fungi are heterotrophic, with nutrition by absorption. Reproductive cycles often include both sexual and asexual phases. Most fungi are haploid, with the zygote the only diploid stage in the life cycle. Some 100,000 species of fungi have been named.

DIVISION

DIVISION CHYTRIDIOMYCOTA: chytrids. Aquatic fungi with a vegetative body, or thallus, usually differentiated into rhizoids and a sporangium. Cell walls contain chitin. Both spores and gametes are flagellated. About 650 species.

DIVISION OOMYCOTA: water molds. Primarily aquatic, with flagellated asexual spores and nonmotile gametes. The mycelium is diploid, and the haploid gametes are produced by meiosis. Their cell walls are composed of glucose polymers, including cellulose. About 475 species.

DIVISION

DIVISION ZYGOMYCOTA: terrestrial fungi, such as black bread mold, with the hyphae septate only during the formation of reproductive bodies. The division includes about 600 described species, of which about 30 occur as components of the endomycorrhizae that are found in about 80 percent of all vascular plants.

DIVISION ASCOMYCOTA: terrestrial and aquatic fungi, including *Neurospora,* powdery mildews, morels, and truffles. The hyphae are septate but the septa perforated; complete septa cut off the reproductive bodies, such as spores or gametangia. Sexual reproduction involves the formation of a characteristic cell, the ascus, in which meiosis takes place and within which spores are formed. The hyphae in many ascomycetes are packed together into complex "fruiting bodies." Yeasts are unicellular ascomycetes that reproduce asexually by budding. About 30,000 species, in addition to the 25,000 species of lichens.

DIVISION BASIDIOMYCOTA: terrestrial fungi, including the mushrooms and toadstools, with the hyphae septate but the septa perforated; complete septa cut off reproductive bodies. Sexual reproduction involves formation of basidia, in which meiosis takes place and on which the spores are borne. There are some 25,000 species.

DIVISION DEUTEROMYCOTA: Fungi Imperfecti. Mainly fungi in which the sexual cycle has not been observed. The deuteromycetes are classified by their asexual spore-bearing structures. There are some 25,000 species, including *Penicillium,* the original source of penicillin, fungi that cause athlete's foot and other skin diseases, and many of the molds that give cheeses, such as Roquefort and Camembert, their special flavor.

KINGDOM PLANTAE

Multicellular photosynthetic eukaryotes, primarily adapted to life on land. The photosynthetic pigment is chlorophyll *a*, with chlorophyll *b* and a number of carotenoids serving as accessory pigments. The cell walls contain cellulose. There is considerable differentiation of tissues and organs. Reproduction is primarily sexual with alternating gametophytic and sporophytic phases; the gametophytic phase has been progressively reduced in the course of evolution. The egg- and sperm-producing structures are multicellular and are surrounded by a sterile (nonreproductive) jacket layer; the zygote develops into an embryo while encased in the archegonium (seedless plants) or embryo sac (seed plants). The living members of the plant kingdom include the bryophytes and nine divisions of vascular plants—plants with complex differentiation of the sporophyte into leaves, roots, and stems and with well-developed strands of conducting tissue for the transport of water and organic materials.

DIVISION

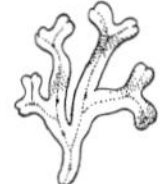

DIVISION BRYOPHYTA: liverworts, hornworts, and mosses. Multicellular plants with photosynthetic pigments and food reserves similar to those of the green algae. They have multicellular gametangia with a sterile jacket one cell layer thick. The sperm are biflagellate and motile. Gametophytes and sporophytes both exhibit complex multicellular patterns of development, but conducting tissues are usually completely absent and not well differentiated when present; true roots, leaves, and stems are absent. Most of the photosynthesis is carried out by the gametophyte, upon which the sporophyte is initially dependent. There are some 16,000 species.

CLASS

Class Hepaticopsida: liverworts. The gametophytes are either thallose (not differentiated into roots, leaves, and stem) or "leafy," and the sporophytes are relatively simple in construction. About 6,000 species.

Class Antherocerotopsida: hornworts. The gametophytes are thallose. The sporophyte grows from a basal meristem for as long as conditions are favorable. Stomata are present on the sporophyte. About 100 species.

Class Muscopsida: mosses. The gametophytes are "leafy." The sporophytes have complex patterns of spore discharge. Stomata are present on the sporophyte. About 9,500 species.

DIVISION

DIVISION PSILOPHYTA: whisk ferns. Homosporous vascular plants with or without microphylls. The sporophytes are extremely simple, and there is no differentiation between root and shoot. The sperm are motile. Two genera and several species.

DIVISION LYCOPHYTA: club mosses. Homosporous and heterosporous vascular plants with microphylls; extremely diverse in appearance. All lycophytes have motile sperm. There are five genera and about 1,000 species.

DIVISION SPHENOPHYTA: horsetails. Homosporous vascular plants with jointed stems marked by conspicuous nodes and elevated siliceous ribs and sporangia borne in a strobilus at the apex of the stem. Leaves are scalelike. Sperm are motile. Although now thought to have evolved from a megaphyll, the leaves of the horsetails are structurally indistinguishable from microphylls. There is one genus, *Equisetum*, with 15 living species.

DIVISION PTEROPHYTA: the ferns. They are mostly homosporous, although some are heterosporous. All possess a megaphyll. The gametophyte is more or less free-living and usually photosynthetic. Multicellular gametangia and free-swimming sperm are present. About 12,000 species.

DIVISION CONIFEROPHYTA: the conifers. Seed plants with active cambial growth and simple leaves, in which the ovules are not enclosed and the sperm are not flagellated. There are some 50 genera and about 550 species; the most familiar group of gymnosperms.

DIVISION CYCADOPHYTA: cycads. Seed plants with sluggish cambial growth and pinnately compound, palmlike or fernlike leaves. The ovules are not enclosed. The sperm are flagellated and motile but are carried to the egg in a pollen tube. Cycads are gymnosperms. There are 10 genera and about 100 species.

DIVISION GINKGOPHYTA: ginkgo. Seed plants with active cambial growth and fan-shaped leaves with open dichotomous venation. The ovules are not enclosed and are fleshy at maturity. Sperm are carried to the egg in a pollen tube but are flagellated and motile. They are gymnosperms. There is one species only.

DIVISION GNETOPHYTA: Seed plants with many angiospermlike features. They are the only gymnosperms in which vessels are present in the xylem. Motile sperm are absent. There are three very distinctive genera, with about 70 species.

DIVISION ANTHOPHYTA: flowering plants (angiosperms). Seed plants in which the ovules are enclosed in a carpel (in all but a very few genera), and the seeds at maturity are borne within fruits. They are extremely diverse vegetatively but are characterized by the flower, which is basically insect-pollinated. Other modes of pollination, such as wind pollination, have been derived in a number of different lines. The gametophytes are much reduced, with the female gametophyte often consisting of only seven cells at maturity. Double fertilization involving the two nonmotile sperm nuclei of the mature male gametophyte gives rise to the zygote and to the primary endosperm nucleus; the former becomes the embryo and the latter a special nutritive tissue, the endosperm. About 235,000 species.

CLASS

Class Monocotyledones: monocots. Flower parts are usually in threes, leaf venation is usually parallel, vascular bundles in the stem are scattered, true secondary growth is not present, and there is one cotyledon. About 65,000 species.

Class Dicotyledones: dicots. Flower parts are usually in fours or fives; leaf venation is usually netlike, pinnate, or palmate; the vascular bundles in the stem are in a ring; there is true secondary growth with vascular cambium commonly present; and there are two cotyledons. About 170,000 species.

KINGDOM ANIMALIA

Eukaryotic multicellular organisms. Their principal mode of nutrition is by ingestion. Many animals are motile, and they generally lack the rigid cell walls characteristic of plants. Considerable cellular migration and reorganization of tissues often occur during the course of embryonic development. Their reproduction is primarily sexual, with male and female diploid organisms producing haploid gametes that fuse to form the zygote. More than a million species have been described and the actual number may be close to 10 million.

PHYLUM

PHYLUM PORIFERA: sponges. Simple multicellular animals, largely marine, with stiff skeletons, and bodies perforated by many pores that admit water containing food particles. All have choanocytes, "collar cells." About 5,000 species.

PHYLUM MESOZOA: extremely simple wormlike animals, all parasites of marine invertebrates. The body consists of 20 to 30 cells, organized in two layers. About 50 species.

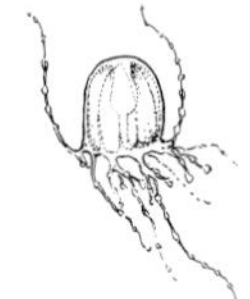

PHYLUM CNIDARIA: polyps and jellyfish. Radially symmetrical animals with a coelenteron and "two-layered" bodies of a jellylike consistency. Reproduction is asexual or sexual. They are the only organisms with cnidocytes, special stinging cells. All are aquatic and most are marine. About 9,000 species.

CLASS

Class Hydrozoa: Hydra, Obelia, and other *Hydra*-like animals. They are often colonial and frequently have a regular alternation of asexual and sexual forms. The polyp form is dominant.

Class Scyphozoa ("cup animals"): marine jellyfishes, including *Aurelia.* The medusa form is dominant. They have true muscle cells.

Class Anthozoa ("flower animals"): sea anemones, colonial corals, and related forms. They have no medusa stage.

PHYLUM

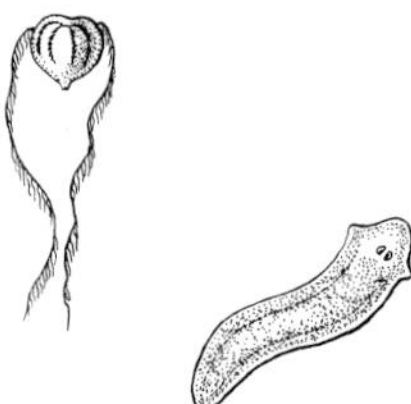

PHYLUM CTENOPHORA: comb jellies and sea walnuts. They are free-swimming, often almost spherical animals, with a coelenteron. They are translucent, gelatinous, delicately colored, and often bioluminescent. They possess eight bands of cilia, for locomotion. About 90 species.

PHYLUM PLATYHELMINTHES: flatworms. Bilaterally symmetrical with three tissue layers. The gut has only one opening. They have no coelom or pseudocoelom and no circulatory system. They have complex hermaphroditic reproductive systems and excrete by means of special flame cells. About 13,000 species.

CLASS

Class Turbellaria: planarians and other nonparasitic flatworms. They are ciliated, carnivorous, and have ocelli ("eyespots").

Class Trematoda: flukes. They are parasitic flatworms with digestive tracts.

Class Cestoda: tapeworms. They are parasitic flatworms with no digestive tracts; they absorb nourishment through body surfaces.

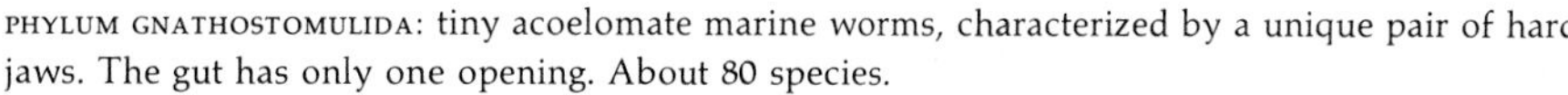

PHYLUM

PHYLUM GNATHOSTOMULIDA: tiny acoelomate marine worms, characterized by a unique pair of hard jaws. The gut has only one opening. About 80 species.

PHYLUM RHYNCHOCOELA: proboscis, nemertine, or ribbon worms. These acoelomate worms are nonparasitic, usually marine, and have a tubelike gut with mouth and anus, a protrusible proboscis armed with a hook for capturing prey, and simple circulatory and reproductive systems. About 650 species.

PHYLUM NEMATODA: roundworms. The phylum includes minute free-living forms, such as vinegar eels, and plant and animal parasites, such as hookworms. They have a pseudocoelom and are characterized by elongated, cylindrical, bilaterally symmetrical bodies. About 12,000 species have been described and named, and there may be as many as 400,000 to 500,000 species.

PHYLUM NEMATOMORPHA: horsehair worms. They are extremely slender, brown or black pseudocoelomate worms up to 1 meter long. Adults are free-living, but the larvae are parasitic in arthropods. About 230 species.

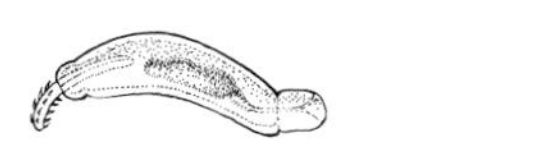

PHYLUM ACANTHOCEPHALA: spiny-headed worms. They are parasitic pseudocoelomate worms with no digestive tract and a head armed with many recurved spines. About 500 species.

PHYLUM KINORHYNCHA: tiny pseudocoelomate worms that burrow in muddy ocean shores. They are short-bodied, covered with spines, and have a spiny retractile proboscis. About 100 species.

PHYLUM GASTROTRICHA: microscopic, wormlike pseudocoelomate animals that move by longitudinal bands of cilia. About 400 species.

PHYLUM ROTIFERA: microscopic, wormlike or spherical pseudocoelomate animals, "wheel animalcules." They have a complete digestive tract, flame cells, and a circle of cilia on the head, the beating of which suggests a wheel; males are minute and either degenerate or unknown in many species. About 1,500 to 2,000 species.

PHYLUM ENTOPROCTA: microscopic stalked, sessile animals that superficially resemble hydrozoans but are much more complex, having three tissue layers, a pseudocoelom, and a complete digestive tract. They were long misclassified with the coelomate bryozoans (ectoprocts), which they also resemble. About 75 species.

PHYLUM MOLLUSCA: unsegmented coelomate animals, with a head, a mantle, and a muscular foot, variously modified. They are mostly aquatic; soft-bodied, often with one or more hard shells, and a three-chambered heart. All mollusks, except bivalves, have a radula (rasplike organ used for scraping or marine drilling). At least 47,000 species, and perhaps many more.

CLASS

Class Aplacophora: solenogasters. Wormlike marine animals, with no clearly defined shell, mantle, or foot. The presence of a radula identifies them as mollusks. About 250 species.

Class Polyplacophora: chitons. The mollusks with the closest resemblance to the hypothetical ancestor, they have an elongated body covered with a mantle in which are embedded eight dorsal shell plates. About 600 species.

Class Monoplacophora: Neopilina. Mostly deep-sea mollusks with a large, single dorsal shell and multiple pairs of gills, nephridia, and retractor muscles. Two genera with eight species.

Class Scaphopoda: tooth or tusk shells. They are marine mollsks with a conical tubular shell. About 350 species.

Class Bivalvia: two-shelled mollusks, including clams, oysters, mussels, scallops. They usually have a hatchet-shaped foot and no distinct head. Generally sessile. At least 7,500 species.

Class Gastropoda: asymmetrical mollusks, including snails, whelks, slugs. They usually have a spiral shell and a head with one or two pairs of tentacles. At least 37,500 species.

Class Cephalopoda: octopuses, squids, *Nautilus.* They are characterized by a "head-foot" with eight or ten arms or many tentacles, mouth with two horny jaws, and well-developed eyes and nervous system. The shell is external *(Nautilus),* internal (squid), or absent (octopus). All except *Nautilus* have ink glands. About 600 species.

PHYLUM

PHYLUM ANNELIDA: segmented worms. They usually have a well-developed coelom, a one-way digestive tract, head, circulatory system, nephridia, and well-defined nervous system. About 9,000 species.

CLASS

Class Oligochaeta: soil, freshwater, and marine annelids, including *Lumbricus* and other earthworms. They have scanty bristles and usually a poorly differentiated head. About 3,000 species.

Class Polychaeta: mainly marine worms, such as *Nereis.* They have a distinct head with tentacles, antennae, and specialized mouthparts. Parapodia are often brightly colored. About 6,000 species.

Class Hirudinea: leeches. They have a posterior sucker and usually an anterior sucker surrounding the mouth. They are freshwater, marine, and terrestrial; either free-living or parasitic. About 300 species.

PHYLUM

PHYLUM SIPUNCULA: peanut worms. Unsegmented marine worms with a stout body, a long, retractile proboscis, and a trochophore larva that resembles those of the polychaete annelids. About 300 species.

PHYLUM ECHIURA: spoon worms. Marine worms with a nonretractile proboscis that contracts to form a structure resembling a spoon. Their embryonic development and trochophore larvae resemble those of the polychaete annelids. About 100 species.

PHYLUM PRIAPULIDA: predatory burrowing marine worms, characterized by a retractile, spine-bearing proboscis. They resemble the pseudocoelomate kinorhynchs, but may have a true coelom. Only nine species.

PHYLUM POGONOPHORA: beard worms. Unsegmented except at the posterior end, these slender marine worms live in long tubes in deep-sea sediments. Although they have no mouth or digestive tract, the anterior region of the body bears a crown of tentacles. About 100 species.

PHYLUM PENTASTOMIDA: wormlike parasites of vertebrate respiratory systems, sometimes with two pairs of short appendages at the anterior end of the body. They lack circulatory, respiratory, and excretory systems, but the nervous system resembles those of annelids and arthropods. About 70 species.

PHYLUM TARDIGRADA: water bears. Tiny segmented animals, with a thin cuticle and four pairs of stubby legs. They are common in fresh water and in the moisture film on mosses; when water is unavailable, they enter a state of suspended animation. About 350 species.

PHYLUM ONYCHOPHORA: *Peripatus*. Caterpillar-like worms with many short, unjointed pairs of legs. Their relatively soft bodies, segmentally arranged nephridia, muscular body walls, and ciliated reproductive tracts resemble those of the annelids; their antennae and eyes resemble those of both the polychaete annelids and the arthropods; their jaws, protective cuticle, brain, and circulatory and respiratory systems resemble those of arthropods. Most give birth to live young, and in some species the embryo is nourished through a placenta. About 70 species.

PHYLUM ARTHROPODA: The largest phylum in the animal kingdom, arthropods are segmented animals with paired, jointed appendages, a hard jointed exoskeleton, a complete digestive tract, reduced coelom, no nephridia, a dorsal brain, and a ventral nerve cord with paired ganglia in each segment. About 850,000 species have been classified to date.

CLASS

Class Merostomata: horseshoe crabs. Aquatic arthropods with chelicerae (pincers or fangs), pedipalps, compound eyes, four pairs of walking legs, and book gills. Only four species.

Class Pycnogonida: sea spiders. Aquatic chelicerates with slender bodies and four or, rarely, five pairs of legs, which are often very long. About 500 species.

Class Arachnida: spiders, mites, ticks, scorpions. Most members are terrestrial, air-breathing; four pairs of walking legs; pedipalps are usually sensory; chelicerae may be pincers or fangs. About 57,000 species.

Class Crustacea: lobsters, crayfish, crabs, shrimps. Crustaceans are mostly aquatic, with compound eyes, two pairs of antennae, one pair of mandibles, and typically two pairs of maxillae. The thoracic segments have appendages, and the abdominal segments are with or without appendages. About 25,000 species.

Class Chilopoda: centipedes. They have a head and 15 to 177 trunk segments, each with one pair of jointed appendages. About 3,000 species.

Class Diplopoda: millipedes. They have a head and a trunk with 20 to 200 body rings, each with two pairs of appendages. About 7,500 species.

Class Pauropoda: tiny, soft-bodied arthropods that resemble millipedes but have only 11 or 12 segments and 9 or 10 pairs of legs. They have branched antennae. About 300 species.

Class Symphyla: garden centipedes and their relatives. Soft-bodied arthropods with one pair of antennae, 12 pairs of jointed legs with claws, and a pair of unjointed posterior appendages. About 130 species.

Class Insecta: insects, including bees, ants, beetles, butterflies, fleas, lice, flies, etc. Most insects are terrestrial and breathe by means of tracheae. The body has three distinct parts (head, thorax, and abdomen); the head bears compound eyes and one pair of antennae; the thorax bears three pairs of legs and usually two pairs of wings. More than 750,000 species.

PHYLUM

PHYLUM PHORONIDA: sedentary, elongated worms that secrete and live in a leathery tube. They have a U-shaped digestive tract and a ring of ciliated tentacles (the lophophore) with which they feed. Marine. Only 18 species.

PHYLUM BRYOZOA (ECTOPROCTA): "moss" animals. These microscopic aquatic organisms are characterized by the lophophore, a U-shaped row of ciliated tentacles, with which they feed. They usually form fixed and branching colonies. Coelomates, they superficially resemble the pseudocoelomate entoprocts. They retain larvae in a special brood pouch. About 4,000 species.

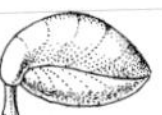

PHYLUM BRACHIOPODA: lamp shells. Marine animals with two hard shells (one orsal and one ventral), they superficially resemble clams. Fixed by a stalk or one shell in adult life, they feed by means of a lophophore. About 250 living species; 30,000 extinct.

PHYLUM ECHINODERMATA: starfish and sea urchins. Echinoderms are radially symmetrical in adult stage, with well-developed coelomic cavities, an endoskeleton of calcareous ossicles and spines, and a unique water vascular system. They have tube feet. About 6,000 species, all marine.

CLASS

Class Crinoidea: sea lilies and feather stars. Sessile animals, they often have a jointed stalk for attachment, and they have 10 arms bearing many slender lateral branches. Most species are fossils.

Class Asteroidea: starfish. They have 5 to 50 arms, an oral surface directed downward, and rows of tube feet on each arm.

Class Ophiuroidea: brittle stars, serpent stars. They have greatly elongated, highly flexible slender arms and rapid horizontal locomotion.

Class Echinoidea: sea urchins and sand dollars. Skeletal plates form rigid external covering that bears many movable spines.

Class Holothuroidea: sea cucumbers. They have a sausage-shaped or wormlike elongated body.

PHYLUM

PHYLUM CHAETOGNATHA: arrow worms. Free-swimming, planktonic marine worms, they have a coelom, a complete digestive tract, and a mouth with strong sickle-shaped hooks on each side. About 60 species.

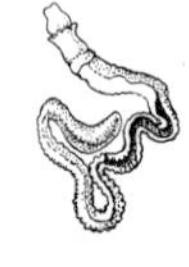

PHYLUM HEMICHORDATA: acorn worms. The body is divided into three regions—proboscis, collar, and trunk. The coelomic cavities provide a hydrostatic skeleton similar to the water vascular system of echinoderms, and their larvae resemble starfish larvae. They have both ventral and dorsal nerve cords, and the anterior portion of the dorsal cord is hollow in some species. They also have a pharynx with gill slits. About 80 species.

PHYLUM CHORDATA: animals having at some stage a notochord, pharyngeal gill slits, a hollow nerve cord on the dorsal side, and a tail. About 43,000 species.

SUBPHYLUM CEPHALOCHORDATA: lancelets. This small subphylum contains only *Branchiostoma* and related forms. They are somewhat fishlike marine animals with a permanent notochord the whole length of the body, a nerve cord, a pharynx with gill slits, and no cartilage or bone. About 28 species.

SUBPHYLUM UROCHORDATA: tunicates. Adults are saclike, usually sessile, often forming branching colonies. They feed by ciliary currents, have gill slits, a reduced nervous system, and no notochord. Larvae are active, with well-developed nervous system and notochord. They are marine. About 1,300 species.

SUBPHYLUM VERTEBRATA: vertebrates. The most important subphylum of Chordata. In the vertebrates the notochord is replaced by cartilage or bone, forming the segmented vertebral column, or backbone. A skull surrounds a well-developed brain. They usually have a tail. About 41,700 species.

CLASS

Class Agnatha: lampreys and hagfish. These are eel-like aquatic vertebrates without limbs, with a jawless sucking mouth, and no bones, scales, or fins. About 60 species.

Class Chondrichthyes: sharks, rays, skates, and other cartilaginous fish. They have complicated copulatory organs, scales, and no air bladders. They are almost exclusively marine. About 625 species.

Class Osteichthyes: bony fish, including nearly all modern freshwater fish, such as sturgeon, trout, perch, anglerfish, lungfish, and some almost extinct groups. They usually have an air bladder or (rarely) a lung. More than 19,000 species.

Class Amphibia: salamanders, frogs, and toads. They usually respire by gills in the larval stage and by lungs in the adult stage. They have incomplete double circulation and a usually naked skin. The limbs are legs. They were the first vertebrates to inhabit the land and were ancestors of the reptiles. Their eggs are unprotected by a shell and embryonic membranes. About 2,500 species.

Class Reptilia: turtles, lizards, snakes, crocodiles; includes extinct species such as the dinosaurs. Reptiles breathe by lungs and have incomplete double circulation. Their skin is usually covered with scales. The four limbs are legs (absent in snakes and some lizards). They are ectotherms. Most live and reproduce on land, although some are aquatic. Their embryo is enclosed in an egg shell and has protective membranes. About 6,000 species.

Class Aves: birds. Birds are homeothermic animals with complete double circulation and a skin covered with feathers. The forelimbs are wings. Their embryo is enclosed in an egg shell with protective membranes. Includes the extinct *Archaeopteryx.* About 8,600 species.

Class Mammalia: mammals. Mammals are homeothermic animals with complete double circulation. Their skin is usually covered with hair. The young are nourished with milk secreted by the mother. They have four limbs, usually legs (forelimbs sometimes arms, wings, or fins), a diaphragm used in respiration, a lower jaw made up of a single pair of bones, three bones in each middle ear connecting eardrum and inner ear, and almost always seven vertebrae in the neck, About 4,500 species.

SUBCLASS

Subclass Prototheria: monotremes. These are the oviparous (egg-laying) mammals with imperfect temperature regulation. There are only three living species: the duckbill platypus and spiny anteaters of Australia and New Guinea.

Subclass Metatheria: marsupials, including kangaroos, opossums, and others. Marsupials are viviparous mammals, usually with a yolk-sac placenta; the young are born in a very immature state and are carried in an external pouch of the mother for some time after birth. They are found chiefly in Australia and South America. About 260 species.

Subclass Eutheria: mammals with a well-developed chorio-allantoic placenta. This subclass comprises the great majority of living mammals. There are 16 orders of Eutheria:

Order

Insectivora: shrews, moles, hedgehogs, etc.
Chiroptera: bats. Aerial mammals with the forelimbs wings.
Dermoptera: "flying" lemurs.
Edentata: toothless mammals—anteaters, sloths, armadillos, etc.
Lagomorpha: rabbits and hares.
Carnivora: carnivorous animals—cats, dogs, bears, weasels, seals, etc.
Tubulidentata: aardvarks.
Pholidota: pangolins.
Rodentia: rodents—rats, mice, squirrels, etc.
Artiodactyla: even-toed ungulates (hoofed mammals)—cattle, deer, camels, hippopotamuses, etc.
Perissodactyla: odd-toed ungulates—horses, zebras, rhinoceroses, etc.
Proboscidea: elephants.
Hyracoidea: hyraxes
Sirenia: manatees, dugongs, and sea cows. Large aquatic mammals with the forelimbs finlike, the hind limbs absent.
Cetacea: whales, dolphins, and porpoises. Aquatic mammals with the forelimbs fins, the hind limbs absent.
Primates: lemurs, monkeys, apes, and humans.

Glossary

This list does not include units of measure or names of taxonomic groups, which can be found in Appendixes A and C, or terms that are used in only one place in the text and defined there. The number in parentheses at the end of each entry is either the text page on which the term is introduced or the page on which a major discussion begins.

abdomen: (1) In vertebrates, the portion of the trunk containing visceral organs other than heart and lungs. (659)
(2) In arthropods, the posterior portion of the body, made up of similar segments and containing the reproductive organs and part of the digestive tract. (528)

abscisic acid (ABA) [L. *ab,* away, off + *scissio,* dividing]: A plant hormone with a variety of inhibitory effects; brings about dormancy in buds, maintains dormancy in seeds, effects stomatal closing, and is involved in the geotropism of roots; also known as the "stress hormone." (638)

abscission [L. *ab,* away, off + *scissio,* dividing]: In plants, the dropping of leaves, flowers, fruits, or stems at the end of a growing season, as the result of formation of a layer of specialized cells (the abscission zone) and the action of a hormone (ethylene). (638)

absorption [L. *absorbere,* to swallow down]: In physiology, the movement of water and dissolved substances into a cell, tissue, or organism. (482)

absorption spectrum: The characteristic pattern of the wavelengths (colors) of light that a particular pigment absorbs. (214)

acetylcholine (a-**sea**-tell-**co**-leen): One of the chemicals (neurotransmitters) responsible for the passing of nerve impulses across synaptic junctions. (774)

acid [L. *acidus,* sour]: A substance that causes an increase in the relative number of hydrogen ions (H^+) in a solution and a decrease in the relative number of hydroxide ions (OH^-); having a pH of less than 7; the opposite of a base. (41)

actin [Gk. *actis,* a ray]: One of the two major proteins of muscle (the other is myosin); the principal constituent of thin filaments. (114)

action potential: A transient change in electric potential across a membrane; in nerve cells, results in transmission of a nerve impulse; in muscle cells, results in contraction. (770)

action spectrum: The characteristic pattern of the wavelenths (colors) of light that elicit a particular reaction or response. (214)

activation energy: The energy that must be possessed by atoms or molecules in order to react. (167)

active site: The region of an enzyme surface that binds the substrate during the reaction catalyzed by the enzyme. (169)

active transport: The energy-requiring transport of a solute across a cellular membrane from a region of lower concentration to a region of higher concentration (that is, against a concentration gradient). (133)

adaptation [L. *adaptare,* to fit]: (1) The evolution of features that make a group of organisms better suited to live and reproduce in their environment. (886)
(2) A peculiarity of structure, physiology, or behavior that aids the organism in its environment. (542)

adaptive radiation: The evolution from a primitive and unspecialized ancestor of several divergent forms, each specialized to fit a distinct niche. (939)

adenosine diphosphate (ADP): A nucleotide consisting of adenine, ribose, and two phosphate groups; formed by the removal of one phosphate from an ATP molecule. (182)

adenosine monophosphate (AMP): A nucleotide consisting of adenine, ribose, and one phosphate group; can be formed by the removal of two phosphate groups from an ATP molecule. *See also* Cyclic AMP. (183)

adenosine triphosphate (ATP): The nucleotide that provides the energy currency for cell metabolism; composed of adenine, ribose, and three phosphate groups. On hydrolysis, ATP loses one phosphate and one hydrogen to become adenosine diphosphate (ADP), releasing energy in the process. ATP is formed from ADP + P_i in an enzymatic reaction that traps energy released by catabolism or energy captured in photosynthesis. (181)

ADH: Abbreviation of antidiuretic hormone. (729)

adhesion [L. *adhaerere,* to stick to]: The holding together of molecules of different substances. (35)

ADP: Abbreviation of adenosine diphosphate. (182)

adrenal gland [L. *ad,* near + *renes,* kidney]: A vertebrate endocrine gland. The cortex (outer surface) is the source of cortisol, aldosterone, and other steroid hormones; the medulla (inner core) secretes adrenaline and noradrenaline. (780)

adrenaline: A hormone produced by the medulla of the adrenal gland, which increases the concentration of sugar in the blood, raises blood pressure and heartbeat rate, and increases muscular power and resistance to fatigue; also a neurotransmitter across synaptic junctions. Also called epinephrine. (785)

adventitious [L. *adventicius,* not properly belonging to]: Referring to a structure arising from an unusual place, such as roots growing from stems or leaves. (586)

aerobic [Gk. *aēr,* air + *bios,* life]: Any biological process that can occur in the presence of molecular oxygen (O_2). (192)

afferent [L. *ad,* near + *ferre,* to carry]: Bringing inward to a central part, applied to nerves and blood vessels. (726)

agar: A gelatinous material prepared from certain red algae that is used to solidify nutrient media for growing microorganisms. (426)

aldosterone [Gk. *aldainō,* to nourish + *stereō,* solid]: A hormone produced by the adrenal cortex that affects the concentration of ions in the blood; it stimulates the reabsorption of sodium and the excretion of potassium by the kidney. (729)

alga, *pl.* **algae** (**al**-gah, **al**-jee): A unicellular or simple multicellular photosynthetic organism lacking multicellular sex organs. (415)

alkaline: Pertaining to substances that increase the relative number of hydroxide ions (OH^-) in water; having a pH greater than 7; basic; opposite of acidic. (41)

allele frequency: The proportion of a particular allele in a population. (893)

alleles (al-**eels**) [Gk. *allelon,* of one another]: Two or more alternative forms of a gene. Alleles occupy the same position (locus) on homologous chromosomes and are separated from each other at meiosis. (241)

allopatric speciation [Gk. *allos,* other + *patra,* fatherland, country]: Speciation that occurs as the result of the geographic separation of a population of organisms. (928)

allosteric interaction [Gk. *allos,* other + *stereos,* solid, shape]: An interaction involving an enzyme that has two binding sites, the active site and a site into which another molecule, an allosteric effector, fits; the binding of the effector changes the shape of the enzyme and activates or inactivates it. (178)

alternation of generations: A sexual life cycle in which a haploid (n) phase alternates with a diploid ($2n$) phase. The gametophyte (n) produces gametes (n) by mitosis. Fertilization of gametes yields zygotes ($2n$). Each zygote develops into a sporophyte ($2n$) that forms haploid spores (n) by meiosis. Each haploid spore forms a new gametophyte, completing the cycle. (250)

altruism: Self-sacrifice for the benefit of others; any form of behavior that increases the fitness of the recipient while reducing the fitness of the altruistic individual. (1017)

alveolus, *pl.* **alveoli** [L. dim. of *alveus,* cavity, hollow]: One of the many small air sacs within the lungs in which the bronchioles terminate. The thin walls of the alveoli contain numerous capillaries and are the site of gas exchange between the air in the alveoli and the blood in the capillaries. (696)

amino acids (am-**ee**-no) [Gk. *Ammon,* referring to the Egyptian sun god, near whose temple ammonium salts were first prepared from camel dung]: Organic molecules containing nitrogen in the form of NH_2 and a carboxyl group, COOH, bonded to the same carbon atom; the "building blocks" of protein molecules. (52)

amnion (**am**-neon) [Gk. dim. of *amnos,* lamb]: Membrane enclosing a fluid-filled space, the amniotic cavity, that surrounds the embryo in reptiles, birds, and mammals. (558)

amniote egg: An egg that is isolated and protected from the environment by a more or less impervious shell during the period of its development and that is completely self-sufficient, requiring only oxygen from the outside. (558)

amoeboid [Gk. *amoibē,* change]: Moving or eating by means of pseudopodia (temporary cytoplasmic protrusions from the cell body). (419)

AMP: Abbreviation of adenosine monophosphate. (183)

amphibian [Gk. *amphibios,* living a double life]: A class of vertebrates intermediate in many characteristics between fish and reptiles, which live part or most of the time on land but must return to water to reproduce because fertilization is external. (557)

anabolism [Gk. *ana,* up + *-bolism* (as in metabolism)]: Within a cell or organism, the sum of all biosynthetic reactions (that is, chemical reactions in which larger molecules are formed from smaller ones). (166)

anaerobe [Gk. *an,* without + *aēr,* air + *bios,* life]: Cell that can live without free oxygen; obligate anaerobes cannot live in the presence of oxygen; facultative anaerobes can live with or without oxygen. (390)

anaerobic [Gk. *an,* without + *aēr,* air + *bios,* life]: Applied to a process that can occur without oxygen, as anaerobic fermentation. (192)

anagenesis [Gk. *ana,* up + *genesis,* origin]: The changes taking place in a single lineage of organisms over a long period of time; one of the principal patterns of organic evolution; also known as phyletic evolution. (938)

analogous [Gk. *analogos,* proportionate]: Applied to structures similar in function but different in evolutionary origin, such as the wing of a bird and the wing of an insect. (378)

anaphase (**anna**-phase) [Gk. *ana,* up + *phasis,* form]: In mitosis and meiosis II, the stage in which the chromatids of each chromosome separate and move to opposite poles; in meiosis I, the stage in which homologous chromosomes separate and move to opposite poles. (145)

androgens [Gk. *andros,* man + *genos,* origin, descent]: Male sex hormones; any chemical with actions similar to those of testosterone. (777)

angiosperms (**an**-jee-o-sperms) [Gk. *angeion,* vessel + *sperma,* seed]: The flowering plants. Literally, a seed borne in a vessel; thus, any plant whose seeds are borne within a matured ovary (fruit). (465)

anisogamy [Gk. *aniso,* unequal + *gamos,* marriage]: Sexual reproduction in which one gamete is larger than the other; both gametes are motile. (424)

annual plant [L. *annus,* year]: A plant that completes its life cycle (from seed germination to seed production) and dies within a single growing season. (588)

antennae: Long, paired sensory appendages on the head of many arthropods. (528)

anterior [L. *ante,* before, toward, in front of]: The front end of an organism; in human anatomy, the ventral surface. (494)

anther [Gk. *anthos,* flower]: In flowering plants, the pollen-bearing portion of a stamen. (473)

antheridium, *pl.* **antheridia:** In bryophytes and some vascular plants, the multicellular sperm-producing organ. (456)

anthropoid [Gk. *anthropos,* man, human]: A higher primate; includes monkeys, apes, and humans. (1042)

antibiotic [Gk. *anti,* against + *bios,* life]: An organic compound formed and secreted by an organism that is inhibitory or toxic to other species. (409)

antibody [Gk. *anti,* against]: A globular protein produced in response to a foreign substance (antigen), with which it combines specifically. (749)

anticodon: In a tRNA molecule, the three-nucleotide sequence that base pairs with the mRNA codon for the amino acid carried by that particular tRNA; the anticodon is complementary to the mRNA codon. (299)

antidiuretic hormone (ADH) [Gk. *anti,* against + *diurgos,* thoroughly wet + *hormaein,* to excite]: A peptide hormone synthesized in the hypothalamus that inhibits urine excretion by inducing the reabsorption of water from the nephrons of the kidneys; also called vasopressin. (729)

antigen [Gk. *anti,* against + *genos,* origin, descent]: A foreign substance, usually a protein or polysaccharide, that stimulates the formation of specific antibodies. (749)

aorta (a-**ore**-ta) [Gk. *aeirein,* to lift, heave]: The major artery in blood-circulating systems; the aorta sends blood to the other body tissues. (709)

apical dominance [L. *apex,* top]: In plants, the influence of a terminal bud in suppressing the growth of lateral buds, sometimes the result of the release of growth-regulating hormones. (633)

apical meristem [L. *apex,* top + Gk. *meristos,* divided]: In vascular plants, the growing point at the tip of the root or stem. (597)

archegonium, *pl.* **archegonia** [Gk. *archegonos,* first of a race]: In bryophytes and some vascular plants, the multicellular egg-producing organ. (456)

archenteron [Gk. *arch,* first, or main + *enteron,* gut]: The main cavity within the early embryo (gastrula) of many animals; lined with endoderm, it opens to the outside by means of the blastopore and ultimately becomes the digestive tract. (853)

artery: A vessel carrying blood from the heart to the tissues; arteries are usually thick-walled, elastic, and muscular. A small artery is known as an arteriole. (705)

arthropod [Gk. *arthron,* joint + *pous, podos,* foot]: An invertebrate animal with jointed appendages; a member of the phylum Arthropoda. (527)

artificial selection: The breeding of selected organisms for the purpose of producing descendants with desired traits. (7)

ascus, *pl.* **asci** (**as**-kus, **as**-i) [Gk. *askos,* wineskin, bladder]: In the fungi of division Ascomycota, a specialized cell within which two haploid nuclei fuse to produce a zygote that immediately divides by meiosis; at maturity, an ascus contains ascospores. (446)

asexual reproduction: Any reproductive process, such as budding or the division of a cell or body into two or more approximately equal parts, that does not involve the union of gametes. (258)

atmospheric pressure [Gk. *atmos,* vapor + *sphaira,* globe]: The weight of the earth's atmosphere over a unit area of the earth's surface. (126)

atom [Gk. *atomos,* indivisible]: The smallest particle into which a chemical element can be divided and still retain the properties characteristic of the element; consists of a central core, the nucleus, containing protons and neutrons, and electrons that move around the nucleus. (20)

atomic number: The number of protons in the nucleus of an atom; equal to the number of electrons in the neutral atom. (20)

atomic weight: The average weight of all the isotopes of an element relative to the weight of an atom of the most common isotope of carbon (^{12}C), which is by convention assigned the integral value of 12; approximately equal to the number of protons plus neutrons in the nucleus of an atom. (20)

ATP: Abbreviation of adenosine triphosphate, the principal energy-carrying compound of the cell. (181)

ATP synthetase: The enzyme complex in the inner membrane of the mitochondrion and the thylakoid membrane of the chloroplast through which protons flow down the gradient established in the first stage of chemiosmosis; the site of formation of ATP from ADP and P_i during oxidative phosphorylation and photophosphorylation. (202)

atrium, *pl.* **atria** (**a**-tree-um) [L., yard, court, hall]: A thin-walled chamber of the heart that receives blood and passes it on to a thick, muscular ventricle. (707)

autonomic [Gk. *autos,* self + *nomos,* usage, law]: Self-controlling, independent of outside influences. (668)

autonomic nervous system [Gk. *autos,* self + *nomos,* usage, law]: In the peripheral nervous system of vertebrates, the motor neurons and ganglia that are not ordinarily under voluntary control; innervates the heart, glands, visceral organs, and smooth muscle; subdivided into the sympathetic and parasympathetic divisions. (765)

autosome [Gk. *autos,* self + *soma,* body]: Any chromosome other than the sex chromosomes. Humans have 22 pairs of autosomes and 1 pair of sex chromosomes. (268)

autotroph [Gk. *autos,* self + *trophos,* feeder]: An organism that is able to synthesize all needed organic molecules from simple inorganic substances (e.g., H_2O, CO_2, NH_3) and some energy source (e.g., sunlight); in contrast to heterotroph. Plants, algae, and some bacteria are autotrophs. (83)

auxin [Gk. *auxein,* to increase + *in,* of, or belonging to]: One of a group of plant hormones with a variety of growth-regulating effects, including promotion of cell elongation. (632)

auxotroph [L. *auxillium,* help + Gk. *trophos,* feeder]: A mutant defective in the synthesis of a particular molecule, which must therefore be supplied for normal growth. (176)

axis: An imaginary line passing through a body or organ around which parts are symmetrically aligned. (486)

axon [Gk. *axon,* axle]: Nerve fiber that conducts impulses away from the cell body of a neuron. (665)

B cell: A type of lymphocyte capable of becoming an antibody-secreting plasma cell; a B lymphocyte. (749)

bacteriophage [L. *bacterium* + Gk. *phagein,* to eat]: A virus that parasitizes a bacterial cell. (281)

bacterium [Gk. dim of *baktron,* staff]: A unicellular prokaryote. (84)

bark: In plants, all tissues outside the vascular cambium in a woody stem. (608)

basal body [Gk. *basis,* foundation]: A cytoplasmic organelle of protozoa and animals from which cilia or flagella arise; identical in structure to the centriole, which is involved in mitosis and meiosis in most protozoa and animals and some plants. (116)

base: A substance that causes an increase in the relative number of hydroxide ions (OH^-) in a solution and a decrease in the relative number of hydrogen ions (H^+); having a pH of more than 7; the opposite of an acid. *See* Alkaline. (41)

base-pairing principle: In the formation of nucleic acids, the requirement that adenine must always pair with thymine (or uracil) and guanine with cytosine. (298)

basidium, *pl.* **basidia** (ba-**sid**-i-um) [L., a little pedestal]: A specialized reproductive cell of the fungi of division Basidiomycota, often club-shaped, in which nuclear fusion and meiosis occur; homologous with the ascus. (447)

biennial [L. *biennium,* a space of two years; *bi,* twice + *annus,* year]: Occurring once in two years; a plant that requires two years to complete its reproductive cycle; vegetative growth occurs in the first year, sexual reproduction and death in the second. (588)

bilateral symmetry [L. *bi,* twice, two + *lateris,* side; Gk. *summetria,* symmetry]: A body form in which the right and left halves of an organism are approximate mirror images of each other. (493)

bile: A yellow secretion of the vertebrate liver, temporarily stored in the gallbladder and composed of organic salts that emulsify fats in the small intestine. (679)

binary fission [L. *binarius,* consisting of two things or parts + *fissus,* split]: Asexual reproduction by division of the cell or body into two equal, or nearly equal, parts. (399)

binomial system [L. *bi,* twice, two + Gk. *nomos,* usage, law]: A system of nomenclature in which the name consists of two parts, with the first designating genus and the second, species; originated by Linnaeus. (372)

biogeochemical cycle [Gk. *bios,* life + *geō,* earth + *chēmeia,* alchemy; *kyklos,* circle, wheel]: The cyclic path of an inorganic substance, such as carbon or nitrogen, through an ecosystem. Its geological components are the atmosphere, the crust of the earth, and the oceans, lakes, and rivers; its biological components are producers, consumers, and detritivores. (986)

biological clock [Gk. *bios,* life + *logos,* discourse]: Proposed internal factor(s) in organisms that governs functions that occur rhythmically in the absence of external stimuli. (649)

biomass [Gk. *bios,* life]: Total weight of all organisms (or some group of organisms) living in a particular habitat or place. (981)

biome: One of the major types of distinctive plant formations; for example, the grassland biome, the tropical rain forest biome, etc. (1000)

biosphere [Gk. *bios,* life + *sphaira,* globe]: The zones of air, land, and water at the surface of the earth occupied by living things. (79)

biosynthesis [Gk. *bios,* life + *synthesis,* a putting together]: Formation by living organisms of organic compounds from elements or simple compounds. (166)

blade: (1) The broad, expanded part of a leaf. (575)
(2) The broad, expanded photosynthetic portion of the thallus of a multicellular alga or a simple plant. (426)

blastodisc [Gk. *blastos,* sprout + *discos,* a round plate]: Disklike area on the surface of a large, yolky egg that undergoes cleavage and gives rise to the embryo. (859)

blastomere: One of many cells produced by division during the cleavage stage of development of the egg. (853)

blastopore [Gk. *blastos,* sprout + *poros,* a way, means, path]: In the gastrula stage of an embryo, the opening that connects the archenteron with the outside; represents the future mouth in some animals (protostomes), the future anus in others (deuterostomes). (506)

blastula [Gk. *blastos,* sprout]: An animal embryo after cleavage and before gastrulation; usually consists of a hollow sphere, the walls of which are composed of a single layer of cells. (853)

bond strength: The strength with which a chemical bond holds two atoms together; conventionally measured in terms of the amount of energy, in kilocalories per mole, required to break the bond. (53)

botany [Gk. *botanikos,* of herbs]: The study of plants. (2)

Bowman's capsule: In the vertebrate kidney, the bulbous unit of the nephron, which surrounds the glomerulus. In filtration, the initial process in urine formation, blood plasma is forced from the glomerular capillaries into Bowman's capsule. (725)

brainstem: The most posterior portion of the vertebrate brain; includes medulla, pons, and midbrain. (817)

bronchus, *pl.* **bronchi** (**bronk**-us, **bronk**-eye) [Gk. *bronchos,* windpipe]: One of a pair of respiratory tubes branching into either lung at the lower end of the trachea; it subdivides into progressively finer passageways, the bronchioles, culminating in the alveoli. (695)

bud: (1) In plants, an embryonic shoot, including rudimentary leaves, often protected by special bud scales. (639)
(2) In animals, an asexually produced outgrowth that develops into a new individual. (258)

buffer: A combination of H^+-donor and H^+-acceptor forms of a weak acid or a weak base; a buffer prevents appreciable changes of pH in solutions to which small amounts of acids or bases are added. (43)

bulb: A modified bud with thickened leaves adapted for underground food storage. (579)

bulk flow: The overall movement of a fluid induced by gravity, pressure, or an interplay of both. (127)

C_3 pathway: *See* Calvin cycle.

C_4 pathway: The set of reactions by which some plants initially fix carbon in the four-carbon compound oxaloacetic acid; the carbon dioxide is later released in the interior of the leaf and enters the Calvin cycle. Also known as the Hatch-Slack pathway. (226)

calorie [L. *calor,* heat]: The amount of energy in the form of heat required to raise the temperature of 1 gram of water 1°C; in making metabolic measurements the kilocalorie (Calorie) is generally used. A Calorie is the amount of heat required to raise the temperature of 1 kilogram of water 1°C. (36)

Calvin cycle: The set of reactions through which carbon dioxide is reduced to carbohydrate during the second stage of photosynthesis. (225)

CAM: *See* Crassulacean acid metabolism.

capillaries [L. *capillaris,* relating to hair]: Smallest thin-walled blood vessels through which exchanges between blood and the tissues occur; connect arteries with veins. (705)

capillary action: The movement of water or any liquid along a surface; results from the combined effect of cohesion and adhesion. (35)

capsule (**kap**-sul) [L. *capsula,* a little chest]: (1) A slimy layer around the cells of certain bacteria. (396)
(2) The sporangium of Bryophyta. (460)
(3) A dehiscent, dry fruit that develops from two or more carpels. (600)

carbohydrate [L. *carbo,* charcoal + *hydro,* water]: An organic compound consisting of a chain or ring of carbon atoms to which hydrogen and oxygen are attached in a ratio of approximately 2:1; carbohydrates include sugars, starch, glycogen, cellulose, etc. (54)

carbon cycle: Worldwide circulation and reutilization of carbon atoms, chiefly due to metabolic processes of living organisms. Inorganic carbon, in the form of carbon dioxide, is incorporated into organic compounds by photosynthetic organisms; when the organic compounds are broken down in respiration, carbon dioxide is released. Large quantities of carbon are "stored" in the seas and the atmosphere, as well as in fossil fuel deposits. (229)

carbon fixation: The second stage of photosynthesis; energy stored in ATP and NADPH by the energy-trapping reactions of the first stage is used to reduce carbon from carbon dioxide to simple sugars. (225)

carnivore [L. *caro, carnis,* flesh + *voro,* to devour]: Predator that obtains its nutrients and energy by eating meat. (983)

carotenoids [L. *carota,* carrot]: A class of pigments that includes the carotenes (yellows, oranges, and reds) and the xanthophylls (yellow); accessory pigments in photosynthesis. (214)

carpel: A leaflike floral structure enclosing the ovule or ovules of angiosperms, typically divided into ovary, style, and stigma; a flower may have one or more carpels, either single or fused. Also known as a pistil. (472)

carrying capacity: In ecology, the largest number of organisms of a particular species that the resources of a particular area can support indefinitely. (953)

cartilage [L. *cartilago,* gristle]: A connective tissue in skeletons of vertebrates; forms much of the skeleton of adult lower vertebrates and immature higher vertebrates. (266)

Casparian strip (after Robert Caspary, German botanist): In plants, a thickened, waxy strip that extends around and seals the walls of endodermal root cells, thus restricting the diffusion of solutes across the endodermis into the vascular tissues of the root. (585)

catabolism [Gk. *katabole,* throwing down]: Within a cell or organism, the sum of all chemical reactions in which large molecules are broken down into smaller parts. (166)

catalyst [Gk. *katalysis,* dissolution]: A substance that lowers the activation energy of a chemical reaction by forming a temporary association with the reacting molecules; as a result, the rate of the reaction is accelerated. Enzymes are catalysts. (167)

cell [L. *cella,* a chamber]: The structural unit of organisms, surrounded by a membrane and composed of cytoplasm and, in eukaryotes, one or more nuclei. In most plants, fungi, and bacteria, there is a cell wall outside the membrane. (76)

cell cycle: A regular, timed sequence of the events of cell growth and division through which dividing cells pass. (141)

cell membrane: The outermost membrane of the cell; also called the plasma membrane. (100)

cell plate: In the dividing cells of most plants (and in some fungi and algae), a flattened structure that forms at the equator of the mitotic spindle in early telophase; gives rise to the middle lamella. (148)

cell theory: All living things are composed of cells; cells arise only from other cells. No exception has been found to these two principles since they were first proposed well over a century ago. (77)

cellulose [L. *cellula,* a little cell]: The chief constituent of the cell wall in all green plants; an insoluble complex carbohydrate formed of microfibrils of glucose molecules. (59)

cell wall: A plastic or rigid structure, produced by the cell and located outside the cell membrane in most plants, algae, fungi, bacteria, and cyanobacteria; in plant cells, it consists mostly of cellulose. (100)

central nervous system: In vertebrates, the brain and spinal cord; in invertebrates it usually consists of one or more cords of nervous tissue plus their associated ganglia. (764)

centriole (sen-tree-ole) [Gk. *kentron,* center]: A cytoplasmic organelle identical in structure to a basal body; flagellated cells and all animal cells, including those without flagella, have centrioles at the spindle poles during division. (116)

centromere (sen-tro-mere) [Gk. *kentron,* center + *meros,* a part]: Region of constriction of chromosome that holds sister chromatids together. (143)

cerebellum [L. dim. of *cerebrum,* brain]: A subdivision of the vertebrate brain that lies above the brainstem and behind the forebrain; functions in coordinating muscular activities and maintaining equilibrium. (817)

cerebral cortex [L. *cerebrum,* brain]: A thin layer of neurons and glial cells forming the upper surface of the cerebrum, well developed only in mammals; the seat of conscious sensations and voluntary muscular activity. (819)

cerebrum [L., brain]: The portion of the vertebrate brain occupying the upper part of the skull, consisting of two cerebral hemispheres united by the corpus callosum; coordinates most activities. (819)

character displacement: A phenomenon in which species that live together in the same environment tend to diverge in those characteristics that overlap; occurs in species whose niches overlap and is also known as niche shift; exemplified by Darwin's finches. (965)

chelicera, *pl.* **chelicerae** [Gk. *cheilos,* the edge, lips + *cheir,* arm]: First pair of appendages in horseshoe crabs, sea spiders, and arachnids; usually take the form of pincers or fangs. (528)

chemical reaction: An interaction among atoms, ions, or molecules that results in the formation of new combinations of atoms, ions, or molecules; the making or breaking of chemical bonds. (28)

chemiosmosis: The mechanism by which ADP is phosphorylated to ATP in mitochondria and chloroplasts. The energy released as electrons pass down an electron transport chain is used to establish a proton gradient across an inner membrane of the organelle; when protons subsequently flow down this electrochemical gradient, the potential energy released is captured in the terminal phosphate bonds of ATP. (199)

chemoautotroph [Gk. *cheimōn,* storm + *autos,* self + *trophos,* feeder]: An autotrophic bacterium that uses the energy released by specific inorganic reactions to power its life processes, including the synthesis of organic molecules. (83)

chemoreceptor: A sensory cell or organ that responds to the presence of a specific chemical stimulus; includes smell and taste receptors. (400)

chemotactic [Gk. *cheimōn,* storm + *taxis,* arrangement, order]: Of an organism, capable of responding to a chemical stimulus by moving toward or away from it. (400)

chiasma, *pl.* **chiasmata (kye-az-ma)** [Gk., a cross]: Connection between paired homologous chromosomes at meiosis; the site at which crossing over and genetic recombination occur. (272)

chitin (kye-tin) [Gk. *chitōn,* a tunic, undergarment]: A tough, resistant, nitrogen-containing polysaccharide present in the exoskeleton of arthropods, the epidermal cuticle or other surface structures of many other invertebrates, and in the cell walls of certain fungi. (60)

chlorophyll [Gk. *chloros,* green + *phyllon,* leaf]: A class of green pigments that are the receptors of light energy in photosynthesis. (214)

chloroplast [Gk. *chloros,* green + *plastos,* formed]: A membrane-bound, chlorophyll-containing organelle in eukaryotes (algae and plants) that is the site of photosynthesis. (112)

chordate [L. *chorda,* cord, string]: Member of the animal phylum Chordata, of which all members possess a notochord, dorsal nerve cord, pharyngeal gill pouches, and a tail, at least at some stage of the life cycle. (554)

chorion (**core**-ee-on) [Gk. *chorion,* skin, leather]: The outermost membrane of the embryos of reptiles, birds, and mammals; in placental mammals it contributes to the structure of the placenta. (861)

chromatid (**crow**-ma-tid) [Gk. *chrōma,* color]: Either of the two strands of a replicated chromosome, which are joined at the centromere. (143)

chromatin [Gk. *chrōma,* color]: The deeply staining complex of DNA and proteins of which eukaryotic chromosomes are composed. (103)

chromosome [Gk. *chrōma,* color + *soma,* body]: One of the bodies in the cell nucleus along which the genes are located; visualized as threads or rods of chromatin, which appear in a contracted form during mitosis and meiosis. (84)

chromosome map: A diagram of the linear order of the genes on a chromosome; determined from the frequency of recombination between pairs of genes. (272)

cilium, *pl.* **cilia** (**silly**-um) [L., eyelash]: A short, thin structure embedded in the surface of some eukaryotic cells, usually in large numbers and arranged in rows; has a highly characteristic internal structure of two inner microtubules surrounded by nine pairs of outer microtubules; involved in locomotion and the movement of substances across the cell surface. (115)

circadian rhythms [L. *circa,* about + *dies,* day]: Regular rhythms of growth or activity that occur on an approximately 24-hour cycle. (648)

cladogenesis [Gk. *clados,* branch + *genesis,* origin]: The splitting of an evolutionary lineage into two or more separate lineages; one of the principal patterns of organic evolution; also known as splitting evolution. (938)

class: A taxonomic grouping of related, similar orders; category above order and below phylum. (374)

cleavage: The successive cell divisions of a fertilized egg to form a multicellular blastula. (853)

cline [Gk. *klinein,* to lean]: A graded series of changes in some characteristic within a species, correlated with some gradual change in climate or another geographic factor. (910)

cloaca [L., sewer]: The exit chamber from the digestive system in lower vertebrates; also may serve as the exit for the reproductive and urinary systems. (840)

clone [Gk. *klon,* twig]: A line of cells, all of which have arisen from the same single cell by mitotic division; a population of individuals derived by asexual reproduction from a single ancestor. (356)

cnidocyte (**ni**-do-site) [Gk. *knide,* nettle + *kytos,* vessel]: A stinging cell containing a nematocyst; characteristic of cnidarians. (487)

coadaptive gene complex: A group of genes that work together in the production of a particular phenotypic trait; when linked together on one chromosome, known as a supergene. (917)

cochlea [Gk. *kochlias,* snail]: Part of the inner ear of mammals; concerned with hearing. (806)

codominance: In genetics, the phenomenon in which the effects of both alleles at a particular locus are apparent in the phenotype of the heterozygote. (262)

codon (**code**-on): Basic unit ("letter") of the genetic code; three adjacent nucleotides in a molecule of DNA or mRNA that form the code for a specific amino acid or for polypeptide chain termination. (297)

coelenteron (see-**len**-t-ron) [Gk. *koilos,* hollow + *enteron,* gut]: A digestive cavity with only one opening, characteristic of the phyla Cnidaria (jellyfish, hydra, corals, etc.) and Ctenophora (comb jellies, sea walnuts). (486)

coelom (**see**-loam) [Gk. *koilos,* a hollow]: A body cavity formed between layers of mesoderm and in which the digestive tract and other internal organs are suspended. (494)

coenocytic (see-no-**sit**-ik) [Gk. *koinos,* shared in common + *kytos,* a hollow vessel]: An organism or part of an organism consisting of many nuclei within a common cytoplasm. (423)

coenzyme [L. *co,* together + Gk. *en,* in + *zyme,* leaven]: A nonprotein organic molecule that plays an accessory role in enzyme-catalyzed processes, often by acting as a donor or acceptor of a substance involved in the reaction. NAD^+, FAD, and coenzyme A are common coenzymes. (173)

coevolution [L. *co,* together + *e-,* out + *volvere,* to roll]: The simultaneous development of adaptations in two or more populations that interact so closely that each is a strong selective force on the other. (924)

cofactor: A nonprotein component that plays an accessory role in enzyme-catalyzed processes; some cofactors are ions, and others are coenzymes. (173)

cohesion [L. *cohaerere,* to stick together]: The attraction or holding together of molecules of the same substance. (35)

cohesion-tension theory: A theory accounting for the upward movement of water in plants. According to this theory, transpiration of a water molecule results in a negative (below 1 atmosphere) pressure in the leaf cells, inducing the entrance from the vascular tissue of another water molecule, which, because of the cohesive property of water, pulls with it a chain of water molecules extending up from the cells of the root tip. (615)

coleoptile (coal-ee-**op**-tile) [Gk. *koleon,* sheath + *ptilon,* feather]: The sheath enclosing the apical meristem and leaf primordia of a germinating monocot. (642)

collagen [Gk. *kolla,* glue]: A fibrous protein in bones, tendons, and other connective tissues. (69)

colony: A group of organisms living together in close association. (422)

commensalism [L. *com,* together + *mensa,* table]: *See* Symbiosis.

community: An association of interacting populations. (962)

competition: Interaction between members of the same population or of two or more populations in order to obtain a mutually required resource available in limited supply. (962)

competitive exclusion: The hypothesis that two species with identical ecological requirements cannot coexist stably in the same locality and the species that is more efficient in utilizing the available scarce resources will exclude the other; also known as Gause's principle, after the Russian biologist G. F. Gause. (962)

compound [L. *componere,* to put together]: A molecule composed of two or more kinds of atoms in definite ratios, held together by chemical bonds. (23)

compound eye: In arthropods, a complex eye composed of many separate elements, each with light-sensitive cells and a lens that can form an image. (542)

condensation [L. *co,* together + *densare,* to make dense]: A type of chemical reaction in which two molecules join to form one larger molecule, simultaneously splitting out a molecule of water. The biosynthetic reactions in which monomers (e.g., monosaccharides, amino acids) are joined to form polymers (e.g., polysaccharides, polypeptides) are condensation reactions. (57)

cone: (1) In plants, the reproductive structure of a conifer. (469)
(2) In vertebrates, a type of light-sensitive neuron in the retina, concerned with the perception of color and with the most acute discrimination of detail. (799)

conifer [Gk. *konos,* cone + *phero,* carry]: One of the cone-bearing plants, such as pines and firs. (464)

conjugation [L. *conjugatio,* a joining, connection]: The sexual process in some unicellular organisms by which genetic material is transferred from one cell to another by cell-to-cell contact. (312)

connective tissues: Supporting or packing tissues that lie between groups of nerves, glands, and muscle cells, and beneath epithelial cells, in which the cells are irregularly distributed through a relatively large amount of intercellular material; include bone, cartilage, blood, and lymph. (662)

consumer, in ecological systems: A heterotroph that derives its energy from living or freshly killed organisms or parts thereof. Primary consumers are herbivores; higher-level consumers are carnivores. (983)

continental drift: The gradual movement of the world's continents that has occurred over hundreds of millions of years. (930)

continuous variation: A gradation of small differences in a particular trait, such as height, within a population; occurs in traits that are controlled by a number of genes. (265)

convergent evolution [L. *convergere,* to turn together; *evolutio,* to unfold]: The independent development of similarities between unrelated groups, such as porpoises and sharks, resulting from adaptation to similar environments. (921)

cork [L. *cortex,* bark]: A secondary tissue that is a major constituent of bark in woody and some herbaceous plants; made up of flattened cells, dead at maturity; restricts gas and water exchange and protects the vascular tissues from injury. (608)

cork cambium [L. *cortex,* bark + *cambium,* exchange]: The lateral meristem that produces cork. (608)

corolla (ko-**role**-a) [L. dim. of *corona,* wreath, crown]: Petals, collectively; usually the conspicuously colored flower parts. (591)

corpus luteum [L., yellowish body]: An ovarian structure that secretes estrogens and progesterone, which maintain the uterus during pregnancy. It develops from the remaining cells of the ruptured follicle following ovulation. (844)

cortex [L., bark]: (1) The outer, as opposed to the inner, part of an organ, as in the adrenal gland. (780)
(2) In a stem or root, the primary tissue bounded externally by the epidermis and internally by the central cylinder of vascular tissue. (583)

cotyledon (cottle-**ee**-don) [Gk. *kotyledon,* a cup-shaped hollow]: A leaflike structure of the embryo of a seed plant; contains stored food used during germination. (470)

countercurrent exchange: An anatomical device for manipulating gradients so as to maximize uptake (or minimize loss) of O_2, heat, etc. (129)

coupled reactions: In cells, the linking of endergonic (energy-requiring) reactions to exergonic (energy-releasing) reactions that provide enough energy to drive the endergonic reactions foward. (182)

covalent bond [L. *con,* together + *valere,* to be strong]: A chemical bond formed as a result of the sharing of one or more pairs of electrons. (26)

Crassulacean acid metabolism: A process by which some species of plants in hot, dry climates take in carbon dioxide during the night, fixing it in organic acids; the carbon dioxide is released during the day and used immediately in the Calvin cycle. (618)

cristae: The "shelves" formed by the intricate folding of the inner membrane of the mitochondrion. (194)

cross-fertilization: Fusion of gametes formed by different individuals; as opposed to self-fertilization. (242)

crossing over: During meiosis, the exchange of genetic material between paired chromatids of homologous chromosomes. (253)

cuticle (ku-tik-l) [L. *cuticula,* dim. of *cutis,* the skin]: (1) In plants, a layer of waxy substance (cutin) on the outer surface of epidermal cell walls. (576)
(2) In animals, the noncellular, outermost layer of many invertebrates. (501)

cyclic AMP: A form of adenosine monophosphate (AMP) in which the atoms of the phosphate group form a ring; functions in chemical communication in slime molds, in regulation of the *lac* operon, and as a "second messenger" for many vertebrate hormones. (120)

cytochromes [Gk. *kytos,* vessel + *chroma,* color]: Heme-containing proteins that participate in electron transport chains; involved in cellular respiration and photosynthesis. (198)

cytokinesis [Gk. *kytos,* vessel + *kinesis,* motion]: Division of the cytoplasm of a cell following nuclear division. (148)

cytokinin [Gk. *kytos,* vessel + *kinesis,* motion]: One of a group of chemically related plant hormones that promote cell division, among other effects. (634)

cytoplasm (**sight**-o-plazm) [Gk. *kytos,* vessel + *plasma,* anything molded]: The living matter within a cell, excluding the genetic material. (104)

cytoskeleton: A network of filamentous protein structures within the cytoplasm that maintains the shape of the cell, anchors its organelles, and is involved in cell motility; includes microtubules, microfilaments, intermediate fibers, and microtrabeculae. (106)

deciduous [L. *decidere,* to fall off]: Refers to plants that shed their leaves at a certain season. (588)

decomposers: Small detritivores, usually bacteria and fungi, that consume such substances as cellulose and nitrogenous waste products. Their metabolic processes release inorganic nutrients, which are then available for reuse by plants and other organisms. (983)

denaturation: The loss of the native configuration of a macromolecule resulting, for instance, from heat treatment, extreme pH changes, chemical treatment, or other denaturing agents. It is usually accompanied by loss of biological activity. (174)

dendrite [Gk. *dendron,* tree]: Nerve fiber, typically branched, that conducts impulses toward the cell body of a neuron. (665)

deoxyribonucleic acid (DNA) (dee-ox-y-rye-bo-new-**klee**-ick): The carrier of genetic information in cells, composed of two complementary chains of nucleotides wound in a double helix; capable of self-replication as well as coding for RNA synthesis. (279)

dermis [Gk. *derma,* skin]: The inner layer of the skin, beneath the epidermis. (667)

desmosome [Gk. *desmos,* bond + *soma,* body]: A type of "spot" junction that holds adjacent cells together in animal tissues. (119)

detritivores [L. *detritus,* worn down, worn away + *voro,* to devour]: Organisms that live on dead and discarded organic matter; include large scavengers, smaller animals such as earthworms and some insects, and decomposers (fungi and bacteria). (983)

deuterostome [Gk. *deuteros,* second + *stoma,* mouth]: An animal in whose embryonic development the anus forms at or near the blastopore and the mouth forms secondarily elsewhere. Deuterostomes are also characterized by radial cleavage during the earliest stages of development and by enterocoelous formation of the coelom. (506)

diaphragm [Gk. *diaphrassein,* to barricade]: In mammals, a sheetlike tissue (tendon and muscle) forming the partition between the abdominal and thoracic cavities; functions in breathing. (659)

dicotyledon (dye-cottle-**ee-**don) [Gk. *di,* double, two + *kotyledon,* a cup-shaped hollow]: A member of the class of flowering plants having two seed leaves, or cotyledons, among other distinguishing features; often abbreviated as dicot. (471)

differentiation: The developmental process by which a relatively unspecialized cell or tissue undergoes a progressive (usually irreversible) change to a more specialized cell or tissue. (119)

diffusion [L. *diffundere,* to pour out]: The net movement of suspended or dissolved particles down a concentration gradient as a result of the random spontaneous movements of individual particles; the process tends to distribute the particles uniformly throughout a medium. (127)

digestion [L. *digestio,* separating out, dividing]: The breakdown of complex, usually insoluble foods into molecules that can be absorbed into the body and used by the cells. (671)

dikaryon (dye-**care-ee**-on) [Gk. *di,* two + *karyon,* kernel]: A cell or organism with paired but not fused nuclei derived from different parents. Found among the higher fungi (divisions Ascomycota and Basidiomycota). (441)

dioecious (dye-**ee**-shus) [Gk. *di,* two + *oikos,* house]: In angiosperms, having the male (staminate) and female (carpellate) flowers on different individuals of the same species. (592)

diploid [Gk. *di,* double, two + *ploion,* vessel]: The condition in which each autosome is represented twice ($2n$); in contrast to haploid (n). (250)

division: A taxonomic grouping of related, similar classes; a high-level category beneath kingdom and above class. Division is generally used in the classification of prokaryotes, algae, fungi, and plants, whereas an equivalent category, phylum, is used in the classification of protozoa and animals. (374)

DNA: Abbreviation of deoxyribonucleic acid. (279)

dominance: (1) In genetics, the ability of one allelic form of a gene to determine the phenotype of a heterozygous individual. The homologous chromosome carries a different allele, which is said to be recessive. (240)
(2) In ecology, the capacity of a group of organisms, by virtue of size, number, or behavior, to exert a controlling influence on their environment and, as a result, to determine what other kinds of organisms exist in that ecosystem. (1000)

dormancy [L. *dormire,* to sleep]: A period during which growth ceases and is resumed only if certain requirements, as of temperature or day length, have been fulfilled. (600)

dorsal [L. *dorsum,* the back]: Pertaining to or situated near the back; opposite of ventral. (493)

double fertilization: A phenomenon unique to the angiosperms, in which the egg and one sperm nucleus fuse (resulting in a $2n$ fertilized egg, the zygote) and simultaneously the second sperm nucleus fuses with the two polar nuclei (resulting in a $3n$ endosperm nucleus). (593)

duodenum (duo-**dee**-num) [L. *duodeni,* twelve each—from its length, about 12 fingers' breadth]: The upper portion of the small intestine in vertebrates, where food is digested into molecules that can be absorbed by intestinal cells. (676)

ecological niche: A description of the roles and associations of a particular species in the community of which it is a part; the way in which an organism interacts with the biotic and abiotic parts of its environment. (964)

ecological pyramid: A graphic representation of the quantitative relationships of number of organisms, biomass, or energy flow between the trophic levels of a food chain. Because large amounts of energy and biomass are dissipated at every trophic level, these diagrams nearly always take the form of pyramids. (986)

ecological succession: The gradual process by which the species composition of a community changes. (992)

ecology [Gk. *oikos,* home + *logos,* a discourse]: The study of the interactions of organisms with their physical environment and with each other and of the results of such interactions. (951)

ecosystem [Gk. *oikos,* home + *systema,* that which is put together]: The organisms in a community plus the associated abiotic factors with which they interact. (980)

ecotype [Gk. *oikos,* home + L. *typus,* image]: A locally adapted variant of a species, differing genetically from other ecotypes of the same species. (910)

ectoderm [Gk. *ecto,* outside + *derma,* skin]: (1) The outermost layer of body tissue. (487)
(2) One of the three embryonic germ layers, it gives rise to the outer epithelium of the body (skin, hair, nails), the sense organs, the brain and spinal cord, etc. (494)

ectotherm [Gk. *ecto,* outside + *therme,* heat]: An organism, such as a reptile, that regulates its body temperature by taking in heat from the environment or giving it off to the environment. *See also* Poikilotherm. (560)

effector [L. *ex,* out of + *facere,* to make]: Cell, tissue, or organ (such as muscle or gland) capable of producing a response to a stimulus. (665)

efferent [L. *ex,* out of + *ferre,* to bear]: Carrying away from a center, applied to nerves and blood vessels. (726)

egg: A female gamete, which usually contains abundant cytoplasm and yolk; nonmotile and often larger than a male gamete. (257)

electric potential: The difference in the amount of electric charge between a region of positive charge and a region of negative charge. The establishment of electric potentials across cellular membranes makes possible a number of phenomena, including the chemiosmotic synthesis of ATP, the transmission of nerve impulses, and muscle contraction. (201)

electron: A subatomic particle with a negative electric charge equal in magnitude to the positive charge of the proton but with a much smaller mass; normally found within orbitals surrounding the atom's positively charged nucleus. (20)

electron acceptor: Substance that accepts or receives electrons in an oxidation-reduction reaction, becoming reduced in the process. (164)

electron carrier: A specialized molecule, such as a cytochrome, that can lose and gain electrons reversibly, alternately becoming oxidized and reduced. (220)

electron donor: Substance that donates or gives up electrons in an oxidation-reduction reaction, becoming oxidized in the process. (403)

electron transport: The movement of electrons down a series of electron-carrier molecules that hold electrons at slightly different energy levels; as electrons move down the chain, the energy released is used to form ATP from ADP and phosphate. Electron transport plays an essential role in the final stage of cellular respiration and in the light-trapping reactions of photosynthesis. (197)

element: A substance composed only of atoms of the same atomic number, which cannot be decomposed by ordinary chemical means; one of about 105 distinct natural or synthetic types of matter that, singly or in combination, compose all materials of the universe. (20)

embryo [Gk. *en,* in + *bryein,* to swell]: The early developmental stage of an organism produced from a fertilized egg; a young organism before it emerges from the seed, egg, or body of its mother. In humans, refers to the first two months of intrauterine life. *See* Fetus. (257)

endergonic [Gk. *endon,* within + *ergon,* work]: Energy-requiring, as in a chemical reaction; applied to an "uphill" process. (160)

endocrine gland [Gk. *endon,* within + *krinein,* to separate]: Ductless gland whose secretions (hormones) are released into the extracellular spaces, from which they diffuse into the circulatory system; in vertebrates, includes pituitary, sex glands, adrenal, thyroid, and others. (777)

endocytosis [Gk. *endon,* within + *kytos,* vessel]: A cellular process in which material to be taken into the cell attaches to special areas on the cell membrane, inducing the membrane to form a vacuole enclosing the material; the vacuole is released into the cytoplasm. (135)

endoderm [Gk. *endon,* within + *derma,* skin]: (1) The innermost layer of body tissue. (487)
(2) One of the three embryonic germ layers, it gives rise to the epithelium that lines certain internal structures, such as most of the digestive tract and its outgrowths, most of the respiratory tract, and the urinary bladder, liver, pancreas, and some endocrine glands. (494)

endodermis [Gk. *endon,* within + *derma,* skin]: In plants, a one-celled layer of specialized cells that lies between the cortex and the vascular tissues in young roots. The Casparian strip of the endodermis prevents diffusion of solutes across the root. (585)

endometrium [Gk. *endon,* within + *metrios,* of the womb]: The glandular lining of the uterus in mammals; thickens in response to secretion of estrogens and progesterone and is sloughed off in menstruation. (842)

endoplasmic reticulum [Gk. *endon,* within + *plasma,* from cytoplasm; L. *reticulum,* network]: An extensive system of membranes present in most eukaryotic cells, dividing the cytoplasm into compartments and channels, often coated with ribosomes. (108)

endorphin: One of a group of small peptides with morphine-like properties produced by the vertebrate brain. (825)

endosperm [Gk. *endon,* within + *sperma,* seed]: In plants, a $3n$ tissue containing stored food that develops from the union of a sperm nucleus and the polar nuclei of the egg; found only in angiosperms. (593)

endotherm [Gk. *endon,* within + *therme,* heat]: An organism, such as a bird or a mammal, that regulates its body temperature internally through metabolic processes. *See also* Homeotherm. (560)

enterocoelous [Gk. *enteron,* gut + *koilos,* a hollow]: Formation of the coelom during embryonic development as cavities within mesoderm originating from outpocketings of the primitive gut; characteristic of deuterostomes. (506)

entropy [Gk. *en,* in + *trope,* turning]: A measure of the randomness or disorder of a system. (161)

enzyme [Gk. *en,* in + *zyme,* leaven]: A globular protein molecule that accelerates a specific chemical reaction. (167)

epidermis [Gk. *epi,* on or over + *derma,* skin]: In plants and animals, the outermost layers of cells. (470)

epinephrine: *See* Adrenaline.

epistasis [Gk., a stopping]: Interaction between two nonallelic genes in which one of them modifies the phenotypic expression of the other. (262)

epithelial tissue [Gk. *epi,* on or over + *thele,* nipple]: In animals, a type of tissue that covers a body or structure or lines a cavity; epithelial cells form one or more regular layers with little intercellular material. (661)

equilibrium [L. *aequus,* equal + *libra,* balance]: The state of a system in which no further net change is occurring; result of counterbalancing forward and backward processes. (40)

equilibrium species [L. *aequus,* equal + *libra,* balance + *species,* kind, sort]: Species characterized by low reproduction rates, long development times, large body size, and long adult life with repeated reproductions; also called prudent or *K*-selected. (956)

erythrocyte (eh-**rith**-ro-site) [Gk. *erythros,* red + *kytos,* vessel]: Red blood cell, the carrier of hemoglobin. (706)

estrogens [Gk. *oistros,* frenzy + *genos,* origin, descent]: A group of steroid hormones, including estradiol, that affect secondary sex characteristics in females, estrus, and, in humans, the menstrual cycle. (780)

estrus [Gk. *oistros,* frenzy]: The mating period in female mammals, characterized by ovulation and intensified sexual activity. (805)

ethology [Gk. *ethos,* habit, custom + *logos,* discourse]: The study of whole patterns of animal behavior in natural environments, stressing adaption and evolution. (1017)

eukaryote (you-**car**-ry-oat) [Gk. *eu,* good + *karyon,* nut, kernel]: A cell having a membrane-bound nucleus, membrane-bound organelles, and chromosomes in which DNA is combined with special proteins; an organism composed of such cells. (84)

eusocial [Gk. *eu,* good + L. *socius,* companion]: Applied to insect societies in which sterile individuals work on behalf of reproductive forms. (1019)

evolution [L. *e-,* out + *volvere,* to roll]: Changes in the gene pool from one generation to the next as a consequence of processes such as natural selection, mutation, and genetic drift. (881)

exergonic [Gk. *ex,* out of + *ergon,* work]: Energy-yielding, as in a chemical reaction; applied to a "downhill" process. (160)

exocrine glands [Gk. *ex,* out of + *krinein,* to separate]: Glands, such as digestive glands and sweat glands, that secrete their products into ducts. (776)

exocytosis [Gk. *ex,* out of + *kytos,* vessel]: A cellular process in which particles are wrapped in a vacuole and transported to the cell surface; there, the membrane of the vacuole fuses with the cell membrane, expelling the vacuole's contents to the outside. (135)

exon: A segment of DNA that is transcribed into RNA and, upon translation, directs the amino acid sequence of part of a polypeptide. (338)

exoskeleton: The outer supporting covering of the body; common in arthropods. (529)

exponential growth: In populations, the increasingly accelerated rate of growth due to the increasing number of individuals being added to the reproductive base. Exponential growth is very seldom approached in natural populations. (952)

extraembryonic membranes: Membranes formed of embryonic tissues that lie outside the embryo, protecting it and aiding metabolism; include amnion, chorion, allantois, and yolk sac. (861)

F_1 (first filial generation): The offspring resulting from the crossing of plants or animals of a parental generation. (240)

F_2 (second filial generation): The offspring resulting from crossing members of the F_1 generation among themselves. (241)

facilitated diffusion: The transport of substances across a cellular membrane from a region of high concentration to a region of low concentration by carrier molecules embedded in the membrane; driven by the concentration gradient. (132)

family: A taxonomic grouping of related, similar genera; the category below order and above genus. (374)

fatty acid: A molecule consisting of a COOH group and a long hydrocarbon chain; fatty acids are components of fats, oils, phospholipids, and waxes. (52)

feedback systems: Control mechanisms whereby an increase or decrease in the level of a particular factor inhibits or stimulates the production, utilization, or release of that factor; important in the regulation of enzyme and hormone levels, ion concentrations, temperature, and many other factors. (180)

fertilization: The fusion of two haploid gamete nuclei to form a diploid zygote nucleus. (249)

fetus [L., pregnant]: An unborn or unhatched vertebrate that has passed through the earliest developmental stages; a developing human from about the second month of gestation until birth. (870)

fibril [L. *fibra,* fiber]: Any minute, threadlike organelle within a cell. (69)

fibrous protein: Insoluble structural protein in which the polypeptide chain is coiled along one dimension; constitutes the main structural elements of many animal tissues. (69)

filament [L. *filare,* to spin]: (1) A chain of cells. (397)
(2) In plants, the stalk of a stamen. (473)

filtration: The first stage of kidney function; blood plasma is forced, under pressure, out of the glomerular capillaries into Bowman's capsule, through which it enters the renal tubule. (726)

fission: *See* Binary fission.

fitness: The genetic contribution of an individual to succeeding generations relative to the contributions of other individuals in the population. (893)

flagellum, *pl.* **flagella** (fla-**jell**-um) [L. *flagellum,* whip]: A long, threadlike organelle found in eukaryotes and used in locomotion and feeding; has an internal structure of nine pairs of microtubules encircling two central microtubules. (115)

flower: The reproductive structure of angiosperms; a complete (perfect) flower includes sepals, petals, stamens (male structures), and carpels (female structures). (472)

food chain: A sequence of organisms related to one another as prey and predator. (981)

food web: A set of interactions among organisms, including producers, herbivores, and carnivores, through which energy and materials move within a community or ecosystem. (968)

fossil [L. *fossilis,* dug up]: The remains of an organism, or direct evidence of its presence (such as tracks). May be an unaltered hard part (tooth or bone), a mold in a rock, petrification (wood or bone), unaltered or partially altered soft parts (a frozen mammoth). (3)

fossil fuels: The remains of once-living organisms that are burned to release energy. Examples are coal, oil, and natural gas. (229)

founder effect: Type of genetic drift that occurs as the result of the founding of a population by a small number of individuals. (896)

fovea [L., pit]: A small area in the center of the retina in which cones are concentrated; the area of sharpest vision. (799)

free energy change: The total energy change that results from a chemical reaction or other process (such as evaporation); takes into account changes in both heat and entropy. (162)

fruit [L. *fructus,* fruit]: In angiosperms, a matured, ripened ovary or group of ovaries and associated structures; contains the seeds. (597)

function [L. *fungor,* to busy oneself]: Characteristic role or action of a structure or process in the nomal metabolism or behavior of an organism. (84)

gametangium, *pl.* **gametangia** [Gk. *gamein,* to marry + L. *tangere,* to touch]: A unicellular or multicellular structure in which gametes are produced. (441)

gamete (*gam*-meet) [Gk., wife]: The haploid reproductive cell whose nucleus fuses with that of another gamete of an opposite sex (fertilization); the resulting cell (zygote) may develop into a new diploid individual or, in some protists and fungi, may undergo meiosis to form haploid somatic cells. (239)

gametophyte: In plants, which have alternation of haploid and diploid generations, the haploid (*n*) gamete-producing generation. (424)

ganglion, *pl.* **ganglia** (**gang**-lee-on) [Gk. *ganglion,* a swelling]: Aggregation of nerve cell bodies; in vertebrates, refers to an aggregation of nerve cell bodies located outside the central nervous system. (494)

gap junction: A junction between adjacent animal cells that allows the passage of materials between the cells. (119)

gastrula [Gk. *gaster,* stomach]: An embryo in the process of gastrulation; the stage of development during which the blastula with its single layer of cells turns into a three-layered embryo, made up of ectoderm, mesoderm, and endoderm, often enclosing an archenteron. (853)

Gause's principle: *See* Competitive exclusion.

gene [Gk. *genos,* birth, race; L. *genus,* birth, race, origin]: A unit of heredity in the chromosome; a sequence of nucleotides in a DNA molecule that performs a specific function, such as coding for an RNA molecule or a polypeptide. (240)

gene flow: The exchange of genes between different populations. (896)

gene frequency: *See* Allele frequency.

gene pool: All the alleles of all the genes of all the individuals in a population. (892)

genetic code: The system of nucleotide triplets in DNA and RNA that carries genetic information; referred to as a code because it determines the amino acid sequence in the enzymes and other protein molecules synthesized by the organism. (296)

genetic drift: Evolution (change in allele frequencies) owing to chance processes. (896)

genetic isolation: The absence of genetic exchange between populations or species as a result of geographic separation or as a result of premating or postmating mechanisms (behavioral, morphological, or physiological) that prevent reproduction. (928)

genome: The complete set of chromosomes, with their associated genes. (249)

genotype (**jean**-o-type): The genetic constitution of an individual cell or organism with reference to a single trait or a set of traits; the sum total of all the genes present in an individual. (242)

genus, *pl.* **genera** (**jean**-us) [L. *genus,* race, origin]: A taxonomic grouping of closely related species. (372)

geologic eras: *See* Table 23–1, page 457.

geotropism [Gk. *ge,* earth + *trope,* turning]: The direction of growth or movement in which the force of gravity is the determining factor; also called gravitropism. (643)

germ cells: Gametes or the cells that give rise directly to gametes. (866)

germination [L. *germinare,* to bud]: In plants, the resumption of growth or the development from seed or spore. (461)

germ layer: A layer of distinctive cells in an embryo; each germ layer always gives rise to certain tissues or structures as the organism develops. The majority of multicellular animals have three embryonic layers: ectoderm, mesoderm, and endoderm. (494)

gibberellins (jibb-e-**rell**-ins) [Fr. *gibberella,* genus of fungi]: A group of chemically related plant growth hormones, whose most characteristic effect is stem elongation in dwarf plants and bolting. (636)

gill: The respiratory organ of aquatic animals, usually a thin-walled projection from some part of the external body surface or, in vertebrates, from some part of the digestive tract. (691)

gland [L. *glans, glandis,* acorn]: An organ composed of modified epithelial cells specialized to produce one or more secretions that are discharged to the outside of the gland. (662)

globular protein [L. dim. of *globus,* a ball]: A polypeptide chain folded into a roughly spherical shape. (68)

glomerulus (glom-**mare**-u-lus) [L. *glomus,* ball]: In the vertebrate kidney, a cluster of capillaries enclosed by Bowman's capsule; blood plasma minus large molecules filters through the walls of the glomerular capillaries into the renal tubule. (725)

glucagon [Gk. *glukus,* sweet + *agō,* to lead toward]: Hormone produced in the pancreas that acts to raise the concentration of blood sugar. (668)

glucose [Gk. *glukus,* sweet]: A six-carbon sugar ($C_6H_{12}O_6$); the most common monosaccharide in animals. (56)

glycogen [Gk. *glukus,* sweet + *genos,* race or descent]: A complex carbohydrate (polysaccharide); one of the main stored food substances of most animals and fungi; it is converted into glucose by hydrolysis. (58)

glycolysis (gly-**coll**-y-sis) [Gk. *glukus,* sweet + *lysis,* loosening]: The process by which a glucose molecule is changed anaerobically to two molecules of pyruvic acid with the liberation of a small amount of useful energy; catalyzed by soluble cytoplasmic enzymes. (189)

Golgi body (**goal**-jee): An organelle present in many eukaryotic cells; consists of flat, disk-shaped sacs, tubules, and vesicles. It functions as a collecting and packaging center for substances that the cell manufactures for export. (109)

gonad [Gk. *gone,* seed]: Gamete-producing organ of multicellular animals; ovary or testis. (836)

granum, *pl.* **grana** [L., grain or seed]: In chloroplasts, stacked membrane-bound disks (thylakoids) that contain chlorophylls and carotenoids and are the sites of the light-trapping reactions of photosynthesis. (218)

guard cells: Specialized epidermal cells surrounding a pore, or stoma, in a leaf or green stem; changes in turgor of a pair of guard cells cause opening and closing of the pore. (577)

gymnosperm [Gk. *gymnos,* naked + *sperma,* seed]: A seed plant in which the seeds are not enclosed in an ovary; the conifers are the most familiar group. (465)

habitat [L. *habitare,* to live in]: The place in which individuals of a particular species can usually be found. (964)

habituation [L. *habitus,* condition]: A response to a repeated stimulus in which the stimulus comes to be ignored and a previous behavior pattern is restored. (434)

half-life: The average time required for the disappearance or decay of one-half of any amount of a given substance. (887)

haploid [Gk. *haploos,* single + *ploion,* vessel]: Having only one set of chromosomes (n), in contrast to diploid ($2n$); characteristic of eukaryotic gametes, of gametophytes in plants, and of some protists and fungi. (250)

Hardy-Weinberg principle: The mathematical expression of the relationship between relative frequencies of two or more alleles in an idealized population; it states that both the allele frequencies and the genotype frequencies will remain constant in a population breeding at random in the absence of evolutionary forces. (893)

heat of vaporization: The amount of heat required to change a given amount of a liquid into a gas; 540 calories are required to change 1 gram of liquid water into vapor. (36)

heme [Gk. *haima,* blood]: The iron-containing group of heme proteins such as hemoglobin and the cytochromes. (70)

hemocoel [Gk. *haima,* blood + *koilos,* a hollow]: A blood-filled space within the tissues; characteristic of animals with an incomplete circulatory system, such as mollusks and arthropods. (509)

hemoglobin [Gk. *haima,* blood + L. *globus,* a ball]: The iron-containing protein in vertebrate blood that carries oxygen. (70)

hemophilia [Gk. *haima,* blood + *philios,* friendly]: A group of hereditary diseases characterized by failure of the blood to clot and consequent excessive bleeding from even minor wounds. (355)

herbaceous (her-**bay**-shus) [L. *herba,* grass]: In plants, nonwoody. (588)

herbivore [L. *herba,* grass + *vorare,* to devour]: A consumer that eats plants or other photosynthetic organisms to obtain its food and energy. (983)

heredity [L. *herres, heredis,* heir]: The transmission of characteristics from parent to offspring. (237)

hermaphrodite [Gk. *Hermes* and *Aphrodite*]: An organism possessing both male and female reproductive organs; hermaphrodites may or may not be self-fertilizing. (486)

heterosis [Gk. *heteros,* other, different]: Hybrid vigor; the overall superiority of the hybrid over either parent. (909)

heterotroph [Gk. *heteros,* other, different + *trophos,* feeder]: An organism that must feed on organic materials formed by other organisms in order to obtain energy and small building-block molecules; in contrast to autotroph. Animals, fungi, and many unicellular organisms are heterotrophs. (83)

heterozygote [Gk. *heteros,* other + *zugōtos,* a pair]: A diploid organism that carries two different alleles at one or more genetic loci. (241)

heterozygote superiority: The greater fitness of an organism heterozygous at a given genetic locus as compared with either homozygote. (909)

hibernation [L. *hiberna,* winter]: A period of dormancy and inactivity, varying in length, depending on the species, and occurring in dry or cold seasons. During hibernation, metabolic processes are greatly slowed and, even in mammals, body temperature may drop to just above freezing. (741)

histones: A group of five relatively small, basic polypeptide molecules found bound to the DNA of eukaryotic cells. (329)

homeostasis (home-e-o-**stay**-sis) [Gk. *homos,* same or similar + *stasis,* standing]: Maintenance of a relatively stable internal physiological environment or internal equilibrium in an organism. (666)

homeotherm [Gk. *homos,* same or similar + *therme,* heat]: An organism, such as a bird or mammal, capable of maintaining a stable body temperature independent of the environment. (736)

hominid [L. *homo,* man]: Humans and closely related primates; includes modern and fossil hominoids but not the apes. (1050)

hominoid [L. *homo,* man]: Hominids and the great apes. (1045)

homologues [Gk. *homologia,* agreement]: Chromosomes that carry corresponding genes and associate in pairs in the first stage of meiosis; each member of the pair is derived from a different parent. (250)

homology [Gk. *homologia,* agreement]: Similarity in structure and/or position, assumed to result from a common ancestry, regardless of function, such as the wing of a bird and the foreleg of a mammal. (378)

homozygote [Gk. *homos,* same or similar + *zugōtos,* a pair]: A diploid organism that carries identical alleles at one or more genetic loci. (241)

hormone [Gk. *hormaein,* to excite]: An organic molecule secreted, usually in minute amounts, in one part of an organism that regulates the function of another tissue or organ. (632, 776)

host: (1) An organism on or in which a parasite lives. (282)
(2) A recipient of grafted tissue. (859)

hybrid [L. *hybrida,* the offspring of a tame sow and a wild boar]:
(1) Offspring of two parents that differ in one or more inheritable characteristics. (240)
(2) Offspring of two different varieties or of two different species. (930)

hydrocarbon [L. *hydro,* water + *carbo,* charcoal]: An organic compound consisting of only carbon and hydrogen. (50)

hydrogen bond: A weak molecular bond linking a hydrogen atom that is covalently bonded to another atom (usually oxygen, nitrogen, or fluorine) to another oxygen, nitrogen, or fluorine atom of the same or another molecule. (34)

hydrolysis [L. *hydro,* water + Gk. *lysis,* loosening]: Splitting of one molecule into two by addition of H^+ and OH^- ions of water. (57)

hydrophilic [L. *hydro,* water + Gk. *philios,* friendly]: Having an affinity for water; applied to polar molecules or polar regions of large molecules. (40)

hydrophobic [L. *hydro,* water + Gk. *phobos,* fearing]: Having no affinity for water; applied to nonpolar molecules or nonpolar regions of molecules. (40)

hypertonic [Gk. *hyper,* above + *tonos,* tension]: Of two solutions of different concentration, the solution that contains the higher concentration of solute particles; water moves across a semipermeable membrane into a hypertonic solution. (130)

hypha [Gk. *hyphe,* web]: A single tubular filament of a fungus; the hyphae together make up the mycelium, the matlike "body" of a fungus. (439)

hypothalamus [Gk. *hypo,* under + *thalamos,* inner room]: The region of the vertebrate brain just below the cerebral hemispheres; responsible for the integration of many basic behavioral patterns that involve correlation of neural and endocrine functions. (667)

hypothesis [Gk. *hypo,* under + *tithenai,* to put]: A temporary working explanation or supposition based on accumulated facts and suggesting some general principle or relation of cause and effect; a postulated solution to a scientific problem that must be tested by experimentation and, if not validated, discarded. (10)

hypotonic [Gk. *hypo,* under + *tonos,* tension]: Of two solutions of different concentration, the solution that contains the lower concentration of solute particles; water moves across a semipermeable membrane from a hypotonic solution. (130)

immune response: A highly specific defensive reaction of the body to invasion by a foreign substance or organism; consists of a primary response in which the invader is recognized as foreign, or "not-self," and eliminated and a secondary response to subsequent attacks by the same invader; is mediated by two types of lymphocytes: B cells, primarily responsible for antibody production, and T cells, responsible for cell-mediated immunity. (748)

inbreeding: The mating of individuals closely related genetically. (237)

inclusive fitness: The total fitness of an individual plus the fitness of its relatives, the latter weighted according to the degree of relatedness. (1027)

incomplete dominance: In genetics, the phenomenon in which the effects of both alleles at a particular locus are apparent in the phenotype of the heterozygote. (262)

independent assortment: *See* Mendel's second law.

induction [L. *inducere,* to induce]: The process in an embryo in which one tissue or body part causes the differentiation of another tissue or body part. (858)

inflammatory response: A nonspecific defensive reaction of the body to invasion by a foreign substance or organism; involves phagocytosis by white blood cells and is often accompanied by accumulation of pus and an increase in the local temperature. (746)

insertion sequences: Relatively short sequences of DNA that can produce copies of themselves that become incorporated at other places in the same chromosome or in other chromosomes. (318)

insulin: A peptide hormone produced by the vertebrate pancreas, the action of which results in the lowering of the concentration of sugar in the blood. (67)

interferon: A protein made by virus-infected cells that inhibits viral multiplication. (747)

intermediate fibers: Fibrous protein filaments that form part of the cytoskeleton; found in greatest density in cells subject to mechanical stress. (106)

interneuron: Neuron that transmits nerve impulses from one neuron to another within the central nervous system; may receive impulses from and transmit impulses to many different neurons. (665)

interphase: The portion of the cell cycle that occurs before mitosis or meiosis can take place; includes the G_1, S, and G_2 phases. (141)

intervening sequence: A segment of DNA that is transcribed into RNA but is removed enzymatically from the RNA molecule before the mRNA enters the cytoplasm and is translated; an intron. (338)

intron: *See* Intervening sequence.

invagination [L. *in,* in + *vagina,* sheath]: The local infolding of a layer of tissue, especially in animal embryos, so as to form a depression or pocket opening to the outside. (863)

inversion: A chromosomal aberration in which a double break occurs and a segment is turned 180° before it is reincorporated into the chromosome. (275)

ion (**eye**-on): Any atom or small molecule containing an unequal number of electrons and protons and therefore carrying a net positive or net negative charge. (25)

ionic bond: A chemical bond formed as a result of the mutual attraction of ions of opposite charge. (25)

isogamy [Gk. *isos,* equal + *gamos,* marriage]: Sexual reproduction in which the gametes are morphologically alike. (424)

isolating mechanisms: Mechanisms that prevent genetic exchange between individuals of different populations or species; may be behavioral, morphological, or physiological. (932)

isotonic [Gk. *isos,* equal + *tonos,* tension]: Having the same concentration of solutes as another solution. If two isotonic solutions are separated by a semipermeable membrane, there will be no net flow of water across the membrane. (129)

isotope [Gk. *isos,* equal + *topos,* place]: Atom of an element that differs from other atoms of the same element in the number of neutrons in the atomic nucleus; isotopes thus differ in atomic weight. Some isotopes are unstable and emit radiation. (21)

karyotype [Gk. *kara,* the head + *typos,* stamp or print]: The general appearance of the chromosomes in a genome with regard to number, size, and shape. (349)

keratin [Gk. *karas,* horn]: One of a group of tough, fibrous proteins formed by certain epidermal tissues and especially abundant in skin, claws, hair, feathers, and hooves. (69)

kidney: In vertebrates, the organ that regulates the balance of water and solutes in the blood and the excretion of nitrogenous wastes in the form of urine. (725)

kinetic energy: Energy of motion. (36)

kinetochore [Gk. *kinetikos,* putting in motion + *choros,* chorus]: Disk-shaped protein structure within the centromere to which the spindle fibers attach during mitosis or meiosis. (143)

kingdom: A taxonomic grouping of related, similar phyla or divisions; the highest-level category in biological classification. (374)

kin selection: The differential reproduction of groups of related individuals, leading to the increase in frequency of alleles shared by members of the group. (1025)

Krebs cycle: Stage of cellular respiration in which pyruvate fragments are broken down into carbon dioxide; molecules reduced in the process can be used in ATP formation. (196)

lamella (lah-**mell**-ah) [L. dim. of *lamina,* plate or leaf]: Layer, thin sheet. (129)

larva [L., ghost]: An immature animal that is morphologically very different from the adult; examples are caterpillars and tadpoles. (486)

lateral meristem [L. *latus, lateris,* side + Gk. *meristos,* divided]: In vascular plants, one of the two rings of tissue (vascular cambium and cork cambium) that produce new cells for secondary growth. (606)

leaching: The dissolving of minerals and other elements in soil or rocks by the downward movement of water. (626)

leucoplast [Gk. *leukos,* white + *plastes,* molder]: In plants, a colorless cell organelle that serves as a starch respository; usually found in cells not exposed to light, such as roots and internal stem tissue. (112)

leukocyte [Gk. *leukos,* white + *kytos,* vessel]: White blood cell. Two principal types are phagocytic cells, involved in the inflammatory response, and lymphocytes, involved in immune reactions. (706)

lichen: Organism composed of an alga and a fungus that are symbiotically associated. (449)

life cycle: The entire span of existence of any organism from time of zygote formation (or asexual reproduction) until it itself reproduces. (250)

limbic system [L. *limbus,* border]: Neuron network forming a loop around the inside of the brain and connecting the hypothalamus to the cerebral cortex; thought to be circuit by which drives and emotions are translated into complex actions. (820)

linkage: The tendency for certain alleles to be inherited together because they are located on the same chromosome. (270)

lipid [Gk. *lipos,* fat]: One of a large variety of organic substances that are insoluble in polar solvents, such as water, but that dissolve readily in nonpolar organic solvents; includes fats, oils, waxes, steroids, phospholipids, and carotenes. (60)

locus, *pl.* **loci** [L., place]: In genetics, the position of a gene in a chromosome. For any given locus, there may be a number of possible alleles (as many as 30, for instance). (272)

loop of Henle (after F. G. J. Henle, German pathologist): A hairpin-shaped portion of the renal tubule found in birds and mammals in which a hypertonic urine is formed by processes of diffusion and active transport. (725)

lymph [L. *lympha,* water]: Colorless fluid derived from blood by filtration through capillary walls in the tissues; carried in special lymph ducts. (714)

lymphatic system: The system through which lymph circulates; consists of lymph capillaries, which begin blindly in the tissues, and a network of progressively larger vessels that empty into the vena cava; also includes the lymph nodes, spleen, thymus, and tonsils. (714)

lymph node: A mass of spongy tissues, separated into compartments; located throughout the lymphatic system, lymph nodes remove dead cells, debris, and foreign particles from the circulation. (748)

lymphocyte [L. *lympha,* water + Gk. *kytos,* vessel]: A type of white blood cell involved in the immune response; B cells differentiate into antibody-producing plasma cells, whereas T cells interact directly with the foreign invader. (749)

lysis [Gk. *lysis,* a loosening]: Disintegration of a cell by rupture of its cell membrane. (110)

lysogenic bacteria (lye-so-**jenn**-ick) [Gk. *lysis,* a loosening + *genos,* race or descent]: Bacteria carrying a bacteriophage integrated into the bacterial chromosome. The virus may subsequently set up an active cycle of infection, causing lysis of the bacterial cells. (316)

lysosome [Gk. *lysis,* loosening + *soma,* body]: A membrane-bound organelle in which hydrolytic enzymes are segregated. (110)

macromolecule [Gk. *makros,* large + L. dim. of *moles,* mass]: An extremely large molecule; refers specifically to proteins, nuclei acids, polysaccharides, and complexes of these. (110)

mandibles [L. *mandibula,* jaw]: In crustaceans, insects, and myriapods, the appendages immediately posterior to the antennae; used to seize, hold, bite, or chew food. (528)

mantle: In mollusks, the outermost layer of the body wall or a soft extension of it; usually secretes a shell. (508)

marine [L. *marini(us),* from *mare,* the sea]: Living in salt water. (486)

marsupial [Gk. *marsypos,* pouch, little bag]: A nonplacental mammal in which the female has a ventral pouch or folds surrounding the nipples; the premature young leave the uterus and crawl into the pouch, where each attaches itself by the mouth to a nipple until development is completed. (562)

matrix: The dense solution in the interior of the mitochondrion, surrounding the cristae; contains enzymes, phosphates, coenzymes, and other molecules involved in cellular respiration. (194)

mechanoreceptor: A sensory cell or organ that receives mechanical stimuli such as those involved in touch, pressure, hearing, and balance. (519)

medulla (med-**dull**-a) [L., the innermost part]: (1) The inner, as opposed to the outer, part of an organ, as in the adrenal gland. (785) (2) The most posterior region of the vertebrate brain; connects with the spinal cord. (818)

medusa: The free-swimming, bell- or umbrella-shaped stage in the life cycle of many cnidarians; a jellyfish. (487)

megaspore [Gk. *megas,* great, large + *spora,* a sowing]: In plants, a haploid (*n*) spore that develops into a female gametophyte. (464)

meiosis (my-**o**-sis) [Gk. *meioun,* to make smaller]: The two successive nuclear divisions in which a single diploid (2*n*) cell forms four haploid (*n*) nuclei, and segregation, crossing over, and reassortment of the alleles occur; gametes or spores may be produced as a result of meiosis. (249)

Mendel's first law: The factors for a pair of alternative characters are separate and only one may be carried in a particular gamete (genetic segregation). In modern form: Alleles segregate in meiosis. (240)

Mendel's second law: The inheritance of a pair of factors for one trait is independent of the simultaneous inheritance of factors for other traits, such factors "assorting independently" as though there were no other factors present (later modified by the discovery of linkage). Modern form: Unlinked genes assort independently. (242)

menstrual cycle [L. *mensis,* month]: In certain primates, the cyclic, hormone-regulated changes in the condition of the uterine lining; marked by the periodic discharge of blood and disintegrated uterine lining through the vagina. Mammals with a menstrual cycle lack a well-defined period of estrus. (844)

meristem [Gk. *meris,* part of, portion + *stamon,* the warp of a loom]: The undifferentiated plant tissue, including a mass of rapidly dividing cells, from which new tissues arise. (597)

mesenteries [Gk. *mesos,* middle + *enteron,* gut]: Double layers of mesoderm that suspend the digestive tract and other internal organs within the coelom. (494)

mesoderm [Gk. *mesos,* middle + *derma,* skin[: One of the three embryonic germ layers, it gives rise to muscle, connective tissue, the circulatory system, and most of the excretory and reproductive systems. (494)

mesophyll [Gk. *mesos,* middle + *phyllon,* leaf]: The internal tissue of a leaf, sandwiched between two layers of epidermal cells; consists of palisade parenchyma and spongy parenchyma cells. (576)

messenger RNA (mRNA): A class of RNA molecules, each of which is complementary to one strand of DNA and which serves to carry the genetic information from the chromosome to the ribosomes, where it is translated into protein. (298)

metabolism [Gk. *metabole,* change]: The sum of all chemical reactions occurring within a cell or organism. (165)

metamere [Gk. *meta,* middle + *meros,* part]: One of a linear series of similar body segments. (516)

metamorphosis [Gk. *metamorphoun,* to transform]: Abrupt transition from larval to adult form, such as the transition from tadpole to adult frog. (539)

metaphase [Gk. *meta,* middle + *phasis,* form]: The stage of mitosis or meiosis during which the chromosomes lie in the equatorial plane of the spindle. (144)

microfilament [Gk. *mikros,* small + L. *filare,* to spin]: A fine protein thread, consisting of molecules of the globular protein actin. Important constituents of the cytoskeleton, microfilaments are involved in cell motility. (106)

micronutrient [Gk. *mikros,* small + L. *nutrire,* to nourish]: A mineral required in only minute amounts for plant growth, such as iron, chlorine, copper, manganese, zinc, molybdenum, and boron. (618)

microspore [Gk. *mikros,* small + *spora,* a sowing of seeds]: In plants, a spore that develops into a male gametophyte; in seed plants, it becomes a pollen grain. (464)

microtrabecula, *pl.* **microtrabeculae** [Gk. *mikros,* small + L. *trabecula,* little beam]: Wisplike fibers visible in the cytoplasm under the high-voltage electron microscope that interconnect all of the other structures within the cell; the enzymes of glycolysis are thought to be located on the microtrabeculae. (107)

microtubule [Gk. *mikros,* small + L. dim. of *tubus,* tube]: An extremely small hollow tube composed of two types of globular protein subunits. Among their many functions, microtubules make up the internal structure of cilia and flagella. (70)

middle lamella: In plants, distinct layer between adjacent cell walls, rich in pectins and other polysaccharides; derived from the cell plate. (100)

mimicry [Gk. *mimos,* mime]: The superficial resemblance in form, color, or behavior of certain organisms (mimics) to other more powerful or more protected ones (models), resulting in protection, concealment, or some other advantage for the mimic. (924)

mineral: A naturally occurring element or inorganic compound. (618)

mitochondrion, *pl.* **mitochondria** [Gk. *mitos,* thread + *chondros,* cartilage or grain]: An organelle bound by a double membrane in which the reactions of the Krebs cycle, terminal electron transport, and oxidative phosphorylation take place, resulting in the formation of CO_2, H_2O, and ATP from acetyl CoA and ADP. Mitochondria are the organelles in which most of the ATP of the eukaryotic cell is produced. (112)

mitosis [Gk. *mitos,* thread]: Nuclear division characterized by chromosome replication and formation of two identical daughter nuclei. (143)

mole [L. *moles,* mass]: The amount of an element equivalent to its atomic weight expressed in grams, or the amount of a substance equivalent to its molecular weight expressed in grams. (42)

molecular weight: The sum of the atomic weights of the constituent atoms in a molecule. (42)

molecule [L. dim. of *moles,* mass]: A particle consisting of two or more atoms held together by chemical bonds; the smallest unit of a compound that displays the properties of the compound. (25)

molting: Shedding of all or part of an organism's outer covering; in arthropods, periodic shedding of the exoskeleton to permit an increase in size. (529)

monocotyledon [Gk. *monos,* single + *kotyledon,* a cup-shaped hollow]: A member of the class of flowering plants having one seed leaf, or cotyledon, among other distinguishing features; often abbreviated as monocot. (471)

monoecious (mo-**nee**-shus) [Gk. *monos,* single + *oikos,* house]: In angiosperms, having the male and female structures (the stamens and the carpels, respectively) on the same individual but on different flowers. (592)

monomer [Gk. *monos,* single + *meros,* part]: A simple, relatively small molecule that can be linked to others to form a polymer. (55)

monosaccharide [Gk. *monos,* single + *sakcharon,* sugar]: A simple sugar, such as glucose, fructose, ribose. (55)

monotreme [Gk. *monos,* single + *trēma,* hole]: A nonplacental mammal, such as the duckbilled platypus, in which the female lays shelled eggs and nurses the young. (562)

morphogenesis [Gk. *morphe,* form + *genesis,* origin]: The development of size, form, and other structural features of organisms. (597)

morphological [Gk. *morphe,* form + *logos,* discourse]: Pertaining to form and structure, at any level of organization. (372)

motor neuron: Neuron that transmits nerve impulses from the central nervous system to an effector, which is typically a muscle or a gland; an efferent neuron. (665)

muscle fiber: Muscle cell; a long, cylindrical, multinucleated cell containing numerous myofibrils, which is capable of contraction when stimulated. (808)

mutagen [L. *mutare,* to change + *genus,* source or origin]: A chemical or physical agent that increases the mutation rate. (264)

mutant [L. *mutare,* to change]: An organism carrying a gene that has undergone a mutation. (264)

mutation [L. *mutare,* to change]: The change of a gene from one allelic form to another; an inheritable change in the DNA sequence of a chromosome. (264)

mutualism [L. *mutuus,* lent, borrowed]: *See* Symbiosis.

mycelium [Gk. *mykes,* fungus]: The mass of hyphae forming the body of a fungus. (439)

mycorrhizae [Gk. *mykes,* fungus + *rhiza,* root]: Symbiotic associations between particular species of fungi and the roots of vascular plants. (451)

myelin sheath [Gk. *myelinos,* full of marrow]: A fatty layer surrounding the long fibers of neurons in the peripheral nervous system of vertebrates; made up of the membranes of Schwann cells. (772)

myofibril [Gk. *mys,* muscle + L. *fibra,* fiber]: Contractile element of a muscle fiber, made up of thick and thin filaments arranged in sarcomeres. (808)

myoglobin [Gk. *mys,* muscle + L. *globus,* a ball]: An oxygen-binding, heme-containing globular protein found in muscles. (700)

myosin [Gk. *mys,* muscle]: One of the principal proteins in muscle; makes up the thick filaments. (114)

NAD: Abbreviation of nicotinamide adenine dinucleotide, a coenzyme that functions as an electron acceptor. (173)

natural selection: A process of interaction between organisms and their environment that results in a differential rate of reproduction of different phenotypes in a population; can result in changes in the relative frequencies of alleles in the gene pool—that is, in evolution. (7)

nectar [Gk. *nektar,* the drink of the gods]: A sugary fluid that attracts insects to plants. (162)

negative feedback: A control mechanism whereby an increase in some substance inhibits the process leading to the increase; also known as feedback inhibition. (668)

nematocyst [Gk. *nema, nematos,* thread + *kyst,* bladder]: A threadlike stinger, containing a poisonous or paralyzing substance, found in the cnidocyte of cnidarians. (488)

nephridium, *pl.* **nephridia** [Gk. *nephros,* kidney]: A tubular excretory structure found in many invertebrates. (510)

nephron [Gk. *nephros,* kidney]: The functional unit of the kidney in reptiles, birds, and mammals; a human kidney contains about 1 million nephrons. (725)

nerve: A group or bundle of nerve fibers with accompanying connective tissue, located in the peripheral nervous system. A bundle of nerve fibers within the central nervous system is known as a tract. (763)

nerve fiber: A filamentous process extending from the cell body of a neuron and conducting the nerve impulse. (763)

nerve impulse: A rapid, transient, self-propagating change in electric potential across the membrane of a nerve fiber. (769)

nervous system: All the nerve cells of an animal; the receptor-conductor-effector system; in humans, the nervous system consists of brain, spinal cord, and all nerves. (487)

net productivity: In a trophic level, a community, or an ecosystem, the amount of energy (in calories) stored in chemical compounds or the increase in biomass (in grams or metric tons) in a particular period of time; it is the difference between gross productivity and the energy used by the organisms in respiration. (980)

neural groove: Dorsal, longitudinal groove that forms in a vertebrate embryo; bordered by two neural folds; preceded by the neural-plate stage and followed by the neural-tube stage. (857)

neural plate: Thickened strip of ectoderm in early vertebrate embryos that forms along the dorsal side of the body and gives rise to the central nervous system. (857)

neural tube: Primitive, hollow, dorsal nervous system of the early vertebrate embryo; formed by fusion of neural folds around the neural groove. (857)

neuron [Gk. *neuron,* nerve]: Nerve cell, including cell body, dendrites, and axon. (665)

neurosecretory cell [Gk. *neuron,* nerve]: A nerve cell that releases one or more hormones into the circulatory system. (541)

neurotransmitter [Gk. *neuron,* nerve + L. *trans,* across + *mittere,* to send]: A chemical agent released by one neuron that acts upon a second neuron or upon a muscle or gland cell and alters its electrical state or activity. (774)

neutron (**new**-tron): An uncharged particle with a mass slightly greater than that of a proton. Found in the atomic nucleus of all elements except hydrogen, in which the nucleus consists of a single proton. (20)

niche: *See* Ecological niche.

niche shift: *See* Character displacement.

nitrification: The oxidation of ammonia or ammonium to nitrites and nitrates, as by nitrifying bacteria. (989)

nitrogen cycle: Worldwide circulation and reutilization of nitrogen atoms, chiefly due to metabolic processes of living organisms; plants take up inorganic nitrogen and convert it into organic compounds (chiefly proteins), which are assimilated into the bodies of one or more animals; excretion and bacterial and fungal action on dead organisms return nitrogen atoms to the inorganic state. (988)

nitrogen fixation: Incorporation of atmospheric nitrogen into inorganic nitrogen compounds available to plants, a process that can be carried out only by some soil bacteria, many free-living and symbiotic cyanobacteria, and certain symbiotic bacteria in association with legumes. (627)

nitrogenous base: A nitrogen-containing molecule having basic properties (tendency to acquire an H^+ ion); a purine or pyrimidine. (53)

node [L. *nodus,* knot]: In plants, a joint of a stem, the place where branches and leaves are joined to the stem. (575)

nondisjunction [L. *non,* not + *disjungere,* to separate]: The failure of chromatids to separate during meiosis, resulting in one or more extra chromosomes in some gametes and correspondingly fewer in others. (351)

noradrenaline: A hormone produced by the medulla of the adrenal gland, which increases the concentration of sugar in the blood, raises blood pressure and heartbeat rate, and increases muscular power and resistance to fatigue; also one of the principal neurotransmitters; also called norepinephrine. (785)

norepinephrine: *See* noradrenaline.

notochord [Gk. *noto,* back + L. *chorda,* cord]: A dorsal rodlike structure that runs the length of the body and serves as the internal skeleton in the embryos of all chordates; in most adult chordates the notochord is replaced by a vertebral column that forms around (but not from) the notochord. (554)

nuclear envelope [L. *nucleus,* a kernel]: The double membrane surrounding the nucleus within a eukaryotic cell. (101)

nucleic acid: A macromolecule consisting of nucleotides; the principal types are deoxyribonucleic acid (DNA) and ribonucleic acid (RNA). (49)

nucleolus (new-**klee**-o-lus) [L. *nucleolus,* a small kernel]: A small, dense body, containing DNA, RNA, and protein, present in the nucleus of eukaryotic cells; site of production of ribosomal RNA. (103)

nucleosome [L. *nucleus,* a kernel + Gk. *soma,* body]: A complex of DNA and histone proteins that forms the fundamental packaging unit of eukaryotic DNA; its structure resembles a bead on a string. (330)

nucleotide [L. *nucleus,* a kernel]: A molecule composed of phosphate, a five-carbon sugar (either ribose or deoxyribose), and a purine or pyrimidine base; nucleotides are the building blocks of nucleic acids. (173)

nucleus [L., a kernel]: (1) The central core of an atom, containing protons and neutrons, around which electrons move. (20)
(2) The membrane-bound structure characteristic of eukaryotic cells that contains the genetic information in the form of DNA organized into chromosomes. (101)
(3) A group of nerve cell bodies within the central nervous system. (763)

ocellus, *pl.* **ocelli** [L. dim. of *oculus,* eye]: A simple light receptor common among invertebrates. (491)

olfactory [L. *olfacere,* to smell]: Pertaining to smell. (803)

ommatidium, *pl.* **ommatidia** [Gk. *ommos,* eye]: The single visual unit in the compound eye of arthropods; contains light-sensitive cells and a lens able to form an image. (542)

omnivore [L. *omnis,* all + *vorare,* to devour]: An organism that "eats everything"; for example, an animal that eats both plants and meat. (512)

oocyte (**o**-uh-sight) [Gk. *oion,* egg + *kytos,* vessel]: A cell that gives rise by meiosis to an ovum. (257)

oogamy (oh-**og**-amy) [Gk. *oion,* egg + *gamos,* marriage]: Sexual reproduction in which one of the gametes, usually the larger, is not motile. (424)

operator: A segment of DNA that interacts with a repressor protein to regulate the transcription of an operon. (305)

operon [L. *opus, operis,* work]: A group of adjacent structural genes transcribed onto a single mRNA molecule, whose functions are related to a particular biochemical pathway and whose transcription is regulated by a single repressor protein. (304)

opportunistic species: Species characterized by high reproduction rates, rapid development, early reproduction, small body size, and uncertain adult survival; also called prodigal or *r*-selected. (956)

opposable thumb: Thumb that rotates at the joint so that the tip of the thumb can be placed opposite the tip of any one of the four fingers. (1040)

orbital [L. *orbis,* circle, disk]: In the current model of atomic structure, the volume of space surrounding the atomic nucleus in which an electron will be found 90 percent of the time. (23)

order: A taxonomic grouping of related, similar families; the category below class and above family. (374)

organ [Gk. *organon,* tool]: A body part composed of several tissues grouped together in a structural and functional unit. (117)

organelle [Gk. *organon,* instrument, tool]: A formed body in the cytoplasm of a cell. (84)

organic [Gk. *organon,* instrument, tool]: Pertaining to (1) organisms or living things generally, or (2) compounds formed by living organisms, or (3) the chemistry of compounds containing carbon. (49)

organism [Gk. *organon,* instrument, tool]: Any living creature, either unicellular or multicellular. (371)

organizer [Gk. *organon,* instrument, tool]: The part of an embryo capable of inducing undifferentiated cells to follow a specific course of development; in particular, the dorsal lip of the blastopore in the amphibian. (858)

osmosis [Gk. *osmos,* impulse, thrust]: The diffusion of water across a selectively permeable membrane (a membrane that permits the free passage of water but prevents or retards the passage of a solute). In the absence of other factors that affect the water potential, the net movement of water is from the side containing a lower concentration of solute to the side containing a higher concentration. (129)

osmotic potential [Gk. *osmos,* impulse, thrust]: The tendency of water to move across a selectively permeable membrane into a solution; it is determined by measuring the pressure required to stop the osmotic movement of water into the solution; the higher the solute concentration, the greater the osmotic potential of the solution. (131)

ovary [L. *ovum,* egg]: (1) In animals, the egg-producing organ. (257)
(2) In flowering plants, the enlarged basal portion of a carpel or a fused carpel, containing the ovule or ovules; the ovary matures to become the fruit. (473)

oviduct [L. *ovum,* egg + *ductus,* duct]: The tube serving to transport the eggs to the outside or to the uterus; also called uterine tube. (496)

ovulation: In animals, release of an egg or eggs from the ovary. (843)

ovule [L. dim. of *ovum,* egg]: In seed plants, a structure composed of a protective outer coat, a tissue specialized for food storage, and a female gametophyte with egg cell; becomes a seed after fertilization. (464)

ovum, *pl.* **ova** [L., egg]: The egg cell; female gamete. (257)

oxidation: Gain of oxygen, loss of hydrogen, or loss of an electron by an atom, ion, or molecule. Oxidation and reduction take place simultaneously, with the electron lost by one reactant being transferred to another reactant. (164)

oxidative phosphorylation: The process by which the energy released as electrons pass down the mitochondrial electron transport chain in the final stage of cellular respiration is used to phosphorylate (add a phosphate group to) ADP molecules, thereby yielding ATP molecules. (198)

pacemaker: Area of the vertebrate heart that initiates the heartbeat; located where the superior vena cava enters the right atrium; sinoatrial node. (709)

paleontology [Gk. *palaios,* old + *onta,* things that exist + *logos,* discourse]: The study of the life of past geologic times, principally by means of fossils. (4)

palisade cells [L. *palus,* stake + *cella,* a chamber]: In plant leaves, the columnar, chloroplast-containing parenchyma cells of the mesophyll. (576)

pancreas (pang-kree-us) [Gk. *pan,* all + *kreas,* meat, flesh]: In vertebrates, a small, complex gland located between the stomach and the duodenum, which produces digestive enzymes and the hormones insulin and glucagon. (682)

parasite [Gk. *para,* beside, akin to + *sitos,* food]: An organism that lives on or in an organism of a different species and derives nutrients from it. (408)

parasitism: *See* Symbiosis.

parasympathetic division [Gk. *para,* beside, akin to]: A subdivision of the autonomic nervous system of vertebrates, with centers located in the brain and in the most anterior part and the most posterior parts of the spinal cord; stimulates digestion; generally inhibits other functions and restores the body to normal following emergencies. Neurotransmitter: acetylcholine. (767)

parenchyma (pah-**renk**-ee-ma) [Gk. *para,* beside, akin to + *en,* in + *chein,* to pour]: A plant tissue composed of living, thin-walled, randomly arranged cells with large vacuoles; usually photosynthetic or storage tissue. (576)

parthenogenesis [Gk. *parthenon,* virgin + *genesis,* birth]: The development of an egg without fertilization. (502)

pecking order: A social ranking in poultry flocks in which each bird dominates (can peck) others lower in rank, but is dominated by birds higher in rank. (1022)

pellicle [L. dim. of *pellis,* skin]: A flexible series of protein strips inside the cell membrane of many protists. (418)

peptide bond [Gk. *pepto,* to soften, digest]: The type of bond formed when two amino acids are joined end to end; the acidic group ($-COOH$) of one amino acid is linked covalently to the basic group ($^{-}NH_2$) of the next, and a molecule of water (H_2O) is removed. (66)

perennial [L. *per,* through + *annus,* year]: A plant that persists in whole or in part from year to year and usually produces reproductive structures in more than one year. (588)

pericycle [Gk. *peri,* around + *kyklos,* circle]: One or more layers of cells completely surrounding the vascular tissues of the root; branch roots arise from the pericycle. (586)

peripheral nervous system [Gk. *peripherein,* to carry around]: All of the neurons and nerve fibers outside the central nervous system, including both motor neurons and sensory neurons; consists of the somatic nervous system and the autonomic nervous system. (763)

peristalsis [Gk. *peristellein,* to wrap around]: Successive waves of muscular contraction in the walls of a tubular structure, such as the digestive tract or an oviduct; moves the contents, such as food or an egg cell, through the tube. (672)

peritoneum [Gk. *peritonos,* stretched over]: A membrane that lines the body cavity and forms the external covering of the visceral organs. (673)

permeable [L. *permeare,* to pass through]: Penetrable by molecules, ions, or atoms; usually applied to membranes that let given solutes pass through. (129)

peroxisome: A membrane-bound organelle in which enzymes catalyzing peroxide-forming and peroxide-destroying reactions are segregated; in plant cells, the site of photorespiration. (111)

petiole (pet-ee-ole) [Fr., from L. *petiolus,* dim. of *pes, pedis,* a foot]: The stalk of a leaf, connecting the blade of the leaf with the branch or stem. (576)

pH: A symbol denoting the relative concentration of hydrogen ions in a solution; pH values range from 0 to 14; the lower the value, the more acidic a solution, that is, the more hydrogen ions it contains; pH 7 is neutral, less than 7 is acidic, more than 7 is alkaline. (42)

phagocytosis [Gk. *phagein,* to eat + *kytos,* vessel]: Cell "eating"; the intake of solid particles by a cell, by flowing over and engulfing them; characteristic of amoebas, the digestive cells of some invertebrates, and vertebrate white blood cells. (135)

phenotype [Gk. *phainein,* to show + *typos,* stamp, print]: Observable properties of an organism, resulting from interactions between the genotype and the environment. (242)

pheromone (**fair**-o-moan) [Gk. *phero,* to bear, carry]: Substance secreted by an animal that influences the behavior or morphological development of other animals of the same species, such as the sex attractants of moths, the odor trail of ants. (547)

phloem (**flow**-em) [Gk. *phloos,* bark]: Vascular tissue of higher plants; conducts sugars and other organic molecules from the leaves to other parts of the plant; in angiosperms, composed of sieve tubes and companion cells, parenchyma, and fibers. (461)

phospholipids: Organic molecules similar in structure to fats, but in which a phosphate group rather than a fatty acid is attached to the third carbon of the glycerol molecule; as a result, the molecule has a hydrophilic "head" and a hydrophobic "tail." Phospholipids form the basic structure of cellular membranes. (62)

phosphorylation: Addition of a phosphate group or groups to a molecule. (183)

photon [Gk. *photos,* light]: The elementary particle of light and other electromagnetic radiations. (213)

photoperiodism [Gk. *photos,* light]: The response to relative day and night length, a mechanism by which organisms measure seasonal change. (644)

photophosphorylation [Gk. *photos,* light + *phosphoros,* bringing light]: The process by which the energy released as electrons pass down the electron transport chain between photosystems II and I during photosynthesis is used to phosphorylate ADP to ATP. (220)

photoreceptor [Gk. *photos,* light]: A cell or organ capable of detecting light. (418)

photosynthesis [Gk. *photos,* light + *syn,* together + *tithenai,* to place]: The conversion of light energy to chemical energy; the synthesis of organic compounds from carbon dioxide and water in the presence of chlorophyll, using light energy. (210)

phototropism [Gk. *photos,* light + *trope,* turning]: Movement in which the direction of the light is the determining factor, such as the growth of a plant toward a light source; turning or bending response to light. (642)

phyletic evolution [Gk. *phylon,* race, tribe + L. *e-,* out + *volvere,* to roll]: The changes taking place in a single lineage of organisms over a long period of time; one of the principal patterns of organic evolution; also known as anagenesis. (938)

phylogeny [Gk. *phylon,* race, tribe]: Evolutionary history of a taxonomic group. Phylogenies are often depicted as "evolutionary trees." (376)

phylum [Gk. *phylon,* tribe, stock]: A taxonomic grouping of related, similar classes; a high-level category beneath kingdom and above class. Phylum is generally used in the classification of protozoa and animals, whereas an equivalent category, division, is used in the classification of prokaryotes, algae, fungi, and plants. (374)

physiology [Gk. *physis,* nature + *logos,* a discourse]: The study of function in cells, organs, or entire organisms; the processes of life. (481)

phytochrome [Gk. *phyton,* plant + *chrōma,* color]: A plant pigment that is a photoreceptor for red or far-red light and is involved with a number of developmental processes, such as flowering, dormancy, leaf formation, and seed germination. (645)

phytoplankton [Gk. *phyton,* plant + *planktos,* wandering]: Aquatic, free-floating, microscopic, photosynthetic organisms. (229)

pigment [L. *pigmentum,* paint]: A colored substance that absorbs light over a narrow band of wavelengths. (214)

pinocytosis [Gk. *pinein,* to drink + *kytos,* vessel]: Cell "drinking"; the intake of fluid droplets by a cell, probably by a mechanism similar to that of phagocytosis but triggered by different stimuli. (135)

pituitary [L. *pituita,* phlegm]: Endocrine gland in vertebrates; the anterior lobe is the source of tropic hormones, growth hormone, and prolactin and is stimulated by secretions of the hypothalamus; the posterior lobe stores and releases oxytocin and ADH produced by the hypothalamus. (782)

placenta [Gk. *plax,* a flat object]: A tissue formed in part from the inner lining of the mammalian uterus and in part from the extraembryonic membranes; serves as the connection through which exchanges of nutrients and wastes occur between the blood of the mother and that of the embryo. (867)

plankton [Gk. *planktos,* wandering]: Small (mostly microscopic) aquatic and marine organisms found in the upper levels of the water, where light is abundant; includes both photosynthetic (phytoplankton) and heterotrophic (zooplankton) forms. (416)

planula [L. dim. of *planus,* a wanderer]: The ciliated, free-swimming type of larva formed by many cnidarians. (488)

plasma [Gk. *plasma,* form or mold]: The clear, colorless fluid component of vertebrate blood, containing dissolved salts and proteins; blood minus the blood cells. (705)

plasma cell: An antibody-producing cell resulting from the multiplication and differentiation of a B lymphocyte that has interacted with an antigen; a mature plasma cell can produce from 3,000 to 30,000 antibody molecules per second. (750)

plasma membrane: The membrane surrounding the cytoplasm of a cell; the cell membrane. (84)

plasmid: An extrachromosomal, independently replicating, small, circular DNA molecule. (311)

plasmodesma, *pl.* **plasmodesmata** [Gk. *plassein,* to mold + *desmos,* band, bond]: In plants, a minute, cytoplasmic thread that extends through pores in cell walls and connects the cytoplasm of adjacent cells. (118)

plastid [Gk. *plastos,* formed or molded]: A cytoplasmic, often pigmented, organelle in plant cells; includes leucoplasts, chromoplasts, and chloroplasts. (112)

platelet (**plate**-let) [Gk. *platus,* flat]: In mammals, a minute, granular body suspended in the blood and involved in the formation of blood clots. (706)

pleiotropy (**plee**-o-trope-ee) [Gk. *pleios,* more + *trope,* a turning]: The capacity of a gene to affect a number of different phenotypic characteristics. (266)

poikilotherm [Gk. *poikilos,* changeable + *therme,* heat]: An organism with a body temperature that varies with that of the environment; an ectotherm. (736)

polar [L. *polus,* end of axis]: Having parts or areas with opposed or contrasting properties, such as positive and negative charges, head and tail. (27)

polar body: Minute, nonfunctioning cell produced during those meiotic divisions that lead to egg cells; contains a nucleus but very little cytoplasm. (257)

polar covalent bond: A covalent bond in which the electrons are shared unequally between the two atoms; the resulting polar molecule has regions of slightly negative and slightly positive charge. (27)

pollen [L., fine dust]: In seed plants, spores consisting of an immature male gametophyte and a protective outer covering. (469)

pollination [L. *pollen,* fine dust]: The transfer of pollen from the anther to a receptive surface of a flower. (240)

polygenic inheritance [Gk. *polus,* many + *genos,* race, descent]: The determination of a given characteristic, such as weight or height, by the interaction of many genes. (265)

polymer [Gk. *polus,* many + *meris,* part or portion]: A large molecule composed of many similar or identical molecular subunits. (55)

polymorphism [Gk. *polus,* many + *morphe,* form]: The presence in a single population of two or more phenotypically distinct forms of a trait. (907)

polyp [Gk. *polus,* many + *pous,* foot]: The sessile stage in the life cycle of cnidarians. (487)

polypeptide [Gk. *polus,* many + *pepto,* to soften, digest]: A molecule consisting of a long chain of amino acids linked together by peptide bonds. (66)

polyploid [Gk. *polus,* many + *ploion,* vessel]: Cell with more than two complete sets of chromosomes per nucleus. (250)

polyribosome: Two or more ribosomes together with a molecule of mRNA that they are simultaneously translating; a polysome. (301)

polysaccharide [Gk. *polus,* many + *sakcharon,* sugar]: A carbohydrate polymer composed of monosaccharide monomers in long chains; includes starch, cellulose. (58)

polysome: *See* Polyribosome.

population: Any group of individuals of one species that occupy a given area at the same time; in genetic terms, an interbreeding group of organisms. (892)

population bottleneck: Type of genetic drift that occurs as the result of a population being drastically reduced in numbers by an event having little to do with the usual forces of natural selection. (897)

posterior: Of or pertaining to the rear end. In humans, the back of the body is said to be posterior. (494)

potential energy: Energy in a potentially usable form that is not, for the moment, being used; often called "energy of position." (22)

predator [L. *praedari,* to prey upon; from *prehendere,* to grasp, seize]: An organism that eats other living organisms. (157)

pressure-flow hypothesis: A hypothesis accounting for sap flow through the phloem system. According to this hypothesis, the solution containing nutrient sugars moves through the sieve tubes by bulk flow, moving into and out of the sieve tubes by active transport. (624)

prey [L. *prehendere,* to grasp, seize]: An organism eaten by another organism. (188)

primary growth: In plants, growth originating in the apical meristem of the shoots and roots, as contrasted with secondary growth; results in an increase in length. (601)

primary structure of a protein: The amino acid sequence of a protein. (66)

primate: A member of the order of mammals that includes anthropoids and prosimians. (563)

primitive [L. *primus,* first]: Not specialized; at an early stage of evolution or development. (9)

primitive streak [L. *primus,* first]: The thickened, dorsal, longitudinal strip of ectoderm and mesoderm in early avian, reptilian, and mammalian embryos; equivalent to blastopore in other forms. (860)

procambium [L. *pro,* before + *cambium,* exchange]: In plants, a primary meristematic tissue; gives rise to vascular tissues of the primary plant body and to the vascular cambium. (596)

producer, in ecological systems: An autotrophic organism, usually a photosynthesizer, that contributes to the net primary productivity of a community. (981)

productivity: A measure of the rate at which energy is assimilated by the organisms in a trophic level, a community, or an ecosystem. (980)

progesterone [L. *progerere,* to carry forth or out + *steiras,* barren]: In mammals, a steroid hormone produced by the corpus luteum that prepares the uterus for implantation of the ovum. (780)

prokaryote [L. *pro,* before + Gk. *karyon,* nut, kernel]: A cell lacking a membrane-bound nucleus and membrane-bound organelles; a bacterium or a cyanobacterium. (84)

promoter: Specific segment of DNA to which RNA polymerase attaches to initiate transcription of mRNA from an operon. (305)

prophage: A bacterial virus (bacteriophage) integrated into a host chromosome. (316)

prophase [Gk. *pro,* before + *phasis,* form]: An early stage in nuclear division, characterized by the condensing of the chromosomes and their movement toward the equator of the spindle. Homologous chromosomes pair up during meiotic prophase. (143)

proprioceptor [L. *proprius,* one's own]: Receptor that senses movements, position of the body, or muscle strength. (545)

prosimian [L. *pro,* before + *simia,* ape]: A lower primate; includes lemurs, lorises, tarsiers, and tree shrews, as well as many fossil forms. (1042)

prostaglandins [Gk. *prostas,* a porch or vestibule + L. *glaus,* acorn]: A group of modified fatty acids that function as chemical messengers; synthesized in most, possibly all, cells of the body. (787)

prostate gland [Gk. *prostas,* a porch or vestibule + L. *glaus,* acorn]: A mass of muscle and glandular tissue surrounding the base of the urethra in male mammals; the vasa deferentia merge with ducts from the seminal vesicles, enter the prostate gland, and there merge with the urethra. The prostate gland secretes an alkaline fluid that has a stimulating effect on the sperm as they are released. (841)

protein [Gk. *proteios,* primary]: A complex organic compound composed of one or more polypeptide chains, each made up of many (about 100 or more) amino acids linked together by peptide bonds. (64)

proton: A subatomic particle with a single positive charge equal in magnitude to the charge of an electron and with a mass slightly less than that of a neutron; a component of every atomic nucleus. (20)

protoplasm [Gk. *protos,* first + *plasma,* anything molded]: Living matter. (427)

protostome [Gk. *protos,* first + *stoma,* mouth]: An animal in whose embryonic development the mouth forms at or near the blastopore. Protostomes are also characterized by spiral cleavage during the earliest stages of development and by schizocoelous formation of the coelom. (506)

pseudocoelom [Gk. *pseudes,* false + *koilos,* a hollow]: A body cavity consisting of a fluid-filled space between the endoderm and the mesoderm; characteristic of the nematodes. (494)

pseudopod [Gk. *pseudes,* false + *pous, pod-,* foot]: A temporary cytoplasmic protrusion from an amoeboid cell, which functions in locomotion or in feeding by phagocytosis. (429)

punctuated equilibrium: A new model of the mechanism of evolutionary change that proposes that long periods of no change ("stasis") are punctuated by periods of rapid speciation, with natural selection acting on species as well as on individuals. (943)

Punnett square: The checkerboard diagram used for analysis of gene segregation. (243)

pupa [L., girl, doll]: A developmental stage of some insects, in which the organism is nonfeeding, immotile, and sometimes encapsulated or in a cocoon; the pupal stage occurs between the larval and adult phases. (541)

purine [Gk. *purinos,* fiery, sparkling]: A nitrogenous base such as adenine or guanine; one of the components of nucleic acids. (281)

pyramid, ecological: *See* Ecological pyramid.

pyramid of energy: A diagram of the energy flow between the trophic levels of a food chain; plants or other autotrophs (at the base of the pyramid) represent the greatest amount of energy, herbivores next, then primary carnivores, secondary carnivores, etc. (986)

pyrimidine: A nitrogenous base such as cytosine, thymine, or uracil; one of the components of nucleic acids. (281)

quaternary structure of a protein: The overall structure of a globular protein molecule that consists of two or more polypeptide chains. (68)

queen: In social insects (ants, termites, and some species of bees and wasps), the fertile, or fully developed, female whose function is to lay eggs. (1019)

radial symmetry [L. *radius,* a spoke of a wheel + Gk. *summetros,* symmetry]: The regular arrangement of parts around a central axis such that any plane passing through the central axis divides the organism into halves that are approximate mirror images; seen in cnidarians, ctenophorans, and adult echinoderms. (486)

radiation [L. *radius,* a spoke of a wheel, hence, a ray]: Energy emitted in the form of waves or particles. (212)

radioactive isotope: An isotope with an unstable nucleus that stabilizes itself by emitting radiation. (21)

recessive allele [L. *recedere,* to recede]: An allele whose phenotypic effect is masked in the heterozygote by that of another, dominant allele. (241)

reciprocal altruism: Performance of an altruistic act with the expectation that the favor will be returned. (1036)

recombinant DNA: DNA formed either naturally or in the laboratory by the joining of segments of DNA from different sources. (310)

recombination: The formation of new gene combinations; in eukaryotes, may be accomplished by new associations of chromosomes produced during sexual reproduction or crossing over; in prokaryotes, may be accomplished through transformation, conjugation, or transduction. (271)

reduction [L. *reducere,* to lead back]: Loss of oxygen, gain of hydrogen, or gain of an electron by an atom, ion, or molecule; oxidation and reduction take place simultaneously, with the electron lost by one reactant being transferred to another. (164)

reflex [L. *reflectere,* to bend back]: Unit of action of the nervous system involving a sensory neuron, often an interneuron or -neurons, and one or more motor neurons. (666)

releasing hormone: One of at least nine small peptide hormones produced by the hypothalamus that stimulate or inhibit the secretion of specific hormones by the anterior pituitary. (782)

renal [L. *renes,* kidneys]: Pertaining to the kidney. (725)

repressor [L. *reprimere,* to press back, keep back]: In genetics, a protein that binds to the operator, preventing RNA polymerase from attaching to the promoter and transcribing the structural genes of the operon. (305)

resolving power [L. *resolvere,* to loosen, unbind]: The ability of a lens to distinguish two lines as separate. (90)

respiration [L. *respirare,* to breathe]: (1) In aerobic organisms, the intake of oxygen and the liberation of carbon dioxide. (688)
(2) In cells, the oxygen-requiring stage in the breakdown and release of energy from fuel molecules. (193)

resting potential: The difference in electric potential (about 70 millivolts) across the membrane of an axon at rest. (770)

restriction enzymes: Enzymes that cleave the DNA double helix at specific sites. (319)

reticular activating system [L. *reticulum,* a network]: A network of fibers and neurons; includes the reticular formation, a core of tissue that runs centrally through the brainstem, and its thalamic extension; involved with alertness and direction of attention to selected events. (820)

retina [L. dim. of *rete,* net]: The photosensitive layer of the vertebrate eye; contains several layers of neurons and light-receptors (rods and cones); receives the image formed by the lens and transmits it to the brain via the optic nerve. (798)

retrovirus [L., turning back]: An RNA virus whose RNA codes for an enzyme, reverse transcriptase, that transcribes the RNA into DNA. (315)

reverse transcriptase: An enzyme that transcribes RNA into DNA; found only in association with retroviruses. (315)

rhizoid [Gk. *rhiza,* root]: Rootlike anchoring structure in fungi and nonvascular plants. (442)

rhizome [Gk. *rhizoma,* mass of roots]: In vascular plants, an underground stem; may be enlarged for storage, or may function in vegetative reproduction. (463)

ribonucleic acid (RNA) (rye-bo-new-**klee**-ick): A class of nucleic acids characterized by the presence of the sugar ribose and the pyrimidine uracil; includes mRNA, tRNA, and rRNA. RNA is the genetic material of many viruses. (297)

ribosomal RNA (rRNA): A class of RNA molecules found, along with characteristic proteins, in ribosomes; transcribed from the DNA of the nucleolus. (299)

ribosome: A small organelle composed of protein and ribonucleic acid; the site of translation in protein synthesis; in eukaryotic cells, often bound to the endoplasmic reticulum. Many ribosomes attached to a single strand of mRNA are called a polyribosome, or polysome. (85)

RNA: Abbreviation of ribonucleic acid. (297)

rod: Light-sensitive nerve cell found in the vertebrate retina; sensitive to very dim light, responsible for "night vision." (799)

root: The descending axis of a plant, normally below ground and serving both to anchor the plant and to take up and conduct water and minerals. (584)

sarcomere [Gk. *sarx,* the flesh + *meris,* part of, portion]: Functional and structural unit of contraction in striated muscle. (809)

schizocoelous [Gk. *schizo,* to split + *koilos,* a hollow]: Formation of the coelom during embryonic development by a splitting of the mesoderm; characteristic of protostomes. (506)

secondary sex characteristics: The many external differences between male and female that are not directly involved in reproduction. (778)

secondary structure of a protein: The simple structure (often a helix, a sheet, or a cable) resulting from the spontaneous folding of a polypeptide chain as it is formed; maintained by hydrogen bonds and other weak forces. (67)

secretion [L. *secermere,* to sever, separate]: (1) Product of any cell, gland, or tissue that is released through the cell membrane and that performs its function outside the cell that produced it. (135)
(2) The stage of kidney function in which, through active transport processes, molecules remaining in the blood plasma are selectively removed from the peritubular capillaries and pumped into the filtrate in the renal tubule. (726)

seed: A complex organ formed by the maturation of the ovule of seed plants following fertilization; upon germination, a seed develops into a new sporophyte; generally consists of seed coat, embryo, and a food reserve. (470)

segregation: *See* Mendel's first law.

selectively permeable [L. *seligere,* to gather apart + *permeare,* to go through]: Applied to membranes that permit passage of water and some solutes but block passage of most solutes; semipermeable. (129)

self-fertilization: The union of egg and sperm produced by a single hermaphroditic organism. (240)

self-pollination: The transfer of pollen from anther to stigma in the same flower or to another flower of the same plant, leading to self-fertilization. (240)

semen [L., seed]: Product of the male reproductive system; includes sperm and the sperm-carrying fluids. (841)

seminal vesicles [L. *semen,* seed + *vesicula,* a little bladder]: In male mammals, small vesicles, the ducts of which merge with the vasa deferentia as they enter the prostate gland; they produce an alkaline, fructose-containing fluid that suspends and nourishes the sperm cells. (841)

sensillum, *pl.* **sensilla:** Sensory-receptor unit on the body surface of arthropods; responsive to touch, smell, taste, and vibration (sound). (543)

sensory neuron: Neuron that transmits nerve impulses from a receptor to the central nervous system or central ganglion; an afferent neuron. (665)

sensory receptor: A cell, tissue, or organ that detects internal or external stimuli. (489)

sessile [L. *sedere,* to sit]: Attached; not free to move about. (484)

sex chromosomes: Chromosomes that are different in the two sexes and that are involved in sex determination. (268)

sex-linked characteristic: A genetic characteristic, such as color blindness, determined by a gene located on a sex chromosome and that therefore shows a different pattern of inheritance in males and females. (269)

sexual reproduction: Reproduction involving meiosis and fertilization. (249)

shoot: The aboveground portions, such as the stem and leaves, of a vascular plant. (575)

sieve cell: A long, slender cell of the phloem of gymnosperms; involved in transport of materials synthesized in the leaves to other parts of the plant. (461)

sieve tube: A series of sugar-conducting cells (sieve-tube members) found in the phloem of angiosperms. (580)

sigmoid growth: A pattern of population growth in which growth is rapid when the population is small, gradually slows as the population approaches some environmental limit, and then oscillates as the population stabilizes at or near its carrying capacity; it is the simplest growth pattern observed for most populations in nature. (953)

sinoatrial node: *See* Pacemaker.

smooth muscle: Nonstriated muscle; lines the walls of internal organs and arteries and is under involuntary control. (662)

social dominance: A hierarchical pattern of social organization involving domination of some members of a group by other members in a relatively orderly and long-lasting pattern. (1022)

society [L. *socius,* companion]: An organization of individuals of the same species in which there are divisions of labor and mutual dependence. (1019)

sociobiology: The study of the biological basis of social behavior. (29)

solution: A homogeneous mixture of the molecules of two or more substances; the substance present in the greatest amount (usually a liquid) is called the solvent, and the substances present in lesser amounts are called solutes. (40)

somatic cells [Gk. *soma,* body]: The differentiated cells composing body tissues of multicellular plants and animals; all body cells except those giving rise to gametes. (250)

somatic nervous system [Gk. *soma,* body]: In vertebrates, the motor and sensory neurons of the peripheral nervous system that control skeletal muscle; the "voluntary" system, as contrasted with the "involuntary," or autonomic, nervous system. (765)

somite: One of the blocks, or segments, of tissue into which the mesoderm is divided during differentiation of the vertebrate embryo. (858)

specialized (1) Of cells, having particular functions in a multicellular organism. (86)
(2) Of organisms, having special adaptations to a particular habitat or mode of life. (918)

speciation: The process by which new species are formed. (928)

species, *pl.* **species** [L., kind, sort]: A group of organisms that actually (or potentially) interbreed in nature and are reproductively isolated from all other such groups; a taxonomic grouping of morphologically similar individuals (the category beneath genus). (371)

species-specific: Characteristic of (and limited to) a particular species. (934)

specific: Unique; for example, the proteins in a given organism, the enzyme catalyzing a given reaction, or the antibody to a given antigen. (70)

specific heat: The amount of heat (in calories) required to raise the temperature of 1 gram of a substance 1°C. The specific heat of water is 1 calorie per gram. (36)

sperm [Gk. *sperma,* seed]: A mature male sex cell, or gamete, usually motile and smaller than the female gamete. (257)

spermatheca [Gk. *sperma,* seed + *theke,* box, coffin, vault]: Receptacle for sperm storage; found in many species of female invertebrates. (520)

spermatid [Gk. *sperma,* seed]: Each of four haploid (*n*) cells resulting from the meiotic divisions of a spermatocyte; each spermatid becomes differentiated into a sperm cell. (257)

spermatocytes [Gk. *sperma,* seed + *kytos,* vessel]: The diploid ($2n$) cells formed by the enlargement of the spermatogonia; they give rise by meiotic division to the spermatids. (257)

spermatogenesis [Gk. *sperma,* seed + *genesis,* origin]: The process by which spermatogonia develop into sperm. (837)

spermatogonia [Gk. *sperma,* seed + *gonos,* a child, the young]: The unspecialized diploid ($2n$) cells on the walls of the testes that, by meiotic division, become spermatocytes, then spermatids, then sperm cells. (257)

spermatozoon, *pl.* **spermatozoa** [Gk. *sperma,* seed + *zoos,* alive, living]: A sperm cell. (838)

spinal cord: Part of the vertebrate central nervous system; consists of a thick, dorsal, longitudinal bundle of nerve fibers extending posteriorly from the brain. (764)

spindle: In dividing cells, the structure formed of microtubules that extends from pole to pole; the spindle fibers appear to maneuver the chromosomes into position during metaphase and to pull the newly separated chromosomes toward the poles during anaphase. (146)

spiracle [L. *spirare,* to breathe]: One of the external openings of the respiratory system in terrestrial arthropods. (530)

splitting evolution: A type of evolutionary change characterized by the splitting, or branching, of phylogenetic lineages; one of the principal patterns of organic evolution; also known as cladogenesis. (938)

sporangiophore (spo-**ran**-ji-o-for) [Gk. *spora,* seed + *phore,* from *phorein,* to bear]: A specialized hypha or a branch bearing one or more sporangia. (441)

sporangium, *pl.* **sporangia** [Gk. *spora,* seed]: A unicellular or multicellular structure in which spores are produced. (441)

spore [Gk. *spora,* seed]: An asexual reproductive cell capable of developing into an adult without fusion with another cell; in contrast to a gamete. (424)

sporophyll [Gk. *spora,* seed + *phyllon,* the leaves]: Spore-bearing leaf. The carpels and stamens of flowers are modified sporophylls. (463)

sporophyte [Gk. *spora,* seed + *phytos,* growing]: In plants, which have alternation of haploid and diploid generations, the diploid ($2n$) spore-producing generation. (424)

stamen [L., a thread]: The male structure of a flower, which produces microspores or pollen; usually consists of a stalk, the filament, bearing a pollen-producing anther at its tip. (473)

starch [M.E. *sterchen,* to stiffen]: A class of complex, insoluble carbohydrates, the chief food-storage substances of plants; composed of 1,000 or more sugar units and readily broken down enzymatically into these units. (58)

statocyst [Gk. *statos,* standing + *kystis,* sac]: An organ of balance, consisting of a vesicle containing granules of sand (statoliths) or some other material that stimulates sensory cells when the organism moves. (491)

stem: The aboveground part of the axis of vascular plants, as well as anatomically similar portions below ground (such as rhizomes). (579)

stereoscopic vision [Gk. *stereos,* solid + *optikos,* pertaining to the eye]: Ability to perceive a single, three-dimensional image from the simultaneous but separate images delivered to the brain by each eye. (797)

steroid: One of a group of lipids having four linked carbon rings and, often, a hydrocarbon tail; cholesterol, sex hormones, and the hormones of the adrenal cortex are steroids. (63)

stigma [Gk. *stigme,* a prick mark, puncture]: In plants, the region of a carpel serving as a receptive surface for pollen grains, which germinate on it. (473)

stimulus [L., goad, incentive]: Any internal or external change or signal that influences the activity of an organism or of part of an organism. (19)

stoma, *pl.* **stomata** [Gk. *stoma,* mouth]: A minute opening bordered by guard cells in the epidermis of leaves and stems through which gases pass. (225)

strategy [Gk. *strategein,* to maneuver]: A group of coadapted traits, often including anatomical, physiological, and behavioral characteristics. (54)

striated muscle [L., from *striare,* to groove]: Skeletal voluntary muscle and cardiac muscle. The name derives from the striped appearance, which reflects the arrangement of contractile elements. (662)

stroma [Gk. *stroma,* a bed, from *stronnymi,* to spread out]: The ground substance that makes up the interior of the chloroplast and surrounds the thylakoids. (218)

structural gene: Any gene that produces a protein; in distinction to regulatory genes. (304)

style [L. *stilus,* stake, stalk]: In angiosperms, the stalk of a carpel, down which the pollen tube grows. (473)

substrate [L. *substratus,* strewn under]: (1) The foundation to which an organism is attached. (484)
(2) A substance on which an enzyme acts. (168)

succession: *See* Ecological succession.

sucrose: Cane sugar; a common disaccharide found in many plants; a molecule of glucose linked to a molecule of fructose. (57)

sugar: Any monosaccharide or disaccharide. (52)

surface tension: A tautness of the surface of a liquid, caused by the cohesion of the molecules of liquid. Water has an extremely high surface tension. (35)

symbiosis [Gk. *syn,* together with + *bioonai,* to live]: An intimate and protracted association between two or more organisms of different species. Includes mutualism, in which the association is beneficial to both; commensalism, in which one benefits and the other is neither harmed nor benefited; and parasitism, in which one benefits and the other is harmed. (408)

sympathetic division: A subdivision of the autonomic nervous system of vertebrates, with centers in the midportion of the spinal cord; slows digestion; generally excites other functions. Neurotransmitters: acetylcholine and noradrenaline. (767)

sympatric speciation [Gk. *syn,* together with + *patra,* fatherland, country]: Speciation that occurs without geographic isolation of a population of organisms; may occur as the result of hybridization accompanied by polyploidy or of disruptive selection. (930)

synapse [Gk. *synapsis,* a union]: A specialized junction between two neurons where the activity in one influences the activity in another; may be chemical or electrical, excitatory or inhibitory. (773)

syngamy (**sin**-gamy) [Gk. *syn,* with + *gamos,* a marriage]: The union of gametes in sexual reproduction; fertilization. (420)

synthesis [Gk. *syntheke,* a putting together]: The formation of a more complex substance from simpler ones. (166)

synthetic theory: The currently prevailing theory of the mechanism of evolutionary change; combines the Darwinian two-step model of variation and selection with the principles of Mendelian genetics. (892)

systematics [Gk. *systema,* that which is put together]: Scientific study of the kinds and diversity of organisms and of the relationships among them. (371)

T cell: A type of lymphocyte arising from precursors in the thymus and, upon maturation, involved in cell-mediated immunity and interactions with B cells; a T lymphocyte. (755)

tagmosis [Gk. *tagma,* arrangement, order + *-osis,* process]: The formation of groups of segments (metameres) into body regions (tagmata) with functional differences. (520)

taxon, *pl.* **taxa** [Gk. *taxis,* arrange, put in order]: A particular group, ranked at a particular categorical level, in a hierarchical classification scheme; for example, *Drosophila* is a taxon at the categorical level of genus. (374)

taxonomy [Gk. *taxis,* arrange, put in order + *nomos,* law]: The study of the classification of organisms, the ordering of organisms into a hierarchy that reflects on their essential similarities and differences. (371)

telophase [Gk. *telos,* end + *phasis,* form]: The last stage in mitosis and meiosis, during which the chromosomes become reorganized into two new nuclei. (145)

temperate bacteriophage: A bacterial virus that may become incorporated into the host-cell chromosome. (316)

template: A pattern or mold guiding the formation of a negative or complement. (292)

tentacles [L. *tentare,* to touch]: Long, flexible protrusions located about the mouth in many invertebrates; usually prehensile or tactile. (487)

territory An area or space occupied and defended by an individual or a group; trespassers are attacked (and usually defeated); may be the site of breeding, nesting, food gathering, or any combination thereof. (1023)

tertiary structure of a protein: A complex structure, usually globular, resulting from further folding of the secondary structure of a protein; forms spontaneously due to attractions and repulsions among amino acids with different charges on their R groups. (67)

testcross: A mating between a phenotypically dominant individual and a homozygous recessive "tester" to determine the genetic constitution of the dominant phenotype, that is, whether it is homozygous or heterozygous for the relevant gene. (242)

testis, *pl.* **testes** [L., witness]: The sperm-producing organ; also the source of male sex hormone. (257)

testosterone [Gk. *testis,* testicle + *steiras,* barren]: A steroid hormone secreted by the testes in higher vertebrates and stimulating the development and maintenance of male sex characteristics and the production of sperm; an androgen. (777)

tetrad [Gk. *tetras,* four]: In genetics, a pair of homologous chromosomes that have replicated and come together in prophase I of meiosis; consists of four chromatids. Also called a bivalent. (253)

thalamus [Gk. *thalamos,* chamber]: A part of the vertebrate forebrain just posterior to the cerebrum; an important intermediary between all other parts of the nervous system and the cerebrum. (818)

thallus [Gk. *thallos,* a young twig]: A simple plant or algal body without true roots, leaves, or stems. (426)

theory [Gk. *theorein,* to look at]: A generalization based on many observations and experiments; a verified hypothesis. (10)

thermodynamics [Gk. *therme,* heat + *dynamis,* power]: The study of transformations of energy. The first law of thermodynamics states that in all processes, the total energy of a system plus its surroundings remains constant. The second law states that all natural processes tend to proceed in such a direction that the disorder or randomness of the system increases. (157)

thorax [Gk., breastplate]: (1) In vertebrates, that portion of the trunk containing the heart and lungs. (659)
(2) In crustaceans and insects, the fused, leg-bearing segments between head and abdomen. (528)

thylakoid [Gk. *thylakos,* a small bag]: A flattened sac, or vesicle, that forms part of the internal membrane structure of the chloroplast; the site of the light-trapping reactions of photosynthesis and of photophosphorylation; stacks of thylakoids collectively form the grana. (217)

thyroid [Gk. *thyra,* a door]: An endocrine gland of vertebrates, located in the neck; source of an iodine-containing hormone (thyroxine) that increases the metabolic rate and affects growth. (781)

tight junction: A junction between adjacent animal cells that prevents materials from leaking through the tissue; for example, intestinal epithelial cells are surrounded by tight junctions. (119)

tissue [L. *texere,* to weave]: A group of similar cells organized into a structural and functional unit. (117)

trachea, *pl.* **tracheae (trake-**ee-a) [Gk. *tracheia,* rough]: An air-conducting tube. (1) In insects and some other terrestrial arthropods, a system of chitin-lined air ducts. (529)
(2) In terrestrial vertebrates, the windpipe. (692)

tracheid (tray-key-idd) [Gk. *tracheia,* rough]: In vascular plants, an elongated, thick-walled conducting and supporting cell of xylem, characterized by tapering ends and pitted walls without true perforations. (461)

transcription [L. *trans,* across + *scribere,* to write]: The enzymatic process by which the genetic information contained in one strand of DNA is used to specify a complementary sequence of bases in an RNA molecule. (297)

transduction [L. *trans,* across + *ducere,* to lead]: The transfer of genetic material (DNA) from one cell to another by a virus. (316)

transfer RNA (tRNA) [L. *trans,* across + *ferre,* to bear or carry]: A class of small RNAs (about 80 nucleotides) with two functional sites; one recognizes a specific activated amino acid; the other carries the nucleotide triplet (anticodon) for that amino acid. Each type of tRNA accepts a specific activated amino acid and transfers it to a growing polypeptide chain as specified by the nucleotide sequence of the messenger RNA being translated. (298)

transformation [L. *trans,* across + *formare,* to shape]: A genetic change produced by the incorporation into a cell of exogenous DNA. (279)

translation [L. *trans,* across + *latus,* that which is carried]: The process by which the genetic information present in a strand of mRNA directs the sequence of amino acids during protein synthesis. (300)

translocation [L. *trans,* across + *locare,* to put or place]: (1) In plants, the transport of the products of photosynthesis from a leaf to another part of the plant. (621)
(2) In genetics, the breaking off of a piece of chromosome with its reattachment to a nonhomologous chromosome. (274)

transpiration [L. *trans,* across + *spirare,* to breathe]: In plants, the loss of water vapor from the stomata. (614)

transposon [L. *transponere,* to change the position of]: A DNA sequence carrying one or more genes and flanked by insertion sequences that confer the ability to move from one DNA molecule to another. (319)

tritium: A radioactive isotope (^3H) of hydrogen with a half-life of 12.5 years. (21)

trophic level [Gk. *trophos,* feeder]: The position of a species in the food web or chain; a step in the movement of biomass or energy through a system. (981)

tropic [Gk. *trope,* a turning]: Pertaining to behavior or action brought about by specific stimuli, for example, phototropic ("light-oriented") motion, gonadotropic ("stimulating the gonads") hormone. (642)

tuber [L. *tuber,* bump, swelling]: A much-enlarged, short, fleshy underground stem, such as that of the potato. (112)

turgor [L. *turgere,* to swell]: The pressure exerted on the inside of a plant cell wall by the fluid contents of the cell; the interior of the cell is hypertonic in relation to the fluids surrounding it and so gains water by osmosis. (131)

urea [Gk. *ouron,* urine]: An organic compound formed in the vertebrate liver; the principal form of disposal of nitrogenous wastes by mammals. (721)

ureter [Gk. from *ourein,* to urinate]: The tube carrying urine from the kidney to the cloaca (in reptiles and birds) or to the bladder (in amphibians and mammals). (726)

urethra [Gk. from *ourein,* to urinate]: The tube carrying urine from the bladder to the exterior of mammals. (726)

uric acid [Gk. *ouron,* urine]: An insoluble nitrogenous waste product that is the principal excretory product in birds, reptiles, and insects. (721)

urine [Gk. *ouron,* urine]: The liquid waste filtered from the blood by the kidney and stored in the bladder pending elimination through the urethra. (726)

uterine tube: *See* Oviduct.

uterus [L., womb]: The muscular, expanded portion of the female reproductive tract modified for the storage of eggs or for housing and nourishing the developing embryo. (842)

vacuole [L. *vacuus,* empty]: A membrane-bound, fluid-filled sac within the cytoplasm of a cell. (107)

vagus nerve [L. *vagus,* wandering]: A nerve arising from the medulla of the vertebrate brain that innervates the heart and visceral organs; carries parasympathetic fibers. (680)

vaporization [L. *vapor,* steam]: The change from a liquid to a gas; evaporation. (36)

vascular [L. *vasculum,* a small vessel]: Containing or concerning vessels that conduct fluid. (456)

vascular bundle: In plants, a group of longitudinal supporting and conducting tissues (xylem and phloem). (461)

vascular cambium [L. *vasculum,* a small vessel + *cambium,* exchange]: A cylindrical sheath of meristematic cells that divide mitotically, producing secondary phloem to one side and secondary xylem to the other, but always with a cambial cell remaining. (606)

vas deferens (vass **deff**-er-ens) [L. *vas,* a vessel + *deferre,* to carry down]: In mammals, the tube carrying sperm from the testes to the urethra. (840)

vein [L. *vena,* a blood vessel]: (1) In plants, a vascular bundle forming a part of the framework of the conducting and supporting tissue of a leaf. (576)
(2) In animals, a blood vessel carrying blood from the tissues to the heart. A small vein is known as a venule. (705)

vena cava (vee-na **cah**-va) [L., blood vessel + hollow]: A large vein that brings blood from the tissues to the right atrium of the four-chambered vertebrate heart. The superior vena cava collects blood from the forelimbs, head, and anterior or upper trunk; the inferior vena cava collects blood from the posterior body region. (708)

ventral [L. *venter,* belly]: Pertaining to the undersurface of an animal that moves on all fours; to the front surface of an animal that holds its body erect. (493)

ventricle [L. *ventriculus,* the stomach]: A muscular chamber of the heart that receives blood from an atrium and pumps blood out of the heart, either to the lungs or to the body tissues. (707)

vertebral column [L. *vertebra,* joint]: The backbone; in nearly all vertebrates, it forms the supporting axis of the body and it protects the spinal cord. (659)

vesicle [L. *vesicula,* a little bladder]: A small, intracellular membrane-bound sac. (107)

vessel [L. *vas,* a vessel]: A tubelike element of the xylem of angiosperms; composed of dead cells (vessel members) arranged end to end. Its function is to conduct water and minerals from the soil. (583)

viable [L. *vita,* life]: Able to live. (351)

villus, *pl.* **villi** [L., a tuft of hair]: In vertebrates, one of the minute, fingerlike projections lining the small intestine that serve to increase the absorptive surface area of the intestine. (676)

virus [L., slimy, liquid, poison]: A submicroscopic, noncellular particle composed of a nucleic acid core and a protein coat; parasitic; reproduces only within a host cell. (405)

viscera [L., internal organs]: The collective term for the internal organs of an animal. (508)

vitamin [L. *vita,* life]: Any of a number of unrelated organic substances that cannot be synthesized by a particular organism and are essential in minute quantities for normal growth and function. (174)

water cycle: Worldwide circulation of water molecules, powered by the sun. Water evaporates from oceans, lakes, rivers, and, in smaller amounts, soil surfaces and bodies of organisms; water returns to the earth in the form of rain and snow. Of the water falling on land, some flows into rivers that pour water back into the oceans and some percolates down through the soil until it reaches a zone where all pores and cracks in the rock are filled with water (groundwater); the deep groundwater eventually reaches the oceans, completing the cycle. (46)

water potential: The potential energy of water molecules; regardless of the reason (e.g., gravity, pressure, concentration of solute particles) for the water potential, water moves from a region where water potential is greater to a region where water potential is lower. (126)

worker: A member of the nonreproductive laboring caste of social insects. (1019)

xanthophyll [Gk. *xanthos,* yellow + *phyllon,* leaf]: In algae and plants, one of a group of yellow pigments; a member of the carotenoid group. (403)

xylem [Gk. *xylon,* wood]: A complex vascular tissue through which most of the water and minerals are conducted from the roots to other parts of the plant; consists of tracheids or vessel elements, parenchyma cells, and fibers; constitutes the wood of trees and shrubs. (461)

yolk: The stored food in egg cells that nourishes the embryo. (558)

zoology [Gk. *zoe,* life + *logos,* a discourse]: The study of animals. (2)

zooplankton [Gk. *zoe,* life + *plankton,* wanderer]: A collective term for the nonphotosynthetic organisms present in plankton. (969)

zygote (zi-got) [Gk. *zygon,* yolk, pair]: The diploid ($2n$) cell resulting from the fusion of male and female gametes (fertilization); a zygote may either develop into a diploid individual by mitotic divisions or may undergo meiosis to form haploid (n) individuals that divide mitotically to form a population of cells. (250)

Illustration Acknowledgments

Frontispiece © John Reader

Page vii Tui De Roy Moore

Page x A. D. Greenwood

Page xi Stanley Falkow

Page xii Tanaka Kojo, Animals Animals Enterprises

Page xiii E. S. Ross

Page xiv George K. Bryce, Animals Animals Enterprises

Page xv Alvin E. Staffan

Page xvi Yvon Le Maho

Page xvii From Lennart Nilsson, *Behold Man,* Little, Brown and Company, Boston, copyright © 1974; photograph courtesy Lennart Nilsson, Bonnier Fakta, Stockholm

Page xviii Bradley Smith, Animals Animals Enterprises

I-1 The Royal College of Surgeons of England

I-2 Tui De Roy Moore, Bruce Coleman

I-3 Nigel Smith, Earth Scenes

I-4 E. R. Degginger, Earth Scenes

I-5 The American Museum of Natural History

I-6 M. P. L. Fogden, Bruce Coleman

I-7 Medical Illustration Unit, The Royal College of Surgeons of England

Page 8 The Royal College of Surgeons of England

Pages 12–13 (a) Oronce Finé, *Théorique de la huitième sphère et sept planètes,* 1528; (b) Ann Ronan Picture Library

1-1 © California Institute of Technology

Pages 18–19 (a), (b) Runk/Schoenberger, Grant Heilman Photography; (c) Grant Heilman; (d) Roberts Rugh and Landrum B. Shettles, M.D., *From Conception to Birth: The Drama of Life's Beginnings,* Harper & Row Publishers, Inc., New York, 1971; (e) Tom Bledsoe, Photo Researchers; (f) D. R. Specker, Animals Animals Enterprises; (g) Stephen Dalton, Photo Researchers

1-2 John Reader, © National Geographic Society

1-5 Elihu Blotnick

1-7 (b) Runk/Schoenberger, Grant Heilman Photography

1-8 Barbara F. Reese and Thomas S. Reese, National Institute of Neurological and Communicative Disorders and Stroke

Page 29 John G. Torrey

2-1 Runk/Schoenberger, Grant Heilman Photography

2-4 Larry Pringle, Photo Researchers

2-5 Jack Dermid

2-7 (b) Grant Heilman

2-11 Jeanne M. Riddle

Page 48 Grant Heilman

Page 50 John M. Sieburth

3-4 After E. J. DuPraw, *Cell and Molecular Biology,* Academic Press, Inc., New York, 1968

3-8 (c) After Albert L. Lehninger, *Biochemistry,* 2d ed., Worth Publishers, Inc., New York, 1975; (d) L. M. Beidler; (e) Don Fawcett

3-9 (c) R. D. Preston

3-10 (b) Larry West

3-12 Russ Kinne, Photo Researchers

3-15 B. E. Juniper

3-20 (b) After E. O. Wilson et al., *Life: Cells, Organisms, Populations,* Sinauer Associates, Inc., Sunderland, Mass., 1977

3-21 (b) Richard Dickerson

3-23 (a) After Bruce Alberts, Dennis Bray, Julian Lewis, Martin Raff, Keith Roberts, and James D. Watson, *Molecular Biology of the Cell,* Garland Publishing Company, New York, 1983; (b) Dan Friend

3-24 (a) Emil Bernstein and Eila Kairinen, Gillette Company Research Institute, Rockville, Md., *Science,* vol. 173, 1971. © 1971 by AAAS. (b) John Mais

3-25 (a) After Gerald Karp, *Cell Biology,* McGraw-Hill Book Company, New York, 1979; (b) Eugene B. Small and Douglas S. Marszalek, *Science,* vol. 163, pages 1064–1065, 1969. © 1969 by AAAS

3-28 Margaret Clark

Page 74 After Albert L. Lehninger, *Principles of Biochemistry,* Worth Publishers, Inc., New York, 1982

4-1 (a) Rare Book Division, The New York Public Library; (b) John Mais

4-2 Pasteur Institute and The Rockefeller University Press

Page 79 National Center for Atmospheric Research

4-3 S. Jonasson, from R. Anderson et al., *Science,* vol. 148, pages 1179–1190, 1965. © 1965 by AAAS

4-5 Sidney W. Fox

4-6 (a) E. S. Barghoorn, *Science,* vol. 152, pages 758–763, 1966. © 1966 by AAAS. (b) J. W. Schopf

4-7 Micrograph by A. Ryter

4-8 Micrograph by Norma J. Lang, *Journal of Phycology,* vol. 1, pages 127–134, 1965

4-9 Micrograph by George Palade

4-11 Micrograph by Michael A. Walsh

4-12 Micrograph by Keith R. Porter

4-13 After Alberts et al., *op. cit.*

4-14 David M. Phillips

4–15 David M. Phillips

4–16 David M. Phillips

4–18 After Alberts et al., *op. cit.*

4–19 *Ibid.*

4–20 Keith Roberts and James Barnett

4–21 T. E. Adams

5–2 Lewis Tilney

5–3 (a) J. David Robertson

5–4 (a) M. C. Ledbetter; (b) Peter Albersham, "The Walls of Growing Plant Cells," *Scientific American,* April 1975

5–5 (a) Daniel Branton; (b) A. C. Fabergé, *Cell and Tissue Research,* vol. 151, pages 403–415, 1974

5–6 Ursula Goodenough

5–17 Mia Tegner and David Epel

5–11 Keith R. Porter

5–12 (a) Keith R. Porter; (b) Karen Anderson

5–13 (a), (b) Don Fawcett, from William Bloom and D. W. Fawcett, *A Textbook of Histology,* 10th ed., W. B. Saunders Company, Philadelphia, 1975

5–14 Don Fawcett

5–15 Micrograph by C. J. Flickinger, *Journal of Cell Biology,* vol. 49, page 221, 1975

5–16 After Stephen Wolfe, *Biology of the Cell,* Wadsworth Publishing Company, Inc., Belmont, Calif., 1972

5–17 (a) Don Fawcett; (b) G. Decker

5–18 David Smith, *Insect Cells: Their Structure and Function,* Oliver & Boyd, Edinburgh, 1968

5–19 M. C. Ledbetter

5–20 David Stetler

5–21 Don Fawcett

5–22 (a) E. Lazarides, *Journal of Cell Biology,* vol. 65, page 549, 1975; (b) K. Weber and U. Groeschel-Stewart, *Proceedings of National Academy of Sciences U.S.,* vol. 171, page 4561, 1974

5–23 Gregory Antipa

5–24 (a) After DuPraw, *op. cit.;* (b) A. V. Grimstone

5–25 Étienne de Harven, *The Nucleus,* Academic Press, Inc., New York, 1968

5–26 J. F. M. Hoeniger, *Journal of General Microbiology,* vol. 40, page 29, 1965

5–27 Ray F. Evert

5–28 (a) Keith R. Porter; (b) L. A. Staehelin and B. E. Hull; (c) D. A. Goodenough, *Journal of Cell Biology,* vol. 68, page 220, 1976

5–29 David Buck

5–30 (a)–(c) K. T. Raper; (d), (e) J. T. Bonner

6–2 Keith R. Porter

6–4 K. Weidmann, National Audubon Society Collection/PR

6–6 (a) G. M. Hughes; (b) after K. Schmidt-Nielsen, *Animal Physiology,* 2d ed., Cambridge University Press, Cambridge, 1979

6–7 (a) Eric V. Gravé; (b), (c) Thomas Eisner

6–8 After Lehninger, *op. cit.,* 1975

6–10 (b) P. Pinto da Silva and D. Branton, *Journal of Cell Biology,* vol. 45, page 598, 1970

6–11 Don Fawcett

6–14 Birgit H. Satir

6–15 Gregory Antipa

7–1 From L. P. Wisniewski, K. Hirschhorn, "A Guide to Human Chromosome Defects," 2d ed., White Plains: March of Dimes Birth Defects Foundation, BD:OAS XVI(6), 1980

7–2 John Mais

7–3 (b) A. Ryter

7–5 General Biological Supply Company

7–7 William Tai

7–8 Andrew S. Bajer

7–9 William Tai

7–10 Eric V. Gravé

7–11 (a) G. Östergren; (b) J. T. Pickett-Heaps

7–12 (a) de Harven, *op. cit.;* (b) Michael Friedlander

7–13 H. W. Beams and R. G. Kessel, *American Scientist,* vol. 64, page 279, 1976

7–14 James Cronshaw

8–1 R. D. Estes

8–2 Georg Gerster, Rapho/PR

Page 159 Lotte Jacobi

8–4 (a) L. L. Rue, Jr., Animals Animals Enterprises; (b) Larry West; (c) Y. Haneda; (d) John Dominis, LIFE © 1971

8–6 J. T. Pickett-Heaps

8–7 After Lehninger, *op. cit.,* 1975

Page 172 The Bettmann Archive

8–9 After Lehninger, *op. cit.,* 1975

8–14 V. Lennard

8–16 After J. D. Watson, *Molecular Biology of the Gene,* 2d ed., The Benjamin/Cummings Publishing Company, Menlo Park, Calif., 1970

Page 179 (a) Ylla, Photo Researchers; (b) Photo Researchers

8–18 After Lehninger, *op. cit.,* 1975

8–19 After Peter H. Raven, Ray F. Evert, and Helena Curtis, *Biology of Plants,* 3d ed., Worth Publishers, Inc., New York, 1981

9–1 (a) Bob Evans, Peter Arnold; (b) Gary Robinson; (c) Richard Bray, *Science,* vol. 200, pages 333–334, 1978. © 1978 by AAAS

9–5 (b) Grant Heilman

9–7 Micrograph by Keith R. Porter

9–8 (b) Lester J. Reed, from P. D. Boyer, ed., *The Enzymes,* vol. 1, Academic Press, Inc., New York, 1970

Page 195 R. H. Kirschner

9–9 After A. J. Vander, J. H. Sherman, and Dorothy Luciano, *Human Physiology,* McGraw-Hill Book Company, New York, 1969

9–10 After Lehninger, *op. cit.,* 1975

9–13 After T. Takano, O. B. Kallai, R. Swanson, and R. E. Dickerson, *Journal of Biological Chemistry,* vol. 248, page 5244, 1973

9–14 Adapted from Lehninger, *op. cit.,* 1982

9–15 After Alberts et al., *op. cit.*

9–16 (a) After Lehninger, *op. cit.,* 1982; (b) John N. Telford

9–18 After Lehninger, *op. cit.,* 1975

Page 206 Charles S. Lieber, "The Metabolism of Alcohol," *Scientific American,* March 1976

9–21 After Lehninger, *op. cit.,* 1975

10–1 J. Robert Waaland

Page 211 N. Pfeiffer

10–7 Micrograph by Oxford Scientific Films, Bruce Coleman

10–9 (a) Daniel Branton; (b) A. D. Greenwood; (c) L. K. Shumway; (d) M. C. Ledbetter

10–13 After Lehninger, *op. cit.,* 1975

Page 224 Walther Stoeckenius, "The Purple Membrane of Salt-Loving Bacteria," *Scientific American,* June 1976

10–16 J. Heslop-Harrison

10–17 After L. Stryer, *Biochemistry,* W. H. Freeman and Company, San Francisco, 1975

10–20 Ray F. Evert

11–1 David M. Phillips

11–2 The Bettmann Archive

11–3 The Bettmann Archive

11–4 The National Library of Medicine

11–6 After Karl von Frisch, *Biology,* translated by Jane Oppenheimer, Harper & Row Publishers, Inc., New York, 1964

11–12 Dr. V. Orel, The Moravian Museum

12–1 (a) Arnold Sparrow, Brookhaven National Laboratory; (b) S. Rannels, Grant Heilman Photography

12–7 B. John

12–8 William Tai

12–9 Mary E. Clutter

12–10 After DuPraw, *op. cit.*

Page 258 After John L. Moore, *Heredity and Development,* Oxford University Press, New York, 1963

12–13 B. John

Page 261 William Tai

13–2 After Francisco J. Ayala and John A. Kiger, Jr., *Modern Genetics,* The Benjamin/Cummings Publishing Company, Menlo Park, Calif., 1980

13–4 C. G. G. J. van Steenis

13–5 (a) After E. D. Merrell, 1964; (b) *Journal of Heredity,* vol. 15, 1914

13–6 Photograph, F. B. Hutt, *Journal of Genetics,* vol. 22, page 126, 1930

13–10 John A. Moore

Page 271 Shirley Baty

13–11 Micrograph by B. John

13–15 B. Kaufmann

14–1 A. K. Kleinschmidt, D. Land, D. Jacherts, and R. K. Zahn, *Biochemica Biophysica Acta,* vol. 61, pages 857–864, 1962

14–2 Robert Austrian, *Journal of Experimental Medicine,* vol. 92, page 21, 1953

14–3 After Derry D. Koob and William E. Boggs, *The Nature of Life,* Addison-Wesley Publishing Company, Inc., Reading, Mass., 1972

14–5 From J. Cairns, G. S. Stent, and J. D. Watson, eds., *Phage and The Origins of Molecular Biology,* Cold Spring Harbor Laboratory of Quantitative Biology, Cold Spring Harbor, N.Y., 1966

14–6 (a) Lee D. Simon

14–7 Lee D. Simon

14–8 From James D. Watson, *The Double Helix,* Atheneum Publishers, New York, 1968

14–10 After Lehninger, *op. cit.,* 1975

14–11 *Ibid.*

Page 289 From Watson, *op. cit.,* 1968

14–13 After Lehninger, *op. cit.,* 1975

14–14 *Ibid.*

14–15 Jack Griffith

Page 294 C. M. Plork, Museum of Comparative Zoology, Harvard University

15–1 After Wendell Stanley and Evan G. Valens, *Viruses and The Nature of Life,* E. P. Dutton & Company, Inc., New York, 1961

15–2 After Linus Pauling et al., *Science,* vol. 110, pages 543–548, 1949

15–4 Jay Hirsh and Robert Schleif

15–6 (b) Sung Hou-Kim

15–10 Hans Ris

15–11 Micrograph by O. L. Miller, Jr., Barbara A. Hamkalo, and C. A. Thomas, Jr., *Science,* vol. 69, pages 392–395, 1970. © 1970 by AAAS

15–13 After Alberts et al., *op. cit.*

15–16 Jack Griffith

16–2 (a) Sunil Palchaudhuri, E. Bell, and M. R. J. Salton, *Infection and Immunity,* vol. 11, page 1141, 1975; (b)–(d) T. Kakefuda

16–3 Judith Carnahan and Charles Brinton, Jr.

16–4 After Ayala and Kiger, *op. cit.*

16–5 *Ibid.*

16–6 Stanley N. Cohen, *Nature,* December 27, 1969

16–7 B. Menge, J. V. D. Brock, H. Wunderli, K. Lickfield, M. Wurtz, and E. Kellenberger

16–10 Based on Alberts et al., *op. cit.*

16–11 After Lehninger, *op. cit.,* 1982

16–12 Stanley Falkow

16–13 (a) Jack Griffith; (b) after Lehninger, *op. cit.,* 1975

16–14 (a) Stanley N. Cohen

16–15 After Lehninger, *op. cit.,* 1982

17–1 E. J. DuPraw

17–2 Ursula Goodenough

17–3 Alexander Rich et al., *Science,* vol. 211, pages 171–176, 1981. © by AAAS

17–4 (a) Donald E. Olins and Ada L. Olins, University of Tennessee, Oak Ridge Graduate School of Biomedical Sciences and Oak Ridge National Laboratory; (b) after Roger D. Kornberg and Aaron Klug, "The Nucleosome," *Scientific American,* February 1981

17–5 After Alberts et al., *op. cit.*

17–6 James German

17–7 George T. Rudkin

17–8 Joseph Gall

17–9 After Lehninger, *op. cit.,* 1982

17–10 H. C. MacGregor

17–11 Ulrich Scheer, W. W. Franke, and M. F. Trendelenberg

17–12 R. Portmann and M. L. Birnsteil

17–13 O. L. Miller, Jr., and Barbara R. Beatty, Biology Division, Oak Ridge National Laboratory

17–14 After Ursula Goodenough, *Genetics,* 3d ed., Saunders College/Holt, Rinehart and Winston, New York, 1983

17–15 After Samuel Rahbar "Abnormalities of Human Hemoglobin," *City of Hope Quarterly,* vol. 11, page 2, Winter 1982

17–16 Micrograph, Pierre Chambon; art, after Pierre Chambon, "Split Genes," *Scientific American,* pages 60–71, May 1981

17–17 After Lehninger, *op. cit.,* 1982

17–18 After Roger Lewin, "Biggest Challenge Since the Double Helix," *Science,* vol. 212, pages 28–32, 1981

17–20 Heather Angel

17–21 Herb Parsons, Cold Spring Harbor Laboratory Research Library Archives

17–22 After Lehninger, *op. cit.,* 1982

17–23 *Ibid.*

17–24 R. D. Goldman

17–25 Department of Plant Pathology, Cornell University

18–1 Gernsheim Collection, Humanities Research Center, University of Texas

18–2 Paediatric Research Unit, Guy's Hospital Medical School

18–3 After J. J. Yunis, *Science,* vol. 191, pages 1268–1270, 1976. © 1976 by AAAS

18–4 (a) Bruce Roberts, Photo Researchers; (b) Paediatric Research Unit, Guy's Hospital Medical School

18–7 J. J. Yunis, University of Minnesota

18–10 After I. Michael Lerner, *Heredity, Evolution, and Society,* W. H. Freeman and Company, San Francisco, 1968

18–11 The Bettmann Archive

18–12 After Stryer, *op. cit.*

18–13 (a) Herbert A. Fischler, Isaac Albert Research Institute of the Kingsbrook Jewish Medical Center; (b) John S. O'Brien

18–14 Margaret Clark

18–16 After *City of Hope Quarterly,* vol. 7, page 2, Winter, 1978

18–17 Kenneth Huttner et al., "DNA-Mediated Gene Transfer without Carrier DNA," *Journal of Cell Biology,* vol. 91, pages 153–156, October 1981

18–18 (a) After Jon W. Gordon and Frank Ruddle; (b) Jon W. Gordon and Frank Ruddle

19–1 E. S. Ross

19–2 (a) Larry West; (b) R. Carr, Bruce Coleman; (c) E. S. Ross

19–3 (a) John H. Gerard, D.P.I.; (b) H. Reinhard, Bruce Coleman

19–4 Burndy Library

19–5 (a) Russ Kinne, Photo Researchers; (b) Stephen J. Krasemann, Photo Researchers; (c) Ken Brate, Photo Researchers; (d) Tanaka Kojo, Animals Animals Enterprises; (e) E. R. Degginger, Animals Animals Enterprises

19–7 After Theodosius Dobzhansky, *Evolution,* W. H. Freeman and Company, San Francisco, 1977

19–8 *Ibid.*

19–9 After Lehninger, *op. cit.,* 1982

19–10 After Ayala and Kiger, *op. cit.*

19–12 (a) T. E. Adams; (b) Eric V. Gravé; (c) E. S. Ross; (d) Les Blacklock; (e) Alvin E. Staffan

20–1 M. E. Edwards

20–2 Eric V. Gravé

20–3 Micrograph by Lee D. Simon

20–4 Victor Lorian and B. Atkinson

20–5 John Swanson

20–6 (a) J. Adler; (b) after M. L. DePamphilis and J. Adler, *Journal of Bacteriology,* vol. 105, page 395, 1971

20–7 J. F. M. Hoeniger

20–8 David Greenwood

20–9 J. F. M. Hoeniger

20–10 (a) Centers for Disease Control; (b), (c) M. A. Listgarten and S. S. Socransky, *Journal of Bacteriology,* vol. 10, pages 127–138, 1964

20–11 Virus Laboratory, Parke, Davis & Co.

20–12 R. S. Wolfe

Page 400 D. L. Balkwill and D. Maratea

20–13 Micrograph by J. F. M. Hoeniger and C. L. Headley, *Journal of Bacteriology,* vol. 96, pages 1835–1847, 1968

20–14 P. L. Grilone and J. Pangborn, *Journal of Bacteriology,* vol. 124, page 1558, 1975

20–15 H. Stolp and M. P. Starr, *Antoine van Leeuwenhoek,* vol. 29, pages 217–248, 1963

20–16 E. Boatman

20–17 Micrograph by M. Jost

20–18 (a), (b) J. D. Almeida and A. F. Howatson, *Journal of Cell Biology,* vol. 16, page 616, 1963; (c) Frederick A. Murphy; (e) Lee D. Simon

20–19 T. Koller and J. M. Sogo, Swiss Federal Institute of Technology, Zurich

20–20 The Ny Carlsberg Glyptotek

21–1 Eric V. Gravé

21–2 (a) William J. Larsen, *Journal of Cell Biology,* vol. 97, page 373, 1970; (b) D. A. Stetler and W. M. Laetsch, *American Journal of Botany,* vol. 56, page 260, 1969

21–3 Richard W. Greene

21–6 D. P. Wilson, Eric and David Hosking Photography

21–7 J. Whatley and R. A. Lewin, *New Phytologist,* vol. 79, pages 303–313, 1977

21–10 H. N. Guttmann

21–11 (a) Eric V. Gravé, Photo Researchers; (b) D. P. Wilson, Eric and David Hosking Photography

21–12 (a) Kent Cambridge Scientific Instruments; (b) G. A. Fryxell

21–13 D. P. Wilson, Eric and David Hosking Photography

21–14 (a) T. E. Adams; (b) M. I. Walker; (c) George Schwartz; (d) Oxford Scientific Films, Animals Animals Enterprises

21–15 (a) Grant Heilman; (b) Oxford Scientific Films, Animals Animals Enterprises; (c) D. P. Wilson, Eric and David Hosking

21–17 Micrograph by Trond Braten, EM Laboratory for Biosciences, Oslo

21–19 D. P. Wilson, Eric and David Hosking Photography

21–20 Larry West

21–21 Eric V. Gravé

21–22 D. S. Marszalek and E. B. Small

21–23 Eric V. Gravé, Photo Researchers

21–24 Eric V. Gravé

21–26 T. E. Adams

Page 432 Micrographs by D. Kubai and H. Ris

21–27 G. A. Horridge and S. L. Tamm, *Science,* vol. 163, pages 817–818, 1969. © 1969 by AAAS

21–29 K. W. Jeon

22–1 Les Blacklock

22–3 John Richardson

22–4 John H. Troughton

22–5 H. C. Hoch and D. P. Maxwell

22–6 John W. Taylor

22–7 Alma W. Barkdale, *Mycologia,* vol. 55, pages 493–501, 1973

22–8 After John S. Niederhauser and William Cobb, *Scientific American,* May 1959

22–9 (c) Carolina Biological Supply

22–10 (a) Alvin E. Staffan; (b) John Shaw

22–11 (a) C. Bracker; (b) drawing by B. O. Dodge, from E. L. Tatum

22–13 American Society for Microbiology

22–14 (a) Agenzia Fotografica, Luisa Ricciarini, Milan; (b) E. S. Ross; (c) Donald Simons

22–15 (a) James R. Leard, National Audubon Society Collection; (b) Jack Dermid; (c) E. S. Ross

22–16 Vernon Ahmadjian

Page 451 David Pramer

22–17 (a) R. J. Molina, J. M. Trappe, and G. S. Strickler, *The Canadian Journal of Botany,* vol. 56, pages 1691–95, 1978

23–1 E. S. Ross

23–2 Ray F. Evert

23–3 R. E. Magill, Botanical Research Institute, Pretoria

23–4 Ray F. Evert

23–5 Alvin E. Staffan

23–7 Turtox

23–8 E. S. Ross

23–9 After Peter Raven, Ray F. Evert, and Helena Curtis, *Biology of Plants,* 2d ed., Worth Publishers, Inc., New York, 1976

23–10 *Ibid.*

23–11 (a), (b) E. S. Ross; (c) R. Carr

23–12 (a) Richard B. Peacock, Photo Researchers; (b) Gene Ahrens, Bruce Coleman

23–13 E. S. Ross

23–14 (a) D. A. Steingraeber; (b) A–Z Collection, 1976, Photo Researchers; (c) C. H. Bornman; (d) William M. Harlow, Photo Researchers

Pages 466–467 (a) After P. Kukuk, *Geologie des Niederheinisch-Westfälischen Steinkohlengebietes,* Verlag von Julius Springer, Berlin, 1938; (b) Field Museum of Natural History

23–16 (a) William M. Harlow; (b) Larry West

23–18 Grant Heilman

23–22 Larry West

23–23 (b) Grant Heilman

23–24 (b) Anita Sabarese

23–26 (a)–(c) E. S. Ross; (d) W. G. Calder; (e) Alan and Joan Root

23–27 (a) H. N. Darrow, Bruce Coleman; (b) Heather Angel

Page 480 R. Borneman, Photo Researchers

24–1 D. P. Wilson, Eric and David Hosking Photography

24–2 Nat Fain

24–3 Nat Fain

24–5 Carleton Ray, Photo Researchers

24–7 The American Museum of Natural History

24–8 D. P. Wilson, Eric and David Hosking Photography

24–9 Robert B. Short

24–14 After Ralph Buchsbaum and Lorus J. Milne, *The Lower Animals: Living Invertebrates of The World,* Doubleday & Company, Inc., Garden City, N.Y., 1962

24–15 (a) D. P. Wilson, Eric and David Hosking Photography; (b) Runk/Schoenberger, Grant Heilman Photography

24–16 Douglas Faulkner

24–18 Steve Earley, Animals Animals Enterprises

Page 492 (a) Allan Power, Bruce Coleman; (b) Ian Took/Biofotos

24–19 Runk/Schoenberger, Grant Heilman Photography

24–22 (a) Maria Wimmer

24–23 Walter E. Harvey, National Audubon Society Collection/PR

24–25 After M. B. V. Roberts, *Biology: A Functional Approach,* The Ronald Press Company, New York, 1971

24–27 (a) British Museum (Natural History); (b) Carolina Biological Supply Company

Page 499 (a) Ming Wong; (b) Klaus D. Francke, Peter Arnold

24–28 (a) W. Sterrer, Bermuda Biological Station; (b) Maria Wimmer

24–29 J. E. Sulston

24–30 (c) P. Emschermann

24–31 (a) R. P. Higgins; (b) M. I. Walker, Photo Researchers; (c) John Gilbert

24–32 P. Emschermann

25–4 (a) After Roberts, *op. cit.;* (b) A. Solem, courtesy Field Museum of Natural History

25–5 (b), (d) Specimens courtesy of Edward I. Coher; (c) after a photograph supplied by Russel H. Jensen, Delaware Museum of Natural History

25–6 Runk/Schoenberger, Grant Heilman Photography

25–7 Peter Ward, Bruce Coleman

25–9 Tom McHugh, Steinhart Aquarium, Photo Researchers

25–10 Jane Burton, Bruce Coleman

25–11 (b) D. P. Wilson, Eric and David Hosking Photography

25–15 Runk/Schoenberger, Grant Heilman Photography

25–16 Maria Wimmer

25–17 Russ Kinne, Photo Researchers

25–18 (a), (b) Mary E. Rice; (c) C. Bradford Calloway

25–19 Meredith Jones, Smithsonian Institution

25–20 J. Teague Self

25–21 Robert O. Schuster

25–22 Runk/Schoenberger, Grant Heilman Photography

25–23 Russell Zimmer

25–24 Heather Angel

25–25 Specimen supplied by Edward I. Coher

26–1 Stephen Dalton, Animals Animals Enterprises

26–2 E. S. Ross

26–3 (a) E. S. Ross; (b) Jack Dermid

26–4 After Wilson et al., *op. cit.*

26–7 Heather Angel

26–9 (a), (c) E. S. Ross; (b) George K. Bryce, Animals Animals Enterprises; (d) W. Clifford Healey, National Audubon Society Collection/PR

26–11 (a) A. Kerstitch, Taurus Photos; (b) T. E. Adams; (c) E. S. Ross

26–12 E. S. Ross

26–13 (a) E. S. Ross; (b) Larry West

26–15 Photograph by Grant Heilman

26–16 (a), (c), (d) E. S. Ross; (b) Larry West

26–17 (a) Oxford Scientific Films, Bruce Coleman; (b)–(d) David Overcash, Bruce Coleman; (e)–(i) E. R. Degginger, Bruce Coleman

26–18 After Malcolm S. Gordon, *Animal Physiology: Principles and Adaptations,* 2d ed., The Macmillan Company, New York, 1972

26–19 John Mais

26–20 After Roberts, *op. cit.*

26–21 Larry West

Page 545 James E. Lloyd

26–23 Ross Hutchins

26–25 E. S. Ross

26–26 Grant Heilman

27–1 Fred Bavendam, Peter Arnold

27–2 Jane Burton, Bruce Coleman

27–4 Heather Angel

27–5 (a) D. P. Wilson, Eric and David Hosking Photography; (b) Heather Angel; (c) Runk/Schoenberger, Grant Heilman Photography

27–6 D. P. Wilson, Eric and David Hosking Photography

27–7 Heather Angel

27–8 (b) Heather Angel

27–9 (a) D. P. Wilson, Eric and David Hosking Photography

27–10 © James Hanken, 1979

27–11 John S. Flannery, Bruce Coleman

27–12 The American Museum of Natural History

27–13 Peter Arnold

27–14 Runk/Schoenberger, Grant Heilman Photography

27–15 Alvin E. Staffan

27–16 After Cleveland P. Hickman and Cleveland P. Hickman, Jr., *Biology of Animals,* C. V. Mosby Company, St. Louis, Mo., 1972

27–17 Alvin E. Staffan

27–18 Lynwood M. Chace, National Audubon Society Collection/PR

27–19 The American Museum of Natural History

27–20 Anthony Mercieca, Photo Researchers

27–21 (a) The New York Zoological Society; (b) Jack Dermid

27–23 (a) Tom McHugh, Photo Researchers; (b) Karl Maslowski, Photo Researchers; (c) Jeff Foote, Bruce Coleman; (d) Mark N. Boulton, National Audubon Society Collection/PR; (e) Catherine Bray, National Audubon Society Collection/PR; (f) Mary M. Thatcher, Photo Researchers

28–1 Jack Dermid

28–5 B. E. Juniper

28–6 Ray F. Evert

Page 578 (a) Alvin E. Staffan; (b) Jack Dermid; (c), (d) Runk/Schoenberger, Grant Heilman Photography

28–7 (a) Grant Heilman; (b) Dan Clark, Grant Heilman Photography; (c) E. S. Ross.

28–8 (a), (b) Myron C. Ledbetter, from M. C. Ledbetter and K. R. Porter, *Introduction*

to The Fine Structure of Plant Cells, Springer-Verlag, New York, 1970; (c) Ray F. Evert

28–9 (a), (b) R. F. Evert, W. Eschrich, and S. E. Eichhorn, *Planta,* vol. 109, pages 193–210, 1973; (c) Ray F. Evert

28–11 (a) J. Heslop-Harrison; (b), (c) H. A. Core, W. A. Côté, and A. C. Day, *Wood: Structure and Identification,* 2d ed., Syracuse University Press, Syracuse, N.Y., 1979

28–12 Ray F. Evert

28–13 Ray F. Evert

28–14 Robert Mitchell, Earth Scenes

28–15 After Peter Ray, *The Living Plant,* 2d ed., Holt, Rinehart and Winston, Inc., New York, 1972

28–16 Ray F. Evert

28–17 Ray F. Evert

28–19 (a) John Colwell, Grant Heilman Photography; (b) Jack Dermid

28–20 (a) Grant Heilman; (b) Earth Scenes; (c) Renee Purse, Photo Researchers

Page 590 Ray F. Evert

29–1 Ray F. Evert

29–2 (a) Allen Rokach, The New York Botanical Garden; (b) L. Anderson; (c) J. F. Skvarla; (d) John Troughton and L. Donaldson

29–3 (b) Myron C. Ledbetter

29–4 (a) Jane Burton, Bruce Coleman; (b), (e) Jack Dermid; (c) E. S. Ross; (d) W. H. Hodge, Peter Arnold; (f) John H. Gerard; (g) Grant Heilman; (h) Alvin E. Staffan

29–5 Micrographs by Ray F. Evert

Page 601 Ray F. Evert

29–12 (a) Ray F. Evert

29–13 Howard Towner

29–14 After Ray, *op. cit.*

29–15 (a) Peter Ray; (b) John Troughton and L. Donaldson

29–17 From Richard Ketchum, *The Secret Life of The Forest,* American Heritage Publishing Company, Inc., New York, 1970

Page 609 (a) U.S. Forest Service; (b) C. W. Ferguson, Laboratory of Tree-Ring Research, University of Arizona

29–19 Jack Dermid

Page 612 Ron Curbow, National Audubon Society Collection/PR

30–2 After A. C. Leopold, *Plant Growth and Development,* McGraw-Hill Book Company, New York, 1964

30–3 George Whiteley, Photo Researchers

30–4 After M. Richardson, *Translocation in Plants,* St. Martin's Press, Inc., New York, 1968 (after Stoat and Hoagland, 1939)

30–5 John H. Troughton

30–7 (a) Grant Heilman; (b) Grant M. Haist, National Audubon Society Collection/PR

Page 621 (a) Grant Heilman; (b) John H. Troughton and L. Donaldson

Page 623 (b), (c) Walter Eschrich and Eberhard Fritz

30–8 (a) Martin H. Zimmermann; (b) George A. Schaefers

30–11 A. W. Ambler, National Audubon Society Collection/PR

30–12 S. A. Wilde

30–13 Raymond C. Valentine

30–14 (a) P. J. Dart; (b), (c) R. R. Hebert, J. G. Griswell, and R. W. F. Hardy, Central Research and Development Department, E. I. duPont de Nemours & Company

31–1 Jack Dermid

31–3 J. P. Nitsch, *American Journal of Botany,* vol. 37, page 3, 1950

31–4 U.S. Department of Agriculture

31–6 H. R. Chen

31–8 S. H. Wittwer and The Michigan Agricultural Experiment Station

31–10 J. E. Varner

31–11 After Wilson et al., *op. cit.*

31–12 (a) Ray F. Evert; (b) Grant Heilman

31–14 Jack Dermid

32–3 B. E. Juniper

32–4 Hugh Spencer, National Audubon Society Collection/PR

Page 647 After Aubrey W. Naylor, "The Control of Flowering," *Scientific American,* May 1952

32–7 Frank Salisbury

32–8 Richard F. Trump, Photo Researchers

32–10 Jack Dermid

32–11 Jerome Wexler, National Audubon Society Collection/PR

32–12 Jack Dermid

33–1 From Nilsson, *op. cit.*; photograph courtesy Lennart Nilsson, Bonnier Fakta, Stockholm

33–5 Micrograph by Keith R. Porter

33–6 (a) M. H. Ross and Edward J. Reith; (b) Don Fawcett

33–10 Keith R. Porter

33–11 After Roger Eckert and David Randall, *Animal Physiology,* W. H. Freeman and Company, San Francisco, 1978

33–13 After Vander et al., *op. cit.*

Page 670 After Eckert and Randall, *op. cit.*

34–2 After W. W. Howells, *Mankind in the Making,* rev. ed., Doubleday & Company, Garden City, N.Y., 1967

34–3 (a) I. Kaufman Arenberg

34–7 (a) Jeanne M. Riddle

34–9 (b) Don Fawcett

Page 681 U. S. Department of Agriculture

34–10 Don Fawcett, from Bloom and Fawcett, *op. cit.*

34–11 G. Decker

34–12 (b) L. L. Rue, III, Photo Researchers

34–13 Keith R. Porter

35–1 Alan Root

35–3 Thomas Eisner

35–4 Russ Kinne, Photo Researchers

35–7 (b) H. R. Duncker, Department of Anatomy, Justus Liebig University, Giessen, Germany

35–10 (a) Keith R. Porter; (b) E. R. Weibel

35–11 After W. S. Beck, *Human Design,* Harcourt Brace Jovanovich, Inc., New York, 1971

Page 697 Myron Melamed

35–14 After Knut Schmidt-Nielsen, *How Animals Work,* Cambridge University Press, New York, 1972

Page 702 Anker Odum, from *Scenic Wonders of Canada,* © 1976 by Reader's Digest Association (Canada) Ltd.

36–1 Thomas Eisner

36–2 Keith R. Porter

36–3 Emil Bernstein, The Gillette Company Research Institute

36–7 Robert Rosenblum

36–8 After Roberts, *op. cit.*

36–9 U.N. Food and Agriculture Organization

36–10 Micrograph by Don Fawcett

36–15 After Roberts, *op. cit.*

36–16 After Vander et al., *op. cit.*

36–18 American Heart Association

37–2 Tom McHugh, Photo Researchers

37–3 From Lewis Carroll, *Alice's Adventures in Wonderland,* illustrated by John Tenniel

37–4 Bob Leatherman, National Audubon Society Collection/PR

37–8 From Richard G. Kessel and Randy H. Kardon, *Tissues and Organs: A Text-Atlas of Scanning Electron Microscopy,* W. H. Freeman and Company, San Francisco, © 1979

37–10 Don Fawcett, from William Bloom and D. W. Fawcett, *Histology,* 9th ed., W. B. Saunders Company, Philadelphia, 1968

38–1 U.S. Fish and Wildlife Service

38–3 (a) Grant Heilman; (b) Jen and Des Bartlett, Photo Researchers; (c) After Gordon, *op. cit.*

38–4 After Gordon, *op. cit.*

38–5 (a) Jack Dermid

38–6 After Schmidt-Nielsen, *op. cit.*

38–7 (a) Ron Garrison, San Diego Zoo; (b) L. L. Rue, III, Bruce Coleman; (c) L. L. Rue, III, National Audubon Society Collection/PR

38–9 (a) Spider, C. D. Marvin Winter; (b) George Daniell, Photo Researchers

38–10 (a) Nicholas Mrosovsky; (b) L. L. Rue, III, National Audubon Society Collection/PR; (c) Yvon Le Maho

38–11 Phyllis McCutcheon Mithassel, Photo Researchers

38–13 Dominique Reger, UNESCO

39–1 From Nilsson, *op. cit.;* photograph courtesy Lennart Nilsson, Bonnier Fakta, Stockholm

39–2 D. W. Fawcett, *The Cell,* 2d ed., W. B. Saunders Company, Philadelphia, 1981/Photo Researchers

39–5 Joseph Feldman

39–6 After Niels Kai Jerne, "The Immune System," *Scientific American,* June 1973

39–8 After Richard A. Lerner and Frank J. Dixon, *Scientific American,* June 1973

Page 755 Thomas Hooker, Woodfin Camp & Associates

39–10 D. W. Fawcett, *op. cit.,* 1981/Photo Researchers

40–6 The American Museum of Natural History

40–7 After Eckert and Randall, *op. cit.*

40–9 (b) David Hubel

40–10 After S. W. Kuffler and J. G. Nicholls, *From Neuron to Brain,* Sinauer Associates, Inc., Sunderland, Mass., 1976

40–11 *Ibid.*

40–12 *Ibid.*

40–13 (b) C. S. Raine

40–14 After C. R. Noback, *The Human Nervous System: Principles of Neurobiology,* McGraw-Hill Book Company, New York, 1975

40–15 R. Perkins, courtesy of Thomas Reese

41–1 Don Fawcett

41–4 After Vander et al., *op. cit.*

41–5 (a) Terence A. Gili, Animals Animals Enterprises; (b) L. L. Rue, III, Photo Researchers; (c) Len Rue, Jr., Animals Animals Enterprises; (d) Jan Lukas, Photo Researchers

41–12 After Wilson et al., *op. cit.,* 1978

41–14 (b) J. T. Bonner

42–1 After W. S. Beck, *op. cit.*

42–2 After Woodburne

42–5 (a) Russ Kinne, Photo Researchers; (b) Gordon S. Smith, Photo Researchers; (c) Jerry Focht, Photo Researchers; (d) A. W. Ambler, National Audubon Society Collection/PR

42–8 E. R. Lewis

42–9 L. L. Rue, III, National Audubon Society Collection/PR

42–10 (b) Keith R. Porter

42–12 Jack Dermid

42–16 Robert Preston, courtesy of J. E. Hawkins, Kresge Hearing Research Institute, The University of Michigan Medical School

42–18 Micrograph by Hugh E. Huxley

42–20 Hugh E. Huxley

42–21 After Vander et al., *op. cit.*

42–22 (b), (c) After Keith R. Porter and Mary Bonneville, *Fine Structure of Cells and Tissues,* Lea and Febiger, Philadelphia, 1968

42–23 From Nilsson, *op. cit.;* photograph courtesy Lennart Nilsson, Bonnier Fakta, Stockholm

42–24 John Heuser

43–1 Manfred Kage, Peter Arnold

43–2 After P. W. Davis and E. P. Solomon, *The World of Biology,* McGraw-Hill Book Company, New York, 1974

43–4 Arthur Jacques

43–5 Photograph by Arthur Jacques

43–12 Jerre Levy, Colwyn Trevarthen, and Roger W. Sperry, *Brain: The Journal of Neurology,* 1972

43–14 Michael J. Kuhar

43–16 National Institutes of Health

43–18 Aquarius Electronics, Mendocino, Calif., © 1971

43–19 W. R. Klemm, *Science, The Brain and Our Future,* Pegasus, New York, 1972

43–20 Arthur Tress, Photo Researchers

43–21 E. R. Lewis

43–23 Lee-Ming Kow and Donald Pfaff, The Rockefeller University

43–24 Joan I. Morrell and Donald Pfaff, The Rockefeller University

44–1 Ken Heyman

44–4 David M. Phillips

44–5 After Roberts, *op. cit.*

44–8 From Nilsson, *op. cit.;* photograph courtesy Lennart Nilsson, Bonnier Fakta, Stockholm

44–10 (b) R. J. Blandau

44–11 Rugh and Shettles, *op. cit.*

44–14 E. S. E. Hafez

45–1 (a) William Byrd; (b) Everett Anderson

45–2 Tryggve Gustafson

45–3 After J. J. W. Baker and G. E. Allen, *The Study of Biology,* 3d ed., Addison-Wesley Publishing Company, Inc., Reading, Mass., 1977

45–4 Tryggve Gustafson

45–5 After E. O. Wilson et al., *Life on Earth,* 2d ed., Sinauer Associates, Inc., Sunderland, Mass., 1978

45–6 After Waddington, 1966

45–8 After Huettner, 1949

45–9 *Ibid.*

45–10 *Ibid.*

45–11 John A. Moore

45–12 After Baker and Allen, *op. cit.*

45–13 After Waddington, 1966

45–14 M. P. Olsen

45–15 After Wilson et al., *op. cit.,* 1978

45–16 *Ibid.*

45–17 After Torrey, 1962

45–19 After Maurice Sussman, *Animal Growth and Development,* 2d ed., Prentice-Hall, Inc., Englewood Cliffs, N.J., 1964

45–20 After Charles W. Bodemer, *Modern Embryology,* Holt, Rinehart and Winston, Inc., New York, 1968

45–21 Runk/Schoenberger, Grant Heilman Photography

45–22 After Sussman, *op. cit.*

45–23 Landrum B. Shettles, M.D.

45–24 (a) Carnegie Institution of Washington; (b) Landrum B. Shettles, M.D.; (c) Russ Kinne, Photo Researchers

45-25 Carnegie Institution of Washington

45-26 Micrograph by Rugh and Shettles, *op. cit.*

45-27 After Roberts, *op. cit.*

45-29 Landrum B. Shettles, M.D.

45-30 From Nilsson, *op. cit.;* photographs courtesy Lennart Nilsson, Bonnier Fakta, Stockholm

45-31 London Scientific Photos

45-32 Rugh and Shettles, *op. cit.*

45-33 From Nilsson, *op. cit.;* photograph courtesy Lennart Nilsson, Bonnier Fakta, Stockholm

45-34 From Lennart Nilsson, *A Child Is Born,* Delacorte Press, New York, copyright © 1978; photograph courtesy Lennart Nilsson, Bonnier Fakta, Stockholm

45-36 Suzanne Arms, Jeroboam

46-1 Peter Ward, Bruce Coleman, Ltd.

46-2 The American Museum of Natural History

46-3 (a) Russ Kinne, Photo Researchers; (b) Heather Angel; (c) Jeanne White, Photo Researchers

46-4 Karl H. Maslowski, Photo Researchers

46-5 Painting by Charles R. Knight, The American Museum of Natural History

46-6 From Roberts Rugh, *Experimental Embryology; Techniques and Procedures,* 7th ed., Burgess Publishing Company, Minneapolis

Page 887 David Archibald, from David Archibald and William Clemens, "Late Cretaceous Extinctions," *American Scientist,* page 379, July/August 1982

46-8 Jeanne White, Photo Researchers

46-9 H. B. D. Kettlewell

47-1 Office of Public Information, Columbia University

47-2 BBC Hulton Picture Library

Page 894 Stephen Dalton, Photo Researchers

47-4 Edmund B. Gerard

47-5 Victor McKusick

47-6 L. L. Rue, III, Photo Researchers

48-1 Baron Wolman

48-2 Eric Hosking

48-4 John Mais

48-5 After Mather and Harrison, *Heredity,* vol. 3, 1949

48-7 George Holton, Photo Researchers

48-8 P. M. Sheppard

48-9 (b) U.N. Food and Agriculture Organization

48-10 From Stephen J. Gould, *Ever Since Darwin,* W. W. Norton & Company, New York, 1977

48-11 After J. Clausen and W. M. Hiesey, Carnegie Institution of Washington, publication 615, 1958

49-1 After Jack R. Harlow, "The Plants and Animals That Nourish Man," *Scientific American,* September 1976

49-2 After Charles W. Brown, Santa Rosa Junior College

49-4 (a) Allan D. Cruickshank, Photo Researchers; (b) Alan Root, Bruce Coleman

49-5 Karu Kurita, *Discover,* © 1982, Time, Inc.

49-6 Tony Gauba, Photo Researchers

49-7 Heather Angel

49-8 Robert Mitchell

49-9 From Lewis Carroll, *Through The Looking Glass*

49-10 (a) Russ Kinne, Photo Researchers; (b) L. L. Rue, III, Bruce Coleman; (c) M. P. Kahl, Photo Researchers; (d) Bradley Smith, Animals Animals Enterprises; (e) George Holton, Photo Researchers

49-11 Department of Indian Affairs and Northern Development

49-12 (a) C. Haagner, Bruce Coleman; (b) Robert Clark, Photo Researchers

49-13 (a) E. S. Ross; (b) Lynn M. Stone, Earth Scenes

49-14 After Theodosius Dobzhansky, Francisco J. Ayala, G. Ledyard Stebbins, and James W. Valentine, *Evolution,* W. H. Freeman and Company, San Francisco, 1977

49-15 (a), (b) Lincoln Brower; (c) John Bova, Photo Researchers; (d) James H. Robinson, Photo Researchers

49-16 Lincoln Brower

49-17 E. S. Ross

50-1 (a) M. Philip Kahl, Bruce Coleman; (b) N. Myers, Bruce Coleman

50-2 After E. O. Wilson and William H. Bossert, *A Primer of Population Biology,* Sinauer Associates, Sunderland, Mass., 1971

Page 931 John Shelton

50-3 After Raven et al., *op. cit.,* 1981

50-6 Jen and Des Bartlett, Bruce Coleman

50-7 After Bruce Wallace and Adrian Srb, *Adaptation,* Prentice-Hall, Inc., Englewood Cliffs, N.J., 1964

Page 935 U.S. Department of Agriculture

50-8 Tui De Roy Moore

50-10 After David Lack, *Darwin's Finches,* Harper & Row, Publishers, Inc., New York, 1961

50-11 After Raven et al., *op. cit.,* 1981

50-12 From George Gaylord Simpson, *Horses,* Oxford University Press, New York, 1951

50-13 *Ibid.*

50-14 After Steven M. Stanley, *Macroevolution: Pattern and Process,* W. H. Freeman and Company, San Francisco, 1979

51-1 Charles Goodnight/Donald R. Perry

51-7 After *Nature,* vol. 298, August 1982

51-8 From Sarukhan and Harper, 1973

52-1 George Holton, Photo Researchers

52-3 After John L. Harper, *Symposia Society for Experimental Biology,* vol. 15, page 25, 1961

52-4 After MacArthur, 1958

52-6 After Robert E. Ricklefs, *Ecology,* 2d ed., Chiron Press, Inc., New York, 1979

52-8 After Henry S. Horn, "Forest Succession," *Scientific American,* May 1975

52-9 Norman Myers, Bruce Coleman

52-12 Australian Department of Lands

52-13 E. S. Ross

52-14 Thomas Eisner

52-15 (a) Ken M. Highfill, Photo Researchers; (c), (d) After M. B. Fenton and James H. Fullard, "Moth Hearing and The Feeding Strategies of Bats," *American Scientist,* vol. 69, pages 266–275, 1981

52-16 Australian News and Information Service

52-17 (a) R. Mariscal, Bruce Coleman; (b) Douglas Faulkner; (c) E. S. Ross; (d) L. L. Rue, III, Bruce Coleman

52-18 Daniel Janzen

52-19 After Ricklefs, *op. cit.*

53-1 Jack Dermid

Page 978 (a) John Porteous, WHO/The Woods Hole Oceanographic Institute; (b) John M. Edmond, from John B. Corliss and Robert D. Ballard, *National Geographic,* vol. 152, page 4, October 1977

53-3 (a) Heather Angel; (b) Tom Brakefield, Bruce Coleman; (c) Jeff Foote, Bruce Coleman

53-4 (a) Tom McHugh, Photo Researchers;

(b) M. A. Chappell, Animals Animals Enterprises; (c) Frank W. Lane, Bruce Coleman

Page 981 Irven deVore, Anthro-Photo

53–10 G. E. Likens

53–11 (b) Michigan Department of Natural Resources

53–12 (a) Larry West; (b) Jack Dermid

53–13 © Roger del Moral

54–1 NASA

54–8 After Karen Arms and Pamela S. Camp, *Biology,* 2d ed., Sanders College Publishing, New York, 1979

54–9 (a), (b) Les Blacklock; (c) L. L. Rue, III, Photo Researchers; (d), (e) K. Schwartz

54–10 (a) Craig Blacklock; (b), (c) Les Blacklock

54–11 (a) Les Blacklock; (b) Alvin E. Staffan; (c) Bill Ruth, Bruce Coleman

54–12 (a) Tom McHugh, Photo Researchers; (b) Patricia Caulfield

54–13 Joe Van Wormer, Bruce Coleman

54–14 (a) Jen and Des Bartlett, Bruce Coleman; (b) Dennis Brokaw; (c) E. S. Ross

54–15 (a) L. L. Rue III, Bruce Coleman; (b) Max Thompson, National Audubon Society Collection/PR; (c) N. Myers, Bruce Coleman; (d) Verna R. Johnston, Photo Researchers

54–16 (a) Bill Ratcliffe; (b) Jen and Des Bartlett, Bruce Coleman; (c) Robert H. Wright, National Audubon Society Collection/PR; (d) Grace Thompson, Photo Researchers; (e) Anthony Mercieca, National Audubon Society Collection/PR; (f) Willis Peterson; (g) Alan Blank, Bruce Coleman

54–17 Karl Weidmann, National Audubon Society Collection/PR

54–18 (a) E. S. Ross; (b) John Markham, Bruce Coleman

54–19 After Arms and Camp, *op. cit.*

55–2 From Jerram Brown, *The Evolution of Behavior,* W. W. Norton & Company, New York, 1975

55–3 Sullivan Rogers, Bruce Coleman

55–4 (a) E. S. Ross; (b) Alan Blank, Bruce Coleman

55–5 (a) Oxford Scientific Films, Bruce Coleman; (b) Treat Davidson, National Audubon Society Collection/PR

55–6 (a) E. S. Ross; (b) Colin G. Butler, Bruce Coleman; (c) Stephen Dalton, Photo Researchers

55–7 Patricia Caulfield

55–8 T. W. Ransom

55–9 (a) L. L. Rue, III, Bruce Coleman; (b) R. R. Pawlaski, Bruce Coleman; (c) S. Gaulin, Anthro-Photo; (d) Douglas Faulkner

55–10 After J. R. Krebs and N. B. Davies, *An Introduction to Behavioral Ecology,* Sinauer Associates, Inc., Sunderland, Mass., 1981

55–11 (a), (c) © Patricia Moehlman; (b) after Krebs and Davies, *op. cit.*

Page 1025 Simon Trevor, Bruce Coleman

55–12 Sarah Blaffer Hrdy, Anthro-Photo

55–13 John S. Crawford, Photo Researchers

Page 1027 (a) Tom McHugh, Photo Researchers; (b) Hans and Judy Beste, Animals Animals Enterprises

55–14 (a) Jeff Foote, Bruce Coleman; (b) E. S. Ross

Page 1029 Joseph Popp, Anthro-Photo

55–15 David Bygott, Anthro-Photo

56–1 © John Reader

56–4 (a) E. S. Ross; (b) H. Albrecht, Bruce Coleman

56–5 After R. E. Leakey and R. E. Lewin, *Origins,* E. P. Dutton & Company, Inc., New York, 1977

56–6 (a) Tom McHugh, Photo Researchers; (b) M. P. L. Fogden, Bruce Coleman

56–7 After Howells, *op. cit.*

Page 1040 C. R. Carpenter

56–8 L. L. Rue, III, Photo Researchers

56–10 Alan Walker

56–11 The American Museum of Natural History

56–12 (a) Donald Johanson; (b) Donald Johanson and the Cleveland Museum of Natural History

56–13 Robert F. Sisson, © National Geographic Society

56–14 The American Museum of Natural History

56–15 After Donald Johanson and Edey Maitland, *Lucy: The Beginnings of Humankind,* Warner Books, New York, 1982

56–16 Tom McHugh, Photo Researchers

56–17 Lee Boltin

Page 1050–1051 (a) Ralph Morse, Time-Life; (b) G. D. Plage, Bruce Coleman, Ltd.; (c)–(e) Peter G. Veit; (f) Jim Moore, Anthro-Photo; (g) Teleki/Baldwin; (h), (i) Richard Wrangham, Anthro-Photo

56–18 Ralph S. Solecki

56–20 (a) After Jacques Bordaz, *Tools of The Old and New Stone Age,* The Natural History Press, Garden City, New York, 1970; (b) Peabody Museum

56–21 Tom McHugh, Photo Researchers

56–22 The American Museum of Natural History

Page 1057 George Herben, Photo Researchers

56–23 International Rice Research Institute

Index

Page numbers in boldface type indicate major text discussions.